The Elements

Name	Symbol	Atomic Number	Atomic Weight	Name	Symbol	Atomic Number	Atomic Weight
Actinium	Ac	89	227.028	Mendelevium	Md	101	(258)
Aluminum	Al	13	26.9815	Mercury	Hg	80	200.59
Americium	Am	95	(243)	Molybdenum	Mo	42	95.94
Antimony	Sb	51	121.76	Neodymium	Nd	60	144.24
Argon	Ar	18	39.948	Neon	Ne	10	20.1797
Arsenic	As	33	74.9216	Neptunium	Np	93	237.048
Astatine	At	85	(210)	Nickel	Ni	28	58.693
Barium	Ba	56	137.327	Niobium	Nb	41	92.9064
Berkelium	Bk	97	(247)	Nitrogen	N	7	14.0067
Beryllium	Be	4	9.01218	Nobelium	No	102	(259)
Bismuth	Bi	83	208.980	Osmium	Os	76	190.23
Bohrium	Bh	107	(262)	Oxygen	O	8	15.9994
Boron	B	5	10.811	Palladium	Pd	46	106.42
Bromine	Br	35	79.904	Phosphorus	P	15	30.9738
Cadmium	Cd	48	112.411	Platinum	Pt	78	195.08
Calcium	Ca	20	40.078	Plutonium	Pu	94	(244)
Californium	Cf	98	(251)	Polonium	Po	84	(209)
Carbon	C	6	12.011	Potassium	K	19	39.0983
Cerium	Ce	58	140.115	Praseodymium	Pr	59	140.908
Cesium	Cs	55	132.905	Promethium	Pm	61	(145)
Chlorine	Cl	17	35.4527	Protactinium	Pa	91	231.036
Chromium	Cr	24	51.9961	Radium	Ra	88	226.025
Cobalt	Co	27	58.9332	Radon	Rn	86	(222)
Copper	Cu	29	63.546	Rhenium	Re	75	186.207
Curium	Cm	96	(247)	Rhodium	Rh	45	102.906
Dubnium	Db	105	(262)	Rubidium	Rb	37	85.4678
Dysprosium	Dy	66	162.50	Ruthenium	Ru	44	101.07
Einsteinium	Es	99	(252)	Rutherfordium	Rf	104	(261)
Erbium	Er	68	167.26	Samarium	Sm	62	150.36
Europium	Eu	63	151.965	Scandium	Sc	21	44.9559
Fermium	Fm	100	(257)	Seaborgium	Sg	106	(263)
Fluorine	F	9	18.9984	Selenium	Se	34	78.96
Francium	Fr	87	(223)	Silicon	Si	14	28.0855
Gadolinium	Gd	64	157.25	Silver	Ag	47	107.868
Gallium	Ga	31	69.723	Sodium	Na	11	22.9898
Germanium	Ge	32	72.61	Strontium	Sr	38	87.62
Gold	Au	79	196.967	Sulfur	S	16	32.066
Hafnium	Hf	72	178.49	Tantalum	Ta	73	180.948
Hassium	Hs	108	(265)	Technetium	Tc	43	(98)
Helium	He	2	4.00260	Tellurium	Te	52	127.60
Holmium	Ho	67	164.930	Terbium	Tb	65	158.925
Hydrogen	H	1	1.00794	Thallium	Tl	81	204.383
Indium	In	49	114.818	Thorium	Th	90	232.038
Iodine	I	53	126.904	Thulium	Tm	69	168.934
Iridium	Ir	77	192.22	Tin	Sn	50	118.710
Iron	Fe	26	55.847	Titanium	Ti	22	47.88
Krypton	Kr	36	83.80	Tungsten	W	74	183.84
Lanthanum	La	57	138.906	Uranium	U	92	238.029
Lawrencium	Lr	103	(260)	Vanadium	V	23	50.9415
Lead	Pb	82	207.2	Xenon	Xe	54	131.29
Lithium	Li	3	6.941	Ytterbium	Yb	70	173.04
Lutetium	Lu	71	174.967	Yttrium	Y	39	88.9059
Magnesium	Mg	12	24.3050	Zinc	Zn	30	65.39
Manganese	Mn	25	54.9381	Zirconium	Zr	40	91.224
Meitnerium	Mt	109	(266)				

General Chemistry

An Integrated Approach

Second Edition

John W. Hill

University of Wisconsin—River Falls

Ralph H. Petrucci

California State University, San Bernardino

PRENTICE HALL
Upper Saddle River, New Jersey 07458

Brief Contents

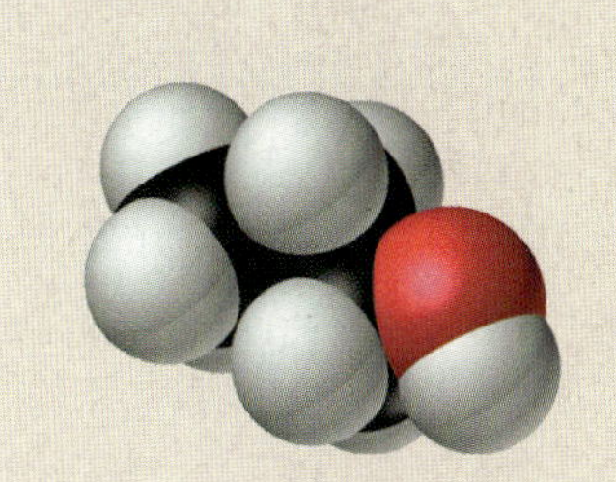

Appendices

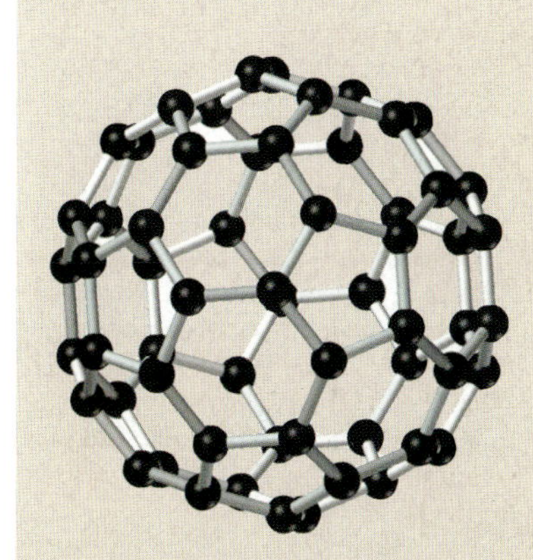

Contents

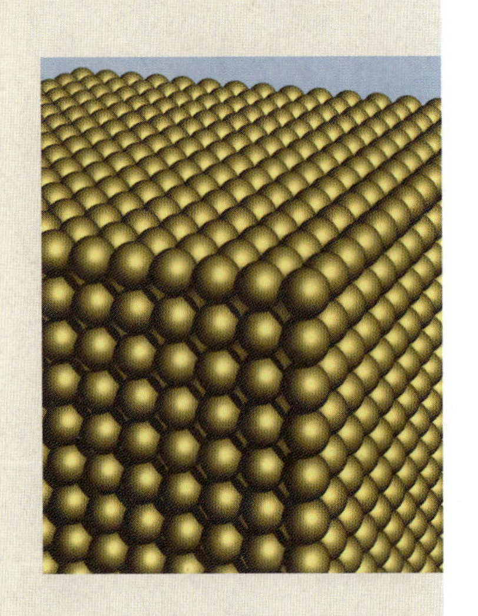

4 Chemical Reactions in Aqueous Solutions 132

5 Gases 177

6 Thermochemistry 226

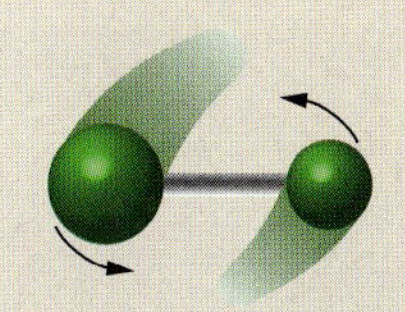

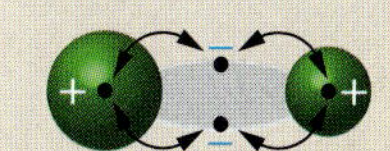

7 Atomic Structure 275

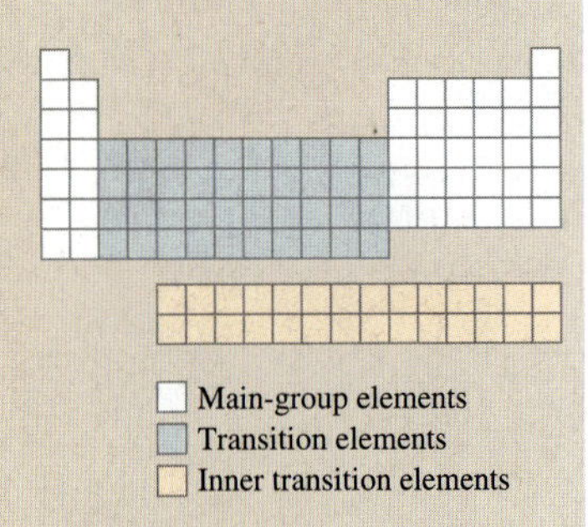
Main-group elements
Transition elements
Inner transition elements

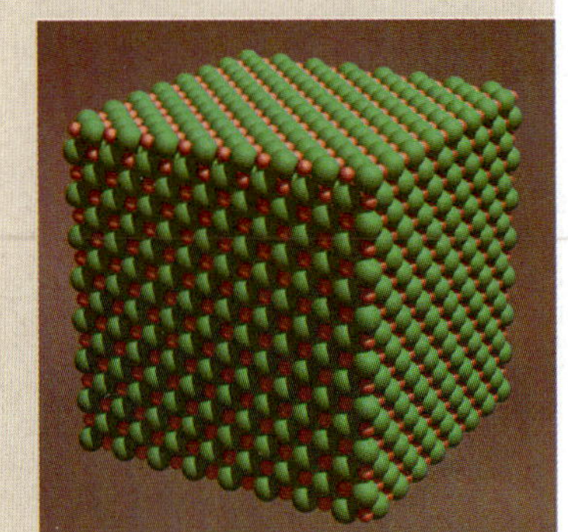

10 Bonding Theory and Molecular Structure 408

VSEPR notation: AX_3E

11 States of Matter and Intermolecular Forces 458

12 Physical Properties of Solutions 509

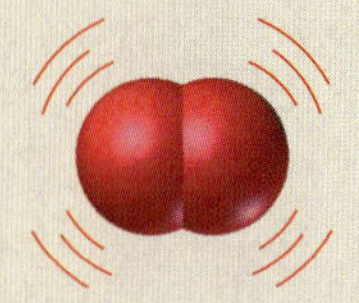

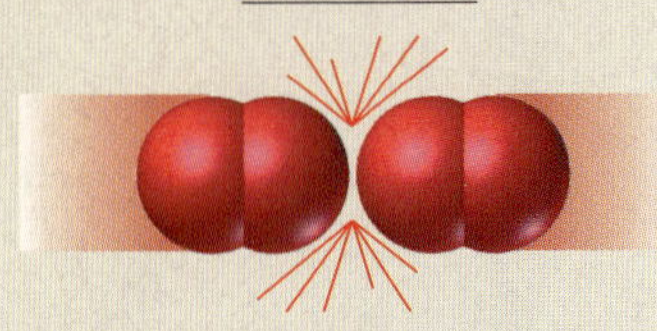

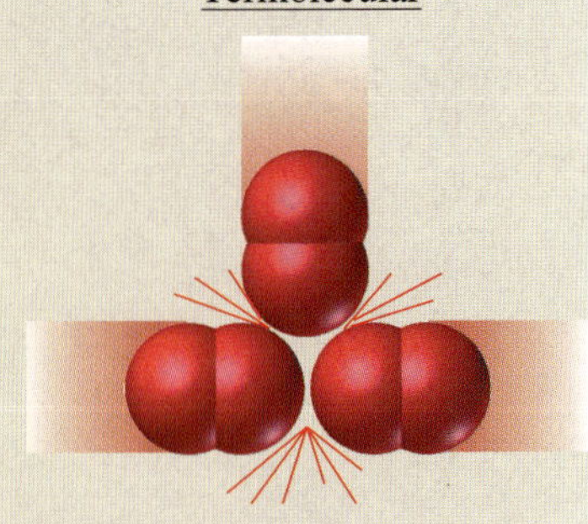

13 Chemical Kinetics: Rates and Mechanisms of Chemical Reactions 552

14 Chemical Equilibrium 594

15 Acids, Bases, and Acid-Base Equilibria 632

19 Nuclear Chemistry 817

20 The *s*-Block Elements 848

21 The *p*-Block Elements 879

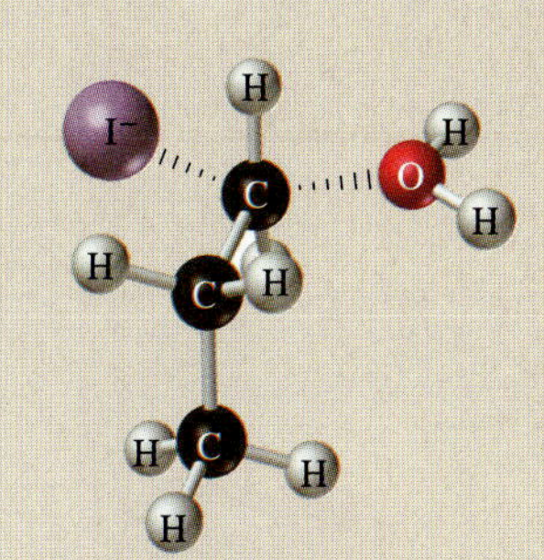

22 The *d*-Block Elements and Coordination Chemistry 921

23 Chemistry and Life: More on Organic, Biological, and Medicinal Chemistry 957

Preface

Students come to a general chemistry course with a variety of backgrounds and interests. Most plan to become scientists, engineers, or professionals in medicine or other areas of the life sciences. Part of the task of the chemistry instructor is convincing students that knowledge of chemistry is essential to a true understanding of fields that range from cell biology to medicine to materials science. Indeed, the chemical properties and principles students learn in this course will pervade almost every aspect of their personal and professional lives. In this text, we have tried to provide students with both the core principles and interesting applications of chemistry, with the belief that such knowledge will both help them in their professions and enrich their everyday lives.

This New Edition: Achieving Balance

In addition to establishing the relevance of chemistry to broader concerns, a textbook can enhance students' success in the study of chemistry by satisfying other needs: the need for background reading and a second voice for students to hear; help in visualizing chemical phenomena, both what students can see with their eyes and what they must learn to see with their minds' eyes; and help in formulating strategies for solving problems, the basis on which their knowledge will so often be tested. In crafting the second edition of this text, we have striven to strike a necessary balance in meeting these basic needs. Following are some of the ways we have revised the previous edition:

- In response to reviewer suggestions, we have expanded coverage of a number of topics while at the same time being sensitive to the common desire for leaner books. In particular, **we have added many more problems**, both mid-level and those of a more challenging nature.
- We have **rewritten explanations of difficult topics to be carefully paced** and appropriate to the background of the typical student. In some instances we have approached the topic in a nontraditional manner, but we have always been careful to adhere to current views, such as in the treatment of expanded valence shells in Lewis structures, the rationalization of Raoult's law, and the use of electrochemical conventions.
- We have **added many new chemical applications** throughout this edition. These include in-text discussions, new boxes, and a new feature called Application Notes. In-text discussions include those of the Alkali Metals and Living Matter (Section 20.7) and the Group 2A Metals and Living Matter (Section 20.11). As in the first edition, our box features focus on applications. New box features introduce topics such as Electron Probabilities and a Close-Up Look at Atoms (p. 308) and NO: A Messenger Molecule (p. 899). Application Notes are marginal notes that highlight interesting applications of key topics. For example, an Application Note in Chapter 5 (Gases, p. 213) describes how methyl mercaptan enables us to detect natural gas leaks. These notes help the student see that we can apply our knowledge of chemistry to solve real-life problems. Another Application Note in Chapter 24 (Chemistry of Materials) explains that the Titanic sank partly because the steel rivets that held the ship together were weakened by the presence of too much slag.
- We have **expanded the teaching of all problem-solving skills** by adding flowcharts for quantitative problem-solving, more Conceptual Examples and

Exercises, and more Estimation Examples and Exercises (which encourage students to develop the habit of asking, "Is this answer reasonable?").

In summary, we have tried to combine our collective experience in teaching and writing for various audiences to produce a textbook that strikes a balance between the principles that give meaning to chemistry and the applications that make it come alive.

Organization

The first 18 chapters of the text emphasize chemical principles, but the principles are illustrated throughout with significant applications and concrete examples from descriptive chemistry. Chapter 20 (The *s*-Block Elements), Chapter 21 (The *p*-Block Elements), and Chapter 22 (The *d*-Block Elements and Coordination Chemistry) provide a systematic treatment of descriptive chemistry, but with an emphasis on how the properties of substances relate to the principles learned earlier in the text. Chapter 19 (Nuclear Chemistry), Chapter 23 (Chemistry and Life), Chapter 24 (Chemistry of Materials), and Chapter 25 (Environmental Chemistry) are fairly independent, free-standing chapters. These chapters can serve as capstones to a general chemistry course, for each revisits the basic principles of earlier chapters to cover topics in which students generally have a strong interest. These chapters can be studied, in whole or in part, in just about any order.

Integrating Organic and Biological Chemistry

A major goal of ours in writing this text has been **to provide a truly general course that integrates all the major areas of chemistry**. Physical principles, inorganic compounds, and analytical techniques are addressed repeatedly. As in the previous edition, organic chemistry is incorporated throughout the text, as students may be ill-served by being sheltered from organic compounds until very late in their study of general chemistry. Thus, some simple organic chemistry is introduced in Chapter 2 and used thereafter to describe physical properties of substances, aspects of chemical bonding, acid-base chemistry, and oxidation-reduction reactions. Biochemistry is introduced in Chapter 6 in a discussion of carbohydrates and fats as fuels for our bodies; it is used frequently in following chapters where appropriate.

New to this edition is Chapter 23, titled "Chemistry and Life: More On Organic, Biological, and Medicinal Chemistry." This chapter brings together the core organic chemistry concepts introduced in earlier chapters, expands on them in those cases where the earlier introduction was necessarily brief, and then discusses the chemistry of selected biomolecules and medicinal compounds. In this way, we have tried to provide a useful set of core material to those who will never take an organic chemistry course, while also offering a broader-than-usual preparation for students who will enroll in organic chemistry courses.

A Balanced Approach to Problem Solving

Problem-solving skills and the ability to think critically are essential for success in today's world. We provide ample opportunities for practicing these skills. For every type of problem we provide *Examples* that are carefully worked out, step-by-step, to guide students in solving similar problems.

Two new problem-solving tools accompany the Examples. *Problem-Solving Notes* provide ready reference and help for students as they study specific Examples:

The notes highlight relevant problem-solving techniques, help students understand and test the assumptions used to solve a worked Example, provide helpful hints, and encourage students to check their answers. Also, in the early chapters, particularly Chapter 3 (Stoichiometry), **the various terms in a series of related calculations may be annotated**. These annotations present a brief rationale for each calculation; we hope they will help students focus on "why" as well as "how."

The Examples are followed by *Exercises* that students can use to practice their understanding of the methods illustrated. In most cases, **two Exercises are given**, labeled *A* and *B*. The goal in an *A* Exercise is to apply to a similar situation the method outlined in the Example. In a *B* Exercise, students often must combine that method with other ideas previously learned. Many of the *B* Exercises provide a context closer to that in which chemical knowledge is applied, and they thus serve as a bridge between the worked Examples and the more challenging problems at the end of the chapter.

The ability to plug numbers into an equation and get an answer, in itself, is seldom enough to attain mastery of a concept. For example, **students should generally be able to judge whether an answer is reasonable**, and in some cases, to obtain a reasonable estimate of an answer without doing a detailed calculation. To assist in the acquisition of these skills, we offer worked-out *Estimation Examples* followed by *Estimation Exercises*. Examples and Exercises of this type are found throughout the text.

Students also need to **develop insights into chemical concepts** that are often best demonstrated by an ability to solve problems of a qualitative nature. To emphasize this aspect of problem-solving, we provide guided *Conceptual Examples* followed by *Conceptual Exercises*.

Through the different types of Examples and Exercises described, students of this text should gain a balanced set of skills in chemical problem-solving. As additional reinforcement, the text offers three kinds of end-of-chapter exercises:

- *Review Questions* are intended to provide a qualitative measure of student understanding of the main ideas introduced in the chapter. Answers to a few of the Review Questions are given in Appendix F (Answers to Selected Problems).
- *Problems* are arranged by topic; they test mastery of the problem-solving techniques discussed in the chapter. The Problems are arranged in matched pairs, with answers to odd-numbered problems given in Appendix F.
- *Additional Problems* are not grouped by type. Some are more challenging than the Problems, often requiring a synthesis of ideas from more than one chapter. Others pursue an idea further than is done in the text, or introduce new ideas. Answers to the odd-numbered Additional Problems are given in Appendix F.

Some of the Problems and Additional Problems are of an estimation or conceptual type, mirroring similar types of exercises within each chapter. We did not specifically label these questions, however; we want to give students experience in recognizing different types of problems as well as solving them.

Improving Students' Visualization Skills

Difficulty seeing the unseeable and imagining things in three dimensions is cited among the top three barriers confronting students in a general chemistry course. (The other two are poor study habits and poor math skills, both of which are addressed by specific print supplements to this text; see p. xvi.) In this book, we use drawings, computer graphics, and photographs to help students visualize chemical phenomena at both the microscopic (molecular) and macroscopic (visible) levels. Users of this text also have available ChemCDX, which includes nearly 70 animations and videos, and a link to the Companion Website that presents almost all the molecules in the book as three-dimensional images that students can examine in detail on their computers.

Supplements

For the Instructor

Annotated Instructor's Edition (0-13-010318-7), with annotations by Leslie Kinsland, University of Southwestern Louisiana. This special edition of the text includes the entire student text plus marginal icons and annotations to aid instructors in preparing their lectures. Included are suggestions for lecture demonstrations, teaching tips, common student misconceptions, and indications of which graphics in the textbook are available as overhead transparencies. The AIE also includes cross-references to all figures, demonstrations, and animations available in electronic form on the *Matter '99* CD-ROM.

Instructor's Resource Manual (0-13-918947-5), prepared by Robert K. Wismer of Millersville University. This book provides chapter-by-chapter lectures outlines, teaching tips, common student misconceptions, background references, and suggested lecture demonstrations for in-class use.

Solutions Manual (0-13-918724-3) by C. Alton Hassell of Baylor University contains worked-out solutions to all in-chapter, end-of-chapter, review, conceptual, and estimation exercises and problems.

Transparencies (0-13-918913-0) Over 200 full-color transparencies chosen from the text put principles into visual perspective and save you time while you are preparing your lectures.

Test Item File (0-13-919051-1), revised and expanded by Michael Mosher, University of Nebraska, Kearney. This printed test bank includes over 1300 questions written exclusively for the Hill & Petrucci text, with all answers section-referenced to the text.

Prentice Hall Custom Test. The computerized version of the Test Item File is available in both a Windows version (0-13-919069-4) and a Mac version (0-13-919077-5). The software available with this database allows you to create and tailor exams to your specific needs.

Matter '99 Visual Presentation Manager CD-ROM is available for Windows (0-13-012503-2) and Macintosh (0-13-012504-0) and contains over 400 pieces of art from the text (in electronic format), 20 lab demonstration video segments, and 50 animations of core concepts. This CD also includes Presentation Manager 3.0, Prentice Hall's software designed specifically for classroom presentation, and Presentation Manager '99, our new PowerPoint-based image and animation gallery. Instructors can select one or the other to display material using a classroom projection system. Instructors can also access a special Website for updated graphics and new images related to their courses.

For the Student

ChemCDX (0-13-012505-9). A free copy of the ChemCDX CD-ROM accompanies each new copy of the textbook. This CD-ROM features 50 chapter-based modules containing 70 animations and videos that allow students to explore, extend, experiment, and expand their understanding of chemistry. Each module includes a brief introduction and is followed by questions that test student understanding of the material. References to relevant reading in the text are provided for students who need additional review to answer questions correctly. ChemCDX also serves as a launching point to the Companion Website for this textbook. ChemCDX and the Companion Website are easy-to-use and offer many ways to integrate media into your course, if you desire.

Study Guide (0-13-918765-0) by D. J. Goss of Hunter College. This book is keyed to the main text and provides further learning material for students: chapter-by-chapter overviews, learning goals, numerous examples and exercises, parallel text material, worked-out solutions, and practice tests with answers. This book serves as an excellent diagnostic tool and also helps sharpen students' skills in test-taking.

Selected Solutions Manual (0-13-918740-5) by C. Alton Hassell of Baylor University, contains worked-out solutions to over half of the text's problems. The answers to these problems also appear in the text as Appendix F, Answers to Selected Problems.

Math Review Toolkit (0-13-919184-4) by Gary Long of Virginia Polytechnic Institute. This brief paperback is engineered for students who find math a significant challenge in this course. The book provides a chapter-by-chapter review of the mathematics used throughout the text; a guide to preparing for a career in chemistry; and a review of some of the special writing requirements often needed in the general chemistry course, focusing particularly on the laboratory notebook. This supplement is free to qualified adopters; please speak with your local Prentice Hall representative.

Interactive Chemistry Journey CD-ROM (0-13-548116-3) by Steven Gammon, University of Idaho, Lynn R. Hunsberger, University of Louisville, and Sharon Hutchison, University of Idaho. This student tutorial CD-ROM is an interactive study tool that fosters understanding of a number of core chemical concepts. In this dynamic, simulation-based environment, students interact with and visualize chemical concepts in ways that are not possible with static learning programs.

Companion Website for *General Chemistry, Second Edition*. This innovative on-line resource center is designed specifically to support and enhance the Hill & Petrucci text. The Website features the following:

- a visualization gallery that includes hundreds of pre-built molecules. Students can manipulate these models in real-time on their computers using Chime or Rasmol.
- an interactive problem-solving center with nearly 1300 unique quiz problems referenced to the text. All quizzes feature hints, scrambled answer choices, and are graded on-line. Student feedback includes readings from the textbook.
- constantly updated links to recently-published articles and other chemistry-related Websites;
- chatrooms and bulletin boards, where students can communicate with you, TAs, or classmates.

Chemistry on the Internet: A Guide for Students (0-13-083977-9) by Thomas Gardner, Tennessee State University. This brief, paperback booklet helps students gain a greater understanding of the Internet while showing them how to access chemical information and learn about chemistry. It includes an overview of the Web and introduces students to search techniques for navigation. This supplement is free to qualified adopters; please speak with your local Prentice Hall representative.

The New York Times/Prentice Hall Themes of the Times. This newspaper-format resource brings together current chemistry-related articles from the award-winning science pages of *The New York Times*. This supplement is free to qualified adopters; please speak with your local Prentice Hall representative.

Acknowledgments

JWH would like to thank his colleagues at the University of Wisconsin-River Falls for so many ideas that made their way into this text, and to Kathy Sumter, Program Aide in the Department of Chemistry. He is especially indebted to Ina Hill and Cynthia Hill for library research, typing, and unfailing support throughout this project, and to Mike Davis for his encouragement and for sharing his love of learning.

RHP would like to thank his colleagues at California State University, San Bernardino for the interest they have shown and the helpful suggestions they have contributed to his textbooks over the years. His greatest debt is to his wife, Ruth, for her love, encouragement, and continuing support of a husband who has been preoccupied and neglectful for so long.

Both of us would like to thank Robert K. Wismer of Millersville University, who is ever ready to offer his excellent advice on the broad range of issues that crop up in the preparation of a new text. We are especially indebted to our students, who have challenged us to be better teachers, and the reviewers of this and our other books, who have challenged us to write more clearly and accurately. We also owe a debt of gratitude to the many creative people at Prentice Hall who have contributed their talents to this edition: John Challice, our chemistry editor, for imaginative guidance throughout the project; Amanda Griffith, editorial assistant, for diligence and patience in managing a myriad of reviews and other correspondence; Shana Ederer, our marvelous development editor, for her innumerable creative contributions; Yvonne Gerin, photo researcher, for her aesthetic contributions to the photographic illustrations; and to the production staff for their diligence and patience in bringing all the parts together to yield a finished work.

This edition has benefited from the thoughtful and careful input of users of the previous edition and reviewers of the new manuscript. To these colleagues we convey our sincere appreciation; we are pleased to acknowledge them below:

Reviewers for the First Edition:

Donald Baird
Florida Atlantic University

Marvin E. Brunch
Corning Community College

Roger Bunting
Illinois State University

Julia Burdge
University of Akron

Jorge Castillo
Chicago State University

Wallace Cordes
University of Arkansas

James Crosthwaite
University of North Carolina

John DeKorte
Glendale Community College

Marian Douglas
University of Arkansas

Kirt Dreyer
Bemidji State University

Michael Eastman
Northern Arizona University

Grover Everett
University of Kansas

Raymond Fort
University of Maine

Robin L. Garrell
University of California, Los Angeles

Elsie Gross
Hillsborough Community College

Anthony Guzzo
University of Wyoming

C. Alton Hassell
Baylor University

Sherman Henzel
Monroe Community College

John Hershberger
North Dakota State University

Bruce Hoffman
Lewis & Clark Community College

Paul W. Hunter
Michigan State University

Roger Hurdlik
Olive-Harvey College

Michael A. Janusa
Nicholls State University

Dale Johnson
University of Arkansas

Philip Kinsey
University of Evansville

Donald Kleinfelter
University of Tennessee

Roy A. Lacey
State University of New York

James Long
University of Oregon

C. Luehrs
Michigan Technological University

Terry McCreary
Murray State University

Paul O'Brien
West Valley College

Don Roach
Miami-Dade Community College

E. Alan Sadorski
Ohio Northern University

Catherine Shea
Mission College

Peter Sheridan
Colgate University

Charles Trapp
University of Louisville

Garth Welch
Weber State University

David White
State University of New York

Reviewers for the Second Edition:

Steven Albrecht
Oregon State University

John J. Alexander
University of Cincinnati

David Ball
Cleveland State University

John E.Bauman
University of Missouri-Columbia

Marcus Bond
Southeast Missouri State University

Cassandra Eagle
Appalachian State University

Charles Drain
Hunter College, CUNY

W. Budd Glenn
Appalachian State University

Yvonne Gindt
University of Nebraska, Kearney

John Henderson
Jackson Community College

W. Vernon Hicks
Northern Kentucky University

Andrew J. Holder
University of Missouri-Kansas City

Werner Horsthemke
Southern Methodist University

Richard W. Kopp
East Tennessee State University

Lois Krause
Clemson University

Robert Kren
University of Michigan, Flint

David Leddy
Michigan Technological University

Massoud Miri
Rochester Institute of Technology

Michael Mosher
University of Nebraska, Kearney

Dean Nelson
University of Wisconsin, Eau Claire

Robert Nelson
Georgia Southern University

Bruce Osterby
University of Wisconsin

Mary Jane Patterson
Brazosport College

Gary F. Riley
St. Louis College of Pharmacy

Paul Reinbold
Southern Nazarene University

Arnold Rheingold
University of Delaware

Edward Shane
Morningside College

Estel Sprague
University of Cincinnati

Anthony Tanner
Austin College

Linda Thomas-Glover
Guilford Technical Community College

Richard Treptow
Chicago State University

Robert K. Wismer
Millersville University

John W. Hill
jwhill@pressenter.com

Ralph H. Petrucci

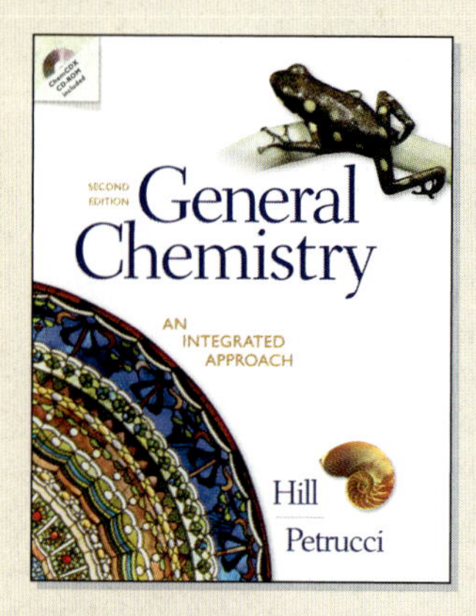

A Guide to Using this Text

You and your classmates come to this course with a variety of backgrounds and interests. Most of you plan to be scientists, engineers, or professionals in medicine or another life science. Knowledge of chemistry is essential to a true understanding of everything from DNA replication to drug discovery to computer chip design and manufacturing. Indeed, the chemical properties and principles you learn in this course will pervade almost every aspect of your private and professional lives. In this text, we provide you with both the principles and applications of chemistry that will help you in your professional practice and enrich your everyday life as well.

This text is rich in pedagogical aids, both within and at the ends of the chapters. We present this "user's guide" to the text to help you get the most out of this book and your course.

3.3 More on the Mole **89**

EXAMPLE 3.4

Calculate (**a**) the mass of a sodium atom, in grams; and (**b**) the number of Cl^- ions present in 1.38 g $MgCl_2$.

SOLUTION

a. The answer must have the unit "grams per sodium atom," that is, "g/Na atom." Thus, if we knew the mass of a certain number of Na atoms, our answer would simply be the mass divided by that number. But we do know these quantities: One mole of Na has a mass of 22.99 g and consists of 6.022×10^{23} Na atoms. As shown, the division (22.99 g/6.022×10^{23} Na atoms) is represented as the product of two factors, one involving the molar mass and the other, Avogadro's number.

$$? \text{ g/Na atom} = \frac{22.99 \text{ g}}{\cancel{1 \text{ mol Na}}} \times \frac{\cancel{1 \text{ mol Na}}}{6.022 \times 10^{23} \text{ Na atoms}} = 3.818 \times 10^{-23} \text{ g/Na atom}$$

Because there is only one naturally-occurring type of sodium atom (^{23}Na), what we have just calculated is truly the mass of a sodium atom. For elements that have two or more isotopes, our calculation would be a *weighted average* mass and not the mass of an atom of any particular isotope.

b. First we need the formula mass of $MgCl_2$ and then the molar mass. Once we have established the amount of $MgCl_2$ in moles, we can use Avogadro's number to determine the number of formula units (f.u.) of $MgCl_2$. Finally, we can use a factor (shown in red) that establishes the number of Cl^- ions per formula unit (f.u.).

$$? \text{ Cl}^- \text{ ions} = 1.38 \cancel{\text{ g MgCl}_2} \times \frac{\cancel{1 \text{ mol MgCl}_2}}{95.21 \cancel{\text{ g MgCl}_2}} \times \frac{6.022 \times 10^{23} \cancel{\text{ f.u.}}}{\cancel{1 \text{ mol MgCl}_2}} \times \frac{2 \text{ Cl}^- \text{ ion}}{1 \cancel{\text{ f.u.}}} = 1.75 \times 10^{22} \text{ Cl}^- \text{ ions}$$

EXERCISE 3.4A

Calculate (**a**) the weighted average mass, in grams, of a bismuth atom; (**b**) the weighted average mass, in grams, of a glycerol molecule, $CH_2OHCHOHCH_2OH$; (**c**) the number of molecules in 0.0100 g of nitrogen gas; and (**d**) the total number of atoms in 215 g of sucrose, $C_{12}H_{22}O_{11}$.

EXERCISE 3.4B

Calculate (**a**) the number of Br_2 molecules in 125 mL of liquid bromine ($d = 3.12$ g/mL); and (**b**) the number of liters of liquid ethanol ($d = 0.789$ g/mL) required to obtain a sample containing 1.00×10^{25} CH_3CH_2OH molecules.

PROBLEM-SOLVING NOTE

The mass of a single atom is tiny—very much less than 1 g. Keep this fact in mind to avoid mistakenly multiplying by Avogadro's number when you should divide.

PROBLEM-SOLVING NOTE

Example 3.4(b) illustrates the common practice of expressing molar mass and Avogadro's number with one more significant figure than the least precisely known quantity (1.38 g $MgCl_2$). This ensures that the precision of the calculated result is limited only by the least precisely known quantity.

◀ Examples

These worked problems help you build your problem-solving skills by showing you how to solve various types of problems. Study the Examples carefully to make sure you understand the model solution. Then start to master the problem-solving process by working the Exercises that follow.

◀ Problem-Solving Notes

These marginal notes highlight good problem-solving practices and warn you of common student misconceptions. These tips and techniques can help you avoid common pitfalls.

◀ Exercise A

Exercise A asks a question similar to that in the Example. By drawing on the model solution in the Example and your growing knowledge of chemistry, you will be able to create a strategy and solve the A Exercise.

▲ Exercise B

To solve Exercise B, you will need all the skills and knowledge derived from the Example and Exercise A. You may also need to apply a problem-solving technique or an idea you learned earlier. These Exercises help prepare you for solving more complex problems, such as those you might face on an exam.

EXAMPLE 11.4—A Conceptual Example

To keep track of how much gas remains in a cylinder, we can weigh the cylinder when it is empty, when it is filled, and after each use. In some cases, though, we can equip the cylinder with a pressure gauge and simply relate the amount of gas to the measured gas pressure. Which method should we use to keep track of the bottled propane, C_3H_8, in a gas barbecue?

SOLUTION

The propane is at a temperature below its T_c (369.8 K) and exists in the cylinder as a mixture of a liquid and vapor. As the fuel is consumed, the volume of liquid in the cylinder decreases, and that of the vapor increases. However, the vapor pressure of the liquid propane does not depend on the amounts of liquid and vapor present. The pressure will remain constant (assuming a constant temperature) as long as some liquid remains. Only after the last of the liquid has vaporized, will the pressure drop. Measuring the pressure doesn't tell us how much propane is left in the cylinder until it's almost gone. We would have to monitor the contents of the cylinder by weighing.

EXERCISE 11.4

Which of the two methods described above could you use if the fuel in the cylinder were methane, CH_4? Explain.

Conceptual Examples and Exercises

Conceptual Examples help you understand the most important ideas in your chemistry course. You will find that they focus on concepts rather than calculations, but they still require careful thought and study. Test your mastery of the material by examining the Conceptual Example and solving the Conceptual Exercise that follows it.

Estimation Examples and Exercises

Have you ever made a mistake using a calculator and come up with the wrong answer? Estimation Examples and Exercises help you learn to estimate the answer to a problem before you perform detailed calculations. This lets you determine at a glance if your calculated answer is a reasonable one.

EXAMPLE 6.10—An Estimation Example

Without doing detailed calculations, determine which of the following is a likely approximate final temperature when 100 g of iron at 100 °C is added to 100 g of water in a Styrofoam® cup calorimeter at 20 °C:

20 °C 30 °C 60 °C 70 °C

SOLUTION

Because we have the same mass of each substance, *if* the water and iron had the same specific heat, the rise in temperature of the water would be the same as the drop in temperature of the iron. The final temperature would be the average of 20 °C and 100 °C, or 60 °C. However, from Table 6.1 we see that the specific heat of water ($4.18\ J\ g^{-1}\ °C^{-1}$) is much greater than that of iron ($0.45\ J\ g^{-1}\ °C^{-1}$). It takes more heat to change the temperature of a given mass of water than of the same mass of iron. The final water temperature must be below 60 °C, but also it has to be above 20 °C, the initial temperature. The only possible approximate temperature of those given is 30 °C.

EXERCISE 6.10

Without doing detailed calculations, determine the final temperature if 200.0 mL of water at 80 °C is added to 100.0 mL of water at 20 °C.

Determining the mass percent of carbon in butane, C_4H_{10}.

1. C_4H_{10} — There are 4 mol C atoms and 10 mol H atoms in 1 mol butane.
2. $(4 \times 12.011)g = 48.044g$ — No. mol C in 1 mol butane; Molar mass of C; We multiply to find out the mass of C in 1 mol butane.
3. $(10 \times 1.0079)g = 10.079g$ — No. mol H in 1 mol butane; Molar mass of H; We find the mass of H in 1 mol butane by the same method.
4. $48.044g + 10.079g = 58.123g$ — Mass of 4 mol C atoms; Mass of 10 mol H atoms; Adding the mass of C and H gives us the mass of 1 mol of C_4H_{10} molecules.
5. $\frac{48.044g\ C}{58.123g\ C_4H_{10}} \times 100\% = 82.66\%\ C$ — Mass percent C; Here we divide the mass of 4 mol C by the mass of 1 mol butane.; We convert our answer to a percentage by multiplying by 100%.

Voice Balloons

Voice balloons help you understand each step in the solution to a problem, as well as why each step is necessary. Make sure you understand each step; don't just memorize them.

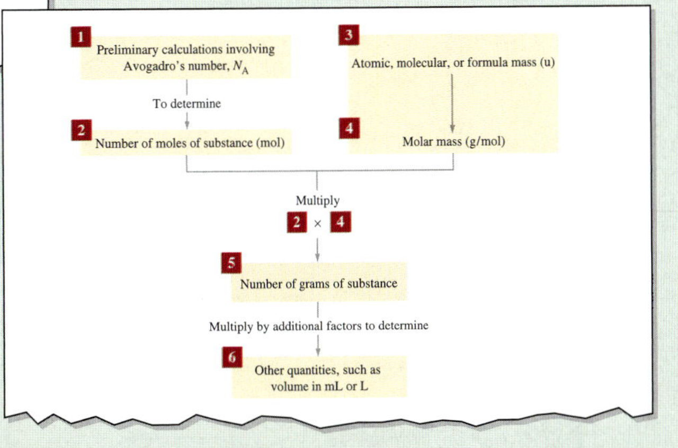

Flow Charts

Flow charts provide a visual outline of the problem-solving process. Use them to plan how to solve a related problem, to check your progress as you solve a problem, or to review your solution after it is complete.

monooxide, and dinitrogen *tetr*oxide, not dinitrogen *tetra*oxide. However, PI_3 is phosphorus *tri*iodide, not phosphorus *tri*odide.

- The names are further altered by adding prefixes such as *mono*, *di*, *tri*, and so on to denote the numbers of atoms of each element in the molecule (see Table 2.3). Thus, P_4S_3 is called *tetra*phosphorus *tri*sulf*ide*. Note that in these examples the prefix *mono-* is treated in a special way. We do not use it for the first-named element, but we do for the second. For example, the name for CO is carbon monoxide, *not* monocarbon monoxide.

EXAMPLE 2.6

Write the formula and name of a compound that has six oxygen atoms and four phosphorus atoms in its molecules.

APPLICATION NOTE

Is P_4S_3 just a curiosity chemical with a tongue twister name? No, indeed. It is the chemical on the head of a strike-anywhere match that is ignited through frictional heat. P_4S_3 does not spontaneously ignite at temperatures below 100 °C.

Application Notes

Application Notes highlight the intriguing ways in which we can put our knowledge of chemistry to work, touching on fields as diverse as medicine, engineering, and agriculture.

Essays

Essays focus on how we apply our chemical knowledge to solve real-life problems, and on historical topics of interest. These readings will help you see how chemistry affects everyday life and how we arrived at our current understanding of chemistry.

What Is a Low-Sodium Diet?

There are about 27.5 million people with hypertension (high blood pressure) in the United States. Physicians usually advise them to follow a low-sodium diet. Just what does that mean? Surely they are not being advised to reduce their consumption of sodium itself. Sodium is an extremely reactive metal that reacts violently with moisture. Sodium metal is not a part of anyone's diet. The concern is really with sodium *ion*, Na^+, but sodium ion taken alone is not a substance. It enters our diet in combination with anions in the form of ionic compounds, the principal one being sodium chloride—common table salt. Some people eat 6 or 7 grams of sodium chloride a day, most of it in prepared foods. Many snack foods, such as potato chips, pretzels, and corn chips, are especially high in salt. Most physicians recommend that people with hypertension restrict their salt intake. Some, but not all, physicians suggest that even people with normal blood pressure should eat less salt.

Ions differ greatly from the atoms from which they are formed. Sodium atoms make up an element that, although quite reactive, can exist by itself. Sodium *ions* are generally unreactive, but they come only in combination with anions. A metal atom and its cation are as different as a whole peach (atom) and a peach pit (ion). Unfortunately, the situation is confused when people talk about diets with too much "sodium" or of taking "calcium" for healthy teeth and bones. They really mean sodium *ions* and calcium *ions*. As scientists, we try to be more precise in our terminology.

Some familiar foods with high Na^+ content.

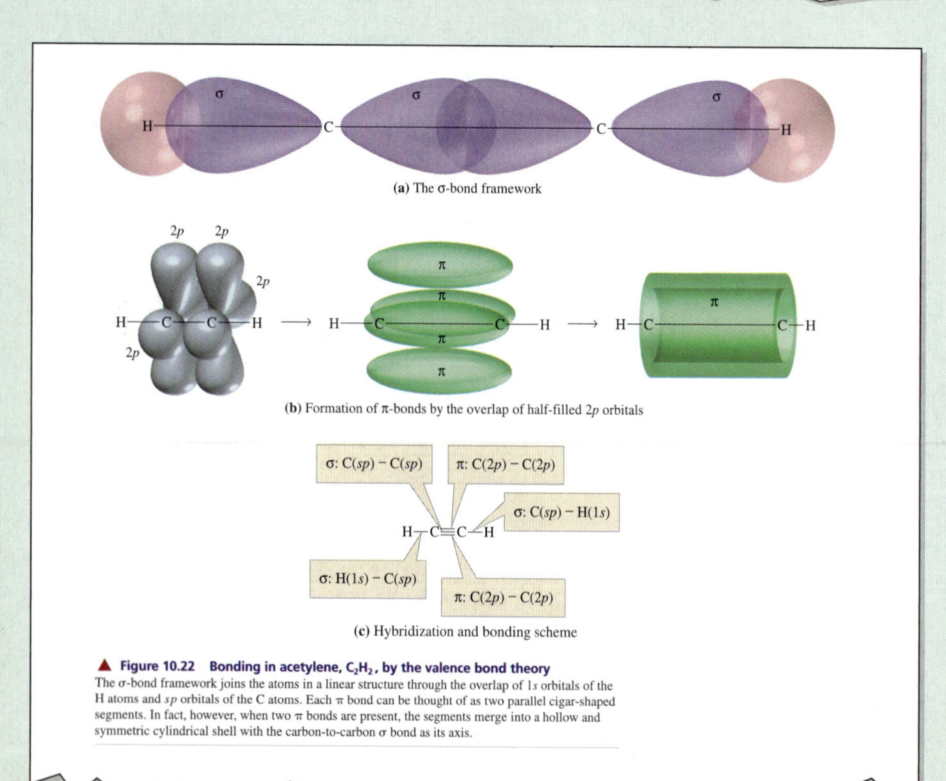

▲ **Figure 10.22 Bonding in acetylene, C_2H_2, by the valence bond theory**
The σ-bond framework joins the atoms in a linear structure through the overlap of 1*s* orbitals of the H atoms and *sp* orbitals of the C atoms. Each π bond can be thought of as two parallel cigar-shaped segments. In fact, however, when two π bonds are present, the segments merge into a hollow and symmetric cylindrical shell with the carbon-to-carbon σ bond as its axis.

Art

Study the art carefully; this will help you to visualize atoms, molecules, and chemical processes that cannot be seen with the unaided (and sometimes even with the aided) eye.

Review Questions

1. Describe what a gas is like at the molecular level.
2. Why don't the molecules of a gas settle to the bottom of their container?
3. Give a kinetic-molecular explanation of the origin of gas pressure.
4. What does a mercury barometer measure? How does it work?
5. Why is mercury (rather than water or another liquid) used as the fluid in barometers?
6. How does a manometer differ from a barometer? How does an open-end manometer differ from a closed-end manometer?

(a) an increase in temperature at constant volume

(b) a decrease in volume at constant temperature

(c) an increase in temperature coupled with a decrease in volume

16. According to the kinetic-molecular theory, **(a)** what change in temperature occurs as the molecules of a gas begin to move more slowly, on average? **(b)** What change in pressure occurs when the walls of the container are struck less often by molecules of the gas?
17. Container A has twice the volume but holds twice as many gas molecules as container B at the same temperature.

Review Questions

These end-of-chapter questions help you to review definitions and key concepts in the chapter without performing detailed calculations. You will find them useful in building your knowledge of chemistry and solidifying your understanding of key ideas.

Problems

Problems test your mastery of the problem-solving techniques presented in the chapter. They are arranged by topic and come in matched pairs (each odd- and even-numbered pair test the same concept). Problems may emphasize estimation skills or conceptual understanding. You can check your answers to odd-numbered problems in Appendix F. Work many of the Problems to develop the strong problem-solving skills that will help you succeed.

Mixtures of Gases

93. Only one of the accompanying sketches is a reasonable molecular view of a mixture of 1.0 g H_2 and 1.0 g He. Which is the correct view, and what is wrong with the other two?

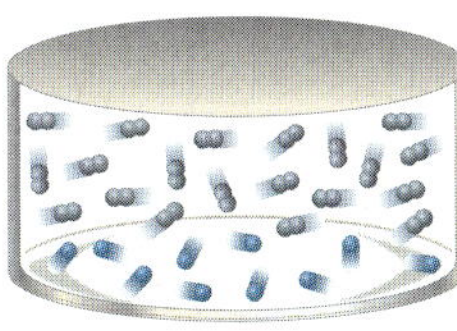

(a)

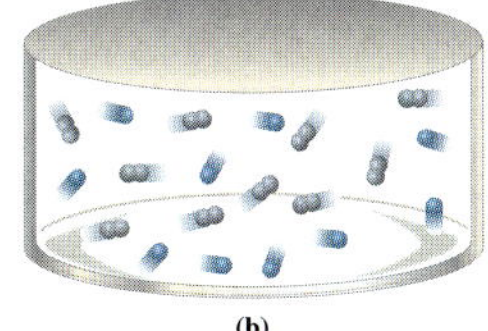

(b)

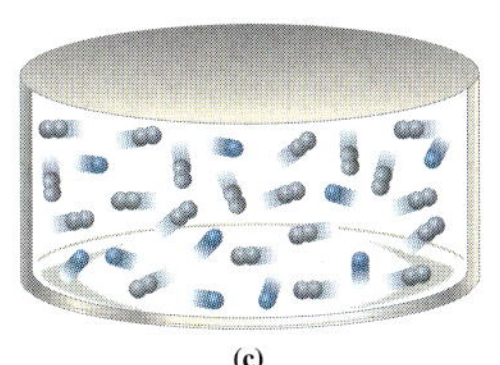

(c)

94. With reference to Problem 93, draw a molecular-level

Torr? [*Hint:* Recall that volume percent is the same as mole percent for ideal gas mixtures.]

97. Mixtures of helium and oxygen are used in scuba diving. What are **(a)** the mole fractions of the two gases, **(b)** their partial pressures, and **(c)** the total pressure in a mixture of 1.96 g He and 60.8 g O_2 confined in a 5.00-L tank at 25.0 °C?
98. A 267-mL sample of a mixture of noble gases at 25.0 °C contains 0.354 g Ar, 0.0521 g Ne, and 0.0049 g Kr. What are **(a)** the mole fractions of the three gases, **(b)** their partial pressures, and **(c)** the total gas pressure?
99. Oxygen is collected over water at 30 °C and a barometric pressure of 742 Torr. What is the partial pressure and mole fraction of $O_2(g)$ in the container?
100. An oxygen-helium gas sample, collected over water at 23 °C, exerts a total pressure of 758 Torr. Calculate the mole fraction of water vapor in the sample.
101. *Elodea* is a green plant that carries out photosynthesis under water.

$$6\ CO_2(g) + 6\ H_2O(l) \longrightarrow C_6H_{12}O_6(aq) + 6\ O_2(g)$$

In an experiment, some *Elodea* produce 122 mL of $O_2(g)$, collected over water at 743 Torr and 21 °C. What mass of oxygen is produced? What mass of glucose ($C_6H_{12}O_6$) is produced concurrently?

102. A 2.02-g sample of aluminum reacts with an excess of HCl(aq), and the liberated hydrogen is collected over water at a temperature of 24 °C. What is the *total* volume of the gas collected?

$$2\ Al(s) + 6\ H^+(aq) \longrightarrow 2\ Al^{3+}(aq) + 3\ H_2(g)$$

103. The reaction between carbon dioxide gas and sodium peroxide, $Na_2O_2(s)$, to produce solid sodium carbonate and oxygen gas, is used in submarines to replace expired

Additional Problems

111. In terms of pressure (P), volume (V), Kelvin temperature (T), and amount of gas (n), and in the manner of Figures 5.6 and 5.8, sketch a graph of each of the following.
    (a) V as a function of P, with T and n held constant
    (b) n as a function of P, with T and V held constant
    (c) T as a function of P, with V and n held constant
    (d) n as a function of T, with P and V held constant
112. Stephen Malaker of Cryodynamics, Inc. has developed a refrigerator that uses compressed helium as a refrigerant gas. A typical system uses 5.00 in^3 of He compressed to 195 psi at 20 °C. What mass of helium, in grams, is needed for one refrigerator?
113. In an attempt to verify Avogadro's hypothesis, small quantities of several different gases were weighed in 100.0-mL syringes. Masses were determined on an analytical balance. The following masses were obtained: 0.0080 g

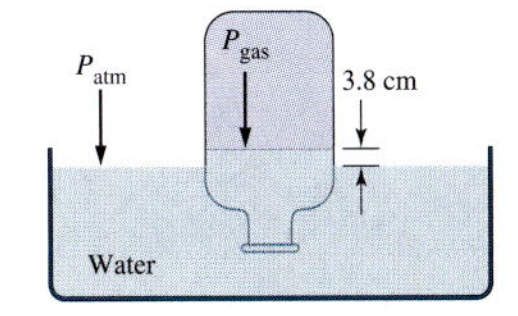

115. A 2.135-g sample of a gaseous chlorofluorocarbon (a type of gas implicated in the depletion of stratospheric ozone) occupies a volume of 315.5 mL at 739.2 mmHg and 26.1 °C. Analysis of the compound shows it to be 14.05% C, 41.48% Cl, and 44.46% F, by mass. What is the molecular formula of this compound?
116. The gaseous hydrocarbon 1,3-butadiene is used to make synthetic rubber. The following measurements were made

Additional Problems

Some of the Additional Problems are more challenging than the Problems, requiring a synthesis of concepts from multiple chapters. Others will help you to attain a stronger mastery of key concepts in this chapter. Additional Problems may also emphasize estimation skills or conceptual understanding. Answers to odd-numbered Additional Problems are given in Appendix F.

Tools to Help You Succeed in this Course

Explore, Expand, Experiment—The Multimedia Program

ChemCDX

The ChemCDX CD-ROM that accompanies the text includes animations of 50 key concepts (processes that are more easily understood in an animation than in a picture), each of which has an introduction and is followed by multiple-choice questions designed to gauge your understanding of the concepts presented. Icons in the margin of the text will refer you to ChemCDX when there is a relevant module for you to study.

its partial pressure. But the ice and liquid water are under atmospheric pressure (760 mmHg). Moreover, at the normal melting point liquid water contains some dissolved air and this affects the equilibrium temperature slightly. To have a true triple point, a system must consist of a *pure* substance existing *only under the pressure of its own vapor*. There can be no extraneous substances present (such as the gases in air).

Module Opening Screen ▶

Every chapter contains modules you can access, each corresponding to a key topic in the text chapter. When reading about a topic, you will always be able to find a related video or figure to help clarify.

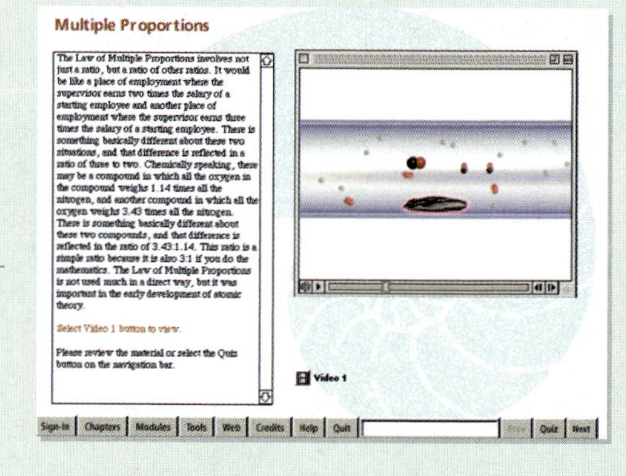

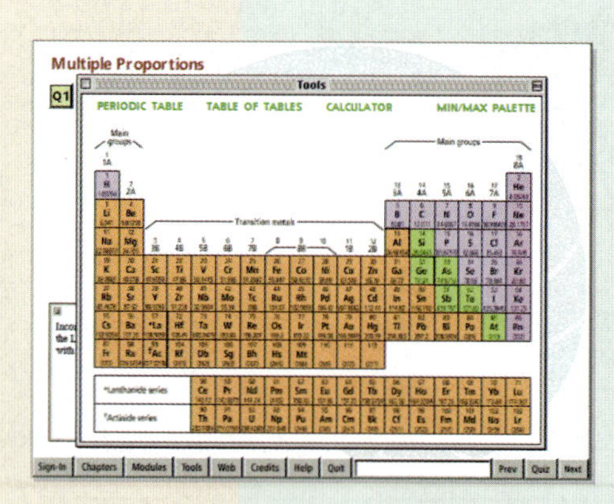

◀ Tools

At any time, you can access one of several tools, including the Periodic Table, a Table of Tables, and a calculator. Each can be used to work through the content and problems on ChemCDX.

Quiz ▶

Each module contains several quiz questions. If your answer is incorrect, you'll receive an explanation and a prompt pointing you to the corresponding section in the text.

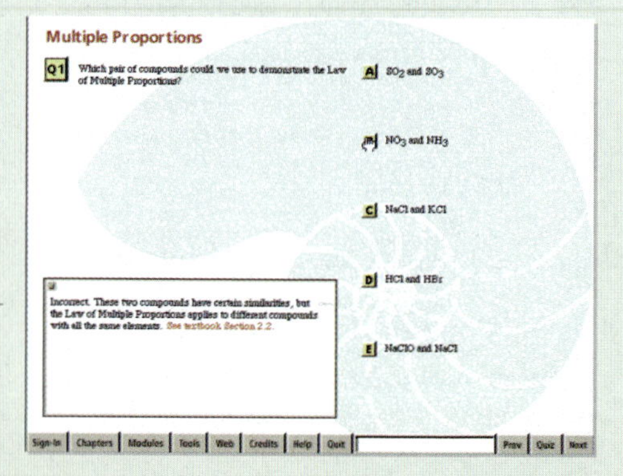

EXPLAIN • EXPLORE • EXPAND • EXAMINE

Companion Website

The interactive student Website that accompanies this text enables you to take multiple-choice quizzes (three different levels) and view a gallery of three-dimensional molecules. Feedback from quizzes includes detailed reading lists from the textbook. Icons in the textbook refer to the student Website and ChemCDX CD-ROM at appropriate points. You can enter the Companion Website directly at www.prenhall.com/~chem or through the ChemCDX CD-ROM.

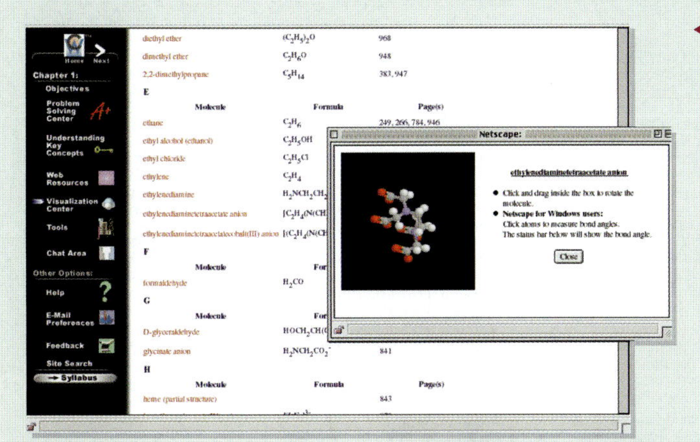

Visualization Center

Engage in tutorials and access an archive of animations tied to the text. Click on a molecule and watch it become "live"; you can re-size and rotate the three-dimensional molecules. All molecules are organized alphabetically and by chapter.

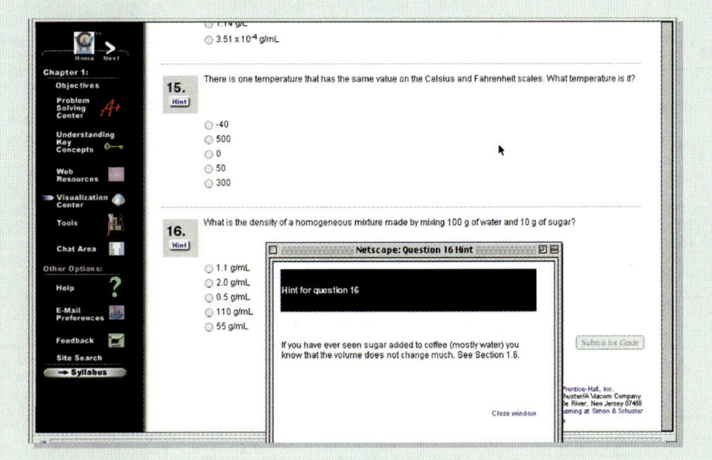

Problem Solving Center

Several quizzes and tests are tied to each chapter in the Problem Solving Center. As you work through these, you can access hints, get immediate feedback, and e-mail results to your instructor or TA.

Other study aids available with Hill/Petrucci's General Chemistry, second edition

Study Guide (0-13-918765-0)

For each chapter, the Study Guide includes a summary of key topics, an overview, worked examples, and expanded self-tests with answers.

Selected Solutions Manual (0-13-918740-5)

The Selected Solutions Manual contains solutions to all in-chapter problems, all conceptual exercises, and selected end-of-chapter problems.

Molecular Model Kit for General and Organic Chemistry (0-13-955444-0)

This model kit comes with a detailed set of instructions that will show you how to build hundreds of ball-and-stick models. You can also use this kit in your organic chemistry course.

If you are interested in purchasing any of these supplements, check with your bookstore to see if these items are in stock. If not, ask if they can be special ordered for you or call Prentice Hall at 1-800-947-7700.

EXTEND • EXCURSION • EXAMINE • EXCITE

Photo Credits

Chapter 1 Page 1: Philippe Plailly/Science Photo Library/ Photo Researchers, Inc. Page 2: Dr. E. R. Degginger; Tony Stone Images. Page 3: Serco International Fire Training Centre, Teesside College. Teesside Airport, Darlington, U.K. Page 5: Carey Van Loon. Page 5: Gregg Addams/Tony Stone Images. Page10: Carey Van Loon. Page 17: Tom Pantages. Page 22: Richard Megna/Fundamental Photographs. Page 23: Yoav Levy/Phototake. NYC Page 32 AP/Wide World Photos.

Chapter 2 Page 36: Stamp from the private collection of Professor C. M. Lang, photography by Gary L. Shulfer, University of Wisconsin, Stevens Point. "1993, Republic of Mali"; Scott Standard Postage Stamp Catalogue, Scott Pub. Co., Sidney, Ohio. Page 37: Dave Bartruff/Stock Boston. Page 38: Corbis-Bettman. Page 45: Science Photo Library/Photo Researchers, Inc. Page 46: Carey Van Loon. Page 54: Richard Megna/Fundamental Photographs. Page 58: Richard Megna/Fundamental Photographs (all). Page 59: Corbis-Bettmann.

Chapter 3 Page 81: Runk/Shoenberger/Grant Heilman Photography, Inc. Page 87: Carey Van Loon. Page 92: Richard Megna/Fundamental Photographs. Page 113: Richard Megna/Fundamental Photographs. Page 113 Carey Van Loon. Page 113: Carey Van Loon/Ralph H. Petrucci. Page 113: Carey Van Loon. Page 117: Carey Van Loon (all). Page 121: Richard Megna/Fundamental Photographs (all). Page 129: Richard Megna/ Fundamental Photographs.

Chapter 4 Page 132: Richard Megna/Fundamental Photographs. Page 134: Richard Megna/Fundamental Photographs; Page 134: Richard Megna/Fundamental Photographs. Page 134: Richard Megna/Fundamental Photographs. Page 141: Richard Megna/Fundamental Photographs. Page 141: Richard Megna/Fundamental Photographs. Page 141: Richard Megna/Fundamental Photographs. Page 143: Richard Megna/Fundamental Photographs. Page 143: Richard Megna/Fundamental Photographs. Page 143: Richard Megna/Fundamental Photographs. Page 146: Carey Van Loon Page 146 Carey Van Loon. Page 147: Richard Megna/Fundamental Photographs. Page 147: Richard Megna/Fundamental Photographs. Page 147: Richard Megna/Fundamental Photographs. Page 148: Richard Megna/Fundamental Photographs. Page 150: Carey Van Loon. Page 151: Dr. E. R. Degginger. Page 151: CLO Research. Page 152: Richard Megna/ Fundamental Photographs. 155: Richard Megna/Fundamental Photographs. Page 156: Richard Megna/Fundamental Photographs. Page 156: Richard Megna/Fundamental Photographs. Page 162: Richard Megna/Fundamental Photographs. Page 162: Richard Megna/Fundamental PhotographsPage Page 162: Richard Megna/Fundamental Photographs. Page 163: Carey Van Loon. Page 163: Carey Van Loon. Page 163: Carey Van Loon. Page 165: Richard Megna/Fundamental Photographs. Page 165: Richard Megna/Fundamental Photographs. Page 165: Richard Megna/Fundamental Photographs. Page 166: Alan Levenson/Tony Stone Images. Page 167: Stephen Frisch/Stock Boston. Page 168: Carey Van Loon. Page 173: Richard Megna/Fundamental Photographs. Page 173: Richard Megna/Fundamental Photographs. Page 173: Richard Megna/Fundamental Photographs. Page 173: Richard Megna/Fundamental Photographs.

Chapter 5 Page 177: Jet Propulsion Laboratory/NASA Headquarters. Page 178: Dr. E. R. Degginger. Page 185: Stamp from the private collection of Professor C. M. Lang, photography by Gary L. Shulfer, University of Wisconsin, Stevens Point. "Ireland;" Scott Standard Postage Stamp Catalogue, Scott Pub. Co., Sidney, Ohio. Page 190: Richard Megna/ Fundamental Photographs. Page 190: Richard Megna/Fundamental Photographs. Page 191: Stamp from the private collection of Professor C. M. Lang, photography by Gary L. Shulfer, University of Wisconsin, Stevens Point. "Italy #714 (1956);" Scott Standard Postage Stamp Catalogue, Scott Pub. Co., Sidney, Ohio. Page 201: Archive/Photo Researchers, Inc. Page 192: Carey Van Loon. Page 219: Richard Megna/Fundamental Photographs.

Chapter 6 Page 226: NASA Headquarters. Page 228: Ian Murphy/Tony Stone Images. Page 228: Richard Megna/Fundamental Photographs. Page 230: Richard Megna/Fundamental Photographs. Page 231: Richard Megna/Fundamental Photographs. Page 232: Richard Megna/Fundamental Photographs. Page 266: Carey Van Loon.

Chapter 7 Page 275: Account Phototake/Phototake NYC. Page 277: Science Photo Library/Photo Researchers, Inc. Page 279: Stamp from the private collection of Professor C. M. Lang, photography by Gary L. Shulfer, University of Wisconsin, Stevens Point. Scott Standard Postage Stamp Catalogue, Scott Pub. Co., Sidney, Ohio. Page 277: Carey Van Loon. Page 281: Stamp from the private collection of Professor C. M. Lang, photography by Gary L. Shulfer, University of Wisconsin, Stevens Point. "New Zealand;" Scott Standard Postage Stamp Catalogue, Scott Pub. Co., Sidney, Ohio. Page 288: UPI/Corbis-Bettman. Page 289: Carey Van Loon. Page 289: Wabash Instrument Corp./Fundamental Photographs. Page 290: Department of Physics, Imperial College/Science Photo Library/Photo Researchers, Inc. Page 290: Carey Van Loon. Page 290: Paul Silverman/Fundamental Photographs. Page 291: Paul Silverman/Fundamental Photographs. Page 295: Science Photo Library/Photo Researchers, Inc. Page 301: M. I. Walker/Photo Researchers, Inc. Page 301: David M. Phillips/Visuals Unlimited. Page 301: UPI/Corbis-Bettman. Page 302: Corbis-Bettman. Page 308: Richard Megna/Fundamental Photographs Page 308: Courtesy of International Business Machines Corporation. Unauthorized use not permitted (both). Page 309: Michael Matthews, Joint Institute for Laboratory Astrophysics, University of Colorado and National Institute for Standards and Technology.

Chapter 8 Page 316: David Nunuk/Science Photo Library/Photo Researchers, Inc. Page 328: From Philosophical Magazine, 1913, with permission of Edward Arnold Publishers, London. Page 348: Richard Megna/Fundamental Photographs. Page 348: Richard Megna/Fundamental Photographs. Page 348: Richard Megna/Fundamental Photographs. Page 348: Richard Megna/Fundamental Photographs. Page 348: Richard Megna/Fundamental Photographs. Page 348: Richard Megna/Fundamental Photographs. Page 348: Carey Van Loon. Page 348: Carey Van Loon. Page 349: Carey Van Loon. Page 349: Dr. E. R. Degginger Page 349: Richard Megna/Fundamental Photographs.

Chapter 9 Page 358: E. R. Degginger/Bruce Coleman, Inc. Page 371: Richard Megna/Fundamental Photographs. Page 360: University of California Berkeley. Page 362: Carey Van Loon. Page 363: Albert Copley/Visuals Unlimited. Page 372: Science Photo Library/Photo Researchers, Inc. Page 398: Richard Megna/Fundamental Photographs. Page 398: Richard Megna/Fundamental Photographs. Page 399: The Proctor & Gamble Company. Page 400: Bob Masini/Phototake NYC.

Chapter 10 Page 408: Richard Megna/Fundamental Photographs. Page 412: Carey Van Loon. Page 412: Carey Van Loon. Page 412: Carey Van Loon. Page 412: Carey Van Loon. Page 412: Carey Van Loon. Page 410: Carey Van Loon Page 421: Dugald Bremner/Tony Stone Images. Page 448: Stamp from the private collection of Professor C. M. Lang, photography by Gary L. Shulfer, University of Wisconsin, Stevens Point. Scott Standard Postage Stamp Catalogue, Scott Pub. Co., Sidney, Ohio. Page 449: Hulton Getty Picture Collection/Tony Stone Images.

Chapter 11 Page 458: C. Haagner/Bruce Coleman, Inc. Page 467: Carey Van Loon. . Page 467: Ulrich Sapountsis/Okapia/Photo Researchers, Inc. Page 471: Carey Van Loon. Page 473: Carey Van Loon. Page 477: Blair Seitz/Photo Researchers, Inc. Page 477: Tom Pantages. Page 478: Carey Van Loon. Page 484: Jim Zuckerman/Westlight. Page 488: Carey Van Loon. Page 488: Herman Eisenbeiss/Photo Researchers, Inc. Page 489: Carey Van Loon. Page 489: Richard Megna/Fundamental Photographs. Page 489: Ralph H. Petrucci. Page 490: Kristen Brochmann/Fundamental Photographs. Page 490: Kristen Brochmann/Fundamental Photographs. Page 496: Carey Van Loon. Page 497: Carey Van Loon. Page 497: Carey Van Loon. Page 498: Carey Van Loon. Page 498: Carey Van Loon. Page 504: Carey Van LoonPage 508: Richard Megna/Fundamental Photographs. Page 508: Richard Megna/Fundamental Photographs.

Chapter 12 Page 509 (All three) David Phillips/Science Source/Photo Researchers, Inc. . Page 510: L.C. Clark, Jr., University of Michigan Medical School. Page 512: Stacy Pick/Stock Boston. Page 513: Richard Megna/Fundamental Photographs. Page 515: Richard Megna/Fundamental Photographs. Page 515: Richard Megna/Fundamental Photographs. Page 521: Richard Megna/Fundamental Photographs. Page 521: Ken Graham/Bruce Coleman, Inc. Page. 525: Richard Megna/Fundamental Photographs. Page 526: Myrleen Ferguson/PhotoEdit. Page 526: Richard Megna/Fundamental Photographs. Page 526: Richard Megna/Fundamental Photographs. Page 526: Richard Megna/Fundamental Photographs. Page 529: Charles D. Winters/Photo Researchers, Inc. Page 530: F. Stuart Westmorland/Photo Researchers, Inc. Page 537: James Lester/Photo Researchers, Inc. Page 539: Tom J. Ulrich/Tony Stone Images. Page 541: Veronika Burmeister/Visuals Unlimited. Page 542: Science Photo Library/Photo Researchers, Inc. Page 543: Omikron/Science Source/Photo Researchers, Inc. Page 544: Stephen Frisch/Stock Boston. Page 545: Carey Van Loon. Page 545: Carey Van Loon. Page 551: Richard Megna/ Fundamental Photographs.

Chapter 13 Page 552: Simon D. Pollard/Photo Researchers, Inc. Page 553: James Prince/Photo Researchers, Inc. Page 553: Richard Megna/Fundamental Photographs. Page 554: Science VU/Visuals Unlimited. Page 576 and Page 582: Peter Pearson/Tony Stone Images. Page 584: Carey Van Loon.

Chapter 14 Page 594: Tom Bochsler/Simon & Schuster/PH College. Page 606: William E. Ferguson. Page 609: Wendell Metzen/Bruce Coleman, Inc. Page 610: Stamp from the private collection of Professor C. M. Lang, photography by Gary J. Shulfer, University of Wisconsin, Stevens Point. Scott Standard Postage Stamp Catalogue, Scott Pub. Co., Sidney, Ohio. Page 610: Stamp from the private collection of Professor C. M. Lang, photography by Gary J. Shulfer, University of Wisconsin, Stevens Point. Scott Standard Postage Stamp Catalogue, Scott Pub. Co., Sidney, Ohio. Page 631: Carey Van Loon. Page 606: Michelangelo (1475-1564) David - p. (testa di profilo). Accademia Firenze. Scala/Art Resource.

Chapter 15 Page 632: Okapia-Frankfurt/Photo Researchers, Inc. Page 633: Carey Van Loon. Page 646: Richard Megna/Fundamental Photographs. Page 650: Richard Megna/Fundamental Photographs. Page 658: Carey Van Loon. Page 660: Carey Van Loon. Page 661: Richard Megna/Fundamental Photographs. Page 661: Richard Megna/Fundamental Photographs. Page 661: Richard Megna/Fundamental Photographs. Page 661: Richard Megna/Fundamental Photographs. Page 664: Carey Van Loon. Page 673: Carey Van Loon. Page 674: Dan McCoy/Rainbow. Page 675: Richard Megna/Fundamental Photographs. Page 676: Richard Megna/Fundamental Photographs. Page 676: Richard Megna/Fundamental Photographs. Page 676: Richard Megna/Fundamental Photographs. Page 676: Richard Megna/Fundamental Photographs.

Chapter 16 Page 695: Carey Van Loon. Page 696: Science Source/Photo Researchers, Inc. Page 703: Richard Megna/Fundamental Photographs. Page 704: Carey Van Loon. Page 704: Carey Van Loon. Page 706: Manfred Kage/Peter Arnold, Inc. Page 708: Richard Megna/Fundamental Photographs. Page 708: Richard Megna/Fundamental Photographs. Page 710: NYC Parks Photo Archive/Fundamental Photographs. Page 710: Kristen Brochmann/ Fundamental Photographs. Page 711: Carey Van Loon. Page 712: Max Listgarten/Visuals Unlimited. Page 713: Richard Megna/Fundamental Photographs. Page 714: Richard Megna/Fundamental Photographs. Page 718: Richard Megna/Fundamental Photographs. Page 720: Richard Megna/Fundamental Photographs. Page 720: Richard Megna/Fundamental Photographs. Page 722: Carey Van Loon. Page 729: Richard Megna/ Fundamental Photographs.

Chapter 17 Page 731: Richard A. Cooke III/Tony Stone Images. Page 732: J. A. Strandberg/Fire Department, City of New York. Page 733: Carey Van Loon. Page 735: Kristen Brochmann/Fundamental Photographs. Page 735: Kristen Brochmann/Fundamental Photographs. Page 735: Kristen Brochmann/Fundamental Photographs. Page 735: Kristen Brochmann/ Fundamental Photographs. Page 737: Carey Van Loon. Page 739: Eric Draper/AP/Wide World Photos. Page 743: Library of Congress. Page 754: Tom Bochsler/Simon & Schuster/PH CollegePage. 765: JPL/NASA Headquarters.

Chapter 18 Page 766: John Mead/Science Photo Library. Page 767: Carey Van Loon. Page 767: Carey Van Loon. Page 773: Carey Van Loon. Page 782: Carey Van Loon. Page 785: Carey Van Loon. Page 790: Science Photo Library/Photo Researchers, Inc. Page 794: Carey Van Loon. Page 797: Alupower, Inc. Page 797: H. Mock/Dr. R. Krauss/Mercedes-Benz of North America, Inc., Photo Library. Page 798: Richard Megna/Fundamental Photographs. Page 799: Robert Erlbacher/Missouri Dry Dock and Repair Company, Inc. Page 800: Carey Van Loon. Page 804: Library of Congress. Page 807: The Stock Market. Page 814: Carey Van Loon.

Chapter 19 Page 817: © The Trustees of the National Museums of Scotland. Page 823: Science VU/NLM/Visuals Unlimited. Page 823: William E. Ferguson. Page 826: Gianni Tortoli/Science Source. Page 828: Corbis-Bettmann. Page 830: AP/Wide World Photos. 19.08U Kevin Schafer/Peter Arnold, Inc. Page 839: Yoav Levy/Phototake NYC. Page 842: Runk/Schoenberger/Grant Heilman Photography, Inc. Page 836: Sylvain Coffie/Tony Stone Images.

Chapter 20 Page 848: Dr. E. R. Degginger. Page 849: Royal Observatory, Edinburgh, Scotland/Anglo-Australian Telescope Board/Science Photo Library/Photo Researchers, Inc. Page 850: Dr. E. R. Degginger. Page 851: Richard Megna/Fundamental Photographs. Page 853: H. Mock/Dr. R. Krauss/Mercedes-Benz of North America, Inc., Photo Library. Page 854: Dr. E. R. Degginger. Page 854: Dr. E. R. Degginger. Page 855: Richard Megna/Fundamental Photographs. Page 856: VU/Mark E. Gibson/Visuals Unlimited. Page 857: Department of Clinical Radiology, Salisbury District Hospital/Science Photo Library/Photo Researchers, Inc. Page 860: Tom Pantages. Page 860: James L. Almos/Peter Arnold, Inc. Page 860: David Cavagnaro/Peter Arnold, Inc. Page 863: E. Nagele/FPG International LLC. Page 866: Richard Megna/Fundamental Photographs. Page 867: Portland Cement Association. Page 868: Michael Collier. Page 869: Howard Sochurek/The Stock Market. Page 870: Richard Megna/Fundamental Photographs. Page 870: Francoise Sauze/Science Photo Library/Photo Researchers, Inc. Page 871: J. Smith, U. of Chicago/Phototake NYC. Page 872: Richard Megna/Fundamental Photographs. Page 872: Richard Megna/Fundamental Photographs.

Chapter 21 Page 879: NASA Headquarters. Page 881: Richard Megna/Fundamental Photographs. Page 883: Photo courtesy of U.S. Borax Inc. Page 884: UPI/Corbis/Bettmann. Page 885: Corbis-Bettmann. Page 886: Dr. E. R. Degginger. Page 887: Carey Van Loon. Page 888: Dorothy Littell Greco/Stock Boston. Page 888: Vaughan Fleming/Science Photo Library/Photo Researchers, Inc. Page 889: Paul Silverman/Fundamental Photographs. Page 889: Russell Munson/The Stock Market. Page 890: J. & L. Weber/Peter Arnold, Inc. Page 892: Robert Essel/The Stock Market. Page 893: M. Angelo/Westlight. Page 894: Paul Silverman/Fundamental Photographs. Page 895: Carey Van Loon. Page 896: Water container. Lead, Roman, 1st-3rd century C.E. Museo Archeologico Nazionale, Naples, Italy. Photograph © Erich Lessing/Art Resource. Page 896: Hank Morgan/ Rainbow. Page 898: Johnson Matthey Plc. Page 898: Grant Heilman/Grant Heilman Photography, Inc. Page 899: Jessica Wecker/Photo Researchers, Inc. Page 900: NASA Headquarters. Page 901: Dr. E. R. Degginger. Page 904: Novosti/Sovfoto/Eastfoto. Page 905: OZ Tech International. Page 906: Space Telescope Science Institute/NASA/Science Photo Library/Photo Researchers, Inc. Page 906: Jeffrey A. Scovil. Page 906: Tom Bochsler/Simon & Schuster/PH College. Page 906: Stephen Frisch/Stock Boston. Page 906: Carey Van Loon. Page 907: Bill Pierce/Rainbow. Page 907: Carey Van Loon. Page 908: Tom Pantages. Page 909: Richard Megna/Fundamental Photographs. Page 913: Argonne National Laboratory. Page 913: Tom Pantages. Page 914: Lili Lakich (American, b. 1944), "And Justice for All," 1990. Aluminum, copper, brass, etched glass, glass lens, and argon with mercury, neon, and helium. 144 in. x 120 in. x 16 in. (366 cm x 305 cm x 41 cm). Commissioned by Ace Parking/The Knoll Co. for Columbia Square, San Diego, California. The Museum of Neon Art, Los Angeles. Photo by Jeff Atherton.

Chapter 22 Page 921: Jan Halaska/Photo Researchers, Inc. Page 925: Phil Degginger/Dr. E. R. Degginger. Page 926: Rich Chisholm/The Stock Market. Page 926: Richard Megna/Fundamental Photographs. Page 927: N.H. (Dan) Cheatham/DRK Photo. Page 927: Carey Van Loon. Page 927: Carey Van Loon. Page 928: Michael Dalton/Fundamental Photographs. Page 928: Carey Van Loon. Page 928: Carey Van Loon. Page 928: Carey Van Loon. Page 929: Carey Van Loon. Page 929: Carey Van Loon. Page 929: Carey Van Loon. Page 930: Science VU/Visuals Unlimited. Page 932: Property of AT&T Archives, reprinted with permission of AT&T. Page 932: Gabe Palmer/The Stock Market. Page 934: Dale Boyer/Photo Researchers, Inc. Page 935: Jeff Greenberg/Omni-Photo Communications, Inc. Page 937: Carey Van Loon. Page 937: Science Photo Library/Photo Researchers, Inc. Page 950: Carey Van Loon. Page 950: Carey Van Loon. Page 951: Carey Van Loon.

Chapter 23 Page 957: Michel Viard/Peter Arnold, Inc. Page 958: Richard Megna/Fundamental Photographs. Page 959: Brownie Harris/The Stock Market. Page 962: PhotoDisc, Inc. Page 966: Phil Degginger/Dr. E. R. Degginger. Page 969: Thomas Kitchin/Tom Stack & Associates. Page 969: Phil Degginger/Dr. E. R. Degginger. Page 970: Ed Bock/The Stock Market. Page 970: Arthur Beck/Photo Researchers, Inc. Page 971: Richard Megna/Fundamental Photographs. Page 972: Philippe Colombi/PhotoDisc, Inc. Page 974: Phil Degginger/Dr. E. R. Degginger. Page 976: Corbis-Bettmann. Page 978: Frank LaBua/Simon & Schuster/PH College. Page 979: Andrew Syred/Science Photo Library/Photo Researchers, Inc. Page 980: Don W. Fawcett/Visuals Unlimited. Page 980: Biophoto Associates/Photo Researchers, Inc. Page 984: A. Barrington Brown. Page 995: Scott Camazine/Photo Researchers, Inc. Page 997: AP/Wide World Photos. Page 998: Dr. Jeremy Burgess/Science Photo Library/Photo Researchers, Inc. Page 1004: Dr. Morley Read/Science Photo Library/Photo Researchers, Inc.

Chapter 24 Page 1013: IBM Research/Peter Arnold, Inc. Page 1014: Jeffrey A. Scovil. Page 1014: Karl Hartmann/Sachs/Phototake NYC. Page 1014: Bruce Cairncross, Rand Afrikaans University. Page 1015: Kennecott Utah Copper Corporation. Page 1016: Kristen Brochmann/Fundamental Photographs. Page 1017: Phyllis Picardi/Stock Boston. Page 1019: Elkay Manufacturing Company. Page 1019: Ken Sherman/Phototake NYC. Page 1020: John Cancalosi/Stock Boston. Page 1021: United McGill Corporation. Page 1023: Robert Semeniuk/The Stock Market. Page 1027: Richard Megna/Fundamental Photographs. Page 1028: IBM Research/Peter Arnold, Inc. Page 1030: Miriam F. Mednick Rothman. Page 1030: Dave Jacobs/Tony Stone Images. Page 1032: Dr. E. R. Degginger. Page 1032: Richard Megna/Fundamental Photographs. Page 1035: Bob Daemmrich/Stock Boston. Page 1036: Richard Megna/Fundamental Photographs. Page 1037: Photograph courtesy of the Hagley Museum and Library, Wilmington, Delaware. Page 1038: Chris Hartlove. Page 1042: Southern Illinois University/Visuals Unlimited. Page 1042: Southern Illinois University/Visuals Unlimited. Page 1042: Smith & Nephew Richards.

Chapter 25 Page 1047: David H. Smith. Page 1050: Norman Owen Tomalin/Bruce Coleman, Inc. Page 1050: David Nunuk/Science Photo Library/Photo Researchers, Inc. Page 1051: Kristen Brochmann/Fundamental Photographs. Page 1052: D. Cavagnaro/Visuals Unlimited. Page 1052: Richard Megna/Fundamental Photographs. Page 1053: Ed Degginger/Bruce Coleman, Inc. Page 1057: Science VU/Visuals Unlimited. Page 1057: Stephanie Maze/National Geographic Society. Page 1058: Delphi Automotive Systems. Page 1058: Marty Corando/DRK Photo. Page 1059: Dr. Gerald L. Fisher/Science Photo Library/Photo Researchers, Inc. Page 1062: NASA Headquarters. Page 1062: M. C. Chamberlain/DRK Photo. Page 1064: Douglas Peebles/Westlight. Page 1069: Mark C. Burnett/Stock Boston. Page 1070: Ford Madox Brown (1821-1893), Dalton Collecting Marsh Fire Gas. Mural at Manchester City Art Galleries, Manchester, England. Photo by John Ingram. Page 1076: Jeff Greenberg/PhotoEdit. Page 1076: Laurie Campbell/Tony Stone Images. Page 1076: Dr. E. R. Degginger. Page 1081: Paul Cherfils/Tony Stone Images. Page 1081: Phil Degginger/Dr. E. R. Degginger. Page 1082: Courtesy of UKAEA. Page 1082: Courtesy of UKAEA.

Appendix Page A12 Richard Megna/Fundamental Photographs. Page A14 Carey Van Loon. Page A14 Manfred Kage/Peter Arnold, Inc.

Chemistry: Matter and Measurement

CHAPTER

1

CONTENTS

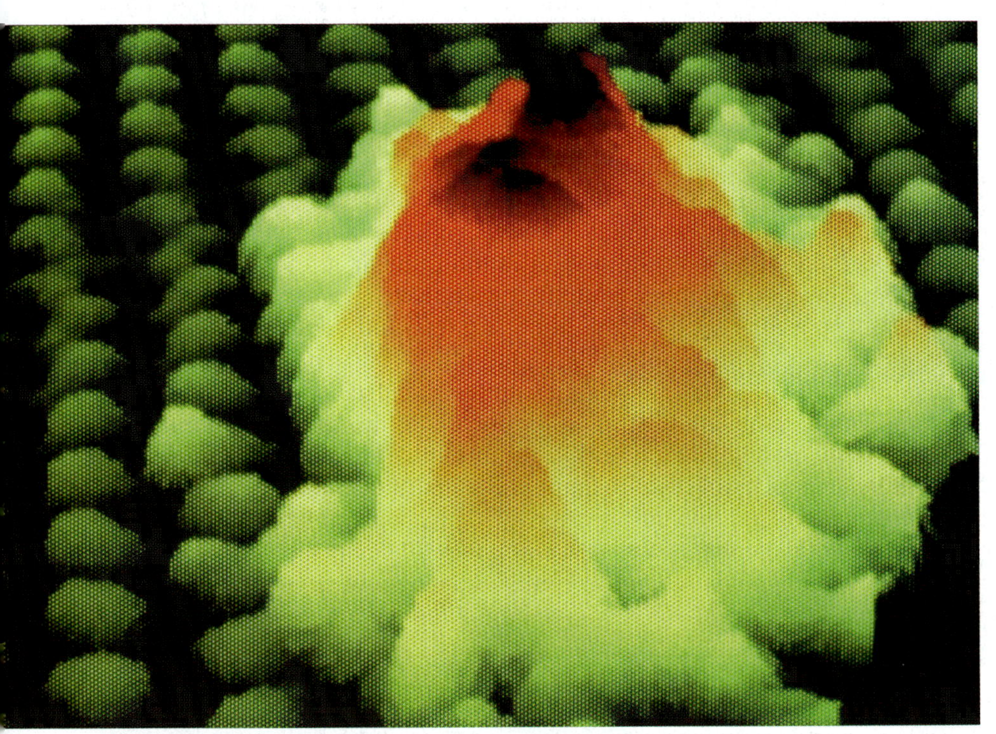

False-color image, obtained by a technique called scanning tunneling microscopy, of a cluster of gold atoms (yellow, orange, and red) on a surface of carbon atoms (green). The idea that all matter is made of atoms is over 2000 years old, but images of atoms have been available for only a few decades.

Today, most people in the industrialized nations have a higher standard of living than the human race has ever known: more nutritious food, better health, greater wealth. Much of this prosperity is due to chemistry. Chemistry enables us to design all sorts of materials: drugs to fight disease; pesticides to protect our health and crops; fertilizers to grow abundant food; fuels for transportation; fibers to provide comfort and variety in clothes; building materials for affordable housing; plastics to package food, replace worn-out body parts, and stop bullets; sports equipment to enrich our leisure time; and much more.

Chemistry also helps us to comprehend the nature of our environment, our universe, and ourselves. It provides essential information about issues such as global warming and ozone depletion over the polar regions. Chemistry plays a vital role in our understanding and treatment of diseases such as cancer and AIDS, and it helps unravel the mysteries of the human mind. In fact, the theories of chemistry

● **ChemCDX**
When you see this icon, there are related animations, demonstrations, and exercises on the CD accompanying this text.

illuminate our understanding of the material world from tiny atoms to giant galaxies. It is a journey we have barely begun, but we can be sure that a knowledge of chemistry will light the path toward a better understanding of our natural world.

1.1 Chemistry: Principles and Applications

Chlorine gas
Chlorine is one of 112 fundamental kinds of matter called chemical *elements*. A chemical *compound* is composed of two or more elements in fixed proportions. Elements and compounds are discussed further in Section 1.2.

The principles of chemistry that we consider in this book have many useful applications. Exploring the practical side of chemistry has stimulated the discovery of many new principles. In chemistry, theory and applications are interwoven like the threads of a fine fabric. To illustrate, let's look briefly at chlorine, one of the more familiar chemical elements.

Chlorine is a pale yellow-green gas at room temperature, but it is a gas that most people will never see. In nature, chlorine is found only combined with other elements, such as with sodium in sodium chloride—common table salt. Seawater is 3% salt, and blood serum is about 0.8% salt. Salt, and hence the chlorine it contains, is essential to life. Although elemental chlorine was discovered in 1774, it did not become an important commercial chemical until late in the nineteenth century when an inexpensive method was developed for obtaining it from salt. Currently, the chemical industry produces some 10,000 chlorine-containing substances that are used in such varied materials as bleaches, flame retardants, pesticides, drugs, solvents, and plastics.

Perhaps the most familiar use of chlorine is in disinfecting water. Water-borne diseases, such as typhoid fever (responsible for 35,000 deaths in the United States in 1900), are all but eliminated when water supplies are treated with chlorine. This use of chlorine does have a drawback, however, in that it converts some dissolved substances into minute amounts of chlorinated compounds that are suspected of causing cancer. These tiny amounts can be detected by only highly sophisticated methods of analysis, and potential problems with the use of chlorine have only recently become apparent.

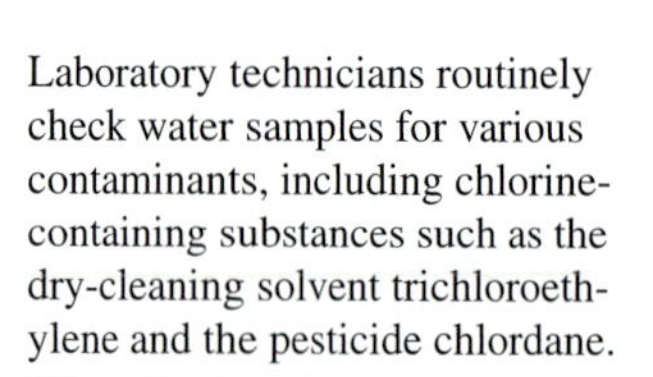

Laboratory technicians routinely check water samples for various contaminants, including chlorine-containing substances such as the dry-cleaning solvent trichloroethylene and the pesticide chlordane.

Trace amounts of toxic substances called dioxins are formed when chlorine-containing plastics are burned and when chlorine is used as a bleach in the pulp and paper industry. Dioxins harm fish and wildlife and perhaps even humans. Still other chlorine compounds that cause environmental concerns are the CFCs (chlorofluorocarbons), used in many refrigerators and air conditioners and once used in making foamed plastics. In the stratosphere, CFCs release chlorine atoms that most atmospheric scientists think react with and thus destroy some of the ozone found there. Ozone, a form of oxygen, absorbs ultraviolet light and protects life on Earth from this harmful component of sunlight. Interestingly, scientists used chemical

Halons, compounds that contain bromine as well as chlorine, have long been used in sprays to extinguish fires, particularly those on aircraft. Halons, like CFCs, have been implicated in the depletion of the ozone layer. To date, no substitutes have been found that have all the desirable properties of the halons: essentially nontoxic, nonconducting, noncorrosive, and leaving no residue.

reactions in the laboratory to *predict* the ozone problem several years before ozone depletion in the stratosphere was actually detected and confirmed.

Scientists and others have carefully thought further about many of chlorine's applications, and certain chlorine-containing substances are no longer used. The once common insecticide DDT has been phased out, and substances known as PCBs (polychlorinated biphenyls) are no longer used in printing inks and electrical transformers. CFCs are being replaced by other refrigerants that are less destructive of stratospheric ozone. Chlorine dioxide and sodium chlorite do not form objectionable chlorinated compounds and are replacing chlorine in the pulp and paper industry. Ozone, which has fewer undesirable side effects, is a potential replacement for chlorine in water treatment.

The total banning of chlorine-containing materials from our lives is neither possible nor even desirable. For example, even though ozone kills microorganisms as well as chlorine, water treated with ozone can become contaminated again as it enters the distribution system. Chlorine, in contrast, has a residual disinfectant action that extends beyond the treatment plant. And for some uses, there are no known replacements for chlorine and its compounds, such as in the cancer-fighting drug cisplatin.

Chemical knowledge was used to make the chlorine products that have provided so many benefits; chemical knowledge played a large role in revealing the negative side of these products; and even more chemical knowledge will be required in the search for suitable substitutes. The applications of chemistry, much like the science itself, undergo constant change.

1.2 Getting Started: Some Key Terms

We will introduce new terms only as we need them in the text. The few we define in this section are those we need to begin our study. Many of them may already be part of your vocabulary.

Chemistry is a study of the composition, structure, and properties of matter and of changes that occur in matter. What is matter? It is the stuff things are made of. Sometimes we study matter at the *macroscopic* level, in which case we deal with quantities large enough to be seen by the unaided eye. At other times, we deal with matter at the *microscopic* level, where particles are so small that they can be viewed only with special instruments such as the scanning tunneling microscope that made the image at the beginning of this chapter. It is easy to see that macroscopic objects of matter occupy space, and that no two objects can occupy the same space at the same time. At the microscopic level, we can also see that no two particles, no matter how tiny, can occupy the same space at the same time.

Wood, sand, people, water, and air are all examples of matter. Heat and light are not matter; they are forms of energy. The amount of iron in the burner on a kitchen range does not change as the burner transfers heat to the water in a teakettle. Nor does the amount of matter in a pane of window glass change as sunlight passes through.

A central concern of chemists is the tiny, microscopic building blocks of matter known as atoms and molecules. We trust that they will become your concern as well, as you continue the study of chemistry. **Atoms** are the smallest distinctive units in a sample of matter, and **molecules** are larger units in which two or more atoms are joined together. What a sample of matter is and how it behaves depend ultimately on the particular atoms that are present and the ways in which they are joined together. Walls built of concrete blocks and of bricks have a different appearance because the building blocks are different. A brick fireplace has a different appearance than a brick wall. The building blocks (bricks) are the same, but they are joined together in different ways.

A model of a water molecule.

Composition refers to the types of atoms and their relative proportions in a sample of matter. For example, at the microscopic level we might describe a molecule of water as consisting of one oxygen atom and two hydrogen atoms. We can conveniently represent this molecule with a drawing or model.

Properties

Suppose we need a beaker for an experiment and find just the right one, but it is filled with a clear, colorless liquid. We can safely pour the liquid down the drain if it is water, but perhaps not if it's something else. How can we determine if the liquid is water? We could smell it. If the liquid has no odor, it could be water—water is odorless, but so are a lot of other liquids. However, if the liquid has a distinctive odor, it can't be water. Thus, we could easily distinguish between water, ethyl alcohol, and acetic acid (vinegar) by odor alone.

A **physical property** is a characteristic displayed by a sample of matter without undergoing any change in its composition. When ethyl alcohol is identified by its odor, there is no change in its composition. Neither do any changes occur in copper when we observe its color or its ability to conduct electric current, or in diamond when we observe its brilliance and hardness. Another way to distinguish between ethyl alcohol and water is that ethyl alcohol burns and water does not. But when ethyl alcohol burns, it is converted to carbon dioxide gas and water. A **chemical property** is a characteristic displayed by a sample of matter as it undergoes a change in composition. Consider the compositions of the carbon dioxide and water, which result from burning ethyl alcohol: They are very different from each other and from that of ethyl alcohol. Carbon dioxide molecules have two oxygen atoms to one carbon atom, water molecules have two hydrogen atoms to one oxygen atom, and ethyl alcohol molecules have six hydrogen atoms to two carbon atoms to one oxygen atom. Thus, we conclude that flammability, an ability to burn, is a chemical property. We will encounter many others in the text, such as an ability to react with acids or an ability to react with chlorine. Figure 1.1 illustrates some physical and chemical properties for copper and ethyl alcohol.

When ice melts, solid water is changed to liquid water. The water undergoes a rather profound change at the macroscopic level—what we see—but not at the microscopic level. Water molecules have two hydrogen atoms for every oxygen atom in both solid and liquid water. In a **physical change**, a sample of matter usually

◀ **Figure 1.1**
Properties of matter
Copper (left) and ethyl alcohol (right) are easily distinguished by their properties. Copper is a solid; ethyl alcohol is a liquid. Copper is opaque and has a red-brown color. Ethyl alcohol is transparent and colorless. Also, ethyl alcohol burns, and copper does not.

undergoes some noticeable change at the macroscopic level, but no change at the microscopic level—no change in composition. By contrast, in a **chemical change**, also called a **chemical reaction**, a sample of matter undergoes a change in composition and/or the structure of its molecules. In many cases, chemical changes also produce effects observable at the macroscopic level, such as the flame observed when ethyl alcohol burns. The cooking of foods and the spoilage of foods are common examples of chemical changes. Figure 1.2 illustrates a situation involving both physical and chemical changes.

● **ChemCDX** Formation of water

◀ **Figure 1.2**
Physical and chemical changes
Propane fuel is stored in the tank as a liquid under pressure. When the tank valve is opened, the liquid vaporizes—a physical change. Propane gas mixes with air and burns—a chemical change. The products of the combustion are carbon dioxide and water.

Classifying Matter

A type of matter that has a definite, or fixed, composition that does not vary from one sample to another is called a **substance**. All substances are either elements or compounds. An **element** is a substance that cannot be broken down into other simpler substances by chemical reactions. Viewed at the microscopic level, an element is made up of atoms of only a single type. (We will specify just what we mean by a "type" of atom in Chapter 2.) At the present time, 112 elements are known. You are most likely already familiar with common elements such as oxygen, nitrogen, carbon, iron, aluminum, copper, silver, and gold, but many of the elements are quite rare.

A **compound** is a substance made up of atoms of two or more elements, with the different kinds of atoms combined in fixed proportions. In the compound water,

TABLE 1.1 Some Elements with Symbols Derived from Latin Names

Usual Name	Latin Name	Symbol
Copper	Cuprum	Cu
Gold	Aurum	Au
Iron	Ferrum	Fe
Lead	Plumbum	Pb
Mercury	Hydrargyrum	Hg
Potassium	Kalium	K
Silver	Argentum	Ag
Sodium	Natrium	Na
Tin	Stannum	Sn

the fundamental units are molecules having two hydrogen atoms joined to an oxygen atom. Carbon dioxide, sodium chloride (table salt), sucrose (cane sugar), and iron oxide (rust) are also compounds. Compounds can be broken down into simpler substances—elements—by chemical reactions. The possible number of compounds is essentially limitless. By 1997, over 16 million compounds had been recorded and over a million new ones are being added each year. If this seems like a huge number of compounds from such a limited number of elements, think of the 26 letters of the English alphabet and the vast number of words that are formed from them.

Because elements and compounds are so fundamental to our study, it is convenient to use symbols to represent them. A **chemical symbol** is a one- or two-letter designation derived from the name of an element. Most symbols are based on English names; a few are based on the Latin name of the element or one of its compounds (see Table 1.1). The first letter of a symbol is capitalized and the second is always lower case. (It makes a difference. For example, Co is the symbol for the element cobalt; CO represents the poisonous compound carbon monoxide.) The names and symbols of all the elements are listed inside the front cover of this book. Compounds are designated by combinations of chemical symbols called *chemical formulas*. Writing chemical formulas is a little more complex than writing symbols. For example, ethyl alcohol has the formula CH_3CH_2OH. We will discuss how to write chemical formulas in Chapter 2.

In the scheme for classifying matter shown in Figure 1.3, we see two broad categories of matter: substances and mixtures. A **mixture** has no fixed composition; its composition may vary over a broad range. Ordinary table salt and water form a mixture, and we can vary the proportions of salt and water from sample to sample.

A mixture that has the same composition and properties throughout is called a **homogeneous mixture**, or a **solution**. A given solution of salt in water, sometimes called a saline solution, has the same "saltiness" throughout the solution. In contrast, a **heterogeneous mixture** varies in composition and/or properties from one part of the mixture to another. Although ice and liquid water have the same composition (both are made of water molecules), the physical properties of the ice and the liquid on which it floats are different. In a sand-water mixture, both the composition and properties vary within the mixture. They are different in the water than in the sand which sinks to the bottom of the water layer. Both homogeneous and heterogeneous mixtures can be separated into their individual components by *physical* changes—chemical reactions are not required. The dissolved salt can be obtained from a saline solution by allowing the water to evaporate away. Sand can be recovered from a sand-water mixture by passing the mixture through filter paper similar to that used in coffee makers. The water passes through the paper, and the sand is held back.

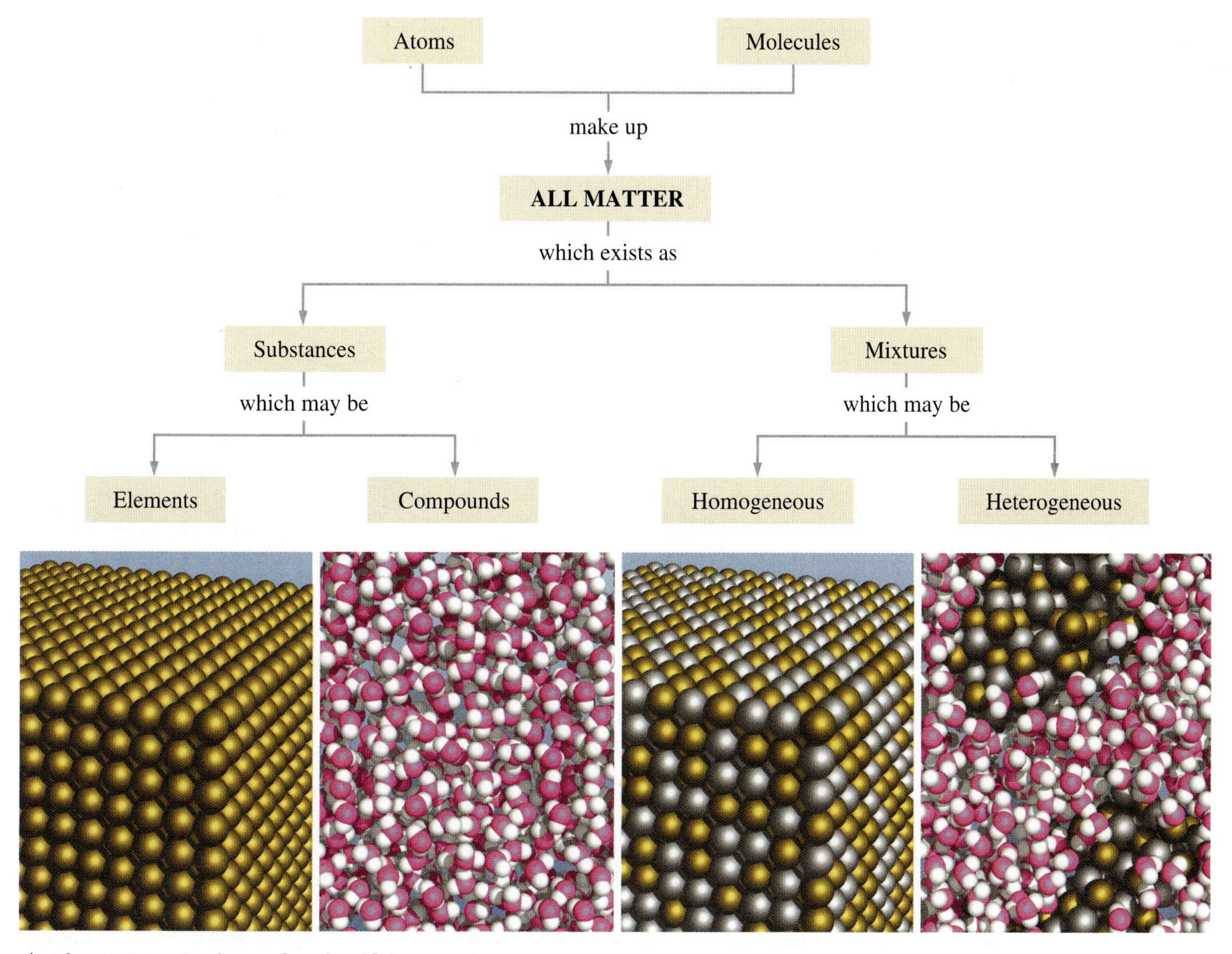

▲ **Figure 1.3 A scheme for classifying matter**
At the "molecular level," gold—an *element*—is made up of only one kind of atom, represented here as spheres. Water—a compound—is made up of only one kind of molecule. Twelve-karat gold—a *homogeneous mixture* of gold and silver—has gold atoms and silver atoms distributed at random. A *heterogeneous mixture* of 12 karat gold in water has clumps of gold and silver atoms surrounded by water molecules.

Scientific Methods

Chemists and other scientists use certain terms to describe the way in which they conduct their studies. We will briefly consider some of these terms, but the key idea to keep in mind is that scientific knowledge is *testable*, *reproducible*, *explanatory*, *predictive*, and *tentative*.

Scientists often begin a study by making observations and then formulating a hypothesis. A **hypothesis** is a *tentative* explanation or prediction concerning some phenomenon. It may be just an educated guess, but it must be a guess that can be tested. Scientists *test* a hypothesis through a carefully controlled procedure called an **experiment**. The facts obtained through careful observation and measurements made during experiments are called scientific **data**. Examples of scientific data are the melting point of iron (1535 °C) and the speed of light (2.99792458×10^8 meters per second). Further experiments may refine these data to some degree, but the basic facts can be verified by other scientists in similar experiments; the data are *reproducible*.

Scientists try to identify patterns in large collections of data and to summarize these patterns in brief statements called **scientific laws**. Many of these laws can be stated mathematically. Scientists also use scientific *models*—tangible items or pictures—to represent invisible processes and explain complicated phenomena. For example, the invisible particles (atoms and molecules) of solids, liquids, and gases can be visualized as billiard balls, marbles, or as dots or circles on paper (Figure 1.4). The ultimate goal of scientists is to formulate theories. A scientific **theory** provides *explanations* of observed natural phenomena and *predictions* that can be tested by further experiments. Theories often serve as a framework for organizing scientific knowledge.

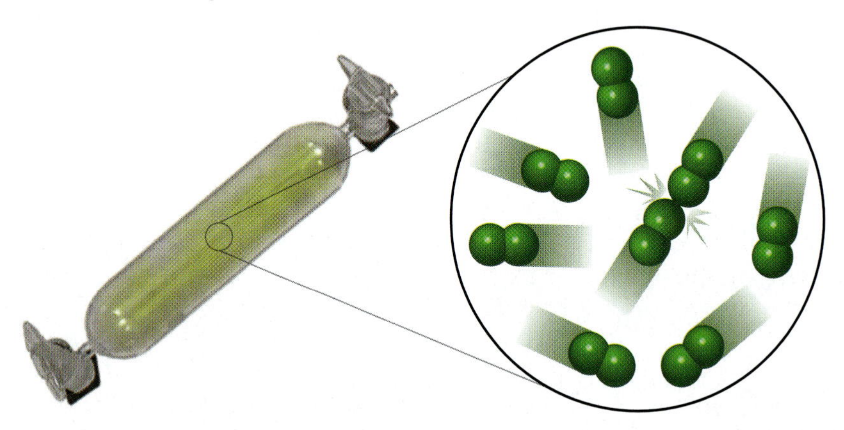

▶ **Figure 1.4**
A scientific model of a gas
The chlorine gas in this container is made up of chlorine molecules. Each molecule is a combination of two chlorine atoms. Chlorine molecules, like all gas molecules, are in constant random, chaotic motion and undergo frequent collisions with each other and with the container walls. This model is used in Chapter 5 to explain several properties of gases.

Contrary to some popular notions, scientific knowledge is not absolute. No hypothesis or theory can ever be proved completely true; it can only be *dis*proved. The most promising hypothesis can be destroyed by one stubborn fact. Thus, the body of scientific knowledge is growing, changing, and never final. Scientists discard old concepts when new tools and new techniques reveal new data and generate new concepts.

There is no single scientific method that serves as a guideline for all to follow. Like everyone, scientists make missteps and sometimes great leaps of the imagination. Often, though, they do proceed rather methodically, checking their ideas with carefully designed experiments. Consider the following as an illustrative but far from unique example.

Observations Felix Hofmann, a chemist working for the Bayer company, had a father who suffered from rheumatoid arthritis. Sodium salicylate, the best drug available for treating his father's chronic rheumatic pain, tasted awful and caused severe stomach problems.

Hypothesis About 1893 Hofmann hit upon the idea that acetylsalicylic acid, if it could be made in a pure form, might be less irritating than sodium salicylate.

Experiments Hofmann searched the literature for information about acetylsalicylic acid, and he did laboratory experiments to find a better way to synthesize the compound than that which had been used to make an impure form of the compound in 1853.

Results With a purer product in hand, he administered acetylsalicylic acid to his father and found that it was indeed less unpleasant to take and less irritating to the stomach than sodium salicylate. It was also just as effective for the relief of pain. This result was widely confirmed by others, and use of acetylsalicylic acid, now known by the brand name aspirin, spread rapidly.

Theories The theories about why aspirin works so well were not developed until the 1970s, and we still have much to learn. Aspirin acts by inhibiting the

APPLICATION NOTE
Aspirin is produced in greater quantity than any other drug in the world. Its synthesis is so simple that it is carried out as an exercise in many school laboratories. Clinical studies over the years have shown aspirin to be effective for relief of pain, relief from fever, reduction of inflammation, and inhibition of blood clotting.

formation of compounds called *prostaglandins* that are involved in the transmittal of pain messages to the brain and in the inflammation of injured tissues.

Prediction These theories subsequently led to the development of many other pain-relieving and anti-inflammatory drugs such as ibuprofen (Motrin®, Advil®), naproxen (Aleve®), and ketoprofen (Orudis KT®).

1.3 Scientific Measurements

It is much easier to gather and check data, activities that are so essential to the methods of science, if all scientists agree to use a common system of measurement. The system agreed upon in 1960 is the International System of Units (SI), a modernized version of the metric system established in France in 1791.

In the original metric system, the standard unit of length was taken as 1/10,000,000th of the distance from the equator to the North Pole measured along the meridian passing through Paris. Other standards were related to the meter. In modern scientific work all measured quantities can be expressed in terms of the seven base units listed in Table 1.2. We will use the first six in this text, and we'll introduce the first four here. An essential part of the SI system is the use of exponential ("powers of ten") notation for numbers. If you are not already familiar with this notation, you will find a discussion of this topic in Appendix A.

Length

The SI base unit of length is the **meter (m)**, a unit about 10% longer than the yard. Units larger and smaller than the base unit are expressed by the use of prefixes (see Table 1.3). For example, to measure lengths much larger than the meter, such as distances along a highway, we often use the kilometer (km).

$$1\ \text{km} = 10^3\ \text{m} = 1000\ \text{m}$$

In the laboratory, lengths smaller than the meter are often more convenient. For example, consider the centimeter (cm)—about as long as the typeset word "lengths" in this sentence—and the millimeter (mm)—about the thickness of the cardboard backing in a notepad.

$$1\ \text{cm} = 10^{-2}\ \text{m} = 0.01\ \text{m}$$
$$1\ \text{mm} = 10^{-3}\ \text{m} = 0.001\ \text{m}$$

For measurements on the microscopic scale, we use the micrometer (μm), the nanometer (nm), and the picometer (pm).

$$1\ \mu\text{m} = 10^{-6}\ \text{m} \qquad 1\ \text{nm} = 10^{-9}\ \text{m} \qquad 1\ \text{pm} = 10^{-12}\ \text{m}$$

TABLE 1.2 The Seven SI Base Units

Physical Quantity	Name of Unit	Symbol of Unit
Length	Meter[a]	m
Mass	Kilogram	kg
Time	Second	s
Temperature	Kelvin	K
Amount of substance	Mole	mol
Electric current	Ampere	A
Luminous intensity	Candela	cd

[a]Spelled *metre* in most countries other than the United States.

TABLE 1.3 Some Common SI Prefixes

Multiple	Prefix
10^9	*giga* (G)
10^6	*mega* (M)
10^3	*kilo* (k)
10^{-1}	*deci* (d)
10^{-2}	*centi* (c)
10^{-3}	*milli* (m)
10^{-6}	*micro* (μ)[a]
10^{-9}	*nano* (n)
10^{-12}	*pico* (p)

[a]The Greek letter μ (spelled "mu" and pronounced "mew").

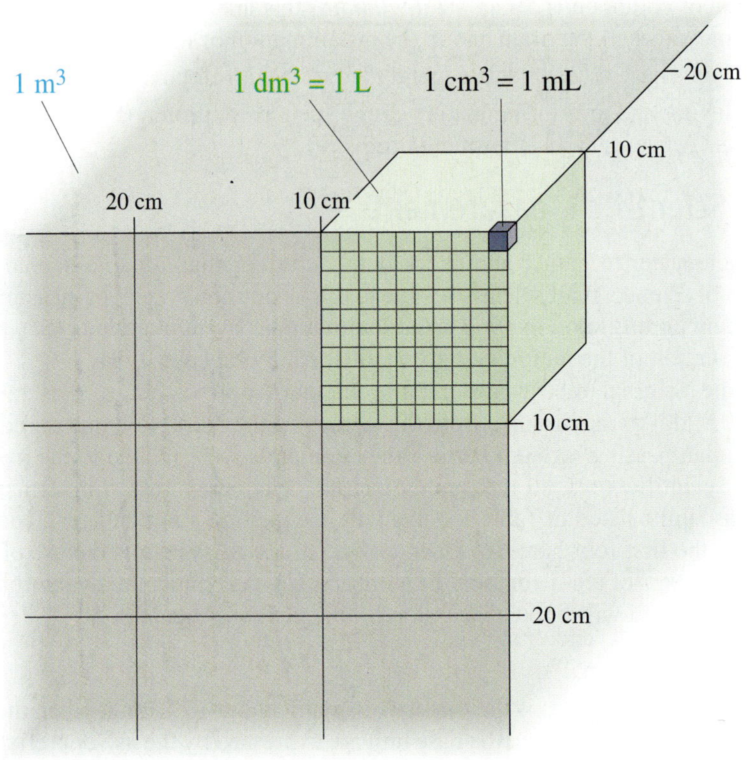

▶ **Figure 1.5**
Some volume units compared
The largest volume, shown in part, is the SI standard of 1 cubic meter (m^3). A cube 10 cm (1 dm) on an edge (green) has a volume of 1000 cm^3 (1 dm^3) and is equal to 1 liter (1 L). The smallest cube is 1 cm on edge (dark blue) and has a volume of 1 cm^3 = 1 mL.

▲ **Figure 1.6**
Measuring mass by weighing
Weight is the force of gravity on an object. This force is proportional to the mass. In the balance shown here, the force of gravity on the object is counterbalanced by a magnetic force, and the magnitude of this force is registered as a mass in the digital readout of the balance. The portion of the metallic cylinder shown has a mass of 999.0 g, and the very thin section, 1.0 g. Their combined mass is 1000.0 g = 1.0000 kg.

For example, a chlorophyll molecule is about 0.1 μm or 100 nm long, and the diameter of a sodium atom is 372 pm.

The units for area and volume are related to the base unit of length. The SI unit of area is the square meter (m^2), although for laboratory work we often find it more convenient to work with square centimeters (cm^2) or square millimeters (mm^2).

$$1\ cm^2 = (10^{-2}\ m)^2 = 10^{-4}\ m^2 \qquad 1\ mm^2 = (10^{-3}\ m)^2 = 10^{-6}\ m^2$$

A square centimeter is easy to picture; it's about the area of a button on many touch-tone telephones. A square millimeter is about the size of a "bullet" (•).

The SI unit of volume is the cubic meter (m^3), but Figure 1.5 pictures two units that are more likely to be used in the laboratory: the cubic centimeter (cm^3)—about the volume of a sugar cube—and the cubic decimeter (dm^3)—slightly larger than 1 quart.

$$1\ cm^3 = (10^{-2})^3\ m^3 = 10^{-6}\ m^3$$
$$1\ dm^3 = (10^{-1})^3\ m^3 = 10^{-3}\ m^3$$

Although it is not an SI unit, the old metric unit *liter* is also commonly used. A **liter (L)** is the same volume as one cubic decimeter, or 1000 cubic centimeters.

$$1\ L = 1000\ mL = 1\ dm^3 = 1000\ cm^3$$

The milliliter (mL) is the same as a cubic centimeter: 1 mL = 1 cm^3.

Mass

Mass is the quantity of matter in an object. Mass can be measured in many ways, but the most common is through weighing (Figure 1.6). The *weight* of an object is the force of Earth's gravity on the object, and this force is directly proportional to the mass of the object. Two objects of the same mass will weigh the same at any given location on Earth. If weighed at different locations, they may have slightly different weights, even though their masses remain equal, because of slight variations in Earth's gravitational pull. Therefore we use mass and not weight as the fundamental measure of a quantity of matter.

The SI base quantity of mass is the **kilogram (kg)**, which has a weight at Earth's surface of about 2.2 pounds. This base quantity is unique in that it already has a prefix. A more convenient mass unit for most laboratory work is the gram (g).

$$1 \text{ kg} = 10^3 \text{ g} = 1000 \text{ g}$$

The milligram (mg) is a suitable unit for small quantities of materials, such as some drug dosages.

$$1 \text{ mg} = 10^{-3} \text{ g}$$

Chemists can now detect masses in the microgram (μm), the nanogram (ng), and even the picogram (pg) range.

Time

The SI base unit for measuring intervals of time is **the second (s)**. Extremely short time periods are expressed through the usual SI prefixes: *milli*seconds, *micro*seconds, *nano*seconds, and *pico*seconds. Long time intervals, in contrast, are usually expressed in traditional, non-SI units: minute (min), hour (h), day (d), and year (y).

EXAMPLE 1.1

Convert the unit of each of the following measurements to a unit that replaces the power of ten by a prefix.

a. 9.56×10^{-3} m **b.** 1.07×10^3 g

SOLUTION

Our goal is to replace each power of ten with the appropriate prefix from Table 1.3. For example, $10^{-3} = 0.001$, leading to: *milli*(unit).

a. 10^{-3} corresponds to the prefix *milli*; 9.56 mm
b. 10^3 corresponds to the prefix *kilo*; 1.07 kg

EXERCISE 1.1

Convert each of the following measurements to a unit that replaces the power of ten by a prefix.

a. 7.42×10^{-3} s
b. 5.41×10^{-6} m
c. 1.19×10^{-9} g
d. 5.98×10^3 m

EXAMPLE 1.2

Use exponential notation to express each of the following measurements in terms of an SI base unit.

a. 1.42 cm **b.** 645 μs

SOLUTION

a. Our goal is to find the power of ten that relates the given unit to the SI base unit. That is, *centi* (base unit) = $10^{-2} \times$ (base unit)

$$1.42 \text{ centimeter} = 1.42 \times 10^{-2} \text{ m}$$

b. To change microsecond to the base unit second, we need to replace the prefix *micro* by 10^{-6}. To get our answer in the conventional exponential form, we also need to replace the coefficient 645 by 6.45×10^2. The result of these two changes is

$$645 \text{ microsecond} = 645 \times 10^{-6} \text{ s} = 6.45 \times 10^2 \times 10^{-6} \text{ s} = 6.45 \times 10^{-4} \text{ s}$$

EXERCISE 1.2

Use exponential notation to express each of the following measurements in terms of an SI base unit.

a. 475 nm **b.** 225 ns **c.** 1415 km **d.** 2.26×10^6 g

Temperature

Temperature is difficult to define. We can say that it is a measure of "hotness," but that isn't very precise. Think about what happens if two objects at different temperatures are brought together: Heat flows from the warmer to the colder object. The temperature of the warmer object drops, and that of the colder object increases until finally the two objects are at the same temperature. Temperature is therefore a property that tells us in what direction heat will flow. For example, if you touch a hot test tube, heat will flow from the tube to your hand. If the tube is hot enough, your hand will be burned. Later in the text, we will present some additional, perhaps more satisfying, ideas about temperature.

The SI base unit of temperature is the **kelvin (K)**. We will define the kelvin temperature scale later on, when we need to do so. For routine laboratory work, we often use the more familiar Celsius temperature scale. On this scale, the freezing point of water is 0 degrees Celsius (°C) and the boiling point is 100 °C. The interval between these two reference points is divided into 100 equal parts, each a degree Celsius. Another temperature scale widely used in the United States, but probably unfamiliar to most people in the world, is the Fahrenheit scale. As pointed out in Figure 1.7, the scales differ in

- their treatment of an important physical property—the freezing point of water: 0 °C and 32 °F;
- the temperature interval called a *degree* (°). A 10-degree temperature interval on the Celsius scale equals an 18-degree interval on the Fahrenheit scale.

These two facts are the basis of two equations that relate temperatures on the two scales. One of these requires multiplying the degrees of Celsius temperature by the factor 1.8 (that is, 18/10) to obtain degrees of Fahrenheit temperature, followed by adding "32" to account for the fact that 0 °C = 32 °F. In the other equation, "32" is subtracted from the Fahrenheit temperature to get the number of degrees Fahrenheit above the freezing point of water. Then this quantity is divided by 1.8.

$$t_F = 1.8\, t_C + 32 \qquad t_C = \frac{(t_F - 32)}{1.8}$$

You will not often need to convert between Celsius and Fahrenheit temperature, but Example 1.3 illustrates a practical situation where this would be necessary.

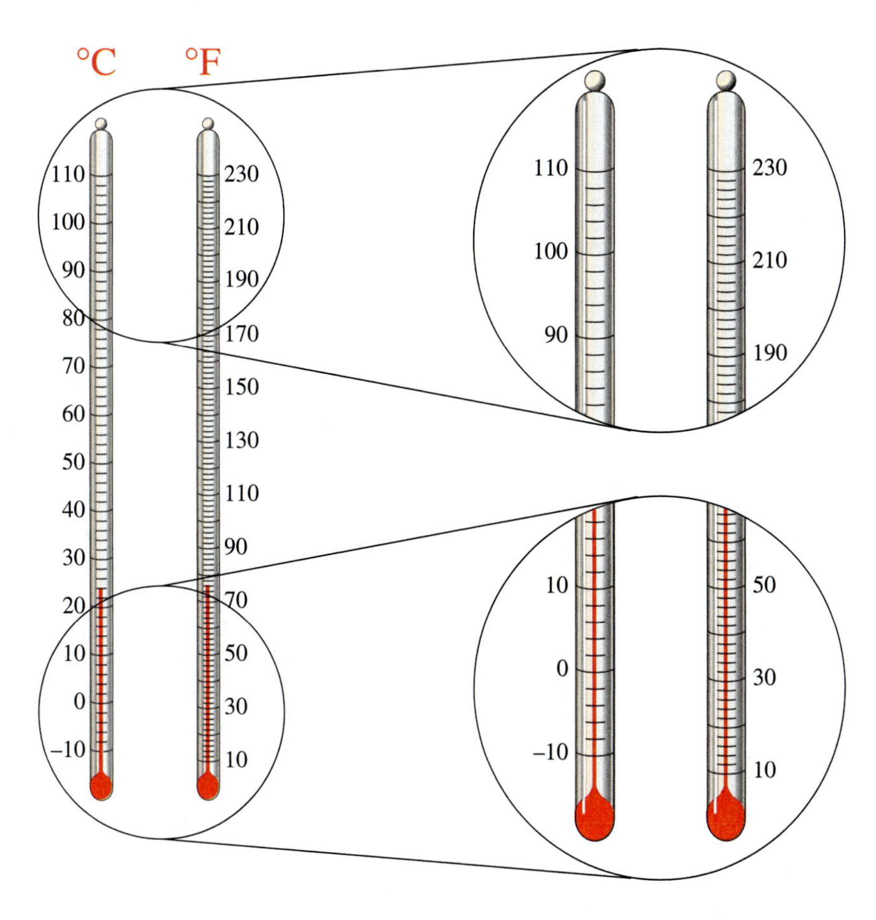

Figure 1.7
The Celsius and Fahrenheit temperature scales compared
The thermometer on the left is marked in degrees Celsius, and the one on the right in degrees Fahrenheit. The freezing point of water is at 0 °C and 32 °F; the boiling point is at 100 °C and 212 °F. Note that for an interval of 10 °C, the corresponding interval on the Fahrenheit scale is 18 °F. This gives rise to the factor 18/10 = 1.8 in the equations that relate the two scales.

EXAMPLE 1.3

A parasite that causes trichinosis is sometimes found in undercooked pork. Trichinosis is characterized by nausea, diarrhea, fever, stiffness and painful swelling of muscles, and swelling of facial tissues. It is rare in industrialized countries, occurring mainly in people who eat undercooked meat of bears or other carnivorous animals, but it is common in developing nations where people eat pork. The last reported incidence in the United States was in Idaho in 1995. It was associated with eating cougar jerky. The parasite that causes trichinosis is killed when meat is cooked to at least 66 °C at the center. If you have only a Fahrenheit meat thermometer, what is the minimum Fahrenheit temperature to which you should heat the center of the pork when cooking it?

SOLUTION

The pork must be heated to a temperature of 66 °C or higher. To get the minimum Fahrenheit temperature, we substitute $t_C = 66\ °C$ in the following equation and solve for t_F.

$$t_F = 1.8\, t_C + 32$$
$$t_F = (1.8 \times 66) + 32 = 151\ °F$$

EXERCISE 1.3

Carry out the following temperature conversions.

a. 85.0 °C to degrees Fahrenheit
b. −12.2 °C to degrees Fahrenheit
c. 355 °F to degrees Celsius
d. −20.8 °F to degrees Celsius

TABLE 1.4 Five Measurements of the Dimensions of a Poster Board

Student	Length, m	Width, m
1	1.827	0.761
2	1.824	0.762
3	1.826	0.763
4	1.828	0.762
5	1.829	0.762
Average:	1.827	0.762

1.4 Precision and Accuracy in Measurements

Counting can be exact: We can count exactly 18 students in a room. Measurements, on the other hand, are subject to error. One source of error is the measuring instruments themselves. An incorrectly calibrated thermometer may consistently yield a result that is 2 °C too low, for example. Other errors may result from the experimenter's lack of skill or care in using measuring instruments.

Suppose you were one of five students asked to measure the dimensions of a poster board, using a meter stick marked off in millimeters. Table 1.4 presents your measurement, along with those of the other students. The **precision** of a set of measurements refers to how closely individual measurements agree with one another. The precision is good (or high) if each of the measurements is close to the average of the series. The precision is poor (or low) if there is a wide deviation from the average value. How would you describe the precision of the data in Table 1.4? Examine the individual data for the height and width, note the average values, and determine how much the individual data differ from the averages. You will find that the maximum deviations from average values are 0.003 m for the length and 0.001 m for the width. Thus, we describe the precision as good.

The **accuracy** of a set of measurements refers to the closeness of the average of the set to the "correct" or most probable value. Measurements of high precision are more likely to be accurate than are those of poor precision, but even highly precise measurements are sometimes inaccurate. For example, what if the meter sticks used to obtain the data in Table 1.4 were actually 1005 mm long, but still carried 1000 millimeter markings? The accuracy of the measurements would be rather poor, even though the precision would remain high. Another comparison of precision and accuracy is made in Figure 1.8.

Sampling Errors

No matter how accurate a measurement is, it will not mean much unless it was performed on valid, representative samples. Consider determining the level of dissolved oxygen in the water of a lake. Dissolved oxygen is vital to fish. Measurements would yield varying results, depending on several factors such as *where* the sample was taken—at the surface or near the bottom, near the mouth of a swiftly flowing entering stream or in a stagnant bay. Results would also depend on *when* the sample was taken—on a warm, still day or when the wind was whipping up whitecap waves. Dissolved oxygen levels also depend on other factors, such as the temperature of the water. There is therefore no one true value for the level of dissolved oxygen in the lake. By extensive random sampling, one could get an average level, but even then, scientists usually record the conditions under which measurements are made.

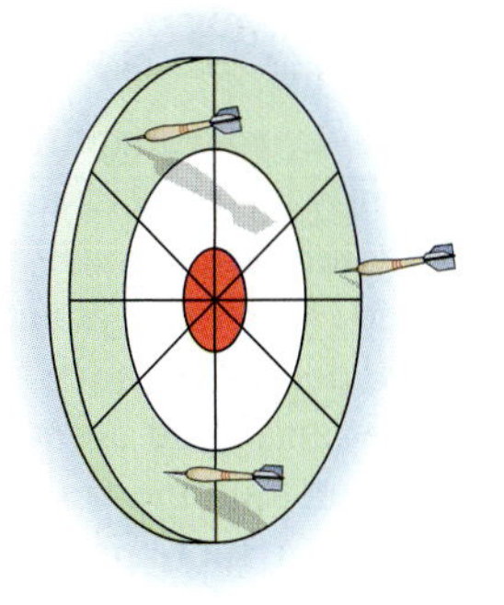

(a) Low accuracy
Low precision

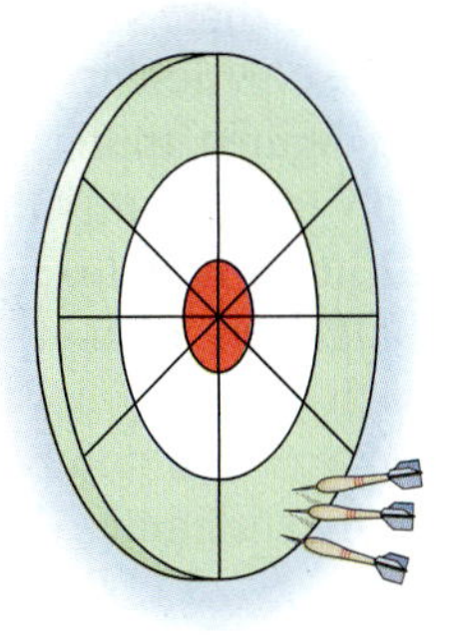

(b) Low accuracy
High precision

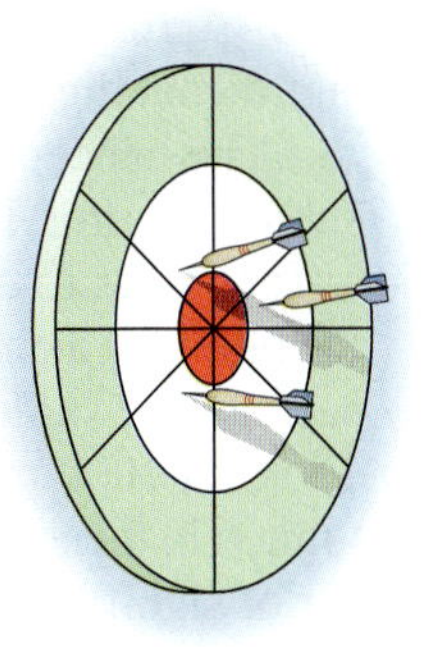

(c) High accuracy
Low precision

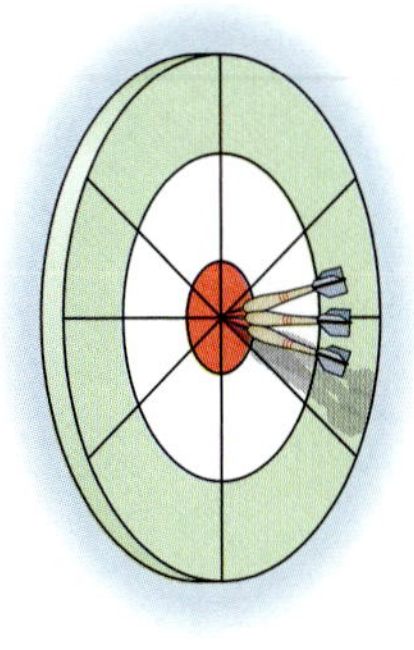

(d) High accuracy
High precision

▲ **Figure 1.8 Comparing precision and accuracy: a dart board analogy**
(a) The darts are both scattered (low precision) and off-center (low accuracy). (b) The darts are in a tight cluster (high precision) but still off-center (low accuracy). (c) The darts are somewhat scattered (low precision) but evenly distributed about the center (high accuracy). (d) The darts are in a tight cluster (high precision) and well centered (high accuracy).

Significant Figures

Look again at Table 1.4. Notice that the five measurements of length agree in the first three digits (1.82); they differ only in the fourth digit. We say the fourth digit is uncertain. All digits known with certainty (three in this case), plus the first uncertain one, are called significant digits, or **significant figures**. We use significant figures to reflect the precision of a measurement—the more significant figures, the more precise the measurement. The length measurements in Table 1.4 have *four* significant figures. In other words, we are quite sure that the length of the board is between 1.82 m and 1.83 m. Our best estimate of the average value, including the uncertain digit, is 1.827 m.

It is rather easy to establish that 1.827 has four significant figures; we simply count the number of digits. In any measurement that is properly reported, all *nonzero* digits are significant. *Zeros* present problems because they can be used in two ways: as a part of the measured value or to position a decimal point.

- Zeros between two other significant digits are significant. Examples: 1107 (four significant figures); 50.002 (five).
- A lone zero preceding a decimal point is written for "cosmetic" purposes. It is not significant. Example: 0.762 (three significant figures).
- Zeros that precede the first nonzero digit and position the decimal point are also *not* significant. Example: 0.000163 (three significant figures).
- Zeros at the end of a number are significant if they are to the *right* of the decimal point. Examples: 0.2000 (four significant figures); 0.050120 (five).

These four situations all conform to the general rule that when we read a number from left to right, all the digits starting with the first nonzero digit are significant. Numbers written without a decimal point and ending in zeros are a special case, however.

- Zeros at the end of a number may or may not be significant if the number is written *without* a decimal point. Example: 400

We do not know whether the number 400 was measured to the nearest unit, ten, or hundred. To avoid this confusion, we can use exponential notation (see Appendix A). In exponential notation, 400 would be recorded as 4×10^2 or 4.0×10^2 or

4.00×10^2 to indicate one, two, or three significant figures, respectively. The only significant digits are those in the coefficient, not in the power of ten.

The concept of significant figures applies only to *measurements*—quantities subject to error. It does not apply to a quantity that is (a) inherently an integer, such as four sides to a square or 12 items in a dozen, (b) inherently a fraction, such as the radius of a circle = one-half ($\frac{1}{2}$) the diameter, or (c) obtained by an exact count, such as 127 students in a class. It also does not apply to *defined* quantities, such as 1 km = 1000 m. In these contexts, the numbers 4, 12, $\frac{1}{2}$, 127, and 1000 are not limited in their numbers of significant figures. In effect, each has an unlimited number of significant figures (4.000 . . ., 12.000 . . ., 0.5000 . . .) or, more properly, each is an *exact* value.

Significant Figures in Calculations: Multiplication and Division

A brief item in a newspaper on December 25, 1997, stated, "As of Jan. 1, 1998 the Census Bureau projects there will be 268,921,733 people living in the United States." Is it possible that all the digits in this number are truly significant and that the projection could account for every single individual? Clearly this projection is only an estimate based on certain data, such as rates of births and deaths and migrations into and out of the country. The reporting of this estimate violates a fundamental principle concerning calculations based on measurements.

A calculated quantity can be no more precise than the data used in the calculation, and the reported result should reflect this fact.

Just as a chain is only as strong as its weakest link, a calculation is only as precise as the least precise measurement that figures into the calculation. A strict application of this principle involves a sophisticated method of analyzing the relevant data that is beyond the scope of this text, but we can do a pretty good job through a practical rule involving significant figures.

In multiplication and division, the reported result should have no more significant figures than the factor with the fewest *significant figures.*

To write a numerical answer with the proper number of significant figures often requires that we round off numbers. In rounding, we drop all digits that are not significant and, if necessary, adjust the last reported digit. We will follow these rules in rounding.

- If the leftmost digit to be dropped is less than 5, leave the final digit *unchanged*. Example: 369.443 rounds to 369.44 if we need five significant figures and to 369.4 if we need four.
- If the leftmost digit to be dropped is 5 or greater, *increase* the final digit by *one*. Example: 538.768 rounds to 538.77 if we need five significant figures and to 538.8 if we need four; 74.397 rounds to 74.40 if we need four significant figures and 74.4 if we need three.

EXAMPLE 1.4

Calculate the area, in square meters, of the poster board whose dimensions are given in Table 1.4. Use the correct number of significant figures in your answer.

SOLUTION

The area of the rectangular poster board is the product of its length and width. In expressing the result of this multiplication, we can show only as many significant figures as found in the least precisely stated dimension: the width (three significant figures).

$$1.827 \text{ m} \times 0.762 \text{ m} = 1.392174 \text{ m}^2 = 1.39 \text{ m}^2$$

When using an electronic calculator, the calculator display often has more digits than are significant (see Figure 1.9). We use the rules on rounding off numbers as the basis for dropping the digits "2174."

EXERCISE 1.4

Calculate the volume, in cubic meters, of the poster board, given that its thickness is 6.4 mm. Use the correct number of significant figures.

PROBLEM-SOLVING NOTE
Note that all three dimensions must be expressed in the same unit.

EXAMPLE 1.5

For a laboratory experiment, a teacher wants to divide all of a 453.6 g sample of sulfur equally among the 21 members of her class. How many grams of sulfur should each student receive?

SOLUTION

Here we need to recognize that the number "21" is a counted number and is an exact number. It is not subject to significant figure rules. The answer should carry *four* significant figures, the same as in 453.6 g.

$$\frac{453.6 \text{ g}}{21} = 21.60 \text{ g}$$

In this calculation, an electronic calculator displays the result "21.6." We need to add the digit "0" to emphasize that the result is precise to four significant figures.

EXERCISE 1.5

The experiment described in Example 1.5 also requires that each student have available 2.04 times the mass of zinc as of sulfur. What is the total mass of zinc that the teacher will need, expressed with the appropriate number of significant figures?

▲ **Figure 1.9**
Significant figures
The "answer" on this electronic calculator is 1.392174, suggesting seven significant figures, but the rule for multiplication tells us there can be only three: 1.39.

Significant Figures in Calculations: Addition and Subtraction

When we add or subtract, our concern is not with the number of significant figures but with digits to the right of the decimal point. If the quantities being added or subtracted have varying numbers of digits to the right of the decimal point, find the one with the *fewest* such digits. The result of the addition or subtraction should contain the same number of digits to the right of *its* decimal point. The idea is that if you are adding several lengths, and one of them is measured only to the nearest *centi*meter, the total length cannot be stated to the nearest *milli*meter, no matter how precise the other measurements are.

We apply this idea in Example 1.6, and illustrate another point as well: In a calculation involving several steps, we should round off only the final result.

EXAMPLE 1.6

Perform the following calculation, and round off the answer to the correct number of significant figures.

$$49.146 \text{ m} + 72.13 \text{ m} - 9.1434 \text{ m} = ?$$

SOLUTION

In this calculation, we must add two numbers and, from their sum, subtract a third. We do this in two ways below. In both cases, we express the answer to two decimal places, the number of decimal places in "72.13 m."

(a)

$$\begin{array}{r} 49.146 \text{ m} \\ +72.13 \text{ m} \\ \hline 121.276 \text{ m} = 121.28 \text{ m} \\ -9.1434 \text{ m} \\ \hline 112.1366 \text{ m} = 112.14 \text{ m} \end{array}$$

(b)

$$\begin{array}{r} 49.146 \text{ m} \\ +72.13 \text{ m} \\ \hline 121.276 \text{ m} \\ -9.1434 \text{ m} \\ \hline 112.1326 \text{ m} = 112.13 \text{ m} \end{array}$$

We will generally use method (b), where we do *not* round off the intermediate result: 121.276. If we use electronic calculators, we generally don't need to write down or otherwise take note of an intermediate result.

EXERCISE 1.6A*

Perform the indicated operations and give answers with the proper number of significant figures.

a. 48.2 m + 3.82 m + 48.4394 m
b. 148 g + 2.39 g + 0.0124 g
c. 451 g − 15.46 g − 20.3 g
d. 15.436 L + 5.3 L − 6.24 L − 8.177 L

EXERCISE 1.6B*

Perform the indicated operations and give answers with the proper number of significant figures.

a. $(51.5 \text{ m} + 2.67 \text{ m}) \times (33.42 \text{ m} - 0.124 \text{ m})$

b. $\dfrac{(125.1 \text{ g} - 1.22 \text{ g})}{(52.5 \text{ mL} + 0.63 \text{ mL})}$

c. $\dfrac{(47.5 \text{ kg} - 1.44 \text{ kg})}{10.5 \text{ m} \times 0.35 \text{ m} \times 0.175 \text{ m}}$

d. $\dfrac{(0.307 \text{ g} - 14.2 \text{ mg} - 3.52 \text{ mg})}{(1.22 \text{ cm} - 0.28 \text{ mm}) \times 0.752 \text{ cm} \times 0.51 \text{ cm}}$

PROBLEM-SOLVING NOTE
In addition and subtraction, all terms must be expressed in the same unit.

1.5 A Problem-Solving Method

We will present many problems in this text. To solve them, you generally will need to apply some basic principles to describe a new situation based on the information given. Many chemistry problems are stated in a way that requires calculations and *quantitative*, or numerical, results. In this section, we will describe a useful method

* We often follow an Example with two Exercises, labeled A and B. The goal in an *A Exercise* is to apply the method outlined in the *Example* to a similar situation. In a *B Exercise*, you often need to combine that method with other ideas previously learned. The B Exercises provide the kind of context in which we usually apply chemical knowledge.

for doing such calculations, and we will present some additional ideas on problem solving in Section 1.6.

The Unit-Conversion Method

Imagine that an American orders a leather belt from a foreign country. The customer knows her waist length in common units but must give it in metric units. Suppose her waist is 26 inches. What is this length in centimeters? A 26-inch waist is a fixed length, whether we express it in inches, feet, centimeters, or millimeters. Thus, when we measure something in one unit and then convert that unit to another one, we must not change the measured quantity in any fundamental way.

In mathematics, multiplying a quantity by "1" does not change its value, and we can therefore use a factor equivalent to "1" to convert between inches and centimeters. We find our factor in the definition of the inch in Table 1.5.

$$1 \text{ in.} = 2.54 \text{ cm}$$

Notice that if we divide both sides of this equation by 1 in. we obtain a ratio of quantities that is equal to 1.

$$1 = \frac{\cancel{1 \text{ in.}}}{\cancel{1 \text{ in.}}} = \frac{2.54 \text{ cm}}{1 \text{ in.}}$$

Or, if we divide both sides of the equation by 2.54 cm, we also obtain a ratio of quantities that is equal to 1.

$$\frac{1 \text{ in.}}{2.54 \text{ cm}} = \frac{2.54 \cancel{\text{cm}}}{2.54 \cancel{\text{cm}}} = 1$$

The two ratios, one printed in red and the other in blue, are conversion factors. A **conversion factor** is a ratio of terms—equivalent to the number one—used to change the unit in which a quantity is expressed.

What happens when we multiply a known quantity by a conversion factor? The original unit used to express the quantity cancels out and is replaced by the desired unit. To put the matter another way, our general approach to using conversion factors is

$$\text{Desired quantity and unit} = \text{given quantity and unit} \times \text{conversion factors}$$

Now let's return to the question about the length of the belt. Its length—the measured quantity—is 26 in. The appropriate conversion factor (in red) must have the desired unit (cm) in the *numerator* and the unit to be replaced (in.) in the *denominator*. In this arrangement, we can cancel the unit *in.* so that only the unit *cm* remains.

$$? \text{ cm} = 26 \cancel{\text{in.}} \times \frac{2.54 \text{ cm}}{1 \cancel{\text{in.}}} = 66 \text{ cm}$$

It's not hard to see why the other conversion factor (blue) won't work. It gives a nonsensical unit.

$$26 \text{ in.} \times \frac{1 \text{ in.}}{2.54 \text{ cm}} = 10 \frac{\text{in.}^2}{\text{cm}}$$

Example 1.7 is a more typical illustration of the *unit-conversion* method of problem solving in that we need to use more than one conversion factor. We also use this example to point out some important practices that we will follow in writing conversion factors and dealing with significant figures.

TABLE 1.5 Some Conversions Between Common (U.S.) and Metric Units

Metric	Common
Mass	
1 kg	= 2.205 lb
453.6 g	= 1 lb
28.35 g	= 1 ounce (oz)
Length	
1 m	= 39.37 in.
1 km	= 0.6214 mi
2.54 cm	= 1 in.[a]
Volume	
1 L	= 1.057 qt
3.785 L	= 1 gal
29.57 mL	= 1 fluid ounce (fl oz)

[a]The U.S. inch is defined as exactly 2.54 cm. The other equivalencies are rounded off.

EXAMPLE 1.7

What is the length, in millimeters, of a 1.25-ft rod?

SOLUTION

There is no relationship between feet and millimeters in Table 1.5, so we cannot make this conversion with a single, simple conversion factor. Rather, we can think of the problem as a series of three conversions:

1. Use the fact that 1 ft = 12 in. to convert from feet to inches.
2. Use data from Table 1.5 to convert from inches to meters.
3. Use a knowledge of prefixes to convert from meters to millimeters.

Although we can solve this problem in three distinct steps by making one conversion in each step, we can just as easily combine three conversion factors into a single setup. This is the procedure sketched below.

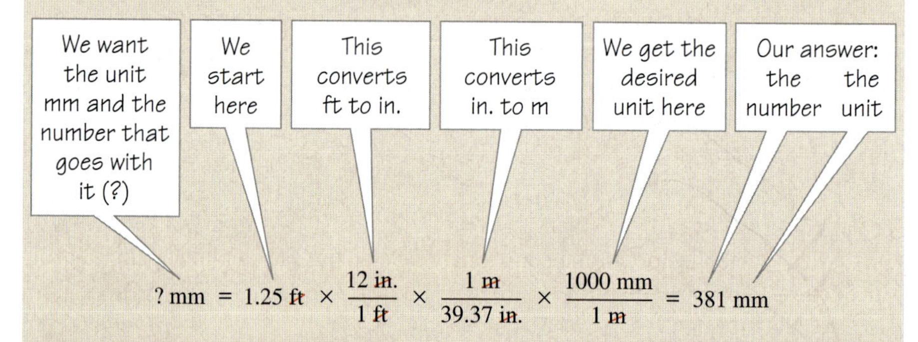

$$? \text{ mm} = 1.25 \text{ ft} \times \frac{12 \text{ in.}}{1 \text{ ft}} \times \frac{1 \text{ m}}{39.37 \text{ in.}} \times \frac{1000 \text{ mm}}{1 \text{ m}} = 381 \text{ mm}$$

PROBLEM-SOLVING NOTE
Express numerical constants and equivalencies (such as 1 m = 39.37 in.) with *more* significant figures than the least precisely measured quantity. In this way, you will not inadvertently make the result less precise.

Now let's look at the use of significant figures in this problem. The measured quantity, 1.25 ft, is given with *three* significant figures. The relationship 12 in. = 1 ft is *exact*, and this will not affect the precision of our calculation. It is understood that the relationship 1 m = 39.37 in. is stated to *four* significant figures; there is no need to write 1 m as 1.000 m. Finally, the relationship 1000 mm = 1 m is *exact*; this defines the way we relate millimeters and meters. Our answer should have *three* significant figures.

EXERCISE 1.7

Carry out the following conversions.

a. 76.3 mm to meters
b. 0.0856 kg to milligrams
c. 0.556 km to feet
d. 48.8 oz to kilograms
e. 3.50 gal to fluid ounces

In Example 1.8, we must replace *two* units in the original measured quantity, one in the numerator and one in the denominator. We will solve the problem in two ways and compare the two approaches.

EXAMPLE 1.8

A sprinter runs a 100.0-m dash in 11.00 s. What is her speed in kilometers per hour?

SOLUTION

First let's identify the measured quantity. It is a speed—the ratio of a distance in meters to a time in seconds.

$$\frac{100.0 \text{ m}}{11.00 \text{ s}}$$

Our goal is to convert this speed to one expressed in kilometers per hour. We must convert from meters to kilometers in the numerator and from seconds to hours in the denominator. We need this set of equivalent values to formulate conversion factors:

$$1 \text{ km} = 1000 \text{ m} \qquad 1 \text{ min} = 60 \text{ s} \qquad 1 \text{ h} = 60 \text{ min}$$

$$? \frac{\text{km}}{\text{h}} = \frac{100.0 \text{ m}}{11.00 \text{ s}} \times \frac{1 \text{ km}}{1000 \text{ m}} \times \frac{60 \text{ s}}{1 \text{ min}} \times \frac{60 \text{ min}}{1 \text{ h}} = 32.73 \text{ km/h}$$

In an alternate approach, we can do the conversions in the numerator and denominator separately and then divide the numerator by the denominator.

$$\text{Numerator: } 100.0 \text{ m} \times \frac{1 \text{ km}}{1000 \text{ m}} = 0.1000 \text{ km}$$

$$\text{Denominator: } 11.00 \text{ s} \times \frac{1 \text{ min}}{60 \text{ s}} \times \frac{1 \text{ h}}{60 \text{ min}} = 3.056 \times 10^{-3} \text{ h}$$

$$\text{Division: } \frac{0.1000 \text{ km}}{3.056 \times 10^{-3} \text{ h}} = 32.72 \text{ km/h}$$

Both of these methods are correct in that they use a sound strategy, even though the answers appear to differ slightly. The source of the difference is in rounding off the intermediate result in the second method (3.0555556×10^{-3} to 3.056×10^{-3}). If we had merely stored this intermediate result in the memory of a calculator rather than rounding it off and writing it down, we would have obtained the very same answer as by the first method.

EXERCISE 1.8

Carry out the following conversions.

a. 90.0 km/h to meters per second
b. 1.39 ft/s to kilometers per hour
c. 4.17 g/s to kilograms per hour.

Because a conversion factor has an intrinsic value of 1 (the numerator and denominator are equivalent), we can raise the factor to a power and it still has an intrinsic value of 1. For example,

$$\frac{2.54 \text{ cm}}{1 \text{ in.}} = 1 \quad \text{and} \quad \left(\frac{2.54 \text{ cm}}{1 \text{ in.}}\right)^2 = 1^2 = 1 \quad \text{and} \quad \left(\frac{2.54 \text{ cm}}{1 \text{ in.}}\right)^3 = 1^3 = 1$$

EXAMPLE 1.9

A box has a volume of 482.2 in.^{3}. What is its volume in cubic centimeters?

SOLUTION

We can base this conversion on a single conversion factor, but the factor itself (2.54 cm/1 in.) must be raised to the third power.

$$? \text{ cm}^3 = 482.2 \text{ in.}^3 \times \left(\frac{2.54 \text{ cm}}{1 \text{ in.}}\right)^3$$

$$= 482.2 \text{ in.}^3 \times \frac{(2.54)^3 \text{ cm}^3}{(1)^3 \text{ in.}^3} = 7902 \text{ cm}^3$$

EXERCISE 1.9A

Carry out the following conversions.

a. 476 cm^2 to square inches **b.** 1.56×10^4 $in.^3$ to cubic meters

EXERCISE 1.9B

The pressure exerted by the atmosphere is found to be 14.70 lb/$in.^2$. What is this pressure expressed in kilograms per square meter?

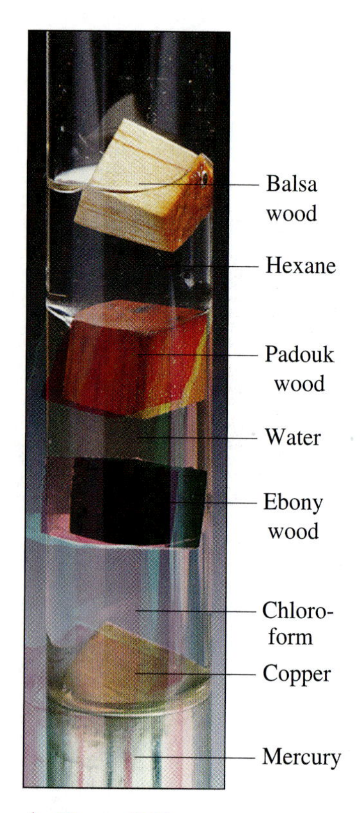

▲ **Figure 1.10**
A comparison of some densities
The liquids are as follows: chloroform (d = 1.48 g/cm^3) floats on mercury (d = 13.6 g/cm^3). Water (d = 1.00 g/cm^3), which does not mix with chloroform, floats on the chloroform. Hexane (d = 0.66 g/cm^3), which does not mix with water, floats on the water. Liquid mercury is so dense that copper (d = 8.94 g/cm^3) will float on it. Wood, a material of variable composition, has a range of densities. The densities listed here are representative for three types of wood. Balsa wood (d = 0.1 g/cm^3) has such a low density that it floats on hexane; padouk wood (d = 0.86 g/cm^3) floats on water but not on hexane; and ebony wood (d = 1.2 g/cm^3) sinks in water but floats on chloroform.

Density: A Physical Property and Conversion Factor

People often say that iron is "heavy" and aluminum is "light." As you know, they don't mean to suggest that a 10-inch iron skillet weighs more than a 16-ft aluminum extension ladder. What they mean is that iron is more *dense* than aluminum. If we compare the masses of *equal volumes* of the two metals, we find that the iron has a greater mass. **Density (d)** is the mass per unit volume of a substance. This important physical property is defined by the following equation.

$$d = \frac{m}{V}$$

The SI unit of density is kilograms per cubic meter (kg/m^3 or kg m^{-3}), but densities are more often measured in grams per cubic centimeter (g/cm^3) or grams per milliliter (g/mL). The density of aluminum is 2.70 g/cm^3, and that of iron is 7.87 g/cm^3.

Although the mass of an object remains constant as temperature is raised, the volume generally increases—the object expands—and therefore its density decreases. As a consequence, it is usually necessary to state the temperature at which a density is measured. Ethylene glycol (the principal ingredient in many antifreeze solutions), water, and ethanol are all colorless liquids. We can distinguish these liquids by their densities. At 20 °C, the density of ethylene glycol is 1.114 g/mL, that of water is 0.998 g/mL, and that of ethanol is 0.789 g/mL. At around room temperature, for most purposes, we can assume the density of water to be 1.00 g/mL, an assumption that you will find useful from time to time in solving problems.

If we combine two liquids that are *immiscible* (do not mix to form solutions), the liquid with lower density will float on top of the liquid of higher density. A solid that does not dissolve in a liquid will float on the liquid if the solid is *less* dense than the liquid; otherwise the solid will sink. A solid that *floats* displaces a *mass* of liquid equal to the mass of the solid. A solid that *sinks* displaces a *volume* of liquid equal to that of the solid. We illustrate some of these ideas with Figure 1.10 and topics presented later in the chapter.

EXAMPLE 1.10

The mass of 325 mL of the liquid methanol (wood alcohol) is found to be 257 g. What is the density of methanol?

Fat Floats: Body Density and Fitness

How fit are you physically? If you are an athlete, or if you simply want to get in better physical condition through exercise, your first step might well be a fitness evaluation. One part of the evaluation is to determine your percent body fat, and this is best done by measuring your body density. The density of human fat is 0.903 g/mL, whereas that of water is 1.00 g/mL. Will fat float or sink if placed in water? Because fat is less dense than water, it floats on water. The higher your proportion of body fat, the more buoyant you are in water. Body density is determined from body mass (weight) and volume. Body volume is determined by weighing a person when submerged in water (Figure 1.11). Once body density is determined, the percent body fat is estimated from a graph or from a table of compiled data. The average percent body fat is about 16% for an adult male and 25% for an adult female. Male athletes in superb condition have less than 7% body fat, and females less than 12% body fat.

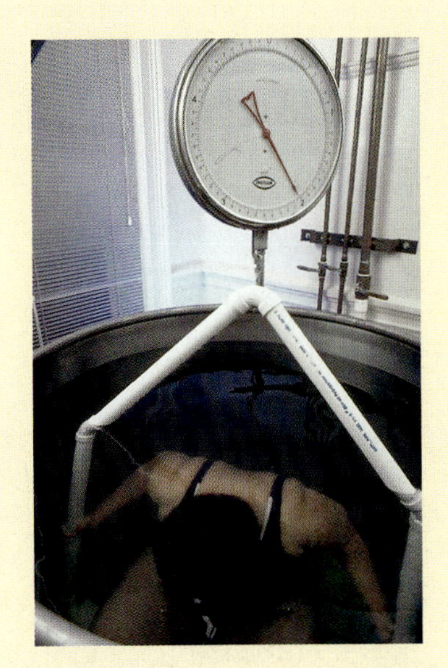

▶ Figure 1.11
Determining body volume
It is easy to measure a person's mass (weight), but what about a person's volume? When submerged in water, a body displaces its own volume of water. The difference in the person's weight in air and when submerged in water equals the mass of displaced water. This mass, divided by the density of water, yields the volume of displaced water and—with an appropriate correction for the volume of air in the lungs and gases in the intestine—the person's volume (see Problem 83).

SOLUTION

We know both mass in grams and the volume in milliliters of the sample of methanol. We can therefore obtain a density in grams per milliliter by simply applying the defining equation.

$$d = \frac{m}{V} = \frac{257\ \text{g}}{325\ \text{mL}} = 0.791\ \text{g/mL}$$

EXERCISE 1.10A

If the block of wood pictured below has a mass of 1.25 kg, what is its density, expressed in grams per cubic centimeter?

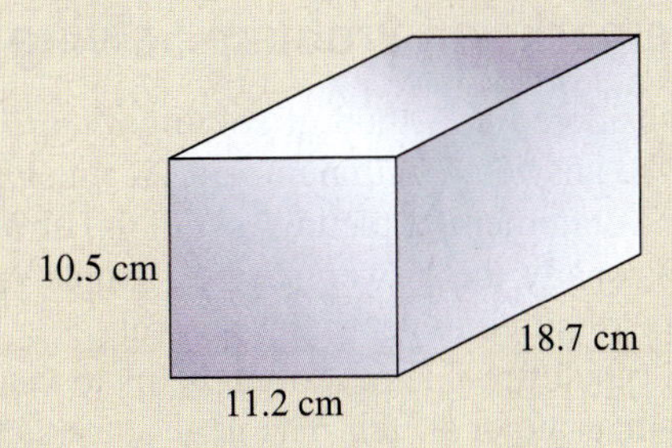

EXERCISE 1.10B

Use appropriate conversion factors from the text to calculate the density of liquid mercury if a 76.0-lb sample is found to occupy a volume of 2.54 L.

In Example 1.10, we found that the density of methanol is 0.791 g/mL. We can also describe this density through the equation that follows.

$$1 \text{ mL methanol} = 0.791 \text{ g methanol}$$

This then allows us to formulate two conversion factors between the mass and volume of methanol.

$$\text{(a)} \quad \frac{0.791 \text{ g methanol}}{1 \text{ mL methanol}} = 1 \quad \text{and} \quad \text{(b)} \quad \frac{1 \text{ mL methanol}}{0.791 \text{ g methanol}} = 1$$

We can use density—factor (a)—as a conversion factor to convert a volume of methanol to a mass. To convert from mass to volume, we can use the inverse of density—that is, factor (b).

EXAMPLE 1.11

How many kilograms of methanol does it take to fill the 15.5-gal fuel tank of an automobile modified to run on methanol?

SOLUTION

Our goal is to determine the mass of a certain volume of methanol. For this, we need to use conversion factor (a) above. Before using the factor, however, we must convert a volume in gallons to one in milliliters. Finally, we need to make a conversion from grams to kilograms of methanol. We can write all of the required factors in the following single setup.

$$? \text{ kg} = 15.5 \cancel{\text{gal}} \times \frac{3.785 \cancel{\text{L}}}{1 \cancel{\text{gal}}} \times \frac{1000 \cancel{\text{mL}}}{1 \cancel{\text{L}}} \times \frac{0.791 \cancel{\text{g}}}{1 \cancel{\text{mL}}} \times \frac{1 \text{ kg}}{1000 \cancel{\text{g}}} = 46.4 \text{ kg}$$

EXERCISE 1.11A

What is the volume in gallons of 10.0 kg of methanol ($d = 0.791$ g/mL)?

EXERCISE 1.11B

What volume of methanol ($d = 0.791$ g/mL) has the same mass as 10.00 gal of gasoline ($d = 0.690$ g/mL)?

1.6 Further Remarks on Problem Solving

In Section 1.5, we considered a method of solving *quantitative* chemistry problems—those that can be answered with numbers. At times, though, we need only a *qualitative* answer: a statement, a picture, symbols, or a diagram to represent what is expected. Occasionally, the answer may be a simple Yes or No, although substantial thought may be required to get to this answer.

Regardless of the type of problem, it often helps to visualize how to solve it, perhaps by drawing a diagram or sketch. The next action is vital: Outline *a* strategy, a step-by-step approach to solving the problem. We emphasize the *a* because often more than one strategy will work. In general, any logical scheme you choose will be valid as long as it leads to an appropriate conclusion. You demonstrate your mastery of the underlying principles by devising a problem-solving strategy that works. On occasion, you may devise a strategy that was not anticipated by the person who posed the problem. In doing so, you may demonstrate a special insight into the principles involved.

To assist you in developing a broad range of problem-solving skills, from time to time we will present two types of examples and exercises that differ somewhat from those introduced earlier. We close this chapter by illustrating both types.

Estimation Examples and Exercises

A calculator will always give you an answer, but the answer may not be correct. You may have punched a wrong number or executed the wrong function. If you learn to make *estimations* of answers, you at least will know whether your answer is a reasonable one. And at times, an approximate answer is quite acceptable because there are not enough data to make a precise calculation. Moreover, the ability to estimate answers is important in everyday life as well as in chemistry.

Keep in mind as you consider the following illustrations that the idea behind an estimation is *not* to do a detailed calculation. Only a rough calculation, if any at all, is required. We will use the designation "An Estimation Example" for in-chapter examples in which estimations form a major part of the solution. However, we will not use such notation for the end-of-chapter problems. When working problems, try to identify from the context of the problem situations in which an estimated answer is appropriate.

EXAMPLE 1.12—An Estimation Example

Which of the following is an approximate room temperature on the Celsius scale?

5 °C 20 °C 50 °C 70 °C

SOLUTION

Two easy points to remember are (1) the freezing point of water is 0 °C = 32 °F, and (2) there are slightly less than two Fahrenheit degrees for every degree Celsius (1.8, to be exact). The first value, 5 °C, is less than 10 °F above the freezing point; it is rather far below room temperature. The second answer, 20 °C, is nearly 40 °F above the freezing point, which makes it about 70 °F; this is a typical room temperature. We can stop here. The other two temperatures are obviously much above room temperature.

EXERCISE 1.12

Without doing detailed calculations, determine which of the following Celsius temperatures is most nearly equal to 900 °F? Explain your reasoning.

300 °C 500 °C 850 °C 1700 °C

EXAMPLE 1.13—An Estimation Example

A small storage tank for liquefied petroleum gas (LPG) appears to be spherical in shape and to have a diameter of about 1 ft. Suppose that some common volumes for LPG tanks are 1 gal, 2 gal, 5 gal, and 10 gal. Which is the most probable volume of this particular tank?

SOLUTION

Draw a sketch similar to Figure 1.12, which shows how the tank will just fit into a cubical box, 12 in. on edge. Each dimension of the box is a little greater than 30 cm (12 in. × 2.54 cm/1 in.). The volume of the box is slightly larger than

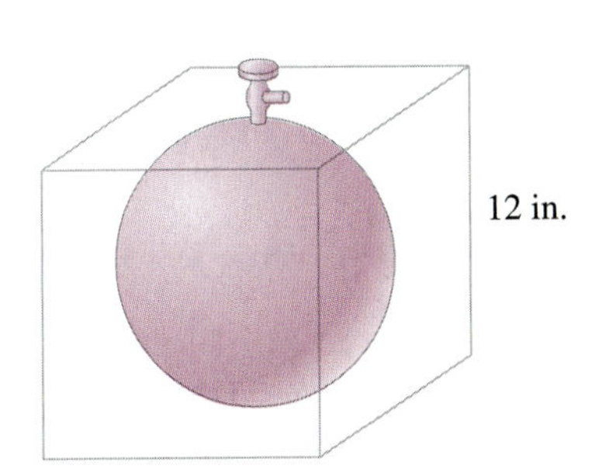

▲ Figure 1.12
Example 1.13 visualized

$V = 30\text{ cm} \times 30\text{ cm} \times 30\text{ cm} = 27 \times 10^3\text{ cm}^3 = 27\text{ L}$. Because there are slightly less than 4 L in one gallon, the volume of the box is about 7 gal. The volume of the tank is clearly less than that of the box, perhaps about one half the volume, say 3.5–4 gal. Of the possibilities given, the one with a volume of 5 gal is correct. (The diameter of the tank is probably about 13 in. rather than 1 ft.)

EXERCISE 1.13

Which of the following is a reasonable estimate of the mass of a 20-qt pail full of water?

5 kg   10 kg   15 kg   20 kg

Conceptual Examples and Exercises

A typical example or exercise in this text illustrates a specific technique or limited idea. Generally, the topic is one currently being discussed, such as significant figures, unit conversions, or density. However, in many cases we need to bring a clear understanding of concepts to bear on a novel situation. We will use a special designation, "A Conceptual Example," for in-chapter Examples in which this is our primary purpose. Conceptual Examples and Exercises may be either qualitative or quantitative; their solutions may require calculations, simple estimates, or verbal explanations.

EXAMPLE 1.14—A Conceptual Example

The fact that iron sinks in water is well known. However, iron is the primary component of steel, a material used to construct oceangoing vessels. How can a material more dense than water be made to float on water?

SOLUTION

If it floats, an intact steel ship displaces a mass of water equal to the mass of the ship. The volume of the ship is not just that of the steel that makes up the ship's hull, but the hull and everything contained within it, including the cargo, the bilge, ballast, empty holds, and so on. Because the ship floats, the mass of a volume of water equal to the volume of the ship must be greater than the mass of the ship. To put the matter in another way: The average density of the total ship, including steel hull, contents, air cavities, is less than the density of water; so the ship floats.

EXERCISE 1.14

The ocean liner *Titanic*, built in 1912, was declared "unsinkable" by its owners, yet it sank when it struck an iceberg on its maiden voyage. How might an ocean liner be constructed to be "unsinkable," and why did the *Titanic* sink despite this?

EXAMPLE 1.15—A Conceptual Example

We want to mail 225 wedding invitations and are unsure whether to use the stamp required for a one-ounce letter or to add additional postage. Of course, we would like to use the minimum postage possible. We don't have a postal scale, but we do have a kitchen scale with a 4-kilogram capacity and marked off in 10-g intervals. The observation we make on weighing one of the invitations is sketched in Figure 1.13. How can we determine the required postage?

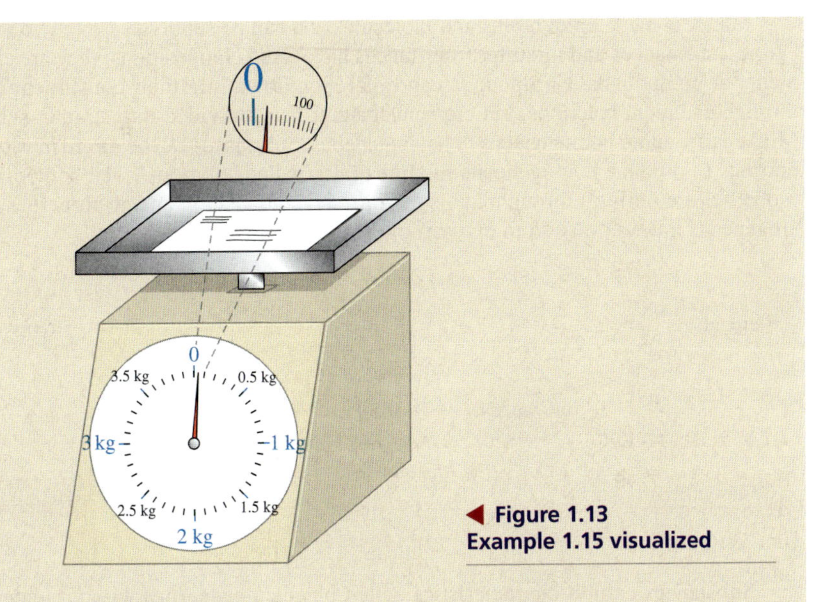

◀ **Figure 1.13 Example 1.15 visualized**

SOLUTION

First, we need the relationship between the ounce (oz) and the gram (g) given in Table 1.5: 1 oz = 28.35 g. From Figure 1.13, it appears that the invitation weighs just about 30 g. Still, we can't be certain that the invitation weighs more than 28.35 g. We can only estimate the mass to the nearest few grams, let's say to the nearest 3 g. This means that the mass of one invitation is 30 ± 3 g; it could be anywhere from 27 g to 33 g.

Now suppose we weigh 100 of the invitations. We expect the mass to be somewhere in the very broad range from 2700 g to 3300 g. To meet the one-stamp requirement, the maximum mass of the 100 invitations is 100 oz × 28.35 g/oz = 2835 g. Thus, if the 100 invitations weigh less than 2830 g, one stamp will probably do; if they weigh more than 2840 g, additional postage is needed.

EXERCISE 1.15

There are a couple of implicit assumptions in the method described in Example 1.15. What are they, and how important do you think they are? When will the method not work?

Summary

Chemistry is the science that deals with the composition and the physical and chemical properties of matter. Some characteristics of a science are that it is testable, reproducible, explanatory, predictive, and tentative. The scientific method involves making observations, forming hypotheses, gathering data by doing experiments, and formulating laws and theories.

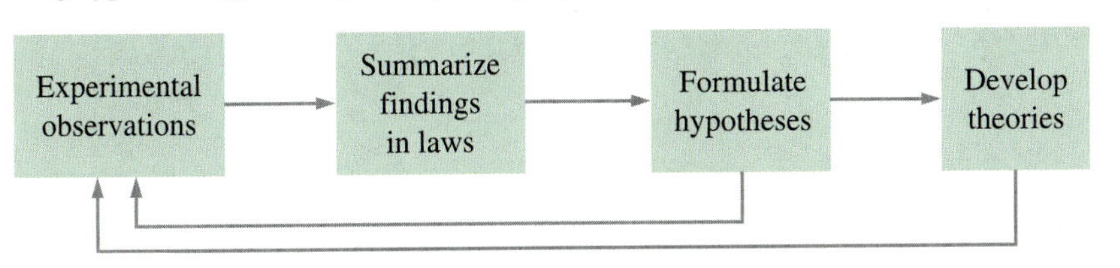

Key Terms

(See Glossary for definitions of these terms.)

accuracy (1.4)
atom (1.2)
chemical change (1.2)
chemical property (1.2)
chemical reaction (1.2)
chemical symbol (1.2)
chemistry (1.2)
composition (1.2)
compound (1.2)
conversion factor (1.5)
data (1.2)
density (*d*) (1.5)
element (1.2)
experiment (1.2)
heterogeneous mixture (1.2)
homogeneous mixture (1.2)
hypothesis (1.2)
kelvin (K) (1.3)
kilogram (kg) (1.3)
liter (L) (1.3)
mass (1.3)
meter (m) (1.3)
mixture (1.2)
molecule (1.2)
physical change (1.2)
physical property (1.2)
precision (1.4)
scientific law (1.2)
second (s) (1.3)
significant figure (1.4)
solution (1.2)
substance (1.2)
theory (1.2)

Matter is made up of atoms and molecules and can be subdivided into two broad categories: substances and mixtures. Substances have fixed compositions; they are either elements or compounds. Compounds can be broken down into their constituent elements through chemical reactions, but elements cannot be subdivided into simpler substances. Mixtures are either homogeneous or heterogeneous. Substances can be mixed in varying proportions to produce homogeneous mixtures (also called solutions). The composition and properties are uniform throughout a solution. The composition and/or properties of a heterogeneous mixture vary from one part of the mixture to another.

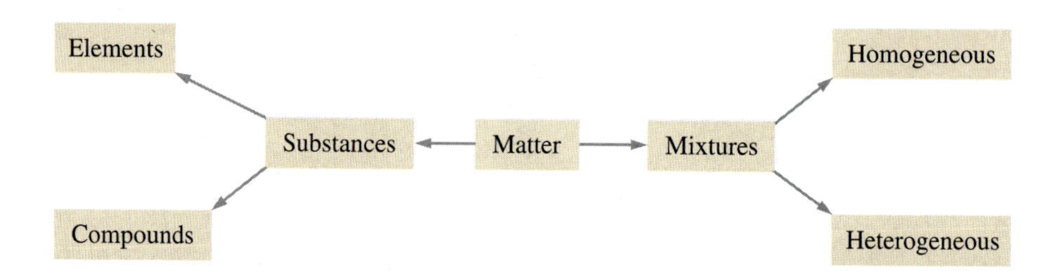

Substances exhibit characteristics called physical properties without undergoing a change in composition. In displaying a chemical property, a substance undergoes a change in composition—new substances are formed. A physical change produces a change in the appearance of a sample of matter but no change in its microscopic structure and composition. In a chemical change, the composition and/or microscopic structure of matter changes.

Four basic physical quantities of measurement are introduced in this chapter: mass, length, time, and temperature. In the SI system, measured quantities may be reported in the base unit or as multiples or submultiples of the base unit. Multiples and submultiples are based on powers of ten and reflected through prefixes in names and abbreviations.

*nano*meter(nm)	*micro*meter (μm)	*milli*meter (mm)	meter(m)	*kilo*meter (km)
10^{-9} m	10^{-6} m	10^{-3} m	m	10^{3} m

The SI base unit of temperature, the kelvin (K), is introduced in Chapter 5, but in this chapter two other temperature scales, Celsius and Fahrenheit, are considered and compared.

$$t_F = 1.8\,t_C + 32 \qquad t_C = \frac{(t_F - 32)}{1.8}$$

To indicate its precision, a measured quantity must be expressed with the proper number of significant figures. Furthermore, special attention must be paid to the concept of significant figures in reporting calculated quantities. Calculations themselves frequently can be done by the unit-conversion method. The physical property of density also serves as an important conversion factor. The density of a material is its mass per unit volume: $d = m/V$. When the volume of a substance or homogeneous mixture (cm^3) is multiplied by its density (g/cm^3), volume is converted to mass. When the mass of a substance or homogeneous mixture (g) is multiplied by the *inverse* of density (cm^3/g), mass is converted to volume.

In this chapter and throughout the text, Examples and Exercises illustrate the ideas, methods, and techniques under current discussion. In addition, Estimation Examples and accompanying Exercises deal with means of obtaining estimated answers with a minimum of calculation. Conceptual Examples and accompanying Exercises apply fundamental concepts to answer questions that are often of a qualitative nature.

Review Questions

1. What is matter?
2. Which of the following are examples of matter?
   (a) iron
   (b) air
   (c) the human body
   (d) red light
   (e) gasoline
   (f) an idea
3. What are some of the requirements of scientific activity, as compared, for example, with artistic activity?
4. What is a hypothesis? How are hypotheses tested?
5. What are scientific data, and what characterizes good data?
6. What is a scientific law? How does it differ from a legislative law?
7. What is a theory? What must one be able to do with a theory? Can a theory be proved?
8. A formerly chubby but now trim person completes a successful diet. Has the person's weight changed? Has the person's mass changed? Explain.
9. Sample A, which is on the moon, has exactly the same mass as sample B on Earth. Do the two samples weigh the same? Explain your answer.
10. How do physical and chemical properties differ?
11. Which of the following is *not* a physical property? Explain.
    (a) Solid iron melts at a temperature of 1535 °C.
    (b) Solid sulfur has a yellow color.
    (c) Natural gas burns.
    (d) Diamond is extremely hard.
12. Which of the following describe a chemical change, and which a physical change?
    (a) Sheep are sheared and the wool is spun into yarn.
    (b) A cake is baked from a mixture of flour, baking powder, sugar, eggs, shortening, and milk.
    (c) Milk turns sour when left out of the refrigerator for many hours.
    (d) Silkworms feed on mulberry leaves and produce silk.
    (e) An overgrown lawn is manicured by mowing it with a lawn mower.
13. Refer to the listing of the elements inside the front cover, and determine which of the following represent elements and which do not. Explain.
    (a) C
    (b) CO
    (c) Cl
    (d) $CaCl_2$
    (e) Na
    (f) KI
14. Which of the following are substances, and which are mixtures? Explain.
    (a) helium gas used to fill a balloon
    (b) the juice squeezed from a lemon
    (c) a premium red wine
    (d) salt used to de-ice roads
15. Which of the following mixtures are homogeneous, and which are heterogeneous?
    (a) gasoline
    (b) raisin pudding
    (c) Italian salad dressing
    (d) a cola drink
16. What are the names and symbols of the SI base units for length, mass, and time?
17. Express the SI units for area, volume, and density in terms of SI *base* units.
18. What is the difference between the precision and the accuracy of a set of measurements?
19. Can a set of measurements be precise without being accurate? Can the average of a set of measurements be accurate even if the individual measurements are imprecise? Explain.
20. How does the number of significant figures in a measured quantity relate to the precision of the measurement? to the accuracy?
21. If the numerical value of a measured quantity includes zeros, are these zeros significant? Explain.
22. What is a conversion factor? How must the numerator and denominator of a conversion factor be related?
23. Why are the terms "light" and "heavy" imprecise when used to refer to low-density and high-density materials?
24. How does temperature generally affect the density of a material? Explain your answer.

Problems

A word of advice: You cannot learn to work problems by reading them or watching your teacher work them, any more than you could become a piano player solely by reading about piano-playing skills or attending a performance. As you work through problems, you will find many opportunities to improve your understanding of the ideas presented in the chapter and also to practice your estimation skills and your ability to synthesize concepts. Plan to work through the great majority of these problems.

Measurement and Unit Conversion

25. Change the unit used to report each of the following measurements by replacing the power of ten by an appropriate SI prefix.
    (a) 4.54×10^{-9} g
    (b) 3.76×10^{3} m
    (c) 6.34×10^{-6} g

26. Use exponential notation to express each of the following measurements in terms of an SI base unit.
(a) 1.09 ng **(b)** 9.01 ms **(c)** 145 pm

27. Carry out the following temperature conversions.
(a) 23.5 °C to °F **(c)** −98.0 °C to °F
(b) 173.9 °F to °C

28. The melting point of a solid is the temperature at which the solid undergoes a physical change to become a liquid. The following melting points are expressed on one temperature scale. Convert them to temperatures on the scale indicated.
(a) Convert to °F: iron, 1535 °C; silver, 961.93 °C; mercury, −38.36 °C
(b) Convert to °C: gold, 1948.57 °F; lead, 621.37 °F; ethyl alcohol, −174.1 °F

29. Make the following temperature conversions.
(a) The high-temperature record for the continent of Africa is 136 °F, recorded at Azizia, Libya, in 1922. What is this temperature in degrees Celsius?
(b) In the Martian winter, the temperature at the poles drops to −120 °C, freezing water vapor and carbon dioxide from the atmosphere. What is this temperature in degrees Fahrenheit?

30. Answer the following temperature scale questions.
(a) A candy recipe calls for heating a sugar mixture to the "soft ball" stage (234 to 240 °F). Can a laboratory thermometer with a range of −10 to 110 °C be used for this measurement?
(b) At what temperature do the Fahrenheit and Celsius readings have the same numerical value? Can they have the same value at more than one temperature?

31. Carry out the following conversions.
(a) 50.0 km to meters
(b) 47.9 mL to liters
(c) 578 μs to milliseconds
(d) 1.55×10^2 kg to milligrams
(e) 87.4 cm^2 to mm^2
(f) 0.0962 km/min to m/s

32. Carry out the following conversions.
(a) 546 mm to meters
(b) 87.6 mg to kilograms
(c) 181 pm to micrometers
(d) 1.00 h to microseconds
(e) 46.3 m^3 to liters
(f) 55 mi/h to km/min

33. *Without doing detailed calculations,* arrange the following in order of increasing length (shortest first): (1) a 1.21-m chain, (2) a 75-in. rope, (3) a 3 ft-5-in. rattlesnake, (4) a yardstick.

34. *Without doing detailed calculations,* arrange the following in order of increasing mass (lightest first): (1) a 5-lb bag of potatoes, (2) a 1.65-kg cabbage, (3) 2500 g sugar.

Significant Figures

35. How many significant figures are there in each of the following measured quantities?
(a) 8008 m **(d)** 6.02×10^5 m
(b) 0.00075 s **(e)** 4.200×10^5 s
(c) 0.049300 g **(f)** 0.1050 °C

36. How many significant figures are there in each of the following measured quantities?
(a) 4051 m **(d)** 5.00×10^9 m
(b) 0.0169 s **(e)** 1.60×10^{-9} s
(c) 0.0430 g **(f)** 0.0150 °C

37. Express each of the following measured quantities in exponential notation.
(a) 2804 m **(c)** 0.00090 cm
(b) 901 s **(d)** 221.0 s

38. Express each of the following measured quantities in exponential notation.
(a) 8352 m **(c)** 0.0885 cm
(b) 300.0 s **(d)** 122.2 s

39. Express each of the following measured quantities in common decimal form (for example, $3.11 \times 10^2 = 311$). Comment on the significance of any terminal zeros that you write.
(a) 5.055×10^2 m **(c)** 6.10×10^{-3} g
(b) 2.12×10^3 s **(d)** 4.00×10^4 mL

40. Express each of the following measured quantities in common decimal form (for example, $6.375 \times 10^3 = 6375$). Comment on the significance of any terminal zeros that you write.
(a) 3.18×10^5 m **(c)** 4.1×10^{-4} s
(b) 7.50×10^2 mL **(d)** 9.200×10^4 m

41. Perform the indicated operations, and give answers with the proper number of significant figures.
(a) 36.5 m − 2.16 m + 3.452 m
(b) 151 g + 4.16 g − 0.0220 g
(c) 15.44 mL − 9.1 mL + 105 mL
(d) 12.52 cm + 5.1 cm − 3.18 cm − 12.02 cm

42. Perform the indicated operations, and give answers in the indicated unit and with the proper number of significant figures.
(a) 13.25 cm + 26 mm − 7.8 cm + 0.186 m (in cm)
(b) 48.834 g + 717 mg − 0.166 g + 1.0251 kg (in kg)

43. Perform the indicated operations, and give answers with the proper number of significant figures.
(a) $73.0 \times 1.340 \times (25.31 - 1.6) = ?$

(b) $\dfrac{33.58 \times 1.007}{0.00705} = ?$

(c) $\dfrac{418.7 \times 31.8}{(19.27 - 18.98)} = ?$

(d) $\dfrac{2.023 - (1.8 \times 10^{-3})}{1.05 \times 10^{4}} = ?$

44. Perform the indicated operations, and give answers with the proper number of significant figures.

a. $265.02 \times 0.000581 \times 12.18 = ?$

b. $\dfrac{22.61 \times 0.0587}{135 \times 28} = ?$

c. $\dfrac{(33.62 + 12.2 - 48.36)}{26.4 \times 12.13} = ?$

d. $\dfrac{(4.6 \times 10^{3} + 2.2 \times 10^{2})}{3.11 \times 10^{4} \times 7.12 \times 10^{-2}} = ?$

Density

45. A 25.0-mL sample of liquid bromine has a mass of 78.0 g. What is the density of the bromine?

46. What is the density of a salt solution if 50.0 mL has a mass of 57.0 g?

47. Some metal chips having a volume of 3.29 cm^3 are placed on a piece of paper and weighed. The combined mass is found to be 18.432 g. The paper itself weighs 1.214 g. Calculate the density of the metal, expressed with the proper number of significant figures.

48. A glass container weighs 48.462 g. A sample of 4.00 mL of antifreeze solution is added, and the container plus the antifreeze weigh 54.513 g. Calculate the density of the antifreeze solution, expressed with the proper number of significant figures.

49. A rectangular block of lead is 1.20 cm $\times$ 2.41 cm $\times$ 1.80 cm and has a mass of 59.01 g. Calculate the density of lead.

50. A rectangular block of gold-colored material measures 3.00 cm $\times$ 1.25 cm $\times$ 1.50 cm and has a mass of 28.12 g. Can the material be gold? The density of gold is 19.3 g/cm^3.

51. What is the mass, in grams, of 30.0 mL of a syrup that has a density of 1.32 g/mL?

52. What is the mass, in kilograms, of 2.75 L of the liquid glycerol, which has a density of 1.26 g/mL?

53. What is the volume of a 898-kg piece of cast iron (d = 7.76 g/cm^3)? How long a cylindrical bar with a base area of 1.50 cm^2 can be made from the iron?

54. What is the volume of 5.79 mg of gold (d = 19.3 g/cm^3)? If the gold is hammered into gold leaf of uniform thickness with an area of 44.6 cm^2, what is the thickness of the gold leaf?

55. A box with a square base, measuring 0.80 m on each side, and having a height of 1.20 m, is filled with 3.2 kg of expanded polystyrene packing material. What is the bulk density, in grams per cubic centimeter, of the packing material? (The bulk density includes the air between the pieces of polystyrene foam.)

56. Hylon VII, a starch-based substitute for polystyrene packing material, has a bulk density of 12.8 kg/m^3. How many grams of the material are needed to fill a volume of 2.00 ft^3?

57. Which of the following items would be most difficult to lift onto the back of a pickup truck: (1) a 100-lb bag of potatoes, (2) a 15-gal plastic bottle filled with water, (3) a 3.0-L flask filled with mercury (d = 13.6 g/cm^3)?

58. Liquid mercury (d = 13.6 g/cm^3) is commonly shipped in iron flasks that contain 76 lb of mercury. Will one of these flasks fit inside a wooden box that has inside dimensions of 4.0 in. $\times$ 4.0 in. $\times$ 8.0 in.?

59. Several irregularly shaped pieces of zinc weighing 30.0 g are dropped into the graduated cylinder below on the left. The water level rises as shown in the right cylinder. Calculate the density of zinc, and express your result with the maximum number of significant figures permitted in this experiment.

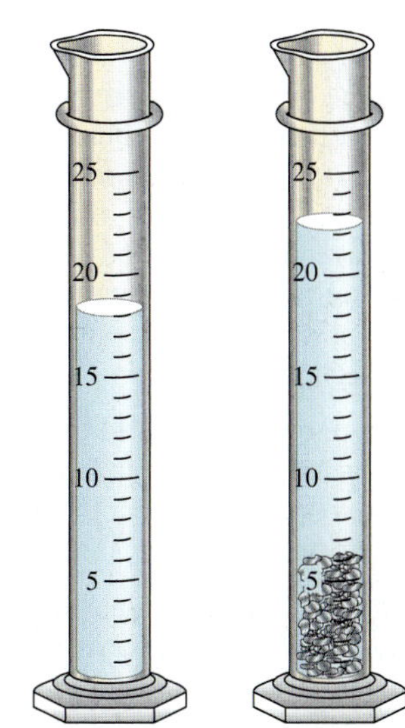

60. The container pictured below on the left is filled with water at 20 °C, just to the overflow spout. A cube of wood with edges of 1.0 in. is floated on the water, and 10.8 mL of water is collected, as shown on the right. Calculate the density of the wood, and express your result with the maximum number of significant figures permitted in this experiment.

Additional Problems

61. Sample B, which is on the moon, has the same mass as does sample A on Earth. Do samples A and B weigh the same? If not, which weighs more? Explain. Sample C, also on the moon weighs the same as does sample A on Earth. Do samples A and C have the same mass? If not, which has the greater mass? Explain.

62. Must it always be true that the result of a calculation can carry no more significant figures than the term in the calculation having the smallest number of significant figures? Explain.

63. If the meter stick used in the measurements in Table 1.4 were actually 1.005 mm long, how much error would this produce in calculating the area of the poster board in Example 1.4?

64. Which device do you think would be more precise for measuring the length of the poster board in Table 1.4 on page 14: a meter stick or a 2-m steel tape measure, each marked in millimeter increments? Which do you think would be more accurate? Explain your choices. Would the situation be the same for measuring the width? Explain.

65. Carl Wunderlich was a German physician who first recognized fever as a symptom of disease. During the 1840s and 1850s, he averaged a large number of human temperature measurements and rounded the value to 37 °C. What is the equivalent temperature in degrees Fahrenheit? In recent years millions of measurements have revealed that average normal body temperature is actually 98.2 °F. What is the equivalent temperature in degrees Celsius? Express both results with the proper number of significant figures.

66. In its nonstop, round-the-world trip in 1986, the aircraft *Voyager* traveled 25,102 mi in 9 days, 3 min, and 44 s. Using the maximum number of significant figures permitted, calculate the average speed of *Voyager* in mi/h.

67. The *furlong* is a unit used in horse racing, and the units *chain* and *link* are used in surveying land. There are 8 furlongs in 1 mi, 10 chains in one furlong, and 100 links in one chain. To three significant figures, calculate the length of one link, in inches. (1 mi = 5280 ft)

68. On July 23, 1983, Air Canada Flight 143 required 22,300 kg of jet fuel to fly from Montreal to Edmonton. The density of jet fuel is 0.803 g/mL, or 1.77 lb/L. The plane had 7682 L of fuel on board in Montreal. The ground crew there multiplied the 7682 L by the factor 1.77 (without units) and concluded that they had 13,597 kg of fuel on board and needed an additional 8703 kg for the trip to Edmonton. They divided 8703 kg by the factor 1.77 (again without units) and concluded that they needed to add 4916 L of fuel. They added 5000 L. On its flight, the plane ran out of fuel and crash-landed near Winnipeg, hundreds of kilometers short of its destination. (There were a few injuries but no fatalities.) What mistake did the ground crew make? How much fuel should they have added before takeoff?

69. In the United States, land area is commonly measured in acres: 640 acre = 1 mi^2. In most of the rest of the world, land area is measured in hectares: 1 hectare = 1 hm^2 [1 hectometer (hm) = 100 m]. Which is the larger area, the acre or the hectare? Write a conversion factor that relates the acre and the hectare. (1 mi = 5280 ft; 1 m = 39.37 in.)

70. In scientific work, densities usually are expressed in grams per cubic centimeter. Engineers often use the unit pounds per cubic foot (lb/ft^3). The density of water at 20 °C is 0.998 g/cm^3. What is its density in pounds per cubic foot?

71. A square of aluminum foil (d = 2.70 g/cm^3) is 5.10 cm on a side and has a mass of 1.762 g. Calculate the thickness of the foil, in millimeters.

72. An empty 3.00-L bottle weighs 1.70 kg. Filled with a homemade wine, the bottle weighs 4.72 kg. The wine contains 11.0% ethyl alcohol by mass. How many ounces of ethyl alcohol are present in 275 mL of this wine?

73. Aerogels consist of a solid framework with most of their volume occupied by air. Silica (silicon dioxide) powder has a density of 2.2 g/cm^3. Silica can be expanded into an aerogel with a bulk density of 0.015 g/cm^3. Calculate the volume of silica aerogel that can be made from 125 cm^3 of silica powder.

74. Calculate the total surface area of the aerogel described in Problem 73, when there is 470 m^2 of surface per gram of silica in the aerogel.

75. Dust from air that is not significantly polluted is deposited ("dustfall") at a typical rate of 10 tons/mi^2 per month. What is this dustfall expressed in milligrams per square meter per hour?

76. The device shown on page 33, called a pycnometer, can be used to determine the densities of liquids very precisely. The pycnometer is weighed empty and again when filled with a liquid of known density, usually water. The data obtained in these two measurements can be used to determine the volume of the pycnometer. Finally, the pycnometer is weighed when filled with a liquid of unknown density. Now the unknown density can be calculated. From the data given, determine the density of benzene at 20 °C.

Empty:
32.105 g

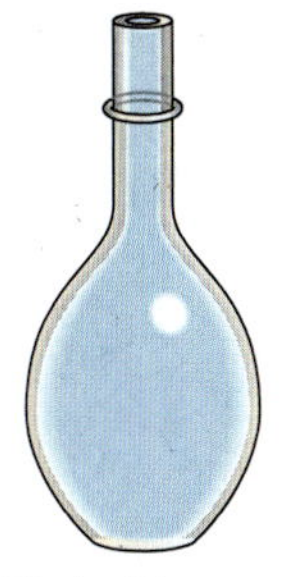

Filled with water
at 20 °C: 42.062 g
Density of water:
0.9982 g/ml

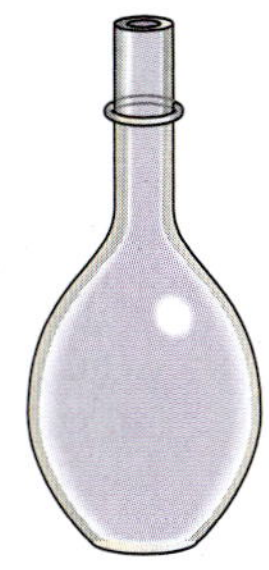

Filled with benzene
at 20 °C: 40.873 g
Density of benzene:
?

77. A pycnometer similar to the one illustrated in Problem 76 is used to determine the density of granular zinc metal. In experiments performed at 20 °C, the mass of the dry, empty pycnometer is found to be 36.2142 g. When the pycnometer is filled with water, its mass is 46.1894 g. When a small sample of zinc is added to the dry, empty pycnometer, the mass is 38.1055 g. Finally, when the pycnometer with the sample of zinc is again filled with water, the measured mass is 47.8161 g. Use these data, together with the density of water at 20 °C—0.99823 g/mL—to determine the density of the zinc.

78. Refer to Figure 1.10. When a block of padouk wood floats on water, what fraction of the block is above the waterline and what fraction is below the waterline? Explain your reasoning.

79. The two vessels shown are completely filled with water. A brass cube 2.0 cm on edge is gently placed on the water in the vessel on the left, and a rectangular block of cork, 5.0 cm × 4.0 cm × 2.0 cm, on the water in the vessel on the right. The density of brass is 8.40 g/cm^3 and that of cork is 0.22 g/cm^3. From which vessel will the greater volume of water overflow?

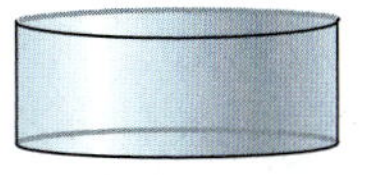

80. The square nut pictured is 14.0 mm on edge, 6.0 mm thick, and has a 7.0-mm diameter hole. The density of the metal used in the nuts is 7.87 g/cm^3. Approximately how many of these nuts are present in a 1.00-lb package?

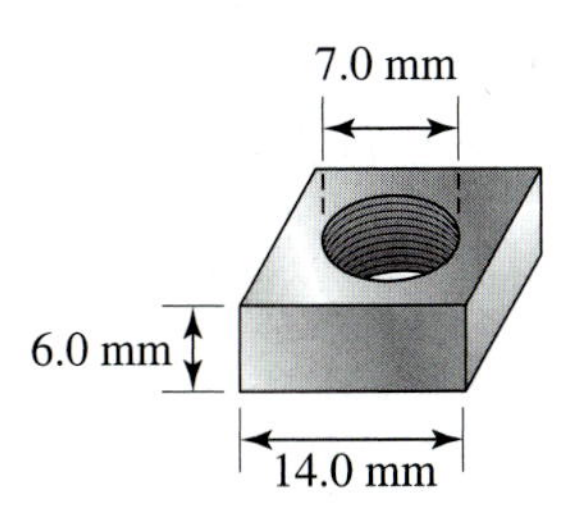

81. The significant figure rules outlined in the chapter give a first approximation of the precision of a calculation based on measurements. A more refined approach looks at the absolute and percentage errors in measured quantities. The data in Table 1.4 on page 14, for example, include a length of 1.827 ± 0.003 m and a width of 0.762 ± 0.001 m. The absolute errors are ± 0.003 m and ± 0.001 m, respectively, and the percent errors are

$$\frac{0.003\ \cancel{m}}{1.827\ \cancel{m}} \times 100\% = 0.2\% \quad \text{and}$$

$$\frac{0.001\ \cancel{m}}{0.762\ \cancel{m}} \times 100\% = 0.1\%$$

In addition and subtraction, the absolute error in the result is the sum of the absolute errors in the measurements. In multiplication and division, the accumulated percent error is the sum of the percent errors.

(a) Use data in Table 1.4 to determine the perimeter of the poster board and the absolute error in the perimeter.

(b) Use data in Table 1.4 to determine the area of the poster board, the percent error, and the absolute error in the area.

(c) Show that by the method outlined here, you are permitted to carry one more significant figure in the area of the poster board than you are by the significant figure rules introduced in the chapter. Do you think that this would be the case in all calculations? Explain.

82. The food section of a newspaper presented a recipe for "Hamburgers Inside Out." The ingredients required for 6 burgers were listed as: $1\frac{1}{2}$ pounds ground beef; 2 to 3 ounces of bleu cheese; vegetable oil; six slices of French-style bread; 4 tablespoons (or so) of butter; salt; black pepper; 1/4 cup minced shallots; 1 cup red wine. The following nutritional information was given for each burger: 1765 calories; 58 g total fat (23 g saturated fat); 62 g protein; 238 g carbohydrates; 128 mg cholesterol; 3117 mg sodium.

Criticize this reporting of nutritional information in relation to the information given about the ingredients. Rewrite the nutritional information in a way that better conforms to the reporting of measured and calculated quantities described in Section 1.4.

83. Following is a table of data used to determine the percent body fat of an individual.

Body density (g/cm^3)	% Body fat
1.010	38.3
1.030	29.5
1.050	21.0
1.070	12.9
1.090	5.07

Make a graph of the data (see Appendix A), and then estimate the percent body fat of the following.

(a) A person with a body density of 1.037 g/cm^3.

(b) A person who weighs 165 lb in air and 14 lb when submerged in water.

(c) Which person is more likely to be a well-trained athlete?

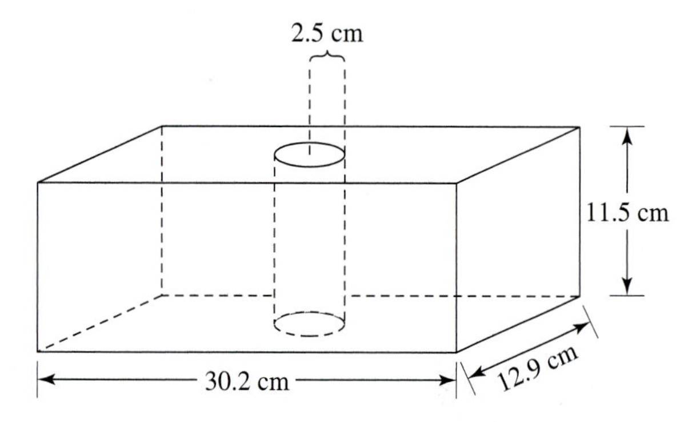

84. We want to determine if the object pictured, which has a mass of 5.15 kg, can be used as a float on pure water.

(a) *Without doing detailed calculations, (that is, by estimation),* show that the object will not float.

(b) Suppose that we completely fill the cylindrical hole with a plug of balsa wood ($d = 0.1 \text{ g/cm}^3$). Show that the object still will not float.

(c) What minimum number of cylindrical holes like that shown would have to be drilled and filled with balsa plugs if the object is to float?

Atoms, Molecules, and Ions

CHAPTER

2

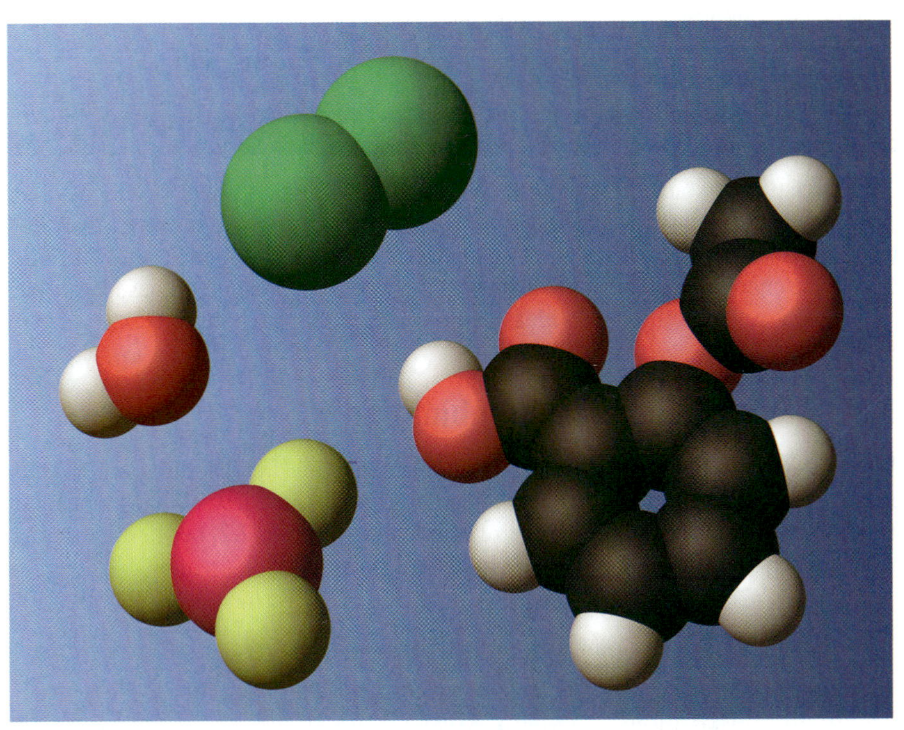

Much of our study of chemistry is focused on molecules, which we often represent by space-filling models. Clockwise from left center are models of water, H_2O; chlorine, Cl_2; acetylsalicyclic acid, $CH_3COOC_6H_4COOH$; and boron trifluoride, BF_3.

CONTENTS

The language of chemistry has a special vocabulary: atoms, ions, molecules, isotopes, acids, bases, salts, saturated hydrocarbons. . . . Chemistry also has a symbolic language. Atoms of elements are represented by *symbols* such as H, C, N, O, Na, and Cl. These symbols are the "alphabet" of chemistry. To represent compounds, symbols are combined into chemical formulas, such as NaCl, H_2O, CO_2, and C_5H_{12}. Formulas are the "words" of chemistry. We will explore both the special vocabulary and the "alphabet" and "words" of the chemist's symbolic language in this chapter. In the next chapter we will extend the symbolic language to include its "sentences"—chemical equations.

In the first part of this chapter we will focus on some basic ideas about atoms and the laws that govern the combination of atoms of the elements into compounds. In the second part we will introduce two broad categories of compounds: molecular and ionic. Our emphasis in that discussion will be on **chemical nomenclature**, the relationship between the names and formulas of chemical compounds.

● **ChemCDX**
When you see this icon, there are related animations, demonstrations, and exercises on the CD accompanying this text.

Antoine Lavoisier (1743–1794) was a chemist, biologist, and economist. He used the return on investments in a much-hated private tax-collecting agency of King Louis XVI to equip a private laboratory and finance his research. Lavoisier's experience as an accountant led him to apply principles of the balance sheet to everything, including chemical reactions. As a royal tax collector, though, he earned the enmity of leaders of the French Revolution and was beheaded on a guillotine. Little did his executioners realize that their victim would later become known as "the father of modern chemistry."

Specifically, we will learn to write the formulas of compounds if we know their names, and the names of compounds if we know their formulas. A working knowledge of nomenclature is useful because the name of a compound often provides clues about its structure, chemical behavior, or both.

In the final part of the chapter we will introduce a general type of compound based on the element carbon and called *organic*. Organic compounds vastly outnumber all other compounds, which collectively are called *inorganic*. Over 95% of the 18 million or so known compounds are organic. Although our introduction will be limited in scope, it should still provide a glimpse of the enormous diversity of organic compounds.

Laws and Theories: A Brief Historical Introduction

Early practitioners accumulated a lot of useful information about chemicals and their behavior, but it wasn't until the laws of chemical combination were firmly established about two centuries ago that chemistry became a scientific study. In this part of the chapter, we look at the historical development of some of these laws. We also look at several other basic ideas that we will use throughout the text.

2.1 Laws of Chemical Combination

We have learned that data obtained by experiment can often be summarized into laws. Scientific theories are then formulated to explain these laws. We present John Dalton's theory of the atomic structure of matter in Section 2.2, but first let's consider two laws that he attempted to explain with his theory.

Lavoisier: The Law of Conservation of Mass

Modern chemistry dates from the eighteenth century, when scientists began to make *quantitative* observations. Antoine Lavoisier, through measurements precise to about 0.0001 g, found that total mass did not change during a chemical reaction. For example, Lavoisier heated the red oxide of mercury, causing it to decompose (break down) to form two new products: mercury metal and oxygen gas. By measuring carefully, he found that the total mass of the products was exactly the same as the mass of the mercury oxide he started with.

Lavoisier summarized his findings through the **law of conservation of mass**.

*The total mass remains constant during a chemical reaction.**

EXAMPLE 2.1—A Conceptual Example

Jan Baptista van Helmont (1579–1644) planted a young willow tree in a weighed bucket of soil. After five years he found that the tree had gained 75 kg in mass, yet the soil had lost only 0.057 kg. He had added only water to the bucket, and so he concluded that the substance of the tree had come from the water. Explain and criticize his conclusion.

*A twentieth-century modification of this law, based on the work of Albert Einstein, allows for the possibility of converting mass to energy (and vice versa). This need not concern us, however, because mass changes corresponding to the energy changes in chemical reactions are too small to detect.

SOLUTION

Van Helmont's explanation anticipated the law of conservation of mass by dismissing the possibility that the mass of the tree could have been created from nothing. His main failure, however, was in not identifying all of the substances involved. He did not know about the role of water vapor and carbon dioxide gas in the growth of plants. In applying the law of conservation of mass we must focus on *all* the substances involved in a chemical reaction.

EXERCISE 2.1

A photographic flashbulb containing magnesium and air weighs 7.500 g. On firing, a brilliant flash of white light is emitted and a white powder is formed inside the bulb. What should be the mass of the bulb after firing? Explain.

Proust: The Law of Definite Proportions

By the end of the eighteenth century, Lavoisier and other scientists had succeeded in causing many compounds to decompose into the elements that form them. One of these scientists, Joseph Proust (1754–1826), did careful quantitative studies by which he established the **law of constant composition** (also called the **law of definite proportions**).

All samples of a compound have the same composition; that is, all samples have the same proportions by mass of the elements present.

Proust based this law on his studies of a substance that we now call basic copper carbonate (Figure 2.1); all samples of this compound have the same composition: 57.48% copper, 5.43% carbon, 0.91% hydrogen, and 36.18% oxygen, by mass.

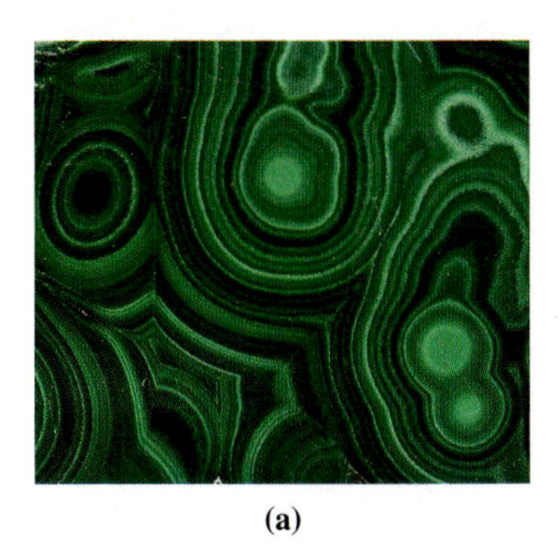

(a)

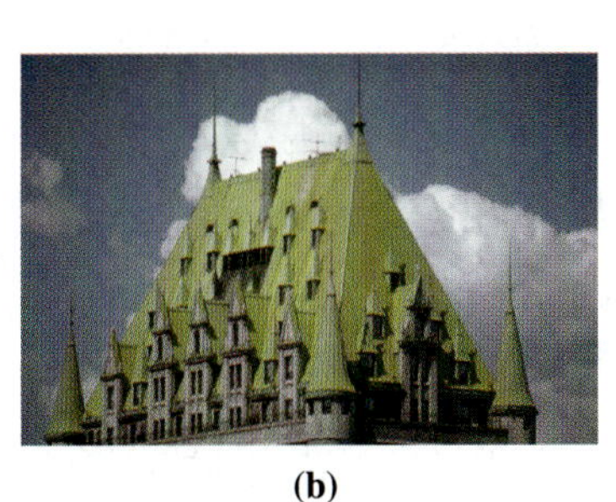

(b)

(c)

▲ **Figure 2.1 The law of definite proportions**
Basic copper carbonate occurs in nature as the mineral *malachite* (a), and it forms as a patina on copper roofs and bronze statues (b). Basic copper carbonate can also be synthesized in the laboratory (c). Regardless of its source, basic copper carbonate has the same composition.

A compound not only has a constant or fixed composition, it also has fixed properties. At 20 °C, 100.0 g of pure water always dissolves a maximum of 35.9 g of sodium chloride (salt). Under normal atmospheric pressure, pure water always freezes at 0 °C and boils at 100 °C. The physical and chemical properties of chemical substances depend on their composition; they are not a matter of chance. We will learn more about how and why this is so in later chapters.

EXAMPLE 2.2

The mass ratio of oxygen to magnesium in the compound magnesium oxide is 0.6583 : 1. What mass of magnesium oxide will form when 2.000 g of magnesium is completely converted to magnesium oxide by burning in pure oxygen gas?

SOLUTION

We usually write mass ratios in a form such as "the ratio of O to Mg is 0.6583 : 1." The first number represents, in this case, a mass of *oxygen*, say 0.6583 g oxygen, and the second number, a mass of *magnesium*. Although written only as "1," we assume that the second number is as precisely known as the first, for example, 1.000 g magnesium. According to the law of constant composition, just the right mass of oxygen must combine with the magnesium so that the mass ratio of oxygen to magnesium in the product is 0.6583 : 1. We can state the mass ratio in the form of a conversion factor, and then determine the required mass of oxygen.

$$? \text{ g oxygen} = 2.000 \cancel{\text{ g magnesium}} \times \frac{0.6583 \text{ g oxygen}}{1.000 \cancel{\text{ g magnesium}}} = 1.317 \text{ g oxygen}$$

Now, according to the law of conservation of mass, the mass of magnesium oxide, the product, must equal the total of the masses of the substances entering into the reaction.

$$\begin{aligned} ? \text{ g magnesium oxide} &= 2.000 \text{ g magnesium} + 1.317 \text{ g oxygen} \\ &= 3.317 \text{ g magnesium oxide} \end{aligned}$$

EXERCISE 2.2A

What mass of magnesium oxide is formed when 1.500 g of oxygen combines with magnesium?

EXERCISE 2.2B

A 3.250-g sample of magnesium is burned in a container of 12.500 g oxygen. What mass of oxygen gas remains *unreacted* after the magnesium has been completely consumed to form magnesium oxide as the sole product?

John Dalton (1766–1844) was not principally an experimenter (perhaps because he was color-blind), but he skillfully used the results of others in formulating his atomic theory. Dalton also kept records of the weather throughout his life. His studies of humidity as measured by dew point temperatures led to his discovery of an important law concerning mixtures of gases (Section 5.10).

2.2 John Dalton and the Atomic Theory of Matter

In 1803, John Dalton proposed a theory to explain the laws of conservation of mass and constant composition. As he developed his atomic theory, Dalton found evidence that required another scientific law that the theory would have to explain. He noted that a given set of elements may be able to combine to produce two or more different compounds, each with a unique composition. Let's illustrate Dalton's reasoning with a simple example, two compounds of the elements carbon and oxygen. In carbon dioxide, the familiar gas produced in respiration and in the burning of fuels, the two elements combine in the ratio of 8.0 g oxygen to 3.0 g carbon. In carbon monoxide, the poisonous gas formed when a fuel is burned in limited air, the elements combine in the ratio of 4.0 g oxygen to 3.0 g carbon. Based on evidence such as this, Dalton formulated the **law of multiple proportions**, which we can state in the following way.

When two or more different compounds of the same two elements are compared, the masses of one element that combine with a fixed mass of the second element are in the ratio of small whole numbers.

Thus for carbon dioxide and carbon monoxide, the masses of oxygen (8.0 g and 4.0 g) that combine with a fixed mass of carbon (3.0 g) are in a small whole-number ratio, 8.0:4.0 = 2:1. Figure 2.2 shows a scheme for establishing this 2:1 ratio.

	Carbon monoxide (CO)	**Carbon dioxide (CO_2)**
The elements	3.0 g carbon (C) + 4.0 g oxygen (O)	3.0 g carbon (C) + 8.0 g oxygen (O)
The compound	7.0 g carbon monoxide (CO)	11.0 g carbon dioxide (CO_2)
Oxygen-to-carbon mass ratio	$\frac{4.0\text{ g oxygen}}{3.0\text{ g carbon}}$	$\frac{8.0\text{ g oxygen}}{3.0\text{ g carbon}}$

Comparing two mass ratios:

Mass ratio for CO_2 / Mass ratio for CO

$$\frac{\frac{8.0\text{ g oxygen}}{3.0\text{ g carbon}}}{\frac{4.0\text{ g oxygen}}{3.0\text{ g carbon}}} = \frac{8.0\text{ g oxygen}}{4.0\text{ g oxygen}} = 2:1$$

Figure 2.2 **The law of multiple proportions** The oxygen-to-carbon mass ratio in carbon dioxide is twice that in carbon monoxide.

Dalton's Atomic Theory

The atomic model Dalton developed to explain the laws of chemical combination is based on four ideas.

● ChemCDX Multiple Proportions

- *All matter is composed of extremely small, indivisible particles called atoms.*
- *All atoms of a given element are alike, in mass and other properties, but atoms of one element differ from the atoms of every other element.*
- *Compounds are formed when atoms of different elements unite in fixed proportions.* [The relative numbers of each kind of atom in a compound form a simple ratio, such as one atom of A to one of B (AB), two atoms of A to one of B (A_2B), and so on.]
- *A chemical reaction involves a rearrangement of atoms. No atoms are created, destroyed, or broken apart in a chemical reaction.*

Explanations Using Dalton's Theory

Let's first explain the law of constant composition and the law of conservation of mass in the way that Dalton might have. For simplicity, we will consider the compound hydrogen fluoride, which was not known in Dalton's time, and use the following reasoning suggested by Figure 2.3.

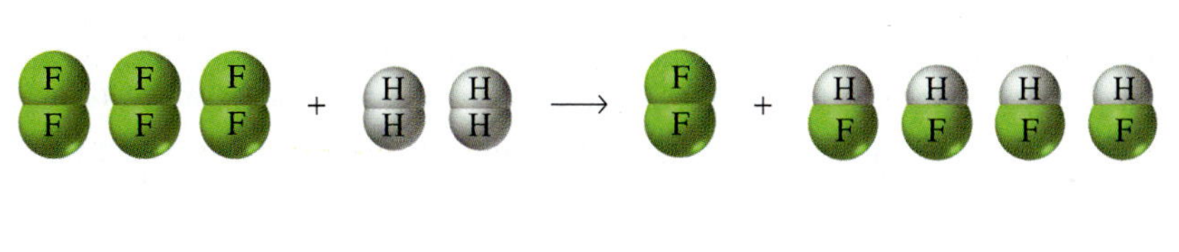

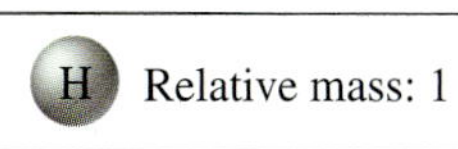

Figure 2.3 **Dalton's theory and the laws of constant composition and conservation of mass** Dalton's theory explained these two basic laws of chemical combination as described in the text.

All hydrogen fluoride molecules are made up of one hydrogen atom and one fluorine atom. Because a fluorine atom has 19 times the mass of a hydrogen atom, 1/20th, or 5.0%, of the mass of hydrogen fluoride is hydrogen; 19/20th, or 95%, is fluorine. This is true whether we consider one hydrogen fluoride molecule, the four molecules shown in Figure 2.3, or many, many molecules. The law of constant composition is confirmed.

The law of conservation of mass is illustrated in this way: Figure 2.3 shows that six fluorine atoms (in the form of three fluorine molecules) and four hydrogen atoms (in the form of two hydrogen molecules) are available. Four of the fluorine atoms combine with the four hydrogen atoms, producing four hydrogen fluoride molecules. Two fluorine atoms do not react and are left unchanged as one fluorine molecule. Thus, after the reaction, there are four hydrogen atoms and six fluorine atoms, just as there were before. The total mass remains unchanged. With Dalton's atomic theory, we can restate the law of conservation of mass in this way.

> *Atoms can neither be created nor destroyed in a chemical reaction, and as a consequence the total mass remains unchanged.*

APPLICATION NOTE

The law of conservation of mass tells us that we cannot create materials from nothing: We can make new materials only by changing the way atoms are combined in existing materials. It also tells us that we cannot make materials disappear: We create as much mass of new matter as of the old matter replaced. Chemistry offers alternatives, however. Through chemical reactions, we can change some kinds of potentially hazardous wastes to less harmful forms. Such transformations of matter are what chemistry is all about.

Dalton's theory also nicely explains the law of multiple proportions. Recall that in carbon dioxide, two oxygen atoms are combined with each carbon atom (CO_2); in carbon monoxide there is one oxygen atom per carbon atom (CO). The ratio of the numbers of oxygen atoms per carbon atom between the two compounds is 2:1. Because all oxygen atoms have the same mass, this 2:1 ratio based on numbers of atoms is the same as the 2:1 ratio based on masses in Figure 2.2.

2.3 The Divisible Atom

Dalton's concept of indivisible atoms fueled scientific work for nearly all of the nineteenth century. Like most scientific theories, however, Dalton's eventually had to be modified in light of later discoveries. About 100 years ago, certain experiments that we will describe later in the text began to reveal that atoms are made up of smaller parts. Of the dozens of subatomic (smaller than atomic) particles now known, three are of special importance in the study of chemistry.

Subatomic Particles

The **proton** has a mass that for now we will call a relative mass of 1. The proton also carries one fundamental unit of positive electric charge, 1+. The **neutron**, as its name implies, is an electrically neutral particle; it has no charge. Although its mass is slightly greater than that of the proton, for many purposes we can consider the neutron also to have a relative mass of 1. The third particle, the **electron**, has a mass that is 1/1836 (or 0.0005447) of the mass of a proton. An electron has the same amount of charge as a proton, but it is a negative charge, that is, 1−. Protons, neutrons, and electrons are *fundamental* particles. This means that all protons are alike, all neutrons are alike, and all electrons are alike in whatever element they are found. These three subatomic particles and their properties are listed in Table 2.1.

TABLE 2.1 Subatomic Particles

Particle	Symbol	Approximate Relative Mass	Relative Charge	Location in Atom
Proton	p^+	1	1+	Inside nucleus
Neutron	n	1	0	Inside nucleus
Electron	e^-	0.000545	1−	Outside nucleus

(If you need to review some fundamental ideas about electric charge and electricity, please refer to Appendix B.)

Protons and neutrons are densely packed into a tiny positively charged core of the atom known as the *nucleus*. The extremely lightweight electrons are widely dispersed around the nucleus. An atom as a whole is electrically neutral: It has no net charge because the negative and positive charges balance each other. *Every atom has the same number of electrons as it does protons.* An atom is mostly empty space. Picture it as something like this: If an entire atom were represented by a room, 5 m × 5 m × 5 m, the nucleus would be only about as big as the period at the end of this sentence. The electrons would be tiny specks, too.

Dalton believed that an element is determined by the *mass* of one of its atoms. We now know that it is not the mass but the *number of protons* in the nucleus that determines the kind of atom and therefore the identity of an element. The **atomic number (Z)** is the number of protons in the nucleus of an atom of a given element. An atom with two protons has $Z = 2$, and it is an atom of helium. An atom with 92 protons has $Z = 92$ and is an atom of uranium. A list of all the known elements, with their symbols and atomic numbers, is located inside the front cover of this book.

Isotopes

Dalton's belief that all atoms of a given element have the same mass isn't quite so. Any two atoms of an element do have the same number of protons (and electrons as well), but they may have different numbers of neutrons. Atoms that have the same number of protons but different numbers of neutrons are called **isotopes**. For example, there are three isotopes of hydrogen. The most abundant isotope, occasionally called protium, has a single proton and no neutrons in its nucleus. About one in every 6700 hydrogen atoms, however, has a neutron as well as a proton. The mass of this kind of hydrogen isotope, called deuterium, is about twice that of protium. A third, very rare isotope of hydrogen, called tritium, has two neutrons and one proton in the nucleus. A tritium atom has about three times the mass of a protium atom.

We cannot determine the mass of an atom by adding together the masses of its protons, neutrons, and electrons. This is because in the formation of a nucleus from protons and neutrons, a tiny amount of mass is released as a quantity of energy called the nuclear binding energy. This mass/energy equivalence is a special matter that we do not need to consider until much later in the text (Chapter 19). For now, let's concentrate on a property closely related to the relative mass of an atom: The **mass number (A)** is an integral number that is the sum of the numbers of protons and neutrons in an atom. An *actual* relative atomic mass, unlike a mass number, is not quite a whole number because of the nuclear binding energy.

Isotopes have the same atomic number but different mass numbers. The three isotopes of hydrogen all have the atomic number 1, but their mass numbers are 1, 2, and 3, respectively. Only the hydrogen isotopes have special names. Other isotopes are usually identified by the name of the element followed by the mass number, such as carbon-14, cobalt-60, and uranium-235. *Isobars* are atoms with the same mass number but different atomic numbers, for example, carbon-14 and nitrogen-14.

For some elements, all the naturally occurring atoms have the same mass number, such as fluorine-19, sodium-23, and phosphorus-31. Most elements, however, have two or more isotopic forms. Tin has the greatest number of naturally occurring isotopes: ten. For all but a few elements, the naturally occurring isotopes always occur in certain precise proportions. In chlorine from natural sources, for example, 75.77% of the atoms are chlorine-35 and 24.23% are chlorine-37.

Chemical symbols for isotopes are commonly written in the form $^{A}_{Z}E$ with A as the mass number and Z as the atomic number of the element E. For the two naturally occurring isotopes of chlorine, we can write $^{35}_{17}Cl$ and $^{37}_{17}Cl$, indicating mass numbers of 35 and 37 for chlorine-35 and chlorine-37, respectively. Because the atomic number of an element is implied by its chemical symbol, we sometimes use simplified forms such as ^{35}Cl and ^{37}Cl. The number of neutrons in an atom is easily calculated from the values of A and Z.

$$\text{Number of neutrons} = A - Z$$

EXAMPLE 2.3

How many protons, neutrons, and electrons are present in a ^{81}Br atom?

SOLUTION

Even though the atomic number is not shown here, we can check the list of elements on the inside front cover, where we find it to be 35. Thus,

$$\text{Number of protons} = \text{number of electrons} = 35$$

Now, knowing that the atomic number is 35 and the mass number is 81, we can write

$$\text{Number of neutrons} = A - Z = 81 - 35 = 46$$

EXERCISE 2.3A

Use the notation $^{A}_{Z}E$ to represent the isotope of tin having 66 neutrons.

EXERCISE 2.3B

Isotones are atoms that have the same number of neutrons but different numbers of protons. What are the numbers of protons and electrons and the atomic and mass numbers of a cadmium atom that is an isotone of a tin-116 atom?

2.4 Atomic Masses

We are familiar with measuring the mass of a macroscopic sample of matter. We can weigh ourselves on a bathroom scale, a bag of apples at the checkout stand in a supermarket, or a letter on a postal scale. But how do we weigh a microscopic sample of matter that we cannot see, specifically, an individual atom?

We can do what John Dalton did: *arbitrarily* assign a mass to one atom and determine the masses of other atoms relative to it. Dalton called these relative masses *atomic weights*, and let's do the same for the time being. Suppose we assign a mass of 1 to the hydrogen atom, just as Dalton did. Now, to establish the mass of an oxygen atom, we can measure the mass ratio of hydrogen to oxygen in water. The best value available in Dalton's time was 1 g H : 7 g O. (The modern value is 1 g H : 8 g O.) At this point, Dalton *assumed* that H and O atoms combined in a 1 : 1 ratio in water. Then, if a given number of oxygen atoms weighed seven times as much as the same number of hydrogen atoms, one oxygen atom must also weigh seven times as much as a hydrogen atom. Thus, Dalton assigned oxygen the atomic weight of 7 (it would have been 8 with modern data). However, as outlined in Figure 2.4, if we apply our current understanding that there are *two* H atoms for every O atom in water, we get a value of 16 for the atomic weight of oxygen. Once chemists figured out how to determine the combining ratios of atoms in compounds, they were able to measure atomic weights rather easily.

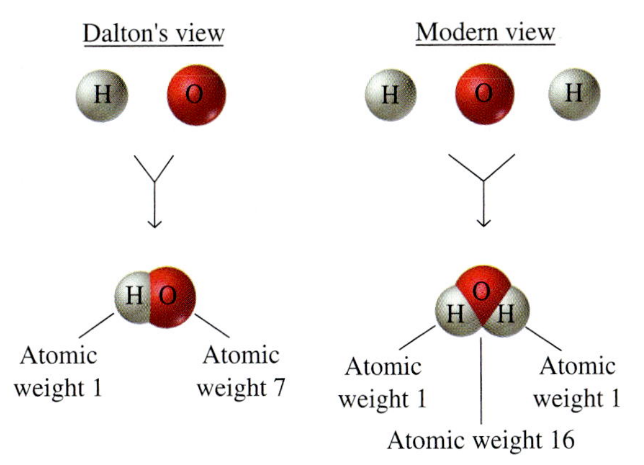

Figure 2.4
Dalton's atomic weight problem Dalton assumed a combining ratio of hydrogen to oxygen atoms of 1 : 1. Data at the time suggested that the mass ratio of hydrogen to oxygen in water was 1 : 7. Taking the atomic weight of hydrogen to be one, that of oxygen was *seven*. Modern data indicate that the combining ratio of hydrogen to oxygen atoms is 2 : 1 and that the mass ratio is 1 : 8 (or 2 : 16). If the atomic weight of hydrogen is taken to be one, that of oxygen must be *16*.

By international agreement, the current atomic mass standard is the *pure* isotope carbon-12, which is assigned a mass of *exactly* 12 atomic mass units (12 u). Based on this standard, we can define an **atomic mass unit** (abbreviated amu and having the unit, u) as exactly one twelfth the mass of a carbon-12 atom. In more familiar units of mass, $1 \text{ u} = 1.66054 \times 10^{-24}$ g. Because we are interested in the masses of atoms and not their weights, we will now shift to the term *atomic mass* in place of the older *atomic weight*,* except in a few cases of historical interest.

Naturally occurring carbon consists of a *mixture* of two isotopes. The much more abundant isotope is ^{12}C. The other is ^{13}C, with a mass of 13.00335 u. Both isotopes are present in substances containing carbon atoms and in proportions that generally do not vary from one carbon-containing substance to another. To describe the atomic mass of carbon, then, we need to use an average value. Because carbon-12 is much more abundant than carbon-13, this average lies much closer to the mass of carbon-12 than to that of carbon-13. We say that it is "weighted" toward the mass of carbon-12. The **atomic mass** of an element is the weighted average of the masses of the naturally occurring isotopes of that element.

To calculate the atomic mass of an element, we need two quantities:

- the atomic masses of the isotopes of the element and
- the naturally occurring fractional abundances of the isotopes.

We will learn how these quantities are obtained experimentally in Chapter 7, but for now let's just describe how they are used. To illustrate, let's return to the atomic mass of carbon.

The masses of the two carbon isotopes are 12.0000 u for carbon-12 (by definition) and 13.00335 u for carbon-13 (by measurement). The fractional abundances of these two isotopes refer to the fraction of all carbon atoms that are carbon-12 and the fraction of carbon atoms that are carbon-13. The two fractions must add up to 1. However, the relative abundances of the isotopes of an element are often expressed as percentages rather than fractions. If this is the case, first we need to convert these percentages into fractions. Percentages are actually fractions expressed on a per-hundred basis, that is,

$$\text{Percent} = \text{fraction} \times 100\%$$

In turn, fractions are percentages divided by 100%.

$$\text{Fraction} = \text{percent} / 100\%$$

*The term *atomic weight* is still widely used, however, as by the *Commission on Atomic Weights* of the International Union of Pure and Applied Chemistry (IUPAC).

The percentage and fractional abundances of the carbon isotopes are

Isotope	Percent abundance	Fractional abundance
Carbon-12	98.892%	0.98892
Carbon-13	1.108%	0.01108

To obtain a weighted average atomic mass, as in Example 2.4, we express the contribution of each isotope to the weighted average as:

$$\text{Contribution of isotope} = \text{fractional abundance} \times \text{mass of isotope.}$$

EXAMPLE 2.4

Use the data cited above to complete the determination of the weighted average atomic mass of carbon.

SOLUTION

As we have already noted, the contribution that each isotope makes to a weighted average atomic mass is given by the following product

$$\text{Contribution of isotope} = \text{fractional abundance} \times \text{mass of isotope}$$

Thus, the contributions of carbon-12 and carbon-13 are

$$\text{Contribution of } ^{12}\text{C} = 0.98892 \times 12.00000 \text{ u} = 11.867 \text{ u}$$

$$\text{Contribution of } ^{13}\text{C} = 0.01108 \times 13.00335 \text{ u} = 0.1441 \text{ u}$$

The weighted average mass is the sum of these two contributions.

$$\text{Atomic mass of carbon} = 11.867 \text{ u} + 0.1441 \text{ u} = 12.011 \text{ u}$$

This is the value listed in a table of atomic masses. As expected, the atomic mass of carbon is much closer to 12 u than to 13 u.

PROBLEM-SOLVING NOTE

Perhaps you are familiar with *grade-point average*. This is a weighted average in which letter grades are converted to points (A = 4.0, B = 3.0, and so on). These grade points are equivalent to the masses of isotopes, and the fraction of the total credit units of enrollment assigned to each course is equivalent to the fractional abundance of an isotope.

EXERCISE 2.4A

The three naturally occurring isotopes of neon, their percent abundances, and their atomic masses are: neon-20, 90.51%, 19.99244 u; neon-21, 0.27%, 20.99395 u; neon-22, 9.22%, 21.99138 u. Calculate the weighted average atomic mass of neon.

EXERCISE 2.4B

The two naturally occurring isotopes of copper are copper-63, with a mass of 62.9298 u, and copper-65, with a mass of 64.9278 u. If the tabulated atomic mass of copper is 63.546 u, what must be the percent natural abundances of the two isotopes?

PROBLEM-SOLVING NOTE

The fractional abundances of the two isotopes are unknowns. Can you see that if one of them is x the other must be $1.000 - x$?

EXAMPLE 2.5—An Estimation Example

Indium has *two* naturally occurring isotopes and a tabulated atomic mass of 114.82 u. One of the isotopes has a mass of 112.9043 u. Which is likely to be the second isotope: ^{111}In, ^{112}In, ^{114}In, or ^{115}In?

SOLUTION

We have seen that the masses of isotopes differ only slightly from whole numbers. The isotope with a mass of 112.9043 u is ^{113}In. To account for the observed weighted average atomic mass of 114.82 u, the second isotope must have a mass number greater than 114. It can only be ^{115}In.

EXERCISE 2.5

The masses of the three naturally occurring isotopes of magnesium are ^{24}Mg, 23.98504 u; ^{25}Mg, 24.98584 u; ^{26}Mg, 25.98259 u. Which of the three is the most abundant isotope? Can you determine which is the second most abundant? Explain.

2.5 The Periodic Table: Elements Organized

In the nineteenth century, chemists discovered elements at a rapid rate. By 1830, 55 elements were known, all with different properties and no apparent pattern among them. Chemists greatly needed a way to organize the growing mass of chemical data by arranging the elements in a meaningful manner, one that would establish categories of elements with similar physical and chemical properties. The first successful arrangement, called a *periodic table*, was published by Dmitri Mendeleev in 1869. In its modern form, the periodic table organizes a vast array of chemical knowledge; it is so important that we will devote most of a later chapter to it. For now, though, we need only consider a few ideas that will help us in naming and writing formulas of chemical compounds.

Dmitri Ivanovich Mendeleev (1834–1907). Mendeleev developed a periodic table while trying to systematize the properties of the elements for presentation in a chemistry textbook. His highly influential text lasted for 13 editions, including five that were published after his death.

Mendeleev's Periodic Table

Mendeleev arranged the elements in the order of increasing atomic weight, from left to right in rows and from top to bottom in groups. In this arrangement, elements that most closely resemble each other tend to fall in the same vertical group. This group similarity recurs *periodically* (once in each row), and the tabular format is known as a **periodic table**. So that there would be no exceptions to the principle that all the elements in a group display similar properties, Mendeleev placed some elements "out of order," that is, not in the strict order of increasing atomic weight. For example, Mendeleev placed tellurium (atomic weight 127.6) ahead of iodine (atomic weight 126.9) so that tellurium would be in the same column as sulfur and selenium, elements that closely resemble it. Mendeleev left gaps in his table. This too was done to avoid exceptions to the idea that elements with similar properties fall in the same vertical group. Instead of seeing these gaps as defects, he boldly predicted the existence of undiscovered elements. Furthermore, because the table was based on patterns of *properties*, he was able to predict some properties of the missing elements. For example, he left a blank space for an undiscovered element that he called "eka-silicon" and used its location between silicon and tin to predict an atomic weight of 72 and other properties. Table 2.2 shows just how accurate his predictions were, when compared to the properties of the actual element germanium discovered 15 years later. The *predictive* nature of Mendeleev's periodic table led to its wide acceptance as a tremendous scientific accomplishment.

The Modern Periodic Table

The modern periodic table shown inside the front cover contains 112 elements. Each entry in the table gives the chemical symbol, atomic number (Z), and atomic mass of an element. Note that the elements are placed in order of increasing *atomic number*, a property that determines the behavior of an element more so than its atomic mass.

TABLE 2.2 Properties of Germanium: Predicted and Observed

Property	Predicted: Eka-silicon[a] (1871)	Observed: Germanium (1886)
Atomic weight	72	72.6
Density, g/cm^3	5.5	5.47
Color	Dirty gray	Grayish white
Density of oxide, g/cm^3	EsO_2: 4.7	GeO_2: 4.703
Boiling point of chloride	$EsCl_4$: below 100 °C	$GeCl_4$: 86 °C
Density of chloride, g/cm^3	$EsCl_4$: 1.9	$GeCl_4$: 1.887

[a]The term "eka" is derived from Sanskrit and means "first." Literally, eka-silicon means, first comes silicon (and then comes the unknown element).

The periodic table is divided into groups and periods. *Groups* (or families) are the vertical columns of elements having similar properties. We place a group number (for example, 1A, 2A, 3B, and so on) at the top of each column. *Periods* are the horizontal rows of elements. The periods vary in length from two elements (the first period) to 32 elements (the sixth and seventh periods). We use a number at the left of the row to identify each period. If we were to include all the elements in a periodic table, it would have an awkward shape. The table would be very long to accommodate the periods with 32 elements and would have much blank space above these long periods. By extracting 14-element series from the sixth and seventh periods and listing them separately at the bottom of the table, we can use a more compact form that limits the width of the table to 18 elements. The sixth period "footnote" is called the *lanthanide* series and that of the seventh period is the *actinide* series.

● **ChemCDX** Reactions of Alkali Metals

Each element is represented by a "box" in the periodic table, and the data typically shown are

26	— Atomic number, Z
Fe	— Chemical symbol
55.847	— Atomic mass (weighted average)

Elements are also divided into two main classes by a heavy stepped, diagonal line. Those to the left of the line, except for hydrogen, are **metals**—elements that have a characteristic luster and are generally good conductors of heat and electricity. Most metals are *malleable*, which means that they can be hammered into thin sheets or foil. Also, most metals are *ductile*; they can be drawn into wires. Except for mercury, a liquid, all the metals are solid at room temperature.

Copper, a metal (left), can be obtained as pellets that can be hammered into a thin foil or drawn into a wire. Sulfur, a nonmetal (right), can be obtained as lumps that crumble into a powder when hammered.

Elements to the right of the stepped line are **nonmetals**, elements that lack metallic properties. For example, nonmetals generally are poor conductors of heat and electricity. At room temperature, several nonmetals, including oxygen, nitrogen, fluorine, and chlorine, are gases. Others, such as carbon, sulfur, phosphorus, and iodine, are brittle solids. Bromine is the only nonmetal that is a liquid at room temperature. Some of the elements bordering the stairstep line resemble metals in some of their properties and nonmetals in others. They are called **metalloids**.

We will consider the theoretical basis of the periodic table in Chapter 8. For example, we will discover that the group numbers and letters (A and B) are related to the way electrons are arranged about the nuclei of atoms. We will also discuss why periods are not all the same length, why the nonmetal hydrogen appears in Group 1A with a group of metals, and other fundamental ideas about the periodic table.

Introduction to Molecular and Ionic Compounds

We have already mentioned chemical compounds several times, but so far we have just scratched the surface of this broad subject. The remainder of this chapter serves as an introduction to chemical compounds and it emphasizes that we will be dealing with this topic throughout the text. In later chapters we will explore the compositions, properties, and reactions of compounds and how real-world, or *macroscopic*, properties are related to the structure of matter at the microscopic level. For example, it is the structures of its atoms—a microscopic property—that determines whether an element displays the macroscopic properties of a metal or a nonmetal. The sections that follow include some useful information about classifying, naming, and writing formulas of compounds. You may want to return to this information from time to time later on; it may be helpful each time you encounter an unfamiliar chemical compound.

Chemists classify compounds in different ways. We already mentioned the two broadest categories on page 36: *organic compounds* and *inorganic compounds*. Although there are borderline cases, most chemical compounds can be placed in one category or the other.

Chemists also use other terms to describe and classify chemical compounds. Some additional classifications that we will introduce and describe in this chapter are:

- molecular compounds and ionic compounds
- acids, bases, and salts

At times we may use more than one term to classify a compound. For example, when we say that sulfuric acid is an *inorganic acid*, we are indicating that it is both an inorganic compound and an acid.

No matter what its class, we can represent any compound symbolically. A **chemical formula** is a symbolic representation of the composition of a compound in terms of its constituent elements. A chemical formula uses symbols to indicate which elements are present, and it uses subscripts to indicate how many atoms of each element are present. For example, the formula B_2O_3 signifies that the compound boron oxide contains two boron atoms for every three oxygen atoms.

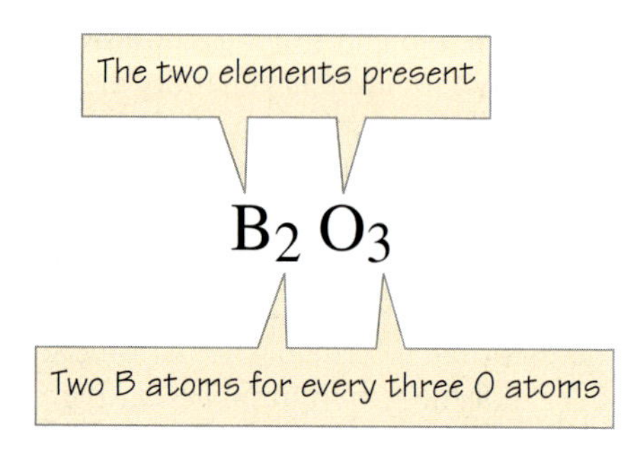

Ammonia has nitrogen and hydrogen atoms in the ratio of 1 to 3, but we write the formula as NH_3, *not* N_1H_3. The *lack* of a subscript number in a formula means that the subscript "1" is implied.

2.6 Molecules and Molecular Compounds

A **molecule** is a group of two or more atoms held together in a definite spatial arrangement by forces called *covalent* bonds.* A **molecular compound** has molecules as its smallest characteristic entities, and these molecules determine the properties of the substance. Considered separately, the component atoms of molecules do not determine the properties of a substance, much as the flavor of zucchini bread does not resemble the taste of zucchini squash.

An **empirical formula** is the simplest formula we can write for a compound. It lists the elements present and indicates the smallest integral (whole number) ratio in which their atoms are combined. The empirical formula CH_2O indicates that the elements C, H, and O are present in the atom ratios 1:2:1, respectively. However, several different compounds have this same empirical formula, for example, acetic acid (found in vinegar) and glucose (a sugar). A **molecular formula** shows the difference between compounds with identical empirical formulas by giving the symbol and the *actual* number of each kind of atom in a molecule. The molecular formula of acetic acid is $C_2H_4O_2$, and that of glucose is $C_6H_{12}O_6$. In the molecular formula, $C_2H_4O_2$, the total number of atoms is *twice* that in the empirical formula, that is, $2 \times (1:2:1) = 2:4:2$. In $C_6H_{12}O_6$, it is *six* times: $6 \times (1:2:1) = 6:12:6$. Water ($H_2O$), ammonia ($NH_3$), and carbon dioxide ($CO_2$) are familiar substances whose molecular formulas are the same as their empirical formulas.

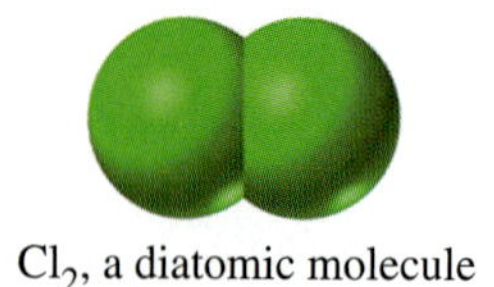
Cl_2, a diatomic molecule

S_8, a polyatomic molecule

Although most of the molecules we encounter will be those of compounds, some of the elements also exist in molecular form. Hydrogen occurs naturally not as individual H atoms but as pairs of atoms joined into H_2 molecules. These are called *diatomic* (two-atom) molecules. Other common elements that form diatomic molecules are nitrogen, N_2, and oxygen, O_2. The Group 7A elements, the halogens, are also diatomic: F_2, Cl_2, Br_2, I_2. A few elements exists as *polyatomic* (many-atom) molecules. Some examples are phosphorus, P_4, and sulfur, S_8. The noble gases of Group 8A exist in *monatomic* (one-atom) form. Thus we represent helium, neon, and argon simply as He, Ne, and Ar.

The atoms in a molecule are not arranged at random. They are attached in a definite order. In water molecules, the order is always H—O—H, never H—H—O. A **structural formula** is a chemical formula that shows how atoms are attached to one another. The structural formulas of ammonia, methane, and acetic acid are shown on the next page.

*Covalent bonds are discussed in Chapters 9 and 10. In this section, we concentrate on the atoms present in molecules and not on the forces between atoms.

```
    H             H             H  O
    |             |             |  ||
  H—N—H         H—C—H         H—C—C—O—H
                  |             |
                  H             H
```

Ammonia (NH_3) Methane (CH_4) Acetic acid (CH_3COOH)

The lines in structural formulas represent the covalent bonds between atoms. A single line represents a single bond; a double line, a double bond; and a triple line, a triple bond. We will discuss these different bond types in Chapters 9 and 10, but for now simply look upon a double bond as being stronger than a single bond and a triple bond as stronger than a double bond.

Molecules have definite shapes, but these are often difficult to represent on paper. Shapes are best represented by molecular models. Figure 2.5 shows two types of models of the acetic acid molecule, and Figure 2.6 presents the color scheme commonly used for atoms in molecular models.

● **ChemCDX** Representing Substances

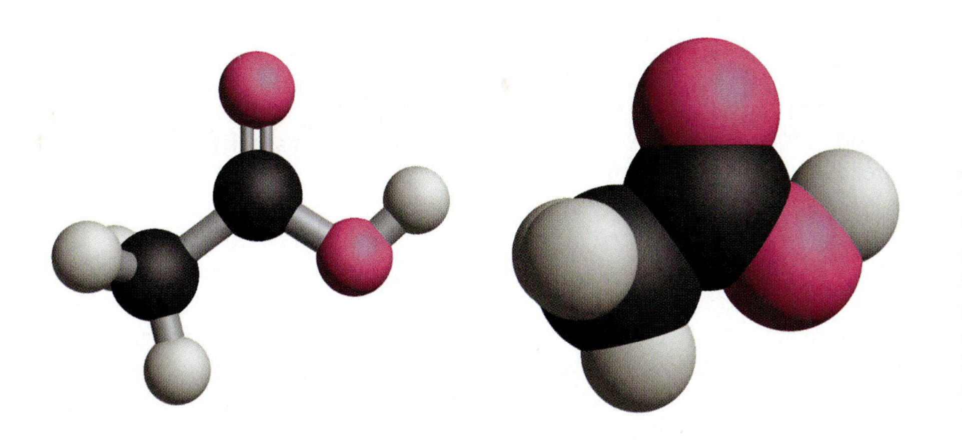

◀ **Figure 2.5**
Two types of molecular models
The ball-and-stick model of acetic acid (left) and the space-filling model (right) conform to the structural formula of the acetic acid molecule (above). However, they are more informative about its three-dimensional shape.

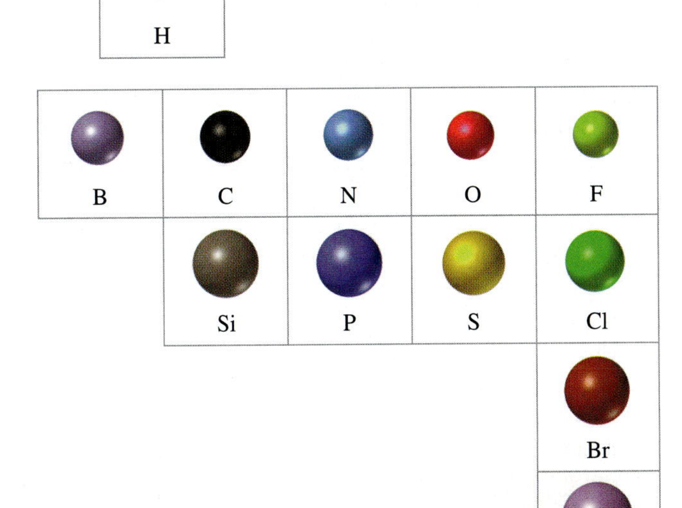

◀ **Figure 2.6**
A common color scheme for atoms in molecular models

The *ball-and-stick* model shows the spatial arrangements of the bonds in a way that the structural formula fails to do. The *space-filling* model shows that atoms in a molecule occupy space and that they are in actual contact with one another.

Writing Formulas and Names of Binary Molecular Compounds

In a *binary* molecular compound, the molecules are made up of *two* elements. Usually both elements are nonmetals, although metalloids may also be involved.

In writing formulas of binary molecular compounds, we must decide (1) which element symbol to write first and (2) what to write as subscripts to these symbols. We can construct the formula in this way:

- *Choosing the first element symbol.* As a general rule, we write first the symbol of the element that lies farthest to the left in its period and/or lowest in its group of the periodic table. Exceptions are required for hydrogen and oxygen and in a few other cases as well. The scheme based on the portion of the periodic table shown in Figure 2.7 will meet most of our current needs. For example, it explains why we show the nitrogen atom first in NH_3 but the hydrogen atom first in H_2Te, and why we write Cl_2O but OF_2.
- *Writing subscripts.* The prefixes *mono*, *di*, *tri*, and so on refer to the numbers of atoms of each element in the molecules of a binary molecular compound. Table 2.3 illustrates the use of prefixes for up to 10 atoms.

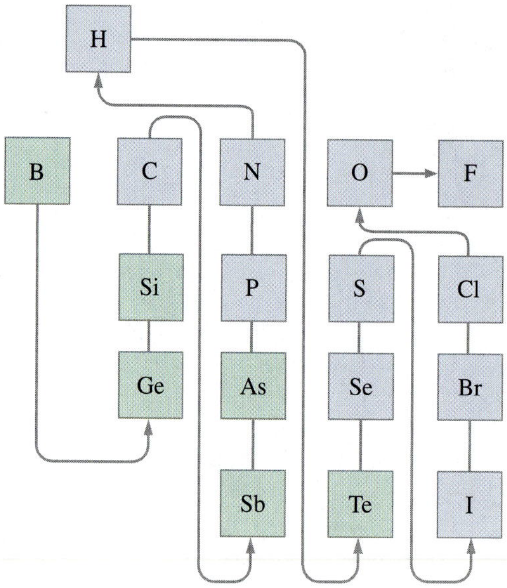

▶ Figure 2.7
A scheme based on the periodic table to assist in writing formulas and names of binary molecular compounds
The lines trace a continuous path from boron (B) to fluorine (F). The element that is generally written first in the formula of a binary molecular compound is the one that is closer to the beginning of this path.

To name a binary molecular compound, we proceed much as in writing its formula but with a few added considerations.

- The name consists of two words, one for each of the elements in the compound.
- The first word is the name of the element that appears first in the formula.
- The second word is an altered version of the name of the second element. This word retains the stem of the element name and replaces the ending by *-ide*. Thus the element name chlor*ine* becomes chlor*ide*.

TABLE 2.3 Naming Binary Molecular Compounds

Number of Atoms	Prefix	Examples[a]
1	mono	NO nitrogen monoxide
2	di	NO_2 nitrogen dioxide
3	tri	N_2O_3 dinitrogen trioxide
4	tetra	N_2O_4 dinitrogen tetroxide
5	penta	N_2O_5 dinitrogen pentoxide
6	hexa	SF_6 sulfur hexafluoride
7	hepta	S_2O_7 disulfur heptoxide
8	octa	
9	nona	
10	deca	

[a]When the prefix ends in "a" or "o" and the element name begins with "a" or "o," the final vowel of the prefix is dropped for ease of pronunciation. For example, nitrogen *mon*oxide and not nitrogen *mono*oxide, and dinitrogen *tetr*oxide, not dinitrogen *tetra*oxide. However, PI_3 is phosphorus *tri*iodide, not phosphorus *tr*iodide.

APPLICATION NOTE

Is P_4S_3 just a curiosity chemical with a tongue twister name? No, indeed. It is the chemical on the head of a strike-anywhere match that is ignited through frictional heat. P_4S_3 does not spontaneously ignite at temperatures below 100 °C.

- The names are further altered by adding prefixes such as *mono*, *di*, *tri*, and so on to denote the numbers of atoms of each element in the molecule (see Table 2.3). Thus, P_4S_3 is called *tetra*phosphorus *tri*sulf*ide*. Note that in these examples the prefix *mono-* is treated in a special way. We do not use it for the first-named element, but we do for the second. For example, the name for CO is carbon monoxide, *not* monocarbon monoxide.

EXAMPLE 2.6

Write the formula and name of a compound that has six oxygen atoms and four phosphorus atoms in its molecules.

SOLUTION

To write the formula, we need to use the chemical symbols of the elements and, as subscripts, the numbers of atoms stated, that is O_6 and P_4. All that remains is to determine which element to place first in the formula. According to the scheme in Figure 2.7, the element O is followed only by F. Phosphorus comes first in the formula; we write P_4O_6.

The name of a compound with four (tetra) P atoms and six (hexa) oxygen atoms in its molecules is *tetra*phosphorus *hex*oxide.

EXERCISE 2.6

Write the formula and name of a compound that has four fluorine atoms and two nitrogen atoms in its molecules.

EXAMPLE 2.7

Write (**a**) the formula of phosphorus pentachloride and (**b**) the name of S_2F_{10}.

SOLUTION

a. *Choosing the first element symbol.* The elements are listed in the same order in the formula as in the name.

Writing subscripts. The lack of a prefix on phosphorus signifies one P atom per molecule. The prefix *penta* indicates five chlorine atoms. The formula is PCl_5.

b. The prefixes indicate two (*di*) sulfur atoms and ten (*deca*) fluorine atoms. The compound is disulfur decafluoride.

EXERCISE 2.7

Write (**a**) the formula of tetraphosphorus decoxide and (**b**) the name of IF_7.

2.7 Ions and Ionic Compounds

An isolated atom has an equal number of protons and electrons and is electrically neutral. But in some chemical reactions, an atom (or group of bonded atoms) may lose or gain one or more electrons, acquire a net electric charge, and become an **ion**. Ions are formed only through the loss or gain of electrons. The formation of ions does *not* involve changes in the number of protons in the nucleus of an atom. If electrons are *lost*, there are more protons than electrons, and the ion has a *positive* charge. If electrons are *gained*, there are more electrons than protons, and the ion has a *negative* charge. *Monatomic* (simple) ions are formed when a single atom loses or gains one or more electrons. If a sodium atom loses one electron, it acquires a charge of 1+ and is represented as Na^+. If a chlorine atom gains one electron, it acquires a charge of 1− and is written as Cl^-. Positively charged ions are called **cations**, and negatively charged ions, **anions**. In general, metal atoms form cations and nonmetal atoms form anions. A grouping of bonded atoms can also lose or gain electrons to form *polyatomic* ions. When a group of one sulfur atom and four oxygen atoms gains two electrons, the resulting ion is SO_4^{2-}.

Ionic compounds are composed of oppositely charged ions (cations and anions) held together in huge clusters by electrostatic attractions. There are no identifiable small units of ionic compounds comparable to the molecules of a molecular compound.

Monatomic Ions

To some extent we can use the periodic table to predict the charges on monatomic ions (Figure 2.8). For metal atoms in the A groups, in most cases the number of electrons given up is equal to the periodic table group number. Thus, the highly reactive metal atoms of Group 1A—the alkali metals—give up one electron to form cations with charges of 1+. Group 2A atoms—the alkaline earth metals—give up two electrons to form cations with 2+ charges. Aluminum, the most common and commercially important metal in Group 3A, forms cations with 3+ charges.

To name monatomic cations, we add the word *ion* to the name of the parent element: sodium ion, magnesium ion, and so on. We write the ionic charge as an Arabic numeral followed by a plus sign. For a 1+ charge, however, we usually omit the 1 and just write +. Thus, we write Mg^{2+} and Al^{3+}, but Na^+ rather than Na^{1+}.

There is no simple way to use the periodic table to determine the most likely charge on cations formed by B-group elements. In a few cases, the magnitude of the charge is equal to the group number, but in most cases it is not. Moreover, you will notice from Figure 2.8 that B-group elements may form ions with different charges, such as Fe^{2+} and Fe^{3+}. We must give these two ions different names, which we can do by using Roman numerals to indicate the charges. Thus we arrive at the names *iron(II)* ion and *iron(III)* ion for Fe^{2+} and Fe^{3+}, respectively. In an older system of nomenclature, no longer much used by chemists but still encountered occasionally, the names of certain ions combine Latin stems and the endings *ous* and *ic*. The *ous* indicates the lower of two possible charges on the ion and *ic*, the higher

1A	2A	3B	4B	5B	6B	7B	8B	8B	8B	1B	2B	3A	4A	5A	6A	7A	8A
Li^+														N^{3-}	O^{2-}	F^-	
Na^+	Mg^{2+}											Al^{3+}		P^{3-}	S^{2-}	Cl^-	
K^+	Ca^{2+}				Cr^{2+} Cr^{3+}	Mn^{2+}	Fe^{2+} Fe^{3+}	Co^{2+} Co^{3+}	Ni^{2+}	Cu^+ Cu^{2+}	Zn^{2+}					Br^-	
Rb^+	Sr^{2+}									Ag^+			Sn^{2+}			I^-	
Cs^+	Ba^{2+}												Pb^{2+}				

▲ **Figure 2.8 Symbols and periodic table locations of some monatomic ions**
In general, (a) the metals of Groups 1A and 2A and aluminum have just one cation, which carries a positive charge equal in magnitude to the A-group number; (b) the metals of the B-group have two or more cations of different charges, though in some cases only one of these cations is commonly encountered; and (c) the nonmetals of Groups 7A and 6A, nitrogen and phosphorus form anions with a charge equal to " the group number minus eight."

charge. In this system Fe^{2+} is called *ferrous* ion and Fe^{3+} is *ferric* ion. For example, ferrous sulfate is listed as the active ingredient in a mineral supplement pill.

When they combine with metal atoms, nonmetal atoms generally gain electrons to form anions with a charge equal to "the periodic table group number minus eight." The nonmetal atoms of Group 7A—the halogens—gain one electron to form anions with the charge $7 - 8 = 1-$, such as F^- and Cl^-. Group 6A atoms gain two electrons to form anions such as O^{2-} and S^{2-}. Nitrogen and phosphorus, in Group 5A, form the anions N^{3-} and P^{3-}, respectively. To name monatomic anions, we use an *-ide* ending on the name of the parent element and add the word *ion*. When a chlor*ine* atom gains one electron it becomes a chlor*ide* ion. An ox*ygen* atom, by gaining two electrons, becomes an ox*ide* ion.

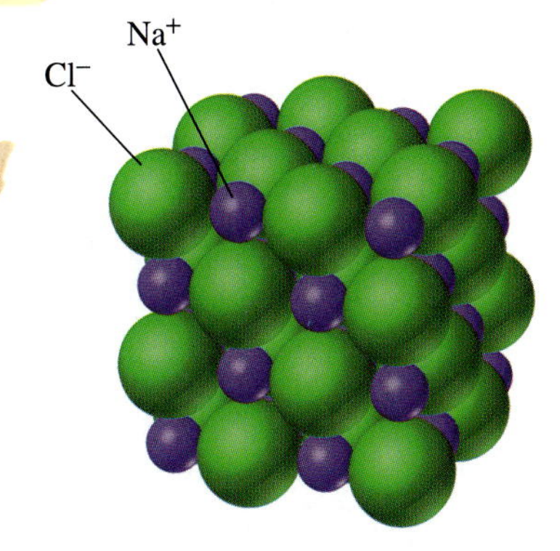

▲ **Figure 2.9**
A formula unit of NaCl
The compound sodium chloride consists of Na^+ and Cl^- ions held together by electrostatic forces of attraction in a very large ordered network called a crystal. The hypothetical combination of the one Na^+ and one Cl^- ion indicated is a formula unit. It is the smallest collection of ions from which we can deduce the formula NaCl.

Formulas and Names for Binary Ionic Compounds

Binary (two-element) ionic compounds are made up of monatomic cations and anions. These combinations of ions must be electrically neutral; they must have no net charge, neither positive nor negative. This requirement dictates how we write formulas of ionic compounds.

We base the formula on the simplest collection of cations and anions that represents an electrically neutral unit. We call this hypothetical collection a **formula unit** of the compound. The formula unit is "hypothetical," because it does not exist as a separate entity. The formula unit of sodium chloride in Figure 2.9 is not hard to imagine; it just consists of one Na^+ and one Cl^- ion. The formula of sodium chloride is NaCl. This example highlights an important general rule: *The name for an ionic compound is a combination of two words—the name of the cation followed by that of the anion.*

Now consider the formula for aluminum oxide. We can't simply combine 1 Al^{3+} and 1 O^{2-} because this would produce AlO^+, a formula unit with a net positive charge. If we base the formula unit on *two* Al^{3+} ions and *three* O^{2-} ions, it is electrically neutral.

$$2(3+) + 3(2-) = +6 - 6 - 0$$

The formula of aluminum oxide is Al_2O_3.

What Is a Low-Sodium Diet?

There are about 27.5 million people with hypertension (high blood pressure) in the United States. Physicians usually advise them to follow a low-sodium diet. Just what does that mean? Surely they are not being advised to reduce their consumption of sodium itself. Sodium is an extremely reactive metal that reacts violently with moisture. Sodium metal is not a part of anyone's diet. The concern is really with sodium *ion*, Na^+, but sodium ion taken alone is not a substance. It enters our diet in combination with anions in the form of ionic compounds, the principal one being sodium chloride—common table salt. Some people eat 6 or 7 grams of sodium chloride a day, most of it in prepared foods. Many snack foods, such as potato chips, pretzels, and corn chips, are especially high in salt. Most physicians recommend that people with hypertension restrict their salt intake. Some, but not all, physicians suggest that even people with normal blood pressure should eat less salt.

Ions differ greatly from the atoms from which they are formed. Sodium atoms make up an element that, although quite reactive, can exist by itself. Sodium *ions* are generally unreactive, but they come only in combination with anions. A metal atom and its cation are as different as a whole peach (atom) and a peach pit (ion). Unfortunately, the situation is confused when people talk about diets with too much "sodium" or of taking "calcium" for healthy teeth and bones. They really mean sodium *ions* and calcium *ions*. As scientists, we try to be more precise in our terminology.

Some familiar foods with high Na^+ content.

Why do we call Al_2O_3 aluminum oxide instead of *di*aluminum *tri*oxide? Simply because we don't use prefixes if we don't need them. If we know the charges on the cations and anions, we can always figure out how many cations and anions are needed to produce an electrically neutral formula unit. Then we can name the formula unit in terms of the cations and anions present, regardless of their relative numbers. You should generally find it easy to write names and formulas of the compounds formed by the ions listed in Figure 2.8.

EXAMPLE 2.8

Determine the formula for (**a**) calcium chloride and (**b**) magnesium oxide.

SOLUTION

a. First we write the symbols for the ions, with the cation first, followed by the anion: Ca^{2+} and Cl^-. The simplest combination of these ions that gives an electrically neutral formula unit is *one* Ca^{2+} ion for every *two* Cl^- ions. The formula is $CaCl_2$.

$$Ca^{2+} + 2\,Cl^- = CaCl_2$$

b. The ions are Mg^{2+} and O^{2-}. The simplest ratio for an electrically neutral formula unit is 1 : 1. The formula of the compound is MgO.

$$Mg^{2+} + O^{2-} = MgO$$

EXERCISE 2.8

Give the formula for each of the following ionic compounds.

a. aluminum fluoride **b.** potassium sulfide
c. calcium nitride **d.** lithium oxide

EXAMPLE 2.9

What are the names of (**a**) MgS and (**b**) $CrCl_3$?

SOLUTION

a. MgS is made up of Mg^{2+} and S^{2-} ions. Its name is magnesium sulfide.
b. The ions are Cr^{3+} and Cl^-. How do we know that the chromium is present as Cr^{3+} and not Cr^{2+}? Because there are three Cl^- ions, the cation must have a charge of 3+, that is, $3 + 3(1-) = 3 - 3 = 0$. The compound is chromium(III) chloride. Notice that we cannot use just "chromium chloride" as the name of $CrCl_3$. From Figure 2.8 we see that there are two simple ions of chromium, Cr^{2+} and Cr^{3+}, and thus two chlorides: $CrCl_2$ and $CrCl_3$. We must assign a different name to each.

EXERCISE 2.9A

Give names for the following compounds:
a. $CaBr_2$ **b.** Li_2S **c.** $FeBr_2$ **d.** CuI

EXERCISE 2.9B

What are the name and formula of the sulfide of copper(I)?

Polyatomic Ions

A **polyatomic ion** is a charged group of bonded atoms, for example, NH_4^+ or SO_4^{2-}. Table 2.4 lists some of the more common polyatomic ions. Notice that polyatomic anions are more common than cations, that the suffixes *-ite* and *-ate* are frequently used, and that the prefixes *hypo-* and *per-* are occasionally used. For any given element, an anion with a name ending in *-ite* generally has one less oxygen atom and the same charge as the anion with a name ending in *-ate*. If an element has more than two polyatomic anions, the prefix *hypo-* represents one less oxygen atom than the *-ite* anion, and the prefix *per-* represents one more oxygen atom than the *-ate* anion. We illustrate the scheme using the oxo (oxygen-containing) anions of chlorine and sulfur.

		Example	Name	Example	Name
Increasing number of oxygen atoms ↓	hypo___ite	ClO^-	hypochlorite ion	—	—
	___ite	ClO_2^-	chlorite ion	SO_3^{2-}	sulfite ion
	___ate	ClO_3^-	chlorate ion	SO_4^{2-}	sulfate ion
	per___ate	ClO_4^-	perchlorate ion	—	—

If a polyatomic anion has hydrogen as a third element, we indicate its presence in the name: HPO_4^{2-} is the *hydrogen* phosphate ion and $H_2PO_4^-$ is the *dihydrogen* phosphate ion.

TABLE 2.4 Some Common Polyatomic Ions

Name	Formula	Typical Compound
Cations		
Ammonium ion	NH_4^+	NH_4Cl
Hydronium ion	H_3O^+	[a]
Anions		
Acetate ion	[b]$C_2H_3O_2^-$	$NaC_2H_3O_2$
Carbonate ion	CO_3^{2-}	Li_2CO_3
Hydrogen carbonate ion (or bicarbonate ion)[c]	HCO_3^-	$NaHCO_3$
Hypochlorite ion	ClO^-	$Ca(ClO)_2$
Chlorite ion	ClO_2^-	$NaClO_2$
Chlorate ion	ClO_3^-	$NaClO_3$
Perchlorate ion	ClO_4^-	$KClO_4$
Chromate ion	CrO_4^{2-}	K_2CrO_4
Dichromate ion	$Cr_2O_7^{2-}$	$(NH_4)_2Cr_2O_7$
Cyanate ion	OCN^-	KOCN
Thiocyanate ion[d]	SCN^-	KSCN
Cyanide ion	CN^-	KCN
Hydroxide ion	OH^-	NaOH
Nitrite ion	NO_2^-	$NaNO_2$
Nitrate ion	NO_3^-	$NaNO_3$
Oxalate ion	$C_2O_4^{2-}$	CaC_2O_4
Permanganate ion	MnO_4^-	$KMnO_4$
Phosphate ion	PO_4^{3-}	Na_3PO_4
Hydrogen phosphate ion	HPO_4^{2-}	Na_2HPO_4
Dihydrogen phosphate ion	$H_2PO_4^-$	NaH_2PO_4
Sulfite ion	SO_3^{2-}	Na_2SO_3
Hydrogen sulfite ion (or bisulfite ion)[c]	HSO_3^-	$NaHSO_3$
Sulfate ion	SO_4^{2-}	Na_2SO_4
Hydrogen sulfate ion (or bisulfate ion)[c]	HSO_4^-	$NaHSO_4$
Thiosulfate ion[d]	$S_2O_3^{2-}$	$Na_2S_2O_3$

[a] In water solution, H^+ associates itself with water molecules and is generally represented as H_3O^+. There are no common compounds containing H_3O^+.
[b] The acetate ion is also represented as CH_3COO^-.
[c] The prefix "bi" means that the ion contains a replaceable H atom. This should not be confused with the prefix "di," which means two (used to represent a doubling of a simpler unit).
[d] The prefix "thio" means that a sulfur atom has replaced an oxygen atom.

We can write formulas and names for compounds containing polyatomic ions by combining information from Figure 2.8 with that from Table 2.4. However, note that we have to use parentheses in some of these formulas. Consider a formula unit of magnesium nitrate, which consists of one Mg^{2+} ion and two NO_3^- ions. We can't simply write a subscript "2" following NO_3; we would get $MgNO_{32}$. Nor can we write MgN_2O_6. Rather, we enclose NO_3 in parentheses, followed by the subscript 2, that is, $(NO_3)_2$. The formula of magnesium nitrate is $Mg(NO_3)_2$.

EXAMPLE 2.10

What are the formulas of each of the following?

a. sodium sulfite **b.** ammonium sulfate

SOLUTION

a. It is possible to identify the sulfite ion without memorizing all the ions in Table 2.4. If you remember the name and formula of one of the sulfur-oxygen polyatomic anions, you should be able to deduce the names of others. Suppose you remember that sulf*ate* is SO_4^{2-}. The *-ite* anion has one less oxygen atom, 3 instead of 4, so it is SO_3^{2-}. The charges of the two ions in a formula unit must balance. The formula unit of sodium sulfite must have Na^+ and SO_3^{2-} in the ratio 2:1. The formula is Na_2SO_3.

b. Ammonium ion is NH_4^+ and sulfate ion is SO_4^{2-}. A formula unit of ammonium sulfate has two NH_4^+ ions and one SO_4^{2-}. To represent the two NH_4^+ ions, we place parentheses around the NH_4, followed by a subscript 2, that is, $(NH_4)_2$. Thus we arrive at the formula $(NH_4)_2SO_4$.

EXERCISE 2.10

What are the formulas of (**a**) ammonium carbonate, (**b**) calcium hypochlorite, and (**c**) chromium(III) sulfate?

EXAMPLE 2.11

What are the names for the following compounds?

a. NaCN **b.** $Mg(ClO_4)_2$

SOLUTION

a. The ions in this compound are Na^+, sodium ion, and CN^-, cyanide ion. The name of the compound is sodium cyanide.

b. The ions present are Mg^{2+}, magnesium ion, and ClO_4^-, perchlorate ion. The name of the compound is magnesium perchlorate.

EXERCISE 2.11

What are the names for the following compounds?

a. $KHCO_3$ **b.** $FePO_4$ **c.** $Mg(H_2PO_4)_2$

Hydrates

If you scan the labels in a chemical storeroom, you may find some that don't match what we have written. For example, a bottle of calcium chloride may carry the label $CaCl_2 \cdot 6H_2O$ instead of $CaCl_2$. This label indicates that the substance is a *hydrate*. A **hydrate** is an ionic compound in which the formula unit includes a fixed number of water molecules together with cations and anions. $CaCl_2 \cdot 6H_2O$ is called calcium chloride *hexa*hydrate. "Calcium chloride" pertains to the Ca^{2+} and two Cl^- ions in the formula unit; "hydrate" signifies the inclusion of H_2O molecules in the formula unit and "hexa," that there are *six* of them (recall Table 2.3). Other examples of hydrates are barium chloride *di*hydrate, $BaCl_2 \cdot 2H_2O$, lithium perchlorate *tri*hydrate, $LiClO_4 \cdot 3H_2O$, and magnesium carbonate *penta*hydrate, $MgCO_3 \cdot 5H_2O$.

A hydrate may form when an anhydrous (nonhydrated) compound is exposed to atmospheric water vapor or, more commonly, when the ionic compound is crystallized from a solution in water. Although hydrates are common, many ionic compounds do not form them, and there is no need to try to learn which ones do. You need only to recognize a hydrate formula when you see one.

APPLICATION NOTE

Anhydrous calcium chloride, $CaCl_2$, is a good *desiccant*—a drying agent. In a tightly closed container it removes water vapor from the air and is successively converted to the hydrates: $CaCl_2 \cdot H_2O$, $CaCl_2 \cdot 2H_2O$, and $CaCl_2 \cdot 6H_2O$. This provides an environment in which other wet substances can be dried without heating.

Anhydrous copper sulfate ($CuSO_4$) is white, but copper sulfate pentahydrate ($CuSO_4 \cdot 5H_2O$) is a brilliant blue.

2.8 Acids, Bases, and Salts

All of us participate in a complicated chemistry, from digesting food and shedding tears to baking bread or taking medicine for an upset stomach. Central to much of this chemistry are two special kinds of compounds called acids and bases. We eat them and drink them, and our bodies produce them. Some common acids and bases used around the home are shown in Figure 2.10.

Historically, acids and bases have been classified according to some distinctive properties. **Acids** are substances that have the following characteristics when dissolved in water.

- Acids taste sour, if diluted with enough water to be tasted safely.
- Acids produce a pricking or stinging sensation on the skin.
- Acids turn the color of litmus, an indicator dye, from blue to red.
- Acids react with many metals, such as magnesium, zinc, and iron, to produce ionic compounds and hydrogen gas.
- Acids also react with bases, thereby losing their acidic properties.

Bases are substances that have the following characteristics when dissolved in water.

- Bases taste bitter if diluted with enough water to be tasted safely.*
- Bases feel slippery or soapy on the skin.
- Bases turn the color of the indicator dye litmus from red to blue.
- Bases react with acids, thereby losing their basic properties.

(a)

(b)

▲ **Figure 2.10**
Some common acids and bases
(a) Some common acids: toilet bowl cleaner, vinegar, aspirin, tomato and fruit juices. (b) Some common bases: ammonia; drain cleaner, antacid tablets, baking soda, washing soda, and oven cleaners.

Acids and Bases: The Arrhenius Concept

In 1887, the Swedish chemist Svante Arrhenius proposed that an **acid** is a molecular compound that ionizes, or breaks up, in water to form a solution containing H^+ ions and anions. He viewed a **base** as a compound that ionizes in water to form a solution containing OH^- ions and cations. Some bases, such as NaOH and KOH, are ionic compounds. When they dissolve in water, the hydroxide ions and cations dissociate from one another. Most bases, however, are not ionic compounds. They do not contain hydroxide ions. The hydroxide ions and cations are formed in a reaction between the base and water.

*According to the biblical account, the Israelites, in their journey from Egypt to Canaan, came upon the bitter waters at Marah (Exodus 15:23). Although the writers may not have meant to, they thus recorded for posterity the existence of a base.

Arrhenius proposed that the essential reaction between an acid and a base, called *neutralization*, is the combination of H^+ ions from the acid and OH^- ions from the base to form water (that is, HOH or H_2O). The cation from the base and the anion from the acid make up an ionic compound called a **salt**. An example of a neutralization reaction is that between HCl (an acid) and NaOH (a base) to form NaCl (a salt) and H_2O (water).

Like many scientific theories, the Arrhenius theory has been supplanted by newer ones that better explain all the available data. We will study more modern acid-base theories in Chapter 15, but for now, this older Arrhenius theory will help us to identify acids and bases and to write names and formulas, and that's all we need.

In addition to his work on acids and bases, Svante Arrhenius (1859–1927) derived a mathematical expression that relates reaction rates to several factors (Chapter 13). Arrhenius was also the first to relate carbon dioxide in the atmosphere to the greenhouse effect (Chapter 25).

Formulas and Names of Acids, Bases, and Salts

Let's begin with the simplest case: salts. Because salts are ionic compounds, we write their formulas and names just as we do for ionic compounds in general. The only new idea is that salts are ionic compounds formed in the reaction between an acid and a base.

Next, consider *ionic* bases. We write their formulas and names like other ionic compounds. In Arrhenius bases, however, the anions will always be hydroxide ions, OH^-. The principal ionic bases are those of the Group 1A and 2A cations. The three most common are

$$\begin{aligned} \text{NaOH} &= \text{sodium hydroxide} \\ \text{KOH} &= \text{potassium hydroxide} \\ \text{Ca(OH)}_2 &= \text{calcium hydroxide} \end{aligned}$$

Most bases are *molecular* compounds and do not contain hydroxide ion. Rather, hydroxide ions are produced when these molecular bases react with water. The principal molecular bases are ammonia (NH_3) and compounds related to ammonia.

The naming of acids is a bit more complex but still not difficult if we take a systematic approach. Let's first consider *binary* acids. These are certain molecular compounds in which hydrogen is combined with a second nonmetallic element. We have learned that HCl is hydrogen chloride, and we will continue to use this name to refer to the pure gaseous compound. When hydrogen chloride dissolves and reacts with water, H^+ and Cl^- ions are formed. We call this *solution* an acid, and we name it by changing *hydrogen* to *hydro* and *chloride* to *chloric*. The name is hydrochloric acid. A few other examples are

Hydrogen bromide	HBr	*hydro*brom*ic* acid
Hydrogen iodide	HI	*hydro*iod*ic* acid
Hydrogen sulfide	H_2S	*hydro*sulfur*ic* acid

Not all binary compounds of hydrogen and a nonmetal are acids by any means—only those that ionize in water to produce H^+ ions and anions. Methane, CH_4, is a binary compound of hydrogen and it is not an acid. We will learn to recognize acids later on. For now, our emphasis is on how to write their formulas and what to call them once they have been identified as acids.

As Table 2.5 suggests, most acids are *ternary* acids. Their molecules are made up of atoms of *three* elements: hydrogen and two other nonmetals. Many of the ternary acids are *oxoacids*. These have atoms of hydrogen, *oxygen*, and a third nonmetal in their molecules. Some nonmetals form a set of ternary oxoacids. We base the names of the members of the set on the number of oxygen atoms per molecule, as illustrated on the next page for ternary oxoacids of chlorine and sulfur.

		Example	Name	Example	Name
Increasing number of oxygen atoms ↓	hypo___ous	HClO	hypochlorous acid	—	—
	___ous	$HClO_2$	chlorous acid	H_2SO_3	sulfurous acid
	___ic	$HClO_3$	chloric acid	H_2SO_4	sulfuric acid
	per___ic	$HClO_4$	perchloric acid	—	—

To name the salts of these acids, we change the anion name endings as follows:

An **acid** with a name ending in		A **salt** with a name ending in
–ous	forms	*–ite*
–ic	forms	*–ate*

As a result, we obtain the same names as on page 56: NaClO is sodium *hypo*chlo*rite*; $NaClO_2$ is sodium chlor*ite*; $NaClO_3$ is sodium chlor*ate*; and $NaClO_4$ is sodium *per*chlor*ate*.

Finally, with a few exceptions such as carbonic acid, H_2CO_3, acids containing carbon are rather different from other acids. Thus, we will categorize them in a different way in Section 2.10.

TABLE 2.5 Formulas and Names of Some Common Acids and Their Salts

Formula of acid	Name of acid	Sodium salt	
		Formula	Name
HCl	*Hydro*chloric acid	NaCl	Sodium chlor*ide*
HClO	*Hypo*chlor*ous* acid	NaClO	Sodium *hypo*chlor*ite*
$HClO_2$	Chlor*ous* acid	$NaClO_2$	Sodium chlor*ite*
$HClO_3$	Chlor*ic* acid	$NaClO_3$	Sodium chlor*ate*
$HClO_4$	*Per*chlor*ic* acid	$NaClO_4$	Sodium *per*chlor*ate*
H_2S	*Hydro*sulfur*ic* acid	Na_2S	Sodium sulf*ide*
H_2SO_3[a]	Sulfur*ous* acid	Na_2SO_3	Sodium sulf*ite*
H_2SO_4[a]	Sulfur*ic* acid	Na_2SO_4	Sodium sulf*ate*
HNO_2	Nitr*ous* acid	$NaNO_2$	Sodium nitr*ite*
HNO_3	Nitr*ic* acid	$NaNO_3$	Sodium nitr*ate*
H_3PO_4[a]	Phosphor*ic* acid	Na_3PO_4	Sodium phosph*ate*
H_2CO_3[a]	Carbon*ic* acid	Na_2CO_3	Sodium carbon*ate*

[a]Table 2.4 lists anions found in some salts of these acids in which not all of the available H atoms are replaced. If one or more H atoms remains unreplaced, formulas and names must be written accordingly, for example, $NaHSO_4$ is sodium hydrogen sulfate, and NaH_2PO_4 is sodium dihydrogen phosphate.

Introduction to Organic Compounds

We now turn our attention to a huge category of compounds that is so important that an entire branch of chemistry—*organic chemistry*—is devoted to their study.

Organic compounds, based on the element carbon, are all around us. The rich red, yellow, and orange colors of watermelons, tomatoes, carrots, and other vegetables and fruits are due to naturally occurring organic compounds. The carbohydrates and fats that fuel our bodies are organic compounds. The gasoline that powers our cars and the fuels that heat our homes are organic compounds. So are synthetic substances such as nylon and acetylsalicyclic acid (aspirin). Carbon-based compounds are vital components of our bodies—indeed of all living creatures. Overall, the diversity of organic compounds is quite astonishing.

What accounts for the extraordinary variety of organic compounds? It can be explained by the ability of carbon atoms to combine readily with atoms of nearly all the other elements and with other carbon atoms. The distinctive feature of organic compounds is that the carbon atoms join together into chains or rings to form a backbone to which other atoms are attached. Carbon compounds containing one or more of the elements H, O, N, and S are especially common. These attributes make possible an almost limitless number of different structures and, therefore, different compounds. Most organic compounds are molecular; only a few are ionic. Some organic compounds also have the attributes of an acid, base, or salt, and we can classify them as such.

So that we will be able to communicate effectively about organic compounds, and recognize them when they appear in this textbook (or in a newspaper article), let's now learn a bit about writing the names and formulas of organic compounds. Often, we can name an organic compound in two ways: We can assign a common, or trivial, name; or we can assign a *systematic* name that conforms to an international convention. We will use mostly common names, but we will introduce a few systematic names as well. Also, we take a more comprehensive look at systematic organic nomenclature in Appendix D.

2.9 Alkanes: Saturated Hydrocarbons

The simplest organic compounds, the *hydrocarbons*, contain only carbon and hydrogen atoms. There are several types of hydrocarbons, and in this section, we will discuss the kind known as alkanes. **Alkanes** are said to be *saturated* hydrocarbons, because their molecules contain the maximum number of hydrogen atoms possible for the number of carbon atoms present. The molecules are "saturated" with H atoms.

The simplest alkane is *methane*, the principal component of natural gas. In methane, four hydrogen atoms are attached to a central carbon atom.

Methane, CH_4

The next member of the series is *ethane*, a minor component of natural gas. In this molecule, two carbon atoms are joined together and three hydrogen atoms are attached to each carbon atom.

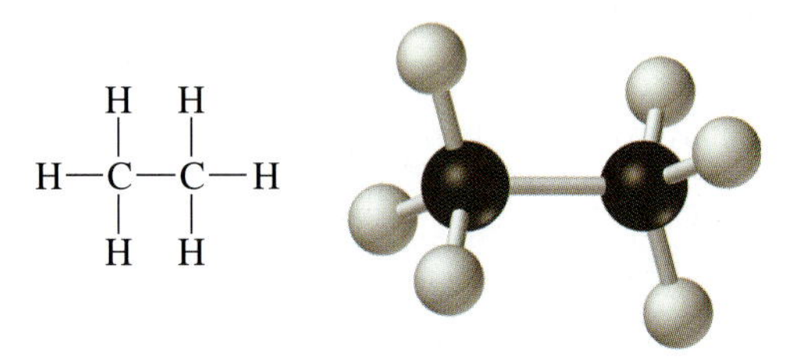

Ethane, C_2H_6

Propane, the familiar bottled gas used as a fuel in portable torches, gas grills and stoves, is the third member of the alkane series.

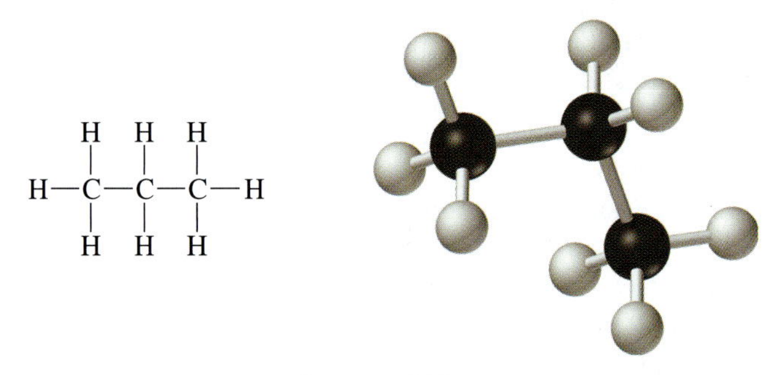

Propane, C_3H_8

TABLE 2.6 Word Stems Indicating the Number of Carbon Atoms in Simple Organic Molecules

Stem	Number of C atoms
meth-	1
eth-	2
prop-	3
but-	4
pent-	5
hex-	6
hept-	7
oct-	8
non-	9
dec-	10

Perhaps you see a pattern emerging: Each member of the alkane series differs from the preceding member by a CH_2 unit, that is, one carbon atom and two hydrogen atoms. Another pattern among the alkanes is that each conforms to the general formula C_nH_{2n+2}, where n is the number of carbon atoms in the molecule. Thus, for propane, with three C atoms, the formula is $C_3H_{(2\times 3)+2} = C_3H_8$.

The names of simple alkanes are composed of two parts. A word stem (Table 2.6) indicates the number of carbon atoms, and the ending *-ane* indicates that the hydrocarbon belongs to the alk*ane* family. Thus, C_5H_{12} is *pent*ane and C_6H_{14} is *hex*ane.

In nearly all carbon compounds, each carbon atom forms *four* bonds. Each hydrogen atom forms *one* bond. We can often decide whether a given structural formula is plausible simply by counting the number of bonds to each C and H atom.

When we consider a possible structural formula of the fourth member of the alkane series, we find that there are *two* possibilities with the formula C_4H_{10}. One has four carbon atoms bonded together in a continuous (straight) chain (left); in the other, a $-CH_3$ group branches off a three-carbon chain (right).

```
    H  H  H  H
    |  |  |  |
 H—C—C—C—C—H
    |  |  |  |
    H  H  H  H
```

Butane, C_4H_{10}

```
          H
          |
       H—C—H
    H     |    H
    |     |    |
 H—C——C——C—H
    |     |    |
    H     H    H
```

Isobutane, C_4H_{10}

Compounds with the same molecular formula but different structural formulas are called **isomers**. When there are only a few isomers, we can use prefixes to give them different names. For example, the prefix *iso-* (meaning "isomer of") indicates that a structural fragment of the type $(CH_3)_2CH—$ is part of the molecule. An example is isobutane, $(CH_3)_2CH—CH_3$. The prefix *neo-* (meaning "new") indicates the presence of the group $(CH_3)_3C—$, as in neopentane, $(CH_3)_3C—CH_3$. However, we need to use a more systematic approach when large numbers of isomers are possible for a given molecular formula. (Imagine trying to learn 18 such individual names for the 18 isomeric alkanes with the formula C_8H_{18}.) A system for naming isomers of organic compounds is outlined in Appendix D.

Isomers are distinctly different compounds. Although both butane and isobutane are gases at room temperature, liquid butane boils at about 0 °C, whereas the boiling point of isobutane is −12 °C. Differences in the structures of isomers are brought out most clearly by models (Figure 2.11), but we can suggest these differences on paper. We can use the structural formulas on page 62, or we can use *condensed* structural formulas, which also bring out the main structural features of molecules. For example,

$$\underset{\text{Butane}}{CH_3CH_2CH_2CH_3} \qquad \underset{\text{Isobutane}}{CH_3\overset{\overset{CH_3}{|}}{C}HCH_3}$$

To save space, we can usually write condensed structures on a single line by using parentheses to set off groups that branch from a longer chain, as in the following representations of isobutane.

$$CH_3CH(CH_3)CH_3 \quad \text{or} \quad CH_3CH(CH_3)_2$$
Isobutane

The number of isomers increases rapidly with the number of carbon atoms: three C_5H_{12} alkanes, five C_6H_{14} alkanes, nine C_7H_{16} alkanes, eighteen C_8H_{18} alkanes, and so on. There are, for example, over 4 billion possible isomers of $C_{30}H_{62}$, few of which have been isolated or synthesized.

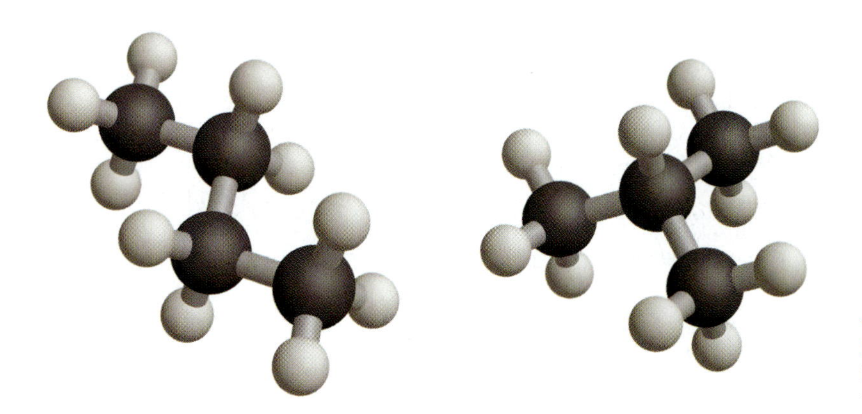

◀ **Figure 2.11**
Ball-and-stick models
Butane (left) and isobutane (right).

Counting Isomers

At times, it may seem difficult to tell whether structures that appear to be different are isomers or merely different representations of the same compound. Consider the question of how many isomers there are with the molecular formula C_5H_{12}. The following structures, in which we have omitted H atoms for simplicity, are all the same. If we think of the interior carbon atoms as being like door hinges, we see that there is only one possibility for a continuous (unbranched) five-carbon chain. That is, structures (1a), (1b), and (1c) are all the same as structure (1).

```
                                                      C—C
                                                        |
                    C       C     C—C   C               C—C
                    |       |        |  |                 |
C—C—C—C—C           C—C—C            C—C                  C
(1) Pentane          (1a)            (1b)               (1c)
```

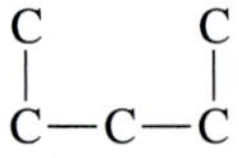

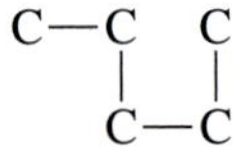

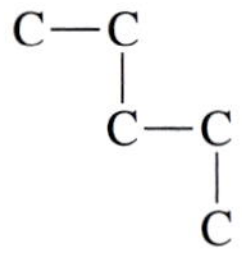

Now consider structures with four carbon atoms in a chain and a one-carbon branch.

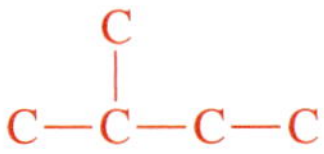

```
      C                    C          C—C—C—C
      |                    |            |
  C—C—C—C            C—C—C—C            C
(2) Isopentane          (2a)           (2b)
```

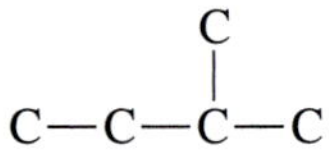

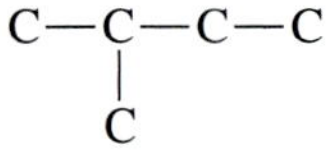

These structures are also identical. Structure (2a), if flipped from left to right, and (2b), if turned top to bottom, are both the same as (2).

Finally, consider a three-carbon chain with two one-carbon branches. There is only one possibility.

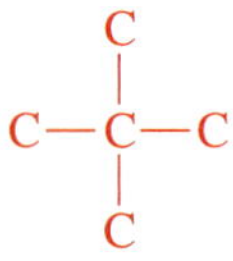

```
      C                    C
      |                    |
  C—C—C              C—C—C
      |                |
      C                C
(3) Neopentane        (3a)
```

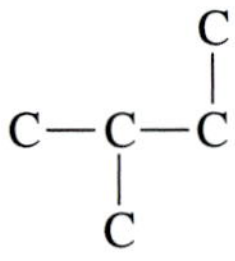

Other structures that we might consider will be identical to one of the three structures shown in color. For example, if you look at structure (3), you might think that moving the top C from the second to the third carbon atom of the chain would give a different structure (3a). However, it would actually have *four* carbons in a continuous chain and one branch; it would be the same as structure (2). There are only three isomers with the formula C_5H_{12}.

EXAMPLE 2.12—A Conceptual Example

Are the following pairs of molecules isomers or not?

a. 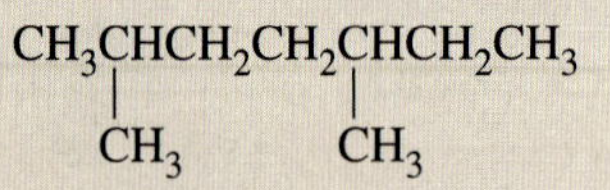
```
CH3CHCH2CH2CHCH2CH3    and    CH3CH2CHCH2CH2CHCH3
   |         |                      |         |
   CH3       CH3                    CH3       CH3
```

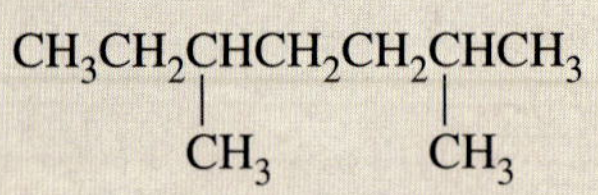

b. 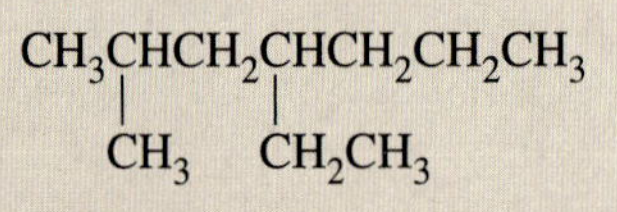
```
CH3CHCH2CHCH2CH2CH3    and    CH3CH2CHCH2CH2CHCH2CH3
   |     |                          |         |
   CH3   CH2CH3                     CH3       CH3
```

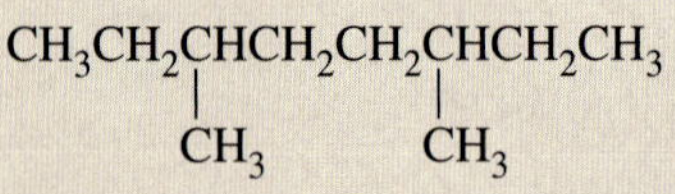

SOLUTION

Check first to see if the structures have the same molecular formula. If they *do not*, they cannot be isomers. If the structures do have the same molecular formula and are otherwise identical, the structures represent the same molecule, not isomers. In short, isomers must have the *same* molecular formula and *different* structures.

a. These two molecules have the same molecular formula, C_9H_{20}. If the second structure is flipped from left to right, we see that it is identical to the first. Each molecule has a CH_3 group on the second carbon atom from one end of the chain and on the third carbon from the other end. The structures represent two molecules of the same substance, not isomers.

b. These two molecules have the same molecular formula, $C_{10}H_{22}$. The first has *seven* C atoms in a chain and two branches. One branch has one C atom and the other branch has two C atoms. The second molecule has *eight* C atoms in a chain and two branches. However, each branch has only one C atom. The two molecules have the same molecular formula but different structures. They are isomers.

EXERCISE 2.12

Are the following pairs of molecules isomers?

```
a. CH3CH2CH2CH2CH(CH3)2    and        CH2
                                     /   \
                                  CH2     CH—CH(CH3)2
                                     \   /
                                      CH2

       CH3       CH3
       |         |
b. CH3CH2CHCH2CH2CHCH3     and   CH3CHCH2CHCH2CH2CH3
                                    |    |
                                    CH3  CH3
```

Cyclic Alkanes

Imagine the two ends of a straight-chain alkane molecule coming close together, the end C atoms each shedding one H atom, and the ends joining to each other. The result would be a ringlike structure with the generic formula C_nH_{2n}. Alkane molecules with ring structures are named with the prefix *cyclo*. The smallest number of carbon atoms in a ring structure is three, as in the compound cyclopropane. Another common cyclic alkane has a six-carbon atom ring and is called cyclohexane. Of the representations shown below, the second one, in which the ring C atoms are not labeled and the H atoms are not shown, is often used because it is easier to write. The ring of C atoms in these representations of cyclohexane appears to be flat, but Figure 2.12 shows that it is not.

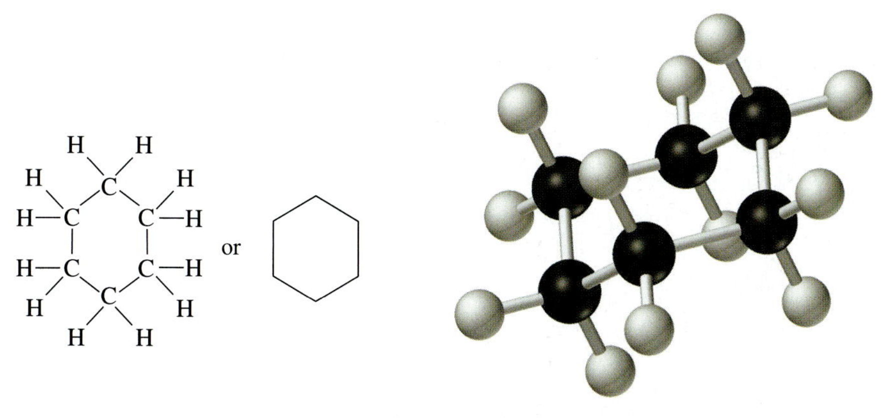

Figure 2.12
Ball-and-stick model of cyclohexane
As this model indicates, the cyclohexane molecule is not planar. Rather, it can assume several different arrangements or conformations. The most stable arrangement, shown here, is called the *chair conformation* because the six carbon atoms outline a structure that somewhat resembles a reclining chair.

Line-angle formulas (also called "stick figures") such as the hexagon for cyclohexane are usually used for cyclic compounds and sometimes for open-chain compounds. In these representations, the lines represent bonds and a carbon atom is assumed to be present wherever two lines meet or where a line begins or ends. We

Pentane: line-angle formula

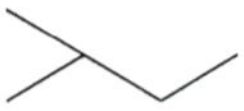

Isopentane: line-angle formula

assume that each carbon atom has enough H atoms to give it a total of four bonds. Thus, we can represent pentane and isopentane (page 64) as in the margin.

Cyclic structures can have other atoms or groups of atoms attached to the ring. Methylcyclopropane, which consists of a three-carbon ring with a $—CH_3$ group (methyl) attached, can be represented with either a partial or complete line-angle formula.

CH_3 or

Methylcyclopropane

Similarly, we can represent 1,2-dimethylcyclopentane and isopropylcyclobutane with the following line-angle formulas.

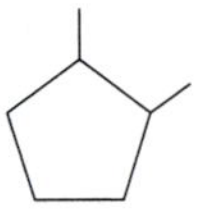

1,2-Dimethylcyclopentane

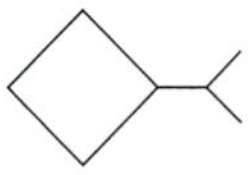

Isopropylcyclobutane

Homology

A series of compounds whose formulas and structures vary in a regular manner also have properties that vary in a predictable manner, a principle called *homology*. For example, both the densities and the boiling points of the straight-chain alkanes increase in a continuous and regular fashion with increasing numbers of carbon atoms in the chain. Homology aids our study of organic chemistry in much the same way that the periodic table provides an organizing principle for the chemistry of the elements. Instead of studying the properties of a bewildering array of individual organic compounds, we can usually study a few members of a homologous series and deduce properties of the others.

EXAMPLE 2.13—An Estimation Example

The boiling points of the straight-chain alkanes pentane, hexane, heptane, and octane are 36.1, 68.7, 98.4, and 125.7 °C, respectively. Estimate the boiling point of the straight-chain alkane *decane*.

SOLUTION

We need to figure out the pattern of increasing boiling point with increasing length of the carbon chain. The data are for carbon-chain lengths of five, six, seven, and eight C atoms. We want to estimate the boiling point of the 10-carbon alkane, *dec*ane.

Alkane formula	C_5H_{12}		C_6H_{14}		C_7H_{16}		C_8H_{18}		C_9H_{20}		$C_{10}H_{22}$
Boiling point, °C	36.1		68.7		98.4		125.7		?		?
Increase per CH_2 unit, °C		32.6		29.7		27.3		≈25		≈23	

We see a trend in boiling point increases per added CH_2 unit—32.6, 29.7, 27.3—that suggests the boiling point of C_9H_{20} will be about 25 °C above that of C_8H_{18}, and that the boiling point of $C_{10}H_{22}$ will be about 23 °C greater than that of C_9H_{20}. We therefore estimate the boiling point of decane to be $\approx$ (125.7 + 25 + 23) °C $\approx$ 174 °C. (The observed boiling point is 174.1 °C.)

EXERCISE 2.13

The kerosene component of petroleum consists of hydrocarbons with boiling points that range from about 200 to 260 °C. What are the formulas of the straight-chain alkanes you would expect to find in kerosene?

2.10 Types of Organic Compounds

In addition to homology, another important organizing principle in organic chemistry is that of functional groups. A **functional group** is an atom or group of atoms attached to or inserted in a hydrocarbon chain or ring that confers characteristic properties to the molecule as a whole. Many simple organic molecules are composed of two parts: a functional group, where most of the reactions of the molecule occur, and a hydrocarbon chain that usually is unreactive. Molecules having the same functional group generally have similar properties. We will briefly consider five functional groups here, and from time to time in the text, we will introduce others. For ready reference, some of the more common functional groups are presented in Table 2.7.

Alcohols

The functional group common to **alcohols** is the *hydroxyl* group, —OH. The simplest alcohol has the —OH group substituted for one of the H atoms in methane.

H—C(H)(H)—O—H or CH_3OH

Methyl alcohol (methanol), CH_3OH

APPLICATION NOTE
Methanol is widely used as a solvent, and because it burns more cleanly than gasoline, methanol is used in some fleet vehicles in smoggy areas. Methanol is fairly toxic. Ingestion of as little as 30 mL can cause blindness and even death.

Both names suggest a relationship to methane. The common name, methyl alcohol, tells us that the methyl group, CH_3—, is the hydrocarbon portion of the molecule, and the family name *alcohol* indicates the presence of the —OH group. In the systematic name, methanol, the stem *meth-* indicates a compound based on methane, and the ending *-ol* signifies that this compound is an alcohol. Note that even though alcohols contain the functional group —OH, they are *not* bases in the Arrhenius sense. Alcohols are not ionic; the hydroxyl group is not present as OH^-, nor is OH^- produced when an alcohol dissolves in water.

The next higher alcohol is based on the two-carbon alkane, ethane.

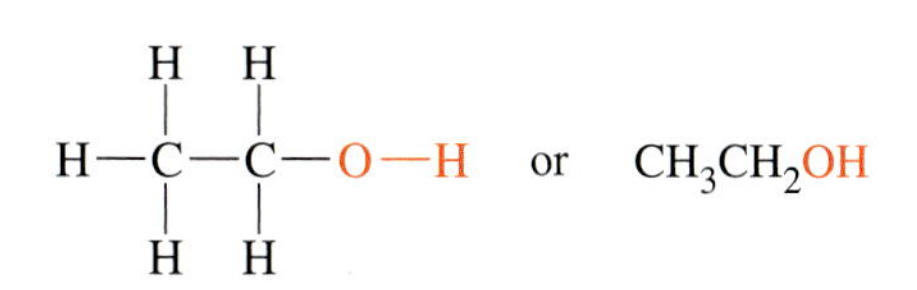

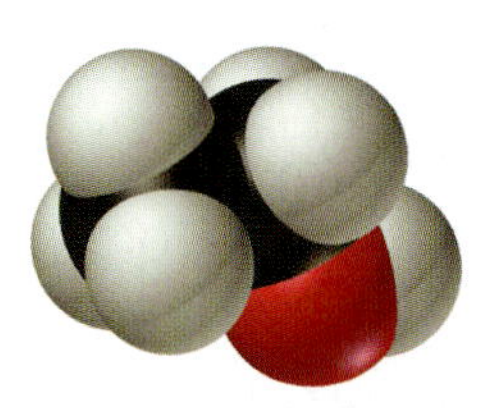

Ethyl alcohol (ethanol), CH_3CH_2OH

APPLICATION NOTE
Ethanol is the alcohol in alcoholic beverages. It is considerably less toxic than methanol. (The lethal dose of ethanol, when rapidly ingested, is about 500 mL.) Like methanol, ethanol is widely used as a solvent. It is also used as a gasoline additive and in some instances even as a gasoline substitute.

There are *two* different three-carbon alcohols, based on propane. They are isomers.

TABLE 2.7 Some Classes of Organic Compounds and Their Functional Groups

Class	General Structural Formula[a]	Example	Name of Example	Cross Reference
Alkane	R—H	$CH_3CH_2CH_2CH_2CH_2CH_3$	hexane	Section 2.9, 6.8, Chap. 23
Alkene	$>C{=}C<$	$CH_2{=}CHCH_2CH_2CH_3$	1-pentene	Section 9.10, Chap. 23
Alkyne	$—C{\equiv}C—$	$CH_3C{\equiv}CCH_2CH_2CH_2CH_2CH_3$	2-octyne	Section 9.10, Chap. 23
Alcohol	R—OH	$CH_3CH_2CH_2CH_2OH$	1-butanol	Section 2.10, Chap. 23
Alkyl halide	R—X [b]	$CH_3CH_2CH_2CH_2CH_2CH_2Br$	1-bromohexane	Chap. 23
Ether	R—O—R	$CH_3–O–CH_2CH_2CH_3$	1-methoxypropane (methyl propyl ether)[c]	Section 2.10
Amine	$R—NH_2$	$CH_3CH_2CH_2–NH_2$	1-aminopropane (propylamine)[c]	Sections 2.10, 4.2, Chap. 15
Aldehyde	R—C(=O)—H	$CH_3CH_2CH_2C(=O)—H$	butanal (butyraldehyde)[c]	Section 4.6, Chap. 23
Ketone	R—C(=O)—R	$CH_3CH_2C(=O)CH_2CH_2CH_3$	3-hexanone (ethyl propyl ketone)[c]	Section 4.6, Chap. 23
Carboxylic acid	R—C(=O)—OH	$CH_3CH_2CH_2C(=O)—OH$	butanoic acid (butyric acid)[c]	Sections 2.10, 4.2, Chap. 15, 23
Ester	R—C(=O)—OR	$CH_3CH_2CH_2C(=O)—OCH_3$	methyl butanoate (methyl butyrate)[c]	Sections 2.10, 6.8 (fats) Chap. 23, Chap. 24 (polymers)
Amide	$R—C(=O)—NH_2$	$CH_3CH_2CH_2C(=O)—NH_2$	butanamide (butyramide)[c]	Section 11.6, Chap. 23, Chap. 24 (polymers)
Arene	Ar—H[d]	$C_6H_5—CH_2CH_3$ (benzene ring)	ethylbenzene	Section 10.8, Chap. 23
Aryl halide	Ar—X[b]	$C_6H_5—Br$ (benzene ring)	bromobenzene	Chap. 23
Phenol	Ar—OH	Cl—C_6H_4—OH (benzene ring)	4-chlorophenol (*p*-chlorophenol)[c]	Section 9.10, Chap. 23

[a] The functional group is shown in red. R stands for an alkyl group.
[b] X stands for a halogen atom—F, Cl, Br, or I.
[c] Common name.
[d] Ar— stands for an aromatic (*aryl*) group such as the benzene ring.

Propyl alcohol (1-propanol), $CH_3CH_2CH_2OH$

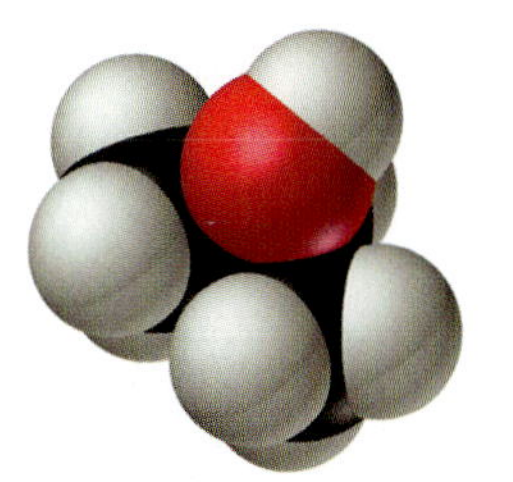

Isopropyl alcohol (2-propanol), $CH_3CHOHCH_3$

In the systematic names, the prefix "1-" indicates that the —OH group is on the first, or end, carbon atom, and the prefix "2-," that the —OH group is on the second carbon from the end.

Sometimes we represent alcohols in general as ROH, where R— stands for any hydrocarbon group. Molecules with more than one —OH group can also be classified as alcohols. A good example is one that is used in cosmetics and known by the trivial names of glycerol and glycerin. The molecular formula of glycerol is $C_3H_8O_3$, and its condensed structural formula is $CH_2(OH)CH(OH)CH_2OH$. Figure 2.13 shows a space-filling model of glycerol and the derivation of its systematic name, 1,2,3-propanetriol.

APPLICATION NOTE
Rubbing alcohol is a solution containing 70% isopropyl alcohol in water.

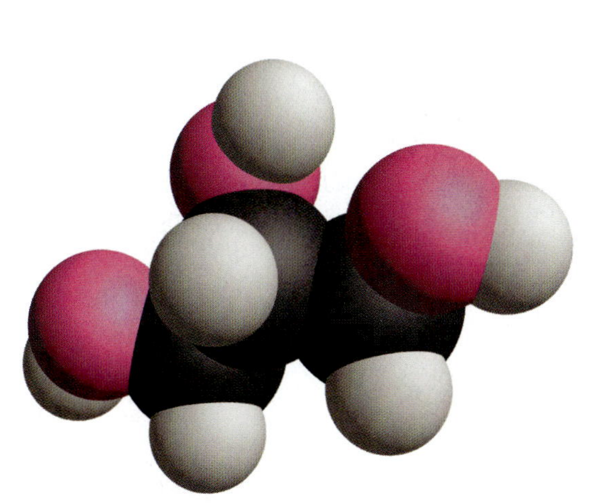

Figure 2.13
The glycerol molecule
The systematic name of glycerol is 1,2,3-propanetriol. In relation to the space-filling molecular model, *propane* signifies a chain of three carbon atoms (black). The ending -*ol* indicates the presence of —OH groups. *Tri*ol means that there are three of these groups, and the numbers "1,2,3" signify that they are attached to the first, second, and third carbon atoms of propane. The system of nomenclature that leads to this name is outlined in Appendix D. (Note that two of the hydrogen atoms are hidden in this model.)

Ethers

Compounds that have two hydrocarbon groups attached to the same oxygen atom are called **ethers**. Ethers in general can be represented as R—O—R (or sometimes R—O—R′ if the two hydrocarbon groups are different). Perhaps the most familiar ether is diethyl ether, $CH_3CH_2OCH_2CH_3$, a substance once used as an anesthetic. Today, diethyl ether is an important solvent; it dissolves a wide variety of organic compounds. It is highly flammable, however, and must be used with great care.

Carboxylic Acids

The functional group that most commonly confers acidic properties to an organic substance is the *carboxyl* group:

$$-\overset{\overset{\displaystyle O}{\|}}{C}-O-H \quad \text{or} \quad -COOH$$

The carbon atom in a carboxyl group forms a *double* bond to one of the oxygen atoms bonded to it, but notice that it still conforms to the requirement of forming four bonds (page 62). In the condensed form (—COOH), the double bond to one of the oxygen atoms is understood.

When a substance with carboxyl groups is dissolved in water, some of the hydrogen atoms of the carboxyl groups become H^+ ions; hydrogen atoms that can split off as H^+ are called *acidic hydrogens*, or *ionizable hydrogens*. Hydrogen atoms attached to carbon atoms do *not* split off as H^+. Thus, it is the presence of a carboxyl group that signifies a molecule is a **carboxylic acid**. The simplest of the carboxylic acids is

$$H-\overset{\overset{\displaystyle O}{\|}}{C}-O-H \quad \text{or} \quad HCOOH$$

Formic acid (methanoic acid), HCOOH

The common name, formic acid, is derived from the Latin word *formica*, meaning "ant." An ant bite hurts because the ant injects formic acid when it bites. In the systematic name, *methan-* indicates one carbon atom, and *-oic acid* tells us that the compound is a carboxylic acid.

Just as with the alkanes and alcohols, there is a homologous series of carboxylic acids. The two-carbon carboxylic acid is

$$H-\underset{\underset{\displaystyle H}{|}}{\overset{\overset{\displaystyle H}{|}}{C}}-\overset{\overset{\displaystyle O}{\|}}{C}-O-H \quad \text{or} \quad CH_3COOH$$

Acetic acid (ethanoic acid)

Molecular models of acetic acid were presented in Figure 2.5. Acetic acid is probably the most frequently used organic acid in chemical laboratories. When an organic acid is neutralized with a base, a salt is produced, just as with an inorganic acid. Thus, acetic acid and sodium hydroxide react to form sodium acetate, CH_3COONa, an ionic compound consisting of Na^+ and CH_3COO^- ions.

The three-carbon carboxylic acid is propionic (propanoic) acid. There are two isomers of the four-carbon acid: butyric and isobutyric acid. As you might expect, the number of possible isomers goes up rapidly as the chain length increases. Carboxylic acids in general are often represented as RCOOH.

APPLICATION NOTE

Acetic acid can be made by the fermentation of cider or wine, producing vinegar, which is about 4 to 10% acetic acid.

Esters

Esters are derived from carboxylic acids and alcohols. The —OH group of a carboxylic acid molecule is replaced by an —OR of an alcohol molecule. In general, then, we can represent an ester as

$$R'-\overset{\overset{\displaystyle O}{\|}}{C}-O-R \quad \text{or} \quad R'COOR$$

where R′ is the hydrocarbon portion of a carboxylic acid, and R is the hydrocarbon group of an alcohol. R and R′ may be the same or different.

Esters are named by indicating the part from the alcohol first and then naming the portion from the carboxylic acid with the name ending in *-ate*. For instance,

$$CH_3-\overset{\overset{\displaystyle O}{\|}}{C}-O-CH_2CH_3$$

is ethyl acet*ate*; it is made from ethyl alcohol and acetic acid.

Many esters are noted for their pleasant odors, and some are used in flavors and fragrances. Pentyl acetate, $CH_3COOCH_2CH_2CH_2CH_2CH_3$, is responsible for most of the odor and flavor of ripe bananas. Many esters are used as flavorings in cakes, candies, and other foods and as ingredients in fragrances, especially those used to perfume household products. Some esters are also used as solvents. Ethyl acetate, for example, is used in some fingernail polish removers: It is a solvent for the resins in the polish.

APPLICATION NOTE
Butyric acid, $CH_3CH_2CH_2COOH$, is one of the most foul-smelling substances known, but turn it into the ester methyl butyrate, $CH_3CH_2CH_2COOCH_3$, and you get the aroma of apples.

Amines

The most common *organic* bases, the amines, are related to ammonia. **Amines** are compounds in which one or more organic groups are substituted for H atoms in NH_3. In these two amines, one of the H atoms has been replaced:

$$H-\underset{\underset{\displaystyle H}{|}}{\overset{\overset{\displaystyle H}{|}}{C}}-\overset{\overset{\displaystyle H}{|}}{N}-H \quad \text{or} \quad CH_3NH_2$$

Methylamine

$$H-\underset{\underset{\displaystyle H}{|}}{\overset{\overset{\displaystyle H}{|}}{C}}-\underset{\underset{\displaystyle H}{|}}{\overset{\overset{\displaystyle H}{|}}{C}}-\overset{\overset{\displaystyle H}{|}}{N}-H \quad \text{or} \quad CH_3CH_2NH_2$$

Ethylamine

The replacement of two and three H atoms, respectively, is seen in dimethylamine $[(CH_3)_2NH]$ and trimethylamine $[(CH_3)_3N]$. In Chapters 4 and 15, we will see that much of what we learn about ammonia as a base applies as well to amines.

APPLICATION NOTE
Amines with one or two carbon atoms per molecule smell much like ammonia. Higher homologs smell like rotting fish. In fact, the foul odors of rotting flesh are due in large part to amines that are given off as the flesh decays.

Summary

The basic laws of chemical combination are the laws of conservation of mass, constant composition, and multiple proportions. Each played an important role in Dalton's development of the atomic theory.

The three components of atoms of most concern to chemists are protons, neutrons, and electrons. Protons and neutrons make up the nucleus, and their combined number is the mass number, *A*, of the atom. The number of protons is the atomic number, *Z*. Electrons, found outside the nucleus, have negative charges equal to the positive charges of the protons. All atoms of an element have the same atomic number, but they may have different mass numbers, giving rise to isotopes.

A chemical formula indicates the relative numbers of atoms of each type in a compound. An empirical formula is the simplest that can be written, and a molecular formula reflects the actual composition of a molecule. Structural and condensed structural formulas describe the arrangement of atoms within molecules. For example, for acetic acid:

Key Terms

acid (2.8)
alcohol (2.10)
alkane (2.9)
amine (2.10)
anion (2.7)
atomic mass (2.4)
atomic mass unit (2.4)
atomic number (*Z*) (2.3)
base (2.8)
carboxylic acid (2.10)
cation (2.7)
chemical formula (p. 47)
chemical nomenclature (p. 35)
electron (2.3)
empirical formula (2.6)
ester (2.10)
ether (2.10)
formula unit (2.7)
functional group (2.10)
hydrate (2.7)
ion (2.7)
ionic compound (2.7)
isomer (2.9)
isotope (2.3)
law of conservation of mass (2.1)
law of constant composition (2.1)
law of definite proportions (2.1)
law of multiple proportions (2.2)
mass number (*A*) (2.3)
metal (2.5)
metalloid (2.5)
molecular compound (2.6)
molecular formula (2.6)
molecule (2.6)
neutron (2.3)
nonmetal (2.5)
periodic table (2.5)
polyatomic ion (2.7)
proton (2.3)
salt (2.8)
structural formula (2.6)

Acetic acid:	CH_2O	$C_2H_4O_2$	H—C(H)(H)—C(=O)—O—H	CH_3COOH
	Empirical formula	Molecular formula	Structural formula	Condensed structural formula

The periodic table is an arrangement of the elements by atomic number that places elements with similar properties into the same vertical groups (families). The periodic table is an important aid in the writing of formulas and names of chemical compounds. A molecular compound consists of molecules; in a binary molecular compound the molecules are made up of atoms of two different elements. In naming these compounds, the numbers of atoms in the molecules are denoted by prefixes; the names also feature *-ide* endings.

Examples: NI_3 = nitrogen *tri*iodide S_2F_4 = *di*sulfur *tetra*fluoride

Ions are formed by the loss or gain of electrons by single atoms or groups of atoms. Positive ions are known as cations and negative ions as anions. An ionic compound is made up of cations and anions, held together by electrostatic forces of attraction. Formulas of ionic compounds are based on an electrically neutral combination of cations and anions called a formula unit. The names of some monatomic cations include Roman numerals to designate the charge on the ion. The names of monatomic anions are those of the nonmetallic elements, modified to an *-ide* ending. For polyatomic anions, the prefixes *hypo-* and *per-* and the endings *-ite* and *-ate* are commonly found.

Examples:
MgF_2 = magnesium fluor*ide* Li_2S = lithium sulf*ide*
Cu_2O = copper(*I*) ox*ide* CuO = copper (*II*) ox*ide*
$Ca(ClO)_2$ = calcium *hypo*chlor*ite* KIO_4 = potassium *per*iod*ate*

Many compounds are classified as acids, bases, or salts. According to the Arrhenius theory, an acid produces H^+ in aqueous (water) solution, and a base produces OH^-. A neutralization reaction between an acid and a base forms water and an ionic compound called a salt. Binary acids have hydrogen and a nonmetal as their constituent elements. Their names feature the prefix *hydro-* and the ending *-ic* attached to the stem of the name of the nonmetal. Ternary oxoacids have oxygen as an additional constituent element, and their names use prefixes (*hypo-* and *per-*) and endings (*-ous* and *-ic*) to indicate the number of O atoms per molecule.

Examples:
HI = *hydro*iod*ic* acid HIO_3 = iod*ic* acid
$HClO_2$ chlor*ous* acid $HClO_4$ = *per*chlor*ic* acid

Organic compounds are based on the element carbon. Hydrocarbons contain only the elements hydrogen and carbon. Alkanes have carbon atoms joined together by single bonds into chains or rings, with hydrogen atoms attached to the carbon atoms. Alkanes with four or more carbon atoms can exist as isomers: molecules with the same molecular formula but different structures and properties.

Functional groups confer distinctive properties to an organic molecule when the groups are substituted for hydrogen atoms in a hydrocarbon. Alcohols feature the hydroxyl group, —OH, and ethers have two hydrocarbon groups joined to the same oxygen atom. Carboxylic acids have a carboxyl group, —COOH. An ester, R′COOR, is derived from a carboxylic acid (R′COOH) and an alcohol (ROH). Amines are compounds in which organic groups are substituted for one or more of the H atoms in ammonia, NH_3.

Review Questions

1. Heptane, a hydrocarbon, is always found to contain 84.0% carbon by mass. What law does this observation illustrate?
2. When 24.3 g of magnesium is burned in 16.0 g of oxygen, 40.3 g of magnesium oxide is formed. When 24.3 g of magnesium is burned in 80.0 g of oxygen, **(a)** what is the total mass of substances present after the reaction? **(b)** What mass of magnesium oxide is formed? **(c)** What law(s) is (are) illustrated by this reaction? **(d)** If 48.6 g of magnesium is burned in 80.0 g of oxygen, what mass of magnesium oxide is formed? Explain.
3. Sulfur and oxygen form two compounds. The mass ratio of oxygen to sulfur in compound A is 1.0:1.0 and that in compound B is 1.5:1.0. What is the ratio of oxygen to sulfur in compound B as compared to that in compound A? Express the ratio in the smallest whole numbers. What law does this observation illustrate?
4. Polychlorinated biphenyls (PCBs) are environmental contaminants composed of carbon, hydrogen, and chlorine atoms. A news report states that a new method of disposal of PCBs converts them completely to carbon dioxide and water. The necessary oxygen atoms come from oxygen in the air. Do you think this method will work? Explain.
5. Outline the main points of Dalton's atomic theory, and use it, together with illustrative examples, to explain **(a)** the law of conservation of mass, **(b)** the law of constant composition, and **(c)** the law of multiple proportions.
6. Are the following findings, expressed to the nearest atomic mass unit, in agreement with Dalton's atomic theory? Explain your answers. **(a)** An atom of calcium has a mass of 40 u and one of vanadium, 50 u. **(b)** An atom of calcium has a mass of 40 u and one of potassium, 40 u. **(c)** One atom of calcium has a mass of 40 u and another calcium atom, 44 u.
7. Give the distinguishing characteristics of the proton, the neutron, and the electron. Why are they called fundamental particles?
8. What is the atomic nucleus? What subatomic particles are found in the nucleus?
9. What are isotopes, and what is meant by the mass number of an isotope?
10. Which of the following pairs of symbols represent isotopes? Which are isobars?
    (a) ${}^{70}_{33}E$ and ${}^{70}_{34}E$
    (b) ${}^{57}_{28}E$ and ${}^{66}_{28}E$
    (c) ${}^{186}_{74}E$ and ${}^{186}_{74}E$
    (d) ${}^{7}_{3}E$ and ${}^{8}_{4}E$
    (e) ${}^{22}_{11}E$ and ${}^{44}_{22}E$
11. Use the symbolism ${}^{A}_{Z}E$ to represent each of the following atoms. You may refer to the periodic table.
    (a) boron-8
    (b) carbon-14
    (c) uranium-235
    (d) cobalt-60
12. What do tabulated atomic mass values, such as those found inside the front cover, represent?
13. List some characteristic properties of metals, and indicate where these elements are located in the periodic table.
14. List some characteristic properties of nonmetals, and indicate where they are located in the periodic table.
15. What is a metalloid? Where are the metalloids located in the periodic table?
16. Explain why a chemist calls the compound $MgCl_2$ magnesium chloride and not magnesium dichloride. Would a chemist similarly call SCl_2 sulfur chloride? Explain.
17. In what two ways might we use assigned names to distinguish between $FeCl_2$ and $FeCl_3$? Which is the systematic and which is the common name?
18. Which of the names listed below refer to actual substances that you might find in containers on a storeroom shelf? What do the other names represent?
    (a) magnesium
    (b) methyl
    (c) chloride
    (d) ammonia
    (e) ammonium
    (f) ethane
19. What type of information is conveyed by each of the following representations of a molecule?
    (a) empirical formula
    (b) molecular formula
    (c) structural formula
    (d) ball-and-stick model
    (e) space-filling model.
20. A substance has the molecular formula $C_4H_8O_2$. **(a)** What is the empirical formula of this substance? **(b)** Can you write a structural formula from the molecular formula given? Explain.
21. According to the Arrhenius theory, all acids have one element in common. What is that element? Are all compounds containing that element acids? Explain.
22. Suggest some ways in which you might determine whether a particular solution contains an acid or a base.
23. Can a substance with the molecular formula C_3H_4 be an *alkane*? Explain.
24. Are hexane and cyclohexane isomers? Explain.
25. For which of the following is the *molecular* formula alone enough to identify the type of compound? For which must you have the *structural* formula?
    (a) an organic compound
    (b) a hydrocarbon
    (c) an alcohol
    (d) an alkane
    (e) a carboxylic acid
26. Explain the difference in meaning between each pair of terms:
    (a) a group and a period of the periodic table
    (b) an ion and an ionic substance
    (c) an acid and a salt
    (d) an isomer and an isotope

Problems

Laws of Chemical Combination

27. A student heats 1.0000 g of zinc and 0.200 g of sulfur in a closed container. She obtains 0.608 g of zinc sulfide and recovers 0.592 g of unreacted zinc. Are these results consistent with the law of conservation of mass? Explain.

28. When 0.2250 g of magnesium is heated with 0.5050 g of nitrogen in a closed container, the magnesium is completely converted to 0.3114 g of magnesium nitride. What mass of unreacted nitrogen must remain?

29. A colorless liquid is thought to be a pure compound of carbon, hydrogen, and oxygen. Analyses of three samples give the following results.

	Mass of sample	Mass of carbon	Mass of hydrogen
Sample 1	1.000 g	0.625 g	0.0419 g
Sample 2	1.549 g	0.968 g	0.0649 g
Sample 3	0.988 g	0.618 g	0.0414 g

Could the material be a pure compound?

30. Azulene, a blue solid, is thought to be a pure compound. Analyses of three samples of the material give the following results.

	Mass of sample	Mass of carbon	Mass of hydrogen
Sample 1	1.000 g	0.937 g	0.0629 g
Sample 2	0.244 g	0.229 g	0.0153 g
Sample 3	0.100 g	0.094 g	0.0063 g

Could the material be a pure compound?

31. Two experiments were performed in which sulfur was burned completely in pure oxygen gas, producing sulfur dioxide and leaving some unreacted oxygen. In the first experiment, 0.312 g of sulfur produced 0.623 g of sulfur dioxide. What mass of sulfur dioxide should have been produced in the second experiment, in which 1.305 g of sulfur was burned?

32. Refer to Example 2.2 (page 38) to determine the mass ratios: mass oxygen/mass magnesium oxide and mass magnesium/mass magnesium oxide. Then, determine the masses of magnesium and oxygen that must combine to form 1.000 g of magnesium oxide.

33. One oxide of nitrogen is found to have an oxygen-to-nitrogen mass ratio of 1.142 : 1 (that is, 1.142 g of oxygen for every 1.000 g of nitrogen). Which of the following oxygen-to-nitrogen mass ratios are possible for different oxides of nitrogen? (*Hint*: Recall the scheme outlined in Figure 2.2.)

(a) 0.571 : 1 **(c)** 2.285 : 1

(b) 1.000 : 1 **(d)** 2.500 : 1

34. A sample of one oxide of tin, SnO, is found to consist of 0.742 g of tin and 0.100 g of oxygen. A sample of a second oxide of tin consists of 0.555 g of tin and 0.150 g of oxygen. What must be the formula of this second oxide?

Atoms and Atomic Masses

35. Indicate how many electrons and how many protons there are in a neutral atom of each of these elements. (You may use the periodic table.)

(a) calcium **(c)** fluorine **(e)** beryllium

(b) sodium **(d)** argon

36. Use the periodic table to determine the number of protons and electrons in a neutral atom of each of these elements.

(a) nitrogen **(c)** cadmium **(e)** silver

(b) iron **(d)** uranium

37. Indicate the numbers of protons and neutrons in the following atoms.

(a) ${}^{62}Zn$ **(b)** ${}^{241}Pu$ **(c)** ${}^{99}Tc$ **(d)** ${}^{99}Mo$

38. Indicate the numbers of electrons and neutrons in the following atoms.

(a) ${}^{11}B$ **(b)** ${}^{154}Sm$ **(c)** ${}^{81}Kr$ **(d)** ${}^{121}Te$

Problems 39 and 40 are based on the table of atomic species shown below.

	#1	#2	#3	#4	#5	#6	#7
No. protons	16	20	19	18	20	22	20
No. neutrons	18	20	21	22	24	26	28
No. electrons	18	20	18	17	18	22	20

39. For each of the numbered species that is a neutral atom, represent its composition in the form ${}^{A}_{Z}E$. Which of these species are isotopes?

40. Use the ${}^{A}_{Z}E$ symbolism to represent each of the ions in the table. Include the net ionic charge as a superscript.

41. The weighted-average atomic mass of bromine, 79.904 u, suggests that the principal isotope of bromine might be ${}^{80}Br$. However, this isotope does not occur naturally. How do you account for the observed weighted average atomic mass?

42. There are two naturally occurring isotopes of silver, existing in approximately equal proportions. One of these is ^{107}Ag. Which of the following must be the other: ^{106}Ag, ^{108}Ag, ^{109}Ag? Explain.

43. Gallium in nature consists of two isotopes, gallium-69, with a mass of 68.926 u and a fractional abundance of 0.601; and gallium-71, with a mass of 70.925 u and a fractional abundance of 0.399. Calculate the weighted average atomic mass of gallium.

44. Europium in nature consists of two isotopes, europium-151, with a mass of 150.92 u and a fractional abundance of 0.478; and europium-153, with a mass of 152.92 u and a fractional abundance of 0.522. Calculate the weighted average atomic mass of europium.

45. Neon in nature consists of the following isotopes.

Isotope	Atomic mass, u	Percent abundance
Neon-20	19.9924	90.51
Neon-21	20.9940	0.27
Neon-22	21.9914	9.22

Calculate the weighted average atomic mass of neon.

46. Natural strontium consists of the following isotopes.

Isotope	Atomic mass, u	Percent abundance
Strontium-84	83.913	0.56
Strontium-86	85.909	9.86
Strontium-87	86.909	7.00
Strontium-88	87.906	82.58

Calculate the weighted average atomic mass of strontium.

47. The two naturally occurring isotopes of rubidium are rubidium-85, with an atomic mass of 84.91179 u, and rubidium-87, with an atomic mass of 86.90919 u. What are the percent natural abundances of these isotopes?

48. The two naturally occurring isotopes of nitrogen are nitrogen-14, with an atomic mass of 14.003074 u, and nitrogen-15, with an atomic mass of 15.000108 u. What are the percent natural abundances of these isotopes?

The Periodic Table

49. Indicate the group and period numbers for each of the following elements, and classify the element as a metal or nonmetal.

(a) C **(b)** Ca **(c)** S **(d)** Sn **(e)** Ti **(f)** Br **(g)** Bi **(h)** In **(i)** Au **(j)** Mo

50. Identify the elements represented by the following group and period numbers. Tell whether each element is a metal or a nonmetal.

(a) Group 3A, period 4
(b) Group 1B, period 4
(c) Group 7A, period 5
(d) Group 1A, period 2
(e) Group 4A, period 2
(f) Group 4B, period 4

Chemical Formulas: Elements

51. Give molecular formulas for the following elements.

(a) oxygen **(b)** bromine **(c)** hydrogen **(d)** nitrogen

52. Give either an atomic symbol or a molecular formula for the following, whichever best represents how the element exists in the natural state.

(a) chlorine **(b)** sulfur **(c)** neon **(d)** phosphorus **(e)** sodium

Chemical Formulas: Binary Molecular Compounds

53. Which of the following are binary molecular compounds? Explain.

(a) HCN; **(b)** ICl; **(c)** KI; **(d)** H_2O; **(e)** ONF

54. Which of the following are binary molecular compounds? Explain.

(a) barium iodide; **(b)** hydrogen bromide; **(c)** chlorofluorocarbons; **(d)** ammonia; **(e)** sodium cyanide.

55. Supply the missing information (formula or name) for the following binary molecular compounds.

(a) dinitrogen monoxide, ____________
(b) sulfur hexafluoride, ____________
(c) tetraphosphorus trisulfide, ____________
(d) ____________, CS_2
(e) ____________, B_2Cl_4
(f) ____________, Cl_2O_7

56. Supply the missing information (name or formula) for the following binary molecular compounds.

(a) ____________, PF_3
(b) ____________, I_2O_5
(c) ____________, P_4S_{10}
(d) phosphorus pentachloride, ____________
(e) sulfur hexafluoride, ____________
(f) dinitrogen pentoxide, ____________

Chemical Symbols and Formulas: Ions

57. Supply the missing information (name or symbol) for each of the following monatomic ions.

(a) ________, K^+

(b) ________, O^{2-}

(c) ________, Cu^{2+}

(d) aluminum ion, ________

(e) nitride ion, ________

(f) chromium(III) ion ________

58. Supply the missing information (symbol or name) for each of the following monatomic ions.

(a) calcium ion, ________

(b) cobalt(II) ion, ________

(c) sulfide ion, ________

(d) ________, Fe^{3+}

(e) ________, Ba^{2+}

(f) ________, Se^{2-}

59. Supply the missing information (formula or name) for each of the following polyatomic ions.

(a) ________, CO_3^{2-}

(b) ________, SO_4^{2-}

(c) ________, OH^-

(d) ________, $H_2PO_4^-$

(e) ammonium ion, ________

(f) nitrite ion, ________

(g) cyanide ion, ________

(h) hydrogen carbonate ion, ________

60. Supply the missing information (formula or name) for each of the following polyatomic ions.

(a) ________, HSO_4^-

(b) ________, NO_3^-

(c) ________, MnO_4^-

(d) ________, CrO_4^{2-}

(e) hydrogen phosphate ion, ________

(f) dichromate ion, ________

(g) perchlorate ion, ________

(h) thiosulfate ion, ________

Chemical Formulas: Ionic Compounds

61. Name the following ionic compounds.

(a) Na_2O (b) $MgBr_2$ (c) $FeCl_2$ (d) Al_2O_3 (e) $LiI \cdot 3H_2O$

(f) K_2S (g) $Ca(OH)_2$ (h) NH_4NO_3 (i) $Cr_2(SO_4)_3$ (j) $NaHSO_3$

(k) $KMnO_4$ (l) $Mg(ClO_4)_2$ (m) $Cu(OH)_2$ (n) $(NH_4)_2C_2O_4$ (o) $FePO_4 \cdot 2H_2O$

62. Name the following ionic compounds.

(a) Li_2S (b) $FeCl_3$ (c) CaS (d) Cr_2O_3 (e) $BaSO_3$ (f) KOH (g) NH_4CN (h) $Cr(NO_3)_3 \cdot 9H_2O$

(i) $Mg(HCO_3)_2$ (j) $Na_2S_2O_3 \cdot 5H_2O$ (k) $K_2Cr_2O_7$ (l) $Ca(ClO_2)_2$ (m) CuI (n) $Mg(H_2PO_4)_2$ (o) $CaC_2O_4 \cdot H_2O$

63. Give formulas for the following ionic compounds.

(a) iron(II) carbonate

(b) barium iodide dihydrate

(c) aluminum sulfate

(d) potassium hydrogen carbonate

(e) sodium bromate

(f) calcium chloride hexahydrate

(g) copper(II) nitrate trihydrate

(h) lithium hydrogen sulfate

(i) magnesium cyanide

(j) iron(III) sulfate

(k) ammonium dichromate

(l) magnesium perchlorate

64. Give formulas for the following ionic compounds.

(a) potassium sulfide

(b) barium carbonate

(c) aluminum bromide hexahydrate

(d) potassium sulfite

(e) copper(I) sulfide

(f) magnesium nitride

(g) cobalt(II) nitrate

(h) magnesium dihydrogen phosphate

(i) potassium nitrite

(j) zinc sulfate heptahydrate

(k) sodium hydrogen phosphate

(l) iron(III) oxide

65. Which of the following is the correct name for the compound $Ca(ClO_2)_2$, and what is wrong with the others?

(a) calcium hypochlorite

(b) calcium chlorite

(c) calcium dichlorite

(d) calcium chlorate

(e) calcium oxychloride

66. Which of the following is a correct formula for the compound chromium(III) sulfite, and what is wrong with the others?

(a) $CrSO_3$ (b) $Cr(SO_3)_3$ (c) $Cr_2(SO_3)_3$ (d) $Cr(HSO_3)_3$ (e) $Cr_2(SO_4)_3$

Acids and Bases

67. Supply the missing information (formula or name) for each of the following acids and bases.

(a) hydroiodic acid, ________________

(b) sulfuric acid, ________________

(c) lithium hydroxide, ________________

(d) nitrous acid, ________________

(e) ____________, HIO_4

(f) ____________, $Ca(OH)_2$

(g) ____________, HBr

(h) ____________, H_3PO_3

68. Supply the missing information (name or formula) for each of the following acids and bases.

(a) ____________, $HClO_2$

(b) ____________, $Ba(OH)_2$

(c) ____________, NH_3

(d) ____________, H_2CO_3

(e) chloric acid, ________________

(f) sulfurous acid, ________________

(g) potassium hydroxide, ________________

(h) hypochlorous acid, ________________

Organic Compounds: Formulas, Structures, and Functional Groups

69. Write structural and condensed structural formulas for each of the following organic compounds.

(a) pentane (e) isobutane
(b) butanoic acid (f) propionic acid
(c) diethylamine (g) dimethyl ether
(d) cyclobutane (h) methyl acetate

70. Write structural and condensed structural formulas for each of the following organic compounds.

(a) hexane (e) 1-butanol
(b) pentanoic acid (f) ethylmethylamine
(c) isopropyl alcohol (g) dipropyl ether
(d) cyclopentane (h) methyl butyrate

71. Write structural and condensed formulas for (**a**) these two acids mentioned in the chapter: butyric acid and isobutyric acid, and (**b**) the ester formed when each of the acids reacts with propyl alcohol.

72. Write structural and condensed formulas for these two amines, not specifically mentioned in the chapter: propylamine and isopropylamine.

73. Which of the structures written below represent isomers of $CH_3CH_2C(CH_3)_2CH_2CH_2OH$?

(a) $CH_3(CH_2)_5CH_2OH$

(b) $CH_3CH(OH)CH_2CH_2CH(CH_3)_2$

(c) $CH_3\underset{\displaystyle CH_3}{\underset{|}{C}}HCH_2CH_2\underset{\displaystyle OH}{\underset{|}{C}}HCH_3$

(d) $CH_3\underset{\displaystyle CH_3}{\underset{|}{C}}HCH_2\underset{\displaystyle CH_3}{\underset{|}{C}}HCH_2CH_2OH$

74. Which of the structures written below represent isomers of $(CH_3)_2CHCH(NH_2)CH_2CH_3$?

(a) $CH_3(CH_2)_3CH_2NHCH_3$

(b) $CH_3CH(NHCH_3)CH_2CH_2CH_2CH_3$

(c) $CH_3\underset{\displaystyle CH_3}{\underset{|}{C}}HCH_2\underset{\displaystyle NH_2}{\underset{|}{C}}HCH_3$

(d) $CH_3CH_2CH(NH_2)\underset{\displaystyle CH_3}{\underset{|}{C}}HCH_3$

75. Write structural formulas for three ether isomers having the molecular formula $C_4H_{10}O$.

76. Write structural formulas for four alcohol isomers having the molecular formula $C_4H_{10}O$.

77. Each of the formulas below belongs to one of the following classes of substances: straight-chain alkane, branched-chain alkane, cyclic alkane, hydrocarbon, alcohol, ether, carboxylic acid, ester, amine, inorganic compound. Match each formula with the term that identifies it *most specifically*.

(a) $CH_3(CH_2)_6CH_3$

(b) $CH_3CH_2CHOHCH_3$

(c) C_5H_{10}

(d) C_2H_2

(e) $CH_3(CH_2)_6COOH$

(f) Na_2CO_3

(g) $CH_3(CH_2)_6\overset{\displaystyle O}{\overset{\|}{C}}OCH_3$

(h) $CH_3CH_2OCH_3$

78. Each of the formulas below belongs to one of the following classes of substances: straight-chain alkane, branched-chain alkane, cyclic alkane, hydrocarbon, alcohol, ether, carboxylic acid, ester, amine, inorganic compound. Match each formula with the term that identifies it *most specifically*.

(a) $CH_3(CH_2)_3\underset{\displaystyle CH_3}{\underset{|}{C}}HCH_2CH_3$

(b) $CH_3CH_2OCH_3$

(c) $KHCO_3$

(d) $CH_3(CH_2)_3\underset{\displaystyle OH}{\underset{|}{C}}HCH_2CH_3$

(e) C_6H_6

(f) $HCOOH$

(g) $CH_3(CH_2)_4CH_3$

(h) C_8H_{16}

(i)
$$CH_3\overset{\overset{\displaystyle O}{\|}}{C}OCH_3$$

(j) $CH_3CH_2NCH_3$ (with CH_3 attached to N)

(k) $CH_3(CH_2)_6OCH_3$

79. Which of the following formulas applies to the molecular model pictured?

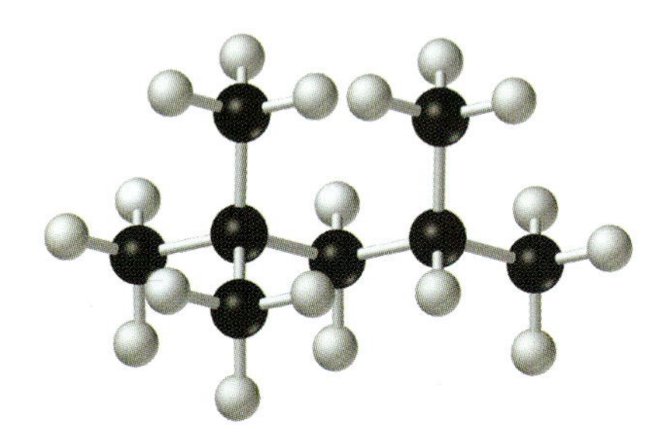

(a) $CH_3(CH_2)_6CH_3$

(b) $CH_3CCH_2CHCH_3$ (with CH_3 above and below the second C, and CH_3 below the fourth C)

(c) $CH_3C—CHCH_3$ (with CH_3 above and below the second C, and CH_3 below the third C)

(d) $CH_3C—CH_2—CCH_3$ (with CH_3 above and below the second C, and CH_3 above and below the fourth C)

80. Write a structural formula for the molecular model of the alcohol pictured.

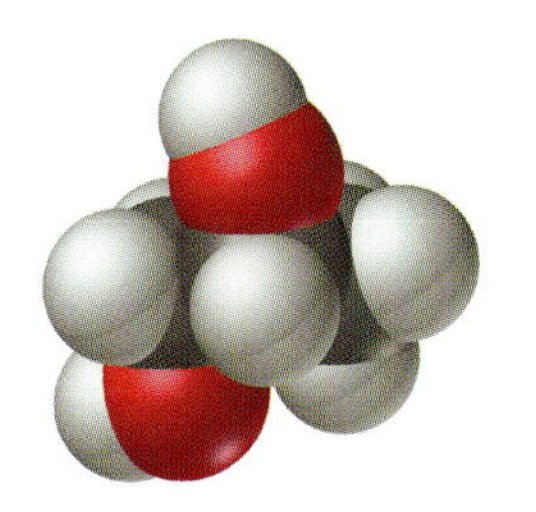

Additional Problems

81. Show that the following experiment is consistent with the law of conservation of mass (within the limits of experimental error): A 10.00-g sample of calcium carbonate was dissolved in 100.0 mL of hydrochloric acid solution (density = 1.148 g/mL). The products were 120.40 g of solution (a mixture of hydrochloric acid and calcium chloride) and 2.22 L of carbon dioxide gas (density = 0.0019769 g/mL).

82. When 3.06 g hydrogen was allowed to react with an excess of oxygen, 27.35 g water was obtained. In a second experiment, electric current is used to break down a sample of water into 1.45 g hydrogen and 11.51 g oxygen. Are these results consistent with the law of constant composition? Demonstrate why or why not.

83. Amino acids have both the carboxylic acid (—COOH) and amino (—NH_2) groups attached to hydrocarbon groups. Analyses of a newly synthesized amino acid yielded the following results. Can these results be used to verify the law of constant composition? Explain.

	Mass of sample	Mass of carbon	Mass of hydrogen
Sample 1	0.2450 g	0.1141 g	0.0216 g
Sample 2	0.3005 g	0.1400 g	0.0264 g
Sample 3	0.1371 g	0.0639 g	0.0121 g

84. In addition to CO and CO_2, another known but uncommon oxide of carbon is the colorless, pungent gas C_3O_2, often called carbon suboxide. Expand Figure 2.2 to include carbon suboxide, and show that it conforms to the law of multiple proportions as do the other two oxides.

85. Mercury and oxygen form two compounds. One contains 96.2% mercury, by mass, and the other, 92.6%. Show that these data conform to the law of multiple proportions.

86. Three oxides of nitrogen, N_2O, NO, and NO_2, were known to Dalton and figured prominently in his proposal of the law of multiple proportions. Explain how he would have been able to use this law to verify the existence of N_2O_5 but not N_2O_4, two oxides of nitrogen that are now well known.

87. One oxide of iron is found to consist of 22.36% oxygen by mass. Which of the following represents a possible mass percent oxygen of a second iron oxide? Explain.
(a) 27.64% **(b)** 44.72% **(c)** 50.00% **(d)** 67.08%

88. William Prout (1815) hypothesized that all other atoms are built up of hydrogen atoms. If so, all elements should have integral (whole number) atomic masses relative to an atomic mass of 1 for hydrogen. However, the hypothesis appeared inconsistent with atomic masses such as 24.3 for magnesium and 35.5 for chlorine. Based on modern knowledge, Prout's hypothesis seems more reasonable. Explain why.

89. Identify the isotope for which the numbers of protons, neutrons, and electrons in its atoms are equal and their *total* is 60.

90. Identify the isotope for which the mass number of its atoms is 234 and its atoms have 60.0% more neutrons than protons.

91. The principal naturally occurring isotope of magnesium has an atomic mass of 23.98504 u and a 78.99% abundance. The other two naturally occurring isotopes have atomic masses of 24.98584 u and 25.98259 u, respectively. What are the percent natural abundances of the three naturally occurring isotopes of magnesium?

92. A newspaper article describing a localized groundwater pollution problem states, "Wells in Redlands, Loma Linda, and Riverside may contain perchlorate, a chemical that can affect the thyroid gland." Elsewhere, the article states that, "Perchlorate, which regulators say came from Lockheed's former aerospace plant in Mentone, is used in manufacturing explosives." Can you identify precisely what substance is being referred to? Do you think a chemist would be able to? Can you write a formula for the substance? Explain.

93. Following are the boiling points of some straight-chain alkanes.

Alkane formula:	$C_{16}H_{34}$	$C_{17}H_{36}$	$C_{18}H_{38}$	$C_{19}H_{40}$	$C_{20}H_{42}$
Boiling point, °C	286.79	301.82	316.12	329.7	342.7

Use this information and data from Example 2.13 (page 66) to estimate the boiling points of $C_{11}H_{24}$, $C_{12}H_{26}$, $C_{13}H_{28}$, $C_{14}H_{30}$, $C_{15}H_{32}$, $C_{21}H_{44}$, and $C_{22}H_{46}$.

94. The following name and formula suggest compounds that do not exist. Explain why they do not.

(a) isoethanol

(b) $CH_3CH_2C(CH_3)_3CH_2CH_3$

95. Are the following pairs of molecules isomers or not? Explain.

(a) $CH_3CH(OH)CH_2CH_3$ and $CH_3CH_2CH(OH)CH_3$ (each with the OH group drawn below the CH carbon)

(b) $CH_3(CH_2)_5CH(CH_3)_2$ and $CH_3(CH_2)_4CH(CH_3)CH_2CH_3$

(c) $CH_3(CH_2)_4CH(CH_3)_2$ and $(CH_3)_2CHCH_2CH_2CH_3$

(d) $CH_3CH_2C(CH_3)_2(CH_2)_3CH_3$ and $C_2H_5C(CH_3)_2C_4H_9$

96. Draw structural formulas of all the possible C_7H_{16} alkane isomers.

97. Write the *general* formula for (**a**) a straight-chain or branched-chain alcohol having one —OH group per molecule, (**b**) a straight-chain or branched-chain carboxylic acid having one —COOH group per molecule, and (**c**) a cyclic alkane alcohol having two —OH groups per molecule.

The molecular models shown below refer to Problems 98–103. The colors used in the models are given in Figure 2.6.

(a)

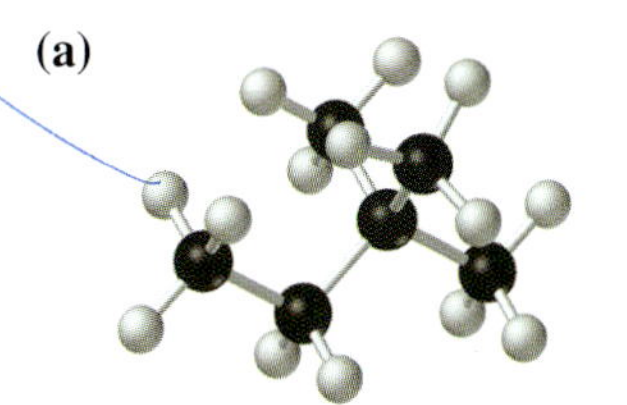

(b)

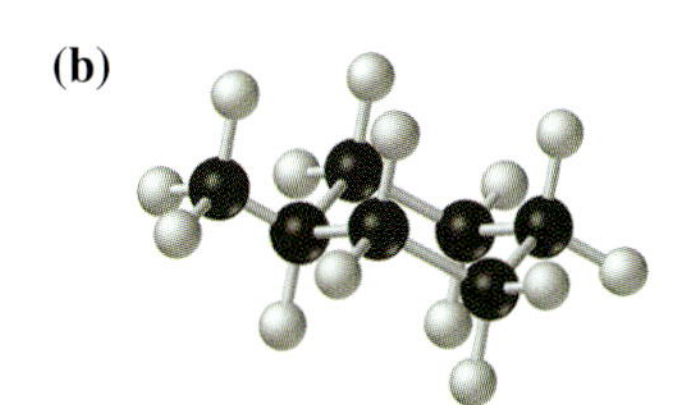

(c)

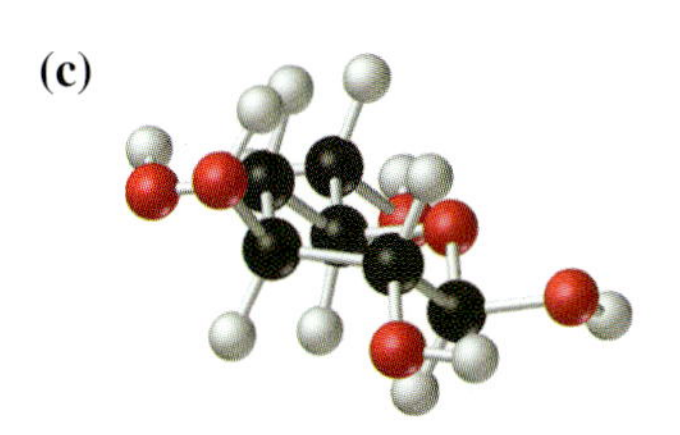

(d)

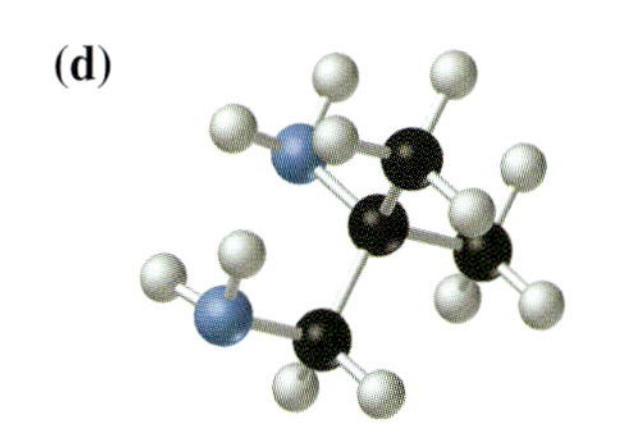

(e)

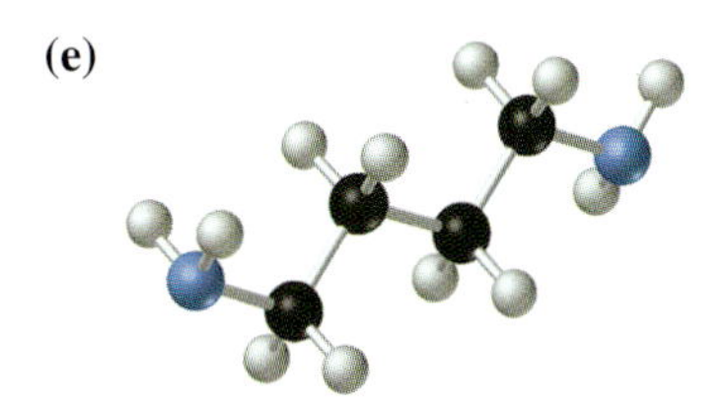

(f)

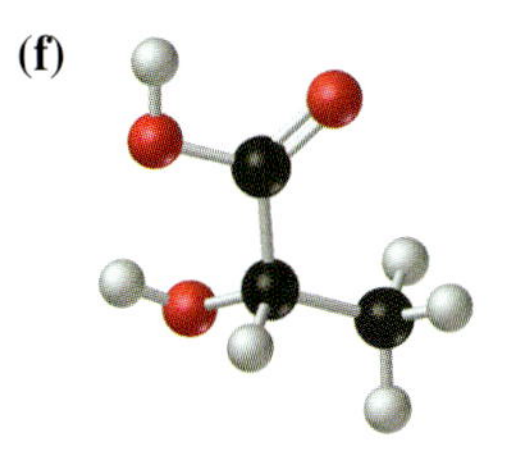

98. Which model represents a cyclic alkane? What is the name of the compound?

99. Which model represents a compound with a carboxylic acid functional group? Write its condensed structural formula.

100. Which pair of models represent isomers? Draw their structural formulas.

101. What are the empirical and molecular formulas of the compound represented by (**e**)?

102. Write structural formulas for two isomers of (**a**).

103. Which model represents a compound which has the same empirical formula as (**c**), but is not an isomer of (**c**)?

104. People often refer to organic food, fertilizer, medicines, cosmetics, and other products. What does *organic* mean in each case? Which of the following is organic? In what sense?

(**a**) cane sugar (sucrose, $C_{12}H_{22}O_{11}$)

(**b**) *E. coli* O157:H7 (a bacterial strain that causes food poisoning)

(**c**) cabbage grown without synthetic fertilizers or pesticides

(**d**) cow manure

(**e**) shampoo made with synthetic detergents

(**f**) cocaine

Explain your answer in each case. You may use an ordinary dictionary and a dictionary of scientific terms.

Stoichiometry: Chemical Calculations

CHAPTER

3

Sodium metal reacts vigorously with water to produce hydrogen gas and a solution of sodium hydroxide. In this chapter, we describe how to write chemical equations that represent chemical reactions and how to do calculations based on chemical formulas and chemical equations.

CONTENTS

You can describe many important physical and chemical properties with words alone: Ammonium nitrate is a solid substance used both as a fertilizer and as an explosive. Chlorine is a poisonous, greenish yellow gas used to disinfect water. Sodium is a soft, silvery metal that reacts violently with water to produce hydrogen gas and sodium hydroxide.

However, chemists often face questions that require *quantitative* answers: How much ammonium nitrate should be used to provide an avocado tree with the desired amount of the element nitrogen? How much chlorine gas is required to establish a level of two parts per million of chlorine in a swimming pool? How much soda ash (Na_2CO_3) is required to neutralize the acidity resulting from the reaction of the chlorine with the pool water? How much hydrogen gas is produced when 1 kg of sodium metal reacts with water?

● **ChemCDX**
When you see this icon, there are related animations, demonstrations, and exercises on the CD accompanying this text.

Quantitative answers require mathematics, and indeed some chemical questions require quite sophisticated mathematics. However, chemical questions like the ones mentioned require only arithmetic or simple algebra. In this chapter, we will consider calculations based on chemical formulas, the symbolic representation of chemical reactions through *chemical equations*, and calculations based on chemical equations. Collectively, these topics constitute a broad study known as **stoichiometry**.

Stoichiometry of Chemical Compounds

In Chapter 2, we stressed writing the names and formulas of chemical compounds and we also learned some of the information that is embodied in a formula: Formulas tell us the ratios in which atoms or ions combine, and structural formulas indicate the order in which atoms join together in molecules. We shall see in the opening sections of this chapter that chemical formulas provide a good deal of additional useful information.

3.1 Molecular Masses and Formula Masses

We learned in Chapter 2 that each element has a characteristic atomic mass. Because chemical compounds are made up of two or more elements, the masses that we associate with molecules or formula units of compounds are combinations of atomic masses. Atomic masses, molecular masses, and formula masses all enter into the calculations we do in this chapter.

Molecular Masses

Recall that we assign masses to the atoms of the different elements that are relative to the mass of a carbon-12 atom. We call these *atomic masses*, or in an older but still widely used terminology, *atomic weights*. For a molecular substance we can similarly assign a **molecular mass** (or molecular weight), which is the mass of a molecule relative to that of a carbon-12 atom. Moreover, we can calculate this molecular mass from atomic masses.

Molecular mass is the sum of the masses of the atoms represented in a molecular formula.

For example, because the formula O_2 specifies two O atoms per molecule of oxygen, the molecular mass of oxygen (O_2) is twice the atomic mass of oxygen.

$$\begin{aligned}\text{Molecular mass of } O_2 &= 2 \times \text{atomic mass of O} \\ &= 2 \times 15.9994\text{ u} = 31.9988\text{ u}\end{aligned}$$

The molecular mass of carbon dioxide (CO_2) is the sum of the atomic mass of carbon and twice the atomic mass of oxygen.

$$\begin{aligned}1 \times \text{atomic mass of C} &= 1 \times 12.011\text{ u} = 12.011\text{ u} \\ 2 \times \text{atomic mass of O} &= 2 \times 15.9994\text{ u} = 31.9988\text{ u} \\ \text{Molecular mass of } CO_2 &= 44.010\text{ u}\end{aligned}$$

Notice that the atomic masses we use in computing a molecular mass are those in the listing on the inside front cover. The isotopes of an element occur in the same proportions in a compound as they do in the free element itself, and usually the calculated molecular mass is an average value.

EXAMPLE 3.1

Calculate the molecular mass of sulfur hexafluoride, SF_6, an unusually stable gas used as an electric insulator in high-voltage equipment.

SOLUTION

The molecular formula indicates that a single molecule of SF_6 contains one atom of sulfur and six atoms of fluorine. We solve the problem by adding the atomic mass of sulfur to six times the atomic mass of fluorine. However, we need only write down the final answer obtained with a calculator, 146.056 u; we need not record the numbers 32.066 and 113.9904.

$$\begin{aligned} 1 \times \text{atomic mass of S} &= 1 \times 32.066\ \text{u} = 32.066\ \text{u} \\ 6 \times \text{atomic mass of F} &= 6 \times 18.9984\ \text{u} = 113.9904\ \text{u} \\ \text{Molecular mass of } SF_6 &= 146.056\ \text{u} \end{aligned}$$

EXERCISE 3.1A

Calculate, to *three* significant figures, the molecular masses of **(a)** S_8, **(b)** N_2H_4, **(c)** H_3PO_4, and **(d)** C_5H_{12}.

EXERCISE 3.1B

Calculate, to *five* significant figures, the molecular masses of **(a)** phosphorus pentafluoride, **(b)** dinitrogen tetroxide **(c)** hexane, C_6H_{14}, and **(d)** propionic acid, CH_3CH_2COOH.

Formula Masses

As we noted in Chapter 2, there are no molecules in sodium chloride, just large clusters of Na^+ and Cl^- ions. The term "molecular mass" is not appropriate for sodium chloride. We base the formula of an ionic compound on a hypothetical entity called a *formula unit*. For an ionic compound or any other compound in which there are no discrete molecules, we use the term **formula mass** (or formula weight); this is the mass of a formula unit relative to that of a carbon-12 atom. And again, we can calculate a formula mass from atomic masses.

> *Formula mass is the sum of the masses of the atoms or ions present in a formula unit.*

Thus, for the ionic compound $BaCl_2$,

$$\begin{aligned} \text{Formula mass of } BaCl_2 &= \text{atomic mass of Ba} + (2 \times \text{atomic mass of Cl}) \\ &= 137.327\ \text{u} \qquad\quad + (2 \times 35.4527\ \text{u}) \\ &= 208.232\ \text{u} \end{aligned}$$

You might think that we should have used masses of the ions Ba^{2+} and Cl^- in the above calculation. It is true that the mass of an atom increases ever so slightly when it gains an electron, but that increase is exactly offset by the very very slight decrease in mass of an atom as it loses an electron. Thus, the formula mass based on atomic masses is the same as if we used masses of the ions.

EXAMPLE 3.2

Calculate the formula mass of ammonium sulfate, a fertilizer commonly used by home gardeners.

SOLUTION

Before we can determine a formula mass, we must first write the correct chemical formula. Ammonium sulfate is an ionic compound consisting of ammonium ions (NH_4^+), and sulfate ions (SO_4^{2-}); its formula is $(NH_4)_2SO_4$. To derive a formula mass from a complex formula, we must make certain that we account for all the atoms in the formula unit. This requires that we note all the subscripts and parentheses in the formula. Let's first note the relevant atomic masses and the way in which they must be combined.

The formula (N H 4) 2 S O 4

The relevant atomic masses (14.0067 u + 1.00794 u x 4) × 2 + 32.066 u + 15.9994 u x 4

Summing the atomic masses

$$(NH_4)_2: \quad 2 \times [14.0067\ u + (4 \times 1.00794\ u)] = 36.0769\ u$$

$$SO_4: \quad [32.066\ u + (4 \times 15.9994\ u)] = 96.064\ u$$

$$\text{Formula mass of } (NH_4)_2SO_4 = 132.141\ u$$

EXERCISE 3.2A

Calculate to *three* significant figures the formula masses of (**a**) Li_2O, (**b**) $Mg(NO_3)_2$, (**c**) $Ca(H_2PO_4)_2$, and (**d**) K_2SbF_5.

EXERCISE 3.2B

Calculate to *five* significant figures the formula masses of (**a**) sodium hydrogen sulfite, (**b**) ammonium perchlorate, (**c**) chromium(III) sulfate, and (**d**) copper(II) sulfate pentahydrate.

3.2 The Mole and Avogadro's Number

We learned in Section 2.2 that when there is a plentiful supply of oxygen, carbon burns to form carbon dioxide. However, when the supply of oxygen is limited, carbon monoxide is formed. How can we determine the minimum quantity of oxygen required to ensure that when carbon burns, it is completely converted to carbon dioxide? The carbon dioxide molecule, CO_2, consists of one C atom and two O atoms. The oxygen molecule, O_2, is *diatomic*; it consists of two O atoms. Thus, we need at least as many O_2 molecules as we have C atoms to ensure that we have two O atoms for every C atom. The problem, though, is that any actual sample of carbon contains so many atoms that we cannot literally count them.

APPLICATION NOTE
Even a sample of carbon as small as a pencil-mark period at the end of a sentence contains about 10^{18} C atoms; that is, about 1,000,000,000,000,000,000 C atoms.

How can we measure a sample of oxygen that has the same number of O_2 molecules as this uncountable number of C atoms? We can use the SI base unit for an amount of substance called the *mole* to deal with questions like this.

*A **mole (mol)** is an amount of substance that contains as many elementary entities as there are atoms in exactly 12 g of the carbon-12 isotope.*

We generally indicate the elementary entities through the symbol or formula of a substance. For example, the elementary entities are *atoms* in samples of carbon, C; sodium, Na; and iron, Fe. They are *molecules* in samples of oxygen, O_2; carbon dioxide, CO_2; and water, H_2O. And they are *formula units* in samples of sodium chloride, NaCl; magnesium hydroxide, $Mg(OH)_2$; and calcium nitrate, $Ca(NO_3)_2$.

Although the mass of 1 mol of carbon-12 is exactly 12 g, that of 1 mol of carbon as it is obtained from natural sources is 12.011 g. This is because the carbon contains a small amount of the carbon-13 isotope (1.108%).

Because naturally occurring oxygen atoms are heavier than carbon-12 atoms by the factor 15.9994 u/12.0000 u, the mass of 1 mol of oxygen atoms is

$$\frac{15.9994 \text{ u}}{12.0000 \text{ u}} \times 12.0000 \text{ g} = 15.9994 \text{ g}$$

And the mass of 1 mol of O_2 molecules is $2 \times 15.9994 \text{ g} = 31.9988 \text{ g}$.

Now, to have a number of O_2 molecules equal to a given number of C atoms, we can take a mole of each. The oxygen and carbon will be in the mass ratio: 31.9988 g O/12.011 g C = 2.6641 g O/g C, as illustrated in Figure 3.1.

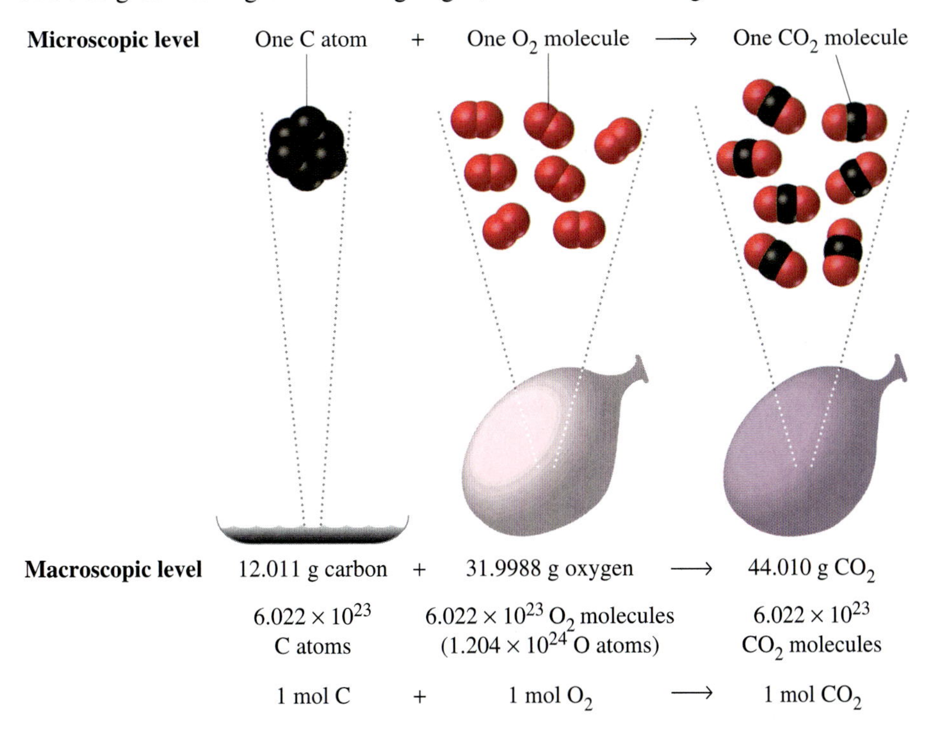

Figure 3.1
Two views of the combination of carbon and oxygen to form carbon dioxide
At the microscopic (molecular) level, chemical reactions occur between atoms and molecules (top), but we can show only a few atoms and molecules to represent the enormous numbers that actually make up the samples. In reality, we usually observe substances at the macroscopic level, and the artist's sketch of the substances at that level suggests one mole each (Avogadro's number) of C atoms, O_2 molecules, and CO_2 molecules.

When we use the *concept* that 1 mole of one substance contains the same number of elementary entities as 1 mole of any other substance, we don't actually need to know what that number is. Sometimes, however, we will need to work with the actual number of elementary entities in a mole of substance. This number is called **Avogadro's number,** N_A, after Amedeo Avogadro, who was the first person to sense the significance of the concept of the mole. Later in the text, we will describe ways in which Avogadro's number can be established, but for now we will just state it.

$$N_A = 6.022137 \times 10^{23} \text{ mol}^{-1}$$

If we don't need all those significant figures in a calculation, we can round off Avogadro's number to $6.022 \times 10^{23} \text{ mol}^{-1}$, or even $6.02 \times 10^{23} \text{ mol}^{-1}$. The unit

"mol^{-1}," which we read as "per mole," signifies that a collection of N_A molecular-level entities (atoms, molecules, formula units) is equivalent to one mole at the macroscopic level. For example, a mole of carbon contains 6.02×10^{23} atoms of C; a mole of oxygen contains 6.02×10^{23} molecules of O_2; and a mole of sodium chloride contains 6.02×10^{23} formula units of NaCl.

Avogadro's number is almost beyond imagination. If you had 6.02×10^{23} dollars, you could spend a billion dollars a second for your entire lifetime and still have used less than 0.001% of your money. If carbon atoms were the size of peas, 6.02×10^{23} of them would cover the entire surface of our planet to a depth of over 100 m.

3.3 More on the Mole

We buy eggs by the dozen (12 eggs), pencils by the gross (144 pencils), and paper by the ream (500 sheets). We don't try to count out very large numbers of objects, however. We could speak of a mole of eggs, but the term would not be very useful. All the hens' eggs laid since the first chicken wouldn't add up to anywhere near one mole. The chemists' mole is useful only when dealing with particles at the microscopic level.

Let us emphasize again that the "elementary entities" that we "count" to add up to one mole are specified by a chemical symbol or formula. They may be atoms, such as C, O, or Pu, or they may be molecules, such as O_2, CO_2, or even $C_{46}H_{65}N_{15}O_{12}S_2$ (vasopressin, a hormone). In ionic compounds, the structural particles are cations and anions, but the elementary entity we choose is the hypothetical formula unit. A formula unit of magnesium chloride, for example, consists of one Mg^{2+} ion and two Cl^- ions. One mole of magnesium chloride is one mole of these formula units, which means one mol Mg^{2+} ions and two mol Cl^- ions.

$$1 \text{ mol } MgCl_2 = 1 \text{ mol } Mg^{2+} + 2 \text{ mol } Cl^-$$

Because bulk matter cannot carry large excesses of either positive or negative charge, we cannot accumulate 1 mol Mg^{2+} or 2 mol Cl^- *separately*, but we can obtain them *together* in 1 mol $MgCl_2$.

Molar Mass

A dozen is the same *number*, whether we have a dozen oranges or a dozen watermelons. However, a dozen oranges and a dozen watermelons do not have the same *mass*. Similarly, a mole of magnesium and a mole of iron contain the same numbers ($N_A = 6.022 \ldots \times 10^{23}$) of atoms but have *different* masses. Figure 3.2 is a photograph of one mole of several different chemical substances.

The **molar mass** of a substance is the mass of 1 mol of that substance. Recall that by definition, 1 mole of carbon-12 has a mass of exactly 12 g. This means that the molar mass of carbon-12 (exactly 12 g) is numerically equal to the atomic mass of carbon-12 (exactly 12 u). For other substances, the molar mass is also numerically equal to the atomic mass, molecular mass, or formula mass but expressed in the unit *grams per mole* (*g/mol*). For example, sodium has an atomic mass of 22.99 u and a molar mass of 22.99 g/mol, the molar mass of CO_2 is 44.01 g/mol, and that of $MgCl_2$ is 95.21 g/mol. We can use the definitions of a mole, Avogadro's number, and molar mass to write the relationships

$$1 \text{ mol Na} = 22.99 \text{ g Na} = 6.022 \times 10^{23} \text{ Na atoms}$$
$$1 \text{ mol } CO_2 = 44.01 \text{ g } CO_2 = 6.022 \times 10^{23}\ CO_2 \text{ molecules}$$
$$1 \text{ mol } MgCl_2 = 95.21 \text{ g } MgCl_2 = 6.022 \times 10^{23}\ MgCl_2 \text{ formula units}$$

Figure 3.2
One mole each of four different elements
The watch glasses contain one mole of copper atoms (left) and one mole of sulfur atoms (right). The 50-mL beaker contains one mole of liquid mercury, and the balloon contains one mole of helium gas.

These relationships supply the conversion factors to make conversions between mass in grams, amount in moles, and number of elementary entities, as we illustrate schematically in Figures 3.3 and 3.4 and demonstrate in Examples 3.3 and 3.4.

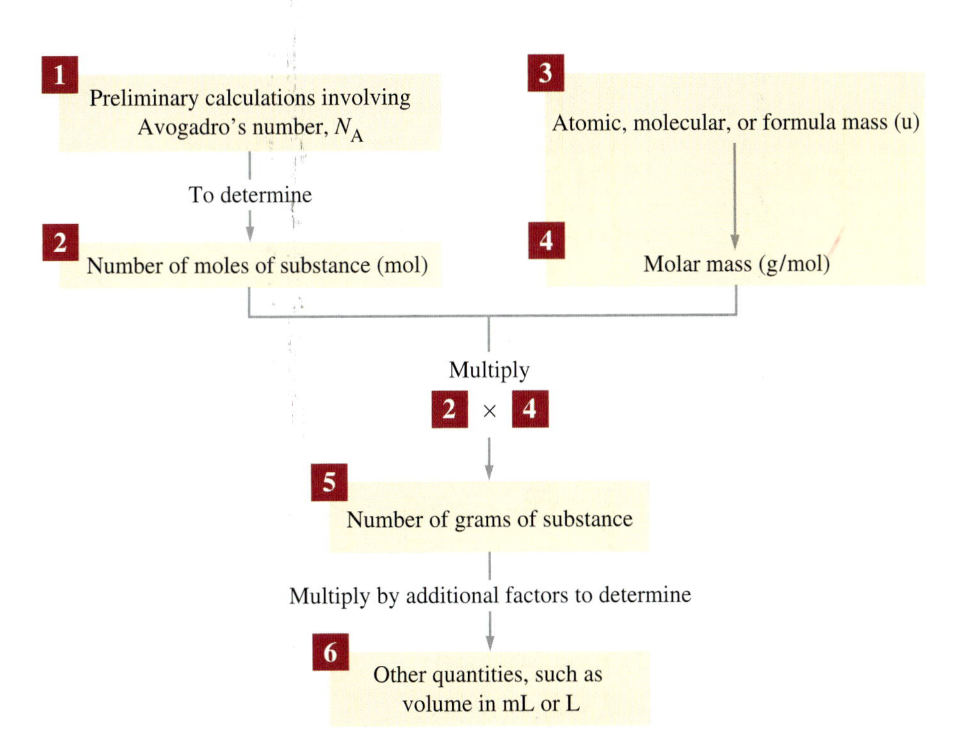

Figure 3.3
Determining mass and related quantities from the number of moles of a substance
Some problems require that you consider only the factors in steps 2–5; others require steps 1 and/or 6 as well.

EXAMPLE 3.3

Calculate (**a**) the mass, in grams, of 0.250 mol Na; and (**b**) the number of moles of CO_2 in a 225-g sample of the gas.

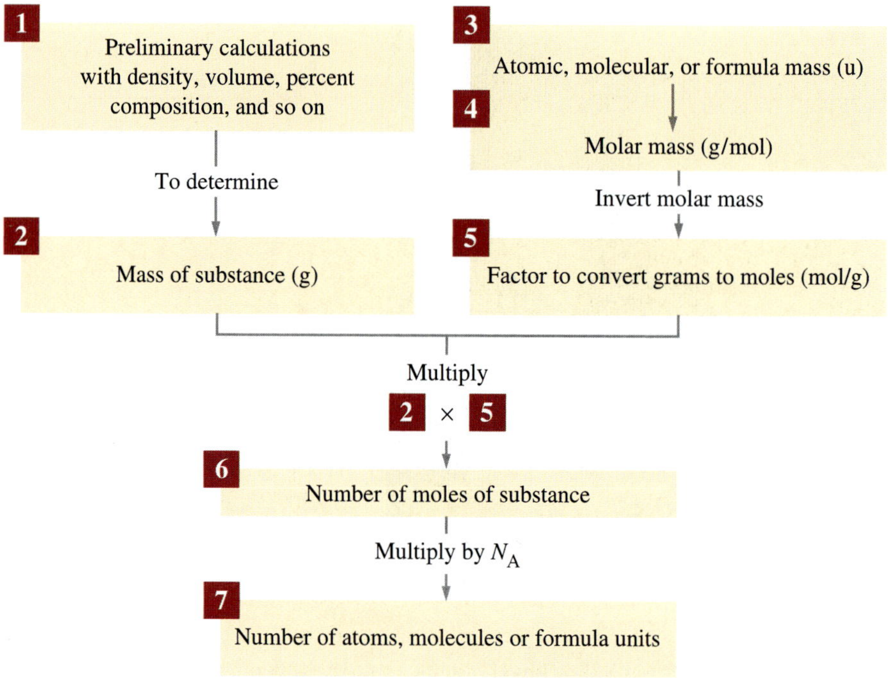

1 Preliminary calculations with density, volume, percent composition, and so on

To determine

2 Mass of substance (g)

3 Atomic, molecular, or formula mass (u)

4 Molar mass (g/mol)

Invert molar mass

5 Factor to convert grams to moles (mol/g)

Multiply

2 × 5

6 Number of moles of substance

Multiply by N_A

7 Number of atoms, molecules or formula units

▶ **Figure 3.4**
Determining number of moles and related quantities from the mass of a substance
Some problems require that you consider only the factors in steps 2–6; others require steps 1 and/or 7 as well.

SOLUTION

a. Sodium has an atomic mass of 22.99 u and a molar mass of 22.99 g/mol. In the setup that follows, notice that we use the molar mass of sodium as a conversion factor from mol Na to g Na.

$$? \text{ g Na} = 0.250 \cancel{\text{ mol Na}} \times \frac{22.99 \text{ g Na}}{1 \cancel{\text{ mol Na}}} = 5.75 \text{ g Na}$$

b. In this case, we need the molar mass of a molecular substance, CO_2. We first determine its molecular mass, 12.011 u + (2 × 15.9994 u) = 44.01 u, and its molar mass is therefore 44.01 g/mol. Also, we see that in converting from a mass in grams to an amount in moles, we must use the *inverse* of the molar mass as a conversion factor to get the proper cancellation of units.

$$? \text{ mol } CO_2 = 225 \cancel{\text{ g } CO_2} \times \frac{1 \text{ mol } CO_2}{44.01 \cancel{\text{ g } CO_2}} = 5.11 \text{ mol } CO_2$$

EXERCISE 3.3A

Calculate (**a**) the mass, in grams, of 0.155 mol C_3H_8; (**b**) the mass, in milligrams, of 2.45×10^{-4} mol of ethane, C_2H_6; (**c**) the number of moles of C_4H_{10} in a 165-kg sample; and (**d**) the number of moles of phosphoric acid in a 76.0-mg sample.

EXERCISE 3.3B

Calculate (**a**) the number of moles of aluminum in a cube of the metal 5.5 cm on an edge ($d = 2.70 \text{ g/cm}^3$); and (**b**) the volume occupied by 1.38 mol of carbon tetrachloride, a liquid with a density of 1.59 g/mL.

PROBLEM-SOLVING NOTE
You will need to use the relationships between mass, volume, and density considered in Section 1.5.

EXAMPLE 3.4

Calculate (**a**) the mass of a sodium atom, in grams; and (**b**) the number of Cl^- ions present in 1.38 g $MgCl_2$.

SOLUTION

a. The answer must have the unit "grams per sodium atom," that is, "g/Na atom." Thus, if we knew the mass of a certain number of Na atoms, our answer would simply be the mass divided by that number. But we do know these quantities: One mole of Na has a mass of 22.99 g and consists of 6.022×10^{23} Na atoms. As shown, the division (22.99 g/6.022×10^{23} Na atoms) is represented as the product of two factors, one involving the molar mass and the other, Avogadro's number.

$$? \text{ g/Na atom} = \frac{22.99 \text{ g}}{\cancel{1 \text{ mol Na}}} \times \frac{\cancel{1 \text{ mol Na}}}{6.022 \times 10^{23} \text{ Na atoms}}$$
$$= 3.818 \times 10^{-23} \text{ g/Na atom}$$

Because there is only one naturally-occurring type of sodium atom (^{23}Na), what we have just calculated is truly the mass of a sodium atom. For elements that have two or more isotopes, our calculation would be a *weighted average* mass and not the mass of an atom of any particular isotope.

b. First we need the formula mass of $MgCl_2$ and then the molar mass. Once we have established the amount of $MgCl_2$ in moles, we can use Avogadro's number to determine the number of formula units (f.u.) of $MgCl_2$. Finally, we can use a factor (shown in red) that establishes the number of Cl^- ions per formula unit (f.u.).

$$? \text{ Cl}^- \text{ ions} = 1.38 \cancel{\text{ g MgCl}_2} \times \frac{1 \cancel{\text{ mol MgCl}_2}}{95.21 \cancel{\text{ g MgCl}_2}} \times \frac{6.022 \times 10^{23} \cancel{\text{ f.u.}}}{1 \cancel{\text{ mol MgCl}_2}} \times \frac{2 \text{ Cl}^- \text{ ion}}{1 \cancel{\text{ f.u.}}}$$
$$= 1.75 \times 10^{22} \text{ Cl}^- \text{ ions}$$

PROBLEM-SOLVING NOTE

The mass of a single atom is tiny—very much less than 1 g. Keep this fact in mind to avoid mistakenly multiplying by Avogadro's number when you should divide.

PROBLEM-SOLVING NOTE

Example 3.4(b) illustrates the common practice of expressing molar mass and Avogadro's number with one more significant figure than the least precisely known quantity (1.38 g $MgCl_2$). This ensures that the precision of the calculated result is limited only by the least precisely known quantity.

EXERCISE 3.4A

Calculate (**a**) the weighted average mass, in grams, of a bismuth atom; (**b**) the weighted average mass, in grams, of a glycerol molecule, $CH_2OHCHOHCH_2OH$; (**c**) the number of molecules in 0.0100 g of nitrogen gas; and (**d**) the total number of atoms in 215 g of sucrose, $C_{12}H_{22}O_{11}$.

EXERCISE 3.4B

Calculate (**a**) the number of Br_2 molecules in 125 mL of liquid bromine ($d = 3.12$ g/mL); and (**b**) the number of liters of liquid ethanol ($d = 0.789$ g/mL) required to obtain a sample containing 1.00×10^{25} CH_3CH_2OH molecules.

EXAMPLE 3.5—An Estimation Example

Which of the following is a reasonable value for the number of atoms in 1.00 g of helium?

a. 0.25 **b.** 4.0 **c.** 4.1×10^{-23} **d.** 1.5×10^{23}

SOLUTION

We could simply calculate the number of atoms and pick the correct answer from among the four. But just by using our knowledge that atoms are extremely small and must appear in extremely large numbers in a 1.00-g sample of matter, we can see that the only possible answer is (d); it is the only large number. To examine the possible responses in more detail, response (a) is the number of *moles* of He atoms in 1.00 g. We certainly cannot have a fraction (0.25) of an atom. Response (b), expressed as 4.0 u, is the atomic mass of He; expressed as 4.0 g/mol He, it is the molar mass. In either case, it is far too small to be the number of atoms for a macroscopic sample. Response (c) is what we get if, in error, we *divide* the number of moles of He, 0.25, by Avogadro's number; and, as in (a), this represents only a fraction of an atom. Response (d) is the correct answer, obtained from the calculation: $0.25 \times N_A$.

EXERCISE 3.5

Which of the following is a reasonable value for the mass of 1.0×10^{23} magnesium atoms? (Try to reason through the problem and avoid using your calculator if possible. Your goal is to find a reasonable answer rather than to calculate a specific number.)

a. 2.4×10^{-22} g **b.** 0.17 g **c.** 2.4 g **d.** 4.0 g

3.4 Mass Percent Composition from Chemical Formulas

APPLICATION NOTE

Carbon dioxide is one of several "greenhouse" gases found in the atmosphere. These gases trap heat radiated by Earth's surface and may contribute to a potential global warming.

To limit the production of carbon dioxide gas in the combustion of a hydrocarbon, we need to know which is the more desirable fuel when comparing equal masses of two fuels. For example, which produces less carbon dioxide when equal masses are burned, methane or butane? There are several ways to answer this question, but we would have an immediate answer if we knew the *percent* by mass of carbon in each of these hydrocarbons. The one with the smaller percent carbon would produce the lesser amount of CO_2. **Mass percent composition** describes the proportions of the constituent elements in a compound as the number of grams of each element per 100 grams of the compound.

We can determine the molar mass of a compound from its chemical formula, as we did in Section 3.3. If we keep track of the contribution that each element makes to the total mass, we can use this information to establish a ratio of the mass of each element to that of the compound as a whole. This gives us the fractional composition of the compound, by mass. If we multiply these fractions by 100%, we obtain the mass percent composition of the compound. For the element carbon in butane, these quantities are

$$\textit{Mass fraction} = \frac{\text{mass carbon}}{\text{mass butane}} = \frac{48.044\text{ g C}}{58.123\text{ g C}_4\text{H}_{10}} = 0.8266$$

$$\textit{Mass percent} = \frac{48.044\text{ g C}}{58.123\text{ g C}_4\text{H}_{10}} \times 100\% = 82.66\%\text{ C}$$

We show how mass percent is obtained in schematic fashion below.

Determining the mass percent of carbon in butane, C_4H_{10}.

1 C_4H_{10} — There are 4 mol C atoms and 10 mol H atoms in 1 mol butane.

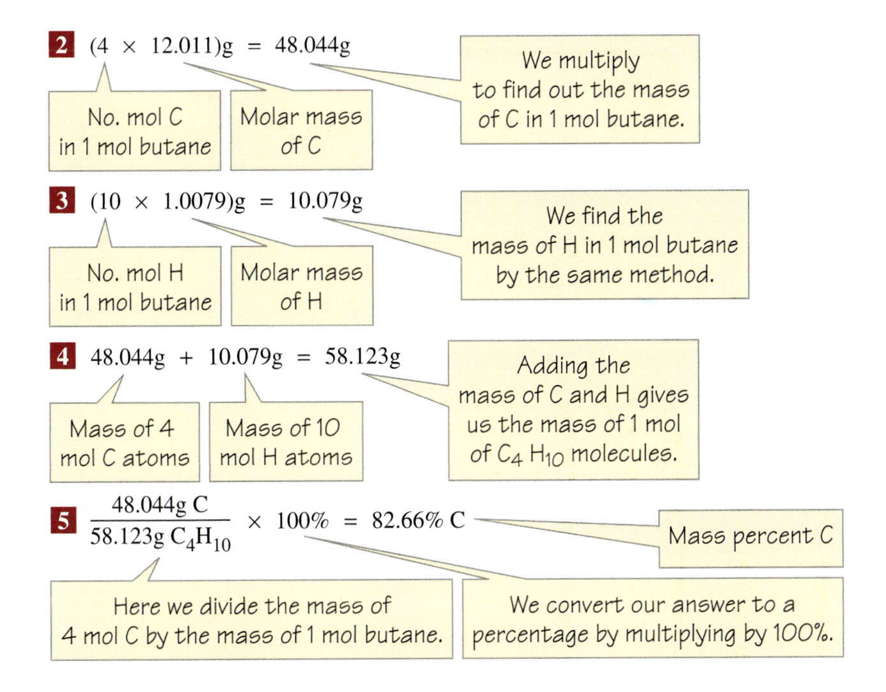

2 (4 × 12.011)g = 48.044g
No. mol C in 1 mol butane
Molar mass of C
We multiply to find out the mass of C in 1 mol butane.
3 (10 × 1.0079)g = 10.079g
No. mol H in 1 mol butane
Molar mass of H
We find the mass of H in 1 mol butane by the same method.
4 48.044g + 10.079g = 58.123g
Mass of 4 mol C atoms
Mass of 10 mol H atoms
Adding the mass of C and H gives us the mass of 1 mol of C4 H10 molecules.
5 48.044g C / 58.123g C4H10 × 100% = 82.66% C
Mass percent C
Here we divide the mass of 4 mol C by the mass of 1 mol butane.
We convert our answer to a percentage by multiplying by 100%.

EXAMPLE 3.6

Calculate the mass percent of each element in ammonium nitrate.

SOLUTION

First, determine the molar mass of ammonium nitrate, based on the formula unit NH_4NO_3.

$$\begin{aligned}
\text{Formula mass} &= (2 \times \text{atomic mass N}) + (4 \times \text{atomic mass H}) \\
&\qquad + (3 \times \text{atomic mass O}) \\
&= (2 \times 14.01)\text{ u} + (4 \times 1.008)\text{ u} + (3 \times 16.00)\text{ u} \\
&= 28.02\text{ u} + 4.032\text{ u} + 48.00\text{ u} = 80.05\text{ u} \\
\text{Molar mass} &= 80.05\text{ g/mol } NH_4NO_3
\end{aligned}$$

Then, for one mole of compound, determine mass ratios and percentages.

$$\%\text{ N} = \frac{28.02\text{ g N}}{80.05\text{ g }NH_4NO_3} \times 100\% = 35.00\%$$

$$\%\text{ H} = \frac{4.032\text{ g H}}{80.05\text{ g }NH_4NO_3} \times 100\% = 5.04\%$$

$$\%\text{ O} = \frac{48.00\text{ g O}}{80.05\text{ g }NH_4NO_3} \times 100\% = 59.96\%$$

To check, add the percentages to ensure that they add up to 100.00%. (Sometimes, the total may differ from 100.00% by ±0.01% due to rounding.)

EXERCISE 3.6A

Calculate the mass percent of each element in **(a)** ammonium sulfate, $(NH_4)_2SO_4$; and **(b)** urea, $CO(NH_2)_2$. Which compound has the greatest mass percent nitrogen: ammonium nitrate (see Example 3.6), ammonium sulfate, or urea?

PROBLEM-SOLVING NOTE

We can also determine the mass percent of one element from the mass percentages of all the others. For example, in NH_4NO_3

$$\begin{aligned}
\%\text{ O} = 100.00\% &- 35.00\%\text{ N} \\
&- 5.04\%\text{ H} = 59.96\%\text{ O}
\end{aligned}$$

When we do this, however, we lose the opportunity to check our result.

EXERCISE 3.6B

Calculate the mass percent N in triethanolamine, $N(CH_2CH_2OH)_3$, a substance used in dry-cleaning agents and household detergents.

We can set up a conversion factor based on mass percent and use it to determine the mass of an element in any sample of a compound. For example, using the result from Example 3.6 that ammonium nitrate is 35.00% N by mass, we can formulate the conversion factor (red) and find the mass of nitrogen in 46.34 g NH_4NO_3.

$$? \text{ g N} = 46.34 \cancel{\text{ g } NH_4NO_3} \times \frac{35.00 \text{ g N}}{100.00 \cancel{\text{ g } NH_4NO_3}} = 16.22 \text{ g N}$$

However, Example 3.7 illustrates a simpler approach. There, we use the chemical formula to formulate factors for the conversion from g NH_4NO_3 to g N; we don't have to evaluate the mass percent N first. Similar conversion factors are compared in Example 3.8, where no detailed calculations are required.

"5-10-5" Fertilizer: What Is It?

A common "5-10-5" fertilizer.

Most gardeners buy *complete fertilizers*, which, despite the name, usually contain compounds of only three main nutrient elements: nitrogen, phosphorus, and potassium. We usually find three numbers on a fertilizer box or bag. The first number represents the mass percent N; the second, the percent P_2O_5; and the third, the percent K_2O. So, "5-10-5" means that a fertilizer contains 5% N, 10% P_2O_5, and 5% K_2O as its active ingredients.

A 100-lb bag of this fertilizer contains 5 lb of the element nitrogen, N. The actual mass of the nitrogen-containing compound, of course, depends on its formula, for example, whether it is NH_4NO_3 (ammonium nitrate) or $CO(NH_2)_2$ (urea). But what about the K_2O and P_2O_5? There are no compounds with these formulas in the fertilizer. These mass percentages based on oxides are a holdover from the way compositions were reported in the early days of analytical chemistry, but we can convert them to actual % P and % K without difficulty. We can show that P_2O_5 is 43.64% P and that K_2O is 83.01% K. Thus, 10% P_2O_5 is the same as $0.10 \times 43.64\% = 4.4\%$ P, and 5% K_2O is about $0.05 \times 83.01\% = 4.2\%$ K. A 100-lb bag of "5-10-5" fertilizer contains 4.4 lb P and 4.2 lb K.

Phosphorus can be supplied by several compounds; $Ca(H_2PO_4)_2$ and $(NH_4)_2HPO_4$ are common ones. The potassium is nearly always supplied as KCl, although any potassium salt will furnish the needed K^+ ion.

Fertilizers greatly increase the production of food and fiber, but they also can cause problems. The fertilizers must be water soluble to be used by plants. When it rains, some of the nutrients from the fertilizers are washed into streams and lakes, where they stimulate blooms of algae. These chemicals, particularly the nitrates, also penetrate the groundwater. In some areas, the nitrates present in well water have reached levels that are toxic to infants.

EXAMPLE 3.7

How many grams of nitrogen are present in 46.34 g ammonium nitrate?

SOLUTION

The central factor (red) in the conversion is based on the chemical formula NH_4NO_3. The other factors in the setup below are based on molar masses.

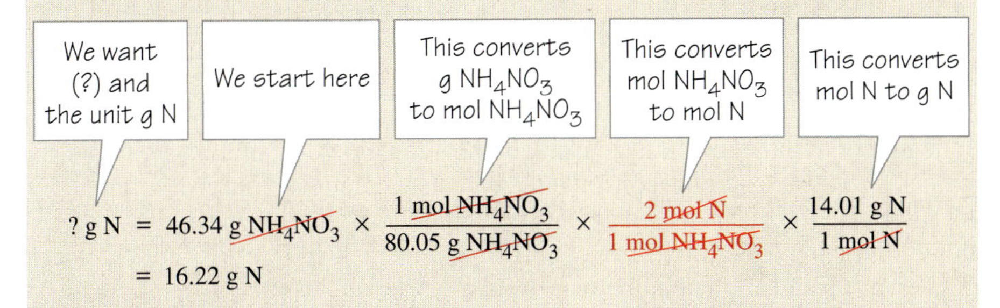

$$? \text{ g N} = 46.34 \text{ g } NH_4NO_3 \times \frac{1 \text{ mol } NH_4NO_3}{80.05 \text{ g } NH_4NO_3} \times \frac{2 \text{ mol N}}{1 \text{ mol } NH_4NO_3} \times \frac{14.01 \text{ g N}}{1 \text{ mol N}}$$
$$= 16.22 \text{ g N}$$

EXERCISE 3.7A

People with hypertension (high blood pressure) are advised to limit the amount of sodium (actually sodium ion, Na^+) in their diets. Baking soda is one of many familiar products that contain sodium ions; these ions are in the compound known as sodium hydrogen carbonate, $NaHCO_3$. Calculate the number of milligrams of Na^+ in 5.00 g of $NaHCO_3$.

EXERCISE 3.7B

A fertilizer mixture contains 12.5% NH_4NO_3 and 35.3% $(NH_4)_2SO_4$, by mass. How many grams of nitrogen are present in a 1.00-kg bag?

EXAMPLE 3.8—An Estimation Example

Without doing detailed calculations, determine which of these compounds contains the greatest mass of sulfur per gram of compound: barium sulfate, lithium sulfate, sodium sulfate, or lead sulfate.

SOLUTION

To make this comparison, we need formulas of the compounds, which we can get from their names.

$$BaSO_4 \qquad Li_2SO_4 \qquad Na_2SO_4 \qquad PbSO_4$$

The compound with the greatest mass of sulfur per gram of compound also has the greatest mass of sulfur per 100 grams of compound—the greatest %S by mass. From the formulas, we see that in one mole of each compound there is one mole of sulfur, 32.066 g S. Thus, the compound with the greatest % S is the one with the *smallest* formula mass. Each formula unit has one SO_4^{2-} ion, so all we have to do is compare some atomic masses: that of barium to twice that of lithium, and so on. With just a glance at an atomic mass table, we see the answer must be lithium sulfate, Li_2SO_4.

EXERCISE 3.8

Without doing detailed calculations, determine which of these compounds has the greatest percent phosphorous by mass: lithium dihydrogen phosphate, calcium dihydrogen phosphate, or ammonium hydrogen phosphate.

3.5 Chemical Formulas from Mass Percent Composition

As we saw in the preceding section, there are practical reasons why we may need to determine the mass percent composition of a compound from its formula. The reverse situation—deducing the formula of a compound from its mass percent composition—is of even more fundamental importance. However, we can obtain only an *empirical* formula from mass composition data, and generally what we actually want is a true *molecular* formula. An important step toward obtaining a molecular formula, however, is first to determine the empirical formula.

Determining Empirical Formulas

When we determine the mass percent composition of a compound by experiment, we deal with *masses* of the constituent elements. To get the empirical formula, we must consider the elements on a relative number basis, that is, on a *mole* basis. An important first step in finding an empirical formula is to convert the mass of each element in a sample of a compound to an amount in moles.

We could base our calculation on a sample of any mass, but we can simplify the task if we choose 100.00 g when the mass percent composition is known. This makes the masses of the elements in the sample numerically equal to their mass percentages. We can use a five-step procedure for converting mass percent composition data to an empirical formula, and we can use butane, C_4H_{10}, as a simple example. We already know its formula and percent composition from page 90.

1 *Convert the percent of each element to a mass.*

Butane is 82.66% C and 17.34% H. A 100.00-g sample consists of

$$82.66 \text{ g C} \quad \text{and} \quad 17.34 \text{ g H}$$

2 *Convert the mass of each element to an amount in moles.*

$$? \text{ mol C} = 82.66 \cancel{\text{g C}} \times \frac{1 \text{ mol C}}{12.011 \cancel{\text{g C}}} = 6.882 \text{ mol C}$$

$$? \text{ mol H} = 17.34 \cancel{\text{g H}} \times \frac{1 \text{ mol H}}{1.0079 \cancel{\text{g H}}} = 17.20 \text{ mol H}$$

3 *Use the numbers of moles of the elements as subscripts in a tentative formula.*

With the numbers of moles from Step 2 as subscripts, we get the formula

$$C_{6.882}H_{17.20}$$

4 *Attempt to get* integers *as subscripts by dividing each of the subscripts in Step 3 by the smallest subscript.*

We divide 6.882 and 17.20 each by 6.882, the smaller of the two.

$$C_{\frac{6.882}{6.882}}H_{\frac{17.20}{6.882}} \longrightarrow CH_{2.500}$$

5 *If any subscripts obtained after Step 4 are fractional quantities, multiply each of the subscripts by the smallest integer that will convert all the subscripts to integers. The result is an* empirical *formula.*

The subscript of H in $CH_{2.500}$ is a fractional quantity: $5/2 = 2.500$.

The smallest multiple that will convert this fractional quantity to an integer is 2.

Empirical formula: $C_{2\times 1}H_{2\times 2.500} = C_2H_5$

You might think that we have not gotten the correct formula for butane, which you know is C_4H_{10}. But C_4H_{10} is a *molecular* formula; the empirical or simplest formula on which it is based is indeed C_2H_5 ($C_{2\times 2}H_{2\times 5} = C_4H_{10}$). Be sure to note once more that the method presented here yields an empirical formula, which may or may not also be a molecular formula.

EXAMPLE 3.9

Phenol, a general disinfectant, has the composition 76.57% C, 6.43% H, and 17.00% O, by mass. Determine its empirical formula.

SOLUTION

1 A 100.00-g sample of phenol contains 76.57 g C, 6.43 g H, and 17.00 g O.

2 Convert the masses of C, H, and O to amounts in moles.

$$76.57 \text{ g C} \times \frac{1 \text{ mol C}}{12.011 \text{ g C}} = 6.375 \text{ mol C}$$

$$6.43 \text{ g H} \times \frac{1 \text{ mol H}}{1.008 \text{ g H}} = 6.38 \text{ mol H}$$

$$17.00 \text{ g O} \times \frac{1 \text{ mol O}}{15.999 \text{ g O}} = 1.063 \text{ mol O}$$

3 Use the numbers of moles in Step 2 as subscripts in a tentative formula.

$$C_{6.375}H_{6.38}O_{1.063}$$

4 Divide each of the subscripts above by the smallest (1.063) to try to get integral subscripts.

$$C_{\frac{6.375}{1.063}}H_{\frac{6.38}{1.063}}O_{\frac{1.063}{1.063}} \longrightarrow C_{5.997}H_{6.00}O_{1.000} \longrightarrow C_6H_6O$$

5 The subscripts in Step 4 are all integers. We need do nothing further. The empirical formula of phenol is C_6H_6O.

PROBLEM-SOLVING NOTE
In rounding off 5.997 to 6, we are following the rule of rounding off to an integer any subscript that is within one or two hundredths of an integral value.

EXERCISE 3.9A

Cyclohexanol, used in the manufacture of plastics, has the mass percent composition: 71.95% C, 12.08% H, and 15.97% O. Determine its empirical formula.

EXERCISE 3.9B

Mebutamate, a diuretic ("water pill") used to treat high blood pressure, has the mass percent composition: 51.70% C, 8.68% H, 12.06% N, and 27.55% O. Determine its empirical formula.

EXAMPLE 3.10

Diethylene glycol, used as an antifreeze, has the composition: 45.27% C, 9.50% H, and 45.23% O, by mass. Determine its empirical formula.

SOLUTION

1 A 100.00-g sample of diethylene glycol contains 45.27 g C, 9.50 g H, and 45.23 g O.

2 Convert the masses of C, H, and O to amounts in moles.

$$45.27\ \cancel{g\,C} \times \frac{1\ \text{mol C}}{12.011\ \cancel{g\,C}} = 3.769\ \text{mol C}$$

$$9.50\ \cancel{g\,H} \times \frac{1\ \text{mol H}}{1.008\ \cancel{g\,H}} = 9.42\ \text{mol H}$$

$$45.23\ \cancel{g\,O} \times \frac{1\ \text{mol O}}{15.999\ \cancel{g\,O}} = 2.827\ \text{mol O}$$

3 Use the numbers of moles in Step 2 as subscripts in a tentative formula.

$$C_{3.769}H_{9.42}O_{2.827}$$

4 Divide each of the subscripts above by the smallest (2.827) in an attempt to get integral subscripts.

$$C_{\frac{3.769}{2.827}}H_{\frac{9.42}{2.827}}O_{\frac{2.827}{2.827}} \longrightarrow C_{1.333}H_{3.33}O_{1.000}$$

5 Multiply each of the subscripts from step 4 by a common factor to convert them all to integers. By recognizing that 1.333 = 4/3 and 3.33 = 10/3, we can see that the common factor we need is *three*.

$$C_{(1.333\times 3)}H_{(3.33\times 3)}O_{(1.000\times 3)} \longrightarrow C_4H_{10}O_3$$

PROBLEM-SOLVING NOTE

Common decimals and their fractional equivalents are 0.500 = 1/2, 0.333 = 1/3, 0.250 = 1/4, 0.200 = 1/5, 0.167 = 1/6, and so on. The factors we must multiply by to convert subscripts with these decimal endings into integers are shown in red.

EXERCISE 3.10A

Anthracene, used in the manufacture of dyes, has the mass percent composition: 94.34% C and 5.66% H. Determine its empirical formula.

EXERCISE 3.10B

Trinitrotoluene (TNT), an explosive, has the mass percent composition: 37.01% C, 2.22% H, 18.50% N, and 42.27% O, by mass. Determine its empirical formula.

Relating Molecular Formulas to Empirical Formulas

For a molecular substance, the subscript integers in the molecular formula are either the same as those in its empirical formula or they are simple multiples of them. For example, in benzene, C_6H_6, the multiplier to convert the subscripts of the empirical formula CH to those of the molecular formula is six.

To establish a molecular formula, we need to know both the molecular mass and the empirical formula mass. The relationship between them can be stated as follows.

$$\text{Empirical formula mass} \times \text{integral factor} = \text{molecular mass}$$

$$\text{Integral factor} = \frac{\text{molecular mass}}{\text{empirical formula mass}}$$

The integral factor is the multiplier that converts the subscripts of the empirical formula to those of the molecular formula. We can obtain the empirical formula mass from the empirical formula, and—as we have just seen—we can establish the empirical formula from mass percent composition data. In later chapters, we will consider several experimental methods of determining molecular masses.

EXAMPLE 3.11

The empirical formula of hydroquinone, a chemical used in photography, is C_3H_3O, and its molecular mass is 110 u. What is its molecular formula?

SOLUTION

The empirical formula mass is $[(3 \times 12.0\ \text{u}) + (3 \times 1.0\ \text{u}) + 16.0\ \text{u}] = 55.0\ \text{u}$. The multiplier we need to convert the subscripts in the empirical formula to those in the molecular formula is the integral factor in the relationship that follows.

$$\text{Integral factor} = \frac{\text{molecular mass}}{\text{empirical formula mass}} = \frac{110\ \cancel{\text{u}}}{55.0\ \cancel{\text{u}}} = 2$$

The molecular formula is $(C_3H_3O)_2$ or $C_6H_6O_2$.

EXERCISE 3.11

Ethylene (molecular mass: 28.0 u), cyclohexane (84.0 u), and 1-pentene (70.0 u) all have the empirical formula CH_2. Give the molecular formula of each compound.

3.6 Elemental Analysis: Experimental Determination of Mass Percent Composition

A chemist has just synthesized a new compound in an organic chemistry research laboratory. Most likely he or she has some idea of the composition and structure of the compound. To publish this discovery in the chemical literature or to obtain a patent covering this work, however, the chemist must first present proof of the identity of the compound. One of the first steps in that proof is likely to be an elemental analysis establishing the mass percent composition of the compound. The methods of analysis vary, depending on the elements present. We will limit our discussion to the analysis of compounds that contain only the elements carbon, hydrogen, and oxygen. However, recall that a vast percentage (over 95%) of all compounds are carbon-based. Nearly all of these contain hydrogen, and many contain oxygen atoms as well.

In the apparatus pictured in Figure 3.5, a weighed sample of a compound in a combustion "boat" is burned in a stream of oxygen gas in a high-temperature furnace. The carbon dioxide gas and water vapor formed in the combustion pass into absorbers. Water vapor is absorbed by a substance such as magnesium perchlorate, and carbon dioxide by sodium hydroxide. The masses of carbon dioxide and water are determined as the differences in masses of the absorbers after and before the combustion.

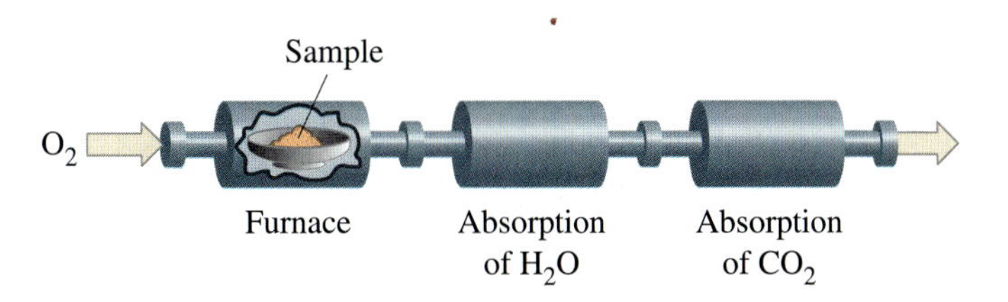

Figure 3.5 Apparatus for combustion analysis
Oxygen gas enters the combustion apparatus and streams over the sample under analysis in a high-temperature furnace. The absorbers collect the water vapor and carbon dioxide that are produced in the combustion.

Consider the combustion of methanol, CH_3OH. By the law of conservation of mass, all the carbon atoms in the compound must appear in carbon dioxide molecules, and all the hydrogen atoms in water molecules. Figure 3.6 depicts this fact

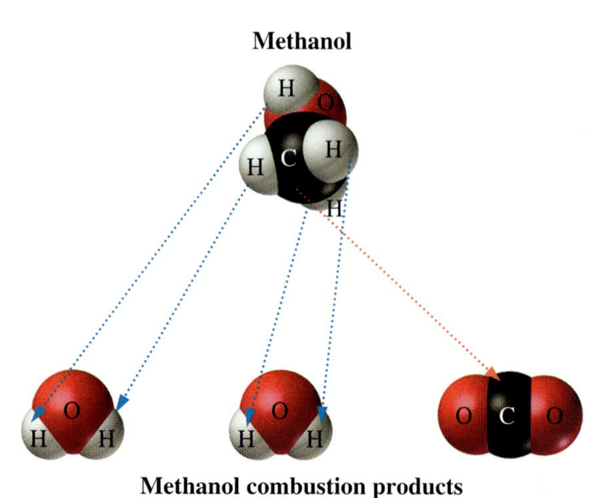

▶ **Figure 3.6**
Tracing C and H atoms in the combustion of methanol, CH_3OH.

in a molecular view, showing that every molecule of CH_3OH is replaced by one molecule of CO_2 and two molecules of H_2O. By determining the mass of carbon in the carbon dioxide and the mass of hydrogen in the water, we actually determine the masses of carbon and hydrogen present in the original sample. From these masses and the mass of the compound itself, we can determine the mass percent of C and H in the compound. Because the oxygen atoms in the combustion products come partly from the methanol and partly from the oxygen used in the combustion, we can only determine the mass percent of O indirectly: % O = 100.00% − % C − % H.

The specific conversions required to establish the mass percent composition of a compound from combustion analysis data are as follows.

$$\text{g } CO_2 \longrightarrow \text{mol } CO_2 \longrightarrow \text{mol C} \longrightarrow \text{g C} \longrightarrow \text{\% C in original sample}$$
$$\text{g } H_2O \longrightarrow \text{mol } H_2O \longrightarrow \text{mol H} \longrightarrow \text{g H} \longrightarrow \text{\% H in original sample}$$

These conversions and a typical application of combustion analysis are illustrated in Example 3.12.

EXAMPLE 3.12

A 0.1000-g sample of a carbon-hydrogen-oxygen compound is burned in oxygen to yield 0.1953 g CO_2 and 0.1000 g H_2O. In a separate experiment, the molecular mass of the compound is found to be 90 u.

a. Calculate the mass percent composition of the compound.
b. What is the empirical formula of the compound?
c. What is its molecular formula?

SOLUTION

a. First, we can do the conversions outlined above to calculate the mass of carbon in the CO_2 produced.

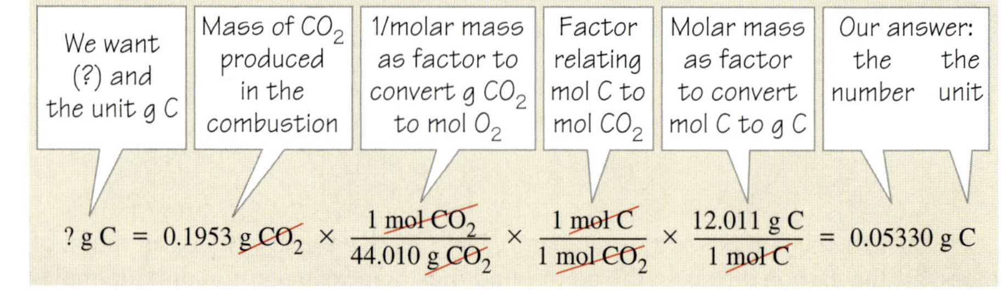

$$? \text{ g C} = 0.1953 \text{ g } CO_2 \times \frac{1 \text{ mol } CO_2}{44.010 \text{ g } CO_2} \times \frac{1 \text{ mol C}}{1 \text{ mol } CO_2} \times \frac{12.011 \text{ g C}}{1 \text{ mol C}} = 0.05330 \text{ g C}$$

This mass of carbon originated from the 0.1000-g sample. Thus, the mass percent carbon in the compound is

$$\%\text{ C} = \frac{0.05330\text{ g C}}{0.1000\text{ g sample}} \times 100\% = 53.30\%\text{ C}$$

With similar calculations, we can determine first the mass of hydrogen and then the mass percent of hydrogen in the compound.

$$?\text{ g H} = 0.1000\ \cancel{\text{g H}_2\text{O}} \times \frac{1\ \cancel{\text{mol H}_2\text{O}}}{18.015\ \cancel{\text{g H}_2\text{O}}} \times \frac{2\ \cancel{\text{mol H}}}{1\ \cancel{\text{mol H}_2\text{O}}} \times \frac{1.0079\text{ g H}}{1\ \cancel{\text{mol H}}}$$

$$= 0.01119\text{ g H}$$

$$\%\text{ H} = \frac{0.01119\text{ g H}}{0.1000\text{ g sample}} \times 100\% = 11.19\%\text{ H}$$

Finally, we can calculate the mass percent oxygen by subtracting the mass percents of C and H from 100.00%.

$$\%\text{ O} = 100.00\% - 53.30\% - 11.19\% = 35.51\%.$$

b. Here, we can again apply the method of Examples 3.9 and 3.10.

1 A 100.00-g sample of the compound contains 53.30 g C, 11.19 g H, and 35.51 g O.

2 Convert the masses of C, H, and O to numbers of moles.

$$53.30\ \cancel{\text{g C}} \times \frac{1\text{ mol C}}{12.011\ \cancel{\text{g C}}} = 4.438\text{ mol C}$$

$$11.19\ \cancel{\text{g H}} \times \frac{1\text{ mol H}}{1.0079\ \cancel{\text{g H}}} = 11.10\text{ mol H}$$

$$35.51\ \cancel{\text{g O}} \times \frac{1\text{ mol O}}{15.999\ \cancel{\text{g O}}} = 2.220\text{ mol O}$$

3 Use the numbers of moles in Step 2 as subscripts in a tentative formula.

$$C_{4.438}H_{11.10}O_{2.220}$$

4 Divide each of the subscripts above by the smallest (2.220) to try to get integral subscripts.

$$C_{\frac{4.438}{2.220}}H_{\frac{11.10}{2.220}}O_{\frac{2.220}{2.220}} \longrightarrow C_{1.999}H_{5.000}O_{1.000} \longrightarrow C_2H_5O$$

c. The empirical formula mass is $(2 \times 12.0)\text{ u} + (5 \times 1.0)\text{ u} + 16.0\text{ u} = 45.0\text{ u}$. The multiplier we need to convert the subscripts in the empirical formula to those in the molecular formula is the integral factor in this relationship.

$$\text{Integral factor} = \frac{\text{molecular mass}}{\text{empirical formula mass}} = \frac{90\ \cancel{\text{u}}}{45.0\ \cancel{\text{u}}} = 2$$

The molecular formula is $(C_2H_5O)_2$ or $C_4H_{10}O_2$.

● **ChemCDX** Representing Substances I

EXERCISE 3.12

A 0.3629-g sample of tetrahydrocannabinol, the principal active component of marijuana, is burned to yield 1.0666 g of carbon dioxide and 0.3120 g of water.

a. Calculate its mass percent composition.
b. Calculate its empirical formula.

Stoichiometry of Chemical Reactions

So far in this chapter, we have considered stoichiometric calculations based on chemical formulas. We can calculate mass percent compositions from formulas, and we can determine empirical formulas from mass percent compositions of chemical compounds. In the following sections, we consider chemical reactions and some of the various stoichiometric calculations that aid our understanding of these processes. We start by considering chemical equations, symbolic representations that provide considerable useful information.

3.7 Writing and Balancing Chemical Equations

A **chemical equation** is a shorthand description of a chemical reaction, using symbols and formulas to represent the elements and compounds involved. Before we can construct this shorthand description, however, someone must have done *experiments* to establish that a reaction has occurred, to identify all the substances involved in the reaction, and at times to establish the formulas of these substances by the methods described earlier in the chapter.

For the reaction of carbon with a plentiful source of oxygen, as we have already noted, the sole product is carbon dioxide. The equation we write for this reaction is

$$C + O_2 \longrightarrow CO_2$$

The plus sign (+) indicates that carbon and oxygen interact in some way, and the arrow (→), usually read as "yields," points to the result of that combination: carbon dioxide. We generally call the starting substances in a reaction the **reactants** and the substances formed in the reaction, the **products**. The reactants appear on the *left* of the arrow in an equation and the products, on the *right*.

At times we need to indicate the physical forms of the reactants and products, and we can do this through the following symbols.

$$(g) = \text{gas};\ (l) = \text{liquid};\ (s) = \text{solid};\ (aq) = \text{aqueous (water) solution}$$

These parenthetical symbols are attached to the reactants and products, as in the example that follows.

$$C(s) + O_2(g) \longrightarrow CO_2(g)$$

If it is necessary to heat a mixture of reactants to bring about a chemical reaction, we sometimes denote this by placing a capital Greek letter delta, Δ, above the yield arrow, that is, $\xrightarrow{\Delta}$. Sometimes the temperature or other conditions under which the reaction is carried out are also noted above the yield arrow.

There are several different ways we can interpret the equation

$$C(s) + O_2(g) \longrightarrow CO_2(g)$$

We can use it as a qualitative description of the reaction, as in "solid carbon and gaseous oxygen react to form gaseous carbon dioxide." We can give a microscopic interpretation, as in "one carbon atom reacts with one oxygen molecule to form one molecule of carbon dioxide gas." However, we must work at the real-life, or macroscopic, level. The most useful interpretation of the equation is based on enormously large numbers of atoms and molecules, specifically the numbers found in a mole of substance: 1 mol (12.0 g) of carbon reacts with 1 mol (32.0 g) of oxygen to produce 1 mol (44.0 g) of carbon dioxide. This interpretation on a molar basis is at the heart of the quantitative calculations that we will do in Section 3.8.

The equation for the reaction of carbon and oxygen to form carbon dioxide is deceptively easy to write. If we try something similar for the reaction of hydrogen and oxygen to form water, we run into a bit of a problem. The following equation does not conform to the law of conservation of mass: It is not balanced.

$$H_2(g) + O_2(g) \longrightarrow H_2O(l) \quad (\textit{not balanced})$$

Two O atoms in the form of an O_2 molecule are shown on the left side of the equation, but there is only one O atom in the H_2O molecule on the right side. An O atom is missing on the product side, but we know that atoms cannot be created or destroyed in a chemical reaction. We went wrong by assuming that H_2 and O_2 molecules react in a 1:1 ratio. As shown in the molecular interpretation in Figure 3.7, they don't. Figure 3.7 illustrates how to balance the equation so that it is in harmony with the law of conservation of mass. However, we don't need to draw molecular pictures to balance equations because we can work directly with the symbolic equation and use stoichiometric coefficients to adjust the ratios of the reactants and products. **Stoichiometric coefficients** are numbers placed in front of formulas in a chemical equation to balance the equation, thereby indicating the combining ratios of the reactants and the ratios in which products are formed. A stoichiometric coefficient multiplies everything in the formula that follows it. If there is no coefficient before a formula, then the coefficient "1" is understood.

To balance the equation

$$H_2(g) + O_2(g) \longrightarrow H_2O(l) \quad (\textit{not balanced})$$

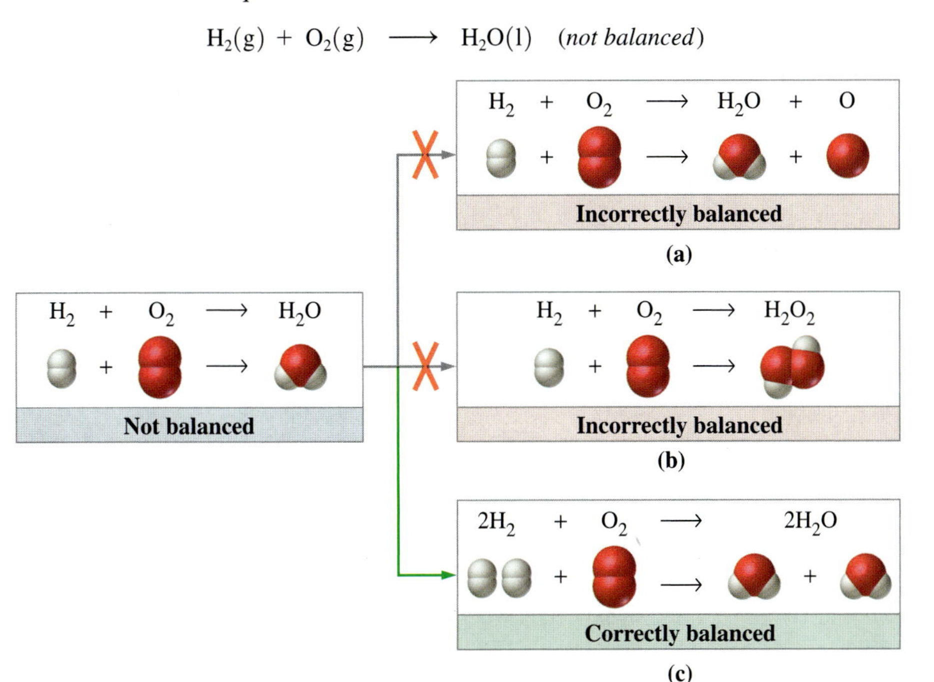

Figure 3.7 Balancing the chemical equation for the reaction of hydrogen with oxygen to form water. (a) *Incorrect*: There is no evidence for the presence of atomic oxygen as a product. *A reactant or product with a different chemical formula cannot be introduced for the purpose of balancing an equation.* (b) *Incorrect*: The product of the reaction is water, H_2O, not hydrogen peroxide, H_2O_2. *A formula cannot be changed in order to balance a chemical equation.* (c) *Correct*: An equation can be balanced only through the use of *correct formulas and coefficients.*

we can begin by placing the coefficient "2" in front of the formula H_2O.

$$H_2(g) + O_2(g) \longrightarrow 2\,H_2O(l) \quad (\textit{O balanced, H not balanced})$$

Now we have two oxygen atoms on each side of the equation. The coefficient "2" increases the number of H atoms on the right to *four* at the same time that it increases the number of O atoms to two. There are only *two* H atoms in the H_2 molecule on the left. We can correct this, however, by placing the coefficient "2" in front of H_2 on the left. The equation is now balanced.

$$2\,H_2(g) + O_2(g) \longrightarrow 2\,H_2O(l) \quad (\textit{balanced})$$

Altogether there are four H atoms and two O atoms on each side of the equation.

We cannot overemphasize the point made in Figure 3.7: The only way to balance an equation is by adjusting coefficients, not by changing formulas. If we add or remove a reactant or product from the equation or change the subscript in a formula, we may achieve a balance, but *we are no longer describing the desired reaction.*

The method we just described is called *balancing by inspection*, but there are some simple strategies that reduce the trial-and-error aspect of the method. For example,

- If an element is present in just one compound on each side of the equation, try balancing that element *first.*
- Balance any reactants or products that exist as the *free* element *last.*
- Some groupings of atoms, such as those in polyatomic ions, remain unchanged in a reaction. Balance these as a unit.

In any case, the most important step in any strategy is to check an equation to ensure that it is indeed balanced: *For each element, the same number of atoms must appear on each side of the equation.* Atoms are conserved in chemical reactions.

EXAMPLE 3.13

Balance the following equation.

$$Fe + O_2 \longrightarrow Fe_2O_3$$

SOLUTION

It seems that the easiest place to begin is with the iron atoms. There is one on the left (Fe) and two on the right (Fe_2O_3), and we can balance them by placing the coefficient *two* on the left.

$$2\,Fe + O_2 \longrightarrow Fe_2O_3 \quad (\textit{Fe balanced, O not balanced})$$

To balance the oxygen atoms, we begin by noting two of them on the left and three on the right. The easiest way to balance them is to use the *fractional* coefficient *3/2* before O_2 on the left (that is, $3/2 \times 2 = 3$).

$$2\,Fe + \tfrac{3}{2}O_2 \longrightarrow Fe_2O_3 \quad (\textit{balanced})$$

Fractional coefficients are perfectly acceptable in equations—sometimes even desirable. If we don't want them, however, we can easily clear the equation of fractional coefficients by multiplying *every* coefficient by the same integer, in this case, 2.

$$2 \times \{2\,Fe + \tfrac{3}{2}O_2 \longrightarrow Fe_2O_3\}$$

becomes

$$4\ Fe + 3\ O_2 \longrightarrow 2\ Fe_2O_3 \quad (\textit{balanced})$$

EXERCISE 3.13

Balance the following equations.

a. $SiCl_4 + H_2O \rightarrow SiO_2 + HCl$
b. $PCl_3 + H_2O \rightarrow H_3PO_3 + HCl$
c. $CaO + P_4O_{10} \rightarrow Ca_3(PO_4)_2$

EXAMPLE 3.14

Balance the following equation.

$$C_2H_6 + O_2 \longrightarrow CO_2 + H_2O$$

SOLUTION

Oxygen appears as the free element on the left, so let's leave it for last and balance the other two elements first. To balance carbon, we place the coefficient 2 in front of CO_2.

$$C_2H_6 + O_2 \longrightarrow 2\ CO_2 + H_2O \quad (\textit{not balanced})$$

To balance hydrogen, we need the coefficient 3 before H_2O.

$$C_2H_6 + O_2 \longrightarrow 2\ CO_2 + 3\ H_2O \quad (\textit{still not balanced})$$

Now, if we count oxygen atoms, we find 2 on the left and 7 on the right. We can get 7 on each side by using the *fractional* coefficient 7/2 on the left ($7/2 \times 2 = 7$).

$$C_2H_6 + \tfrac{7}{2}O_2 \longrightarrow 2\ CO_2 + 3\ H_2O \quad (\textit{balanced})$$

To obtain an equation with only integral (whole number) coefficients, we can multiply each coefficient by 2. That is,

$$2 \times \{C_2H_6 + \tfrac{7}{2}O_2 \longrightarrow 2\ CO_2 + 3\ H_2O\}$$

leads to

$$2\ C_2H_6 + 7\ O_2 \longrightarrow 4\ CO_2 + 6\ H_2O \quad (\textit{balanced})$$

PROBLEM-SOLVING NOTE
Always make a final check to ensure that an equation is balanced. Here we count four C atoms, 12 H atoms, and 14 O atoms on each side.

EXERCISE 3.14

Balance the following equations.

a. $C_4H_{10} + O_2 \rightarrow CO_2 + H_2O$
b. $CH_3CH_2CH_2CH(OH)CH_2OH + O_2 \rightarrow CO_2 + H_2O$

EXAMPLE 3.15

Balance the following equation.

$$H_3PO_4 + NaCN \longrightarrow HCN + Na_3PO_4$$

SOLUTION

Notice that the PO_4 and CN *groups* remain unchanged in the reaction. For purposes of balancing equations, we can often treat such groups as a whole rather than breaking them down into their constituent atoms. To balance hydrogen atoms, we place a 3 before HCN.

$$H_3PO_4 + NaCN \longrightarrow 3\ HCN + Na_3PO_4 \quad (\textit{not balanced})$$

To balance sodium atoms, we put a 3 in front of the NaCN.

$$H_3PO_4 + 3\ NaCN \longrightarrow 3\ HCN + Na_3PO_4 \quad (\textit{balanced})$$

Note that in the final balanced equation, there is one PO_4 group and three CN groups on each side of the equation.

EXERCISE 3.15

Balance the following equations.

a. $FeCl_3 + NaOH \rightarrow Fe(OH)_3 + NaCl$
b. $Ba(NO_3)_2 + Al_2(SO_4)_3 \rightarrow BaSO_4 + Al(NO_3)_3$
c. $Ca(OH)_2 + H_3PO_4 \rightarrow Ca_3(PO_4)_2 + H_2O$

EXAMPLE 3.16—A Conceptual Example

Write a plausible chemical equation for the reaction of a liquid chloride of phosphorus (77.45% Cl by mass) with water, forming an aqueous solution of hydrochloric and phosphorous acids.

SOLUTION

The concept that we have just been exploring—balancing a chemical equation—is the last step to this problem. First we must apply some earlier ideas:

Establishing the formula of the chloride of phosphorus.
We can apply the method of Examples 3.9 and 3.10 to a compound that is 22.55% P and 77.45% Cl. A 100.00-g sample of the compound consists of 22.55 g P and 77.45 g Cl, corresponding to 0.728 mol P and 2.185 mol Cl. The formula: $P_{0.728}Cl_{2.185}$ reduces to PCl_3.

Establishing the formulas of hydrochloric and phosphorous acids.
The relationship between names and formulas was described on pages 59–60. Hydrochloric acid, a binary acid, has the formula HCl. Table 2.5 gives H_3PO_4 as the formula of phosphor*ic* acid. Phosphor*ous* acid should have one less O atom per molecule than the "*ic*" acid, giving the formula: H_3PO_3.

Writing and balancing the equation.
The unbalanced equation, including an indication of the physical states, is

$$PCl_3(l) + H_2O(l) \longrightarrow HCl(aq) + H_3PO_3(aq)$$

The balanced equation requires the coefficient 3 for $H_2O(l)$ and for $HCl(aq)$.

$$PCl_3(l) + 3\,H_2O(l) \longrightarrow 3\ HCl(aq) + H_3PO_3(aq)$$

EXERCISE 3.16

Write a plausible chemical equation for the combustion of liquid triethylene glycol in an abundant supply of oxygen gas. Triethylene glycol is 47.99% C, 9.40% H, and 42.62% O by mass and has a molecular mass of 150.2 u.

PROBLEM-SOLVING NOTE
What products would you expect from the combustion of this compound with a plentiful supply of oxygen?

Our main interest in this section has been simply balancing equations. It is much more important, however, to be able to *predict* whether a chemical reaction will occur and then to write an equation to represent it. We will introduce certain new ideas in later chapters to show how to make such predictions.

3.8 Stoichiometric Equivalence and Reaction Stoichiometry

Whether making medicines, obtaining metals from their ores, studying the combustion of a rocket fuel, synthesizing new compounds, or simply testing a hypothesis, chemists need to consider mole and mass relationships in chemical reactions. These relationships are derived from *chemical equations*.

Consider the reaction of carbon monoxide and hydrogen to form methanol.

$$CO + 2\,H_2 \longrightarrow CH_3OH$$

At the microscopic level, the stoichiometric coefficients mean that for every *one* molecule of CO that reacts, *two* molecules of H_2 react and *one* molecule of CH_3OH is formed. In the reaction of *10* molecules of CO, *20* molecules of H_2 also react and *20* molecules of CH_3OH are formed. The reactants and product retain this 1 : 2 : 1 ratio no matter how large the numbers of molecules we choose. If we work in the range of Avogadro's number ($N_A = 6.022 \times 10^{23}$), we are at the macroscopic level and can switch to a molar basis: *One* mole of CO reacts with *two* moles of H_2 to produce *one* mole of CH_3OH.

The amounts of CO, H_2, and CH_3OH in the previous sentence are said to be *stoichiometrically equivalent*, which we can represent in the following way.

$$1 \text{ mol CO} \Leftrightarrow 2 \text{ mol } H_2$$
$$1 \text{ mol CO} \Leftrightarrow 1 \text{ mol } CH_3OH$$
$$2 \text{ mol } H_2 \Leftrightarrow 1 \text{ mol } CH_3OH$$

The symbol $\Leftrightarrow$ means "is chemically equivalent to." To say that 1 mol CO is chemically equivalent to 2 mol H_2 is not the same as saying that they are equal. There is no way that CO and H_2 can be thought of as equal or identical—they are two completely different substances. Nevertheless, because *in the reaction to form CH_3OH* the two react in the ratio, 1 mol CO : 2 mol H_2, we can form conversion factors for use in problem solving that have the usual form, such as

$$\frac{2 \text{ mol } H_2}{1 \text{ mol CO}}$$

Conversion factors formed from the stoichiometric coefficients in a chemical equation are called **stoichiometric factors** or (sometimes) *mole ratios*.

It is necessary to keep in mind that the stoichiometric factor linking H_2 and CO depends on the particular reaction. For another reaction between these two same substances

$$CO + 3\,H_2 \longrightarrow CH_4 + H_2O$$

the corresponding stoichiometric factor is

$$\frac{3 \text{ mol } H_2}{1 \text{ mol CO}}$$

APPLICATION NOTE

Alkanes with more carbon atoms than CH_4, such as those in gasoline, can be synthesized by using different proportions of CO and H_2. These CO-H_2 mixtures, called *synthesis gas*, can be produced by a reaction between coal and steam.

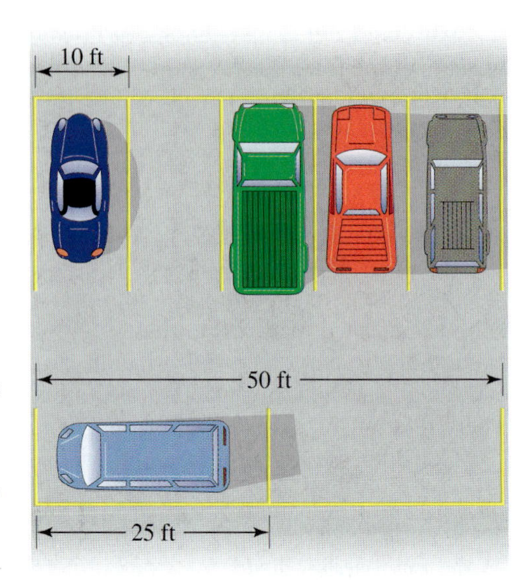

▶ **Figure 3.8**
The concept of equivalence
With parallel parking, each automobile *is equivalent* to 25 ft of curb space; that is, 1 automobile ≎ 25 ft. The number of automobiles that can be parked is

$$\text{Number of automobiles} = 50 \text{ ft} \times \frac{1 \text{ automobile}}{25 \text{ ft}} = 2 \text{ automobiles}$$

With perpendicular parking, each automobile *is equivalent* to 10 ft of curb space. How many automobiles can be parked along the 50-ft section of curb?

Figure 3.8 illustrates a commonplace example of an equivalence—that encountered when parking automobiles. The equivalence between automobile and curb length is different depending on whether the parking arrangement is parallel or perpendicular.

EXAMPLE 3.17

When 0.105 mol propane is burned in an excess of oxygen, how many moles of oxygen are consumed?

$$C_3H_8 + 5\,O_2 \longrightarrow 3\,CO_2 + 4\,H_2O$$

SOLUTION

The stoichiometric coefficients in the equation allow us to write the following equivalence.

$$1 \text{ mol } C_3H_8 \Leftrightarrow 5 \text{ mol } O_2$$

From this equivalence we can derive two conversion factors.

$$\frac{5 \text{ mol } O_2}{1 \text{ mol } C_3H_8} \quad \text{and} \quad \frac{1 \text{ mol } C_3H_8}{5 \text{ mol } O_2}$$

Because we are given the number of moles of propane and are seeking the number of moles of O_2 consumed, we need the factor that has the unit "mol O_2" in the numerator and the unit "mol C_3H_8" in the denominator. This is the factor in red.

$$? \text{ mol } O_2 = 0.105 \cancel{\text{mol } C_3H_8} \times \frac{5 \text{ mol } O_2}{\cancel{1 \text{ mol } C_3H_8}} = 0.525 \text{ mol } O_2$$

EXERCISE 3.17

For the combustion of propane in Example 3.17:

a. How many moles of carbon dioxide are formed when 0.529 mol C_3H_8 is burned?
b. How many moles of water are produced when 76.2 mol C_3H_8 is burned?
c. How many moles of carbon dioxide are produced when 1.010 mol O_2 is consumed?

Although the mole is essential in calculations based on chemical equations, we cannot measure out molar amounts directly. We have to relate them to quantities that we can measure: mass in grams or kilograms, volume in milliliters or liters, and so on. Figure 3.9 outlines a general problem-solving approach for stoichiometric calculations based on chemical equations. The most important step is deriving a stoichiometric factor from the balanced equation (Box 5), but note also that several aspects of the approach resemble calculations based on the chemical formula (see Figures 3.3 and 3.4). In Example 3.17 we used only a portion of Figure 3.9. In other examples that follow we will deal with additional portions of the outline and will refer to them when we do.

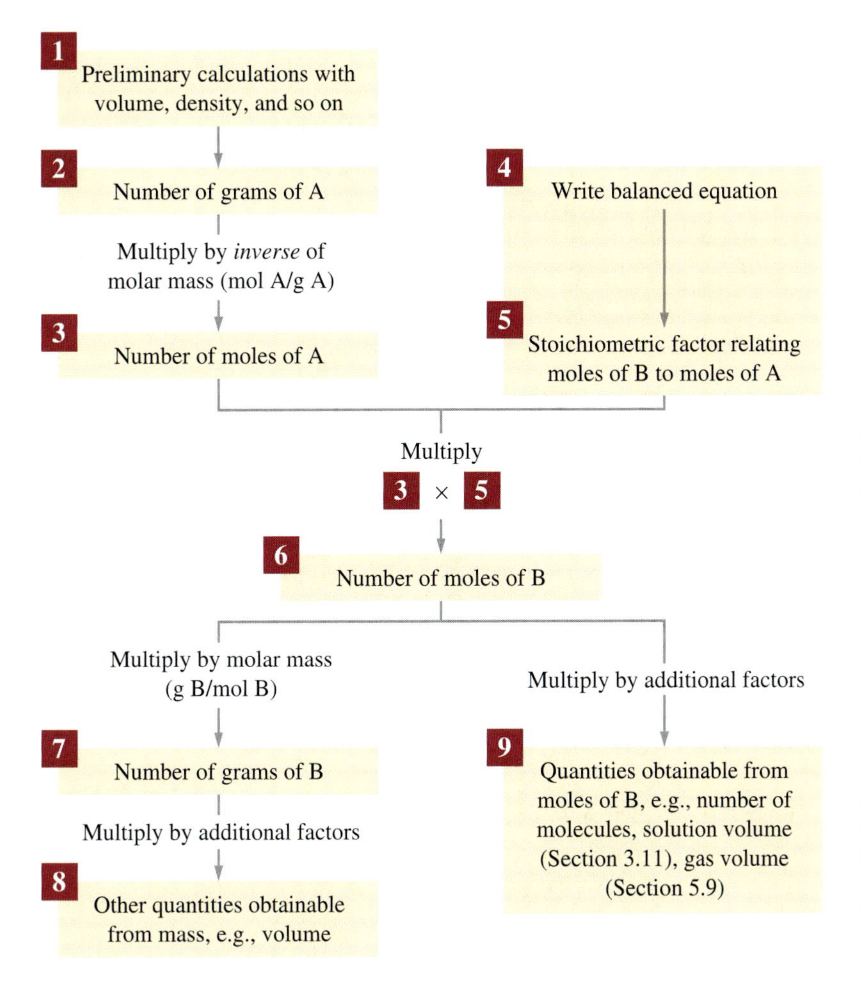

PROBLEM-SOLVING NOTE

The heart of a reaction stoichiometry problem is converting moles A to moles B: 3 × 5

◀ **Figure 3.9 Outline of reaction stoichiometry**

The substances A and B are two reactants, a reactant and product, or two products of a chemical reaction.

EXAMPLE 3.18

The final step in the production of nitric acid involves the reaction of nitrogen dioxide with water; also produced is nitrogen monoxide. How many grams of nitric acid are produced for every 100.0 g of nitrogen dioxide that reacts?

SOLUTION

Because no equation is given, we must first write a chemical equation from the description of the reaction. Recall that we related the names and formulas of these reactants and products in Chapter 2.

PROBLEM-SOLVING NOTE

Writing a balanced equation is noted as 4 in Figure 3.9.

$$NO_2 + H_2O \longrightarrow HNO_3 + NO \quad (\textit{not balanced})$$

To balance the equation, let's first balance H atoms because H appears in one reactant and one product. Then we can balance N atoms, which on the reactant side appear only in NO_2. At this point, O atoms will also be balanced.

$$3\ NO_2 + H_2O \longrightarrow 2\ HNO_3 + NO \quad (\textit{balanced})$$

Now we convert the mass of the given substance, NO_2, to an amount in moles.

PROBLEM-SOLVING NOTE
This is the conversion 2 → 3 in Figure 3.9.

$$?\ \text{mol } NO_2 = 100.0\ \cancel{\text{g } NO_2} \times \frac{1\ \text{mol } NO_2}{46.006\ \cancel{\text{g } NO_2}} = 2.174\ \text{mol } NO_2$$

Next, we use coefficients from the balanced equation to establish the stoichiometric equivalence of NO_2 and HNO_3.

$$3\ \text{mol } NO_2 \approx 2\ \text{mol } HNO_3$$

PROBLEM-SOLVING NOTE
Writing a stoichiometric factor based on a balanced equation is represented as 5 in Figure 3.9.

Because we need to convert from moles of NO_2 to moles of HNO_3, we should use this equivalence to write the stoichiometric factor.

$$\frac{2\ \text{mol } HNO_3}{3\ \text{mol } NO_2}$$

PROBLEM-SOLVING NOTE
This is the step labeled 3 × 5 in Figure 3.9.

Now we can use this stoichiometric factor in the conversion

$$?\ \text{mol } HNO_3 = 2.174\ \cancel{\text{mol } NO_2} \times \frac{2\ \text{mol } HNO_3}{3\ \cancel{\text{mol } NO_2}} = 1.449\ \text{mol } HNO_3$$

PROBLEM-SOLVING NOTE
This is the conversion 6 → 7 in Figure 3.9.

Finally, we can convert from moles of HNO_3 to grams of HNO_3.

$$?\ \text{g } HNO_3 = 1.449\ \cancel{\text{mol } HNO_3} \times \frac{63.013\ \text{g } HNO_3}{1\ \cancel{\text{mol } HNO_3}} = 91.31\ \text{g } HNO_3$$

Combined Setup. We can combine all of the individual steps into a single setup.

$$?\ \text{g } HNO_3 = 100.0\ \cancel{\text{g } NO_2} \times \frac{1\ \cancel{\text{mol } NO_2}}{46.006\ \cancel{\text{g } NO_2}} \times \frac{2\ \cancel{\text{mol } HNO_3}}{3\ \cancel{\text{mol } NO_2}} \times \frac{63.013\ \text{g } HNO_3}{1\ \cancel{\text{mol } HNO_3}}$$
$$= 91.31\ \text{g } HNO_3$$

EXERCISE 3.18A

How many grams of magnesium are required to convert 83.6 g $TiCl_4$ to titanium metal?

$$TiCl_4 + 2\ Mg \xrightarrow{\Delta} Ti + 2\ MgCl_2$$

EXERCISE 3.18B

Upon being strongly heated or subjected to severe mechanical shock, ammonium nitrate decomposes into nitrogen and oxygen gases and water vapor. If 75.5 grams of ammonium nitrate decomposes in this way, how many grams of nitrogen and how many grams of oxygen are produced?

EXAMPLE 3.19

Ammonium sulfate, a common fertilizer used by gardeners, is produced commercially by passing gaseous ammonia into an aqueous solution that is 65% H_2SO_4 by mass and has a density of 1.55 g/mL. What volume of this sulfuric acid solution is required to convert 1.00 kg NH_3 to $(NH_4)_2SO_4$?

SOLUTION

First we must write the balanced equation for the reaction.

$$2\,NH_3(g) + H_2SO_4(aq) \longrightarrow (NH_4)_2SO_4(aq)$$

The required setup has 1.00 kg NH_3—the given substance—as its starting point. Because we are seeking a *volume* of $H_2SO_4(aq)$, the setup has the general form.

$$?\ \text{mL } H_2SO_4(aq) = 1.00\ \text{kg } NH_3 \times \text{conversions factors}$$

The required series of conversions is indicated below. The upper set of numbers refer to boxes in the stoichiometry outline in Figure 3.9. The lower set of conversion factors are needed to complete the "preliminary calculations" and "additional factors" referred to in Figure 3.9.

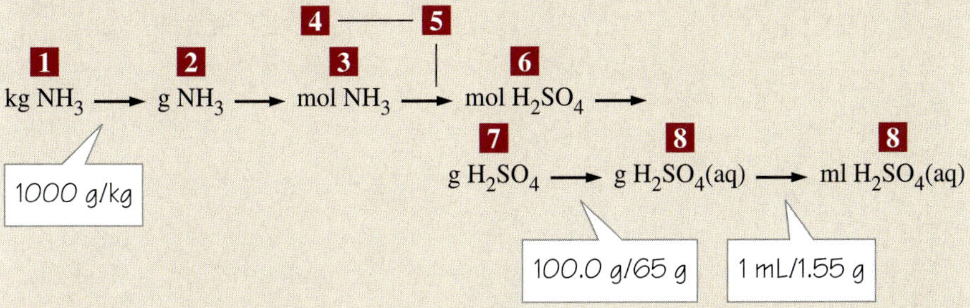

4 — 5

1 kg NH_3 ⟶ 2 g NH_3 ⟶ 3 mol NH_3 ⟶ 6 mol H_2SO_4 ⟶

7 g H_2SO_4 ⟶ 8 g $H_2SO_4(aq)$ ⟶ 8 ml $H_2SO_4(aq)$

1000 g/kg

100.0 g/65 g

1 mL/1.55 g

In the setup below, the conversions are done in the same order as they are sketched out above.

$$?\ \text{mL } H_2SO_4(aq) = 1.00\ \text{kg } NH_3 \times \frac{1000\ \text{g } NH_3}{1\ \text{kg } NH_3} \times \frac{1\ \text{mol } NH_3}{17.03\ \text{g } NH_3} \times \frac{1\ \text{mol } H_2SO_4}{2\ \text{mol } NH_3}$$

$$\times \frac{98.08\ \text{g } H_2SO_4}{1\ \text{mol } H_2SO_4} \times \frac{100.0\ \text{g } H_2SO_4(aq)}{65\ \text{g } H_2SO_4} \times \frac{1\ \text{mL } H_2SO_4(aq)}{1.55\ \text{g } H_2SO_4(aq)}$$

$$= 2.9 \times 10^3\ \text{mL } H_2SO_4(aq)\ [2.9\ \text{L } H_2SO_4(aq)]$$

PROBLEM-SOLVING NOTE

Note that the percent composition (65% H_2SO_4) and density (1.55 g/mL) are written in the *inverted* form. This provides for the proper cancellation of units, and it makes sense. The mass of $H_2SO_4(aq)$ should be greater than that of the pure H_2SO_4 represented in 7, and the volume of $H_2SO_4(aq)$ should be a smaller number than its mass because the density of the solution is greater than 1.

EXERCISE 3.19

How many milliliters of liquid water should be produced by the combustion in abundant oxygen of 775 mL of octane, $C_8H_{18}(l)$? Assume that the volumes of both liquids are measured at 20 °C, where the densities are 0.7025 g/mL for octane and 0.9982 g/mL for water.

$$2\,C_8H_{18}(l) + 25\,O_2(g) \longrightarrow 16\,CO_2(g) + 18\,H_2O(l)$$

3.9 Limiting Reactants

Suppose we carry out a reaction by using numbers of moles of reactants that are in the same ratio as the stoichiometric coefficients in the balanced equation. In this case, we say that the reactants are in **stoichiometric proportions**, and we find that if the reaction goes to completion, the initial reactants are totally consumed. In practice, however, we often carry out reactions with a *limited* amount of one reactant and *plentiful* amounts of the others. We may do this, for example, because one reactant is much more expensive than the others.

The reactant that is completely consumed in a chemical reaction limits the amounts of products formed and is called the **limiting reactant**, or *limiting reagent*. (Reagent is a general term for a chemical.) In the combustion of octane described in Exercise 3.19,

$$2\ C_8H_{18}(l) + 25\ O_2(g) \longrightarrow 16\ CO_2(g) + 18\ H_2O(l)$$

if we allow 2 mol C_8H_{18} to react with 25 mol O_2, the reactants are in stoichiometric proportions. On the other hand, if we allow the 2 mol C_8H_{18} to burn in a plentiful supply of $O_2(g)$—more than 25 moles—then the C_8H_{18} is the limiting reactant. The octane is completely consumed and some unreacted O_2 remains; the O_2 is a reactant present in excess. Figure 3.10 is a microscopic view of the limiting and excess reactants in a reaction.

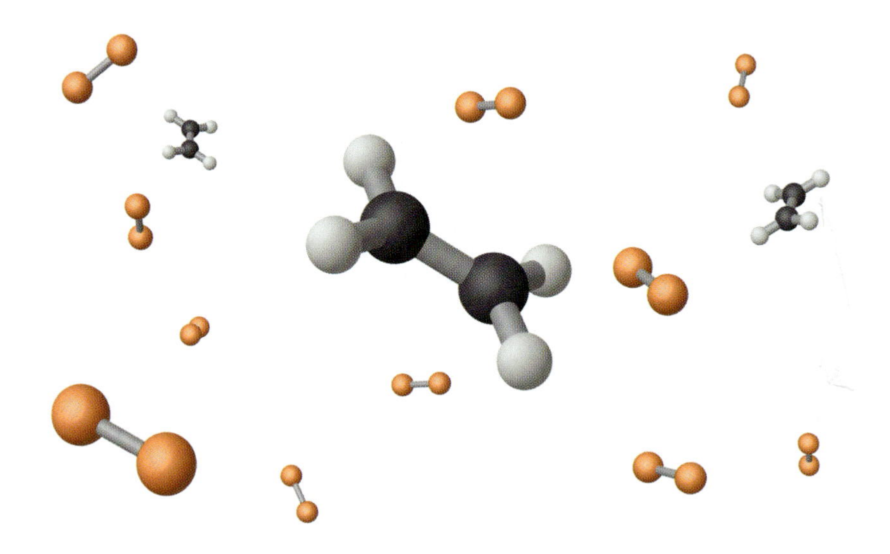

Figure 3.10
A molecular view of the reactants in the reaction between ethylene and bromine
Ethylene (shown by the black and gray model) and bromine (orange) react in a 1 : 1 mole ratio.

$$C_2H_4(g) + Br_2(g) \longrightarrow C_2H_4Br_2(g)$$

Here bromine molecules outnumber ethylene molecules. Thus bromine is present in excess, and ethylene is the limiting reactant.

A limiting reactant problem is much like the task illustrated in Figure 3.11, that of packaging a snack meal for airline passengers. Each package consists of a sandwich, two cookies, and an orange.

$$1 \text{ sandwich} + 2 \text{ cookies} + 1 \text{ orange} \longrightarrow 1 \text{ packaged snack}$$

The "stoichiometric proportions" of the "reactants" are 1 : 2 : 1. If we have 100 sandwiches, 200 cookies, and 100 oranges, we can prepare 100 snack meals and have nothing left over—the components needed for the packages are in "stoichiometric proportions." Suppose we have available 105 sandwiches, 202 cookies, and 107 oranges. How many of the packaged meals can we prepare? One way to answer the question is to consider each of the components separately and determine how many of the meals can be made from the available quantity of that component, assuming for the moment that there is enough of the other components available.

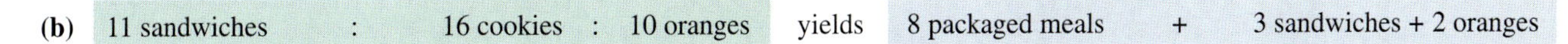

(b)
11 sandwiches : 16 cookies : 10 oranges
yields
8 packaged meals + 3 sandwiches + 2 oranges

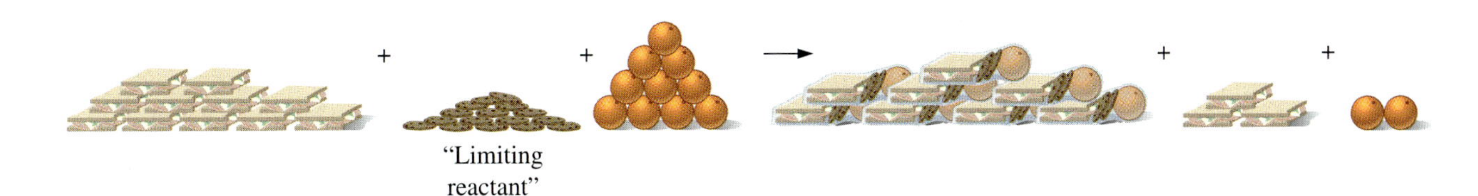

▲ FIGURE 3.11 An analogy to determining a limiting reactant
(a) To package the snack meal shown on the right requires sandwiches, cookies, and oranges in the ratio 1 : 2 : 1. Cookies are the "limiting reactant" for the situation shown in (b).

$$? \text{ Meals} = 105 \cancel{\text{ sandwiches}} \times \frac{1 \text{ meal}}{1 \cancel{\text{ sandwich}}} = 105 \text{ meals}$$

$$? \text{ Meals} = 202 \cancel{\text{ cookies}} \times \frac{1 \text{ meal}}{2 \cancel{\text{ cookies}}} = 101 \text{ meals}$$

$$? \text{ Meals} = 107 \cancel{\text{ oranges}} \times \frac{1 \text{ meal}}{1 \cancel{\text{ orange}}} = 107 \text{ meals}$$

Only one of these "answers" can be correct. It must always be the *smallest*. We have enough sandwiches to make 105 meals and enough oranges for 107 meals, but only enough cookies for 101 meals. When we have prepared 101 meals we will have used up all the cookies and have some sandwiches and oranges left over. The cookies are the "limiting reactant," and the sandwiches and oranges are in excess. A similar situation is shown in Figure 3.11(b).

Be sure to keep in mind that the limiting reactant is not necessarily the one present in smallest quantity. In the snack meal analogy, we had more cookies than sandwiches or oranges, but the situation required twice as many cookies as sandwiches and oranges, and we didn't have that many.

EXAMPLE 3.20

Magnesium nitride can be formed by the reaction of magnesium metal with nitrogen gas. **(a)** How many grams of magnesium nitride can be made through the reaction of 35.00 g of magnesium and 15.00 g of nitrogen? **(b)** How many grams of the excess reactant remain after the reaction?

SOLUTION

a. As usual, we must first write a balanced equation, and we can do this by using ideas from this chapter and Chapter 2.

$$3\,\text{Mg(s)} + \text{N}_2\text{(g)} \longrightarrow \text{Mg}_3\text{N}_2\text{(s)}$$

To identify the limiting reactant, we can do two simple calculations: the number of moles of $Mg_3N_2(s)$ produced based on two different assumptions.

Assuming that Mg is the limiting reactant and N_2 *is in excess*:

$$? \text{ mol } Mg_3N_2 = 35.00 \cancel{\text{ g Mg}} \times \frac{1 \cancel{\text{ mol Mg}}}{24.305 \cancel{\text{ g Mg}}} \times \frac{1 \text{ mol } Mg_3N_2}{3 \cancel{\text{ mol Mg}}}$$

$$= 0.4800 \text{ mol } Mg_3N_2$$

Assuming that N_2 *is the limiting reactant and Mg is in excess*:

$$? \text{ mol } Mg_3N_2 = 15.00 \cancel{\text{ g } N_2} \times \frac{1 \text{ mol } N_2}{28.013 \cancel{\text{ g } N_2}} = 0.5355 \text{ mol } Mg_3N_2$$

Because the amount of product in the first calculation (0.4800 mol Mg_3N_2) is smaller than that in the second (0.5355 mol Mg_3N_2), we know that magnesium is the limiting reactant. When 0.4800 mol Mg_3N_2 has been formed, the Mg is completely consumed and the reaction stops. The mass of this amount of Mg_3N_2 is

$$? \text{ g } Mg_3N_2 = 0.4800 \cancel{\text{ mol } Mg_3N_2} \times \frac{100.93 \text{ g } Mg_3N_2}{1 \cancel{\text{ mol } Mg_3N_2}} = 48.45 \text{ g } Mg_3N_2$$

b. Having found that the amount of product is 0.4800 mol Mg_3N_2, we can now calculate how much N_2 must have been consumed.

$$? \text{ g } N_2 = 0.4800 \cancel{\text{ mol } Mg_3N_2} \times \frac{1 \cancel{\text{ mol } N_2}}{1 \cancel{\text{ mol } Mg_3N_2}} \times \frac{28.013 \text{ g } N_2}{1 \cancel{\text{ mol } N_2}} = 13.45 \text{ g } N_2$$

The mass of N_2 present in excess is

$$15.00 \text{ g } N_{2\,\text{(initially)}} - 13.45 \text{ g } N_{2\,\text{(consumed)}} = 1.55 \text{ g } N_{2\,\text{(excess)}}$$

PROBLEM-SOLVING NOTE

Comparing the starting *masses*—35.00 g Mg and 15.00 g N_2—is of no use and can actually be misleading. We need to work on a molar basis and with stoichiometric factors as well. The substance present in greater mass, Mg, is actually the limiting reactant in this instance.

EXERCISE 3.20A

One way to produce hydrogen sulfide gas is by the reaction of iron(II) sulfide with hydrochloric acid.

$$FeS(s) + 2\,HCl(aq) \longrightarrow FeCl_2(aq) + H_2S(g)$$

If 10.2 g HCl is added to 13.2 g FeS, how many grams of H_2S can be formed? What is the mass of the excess reactant remaining?

EXERCISE 3.20B

A convenient laboratory source of hydrogen gas is the reaction of an aqueous hydrochloric acid solution with aluminum metal. An aqueous solution of aluminum chloride is the other product of the reaction. How many grams of hydrogen are produced in the reaction between 12.5 g of aluminum and 250.0 mL of an aqueous hydrochloric acid solution that is 25.6% HCl by mass and has a density of 1.13 g/mL?

● **ChemCDX** Limiting Reactant

3.10 Yields of Chemical Reactions

The calculated quantity of product in a reaction is called the **theoretical yield** of the reaction. In Example 3.20 the calculated quantity of product is 48.45 g Mg_3N_2. This is the theoretical yield of magnesium nitride. The *measured* mass of magnesium nitride formed in the reaction described in Example 3.20—the **actual yield**—might well be *less* than the theoretical yield of 48.45 g Mg_3N_2. Actual yields of chemical reactions are often less than theoretical yields for a variety of reasons (see Figure 3.12). The starting materials may not be pure, meaning that the actual quantities of reactants are less than what we weighed out. Some of the product may be left behind during the process of separating it from excess reactants. Side reactions may occur in addition to the main reaction, converting some of the original reactants into products other than the desired one. (In this reaction, for example, if there is oxygen present, some of the Mg is converted to by-product MgO, reducing the yield of Mg_3N_2.).

(a) (b) (c)

Figure 3.12
A reaction that has less than 100% yield

$$8\ Zn(s) + S_8(s) \longrightarrow 8\ ZnS(s)$$

The actual yield of ZnS(s) obtained (c) is less than that calculated for the starting mixture (a) for several reasons.

- Neither the powdered zinc nor the powdered sulfur is pure.
- The Zn(s) can combine with $O_2(g)$ in air to produce ZnO(s), and some of the sulfur burns in air to produce $SO_2(g)$.
- As suggested in (b), some of the product escapes from the reaction mixture as small lumps and as a fine dust.

Yields in an Organic Reaction

In some areas of chemistry—for example, quantitative analytical chemistry—only those reactions with 100% yield are useful. In other areas, 100% yield is rare, and improving the percent yield of a reaction can be an important consideration. Yield calculations are almost always important in synthesis reactions, especially those in organic chemistry. Consider, for example, the reaction by which ethanol is converted to diethyl ether (Section 2.10). The reaction is carried out in the presence of sulfuric acid, a fact that we indicate by writing the formula H_2SO_4 above the yield arrow.

$$\underset{\text{Ethanol}}{2\ CH_3CH_2OH} \xrightarrow{H_2SO_4} \underset{\text{Diethyl ether}}{CH_3CH_2OCH_2CH_3} + H_2O$$

On paper, the reaction seems to be straightforward, but in the laboratory there are several complications. There is an important side reaction: Some of the ethanol is converted to ethylene, a hydrocarbon with a double bond (described in Chapter 9).

$$\underset{\text{Ethanol}}{CH_3CH_2OH} \longrightarrow \underset{\text{Ethylene}}{CH_2{=}CH_2} + H_2O$$

Any molecules of ethanol that are converted to ethylene obviously cannot also form diethyl ether, and the yield of diethyl ether is accordingly diminished.

There are practical problems also. For example, the diethyl ether is purified by distilling it from the reaction mixture, and some will always remain in the distillation glassware. Also, some of the ethanol may distill with the ether, effectively removing it as a reactant. Even under the best of conditions, yields above 80–85% are difficult to achieve. Chemists often have to settle for 50%—and sometimes even less than that.

Organic reactions often have less than 100% yield. Here, reaction of isopropyl alcohol, $CH_3CHOHCH_3$, with potassium dichromate produces acetone, CH_3COCH_3, with chromium(III) compounds (the gray-green solid) as by-products. To obtain pure acetone the crude product would have to be "worked up" (purified), and some would be left behind at each stage of the process.

Yields usually are expressed as percentages. The **percent yield** is the ratio of the actual yield to the theoretical yield times 100%.

$$\text{Percent yield} = \frac{\text{actual yield}}{\text{theoretical yield}} \times 100\%$$

If the actual yield of Mg_3N_2 in Example 3.20 had been 47.87 g, the percent yield would have been

$$\text{Percent yield} = \frac{47.87 \text{ g}}{48.45 \text{ g}} \times 100\% = 98.80\%$$

EXAMPLE 3.21

The ester ethyl acetate (Section 2.10) is a solvent used as fingernail polish remover. What mass of acetic acid is needed to prepare 252 g ethyl acetate if the expected percent yield is 85.0%? Assume that the other reactant, ethanol, is present in excess. The equation for the reaction, carried out in the presence of H_2SO_4, is

$$\underset{\text{Acetic acid}}{CH_3COOH} + \underset{\text{Ethanol}}{HOCH_2CH_3} \xrightarrow{H_2SO_4} \underset{\text{Ethyl acetate}}{CH_3COOCH_2CH_3} + H_2O$$

SOLUTION

The mass we are given in this problem—252 g ethyl acetate—is an *actual* yield, but we must use the *theoretical* yield in the stoichiometric approach outlined in Figure 3.9 (box 2). Therefore, we must first calculate the theoretical yield of the reaction. The actual and theoretical yields are related through the percent yield.

$$\text{Percent yield} = \frac{\text{actual yield}}{\text{theoretical yield}} \times 100\%$$

We can solve this equation for the theoretical yield, and substitute known quantities for the actual and percent yields.

$$\text{Theoretical yield} = \frac{\text{actual yield}}{\text{percent yield}} \times 100\%$$

$$= \frac{252 \text{ g ethyl acetate}}{85.0 \cancel{\%}} \times 100 \cancel{\%} = 296 \text{ g ethyl acetate.}$$

PROBLEM-SOLVING NOTE
We need to form enough product so that 85.0% of it will be our desired 252 g ethyl acetate. The required amount of product must be more than 252 g, by the factor 100/85.0. This is why we base the calculation on 252 × 100/85.0 = 296 g ethyl acetate.

We can now determine the mass of acetic acid required to produce 296 g ethyl acetate.

$$? \text{ g } CH_3COOH = 296 \cancel{\text{ g } CH_3COOCH_2CH_3} \times \frac{1 \cancel{\text{ mol } CH_3COOCH_2CH_3}}{88.11 \cancel{\text{ g } CH_3COOCH_2CH_3}}$$

$$\times \frac{1 \cancel{\text{ mol } CH_3COOH}}{1 \cancel{\text{ mol } CH_3COOCH_2CH_3}} \times \frac{60.05 \text{ g } CH_3COOH}{1 \cancel{\text{ mol } CH_3COOH}}$$

$$= 202 \text{ g } CH_3COOH$$

EXERCISE 3.21A

Isopentyl acetate is the main component of banana flavoring. Calculate the theoretical yield of isopentyl acetate that can be made from 20.0 g isopentyl alcohol and 25.0 g acetic acid.

$$\underset{\text{Acetic acid}}{CH_3COOH} + \underset{\text{Isopentyl alcohol}}{HOCH_2CH_2CH(CH_3)_2} \longrightarrow \underset{\text{Isopentyl acetate}}{CH_3COOCH_2CH_2CH(CH_3)_2} + H_2O$$

If the percent yield of the reaction is 90.0%, what is the actual yield of isopentyl acetate?

EXERCISE 3.21B

How many grams of isopentyl alcohol are needed to make 433 g isopentyl acetate in the reaction described in Exercise 3.21A if the expected yield is 78.5%? Assume the acetic acid is in excess.

EXAMPLE 3.22—A Conceptual Example

What is the maximum yield of CO(g) obtainable from 725 g of $C_6H_{14}(l)$, regardless of the reaction(s) used, assuming no other carbon-containing reactant?

SOLUTION

If the maximum yield is independent of the reaction(s) used, then we ought to be able to determine this quantity without writing any chemical equations and without using stoichiometric factors based on chemical equations. While at first this may not seem feasible, recall that in Section 3.6 we dealt with a similar idea. There we needed to relate the mass of CO_2 produced in combustion analysis to the mass of a carbon-containing compound from which it formed. According to the law of conservation of mass, all the C atoms in the C_6H_{14} have to be accounted for, and the maximum yield results if all the C atoms end up in CO.

Let's rephrase the original question as: What mass of CO contains the same number of C atoms as 725 g C_6H_{14}? We can start the calculation in the manner outlined in Figure 3.4 for converting the number of grams to the number of moles of C_6H_{14}, and we can end it in the manner outlined in Figure 3.3 for converting the number of moles to the number of grams of CO. The critical link between these two "ends" of the calculation are the factors that relate moles of C to moles of C_6H_{14} on the one hand and moles of CO to moles of C on the other. These are the factors shown in red in the following equation.

$$? \text{ g CO} = 725 \cancel{\text{g } C_6H_{14}} \times \frac{1 \cancel{\text{mol } C_6H_{14}}}{86.18 \cancel{\text{g } C_6H_{14}}} \times \frac{6 \cancel{\text{mol C}}}{1 \cancel{\text{mol } C_6H_{14}}} \times \frac{1 \cancel{\text{mol CO}}}{1 \cancel{\text{mol C}}} \times \frac{28.01 \text{ g CO}}{1 \cancel{\text{mol CO}}}$$

$$= 1.41 \times 10^3 \text{ g CO}$$

EXERCISE 3.22

What is the maximum yield of the important commercial fertilizer ammonium hydrogen phosphate that can be obtained per kilogram of phosphoric acid?

PROBLEM-SOLVING NOTE

Can you write the formulas for these two compounds? If not, review Tables 2.4 and 2.5.

3.11 Solutions and Solution Stoichiometry

Because a reaction involving solids generally is quite slow, chemists usually dissolve the solids in a liquid and carry out the reaction in a liquid medium. Most of the reactions in our bodies occur in aqueous solutions.

In Chapter 1, we noted that a solution is a homogeneous mixture of two or more substances. A solution of sugar in water does not consist of tiny particles of solid sugar dispersed among droplets of liquid water. Rather, individual sugar molecules are randomly distributed among water molecules in a uniform liquid medium. A solution is *homogeneous* right down to the molecular level. The composition and physical and chemical properties are identical in all portions of a solution.

The components of a solution are the **solute**(s)—the substance(s) being dissolved—and the **solvent**—the substance doing the dissolving. The solutes are usually the components present in lesser amounts, and the solvent is usually present in the greatest amount. There are many common solvents. *Hexane* dissolves grease. *Ethanol* is the solvent for many medicines. *Isopentyl acetate*, a component of banana oil, is a solvent for the glue used in making model airplanes. *Water* is the most familiar solvent, dissolving as it does many common substances such as sugar, salt, and ethanol. Solutions in which water is the solvent are called *aqueous* solutions.

The *concentration* of a solution refers to the quantity of solute in a given quantity of solvent or of solution. A *dilute* solution is one that contains relatively little solute in a large quantity of solvent. A *concentrated* solution contains a relatively large amount of solute in a given quantity of solvent. These terms are imprecise, however. For example, a dilute sugar solution tastes faintly sweet and a concentrated solution has a sickeningly sweet taste, but the terms "dilute" and "concentrated" do not in themselves tell us the precise proportions of sugar and water.

For commercially available acids and bases, the term *concentrated* generally signifies the highest concentration, usually expressed as a mass percent, that is commonly available. Commercial concentrated hydrochloric acid is about 38% HCl by mass; the rest is water. Commercial concentrated sulfuric acid contains 93–98% H_2SO_4 by mass, with the remainder water.

Molar Concentration

Mass percent composition, as in 38% HCl by mass, is one way to describe the concentration of a solution, but there are several other ways as well. Chemists generally prefer to work with the concentration unit *molarity* for two reasons.

- Substances enter into chemical reactions according to certain *molar* ratios.
- Volumes of solutions are more convenient to measure than their masses.

The **molarity (M)** or **molar concentration** is the amount of solute, in moles, per liter of solution.*

$$\text{Molarity (M)} = \frac{\text{moles of solute}}{\text{liters of solution}}$$

The molarity of a solution made by dissolving 3.50 mol NaCl in enough water to produce 2.00 L of solution is

$$\text{Molarity} = \frac{\text{moles of solute}}{\text{liters of solution}} = \frac{3.50 \text{ mol NaCl}}{2.00 \text{ L solution}} = 1.75 \text{ M NaCl}$$

We read "1.75 M NaCl" as "1.75 molar NaCl."

Keep in mind that molarity signifies moles of solute per liter of *solution*, not per liter of solvent. To make 0.01000 M $KMnO_4$, we therefore weigh out 1.580 g $KMnO_4$ and add it to a 1.000-L flask partially filled with water. After the solid has completely dissolved, we add more water to bring the volume up to the mark that indicates a solution volume of 1.000 L. A flask designed to hold a precise volume of solution is called a *volumetric flask*. Figure 3.13 depicts a 1.000-L volumetric flask and its use to prepare the 0.01000 M $KMnO_4$ solution just described. Volumetric flasks of different volumes are available. They can be used with proportionate quantities of solute when preparing solutions.

*Recall from Chapter 1 that a liter is the same as a cubic decimeter: 1 L = 1 dm^3. The derived SI unit for molarity is moles per cubic decimeter (mol/dm^3). The liter isn't an authorized SI unit, but the unit mol/L is still widely used.

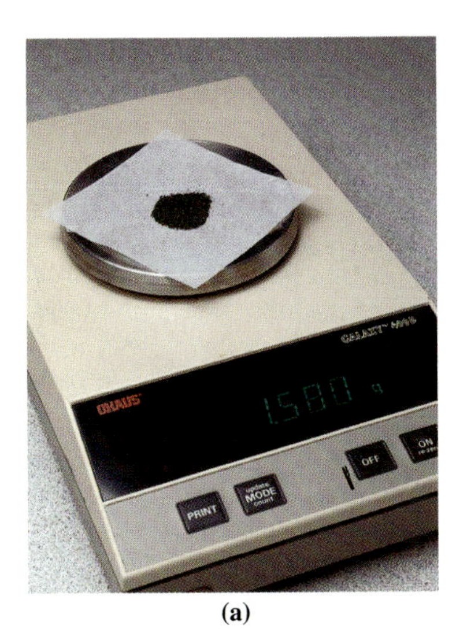

(a)

(b)

(c)

◀ **Figure 3.13**
Preparation of 0.01000 M $KMnO_4$
In a step not shown, the balance is set to zero (tared) with just the weighing paper present. (a) The sample of $KMnO_4$ has a mass of 1.580 g (0.01000 mol). (b) The $KMnO_4$ is dissolved in water in the partially filled 1.000-L volumetric flask. More water is added (c), and the solution is thoroughly mixed. Finally, the flask is filled to the mark by adding a small quantity of water one drop at a time.

EXAMPLE 3.23

What is the molarity of a solution in which 333 g potassium hydrogen carbonate is dissolved in enough water to make 10.0 L of solution?

SOLUTION

First, let's prepare the setup that converts mass to number of moles of $KHCO_3$.

$$333\ \cancel{\text{g KHCO}_3} \times \frac{1\ \text{mole KHCO}_3}{100.1\ \cancel{\text{g KHCO}_3}}$$

Now, without solving this expression, let's use it as the *numerator* in the defining equation for molarity. The solution volume, 10.0 L, is the *denominator*.

$$\text{Molarity} = \frac{333\ \cancel{\text{g KHCO}_3} \times \dfrac{1\ \text{mole KHCO}_3}{100.1\ \cancel{\text{g KHCO}_3}}}{10.0\ \text{L solution}} = 0.333\ \text{M KHCO}_3$$

EXERCISE 3.23A

Calculate the molarity of solute in each of the following solutions.

a. 3.00 mol KI in 2.39 L of solution
b. 0.522 g HCl in 0.592 L of solution
c. 2.69 g $C_{12}H_{22}O_{11}$ in 225 mL of solution

EXERCISE 3.23B

Calculate the molarity of (**a**) glucose, $C_6H_{12}O_6$, in a solution containing 126 mg of glucose per 100.0 mL, (**b**) ethanol, CH_3CH_2OH, in a solution containing 10.5 mL of ethanol ($d = 0.789$ g/mL) in 25.0 mL of solution, and (**c**) urea, $CO(NH_2)_2$, in a solution of urea whose concentration is expressed as 9.5 mg N/mL solution.

PROBLEM-SOLVING NOTE
In (c), you need to determine the mass of urea having a nitrogen content of 9.5 mg. You will then have the concentration in mg urea/mL solution, which you must convert to mol urea/L solution.

At times we need to determine the number of *moles of solute* required to prepare a given volume of solution of a specified molarity. Other times we want to determine the *volume of solution* of a specified molarity we need to obtain a given number of moles of solute. We illustrate both of these types of calculations in Example 3.24, where the central conversion factors are those derived from the following equivalency for 6.68 M NaOH.

$$6.68 \text{ mol NaOH} \approx 1 \text{ L soln}$$

When we recast this equivalency into the usual *two* conversion factors, we find that one is simply the definition of the molarity and the other is its inverse.

$$\frac{6.68 \text{ mol NaOH}}{1 \text{ L soln}} \quad \text{and} \quad \frac{1 \text{ L soln}}{6.68 \text{ mol NaOH}}$$

EXAMPLE 3.24

We want to prepare a 6.68 molar solution of NaOH, that is, 6.68 M NaOH.

a. How many moles of NaOH are required to prepare 0.500 L of 6.68 M NaOH?
b. How many liters of 6.68 M NaOH can we prepare with 2.35 kg NaOH?

SOLUTION

a. This calculation requires only a one-step conversion from liters of solution to moles of NaOH, with the molarity of the solution as the conversion factor.

$$? \text{ mol NaOH} = 0.500 \text{ L soln} \times \frac{6.68 \text{ mol NaOH}}{1 \text{ L soln}} = 3.34 \text{ mol NaOH}$$

b. The central conversion factors in this calculation are the *inverse* of the molar mass of NaOH to convert from grams to moles of NaOH and the *inverse* of the molarity—1 L soln/6.68 mol NaOH—to convert from moles of NaOH to liters of solution. We must also convert from kilograms to grams of NaOH. In all, the required conversions are

$$\text{kg NaOH} \longrightarrow \text{g NaOH} \longrightarrow \text{mol NaOH} \longrightarrow \text{L soln}$$

which are set up as follows.

$$? \text{ L soln} = 2.35 \text{ kg NaOH} \times \frac{1000 \text{ g NaOH}}{1 \text{ kg NaOH}} \times \frac{1 \text{ mol NaOH}}{40.00 \text{ g NaOH}} \times \frac{1 \text{ L soln}}{6.68 \text{ mol NaOH}}$$

$$= 8.79 \text{ L soln}$$

EXERCISE 3.24A

How many grams of potassium hydroxide are required to prepare each of the following solutions?

a. 2.00 L of 6.00 M KOH **b.** 10.0 mL of 0.100 M KOH
c. 35.0 mL of 2.50 M KOH

● **ChemCDX** Solution Formation from a Solid

EXERCISE 3.24B

How many mL of 1-butanol, $CH_3CH_2CH_2CH_2OH$, ($d = 0.810$ g/mL) are required to prepare 725 mL of a 0.350 molar solution of this solute?

Labels on bottles of stock solutions of acids and bases often indicate concentrations only in percent solute by mass. If we want to know the molarity of such a solution, we must either know or measure the density of the solution. Density provides the conversion factor from mass of solution (g) to volume of solution (mL). We need to convert from milliliters to liters of solution, and we also need conversion factors based on mass percent and molar mass. We illustrate how all of these factors enter into the calculation in Example 3.25.

EXAMPLE 3.25

The label of a stock bottle of aqueous ammonia indicates that the solution is 28.0% NH_3 by mass and has a density of 0.898 g/mL. Calculate the molarity of the solution.

SOLUTION

We will find it most convenient to base the calculation on a 1.00-L volume of solution. When we have found the number of moles of NH_3 in this 1.00-L solution we will have found the molarity. The factors cited above enter into the calculation as follows.

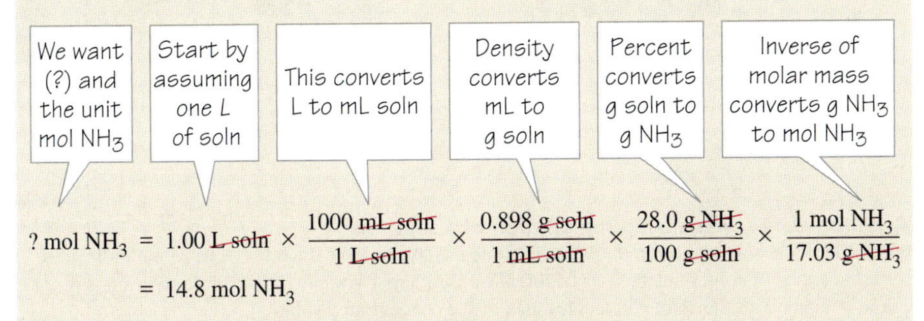

$$? \text{ mol NH}_3 = 1.00 \text{ L soln} \times \frac{1000 \text{ mL soln}}{1 \text{ L soln}} \times \frac{0.898 \text{ g soln}}{1 \text{ mL soln}} \times \frac{28.0 \text{ g NH}_3}{100 \text{ g soln}} \times \frac{1 \text{ mol NH}_3}{17.03 \text{ g NH}_3}$$

$$= 14.8 \text{ mol NH}_3$$

Because the 14.8 mol NH_3 is present in 1.00 L, the solution is 14.8 molar. That is,

$$\text{Molarity} = \frac{14.8 \text{ mol NH}_3}{1.00 \text{ L soln}} = 14.8 \text{ M NH}_3$$

EXERCISE 3.25A

A stock bottle of aqueous formic acid indicates that the solution is 90.0% HCOOH by mass and has a density of 1.20 g/mL. Calculate the molarity of the solution.

EXERCISE 3.25B

A concentrated solution of perchloric acid, $HClO_4$, is 11.7 molar and has a density of 1.67 g/mL. What is the mass percent perchloric acid in this solution?

PROBLEM-SOLVING NOTE

What is the mass of 1.00 L of the perchloric acid solution? What is the mass of perchloric acid in the 1.00 L of solution?

Dilution of Solutions

We can generally find concentrated solutions, often ones that are commercially available, in a chemical storeroom. When we need solutions of lower concentrations, we can prepare them from the concentrated solutions. The process of preparing a more dilute solution by adding solvent to a more concentrated one is called **dilution**. The following basic principle of dilution is suggested by Figure 3.14.

Addition of solvent does not change the amount of solute in a solution but does change the solution concentration.

Suppose we let M_{conc} and M_{dil} represent the molar concentrations and V_{conc} and V_{dil}, the volumes of a concentrated and a diluted solution. Because the product of a molarity (mol/L) and a volume (L) is the number of moles of solute in a solution, and *because the amount of solute doesn't change during dilution*, we can write the following simple equation.

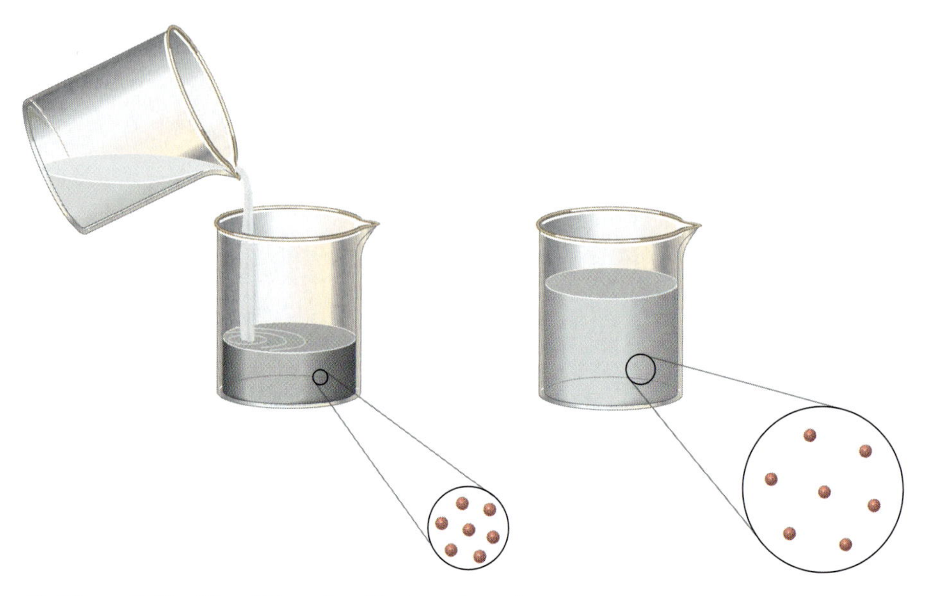

▲ **Figure 3.14 Visualizing the dilution of a solution**
When additional solvent is added to a concentrated solution (left), the number of solute molecules *per unit volume* of solution decreases (right). That is, a larger volume of the dilute solution is required to contain all the solute molecules than of the concentrated solution.

$$M_{conc} \times V_{conc} = \text{moles of solute} = M_{dil} \times V_{dil} \quad \text{and} \quad M_{conc} \times V_{conc} = M_{dil} \times V_{dil}$$

In Example 3.26 we will consider a problem involving dilution in two ways: first by considering it as an illustration of the principle of dilution, and then by applying the dilution equation. Of course, we will get the same answer either way.

EXAMPLE 3.26

How many milliliters of a 2.00 M $CuSO_4$ stock solution are needed to prepare 0.250 L of 0.400 M $CuSO_4$?

SOLUTION

Applying the principle of dilution:
The key is to note that all the solute in the unknown volume of the stock solution appears in the 0.250 L of 0.400 M $CuSO_4$. First, let's calculate that amount of solute.

$$? \text{ mol } CuSO_4 = 0.250 \text{ L} \times \frac{0.400 \text{ mol } CuSO_4}{1 \text{ L}} = 0.100 \text{ mol } CuSO_4$$

Now we need to answer the question, "What volume of 2.00 M $CuSO_4$ contains 0.100 mol $CuSO_4$?" In doing so, we will have answered the original question.

$$? \text{ mL} = 0.100 \text{ mol } CuSO_4 \times \frac{1 \text{ L}}{2.00 \text{ mol } CuSO_4} \times \frac{1000 \text{ mL}}{1 \text{ L}} = 50.0 \text{ mL}$$

Of course, we could have done all of this in a single setup.

$$? \text{ mL} = 0.250 \text{ L} \times \frac{0.400 \text{ mol } CuSO_4}{1 \text{ L}} \times \frac{1 \text{ L}}{2.00 \text{ mol } CuSO_4} \times \frac{1000 \text{ mL}}{1 \text{ L}} = 50.0 \text{ mL}$$

Using the dilution equation:
First we can identify the terms we need for the equation.

$$M_{conc} = 2.00\text{ M};\ V_{conc} = ?;\ M_{dil} = 0.400\text{ M};\ V_{dil} = 0.250\text{ L}$$

Then we can substitute these terms into the equation.

$$M_{conc} \times V_{conc} = M_{dil} \times V_{dil}$$
$$2.00\text{ M} \times V_{conc} = 0.400\text{ M} \times 0.250\text{ L}$$
$$V_{conc} = \frac{0.400\ \cancel{\text{M}}}{2.00\ \cancel{\text{M}}} \times 0.250\text{ L} = 0.0500\text{ L}$$
$$V_{conc} = 0.0500\ \cancel{\text{L}} \times \frac{1000\text{ mL}}{1\ \cancel{\text{L}}} = 50.0\text{ mL}$$

To make the dilute solution, we measure out 50.0 mL of 2.00 M $CuSO_4$ and add it to enough water to make 0.250 L of solution, as illustrated in Figure 3.15.

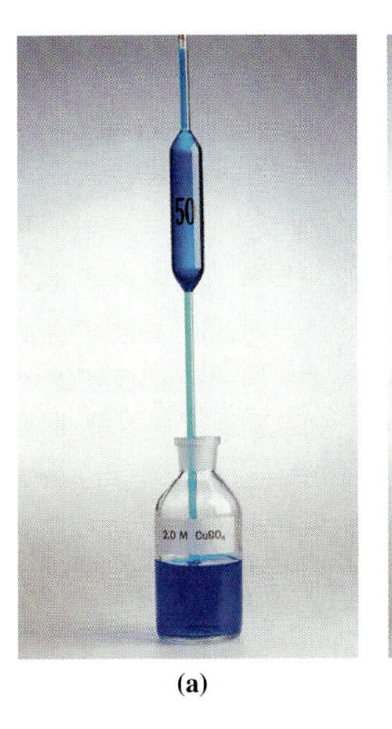

(a)

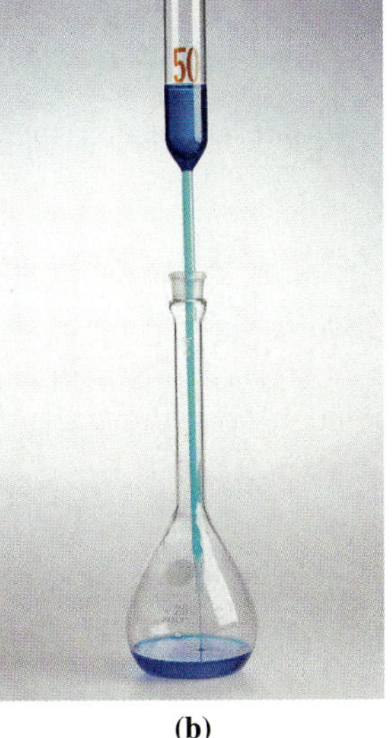

(b)

(c)

◀ **Figure 3.15 Dilution of a copper(II) sulfate solution: Example 3.26 illustrated.**
(a) The pipet is being filled with 50.0 mL of 2.00 M $CuSO_4$. The amount of $CuSO_4$ in the filled pipet will be 0.100 mol. (b) The 50.0 mL of 2.00 M $CuSO_4$ solution is transferred to a 250.0-mL volumetric flask, water is added, and the solution is thoroughly mixed. (c) Finally, the flask is filled to the mark by adding the remaining water dropwise.

EXERCISE 3.26A

How many milliliters of a 10.15 M NaOH stock solution are needed to prepare 15.0 L of 0.315 M NaOH?

EXERCISE 3.26B

How many milliliters of a 5.15 M CH_3OH stock solution are needed to prepare 375 mL of a solution having 7.50 mg of methanol per mL of solution?

● **ChemCDX** Solution Formation by Dilution

Solutions in Chemical Reactions

Molarity concentration provides an additional tool in stoichiometric calculations based on the chemical equation. Specifically, it gives us two conversion factors: one to convert from liters of solution to moles of solute and another to convert from moles of solute to liters of solution. We use these conversion factors in the early and/or late stages of the setup of a stoichiometric calculation. The heart of the calculation, however, is still a stoichiometric factor derived from the chemical equation. We illustrate these points in Example 3.27, and we will explore reaction stoichiometry in solutions in greater detail in Chapter 4.

EXAMPLE 3.27

A chemical reaction familiar to geologists is that used to identify limestone. The reaction of hydrochloric acid with limestone, which is largely calcium carbonate, is seen through an effervescence—a bubbling due to the liberation of carbon dioxide gas.

$$CaCO_3(s) + 2\,HCl(aq) \longrightarrow CaCl_2(aq) + H_2O(l) + CO_2(g)$$

a. How many moles of $CaCO_3(s)$ are consumed in a reaction with 225 mL of 3.25 M HCl?

b. How many milliliters of 3.25 M HCl are consumed in a reaction with excess $CaCO_3$ to produce 1.00 mol CO_2?

c. How many moles of $CO_2(g)$ are produced in the reaction of 175 mL of 3.25 M HCl with 45.0 g $CaCO_3(s)$?

SOLUTION

a. To relate the amount of $CaCO_3$ to that of HCl, we need to express the amount of HCl in moles and multiply by the stoichiometric factor: 1 mol $CaCO_3$/2 mol HCl. To get the number of moles of HCl, all we have to do is multiply the volume of HCl solution (in liters) by its molarity. Thus, in this problem we use molarity as a conversion factor before introducing the stoichiometric factor.

$$? \text{ mol } CaCO_3 = 225 \text{ mL soln} \times \frac{1 \text{ L soln}}{1000 \text{ mL soln}} \times \frac{3.25 \text{ mol HCl}}{1 \text{ L soln}} \times \frac{1 \text{ mol } CaCO_3}{2 \text{ mol HCl}}$$

$$= 0.366 \text{ mol } CaCO_3$$

b. Here we can use a stoichiometric factor from the equation to establish the number of moles of HCl required. Then we use a conversion factor that is the *inverse* of the molarity.

$$? \text{ mL soln} = 1.00 \text{ mol } CO_2 \times \frac{2 \text{ mol HCl}}{1 \text{ mol } CO_2} \times \frac{1 \text{ L soln}}{3.25 \text{ mol HCl}} \times \frac{1000 \text{ mL soln}}{1 \text{ L soln}}$$

$$= 615 \text{ mL soln}$$

c. We have information about both reactants, but don't know which one is in excess. We can proceed as in Example 3.20 to determine the limiting reactant, and then base the calculation of the amount of product on that reactant.

Assuming $CaCO_3$ is the limiting reactant:

$$? \text{ mol } CO_2 = 45.0 \text{ g } CaCO_3 \times \frac{1 \text{ mol } CaCO_3}{100.1 \text{ g } CaCO_3} \times \frac{1 \text{ mol } CO_2}{1 \text{ mol } CaCO_3}$$

$$= 0.450 \text{ mol } CO_2$$

Assuming HCl is the limiting reactant:

$$? \text{ mol } CO_2 = 175 \text{ mL soln} \times \frac{1 \text{ L soln}}{1000 \text{ mL soln}} \times \frac{3.25 \text{ mol HCl}}{1 \text{ L soln}} \times \frac{1 \text{ mol } CO_2}{2 \text{ mol HCl}}$$

$$= 0.284 \text{ mol } CO_2$$

We see that HCl is the limiting reactant and the amount of $CO_2(g)$ produced is 0.284 mol.

EXERCISE 3.27A

How many milliliters of 0.100 M $AgNO_3(aq)$ are required to react completely with 750.0 mL of 0.0250 M $Na_2CrO_4(aq)$?

$$2\ AgNO_3(aq) + Na_2CrO_4(aq) \longrightarrow Ag_2CrO_4(s) + 2\ NaNO_3(aq)$$

EXERCISE 3.27B

In a reaction similar to that of baking soda in the neutralization of stomach acid, 175 mL of 1.55 M $NaHCO_3$ is added to 235 mL of 1.22 M HCl.

$$NaHCO_3(aq) + HCl(aq) \longrightarrow NaCl(aq) + H_2O(l) + CO_2(g)$$

a. How many grams of CO_2 are liberated?
b. What is the molarity of the NaCl(aq) produced? Assume that the solution volume is 175 mL + 235 mL = 410 mL.

Summary

Molecular and formula masses relate to the masses of molecules and formula units. Molecular mass applies to molecular compounds, but only formula mass is appropriate for ionic compounds.

A mole is an amount of substance containing a number of elementary entities equal to the number of atoms in exactly 12 g of carbon-12. This number, called Avogadro's number, is $N_A = 6.022 \times 10^{23}$. The mass, in grams, of one mole of substance is called the molar mass and is numerically equal to an atomic, molecular, or formula mass. Conversions between number of moles and number of grams of a substance require molar mass as a conversion factor; conversions between number of grams and number of moles require the inverse of molar mass. Other calculations involving volume, density, number of atoms or molecules, and so on may be required prior to or following the gram/mole conversion. That is,

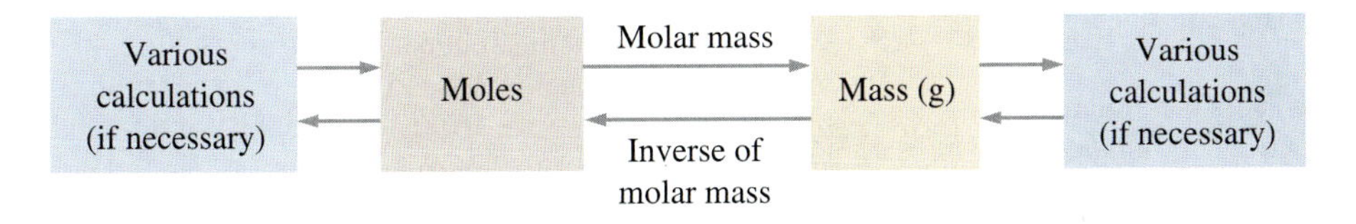

Formulas and molar masses can be used to calculate the mass percent compositions of compounds. And conversely, an *empirical* formula can be established from the mass percent composition of a compound; to establish a *molecular* formula, we must also know the molecular mass. The mass percents of carbon, hydrogen, and oxygen in organic compounds can be determined by combustion analysis.

A chemical equation uses symbols and formulas for the elements and/or compounds involved in a reaction. Stoichiometric coefficients are used in the equation to reflect that a chemical reaction obeys the law of conservation of mass.

Calculations concerning reactions use conversion factors, called stoichiometric factors, that are based on stoichiometric coefficients in the balanced equation. Also required are molar masses and often other quantities such as volume, density, and percent composition. The general format of a reaction stoichiometry calculation is

Key Terms

actual yield (3.10)
Avogadro's number, N_A (3.2)
chemical equation (3.7)
dilution (3.11)
formula mass (3.1)
limiting reactant (3.9)
mass percent composition (3.4)
molar concentration (3.11)
molarity, M (3.11)
molar mass (3.3)
mole (3.2)
molecular mass (3.1)
percent yield (3.10)
product (3.7)
reactant (3.7)
solute (3.11)
solvent (3.11)
stoichiometric coefficient (3.7)
stoichiometric factor (3.8)
stoichiometric proportions (3.9)
stoichiometry (page 82)
theoretical yield (3.10)

Various calculations (if necessary) ⇄ Moles A ⇄ Moles B ⇄ Various calculations (if necessary)

$$\text{Moles A} \xrightarrow{\frac{\text{no. mol B}}{\text{no. mol A}}} \text{Moles B} \qquad \text{Moles B} \xrightarrow{\frac{\text{no. mol A}}{\text{no. mol B}}} \text{Moles A}$$

The limiting reactant determines the amounts of products in a reaction. The calculated quantity of a product is the theoretical yield of a reaction. The quantity obtained, called the actual yield, is often less. It is commonly expressed as a percentage of the theoretical yield, known as the percent yield. The relationship involving theoretical, actual, and percent yield is

$$\text{Percent yield} = \frac{\text{actual yield}}{\text{theoretical yield}} \times 100\%$$

The molarity of a solution is the number of moles of solute per liter of solution. Common calculations include relating an amount of solute to solution volume and molarity. Solutions of a desired concentration are often prepared from more concentrated solutions by dilution. The principle of dilution is that the volume of a solution increases as it is diluted, but the amount of solute is unchanged. As a consequence, the amount of solute per unit volume—the concentration—decreases. A useful equation describing the process of dilution is

$$M_{conc} \times V_{conc} = M_{dil} \times V_{dil}$$

In addition to other conversion factors, stoichiometric calculations for reactions in solution use molarity or its inverse as a conversion factor.

Review Questions

1. Explain the difference between the *atomic mass* of oxygen and the *molecular mass* of oxygen. Explain how each is determined from data in the periodic table.
2. What is Avogadro's number, and how is it related to the quantity called one mole?
3. How many oxygen molecules and how many oxygen atoms are in 1.00 mol O_2?
4. How many calcium ions and how many chloride ions are in 1.00 mol $CaCl_2$?
5. What is the molecular mass, and what is the molar mass of carbon dioxide? Explain how each is determined from the formula, CO_2.
6. Describe how the mass percent composition of a compound is established from its formula.
7. Describe how the empirical formula of a compound is determined from its mass percent composition.
8. What are the empirical formulas of the compounds with the following molecular formulas?
   (a) H_2O_2 **(b)** C_8H_{16} **(c)** $C_{10}H_8$ **(d)** $C_6H_{16}O$
9. Describe how the empirical formula of a compound that contains carbon, hydrogen, and oxygen is determined by combustion analysis.
10. What is the purpose of balancing a chemical equation?
11. Explain the meaning of the equation

$$CH_4 + 2\,O_2 \longrightarrow CO_2 + 2\,H_2O$$

at the molecular level. Interpret the equation in terms of moles. State the mass relationships conveyed by the equation.

12. Translate the following chemical equations into words:
    (a) $2\,H_2(g) + O_2(g) \rightarrow 2\,H_2O(l)$
    (b) $2\,KClO_3(s) \rightarrow 2\,KCl(s) + 3\,O_2(g)$
    (c) $2\,Al(s) + 6\,HCl(aq) \rightarrow 2\,AlCl_3(aq) + 3\,H_2(g)$
13. Write balanced chemical equations to represent **(a)** the reaction of solid magnesium and gaseous oxygen to form solid magnesium oxide, **(b)** the decomposition of solid ammonium nitrate into dinitrogen monoxide gas and liquid water, and **(c)** the combustion of liquid heptane, C_7H_{16}, in oxygen gas to produce carbon dioxide gas and liquid water as the sole products.
14. What is meant by the limiting reactant in a chemical reaction? Under what circumstances might we say that a reaction has two limiting reactants? Explain.
15. Why are the actual yields of products often less than the theoretical yields? Can actual yields ever be greater than theoretical yields? Explain.
16. Define each of the following terms.
    (a) solution **(d)** molarity
    (b) solvent **(e)** dilute solution
    (c) solute **(f)** concentrated solution

17. Is the volume of a solution changed by dilution? Is the concentration? Is the number of moles of solute? Explain.

18. Some handbooks list the concentration of solute in an aqueous solution in the unit, g solute/100 mL H_2O, and others use g solute/100 mL solution. Are these concentration units the same? Explain.

19. For some applications, solution concentrations are expressed as grams of solute per liter of solution, and for some, as milligrams of solute per milliliter of solution. How are these two concentrations units related? Explain.

20. Explain why a stoichiometric calculation gives the same result, regardless of the coefficients used in the equation for a chemical reaction, as long as the equation is balanced.

Problems

Molecular and Formula Masses

21. Calculate the molecular or formula mass of each of the following. Specify whether the answer is a molecular or formula mass.

(a) C_6H_5Br **(d)** potassium dichromate

(b) $Ca(HCO_3)_2$ **(e)** $Al_2(SO_4)_3 \cdot 18H_2O$

(c) phosphoric acid **(f)** $Na_2[Pt(CN)_4] \cdot 3H_2O$

22. Calculate the molecular or formula mass of each of the following. Specify whether the answer is a molecular or formula mass.

(a) $C_2H_5NO_2$ **(d)** $K_3[Co(NO_2)_6]$

(b) $Na_2S_2O_3$ **(e)** chlorous acid

(c) $Fe(NO_3)_3 \cdot 9H_2O$ **(f)** ammonium hydrogen phosphate

23. Calculate the molecular mass of **(a)** the substance fensulfothion, an insecticide, with the condensed structural formula $(CH_3CH_2O)_2PSOC_6H_4SOOCH_3$, and **(b)** methyl salicylate represented at the upper right. (*Hint:* Refer to the color code in Figure 2.6.)

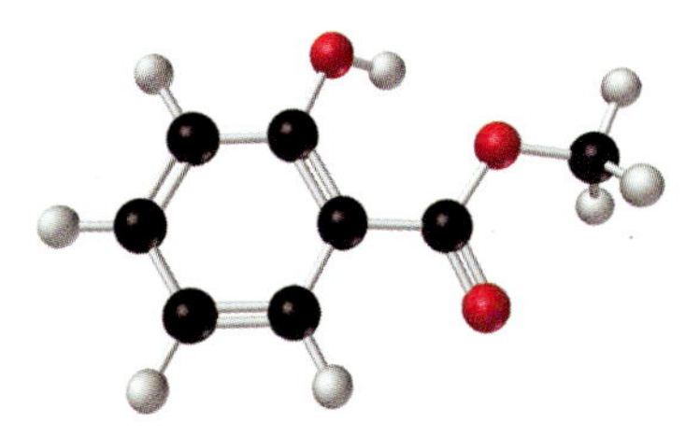

Methyl salicylate

24. Calculate the molecular mass of **(a)** the substance trimethobenzamide, used to suppress nausea and vomiting, which has the condensed structural formula $(CH_3)_2NCH_2CH_2OC_6H_4CH_2NHCOC_6H_2(OCH_3)_3$, and **(b)** the substance represented by the space-filling model below. (*Hint:* Refer to the color code in Figure 2.6.)

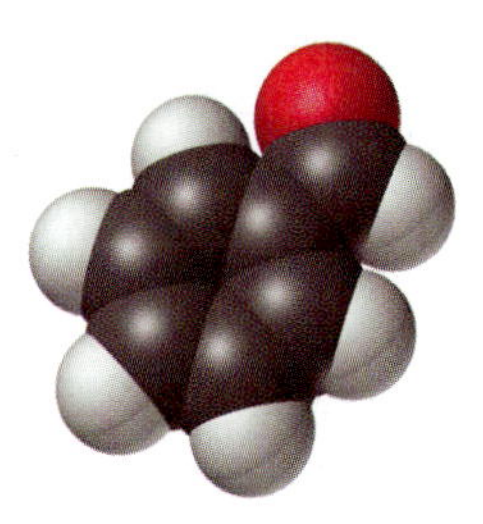

Avogadro's Number and Molar Masses

25. Calculate the mass, in grams, of **(a)** 1.12 mol CaH_2; **(b)** 0.250 mol $(CH_3)_2CHCOOH$; and **(c)** 0.158 mol of iodine pentafluoride.

26. Calculate the mass, in grams, of **(a)** 4.61 mol $AlCl_3$; **(b)** 0.314 mol $HOCH_2(CH_2)_4CH_2OH$; and **(c)** 0.615 mol of chromium(III) oxide.

27. Calculate the amount in moles of **(a)** 98.6 g HNO_3; **(b)** 16.3 g of sulfur hexafluoride; and **(c)** 35.6 g of iron(III) sulfate heptahydrate; and **(d)** 218 mg methanol, CH_3OH.

28. Calculate the amount in moles of **(a)** 24.2 g H_2SO_4; **(b)** 198.2 g dinitrogen pentoxide; **(c)** 65.2 g lithium nitrate trihydrate; **(d)** 746 mg hexane, C_6H_{14}.

29. Calculate **(a)** the number of molecules in 4.68 mol H_2O; **(b)** the number of sulfate ions in 86.2 g of aluminum sulfate; and **(c)** the average mass of an atom of copper.

30. Calculate **(a)** the total number of ions in 1.75 mol of magnesium chloride; **(b)** the number of molecules in 37.0 mL of ethanol, CH_3CH_2OH, ($d = 0.789$ g/mL); **(c)** the average mass of a glucose molecule, $C_6H_{12}O_6$.

31. *Without doing detailed calculations*, determine which of the following samples contains the greatest number of atoms, and explain your choice. **(a)** 1.0 mol N_2; **(b)** 50.0 g Na; **(c)** 17.0 mL H_2O; or **(d)** 1.2×10^{24} Mg atoms.

32. *Without doing detailed calculations*, determine which of the following samples has the greatest mass, and explain your choice. **(a)** 0.80 mol Fe; **(b)** 1.1×10^{24} S atoms; **(c)** 50.0 mL H_2O; or **(d)** 41.0 g $Al_2(SO_4)_3$.

Percent Composition and Empirical Formulas

33. Calculate the mass percent of each element in (**a**) $BaSiO_3$; (**b**) $C_6H_5NO_2$; (**c**) $Mg(HCO_3)_2$; and (**d**) $Al(BrO_3)_3 \cdot 9H_2O$.

34. Calculate the mass percent of each element in (**a**) C_3H_8O; (**b**) $(NH_4)_2SO_4$; (**c**) $C_6H_8N_2$; and (**d**) $Ba(ClO_4)_2 \cdot 3H_2O$.

35. What is the mass percent oxygen in the compound having the condensed structural formula $HOOCCH_2CH(CH_3)COOH$?

36. What is the mass percent nitrogen in the compound having the condensed structural formula $CH_3CH_2CH(CH_3)CONH_2$?

37. What is the mass percent of beryllium in the mineral beryl, $Be_3Al_2Si_6O_{18}$? What is the maximum mass of beryllium obtainable from 1.00 kg of beryl?

38. What is the mass percent uranium in the mineral carnotite, $K_2(UO_2)_2(VO_4)_2 \cdot 3H_2O$? What is the smallest mass of carnotite from which 1.00 kg of uranium can be obtained?

39. The empirical formula of *para*-dichlorobenzene, used as a moth repellent, is C_3H_2Cl. The molecular mass of the compound is 147 u. What is the molecular formula?

40. The empirical formula of apigenin, a yellow dye for wool, is C_3H_2O. The molecular mass of the compound is 270 u. What is the molecular formula?

41. The pain killer codeine has the mass percent composition: 72.22% C, 7.07% H, 4.68% N, and 16.03% O. Determine its empirical formula.

42. Urea, used as a fertilizer and in the manufacture of plastics, is 20.00% C, 6.71% H, 46.65% N, and 26.64% O by mass. Determine its empirical formula.

43. Resorcinol, used in manufacturing resins, drugs, and other products, is 65.44% C, 5.49% H, and 29.06% O by mass. Its molecular mass is 110 u. What is its molecular formula?

44. Sodium tetrathionate, an ionic compound formed when sodium thiosulfate reacts with iodine, is 17.01% Na, 47.46% S, and 35.52% O by mass, and has a formula mass of 270 u. What is its formula?

45. A hydrate is found to have the mass percent composition: 4.33% Li, 22.10% Cl, 39.89% O, and 33.69% H_2O by mass. What is its formula?

46. A hydrate of magnesium bromide is found to contain 37% H_2O by mass. What is its formula?

47. *Without doing detailed calculations*, determine which of these compounds has the greatest mass percent nitrogen: $(NH_4)_2SO_4$, NH_4NO_2, NH_4NO_3, NH_4Cl.

48. *Without doing detailed calculations*, determine which of the following hydrates has the greatest mass percent of water: $LiNO_3 \cdot 3H_2O$; $MgCl_2 \cdot 6H_2O$; $ZnSO_4 \cdot 7H_2O$; $Al(NO_3)_3 \cdot 9H_2O$.

49. An 0.1204-g sample of a carboxylic acid is burned in oxygen to yield 0.2147 g CO_2 and 0.0884 g H_2O. Calculate the mass percent composition and empirical formula of the compound.

50. An 0.0989-g sample of an alcohol is burned in oxygen to yield 0.2160 g CO_2 and 0.1194 g H_2O. Calculate the mass percent composition and empirical formula of the compound.

51. Thiophene is a carbon-hydrogen-sulfur compound used in the manufacture of pharmaceuticals. When burned completely in excess oxygen, its combustion products are CO_2, H_2O, and SO_2. Combustion of a 0.535-g sample yields 1.119 g CO_2, 0.229 g H_2O, and 0.407 g SO_2. What is the empirical formula of thiophene?

52. Dimethylhydrazine is a carbon-hydrogen-nitrogen compound used in rocket fuels. When burned completely in excess oxygen, a 0.312-g sample produces 0.458 g CO_2 and 0.374 g H_2O. The nitrogen content of a separate 0.525-g sample is converted to 0.244 g N_2. What is the empirical formula of dimethylhydrazine?

53. To decrease carbon monoxide emissions from automobiles in some geographic areas, gasoline is required to include oxygen-containing additives ("oxygenates") such as methanol (CH_3OH), ethanol (CH_3CH_2OH), and methyl *tertiary*-butyl ether (MTBE), $CH_3OC(CH_3)_3$.

(**a**) *Without doing detailed calculations*, arrange these three compounds in the order of *increasing* mass percent oxygen.

(**b**) The 1990 U.S. Clean Air Amendment requires fuels to contain 2.7% O. Does a fuel that contains 10.5% methanol by mass as its only oxygen-containing component meet this standard?

(**c**) What mass percent of MTBE should be present in gasoline if the gasoline is to contain 2.7% O? Assume that the gasoline has no other oxygen-containing components.

54. Morton Lite Salt® has 290 mg of sodium (as Na^+) and 340 mg of potassium (as K^+) per 0.25 teaspoon. Assume that 1.0 teaspoon of the Lite Salt has a mass of 6.0 g and that the Na^+ comes from NaCl and the K^+ from KCl. Calculate the mass percent of NaCl and of KCl in Lite Salt.

Chemical Equations

55. Balance the following equations.

(**a**) $Cl_2O_5 + H_2O \rightarrow HClO_3$

(**b**) $V_2O_5 + H_2 \rightarrow V_2O_3 + H_2O$

(**c**) $Al + O_2 \rightarrow Al_2O_3$

(**d**) $C_4H_{10} + O_2 \rightarrow CO_2 + H_2O$

(**e**) $Sn + NaOH \rightarrow Na_2SnO_2 + H_2$

(**f**) $PCl_5 + H_2O \rightarrow H_3PO_4 + HCl$

(**g**) $CH_3OH + O_2 \rightarrow CO_2 + H_2O$

(**h**) $Zn(OH)_2 + H_3PO_4 \rightarrow Zn_3(PO_4)_2 + H_2O$

56. Balance the following equations.

(**a**) $TiCl_4 + H_2O \rightarrow TiO_2 + HCl$

(**b**) $WO_3 + H_2 \rightarrow W + H_2O$

(**c**) $C_5H_{12} + O_2 \rightarrow CO_2 + H_2O$

(**d**) $Al_4C_3 + H_2O \rightarrow Al(OH)_3 + CH_4$

(e) $Al_2(SO_4)_3 + NaOH \rightarrow Al(OH)_3 + Na_2SO_4$

(f) $Ca_3P_2 + H_2O \rightarrow Ca(OH)_2 + PH_3$

(g) $Cl_2O_7 + H_2O \rightarrow HClO_4$

(h) $MnO_2 + HCl \rightarrow MnCl_2 + Cl_2 + H_2O$

57. Write a balanced equation to represent (**a**) the reaction of the gases carbon monoxide and nitrogen monoxide to form the gases carbon dioxide and nitrogen; (**b**) the reaction of the gases propane, C_3H_8, and water to form the gases carbon monoxide and hydrogen; (**c**) the reaction of solid magnesium nitride and liquid water to form solid magnesium hydroxide and gaseous ammonia; and (**d**) the reaction that produces electricity in a lead-acid storage battery: The solids lead and lead(IV) oxide react with an aqueous solution of sulfuric acid to produce solid lead(II) sulfate and liquid water.

58. Write a balanced equation to represent (**a**) the decomposition of solid potassium chlorate upon heating to form solid potassium chloride and oxygen gas; (**b**) the combustion of liquid 2-butanol, $CH_3CHOHCH_2CH_3$, in oxygen to produce gaseous carbon dioxide and liquid water; (**c**) the reaction of the gases ammonia and oxygen to produce the gases nitrogen monoxide and water; and (**d**) the reaction of the gases chlorine and ammonia with an aqueous solution of sodium hydroxide to form water and an aqueous solution containing sodium chloride and hydrazine, N_2H_4 (a chemical used in the synthesis of pesticides).

59. At 400 °C, hydrogen gas is passed over iron(III) oxide. Water vapor is formed, together with a black residue—a compound that is 72.3% Fe and 27.7% O by mass. Write a balanced equation for this reaction.

60. An aluminum-carbon compound that is 74.97% Al and 25.03% C by mass, reacts with water to produce aluminum hydroxide and a hydrocarbon that is 74.87% C and 25.13% H by mass. Write a balanced equation for this reaction.

Stoichiometry of Chemical Reactions

61. Consider the combustion in excess oxygen of octane, a major component of gasoline.

$$2\,C_8H_{18} + 25\,O_2 \longrightarrow 16\,CO_2 + 18\,H_2O$$

(**a**) How many moles of CO_2 are produced when 1.8×10^4 mol C_8H_{18} is burned?

(**b**) How many moles of oxygen are consumed in the combustion of 4.4×10^4 mol C_8H_{18}?

62. Use the equation in Problem 61 to determine (**a**) how many moles of H_2O are produced when 7.6×10^3 mol C_8H_{18} is burned, and (**b**) how many moles of CO_2 are produced when 2.2×10^4 mol O_2 is consumed?

63. Lead(II) oxide reacts with ammonia as follows:

$$PbO(s) + NH_3(g) \longrightarrow Pb(s) + N_2(g) + H_2O(l)$$

(not balanced)

(**a**) How many grams of NH_3 are consumed in the reaction of 75.0 g PbO?

(**b**) If 56.4 g Pb(s) are produced in this reaction, how many grams of nitrogen are also formed?

64. Toluene and nitric acid are used in the production of trinitrotoluene (TNT), an explosive.

$$\underset{\text{Toluene}}{C_7H_8} + HNO_3 \longrightarrow \underset{\text{TNT}}{C_7H_5N_3O_6} + H_2O \quad \textit{(not balanced)}$$

(**a**) How many grams of nitric acid are required to react with 454 g C_7H_8?

(**b**) How many grams of TNT can be made when 829 g C_7H_8 reacts with excess nitric acid?

65. The two solids, calcium cyanamide and magnesium nitride, both react with water to produce ammonia gas. *Without doing detailed calculations*, determine which of the two produces the greater amount of ammonia per kilogram of solid when it reacts with an excess of water.

$$CaCN_2(s) + H_2O(l) \longrightarrow CaCO_3(s) + NH_3(g)$$

(not balanced)

$$Mg_3N_2(s) + H_2O(l) \longrightarrow Mg(OH)_2(s) + NH_3(g)$$

(not balanced)

66. The two solids ammonium nitrate and potassium chlorate both decompose on heating to form oxygen gas. *Without doing detailed calculations*, determine which of the two yields the greater (a) number of moles of O_2 per mole of solid and (b) number of grams of O_2 per gram of solid.

$$NH_4NO_3(s) \longrightarrow N_2(g) + O_2(g) + H_2O(g)$$

(not balanced)

$$KClO_3(s) \longrightarrow KCl(s) + O_2(g) \quad \textit{(not balanced)}$$

67. Kerosene is a mixture of hydrocarbons used in domestic heating and as a jet fuel. Assume that kerosene can be represented as $C_{14}H_{30}$ and that it has a density of 0.763 g/mL.

$$C_{14}H_{30}(l) + O_2(g) \longrightarrow CO_2(g) + H_2O(l)$$

(not balanced)

How many grams of CO_2 are produced by the combustion of 1.00 gal (3.785 L) of kerosene?

68. Acetaldehyde, CH_3CHO, ($d = 0.788$ g/mL), a liquid used in the manufacture of perfumes, flavors, dyes, and plastics, can be produced by the reaction of ethanol with oxygen.

$$CH_3CH_2OH + O_2 \longrightarrow CH_3CHO + H_2O$$

(not balanced)

How many liters of liquid ethanol ($d = 0.789$ g/mL) must be consumed to produce 25.0 L of acetaldehyde?

69. Ordinary chalkboard chalk is a solid mixture, with limestone ($CaCO_3$) and gypsum ($CaSO_4$) as its principal ingredients. The limestone dissolves in dilute HCl(aq) but the gypsum does not.

9.5% → 9.5g

$$CaCO_3(s) + HCl(aq) \longrightarrow CaCl_2(aq) + CO_2(g) \quad \textit{(not balanced)}$$

If a 5.05-g piece of chalk that is 72.0% $CaCO_3$ is dissolved in excess HCl(aq), what mass of CO_2(g) will be produced?

70. Refer to Problem 69 and determine the mass percent of $CaCO_3$ in a 4.38-g piece of chalk that yields 1.31 g CO_2 when it reacts with excess HCl(aq).

71. How many milliliters of dilute HCl(aq) (d = 1.045 g/mL) that is 9.50% HCl by mass are required to react completely with 0.858 g Al?

$$Al(s) + HCl(aq) \longrightarrow AlCl_3(aq) + H_2(g) \quad \textit{(not balanced)}$$

72. Refer to Problem 71, and determine how many grams of hydrogen can be produced by the reaction of 125 mL of the HCl(aq) with an excess of aluminum.

Limiting Reactant and Yield Calculations

73. Lithium hydroxide absorbs carbon dioxide to form lithium carbonate and water.

$$2\ LiOH + CO_2 \longrightarrow Li_2CO_3 + H_2O$$

If a reaction vessel contains 0.150 mol LiOH and 0.080 mol CO_2, which compound is the limiting reactant? How many moles of Li_2CO_3 can be produced?

74. Boron trifluoride reacts with water to produce boric acid (H_3BO_3) and fluoroboric acid (HBF_4).

$$4\ BF_3 + 3\ H_2O \longrightarrow H_3BO_3 + 3\ HBF_4$$

If a reaction vessel contains 0.496 mol BF_3 and 0.313 mol H_2O, which compound is the limiting reactant? How many moles of HBF_4 can be produced?

75. The reaction of liquid mercury and oxygen gas is represented by the equation

$$Hg(l) + O_2(g) \longrightarrow HgO(s) \quad \textit{(not balanced)}$$

Without doing detailed calculations, explain which of the following outcomes should result from the reaction of 0.200 mol Hg(l) and 4.00 g O_2(g).

(a) 4.00 g HgO(s) and 0.200 mol Hg(l)

(b) 0.100 mol HgO(s), 0.100 mol Hg(l), and 2.40 g O_2

(c) 0.200 mol HgO(s) and no O_2(g)

(d) 0.200 mol HgO(s) and 0.80 g O_2(g)

76. The purification of titanium dioxide, an important step in the commercial production of titanium metal, involves the reaction

$$TiO_2(s) + C(s) + Cl_2(g) \longrightarrow TiCl_4(g) + CO(g) \quad \textit{(not balanced)}$$

Without doing detailed calculations, determine which of the following initial conditions will result in the production of the maximum amount of $TiCl_4$(g). Explain your reasoning.

(a) 1.5 mol of TiO_2, 2.1 mol C, and 4.4 mol Cl_2

(b) 1.6 mol TiO_2, 2.5 mol C, and 3.6 mol Cl_2

(c) 2.0 mol each of TiO_2, C, and Cl_2

(d) 3.0 mol each of TiO_2, C, and Cl_2

77. Potassium iodide, used as a dietary supplement to prevent the iodine-deficiency disease, goiter, is prepared by the reaction of hydroiodic acid and potassium hydrogen carbonate. Water and carbon dioxide are also produced. In the reaction of 481 g HI and 318 g $KHCO_3$, **(a)** how many grams of KI are produced; and **(b)** which reactant is in excess, and how many grams of it remain after the reaction?

78. Sodium nitrite, used as a preservative in meat processing (to prevent botulism) is prepared by passing nitrogen monoxide and oxygen gases into an aqueous solution of sodium carbonate. Carbon dioxide gas is another product of the reaction. In the reaction of 154 g Na_2CO_3, 105 g NO, and 75.0 g O_2(g), **(a)** how many grams of $NaNO_2$ are produced, and **(b)** which reactants are in excess, and how many grams of each remain after the reaction?

79. Calculate the theoretical yield of ZnS, in grams, that can be made from 0.488 g Zn and 0.503 g S_8.

$$8\ Zn + S_8 \longrightarrow 8\ ZnS$$

If the actual yield is 0.606 g ZnS, what is the percent yield?

80. Calculate the theoretical yield of CH_3CH_2Cl, in grams, that can be made from 11.3 g of ethanol and 3.48 g PCl_3.

$$3\ CH_3CH_2OH + PCl_3 \longrightarrow CH_3CH_2Cl$$

If the actual yield is 12.4 g CH_3CH_2Cl, what is the percent yield?

81. A student prepares ammonium bicarbonate by the reaction

$$NH_3 + CO_2 + H_2O \longrightarrow NH_4HCO_3$$

She uses 14.8 g NH_3 and 41.3 g CO_2. Water is present in excess. What is her actual yield of ammonium bicarbonate if she obtains a 74.7% yield in the reaction?

82. A student needs 625 g of zinc sulfide, a white pigment, for an art project. He can synthesize it using the reaction

$$Na_2S(aq) + Zn(NO_3)_2(aq) \longrightarrow ZnS(s) + 2\ NaNO_3(aq)$$

How many grams of zinc nitrate will he need if he can make the zinc sulfide in 85.0% yield? Assume that he has plenty of sodium sulfide.

Solution Stoichiometry

83. Calculate the molarity of each of the following aqueous solutions.
    (a) 6.00 mol HCl in 2.50 L of solution
    (b) 0.000700 mol Li_2CO_3 in 10.0 mL of solution
    (c) 8.905 g H_2SO_4 in 100.0 mL of solution
    (d) 439 g $C_6H_{12}O_6$ in 1.25 L of solution
    (e) 15.50 mL glycerol, $C_3H_8O_3$ (d = 1.265 g/mL) in 225.0 mL of solution
    (f) 35.0 mL 2-propanol, $CH_3CHOHCH_3$, (d = 0.786 g/mL) in 250 mL of solution
84. Calculate the molarity of each of the following aqueous solutions.
    (a) 2.50 mol H_2SO_4 in 5.00 L of solution
    (b) 0.200 mol C_2H_5OH in 35.0 mL of solution
    (c) 44.35 g KOH in 125.0 mL of solution
    (d) 2.46 g $H_2C_2O_4$ in 750.0 mL of solution
    (e) 22.00 mL triethylene glycol, $(CH_2OCH_2CH_2OH)_2$, (d = 1.127 g/mL) in 2.125 L of solution
    (f) 15.0 mL isopropylamine, $CH_3CH(NH_2)CH_3$, (d = 0.694 g/mL) in 225 mL of solution
85. How much solute is required to prepare each of the following solutions?
    (a) moles NaOH for 1.25 L of 0.0235 M NaOH
    (b) grams $C_6H_{12}O_6$ for 10.0 mL of 4.25 M $C_6H_{12}O_6$
    (c) milliliters triethanolamine, $N(CH_2CH_2OH)_3$, (d = 1.0985 g/mL) for 2.225 L of 0.2500 M $N(CH_2CH_2OH)_3$
    (d) milliliters 2-butanol, $CH_3CHOHCH_2CH_3$, (d = 0.808 g/mL) for 715 mL of a 1.34 molar solution
86. How much solute is required to prepare each of the following solutions?
    (a) moles $K_2Cr_2O_7$ for 315 mL of 2.50 M $K_2Cr_2O_7$
    (b) grams $KMnO_4$ for 20.0 mL of 0.0100 M $KMnO_4$
    (c) milliliters 1,4-butanediol, $HOCH_2(CH_2)_2CH_2OH$, (d = 1.020 g/mL) for 175.0 mL of 1.305 M $HOCH_2(CH_2)_2CH_2OH$
    (d) milliliters ethyl acetate, $CH_3COOCH_2CH_3$, (d = 0.902 g/mL) for 315 mL of a 0.0150 molar solution
87. How many milliliters of a 0.215 molar solution are required to contain (**a**) 0.0867 mol NaBr; (**b**) 32.1 g $CO(NH_2)_2$; and (**c**) 715 mg methanol, CH_3OH?
88. How many milliliters of an 0.0886 molar solution are required to contain (**a**) 3.52×10^{-2} mol $Al_2(SO_4)_3$; (**b**) 15.6 g $C_{12}H_{22}O_{11}$ and (**c**) 35.4 mg diethyl ether, $CH_3CH_2OCH_2CH_3$?
89. A stock bottle of nitric acid indicates that the solution is 67.0% HNO_3 by mass and has a density of 1.40 g/mL. Calculate the molarity of the solution.
90. A stock bottle of potassium hydroxide solution indicates that the solution is 50.0% KOH by mass and has a density of 1.52 g/mL. Calculate the molarity of the solution.
91. If 25.00 mL of 1.04 M Na_2CO_3 is diluted to 0.500 L, what is the molarity of Na_2CO_3 in the diluted solution?
92. If 50.00 mL of 19.1 M NaOH is diluted to 2.00 L, what is the molarity of NaOH in the diluted solution?
93. How many milliliters of 6.052 M HCl are required to make (**a**) 2.000 L of 0.5000 M HCl; and (**b**) 500.0 mL of a solution containing 7.150 mg HCl per milliliter?
94. How many milliliters of 3.124 M KOH are required to make (**a**) 250.0 mL of 1.200 M KOH; and (**b**) 15.0 L of a solution containing 0.245 g KOH per liter?
95. *Without doing detailed calculations*, determine which of the following is the most likely concentration of an aqueous solution obtained by mixing 0.100 L of 0.100 M NH_3 and 0.200 L of 0.200 M NH_3. Explain your reasoning.
    (a) 0.13 M NH_3
    (b) 0.15 M NH_3
    (c) 0.17 M NH_3
    (d) 0.30 M NH_3
96. The "Acculute" solution in the vial in the photograph, when diluted with water to 1.000 L, produces 0.1000 M HCl. *Without doing detailed calculations*, indicate which of the following is a plausible approximate concentration of the solution in the vial. Explain your reasoning.
    (a) 0.5 M HCl
    (b) 2.0 M HCl
    (c) 10.0 M HCl
    (d) 12.0 M HCl

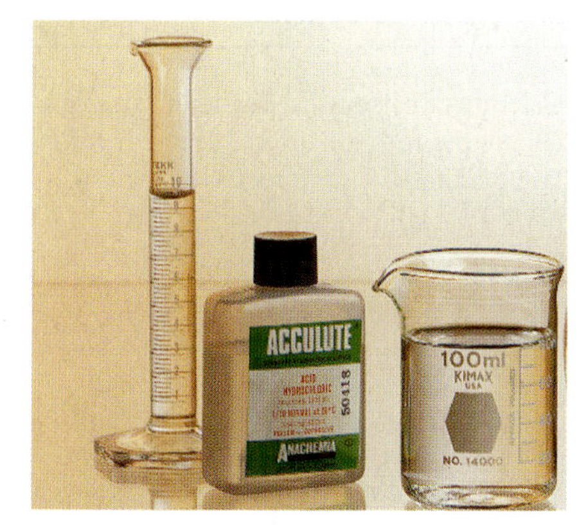

The contents of the vial, when diluted to 1.000 L with water, produce a solution that is 0.1000 M HCl.

97. Suppose you need about 80 mL of 0.100 M $AgNO_3$. You have available about 150 mL of 0.04000 M $AgNO_3$ and also about 1.0 g of solid $AgNO_3$. Assume that you have available standard laboratory equipment such as an analytical balance, 10.00-mL and 25.00-mL pipets, 100.0-mL and 250.0-mL volumetric flasks, and so on. Describe how you would prepare the desired $AgNO_3$ solution, including actual masses or volumes required.
98. Two sucrose solutions, 125 mL of 1.50 M $C_{12}H_{22}O_{11}$ and 275 mL of 1.25 M $C_{12}H_{22}O_{11}$, are mixed. Assuming a final solution volume of 400 mL, what is the molarity of $C_{12}H_{22}O_{11}$ in the final solution?
99. How many grams of $BaSO_4(s)$ are formed when an excess of $BaCl_2(aq)$ is added to 635 mL of 0.314 M $Na_2SO_4(aq)$?

$$BaCl_2(aq) + Na_2SO_4(aq) \longrightarrow BaSO_4(s) + 2\,NaCl(aq)$$

100. How many milliliters of 3.84 M HCl are required to consume 4.12 grams of zinc in the reaction

$$Zn(s) + 2\,HCl(aq) \longrightarrow ZnCl_2(aq) + H_2(g)$$

101. After 2.02 g Al has reacted completely with 0.400 L of 2.75 M HCl (the excess reactant), what is the molarity of the remaining HCl(aq)?

$$2\,Al(s) + 6\,HCl(aq) \longrightarrow 2\,AlCl_3(aq) + 3\,H_2(g)$$

102. How many grams of aluminum should be added to 0.400 L of 2.75 M HCl to reduce the concentration of the acid to 2.50 M HCl as a result of the reaction in Problem 101.

103. The reaction of calcium carbonate and hydrochloric acid produces calcium chloride, carbon dioxide and water. How many grams of carbon dioxide are produced when 4.35 g of calcium carbonate is added to 75.0 mL of 1.50 M hydrochloric acid?

104. In aqueous solution, the reaction of silver nitrate and potassium chromate yields solid silver chromate and aqueous potassium nitrate. How many grams of silver chromate are produced when 37.5 mL of a 0.625 M potassium chromate solution is added to 145 mL of 0.0525 M silver(I) nitrate solution?

105. A drop (0.05 mL) of 12.0 M HCl is spread over a sheet of thin aluminum foil (see photograph). What will be the maximum area, in cm^2, of the hole produced? The thickness of the foil is 0.10 mm, and the density of aluminum is 2.70 g/cm^3.

$$Al(s) + HCl(aq) \longrightarrow AlCl_3(aq) + H_2(g)$$

(not balanced)

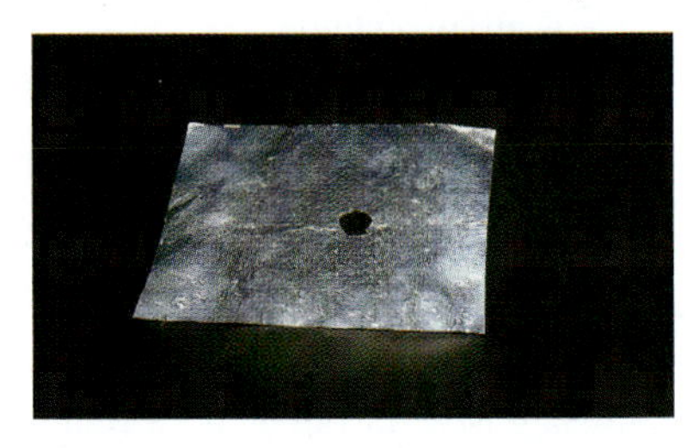

106. Refer to Problem 105 and the accompanying photograph. Suppose the foil is 0.065 mm thick and is made of copper (d = 8.96 g/cm^3) rather than aluminum. What is the minimum number of drops (1 drop = 0.05 mL) of 6.0 M HNO_3 needed to dissolve a 1.50-cm^2 hole in the metal?

Additional Problems

107. Calcium tablets for use as dietary supplements are available in the form of several different compounds. Calculate the mass of each required to furnish 875 mg Ca^{2+}.

(a) calcium carbonate, $CaCO_3$

(b) calcium lactate, $Ca(C_3H_5O_3)_2$

(c) calcium gluconate, $Ca(C_6H_{11}O_7)_2$

(d) calcium citrate, $Ca_3(C_6H_5O_7)_2$

108. Iron, as Fe^{2+}, is an essential nutrient. Pregnant women often take 325-mg ferrous sulfate ($FeSO_4$) tablets as a dietary supplement. Yet iron tablets are the leading cause of poisoning deaths in children. As little as 550 mg Fe^{2+} can be fatal to a 22-lb child. How many 325-mg ferrous sulfate tablets would it take to constitute a lethal dose to a 22-lb child?

109. Chlorophyll, found in plant cells and essential to the process of photosynthesis, contains 2.72% Mg by mass. Assuming one magnesium atom per chlorophyll molecule, calculate the molecular mass of chlorophyll.

110. A pentafluoride of an element is found to contain 55.91% F by mass. What must the element be?

111. A 0.8150-g sample of the compound MCl_2 reacts with an excess of $AgNO_3(aq)$ to produce 1.8431 g AgCl(s) and $M(NO_3)_2(aq)$. What is the atomic mass of the element M, and what is the element M?

112. When burned in oxygen in combustion analysis, a 0.1888-g sample of a hydrocarbon produced 0.6260 g CO_2 and 0.1602 g H_2O. The molecular mass of the compound is 106 u. Calculate **(a)** the mass percent composition, **(b)** the empirical formula, and **(c)** the molecular formula of the hydrocarbon.

113. The combustion in oxygen of 1.525 g of a compound of carbon, hydrogen, and oxygen derived from an alkane yields 3.047 g CO_2 and 1.247 g H_2O. The molecular mass of this compound is 88.1 u. Draw a plausible structural formula for this compound. Is there more than one possibility? Explain.

114. Explain whether in the combustion analysis of an alkane, **(a)** the mass of CO_2 obtained can be greater than the mass of H_2O; **(b)** the mass of CO_2 obtained can be less than the mass of H_2O; and **(c)** the masses of CO_2 and H_2O obtained can be equal.

115. At temperatures above 300 °C, silver oxide, Ag_2O, decomposes to give metallic silver and oxygen gas. A 2.95-g sample of *impure* silver oxide yields 0.183 g of oxygen. Assuming that $Ag_2O(s)$ is the only source of oxygen, what is the mass percent of Ag_2O in the sample?

116. A sheet of iron with a surface area of 525 cm^2 is covered with a coating of rust that has an average thickness of 0.0021 cm. What minimum volume of an HCl solution having a density of 1.07 g/mL and consisting of 14% HCl by mass is required to clean the surface of the metal by reacting with the rust? Assume that the rust is $Fe_2O_3(s)$, that it has a density of 5.2 g/cm^3, and that the reaction is

$$Fe_2O_3(s) + 6\,HCl(aq) \longrightarrow 2\,FeCl_3(aq) + 3\,H_2O(l)$$

117. A laboratory manual calls for 13.0 g of 1-butanol, 21.6 g of sodium bromide, and 33.8 g H_2SO_4 as reactants in this reaction.

$$C_4H_9OH + NaBr + H_2SO_4 \longrightarrow C_4H_9Br + NaHSO_4 + H_2O$$

A student following these directions obtains 16.8 g of butyl bromide (C_4H_9Br). What are the theoretical yield and the percent yield of this reaction?

118. An environmental newsletter states that 11 billion gallons of a sulfuric acid solution that is 70% H_2SO_4 by mass is scheduled for shipment to the White Pine Mine over a one-year period. The current annual production of pure H_2SO_4 in the United States is just under 100 billion pounds. Do you think the statement in the newsletter is accurate? Explain. Assume a density of 1.61 g/mL for the sulfuric acid.

119. How many grams of sodium metal must react with 250.0 mL of water to produce a solution that is 0.315 M NaOH. (Assume the final solution volume is 250.0 mL.)

$$Na(s) + H_2O(l) \longrightarrow NaOH(aq) + H_2(g)$$
(not balanced)

120. In Example 3.20 (page 111), we described the reaction of magnesium and nitrogen to form magnesium nitride. Magnesium reacts even more readily with oxygen to form magnesium oxide. The noble gas argon, on the other hand, is inert; it does not react with magnesium. Suppose that for the conditions described in the example—35.00 g Mg and 15.00 g N_2—the nitrogen gas was not pure. Calculate the number of grams of product you would expect if the nitrogen contained (**a**) 5% argon by mass, (**b**) 15% argon by mass, and (**c**) 25% oxygen by mass.

121. The urea [$CO(NH_2)_2$] solutions pictured are mixed, and the resulting solution is evaporated to a final volume of 825 mL. What is the molarity of $CO(NH_2)_2$ in this final solution?

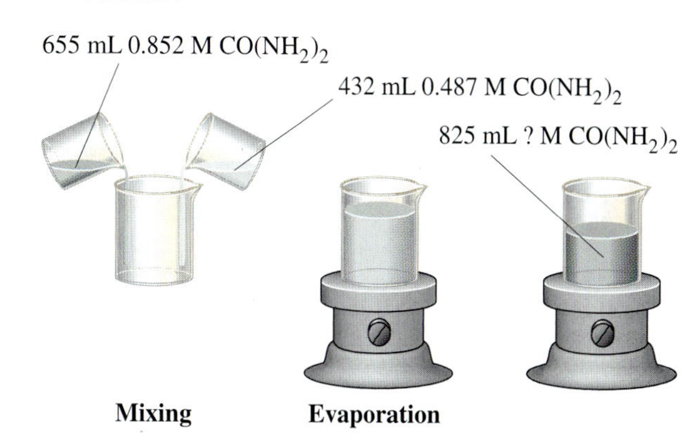

122. A certain commercial fertilizer consists of a mixture of two compounds in a mole ratio of 1 : 1. The fertilizer carries a rating of about 10-53-18. What combination of two of the following compounds might this be: KH_2PO_4, K_2HPO_4, K_3PO_4, $NH_4H_2PO_4$, $(NH_4)_2HPO_4$, $(NH_4)_3PO_4$? Use information from the box feature on page 92.

CHAPTER

4

Chemical Reactions in Aqueous Solutions

CONTENTS

The reaction of copper with nitric acid is an oxidation-reduction reaction, one of the three types of chemical reactions considered in this chapter.

Three quarters of Earth's surface is covered with water, but this natural water is not chemically pure. It contains dissolved substances. Solutions are formed by the contact of liquid water with gases in the atmosphere, with materials in Earth's solid crust, and with materials released to the environment by human activities. Water in living organisms is also in the form of aqueous solutions.

Water is a medium that chemists frequently choose for carrying out chemical reactions in the laboratory. Reactions in aqueous solutions fall into a few basic categories, all of which have important applications, as we will see in this chapter. In exploring these reactions, we will find more applications of ideas we learned in the previous two chapters. Later in the text, we will examine each of the reaction types in more detail.

● **ChemCDX**
When you see this icon, there are related animations, demonstrations, and exercises on the CD accompanying this text.

4.1 Some Electrical Properties of Aqueous Solutions

Water is such a good solvent for so many simple ionic and molecular substances that it has been called the "universal" solvent. The outward appearance of a solution doesn't tell us what is present at the microscopic level, that is, whether the solute particles are ions or molecules or a mixture of the two. Nevertheless, some of the earliest insights into the microscopic nature of aqueous solutions came through macroscopic observations on the abilities of solutions to conduct electricity. To understand why, let's first look at some significant early discoveries about electricity.

Static electricity, such as that produced by running a comb through your hair, has been known from ancient times. By the end of the eighteenth century, two types of electric charge (*positive* and *negative*) had been identified, and the interactions between positively and negatively charged objects were well understood (Figure 4.1).

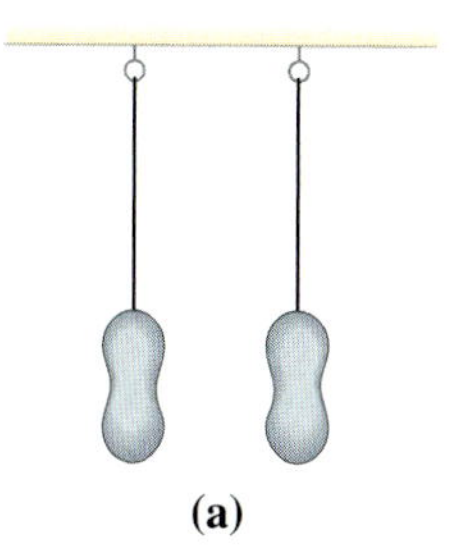

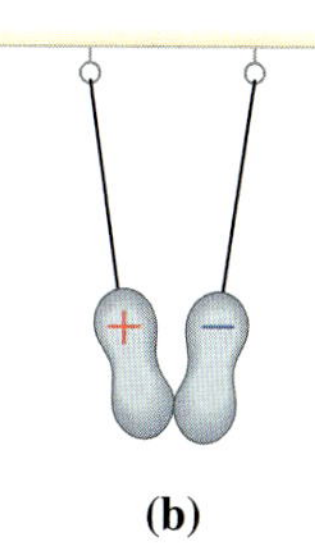

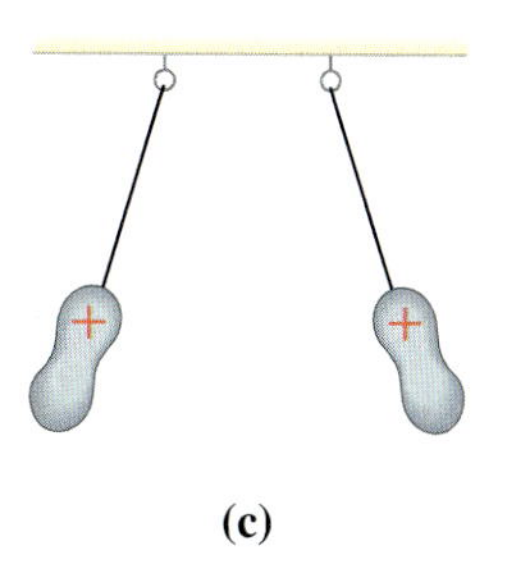

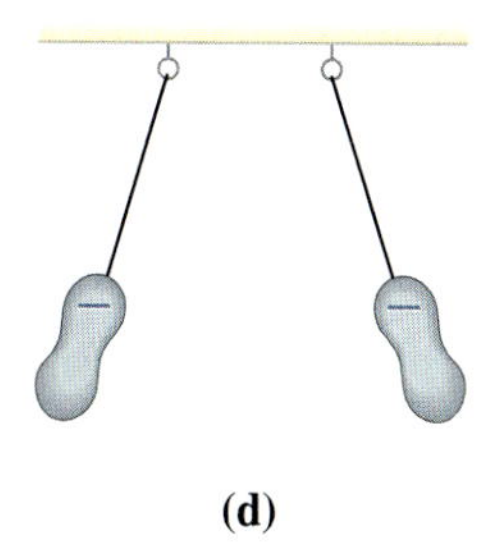

▲ Figure 4.1 Electrostatic forces
The objects pictured here are Styrofoam® "peanuts" used as a packing material. (a) The "peanuts" carry no charge, and no electrostatic force acts between them. (b) The "peanuts" are oppositely charged and attract one another. One has a positive charge, the other, a negative charge. In both (c) and (d), the "peanuts" carry like charges and repel one another.

At the beginning of the nineteenth century, the familiar electricity that flows through metal wires—current electricity—was discovered. We now know that an electric current is a *flow of charged particles*. In solid and liquid metals, the charged particles that flow are *electrons*. We say that metals *conduct* electricity, or that they are good electrical conductors. Electrons can be set in motion in different ways, for example, by an electric generator in a power plant or by a chemical reaction occurring in an electric battery.

Molten (liquid) ionic compounds and aqueous solutions of ionic compounds are also good electrical conductors, but in these cases the charged particles that flow are *ions*. Michael Faraday (1791–1867) did much of the early work on the conduction of electricity through solutions, the topic of interest in this section. Faraday coined the following terms, which are still used by chemists.

- *Electrodes* are electrical conductors (wires or metal plates) partially immersed in a solution and connected to a source of electricity. The *anode* is the electrode connected to the *positive* pole of the source of electricity, and the *cathode* is the electrode connected to the *negative* pole.
- An *ion* is a carrier of electricity through a solution. (*Ion* is derived from Greek and means "wanderer.") *Anions*, being negatively charged ions (−) are attracted to the anode (+); *cations* carry a positive charge (+) and are attracted to the cathode (−).

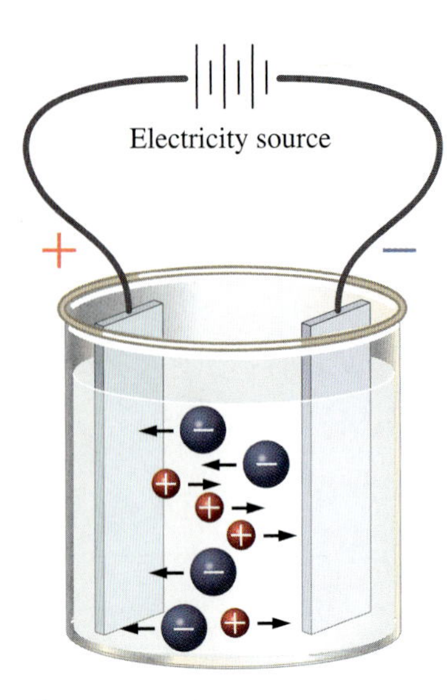

▲ **Figure 4.2**
Conduction of electric current through a solution
The electricity source directs electrons through wires from the anode to the cathode. Cations (+) are attracted to the cathode (−), and anions (−) are attracted to the anode (+). This migration of ions is the flow of electricity through the solution.

Figure 4.2 suggests how electricity passes through a solution. The external source of electricity is the driving force behind the electric current. It forces electrons through wires from the anode (+) to the cathode (−). The ions in solution then migrate to the electrodes. In this way they carry electricity through the solution and complete the electric circuit.

Arrhenius's Theory of Electrolytic Dissociation

Faraday did not speculate about how the ions necessary to conduct electricity are formed in a solution. Other scientists at the time thought that solute molecules break down, or dissociate, into cations and anions only as electric current enters a solution.

In his doctoral dissertation in 1884, Svante Arrhenius presented the hypothesis that certain substances, such as NaCl and HCl, dissociate into cations and anions *when they dissolve in water*. Electricity does not produce ions in an aqeuous solution; rather, it is the ions that pre-exist in solution that allow electricity to flow. We use the term *electrolyte* to describe a solute that produces enough ions to make a solution an electrical conductor. Arrhenius's ideas are now referred to as the *theory of electrolytic dissociation*.

Figure 4.3 illustrates a way to demonstrate the relative abilities of solutions to conduct electric current. We place two electrodes made of graphite (the material in pencil "lead") in the solution to be tested and connect them by wires to a source of electric current. We also place an electric lightbulb in the circuit. Electrons can pass freely through the wires and the bulb, but they cannot pass through the solution. In order for the electric circuit to be completed and the lightbulb to light up, there must be enough ions present to carry electric charge through the solution. Suppose that the solutions in the beakers are all one molar (1 M) in a solute. No matter what the solute is, we will always observe one of the following three possibilities.

- Figure 4.3(a): *The bulb lights up brightly*. This signals the presence of a large number of ions in the solution. In NaCl(aq), as in NaCl(s), there are no NaCl molecules but only separated Na^+ and Cl^- ions. We say that NaCl is a strong electrolyte.
  A ***strong electrolyte*** *is a solute that is present in solution almost exclusively as ions. A strong electrolyte solution is a good electrical conductor.*
  Hydrogen chloride is also a strong electrolyte in an aqueous solution. In this case, however, HCl is a molecular substance in the pure state. It forms ions (H^+ and Cl^-) when dissolved in water.

- Figure 4.3(b): *The bulb does not light up*. This observation signifies that there are essentially no ions in the solution, certainly not nearly enough to carry a significant amount of charge through the solution. This is what we see with *pure* water between the electrodes,* or an aqueous solution of a molecular substance that does not ionize, such as ethanol, CH_3CH_2OH.
  A ***nonelectrolyte*** *is a solute that is present in solution almost exclusively as molecules. A nonelectrolyte solution is a nonconductor of electricity.*

- Figure 4.3(c): *The bulb lights up but glows only dimly*. This is what we expect if a solute exists partly in molecular and partly in ionic form in solution. Acetic acid, CH_3COOH, behaves this way.
  A ***weak electrolyte*** *is a solute that is only partly ionized in solution. A weak electrolyte solution is a poor conductor of electricity.*

It is important to understand that the terms "strong" and "weak" refer to the extent to which an electrolyte produces ions in solution. If it exists almost exclusively as ions in solution, the electrolyte is *strong*; if much of it remains in molecular

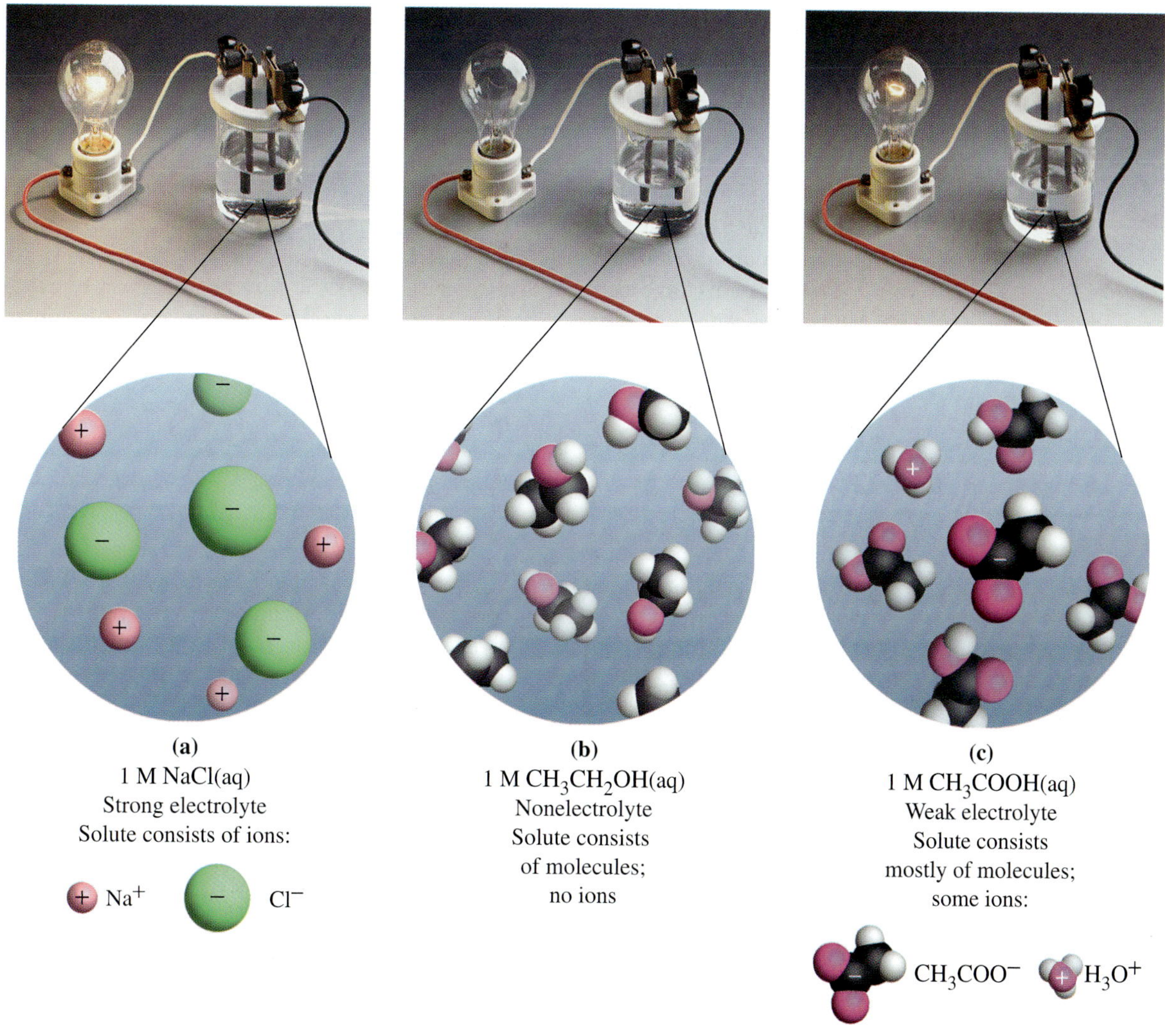

▲ **Figure 4.3 Electrolytic properties of aqueous solutions**
If electric current is to flow, cations and anions must be present and be free to migrate between the two graphite electrodes. In the microscopic view of each solution, solvent molecules are not shown because we want to focus on the solutes; the blue background suggests the presence of water as the solvent. In (a) all the solute particles are ions: Na^+ and Cl^-. In (b) there are no ions. In (c) about one in every 200 molecules ionizes, producing an acetate ion, CH_3COO^-, and a hydrogen ion, which attaches itself to a water molecule, forming H_3O^+.

form, the electrolyte is *weak*. Thus, even though an extremely dilute solution of sodium chloride might not conduct electricity as well as a somewhat more concentrated solution of acetic acid, NaCl is still a strong electrolyte and CH_3COOH is a weak electrolyte. We can use the following generalizations to classify water-soluble solutes according to their electrolytic properties.

- With only a few exceptions, ionic compounds are *strong electrolytes*.
- A few molecular compounds (mainly acids such as HCl) are *strong electrolytes*.

* Ordinary tap water is not pure. It contains enough dissolved ionic compounds to conduct electricity to a limited extent. This is why we must exercise great care when handling electrical equipment near pools of water.

- Most molecular compounds are either *nonelectrolytes* or *weak electrolytes*.
- Most organic compounds are molecular and *nonelectrolytes;* carboxylic acids and amines are *weak electrolytes*.

● **ChemCDX** Electrolytes and Nonelectrolytes

We can use these generalizations to help us decide how best to represent the solute in an aqueous solution. Because a solute that is a *nonelectrolyte* exists only in molecular form, we use its molecular formula. For example, we represent an aqueous solution of ethanol as $CH_3CH_2OH(aq)$.

When we dissolve a crystal of a water-soluble ionic compound—a *strong electrolyte*—the cations and ions dissociate from one another and appear in solution as independent solute particles. We can represent the dissociation of NaCl as

$$NaCl(s) \xrightarrow{H_2O} Na^+(aq) + Cl^-(aq)$$

We show the formula H_2O above the arrow to signify that the dissociation occurs when NaCl(s) dissolves in water, but H_2O is not a reactant in the usual sense. In many cases, we can continue to represent aqueous solutions of ionic compounds as we did in Chapter 3. For example, an aqueous solution of sodium chloride is represented as NaCl(aq). In other situations, however, the ionic form—$Na^+(aq) + Cl^-(aq)$—gives a better representation of the matter at hand, as we shall see next.

Calculating Ion Concentrations in Solution

We know that to prepare 0.010 M $Na_2SO_4(aq)$, we must dissolve 0.010 mol Na_2SO_4 per liter of solution. We also know that Na_2SO_4 is an ionic compound and a strong electrolyte. The dissolved particles in 0.010 M $Na_2SO_4(aq)$ are Na^+ and SO_4^{2-} ions, and the concentrations of these ions determine certain solution properties, for example, the freezing point. We will relate physical properties of solutions to their concentrations in Chapter 12. For that and other purposes, we must be able to calculate ion concentrations.

We use the bracket symbol, [ ], to represent the molar concentrations of ions or molecules in solution. Because the dissociation of Na_2SO_4 produces *two* Na^+ and *one* SO_4^{2-} ions per formula unit, in 0.010 M Na_2SO_4 we find (2×0.010) mol Na^+ per liter and 0.010 mol SO_4^{2-} per liter. That is, $[Na^+] = 0.020$ M and $[SO_4^{2-}] = 0.010$ M. Note that there is only one concentration for any given ion in a solution, even if the ion has two or more sources. We apply this principle in the calculation in Example 4.1.

EXAMPLE 4.1

Calculate the molarity of each type of ion present in an aqueous solution that is both 0.00384 M Na_2SO_4 and 0.00202 M $Al_2(SO_4)_3$.

SOLUTION

With practice, you can no doubt relate the concentrations of ions to that of the solutes that produce them, without writing equations. For this example, however, let's begin with equations to represent the dissociations.

$$\underset{\text{Salt}}{Na_2SO_4(s)} \xrightarrow{H_2O} \underset{\text{Cation}}{2\,Na^+(aq)} + \underset{\text{Anion}}{SO_4^{2-}(aq)}$$

$$\underset{\text{Salt}}{Al_2(SO_4)_3(s)} \xrightarrow{H_2O} \underset{\text{Cation}}{2\,Al^{3+}(aq)} + \underset{\text{Anion}}{3\,SO_4^{2-}(aq)}$$

Thus, we need to calculate molarities for three different types of ions: Na^+, Al^{3+}, and SO_4^{2-}. The dissociation equations are the basis for relevant conversion factors. We can use the conversion factors shown in red to establish the molarities of the ions.

Ions derived from $Na_2SO_4(s)$:

$$[Na^+] = \frac{0.00384\ \cancel{mol\ Na_2SO_4}}{1\ L} \times \frac{2\ mol\ Na^+}{1\ \cancel{mol\ Na_2SO_4}} = \frac{0.00768\ mol\ Na^+}{1\ L} = 0.00768\ M$$

$$[SO_4^{2-}] = \frac{0.00384\ \cancel{mol\ Na_2SO_4}}{1\ L} \times \frac{1 mol\ SO_4^{2-}}{1\ \cancel{mol\ Na_2SO_4}} = \frac{0.00384\ mol\ SO_4^{2-}}{1\ L}$$

$$= 0.00384\ M$$

Ions derived from $Al_2(SO_4)_3(s)$:

$$[Al^{3+}] = \frac{0.00202\ \cancel{mol\ Al_2(SO_4)_3}}{1\ L} \times \frac{2\ mol\ Al^{3+}}{1\ \cancel{mol\ Al_2(SO_4)_3}} = \frac{0.00404\ mol\ Al^{3+}}{1\ L}$$

$$= 0.00404\ M$$

$$[SO_4^{2-}] = \frac{0.00202\ \cancel{mol\ Al_2(SO_4)_3}}{1\ L} \times \frac{3\ mol\ SO_4^{2-}}{1\ \cancel{mol\ Al_2(SO_4)_3}} = \frac{0.00606\ mol\ SO_4^{2-}}{1\ L}$$

$$= 0.00606\ M$$

The SO_4^{2-} ions derived from $Al_2(SO_4)_3$ are identical to those derived from Na_2SO_4. To obtain the total sulfate ion concentration, we add the sulfate ion concentration from the two sources. The cations Na^+ and Al^{3+} each has only one source.

$$[Na^+] = 0.00768\ M$$

$$[Al^{3+}] = 0.00404\ M$$

$$[SO_4^{2-}] = 0.00384\ M + 0.00606\ M = 0.00990\ M$$

EXERCISE 4.1A

Seawater is essentially 0.438 M NaCl and 0.0512 M $MgCl_2$, together with several other minor solutes. What are $[Na^+]$, $[Mg^{2+}]$, and $[Cl^-]$ in seawater?

EXERCISE 4.1B

Each year, oral rehydration therapy (ORT)—feeding of an electrolyte solution—saves the lives of a million children worldwide who become dehydrated from diarrhea. The solution contains 3.5 g sodium chloride, 1.5 g potassium chloride, 2.9 g sodium citrate ($Na_3C_6H_5O_7$), and 20.0 g glucose ($C_6H_{12}O_6$) per liter. Calculate the molarity of each of the species present in the solution.

[*Hint:* Sodium citrate is a strong electrolyte, and glucose is a nonelectrolyte.]

4.2 Reactions of Acids and Bases

We first encountered Arrhenius's theory of acids and bases in Chapter 2, where we focused on names and formulas, and we used acids and/or bases as reactants in a few of the reactions presented in Chapter 3. In this section, we will look at acid-base reactions in more detail.

Strong and Weak Acids

We will adopt a more general acid-base theory in Chapter 15, but for now, we will continue to use the Arrhenius view that an acid is a substance that produces hydrogen ion (H^+)* in aqueous solution. And we will expand the definition of an acid to incorporate information about the ability of aqueous acid solutions to conduct electricity.

Acids that are completely ionized in water produce aqueous solutions that are good electrical conductors. These acids are *strong* electrolytes and are called **strong acids**. To represent the ionization of a strong acid, such as hydrochloric acid, we can write the equation

$$HCl(g) \xrightarrow{H_2O} H^+(aq) + Cl^-(aq)$$

We can therefore represent an aqueous solution of hydrochloric acid either as $HCl(aq)$ or $H^+(aq) + Cl^-(aq)$, and we can calculate the concentrations of cations and anions in the same way as we did for ionic compounds in Example 4.1. For example, the ionization equation tells us that 1 mol $HCl(g)$ dissociates to yield 1 mol $H^+(aq)$ and 1 mol $Cl^-(aq)$. Thus, in 0.0010 M HCl,

$$[H^+] = 0.0010 \text{ M} \quad \text{and} \quad [Cl^-] = 0.0010 \text{ M}$$

and, because essentially all the HCl molecules have ionized, we take the concentration of HCl *molecules* to be zero: $[HCl] = 0$.

The vast majority of acids are *weak* electrolytes. Only some of their molecules ionize in aqueous solution; the rest remain as intact molecules. These acids are called **weak acids**. Most carboxylic acids (Section 2.10) with one to four carbon atoms are soluble in water, and almost all of them are weak acids.

A ball-and-stick model of acetic acid, CH_3COOH, a weak acid.

In 1.0 M CH_3COOH, only about 0.5% of the molecules ionize. Because most weak acids remain largely in molecular form, we represent them through a molecular formula, such as $CH_3COOH(aq)$. To represent the limited ionization that does occur, we might write

$$\textit{Ionization:} \quad \underset{\text{Acetic acid}}{CH_3COOH(aq)} \longrightarrow H^+(aq) + \underset{\text{Acetate ion}}{CH_3COO^-(aq)}$$

This equation alone is misleading because it falsely implies that the ionization goes to completion. It doesn't because the ions are able to recombine to form neutral molecules in a reverse reaction.

$$\textit{Recombination of ions:} \quad H^+(aq) + CH_3COO^-(aq) \longrightarrow CH_3COOH(aq)$$

The best way to represent the limited ionization of a weak acid is to write a single equation but with a *double arrow* ($\rightleftharpoons$).

$$CH_3COOH(aq) \rightleftharpoons H^+(aq) + CH_3COO^-(aq)$$

A double arrow in a chemical equation signifies that the reaction is *reversible:* a forward and reverse reaction occur simultaneously. Instead of going to completion, a reversible reaction reaches a state of *equilibrium* in which concentrations of

*The simple hydrogen ion, H^+, does not exist in aqueous solutions. Instead, it is associated with several H_2O molecules; that is, it is present as $H(H_2O)_n^+$, where n is an integer. The most abundant of these hydrated hydrogen ions has $n = 1$. It is H_3O^+ and is called *hydronium ion*. We will use the abbreviation H^+ here and switch to H_3O^+ in Chapter 15.

reactants and products remain constant with time. We will discuss reversible reactions and the nature of equilibrium in more detail in Chapter 14, but for now, we will simply note the significance of the double arrow.

Some acids can produce two or more hydrogen ions per molecule of the acid. Sulfuric acid, H_2SO_4, and phosphoric acid, H_3PO_4, are two common examples. Sulfuric acid is interesting in that it is a *strong* acid in its first ionization and a *weak* acid in its second.

A space-filling model of sulfuric acid, a strong acid in its first ionization and a weak acid in its second.

First ionization: $H_2SO_4(aq) \longrightarrow H^+(aq) + HSO_4^-(aq)$

Second ionization: $HSO_4^-(aq) \rightleftharpoons H^+(aq) + SO_4^{2-}(aq)$

Phosphoric acid, on the other hand, is a *weak* acid in each of its three ionization steps.

$$H_3PO_4(aq) \rightleftharpoons H^+(aq) + H_2PO_4^-(aq)$$
$$H_2PO_4^-(aq) \rightleftharpoons H^+(aq) + HPO_4^{2-}(aq)$$
$$HPO_4^{2-}(aq) \rightleftharpoons H^+(aq) + PO_4^{3-}(aq)$$

We will discuss these acids further in Chapter 15.

Can we look at a chemical formula and tell if a substance is an acid or not? And, if it is, can we tell whether it is a strong or weak acid? In many cases, we can. Chemists generally identify *ionizable* hydrogen atoms by writing formulas in one of two ways.

1. *A molecular formula of a compound with ionizable H atoms has the ionizable H atoms written first.* HNO_3, H_2SO_4, and H_3PO_4 are acids with one, two, and three ionizable hydrogen atoms, respectively. Methane, CH_4, has four H atoms, but they are *not* ionizable; CH_4 is *not* an acid. From its name, we know that acetic acid is an acid. When its formula is written $HC_2H_3O_2$, we see that it has four H atoms, but that only one is ionizable—the H atom that is written first.
2. *A condensed structural formula shows where H atoms are found in a molecule and which of them are ionizable.* The condensed structural formula of acetic acid, CH_3COOH, shows that three of the H atoms are bonded to a C atom. Just like the four H atoms in CH_4, the three in the $-CH_3$ group are *not* ionizable. Only the H atom bonded to an O atom in the carboxylic acid group ($-COOH$) is *ionizable*.

The simplest way to tell whether an acid is strong or weak is to note that there aren't many strong acids; the most common ones are those in Table 4.1. Unless you have information to the contrary, you can assume that any other acid is a weak acid.

● **ChemCDX** Introduction to Aqueous Acids

TABLE 4.1 Common Strong Acids and Strong Bases

Acids		Bases	
Group 7A hydrides	*Oxoacids*	*Group 1A hydroxides*	*Group 2A hydroxides*
HCl	HNO_3	LiOH	$Mg(OH)_2$
HBr	H_2SO_4[a]	NaOH	$Ca(OH)_2$
HI	$HClO_4$	KOH	$Sr(OH)_2$
		RbOH	$Ba(OH)_2$
		CsOH	

[a] H_2SO_4 is a strong acid in its first ionization step but weak in its second ionization step.

Strong and Weak Bases

By the Arrhenius definition, a base is a substance that produces hydroxide ions, OH^-, in aqueous solution. Moreover, because ionic compounds are *strong* electrolytes, ionic hydroxides such as NaOH are **strong bases.**

$$NaOH(s) \xrightarrow{H_2O} Na^+(aq) + OH^-(aq)$$

Some compounds produce OH^- ions by *reacting* with water, not just by dissolving in it. Such substances are also bases. Gaseous ammonia dissolves in water, and then $NH_3(aq)$ reacts with water to produce some ions, but most remains as molecules.

$$NH_3(aq) + H_2O(l) \rightleftharpoons NH_4^+(aq) + OH^-(aq)$$

As in the case of acetic acid, most of the NH_3 molecules remain nonionized at equilibrium, and ammonia is therefore a **weak base.*** Most molecular substances that act as bases are *weak* bases.

As we saw in Chapter 2, *amines* are compounds consisting of molecules in which one or more of the H atoms in NH_3 is replaced by a hydrocarbon group. Amines with molecules having one to four carbon atoms are water soluble, and like NH_3 they are *weak* bases. For example,

$$\underset{\text{Methylamine}}{CH_3NH_2(aq)} + H_2O(l) \rightleftharpoons \underset{\text{Methylammonium ion}}{CH_3NH_3^+(aq)} + OH^-(aq)$$

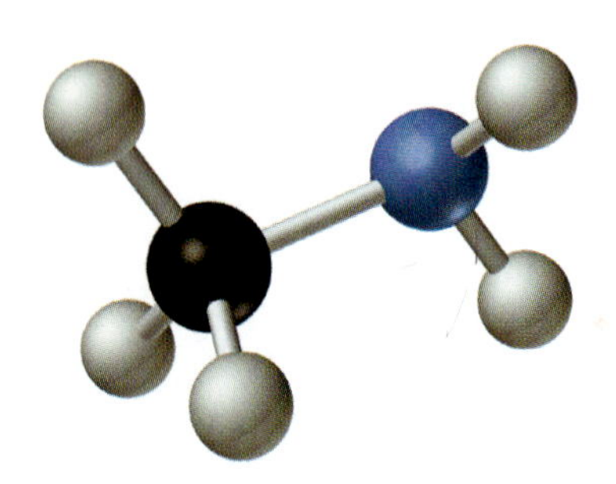

A ball-and-stick model of methylamine, CH_3NH_2, a weak base.

Can we look at a chemical formula and tell if a substance is a base or not? And if it is, can we tell whether the base is weak or strong? In many cases, we can. For example, if the formula indicates an *ionic* compound containing OH^- ions, we expect it to be a strong base. NaOH and KOH are strong bases. Table 4.1 lists the common strong bases. In contrast, methanol, CH_3OH, is *not* a base. There are only nonmetals in the compound—no metals—and therefore CH_3OH is a *molecular* compound. The OH group is covalently bonded to the C atom, *not* present as OH^-. Similarly, acetic acid (CH_3COOH) is not a base; it produces H^+ and is therefore an *acid.* To identify a weak base, you usually need a chemical equation for the ionization reaction. However, you can identify many weak bases by using these facts:

There are only a few common strong acids and strong bases. The most common weak bases are ammonia and the amines.

● **ChemCDX** Introduction to Aqueous Bases

Acid-Base Reactions: Neutralization

In the reaction of an acid and a base, called **neutralization,** the identifying characteristics of the acid and base (page 58) cancel out, or neutralize, each other. The acid and base are converted to an aqueous solution of an ionic compound called a **salt.** Generally, we don't see anything happen in a neutralization reaction. Both the acid and base are usually colorless, as is the salt that replaces them. How do we know when a solution is *neutral,* that is, neither acidic nor basic? One of the characteristics of acids and bases mentioned in Chapter 2 (page 58) is their ability to af-

* In Arrhenius's time, chemists generally believed that a substance must *contain* OH groups to be a base. For example, $NH_3(aq)$ was thought to be NH_4OH (ammonium hydroxide). Even though there is no compelling evidence for the existence of NH_4OH molecules, this formula is still often seen as a representation of $NH_3(aq)$.

(a)

(b)

(c)

Figure 4.4
Phenol red—an acid-base indicator
Phenol red is (a) *yellow* in an acidic solution (0.10 M HCl), (b) *orange* in a neutral solution (0.10 M NaCl), and (c) *red* in a basic solution (0.10 M NaOH).

fect the color of litmus (a natural product obtained from certain lichens). An *acid-base indicator* is a chemical that changes color as a solution passes from being acidic to basic. For example, as shown in Figure 4.4, the indicator phenol red is yellow in acidic solutions, orange when a solution is neutral, and red in basic solutions.

If we use conventional formulas for the acid and base, we can write what we might call a "complete formula" equation* for a neutralization reaction as follows:

$$\underset{\text{Acid}}{HCl(aq)} + \underset{\text{Base}}{NaOH(aq)} \longrightarrow \underset{\text{Salt}}{NaCl(aq)} + \underset{\text{Water}}{H_2O(l)}$$

But this complete formula equation is not always the best way to show what happens in the neutralization. To show that the reactants and one of the products of the reaction exist in solution as ions, we can write the equation in *ionic* form.

$$\underbrace{H^+(aq) + Cl^-(aq)}_{\text{Acid}} + \underbrace{Na^+(aq) + OH^-(aq)}_{\text{Base}} \longrightarrow \underbrace{Na^+(aq) + Cl^-(aq)}_{\text{Salt}} + \underset{\text{Water}}{H_2O(l)}$$

When we eliminate "spectator" ions—those that just "look on" and that appear unchanged on both sides of the ionic equation—we find that the equation reduces to a simple form called a **net ionic equation.** In the equation above, Na^+ and Cl^- are spectator ions, and the net ionic equation is

$$H^+(aq) + OH^-(aq) \longrightarrow H_2O(l)$$

The net ionic equation conveys the essence of a neutralization: H^+ and OH^- ions combine to form water. In general, net ionic equations represent the gist of a reaction. If the spectator ions in a neutralization reaction are those of a soluble salt, they remain in solution. If we evaporate this solution to dryness we get the pure salt, for example, NaCl (table salt).

APPLICATION NOTE

Phenol red is familiar to people who maintain a home swimming pool. It is the indicator commonly used to ensure that the pool water is kept about neutral, that is, neither acidic nor basic.

EXAMPLE 4.2

Barium nitrate is used to produce a green color in fireworks. It can be made by the reaction of nitric acid with barium hydroxide. Write an **(a)** "complete formula" equation, **(b)** ionic equation, and **(c)** net ionic equation for this neutralization reaction.

* The "complete formula" equation is often called a molecular equation, but this term is misleading. Many of the formulas written in such an equation, for example, NaCl(aq), represent formula units of ionic compounds, not actual molecules.

SOLUTION

a. Write chemical formulas for the substances involved in the reaction and balance the equation.

Nitric acid		*Barium hydroxide*		*Barium nitrate*		*Water*	

$$HNO_3(aq) + Ba(OH)_2(aq) \longrightarrow Ba(NO_3)_2(aq) + H_2O(l) \quad \textit{(not balanced)}$$

$$2\,HNO_3(aq) + Ba(OH)_2(aq) \longrightarrow Ba(NO_3)_2(aq) + 2\,H_2O(l) \quad \textit{(balanced)}$$

b. Now, represent the strong electrolytes by the formulas of their ions and the nonelectrolyte water by its molecular formula.

$$\underbrace{2\,H^+(aq) + 2\,NO_3^-(aq)}_{\text{A strong acid}} + \underbrace{Ba^{2+}(aq) + 2\,OH^-(aq)}_{\text{A strong base}} \longrightarrow \underbrace{Ba^{2+}(aq) + 2\,NO_3^-(aq)}_{\text{A salt}} + \underbrace{2\,H_2O(l)}_{\text{Water}}$$

c. Cancel the spectator ions (Ba^{2+} and NO_3^-) in the ionic equation.

$$2\,H^+(aq) + \cancel{2\,NO_3^-(aq)} + \cancel{Ba^{2+}(aq)} + 2\,OH^-(aq) \longrightarrow \cancel{Ba^{2+}(aq)} + \cancel{2\,NO_3^-(aq)} + 2\,H_2O(l)$$

This gives the equation

$$2\,H^+(aq) + 2\,OH^-(aq) \longrightarrow 2\,H_2O(l)$$

or, more simply,

$$H^+(aq) + OH^-(aq) \longrightarrow H_2O(l)$$

EXERCISE 4.2

Calcium hydroxide is used to neutralize a waste stream of hydrochloric acid. Write an **(a)** "complete formula" equation, **(b)** ionic equation, and **(c)** net ionic equation for this neutralization reaction.

Acid-Base Reactions: Titrations

In most of the stoichiometry problems in Chapter 3, we focused on calculating the amount of product formed from one limiting reactant and an excess of other reactants. In neutralization reactions, in contrast, we usually are not interested in the product of the reaction but rather in mixing the reactants—an acid and a base—in *stoichiometric* proportions. A technique in which two reactants in solution are made to combine in their stoichiometric proportions is called a **titration.**

We can use titrations to determine the concentrations of acids or bases in "unknown" solutions, such as the concentration of H_2SO_4 in a battery acid. We can also use titrations to determine the concentrations of solutions that we cannot precisely prepare. For example, we cannot simply dissolve 1.000 mol NaOH (40.00 g NaOH) in 1.000 L of aqueous solution to prepare 1.000 M NaOH. As it is usually obtained, solid sodium hydroxide generally is not quite pure, and it has some absorbed water.

APPLICATION NOTE
Vinegar can contain from 4 to 10% acetic acid. The exact amount of acetic acid in a particular vinegar can be determined by titration with a base of known concentration.

The central piece of equipment in a titration is a **buret**, a graduated, long glass tube constructed to deliver precise volumes of a solution through a stopcock valve. In a titration, we place a solution of one of the reactants, called the **titrant**, in the buret. We place a solution of the other reactant in a small flask to which we also add

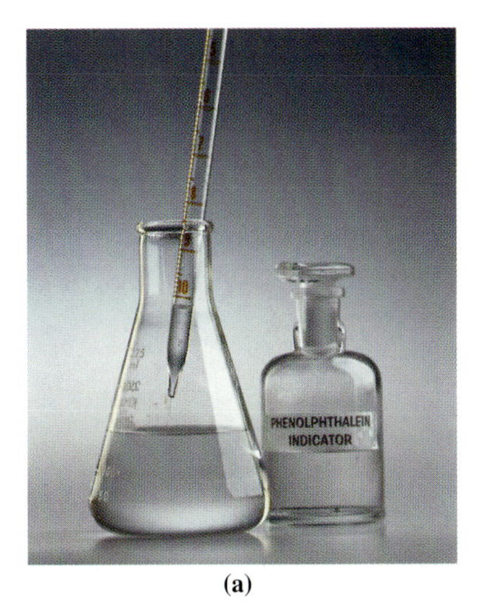

(a)

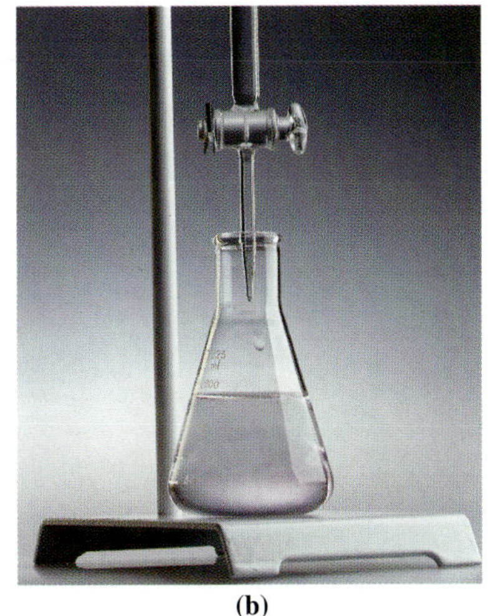

(b)

(c)

▲ **Figure 4.5 The technique of titration**
The steps in the technique corresponding to the photographs (a), (b), and (c) are described in the text.

a few drops of an acid-base indicator solution. Titrant is delivered from the buret into the flask, where a chemical reaction occurs. Consider the titration pictured in Figure 4.5.

(a) A precisely measured volume of HCl(aq) is discharged from a pipet into a quantity of water in a small flask. Then a few drops of phenolphthalein indicator solution are added. The solution is colorless, indicating that it is acidic.

(b) NaOH(aq) is slowly added from a buret into the flask. Until all of the HCl has been neutralized, the HCl is in excess and the NaOH is the limiting reactant. The solution remains colorless, indicating that it is still acidic.

(c) At the point where the acid has just been completely neutralized, called the *equivalence* point, HCl and NaOH are in stoichiometric proportions. An additional drop of NaOH(aq) beyond this point makes the solution slightly basic, and the indicator turns to a light pink color. The titration is stopped, and the volume of solution delivered from the buret is recorded.

To perform a successful titration, we must use an indicator that changes color at the equivalence point. We will discuss how to do this in Chapter 15.

Let's now explore how to combine the data collected in a titration with the equation for a neutralization reaction to do some calculations.

EXAMPLE 4.3

What volume of 0.2010 M NaOH is required to neutralize 20.00 mL of 0.1030 M HCl in an acid-base titration?

SOLUTION

We need to do four things to solve this problem:

1. Write an equation describing the neutralization, and obtain a stoichiometric factor relating moles of NaOH and HCl.
2. Determine how many moles of HCl are to be neutralized.
3. Find the number of moles of NaOH required in the neutralization.
4. Determine the volume of solution containing this number of moles of NaOH.

Let's now apply these steps.

$M = \frac{mol}{L}$

1 The complete formula equation for the reaction is

$$NaOH(aq) + HCl(aq) \longrightarrow NaCl(aq) + H_2O(l)$$

From this equation, we see that the stoichiometric equivalence between the reactants is

$$1 \text{ mol NaOH} \approx 1 \text{ mol HCl}$$

and the needed stoichiometric factor is

$$\frac{1 \text{ mol NaOH}}{1 \text{ mol HCl}}$$

2 The number of moles of HCl to be titrated is the product of the volume (0.02000 L) and the molarity of the HCl(aq) (0.1030 mol HCl/L).

$$? \text{ mol HCl} = 0.02000 \cancel{\text{L HCl(aq)}} \times \frac{0.1030 \text{ mol HCl}}{1 \cancel{\text{L HCl(aq)}}} = 0.002060 \text{ mol HCl}$$

3 Use the stoichiometric factor (in red) to convert moles of HCl (0.002060 mol) to moles of NaOH required for the titration.

$$? \text{ mol NaOH} = 0.002060 \cancel{\text{mol HCl}} \times \frac{1 \text{ mol NaOH}}{1 \cancel{\text{mol HCl}}} = 0.002060 \text{ mol NaOH}$$

4 Use the *inverse* of the molarity of the NaOH(aq)—that is, 1 L/0.2010 mol NaOH—to convert from moles of NaOH to liters of NaOH(aq). Follow this by a conversion of liters to milliliters of NaOH(aq).

$$? \text{ mL NaOH(aq)} = 0.002060 \cancel{\text{mol NaOH}} \times \frac{1 \cancel{\text{L NaOH(aq)}}}{0.2010 \cancel{\text{mol NaOH}}} \times \frac{1000 \text{ mL NaOH(aq)}}{1 \cancel{\text{L NaOH(aq)}}} = 10.25 \text{ mL NaOH(aq)}$$

As in most stoichiometric calculations, we can combine these steps into a single setup.

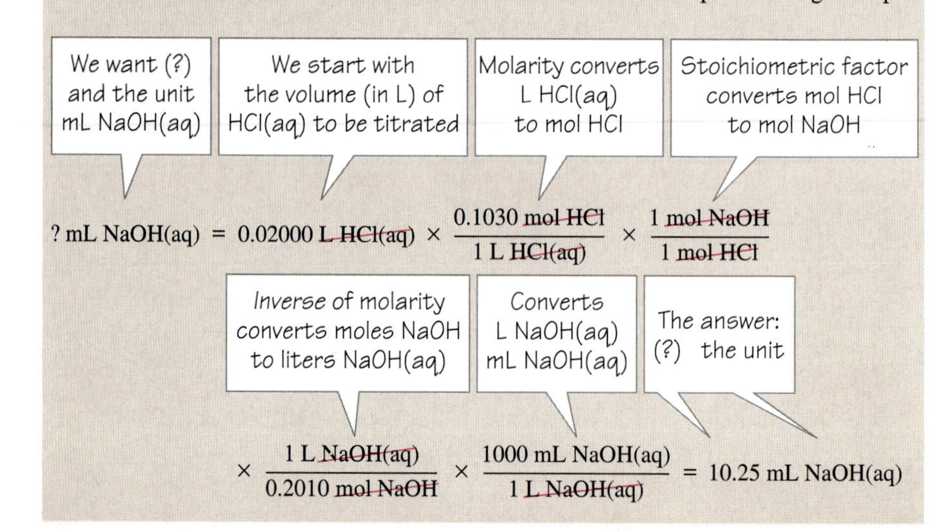

EXERCISE 4.3A

What volume of 0.01060 M HBr(aq) is required to neutralize 25.00 mL of 0.01580 M $Ba(OH)_2$?

$$2\ HBr(aq) + Ba(OH)_2(aq) \longrightarrow BaBr_2(aq) + 2\ H_2O(l)$$

EXERCISE 4.3B

A 2.000-g sample of a sulfuric acid solution that is 96.5% H_2SO_4 by mass is dissolved in a quantity of water and titrated. What volume of 0.3580 M KOH(aq) is required for the titration? Assume that at the equivalence point of the titration the solution is $K_2SO_4(aq)$.

EXAMPLE 4.4

A 10.00-mL sample of an aqueous solution of calcium hydroxide requires 23.30 mL of 0.02000 M $HNO_3(aq)$ for its neutralization. What is the molarity of the calcium hydroxide solution?

SOLUTION

Let's start with a balanced equation for this reaction and then decide how to proceed.

$$\underset{\text{Nitric acid}}{2\ HNO_3(aq)} + \underset{\text{Calcium hydroxide}}{Ca(OH)_2(aq)} \longrightarrow \underset{\text{Calcium nitrate}}{Ca(NO_3)_2(aq)} + \underset{\text{Water}}{2\ H_2O(l)}$$

To determine the molarity of the $Ca(OH)_2(aq)$, we need to determine how many moles of $Ca(OH)_2$ are consumed in the titration and divide this number by the 0.01000-L (10.00-mL) volume of the sample. And to determine the number of moles of $Ca(OH)_2$ we can proceed through much of the problem as we did in Example 4.3 by making the conversions outlined below.

$$\text{mL } HNO_3(aq) \longrightarrow \text{L } HNO_3(aq) \longrightarrow \text{mol } HNO_3 \longrightarrow \text{mol } Ca(OH)_2$$

We want (?) and the unit mol $Ca(OH)_2$

mL HNO_3 required in titration

Converts mL to L

Molarity converts L acid to mol HNO_3

Stoichiometric factor converts mol HNO_3 to mol $Ca(OH)_2$

$$?\ \text{mol } Ca(OH)_2 = 23.30\ \cancel{\text{mL}}\ HNO_3 \times \frac{1\ \cancel{\text{L}}}{1000\ \cancel{\text{mL}}} \times \frac{0.02000\ \cancel{\text{mol } HNO_3}}{1\ \cancel{\text{L}}} \times \frac{1\ \text{mol } Ca(OH)_2}{2\ \cancel{\text{mol } HNO_3}}$$

The answer: (?) the unit

$$= 2.330 \times 10^{-4}\ \text{mol } Ca(OH)_2$$

We have just calculated the amount of $Ca(OH)_2$ found in a 10.00-mL (0.01000 L) sample. We can therefore use the definition of molarity to write

$$\text{Molarity} = \frac{2.33 \times 10^{-4}\ \text{mol } Ca(OH)_2}{0.01000\ \text{L}} = 0.02330\ \text{M } Ca(OH)_2$$

EXERCISE 4.4A

The titration of 25.00 mL of $H_2SO_4(aq)$ of an unknown concentration requires 25.20 mL of 0.1000 M NaOH(aq). What is the molarity of the $H_2SO_4(aq)$?

$$2\,NaOH(aq) + H_2SO_4(aq) \longrightarrow Na_2SO_4(aq) + 2\,H_2O(l)$$

EXERCISE 4.4B

What volume of 0.550 M NaOH(aq) is required to titrate a 10.00-mL sample of vinegar that is 4.12% by mass of acetic acid, CH_3COOH? Assume that the vinegar has a density of 1.01 g/mL.

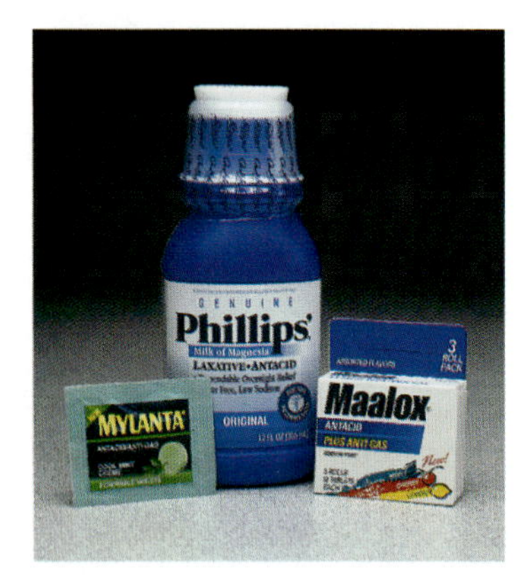

Water-insoluble hydroxides such as $Mg(OH)_2(s)$ (milk of magnesia) and $Al(OH)_3(s)$ are used as antacids. Using ionic hydroxides that are soluble in water would be quite dangerous. In high concentrations, $OH^-(aq)$ is highly basic and causes severe burning and scarring of tissue.

More Acid-Base Reactions

Magnesium hydroxide is a strong base, but it is only slightly soluble in water. In a net ionic equation that shows how a slurry of $Mg(OH)_2(s)$ in water neutralizes excess stomach acid, we should use its complete formula to represent the magnesium hydroxide. Nevertheless, OH^- from $Mg(OH)_2$ combines with H^+ of the stomach acid to form water.

$$Mg(OH)_2(s) + 2\,H^+(aq) \longrightarrow Mg^{2+}(aq) + 2\,H_2O(l)$$

The net ionic equation for the neutralization of HCl(aq) with $NH_3(aq)$ looks rather different from that of a strong acid and strong base for two reasons: (1) NH_3 is a weak base, and we should therefore write its complete formula, and (2) the NH_3 molecule contains no OH^-, which means that it cannot be satisfactorily described by the Arrhenius theory (see again, the footnote on page 140). Instead of the net ionic equation $H^+(aq) + OH^-(aq) \longrightarrow H_2O(l)$, we have

$$H^+(aq) + NH_3(aq) \longrightarrow NH_4^+(aq)$$

That is, think of H^+ from the acid as combining directly with NH_3 molecules in solution.

In the same way that H^+ can combine with OH^- to form H_2O and with NH_3 to form NH_4^+, it can combine with certain other anions to produce weak electrolytes or nonelectrolytes. In a broad sense, these reactions are also acid-base reactions.

In baking, carbon dioxide gas causes the dough to rise. The basis of the process is the action of an acidic ingredient of the dough on baking soda, $NaHCO_3$ (see Figure 4.6). Specifically, H^+ from a weak acid (let's call it HA) reacts with hydrogen carbonate (bicarbonate) ion, HCO_3^-, to form the weak acid carbonic acid, H_2CO_3. Carbonic acid is unstable and decomposes to $H_2O(l)$ and $CO_2(g)$. Equations

▶ Figure 4.6 The leavening action of baking soda
When acidified, here with citric acid ($H_3C_6H_5O_7$) from a lemon, baking soda ($NaHCO_3$) reacts to produce carbonic acid (H_2CO_3), which decomposes to carbon dioxide and water. The carbon dioxide gas produces a "lift" in the dough being baked.

TABLE 4.2 Some Common Gas-Forming Reactions

Anion	Reaction with H^+
HCO_3^-	$HCO_3^- + H^+ \longrightarrow CO_2(g) + H_2O(l)$
CO_3^{2-}	$CO_3^{2-} + 2\,H^+ \longrightarrow CO_2(g) + H_2O(l)$
HSO_3^-	$HSO_3^- + H^+ \longrightarrow SO_2(g) + H_2O(l)$
SO_3^{2-}	$SO_3^{2-} + 2\,H^+ \longrightarrow SO_2(g) + H_2O(l)$
HS^-	$HS^- + H^+ \longrightarrow H_2S(g)$
S^{2-}	$S^{2-} + 2\,H^+ \longrightarrow H_2S(g)$

for these two reactions and the sum of the equations, which represents the net ionic equation, are shown below.

$$HA(aq) + HCO_3^-(aq) \longrightarrow \cancel{H_2CO_3(aq)} + A^-(aq)$$

$$\cancel{H_2CO_3(aq)} \longrightarrow H_2O(l) + CO_2(g)$$

$$\textit{Net}: \quad HA(aq) + HCO_3^-(aq) \longrightarrow A^-(aq) + H_2O(l) + CO_2(g)$$

The reaction of carbonate ion (CO_3^{2-}) with an acid also produces unstable carbonic acid that decomposes to $H_2O(l)$ and $CO_2(g)$. The reactions of sulfite ion (SO_3^{2-}) and hydrogen sulfite ion (HSO_3^-) with an acid produce unstable sulfurous acid (H_2SO_3). It decomposes to $H_2O(l)$ and $SO_2(g)$. Net ionic equations for a few gas-forming reactions are summarized in Table 4.2.

EXAMPLE 4.5—A Conceptual Example

Explain the observations illustrated in Figure 4.7.

SOLUTION

The bulb in Figure 4.7(a) is only dimly lit because acetic acid is a weak acid, and therefore a weak electrolyte. The situation in Figure 4.7(b) is similar, here because ammonia is a weak base. In both solutions, only a small fraction of all the molecules are ionized. When the solutions are mixed (c), the H^+ ions from the CH_3COOH readily combine with NH_3 molecules to form NH_4^+ ions.

$$H^+(aq) + NH_3(aq) \longrightarrow NH_4^+(aq)$$

The removal of H^+ by this reaction causes more CH_3COOH molecules to ionize, producing more H^+ to react with more NH_3, and so on. Soon all the CH_3COOH molecules ionize, and the neutralization goes to completion.

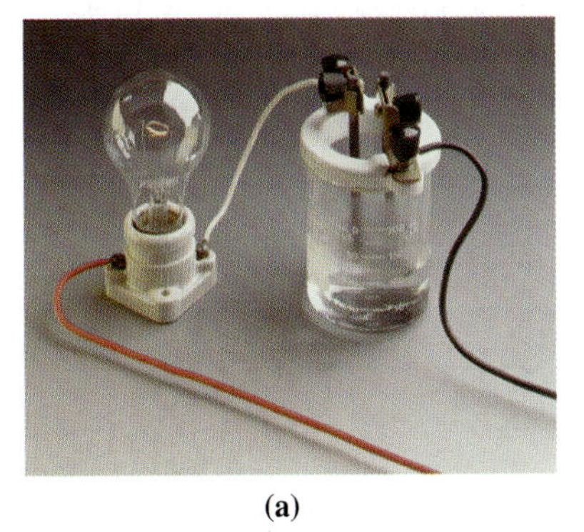

(a)

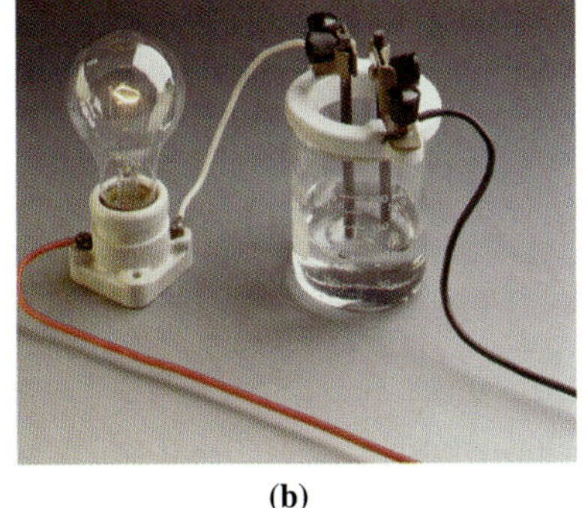

(b)

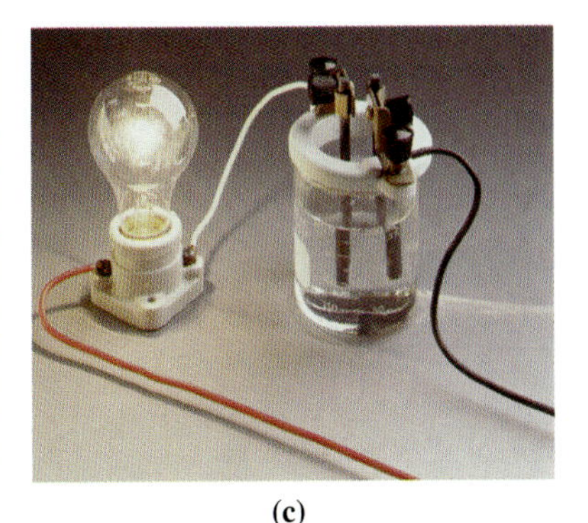

(c)

▲ **Figure 4.7 Change in electrical conductivity in a chemical reaction**
The change in electrical conductivity on mixing (a) 1 M $CH_3COOH(aq)$ and (b) 1 M NH_3 is described in Example 4.5.

$$\underset{\text{Weak acid}}{CH_3COOH(aq)} + \underset{\text{Weak base}}{NH_3(aq)} \longrightarrow \underbrace{NH_4^+(aq) + CH_3COO^-(aq)}_{\text{Salt}}$$

The original weak acid and weak base are replaced by an aqueous solution of a salt—an ionic compound and strong electrolyte. The solution is now a good electrical conductor, as seen in Figure 4.7(c).

EXERCISE 4.5

In a situation similar to that in Figure 4.7, describe the observations you would expect to make if the original solutions were $CH_3NH_2(aq)$ and $HNO_3(aq)$.

4.3 Reactions that Form Precipitates

Earlier in the chapter, we noted that many ionic compounds dissolve in water, but there is a limit to the amount of such solutes that will dissolve in a given quantity of water. For NaCl(aq) at 25 °C, this limit corresponds to a concentration of about 5.47 M NaCl. Because the maximum solute concentration for NaCl is relatively high, we say that NaCl(s) is readily soluble in water. If the maximum concentration of solute that we can obtain is less than about 0.01 M, however, we generally refer to the solute as *insoluble* in water.

Combinations of certain cations and anions, then, yield ionic compounds that are insoluble in water. If these ions are taken from separate sources and brought together in an aqueous solution, the insoluble ionic compound deposits, or precipitates, from the solution. This insoluble ionic compound is a solid known as a **precipitate**. A chemical reaction between ions that produces a precipitate is called a **precipitation reaction**. Figure 4.8 shows that a precipitate of insoluble silver iodide forms when mixing solutions of the soluble compounds silver nitrate and potassium iodide.

▲ **Figure 4.8**
The precipitation of silver iodide, AgI(s)
When a clear, colorless aqueous solution of silver nitrate is added to one of potassium iodide, a yellow precipitate of silver iodide is formed.

$$Ag^+(aq) + I^-(aq) \longrightarrow AgI(s)$$

Predicting Precipitation Reactions

Let's now consider how we might have *predicted* that a precipitate of silver iodide should form in the reaction illustrated in Figure 4.8. When asked to *predict* a chemical reaction, you will have information about the reactants—the left side of an equation—and need to provide information about the products—the right side of the equation. Predicting chemical reactions also includes recognizing when no reaction occurs, noted by writing "no reaction" on the right side of the equation. Now let's turn to the case in Figure 4.8: $AgNO_3(aq) + KI(aq) \longrightarrow$? Generally, to predict precipitation reactions, it helps to write the equation in its *ionic* form.

$$Ag^+(aq) + NO_3^-(aq) + K^+(aq) + I^-(aq) \longrightarrow ?$$

The only simple compounds that could form by combinations of these ions that are *different* from the starting reactants are KNO_3 and AgI. A precipitation reaction will occur *only* if a potential product is insoluble; that is, forms a *precipitate*.

In short, to make predictions, we need to know which ionic compounds are soluble in water and which are not. We can either look up solubility data in handbooks or we can memorize solute solubilities. Memorizing solute solubilities is easier than it might first seem, because we can summarize a lot of data for common ionic compounds in the few **solubility rules** listed in Table 4.3.

From Table 4.3, we conclude that KNO_3 is soluble (all nitrates are soluble) and that AgI is not (all chlorides, bromides, and iodides are soluble *except* those of Pb^{2+}, Ag^+, and Hg_2^{2+}). Thus, we conclude that AgI is the insoluble compound that

TABLE 4.3 General Rules for the Water Solubilities of Common Ionic Compounds

Compounds that are *soluble:*

Nitrates, acetates, and perchlorates
Group 1A metal salts and ammonium salts

Compounds that are *mostly soluble:*

Chlorides, bromides, and iodides, *except* for those of Pb^{2+}, Ag^{+}, and Hg_2^{2+}.
Sulfates, *except* for those of Sr^{2+}, Ba^{2+}, Pb^{2+}, and Hg_2^{2+}. ($CaSO_4$ is slightly soluble.)

Compounds that are *mostly insoluble:*

Carbonates, hydroxides, phosphates, and sulfides, *except* for ammonium compounds and those of the Group 1A metals. (The hydroxides and sulfides of Ca^{2+}, Sr^{2+}, and Ba^{2+} are slightly to moderately soluble.)

precipitates. We can complete the equation by showing the AgI precipitate as a solid and the KNO_3 as dissociated into ions.

$$Ag^{+}(aq) + NO_3^{-}(aq) + K^{+}(aq) + I^{-}(aq) \longrightarrow AgI(s) + K^{+}(aq) + NO_3^{-}(aq)$$

Usually our interest is only in the net ionic equation, so we can eliminate the spectator ions and write

$$Ag^{+}(aq) + I^{-}(aq) \longrightarrow AgI(s)$$

This equation tells us that regardless of their sources, if Ag^{+} and I^{-} are placed in the same solution, they will form a precipitate of AgI(s).

EXAMPLE 4.6

Predict whether a precipitation reaction will occur in each of the following cases. If so, write a net ionic equation for the reaction.

a. $Na_2SO_4(aq) + MgCl_2(aq) \longrightarrow ?$

b. $(NH_4)_2S(aq) + Cu(NO_3)_2(aq) \longrightarrow ?$

c. $K_2CO_3(aq) + ZnCl_2(aq) \longrightarrow ?$

SOLUTION

a. The initial reactants are in aqueous solution. We can begin by writing the equation in ionic form:

$$2\,Na^{+}(aq) + SO_4^{2-}(aq) + Mg^{2+}(aq) + 2\,Cl^{-}(aq) \longrightarrow ?$$

Then we can identify the possible products: NaCl and $MgSO_4$. The reaction will occur only if one or both of these products are insoluble. However, from Table 4.3 we know that both products are water soluble; all common sodium compounds are soluble, as are most sulfates, including that of magnesium. Thus, we conclude

$$Na_2SO_4(aq) + MgCl_2(aq) \longrightarrow \textit{no precipitate} \quad \textit{(no reaction)}$$

b. We can represent the reactants in solution by an ionic equation:

$$2\,NH_4^{+}(aq) + S^{2-}(aq) + Cu^{2+}(aq) + 2\,NO_3^{-}(aq) \longrightarrow ?$$

Here the possible products are NH_4NO_3 and CuS. According to the solubilities rules, all nitrates are water soluble, but most sulfides are not, including CuS. Thus, CuS will form a precipitate; the reaction that occurs is

$$\cancel{2\,NH_4^+(aq)} + S^{2-}(aq) + Cu^{2+}(aq) + \cancel{2\,NO_3^-(aq)} \longrightarrow CuS(s) + \cancel{2\,NH_4^+(aq)} + \cancel{2\,NO_3^-(aq)}$$

NH_4^+ and NO_3^- are spectator ions and can be canceled out, yielding the net ionic equation

$$S^{2-}(aq) + Cu^{2+}(aq) \longrightarrow CuS(s)$$

c. The ionic equation is

$$2\,K^+(aq) + CO_3^{2-}(aq) + Zn^{2+}(aq) + 2\,Cl^-(aq) \longrightarrow ?$$

and the potential products are KCl and $ZnCO_3$. All common potassium compounds are soluble, and among carbonates, only those of the Group 1A metals are soluble. We should expect a precipitate of $ZnCO_3(s)$ and an aqueous solution containing K^+ and Cl^- ions. The net ionic equation is

$$CO_3^{2-}(aq) + Zn^{2+}(aq) \longrightarrow ZnCO_3(s)$$

EXERCISE 4.6

Predict whether a reaction will occur in each of the following cases. If so, write a net ionic equation for the reaction.

a. $MgSO_4(aq) + KOH(aq) \longrightarrow ?$

b. $FeCl_3(aq) + Na_2S(aq) \longrightarrow ?$

c. $CaCO_3(s) + NaCl(aq) \longrightarrow ?$

▲ **Figure 4.9**
Predicting the product of a precipitation reaction
Addition of $NH_3(aq)$ to $FeCl_3(aq)$ produces a precipitate. What is it?

EXAMPLE 4.7—A Conceptual Example

Figure 4.9 shows that the dropwise addition of $NH_3(aq)$ to $FeCl_3(aq)$ produces a precipitate. What is the precipitate?

SOLUTION

The initial reactants, NH_3 and $FeCl_3$, are both water soluble. All common ammonium compounds are water soluble, so the precipitate is not likely to contain NH_4^+. It must therefore contain Fe^{3+}. But what is the anion? Recall that NH_3 is a weak base, and that it produces OH^- ions in aqueous solution.

$$\text{(a) } NH_3(aq) + H_2O(l) \rightleftharpoons NH_4^+(aq) + OH^-(aq)$$

From the solubilities rules, we expect $OH^-(aq)$ to combine with $Fe^{3+}(aq)$ to form *insoluble* $Fe(OH)_3(s)$.

$$\text{(b) } Fe^{3+}(aq) + 3\,OH^-(aq) \longrightarrow Fe(OH)_3(s)$$

To get the complete net ionic equation for the precipitation reaction, we need to multiply equation (a) by *three* to get the three OH^- ions needed in equation (b). Then we can add equation (b) to 3 × equation (a). That is,

$$3\,NH_3(aq) + 3\,H_2O(l) \rightleftharpoons 3\,NH_4^+(aq) + \cancel{3\,OH^-(aq)}$$

$$Fe^{3+}(aq) + \cancel{3\,OH^-(aq)} \longrightarrow Fe(OH)_3(s)$$

$$\textit{Net:}\quad Fe^{3+}(aq) + 3\,NH_3(aq) + 3\,H_2O(l) \longrightarrow 3\,NH_4^+(aq) + Fe(OH)_3(s)$$

The cancellation slashes show that the hydroxide ions formed in the ionization of NH_3 are consumed in the precipitation of $Fe(OH)_3(s)$ and that no free $OH^-(aq)$ appears in the net ionic equation.

EXERCISE 4.7

Suppose that after all the $Fe(OH)_3(s)$ is precipitated, a large quantity of HCl(aq) is added to the beaker in Figure 4.9. Describe what you would expect to see, and write a net ionic equation for this change.

Some Applications of Precipitation Reactions

In many cases, precipitation reactions are an attractive method of preparing chemical substances because they generally can be done with simple equipment and give a high percent yield of product. Table 4.4 lists a few industrially important precipitation reactions, and Figure 4.10 pictures one of these.

TABLE 4.4 Some Precipitation Reactions of Practical Importance

Reaction in Aqueous Solution	Application
$Al^{3+} + 3\,OH^- \longrightarrow Al(OH)_3(s)$	Water purification. (The gelatinous precipitate carries down suspended matter.)
$Al^{3+} + PO_4^{3-} \longrightarrow AlPO_4(s)$	Removal of phosphates from wastewater in sewage treatment.
$Mg^{2+} + 2\,OH^- \longrightarrow Mg(OH)_2(s)$	Precipitation of magnesium ion from seawater. (First step in the Dow process for extracting magnesium from seawater.)
$Ag^+ + Br^- \longrightarrow AgBr(s)$	Preparation of AgBr for use in photographic film.
$Zn^{2+} + SO_4^{2-} + Ba^{2+} + S^{2-} \longrightarrow ZnS(s) + BaSO_4(s)$	Production of *lithopone*, a mixture used as a white pigment in both water paints and oil paints.
$H_3PO_4(aq) + Ca(OH)_2(aq) \longrightarrow CaHPO_4 \cdot 2H_2O(s)$	Preparation of calcium hydrogen phosphate dihydrate, used as a polishing agent in toothpastes.

Figure 4.10 An important industrial precipitation reaction
The precipitation of $Mg(OH)_2$ from seawater is carried out in huge vats. This is the first step in the Dow process for extracting magnesium from seawater.

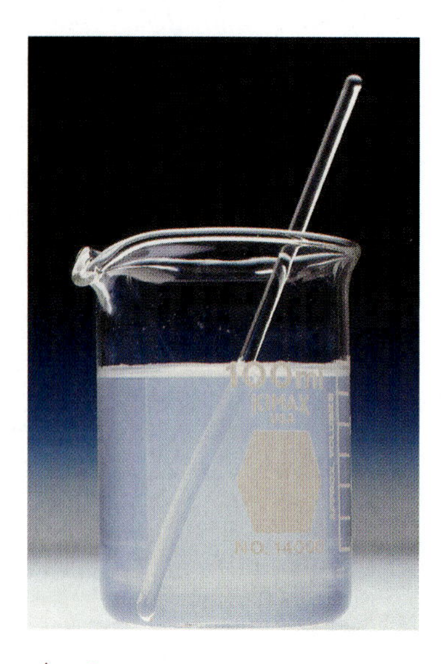

▲ **Figure 4.11**
Chloride ion in tap water
When a solution of silver nitrate [$AgNO_3(aq)$] is added to tap water, the solution turns cloudy. Eventually a precipitate of AgCl(s) settles to the bottom of the beaker. The formation of AgCl(s) is a qualitative test for Cl^-. Here it indicates the presence of Cl^- in the tap water.

Analysis of a sample of matter to find out what it contains, but with no concern for how much, is called a *qualitative* analysis. For example, as shown in Figure 4.11, a municipal water sample typically becomes cloudy when $AgNO_3(aq)$ is added to it, signaling that chloride ion is probably present in the water.

$$Cl^-\text{(from sample under analysis)} + Ag^+(aq) \longrightarrow AgCl(s)$$

Within the precision of the measurements used, the actual yield of a precipitation reaction often is equal to the theoretical yield. In this case, the reaction can be used for a *quantitative* analysis. For example, if the source of the chloride ion in a water-soluble material is NaCl, we can determine the actual quantity of NaCl present, as shown in Example 4.8.

EXAMPLE 4.8

One cup (about 240 g) of a particular clear chicken broth yields 4.302 g AgCl when excess $AgNO_3(aq)$ is added to it. Assuming that all the Cl^- is derived from NaCl, what is the mass of NaCl in this sample of broth?

SOLUTION

The equation for the precipitation reaction is

$$Ag^+(aq) + Cl^-(aq) \longrightarrow AgCl(s)$$

The stoichiometric equivalence from which we derive a stoichiometric factor is

$$1\text{ mol } Cl^-(aq) \approx 1\text{ mol AgCl(s)}$$

and the series of conversions required in the stoichiometric calculation is

$$\text{g AgCl} \longrightarrow \text{mol AgCl} \longrightarrow \text{mol } Cl^- \longrightarrow \text{mol NaCl} \longrightarrow \text{g NaCl}$$

In the usual manner, we can combine these conversions into a single setup.

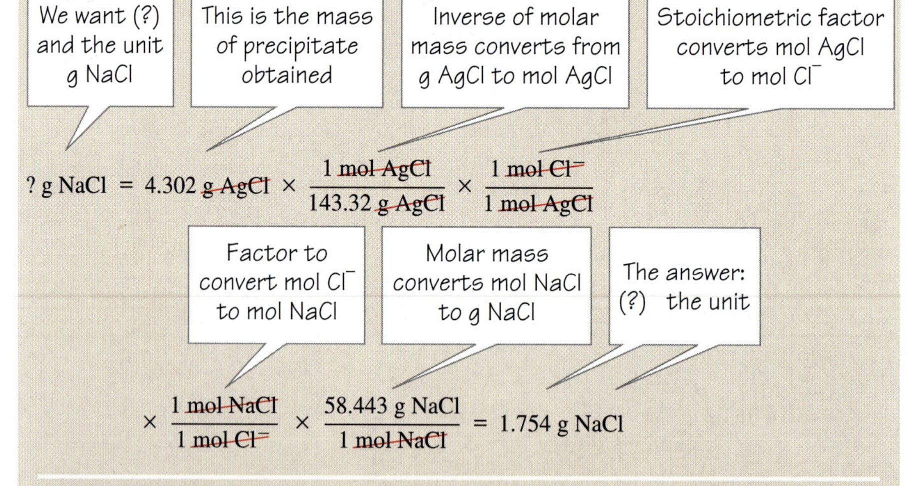

$$?\text{ g NaCl} = 4.302\ \cancel{\text{g AgCl}} \times \frac{1\ \cancel{\text{mol AgCl}}}{143.32\ \cancel{\text{g AgCl}}} \times \frac{1\ \cancel{\text{mol Cl}^-}}{1\ \cancel{\text{mol AgCl}}} \times \frac{1\ \cancel{\text{mol NaCl}}}{1\ \cancel{\text{mol Cl}^-}} \times \frac{58.443\text{ g NaCl}}{1\ \cancel{\text{mol NaCl}}} = 1.754\text{ g NaCl}$$

EXERCISE 4.8A

What is the mass percent NaCl in a mixture of sodium chloride and sodium nitrate if a 0.9056-g sample of the mixture yields 0.9372 g AgCl(s) when allowed to react with excess $AgNO_3(aq)$?

EXERCISE 4.8B

Consider the seawater sample described in Exercise 4.1A (page 137). How many grams of precipitate would you expect to get by adding **(a)** an excess of $AgNO_3$(aq) to 225 mL of the seawater; **(b)** an excess of NaOH(aq) to 5.00 L of the seawater? What are the precipitates?

4.4 Oxidation-Reduction

The third category of chemical reactions that we consider in this chapter, known as *oxidation-reduction* reactions, is perhaps the largest of all. It includes all combustion processes, most metabolic reactions in living organisms, the extraction of metals from their ores, the manufacture of countless chemicals, and many of the reactions occurring in our natural environment. Chemists often refer to oxidation-reduction reactions by turning the words around and abbreviating them to "redox" reactions; redox is a little easier to say than "oxred."

The term "oxidation" was originally used to describe reactions in which a substance combines with oxygen. The opposite process, the removal of oxygen, was described by the term "reduction." To encompass the wide range of reactions described above, however, we need a much broader definition of oxidation and reduction. First, let's introduce a concept that assists us in formulating this broader definition.

Oxidation Numbers

As an aid to understanding, we have tried whenever possible to relate observations at the macroscopic level to behavior at the molecular level. Oxidation numbers, however, don't precisely reflect anything at the microscopic or molecular level. They are an arbitrary construction that chemists find useful when dealing with a variety of matters concerning oxidation and reduction.

Oxidation numbers are easier to illustrate than to define, but after we have considered some practical examples, you should be able to relate them to this definition: An **oxidation number** represents the actual charge on a monatomic ion or a *hypothetical* charge assigned by a set of conventions to an atom in a molecule or in a polyatomic ion. For example, in the formation of sodium chloride each Na atom loses an electron and each Cl atom gains one. The compound is made up of Na^+ and Cl^- ions. We say that Na has the oxidation number +1 and Cl has the oxidation number −1. Note that we place the plus or minus sign *in front* of the numeral. This distinguishes the oxidation number from the electronic charge, where the plus or minus sign follows the numeral, as in 1+, 1−, and so on. The related term *oxidation state* refers to the actual state or condition corresponding to a given oxidation number. For example, Cl^- ion is an oxidation state of chlorine with the oxidation number of −1. The terms oxidation number and oxidation state are often used interchangeably.

● **ChemCDX** Reaction of Na and Cl_2

In the ionic compound $CaCl_2$, chlorine also has the oxidation number −1, existing as Cl^- ions. The oxidation number of calcium, however, is +2; it is present as Ca^{2+} ions. The total of the oxidation numbers of the atoms (ions) in a formula unit of $CaCl_2$ is $+2 - 1 - 1 = 0$.

In the formation of a molecule, no electrons are transferred; they are shared. We can, however, *arbitrarily* assign oxidation numbers *as if* electrons were transferred. For example, in the molecule H_2O, we assign each H atom an oxidation number of +1. If we also require that the total of the oxidation numbers for the three atoms in the molecule be *zero*, then we must assign the O atom an oxidation number of −2, because $+1 + 1 - 2 = 0$.

In the H_2 molecule, the H atoms are identical and we should assign them the same oxidation number. If we require the sum of these oxidation numbers to be *zero*, then the oxidation number of each H atom must also be 0.

From these examples you can see that we must assign oxidation numbers in a systematic way. We can deal with the great majority of compounds with the following rules. Important exceptions are listed in notes in the margin. The rules are listed below *by priority*; if two rules contradict one another, use the one with the higher priority, and this generally takes care of exceptions. For each rule, we have provided some examples. The complete set is illustrated in Example 4.9.

Because all the atoms in a molecule of an element are alike, each Cl atom in Cl_2 and each S atom in S_8 has an oxidation number of 0.

1. *For the atoms in a neutral species—an isolated atom, a molecule, or a formula unit—the total of all the oxidation numbers is 0.*
   [Examples: The oxidation number of an uncombined Fe atom is 0. The sum of the oxidation numbers of all the atoms in Cl_2, S_8, and $C_6H_{12}O_6$ is 0. The sum of the oxidation numbers of the ions in $MgBr_2$ is 0.]
   For the atoms in an ion, the total of the oxidation numbers is equal to the charge on the ion.
   [Examples: The oxidation number of Cr in the Cr^{3+} ion is +3. The sum of the oxidation numbers in PO_4^{3-} is −3, and the sum in NH_4^+ is +1.]
2. *In their compounds, the Group 1A metals all have an oxidation number of +1 and the Group 2A metals have an oxidation number of +2.*
   [Examples: The oxidation number of Na in Na_2SO_4 is +1 and that of Ca in $Ca_3(PO_4)_2$ is +2.]
3. *In its compounds, the oxidation number of fluorine is −1.*
   [Examples: The oxidation number of F is −1 in HF, ClF_3, and SO_2F_2.]
4. *In its compounds, hydrogen has an oxidation number of +1.*
   [Examples: The oxidation number of H is +1 in HCl, H_2O, NH_3, CH_4.]
5. *In most of its compounds, oxygen has an oxidation number of −2.*
   [Examples: The oxidation number of O is −2 in CO, CH_3OH, $C_6H_{12}O_6$, and ClO_4^-.]
6. *In their binary (two-element) compounds with metals, Group 7A elements have an oxidation number of −1, Group 6A elements have an oxidation number of −2, and Group 5A elements have an oxidation number of −3.*
   [Examples: The oxidation number of Br is −1 in $CaBr_2$, that of S is −2 in Na_2S, and that of N is −3 in Mg_3N_2.)

The principal exception to rule 4 is when H is bonded to a metal, as in compounds called metal hydrides, where H has an oxidation number of −1. Examples are NaH and CaH_2.

The principal exceptions to rule 5 are when O atoms are bonded to one another, as in *peroxides* (for example, in H_2O_2 the oxidation number of O is −1) and in *superoxides* (for example, in KO_2 the oxidation number of O is −1/2).

EXAMPLE 4.9

What are the oxidation numbers assigned to the atoms of each element in the following?

a. $KClO_4$ **b.** $Cr_2O_7^{2-}$ **c.** CaH_2 **d.** Na_2O_2 **e.** Fe_3O_4

SOLUTION

a. The oxidation number of K is +1 (rule 2) and that of O is −2 (rule 5). The total for *four* O atoms is −8. For these two elements the total is $+1 - 8 = -7$. The oxidation number of the Cl atom must be +7, so that the total for all atoms in the formula unit is 0 (rule 1). The oxidation numbers are +1 for K, +7 for Cl, and −2 for O.

b. The oxidation number of O is −2 (rule 5), and the total for *seven* O atoms is −14. The total of the oxidation numbers for all atoms in this *ion* must be −2 (rule 1). This means that the total of the oxidation numbers of *two* Cr atoms is +12, and that of *one* Cr atom is +6. The oxidation numbers are +6 for Cr and −2 for O.

c. The oxidation number of Ca is +2 (rule 2). The total for the formula unit must be 0 (rule 1). Even though the oxidation number of H is usually +1 (rule 4), here it must be −1, so that the total for the *two* H atoms is −2. Rule 2 takes priority over rule 4. The oxidation numbers are +2 for Ca and −1 for H.

d. The oxidation number of Na is +1 (rule 2), and for the *two* Na atoms, +2. The total for the formula unit must be 0 (rule 1). Even though the oxidation number of O is usually −2 (rule 5), here it must be −1, so that the total for the *two* O atoms is −2. Rule 2 takes priority over rule 5. The oxidation numbers are +1 for Na and −1 for O.

e. The oxidation number of O is −2 (rule 5). For *four* O atoms, the total is −8. The total for the formula unit must be 0 (rule 1). The total for *three* Fe atoms must be +8, and for each Fe atom, +8/3. The oxidation numbers are $+2\frac{2}{3}$ for Fe and −2 for O.

Usually fractional oxidation numbers signify an *average*. The compound Fe_3O_4 is actually $Fe_2O_3 \cdot FeO$. Two of the Fe atoms have oxidation numbers of +3 and one has an oxidation number of +2. The average is (3 + 3 + 2)/3 = +8/3.

EXERCISE 4.9

What are the oxidation numbers assigned to the atoms of each element in the following?

a. Al_2O_3	**d.** ClO^-	**g.** CsO_2	**j.** CH_3COOH
b. P_4	**e.** $HAsO_4^{2-}$	**h.** CH_3F	
c. $NaMnO_4$	**f.** $HSbF_6$	**i.** $CHCl_3$	

PROBLEM-SOLVING NOTE
The compounds CH_3F, $CHCl_3$, and CH_3COOH demonstrate the variability of the oxidation numbers of carbon atoms in organic compounds.

Identifying Oxidation-Reduction Reactions

The spectacular reaction pictured in Figure 4.12, called the *thermite* reaction, is used to produce liquid iron for welding large iron objects.

$$2\,Al(s) + Fe_2O_3(s) \longrightarrow 2\,Fe(l) + Al_2O_3(s)$$

Even by the limited definitions we gave at the start of this section, we can call this an oxidation-reduction reaction. Al is *oxidized* to Al_2O_3; aluminum atoms take on or *gain* oxygen atoms. Fe_2O_3 is *reduced* to Fe; iron(III) oxide *loses* oxygen atoms.

We can use oxidation numbers to identify an oxidation-reduction reaction. Take the thermite reaction as an example. In the equation below, the oxidation numbers of the Al, Fe, and O atoms are assigned according to the conventions just established and written as small numbers above the chemical symbols.

$$\overset{0}{2\,Al(s)} + \overset{+3\,-2}{Fe_2O_3(s)} \longrightarrow \overset{0}{2\,Fe(l)} + \overset{+3\,-2}{Al_2O_3(s)}$$

In the thermite reaction, the oxidation number of Al atoms *increases* from 0 to +3, and the oxidation number of Fe atoms *decreases* from +3 to 0, illustrating this oxidation-number view of oxidation-reduction:

> *In an* oxidation-reduction *reaction, the oxidation number of one or more elements increases—an* **oxidation** *process—and the oxidation number of one or more elements decreases—a* **reduction**.

The reaction pictured in Figure 4.13 differs strikingly from the thermite reaction, but the expanded definition identifies this also as an oxidation-reduction reaction. Oxidation numbers are noted in the equation that follows.

● ChemCDX Reduction of Fe_2O_3 by Al

▲ **Figure 4.12**
The thermite reaction:
$2\,Al(s) + Fe_2O_3(s) \longrightarrow 2\,Fe(l) + Al_2O_3(s)$.

Figure 4.13
An oxidation-reduction reaction:

$$Mg(s) + Cu^{2+}(aq) \longrightarrow Mg^{+}(aq) + Cu(s)$$

In the photograph on the left, a coil of magnesium ribbon is added to a solution of $CuSO_4(aq)$. After several hours, all of the $Cu^{2+}(aq)$ has been displaced from solution, leaving a deposit of red-brown copper metal, some unreacted magnesium, and clear, colorless $MgSO_4(aq)$.

● **ChemCDX** Reduction of $CuSO_4$ by Fe

● **ChemCDX** Reduction of CuO by C

● **ChemCDX** Reaction of Br_2 and Al

$$\overset{0}{Mg(s)} + \overset{+2}{Cu^{2+}(aq)} \longrightarrow \overset{+2}{Mg^{2+}(aq)} + \overset{0}{Cu(s)}$$

Mg(s) is *oxidized* to $Mg^{2+}(aq)$, and $Cu^{2+}(aq)$ is *reduced* to Cu(s).

The simple reaction in Figure 4.13 actually suggests the most fundamental definition of oxidation-reduction reactions. In this reaction, two things appear to happen simultaneously, that is,

Oxidation: $\quad Mg(s) \longrightarrow Mg^{2+}(aq) + 2\,e^-$

Reduction: $\quad Cu^{2+}(aq) + 2\,e^- \longrightarrow Cu(s)$

The oxidation number of Mg increases from 0 to +2—an oxidation—as the Mg atom loses two electrons. The Cu^{2+} ion gains the two electrons, and its oxidation number decreases from +2 to 0—a reduction. Because electrons—particles of matter—can be neither created nor destroyed, oxidation and reduction must always occur together. This fundamental definition of oxidation and reduction reflects these ideas:

> *An oxidation-reduction reaction consists of two processes that occur simultaneously. In one process, electrons are lost—oxidation—and in the other they are gained—reduction. Oxidation and reduction must always occur together.*

● **ChemCDX** Oxidation-Reduction Reactions I

Writing and Balancing Oxidation-Reduction Equations

The oxidation-reduction concept gives us a new perspective from which to consider chemical equations. Suppose, for example, we were asked to balance the following equation:

$$MnO_2(s) + O_2(g) + H^+(aq) \longrightarrow Mn^{2+}(aq) + H_2O \quad (\textit{not balanced})$$

To balance the four O atoms on the left, we need "4 H_2O" on the right, and to balance the eight H atoms on the right we need "8 H^+" on the left.

$$MnO_2(s) + O_2(g) + 8\,H^+(aq) \longrightarrow Mn^{2+} + 4\,H_2O$$

(*charge not balanced* and *reaction not possible*)

Why have we labeled this equation both "not balanced" and "not possible"? It is not balanced because the equation must reflect the fact that *electric charge cannot be created or destroyed in a chemical reaction.* The net electric charges associated with the reactants and with the products must be the same. The above equation shows *8* units of positive charge on the left (8 × 1+) but only *2* units on the right (2+). This violates the principle of conservation of electric charge.

To see that the reaction is not even possible, we can look at oxidation numbers.

$$\overset{+4\,-2}{MnO_2} + \overset{0}{O_2} + \overset{+1}{H^+} \longrightarrow \overset{+2}{Mn^{2+}} + \overset{+1\,-2}{H_2O}$$

The oxidation number of the Mn *decreases* from +4 to +2, a reduction. The oxidation number of O *decreases* from 0 in O_2 to −2 in H_2O. This is also a reduction. The oxidation number of H does not change. The reaction is impossible because reductions cannot occur without an accompanying oxidation.

We can continue to balance oxidation-reduction equations by inspection as long as we keep the two points just cited clearly in mind. Some, however, are quite difficult to balance this way. In our discussion of electrochemistry in Chapter 18, we will develop a systematic method based on the transfer of electrons in redox reactions. For the present, we will focus on a method that uses oxidation numbers and their changes.

In the *change in oxidation number* method of balancing a redox equation, we start by identifying the elements whose atoms undergo a *change* in oxidation number. At least one of these changes must involve an *increase* in oxidation number—an *oxidation* process—and one, a *decrease* in oxidation number—a *reduction* process. Thus, in the following redox reaction, which converts the noxious air pollutant NO to harmless N_2, we begin the task of equation balancing by assigning oxidation numbers.

$$\overset{-3\,+1}{NH_3} + \overset{+2\,-2}{NO} \longrightarrow \overset{0}{N_2} + \overset{+1\,-2}{H_2O} \quad (\textit{not balanced})$$

We note that N atoms in NH_3 undergo an *increase* in oxidation number (from −3 to 0) and that N atoms in NO undergo a *decrease* in oxidation number (from +2 to 0), and we summarize these changes in a diagrammatic form.

$$\overset{-3}{NH_3} + \overset{+2}{NO} \longrightarrow \overset{0}{N_2} + H_2O \quad (\textit{not balanced})$$

Oxidation number increase: 3/N atom

Oxidation number decrease: 2/N atom

The requirement:

total increase in oxidation number	=	total decrease in oxidation number
Oxidation		Reduction

is equivalent to saying that the total loss and gain of electrons are equal in a redox reaction.

The next step is a key one in the method: We adjust the coefficients of the reactants to make the *total increase* in oxidation numbers in the oxidation process equal to the *total decrease* in oxidation numbers in the reduction process. The necessary adjustments are "2 NH_3" (for a total oxidation number *increase* of 6) and "3 NO" (for a total oxidation number *decrease* of 6):

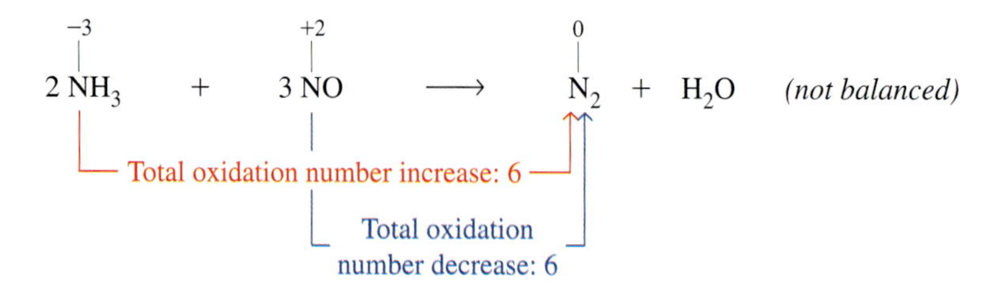

We now use familiar steps to complete the balancing: To balance the N atoms, we need a total of 5 N atoms on each side of the equation. This requires the term "5/2 N_2" on the right. We also need a total of 3 O atoms on each side, and this requires the term "3 H_2O" on the right.

$$2\ NH_3 + 3\ NO \longrightarrow 5/2\ N_2 + 3\ H_2O \quad (balanced)$$

We can further verify that the equation is balanced by noting that there are now 6 H atoms on each side. Finally, if desired, we can remove the fractional coefficient by multiplying all coefficients by *two*.

$$2 \times \{2\ NH_3 + 3\ NO \longrightarrow 5/2\ N_2 + 3\ H_2O\}$$

$$4\ NH_3 + 6\ NO \longrightarrow 5\ N_2 + 6\ H_2O \quad (balanced)$$

In Example 4.10 we apply the *change in oxidation number* method to a net ionic equation for a reaction in aqueous solution. The example brings out these points:

- In a reaction in aqueous solution, H_2O is often a reactant or product, as well as the medium in which the reaction occurs.
- If a reaction occurs in *acidic* solution, H^+ will generally appear as a reactant or product.

In Exercise 4.10B, the same reactant (Cl_2) undergoes both oxidation and reduction, and this type of redox reaction is called a **disproportionation reaction**. Also, the reaction occurs in basic solution.

- If a reaction occurs in *basic* solution, OH^- will generally appear as a reactant or product.
- In a *disproportionation* reaction, a portion of a reactant is oxidized and a portion of the same reactant is simultaneously reduced.

EXAMPLE 4.10

Balance the following redox equation.

$$Fe^{2+}(aq) + Cr_2O_7^{2-}(aq) + H^+(aq) \longrightarrow Fe^{3+}(aq) + Cr^{3+}(aq) + H_2O(l)$$

SOLUTION

1 *Identify the elements whose oxidation numbers change in the reaction.* The oxidation number of Fe increases from +2 in Fe^{2+} to +3 in Fe^{3+}, an oxidation process. The oxidation number of Cr decreases from +6 in $Cr_2O_7^{2-}$ to +3 in Cr^{3+}, a reduction process. The oxidation numbers of O and H remain unchanged at −2 and +1, respectively.

$$\overset{+2}{Fe^{2+}} + \overset{+6}{Cr_2O_7^{2-}} + H^+ \longrightarrow \overset{+3}{Fe^{3+}} + \overset{+3}{Cr^{3+}} + H_2O$$

2 *Determine the oxidation number changes per atom for elements undergoing change.*

$$\overset{+2}{Fe^{2+}} + \overset{+6}{Cr_2O_7^{2-}} + H^+ \longrightarrow \overset{+3}{Fe^{3+}} + \overset{+3}{Cr^{3+}} + H_2O$$

Oxidation number increase: 1/Fe atom

Oxidation number decrease: 3/Cr atom

3 *Adjust coefficients so total increase in oxidation number equals total decrease.* Because the Cr atoms come in pairs in the $Cr_2O_7^{2-}$, let's base these totals on *two* Cr atoms (total decrease in oxidation number: 6) and *six* Fe atoms (total increase in oxidation number: 6).

$$6\,\overset{+2}{Fe^{2+}} + \overset{+6}{Cr_2O_7^{2-}} + H^+ \longrightarrow 6\,\overset{+3}{Fe^{3+}} + 2\,\overset{+3}{Cr^{3+}} + H_2O$$

Total oxidation number increase: 6

Total oxidation number decrease: 6

4 *Adjust the remaining coefficients by inspection.* To balance the *seven* O atoms on the left, we need the term "7 H_2O" on the right, and to balance the *14* H atoms on the right, we need the term "14 H^+" on the left.

$$6\,Fe^{2+} + Cr_2O_7^{2-} + 14\,H^+ \longrightarrow 6\,Fe^{3+} + 2\,Cr^{3+} + 7\,H_2O \quad (\textit{balanced})$$

5 *Verify by checking the balance of electric charge.* We have balanced the equation atomically, and we can see that it is also balanced electrically by totaling the electric charges represented on each side of the equation.

$$(6 \times 2+) + (2-) + (14 \times 1+) = (6 \times 3+) + (2 \times 3+)$$
$$+12 \qquad -2 \qquad +14 \quad = \quad +18 \qquad +6$$
$$+24 = +24$$

EXERCISE 4.10A

Balance the following redox equation:

$$MnO_4^-(aq) + C_2O_4^{2-}(aq) + H^+(aq) \longrightarrow Mn^{2+}(aq) + H_2O(l) + CO_2(g)$$

EXERCISE 4.10B

Balance the following redox equation:

$$Cl_2(g) + OH^-(aq) \longrightarrow ClO_3^-(aq) + Cl^-(aq) + H_2O(l)$$

4.5 Oxidizing and Reducing Agents

An inorganic chemistry treatise lists dinitrogen tetroxide, N_2O_4, as a "fairly strong oxidizing agent" and hydrazine, N_2H_4, as a "powerful reducing agent." Such terms describe the way substances participate in redox reactions, and chemists generally understand their meanings.

In an oxidation-reduction reaction, the substance that is *oxidized*—because it causes some other substance to be reduced—is called a **reducing agent**. Similarly, the substance that is *reduced* is called an **oxidizing agent** because it causes another substance to be oxidized. By analogy, a lender lends money to a borrower. The borrower becomes more moneyed (is oxidized) and the lender becomes less moneyed (is reduced). The lender is like an oxidizing agent and the borrower is like a reducing agent. The borrower cannot borrow money unless there is a lender, and the lender cannot lend money unless there is a borrower. Likewise, both a reducing agent and an oxidizing agent are needed for a redox reaction.

We might well predict that nitrogen tetroxide, a "fairly strong oxidizing agent," and hydrazine, a "powerful reducing agent," should react with one another. They do indeed. The following reaction, which is accompanied by the release of large quantities of heat, is the basis of a rocket propulsion system.

$$N_2O_4(l) + 2\,N_2H_4(l) \longrightarrow 3\,N_2(g) + 4\,H_2O(g)$$

In the reaction, N_2O_4 is *reduced* to N_2 (oxidation number of N decreases from +4 to 0); N_2O_4 is the *oxidizing agent*. N_2H_4 is *oxidized* to N_2 (oxidation number of N increases from −2 to 0); N_2H_4 is the *reducing agent*. Note that although the changes in oxidation numbers occur in N atoms, we don't refer to the *atoms* as oxidizing or reducing agents. It is the *compounds* in which these atoms are found (N_2O_4 and N_2H_4) that are the oxidizing and reducing agents, respectively.

Oxidation Numbers of Nonmetals

Some compounds and ions that contain the nonmetallic elements nitrogen, sulfur, or chlorine are listed in Figure 4.14. They are arranged in order of decreasing oxidation number of the nonmetal atoms and in columns that correspond to the periodic table. We can use this figure to illustrate some additional ideas about oxidizing and reducing agents.

- The *maximum* oxidation number for a nonmetal atom is equal to the number of the group in the periodic table in which it is found: +5 for Group 5A atoms, +6 for Group 6A, and +7 for Group 7A. Oxygen and fluorine are exceptions (recall the conventions stated on page 154).
- The *minimum* oxidation number for a nonmetal atom is equal to the group number minus eight: −3 for Group 5A atoms, −2 for Group 6A, and −1 for Group 7A.
- Species in which a nonmetal atom has its maximum oxidation number are invariably oxidizing agents. The oxidation number of the nonmetal atom in these species can only *decrease* in a redox reaction. Thus in a redox reaction NO_3^- can only be an oxidizing agent.
- Species in which a nonmetal atom has its minimum oxidation number are reducing agents. Thus, in a redox reaction H_2S can only be a reducing agent.
- In principle, species in which a nonmetal atom has an intermediate oxidation number can be either an oxidizing or a reducing agent, depending on the particular reaction. In practice, one role or the other is generally more common. For example, N_2O_4, with the oxidation number +4 for N, is almost always an

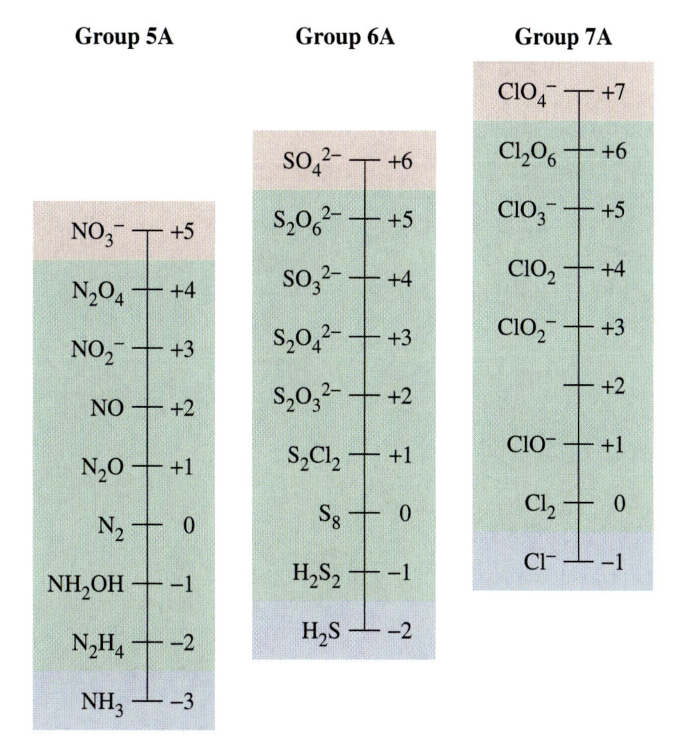

Figure 4.14
Oxidation numbers in some nitrogen-, sulfur-, and chlorine-containing species
The species in red can act only as oxidizing agents; those in blue, only as reducing agents. Those in between can act as either, depending on the reaction.

oxidizing agent; N_2H_4, with the oxidation number −2 for N, a reducing agent. Even though it is high on the oxidation-number scale for sulfur, SO_3^{2-}, usually acts as a reducing agent; Cl_2, though low on the oxidation number scale for chlorine, is generally an oxidizing agent.

Metals as Reducing Agents

In all common metallic compounds, the metal atom has a positive oxidation number. Elemental metals, of course, have atoms with oxidation number 0, their lowest common oxidation state. Metals are *reducing* agents, but their strengths as reducing agents vary widely. The atoms of some metals, such as those of Groups 1A and 2A, lose electrons easily. The metal atoms are readily oxidized to metal cations and are therefore powerful reducing agents. Other metals, such as silver and gold, are oxidized with great difficulty. They are exceptionally poor reducing agents. Figure 4.15 is a listing of some common metals in a form called the **activity series of the metals.**

> *A metal will displace from solution the ions of any metal that lies* below it *in the activity series.*

For example, with the activity series we could have predicted the reaction pictured in Figure 4.13.

$$Mg(s) + Cu^{2+}(aq) \longrightarrow Mg^{2+}(aq) + Cu(s)$$

Mg lies *above* Cu in the activity series. Therefore Mg(s), a good reducing agent, reduces Cu^{2+} to Cu(s), and the Mg(s) is itself oxidized to $Mg^{2+}(aq)$. Because Mg^{2+} takes the place of Cu^{2+} in the solution and Cu replaces Mg as the solid, we say that magnesium *displaces* copper(II) ion from solution.

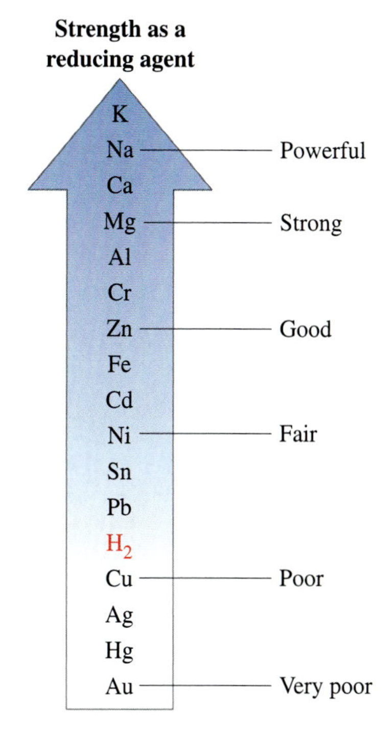

Figure 4.15
Activity series of some metals

With the activity series, we can also confidently predict that if we add Ag(s) rather than Mg(s) to Cu^{2+}(aq), there is no reaction.

$$Ag(s) + Cu^{2+}(aq) \longrightarrow \text{no reaction}$$

Ag lies *below* Cu in the activity series. Silver is unable to reduce Cu^{2+}(aq) to Cu(s).

Including H_2 in the activity series of the metals greatly increases its usefulness. This permits us to say that any metal *above* hydrogen in the series can react with an acid to produce H_2(g). For example,

$$2\,Al(s) + 6\,H^+(aq) \longrightarrow 2\,Al^{3+}(aq) + 3\,H_2(g)$$

And any metal *below* hydrogen *cannot* react with an acid to produce H_2(g). For example,

$$Ag(s) + H^+(aq) \longrightarrow \text{no reaction}$$

Finally, there are a few circumstances in which a metal that lies below hydrogen in the activity series may still undergo reaction in an acidic solution. One is described in Example 4.11.

● **ChemCDX** Oxidation-Reduction Reactions II

● **ChemCDX** Reduction of $SnCl_2$ by Zn

● **ChemCDX** Explosive Reaction of Hydrogen in Air

EXAMPLE 4.11—A Conceptual Example

Explain the difference in the behavior of a copper penny toward hydrochloric and nitric acids pictured in Figure 4.16. Write a net ionic equation for any probable reaction(s) that occurs.

SOLUTION

Copper lies below hydrogen in the activity series of the metals. Cu(s) is unable to reduce H^+(aq) to H_2(g) and be oxidized to Cu^{2+}(aq). Or, looking at it the other way, H^+ is not an oxidizing agent strong enough to oxidize Cu(s) to Cu^{2+}(aq). Chloride ion in HCl(aq) can only be a *reducing* agent. As neither H^+ nor Cl^- can oxidize Cu, we expect no reaction between Cu(s) and HCl(aq).

In the case of nitric acid, there are *two* potential oxidizing agents: H^+(aq) *and* NO_3^-(aq). From Figure 4.14, we see that NO_3^- is an oxidizing agent because the N atom has its highest possible oxidization number. Figure 4.14 also suggests that NO_3^- might be reduced to any one of several products. The red-brown gas in Figure 4.16 is nitrogen dioxide.

We can balance the equation for this reaction without employing the full *change in oxidation number* method. Notice that when HNO_3 is reduced to NO_2, the oxidation number of N decreases by 1, that is, from +5 to +4. When Cu is oxidized to Cu^{2+}, the

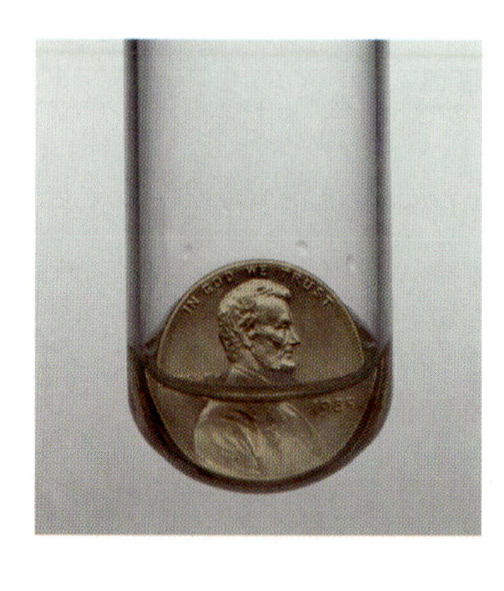

▶ Figure 4.16
The action of HCl(aq) and HNO_3(aq) on copper
As seen in these photographs, a copper penny does not react with hydrochloric acid (left), but it does react with concentrated nitric acid, producing red-brown fumes and a green-blue solution (right).

oxidation number increases by 2. This requires the terms "Cu", "Cu^{2+}", "2 NO_3^-", and "2 NO_2" in the equation. The remainder of the equation is easily balanced by inspection.

$$Cu(s) + 4\,H^+(aq) + 2\,NO_3^-(aq) \longrightarrow Cu^{2+}(aq) + 2\,NO_2(g) + 2\,H_2O(l)$$

EXERCISE 4.11

Potassium dichromate, $K_2Cr_2O_7(s)$, also exhibits a difference in behavior when heated with hydrochloric and nitric acids. With one of the acids, a gas is evolved and the solution color changes from red-orange to green. With the other acid, the red-orange color remains; that is, no reaction occurs. Explain this difference in behavior.

4.6 Some Practical Applications of Oxidation and Reduction

We conclude this chapter by considering some applications of oxidation-reduction reactions in a few of the settings where they are commonly encountered.

In Analytical Chemistry

Permanganate ion, usually from $KMnO_4$, is one of the most commonly used oxidizing agents in the chemical laboratory. For example, it can be used to oxidize Fe^{2+} to Fe^{3+} in an acidic solution in a titration for determining the percent iron in an iron ore.

$$5\,Fe^{2+}(aq) + MnO_4^-(aq) + 8\,H^+(aq) \longrightarrow 5\,Fe^{3+}(aq) + Mn^{2+}(aq) + 4\,H_2O(l)$$

The stoichiometric equivalence between the reactants is

$$5 \text{ mol } Fe^{2+} \approx 1 \text{ mol } MnO_4^-$$

The titration is illustrated in Figure 4.17, and some sample data are presented in Example 4.12.

Analytical chemistry deals with determining the composition of substances and mixtures. In Figure 4.11 we gave an example of a *qualitative* analysis: testing for the presence of Cl^- in a water sample. In Example 4.8 we had an example of a *quantitative* analysis: determing the quantity of NaCl in a sample of soup. An acid-base titration is a *volumetric* analysis: an analysis based on the measurement of solution volumes. We can also use redox reactions as the basis of titrations.

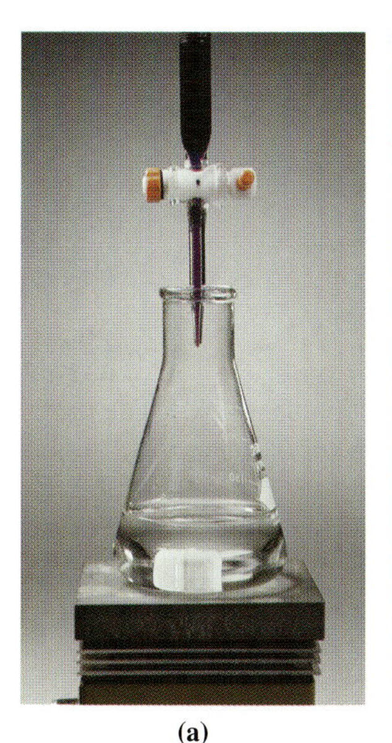
(a)

(b)

(c)

Figure 4.17 A redox titration using permanganate ion as an oxidizing agent
(a) The acidic solution in the flask contains an unknown amount of Fe^{2+}, and the buret contains $KMnO_4(aq)$ of a known concentration. (b) The solution is immediately decolorized as MnO_4^- reacts with Fe^{2+}. (c) When all the Fe^{2+} has been oxidized to Fe^{3+}, the next drop of $KMnO_4(aq)$ produces a lasting pink coloration of the solution.

EXAMPLE 4.12

A 0.2865-g sample of an iron ore is dissolved in acid and the iron obtained as $Fe^{2+}(aq)$. To titrate the solution, 0.02645 L of 0.02250 M $KMnO_4(aq)$ is required. What is the mass percent of iron in the ore?

SOLUTION

Let's do this problem in two parts. The first requires finding the number of grams of Fe in the 0.2865-g sample of iron ore from the titration data. This calculation is similar to what we have already seen in acid-base titrations. The second part is a straightforward calculation of a percentage.

The purpose of each factor in the calculation involving titration data is indicated in the outline below.

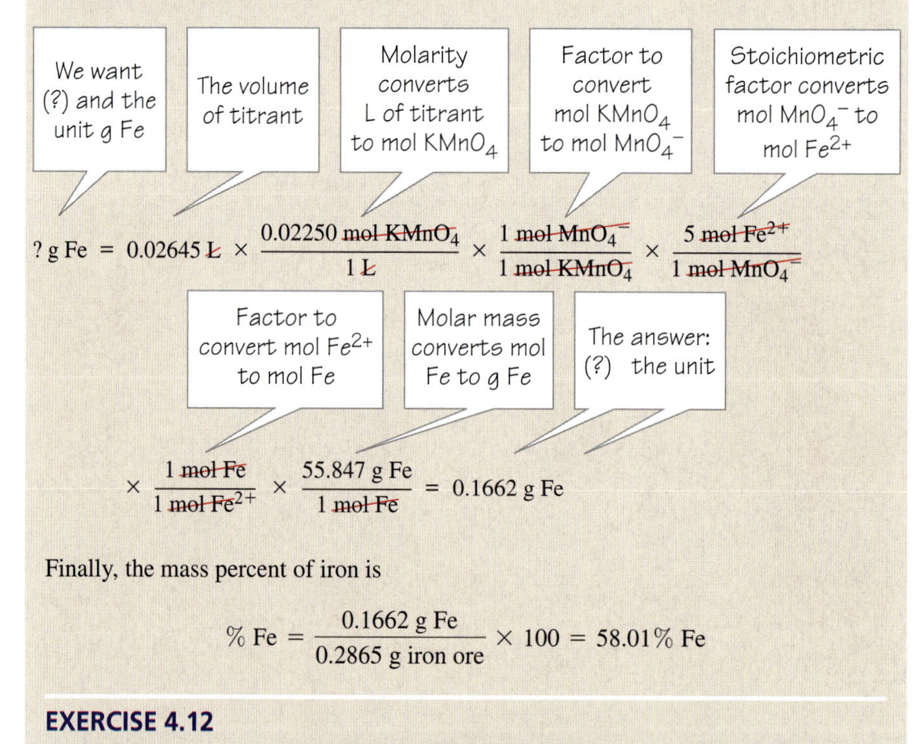

$$? \text{ g Fe} = 0.02645 \text{ L} \times \frac{0.02250 \text{ mol KMnO}_4}{1 \text{ L}} \times \frac{1 \text{ mol MnO}_4^-}{1 \text{ mol KMnO}_4} \times \frac{5 \text{ mol Fe}^{2+}}{1 \text{ mol MnO}_4^-}$$

$$\times \frac{1 \text{ mol Fe}}{1 \text{ mol Fe}^{2+}} \times \frac{55.847 \text{ g Fe}}{1 \text{ mol Fe}} = 0.1662 \text{ g Fe}$$

Finally, the mass percent of iron is

$$\% \text{ Fe} = \frac{0.1662 \text{ g Fe}}{0.2865 \text{ g iron ore}} \times 100 = 58.01\% \text{ Fe}$$

EXERCISE 4.12

Suppose the titration in Example 4.12 were carried out with 0.02250 M $K_2Cr_2O_7(aq)$ rather than $KMnO_4(aq)$. What volume of the $K_2Cr_2O_7(aq)$ would be required?

$$6\,Fe^{2+}(aq) + Cr_2O_7^{2-}(aq) + 14\,H^+ \longrightarrow 6\,Fe^{3+}(aq) + 2\,Cr^{3+}(aq) + 7\,H_2O$$

In Organic Chemistry

In Exercise 4.12, we noted the use of dichromate ion, $Cr_2O_7^{2-}$, as an oxidizing agent in analytical chemistry. It is also used to oxidize alcohols (ROH). The product of the reaction depends on the location of the OH group in the alcohol molecule, the relative proportions of alcohol and dichromate ion, and reaction conditions such as temperature. If the OH group is on a terminal carbon atom and the product is distilled off as formed, the product is an **aldehyde**. The characteristic functional group of an aldehyde is shown below.

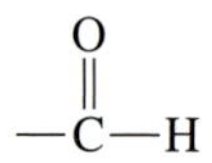

It is often written on one line as —CHO, in which case the carbon-oxygen double bond is understood. As an example, consider the following reaction.

$$3\ CH_3CH_2OH + Cr_2O_7^{2-} + 8\ H^+ \longrightarrow 3\ CH_3CHO + 2\ Cr^{3+} + 7\ H_2O$$

Ethanol, Boiling point 78 °C — Acetaldehyde, Boiling point 21 °C

Because it has a lower boiling point, the acetaldehyde can be boiled off from the reaction mixture, whereas the ethanol is retained. If the acetaldehyde is not removed as it forms, it is further oxidized to acetic acid. In this case, the overall reaction is

$$3\ CH_3CH_2OH + 2\ Cr_2O_7^{2-} + 16\ H^+ \longrightarrow 3\ CH_3COOH + 4\ Cr^{3+} + 11\ H_2O$$

From the oxidation number of Cr in $Cr_2O_7^{2-}$ (+6) and in Cr^{3+} (+3), it's not hard to see that $Cr_2O_7^{2-}$ is *reduced* in this reaction. This means that the ethanol is *oxidized* to acetaldehyde in the first reaction and to acetic acid in the second. We can reason that the conversion of acetaldehyde to acetic acid is also an oxidation in this way: Compared to acetaldehyde (CH_3CHO), acetic acid (CH_3COOH) has the same number of H atoms but one more O atom. Because the oxidation number of O is −2, the average of the oxidation numbers of the two C atoms must be *higher* in acetic acid than in acetaldehyde. An oxidation number increase occurs in oxidation. The oxidations described here are pictured in Figure 4.18.

APPLICATION NOTE

When we drink an alcoholic beverage, enzymes in our livers cause the ethanol to be oxidized to acetaldehyde. If we drink moderately, the acetaldehyde is further oxidized to acetic acid and then to carbon dioxide and water. The liver can handle about 1 oz of ethanol an hour, the quantity in one average drink. If we drink more than that, the acetaldehyde concentration builds up and we get intoxicated. Acetaldehyde is thought to be responsible for many of the harmful effects of ethanol, such as the hangover the next day and the fetal alcohol syndrome in babies born to women who drink heavily while pregnant.

(a)

(b)

(c)

◀ **Figure 4.18 The oxidation of ethanol by dichromate ion in acidic solution** (a) Orange $K_2Cr_2O_7(aq)$ about to be added to colorless $CH_3CH_2OH(aq)$ that has been acidified with H_2SO_4. (b) The ethanol solution becomes colored due to the $Cr_2O_7^{2-}(aq)$. (c) After the reaction, the solution becomes a pale violet color, signifying that the $Cr_2O_7^{2-}$ is gone and that $Cr^{3+}(aq)$ is now present.

When the —OH group of an alcohol is bonded to an interior carbon atom, the product of its oxidation is called a **ketone** and has a functional group known as the *carbonyl* group.

$$-\overset{\overset{\displaystyle O}{\|}}{C}-$$

The simplest ketone is derived from 2-propanol. It is the common solvent *acetone*, $(CH_3)_2CO$, used in applications as varied as varnishes and lacquers, rubber cement, and nail polish remover.

$$3\ CH_3\overset{\overset{\displaystyle OH}{|}}{C}HCH_3 + Cr_2O_7^{2-} + 8\ H^+ \longrightarrow 3\ CH_3\overset{\overset{\displaystyle O}{\|}}{C}CH_3 + 2\ Cr^{3+} + 7\ H_2O$$

Isopropyl alcohol, 2-Propanol — Acetone, 2-Propanone

As we have just seen, aldehydes and ketones can be formed by the oxidation of alcohols. Conversely, aldehydes and ketones can be *reduced* to alcohols. Reduction of the carbonyl group is important in living organisms. For example, in *anaerobic* exercise (exercise in which the supply of oxygen is limited), pyruvic acid is reduced to lactic acid in the muscles.

$$\underset{\text{Pyruvic acid}}{CH_3-\overset{\overset{\displaystyle O}{\|}}{C}-\overset{\overset{\displaystyle O}{\|}}{C}-OH} \xrightarrow{\text{reduction}} \underset{\text{Lactic acid}}{CH_3-\overset{\overset{\displaystyle OH}{|}}{CH}-\overset{\overset{\displaystyle O}{\|}}{C}-OH}$$

(Note that pyruvic acid is both a carboxylic acid and a ketone; it is the ketone group that is reduced.) The buildup of lactic acid during vigorous exercise is responsible in large part for the fatigue and the sore muscles that we experience later.

In Industry

The most widely used *oxidizing* agent in industrial processes, and certainly the cheapest and least objectionable environmentally, is oxygen itself. In the first step of the conversion of iron to steel, high-pressure oxygen gas is blown over molten impure iron (pig iron). This process burns off carbon and sulfur as gaseous oxides. The oxygen also converts the elements Si, P, and Mn to oxides.

Oxygen is utilized to oxidize hydrogen or acetylene in welding and cutting metals. The heat released in such reactions provides the high temperatures needed to melt metals. In the oxyacetylene torch, for example, the reaction is

$$2\,C_2H_2(g) + 5\,O_2(g) \longrightarrow 4\,CO_2(g) + 2\,H_2O(g)$$

An oxyacetylene torch, welding air-conditioning equipment in Malaysia.

Another group of important industrial *oxidizing* agents consists of chlorine gas and chlorine compounds in which the chlorine atoms have positive oxidation numbers. Chlorine gas and solutions containing hypochlorite ion, OCl^-, are used in water-treatment plants to kill pathogenic (disease-causing) microorganisms and in the paper and textile industries for bleaching.

The principal industrial *reducing* agents are carbon and hydrogen. Carbon is extensively used in a form known as *coke*. Coke is produced by driving off the volatile matter from coal. In the blast furnace method for the manufacture of iron from iron ore, the actual reducing agent is carbon monoxide, which is produced from coke.

$$C(s) + O_2(g) \longrightarrow CO_2(g)$$

$$C(s) + CO_2(g) \longrightarrow 2\,CO(g)$$

The CO(g) then reduces the iron ore to the metal.

$$Fe_2O_3(s) + 3\,CO(g) \longrightarrow 2\,Fe(l) + 3\,CO_2(g)$$

APPLICATION NOTE

Tungsten has a high electrical conductivity and the highest melting point of all the metals (3410 °C). These properties make it ideal for use as the light-emitting filaments in electric lightbulbs.

Hydrogen is used as a reducing agent in smaller-scale processes and in cases where a metal might react with carbon to form an objectionable metal carbide. As an example, tungsten metal is produced by passing a stream of $H_2(g)$ over the oxide WO_3 at 1200 °C.

$$WO_3(s) + 3\,H_2(g) \longrightarrow W(s) + 3\,H_2O(g)$$

In Everyday Life

Oxygen is undoubtedly the most important oxidizing agent in all aspects of life. We use it to oxidize fuels in heating our homes and in propelling automobiles. Oxygen corrodes metals, that is, oxidizes them to positive oxidation numbers, as in the rusting of iron. Oxygen even "burns" the foods we eat, to release the energy we need for all mental and physical activity.

A common household oxidizing agent is hydrogen peroxide, H_2O_2, usually in the form of an aqueous solution with 3% H_2O_2. An advantage of hydrogen peroxide over other oxidizing agents is that in most reactions, it is converted to water, an innocuous product. The 3% hydrogen peroxide solution is used in medicine as an *antiseptic* to treat minor cuts and abrasions. An enzyme in blood catalyzes the decomposition of hydrogen peroxide, a *disproportionation* reaction.

$$2\ H_2O_2(aq) \longrightarrow 2\ H_2O(l) + O_2(g)$$

The escaping bubbles of oxygen gas also help to carry dirt and germs out of a wound.

Benzoyl peroxide $(C_6H_5COO)_2$ is a powerful oxidizing agent that has long been used at 5% and 10% concentrations for treating acne. In addition to its antibacterial action, benzoyl peroxide acts as a skin irritant, causing the old skin to flake off and be replaced by newer, fresher-looking skin. When used on areas exposed to sunlight, however, benzoyl peroxide is thought to promote skin cancer; such use is therefore discouraged.

Chlorine and its compounds are commonly encountered as oxidizing agents in daily life. Cl_2 is used to kill microorganisms, both in drinking water and in wastewater treatment. Swimming pools are usually disinfected by "chlorination." In large pools, the chlorine is often introduced from steel cylinders in which it is stored as a liquid. The chlorine *disproportionates* in water to form hypochlorous acid, the actual sanitizing agent, and HCl(aq).

$$Cl_2(g) + H_2O(l) \longrightarrow H^+(aq) + Cl^-(aq) + HOCl(aq)$$

The HCl(aq) soon makes the pool water too acidic, and it must be neutralized by adding a base such as sodium carbonate.

Small swimming pools are often chlorinated with sodium hypochlorite, NaOCl, dissolved in NaOH(aq). The added NaOH causes these pools to become too basic, and the excess basicity must be neutralized by adding hydrochloric acid. Calcium hypochlorite, $Ca(OCl)_2$, is used to disinfect clothing and bedding in hospitals and nursing homes.

Nearly any oxidizing agent can be used as a bleach to remove unwanted color from fabrics, hair, or other materials. However, some are too expensive, some harm fabrics, some produce undesirable products, and some are simply unsafe.

Swimming pool "chlorine" used to treat small home pools is actually an alkaline solution of sodium hypochlorite, formed by the reaction of chlorine gas with aqueous sodium hydroxide.

$$Cl_2(g) + 2\ NaOH(aq) \longrightarrow NaOCl(aq) + NaCl(aq) + H_2O(l)$$

In Foods and Nutrition

In food chemistry, the substances known as *antioxidants* are reducing agents. Ascorbic acid (vitamin C), which is water soluble, is thought to retard potentially damaging oxidation of living cells. Tocopherol (vitamin E) is a fat-soluble antioxidant. In the body, vitamin E is thought to act by scavenging harmful by-products of metabolism such as the highly reactive molecular fragments called free radicals. In foods, vitamin E acts to prevent fats from being oxidized and thus becoming rancid.

When it acts as an antioxidant, vitamin C $(C_6H_8O_6)$, is oxidized to dehydroascorbic acid, $C_6H_6O_6$. For example, when nitrite ions from foods get into the

Oxidation-Reduction in Bleaching and in Stain Removal

Many colored organic materials are made up of large molecules with alternate double and single carbon-to-carbon bonds. An example is lycopene, the compound that gives tomatoes their bright red color.

$$\left[CH_3{-}\underset{|}{\overset{CH_3}{C}}{=}CH{-}CH_2{-}CH_2{-}\overset{CH_3}{C}{=}CH{-}CH{=}CH{-}\overset{CH_3}{C}{=}CH{-}CH{=}CH{-}\overset{CH_3}{C}{=}CH{-}CH{=} \right]_2$$

The brackets and subscript "2" mean that the formula within the brackets is "doubled." Think of the structure shown here as the complete formula folded over from right to left; the right hand bracket is like a hinge. When you "open up" the hinge, you get the complete formula. Chemists often use space-saving devices like this to represent complex formulas. The interesting feature of the lycopene molecule is the series of alternating double and single bonds that are shown in color. This feature is called a *conjugated system*, and such systems often confer the property of color to organic compounds.

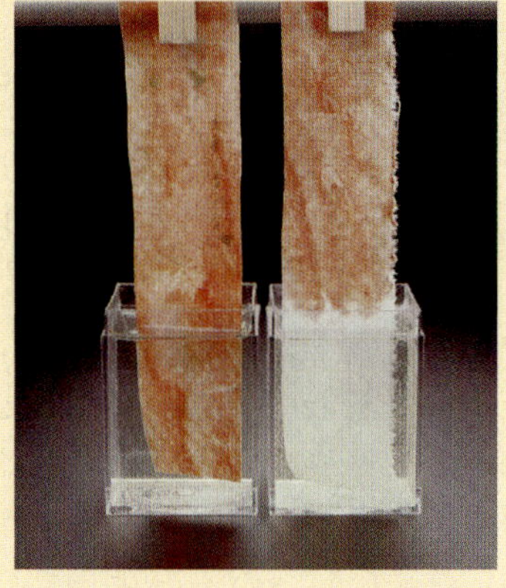

Pure water (left) has little ability to remove a dried tomato sauce stain. Sodium hypochlorite, NaOCl(aq) (right), easily bleaches the stain away by oxidizing the colored pigments of the sauce to colorless products.

Bleaching agents oxidize colored molecules such as lycopene by destroying the conjugated system. Thus, lycopene is oxidized by aqueous hypochlorite ions to colorless organic compounds, and the hypochlorite ion is reduced to chloride ion.

$$C_{40}H_{56} + OCl^-(aq) \longrightarrow Cl^-(aq) + \text{colorless organic products}$$

Hypochlorite bleaches are safe and effective for cotton and linen fabrics because they do not oxidize the cellulose that makes up the fabric. However, hypochlorites do oxidize the protein and proteinlike molecules that make up wool, silk, and nylon, and should not be used on those fabrics. Chlorine dioxide (ClO_2) and sodium chlorite ($NaClO_2$) are equally good oxidizing agents that do less damage to fabrics than do chlorine and hypochlorites.

Other bleaching agents include hydrogen peroxide, sodium perborate (often represented as $NaBO_2 \cdot H_2O_2$ to indicate an association of $NaBO_2$ and H_2O_2), and a variety of chlorine-containing organic compounds that release Cl_2 in water.

Stain removal is not nearly so simple a process as bleaching. A few stain removers are oxidizing agents or reducing agents; others have quite different chemical natures. Nearly all stains require rather specific stain removers.

Hydrogen peroxide in cold water removes blood stains from cotton and linen fabrics. Potassium permanganate can be used to remove most stains from white fabrics (except rayon). The purple permanganate stain then can be removed in a redox reaction with oxalic acid.

$$5\,H_2C_2O_4(aq) + 2\,MnO_4^- + 6\,H^+(aq) \longrightarrow 2\,Mn^{2+}(aq) + 8\,H_2O + 10\,CO_2(g)$$

Iodine, used as a disinfectant, often stains clothing in contact with the treated area. The iodine stain is readily removed in a redox reaction with sodium thiosulfate.

$$I_2 + 2\,S_2O_3^{2-}(aq) \longrightarrow 2\,I^-(aq) + S_4O_6^{2-}(aq)$$

bloodstream, they oxidize iron in hemoglobin, destroying its ability to carry oxygen. In the stomach, however, ascorbic acid reduces nitrite ion to NO(g).

$$\underset{\text{Ascorbic acid (Vitamin C)}}{C_6H_8O_6(aq)} + 2\,H^+(aq) + 2\,NO_2^-(aq) \longrightarrow \underset{\text{Dehydroascorbic acid}}{C_6H_6O_6(aq)} + 2\,H_2O(l) + 2\,NO(g)$$

Green plants carry out the redox reaction that makes possible almost all life on Earth. They do this through a process called photosynthesis, in which carbon dioxide and water are converted to glucose, a simple sugar. The synthesis of glucose requires a variety of proteins called enzymes and a green pigment called chlorophyll that converts sunlight into chemical energy. The overall change that occurs is

$$6\,CO_2 + 6\,H_2O \longrightarrow C_6H_{12}O_6 + 6\,O_2$$

In this reaction, CO_2 is *reduced* to glucose and H_2O is *oxidized* to oxygen gas. Other reactions convert the simple sugar to more complex carbohydrates and to plant proteins and oils. Animals that feed on plants are secondary sources of fats and proteins.

Summary

Soluble ionic compounds are strong electrolytes. They are completely dissociated into ions in aqueous solution. A few water-soluble molecular compounds are completely ionized in aqueous solution and are also strong electrolytes. Most molecular compounds, however, exist in solution only as molecules (nonelectrolytes) or as a mixture of molecules and ions (weak electrolytes).

A small number of acids are strong acids; they are strong electrolytes. The majority of acids, however, are weak acids. Only a small fraction of their molecules ionize in aqueous solution. The common strong bases are water-soluble ionic hydroxides. Weak bases, like the weak acids, are molecular compounds that exist as a mixture of molecules and ions in aqueous solution. Neutralization reactions between acids and bases are conveniently represented by ionic and net ionic equations. An acid-base titration is a neutralization reaction carried out in a way that brings together the acid and base in their exact stoichiometric proportions.

Another important type of reaction in solution is one in which ions combine to form an insoluble solid—a precipitate. Precipitation reactions can often be predicted with the solubility rules. Precipitation reactions are extensively used in qualitative and quantitative chemical analyses and in industry.

The concept of oxidation numbers is used to deal with a third reaction type known as oxidation-reduction (redox). Oxidation entails an increase in oxidation number, and reduction entails a decrease. Oxidation and reduction always occur simultaneously in an oxidation-reduction reaction. The *change in oxidation number* method offers a special approach to balancing redox equations.

In oxidation-reduction, the reactant that undergoes reduction is the oxidizing agent. The reactant that undergoes oxidation is the reducing agent. In a disproportionation reaction, the same substance acts both as the oxidizing and the reducing agent. Among the strong oxidizing agents are a few of the free nonmetals and polyatomic anions or compounds having atoms with high oxidation numbers. Among the strong reducing agents are active metals and certain polyatomic ions or compounds having atoms with low oxidation numbers. The activity series is a ranking of metals by their relative strengths as reducing agents. The series can be used to predict reactions between a given metal and other metal ions in solution.

Permanganate ion and dichromate ion are important oxidizing agents in the laboratory. For example, dichromate ion in acidic solution oxidizes certain alcohols to aldehydes and others to ketones. Also, it oxidizes aldehydes to carboxylic acids. Some commonly used oxidizing agents in industry are oxygen, chlorine, and hypochlorite ion. Carbon (coke) and hydrogen are common industrial reducing agents.

Key Terms

activity series of the metals (4.5)
aldehyde (4.6)
amine (4.2)
analytical chemistry (4.6)
buret (4.2)
disproportionation reaction (4.4)
ketone (4.6)
net ionic equation (4.2)
neutralization (4.2)
nonelectrolyte (4.1)
oxidation (4.4)
oxidation number (4.4)
oxidizing agent (4.5)
precipitate (4.3)
precipitation reaction (4.3)
reducing agent (4.5)
reduction (4.4)
salt (4.2)
solubility rules (4.3)
strong acid (4.2)
strong base (4.2)
strong electrolyte (4.1)
titrant (4.2)
titration (4.2)
weak acid (4.2)
weak base (4.2)
weak electrolyte (4.1)

Review Questions

1. In aqueous solution, which of the following substances are *strong* electrolytes, which are *weak* electrolytes, and which are *non*electrolytes?

(a) CH_3OH **(b)** KCl **(c)** HI **(d)** HCOOH **(e)** NaOH **(f)** HNO_2 **(g)** HBr **(h)** $CH_2OHCHOHCH_2OH$

2. Identify each of the following substances as either a strong acid, a weak acid, a strong base, a weak base, or a salt.

(a) Na_2SO_4 **(b)** KOH **(c)** $CaCl_2$ **(d)** CH_3CH_2COOH **(e)** HBr **(f)** $CH_3CH_2NH_2$ **(g)** NH_4I **(h)** $Ca(OH)_2$

3. Which of the following solutions has the *highest* and which has the *lowest* concentration of NO_3^-?

(a) 0.10 M KNO_3 **(b)** 0.040 M $Ca(NO_3)_2$ **(c)** 0.040 M $Al(NO_3)_3$ **(d)** 0.050 M $Mg(NO_3)_2$

4. Which of the following solutions has the highest *total* concentration of ions?

(a) 0.012 M $Al_2(SO_4)_3$ **(b)** 0.030 M KCl **(c)** 0.022 M $CaCl_2$ **(d)** 0.025 M K_2SO_4

5. Which of the following aqueous solutions is the *best* electrical conductor? Explain.

(a) 0.10 M NaCl **(b)** 0.10 M CH_3CH_2OH **(c)** 0.10 M CH_3COOH **(d)** 0.10 M $C_6H_{12}O_6$

6. Which of the following aqueous solutions has the highest concentration of H^+ ion? Explain.

(a) 0.10 M HCl **(b)** 0.10 M H_2SO_4 **(c)** 0.10 M CH_3COOH **(d)** 0.10 M NH_3

7. According to the Arrhenius theory, are all hydrogen-containing compounds acids? Explain.

8. According to the Arrhenius theory, are all compounds containing OH groups bases? Explain.

9. A reaction occurs in one of these mixtures but not the other. Explain why this is so, and write a net ionic equation for the reaction that occurs.

$$HCl(aq) + CH_3CH_2NH_2(aq) \longrightarrow$$

$$NaOH(aq) + CH_3NH_2 \longrightarrow$$

10. Slaked lime [$Ca(OH)_2(s)$] can be used to neutralize excess acidity in natural waters, such as lakes. Write a net ionic equation for the reaction that occurs.

11. What is the equivalence point in an acid-base titration? What is the function of the indicator in the titration?

12. In the acid-base titration pictured in Figure 4.5 (page 143), is it necessary to know the volume of water initially present in the small flask in (a)? Explain.

13. What observations would you expect to make regarding the photographs in Figure 4.7 (page 147) if the solution in (a) were 1.00 M $CH_3CH_2CH_2COOH$ and that in (b) 1.00 M KOH?

14. A reaction occurs in one of these mixtures but not the other. Explain why this is so, and write a net ionic equation for the reaction that occurs.

$$ZnCl_2(aq) + MgSO_4(aq) \longrightarrow$$

$$ZnCl_2(aq) + KOH(aq) \longrightarrow$$

15. Only one of the following compounds is *insoluble* in water. Which one must it be? Explain.

(a) $Ba(NO_3)_2$ **(b)** $ZnCl_2$ **(c)** $CuSO_4$ **(d)** $PbCrO_4$

16. Which of the following compounds reacts to precipitate Mg^{2+} from an aqueous solution of $MgCl_2$? Write an equation for the reaction.

(a) Na_2S **(b)** NaI **(c)** Na_2CO_3 **(d)** $NaNO_3$

17. What simple chemical test can you perform to determine whether a particular barium compound is $BaSO_4(s)$ or $BaCO_3(s)$?

18. What is the usual oxidation number of hydrogen atoms in compounds? What is that of oxygen atoms in compounds? What are some exceptions?

19. What happens to the oxidation number of one of its elements when a compound is oxidized, and when it is reduced?

20. Is it possible for the same compound to be both an oxidizing agent and a reducing agent? Can this occur in the same reaction? Explain.

21. Indicate the oxidation number of the underlined atom in each of the following.

(a) $\underline{Cr}$ **(b)** $\underline{Cl}O_2^-$ **(c)** $K_2\underline{Se}$ **(d)** $\underline{Te}F_6$ **(e)** $\underline{P}H_4^+$ **(f)** $Ca\underline{Ru}O_3$ **(g)** $Sr\underline{Ti}O_3$ **(h)** $\underline{P}_2O_7^{4-}$ **(i)** $\underline{S}_4O_6^{2-}$ **(j)** $\underline{N}H_2OH$

22. Use the conventions on page 154 to determine the oxidation numbers of the carbon atoms in the following organic compounds.

(a) C_2H_6
(b) CH_2O_2
(c) $C_2H_2O_2$
(d) C_2H_6O
(e) $C_2H_2O_4$

23. In the rection

$$Cu(s) + 2\,H_2SO_4(aq) \longrightarrow CuSO_4(aq) + 2\,H_2O(l) + SO_2(g)$$

is the H_2SO_4(aq) oxidized or reduced or neither? Explain.

24. Show that according to the definitions of oxidation and reduction given in the text, the combustion of a hydrocarbon is an oxidation-reduction reaction. What is the oxidizing agent, and what is the reducing agent in the reaction?

25. A reaction occurs in one of these mixtures but not the other. Explain why this is so, and write a net ionic equation for the reaction that occurs.

$$Zn(s) + CH_3COOH(aq) \longrightarrow$$

$$Au(s) + HCl(aq) \longrightarrow$$

26. Both magnesium and aluminum react with an acidic solution to produce hydrogen. Why is it that only one of the following equations correctly describes the reaction?

$$Mg(s) + 2\,H^+(aq) \longrightarrow Mg^{2+}(aq) + H_2(g)$$

$$Al(s) + 2\,H^+(aq) \longrightarrow Al^{3+}(aq) + H_2(g)$$

Problems

Ion Molarities

27. Determine the molarity of each of the following.

(a) Li^+ in 0.0385 M $LiNO_3$
(b) Cl^- in 0.035 M $CaCl_2$
(c) Al^{3+} in 0.0112 M $Al_2(SO_4)_3$
(d) Na^+ in 0.12 M Na_2HPO_4

28. Determine the molarity of each of the following.

(a) I^- in 0.0185 M KI
(b) CH_3COO^- in 1.04 M $Ca(CH_3COO)_2$
(c) Li^+ in 0.053 M Li_2SO_4
(d) NO_3^- in 0.0205 M $Al(NO_3)_3$

29. A solution is 0.0554 M NaCl and 0.0145 M Na_2SO_4. What are $[Na^+]$, $[Cl^-]$, and $[SO_4^-]$ in this solution?

30. A solution is 0.015 M each in LiCl, MgI_2, Li_2SO_4, and $AlCl_3$. What is the molarity of each ion in this solution?

31. An aqueous solution is prepared by dissolving 0.112 g $Mg(NO_3)_2 \cdot 6H_2O$ in 125 mL of solution. What is $[NO_3^-]$ in this solution?

32. A solution has 25.0 mg K_2SO_4/mL of solution. What is $[K^+]$ in this solution?

33. The components of seawater are sometimes listed in the unit, milligrams per liter (mg/L). Use the description of seawater given in Exercise 4.1A to determine the chloride ion content of seawater in mg Cl^-/L seawater.

34. A unit commonly used to describe low concentrations of a solute is *ppm* (parts per million, meaning, for example, grams solute per million grams of solution). If the concentration of chloride ion in a municipal water supply is given as 30.6 ppm Cl^-, what is the molarity of Cl^- in the water? (Assume the density of the water is 1.00 g/mL.)

35. What volume of 0.0250 M $MgCl_2$ should be diluted to 250.0 mL to obtain a solution with $[Cl^-] = 0.0135$ M?

36. A solution is prepared by mixing 100.0 mL 0.438 M NaCl, 100.0 mL 0.0512 M $MgCl_2$, and 250.0 mL of water. What are $[Na^+]$, $[Mg^{2+}]$, and $[Cl^-]$ in the resulting solution?

37. *Without doing detailed calculations*, determine which of the following has the greatest $[Cl^-]$.

(a) 0.00105 molar aluminum chloride
(b) an ammonium chloride solution containing 50 mg Cl^-/L
(c) a 0.10% by mass magnesium chloride solution (d = 1.00 g/mL)

38. *Without doing detailed calculations*, determine which of the following contains the greatest mass of the element nitrogen.

(a) 1.00 L of 0.0020 molar aluminum nitrate
(b) 500 mL of an ammonium nitrate solution containing 80 mg N/L
(c) 100 mL of a 0.10% by mass magnesium nitrate solution (d = 1.00 g/mL)

Acid-Base Reactions

39. Write equations to show the ionization of the following acids and bases.

(a) $HI(aq)$
(b) $KOH(aq)$
(c) $HNO_2(aq)$
(d) $H_2PO_4^-(aq)$
(e) $CH_3NH_2(aq)$
(f) $CH_3CH_2COOH(aq)$

40. Write equations to show the ionization of the following acids and bases.

(a) $HNO_3(aq)$
(b) $CH_3(CH_2)_2COOH(aq)$
(c) $Ba(OH)_2(aq)$
(d) $HClO_2(aq)$
(e) $HC_2O_4^-(aq)$
(f) $(CH_3)_2NH(aq)$

41. Rubidium chloride has been used in medical studies as an antidepressant. It can be made by the reaction of an aqueous solution of rubidium hydroxide and hydrochloric acid. Write **(a)** complete formula, **(b)** ionic, and **(c)** net ionic equations for this reaction.

42. Strontium iodide can be made by the reaction of solid strontium carbonate with hydroiodic acid. Write **(a)** complete formula, **(b)** ionic, and **(c)** net ionic equations for this reaction.

43. Automatic coffeemakers often develop a mineral deposit ($CaCO_3$). The manufacturer's instructions generally call for removing the deposit by treatment with vinegar. Write a net ionic equation for the reaction that occurs. (*Hint:* Recall that vinegar contains acetic acid, CH_3COOH.)

44. A paste of sodium hydrogen carbonate (sodium bicarbonate) and water can be used to relieve the pain of an ant bite. The irritant in the ant bite is formic acid (HCOOH). Write a net ionic equation for the reaction that occurs.

45. How many milliliters of 0.0195 M HCl are required to titrate **(a)** 25.00 mL 0.0365 M KOH(aq); **(b)** 10.00 mL 0.0116 M $Ca(OH)_2(aq)$; **(c)** 20.00 mL 0.0225 M $NH_3(aq)$?

46. How many milliliters of 0.0108 M $Ba(OH)_2(aq)$ are required to titrate **(a)** 20.00 mL 0.0265 M $H_2SO_4(aq)$; **(b)** 25.00 mL 0.0213 M HCl(aq); **(c)** 10.00 mL 0.0868 M $CH_3COOH(aq)$?

47. Vinegar is an aqueous solution of acetic acid, CH_3COOH. A 10.00-mL sample of a particular vinegar requires 31.45 mL of 0.2560 M KOH for its titration. What is the molarity of acetic acid in the vinegar?

48. Most window cleaners are aqueous solutions of ammonia. A 10.00-mL sample of a particular window cleaner requires 39.95 mL of 0.1008 M HCl for its titration. What is the molarity of ammonia in the window cleaner?

49. A 10.00-mL sample of a sulfuric acid solution used as a battery acid (d = 1.265 g/mL) is diluted to 100.0 mL with water. A 10.00-mL sample of the diluted acid requires 38.70 mL of 0.238 M KOH for its titration to the equivalence point, where the solution is $K_2SO_4(aq)$. What is **(a)** the molarity and **(b)** the mass percent H_2SO_4 in the battery acid?

50. A 5.00% NaOH solution by mass has a density of 1.054 g/mL. What is the minimum molarity of an HCl(aq) solution that can be used to titrate a 5.00-mL sample of the NaOH(aq) if the titration is to be accomplished without having to refill a 50.00-mL buret used in the titration?

51. *Without doing detailed calculations*, determine which of the following popular antacids is able to neutralize more stomach acid [dilute HCl(aq)] if equal masses are compared: Alka-Seltzer® (sodium hydrogen carbonate) or Tums® (calcium carbonate).

52. *Without doing detailed calculations*, determine which of the following is more effective in reducing the acidity of a home swimming pool if equal masses are compared: caustic soda (sodium hydroxide) or soda ash (sodium carbonate).

53. Which of the following points in the titration of 10.00 mL of 1.00 M CH_3COOH with 0.500 M NaOH would produce a solution with the "molecular view" like the figure? After the volume of 0.5000 M NaOH added is (a) 0.00 mL; (b) 5.00 mL; (c) 20.00 mL; (d) 22.00 mL; (e) 30.00 mL. Explain.

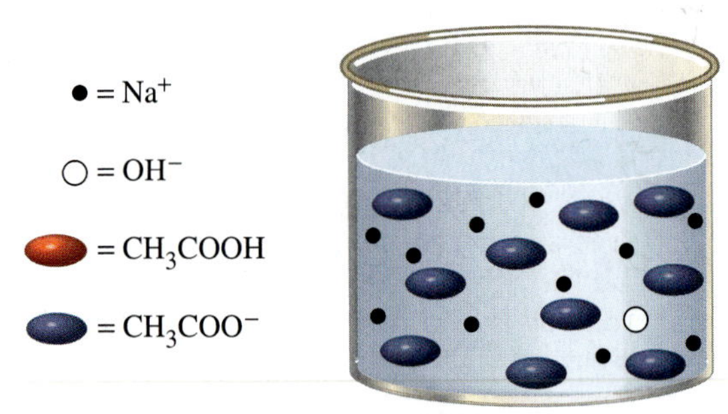

54. Refer to Problem 53. In a similar manner to the figure shown, draw a sketch to represent the titration mixture after 10.00 mL of 0.5000 M NaOH has been added.

Ionic Equations

55. Complete each of the following as a *net ionic equation*. If no reaction occurs, say so.

(a) $K^+(aq) + I^-(aq) + Pb^{2+}(aq) + 2\,NO_3^-(aq) \longrightarrow$

(b) $Mg^{2+}(aq) + 2\,Br^-(aq) + Zn^{2+}(aq) + SO_4^{2-}(aq) \longrightarrow$

(c) $Cr^{3+}(aq) + 3\,Cl^-(aq) + Li^+(aq) + OH^-(aq) \longrightarrow$

(d) $H^+(aq) + Cl^-(aq) + CH_3COOH(aq) \longrightarrow$

(e) $Ba^{2+}(aq) + 2\,OH^-(aq) + H^+(aq) + I^-(aq) \longrightarrow$

(f) $K^+(aq) + HSO_4^-(aq) + Na^+(aq) + OH^-(aq) \longrightarrow$

56. Complete each of the following as a *net ionic equation*. If no reaction occurs, say so.

(a) $Ba^{2+}(aq) + 2\ Cl^-(aq) + 2\ Na^+(aq) + CO_3^{2-}(aq) \longrightarrow$

(b) $Pb^{2+}(aq) + 2\ NO_3^-(aq) + Mg^{2+}(aq) + SO_4^{2-}(aq) \longrightarrow$

(c) $2\ Na^+(aq) + SO_4^{2-}(aq) + Cu^{2+}(aq) + 2\ Cl^-(aq) \longrightarrow$

(d) $K^+(aq) + OH^-(aq) + Na^+(aq) + HSO_4^-(aq) \longrightarrow$

(e) $Na^+(aq) + OH^-(aq) + Mg^{2+}(aq) + 2\ Cl^-(aq) \longrightarrow$

(f) $CH_3CH_2COOH(aq) + Ba^{2+}(aq) + 2\ OH^-(aq) \longrightarrow$

57. Predict whether a reaction is likely to occur in each of the following cases. If so, write a *net ionic equation* for the reaction.

(a) $MgO(s) + HI(aq) \longrightarrow$

(b) $HCOOH(aq) + NH_3(aq) \longrightarrow$

(c) $CH_3COOH(aq) + H_2SO_4(aq) \longrightarrow$

(d) $CuSO_4(aq) + Na_2CO_3(aq) \longrightarrow$

(e) $KBr(aq) + Zn(NO_3)_2 \longrightarrow$

58. Predict whether a reaction is likely to occur in each of the following cases. If so, write a net ionic equation for the reaction.

(a) $BaS(aq) + CuSO_4(aq) \longrightarrow$

(b) $Cr(OH)_3(s) + HBr(aq) \longrightarrow$

(c) $NH_3(aq) + H_2SO_4(aq) \longrightarrow$

(d) $MgBr_2(aq) + ZnSO_4(aq) \longrightarrow$

(e) $NaOH(aq) + Mg(NO_3)_2(aq) \longrightarrow$

Solubility Rules and Precipitation Reactions

59. You suspect that a certain white powder is either $MgSO_4(s)$ or $Mg(OH)_2(s)$. You add dilute $HCl(aq)$ and obtain the result shown. Does the test indicate what the powder is? If not, what test would you perform instead? Explain.

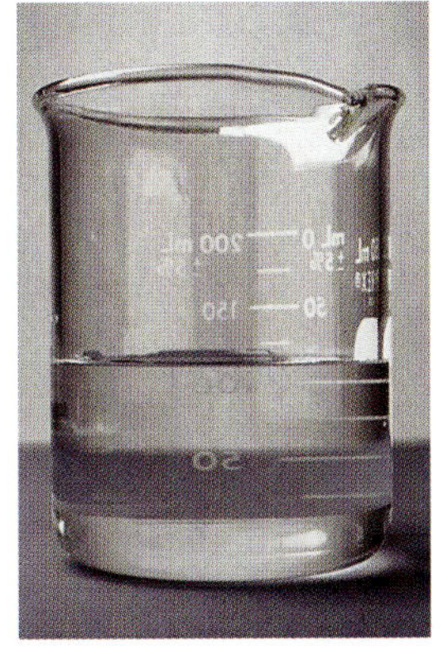

60. You suspect that a particular solution is either $CuCl_2(aq)$ or $Cu(NO_3)_2(aq)$. You add dilute $KOH(aq)$ and obtain the result shown. Does the test indicate what the solution is? If not, what test would you perform instead? Explain.

61. When aqueous solutions of copper(II) nitrate and potassium carbonate are mixed, a precipitate forms. Write the net ionic equation for this reaction.

62. When aqueous solutions of iron(III) chloride and sodium sulfide are mixed, a precipitate forms. Write the net ionic equation for this reaction.

63. You suspect that a particular unlabeled aqueous solution is one of the following: $Na_2SO_4(aq)$, $NH_3(aq)$, or $Ba(NO_3)_2(aq)$. Explain how you can use precipitation reactions on small test samples of the solution to determine its identity. You have available the variety of aqueous solutions usually found in a general chemistry laboratory.

64. The addition of $MgCl_2(g)$ to an aqueous solution containing a single unknown ionic compound as a solute produces a white precipitate. List three compounds that the unknown might be, and three that it cannot be. Explain your choices.

Oxidation-Reduction Reactions

65. Indicate whether the first named substance in each change undergoes an oxidation, a reduction, or neither. Explain your reasoning.

(a) Blue $CrCl_2(aq)$ changes to green $CrCl_3(aq)$ when exposed to air.

(b) Yellow $K_2CrO_4(aq)$ changes to orange $K_2Cr_2O_7(aq)$ when acidified.

(c) Nitrogen pentoxide produces nitric acid when it reacts with water.

66. Indicate whether the first named substance in each change undergoes an oxidation, a reduction, or neither. Explain your reasoning.

(a) Sulfur trioxide gas produces sulfuric acid when passed into water.

(b) Nitrogen dioxide converts to dinitrogen tetroxide when cooled.

(c) Carbon monoxide is converted to methane in the presence of hydrogen.

67. Balance the following redox equations.

(a) $HCl + O_2 \longrightarrow Cl_2 + H_2O$

(b) $NO + H_2 \longrightarrow NH_3 + H_2O$

(c) $CH_4 + NO \longrightarrow N_2 + CO_2 + H_2O$

68. Balance the following redox equations.

(a) $NH_3 + O_2 \longrightarrow NO + H_2O$

(b) $CH_4 + NO_2 \longrightarrow N_2 + CO_2 + H_2O$

(c) $Ca(ClO)_2 + HCl \longrightarrow CaCl_2 + H_2O + Cl_2$

69. Balance the following redox equations, except in any case where the reaction is not possible.

(a) $Cu^{2+} + I_2 \longrightarrow Cu^+ + I^-$

(b) $Ag + H^+ + NO_3^- \longrightarrow Ag^+ + H_2O + NO$

(c) $H_2O_2 + MnO_4^- + H^+ \longrightarrow Mn^{2+} + H_2O + O_2$

(d) $PbO + V^{3+} + H_2O \longrightarrow PbO_2 + VO^{2+} + H^+$

(e) $IO_4^- + I^- + H^+ \longrightarrow I_2 + H_2O$

70. Balance the following redox equations, except in those cases where the reaction is not possible.

(a) $Mn^{2+} + ClO_3^- + H_2O \longrightarrow MnO_2(s) + Cl^- + H^+$

(b) $O_2(g) + NO_3^- + H^+ \longrightarrow H_2O_2 + H_2O + NO(g)$

(c) $S_8(s) + O_2(g) + H_2O \longrightarrow H_2SO_4 + H^+$

(d) $I^- + MnO_4^- + H^+ \longrightarrow Mn^{2+} + I_2 + H_2O$

(e) $Ce^{3+} + I^- + H_2O \longrightarrow CeO_2 + I_2 + H^+$

71. Balance the following redox equations.

(a) $Zn + Cr_2O_7^{2-} + H^+ \longrightarrow Zn^{2+} + Cr^{3+} + H_2O$

(b) $SeO_3^{2-} + I^- + H^+ \longrightarrow Se + I_2 + H_2O$

(c) $Mn^{2+} + S_2O_8^{2-} + H_2O \longrightarrow MnO_4^- + SO_4^{2-} + H^+$

(d) $CaC_2O_4 + MnO_4^- + H^+ \longrightarrow Ca^{2+} + Mn^{2+} + H_2O + CO_2$

(e) $CrO_4^{2-} + AsH_3 + H_2O \longrightarrow Cr(OH)_3 + As + OH^-$

(f) $S_2O_4^{2-} + CrO_4^{2-} + H_2O + OH^- \longrightarrow Cr(OH)_3 + SO_3^{2-}$

(g) $P_4 + H_2O + OH^- \longrightarrow H_2PO_2^- + PH_3$

72. Balance the following redox equations.

(a) $Fe^{2+} + NO_3^- + H^+ \longrightarrow Fe^{3+} + H_2O + NO$

(b) $Mn^{2+} + MnO_4^- + H_2O \longrightarrow MnO_2 + H^+$

(c) $BiO_3^- + Mn^{2+} + H^+ \longrightarrow Bi^{3+} + MnO_4^- + H_2O$

(d) $BaCrO_4 + Fe^{2+} + H^+ \longrightarrow Ba^{2+} + Cr^{3+} + Fe^{3+} + H_2O$

(e) $S_8 + OH^- \longrightarrow S_2O_3^{2-} + S^{2-} + H_2O$

(f) $CH_3OH + MnO_4^- \longrightarrow HCOO^- + MnO_2 + OH^- + H_2O$

(g) $Fe_2S_3 + O_2 + H_2O \longrightarrow Fe(OH)_3 + S_8$

73. Write balanced net redox equations for the following reactions.

(a) The action of potassium permanganate on oxalic acid (HOOCCOOH) in acidic solution. The products are $Mn^{2+}(aq)$ and carbon dioxide gas.

(b) The action of permanganate ion on acetaldehyde (CH_3CHO) in basic solution. The products are manganese(IV) oxide and acetate ion (CH_3COO^-).

(c) The oxidation of thiosulfate ion to hydrogen sulfate ion in an acidic solution by chlorine gas, which is reduced to chloride ion.

74. Write balanced net redox equations for the following reactions.

(a) The action of concentrated nitric acid on zinc metal. The products are zinc(II) nitrate and nitrogen dioxide gas.

(b) The action of nitrate ion on zinc metal in basic solution. The products are $Zn^{2+}(aq)$ and $NH_3(g)$.

(c) The disproportionation of liquid bromine in a basic solution to produce bromate and bromide ions.

75. Identify the oxidizing and reducing agents in Problem 73.

76. Identify the oxidizing and reducing agents in Problem 74.

77. Use the activity series of the metals to predict chemical reactions in the following cases. Write a plausible balanced equation for the reactions that do occur and "no reaction" for those that do not.

(a) $Zn(s) + H^+(aq) \longrightarrow$

(b) $Cu(s) + Zn^{2+}(aq) \longrightarrow$

(c) $Fe(s) + Ag^+(aq) \longrightarrow$

(d) $Au(s) + H^+(aq) \longrightarrow$

78. An "unknown" metal M gives the following results in some laboratory tests. Use these data to find the approximate location of M in the activity series on page 161.

$$M(s) + 2\,H^+(aq) \longrightarrow M^{2+}(aq) + H_2(g)$$
$$M(s) + Cu^{2+}(aq) \longrightarrow M^{2+}(aq) + Cu(s)$$
$$M(s) + Fe^{2+}(aq) \longrightarrow M^{2+}(aq) + Fe(s)$$
$$2\,Al(s) + 3\,M^{2+}(aq) \longrightarrow 2\,Al^{3+}(aq) + 3\,M(s)$$
$$M(s) + Zn^{2+}(aq) \longrightarrow \text{no reaction}$$

79. How many milliliters of 0.1050 M $KMnO_4(aq)$ are required for the titration of **(a)** 20.00 mL of 0.3252 M $Fe^{2+}(aq)$, and **(b)** a 1.065-g sample of KNO_2?

$$5\,Fe^{2+} + MnO_4^- + 8\,H^+ \longrightarrow 5\,Fe^{3+} + Mn^{2+} + 4\,H_2O$$

$$5\,NO_2^- + 2\,MnO_4^- + 6\,H^+ \longrightarrow 2\,Mn^{2+} + 5\,NO_3^- + 3\,H_2O$$

80. For the reactions described in Problem 79, what is the molarity of $KMnO_4(aq)$ if **(a)** 22.55 mL $KMnO_4(aq)$ is required to titrate 10.00 mL of 0.2434 M $FeSO_4(aq)$, and **(b)** if a 567.4 mg sample of KNO_2 requires 31.61 mL $KMnO_4(aq)$ for its titration?

81. $Mn^{2+}(aq)$ can be determined by titration with $MnO_4^-(aq)$ in basic solution.

$$Mn^{2+} + MnO_4^- + OH^- \longrightarrow MnO_2(s) + H_2O \text{ (not balanced)}$$

A 25.00-mL sample of $Mn^{2+}(aq)$ requires 34.77 mL of 0.05876 M $KMnO_4(aq)$ for its titration. What is the molarity of the $Mn^{2+}(aq)$?

82. The concentration of $KMnO_4(aq)$ is to be determined by titration against $As_2O_3(s)$. A 0.1156-g sample of $As_2O_3(s)$ requires 27.08 mL of the $KMnO_4(aq)$ for its titration. What is the molarity of the $KMnO_4(aq)$?

$$As_2O_3 + MnO_4^- + H_2O + H^+ \longrightarrow H_3AsO_4 + Mn^{2+} \text{ (not balanced)}$$

83. An iron ore sample weighing 0.8765 g is dissolved in HCl(aq), and the iron is obtained as $Fe^{2+}(aq)$. This solution is then titrated with 29.43 mL of 0.04212 M $K_2Cr_2O_7(aq)$. What is the % Fe, by mass, in the ore sample?

$$Fe^{2+} + Cr_2O_7^{2-} + H^+ \longrightarrow Fe^{3+} + Cr^{3+} + H_2O \quad \text{(not balanced)}$$

84. To titrate a 5.00-mL sample of a saturated aqueous solution of sodium oxalate, $Na_2C_2O_4$, requires 25.82 mL of 0.02140 M $KMnO_4(aq)$. How many grams of $Na_2C_2O_4$ would be present in 250.0 mL of the saturated solution?

$$C_2O_4^{2-}(aq) + MnO_4^-(aq) + H^+(aq) \longrightarrow Mn^{2+}(aq) + H_2O(l) + CO_2(g) \quad \text{(not balanced)}$$

Additional Problems

85. A sample of ordinary table salt is 98.8% NaCl and 1.2% $MgCl_2$ by mass. What is $[Cl^-]$ if 6.85 g of this mixture is dissolved in 500.0 mL of an aqueous solution?

86. A solution is 0.0240 M KI and 0.0146 M MgI_2. What volume of water should be added to 100.0 mL of this solution to produce a solution with $[I^-] = 0.0500$ M?

87. A white solid is known to be either $MgCl_2$, $MgSO_4$, or $Mg(NO_3)_2$. A water solution prepared from the solid yields a white precipitate when treated with $Ba(NO_3)_2$. What must the solid be?

88. What reagent solution (including pure water) would you use to separate the cations in the following pairs, that is, with one cation appearing in solution and the other in a precipitate?

(a) $BaCl_2(s)$ and $NaCl(s)$

(b) $MgCO_3(s)$ and $Na_2CO_3(s)$

(c) $AgNO_3(s)$ and $KNO_3(s)$

(d) $PbSO_4(s)$ and $CuCO_3(s)$

(e) $Mg(OH)_2(s)$ and $BaSO_4(s)$

89. A railroad tank car carrying 1.5×10^3 kg of concentrated sulfuric acid derails and spills its load. The acid is 93.2% H_2SO_4 and has a density of 1.84 g/mL. How many kilograms of sodium carbonate (soda ash) are needed to neutralize the acid? (*Hint:* What is the neutralization reaction?)

90. To 125 mL of 1.05 M $Na_2CO_3(aq)$ is added 75 mL of 4.5 M HCl(aq). Then the solution is evaporated to dryness. What mass of NaCl(s) is obtained?

91. A home swimming pool is disinfected by the daily addition of 0.50 gal of a "chlorine" solution—NaOCl in NaOH(aq). To maintain the proper acidity in the pool, the basic components in the "chlorine" solution must be neutralized. By experiment, it is found that about 220 mL of an HCl(aq) solution that is 31.4% HCl by mass ($d = 1.16$ g/mL) is required to neutralize 0.50 gal of the "chlorine" solution. What is the $[OH^-]$ of the "chlorine" solution? (1 gal = 3.785 L)

92. What is the molarity of OH^- in the solution that results from mixing 25.10 mL of 0.2455 M NaOH and 35.05 mL of 0.1524 M HNO_3?

93. A sample of battery acid is to be analyzed for its sulfuric acid content. A 1.00-mL sample has a mass of 1.239 g. This 1.00-mL sample is diluted to 250.0 mL with water, and 10.00 mL of the diluted acid requires 32.44 mL of 0.00986 M NaOH for its titration. What is the mass percent H_2SO_4 in the battery acid?

94. The compound $NdCaMn_2O_6$ has half of its Mn atoms with oxidation number +3 and half with oxidation number +4. What is the oxidation number of the neodymium?

95. Balance the following redox equations.

(a) $CrI_3 + H_2O_2 + OH^- \longrightarrow CrO_4^{2-} + IO_4^- + H_2O$

(b) $As_2S_3 + H_2O_2 + OH^- \longrightarrow AsO_4^{3-} + SO_4^{2-} + H_2O$

(c) $I_2 + H_5IO_6 \longrightarrow IO_3^- + H_2O + H^+$

(d) $XeF_6 + OH^- \longrightarrow XeO_6^{4-} + Xe + O_2 + F^- + H_2O$

(e) $As_2S_3 + H^+ + NO_3^- + H_2O \longrightarrow H_3AsO_4 + S_8 + NO$

96. Biochemists sometimes describe reduction as the gain of H atoms. Use examples from the text to show that this definition conforms to that based on oxidation numbers.

97. The following reactions are associated with wood-eating termites: **(a)** Certain bacteria that live in the hindgut of the termites convert glucose ($C_6H_{12}O_6$) to acetic acid (CH_3COOH), carbon dioxide, and hydrogen. **(b)** Other bacteria convert part of the carbon dioxide and hydrogen to more acetic acid. **(c)** The termite then meets its respiratory energy requirements through the reaction of acetic acid and oxygen to produce carbon dioxide and water. Write balanced equations for these three reactions and identify the oxidizing and reducing agents.

98. Incineration of a chlorine-containing toxic waste such as a polychlorinated biphenyl (PCB) produces CO_2 and HCl.

$$C_{12}H_4Cl_6 + O_2 \longrightarrow CO_2 + HCl$$

Balance the equation for this combustion reaction. Comment on the advantages and disadvantages of incineration as a method of disposal of such wastes.

99. Methanol (CH_3OH) can reduce chlorate ion to chlorine dioxide in an acidic solution. The methanol is oxidized to carbon dioxide. What volume of methanol (d = 0.791 g/mL), in mL, is needed to produce 125 kg ClO_2(g)?

100. A 10.00-mL sample of an aqueous solution of H_2O_2 is treated with an excess of KI(aq). The liberated I_2 requires 28.91 mL of 0.1522 M $Na_2S_2O_3$ for its titration. Is the H_2O_2(aq) up to full strength (3% H_2O_2 by mass) as an antiseptic solution? Assume that the density of the H_2O_2(aq) is 1.00 g/mL.

$H_2O_2(aq) + H^+(aq) + I^-(aq) \longrightarrow H_2O(l) + I_2(s)$ *(not balanced)*

$I_2(s) + S_2O_3^{2-}(aq) \longrightarrow S_4O_6^{2-}(aq) + I^-(aq)$ *(not balanced)*

101. The exact concentration of an aqueous solution of oxalic acid (HOOCCOOH, that is, $H_2C_2O_4$) is determined by an acid-base titration. Then the oxalic acid solution is used to determine the concentration of $KMnO_4$(aq) by a redox titration in acidic solution. The titration of 25.00-mL samples of the oxalic acid solution requires 32.15 mL of 0.1050 M NaOH and 28.12 mL of the $KMnO_4$(aq). What is the molarity of the $KMnO_4$(aq)? (*Hint:* Oxalic acid is oxidized to carbon dioxide.)

102. Although Figure 4.7 (page 147) gives a dramatic picture, chemists use precise measurements to determine how well a solution conducts electric current and how the conductance changes as the result of a chemical reaction. For our purposes, we don't need the details of the method used; we just need to note the ideas that follow.

- The conductivity of a solution depends on the total concentration of ions.
- Ions differ in their individual abilities to carry electric current.
- H^+ and OH^- are much better electrical conductors than other ions.

Shown below are four idealized graphs of the electrical conductance of a solution during the course of a titration. For example, (a) represents the titration of HCl(aq) with NaOH(aq).

$$H^+ + Cl^- + Na^+ + OH^- \longrightarrow Na^+ + Cl^- + H_2O$$

The conductance starts high because $[H^+]$ is greatest at the start. The conductance decreases during the titration because H^+ reacts with OH^- to form the nonelectrolyte H_2O, and it is replaced by Na^+, which does not conduct current nearly as well as H^+. The conductance falls to a minimum at the equivalence point, where the solution is simply NaCl(aq). Beyond the equivalence point, the conductance again rises because of the accumulation of excess OH^-, which is a very good electrical conductor.

Match each of the following titrations to the appropriate remaining graph: (b), (c), or (d). In the manner used above for graph (a), explain the general shape of each graph.

NH_3(aq) with HCl(aq) as the titrant

CH_3COOH(aq) with NH_3(aq) as the titrant

$Ba(OH)_2$ with $(NH_4)_2SO_4$ as the titrant

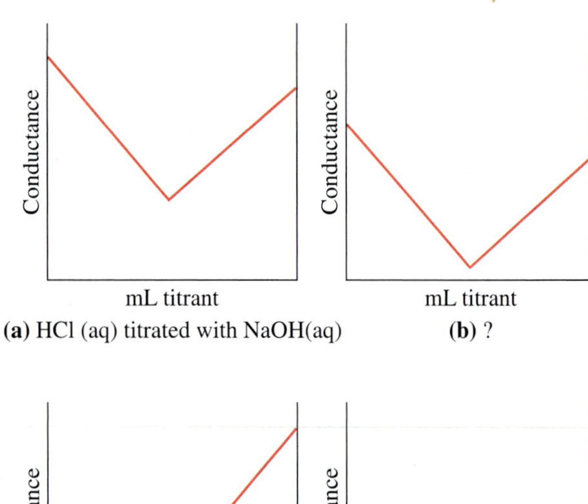

(a) HCl (aq) titrated with NaOH(aq) **(b)** ?

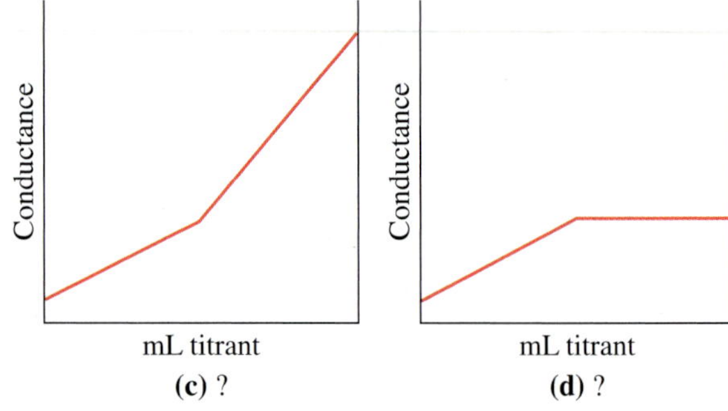

(c) ? **(d)** ?

Gases

CHAPTER

5

Air bags can protect passengers in automobile crashes or delicate instruments as they land on the surface of a planet. These giant bags protect the Pathfinder lander as it lands on the surface of Mars on July 4, 1997. In order to design an effective air bag, scientists must understand the gas laws, reaction stoichiometry of gases, and much more. See Example 5.17, page 203.

Earth's life-support system relies on a thin blanket of gases that we call air. Air is so insubstantial that it is difficult to think of it as matter. But air is matter; it is matter in the gaseous state. Gases have mass and occupy space.

In the macroscopic world, we probably have a more intuitive feeling for liquids and solids than for gases. In the microscopic realm, however, gases are much easier to conceive than are liquids and solids. This is because the behavior of liquids and solids is critically dependent on intermolecular forces, which in turn are closely related to molecular structure. In contrast, we don't need to know much about molecular structure or intermolecular forces to describe gases. The simple laws governing their behavior are based on the *absence* of intermolecular forces.

Among the most successful of scientific theories is one that we will introduce in this chapter—the kinetic-molecular theory. In a later chapter (Chapter 13), we will find that the kinetic-molecular theory not only explains the behavior of gases, but also provides important insights into how molecules enter into chemical reactions.

CONTENTS

● **ChemCDX**
When you see this icon, there are related animations, demonstrations, and exercises on the CD accompanying this text.

Several gases are among the top industrial chemicals and are often shipped as liquids in tanker trucks. Among all the industrial chemicals produced in the United States, typical annual rankings, by mass, have nitrogen in the No. 2 position; oxygen, No. 3; ethylene, No. 4; ammonia, No. 6; and chlorine, No. 10. The tanker truck here is being filled with liquefied natural gas.

5.1 Gases: What Are They Like?

Gases are composed of widely separated particles (molecules in most cases, atoms in some) in constant, random motion. Gases flow readily and occupy the entire volume of their container, regardless of its shape. If we release ammonia in one part of a room, we can soon smell it throughout the room.

Unlike a liquid or solid, in which molecules or atoms are already quite close to one another, a gas with widely separated particles is readily compressed by the application of pressure. It is possible to compress enough air for an hour or more of breathing into a small portable tank for underwater diving. When air is compressed, the particles are forced closer together, but they are still far apart compared to those of a liquid or solid.

At room temperature, nearly all *ionic* substances, even those of low molar mass, are solids (for example, NaCl: 58.44 g/mol). In contrast, most *molecular* substances of low molar mass are gases or liquids that are easily vaporized. For example, nitrogen is a gas (N_2: 28.01 g/mol) and ethanol is a liquid (CH_3CH_2OH: 46.07 g/mol). **Vapor** is a term used to denote the gaseous state of a substance that is more commonly encountered as a liquid (for example, water or ethanol) or a solid (for example, *para*-dichlorobenzene, used as a moth repellent and room deodorizer).

Table 5.1 lists several common gases. All of these gases and many others are commercially available in tanks, either as highly compressed gases or as liquids that vaporize when the pressure on them is released. Perhaps the most familiar and important gases are those found mixed together in ordinary air—N_2, O_2, Ar, and CO_2. The gases that make up the air we breathe are so significant that we will devote considerable attention to a study of the atmosphere in a later chapter (Chapter 25).

TABLE 5.1 Some Common Gases[a]

Substance	Formula	Typical Use(s)
Acetylene	C_2H_2	Welding of metals
Ammonia	NH_3	Fertilizer, manufacture of plastics
Argon	Ar	Filling gas for specialized lightbulbs
Butane	C_4H_{10}	Fuel for heating (LPG)
Carbon dioxide	CO_2	Carbonation of beverages
Carbon monoxide	CO	Obtaining metals from ores
Chlorine	Cl_2	Disinfectant, bleach
Ethylene	C_2H_4	Manufacture of plastics
Helium	He	Lifting gas for balloons
Hydrogen	H_2	Chemical reagent
Hydrogen sulfide	H_2S	Chemical reagent
Methane	CH_4	Fuel, manufacture of hydrogen
Nitrogen	N_2	Manufacture of ammonia
Nitrous oxide	N_2O	Anesthetic
Oxygen	O_2	Support of combustion; respiration
Propane	C_3H_8	Fuel for heating (LPG)
Sulfur dioxide	SO_2	Preservative, disinfectant, bleach

[a] All of these substances are gases at room temperature (about 25 °C) and at pressures comparable to that of the atmosphere, but they can be converted to liquids and solids by a sufficient lowering of the temperature and/or increase of the pressure.

5.2 The Kinetic-Molecular Theory: An Introduction

The **kinetic-molecular theory** was developed in the mid-nineteenth century. It provides a model for gases at the *microscopic* level that explains their physical properties, which we observe at the macroscopic level. The theory treats gases as collections of particles in rapid, random motion. In the name "kinetic-molecular" theory, "kinetic" conveys the idea of motion, and "molecular" indicates that the motion is that of molecules. The theory refers to particles of a gas as "molecules," even if the particles are monatomic in some cases. The particles are molecules (N_2) in nitrogen gas, for instance, and atoms (Ar) in argon gas.

The molecules of a gas are in such rapid motion that they seem to resist the force of gravity. They do not fall and collect at the bottom of a container, as do the molecules in a liquid. Gases fill their containers completely. *Filled* does not mean that the gas molecules are tightly packed, but rather that they are distributed uniformly throughout the entire volume of a container. In fact, because the distances between gas molecules are generally much greater than the dimensions of the molecules themselves, the typical gas is mostly empty space. For example, on average, a molecule of N_2 in room temperature air travels a distance about 200 times its molecular diameter between collisions.

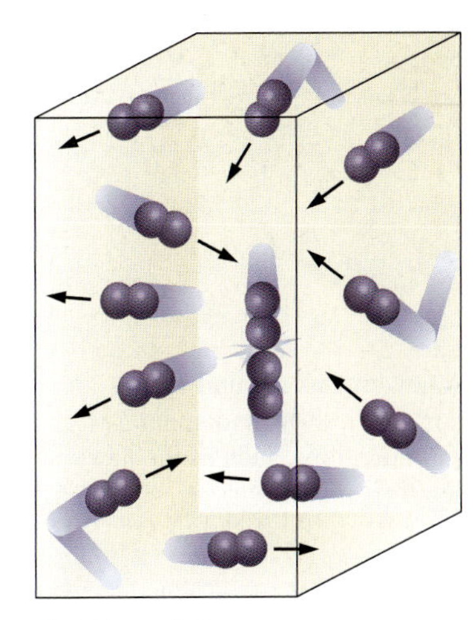

▲ **Figure 5.1**
Visualizing molecular motion in a gas
Molecules of a gas are in constant, random, straight-line motion. They undergo elastic collisions with each other and with the walls of the container.

The movement of gaseous molecules through three-dimensional space, called *translational* motion, is *random*. That is, we find it impossible to predict the speed and direction of motion of any given molecule at any given time. The speed can vary widely, and all directions of motion are equally probable. A molecule of a gas moves along a straight-line path unless it strikes another molecule of the gas or hits the container wall. Then it bounces off and travels along a new straight-line path until its next collision (see Figure 5.1). Some molecules lose energy and are slowed down by collisions, but others gain energy and are speeded up. However, there is no *net* loss or gain of energy in a collision. Therefore, the total translational kinetic energy of the molecules of a gas is not changed by collisions. Because of this conservation of kinetic energy, we say that the collisions of gas molecules are *elastic*.

● **ChemCDX** Kinetic Energy in a Gas

The kinetic-molecular theory explains what we measure when we measure temperature. According to the theory, temperature is a measure of the average translational kinetic energies of the molecules of a sample of gas. The higher the average translational kinetic energy (signifying the faster, on average, the molecules are moving), the higher is the temperature of the gas. On average, molecules of a cold sample of a gas are moving more slowly than those of a hot sample of the same gas. We must relate temperature to the *average* translational energy because the molecules of a gas, moving at different speeds, have different kinetic energies.

The kinetic-molecular theory also explains the origin of gas pressure, an important property of gases that we explore in the next section. Consider a gas-filled balloon: When a molecule of the gas strikes the wall of the balloon, it gives the wall a little push. When we measure gas pressure, we assess these "molecular pushes." We will return to a more quantitative discussion of the kinetic-molecular theory in Section 5.11 after we have studied some of the natural laws that describe gas behavior.

5.3 Gas Pressure

The "molecular push" exerted by an individual molecule of air as it bounces off our skin is too tiny for us to feel. However, we can perceive the collective effect of very large numbers of gas molecules by measuring the pressure they exert. **Pressure** is defined as force per unit area, that is, a force divided by the area over which the force is exerted.

$$\text{Pressure} = \frac{\text{force}}{\text{area}} = \frac{F}{A}$$

In SI, force is expressed in *newtons* (N) and area in square meters (m^2). Therefore the derived SI unit for pressure is newton per square meter, also called a **pascal (Pa)**.*

A newton has the unit kg m s^{-2}. For a review of fundamental physical quantities, such as the newton, see Appendix B.

$$1\ \text{Pa} = 1\ \text{N/m}^2$$

The pascal is such a small unit that the **kilopascal (kPa)** is often used instead. As we shall see next, the easiest way to measure a gas pressure is to compare it to the pressure exerted by a column of liquid. Units of pressure stated in terms of liquid pressures are still widely used.

Measuring Atmospheric Pressure: Barometers

The pressure of the atmosphere is measured by a device called a **barometer**. In the mercury barometer, invented in 1643 by Evangelista Torricelli (1608–1647), a long glass tube, closed at one end, is filled with mercury and inverted in a shallow dish that also contains mercury (Figure 5.2). Suppose the tube is 1 m long. Some of the mercury in the tube drains into the dish, but *not all* of it. The mercury drains out only until the pressure exerted by the mercury remaining in the tube exactly balances the pressure exerted by the atmosphere on the surface of the mercury in the dish. The mercury in the tube tries to push its way out under the influence of gravity, and the air pressure tries to push the mercury back in. At some point these two opposing tendencies reach a stalemate, or equilibrium.

Mercury is a dense liquid. At sea level, a column of mercury about 760 mm high will balance the push of a column of air many kilometers high, a column of air that is measured from Earth's surface to a distance where the atmosphere gives way to outer space. The pressure that is exerted by a column of mercury that is *exactly* 760 mm high is called **1 atmosphere (atm)**.[†] The pressure unit, one **millimeter of mercury (mmHg)** is often called a **Torr** (after Torricelli). That is,

$$1\ \text{atm} = 760\ \text{mmHg} = 760\ \text{Torr}$$

These are the pressure units that we will use for the most part in this text.

The relationship of the atmosphere unit to SI units is

$$1\ \text{atm} = 101{,}325\ \text{Pa} = 101.325\ \text{kPa}$$

For approximate work, it is helpful to remember that 1 atm is about 100 kPa. Several other pressure units are widely used. Weather reports in the United States often include atmospheric pressure in *millibars* (*mb*) or *inches of mercury* (*in. Hg*).

$$1\ \text{atm} = 29.921\ \text{in. Hg} = 1.01325\ \text{bar} = 1013.25\ \text{mb}$$

* Units that are named after individuals are not capitalized, but the symbols for such units are. Thus, the pressure unit named for Blaise Pascal (1623–1662) is called a "pascal," but it is represented by the symbol "Pa."

[†] For the pressure to be exactly 1 atm, the 760-mm column of mercury must be at 0 °C (d = 13.59508 g/cm^3) and at a location where the acceleration due to gravity (g) is 9.80665 m/s^2.

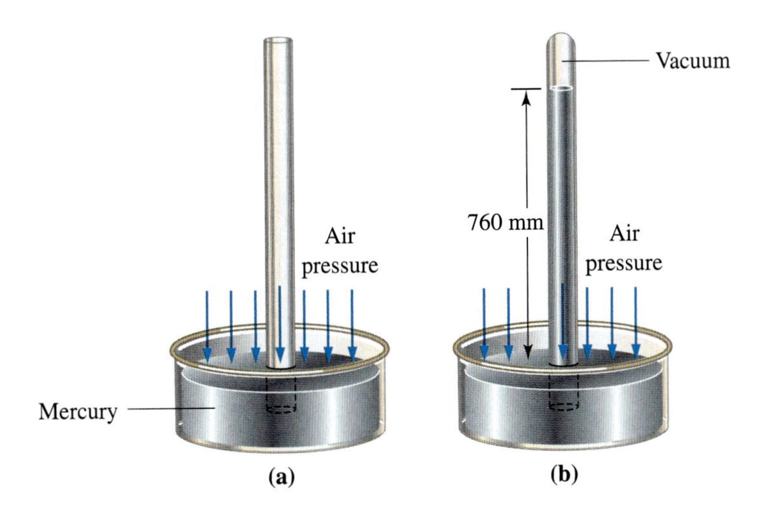

Figure 5.2
Measurement of air pressure with a mercury barometer
(a) The mercury levels are equal inside and outside the *open-end* tube because the tube is open to the atmosphere and filled with air. (b) A column of mercury 760 mm high is maintained in the *closed-end* tube for standard atmospheric pressure. The space above the mercury is devoid of air (a vacuum), containing only a tiny trace of mercury vapor.

Engineers generally use *pounds per square inch* ($lb/in.^2$) and the symbol *psi* for practical applications like steam pressure in boilers and turbines.

$$1 \text{ atm} = 14.696 \text{ lb/in.}^2 = 14.696 \text{ psi}$$

EXAMPLE 5.1

A Canadian weather report gives the atmospheric pressure as 100.2 kPa. What is the pressure in Torr?

SOLUTION

To convert a pressure from one unit to another, we need a relationship between the two units. Such relationships are suggested by Table 5.2, which expresses the standard atmosphere of pressure in several different units. The relationships we need here are 1 atm = 760 Torr = 101.325 kPa, from which we can write

$$\frac{760 \text{ Torr}}{101.325 \text{ kPa}} = 1$$

With this conversion factor, we find that

$$? \text{ Torr} = 100.2 \cancel{\text{kPa}} \times \frac{760 \text{ Torr}}{101.325 \cancel{\text{kPa}}} = 751.6 \text{ Torr}$$

EXERCISE 5.1A

Carry out the following conversions of pressure units.

a. 0.947 atm to mmHg
b. 98.2 kPa to Torr
c. 29.95 in. Hg to Torr
d. 768 Torr to atm.

EXERCISE 5.1B

What is the pressure, in kilopascals and in atmospheres, if a force of 1.00×10^2N is exerted on an area of 5.00 cm^2?

TABLE 5.2 The Standard Atmosphere of Pressure in Different Units

1 atm	=	760 mmHg
		760 Torr
		14.696 $lb/in.^2$
		101,325 N/m^2
		101,325 Pa
		101.325 kPa
		1.01325 bar
		1013.25 mb

Although gas molecules do not settle under the force of gravity in laboratory-size containers, they do settle in that huge container that we call the atmosphere. As a result, the density of air—the mass per unit volume—decreases rapidly with increased height above Earth's surface. Because the atmospheric pressure falls off rapidly with altitude, we are required to define the standard atmosphere in terms of a mercury column at sea level.

When driving up a mountain or riding an express elevator to the top of a tall building, our ears may "pop." This popping sensation is caused by an unequal air pressure on the two sides of our eardrums. The popping stops as soon as the higher pressure low-altitude air inside the ear is replaced by the lower pressure high-altitude air.

Measuring Other Gas Pressures: Manometers

A mercury barometer is fine for measuring the pressure of the atmosphere, but we often need to know the pressure of a gas in a small closed container. One common device for measuring gas pressures is called a **manometer**.

The manometer in Figure 5.3, known as a *closed-end* manometer, is like a barometer that has been bent into a U-shape. When the bulb connected to the manometer is evacuated of gases, the mercury levels in the two arms are equal and the gas pressure is essentially zero. When the bulb is filled with a gas, the gas pressure pushes a column of mercury up the closed-end arm. The gas pressure is the *difference* in height of the mercury in the two arms.

With an *open-end* manometer (see Figure 5.4), the gas pressure, P_{gas}, is not zero when the mercury column heights are equal. Instead, P_{gas} is equal to the prevailing atmospheric pressure, as measured with a barometer and called *barometric pressure*, P_{bar}. When there is a difference in the heights of the mercury columns,

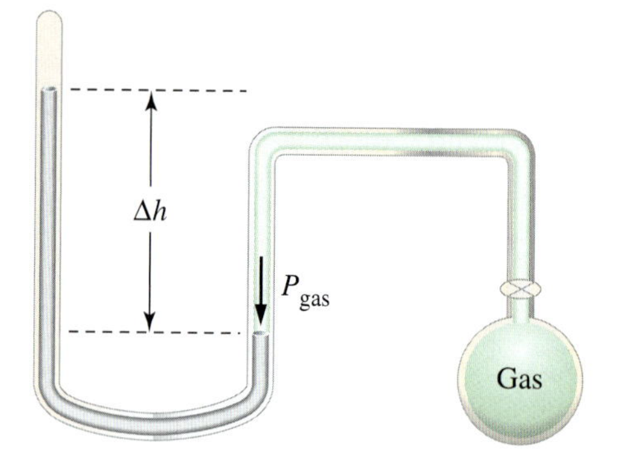

▶ **Figure 5.3**
Measuring gas pressure with a closed-end manometer
The gas pressure is equal to the *difference* in height (Δh) of the mercury column in the two arms of the manometer.

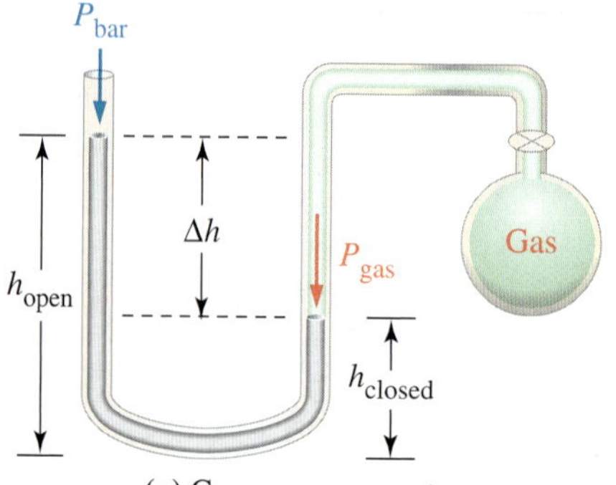

(a) Gas pressure greater than barometric pressure

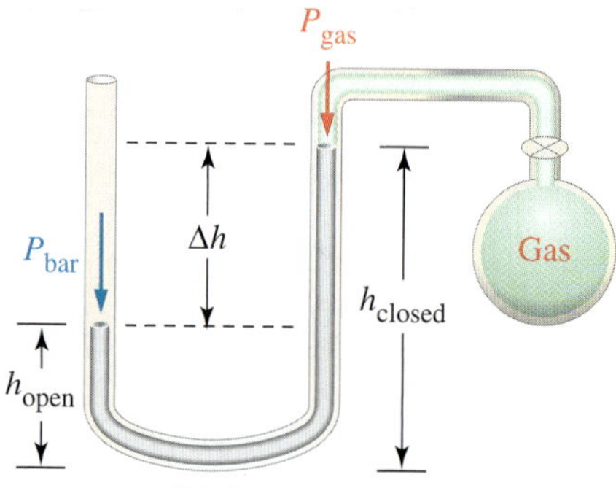

(b) Gas pressure less than barometric pressure

▶ **Figure 5.4**
Measuring gas pressure with an open-end manometer
The difference in mercury levels (Δh) between the two arms of the manometer gives the difference between barometric pressure and the gas pressure.

P_{gas} differs from P_{bar}. In Figure 5.4(a), the greater mercury height is in the arm open to the atmosphere, and P_{gas} is *greater* than P_{bar}. In Figure 5.4(b), the greater mercury height is in the closed arm, and P_{gas} is *smaller* than P_{bar}. Suppose we define the difference in height of the two mercury columns (Δh) as follows.

$$\Delta h = h_{open} - h_{closed}$$

Now, by noting that Δh may be either positive or negative, we see that the gas pressure, in mmHg, is always given by the following equation.

$$P_{gas} = P_{bar} + \Delta h$$

Let's extend the idea of the pressures exerted by liquid columns. First, we note that force is the product of mass and acceleration. If the acceleration is that due to gravity ($g = 9.807 \text{ m s}^{-2}$), we can write

$$P = \frac{F}{A} = \frac{g \cdot m}{A}$$

The mass (m) of a liquid column is the product of its density (d) and volume (V). In turn, the volume is equal to the product of the area at the base of the liquid column (A) and the height of the column (h). With these ideas we can obtain the general expression below.

$$P = \frac{g \cdot m}{A} = \frac{g \cdot d \cdot V}{A} = \frac{g \cdot d \cdot h \cdot \cancel{A}}{\cancel{A}} = g \cdot d \cdot h$$

APPLICATION NOTE

For gas pressures that are close to atmospheric pressure, liquids much less dense than mercury are used in manometers. Respiration therapists, for example, measure pressure differences in the unit *centimeters of water*.

EXAMPLE 5.2

Calculate the height of a column of water that would exert the same pressure as a column of mercury that is 760 mm high. (Use densities of 13.6 g/cm^3 for mercury and 1.00 g/cm^3 for water.)

SOLUTION

Let's use the general expression that relates a liquid pressure to the height of a column and the density of the liquid it contains. For the mercury column, we can write

$$P_{Hg} = g\, d_{Hg}\, h_{Hg}$$

and for the water column,

$$P_{H_2O} = g\, d_{H_2O}\, h_{H_2O}$$

Because we are interested in the situation in which the liquid pressures are equal to one another,

$$\cancel{g}\, d_{Hg} h_{Hg} = \cancel{g}\, d_{H_2O} h_{H_2O}$$

Canceling g and substituting numerical values, we have

$$13.6 \text{ g/cm}^3 \times 760 \text{ mm} = 1.00 \text{ g/cm}^3 \times h_{H_2O}$$

Solving the expression for h_{H_2O}, we get

APPLICATION NOTE
Atmospheric pressure can easily push a beverage up the 20-cm length of a straw when we draw some of the air from the straw. In contrast, atmospheric pressure cannot push water to a height of more than 10.3 m, even if the suction pump in a well were able to remove nearly all the air, creating a near-perfect vacuum.

$$h_{H_2O} = \frac{13.6 \text{ g/cm}^3 \times 760 \text{ mm}}{1.00 \text{ g/cm}^3} = 10{,}300 \text{ mm} = 10.3 \text{ m}$$

One of the implications of this result is that to measure normal atmospheric pressure with a water-filled barometer, we would have to use one that is over 10 m tall—as tall as a three-story building!

EXERCISE 5.2A

Calculate the height of a column of carbon tetrachloride, CCl_4, that would exert the same pressure as a column of mercury that is 760 mm high. The density of CCl_4 is 1.59 g/cm^3 and that of Hg is 13.6 g/cm^3.

EXERCISE 5.2B

A diver reaches a depth of 30.0 m. What is the pressure in atmospheres that is exerted by this depth of water? What is the *total* pressure that the diver experiences at this depth? Explain.

EXAMPLE 5.3—A Conceptual Example

Without doing calculations, arrange the drawings in Figure 5.5 so that the pressures denoted in red are in *increasing* order.

SOLUTION

The pressure in (a) is expressed as the depth of the liquid mercury, that is, as 745 mmHg. The pressure of helium in the open-end manometer (b) is slightly *greater* than P_{bar}, which is 762 Torr, and therefore greater than the pressure in (a). Although there is much less mercury in (c) than in (a), the *pressure* of the mercury column in (c) is greater than in (a) because of its greater height: 75.0 cm = 750 mm. Finally, in the closed-end manometer in (d), the difference in the heights of the two mercury columns, 735 mm, is the actual gas pressure.

$$\begin{array}{ccccccc} (d) & < & (a) & < & (c) & < & (b) \\ P_{Ne} & < & P_{Hg}(l) & < & P_{atm} & < & P_{He} \end{array}$$

735 mmHg < 745 mmHg < 750 mmHg < above 762 mmHg

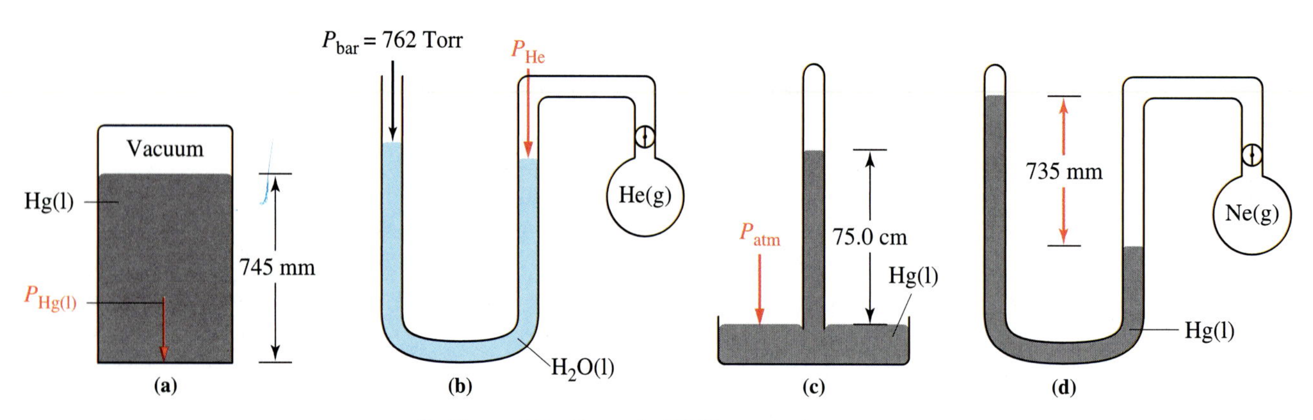

▲ **Figure 5.5 Example 5.3 illustrated.**

EXERCISE 5.3

Place a barometric pressure of (e) 101 kPa into the order of increasing pressures established in Example 5.3. If possible, also place a barometric pressure of (f) 103 kPa in the order. If it is not possible, explain why.

5.4 Boyle's Law: The Pressure–Volume Relationship

We use four variables to specify a sample of gas in calculations: its amount in moles (n), its volume (V), its temperature (T), and its pressure (P). These variables are related through several simple gas laws that show how one of the variables (for example, V) changes as a second variable (for example, P) changes and the other two (for example, n and T) remain constant.

The first of these simple gas laws, discovered by Robert Boyle in 1662, concerns the relationship between the pressure and volume of a gas. **Boyle's law** is expressed below.

For a given amount of gas at a constant temperature, the volume of the gas varies inversely with its pressure.

That is, when the pressure increases, the volume decreases; when the pressure decreases, the volume increases.

Let's think of a gas in terms of the kinetic-molecular theory, as is suggested in the upper portion of Figure 5.6. The molecules bounce off the container walls with

Robert Boyle's (1627–1691) experiments on air and his textbook, *The Sceptical Chemist,* helped to establish modern chemistry.

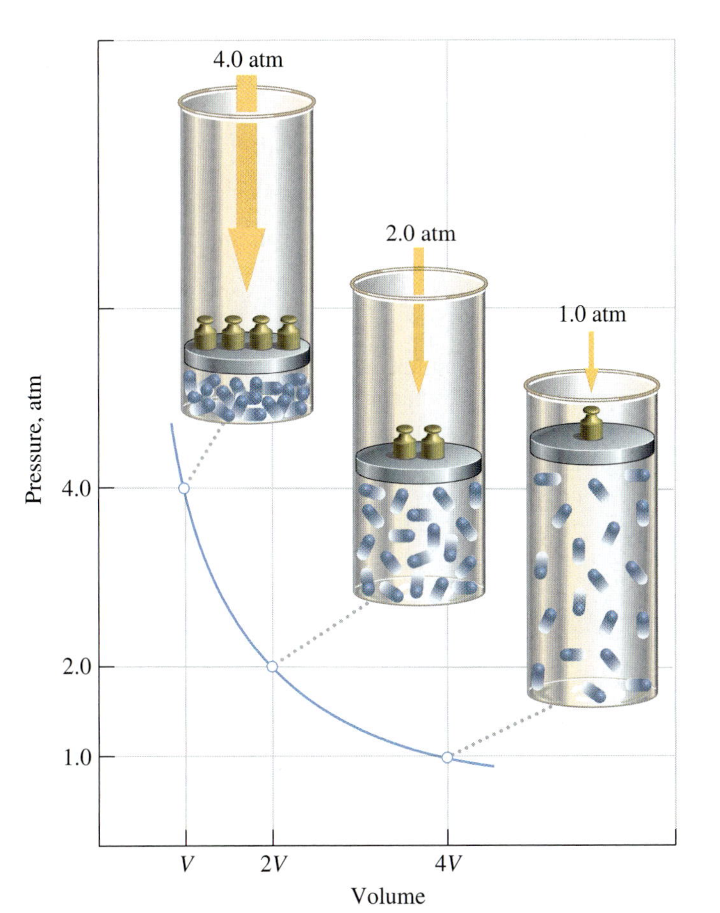

Figure 5.6
Boyle's law: A kinetic-theory view and a graphical representation
As the pressure is successively reduced from 4.00 atm to 2.00 atm and then to 1.00 atm, the volume doubles and then doubles again. The volume is *inversely* proportional to the pressure, and the pressure-volume product is a constant ($PV = a$).

a certain frequency, creating a particular pressure. If we expand the volume of the container, but the amount of gas inside the container remains unchanged, the number of molecules per unit volume of gas decreases. The frequency with which molecules strike a unit area of the container walls decreases, and the gas pressure decreases. As the volume of a gas is increased, its pressure therefore decreases.

Mathematically, for a given amount of gas at a constant temperature, we can express Boyle's law as

$$V \propto \frac{1}{P}$$

where the symbol $\propto$ means "is proportional to." However, we can't use a proportionality for quantitative calculations. For example, the time required for a trip on a freeway is inversely proportional to the speed. The faster you drive, the shorter your driving time is. But this doesn't allow you to calculate an actual time. You need to change the proportionality to an equation by introducing a proportionality constant—the distance to be driven. Let's denote the proportionality constant in Boyle's law by the symbol, a, and then write the equation

$$V = \frac{a}{P}$$

Multiplying both sides of the equation by P, we get

$$PV = a$$

Another way to state Boyle's law, then, is that

For a given amount of gas at a constant temperature, the product of the pressure and volume is a constant.

This is an elegant and precise way of summarizing a lot of experimental data. If the product $P \times V$ is to be constant, then if V increases, P must decrease, and vice versa. This relationship is demonstrated in the pressure–volume graph in the lower part of Figure 5.6.

Boyle's law has a number of practical applications perhaps best illustrated by some examples. Note that in these applications, both the amount of gas and the temperature must be held constant, and that any units can be used for pressure and volume, as long as the same units are used throughout a calculation. Also note that in the typical problem, we deal with the same gas under two different conditions, which we might call the "initial" and "final" conditions. The pressure–volume product of the gas is a constant under these two conditions, leading to the expression

APPLICATION NOTE
Gases are usually stored under high pressure even though they will be used at atmospheric pressure. This allows a large amount of gas to be stored in a small volume.

$$P_{\text{initial}} \times V_{\text{initial}} = a = P_{\text{final}} \times V_{\text{final}}$$

or

$$P_{\text{initial}} \times V_{\text{initial}} = P_{\text{final}} \times V_{\text{final}}$$

EXAMPLE 5.4

A party balloon is filled to a volume of 4.50 L with helium at a location near sea level, where the atmospheric pressure is measured at 748 Torr. Assuming that the temperature remains constant, what will be the volume of the balloon when it is taken to a mountain resort at 2500 m? The atmospheric pressure at the resort is 557 Torr.

SOLUTION

First, we can solve the mathematical expression of Boyle's law,

$$P_{\text{initial}} \times V_{\text{initial}} = P_{\text{final}} \times V_{\text{final}}$$

for V_{final}. Then, we can substitute known values for the three variables on the right side of the equation.

$$V_{\text{final}} = V_{\text{initial}} \times \frac{P_{\text{initial}}}{P_{\text{final}}} = 4.50\text{ L} \times \frac{748\text{ Torr}}{557\text{ Torr}} = 6.04\text{ L}$$

PROBLEM-SOLVING NOTE

We have an answer, but is it reasonable? Because the final pressure in Example 5.4 is *smaller than* the initial pressure, we expect the final volume to be *greater than* the initial volume; and it is. You should always check an answer by qualitative reasoning when possible.

EXERCISE 5.4A

A sample of helium occupies 535 mL at 988 Torr and 25 °C. If the sample is transferred to a 1.05-L flask at 25 °C, what will be the gas pressure in the flask?

EXERCISE 5.4B

A sample of air occupies 73.3 mL at 98.7 kPa and 0 °C. What volume will the air occupy at 4.02 atm and 0 °C?

EXAMPLE 5.5—An Estimation Example

A gas is enclosed in a cylinder fitted with a piston. The volume of the gas is 2.00 L at 398 Torr. The piston is moved to increase the gas pressure to 5.15 atm. Which of the following is a reasonable value for the volume of the gas at the greater pressure?

0.20 L   0.40 L   1.00 L   16.0 L

SOLUTION

The initial pressure (398 Torr) is about 0.5 atm. An increase in pressure to 5.15 atm represents about a tenfold increase. As a result, the volume should drop to about one tenth of its initial value of 2.00 L. A final volume of 0.20 L seems like the most reasonable estimate. (The calculated value is 0.203 L.)

● **ChemCDX** Pressure-Volume Relationships

EXERCISE 5.5

A gas is enclosed in a 10.2-L tank at 1208 Torr. Which of the following is a reasonable value for the pressure when the gas is transferred to a 30.0-L tank?

0.40 atm   25 lb/in^2 400 mmHg   3600 Torr

5.5 Charles's Law: The Temperature–Volume Relationship

In 1787, Jacques Charles (1746–1823) studied the relationship between the volume and the temperature of gases. He found that when a fixed mass of gas is cooled at constant pressure, its volume decreases. When the fixed mass of gas is heated, its volume increases. Temperature and volume are directly proportional; that is, they rise or fall together. However, the relationship between volume and temperature is not as tidy as it may seem on first impression. For example, consider the following: If a quantity of gas that occupies 1.00 L is heated from 100 °C to 200 °C

Boyle's Law and Breathing

The pressure–volume relationship for gases helps to explain the mechanics of breathing. When we breathe in (inspire), the diaphragm is lowered and the chest wall is expanded, increasing the volume of the chest cavity (Figure 5.7). Boyle's law tells us that the pressure inside the cavity must decrease. Outside air enters the lungs because it is at a higher pressure than the air in the chest cavity. When we breathe out (expire), the diaphragm rises and the chest wall contracts, decreasing the volume of the chest cavity. The pressure is increased, and some air is forced out.

During normal inspiration, the pressure inside the lungs drops about 3 Torr below atmospheric pressure. During expiration, the internal pressure is about 3 Torr above atmospheric pressure. About one-half liter of air is moved in and out of the lungs in this process, and this normal breathing volume is referred to as the *tidal volume*. The *vital capacity* is the maximum volume of air that can be forced from the lungs and ranges from 3 to 7 L, depending on the individual. A pressure inside the lungs 100 Torr greater than the external pressure is not unusual during such a maximum expiration.

The lungs are never emptied completely, however. The space around the lungs is maintained at a slightly lower pressure than are the lungs themselves, causing the lungs to be kept partially inflated by the higher pressure within them. If a lung, the diaphragm, or the chest wall is punctured, allowing the two pressures to equalize, the lung will collapse. Sometimes a medical doctor will collapse a patient's damaged lung intentionally to give it time to heal. Closing the opening to the lung reinflates it.

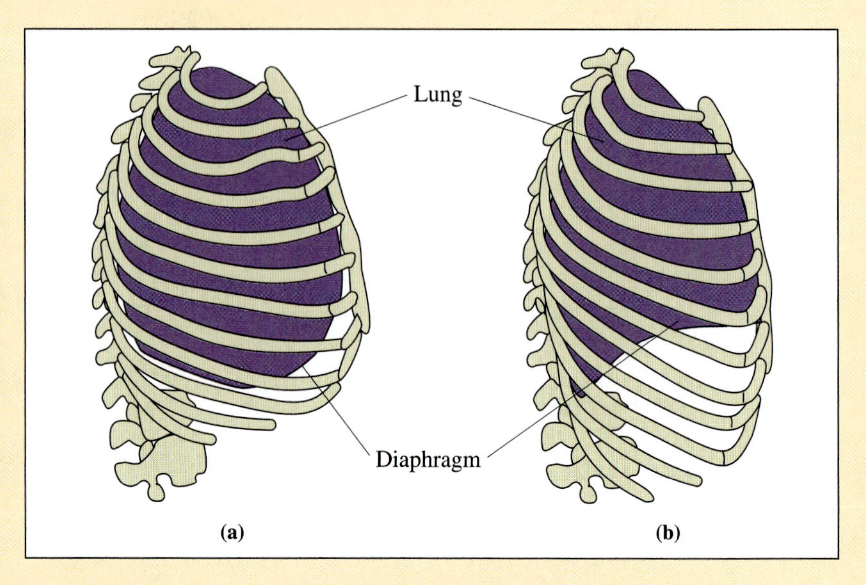

▲ **Figure 5.7**
The mechanics of breathing
(a) Inspiration. The diaphragm is pulled down and the rib cage is lifted up and out, increasing the volume of the chest cavity. (b) Expiration. The diaphragm is relaxed, the rib cage is down, and the volume of the chest cavity decreases.

at constant pressure, the volume does not double; instead, it increases to only about 1.27 L.

Zero pressure or zero volume really means zero—no pressure or volume to be measured. Zero degrees Celsius (0 °C) signifies only the freezing point of water. This zero point is arbitrarily set, much like mean sea level is the arbitrary zero used to measure altitudes on Earth. Negative pressures and volumes are not possible, but temperatures below 0 °C are often encountered, as are altitudes below sea level.

Charles noted that for each Celsius degree rise in temperature, the volume of a gas expands by 1/273 of its volume at 0 °C. For each Celsius degree drop in temperature, the volume of a gas decreases by 1/273 of its value at 0 °C. If we plot the volume of a fixed amount of gas at constant pressure against its temperature, we get the straight line shown in Figure 5.8. We can *extrapolate* the line beyond the range of measured temperatures to the temperature at which the volume of the gas *appears* to become zero. Before a gas ever reaches this temperature, however, it liquefies, and then the liquid freezes, so this an exercise for the imagination.

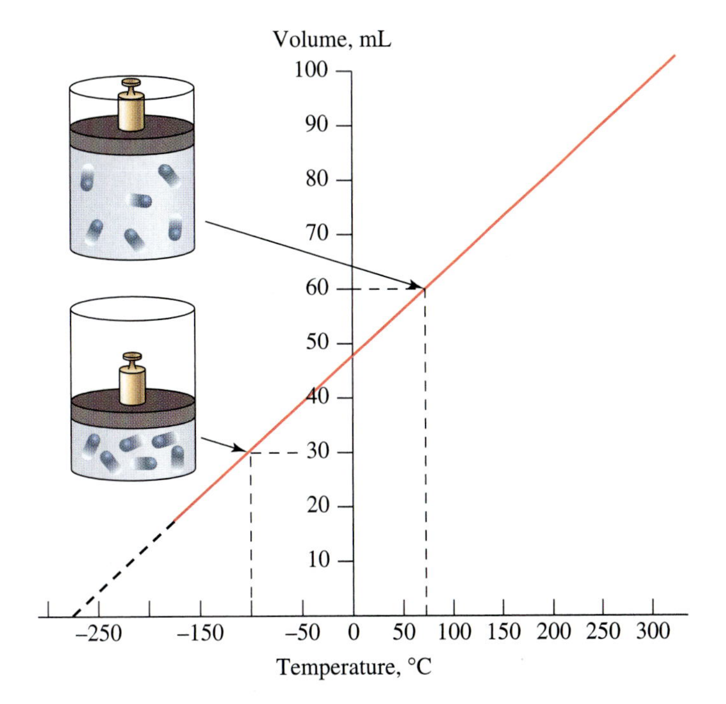

Figure 5.8
Charles's law: Gas volume as a function of temperature
The gas shown has a volume of 60 mL at about 70 °C. When the gas is cooled through the temperature interval from about 70 °C to −100 °C, its volume drops to 30 mL. The gas volume continues to decrease in a linear fashion as the temperature is lowered. The extrapolated line intersects the temperature axis (corresponding to a volume of zero) at about −270 °C.

The temperature obtained by extrapolation to "zero volume" is −273.15 °C. In 1848, William Thomson (Lord Kelvin) made this temperature the **absolute zero** on a temperature scale now called the **Kelvin scale**. The unit of temperature on this scale is the **kelvin (K)**, which is equal to a degree of Celsius temperature. Note that the degree sign is *not* used for the Kelvin. Temperatures on the Kelvin scale are 273.15 degrees higher than on the Celsius scale, a fact that we can express through the equation

$$T(\text{K}) = T(°\text{C}) + 273.15$$

A modern statement of **Charles's law** reflects the significance of the Kelvin temperature scale and the absolute zero of temperature on which it is based.

The volume of a fixed amount of a gas at a constant pressure is directly proportional to its Kelvin temperature.

Mathematical statements of Charles's law are of the form

$$V \propto T \quad \text{and} \quad V = bT \quad (\text{where } b \text{ is a constant})$$

Similar to what we did with Boyle's law, we can express Charles's law for an "initial" and "final" condition of a sample of gas at constant pressure, leading to the expressions

$$\frac{V_{\text{initial}}}{T_{\text{initial}}} = b = \frac{V_{\text{final}}}{T_{\text{final}}}$$

and

$$\frac{V_{\text{initial}}}{T_{\text{initial}}} = \frac{V_{\text{final}}}{T_{\text{final}}}$$

The kinetic–molecular model easily accounts for the relationship between gas volume and temperature. When we heat a gas, we supply the gas molecules with energy and they begin to move faster. These speedier molecules strike the walls of their container harder and more often. For the pressure to stay the same, the volume of the container must increase so that the increased molecular motion will be distributed over a greater space. In this way, the pressure exerted by the faster molecules in the larger volume (high temperature) is the same as that of the slower moving molecules in the smaller volume (low temperature). Figure 5.9 illustrates the dramatic change in gas volume that occurs over a wide range of temperatures.

(a)

(b)

▶ Figure 5.9
Charles's law: A dramatic illustration
(a) Liquid nitrogen (boiling point, −196 °C) cools the balloon and its contents to a temperature far below room temperature, and the balloon collapses. (b) As the balloon warms to room temperature, it reinflates. The air in the balloon regains the volume it previously had at room temperature.

EXAMPLE 5.6

A balloon indoors, where the temperature is 27 °C, has a volume of 2.00 L. What will its volume be outdoors, where the temperature is −23 °C? (Assume no change in the gas pressure.)

SOLUTION

First, we must convert both temperatures to the Kelvin scale, using the relationship

$$T(\text{K}) = T(°\text{C}) + 273.15$$

The initial temperature is

$$T_{\text{initial}} = 27 + 273.15 = 300 \text{ K}$$

and the final temperature is

$$T_{\text{final}} = -23 + 273.15 = 250 \text{ K}$$

Now let's apply Charles's law in the form

$$\frac{V_{\text{final}}}{T_{\text{final}}} = \frac{V_{\text{initial}}}{T_{\text{initial}}}$$

and solve for V_{final}.

$$V_{\text{final}} = V_{\text{initial}} \times \frac{T_{\text{final}}}{T_{\text{initial}}} = 2.00\text{ L} \times \frac{250\text{ K}}{300\text{ K}} = 1.67\text{ L}$$

PROBLEM-SOLVING NOTE

Because the final temperature is *lower than* the initial temperature, we expect the final volume to be *smaller than* the initial volume; and it is.

EXERCISE 5.6A

A sample of hydrogen gas occupies 692 L at 602 °C. If the pressure is held constant, what volume will the gas occupy after cooling to 23 °C?

EXERCISE 5.6B

The balloon described in Example 5.6 needs to be maintained at a particular temperature in order to have a volume of 2.25 L. Assuming that the pressure remains constant, find this temperature in kelvins and in degrees Celsius.

EXAMPLE 5.7—An Estimation Example

A sample of nitrogen occupies a volume of 2.50 L at −120 °C and 1 atm pressure. To which of the following approximate temperatures should the gas be heated in order to double its volume while maintaining a constant pressure?

−240 °C −60 °C −12 °C 30 °C

SOLUTION

The initial temperature is about 150 K (−120 + 273.15 ≈ 150). The final temperature must be twice the initial temperature—about 300 K—if the volume is to double. The Celsius temperature corresponding to 300 K is about 30 °C (30 + 273.15 ≈ 300). Notice that the four temperatures from which to choose are all multiples or fractions of −120 °C, but this has *nothing* to do with our answer because the relationship between volume and temperature must be based on the *Kelvin* scale.

EXERCISE 5.7

If the gas in Example 5.7—initially 2.50 L at −120 °C and 1 atm—is brought to a temperature of 180 °C while a constant pressure is maintained, which of the following is the approximate final volume of the gas?

3.75 L 5.0 L 7.5 L 10.0 L

5.6 Avogadro's Law: The Mole–Volume Relationship

As we will describe in Section 5.9, Amedeo Avogadro proposed an important hypothesis in 1811 to explain some observations about the ratios in which volumes of gases combine during chemical reactions. **Avogadro's hypothesis** states that equal numbers of molecules of different gases, when compared at the same temperature and pressure, occupy equal volumes. Thus, Avogadro's hypothesis relates an amount of gas (numbers of molecules) and gas volume when temperature and pressure remain constant. We call the simple gas law implied by this statement **Avogadro's law**.

At a fixed temperature and pressure, the volume of a gas is directly proportional to the amount of gas (that is, to the number of moles of gas, n, or to the number of molecules of gas).

Amedeo Avogadro (1776–1856) did not live to see his ideas accepted by the scientific community. This acceptance finally came in 1860 at an international conference where Stanislao Cannizzaro (1826–1910) effectively put forth Avogadro's ideas from five decades before.

If we double the number of moles of gas at a fixed temperature and pressure, the volume of the gas doubles. Because the mass of a gas is proportional to the number of moles, doubling the *mass* of a gas also doubles its volume. Mathematically, we can state Avogadro's law as

$$V \propto n \quad \text{or} \quad V = cn \quad \text{(where } c \text{ is a constant)}$$

When we use Avogadro's hypothesis to compare different gases, the gases must be at the same temperature and pressure. As we will illustrate next, a convenient temperature/pressure combination for such comparisons is 0 °C (273.15 K) and 1 atm (760 Torr), known as the **standard temperature and pressure (STP)**.

Molar Volume of a Gas

Suppose that in comparing different gases at STP, we use Avogadro's number as the number of molecules present. Avogadro's hypothesis states that under these conditions, 1 mol (6.022×10^{23} molecules) of *all* gases should occupy the same volume. The **molar volume of a gas** is the volume occupied by one mole of the gas. By experiment, the molar volumes at STP of some common gases are

$$22.428 \text{ L of } H_2;\ 22.404 \text{ L of } N_2;\ 22.394 \text{ L of } O_2;\ \text{and } 22.360 \text{ L } CH_4$$

So, to *three* significant figures, we can state that for any gas,

$$1 \text{ mol gas} = 22.4 \text{ L gas (at STP)}$$

Figure 5.10 pictures a volume of 22.4 L and relates it to some familiar objects. The 22.4-L container would hold 2.02 g H_2, 28.0 g N_2, 32.0 g O_2, or 44.0 g CO_2.

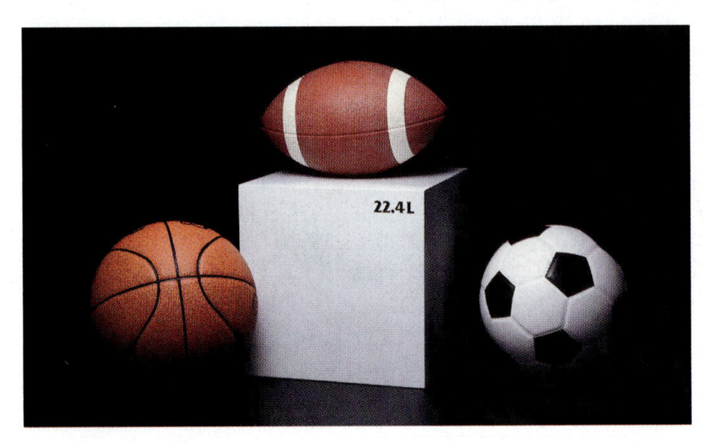

▶ **Figure 5.10**
Molar volume of a gas visualized
The wooden cube has a volume of 22.4 L, the same volume as 1 mol of gas at STP. The familiar objects offer a contrast to this volume.

EXAMPLE 5.8

Calculate the volume occupied by 4.11 kg of methane gas, $CH_4(g)$, at STP.

SOLUTION

We must convert the mass of gas to an amount in moles, and then use the molar volume (in red) as a conversion factor to go from the amount of gas to its volume at STP. We can do all of this in a single setup.

$$? \text{ L } CH_4 = 4.11 \text{ kg } CH_4 \times \frac{1000 \text{ g } CH_4}{1 \text{ kg } CH_4} \times \frac{1 \text{ mol } CH_4}{16.04 \text{ g } CH_4} \times \frac{22.4 \text{ L } CH_4}{1 \text{ mol } CH_4}$$

$$= 5.74 \times 10^3 \text{ L } CH_4$$

EXERCISE 5.8A

What is the mass of propane, C_3H_8, in a 50.0-L container of the gas at STP?

EXERCISE 5.8B

Solid carbon dioxide, called "dry ice," is useful in maintaining frozen foods because it vaporizes to $CO_2(g)$ rather than melting to a liquid. How many liters of $CO_2(g)$, measured at STP, will be produced by the vaporization of a block of dry ice that measures 12.0 in. × 12.0 in. × 2.0 in.? The density of the dry ice is 1.56 g/cm³.

5.7 The Combined Gas Law

From the three simple gas laws,

$$V = \frac{a}{P}; \quad V = bT; \quad \text{and} \quad V = cn$$

it seems reasonable that the volume of a gas (V) should be *directly* proportional to the Kelvin temperature (T) and to the amount of gas (n), and *inversely* proportional to the pressure (P). That is indeed the case:

$$V \propto \frac{nT}{P}$$

Or, expressed as an equation rather than a proportionality,

$$\frac{PV}{nT} = \text{constant}$$

This *combined* gas law is most useful when we want to describe the *final* conditions for a gas from a knowledge of the *initial* conditions and the changes to which the gas is subjected. In these cases, we write,

$$\begin{array}{ccc} \text{Initial} & & \text{Final} \\ \dfrac{P_1V_1}{n_1T_1} & = \text{constant} = & \dfrac{P_2V_2}{n_2T_2} \end{array}$$

The subscript "1" represents the initial condition of the gas with respect to pressure (P_1), volume (V_1), amount (n_1), and temperature (T_1); the subscript "2" represents the final condition. If one or more of the gas properties remain constant during a change from initial to final conditions, we can simplify this expression by canceling out these constant terms. To derive Boyle's law from the combined gas law, we would fix n and T (that is, assume that the final amount and temperature of the gas were identical to the initial amount and temperature). To derive Charles's law, we would fix n and P. Now, let's consider the case of a fixed amount of gas ($n_1 = n_2$) in a fixed volume ($V_1 = V_2$). The combined gas equation

$$\frac{P_1V_1}{n_1T_1} = \frac{P_2V_2}{n_2T_2}$$

simplifies to

$$\frac{P_1}{T_1} = \frac{P_2}{T_2}$$

This equation shows that the pressure of a fixed amount of gas in a constant volume is proportional to its Kelvin temperature. This relationship, a simple gas law sometimes called *Amontons'* law, certainly seems reasonable from the standpoint of the kinetic-molecular theory, as is illustrated in Figure 5.11. Amontons' and Charles's law both deal with the effect of temperature changes on a gas. Charles's law relates the volume and temperature if the amount of gas and its pressure are held constant. Amontons' law relates the pressure and temperature if the amount of gas and its volume are held constant.

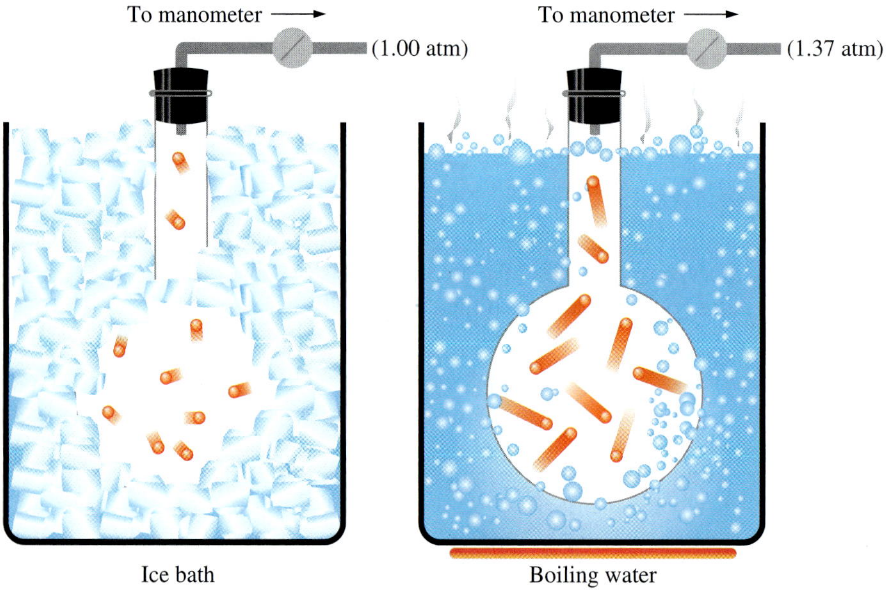

To manometer → (1.00 atm)

To manometer → (1.37 atm)

Ice bath

Boiling water

▶ Figure 5.11
The effect of temperature on the pressure of a fixed amount of gas in a constant volume: A kinetic theory interpretation
The amount of gas and its volume are the same in either case, but if the gas in the ice bath (0 °C) exerts a pressure of 1 atm, the gas in the boiling-water bath (100 °C) exerts a pressure of 1.37 atm. The frequency and the force of the molecular collisions with the container walls are greater at the higher temperature.

EXAMPLE 5.9

The flask pictured in Figure 5.11 contains $O_2(g)$, first at STP and then at 100 °C. What is the pressure at 100 °C?

SOLUTION

We start with the expression that we just derived,

$$\frac{P_1}{T_1} = \frac{P_2}{T_2}$$

where the subscript "1" represents the initial condition (STP) and "2," the final condition. Then we solve it for the final pressure, P_2.

$$P_2 = P_1 \times \frac{T_2}{T_1} = 1.00 \text{ atm} \times \frac{(100 + 273)\cancel{\text{K}}}{273\ \cancel{\text{K}}} = 1.37 \text{ atm}$$

PROBLEM-SOLVING NOTE
Note that in this calculation we need to know that the amount and volume of $O_2(g)$ remain constant, but we don't need to know their actual values.

EXERCISE 5.9A

Aerosol containers often carry the warning that they should not be heated. Suppose such a container were filled with a gas at 2.5 atm and 22 °C, and suppose that the container may rupture if the pressure exceeds 8.0 atm. At what temperature is that rupture likely to occur?

EXERCISE 5.9B

In the manner we used to derive the simple gas law relating pressure and temperature (Amontons' law), derive an expression relating the pressure and amount of gas when both the temperature and volume remain constant. Rationalize your result in terms of the kinetic-molecular theory and a sketch in the manner of Figure 5.11.

5.8 The Ideal Gas Law

We have seen how three simple gas laws—Boyle's, Charles's, and Avogadro's—serve as the basis of a more general gas law—the combined gas law. In turn, this combined law can be used to establish other simple relationships, such as Amontons' law. There is another expression, however, that is more useful than the combined gas law. In this equation, we replace the unspecified constant in the combined gas law with an actual numerical value. Let's use the symbol R for the constant term.

$$\frac{PV}{nT} = \text{constant} = R$$

Then, let's substitute data into the left side of the equation that will allow us to obtain the value of R. For this purpose, we can use the data associated with the molar volume of a gas at STP. The "best" value we can use for the molar volume is that for an *ideal gas*. An **ideal gas** is a gas that *strictly* obeys all the simple gas laws and has a molar volume at STP of 22.4141 L. By strictly obeys, we mean that the behavior of the ideal gas is accurately described by the simple gas laws. We can obtain the three different values for R listed in Table 5.3 depending on the units we use to express pressure. If we choose pressure in atmospheres,

$$R = \frac{PV}{nT} = \frac{1\ \text{atm} \times 22.4141\ \text{L}}{1\ \text{mol} \times 273.15\ \text{K}} = 0.082058\ \frac{\text{L atm}}{\text{mol K}}$$

R is called the **universal gas constant**, or simply the *ideal gas constant*. We often round off the value of R to three significant figures and use negative exponents to indicate the units in the denominator, that is, $R = 0.0821\ \text{L atm mol}^{-1}\ \text{K}^{-1}$.

The universal gas constant, R, is very important in chemistry. We will use it often, usually in the form $8.3145\ \text{J mol}^{-1}\ \text{K}^{-1}$.

The equation, called the **ideal gas law** or the **ideal gas equation**, is generally written in the form

$$PV = nRT$$

An ideal gas is a *hypothetical* gas. With *real* gases we need to know how closely they behave like an ideal gas. At the end of this chapter (Section 5.12), we will examine why real gases may fail to behave like ideal gases. For now, we will avoid

TABLE 5.3 Units for the Universal Gas Constant, *R*

***R* has the value**	**When**
$0.082058\ \text{L atm mol}^{-1}\ \text{K}^{-1}$	P is in atmospheres
$62.364\ \text{L Torr mol}^{-1}\ \text{K}^{-1}$	P is in Torr
$8.3145\ \text{J mol}^{-1}\ \text{K}^{-1}$	P is in Pa; V is in m^3

conditions (generally, high pressures and low temperatures) where real gases depart most seriously from ideal gas behavior, so that we can describe them with the ideal gas law.

The typical calculation that we'll make is to determine any one of the four quantities—*P*, *V*, *n*, or *T*—if we know the other three. As indicated in the following examples, we begin each calculation by solving the ideal gas equation for the unknown variable, and we also make certain to express quantities in the units needed to match the units of *R*, for example: L (volume), atm (pressure), mol (amount of gas), and K (temperature) to match $R = 0.0821\ \text{L atm mol}^{-1}\ \text{K}^{-1}$.

EXAMPLE 5.10

What is the pressure exerted by 0.508 mol O_2 in a 15.0-L container at 303 K?

SOLUTION

First, we need to solve the ideal gas equation for pressure, *P*. We accomplish this by dividing both sides of the equation by the volume, *V*, and then canceling *V* on the left side of the equation.

$$\frac{P\cancel{V}}{\cancel{V}} = \frac{nRT}{V}$$

$$P = \frac{nRT}{V}$$

Because data are given in the units mol, L, and K, we can use $R = 0.0821$ L atm $\text{mol}^{-1}\ \text{K}^{-1}$ and substitute the data directly into the above equation. Our result will be a pressure in atmospheres.

$$P = \frac{nRT}{V} = \frac{0.508\ \cancel{\text{mol}} \times 0.0821\ \cancel{\text{L}}\ \text{atm} \times 303\ \cancel{\text{K}}}{15.0\ \cancel{\text{L}}\ \cancel{\text{mol}}\ \cancel{\text{K}}} = 0.842\ \text{atm}$$

EXERCISE 5.10

How many moles of nitrogen, N_2, are there in a sample that occupies 35.0 L at a pressure of 3.15 atm and a temperature of 852 K?

EXAMPLE 5.11

What is the volume occupied by 16.0 g ethane gas (C_2H_6) at 720 Torr and 18 °C?

SOLUTION

In addition to solving the ideal gas equation for volume, *V* (by dividing both sides of the equation by the pressure, *P*), we need to use an appropriate unit for each variable in the equation. We can use the units that are consistent with the value of $R = 0.0821$ L atm $\text{mol}^{-1}\ \text{K}^{-1}$ as shown below.

$$n = 16.0\ \cancel{\text{g}\ C_2H_6} \times \frac{1\ \text{mol}\ C_2H_6}{30.07\ \cancel{\text{g}\ C_2H_6}} = 0.532\ \text{mol}\ C_2H_6$$

$$T = (273.15 + 18)\ \text{K} = 291\ \text{K}$$

$$P = 720\ \cancel{\text{Torr}} \times \frac{1\ \text{atm}}{760\ \cancel{\text{Torr}}} = 0.947\ \text{atm}$$

Finally, let's substitute the converted data for the variables n, T, and P into the ideal gas equation, which we have solved for V.

$$V = \frac{nRT}{P} = \frac{0.532 \cancel{\text{mol}} \times 0.0821 \text{ L} \cancel{\text{atm}} \times 291 \cancel{\text{K}}}{0.947 \cancel{\text{atm}} \cancel{\text{mol}} \cancel{\text{K}}} = 13.4 \text{ L}$$

PROBLEM-SOLVING NOTE

An alternative approach is to leave the pressure in Torr and to use $R = 62.4$ L Torr/mol K.

EXERCISE 5.11A

What is the temperature, in degrees Celsius, at which 15.0 g O_2 will exert a pressure of 785 Torr in a volume of 5.00 L?

EXERCISE 5.11B

How many grams of $N_2(g)$, at 25.0 °C and 734 Torr, will occupy the same volume as does 25.0 g $O_2(g)$ at 30.0 °C and 755 Torr?

PROBLEM-SOLVING NOTE

You need to use the ideal gas law twice in this calculation.

Molecular Mass Determination

Suppose we have a *fixed quantity* of gas in a known volume at a known temperature and pressure. By the methods used in Examples 5.10 and 5.11, we can readily calculate the number of moles of the gas, n.

$$n = \frac{PV}{RT}$$

Suppose we also measure the mass, m, of this *fixed quantity* of gas. We now know the quantity of a sample of substance both in grams and in moles. Recall that many times in Chapter 3 we had to convert from moles to grams of a substance using molar mass, M, as a conversion factor. That is,

$$\text{Mass in grams} = \text{number of moles} \times \text{molar mass in g/mol}$$

or

$$m = n \times M$$

When we know both the mass and the number of moles of a substance, we can calculate the molar mass.

$$M = \frac{m}{n}$$

Finally, recall that molar mass and molecular mass are numerically equal, although having different units. For example, the molar mass of CO_2 is 44.01 g/mol and the molecular mass of CO_2 is 44.01 u. What we have outlined here, then, is a way to use ideal gas law measurements to obtain unknown molecular masses. This method works for any gas under conditions where its behavior is essentially that of an ideal gas. It usually works well for the vapor obtained from an easily vaporized liquid.

EXAMPLE 5.12

Calculate the molecular mass of a gas if 0.550 g of the gas occupies 0.200 L at 0.968 atm and 289 K.

SOLUTION

First, let's calculate the amount of gas, in moles, from the ideal gas equation.

$$n = \frac{PV}{RT} = \frac{0.968 \text{ atm} \times 0.200 \text{ L}}{0.0821 \text{ L atm mol}^{-1} \text{ K}^{-1} \times 289 \text{ K}} = 0.00816 \text{ mol}$$

Now we can use the known mass and the calculated number of moles to determine the mass per mole.

$$\text{Molar mass} = \frac{0.550 \text{ g}}{0.00816 \text{ mol}} = 67.4 \text{ g/mol}$$

The molar mass of the gas is 67.4 g/mol. The molecular mass, therefore, is 67.4 u.

EXERCISE 5.12

Calculate the molar mass of a gas if 0.440 g occupies 179 mL at 741 mmHg and 86 °C.

An alternative approach to the one in Example 5.12 is to combine the two equations

$$n = \frac{PV}{RT} \quad \text{and} \quad M = \frac{m}{n}$$

into a variation of the ideal gas equation

$$n = \frac{m}{M} = \frac{PV}{RT}; \; m\,RT = M\,PV;$$

and

$$M = \frac{mRT}{PV}$$

We illustrate this alternative approach in Example 5.13.

In Chapter 3, we learned how to establish the *empirical* formula of a compound from its experimentally determined mass percent composition. And if we also knew the molecular mass of the compound, we were able to determine its true *molecular* formula. We can combine these earlier ideas with what we have learned here to determine the molecular formula of a compound obtainable as a gas. This is what you will need to do in Exercise 5.13B.

EXAMPLE 5.13

Calculate the molecular mass of a liquid that, when vaporized at 100 °C and 755 Torr, yields 185 mL of vapor with a mass of 0.523 g.

SOLUTION

We must convert from mL to L and from °C to K, but we can leave the pressure in Torr and use the corresponding value of R, that is, 62.364 L Torr mol^{-1} K^{-1}.

$$V = 185 \text{ mL} \times \frac{1 \text{ L}}{1000 \text{ mL}} = 0.185 \text{ L}$$

$$T = (273 + 100) \text{ K} = 373 \text{ K}$$

Now we can substitute these data into the equation

$$M = \frac{mRT}{PV} = \frac{0.523\ \text{g} \times 62.364\ \cancel{\text{L}}\ \cancel{\text{Torr}}\ \text{mol}^{-1}\ \cancel{\text{K}}^{-1} \times 373\ \cancel{\text{K}}}{755\ \cancel{\text{Torr}} \times 0.185\ \cancel{\text{L}}}$$

$$= 87.1\ \text{g mol}^{-1}$$

The molar mass of the gas is 87.1 g/mol, and the molecular mass is 87.1 u.

EXERCISE 5.13A

Calculate the molecular mass of a liquid which, when vaporized at 98 °C and 715 mmHg, yields 121 mL of vapor with a mass of 0.471 g.

EXERCISE 5.13B

Diacetyl, a substance contributing to the characteristic flavor and aroma of butter, consists of 55.80% C, 7.03% H, and 37.17% O by mass. In the gaseous state at 100 °C and 747 mmHg, a 0.3060-g sample occupies a volume of 111 mL. Establish the molecular formula of diacetyl.

Gas Densities

Gases are much less dense than liquids and solids, and we generally express their densities in g/L rather than g/mL. To establish the density of $O_2(g)$ at STP is easy enough; we simply divide the molar mass of O_2 by its molar volume at STP.

$$d = \frac{m}{V} = \frac{32.00\ \text{g}\ O_2}{22.4\ \text{L}} = 1.43\ \text{g}\ O_2/\text{L}$$

However, this value does not apply under conditions other than STP. Much more so than the densities of liquids and solids, the densities of gases are critically dependent on both temperature and pressure. This is because the volume of a fixed mass of gas depends so strongly on temperature and pressure. We can best see the effects of different variables on gas density by another rearrangement of the ideal gas equation. First, we substitute $n = m/M$ into the ideal gas equation, $PV = nRT$, to obtain

$$PV = \frac{mRT}{M}$$

which rearranges to

$$MPV = mRT$$

Solve this equation for m,

$$m = \frac{MPV}{RT}$$

and divide both sides of the equation by V. The gas density $d = m/V$.

$$d = \frac{m}{V} = \frac{MP}{RT}$$

Thus, we see that the density of a gas is *directly* proportional to its molar mass (M) and pressure (P) and *inversely* proportional to its Kelvin temperature (T). You can think of $1/R$ as the proportionality constant that allows these proportionalities to be expressed in an equation.

EXAMPLE 5.14

Calculate the density of methane gas, CH_4, in grams per liter, at 25 °C and 0.978 atm.

SOLUTION

We could do this as a two-step problem in which we (1) determine the number of moles of CH_4 in 1.00 L of the gas at the stated temperature and pressure, and (2) determine the mass of this gas. To do so, however, is really the same as solving the gas density equation derived on page 199.

$$d = \frac{M \times P}{R \times T} = \frac{16.04 \text{ g CH}_4 \cancel{\text{mol}^{-1}} \times 0.978 \cancel{\text{atm}}}{0.0821 \text{ L} \cancel{\text{atm}} \cancel{\text{mol}^{-1}} \cancel{\text{K}^{-1}} \times 298 \cancel{\text{K}}}$$

$$= 0.641 \text{ g CH}_4/\text{L}$$

EXERCISE 5.14

Calculate the density of ethane gas (C_2H_6), in grams per liter, at 15 °C and 748 Torr.

EXAMPLE 5.15

Under what pressure must $O_2(g)$ be maintained at 25 °C to have a density of 1.50 g/L?

SOLUTION

We can use the equation $d = MP/RT$ to solve for any one of the five quantities if the other four are known. Here we must solve it for the unknown pressure

$$P = \frac{dRT}{M}$$

and then substitute the known quantities into the right side.

$$P = \frac{1.50 \cancel{\text{g}} \cancel{\text{L}^{-1}} \times 0.0821 \cancel{\text{L}} \text{ atm} \cancel{\text{mol}^{-1}} \cancel{\text{K}^{-1}} \times 298 \cancel{\text{K}}}{32.00 \cancel{\text{g}} \cancel{\text{mol}^{-1}}}$$

$$= 1.15 \text{ atm}$$

PROBLEM-SOLVING NOTE
To ensure that you have used a correct variation of the ideal gas equation, you should use the appropriate cancellation of units as a check . Here, all units cancel except for the desired "atm."

EXERCISE 5.15A

To what temperature must propane gas (C_3H_8) at 785 Torr be heated to have a density of 1.51 g/L?

EXERCISE 5.15B

At what temperature will a sample of $O_2(g)$ at 725 Torr have the same density as $NH_3(g)$ does at 22.5 °C and 1.45 atm?

Connections: Balloons, the Gas Laws, and Chemistry

Early knowledge of the behavior of gases was stimulated by an interesting development—human flight using balloons. In June 1782, two brothers, Joseph Michel and Jacques Étienne Montgolfier, launched the first such balloon. They lit a fire under an opening in a large bag. The less-dense hot air allowed the balloon to rise slowly through the denser, cooler atmosphere. By August of that year, Jacques Charles filled a balloon with hydrogen gas, discovered 16 years earlier by Henry Cavendish. Charles made hydrogen on a scale never before attempted, using the reaction of about 500 kg of iron with acid.

$$Fe(s) + 2\ HCl(aq) \rightarrow FeCl_2(aq) + H_2(g)$$

When they are compared at the same temperature and pressure, hydrogen is only 1/14th as dense as air. Each kilogram of hydrogen can carry aloft a payload of 13 kg. In December 1782, Charles and a companion flew for 25 km from Paris in a balloon filled with hydrogen. They landed in a small village where they were attacked by frightened farmers who tore the balloon with pitchforks.

By 1804, Joseph Gay-Lussac had ascended to an altitude of 7 km and brought back samples of the rarefied air at that altitude. Within just a few years, scientists had used balloons to learn a great deal about the nature of gases, about Earth's atmosphere, and about weather.

A contemporary drawing depicting the first ascent of the Montgolfier brothers in a hot air balloon in 1782.

5.9 Gases in Reaction Stoichiometry

With what we have learned about gases, we can greatly expand the scope of stoichiometry calculations, as we shall see in this section.

The Law of Combining Volumes

In 1809, Joseph Gay-Lussac published an experimental result known as the **law of combining volumes**.

> *When gases measured at the same temperature and pressure are allowed to react, the volumes of gaseous reactants and products are in small whole-number ratios.*

For example, at 100 °C two volumes of hydrogen unite with one volume of oxygen to produce two volumes of steam (water vapor), as suggested by Figure 5.12. The combining ratio for volume is 2:1:2.

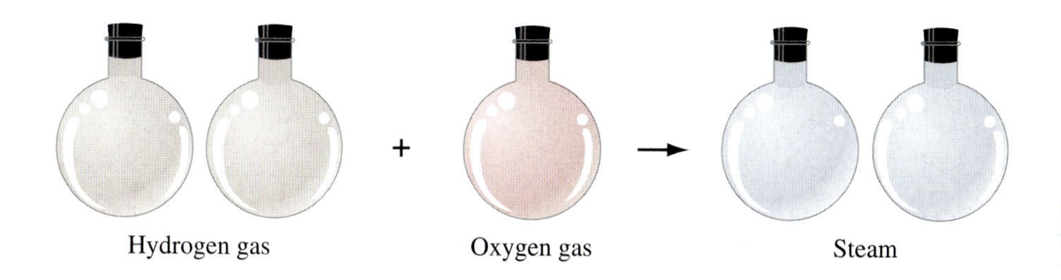

◀ **Figure 5.12**
Gay-Lussac's law of combining volumes

One hypothesis that was successful in explaining a few of Gay-Lussac's results was that equal volumes of different gases at the same temperature and pressure contained equal numbers of atoms. John Dalton rejected this idea, however. If two volumes of hydrogen did react with one of oxygen, he reasoned that only *one* volume of steam should have formed, not two. That is, there could be no more combinations of H and O atoms in water than there were O atoms to start with. Dalton's reasoning was based on the following *incorrect* equation.

$$2\,H(g) + O(g) \longrightarrow H_2O(g)$$

Avogadro gave the correct explanation of Gay-Lussac's law in 1811. In addition to accepting the "equal volumes-equal numbers" hypothesis, Avogadro proposed that gases may exist in *molecular* form. By postulating the existence of the molecules H_2 and O_2, he explained the reaction in the same way that we do today. *Two molecules* of H_2 react with *one molecule* of O_2 to form *two molecules* of water. The combining ratios by number of molecules, number of moles, and volumes are all 2:1:2.

$$2\,H_2(g) + O_2(g) \longrightarrow 2\,H_2O(g)$$

Figure 5.13 suggests Avogadro's line of reasoning.

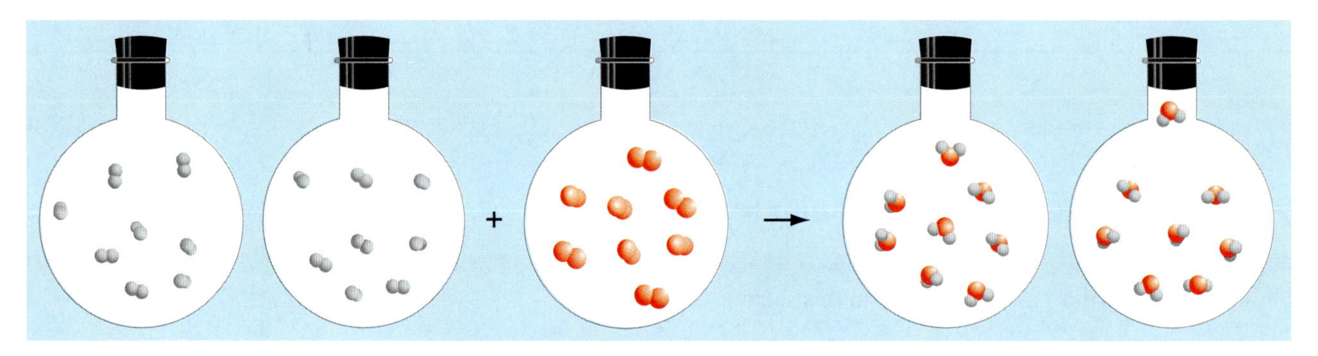

▲ **Figure 5.13 Avogadro's explanation of Gay-Lussac's law of combining volumes** When the gases are measured at the same temperature and pressure, each of the identical flasks contains the same number of molecules. Notice how the combining ratio: 2 volumes H_2 to 1 volume O_2 to 2 volumes H_2O leads to a result in which all the atoms present initially are accounted for in the product.

In applying the law of combining volumes in Example 5.16, we will also use the following ideas:

1. The law of combining volumes relates the volumes of any two *gaseous* species in a reaction, even if some reactants or products are liquid or solid.
2. It is not necessary that a reaction be carried out at the particular temperature and/or pressure at which the gaseous species are compared.
3. The actual conditions under which the gaseous species are compared need not be known, as long as the individual gases are measured at the *same* temperature and the *same* pressure.

EXAMPLE 5.16

How many liters of $O_2(g)$ are consumed for every 10.0 L of $CO_2(g)$ produced in the combustion of liquid pentane, C_5H_{12}, if each gas is measured at STP?

SOLUTION

We begin by writing a balanced equation for the combustion reaction.

$$C_5H_{12}(l) + 8\,O_2(g) \longrightarrow 5\,CO_2(g) + 6\,H_2O(l)$$

The fact that pentane and water are in the liquid state does not affect our calculation, because the substances of interest, CO_2 and O_2, are both gases. The actual temperatures and pressures before, during, and after the combustion reaction are not relevant, as long as each gas is *measured* at STP. The condition of STP is relevant only in establishing that the two gases are compared at the same temperature and pressure.

The fundamental equivalence in the reaction is

$$8 \text{ mol } O_2 \Leftrightarrow 5 \text{ mol } CO_2$$

but because of the "equal numbers-equal volumes" hypothesis, we can substitute a volume unit for the unit mol.

$$8 \text{ L } O_2 \Leftrightarrow 5 \text{ L } CO_2$$

This equivalence provides the stoichiometric factor we need. Thus,

$$? \text{ L } O_2(g) = 10.0 \text{ L } \cancel{CO_2(g)} \times \frac{8 \text{ L } O_2(g)}{5 \text{ L } \cancel{CO_2(g)}} = 16.0 \text{ L } O_2(g)$$

EXERCISE 5.16A

What volume of oxygen gas is required to burn 0.556 L of propane gas, C_3H_8, if both gases are measured at the same temperature and the same pressure?

EXERCISE 5.16B

What volume of $O_2(g)$ is consumed in the combustion of 125 g of gaseous dimethyl ether, $(CH_3)_2O$, if both gases are measured at the temperature and pressure at which the density of dimethyl ether is 1.81 g/L?

The Ideal Gas Equation in Reaction Stoichiometry

We can substitute ratios of gas volumes for mole ratios *only* for *gases* and *only* if the gases are at *the same temperature and pressure*. Often the amount of a gaseous reactant or product needs to be related to that of a *solid* or *liquid*. In these cases, we must work with mole ratios, but the ideal gas equation allows us to relate moles of gas to other gas properties.

EXAMPLE 5.17

In the chemical reaction used in automotive air-bag safety systems, $N_2(g)$ is produced by the decomposition of sodium azide, NaN_3.

$$2\,NaN_3(s) \longrightarrow 2\,Na(l) + 3\,N_2(g)$$

What volume of $N_2(g)$, measured at 25 °C and 0.980 atm, is produced by the decomposition of 62.5 g NaN_3?

SOLUTION

First, we can determine the amount of $N_2(g)$ that is produced when 62.5 g NaN_3 is decomposed. Here we use the molar mass (1 mol NaN_3/65.01 g NaN_3) and a stoichiometric factor from the balanced equation (3 mol N_2/2 mol NaN_3) as conversion factors.

PROBLEM-SOLVING NOTE
Although we have solved this problem in two steps, there is no need to reenter the intermediate result (1.44 mol N_2) into your calculator. The number displayed at the end of the first step simply becomes the first entry of the second step.

$$? \text{ mol } N_2 = 62.5 \text{ g NaN}_3 \times \frac{1 \text{ mol NaN}_3}{65.01 \text{ g NaN}_3} \times \frac{3 \text{ mol } N_2}{2 \text{ mol NaN}_3} = 1.44 \text{ mol } N_2$$

Then, we can use the ideal gas equation to determine the volume of 1.44 mol $N_2(g)$ under the stated conditions.

$$V = \frac{nRT}{P} = \frac{1.44 \text{ mol} \times 0.0821 \text{ L atm mol}^{-1} \text{ K}^{-1} \times (273 + 25) \text{ K}}{0.980 \text{ atm}}$$

$$= 35.9 \text{ L } N_2(g)$$

EXERCISE 5.17A

Quicklime (CaO) for use in the construction industry is manufactured by decomposing limestone ($CaCO_3$) by heating.

$$CaCO_3(s) \xrightarrow{\Delta} CaO(s) + CO_2(g)$$

How many liters of $CO_2(g)$ at 825 °C and 754 Torr are produced in the decomposition of 45.8 kg $CaCO_3(s)$?

● **ChemCDX** Airbags

EXERCISE 5.17B

How many liters of cyclopentane, C_5H_{10} (d = 0.7445 g/mL), must be burned in excess $O_2(g)$ to produce 1.00×10^6 L of $CO_2(g)$ measured at 25.0 °C and 736 Torr?

5.10 Mixtures of Gases: Dalton's Law of Partial Pressures

Early experimenters did not make a strong distinction between air and other gases. Carbon dioxide was first called "fixed air"; oxygen, "dephlogisticated air"; and hydrogen, "flammable air." We now know that air is a *mixture* of gases, but fortunately the ideal gas equation applies to mixtures of gases about as well as it does to individual gases. Let's begin by considering Dalton's views on gaseous mixtures.

Partial Pressures and Mole Fractions

Although best known for his atomic theory, Dalton had wide-ranging interests, one of which was meteorology. He performed several experiments on atmospheric gases in an attempt to understand weather. In fact, he formulated his atomic theory to explain the results of those meteorological experiments. In one experiment, he found that if he added water vapor at a certain pressure to dry air, the pressure exerted by the air increased by an amount equal to the pressure of the water vapor. Based on this and other experiments, Dalton concluded that each gas in a mixture behaves independently of the other gases. **Dalton's law of partial pressures** states that in a mixture of gases, each gas expands to fill the container and exerts its own pressure; this individual gas pressure is called a **partial pressure**. Moreover,

> *The total pressure exerted by a mixture of gases is equal to the sum of the partial pressures exerted by the separate gases.*

The meaning of this statement at both the macroscopic and microscopic levels is suggested pictorially in Figure 5.14. The 5.0-L gas samples in Figure 5.14a and 5.14b are forced into the same 5.0-L volume, forming the mixture of gases in Figure 5.14c. The number of moles of gas in the mixture is the sum of the number of

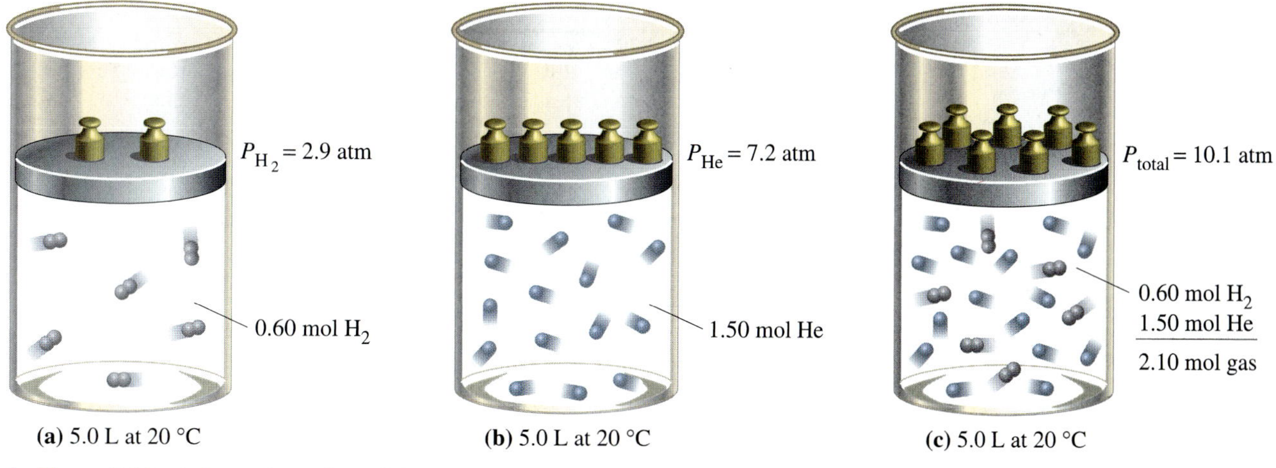

▲ **Figure 5.14 Dalton's law of partial pressures illustrated**
Each gas expands to fill the container and exerts a pressure that is readily calculated using the ideal gas equation. The total pressure of the mixture of gases is equal to the sum of the partial pressures of the individual gases.

moles of each gas. Likewise, the total pressure in the mixture (10.1 atm) is the sum of the individual gas pressures. Thus, we can state Dalton's law mathematically as

$$P_{total} = P_1 + P_2 + P_3 + \dots$$

where the terms on the right side refer to the partial pressures of gases 1, 2, 3, . . . (The series of three dots signifies that there may be more gases in the mixture than the three noted.) Applying this equation to the mixture of gases in Figure 15.14, we find

$$P_{total} = P_{H_2} + P_{He} = 2.9 \text{ atm} + 7.2 \text{ atm} = 10.1 \text{ atm}$$

To obtain the individual partial pressures we assume that each gas can be described by the ideal gas law. For gas 1, of which there are n_1 moles, gas 2, of which there are n_2 moles, and so on,

$$P_1 = \frac{n_1 RT}{V}; \quad P_2 = \frac{n_2 RT}{V}; \text{ and so on.}$$

Example 5.18 illustrates the idea that a component (N_2) of a mixture of gases (air) expands to fill its container and exerts a distinctive partial pressure.

EXAMPLE 5.18

A 1.00-L sample of dry air at 25 °C contains 0.0319 mol N_2, 0.00856 mol O_2, 0.000381 mol Ar, and 0.00002 mol CO_2. Calculate the partial pressure of $N_2(g)$ in the mixture.

SOLUTION

From the ideal gas equation,

$$P_{N_2} = \frac{n_{N_2} RT}{V}$$

$$= \frac{0.0319 \text{ mol} \times 0.0821 \text{ L atm mol}^{-1} \text{ K}^{-1} \times 298 \text{ K}}{1.00 \text{ L}} = 0.780 \text{ atm}$$

PROBLEM-SOLVING NOTE

Of course, for a mixture of gases, $P_{total} \times V = n_{total} \times RT$.

EXERCISE 5.18A

Calculate the partial pressure of each of the other components of the air sample in Example 5.18. What is the total pressure exerted by all the gases in the sample?

EXERCISE 5.18B

What is the total pressure exerted by a mixture of 4.05 g N_2, 3.15 g H_2, and 6.05 g He when confined to a 6.10-L container at 25 °C?

We can obtain another set of useful expressions from the equations we discussed earlier. Let's designate one of the components of a gaseous mixture as "1," and then take the *ratio* of the partial pressure of component "1" to the total gas pressure.

$$\frac{P_1}{P_{total}} = \frac{P_1}{P_1 + P_2 + \cdots}$$

$$\frac{P_1}{P_{total}} = \frac{\dfrac{n_1 \cancel{RT}}{\cancel{V}}}{\dfrac{n_1 \cancel{RT}}{\cancel{V}} + \dfrac{n_2 \cancel{RT}}{\cancel{V}} + \cdots} = \frac{n_1}{n_1 + n_2 + \cdots} = \frac{n_1}{n_{total}} = x_1$$

We can write similar expressions for all the other components in the gaseous mixture. The ratio, n_1/n_{total}, has a special name and symbol. It is called the *mole fraction* of component "1" in the mixture and is represented as x_1. The **mole fraction** is the fraction of all the molecules in a mixture that are of a given type. The sum of the mole fractions of all the components in a gaseous mixture is one. That is,

$$\underbrace{\frac{n_1}{n_1 + n_2 + \cdots}}_{x_1} + \underbrace{\frac{n_2}{n_1 + n_2 + \cdots}}_{x_2} + \cdots = \frac{n_1 + n_2 + \cdots}{n_1 + n_2 + \cdots} = 1$$

As with other fractional parts of the whole, we can multiply mole fractions by 100% to obtain *mole percents*.

Expressions of the type

$$\frac{P_1}{P_{total}} = \frac{n_1}{n_{total}} = x_1$$

have a special significance: They allow us to relate the partial pressures of the components of a gaseous mixture to the total gas pressure. That is,

$$P_1 = x_1 \times P_{total}$$

The compositions of gaseous mixtures are often given in percent by volume, but this is actually a mole percent because in a gas at a fixed temperature and pressure, the volume and moles of a gas are directly proportional to one another. In Example 5.19, we calculate the partial pressures of the components of air from its volume percent composition.

EXAMPLE 5.19

The main components of dry air, by volume, are N_2, 78.08%; O_2, 20.95%; Ar, 0.93%; and CO_2, 0.04%. What are the partial pressures of each of the four gases in a sample of air at 1.000 atm?

SOLUTION

Volume percent is the same as mole percent, and from the mole percents, we can write the mole fractions. Thus, 78.08 volume percent N_2 is the same as 78.08 mole percent N_2, which, in turn, is the same as a 0.7808 mole fraction of N_2. We can apply this reasoning to other gases in the mixture, too: The mole fraction for O_2 is 0.2095; for Ar, 0.0093; and for CO_2, 0.0004. Each partial pressure can be obtained from the expression

$$\text{Partial pressure} = \text{mole fraction} \times \text{total pressure}$$

and,

$$P_{N_2} = 0.7808 \times 1.000 \text{ atm} = 0.7808 \text{ atm}$$

$$P_{O_2} = 0.2095 \times 1.000 \text{ atm} = 0.2095 \text{ atm}$$

$$P_{Ar} = 0.0093 \times 1.000 \text{ atm} = 0.0093 \text{ atm}$$

$$P_{CO_2} = 0.0004 \times 1.000 \text{ atm} = 0.0004 \text{ atm}$$

EXERCISE 5.19A

A sample of expired air (air that has been exhaled) is composed, by volume, of the following main gaseous components: N_2, 74.1%; O_2, 15.0%; H_2O, 6.0%; Ar, 0.9%; and CO_2, 4.0%. What are the partial pressures of each of the five gases in the expired air at 37 °C and 1.000 atm?

EXERCISE 5.19B

A 75.0-L sample of natural gas at 23.5 °C contains methane (CH_4) at a partial pressure of 505 Torr, ethane (C_2H_6) at 201 Torr, propane (C_3H_8) at 43 Torr, and 10.5 g of butane (C_4H_{10}). Calculate the mole fraction of each of the four components of this natural gas.

EXAMPLE 5.20—A Conceptual Example

Describe what we would need to do to get from condition (a) to condition (b) in Figure 5.15.

SOLUTION

First, we can assess the situation in (a): Because the chosen volume is 22.4 L, the temperature is 273 K (0 °C), and the total amount of gas is 1.00 mol, the total pressure must be 1.00 atm. This condition is the familiar molar volume of an ideal gas at STP (22.4 L).

Now let's appraise the situation in (b): The volume and temperature remain as they were in (a), and the pressure increases to 3.00 atm. This tripling of the pressure while

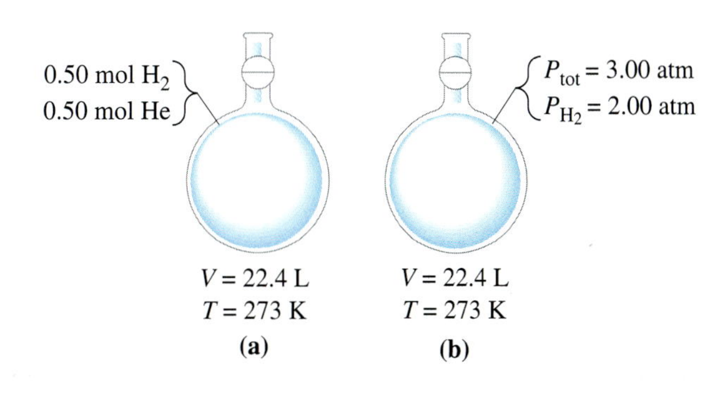

Figure 5.15 Example 5.20 illustrated

the volume and temperature are fixed requires a tripling of the amount of gas. There must be 3.00 mol gas in (b) compared to 1.00 mol in (a). P_{H_2} is given as 2.00 atm; the partial pressure(s) of the other gas or gases present must be 1.00 atm. The corresponding mole fractions are 2/3 for H_2 and 1/3 for the other gas or gases. The 3.00-mol mixture of gases in (b) must consist of 2.00 mol H_2 and 1.00 mol of another gas or gases.

So to get from (a)—0.50 mol H_2 and 0.50 mol He—to (b)—2.00 mol H_2 and 1.00 mol of other gases—we must add 1.50 mol H_2 and 0.50 mol of some other gas or gases to the flask in (a). Note that this 0.50 mol *can be* 0.50 mol He, but it could also be any single gas or combination of gases, *other than* hydrogen.

EXERCISE 5.20

Why can't the change between (a) and (b) in Figure 5.15 be achieved by adding hydrogen gas alone? Why can't it be achieved by adding helium gas alone? Why is it necessary that some hydrogen be added but not necessary that any helium be added?

Collection of Gases over Water

If a gas is highly soluble in water, molecules of the gas intermix with molecules of water and the aqueous solution of gas can be accommodated in essentially the same volume as was the water itself. If a gas is only slightly soluble in water, as are such common gases as oxygen, nitrogen, and hydrogen, the situation is different. As the essentially insoluble gas is passed into a container of water like that pictured in Figure 5.16, the gas rises because its density is much less than that of water and water must be displaced or pushed out of the collection bottle. This simple technique is called the collection of a gas over water. As a gas is collected by displacement, water vapor produced by evaporation gathers in the collection vessel as well. The total pressure in the collection vessel is that of the gas (P_{gas}) plus that of water vapor $\left(P_{H_2O(g)}\right)$.

$$P_{total} = P_{gas} + P_{H_2O(g)}$$

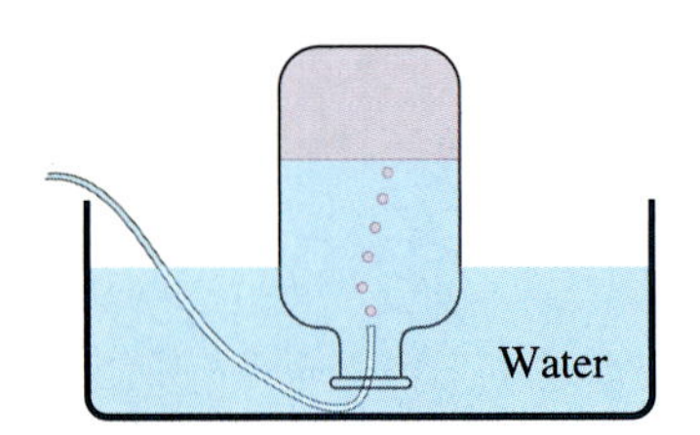

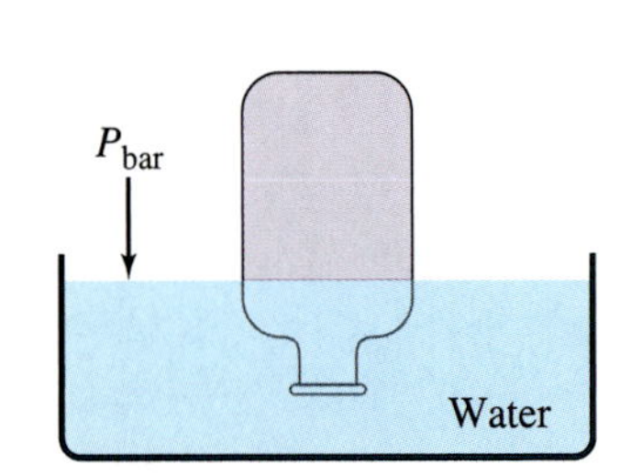

▲ **Figure 5.16**
Collection of a gas over water
To make the total pressure of the gaseous mixture in the bottle equal to the atmospheric pressure as measured with a barometer, it is necessary to adjust the position of the bottle so that the water levels inside and outside the bottle are the same. When this is done, the partial pressure of the gas is given by

$$P_{gas} = P_{bar.} - P_{H_2O}$$

P_{bar} is the barometer reading and P_{H_2O}, the vapor pressure of water, is obtained from tabulated data such as Table 5.4.

Assuming the gas is saturated with water vapor, the partial pressure of the water vapor is the *vapor pressure* of the water. The vapor pressure depends only on the temperature: The warmer the water, the higher the vapor pressure is (Table 5.4). To find the partial pressure of the collected gas, we need only subtract the vapor pressure of the water from the total pressure within the collection vessel, which is made equal to the measured atmospheric pressure, P_{bar}. That is,

$$P_{gas} = P_{total} - P_{H_2O(g)}$$
$$= P_{bar} - P_{H_2O(g)}$$

Thus, if the barometric pressure is 735 Torr and the water temperature is 22 °C, the water vapor pressure is 19.8 Torr (from Table 5.4), and the gas pressure is 735 Torr − 19.8 Torr = 715 Torr.

EXAMPLE 5.21

Hydrogen produced in the following reaction is collected over water at 23 °C and 742 Torr barometric pressure.

$$2\,Al(s) + 6\,HCl(aq) \longrightarrow 2\,AlCl_3(aq) + 3\,H_2(g)$$

What volume of the "wet" gas will be collected in the reaction of 1.50 g Al(s) with excess HCl(aq)?

SOLUTION

The "wet" gas is the mixture of hydrogen and water vapor. However, because the two gases are found in the same collection vessel, they occupy the same volume. We need to calculate the volume of only one of them. We don't have enough data to calculate the volume of the water vapor. We can find the pressure of the water vapor (21.1 Torr) in Table 5.4, but this partial pressure and the temperature are not enough; we also need to know the number of moles of water vapor, and we don't have a simple way of getting this.

However, we can gather the necessary data to determine the volume of hydrogen. To begin, we know the temperature, 23 °C, and we can get the partial pressure of hydrogen by noting these facts.

$$P_{total} = P_{H_2} + P_{H_2O}$$

$$742 \text{ Torr} = P_{H_2} + 21.1 \text{ Torr}$$

$$P_{H_2} = 742 \text{ Torr} - 21.1 \text{ Torr} = 721 \text{ Torr [or } (721/760) \text{ atm]}$$

And we can determine the number of moles of hydrogen from the stoichiometry of the reaction.

$$? \text{ mol } H_2 = 1.50 \cancel{\text{g Al}} \times \frac{1 \cancel{\text{mol Al}}}{26.98 \cancel{\text{g Al}}} \times \frac{3 \text{ mol } H_2}{2 \cancel{\text{mol Al}}} = 0.0834 \text{ mol } H_2$$

So, now we can use the ideal gas equation to calculate the volume of hydrogen, and, therefore, that of the wet gas.

$$V = \frac{nRT}{P} = \frac{0.0834 \cancel{\text{mol}} \times 0.0821 \text{ L} \cancel{\text{atm}} \cancel{\text{mol}^{-1}} \cancel{\text{K}^{-1}} \times (273 + 23) \cancel{\text{K}}}{(721/760) \cancel{\text{atm}}}$$

$$= 2.14 \text{ L}$$

EXERCISE 5.21A

Hydrogen gas is collected over water at 18 °C. The total pressure inside the collection jar is set at the barometric pressure of 738 Torr. If the volume of the gas is 246 mL, what mass of hydrogen is collected? What is the mass of the wet gas?

(*Hint:* What is the mass of water vapor?)

EXERCISE 5.21B

A sample of $KClO_3$ was decomposed to potassium chloride and oxygen by heating. The O_2 was collected over water at 21 °C and a barometric pressure of 746 mmHg. A 155-mL volume of the gaseous mixture was obtained. What mass of $KClO_3$ was decomposed?

TABLE 5.4 Vapor Pressure of Water as a Function of Temperature

Temp. (°C)	Pressure (mmHg)
15	12.8
16	13.6
17	14.5
18	15.5
19	16.5
20	17.5
21	18.7
22	19.8
23	21.1
24	22.4
25	23.8
30	31.8
40	55.3

5.11 The Kinetic-Molecular Theory: Some Quantitative Aspects

We introduced the kinetic-molecular theory in Section 5.2. We now examine the theory in a more quantitative fashion. Recall that a theory is invented to explain data, but that successful theories also allow for predictions. To begin, let's restate the principal assumptions of the kinetic-molecular theory.

1. A gas is made up of molecules that are in constant, random, straight-line motion.
2. Molecules of a gas are far apart; a gas is mostly empty space.
3. There are no forces between molecules except during the instant of collision. Each molecule acts independently of all the others and is unaffected by their presence, except during collisions.
4. Individual molecules may gain or lose energy as a result of collisions; however, in a collection of molecules at constant temperature, *the total energy remains constant.*

With these assumptions, it is possible to calculate the pressure exerted by a collection of molecules when confined to a given volume at a constant temperature. The derivation is too complex for us to consider in detail, but the final result is the equation below.

$$P = \frac{1}{3} \cdot \frac{N}{V} \cdot m\overline{u^2}$$

Here, V is the volume containing N molecules each having mass m, $\overline{u^2}$ is the *average* of the *squares* of their speeds (the bar above a term indicates an average value), and P is the pressure exerted by the gas. The important conclusions of the kinetic-molecular theory stem from this equation.

The Kinetic-Molecular Theory and Temperature

The kinetic-molecular theory gives us an important insight into the meaning of temperature. Suppose we modify the basic equation by multiplying both sides by V and replacing the fraction 1/3 by the product $2/3 \times 1/2$.

$$PV = \frac{1}{3} \cdot N \cdot m\overline{u^2} = \frac{2}{3} \cdot N \cdot \left(\frac{1}{2} m\overline{u^2}\right)$$

We do this to be able to isolate the term $1/2\ m\overline{u^2}$, which represents the average translational kinetic energy, e_k, of the gas molecules. Recall from page 179 that the translational kinetic energy of a molecule is that associated with the movement of the molecule as a whole in three-dimensional space (in mathematical terms, its movement in the x, y, and z directions).

Now let's assume we have 1 mol of gas, which means N will be Avogadro's number, N_A. We can then use the ideal gas equation for one mole of gas.

$$PV = nRT \quad (\text{where } n = 1 \text{ mol})$$

This allows us to substitute RT for PV and e_k for $1/2\ m\overline{u^2}$, and to write the following equation.

$$RT = \frac{2}{3} \cdot N_A \cdot e_k$$

Finally, we can solve for e_k.

$$e_k = \frac{3}{2} \cdot \frac{R}{N_A} \cdot T$$

Because 3/2, R, and N_A are all fixed quantities, this equation is equivalent to the expression that follows.

$$e_k = \text{constant} \times T$$

That is,

> *The average translational kinetic energy of the molecules of a gas, e_k, is directly proportional to the Kelvin temperature.*

You will find some basic information about kinetic energy in Appendix B.

Thus, the molecules of *any* gas at a given temperature have the same average translational kinetic energy as the molecules of *any other* gas at the same temperature.

When we heat a gas, we increase the average translational kinetic energy of the gas molecules and the temperature increases. If we were to cool a gas to 0 K, the average translational kinetic energy of the molecules would drop to zero. This, then, is the kinetic-molecular interpretation of the absolute zero of temperature: At 0 K, translational molecular motion ceases; molecules stop moving around. Absolute zero temperature is unattainable, but recent attempts have produced temperatures as low as a few nanokelvins (nK).

Molecular Speeds: Faster than a Speeding Bullet

Let's explore the notion of average molecular speeds that is required in the equations of the kinetic-molecular theory. The average speed $\bar{u}$ is defined as

$$\bar{u} = \frac{\text{sum of the speeds of all the molecules}}{\text{total number of molecules}} = \frac{u_1 + u_2 + u_3 + \ldots + u_n}{N}$$

For most applications, however, the root-mean-square speed is a more significant expression of molecular speed than is the average speed. The **root-mean-square speed, u_{rms}**, is the *square root* of the *average* of the *squares* of the molecular speeds.

$$u_{rms} = \sqrt{\overline{u^2}} = \sqrt{\frac{u_1^2 + u_2^2 + \ldots + u_n^2}{N}}$$

Because we can't measure the speeds of individual molecules, we can't use the equations above to calculate either the average molecular speed or the root-mean-square speed. One of the triumphs of the kinetic-molecular theory is that it permits us to derive equations for calculating average speeds. We will not attempt

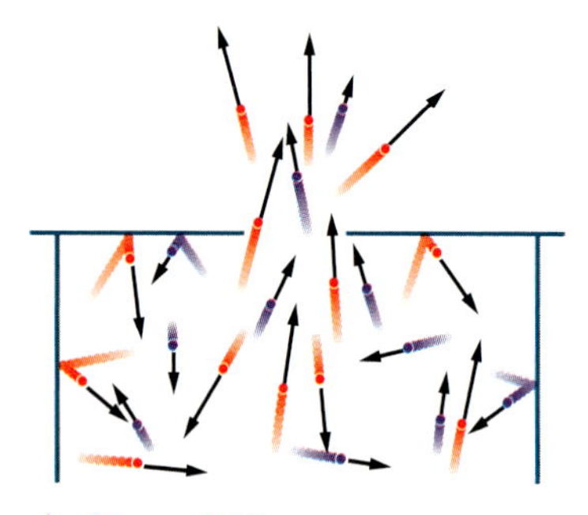

▲ **Figure 5.18**
Effusion of gases through an orifice
Average speeds of the two different types of molecules are suggested by the lengths of the arrows. The faster molecules (red) effuse more rapidly.

about 6×10^{-8} m. Although the rate of diffusion of a gas does depend on the average molecular speed of the gas molecules, frequent collisions make precise calculations quite complicated. Let's look at a somewhat simpler process for some insight into the process.

Effusion is a process in which a gas escapes from its container through a tiny hole (called an orifice). As suggested by Figure 5.18, because lighter molecules have greater speeds, we expect them to escape more quickly through an orifice than heavier molecules do. Although the situation is more complicated, we can roughly relate the rates at which gases effuse to their root-mean-square speeds. The following expression applies to two different gases, designated by the subscripts 1 and 2, whose molar masses are M_1 and M_2.

$$\frac{(\text{rate})_1}{(\text{rate})_2} = \frac{\sqrt{\dfrac{\cancel{3RT}}{M_1}}}{\sqrt{\dfrac{\cancel{3RT}}{M_2}}} = \sqrt{\frac{M_2}{M_1}}$$

This equation is a kinetic-molecular description of a nineteenth-century law proposed by Thomas Graham (1805–1869) and known as *Graham's law of effusion*.*

> *At a given temperature, the rates of effusion of gas molecules are inversely proportional to the square roots of their molar masses.*

In Example 5.23, we compare the rates or speeds of effusion of two different gases. In Example 5.24, instead of working with rates of effusion, we work with effusion *times*. Effusion times and rates are *inversely* related. That is,

$$\text{Effusion time} \propto \frac{1}{\text{effusion rate}}$$

This relationship should seem reasonable. It's quite similar to the example of an inverse proportionality that we gave on page 186 when considering Boyle's law. For example, if a car traveling at 100 km/h makes a particular trip in 1.00 h, a car traveling 50 km/h will require 2.00 h for the same trip.

EXAMPLE 5.23

If they are compared under the same conditions, how much faster than helium does hydrogen effuse through a tiny hole?

SOLUTION

We can answer this question by simply setting up a ratio of the two rates of effusion in accordance with Graham's law. Thus, if we place the rate of effusion of hydrogen, r_{H_2}, in the numerator on the left, its molar mass, M_{H_2}, must go into the denominator under the square root sign on the right. The situation is reversed for helium.

$$\frac{r_{H_2}}{r_{He}} = \sqrt{\frac{M_{He}}{M_{H_2}}} = \sqrt{\frac{4.00}{2.02}} = 1.41$$

* Graham's law adequately describes effusion only if the gas pressure is very low, so that molecules escape because of their random motion and not as a jet of gas. Also, the orifice must be tiny, so that no molecular collisions occur in the orifice as the gas effuses.

The ratio of the two rates is 1.41, and $r_{H_2} = 1.41 \times r_{He}$. Hydrogen effuses 1.41 times faster than helium.

EXERCISE 5.23A

Which effuses faster when the two gases are compared under the same conditions: N_2 or Ar? How much faster?

EXERCISE 5.23B

Which effuses faster when the two gases are compared under the same conditions: nitrogen monoxide or dimethyl ether, $(CH_3)_2O$? How much faster, expressed as a percentage?

EXAMPLE 5.24

One percent of a measured amount of Ar(g) escapes through a tiny hole in 77.3 s. One percent of the same amount of an unknown gas escapes under the same conditions in 97.6 s. Calculate the molar mass of the unknown gas.

SOLUTION

First, let's get the appropriate form for an expression relating effusion times and molar masses. To do this, we need to recall that "effusion time" and "effusion rate" are inversely related. Thus we can write

$$\frac{(\text{effusion rate})_{\text{Ar}}}{(\text{effusion rate})_{\text{unk}}} = \frac{1/(\text{effusion time})_{\text{Ar}}}{1/(\text{effusion time})_{\text{unk}}} = \frac{(\text{effusion time})_{\text{unk}}}{(\text{effusion time})_{\text{Ar}}} = \sqrt{\frac{M_{\text{unk}}}{M_{\text{Ar}}}}$$

$$\frac{97.6\ \text{s}}{77.3\ \text{s}} = \sqrt{\frac{M_{\text{unk}}}{39.95\ \text{g/mol}}} = 1.26$$

If we square both sides of the final equation and solve for M_{unk}, we get

$$\left(\sqrt{\frac{M_{\text{unk}}}{39.95\ \text{g/mol}}}\right)^2 = (1.26)^2$$

$$M_{\text{unk}} = 39.95\ \text{g/mol} \times (1.26)^2$$

The molar mass of the unknown gas is

$$M_{\text{unk}} = 63.4\ \text{g/mol}$$

PROBLEM-SOLVING NOTE

We can see that this is a reasonable answer. Because the unknown gas effuses more *slowly* than does Ar, its molar mass must be *greater* than that of Ar.

EXERCISE 5.24A

Two percent of a sample of N_2(g) effuses from a tiny opening in 57 s. Two percent of the same amount of an unknown gas escapes under the same conditions in 83 s. Calculate the molar mass of the unknown gas.

Separation of Uranium Isotopes

Naturally occurring uranium is a mixture of isotopes, with only about 0.7% being the fissionable uranium-235 isotope. To make fuel for a nuclear power plant or material for a nuclear bomb, the ^{235}U must be separated from the more abundant ^{238}U. During World War II, the United States undertook a massive program, code-named the "Manhattan Project," to develop a nuclear bomb. Project scientists developed a technique of separating the uranium isotopes by gaseous diffusion, a method still used to produce ^{235}U-enriched fuels for nuclear power plants.

One of the few uranium compounds that can be converted to a gas at moderate temperatures is uranium hexafluoride, UF_6. When high-pressure $UF_6(g)$ is forced through a porous barrier, molecules of the lighter ^{235}U isotope move through the barrier a tiny bit faster than those of ^{238}U. The portion of $UF_6(g)$ that has passed through the barrier has a slightly higher ratio of ^{235}U to ^{238}U than the original gas; it is said to be enriched in ^{235}U. The more times the process is repeated, the greater the concentration of ^{235}U becomes. Uranium must be enriched to 3% or 4% ^{235}U for use in typical nuclear power plants and to about 90% to make a nuclear bomb.

Uranium-235 and uranium-238 isotopes are separated by the diffusion of $UF_6(g)$ through porous barriers in these cylinders.

EXERCISE 5.24B

Five percent of a sample of $O_2(g)$ effuses from an orifice in 123 s. How long should it take five percent of the same amount of butane, C_4H_{10}, to effuse under the same conditions?

5.12 Real Gases

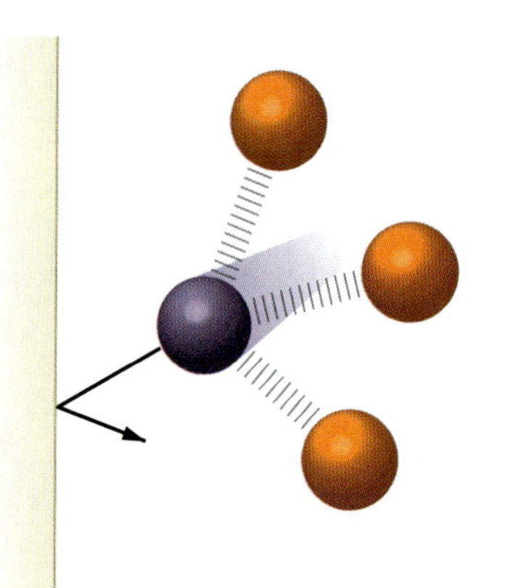

▲ **Figure 5.19 Intermolecular forces of attractions**
Attractive forces of the orange molecules for the purple molecule cause the purple molecule to exert less force when it collides with the wall than it would if these attractive forces did not exist.

Under many conditions, real gases do not follow the ideal gas law. We can account for their deviations from ideal gas behavior by reexamining the assumptions of the kinetic-molecular theory. We can safely assume minimal forces of attraction between molecules when they are far apart. However, at high pressures, molecules are closer together, and at low temperatures, they are both more closely spaced and pass by one another more slowly. Under these conditions we must reckon with two factors: (1) Intermolecular forces of attraction cause the measured pressure of a real gas to be less than we would expect, as suggested by Figure 5.19. (2) When molecules are close together, the volume of the molecules themselves becomes a significant fraction of the total volume of a gas.

The ideal gas law is based on the assumptions of *no intermolecular forces* and molecules as point masses, that is, molecules having mass but *no volume*. One way to obtain an equation to describe real gases under nonideal conditions is to adjust the ideal gas equation to account for the two factors just cited. The adjustments must be of the following sort.

Ideal gas equation: $P \times V = nRT$

"Adjusted" ideal gas equation: $(P_{\text{measured}} + ?)(V_{\text{measured}} - ?) = nRT$

TABLE 5.5 van der Waals Constants for Selected Gases		
Substance	***a*** **(L^2 atm mol^{-2})**	***b*** **(L mol^{-1})**
He	0.0341	0.02370
Ar	1.34	0.0322
H_2	0.244	0.0266
O_2	1.36	0.0318
CO_2	3.59	0.0427
CCl_4	20.4	0.1383

The adjusted equation acknowledges that the measured pressure is too low and must be increased by a certain amount (?) to compensate for intermolecular forces of attraction. The measured volume, on the other hand, is too high and must be decreased by a certain amount (?) to obtain the volume of empty space in the gas.

Perhaps the best known of the equations to describe the behavior of real gases is the one proposed by Johannes van der Waals in 1873. In the *van der Waals equation* written below, the term $n^2 a/V^2$ makes the adjustment for intermolecular forces of attraction and the term nb, for the volume of the gas molecules.

$$\left(P + \frac{n^2 a}{V^2}\right)(V - nb) = nRT$$

Whereas the ideal gas equation is perfectly general and can be applied without regard to the particular gas, the van der Waals equation includes two parameters, a and b, which are different for different gases and must be determined by experiment (Table 5.5).

For a real gas to approach ideal gas behavior, the term n^2a/V^2 should be small compared to P, and nb should be small compared to V. These conditions are most likely to be met when a small amount of gas (n) is found in a large volume (V) of gas, or, generally speaking, when a gas is at a *high temperature* and a *low pressure*. At room temperature or above and at pressures less than a few atmospheres, most gases obey the ideal gas equation reasonably well.

Summary

Gases consist of widely separated molecules, moving constantly and randomly throughout their containers. Molecular collisions with the container walls create a gas pressure. Atmospheric pressure can be measured with a barometer, and other gas pressures, with a manometer. The SI unit of pressure is the pascal (Pa), but more commonly used are the standard atmosphere (atm), the millimeter of mercury (mmHg), and the Torr: 1 atm = 760 mmHg = 760 Torr.

For a given amount of gas at a constant temperature, the volume and pressure are *inversely* proportional: $V = a/P$. For a given amount of gas at a constant pressure, the volume and temperature are *directly* proportional: $V = bT$. The temperature in this relationship is an absolute temperature measured on the Kelvin scale, with 0 K = −273.15 °C. At a fixed temperature and pressure, the volume and amount of a gas are *directly* proportional: $V = c\,n$. At 0 °C and 1 atm pressure (STP), the molar volume of a gas is about 22.4 L. The three simple gas laws can be consolidated into the combined gas law: $P_1V_1/n_1T_1 = P_2V_2/n_2T_2$.

If a gas has properties that conform to the simple gas laws, they are said to be ideal gases. The ideal gas law is stated through the equation, $PV = nRT$. The ideal gas equation can be used in place of the simple or combined gas laws, especially in molecular mass and gas density determinations. Of particular importance is its use in stoichiometric calculations for reactions involving gases.

Each gas in a mixture expands to fill the container and exerts its own *partial* pressure. Dalton's law of partial pressures states that the total pressure in a gaseous mixture is the sum

Key Terms

absolute zero (5.5)
atmosphere (atm) (5.3)
Avogadro's hypothesis (5.6)
Avogadro's law (5.6)
barometer (5.3)
Boyle's law (5.4)
Charles's law (5.5)
Dalton's law of partial pressures (5.10)
diffusion (5.11)
effusion (5.11)
ideal gas (5.8)
ideal gas law (ideal gas equation) (5.8)
kelvin (K) (5.5)
Kelvin scale (5.5)
kilopascal (kPa) (5.3)
kinetic-molecular theory of gases (5.2, 5.11)
law of combining volumes (5.9)
manometer (5.3)
millimeter of mercury (mmHg) (5.3)
molar volume of a gas (5.6)
partial pressure (5.10)
pascal (Pa) (5.3)
pressure (5.3)
root-mean-square speed (u_{rms}) (5.11)
standard temperature and pressure (STP) (5.6)
Torr (5.3)
universal gas constant (R) (5.8)
vapor (5.1)

of the partial pressures. A gas collected over water is a mixture. It also contains water vapor at a partial pressure equal to the vapor pressure of water. Stoichiometric calculations involving gases collected over water use stoichiometric factors from the balanced equation, Dalton's law of partial pressures, and the ideal gas equation.

The postulates of the kinetic-molecular theory lead to an equation for gas pressure and a relationship between Kelvin temperature and the average translational kinetic energy of gas molecules. The theory also provides a basis for describing the diffusion and effusion of gases.

Real gases are most likely to exhibit ideal behavior at high temperatures and low pressures. When the ideal gas equation fails, it generally can be replaced by another equation, such as that of van der Waals.

Review Questions

1. Describe what a gas is like at the molecular level.

2. Why don't the molecules of a gas settle to the bottom of their container?

3. Give a kinetic-molecular explanation of the origin of gas pressure.

4. What does a mercury barometer measure? How does it work?

5. Why is mercury (rather than water or another liquid) used as the fluid in barometers?

6. How does a manometer differ from a barometer? How does an open-end manometer differ from a closed-end manometer?

7. State Boyle's law in words and as a mathematical equation.

8. Use the kinetic-molecular theory to explain Boyle's law.

9. What is the advantage of storing gases under high pressure, for example, oxygen used in respiration therapy?

10. State Charles's law in words and in the form of a mathematical equation.

11. Use the kinetic-molecular theory to explain Charles's law.

12. Why must an absolute temperature scale rather than the Celsius scale be used for Charles's law calculations?

13. How is the Kelvin scale related to the Celsius temperature scale?

14. What effect will the following changes have on the volume of a fixed amount of gas?

(a) an increase in pressure at constant temperature

(b) a decrease in temperature at constant pressure

(c) a decrease in pressure coupled with an increase in temperature

15. What effect will the following changes have on the pressure of a fixed amount of a gas?

(a) an increase in temperature at constant volume

(b) a decrease in volume at constant temperature

(c) an increase in temperature coupled with a decrease in volume

16. According to the kinetic-molecular theory, **(a)** what change in temperature occurs as the molecules of a gas begin to move more slowly, on average? **(b)** What change in pressure occurs when the walls of the container are struck less often by molecules of the gas?

17. Container A has twice the volume but holds twice as many gas molecules as container B at the same temperature. Use the kinetic-molecular theory to compare the pressures in the two containers.

18. What are the standard conditions of temperature and pressure for gases? Why is it useful to define such conditions?

19. What is meant by the *molar volume* of a gas? What is the approximate value of the molar volume of a gas at STP?

20. State the ideal gas law in words and in the form of a mathematical equation.

21. Use the kinetic-molecular theory to explain the dependence of gas density on molar mass, temperature, and pressure referred to on page 199.

22. For each of the following, indicate whether a given gas would have the same or different densities in the two containers. If the densities are different, in which container is the density greater?

(a) Containers A and B have the same volume and are at the same temperature, but the gas in A is at a higher pressure.

(b) Containers A and B are at the same pressure and temperature, but the volume of A is greater than that of B.

(c) Containers A and B are at the same pressure and volume, but the gas in A is at a higher temperature.

23. State Dalton's law of partial pressures in words and in the form of a mathematical equation.

24. Describe how Dalton's law of partial pressures is applied to gases collected over water.

25. State the definition of mole fraction in the form of a mathematical equation, and then restate it in terms of partial pressures of gases.

26. Use the kinetic-molecular theory to explain what we mean by temperature.

27. What is meant by u_{rms}, the root-mean-square speed of molecules of a gas? What properties of a gas determine the value of u_{rms}?

28. State Graham's law of effusion, and indicate whether $O_2(g)$ or $N_2(g)$ has the greater effusion rate when the gases are compared under identical conditions.

29. Describe the primary factors that cause a gas not to conform to the ideal gas equation.

30. Under what conditions do the properties of real gases differ substantially from those of ideal gases?

Problems

(Assume that the simple and ideal gas laws apply in all cases, unless the problem states otherwise.)

Pressure

31. Carry out the following conversions between pressure units.
    (a) 0.985 atm to mmHg
    (b) 849 Torr to atm
    (c) 642 Torr to kPa
    (d) 15.5 lb/in.^{2} to mmHg

32. Carry out the following conversions between pressure units.
    (a) 4.00 atm to Torr
    (b) 721 Torr to mmHg
    (c) 1050 mb to kPa
    (d) 30.10 in. Hg to atm

33. Elephant seals dive to depths as great as 1250 m. What is the pressure, in atm, exerted by water at that depth? The densities of water and mercury are 1.00 g/mL and 13.6 g/mL, respectively.

34. City fire codes specify the water pressure that must be maintained at the end of a water line—for example, 35.5 lb/in.^{2}. Calculate the height of a water column that exerts a pressure of 35.5 lb/in.^{2}. The densities of water and mercury are 1.00 g/mL and 13.6 g/mL, respectively.

35. Calculate the *height*, in meters, of each liquid column.
    (a) a column of Hg(l) (d = 13.6 g/mL) that exerts a pressure of 4.36 atm
    (b) a column of CCl_4(l) (d = 1.59 g/mL) that exerts the same pressure as a 25.0-cm column of Hg(l) (d = 13.6 g/mL)

36. Calculate the *pressure*, in atmospheres, exerted by the following columns of liquid. Note that for Hg(l), d = 13.6 g/mL.
    (a) a 35.0-ft column of H_2O(l)
    (b) a 2.50-m column of carbon disulfide having a volume of 1.25 L and found to weigh 1.58 kg

37. The open-end manometer pictured in Figure 5.4(a) on page 182 is filled with an oil that has a density of 0.789 g/mL. Calculate the pressure, in Torr, of the gas in the flask if barometric pressure is 755 Torr and the oil level on the right side, h_{closed}, is 44 mm below that on the left side.

38. The open-end manometer pictured in Figure 5.4(b) on page 182 is filled with glycerol (d = 1.261 g/mL). Calculate the pressure, in Torr, of the gas in the flask if barometric pressure is 738 Torr and the glycerol level on the left side, h_{open}, is 12.7 cm below that on the right side.

39. We have noted that to measure atmospheric pressure, a mercury barometer must be nearly one meter in height. What is the height requirement for the closed-end mercury manometer pictured in Figure 5.3 on page 182? Explain.

40. How would you explain the "trick" suggested by the accompanying photograph to a young school child?

Boyle's Law

41. A sample of helium occupies 521 mL at 1752 Torr. Assume that the temperature is held constant, and determine the volume of the helium at **(a)** 752 Torr; **(b)** 3.55 atm; **(c)** 125 kPa.

42. A sample of nitrogen occupies 4.35 L at 732 mmHg. Assume that the temperature is held constant, and determine the pressure of the nitrogen if the volume is changed to **(a)** 12.5 L; **(b)** 435 mL; **(c)** 0.150 m^3.

43. A decompression chamber used by deep-sea divers has a volume of 10.3 m^3 and operates at an internal pressure of 4.50 atm. What volume, in m^3, would the air in the chamber occupy if it were at normal atmospheric pressure, assuming no temperature change?

44. A novel energy storage system involves storing air under high pressure. (Energy is released when the air is allowed to expand.) How many cubic feet of air, measured at standard atmospheric pressure of 14.7 pounds per square inch (psi), can be compressed into a 19-million-ft^3 underground cavern at a pressure of 1070 psi?

45. A 56.0-mL sample of neon gas is at the same pressure as the prevailing barometric pressure. If the pressure of the gas is increased to 100.0 Torr above the barometric pressure while the temperature is held constant, the volume is observed to change to 49.4 mL. What is the prevailing barometric pressure?

46. A 60.0-mL sample of hydrogen gas is obtained at a pressure of 762 Torr. How much additional pressure is required to compress the sample into a volume of 55.0 mL?

47. Oxygen used in respiration therapy is stored at room temperature under a pressure of 1.50×10^2 atm in gas cylinders with a volume of 60.0 L.

(a) What volume would the gas in one of these cylinders occupy at a pressure of 750.0 Torr? Assume no temperature change.

(b) If the oxygen flow to the patient is adjusted to 8.00 L per minute at room temperature and 750.0 Torr, how long will the tank of gas last?

48. The figure shows a cylinder with 1.20 L of a gas at the standard temperature and pressure (STP): 0 °C and 760 Torr. What will the gas pressure be if the piston is depressed 5.25 cm further into the cylinder?

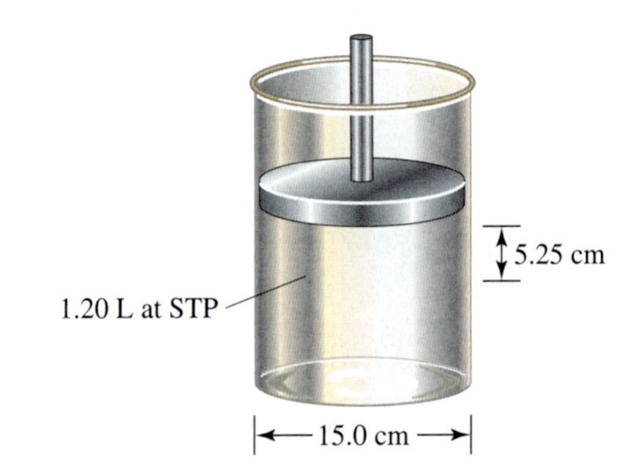

Charles's Law

49. A gas at a temperature of 99.8 °C occupies a volume of 154 mL. What will the volume be at a temperature of 10.0 °C, assuming no change in pressure?

50. A balloon is filled with helium. Its volume is 5.90 L at 26 °C. What will be its volume at −78 °C, assuming no pressure change?

51. A 567-mL sample of a gas at 305 °C and 1.20 atm is to be cooled at constant pressure until its volume becomes 425 mL. What will be the new gas temperature?

52. A sample of gas at 0 °C and 760 torr is to be heated at constant pressure until its volume triples. What will be the new gas temperature?

53. If a 15 °C temperature increase caused a 10.0% increase in the volume of a 95.0-mL sample of He(g) while the gas pressure was held constant, what was the original temperature?

54. What increase in the Celsius temperature will produce a 5.0% increase in the volume of a sample of gas originally at 25.0 °C if the gas pressure is held constant?

Avogadro's Law and Molar Volume at STP

55. What is the mass, in kilograms, of 4.55×10^3 L of neon gas at STP?

56. What is the volume, in milliliters, of 837 mg of xenon gas at STP?

57. If 125 mg of Ar(g) is added to a 505-mL sample of Ar(g) at STP, what volume will the sample occupy when the conditions of STP are restored?

58. How many grams of gas must be released from a 45.2-L sample of N_2(g) at STP to reduce the volume to 45.0 L at STP?

59. *Without doing detailed calculations*, determine which of the following samples contains the greatest number of molecules.

(a) 5.0 g H_2

(b) 50 L SF_6(g) at STP

(c) 1.0×10^{24} molecules of CO_2

(d) 67 L of a gaseous hydrocarbon at STP

60. *Without doing detailed calculations*, determine which of the following samples occupies the greatest volume *when measured at STP*.

(a) 30.0 g $O_2(g)$

(b) 1.10 mol $SO_2(g)$

(c) 24.0 L CO(g) at 22 °C and 745 mmHg

(d) 7.2×10^{23} molecules of $Cl_2(g)$

The Combined Gas Law

61. A sealed can with an internal pressure of 721 Torr at 25 °C is thrown into an incinerator operating at 755 °C. What will be the pressure inside the heated can, assuming the container remains intact during incineration?

62. A fixed amount of He exerts a pressure of 775 mmHg in a 1.05-L container at 26 °C. To what value must the temperature be changed to change the gas pressure to 725 mmHg? Assume the volume of gas remains constant.

63. Suppose we wish to contain a 1.00-mol sample of gas at 1.00 atm pressure and 25 °C. What volume container would we need?

64. At 25 °C, the pressure in a gas cylinder containing 8.00 mol O_2 is 5.05 atm. To maintain a constant pressure of 5.05 atm, how many moles of $O_2(g)$ should be released when the temperature is raised to 235 °C?

65. If a fixed amount of gas occupies 2.53 m^3 at a temperature of −15 °C and 191 Torr, what volume will it occupy at 25 °C and 1142 Torr?

66. What is the molar volume of a gas at **(a)** 25.0 °C and a pressure of 55.0 psi, and **(b)** at −78 °C and a pressure of 100.0 kPa?

The Ideal Gas Law

67. Calculate **(a)** the volume, in liters, of 1.12 mol of an ideal gas at 62 °C and 1.38 atm; **(b)** the pressure, in atmospheres, of 125 g CO(g) in a 3.96-L tank at 29 °C; **(c)** the mass, in milligrams, of 34.5 mL $H_2(g)$ at 725 Torr and −12 °C; and **(d)** the pressure, in kilopascals, of 173 g $N_2(g)$ in an 8.35-L flask at 0 °C.

68. Calculate **(a)** the pressure, in Torr, of 1.57×10^{-3} mol of an ideal gas in a volume of 225 mL at 17 °C; **(b)** the volume, in milliliters, of 42.4 mg $H_2(g)$ at 1.35 atm and −45 °C; **(c)** the mass, in kilograms, of 652 m^3 of $O_2(g)$ at 712 Torr and 6.5 °C; and **(d)** the pressure, in pounds per square inch (psi), of 108 mg Ar(g) in a 265-mL flask at 37 °C.

69. If the gas present in 4.65 L at STP is changed to a temperature of 15 °C and a pressure of 756 Torr, what will be the new volume?

70. If 7.75 g $SO_2(g)$ is added, at constant temperature and volume, to 5.87 L of $SO_2(g)$ at 705 mmHg and 26.5 °C, what will be the new gas pressure?

71. The interior volume of the Hubert H. Humphrey Metrodome in Minneapolis is 1.70×10^{10} L. The Teflon™-coated fiberglass roof is supported by air pressure provided by 20 huge electric fans. How many moles of air are present in the dome if the pressure is 1.02 atm at 18 °C?

72. A hyperbaric chamber is an enclosure containing oxygen at higher than normal pressures used in the treatment of certain heart and circulatory conditions. What volume of $O_2(g)$ from a cylinder at 25 °C and 151 atm is required to fill a 4.20×10^3-L hyperbaric chamber to a pressure of 3.00 atm at 17 °C?

73. The best laboratory vacuum systems can pump down to as few as 1.0×10^9 molecules per cubic meter of gas. Calculate the corresponding pressure, in atmospheres, assuming a temperature of 25 °C.

74. The atmosphere at the surface of Venus has an average temperature of 725 K and a pressure of 95 atm. At the surface of Mars, the atmosphere has an average temperature of 225 K and a pressure of about 8 mmHg. Calculate the molar volume of the atmospheric gases at the surface of each planet.

Molecular Mass Determinations

75. Calculate the molecular mass of a liquid that when vaporized at 98 °C and 756 Torr gave 125 mL of vapor with a mass of 0.625 g.

76. Calculate the molecular mass of a liquid that when vaporized at 99 °C and 716 Torr gave 225 mL of vapor with a mass of 0.773 g.

77. A liquid hydrocarbon is found to be 8.75% H by mass. A 1.261-g vaporized sample of the hydrocarbon has a 435-mL volume at 115 °C and 761 Torr. What is the molecular formula of this hydrocarbon?

78. A liquid hydrocarbon is found to be 16.37% H by mass. A 1.158-g vaporized sample of the hydrocarbon has a 385-mL volume at 71.0 °C and 749 mmHg. What is the molecular formula of this hydrocarbon?

Gas Densities

79. Calculate the density, in grams per liter, of each of the following.

(a) $CO(g)$ at STP

(b) $Ar(g)$ at 1.26 atm and 325 °C

80. Calculate the density, in grams per liter, of each of the following.

(a) $AsH_3(g)$ at STP

(b) $N_2(g)$ at 715 Torr and 98 °C

81. At what temperature will $O_2(g)$ at 0.982 atm pressure have a density of 1.05 g/L?

82. At what pressure will $N_2(g)$ have a density of 0.985 g/L at 25 °C?

83. The density of sulfur vapor at 445 °C and 755 mmHg is 4.33 g/L. What is the molecular formula of sulfur vapor?

84. A gaseous hydrocarbon contains 14.37% H by mass and has a density of 1.69 g/L at 24 °C and 743 mmHg. What is the molecular formula of this hydrocarbon?

85. *Without doing detailed calculations*, determine which of the following gases has the greatest density.

(a) $H_2(g)$ at −15 °C and 745 Torr

(b) $He(g)$ at STP

(c) $CH_4(g)$ at −10 °C and 1.15 atm

(d) $C_2H_6(g)$ at 50 °C and 435 Torr

86. *Without doing detailed calculations*, determine which of the following gases is (are) *more* dense than $O_2(g)$ at STP.

(a) N_2 at STP

(b) $CO(g)$ at 0 °C and 1100 Torr

(c) $SO_2(g)$ at 300 °C and 750 Torr

(d) $H_2(g)$ at 25 °C and 10 atm

Gases and the Stoichiometry of Reactions

87. How many liters of $SO_3(g)$ can be produced by the reaction of 1.15 L $SO_2(g)$ and 0.65 L $O_2(g)$ if all three gases are measured at the same temperature and pressure?

$$2\,SO_2(g) + O_2(g) \longrightarrow 2\,SO_3(g)$$

88. How many liters of $CO_2(g)$ can be produced in the reaction of 3.06 L $CO(g)$ and 1.48 L $O_2(g)$ if all three gases are measured at the same temperature and pressure?

89. How many liters of $CO_2(g)$ measured at 26 °C and 767 Torr are produced in the complete combustion of 125 mL of 1-propanol (d = 0.804 g/mL)?

$$CH_3CH_2CH_2OH(l) + O_2(g) \longrightarrow CO_2(g) + H_2O(l) \text{ (not balanced)}$$

90. How many liters of $O_2(g)$ measured at 22 °C and 763 Torr are consumed in the complete combustion of 2.55 L dimethyl ether measured at 25 °C and 748 Torr?

$$CH_3OCH_3(g) + O_2(g) \longrightarrow CO_2(g) + H_2O(l) \text{ (not balanced)}$$

91. How many milligrams of magnesium metal must react with excess $HCl(aq)$ to produce 28.50 mL of $H_2(g)$, measured at 26 °C and 758 Torr?

$$Mg(s) + 2\,HCl(aq) \longrightarrow MgCl_2(aq) + H_2(g)$$

92. A 100.0-g sample of aqueous hydrogen peroxide solution decomposes over time, producing 2.17 L $O_2(g)$ at 25 °C and 755 Torr.

$$2\,H_2O_2(aq) \longrightarrow 2\,H_2O(l) + O_2(g)$$

What must have been the mass percent H_2O_2 in the solution?

Mixtures of Gases

93. Only one of the accompanying sketches is a reasonable molecular view of a mixture of 1.0 g H_2 and 1.0 g He. Which is the correct view, and what is wrong with the other two?

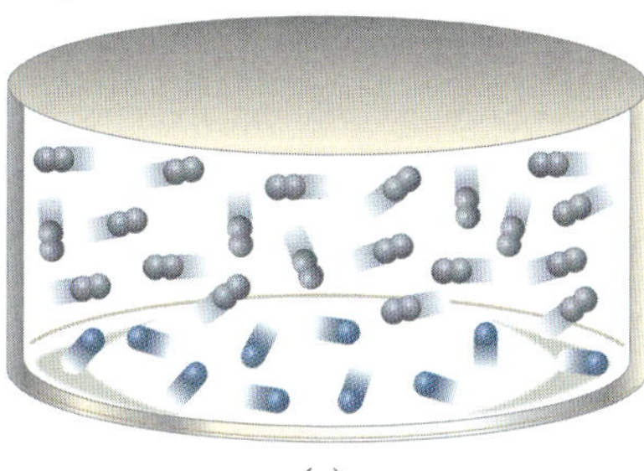

(a)

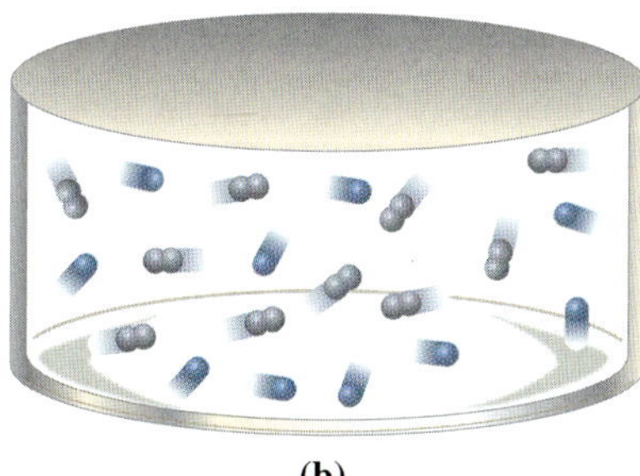

(b)

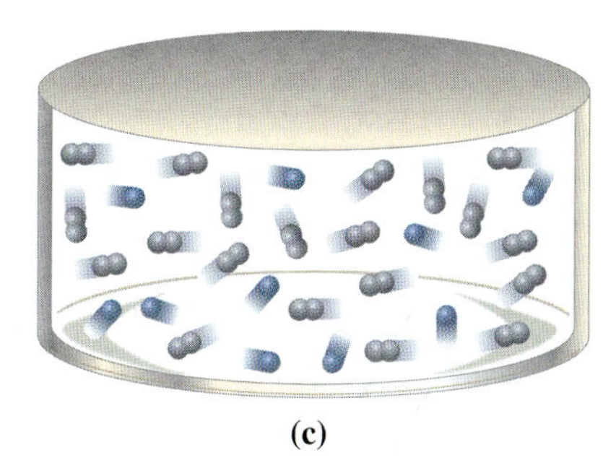

(c)

94. With reference to Problem 93, draw a molecular-level sketch of the gaseous mixture after the addition of 7.5 g $C_2H_6(g)$ to the mixture.

95. A gas sample has 76.8 mole percent N_2, 20.1 mole percent O_2, and 3.1 mol percent CO_2. If the total pressure is 762 mmHg, what are the partial pressures of the three gases?

96. A sample of intestinal gas was collected and found to consist of 44% CO_2, 38% H_2, 17% N_2, 1.3% O_2, and 0.003% CH_4, by volume. (The percentages do not total 100% because of rounding.) What is the partial pressure of each gas if the total pressure in the intestine is 818 Torr? [*Hint:* Recall that volume percent is the same as mole percent for ideal gas mixtures.]

97. Mixtures of helium and oxygen are used in scuba diving. What are **(a)** the mole fractions of the two gases, **(b)** their partial pressures, and **(c)** the total pressure in a mixture of 1.96 g He and 60.8 g O_2 confined in a 5.00-L tank at 25.0 °C?

98. A 267-mL sample of a mixture of noble gases at 25.0 °C contains 0.354 g Ar, 0.0521 g Ne, and 0.0049 g Kr. What are **(a)** the mole fractions of the three gases, **(b)** their partial pressures, and **(c)** the total gas pressure?

99. Oxygen is collected over water at 30 °C and a barometric pressure of 742 Torr. What is the partial pressure and mole fraction of $O_2(g)$ in the container?

100. An oxygen-helium gas sample, collected over water at 23 °C, exerts a total pressure of 758 Torr. Calculate the mole fraction of water vapor in the sample.

101. *Elodea* is a green plant that carries out photosynthesis under water.

$$6\ CO_2(g) + 6\ H_2O(l) \longrightarrow C_6H_{12}O_6(aq) + 6\ O_2(g)$$

In an experiment, some *Elodea* produce 122 mL of $O_2(g)$, collected over water at 743 Torr and 21 °C. What mass of oxygen is produced? What mass of glucose ($C_6H_{12}O_6$) is produced concurrently?

102. A 2.02-g sample of aluminum reacts with an excess of HCl(aq), and the liberated hydrogen is collected over water at a temperature of 24 °C. What is the *total* volume of the gas collected?

$$2\ Al(s) + 6\ H^+(aq) \longrightarrow 2\ Al^{3+}(aq) + 3\ H_2(g)$$

103. The reaction between carbon dioxide gas and sodium peroxide, $Na_2O_2(s)$, to produce solid sodium carbonate and oxygen gas, is used in submarines to replace expired carbon dioxide with fresh oxygen. How many liters of oxygen can be collected over water at 23 °C and a barometric pressure of 758 Torr in the reaction of 1.45 kg $Na_2O_2(s)$ with an excess of carbon dioxide?

104. A 9.90-g sample of potassium chlorate is decomposed by heating, yielding potassium chloride and oxygen. The $O_2(g)$ is collected over water at 18 °C. What are the partial pressures of the two gases (oxygen and water vapor) when contained in a 1.00-L container at 22 °C? What is the total pressure?

Kinetic-Molecular Theory

105. A gaseous mixture with equal numbers of molecules of H_2 and He is allowed to effuse through an orifice for a certain period of time. Which of the conditions pictured (a, b, c, or d) is most likely to result?

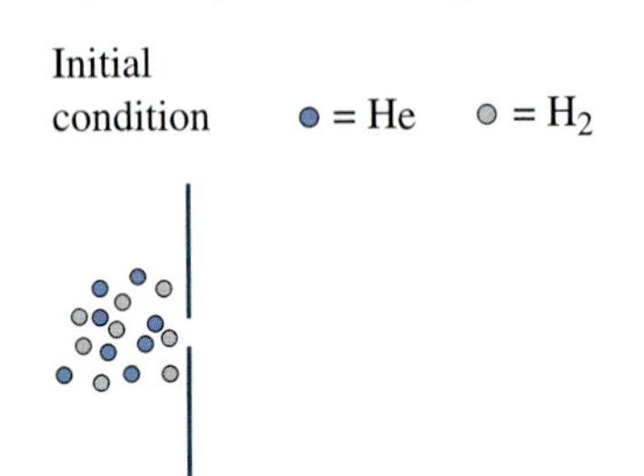

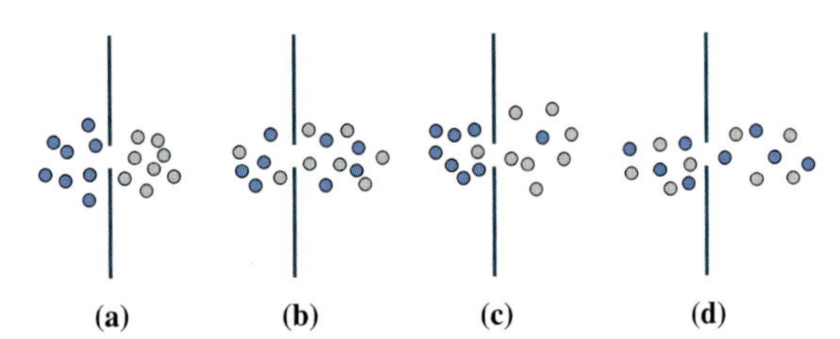

106. Concerning the molecular-level sketches in Problem 105, propose two gases that could be substituted for the He and H_2 that would yield the result pictured in sketch (d). Explain your reasoning.

107. Explain why two different gases in the same container *may* exert different pressures but *may not* have different temperatures or occupy different volumes.

108. Explain how Avogadro's "equal volumes-equal numbers of molecules" hypothesis for gases can be rationalized by using the basic equation of the kinetic-molecular theory:

$$P = \frac{1}{3} \cdot \frac{N}{V} \cdot m\overline{u^2}$$

109. It takes 44 s for a sample of $N_2(g)$ to effuse through a tiny orifice. Determine the molecular masses of gases whose effusion time under exactly the same conditions are **(a)** 75 s and **(b)** 42 s.

110. At a certain temperature, the root-mean-square speed of CH_4 molecules is 1610 km/h. What is the root-mean-square speed of CO_2 molecules at the same temperature?

Additional Problems

111. In terms of pressure (P), volume (V), Kelvin temperature (T), and amount of gas (n), and in the manner of Figures 5.6 and 5.8, sketch a graph of each of the following.

(a) V as a function of P, with T and n held constant

(b) n as a function of P, with T and V held constant

(c) T as a function of P, with V and n held constant

(d) n as a function of T, with P and V held constant

112. Stephen Malaker of Cryodynamics, Inc. has developed a refrigerator that uses compressed helium as a refrigerant gas. A typical system uses 5.00 in^3 of He compressed to 195 psi at 20 °C. What mass of helium, in grams, is needed for one refrigerator?

113. In an attempt to verify Avogadro's hypothesis, small quantities of several different gases were weighed in 100.0-mL syringes. Masses were determined on an analytical balance. The following masses were obtained: 0.0080 g H_2, 0.1112 g N_2, 0.1281 g O_2, 0.1770 g CO_2, 0.2320 g C_4H_{10}, and 0.4824 g CCl_2F_2. Within 1%, are these results consistent with Avogadro's hypothesis? Explain.

114. The $O_2(g)$ produced in the decomposition of a 3.275-g mixture of potassium chlorate and potassium chloride with 65.82% $KClO_3$ is collected over water at 21 °C and 753.5 mmHg atmospheric pressure.

$$2\,KClO_3(s) \longrightarrow 2\,KCl(s) + 3\,O_2(g)$$

How many milliliters of the gas are collected for the conditions shown in the figure at right?

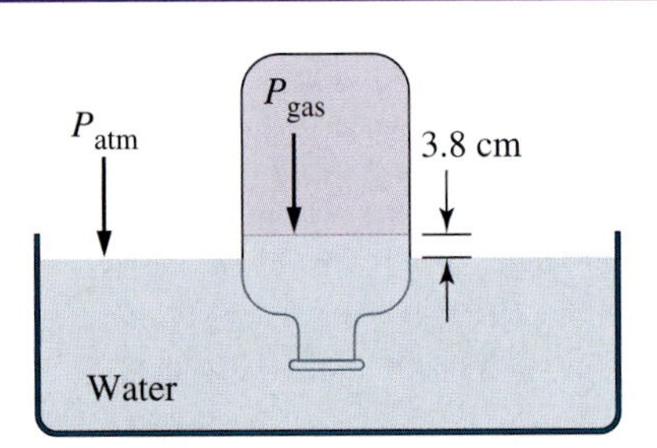

115. A 2.135-g sample of a gaseous chlorofluorocarbon (a type of gas implicated in the depletion of stratospheric ozone) occupies a volume of 315.5 mL at 739.2 mmHg and 26.1 °C. Analysis of the compound shows it to be 14.05% C, 41.48% Cl, and 44.46% F, by mass. What is the molecular formula of this compound?

116. The gaseous hydrocarbon 1,3-butadiene is used to make synthetic rubber. The following measurements were made to determine its molecular mass: A glass container weighed 45.0143 g when evacuated; 192.8273 g when filled with Freon-113™, a liquid with a density of 1.576 g/mL, and 45.2217 g when filled with butadiene at 751.2 mmHg and 21.48 °C. What is the molecular mass of 1,3-butadiene?

117. A 1.405-g sample of an alkane hydrocarbon yields 4.305 g CO_2 and 2.056 g H_2O on combustion. A 0.403-g sample of the gaseous hydrocarbon occupies a volume of 145 mL at 99.8 °C and 749 Torr. Another type of analysis reveals that one of the carbon atoms is in a methyl group, $-CH_3$, attached to the main hydrocarbon chain.

(a) What is the molecular formula of this hydrocarbon?

(b) Draw structural formulas for all the possible isomers that fit this description.

118. Calculate the mass of Earth's atmosphere, in tons, given that the surface area of Earth is 1.95×10^8 mi^2. (*Hint:* What is normal atmospheric pressure in pounds per square inch?)

119. An inflatable air bag made of steel-reinforced rubber is used by rescue squads to extricate people trapped under heavy objects, such as automobiles and trains. A square bag 3.0-ft on a side can lift a 73-ton object. What must be the air pressure in the bag, in pounds per square inch, to make this possible?

120. Calculate the volume of $H_2(g)$ required to react with 15.0 L CO(g) in the reaction

$$3\ CO(g) + 7\ H_2(g) \longrightarrow C_3H_8 + 3\ H_2O(l)$$

(a) if both gases are measured at STP; **(b)** if the CO(g) is measured at STP, and the $H_2(g)$ at 22 °C and 745 mmHg; **(c)** if both gases are measured at 22 °C and 745 mmHg; and **(d)** if the CO(g) is measured at 25 °C and 757 mmHg, and the $H_2(g)$ at 22 °C and 745 mmHg.

121. Use C_8H_{18} as the "formula" of gasoline and 0.71 g/mL for its density. If a car gets 31.2 mi/gal (1 gal = 3.785 L), what volume of $CO_2(g)$ measured at 28 °C and 732 mmHg is produced in a trip of 265 mi? Assume complete combustion of the gasoline.

122. What volume of *air*, measured at 23 °C and 741 mmHg, is required for the complete combustion of 1.00×10^3 L of a particular natural gas, measured at STP? The composition of the natural gas is 77.3% CH_4, 11.2% C_2H_6, 5.8% C_3H_8, 2.3% C_4H_{10}, and 3.4% noncombustible gases, by volume. Use the composition of air given in Example 5.19 on page 206.

123. Use the definitions of $\bar{u}$ and u_{rms} on page 211 to calculate $\bar{u}$ and u_{rms} for a group of six particles with the speeds: 9.83×10^3, 9.05×10^3, 8.33×10^3, 6.48×10^3, 3.67×10^3, and 1.75×10^3 m/s, respectively.

124. Calculate the root-mean-square speed of SO_2 molecules at 27 °C. [*Hint:* Use $R = 8.3145$ J mol^{-1} K^{-1} and 1 J = 1 kg m^2 s^{-2}.]

125. Calculate the pressure exerted by 1.00 mol $CO_2(g)$ when it is confined to a volume of 2.50 L at 298 K by using **(a)** the ideal gas equation, **(b)** the van der Waals equation (page 217). Use data from Table 5.5. **(c)** Compare the two results, and comment on the reason(s) for the difference between them.

126. Use other appropriate equations from the text to derive the kinetic-molecular theory equation for the root-mean-square speed of a gas given on page 212.

127. Typically, when a person coughs, he or she first inhales about 2.0 L of air at 1.0 atm and 25 °C. The epiglottis and the vocal cords then shut, trapping the air in the lungs, where it is warmed to 37 °C and compressed to a volume of about 1.7 L by the action of the diaphragm and chest muscles. The sudden opening of the epiglottis and vocal cords releases this air explosively. Just prior to this release, what is the approximate pressure of the gas inside the lungs?

128. Use the composition of air presented in Example 5.19 on page 206 to determine **(a)** the *apparent molar mass* of air (this is a weighted average of the molar masses of the different components of air in the same sense that a tabulated atomic mass of an element is the weighted average of the masses of the different naturally occurring isotopes of the element), and **(b)** how long it takes a person at rest to breathe "one mole of air" if the person breathes 80 mL per second of air that is measured at 25 °C and 755 mmHg.

129. A sounding balloon is a bag filled with $H_2(g)$ that carries a set of instruments (the payload). Because this combination of bag, gas, and payload has a smaller mass than a corresponding volume of air, the balloon rises and expands as it does so. From the following data, estimate the maximum height to which the balloon can rise: mass of empty balloon, 1200 g; mass of payload, 1700 g; quantity of $H_2(g)$ in balloon, 120 ft^3 at STP; diameter of spherical balloon at maximum height, 25 ft. Air pressure and temperature as a function of altitude are

0 km, 1.0×10^3 mb, 290 K

5 km, 5.4×10^2 mb, 266 K

10 km, 2.7×10^2 mb, 235 K

20 km, 5.5×10^1 mb, 217 K

30 km, 1.2×10^1 mb, 239 K

40 km, 2.9×10^0 mb, 267 K

50 km, 8.1×10^{-1} mb, 280 K

60 km, 2.3×10^{-1} mb, 260 K

Treat air as if it were a single gas with a molar mass of 28.96 g/mol air (referred to as the *apparent* molar mass of air).

CHAPTER

6

Thermochemistry

CONTENTS

Combustion in this Delta II rocket engine converts chemical energy of the fuel to heat and work, lifting the Global Surveyor from the pad and sending it on its way to Mars. Thermochemistry is a study of the heat and work that accompany chemical reactions.

We carry out many chemical reactions just for the associated energy changes, not to obtain products. In fact, we sometimes use energy-producing reactions *in spite of* the products. Consider the combustion of methane, the principal component of natural gas.

$$CH_4(g) + 2\,O_2(g) \longrightarrow CO_2(g) + 2\,H_2O(l)$$

We burn methane for cooking and to warm our homes, using the heat that is released when methane burns.

Modern industrialized nations run mainly on fossil fuels: natural gas, petroleum, and coal. Unfortunately, burning these fuels produces more than heat. Their combustion is accompanied by the formation of air pollutants such as carbon monoxide, oxides of nitrogen, and particulate matter. Even carbon dioxide, although generally not considered an air pollutant, may present future problems. It is the main one of several "greenhouse" gases that trap heat in Earth's atmosphere,

● **ChemCDX**
When you see this icon, there are related animations, demonstrations, and exercises on the CD accompanying this text.

and most atmospheric scientists think that a buildup of these gases will lead to global warming, an important environmental issue.

In this chapter, after first examining a few fundamental concepts, we will look at some quantitative relationships involving chemical reactions and energy changes. We will conclude the chapter with a discussion of the foods that sustain life and the fuels that enrich it.

6.1 Energy

Because we will use the concept of energy and energy changes in nearly everything we do in this chapter, let's briefly consider some basic ideas about energy here. If you would like more detail on some of these matters, refer to Appendix B.

The term *energy* is derived from Greek and means, literally, "work within." As we learn more about work, however, we will see that no object of matter *contains* work. Work is done when matter is displaced or moved. We might say, then, that energy is the capacity to do work. Objects that can do work because of their composition or position possess **potential energy**. Moving objects also have the capacity to do work; they possess **kinetic energy**. In Chapter 5 we discussed the kinetic energy associated with the motion of gas molecules.

Let's compare these two types of energy with the practical example in Figure 6.1. Water at the top of a dam has *potential* energy due its gravitational attraction to Earth's center. Thus, the water has the capacity to do work, but as long as it remains behind the dam, it does none. When the water is allowed to flow through a pipe to a lower level, some of its potential energy is converted to *kinetic* energy. Water rushing through the pipe can be made to turn the blades of a turbine (a waterwheel), which in turn can rotate a coil of wire in an electrical generator, producing electricity. The net result is that some of the potential energy originally stored in the water is converted to electrical work.

To get an idea of the magnitude and units of quantities of energy, let's use the mathematical definition of kinetic energy (E_k) that we introduced in Section 5.11.

$$E_k = \frac{1}{2}mv^2 \text{ *}$$

If we express mass (m) in kilograms and speed (v) in meters per second, the units of kinetic energy are

$$(\text{kg}) \times (\text{m/s})^2 = \text{kg m}^2 \text{ s}^{-2}$$

For example, a 2-kg object moving at a speed of 1 m/s has the kinetic energy

$$E_k = \frac{1}{2} \times 2 \text{ kg} \times (1 \text{ m/s})^2 = 1 \text{ kg m}^2 \text{ s}^{-2} = 1 \text{ J}$$

The quantity 1 kg m^2 s^{-2} is the SI unit of energy called **1 joule (J)**.

From a macroscopic point of view, the joule is a small energy unit. A bowling ball rolling slowly down a bowling alley has a kinetic energy of a few joules. A 60-watt lightbulb uses 60 J of energy per second. To describe an energy change during a chemical reaction, the term kilojoule (kJ) is commonly used: 1 kJ = 1000 J.

* In Section 5.11, we used the symbol u when describing the speed of an object at the molecular level. The symbol v is more commonly used to represent the speed of a macroscopic object.

Figure 6.1
Energy conversion
Potential energy of water stored behind this dam is converted to kinetic energy as the water falls through conduits called penstocks. They deliver the flowing water to turbines. The water turns the rotors in turbines and generators, thereby producing electricity.

Before we can establish the relationship between energy and work, we need a definition of work. The **work** required to move an object is the product of a force in the direction of motion and the distance the object is moved.

$$\text{Work} = \text{force} \times \text{distance}$$

The work done by a force of 1 newton (N) acting over a distance of 1 meter is 1 newton-meter. In terms of the SI base quantities of mass, length, and time,

$$1\text{ N} \times 1\text{ m} = 1\text{ kg m}^2\text{ s}^{-2} = 1\text{ J}$$

Thus, work has the same unit as energy, the joule.

To further illustrate the relationship between energy and work, let's consider the bouncing tennis ball in Figure 6.2. First, we have to do work to raise the ball to its starting position. That is, we have to apply an upward force on the ball to overcome the force of gravity that tends to pull the ball straight down. This work of lifting is "stored" in the ball as potential energy.

When we release the ball, it falls as it is pulled toward Earth's center by the force of gravity. During this fall, potential energy is converted to kinetic energy, and the kinetic energy reaches a maximum just as the ball strikes the surface. As the ball rebounds, its kinetic energy decreases (the ball slows down) as its potential energy increases (the ball rises).

If its collisions with the immovable surface were perfectly *elastic* (like those of gas molecules according to the kinetic-molecular theory), the ball would always rebound with the same energy it had before the collision. At every point on its path,

Figure 6.2
Potential energy and kinetic energy
A bouncing ball illustrates the interconversion of potential energy and kinetic energy.

the *sum* of the potential and kinetic energies would be constant: The ball would return to its original highest position on each bounce, and the ball would go on bouncing forever. We know, however, that a real ball behaves otherwise.

In reality, the maximum height on each bounce is less than on the previous one and the ball eventually comes to rest on the surface. The potential energy invested in lifting the ball is lost, but it does not appear as kinetic energy of the ball itself. Instead, it appears as additional kinetic energy in the atoms and molecules that make up the ball, the surface, and the surrounding air. The temperature of the ball and its surroundings increases slightly.

Thus, the situation is a bit more complicated than it might first appear, and we need to explore a few basic ideas in more detail. We do so in the next section.

6.2 Thermochemistry: Some Basic Terms

Thermochemistry is the study of energy changes that occur during chemical reactions. Usually we assess these energy changes in terms of something losing energy and something else gaining it. When we warm our cold hands over a campfire, the burning wood gives off energy (as heat) and our hands gain energy, raising their temperature. For scientific purposes, however, we need to describe the two "somethings" that exchange energy a little more precisely.

System and Surroundings

We define the **system** as that part of the universe that we are studying. The system may be a solution in a beaker, a gas in a cylinder, or a block of frozen spinach in a dish. The system may be as complicated as a polluted lake or even Earth's atmosphere. The **surroundings** are the rest of the universe, although we can typically limit our concern only to the parts that interact with the system. *Interactions* refer to the exchange of energy and/or matter between a system and its surroundings. The surroundings of a solution in a beaker, for example, might be the beaker itself, the surface on which the beaker is placed, and that part of the surrounding air with which the solution effectively exchanges heat, matter, or both.

There are three kinds of systems. An *open system* interacts readily with its surroundings, exchanging matter and energy, as does a steaming hot cup of coffee. The coffee loses heat and water vapor to its surroundings, as seen by its gradual cooling and the appearance of condensed water vapor (steam) in the cooler air above the coffee. A *closed system* exchanges energy but not matter with the surroundings. An example is a capped bottle of water and tea bags placed in sunlight to make sun tea. Only energy (as heat) is transferred from the surroundings into the system. Figure 6.3 is a chemist's abstraction to represent the formation of sun tea. An *isolated system* exchanges neither matter nor energy with its surroundings. A tightly stoppered thermos flask with hot coffee is, at least for a short period of time, a good approximation of an isolated system. In our study of thermochemistry, we will try to keep systems simple, often considering the system to be isolated from its surroundings. We know from experience, however, that we cannot isolate a system completely. The stoppered thermos flask, for example, is not a truly isolated system; the contents of the flask slowly cool, indicating that heat is being lost to air surrounding the flask.

Internal Energy

The **internal energy** (*U*) of a system is the total energy contained within the system, partly as kinetic energy and partly as potential energy. As Figure 6.4 suggests, the kinetic energy component comes from various types of motion at the molecular level. Such movements include straight-line motion, or *translational motion*, of

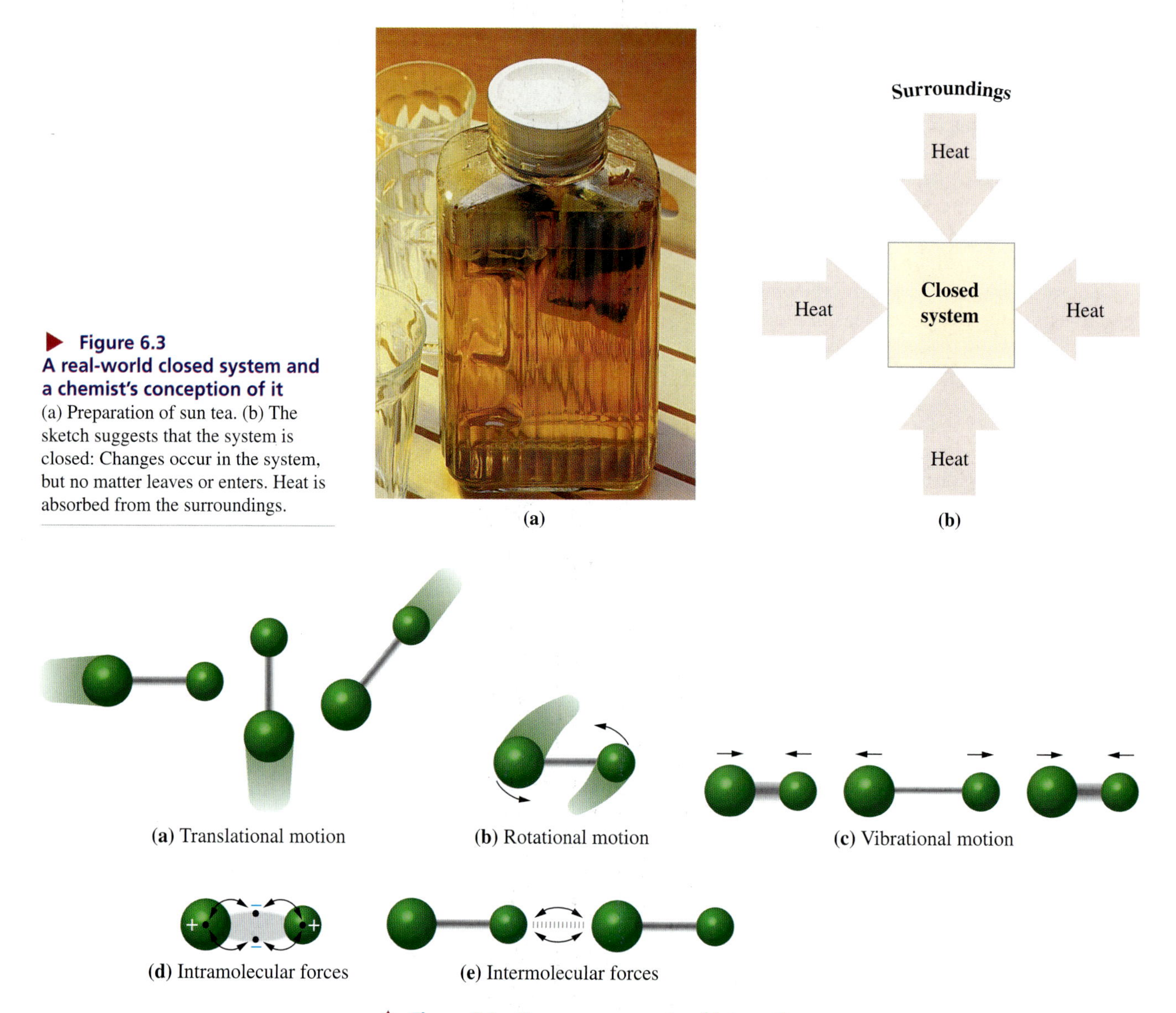

Figure 6.3 A real-world closed system and a chemist's conception of it
(a) Preparation of sun tea. (b) The sketch suggests that the system is closed: Changes occur in the system, but no matter leaves or enters. Heat is absorbed from the surroundings.

Figure 6.4 Some components of internal energy
The system suggested here is a gas made up of diatomic molecules. Most actual systems are more complex. Other important contributions to internal energy not shown here involve the attractions between electrons and protons within atoms and those between protons and neutrons in atomic nuclei.

molecules (Figure 6.4a), the spinning, or *rotational motion* of molecules (Figure 6.4b), and the displacements of atoms within molecules, called *vibrational motion* (Figure 6.4c). Collectively, the kinetic energy contributions to internal energy are sometimes called *thermal energy*. The potential energy component of internal energy comes from interactions between particles of matter. Some of these interactions, such as those that hold protons and neutrons together in atomic nuclei, remain unchanged in chemical reactions and we do not concern ourselves with them. The most important interactions that chemists consider are the electrostatic attractions that produce chemical bonds between atoms. These are known as *intramolecular forces* (Figure 6.4d). Also important are electrostatic attractions between molecules, called *intermolecular forces* (Figure 6.4e). Collectively, these potential energy contributions to internal energy are sometimes referred to as *chemical energy*.

Heat

Heat (q) is an energy *transfer* between a system and its surroundings, caused by a difference in temperature between them. Heat passes spontaneously from the region of higher temperature to the region of lower temperature. Heat transfer stops when the system and surroundings reach the same temperature and the system and surroundings are at *thermal equilibrium*. Although we often use expressions such as "heat flows," "heat is lost," and "heat is gained," *a system does not contain heat*. Rather, it contains energy—internal energy.

Consider heat transfer at the molecular level. As we noted on page 211, temperature is a measure of the average kinetic energy of translational motion of molecules. As suggested in Figure 6.5, kinetic energy is transferred during molecular collisions at the interface between a system and its surroundings. The more energetic molecules (the ones at the higher temperature) lose energy to the less energetic molecules (the ones at the lower temperature). In effect, heat is simply a transfer of molecular kinetic energy that brings a system and its surroundings to the same temperature.

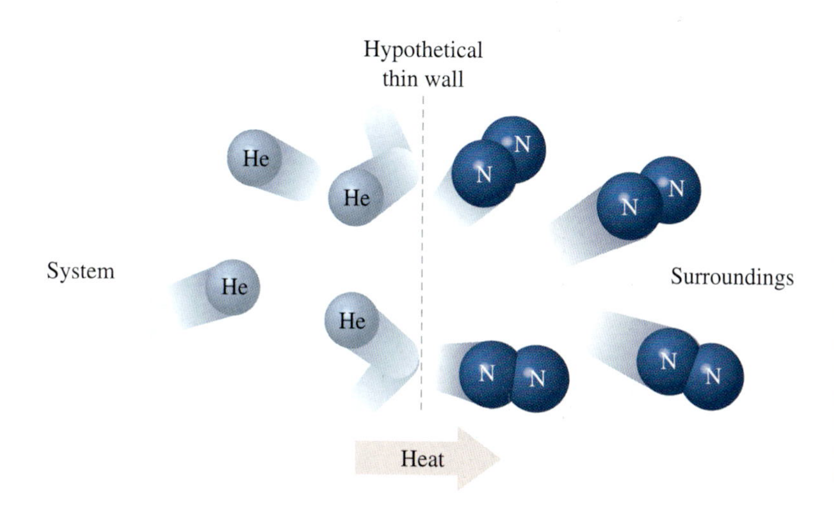

Figure 6.5
Heat transfer between a system and surroundings
In their collisions with the hypothetical thin wall, the more energetic gaseous helium atoms of the system transfer some kinetic energy to the less energetic nitrogen molecules in the cooler surroundings. The average translational kinetic energy of the helium atoms decreases and that of the nitrogen molecules increases until the average translational kinetic energies of the two gases become equal. At this point of thermal equilibrium, the temperatures of the system and surroundings have become equal and heat no longer flows.

Now let's consider a system consisting of methane gas burning in oxygen. This is the principal reaction that takes place in a laboratory Bunsen burner (see Figure 6.6). We will define the methane and oxygen gases and the reaction products as the system. Likewise, we consider the burner and everything else around it to be the

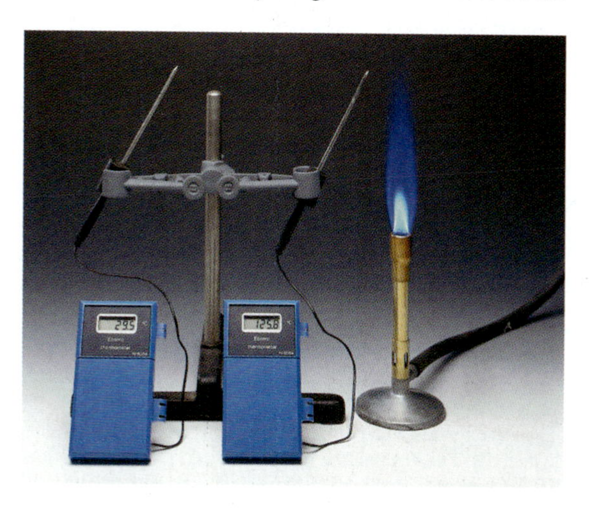

Figure 6.6
A reaction evolving heat
In the combustion of methane (natural gas), heat is given off by the system—the burning gas—to the surroundings—the air around the flame, the burner, the bench top, and so on. The thermometer closer to the system is at a higher temperature (125.8 °C) than is the one farther away (29.5 °C).

surroundings. When the methane burns, some of the energy associated with chemical bonds in the methane and oxygen molecules—chemical energy—is converted to thermal energy and the temperature of the system rises. Because the temperature of the system is now above that of the surroundings, heat is transferred from the system to the surroundings through collisions between the gaseous molecules of the system and the molecules in the surrounding air.

Next, consider a system composed of water, initially in the form of a block of ice. When the ice is placed in surroundings above 0 °C, heat passes from the surroundings into the ice (see Figure 6.7). The heat absorbed by the system is used to melt the ice, which remains at 0 °C while melting occurs.

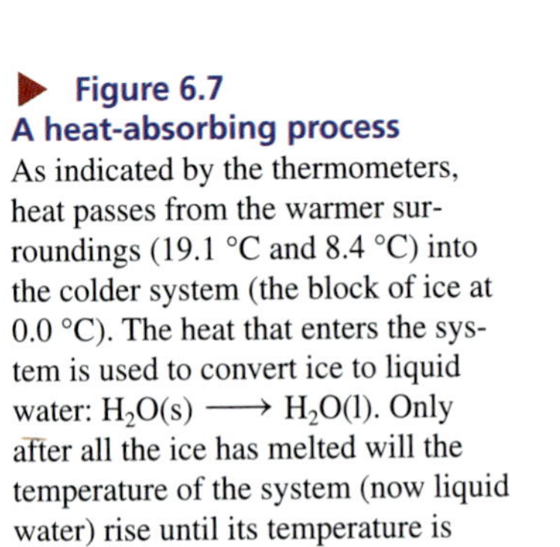

Figure 6.7
A heat-absorbing process
As indicated by the thermometers, heat passes from the warmer surroundings (19.1 °C and 8.4 °C) into the colder system (the block of ice at 0.0 °C). The heat that enters the system is used to convert ice to liquid water: $H_2O(s) \longrightarrow H_2O(l)$. Only after all the ice has melted will the temperature of the system (now liquid water) rise until its temperature is equal to that of the surroundings.

Work

Work (w), like heat, is an energy transfer between a system and its surroundings; *a system does not contain work*. There are several types of work, but for now we will consider only *pressure-volume work*—that done when gases are compressed or when they expand. Much later in the text we will discuss another type—electrical work.

In Figure 6.8, a gas is confined in a cylinder by a freely moving, weightless piston. At first, the gas is maintained in a fixed volume by the two identical weights.

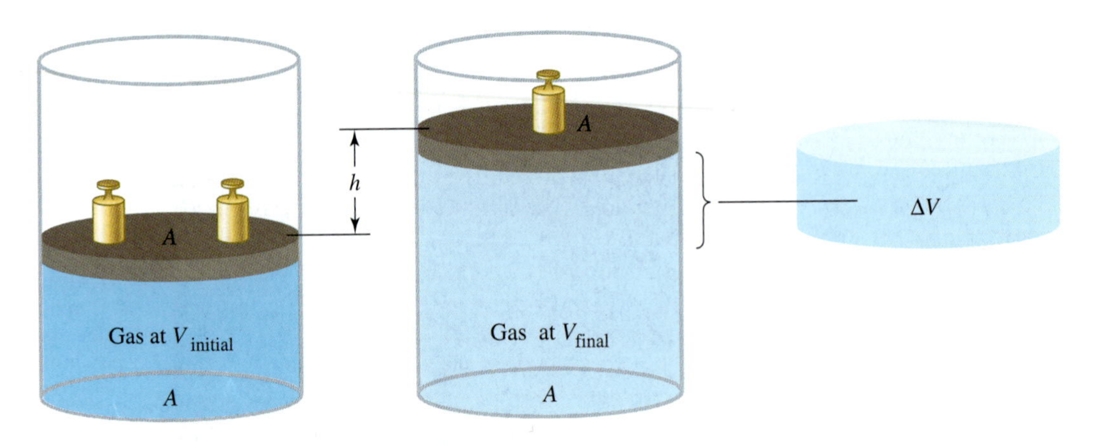

Figure 6.8 Pressure-volume work
The gas expands when one of the weights is removed. The remaining weight is lifted through the distance h. The gas volume increases by ΔV, and the gas does work.

When one of these weights is removed, the gas expands and the remaining weight is lifted through the distance h. At this point, the volume has doubled and the piston no longer moves. To raise the weight, the gas exerts the force necessary to overcome the force of gravity on the weight. This force acts through the distance h, and the product of a force and a distance is an amount of work.

In Chapter 5, we described forces exerted by gas molecules in another way—through the concept of pressure. Pressure is a force divided by the area over which the force is distributed: $P = F/A$. The force exerted by the gas, then, is given by $F = P \times A$. We can combine these ideas into the following expression.

$$\text{Work } (w) = \text{force } (F) \times \text{distance } (h) = P \times A \times h$$

The product of the cross-sectional area of the cylinder A and the height h represents the *change* in volume of the gas, which is designated by the symbol ΔV. (The Greek letter *delta*, Δ, represents a *change* in some quantity.) Mathematically, we can represent a change as some *final* value *minus* an initial value: $\Delta V = V_{\text{final}} - V_{\text{initial}}$. By replacing $A \times h$ with its equivalent, ΔV, we see that the work associated with a gas that expands at constant pressure is described by the following equation.

$$\text{Work } (w) = \text{pressure } (P) \times \text{change in volume } (\Delta V) = P\,\Delta V$$

This product of pressure and a change in volume is usually called "pressure-volume" work. In Figure 6.8, the constant pressure is that needed to support the single weight, and the change in volume (ΔV) is indicated.

As a final adjustment to the equation for pressure-volume work, we need to add a negative sign. That is,

$$\text{Work } (w) = -P\,\Delta V$$

Thus, when a gas expands, ΔV is positive and the work is negative. A negative quantity of work signifies that the system loses energy, that is, that energy passes from the system to the surroundings. When a gas is compressed by its surroundings, ΔV is negative, the quantity of work is positive and energy is gained by the system. We'll say more about this and other sign conventions in the next section.

● **ChemCDX** Work of Gas Expansion

6.3 Internal Energy (*U*), State Functions, and the First Law of Thermodynamics

Earlier (page 230), we described some of the kinds of energy that collectively make up the internal energy of a system, but—for a very good reason—we didn't say how one goes about determining an actual value of this quantity: We can't measure the absolute value of the internal energy of a system because we cannot account exactly for all the energy quantities. Here, we encounter for the first time a property that seems to have fundamental significance, yet we can't measure it.

Fortunately, internal energy has one crucial property that enables us to use it successfully even though we can't measure it. *Internal energy is a function of state.* The **state** of a system refers to its exact condition, determined by the kinds and amounts of matter present, the structure of this matter at the molecular level, and the prevailing temperature and pressure. A **function of state**, also called a **state function**, is *a property that has a unique value that depends only on the present state of a system and* not *on how that state was reached, that is,* not *on the previous history of the system.*

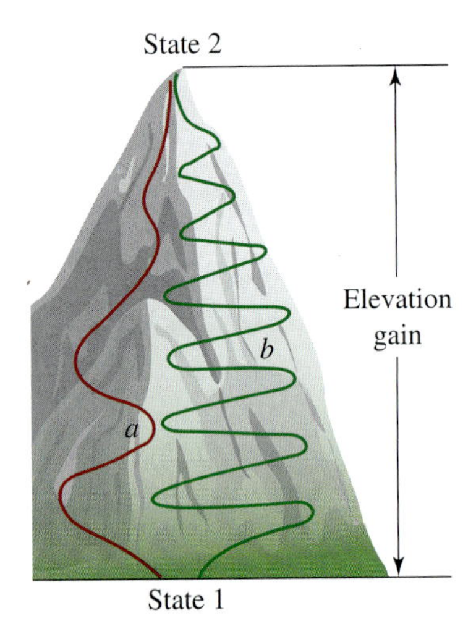

▲ **Figure 6.9**
Mountain-climbing analogy to a function of state
Path *a* (red) is shorter than path *b* (green), so a shorter distance is walked along path *a*. The gain in elevation, however, is independent of which path is chosen. The internal energy of a system is also independent of the path by which a change occurs. Internal energy, *U*, is a *function of state*.

Interestingly, heat, q, and work, w, are *not* functions of state. Because they are not contained in a system, they are quantities of energy that we observe *only when a system changes from one state to another*. Consequently, the values of q and w depend on the way in which a change is brought about—they depend on the *path* chosen.

Perhaps we can better understand state functions by considering the analogy to climbing a mountain, presented in Figure 6.9. Think of the base of the mountain as the initial state and the summit as the final state. Climbing the mountain is like a thermodynamic process that proceeds from an initial to a final state. The *elevation gain*, the difference in elevation between the base and summit of the mountain, is analogous to the *change* in internal energy, ΔU. We can measure the elevation gain without knowing the actual elevation of either the base or the summit of the mountain. It has a unique value regardless of the path we take in our climb. Similarly, because both the internal energies of the initial and final states of a system have unique values, so does the *change* in internal energy of the system ΔU between the two states. That is,

$$\text{Initial state } (U_{\text{initial}}) \longrightarrow \text{final state } (U_{\text{final}})$$

$$\Delta U = U_{\text{final}} - U_{\text{initial}}$$

In the mountain-climbing analogy, when we return to the base of the mountain from the summit, our net elevation gain for the round-trip is *zero*. The positive elevation gain we make in climbing to the summit is wiped out by the negative elevation gain when we return to the base. This illustrates an important property of state functions: *A state function returns to its initial value if a process is reversed and a system is returned to its initial state.*

We have described why a change in internal energy—the difference in internal energy between two states of a system—has a unique value. But how can we measure this difference? Here, another fundamental scientific law comes to our aid: the **law of conservation of energy**.

> *In a physical or chemical change, energy can be exchanged between a system and its surroundings, but no energy can be created or destroyed.*

This law implies that if we can account for all the energy exchanges between a system and its surroundings, we should be able to determine how the internal energy of a system changes in a physical or chemical process. That is, the *change* in internal energy of a system, ΔU, must be related to the energy exchanges that occur as heat, q, and work, w. When the law of conservation of energy is recast in this way, we call it the **first law of thermodynamics** and express it mathematically as follows.

$$\Delta U = q + w$$

We can see from this equation that in an *isolated* system the internal energy remains constant. That is, if a system can exchange neither heat nor work with its surroundings, q and w are both *zero* and ΔU for the system is also *zero*. Intuitively, it would seem that if a system *gains* energy as heat but *loses* an equal quantity as work, ΔU for the system should also be *zero*. However, the defining equation gives this result only if one of the energy exchanges has a positive sign and the other a negative sign. This suggests that we need to establish some sign conventions to be used with the first law of thermodynamics. The conventions are as follows.

- Energy *entering* a system carries a *positive* sign: If heat is *absorbed* by the system, $q > 0$. If work is done *on* a system, $w > 0$.
- Energy *leaving* a system carries a *negative* sign: If heat is *given off* by a system, $q < 0$. If work is done *by* a system, $w < 0$.

Accounting for energy exchanges between a system and its surroundings is much like balancing a checking account. You keep track of the activity in the account (analogous to ΔU of the system) by recording all the deposits and withdrawals. This can be rather simple, but some checking accounts present more of a challenge, as suggested by Figure 6.10. We will work only with thermodynamic systems that are like simple checking accounts, though scientists often face systems that are quite complex.

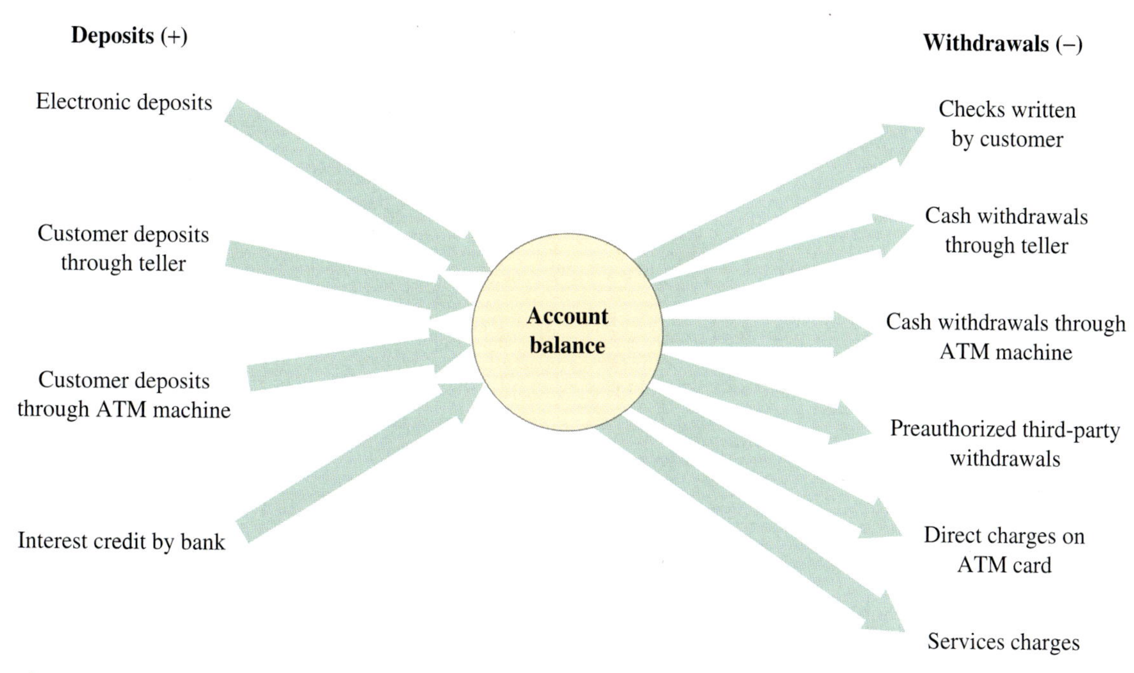

▲ **Figure 6.10 Checking account analogy to internal energy**
In a modern interest-bearing checking account, there are many ways in which funds can be deposited or withdrawn. Nevertheless, we can accurately determine the account balance if all the transactions are known. Similarly, we can evaluate the change in internal energy in a system by accounting for all the interactions that involve the transfer of energy between a system and its surroundings.

EXAMPLE 6.1

A gas does 135 J of work while expanding, and at the same time, it absorbs 156 J of heat. What is the change in internal energy?

SOLUTION

We are given the quantities w and q, and we seek ΔU. The key is to assign the correct signs to the values of w and q, as indicated in parentheses below. Note specifically that heat is *absorbed by* the system (a positive quantity, +156 J) and work is *done by* the system (a negative quantity, −135 J). Because more heat is absorbed than work done, the internal energy increases.

$$\Delta U = q + w = (+156 \text{ J}) + (-135 \text{ J}) = +21 \text{ J}$$

EXERCISE 6.1A

In a process in which 89 J of work is done on a system, 567 J of heat is given off. What is ΔU of the system?

EXERCISE 6.1B

In a particular process, the internal energy of a system increases by 41.4 J and the quantity of work the system does on its surroundings is 81.2 J. Is heat absorbed or given off by the system? What is the value of q?

EXAMPLE 6.2—A Conceptual Example

The internal energy of a fixed quantity of an ideal gas depends only on its temperature. If a sample of an ideal gas is allowed to expand against a *constant* pressure at a *constant* temperature, **(a)** what is ΔU for the gas? **(b)** does the gas do work? **(c)** is any heat exchanged with the surroundings?

SOLUTION

a. If the expansion of the ideal gas occurs at a constant temperature, the expanded gas (state 2) is at a lower pressure than the compressed gas (state 1) but the temperature is *unchanged*. Because the internal energy of the ideal gas depends only on the temperature, $U_2 = U_1$ and

$$\Delta U = U_2 - U_1 = 0$$

b. The gas does work in expanding against the confining pressure, P. The pressure-volume work is $w = -P\,\Delta V$, as was illustrated in Figure 6.8.

c. The work done by the gas represents energy leaving the system. If this were the only energy exchange between the system and its surroundings, the internal energy of the system would decrease, and so would the temperature. However, because the temperature remains constant, the internal energy does not change. This means that the gas must absorb enough heat from the surroundings to compensate for the work that it does in expanding: $q = -w$. And, according to the first law of thermodynamics

$$\Delta U = q + w = -w + w = 0$$

EXERCISE 6.2

In an *adiabatic* process, a system is thermally insulated from its surroundings so there is no exchange of heat ($q = 0$). If an ideal gas undergoes an adiabatic expansion against a constant pressure, **(a)** does the gas do work? **(b)** does the internal energy of the gas increase, decrease, or remain unchanged? **(c)** what happens to the temperature?

We have now examined in some detail the three quantities that are related by the first law of thermodynamics: $\Delta U = q + w$. Of these three, the easiest to measure experimentally is q. When the thermodynamic process is a chemical reaction, the heat involved, q, is called the *heat of reaction*. We will consider heats of reaction in the next section and also introduce an important new thermodynamic function called *enthalpy*.

6.4 Heats of Reaction and Enthalpy Change, ΔH

We can think of a chemical reaction as a process in which a thermodynamic system changes from an *initial* state (the reactants) to a *final* state (the products). Almost always a quantity of heat is associated with the change, and sometimes work as well. An important way of classifying thermochemical processes is by the descriptive terms *exothermic* and *endothermic*. First we will focus on how exothermic and endothermic reactions relate to the first law of thermodynamics.

During an **exothermic reaction**, chemical energy in a system is converted to thermal energy. Changes characteristic of exothermic reactions are as follows.

- An exothermic reaction in an *isolated* system produces a temperature increase in the system.
- An exothermic reaction in a nonisolated (*open* or *closed*) system causes heat to be *given off* to the surroundings. Thus, $q < 0$—a *negative* quantity.

These situations are pictured in Figure 6.11. The temperature of the system rises to the greatest extent possible for an exothermic reaction occurring in an isolated system (Figure 6.11a). In a nonisolated system, the temperature does not change at all if the evolved heat escapes instantaneously (Figure 6.11b). What is commonly observed, however, is that the temperature of the nonisolated system does rise to some extent initially because the exothermic heat of reaction cannot escape fast enough (Figure 6.11c).

APPLICATION NOTE
Solid NaOH, sold in markets as *lye*, is a common drain cleaner. The dissolving of lye in water is an *exothermic* process, and the liberated heat may melt congealed grease, allowing it to be flushed from a clogged drainpipe.

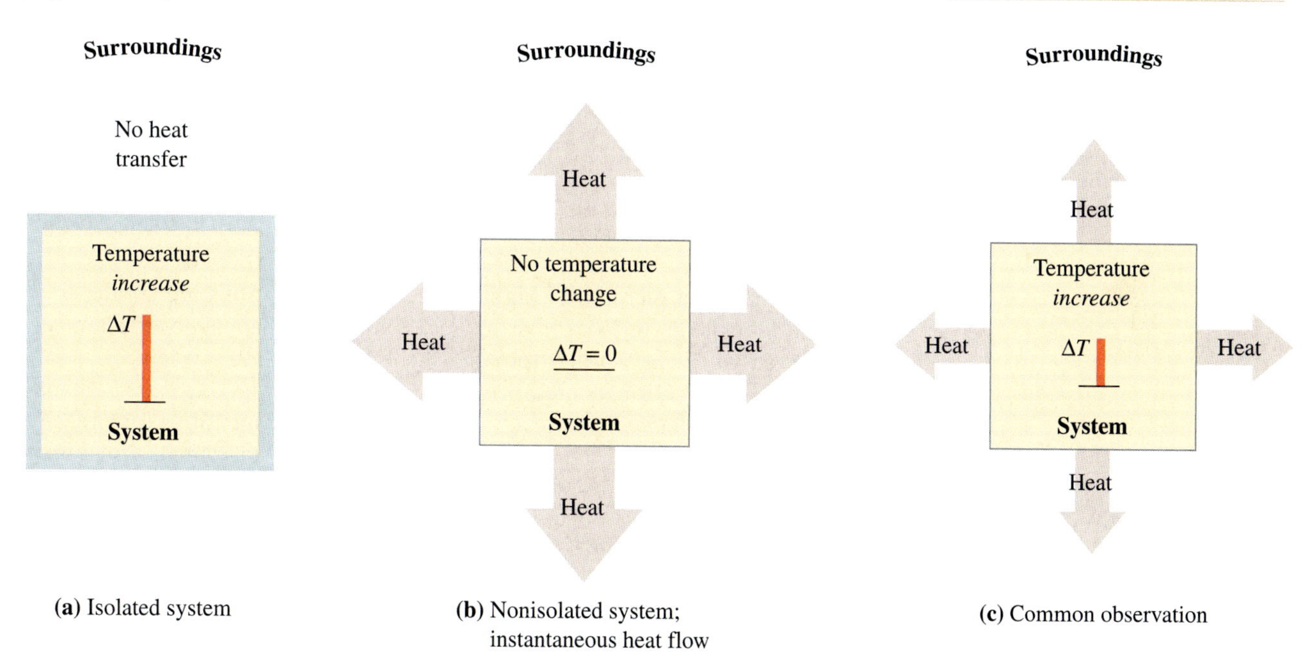

▲ **Figure 6.11 Conceptualizing an exothermic reaction**
In an exothermic reaction, chemical energy in a system is converted to thermal energy. (a) The thermal energy is retained in the isolated system. A maximum temperature increase occurs. (b) The thermal energy is released as heat to the surroundings as rapidly as it is formed. There is no temperature change in the system. (c) The system is not completely isolated from its surroundings, nor is the flow of heat instantaneous. There is both an initial increase in temperature in the system *and* a flow of heat to the surroundings.

In an **endothermic reaction**, thermal energy in a system is converted to chemical energy. Changes characteristic of endothermic reactions are as follows.

- An endothermic reaction in an *isolated* system produces a temperature decrease in the system.
- An endothermic reaction in a nonisolated (*open* or *closed*) system causes heat to be *absorbed* from the surroundings. Thus, $q > 0$—a *positive* quantity.

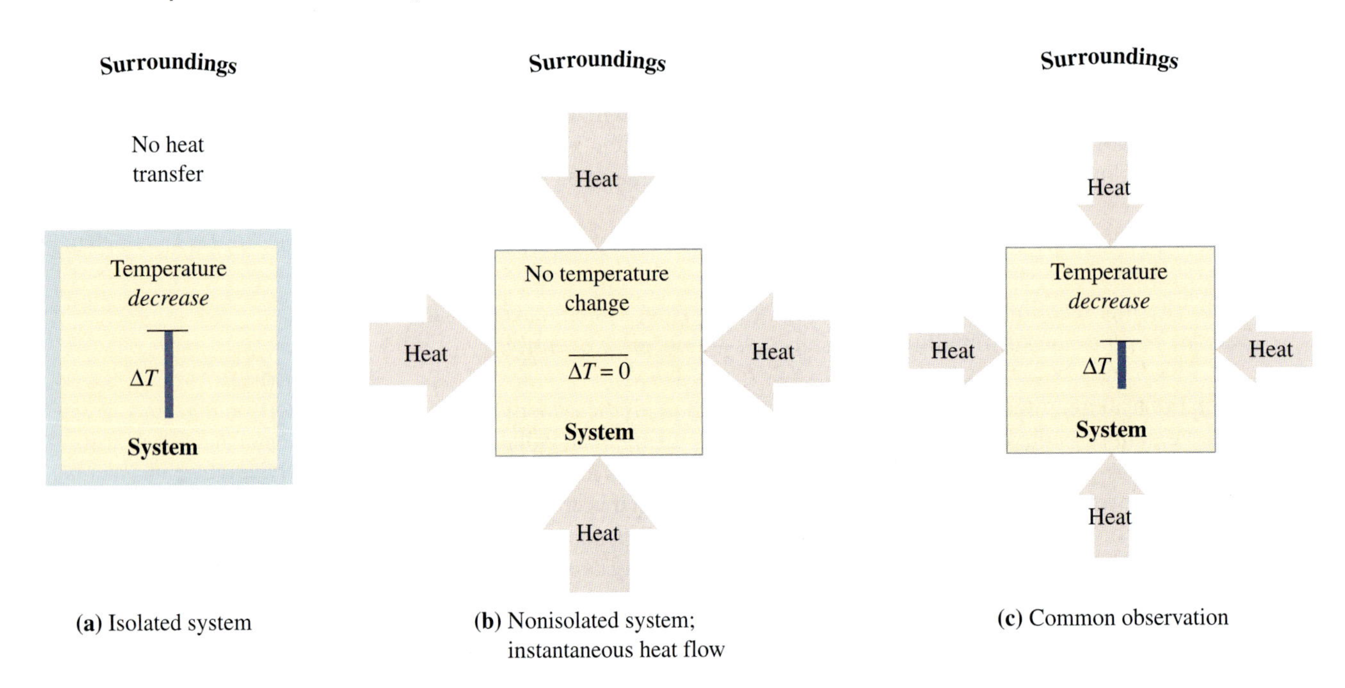

(a) Isolated system

(b) Nonisolated system; instantaneous heat flow

(c) Common observation

▲ **Figure 6.12 Conceptualizing an endothermic reaction**
In an endothermic reaction, thermal energy in a system is converted to chemical energy. (a) The loss of thermal energy is uncompensated in the isolated system. A maximum temperature decrease occurs. (b) The loss of thermal energy in the system is immediately compensated for by the absorption of heat from the surroundings. There is no temperature change in the system. (c) This situation is similar to that described in Figure 6.11(c), except that it involves a temperature decrease.

APPLICATION NOTE
Instant cold packs, used to treat sports injuries, have separate compartments of water and solid ammonium nitrate in a plastic bag. When the barrier between the two is broken, the NH_4NO_3 dissolves in the water endothermically. Heat is absorbed from the surroundings—including the injured area of the athlete's body.

These situations are pictured in Figure 6.12. Our observations are similar to the three cases described for an exothermic reaction, except that we observe temperature decreases and heat passing from the surroundings into the system.

The **heat of reaction**, sometimes denoted by the symbol q_{rxn}, is the quantity of heat exchanged between the reaction system and its surroundings for a reaction occurring *at a fixed temperature*. This definition implies measuring the quantity of heat represented in Figure 6.11b for an exothermic reaction and Figure 16.12b for an endothermic reaction. We will see how to assess these quantities of heat when we describe the experimental measurement of heats of reactions.

A reaction can produce work in several different forms, but most commonly this is the pressure-volume work associated with the expansion or compression of gaseous reactants and products.

Before proceeding further with our discussion of heats of reaction, let's pause to summarize a few ideas. We can treat a chemical reaction as a thermodynamic system in which the reactants represent the initial state (state 1) and the products represent the final state (state 2).

$$\begin{array}{ccc} \text{Reactants} & \longrightarrow & \text{products} \\ \text{State 1: } U_1 & & \text{State 2: } U_2 \end{array}$$

The internal energies of these two states have unique values, as does their difference.

$$\Delta U = U_2 - U_1$$

The change in internal energy of the system is related to the exchanges of heat and work between the system and its surroundings, as given by the first law of thermodynamics.

$$\Delta U = q + w$$

Even though q and w depend on the path chosen, ΔU has a unique value for a given reaction; that is, the value of ΔU does not depend on the method used to carry out the reaction. Recall also that pressure-volume work is given by the following expression.

$$w = -P\,\Delta V$$

To continue, most reactions are carried out either at *constant volume* or at *constant pressure*. If we carry out a reaction in a system of *constant volume*, the volume cannot change during the reaction and $\Delta V = 0$. But if $\Delta V = 0$, so too must $P\,\Delta V = 0$. This means that no pressure-volume work is done. And, if no other type of work takes place during the reaction—as is often the case—then $w = 0$. Now notice the implication of this fact on the heat of reaction, q.

$$\Delta U = q + w$$

$$\Delta U = q + 0$$

The heat of reaction is equal to ΔU for this reaction. A heat of reaction *at constant volume* is indicated as q_V (the subscript V indicates that the volume remains constant).

$$\Delta U = q_V$$

Now consider a reaction carried out at *constant pressure* and with only pressure-volume work. In this case, we can use the subscript P to represent a heat of reaction at constant pressure, that is, q_P. Then when we substitute $-P\,\Delta V$ for w, the first law of thermodynamics becomes

$$\Delta U = q + w = q_P - P\Delta V$$

Finally, let's rearrange the equation to isolate the term q_P.

$$q_P = \Delta U + P\Delta V$$

Let's use Figure 6.13 to help us visualize two ways of carrying out an *exothermic* reaction. When the reaction is carried out *at constant volume*, all the thermal

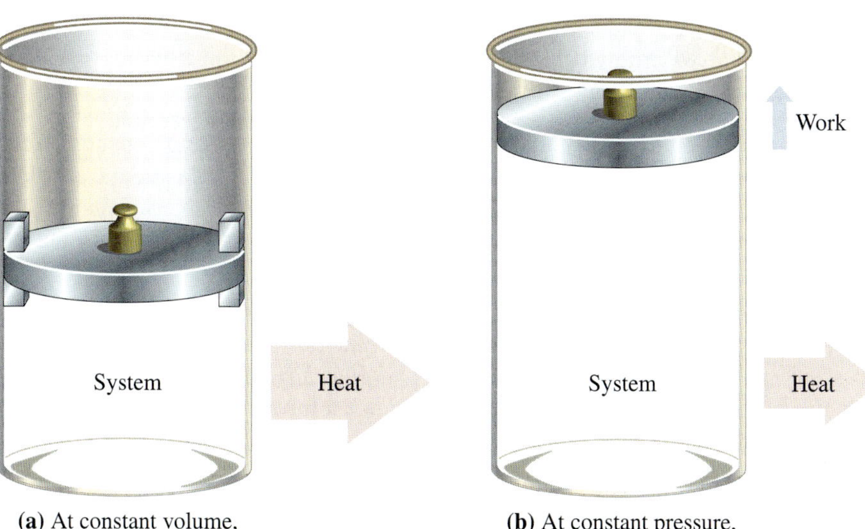

Figure 6.13 Comparing the heats of reaction at constant volume and constant pressure for an exothermic reaction

(a) The stops above and below the piston prevent the system volume from changing. The heat evolved, q_V, is equal to ΔU. (b) The system is allowed to expand against the constant pressure of the weight. A small amount of work, w, and a quantity of heat, q_P, are evolved. The magnitude of q_P is slightly less than that of q_V.

▲ **Figure 6.14**
Visualizing a reaction carried out under the pressure of the atmosphere
In the reaction

$$Mg(s) + 2\,HCl(aq) \longrightarrow MgCl_2(aq) + H_2(g)$$

the evolved $H_2(g)$ pushes air from the flask and surroundings. The expanding gas does work, represented by the blue arrows.

energy produced by conversion from chemical energy is released as heat. That is, $q_V = \Delta U$. And, because the reaction is exothermic, both q_V and ΔU are *negative* quantities.

When the reaction is carried out *at constant pressure*, most of the thermal energy is also released as heat, but a small amount of it appears as the work required to expand the system against the surroundings. The quantity of heat liberated is somewhat less than in the constant-volume case. Notice how the equation $q_P = \Delta U + P\,\Delta V$ leads us to the same conclusion: ΔU is negative, the $P\,\Delta V$ term is positive, and their sum, q_P, is not quite as negative as is ΔU.

Enthalpy

Determining ΔU for a reaction carried out at constant volume is thus relatively simple: Just by measuring q_V, we can find the value for ΔU. However, the vast majority of chemical reactions are carried out at constant pressure, not at constant volume. They take place in vessels that are open to the atmosphere, and the pressure in the reaction system is the prevailing atmospheric pressure. Figure 6.14 shows how we might think of the reaction of magnesium with hydrochloric acid as it is carried out in a chemistry laboratory.

$$Mg(s) + 2\,HCl(aq) \longrightarrow MgCl_2(aq) + H_2(g)$$

Because the vast majority of heats of reaction that chemists measure are q_P values, it is useful to have a thermodynamic function whose change in a chemical reaction is exactly equal to q_P. Chemists have defined a function called *enthalpy* for just this purpose. **Enthalpy (H)** is defined as the sum of the internal energy and the pressure-volume product of a system.

$$H = U + PV$$

From this definition we see that the heat of reaction q_P is just the **enthalpy change (ΔH)** for a process carried out at constant temperature and pressure and with work limited to pressure-volume work.

$$q_P = \Delta H = \Delta U + P\,\Delta V$$

Some Properties of Enthalpy

Enthalpy (H) is especially useful because of certain properties that it shares with internal energy (U) and some other thermodynamic functions.

1. **Enthalpy is an *extensive* property.** The enthalpy of a system depends on the quantities of substances present. Even though we cannot measure the absolute enthalpy of a substance any more than we can measure its internal energy, we can say, for example, that the enthalpy of 2.00 mol CO_2 is exactly twice the enthalpy of 1.00 mol CO_2.
2. **Enthalpy is a function of state.** The enthalpy of a system depends only on its present state or condition, and *not* on the route by which it got there. Because U, P, and V are all functions of state, and H is a function of only these variables (that is, $H = U + PV$), then H is also a function of state.
3. **Enthalpy changes have unique values.** Because the enthalpy in each of two states of a system has a unique value, the difference in enthalpy between the two states—the enthalpy *change*, ΔH—also has a unique value. And this change in enthalpy is equal to the heat of reaction at constant pressure: $\Delta H = q_P$.

Now let's examine some practical consequences of these statements about enthalpy and enthalpy change.

Representing ΔH for a Chemical Reaction

First, we can expand our symbolic description of a chemical reaction, in the manner shown below for the combustion of methane at 25 °C.

$$CH_4(g) + 2\,O_2(g) \longrightarrow CO_2(g) + 2\,H_2O(l) \qquad \Delta H = -890.3\text{ kJ}$$

We say that the enthalpy change in the combustion of 1 mol $CH_4(g)$ at 25 °C is −890.3 kJ. ΔH is equal to q_P, and its negative value shows that the combustion of methane at constant pressure is an *exothermic* reaction.

In the equation representing the decomposition of mercury(II) oxide at 25 °C, the positive value of ΔH shows this to be an *endothermic* reaction.

$$2\,HgO(s) \longrightarrow 2\,Hg(l) + O_2(g) \qquad \Delta H = +181.66\text{ kJ}$$

APPLICATION NOTE
Lavoisier used this reaction in the discovery of the law of conservation of mass (page 36). To supply the necessary heat, he focused sunlight on HgO(s) in sealed glass containers by using a magnifying glass.

Because enthalpies and enthalpy changes are extensive properties, the decomposition of *one* mole of HgO(s) is accompanied by exactly one half the enthalpy change noted above.

$$HgO(s) \longrightarrow Hg(l) + \frac{1}{2}O_2(g) \qquad \Delta H = \frac{1}{2} \times 181.66\text{ kJ} = +90.83\text{ kJ}$$

We can represent enthalpy change through a chemical equation, but we can also do so through a graphical representation known as an **enthalpy diagram**, as illustrated in Figure 6.15.

If a process is carried out first in one direction and then brought back to its initial state, the total enthalpy change is *zero* because enthalpies are state functions. In any such *cyclic* process, the initial and final states are the same. This means that

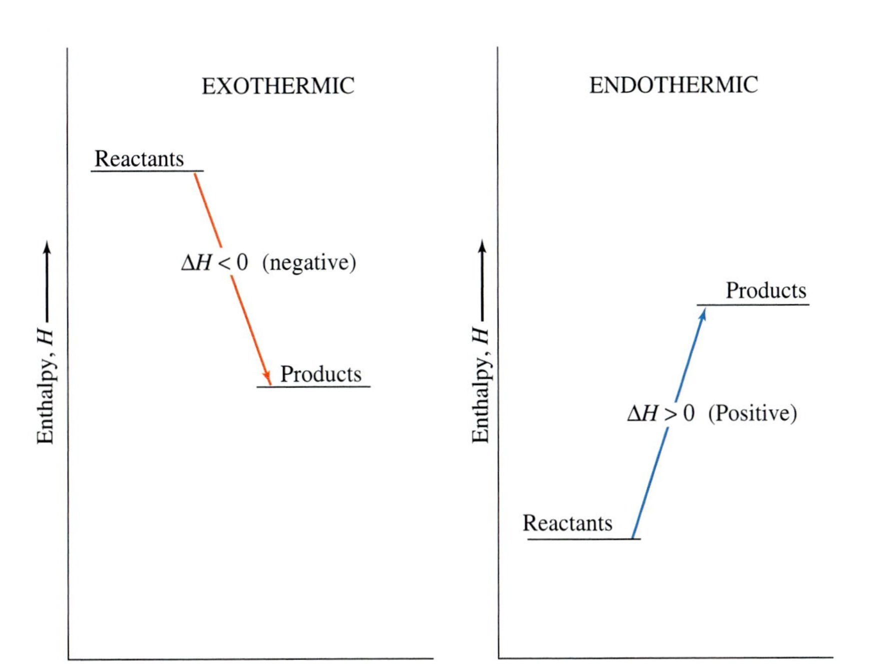

◀ **Figure 6.15**
Enthalpy diagrams
No numbers are shown on the enthalpy axis because absolute values of enthalpy cannot be measured. For an *exothermic* reaction (red), the products have a lower enthalpy than the reactants and ΔH is *negative* ($\Delta H < 0$). For an *endothermic* reaction (blue), the products have a higher enthalpy than the reactants and ΔH is *positive* ($\Delta H > 0$).

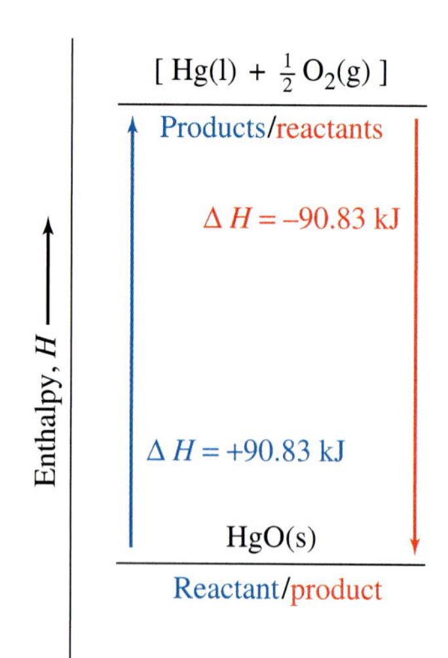

▲ **Figure 6.16**
Reversing a chemical reaction
The *forward* reaction (shown in blue) is the dissociation at 25 °C of 1.00 mol HgO(s) into its elements and is accompanied by an increase in enthalpy of 90.83 kJ; $\Delta H = +90.83$ kJ. In the *reverse* reaction (red), 1.00 mol HgO(s) is formed from its elements. This reaction is accompanied by a *decrease* in enthalpy of 90.83 kJ; $\Delta H = -90.83$ kJ. When a reaction is reversed, the magnitude of ΔH remains the same, but its sign changes.

ΔH changes sign when a process is reversed (see Figure 6.16). Thus, for the formation of HgO(s) from its elements at 25 °C,

$$\text{Hg(l)} + \frac{1}{2}\,\text{O}_2\text{(g)} \longrightarrow \text{HgO(s)} \qquad \Delta H = -90.83\text{ kJ}$$

EXAMPLE 6.3

Given the equation:

$$\text{H}_2\text{(g)} + \text{I}_2\text{(s)} \longrightarrow 2\,\text{HI(g)} \qquad \Delta H = +52.96\text{ kJ},$$

calculate ΔH for the reaction:

$$\text{HI(g)} \longrightarrow \frac{1}{2}\,\text{H}_2\text{(g)} + \frac{1}{2}\,\text{I}_2\text{(s)}.$$

SOLUTION

HI(g) is on the *right* in the given equation but on the *left* in the equation of interest. We have to reverse the given equation and change the sign of the ΔH value.

$$2\,\text{HI(g)} \longrightarrow \text{H}_2\text{(g)} + \text{I}_2\text{(s)} \qquad \Delta H = -52.96\text{ kJ}$$

Because the coefficient of HI(g) is "2" in the above equation but only "1" in the equation we are seeking, we must multiply the enthalpy change for the reaction by one half.

$$\text{HI(g)} \longrightarrow \frac{1}{2}\,\text{H}_2\text{(g)} + \frac{1}{2}\,\text{I}_2\text{(s)} \qquad \Delta H = \frac{1}{2} \times (-52.96) = -26.48\text{ kJ}$$

EXERCISE 6.3A

Given the equation:

$$3\,\text{O}_2\text{(g)} \longrightarrow 2\,\text{O}_3\text{(g)} \qquad \Delta H = +285.4\text{ kJ},$$

calculate ΔH for the following reaction.

$$\frac{3}{2}\,\text{O}_2\text{(g)} \longrightarrow \text{O}_3\text{(g)}.$$

EXERCISE 6.3B

Given the equation:

$$2\,\text{Ag}_2\text{S(s)} + 2\,\text{H}_2\text{O(l)} \longrightarrow 4\,\text{Ag(s)} + 2\,\text{H}_2\text{S(g)} + \text{O}_2\text{(g)} \qquad \Delta H = +595.5\text{ kJ}$$

calculate ΔH for the following reaction.

$$\text{Ag(s)} + \frac{1}{2}\,\text{H}_2\text{S(g)} + \frac{1}{4}\,\text{O}_2\text{(g)} \longrightarrow \frac{1}{2}\,\text{Ag}_2\text{S(s)} + \frac{1}{2}\,\text{H}_2\text{O(l)}$$

EXAMPLE 6.4

The complete combustion of liquid octane, C_8H_{18}, to produce gaseous carbon dioxide and liquid water at 25 °C and at a constant pressure gives off 47.9 kJ per gram of octane. Write a chemical equation to represent this information.

SOLUTION

Because the reaction is carried out at constant pressure, the heat of reaction given is a q_P value, and therefore an enthalpy change, ΔH. Moreover, the fact that heat is given off is a clue that the reaction is *exothermic* and that ΔH is a *negative* quantity. The enthalpy change, however, is given for one gram of reactant, and we want to enter the enthalpy change into an equation in which the coefficients represent *moles* of substances. We need to convert the enthalpy change to a molar basis.

$$? \text{ kJ/mol } C_8H_{18} = \frac{-47.9 \text{ kJ}}{1 \text{ g } C_8H_{18}} \times \frac{114.2 \text{ g } C_8H_{18}}{1 \text{ mol } C_8H_{18}} = -5.47 \times 10^3 \text{ kJ/mol } C_8H_{18}$$

Now we can write a balanced equation for the reaction and enter the ΔH value.

$$C_8H_{18}(l) + \frac{25}{2} O_2(g) \longrightarrow 8\, CO_2(g) + 9\, H_2O(l) \quad \Delta H = -5.47 \times 10^3 \text{ kJ}$$

Note that if we remove the fractional coefficient by multiplying all the coefficients by 2, as we so often do in balancing an equation, we must also multiply the ΔH value by 2.

$$2\, C_8H_{18}(l) + 25\, O_2(g) \longrightarrow 16\, CO_2(g) + 18\, H_2O(l) \quad \Delta H = -1.09 \times 10^4 \text{ kJ}$$

EXERCISE 6.4

Express the following information as a chemical equation. At 25 °C and at a constant pressure, dinitrogen trioxide gas decomposes to nitrogen monoxide and nitrogen dioxide gases with the absorption of 0.533 kJ of heat for every gram of dinitrogen trioxide that decomposes.

ΔH in Stoichiometric Calculations

We can calculate the quantity of heat involved in a chemical reaction in much the same way that we calculate masses of reactants and products. That is, we can think of heat absorbed in an endothermic reaction as being like a reactant and heat evolved in an exothermic reaction like a product. The ΔH value then becomes the basis for a conversion factor. Consider, for example, this exothermic reaction.

$$H_2(g) + Cl_2(g) \longrightarrow 2\, HCl(g) \qquad \Delta H = -184.6 \text{ kJ}$$

This equation indicates that 184.6 kJ of heat is released by the reaction system to the surroundings for every one mol H_2 consumed. We can use such reasoning to write the conversion factors

$$\frac{-184.6 \text{ kJ}}{1 \text{ mol } H_2} \qquad \frac{-184.6 \text{ kJ}}{1 \text{ mol } Cl_2} \qquad \frac{-184.6 \text{ kJ}}{2 \text{ mol HCl}}$$

We can use such conversion factors to determine the enthalpy change for a reaction, given a specific quantity of a product or reactant. (See Example 6.5 and Exercise 6.5A.)

Another typical situation is presented in Exercise 6.5B. There, we need a conversion factor to relate the amount of a reactant to a quantity of heat through a ΔH value. Then we use the ideal gas equation to convert the amount of the gaseous reactant to a gas volume at a given temperature and pressure.

EXAMPLE 6.5

What is the enthalpy change associated with the formation of 5.67 mol HCl(g) in the following reaction?

$$H_2(g) + Cl_2(g) \longrightarrow 2\,HCl(g) \qquad \Delta H = -184.6\text{ kJ}$$

SOLUTION

The conversion factor we need from the chemical equation is shown in color below.

$$5.67\ \cancel{\text{mol HCl}} \times \frac{-184.6\text{ kJ}}{2\ \cancel{\text{mol HCl}}} = -523\text{ kJ}$$

The formation of 5.67 mol HCl in this *exothermic* reaction results in the release of 523 kJ of heat to the surroundings.

EXERCISE 6.5A

What is the enthalpy change when 12.8 g $H_2(g)$ reacts with excess $Cl_2(g)$ to form HCl(g)?

$$H_2(g) + Cl_2(g) \longrightarrow 2\,HCl(g) \qquad \Delta H = -184.6\text{ kJ}$$

EXERCISE 6.5B

What volume of $CH_4(g)$, measured at 25 °C and 745 Torr, must be burned in excess oxygen to release 1.00×10^6 kJ of heat to the surroundings?

$$CH_4(g) + 2\,O_2(g) \longrightarrow CO_2(g) + 2\,H_2O(l) \qquad \Delta H = -890.3\text{ kJ}$$

6.5 Calorimetry: Measuring Quantities of Heat

We have now seen how ΔU and ΔH can be related to heats of reaction, how these quantities—generally ΔH—can be incorporated into chemical equations, and how we can include quantities of heat in stoichiometric calculations. But we also need to consider another question: How do we *measure* a heat of reaction? We do so by measuring quantities of heat.

The measurement of a quantity of heat is called *calorimetry*, and the device in which the measurement is made is called a **calorimeter.** Calorimetry is based on the *law of conservation of energy*: Whatever heat is *lost* by a system must be *gained* by its surroundings (or vice versa). Successful calorimetry depends on our ability to measure the quantities of such heat accurately.

To do calorimetry, we need to understand the concepts of heat capacity and specific heat. We begin by introducing and illustrating these terms.

Heat Capacity

When some systems absorb a given quantity of heat from the surroundings, they experience a large increase in temperature; they are said to have a low heat capacity. Other systems may absorb the same quantity of heat but experience a smaller temperature increase; they are said to have a higher heat capacity. The *heat capacity* (C) of a system is the quantity of heat required to change the temperature of the system by 1 °C (or by 1 K). We can determine a heat capacity by dividing the quantity of heat (q) by the temperature change it produces (ΔT).

$$C = \frac{q}{\Delta T}$$

The units of heat capacity are joules per degree Celsius (J/°C) or joules per kelvin (J/K). Because one kelvin is equal to a change of one degree Celsius, heat capacity has the same numerical value in either case.

EXAMPLE 6.6

Calculate the heat capacity of an aluminum block that must absorb 629 J of heat from its surroundings for its temperature to rise from 22 °C to 145 °C.

SOLUTION

We need the ratio of q to ΔT. The value of $q = 629$ J, and ΔT is the final temperature minus the initial temperature: $\Delta T = T_f - T_i = 145\ °C - 22\ °C$.

$$C = \frac{q}{\Delta T} = \frac{629\ \text{J}}{(145 - 22)\ °\text{C}} = \frac{629\ \text{J}}{123\ °\text{C}} = 5.11\ \text{J/°C}$$

EXERCISE 6.6A

Calculate the heat capacity of a sample of brake fluid if the sample must absorb 911 J of heat in order for its temperature to rise from 15 °C to 100 °C.

EXERCISE 6.6B

A burner on an electric range has a heat capacity of 345 J/K. What is the value of q, in kilojoules, as the burner cools from a temperature of 467 °C to a room temperature of 23 °C?

Specific Heats

The heat capacity of a system depends on the quantity and type(s) of matter in the system. A large block of aluminum has a higher heat capacity than does a small piece of the metal. **Molar heat capacity** is the heat capacity of one mole of a substance. **Specific heat** is the heat capacity of a one-gram sample; it is the quantity of heat required to change the temperature of *one gram* of the substance by 1 K (or by 1 °C). We get the specific heat when we divide the heat capacity of a substance by its mass.

$$\text{Specific heat} = \frac{\text{heat capacity}}{\text{mass}} = \frac{C}{m}$$

Because the heat capacity itself is $C = q/\Delta T$,

$$\text{Specific heat} = \frac{q}{m \times \Delta T}$$

If the aluminum block described in Example 6.6 has a mass of 5.7 g, then the specific heat of aluminum is

$$\text{Specific heat} = \frac{629\ \text{J}}{5.7\ \text{g} \times 123\ °\text{C}} = 0.90\ \text{J g}^{-1}\ °\text{C}^{-1}$$

The molar heat capacity of aluminum is

$$\text{Specific heat} \times \text{molar mass} = 0.90\ \text{J}\ \cancel{\text{g}^{-1}}\ °\text{C}^{-1} \times 27.0\ \cancel{\text{g}}\ \text{mol}^{-1} = 24\ \text{J mol}^{-1}\ °\text{C}^{-1}$$

The specific heats of several familiar substances are given in Table 6.1.

APPLICATION NOTE

The high specific heat of water compared to other substances helps to account for the observation that the climate near large bodies of water, such as oceans or large lakes, is generally more moderate than in interior locations. The daily temperature rises and falls more slowly, for example, than in a desert region.

TABLE 6.1 Specific Heats of Several Substances at 25 °C

Substance	Specific heat, J g^{-1} °C^{-1}
Aluminum (Al)	0.902
Copper (Cu)	0.385
Ethanol (CH_3CH_2OH)	2.46
Iron (Fe)	0.449
Lead (Pb)	0.128
Mercury (Hg)	0.139
Silver (Ag)	0.235
Sulfur (S)	0.706
Water (H_2O)	4.182

In a typical calorimetric calculation, we relate a quantity of heat, a temperature change, and the mass and specific heat of a substance. For this purpose, we can rearrange the specific-heat equation into the following form.

$$q = \text{mass} \times \text{specific heat} \times \Delta T$$

When we raise the temperature of a system, the final temperature, T_f, is higher than the initial temperature, T_i. The temperature change, which is $\Delta T = (T_f - T_i)$, is *positive*, and so is q. Heat is *gained* by the system. Lowering the temperature of a system means that T_f is smaller than T_i, ΔT is *negative*, q is also *negative*, and heat is *lost* by the system.

The specific heat of a substance varies with temperature, and some of our calculations will yield only approximate answers. For example, the specific heat of silver is 0.235 J g^{-1} K^{-1} at 298 K (25 °C), but it increases steadily to a value of 0.278 J g^{-1} K^{-1} at 1000 K. The most important specific heat value that we will use in calculations is that of water. At about room temperature, the specific heat of water is 4.18 J g^{-1} °C^{-1}, and over the temperature range from 0 °C to 100 °C, it remains within 1% of this value.

The specific heat of water at 15 °C is 4.184 J g^{-1} °C. At one time, the quantity of heat required to raise the temperature of one gram of water from 14.5 °C to 15.5 °C was defined as one **calorie (cal)**. The calorie is not an SI unit, but it was widely used in the past and is still used to some extent. Now we simply *define* the calorie through the expression: 1 cal = 4.184 J. Like the joule, the calorie is a relatively small energy unit, and the *kilo*calorie is also used (see, for example, p. 266).

EXAMPLE 6.7

How much heat, in joules and kilojoules, does it take to raise the temperature of 225 g of water from 25.0 °C to 100.0 °C?

SOLUTION

The specific heat of water is 4.18 J g^{-1} °C^{-1}. The temperature change is (100.0 − 25.0) °C = 75.0 °C, and the quantity of water to be heated is 225 g.

$$q = \text{mass} \times \text{specific heat} \times \Delta T$$

$$q = 225\ \cancel{\text{g H}_2\text{O}} \times \frac{4.18\ \text{J}}{\cancel{\text{g H}_2\text{O}}\ \cancel{°\text{C}}} \times (100.0 - 25.0)\ \cancel{°\text{C}}$$

$$= 7.05 \times 10^4\ \text{J} = 70.5\ \text{kJ}$$

EXERCISE 6.7A

How much heat, in calories and kilocalories, does it take to raise the temperature of 814 g of water from 18.0 °C to 100.0 °C?

EXERCISE 6.7B

What mass of water, in kilograms, can be heated from 5.5 °C to 55.0 °C by 9.09×10^{10} J of heat?

EXAMPLE 6.8

What will be the final temperature if a 5.00-g silver ring at 37.0 °C gives off 25.0 J of heat to its surroundings? Use the specific heat of silver listed in Table 6.1.

SOLUTION

The loss of heat from the system—the silver—means that q is negative: $q = -25.0$ J. This loss is not compensated for by any energy input into the system, and so the temperature must fall, $\Delta T < 0$. We solve the specific-heat equation for the temperature change ($T_f - T_i$), which will be a *negative* quantity. We note that $T_i = 37.0$ °C, and we solve for the final temperature, T_f.

$$\Delta T = (T_f - T_i) = \frac{q}{\text{mass} \times \text{specific heat}} = \frac{-25.0\ \text{J}}{5.00\ \text{g} \times 0.235\ \text{J g}^{-1}\ °\text{C}^{-1}}$$

$$= -21.3\ °\text{C}$$

$$T_f - T_i = -21.3\ °\text{C}$$

$$T_f = T_i - 21.3\ °\text{C} = 37.0\ °\text{C} - 21.3\ °\text{C} = 15.7\ °\text{C}$$

EXERCISE 6.8A

A 454-g block of lead is at an initial temperature of 22.5 °C. What will be the temperature of the lead after it absorbs 4.22 kJ of heat from its surroundings?

EXERCISE 6.8B

How many grams of copper can be heated from 22.5 °C to 35.0 °C by the same quantity of heat that is capable of raising the temperature of 145 g H_2O from 22.5 °C to 35.0 °C?

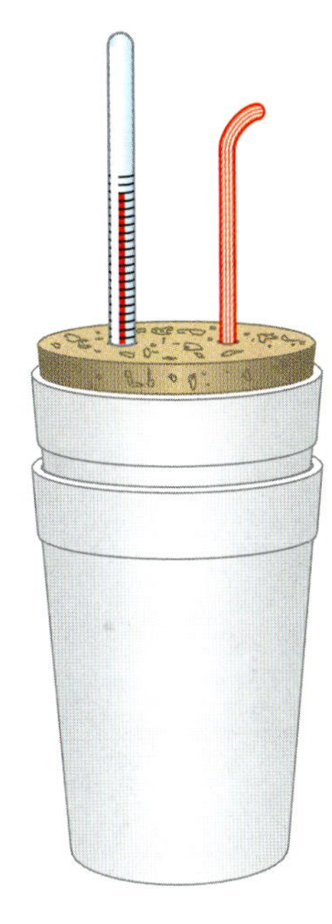

▲ **Figure 6.17**
A simple calorimeter made from Styrofoam® cups
The inner cup is closed off with a cork stopper through which a thermometer and stirrer are immersed into the calorimeter. The outer cup provides additional thermal insulation from the surroundings.

Measuring Specific Heats

The simple calorimeter pictured in Figure 6.17 is made from Styrofoam® cups. It is quite suitable for determining the specific heats of solid substances that are insoluble in water. One approach is to consider the solid as the system and the water as the surroundings.

A measured mass of the solid is heated to a given temperature and the solid is dropped into the calorimeter, which contains a fixed mass of water at a known temperature. Heat is transferred from the hot solid into the cooler water. The temperature of the solid falls and that of the water rises until the solid and water reach the same final temperature.

The basic principle of the method is that

All the heat lost by the hot solids is gained by the water in the cup.

Styrofoam® is an ideal material for the simple calorimeter in Figure 16.17 for two reasons. (1) Its very low heat capacity allows us to neglect the small amount

APPLICATION NOTE

An old-time application of this technique is to heat a poker over an open fire and then plunge it into a mug of cider. The result is instant hot cider.

of heat gained by the Styrofoam®, and (2) the foam is a good thermal insulator that permits very little heat to escape from the calorimeter.

From the mass and temperature change of the water, we can determine the quantity of heat transferred. We then use this quantity of heat and the mass and temperature change of the solid to determine its specific heat. The method is illustrated in Example 6.9.

EXAMPLE 6.9

A 15.5-g sample of a metal alloy is heated to 98.9 °C and then dropped into 25.0 g of water in a calorimeter. The temperature of the water rises from 22.5 °C to 25.7 °C. Calculate the specific heat of the alloy.

SOLUTION

First, let's calculate the quantity of heat absorbed by the water (the surroundings).

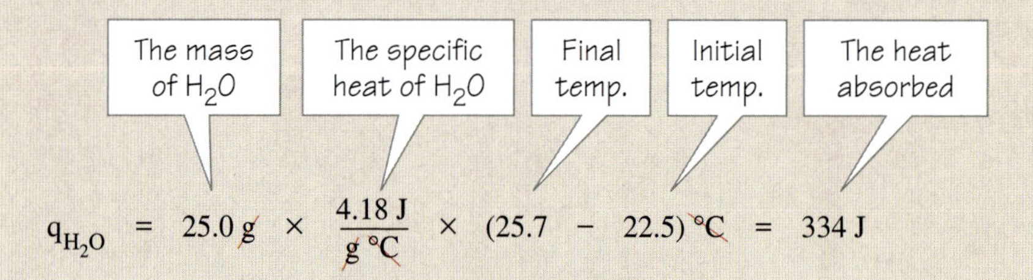

$$q_{H_2O} = 25.0\ \text{g} \times \frac{4.18\ \text{J}}{\text{g}\ ^\circ\text{C}} \times (25.7 - 22.5)\ ^\circ\text{C} = 334\ \text{J}$$

We assume that only the water (and not the container) absorbs heat and the only source of this heat is the alloy (the system) as it cools. Because the alloy *loses* heat, we write

$$q_{alloy} = -334\ \text{J}$$

Now we can calculate the specific heat of the alloy.

$$\text{Specific heat}_{alloy} = \frac{q_{alloy}}{\text{mass} \times \Delta T} = \frac{-334\ \text{J}}{15.5\ \text{g} \times (25.7 - 98.9\ ^\circ\text{C})}$$
$$= 0.29\ \text{J g}^{-1}\ ^\circ\text{C}^{-1}$$

We really aren't justified in writing the intermediate result (334 J) to three significant figures. Ordinarily we would not record this number and would simply round off to the appropriate two significant figures only once, in the final answer.

EXERCISE 6.9A

A 23.9-g sample of iridium is heated to 89.7 °C and then dropped into 20.0 g of water in a calorimeter. The temperature of the water rises from 20.1 °C to 22.6 °C. Calculate the specific heat of iridium.

EXERCISE 6.9B

A 135-g piece of iron (specific heat, 0.449 J g^{-1} °C^{-1}) is heated to 225 °C in an oven and then dropped into a Styrofoam® cup calorimeter containing 250.0 mL of glycerol (d = 1.261 g/mL) at 23.5 °C. The temperature of the glycerol is found to rise to a maximum value of 44.7 °C. Use these data to determine the specific heat of glycerol.

EXAMPLE 6.10—An Estimation Example

Without doing detailed calculations, determine which of the following is a likely approximate final temperature when 100 g of iron at 100 °C is added to 100 g of water in a Styrofoam® cup calorimeter at 20 °C:

20 °C 30 °C 60 °C 70 °C

SOLUTION

Because we have the same mass of each substance, *if* the water and iron had the same specific heat, the rise in temperature of the water would be the same as the drop in temperature of the iron. The final temperature would be the average of 20 °C and 100 °C, or 60 °C. However, from Table 6.1 we see that the specific heat of water ($4.18\ J\ g^{-1}\ °C^{-1}$) is much greater than that of iron ($0.45\ J\ g^{-1}\ °C^{-1}$). It takes more heat to change the temperature of a given mass of water than of the same mass of iron. The final water temperature must be below 60 °C, but also it has to be above 20 °C, the initial temperature. The only possible approximate temperature of those given is 30 °C.

EXERCISE 6.10

Without doing detailed calculations, determine the final temperature if 200.0 mL of water at 80 °C is added to 100.0 mL of water at 20 °C.

Measuring Enthalpy Changes for Chemical Reactions

By our definition of a heat of reaction, the reactants and products must be at the same temperature. We pictured constant-temperature exothermic and endothermic reactions in a schematic fashion in Figures 6.11(b) and 6.12(b). But we also noted that what we commonly observe are the results pictured in Figures 6.11(c) and 6.12(c): The temperature does not remain constant as heat leaves or enters the system in which a chemical reaction occurs. If we measure a heat of reaction conducted in an *isolated* system, as pictured in Figures 6.11(a) and 6.12(a), the change in chemical energy will appear as a change in thermal energy of the system. We will observe a corresponding temperature increase if the reaction is exothermic or a temperature decrease if the reaction is endothermic. Then we can use measured masses of substances, specific heats, and a temperature change to calculate the heat of reaction.

Consider the exothermic neutralization reaction in which the acid HCl(aq) neutralizes the base NaOH(aq) to produce the salt NaCl(aq) and $H_2O(l)$.

$$HCl(aq) + NaOH(aq) \longrightarrow NaCl(aq) + H_2O(l)$$

Suppose we allow stoichiometric proportions of the acid and base at the temperature T_i to react in an *isolated* system, such as a Styrofoam® cup calorimeter. After the reaction, an aqueous solution of NaCl is all that is present. The heat of the reaction remains in the system and raises the temperature to T_f. The heat of reaction (q_{rxn}) is the quantity of heat that would be given off by the NaCl(aq) if we allowed it to *cool* back to the initial temperature (T_i) by giving off heat to the surroundings. But this is just the *negative* of the quantity of heat that was required to *raise* the temperature in the calorimeter from T_i to T_f in the first place. And this is just the product of the mass of the NaCl(aq) formed in the reaction, the specific heat of the solution, and the temperature change ($\Delta T = T_f - T_i$). Let's call this quantity of heat $q_{calorim}$.

$$q_{calorim} = \text{mass of NaCl(aq)} \times \text{specific heat of NaCl(aq)} \times \Delta T$$

The heat of the reaction, then, is the *negative* of $q_{calorim}$, that is,

$$q_{rxn} = -q_{calorim}$$

For reactions carried out in systems under the constant pressure of the atmosphere, the quantity we measure is a heat of reaction at constant pressure, q_P. Thus we can write that $q_{rxn} = q_P = \Delta H$, and

$$\Delta H = -q_{calorim}$$

In Example 6.11 we carry out the calculation outlined above for the reaction of HCl(aq) and NaOH(aq).

EXAMPLE 6.11

A 50.0-mL sample of 0.250 M HCl at 19.50 °C is added to 50.0 mL of 0.250 M NaOH, also at 19.50 °C, in a Styrofoam® cup calorimeter. After mixing, the solution temperature rises to 21.21 °C. Calculate the heat of this reaction.

SOLUTION

As outlined in the preceding paragraphs, we are considering a neutralization reaction in which the acid and base are present in their stoichiometric proportions. The product of the reaction is NaCl(aq).

$$\text{HCl(aq)} + \text{NaOH(aq)} \longrightarrow \text{NaCl(aq)} + \text{H}_2\text{O(l)}$$

Now we make four assumptions to keep the calculation simple.

(1) The solution volumes are additive. The volume of NaCl(aq) that forms is equal to 50.0 mL + 50.0 mL = 100.0 mL.

(2) The NaCl(aq) is sufficiently dilute that its density and specific heat are about the same as that of pure water: 1.00 g/mL and 4.18 J g^{-1} °C^{-1}, respectively.

(3) The system is completely *isolated*. No heat escapes from the calorimeter.

(4) The heat required to warm any part of the calorimeter other than the NaCl(aq) is negligible.

PROBLEM-SOLVING NOTE
The first assumption allows us to work with the volume of NaCl(aq) and its density rather than having to measure its mass.

PROBLEM-SOLVING NOTE
The second assumption allows us to avoid having to measure the density and specific heat of the NaCl(aq) or having to find these data somewhere in the chemical literature.

The heat produced by the reaction and retained in the calorimeter is

$$q_{calorim} = \text{mass} \times \text{specific heat} \times \Delta T$$

$$= 100.0\ \cancel{\text{mL}} \times \frac{1.00\ \cancel{\text{g}}}{\cancel{\text{mL}}} \times \frac{4.18\ \text{J}}{\cancel{\text{g}}\ \cancel{°\text{C}}} \times (21.21 - 19.50)\ \cancel{°\text{C}} = 715\ \text{J}$$

$$q_{rxn} = q_P = -q_{calorim} = -715\ \text{J}$$

EXERCISE 6.11

A 100.0-mL portion of 0.500 M HBr at 20.29 °C is added to 100.0 mL of 0.500 M KOH, also at 20.29 °C, in a foam cup calorimeter. After mixing, the temperature rises to 23.65 °C. Calculate the heat of this reaction.

EXAMPLE 6.12

Express the result of Example 6.11 for *molar* amounts of the reactants and products. That is, determine the value ΔH that should be written in the equation for the neutralization reaction.

$$\text{HCl(aq)} + \text{NaOH(aq)} \longrightarrow \text{NaCl(aq)} + \text{H}_2\text{O(l)} \quad \Delta H = ?$$

SOLUTION

We begin by determining just how many moles of reactants and products were involved in Example 6.11.

Reactants:

$$? \text{ mol HCl} = 0.0500 \cancel{\text{L HCl(aq)}} \times \frac{0.250 \text{ mol HCl}}{1 \cancel{\text{L HCl(aq)}}} = 0.0125 \text{ mol HCl}$$

$$? \text{ mol NaOH} = 0.0500 \cancel{\text{L NaOH(aq)}} \times \frac{0.250 \text{ mol NaOH}}{1 \cancel{\text{L NaOH(aq)}}} = 0.0125 \text{ mol NaOH}$$

Products:

The numbers of moles of NaCl and H_2O formed are the same as the numbers of moles of HCl and NaOH that react. For example,

$$? \text{ mol H}_2\text{O} = 0.0125 \cancel{\text{mol HCl}} \times \frac{1 \text{ mol H}_2\text{O}}{1 \cancel{\text{mol HCl}}} = 0.0125 \text{ mol H}_2\text{O}$$

Thus the enthalpy change that we found in Example 6.11 was for the formation of 0.0125 mol H_2O, that is, -715 J/0.0125 mol H_2O. To calculate the enthalpy change for the formation of 1.00 mol H_2O, we can write

$$\Delta H = 1.00 \cancel{\text{mol H}_2\text{O}} \times \frac{-715 \text{ J}}{0.0125 \cancel{\text{mol H}_2\text{O}}} = -5.72 \times 10^4 \text{ J } (-57.2 \text{ kJ})$$

PROBLEM-SOLVING NOTE

Because we have equal numbers of moles of HCl and NaOH and they react in a 1:1 mole ratio, we can say that the reactants are in *stoichiometric proportions*. Neither reactant is present in excess.

EXERCISE 6.12A

Express the result of Exercise 6.11 for molar amounts of reactants and products.

$$\text{HBr(aq)} + \text{KOH(aq)} \longrightarrow \text{KBr(aq)} + \text{H}_2\text{O(l)} \quad \Delta H = ?$$

EXERCISE 6.12B

A 125-mL sample of 1.33 M HCl and 225 mL of 0.625 M NaOH, both initially at 24.4 °C, are allowed to react in a calorimeter.

What is the final temperature that would be observed in the calorimeter? (Use the type of assumptions stated in Example 6.11 and the value of ΔH obtained in Example 6.12 to determine the final temperature.)

PROBLEM-SOLVING NOTE

The reactants are not in stoichiometric proportions. One is the limiting reactant and one is in excess.

Bomb Calorimetry: Reactions at Constant Volume

Reactions carried out in Styrofoam® cup calorimeters and in most other laboratory vessels are open to the atmosphere, that is, under constant pressure. (The lid on the calorimeter is there to minimize heat loss; it is not airtight.) We cannot carry out reactions involving gases in these calorimeters because the gases would escape; the system would no longer be isolated.

For reactions involving gases, such as combustion reactions, we ordinarily use a device like the one illustrated in Figure 6.18: a *bomb calorimeter*. The heart of this calorimeter is a container with strong steel walls in which an exothermic reaction is carried out at a near-explosive rate and the reactants and products are confined to a constant volume. Perhaps appropriately, this container is called a "bomb."

To perform a bomb calorimetry experiment, we place a small sample of known mass in a metal cup in the bomb. We then fill the bomb with oxygen at a pressure of about 30 atm, and place the bomb in an insulated container filled with a known quantity of water. We initiate the reaction with an electric ignition wire, and record the highest water temperature reached.

From the temperature rise and the heat capacity of the calorimeter (established in a separate experiment), we can determine the increase in thermal energy of the calorimeter contents, $q_{calorim}$; that is,

$$q_{calorim} = \text{heat capacity of calorimeter} \times \Delta T$$

Notice that, unlike with foam cup calorimeters where we worked with the product (mass × specific heat), we use the heat capacity of the calorimeter as a whole in bomb calorimetry. A bomb calorimeter has too many component parts for us to determine their separate masses and specific heats. We treat the contents of the calorimeter as an *isolated* system. The system is similar to that pictured schematically in Figure 6.11(a) on page 237. As in Example 6.11, we can write this equation.

$$q_{rxn} = -q_{calorim}$$

The pressure in the bomb does not necessarily remain constant during the reaction, but the *volume* does. This means that $\Delta V = 0$ and $w = -P\,\Delta V = 0$. The heat of reaction we measure is q_V, and $q_V = \Delta U$. Thus, for a reaction carried out in a bomb calorimeter,

$$\Delta U = q_V = q_{rxn} = -q_{calorim}$$

Even though we measure heats of combustion in bomb calorimeters, we usually carry out combustion reactions in the open atmosphere, such as in a natural-gas heater or a propane torch. The more useful property of combustion reactions is therefore ΔH (the heat of reaction at constant pressure), not ΔU (the heat of reaction at constant volume). Although it is not difficult to calculate a ΔH value from a ΔU value, we will not do so here. Actually, in some cases, ΔH and ΔU are equal because $P\Delta V$ is zero, and in many others ΔH and ΔU are nearly equal because the $P\Delta V$ term is quite small. The principal exceptions are reactions that involve a change in the number of moles of gas.

In examples and exercises like those that follow, we will treat the results of bomb calorimetry experiments as if they were ΔH values. To do a bomb calorimetry calculation, we must know the heat capacity of the calorimeter assembly, and Exercise 6.13B suggests how this quantity can be determined by experiment.

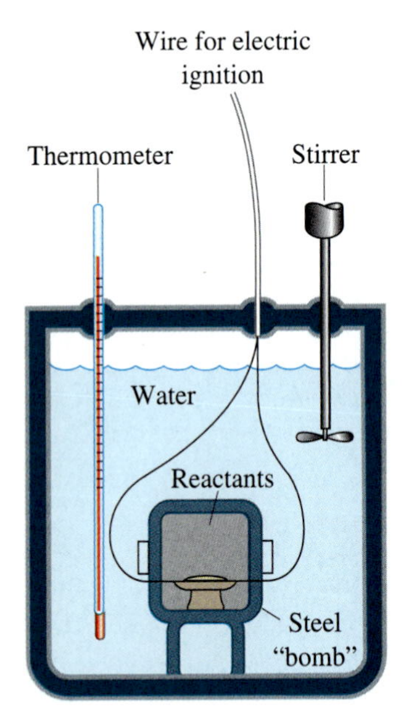

▲ **Figure 6.18**
A bomb calorimeter
The sample to be burned is placed in a small cup in the steel bomb, which is then filled with oxygen. The sample is ignited by an electric current. The heat of the reaction is determined from the temperature rise in the water that surrounds the bomb. The steel bomb confines the reactants and products to a constant volume.

EXAMPLE 6.13

In a preliminary experiment, the heat capacity of a bomb calorimeter assembly is found to be 5.15 kJ/°C. In a second experiment, a 0.480-g sample of graphite (carbon) is placed in the bomb with an excess of oxygen. The water, bomb, and other contents of the calorimeter are in thermal equilibrium at 25.00 °C. The graphite is ignited and burned, and the water temperature rises to 28.05 °C. Calculate ΔH for the reaction

$$\text{C(graphite)} + \text{O}_2\text{(g)} \longrightarrow \text{CO}_2\text{(g)} \quad \Delta H = ?$$

SOLUTION

First, we can calculate $q_{calorim}$ from the heat capacity of the calorimeter and the temperature change.

$$q_{calorim} = 5.15 \text{ kJ °C}^{-1} \times (28.05 - 25.00) \text{ °C} = 15.7 \text{ kJ}$$

Then, we can establish q_{rxn}.

$$q_{rxn} = -q_{calorim} = -15.7 \text{ kJ}$$

To get ΔH, we need to determine the heat released by the combustion of one mole (12.01 g) of graphite.

$$\Delta H = \frac{12.01 \text{ g C}}{1 \text{ mol C}} \times \frac{-15.7 \text{ kJ}}{0.480 \text{ g C}} = -393 \text{ kJ/mol C}$$

$$\text{C(graphite)} + \text{O}_2\text{(g)} \longrightarrow \text{CO}_2\text{(g)} \quad \Delta H = -393 \text{ kJ}$$

EXERCISE 6.13A

A 0.250-g sample of diamond (another form of carbon) is burned with an excess of O_2(g) in a bomb calorimeter having a heat capacity of 6.52 kJ/°C. The calorimeter temperature rises from 20.00 °C to 21.26 °C. Calculate ΔH for the reaction

$$\text{C(diamond)} + \text{O}_2\text{(g)} \longrightarrow \text{CO}_2\text{(g)} \qquad \Delta H = ?$$

PROBLEM-SOLVING NOTE:
You will find that the result in this case is slightly different from that in Example 6.13 because the initial states of the systems differ. [C(diamond) is different from C(graphite).]

EXERCISE 6.13B

A 0.8082-g sample of glucose ($C_6H_{12}O_6$) is burned in a bomb calorimeter assembly, and the temperature is noted to rise from 25.11 °C to 27.21 °C.

Determine the heat capacity of the bomb calorimeter assembly.

$$\text{C}_6\text{H}_{12}\text{O}_6\text{(s)} + 6\,\text{O}_2\text{(g)} \longrightarrow 6\,\text{CO}_2\text{(g)} + 6\,\text{H}_2\text{O(l)} \qquad \Delta H = -2803 \text{ kJ}$$

PROBLEM-SOLVING NOTE:
How much heat is given off in the combustion reaction?

6.6 Hess's Law of Constant Heat Summation

What if we are interested in the enthalpy change for the combustion of carbon to carbon monoxide (CO)? We *cannot* use bomb calorimetry in the same way that we did for the combustion of carbon to carbon dioxide in Example 6.13 because CO(g), once formed, will undergo combustion to CO_2(g). Carbon monoxide is an intermediate in the combustion of carbon to CO_2(g), but we cannot stop the reaction at that point. However, if we start with pure CO(g), we can carry out its combustion to CO_2(g) in a bomb calorimeter. Below, we summarize what we can and cannot measure directly, and the values we would obtain, expressed to four significant figures.

(a) $$\text{C(graphite)} + \frac{1}{2}\text{O}_2\text{(g)} \longrightarrow \text{CO(g)} \quad \Delta H_{(a)} = ?$$

(b) $$\text{CO(g)} + \frac{1}{2}\text{O}_2\text{(g)} \longrightarrow \text{CO}_2\text{(g)} \quad \Delta H_{(b)} = -283.0 \text{ kJ}$$

(c) $$\text{C(graphite)} + \text{O}_2\text{(g)} \longrightarrow \text{CO}_2\text{(g)} \quad \Delta H_{(c)} = -393.5 \text{ kJ}$$

But suppose we look upon these three situations as different states that appear on the enthalpy diagram of Figure 6.19. The three enthalpy changes are represented by arrows, and we see that the red arrow, representing $\Delta H_{(a)}$ is the difference between the red arrows representing $\Delta H_{(c)}$ and $\Delta H_{(b)}$.

$$\Delta H_{(a)} = \Delta H_{(c)} - \Delta H_{(b)}$$

$$\Delta H_{(a)} = -393.5 \text{ kJ} - (-283.0 \text{ kJ}) = -110.5 \text{ kJ}$$

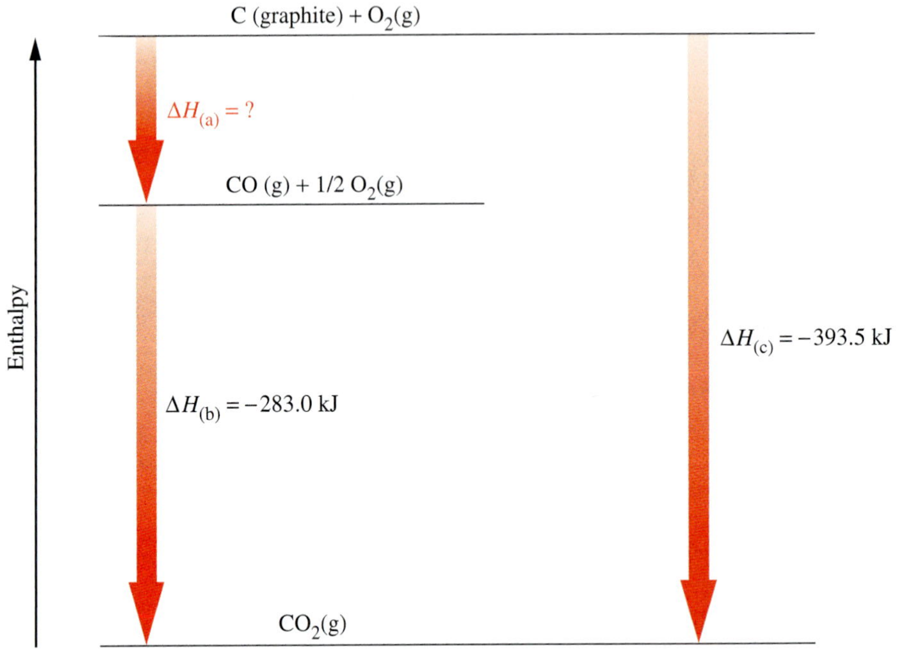

C (graphite) + $O_2(g)$

$\Delta H_{(a)} = ?$

CO (g) + 1/2 $O_2(g)$

Enthalpy

$\Delta H_{(b)} = -283.0$ kJ

$\Delta H_{(c)} = -393.5$ kJ

$CO_2(g)$

▶ **Figure 6.19**
Determining an unknown enthalpy change through an enthalpy diagram
This diagram brings out the relationship of the unknown $\Delta H_{(a)}$ to the known values of $\Delta H_{(b)}$ and $\Delta H_{(c)}$.

$$\begin{aligned}\Delta H_{(a)} &= \Delta H_{(c)} - \Delta H_{(b)} \\ &= -393.5 \text{ kJ} - (-283.0 \text{ kJ}) \\ &= -110.5 \text{ kJ.}\end{aligned}$$

We could use this graphical method to evaluate the enthalpy change for almost any reaction that we cannot measure directly by relating it to enthalpy changes that we can measure. After doing this a few times, we would probably discover that we really don't have to draw any graphs at all. In this way, we would repeat the discovery of a principle first stated by Germain Hess (1802–1850). According to **Hess's law**,

> *The heat of a reaction is constant, whether the reaction is carried out directly in one step or indirectly through a number of steps.*

The basic principle in using Hess's law is to express the equation for the reaction of interest as the sum of two or more other equations. At the same time that we add the equations, we add their ΔH values to obtain the unknown ΔH. Thus, referring to the equations on page 253 to get equation (a) and its ΔH value, we must *reverse* equation (b) and change the sign of its ΔH value. Then we can add it to equation (c).

$$\begin{array}{lll} -\text{(b)} \quad \cancel{CO_2(g)} \longrightarrow CO(g) + \frac{1}{2}O_2(g) & \Delta H_{(b)} = -(-283.0 \text{ kJ}) \\ \text{(c)} \quad \text{C(graphite)} + O_2(g) \longrightarrow \cancel{CO_2(g)} & \Delta H_{(c)} = -393.5 \text{ kJ} \\ \hline \text{(a)} \quad \text{C(graphite)} + \frac{1}{2}O_2(g) \longrightarrow CO(g) & \Delta H_{(a)} = -110.5 \text{ kJ} \end{array}$$

Note that in adding the equations, $CO_2(g)$ cancels out and the sum of $\frac{1}{2}\,O_2(g)$ on the right of the equation labeled $-$(b) and "1" $O_2(g)$ on the left of equation (c) leaves a net of $\frac{1}{2}\,O_2(g)$ on the left of equation (a).

A strategy for combining two or more chemical equations and their ΔH values to obtain the equation of interest and its ΔH value is outlined in Example 6.14. As part of this strategy, you can generally expect to do the following.

- *Reverse* certain equations and change the signs of their ΔH values.
- *Multiply* certain equations and their ΔH values by appropriate factors. The factors may be whole numbers (2, 3, ...) or fractions $\left(\frac{1}{4}, \frac{1}{2}, \ldots\right)$.

EXAMPLE 6.14

Calculate the enthalpy change for the reaction

(a) $2\,C(graphite) + 2\,H_2(g) \longrightarrow C_2H_4(g) \qquad \Delta H = ?$

given the following data.

(b) $C(graphite) + O_2(g) \longrightarrow CO_2(g) \qquad \Delta H = -393.5\ kJ$

(c) $C_2H_4(g) + 3\,O_2(g) \longrightarrow 2\,CO_2(g) + 2\,H_2O(l) \qquad \Delta H = -1410.9\ kJ$

(d) $H_2(g) + \frac{1}{2}O_2(g) \longrightarrow H_2O(l) \qquad \Delta H = -285.8\ kJ$

SOLUTION

Let's begin by looking for an equation that will place the term "2 C(graphite)" on the *left*, to match this term in equation (a). Only equation (b) has "C(graphite)" on the left, and to get "2 C(graphite)" we need to *double* equation (b) and its ΔH value.

2(b) $2\,C(graphite) + 2\,O_2(g) \longrightarrow 2\,CO_2(g) \qquad \Delta H = 2 \times (-393.5)\ kJ = -787.0\ kJ$

Next, let's identify an equation that will place the term "$C_2H_4(g)$" on the *right*. For this, we need equation (c), and we need to *reverse* it and *change the sign* of its ΔH value.

−(c) $2\,CO_2(g) + 2\,H_2O(l) \longrightarrow C_2H_4(g) + 3\,O_2(g)$

$\Delta H = -(-1410.9)\ kJ = 1410.9\ kJ$

Finally, we see that equation (a) has the term "$2\,H_2(g)$" on the *left*. To get "$2\,H_2(g)$" into our final equation, we need to *double* equation (d) and its ΔH value.

2(d) $2\,H_2(g) + O_2(g) \longrightarrow 2\,H_2O(l) \qquad \Delta H = 2 \times (-285.8\ kJ) = -571.6\ kJ$

We have used all the information given to us in order to correctly account for the "C(graphite)" and "$2\,H_2(g)$" on the left and the "C_2H_4" on the right. If we now add together the equations 2(b), −(c), and 2(d), we expect $O_2(g)$, $CO_2(g)$, and $H_2O(l)$ to cancel out in the summation. They do.

2(b) $2\,C(graphite) + \cancel{2\,O_2(g)} \longrightarrow \cancel{2\,CO_2(g)}$

$\Delta H = 2 \times (-393.5)\ kJ = -787.0 kJ$

−(c) $\cancel{2\,CO_2(g)} + \cancel{2\,H_2O(l)} \longrightarrow C_2H_4(g) + \cancel{3\,O_2(g)}$

$\Delta H = -(-1410.9)\ kJ = 1410.9\ kJ$

2(d) $2H_2(g) + \cancel{O_2(g)} \longrightarrow \cancel{2\,H_2O(l)}$

$\Delta H = 2 \times (-285.8\ kJ) = -571.6\ kJ$

(a) $2C(graphite) + 2\,H_2(g) \longrightarrow C_2H_4(g)$

$\Delta H = -787.0\ kJ + 1410.9\ kJ - 571.6\ kJ = 52.3\ kJ$

PROBLEM-SOLVING NOTE

It doesn't matter in which order we adjust the equations before combining them into the required final equation as long as each adjustment is a step toward our goal of getting the correct terms on each side of the equation.

EXERCISE 6.14A

Calculate the enthalpy change for the reaction

$$2\,CH_4(g) + 3\,O_2(g) \longrightarrow 2\,CO(g) + 4\,H_2O(l) \qquad \Delta H = ?$$

from the following data.

$$CO(g) + \frac{1}{2}O_2(g) \longrightarrow CO_2(g) \qquad \Delta H = -283.0\text{ kJ}$$

$$CH_4(g) + 2\,O_2(g) \longrightarrow CO_2(g) + 2\,H_2O(l) \qquad \Delta H = -890.3\text{ kJ}$$

EXERCISE 6.14B

Calculate the enthalpy change for the reaction

$$C_2H_4(g) + H_2(g) \longrightarrow C_2H_6(g) \qquad \Delta H = ?$$

given the following data.

$$C_2H_4(g) + 3\,O_2(g) \longrightarrow 2\,CO_2(g) + 2\,H_2O(l) \qquad \Delta H = -1410.9\text{ kJ}$$

$$2\,C_2H_6(g) + 7\,O_2(g) \longrightarrow 4\,CO_2(g) + 6\,H_2O(l) \qquad \Delta H = -3119.4\text{ kJ}$$

$$2\,H_2(g) + O_2(g) \longrightarrow 2\,H_2O(l) \qquad \Delta H = -571.6\text{ kJ}$$

6.7 Standard Enthalpies of Formation

We can determine the enthalpy changes for some reactions by calorimetry. Other enthalpy changes we can *calculate* with Hess's law, using ΔH values that can be measured. Of course, it would be best if we could simply *calculate* ΔH values directly, using the following relationship.

$$\Delta H = H_{\text{products}} - H_{\text{reactants}}$$

We can't do this, however, because we can't obtain *absolute* values of enthalpies. However, this situation is similar to that of determining the gain in elevation in climbing a mountain. We can express the elevation gain in a climb as

$$\text{Elevation gain } (\Delta h) = \text{elevation}_{\text{top}} - \text{elevation}_{\text{base}}$$

Yet we don't have *absolute* values for the elevations either. We use instead the elevations that are *relative* to mean sea level.

Chemists have devised a similar scale of *relative* enthalpies called *enthalpies of formation*. To understand how this scale works, we need first to describe the *standard* states for elements and compounds. The **standard state** of a solid or liquid substance is the pure element or compound at 1 atm pressure* and the temperature of interest. For a gaseous substance, the standard state is the (hypothetical) pure gas behaving as an ideal gas at 1 atm pressure* and at the temperature of interest. The **standard enthalpy of reaction** ($\Delta H°$) is the enthalpy *change* for

* The International Union of Pure and Applied Chemistry (IUPAC) has recommended that the standard state pressure be changed from 1 atm (101,325 Pa) to 1 bar (1×10^5 Pa). The effects of this change are quite small, and we will continue to use the standard state pressure of 1 atm in this text.

a reaction in which the reactants in their standard states yield products in their standard states; we denote a standard enthalpy change with a superscript degree symbol (°). Most of the enthalpy changes that we have used to this point are in fact *standard* enthalpies of reaction, and henceforth we will use the symbol $\Delta H°$ where appropriate.

The **standard enthalpy of formation** ($\Delta H_f°$) of a substance is the enthalpy *change* that occurs in the formation of 1 mol of the substance from its elements when both products and reactants are in their standard states. Moreover, the elements must be in their *reference* forms. With only a few of exceptions,* the reference form of an element is the most stable form of the element at the given temperature and 1 atm pressure.

The superscript degree sign labels the enthalpy change as a standard enthalpy change, and the subscript f refers to the reaction in which a compound is *formed* from its elements. The standard enthalpy of formation is often called the standard heat of formation, or more simply, the *heat of formation*.

What should we use for the standard enthalpy of formation for the reference form of an element? Because formation of the reference form of an element from itself is not an actual change, its $\Delta H_f°$ value is *zero*. That is,

The standard enthalpy of formation of a pure element in its reference form is 0.

Extensive tables of standard enthalpies of formation have been compiled that include both compounds and elements; these values are usually given for 298 K (25 °C) and expressed in kJ/mol. You will find a short list in Table 6.2 and a longer tabulation in Appendix C.

Two forms of carbon, graphite and diamond, are readily attainable at 25 °C and 1 atm pressure. The two forms are different states of carbon and must have different enthalpies of formation. As we demonstrated in Example 6.13 and Exercise 6.13A, they have different enthalpies of combustion, which are expressed below to four significant figures.

$$\text{C(graphite)} + \text{O}_2\text{(g)} \longrightarrow \text{CO}_2\text{(g)} \quad \Delta H° = -393.5 \text{ kJ}$$

$$\text{C(diamond)} + \text{O}_2\text{(g)} \longrightarrow \text{CO}_2\text{(g)} \quad \Delta H° = -395.4 \text{ kJ}$$

If we reverse the second of these equations and add it to the first, then by Hess's law we get this equation.

$$\text{C(graphite)} \longrightarrow \text{C(diamond)} \quad \Delta H° = -393.5 \text{ kJ} - (-395.4 \text{ kJ}) = 1.9 \text{ kJ}$$

Graphite has been chosen as the reference form of carbon at 298 °C, so $\Delta H_f°$ [C(graphite)] = 0. When graphite is converted to diamond, there is an increase in enthalpy of 1.9 kJ/mol, so $\Delta H_f°$[C(diamond)] = 1.9 kJ/mol.

Calculations Based on Standard Enthalpies of Formation

Now let's explore how we can use standard enthalpies of formation to *calculate* standard enthalpy changes of chemical reactions. Suppose we need to know the standard enthalpy change for the conversion of nitrogen dioxide to dinitrogen tetroxide at 25 °C and 1 atm.

$$2\ \text{NO}_2\text{(g)} \longrightarrow \text{N}_2\text{O}_4\text{(g)} \quad \Delta H° = ?$$

* A notable exception is that of phosphorus, for which the reference form at 25 °C and 1 atm pressure is solid *white* phosphorus, even though solid *red* phosphorus is the more stable of the two forms. Over a long period of time, a sample of white phosphorus slowly converts to the more stable red phosphorus.

TABLE 6.2 Some Standard Enthalpies of Formation at 25 °C

Substance	ΔH_f°, kJ/mol	Substance	ΔH_f°, kJ/mol
$CO(g)$	−110.5	$HCl(g)$	−92.31
$CO_2(g)$	−393.5	$HF(g)$	−271.1
$CH_4(g)$	−74.81	$HI(g)$	26.48
$C_2H_2(g)$	226.7	$H_2O(g)$	−241.8
$C_2H_4(g)$	52.26	$H_2O(l)$	−285.8
$C_2H_6(g)$	−84.68	$NH_3(g)$	−46.11
$C_3H_8(g)$	−103.8	$NO(g)$	90.25
$C_4H_{10}(g)$	−125.7	$N_2O(g)$	82.05
$C_6H_6(l)$	48.99	$NO_2(g)$	33.18
$CH_3OH(l)$	−238.7	$N_2O_4(g)$	9.16
$CH_3CH_2OH(l)$	−277.7	$SO_2(g)$	−296.8
$HBr(g)$	−36.40	$SO_3(g)$	−395.7

From Table 6.2, we can find the enthalpies of formation of these two gases.

$$\text{(a)} \quad 1/2\,N_2(g) + O_2(g) \longrightarrow NO_2(g) \quad \Delta H^\circ = \Delta H_f^\circ[NO_2(g)] = 33.18 \text{ kJ}$$

$$\text{(b)} \quad N_2(g) + 2\,O_2(g) \longrightarrow N_2O_4(g) \quad \Delta H^\circ = \Delta H_f^\circ[N_2O_4(g)] = 9.16 \text{ kJ}$$

With Hess's law, we can combine these two equations and arrive at the equation of interest to us. To do this, we must first *reverse* and then *double* equation (a).

$$-2 \times \text{Eq(a)} \quad 2\,NO_2(g) \longrightarrow \cancel{N_2(g)} + \cancel{2\,O_2(g)}$$

$$\Delta H^\circ = -2\,\Delta H_f^\circ[NO_2(g)] = -2 \times 33.18 = -66.36 \text{ kJ}$$

$$\text{Eq(b)} \quad \cancel{N_2(g)} + \cancel{2\,O_2(g)} \longrightarrow N_2O_4(g)$$

$$\Delta H^\circ = \Delta H_f^\circ[N_2O_4(g)] = 9.16 \text{ kJ}$$

$$2\,NO_2(g) \longrightarrow N_2O_4(g)$$

$$\Delta H^\circ = -66.36 \text{ kJ} + 9.16 \text{ kJ} = -57.20 \text{ kJ}$$

If we think about this calculation for a moment, we see that what we have done is equivalent to writing this equation.

$$\Delta H^\circ = \Delta H_f^\circ[N_2O_4(g)] - 2 \times \Delta H_f^\circ[NO_2(g)] = 9.16 \text{ kJ} - (2 \times 33.18) \text{ kJ} = -57.20 \text{ kJ}$$

This is just an example of a more general expression that allows us to relate a standard enthalpy of reaction and standard enthalpies of formation *without* formally going through Hess's law.

$$\Delta H^\circ = \sum \nu_p \times \Delta H_f^\circ(\text{products}) - \sum \nu_r \times \Delta H_f^\circ(\text{reactants})$$

The symbol $\sum$ signifies that we take a sum of several terms. The symbol ν (Greek letter "nu") is the stoichiometric coefficient used in front of a symbol or formula in a chemical equation. We write one term for each substance in the chemical reaction, and we obtain this term by multiplying the stoichiometric coefficient by the standard enthalpy of formation of the substance. First we take the sum of terms for the *products* of a reaction, and then we *subtract* the sum of terms for the *reactants*. The result is the standard enthalpy change, ΔH°.

Example 6.15 shows how we can use this equation to calculate a standard enthalpy of reaction, $\Delta H°$, from tabulated standard enthalpies of formation, $\Delta H_f°$. Example 6.16 illustrates the reverse: how we can establish an unknown standard enthalpy of formation from a measured standard enthalpy of reaction. With organic compounds, the measured $\Delta H°$ is often the standard enthalpy of combustion, $\Delta H°_{comb}$.

EXAMPLE 6.15

Synthesis gas is a mixture of carbon monoxide and hydrogen that is used to synthesize a variety of organic compounds, such as methanol. One reaction for producing synthesis gas is shown here.

$$3\ CH_4(g) + 2\ H_2O(l) + CO_2(g) \longrightarrow 4\ CO(g) + 8\ H_2(g) \quad \Delta H° = ?$$

Use standard enthalpies of formation from Table 6.2 to calculate the standard enthalpy change for this reaction.

SOLUTION

A convenient way to begin this kind of calculation is to list $\Delta H_f°$ under the formula of each substance in the equation.

	$3\ CH_4(g)$	$+\ 2\ H_2O(l)$	$+\ CO_2(g)$	$\longrightarrow 4\ CO(g)$	$+\ 8\ H_2(g)$
$\Delta H_f°$, kJ/mol:	−74.81	−285.8	−393.5	−110.5	0

Now we can multiply these $\Delta H_f°$ values by the numbers of moles given by the coefficients in the equation. In substituting these values into the general equation for $\Delta H°$, remember that we must *subtract* the sum of the terms for the reactants from the sum of the terms for the products.

$$\begin{aligned}
\Delta H° &= \sum \nu_p \times \Delta H_f°(\text{products}) - \sum \nu_r \times \Delta H_f°(\text{reactants}) \\
&= \left[4\ \cancel{\text{mol CO}} \times \frac{-110.5\ \text{kJ}}{1\ \cancel{\text{mol CO}}} + 8\ \cancel{\text{mol H}_2} \times \frac{0\ \text{kJ}}{1\ \cancel{\text{mol H}_2}}\right] \\
&\quad - \left[3\ \cancel{\text{mol CH}_4} \times \frac{-74.81\ \text{kJ}}{1\ \cancel{\text{mol CH}_4}} + 2\ \cancel{\text{mol H}_2\text{O}} \times \frac{-285.8\ \text{kJ}}{1\ \cancel{\text{mol H}_2\text{O}}} \right. \\
&\quad \left. + 1\ \cancel{\text{mol CO}_2} \times \frac{-393.5\ \text{kJ}}{1\ \cancel{\text{mol CO}_2}}\right] \\
&= -442.0\ \text{kJ} - [-224.4\ \text{kJ} - 571.6\ \text{kJ} - 393.5\ \text{kJ}] \\
&= -442.0\ \text{kJ} - [-1189.5\ \text{kJ}] \\
&= -442.0\ \text{kJ} + 1189.5\ \text{kJ} = +747.5\ \text{kJ}
\end{aligned}$$

EXERCISE 6.15A

Ethylene, derived from petroleum, is used to make ethanol for use as a fuel or solvent. The reaction is

$$C_2H_4(g) + H_2O(l) \longrightarrow CH_3CH_2OH(l)$$

Use data from Table 6.2 to calculate $\Delta H°$ for this reaction.

EXERCISE 6.15B

Use data from Table 6.2 to calculate $\Delta H°$ for the combustion of butane gas, C_4H_{10}, to produce gaseous carbon dioxide and liquid water.

APPLICATION NOTE

A chemical equation simply notes the initial and final conditions, but often there are rather specific requirements of temperature, pressure, and the use of catalysts for bringing about a chemical reaction. This reaction certainly will not occur just by passing methane and carbon dioxide gases into water.

EXAMPLE 6.16

The combustion of isopropyl alcohol, common rubbing alcohol, is represented by this equation.

$$2\,(CH_3)_2CHOH(l) + 9\,O_2(g) \longrightarrow 6\,CO_2(g) + 8\,H_2O(l) \qquad \Delta H° = -4011\text{ kJ}$$

Use this equation and data from Table 6.2 to establish the standard enthalpy of formation for isopropyl alcohol.

SOLUTION

We can proceed as in Example 6.15, but here we must solve for an unknown standard enthalpy of formation, ΔH_f°.

$$\begin{array}{lcccccc} & 2\,(CH_3)_2CHOH(l) & + & 9\,O_2(g) & \longrightarrow & 6\,CO_2(g) & + & 8\,H_2O(l) \\ \Delta H_f^\circ\text{, kJ/mol:} & \Delta H_f^\circ & & 0 & & -393.5 & & -285.8 \end{array}$$

$$\Delta H° = \Sigma \nu_p \times \Delta H_f^\circ(\text{products}) - \Sigma \nu_r \times \Delta H_f^\circ(\text{reactants})$$

$$-4011\text{ kJ} = \left[6\text{ mol CO}_2 \times \frac{-393.5\text{ kJ}}{1\text{ mol CO}_2} + 8\text{ mol H}_2\text{O} \times \frac{-285.8\text{ kJ}}{1\text{ mol H}_2\text{O}}\right]$$

$$- \left[2\text{ mol (CH}_3)_2\text{CHOH} \times \Delta H_f^\circ + 9\text{ mol O}_2 \times \frac{0\text{ kJ}}{1\text{ mol O}_2}\right]$$

$$= [-2361\text{ kJ} - 2286\text{ kJ}] - 2\text{ mol (CH}_3)_2\text{CHOH} \times \Delta H_f^\circ$$

$$2\text{ mol (CH}_3)_2\text{CHOH} \times \Delta H_f^\circ = 4011\text{ kJ} - 2361\text{ kJ} - 2286\text{ kJ} = -636\text{ kJ}$$

$$\Delta H_f^\circ = \frac{-636\text{ kJ}}{2\text{ mol (CH}_3)_2\text{CHOH(l)}}$$

$$= -318\text{ kJ/mol (CH}_3)_2\text{CHOH(l)}$$

EXERCISE 6.16A

Tetrachloroethylene, a degreasing solvent, is produced by the reaction

$$C_2H_4(g) + 4\,HCl(g) + 2\,O_2(g) \longrightarrow C_2Cl_4(l) + 4\,H_2O(l) \quad \Delta H° = -878.5\text{ kJ}$$

Use data from Table 6.2 to establish ΔH_f° for $C_2Cl_4(l)$.

EXERCISE 6.16B

The combustion of thiophene, $C_4H_4S(l)$, a compound used in the manufacture of pharmaceuticals, produces carbon dioxide and sulfur dioxide gases and liquid water. The enthalpy change in the combustion of one mole of $C_4H_4S(l)$ is found to be −2523 kJ. Use this information and data from Table 6.2 to establish ΔH_f° for $C_4H_4S(l)$.

EXAMPLE 6.17—A Conceptual Example

Without performing a calculation, determine which of these two substances should yield the greater quantity of heat upon complete combustion, on a *per mole* basis: ethane, $C_2H_6(g)$, or ethanol, $CH_3CH_2OH(l)$.

SOLUTION

Each combustion has precisely the same final state: 2 mol $CO_2(g)$ + 3 mol $H_2O(l)$.

$$C_2H_6(g) + 7/2\,O_2(g) \longrightarrow 2\,CO_2(g) + 3\,H_2O(l)$$
$$CH_3CH_2OH(l) + 3\,O_2(g) \longrightarrow 2\,CO_2(g) + 3\,H_2O(l)$$

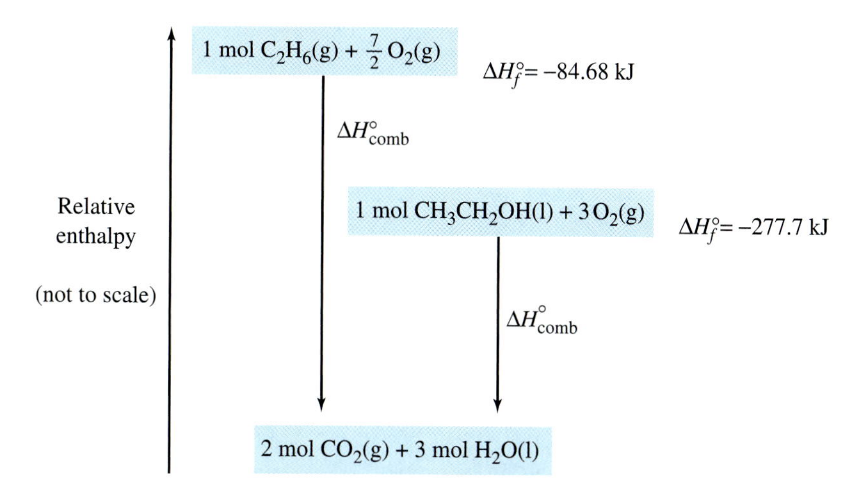

Figure 6.20 Example 6.17 illustrated

The initial state is one mole of the compound (ethane or ethanol) and enough oxygen to complete the combustion. Because $\Delta H_f^\circ[O_2(g)] = 0$, the enthalpy of the initial state is simply the standard enthalpy of formation of one mole of ethane in one case and one mole of ethanol in the other. The combustion reaction that releases more heat is the one with the more negative ΔH°_{comb}. And this will be the reaction that has in its initial state the compound with the higher standard enthalpy of formation. The standard enthalpies of formation are -84.68 kJ/mol C_2H_6 and -277.7 kJ/mol CH_3CH_2OH. Ethane has the higher (that is, less negative) stnadard enthalpy of formation, and its combustion will liberate more heat than the combustion of ethanol. The sketch in Figure 6.20 should help you to visualize this conclusion.

EXERCISE 6.17

Without doing a detailed calculation, determine which alcohol gives off the most heat upon combustion, on a per mole basis: $CH_3OH(l)$ or $C_2H_5OH(l)$? Use data from Table 6.2.

Ionic Reactions in Solutions

In Chapter 4, we saw the importance of reactions involving ions, especially in aqueous solutions. How can we apply thermochemical concepts to such reactions? Consider the reaction between HCl(aq) and NaOH(aq) in Example 6.11. It is actually an ionic reaction, expressed here through a net ionic equation and a somewhat more accurate ΔH° value.

$$H^+(aq) + OH^-(aq) \longrightarrow H_2O(l) \qquad \Delta H^\circ = -55.8 \text{ kJ}$$

There is a slight difficulty in extending the concept of enthalpy of formation to ions: There is no reaction in aqueous solution by which we can form cations alone or anions alone; we always obtain both. We resolve the difficulty by arbitrarily assigning $H^+(aq)$ an enthalpy of formation of *zero*. Example 6.18 illustrates how we can use this assignment to establish ΔH_f° of $OH^-(aq)$. By similar calculations, we can tabulate data such as those in Table 6.3.

TABLE 6.3 Some Standard Enthalpies of Formation of Ions in Aqueous Solution at 25 °C

Ion	ΔH_f°, kJ/mol
H^+	0
Na^+	−240.1
K^+	−252.4
NH_4^+	−132.5
Ag^+	+105.6
Mg^{2+}	−466.9
Ca^{2+}	−542.8
Ba^{2+}	−537.6
OH^-	−230.0
Cl^-	−167.2
NO_3^-	−205.0
CO_3^{2-}	−677.1
SO_4^{2-}	−909.3

EXAMPLE 6.18

Use the net ionic equation given above, together with $\Delta H_f^\circ = 0$ for $H^+(aq)$, to obtain ΔH_f° for $OH^-(aq)$.

SOLUTION

By proceeding exactly as we did in Example 6.16, we can write

$$\mathrm{H^+(aq) + OH^-(aq) \longrightarrow H_2O(l)} \qquad \Delta H^\circ = -55.8\ \mathrm{kJ}$$

$$\Delta H^\circ_f,\ \mathrm{kJ/mol:} \qquad 0 \qquad \Delta H^\circ_f \qquad -285.8$$

$$-55.8\ \mathrm{kJ} = 1\ \cancel{\mathrm{mol\ H_2O}} \times (-285.8\ \mathrm{kJ}/\cancel{\mathrm{mol\ H_2O}})$$
$$- [1\ \cancel{\mathrm{mol\ H^+}} \times 0\ \mathrm{kJ}/\cancel{\mathrm{mol\ H^+}} + 1\ \mathrm{mol\ OH^-} \times \Delta H^\circ_f]$$

$$-55.8\ \mathrm{kJ} = -285.8\ \mathrm{kJ} - 1\ \mathrm{mol\ OH^-} \times \Delta H^\circ_f$$

$$\Delta H^\circ_f = \frac{-285.8\ \mathrm{kJ} + 55.8\ \mathrm{kJ}}{1\ \mathrm{mol\ OH^-}} = -230.0\ \mathrm{kJ/mol\ OH^-}$$

EXERCISE 6.18A

Use data from Table 6.3 and the following data about the precipitation of $BaSO_4(s)$ to determine $\Delta H^\circ_f[BaSO_4(s)]$.

$$\mathrm{Ba^{2+}(aq) + SO_4^{2-}(aq) \longrightarrow BaSO_4(s)} \qquad \Delta H^\circ = -26\ \mathrm{kJ}$$

EXERCISE 6.18B

Given that $\Delta H^\circ_f[Mg(OH)_2(s)] = -924.5$ kJ/mol, what is the standard enthalpy change, ΔH°, for the reaction of aqueous solutions of magnesium chloride and potassium hydroxide?

Looking Ahead

A reaction that occurs simply by bringing the reactants together under the appropriate conditions is said to be *spontaneous*. Most exothermic reactions are spontaneous. In exothermic reactions, enthalpy decreases. The tendency for the enthalpy of a system to be lowered is a powerful driving force, much like the decrease in the potential energy of water as it flows downhill. However, there is more to the matter. Some endothermic processes, like the dissolving of many ionic solids in water, also occur spontaneously, even though they involve an enthalpy increase. We need to consider a second factor, called *entropy*, before we can fully answer the question of why some processes occur spontaneously and others do not. We will do this in Chapter 17.

In describing spontaneous change, we will also have to distinguish between the *tendency* of a reaction to occur and the *rate* at which a reaction occurs. Some reactions that we expect to occur from thermodynamic considerations take place too slowly to be observed. We will discuss rates of chemical reactions in Chapter 13.

6.8 Combustion and Respiration: Fuels and Foods

Two types of exothermic reactions are of utmost importance: (1) The combustion of fuels provides the energy that sustains our modern way of life. (2) The foods we eat are "burned" in the respiratory process to give us the energy to stay alive and perform our many activities.

Fossil Fuels: Coal, Natural Gas, and Petroleum

A *fuel* is a substance that burns with the release of heat. In much of the world today, wood is the principal fuel for cooking food and keeping warm, just as it was in ancient times. The modern industrial world, however, is powered mainly by the fossil fuels coal, petroleum, and natural gas. These materials were formed over a period of millions of years from organic matter that became buried and compressed under mud and water. It is inevitable that Earth's fossil fuel reserves will be exhausted some day because we are using them at a rate 50,000 times as fast as they were formed. Renewable energy sources, such as solar energy and wind power, are likely to become increasingly important in the future.

Coal is a complex organic material with C as the primary element and also significant quantities of H, O, N, and S. Several varieties of coal are listed in Table 6.4. The fuel value of coal lies mainly in its carbon content. Complete combustion of carbon produces carbon dioxide.

$$C(s) + O_2(g) \longrightarrow CO_2(g)$$

High-grade coal yields about 30 kJ per gram. The burning of coal also produces carbon monoxide and soot (unburned carbon) from incomplete combustion. Noncombustible matter is left unburned as ashes and enters the atmosphere as fly ash. Sulfur oxides are formed from sulfur compounds in the coal, and nitrogen oxides are produced from air by the reaction of nitrogen and oxygen at the high temperatures at which coal burns.

TABLE 6.4 Mass Percent Carbon in Some Typical Coal Samples

Grade of Coal	Mass Percent C
Lignite	41.2
Subbituminous	53.9
Bituminous	76.7
Semianthracite	78.3
Anthracite	79.8

Natural gas is mainly methane (CH_4), with some ethane (C_2H_6), propane (C_3H_8), and butane (C_4H_{10}). Complete combustion of methane gives carbon dioxide and water.

$$CH_4(g) + 2\,O_2(g) \longrightarrow CO_2(g) + 2\,H_2O(l)$$

The enthalpy change in the combustion of methane (the heat of combustion) is −890.3 kJ/mol, or −55.49 kJ/g. Natural gas has a heat of combustion on a mass basis that is almost twice as great as that of coal. With far fewer impurities than coal, natural gas also burns more cleanly. Some carbon monoxide, soot, and nitrogen oxides are formed, but there are only traces of sulfur oxides and no ash.

Petroleum is an exceedingly complex mixture of hydrocarbons. Petroleum products such as gasoline burn more cleanly than coal, although not as cleanly as natural gas. Like petroleum itself, gasoline is a mixture of hydrocarbons with molecules having five to 12 carbon atoms (see Figure 6.21). Octane, C_8H_{18}, is often used as a representative gasoline molecule. Complete combustion of octane yields carbon dioxide and water.

$$C_8H_{18}(l) + 25/2\,O_2(g) \longrightarrow 8\,CO_2(g) + 9\,H_2O(l) \qquad \Delta H^\circ = -5450 \text{ kJ}$$

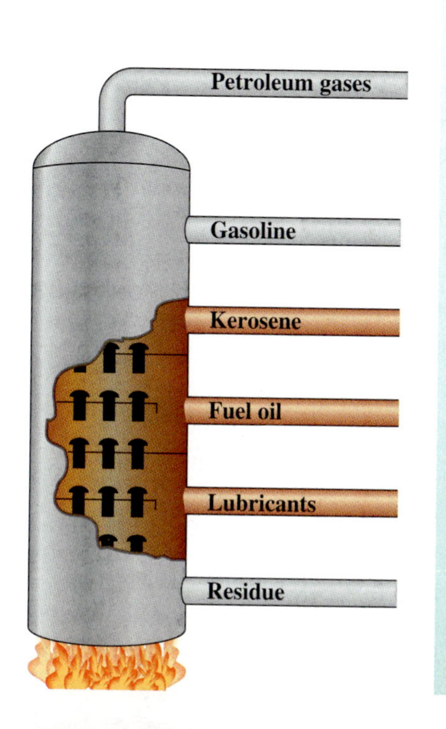

Shorter carbon chains	Lower boiling points
C_1–C_4	< 40 °C
C_5–C_{12}	40–200 °C
C_{12}–C_{16}	175–275 °C
C_{15}–C_{18}	250–400 °C
C_{17}–up	> 300 °C
C_{20}–up	> 350 °C
Longer carbon chains	**Higher boiling points**

Figure 6.21 Petroleum distillation and gasoline production
Crude oil is vaporized, and the fractional distillation column separates the components of the vapor according to their boiling temperatures. The lower-boiling components come off at the top of the column, and higher-boiling components come off lower in the column. A residue that does not boil collects at the bottom. The residue may be paraffin or asphalt, depending on the source of the crude oil.

Alkanes: Our Principal Fuels

The modern industrial world runs mainly on alkanes. Petroleum is a complex mixture of hydrocarbons, mostly alkanes. But in the form in which it comes from the ground—a thick, sticky, smelly liquid made up of hundreds of compounds—it is of limited use. To better suit our needs, petroleum must be refined.

In a key step in its refining, petroleum is boiled in a device called a distillation column (Figure 6.21) and the petroleum vapor is separated into portions, called *fractions*, having different boiling-point ranges. The molecules of lowest mass, those with one to four carbon atoms, are found in the fraction that comes off at the top of the column. The gasoline fraction comes off just below this, and it contains mainly molecules having five to twelve carbon atoms. The fractions in still lower regions of the column contain molecules with still longer carbon-atom chains; the residue obtained from the bottom of the column contains molecules with twenty or more carbon atoms.

Natural gas is about 80% methane, 10% ethane, and 10% a mixture of higher alkanes. Natural gas is used as a cooking fuel—and in Bunsen burners. It is the cleanest of the fossil fuels because it contains the smallest quantity of sulfur compounds. Propane and the butanes are familiar fuels. Although they are gases at ordinary temperatures and pressures, they liquefy under high pressure. In liquid form, they are known as liquefied petroleum gas (LPG).

Gasoline, like the petroleum from which it is derived, is a mixture of hydrocarbons. Alkanes typically found in gasoline have formulas ranging from C_5H_{12} to $C_{12}H_{26}$. Small amounts of other kinds of hydrocarbons are also present, and there are even some sulfur- and nitrogen-containing compounds. Some gasoline components burn unevenly, creating an engine "knock." Others burn more smoothly. Isooctane is an excellent engine fuel and is assigned an *octane number* of 100. Heptane is an exceptionally poor fuel with an octane number of 0.

$$\begin{array}{ccccccccc} & & CH_3 & & & & CH_3 & & \\ & & | & & & & | & & \\ CH_3 & - & C & - & CH_2 & - & C & - & CH_3 \\ & & | & & & & | & & \\ & & CH_3 & & & & H & & \end{array}$$

Isooctane
(2,2,4-trimethylpentane)
Octane number: 100

$$CH_3CH_2CH_2CH_2CH_2CH_2CH_3$$

Heptane
Octane number: 0

Gasolines are rated against these two standards. For example, a gasoline with an octane number of 87 gives the same engine performance as a mixture of 87% isooctane and 13% heptane. Methanol and ethanol raise the octane number of gasoline, as do other oxygen-containing (oxygenated) hydrocarbons such as methyl *tert*-butyl ether (MTBE).

Because gasoline is generally the petroleum fraction in greatest demand, there is often an excess of the fractions with higher boiling points. These can be converted to gasoline by heating them in the absence of air. This process, called *cracking*, breaks the big molecules apart. Cracking converts higher alkanes into those in the gasoline range, but it also affords a variety of useful by-product chemicals from which chemists can synthesize a remarkable array of substances: plastics, painkillers, antibiotics, stimulants, depressants, and detergents, to name just a few. Any future shortages of petroleum could mean a great deal more than just scarce, high-priced gasoline.

Foods: Fuels for the Body

Our bodies require food for the growth and repair of tissue and as a source of energy. The three principal classes of foods are carbohydrates, fats, and proteins. The body's preferred fuel is carbohydrates (starches and sugars). During digestion, starches and sugars are converted to a simple sugar: glucose, $C_6H_{12}O_6$.

If glucose is burned in a calorimeter, its heat of combustion at 25 °C is found to be −2803 kJ/mol, or −15.56 kJ/g.

$$C_6H_{12}O_6(s) + 6\,O_2(g) \longrightarrow 6\,CO_2(g) + 6\,H_2O(l) \qquad \Delta H° = -2803\text{ kJ}$$

When it fuels our bodies, glucose produces the same products and yields about the same quantity of energy. It isn't exactly the same because the reactions occur at body temperature, about 37 °C, rather than 25 °C. Rather than being burned rapidly, the glucose is metabolized through a sequence of many reaction steps. Still, Hess's law applies. The overall enthalpy change is the same: 2803 kJ are released with the complete oxidation of each mole of glucose.

Fats are formed by the reaction of the three-carbon alcohol, glycerol, with a long-chain carboxylic acid (a *fatty acid*). Fats are *esters*. A typical fat, glyceryl trilaurate, has the molecular formula $C_{39}H_{74}O_6$ and the structure shown in Figure 6.22. Complete combustion of glyceryl trilaurate gives carbon dioxide and water.

$$C_{39}H_{74}O_6(s) + 54.5\,O_2(g) \longrightarrow 39\,CO_2(l) + 37\,H_2O(l) \qquad \Delta H° = -2.39 \times 10^4\text{ kJ}$$

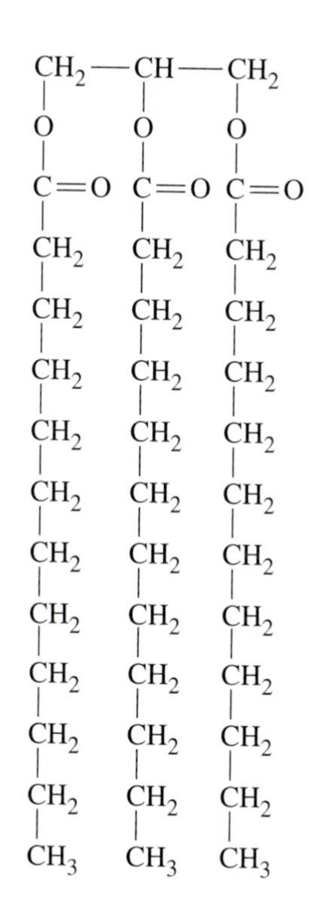

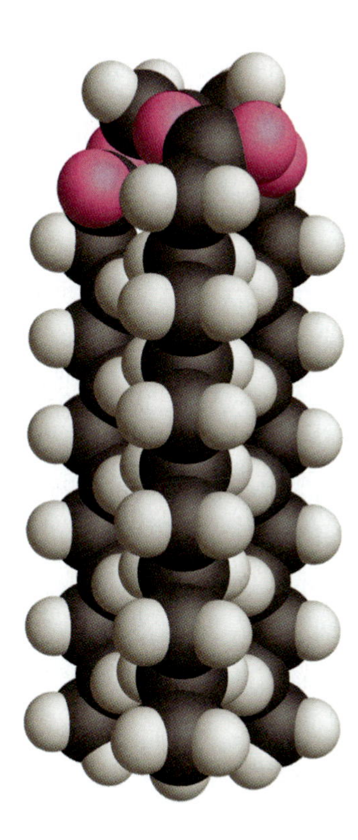

Figure 6.22
Glyceryl trilaurate, a typical fat
Glycerol provides the three-carbon backbone (at the top), and lauric acid molecules contribute the three tails.

The Food Calorie

In nutrition, the energy value of foods is measured in kilocalories.

$$1 \text{ kcal} = 1000 \text{ cal} = 4.184 \text{ kJ}$$

The energy value of glucose is 3.72 kcal/g, and that of glyceryl trilaurate is 8.94 kcal/g. In everyday life, people refer to the kilocalorie as the Calorie.

$$1 \text{ kcal} = 1 \text{ Cal} = 1000 \text{ cal}$$

The capital C is sometimes used to distinguish the food Calorie (a kilocalorie) from the calorie used in science. The term "large calorie" has also been used to designate the food calorie. People are often not careful to make a clear distinction between the two different meanings and we sometimes need to decide what is intended by the context in which the word is used.

In general, the energy value of carbohydrates and proteins is about 4.0 kcal/g, and that of fats is more than twice as high, about 9.0 kcal/g.

A nutrition label from a box of pasta. Can you verify, per serving, the total calories and the calories from fat shown on the label? Use other information in the label and in the text.

The heat of combustion is -2.39×10^4 kJ/mol $C_{39}H_{74}O_6$, or -37.4 kJ/g $C_{39}H_{74}O_6$. Note that, per gram, this heat of combustion is more than twice that of glucose. This is true in general: fats yield over twice as much energy per gram as carbohydrates. Fats are the body's preferred material for energy storage. More energy can be packed into a given mass of fat than of carbohydrate. The body can store at most about 500 g carbohydrate as glycogen, a form of starch. The ability to store fat appears to be almost without limit.

Proteins are found mainly in structural materials such as muscle, connective tissue, skin, and hair. Also, enzymes that regulate the many chemical reactions that occur in living cells are proteins. Proteins eaten in excess of the body's need for growth, repair, and replacement can be used for energy. Proteins have about the same energy value as carbohydrates.

Summary

Thermochemistry concerns energy changes in physical processes or chemical reactions. Among its basic ideas are the notion of a system and its surroundings; the concepts of kinetic energy, potential energy, and internal energy; and the distinction between two types of energy exchanges, heat (q) and work (w).

Internal energy (U) is a function of state; it has a unique value once the state or condition of a system is defined. The heat associated with a reaction is not a unique quantity. If a reaction is carried out at constant volume, the heat of reaction (q_V) is equal to the change in internal energy (ΔU). At a constant pressure and with work limited to pressure-volume work, the heat of reaction (q_P) is equal to the change in enthalpy. Enthalpy (H) is a function based on internal energy but with a modification that makes it especially useful for constant-pressure processes ($H = U + PV$). Like internal energy, enthalpy is a function of state. The first law of thermodynamics relates the heat and work exchanged between a system and its surroundings to changes in the internal energy of a system. This law is the basis for many of the calculations of thermochemistry.

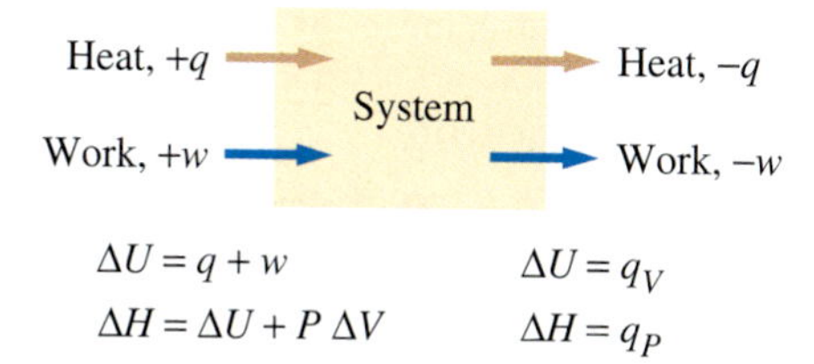

Enthalpy changes (ΔH) can be written into chemical equations and incorporated into conversion factors that relate amounts of substances with quantities of heat released or absorbed in chemical reactions. In an *exothermic* reaction, enthalpy decreases and heat is given off to the surroundings. In an *endothermic* reaction, enthalpy decreases and heat is absorbed from the surroundings.

A calorimeter is used to measure quantities of heat. The actual measurements are of masses and temperature changes. Other data required are heat capacities and/or specific heats. Many reactions can be carried out under the constant pressure of the atmosphere in the Styrofoam® calorimeter. The results obtained are q_P or ΔH, values. Combustion reactions are generally carried out under constant-volume conditions in a bomb calorimeter. Here, the results are q_V or ΔU values; ΔH values can be calculated from ΔU values when necessary.

Because internal energy and enthalpy are state functions, the values of ΔU and ΔH do not depend on how a reaction is carried out (for example, in one step or several). This makes it possible to use Hess's law to evaluate ΔU or ΔH for one reaction from a knowledge of their values for other reactions.

The concepts of the standard state, a standard enthalpy change, ΔH°, and a standard enthalpy of formation (heat of formation), ΔH_f°, are important in thermochemical calculations. The standard enthalpy of formation of an element in its reference form has a value of zero. The standard enthalpy of formation of a compound is the standard enthalpy change of a reaction in which the compound is formed from the reference forms of its elements. Standard enthalpies of formation (usually listed at 25 °C) can be used to *calculate* standard enthalpies of reaction.

$$\Delta H^\circ = \sum \nu \times \Delta H_f^\circ(\text{products}) - \sum \nu \times \Delta H_f^\circ(\text{reactants})$$

Some practical applications of thermochemistry deal with the heats of combustion of fossil fuels—coal, natural gas, and petroleum—and the energy content of foods: carbohydrates, fats, and proteins.

Key Terms

calorie (cal) (6.5)
calorimeter (6.5)
endothermic reaction (6.4)
enthalpy (H) (6.4)
enthalpy change (ΔH) (6.4)
enthalpy diagram (6.4)
exothermic reaction (6.4)
first law of thermodynamics (6.3)
heat (q) (6.2)
heat of reaction (q_{rxn}) (6.4)
Hess's law (6.6)
internal energy (U) (6.2)
joule (J) (6.1)
kinetic energy (6.1)
law of conservation of energy (6.3)
molar heat capacity (6.5)
potential energy (6.1)
specific heat (6.5)
standard enthalpy of formation (ΔH_f°) (6.7)
standard enthalpy of reaction (ΔH°) (6.7)
standard state (6.7)
state (6.3)
state function (function of state) (6.3)
surroundings (6.2)
system (6.2)
thermochemistry (6.2)
work (6.1, 6.2)

Review Questions

1. What is a thermodynamic system? What are the surroundings?
2. Describe the three types of thermodynamic systems and give an example of each.
3. An aqueous solution of copper(II) sulfate is a thermodynamic system under study. What type of system is this if the solution is contained in an open beaker on a laboratory table?
4. Describe the transfer of heat between a system and its surroundings. What is meant by thermal equilibrium?
5. Define the term exothermic reaction. Give an example, and indicate the sign of ΔH.
6. Define the term endothermic reaction. Give an example, and indicate the sign of ΔH.
7. Why must we indicate the physical state of all reactants and products when designating the enthalpy change for a reaction?
8. How is the enthalpy of a system related to its internal energy?

9. Under what conditions is the heat of reaction equal to the change in internal energy in a reaction?

10. Under what conditions is the heat of reaction equal to the change in enthalpy in a reaction?

11. What is calorimetry? How does a calorimeter work?

12. Define heat capacity and specific heat, and give some typical units used to express these quantities.

13. Describe how the specific heat of a substance can be determined.

14. Describe how enthalpy changes for chemical reactions are determined in a Styrofoam® cup calorimeter.

15. Describe how a bomb calorimeter works. What kind of reactions are carried out in bomb calorimeters?

16. State Hess's law of constant heat summation, and explain its significance.

17. What do we mean when we say a substance is in its standard state? when we say a substance is in its reference form?

18. What is meant by the standard molar enthalpy of formation of a substance? the standard enthalpy change for a reaction?

19. Write the equation for a reaction that has as its heat of reaction the standard enthalpy of formation of $Fe_2O_3(s)$.

20. Describe how a standard enthalpy of reaction can be calculated from tabulated standard enthalpies of formation. Describe how an unknown standard enthalpy of formation can be determined from a measured standard enthalpy of reaction and tabulated standard enthalpies of formation.

21. Explain why it is necessary to assign standard enthalpies of formation of *zero* both to $H_2(g)$ and to $H^+(aq)$. Is the same true for $Cl_2(g)$ and $Cl^-(aq)$? Explain.

22. What is a fuel? List the three principal fossil fuels and some advantages of gaseous and liquid fuels over solid fuels.

23. How is the food Calorie related to the calorie used in science?

24. List the three major classes of foods. Which type has the highest energy value per gram?

Problems

First Law of Thermodynamics

25. What is the change in internal energy of a system that absorbs 455 J of heat and does 325 J of work?

26. What is the change in internal energy of a system that has 625 J of work done on it and gives off 515 J of heat?

27. How much work, in joules, must be involved in a process in which a system gives off 58 cal of heat if the internal energy of the system is to remain unchanged?

28. If ΔU for a system is 217 J in a process in which the system absorbs 185 cal of heat, how much work, in joules, must have been involved?

29. Can a system do work and absorb heat at the same time? Can it do so while maintaining a constant internal energy?

30. Can a system do work and give off heat at the same time? Can it do so while maintaining a constant internal energy?

Enthalpy and Enthalpy Changes

31. At high temperatures, water is decomposed to hydrogen and oxygen.

$$H_2O(g) \longrightarrow H_2(g) + 1/2\ O_2(g)$$

Decomposition of 10.0 g H_2O at constant pressure requires that 134 kJ of heat be absorbed by the system. Is the reaction endothermic or exothermic? What is the value of q for the reaction, per mole of water? Is the value of q equal to ΔU or ΔH? Explain.

32. Combustion of 0.144 g of sucrose (table sugar, $C_{12}H_{22}O_{11}$) in the open air results in the release of 2.38 kJ of heat.

$$C_{12}H_{22}O_{11}(s) + 12\ O_2(g) \longrightarrow 12\ CO_2(g) + 11\ H_2O(l)$$

Is the reaction endothermic or exothermic? What is the value of q for the reaction, per mole of sucrose? Is this a value of ΔU or ΔH? How would you expect the values of ΔU and ΔH to compare for this reaction? Explain.

33. When 0.0500 mol of solid calcium carbonate is heated in air, it decomposes to solid calcium oxide and carbon dioxide gas; 8.90 kJ of heat is absorbed. Write a chemical equation for the decomposition of 1 mol of calcium carbonate, including the physical states of all the substances and the value of ΔH.

34. When 0.200 mol of ethane gas, C_2H_6, is burned in excess oxygen, 312 kJ of heat is evolved. Write a chemical equation for the combustion of 1 mol of ethane, including the physical states of all the substances and the value of ΔH. (*Hint*: What are the products of the reaction, and in what forms do they appear at 25 °C and 1 atm pressure?)

35. Given the reaction

$$3\ Fe_2O_3(s) + CO(g) \longrightarrow 2\ Fe_3O_4(s) + CO_2(g) \qquad \Delta H = -46\text{ kJ}$$

determine ΔH for the following reactions.

(a) $Fe_2O_3(s) + 1/3\ CO(g) \longrightarrow 2/3\ Fe_3O_4(s) + 1/3\ CO_2(g)$

(b) $Fe_3O_4(s) + 1/2\ CO_2(g) \longrightarrow 3/2\ Fe_2O_3(s) + 1/2\ CO(g)$

36. Given the reaction

$$2\ Na_2O_2(s) + 2\ H_2O(l) \longrightarrow 4\ NaOH(s) + O_2(g) \qquad \Delta H = -109\ kJ$$

determine ΔH for the following reactions.

(a) $Na_2O_2(s) + H_2O(l) \longrightarrow 2\ NaOH(s) + 1/2\ O_2(g)$

(b) $NaOH(s) + 1/4\ O_2(g) \longrightarrow 1/2\ Na_2O_2(s) + 1/2\ H_2O(l)$

37. The reaction of solid white phosphorus (P_4) with chlorine gas produces phosphorus trichloride as the sole product. The reaction of 1.00 g P_4 with an excess of $Cl_2(g)$ gives off 9.27 kJ of heat. Write the equation representing the formation of one mole of phosphorus trichloride by this reaction and the accompanying ΔH value.

38. A handbook gives the heat of combustion of liquid carbon disulfide as −3.24 kcal per gram of the liquid. Write the equation representing the combustion of one mole of carbon disulfide and the accompanying ΔH value. (*Hint*: The sulfur-containing combustion product is sulfur dioxide.)

Heats of Reaction and Reaction Stoichiometry

39. Calcium oxide (lime) reacts with water to form calcium hydroxide (slaked lime).

$$CaO(s) + H_2O(l) \longrightarrow Ca(OH)_2(s) \qquad \Delta H = -65.2\ kJ$$

How many kilojoules of heat are evolved in the reaction of 0.500 kg CaO(s) with an excess of water?

40. Calcium carbide (CaC_2) can be made by heating calcium oxide (lime) with carbon (charcoal).

$$CaO(s) + 3\ C(s) \xrightarrow{\Delta} CaC_2(s) + CO(g) \qquad \Delta H = +464.8\ kJ$$

How many kilojoules of heat are absorbed in a reaction in which 76.5 g C(s) is consumed?

41. The reaction of sodium peroxide and water is a source of $O_2(g)$.

$$2\ Na_2O_2(s) + 2\ H_2O(l) \longrightarrow 4\ NaOH(aq) + O_2(g) \qquad \Delta H = -287\ kJ$$

How many kilojoules of heat are evolved in the reaction of 150.0 g Na_2O_2 with 50.0 mL H_2O?

42. Calcium carbide (CaC_2) reacts with water to form acetylene (C_2H_2), a gas used as a fuel in welding.

$$CaC_2(s) + 2\ H_2O(l) \longrightarrow C_2H_2(g) + Ca(OH)_2(s) \qquad \Delta H = -128.0\ kJ$$

How many kilojoules of heat are evolved in the reaction of 3.50 kg CaC_2 with 1.25 L H_2O?

43. How many liters of ethane, measured at 23 °C and 751 Torr, are required to give off 1.00×10^6 kJ of heat on burning?

$$2\ C_2H_6(g) + 7\ O_2(g) \longrightarrow 4\ CO_2(g) + 6\ H_2O(l) \qquad \Delta H = -3.12 \times 10^3\ kJ$$

44. How many liters of $CO_2(g)$, measured at 26 °C and 764 Torr, are produced when 1.50×10^4 kJ of heat is evolved in the burning of butane?

$$2\ C_4H_{10}(g) + 13\ O_2(g) \longrightarrow 8\ CO_2(g) + 10\ H_2O(l) \qquad \Delta H = -5.76 \times 10^3\ kJ$$

45. How many grams of CaO(s) must react with an excess of water to liberate the same quantity of heat as does the combustion of 1.00 L of $CH_4(g)$, measured at 21.5 °C and 772 Torr?

$$CaO(s) + H_2O(l) \longrightarrow Ca(OH)_2(s) \qquad \Delta H = -65.2\ kJ$$

$$CH_4(g) + 2\ O_2(g) \longrightarrow CO_2(g) + 2\ H_2O(l) \qquad \Delta H = -890.3\ kJ$$

46. A reaction mixture of $N_2O_4(g)$ and $NO_2(g)$ absorbs the heat given off in the combustion of 0.100 L $CH_4(g)$, measured at 24.7 °C and 762 Torr. How many moles of N_2O_4 can be converted to NO_2 as a result?

$$CH_4(g) + 2\ O_2(g) \longrightarrow CO_2(g) + 2\ H_2O(l) \qquad \Delta H = -890.3\ kJ$$

$$N_2O_4(g) \longrightarrow 2\ NO_2(g) \qquad \Delta H = +57.20\ kJ$$

47. *Without doing detailed calculations*, determine whether ethane or butane gives off the greater quantity of heat on combustion, when the comparison is made on an equal mass basis. The combustion reactions and their enthalpy changes are given in Problems 43 and 44.

48. *Without doing detailed calculations*, determine whether ethane or butane produces the *lesser* mass of $CO_2(g)$ on combustion when the comparison is made on the basis of an equal quantity of heat liberated. The combustion reactions and their enthalpy changes are given in Problems 43 and 44.

Calorimetry: Heat Capacity and Specific Heat

(Use data from Table 6.1, as necessary.)

49. Calculate the heat capacity of a piece of iron if a temperature rise from 18 °C to 45 °C requires 112 J of heat.

50. Calculate the heat capacity of a sample of radiator coolant if a temperature rise from −5 °C to 142 °C requires 932 J of heat.

51. How much heat, in kilojoules, is required to raise the temperature of **(a)** 20.0 g water from 20.0 °C to 96.0 °C; and **(b)** 120.0 g ethanol from −10.5 °C to 44.5 °C?

52. How much heat, in kilojoules, is released when the temperature of **(a)** 47.0 g water drops from 45.4 °C to 10.0 °C; and **(b)** 209 g iron drops from 400.0 °C to 22.6 °C?

53. A 48.7-g block of lead initially at 27.0°C absorbs 93.5 J of heat. What is the final temperature of the lead?

54. A 454-g iron block initially at 16 °C absorbs 63.9 kJ of heat. What is the final temperature of the iron?

55. A 10.25-g sample of a metal alloy is heated to 99.10 °C and is then dropped into 20.0 g water in a calorimeter. The water temperature rises from 18.51 °C to 22.03 °C. Calculate the specific heat of the alloy.

56. A 2.05-g sample of a metal alloy is heated to 98.88 °C. It is then dropped into 28.0 g water in a calorimeter. The water temperature rises from 19.73 °C to 21.23 °C. Calculate the specific heat of the alloy.

57. A 1.35-kg piece of iron (sp. heat = 0.449 J g^{-1} $°C^{-1}$) is dropped into 0.817 kg of water, and the water temperature rises from 23.3 °C to 39.6 °C. What must have been the initial temperature of the iron?

58. A piece of stainless steel (sp. heat = 0.50 J g^{-1} $°C^{-1}$) is taken from an oven at 178 °C and immersed in 225 mL of water at 25.9 °C. The water temperature rises to 42.4 °C. What is the mass of the piece of steel? How precise is this method of mass determination? Explain.

59. *Without doing detailed calculations*, determine which of the following metal samples must absorb the most heat if it is to be brought from room temperature to the boiling point of water: 2.25 g copper, 4.50 g lead, or 1.87 g silver.

60. *Without doing detailed calculations*, determine which of the following metal samples will be raised to the highest temperature when a 100.0-g sample at 22 °C absorbs 1.00 kJ of heat: aluminum, iron, or silver.

Calorimetry: Measuring Heats of Reaction

61. A 500.0-mL sample of 0.500 M NaOH at 20.00 °C is mixed with an equal volume of 0.500 M HCl at the same temperature in a Styrofoam® cup calorimeter. The reaction

$$HCl(aq) + NaOH(aq) \longrightarrow NaCl(aq) + H_2O(l) \quad \Delta H = ?$$

takes place, and the temperature rises to 23.21 °C. Calculate ΔH for the reaction. (You may make the same assumptions as in Example 6.11.)

62. A 65.0-mL sample of 0.600 M HI at 18.46 °C is mixed with 84.0 mL of a solution containing excess potassium hydroxide, at 18.46 °C in a Styrofoam® cup calorimeter. The reaction

$$HI(aq) + KOH(aq) \longrightarrow KI(aq) + H_2O(l) \quad \Delta H = ?$$

takes place, and the temperature rises to 21.96 °C. Calculate ΔH for the reaction. (You may make the same assumptions as in Example 6.11.)

63. A 1.50-g sample of $NH_4NO_3(s)$ is added to 35.0 g of water in a foam cup and stirred until it dissolves. The temperature of the solution drops from 22.7 to 19.4 °C. What is the heat of solution of NH_4NO_3, expressed in kJ/mol NH_4NO_3, that is, what is ΔH for the process

$$NH_4NO_3(s) \xrightarrow{H_2O} NH_4NO_3(aq) \quad \Delta H = ?$$

64. What is the maximum temperature that can be reached in a Styrofoam® cup containing 105 mL of water at 20.12 °C following the addition and dissolving of a small pellet of KOH(s) weighing 0.215 g?

$$KOH(s) \xrightarrow{H_2O} KOH(aq) \quad \Delta H = -57.6 \text{ kJ}$$

65. A 0.309-g sample of coal is burned in a bomb calorimeter with a heat capacity of 4.62 kJ/°C. The temperature in the calorimeter rises from 20.45 °C to 22.28 °C. Calculate the heat of combustion of the coal, in kilojoules per gram.

66. A 0.196-g sample of gasoline is burned in a bomb calorimeter with a heat capacity of 5.01 kJ/°C. The temperature in the calorimeter rises from 22.75 °C to 24.50 °C. Calculate the heat of combustion of the gasoline, in kilojoules per gram of gasoline.

67. A pure substance of known heat of combustion can be used to determine the heat capacity of a bomb calorimeter. When 2.00 g sucrose is burned in a particular calorimeter, the temperature rises from 22.83 °C to 25.67

°C. What is the heat capacity of the calorimeter? The heat of combustion of sucrose is −16.5 kJ/g.

68. Benzoic acid (C_6H_5COOH) is sometimes used as a standard to determine the heat capacity of a bomb calorimeter. When 1.22 g C_6H_5COOH is burned in a calorimeter that is being calibrated, the temperature rises from 21.13 °C to 22.93 °C. What is the heat capacity of the calorimeter? The heat of combustion of benzoic acid is −26.42 kJ/g.

69. A 1.108-g sample of naphthalene, $C_{10}H_8(s)$, is burned in a bomb calorimeter assembly and a temperature increase of 5.92 °C is noted. When a 1.351-g sample of thymol, $C_{10}H_{14}O(s)$ (a preservative and mold and mildew inhibitor), is burned in the same calorimeter assembly, the temperature increase is 6.74 °C. If the heat of combustion of naphthalene is −5153.5 kJ/mol $C_{10}H_8$, what is the heat of combustion of thymol, in kJ/mol $C_{10}H_{14}O$?

70. A 1.148-g sample of benzoic acid is burned in an excess of oxygen in a bomb calorimeter. The temperature of the water rises from 24.96 °C to 30.25 °C. The heat of combustion of benzoic acid is −26.42 kJ/g. In a second experiment, a 0.895-g powdered coal sample is burned in the same calorimeter assembly. The temperature of the water rises from 24.98 °C to 29.73 °C. How many kilograms of this coal would have to be burned to liberate 1.00×10^9 kJ of heat?

Hess's Law of Constant Heat Summation

71. Use the following equations

$$N_2(g) + 2\,O_2(g) \longrightarrow N_2O_4(g) \qquad \Delta H = +9.2\text{ kJ}$$
$$N_2(g) + 2\,O_2(g) \longrightarrow 2\,NO_2(g) \qquad \Delta H = +33.2\text{ kJ}$$

to calculate the enthalpy change for the reaction

$$2\,NO_2(g) \longrightarrow N_2O_4(g) \qquad \Delta H = ?$$

72. Use the following equations

$$C(\text{graphite}) + O_2(g) \longrightarrow CO_2(g) \qquad \Delta H = -393.5\text{ kJ}$$
$$2\,CO(g) + O_2(g) \longrightarrow 2\,CO_2(g) \qquad \Delta H = -566.0\text{ kJ}$$

to calculate the enthalpy change for the reaction

$$2\,C(\text{graphite}) + O_2(g) \longrightarrow 2\,CO(g) \qquad \Delta H = ?$$

73. Use the following equations

$$C_3H_8(g) + 5\,O_2(g) \longrightarrow 3\,CO_2(g) + 4\,H_2O(l) \qquad \Delta H = -2219.9\text{ kJ}$$
$$CO(g) + 1/2\,O_2(g) \longrightarrow CO_2(g) \qquad \Delta H = -283.0\text{ kJ}$$

to calculate the enthalpy change for the reaction

$$C_3H_8(g) + 7/2\,O_2(g) \longrightarrow 3\,CO(g) + 4\,H_2O(l) \qquad \Delta H = ?$$

74. Use the following equations

$$N_2H_4(l) + O_2(g) \longrightarrow N_2(g) + 2\,H_2O(l) \qquad \Delta H = -622.2\text{ kJ}$$
$$H_2(g) + 1/2\,O_2(g) \longrightarrow H_2O(l) \qquad \Delta H = -285.8\text{ kJ}$$
$$H_2(g) + O_2(g) \longrightarrow H_2O_2(l) \qquad \Delta H = -187.8\text{ kJ}$$

to calculate the enthalpy change for the reaction

$$N_2H_4(l) + 2\,H_2O_2(l) \longrightarrow N_2(g) + 4\,H_2O(l) \qquad \Delta H = ?$$

75. Determine the enthalpy change for the oxidation of ammonia

$$4\,NH_3(g) + 5\,O_2(g) \longrightarrow 4\,NO(g) + 6\,H_2O(l) \qquad \Delta H = ?$$

from the following data:

$$N_2(g) + 3\,H_2(g) \longrightarrow 2\,NH_3(g) \qquad \Delta H = -92.22\text{ kJ}$$
$$N_2(g) + O_2(g) \longrightarrow 2\,NO(g) \qquad \Delta H = +180.5\text{ kJ}$$
$$2\,H_2(g) + O_2(g) \longrightarrow 2\,H_2O(l) \qquad \Delta H = -571.6\text{ kJ}$$

76. Calculate the enthalpy change for the reaction

$$BrCl(g) \longrightarrow Br(g) + Cl(g) \qquad \Delta H = ?$$

by using the following data:

$$Br_2(l) \longrightarrow Br_2(g) \qquad \Delta H = +30.91\text{ kJ}$$
$$Br_2(g) \longrightarrow 2\,Br(g) \qquad \Delta H = +192.9\text{ kJ}$$
$$Cl_2(g) \longrightarrow 2\,Cl(g) \qquad \Delta H = +243.4\text{ kJ}$$
$$Br_2(l) + Cl_2(g) \longrightarrow 2\,BrCl(g) \qquad \Delta H = +29.2\text{ kJ}$$

Standard Enthalpies of Formation and Standard Enthalpies of Reaction

77. Use standard enthalpies of formation from Appendix C to calculate the standard enthalpy change for each of the following reactions.

(a) $NH_3(g) + HCl(g) \longrightarrow NH_4Cl(s)$
(b) $NH_3(g) + HNO_3(l) \longrightarrow NH_4NO_3(s)$
(c) $MgCl_2(s) + Ca(s) \longrightarrow Mg(s) + CaCl_2(s)$
(d) $FeO(s) + CO(g) \longrightarrow Fe(s) + CO_2(g)$

78. Use standard enthalpies of formation from Appendix C to calculate the standard enthalpy change for each of the following reactions.

(a) $Cl_2(g) + I_2(s) \longrightarrow 2\ ICl(g)$
(b) $NO(g) + O_3(g) \longrightarrow NO_2(g) + O_2(g)$
(c) $Zn(s) + 2\ HCl(g) \longrightarrow ZnCl_2(s) + H_2(g)$
(d) $3\ C_2H_2(g) \longrightarrow C_6H_6(l)$
(e) $2\ C_2H_5OH(l) + O_2(g) \longrightarrow 2\ CH_3CHO(g) + 2\ H_2O(l)$

79. Use the following equation and data from Appendix C to calculate the enthalpy of formation, per mole, of ZnS(s).

$$2\ ZnS(s) + 3\ O_2(g) \longrightarrow 2\ ZnO(s) + 2\ SO_2(g) \qquad \Delta H° = -878.2\ kJ$$

80. Use the following equation and data from Appendix C to calculate the enthalpy of formation, per mole, of sucrose, $C_{12}H_{22}O_{11}(s)$.

$$C_{12}H_{22}O_{11}(s) + 12\ O_2(g) \longrightarrow 12\ CO_2(g) + 11\ H_2O(l) \qquad \Delta H° = -5.65 \times 10^3\ kJ$$

81. When it undergoes complete combustion in oxygen, a 1.050-g sample of the industrial solvent, diethylene glycol, $C_4H_{10}O_3$, gives off 23.50 kJ of heat to the surroundings. Calculate the standard enthalpy of formation of liquid diethylene glycol. Assume that the initial reactants and the products of the combustion are at 25 °C and 1 atm pressure.

82. When it undergoes complete combustion in oxygen, a 658.0-mg sample of adipic acid, $HOOC(CH_2)_4COOH$, a substance used in the manufacture of nylon, gives off 12.63 kJ of heat to the surroundings. Calculate the standard enthalpy of formation of solid adipic acid. Assume that the reactants and the products of the combustion are at 25 °C and 1 atm pressure.

83. Use data from Appendix C to determine the standard enthalpy change at 25 °C for the reaction

$$NH_4^+(aq) + OH^-(aq) \longrightarrow NH_3(g) + H_2O(l)$$

84. Use data from Appendix C to determine how many kilojoules of heat are given off per gram of aluminum consumed in the reaction that follows.

$$2\ Al(s) + 6\ H^+(aq) \longrightarrow 2\ Al^{3+}(aq) + 3\ H_2(g)$$

Fuels and Foods

85. The heats of combustion, in kilojoules per mole, of the first eight, straight-chain alkanes are: methane (−890); ethane (−1560); propane (−2220); butane (−2879); pentane (−3536); hexane (−4163); heptane (−4811); octane (−5450).

(a) Use these data to make a graph of the heat of combustion, *per mol C*, versus the number of C atoms per molecule for these eight alkanes.

(b) Use this graph to estimate the heats of combustion of decane ($C_{10}H_{22}$) and dodecane ($C_{12}H_{26}$).

86. *Without doing detailed calculations*, determine which of the alkanes listed in Problem 85 will evolve the greatest quantity of heat upon combustion **(a)** per mole of alkane; **(b)** per gram of alkane; **(c)** per gram of carbon content; and **(d)** per gram of $CO_2(g)$ produced.

87. A person on a 2500-Cal low-fat diet attempts to maintain a fat intake of no more than 20% of food Calories from fat. What percent of the daily fat allowance is represented by a tablespoon of peanut butter (15.0 g and 50.1% fat)? Recall that fat has a food value of about 9.0 Cal/g.

88. Verify the claim by sugar manufacturers that a teaspoon (about 4.8 g) of sugar (sucrose, $C_{12}H_{22}O_{11}$) contains only 19 Cal. Use the heat of combustion from Problem 80, and recall the definition of a food Calorie.

Additional Problems

89. Which of the following quantities of energy is the largest: (a) the kinetic energy of a hydrogen molecule at STP; (b) the kinetic energy of a 1.0-g BB shot traveling with a speed of 100 m/s; or (c) the heat required to raise the temperature of 10 mL of water from 20 °C to 21 °C?

90. For the melting of 1 mol of ice at 0 °C, we can write the expression

$$H_2O(s, d = 0.917\ g/cm^3) \longrightarrow H_2O(l, d = 0.9998\ g/mL) \qquad \Delta H = +6.01\ kJ$$

Would you expect ΔU for the melting of ice at 0 °C to be greater, less than, or equal to +6.01 kJ/mol? Explain.

91. While on his honeymoon in Switzerland, James Joule (whose discoveries led to the first law of thermodynamics) measured the temperature of the water at the top of a waterfall and again at the bottom. What observation do you think he made? Explain.

92. Calculate the final temperature for the situation described in Example 6.10 of page 249.

93. The sample of water in the smaller Styrofoam® cup is added to that in the larger cup. What is the water temperature when thermal equilibrium is reached?

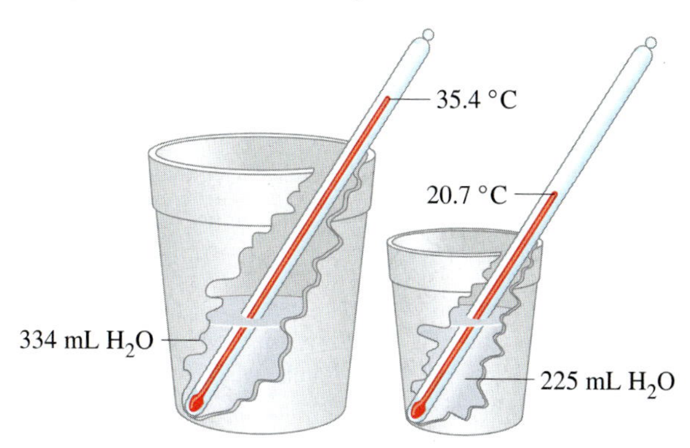

94. The sketch shows a sample of hot water in a foam cup. Also shown are three objects, all at the same room temperature of 25.0 °C: (a) a steel ball, 2.20 cm in diameter (d = 7.83 g/cm^3, specific heat = 0.45 J g^{-1} °C^{-1}); (b) a tightly wound roll of aluminum foil, 2.00 m long, 5.0 cm wide, and 0.10 mm thick (d = 2.70 g/cm^3, specific heat = 0.902 J g^{-1} °C^{-1}); and (c) a graduated cylinder with 5.0 mL of water. Which would you expect to produce the greatest lowering of the temperature of the water in the cup: submerging the steel ball, submerging the roll of aluminum foil, or adding the 5.0 mL of water? Explain. (*Hint*: Do you have to calculate a final temperature?)

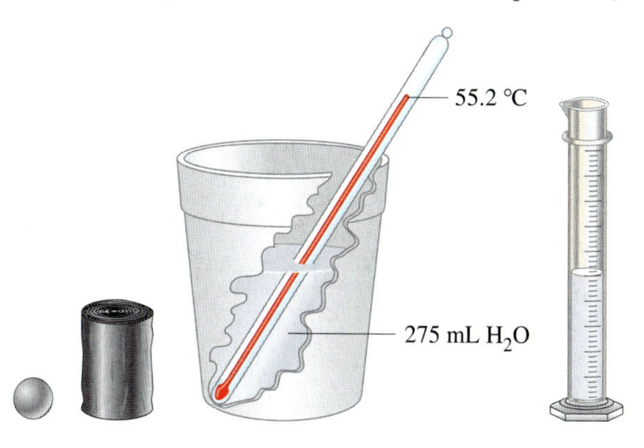

95. In 1818, Dulong and Petit observed that the molar heat capacities of the elements in their solid states were approximately constant. A modern statement of their law is

$$\text{molar mass} \times \text{specific heat (J g}^{-1}\ {}^\circ\text{C}^{-1}) = \text{constant.}$$

(a) Use data from Table 6.1 to establish a value of the constant in the above equation.

(b) What form of a graph of molar masses and specific heats would you plot to obtain a *straight line*? Explain.

(c) The element cobalt (specific heat = 0.421 J g^{-1} °C^{-1}) was known at the time of Dulong and Petit. What approximate value would they have obtained for the atomic weight of cobalt with their equation?

(d) When a 25.5-g sample of titanium at 99.7 °C was added to a quantity of water in a foam cup calorimeter, the water temperature was found to increase from 24.6 °C to 27.4 °C. What must have been the approximate volume of the water?

96. A British thermal unit (Btu) is defined as the quantity of heat required to change the temperature of 1 lb of water by 1 °F. Assuming the specific heat of water to be independent of temperature, how much heat is required to raise the temperature of the water in a 30-gal water heater from 66 to 145 °F: **(a)** in Btu; **(b)** in kcal; **(c)** in kJ?

97. The combustion of methane is represented by the equation

$$CH_4(g) + 2\,O_2(g) \longrightarrow CO_2(g) + 2\,H_2O(l) \qquad \Delta H = -890.3\ \text{kJ}$$

(a) What mass of $CH_4(g)$ must be burned to give off 1.00×10^5 kJ of heat?

(b) What quantity of heat is given off by the combustion of 105 L $CH_4(g)$, measured at 23 °C and 746 mmHg?

98. The combustion of hydrogen-oxygen mixtures is used to produce very high temperatures (about 2500 °C) needed for certain types of welding. Consider the combustion reaction to be

$$H_2(g) + 1/2\,O_2(g) \longrightarrow H_2O(g) \qquad \Delta H^\circ = -241.8\ \text{kJ}$$

How much heat is evolved when a 100.0-g mixture containing 15.5% $H_2(g)$, by mass, is burned?

99. The thermite reaction is highly exothermic.

$$Fe_2O_3(s) + 2\,Al(s) \longrightarrow Al_2O_3(s) + 2\,Fe(s) \qquad \Delta H^\circ = -852\ \text{kJ}$$

The reaction is started in a room-temperature (25 °C) mixture of 1.00 mol $Fe_2O_3(s)$ and 2.00 mol Al(s). The liberated heat is retained within the products, whose combined specific heat over a broad temperature range is about 0.8 J g^{-1} °C^{-1}. Show that the quantity of heat liberated is sufficient to raise the temperature of the products to the melting point of iron (1530 °C).

100. The composition of a particular natural gas, expressed on a mole fraction basis, is CH_4, 0.830; C_2H_6, 0.112; C_3H_8, 0.058. A 215-L sample of this natural gas, measured at 24.5 °C and 744 mmHg, is burned in an excess of oxygen.

How much heat is evolved in the combustion? (*Hint*: What are the heats of combustion of the individual gases?)

101. Care must be taken in preparing solutions of solutes that liberate heat on dissolving. The heat of solution of NaOH is −42 kJ/mol NaOH. What should be the approximate maximum temperature reached in the preparation of 500.0 mL of 6.0 M NaOH from NaOH(s) and water at 20 °C?

102. A particular gasohol fuel is 85% gasoline and 15% ethanol by mass. If we approximate gasoline by the formula for octane C_8H_{18}, determine which gives off more heat on combustion, straight gasoline or the gasohol mixture. (*Hint*: Use data from Table 6.2 and Problem 85.)

103. The heat of neutralization of HCl(aq) by NaOH(aq) is −55.90 kJ/mol H_2O produced. If 50.00 mL of 1.16 M NaOH at 25.15 °C is added to 25.00 mL of 1.79 M HCl at 26.34 °C in a Styrofoam® cup calorimeter, what will be the solution temperature be immediately after the neutralization reaction has occurred. You can make the same assumptions as in Example 6.11 on page 250.

104. Write two equations to represent the combustion of dodecane, $C_{12}H_{26}(l)$, in an excess of oxygen gas. In each case, the products are carbon dioxide gas and water, but in one reaction, the water is obtained as a liquid and in the other, as a gas. *Without doing detailed calculations*, determine which of these two combustion reactions gives off the greater quantity of heat. Also determine the *difference* in the two heats of combustion, per mole of dodecane burned.

105. Under the entry "H_2SO_4", a handbook lists several values for the enthalpy of formation, ΔH_f. For example, for pure $H_2SO_4(l)$, $\Delta H_f = -814$ kJ/mol; for 1.0 M $H_2SO_4(aq)$, $\Delta H_f = -888$ kJ/mol; for 0.25 M $H_2SO_4(aq)$, $\Delta H_f = -890$ kJ/mol; for 0.020 M $H_2SO_4(aq)$, $\Delta H_f = -897$ kJ/mol.

(a) Explain why these values are not all the same.

(b) When concentrated aqueous solutions are diluted, does the solution temperature increase or decrease? Explain.

106. Substitute natural gas (SNG) is a gaseous mixture containing $CH_4(g)$ that can be used as a fuel. One reaction for the production of SNG is

$$4\,CO(g) + 8\,H_2(g) \longrightarrow 3\,CH_4(g) + CO_2(g) + 2\,H_2O(l) \qquad \Delta H = ?$$

Show how the following data, as necessary, can be used to determine $\Delta H°$ for this SNG reaction.

$$C(\text{graphite}) + 1/2\,O_2(g) \longrightarrow CO(g) \qquad \Delta H = -110.5\text{ kJ}$$
$$CO(g) + 1/2\,O_2(g) \longrightarrow CO_2(g) \qquad \Delta H = -283.0\text{ kJ}$$
$$H_2(g) + 1/2\,O_2(g) \longrightarrow H_2O(l) \qquad \Delta H = -285.8\text{ kJ}$$
$$C(\text{graphite}) + 2\,H_2(g) \longrightarrow CH_4(g) \qquad \Delta H = -74.81\text{ kJ}$$

Compare this result with that obtained by using enthalpy of formation data directly on the SNG reaction.

107. The sketch below is an enthalpy diagram based on these reactions.

$$3\,H_2(g) + 3/2\,O_2(g) \longrightarrow 3\,H_2O(l)$$
$$N_2(g) + O_2(g) \longrightarrow 2\,NO(g)$$
$$N_2(g) + 3\,H_2(g) \longrightarrow 2\,NH_3(g)$$
$$2\,NH_3(g) + 5/2\,O_2(g) \longrightarrow 2\,NO(g) + 3\,H_2O(l)$$

Determine the enthalpy changes for these reactions, using data from Table 6.2, and enter these values in the appropriate red blanks on the sketch. Also, write on the blue lines the substances present, starting with the information provided on one of those lines.

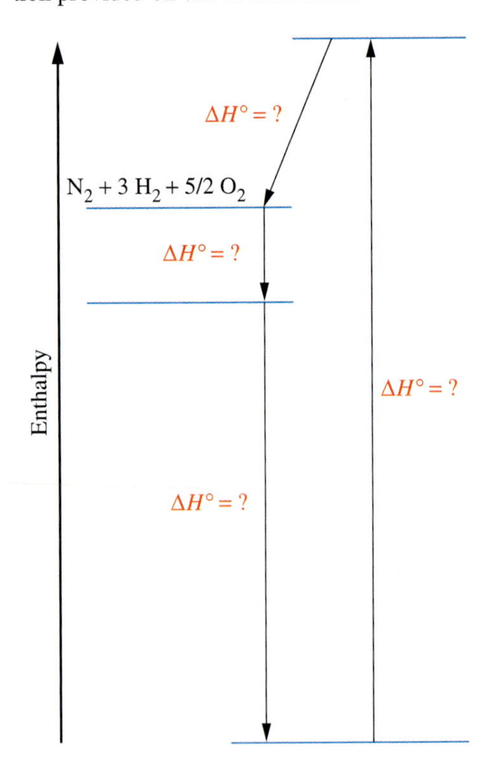

Atomic Structure

CHAPTER

7

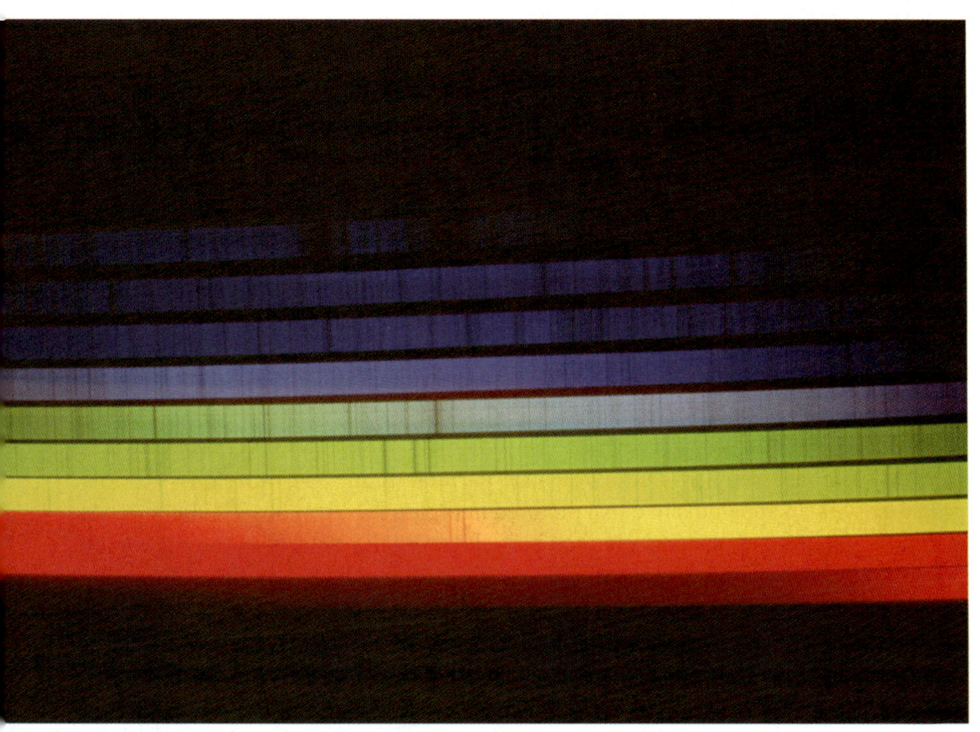

The dark lines in this solar absorption spectrum, taken from Sacramento Peak, reflect the electronic structure of atoms in the sun's atmosphere.

Some discoveries in chemistry have had a great impact on other disciplines. Using the structures of proteins and other biochemical substances determined by chemists, biologists are able to study life processes at the molecular level. For example, progress in genetics is now made not solely by studying patterns of inheritance in populations, but in a more fundamental way by determining the structure of DNA molecules in minute detail. By the same token, discoveries in other sciences have had dramatic effects on chemistry, changing forever the way chemists look at the world. In this chapter, we will consider developments leading to our current knowledge of the structure of atoms.

We began the story of atomic structure in Chapter 2 with a brief introduction. There, we focused on the atomic nucleus with its protons and neutrons. This focus included obtaining atomic masses, and atomic masses and the idea of the mole are at the heart of stoichiometry. So far, we have limited our interest in electrons mainly to their loss or gain in forming ions. In contrast to Chapter 2, we focus mostly on electrons in much of this chapter. Electrons are at the heart of our modern view of atomic structure.

CONTENTS

● **ChemCDX**
When you see this icon, there are related animations, demonstrations, and exercises on the CD accompanying this text.

The chapter consists of three parts. The first provides experimental evidence for the picture of the atom we adopted in Chapter 2, with emphasis on the discovery of the electron and determination of its properties. At first glance, the second part—which is mainly about light—might seem unrelated to atomic structure. However, a complete understanding of the nature of light requires some totally new ideas. And it is these ideas, when used to explain the behavior of electrons in atoms, that lead to the modern view of atomic structure presented in the third part of the chapter.

Classical View of Atomic Structure

The classical view of atomic structure was constructed with the body of knowledge accumulated in physics over several centuries preceding the twentieth century. This body of knowledge is often called "classical physics."

7.1 The Electron: Experiments of Thomson and Millikan

We begin our story of the discovery of the electron at about the middle of the nineteenth century. By this time, static electricity, current electricity, and magnetism were rather well understood (recall Section 4.1, and see Appendix B).

Cathode Rays

In 1838, Michael Faraday studied electric currents in air and other gases at low pressures with an apparatus like that pictured in Figure 7.1. Air does not contain ions in significant number, and any charged particles from the electrodes are not able to penetrate through the molecules of gas. As a result, no electricity flows when the tube is filled with air. However, if the tube is evacuated, that is, if most of the air is removed with a vacuum pump, electricity does flow through the tube. As long as electric current flows through the tube, the residual air gives off a purple light. About 20 years after Faraday's work, better vacuum pumps became available and scientists made some interesting new observations: The purple glow inside the tube disappeared, but portions of the tube itself began to fluoresce. Fluorescence is a light emission similar to that which we see when electricity passes through a fluorescent tube in a lighting fixture. The mysterious, invisible carriers of electric current from cathode to anode in these tubes were called **cathode rays**. For the rest of the nineteenth century, cathode rays were the object of many scientific studies.

APPLICATION NOTE
If the air in the tube in Figure 7.1 is replaced by neon, the glow is a familiar orange-red color. Faraday's electric discharge tube was much like a modern neon sign.

Figure 7.2 shows an electric discharge tube of the sort used in 1879 by William Crookes. In the Crookes tube, cathode rays are concentrated into a narrow beam as they pass through a metal slit. The beam of cathode rays then passes along a screen

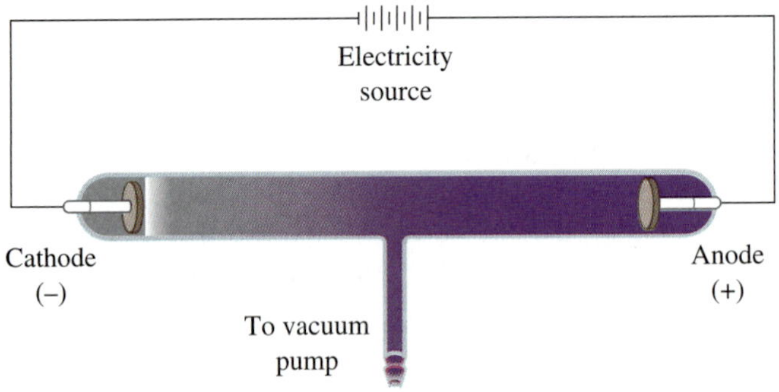

Electricity source

Cathode (−)

Anode (+)

To vacuum pump

Figure 7.1
Electric discharge in an evacuated tube

Figure 7.2 Cathode rays and their deflection in a magnetic field
The beam of cathode rays, made visible by the green fluorescence of a zinc sulfide-coated screen, originates at the cathode on the left and is deflected by the magnet located slightly behind the anode.

that gives off a green fluorescence, thereby marking the path of the otherwise invisible cathode rays. The photograph also shows that when cathode rays pass near a magnet, they are deflected from their usual straight-line path. The magnet exerts a force on the cathode rays in much the same way that a strong crosswind does in deflecting a tennis ball from its straight path.

By the 1890s, scientists had compiled many observations about cathode rays, including the following.

1. The rays are emitted from the cathode when electricity is passed through an evacuated tube.
2. The rays are emitted in a direction perpendicular to the cathode surface.
3. Cathode rays travel in straight lines.
4. Cathode rays cause glass and other materials to give off a fluorescent glow.
5. Cathode rays are deflected in a magnetic field in the way expected for negatively charged particles.*
6. The properties of cathode rays *do not* depend on the composition of the cathode. For example, the cathode rays from an aluminum cathode are the same as those from a silver cathode.

Research on cathode rays was carried out mainly in Germany and in Great Britain. Although German and British scientists agreed on the facts, they developed strikingly different hypotheses about cathode rays. Most of the German scientists thought that the rays were a form of electromagnetic radiation, much like light. Most British scientists thought that the rays were particles of matter—probably residual gas molecules that had acquired a negative charge from the cathode. When J. J. Thomson assessed all the data on cathode rays available in 1897, he leaned heavily toward the particle hypothesis. Thomson then settled the question unequivocally in a landmark series of experiments.

The Experiments of J. J. Thomson

In his first experiments, Thomson confirmed that cathode rays are negatively charged particles by measuring their deflections in magnetic *and* electric fields. In another crucial set of experiments, he showed that the magnetic deflections were the same

* *Electromagnetism*, further described in Appendix B, refers to phenomena in which electricity and magnetism are interrelated.

APPLICATION NOTE
Television picture tubes and computer monitor tubes are cathode ray tubes (CRT). Cathode rays are deflected by magnetic fields to appropriate spots on a screen, where the fluorescence they cause creates an image.

J. J. Thomson (1856–1940) entered Cambridge University as an undergraduate in 1875, and a scant eight years later, was named the Cavendish Professor. Thomson's scientific reputation was greatly enhanced with the publication in 1893 of the book, *Recent Researches in Electricity and Magnetism*, and he soon attracted outstanding students from throughout the world. Thomson won the Nobel Prize in 1906 for characterizing the electron.

no matter what the residual gas in the cathode ray tube—whether it was hydrogen, air, carbon dioxide, or other gases. This observation strongly suggested that cathode rays are not ions formed from gaseous atoms or molecules; instead, they are negatively charged particles *found in all matter*. To strengthen his argument, Thomson designed an experiment to obtain an easily measured property of cathode rays: the ratio of their mass (m_e) to charge (e).

$$\frac{\text{Mass of cathode ray particle}}{\text{Charge on cathode ray particle}} = \frac{m_e}{e}$$

In the apparatus outlined in Figure 7.3, a beam of cathode rays passes through a hole in the anode, and then between the oppositely charged plates of an electrical condenser. Superimposed on the electric field between the condenser plates is a magnetic field; one pole of the magnet is in front of the tube and the other is behind it. The cathode rays are bent upward toward the positive plate by the electric field, and downward by the magnetic field. The electric and magnetic fields can be adjusted so the cathode ray beam strikes the fluorescent screen undeflected. The strengths of the two fields producing this condition are used to calculate the mass-to-charge ratio of the cathode rays. The best available measurements give the following value.

$$m_e/e = -5.686 \times 10^{-12} \text{ kilograms per coulomb (kg/C)}$$

[The coulomb (C) is the SI unit of electric charge (see Appendix B).]

This value of m_e/e is about 2000 times *smaller* than the smallest previously known value—the mass-to-charge ratio of hydrogen ions in the electrolysis of water ($+1.045 \times 10^{-8}$ kg/C). This observation suggests the following.

1. If the charge on a cathode ray particle is comparable to that on a H^+ ion, the mass of a cathode ray particle (m_e) is much smaller than the mass of H^+; or

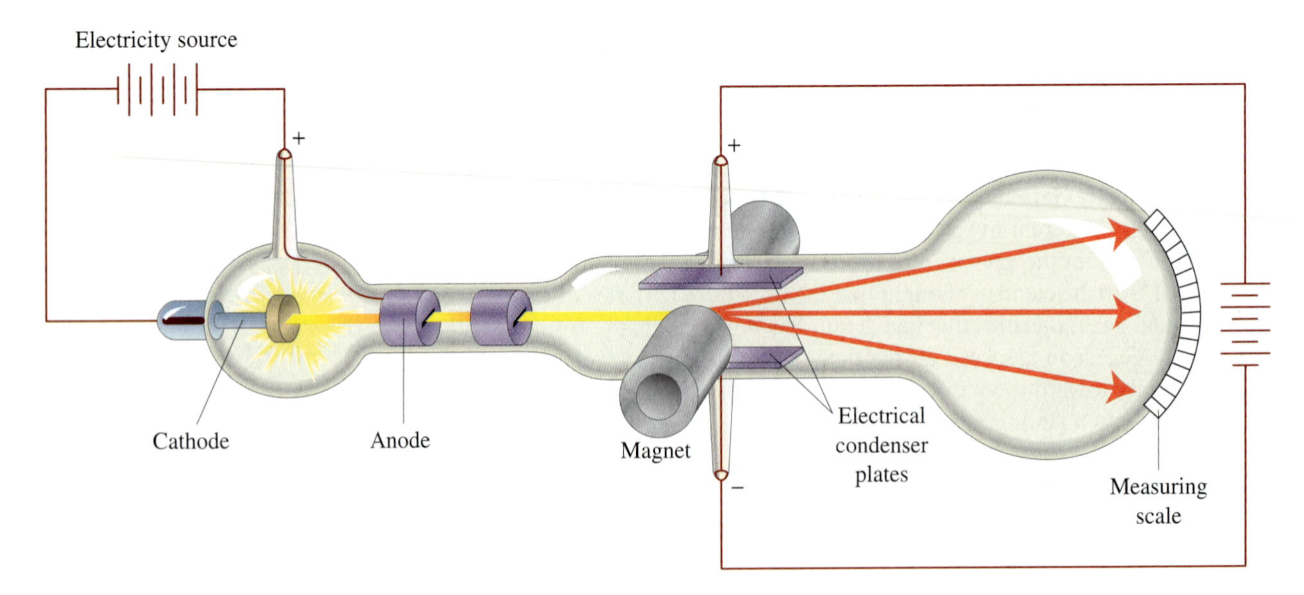

▲ **Figure 7.3**
Thomson's apparatus for determining the mass-to-charge ratio, m_e/e, of cathode rays

2. If the mass of a cathode ray particle is comparable to that of an H^+ ion, the charge of a cathode ray particle (e) is much larger than the charge on H^+; or
3. The situation is somewhere between the extremes described in the first two statements.

Thomson strongly believed that the first statement would prove to be true, but he had no precise way to determine either m_e or e. Once either of these could be measured, the other could be calculated through the value of m_e/e.

Based on Faraday's early work on electrolysis, the Irish physicist George Stoney proposed, in 1874, the existence of a fundamental unit of electric charge that he later called an *electron*. After Thomson's work was published, electron became the commonly accepted term for cathode ray particles. Next we will consider how e, the charge on the electron, was established.

"Thus we have in the cathode rays matter in ... a state in which the subdivision of matter is carried very much further than in the ordinary gaseous state: a state in which all matter ... is of one and the same kind."

J. J. Thomson

Electron Charge: Millikan's Oil-Drop Experiment

In 1909, Robert Millikan devised an elegant experiment to determine the charge on an electron. A version of his apparatus is shown in Figure 7.4. The idea is to produce tiny oil droplets, have them acquire an electric charge, and measure the velocity of a falling droplet both in the absence of an electric field and in the presence of an electric field. The velocity is measured by observing the time required for the droplet to fall between two fine lines in the eyepiece of an observing telescope.

Without the electric field, the droplet falls under the force of gravity and quickly reaches a constant terminal velocity, v_g. This is analogous to the terminal velocity reached by a parachutist. When an electric field is imposed, the velocity of an electrically charged droplet changes to a new value, v_e. Consider the situation in Figure 7.4. A droplet carrying a negative charge would be attracted to the positively charged plate A and be slowed down ($v_e < v_g$), just as the descent of a parachutist is slowed if the parachute is caught in an updraft.

By analyzing the data from hundreds of experiments, Millikan found that droplets carried an electric charge equal to a fundamental unit of charge, e, or some multiple of e. The fundamental unit of *negative* charge is that carried by an electron, and thus by an ion with a single negative charge. The charge on an electron is

$$e = -1.602 \times 10^{-19} \text{ coulomb (C)}$$

With this value and Thomson's value of m_e/e, we can calculate the mass of an electron, m_e.

$$m_e = \frac{m_e}{\cancel{e}} \times \cancel{e} = -5.686 \times 10^{-12}\ \text{kg}/\cancel{\text{C}} \times \frac{-1.602 \times 10^{-19}\ \cancel{\text{C}}}{1\ \text{electron}}$$

$$= 9.109 \times 10^{-31}\ \text{kg/electron}$$

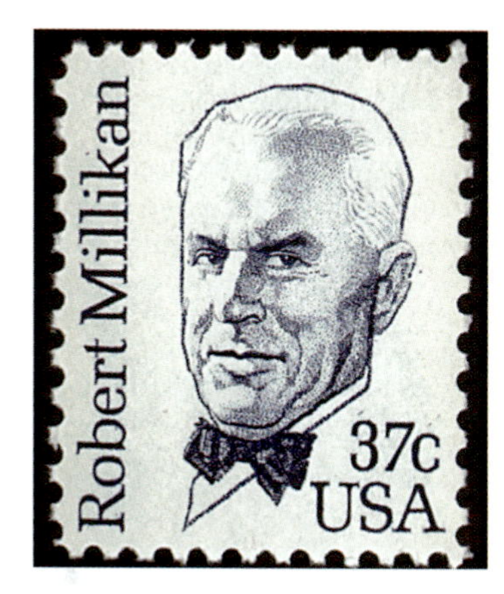

Robert A. Millikan (1868–1953) was a professor at the University of Chicago at the time of his famous experiments on the electronic charge (1909–1913). He won the Nobel Prize in 1923 for these experiments and other experiments he did to confirm Einstein's explanation of the photoelectric effect (page 293). In 1921 he moved to the California Institute of Technology. There, he developed an interest in a newly discovered type of radiation that enters Earth's atmosphere from outer space, for which he coined the term "cosmic radiation."

"Here, then, is direct, unimpeachable proof that ... the electrical charges found on ions all have either exactly the same value or else small exact multiples of that value."

Robert Millikan

7.2 Atomic Models: J. J. Thomson and Ernest Rutherford

Once he established electrons as fundamental particles of all matter, J. J. Thomson became keenly aware that an atom could not consist of electrons alone. The negative charges of the electrons had to be neutralized by an equivalent amount of positive charge, and the electrons had to be organized into some kind of a stable configuration, or else they would fly apart.

● **ChemCDX** Millikan Oil Drop Experiment

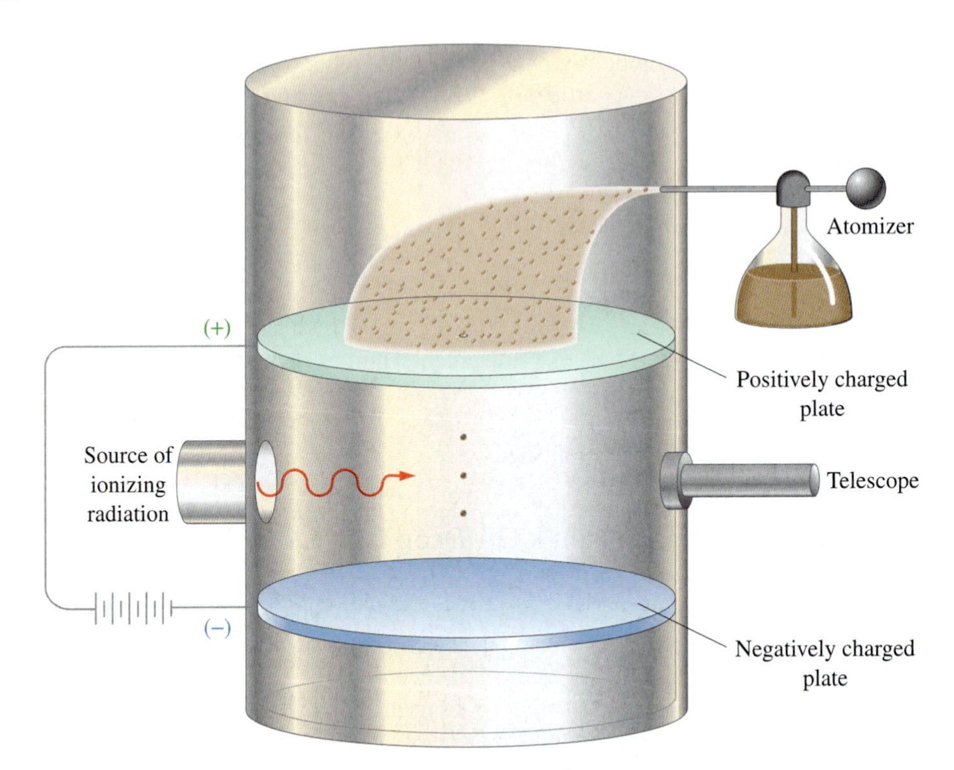

Figure 7.4
Millikan's oil-drop experiment
Oil droplets from an atomizer enter the apparatus through a tiny hole. Some droplets acquire a frictional electric charge as they escape from the atomizer. A source of ionizing radiation, such as X rays, also produces ions. The absorption of these ions sporadically changes the electric charges on the droplets. The droplets are observed through a telescope with a measuring scale in the eyepiece.

Thomson's "Raisin Pudding" Atomic Model

Thomson didn't know exactly how positive electric charge is distributed in an atom, so he considered the case that was easiest to describe mathematically. He developed a model in which the positive charge is uniformly distributed in a sphere and the electrons are imbedded in the sphere in such a way that their attraction for the positive charge just offsets the repulsions among the electrons. This arrangement somewhat resembles a raisin pudding.

For a hydrogen atom, Thomson proposed one electron at the exact center of the sphere. For an atom with two electrons (helium), he proposed two electrons along a straight line through the center, with each electron halfway between the center and the outer surface of the sphere (see Figure 7.5). Thomson applied this kind of analysis to atoms with up to 100 electrons.

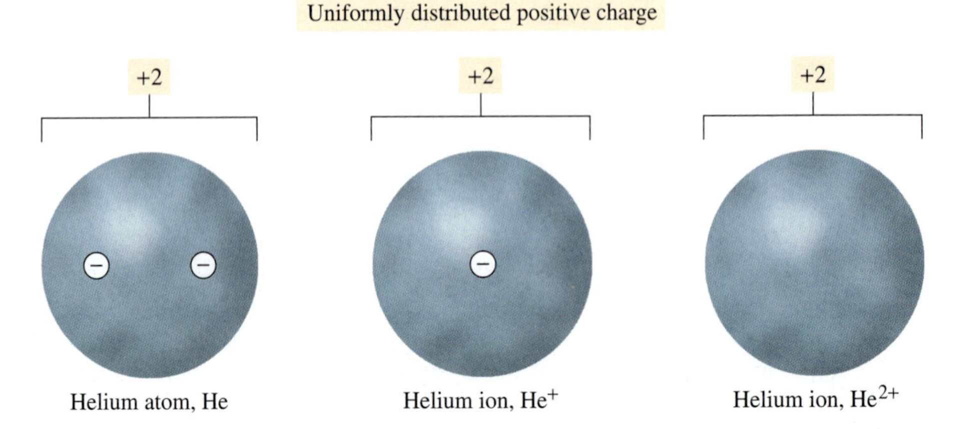

Figure 7.5
Thomson's "raisin pudding" model of the atom
For a helium atom, the model proposes a large spherical cloud with two units of positive charge. The two electrons lie on a line through the center of the cloud. The loss of one electron produces the He^+ ion, with the remaining electron at the center of the cloud. The loss of a second electron produces He^{2+}, in which there is just a cloud of positive charge.

Rutherford's Nuclear Model of the Atom

Ernest Rutherford was a pioneer in the study of radioactivity, a phenomenon in which unstable heavy atoms give off radiation as they disintegrate. Rutherford characterized one type of radiation, called *alpha* (α) particles, as identical to doubly ionized helium atoms, He^{2+}. We will study radioactivity in Chapter 19, but for now we want only to describe how Rutherford used α particles to probe the structure of matter.

Ernest Rutherford (1871–1937) is said to have been digging potatoes on his father's farm in New Zealand when notified of a scholarship to work with J. J. Thomson at Cambridge. He threw down his spade, proclaiming, "That's the last potato I'll dig." Although a physicist, Rutherford was awarded the Nobel Prize in chemistry in 1908 for his pioneering work in radioactivity.

Based on Thomson's model of the atom, Rutherford expected that most α particles would pass through atoms undeflected. However, he also expected that any of the positively charged α particles that came close to an electron should be deflected to some extent. By measuring such deflections, he hoped to gain information about the distribution of electrons in an atom.

Rutherford assigned the experiment illustrated in Figure 7.6 to his assistant, Hans Geiger, and an undergraduate student, Ernest Marsden. When they bombarded very thin foils of metals such as gold, silver, and platinum with α particles, they found that most of the particles went right through the foil undeflected or deflected only slightly. This is just what Rutherford had expected. However, much to Rutherford's surprise, a few particles were deflected sharply, and once in a while, an α particle would bounce right back toward the source.

"It is about as incredible as if you had fired a 15-inch shell at a piece of tissue paper and it came back and hit you."
Ernest Rutherford

These alpha-particle scattering experiments could not be explained by the Thomson model, in which positive charge is spread throughout an atom. Thus, Rutherford concluded that all the positive charge of an atom is concentrated at the center of an atom in a tiny core called the *nucleus*. When a positively charged α particle approaches a positively charged nucleus, it is strongly repelled and therefore sharply deflected. Because only a few α particles were deflected, Rutherford concluded that the nucleus occupies only a tiny fraction of the volume of an atom. Most of the α particles passed right through because, except for the tiny nucleus and a small number of electrons outside, an atom is empty space.

● **ChemCDX** Rutherford Experiment: Nuclear Atom

To picture Rutherford's model of a nuclear atom, visualize the atom as a giant indoor football stadium. A pea at the middle of the structure represents the nucleus. A few houseflies flitting here and there throughout the stadium represent the electrons. The roof of the stadium prevents the flies from leaving, but the atom has no comparable covering. Electrons remain in the atom because they are strongly attracted to the positively charged nucleus.

7.3 Protons and Neutrons

The experiments that led to the idea of a nuclear atom also provided data that could be used to determine the amount of positive nuclear charge. Rutherford believed that this positive charge was carried by particles called *protons*, that the proton charge was the fundamental unit of positive charge, and that the nucleus of a hydrogen atom consisted of a single proton. Rutherford's ideas were proved correct some years later through experiments in which protons were ejected from atomic nuclei and found to be identical to the nuclei of hydrogen atoms.

In the early 1900s, scientists commonly used the notion of an *atomic number*, but it seemed to have no underlying fundamental significance. It was just a numerical ranking of the elements in order of increasing atomic mass: hydrogen first, helium second, and so on. By 1914, however, experiments that we will describe in the next chapter (page 328) showed that the atomic number is equal to the number of units of positive charge on the nucleus. And, when protons were shown to be the particles responsible for the positive charge on the nucleus, the atomic number was seen to be equal to the number of protons in the nucleus.

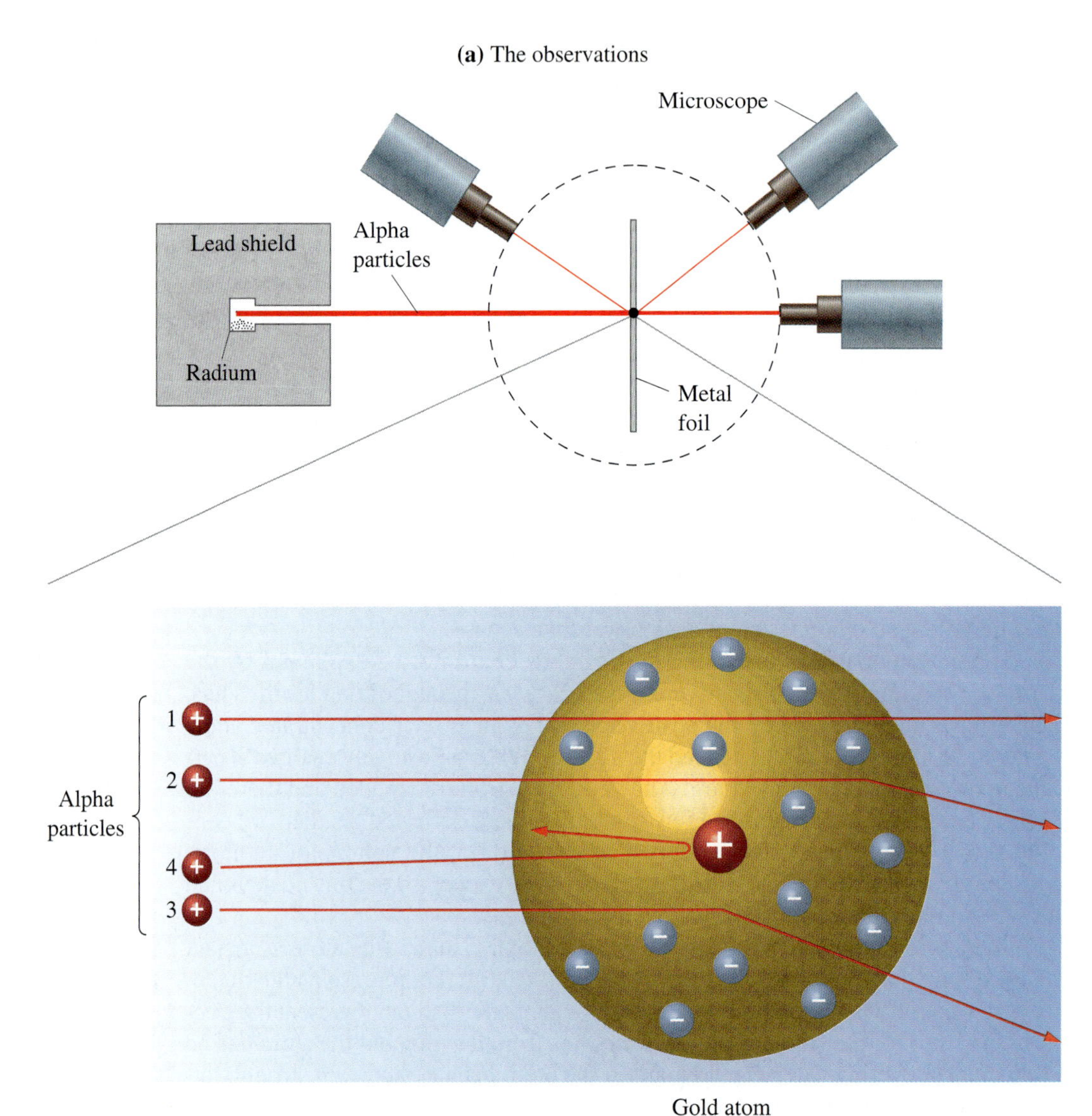

▲ **Figure 7.6 The scattering of alpha (α) particles by a thin metal foil**
(a) The observations: (1) Most of the α particles pass through the foil undeflected. (2) Some α particles are deflected slightly as they penetrate the foil. (3) A few (about 1 in 20,000) are greatly deflected. (4) A similar small number do not penetrate the foil at all, but are reflected back toward the source. (b) Rutherford's interpretation: If the atoms of the foil have a massive, positively charged nucleus and light electrons outside the nucleus, one can explain how: (1) an α particle passes through the atom undeflected (a fate shared by most of the particles); (2) an α particle is deflected slightly as is passes near an electron; (3) an α particle is strongly deflected by passing close to the atomic nucleus; and (4) an α particle bounces back as it approaches the nucleus head-on.

If we assume that all protons have the same mass, then, except for hydrogen, an atom does not have enough protons to account for the mass of the atom. Electrons contribute so little to the mass that they can't make up the difference. Where does the rest of the mass come from? One hypothesis was that the atomic nucleus contains additional particles with masses comparable to protons but *no electric charge*. James Chadwick discovered these particles in 1932 as a form of radiation produced in the bombardment of beryllium atoms with alpha particles. These particles, called *neutrons*, did indeed prove to have about the same mass as protons and no electric charge.

7.4 Positive Ions and Mass Spectrometry

We have seen that some early investigators believed cathode rays to be negatively charged ions, but their mass-to-charge ratio showed them to be simply electrons. Research on cathode rays also revealed the existence of particles that are deflected in the opposite direction from cathode rays in magnetic and electric fields. These *positively charged* particles *were* shown to be ions, because their mass-to-charge ratios are (a) significantly larger than of cathode rays, and (b) different from one type of gas to another. The ions are produced when atoms or molecules of the residual gas in a cathode ray tube are struck by cathode rays (electrons). In such collisions, one or more electrons may be dislodged from a neutral atom or molecule, leaving a positive ion. In Figure 7.3, we saw how Thomson obtained a beam of cathode rays by using a perforated *anode* in a cathode ray tube. We can obtain beams of positive ions by using a perforated *cathode*, and we can study these beams in an experimental technique known as *mass spectrometry*.

A **mass spectrometer** is a device that separates positive gaseous ions according to their mass-to-charge ratios. Actually, because most of the ions in a beam carry a 1+ charge, we can think of the separation as being based just on mass (that is $m/e = m/1 = m$). The mass spectrometer illustrated in Figure 7.7 is of an early design but serves to illustrate the principle involved. If a stream of positive ions having equal velocities is brought into a magnetic field, (a) all the ions experience a force that deflects them from straight-line paths into circular paths, and (b) the lightest ions are deflected the most (into the tightest circle) and the heaviest ones are deflected the least. Think of this analogous situation: A Ping-Pong® ball, a tennis ball, and a baseball are all thrown with the same speed in the same direction but into a strong crosswind. In making a certain forward progress the Ping-Pong® ball, the lightest object, is most strongly deflected. Next comes the tennis ball. The heaviest object, the baseball, makes the same forward progress with the least deflection.

Each circular path in the mass spectrometer of Figure 7.7 represents ions with a specific mass-to-charge ratio. The point where the ions strike the photographic

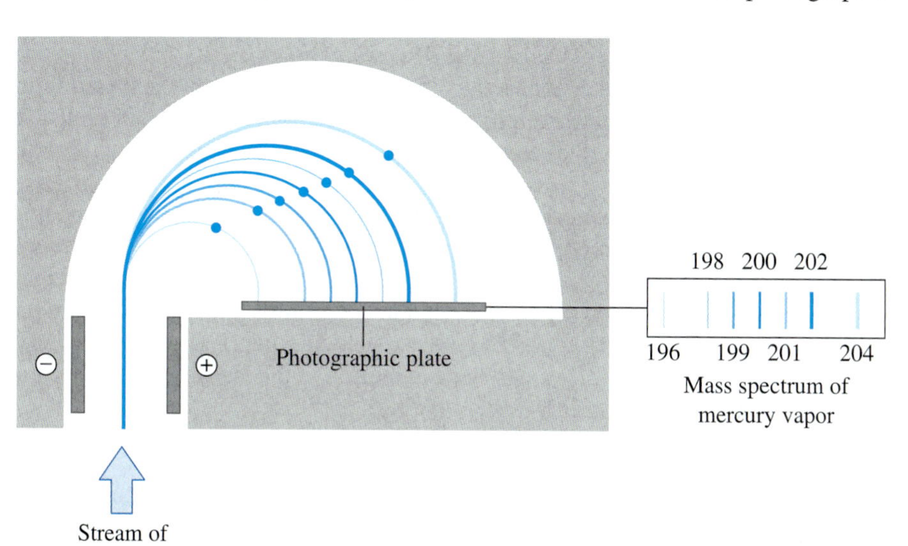

Figure 7.7
A mass spectrometer
A gaseous sample is ionized by bombarding it with electrons in the lower part of the apparatus (not shown), producing positive ions. The ions pass through an electric field in which they are brought to a particular velocity. The ions then pass through a narrow slit into a curved chamber. A magnetic field is applied perpendicular to the beam of ions (perpendicular to the page). All the ions with the same mass-to-charge ratio are deflected into the same circular path. (In most cases, the ionic charge is 1+ and the mass-to-charge ratio is the same as the mass.) Modern spectrometers use electronic detection devices rather than photographic plates or film to establish mass-to-charge ratios and relative numbers of ions.

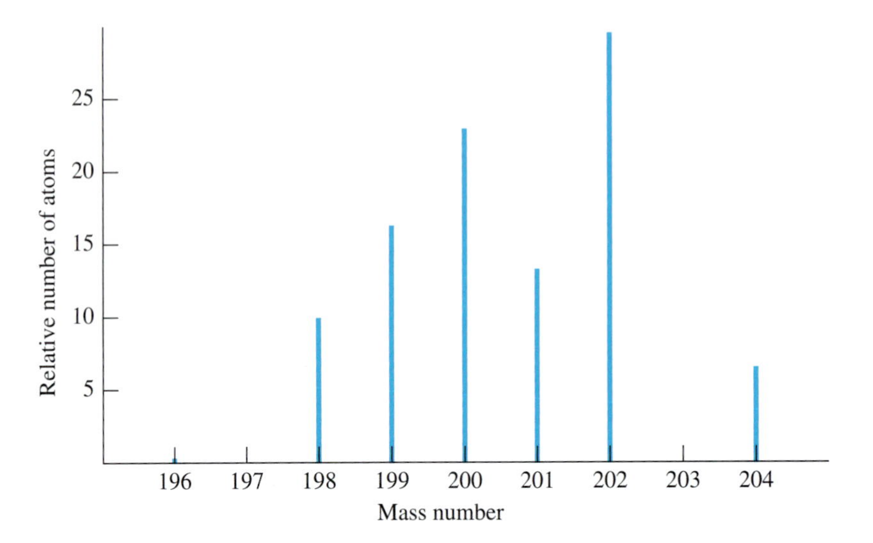

Figure 7.8
Mass spectrum for mercury
The photographic record of Figure 7.7 has been converted to a scale of relative numbers of atoms. The percent natural abundances for the mercury isotopes are: ^{196}Hg, 0.146%; ^{198}Hg, 10.02%; ^{199}Hg, 16.84%; ^{200}Hg, 23.13%; ^{201}Hg, 13.22%; ^{202}Hg, 29.80%; ^{204}Hg, 6.85%.

plate is related to the mass-to-charge ratio. The density (or darkness) of the image on the photographic plate is related to the relative number of ions with this ratio. A record of the separation of ions in a mass spectrometer is called a *mass spectrum*. Figure 7.8 shows the form in which a mass spectrum is usually recorded by modern instruments. It is mass spectral data such as these that we used in Chapter 2 to calculate weighted average atomic masses (page 44).

Light and the Quantum Theory

Space is vast and dark. Here and there, a star lights a small segment of the void. Our sun is one such star; it lights Earth and the other parts of the solar system. Our eyes perceive sunlight to be white, but under proper conditions, we can separate white light into a rainbow of colors. Later in the chapter, we will see that to describe how electrons are arranged outside the nuclei of atoms, we need to analyze the light emitted by energetic atoms. To do this analysis, we need to know something about the physical nature of light.

7.5 The Wave Nature of Light

Many types of waves exist in nature. Waves of water break on the shores of oceans and lakes. A concentric pattern of waves is formed when a pebble is tossed into a pond. Earthquakes send waves through Earth's crust. Sound waves carry music to our ears. With microwaves, we can thaw a TV dinner or heat a cup of soup. But just what is a wave?

A **wave** is a progressive, repeating disturbance that spreads through a medium from a point of origin to more distant points. There is little movement of the material that is in the direction of the disturbance. It is rather like a whispered message ("disturbance") passed along by seated people ("little movement") from one end of a row ("point of origin") to the other ("distant point"). A person holding a long rope fastened to a wall can start a wave motion in the rope by jerking the end of the rope up and down (Figure 7.9).

Electromagnetic waves originate from the movement of electric charges. This movement produces oscillations (fluctuations) in electric and magnetic fields that are propagated over distances. Unlike water waves and sound waves, electromagnetic waves require no medium for their propagation. It is the ability of electromagnetic radiation to travel through empty space that makes it possible for the sun

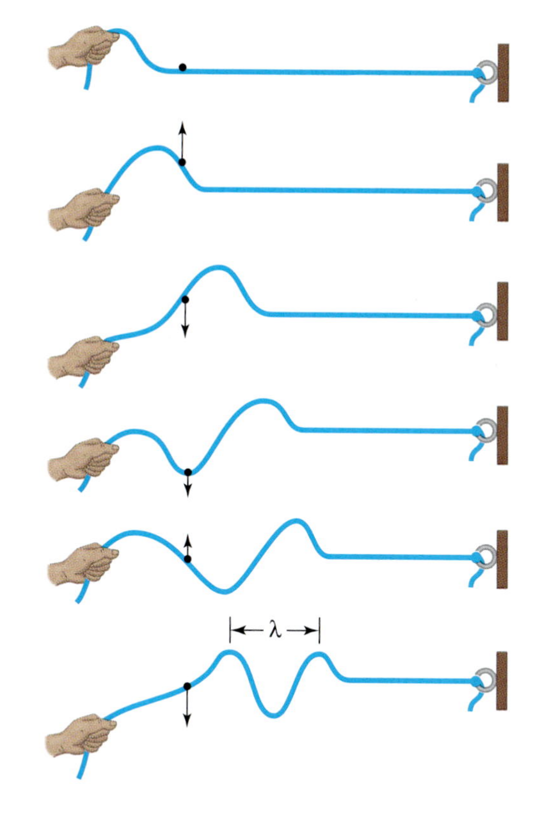

Figure 7.9
The simplest wave motion: A traveling wave in a rope
Imagine an infinitely long rope. Up-and-down hand motion (top to bottom) causes waves to pass along the rope from left to right. The up-and-down motion of a typical point (dot) on the rope is also shown. This one-directional moving wave is called a traveling wave.

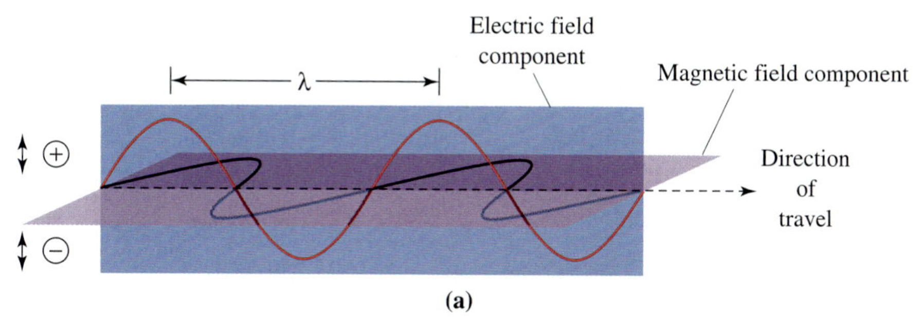

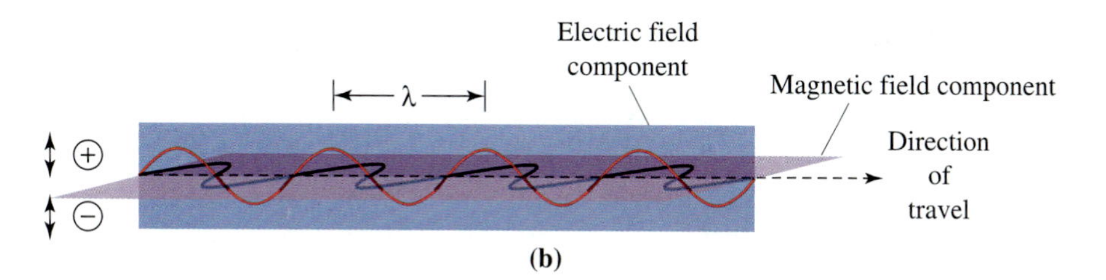

Figure 7.10
An electromagnetic wave
This artist's rendering visualizes an electromagnetic wave as a superposition of two wave motions, that of an alternating electric field and, perpendicular to it, an alternating magnetic field. These alternating fields are generated by the relative motion of electric charges. To travel the same distance requires four cycles of the wave in (b) and only two cycles of the wave in (a). The wave in (b) has a *higher* frequency and a *shorter* wavelength, λ, than the wave in (a). The wave in (a) is shown with a greater amplitude (intensity) than in (b), but the amplitude of a wave is independent of frequency and wavelength.

to transmit some of its energy to Earth as sunlight or for scientists to send a command from Earth to a computerized robot on the surface of Mars.

Electromagnetic radiation is characterized by its wavelength, frequency, and amplitude (Figure 7.10). The **wavelength** is the distance between any two identical points in consecutive cycles; we usually choose peaks or crests for convenience. Wavelength is denoted by the Greek letter λ (lambda). A common SI unit of wavelength, especially for visible light, is the nanometer (1 nm $= 10^{-9}$ m). A non-SI unit still used to some extent in scientific work is the angstrom unit, Å (1 Å $= 10^{-10}$ m).

The **frequency** of a wave is the number of cycles of the wave that pass through a point in a unit of time. For example, if some point through which the wave passes is at the crest of the wave 60 times in one second (and also at the bottom or trough of the wave 60 times), we say that 60 cycles of the wave have passed in one second. Frequency is denoted by the Greek letter ν (nu, pronounced the same as "new"). The SI unit of frequency is the *hertz* (*Hz*). A hertz is one cycle per second, but generally the word cycle is understood. That is, $1\ \text{Hz} = 1\ \text{s}^{-1}$.

The *amplitude* of a wave is its height: the distance from a line of no disturbance through the center of the wave to a peak (crest). Imagine yourself on a raft at sea. The number of times you bob up and down per unit time is the frequency of the ocean waves. How far up or down you go relative to a calm sea is the amplitude of the waves.

Consider again the standing wave in a rope shown in Figure 7.9. If ν cycles of the wave pass through a point on the rope each second, and if the length of each cycle is the wavelength λ, the total distance the wave front travels in one second is

$$\nu\lambda = c$$

If we express the frequency ν in s^{-1} and the wavelength λ in meters, we see that the unit for c is $\text{s}^{-1} \times \text{m} = \text{m s}^{-1}$, a distance per unit time. Thus c is a *speed*—the speed of the wave.

In a vacuum, light travels at a *constant* speed of 2.99792458×10^8 m/s (often rounded to 3.00×10^8 m/s). In air, the speed of light is slightly less, but in calculations, we will generally assume that the light waves travel in a vacuum. The speed of light is the fastest speed attainable. A beam of light would travel a distance equal to that from London to San Francisco in about 0.03 s and from Earth to Moon in 1.28 s. Because the speed of light in any given medium is constant, wavelength and frequency are inversely related. The longer the wavelength is, the lower the frequency; or, the shorter the wavelength, the greater the frequency. The amplitude is unrelated to wavelength and frequency. The amplitude indicates the intensity of the wave. For example, the intensity or brightness of a searchlight beam is greater than that of a reading lamp.

EXAMPLE 7.1

Calculate the frequency, in s^{-1}, of an X ray that has a wavelength of 8.21 nm.

SOLUTION

Because we will need to use the relationship, $c = \nu\lambda$, and we know that the units of c are m s^{-1}, we can express the wavelength in meters.

$$\lambda = 8.21\ \cancel{\text{nm}} \times \frac{1 \times 10^{-9}\ \text{m}}{1\ \cancel{\text{nm}}} = 8.21 \times 10^{-9}\ \text{m}$$

Next we can solve the expression: $\nu\lambda = c$ for the frequency ν.

$$\nu = \frac{c}{\lambda}$$

Now we can substitute the known values of c and λ.

$$\nu = \frac{3.00 \times 10^8\ \cancel{\text{m}}\ \text{s}^{-1}}{8.21 \times 10^{-9}\ \cancel{\text{m}}} = 3.65 \times 10^{16}\ \text{s}^{-1}$$

EXERCISE 7.1A

Calculate the frequency, in hertz, of a microwave that has a wavelength of 1.07 mm.

EXERCISE 7.1B

Calculate the wavelength, in nanometers, of infrared radiation that has a frequency of 9.76×10^{13} Hz.

The Electromagnetic Spectrum

The types of electromagnetic radiation range from the extremely short-wavelength, high-frequency *gamma* (γ) rays, which are produced in certain radioactive decay processes, to the very long-wavelength, extra low-frequency waves emitted by electric power transmission lines. This complete range of wavelengths and frequencies is known as the **electromagnetic spectrum** and is illustrated in Figure 7.11.

The electromagnetic spectrum is largely *invisible*; we see only a tiny portion near the middle of the range. The visible spectrum ranges from about 760 nm, where red light blends into the infrared, to about 390 nm, where violet fades into the ultraviolet. Our senses can detect some radiation other than visible. We experience the energy content of infrared radiation by the warmth we feel when this radiation comes through a car window on a cold winter day. Sunburned skin is a sign that we have been exposed too long to ultraviolet radiation. We can detect the different forms of electromagnetic radiation with appropriate devices: radios to detect radio waves, heat sensors to respond to infrared radiation, ordinary photographic film to record images of objects in visible light, and so on.

Materials vary in their abilities either to absorb or let pass (transmit) different parts of the electromagnetic spectrum. Our bodies absorb visible light but transmit most X rays; that is, our bodies are largely transparent to X rays. Ordinary window

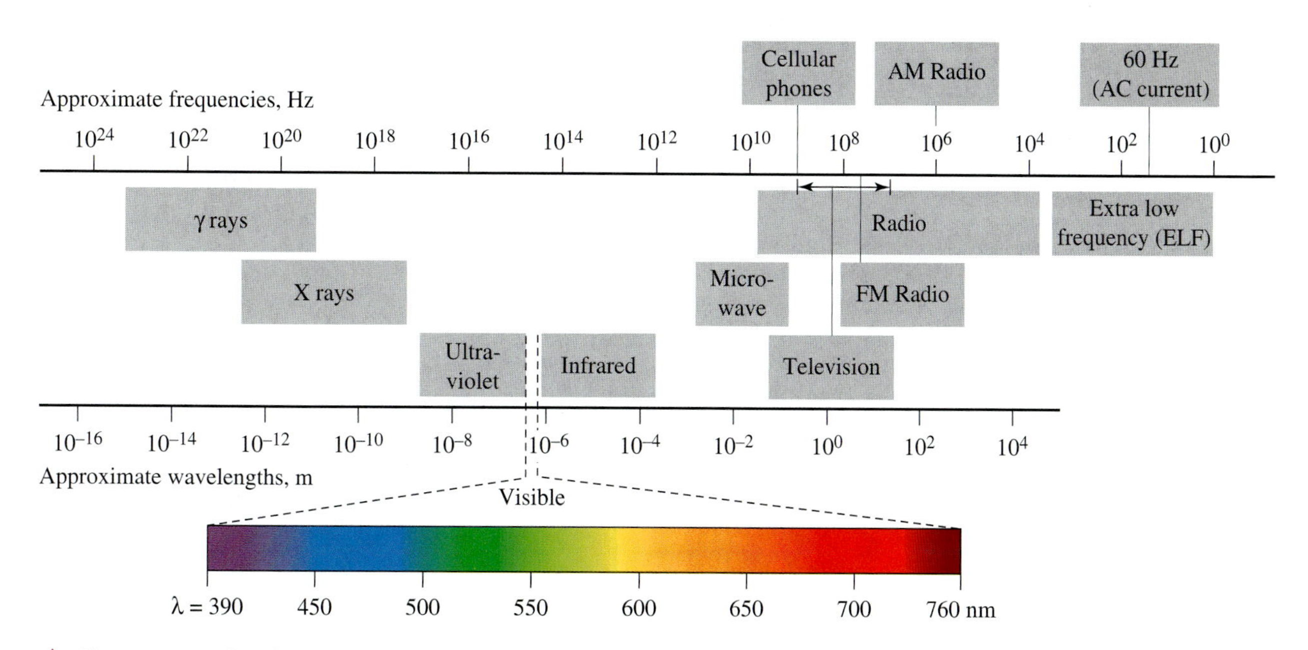

▲ **Figure 7.11 The electromagnetic spectrum**
The visible region, which extends from red at the longest wavelength to violet at the shortest wavelength, is only a small portion of the entire spectrum. The approximate wavelength, frequency ranges, and uses of some other forms of electromagnetic radiation are also indicated.

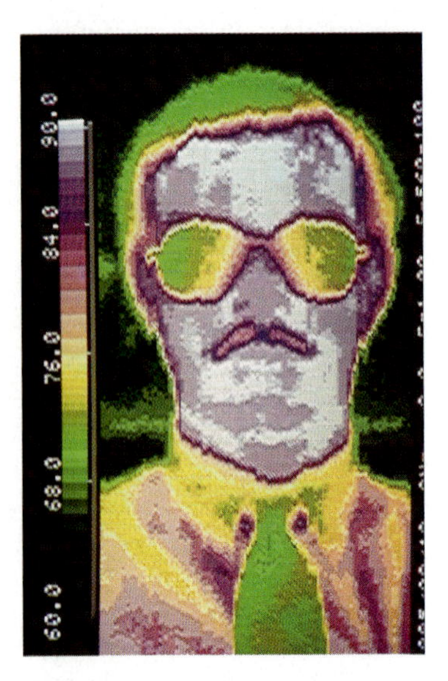

Thermography, a photographic technique that uses infrared radiation to make an image, is used to make a temperature profile of a human face.

glass is transparent to visible light and some infrared radiation, but it effectively blocks, or absorbs, most ultraviolet radiation. (You can't get a suntan from sunlight coming through window glass.) Thus, we use some materials to shield us from harmful rays and others to transmit beneficial rays.

EXAMPLE 7.2—A Conceptual Example

Which light has the higher frequency: the bright red brake light of an automobile or the faint green light of a distant traffic signal?

SOLUTION

The difference in brightness of the two lights is immaterial because, as we have noted, the amplitude of a light wave has no bearing on the frequency or wavelength. From Figure 7.11 we see that there is a relationship between the color of light and its frequency and wavelength. The order of *increasing* frequency in the visible spectrum proceeds from red to violet.

$$\text{Red} < \text{orange} < \text{yellow} < \text{green} < \text{blue} < \text{violet}$$

The green light has a higher frequency than the red light.

EXERCISE 7.2

Which source produces electromagnetic radiation of the longer wavelength: a microwave oven or the fluorescent screen of a color television set?

Continuous and Line Spectra

When white light from an incandescent lamp passes through a narrow slit and then through a glass prism, it separates into a *spectrum*; that is, the various components of white light are spread out into a rainbow of colors (Figure 7.12). The spectrum is *continuous* in that each color merges into the next without a break. Sunlight interacting with raindrops forms a rainbow in the sky. The different colors of light correspond to different wavelengths and frequencies, as we saw in Figure 7.11.

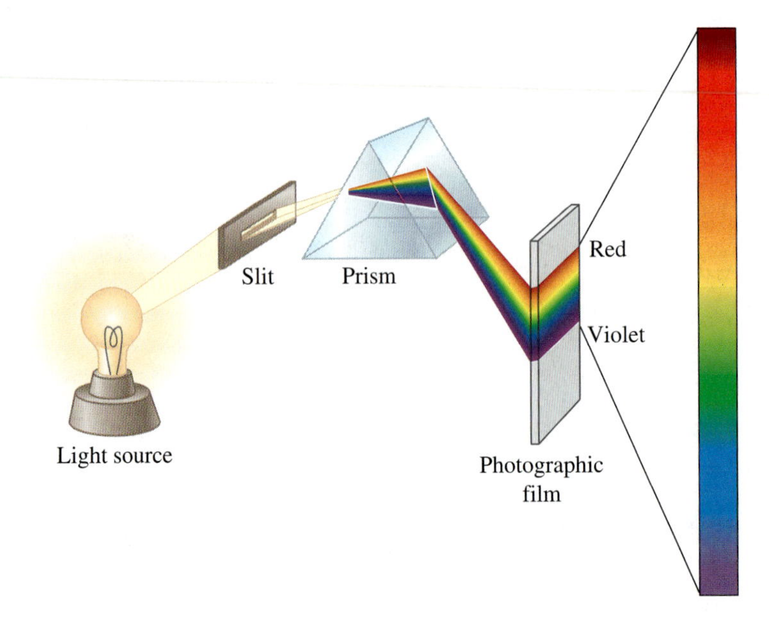

Figure 7.12 The spectrum of ordinary ("white") light
Red light is bent the least and violet light the most when white light is passed through a glass prism. The other colors of the visible spectrum are found between red and violet.

Now consider light from a hydrogen lamp, essentially a cathode ray tube with hydrogen at low pressure as the residual gas. When an electric discharge passes through the tube, some of the hydrogen atoms are energized or excited through collisions with cathode rays. These excited hydrogen atoms give off energy as light. If we pass light from a hydrogen lamp through a slit and prism, we observe only a few images of the slit. These images appear as narrow colored lines separated by dark regions (Figure 7.13); this is a *discontinuous* spectrum. Each line corresponds to electromagnetic radiation of a specific frequency and wavelength. The pattern of lines produced by the light emitted by excited atoms of an element is called a **line spectrum** (Figure 7.14). The line spectrum of an element is characteristic of that element and can be used to identify it. However, many of the lines in a line spectrum are not directly visible; they appear in the ultraviolet or infrared regions of the electromagnetic spectrum.

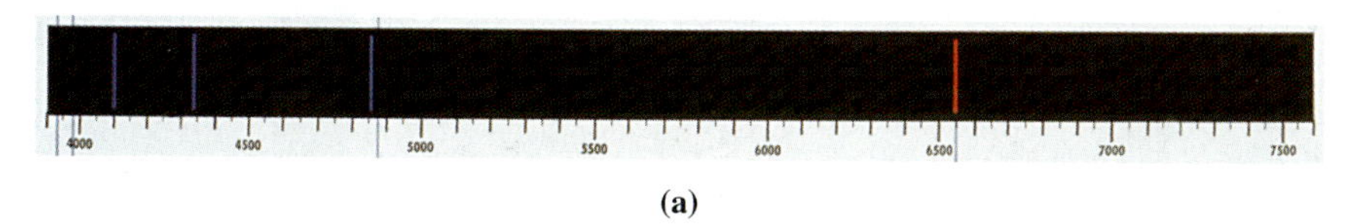

(a)

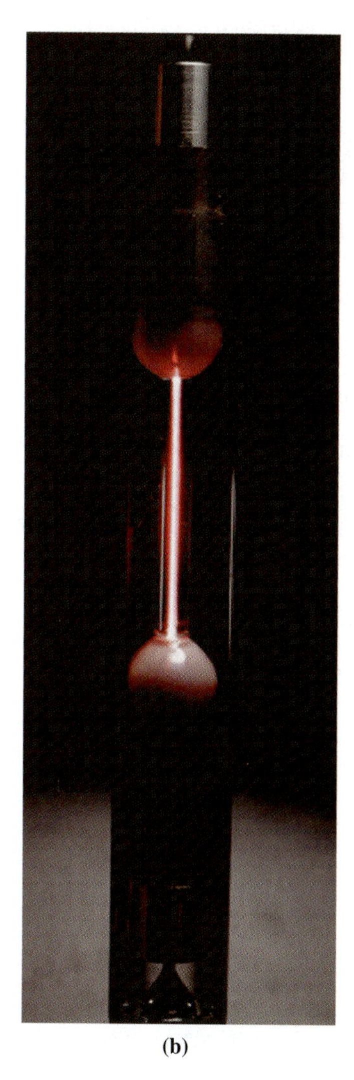

(b)

▲ **Figure 7.13 The visible spectrum of hydrogen**
(a) The Balmer series of the hydrogen spectrum. The Balmer series consists of four lines that appear in the visible region and a much larger number of closely spaced lines in the near ultraviolet that are invisible to the eye. (b) Light from a hydrogen lamp as it appears to the unaided eye.

Emission spectroscopy refers to an analysis of the light emitted when an element is strongly heated or energized by an electric spark or in an electric discharge tube (see Figure 7.15, p. 290). The emitted light is dispersed into individual wavelength components. A photograph or other record of the emitted light is called the **emission spectrum** of the element. Every element has a unique emission spectrum. The element can be identified by the frequencies or wavelengths of its spectral lines. Thus, the emission spectrum is sort of an "atomic fingerprint."

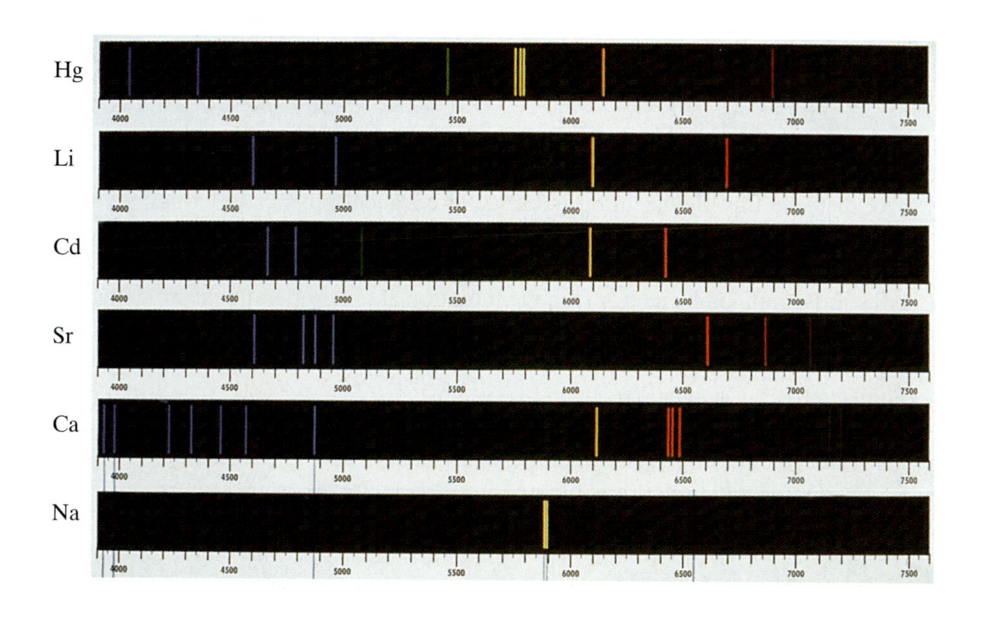

◀ **Figure 7.14 Line spectra of selected elements**
Atomic spectra of several elements are shown here. Wavelengths are given in angstrom units (1 Å $= 10^{-10}$ m). Each element has its own distinctive spectrum which can be used to identify the element. In addition to their practical use in analyzing matter, atomic spectra have led to many of the ideas concerning atomic structure.

Atoms of certain elements—Li, Na, K, Rb, Cs, Ca, Sr, and Ba, for example—emit light when the elements or their compounds are heated in a gas flame. The flame takes on a distinctive color determined by the particular element, and the colored flame is referred to as a *flame test* for the element. The emission spectrum of sodium, for example, is dominated by two closely spaced, exceptionally bright lines in the yellow region of the spectrum (see Figure 7.14). The flame test for sodium is a yellow flame color that appears when sodium metal or sodium compounds are heated. Figure 7.16 shows this color as emitted (a) by a sodium vapor lamp and (b) when a piece of glass tubing is heated in a gas flame. (The raw materials used to make ordinary soda-lime glass are sodium carbonate, calcium carbonate, and silicon dioxide.)

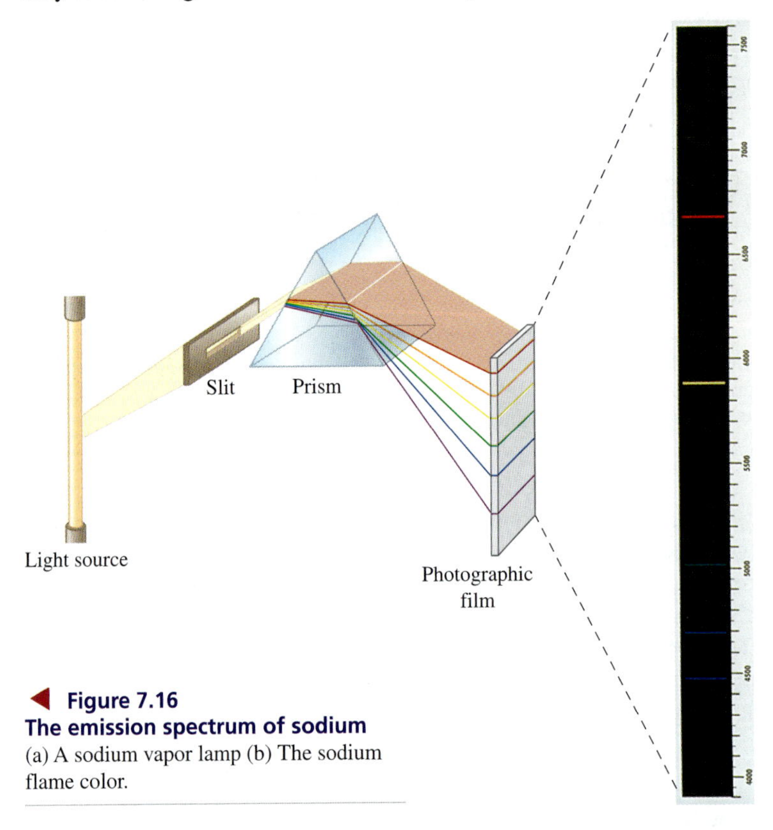

Figure 7.15 The emission spectrum of helium
The light source is a helium lamp. The visible spectrum of helium consists of six colored lines in the visible portion of the electromagnetic spectrum. The quantity of the element is determined from the *intensities* of the lines; the more intense the line, the greater the amount of the element.

(a)

(b)

Figure 7.16 The emission spectrum of sodium
(a) A sodium vapor lamp (b) The sodium flame color.

7.6 Photons: Energy by the Quantum

Toward the end of the nineteenth century, some scientists seemed to feel that classical physics was capable of dealing with all scientific questions. Others, however, were disturbed by the seeming inability of classical physics to explain some rather fundamental natural phenomena, such as the existence of line spectra. Let's consider another phenomenon that classical physics could not explain.

All solids emit some electromagnetic radiation at all temperatures, but in most cases it is invisible infrared radiation. This is the type of radiation, emitted by Earth's surface and trapped by CO_2 and certain atmospheric gases, that causes the greenhouse effect (Chapter 25).

At high temperatures, solids emit radiation that we can generally see; the wavelengths of such radiation are in the visible range. At about 750 °C, for example, a solid emits considerable red light (think of a red-hot poker). As the temperature rises further, more light in the yellow and blue portions of the spectrum blends with the red, until at about 1200 °C, the solid glows white (hence, the term white-hot). This radiation, which depends on the temperature of a solid and *not* on the particular ele-

to explain emission spectra, black-body radiation, and the photoelectric effect. A satisfactory explanation requires the quantum theory.

7.7 Bohr's Hydrogen Atom: A Planetary Model

In 1913, Niels Bohr combined ideas from classical physics and the new quantum theory to explain the structure of the hydrogen atom. In doing so, he was also able to explain the spectrum of light emitted by hydrogen atoms.

Basing his work on that of Planck and Einstein, Bohr assumed that a particular property of the electron (angular momentum) was quantized, that is, that it could have only certain specified values. With this basic assumption, Bohr was then able to use classical physics to calculate other properties. In particular, he found that the electron energy (E_n) was also quantized. Each specified energy value (E_1, E_2, $E_3, \ldots$) is called an **energy level** of the atom, and the only allowable values are given by the following equation.

$$E_n = \frac{-B}{n^2}$$

In this equation, n is an *integer* (that is, $n = 1, 2, 3, \ldots$) and B is a constant based on quantities such as Planck's constant and the mass and charge of an electron: $B = 2.179 \times 10^{-18}$ J. The energy is *zero* when the electron is located infinitely far from the nucleus. Energies associated with forces of attraction are taken to be *negative*, accounting for the negative sign in the equation.

An especially important part of Bohr's theory was the postulate that as long as an electron remains in a given energy level, it cannot emit energy as electromagnetic radiation. This prevents the electron in a hydrogen atom from spiraling into the nucleus. Bohr imagined the electron to orbit about the nucleus much as planets orbit the sun. According to this model, different energy levels for the electrons correspond to different orbits, and only a discrete set of energy levels, or orbits, is possible (Figure 7.19). The lowest (most negative) energy level has $n = 1$ and represents the orbit closest to the nucleus; the next higher energy level has $n = 2$; and so on.

Niels Bohr (1885–1962) proposed his theory of the hydrogen atom early in his career, and he received the Nobel Prize for this work in 1922. Later, he directed the Institute of Theoretical Physics in Copenhagen, a center of attraction for theoretical physicists in the 1920s and 1930s. He worked on the atomic bomb project in World War II, but after the war, he became one of the strongest proponents of peaceful uses of atomic energy.

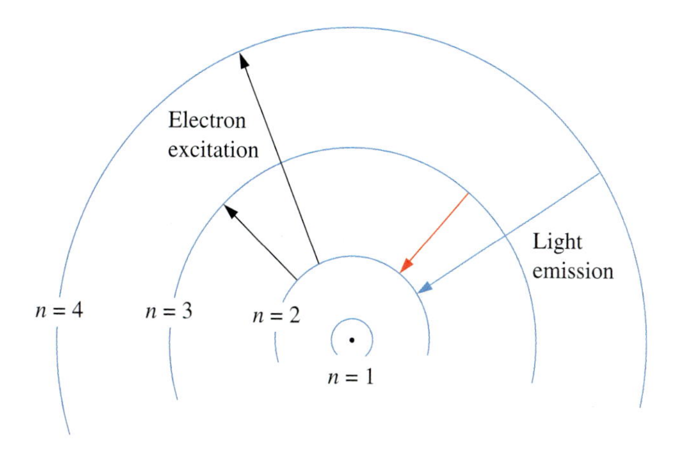

Figure 7.19
The Bohr model of the hydrogen atom
A portion of the hydrogen atom model is shown with the nucleus at the center of the atom and with the electron in one of a set of discrete orbits $n = 1, 2, 3, 4, \ldots$. When the atom is excited, the electron moves to a higher orbit (black arrows). Transitions in which an electron falls to a lower level are accompanied by the emission of light. Two such transitions are shown in colors similar to those of the spectral lines they produce in the Balmer series (see Figure 7.13).

EXAMPLE 7.5

Calculate the energy of an electron in the second energy level of a hydrogen atom.

SOLUTION

Use the Bohr equation for the hydrogen atom, with $B = 2.179 \times 10^{-18}$ J and $n = 2$.

$$E_2 = \frac{-B}{n^2} = \frac{-2.179 \times 10^{-18}\ \text{J}}{2^2}$$

$$= \frac{-2.179 \times 10^{-18}\ \text{J}}{4} = -5.448 \times 10^{-19}\ \text{J}$$

EXERCISE 7.5

Calculate the energy of an electron in the energy level $n = 6$ of a hydrogen atom.

Bohr's Explanation of Line Spectra

When we use the Bohr equation, we are generally interested in the energy change (ΔE_{level}) that accompanies the leap of an electron from one energy level to another in the hydrogen atom. We will define ΔE_{level} as the energy in the final level (E_f) minus the energy in the initial level (E_i).

$$\Delta E_{\text{level}} = E_f - E_i$$

For the final and initial levels we can write

$$E_f = \frac{-B}{n_f^2} \quad E_i = \frac{-B}{n_i^2}$$

The energy difference between n_f and n_i is

$$\Delta E_{\text{level}} = \frac{-B}{n_f^2} - \frac{-B}{n_i^2} = B\left(\frac{1}{n_i^2} - \frac{1}{n_f^2}\right)$$

If $n_f > n_i$, the electron *absorbs* a quantum of energy and moves farther away from the nucleus, that is, from the energy level n_i to the higher level n_f and ΔE_{level} is positive. This energy absorption can occur, for example, during an electric discharge through a gas at low pressure. If $n_f < n_i$, the electron drops from a higher energy level n_i to a lower energy level n_f. The electron moves closer to the nucleus, and a quantum of energy is *emitted* as a photon of light. In this case, ΔE_{level} is negative. In all changes in energy level—called transitions—the electron makes a single jump from one level to another; there is no stopping between levels. Every hydrogen atom in which an electron makes the same transition emits a photon of the same energy. Together, all the photons of this energy produce one spectral line. The collection of lines for all the possible transitions is the observed emission spectrum.

Consider the analogy of a person on a ladder. The person can stand on the first rung, the second rung, the third rung, and so on, but is unable to stand between rungs. As the person goes from one rung to another, the potential energy (energy due to position) changes by definite amounts, or quanta. For an electron, its total energy (both potential and kinetic) changes as it moves from one energy level to another.

EXAMPLE 7.6

Calculate the energy change, in joules, that occurs when an electron falls from the $n_i = 5$ to the $n_f = 3$ energy level in a hydrogen atom.

SOLUTION

To determine the energy change during an electron transition, we need to identify the initial and final energy levels of the electron. We find them in the statement of the problem. We then substitute the numbers of the levels ($n_i = 5$ and $n_f = 3$) and the value of the constant B (2.179×10^{-18} J) into the following equation.

$$\Delta E_{\text{level}} = B\left(\frac{1}{n_i^2} - \frac{1}{n_f^2}\right)$$

$$= 2.179 \times 10^{-18}\left(\frac{1}{5^2} - \frac{1}{3^2}\right) = 2.179 \times 10^{-18}\left(\frac{1}{25} - \frac{1}{9}\right)$$

$$= 2.179 \times 10^{-18}(0.04000 - 0.1111) = -1.550 \times 10^{-19}\text{ J}$$

The negative sign indicates that the atom has given up energy as a photon of light.

EXERCISE 7.6

Calculate the energy change that occurs when an electron is raised from the $n_i = 2$ to the $n_f = 4$ energy level of a hydrogen atom.

We can easily calculate the frequency and wavelength of the photons released when an electron drops from one energy level to a lower one. We first calculate the energy change, as in Example 7.6. This energy change, ΔE_{level}, becomes the E value that we use in Planck's equation.

$$\Delta E_{\text{level}} = h\nu$$

Now we can determine the frequency of the light corresponding to this energy. Finally, if necessary, we can use the relationship $c = \nu\lambda$ to calculate the wavelength of the light.

EXAMPLE 7.7

Calculate the frequency of the radiation released by the transition of an electron in a hydrogen atom from the energy level $n = 5$ to the level $n = 3$.

SOLUTION

We have already determined the energy of a photon of the radiation: ΔE_{level} in Example 7.6 (1.550×10^{-19} J). We can use this ΔE_{level} as the value of E in Planck's equation, $E = h\nu$. Then we can solve the equation for ν.

$$\nu = \frac{E}{h} = \frac{1.550 \times 10^{-19}\text{ J}}{6.626 \times 10^{-34}\text{ J s}} = 2.339 \times 10^{14}\text{ s}^{-1}$$

EXERCISE 7.7A

Calculate the frequency of the radiation emitted during the transition of an electron from the energy level $n = 4$ to $n = 1$ in a hydrogen atom.

EXERCISE 7.7B

Calculate the wavelength, in nanometers, of the radiation emitted by the electron transition from $n_i = 5$ to $n_f = 2$ in a hydrogen atom. In what region of the electromagnetic spectrum is the spectral line produced by this radiation?

The Line Spectrum of Hydrogen

The emission spectrum of hydrogen consists of several series of lines. The most common series are in the ultraviolet, the visible, and the near infrared portions of the electromagnetic spectrum. The electron transitions that produce these spectral series are shown in Figure 7.20. The spectral series in which each electron transi-

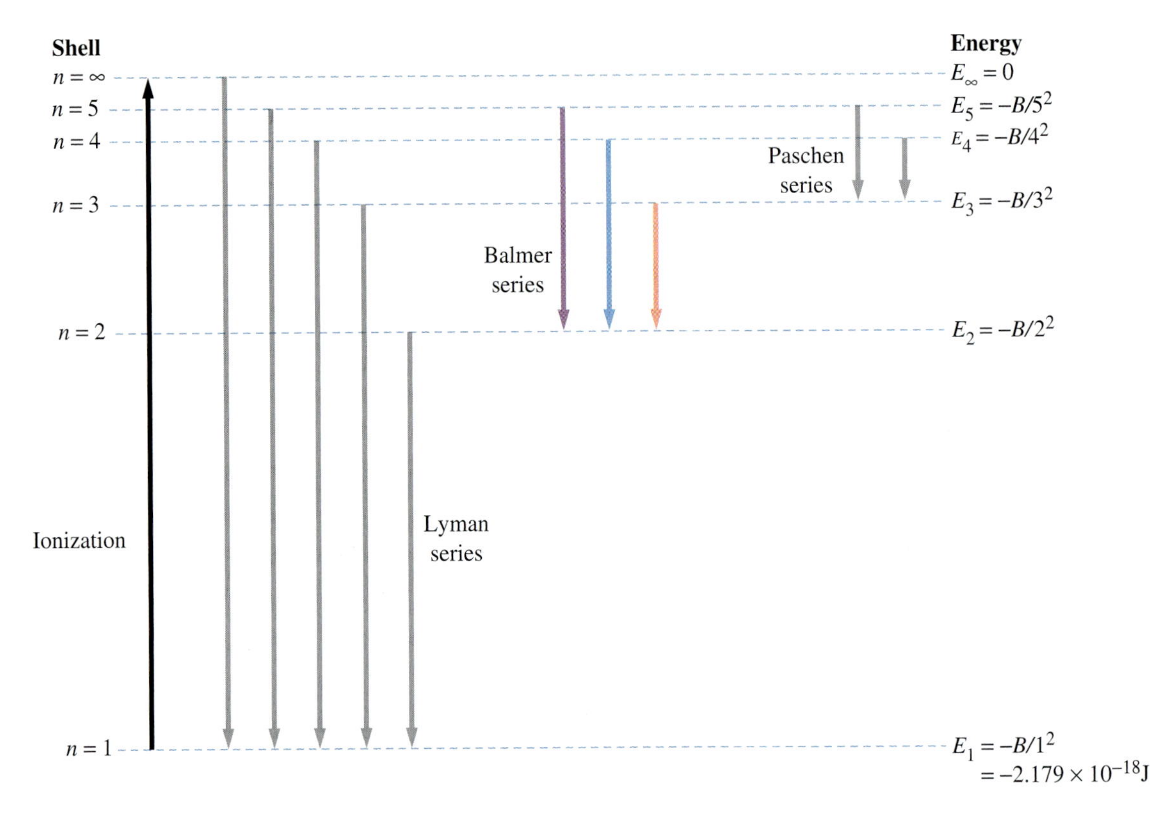

▲ **Figure 7.20 Energy levels and spectral lines for hydrogen**
The distance between energy levels is not to scale. Three of the four visible lines in the Balmer series are shown. Each series is named for the scientist who discovered or characterized it.

tion terminates at $n = 1$, called the Lyman series, is in the ultraviolet range. Four of the lines in the Balmer series, in which an electron drops to the $n = 2$ energy level, are in the visible region (recall Figure 7.13); the other lines for this series are in the ultraviolet region. In the Paschen series, electron transitions terminate at $n = 3$. This series is in the infrared range.

We can also use Figure 7.20 to explain ionization of the normal hydrogen atom, which requires complete removal of the electron. In ionization, the electron is boosted from the level $n = 1$ to a level having an infinitely large value of n, that is, to the level $n = \infty$. The zero value for energy (E_∞) corresponds to the completely ionized atom ($n = \infty$).

Ground States and Excited States

The electron in a hydrogen atom is usually in the lowest energy level (the orbit closest to the nucleus). When an atom has its electrons in their lowest possible energy levels, we say that it is in its **ground state**. When an electric discharge, a flame, or some other source has supplied the energy to promote an electron from the lowest possible level to a higher level, we say the atom is in an **excited state**. An atom in an excited state eventually emits energy in the form of photons as the electron drops back to one of the lower energy levels and ultimately reaches the ground state.

Bohr's theory introduced the important idea of energy levels for electrons in atoms and was a great success in explaining the line spectra of the hydrogen atom. It also gives good results for other one-electron species, such as He^+ and Li^{2+}. However, the theory does not work for atoms with several electrons. Further progress in the study of atomic structure required the introduction of new ideas into quantum theory. We introduce these ideas and the "new" quantum mechanics based on them in the next section.

EXAMPLE 7.8—A Conceptual Example

Without doing detailed calculations, determine which of the four electron transitions shown in Figure 7.21 produces the *shortest wavelength* line in the hydrogen emission spectrum.

SOLUTION

First, we should recognize that transition (a), even though it involves the greatest span of energy levels, requires energy *absorption*, not emission. We can therefore eliminate (a). The other three transitions do involve light emission. Let's compare the energy changes in these transitions, recognizing that the larger the energy change, the *greater* the frequency and the *shorter* the wavelength of the spectral line produced. Transitions (b) and (d) both terminate at the same level, $n = 2$, but the drop in energy in (b) is greater than in (d). Transition (c) begins at the same level as transition (b), but terminates at a higher energy level, $n = 3$. The energy change for transition (c) is therefore less than that for transition (b). Thus, the shortest wavelength spectral line of the three is that produced by transition (b).

EXERCISE 7.8

Without doing detailed calculations, determine which of the following electron transitions in a hydrogen atom requires absorption of the greatest amount of energy (a) from $n = 1$ to $n = 2$; (b) from $n = 3$ to $n = \infty$; (c) from $n = 4$ to $n = 1$; (d) from $n = 2$ to $n = 3$.

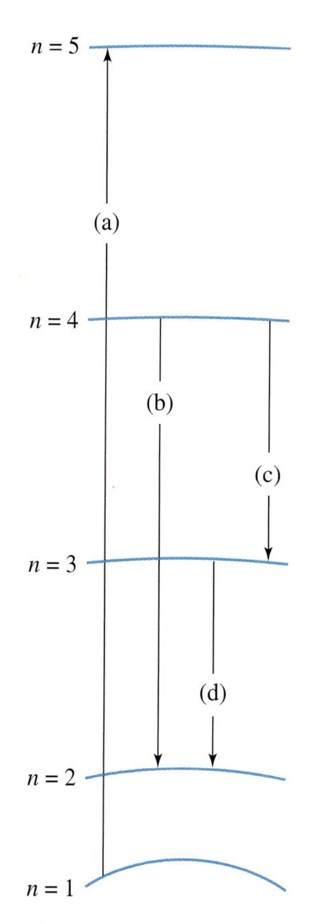

▲ **Figure 7.21**
Several electronic transitions in a hydrogen atom: Example 7.8 illustrated

7.8 Wave Mechanics: Matter as Waves

As we have noted, light seems to be able to act as either waves or particles; we call this the *wave-particle duality* of light. The wave nature is evident when light is dispersed into a spectrum by a prism. The particle nature of light is displayed when photons displace electrons from a metal in the photoelectric effect. We usually regard matter as consisting of particles. Is it possible that under the proper circumstances, matter can behave as waves? Such speculation led Louis de Broglie to propose a startling new theory in 1923 as part of his doctoral thesis. De Broglie's theory, in turn, led to a new mathematical description of atoms that has broad application in modern chemistry.

De Broglie's Equation

De Broglie proposed that a particle with a mass m moving at a speed v will have a wave nature consistent with a wavelength given by the following equation.

$$\lambda = \frac{h}{mv}$$

The symbol h, once again, is Planck's constant.

Atomic Absorption Spectroscopy

On page 289, we described emission spectroscopy, in which light emitted by excited atoms is dispersed into line spectra, and these spectra are analyzed. Now we know that the atoms become excited by absorbing quanta with just the right energy to promote ground-state atoms to an excited state.

An effective way to produce an excited state is to strike the atoms in a high-temperature gaseous sample with a beam of light of the same frequency that the excited gas atoms themselves would emit. Ground-state atoms *absorb* photons from the light beam, and the beam leaves the high-temperature region with a reduced intensity. The more atoms present of the type that can absorb the photons, the lower is the intensity of the transmitted light. In *atomic absorption spectroscopy*, we analyze light that is absorbed rather than emitted.

Suppose we gradually increase the wavelength of a light beam passing through gaseous sodium. When we get to the wavelength 589.00 nm, a strong *absorption* of light occurs. At 589.59 nm, we see another strong absorption. These absorptions occur at the same wavelengths as the two bright yellow lines in the sodium emission spectrum. The transition producing a line in an emission spectrum *emits* the same amount of energy as is *absorbed* in the transition producing the corresponding line in the absorption spectrum.

Over 70 elements can be detected and their amounts determined by atomic absorption spectroscopy, even if their concentrations are as low as a few parts per billion in some cases. Atomic absorption spectroscopy is routinely used in the food industry to test for metals such as calcium, copper, and iron; in environmental analyses for metals such as cadmium, lead, and mercury; and in astronomy to study the chemical makeup of the stars and the atmospheres of the planets.

Sunlight is a white light because the black-body radiation from the extremely hot surface of the sun contains all the wavelength components of the visible electromagnetic spectrum. However, some gaseous atoms in the cooler outer atmosphere of the sun absorb portions of this radiation, creating dark lines in the solar spectrum called the Fraunhofer lines. Figure 7.22 represents the absorption spectrum of sunlight; the elements responsible for the dark lines are identified.

During the solar eclipse of 1868, a line was observed at 587.6 nm. Although quite close to the two sodium lines, this new line was not in the sodium spectrum. It corresponded to no known element on Earth. When helium was discovered in 1895, its emission spectrum produced the same line as the unknown line seen on the sun. This spectral line is the yellow line in Figure 7.15 on page 290.

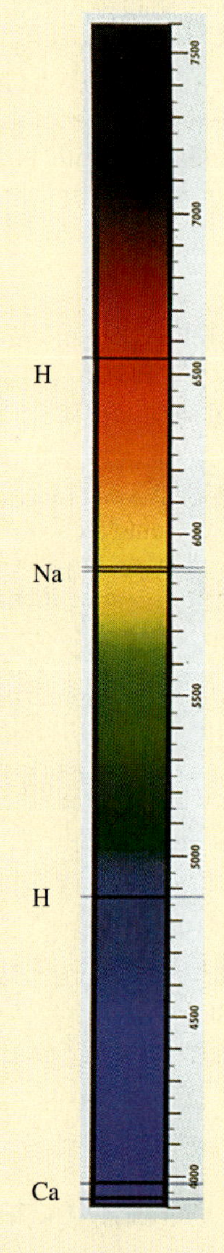

▲ **Figure 7.22**
The absorption spectrum of sunlight

Even large objects presumably have wave properties, but their associated wavelengths are so short that we cannot observe the waves. For example, a 1000-kg car moving at 100 km/h has an associated wavelength of 2.39×10^{-38} m. How short is this wavelength? From Figure 7.11 on page 287 we see that this wavelength is far shorter than anything in the electromagnetic spectrum. It is undetectable. The wave properties of subatomic particles, in contrast, are readily observable.

De Broglie's prediction of matter waves was verified six years later, and his work soon led to the development of the electron microscope. This instrument makes use of the wave nature of electrons, and it is now a standard piece of equipment in many scientific laboratories. An electron microscope can make images of objects as tiny as a few hundred picometers in size (1 pm $= 10^{-12}$ m).

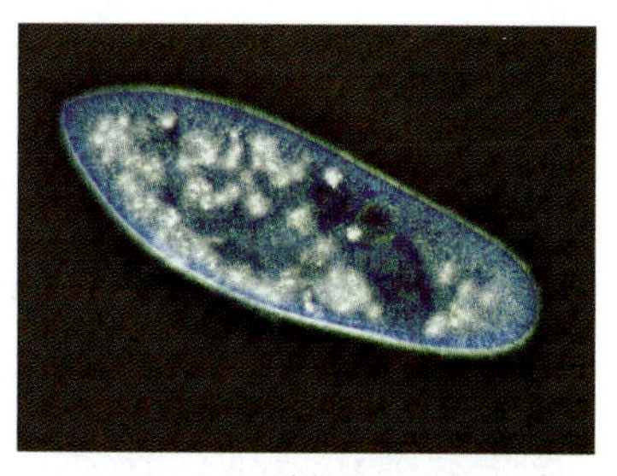

(a)

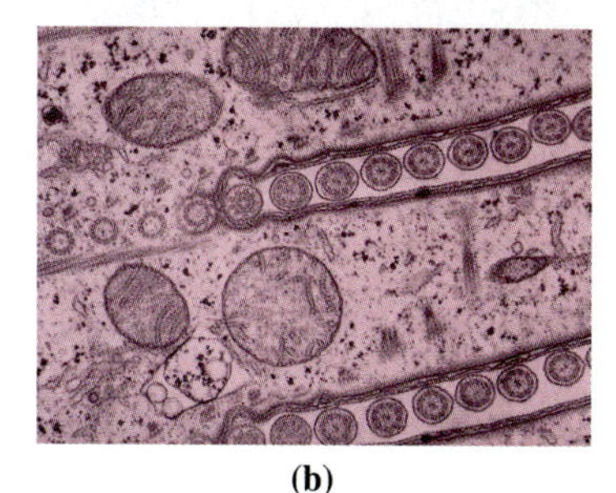

(b)

In an electron microscope, electric and magnetic fields direct and focus electron beams, much as lenses and prisms focus light in optical microscopes. Because of the short wavelengths of electrons, however, the resolving power of an electron microscope is thousands of times greater than an optical microscope. (a) A paramecium as seen with an optical microscope. (b) An image made by an electron microscope shows details of the structure of the paramecium.

EXAMPLE 7.9

Calculate the wavelength, in meters and nanometers, of an electron moving at a speed of 2.74×10^6 m/s. The mass of an electron is 9.11×10^{-31} kg, and 1 J $=$ 1 kg m^2 s^{-2}.

SOLUTION

To use de Broglie's equation, we must first substitute the units kg m^2 s^{-2} for the energy unit J, thus changing the units of Planck's constant, h, from J s to kg m^2 s^{-2} s. Then we must make sure that the particle mass is in kilograms and the particle velocity is in meters per second. These required units are used in the statement of the problem, and we can substitute the given data directly into the de Broglie equation.

$$\lambda = \frac{h}{mv} = \frac{6.626 \times 10^{-34}\ \text{kg m}^2\ \text{s}^{-2}\ \text{s}}{9.11 \times 10^{-31}\ \text{kg} \times 2.74 \times 10^{6}\ \text{m s}^{-1}}$$

$$= 2.65 \times 10^{-10}\ \text{m}$$

The wavelength of the electron is 2.65×10^{-10} m or 0.265 nm.

EXERCISE 7.9

Calculate the wavelength, in nanometers, of a proton moving at a speed of 3.79×10^3 m/s. The mass of a proton is 1.67×10^{-27} kg.

Wave Functions

Although the Bohr model of the hydrogen atom invoked quantum theory, it was still primarily based on classical physics, and we could picture the atom in fairly concrete terms: An electron orbiting around a nucleus is not unlike Earth orbiting the sun. The treatment of atomic structure through the wavelike properties of the electron is called **quantum mechanics**, or *wave mechanics*. In 1926, Erwin Schrödinger developed an equation, called a wave equation, to describe the hydrogen atom. An acceptable solution to Schrödinger's wave equation is called a *wave function*. A wave function, denoted by the Greek letter ψ (psi, pronounced "sigh"), represents an energy state of the atom.

In contrast to the precise planetary orbits of the Bohr atom, wave mechanics provides a less certain picture of the hydrogen atom. Instead of talking about the exact location of the electron, we can only speak of the *probability* of the electron's being found in certain regions of the atom. A useful interpretation of a wave function is the one first proposed by Max Born (1882–1970).

The square of a wave function (ψ^2) gives the probability of finding an electron in a particular volume of space in an atom.

Erwin Schrödinger (1887–1961) was at the University of Zurich when he promulgated his wave equation (1926). He moved to the University of Berlin in 1928 to succeed Max Planck. Although he was not Jewish, Schrödinger left Nazi Germany in 1933, the same year that he received the Nobel Prize in physics. In his later years, Schrödinger's interest turned to biology, expressed in his famous 1944 book, *What Is Life?*

Werner Heisenberg (1901–1976) devised an approach to atomic structure while on a North Sea vacation in 1927. His approach, called matrix mechanics, was purely mathematical but proved to be equivalent to Schrödinger's wave equation. Later in that year, he deduced his famous uncertainty principle. Unlike Schrödinger, Heisenberg remained in Nazi Germany during World War II, directing an unsuccessful atomic bomb project.

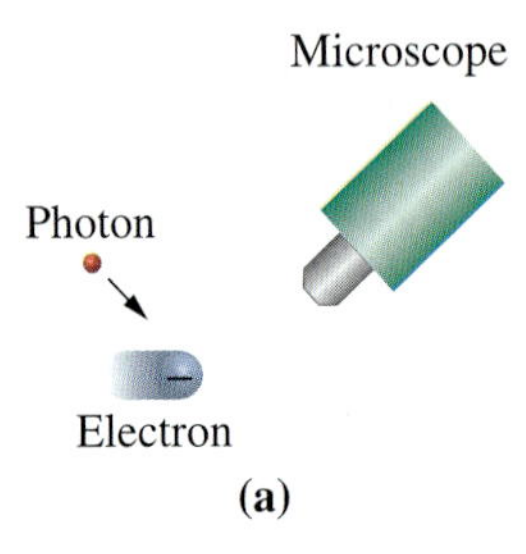

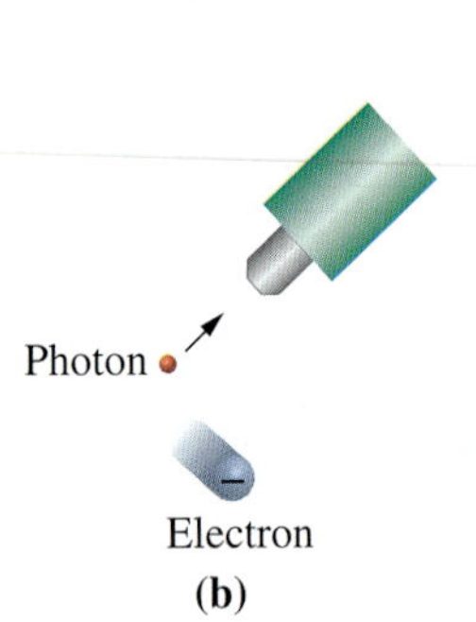

▲ **Figure 7.23**
The uncertainty principle
A free electron moves into the focus of a hypothetical microscope (a) and is struck by a photon of light; the photon transfers momentum to the electron. The reflected photon is seen in the microscope (b), but the electron has moved out of focus. The electron is not where it appears to be.

If we adopt the view that the electron is a cloud of negative electric charge rather than a particle, we can only speak of the charge densities in various parts of the atom. This "fuzzier" atomic model is the only type acceptable according to an important principle of science established in 1927 by Werner Heisenberg.

Heisenberg's **uncertainty principle** essentially states that we can't simultaneously know exactly where a tiny particle like an electron is and exactly how it is moving. One way to visualize the principle is to recognize that when we try to make measurements on an extremely small particle, the act of measuring interferes with the particle we are measuring. That is, we produce changes in one measured quantity just in the act of measuring some other quantity (see Figure 7.23).

The momentum (p) of a particle is the product of its mass (m) and its velocity (v): $p = mv$. If we focus on measuring a particle's momentum, we lose a sense of its exact position (x); and if we focus on measuring the particle's position, we lose a sense of its exact momentum (p). Heisenberg related the product of the *uncertainties* in position (Δx) and momentum (Δp) to Planck's constant, h.

$$\Delta x \Delta p \geqq \frac{h}{4\pi}$$

The term $h/4\pi$ has the magnitude 5.3×10^{-35}, an extremely small number. No matter how precisely we measure the position and momentum of a large object, the product of the uncertainties in the two measurements far exceeds 5.3×10^{-35}, just as Heisenberg's principle states that it should. This means, for example, that we can readily and accurately describe the orbit of Earth about the sun or of the smallest artificial satellite about Earth. The situation is entirely different when we deal with a particle as small as an electron, whose mass is only 9.11×10^{-31} kg. Even the smallest uncertainties in position and momentum that meet the requirements of the uncertainty principle are *relatively* large. We cannot measure both these properties with great precision. As a result, the uncertainty in the position of an electron in an atom might be about as large as the atom itself.

In light of the uncertainty principle, Bohr's model of the hydrogen atom fails, in part, because it tells us *more* than we can know with certainty. It gives us a precise location for the electron—the particular orbit—and it enables us to make a precise calculation of the electron speed (and its momentum, mv) in this orbit.

7.9 Quantum Numbers and Atomic Orbitals

The wave functions for the hydrogen atom contain three parameters to which we must assign specific integral values called **quantum numbers**. A wave function with a given set of these three quantum numbers is called an **atomic orbital**. Atomic orbitals, although they are only mathematical expressions, allow us to visualize a three-dimensional region in an atom in which there is a significant probability of finding electrons. Consequently, we tend to think of orbitals as geometric regions as well as mathematical expressions. Because orbitals are characterized by quantum numbers, we first say a bit more about quantum numbers.

Quantum Numbers

Let's consider the three quantum numbers and their possible values. When we specify a set of values for the three quantum numbers, we define a specific atomic orbital. We might think of the three quantum numbers as similar to a residential address. One part of the address locates the state; a second, the city; and a third, the street address. Table 7.1 summarizes the discussion that follows, and you may find it helpful to refer to the table as we proceed.

TABLE 7.1 Electronic Shells, Orbitals, and Quantum Numbers.

Principal Shell	**1st**	**2nd**				**3rd**									
n	1	2	2	2	2	3	3	3	3	3	3	3	3	3
l	0	0	1	1	1	0	1	1	1	2	2	2	2	2
m_l	0	0	−1	0	+1	0	−1	0	+1	−2	−1	0	+1	+2
Subshell and orbital designation	1s	2s	2p	2p	2p	3s	3p	3p	3p	3d	3d	3d	3d	3d
Number of orbitals in the subshell	1	1		3		1		3				5		

1. The **principal quantum number (n)** is assigned first because the allowable values of the other two quantum numbers depend on the value assigned to n. The value of n can only be a *positive integer* (whole number).

$$n = 1, 2, 3, 4, 5, 6, 7, \ldots$$

The quantum number n is analogous to n in Bohr's planetary model. The size of an orbital and its electron energy depend mainly on the n quantum number. As n increases, the electron is at a higher energy level and spends more time farther from the nucleus. Orbitals with the same value of n are said to be in the same **principal shell**.

2. The **orbital angular momentum quantum number (l)** determines the *shape* of an orbital. This quantum number can have positive integral values from zero to one less than the value of n.

$$l = 0, 1, 2, \ldots, (n - 1)$$

All orbitals having the same value of n and the same value of l are said to be in the same principal shell and in the same **subshell**. Orbitals and subshells are also designated by a letter (s, p, d, f) according to the following scheme.

Value of l	0	1	2	3
Orbital and subshell designations	s	p	d	f

Each of these four orbital types describes a region of space with a different shape. Also note that the number of different kinds of orbitals and subshells in a principal energy shell is equal to the principal quantum number n. For example, the third principal shell ($n = 3$) has three subshells and three kinds of orbitals: s, p, and d, corresponding to l values of 0, 1, and 2, respectively.

3. The **magnetic quantum number (m_l)** determines the orientation in space of the orbitals of a given type in a subshell. This quantum number can have any integral value ranging from $-l$ to $+l$, including 0.

$$m_l = 0, \pm1, \pm2, \ldots \pm l$$

Thus, if $l = 0$, the only permitted value of m_l is also 0; if $l = 1$, m_l may have any one of three values: -1, 0, or $+1$; and so on. In all, the number of possible values for $m_l = 2l + 1$, and this determines the number of orbitals in a subshell.

Let's summarize the relationship between the l and m_l quantum numbers.

s orbitals	*p* orbitals	*d* orbitals	*f* orbitals
$l = 0$	$l = 1$	$l = 2$	$l = 3$
$m_l = 0$	$m_l = 0, \pm 1$	$m_l = 0, \pm 1, \pm 2$	$m_l = 0, \pm 1, \pm 2, \pm 3$
one s orbital in an *s* subshell	*three p* orbitals in a *p* subshell	*five d* orbitals in a *d* subshell	*seven f* orbitals in an *f* subshell

To indicate the principal shell in which an orbital or subshell is found, we combine the principal quantum number of that shell with the orbital or subshell designation. Thus, the symbol 2*p* represents either the *p* subshell of the second principal shell or one of the three *p* orbitals that make up the subshell.

Example 7.10 tests your understanding of relationships among the quantum numbers. Example 7.11 applies this understanding to questions about principal shells, subshells, and atomic orbitals.

EXAMPLE 7.10

Considering the limitations on values for the various quantum numbers, state whether an electron can be described by each of the following sets. If a set is not possible, state why it is not.

a. $n = 2, l = 1, m_l = -1$

b. $n = 1, l = 1, m_l = +1$

c. $n = 7, l = 3, m_l = +3$

d. $n = 3, l = 1, m_l = -3$

SOLUTION

a. All the quantum numbers are allowed values.

b. Not possible. The value of l must be less than n.

c. All the quantum numbers are allowed values.

d. Not possible. The value of m_l must be in the range $-l$ to $+l$ (in this case, -1, 0, or $+1$).

EXERCISE 7.10

Considering the limitations on values for the various quantum numbers, state whether an electron can be described by each of the following sets. If a set is not possible, state why it is not.

a. $n = 2, l = 1, m_l = -2$

b. $n = 3, l = 2, m_l = +2;$

c. $n = 4, l = 3, m_l = +3$

d. $n = 5, l = 2, m_l = +3$

EXAMPLE 7.11

Consider the relationship among quantum numbers and orbitals, subshells, and principal shells to answer the following: **(a)** How many orbitals are there in the 4*d* subshell? **(b)** What is the first principal shell in which *f* orbitals can be found? **(c)** Can an atom have a 2*d* subshell? **(d)** Can a hydrogen atom have a 3*p* subshell?

SOLUTION

a. The *d* subshell corresponds to a value of $l = 2$. With $l = 2$, there are five possible values of m_l: $-2, -1, 0, +1$, and $+2$. There are *five d* orbitals in the 4*d* subshell.

b. The *f* orbital corresponds to a value of $l = 3$. Since the maximum value of l is $n - 1$, the allowable values of n are 4, 5, 6, and so on. The first shell to contain *f* orbitals is the fourth principal shell ($n = 4$).

c. No. For a *d* subshell, $l = 2$. The maximum value of l is $n - 1$, so if $n = 2$, l cannot be 2. There cannot be a 2*d* subshell.

d. Yes. Although the electron in a hydrogen atom is usually in the 1*s* orbital, it can be excited to one of the higher energy states, such as one of the 3*p* orbitals.

EXERCISE 7.11

Consider the relationship among quantum numbers and orbitals, subshells, and principal shells to answer the following: **(a)** How many orbitals are there in the 5*p* subshell? **(b)** What is the *total* number of orbitals in the principal shell $n = 4$? **(c)** What subshells of the hydrogen atom consist of a total of seven orbitals?

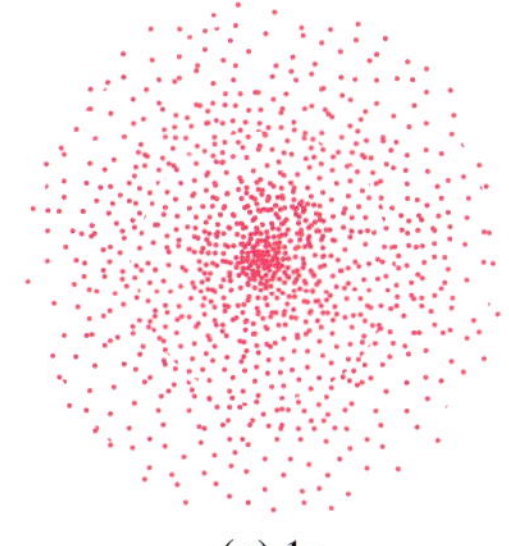

(a) 1*s*

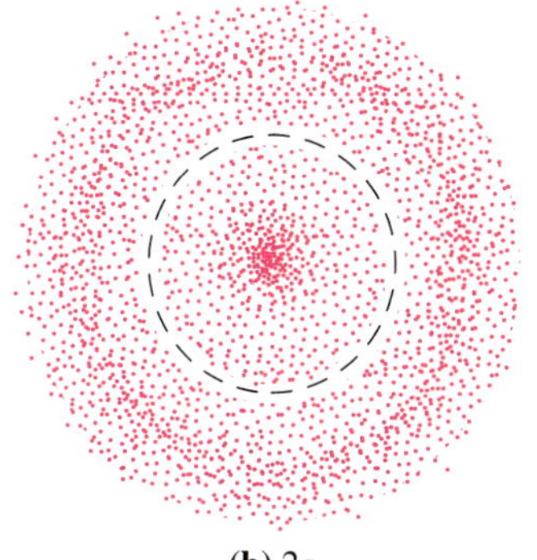

(b) 2*s*

▲ **Figure 7.24**
The 1*s* and 2*s* orbitals
The geometric regions defined by *s* orbitals are *spheres*. Shown here are circular *cross-sections* obtained by passing a plane through the center of each of the spherical orbitals, where we imagine the nucleus to be. (a) The close spacing of the dots in the center of the 1*s* orbital signifies that the highest probability of finding an electron is in the vicinity of the nucleus. The probability falls off with distance from the nucleus. (b) The pattern of dots in the 2*s* orbital covers a larger region than that in a 1*s* orbital. The broken circle outlines a region (called a *node*) in which there is no probability of finding an electron. The most probable location of the electron is in a spherical shell found in the second ring of dots.

Electron Probabilities and the Shapes of Orbitals

What do the regions of high electron probabilities described by atomic orbitals look like? We can't see these regions, of course, but we can translate the mathematical forms of the orbitals into geometric pictures. Strictly speaking, an atomic orbital must have an infinite size to guarantee a 100% certainty of finding an electron in it, and we can't picture that. For practical purposes, though, we can outline a region in which there is a certain probability (say, a 90% probability) that the electron will be found within the region. Alternatively, if we think of the orbital as a "charge cloud," we can say that this region contains a certain percentage of the electron's charge (say, 90%).

The wave mechanical picture of an electron in an *s* orbital looks like a fuzzy ball; that is, the orbital has *spherical* symmetry. The size of the orbital and the specific electron probability distribution depend on the value of *n*. Figure 7.24 suggests that the greatest probability of finding an electron in a 1*s* orbital is at the nucleus itself, as suggested by the heavy concentration of dots at the center of the sphere. Another useful quantity is the total probability of an electron's being at a given distance from nucleus. Finding this probability is like adding up all the dots that are a given distance from the nucleus in Figure 7.24a. These dots would lie on the circumference of a circle with the nucleus at its center. In the three-dimensional picture, this means summing the probabilities at all the points that lie in a very thin spherical shell, with the nucleus at its center. Of all the possible shells we might choose, the one with the greatest probability of containing an electron has a radius of 52.9 pm, a distance identical to the first Bohr orbit of hydrogen. The dartboard in Figure 7.25 is a two-dimensional analogy to this description of a three dimensional 1*s* orbital.

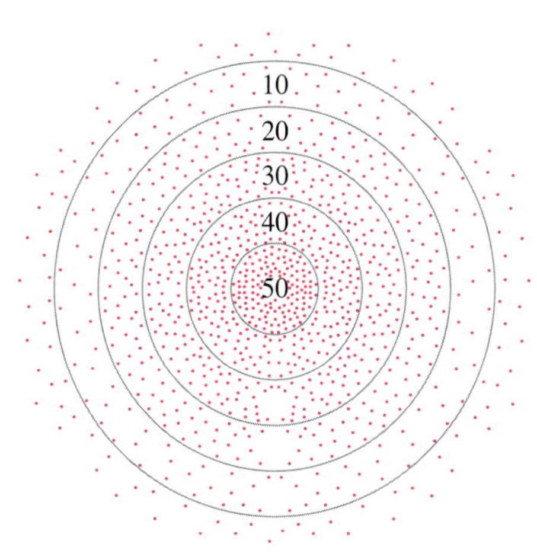

◀ **Figure 7.25**
An analogy to a 1*s* orbital
Imagine that the hole left by each dart throw represents the probability of an electron's being at that point. If 900 of 1000 dart throws hit the board, the dartboard is analogous to the 90% outline of a 1*s* orbital. The greatest number of holes per unit area (say, per 1 cm^2) is in the "50" region—that's what the dart thrower aims at. But what is the most likely score that the average dart thrower will make on a given throw? Can you see that the answer is "30"? Although the number of holes per square centimeter in the "30" region is smaller than in the "50" region, the total area of the "30" region is much greater. The "30" ring of the dartboard is analogous to the 52.9-pm spherical shell of the 1*s* orbital in the hydrogen atom.

The 2s orbital has two regions of high electron probability; both are spherical. The region near the nucleus is separated from the outer region by a spherical *node*—a spherical shell in which the electron probability is zero. The 2s electron is more likely to be found in the outer region.

The second principal shell consists of two subshells, 2s and 2p. The 2s subshell has only one orbital, the 2s orbital just described. The 2p subshell consists of three 2p orbitals, corresponding to $l = 1$ and three possibilities for m_l. The 2p orbitals describe dumbbell-shaped regions (Figure 7.26); two lobes lie along a line with the nucleus at the center between the lobes. In a 2p orbital, there is a planar node between the two "halves" of the orbital. To distinguish among the different 2p orbitals, we sometimes refer to them as the $2p_x$, $2p_y$, and $2p_z$ orbitals because they are perpendicular to one another and can be drawn along the x, y, and z coordinate axes (see Figure 7.26). Higher p orbitals—that is, 3p, 4p, and so on—have similar overall shapes, but they have spherical and planar nodes that separate regions of higher electron density. Their sizes increase with the value of n.

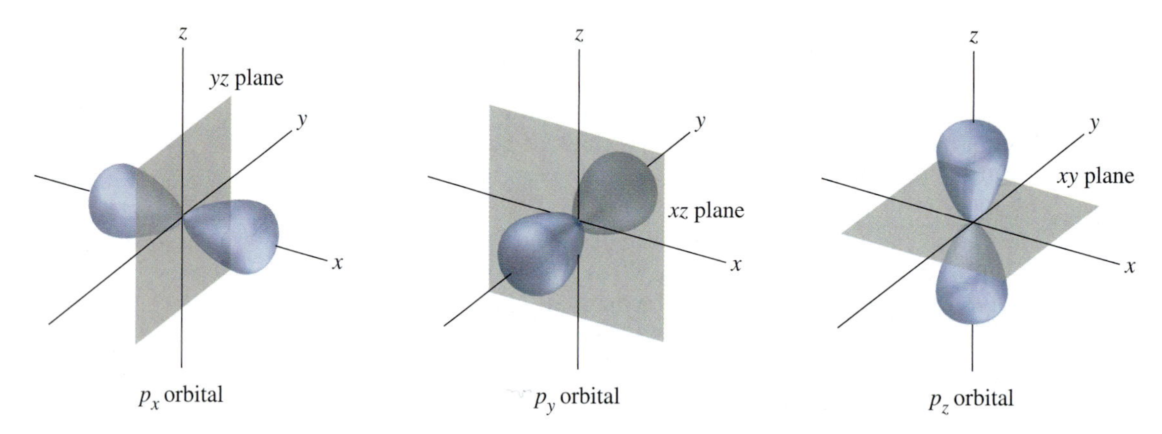

▲ **Figure 7.26 The three 2*p* orbitals**
The regions of high electron probabilities are dumbbell-shaped and oriented along the x, y, and z axes, respectively. Each of the orbitals has a *nodal plane*, a planar region of zero electron probability that passes through the nucleus of the atom.

● **ChemCDX** Radial Electron Distribution

The third principal energy shell is divided into three subshells: one 3s orbital, three 3p orbitals, and five 3d orbitals. The d orbitals are pictured in Figure 7.27.

The fourth principal energy shell is divided into four subshells: a 4s subshell with one orbital, a 4p subshell with three orbitals, a 4d subshell with five orbitals, and a 4f subshell with seven orbitals. The seven f orbitals have complex shapes that we will not consider here.

Electron Spin: A Fourth Quantum Number

The quantum numbers n, l, and m_l arise from Schrödinger's treatment of the electron in a hydrogen atom as a matter wave. This concept can be extended to electrons in other atoms as well. Although these three quantum numbers fully characterize the orbitals in an atom, we need an additional quantum number to describe the electrons that occupy these orbitals. This fourth quantum number, the **electron spin quantum number (m_s)**, was proposed in 1925 by Samuel Goudsmit and George Uhlenbeck to explain some of the finer features of atomic emission spectra. The two possible values of the spin quantum number are $+1/2$ (also represented by the arrow ↑) and $-1/2$ (↓).

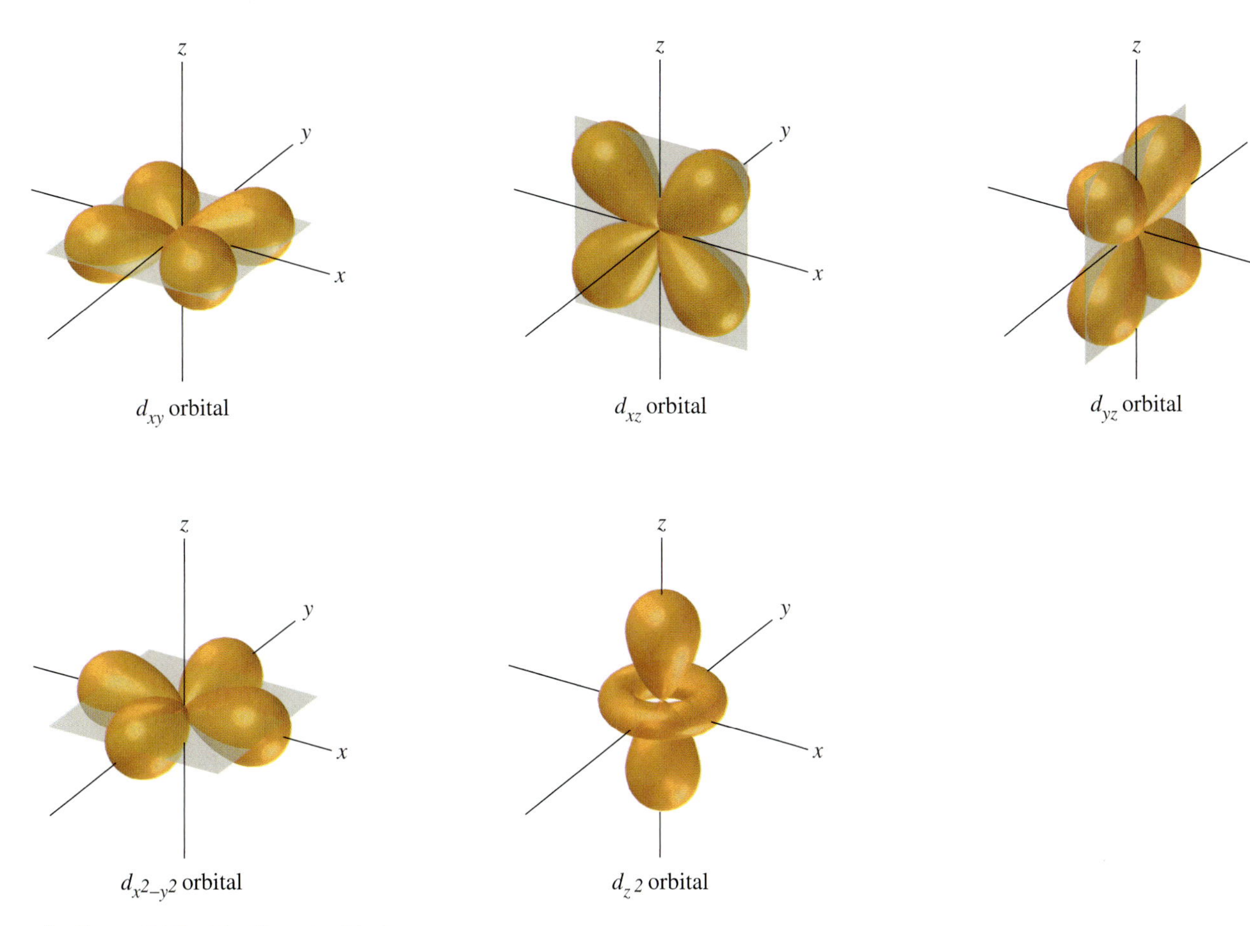

Figure 7.27 The five *d* orbitals
The designations xy, xz, yz, $x^2 - y^2$, and z^2 refer to the directional characteristics of certain combinations of the orbitals having the allowed values of m_l when $l = 2$.

The name, electron spin quantum number, suggests that electrons have a spinning motion, as suggested in Figure 7.28. However, we can't attach a precise physical reality to electron spin. All we know is that certain observations made in *experiments* can best be explained in terms of electron spin.

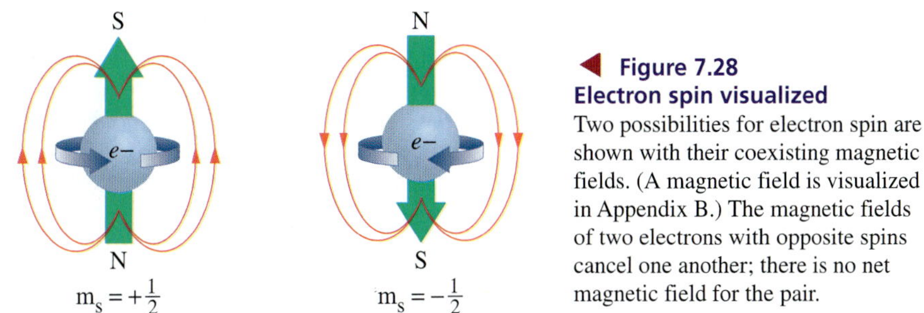

Figure 7.28 Electron spin visualized
Two possibilities for electron spin are shown with their coexisting magnetic fields. (A magnetic field is visualized in Appendix B.) The magnetic fields of two electrons with opposite spins cancel one another; there is no net magnetic field for the pair.

In 1921, Otto Stern and Walter Gerlach performed the experiment illustrated in Figure 7.29. When they shot a beam of gaseous silver atoms into a powerful magnetic field, the beam was split into two by the field, indicating that the atoms themselves act like small magnets.

It has long been known that a moving electric charge induces a magnetic field about itself, so a spinning electron should have an associated magnetic field. The

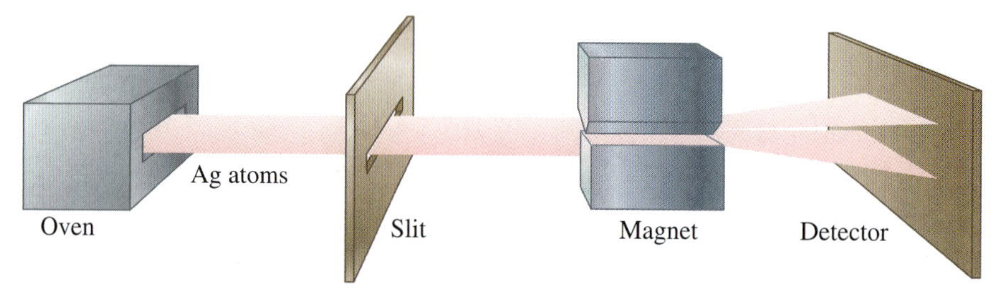

▲ **Figure 7.29 The Stern-Gerlach experiment: Demonstration of electron spin**
Silver atoms vaporized in the oven are shaped into a beam by the slit, and the beam is passed through a nonuniform magnetic field. The beam splits in two. (The beam of atoms would not experience a force if the magnetic field were uniform. The field strength must be greater in certain directions than in others.)

magnetic field produced by an electron with $m_s = +1/2$ is in the opposite direction to the field caused by an electron with $m_s = -1/2$. The magnetic effect is canceled for a pair of electrons with opposing spins.

Here is a way we can explain the Stern-Gerlach experiment: Of the 47 electrons in a silver atom, 23 have a spin in one direction and 24 have the opposite spin. If 46 of the electrons have their spins paired, their magnetic fields cancel. The direction of deflection of a given silver atom depends on the type of spin on the 47th elec-

Electron Probabilities and a Close-Up Look at Atoms

In the Bohr atom, we considered the energy requirement to move an electron from one orbit to another orbit more distant from the positively charged nucleus. In wave mechanics, we replace definite orbits with the probabilities of finding an electron in different regions outside the nucleus. The wave-mechanical interpretation even includes the extremely small but nonzero possibility that an atom may transfer an electron to an adjacent atom without first ionizing. This can occur when an electron has a significant probability of being closer to the nucleus of another atom than to the nucleus of its "parent" atom. This transfer is called *tunneling* and requires much less energy than ionization.

The tunneling effect was put to good effect in the early 1980s when Gerd Binnig and Heinrich Rohrer developed the technique of *scanning tunneling microscopy* to obtain the images of individual atoms on the surface of a solid. A scanning tunneling microscope (STM) uses a tungsten probe with an extraordinarily sharp tip that is carefully placed only about 0.5 nm from the surface being studied. A minute electric potential is applied across the gap separating the tip from the surface, increasing the probability of electrons tunneling from the probe to the surface. The flow of these electrons creates a small electric current. The instrument is designed to keep this current constant, which in turn requires that the probe move up and down as it follows the contours of the atoms on the surface. The surface is scanned repeatedly, and the results are processed by a computer to give a three-dimensional map of the surface. These maps tell us how atoms are arranged on a surface. However, STM images tell us nothing about the internal structure of atoms. For that, we still use indirect evidence such as atomic spectra.

In this image made by a scanning tunneling microscope (STM), individual silicon atoms on the surface of a silicon crystal are seen at a magnification of 10 million.

tron. In a collection of silver atoms, there is an equal chance that the "odd" electron will have a spin of $+1/2$ (↑) or $-1/2$ (↓). The beam of silver atoms is thus split into two beams. A beam of hydrogen atoms would be similarly split in a magnetic field, indicating that half the electrons have a spin of $+1/2$ and half have a spin of $-1/2$.

Looking Ahead

We have now described the four quantum numbers that are used to characterize electrons in atoms. In the next chapter, we will see how these quantum numbers, together with a few simple rules, allow us to establish which orbitals in atoms are occupied by electrons. We will also discover a close relationship between the electronic structures of atoms and the periodic table. In Chapters 9 and 10, we will see how the electronic structures of atoms in turn determine the nature of chemical bonds between atoms and the structures of molecules.

The Stillness at 0 K: Temperature, Uncertainty, and Superconductivity

Science is not just a collection of facts, but rather it is a harmonious whole that helps us to understand the entire universe. Understanding gained in one area can often be applied in another seemingly unrelated area. Science is knowledge built to a large degree on mathematics. Consider the following case.

At room temperature, the kinetic molecular theory predicts a typical speed of gaseous atoms of the order of 1000 mi/h. At temperatures of about 20 nanokelvins (0.000000020 K), their speed should be of the order of a few *feet* per hour. In 1995, scientists succeeded in "chilling" rubidium atoms to this extremely low temperature by using laser beams to slow down the atoms and nudge them together. This is much like using sheep dogs to turn stray sheep back into the flock. The speed of the rubidium atoms, however, proved to be much slower than expected; in fact, too slow to measure. The atoms remained in a closely knit group. This behavior conformed to Albert Einstein's speculations on some mathematics developed by the Indian physicist, Satyendra Bose, in 1924. This clustering of atoms at near 0 K is now called a Bose-Einstein condensate (BEC).

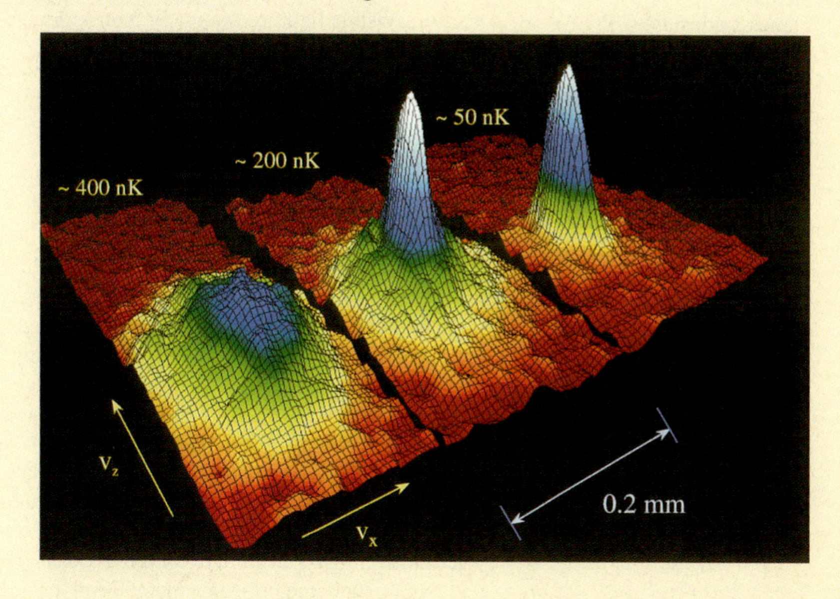

This image shows the speed distribution of atoms that have formed a Bose-Einstein condensate.

Now consider Heisenberg's uncertainty principle. Since the atoms in the BEC are hardly moving, their *momentum* is very near 0. Since we know the momentum quite accurately, the uncertainty principle tells us that we cannot know their *position* very well. Indeed the atoms act as a group; individual atoms cannot be distinguished.

BEC properties are related to superconductivity (Chapter 24). Scientists hope that study of BEC samples will shed some light on that phenomenon.

Key Terms

atomic orbital (7.9)
cathode ray (7.1)
electromagnetic spectrum (7.5)
electromagnetic wave (7.5)
electron spin quantum number (m_s) (7.9)
emission spectrum (7.5)
energy level (7.7)
excited state (7.7)
frequency (ν) (7.5)
ground state (7.7)
line spectrum (7.5)
magnetic quantum number (m_l) (7.9)
mass spectrometer (7.4)
orbital angular momentum quantum number (l) (7.9)
photon (7.6)
Planck's constant (h) (7.6)
principal quantum number (n) (7.9)
principal shell (7.9)
quantum (7.6)
quantum (wave) mechanics (7.8)
quantum number (7.9)
subshell (7.9)
uncertainty principle (7.8)
wave (7.5)
wavelength (λ) (7.5)

Summary

Cathode rays, formed when electricity passes through gases at very low pressures, can be detected through the fluorescence they produce when they strike surfaces. Studies on their deflection in electric and magnetic fields yielded their mass-to-charge ratio and the suggestion that they are negatively charged fundamental particles of matter. The electric charge on cathode ray particles, now called electrons, is the fundamental unit of negative electric charge.

Thomson's "raisin pudding" model viewed an atom as electrons ("raisins") embedded in a positively charged "pudding." Rutherford's atomic model is that of a tiny positively charged nucleus and extranuclear electrons. The nucleus consists of protons and neutrons and contains practically all the mass of an atom. Atomic masses and relative abundances of the isotopes of an element can be established by mass spectrometry.

Electromagnetic radiation is an energy transmission in the form of oscillating electric and magnetic fields. The oscillations produce waves that are characterized by their frequencies (ν), wavelengths (λ), and velocity (c). The complete span of possibilities for frequency and wavelength is described as the electromagnetic spectrum, of which visible light is only a very small part.

$$c = \nu\lambda$$

$$c = 3.00 \times 10^8 \text{ m s}^{-1} \text{ in vacuum}$$

Visible light: red ($\lambda < 760$ nm), orange, yellow, green, blue, violet ($\lambda > 390$ nm)

Light consisting of essentially all wavelength components in a region of the electromagnetic spectrum is a continuous spectrum (for example, the "rainbow" of colors). Energetically excited atoms typically emit only a discrete set of wavelength components in the form of a line spectrum.

The energy of electromagnetic radiation is limited to integral multiples of a fundamental quantity called a *quantum* and having a value of $E = h\nu$. The photoelectric effect is explained by thinking of quanta of energy as concentrated into "particles" of light called *photons*.

Bohr's theory requires the electron in a hydrogen atom to be in one of a discrete set of energy levels. The fall of an electron from a higher to a lower energy level releases a discrete amount of energy as a photon of light with a characteristic frequency. Bohr's theory accounts for the observed atomic spectrum of hydrogen.

The electron in a hydrogen atom can be viewed as a matter wave enveloping the nucleus. The matter wave is represented by a wave equation, and solutions of the wave equation are wave functions. Wave functions require the assignment of three quantum numbers: the principal quantum number, n, the orbital angular momentum quantum number, l, and the magnetic quantum number, m_l. Wave functions with acceptable values of the three quantum numbers are called atomic orbitals. Orbitals describe regions in an atom that have a high probability of containing an electron or a high electronic charge density. Orbitals with the same value of n are in the same principal energy level, or principal shell. Those with the same value of n and of l are in the same subshell, or sublevel. The shapes associated with orbitals depend on the value of l. Thus, an s orbital ($l = 0$) is spherical and a p orbital ($l = 1$) is dumbbell-shaped.

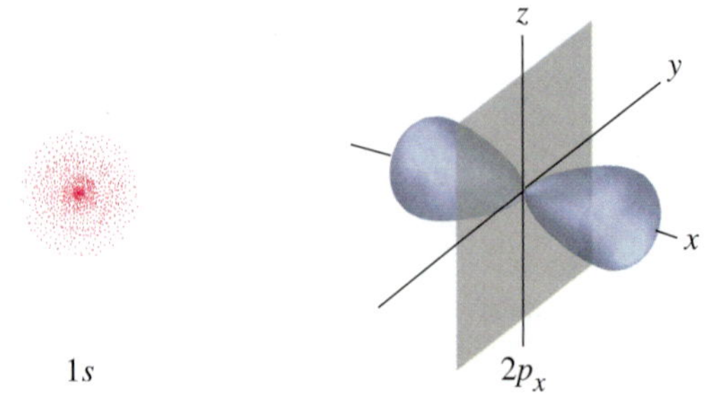

The n, l, and m_l quantum numbers define an orbital, but a fourth quantum number is also required to characterize an electron in an orbital—the spin quantum number, m_s. This quantum number may have either of two values: $+1/2$ or $-1/2$.

Review Questions

1. What are cathode rays? State some distinctive properties of cathode rays. Which property of cathode rays gives strongest support to the notion that they are negatively charged particles?

2. Describe J. J. Thomson's experiment that determined the mass-to-charge ratio of the electron.

3. Millikan's experiments hinged on the absorption of positive and negative ions by oil droplets. How did his observations of these droplets translate into a determination of the fundamental unit of electric charge, and specifically the charge on an electron?

4. How did Rutherford's interpretation of the Geiger-Marsden metal-foil experiments contradict Thomson's "raisin pudding" model of the atom?

5. What do we mean by the term nuclear atom? What particles are found in the nucleus of an atom? Must the nucleus of an atom be electrically neutral? Explain.

6. Why did Rutherford's model of the atom require that electrons be in motion, whereas Thomson's model did not?

7. Is there any difference in the concept of atomic number as (a) an ordering of the elements by increasing atomic mass and (b) equal to the number of protons in the nucleus of an atom? Explain.

8. Both negative and positive particles were discovered in the passage of electric discharges through gases at low pressure. Explain why the negative particles proved to be fundamental particles of matter but the positive particles did not.

9. Describe how a mass spectrometer works.

10. What do each of the following terms mean in reference to an electromagnetic wave?

    (a) wavelength
    (b) frequency
    (c) speed
    (d) amplitude

11. What do we mean by **(a)** the electromagnetic spectrum? **(b)** the visible spectrum? **(c)** a continuous spectrum? **(d)** a line spectrum?

12. What is black-body radiation? Can we detect it with our senses? Explain.

13. Explain the meaning of **(a)** a quantum of energy, **(b)** a photon of light, and **(c)** Planck's equation.

14. Why is the quantum theory of such importance in describing behavior at the microscopic level? Why is it of little importance in describing events at the macroscopic level, for example, the speed and trajectory of a bullet fired from a rifle?

15. Describe the photoelectric effect. How did Einstein explain it?

16. What is meant by the wave-particle duality of light?

17. How was Rutherford's nuclear model of the atom refined by Bohr?

18. Explain what change has occurred within an atom when **(a)** light has been emitted; **(b)** light has been absorbed.

19. Explain why it is reasonable to expect the energy levels of the Bohr hydrogen atom to be given by an equation of the form $E_n = -B/n^2$, rather than $E_n = -B \times n^2$. What is the significance of the negative sign?

20. Which hydrogen atom has absorbed more energy: one in which the electron moved from the first to third energy levels or one in which the electron moved from the second to the fourth energy levels? Explain.

21. Explain why Heisenberg's uncertainty principle places a limitation on our ability to learn about the structure of an atom.

22. The text states that the Bohr hydrogen atom violates the Heisenberg uncertainty principle whereas the Schrödinger hydrogen atom does not. Yet, both models seem to give the same distance (52.9 pm) for the electron from the nucleus. How do you explain this?

23. What is a matter wave? Is there any physical evidence to support the existence of matter waves? What is the significance of matter waves in describing atomic structure?

24. What is an atomic orbital? What information does it give about the electronic structure of an atom?

25. Describe the three quantum numbers that arise from the solution of Schrödinger's wave equation and list their possible values.

26. Give the subshell notation for each of the following sets of quantum numbers.

    (a) $n = 3$ and $l = 2$
    (b) $n = 2$ and $l = 0$
    (c) $n = 4$ and $l = 1$
    (d) $n = 4$ and $l = 3$

27. How many subshells are there in a principal shell with **(a)** $n = 3$, **(b)** $n = 2$, and **(c)** $n = 4$?

28. What is the shape of the geometric region described by an *s* orbital? How does a 2*s* orbital differ from a 1*s* orbital?

29. What is the shape of the geometric region described by a *p* orbital?

30. Describe the property called electron spin and an experiment that can be explained in terms of electron spin.

Problems

Subatomic Particles

31. The mass-to-charge ratio of the proton is found to be 1.044×10^{-8} kg/C. What is the charge on a proton? What is its mass?

32. The mass-to-charge ratio of the positron (a product of the radioactive decay of certain isotopes) is 5.686×10^{-12} kg/C. The charge on a positron is the same as that on a proton. Calculate the mass of a positron. Does any other fundamental particle have this same mass? Explain.

33. An oil-drop experiment yielded the following charges on an oil droplet at various times: 6.4×10^{-19} C, 3.2×10^{-19} C, 8.0×10^{-19} C, 4.8×10^{-19} C. What value for the electronic charge can you deduce from these data?

34. An oil-drop experiment yielded the following charges on an oil droplet at various times: 9.6×10^{-19} C, 3.2×10^{-19} C, 16.0×10^{-19} C, 6.4×10^{-19} C. What value of the electronic charge can you deduce from these data?

35. Determine approximate mass-to-charge ratios for the following ions.

(a) $^{80}Br^-$ **(b)** $^{18}O^{2-}$ **(c)** $^{40}Ar^+$

Why are these values only approximate? (*Hint:* What are the masses of these ions?)

36. Which of the following ions should exhibit the same mass-to-charge ratios: $^{40}Ca^{2+}$, $^{10}B^+$, $^{60}Co^{2+}$, $^{20}Ne^+$, $^{120}Sn^{2+}$, $^{80}Br^-$? Would you expect the ratios to be exactly the same or only approximately so? Explain.

Mass Spectrometry

37. The atomic masses of the five naturally occurring isotopes of germanium are germanium-70, 69.9243 u; germanium-72, 71.9217 u; germanium-73, 72.9234 u; germanium-74, 73.9219 u; germanium-76, 75.9214 u. Use these values and data from the accompanying bar graph mass spectrum of germanium to determine a weighted average atomic mass of germanium. (*Hint:* Recall Example 2.4 on page 44.)

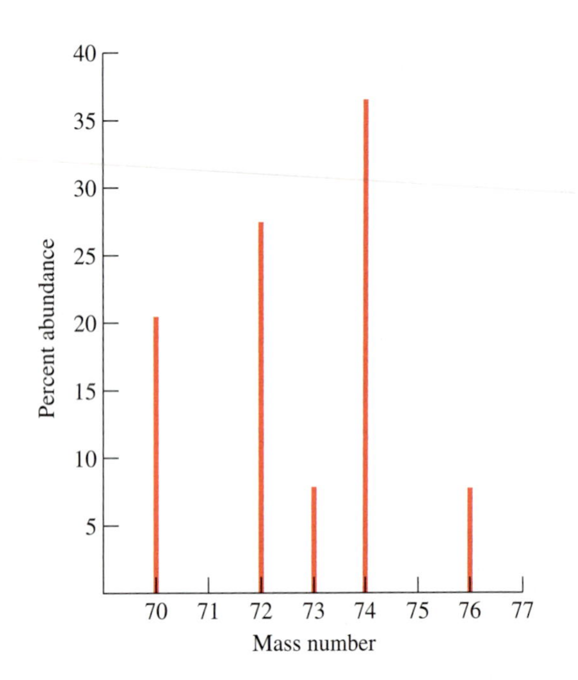

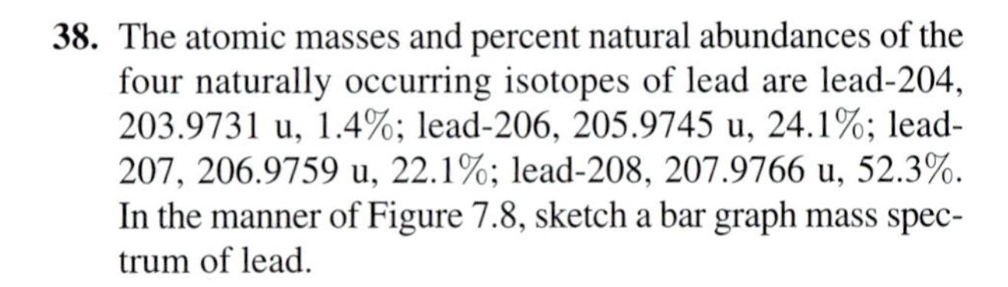

38. The atomic masses and percent natural abundances of the four naturally occurring isotopes of lead are lead-204, 203.9731 u, 1.4%; lead-206, 205.9745 u, 24.1%; lead-207, 206.9759 u, 22.1%; lead-208, 207.9766 u, 52.3%. In the manner of Figure 7.8, sketch a bar graph mass spectrum of lead.

Electromagnetic Radiation

39. When the Voyager space probe radioed information back from Neptune to Earth, the electromagnetic signals had to travel about 4.4×10^9 km. How long did it take for these signals to reach Earth?

40. It took 11 min for a command from the controllers on Earth to reach the Pathfinder space vehicle on the surface of Mars on July 4, 1997. How many miles did the message have to travel?

41. Radio station XOOX operates at a frequency of 992 kHz on the AM band. What is the wavelength, in meters, of the radio waves?

42. Nuclear magnetic resonance (NMR) spectrometers use radio frequency waves to obtain information from which the structures of molecules can be deduced. One kind of NMR operates at a frequency of 200 MHz. What is the wavelength, in meters, of this radiation?

43. What are the frequency and color of light that has a wavelength of 650 nm?

44. What are the wavelength and color of light that has a frequency of $5.2 \times 10^{14}\ s^{-1}$?

45. The accompanying sketch suggests a particular light wave. What are the wavelength and frequency of this light? What color light is represented by this wave?

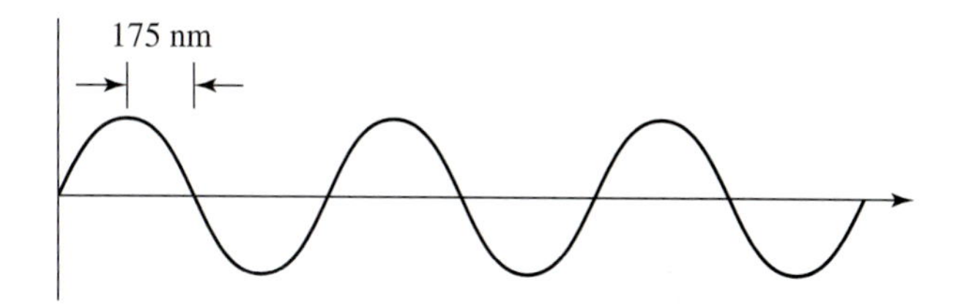

46. On the same scale as the light wave in Problem 45, sketch a reasonable representation of **(a)** a wave of yellow light and **(b)** an infrared light wave with a frequency of $1.0 \times 10^{14}\ s^{-1}$.

Photons and the Photoelectric Effect

47. Calculate the energy, in joules, of a photon of violet light that has a frequency of $7.42 \times 10^{14}\ s^{-1}$. Is this energy greater or less than that of light of 655-nm wavelength? Explain.

48. Calculate the energy, in joules, of a photon of red light that has a frequency of $3.73 \times 10^{14}\ s^{-1}$. Is this energy greater or less than that of light of 530-nm wavelength? Explain.

49. The laser in a compact disc player uses light with a wavelength of 780 nm. Calculate the energy of this radiation, in joules per photon and in kilojoules per mole of photons.

50. A microwave oven operates with radiation having a wavelength of 12.0 cm. Calculate the energy of this radiation, in joules per photon and in kilojoules per mole of photons.

51. The minimum energy necessary to knock an electron from an iron atom is 7.21×10^{-19} J. What is the maximum wavelength of light, in nanometers, that will show a photoelectric effect with iron?

52. The minimum energy necessary to knock an electron from a lithium atom is 4.65×10^{-19} J. What is the maximum wavelength of light, in nanometers, that will show a photoelectric effect with lithium?

53. What is the approximate color of light that has an energy content of 191 kJ/mol photons?

54. What is the wavelength, in nanometers, of light that has an energy content of 487 kJ/mol photons? In what portion of the electromagnetic spectrum will this light be found?

The Bohr Model of the Hydrogen Atom

55. What is the energy of the photon emitted when the electron in a hydrogen atom drops from the energy level $n = 5$ to **(a)** the energy level $n = 1$? **(b)** the energy level $n = 4$?

56. How much energy must a hydrogen atom absorb to raise its electron from the energy level $n = 1$ to **(a)** the energy level $n = 3$? **(b)** the energy level $n = 6$?

57. Calculate the frequency, in s^{-1}, of the electromagnetic radiation emitted by a hydrogen atom when its electron drops from **(a)** the $n = 3$ to the $n = 2$ energy level, and **(b)** from the $n = 4$ to the $n = 1$ energy level.

58. Calculate the wavelength, in nanometers, of the electromagnetic radiation emitted by a hydrogen atom when its electron drops from **(a)** the $n = 5$ to the $n = 2$ energy level, and **(b)** from the $n = 3$ to the $n = 1$ energy level.

59. *Without doing detailed calculations*, determine whether it is likely that one of the energy levels in the Bohr hydrogen atom has $E_n = -2.179 \times 10^{-21}$ J. Explain your reasoning.

60. *Without doing detailed calculations*, determine whether there can be an energy level in the Bohr hydrogen atom with $E_n = -1.00 \times 10^{-17}$ J. Explain your reasoning.

61. The energy required to raise the energy of a hydrogen atom from its ground state to the state E_n in which $n = \infty$ is called the *ionization energy*. What is the ionization energy of hydrogen atoms, expressed in kilojoules per mole?

62. Would you expect to find energy states of the hydrogen atom in which the energy is a *positive* quantity? Explain.

Matter as Waves

63. Calculate the wavelength, in nanometers, associated with a proton traveling at a speed of 2.55×10^6 m s^{-1}. Use 1.67×10^{-27} kg as the proton mass.

64. Calculate the wavelength, in nanometers, associated with an electron traveling 90.0% of the speed of light.

65. Electrons have an associated wavelength of 84.4 nm. How many meters per second are the electrons traveling?

66. To what speed, in meters per second, must electrons be accelerated to have an associated wavelength of 174 pm?

Quantum Numbers and Atomic Orbitals

67. What is the lowest numbered principal shell in which **(a)** p orbitals are found, and **(b)** the f subshell can be found?

68. Is there a $3d$ subshell in a hydrogen atom? $3f$ orbitals? Explain.

69. If $n = 5$, what are the possible values of l? If $l = 3$, what are the possible values of m_l?

70. If $n = 4$, what are the possible values of l? If $l = 1$, what are the possible values of m_l?

71. Indicate whether each of the following is a permissible set of quantum numbers. If the set is not permissible, state why it is not.

(a) $n = 2, l = 1, m_l = +1$

(b) $n = 3, l = 3, m_l = -3$

(c) $n = 3, l = 2, m_l = -2$

(d) $n = 0, l = 0, m_l = 0$

72. Indicate whether each of the following is a permissible set of quantum numbers. If the set is not permissible, state why it is not.

(a) $n = 3, l = 1, m_l = +2$

(b) $n = 4, l = 3, m_l = -3$

(c) $n = 3, l = 2, m_l = +2$

(d) $n = 5, l = 0, m_l = 0$

73. Consider the electronic structure of an atom: **(a)** What are the n, l, and m_l quantum numbers corresponding to the $3s$ orbital? **(b)** List all the possible quantum number values for an orbital in the $5f$ subshell. **(c)** In what specific subshell will an electron having the quantum numbers $n = 3, l = 1$, and $m_l = -1$ be found?

74. Consider the electronic structure of an atom: **(a)** What are the n, l, and m_l quantum numbers corresponding to the $3p$ subshell? **(b)** List all the possible quantum number values for an orbital in the $4d$ subshell. **(c)** In what specific subshell will an electron having the quantum numbers $n = 4, l = 2$, and $m_l = 0$ be found?

75. In the following assignments of the four quantum numbers, write acceptable values for the missing ones. What are the orbitals corresponding to these quantum numbers.

(a) $n = ?, l = 2, m_l = 0, m_s = ?$

(b) $n = 2, l = ?, m_l = -1, m_s = -\frac{1}{2}$

(c) $n = 4, l = ?, m_l = 2, m_s = ?$

(d) $n = ?, l = 0, m_l = ?, m_s = \frac{1}{2}$

76. Which of the following statements is (are) correct? An electron having the quantum numbers $n = 5$ and $m_l = -3$?: **(a)** may be in d subshell; **(b)** must have $m_s = +\frac{1}{2}$, **(c)** must have $l = 4$; **(d)** may have $l = 0, 1, 2$, or 3; **(e)** may have $m_s = +\frac{1}{2}$ or $m_s = -\frac{1}{2}$; **(f)** cannot exist.

Additional Problems

77. Use the data presented in Figure 7.8 to establish an approximate weighted average atomic mass of mercury. Why is the result only approximately correct?

78. In Problems 33 and 34, data from oil-drop experiments are used to deduce the value of the electronic charge. The values we get in these two problems are different. Can they both be correct? If not, is one of them correct? Explain.

79. The *period* of a wave is the time required for one cycle. Household electricity has a frequency of 60 s^{-1}. What is the period of this electricity? How are period and frequency related?

80. One photon of ultraviolet light can dislodge an electron from the surface of a metal. Two photons of red light with the same total energy as the one photon of ultraviolet do not produce any photoelectrons. Explain why this is so.

81. In everyday usage the term "quantum jump" describes a change of significant magnitude as compared to more gradual, incremental changes. Does the term have the same meaning when used by a scientist to describe events in the microscopic world? Explain.

82. The *work function* (ϕ) of a photoelectric material is the energy that a photon of light must possess to just expel an electron from the surface of the material. The corresponding frequency of the light is the *threshold frequency* (ν_{th}), and $\phi = h\nu_{th}$. The higher the energy of the incident radiation, the more kinetic energy that ejected electrons have in moving away from the surface. The work function for cesium is 3.42×10^{-19} J.

(a) What is the threshold frequency for the photoelectric effect for cesium?

(b) Will 1000-nm infrared radiation produce the photoelectric effect on cesium? Explain.

(c) A 425-nm blue light shines on a piece of cesium. What should be the speed of the electrons emitted from the metal. (*Hint:* Think of the energy of the photons as used partly to overcome the work function and partly to give the photoelectrons kinetic energy.)

83. The energy difference between the first and second energy levels of the Bohr hydrogen atom is 3/4 B. Which of the following represents the energy difference between the first and third energy levels ? Explain your reasoning.

(a) $3/2 \times (3/4\ \mathrm{B})$ **(c)** $32/27 \times (3/4\ \mathrm{B})$

(b) $2/3 \times (3/4\ \mathrm{B})$ **(d)** $(3/4\ \mathrm{B}) + (4/9\ \mathrm{B})$

84. Show that there should be only *four* lines in the visible spectrum of hydrogen and at the wavelengths indicated in Figure 7.13.

85. What are the maximum and minimum wavelengths of lines in the Lyman series of the hydrogen spectrum (recall Figure 7.20)?

86. From what energy level must an electron fall to produce a line at 1094 nm in the Paschen series of the hydrogen spectrum (recall Figure 7.20)?

87. Between which two energy levels of a hydrogen atom must an electron fall to account for a spectral line at 486.1 nm?

88. Show that if an electron in a hydrogen atom drops from the energy level $n = 3$ to $n = 2$, followed by a drop from $n = 2$ to $n = 1$, the total energy emitted is the same as if the electron had fallen directly from the energy level $n = 3$ to $n = 1$.

89. The Bohr theory can be extended to one-electron species such as He^+ and Li^{2+}, in which case the energy levels are $E_n = -Z^2\ B/n^2$, where Z is the atomic number. How much energy, in joules, is required to ionize completely the electron from a ground-state He^+ ion?

90. Calculate the wavelength, in meters, associated with a bullet with a mass of 25.0 g traveling with a speed of 110 m/s.

91. Einstein's equation for the equivalence of matter and energy is $E = mc^2$. Use this equation, together with Planck's equation, to derive a relationship similar to de Broglie's equation.

92. Which must have the greater velocity to produce matter waves of the same wavelength (for example, 1 nm), protons or electrons? Explain your reasoning.

93. What must be the speed of electrons if the matter wave associated with them is to have the same wavelength as the spectral line in the Lyman series of the hydrogen spectrum in which the electron transition is from $n = 5$ to $n = 1$?

94. On page 302, we make the statement that the Heisenberg uncertainty principle has no practical importance when dealing with large objects. Show that this should be the case for a 1000-kg automobile traveling at 100 km/h. That is, show that we could measure the mass and speed of this automobile with very high precision and still be able to locate its position precisely.

95. Radio signals from Voyager 1 spacecraft on its trip to Jupiter in the late 1970s were broadcast at a frequency of $8.4 \times 10^9\ \mathrm{s^{-1}}$. These signals were intercepted on Earth by an antenna capable of detecting 4×10^{-21} watt (1 watt = 1 J/s). At a minimum, approximately how many photons per second of electromagnetic radiation was the antenna capable of intercepting?

96. A typical microwave oven uses radiation with a 12.2-cm wavelength.

(a) How many moles of photons of this radiation are required to raise the temperature of 345 g water from 26.5 °C to 99.8 °C?

(b) The *watt* is a unit of power—the rate at which energy is delivered or consumed: $1\ \mathrm{W} = 1\ \mathrm{J\ s^{-1}}$. Assume that all the energy of a 700-W microwave oven is delivered to the heating of water described in (a). How long will it take to heat the water?

97. The greatest probability of finding the electron in a small volume element of the 1*s* orbital of the hydrogen atom is at the nucleus. Yet, the most probable distance of the electron from the nucleus is 52.9 pm. How can you reconcile these two statements?

CHAPTER

8

Electron Configurations, Atomic Properties, and the Periodic Table

CONTENTS

● **ChemCDX**
When you see this icon, there are related animations, demonstrations, and exercises on the CD accompanying this text.

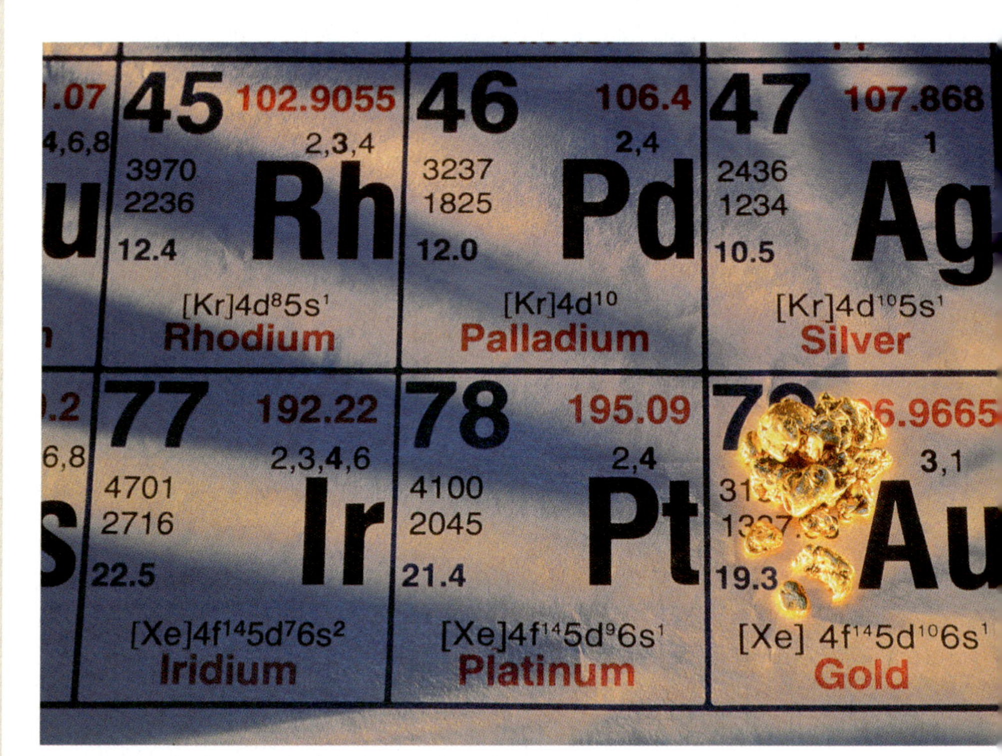

The periodic table is a systematic listing of all the elements. The element gold, shown here, has been known since ancient times.

In Chapter 2, we described how Dmitri Mendeleev constructed the periodic table in 1869 by arranging elements with similar properties into groups. Recall that to make his classification scheme work, Mendeleev corrected some atomic mass values and boldly left gaps in his table for undiscovered elements. Mendeleev's approach to the periodic table was *empirical*; he based his classification scheme on the observed facts. In this chapter, we will learn that Mendeleev's scheme also makes sense from a *theoretical* standpoint. We will first learn how electrons are distributed among the regions of an atom described by atomic orbitals, a description called the *electron configuration* of the atom. Then we will explore electron configurations as a basis for the periodic table. We will also introduce some properties of individual atoms—atomic properties—and see how the periodic table helps us to summarize trends in these properties among the elements.

8.1 Multielectron Atoms

Schrödinger's wave-mechanical treatment of atomic structure, which we described in Chapter 7, applies exactly only to the hydrogen atom or to other species consisting of a positively charged nucleus and a *single* electron; that is, to species such as H, He^+, Li^{2+}, and so on. Wave mechanics quickly gets complicated as we consider *multielectron* species.

Multielectron atoms are especially complex because several electrons are attracted to the nucleus while simultaneously repelling one another. It is possible to write a wave equation that accounts for all the interactions in a multielectron atom—repulsions as well as attractions—but we cannot solve the equation exactly. The usual approach is to assume that, with appropriate adjustments, all atoms can be described through *hydrogen-like* orbitals. An important difference between the orbitals of hydrogen and multielectron atoms is in their orbital energies. Figure 8.1 presents one orbital energy level diagram for hydrogen and another for a typical but unspecified multielectron atom. Orbital energy diagrams can be obtained experimentally. Figure 8.1 illustrates four points that are relevant to the discussion in several sections of this chapter.

1. *In the hydrogen atom, all subshells of a principal shell are at the same energy level.* The orbital energies in the hydrogen atom depend only on the principal quantum number, n. This is what we predict with the Bohr model, that is, $E_n = -B/n^2$.
2. *Orbital energies are lower in multielectron atoms than in the hydrogen atom.* All energy states associated with forces of attraction are negative. Because the

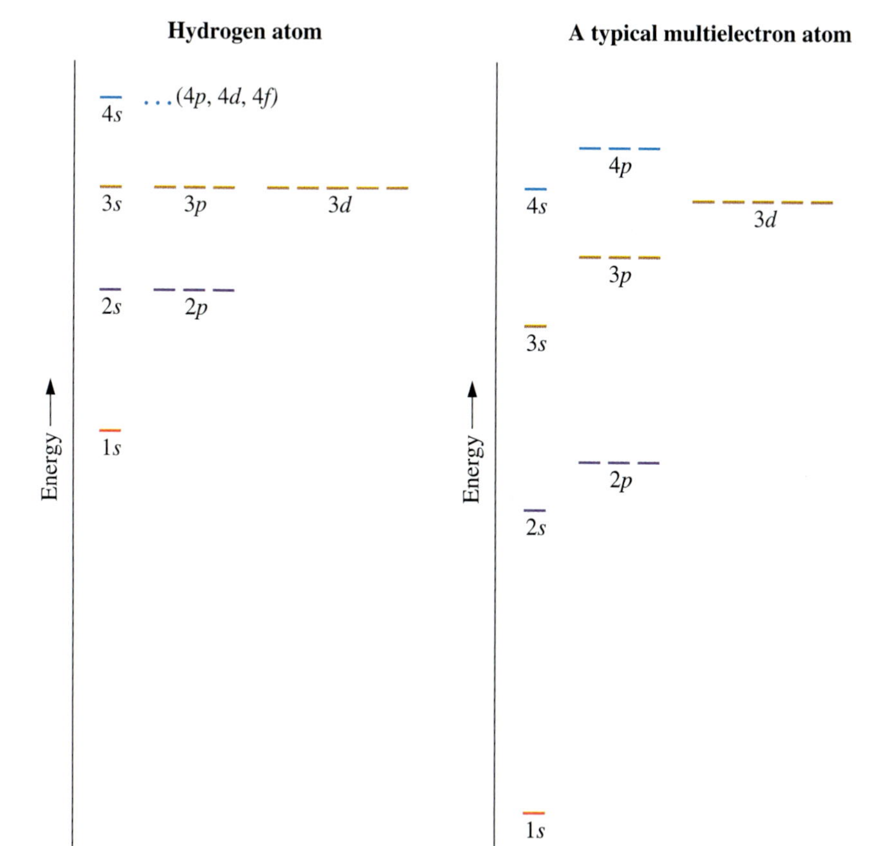

Figure 8.1
Orbital energy diagrams
Some orbital energies of the first four principal shells are compared for a hydrogen atom and a typical multielectron atom. The energies are not plotted to scale.

attractive force between the nucleus and an electron in any orbital increases with increasing nuclear charge, the orbital energy becomes more negative (lower) as the atomic number increases. Thus, the energies of the 1*s*, 2*s*, 3*s*, and 4*s* orbitals in Figure 8.1 are lower in all multielectron atoms than in the hydrogen atom.

3. *In a multielectron atom the various subshells of a principal shell are at different energy levels, but all orbitals within a subshell are at the same energy level.* One way to think about the repulsive forces between electrons in a multielectron atom is that other electrons between a given electron and the nucleus *screen* or *shield* that electron from the full attractive force of the nucleus. The effectiveness of the shielding depends on the type of orbital in which the shielded electron is found.

There is a probability of finding an *s* electron from any principal shell near the nucleus (Figure 7.24, page 305). We say that an *s* electron *penetrates* through inner-shell electrons to approach the nucleus. In contrast, the two lobes of a *p* orbital are separated by a nodal plane and the electron probability falls to zero at the nucleus (Figure 7.26, page 306). A *p* electron is less penetrating than an *s* electron. Electrons in *d* and *f* orbitals are less penetrating still. Highly penetrating electrons are less effectively shielded by inner-shell electrons, more strongly attracted to the nucleus, and at lower energies than less penetrating electrons. The increasing order of subshell energies is

$$s < p < d < f$$

In Figure 8.1, we see this pattern for the multielectron atom. Of the subshells shown, the order of increasing energy is

$$2s < 2p;\ 3s < 3p < 3d;\ 4s < 4p.$$

In an isolated multielectron atom, energy levels within a subshell are not split further. For example, the three 3*p* orbitals are at the same energy level. Orbitals at the same energy level are called **degenerate orbitals**. Thus, the five 3*d* orbitals are degenerate; they all have the same energy. However, their energy is higher than that of the 3*p* orbitals.

In terms of quantum numbers, orbital energy is determined primarily by n: the lower the value of n, the lower the orbital energy. Orbital energy is further affected by l: In any given shell, the lower the value of l, the lower the orbital energy is. The orbital energy does not depend on the value of m_l.

4. *In higher numbered principal shells of a multielectron atom, some subshells of different principal shells have nearly identical energies.* In Figure 8.1, we see that the energy-level difference between 3*d* and 4*s* is quite small. This small energy difference has important consequences that we explore in Section 8.4.

8.2 An Introduction to Electron Configurations

The scientific view of where electrons are found in atoms has evolved since the discovery of the electron in 1897. J. J. Thomson embedded them in a cloud of positive charge. Rutherford placed them in motion outside a positively charged atomic nucleus. Bohr placed them in motion in discrete circular orbits around the nucleus. Schrödinger, Heisenberg, and others referred to regions of high electron probability, or electron charge density, defined by mathematical functions called atomic orbitals.

Although we often say that an electron is "in a 1*s* orbital" or "in a 2*p* orbital" and so on, remember from Chapter 7 that the orbitals are not *actual* regions in an atom. They are mathematical expressions related to the probabilities of finding an electron in various regions of an atom. It is this view of orbitals we have in mind when we say the **electron configuration** of an atom describes the distribution of electrons among the various orbitals in the atom. We can represent electron configurations in two quite similar ways.

The ***spdf* notation** uses numbers to designate a principal shell and the letters *s*, *p*, *d*, or *f* to identify a subshell. A superscript number following the letter indicates the number of electrons in the designated subshell. Empty subshells are *not* included in the notation, that is, we do not include terms such as $3d^0$. Thus, the notation $1s^2 2s^2 2p^3$ indicates an atom with two electrons in the 1*s* subshell, two in the 2*s* subshell, and three in the 2*p* subshell. The atom having this electron configuration has an atomic number of seven; it is an atom of the element *nitrogen*.

This subshell notation leaves an unanswered question: How are the three 2*p* electrons distributed among the three orbitals in the 2*p* subshell? We can show this through an *expanded spdf notation*.

$$1s^2 2s^2 2p_x^1 2p_y^1 2p_z^1$$

This representation shows that each of the three 2*p* orbitals contains a single electron.

Another way to represent an electron configuration is with an **orbital diagram**, which uses boxes to represent orbitals within subshells and arrows to represent electrons. As described on page 306, the directions of the arrows represent the sign of the spin quantum number. The orbital diagram for nitrogen is

↑↓	↑↓	↑	↑	↑
1*s*	2*s*		2*p*	

It indicates that nitrogen has two electrons of opposite or opposing spins in the 1*s* subshell and two more with opposing spins in the 2*s* subshell. Electrons with opposing spins are *paired*. Each of the orbitals of the 2*p* subshell has one electron, and all three electrons have spins in the same direction; they are said to have *parallel* spins.

Several questions arise when we look carefully at the electron configurations we have written: Why are there never more than two electrons in an atomic orbital? When there are two electrons in an orbital, why do they always have opposing spins? Why are orbitals occupied singly before pairing of electrons occurs? Why do the electrons in singly occupied orbitals have parallel spins?

8.3 The Rules for Electron Configurations

To answer the questions just posed and to lay the groundwork for *predicting* electron configurations of the elements, we consider three basic principles that govern the distribution of electrons among atomic orbitals.

1. *Electrons occupy orbitals of the lowest energy available.*

For a given multielectron atom, the order in which orbitals of the first three principal shells fill is suggested by Figure 8.1: The 1*s* orbital is at a lower energy level than the 2*s*, so electrons enter it before the 2*s*; electrons enter the three orbitals in the 2*p* subshell before the 3*s* orbital; and so on. With some exceptions, the order in which the subshells of atoms are occupied by electrons is

$1s, 2s, 2p, 3s, 3p, 4s, 3d, 4p, 5s, 4d, 5p, 6s, 4f, 5d, 6p, 7s, 5f, 6d, 7p.$

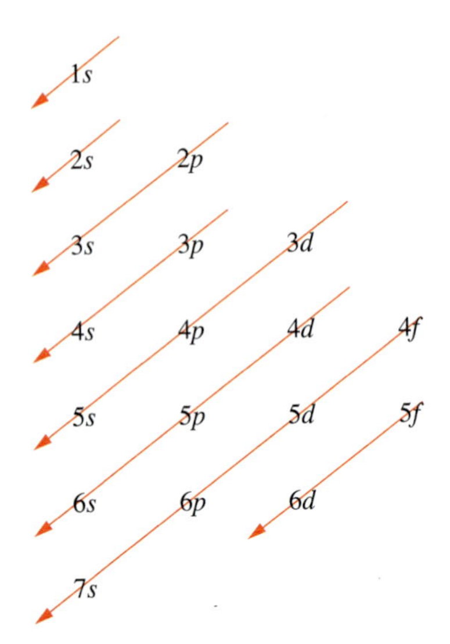

▲ **Figure 8.2**
The order in which subshells are filled with electrons
Start at the upper left of the grid and follow the arrows. The order in which the arrows slice through the subshell designations is the order in which the subshells fill. The blank space at the upper right corresponds to nonexistent orbitals ($1p$, $1d$, $2d$, and so on). The blank space at the bottom corresponds to orbitals that are not filled in the known elements ($7p$, $8s$, and so on).

The scheme shown in Figure 8.2 may be easier to use than this list. Later in the chapter, we will learn to relate electron configurations to the periodic table, which is probably the best guide to the order of filling of subshells.

Exceptions to the above order of filling are mostly due to the overlapping of energy levels for certain orbitals in the higher numbered principal shells. However, electrons occupy orbitals not just according to orbital energies. The orbitals are occupied in a way that leads to the lowest possible energy state for the atom when all electrons are in place. With quantum mechanics, we can determine the lowest energy state for an atom among possible alternatives. However, in the final analysis, the order of subshell filling and other details of electron configurations are established *by experiments*, such as spectroscopic and magnetic studies.

2. *No two electrons in the same atom may have all four quantum numbers alike.*

   This is a statement of the **Pauli exclusion principle**, made by Wolfgang Pauli in 1926 to explain the complex features of emission spectra in a magnetic field. This principle has an important consequence for electron configurations: Because electrons in a given orbital must have the same values of n, l, and m_l (for example, $n = 3, l = 0, m_l = 0$ in the $3s$ orbital), they must have different values of m_s. Only two values of m_s are possible: $+\frac{1}{2}$ and $-\frac{1}{2}$. As a result, we conclude that

   An atomic orbital can accommodate only two electrons, and these electrons must have opposing spins.

   Each principal shell consists of a given number of subshells, and each subshell contains a given number of orbitals. Pauli's exclusion principle therefore limits the number of electrons that can be found in individual orbitals, subshells, and principal shells. These limitations are set forth in Table 8.1.

3. *Of a group of orbitals of identical energy, electrons enter empty orbitals whenever possible. Electrons in half-filled orbitals have the same spins, that is, parallel spins.*

   The first part of statement 3 is known as **Hund's rule** (after the quantum physicist H. Friedrich Hund). We can rationalize Hund's rule as follows: Two electrons carry identical charges and therefore repel each other; they tend not

TABLE 8.1 Maximum Capacities of Subshells and Principal Shells

n	1	2		3			4				$\dots n$
l	0	0	1	0	1	2	0	1	2	3	
Subshell designation	s	s	p	s	p	d	s	p	d	f	
Orbitals in subshell	1	1	3	1	3	5	1	3	5	7	
Subshell capacity	2	2	6	2	6	10	2	6	10	14	
Principal shell capacity	2	8		18			32				$\dots 2n^2$

to occupy the same region of space. As a result, electrons go into separate orbitals as long as empty orbitals of the same energy are available. This is rather like the inclination of bus passengers to sit singly, which gives each person maximum seat space. Only when all seats have one passenger, do passengers begin to sit two to a seat.

It is Hund's rule that leads us to write the electron configuration of nitrogen as

↑↓ ↑↓ ↑ ↑ ↑

1*s* 2*s* 2*p*

The following electron configuration for nitrogen is in conflict with experimental evidence and Hund's rule.

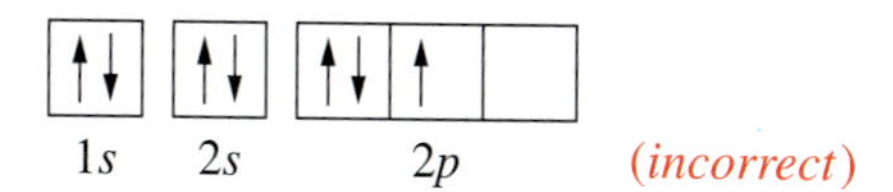

It is not so easy to explain why electrons in singly occupied orbitals have *parallel* spins. However, both experiment and theory indicate that an electron configuration in which all unpaired electrons have parallel spins represents a lower energy state of an atom than any other electron configuration that we can write. This fact tells us, for example, that we should *not* write the electron configuration of nitrogen as

↑↓ ↑↓ ↑ ↓ ↑

1*s* 2*s* 2*p* (*incorrect*)

8.4 Electron Configurations: The Aufbau Principle

The electron configurations that we write in this section are for *gaseous* atoms in their lowest energy or *ground state*. To write these *ground-state* electron configurations, we will use the rules of Section 8.3 and an idea known as the *aufbau principle*. (Aufbau is a German word that means "building up.") The **aufbau principle** describes a *hypothetical* process (we can't actually do this) in which we *think about* building up each atom from the one that precedes it in atomic number. To do so, we imagine adding a proton and some neutrons to the nucleus and one electron to an atomic orbital. That is, we think about building a helium atom from a hydrogen atom; a lithium atom from a helium atom; and so on. In this process, we focus on the particular atomic orbital into which the added electron goes to produce a *ground-state* atom, an atom in its lowest energy state.

Let's illustrate the aufbau principle by "building up" a few atoms. A hydrogen atom has the atomic number 1 ($Z = 1$) and thus only one electron. That electron goes into the orbital with the lowest possible energy, the 1*s* orbital.

$$(Z = 1)\ \mathrm{H}\ 1s^1$$

Based on the order of filling of subshells, the electron we add in building up helium from hydrogen goes into the lowest energy orbital, the 1*s*. The electron configuration of helium is

$$(Z = 2)\ \mathrm{He}\ 1s^2$$

In the lithium atom, the first two electrons are in the $1s$ orbital, just as in the helium atom. The Pauli exclusion principle tells us that the third electron *cannot* enter the fully occupied $1s$ orbital. The added electron must go into the $2s$ orbital, the vacant orbital next lowest in energy. The first principal shell is filled, and the second begins to fill.

$$(Z = 3)\ \text{Li}\ 1s^2 2s^1$$

In the beryllium atom, the added electron also goes into the $2s$ orbital.

$$(Z = 4)\ \text{Be}\ 1s^2 2s^2$$

With the boron atom, the Pauli exclusion principle dictates that the added electron enter a $2p$ orbital.

$$(Z = 5)\ \text{B}\ 1s^2 2s^2 2p^1$$

The $2p$ subshell fills as the atomic number increases from $Z = 6$ to $Z = 10$ in the series: carbon, nitrogen, oxygen, fluorine, neon. In these atoms, Hund's rule requires that electrons enter $2p$ orbitals singly and with parallel spins before electrons pair up. This fact is not explicitly shown in the usual *spdf* notation, but it is clearly shown in orbital diagrams.

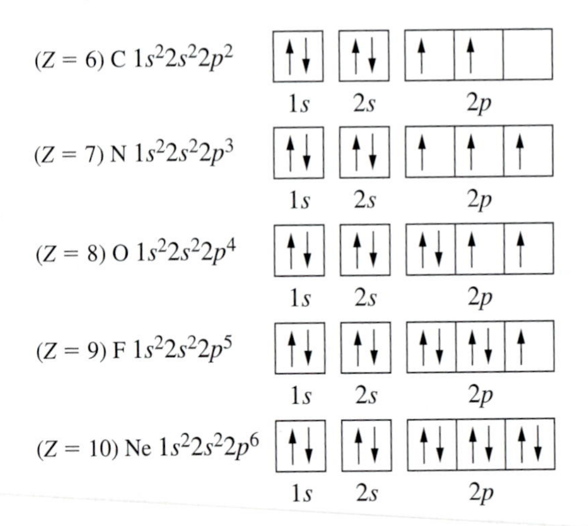

In a *noble-gas-core abbreviated* electron configuration, we replace the portion that corresponds to the electron configuration of a noble gas by a bracketed chemical symbol. Thus, [He] replaces the configuration $1s^2$; so the electron configuration of lithium, $1s^2 2s^1$, is replaced by [He]$2s^1$. Likewise, [He]$2s^2 2p^3$ is equivalent to $1s^2 2s^2 2p^3$, the electron configuration of nitrogen.

The electron configuration of neon has both the first and second principal shells filled to their maximum capacities (recall Table 8.1). In extending the aufbau process from neon to sodium, the added electron goes into the available orbital of lowest energy, the $3s$.

$$(Z = 11)\ \text{Na}\ 1s^2 2s^2 2p^6 3s^1 \quad \textit{or} \quad (Z = 11)\ \text{Na}\ [\text{Ne}]3s^1$$

Other atoms with electrons entering the third principal shell are featured in Example 8.1 and Exercise 8.1, where we illustrate the following general procedure for writing electron configurations.

1 *Determine the number of electrons to appear in the electron configuration.* This is simply the atomic number of the element.

2 *Add electrons to subshells in order of increasing subshell energy.* This will be the order of filling of subshells outlined on page 320.

3 *Observe the Pauli exclusion principle.* There can be no more than two electrons in an orbital, and they must have opposing spins.

4 *Observe Hund's rule.* Orbitals in a subshell are singly occupied whenever possible. Also, all unpaired electrons have the same spin direction (parallel spins).

EXAMPLE 8.1

Write out the electron configuration for sulfur, using both the *spdf* notation and an orbital diagram.

SOLUTION

We can follow the steps just outlined.

1 *Determine the number of electrons to appear in the electron configuration.* The sulfur atom, with atomic number 16, has 16 electrons. All 16 must appear in the electron configuration.

2 *Add electrons to subshells in order of increasing subshell energy.* We must place the electrons into the lowest-energy subshells available, using the order of filling for subshells (Figure 8.2). *Two* electrons go into the $1s$ subshell, *two* into the $2s$ subshell, *six* into the $2p$ subshell, and *two* into the $3s$ subshell. That leaves four electrons to be placed into the $3p$ subshell.

3 *Observe the Pauli exclusion principle.* In accordance with the Pauli exclusion principle, each of the orbitals in the subshells $1s$, $2s$, $2p$, and $3s$ is filled to its limit of two electrons with opposing spins. This accounts for a total of 12 electrons. The remaining four electrons go into the $3p$ subshell.

4 *Observe Hund's rule.* In adding *four* electrons to the *three* orbitals in the $3p$ subshell, *two* of the orbitals have a single electron and *one* has a pair of electrons. The unpaired electrons have parallel spins.

$$(Z = 16)\ \text{S}\ \ 1s^2 2s^2 2p^6 3s^2 3p^4$$

(Z = 16) S [↑↓] [↑↓] [↑↓|↑↓|↑↓] [↑↓] [↑↓|↑|↑]
1s 2s 2p 3s 3p

In the noble-gas-core abbreviated electron configuration, we substitute the symbol [Ne] for $1s^2 2s^2 2p^6$, the portion of the electron configuration that corresponds to that of neon.

$$(Z = 16)\ \text{S [Ne]}\ \ 3s^2 3p^4$$

(Z = 16) S [Ne] [↑↓] [↑↓|↑|↑]
3s 3p

EXERCISE 8.1

Write out the electron configuration for **(a)** phosphorus and **(b)** chlorine, using the *spdf* notation, the noble-gas-core abbreviated electron configuration, and an orbital diagram.

PROBLEM-SOLVING NOTE

It is immaterial whether we represent a pair of electrons with opposing spins as ↑↓ or ↓↑. It is also immaterial whether we designate the parallel spins of unpaired electrons, such as those in the nitrogen atom, as ↑↑↑ or ↓↓↓.

The electron configurations predicted by the aufbau principle for the rest of the third period elements follow a regular pattern through argon, in which the $3p$ subshell fills.

$$(Z = 18)\ \text{Ar}\ 1s^2 2s^2 2p^6 3s^2 3p^6 \quad or \quad (Z = 18)\ \text{Ar [Ne]}3s^2 3p^6$$

Now we need to be careful. In proceeding to potassium ($Z = 19$), we must be guided by the order of filling of subshells (see Figure 8.2), rather than by Table 8.1, which states that the maximum capacity of the third principal shell is 18. That is, the 19th electron goes into the $4s$ subshell, not the $3d$.

$$(Z = 19)\ \mathrm{K}\ 1s^2 2s^2 2p^6 3s^2 3p^6 4s^1 \quad or \quad (Z = 19)\ \mathrm{K}\ [\mathrm{Ar}]4s^1$$

In calcium, the 20th electron pairs up with the 19th electron in the $4s$ orbital.

$$(Z = 20)\ \mathrm{Ca}\ 1s^2 2s^2 2p^6 3s^2 3p^6 4s^2 \quad or \quad (Z = 20)\ \mathrm{Ca}\ [\mathrm{Ar}]4s^2$$

Main-Group and Transition Elements

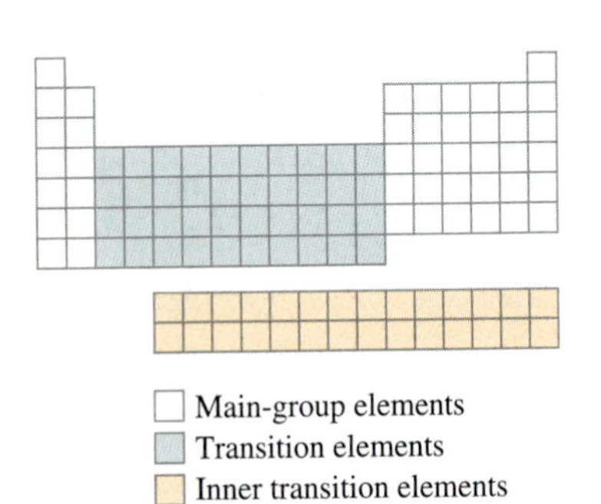

Elements in which the orbitals being filled in the aufbau process are either s or p orbitals of the outermost shell (the shell with the highest principal quantum number) are called **main-group elements**, or *representative elements*. The first 20 elements are all main-group elements. Scandium ($Z = 21$) is the first of the *transition elements*. In **transition elements**, the subshell being filled in the aufbau process is in an inner principal shell. For example, in the fourth period transition elements, the d subshell of the *third* principal shell ($3d$) fills. These elements have the *fourth* principal shell as their outermost shell.

The electron configuration of scandium is shown below in two commonly used representations.

$$\text{(a) Sc } 1s^2 2s^2 2p^6 3s^2 3p^6 3d^1 4s^2 \qquad \text{(b) Sc } 1s^2 2s^2 2p^6 3s^2 3p^6 4s^2 3d^1$$

In method (a), all subshells in the same principal shell are grouped together, and the principal shells are arranged according to increasing principal quantum number. In method (b), the subshells are arranged in the order in which they appear to fill. We will generally use method (a).

Electron configurations of the elements from scandium through zinc, known as the first transition series, are summarized in Table 8.2. Following zinc, we return to a series of main-group elements. Because the third principal shell is filled to its maximum capacity of 18 in zinc, electrons now enter the lowest energy subshell available, the $4p$ subshell. The six atoms in which this subshell fills range from gallium to krypton:

$$(Z = 31)\ \mathrm{Ga}\ [\mathrm{Ar}]3d^{10} 4s^2 4p^1 \quad \text{to} \quad (Z = 36)\ \mathrm{Kr}\ [\mathrm{Ar}]3d^{10} 4s^2 4p^6$$

Exceptions to the Aufbau Principle

Notice the electron configurations of chromium (Cr) and copper (Cu) in Table 8.2. The configurations that we might expect for them, based on the order of filling of subshells in Figure 8.2 and other aspects of the aufbau process, are not the ones that have been established by experiment. Measurements involving the emission spectra and magnetic properties of these elements indicate the electron configurations shown below at the right.

	Expected	*Observed*
Cr ($Z = 24$)	$[\mathrm{Ar}]3d^4 4s^2$	$[\mathrm{Ar}]3d^5 4s^1$
Cu ($Z = 29$)	$[\mathrm{Ar}]3d^9 4s^2$	$[\mathrm{Ar}]3d^{10} 4s^1$

TABLE 8.2 Electron Configurations of the First Transition Series Elements

		3d					4s	
Sc	[Ar]	↑					↑↓	[Ar]$3d^1 4s^2$
Ti	[Ar]	↑	↑				↑↓	[Ar]$3d^2 4s^2$
V	[Ar]	↑	↑	↑			↑↓	[Ar]$3d^3 4s^2$
Cr	[Ar]	↑	↑	↑	↑	↑	↑	[Ar]$3d^5 4s^1$
Mn	[Ar]	↑	↑	↑	↑	↑	↑↓	[Ar]$3d^5 4s^2$
Fe	[Ar]	↑↓	↑	↑	↑	↑	↑↓	[Ar]$3d^6 4s^2$
Co	[Ar]	↑↓	↑↓	↑	↑	↑	↑↓	[Ar]$3d^7 4s^2$
Ni	[Ar]	↑↓	↑↓	↑↓	↑	↑	↑↓	[Ar]$3d^8 4s^2$
Cu	[Ar]	↑↓	↑↓	↑↓	↑↓	↑↓	↑	[Ar]$3d^{10} 4s^1$
Zn	[Ar]	↑↓	↑↓	↑↓	↑↓	↑↓	↑↓	[Ar]$3d^{10} 4s^2$

An electron configuration with a half-filled $3d$ subshell ($3d^5$) and a half-filled $4s$ subshell ($4s^1$) represents a lower energy state for chromium than does the configuration based on a strict application of the aufbau principle. And in copper, the lower energy state corresponds to a filled $3d$ subshell ($3d^{10}$) and half-filled $4s$ subshell ($4s^1$). Actually, because there is little difference between the $4s$ and $3d$ orbital energies, the expected and observed electron configurations are quite close in energy.

At higher principal quantum numbers, the energy difference between certain subshells is even smaller than that between the $3d$ and $4s$ subshells. As a result, there are additional exceptions to the aufbau principle among the heavier transition elements.

8.5 Electron Configurations: Periodic Relationships

As we have just seen, we can use the aufbau principle to assign electron configurations to the elements in order of increasing atomic number. In the modern periodic table, the elements are classified by similarities in physical and chemical properties in an arrangement based on atomic number. In Figure 8.3, where we show the electron configurations of the elements arranged in the periodic table format, it is easy to spot some distinctive patterns.

Let's focus on the portion of the electron configuration describing the principal shell with the highest principal quantum number, n, that contains electrons. This outermost occupied shell is called the **valence shell**. For example, the valence shell of the chlorine atom has the electron configuration $3s^2 3p^5$. From Figure 8.3, we find that, with few exceptions, all the elements in a vertical group of the periodic table have the same number of electrons in the valence shells of their atoms. Moreover, we can make the following generalization.

1A	2A	3B	4B	5B	6B	7B	8B	8B	8B	1B	2B	3A	4A	5A	6A	7A	8A
1 **H** $1s^1$																	2 **He** $1s^2$
3 **Li** $2s^1$	4 **Be** $2s^2$											5 **B** $2s^22p^1$	6 **C** $2s^22p^2$	7 **N** $2s^22p^3$	8 **O** $2s^22p^4$	9 **F** $2s^22p^5$	10 **Ne** $2s^22p^6$
11 **Na** $3s^1$	12 **Mg** $3s^2$											13 **Al** $3s^23p^1$	14 **Si** $3s^23p^2$	15 **P** $3s^23p^3$	16 **S** $3s^23p^4$	17 **Cl** $3s^23p^5$	18 **Ar** $3s^23p^6$
19 **K** $4s^1$	20 **Ca** $4s^2$	21 **Sc** $3d^14s^2$	22 **Ti** $3d^24s^2$	23 **V** $3d^34s^2$	24 **Cr** $3d^54s^1$	25 **Mn** $3d^54s^2$	26 **Fe** $3d^64s^2$	27 **Co** $3d^74s^2$	28 **Ni** $3d^84s^2$	29 **Cu** $3d^{10}4s^1$	30 **Zn** $3d^{10}4s^2$	31 **Ga** $4s^24p^1$	32 **Ge** $4s^24p^2$	33 **As** $4s^24p^3$	34 **Se** $4s^24p^4$	35 **Br** $4s^24p^5$	36 **Kr** $4s^24p^6$
37 **Rb** $5s^1$	38 **Sr** $5s^2$	39 **Y** $4d^15s^2$	40 **Zr** $4d^25s^2$	41 **Nb** $4d^45s^1$	42 **Mo** $4d^55s^1$	43 **Tc** $4d^55s^2$	44 **Ru** $4d^75s^1$	45 **Rh** $4d^85s^1$	46 **Pd** $4d^{10}$	47 **Ag** $4d^{10}5s^1$	48 **Cd** $4d^{10}5s^2$	49 **In** $5s^25p^1$	50 **Sn** $5s^25p^2$	51 **Sb** $5s^25p^3$	52 **Te** $5s^25p^4$	53 **I** $5s^25p^5$	54 **Xe** $5s^25p^6$
55 **Cs** $6s^1$	56 **Ba** $6s^2$	57 ***La** $5d^16s^2$	72 **Hf** $5d^26s^2$	73 **Ta** $5d^36s^2$	74 **W** $5d^46s^2$	75 **Re** $5d^56s^2$	76 **Os** $5d^66s^2$	77 **Ir** $5d^76s^2$	78 **Pt** $5d^96s^1$	79 **Au** $5d^{10}6s^1$	80 **Hg** $5d^{10}6s^2$	81 **Tl** $6s^26p^1$	82 **Pb** $6s^26p^2$	83 **Bi** $6s^26p^3$	84 **Po** $6s^26p^4$	85 **At** $6s^26p^5$	86 **Rn** $6s^26p^6$
87 **Fr** $7s^1$	88 **Ra** $7s^2$	89 **†Ac** $6d^17s^2$	104 **Rf** $6d^27s^2$	105 **Db** $6d^37s^2$	106 **Sg** $6d^47s^2$	107 **Bh**	108 **Hs**	109 **Mt**	110	111	112						

*	58 **Ce** $4f^26s^2$	59 **Pr** $4f^36s^2$	60 **Nd** $4f^46s^2$	61 **Pm** $4f^56s^2$	62 **Sm** $4f^66s^2$	63 **Eu** $4f^76s^2$	64 **Gd** $4f^75d^16s^2$	65 **Tb** $4f^96s^2$	66 **Dy** $4f^{10}6s^2$	67 **Ho** $4f^{11}6s^2$	68 **Er** $4f^{12}6s^2$	69 **Tm** $4f^{13}6s^2$	70 **Yb** $4f^{14}6s^2$	71 **Lu** $4f^{14}5d^16s^2$
†	90 **Th** $6d^27s^2$	91 **Pa** $5f^26d^17s^2$	92 **U** $5f^36d^17s^2$	93 **Np** $5f^46d^17s^2$	94 **Pu** $5f^67s^2$	95 **Am** $5f^77s^2$	96 **Cm** $5f^76d^17s^2$	97 **Bk** $5f^97s^2$	98 **Cf** $5f^{10}7s^2$	99 **Es** $5f^{11}7s^2$	100 **Fm** $5f^{12}7s^2$	101 **Md** $5f^{13}7s^2$	102 **No** $5f^{14}7s^2$	103 **Lr** $5f^{14}6d^17s^2$

s-block elements

d-block elements

p-block elements

f-block elements

▲ **Figure 8.3 Electron configurations and the periodic table**
Subshells that lie at lower energies than those listed are filled with electrons. For example, the electron configuration of arsenic is $[Ar]3d^{10}4s^24p^3$, that of iodine is $[Kr]4d^{10}5s^25p^5$, that of lead is $[Xe]4f^{14}5d^{10}6s^26p^2$, and that of uranium is $[Rn]5f^36d^17s^2$.

tomic Number:
ne Work of Henry G. J. Moseley

Henry G. J. Moseley (1887–1915) was one of a group of brilliant students who studied
ı Ernest Rutherford just before World
r I. Moseley's most important work
in finding a relationship between the
'elengths of X rays and the elements
ıng rise to them. X rays are a form
·lectromagnetic radiation produced
:n cathode rays strike a metal anode
ed a *target* (Figure 8.4).

Moseley reasoned that X rays are
tted in the same way as other emis-
ı spectra—the dropping of electrons
xcited atoms to lower energy levels.
: energy levels of the electrons de-
d primarily on the magnitude of the
itive charge on the nucleus of an
n, as do differences between energy
:ls, ΔE_{level}. The energy, frequency,
wavelength of the emitted X rays
ıld also depend on the magnitude of
nuclear charge ($\Delta E_{\text{level}} = h\nu$).
seley used different metals as the tar-
; in an X ray tube and determined the
'elengths of the emitted X rays.

Moseley's startling results are seen in the regular pattern in the photographic images produced by the X rays in Figure 8.5. The two lines shown for each metal represent two different X ray emissions from the target in the X ray tube. Starting with calcium (Ca), the pairs of lines are displaced progressively to the left, toward shorter wavelengths and higher frequencies. Moseley derived an equation that linked each X ray frequency to the metal producing it, by assigning to the element a number equal to the positive charge on the atomic nucleus—the *atomic number*. In this way, Moseley accounted for all the atomic numbers from 13 (aluminum) to 79 (gold). This included predicting three new elements at $Z = 43$ (Tc), $Z = 61$ (Pm), and $Z = 75$ (Re). He also demonstrated that there could be no other elements in this portion of the periodic table.

Moseley was only in his mid-twenties when he performed this work. Within a year of its conclusion, he was killed in battle at Gallipoli in Turkey in World War I.

▲ **Figure 8.5**
Moseley's X-ray spectra of several metals
Each pure metal displays a pair of lines. The line at the left is the shorter wavelength line, and the one at the right, the longer wavelength line. The spectrum of cobalt has four lines, but only two are produced by cobalt; one of the remaining two lines corresponds to iron as an impurity, and the other line to nickel. Brass is an alloy of copper and zinc. Two of the four lines in its spectrum are identical to those in the copper spectrum shown above it. The other two lines are produced by zinc.

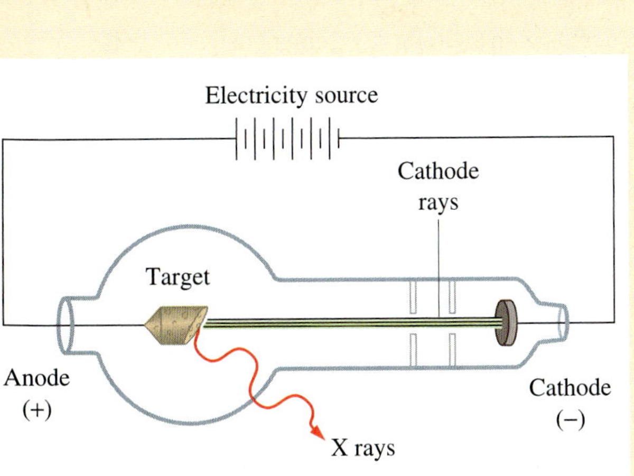

Figure 8.4
production of X rays

For main-group elements (A-group elements), the number of valence-shell electrons is the same as the periodic table group number.

Thus, except for helium, which has only two electrons, all the noble gases (Group 8A) have *eight* valence-shell electrons. These electrons are in the valence-shell configuration ns^2np^6, that is, $2s^22p^6$ for neon, $3s^23p^6$ for argon, and so on. As we shall see later in the chapter, this valence-shell electron configuration seems to impart to the noble gases a stability, and hence lack of reactivity, that is not found in any other group of elements.

The atoms in Group 1A (the alkali metals) have a *single* electron in an s orbital of the valence shell: $2s^1$ for Li, $3s^1$ for Na, or, in general, ns^1. Similarly, the Group 2A atoms have two electrons in valence-shell electron configuration ns^2. As a further example, the seven valence-shell electrons of the Group 7A atoms are in the configuration ns^2np^5. These valence-shell electron configurations relate to the kinds of ions formed by the Group 1A, 2A, and 7A elements, as we will see shortly.

The correlation between an A-group number and the number of valence-shell electrons does *not* extend to the B-group elements, the *transition* elements. Atoms of the transition elements typically have only *two* valence shell electrons (ns^2); some have only one (ns^1).*

Another fundamental relationship that we can infer from Figure 8.3 and the periodic table inside the front cover is:

The period number is the same as the principal quantum number, n, *of the electrons in the valence shell.**

All elements in the fourth period, for example, have one or more electrons with $n = 4$ and no electrons with a higher value of n. This period begins with potassium (K), which has the electron configuration [Ar]$4s^1$, and ends with krypton: Kr ([Ar]$3d^{10}4s^24p^6$). The next subshell to fill after $4p$ is $5s$, so the element rubidium (Rb), with the electron configuration [Kr]$5s^1$, begins the fifth period.

Each period of the periodic table begins with a Group 1A element and ends with a Group 8A element. The periods differ in length because the number of subshells that must fill to get from the electron configuration ns^1 to ns^2np^6 differs. In the two-member first period, we go directly from $1s^1$ to $1s^2$ (there is no p subshell in the first principal shell). In the eight-member second period, the $2s$ and $2p$ subshells fill; in the eight-member third period, the $3s$ and $3p$ subshells fill. The fourth and fifth periods have 18 members each. As we progress to higher atomic numbers, electrons fill the $4s$, $3d$, and $4p$ subshells in the fourth period, and the $5s$, $4d$, and $5p$ subshells in the fifth period. Finally, the 32-member sixth and seventh periods require the filling of the $6s$, $4f$, $5d$, and $6p$, and the $7s$, $5f$, $6d$, and $7p$ subshells, respectively.

Using the Periodic Table to Write Electron Configurations

We don't have to memorize an order of filling of orbitals (Figure 8.2) or use an orbital energy diagram (Figure 8.1) to write probable electron configurations. We can deduce the general form of the electron configurations given in Figure 8.3 (but without some of the details) directly from the periodic table. All we need to know

* Palladium ($Z = 46$) has no electrons in its $5s$ subshell. Thus, it has only four shells of electrons, even though it is in the fifth period.

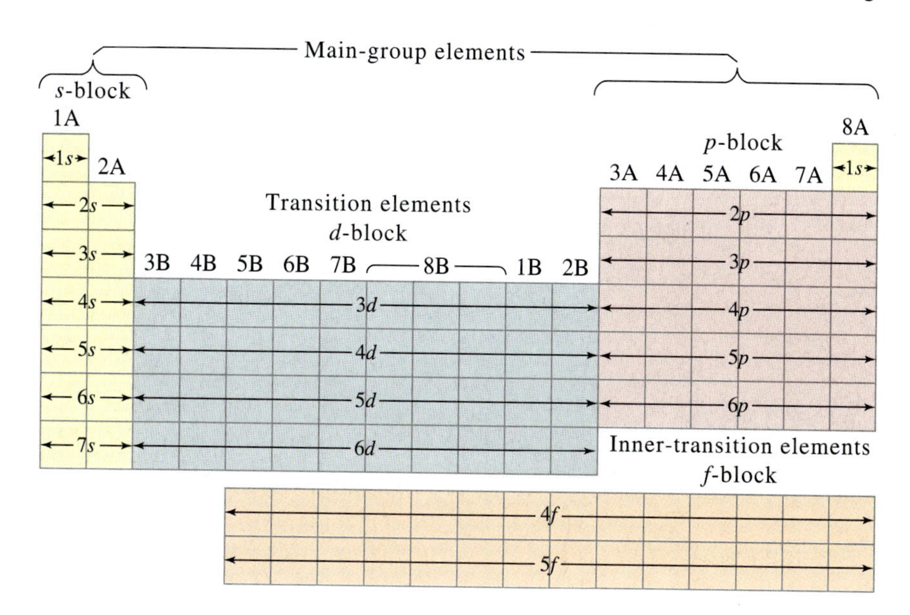

◀ **Figure 8.6**
The periodic table and the order of filling of subshells
Read through this periodic table, starting at the upper left, and you will discover the same order of filling of subshells as shown in Figure 8.2. Note that helium ($Z = 2$) is an *s*-block element, but it is grouped with the *p*-block elements because we place it in Group 8A with the other noble gas elements that it so strongly resembles.

is which subshells fill in different regions of the periodic table. To help make such deductions, we refer to the four blocks of elements shown in Figure 8.6.

***s*-block:** The *ns* subshell (the *s* subshell of the valence shell) fills by the aufbau process. These are *main-group* elements.

***p*-block:** The *np* subshell (the *p* subshell of the valence shell) fills. These are also *main-group* elements.

***d*-block:** The $(n - 1)d$ subshell (the *d* subshell of the next-to-outermost shell) fills. These are *transition* elements found in the main body of the periodic table.

***f*-block:** The $(n - 2)f$ subshell (the *f* subshell of the second from outermost shell) fills. To keep the periodic table at a maximum width of 18 members, these elements are placed below the main body of the table. The $4f$ subshell fills in the **lanthanide** series, and the $5f$ subshell fills in the **actinide** series. Because they fall within series of *d*-block elements, the *f*-block elements are sometimes called the *inner-transition* elements. The lanthanide series follows lanthanum ($Z = 57$) in the periodic table, and the actinide series follows actinium ($Z = 89$).

EXAMPLE 8.2

Give the complete ground-state electron configuration of a strontium atom **(a)** in the *spdf* notation and **(b)** in the noble-gas-core abbreviated notation.

SOLUTION

a. We proceed through Figure 8.6 until we reach Sr ($Z = 38$). The 1*s* subshell fills as we go through the first period. As we continue through the second period (Li through Ne), the 2*s* and 2*p* subshells fill. Moving through the third period, the 3*s* and 3*p* subshells fill. In the fourth period, the 4*s*, 3*d*, and 4*p* subshells fill in succession. In the fifth period, we find Sr in Group 2A, indicating two electrons in the 5*s* orbital. We can write the electron configuration, in the order of filling

$$1s^2 2s^2 2p^6 3s^2 3p^6 4s^2 3d^{10} 4p^6 5s^2$$

We can also group the subshells into principal shells.

$$1s^2 2s^2 2p^6 3s^2 3p^6 3d^{10} 4s^2 4p^6 5s^2$$

b. The noble-gas-core abbreviated notation of the electron configuration is $[Kr]5s^2$.

● **ChemCDX** Electron Configurations

EXERCISE 8.2A

Refer to Figure 8.6 and give the ground-state electron configuration, in the *spdf* notation and in the noble-gas-core abbreviated *spdf* notation, for **(a)** Mo and **(b)** Bi.

EXERCISE 8.2B

Refer only to the periodic table on the inside front cover and give the ground-state electron configuration **(a)** for Sn, using the noble-gas-core abbreviated *spdf* notation, and **(b)** for Zr, using an orbital diagram.

Valence Electrons and Core Electrons

Valence electrons are those that occupy the outermost principal shell of an atom.* They are the electrons with the highest principal quantum number, n. Electrons in inner shells are called **core electrons**; their principal quantum numbers are less than n. Thus, in a calcium atom, which has the electron configuration $[Ar]4s^2$, the 4*s* electrons are valence electrons and those in the [Ar] configuration are core electrons. Bromine, with the electron configuration $[Ar]3d^{10}4s^2 4p^5$, has seven valence electrons; the core electrons are in the configuration $[Ar]3d^{10}$.

Electron Configurations of Ions

To obtain the electron configuration of an *anion* by the aufbau process, we introduce additional electrons to the *valence shell* of the neutral nonmetal atom, *without* adding protons or neutrons to the nucleus. The number of electrons gained is usually the number needed to complete the valence shell of the atom. Thus, a nonmetal atom usually gains one or two electrons (sometimes three) and thereby attains the electron configuration of a noble gas atom. Some examples follow.

$$\text{Br } ([Ar]3d^{10}4s^2 4p^5) + e^- \longrightarrow \text{Br}^- \ ([Ar]3d^{10}4s^2 4p^6)$$
$$\text{S } ([Ne]3s^2 3p^4) + 2\,e^- \longrightarrow \text{S}^{2-} \ ([Ne]3s^2 3p^6)$$
$$\text{N } ([He]2s^2 2p^3) + 3\,e^- \longrightarrow \text{N}^{3-} \ ([He]2s^2 2p^6)$$

A metal atom loses one or more electrons in forming a *cation*. The p valence electrons (if there are any) tend to be lost first, and then s valence electrons; in some cases, d electrons of the next to outermost shell follow. For cations of the

* Some chemists view valence electrons as those involved in chemical reactions and core electrons as those that are not. From this perspective, some inner-shell electrons of the transition elements might be considered valence electrons. However, we will limit our use of the term valence electrons to those that occupy the outermost principal shell of an atom.

main-group metals, we essentially reverse the aufbau process by removing electrons. In many cases, this produces the electron configuration of a noble-gas atom—as in the following examples.

$$Na\ ([Ne]3s^1) \longrightarrow Na^+\ ([Ne]) + e^-$$

$$Mg\ ([Ne]3s^2) \longrightarrow Mg^{2+}\ ([Ne]) + 2\ e^-$$

$$Al\ ([Ne]3s^2 3p^1) \longrightarrow Al^{3+}\ ([Ne]) + 3\ e^-$$

However, in other cases, the electron configuration is not that of a noble gas.

$$Ga\ ([Ar]3d^{10} 4s^2 4p^1) \longrightarrow Ga^{3+}\ ([Ar]3d^{10}) + 3\ e^-$$

$$Sn\ ([Kr]4d^{10} 5s^2 5p^2) \longrightarrow Sn^{2+}\ ([Kr]4d^{10} 5s^2) + 2\ e^-$$

For cations formed from transition metal atoms, we do *not* simply reverse the aufbau process for atoms. That is, the first electrons lost are those of the highest principal quantum number—the s valence electrons—and not the ones added last in the aufbau process. Thus, in the formation of Fe^{2+} from Fe, the $4s^2$ electrons are lost.

$$Fe([Ar]3d^6 4s^2) \longrightarrow Fe^{2+}([Ar]3d^6) + 2\ e^-$$

In the formation of Fe^{3+} from Fe, one $3d$ electron is lost as well as $4s^2$.

$$Fe([Ar]3d^6 4s^2) \longrightarrow Fe^{3+}([Ar]3d^5) + 3\ e^-$$

Notice that in neither case does the ion have a noble-gas electron configuration. The possible types of electron configurations for cations are summarized in Table 8.3.

EXAMPLE 8.3

Write the electron configuration of the ion, Co^{3+}, in a noble-gas-core abbreviated *spdf* notation.

SOLUTION

One approach is to start with the electron configuration of the cobalt *atom*, which we can write with the aid of Figure 8.6. The first group of subshells that fill progressively

TABLE 8.3 Electron Configurations of Some Metal Ions

Noble Gas			Pseudo-Noble Gas[a]		18 + 2[b]	Various
Li^+	Be^{2+}	Al^{3+}	Cu^+	Zn^{2+}	In^+	Cr^{2+}: $[Ar]3d^4$
Na^+	Mg^{2+}		Ag^+	Cd^{2+}	Tl^+	Cr^{3+}: $[Ar]3d^3$
K^+	Ca^{2+}		Au^+	Hg^{2+}	Sn^{2+}	Mn^{2+}: $[Ar]3d^5$
Rb^+	Sr^{2+}				Pb^{2+}	Mn^{3+}: $[Ar]3d^4$
Cs^+	Ba^{2+}				Sb^{3+}	Fe^{2+}: $[Ar]3d^6$
					Bi^{3+}	Fe^{3+}: $[Ar]3d^5$
						Co^{2+}: $[Ar]3d^7$
						Co^{3+}: $[Ar]3d^6$
						Ni^{2+}: $[Ar]3d^8$

[a] In the pseudo-noble gas configuration, all valence electrons are lost and the remaining $(n-1)$ shell has 18 electrons in the configuration $(n-1)s^2(n-1)p^6(n-1)d^{10}$.

[b] In the 18 + 2 configuration, $(n-1)s^2(n-1)p^6(n-1)d^{10}ns^2$, two valence electrons remain.

Unsettled Issues Concerning the Periodic Table

The periodic table has been around for well over a century, and in that time, many variations of the table have been proposed. Yet, in all that time, no single version of the table has gained acceptance by all chemists.

A particular state of confusion results from differing use of the letters A and B. In the United States, A is used to designate the main-group elements and B, the transition elements. In Europe, the groups to the left of Fe/Ru/Os have typically been labeled as A groups, and those to the right of Ni/Pd/Pt, as B groups.

In order to eliminate confusion over the use of A and B, the International Union of Pure and Applied Chemistry (IUPAC) has recommended that the groups simply be numbered from 1 to 18. On the other hand, the A/B system that we use in this text is helpful in that the A-group numbers are equal to the numbers of valence-shell electrons in atoms of the main-group elements.

Placing hydrogen properly is difficult. Although hydrogen is a member of Group 1A, it is a gaseous nonmetal and not at all like an alkali metal. Most periodic tables put hydrogen in Group 1A because it has the electron configuration $1s^1$. But we could also place it in Group 7A because, like the halogen atoms, it is one electron short of a noble-gas electron configuration. Some periodic tables place it in *both* Groups 1A and 7A, and some place it in neither group, but alone at the top of the table near the center.

Although chemists may disagree over just what form of the periodic table is *most* useful, all agree that the table is tremendously useful.

in the aufbau process are $1s$, $2s$, $2p$, $3s$, $3p$, and $4s$. This accounts for 20 electrons (18 in the argon core and an additional two $4s$ electrons). The atomic number of cobalt is 27, and so the partially filled $3d$ subshell must have seven electrons.

$$\text{Co: [Ar]}3d^7 4s^2$$

We must remove three electrons to obtain Co^{3+} from Co. These are the $4s^2$ pair and one $3d$ electron. The noble-gas-core abbreviated *spdf* notation of Co^{3+} is $[Ar]3d^6$.

EXERCISE 8.3A

Write electron configurations in the noble-gas-core abbreviated *spdf* notation for the ions: Se^{2-} and Pb^{2+}.

EXERCISE 8.3B

Write noble-gas-core abbreviated orbital diagrams for the ions: I^- and Cr^{3+}.

8.6 Magnetic Properties: Paired and Unpaired Electrons

When considering electron spin in Chapter 7, we noted that each electron in an atom creates a magnetic field—each electron acts like a tiny magnet. For a *pair* of electrons with opposite spins, the magnetic fields cancel one another and the pair shows no resultant magnetism. Most substances exhibit no magnetic properties until they are placed in a magnetic field. Then we find two possible types of behavior.

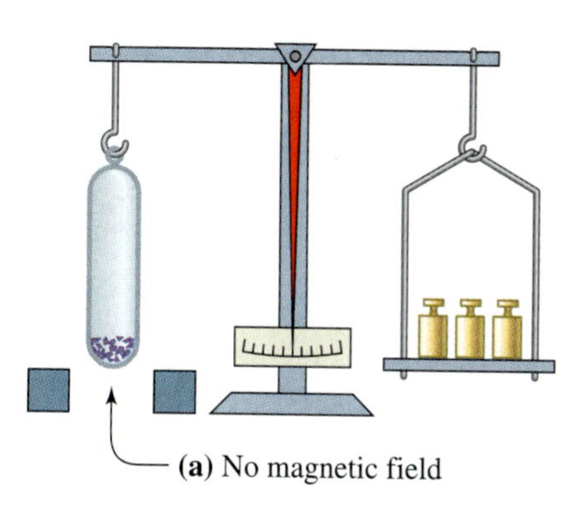

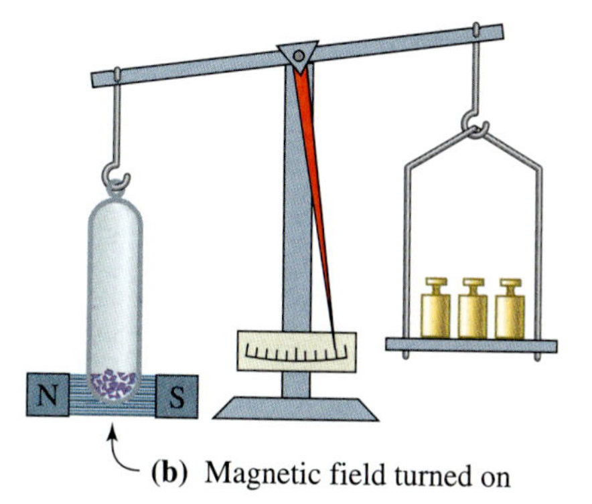

Figure 8.7
Paramagnetism illustrated
(a) A sample (left side of balance) is weighed in the absence of a magnetic field. (b) When the field is turned on, the balanced condition is upset. The sample appears to gain weight. This is because it is now subjected to two attractive forces, the force of gravity *and* the force exerted by the magnetic field.

- If all the electrons in the atoms, ions, or molecules of a substance are paired, the substance is *weakly repelled* by a magnetic field. This weak repulsion associated with paired electrons is called **diamagnetism**.
- Atoms, ions, or molecules having *unpaired* electrons are *attracted* into a magnetic field. This attraction associated with unpaired electrons is called **paramagnetism**. The paramagnetism of unpaired electrons is a much stronger effect than the weak diamagnetism of paired electrons.

The exceptionally strong attractions of a magnetic field for iron and a few other substances is a third type of magnetic behavior, called *ferromagnetism*, which we will consider in Chapter 22.

The magnetic properties of a substance can be determined by weighing the substance in the absence and in the presence of a magnetic field, as indicated in Figure 8.7. Such measurements would show, for example, that the number of unpaired electrons in Fe^{3+} is greater than in Fe^{2+}. This is experimental verification for the orbital diagrams for these two ions.

Fe^{2+} [Ar] [↑↓][↑][↑][↑][↑] (3*d*) and Fe^{3+} [Ar] [↑][↑][↑][↑][↑] (3*d*)

EXAMPLE 8.4

A sample of chlorine gas is found to be diamagnetic. Can this gaseous sample be composed of individual Cl atoms?

SOLUTION

We can deduce the electron configuration of Cl from its position in the periodic table and represent it by an orbital diagram.

(Z = 17) Cl [↑↓] (1*s*) [↑↓] (2*s*) [↑↓][↑↓][↑↓] (2*p*) [↑↓] (3*s*) [↑↓][↑↓][↑] (3*p*)

The diagram shows one unpaired electron per atom. We would expect atomic chlorine (Cl) to be paramagnetic. Because the gas is diamagnetic, it cannot be composed of individual Cl atoms. (In fact, chlorine gas consists of diatomic molecules, Cl_2, in which all electrons are paired.)

EXERCISE 8.4

Which of the following would you expect to exhibit paramagnetism? Explain.

a. a K atom
b. a Hg atom
c. a Ba^{2+} ion
d. a N atom
e. a F^- ion
f. a Ti^{2+} ion
g. a Cu^{2+} ion

8.7 Periodic Atomic Properties of the Elements

In developing his periodic table, Mendeleev stressed the similarities in chemical properties for groups of elements, such as those elements whose oxides and hydrides had similar formulas. Lothar Meyer, a contemporary of Mendeleev's, developed a classification scheme based primarily on similarities in physical properties, such as density. The underlying principle in both cases is the **periodic law**, which in its modern form states that certain sets of physical and chemical properties recur at regular intervals (periodically) when the elements are arranged according to increasing atomic number.

Some physical properties, such as thermal and electrical conductivity, hardness, and melting point, are associated only with *bulk* matter, that is, aggregations of atoms large enough to see and measure at the macroscopic level. Other properties, called *atomic properties*, are associated with individual atoms. We have already discussed one atomic property, electron configuration, at some length. In this section, we examine three other atomic properties: atomic radii, ionization energies, and electron affinities.

Atomic Radii

We cannot measure the exact size of an isolated atom because its outermost electrons have a remote chance of being found quite far from the nucleus. What we can measure is the distance between the nuclei of two atoms, and we can derive a property called the **atomic radius** from this distance. However, internuclear distances depend on the particular environment in which the atoms are found. Thus, there can be more than one value for the "atomic" radius of an element. Let's consider two specific cases.

The **covalent radius** of an atom is one half the distance between the nuclei of two identical atoms joined into a molecule. Consider, for example, the I_2 molecule (Figure 8.8). The distance between the two iodine nuclei is 266 pm, and the covalent radius of iodine is one half of this distance, or 133 pm.* The **metallic radius** of an atom is half the distance between the nuclei of adjacent atoms in a solid metal. In this text, whenever we use the simple expression "atomic radius," you can assume we mean *covalent* radius for *nonmetals* and *metallic* radius for *metals*.

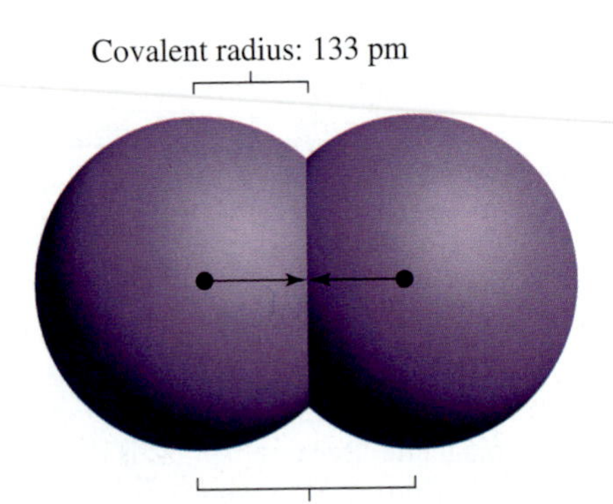

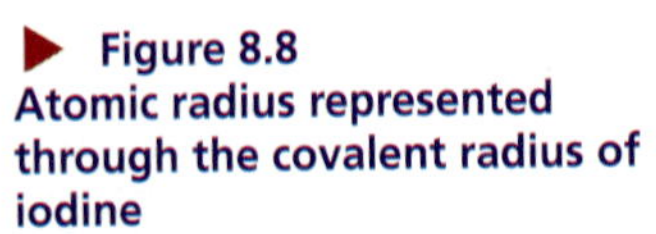

▶ Figure 8.8 Atomic radius represented through the covalent radius of iodine
The covalent radius is half the distance between the nuclei of the two I atoms in the I_2 molecule.

* The value of a covalent radius depends on whether the bond between the two atoms is single, double, or triple. The covalent radii used in this chapter are single-bond covalent radii.

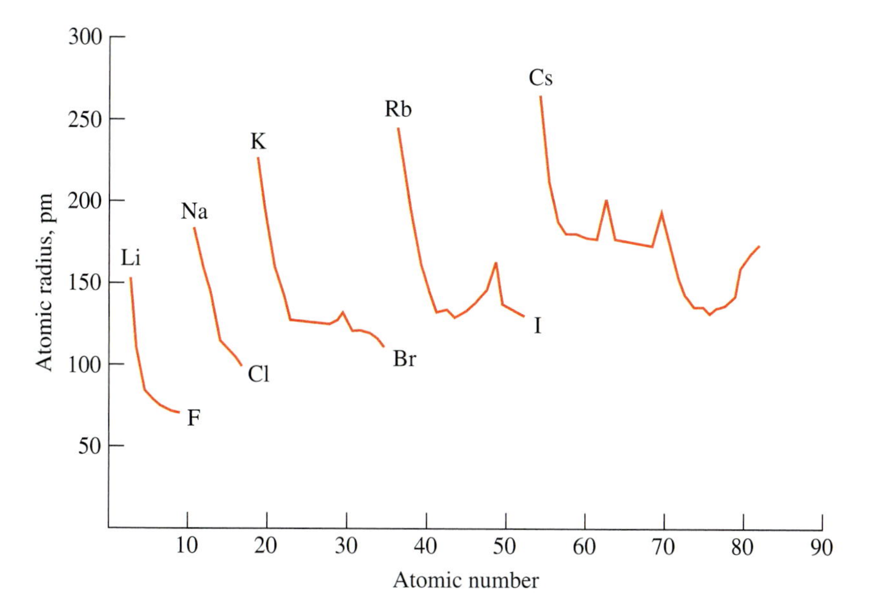

Figure 8.9
Atomic radii of the elements
The values shown here, in picometers (pm), are metallic radii for metals and covalent radii for nonmetals. Data are not included for the noble gases because it is difficult to assess their covalent radii (only Kr and Xe compounds are known). Explanations have been offered for the small peaks in the middle of some periods and for other irregularities, but they are beyond the scope of this book.

Atomic radii are sometimes expressed in *angstroms* (Å), a non-SI unit, but we will use picometers (1 pm $= 10^{-12}$ m).

$$1\ \text{Å} = 10^{-10}\ \text{m} = 100\ \text{pm}$$

Figure 8.9 shows graphically that the atomic radius is an excellent example of a periodic atomic property of the elements. Each red line plots the atomic radii for elements within a single period. Notice that the first member of each period, a Group 1A metal, has the largest atomic radius for the period (Li for period 2, Na for period 3, and so on). The atomic radii then generally decline through the period reaching a low value with the Group 7A nonmetal at the end of the period. To explain these trends, let's think of an atomic radius as roughly equal to the distance from the nucleus to the outer-shell (valence) electrons. Any factor that causes this distance to increase makes for a larger atomic radius.

Within a vertical *group* of the periodic table, each succeeding member has one more principal shell with electrons than its immediate predecessor. Thus, the sodium atom has electrons in the principal shells, $n = 1, 2,$ and 3, whereas the lithium atom has electrons only in $n = 1$ and 2. The electrons in the outer shell are farther from the nucleus as n increases, and therefore

Atomic radii increase from top to bottom within a group of the periodic table.

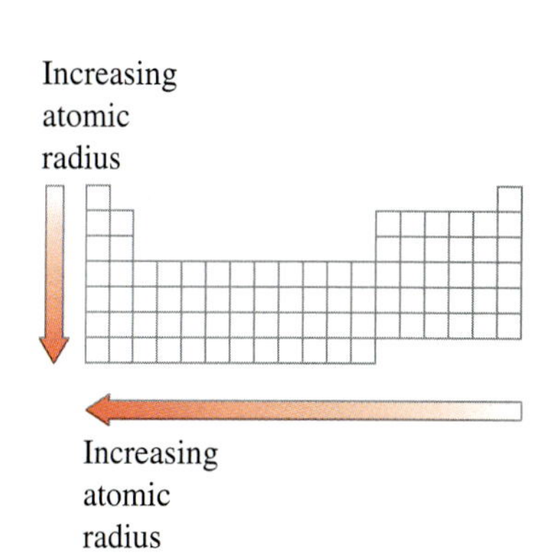

To explain the trend in atomic radii within a *period* of the periodic table, we find a new concept helpful: The **effective nuclear charge (Z_{eff})** acting on an electron is the actual nuclear charge less the screening effect of other electrons in the atom.

First, as an oversimplified example, consider a sodium atom. *If* the 3s valence electron were at all times completely outside the region in which the ten core electrons ($1s^2 2s^2 2p^6$) are found, the 3s electron would be perfectly screened from the positively charged nucleus. Thus, the valence electron would experience an attraction to a net positive charge of only $+11 - 10 = +1$. Similarly, for a magnesium

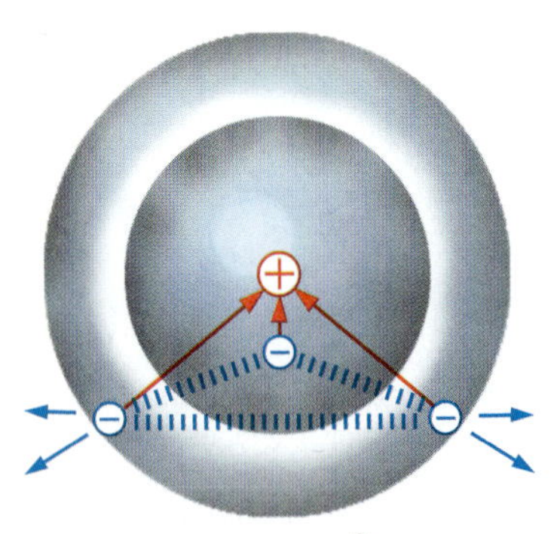

▲ **Figure 8.10**
Shielding effect and effective nuclear charge
This simplified sketch of a magnesium atom shows the two valence electrons and one core electron as discrete particles. The remaining 9 core electrons are shown as a circular cloud of negative electric charge. Interactions (discussed in the text) between valence electrons, core electrons, and the atomic nucleus determine an effective nuclear charge. Attractions are indicated in red and repulsions in blue.

atom pictured in Figure 8.10, there would be two $3s$ electrons outside the neon core and a net positive charge of $+12 - 10 = +2$ acting on each $3s$ electron. Proceeding in this fashion, we would find that the net positive charge acting on the valence electrons would progressively increase across the third period.

Evaluation of an effective nuclear charge also requires the following ideas: (1) Valence electrons can *penetrate* inner shells and approach the nucleus. On page 318, we described how the difference in the penetrating abilities of *s*, *p*, *d*, and *f* electrons cause a splitting of subshell energies. (2) Core electrons are not all equally effective in shielding valence electrons. (3) To some extent, valence electrons shield one another from the nuclear charge. By taking these three factors into account, we can get much better estimates of effective nuclear charges than in our crude approximations. However, for the purpose of explaining trends in atomic sizes, the crude estimates and the more precise ones both lead to the same conclusion: The effective nuclear charge increases from left to right in a period of main-group elements. Because the effective nuclear charge increases, valence electrons are pulled in toward the nucleus and held more tightly.

Atomic radii of the A-group elements decrease from left to right in a period of the periodic table.

● **ChemCDX** Effective Nuclear Charge

The restriction to A-group elements is important because in a series of B-group elements (transition elements) electrons enter an *inner* electron shell. The effective nuclear charge therefore remains essentially constant instead of increasing. For example, consider the effective nuclear charges, Z_{eff}, of iron, cobalt and nickel, by the crude method outlined above (core electrons are shown in blue).

Fe: $[Ar]3d^6 4s^2$ Co: $[Ar]3d^7 4s^2$ Ni: $[Ar]3d^8 4s^2$

$Z_{eff} \approx 26 - 24 \approx 2$ $Z_{eff} \approx 27 - 25 \approx 2$ $Z_{eff} \approx 28 - 26 \approx 2$

● **ChemCDX** Periodic Trends: Atomic Radii

Because the effective nuclear charges are about the same, we conclude that the radii should also be about the same. The actual values are 124, 124, and 125 pm, respectively, for Fe, Co, and Ni. The difference in trends in atomic radii in the main-group and transition-element regions of a period of elements is clearly illustrated in Figure 8.9.

EXAMPLE 8.5

Arrange each set of elements in order of increasing atomic radius, that is, from smallest to largest.

a. Mg, S, Si   **b.** As, N, P   **c.** As, Sb, Se

SOLUTION

a. All three are main-group elements in the same period (third). Atomic radii within a period decrease from left to right. The order of *increasing* radius is:

S (smallest) $<$ Si $<$ Mg (largest).

b. All three elements are in the same group (5A). Atomic radii increase from top to bottom. The order of *increasing* radius is:

N (smallest) $<$ P $<$ As (largest).

c. Of this set, As and Se are in the same period (fourth); As is to the left and therefore larger than Se. Because Sb is below As in the same group (5A), it is larger than As. The overall order is:

Se (smallest) < As < Sb (largest).

EXERCISE 8.5

Arrange each set of elements in order of increasing atomic radius.

a. Be, F, N **b.** Ba, Be, Ca **c.** Cl, F, S **d.** Ca, K, Mg

Ionic Radii

Like atomic radii, ionic radii are based on an internuclear distance—in this case, the distance between the nuclei of two ions. Figure 8.11 shows a Mg^{2+} and an O^{2-} ion in contact. The **ionic radii** of the two ions are the portions of the distance between the nuclei occupied by each ion. Ionic radii are determined by crystal structure studies of the type described in Chapter 11. When atoms of metals react, they often lose all their valence-shell electrons. With one fewer electronic shell, a metal ion is smaller than the atom from which it comes. Also, because the nuclear charge is now greater than the number of electrons, the nucleus attracts the remaining electrons more strongly and holds them more closely in a cation than in the corresponding atom.

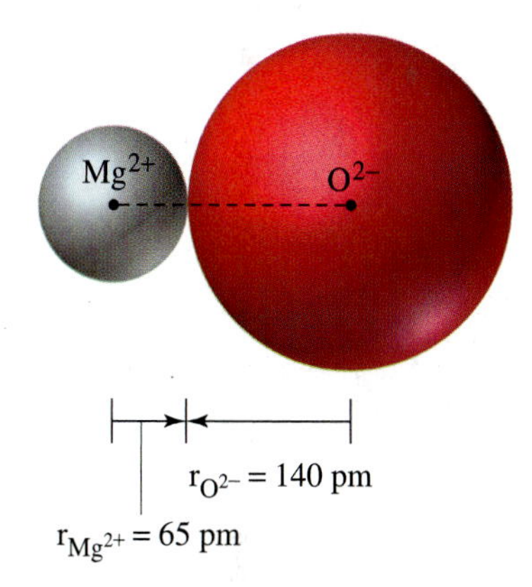

▲ **Figure 8.11**
The ionic radii of Mg^{2+} and O^{2-}
The distance between the centers of the two ions (205 pm) is apportioned between the Mg^{2+} (65 pm) and O^{2-} (140 pm) ions. The sizes of other cations can be related to their internuclear distance to the oxide ion.

Cations are smaller *than the atoms from which they are formed.*

Figure 8.12 compares the radii of five species: a sodium atom (Na), a magnesium atom (Mg), a neon atom (Ne), a sodium ion (Na^+), and a magnesium ion (Mg^{2+}). The Mg atom is smaller than the Na atom, and the ions are smaller than the corresponding atoms. The species Ne, Na^+, and Mg^{2+} are **isoelectronic**; they all have the same number (10) of electrons. They also have identical electron configurations ($1s^2 2s^2 2p^6$). Neon has a nuclear charge of +10. Because the sodium ion has a nuclear charge of +11, Na^+ is *smaller* than Ne. Because its nuclear charge is +12, Mg^{2+} is smaller still.

When a nonmetal atom gains an electron to form an anion, the positive nuclear charge remains constant and repulsions among the negatively charged electrons increase. Thus, the electrons spread out more, and the size increases. The formation of two Cl^- ions from a Cl_2 molecule and the large increase from the covalent atomic radius in Cl_2 to the anionic radius in Cl^- are suggested in Figure 8.13. In general, we can say that

- *Anions are* larger *than the atoms from which they are formed.*
- *For a series of isoelectronic ions with the same electron configuration, the greater the nuclear charge, the smaller the ion is.*

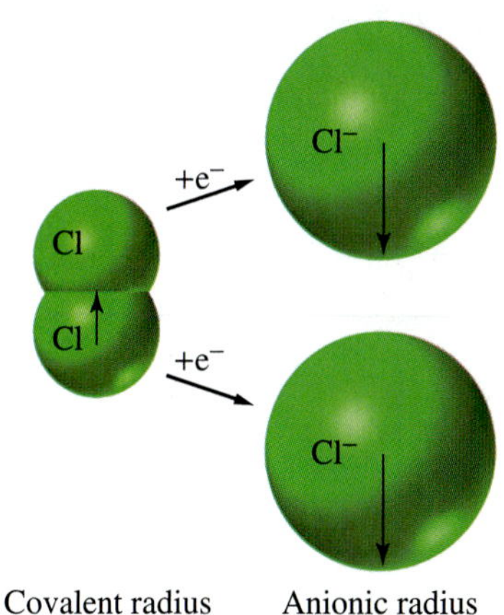

▲ **Figure 8.13**
A comparison of atomic (covalent) and anionic radii
The two chlorine atoms in a Cl_2 molecule gain one electron each to form two chloride ions (2 Cl^-).

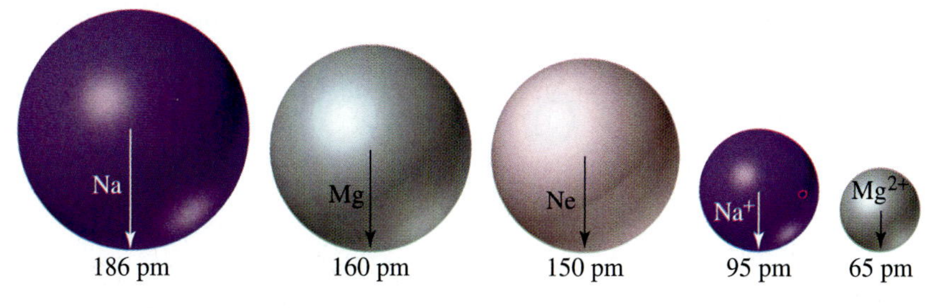

▲ **Figure 8.12 A comparison of atomic and cationic radii**
Metallic radii are shown for Na and Mg, and ionic radii for Na^+ and Mg^{2+}. The radius for Ne is for a nonbonded atom.

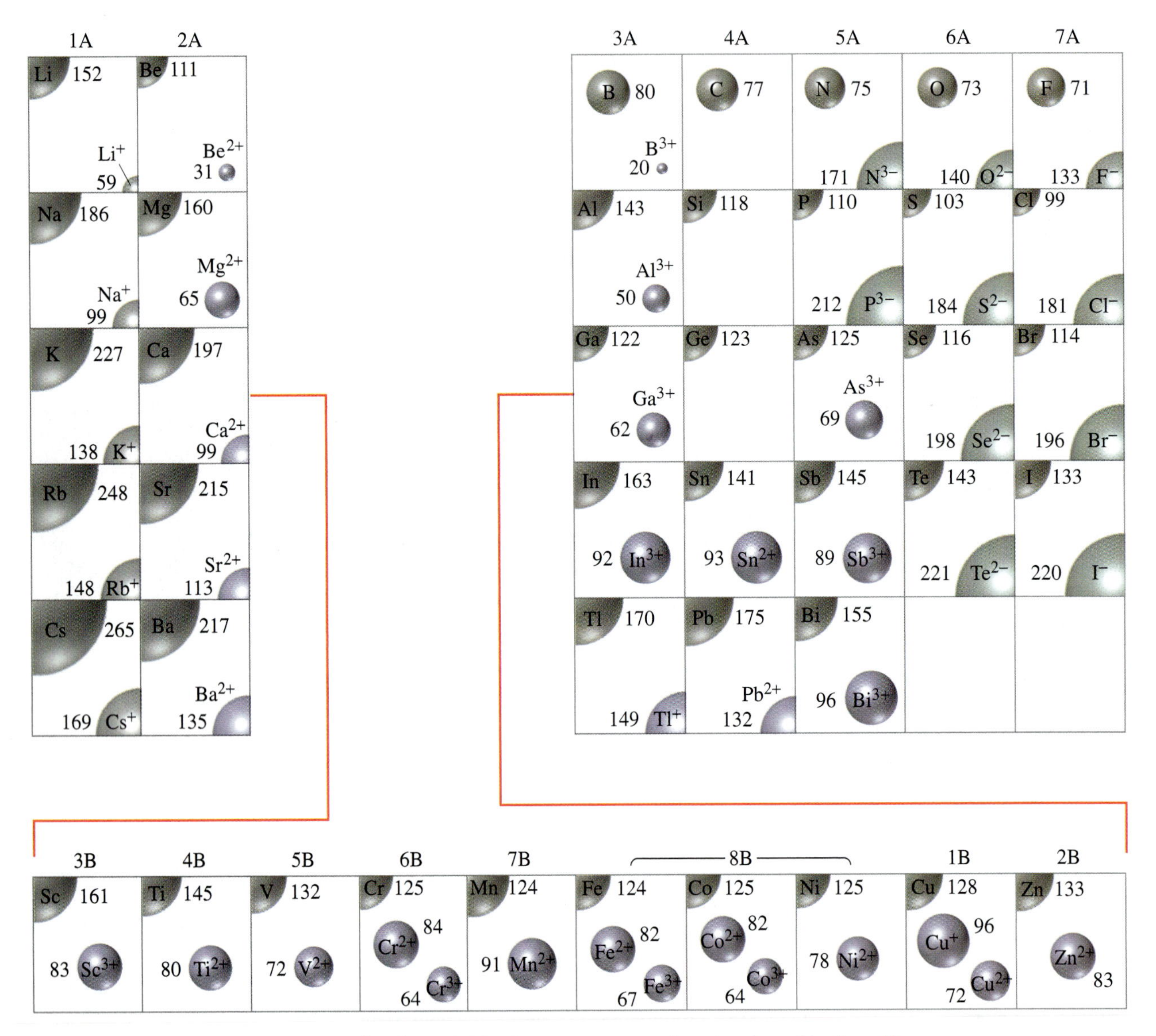

▲ **Figure 8.14 Some representative atomic and ionic radii**
The values, in picometers (pm), are metallic radii for metals, single covalent radii for nonmetals, and ionic radii for the ions indicated.

● **ChemCDX** Gain and Loss of Electrons

Figure 8.14 summarizes much of our discussion of atomic and ionic radii: how atomic radii vary with the position in a group and period of the periodic table; how the radii of cations and anions are related to the radii of the atoms from which they are formed; how the magnitude of the charge affects the radii of cations and anions; and so on. You may need to refer to this figure from time to time.

EXAMPLE 8.6

Refer to a periodic table, but not to Figure 8.14, and arrange the following species in the expected order of increasing radius: Ca^{2+}, Fe^{3+}, K^{+}, S^{2-}, Se^{2-}.

SOLUTION

Three of these (Ca^{2+}, K^{+}, and S^{2-}) are isoelectronic; they have the same electron configuration ($1s^2 2s^2 2p^6 3s^2 3p^6$). According to our generalization, their comparative radii should be $Ca^{2+} < K^{+} < S^{2-}$. Because Fe is in the same period as Ca and farther to the right, we expect the Fe atom to be smaller than the Ca atom. Also, we expect the loss of *three* electrons (two 4*s* and one 3*d*) by an Fe atom to produce a smaller ion than the loss of *two* 4*s* electrons by a Ca atom. The ion Fe^{3+} should therefore be smaller than Ca^{2+}. To get a feel for the size of the Se^{2-} ion, note that it is in the same group as S^{2-}. Because Se^{2-} has the same charge but more electron shells than S^{2-}, we expect Se^{2-} to be larger than S^{2-}. We conclude that the overall order is as shown below.

$$Fe^{3+} < Ca^{2+} < K^{+} < S^{2-} < Se^{2-}$$

EXERCISE 8.6A

Refer to a periodic table, but not to Figure 8.14, and arrange the following species in the expected order of increasing radius: Br^{-}, Rb^{+}, Se^{2-}, Sr^{2+}, Y^{3+}.

EXERCISE 8.6B

Without reference to Figure 8.14, arrange the following species in the expected order of increasing radius: Ca^{2+}, Cr^{2+}, Cs^{+}, Cl^{-}, Cr^{3+}, K^{+}.

Ionization Energy

When they undergo chemical reactions, metal atoms tend to lose valence electrons. However, isolated atoms do not eject electrons spontaneously. Work must be done to remove an electron from an atom, and the amount of work depends on the size of the atom.

The **ionization energy** is the energy required to remove an electron from a ground-state atom (or ion) in the gaseous state. The quantity of energy is usually expressed in terms of a mole of atoms. Atoms with more than one electron can ionize in successive steps. Consider the boron atom, which has five electrons—two in an inner core ($1s^2$) and three valence electrons ($2s^2 2p^1$). The five steps and their successive ionization energies, I_1 through I_5, follow.

$$B(g) \longrightarrow B^{+}(g) + e^{-} \qquad I_1 = 801 \text{ kJ/mol}$$

$$B^{+}(g) \longrightarrow B^{2+}(g) + e^{-} \qquad I_2 = 2{,}427 \text{ kJ/mol}$$

$$B^{2+}(g) \longrightarrow B^{3+}(g) + e^{-} \qquad I_3 = 3{,}660 \text{ kJ/mol}$$

$$B^{3+}(g) \longrightarrow B^{4+}(g) + e^{-} \qquad I_4 = 25{,}025 \text{ kJ/mol}$$

$$B^{4+}(g) \longrightarrow B^{5+}(g) + e^{-} \qquad I_5 = 32{,}822 \text{ kJ/mol}$$

The first electron to be removed is from the highest energy subshell (2*p*). It is by far the easiest to remove, and the energy required is called the *first* ionization energy, I_1. The *second* ionization energy, I_2, is more than three times greater than the first. Why is this so? The first electron is removed from the 2*p* orbital of a *neutral* B atom. The second electron is removed from the 2*s* orbital of a B^{+} *ion*. The second ionization energy is larger than the first, partly because the 2*s* orbital is at a lower energy than the 2*p*, but mostly because the second electron must be stripped from a positive ion, to which it is strongly attracted. Removal of the third electron must occur from a more highly charged ion, B^{2+}, making I_3 larger still.

Compared to the first three, the fourth and fifth successive ionization energies, I_4 and I_5, of boron are extremely large. Note that the first three electrons are *valence* electrons and the last two are *core* electrons; they have a lower value of *n*, the principal quantum number. For the main-group elements, removing a core electron

TABLE 8.4 Some Selected Ionization Energies, kJ/mol

	1A	2A	3A	4A	5A	6A	7A	8A
	Li	Be	B	C	N	O	F	Ne
I_1	520	900	801	1086	1402	1314	1681	2081
I_2	7298	1757						
	Na	Mg						
I_1	496	738						
I_2	4562	1451						
	K	Ca						
I_1	419	590						
I_2	3051	1145						
	Rb	Sr						
I_1	403	550						
I_2	2633	1064						
	Cs	Ba						
I_1	376	503						
I_2	2230	965						

takes *much* more work than removing an additional valence electron. This is consistent with the observation that only valence electrons are associated with the chemical reactivity of the main-group elements.

Table 8.4 lists some ionization energies for main-group elements, and Figure 8.15 is a graph of first ionization energy versus atomic number. These two sources are the basis of some useful generalizations.

- The first ionization energy of an atom is its lowest. Compare I_1 and I_2 values for the members of Group 2A in Table 8.4.

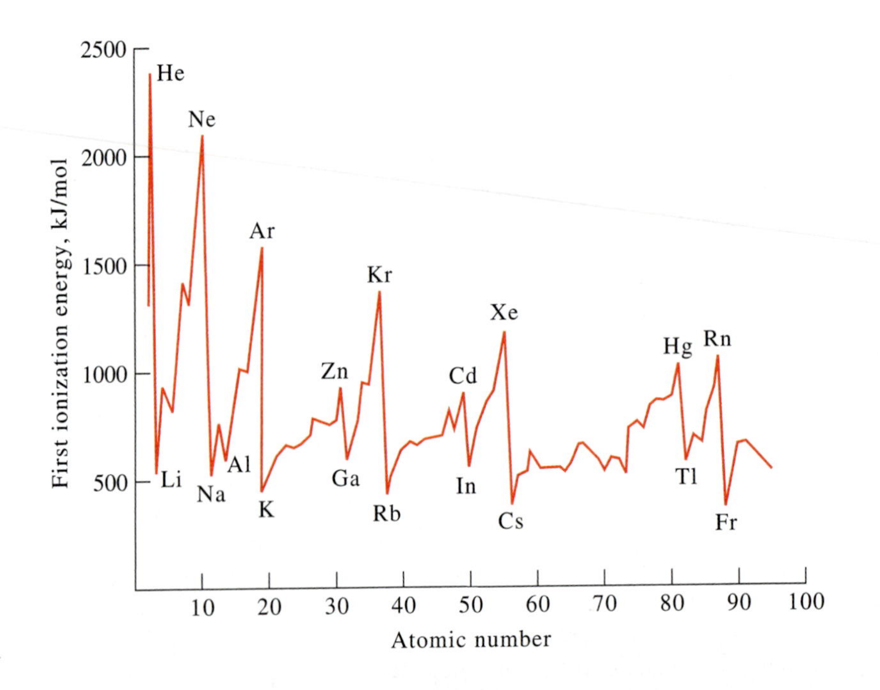

Figure 8.15
First ionization energy as a function of atomic number

- A large increase in successive ionization energies occurs between the removal of the last valence electron and the removal of the first core electron. Compare I_1 and I_2 values for the members of Group 1A in Table 8.4.
- Ionization energies *decrease* down a vertical group in the periodic table, that is, from lower to higher atomic numbers. Compare I_1 values for different members of Group 1A and of Group 2A in Table 8.4. Also, notice the gradual decrease in the minimum point of the graph of Figure 8.15 each time a minimum recurs.
- In general, ionization energies *increase* in going from left to right through a period in the periodic table. Compare the I_1 values across the top row of Table 8.4. In Figure 8.15, this is seen in the steady rise in I_1 values (with some notable exceptions) from the minima for Group 1A (alkali metal) elements to the maxima for Group 8A (noble gas) elements.

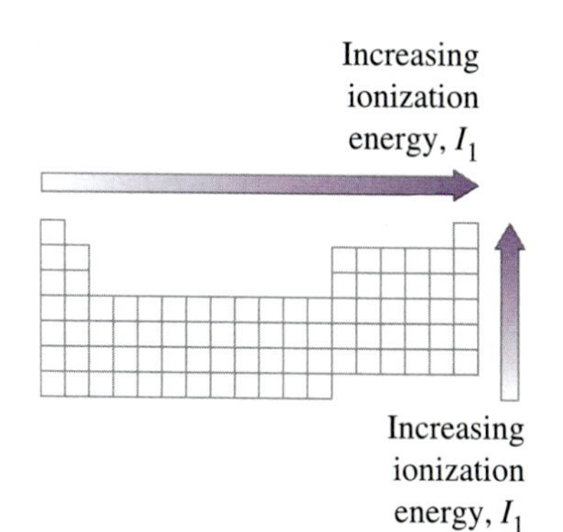

Other factors contribute to these patterns, but we can explain general trends most easily in terms of atomic sizes. The greater the distance between the atomic nucleus and the electron to be removed, the less tightly that electron is held to the nucleus and the more readily ionization occurs. As we have already seen, the general trends in atomic radii are an increase down a group and a decrease across a period of the periodic table.

● **ChemCDX** Ionization Energy

The irregularity between Groups 2A and 3A in Table 8.4 can be explained in this way. A $2p$ electron is more easily removed from the $1s^2 2s^2 2p^1$ electron configuration of a boron atom than is a $2s$ electron from the $1s^2 2s^2$ configuration of a beryllium atom. The $2p$ electron is at a higher energy than a $2s$ electron. As a result, I_1 is smaller for B (801 kJ/mol) than for Be (900 kJ/mol).

To explain the irregularity between Groups 5A and 6A in Table 8.4, we can consider repulsions between electrons. This repulsive tendency makes it easier to remove one of the *paired* electrons in a filled $2p$ orbital of an oxygen atom ($[\text{He}]2s^2 2p_x^2 2p_y^1 2p_z^1$) than an *unpaired* electron from a half-filled $2p$ orbital of a nitrogen atom ($[\text{He}]2s^2 2p_x^1 2p_y^1 2p_z^1$). The I_1 values are 1314 kJ/mol for O and 1402 kJ/mol for N.

● **ChemCDX** Periodic Trends: Ionization Energy

EXAMPLE 8.7

Without reference to Figure 8.15, arrange each set of elements in order of increasing first ionization energy.

a. Mg, S, Si   **b.** As, N, P   **c.** As, Ge, P

SOLUTION

a. All three elements are in the same period. Within a period, I_1 *increases* from left to right as the atoms become smaller. The order of increasing I_1 is:

Mg (lowest) $<$ Si $<$ S (highest)

b. All three elements are in the same group. Within a group, I_1 *decreases* from top to bottom as atoms become larger. The order of increasing I_1 is:

As (lowest) $<$ P $<$ N (highest)

c. Of this set, As and Ge are in the same period; Ge is to the left of As and has a lower I_1 than As. Because it is below P in the same group, As has a lower I_1 than P. The order of increasing I_1 is:

Ge (lowest) $<$ As $<$ P (highest)

EXERCISE 8.7

Without reference to Figure 8.15, arrange each set of elements in order of increasing first ionization energy.

a. Be, F, N **b.** Ba, Be, Ca **c.** F, P, S **d.** Ca, K, Mg

Electron Affinity

Ionization energy relates to forming a gaseous *positive* ion from a gaseous atom. An analogous atomic property for the formation of a gaseous *negative* ion is **electron affinity**, the energy change that occurs when an electron is added to a gaseous atom.

An electron approaching a neutral atom experiences an attraction for the positively charged nucleus. Repulsion of the incoming electron by electrons in the atom tends to offset this attraction. Still, in many cases, the incoming electron is absorbed by the atom and energy is given off, as in the process

$$F(g) + e^- \longrightarrow F^-(g) \qquad EA = -328 \text{ kJ/mol}$$

When a fluorine atom gains an electron, energy is given off. The process is *exothermic*, and we represent the electron affinity (EA) as a negative quantity.*

Table 8.5 lists several electron affinities, but we see fewer clear-cut trends and more irregularities than in the ionization energies in Table 8.4. Some of the data suggest a rough correlation between electron affinity and atomic size: Smaller atoms have more negative electron affinities. It seems reasonable that the smaller the atom, the closer an added electron can approach the atomic nucleus and the more strongly it is attracted to the nucleus. This certainly seems to be the case for the Group 1A elements, for Group 6A from S to Po, and for Group 7A from Cl to At. The second row elements present some problems, though. The electron affinity of O is not as negative as that of S, nor is that of F as negative as that of Cl. It may be that electron repulsions in the small compact atoms keep the added electron from being as strongly attracted as we might expect.

More negative electron affinity

More negative electron affinity

In most cases in Table 8.5, the added electron goes into a subshell that is already partly filled. For the Group 2A and 8A atoms, however, the added electron would be at a significantly higher energy level. For Group 2A atoms, the added electron enters the *np* subshell; for Group 8A atoms, the electron enters the *s* orbital of the next principal shell. The electron affinities of the atoms in Groups 2A and 8A have positive values, and thus the atoms do not form stable anions.

Similar to the stepwise loss of electrons in the formation of positive ions, we can consider a stepwise addition of electrons in anion formation. And we can write a separate electron affinity for each step. The first and second electron affinities for the oxygen atom are

$$O(g) + e^- \longrightarrow O^-(g) \qquad EA_1 = -141 \text{ kJ/mol}$$

$$O^-(g) + e^- \longrightarrow O^{2-}(g) \qquad EA_2 = +744 \text{ kJ/mol}$$

● **ChemCDX** Electron Affinity

* Some scientists define electron affinity as the reverse of the process shown above. That is, as

$$F^-(g) \longrightarrow F(g) + e^- \; EA = +328 \text{ kJ/mol.}$$

As we learned in Chapter 6, when we reverse a process, we change the sign of thermodynamic functions. Energy must be absorbed to remove an electron from a gaseous fluoride ion. You may see electron affinities written in both ways in the chemical literature.

TABLE 8.5 Some Selected First Electron Affinities, kJ/mol

1A	2A	3A	4A	5A	6A	7A	8A
Li −60	Be	B −27	C −154	N −7	O −140	F −328	Ne
Na −53					S −200	Cl −349	
K −48					Se −195	Br −325	
Rb −47					Te −190	I −295	
Cs −46					Po −183	At −270	

It's not hard to see why the second electron affinity is a positive quantity. Here an electron approaches an ion with a net charge of −1. The electron is strongly repelled, and work must be done to force the extra electron onto the $O^-(g)$ ion. In fact, the O^{2-} ion forms only in situations where other energetically favorable processes offset the large energy expense to create it. This happens, for example, in the formation of ionic oxides such as MgO.

EXAMPLE 8.8—A Conceptual Example

Which of the values given is a reasonable estimate of the second electron affinity (EA_2) for sulfur?

$$S^-(g) + e^- \longrightarrow S^{2-}(g) \qquad EA_2 = ?$$

−200 kJ/mol   +450 kJ/mol   +800 kJ/mol   +1200 kJ/mol

SOLUTION

Because of the repulsion between the $S^-(g)$ anion and the approaching electron, EA_2 must be *positive* (the process is endothermic). This eliminates −200 kJ/mol as a possibility. To choose among the other three values, consider the value of EA_2 for oxygen: +744 kJ/mol (page 342). Because of its *smaller* size, we should expect the $O^-(g)$ anion to exert a stronger repulsion for the incoming electron than would the $S^-(g)$ anion. The EA_2 value for $S^-(g)$ should be less than +744 kJ/mol. The only reasonable estimate is +450 kJ/mol.

EXERCISE 8.8

Based on the result of Example 8.8, which of the values given is a reasonable estimate of the second electron affinity for this process?

$$Se^-(g) + e^- \longrightarrow Se^{2-}(g) \qquad EA_2 = ?$$

−390 kJ/mol   +400 kJ/mol   +460 kJ/mol   +500 kJ/mol

8.8 Metals, Nonmetals, Metalloids, and Noble Gases

In Section 2.5, we distinguished between metals, metalloids, and nonmetals in terms of general appearance and bulk physical properties. Now we can consider them in relation to atomic properties and positions in the periodic table.

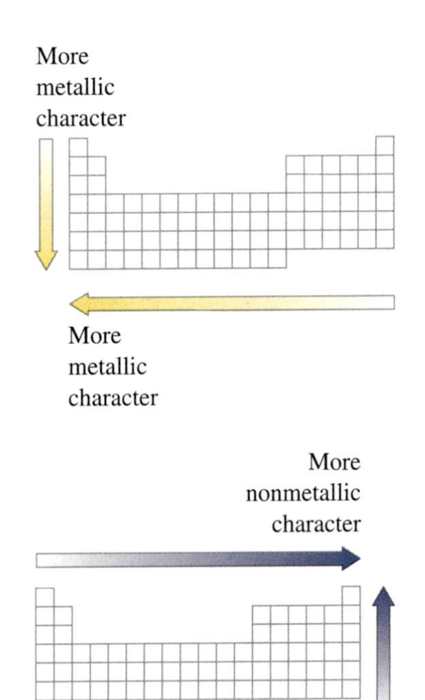

Generally speaking, **metal** atoms have small numbers of electrons in their valence shells and tend to form positive ions. For example, an aluminum atom loses the three valence electrons in the electron configuration $[Ne]3s^2 3p^1$ when forming the ion Al^{3+}. Except for hydrogen and helium, all s-block elements are metals. All d- and f-block elements are metals, and a few p-block elements also are metals. **Nonmetal** atoms generally have larger numbers of electrons in their valence shell than do metals, and many tend to form negative ions. Except for the special cases of hydrogen and helium, nonmetals are all p-block elements. The metals and nonmetals are separated by a heavy, stepped diagonal line in the periodic table. Most of the elements along this line have the physical appearance of metals, but have some nonmetallic properties as well; these "borderline" elements are sometimes called **metalloids.** The gaseous nonmetal elements in the column at the extreme right of the periodic table are often singled out as a special group, the **noble gases.** Each period of the periodic table ends with a noble-gas element. This information is summarized in the format of the periodic table in Figure 8.16.

Metallic character is closely related to atomic radius and ionization energy. The easier it is to remove electrons from an atom, the more metallic the element. Easy removal of electrons corresponds to *large* atomic radius and *low* ionization energy.

Metallic character increases *from top to bottom in a group and* decreases *from left to right in a period of the periodic table.*

The more readily an atom gains an electron, the more nonmetallic the character of the atom. A strong tendency to gain electrons corresponds to a large negative electron affinity, a property found among the smaller nonmetal atoms.

Nonmetallic character decreases *from top to bottom in a group and* increases *from left to right in a period of the periodic table.*

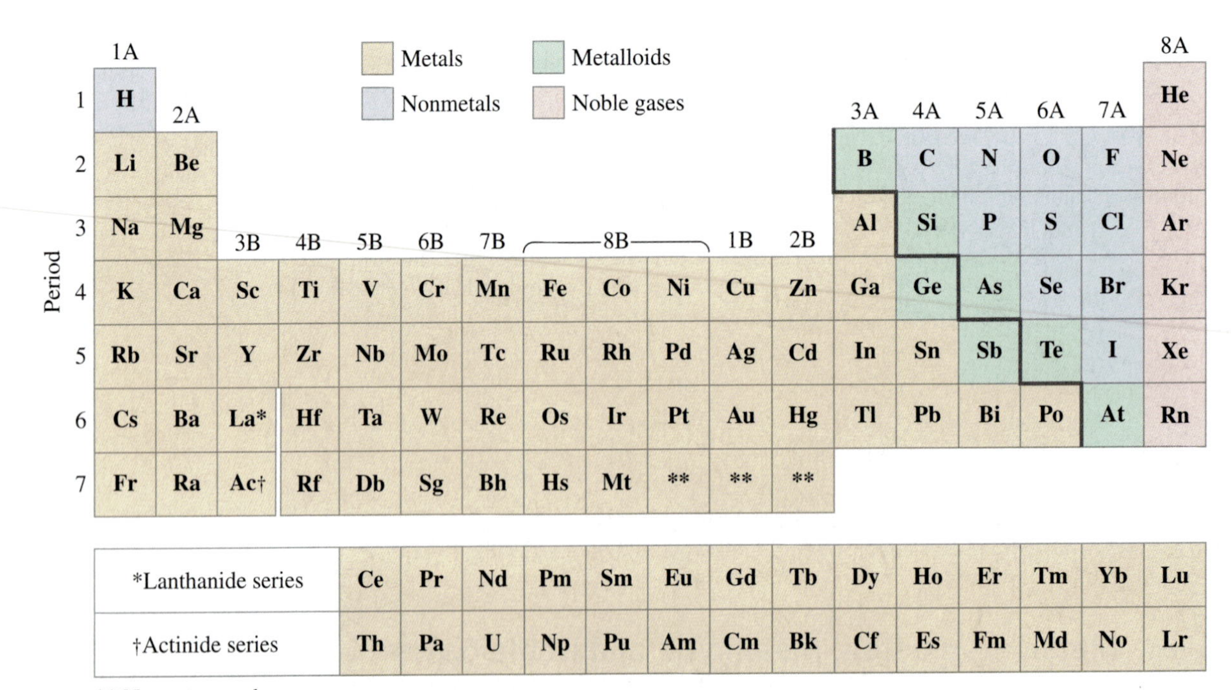

▲ **Figure 8.16 Metals, nonmetals, metalloids, and noble gases**

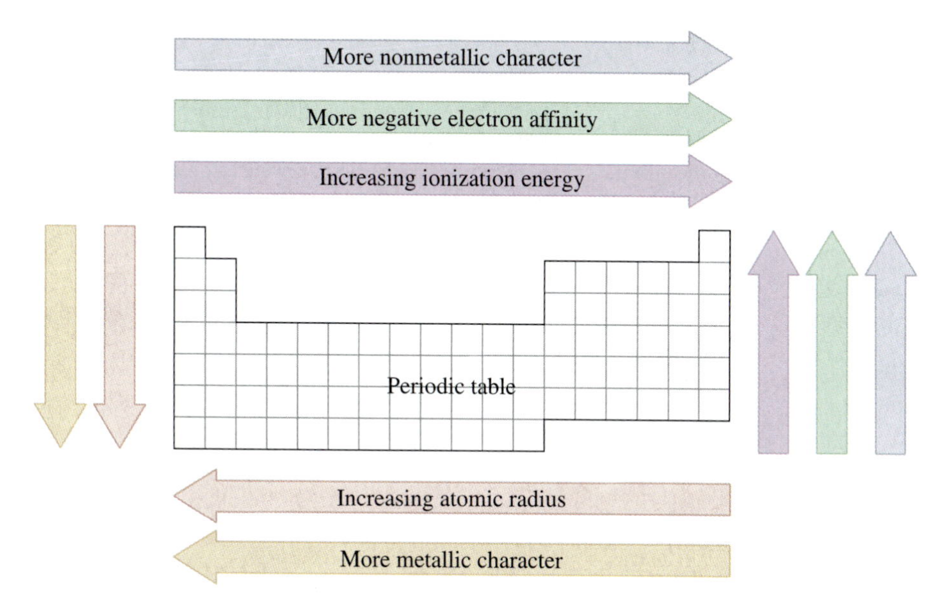

Figure 8.17
Atomic properties—A summary of trends in the periodic table
This figure summarizes the trends noted in the margins of some of the preceding pages. Vertical arrows point in the direction of a trend within a group. Horizontal arrows point in the direction of a trend within a period.

From the first generalization, we can identify the alkali metals (Group 1A) as highly metallic, and from the second generalization we identify the halogens (Group 7A) as highly nonmetallic. In the middle of the periodic table, we see both metallic and nonmetallic behavior. Group 4A has carbon, a nonmetal, at the top and two metals, tin and lead, at the bottom. In between are two metalloids, silicon and germanium. Figure 8.17 summarizes trends in atomic properties and metallic/nonmetallic behavior in the periodic table.

EXAMPLE 8.9

In each set, indicate which is the more metallic element.

a. Ba, Ca **b.** Sb, Sn **c.** Ge, S

● **ChemCDX** Periodic Properties

SOLUTION

a. Ba is below Ca in Group 2A. The Ba atom is larger than the Ca atom, and its first and second ionization energies are lower than those of Ca. As a result, Ba is more metallic than Ca.

b. Sn is to the left of Sb in the fifth period. We expect it to be a larger atom with lower ionization energies than those of Sb. Sn is more metallic than Sb.

c. Ge is to the left of S and in the following period. Because of its larger atoms, we expect Ge to be more metallic than S. (Actually, Ge is a metalloid and S is a nonmetal.)

EXERCISE 8.9

In each set, indicate which is the more nonmetallic element.

a. O, P **b.** As, S **c.** P, F

EXAMPLE 8.10—A Conceptual Example

Without reference to tables or figures in the text, enter into the proper position in the blank periodic table of Figure 8.18 the atomic number of **(a)** the element that has the electron configuration $4s^2 4p^6 4d^5 5s^1$ for its fourth and fifth principal shells, and **(b)** the most metallic of the fifth-period p-block elements.

SOLUTION

a. The $4d^5 5s^1$ part of the electron configuration identifies this as a transition element in the d-block (the $4d$ subshell is being filled) and in the fifth period ($n = 5$ is the highest principal quantum number). The underlying noble-gas electron configuration is that of krypton ($Z = 36$): $1s^2 2s^2 2p^6 3s^2 3p^6 3d^{10} 4s^2 4p^6$. The complete noble-gas-core abbreviated electron configuration is $[Kr]4d^5 5s^1$, so this is the element with the atomic number $Z = 42$ (molybdenum). It is the sixth element in the fifth period. It is in Group 6B, and its position in the table is indicated by this atomic number: 42.

b. In general, the more metallic elements are those with large atomic radii and low ionization energies. These elements tend to be found toward the *left* of their respective periods. The elements in Group 3A are farthest to the left in the p-block, and therefore the most metallic of the fifth-period p-block elements has the valence-shell electron configuration $5s^2 5p^1$. Its noble-gas-abbreviated electron configuration is $[Kr]4d^{10} 5s^2 5p^1$. This is the element with atomic number $Z = 49$ (In). (The I_1 values of In, Sn, and Sb are found to be 558 kJ/mol, 709 kJ/mol, and 834 kJ/mol, respectively.)

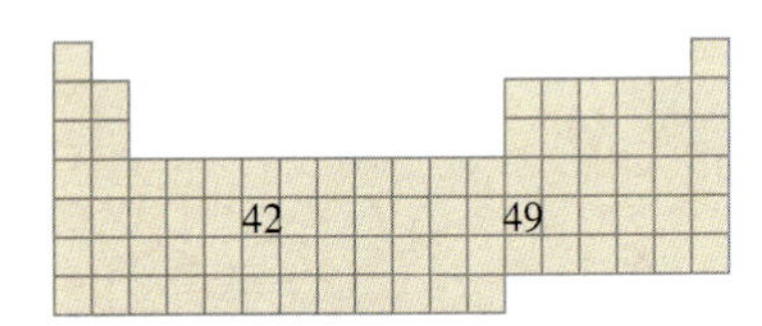

▲ **Figure 8.18**

EXERCISE 8.10

Without reference to tables or figures in the text, enter into the proper position in the blank periodic table of Figure 8.18, the following information.

a. The atomic number of the element having $4s^2 4p^6 4d^{10} 5s^2 5p^4$ as the electron configuration of its fourth and fifth principal shells.

b. The atomic number of the largest atom in the first transition series.

c. The atomic numbers of the d-block elements of the fifth period having the lowest and highest I_1 values.

d. The symbol of the most *nonmetallic* element of Group 5A.

The Noble Gases

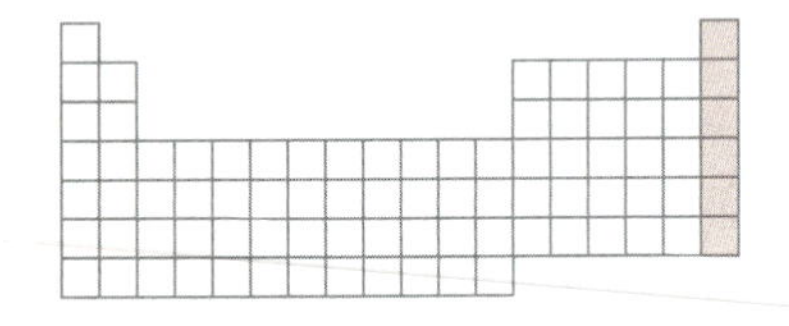

Noble gases

In the last decade of the nineteenth century, a group of elements was discovered that made up an entirely new family, one completely unexpected by Mendeleev and his contemporaries. This new group, called the noble gases, was placed between the highly active nonmetals of Group 7A and the very reactive alkali metals (Group 1A). In modern periodic tables, the noble gases are placed to the far right as Group 8A.

The six noble gases are helium, neon, argon, krypton, xenon, and radon. All are found to some extent in the atmosphere. Argon is abundant, making up nearly 1% of the atmosphere by volume. Xenon, on the other hand, is quite rare, accounting for only 91 parts per billion of the atmosphere (that is, having a mole fraction of 0.000000091). Radon, a radioactive element produced by the radioactive decay of heavier elements such as uranium, makes up only an inconsequential portion of the atmosphere. Radon is a potential health problem, however, when it seeps from the ground into a poorly ventilated building.

The noble gases rarely enter into chemical reactions. This lack of reactivity is a reflection of their electron configurations, ionization energies, and electron affinities. For example, the helium atom has a filled first principal shell ($1s^2$) and an exceptionally high first ionization energy (2372 kJ/mol). Helium gives up an electron only with great difficulty. The helium atom has no affinity for an additional electron. The added electron would have to be accommodated in the $2s$ orbital at a much higher energy than the filled $1s$ orbital. Thus, helium does not form an anion. The other noble gas atoms have the valence-shell electron configuration ns^2np^6 and show a similar aversion to both losing and gaining an electron.

Because of their lack of reactivity, noble-gas atoms tend not to combine with other atoms, even of their own kind. Therefore the noble gases occur naturally only in elemental form and only as monatomic species. Since 1962, a few compounds of the heavier noble gases have been prepared, but as yet, no compounds have been made of the lighter noble gases, helium, neon, and argon. While this family of elements is no longer called "inert," as it once was, its nobility is unquestioned.

8.9 Explaining the Behavior of the Elements Through Atomic Properties and the Periodic Table

We conclude this chapter by looking again at a few matters discussed earlier in the text, but now with the added insights gained through the concepts presented in this chapter.

Flame Colors

As we noted in Chapter 7, atoms absorb electromagnetic energy when electrons move from lower to higher energy states. Atoms also emit energy when electrons fall from higher to lower energy states. If the energy change in the transition is in the range of visible light, the emitted light is colored. A gas flame is not a very high energy source, and only atoms of elements with low first ionization energies are excited in a Bunsen burner flame. The lowest ionization energies are those of the Group 1A metals and the heavier Group 2A metals, and all these elements exhibit flame colors. Beryllium and magnesium have higher first ionization energies and do not give colored flames. The flame test for sodium, for example, results when the valence electron in excited atoms drops back to the 3*s* subshell. The transition from the 3*p* to the 3*s* orbital produces a yellow color.

$$\underset{\text{Excited state}}{\mathrm{Na([Ne]}3p^1)} \longrightarrow \underset{\text{Ground state}}{\mathrm{Na([Ne]}3s^1)}$$

The colors exhibited by three Group 1A and three Group 2A elements are shown in Figure 8.19.

The Halogens as Oxidizing Agents

In general, the halogens (Group 7A) are good oxidizing agents, but their strength decreases sharply from F_2 down the group to I_2. Their relative strengths are illustrated in a series of reactions in aqueous solution. For example, when $Cl_2(g)$ is bubbled into a solution containing iodide ions (Figure 8.20), chlorine acts as an *oxidizing agent;* it oxidizes I^- to I_2. Chlorine atoms gain electrons and are *reduced* to chloride ions. The oxidation number of the chlorine drops from 0 in Cl_2 to -1 in Cl^-.

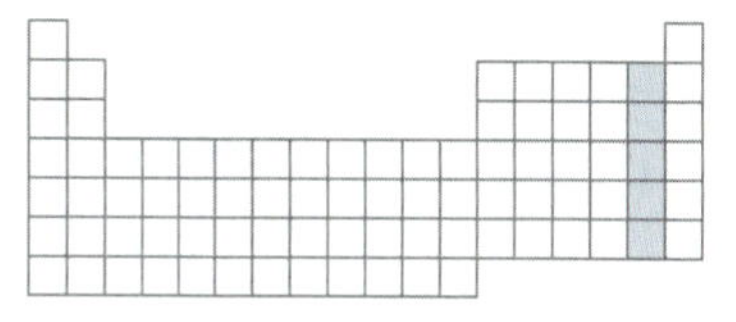
Group 7A: the halogens

$$\mathrm{Cl_2(g)} + 2\,\mathrm{I^-(aq)} \longrightarrow 2\,\mathrm{Cl^-(aq)} + \mathrm{I_2(aq)}$$

This redox reaction should seem reasonable because the Cl atom has a more negative electron affinity (-349 kJ/mol) than an I atom (-295 kJ/mol). In the competition for electrons, the Cl atoms in Cl_2 extract electrons from the I^- ions. Reasoning in this same way, we would predict that $Cl_2(g)$ also reacts with $Br^-(aq)$, and that $Br_2(l)$ reacts with $I^-(aq)$.

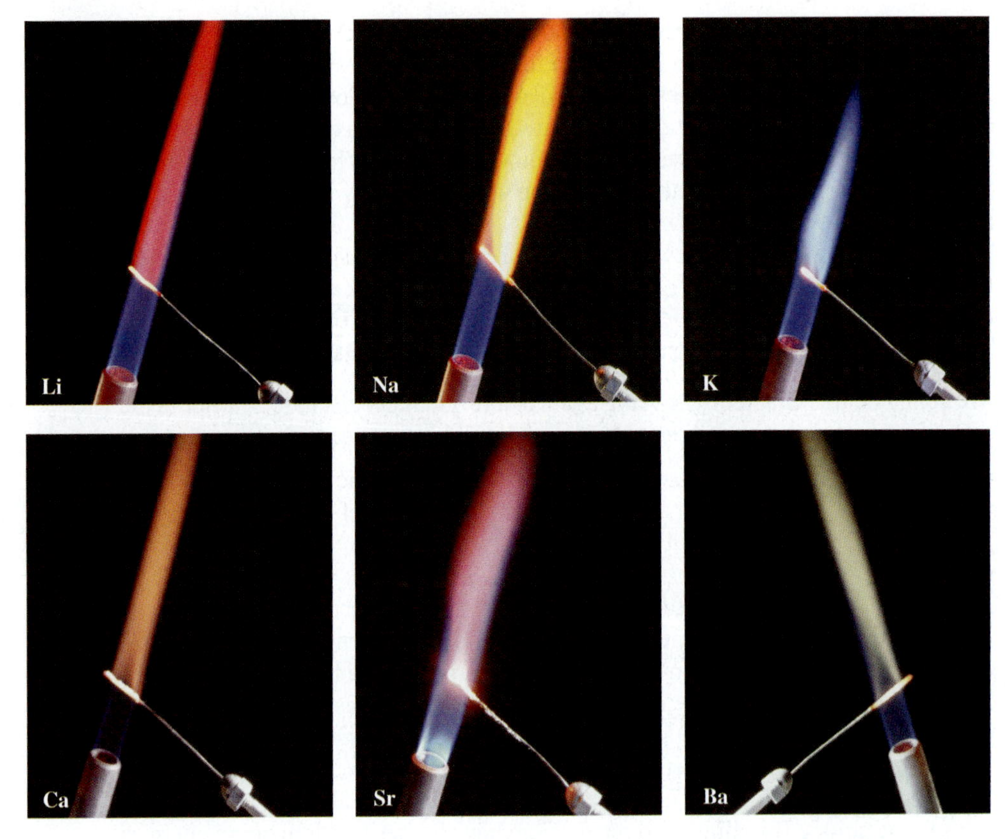

▲ **Figure 8.19 Flame colors of some alkali metals (Group 1A) and alkaline earth metals (Group 2A)**

▲ **Figure 8.20**
Displacement of I^-(aq) by Cl_2(g)
(a) Chlorine gas is bubbled through a colorless aqueous solution containing iodide ions. (b) Cl_2(g) is reduced to Cl^-(aq), and I^-(aq) is oxidized to I_2. Dense CCl_4(l) is added, and the I_2 concentrates in the CCl_4 (purple layer), in which it is much more soluble than in water.

$$Br_2(l) + 2\,I^-(aq) \longrightarrow 2\,Br^-(aq) + I_2(s)$$

In contrast, the reverse of the above reaction does not occur. When iodine is added to a solution of bromide ions, the atom with the greater affinity for an additional electron—bromine—already has the electron.

$$I_2(s) + Br^-(aq) \longrightarrow \text{no reaction}$$

We might think that F_2, a stronger oxidizing agent than Cl_2, would react with Cl^- in aqueous solution to form F^- and Cl_2. However, F_2 is the most powerful of all oxidizing agents, and in aqueous solutions it oxidizes *water* to oxygen gas (rather than Cl^- to Cl_2).

$$2\,F_2(g) + 2\,H_2O(l) \longrightarrow 4\,HF(aq) + O_2(g)$$

Electron affinities do correlate somewhat with the oxidizing power of the halogens, but other factors are also involved. We will take another look at the halogens as oxidizing agents later in the text.

The *s*-Block Metals as Reducing Agents

In Chapter 4 (page 161), we related the activity series of the metals to the tendency of metal atoms in the solid state to become cations in aqueous solution, that is, to undergo *oxidation*, as does Mg in the reaction

$$Mg(s) + 2\,H^+(aq) \longrightarrow Mg^{2+}(aq) + H_2(g)$$

As Mg(s) is oxidized, it reduces H^+(aq) to H_2(g). By undergoing oxidation, a metal acts as a *reducing agent.*

s-block metals

Ionization energy deals with the tendency of gaseous atoms to lose electrons to form gaseous ions, rather than ions in solution. Even so, ionization energy is roughly related to the oxidation tendencies expressed by the activity series. Thus, the *lower* the ionization energy of a metal atom, the more easily it is oxidized and the *better* reducing agent we expect it to be. All the *s*-block elements except hydrogen and helium are metals, and all these metals are above hydrogen in the activity series (page 161), meaning that they are strong enough as reducing agents to displace H_2(g) from an acidic solution.

In fact, the *s*-block elements are among the most powerful reducing agents known. All the alkali metals (Group 1A) and the heavier alkaline earth metals (Group 2A) are able to displace H_2(g), not only from acidic solutions, but even from neutral or basic solutions, in which the concentration of H^+(aq) is extremely low. For example, calcium and potassium react as follows.

$$2\,K(s) + 2\,H_2O(l) \longrightarrow 2\,K^+(aq) + 2\,OH^-(aq) + H_2(g)$$

$$Ca(s) + 2\,H_2O(l) \longrightarrow Ca^{2+}(aq) + 2\,OH^-(aq) + H_2(g)$$

These two reactions are pictured in Figure 8.21, which also shows that magnesium apparently does not react with cold water. There is a rough correlation of the vigor (or lack of vigor) of these reactions to ionization energies: for K ($I_1 = 419$ kJ/mol); for Ca ($I_1 = 590$, $I_2 = 1145$); for Mg ($I_1 = 738$, $I_2 = 1451$). Other factors are also involved, and we will consider reactions of this sort again in later chapters.

Acidic, Basic, and Amphoteric Oxides

The name *oxygen* was coined by Lavoisier because he thought all acids contained this element; oxygen means "acid former" in Greek. We now know that hydrogen, not oxygen, is the element common to acids. Nevertheless, the vast majority of acids do contain oxygen, and some acids form simply by allowing an element oxide to react with water. Oxides that produce acids in this way are **acidic oxides**. They are molecular substances and are generally the oxides of *nonmetals*, for example, SO_3 and P_4O_{10}.

(a)

(b)

(c)

Figure 8.21 The behaviors of potassium, calcium, and magnesium with cold water
(a) Potassium, being less dense than water, floats on the surface as it enters into a vigorous exothermic reaction. The evolved hydrogen gas spontaneously ignites. (b) Calcium, being more dense than water, sinks to the bottom of the test tube and reacts more slowly with the water. The evolved hydrogen gas bubbles to the surface. In this case, a few drops of phenolphthalein indicator have been added to the water to show that the solution is basic. (c) Magnesium metal shows no evidence of a reaction in cold water.

$$SO_3(g) + H_2O(l) \longrightarrow H_2SO_4(aq)$$

$$P_4O_{10}(s) + 6\,H_2O(l) \longrightarrow 4\,H_3PO_4(aq)$$

Like acids, acidic oxides can react directly with bases in neutralization reactions, for example

$$SO_2(g) + 2\,NaOH(aq) \longrightarrow Na_2SO_3(aq) + H_2O(l)$$

In contrast to the oxides of nonmetals, when *metal* oxides—ionic oxides—react with water, they typically form bases; they are **basic oxides**. Two examples are

$$Li_2O(s) + H_2O(l) \longrightarrow 2\,LiOH(aq)$$

$$BaO(s) + H_2O(l) \longrightarrow Ba(OH)_2(aq)$$

Basic oxides can also react directly with acids in neutralization reactions, for example,

$$MgO(s) + 2\,HCl(aq) \longrightarrow MgCl_2(aq) + H_2O(l)$$

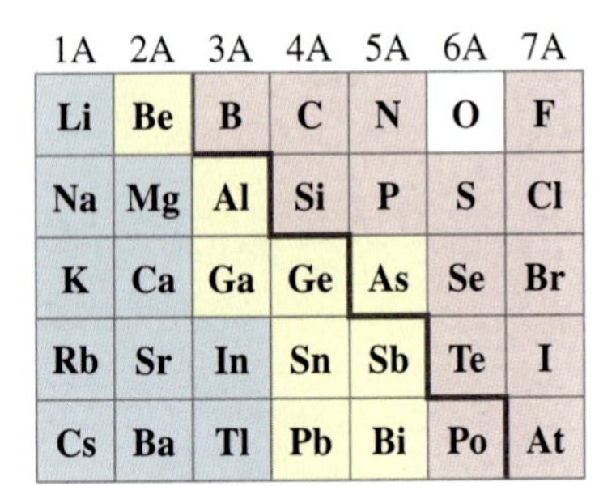

▲ **Figure 8.22 Acidic, basic, and amphoteric oxides of the main-group elements**

Now, let's assess the character of the element oxides in the third period of the periodic table. Starting at the left, the oxides of the Group 1A and 2A members, Na_2O and MgO, are both *basic*. Starting at the right, Ar does not form an oxide but those of Cl, S, P, and Si are all *acidic*. The oxide of the Group 3A member, Al_2O_3, is an interesting case. Consider the following reactions of Al_2O_3, first with the acid HCl(aq) and then with the base NaOH(aq).

$$\underset{\text{Base}}{Al_2O_3(s)} + \underset{\text{Acid}}{6\,HCl(aq)} \longrightarrow 2\,AlCl_3(aq) + 3\,H_2O(l)$$

$$\underset{\text{Acid}}{Al_2O_3(s)} + \underset{\text{Base}}{2\,NaOH(aq)} + 3\,H_2O(l) \longrightarrow \underset{\text{Sodium aluminate}}{2\,Na[Al(OH)_4](aq)}$$

In the first reaction, Al_2O_3 acts like a base and the aluminum appears in solution as a *cation* (Al^{3+}). In the second reaction, Al_2O_3 acts like an acid and the aluminum appears in solution in an *anion* ($[Al(OH)_4]^-$).* An oxide like Al_2O_3 that can react with either an acid or a base is said to be **amphoteric**.

Figure 8.22 categorizes the main-group elements according to the acidic, basic, or amphoteric nature of their oxides. Not unexpectedly, we see that amphoterism is found in a belt of elements that overlaps the stairstep diagonal line separating the metals and nonmetals.

APPLICATION NOTE
The amphoterism of $Al_2O_3(s)$ is not just a laboratory curiosity. It plays an important role in the commercial extraction of aluminum metal from *bauxite* ore (Section 21.3).

● **ChemCDX** Periodic Trends: Acid/Base Behavior of Oxides

* The anion $[Al(OH)_4]^-$ is an example of a *complex ion*. We will describe this and other complex ions in Chapters 16 and 22.

Summary

The wave-mechanical treatment of the hydrogen atom can be extended to multielectron atoms, but with this essential difference: Energy levels are lower in multielectron atoms than in the hydrogen atom, and the energy levels are split, that is, the different subshells of a principal shell have different energies. The order of increasing subshell energy in a principal shell is $s < p < d < f$. All the orbitals *within* a subshell, however, have the same energy; they are *degenerate*.

Electron configuration is the distribution of electrons in orbitals among the subshells and principal shells in an atom. Two types of notation of electron configuration are the *spdf* notation and the orbital diagram. Key ideas required to write a probable electron configuration are that (a) electrons tend to occupy the lowest energy orbitals available, (b) no two electrons can have all four quantum numbers alike (Pauli exclusion principle), and (c) where possible, electrons occupy orbitals singly, and with parallel spins, rather than in pairs (Hund's rule).

N $1s^2 2s^2 2p^3$	N [He]$2s^2 2p^3$	N [↑↓] [↑↓] [↑ | ↑ | ↑] $1s$ $2s$ $2p$
spdf notation	Noble-gas core abbreviated notation	Orbital diagram

The aufbau principle describes a process of hypothetically building up one atom from the atom of the preceding atomic number. With this principle and the ideas cited above, it is possible to predict probable electron configurations for many of the elements. In the aufbau process, for the *main-group* elements electrons are added to the s or p subshell of the valence shell, the principal shell of highest principal quantum number (n). In *transition* elements electrons go into the d subshell of the next to outermost ($n - 1$) shell. In the *inner-transition* elements electrons enter the f subshell of the second from outermost shell ($n - 2$).

Elements with similar electron configurations fall in the same group of the periodic table. For A-group elements, the group number corresponds to the number of electrons in the principal shell of highest quantum number. The period number is the same as the principal quantum number of the valence shell. The division of the periodic table into s, p, d, and f blocks assists in the assignment of probable electron configurations.

An atom with all electrons paired is diamagnetic, and an atom with one or more unpaired electrons is paramagnetic. Experimentally determined magnetic properties can be used to verify electron configurations.

Certain atomic properties recur periodically when atoms are considered in terms of increasing atomic number. The properties and trends considered in this chapter are those of atomic radius, ionic radius, ionization energy, and electron affinity.

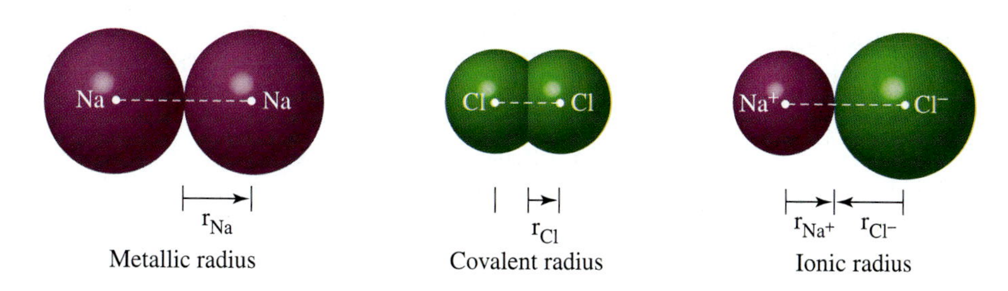

The regions of the periodic table ascribed to metals, nonmetals, metalloids, and the noble gases are related to the values of atomic properties. In general, metallic properties are associated with the ease with which atoms lose electrons and nonmetallic properties, with the ease of gaining electrons.

Key Terms

acidic oxide (8.9)
actinide (8.5)
amphoteric oxide (8.9)
atomic radius (8.7)
aufbau principle (8.4)
basic oxide (8.9)
core electrons (8.5)
covalent radius (8.7)
d-block (8.5)
degenerate orbitals (8.1)
diamagnetism (8.6)
effective nuclear charge (Z_{eff}) (8.7)
electron affinity (8.7)
electron configuration (8.2)
f-block (8.5)
Hund's rule (8.3)
ionic radii (8.7)
ionization energy (8.7)
isoelectronic (8.7)
lanthanide (8.5)
main-group element (8.4)
metal (8.8)
metallic radius (8.7)
metalloid (8.8)
noble gas (8.8)
nonmetal (8.8)
orbital diagram (8.2)
paramagnetism (8.6)
Pauli exclusion principle (8.3)
p-block (8.5)
periodic law (8.7)
s-block (8.5)
spdf notation (8.2)
transition element (8.4)
valence electrons (8.5)
valence shell (8.5)

Review Questions

1. In what way do the orbitals of multielectron atoms resemble those of the hydrogen atom, and in what way(s) do they differ?

2. What is meant by the term degenerate orbitals?

3. Which of the following hydrogen orbitals have identical energies? How would your answer change for an atom other than hydrogen?

(a) $1s, 2s, 2p$ **(c)** $3p_x, 3p_y, 3p_z$

(b) $3s, 3p, 3d$

4. State Pauli's exclusion principle. What restrictions does it place on the number of electrons that can be found in an atomic orbital, a subshell, and a principal shell?

5. State Hund's rule. How does it help us to select the correct electron configuration for the carbon atom from the two shown below?

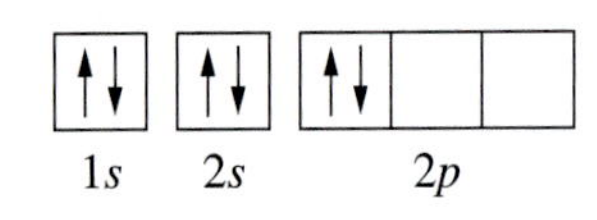

6. Tell in words what is meant by each of the following electron configuration notations. Which element corresponds to each configuration?

(a) $1s^2 2s^2 2p^5$

(b) $1s^2 2s^2 2p^6 3s^2 3p^6 3d^{10} 4s^1$

(c) $1s^2 2s^2 2p_x^1 2p_y^1 2p_z^1$

(d) $[Ne]3s^1$

(e) 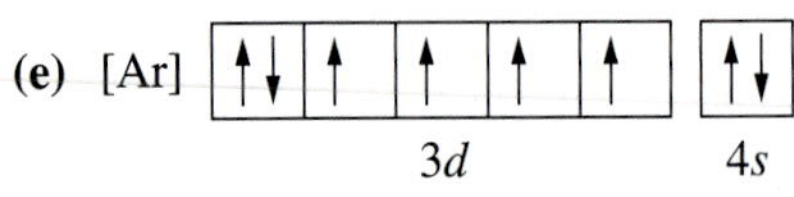

7. State what is meant by a ground-state and an excited-state electron configuration.

8. What is meant by the aufbau process? How is it used?

9. What are the valence electrons in an atom, and what are the core electrons?

10. What subshell(s) is(are) being filled in each of the following regions of the periodic table?

(a) Groups 1A and 2A

(b) Groups 3A through 7A

(c) the transition elements

(d) the lanthanides and actinides

11. What similarity in electron configuration is shared by lithium, sodium, and potassium? by beryllium, magnesium and calcium?

12. How many valence electrons are in an atom of each of the following?

(a) C **(c)** F **(e)** Mg

(b) Ne **(d)** Al

13. Referring only to the periodic table inside the front cover, indicate what similarity in electron configuration is shared by fluorine and chlorine, and by carbon and silicon. What is the difference in the electron configurations of each pair? What is the difference in the electron configurations of oxygen and fluorine?

14. Do you think it possible that someone might discover **(a)** a new element with atomic number 113; **(b)** a new element that would fit between magnesium and aluminum in the periodic table? Explain.

15. What is the configuration of the electrons in the valence shell in Group 4A? in Group 6A?

16. How many electrons are described by the notation $2p^6$? What is the general shape of the orbitals described in the notation? How many orbitals are included in the notation?

17. What are the differences in electron configurations that distinguish **(a)** main-group and transition elements; **(b)** paramagnetic and diamagnetic elements?

18. Must all atoms having an odd atomic number be paramagnetic? Must all atoms having an even atomic number be diamagnetic? Explain.

19. State the periodic law in its modern form.

20. Explain why the several periods of the periodic table do not all have the same number of members.

21. Explain why the sizes of atoms do not simply increase uniformly with increasing atomic number.

22. Why are cations smaller than the atoms from which they are formed, whereas anions are larger?

23. Why is it that isoelectronic ions in the same electron configuration do not have the same ionic radii?

24. What is the general trend in first ionization energies within a period? within a group? Explain each trend.

25. What are the characteristics of the atomic properties (electron configuration, atomic radius, ionization energy, electron affinity) associated with **(a)** metals and **(b)** nonmetals?

26. What are the characteristics of the atomic properties (electron configuration, atomic radius, ionization energy, electron affinity) associated with **(a)** metalloids and **(b)** noble gases?

27. Explain why all ionization energies are positive quantities whereas both positive and negative electron affinities are possible.

28. What is the characteristic behavior of an oxide that permits us to call it **(a)** acidic, **(b)** basic, and **(c)** amphoteric?

Problems

The Rules for Electron Configurations

29. Use the Pauli exclusion principle and Hund's rule to determine which of the following orbital diagrams are possible for a ground-state electron configuration and which are not. If the orbital diagram is not allowed, state why it is not.

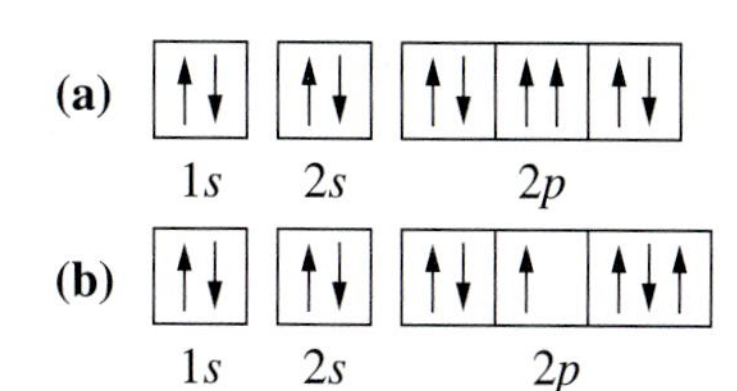

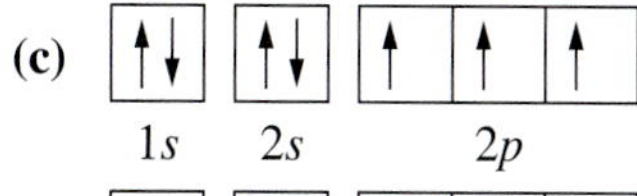

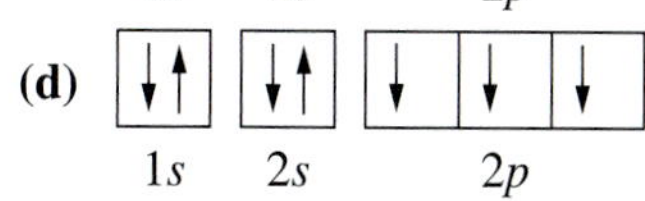

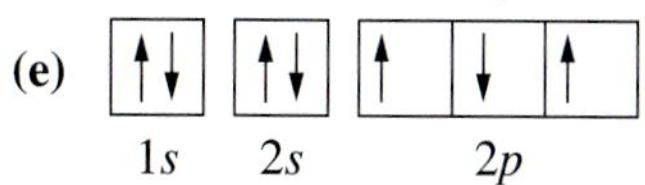

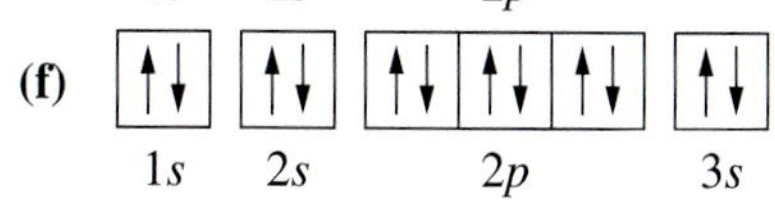

30. Use the Pauli exclusion principle and Hund's rule to determine which of the following orbital diagrams are possible for a ground-state electron configuration and which are not. If the orbital diagram is not allowed, state why it is not.

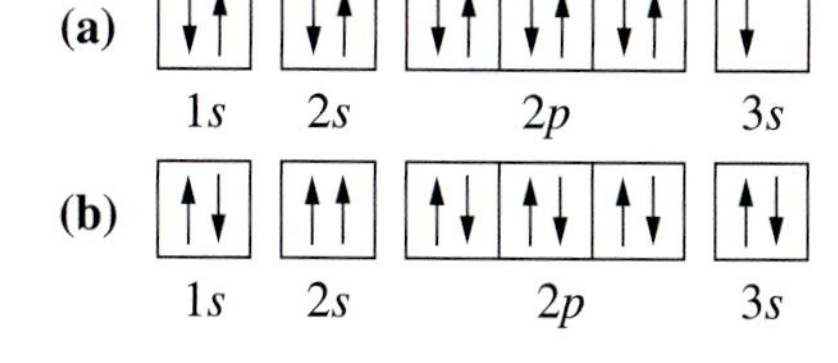

(c) 1*s* 2*s* 2*p* 3*s*

(d) 1*s* 2*s* 2*p* 3*s* 3*p*

(e) 1*s* 2*s* 2*p* 3*s* 3*p*

(f) 1*s* 2*s* 2*p* 3*s* 3*p*

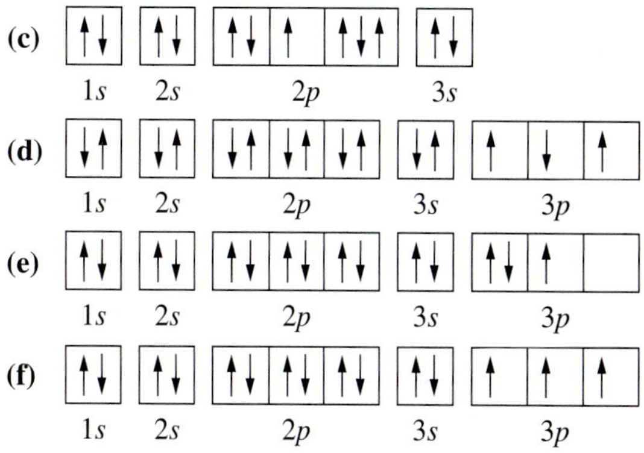

31. None of the following electron configurations is reasonable for a ground-state atom. In each case, explain why.

(a) $1s^2 2s^2 3s^2$ **(c)** $1s^2 2s^2 2p^6 2d^5$

(b) $1s^2 2s^2 2p^2 3s^1$

32. None of the following electron configurations is reasonable for a ground-state atom. In each case, explain why.

(a) $1s^2 2s^2 2p^6 3s^1 3p^1$

(b) $1s^2 2s^2 2p^6 3s^2 3p^6 3d^1$

(c) $1s^2 2s^2 2p^6 3s^2 3p^6 3d^8 4s^2 4p^1$

33. Explain the principle(s) or rule(s) that each of the following electron configurations violates.

(a) $1s^2 2s^6 3s^2$ **(c)** $1s^2 2s^2 2p^6 2d^3$

(b) $1s^2 2s^2 2p^7 3s^1$

34. Explain the principle(s) or rule(s) that each of the following electron configurations violates.

(a) $[Ar]2d^{10}$ **(c)** $[Kr]4d^{10} 4f^{14} 5s^2$

(b) $[Ar]3f^3 4s^2$

Electron Configurations: The Aufbau Principle

35. Using *spdf* notation and referring only to the periodic table inside the front cover, write out the ground-state electron configuration of each of the following:

(a) Al **(d)** B **(g)** C

(b) Cl **(e)** He **(h)** Li

(c) Na **(f)** O **(i)** Si.

36. Using *spdf* notation and referring only to the periodic table inside the front cover, write out the ground-state electron configuration of each of the following.

(a) Ar **(d)** Be **(g)** Ca

(b) H **(e)** K **(h)** Mg

(c) Ne **(f)** P **(i)** Br.

37. Using a noble-gas-core abbreviated *spdf* notation and referring only to the periodic table inside the front cover, write out the ground-state electron configuration for each of the following.

(a) Ba **(c)** As **(e)** Se

(b) Rb **(d)** F **(f)** Sn.

38. Using a noble-gas-core abbreviated *spdf* notation and referring only to the periodic table inside the front cover, write out the ground-state electron configuration for each of the following.

(a) Ga **(c)** I **(e)** Sb
(b) Te **(d)** Cs **(f)** Sr.

39. Give orbital diagrams for the ground-state electron configuration of each of the following.

(a) C **(c)** K **(e)** S
(b) O **(d)** Al **(f)** Mg.

40. Give orbital diagrams for the ground-state electron configuration of each of the following.

(a) N **(c)** Si **(e)** Cl
(b) B **(d)** Ca **(f)** Sc.

41. Give the orbital diagram for the electrons beyond the xenon core of the hafnium (Hf) atom. Relate this electron configuration to the position of hafnium in the periodic table.

42. Give the orbital diagram for the electrons beyond the xenon core of the mercury (Hg) atom. Relate this electron configuration to the position of mercury in the periodic table.

43. Use orbital diagrams to represent the electron configurations of the following ions. (Part of your diagram may be a noble-gas-core abbreviation.)

(a) Br^- **(c)** Sb^{3+}
(b) Ni^{2+} **(d)** Te^{2-}

44. Use orbital diagrams to represent the electron configurations of the following ions. (Part of your diagram may be a noble-gas-core abbreviation.)

(a) Ga^{3+} **(c)** I^-
(b) V^{3+} **(d)** Pb^{2+}

Electron Configurations and the Periodic Table

45. Give the period number and group number for the element whose atoms have the electron configuration

(a) $1s^2 2s^2 2p^6$ **(d)** $1s^2 2s^2$
(b) $1s^2 2s^2 2p^6 3s^2 3p^2$ **(e)** $1s^2 2s^2 2p^3$
(c) $1s^2 2s^2 2p^6 3s^1$ **(f)** $1s^2 2s^2 2p^6 3s^2 3p^1$

46. Write the noble-gas-core abbreviated electron configuration for the element in **(a)** Group 5A and Period 4, **(b)** Group 3A and Period 6, and **(c)** Group 3B and Period 5.

47. Based on the relationship between electron configurations and the periodic table, give the number of **(a)** valence-shell electrons in an atom of Bi; **(b)** electrons in the *fourth* principal shell of Au; **(c)** elements whose atoms have five valence-shell electrons; **(d)** unpaired electrons in an atom of Se; and **(e)** transition elements in the fifth period.

48. On the basis of the periodic table and rules for electron configurations, indicate the number of **(a)** $3p$ electrons in an atom of P; **(b)** $4s$ electrons in an atom of Cs; **(c)** $4d$ electrons in an atom of Se; **(d)** $4f$ electrons in an atom of Bi; **(e)** unpaired electrons in an atom of Ga; **(f)** elements in Group VA of the periodic table; **(g)** elements in the sixth period of the periodic table.

Magnetic Properties

49. Determine which of the following species are diamagnetic and which are paramagnetic.

(a) a S atom **(d)** an O^{2-} ion
(b) a Ba atom **(e)** an Ag atom
(c) a V^{2+} ion

50. Determine which of the following species are diamagnetic and which are paramagnetic.

(a) a Ra^{2+} ion **(d)** an O atom
(b) an I^- ion **(e)** a Co atom
(c) a Sn^{2+} ion

Atomic and Ionic Radii

51. By consulting only the periodic table on the inside front cover, determine which member of the following pairs has the *larger* radius. Explain.

(a) Cl or S **(c)** Al or Mg
(b) Cl^- or S^{2-} **(d)** Mg^{2+} or F^-

52. By consulting only the periodic table on the inside front cover, determine which member of the following pairs has the *smaller* radius. Explain.

(a) Ca or Rb **(c)** N or S
(b) Mg^{2+} or Fe^{2+} **(d)** V^{2+} or Co^{3+}

53. By consulting only the periodic table, arrange each set of the following elements in order of increasing atomic radius, and explain the basis for this order.

(a) Al, Mg, Na **(b)** Ca, Mg, Sr

54. By consulting only the periodic table, arrange each set of the following elements in order of increasing atomic radius, and explain the basis for this order.

(a) Ca, Rb, Sr **(b)** Al, C, Si

55. In what location of the periodic table would you expect to find the two or three elements having the largest atoms? Explain.

56. Explain why there is a large difference in atomic radius between the elements $Z = 11$ (Na; 186 pm) and $Z = 12$ (Mg; 160 pm), whereas that between $Z = 24$ (Cr; 125 pm) and $Z = 25$ (Mn; 124 pm) is quite small.

57. Without reference to a table of data, should you be able to predict **(a)** whether atoms of Ca are larger than atoms of Cl and **(b)** whether the ionic radius of K^+ is greater than that of F^-? Explain your reasoning.

58. Without reference to a table of data, should you be able to predict **(a)** whether atoms of I are larger than those of Li and **(b)** whether the ionic radius of Cl^- is greater than that of Ca^{2+}? Explain your reasoning.

Ionization Energy

59. Arrange each set of the following elements in order of increasing first ionization energy and explain the basis for this order.

(a) Ca, Mg, Ba **(c)** F, Na, Fe, Cl, Ne

(b) P, Cl, Al

60. Arrange each set of the following elements in order of increasing first ionization energy and explain the basis for this order.

(a) Ca, Na, As **(c)** Kr, Ba, Zn, Sc, Al, Br

(b) S, As, Sn

61. Describe the trend in successive ionization energies as electrons are removed one at a time from an aluminum atom. Why is there a big jump between I_3 and I_4?

62. Why does sulfur have a lower first ionization energy than phosphorus?

Electron Affinity

63. Which group of elements has electron affinities with the largest negative values? Explain.

64. Which of the main groups of elements do not form stable negative ions? Use electron configurations to explain this behavior.

65. Silicon has an electron affinity of −134 kJ/mol. The electron affinity of phosphorus is −72 kJ/mol. Give a plausible reason for this difference.

66. Lithium has an electron affinity of −60 kJ/mol. That of boron is −27 kJ/mol. Give a plausible reason for this difference.

Atomic Properties and the Periodic Table

67. One of the first periodic properties studied was that of *atomic volume*, the atomic mass of an element divided by its density. Draw a graph to show that atomic volume is a periodic property of the following elements. Densities, in g/cm^3, are: Na, 0.971; Mg, 1.74; Al, 2.70; Si, 2.33; P, 2.20; S, 2.07; Cl, 2.03; Ar, 1.66; K, 0.862; Ca, 1.55; Sc, 2.99; Cr, 7.19; Co, 8.90; Zn, 7.13; Ga, 5.91; As, 4.70; Br, 4.05; Kr, 2.82; Rb, 1.53; Sr, 2.54. To what atomic property described in this chapter does the atomic volume seem most closely related? Explain.

68. Draw a graph to show that melting point is a periodic property of these elements (melting points are in °C): Al, 660; Ar, −189; Be, 1278; B, 2300; C, 3350; Cl −101; F, −220; Li, 179; Mg, 651; Ne, −249; N, −210; O, −218; P, 590; Si, 1410; Na, 98; S, 119. For this purpose, does it matter whether melting points are expressed on the Celsius scale or the Kelvin scale? Explain.

69. What properties could you use to assess the degree of metallic character in an element? Arrange the following elements in the expected order of *increasing* metallic character: K, P, Al, Rb, Bi, Ca, Ge, and explain your arrangement.

70. What properties could you use to assess the degree of nonmetallic character in an element? Arrange the following elements in the expected order of *increasing* nonmetallic character: Pb, Sb, N, Br, As, F, O, Si, and explain your arrangement.

71. Listed below are the periodic table locations of certain elements. Arrange them in the expected order of *increasing* ionization energy, I_1. **(a)** Group 4A, Period 4;

(b) Group 6A, Period 3; **(c)** Group 3A, Period 6; **(d)** Group 8A, Period 2; **(e)** Group 6A, Period 4.

72. Classify each of the following substances as an acidic oxide, a basic oxide, or neither, and explain your classification: **(a)** CO_2, **(b)** O_2, **(c)** SrO, **(d)** HCOOH, **(e)** P_4O_6, **(f)** $Ba(OH)_2$.

73. Complete and balance the following equations. If no reaction occurs, so state.

(a) $Cl_2(g) + Br^-(aq) \longrightarrow$

(b) $I_2(s) + F^-(aq) \longrightarrow$

(c) $Br_2(l) + I^-(aq) \longrightarrow$

74. Complete and balance the following equations. If no reaction occurs, so state.

(a) $K(s) + H_2O(l) \longrightarrow$

(b) $Ca(s) + H^+(aq) \longrightarrow$

(c) $Be(s) + H_2O(l) \longrightarrow$

75. Complete and balance the following equations.

(a) $N_2O_5(s) + H_2O(l) \longrightarrow$

(b) $MgO(s) + CH_3COOH(aq) \longrightarrow$

(c) $Li_2O(s) + H_2O(l) \longrightarrow$

76. Complete and balance the following equations.

(a) $SO_3(g) + KOH(aq) \longrightarrow$

(b) $Al_2O_3(s) + H^+(aq) \longrightarrow$

(c) $CaO(s) + H_2O(l) \longrightarrow$

Additional Problems

77. Without referring to any tables or listing in the text, mark an appropriate location for each of the following in the blank periodic table provided: **(a)** the fourth period noble gas; **(b)** a fifth period element whose atoms have three unpaired electrons; **(c)** the *d*-block element having one 3*d* electron; **(d)** a *p*-block element that is a metalloid; **(e)** a metal that forms the oxide M_2O_3.

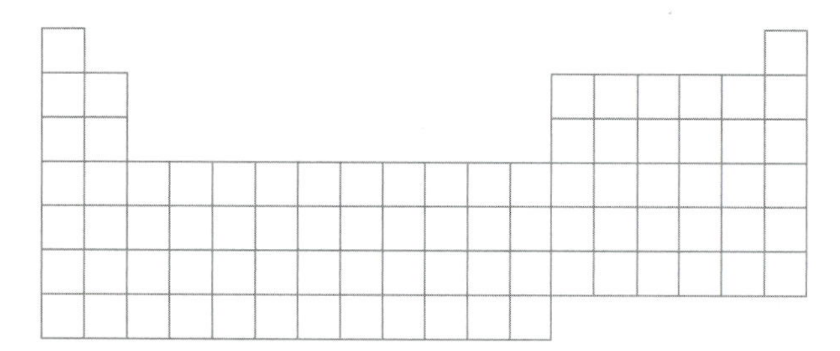

78. Propose a probable electron configuration for the unknown and as yet undiscovered element with $Z = 114$.

79. Use ideas presented in this chapter to indicate **(a)** three metals that you would expect to exhibit the photoelectric effect with visible light, and three that you would not; **(b)** the approximate first ionization energy of fermium ($Z = 100$); **(c)** the approximate atomic radius of francium.

80. Arrange the following ionization energies in the most probable order of increasing value and explain your reasoning: I_1 for B, I_1 for Cs, I_2 for In, I_2 for Sr, I_2 for Xe, I_3 for Ca.

81. If all other rules governing electron configurations were valid, what would be the electron configuration of rubidium if **(a)** there were *three* values of m_s instead of two and **(b)** the quantum number l could have the value n, as well as its other values. Would sodium and rubidium fall in the same group of elements in a periodic table based on electron configurations in either case? Explain.

82. In a hydrogen atom, the 1*s* subshell is at an energy of -1.31×10^3 kJ/mol, whereas in a helium atom it is at -2.37×10^3 kJ/mol. Explain why the level is not the same in the two cases. For hydrogen-like atoms, Bohr's equation can be written as $E_n = -B\,Z^2/n^2$ (see Problem 89 on page 315). Calculate the energy of the 1*s* subshell for He^+ using this equation, and explain why the value you get is not the same as that given above for helium.

83. Calculate I_1 for hydrogen **(a)** using Bohr's equation for the energy levels of the hydrogen atom, $E_n = -B/n^2$, and **(b)** using the shortest wavelength line in the Balmer series of the hydrogen spectrum and the longest wavelength line in the Lyman series. Explain why they give the same result.

84. The trend in I_1 values and explanations for small irregularities were presented in the text for the second period elements (page 341). In a similar way, discuss the trend and irregularities in the I_2 values for the second period elements: Li (I_2 = 7298 kJ/mol), Be (1757), B (2427), C (2352), N (2856), O (3388), F (3374), Ne (3952).

85. The production of gaseous chloride ions from chlorine molecules can be considered a two-step process in which the first step is as follows:

$$Cl_2(g) \longrightarrow 2\,Cl(g) \quad \Delta H = +242.8 \text{ kJ}$$

What is the second step? Is the overall process endothermic or exothermic?

86. Use ionization energies and electron affinities to determine whether the following reaction is endothermic or exothermic.

$$Mg(g) + 2\,Cl(g) \longrightarrow Mg^{2+}(g) + 2\,Cl^-(g)$$

87. Show how you might use the molar mass and density of a solid metallic element, together with Avogadro's number, to make a rough estimate of the volume of an atom of the metal. How would you estimate the metallic radius from this volume? Why is this estimate not exact, even if the molar mass, the density, and Avogadro's number are all known with considerable accuracy? Is the estimated atomic radius likely to be too high or too low? Explain. [*Hint:* Use sodium as a specific example ($d = 0.968$ g/cm^3), and compare the estimated metallic radius with that shown in Figure 8.14].

88. In this text, ionization energies are given in the unit kJ/mol. Another way of expressing these quantities is in terms of *single* atom rather than a mole of atoms, using the unit electron volts per atom, eV/atom. Use physical constants and other data from the appendices to show that 1 eV/atom = 96.49 kJ/mol.

89. Gaseous atoms can be ionized when struck by sufficiently energetic photons of electromagnetic radiation. At photon energies just corresponding to I, electrons are ejected with no energy to spare. At higher photon energies, the ejected electrons acquire some kinetic energy. The relationship between photon energy ($h\nu$), ionization energy (I), and the kinetic energies ($\frac{1}{2}$ mv^2) of the ejected electrons is

$$h\nu = I + \frac{1}{2}\text{mv}^2$$

By knowing the frequency of the radiation used and measuring the kinetic energy of the ejected electrons, we can calculate the ionization energy.

In the above equation, we have written I rather than I_1 because the method is not limited to ejecting an electron from the highest energy subshell. A graph of the number of electrons ejected (or a property related to it) versus the calculated ionization energies consists of a series of peaks corresponding to ionization from different subshells. The technique of producing such graphs is called *photoelectron spectroscopy*. The graph shown is for the element boron.

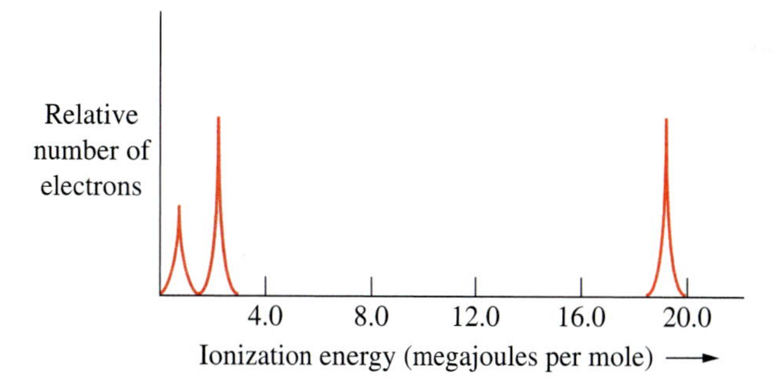

(a) What subshell is represented by each peak in the graph?

(b) Why are two of the peaks similar in size but one peak smaller than the other two?

(c) Is there a relationship between the ionization energies shown here and the successive ionization energies for boron on page 339? Explain.

(d) What is the maximum wavelength of radiation that can be used to obtain the complete photoelectron spectrum of boron?

(e) Sketch a graph similar to the one above (but not to scale) for the expected photoelectron spectrum of aluminum.

CHAPTER

9

Chemical Bonds

CONTENTS

● ChemCDX
When you see this icon, there are related animations, demonstrations, and exercises on the CD accompanying this text.

These cubic crystals of sodium chloride—common table salt—reflect the microscopic arrangement of sodium ions (Na^+) and chloride ions (Cl^-). The bonds that hold the ions in this particular arrangement are discussed in Section 9.3.

The forces that hold atoms together in molecules and keep ions in place in solid ionic compounds are called **chemical bonds.** Most properties of a material are determined by the nature of its chemical bonds. We can use the concepts of chemical bonding described in this and the next two chapters to answer such important questions as the following.

- Why are some compounds solids that melt only at high temperatures, and others liquids or even gases at room temperature?
- Why are carbon atoms the key atoms in fats, proteins, carbohydrates, and nucleic acids—the molecules of living things?
- Both carbon monoxide and oxygen form bonds to hemoglobin. But why does carbon monoxide kill and oxygen sustain life?
- How is the energy required to maintain a heartbeat, breathing, and other physiological functions stored in chemical bonds?

9.1 Chemical Bonds: A Preview

Chemical bonds are electrical forces; they reflect a balance in the forces of attraction and repulsion between electrically charged particles. In Figure 9.1a, we represent the attractions and repulsions in the molecular ion H_2^+. The two nuclei (protons) repel one another, whereas the single electron is simultaneously attracted to both nuclei. Figure 9.1b represents the attractions and repulsions in the H_2 molecule. Here the two nuclei also repel each other, and each of the two electrons is simultaneously attracted to both nuclei. In addition, the two electrons repel each other.

We can assess the relative importance of the attractive and repulsive forces represented in Figure 9.1b by examining the changes in potential energy that occur when two H atoms are allowed to come together. This process is suggested in Figure 9.2. We take the zero of energy as two unbonded H atoms so far apart that no

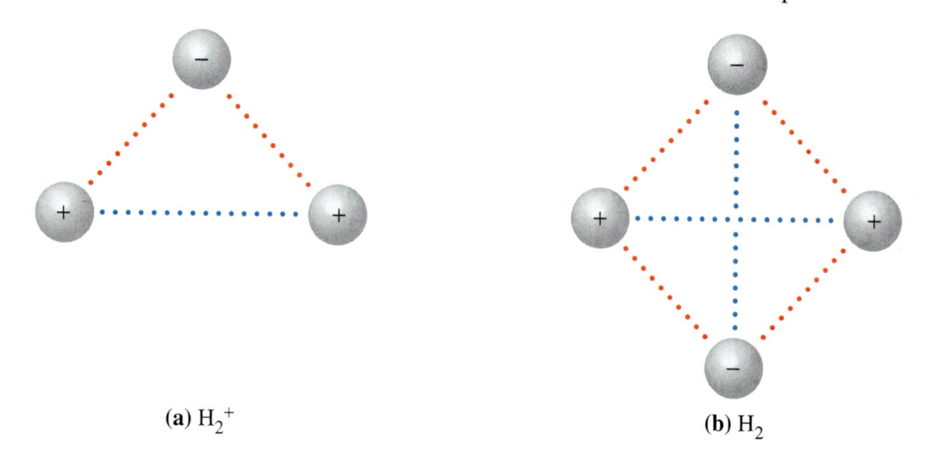

Figure 9.1 Electrostatic attractions and repulsions in two simple molecular species
The electrostatic attractions (red) and repulsions (blue) in H_2^+ and H_2 are described in the text.

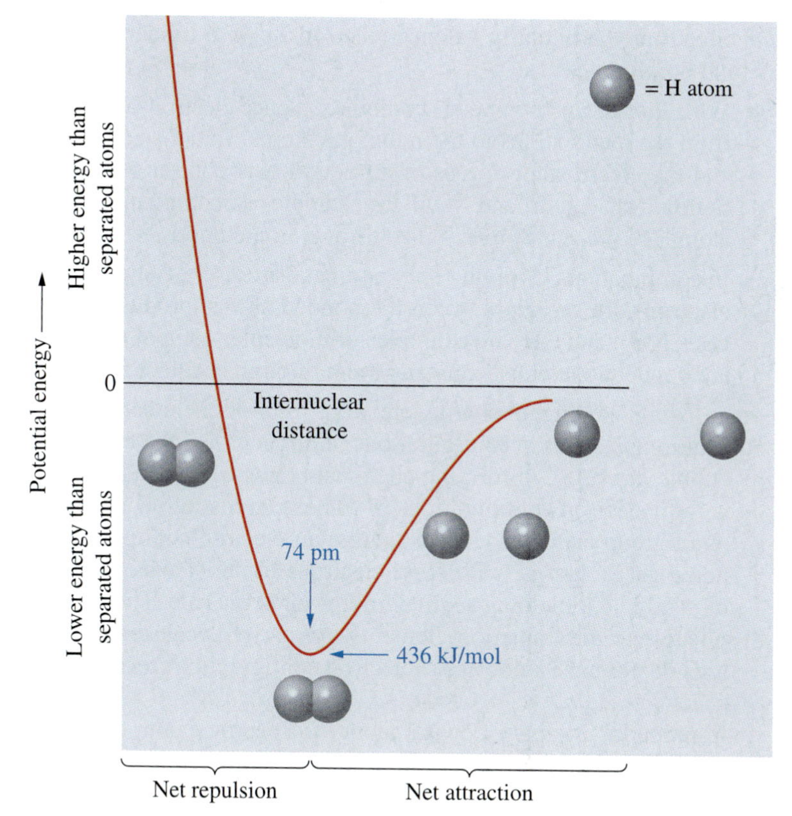

Figure 9.2 The energy of interaction of two hydrogen atoms
The molecular models and graph together show that the lowest energy for two H atoms comes at the internuclear distance in the H_2 molecule, 74 pm.

forces of attraction or repulsion exist between them. The curve representing the potential energy of the system of two H atoms begins at the far right of Figure 9.2, where the internuclear distance is large. The potential energy is slightly below zero. The attractive forces of the electrons to the atomic nuclei slightly outweigh repulsive forces. Moving to the left along the curve, as the internuclear distance decreases, the potential energy also decreases. This decrease in potential energy continues as long as attractive forces outweigh repulsive forces. When the internuclear distance reaches 74 pm, the attractive forces and repulsive forces become equal. At shorter internuclear distances repulsive forces exceed attractive forces and the atoms are pushed back to the equilibrium distance of 74 pm. Two hydrogen atoms at an internuclear distance of 74 pm and a potential energy of −436 kJ/mol correspond to a H_2 molecule in its lowest energy state.

Scientists can determine internuclear distances that correspond to the lowest energy states of molecules experimentally. Then they can use quantum mechanical calculations to develop a theoretical model that fits the experimental measurements. We will explore descriptions of chemical bonding based on quantum mechanics in the next chapter. For most of this chapter, however, we will use a simpler approach to chemical bonding that predates quantum mechanics but is still extensively used by chemists.

● **ChemCDX** H_2 Bond Formation

9.2 The Lewis Theory of Chemical Bonding: An Overview

In the years during and following World War I, two Americans, G. N. Lewis and Irving Langmuir, and a German, Walther Kossel, published similar ideas about chemical bonding. These ideas, presented here in modern form, are generally referred to as the *Lewis theory*.

Gilbert Newton Lewis (1875–1946) was one of the foremost American chemists of the first half of the twentieth century. In addition to his pioneer work in describing chemical bonding, Lewis was a driving force in bringing thermodynamics into mainstream chemistry; he also made important contributions to acid-base theory.

- Electrons, particularly valence electrons, play a fundamental role in chemical bonding.
- When metals and nonmetals combine, valence electrons usually are transferred from the metal atoms to the nonmetal atoms. Cations and anions are formed, and the electrostatic forces of attraction between the ions give rise to **ionic bonds**. NaCl, KBr, and MgO are examples of compounds whose constituent atoms are present as ions joined through ionic bonds.
- In combinations involving only nonmetal atoms, one or more *pairs* of valence electrons are *shared* between the bonded atoms, producing **covalent bonds**. H_2O, NH_3, and CH_4 are examples of molecular compounds in which H atoms are bonded to another nonmetal atom through covalent bonds.
- In losing, gaining, or sharing electrons to form chemical bonds, atoms tend to acquire the electron configurations of noble gases. We might refer to this as a "noble-gas rule." In forming chemical bonds, H, Li, and Be tend to acquire a helium electron configuration, $1s^2$. These atoms follow a *duet* rule. The other main-group elements tend to acquire an electron configuration like the other noble gases, that is, with *eight* electrons in the valence shell: ns^2np^6 (where $n = 2, 3, \ldots$). In doing so, they follow an **octet rule**. The octet rule is less applicable to the transition elements, primarily because most transition metal ions do not have a noble-gas electron configuration (recall Table 8.3).

In preparing to apply Lewis's theory to specific examples of ionic and covalent bonding, we will next consider a special set of symbols Lewis used to represent his ideas.

Lewis Symbols

A **Lewis symbol** uses the chemical symbol for an atom to represent the nucleus and core electrons and dots around the symbol to represent the *valence* electrons. The Lewis theory is closely linked to noble-gas electron configurations. We generally write Lewis symbols only for the elements that acquire such configurations when they form bonds. For the most part, these are main-group elements, and their numbers of valence electrons are equal to periodic table group numbers. For the second-row elements, for example,

1A	2A	3A	4A	5A	6A	7A	8A
Li·	Be·	·B·	·C·	·N:	·O:	:F:	:Ne:

In writing Lewis symbols, we will follow Lewis's practice of placing the first four bold dots separately on the four sides of the chemical symbol and pairing the dots as we add the next four. Lewis symbols do *not* specifically reflect electron pairing; the idea of electron spin had not yet been established when he proposed the theory. Thus, although the atoms Be, B, and C have zero, one, and two unpaired electrons, respectively, the Lewis symbols of Be, B, and C have two, three, and four separate dots. Interestingly, as we will see in Chapter 10, in some cases, Lewis symbols lead to a better prediction of chemical bonding than do ground-state electron configurations.

EXAMPLE 9.1

Give Lewis symbols for magnesium, silicon, and phosphorus.

SOLUTION

Mg, Si, and P are all in the third *period*, but the Lewis symbol depends on the *group* number. Consequently, each element has the same valence-shell electron configuration and distribution of dots in its Lewis symbol as the second-period element preceding it in the same group. Thus, the Lewis symbols of Mg (Group 2A), silicon (Group 4A), and phosphorus (Group 5A) resemble those of Be, C, and N, respectively.

Mg·   ·Si·   ·P:

EXERCISE 9.1

Give Lewis symbols for each of the following atoms.

a. Ar **b.** Ca **c.** Br **d.** As **e.** K **f.** Se

PROBLEM-SOLVING NOTE

It makes no difference on which side of the symbol we begin the process of adding dots. Here we have placed them around the symbol in a clockwise direction, starting on the right side.

Ionic Bonding

Figure 9.3 shows the reaction between sodium, a soft, low-density, silvery *metal*, and chlorine, a yellow-green, toxic gaseous *nonmetal*, to produce white crystalline, solid sodium chloride. One way to demonstrate that sodium chloride is an *ionic* compound is to measure the electrical conductivity of its aqueous solutions. In Chapter 4, we saw that NaCl is completely dissociated into ions in aqueous solutions. NaCl(aq) is a good electrical conductor—a strong electrolyte. We begin our discussion of chemical bonding with ionic bonding because conceptually this is the easiest type to describe.

● **ChemCDX** Reaction of Sodium and Chlorine

▶ **Figure 9.3**
The reaction of sodium and chlorine
Sodium metal and chlorine gas provide striking visual evidence of their reaction to produce the solid ionic substance sodium chloride.

9.3 Ionic Bonds and Ionic Crystals

Let's use the electron configurations of Na and Cl atoms to interpret the reaction between sodium and chlorine. A sodium atom, Na, by *losing* an electron, forms the cation Na^+, which has the same electron configuration as the noble gas neon.

$$\begin{array}{llcll} & \text{Na} & \longrightarrow & \text{Na}^+ & + \ e^- \\ \text{Electron configurations:} & 1s^2 2s^2 2p^6 3s^1 & & 1s^2 2s^2 2p^6 = [\text{Ne}] & \end{array}$$

A chlorine atom, Cl, by *gaining* an electron, forms the anion Cl^-, which has the same electron configuration as the noble gas argon.

$$\begin{array}{lllcl} & \text{Cl} & + \ e^- & \longrightarrow & \text{Cl}^- \\ \text{Electron configurations:} & [\text{Ne}]3s^2 3p^5 & & & [\text{Ne}]3s^2 3p^6 = [\text{Ar}] \end{array}$$

These two processes occur together in the reaction in Figure 9.3. That is, the sodium atoms lose electrons, and the chlorine atoms gain them. Moreover, although we can represent the Na atoms in solid sodium as Na(s), the Cl atoms in gaseous chlorine are present in diatomic molecules, $Cl_2(g)$. We can write an equation that shows the transfer of an electron from each of two Na atoms to each of two Cl atoms.

$$2\ \text{Na(s)} + \text{Cl}_2\text{(g)} \longrightarrow 2\ \text{Na}^+\text{Cl}^-\text{(s)}$$

We must emphasize that in giving up an electron, a sodium atom does not become a neon atom. The sodium ion and neon atom do have the same electron configuration, but the sodium ion has 11 protons in its nucleus and a charge of 1+, whereas the neon atom has 10 protons in its nucleus and is electrically neutral. Similarly, a chlorine atom does not become an argon atom.

Because the two ions formed in the reaction between a sodium atom and a chlorine atom have opposite charges, they are strongly attracted to one another and form an *ion pair* (Na^+Cl^-). However, the attractive force of a given sodium ion in solid sodium chloride is not limited to one chloride ion. Each sodium ion most strongly attracts (and is attracted by) six neighboring chloride ions. It also has a much weaker attraction for more distant chloride ions. Each chloride ion most strongly attracts (and is attracted by) six neighboring sodium ions, and attracts the more distant sodium ions much less strongly. Moreover, ions of like charge repel one another. The attractive and repulsive forces counteract one another to some extent, but the net effect of all the interactions is to produce an extensive cluster of ions arranged in a regular pattern of alternating cations and anions. The net attractive electrostatic forces that hold the cations and anions together are **ionic bonds**, and the highly ordered solid collection of ions is called an *ionic crystal*. In general, a **crystal** is a distinctive repeating pattern of particles at the *microscopic* level that produces a solid structure characterized by plane surfaces, sharp edges, and a regular geometric shape at the *macroscopic* level. Figure 9.4 suggests the progressive stages by which an ionic crystal of sodium chloride is formed from individual ion pairs. We have illustrated ionic bonding and crystal formation with sodium chloride, but these processes are relevant to ionic compounds in general.

9.4 Using Lewis Symbols to Represent Ionic Bonding

Lewis developed his bonding theory primarily to describe covalent bonding, but we can also use it to represent ionic bonding. However, because we limit our use of Lewis symbols to atoms that can acquire noble-gas electron configurations, we will

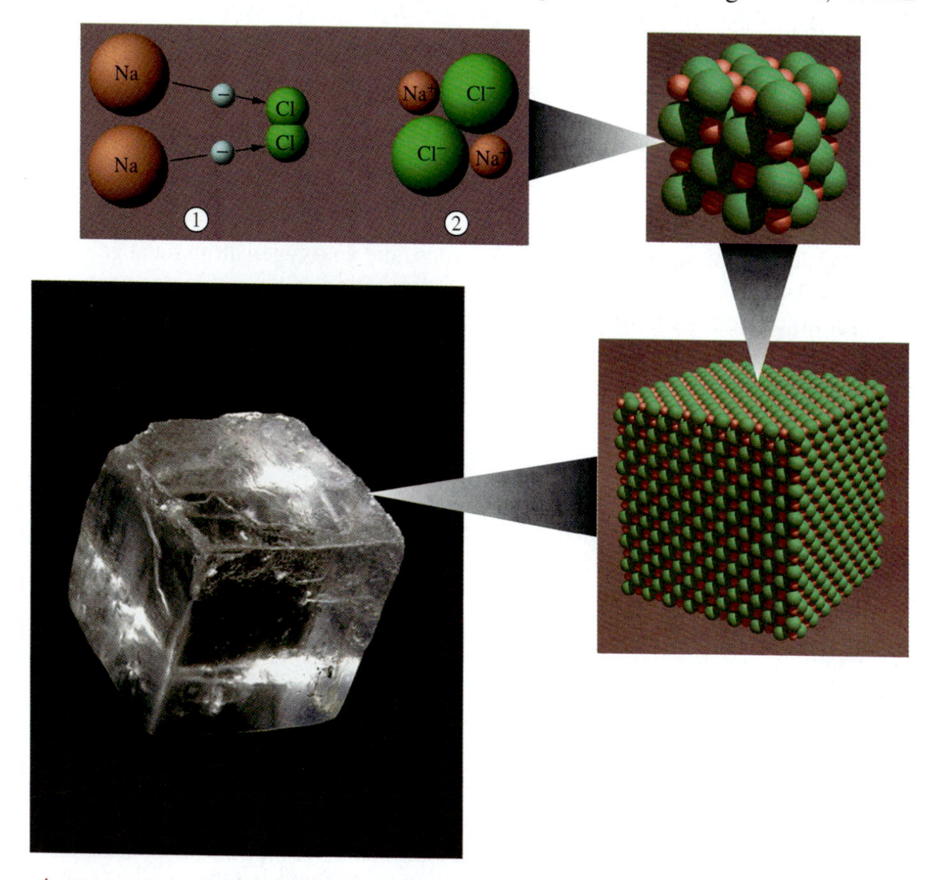

▲ **Figure 9.4 Formation of a crystal of sodium chloride**
The formation of two ion pairs from two atoms of sodium and a molecule of chlorine is depicted at the upper left (steps 1 and 2). In the formation of crystalline NaCl, each Na^+ ion (small sphere) is surrounded by six Cl^- ions (large spheres). In turn, each Cl^- ion is surrounded by six Na^+ ions. This arrangement repeats itself many many times over, ultimately resulting in a crystal of sodium chloride (lower left).

use Lewis symbols to represent only ionic bonding between nonmetals and the *s*-block metals, a few *d*-block metals, and the *p*-block metal aluminum.

Suppose that instead of using complete electron configurations to represent the loss and gain of electrons, we use Lewis symbols.

$$\text{Na}\cdot \longrightarrow \text{Na}^+ + e^-$$

$$:\dot{\underset{\cdot\cdot}{\text{Cl}}}: + e^- \longrightarrow :\ddot{\underset{\cdot\cdot}{\text{Cl}}}:^-$$

Because these processes occur together, we can use an equation to represent the net result.

$$\text{Na}\cdot + :\dot{\underset{\cdot\cdot}{\text{Cl}}}: \longrightarrow \text{Na}^+ + :\ddot{\underset{\cdot\cdot}{\text{Cl}}}:^-$$

As another example, consider the reaction of magnesium, a Group 2A metal, with oxygen, a Group 6A element, to form a stable white crystalline solid, magnesium oxide (MgO).

$$\text{Mg}: + \cdot\dot{\underset{\cdot\cdot}{\text{O}}}: \longrightarrow \text{Mg}^{2+} + :\ddot{\underset{\cdot\cdot}{\text{O}}}:^{2-}$$

A magnesium atom must give up two electrons, and an oxygen atom must gain two electrons for each to acquire the electron configuration of the noble gas neon.

An atom such as oxygen, which needs two electrons to complete an octet, may react with lithium atoms, which have only one electron to give. In this case, *two* atoms of lithium are needed for each oxygen atom. The product is lithium oxide, Li_2O.

$$\begin{matrix}\text{Li}\cdot \\ \text{Li}\cdot\end{matrix} + \cdot\dot{\underset{\cdot\cdot}{\text{O}}}: \longrightarrow \begin{matrix}\text{Li}^+ \\ \text{Li}^+\end{matrix} + :\ddot{\underset{\cdot\cdot}{\text{O}}}:^{2-} \quad \text{or} \quad 2\,\text{Li}\cdot + \cdot\dot{\underset{\cdot\cdot}{\text{O}}}: \longrightarrow 2\,\text{Li}^+ + :\ddot{\underset{\cdot\cdot}{\text{O}}}:^{2-}$$

Lithium atoms have only three electrons. They lose one electron to become Li^+, and in doing so they acquire the electron configuration of helium, $1s^2$.

EXAMPLE 9.2

Use Lewis symbols to show the formation of ionic bonds between magnesium and nitrogen. What are the name and formula of the compound that results?

SOLUTION

To acquire noble-gas electron configurations, Mg atoms (Group 2A) must lose their two valence electrons and N atoms (Group 5A) must gain three additional valence electrons. All the electrons lost by Mg atoms must be gained by N atoms to produce an electrically neutral formula unit, so *three* Mg atoms must lose a total of *six* electrons and *two* N atoms must gain a total of *six*.

$$3\,Mg\cdot + 2\,\cdot\ddot{N}\cdot \longrightarrow 3\,Mg^{2+} + 2\,:\ddot{N}:^{3-}$$

The compound is magnesium nitride, Mg_3N_2.

EXERCISE 9.2A

Use Lewis symbols to show the formation of ionic bonds between barium and iodine. What are the name and formula of the compound that results?

EXERCISE 9.2B

Use Lewis symbols to show the formation of ionic bonds in aluminum oxide.

● **ChemCDX** Periodic Trends: Common Oxidation States

9.5 Energy Changes in Ionic Compound Formation

In Figure 9.2, we illustrated the tendency for two H atoms to reach a lower energy state by forming the molecule H_2. Metal and nonmetal atoms should show a similar tendency to reach a lower energy state by forming ionic bonds. Let's see if this is the case with sodium chloride.

The energy needed to remove the valence electron from a gaseous sodium atom is its *first ionization energy* (I_1). The energy released when an electron is added to a gaseous chlorine atom is the *electron affinity* (*EA*) of chlorine.

$$Na(g) \longrightarrow Na^+(g) + e^- \qquad I_1 = +496\ kJ/mol$$

$$Cl(g) + e^- \longrightarrow Cl^-(g) \qquad EA = -349\ kJ/mol$$

The net energy change for the transfer of an electron from an isolated sodium atom to a lone chlorine atom is $(496 - 349)$ kJ/mol $= +147$ kJ/mol. Based on this calculation alone, the simultaneous formation of separate sodium cations and chloride anions from gaseous atoms is not energetically favorable. However, there are several other issues to consider.

To begin, the reaction in Figure 9.3 is between *solid* sodium, Na(s), and *gaseous* chlorine, $Cl_2(g)$, to form *solid* sodium chloride, NaCl(s). The enthalpy change in the reaction is the *enthalpy of formation* of NaCl(s).

$$Na(s) + \frac{1}{2}Cl_2(g) \longrightarrow NaCl(s) \qquad \Delta H^\circ_f = -411\ kJ$$

This negative enthalpy change indicates that the reaction is energetically favorable. That is, the system loses energy, and the ionic compound NaCl(s) is in a lower energy state than the elements. We can reach the same conclusion by considering a *hypothetical* multistep process, known as a *Born-Haber cycle*.

We need to imagine a way of converting Na(s) and $Cl_2(g)$ into NaCl(s) in a series of steps. Each step has to be one for which we can obtain a ΔH value. The combination of the steps is the overall reaction for the formation of one mole of

NaCl(s) from one mole of Na(s) and one-half mole of $Cl_2(g)$. The sum of ΔH values for the individual steps is $\Delta H°_f$ for NaCl(s), in accordance with Hess's law (Section 6.6).

In the outline below, each step is described in the column on the left. An equation for that step is written in the center column. The energy change for the step, written as an enthalpy change (ΔH), is given in the column on the right. The process is also represented schematically by the enthalpy diagram in Figure 9.5.

Starting Point: 1 mol Na(s) and $\frac{1}{2}$ mol $Cl_2(g)$

1. *Conversion of solid to gaseous Na atoms*
   The energy required to convert one mole of a solid to a gas is the *enthalpy of sublimation*. For 1 mol Na(s), this is 107 kJ.

   $Na(s) \longrightarrow Na(g)$ $\qquad \Delta H_1 = +107\ kJ$

2. *Dissociation of Cl_2 molecules into Cl atoms*
   The energy required to break one mole of bonds is the *bond-dissociation energy*. For $Cl_2(g)$, this is 243 kJ/mol. To break the bonds in $\frac{1}{2}$ mol $Cl_2(g)$ requires $\frac{1}{2}$ mol $\times$ 243 kJ/mol = 122 kJ

   $\frac{1}{2} Cl_2(g) \longrightarrow Cl(g)$ $\qquad \Delta H_2 = +122\ kJ$

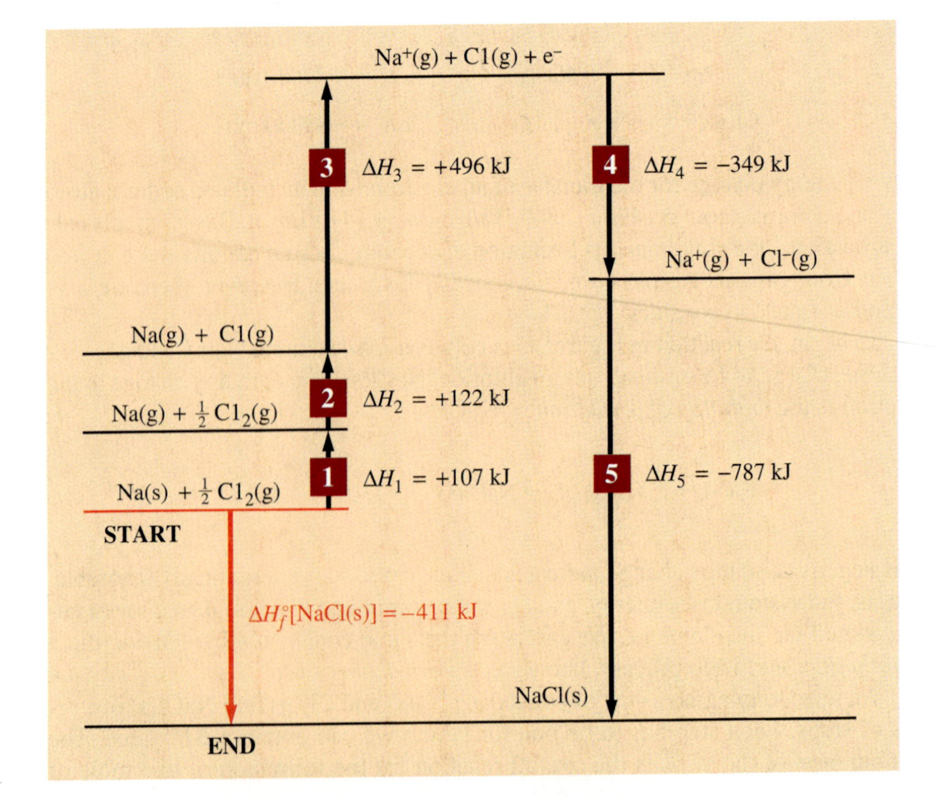

Figure 9.5
Born-Haber cycle for 1 mol of sodium chloride
The starting point and the five steps of the cycle are shown. (ΔH values are not to scale.) The sum of the five enthalpy changes (ΔH_1 to ΔH_5) gives $\Delta H°_f$. The equivalent one-step reaction for the formation of NaCl(s) directly from Na(s) and 1/2 $Cl_2(g)$ is shown in red.

3. *Ionization of Na(g) atoms to form $Na^+(g)$ ions*
   1 mol Na(g) is converted to 1 mol $Na^+(g)$. The energy requirement is the first ionization energy, $I_1 = 496$ kJ/mol. $\cancel{Na(g)} \longrightarrow \cancel{Na^+(g)} + \cancel{e^-}$ $\Delta H_3 = +496$ kJ
4. *Conversion of Cl(g) atoms to $Cl^-(g)$ ions*
   1 mol Cl(g) is converted to 1 mol $Cl^-(g)$. The energy requirement is the electron affinity, EA = −349 kJ/mol $\cancel{Cl(g)} + \cancel{e^-} \longrightarrow \cancel{Cl^-(g)}$ $\Delta H_4 = -349$ kJ
5. *Assembling of Na^+ and Cl^- ions into a crystal*
   1 mol NaCl(s) is formed from 1 mol $Na^+(g)$ and 1 mol $Cl^-(g)$. The energy of formation of one mole of an ionic solid from its separated gaseous ions is the **lattice energy**. For NaCl(s), this is −787 kJ. $\cancel{Na^+(g)} + \cancel{Cl^-(g)} \longrightarrow NaCl(s)$ $\Delta H_5 = -787$ kJ

Ending Point: 1 mol NaCl(s)

The Overall Reaction:

$$Na(s) + \tfrac{1}{2}\,Cl_2(g) \longrightarrow NaCl(s)$$

$$\Delta H_f^\circ = \Delta H_1 + \Delta H_2 + \Delta H_3 + \Delta H_4 + \Delta H_5 = -411 \text{ kJ}$$

From this analysis, we see that the large negative value of the lattice energy (ΔH_5) is the major factor that makes ionic compound formation an energetically favorable process. In actual practice, we cannot establish lattice energies by a direct measurement. In fact, we generally use the Born-Haber cycle to calculate a lattice energy from other measured quantities, as illustrated in Example 9.3.

EXAMPLE 9.3

Determine the lattice energy of $MgF_2(s)$, using the following data: enthalpy of sublimation of magnesium, +146 kJ/mol; I_1 for Mg, +738 kJ/mol; I_2 for Mg, +1451 kJ; bond-dissociation energy of $F_2(g)$, +159 kJ/mol F_2; electron affinity of F, −328 kJ/mol F; enthalpy of formation of $MgF_2(s)$, −1124 kJ/mol.

SOLUTION

We need a setup here that differs in three ways from the one we used for NaCl(s). (1) The compound $MgF_2(s)$ has *two* anions for each cation in the crystal. In the step where $F_2(g)$ dissociates, we will use the bond-dissociation energy based on *one* mole of $F_2(g)$ in order to get *two* moles of F(g). Similarly, in the electron affinity step, we must produce *two* moles of $F^-(g)$ rather than one. (2) Because the magnesium cation carries a 2+ charge, we have to include *two* ionization steps and both I_1 and I_2 in our setup. (3) The enthalpy of formation of $MgF_2(s)$ is given, and our unknown is the lattice energy. Thus, we need the setup that follows.

PROBLEM-SOLVING NOTE
In a Born-Haber cycle, we can use cancellation marks to help ensure that we have done all the appropriate steps.

$$\begin{array}{rcll}
Mg(s) & \longrightarrow & \cancel{Mg(g)} & \Delta H_1 = +146 \text{ kJ} \\
F_2(g) & \longrightarrow & \cancel{2\,F(g)} & \Delta H_2 = +159 \text{ kJ} \\
\cancel{Mg(g)} & \longrightarrow & \cancel{Mg^+(g)} + \cancel{e^-} & \Delta H_3 = +738 \text{ kJ} \\
\cancel{Mg^+(g)} & \longrightarrow & \cancel{Mg^{2+}(g)} + \cancel{e^-} & \Delta H_4 = +1451 \text{ kJ} \\
\cancel{2\,F(g)} + \cancel{2\,e^-} & \longrightarrow & \cancel{2\,F^-(g)} & \Delta H_5 = -2 \times 328 \text{ kJ} \\
\cancel{Mg^{2+}(g)} + \cancel{2\,F^-(g)} & \longrightarrow & MgF_2(s) & \Delta H_6 = \text{Lattice energy} \\
\hline
\textit{Overall: } Mg(s) + F_2(g) & \longrightarrow & MgF_2(s) & \Delta H_f^\circ = -1124 \text{ kJ}
\end{array}$$

$$\Delta H_f^\circ = -1124 \text{ kJ} = (146 + 159 + 738 + 1451 - 656) \text{ kJ} + \text{lattice energy}$$

$$\text{Lattice energy} = (-1124 - 146 - 159 - 738 - 1451 + 656) \text{ kJ} = -2962 \text{ kJ}$$

The lattice energy is -2962 kJ/mol $MgF_2(s)$.

EXERCISE 9.3A

For lithium, the enthalpy of sublimation is +161 kJ/mol and the first ionization energy is +520 kJ/mol. The bond-dissociation energy of fluorine is +159 kJ/mol F_2, and the electron affinity of fluorine is -328 kJ/mol. The lattice energy of LiF is -1047 kJ/mol. Calculate the overall enthalpy change for the reaction

$$Li(s) + \frac{1}{2}F_2(g) \longrightarrow LiF(s) \quad \Delta H_f^\circ = ?$$

● **ChemCDX** Born-Haber Cycle

EXERCISE 9.3B

Use the data provided in this section and the enthalpy of formation of lithium chloride, $\Delta H_f^\circ = -409$ kJ/mol LiCl(s), to determine the lattice energy of LiCl.

Covalent Bonding

We showed in Figure 9.2 that the energy state of two H atoms joined by a chemical bond in H_2 is lower than that of two isolated H atoms. But the bond can't be ionic. One hydrogen atom cannot accept an electron from another, because all hydrogen atoms have an equal electron affinity. They also have a high ionization energy (I_1 = 1312 kJ/mol) indicating that hydrogen atoms lose electrons only with great difficulty. Lewis proposed that in such cases, the chemical bond is a pair of electrons *shared* between the bonded atoms; it is a **covalent bond**.

$$H\cdot + \cdot H \longrightarrow H{:}H$$

Covalent bond (shared pair of electrons)

9.6 Lewis Structures of Some Simple Molecules

A representation such as H:H for the H_2 molecule is called a Lewis structure. A **Lewis structure** is a combination of Lewis symbols that represents the formation of covalent bonds between atoms. In addition,

- A Lewis structure indicates the proportions in which atoms combine.
- In most cases, a Lewis structure shows that the bonded atoms acquire the electron configuration of a noble gas, that is, the atoms obey the *octet* rule. (H atoms obey the *duet* rule.)

Notice that by "double counting" the shared electrons in the Lewis structure for H_2, each H atom appears to have two electrons in its valence shell, the electron configuration of helium.

Consider next the case of chlorine, which also exists in the diatomic form, Cl_2. Chlorine atoms, too, are joined by a covalent bond.

$$:\ddot{\underset{..}{Cl}}\cdot \quad + \quad \cdot\ddot{\underset{..}{Cl}}: \quad \longrightarrow \quad :\ddot{\underset{..}{Cl}}:\ddot{\underset{..}{Cl}}:$$

Again, by double counting the electrons shared by the two atoms, the number of electrons around each Cl atom in the Lewis structure of Cl_2 is *eight*; each Cl atom follows to the octet rule.

The shared pairs of electrons in a molecule are called **bonding pairs.** The other electron pairs, which stay with one atom and are not shared, are called *nonbonding pairs*, or **lone pairs.** Bonding pairs (:) and lone pairs (:) in the Cl_2 molecule are shown below. We also illustrate the common practice of representing a bonding pair by a dash (—).

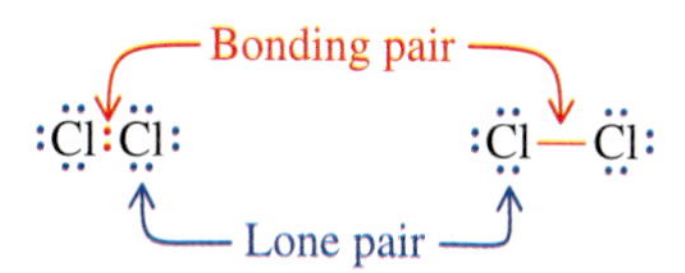

The nonmetals of the *second period* (except boron) tend to form a number of covalent bonds equal to *eight minus the group number*. Thus, fluorine (Group 7A) forms *one* bond; oxygen (Group 6A), *two*; nitrogen (Group 5A), *three*; and carbon (Group 4A), *four*. This idea is illustrated for the molecules HF, H_2O, NH_3, and CH_4 in Figure 9.6. Although the Lewis structures in Figure 9.6 are written in a way that suggests the shapes of the molecules, Lewis structures in themselves do *not* predict molecular shapes. A straight-line Lewis structure for water ($H-\ddot{\underset{..}{O}}-H$) is as acceptable as the angular one shown. We will consider how to predict molecular shapes in Chapter 10.

Coordinate Covalent Bonds

In the molecules we have considered thus far, H_2, Cl_2, HF, H_2O, NH_3, CH_4, each of the bonded atoms contributes one electron to a shared pair of electrons. In some cases, though, one atom can provide *both* electrons of the shared pair to form a **coordinate covalent bond**.

Consider what happens when an acid (a substance that ionizes to produce H^+) is added to water. Oxygen atoms in some of the water molecules use a lone pair of electrons to form a covalent bond to a H^+ ion. The H^+ ion has *no* electrons and cannot contribute to the bond. The bond is a coordinate covalent bond.

$$H^+ \quad + \quad :\ddot{\underset{\underset{H}{..}}{O}}:H \quad \longrightarrow \quad \left[H:\ddot{\underset{\underset{H}{..}}{O}}:H\right]^+ \quad \text{or} \quad \left[H-\ddot{\underset{\underset{H}{|}}{O}}-H\right]^+$$

The positive charge of the H^+ ion is now associated with the entire H_3O^+ ion, not with any particular H atom. Once they have formed, we cannot distinguish among the three O—H bonds. They are identical, and we can't pinpoint one of them as the coordinate covalent bond. We will discuss the H_3O^+ ion, called *hydronium ion*, in Chapter 15, but until then, we will continue to represent acidic solutions as $H^+(aq)$.

In a manner similar to the formation of H_3O^+, the nitrogen atom of ammonia can use its lone pair of electrons to form a fourth bond to a hydrogen ion, H^+. this fourth bond is a coordinate covalent bond, and the product is the *ammonium ion*.

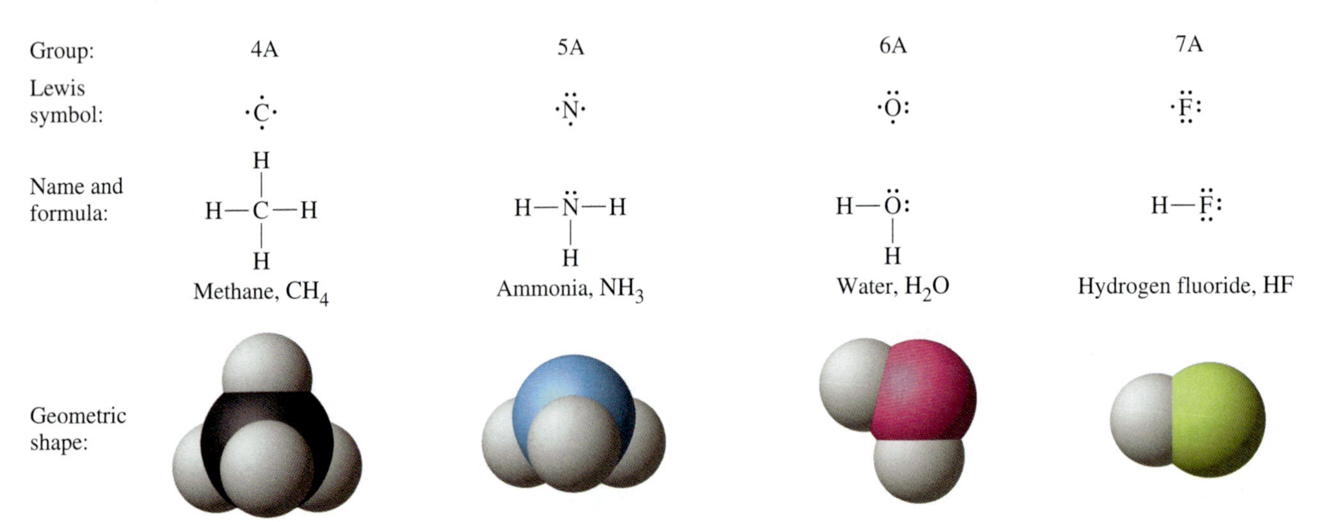

▲ **Figure 9.6 Four hydrogen compounds of second-period nonmetals**
The formulas of the four molecules can be deduced from Lewis structures, but their geometrical shapes cannot. Geometric shapes of molecules can be predicted by methods presented in Chapter 10, and they can be precisely established by experiment.

$$H^+ + :\underset{\cdot\cdot}{\overset{H}{N}}:H \longrightarrow \left[H:\underset{\underset{H}{\cdot\cdot}}{\overset{H}{N}}:H\right]^+ \quad \text{or} \quad \left[H-\underset{\underset{H}{|}}{\overset{\overset{H}{|}}{N}}-H\right]^+$$

Again, however, once the ion has formed, all four N—H bonds in NH_4^+ are identical; we cannot single out one bond as the coordinate covalent bond.

Multiple Covalent Bonds

All the covalent bonds we have considered to this point have involved one shared pair of electrons, a bond type called a **single bond**. Two bonded atoms can also share more than one pair of electrons between them, resulting in a **multiple bond**. A **double bond** is the covalent bond formed when bonded atoms share *two* pairs of electrons, and a **triple bond** involves the sharing of *three* pairs of electrons.

Let's see how to write Lewis structures involving multiple bonds. Our first attempt at a Lewis structure for carbon dioxide, CO_2, might look like this.

$$:\dot{\underset{\cdot\cdot}{O}}\cdot + \cdot\dot{\underset{\cdot}{C}}\cdot + \cdot\ddot{O}: \longrightarrow :\dot{\underset{\cdot\cdot}{O}}:\dot{\underset{\cdot}{C}}:\ddot{O}: \quad \text{or} \quad :\dot{\underset{\cdot\cdot}{O}}-\dot{\underset{\cdot}{C}}-\ddot{O}: \quad (\textit{not correct})$$

However, this structure is unsatisfactory because none of the atoms has acquired a valence-shell octet. By shifting the four unpaired electrons into the regions between the C and O atoms, however, we do obtain a satisfactory structure. Each atom has a valence-shell octet.

$$:\dot{\underset{\cdot\cdot}{O}}-\dot{\underset{\cdot}{C}}-\ddot{O}: \longrightarrow :\underset{\cdot\cdot}{O}=C=\ddot{O}:$$

Each O atom is joined to the C atom by a double bond. We can represent each electron pair by a single dash; thus, two parallel dashes represent a double bond.

Our first attempt at a Lewis structure for the nitrogen molecule, N_2, looks equally bad: Neither of the nitrogen atoms has eight electrons in its valence shell.

$$:\dot{\underset{\cdot}{N}}\cdot \;+\; \cdot\dot{\underset{\cdot}{N}}: \longrightarrow :\dot{\underset{\cdot}{N}}:\dot{\underset{\cdot}{N}}: \quad \text{or} \quad :\dot{\underset{\cdot}{N}}-\dot{\underset{\cdot}{N}}: \qquad (\textit{incorrect})$$

Here we can produce an octet for each nitrogen atom by shifting all the unpaired electrons into the region between the two N atoms.

$$:\dot{\underset{\cdot}{N}}\,\dot{\underset{\cdot}{N}}: \longrightarrow :N{\equiv}N:$$

In doing so, we obtain a triple bond. As in the single and double bond, each electron pair is represented by a dash.

The Importance of Experimental Evidence

We can write what looks like a good Lewis structure for the O_2 molecule, using the same approach that we used for CO_2 and N_2. Doing so, we obtain a structure with a double bond.

$$:\dot{\underset{\cdot\cdot}{O}}\cdot \;+\; \cdot\dot{\underset{\cdot\cdot}{O}}: \longrightarrow :\underset{\cdot\cdot}{O}{=}\underset{\cdot\cdot}{O}: \qquad (\textit{incorrect})$$

This Lewis structure conforms to the octet rule (each atom has eight valence electrons), but it is *not* consistent with an important property of the O_2 molecule. As shown in Figure 9.7, oxygen, is *paramagnetic*; the O_2 molecule must contain unpaired electrons. It is important to note, then, that no matter how plausible a Lewis structure may be, we cannot accept it as the true structure unless it conforms to all

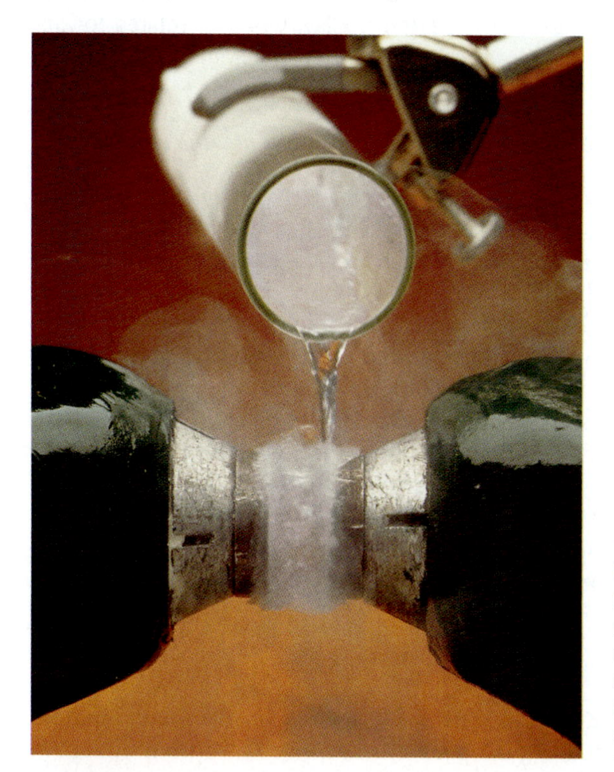

Figure 9.7
Paramagnetism of oxygen
Liquid oxygen is attracted into a magnetic field, the region between the poles of this large magnet. This is the expected behavior for paramagnetic substances (recall page 333).

Linus Pauling (1901–1994) used quantum theory to extend Lewis's theory. Pauling's work was summarized in the 1939 text, *The Nature of the Chemical Bond*, and he was awarded the Nobel Prize in chemistry in 1954. Pauling also won the Nobel Peace Prize in 1962 for his fight to control nuclear weapons. His efforts were influential in the establishment of the 1963 nuclear test ban treaty. Later in life, Pauling became interested in the medical value of megadoses of vitamins, particularly vitamin C. His ideas on vitamins are controversial, but they have stimulated continuing research.

the available *experimental* evidence. We will present a more satisfactory structure of the O_2 molecule in Chapter 10.

In the next section, we consider bonding based on the *unequal* sharing of an electron pair between atoms. This possibility gives rise to an important idea, electronegativity, that we will use when we return to writing Lewis structures in Section 9.8.

9.7 Polar Covalent Bonds and Electronegativity

Metal atoms can transfer electrons to nonmetal atoms to form ionic bonds. Identical atoms combine by sharing pairs of electrons to form covalent bonds. What kind of bond forms between atoms that are different, but not different enough to form ionic bonds? Let's consider the bond between a hydrogen atom and a chlorine atom as an example.

Electronegativity

As shown in the following Lewis structure, a hydrogen atom and a chlorine atom share a pair of electrons in the HCl molecule.

$$\mathrm{H\cdot} + \mathrm{\cdot\ddot{\underset{..}{Cl}}:} \longrightarrow \mathrm{H:\ddot{\underset{..}{Cl}}:} \quad \text{or} \quad \mathrm{H{-}\ddot{\underset{..}{Cl}}:}$$

What the Lewis structure does not show, however, is that the atoms do not share the electrons equally. The chlorine atom has a greater attraction for the electrons than does the hydrogen atom.

In Chapter 8, we encountered two atomic properties that relate in a way to an attraction for electrons: ionization energy and electron affinity. The greater the ionization energy of an atom, the more inclined the atom is to retain its electrons. The more negative its electron affinity, the more inclined it is to acquire an additional electron. However, these properties apply only to isolated gaseous atoms and not directly to atoms in molecules. **Electronegativity (EN)**, which is related to ionization energy and electron affinity, describes the ability of an atom to attract *bonding* electrons to itself when the atom is in a molecule.

> *The greater the electronegativity of an atom in a molecule, the more strongly it attracts the electrons in a covalent bond.*

The electronegativity of a chlorine atom is greater than that of a hydrogen atom.

Atoms of the elements in the upper right of the periodic table—relatively small, nonmetal atoms—attract bonding electrons most strongly: They have the greatest electronegativities. Atoms of the elements to the lower left of the table—relatively large, metal atoms—have a weaker hold on electrons: They have the smallest electronegativities. Several electronegativity scales have been devised, each defined a little differently from the others. As a consequence, the same element might have somewhat different electronegativity values on the different scales. On the scale devised by Linus Pauling and shown in Figure 9.8, the most nonmetallic and most electronegative element, fluorine, is assigned a value of 4.0. The most metallic elements have electronegativities of about 1.0 or less.

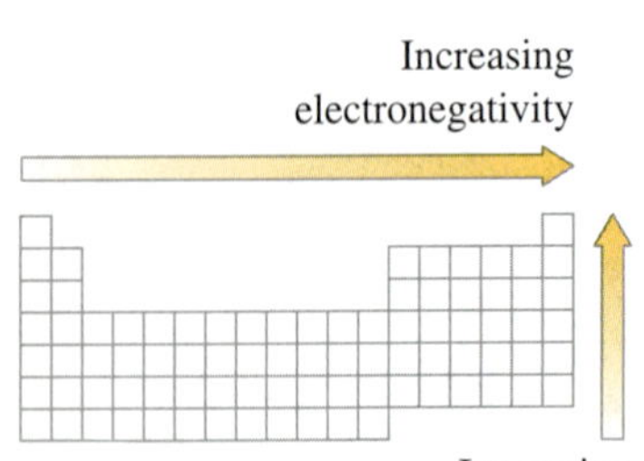

Regardless of the electronegativity scale used, there are two general trends among the elements:

> *Within a period, electronegativity generally increases from left to right.*

In Figure 9.8, we see that the trend is completely regular for the second period. Electronegativity increases by about 0.5 per element as we move from lithium at the left to fluorine at the right. In other periods, the trend is less regular.

Key	
Below 1.0	2.0–2.4
1.0–1.4	2.5–2.9
1.5–1.9	3.0–4.0

Period	1A	2A	3B	4B	5B	6B	7B	8B	8B	8B	1B	2B	3A	4A	5A	6A	7A
1	**H** 2.1																
2	**Li** 1.0	**Be** 1.5											**B** 2.0	**C** 2.5	**N** 3.0	**O** 3.5	**F** 4.0
3	**Na** 0.9	**Mg** 1.2											**Al** 1.5	**Si** 1.8	**P** 2.1	**S** 2.5	**Cl** 3.0
4	**K** 0.8	**Ca** 1.0	**Sc** 1.3	**Ti** 1.5	**V** 1.6	**Cr** 1.6	**Mn** 1.5	**Fe** 1.8	**Co** 1.8	**Ni** 1.8	**Cu** 1.9	**Zn** 1.7	**Ga** 1.6	**Ge** 1.8	**As** 2.0	**Se** 2.4	**Br** 2.8
5	**Rb** 0.8	**Sr** 1.0	**Y** 1.2	**Zr** 1.4	**Nb** 1.6	**Mo** 1.8	**Tc** 1.9	**Ru** 2.2	**Rh** 2.2	**Pd** 2.2	**Ag** 1.9	**Cd** 1.7	**In** 1.7	**Sn** 1.8	**Sb** 1.9	**Te** 2.1	**I** 2.5
6	**Cs** 0.7	**Ba** 0.9	**La*** 1.1	**Hf** 1.3	**Ta** 1.5	**W** 1.7	**Re** 1.9	**Os** 2.2	**Ir** 2.2	**Pt** 2.2	**Au** 2.4	**Hg** 1.9	**Tl** 1.8	**Pb** 1.8	**Bi** 1.9	**Po** 2.0	**At** 2.2
7	**Fr** 0.7	**Ra** 0.9	**Ac†** 1.1														

*Lanthanides: 1.1–1.3
†Actinides: 1.3–1.5

▲ **Figure 9.8 Pauling's electronegativities of the elements**
Values are from L. Pauling, *The Nature of the Chemical Bond*, 3rd edition, Cornell University, Ithaca, NY, 1960, p. 93. (Some values have been modified by later investigators.) Because noble-gas compounds are limited to just a few compounds of Kr and Xe, electronegativities for the Group 8A elements are not included.

Within a group, electronegativity generally increases from the bottom to the top.

Chlorine is less electronegative than fluorine, and sulfur is less electronegative than oxygen. A comparison of electronegativities is not quite so straightforward when we consider two elements that are neither in the same period nor in the same group. As we suggest in Figure 9.9, generally, the element above or to the right (or both) is more electronegative than one below or to the left (or both).

PROBLEM-SOLVING NOTE
In Chapter 2, we used a portion of the periodic table (Figure 2.7) to determine which element symbol to write first when we write the formula of a compound. We can now rationalize the rule: We generally write the symbol of the element of lower electronegativity first.

● **ChemCDX** Electronegativity

EXAMPLE 9.4

Refer only to the periodic table inside the front cover, and arrange the following sets of atoms in the expected order of increasing electronegativity, that is, from the lowest to highest electronegativity value.

a. Cl, Mg, Si **b.** As, N, Sb **c.** As, Se, Sb

SOLUTION

a. The order of increasing electronegativity within a period (the third, in this case) is from left to right: Mg < Si < Cl.

b. The order of increasing electronegativity within a group (5A) is from bottom to top: Sb < As < N.

c. As and Se are in the same period; Se is to the right of As, and we expect it to be the more electronegative. As and Sb are in the same group; As is above Sb, and we expect it to be the more electronegative. Thus, As is in the center of the set of three, and the order of increasing electronegativity is Sb < As < Se.

EXERCISE 9.4

Refer only to the periodic table inside the front cover, and arrange the following sets of atoms in the expected order of increasing electronegativity.

a. Ba, Be, Ca **b.** Ga, Ge, Se **c.** Cl, S, Te **d.** Bi, P, S

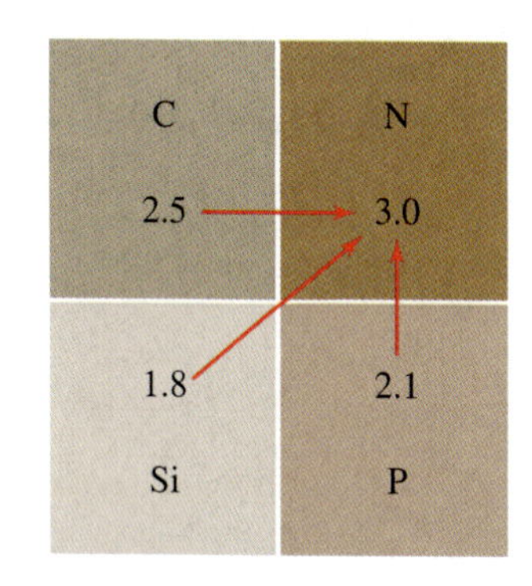

▲ **Figure 9.9 Electronegativities in relation to position in the periodic table**
In general, electronegativities increase in the directions of the colored arrows.

Electronegativity Difference and Bond Type

It is the *difference* in electronegativity of bonded atoms that is relevant to chemical bonding. These electronegativity differences allow us to characterize covalent bonds as *nonpolar* or *polar*, and the polarity of bonds can greatly influence properties of molecular substances. Two identical atoms have the same electronegativity and share a bonding electron pair equally. That is, the bonding electrons are not drawn any closer to one atom than to the other, and the bond is a **nonpolar** covalent bond. The covalent bonds in H—H and Cl—Cl are nonpolar. Even if the atoms are not identical, the bonds in a molecule may be essentially nonpolar if the electronegativity difference is quite small. In CH_4, for example, the electronegativity difference between C (EN = 2.5) and H (EN = 2.1) is only 0.4, and C—H bonds are essentially nonpolar.

In covalent bonds between atoms with somewhat larger electronegativity differences, electron pairs are shared unequally. The electrons are drawn closer to the atom of higher electronegativity, and the bond is a **polar** covalent bond. The H—Cl bond is a polar covalent bond; the electronegativity difference between H (EN = 2.1) and Cl (EN = 3.0) is 0.9. With still larger differences in electronegativity, electrons may be completely transferred from metal to nonmetal atoms to form *ionic* bonds.

Figure 9.10 suggests that there is no distinct boundary between covalent and ionic bonds. Some bonds are clearly nonpolar covalent, and some are essentially 100% ionic, but many bonds have an intermediate character—they are *polar covalent* bonds.

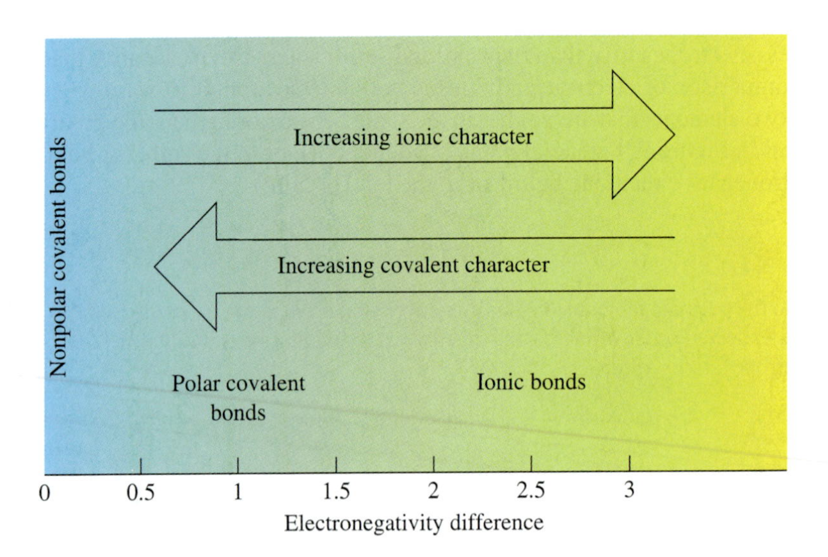

▶ **Figure 9.10 Electronegativity and bond type**

EXAMPLE 9.5

Use electronegativity values to arrange the following bonds in order of increasing polarity: Br—Cl, Cl—Cl, Cl—F, H—Cl, I—Cl.

SOLUTION

The polar character of a bond is determined by an electronegativity *difference*, regardless of what the two electronegativity values themselves happen to be. For the five bonds in question, the electronegativities (EN) of the atoms and the electronegativity *differences* (ΔEN) are as follows.

EN:	2.8 3.0	3.0 3.0	3.0 4.0	2.1 3.0	2.5 3.0
	Br—Cl	Cl—Cl	Cl—F	H—Cl	I—Cl
ΔEN:	0.2	0.0	1.0	0.9	0.5

The order of increasing polarity is:
Cl—Cl < Br—Cl < I—Cl < H—Cl < Cl—F.

EXERCISE 9.5

Use electronegativity values to arrange the following bonds in order of increasing polarity: C—Cl, C—H, C—Mg, C—O, C—S

Depicting Polar Covalent Bonds

According to modern quantum theory, which we will discuss in Chapter 10, we can picture the electron-pair bond between two atoms as a cloud of negative electric charge that encompasses both atoms. In a *nonpolar* covalent bond such as H—H, the contribution of the electron-pair bond to the overall negative charge density in the molecule is greater between the bonded atoms than elsewhere, but otherwise the charge is uniformly distributed (Figure 9.11). In a *polar* covalent bond such as H—Cl, the contribution of the electron-pair bond to the overall negative charge density is strongly displaced toward the more electronegative chlorine atom.

Two methods are widely used to indicate the polar nature of a bond. One method uses the lower case Greek letter delta (δ) to indicate partial charges.

$$\overset{\delta+}{\text{H}}-\overset{\delta-}{\text{Cl}}$$

The δ+ and δ− (read "delta plus" and "delta minus") signify that one end (H) is partially positive and the other end (Cl) is partially negative. The term *partial charge* indicates something less than the full charges of the ions that would result from complete electron transfer. The second method uses a cross-based arrow.

$$\overset{+\!\!\!\longrightarrow}{\text{H}-\text{Cl}}$$

This arrow indicates the direction of displacement of negative charge, from the less electronegative element to the more electronegative element. Note that the "+" of the δ+ and the "+" on the tail of the cross-based arrow always indicate the *positive* end of the bond, that is, the less electronegative element.

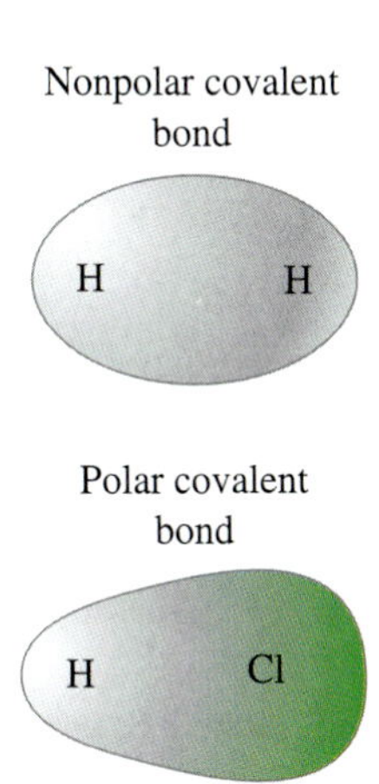

▲ **Figure 9.11**
Nonpolar and polar covalent bonds
In the H—H molecule, there is an even distribution of electron charge density between the atoms. In the H—Cl molecule, electron charge density is displaced toward the Cl atom.

9.8 Strategies for Writing Lewis Structures

Now let's use some of the concepts of covalent bonding we have developed—especially the octet rule, multiple bonds, and electronegativity—together with one or two new concepts to extend the range of Lewis structures that we can write.

When possible, we should check a Lewis structure to see if it is consistent with *experimental evidence*. Lacking such evidence, we can still use the strategy developed over the next several pages to write a *plausible* Lewis structure.

Skeletal Structures

We can use the duet and octet rules alone to conclude that the Lewis structure of H_2O is $\text{H}-\underset{\cdot\cdot}{\ddot{\text{O}}}-\text{H}$ (not $\text{H}-\text{H}-\ddot{\text{O}}\text{:}$). In the correct structure, each H atom has a valence shell *duet*, and the O atom has an *octet*. In the second (incorrect) structure, the H atom on the left has a duet, but the other H atom has *four* valence-shell electrons. Also, the O atom has only *six* valence-shell electrons, *not* an octet. In many cases, however, the way to arrange atoms within a structure may not be as evident. We need some sort of strategy to deal with these cases.

We refer to the arrangement of atoms within a molecule or polyatomic ion as the **skeletal structure**; it shows the order in which the atoms are attached to one another. The skeletal structure consists of one or more central atoms and terminal atoms. A **central atom** is bonded to two or more atoms in the structure, and a **terminal atom** is bonded only to one other atom. The central and terminal atoms in NH_3 are identified below.

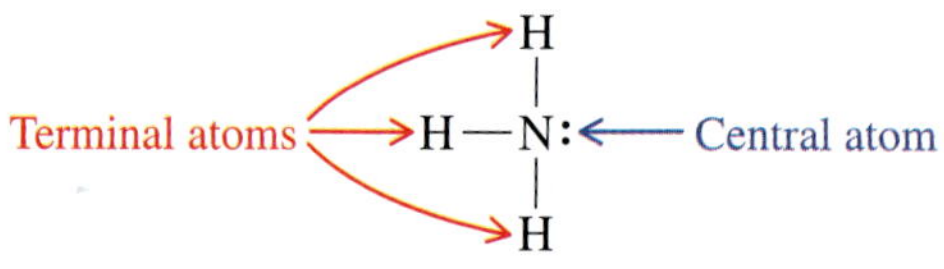

Terminal atoms → H—N: ← Central atom (with H above and H below N)

In writing a skeletal structure, we must attach every atom to the rest of the structure by at least one bond. We do this by joining the bonded atoms by single dashes and making no attempt to account for all the valence electrons. The skeletal structure is *not* a Lewis structure; it is the *first step* in deducing a plausible Lewis structure. In the absence of specific information about a skeletal structure, we can use the following observations to devise a likely one.

1. *Hydrogen atoms are terminal atoms.* This is because a bonded H atom can have only two electrons in its valence shell and thus form only one bond. Exceptions to this rule are rare, and we will not encounter any in this chapter. As a typical example, the ethane molecule, C_2H_6, or CH_3CH_3, has two C atoms as central atoms and six H atoms as terminal atoms.

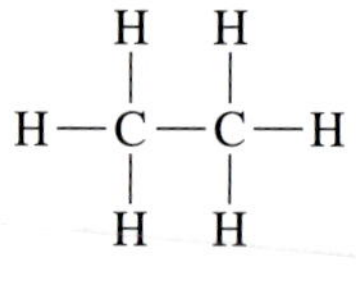

```
    H   H
    |   |
H — C — C — H
    |   |
    H   H
```

2. *The central atom(s) of a structure usually has the* lowest *electronegativity, and the terminal atoms generally have* higher *electronegativities.* Because hydrogen atoms must always be terminal atoms, they will often be an exception to this rule (as they are in H_2O, NH_3, and C_2H_6). Also, because it has a higher electronegativity than any other element, we expect fluorine always to be a terminal atom. We will consider the basis for this rule later in the section, but for now we can see its application to the skeletal structure of the poisonous gas phosgene, $COCl_2$, used in the manufacture of plastics.

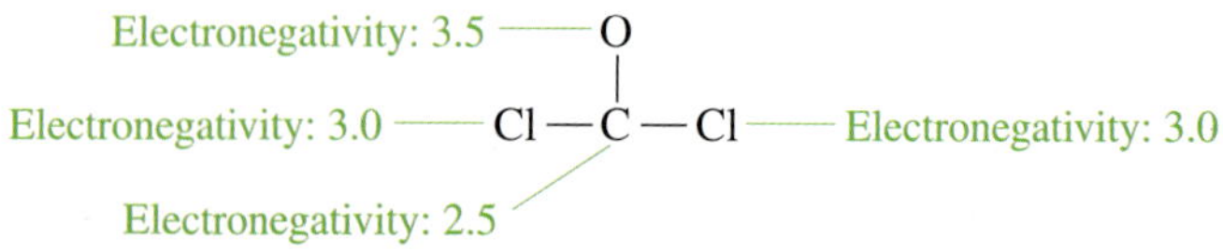

Electronegativity: 3.5 — O
Electronegativity: 3.0 — Cl—C—Cl — Electronegativity: 3.0
Electronegativity: 2.5 (C)

3. *In oxoacids, (Section 2.8), hydrogen atoms are usually bonded to oxygen atoms.*

```
      O
      |
H—O—Cl—O
```
Chloric acid ($HOClO_2$)

```
        O
        |
H—O—S—O—H
        |
        O
```
Sulfuric acid [$(HO)_2SO_2$]

4. *Molecules and polyatomic ions usually have compact, symmetrical structures.* A typical application of this idea is to the molecule SO_2F_2.

```
    O
    |
F—S—F    not    F—O—S—O—F
    |
    O
```
Sulfuryl fluoride (SO_2F_2)

Organic compounds, which can be based on long chains of carbon atoms, are a major exception to this idea.

A Method for Writing Lewis Structures

For the Lewis structures that we write in the remainder of the chapter, we will follow a systematic approach that minimizes false starts and gets us directly to a plausible structure. The following five-step procedure is designed for that purpose.

1 *Determine the total number of valence electrons; these and no other electrons must appear in the Lewis structure.* The total number of electrons in the Lewis structure of a molecule is the sum of the valence electrons for each atom. For a polyatomic anion, *add* to the sum of the valence electrons one electron for each unit of negative charge. For a polyatomic cation, *subtract* from the sum of the valence electrons one electron for each unit of positive charge. Examples: N_2O_4 has $(2 \times 5) + (4 \times 6) = 34$ valence electrons; NO_3^- has $5 + (3 \times 6) + 1 = 24$ valence electrons; and NH_4^+ has $5 + (4 \times 1) - 1 = 8$ valence electrons.

PROBLEM-SOLVING NOTE

Because electrons in Lewis structures are almost always placed in pairs, you may find it helpful to divide the valence electrons into pairs in step 1. Thus, in N_2O_4 there are *17* pairs of valence electrons, and in NO_3^- and NH_4^+ there are *12* and *4* pairs, respectively.

2 *Use the ideas previously listed to write a skeletal structure; connect the bonded atoms by dashes (single covalent bonds).*

3 *Place pairs of electrons as lone pairs around the terminal atoms to give each terminal atom (except hydrogen) an octet.*

4 *Assign any remaining electrons as lone pairs around the central atom(s).*

5 *If necessary, move one or more lone pairs of electrons from a terminal atom(s) to form a multiple bond to a central atom(s).* If the number of valence electrons is just sufficient that all atoms in the structure in step 4 have an octet (duet for H), the structure has only single bonds. If there are not enough electrons to form octets, it is necessary to form one or more multiple bonds. The atoms most commonly involved in double bonds are *carbon*, *nitrogen*, *oxygen*, and *sulfur*; in triple bonds, *carbon* and *nitrogen*.

EXAMPLE 9.6

Write the Lewis structure of nitrogen trifluoride, NF_3.

SOLUTION

1 *Determine number of valence electrons.* The number of valence electrons in *one* N atom (Group 5A) and *three* F atoms (Group 7A) is $5 + (3 \times 7) = 26$.

2 *Write a skeletal structure.* The electronegativity of N is 3.0; that of F is 4.0. We expect a skeletal structure with N as a central atom and F as terminal atoms. The three nitrogen-to-fluorine bonds in this structure account for *six* electrons.

```
F—N—F
  |
  F
```

3 *Complete octets of terminals atoms.* We complete the octets of the F atoms by placing *three* lone pairs of electrons around each. This accounts for *18* additional electrons.

```
 ..     ..
:F—N—F:
 ''  |  ''
    :F:
     ''
```

4 *Assign lone pairs electrons to central atom(s).* We have now assigned $6 + 18 = 24$ electrons to the Lewis structure. Place the remaining two as a lone pair on the N atom.

```
 ..  ..  ..
:F—N—F:
 ''  |  ''
    :F:
     ''
```

Because each atom now has an octet, this is the Lewis structure of NF_3. We do not need to proceed further.

EXERCISE 9.6A

Write the Lewis structure of hydrazine, N_2H_4.

EXERCISE 9.6B

Write the Lewis structure of ethyl chloride, C_2H_5Cl.

EXAMPLE 9.7

Write a plausible Lewis structure for phosgene, $COCl_2$, (page 376).

SOLUTION

1 *Determine number of valence electrons.* The number of valence electrons in *one* C atom (Group 4A), *one* O atom (Group 6A) and *two* Cl atoms (Group 7A) is $4 + 6 + (2 \times 7) = 24$.

2 *Write a skeletal structure.* The electronegativities are C (2.5), O (3.5), and Cl(3.0). We expect C, the least electronegative atom, to be the central atom, and O and Cl to be terminal atoms attached to it. The skeletal structure accounts for *six* of the valence electrons.

```
      O
      |
Cl—C—Cl
```

3 *Complete octets of terminal atoms.* Place *three* lone pairs of electrons around the O atom and the two Cl atoms, for a total of 18 electrons. This completes the octets of the terminal atoms.

```
       ..
      :O:
       |
 ..        ..
:Cl—C—Cl:
 ''        ''
```

4 *Assign lone pair electrons to central atom(s).* We have now assigned 6 + 18 = 24 electrons to the Lewis structure. This accounts for all the valence electrons, and there are none available to complete the octet of the central C atom.

5 *Form multiple bonds to complete octets of central atom(s).* Complete the octet of the C atom by shifting a lone pair of electrons from the O atom to form a carbon-to-oxygen double bond. (C and O are two atoms that are able to form a double bond. In Example 9.9 we will see why the double bond is not carbon-to-chlorine.)

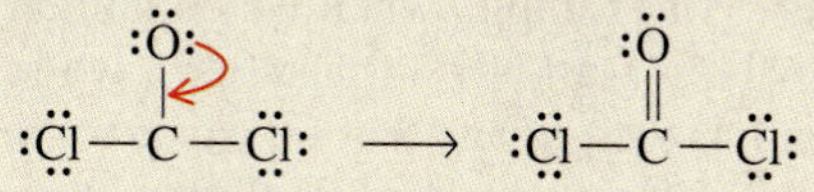

:Ö: (with Cl—C—Cl) ⟶ :Ö=C (Cl—C—Cl)

EXERCISE 9.7A

Write a plausible Lewis structure for carbonyl sulfide, COS.

EXERCISE 9.7B

Write the Lewis structure of nitryl fluoride, NO_2F.

EXAMPLE 9.8

Write a plausible Lewis structure for the chlorate ion, ClO_3^-.

SOLUTION

1 *Determine number of valence electrons.* For this polyatomic *anion*, the number of valence electrons that must appear in the Lewis structure are those for *one* Cl atom (Group 7A), *three* O atoms (Group 6A), and *one* additional electron to convey the −1 ionic charge: 7 + (3 × 6) + 1 = 26.

2 *Write a skeletal structure.* The electronegativities of the elements are Cl (3.0) and O (3.5). We expect the *less* electronegative chlorine atom to be the central atom and the skeletal structure to be

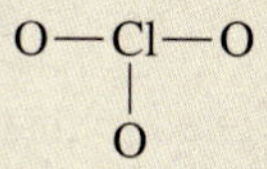

O—Cl—O
|
O

3 *Complete octets of terminal atoms.* Complete the octets of the oxygen atoms by placing three lone pairs of electrons on each.

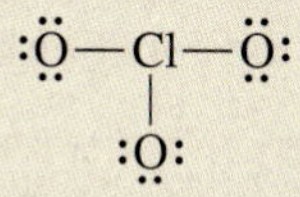

:Ö—Cl—Ö:
|
:O:

4 *Assign lone pair electrons to central atom(s).* We have now assigned 6 + 18 = 24 of the 26 valence electrons. Place the remaining two electrons as a lone pair on the central Cl atom.

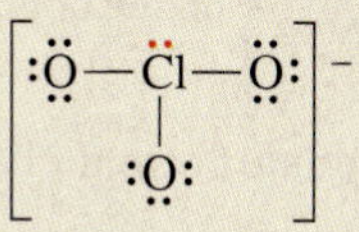

[:Ö—C̈l—Ö: | :O:]⁻

Each atom now has an octet. This is a plausible Lewis structure of ClO_3^-. As in Example 9.6, we don't need to go any further.

EXERCISE 9.8A

Write a plausible Lewis structure for the cyanide ion.

EXERCISE 9.8B

Write a plausible Lewis structure for the phosphonium ion, PH_4^+.

Formal Charge

You may be wondering about some of the ideas that we have used in our discussion of Lewis structures. For example, when we combined Lewis symbols to obtain Lewis structures of H_3O^+ and NH_4^+ (page 369), we introduced the idea of a coordinate covalent bond. Is there a way that we can recognize coordinate covalent bonding when we write a Lewis structure? Is there a reason why we choose the atom(s) of lowest electronegativity for the central atom(s) of a skeletal structure? Also, you may recall that in writing the Lewis structure of $COCl_2$ in Example 9.7, we had to form one double bond. Is there a reason why we made this a carbon-to-oxygen bond rather than a carbon-to-chlorine double bond?

We can use the concept of *formal charge* to shed some light on these issues and also to develop an additional tool to use in writing Lewis structures. To evaluate formal charges, we use a form of "electron bookkeeping" that goes beyond just accounting for the total number of valence electrons in a Lewis structure. **Formal charge** is the *difference* between the number of valence electrons in a free (uncombined) atom and the number of electrons assigned to that atom when bonded to others in a Lewis structure.

$$\text{Formal charge (FC)} = \begin{pmatrix}\text{number of valence}\\ \text{electrons in an}\\ \text{uncombined atom}\end{pmatrix} - \begin{pmatrix}\text{number of valence electrons}\\ \text{assigned to the bound atom}\\ \text{in a Lewis structure}\end{pmatrix}$$

We can get the first term on the right easily enough because the number of valence electrons for a main-group element is equal to its group number. To get the second term, we use the following simple rules to assign electrons in a Lewis structure to a given atom.

- All *lone-pair* electrons are assigned to that atom.
- Electrons in a bond are assigned equally to the two bonded atoms; half are assigned to one atom and half to the other.

We can now write these rules into the equation for formal charge.

$$\text{Formal charge (FC)} = \begin{pmatrix}\text{number of valence}\\ \text{electrons in the}\\ \text{uncombined atom}\end{pmatrix} - \begin{pmatrix}\text{number of lone-}\\ \text{pair electrons}\\ \text{on the bound atom}\end{pmatrix} - \frac{1}{2}\begin{pmatrix}\text{number of}\\ \text{electrons in}\\ \text{bonds to the atom}\end{pmatrix}$$

You can use this equation to evaluate formal charges, or you can use the simple method outlined in Figure 9.12.

Where we do find formal charges, we can designate them as shown below for structure (b) in Figure 9.12.

$$\overset{\ominus 2}{:\underset{\cdot\cdot}{C}}=\overset{\oplus 2}{S}=\underset{\cdot\cdot}{S}:$$

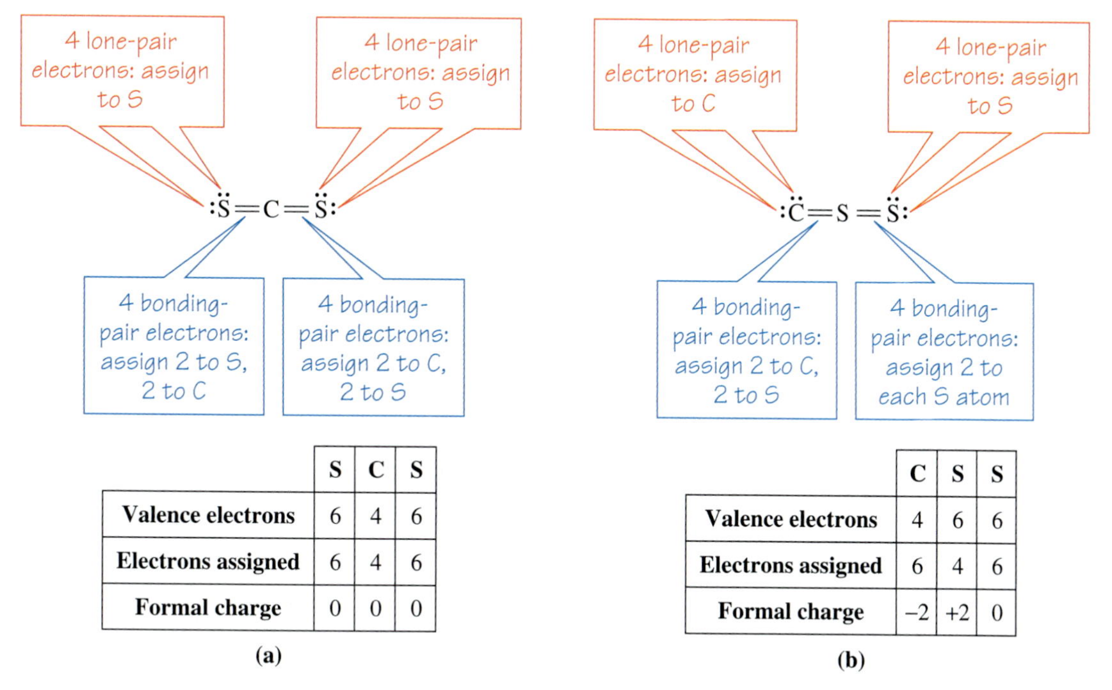

	S	C	S
Valence electrons	6	4	6
Electrons assigned	6	4	6
Formal charge	0	0	0

(a)

	C	S	S
Valence electrons	4	6	6
Electrons assigned	6	4	6
Formal charge	−2	+2	0

(b)

▲ **Figure 9.12 The concept of formal charge illustrated**
Lewis structure (a) is more plausible than (b) because it has no formal charges.

We use small encircled numbers for formal charges to distinguish them from authentic ionic charges. Formal charges are *hypothetical*; the individual atoms in a covalent molecule do not actually carry these charges.

● **ChemCDX** Representing Substances II

If each atom in a Lewis structure contributes half of the electrons to the bonds it forms, none of the atoms will have a formal charge. The existence of one or more formal charges in a Lewis structure indicates that one or more coordinate covalent bonds are involved. Often we can write two or more Lewis structures for a molecule or polyatomic ion, and we look for the Lewis structure with an optimal set of formal charges, which we can find with the help of the following ideas.

- Usually, the most plausible Lewis structure is one with no formal charges (that is, with formal charges of zero on all atoms).
- Where formal charges are required, they should be as small as possible, and negative formal charges should appear on the most electronegative atoms.
- Adjacent atoms in a structure should not carry formal charges of the same sign.
- The total of formal charges on the atoms in a Lewis structure must be *zero* for a neutral molecule and must equal the net charge for a polyatomic ion.

When we apply the formal charge concept to the $COCl_2$ molecule (Example 9.9), we confirm that the most likely Lewis structure has carbon, the least electronegative atom, as its central atom and that the double bond in the structure is carbon-to-oxygen and not carbon-to-chlorine.

EXAMPLE 9.9

In Example 9.7, we wrote the Lewis structure for the molecule $COCl_2$. Show that structure (a) is more plausible than (b) or (c).

(a)	(b)	(c)
:Ö=C—C̤̈l: with :C̤l: bonded to C	:Ö̤—C=C̤l: with :C̤l: bonded to C	:C̈=O—C̤̈l: with :C̤l: bonded to O

SOLUTION

We can use the equation on page 380 to assign formal charges to the atoms in each structure, and then we can apply the ideas listed above to evaluate the results.

Formal charge (FC) =

$$\begin{pmatrix}\text{number of valence}\\ \text{electrons in the}\\ \text{uncombined atom}\end{pmatrix} - \begin{pmatrix}\text{number of lone-}\\ \text{pair electrons}\\ \text{on the bound atom}\end{pmatrix} - \frac{1}{2}\begin{pmatrix}\text{number of}\\ \text{electrons in}\\ \text{bonds to the atom}\end{pmatrix}$$

Evaluating Structure (a)

For the O atom: FC = 6 valence e^- − 4 lone pair e^- − (1/2 × 4 bond pair e^-) = 0

For the C atom: FC = 4 valence e^- − 0 lone pair e^- − (1/2 × 8 bond pair e^-) = 0

For each Cl atom: FC = 7 valence e^- − 6 lone pair e^- − (1/2 × 2 bond pair e^-) = 0

Because none of the atoms has a formal charge, structure (a) is entirely plausible for $COCl_2$.

Evaluating Structure (b)

For the O atom: FC = 6 valence e^- − 6 lone pair e^- − (1/2 × 2 bond pair e^-) = −1

For the C atom: FC = 4 valence e^- − 0 lone pair e^- − (1/2 × 8 bond pair e^-) = 0

For the —Cl atom: FC = 7 valence e^- − 6 lone pair e^- − (1/2 × 2 bond pair e^-) = 0

For the =Cl atom: FC = 7 valence e^- − 4 lone pair e^- − (1/2 × 4 bond pair e^-) = +1

The formal charges in this structure, indicated below, add up to zero, as they should. However, a Cl atom has a +1 formal charge, even though it is more electronegative than a C atom, which has no formal charge. Structure (b) is not as plausible as structure (a).

(b) :Ö̤—C=C̤l: with :C̤l: bonded to C; formal charges: (−1) on O, (+1) on =Cl

Evaluating Structure (c)

For the C atom: FC = 4 valence e^- − 4 lone pair e^- − (1/2 × 4 bond pair e^-) = −2

For the O atom: FC = 6 valence e^- − 0 lone pair e^- − (1/2 × 8 bond pair e^-) = +2

For each Cl atom: FC = 7 valence e^- − 6 lone pair e^- − (1/2 × 2 bond pair e^-) = 0

This structure, shown next, is the least plausible of the three. It has the largest formal charges, and the O atom, with the highest electronegativity, has a positive rather than a negative formal charge.

(c) $\overset{\ominus 2}{:\ddot{C}}=\overset{\oplus 2}{\ddot{O}}-\ddot{\underset{..}{Cl}}:$ with $:\underset{..}{Cl}:$ bonded below O

Again, structure (a), which has no formal charges, is the most plausible Lewis structure for $COCl_2$.

EXERCISE 9.9A

Nitrosyl chloride, NOCl, is present in *aqua regia*, a mixture of concentrated nitric and hydrochloric acids capable of reacting with gold. Write the best Lewis structure that you can for NOCl.

EXERCISE 9.9B

Methyl acetate, CH_3COOCH_3, is used as a paint remover, a solvent for lacquers, and as an artificial flavoring. Write the best Lewis structure that you can for methyl acetate.

Resonance: Delocalized Bonding

When we apply the general strategy for Lewis structures to the ozone molecule, O_3, we get the following structure.

$$\overset{-1}{:\ddot{\underset{..}{O}}}-\overset{+1}{\ddot{O}}=\underset{..}{O}:$$

This structure implies that one of the oxygen-to-oxygen bonds should be a single bond and the other a double bond. However, experimental data show that the two oxygen-to-oxygen bonds are identical. The bonds are intermediate between a single and a double bond, and we cannot write a single Lewis structure consistent with this fact. The best we can do is to write some plausible Lewis structures and try to "average" these in our minds into a composite, or hybrid.

This description of O_3 is based on a theory called *resonance*. The **resonance** theory states that whenever a molecule or ion can be represented by two or more plausible Lewis structures that differ *only in the distribution of electrons*, the true structure is a composite, or hybrid, of them. The different plausible structures are called **resonance structures**. All the atoms are located in exactly the same place in each resonance structure; the only difference between them is the distribution of electrons among the atoms. The actual molecule or ion that is a hybrid of the resonance structures is called a **resonance hybrid**. One way to represent the resonance hybrid is to write each of the resonance structures and join them by a double-headed arrow. The resonance structures of ozone are shown below.

$$\overset{-1}{:\ddot{\underset{..}{O}}}-\overset{+1}{\ddot{O}}=\underset{..}{O}: \longleftrightarrow :\underset{..}{O}=\overset{+1}{\ddot{O}}-\overset{-1}{\ddot{\underset{..}{O}}}:$$

Although we might take the double-headed arrow to suggest that the structure of the O_3 molecule shifts back and forth from one resonance structure to the other, this is *not* the case. There is a single, actual structure—the resonance hybrid—which is a blend of the two resonance structures.

We can find an analogy to a resonance hybrid in the printing of this page. No *purple* ink is used in the printing, but we can get *purple* by blending *magenta* and *cyan*. The page is run through the press twice, first in magenta and then in cyan. Magenta and cyan are like resonance structures, and purple is the resonance hybrid.

APPLICATION NOTE

Near ground level, ozone, O_3, is an environmental pollutant found in smog. In the stratosphere, it provides a crucial shield against harmful ultraviolet radiation.

In molecules such as H_2O, NH_3, and CH_4, in which resonance is not involved, bonding electron pairs are considered to exist in fairly well defined regions between two atoms; the electrons are *localized*. In O_3, to produce oxygen-to-oxygen bonds that are intermediate between single and double bonds, some of the electrons in the resonance hybrid are *delocalized*. Delocalized electrons are bonding electrons that are spread out over several atoms. We can attempt to picture a resonance hybrid by using dotted lines to represent delocalized electrons. In the resonance hybrid of O_3, suggested below, the dotted line across the top of the structure represents four electrons, with an average of one electron on each terminal O atom and two on the central O atom.

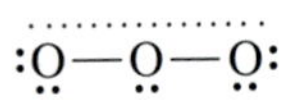

:O—O—O:

Resonance hybrid

EXAMPLE 9.10

Write three equivalent Lewis structures for the SO_3 molecule that conform to the octet rule, and describe how the resonance hybrid is related to the three structures.

SOLUTION

Whether resonance is involved or not, in writing a plausible Lewis structure we follow the five-step procedure introduced earlier. In fact, we often don't realize that resonance is involved until we have examined one of the resonance structures.

1 *Determine number of valence electrons.* There are $6 + (3 \times 6) = 24$ valence electrons.

2 *Write a skeletal structure.* The skeletal structure has sulfur, the atom of lower electronegativity, as the central atom.

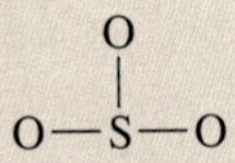

O
|
O—S—O

3 *Complete octets of terminal atoms.* Place three lone pairs of electrons on each O atom to complete its octet.

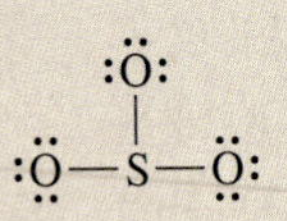

:Ö:
|
:Ö—S—Ö:

4 *Assign lone pair electrons to central atom(s).* Through step 3, we have already assigned all 24 valence electrons. We must go on to step 5.

5 *Form multiple bonds to complete octets of central atom(s).* The central atom shows only six valence electrons. Move a lone pair of electrons from a terminal O atom to form a double bond to the central S atom. Because the double bond can go to any one of the three terminal O atoms, we get three structures that differ only in the position of the double bond.

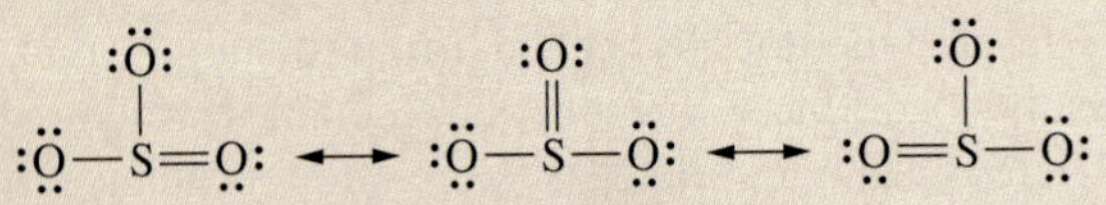

:Ö: | :Ö—S=O: ⟷ :O: ‖ :Ö—S—Ö: ⟷ :Ö: | :O=S—Ö:

Resonance structures

The SO_3 molecule is the resonance hybrid of these three resonance structures. The three sulfur-to-oxygen bonds are identical; each is intermediate between a single and a double bond.

EXERCISE 9.10

Write three Lewis structures for the nitrate ion, NO_3^-, and describe how the resonance hybrid is related to the three structures.

9.9 Molecules that Don't Follow the Octet Rule

Molecules made of atoms of the main-group elements generally have Lewis structures that follow the octet rule, but there are exceptions. These fall into three categories, each category readily identified by some structural characteristic.

Molecules with an Odd Number of Valence Electrons

In all Lewis structures so far, we have dealt with electron *pairs*, either as bonding pairs or lone pairs. However, in Lewis structures with an *odd* number of valence electrons, it is not possible to have all the electrons in pairs, nor is it possible to satisfy the octet rule for all of the atoms. Three examples are given below. They are nitrogen monoxide, NO, with $5 + 6 = 11$ valence electrons; nitrogen dioxide, NO_2, with $5 + 6 + 6 = 17$ valence electrons; and chlorine dioxide, ClO_2, with $7 + 6 + 6 = 19$ valence electrons.

$$\cdot\ddot{N}{=}\ddot{O}: \qquad :\ddot{O}{-}\dot{N}{=}\ddot{O}: \longleftrightarrow :\ddot{O}{=}\dot{N}{-}\ddot{O}: \qquad \cdot\ddot{O}{-}\ddot{Cl}{-}\ddot{O}: \longleftrightarrow :\ddot{O}{-}\ddot{Cl}{-}\ddot{O}\cdot$$

APPLICATION NOTE

NO and NO_2 are major components of smog. ClO_2 is made in multiton quantities and is used for bleaching flour and paper.

There are relatively few stable molecules with odd numbers of electrons. Most of the other species with odd numbers of electrons are fragments of molecules called **free radicals.** Generally, free radicals are highly reactive and have only a fleeting existence as intermediates in chemical reactions. An important free radical in the atmosphere is the *hydroxyl* radical ($\cdot OH$). We usually use a bold dot ($\cdot$) to represent an unpaired electron. The equation below shows the reaction of a hydroxyl radical with a methane molecule to produce a *methyl* radical.

$$\cdot OH(g) + CH_4(g) \longrightarrow \cdot CH_3(g) + H_2O(g)$$

APPLICATION NOTE

The free radicals $H\cdot$, $O\cdot$, $HO\cdot$, and $HO_2\cdot$ are all believed to be present as intermediates in the explosive reaction of hydrogen and oxygen gases: $2\,H_2 + O_2 \longrightarrow 2\,H_2O$

Molecules with Incomplete Octets

Sometimes when we write a Lewis structure, we don't seem to have enough electrons to give every atom a valence-shell octet. Such electron-deficient molecules generally have some unusual bonding characteristics and are often quite reactive. You can expect to encounter electron deficient molecules when the central atom is Be, B, or Al, but not much otherwise. As an example, let's consider the bonding of boron and fluorine. Boron atoms have three valence electrons, and fluorine atoms have seven. In the molecule boron trifluoride, the central boron atom shares its three valence electrons with three fluorine atoms.

$$\begin{array}{l} :\ddot{F}: \\ \;| \\ B{-}\ddot{F}: \\ \;| \\ :\ddot{F}: \end{array}$$

This structure accounts for all 24 valence electrons, but the central B atom is left with only six electrons, two short of a complete octet. Nevertheless, this structure is quite acceptable according to the formal-charge rules; all atoms have a formal charge of 0. We can also write a Lewis structure that conforms to the octet rule by using a boron-to-fluorine double bond.

```
        :F:
         |
   (-1) B=F:(+1)
         |
        :F:
```

In fact, there are three equivalent structures with boron-to-fluorine double bonds. We might strongly object to these structures because they place a positive formal charge on a fluorine atom, the most electronegative of all atoms. What, then, is the real structure of the BF_3 molecule? Experimental evidence suggests that the best representation is a hybrid of four resonance structures.

```
:F:             :F:             :F:             :F:
 |               |               |               ||
 B—F:   <—>      B=F:   <—>      B—F:   <—>      B—F:
 |               |               ||              |
:F:             :F:             :F:             :F:
```

The fact that the central boron atom is deficient in electrons is consistent with the observed high chemical reactivity of BF_3. For example, BF_3 readily reacts to form a coordinate covalent bond with a lone pair of electrons, such as those provided by a fluoride ion.

APPLICATION NOTE

BF_3 is an important industrial chemical, but its main uses are related to its electron-deficient nature, not to its boron or fluorine content. It is an important catalyst in organic chemical reactions.

```
                :F:              [      :F:      ]-
                 |               [       |       ]
:F:-     +       B—F:    —>      [  :F—B—F:      ]
                 |               [       |       ]
                :F:              [      :F:      ]
```

The measured boron-to-fluorine internuclear distance in BF_3 is much shorter than in the single-bonded BF_4^- ion (130 pm compared to 145 pm). As we shall see in Section 9.10, this shorter distance is consistent with some double bond character in the boron-to-fluorine bonds.

In conclusion, we might say that BF_3 exists as the resonance hybrid of the four resonance structures shown above, with perhaps the major contributor being the structure with an incomplete octet on the boron atom. This conclusion highlights two other important facets of resonance theory: (1) Structures that contribute to a resonance hybrid need not be completely equivalent (as they are in O_3 and SO_3), and (2) experimental evidence may indicate that one or more resonance structures make a greater contribution to the resonance hybrid structure than others do.

Structures with Expanded Valence Shells

The second-period elements carbon, nitrogen, oxygen, and fluorine nearly always obey the octet rule. Electron-deficient and odd-electron molecules are obvious exceptions. Because the valence shell of the second-period elements holds a maximum of eight electrons ($2s^2 2p^6$), these elements do not exceed an octet. However, for elements in the *third* and higher-numbered periods, the situation is different. Although the third period ends with argon ($3s^2 3p^6$), the third principal *shell* can hold up to 18 electrons ($3s^2 3p^6 3d^{10}$). We cannot write a Lewis structure based on the octet rule for the molecules PCl_5 and SF_6, because the octet rule allows for a maximum of only four bonds between a central and terminal atoms. In PCl_5, we need

10 electrons in five bonds and in SF_6, 12 electrons in six bonds. To accommodate more than 8 valence electrons for a central atom, we can use **expanded valence shells**, as in the following structures.

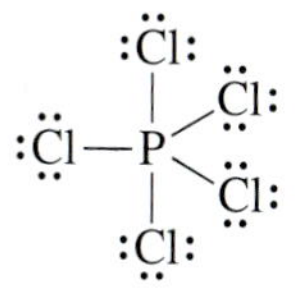

PCl_5 SF_6

Phosphorus pentachloride 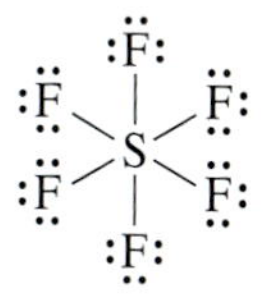   Sulfur hexafluoride

In some cases, even though we can write Lewis structures that conform to the octet rule, structures based on expanded valence shells appear to agree better with experimental data than do those based on the octet rule. Consider the cases of dichlorine monoxide, Cl_2O, and perchlorate ion, ClO_4^-. A Lewis structure of Cl_2O with O as the central atom has valence shell octets for all three atoms and no formal charges.

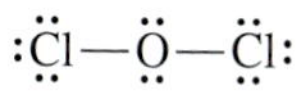

:Cl—O—Cl:

Let's assume that this plausible Lewis structure for Cl_2O corresponds to the normal chlorine-to-oxygen internuclear distance in a Cl—O single bond. Experimental evidence indicates this distance is 170 pm.

The Lewis structure for the perchlorate ion shown below conforms to the octet rule but has formal charges. The positive formal charge on the Cl atom is particularly high.

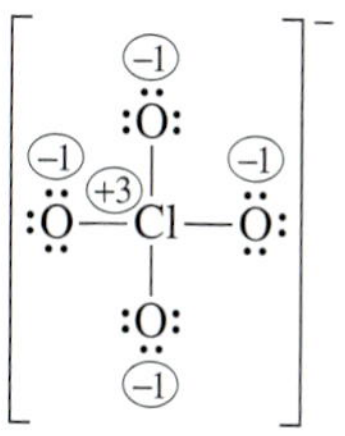

$[O^{(-1)}\text{—}Cl^{(+3)}(\text{—}O^{(-1)})_3]^-$

Experimental evidence indicates that the chlorine-to-oxygen internuclear distance in ClO_4^- is 144 pm. These shorter distances suggest that the Cl—O bond has some multiple bond character. With the concept of expanded valence shells, we can write a number of Lewis structures having double bonds and reduced formal charges. The true structure is a resonance hybrid of many structures, not all of which are equally important. One example of a possible contributing structure is shown below.

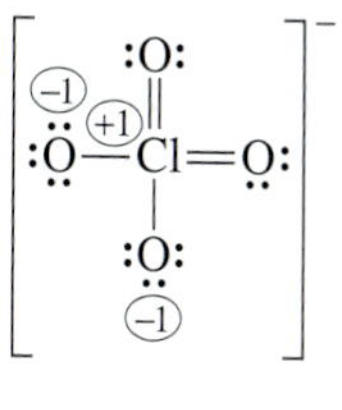

$[O^{(-1)}\text{—}Cl^{(+1)}(=O)_2\text{—}O^{(-1)}]^-$

Although we can make a good case for using formal charge and expanded valence shells to obtain the "best" Lewis structures when the central atoms are in the third period, there is evidence that simpler Lewis structures based on the octet rule may in fact be better. With the aid of high-speed computers, chemists can do calculations that assess the relative importance of various resonance structures to a resonance hybrid. A recent study* suggests that the octet structure with the high formal charge on Cl is the most important structure contributing to the resonance hybrid of ClO_4^-. In contrast, structures with expanded valence shells are of negligible importance. The study suggests that most of the remaining contributions to the resonance hybrid are due to "ionic" structures of the type shown here.

$$\left[\begin{array}{c} :\!O\!: \\ \| \\ :\!\ddot{O}\!-\!Cl^+ \quad :\!\ddot{O}\!:^{2-} \\ | \\ :\!\ddot{O}\!: \end{array} \right]^-$$

These ionic structures suggest a partial ionic character in the Cl—O bonds that can account for the shorter bond lengths. The chlorine-to-oxygen internuclear distance in ClO_4^- predicted by the computer model is 145 pm.

Our approach to these conflicting views will be to use expanded valence shells in cases where we are unable to write a Lewis structure without them (as with PCl_5 and SF_6). Otherwise, we will continue to rely mostly on the octet rule. In Chapter 10, we will find that this approach works quite well when it comes to predicting the shapes of molecules.

In Example 9.11, we find that an expanded valence shell may need to accommodate lone-pair electrons as well as bonding pairs. In Exercise 9.11B, you can't tell from the name and formula alone whether to use an expanded valence shell. However, you will find that you need to do so as you apply the general approach to writing a Lewis structure.

EXAMPLE 9.11

Write the Lewis structure for bromine pentafluoride, BrF_5.

SOLUTION

The formula BrF_5 tells us that we must use an expanded valence shell because five Br—F single bonds require a minimum of 10 electrons in the valence shell of Br.

1 *Determine number of valence electrons.* Both Br and F are in Group 7A. All the atoms in the structure have 7 valence electrons. The total number of valence electrons that must appear in the Lewis structure is $7 + (5 \times 7) = 42$.

2 *Write a skeletal structure.* The skeletal structure has Br, the atom of lowest electronegativity, as the central atom.

$$\begin{array}{c} F \\ | \\ F\!-\!Br\!\begin{array}{l} \diagup F \\ \diagdown F \end{array} \\ | \\ F \end{array}$$

3 *Complete octets of terminal atoms.* To complete the octets of the F atoms, we place three lone pairs of electrons around each one.

$$\begin{array}{c} :\!\ddot{F}\!: \\ | \\ :\!\ddot{F}\!-\!Br\!\begin{array}{l} \diagup \ddot{F}\!: \\ \diagdown \ddot{F}\!: \end{array} \\ | \\ :\!\ddot{F}\!: \end{array}$$

*L. Suidan, *et. al. J. Chem. Educ.* **72**, 583 (1995).

4 *Assign lone pair electrons to central atom(s).* Through step 3, we have assigned 40 electrons. Two remain to be placed. They can be shown as a lone pair on the bromine atom. In this representation, the Br atom has an expanded valence shell with 12 electrons.

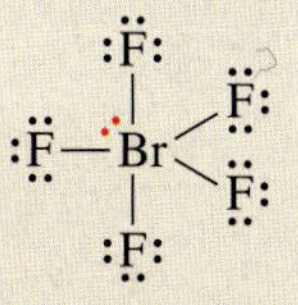

:F̤̈:
|
:F̤̈—B̈r(—F̤̈:)(—F̤̈:)
|
:F̤̈:

There is no need to proceed further. All electrons were placed in Step 4 in a plausible structure having no formal charges.

EXERCISE 9.11A

Write the Lewis structure of phosphorus trichloride.

EXERCISE 9.11B

Write the Lewis structure of **(a)** chlorine trifluoride and **(b)** sulfur tetrafluoride.

EXAMPLE 9.12—A Conceptual Example

Indicate the error involved in each of the following Lewis structures. Replace each by a more acceptable structure(s).

a. :C≡N: **b.** 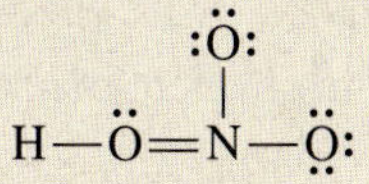 H—Ö=N(—Ö:)—Ö̤:

SOLUTION

a. This structure conforms to the octet rule, but let's back up to the first step in the general strategy for writing Lewis structures: assessing the total number of valence electrons. This number is 4 (from C) + 5 (from N) = 9, not the 10 electrons shown. The structure shown here is not for a molecular species but for the cyanide *ion*, CN^-. It should be written as $[:C{\equiv}N:]^-$.

b. The number of valence electrons required in the structure is $1 + 5 + (3 \times 6) = 24$. If we count the number of valence electrons represented by the given structure, we do indeed find 24. So far, there is no problem. When we assign formal charges to the atoms in the structure, we find them on all the atoms except H.

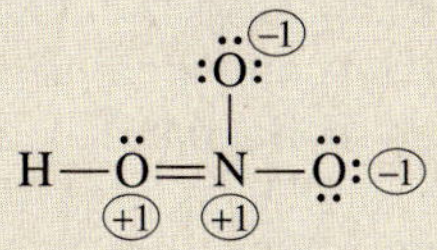

H—Ö(+1)=N(+1)(—Ö:(−1))—Ö̤:(−1)

Further, we have formal charges of the same sign (+1) on adjacent atoms, the N atom and one of the O atoms. We should be able to write a better structure. To do so, we need to have the nitrogen-to-oxygen double bond to a *terminal* O atom rather than another central atom. When we do this, we discover that there are *two equivalent structures*, each of which has only two atoms with formal charges. The correct structure is a resonance hybrid of the following resonance structures.

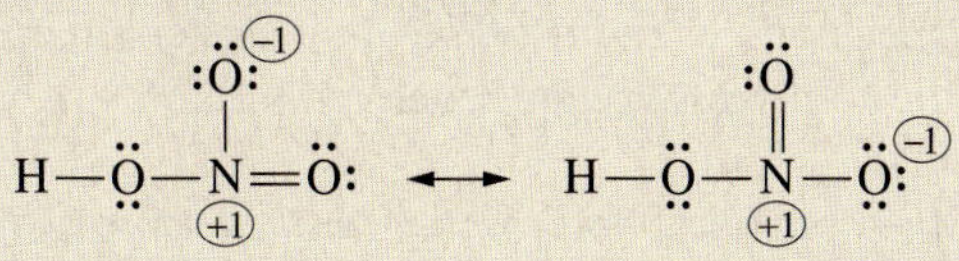

H—Ö̤—N(+1)(—Ö:(−1))=Ö: ⟷ H—Ö̤—N(+1)(=Ö:)—Ö̤:(−1)

EXERCISE 9.12

Only one of the following Lewis structures is correct. Identify that one, and indicate the error(s) in the others.

a. chlorine dioxide, :Ö—C̈l—Ö: (with lone pairs below each atom)

b. hydrogen peroxide, H—Ö—Ö—H (with lone pairs below each O)

c. dinitrogen difluoride, :F̈—N̈—N̈—F̈: (with lone pairs below each F)

9.10 Bond Lengths and Bond Energies

The electron charge density associated with shared electron pairs in covalent bonds is concentrated in the region between the nuclei of the bonded atoms. The greater the attractive forces between the positively charged nuclei and the bonding electrons, the more tightly the atoms are joined. Thus, shared electrons are the "glue" that binds atoms together in molecules and polyatomic ions. Generally, bond strengths vary for different combinations of atoms and for different numbers of shared electrons. The usual pattern for a bond between two particular atoms is that *the more electrons in the bond, the more tightly the atoms are held together*. Think of the electron charge between bonded atoms as partially neutralizing the repulsion between the positively charged nuclei. The greater the electron charge density in the bond, the closer the nuclei can approach each other.

The term **bond order** indicates whether a covalent bond is *single* (bond order = 1), *double* (bond order = 2), or *triple* (bond order = 3). When writing a Lewis structure, we need to take into account any experimental evidence that relates to bond order. Even without such information, we can often figure out probable bond orders from a plausible Lewis structure. Bond length and bond energy are related to bond order.

Bond length is the distance between the nuclei of two atoms joined by a covalent bond. Bond length depends on the particular atoms in the bond and on the bond order. For a given pair of atoms, there is little variation in bond length from one molecule to another. Thus, in the alkane hydrocarbons ethane, CH_3CH_3; propane, $CH_3CH_2CH_3$; and butane, $CH_3CH_2CH_2CH_3$, all the C—C single bonds have a length of 154 pm. Because the atoms in a double bond are more tightly bound than in a single bond, a double bond between two atoms is shorter than a single bond between the same two atoms. For example, the C=C bond length in ethylene, $H_2C{=}CH_2$, is 134 pm, compared to the 154-pm bond length in $H_3C{-}CH_3$. A triple bond is even shorter than a double bond between the same two atoms. The C≡C bond length in HC≡CH is 120 pm. Several single, double, and triple bond lengths are listed in Table 9.1.

Recall that we introduced the *single covalent radius*, a particular kind of atomic radius, in Section 8.7. Now you can appreciate its meaning: Because it is one-half the distance between the nuclei of identical atoms when they are joined by a *single* covalent bond, the single covalent radius is *one-half the bond length*. If you look again at Figure 8.8, you will see that the single covalent radius of an iodine atom is 133 pm. That value is one-half the bond length given in Table 9.1, that is, $1/2 \times 266$ pm. Furthermore, even for *unlike* atoms of comparable electronegativities, we can make the following *rough* generalization.

> *The length of the covalent bond joining unlike atoms is the sum of the covalent radii of the two atoms.*

TABLE 9.1 Some Representative Bond Lengths and Bond Energies

Bond	Bond Length, pm	Bond Energy, kJ/mol	Bond	Bond Length, pm	Bond Energy, kJ/mol
H—H	74	436	C—O	143	360
H—C	110	414	C=O	120	736[a]
H—N	100	389	C—Cl	178	339
H—O	97	464	N—N	145	163
H—S	132	368	N=N	123	418
H—F	92	565	N≡N	110	946
H—Cl	127	431	N—O	136	222
H—Br	141	364	N=O	120	590
H—I	161	297	O—O	145	142
C—C	154	347	O=O	121	498
C=C	134	611	F—F	143	159
C≡C	120	837	Cl—Cl	199	243
C—N	147	305	Br—Br	228	193
C=N	128	615	I—I	266	151
C≡N	116	891			

[a] The value for the C—O bond in CO_2 is considerably different: 799 kJ/mol.

When there is a significant electronegativity difference between the bonded atoms, we are likely to find exceptions to this generalization. The partial ionic character of polar covalent bonds, for example, produces greater bond strengths and *shorter* bond lengths than would otherwise be expected. Thus, the H—Cl bond length that we calculate from the H—H and Cl—Cl bond lengths is $\left[\left(\frac{1}{2} \times 74\right) + \left(\frac{1}{2} \times 199\right)\right]$ pm = 137 pm, but the measured bond length of the polar covalent bond in HCl is only 127.4 pm.

At times, experimentally determined bond lengths can help us to choose the best Lewis structure for a molecule. But also, as illustrated in Example 9.13, we can estimate a bond length from a plausible Lewis structure.

EXAMPLE 9.13

Estimate the indicated bond lengths.

a. The nitrogen-to-nitrogen bond in N_2H_4.

b. The Br-to-Cl bond in BrCl.

SOLUTION

a. Table 9.1 lists three nitrogen-to-nitrogen bond lengths. Which one do we use? First, let's draw a plausible Lewis structure for the molecule to see whether the nitrogen-to-nitrogen bond is single, double, or triple. Then we can select the appropriate value from Table 9.1.

```
   ..  ..
H—N—N—H
   |   |
   H   H
```

The nitrogen-to-nitrogen bond is a single bond, and therefore we expect its bond length to be 145 pm.

b. The bromine-to-chlorine bond length is not given in Table 9.1. Here, we can use the approximation that the covalent bond length is the sum of the covalent radii of the

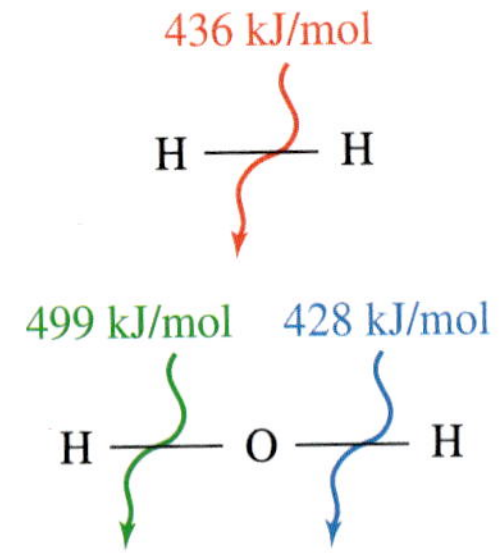

436 kJ/mol

H — H

499 kJ/mol 428 kJ/mol

H — O — H

▲ **Figure 9.13**
Some bond dissociation energies compared
The same quantity of energy (436 kJ/mol) is required to break all H—H bonds. In water, more energy is required to break the first O—H bond (499 kJ/mol) than to break the second (428 kJ/mol). The value of the O—H bond energy in Table 9.1 is an *average* value based on water and other compounds that contain the O—H bond.

atoms. To get the Lewis structure of BrCl, we imagine substituting one Br atom for one Cl atom in the Lewis structure of Cl_2.

$$:\ddot{\underset{..}{Br}}-\ddot{\underset{..}{Cl}}:$$

The BrCl molecule therefore contains a Br—Cl single bond. The length of the Br—Cl bond is approximately one-half the Cl—Cl bond length *plus* one-half the Br—Br bond length. From Table 9.1, we get $[(\frac{1}{2} \times 199) + (\frac{1}{2} \times 228)] =$ 214 pm. This estimated value is in excellent agreement with the measured value of 213.8 pm, suggesting that the small difference in electronegativity (0.2) between Br (2.8) and Cl(3.0) is not an important consideration.

EXERCISE 9.13A

Estimate the oxygen-to-fluorine bond length in OF_2.

EXERCISE 9.13B

Estimate the nitrogen-to-nitrogen bond length in nitramide, NH_2NO_2.

Bond Energy

Energy must be *absorbed* to break a covalent bond. We define **bond-dissociation energy (*D*)** as the quantity of energy required to break one mole of covalent bonds between atoms *in the gas phase.* These energies are generally expressed in kilojoules per mole of bonds (kJ/mol) and compiled in tables such as Table 9.1.

It is easy to understand bond-dissociation energies for *diatomic* molecules. Because there is only one bond (be it single, double, or triple) per molecule, we can represent bond-dissociation energy as an enthalpy change or a heat of reaction. The enthalpy change for the reverse reaction, in which a bond is formed, is the *negative* of the bond dissociation energy. For example,

$$\textit{Bond breaking} \quad Cl_2(g) \longrightarrow 2\,Cl(g) \quad \Delta H = D(Cl-Cl) = +243\text{ kJ/mol}$$

$$\textit{Bond forming} \quad 2\,Cl(g) \longrightarrow Cl_2(g) \quad \Delta H = -D(Cl-Cl) = -243\text{ kJ/mol}$$

With a polyatomic molecule, such as H_2O, the situation is different. As shown in Figure 9.13, the energy required to dissociate one mole of H atoms by breaking one O—H bond per $H_2O(g)$ molecule is different from the energy required to dissociate a second mole of H atoms by breaking the remaining O—H bond in OH(g). (The colored arrows in Figure 9.13 represent the breaking of a bond.)

$$H-OH(g) \longrightarrow H(g) + OH(g) \quad \Delta H = +499\text{ kJ/mol}$$

$$O-H(g) \longrightarrow H(g) + O(g) \quad \Delta H = +428\text{ kJ/mol}$$

The value for the O—H bond energy listed in Table 9.1 is the *average* of these two.

As we can readily see from Table 9.1, there is a relationship between bond order and bond-dissociation energy. A double bond between a given pair of atoms has a higher bond-dissociation energy than does a single bond between the same atoms, and a triple bond has a still higher bond-dissociation energy. Generally, the higher the bond order, the higher is the bond-dissociation energy for bonds between a given pair of atoms. However, Table 9.1 does not reflect another important factor. Bond-dissociation energies also depend on the *environment* of a given bond; that is, on the other atoms in the molecule that are near the bond. For example, the bond-dissociation energy of the O—H bond in H—O—H is somewhat different from that in H—O—O—H or in H_3C—O—H. For this reason, it is customary to use *average* values. An **average bond energy** is the average of the bond-dissociation energies for a number of different molecules containing the particular bond. In Table 9.1, the values for bonds in stable diatomic molecules, such as H—H, H—Cl, and Cl—Cl, are *bond-dissociation energies*. For other bonds, such as H—C, H—N, and H—O, the values are *average bond energies*. We will use average bond energies in the remainder of this discussion.

Calculations Involving Bond Energies

Let's imagine carrying out a gas-phase reaction in the following way. First, while keeping track of the energy required, we break all the bonds in the reactant molecules to produce a gas comprised of free, uncombined atoms. Then, we recombine these atoms into the product molecules and note how much energy is released.

$$\text{Gaseous reactants} \longrightarrow \text{gaseous atoms} \longrightarrow \text{gaseous products}$$

The sum of the enthalpy changes for breaking the old bonds and forming the new bonds is the enthalpy change, ΔH, for the reaction.

$$\Delta H = \Delta H_{\text{bonds broken}} + \Delta H_{\text{bonds formed}}$$

Because some of the bond energies (BE)* that we use in calculations are *average* bond energies rather than the more precise bond dissociation energies, we expect the calculated enthalpy changes to be only approximately correct. That is, the calculations are based on the following relationships.

$$\Delta H_{\text{bonds broken}} \approx \text{BE(reactants)}$$

$$\Delta H_{\text{bonds formed}} \approx -\text{BE(products)}$$

$$\Delta H \approx \text{BE(reactants)} - \text{BE(products)}$$

In Figure 9.14, we illustrate these ideas from a molecular viewpoint by considering the formation of gaseous hydrazine, N_2H_4, from nitrogen and hydrogen gases. In Example 9.14, we use bond energies to estimate the enthalpy of formation for $N_2H_4(g)$.

EXAMPLE 9.14

Use bond energies from Table 9.1 to estimate the enthalpy of formation of gaseous hydrazine. Compare the result with the value of $\Delta H°_f[N_2H_4(g)]$ from Appendix C.

SOLUTION

To choose the proper bond energies from Table 9.1, we need to write Lewis structures for the substances involved in the reaction.

```
                              H   H
                              |   |
:N≡N:  +  2 H—H   ⟶          :N—N:
                              |   |
                              H   H
```

Now we can assess ΔH for bonds broken and bonds formed.

*These energies are more properly called *bond enthalpies*, but we will use the more familiar term *bond energies*.

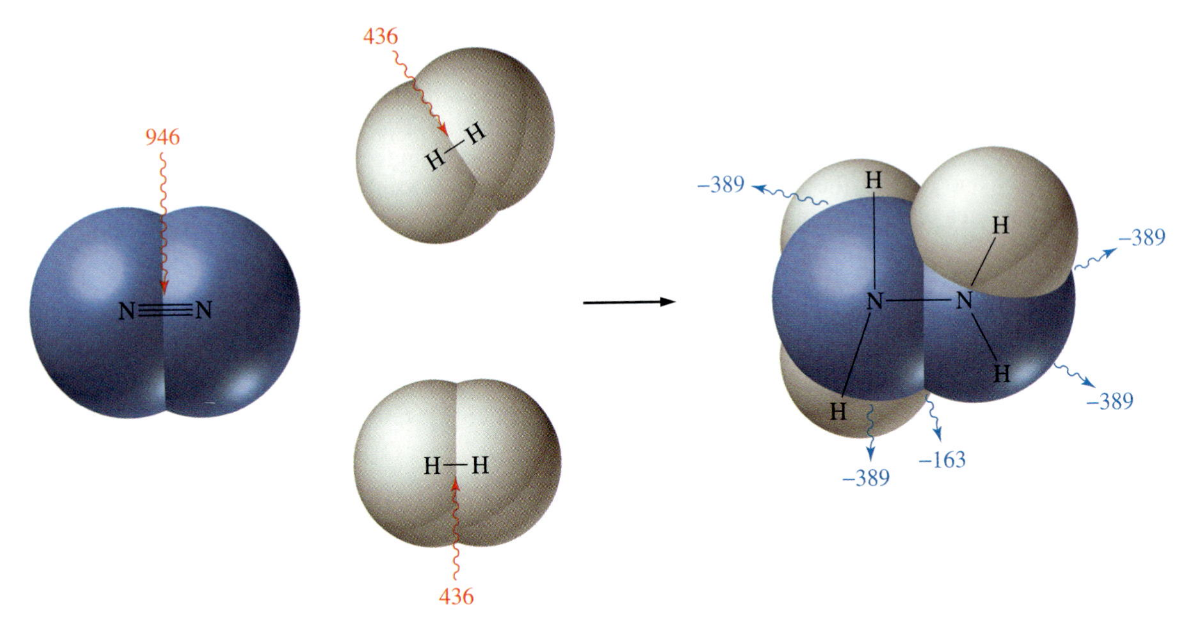

▲ **Figure 9.14 Visualizing bond breakage and bond formation in a reaction**
The values shown, in kilojoules per mole, are the amounts of energy absorbed in breaking bonds in the reactants, N_2 and H_2, and released in forming bonds in the product, N_2H_4.

ΔH for bond breakage:

$$\begin{aligned} 1 \text{ mol N}\equiv\text{N bonds} &= 946 \text{ kJ} \\ 2 \text{ mol H}-\text{H bonds} &= (2 \times 436) \text{ kJ} = 872 \text{ kJ} \\ \text{Total for bonds broken} &= 1818 \text{ kJ} \end{aligned}$$

ΔH for bond formation:

$$\begin{aligned} 1 \text{ mol N}-\text{N bonds} &= -163 \text{ kJ} \\ 4 \text{ mol N}-\text{H bonds} &= 4 \times (-389) \text{ kJ} = -1556 \text{ kJ} \\ \text{Total for bonds formed} &= -1719 \text{ kJ} \end{aligned}$$

$$\Delta H = \Delta H_{\text{bonds broken}} + \Delta H_{\text{bonds formed}}$$

$$\Delta H = 1818 \text{ kJ} - 1719 \text{ kJ} = 99 \text{ kJ}$$

For the reaction in which gaseous hydrazine is formed from its elements, the estimate that $\Delta H°_f[N_2H_4(g)] = 99$ kJ/mol $N_2H_4(g)$ agrees reasonably well with the value of 95.40 kJ/mol listed in Appendix C.

EXERCISE 9.14

Estimate ΔH for the reaction

$$C_2H_6(g) + Cl_2(g) \longrightarrow C_2H_5Cl(g) + HCl(g)$$

The result of Example 9.14 suggests that, as expected, enthalpy changes calculated from average bond energies are not as accurate as those calculated from standard enthalpies of formation (Section 6.7). We use average bond energies mostly in applications where the necessary thermodynamic data are not available or where rough estimates will suffice. Consider the reaction of hydroxyl radicals, $\cdot OH$, with methane in the atmosphere.

$$\cdot OH + CH_4 \longrightarrow \cdot CH_3 + H_2O \quad \Delta H = ?$$

To assess ΔH for this reaction, let's simplify the method of Example 9.14 a bit. Because *four* C—H bonds would be broken in CH_4 and *three* would be formed in $\cdot CH_3$, the net effect would be to *break one* C—H bond. Because *one* O—H bond would be broken in $\cdot OH$ and *two* would be formed in H_2O, the net effect would be to *form one* O—H bond. The enthalpy change for the reaction, then, is simply

$$\Delta H = \text{BE(C—H)} - \text{BE(O—H)} = +414 \text{ kJ} - 464 \text{ kJ} = -50 \text{ kJ}.$$

Unsaturated Hydrocarbons

Some of the most representative examples of covalent bonding are found in organic compounds. In Chapter 2, we learned to write structural formulas for the alkanes. In these compounds, which have the general formula C_nH_{2n+2}, all the bonds are single bonds, and there are no lone-pair electrons. The Lewis structures of the alkanes are the same as their structural formulas, as in the following representation of ethane.

```
     H   H
     |   |
  H—C—C—H
     |   |
     H   H
```

Because alkane molecules are unable to form bonds to additional atoms, the alkanes are called *saturated* hydrocarbons.

Multiple bonds are common in organic compounds. Collectively, hydrocarbons with double or triple bonds between carbon atoms are called **unsaturated hydrocarbons**. In contrast to the saturated hydrocarbons, unsaturated hydrocarbons are able to bond to additional atoms under the appropriate conditions. Carbon atoms can also form double bonds with nitrogen, oxygen, and sulfur atoms, and a triple bond with nitrogen atoms.

9.11 Alkenes and Alkynes

The **alkenes** (note the *-ene* ending) are hydrocarbons with molecules having carbon-to-carbon double bonds. *Simple* alkenes have just one double bond in their molecules; they have the general formula C_nH_{2n}. Hydrocarbons with this formula make up a homologous series parallel to the straight-chain and branched chain alkanes, C_nH_{2n+2}. The simplest alkene is *ethene*, C_2H_4. When we construct its Lewis structure from Lewis symbols, we can see that the ethene molecule must have a carbon-to-carbon double bond.

$$2\,\cdot\dot{\underset{\cdot}{C}}\cdot \; + \; 4\,H\cdot \longrightarrow \begin{matrix} & H & H & \\ & | & | & \\ H- & C & C & -H \end{matrix} \longrightarrow \begin{matrix} & H & & H & \\ & | & & | & \\ H- & C & = & C & -H \end{matrix}$$

We can also use a structural formula or condensed structural formula to represent ethene. The structural formula, which is the same as the Lewis structure, shows that the double bond is shared by the two carbon atoms and does not involve the hydrogen atoms. The condensed structural formula does not make this point quite as obvious, but it is the easier formula to set in type, requires less space on a printed page, and is consequently more widely used.

▲ **Figure 9.15**
Molecular model of ethene (ethylene)

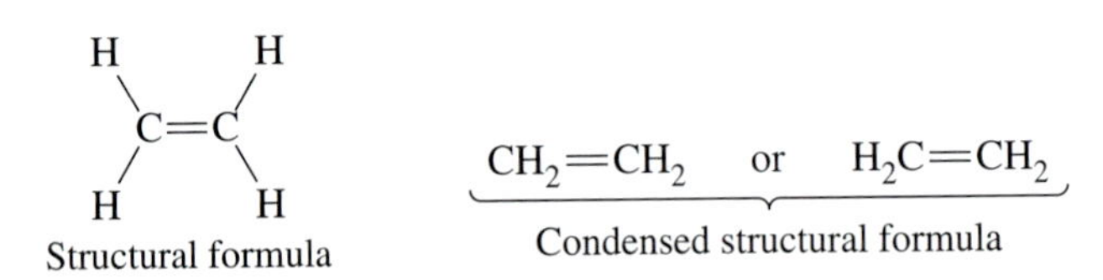

A space-filling molecular model of ethene is shown in Figure 9.15.

Many simple alkenes have common, nonsystematic names. Ethene, the two-carbon alkene, is most often called *ethylene*. The three-carbon alkene, propene, $CH_3CH{=}CH_2$, is frequently called *propylene*. Four different compounds (isomers) have the molecular formula C_4H_8, and the number of isomers increases rapidly with the number of carbon atoms. Whenever there are large numbers of isomers, it is difficult to remember the many possible common names, and chemists resort to the IUPAC system described in Appendix C.

APPLICATION NOTE

Ethylene is produced in greater quantity than any other organic chemical in the United States; propylene is second in production. Both are used in the manufacture of polymers (see page 397).

Hydrocarbon molecules with double bonds—*unsaturated* molecules—are chemically reactive. They can use two electrons from one of the electron pairs in a double bond to form bonds to two other atoms. In the resulting product, all the covalent bonds are single bonds and the molecules are *saturated*. A commercially important reaction in which the H atoms of hydrogen are added to the double-bonded carbon atoms in unsaturated molecules is called a *hydrogenation* reaction. The reaction requires a *catalyst*, a substance that greatly increases the speed of a reaction while remaining unchanged by the reaction. The catalysts most often used in hydrogenation reactions are nickel, platinum, and palladium. In the hydrogenation of ethene represented below, the nickel catalyst is indicated above the yield arrow. The formation of the additional carbon-to-hydrogen bonds is suggested by the red bond lines.

$$\underset{\text{Ethene}}{\begin{matrix} H & & H \\ & C{=}C & \\ H & & H \end{matrix}} \; + \; H{-}H \xrightarrow{\text{Ni}} \underset{\text{Ethane}}{\begin{matrix} & H & H & \\ & | & | & \\ H- & C & -C & -H \\ & | & | & \\ & H & H & \end{matrix}}$$

The product of this reaction, ethane, is an alkane with the same carbon skeletal structure as the original alkene, ethene.

Alkynes

In the double bond of an alkene molecule, carbon atoms share two pairs of electrons. Carbon atoms can also share three pairs of electrons, forming triple bonds. Hydrocarbons with molecules containing carbon-to-carbon triple bonds are called **alkynes.** The simplest alkyne is

$$\mathrm{H{:}C{:}{:}{:}C{:}H} \quad \text{or} \quad \mathrm{H{-}C{\equiv}C{-}H}$$

▲ **Figure 9.16**
Molecular model of ethyne (acetylene)

The common name of this compound is *acetylene* (Figure 9.16). The IUPAC nomenclature parallels that for alkenes, except that the family ending is *-yne* rather than *-ene*. The official name for acetylene is *ethyne*.

Alkyne molecules are unsaturated; they can add atoms and form additional bonds just as alkene molecules can. However, because either one or two pairs of electrons in the triple bond can participate, it is possible to add twice as much of a reagent to an alkyne as to an alkene. For example, in hydrogenation under carefully controlled conditions, acetylene can react with one molecule of hydrogen to form ethene.

$$\mathrm{H{-}C{\equiv}C{-}H} + \mathrm{H{-}H} \xrightarrow{\mathrm{Ni}} \mathrm{H_2C{=}CH_2}$$

More likely, however, an acetylene molecule will react with two molecules of hydrogen to form first ethylene and then ethane. The overall reaction is

$$\mathrm{H{-}C{\equiv}C{-}H} + 2\,\mathrm{H{-}H} \xrightarrow{\mathrm{Ni}} \mathrm{H{-}CH_2{-}CH_2{-}H}$$

9.12 Polymers

If we attempt to write a Lewis structure for ethylene by simply bringing together two Lewis symbols of carbon, $\cdot\dot{\underset{\cdot}{\mathrm{C}}}\cdot$, and four of hydrogen, H·, this is what we might get as a first attempt.

$$\cdot\mathrm{CH_2{-}CH_2}\cdot$$

Of course, we would reject this structure because the C atoms have incomplete octets. Suppose, however, that instead of creating a second bond between the carbon atoms, we consider joining two more such structures (red) to the one above. We would get this.

$$\cdot\mathrm{CH_2{-}CH_2{-}CH_2{-}CH_2{-}CH_2{-}CH_2}\cdot$$

Fats and Oils: Hydrogenation

Chemically, fats are esters. Fats are called **triglycerides** because a molecule of fat is made by the reaction of three molecules of long-chain carboxylic acids called *fatty acids* (RCOOH) with a molecule of the *trihydroxy alcohol* glycerol, $CH_2OHCHOHCH_2OH$. You can find a structural formula and a space-filling model of a typical triglyceride on page 265. In general, a triglyceride is called a *fat* if it is a solid at 25 °C and an *oil* if it is a liquid at that temperature. An important factor in these differences in melting points is the degree of *un*saturation—that is, the number of carbon-to-carbon double bonds—of the constituent fatty acids. The more double bonds a molecule has, the more unsaturated it is.

Saturated fatty acids contain no carbon-to-carbon double bonds; stearic acid, $C_{17}H_{35}COOH$, is a typical example.

$$\mathrm{CH_3CH_2CH_2CH_2CH_2CH_2CH_2CH_2CH_2CH_2CH_2CH_2CH_2CH_2CH_2CH_2CH_2COOH}$$

Stearic acid (saturated)

Monounsaturated fatty acids contain one carbon-to-carbon double bond per molecule; oleic acid, $C_{17}H_{33}COOH$, is a common one.

$$\mathrm{CH_3CH_2CH_2CH_2CH_2CH_2CH_2CH_2CH{=}CHCH_2CH_2CH_2CH_2CH_2CH_2CH_2COOH}$$

Oleic acid (monounsaturated)

Polyunsaturated fatty acids are those that have two or more carbon-to-carbon double bonds per molecule. Linolenic acid, $C_{17}H_{29}COOH$, with three such bonds, is a typical example.

$$\mathrm{CH_3CH_2CH{=}CHCH_2CH{=}CHCH_2CH{=}CHCH_2CH_2CH_2CH_2CH_2CH_2CH_2COOH}$$

Linolenic acid (polyunsaturated)

Saturated fats contain a high portion of saturated fatty acids; the fat molecules have relatively few carbon-to-carbon double bonds. *Polyunsaturated fats* (oils) incorporate mainly unsaturated fatty acids; these fat molecules have many carbon-to-carbon double bonds. Triglycerides obtained from animal sources are usually solids, whereas those of plant origin are generally oils. Therefore, we speak of *animal fats* and *vegetable oils*.

Unsaturated oils can be converted to more saturated ones by hydrogenation. Margarine, a butter substitute, and vegetable shortening, a lard substitute, consist of vegetable oils that have been partially hydrogenated. (If all the bonds were hydrogenated, the product would become hard and brittle like tallow, a fat used to make soaps and candles.) If reaction conditions are properly controlled, it is possible to prepare a fat with a desirable physical consistency (soft and pliable). In this manner, inexpensive and abundant vegetable oils (cottonseed, corn, soybean) are converted into oleomargarine and cooking fats (Crisco®, for example).

The reddish brown color of bromine vapor (left) is removed when a strip of uncooked bacon is a placed in the beaker (right). Bromine molecules react with unsaturated fats in the bacon in a reaction similar to a hydrogenation reaction.

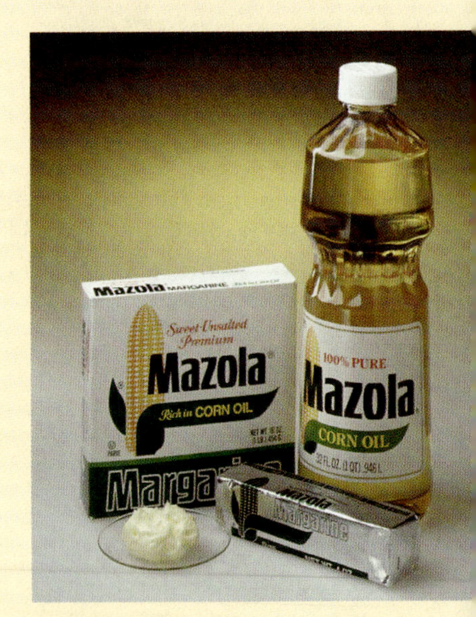

About 85% of the fatty acid residues in corn oil are unsaturated. To convert the oil to a solid (margarine), it is hydrogenated. The process converts some of the unsaturated molecules in the oil to saturated molecules in the product.

Fats and oils are broken down in the digestive tract to smaller molecules that can be absorbed into the body and used for energy. A typical fat yields about 37 kJ per gram. Like nearly all biochemical reactions, fat digestion is made possible by biological catalysts called *enzymes*. Without the enzymes, the reactions would not take place.

Now, it is even possible to get "caloric-free" fats. Olestra, an ester made from fatty acids and sucrose (a sugar), is one such fake fat. Sold by Procter and Gamble under the trade name OLEAN®, olestra is used as a substitute for fats and oils in cooking potato chips and other snack foods. If oxidized in a bomb calorimeter (page 251), olestra would yield about the same quantity of energy per gram as an ordinary fat. However, unlike ordinary fats, the olestra molecules are not broken down by human enzymes and are therefore not absorbed into the body. Because it is not digested, olestra contributes no calories to foods cooked in it. It provides the rich taste and sensory effect of fat-fried foods without the calories of those fried in real fat. However, olestra in large quantities can cause gastrointestinal discomfort and diarrhea in some people. Also, because it is not absorbed, it carries fat-soluble vitamins A, D, E, and K right through the digestive tract, lessening their absorption. Commercial olestra has enough of these vitamins added to offset their lower absorption.

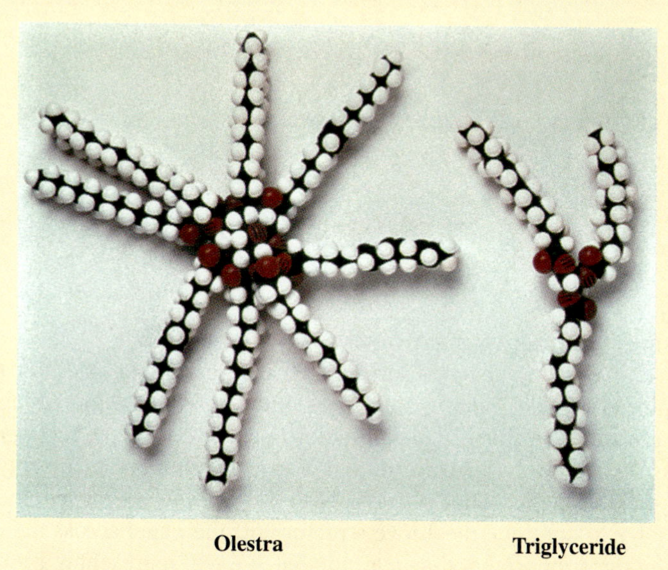

Olestra **Triglyceride**

Sucrose polyester (olestra) is an ester of the alcohol groups on a sucrose molecule with six, seven, or eight fatty acid molecules. Unlike ordinary fats (triglycerides), which are broken down by digestive enzymes, the olestra molecules are not. Olestra is not digested or absorbed; it contributes no calories to the diet.

▶ **Figure 9.17**
The formation of polyethylene
In the synthesis of polyethylene, many monomer units ($CH_2{=}CH_2$) join together to form giant polymer molecules. In this computer-generated representation, the yellow dots indicate a new bond forming as an ethylene molecule (upper left) is added to the growing polymer chain.

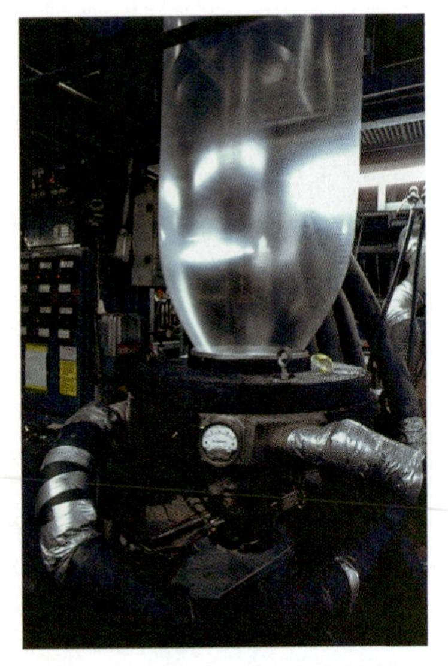

A giant bubble of a tough, transparent plastic film emerges from the die of an extruding machine. The film is used in packaging, consumer products, and food-service operations.

And if we add two more structures (blue), we get

$$\cdot\overset{\displaystyle H}{\underset{\displaystyle H}{\overset{|}{\underset{|}{C}}}}-\overset{\displaystyle H}{\underset{\displaystyle H}{\overset{|}{\underset{|}{C}}}}-\overset{\displaystyle H}{\underset{\displaystyle H}{\overset{|}{\underset{|}{C}}}}-\overset{\displaystyle H}{\underset{\displaystyle H}{\overset{|}{\underset{|}{C}}}}-\overset{\displaystyle H}{\underset{\displaystyle H}{\overset{|}{\underset{|}{C}}}}-\overset{\displaystyle H}{\underset{\displaystyle H}{\overset{|}{\underset{|}{C}}}}-\overset{\displaystyle H}{\underset{\displaystyle H}{\overset{|}{\underset{|}{C}}}}-\overset{\displaystyle H}{\underset{\displaystyle H}{\overset{|}{\underset{|}{C}}}}-\overset{\displaystyle H}{\underset{\displaystyle H}{\overset{|}{\underset{|}{C}}}}-\overset{\displaystyle H}{\underset{\displaystyle H}{\overset{|}{\underset{|}{C}}}}\cdot$$

We could repeat the process many times over (Figure 9.17).

We have just described a process in which a large number of simple molecules called **monomers** (Greek *monos*, single; and *meros*, parts) are joined together into a giant, long-chain molecule called a **polymer** (Greek *poly*, many, and *meros*, parts). The process is called **polymerization**. In this case, the monomers are ethylene molecules and the polymer is called *polyethylene*. To accurately represent the polymer, we would have to continue the above structure off the page, across the room, and (perhaps) out the door. Rather than attempt that, we will represent it through the repeating unit

$$\left[-\overset{\displaystyle H}{\underset{\displaystyle H}{\overset{|}{\underset{|}{C}}}}-\overset{\displaystyle H}{\underset{\displaystyle H}{\overset{|}{\underset{|}{C}}}}-\right]_n$$

Here n indicates the number of monomer units joined together in the polymer; typically n has a value from several hundred to more than one thousand for polyethylene. Even though it is called polyethylene, the polymer is *not* an alkene. The double bonds of the alkene starting material are converted to single bonds.

A few common alkene monomers and the polymers made from them are listed in Table 9.2. In Chapter 24, we will provide some additional details about poly-

TABLE 9.2 A Selection of Addition Polymers

Monomer	Polymer	Polymer Name	Some Uses
$H_2C{=}CH_2$	$[-CH_2-CH_2-]_n$	Polyethylene	Plastic bags, bottles, toys, electrical insulation
$H_2C{=}CH{-}CH_3$	$[-CH_2-CH(CH_3)-]_n$	Polypropylene	Indoor-outdoor carpeting, bottles
$H_2C{=}CH{-}Cl$	$[-CH_2-CHCl-]_n$	Poly(vinyl chloride), PVC	Plastic wrap, simulated leather (Naugahyde®), compact discs, garden hoses
$F_2C{=}CF_2$	$[-CF_2-CF_2-]_n$	Polytetrafluoroethylene, Teflon®	Nonstick coating for cooking utensils, electrical insulation

merization, such as how the double bond in ethylene "opens up" to get the polymerization started and how unpaired electrons are eliminated to terminate a polymer chain.

Summary

Lewis symbols of main-group elements are related to their locations in the periodic table. An ionic bond can be represented by combining the Lewis symbols of cations and anions.

The net energy decrease in the formation of an ionic crystal from its gaseous ions is the lattice energy. The lattice energy and enthalpy of formation of an ionic compound, together with other atomic and molecular properties, can be related through a Born-Haber cycle.

A covalent bond forms by the sharing of an electron pair between atoms. In a Lewis structure of a molecule, electron pairs are either bonding pairs or lone pairs. Usually, each atom in a structure acquires a noble-gas electron configuration, which for most atoms is a valence-shell octet of electrons.

In a covalent bond between atoms of different electronegativity (EN), electrons are displaced toward the atom with the higher EN. Electronegativity values are related to positions of the elements in the periodic table. In terms of electronegativity differences, chemical bonds vary from nonpolar (zero or very small EN difference) to polar covalent to ionic (large EN difference).

In some cases of covalent bonding, one atom appears to provide both electrons in the bonding pair; the bond is coordinate covalent. In some cases, bonded atoms may share more than one pair of electrons between them, giving rise to *multiple* covalent bonding.

One strategy for writing plausible Lewis structures for molecules or polyatomic ions involves two general tasks: (1) writing the skeletal structure and (2) distributing the valence electrons of the bonded atoms in a particular way. It may also be necessary to use the system of "electron bookkeeping" known as *formal charge*.

Key Terms

alkene (9.11)
alkyne (9.11)
average bond energy (9.10)
bond-dissociation energy, D (9.10)
bonding pair (9.6)
bond length (9.10)
bond order (9.10)
central atom (9.8)
chemical bond (page 360)
coordinate covalent bond (9.6)
covalent bond (9.2, 9.6)
crystal (9.3)
double bond (9.6)
electronegativity (EN) (9.7)
expanded valence shell (9.9)
formal charge (9.8)
free radical (9.9)
ionic bond (9.2, 9.3)
lattice energy (9.5)
Lewis structure (9.6)
Lewis symbol (9.2)
lone pair (9.6)
monomer (9.12)
multiple bond (9.6)
nonpolar (9.7)
octet rule (9.2)
polar (9.7)
polymer (9.12)
polymerization (9.12)
resonance (9.8)
resonance hybrid (9.8)
resonance structure (9.8)
single bond (9.6)
skeletal structure (9.8)
terminal atom (9.8)
triglyceride (9.11)
triple bond (9.6)
unsaturated hydrocarbon (page 395)

In the phenomenon of resonance, two or more Lewis structures have the same skeletal structure but different bonding arrangements. The best description of the actual structure, the resonance hybrid, is a combination of plausible resonance structures. The resonance hybrid should conform to experimental data, such as bond lengths and/or bond energies, if these data are available.

Exceptions to the octet rule are found in odd-electron molecules and molecular fragments called free radicals. A few structures appear to have too few electrons to complete all the octets. Some structures appear to have too many. In the latter case, a central atom may employ an expanded valence shell with five, six, or even seven electron pairs.

Bond order is the number of bonds between a pair of atoms: single, double, or triple. Bond length, which in some cases can be related to the appropriate atomic radii, is the internuclear distance between two bonded atoms. Bond-dissociation energy is the energy required to break one mole of bonds in a gaseous species; its value depends on (1) the atoms in the bond, (2) the bond order, and (3) the particular molecule in which the bond is found. Tabulated values are often averaged over a number of molecules containing the same bond. Bond-dissociation energies and average bond energies can be used to estimate enthalpy changes of reactions.

Unsaturated hydrocarbon molecules have one or more multiple bonds between carbon atoms. Alkenes have double bonds, and alkynes triple bonds. A characteristic reaction of alkenes and alkynes is the ability to add atoms by converting multiple bonds into single bonds. Some molecules with multiple bonds undergo polymerization, a reaction in which small molecules (monomers) join together in large numbers to produce a giant molecule (polymer).

Review Questions

1. What are the essential propositions of the Lewis theory of chemical bonding?

2. What is the *octet* rule in Lewis's theory? Why do some elements follow a *duet* rule instead?

3. According to the Lewis theory, why do helium and neon not form chemical bonds.

4. What is the structural difference between a sodium ion and a neon atom? What is the similarity between them?

5. What are the structural differences between chlorine atoms, chlorine molecules, and chloride ions?

6. Write the Lewis symbol for an atom of each of the following elements. You may use a periodic table.

(a) sodium **(d)** bromine
(b) oxygen **(e)** calcium
(c) silicon **(f)** arsenic

7. Explain how Lewis symbols, *spdf* notation, and orbital diagrams differ in their representation of electron spin.

8. Use *spdf* notation *and* Lewis symbols to represent the electron configuration of each of the following.

(a) K^+ **(d)** Cl^-
(b) S^{2-} **(e)** Mg^{2+}
(c) Al^{3+} **(f)** N^{3-}

9. Use Lewis symbols to represent formation of the ionic compound between

(a) calcium and bromine atoms
(b) barium and oxygen atoms
(c) aluminum and sulfur atoms

10. Use Lewis symbols to represent an ionic compound of a Group 1A element in which all the ions obey the *duet* rule.

11. What is meant by the lattice energy of an ionic compound? Describe how lattice energies can be evaluated indirectly from other measured properties.

12. Use Lewis symbols to show the sharing of electrons between iodine atoms to form an iodine molecule. Label all electron pairs as bonding pairs or lone pairs.

13. Use Lewis symbols to show the sharing of electrons between a hydrogen atom and a fluorine atom. Label ends of the molecule with symbols that indicate polarity.

14. What is the electronegativity of an atom? How does it differ from electron affinity?

15. If atoms of the two elements in each set below are joined by a covalent bond, which atom will more strongly attract the electrons in the bond?

(a) N and S **(c)** As and F
(b) B and Cl **(d)** S and O

16. Using only the periodic table (inside front cover), indicate which element in each set is more electronegative.

(a) Br or F **(c)** Cl or As
(b) Br or Se **(d)** N or H

17. Classify the following bonds as ionic or covalent. For those bonds that are covalent, indicate whether they are polar or nonpolar.

(a) KF
(b) IBr
(c) MgS
(d) NO
(e) CaO
(f) NaBr
(g) Br_2
(h) F_2
(i) HCl

18. How many single covalent bonds do each of the following atoms usually form in molecules that have only single covalent bonds? You may refer to the periodic table.

(a) H
(b) C
(c) O
(d) F
(e) N
(f) Br

19. What is a coordinate covalent bond? Use Lewis structures to show the formation of a coordinate covalent bond between BF_3 and F^-.

20. Name four elements that readily form double covalent bonds and two that form triple covalent bonds.

21. List the three main types of compounds that are exceptions to the octet rule. Given an example of each.

22. Where in the periodic table do you expect to find the elements whose atoms can exhibit expanded valence shells when forming covalent bonds?

23. Describe the phenomenon of resonance. What is the difference between a resonance structure and a resonance hybrid?

24. Are all carbon-to-hydrogen bonds of the same strength? Is the same amount of energy required to break the first and second of these bonds in methane, CH_4? Explain.

25. What are unsaturated hydrocarbons? Which of the following compounds are unsaturated hydrocarbons?

(a) $CHCCH_3$
(b) $CH_3CH_2CHCH_2$
(c) $CHCCH_2CH_3$
(d) $CH_3(CH_2)_3CH_3$

26. Which of the compounds in Question 25 are alkenes? Which are alkynes?

27. Define each of the following:

(a) monomer
(b) polymer
(c) polymerization

28. What is the essential structural feature of the alkene monomer molecules that enables them to polymerize?

Problems

Ions and Ionic Bonding

29. Which of the following ions have noble-gas electron configurations? What are the electron configurations of the ions that do not?

(a) Cr^{3+}
(b) Sc^{3+}
(c) Zn^{2+}
(d) Te^{2-}
(e) Zr^{4+}
(f) Cu^+

30. Use noble-gas-core abbreviated *spdf* notation to give the electron configuration for the most likely simple ion formed by each of the following elements.

(a) Ba
(b) K
(c) Se
(d) I
(e) N
(f) Te

31. Use Lewis symbols to represent the following ionic compounds.

(a) KI
(b) barium fluoride
(c) Rb_2S
(d) aluminum oxide

32. Use Lewis symbols to represent the following ionic compounds.

(a) CaO
(b) potassium bromide
(c) $BaCl_2$
(d) strontium nitride

Energetics of Ionic Bond Formation

33. The lattice energy of sodium fluoride is −914 kJ/mol NaF. Use this value and values from Section 9.5 (including Example 9.3) to determine the enthalpy of formation of NaF(s). Compare your result with the value listed in Appendix C.

34. The lattice energy of potassium chloride is −701 kJ/mol KCl. The enthalpy of sublimation of K(s) is 89.24 kJ/mol, and the first ionization energy of K(g) is 419 kJ/mol. Use these values and values from Section 9.5 to determine the enthalpy of formation of KCl(s). Compare your result with the value listed in Appendix C.

35. The enthalpy of formation of cesium bromide is

$$Cs(s) + \frac{1}{2}Br_2(l) \longrightarrow CsBr(s) \quad \Delta H_f^\circ = -405.8 \text{ kJ/mol}$$

The enthalpy of sublimation of cesium is 76.1 kJ/mol, and the enthalpy of vaporization of liquid bromine is

$$Br_2(l) \longrightarrow Br_2(g) \quad \Delta H^\circ = 30.9 \text{ kJ/mol}$$

Use these data and data from Tables 8.4, 8.5, and Appendix C to calculate the lattice energy of CsBr(s).

36. The enthalpy of formation of sodium iodide is

$$Na(s) + 1/2\ I_2(s) \longrightarrow NaI(s) \quad \Delta H_f^\circ = -288\ kJ/mol$$

The enthalpy of sublimation of iodine is

$$I_2(s) \longrightarrow I_2(g) \quad \Delta H^\circ = +62\ kJ/mol$$

Use these data and data from Tables 8.4, 8.5, and Appendix C to calculate the lattice energy of NaI(s).

37. Use data from this chapter, tables in Chapter 8, and Appendix C, together with an enthalpy of sublimation of lithium of 159.4 kJ/mol, to calculate the lattice energy of lithium oxide.

38. Use data from this chapter, tables in Chapter 8, and Appendix C, together with an enthalpy of sublimation of calcium of 178.2 kJ/mol and an enthalpy of vaporization of liquid bromine

$$Br_2(l) \longrightarrow Br_2(g) \quad \Delta H^\circ = 30.9\ kJ/mol$$

to calculate the lattice energy of calcium bromide.

Lewis Structures

39. Write Lewis structures for the simplest covalent molecules formed by the following atoms, assuming that the octet rule (duet rule for hydrogen) is followed in each case.
(a) P and H **(b)** C and F

40. Write Lewis structures for the simplest covalent molecules formed by the following atoms, assuming that the octet rule (duet rule for hydrogen) is followed in each case.
(a) Si and H **(b)** N and Cl

41. Write plausible Lewis structures for the following covalent molecules:
(a) CH_3OH **(d)** N_2H_4
(b) CH_2O **(e)** COF_2
(c) NH_2OH **(f)** PCl_3

42. Write plausible Lewis structures for the following covalent molecules:
(a) NF_3 **(d)** CH_3NH_2
(b) C_2H_2 **(e)** H_2SiO_3
(c) C_2H_4 **(f)** HCN

Electronegativity: Polar Covalent Bonds

43. Without referring to figures or tables in the text (other than the periodic table), arrange each of the following sets of atoms in their expected order of increasing electronegativity.
(a) B, F, N **(c)** C, O, Ga
(b) As, Br, Ca

44. Without referring to figures or tables in the text (other than the periodic table), arrange each of the following sets of atoms in their expected order of increasing electronegativity.
(a) I, Rb, Sb **(c)** Cl, P, Sb
(b) Cs, Li, Na

45. Use differences in electronegativity values to arrange each of the following sets of bonds in order of increasing polarity. Use the symbols δ+ and δ− to indicate partial charges, if any, in the bonds.
(a) Cl—F, F—F, Br—F, H—F, I—F
(b) H—Br, H—Cl, H—F, H—H, H—I

46. Use differences in electronegativity values to arrange each of the following sets of bonds in order of increasing polarity. Use the symbols δ+ and δ− to indicate partial charges, if any, in the bonds.
(a) H—C, H—F, H—H, H—N, H—O
(b) C—Br, C—C, C—Cl, C—F, C—I

Formal Charges

47. Assign formal charges to each atom in the following structures.

(a) 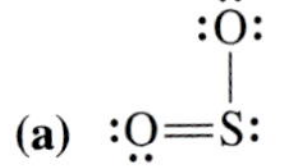

(c) [H—Ö—Ö:]⁻

(b) :C≡O:

(d) H—C≡N:

48. Assign formal charges to each atom in the four structures to the right.

(a) $[:\ddot{O}-\ddot{S}(-\ddot{O}:)-\ddot{O}:]^{2-}$

(b) [:N̈=N=N̈:]⁻

(c) $[BF_4]^-$ (:F̈—B—F̈: with :F̈: above and below B)

(d) ·N=Ö: with :Ö: bonded above N

49. Assign formal charges to each atom in the following structures. Based on these formal charges, which is the preferred Lewis structure for this ion, the cyanate ion?

(a) [:C̈=N=Ö:]$^-$ **(b)** [:N≡C—Ö̤:]$^-$

50. Assign formal charges to each atom in the following structures. Based on these formal charges, which is the preferred Lewis structure for this molecule, the dinitrogen monoxide molecule?

(a) :N≡N—Ö̤: **(b)** :N̈=N=Ö:

Resonance Structures

51. Write two equivalent resonance structures for the bicarbonate ion, $HOCO_2^-$. Describe the resulting resonance hybrid structure for the bicarbonate ion.

52. Write three equivalent resonance structures for the carbonate ion, CO_3^{2-}.Describe the resulting resonance hybrid structure for the carbonate ion.

53. Write Lewis structures for **(a)** nitrous acid and **(b)** nitric acid. In which of these substances is resonance more important? Explain.

54. Write Lewis structures for **(a)** acetic acid and **(b)** acetate ion. In which of these substances is resonance more important? Explain.

More Lewis Structures

55. Write the simplest Lewis structure for each of the following molecules. Comment on any unusual features of the structures.

(a) NO **(c)** BCl_3

(b) ClF_3 **(d)** SeF_4

56. Write the simplest Lewis structure for each of the following molecules. Comment on any unusual features of the structures.

(a) ClO_2 **(c)** $Be(CH_3)_2$

(b) IF_5 **(d)** XeF_6

57. Write Lewis structures for the following molecules and anions. Where appropriate, use the concepts of formal charge, expanded valence shells, and resonance to choose the most likely structure(s).

(a) SSF_2 **(d)** CN^-

(b) I_3^- **(e)** SF_5^-

(c) H_2CO_3 **(f)** BrO_3^-

58. Write Lewis structures for the following molecules and anions. Where appropriate, use the concepts of formal charge, expanded valence shells, and resonance to choose the most likely structure(s).

(a) HNO_3 **(d)** ICl_2^-

(b) IF_4^- **(e)** CH_2SF_4

(c) XeO_4 **(f)** IO_4^-

59. Write Lewis structures for the following organic molecules.

(a) isopropyl alcohol

(b) methanoic (formic) acid

(c) dimethyl ether

60. Write Lewis structures for the following organic molecules.

(a) methylacetylene

(b) dimethylamine

(c) propionaldehyde

61. Indicate what is wrong with each of the following Lewis structures. Replace each with a more acceptable structure.

(a) [:S̤̈—C=N̈:]$^-$

(b) :Ö=N=Ö:

(c) :C̤̈l—N̈=C̈l: (with :C̤̈l: bonded below N)

62. Indicate what is wrong with each of the following Lewis structures. Replace each with a more acceptable structure.

(a) [:Ö̤—N(—:Ö:)—Ö̤:]$^-$

(b) H—N̤(—H)=C=N:

(c) H—N̤̈—C≡O:

63. A carbon-hydrogen-oxygen compound is 53.31% C, 11.18% H, and 35.51% O. Write a Lewis structure based on the empirical formula of this compound. Is it a satisfactory structure? If not, substitute one that is.

64. A carbon-hydrogen-nitrogen compound is 39.97% C, 13.42% H, and 46.61% N. Write a Lewis structure based on the empirical formula of this compound. Is it a satisfactory structure? If not, substitute one that is.

Bond Lengths and Bond Energies

65. Use data from Table 9.1 to estimate bond lengths for the following bonds. Do you think your estimate is likely to be too high or too low? Explain.

(a) I—Cl (b) C—F

66. Use data from Table 9.1 to estimate bond lengths for the following bonds. Do you think your estimate is likely to be too high or too low? Explain.

(a) O—Cl (b) N—I

67. On page 391 we compared the experimentally determined bond length of HCl with a value determined from atomic radii. For which of the hydrogen halides would you expect the *smallest* percent difference between the calculated and experimental values of the bond length: HF, HCl, HBr, or HI? Explain your reasoning.

68. A diatomic interhalogen molecule XZ has a bond between two different halogen elements (F, Cl, Br, I). In a manner similar to that of Problem 67, for which of the interhalogen compounds XZ would you expect to find the *greatest* percent difference between the calculated and experimental values of the bond length? Explain.

69. Dinitrogen difluoride has a nitrogen-to-nitrogen bond length of 123 pm and nitrogen-to-fluorine bond lengths of 141 pm. Write a Lewis structure consistent with these data.

70. Hydrogen peroxide has an oxygen-to-oxygen bond length of 147.5 pm and two oxygen-to-hydrogen bonds with lengths of 95.0 pm. Write a Lewis structure consistent with these data.

71. Use bond energies from Table 9.1 to estimate the enthalpy change (ΔH) for the reaction

$$H_2(g) + F_2(g) \longrightarrow 2\,HF(g)$$

72. Use bond energies from Table 9.1 to estimate the enthalpy change (ΔH) for the reaction

$$CH_4(g) + Cl_2(g) \longrightarrow CH_3Cl(g) + HCl(g)$$

73. *Without doing detailed calculations*, indicate whether you expect the following reaction to be endothermic or exothermic.

$$C_3H_8(g) + Cl_2(g) \longrightarrow C_3H_7Cl(g) + HCl(g)$$

74. *Without doing detailed calculations*, indicate whether you expect the following reaction to be endothermic or exothermic.

$$N_2H_4(g) + H_2(g) \longrightarrow 2\,NH_3(g)$$

75. A reaction occurring in the stratosphere that helps to maintain the heat balance on Earth is that of ozone and atomic oxygen.

$$O_3(g) + O(g) \longrightarrow 2\,O_2(g) \quad \Delta H = -391.9\text{ kJ}$$

Use this information and data from Table 9.1 to obtain an estimate of the oxygen-to-oxygen bond energy in the O_3 molecule.

76. The oxidation of carbon monoxide to carbon dioxide by oxygen can be represented by the equation

$$2\,CO(g) + O_2(g) \longrightarrow 2\,CO_2(g) \quad \Delta H = -566\text{ kJ}$$

Use this information, together with Table 9.1 and its footnote, to obtain a value for the bond dissociation energy of the CO molecule.

77. Use bond energies from Table 9.1 and an enthalpy of formation from Appendix C to determine the dissociation energy of the bond in NO(g).

78. Use an enthalpy of formation from Appendix C, two average bond energies from Table 9.1, and Hess's law to obtain a value for the enthalpy of formation of the ethyl radical, $\cdot C_2H_5$.

$$C(\text{graphite}) + 5/2\,H_2(g) \longrightarrow \cdot C_2H_5(g) \quad \Delta H^\circ_f = ?$$

Alkenes and Alkynes

79. Give the condensed structural formula for each of the following.

(a) ethene (c) propyne

(b) 1-butene (d) 2-pentyne

80. Give the condensed structural formula for each of the following:

(a) ethyne (c) propene

(b) 1-butyne (d) 3-hexyne

81. Write equations to represent the reaction of **(a)** hydrogen with ethene and **(b)** two molecules of hydrogen with ethyne.

82. Write equations to represent each of the following reactions.

(a) hydrogen with 1-butene

(b) one molecule of hydrogen with ethyne

83. Determine the enthalpy change for the following reaction by using **(a)** average bond energies and **(b)** enthalpy of formation data. Compare the results.

$$C_2H_4(g) + H_2(g) \longrightarrow C_2H_6(g) \quad \Delta H = ?$$

84. Determine the enthalpy change for the following reaction by using **(a)** average bond energies and **(b)** enthalpy of formation data. Compare the results.

$$C_2H_4(g) + H_2O(g) \longrightarrow CH_3CH_2OH(g) \quad \Delta H = ?$$

Polymers

85. With the aid of Table 9.2, draw a section of polymer chain that is at least eight carbon atoms long for each of the following.
 (a) polyethylene
 (b) poly(vinyl chloride)
 (c) polypropylene

86. Draw four-monomer-unit sections of the polymers formed from the following monomers.
 (a) tetrafluoroethylene, $CF_2{=}CF_2$
 (b) vinylidene chloride, $H_2C{=}CCl_2$
 (c) acrylonitrile, $H_2C{=}CH{-}CN$

Additional Problems

87. The Lewis structures and structural formulas of the alkane hydrocarbons are identical. Explain why this is the case, but why the same statement is not true for all organic compounds.

88. Both the oxidation state and formal charge concepts deal with the *arbitrary* assignment of electrons between bonded atoms. Describe several ways in which these concepts differ. Can you think of any ways in which they are similar? Explain.

89. Two different molecules have the formula C_2H_6O. Write Lewis structures for the two molecules.

90. Draw four resonance structures for the oxalate ion, $C_2O_4^{2-}$.

91. Draw the structural formulas of five hydrocarbons that have the formula C_4H_6. (*Hint:* Use multiple bonds, cyclic structures, and combinations of these.)

92. The aldehyde propynal has the formula HCCCHO. Write a Lewis structure for propynal and estimate all the bond lengths in the molecule.

93. Hydrogen azide, HN_3, is an acid in aqueous solutions, and its salts are quite reactive (sodium azide, NaN_3, is used in air-bag systems, and lead azide, $Pb(N_3)_2$, is a detonator). The structure of hydrogen azide is indicated below. Write a Lewis structure(s) consistent with the data given.

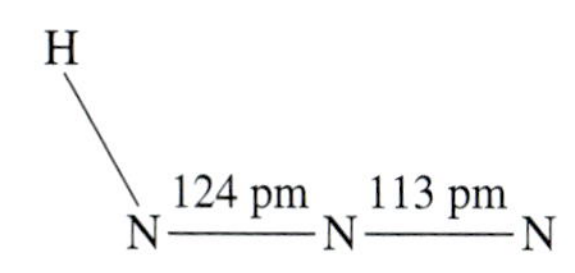

94. Estimate the nitrogen-to-fluorine and nitrogen-to-oxygen bond lengths in the molecule nitryl fluoride, NO_2F. To do so, draw plausible resonance structures for this molecule and use data from Table 9.1.

95. A 0.507-g sample of a gaseous oxide of carbon having 47.04% O by mass occupies a volume of 184 mL at 25 °C and 752 mmHg. The sample of oxide reacts with water to produce as the sole product an acid having a molar mass of 104 g/mol and requiring 29.52 mL of 0.5050 M NaOH for its complete neutralization. **(a)** Write an equation for the reaction of the oxide with water, and **(b)** write plausible Lewis structures for the reactants and products of this reaction.

96. The free radical $\cdot$ClO is intimately involved in reactions that lead to the destruction of ozone in the stratosphere. Use an enthalpy of formation from Appendix C, a bond energy from Table 9.1, a chlorine-to-oxygen bond energy of 243 kJ/mol, and Hess's law to obtain a value for the enthalpy of formation of $\cdot$ClO.

97. The following reaction has been observed in the atmosphere.

$$CO(g) + O_3(g) \longrightarrow CO_2(g) + O_2(g)$$

Use a bond-dissociation energy of 1072 kJ/mol for CO(g), data from Table 9.1 and its footnote, and data from Appendix C to obtain a value of the O-to-O bond energy in the O_3 molecule. Also, obtain an estimate based on the Lewis structure(s) of O_3 and data from Table 9.1. How well do the two values agree?

98. The enthalpy of vaporization of carbon is

$$C(graphite) \longrightarrow C(g) \qquad \Delta H^\circ = 717 \text{ kJ}$$

Use this value, the enthalpy of formation of methane, and the bond-dissociation energy of hydrogen to calculate a value for the carbon-to-hydrogen single bond energy. Compare your result with the value listed in Table 9.1.

99. Calculate ΔH_f° for the hypothetical compound MgCl(s) by using the following data:

Enthalpy of sublimation of Mg:	+150 kJ/mol
First ionization energy of Mg(g):	+738 kJ/mol
Enthalpy of dissociation of $Cl_2(g)$:	+243 kJ/mol
Electron affinity of Cl(g):	−349 kJ/mol
Lattice energy of MgCl(s):	−676 kJ/mol

100. Use the data of Problem 99, I_2 of Mg(g) (1451 kJ/mol), and the lattice energy of $MgCl_2(s)$ (−2500 kJ/mol) to calculate ΔH_f° for $MgCl_2(s)$. Explain why you would expect $MgCl_2(s)$ to be more stable than MgCl(s).

101. Hydrogen forms ionic compounds with certain metals in which it is present as hydride ion, H^-. Determine the electron affinity of H through a Born-Haber calculation, using a lattice energy of −812 kJ/mol NaH and data from page 366, Table 9.1, and Appendix C.

CHAPTER

10

Bonding Theory and Molecular Structure

CONTENTS

● **ChemCDX**
When you see this icon, there are related animations, demonstrations, and exercises on the CD accompanying this text.

The paramagnetism of O_2 is clearly evident in this photograph. We were unable to explain this property of oxygen with the Lewis theory. In this chapter, we consider a theory that does explain it.

In this chapter, we will use quantum mechanics to extend our knowledge of chemical bonding and molecular structure beyond what we learned using the Lewis theory (Chapter 9). Recall that in Chapter 7 we used modern quantum mechanics to extend our knowledge of atomic structure beyond the Bohr model.

With added insights into chemical bonding, we will be able to explain some matters that we can't explain with simple Lewis structures. For example, we will be able to give plausible explanations of the following observations.

- The three atoms in a CO_2 molecule are arranged in a straight line, whereas the three atoms in H_2O form an angular or bent molecule.
- Chloroform ($CHCl_3$) molecules are polar, whereas carbon tetrachloride (CCl_4) molecules are nonpolar.
- There is only one butane molecule $(CH_3CH_2—CH_2CH_3)$, whereas there are two isomers of 2-butene, $CH_3CH{=}CHCH_3$.
- The Ne_2 molecule does not exist, but the F_2 molecule and the Na_2 molecule do exist.

- The O_2 molecule has a double covalent bond, yet has unpaired electrons that account for its paramagnetism (recall page 371).

Molecular Geometry

Let's start by clarifying just what we mean by the shape of a molecule. Like every sample of matter, a molecule occupies space and is, therefore, three-dimensional in shape. A simple verbal description of the overall shape of a molecule can be difficult to formulate and is actually not the most useful one. Instead, the **molecular geometry**, or the shape, of a molecule is described by the geometric figure formed when the *atomic nuclei* of the molecule are joined by the appropriate straight lines.

Diatomic molecules like O_2 have only two nuclei. Because the two points representing the nuclei determine a straight line, we describe the molecular geometry as *linear*. If the three nuclei of a *triatomic* molecule fall along a straight line, as they do in CO_2, the molecular geometry is also linear. If the three nuclei are *not* in a straight line, the molecular geometry is *angular*, or *bent*. The water molecule is a good example of an angular molecule. Figure 10.1 shows the framework of the H_2O molecule. It is set up to emphasize two key features of the geometric structure of a molecule: bond lengths and bond angles. We have this framework in mind when we say that the molecule is angular or bent. The space-filling molecular model is superimposed on the framework model to show how the molecule actually occupies space. Most polyatomic molecules have molecular geometries that are more complex than linear or angular, as we shall see shortly.

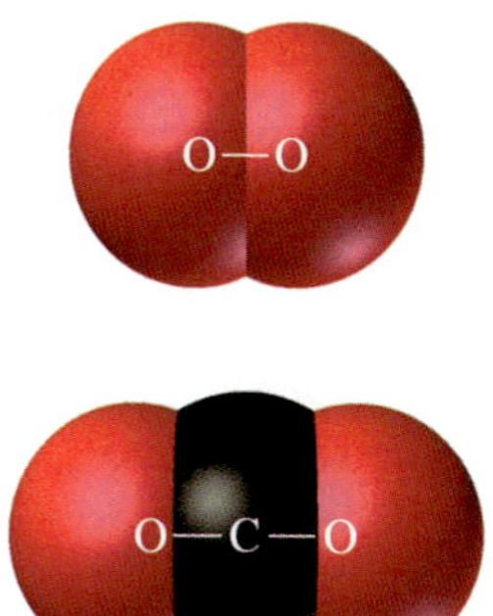

Actual molecular geometry can be determined only by experiment, but in many cases, we can make fairly good predictions of the shapes of molecules and polyatomic ions. We now consider a method for making these predictions.

10.1 Valence-Shell Electron-Pair Repulsion (VSEPR) Method

As the name implies, the **valence-shell electron-pair repulsion**, or **VSEPR**, method is based on the idea that pairs of valence electrons in bonded atoms repel one another. These mutual repulsions push electron pairs as far from one another as possible. This minimizes the energy of repulsion and represents the lowest energy configuration of the molecule or polyatomic ion. Depending on the number of valence electron pairs and whether they are bonding pairs or lone pairs, the terminal atoms adopt specific orientations about the central atom to which they are bonded. This gives the molecule or polyatomic ion a distinctive shape.

Electron-Group Geometries

We can predict the shapes of molecules and polyatomic ions by the VSEPR method. Before we begin to do so, however, we need to broaden our view of repulsions to involve *groups* of valence electrons rather than just pairs. An **electron group** is any collection of valence electrons localized in a region around a central atom that exerts repulsions on other groups of valence electrons. An electron group can be any of the following.

- A single unpaired electron
- A lone pair of electrons
- *One* bonding pair of electrons in a single covalent bond
- *Two* bonding pairs of electrons in a double covalent bond
- *Three* bonding pairs of electrons in a triple covalent bond

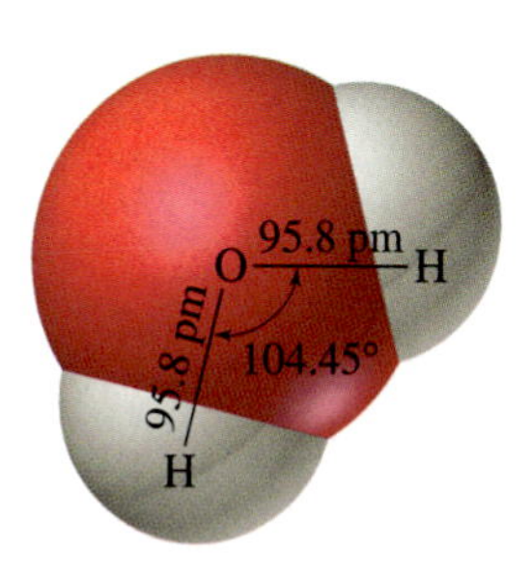

▲ **Figure 10.1**
Molecular geometry and a space-filling model of H_2O
Each of the two O—H bonds has a length of 95.8 pm, and the angle between the two bonds is 104.45°. The molecular geometry of H_2O is angular or bent.

Most commonly there are either 2, 3, 4, 5, or 6 electron groups about a central atom. The mutual repulsions among electron groups lead to an orientation of the groups that we call the **electron-group geometry**. The electron-group geometries are shown below.

- 2 electron groups: *linear*
- 3 electron groups: *trigonal planar*
- 4 electron groups: *tetrahedral*
- 5 electron groups: *trigonal bipyramidal*
- 6 electron groups: *octahedral*

Figure 10.2 offers an analogy that may help you to visualize electron-group repulsions. When balloons are twisted together, the lobes of the balloons seek positions that cause the least interference from other lobes, just as electron groups tend to remain as far apart as possible.

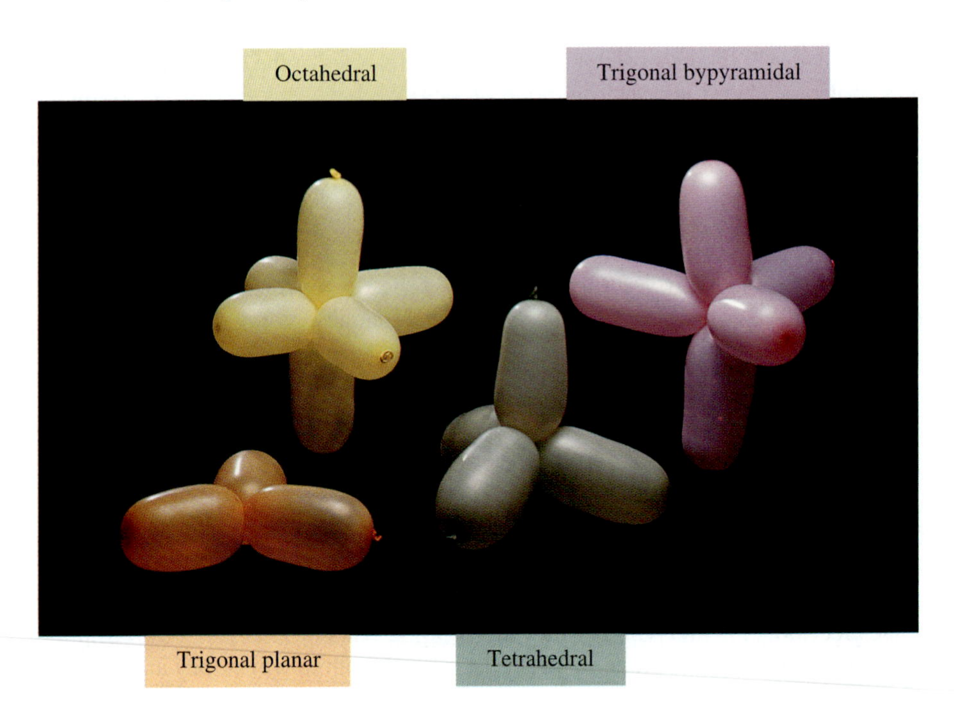

Figure 10.2
Balloon analogy to electron-group geometries for 3, 4, 5, and 6 electron groups
Each lobe represents an electron group. The electron-group geometries are trigonal planar (orange), tetrahedral (green), trigonal bipyramidal (lavender), and octahedral (yellow).

VSEPR Notation and Molecular Geometry

We use a special VSEPR notation to describe molecular geometries. We denote the central atom in a structure as "A," terminal atoms as "X," and lone pairs of electrons as "E." In this notation, AX_2E_2 describes a structure with *two* terminal atoms and *two* lone pairs of electrons around a central atom. For example, the water molecule, H—Ö—H, is described by the AX_2E_2 notation.

Let's pause for a moment to clarify the difference between *electron-group geometry* and *molecular geometry*. An electron-group geometry describes how, as a result of their mutual repulsions, groups of valence electrons are arranged about a central atom. A molecular geometry describes how bonded atoms are arranged about the same central atom. If all the electron groups are bonding groups, the orientation of bonded atoms and electron groups will be the same. Thus, for structures with *no lone-pair electrons* (AX_n) the molecular geometry and electron-group geometry are the same. If there are lone-pair electrons, there will be one or more positions around the central atom where an electron group is found *but no atom*. In these cases, the molecular geometry is related to and determined by the electron-group geometry, but the two are *not the same*.

Table 10.1 summarizes various possibilities for molecular geometries in relation to electron-group geometries. Over the next few pages, we will describe several of the entries in Table 10.1.

Structures with No Lone-Pair Electrons, AX_n

AX_2: The molecules $BeCl_2$ and CO_2 are of the type AX_2. From its Lewis structure, we see that in electron-deficient $BeCl_2$ the two electron groups around the Be atom are both electron pairs in single covalent bonds. In CO_2, both electron groups around the C atom consist of *two* electron pairs: They are double covalent bonds.

:C̤̈l:Be:C̤̈l: :Ö::C::Ö:

Both the electron-group geometry and the molecular geometry for two electron groups is *linear*.

AX_3: The molecules BF_3 and SO_3 are of the type AX_3. In the electron-deficient BF_3, the three electron groups are electron pairs in single covalent bonds. In the Lewis structure below, the F atoms are distributed around the B atom in the same *trigonal planar* geometry as the electron-group orientation.

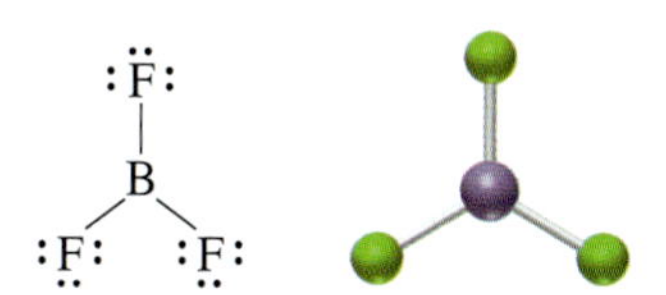

There are several possible Lewis structures for the molecule SO_3; we show two of them. The first is one of three resonance structures that conform to the octet rule. The second is one of four possible structures based on an expanded valence shell. The true structure is a resonance hybrid.

:Ö:
|
S
:O: :O:

:O:
||
S
:O: :O:

These structures illustrate an important point:

All the plausible Lewis structures that we can write for a molecule or polyatomic ion should give the same electron-group geometry.

Both of the structures above indicate a *trigonal planar* distribution of three electron groups and hence of the three O atoms.

TABLE 10.1 VSEPR Notations, Electron-Group Geometry, and Molecular Geometry

Number of Electron Groups	Electron-Group Geometry	Number of Lone Pairs	VSEPR Notation	Molecular Geometry	Ideal Bond Angles	Example
2	Linear	0	AX_2	X—A—X Linear	180°	$BeCl_2$ 
3	Trigonal planar	0	AX_3	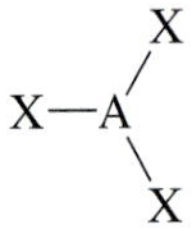 Trigonal planar	120°	 BF_3
3	Trigonal planar	1	AX_2E	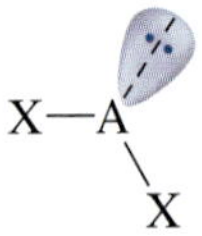 Angular	120°	SO_2
4	Tetrahedral	0	AX_4	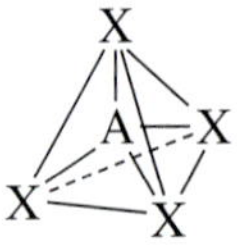 Tetrahedral	109.5°	 CH_4
4	Tetrahedral	1	AX_3E	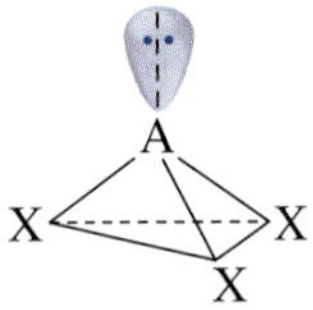 Trigonal pyramidal	109.5°	NH_3
4	Tetrahedral	2	AX_2E_2	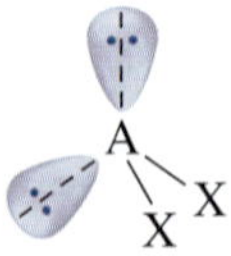 Angular	109.5°	OH_2
5	Trigonal bipyramidal	0	AX_5	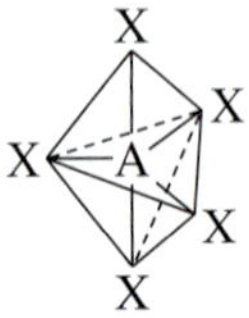 Trigonal bipyramidal	90°, 120°, 180°	PCl_5 

($BeCl_2$)

(BF_3)

(CH_4)

(PCl_5)

TABLE 10.1 (continued)

Number of Electron Groups	Electron-Group Geometry	Number of Lone Pairs	VSEPR Notation	Molecular Geometry	Ideal Bond Angles	Example
5	Trigonal bipyramidal	1	AX_4E	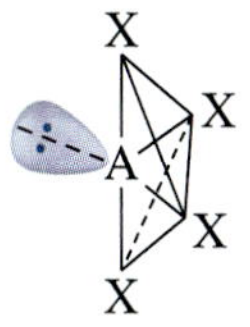 Seesaw	90°, 120°, 180°	SF_4
5	Trigonal bipyramidal	2	AX_3E_2	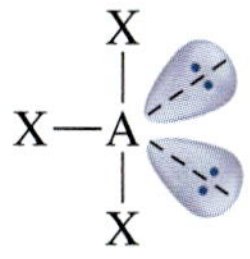 T-shaped	90°, 180°	ClF_3
5	Trigonal bipyramidal	3	AX_2E_3	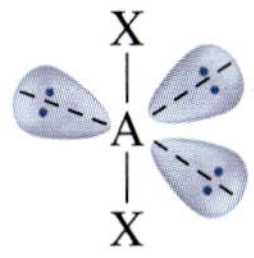 Linear	180°	XeF_2
6	Octahedral	0	AX_6	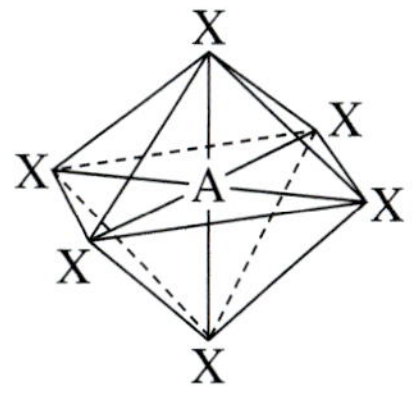 Octahedral	90°, 180°	SF_6
6	Octahedral	1	AX_5E	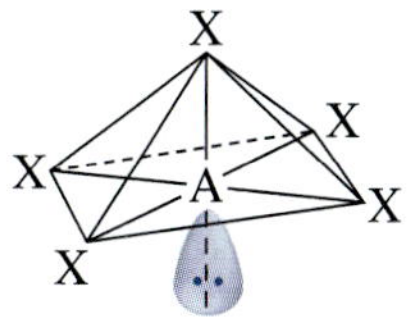 Square pyramidal	90°	BrF_5
6	Octahedral	2	AX_4E_2	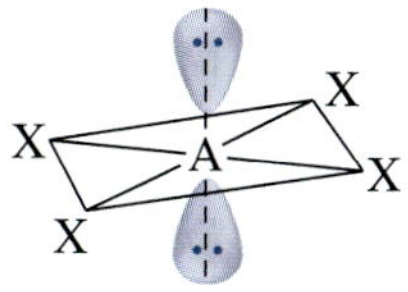 Square planar	90°	XeF_4

(SF_6)

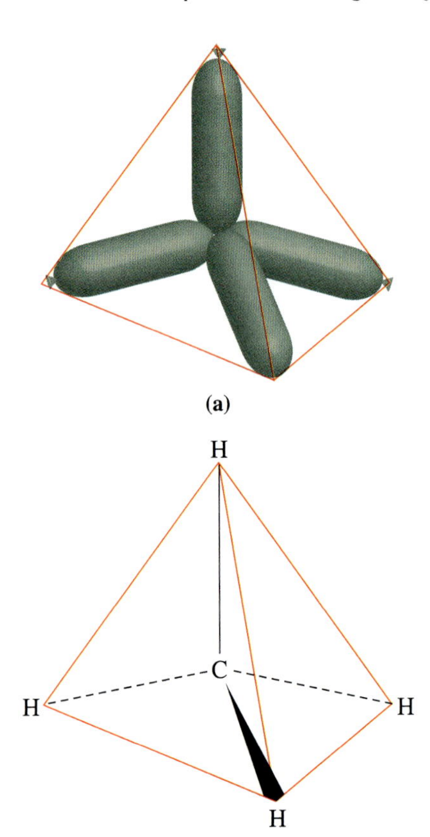

▲ **Figure 10.3**
The electron-group geometry and molecular geometry of methane
(a) The electron-group geometry of methane represented by balloons. Two balloons are twisted together and separate into four lobes. The four lobes are directed to the corners of a hypothetical tetrahedron. Each lobe represents an electron pair in a C—H bond. (b) The molecular geometry of methane. The four H atoms in CH_4 are situated at the corners of the tetrahedron. The C—H bonds are represented by black lines, and the edges of the tetrahedron by red lines.

The distinguishing features of *trigonal planar* geometry are that the central atom and three terminal atoms all lie in the same plane and the bond angles are 120°.

AX_4: A prime example of a molecule of the type AX_4 is CH_4, methane.

```
    H                 H
    ..                |
H : C : H   or   H — C — H
    ..                |
    H                 H
```

The energy of the molecule is at a minimum when the four bonding pairs of electrons are as far from each other as possible. The carbon atom is at the center of a *regular tetrahedron*, and the electron pairs are directed to the four corners where the hydrogen atoms reside. A regular tetrahedron has four faces, each an equilateral triangle. Both the predicted and experimentally determined H—C—H bond angles in CH_4 are 109.5°.

Figure 10.3 illustrates that both the electron-group and molecular geometries in CH_4 are *tetrahedral*. In addition, we portray the molecular geometry in a way often used by organic chemists: A solid line represents a bond in the plane of the page; broken lines are used for bonds that extend *behind* the plane; "wedges" are used for bonds projecting *out* from the plane of the page.

AX_5 and AX_6: In Table 10.1, PCl_5 is of the type AX_5, and SF_6 is of the type AX_6. The AX_5 and AX_6 structures require that the central atom be surrounded by more than 4 electron pairs. They require that the central atom employ an expanded valence shell, and therefore we encounter such structures only when the central atom is a third-period element or higher.

The *trigonal bipyramidal* electron-group geometry of PCl_5 has three electron pairs directed to the vertices of an equilateral triangle. The two remaining pairs are directed along a line perpendicular to the triangle, one pair above and one below the plane of the triangle. The *trigonal bipyramidal* molecular geometry of PCl_5 has a P atom at the center of the equilateral triangle and Cl atoms at the corners of the trigonal bipyramid. (A trigonal bipyramid is a geometric figure with six faces, each face a triangle.)

The *octahedral* electron-group geometry of SF_6 has four electron pairs directed to the vertices of a square. The remaining pairs are directed along a line perpendicular to the square at its center. One pair is above the plane of the square and the other below. The *octahedral* molecular geometry of SF_6 has a S atom at the center of the square and F atoms at the corners of the octahedron. (A regular octahedron is a geometric figure with eight faces, each face an equilateral triangle.)

A Strategy for Applying the VSEPR Method

To predict the shape of a molecule or polyatomic ion, we can generally follow the four-point strategy described below and illustrated in the margin for sulfur trioxide, a molecule whose molecular geometry we have already described.

Name
Sulfur trioxide

Formula
SO_3

1 *Draw a Lewis structure of the molecule or polyatomic ion.* The structure must be plausible but it does not need to be the "best" one. That is, it may have formal charges and may be only one of several contributing structures to a resonance hybrid.

Lewis structure

```
      :O:
       ‖
:Ö — S — Ö:
```

Electron groups
Three; all bonding groups

2 *Determine the number of electron groups around the central atom, and identify each either as a bonding group or a lone pair of electrons.* Keep in mind that a bonding group can be a double bond or a triple bond.

Electron-group geometry
Trigonal planar

3 *Establish the electron-group geometry.* This may be linear, trigonal planar, tetrahedral, trigonal bipyramidal, or octahedral, corresponding to 2, 3, 4, 5, and 6 electron groups, respectively.

Molecular geometry
Trigonal planar: S atom at center;
120° O-S-O bond angles

4 *Describe the molecular geometry.* This is based on the positions around the central atom that are occupied by other atoms (*not by lone-pair electrons*). Use information from Table 10.1 if necessary.

EXAMPLE 10.1

Use the VSEPR method to predict the shape of the nitrate ion.

SOLUTION

We can begin applying the four-step method outlined above after we have written the formula for the nitrate ion—NO_3^-. To write a plausible Lewis structure of NO_3^-, we could first go through the systematic approach for writing Lewis structures that we introduced in Chapter 9 (page 377). Instead, let's just summarize the thinking required by that approach.

1 *Write a plausible Lewis structure.* The number of valence electrons in NO_3^- is $5 + (3 \times 6) + 1 = 24$. The structure has N as the central atom and three O atoms as terminal atoms. The Lewis structure obtained by joining the three O atoms to the N atom and distributing lone-pair electrons around all the atoms accounts for all the valence electrons; it is shown below.

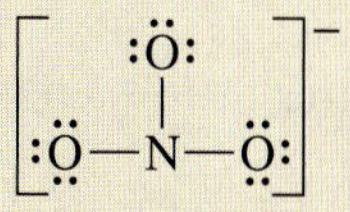

[:Ö: | :Ö—N—Ö:]⁻

This structure is unsatisfactory; the N atom does not have an octet. We can remedy this by forming a nitrogen-to-oxygen double bond. There are three resonance structures, and the true structure is the resonance hybrid of the three. However, we can base a prediction on just one of these resonance structures.

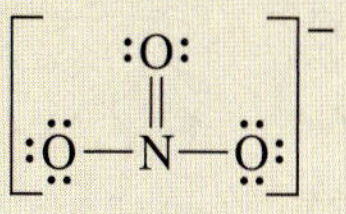

[:O: ‖ :Ö—N—Ö:]⁻

2 *Determine the number of electron groups.* There are *three* electron groups around the N atom; all are bonding groups. The VSEPR notation is AX_3.

3 *Determine the electron-group geometry.* The distribution of the three electron groups is *trigonal planar*.

4 *Describe the molecular geometry.* For the structure AX_3, the molecular geometry is the same as the electron-group geometry: *trigonal planar*. The N and O atoms are all in the same plane; the O—N—O bond angles should be 120° (see Figure 10.4).

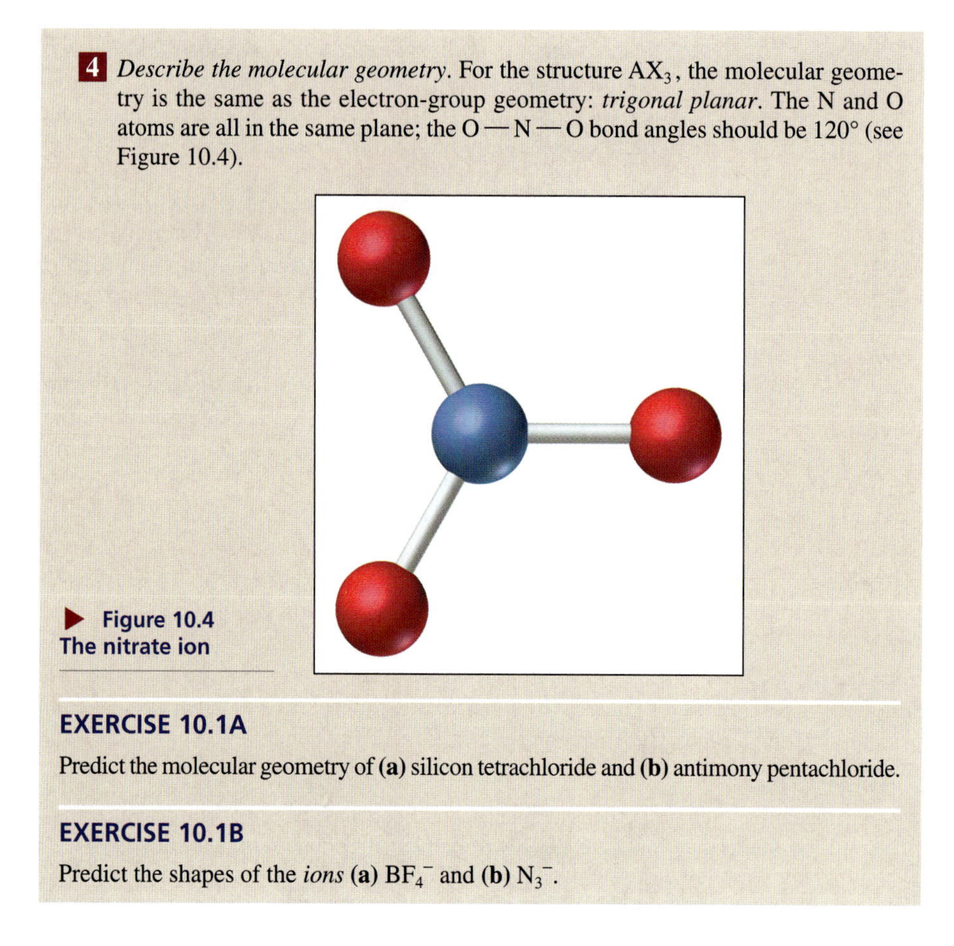

▶ **Figure 10.4**
The nitrate ion

EXERCISE 10.1A

Predict the molecular geometry of **(a)** silicon tetrachloride and **(b)** antimony pentachloride.

EXERCISE 10.1B

Predict the shapes of the *ions* **(a)** BF_4^- and **(b)** N_3^-.

Structures with Lone-Pair Electrons, AX_nE_m

As we mentioned earlier, in structures that have lone-pair electrons (AX_nE_m) the molecular geometry differs from the electron-pair geometry. Let's examine some of the differences.

AX_2E: Consider two resonance structures of the SO_2 molecule.

:Ö::S̈:Ö: ⟷ :Ö:S̈::Ö:

We find three electron groups around the central atom. The electron-group geometry is *trigonal planar*. Oxygen atoms form bonds with two of the electron groups but the third electron group remains as a lone pair of electrons. Because molecular geometry must be based on the geometric arrangement of *atoms*, not electron groups, SO_2 is not trigonal planar. It is angular or bent. We predict a bond angle of 120°, and the observed bond angle is 119°.

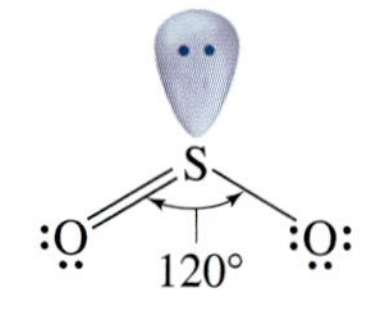

AX_3E: In the NH_3 molecule, three electron pairs in the valence shell around the N atom are bonding pairs, and the fourth is a lone pair. The electron-group geometry is *tetrahedral*, but one of the corners of the tetrahedron is occupied by the lone-pair electrons. The molecular geometry pictured in Figure 10.5 is called *trigonal pyramidal*. The N atom is situated at the top of the pyramid, and the three H atoms are at the vertices of the triangular base. We predict H—N—H bond angles of 109.5° (the same as in CH_4 in Figure 10.3). The actual angle is found to be 107°. Repulsions between the lone-pair electrons and the bonding electrons push the bonding pairs closer together, reducing the bond angle slightly from the regular tetrahedral angle.

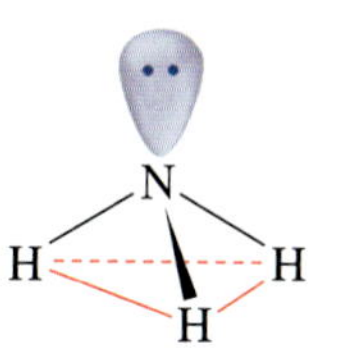

VSEPR notation: AX_3E

▲ **Figure 10.5**
Molecular geometry of ammonia
The bonds in NH_3 are represented by the solid black lines. The molecular geometry—a trigonal pyramid—is outlined by these bond lines and the red lines. The lone pair of electrons (shown in blue) is directed to a corner of the tetrahedron that represents the electron-group geometry. (The tetrahedron is not shown.)

AX_2E_2: The VSEPR notation for the H_2O molecule is AX_2E_2, as we see from the Lewis structure

$$H:\ddot{\underset{\cdot\cdot}{O}}:H$$

Again, the electron-group symmetry for four electron pairs is *tetrahedral*, but only two of the tetrahedral positions around the central atom (O) are occupied by terminal atoms (H). This triatomic molecule has an angular, or bent, shape, as shown in Figure 10.6. Recall also the representation of the H_2O molecule in Figure 10.1. There, we gave the H—O—H bond angle as 104.45°. This angle deviates from the tetrahedral angle (109.5°) somewhat more than does the H—N—H angle in NH_3. The reason for the deviation is basically the same as in NH_3, but the *two* lone pairs of electrons on the O atom of H_2O exert an even stronger repulsion of the bonding electrons than does the *one* lone pair on the N atom of NH_3.

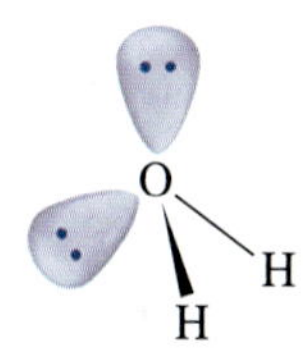

VSEPR notation: AX_2E_2

▲ **Figure 10.6**
Molecular geometry of water
The bonds in H_2O are represented by solid black lines and outline the molecular geometry—angular or bent. The two lone pairs of electrons (shown in blue) are directed to corners of the tetrahedron that represents the electron-group geometry. (The tetrahedron is not shown.)

Other structures: By considering some of the remaining structures in Table 10.1 and those in Example 10.2 and Exercises 10.2A and 10.2B, we can expand our view of electron-pair repulsions to include the following ideas.

- *The closer together two groups of electrons are, the stronger the repulsion between them.* The force of repulsion between two electron groups increases dramatically as the groups are forced very close together. Thus, repulsions between two groups of bonding electrons *increase* significantly when a bond angle is reduced from 180° to 120° to 90°.
- *Lone-pair (LP) electrons spread out more than do bonding-pair (BP) electrons.* Because bonding-pair electrons are simultaneously attracted to two nuclei, the charge cloud associated with them is pulled into a compact shape. In contrast, because lone-pair electrons are associated with just one nucleus, their charge cloud is much more spread out. As a result, the repulsion of one lone pair of electrons for another lone pair is greater than the repulsion between two bonding pairs. In general, the strength of repulsive forces, from strongest to weakest, is

Lone-pair–lone-pair repulsions > Lone-pair–bonding-pair repulsions

> Bonding-pair–bonding-pair repulsions

EXAMPLE 10.2

Use the VSEPR method to predict the molecular geometry of XeF_2.

SOLUTION

We can try the four-step method outlined on page 414 to see if it leads us directly to the molecular geometry. If it does not, we will need to consider additional steps.

1 *Write a plausible Lewis structure.* There are $8 + (2 \times 7) = 22$ valence electrons in XeF_2. We need only 20 electrons to attach two F atoms to a central Xe atom and to provide each of the three atoms an octet. We must place the additional pair in an expanded valence shell on the Xe atom. The Lewis structure of XeF_2 is

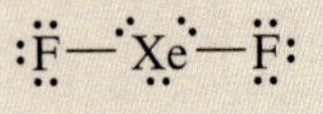

:F̤̈—Ẍe̤—F̤̈:

2 *Determine the number of electron groups.* There are *five* electron groups around the Xe atom, *two* bonding electron pairs and *three* lone pairs. The VSEPR notation is AX_2E_3.

3 *Determine the electron-group geometry.* The distribution of the electron groups is *trigonal bipyramidal.*

4 *Describe the molecular geometry.* We cannot decide on the molecular geometry at this point. There seem to be three possibilities, but only one can be correct.

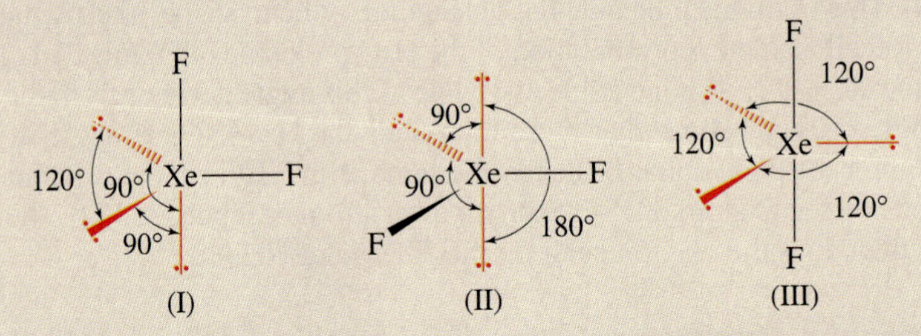

The first possibility, (I), has *one 120°* lone-pair–lone-pair repulsion and *two 90°* lone-pair–lone-pair repulsions. Structure II has *one 180°* lone-pair–lone-pair repulsion and *two 90°* lone-pair–lone-pair repulsions. In contrast, structure III has *no 90°* lone-pair–lone-pair repulsions; there are *three 120°* lone-pair–lone-pair repulsions. Based on the strong repulsions for a 90° bond angle and decreasing repulsions with increasing angle of interaction (90° > 120° > 180°), we predict that the arrangement of bonded atoms and lone-pair electrons in structure III represents the lowest energy configuration. Thus, we expect the Xe and two F atoms to lie in a straight line. The predicted *linear* molecular geometry is also what is observed experimentally.

EXERCISE 10.2A

Use the VSEPR method to explain why SF_4 has the seesaw molecular geometry shown in Table 10.1 rather than the one shown below.

● ChemCDX VSEPR

F
F—S—F
F

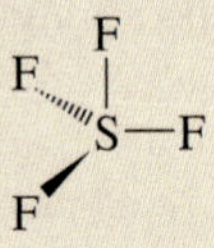

EXERCISE 10.2B

Use the VSEPR method to explain why ClF_3 has the T-shaped molecular geometry shown in Table 10.1 rather than the one shown below.

F
Cl—F
F

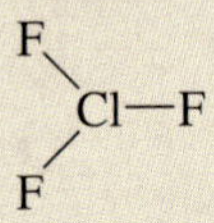

VSEPR Treatment of Structures with More than One Central Atom

So far, we have used the VSEPR method only in structures with one central atom, such as CH_4, NH_3, H_2O, NO_3^-, and XeF_2. However, we already have seen many molecules and polyatomic ions that have more than one central atom, for example, C_2H_6, C_2H_4, CH_3OH, and H_2O_2. We can use the VSEPR method to describe their molecular geometries by first working out the orientation of other atoms or groups of atoms around each central atom and then combining the results to give an overall description of the molecular geometry.

EXAMPLE 10.3

Use the VSEPR method to describe, as well as possible, the molecular geometry of the nitric acid molecule, HNO_3.

SOLUTION

We will use the general approach of page 414, starting with a plausible Lewis structure that will enable us to identify the central atoms. In the next three steps, we will describe the geometry about each of the central atoms. Finally, we will formulate a description of the overall molecular geometry.

1 *Write a plausible Lewis structure.* There are $1 + 5 + (3 \times 6) = 24$ valence electrons in HNO_3. Of the two structures shown below, only the first is plausible. The second, which has only N as a central atom, would not have enough electrons to give each atom an octet.

```
           :O:                       ..
           ||                        :O:
      ..        ..                 ..  |    ..
   H—O—N—O: (correct)           :O—N—O: (incorrect)
      ..        ..                 ..  |    ..
                                       H
```

Thus, we see that there are *two* central atoms: the N atom and the O atom bonded to the H atom.

2 *Determine the number of electron groups around each central atom. Central N atom*: This atom has *three* electron groups involved in two single bonds and one double bond. Its VSEPR notation is AX_3. *Central O atom*: This atom has *four* electron groups. Two of the groups are bonding pairs of electrons and the other two are lone-pair electrons. The VSEPR notation for the O atom is AX_2E_2.

3 *Determine the electron-group geometry for each central atom. Central N atom*: trigonal planar. *Central O atom*: tetrahedral

4 *Describe the molecular geometry around each central atom. Central N atom*: Because there are no lone-pair electrons, the molecular geometry around this atom is the same as the electron-group geometry: *trigonal planar*. *Central O atom*: The atoms in an AX_2E_2 structure take on an orientation that is angular or bent. The H—O—N portion of the molecule has the same general shape as the H—O—H molecule, although not necessarily the same bond angle.

As suggested in Figure 10.7, our description of the overall molecular geometry of HNO_3 has the N and three O atoms in the same plane and O—N—O bond angles of about 120°. We know that the H atom is situated to produce a H—O—N angle of about 109°, but we can't say whether the H atom lies above, below, or in the plane of the other atoms. (Experimental evidence indicates that the H atom is in about the same plane as the other atoms.)

EXERCISE 10.3A

Use the VSEPR method to describe the molecular geometry of dimethyl ether, $(CH_3)_2O$, as completely as you can.

PROBLEM-SOLVING NOTE

Recall that in an oxoacid hydrogen atoms are almost always bonded to an O atom. We could use this fact at the outset to reject the incorrect Lewis structure.

PROBLEM-SOLVING NOTE

Remember that the electrons in a double or triple bond are treated as a single electron group (page 409).

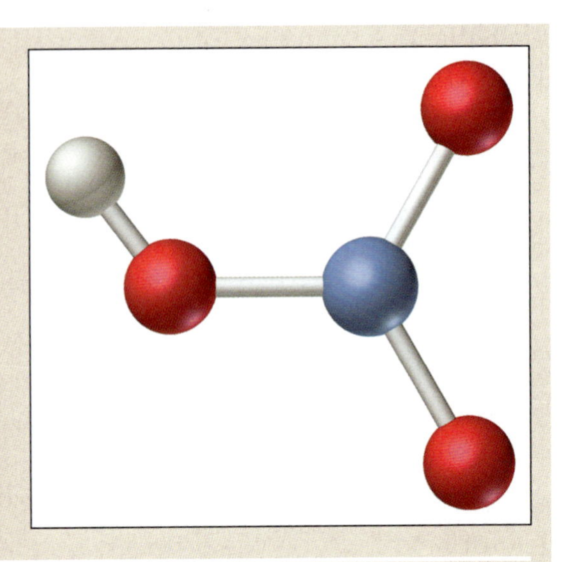

▶ Figure 10.7 Predicted molecular geometry of the HNO_3 molecule by the VSEPR method
The prediction is that the N and O atoms lie in the same plane. The trigonal planar electron-group geometry around the central N atom leads us to predict O—N—O bond angles of about 120°. The tetrahedral electron-group geometry around the O atom should lead to a H—O—N bond angle of about 109°.

EXERCISE 10.3B

Use the VSEPR method to describe the molecular geometry of ethanol, CH_3CH_2OH, as completely as you can.

10.2 Polar Molecules and Dipole Moments

In Section 9.7, we saw that in most covalent bonds the electronegativities of the bonded atoms differ. The opposite ends of the bond therefore acquire slight charges, indicated in Lewis structures by δ+ and δ−; the bond is *polar covalent*. We can represent the polar covalent bond in the HCl molecule as

$$\overset{\delta+}{\mathrm{H}}:\overset{\delta-}{\ddot{\underset{\cdot\cdot}{\mathrm{Cl}}}}:$$

A molecule with separate centers of positive and negative charge is called a **polar molecule**. A quantity called the dipole moment describes the significance of this charge separation. The **dipole moment** (μ) of a molecule is the product of the magnitude of the charge (δ) and the distance (*d*) that separates the centers of positive and negative charge.

$$\mu = \delta\, d$$

Thus, a polar molecule is one having a *nonzero* dipole moment, and a *nonpolar* molecule is one with dipole moment $\mu = 0$.

Dipole moments have units that reflect the electric charge times the distance between the two charge centers, coulomb × meter (C m). Dipole moments are generally expressed in a quantity called a *debye*. One **debye (D)** is equal to 3.34×10^{-30} C m. For example, the measured dipole moment of HCl is $\mu = 1.07$ D. Figure 10.8 suggests a way to measure dipole moments. The metal plates in the figure are separated by a *nonconducting* medium, and small quantities of electric charge are stored, so that one plate acquires a negative charge and the other a positive charge. When polar molecules are introduced, the molecules align themselves as shown. The polar molecules increase the charge-storing capacity of the plates to an extent that is related to their dipole moment.

Molecular Shape and Drug Action

Most drugs act on protein molecules in the tissues or organs they target. They don't act just anywhere on the molecules, however. They must attach to specific locations called *receptor sites*. Receptor sites have three-dimensional shapes designed to accommodate certain molecules produced naturally in the body. An effective drug molecule, by virtue of its size and shape and bond polarities, also fits a receptor site. The drug acts either by replacing natural molecules on the site or by blocking the site from the natural molecules.

Drug molecules that fit a receptor site and cause their own characteristic response are called *agonists*. Those that bind to the site and prevent the action of the intended molecules are called *antagonists*. If an agonist and an antagonist both are present, they compete for the receptor sites. In general, an antagonist binds more tightly than an agonist, and a small quantity of an antagonist can block the action of a larger quantity of an agonist. For example, the human brain has receptor sites for morphine, the principal narcotic found in opium poppies. A 1-mg intravenous dose of naloxone, a morphine antagonist, can block the action of 25 mg of heroin, a morphine-related agonist.

Substances produced in the body during strenuous exercise may create the "runner's high" by activating receptor sites on nerve cells.

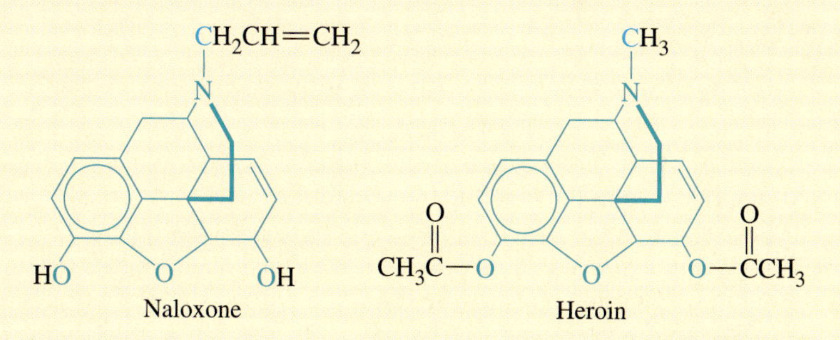

Why should the human brain have receptors for a plant-derived drug like morphine? The answer seems to lie in several substances formed naturally by the human body, called *endorphins*. Endorphins are peptides. β-Endorphin, the most potent human form, is composed of a chain of 30 amino acid residues. Endorphin is a contraction of endogenous morphine, (Greek, *endon*, within; *genes*, born, produced). For example, the production of these compounds during strenuous athletic activity may account for the fact that athletes can sometimes continue to compete after an injury. They don't feel the pain until the event is over. The so-called runner's high experienced by some marathon runners and joggers may also be due to endorphin production.

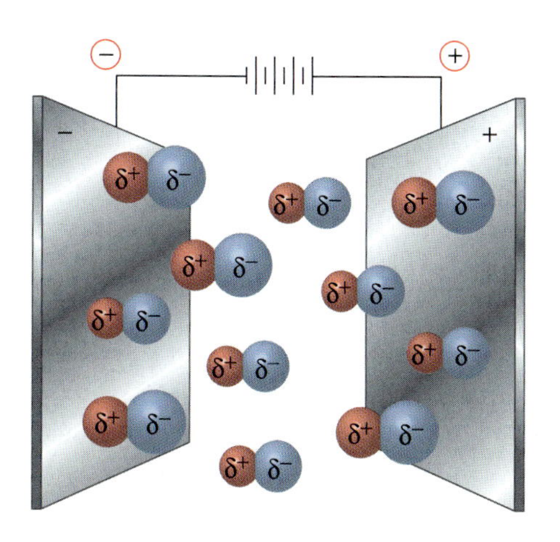

Figure 10.8
Behavior of polar molecules in an electric field
In the absence of an electric field, the polar molecules are oriented randomly. When the metal plates are electrically charged, the polar molecules orient themselves in the manner shown.

Bond Dipoles and Molecular Dipoles

In our discussion of polar molecules and dipole moments, we need to distinguish between a bond dipole and a molecular dipole. A *bond dipole*, found in all polar covalent bonds, describes a separation of positive and negative charges in an individual bond. A *molecular dipole* describes a charge separation in the molecule as a whole, considering all the bonds.

In *diatomic* molecules formed by identical atoms (*homonuclear* molecules), there is no electronegativity difference and no separation of charge. There is neither a bond dipole nor a molecular dipole. The bond is nonpolar covalent and the molecule is *nonpolar*. In *diatomic* molecules formed by different atoms (*heteronuclear* molecules), there is generally an electronegativity difference and a charge separation in the bond. However, the bond dipole and molecular dipole are the same because there is only one bond. The molecule is *polar* and has a dipole moment, as we saw with HCl.

In *polyatomic* molecules, the situation is more complex. Consider the bonds in the CO_2 molecule indicated below.

$$\overset{\longleftarrow\!\!+}{\text{O—}}\overset{+\!\!\longrightarrow}{\text{C—O}}$$
$$\mu = 0$$

The cross-based arrows (recall page 375) point from the positive to the negative end of a bond dipole, here from the C to the more electronegative O atoms. There is a bond dipole in each carbon-oxygen bond. Bond dipoles are *vector* quantities because they have both a *magnitude* and a *direction*. In CO_2, the bond dipoles are equal in magnitude and they point in opposite directions, leading to a cancellation of their effects. There is no molecular dipole ($\mu = 0$), and the CO_2 molecule is *nonpolar*. The situation is analogous to a tug-of-war in which two evenly matched teams (the O atoms) pull on a rope equally hard. A knot in the center of the rope (the C atom), though subjected to strong pulls, doesn't move at all.

If we start with the experimental fact that water has the dipole moment, $\mu = 1.84$ D, we must conclude that H_2O cannot be a linear molecule. The O—H bond dipoles must combine in such a way as not to cancel. The following structure is consistent with the experimental evidence.

O—H
H
104.5°

In this structure, the black cross-based arrows represent the bond dipoles, and the red cross-based arrow depicts the molecular dipole that results by combining the bond dipoles.

Molecular Shapes and Dipole Moments

To determine whether a molecule is polar or nonpolar, we can generally use the three-step approach outlined below and illustrated in Example 10.4.

1. *Use electronegativity values to predict bond dipoles.*
2. *Use the VSEPR method to predict the molecular shape.*
3. *From the molecular shape, determine whether bond dipoles cancel to give a nonpolar molecule or combine to produce a resultant dipole moment for the molecule.*

EXAMPLE 10.4

Explain whether you expect the following molecules to be polar or nonpolar.

a. CCl_4 **b.** $CHCl_3$

SOLUTION

1 *Use electronegativity values to predict bond dipoles.* There are electronegativity differences between C and H atoms and between C and Cl atoms. As a result, there are bond dipoles in all the bonds in CCl_4 and in $CHCl_3$.

2 *Use the VSEPR method to predict the molecular shape.* The Lewis structures of the two molecules are

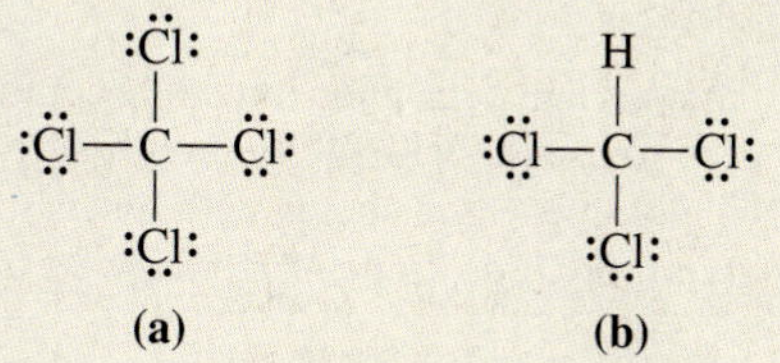

(a) (b)

Both molecules have *tetrahedral* electron-group geometry *and* molecular geometry (VSEPR notation: AX_4).

3 *Determine whether bond dipoles cancel or combine to produce a resultant dipole moment.* The distribution of bond dipoles in CCl_4 is shown in Figure 10.9. The net pull in the downward direction exerted by the three bond dipoles directed down and away from the center of the structure is just matched by the straight upward pull of the bond dipole at the top. There is no net displacement of electrons in the structure and no dipole moment; CCl_4 is a *nonpolar* molecule.

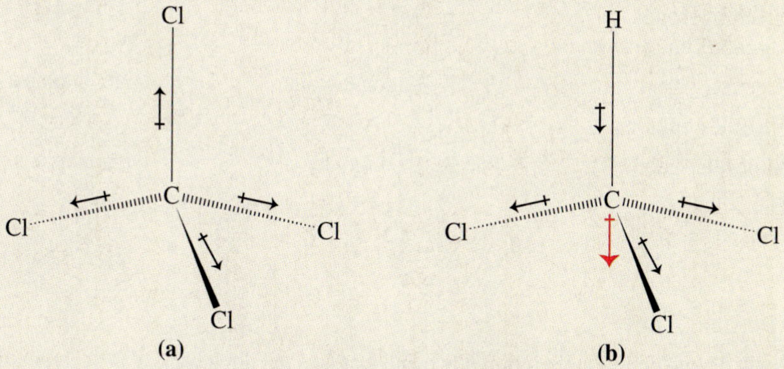

(a) (b)

Things are dramatically different when we substitute a H atom for the Cl atom at the top of the structure. The net downward pull by the three Cl atoms at the bottom remains the same, but there is no counterbalancing upward pull. Instead, there is an additional slight downward pull on the electrons because the C atom is more electronegative than H. The bond dipoles combine to form a molecular dipole pointed toward the bottom of the structure. $CHCl_3$ is a *polar* molecule. (Its dipole moment is found to be 1.01 D.)

EXERCISE 10.4A

Explain whether you expect each of the following molecules to be polar or nonpolar: BF_3, SO_2, BrCl, N_2.

PROBLEM-SOLVING NOTE

Obtain electronegativities from Figure 9.8. The electronegativity for C is 2.5, whereas that for H is 2.1, and that for Cl is 3.0.

Figure 10.9 Molecular geometry and dipole moments

The bond dipoles are represented by black cross-based arrows (⟻). (a) The individual bond dipoles cancel, and there is no dipole moment for the molecule as a whole. (b) All the bond dipoles point downward and combine to form a molecular dipole (red arrow).

EXERCISE 10.4B

Explain whether you expect each of the following molecules to be polar or nonpolar: SO_3, SO_2Cl_2, ClF_3, BrF_5.

EXAMPLE 10.5—A Conceptual Example

Of the two compounds NOF and NO_2F, one has a resultant dipole moment of μ = 1.81 D, and the other, μ = 0.47 D. Which dipole moment do you predict for each compound? Explain.

SOLUTION

We need to identify bond dipoles and determine how they combine to produce a dipole moment. To do this, we need to use the VSEPR method first to get a sense of the geometric structures of the two molecules. Let's begin with Lewis structures. Because it has the lowest electronegativity of the three kinds of atoms, we expect N to be the central atom in both structures.

	NOF	NO_2F
Valence electrons	18	24
Lewis structures	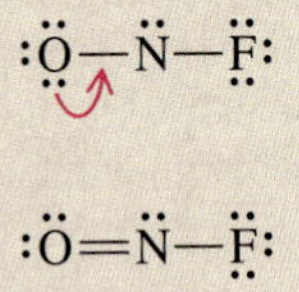 $:\ddot{O}-\ddot{N}-\ddot{F}:$ (with lone pairs on O, N, F) → $:\ddot{O}=\ddot{N}-\ddot{F}:$	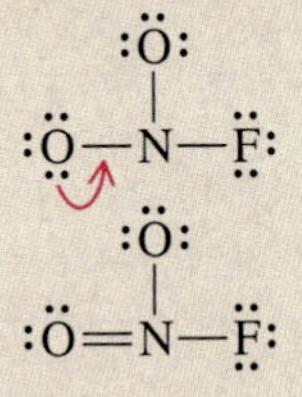 $:\ddot{O}-N(-\ddot{O}:)-\ddot{F}:$ → $:\ddot{O}=N(-\ddot{O}:)-\ddot{F}:$
Electron groups around central atom	3	3
Electron-group geometry	Trigonal planar	Trigonal planar
VSEPR notation	AX_2E	AX_3
Molecular geometry	Angular	Trigonal planar
Bond dipoles (electronegativities F > O > N)	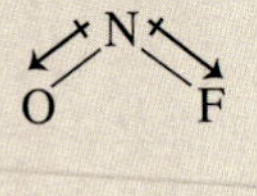 O ← N → F	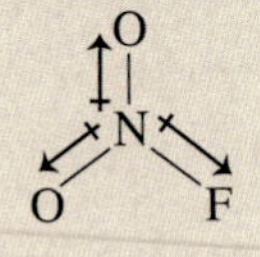 O ↑ N; O ← N → F

In NOF, two bond dipoles point down and lead to a net downward displacement of electron charge density. In NO_2F, one of the bond dipoles opposes the other two, and we expect a smaller net displacement of electron charge density. Our prediction is therefore NOF, μ = 1.81 D, and NO_2F, μ = 0.47 D.

EXERCISE 10.5

Of the two molecules NOF and NO_2F, one has a measured F—N—O bond angle of 110°, and the other, 118°. Which bond angle do you predict for each molecule? Explain.

We have stressed how electronegativity differences lead to bond dipoles and how bond dipoles may combine to produce a resultant dipole moment in a molecule. Lone-pair electrons may also make a contribution to dipole moments. Consider the molecule NF_3. Electrons in the N—F bonds are displaced toward the F atoms and *away* from the N atom. However, to a considerable extent, this displacement is counteracted by the spreading out of the charge cloud of the lone-pair electrons on the N atom. As a result, the dipole moment of NF_3 is rather small: $\mu = 0.24$ D. In contrast, the NH_3 molecule has the same shape as NF_3 but the displacement of bonding-pair electrons is *toward* the N atom. Because this displacement is in the same general direction as that produced by the lone-pair electrons, NH_3 has a much larger dipole moment: $\mu = 1.47$ D.

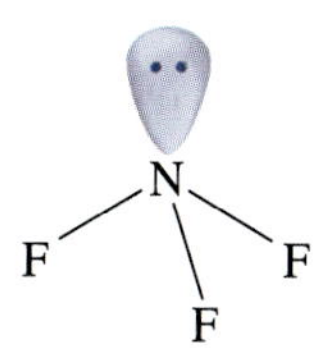

Valence Bond Theory

Soon after Schrödinger applied wave mechanics to the hydrogen *atom*, other scientists began to apply quantum mechanics to the structures of *molecules*, beginning with the H_2 molecule. We will consider two quantum mechanical approaches to molecular structure. In the first, the *valence bond theory*, we continue to think about the atoms in a molecule in terms of their atomic orbitals, and we focus on the particular orbitals involved in covalent bond formation. We will discuss the second approach later in the chapter.

10.3 Atomic Orbital Overlap

Imagine two hydrogen atoms approaching one another. Each atom has a single electron in its 1*s* orbital. As the atoms get close, the electron charge clouds represented by the 1*s* orbitals begin to merge. We describe this intermingling as the *overlap* of the 1*s* orbitals of the two atoms. This region of overlap now accommodates two electrons, one from each atom, and the electrons must have opposing spins. The atomic orbital overlap results in an increased electron charge density in the region between the atomic nuclei. The increased density of negative charge serves to hold the two positively charged atomic nuclei together. Thus, the **valence bond (VB) theory** views a covalent bond in this way:

> *A covalent bond is formed by the pairing of two electrons with opposing spins in the region of overlap of atomic orbitals between two atoms. This overlap region has a high electron charge density.*

In general, the more extensive the overlap between two orbitals, the stronger is the bond between two atoms. If two atoms are forced more closely together, however, the repulsion of the atomic nuclei becomes more important than the attractions between electrons and nuclei, and the bond becomes unstable. For each bond, then, there is a condition of optimal orbital overlap that leads to a maximum bond strength (bond energy) at a particular internuclear distance (bond length). The valence bond theory attempts to find the best approximation of this condition for all the bonds in a molecule.

Figure 10.10 depicts the bonding of two hydrogen atoms into a hydrogen molecule through the overlap of their 1*s* orbitals. Next, let's consider H_2S, a more representative example of a molecule that can be described by the valence bond theory. First, let's focus on the isolated atoms in Figure 10.11a. The orbitals that contain an unpaired electron are those that will overlap to form bonds. For the hydrogen atoms, this is the 1*s* orbital (red). For the sulfur atom, we need consider only the *valence* shell (the third). As seen in the orbital diagram, the sulfur orbitals with a single electron are the $3p_y$ and $3p_z$ (gray); the $3p_x$ is filled (blue). The 3*s* orbital, also filled, is not shown. Thus, the 1*s* orbitals of the two H atoms overlap with the $3p_y$ and $3p_z$ orbitals of the sulfur atom to form the molecule H_2S. Several important points are brought out by this example.

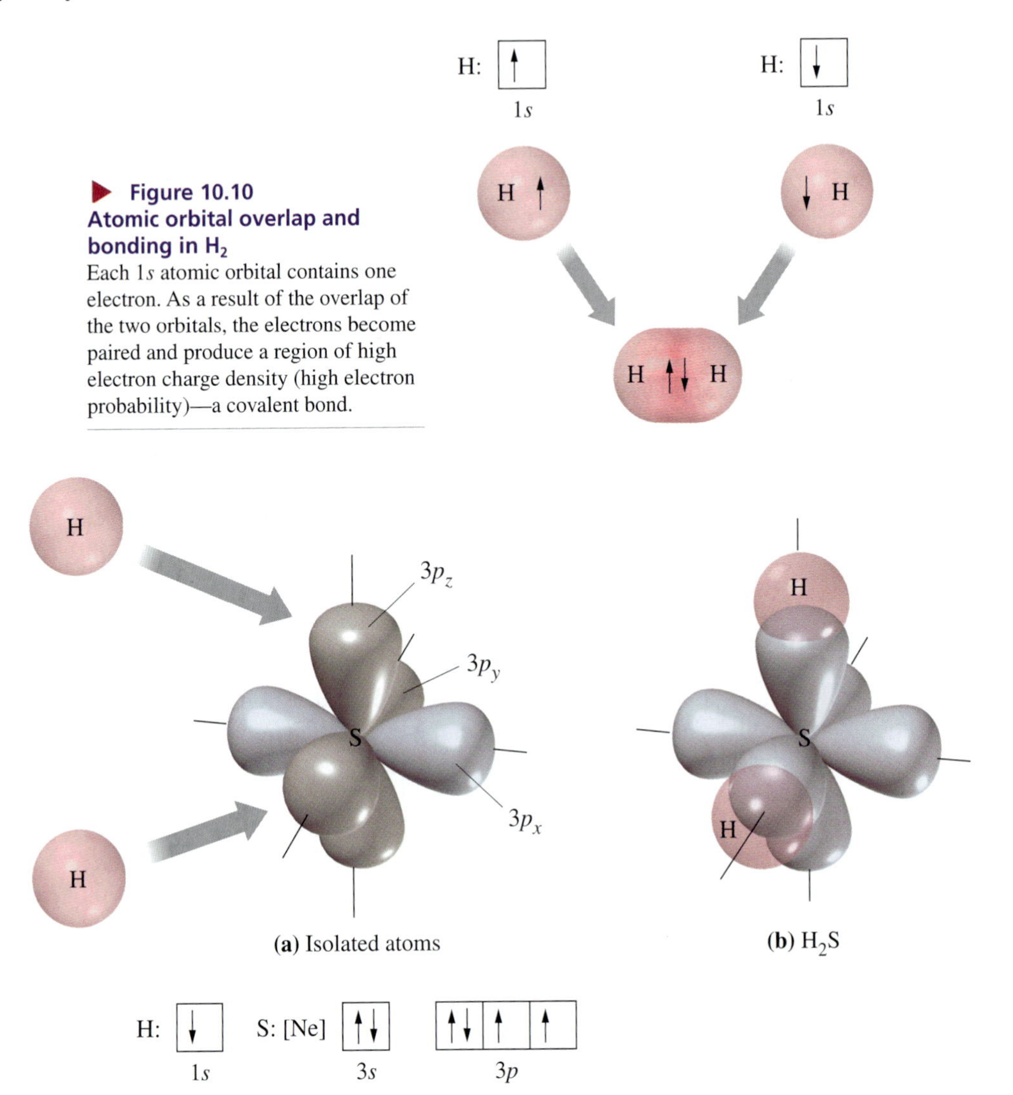

Figure 10.10 Atomic orbital overlap and bonding in H_2
Each 1*s* atomic orbital contains one electron. As a result of the overlap of the two orbitals, the electrons become paired and produce a region of high electron charge density (high electron probability)—a covalent bond.

Figure 10.11 Atomic orbital overlap and bonding in H_2S
(a) Orbitals with a single electron are shown in gray, and those with an electron pair in blue. For S, only 3*p* orbitals are shown. (b) The 1*s* orbitals of the two H atoms overlap with the $3p_y$ and $3p_z$ orbitals of the S atom, producing an H_2S molecule with a predicted bond angle of 90°.

- Most of the electrons in a molecule remain in the same orbital locations that they occupied in the separated atoms.
- Bonding electrons are *localized* in the region of atomic orbital overlap; the greatest probability of finding the bonding electron pair is in this region.
- For orbitals with directional lobes, maximum overlap occurs when atomic orbitals overlap end to end; that is, a hypothetical line joining the nuclei of the bonded atoms passes through the region of maximum overlap. (Recall that *p* orbitals are directed along perpendicular axes through the nucleus of an atom, but *s* orbitals are spherically symmetric—they have no directional lobes.)
- The molecular geometry depends on the geometric relationships among the atomic orbitals of the central atom that participate in bonding. The two 3*p* orbitals of the S atom that overlap with 1*s* orbitals of the H atoms are perpendicular to one another. Thus, the predicted H—S—H bond angle in H_2S is 90° (Figure 10.11b).

As a first approximation, the VSEPR method predicts a tetrahedral bond angle (109.5°) in H_2S. However, by taking into account the strong repulsions between the lone pairs of electrons on the S atom and the bonding pairs, we expect the bonds in H_2S to be forced into a smaller angle. The measured bond angle in the H_2S molecule is 92.1°, suggesting that the valence bond theory describes the covalent bonding in H_2S better than the VSEPR method does. However, using the valence bond theory with unmodified atomic orbitals produces superior results for relatively few molecules. We will be most successful in describing molecular geometries by using a combination of the VSEPR method and an extension of the valence bond theory.

10.4 Hybridization of Atomic Orbitals

Let's use the valence bond theory to describe the simplest hydrocarbon molecule possible. We start with the ground-state electron configuration of carbon and consider only the valence-shell orbitals.

Ground-state electron configuration of C: [He] [↑↓] [↑][↑][ ]

2s 2p

We see two unpaired electrons in the $2p$ subshell and might *predict* the simplest hydrocarbon molecule to be CH_2 with a bond angle of 90°. We really wouldn't expect this molecule to be stable, though, because it doesn't follow the octet rule; the central C atom is surrounded by only six electrons. Experiment shows that CH_2 is not a stable molecule. The simplest *stable* hydrocarbon is methane, CH_4. To account for the *four* covalent bonds in this molecule, we need an orbital diagram for carbon with *four* unpaired electrons in the valence shell and each electron in a separate orbital. To get such a diagram, we imagine that one of the $2s$ electrons is *promoted* to the empty $2p$ orbital. To boost the $2s$ electron to a higher energy subshell requires that energy be absorbed. The resulting electron configuration is that of an *excited state.*

Excited state electron configuration of C: [He] [↑] [↑][↑][↑]

2s 2p

The three mutually perpendicular $2p$ orbitals of this excited state configuration would lead us to predict a molecule with three C—H bonds at angles of 90°. The fourth C—H bond would be based on the overlap of the spherical $2s$ orbital for carbon with the spherical $1s$ orbital for hydrogen. This bond would be oriented in a direction that interfered least with the other three C—H bonds. *By experiment,* however, we find that all four C—H bonds have the same bond lengths and bond energies, and the four H—C—H bond angles are the same; they are the tetrahedral angles of 109.5° (see again Figure 10.3). This tetrahedral molecular geometry is the one predicted by the VSEPR method. Thus, the excited state electron configuration of carbon allows only for the correct number of carbon-to-hydrogen bonds, but not for the correct bond lengths, bond energies, and bond angles.

The problem with this analysis of bonding in CH_4 is that we have assumed *bonded* atoms have the same kinds of orbitals ($s, p, \ldots$) as isolated free atoms. Quite often, this seems not to be the case, and there is a way to get around the problem.

sp^3 Hybridization

Let's consider again the excited state electron configuration of the carbon atom in CH_4. Suppose we "blend" together the one $2s$ and three $2p$ orbitals of the carbon atom to produce four new orbitals that are equivalent to each other in energy and shape and point to the four corners of a tetrahedron. This blending is called **hybridization**, and it is a *hypothetical* process rather than an observed one. We can carry out hybridization only as a quantum mathematical calculation. Figure 10.12 pictures the hybridization of an *s* orbital and three *p* orbitals into the new set of four **hybrid orbitals**. The new orbitals are called sp^3 hybrid orbitals. The notation for a hybridized orbital is based on the kinds and numbers of atomic orbitals used to form the hybrids. For example, sp^3 signifies the hybridization of *one s* orbital and *three p* orbitals.

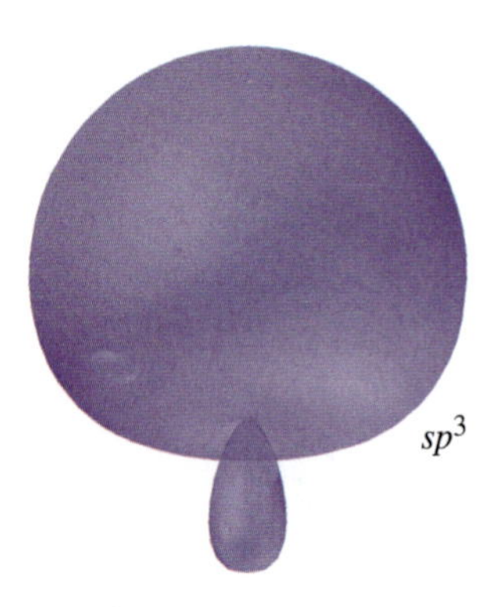

An sp^3 hybrid orbital has a small lobe pointing toward the nucleus and a large lobe pointing away from the nucleus. To add clarity to drawings of sets of hybrid orbitals, we usually omit the small lobe and elongate the large lobe, as in Figure 10.12.

When we find that a molecular structure is best described by a hybridization scheme in the valence bond theory, the following points are important to keep in mind.

- Hybridization schemes can be used rather generally for atoms in covalent bonds, but they are most often employed for central atoms only.
- The number of hybrid orbitals produced in a hybridization scheme is equal to the total number of atomic orbitals combined.
- In forming covalent bonds, hybrid orbitals may overlap with pure atomic orbitals or with other hybrid orbitals.
- Molecular geometry is determined by the shapes and orientations of hybrid orbitals, which are different from the shapes and orientations of atomic orbitals.

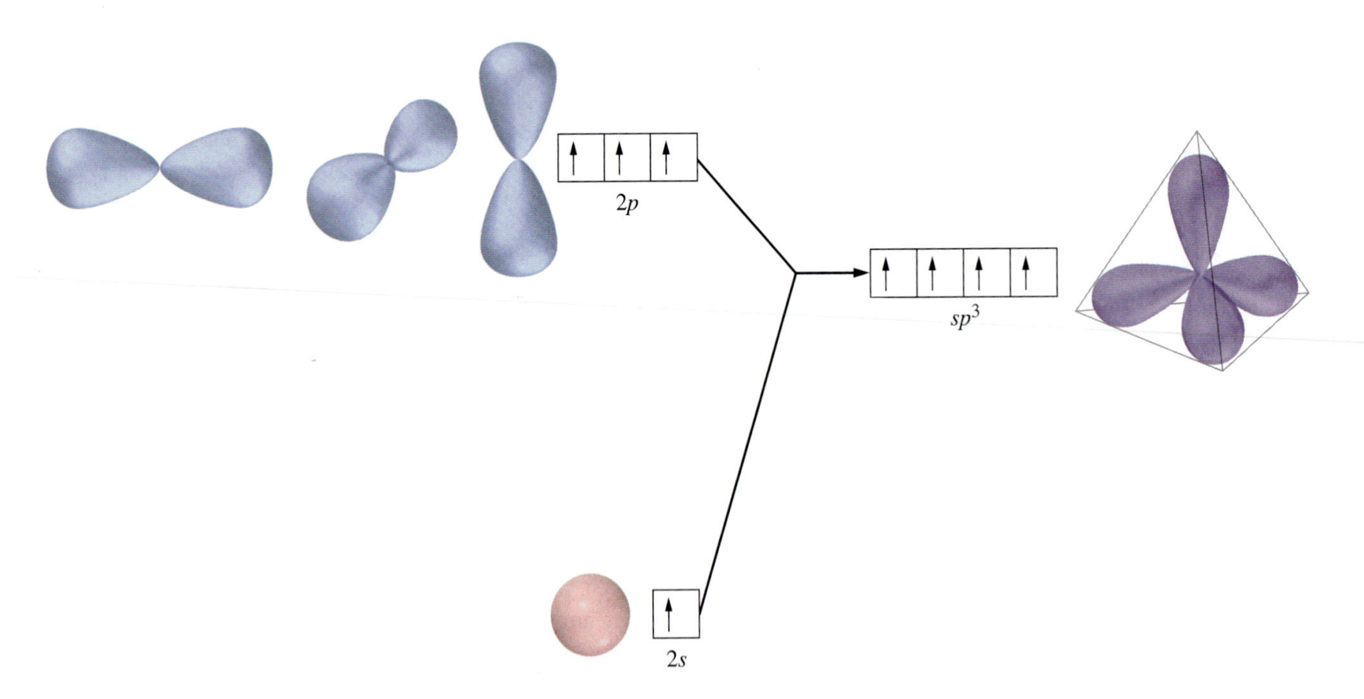

▲ **Figure 10.12 The sp^3 hybridization scheme for C**
The $2s$ and $2p$ orbitals shown on the left combine to yield the four sp^3 orbitals on the right. The placement of the valence-shell orbital diagrams suggests that the $2s$ orbital makes about a 25% contribution and the $2p$ orbitals contribute about 75% to the hybridized orbitals.

The most important point, perhaps, is that hybridization is a way to rationalize a particular molecular structure that has been determined *by experiment.* The hybridization scheme must assess the energy changes in a series of hypothetical processes: promoting an electron from the ground state to an excited state of an atom, hybridizing orbitals in the excited state, and using the hybrid orbitals to form bonds. A successful hybridization scheme minimizes the energy of the molecular structure and satisfactorily accounts for the observed molecular geometry.

To illustrate the role of sp^3 hybrid orbitals in bond formation in methane, we can show hybridization in the central carbon atom with an orbital diagram and show orbital overlaps as illustrated in Figure 10.13.

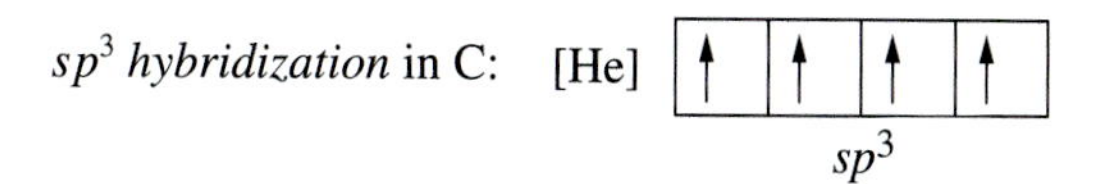

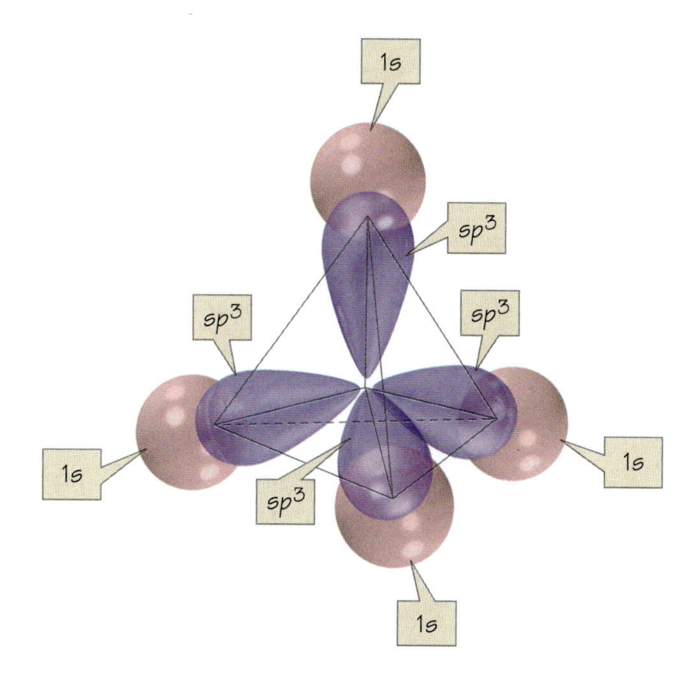

Figure 10.13
sp^3 hybrid orbitals and bonding in CH_4
The four sp^3 hybrid orbitals of the C atom (purple) have been modified to eliminate the small lobes directed toward the center of the structure; they are not involved in orbital overlaps. The hydrogen orbitals are 1*s* (red). The molecular geometry is tetrahedral; the H—C—H bond angles are 109.5°.

We might expect to use sp^3 hybridization not only for structures of the type AX_4 (as in CH_4) but also for AX_3E (as in NH_3) and AX_2E_2 (as in H_2O). For example, suppose we form sp^3 hybrid orbitals from the valence-shell atomic orbitals of a central N atom in NH_3. When we assign the *five* valence electrons to the *four* hybrid orbitals, we have a lone pair of electrons in one of the orbitals and unpaired electrons in the other three (Figure 10.14).

sp^3 *hybridization* in N: [He] [↑↓][↑][↑][↑] sp^3

The hybrid orbitals with the unpaired electrons overlap with the 1*s* orbitals of H atoms to form the three N—H bonds. The predicted H—N—H bond angles of

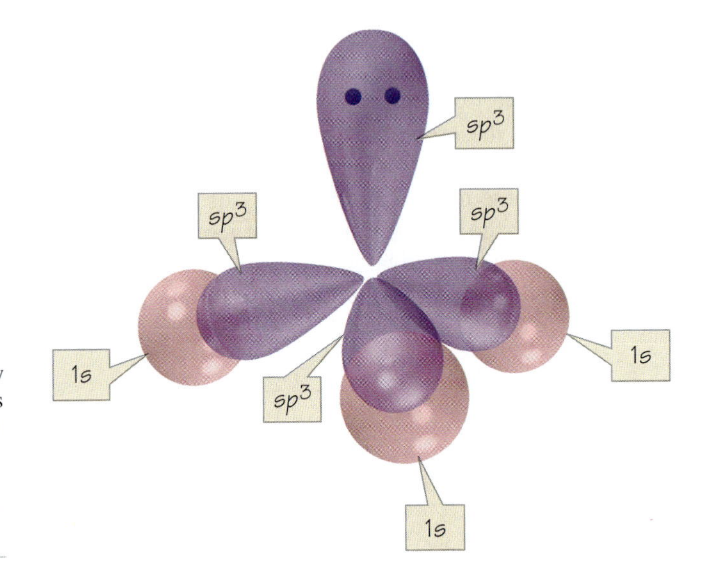

▶ **Figure 10.14**
***sp*^{3} hybrid orbitals and bonding in NH_3**

The three N — H bonds are formed by the overlap of three sp^3 hybrid orbitals of the N atom and 1*s* orbitals of the H atoms. The molecular geometry of NH_3 is trigonal pyramidal. The lone-pair electrons on the N atom reside in the fourth sp^3 orbital.

109.5° are close to the experimentally observed angles of 107°. A similar hybridization scheme for H_2O accounts for the formation of two O — H bonds and the presence of two lone pairs of electrons on the O atom.

sp^3 *hybridization* in O: [He] [↑↓][↑↓][↑][↑]
sp^3

The predicted H — O — H bond angle of 109.5° is also reasonably close to the observed 104.5° angle. As in our discussion of the VSEPR method, we can explain the somewhat smaller than tetrahedral bond angles in NH_3 and H_2O through repulsions involving lone-pair electrons.

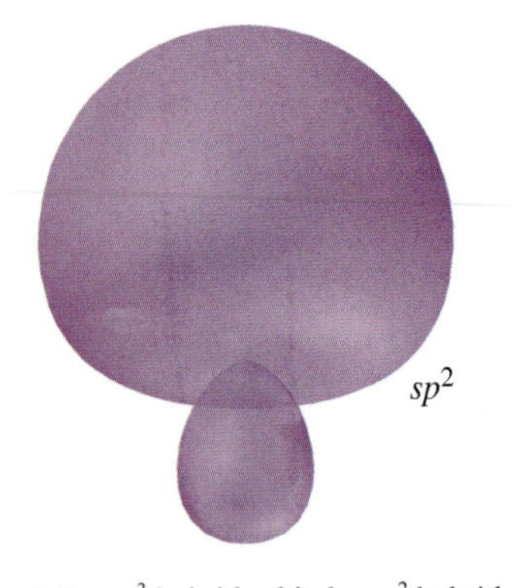

Like sp^3 hybrid orbitals, sp^2 hybrid orbitals have a small lobe pointing toward the nucleus and a large lobe pointing away from the nucleus. The small lobe is larger here, indicating more *s* character in an sp^2 orbital (about 33%) than in an sp^3 orbital (about 25%).

sp^2 Hybrid Orbitals

Next, let's consider sp^2 hybridization. The sp^2 hybridization scheme is especially useful in describing double covalent bonds, as we shall see in Section 10.5. For now, though, we look at a simpler application of its use in describing boron compounds. The first step in the sp^2 hybridization scheme involves promoting a 2*s* electron to an empty 2*p* orbital.

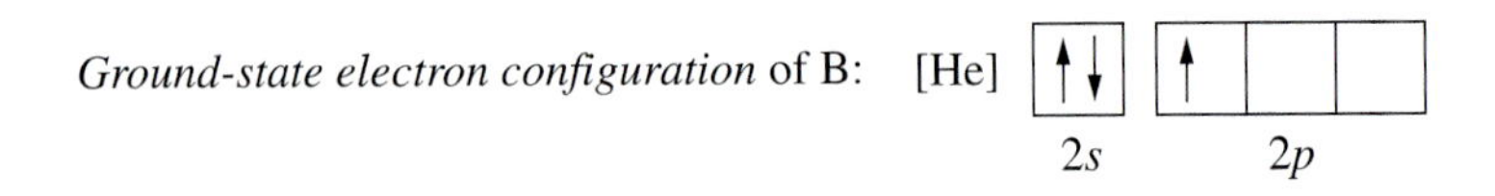

Excited–state electron configuration of B: [He] [↑] [↑][↑][ ]
2*s* 2*p*

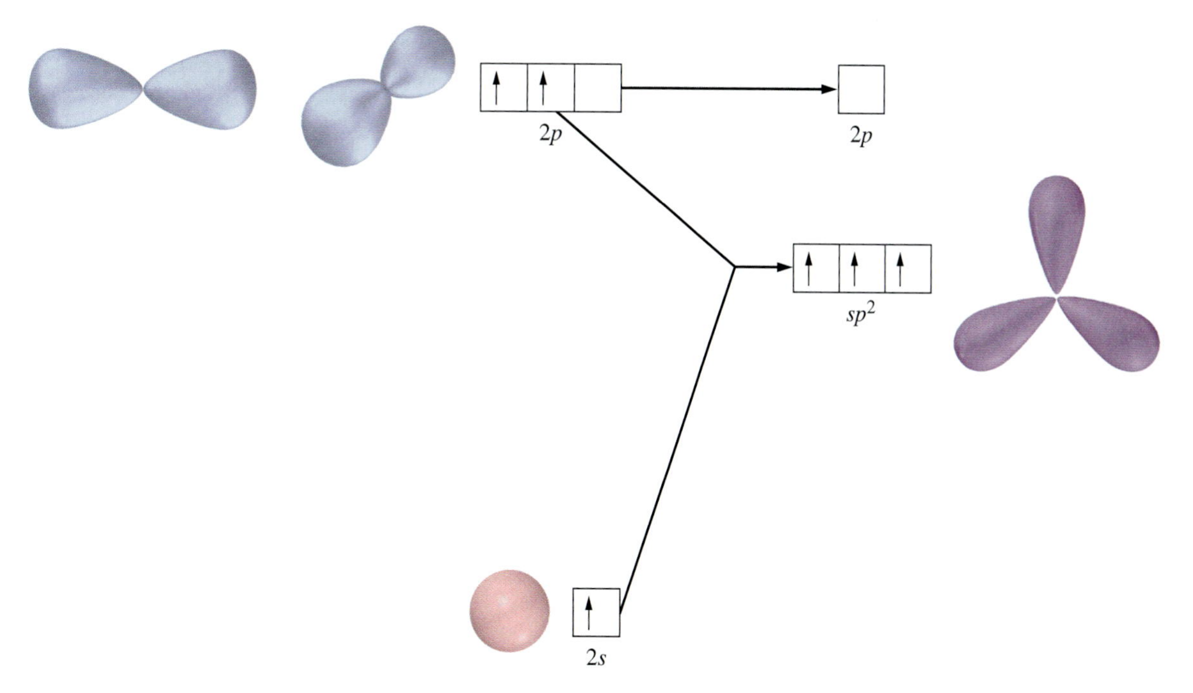

▲ **Figure 10.15 The sp^2 hybridization scheme for boron**
The placement of the valence-shell orbital diagrams suggests that the $2s$ orbital makes about a 33% contribution and that the $2p$ orbitals contribute about 67% to the hybridized orbitals.

As shown in Figure 10.15, the one $2s$ orbital and the two $2p$ orbitals occupied by unpaired electrons are hybridized into three $\boldsymbol{sp^2}$ hybrid orbitals, and the empty $2p$ orbital remains unhybridized.

sp^2 hybridization in B: [He] [↑][↑][↑] sp^2 [ ] $2p$

The geometric distribution of the three sp^2 hybrid orbitals is within a plane, directed at 120° angles. The valence bond theory predicts that BF_3 is a trigonal planar molecule with 120° F-B-F bond angles, exactly as is observed experimentally.

sp Hybrid Orbitals

The *sp* hybridization scheme is especially useful in describing triple covalent bonds (Section 10.5). However, let's look first at a simpler compound. Beryllium and chlorine form the triatomic molecule, $BeCl_2$, which is present in gaseous $BeCl_2$ at high temperatures. To describe bonding in this molecule, again, the first step is to promote a $2s$ electron to a $2p$ orbital, followed by hybridization of the orbitals in the excited-state atom.

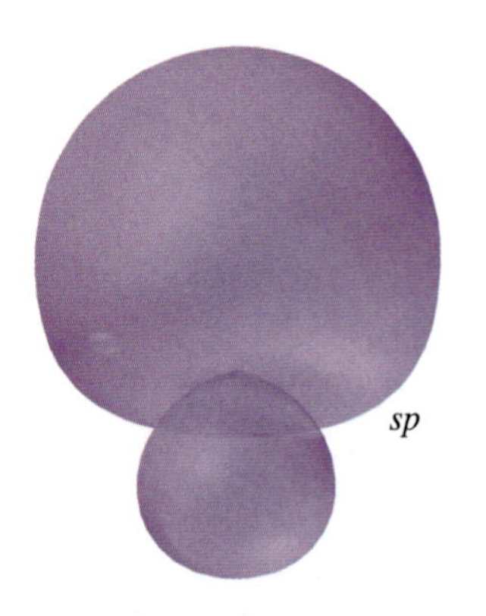

Like sp^3 and sp^2 hybrid orbitals, *sp* hybrid orbitals have a small lobe pointing toward the nucleus and a large lobe pointing away from the nucleus. The small lobe is still larger here, indicating more *s* character in an *sp* orbital (about 50%) than in an sp^2 orbital (about 33%) or in an sp^3 orbital (about 25%).

Ground-state electron configuration of Be: [He] [↑↓] 2s [ ][ ][ ] 2p

Excited–state electron configuration of Be: [He] [↑] 2s [↑][ ][ ] 2p

As shown in Figure 10.16, the 2*s* orbital and the 2*p* orbital occupied by an unpaired electron are hybridized into two ***sp*** hybrid orbitals, and the two empty 2*p* orbitals remain unhybridized.

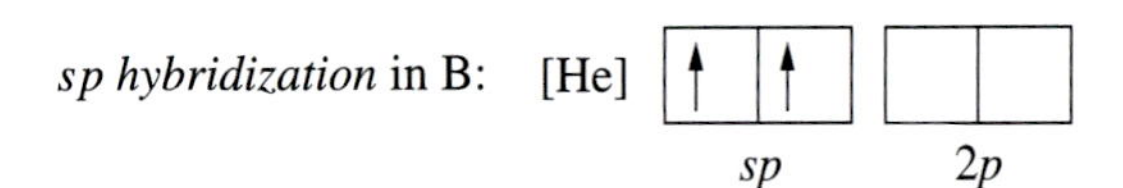

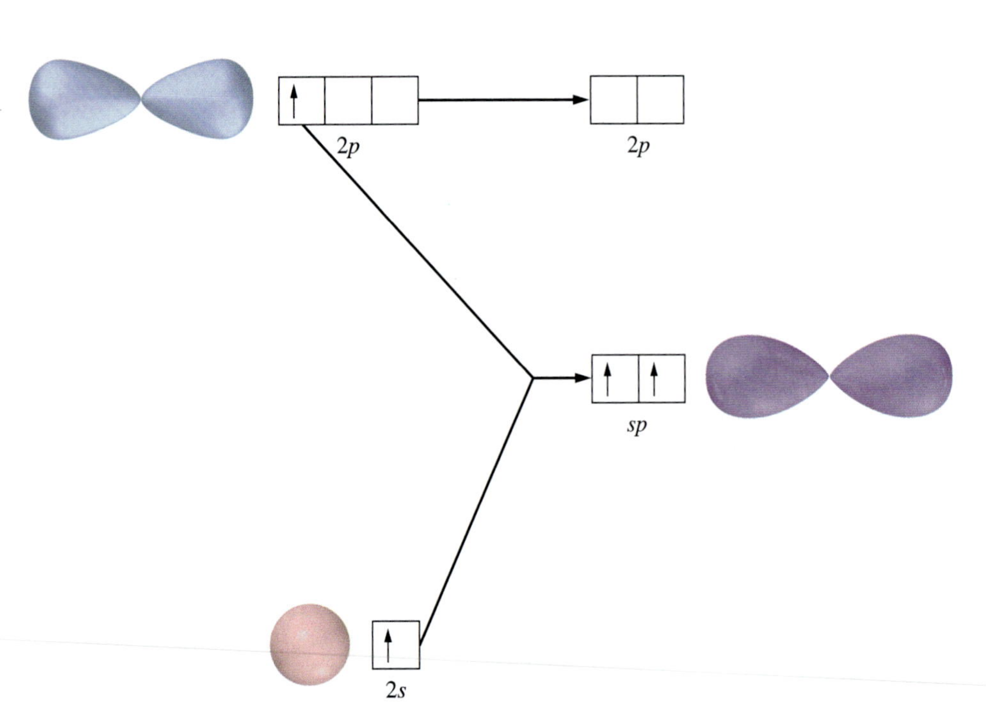

▲ Figure 10.16 The *sp* hybridization scheme for beryllium
The placement of the valence-shell orbital diagrams suggests that the 2*s* and 2*p* orbitals make about equal contributions to the hybridized orbitals.

The geometric distribution of the two *sp* hybrid orbitals is along a line through the Be atom, with a 180° angle between them. We predict that the $BeCl_2$ molecule should be linear, and this is confirmed by experimental evidence.

Hybrid Orbitals Involving *d* Subshells

Any hybridization scheme that involves only *s* and *p* orbitals can accommodate a maximum of *eight* valence electrons, that is, a valence-shell octet. To apply a hybridization scheme to structures involving *expanded* valence shells, we need additional orbitals, and we can find these in the *d* subshell.

For example, we need *five* hybrid orbitals to describe bonding in PCl_5. We get these by combining *one s*, *three p*, and *one d* orbital from the central phosphorus atom, as suggested by the following orbital diagrams.

Ground-state electron configuration of P:

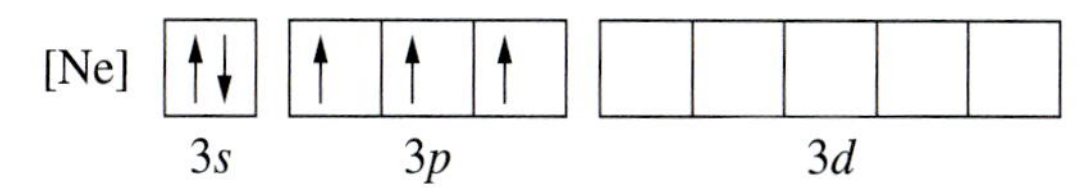

Excited–state electron configuration of P:

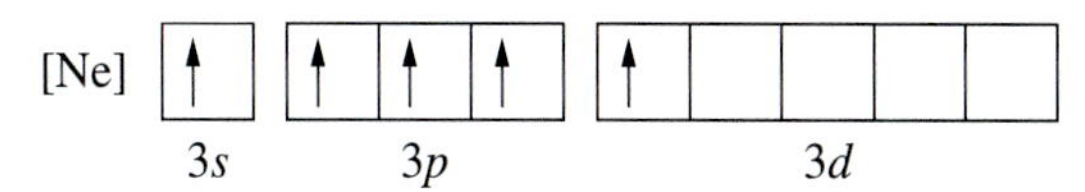

sp^3 d hybridization in P:

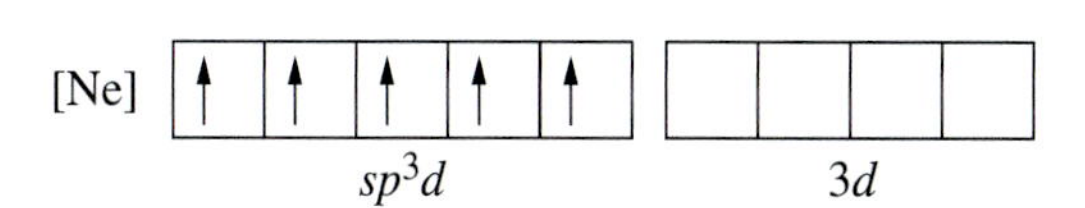

The ***sp^3d*** hybrid orbitals and their trigonal bipyramidal orientation are shown in Figure 10.17. The orbital overlap in each bond in PCl_5 involves an sp^3d hybrid orbital of the P atom and a 3*p* orbital of a Cl atom.

▲ **Figure 10.17**
The *sp^3d* hybrid orbitals
These sp^3d hybrid orbitals in a trigonal bipyramidal arrangement are deployed by a phosphorus atom in the molecule PCl_5 (see p. 412).

Another molecule featuring an expanded valence shell is SF_6. Here, *six* hybrid orbitals for the central sulfur atom are required to describe bonding. We obtain them through the hybridization scheme sp^3d^2, represented through the following orbital diagrams.

Ground-state electron configuration of S:

[Ne] 3s 3p 3d

Excited–state electron configuration of S:

[Ne] 3s 3p 3d

sp^3d^2 hybridization in S: [Ne] sp^3d^2 3d

The ***sp^3d^2* hybrid orbitals** and their octahedral orientation around the central sulfur atom are shown in Figure 10.18. The orbital overlap in each bond in SF_6 involves an sp^3d^2 hybrid orbital of the S atom and a 2*p* orbital of an F atom.

▲ **Figure 10.18**
The *sp^3d^2* hybrid orbitals
The sp^3d^2 hybrid orbitals in an octahedral arrangement are deployed by a sulfur atom in the molecule SF_6 (see p. 413).

Although chemists generally have described bonding in structures with expanded valence shells as involving hybrid orbitals with *d*-orbital contributions, experimental evidence does not provide strong support for these descriptions. The situation is similar to that of expanded-valence-shell Lewis structures in Chapter 9 (page 388). The energy cost in promoting electrons to *d* orbitals in hybridization schemes may not be offset by the lowering of energy produced by orbital overlaps. Here, our approach will be the same as in Chapter 9: We will describe bonding through expanded valence shells and the hybridization schemes that produce them, such as sp^3d and sp^3d^2, when we do not have a simple alternative. Fortunately, the evidence for hybridization schemes using *s* and *p* orbitals is much stronger, and sp, sp^2, and sp^3 hybrid orbitals are the ones that we will encounter most frequently. In Section 10.5, for example, these are the only types of hybrid orbitals that we will use.

Predicting Hybridization Schemes

In hybridization schemes, one hybrid orbital is produced for every simple atomic orbital involved. In a molecule, each of the hybrid orbitals of the central atom acquires an electron pair, either a bonding pair or a lone pair. Also, the hybrid orbitals have the same symmetrical orientations as the electron-group geometries predicted by the VSEPR method and summarized in Table 10.2.

TABLE 10.2 Hybrid Orbitals and Their Geometric Orientations

Hybrid Orbitals	Geometric Orientation	Example
sp	Linear	$BeCl_2$
sp^2	Trigonal planar	BF_3
sp^3	Tetrahedral	CH_4
sp^3d	Trigonal bipyramidal	PCl_5
sp^3d^2	Octahedral	SF_6

If we have *experimental* evidence relating to a molecular structure, the hybridization scheme we choose to describe bonding in the structure must conform to the evidence. Often, however, we don't have experimental evidence at hand and need to *predict* a probable hybridization scheme. In these cases, the following four-step approach generally works.

1. *Write a plausible Lewis structure for the molecule or ion.*
2. *Use the VSEPR method to predict the electron-group geometry of the central atom.*
3. *Select the hybridization scheme that corresponds to the VSEPR prediction.*
4. *Describe the orbital overlap and molecular geometry.*

EXAMPLE 10.6

Iodine pentafluoride, IF_5, is used commercially as a fluorinating agent—a substance that introduces fluorine into other compounds in its chemical reactions. Describe a hybridization scheme for the central atom, and sketch the molecular geometry of the IF_5 molecule.

SOLUTION

Before we even begin with the four-step approach outlined above, can you see that the hybridization scheme cannot be *sp*, sp^2, or sp^3? The central iodine atom forms *five* bonds, and this requires an expanded valence shell.

1 *Write a plausible Lewis structure.* Because all six atoms are in Group 7A of the periodic table, there must be (6 × 7) = 42 valence-shell electrons in the Lewis structure. In the structure below, we draw the five I—F bonds and then complete the octets of the F atoms.

```
       :F:
        |
   :F—I—F:
       / \
    :F:  :F:
```

This accounts for 40 of the 42 valence electrons. The remaining electron pair (red) must go on the central iodine atom.

```
       :F:
        |
   :F—I:—F:
       / \
    :F:  :F:
```

2 *Use the VSEPR method to predict the electron-group geometry of the central atom.* The electron-group geometry for six electron pairs is *octahedral*; the six electron pairs are directed to the corners of an octahedron.

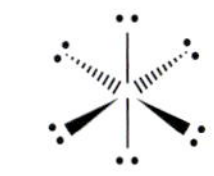

3 *Select the hybridization scheme that corresponds to the VSEPR prediction.* The hybridization scheme that gives an octahedral distribution of hybridized orbitals is sp^3d^2.

4 *Describe the orbital overlap and molecular geometry.* To form five I—F single covalent bonds, we need one electron each from the I atom. The lone-pair electrons belong entirely to the I atom. The assignment of (5 + 2) = 7 valence electrons to the six sp^3d^2 orbitals of iodine in shown in the orbital diagram below. (The lone-pair electrons are in red.)

sp^3d^2 *hybridization* in I: [↑↓][↑][↑][↑][↑][↑] [ ][ ][ ]

sp^3d^2 5*d*

As shown in Figure 10.19, five of the hybrid orbitals for iodine overlap with 2*p* orbitals of the fluorine atoms. Because the six hybrid orbitals are equivalent, we can place the lone-pair electrons in any one of them. The molecular geometry that corresponds to this distribution of bonding and lone-pair electrons is that of a *square pyramid.* We expect F—I—F bond angles of about 90°.

EXERCISE 10.6A

Describe a hybridization scheme for the central atoms and the molecular geometry of the silicon tetrachloride molecule.

PROBLEM-SOLVING NOTE

Note that because of the similarity in formulas, we might expect IF_5 to have the same structure as PCl_5, trigonal bipyramidal. It doesn't, however, because PCl_5 is of the VSEPR type AX_5, whereas IF_5 is AX_5E.

EXERCISE 10.6B

Describe a hybridization scheme for the central atom and the molecular geometry of the triiodide ion, I_3^-.

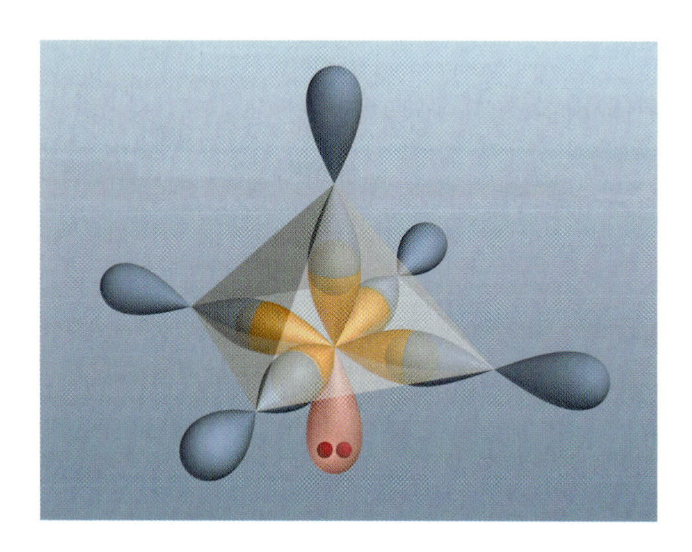

Figure 10.19 Bonding scheme for iodine pentafluoride, IF_5
The central I atom is hybridized sp^3d^2. One of the hybrid orbitals (red) is occupied by lone-pair electrons. The other orbitals are the bonding orbitals. Each bond involves the overlap of an sp^3d^2 orbital with a $2p$ orbital of a terminal F atom. Because of repulsions between the lone-pair electrons and the I—F bonding pairs, the plane of the four F atoms at the base is raised slightly above the I atom.

10.5 Hybrid Orbitals and Multiple Covalent Bonds

We already know how to predict the geometric structures of molecules and polyatomic ions with double and triple covalent bonds. When we combine this knowledge with the valence bond theory, we gain additional insight into some of the essential characteristics of multiple covalent bonds, such as their bond energies.

In Chapter 9, we found the Lewis structure of ethylene, C_2H_4, to have a double bond between the two carbon atoms.

```
   H  H
   |  |
H—C=C—H
```

With the VSEPR method, we would predict that the electron-group geometry around each C atom is *trigonal planar*. Thus, each CH_2 group lies in a plane with a H—C—H bond angle of 120°. Each H—C—C bond is also 120°. But as we can see from Figure 10.20, the VSEPR method doesn't tell us how the two CH_2 groups are oriented with respect to one another. Are they both in the same plane, in planes perpendicular to one another, or at some other angle? To see how the valence bond theory accounts for both the 120° bond angles *and* the orientation of the CH_2 groups, let's write orbital diagrams for sp^2 hybridization of the two C atoms.

sp^2 hybridization in first C: [He] [↑][↑][↑] (sp^2) [↑] ($2p$)

sp^2 hybridization in second C: [He] [↑][↑][↑] (sp^2) [↑] ($2p$)

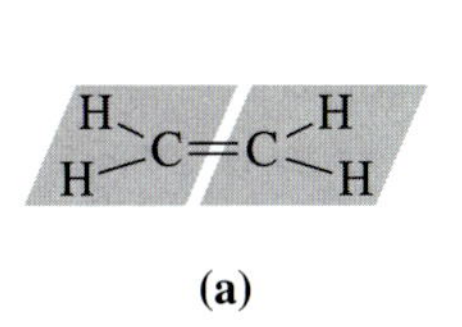

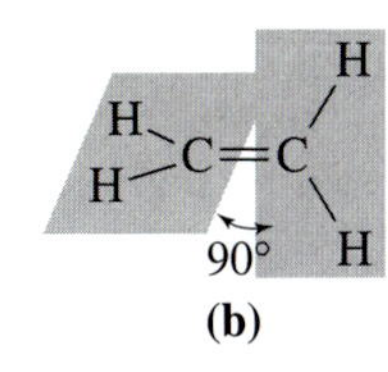

▲ **Figure 10.20 Lewis structure and VSEPR description of C_2H_4**
Lewis theory predicts a double bond between C atoms in ethylene. The VSEPR method predicts that the three atoms in a CH_2 group lie in the same plane, but it doesn't indicate how the two planes are oriented with respect to one another. That is, we don't know whether the two CH_2 groups are (a) coplanar, (b) perpendicular to one another, or (c) at some angle α.

As we can see from Figure 10.21, all the C—H bonds in C_2H_4 are formed by the overlap of sp^2 hybrid orbitals of the C atoms with $1s$ orbitals of the H atoms. The region of maximum overlap occurs along hypothetical lines drawn between the nuclei of the bonded atoms. We say that the orbitals overlap "end to end." Covalent bonds formed by the end-to-end overlap of orbitals, regardless of orbital type, are called **sigma (σ) bonds**. All single covalent bonds are σ bonds. The double bond between the two C atoms in C_2H_4, in contrast, has *two* components: One of the bonds involves sp^2 orbitals overlapping along the line joining the two carbon nuclei; it, like the C—H bonds, is a σ bond. The other bond between the two C atoms results from the overlap of the half-filled, unhybridized $2p$ orbitals that extend above and below the plane of the C and H atoms. These orbitals overlap in a "parallel," or "side-by-side," fashion, not along a line joining the carbon nuclei. A bond formed by this type of orbital overlap is called a **pi (π) bond**. A double covalent bond consists of *one* σ and *one* π bond. Valence bond theory predicts that all six atoms in C_2H_4 lie in the same plane, as shown in Figure 10.21. This is the arrangement that allows for the *maximum* overlap of $2p$ orbitals and the *strongest* π bond. If one CH_2 group were twisted out of the plane of the other, the extent of overlap would be lessened, the π bond would be weakened, and the molecule would be less stable. The point of *minimum* overlap would occur if the planes of the two CH_2 groups were mutually perpendicular.

The positions of terminal atoms with respect to central atoms is determined by the end-to-end orbital overlaps, the σ bonds. Collectively, the σ bonds in a structure constitute the *σ-bond framework*, and the σ-bond framework outlines the molecular geometry. Electron pairs in π bonds *do not* affect the positions of bonded atoms. In the VSEPR method, when we treat all the electrons in a multiple covalent bond as a single electron group, we are in effect constructing an electron-group geometry that is the same as the σ-bond framework of the valence bond theory.

We can describe a triple covalent bond in much the same way as a double bond. Consider the acetylene molecule, C_2H_2, with the Lewis structure H—C≡C—H. The molecule is *linear*, with 180° H—C—C bond angles, as predicted by the VSEPR method and confirmed by experiment. To account for these bond angles with the valence bond theory, we assume *sp* hybridization of the valence-shell orbitals of the two C atoms.

sp hybridization in first C atom: [He] 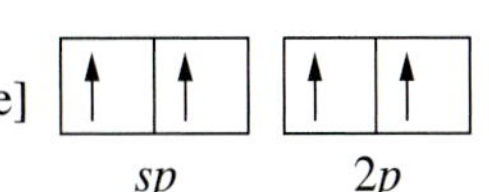

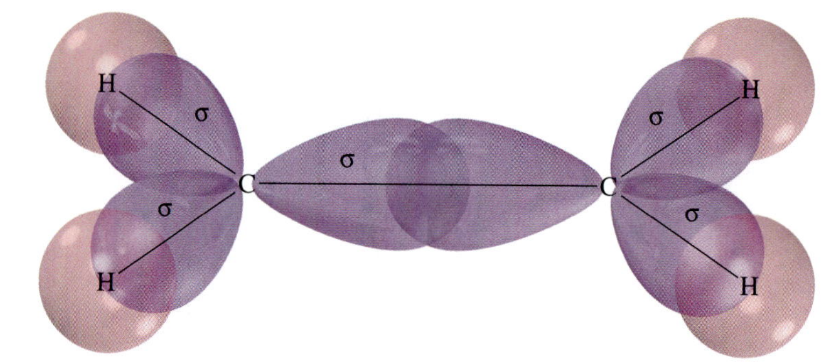

(a) The σ-bond framework

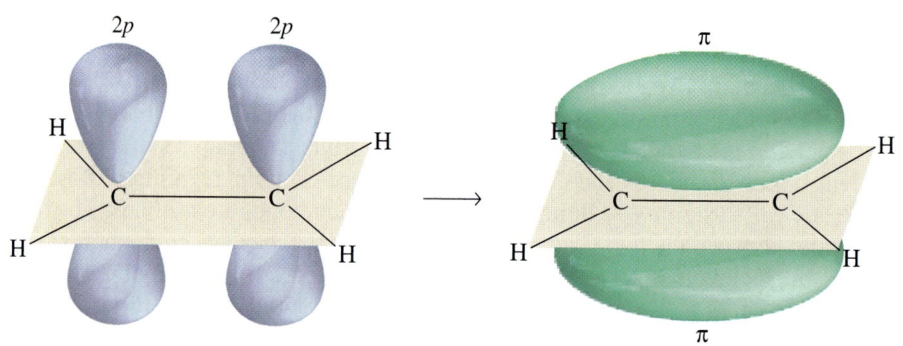

(b) The formation of a π-bond by the overlap of the half-filled 2*p* orbitals

π: C(2*p*) − C(2*p*)

σ: H(1*s*) − C(sp^2)

σ: H(1*s*) − C(sp^2)

σ: C(sp^2) − C(sp^2)

(c) Hybridization and bonding scheme

▲ **Figure 10.21 Bonding in ethylene, C_2H_4, by the valence bond theory**
(a) The σ-bond framework, (b) formation of π bonds by the overlap of half-filled 2*p* orbitals, and (c) hybridization and bonding scheme

sp hybridization in second C atom: [He] ↑ ↑ (*sp*) ↑ ↑ (2*p*)

In the C≡C bond in C_2H_2, as in all triple bonds, *one* bond is a σ bond and *two* are π bonds. The bonding scheme in acetylene is illustrated in Figure 10.22.

EXAMPLE 10.7

Formic acid, HCOOH, is the irritant released in an ant bite (Latin, *formica*, ant).

a. Predict a plausible molecular geometry for this molecule.

(a) The σ-bond framework

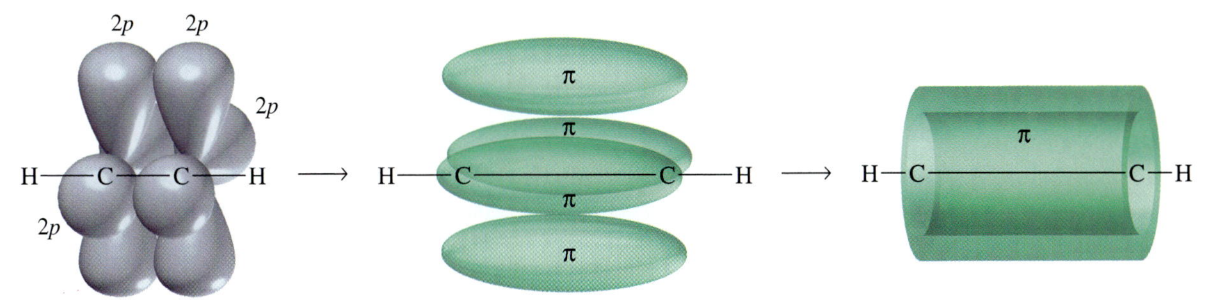

(b) Formation of π-bonds by the overlap of half-filled 2p orbitals

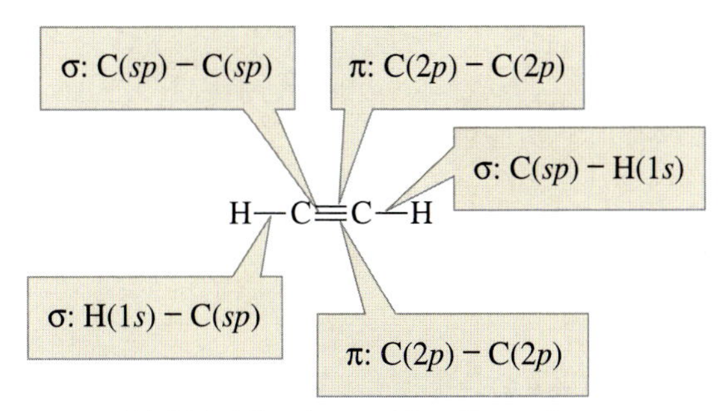

(c) Hybridization and bonding scheme

▲ **Figure 10.22 Bonding in acetylene, C_2H_2, by the valence bond theory**
The σ-bond framework joins the atoms in a linear structure through the overlap of 1s orbitals of the H atoms and sp orbitals of the C atoms. Each π bond can be thought of as two parallel cigar-shaped segments. In fact, however, when two π bonds are present, the segments merge into a hollow and symmetric cylindrical shell with the carbon-to-carbon σ bond as its axis.

b. Propose a hybridization scheme for the central atoms that is consistent with that geometry.

c. Sketch a bonding scheme for the molecule.

SOLUTION

a. We can begin by writing a plausible Lewis structure. There are $(2 \times 1) + (2 \times 6) + 4 = 18$ valence electrons. We need all 18 electrons for the skeletal structure and to provide valence-shell octets for the two O atoms. To complete the octet for C, we then form a carbon-to-oxygen double bond.

$$\mathrm{H{-}C({-}\ddot{\underset{..}{O}}{:}){-}\ddot{\underset{..}{O}}{-}H \longrightarrow H{-}C({=}\ddot{O}{:}){-}\ddot{\underset{..}{O}}{-}H}$$

Then, with the VSEPR method, we can describe the electron-group geometry about the central atoms, C and O. The orientation of *three* electron groups about the C atom—two single bonds and a double bond—is *trigonal planar*. The orientation of *four* electron groups about the central O atom—two bonding pairs and two lone pairs—is *tetrahedral*.

Because the three electron groups in the valence shell of the C atom are in bonds, the molecular geometry around the C atom is the same as the electron-group geometry (*trigonal planar*). We expect the H—C—O and the O—C—O bond angles to be about 120°.

The molecular geometry around the central O atom is *angular*, or *bent*, and we therefore expect a C—O—H bond angle of about 109.5° (based on the VSEPR notation AX_2E_2 and the tetrahedral bond angle).

b. The hybridization schemes corresponding to the electron-group geometries described in (a) are sp^2 for the central C atom and sp^3 for the central O atom.

c. One of the bonds in the C=O double bond is a π bond. All the other bonds in the molecule are σ bonds. These bonds and the orbital overlaps producing them are suggested by the schematic diagram below.

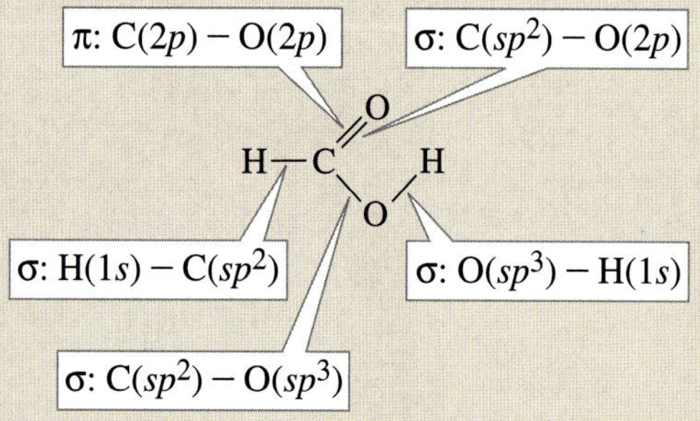

π: C(2p) – O(2p)
σ: C(sp^2) – O(2p)
σ: H(1s) – C(sp^2)
σ: O(sp^3) – H(1s)
σ: C(sp^2) – O(sp^3)

EXERCISE 10.7A

Methanol, CH_3OH, the simplest alcohol, shows promise as a future motor fuel.

a. Predict a plausible molecular geometry for this molecule.

b. Propose a hybridization scheme for the central atoms that is consistent with that geometry.

c. Sketch a bonding scheme for the molecule.

EXERCISE 10.7B

Cyanogen, C_2N_2, is a highly toxic gas used as a fumigant and in synthesizing organic compounds.

a. Predict a plausible molecular geometry for this molecule.

b. Propose a hybridization scheme for the central atoms that is consistent with the predicted geometry.

c. Sketch a bonding scheme for the molecule.

Geometric Isomerism

In discussing Figure 10.20, we concluded that the two CH_2 groups in ethylene, $CH_2{=}CH_2$, must lie in the same plane to produce maximum overlap of the $2p$ orbitals in the π bond between the carbon atoms. Also, to maintain maximum orbital overlap in the π bond, one CH_2 group is prevented from rotating (spinning) with respect to the other. As we will now see, this restricted rotation about a double bond has an important consequence.

There are four C_4H_8 alkene isomers. 1-Butene is related to butane, and isobutylene is related to isobutane.

$$\underset{\text{1-Butene}}{CH_2{=}CHCH_2CH_3} \qquad \underset{\text{Isobutylene}}{CH_2{=}\overset{\overset{\displaystyle CH_3}{|}}{C}{-}CH_3}$$

A third structure readily comes to mind, also related to butane, in which the double bond connects the second and third carbon atoms.

$$\underset{\text{"2-Butene"}}{CH_3CH{=}CHCH_3}$$

What is not immediately apparent is that there are *two* isomers for this structural formula: There are two different 2-butenes.

The two 2-butenes are shown in Figure 10.23. At first, it appears that we should be able to convert one structure to the other just by holding one end of the molecule in a fixed position and rotating the other end by 180°. However, because this rotation about the bond axis would decrease the overlap of $2p$ orbitals on the C atoms and effectively break the π bond, this conversion doesn't occur. The two isomers of 2-butene are *two distinctly different compounds.* To distinguish between them, we call one *cis*-2-butene and the other *trans*-2-butene.

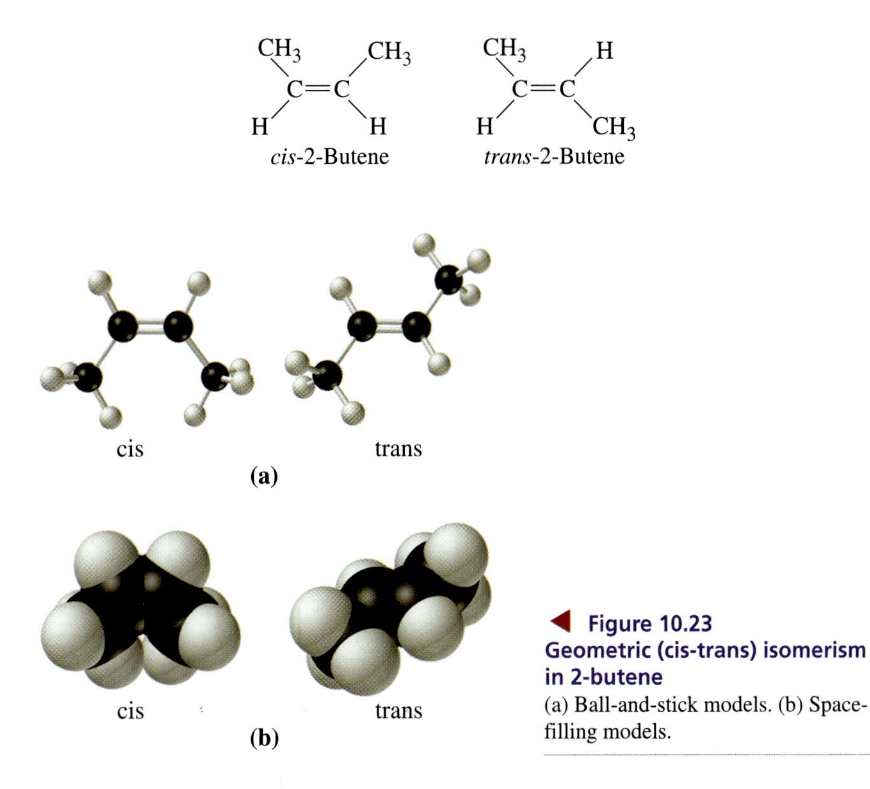

Figure 10.23 Geometric (cis-trans) isomerism in 2-butene
(a) Ball-and-stick models. (b) Space-filling models.

The **cis** isomer has both methyl groups (CH_3) on the same side of the molecule. In the **trans** isomer, the methyl groups are on opposite sides or across the double bond from one another. (Draw a straight line passing through the two carbon atoms in the double bond. If the methyl groups fall on the same side of the line, the compound is *cis*. If they fall on opposite sides, the compound is *trans*.) Cis and trans isomers

In Latin, *cis* means "on this side," and *trans* means "across." With these prefixes, it is not difficult to figure out the meanings of the adjectives cisatlantic and transatlantic.

differ only in the geometric arrangement of certain substituent groups; they are called **geometric isomers**.

If either carbon atom in a double bond has two identical atoms or groups bonded to it, cis-trans isomerism is not possible. Propylene has two hydrogen atoms bonded to one of the carbon atoms in the double bond; there are no cis-trans isomers of propylene. Although we can draw two structural formulas, structure (II) is really the same as structure (I).

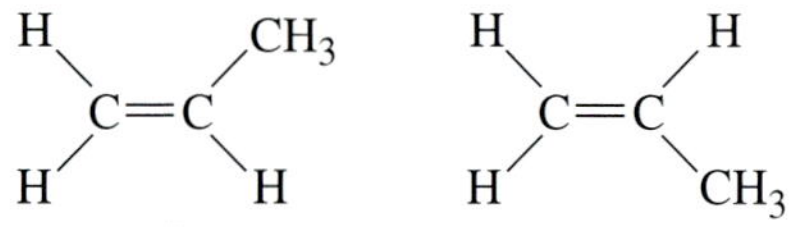

H CH₃ H H
C=C C=C
H H H CH₃
(I) (II)

You need only to imagine picking structure (II) up from the page and flipping it over to see that the two structures are identical.

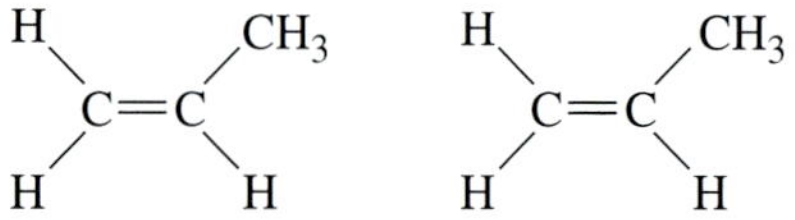

H CH₃ H CH₃
C=C C=C
H H H H
(I) (II: flipped over)

EXAMPLE 10.8—A Conceptual Example

A molecule having the formula $C_2H_2Cl_2$ is found to be nonpolar. Can we use this information to write the structural formula of the molecule?

SOLUTION

We have written structural formulas based on molecular formulas before. If only one structural formula corresponds to $C_2H_2Cl_2$, our problem is solved immediately. If isomers are possible, the answer is not so clear cut. The molecule in question is a substituted *ethene*, and we see that there are *three* possible isomers.

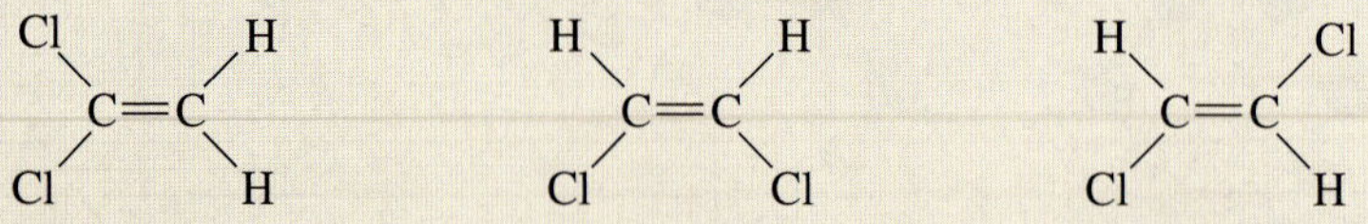

Cl H H H H Cl
C=C C=C C=C
Cl H Cl Cl Cl H
(I) (II) (III)
1,1-Dichloroethene *cis*-1,2-Dichloroethene *trans*-1,2-Dichloroethene

In structure (I), with two Cl atoms on the same C atom, we expect a net displacement of electrons toward the Cl atoms, creating a molecular dipole. In structure (II), a cis isomer, both Cl atoms are on the same side of the double bond, again giving rise to a molecular dipole. In structure (III), the bond dipoles associated with the C—Cl bonds cancel, as do the small dipoles associated with the C—H bonds. Structure (III) represents the only *nonpolar* molecule of the three, and it is the one that we should write. (Can you see that if the molecule we had been asked to identify were *polar* we would not have been able to come up with a definite answer?)

EXERCISE 10.8

Which of the following molecules would you expect to be polar: **(a)** fluoroethene; **(b)** *trans*-2-butene; **(c)** acetylene; or **(d)** 2,3-dichloro-*cis*-2-butene? Explain your reasoning.

Geometric Isomerism and Vision

Our vision requires that light convert one geometric isomer to another. The molecule that undergoes conversion is a light-sensitive pigment known as rhodopsin, which is found in the receptor cells of the retina of the eye. Rhodopsin is a complex of a compound called 11-*cis*-retinal and a protein called opsin. When light strikes rhodopsin, a reaction called *isomerization* occurs. The cis isomer is converted to the trans isomer.

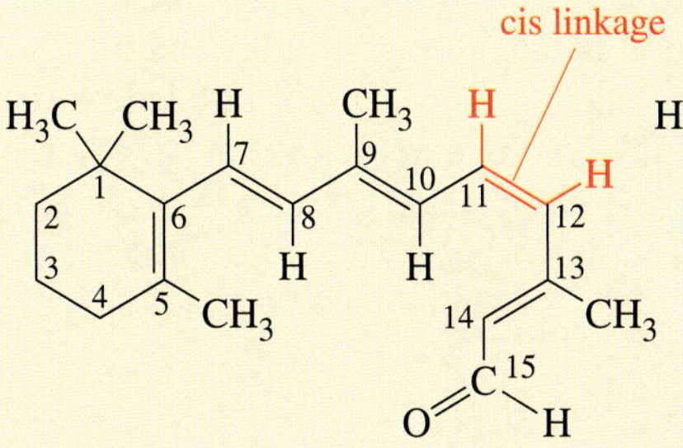

cis linkage

11-*cis*-Retinal

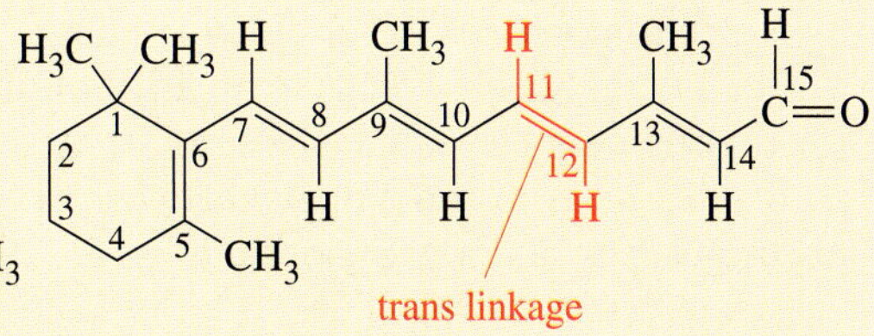

trans linkage

11-*trans*-Retinal

Cis to trans isomerization of 11-*cis*-retinal initiates an electrical impulse that is transmitted to the brain through the optic nerve. The brain translates the nerve impulse into an image, and we see. Following transmission of the impulse, the opsin protein releases the 11-*trans*-retinal, and another enzyme converts 11-*trans*-retinal back to 11-*cis*-retinal. The 11-*cis*-retinal again complexes with opsin and is primed for the next pulse of light.

Some retinal is lost during the regeneration of opsin from rhodopsin and must be replaced by vitamin A from the bloodstream, making vitamin A an essential vitamin for proper vision. Vitamin A differs from 11-*trans*-retinal only in that it has a terminal alcohol group, $-CH_2OH$, instead of the aldehyde group, $-CHO$.

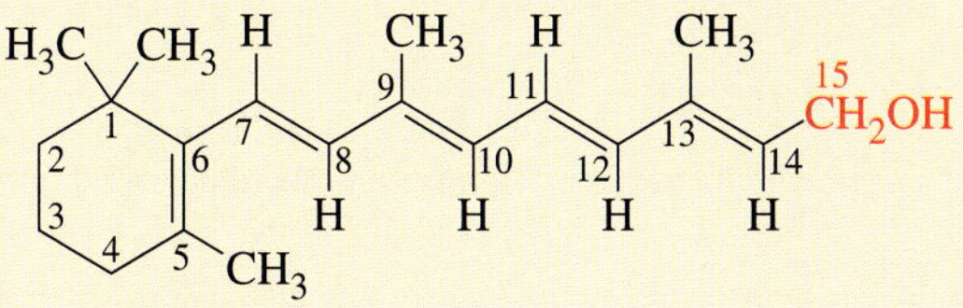

Vitamin A

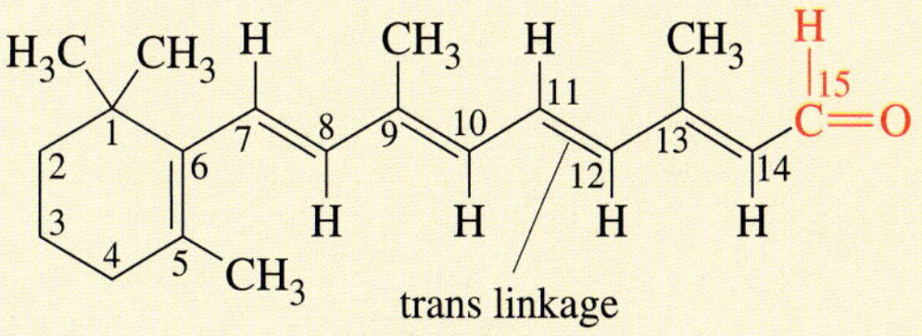

trans linkage

11-*trans*-Retinal

Molecular Orbital Theory

In the valence bond theory, some of the features of isolated atoms are retained when the atoms are bonded together. Only the valence orbitals involved in the bonding are modified. There is an alternative quantum mechanical approach in which we start from scratch. In **molecular orbital theory**, we use an arrangement of appropriately placed atomic nuclei and place electrons in *molecular orbitals* in a way that leads to an energetically favorable, stable molecule.

10.6 Characteristics of Molecular Orbitals

Molecular orbitals (MOs) are mathematical equations that describe the regions in a molecule where there is a high probability of finding electrons. In this regard, they are like the atomic orbitals used to describe the electrons in atoms. In fact, one way to derive molecular orbitals is by an appropriate combination of the atomic orbitals of the atoms being united into a molecule.

Figure 10.24 describes the formation of molecular orbitals by the combination of two 1*s* atomic orbitals. One combination leads to a **bonding molecular orbital**, σ_{1s}, which is at a lower energy level than the separate atomic orbitals and places a high electron probability, or electron charge density, *between* the bonded atoms. The other combination leads to an **antibonding molecular orbital**, σ^*_{1s}, which is at a higher energy level than the separate atomic orbitals and places a high electron probability *away from* the region between the bonded atoms. We use an asterisk (*) to designate an antibonding orbital.

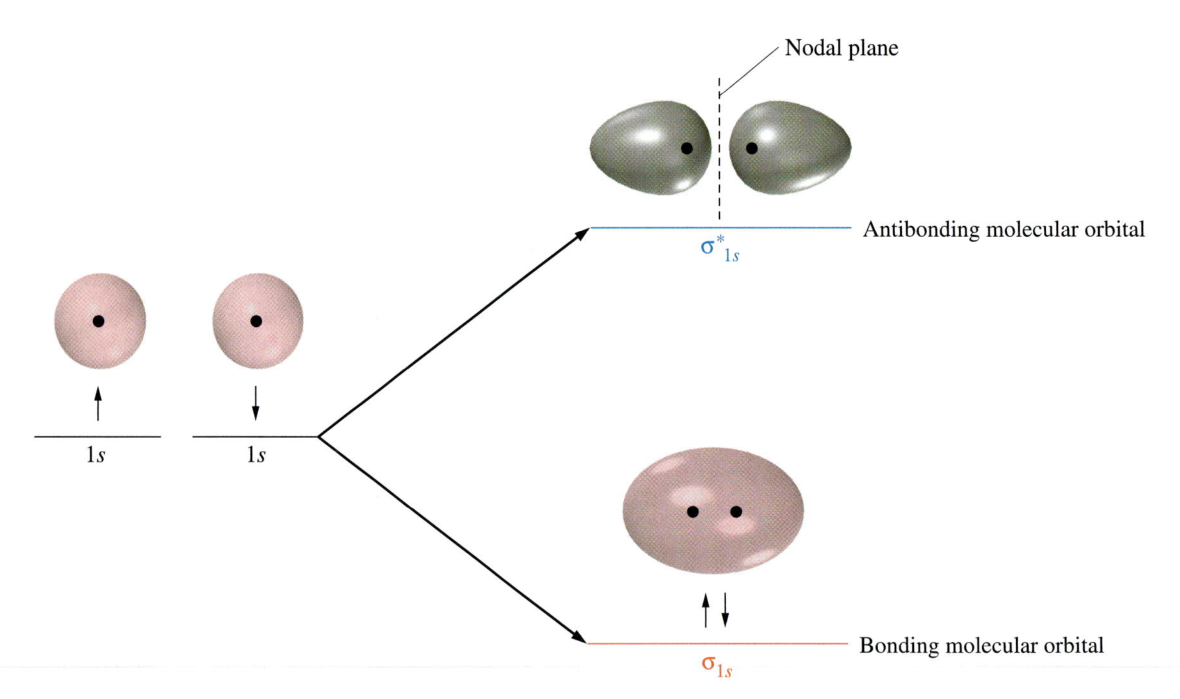

Figure 10.24 Molecular orbitals and bonding in the H_2 molecule
The short horizontal lines represent the relative energy levels of the atomic and molecular orbitals (MOs). The small arrows represent electrons in the hydrogen atoms and in the hydrogen molecule. The figures above the energy levels represent the distribution of electron probability or electron charge density. Note that this quantity falls to zero in the nodal plane between the atomic nuclei (black dots) in the antibonding orbital.

Figure 10.24 shows the assignment of the two electrons in the H_2 molecule to the σ_{1s} bonding molecular orbital. This assignment conforms to an aufbau process similar to that used in writing electron configurations of atoms and subject to these rules.

- Electrons seek the lowest energy molecular orbitals available to them.
- A maximum of two electrons can be present in a molecular orbital (Pauli exclusion principle).
- Electrons enter molecular orbitals of identical energies singly with parallel spins before they pair up (Hund's rule).

Electrons in bonding molecular orbitals *contribute to* the strength of a bond between atoms. Electrons in antibonding molecular orbitals *detract from* the strength of the bond. (In some cases, there is a third type of molecular orbital, a *nonbonding* molecular orbital, that neither contributes to nor detracts from bonding.) The **bond order** in a molecule is one-half the difference between the number of electrons in bonding molecular orbitals and the number in antibonding molecular orbitals.

$$\text{Bond order} = \frac{(\text{number of e}^- \text{ in bonding MOs}) - (\text{number of e}^- \text{ in antibonding MOs})}{2}$$

EXAMPLE 10.9—A Conceptual Example

Molecular orbital theory accounts for species with a "one-electron" bond. What does this term signify? Cite an example of such a bond.

SOLUTION

From the definition of bond order, we can think of a one-electron bond as a bond that has one more electron in bonding MOs than in antibonding MOs; the bond order = 1/2. The simplest example has one electron in the σ_{1s} bonding MO and an empty antibonding σ^*_{1s} MO. It is the hydrogen molecule ion, H_2^+, formed by combining a hydrogen atom and a hydrogen ion. The formation of H_2^+ is suggested by the energy-level diagram of Figure 10.25.

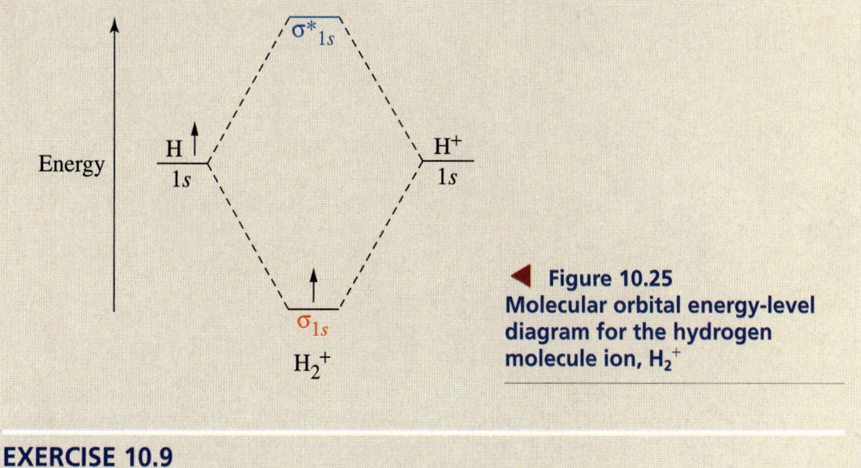

Figure 10.25 Molecular orbital energy-level diagram for the hydrogen molecule ion, H_2^+

EXERCISE 10.9

Would you expect the ion H_2^- to exist as a stable species? Explain.

The scheme for combining 1*s* atomic orbitals in Figure 10.24 and the energy-level diagram shown in Figure 10.25 can be used to describe various diatomic species involving the first-period elements, H and He. Let's consider next molecular orbitals associated with the second-period elements.

10.7 Homonuclear Diatomic Molecules of the Second-Period Elements

We can treat the diatomic species of the first two members of the second period—Li and Be—in a similar fashion to what we did in the first period. First, we can fill the σ_{1s} and σ^*_{1s} molecular orbitals with electrons and then concentrate on the molecular orbitals formed from the valence-shell atomic orbitals, the 2*s*. We obtain another pair of molecular orbitals, σ_{2s} and σ^*_{2s}, at a higher energy than the first-shell molecular orbitals.

To continue with the second period, proceeding from boron to neon, we need to consider molecular orbitals that are formed by combining $2p$ atomic orbitals. These molecular orbitals are shown in Figure 10.26. They share two features with those that we first considered (Figure 10.24).

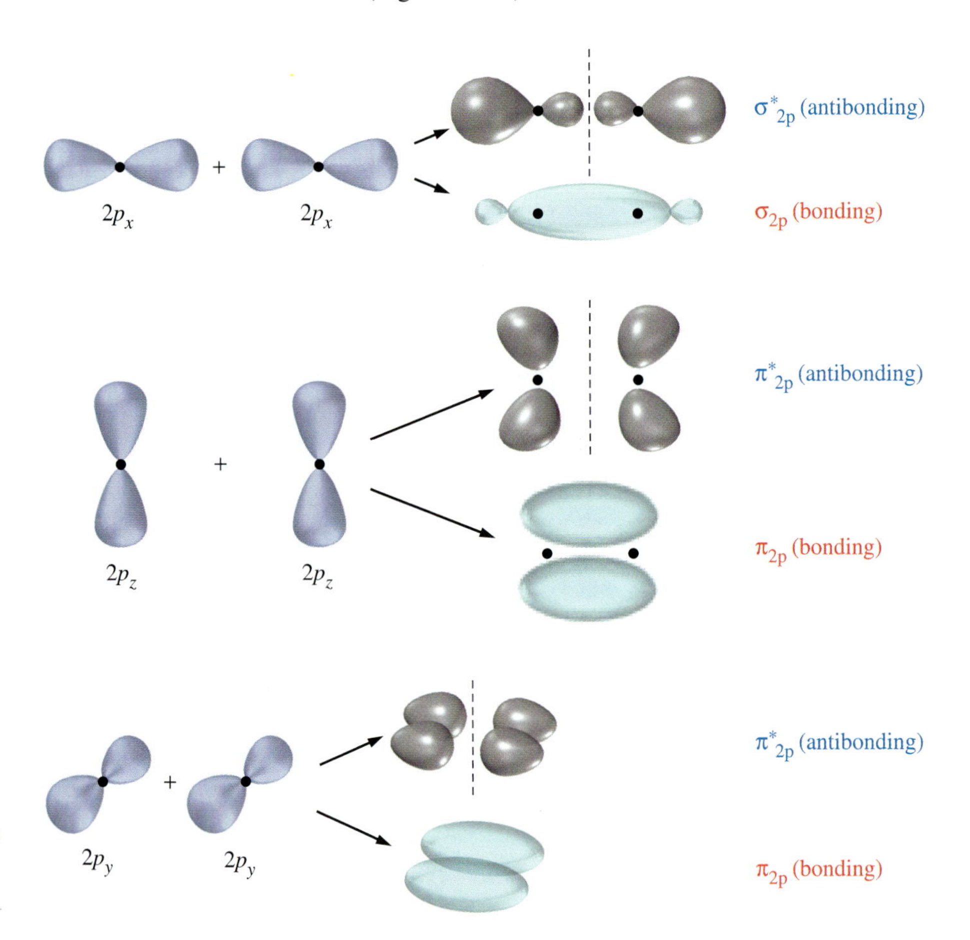

Figure 10.26 Molecular orbitals formed by combining 2*p* atomic orbitals
These figures suggest the distribution of electron probability or electron charge density for the several molecular orbitals. The broken lines represent nodal planes in the antibonding orbitals.

- For every *two* atomic orbitals that are combined, *two* molecular orbitals result. Six molecular orbitals are formed altogether from the $2p$ atomic orbitals.
- Of each pair of molecular orbitals, one is a bonding molecular orbital at a lower energy than the separate atomic orbitals, and one is an antibonding orbital at a higher energy.

A new feature, however, is that we obtain two types of molecular orbitals from $2p$ atomic orbitals. One pair of molecular orbitals, designated σ_{2p} and σ^*_{2p}, corresponds to what amounts to an end-to-end overlap of atomic orbitals. Two pairs resemble the side-by-side overlap of atomic orbitals, leading to two degenerate π_{2p} bonding molecular orbitals and two degenerate π^*_{2p} antibonding molecular orbitals. (Recall that degenerate means having the same energy.)

We have already remarked that the σ_{2s} and σ^*_{2s} molecular orbitals are at higher energies than the underlying σ_{1s} and σ^*_{1s} orbitals. Figure 10.27 suggests the energy levels of all the valence-shell molecular orbitals of the homonuclear diatomic molecules of the second period—Li_2, Be_2, B_2, and so on. For each of the molecules, the energy levels are arranged according to increasing energy, but the drawing is not to scale. For example, the continuous decline in energy levels with increasing nuclear charge of the atoms (recall Figure 8.1) is not brought out in Figure 10.27.

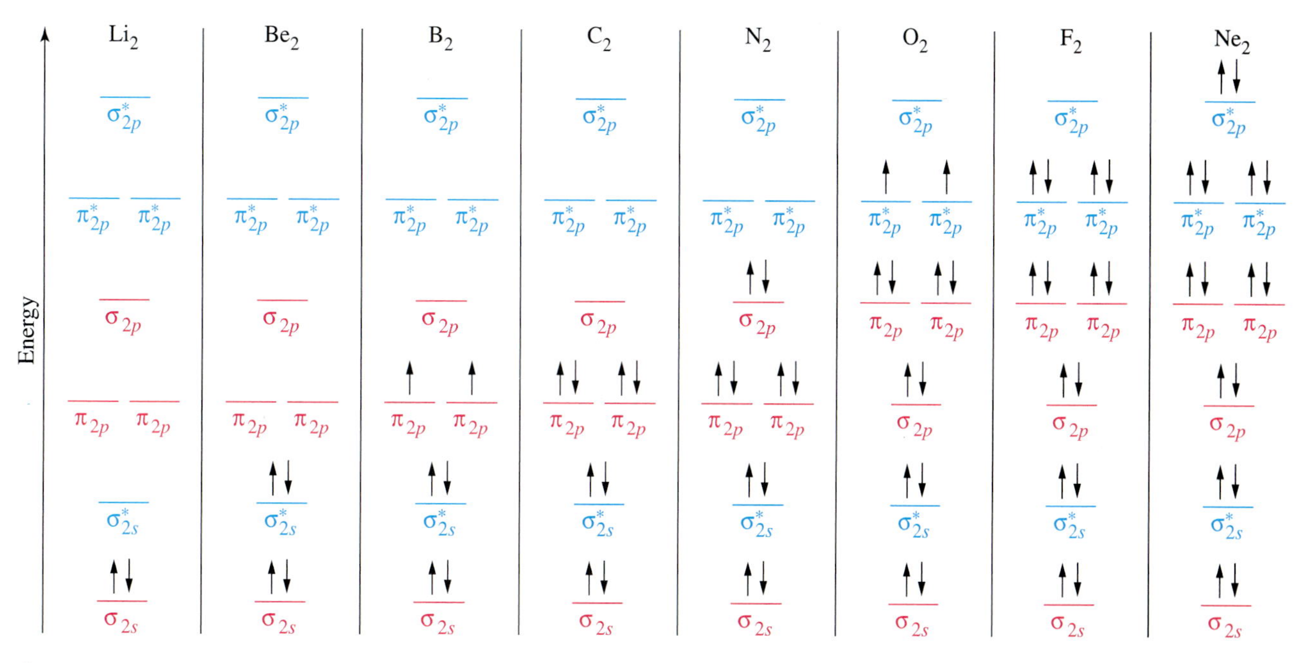

▲ Figure 10.27 Molecular orbitals obtained from 2*s* and 2*p* atomic orbitals, and some actual and hypothetical diatomic molecules
The σ_{1s} and σ^*_{1s} molecular orbitals are filled, and the second-shell orbitals are occupied as shown here. The crossover in the energy levels of the σ_{2p} and π_{2p} bonding molecular orbitals that occurs between N_2 and O_2 is predicted by molecular orbital theory and has been confirmed by experiment.

We can, however, draw some important conclusions from the molecular orbital diagrams in Figure 10.27. One is that we can use the definition of bond order to show that the molecules Be_2 and Ne_2 are unlikely to exist. Each has the same number of electrons in antibonding molecular orbitals as in bonding orbitals. The bond orders are zero, and there is no bond at all.

Another interesting conclusion has to do with the O_2 molecule. Written below is a molecular orbital diagram for the second shell arranged in a manner similar to that used for electron configurations of atoms. We have to assign 12 valence electrons to this diagram, following the rules of the aufbau principle.

O_2	↑↓	↑↓	↑↓	↑↓ ↑↓	↑ ↑	
	σ_{2s}	σ^*_{2s}	σ_{2p}	π_{2p}	π^*_{2p}	σ^*_{2p}

To assess the bond order in O_2, we note that there are *eight* electrons in bonding MOs: *two* in σ_{2s}, *two* in σ_{2p}, and *four* in π_{2p}; and *four* electrons in antibonding MOs: *two* in σ^*_{2s} and *two* in π^*_{2p}. The bond order is $\frac{1}{2}(8 - 4) = 2$, so we predict an oxygen-to-oxygen double bond. Moreover, the presence of two unpaired electrons in the π^*_{2p} MOs indicates that the O_2 molecule should be *paramagnetic*. Recall that in Chapter 9 we were able to account for the double bond with the Lewis structure, $:\ddot{O}=\ddot{O}:$, but we were unable also to account for the paramagnetism.

EXAMPLE 10.10—A Conceptual Example

When an electron is removed from a N_2 molecule, forming an N_2^+ ion, the bond between the N atoms is weakened. When an O_2 molecule is similarly ionized to O_2^+, the bond between the O atoms is strengthened. Explain this difference.

SOLUTION

To explain the difference, we need to identify the electron lost in the ionization in each case. We expect this to be an electron from the highest-energy molecular orbital that is occupied. Electrons from such orbitals should be lost most easily. If we look at the molecular orbital diagrams in Figure 10.27, we see that for N_2, the ionized electron is from the *bonding* molecular orbital, σ_{2p}. The loss of this *bonding* electron *reduces* the bond order; the bond becomes weaker. In contrast, with O_2, the ionized electron is from an *antibonding* orbital, π_{2p}^*. The loss of this *antibonding* electron *increases* the bond order; the bond becomes stronger.

EXERCISE 10.10

Predict the bond order in nitrogen monoxide, NO.

[*Hint:* Which molecular orbital diagram(s) in Figure 10.27 will be useful in making this prediction?]

Friedrich August Kekulé (1829–1896) claimed to have discovered the cyclic structure of benzene while dozing by the fire. (In some versions of the story, he was dozing on an omnibus.) He dreamed of atoms and molecules as snakes. Suddenly one of the twisting snakes seized its own tail, forming a ring. Kekulé's other contributions included the idea that carbon is tetravalent (1858) and an influential textbook in which he defined organic chemistry as the chemistry of carbon compounds.

10.8 Bonding in Benzene

Benzene, C_6H_6, a liquid that smells somewhat like gasoline, was discovered by Michael Faraday in 1825. In 1865, F. A. Kekulé proposed a *cyclic* structure for the benzene molecule that features six carbon atoms in a hexagonal ring. He proposed that a hydrogen atom was attached to each carbon atom and that alternating single and double bonds joined the carbon atoms together.

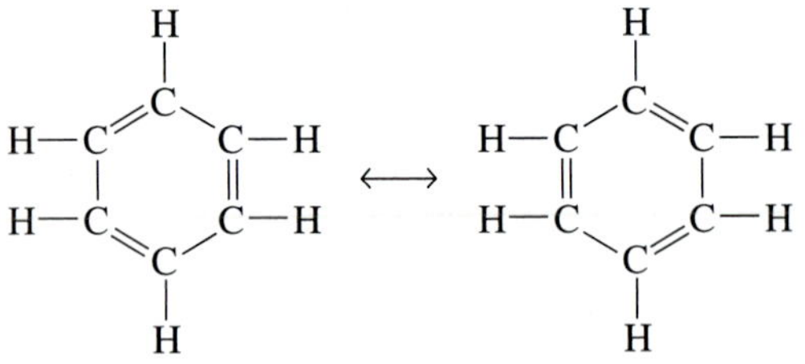

These Kekulé structures are often abbreviated as shown below, where the symbols C and H are not written but understood. (If atoms of other elements substitute for H atoms in C_6H_6, we do write their symbols.)

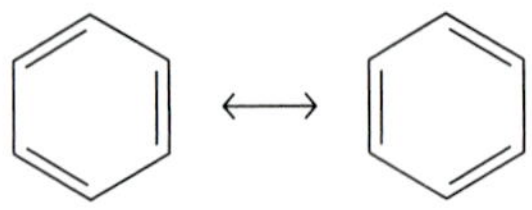

Kekulé structures of C_6H_6

It was not until more than 100 years after the discovery of benzene that its true structure was recognized as a resonance hybrid to which the Kekulé structures are the most important contributors. The carbon-to-carbon bonds in the benzene molecule are all equivalent and intermediate in length and strength between single and double bonds. However, even with its partial double bond character, the benzene molecule does not behave like unsaturated hydrocarbon molecules. For one thing,

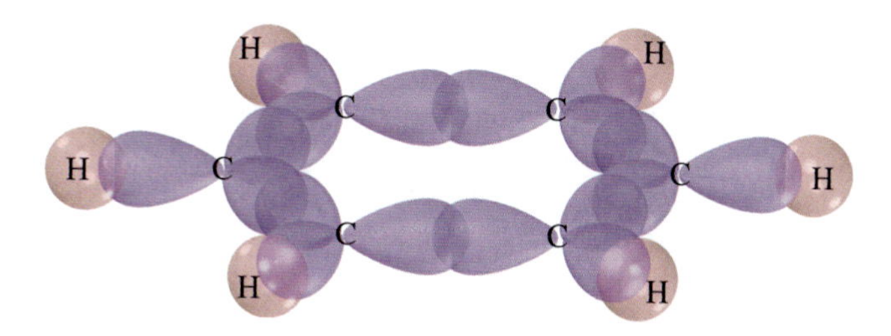

Figure 10.28
The σ-bond framework for benzene, C_6H_6
The carbon atoms in benzene use sp^2 hybrid orbitals to form σ bonds. Each carbon atom bonds to two other carbon atoms and to a hydrogen atom. All the atoms in the σ-bond framework are in the same plane.

it doesn't undergo a hydrogenation reaction (recall page 396). The six electrons that we ordinarily would place between carbon atoms to make three double bonds are spread out over all six carbon atoms; they are *delocalized*. The modern symbol for benzene—a hexagon with an inscribed circle—incorporates this idea.

Resonance hybrid of C_6H_6

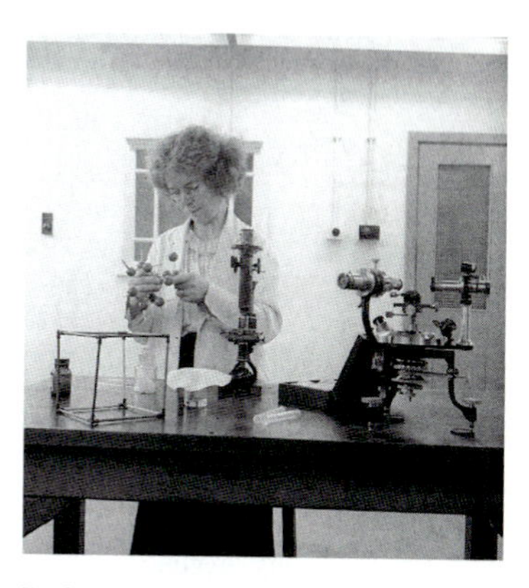

In the 1920s, Kathleen Londsdale (1903–1971) used a technique called X-ray diffraction to prove that the benzene ring is planar and not puckered. She also determined the three-dimensional structures of diamond and graphite (Chapter 11).

Now let's use features of both the valence bond theory and the molecular orbital theory to describe the benzene molecule. To account for the planar shape of the molecule and the 120° bond angles, we presume that all six carbon atoms employ an sp^2 hybridization scheme.

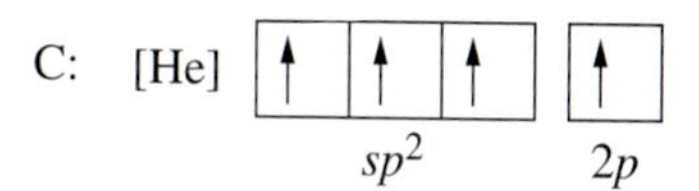

The sp^2 hybrid orbitals overlap to produce the σ-bond framework shown in Figure 10.28. The six half-filled $2p$ orbitals combine to form six π molecular orbitals. As shown in Figure 10.29, three of these molecular orbitals are bonding and three are antibonding. All six $2p$ electrons are found in the bonding MOs. These six electrons produce a total of three π bonds: $\left[(6 - 0)/2\right] = 3$. The molecule also has six carbon-to-carbon σ bonds. Thus, the total number of bonds joining the six carbon atoms is nine, and the average bond order of the carbon-to-carbon bonds is $9/6 = 1.5$. If we "average" the two Kekulé structures, we see that each carbon-to-carbon bond is half way between a single and a double bond, also a bond order of 1.5. Figure 10.30 is a computer-generated model of the benzene molecule that emphasizes the delocalized electrons in π molecular orbitals.

Antibonding MOs

Bonding MOs

Figure 10.29
π Molecular orbital diagram for benzene, C_6H_6

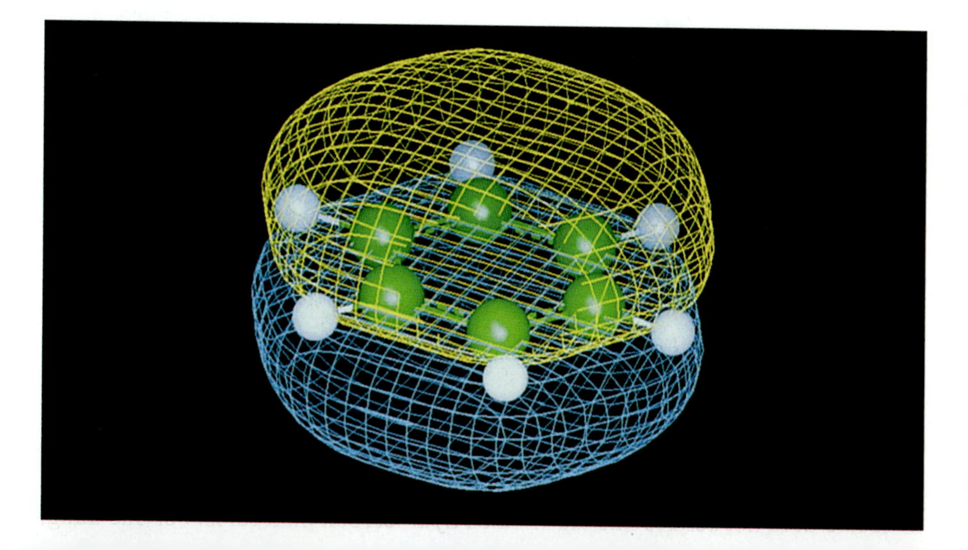

Figure 10.30
Representation of bonding in benzene, C_6H_6

10.9 Aromatic Compounds

Many of the first benzenelike compounds discovered had pleasant odors and hence acquired the name *aromatic*. In modern chemistry, the term **aromatic compounds** simply refers to substances with ring structures and with bonding characteristics and properties related to those of benzene, C_6H_6. Structurally, aromatic compounds have some delocalized bonding. Alkanes, alkenes, alkynes, and their cyclic analogues are called *aliphatic hydrocarbons*, and all organic compounds that are *not* aromatic are called **aliphatic compounds**.

More than 7 billion kg of benzene are produced in the United States each year, most of it from petroleum. Benzene is used as a solvent and as a starting material for the synthesis of many other organic compounds, including drugs, detergents, dyes, pesticides, and plastics. It is also used as a component of some unleaded gasolines.

Table 10.3 lists some important aromatic compounds related to benzene. Methylbenzene, $C_6H_5CH_3$, more commonly called *toluene*, is both aromatic (the ring) and aliphatic (the CH_3 group that substitutes for one of the H atoms of C_6H_6). More than 3 billion kg of toluene are produced in the United States each year. About 6 billion kg of ethylbenzene, $C_6H_5CH_2CH_3$, are also produced annually, nearly all of which is converted to styrene, $C_6H_5CH{=}CH_2$. Styrene is a monomer used to produce the common plastic, polystyrene. You can think of this polymerization as proceeding just like that of ethylene to polyethylene (Section 9.12). Simply replace one of the H atoms of ethylene, $CH_2{=}CH_2$, with the *phenyl* group, $-C_6H_5$.

APPLICATION NOTE

Many aromatic compounds, including benzene, are carcinogenic (cancer causing). Modern workplace regulations severely restrict exposures to many aromatic compounds.

TABLE 10.3 Some Representative Aromatic Compounds

Name	Structure	Typical Use(s)
Aniline	$C_6H_5{-}NH_2$	Starting material for the synthesis of dyes, drugs, resins, varnishes, perfumes; solvent; vulcanizing rubber
Benzoic acid	$C_6H_5{-}COOH$	Food preservative; starting material for the synthesis of dyes and other organic compounds; curing of tobacco
Bromobenzene	$C_6H_5{-}Br$	Starting material for the synthesis of many other aromatic compounds; solvent; motor oil additive
Nitrobenzene	$C_6H_5{-}NO_2$	Starting material for the synthesis of aniline; solvent for cellulose nitrate; in soaps and shoe polish
Phenol	$C_6H_5{-}OH$	Disinfectant; starting material for the synthesis of resins, drugs, and other organic compounds
Toluene	$C_6H_5{-}CH_3$	Solvent; gasoline octane booster; starting material for the synthesis of benzoic acid, benzaldehyde, and many other organic compounds

Halogen atoms may also be substituted for one of the hydrogen atoms on a benzene ring. These compounds, known as *halogenated benzenes*, are important intermediates for the synthesis of many other aromatic compounds. We name them by using prefixes to indicate the halogen, as with *bromo*benzene, C_6H_5Br, which is listed in Table 10.3. No number is needed to indicate the position of the substituent in *mono*substituted benzenes because the hexagonal ring is symmetrical. All the positions are equivalent.

When there are two or more substituents, positions are no longer equivalent, and we must designate relative positions. There are three possible *di*substituted benzenes, and we can use numbers to distinguish them. For example, the dichlorobenzenes are as follows.

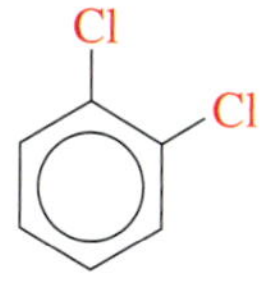

1,2-Dichlorobenzene (*o*-Dichlorobenzene) | 1,3-Dichlorobenzene (*m*-Dichlorobenzene) | 1,4-Dichlorobenzene (*p*-Dichlorobenzene)

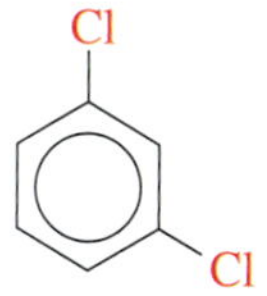

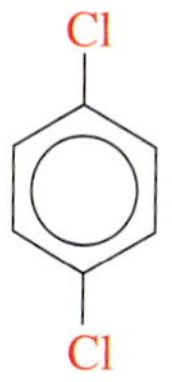

Common names are also used: *ortho*- (*o*-) designates a 1,2 substitution, *meta*- (*m*-) indicates a 1,3 substitution, and *para*- (*p*-) indicates a 1,4 arrangement. *p*-Dichlorobenzene is used as an insecticide, especially as a moth repellent, and as a bathroom deodorant.

With three or more substituents, numbers are almost always used to indicate the location of substituents.

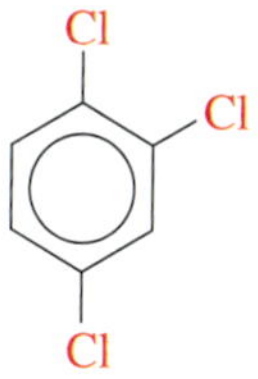

1,2,4-Trichlorobenzene

The compounds toluene, aniline, and phenol have structures in which the groups $—CH_3$, $—NH_2$, and $—OH$, respectively, substitute for a H atom in C_6H_6 (Table 10.3). These groups are given the position "1" on the benzene ring. Substances derived from toluene, aniline, and phenol are named in the manner illustrated below.

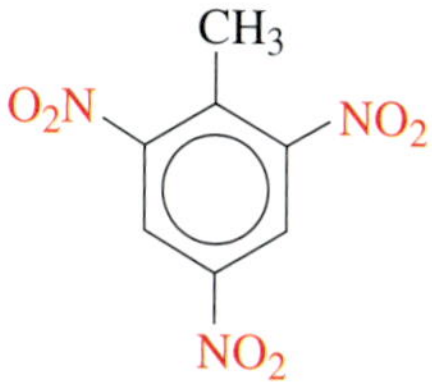

2,4,6-Trinitrotoluene (TNT, an explosive) | 2,4-Dinitroaniline (used to make dyes) | 2,4,5-Trichlorophenol (a fungicide)

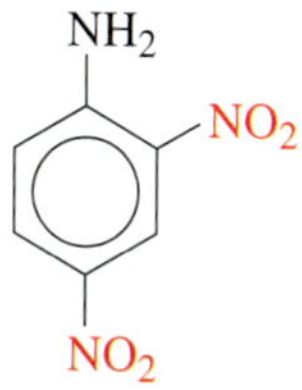

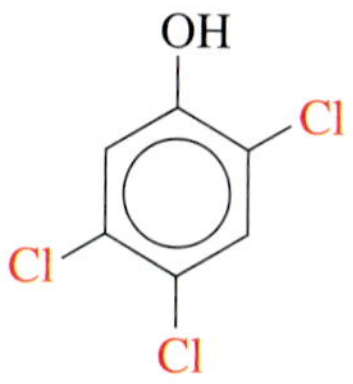

There are many organic compounds that have atoms other than carbon in an aromatic ring. These *heterocyclic* compounds include amines related to purine and pyrimidine that are important constituents of the nucleic acids DNA and RNA.

Key Terms

aliphatic compound (10.8)
antibonding molecular orbital (10.6)
aromatic compound (10.9)
bond order (10.6)
bonding molecular orbital (10.6)
cis (10.5)
debye, D (10.2)
dipole moment, μ (10.2)
electron group (10.1)
electron-group geometry (10.1)
geometric isomers (10.5)
hybridization (10.4)
hybrid orbital (10.4)
molecular geometry (page 409)
molecular orbital (10.6)
pi (π) bond (10.5)
polar molecule (10.2)
sigma (σ) bond (10.5)
sp (10.4)
sp^2 (10.4)
sp^3 (10.4)
sp^3d (10.4)
sp^3d^2 (10.4)
trans (10.5)
valence bond (VB) theory (10.3)
valence-shell electron-pair repulsion (VSEPR) method (10.1)

Summary

The VSEPR method is used to predict the shapes of molecules and polyatomic ions based on the mutual repulsions among valence-shell electron groups. This requires determining the number of valence-shell electron groups for each central atom and assessing the geometric distribution of these electron groups. If all electron groups are bonding groups, the molecular geometry is the same as the electron-group geometry. If some of the electron groups are lone pairs, the molecular and electron-group geometry are related but not identical.

A polar covalent bond has separate centers of positive and negative charge, creating a bond dipole. Whether a molecule as a whole is polar, that is, whether there is a molecular dipole with a dipole moment, depends on bond dipoles *and* molecular geometry. A symmetrical distribution of identical bond dipoles about a central atom can lead to cancellation of all bond dipoles, resulting in a nonpolar molecule.

In the valence bond theory, a covalent bond is viewed as the overlap of atomic orbitals of the bonded atoms in a region between the atomic nuclei. Molecular geometry is determined by the spatial orientations of the atomic orbitals involved in bonding. The theory often requires that bonding atomic orbitals be hybridized. A hybridized orbital is some combination of "pure" *s*, *p*, *d*, and *f* orbitals, such as *sp*; sp^2; sp^3; sp^3d; and sp^3d^2. The geometric distribution of the hybridized orbitals in the valence bond theory is the same as the electron-group geometry predicted by the VSEPR method.

To apply the valence bond theory to structures containing multiple bonds, hybridization schemes must leave some orbitals unhybridized, as in the set $sp^2 + p$. Hybrid orbitals overlap in the usual way (end-to-end) to form σ bonds. Unhybridized *p* orbitals overlap in a side-by-side manner to form π bonds. A double bond consists of one σ bond and one π bond, and a triple bond consists of one σ bond and two π bonds.

In molecular orbital theory, atomic orbitals of separated atoms are combined into molecular orbitals. A pair of molecular orbitals is formed for every pair of atomic orbitals combined. One is a bonding molecular orbital, and one is antibonding. The bond order is one-half the difference between the number of electrons in bonding molecular orbitals and in antibonding molecular orbitals. Valence electrons are assigned to the molecular orbitals in an aufbau procedure similar to that used for the electron configurations of atoms. Bonding can be represented through molecular orbital diagrams.

The benzene molecule can be represented through two Kekulé structures, which are resonance structures of a resonance hybrid. A symbolic representation of the resonance hybrid is also possible. Bonding in benzene can most easily be explained by a combination of valence bond theory for the σ-bond framework and molecular orbital theory for the π bonds. Compounds whose molecules have ring structures based on the benzene molecule are called aromatic compounds. All open-chain organic compounds and all ring compounds that are not aromatic are called aliphatic compounds.

Review Questions

1. Explain why we describe all diatomic molecules as linear.

2. Is it possible for a molecule consisting of three or more atoms to be linear? Explain.

3. In the VSEPR method, what is the distinction between the terms *electron-group geometry* and *molecular geometry*?

4. In the VSEPR method, electron pairs are designated as bonding pairs or lone pairs. Which of the electron-group repulsions are the strongest? Explain.

5. In the VSEPR method, what is the meaning of the notation AX_2E_2?

6. What approximate bond angles would you expect in triatomic molecules having the following electron-group geometries about the central atom?

(a) linear
(b) trigonal planar
(c) tetrahedral

7. Is it possible to have a *linear* molecule in which the electron-group geometry is *tetrahedral*? If so, give an example.

8. Explain why it is not necessary to find the "best" Lewis structure when applying the VSEPR method.
9. Explain how the VSEPR method can be applied to a molecule or ion containing more than one central atom.
10. Must every chemical bond have a bond dipole? Explain.
11. Must every molecule have a dipole moment? Explain.
12. Explain why SO_2 is a polar molecule whereas SO_3 is not.
13. Water has a dipole moment of 1.84 D. Explain why this fact proves that the H_2O molecule has a bent shape.
14. Explain how the valence bond theory is able to predict a 90° bond angle in H_2S, whereas the VSEPR prediction is 109.5°.
15. Why is it necessary to introduce the idea of *hybridized* atomic orbitals in the valence bond approach to chemical bonding?
16. In valence bond theory, what is the meaning of the designation sp^3?
17. Which hybridization scheme in valence bond theory accounts for trigonal planar electron-group geometry? For octahedral electron-group geometry?
18. Explain the difference between a σ and a π bond.
19. Why does the valence bond theory account for σ and π bonds, whereas Lewis structures do not?
20. Is the following a valid statement? "All bonds in an alkanes are σ bonds." Explain.
21. Is the following a valid statement? "All bonds in an alkenes are π bonds." Explain.
22. What are cis and trans isomers? What are the structural features of a molecule that lead to cis-trans isomerism?
23. Describe the differences between an *atomic* orbital, a *hybridized* atomic orbital, and a *molecular* orbital.
24. How do *antibonding* molecular orbitals differ from *bonding* molecular orbitals?
25. How is the term *bond order* defined in molecular orbital theory? What are the conditions under which a bond has the order 1/2? 3/2?
26. What is the important difference between *localized* and *delocalized* electrons in the formation of covalent bonds? Give an example of each type.
27. What are the Kekulé structures of benzene? Why is a single Kekulé structure inadequate for describing the benzene molecule?
28. What are the similarities in the covalent bonding in ethene, C_2H_4, and in benzene, C_6H_6? What are the differences?
29. What is the chemical distinction between the terms *aliphatic* compound and *aromatic* compound?
30. Chlorobenzene (C_6H_5Cl) is derived from benzene by replacing one hydrogen atom with a chlorine atom. Draw the two Kekulé structures for chlorobenzene. Draw a single structure to represent the resonance hybrid of the two Kekulé structures.

Problems

VSEPR Method

31. What is the VSEPR notation of the central atom in each of the following?
    (a) H_2O (b) OCl^- (c) I_3^-
32. What is the VSEPR notation of the central atom in each of the following?
    (a) NI_3 (b) PCl_4^+ (c) H_2CO
33. Predict whether each of the following species is probable. For those that seem improbable, tell why.
    (a) a linear H_2O molecule (b) a planar SO_3 molecule (c) a planar PH_3 molecule
34. Predict whether each of the following species is probable. For those that seem improbable, tell why.
    (a) a tetrahedral $GeCl_4$ molecule
    (b) a trigonal bipyramidal NCl_5 molecule
    (c) a bent HCN molecule
35. Explain why the molecule BF_3 is trigonal planar, whereas a molecule with a similar formula, ClF_3, is T-shaped.
36. Explain why the ion ICl_4^- is square planar, whereas an ion with a similar formula, BF_4^-, is tetrahedral.
37. Predict the molecular geometry of each of the following.
    (a) PCl_3 (b) ClO_4^- (c) XeF_4 (d) OCN^- (e) SF_5^-
38. Predict the molecular geometry of each of the following.
    (a) PH_4^+ (b) NI_3 (c) Cl_2CO (d) NSF (e) ICl_4^-
39. Describe the shape of each of the following.
    (a) H_2O_2 (b) C_3O_2 (c) OSF_4 (d) N_2O
40. Describe the shape of each of the following.
    (a) $HClO_3$ (b) N_2O_4 (c) CH_3CN (d) SO_2Cl_2
41. In which of the molecules, CH_4 or $COCl_2$, do you think the actual bond angles are closer to those predicted by the VSEPR method? Explain.
42. In which of these molecules, BF_3 or SF_4, do you think the actual bond angles are closer to those predicted by the VSEPR method? Explain.

Molecular Shape and Dipole Moments

43. Indicate which of the following molecules you would expect to be polar, and explain your reasoning.

(a) CS_2 (c) XeF_4

(b) NO_2 (d) ClF_3

44. Indicate which of the following molecules you would expect to be polar, and explain your reasoning.

(a) C_2H_4 (c) HCN

(b) $COCl_2$ (d) SF_6

45. Which of the molecules, H_2O or OF_2, would you expect to have the larger dipole moment:? Explain your reasoning.

46. Which of the molecules, NO or NO_2, would you expect to have the larger dipole moment? Explain your reasoning.

47. Draw structural formulas and use cross-based arrows to represent bond dipoles and any resultant molecular dipole in molecules of **(a)** NF_3 and **(b)** SF_4.

48. Draw structural formulas and use cross-based arrows to represent bond dipoles and any resultant molecular dipole in molecules of **(a)** NH_3 and **(b)** $GeCl_4$.

Valence Bond Theory

49. Describe bonding in the molecules $Li_2(g)$ and $F_2(g)$ by the valence bond theory. How can the theory account for the difference in bond energies, 106 kJ/mol for the Li—Li bond and 157 kJ/mol for the F—F bond?

50. The molecule phosphine, PH_3, has H—P—H bond angles of 93.6°. Use the valence bond theory to describe the bonding in this molecule.

51. Indicate the hybridization scheme expected for the central atom in each of the following.

(a) OF_2 (c) CO_2

(b) NH_4^+ (d) $COCl_2$

52. Indicate the hybridization scheme expected for the central atom in each of the following.

(a) BF_4^- (c) NO_2^-

(b) SO_3 (d) XeF_4

53. Indicate the hybridization schemes expected for the central atoms in each of the following

(a) C_2N_2 (c) NH_2OH

(b) HNCO (d) CH_3COOH

54. Indicate the hybridization schemes expected for the central atoms in each of the following

(a) CH_3CN (c) CH_3CCCH_3

(b) CH_3NH_2 (d) CH_3NCO

55. Propose a simple Lewis structure (or resonance structures), the molecular geometry, and a bonding scheme of the type used in Example 10.7 for each of the following.

(a) $ClNO_2$ (b) OF_2 (c) CO_3^{2-}

56. Propose a simple Lewis structure (or resonance structures), the molecular geometry, and a bonding scheme of the type used in Example 10.7 for each of the following.

(a) HNO_3 (b) AsF_6^- (c) CH_3CCH.

57. In both of the ions, ICl_2^+ and ICl_2^-, an iodine atom is bonded to two Cl atoms. Do you expect that the same hybridization scheme for the central I atom applies in each case? Explain.

58. In both the molecule OSF_4 and the ion SF_5^-, a sulfur atom is bonded to five other atoms. Do you expect that the same hybridization scheme for the central S atom applies in each case? Explain.

59. The structure of oxalate ion, $C_2O_4^{2-}$, is represented below.

O O
157 pm
126° C—C
125 pm
O O

Propose hybridization and bonding schemes consistent with this structure. (*Hint:* Use data from Table 9.1.)

60. The structure of hydrazoic acid, HN_3, is indicated below.

124 pm 113 pm
N—N—N
113°
H

Propose hybridization and bonding schemes consistent with this structure. (*Hint:* See Table 9.1.)

61. Draw hybridization and bonding schemes like those in Example 10.7 that are consistent with the space-filling molecular models shown below (white = H, red = O, blue = N, and black = C).

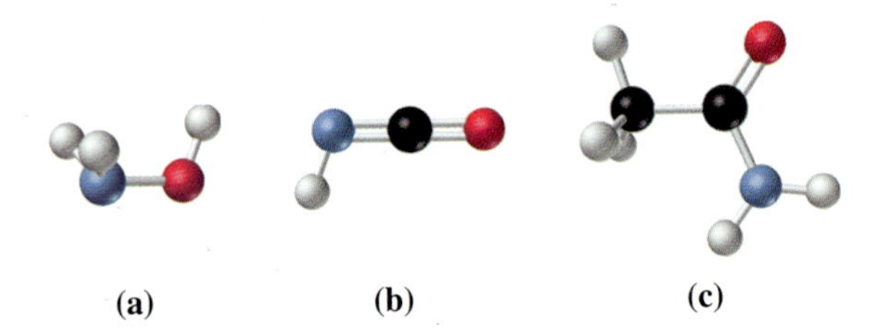

(a) (b) (c)

62. Draw hybridization and bonding schemes like those in Example 10.7 that are consistent with the space-filling molecular models shown below (white = H, red = O, blue = N, black = C, and green = Cl).

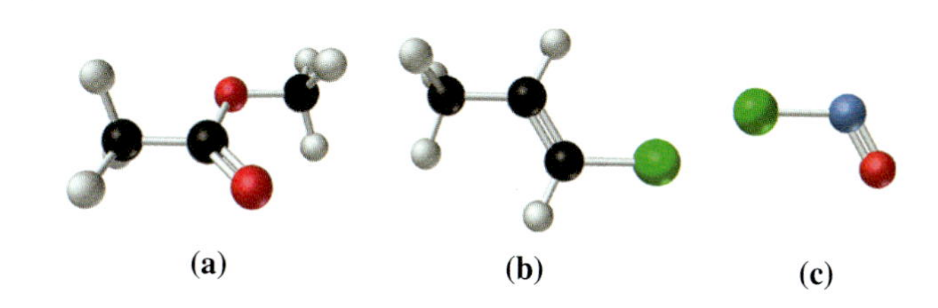

(a) (b) (c)

Geometric Isomerism

63. Which of the following compounds can exist as cis-trans isomers? Explain.
   (a) $CH_3CH_2CH{=}CHCH_2CH_3$
   (b) $CH_2{=}CHCH_2CH_2CH_3$
   (c) $CH_3CH_2CH{=}CHCH_3$

64. Which of the following compounds can exist as cis-trans isomers? Explain.
   (a) $CH_3CH{=}CHCH_2CH_3$ (with a CH_3 group attached to the first CH)
   (b) $CH_3CH{=}CHCH_2CH_2CH_3$
   (c) $CH_3CHCH{=}CHCH_3$ (with a CH_3 group attached to the second carbon)

65. Refer to the structures of 1-butene and isobutylene on page 443. Would you expect either of these substances to exhibit cis-trans isomerism? Explain. Imagine substituting a Cl atom for one H atom on the first carbon atom in each structure. Would the resulting substances exhibit cis-trans isomerism? Explain.

66. Arrange the following molecules in the expected order of increasing dipole moment: **(a)** chloroethane, **(b)** *cis*-2-butene, **(c)** *cis*-1,2-dichloroethene, **(d)** *trans*-1,2-dibromoethene. Explain your arrangement.

Molecular Orbital Theory

67. Which ion would you expect to have the greater fluorine-to-fluorine bond energy, F_2^+ or F_2^-?

68. The text mentions that Be_2 is not likely to exist as a stable molecule. Would you expect the ions Be_2^+ or Be_2^- to be any more stable? Explain.

69. Compare the bond order of the carbon-to-carbon bond in the diatomic molecule C_2 that you would establish from a Lewis structure and from a molecular orbital diagram. Why are they not the same?

70. In contrast to the situation in Problem 69, show that both a Lewis structure and a molecular orbital diagram give the same bond order for the carbide ion, C_2^{2-}.

71. Assume that the energy-level diagrams of Figure 10.27 apply to the ions CN^+ and CN^-.
   (a) Which ion has the stronger carbon-to-nitrogen bond? Explain.
   (b) Is either of these ions paramagnetic? Explain.

72. Assume that the energy-level diagrams of Figure 10.27 apply to the diatomic molecules BN and CN.
   (a) Which molecule has the stronger bond between the atoms? Explain.
   (b) Is either of these molecules paramagnetic? Explain.

73. These two ions of O_2 are commonly encountered in the oxides of the Group 1A metals: *peroxide* ion, O_2^{2-}, and *superoxide* ion, O_2^-. Compare these ions with respect to bond order and magnetic properties by using molecular orbital theory.

74. Explain how molecular orbital theory makes it possible to describe a structure of benzene without the use of resonance structures and a resonance hybrid.

Aromatic Compounds

75. Draw structural formulas for **(a)** *p*-bromotoluene, **(b)** *m*-diethylbenzene, and **(c)** 2,4-diiodotoluene.

76. Draw structural formulas for **(a)** *o*-dinitrobenzene, **(b)** 2,4-dibromophenol, and **(c)** 1,2,4-trimethylbenzene.

77. Name the following compounds by the IUPAC system.

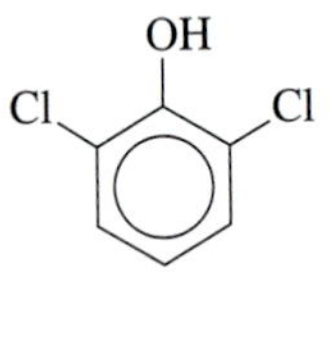

(a) OH, Cl, Cl

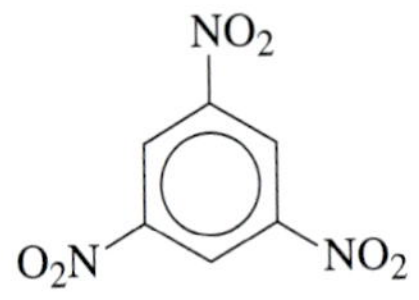

(b) NO_2, O_2N, NO_2

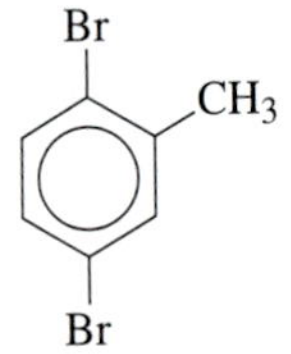

(c) Br, CH_3, Br

78. Name the following compounds by the IUPAC system.

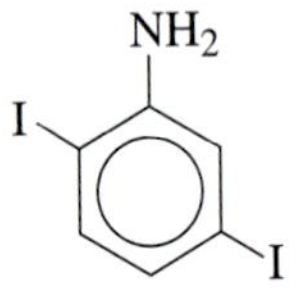

(a) NH_2, I, I

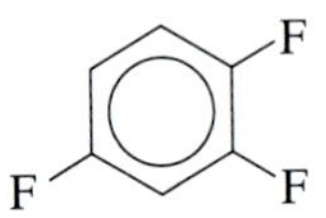

(b) F, F, F

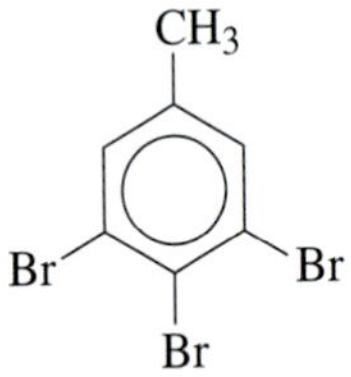

(c) CH_3, Br, Br, Br

79. The aromatic compound azobenzene, $(C_6H_5N)_2$, is an important starting material in the production of dyes (called azo dyes).

(a) Write two Lewis structures for azobenzene that demonstrate a type of cis-trans isomerism in this substance.

(b) Propose hybridization schemes for the C and N atoms in these structures.

(c) What are the bond angles in these structures?

80. Toluene-2,4-diisocyanate is used in the manufacture of polyurethane foam. Its structure is suggested by the following formula.

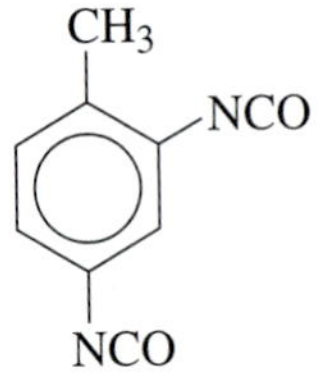

CH_3, NCO, NCO

(a) How many different hybridization schemes must be used for the C atoms in this structure? What are they? Explain.

(b) What is the hybridization scheme for the two N atoms?

(c) What are the N—C—O bond angles? What are the C—N—C bond angles?

Additional Problems

81. Do you think the following is a valid statement? "The greater the electronegativity difference between the atoms in a molecule, the greater the dipole moment of that molecule is." Explain.

82. We examined three possible structures for the XeF_2 molecule in Example 10.2 and then selected the correct one. Why is the actual structure not just a resonance hybrid of the three structures?

83. The NO_2F molecule depicted in Example 10.5 is a symmetrical molecule with bond angles of about 120°. Why doesn't it have a dipole moment of 0?

84. Apply the VSEPR method to predict the shape of the BrF_4^+ ion.

85. Apply the VSEPR method to predict the shape of the XeF_5^+ ion.

86. Draw a Lewis structure for the N_2 molecule and then predict a hybridization scheme consistent with this structure. Could bonding in N_2 be described by the use of pure (unhybridized) atomic orbitals? Explain.

87. Phosphorus pentachloride is molecular in the gas phase, but in the solid phase, it is an ionic compound consisting of the ions $[PCl_4]^+$ and $[PCl_6]^-$. Propose hybridization and bonding schemes to represent these molecular and ionic forms of phosphorus pentachloride.

88. Nitrogen and phosphorus are both in Group 5A of the periodic table. Apparently, a phosphorus atom can employ sp^3d hybridization in forming covalent bonds. Can a nitrogen atom do likewise? Explain.

89. Methyl isocyanate, CH_3CNO, is used in the manufacture of pesticides.

(a) Write a plausible Lewis structure for this molecule.

(b) Indicate a hybridization scheme for the central atoms in this structure.

(c) Indicate the approximate bond angles you would expect to find in the molecule.

(d) Make a rough sketch of the molecule.

90. Dicyandiamide, $NCNC(NH_2)_2$, is used in the manufacture of melamine plastics.

(a) Write a plausible Lewis structure for this molecule.

(b) Indicate a hybridization scheme for the central atoms in this structure.

(c) Indicate the approximate bond angles you would expect to find in the molecule.

(d) Make a rough sketch of the molecule.

91. Carbon suboxide, C_3O_2, is a foul-smelling gas. The C—O bond lengths in the molecule are 116 pm and the C—C bond lengths are 128 pm. Draw a plausible Lewis structure for this molecule, propose hybridization and bonding schemes, and predict its geometric shape. (*Hint:* Use data from Table 9.1.)

92. The structure of the allene molecule, CH_2CCH_2, is indicated below. Propose hybridization schemes for the C atoms in this molecule.

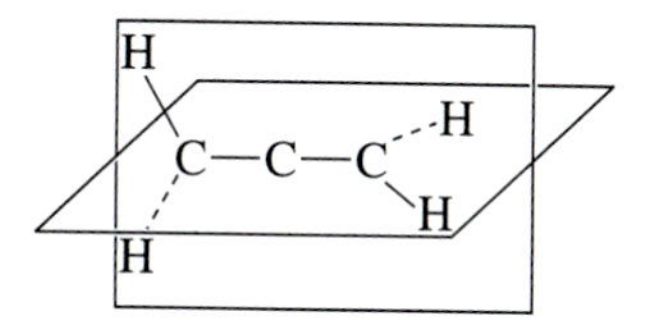

93. The molecule disulfur tetrafluoride has the skeletal structure F_3SSF. Two of the F—S—F bond angles are 90°, and one is 180°. Write a Lewis structure for the molecule, and propose hybridization schemes for the sulfur atoms that are consistent with this structure. Make a rough sketch of the molecule.

94. To reduce the risk of heart disease, people are advised to limit their intake of total and saturated fats. Nearly all naturally occurring *unsaturated* fatty acids are cis isomers, and these are not associated with increased risk of heart disease. When vegetable oils are hydrogenated to make margarine, some of the cis fatty acids are converted to saturated fatty acids, but others are converted to *trans* unsaturated fatty acids. Margarines have 11% to 49% trans fatty acids. Fried foods and pastries may have 35 % to 38% trans fatty acids. Shown below are ball-and-stick models of three fatty acids. **(a)** Identify the saturated, cis unsaturated, and trans unsaturated molecules. **(b)** Why might trans fatty acids have the same effect as saturated fatty acids in the development of heart disease?

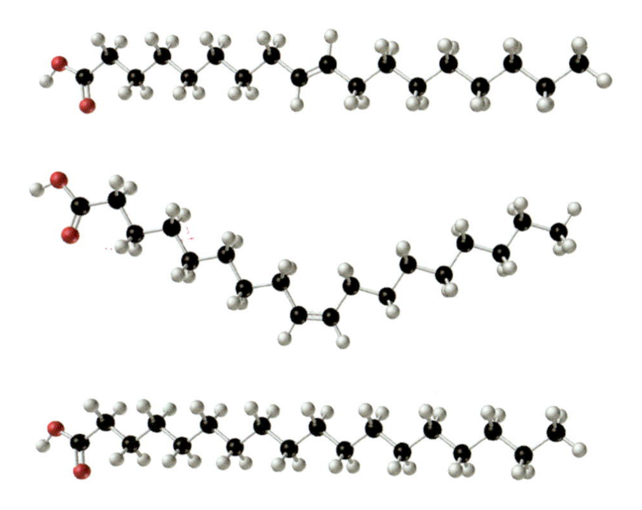

CHAPTER

11

States of Matter and Intermolecular Forces

CONTENTS

● ChemCDX
When you see this icon, there are related animations, demonstrations, and exercises on the CD accompanying this text.

Water exists in three physical states in this view of Paradise Bay in Antarctica. Water vapor is invisible, but when condensed into liquid droplets, water is visible in clouds. The liquid water in the bay and the solid water in the snow and ice on the mountains and in the glacier represent the two condensed states that are the focus of this chapter.

As we noted in Chapter 5, we can describe many gases with a single, general equation—the ideal gas equation—because intermolecular forces between gas molecules are often small enough to ignore. With liquids and solids, however, intermolecular forces are of prime importance. Moreover, because these forces differ in kind and strength from one substance to another, we can't use general equations to describe liquids and solids. Instead of quantitative calculations, then, our goal in this chapter will be to achieve a *qualitative* understanding of natural phenomena. That is, we seek to answer many questions, including the following.

- Why does ice float on liquid water?
- Why does dry ice (solid CO_2) fail to melt under normal conditions?
- Why does it take longer to boil an egg on a mountain top than at the seashore?
- Why is diamond, one form of carbon, so hard that it can scratch glass, but graphite, another form of carbon, soft enough to use in pencils?

11.1 Intermolecular Forces and the States of Matter: A Chapter Preview

In the previous two chapters, we focused on the forces that bind atoms to one another *within* molecules. We called these forces chemical bonds, but we could also have called them *intra*molecular forces. Intramolecular forces determine such molecular properties as molecular geometries and dipole moments. Forces that exist *between* molecules—*inter*molecular forces—determine the macroscopic *physical* properties of liquids and solids. In fact, if there were no intermolecular attractive forces, there would be no liquids or solids—everything would be in a gaseous state. Moreover, we will see that intermolecular forces are themselves related to the kinds of chemical bonds found in substances. Figure 11.1 compares *inter*molecular and *intra*molecular forces.

Latin, *intra*, within; *inter*, between. Notice that these are the meanings conveyed by the terms *intra*mural sports and *inter*collegiate athletics.

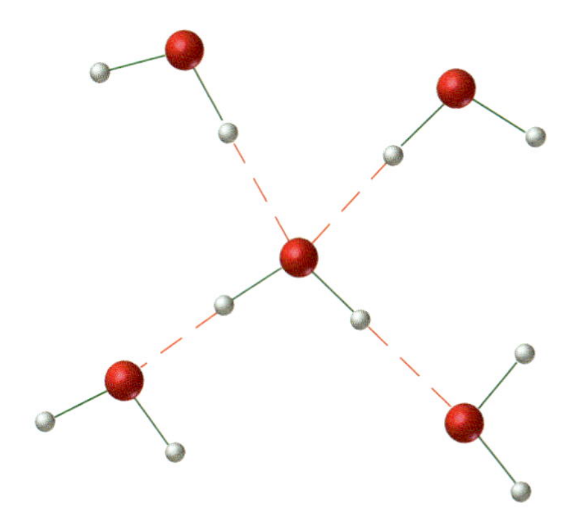

Figure 11.1 *Inter*molecular and *intra*molecular forces compared In the hypothetical situation described here, *intra*molecular forces—chemical bonds—are represented by the green lines. Forces between molecules—*inter*molecular forces—are suggested by the broken red lines.

Throughout this text, we have encountered substances in one of three physical forms called the **states of matter**: solid, liquid, and gas. Let's briefly compare these three states, first at the macroscopic level.

- A gas expands to fill its container, has neither a fixed volume nor shape, and is easily compressible.
- A liquid has a fixed volume, flows to cover the bottom and assume the shape of its container, and is only slightly compressible.
- A solid has a fixed volume, maintains a definite shape, and is even more difficult to compress than a liquid.

We find the best explanations for these observations by considering the behavior of particles at the microscopic level.

- In a gas, we visualize speedy, energetic, widely spaced atoms or molecules undergoing frequent collisions but never coming to rest or clumping together.
- In a liquid, atoms or molecules are close together, and intermolecular forces are strong enough to hold them in a fixed volume but not a definite shape.
- In a solid, the structural particles (atoms, ions, or molecules) are in direct contact, and intermolecular forces hold them into a fixed volume and definite shape.

The microscopic view in Figure 11.2 suggests that in each of the three states of matter, the atoms, ions, or molecules move. In gases and liquids, this motion includes a translational component; the structural particles move from point to point in three-dimensional space. In solids, the motion is mainly vibrational; the structural units move back and forth about fixed points.

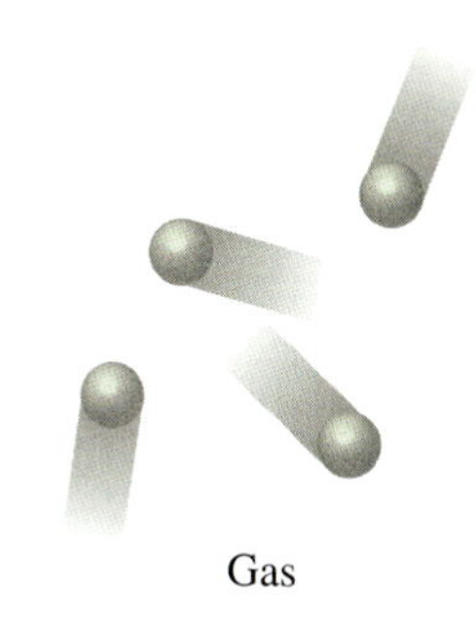

Gas

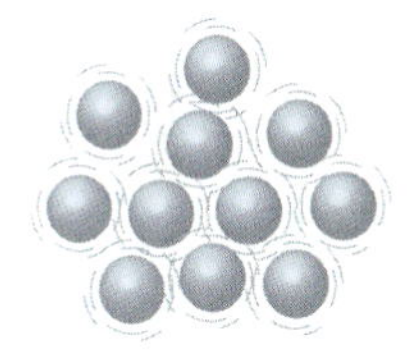

Liquid

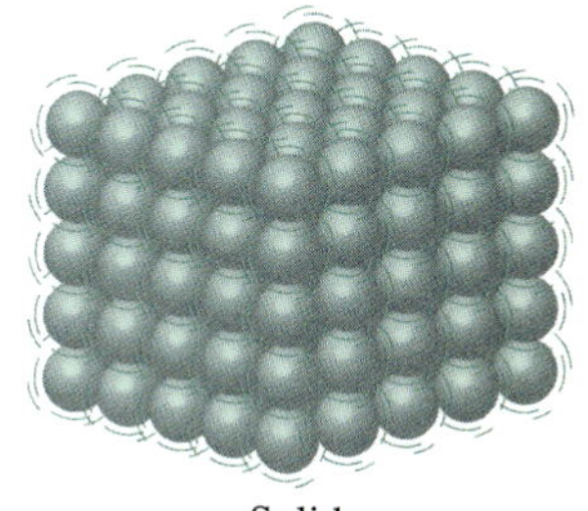

Solid

Figure 11.2 Some comparisons of the states of matter In the gaseous state, molecules are widely spaced, motion is chaotic, and disorder is at a maximum. In the liquid state, molecules continue to undergo translational motion but are in much closer proximity. In the solid state, there is a high degree of order among the structural particles; the motion of the particles is vibrational.

In this chapter, we will first describe changes from one form of matter to another. Then we will explore the types of intermolecular forces that underlie the physical properties of substances. Finally, we will explore the structures of solids in some detail.

Phase Changes

When ice melts, water undergoes a change in state from a solid to a liquid. When a block of dry ice sublimes, carbon dioxide changes from a solid to a gas. Changes in state such as these belong to a larger category of physical changes called **phase changes**. We will consider six main types of phase changes.

- *Melting* solid $\longrightarrow$ liquid
- *Freezing* liquid $\longrightarrow$ solid
- *Vaporization* liquid $\longrightarrow$ gas
- *Condensation* gas $\longrightarrow$ liquid
- *Sublimation* solid $\longrightarrow$ gas
- *Deposition* gas $\longrightarrow$ solid

We will first describe each of these phase changes, and then we will learn how to represent them collectively in a graph called a *phase diagram*.

11.2 Vaporization and Vapor Pressure

To clean up a water spill on a tile floor, we use a mop to spread the water into a thin film. The floor dries as the water evaporates. Evaporation, or **vaporization,** is the conversion of a liquid to a gas. As we noted before (page 178), the term vapor implies the gaseous state of a substance that is more commonly encountered in the liquid or solid state.

Let's try to picture the process of vaporization at the molecular level. The molecules of a liquid have a range of speeds and kinetic energies. The *average* kinetic energy of the molecules is determined by the temperature of the liquid. Surface molecules having enough kinetic energy in excess of this average overcome the intermolecular attractive forces of neighboring molecules in and below the surface. They fly off into the gaseous state: They *vaporize*.

As more energetic molecules leave the surface of a liquid through vaporization, the average kinetic energy of the molecules remaining in the liquid decreases. The temperature of the liquid *falls*, as in the familiar cooling sensation on the skin when a volatile (readily vaporized) liquid such as rubbing alcohol (isopropyl alcohol) evaporates.

Enthalpy (Heat) of Vaporization

To maintain a *constant* temperature while it vaporizes, a liquid must absorb heat to replace the excess kinetic energy carried away by the vaporizing molecules. Vaporization is an *endothermic process*. The **enthalpy (heat) of vaporization** is the quantity of heat that must be absorbed to vaporize a given amount of liquid at a constant temperature. We usually express this quantity in kilojoules per mole (kJ/mol), often in conjunction with an equation like this one for the vaporization of water at 298 K.

$$H_2O(l) \longrightarrow H_2O(g) \qquad \Delta H = 44.0 \text{ kJ/mol}$$

To signify that an enthalpy change is for a vaporization process, we can attach the subscript "vapn" to the ΔH symbol. Several ΔH_{vapn} values are listed in Table 11.1.

TABLE 11.1 Some Enthalpies (Heats) of Vaporization at 298 K[a]

Liquid	ΔH_{vapn}, kJ/mol
Carbon disulfide, CS_2	27.4
Carbon tetrachloride, CCl_4	37.0
Methanol, CH_3OH	38.0
Octane, C_8H_{18}	41.5
Ethanol, CH_3CH_2OH	43.3
Water, H_2O	44.0
Aniline, $C_6H_5NH_2$	52.3

[a] ΔH_{vapn} values are somewhat temperature dependent.

The conversion of a gas to a liquid—the reverse of vaporization—is called **condensation**. A familiar example is the condensation of water on the mirror in a steamy bathroom. Because enthalpy is a function of state, if we vaporize a quantity of liquid and then condense the vapor back to liquid at the same temperature, the total enthalpy change must be *zero*. As a consequence,

$$\Delta H_{vapn} + \Delta H_{condn} = 0$$

$$\Delta H_{condn} = -\Delta H_{vapn}$$

$$H_2O(g) \longrightarrow H_2O(l) \qquad \Delta H = -44.0 \text{ kJ/mol (at 298 K)}$$

In contrast to vaporization, which is an *endothermic* process, condensation is an *exothermic* one.

Most refrigeration and air-conditioning systems are based on a repeated cycle of vaporization and condensation. The refrigerant, a volatile liquid, is allowed to vaporize in a closed system. Part of the surroundings—the interior of a refrigerator, a room, or an entire building—becomes colder by giving up the heat required to vaporize the refrigerant. When the gaseous refrigerant is compressed and condensed back to a liquid, the heat of condensation of the refrigerant is expelled into other parts of the surroundings. In the operation of a refrigerator, the contents of the refrigerator are cooled, and the room in which the refrigerator is kept warms up. In an air-conditioning system, the heat of condensation is expelled outdoors.

APPLICATION NOTE

Common refrigerants include the chlorofluorocarbons (CFCs), such as CCl_3F (b.p. 24 °C). Some of these compounds have been implicated in the depletion of stratospheric ozone discussed in Chapter 25.

EXAMPLE 11.1

How many kilojoules of heat are required to vaporize 175 g methanol, CH_3OH, at 25 °C?

SOLUTION

We can find a value of ΔH_{vapn} for CH_3OH at 298 K in Table 11.1, but notice that it is for one *mole*. We must convert from grams to moles of methanol. Then we can multiply the number of moles of methanol by the ΔH_{vapn} value and obtain the required amount of heat.

$$? \text{ kJ} = 175 \text{ g } CH_3OH \times \frac{1 \text{ mol } CH_3OH}{32.04 \text{ g } CH_3OH} \times \frac{38.0 \text{ kJ}}{1 \text{ mol } CH_3OH} = 208 \text{ kJ}$$

EXERCISE 11.1A

To vaporize a 1.50-g sample of liquid benzene, C_6H_6, requires 652 J of heat. What is ΔH_{vapn} of benzene in kilojoules per mole?

PROBLEM-SOLVING NOTE

You will need ideas from Chapter 1 (relating mass to volume), Chapter 3 (relating grams and moles), Chapter 6 (relating a quantity of heat to a mass and specific heat), together with data from Tables 6.1 and 11.1.

EXERCISE 11.1B

How many kilojoules of heat are required to raise the temperature of 0.750 L of ethanol from 0.0 to 25.0 °C and then to vaporize 10% of the sample? The density (d) of ethanol is 0.789 g/mL, and its specific heat is 2.46 J g^{-1} °C^{-1}.

EXAMPLE 11.2—An Estimation Example

Without doing detailed calculations, determine which liquid in Table 11.1 requires the *greatest* quantity of heat for the vaporization of *one kilogram*.

SOLUTION

The liquid requiring the greatest quantity of heat, *per mole*, is the one with the largest ΔH_{vapn}: aniline, $C_6H_5NH_2$, 52.3 kJ/mol. Next greatest on a per mole basis is water, H_2O, with $\Delta H_{vapn} = 44.0$ kJ/mol. However, we need to consider the heat of vaporization on a mass basis. Let's think about the molar mass of each substance. The molar mass of H_2O is approximately 18 g/mol; the molar mass of $C_6H_5NH_2$ is approximately 93 g/mol. By comparing molar masses, we see that there are about *five times as many* moles of H_2O as $C_6H_5NH_2$ in samples of equal mass. (That is, a 93-g sample is 1 mol of $C_6H_5NH_2$, but $93/18 \approx 5$ mol H_2O.) Therefore, much more heat is required to vaporize 1 kg of H_2O than 1 kg of $C_6H_5NH_2$. Of the remaining liquids in Table 11.1, none has a ΔH_{vapn} as great as that of H_2O, or a molar mass as small. Among the substances shown, water requires the greatest quantity of heat for the vaporization of 1 kg.

EXERCISE 11.2

Without doing detailed calculations, determine which liquid in Table 11.1 requires the *smallest* quantity of heat for the vaporization of *one kilogram*.

Vapor Pressure

When we place a liquid in an *open* container, eventually all the liquid molecules escape into the gaseous state and are dispersed in the atmosphere. When we place a liquid in a *closed* container, at first the volume of liquid decreases, even if only slightly, as some of the liquid is converted to gas. Then evaporation seems to stop, and the volume of liquid remains constant. It appears that nothing further is happening. To explain what really happens, let's see what's going on at the molecular level (Figure 11.3). Also, let's call the gaseous state in contact with the liquid state a vapor, as is commonly done.

In Figure 11.3(a), vaporization of a liquid begins. However, as soon as molecules appear in the vapor state, some of the vapor molecules strike the liquid surface, are captured, and return to the liquid state (Figure 11.3b). Condensation and vaporization occur simultaneously. We can represent these two concurrent and opposing processes by arrows pointing in opposite directions.

$$\text{Liquid} \underset{\text{condensation}}{\overset{\text{vaporization}}{\rightleftharpoons}} \text{Vapor}$$

At first, many more molecules pass from the liquid to the vapor than in the reverse direction; the rate of vaporization is greater than the rate of condensation. As the number of vapor molecules increases, however, so does the rate of condensation. Eventually the rate of condensation becomes equal to the rate of vaporization (Figure 11.3c). At this point, the maximum number of molecules that can be accommodated in the vapor state has been reached. When this happens, the number of molecules per unit volume in the vapor state remains constant with time, and the liquid and vapor are in *dynamic equilibrium*.

Molecules undergoing vaporization — Molecules undergoing condensation

(a) Vaporization

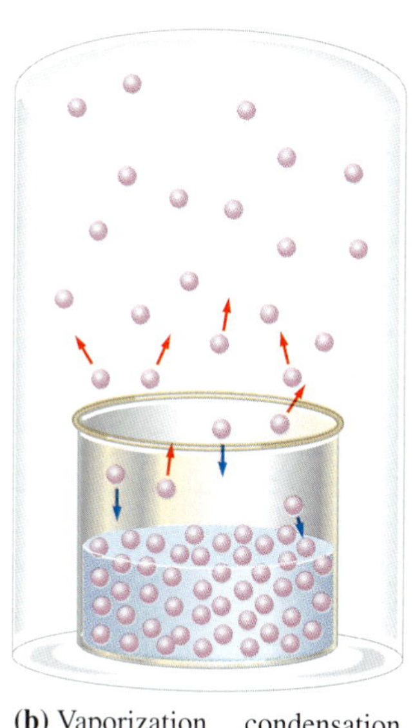

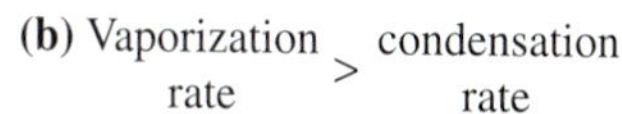

(b) Vaporization rate > condensation rate

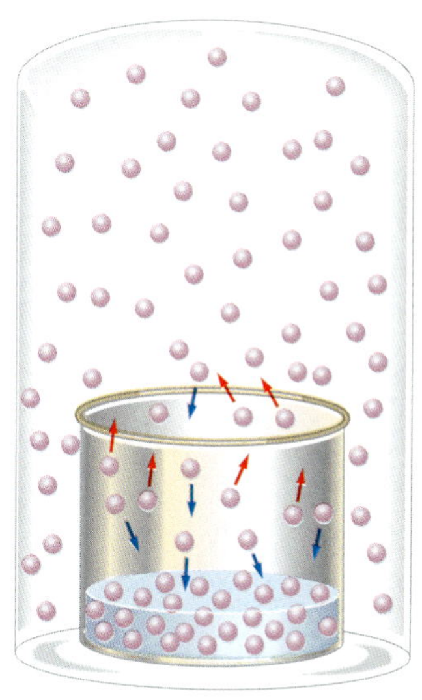

(c) Vaporization rate = condensation rate

▲ **Figure 11.3 Liquid-vapor equilibrium and vapor pressure**
(a) Vaporization of a liquid begins. (b) Condensation begins as soon as the first vapor molecules appear; here, the rate of condensation is still less than the rate of vaporization. (c) The rates of vaporization and condensation have become equal. The partial pressure exerted by vapor molecules on the container walls—the vapor pressure of the liquid—remains constant.

The **vapor pressure** of a liquid is the partial pressure exerted by the vapor when it is in dynamic equilibrium with a liquid at a constant temperature. **Dynamic equilibrium** occurs whenever two opposing processes take place at exactly the same rate. At the macroscopic level, it appears as if no change takes place in an equilibrium condition. However, things keep happening at the molecular level. This is what we mean by *dynamic*. In liquid-vapor equilibrium, molecules of liquid continue to vaporize, and molecules of vapor continue to condense, as suggested in Figure 11.3(c).

The time required to reach liquid-vapor equilibrium at a particular temperature depends on several factors. For example, the smaller the vapor volume and the greater the surface area of the liquid, the faster equilibrium is reached. The equilibrium vapor pressure, in contrast, depends only on what the liquid is and on the equilibrium temperature.

Look again at Figure 11.3. What happens as the temperature is raised? As the average kinetic energy of the molecules in the liquid increases, more molecules have enough kinetic energy to escape from the liquid state. The rate of vaporization increases. At equilibrium, the rates of vaporization and condensation are again equal, but the pressure exerted by the vapor is higher at the higher temperature. *The vapor pressures of liquids increase with temperature.*

Table 11.2 lists the vapor pressure of water at various temperatures. Recall that we used data from a table like this when we applied Dalton's law of partial pressures to the collection of gases over water (page 209). In Example 11.3, we show how a vapor pressure may enter into an ideal gas law calculation.

TABLE 11.2 Vapor Pressure of Water at Various Temperatures

Temperature, °C	Pressure, mmHg	Temperature, °C	Pressure, mmHg	Temperature, °C	Pressure, mmHg
0.0	4.6	29.0	30.0	93.0	588.6
10.0	9.2	30.0	31.8	94.0	610.9
20.0	17.5	40.0	55.3	95.0	633.9
21.0	18.7	50.0	92.5	96.0	657.6
22.0	19.8	60.0	149.4	97.0	682.1
23.0	21.1	70.0	233.7	98.0	707.3
24.0	22.4	80.0	355.1	99.0	733.2
25.0	23.8	90.0	525.8	100.0	760.0
26.0	25.2	91.0	546.0	110.0	1074.6
27.0	26.7	92.0	567.0	120.0	1489.1
28.0	28.3				

EXAMPLE 11.3

Suppose that the equilibrium illustrated in Figure 11.3 is between liquid hexane, C_6H_{14}, and its vapor at 298.15 K. A sample of the equilibrium vapor is found to have a density of 0.701 g/L. What is the vapor pressure of hexane at 298.15 K, expressed in mmHg?

SOLUTION

We will continue to assume that intermolecular forces are negligible in *gases* unless given specific information to the contrary. This means that we can use the ideal gas equation, $PV = nRT$, to describe hexane vapor. Our first step will be to rearrange the equation to isolate the variable, P.

$$P = \frac{nRT}{V}$$

In the above equation, we know two quantities: the temperature (T) and the value of the gas constant (R). We need values of the volume (V) and number of moles of hexane vapor (n). Let's base our calculation on 1.00 L of the equilibrium vapor ($V = 1.00$ L). Because the density of the hexane vapor is 0.701 g/L, the *mass* of 1.00 L of gaseous hexane is 0.701 g. Next, we can use the molar mass of hexane to convert from grams to moles of hexane.

$$n = 0.701\ \cancel{\text{g C}_6\text{H}_{14}} \times \frac{1\ \text{mol C}_6\text{H}_{14}}{86.18\ \cancel{\text{g C}_6\text{H}_{14}}} = 0.00813\ \text{mol}$$

Now we have all of the quantities that we need to solve for vapor pressure.

$$n = 0.00813\ \text{mol}$$
$$V = 1.00\ \text{L}$$
$$R = 0.0821\ \text{L atm mol}^{-1}\ \text{K}^{-1}$$
$$T = 298.15\ \text{K}$$

We substitute these data into the equation that we solved for P.

$$P = \frac{nRT}{V}$$
$$= \frac{0.00813\ \cancel{\text{mol}} \times 0.0821\ \cancel{\text{L}}\ \text{atm}\ \cancel{\text{mol}^{-1}}\ \cancel{\text{K}^{-1}} \times 298.15\ \cancel{\text{K}}}{1.00\ \cancel{\text{L}}}$$
$$= 0.199\ \text{atm}$$

To convert this vapor pressure to the unit mmHg, recall that 1 atm = 760 mmHg (page 180).

$$P = 0.199\ \cancel{\text{atm}} \times \frac{760\ \text{mmHg}}{1\ \cancel{\text{atm}}} = 151\ \text{mmHg}$$

EXERCISE 11.3A

A 335-mL sample of Br_2 vapor is in equilibrium with liquid bromine at 39 °C. The vapor sample has a mass of 1.100 g. Calculate the vapor pressure of bromine at 39 °C.

EXERCISE 11.3B

Suppose that the equilibrium illustrated in Figure 11.3 is between liquid water and its vapor at 22 °C. What mass of water vapor would be present in a vapor volume of 275 mL at 22 °C? Use data from Table 11.2.

A graph of vapor pressure as a function of temperature is called a **vapor pressure curve** (see Figure 11.4). Liquid and vapor coexist in equilibrium at all points lying on the curve. Each point on a vapor pressure curve represents the vapor pressure of the liquid at a given temperature.

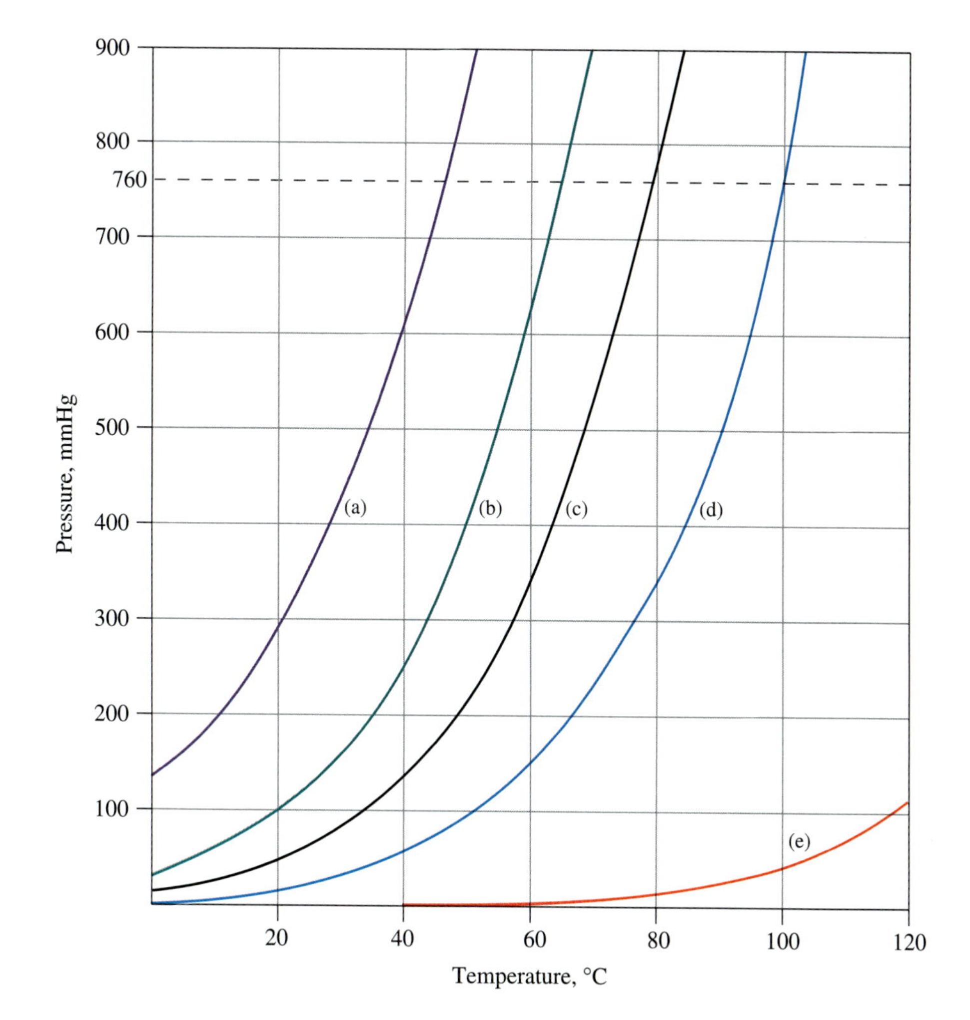

● **ChemCDX** Vapor Pressure vs. Temperature

Figure 11.4
Vapor pressure curves of several liquids
(a) carbon disulfide, CS_2; (b) methanol, CH_3OH; (c) ethanol, CH_3CH_2OH; (d) water, H_2O; (e) aniline, $C_6H_5NH_2$. The temperature of the intersection of the line $P = 760$ mmHg with a vapor pressure curve is the normal boiling point.

If intermolecular forces within a liquid are weak, many molecules escape from the liquid surface before equilibrium is established. The resulting vapor pressure is high, and we say that the liquid is *volatile*. If the intermolecular forces are strong, the equilibrium vapor pressure is low, and we say that the liquid is *nonvolatile*. Diethyl ether, with a vapor pressure of 534 mmHg at 25 °C, is highly volatile. At 25 °C, water has a vapor pressure of 23.8 mmHg; it is moderately volatile. Compared to diethyl ether and water, mercury would be considered *nonvolatile*. Nevertheless, mercury vapor in equilibrium with liquid mercury is present at a high enough partial pressure in air to be toxic for exposures over an extended time. Gasoline is a mixture of hydrocarbons, some of which are quite volatile and others less so. Gasoline for use in cold weather is formulated to have a higher proportion of volatile components so that the vapor produced by the cold gasoline can be ignited more readily.

Vapor Pressure Equations

Vapor pressure curves help us to understand the concept of vapor pressure, but you are not likely to find them in chemistry reference books. Nor are you likely to find many tables of vapor pressure. Instead you will find mathematical equations that summarize experimental vapor pressure data. In this way, vapor pressure data for hundreds of liquids can be presented in just a page or two. Moreover, data can be calculated from these equations more precisely than they can be read from a graph. In Chapter 17, we will see that a particular equation relating vapor pressures and temperature (the Clausius-Clapeyron equation) is just a specific example of a more general expression that we will work with at that time.

Boiling Point

When we heat a liquid in an open container, we observe a special vaporization phenomenon. At a particular temperature, vaporization occurs not only at the surface but *throughout the liquid*. Vapor produced in the interior of the liquid forms bubbles that rise to the surface and escape; the liquid *boils*.

The **boiling point** of a liquid is the temperature at which its vapor pressure becomes equal to prevailing atmospheric pressure. Why can't boiling occur at lower vapor pressures? No boiling occurs because the pressure of the atmosphere would cause any incipient vapor bubbles in the liquid to collapse. That is, the pressure inside the bubbles—the vapor pressure of the liquid—would be less than that exerted by the surrounding atmosphere. Once boiling begins, the temperature of a liquid cannot rise above the boiling point. As long as some liquid remains, the energy provided by continued heating goes into converting more liquid to vapor, not into raising the temperature.

The **normal boiling point** is the temperature at which a liquid boils when the atmospheric pressure is 1 atm, that is, the temperature at which the vapor pressure of the liquid is exactly 1 atm. We can establish the normal boiling point of a liquid from a vapor pressure curve. It is the temperature at which a line at $P = 1$ atm (760 mmHg) intersects the vapor pressure curve. From Figure 11.4 we estimate the normal boiling point of carbon disulfide to be about 46 °C and that of methanol, about 65 °C.

The boiling point is a useful property in identifying liquids. For example, if we find that an unknown liquid boils at about 78 °C, we could refer to a reference book (or Figure 11.4) and conclude that the liquid *could* be ethanol. We would have to do further tests to be sure. However, we could be sure that the liquid is *not* carbon disulfide, methanol, water, or aniline.

Air is less dense at high altitudes than at sea level, and atmospheric pressure therefore decreases with increasing altitude. In turn, the temperature at which a liq-

APPLICATION NOTE

Some liquids undergo decomposition at their normal boiling point, and their boiling points may be recorded at lower pressures. For example, the boiling point of lauric acid, a fatty acid component of vegetable oils, may be listed as $225^{100\text{mm}}$: The liquid boils under a pressure of 100 mmHg at 225 °C.

uid boils decreases as the atmospheric pressure decreases. Thus, although the boiling point of water is 100 °C at normal atmospheric pressure, in the mile-high city of Denver, Colorado (elevation: 1609 m), where the prevailing air pressure is about 630 mmHg, the boiling point of water is 95 °C (203 °F).

The cooking of food involves chemical reactions, and the rates of these reactions depend on the temperature. The chemical reactions go faster with even a small temperature increase, and slower with a temperature decrease. When water boils at 100 °C, an egg can be soft-boiled in three minutes. In Denver, it might take five to six minutes to boil an egg to the same condition. On top of Mount Everest (8848 m), it would take much longer still. Figure 11.5 illustrates an extreme case of the effect of a reduced atmospheric pressure on the boiling point of water. We can easily achieve this effect in the laboratory.

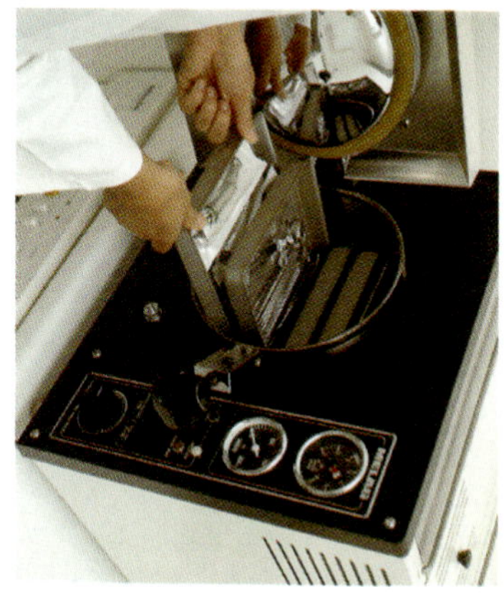

In an autoclave, bacteria (even resistant spores) are killed more rapidly than in boiling water, not directly by the increased pressure but by the higher temperatures attained.

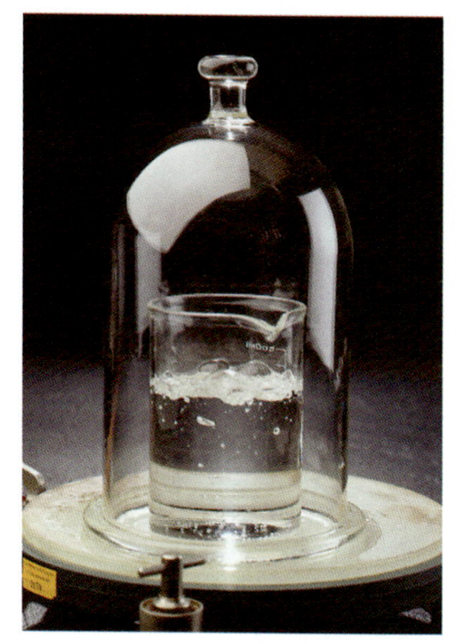

◀ **Figure 11.5**
The effect of pressure on the boiling point of water
Reducing the air pressure in the bell jar can cause water to boil at room temperature.

We exploit the increase in boiling point that is caused by increased external pressure when we use a pressure cooker. In a pressure cooker, water boils at higher than normal temperatures. At such temperatures, the chemical reactions involved in the cooking of potatoes, beets, or a tough piece of meat proceed faster than normal.

The Critical Point

If we heat a liquid in a *closed* container, boiling does not occur. The gas pressure above the liquid increases continuously as vapor accumulates above the liquid, and incipient bubbles of vapor in the liquid cannot overcome this pressure. What we observe instead of boiling as we raise the temperature (Figure 11.6) is that the density of the liquid decreases, the density of the vapor increases, and the boundary (meniscus) between the liquid and vapor becomes blurred and disappears. Finally, the liquid and gaseous states become indistinguishable. The **critical temperature,** T_c, is the highest temperature at which a liquid and vapor can coexist in equilibrium as physically distinct states of matter. The vapor pressure at this temperature is called the **critical pressure,** P_c. The condition corresponding to a temperature of T_c and a pressure of P_c is called the **critical point**. The critical point is the final point on the vapor pressure curve.

Another way to describe the critical temperature is that it is the highest temperature at which a gas can be condensed into a liquid solely by increasing the pressure of the gas. If a gaseous substance has its T_c *above* room temperature, we can

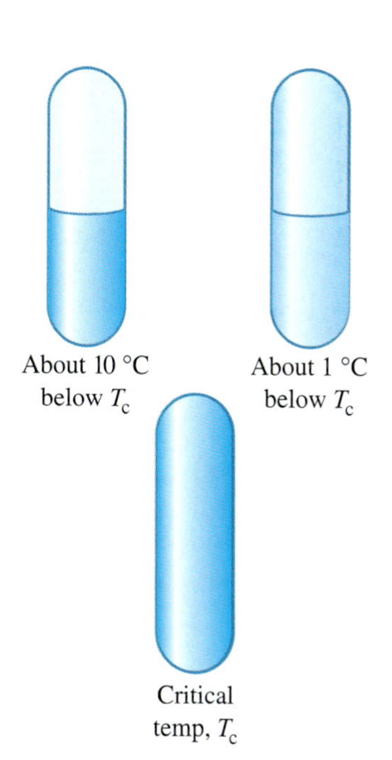

▲ **Figure 11.6**
The critical point
The meniscus separating a liquid (bottom) from its vapor (top) disappears at the critical point. The liquid state does not exist above the critical temperature, regardless of the pressure that might be applied to the substance.

liquefy it at room temperature just by applying enough pressure to the gas. If the T_c is *below* room temperature, however, we must both apply pressure *and* lower the temperature to a value below T_c. Sometimes we use the term *vapor* to refer to the gaseous state of a substance below its T_c and *gas* to the gaseous state above T_c. In these terms, a vapor can be condensed to a liquid simply by applying pressure; a gas cannot. Table 11.3 lists critical temperatures and pressures for several substances.

TABLE 11.3 Critical Temperature and Pressure of Various Substances

Substance	T_c, K	P_c, atm
H_2	33.0	12.8
N_2	126.3	33.5
O_2	154.8	50.1
CH_4	190.6	45.4
CO_2	304.2	72.9
C_2H_6	305.4	48.2
HCl	324.6	81.5
C_3H_8	369.8	41.9
NH_3	405.6	111.3
SO_2	430.6	77.9
H_2O	647.3	218.3

EXAMPLE 11.4—A Conceptual Example

To keep track of how much gas remains in a cylinder, we can weigh the cylinder when it is empty, when it is filled, and after each use. In some cases, though, we can equip the cylinder with a pressure gauge and simply relate the amount of gas to the measured gas pressure. Which method should we use to keep track of the bottled propane, C_3H_8, in a gas barbecue?

SOLUTION

The propane is at a temperature below its T_c (369.8 K) and exists in the cylinder as a mixture of a liquid and vapor. As the fuel is consumed, the volume of liquid in the cylinder decreases, and that of the vapor increases. However, the vapor pressure of the liquid propane does not depend on the amounts of liquid and vapor present. The pressure will remain constant (assuming a constant temperature) as long as some liquid remains. Only after the last of the liquid has vaporized will the pressure drop. Measuring the pressure doesn't tell us how much propane is left in the cylinder until it's almost gone. We would have to monitor the contents of the cylinder by weighing.

EXERCISE 11.4

Which of the two methods described above could you use if the fuel in the cylinder were methane, CH_4? Explain.

Supercritical Fluids

What lies beyond the critical point? It is hard to say exactly. Because liquids and gases both flow readily—they are *fluids*—and because the liquid and gaseous states become indistinguishable at the critical point and remain so somewhat beyond this point, the term *supercritical fluid* (SCF) is commonly used. A **supercritical fluid** is a fluid at temperatures above its critical temperature and pressures above its critical pressure. Although supercritical fluids have been studied in depth only in recent years, they are already being used in many practical applications.

Supercritical Fluids in the Food-Processing Industry

Supercritical fluids are versatile solvents because their properties vary significantly with changes in temperature and pressure. Thus, by an appropriate choice of temperature and pressure, a supercritical fluid can be made to dissolve or extract one component of a mixture while leaving others unaffected. Supercritical fluids are now widely used to perform extractions in laboratories and in industries. Supercritical fluid extraction can be used to measure the fat content of foods, to remove environmental contaminants such as diesel fuel or polychlorinated biphenyls (PCBs) from soil, and to prepare samples (particularly nonvolatile ones such as polymers) for analysis.

Supercritical carbon dioxide is particularly attractive as a solvent in the food-processing industry. A decade or so ago, decaffeinated coffee was made by extracting the caffeine with solvents such as methylene chloride (dichloromethane, CH_2Cl_2). At high concentrations, CH_2Cl_2 has been shown to cause several health problems. Even though no CH_2Cl_2 has been found in the beverage brewed from decaffeinated coffee beans, dichloromethane can be a workplace hazard, and some companies have shifted to supercritical CO_2 to dissolve and carry off the caffeine.

Supercritical fluids also are used to extract cholesterol from eggs, butter, lard, and other fatty foods, making them more suitable for low-cholesterol diets. Potato chips, made by cooking in oils, can be treated with supercritical CO_2 to reduce their normally high fat content. This change improves both the nutritional value and the shelf life of potato chips. Supercritical fluids can also be used to extract the chemical compounds that impart flavor and fragrance to such products as lemons, black pepper, almonds, and nutmeg. These extracts can be used to flavor other foods or give a fragrance to household products.

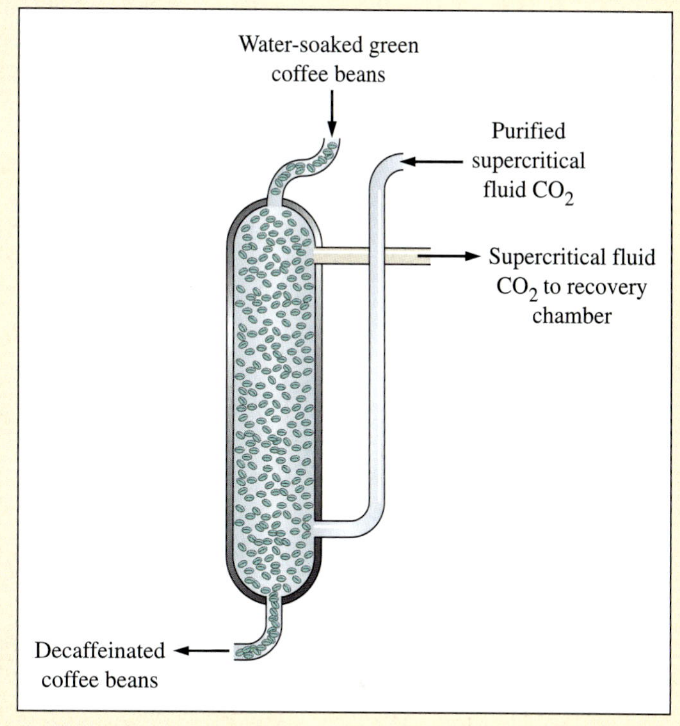

Water-soaked coffee beans enter the vessel from the top and slowly drop through supercritical fluid carbon dioxide at 160 to 220 atm pressure, which enters from the bottom. The caffeine content of the beans drops from an initial 1 to 3% before processing to 0.02%. The CO_2 passes into an absorption chamber where a spray of water leaches the caffeine from the supercritical fluid. The purified CO_2 is recycled.

11.3 Phase Changes Involving Solids

Now let's consider phase changes that involve solids. In particular, we will emphasize energy changes that accompany the transitions of solids to liquids and to vapors.

Melting, Melting Point, and Heat of Fusion

The structural units (atoms, ions, or molecules) of a solid exhibit little motion other than vibrations about fixed points. As the temperature is raised, these vibrations become more vigorous. Finally, the vibrations are strong enough to cause the solid structure to break apart, and a liquid is formed. The conversion of a solid to a liquid is called *melting*, or **fusion**. The temperature at which a solid melts is called its

melting point. The reverse of melting—the conversion of a liquid to a solid—is called *freezing*; the temperature at which it occurs is the **freezing point**. Whether we think of it as a melting solid or a freezing liquid, solid and liquid coexist at equilibrium, and the freezing point and melting point are identical for pure crystalline solids. The freezing point of water and the melting point of ice are the same temperature, 0 °C. Melting points vary slightly with pressure. The data in tables of melting points are usually **normal melting points**, the temperatures at which melting occurs when the pressure on the solid and liquid states is 1 atm.

The quantity of heat required to melt a given amount of solid is called the **enthalpy (heat) of fusion.** As shown below for the melting of one mole of ice, melting is an *endothermic process.*

$$H_2O(s) \longrightarrow H_2O(l) \qquad \Delta H = +6.01 \text{ kJ/mol}$$

Freezing, the reverse of melting, is an *exothermic* process, and the enthalpy of freezing is the negative of the enthalpy of fusion.

$$H_2O(l) \longrightarrow H_2O(s) \qquad \Delta H = -6.01 \text{ kJ/mol}$$

Table 11.4 lists some typical enthalpies of fusion, ΔH_{fusion}.

TABLE 11.4 Some Enthalpies (Heats) of Fusion

Substance	Melting point, °C	ΔH_{fusion}, kJ/mol
Mercury, Hg	−38.9	2.30
Ethanol, CH_3CH_2OH	−114	5.01
Water, H_2O	0.0	6.01
Benzene, C_6H_6	5.5	9.87
Silver, Ag	960.2	11.95
Iron, Fe	1537	15.19

We can determine the freezing point of a liquid by measuring its temperature as it slowly cools. The temperature falls regularly with time, but when freezing begins, the temperature remains constant until all the liquid has frozen. Then the temperature falls regularly as the solid cools. If we plot temperature versus time with data obtained in this way, we get a graph called a **cooling curve.** A typical cooling curve is shown in Figure 11.7. The freezing point of a liquid is the temperature along the straight-line, constant-temperature portion of the cooling curve. Actually, if our interest is only in the freezing-point temperature, we don't need to plot an entire cooling curve. We just need to observe the liquid as it cools and record the constant temperature as it freezes.

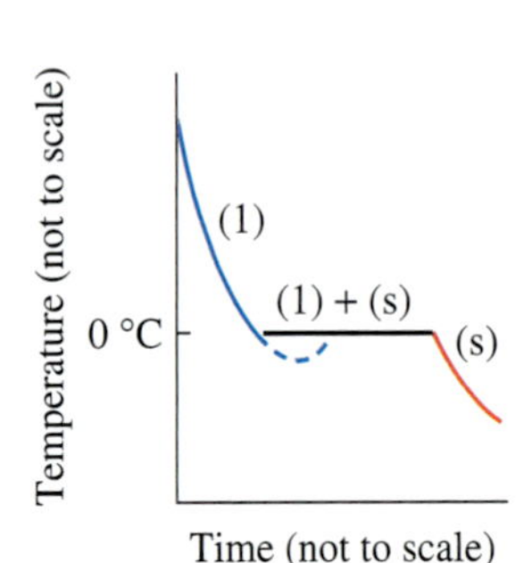

Figure 11.7
Cooling curve for water
The dashed line portion represents the phenomenon of supercooling described in the text, and (l) = liquid; (s) = solid.

Sometimes the temperature of a cooling liquid may drop below the freezing point without a solid appearing, a condition called **supercooling.** This condition may occur if a liquid is quickly cooled to below the freezing point. Formation of a solid requires centers, or sites, on which crystals can grow, and supercooling is more likely to occur if a liquid is kept free of such centers, for example, dust particles. A supercooled liquid is unstable; it cannot persist indefinitely and may begin to freeze at any time. When freezing begins, the temperature of the liquid rises to the freezing point and remains there until freezing is completed. (This process is suggested by the dashed line in Figure 11.7.)

To determine a melting point, we start with the pure solid and add heat. The temperature rises until melting begins, remains constant until melting is completed, and then rises again. A plot of temperature versus time is called a **heating curve.** In appearance, the heating curve of Figure 11.8 resembles the cooling curve of Figure 11.7 but flipped over from right to left. As expected, the freezing point obtained from a cooling curve and the melting point from a heating curve are identical.

APPLICATION NOTE

The droplets in high clouds are often supercooled water. When an airplane flies through these clouds, the surface of the aircraft provides sites for crystal formation, and the airplane suffers from a phenomenon called *rime icing*. If allowed to accumulate, the ice can cause the airplane to crash. Rime icing caused the crash of an American Eagle aircraft in Indiana in 1994, killing all 68 people aboard.

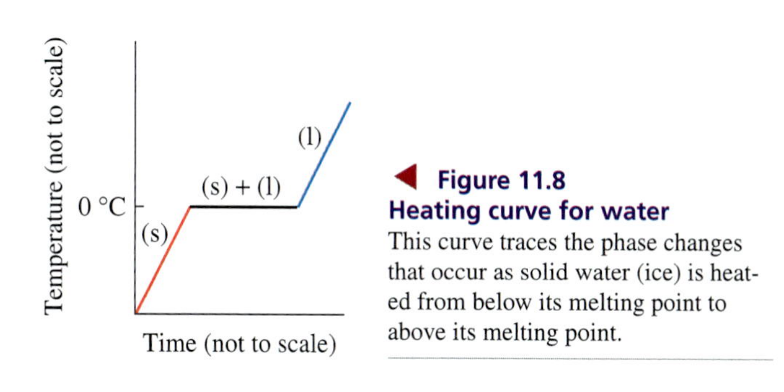

Figure 11.8
Heating curve for water
This curve traces the phase changes that occur as solid water (ice) is heated from below its melting point to above its melting point.

Sublimation

At room temperature, few solids are as volatile as liquids such as ethanol, diethyl ether, and gasoline, but some solids do vaporize. Mothballs, solid-stick room deodorizers, and dry ice are three familiar examples. The passage of molecules directly from the solid to the vapor state is called **sublimation**. The reverse process, the condensation of a vapor to a solid, is generally called *deposition*. A dynamic equilibrium is reached when the rates of sublimation and deposition become equal. Just as the vapor in equilibrium with a liquid does, the vapor in equilibrium with a solid exerts a characteristic vapor pressure, often called the *sublimation pressure*. The sublimation of solid iodine and the deposition of iodine vapor are pictured in Figure 11.9. A plot of the vapor pressure of a solid versus temperature is called a **sublimation curve.**

Although no liquid phase is involved when a solid sublimes, it is sometimes useful to think of sublimation as equivalent to a two-step process: melting followed by vaporization. This assumption and Hess's law help us to understand that the **enthalpy (heat) of sublimation** is simply the sum of the enthalpies of fusion and vaporization.

$$\Delta H_{\text{subln}} = \Delta H_{\text{fusion}} + \Delta H_{\text{vapn}}$$

People living in cold climates are especially familiar with the phenomenon of sublimation. Snow disappears from the ground and ice from the windshield of an automobile even though the temperature stays below 0 °C. This occurs through sublimation, not melting; there is no liquid water at any point in the process. The vapor pressure of ice at 0 °C is 4.58 mmHg.

Figure 11.9
Sublimation of iodine
Even at 70 °C, well below the melting point of iodine (114 °C), solid iodine has a high vapor pressure; it sublimes. The purple iodine vapor deposits as $I_2(s)$ on the colder walls of the flask.

APPLICATION NOTE
Foods and other materials are often preserved by freeze-drying. The material is frozen under a high vacuum so that the water is rapidly removed by sublimation. Harmful microorganisms cannot grow without water. Freeze-dried foods are especially favored for backpacking and camping trips; they are rendered light in weight but retain their nutritive value.

● **ChemCDX** Changes of State

The Triple Point

We have now discussed all the phase changes listed on page 460, but we should consider one additional situation, that in which solid, liquid, and vapor exist together at equilibrium. It is represented by the point at which the vapor pressure curve (extending from higher temperatures) and the sublimation curve (extending from lower temperatures) meet. This point is called the **triple point**; it defines the *only* temperature and pressure at which the *three* states of matter—solid, liquid, and gas—can coexist.

The triple point for water is at 0.0098 °C and 4.58 mmHg. Why isn't the triple point temperature for water exactly 0 °C? Can't solid water (ice), liquid water, and water vapor exist together in the open atmosphere at 0 °C? They can, but this system involves two different pressures. The vapor exists at a pressure of 4.58 mmHg, its partial pressure. But the ice and liquid water are under atmospheric pressure (760 mmHg). Moreover, at the normal melting point liquid water contains some dissolved air and this affects the equilibrium temperature slightly. To have a true triple point, a system must consist of a *pure* substance existing *only under the pressure of its own vapor*. There can be no extraneous substances present (such as the gases in air).

11.4 Phase Diagrams

We can use a *phase diagram* to summarize information about phase changes. A **phase diagram** is a graphical representation of the conditions of temperature and pressure under which a substance exists as a solid, a liquid, a gas, or some combination of these in equilibrium. Figure 11.10 is a hypothetical phase diagram of the simplest sort possible. The colored, labeled areas denote the range of temperatures and pressures in which a substance exists as a solid (green), a liquid (blue), or a gas (cream). Supercritical fluid is represented in the gray area at temperatures above T_c and pressures above P_c. The boundaries of this area are blurred because there is no sharp change in properties in passing from either the liquid or gas area into that of the supercritical fluid.

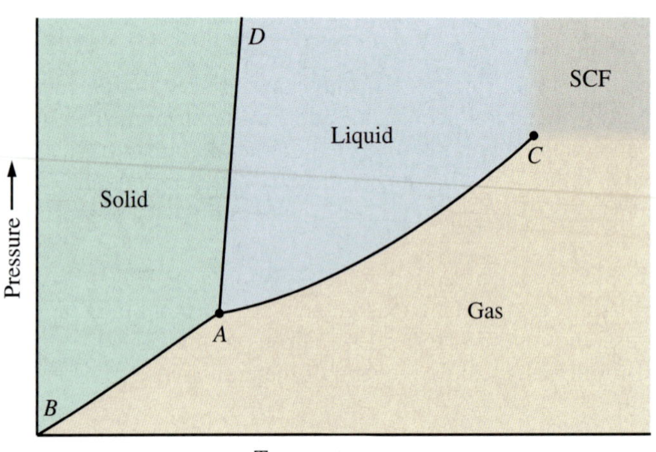

Figure 11.10 A generalized phase diagram: Representing temperatures, pressures, and states of matter of a substance
Because of the extreme range of pressures and temperatures involved, a phase diagram is usually not drawn to scale.

The curves that separate adjoining areas in the diagram represent conditions of temperature and pressure in which two states of matter are in equilibrium. We can summarize the significance of the points and curves in Figure 11.10 as follows:

- Point *A* is the triple point (solid + liquid + gas).
- Curve *AD* is the fusion curve (solid + liquid).

- Curve *AB* is the sublimation curve (solid + gas).
- Curve *AC* is the vapor pressure curve (liquid + gas).
- Point *C* is the critical point (liquid and gas become indistinguishable).

The boundaries of the supercritical fluid area are *not* transition curves, as the fusion, vapor pressure, and sublimation curves are.

We present the phase diagrams of three specific substances in this section: mercuric iodide, carbon dioxide, and water. Each adds important new ideas to what we have outlined here.

Mercury(II) Iodide, HgI_2

The phase diagram of mercury(II) iodide (Figure 11.11) illustrates **polymorphism**, a phenomenon in which a solid exists in two or more different forms. Red $HgI_2(s)$, the form we see in a storeroom bottle of mercury(II) iodide, persists up to 127 °C. At this temperature, red $HgI_2(s)$ converts to yellow $HgI_2(s)$, and only the yellow form is stable up to 259 °C. Red and yellow $HgI_2(s)$ are two *phases* of solid mercury(II) iodide.

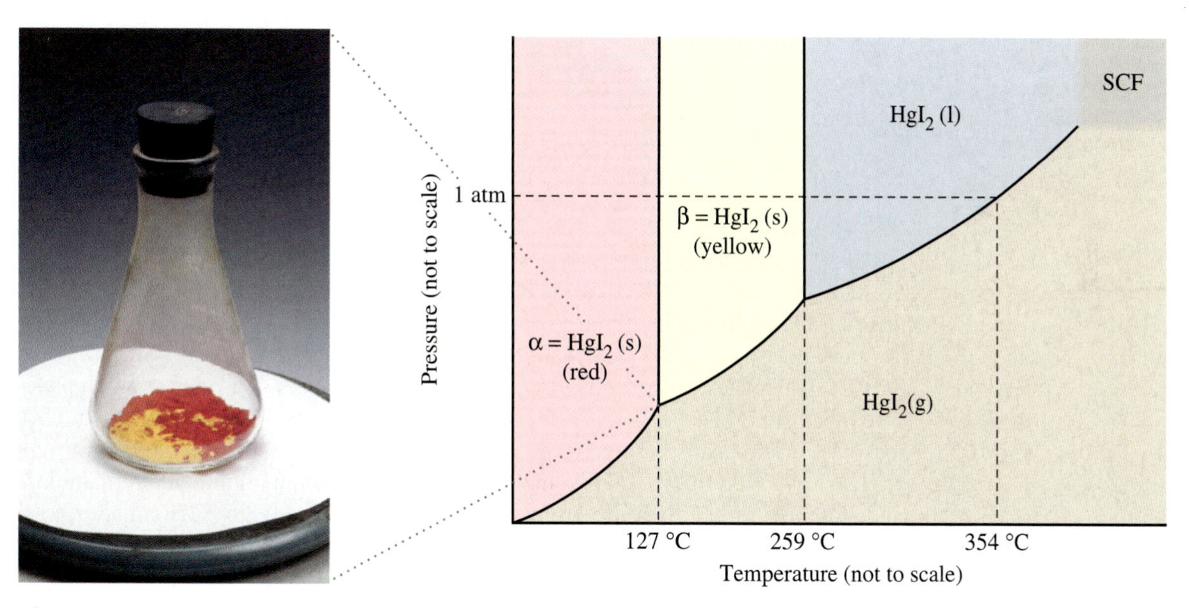

▲ **Figure 11.11 The phase diagram of mercury(II) iodide, HgI_2**
The photograph shows equilibrium between the red and yellow solid phases and gaseous HgI_2 at 127 °C. Also noted in the diagram are the normal melting point of yellow $HgI_2(s)$ at 259 °C and the boiling point of $HgI_2(l)$ at 354 °C.

There can be only one triple point involving the three *states* of matter in a phase diagram. For mercury(II) iodide, this is the equilibrium between yellow $HgI_2(s)$, liquid, and gas at 259 °C. However, there can be additional triple points that involve three *phases* of matter. For mercury(II) iodide the additional triple point, pictured in the photograph in Figure 11.11, has two solid phases and a gas: red $HgI_2(s)$, yellow $HgI_2(s)$, and gas at 127 °C. The deposition of a mixture of the red and yellow solids on the colder walls of the flask indicates that HgI_2 vapor is present.

Carbon Dioxide, CO_2

Now let's examine the phase diagram for carbon dioxide (Figure 11.12). First, the diagram suggests that if a solid-liquid fusion curve, is followed to sufficiently high pressures, the curve extends beyond the critical temperature. (For CO_2, a pressure of several thousand atm is required.) At first, this observation may seem unlikely. However, if we think about how closely together molecules are forced at extremely high pressures, the solid state is the one we should expect. Furthermore, although the critical temperature may seem like a high temperature, T_c for carbon dioxide (304.2 K) is only a little above room temperature (about 293 K).

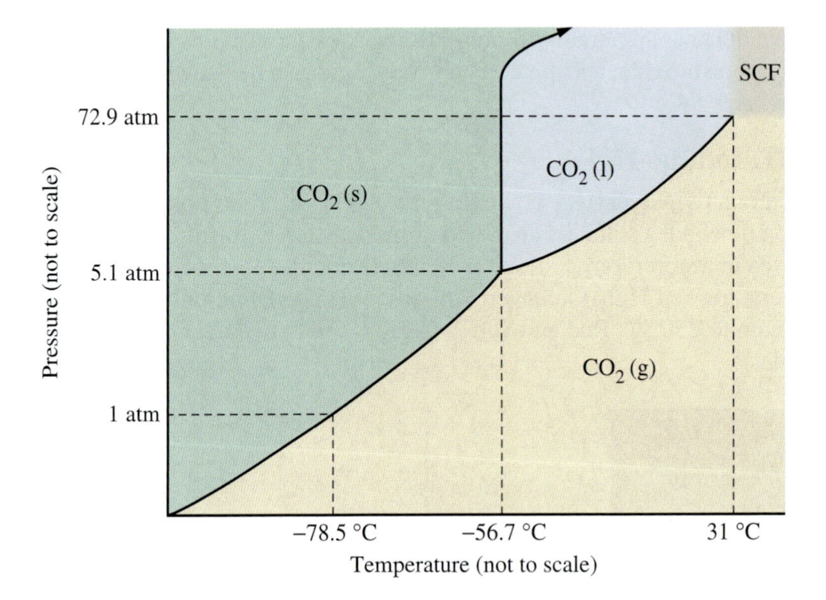

▶ Figure 11.12
Phase diagram of carbon dioxide, CO_2
Noted in the diagram are the normal sublimation temperature of CO_2(s), −78.5 °C; the triple point, −56.7 °C, 5.1 atm; and the critical point, 31 °C, 72.9 atm. The fusion curve slopes away from the pressure axis and at very high pressures reaches temperatures above the critical temperature.

Another interesting feature of the CO_2 phase diagram is that the triple point pressure, 5.1 atm, is well above usual atmospheric pressures. Liquid carbon dioxide can be maintained only at pressures greater than 5.1 atm, and certainly not at normal atmospheric pressure. Liquid CO_2 does not have a normal boiling point. Solid CO_2 also lacks a normal melting point. When solid CO_2 is heated at 1 atm pressure, it sublimes at −78.5 °C. Solid CO_2, called dry ice, is a useful coolant for two reasons. First, because no melting occurs when solid CO_2 is maintained at normal atmospheric pressure, the chilled materials do not contact any liquid and remain dry. Second, the temperature of the remaining dry ice stays at −78.5 °C, no matter how much dry ice has already sublimed. Thus, the cooling effect of the dry ice can be maintained for a relatively long time.

Water, H_2O

The notable feature of the phase diagram for water (Figure 11.13) is that the solid-liquid equilibrium curve (the fusion curve) has a negative slope; it tilts to the left, toward the pressure axis. The normal melting point of ice (1 atm, 0 °C) is represented by the broken black lines. Now notice the pressure-temperature condition represented by the broken blue lines. If solid water (ice) is maintained at a pressure greater than 1 atm, the melting point is below 0 °C. The melting point of solid water (ice) *decreases* as the pressure *increases*. Water is unusual in this respect. In the phase diagrams for almost all other substances, the fusion curve has a positive slope; it tilts to the right, away from the pressure axis. The melting points of most solids *increase* as the pressure *increases*.

Let's use these ideas about pressure and melting point to examine some hypotheses about a common activity: ice-skating. People have claimed that ice-skating is possible because the high pressure exerted by the skate blades causes the ice to melt. Thus, the skater may skim along on a thin film of liquid water from the melted ice. This explanation is probably not correct, however. The last point (*D*)

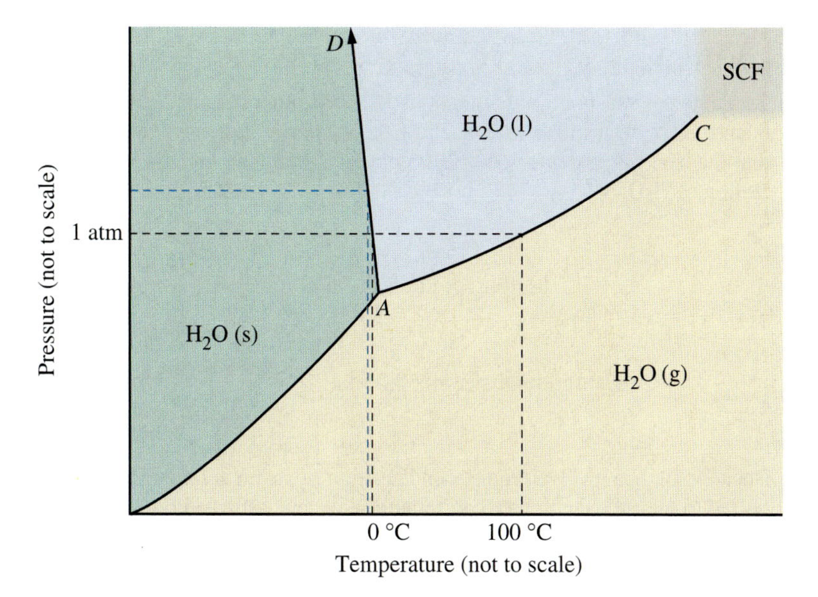

Figure 11.13
Phase diagram of water, H_2O
The triple point, *A*, is at +0.0098 °C and 4.58 mmHg. The critical point, *C*, is at 374.1 °C and 218.2 atm. The negative slope of the fusion curve *AD* is greatly exaggerated. Not shown are the several polymorphic forms of ice that exist at high pressures (in excess of 2045 atm). We describe the significance of the broken blue line in the text.

on the fusion curve (*AD*) is at 2045 atm and a temperature of −22.0 °C (−8 °F). At higher pressures and lower temperatures, ice exists in several different polymorphic forms, and liquid water does not exist at all. Skaters cannot produce pressures nearly as high as 2045 atm, and moreover, it is possible to skate at temperatures below −22 °C.

A second hypothesis is that the frictional resistance of the ice to the skate blades causes an increase in temperature, which melts some of the ice. However, a more recent proposal is backed by experimental measurements. This model suggests that H_2O molecules on the surface of ice retain the vibrational freedom found in liquid water. The vibrational freedom of these molecules gives the surface of the ice a liquidlike quality.

EXAMPLE 11.5—A Conceptual Example

In Figure 11.14, 50.0 mol $H_2O(g)$ (steam) at 100.0 °C and 1.00 atm is added to an insulated cylinder with 5.00 mol $H_2O(s)$ (ice) at 0 °C. *With a minimum of calculation*, and the data provided, determine which of the following will describe the final equilibrium condition: (a) ice and liquid water at 0 °C; (b) liquid water at 50 °C; (c) a mixture of steam and liquid water at 100 °C; or (d) steam at 100 °C. $\Delta H_{\text{fusion}} = 6.01$ kJ/mol; $\Delta H_{\text{vapn}} = 40.6$ kJ/mol (at 100 °C); molar heat capacity of $H_2O(l) = 76$ J mol^{-1} °C^{-1}.

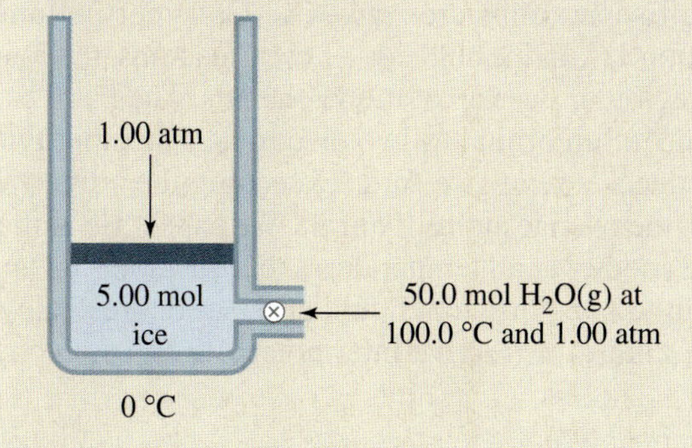

Figure 11.14
Visualizing Example 11.5

SOLUTION

Two changes must occur: Ice must melt and steam must condense. Let's calculate the quantities of heat involved if each of these changes were to go to completion. Note that the heat of condensation of the steam is simply the *negative* of the heat of vaporization of $H_2O(l)$.

$$\text{Melting of ice: } 5.00 \text{ mol} \times \frac{+6.01 \text{ kJ}}{1 \text{ mol}} = +30.05 \text{ kJ}$$

$$\text{Condensation of steam: } 50.0 \text{ mol} \times \frac{-40.6 \text{ kJ}}{1 \text{ mol}} = -2030 \text{ kJ}$$

Because the heat evolved in the condensation of steam is far and away greater than required to melt the ice, no ice can remain. We can eliminate condition (a).

Suppose that the water from the melted ice is heated to 50.0 °C, the temperature for condition (b). The amount of heat required is

$$? \text{ kJ} = 5.00 \text{ mol} \times \frac{76 \text{ J}}{\text{mol } ^\circ\text{C}} \times (50.0 - 0.0) \ ^\circ\text{C} \times \frac{1 \text{ kJ}}{1000 \text{ J}} = 19 \text{ kJ}$$

Far more heat is available from the condensation of steam than needed to melt the ice *and* raise its temperature to 50.0 °C. We can eliminate condition (b).

Because the condensation of steam produces $H_2O(l)$, liquid water must be present; this eliminates condition (d). The final condition, then, must be one of liquid *and* gaseous water at 100 °C; condition (c). (See also Problem 91.)

EXERCISE 11.5

A 1.05-mol sample of $H_2O(g)$ is compressed into a 2.61-L flask at 30.0 °C. Use a point on the phase diagram of Figure 11.13 to indicate the final condition reached.

Intermolecular Forces

An **intermolecular force** is a force that exists between molecules, apart from the forces that hold atoms together through covalent bonds. Intermolecular forces between molecules at very small separations are *repulsive.* Consider helium, which has the electron configuration $1s^2$. In the collision of two He atoms, the four electrons cannot all occupy the region of overlap of the $1s$ orbitals. The Pauli exclusion principle limits occupancy of an orbital to two electrons. The other two are forced into a region that produces a repulsive force exceeding the attractive force of the orbital overlap. Thus, the two He atoms fly apart. We expect He atoms to "keep their distance" from one another at all temperatures, but He atoms do approach each other closely enough to condense to a liquid at temperatures below about 5 K. There must be some kind of *attractive* intermolecular force among He atoms.

Liquid Crystals

Cholesteryl benzoate is a cholesterol derivative. It melts sharply at 145.5 °C to produce a milky fluid. Heating the fluid to 178.5 °C causes it to change abruptly to a clear liquid. Between 145.5 and 178.5 °C, cholesteryl benzoate has the fluid properties of a liquid, the optical properties of a crystalline solid, and some unique properties of its own. It is in a form commonly called *liquid crystal*. Long regarded as mere laboratory curiosities, liquid crystals are now known to almost everyone. We use them in liquid crystal display (LCD) devices such as digital watches, calculators, thermometers, and computer screens.

Liquid crystal formation is relatively common among organic compounds that consist of rodlike molecules and have molar masses in the range of a few hundred grams per mole. Three possibilities for the orientation of these rodlike molecules in liquid crystals are suggested in Figure 11.15. In *nematic* (threadlike) liquid crystals, molecules are arranged parallel to one another. The molecules can move in any direction, and they can rotate on their long axes in the same way that a single pencil can be moved in a loosely packed box of pencils. In *smectic* (greaselike) liquid crystals, molecules are arranged in layers, with the long axes of the molecules perpendicular to the planes of the layers. The molecules can rotate on their long axes and move within the layers.

Cholesteric liquid crystals are related to the smectic form, but the orientation of the molecules in each layer is different from that in the layer above and below. A collection of layers forms a repeating sequence of orientations. The distance between two layers with the same orientation of molecules is a distinctive property of a cholesteric liquid crystal. When a beam of white light strikes a film of cholesteric liquid crystal, the color of the reflected light depends on this characteristic distance. Because this distance changes with temperature, so does the color of the reflected light. Some liquid crystal temperature-sensing devices exhibit color changes for temperature changes as small as 0.01 °C.

The orientation of molecules in a thin film of a nematic liquid crystal changes in the presence of an electric field. Liquid crystal displays (LCDs) exploit this change in orientation, which produces changes in optical properties of the film. In a digital watch or calculator display, electrodes coated with films of liquid crystals are arranged in patterns in the shape of numbers. When an electric field is imposed on the electrodes, the patterns of the electrodes (the numbers) become visible.

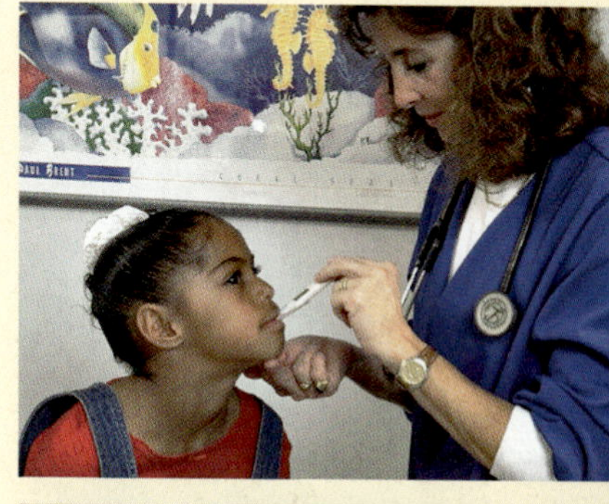

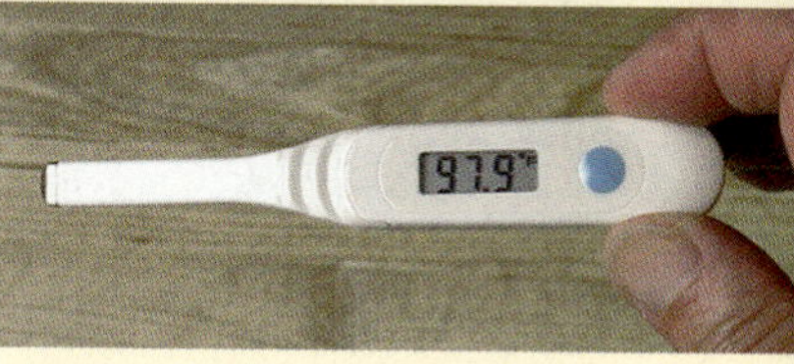

The young patient's temperature is indicated by the liquid crystal display of the thermometer shown here.

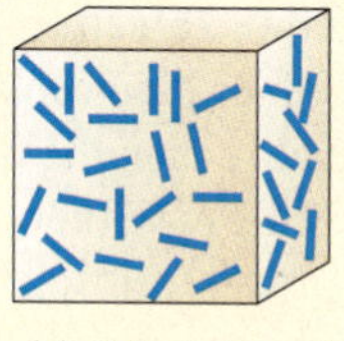

(a) Orientation of molecules in liquid

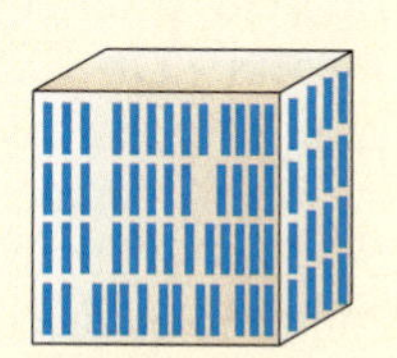

Smectic liquid crystal

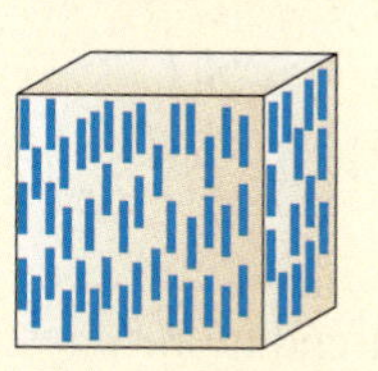

Nematic liquid crystal

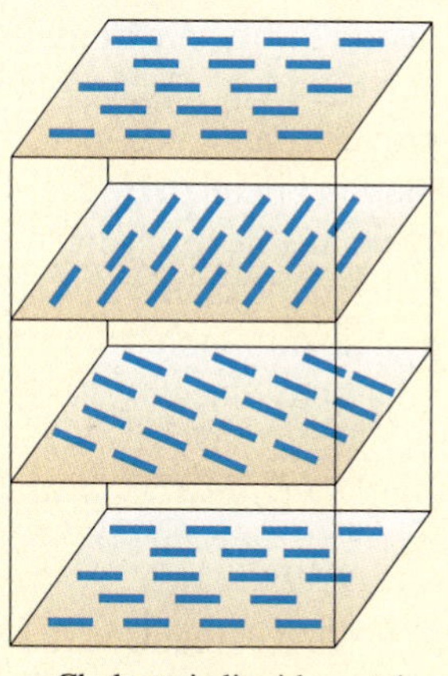

Cholesteric liquid crystal

(b) Orientation of molecules in liquid crystals

◀ **Figure 11.15 The liquid crystalline state**

11.5 Intermolecular Forces of the van der Waals Type

In this section, we will describe two types of intermolecular forces. They belong to a class called *van der Waals forces* because they are the kind of forces we take into account in modifying the ideal gas equation into the van der Waals equation (page 217).

Dispersion Forces

We can visualize an *attractive* molecular force among helium atoms if we keep in mind that the electron charge density pictures we have used since Chapter 7 represent *average* situations only. For example, on average, the electron charge density associated with helium's two 1*s* electrons is evenly distributed in a spherical region about the nucleus. However, at any given instant, the actual location of the two electrons relative to the nucleus can produce an *instantaneous* dipole. This transitory dipole, in turn, can influence the distribution of electrons in neighboring helium atoms, converting those atoms into *induced* dipoles. Figure 11.16 illustrates the formation of an instantaneous dipole and an induced dipole. Figure 11.17 shows a familiar example of the induction of electric charge from the everyday world. The force of attraction between an instantaneous dipole and an induced dipole is known as a **dispersion force** (also called a *London* force, named for Fritz London, who offered a theoretical explanation of these forces in 1928.)

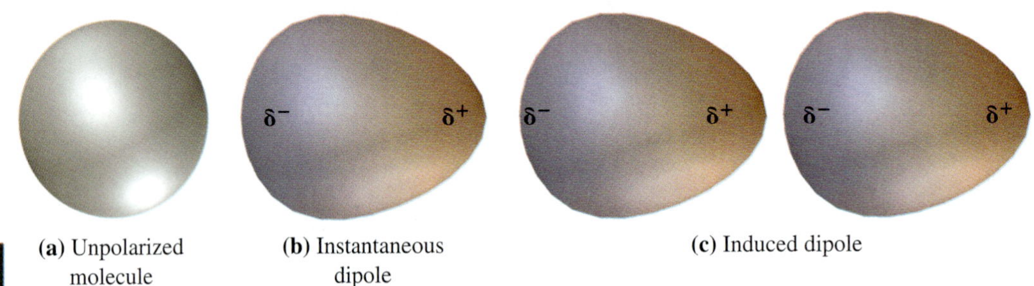

(a) Unpolarized molecule **(b)** Instantaneous dipole **(c)** Induced dipole

▲ **Figure 11.16 Dispersion forces**
(a) *Unpolarized molecule.* The electron charge distribution is symmetrical. (b) *Instantaneous dipole.* A displacement of electron charge density (to the left) produces an instantaneous dipole. (c) *Induced dipole.* The instantaneous dipole on the left induces charge separation in the molecule on the right, making it also a dipole. The attraction between the two dipoles constitutes an intermolecular force.

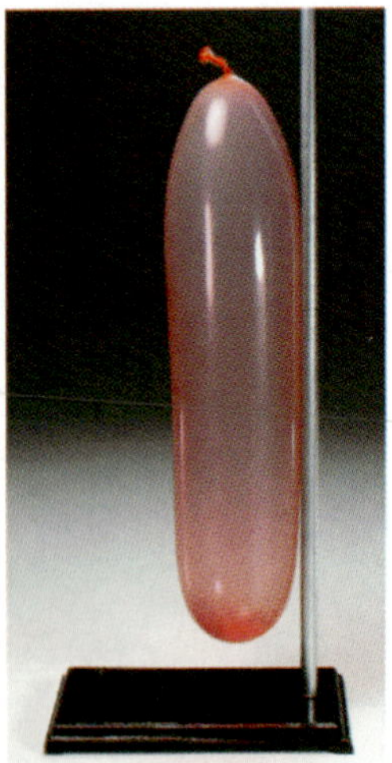

▲ **Figure 11.17 The phenomenon of induction**
The balloon develops a static electric charge when it is rubbed with a cloth. When brought close to a surface, the charged balloon induces an electric charge on the surface with the sign opposite to that on the balloon. The balloon is then attracted to the surface and maintained there by an electrostatic force of attraction.

Polarizability is a measure of the ease with which electron charge density is distorted by an external electric field; it reflects the facility with which a dipole can be induced in an atom or molecule. Large atoms have more electrons and larger electron clouds than small atoms. In large atoms, the outer electrons are more loosely bound; they can shift toward another atom more readily than the more tightly bound electrons in small atoms. Large atoms and molecules are therefore more polarizable than small ones. Atomic and molecular sizes are also closely related to atomic and molecular masses, which means that polarizability increases with increased molecular mass.

The greater the polarizability of molecules, the stronger the intermolecular forces between them.

We can readily note this trend in the physical properties of the Group 7A elements, the halogens. All are nonpolar. The first member, fluorine (F_2), is a *gas* at room temperature (its boiling point is −188 °C). The second member, chlorine (Cl_2), is also a *gas* (boiling point −34 °C), but it is more easily liquefied than is fluorine. At room temperature, bromine (Br_2) is a *liquid* (boiling point 58.8 °C) and iodine (I_2) is a *solid* (melting point 184 °C). Because large molecules are easily polarizable, intermolecular forces between them are strong enough to form liquids or even solids.

Another factor that affects the strength of dispersion forces is molecular shape. Elongated molecules make contact with neighboring molecules over a greater surface than do more compact molecules. As a result, the dispersion forces among the elongated molecules are greater than among the more compact molecules. Figure 11.18 describes two isomers of octane found in gasoline, octane, and isooctane. They have identical molecular masses but different molecular shapes. Intermolecular forces are stronger among the more elongated octane molecules than among the more compact isooctane molecules. As a result, octane has a higher melting point and boiling point than isooctane.

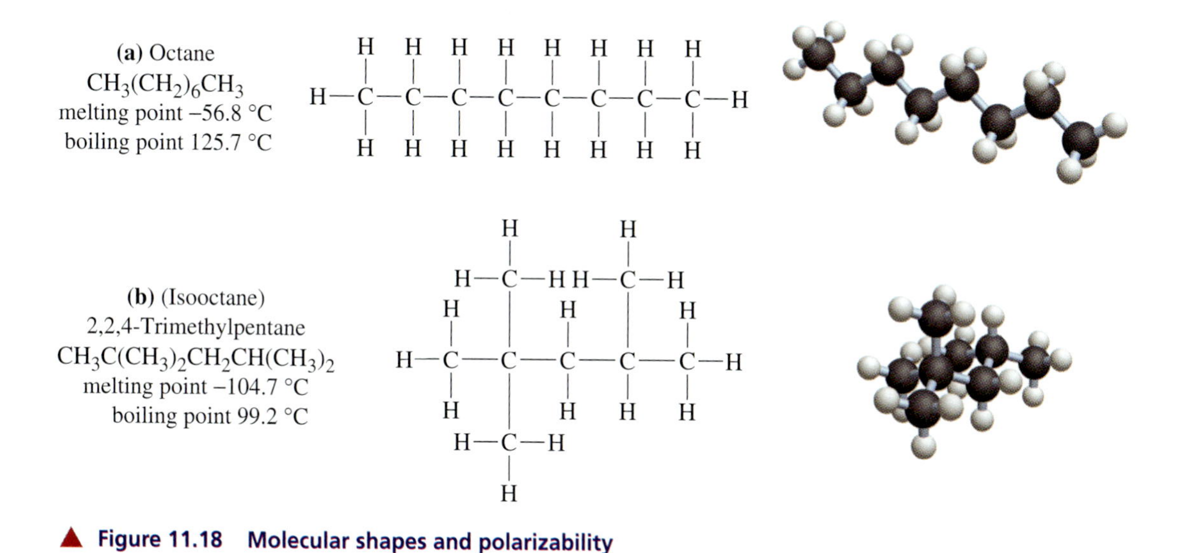

▲ **Figure 11.18 Molecular shapes and polarizability**

Dipole–Dipole Forces

We have just shown that instantaneous and induced dipoles can form in a nonpolar substance. Recall that a *polar* substance, because of its molecular shape and differences in electronegativities between bonded atoms, has *permanent* dipoles. As shown in Figure 11.19, permanent dipoles attempt to align themselves with the positive end of one dipole directed toward the negative ends of neighboring dipoles, giving rise to dipole–dipole forces. The preferred alignment of permanent dipoles is partially offset by the random thermal motion of molecules, and this occurs more so in liquids than in solids. Also, when molecules come close to one another, repulsions occur between like charged regions of dipoles. However, a permanent dipole in one molecule can induce a dipole in a neighboring molecule, giving rise to a dipole-induced dipole force. And this type of attraction exists even when the permanent dipoles are not perfectly aligned.

On balance, there are net attractive forces in a collection of polar molecules. Moreover, the dipole–dipole and dipole-induced dipole forces are *in addition to* the dispersion forces found between all molecules. As a result, the sum of the intermolecular forces is greater in a polar substance than in a nonpolar substance of about the same molar mass. Let's compare nitrogen, nitrogen monoxide, and oxygen to see how polarity affects intermolecular forces and hence the boiling points of these substances.

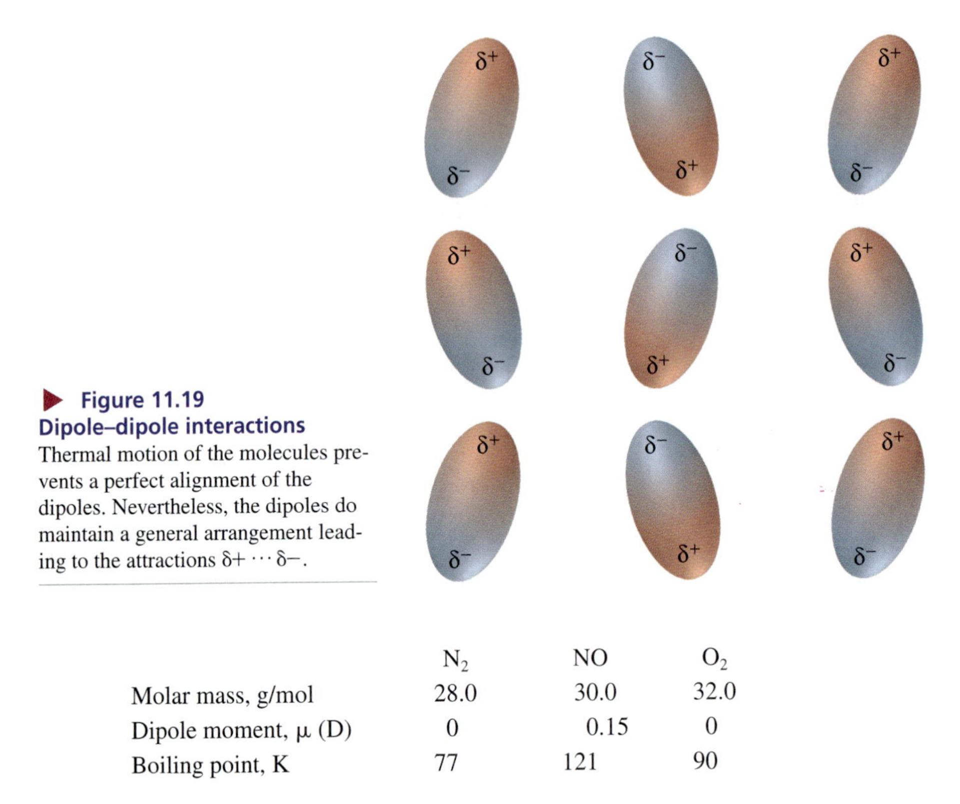

Figure 11.19
Dipole–dipole interactions
Thermal motion of the molecules prevents a perfect alignment of the dipoles. Nevertheless, the dipoles do maintain a general arrangement leading to the attractions δ+ ··· δ−.

	N_2	NO	O_2
Molar mass, g/mol	28.0	30.0	32.0
Dipole moment, μ (D)	0	0.15	0
Boiling point, K	77	121	90

There are stronger intermolecular forces present in NO(l) than in N_2(l) or O_2(l); liquid NO must therefore be warmed to a higher temperature than the other two before it boils.

The more polar a molecule, that is, the greater its dipole moment, the more pronounced is the effect of dipole–dipole forces on physical properties. We can see this by comparing two substances of nearly identical molar masses: *propane*, C_3H_8 (44.10 g/mol) and *acetaldehyde*, CH_3CHO (44.05 g/mol). The electronegativity difference between C and H atoms is very small, and propane is a nonpolar substance. In acetaldehyde, the electronegativity difference between C and O is large. This produces a bond dipole, not offset by other bond dipoles, and a large resultant dipole moment (μ = 2.69 D). As we expect, acetaldehyde has a considerably higher boiling point (20.2 °C) than propane (−42.1 °C).

Predicting Physical Properties of Molecular Substances

We can make predictions about trends in such properties as melting points, boiling points, and enthalpies of vaporization by assessing the impact of the intermolecular forces described in this section. The following summary is helpful in this respect.

- Dispersion forces become stronger with increasing molar mass and elongation of molecules. *In comparing nonpolar substances, molar mass and molecular shape are the essential factors to consider.*
- Dipole–dipole and dipole-induced dipole forces are found in polar substances. *In comparing polar substances to nonpolar substances of comparable molar masses, intermolecular forces are usually stronger in the polar substances. The more polar the substance,* that is, the larger its dipole moment (μ), *the greater we expect the intermolecular force to be.*
- Because they occur in *all* molecular substances, dispersion forces must always be considered. Often they predominate.

EXAMPLE 11.6

Arrange the following substances in the expected order of increasing boiling points: carbon tetrabromide, CBr_4; butane, $CH_3CH_2CH_2CH_3$; fluorine, F_2; acetaldehyde, CH_3CHO.

SOLUTION

The first three substances are *nonpolar*. F_2 is nonpolar because the atoms in the molecules are identical. CBr_4 is nonpolar because of the symmetrical tetrahedral molecular structure ($\mu = 0$), and butane is nonpolar because the electronegativities of the C and H atoms are so nearly alike. We expect the boiling points of the three to increase with increasing molar mass:

$$F_2 \text{ (38.00 g/mol)} < CH_3CH_2CH_2CH_3 \text{ (58.12 g/mol)} < CBr_4 \text{ (331.6 g/mol)}$$

As we noted on page 480, the carbon-to-oxygen bond dipole in acetaldehyde is not offset by any other bond dipoles. Even though the molar mass of acetaldehyde (44.05 g/mol) is somewhat smaller than that of butane, because acetaldehyde is polar ($\mu = 2.69$ D), we expect it to have a higher boiling point than butane.

Comparing acetaldehyde and CBr_4 is more difficult. The polar character of acetaldehyde suggests that it should have the higher boiling point. But because the molar mass of CBr_4 is so much greater (almost 300 g/mol greater) than that of acetaldehyde, we should expect CBr_4 to have the higher boiling point. We predict the following order of increasing boiling points.

$$F_2 < CH_3CH_2CH_2CH_3 < CH_3CHO < CBr_4$$

(The observed boiling points, −188.1, −0.50, 20.2, and 189.5 °C, respectively, support our prediction.)

PROBLEM-SOLVING NOTE

We can't say precisely how large a difference in molar mass is required for a nonpolar substance to have a higher boiling point than a polar one. But as we see here, a difference of several hundred g/mol is more than enough, and a difference of 10 or 20 g/mol is not enough.

EXERCISE 11.6A

Consider the two substances BrCl and IBr. One is a solid at room temperature, and one is a gas. Which substance is a gas? a solid? Explain.

EXERCISE 11.6B

Which member of each of the following pairs would you expect to have the *lower* boiling point? Explain.

a. toluene, $C_6H_5CH_3$, or aniline, $C_6H_5NH_2$

b. *cis*-1,2-dichloroethene or *trans*-1,2-dichloroethene (see page 442)

c. *ortho*-dichlorobenzene or *para*-dichlorobenzene (see page 451)

11.6 Hydrogen Bonds

Suppose we are asked to predict the boiling points of water and acetaldehyde. If we reason as we did in Example 11.7, we would conclude that acetaldehyde, CH_3CHO ($\mu = 2.69$ D), has a higher boiling point than water, H_2O ($\mu = 1.84$ D). Both are polar molecules, but acetaldehyde has a larger dipole moment and also a higher molar mass (44.05 g/mol) than water (18.02 g/mol). However, our prediction would be wrong. Water boils at 100 °C, whereas acetaldehyde boils at 20.2 °C. How can we explain why our prediction is incorrect? The observed boiling points suggests that there is an *additional* kind of intermolecular force that is either not found in acetaldehyde or is not nearly so important in acetaldehyde as it is in water. There is indeed such a force, a kind known as a *hydrogen bond*.

A **hydrogen bond** between molecules is an intermolecular force in which a hydrogen atom covalently bonded to a nonmetal atom in one molecule is *simultaneously* attracted to a nonmetal atom of a neighboring molecule. Although in rare cases the nonmetal atoms may include chlorine and sulfur, the strongest hydrogen bonds are formed if the nonmetal atoms are *small* and *highly electronegative*. Only nitrogen, oxygen, and fluorine atoms can significantly participate in hydrogen bonding.

Think of a hydrogen bond in this way: In a covalent bond, an electron cloud joins a hydrogen atom to another atom, for example, oxygen. The electron cloud is much denser (the electron charge density is greater) at the oxygen end of the bond. The bond is polar, with δ^- on the O atom and δ^+ on the H atom. This leaves the hydrogen nucleus somewhat exposed. As a result, an oxygen atom of a neighboring molecule can approach that hydrogen nucleus rather closely and share some of its "electron wealth" with the hydrogen atom. Hydrogen bonding in water is suggested in Figure 11.20, where we follow the common convention of using dotted lines to represent hydrogen bonds.

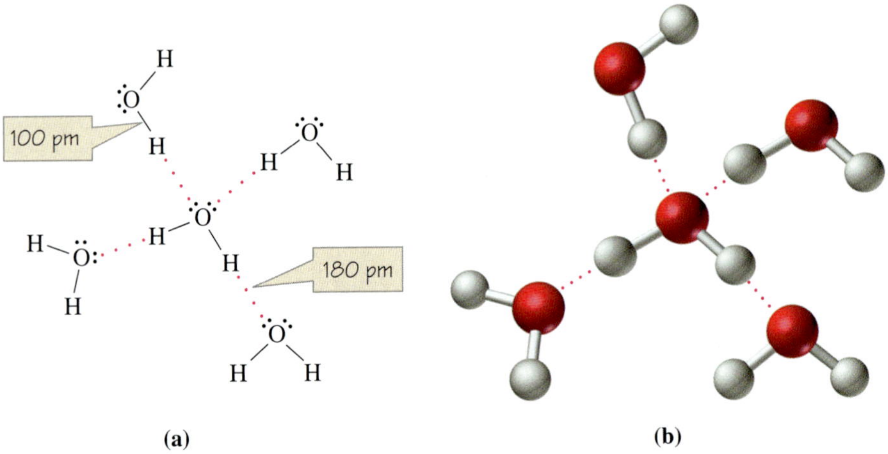

100 pm

180 pm

(a)

(b)

▶ Figure 11.20 Hydrogen bonds in water
As suggested through (a) Lewis structures and (b) ball-and-stick models, each water molecule is linked to four others through hydrogen bonds. Each H atom lies along a line that joins two O atoms. The shorter distances (100 pm) correspond to O—H covalent bonds, and the longer distances (180 pm) to the hydrogen bonds.

EXAMPLE 11.7

For each of the following substances, comment on whether hydrogen bonding is an important intermolecular force: N_2, HI, HF, CH_3CHO, CH_3OH.

SOLUTION

Let's examine each of the molecules from the standpoint that strong hydrogen bonding is expected for molecules in which H atoms are bonded to *small* and *highly electronegative* nonmetal atoms—N, O, or F.

N_2: Nitrogen atoms are small and highly electronegative, but we can't have hydrogen bonds without hydrogen atoms. *Hydrogen bonding does not occur.*

HI: Hydrogen atoms are present, but iodine atoms are large and only moderately electronegative. *Hydrogen bonding does not occur.*

HF: Hydrogen atoms are bonded to small, highly electronegative nonmetal atoms (fluorine). *Hydrogen bonding is a significant intermolecular force.*

CH_3CHO: Both hydrogen atoms and highly electronegative nonmetal atoms (oxygen) are present, but as we see from the structural formula below, the hydrogen atoms are bonded to *carbon*, not oxygen.

```
    H   O
    |   ||
H—C—C—H
    |
    H
```

Hydrogen bonding is not a significant intermolecular force.

CH_3OH: Again, both hydrogen atoms and a small, highly electronegative nonmetal oxygen atom are present, but in this case one of the hydrogen atoms is bonded to the oxygen atom.

```
    H
    |
H—C—O—H
    |
    H
```

Hydrogen bonding is a significant intermolecular force.

EXERCISE 11.7

For each of the following substances, comment on whether hydrogen bonding is an important intermolecular force.

NH_3 CH_4 C_6H_5OH CH_3COOH H_2S H_2O_2

Water: Some Unusual Properties

The model for ice in Figure 11.21 shows how hydrogen bonds hold water molecules in a rigid but open structure. As ice melts, some of the hydrogen bonds break, and water molecules move into the "holes" that were in the ice structure. As a result, the H_2O molecules are closer together in liquid water than in ice. When ice melts, there is about a 10% decrease in volume and a corresponding increase in density. Liquid water at 0 °C is *more dense* than ice, but water is most unusual in this regard. For the vast majority of substances, the liquid state is *less dense* than the solid. High pressures facilitate the disruption of hydrogen bonding and the decrease in volume that accompany the melting of ice. Thus, the greater the pressure, the lower the temperature at which ice can melt. This accounts for the *negative* slope of the fusion curve in the phase diagram of water (Figure 11.13).

If we continue to heat liquid water just above the melting point, more hydrogen bonds break. The molecules become still more closely packed, and the density of liquid water increases to a maximum density at 3.98 °C. Above 3.98 °C, the density of water decreases with temperature, as we normally expect for a liquid. These density phenomena explain why a freshwater lake freezes from the top down. In winter, when the water temperature falls below 4 °C, the more dense water sinks to the bottom of the lake. The colder surface water freezes first. Because ice is less dense than water, the water that does freeze is soon covered with a layer of ice. The ice at the top of the lake then insulates the water below the ice from further heat loss. Except for relatively shallow lakes in extremely cold climates, lakes generally do not freeze solid in the winter time.

The high boiling point of water (100 °C) is also unusual. In fact, a number of substances having considerably higher molar masses than water are *gases* at room

▲ **Figure 11.21 Hydrogen bonds in ice**
(a) Oxygen atoms are arranged in layers of distorted hexagonal rings. Hydrogen atoms lie between pairs of O atoms, closer to one (covalent bond) than to the other (hydrogen bond). (b) At the macroscopic level, this structural pattern is revealed in the hexagonal shapes of snowflakes.

temperature. Two such gases are the nonpolar substances CO_2 and SO_3. The only intermolecular forces in these gases are dispersion forces. Moreover, even some polar substances, such as SO_2, are gases at room temperature. In contrast, methanol, CH_3OH, like water, has a low molar mass (32.04 g/mol), but it is a liquid at room temperature because of strong hydrogen bonding between molecules.

Hydrogen Bonding and Life Processes

The hydrogen bond may seem to be merely an interesting piece of chemical theory, but we cannot overstate its importance to life and health. The structure of proteins, substances essential to life, is determined partly by hydrogen bonding. The action of enzymes, the protein molecules that catalyze the reactions that sustain life, depends in part on the forming and breaking of hydrogen bonds. The heredity that one generation passes on to the next is carried in nucleic acids joined through hydrogen bonds into an elegant arrangement. In DNA and proteins, certain bonds must be easy to break and reform. Of all the kinds of intermolecular forces, only hydrogen bonds have exactly the right amount of energy for this—from about 15 to 40 kJ/mol. In contrast, covalent chemical bonds have energies that range from about 150 to several hundred kJ/mol, and van der Waals forces have energies in the range of only 2 to 20 kJ/mol.

Hydrogen Bonding in Organic Substances

Hydrocarbons form no hydrogen bonds because carbon atoms are not highly electronegative. Many other organic compounds contain oxygen or nitrogen, however, so hydrogen bonding is common in much of organic chemistry.

Acetic acid, CH_3COOH, meets the requirements that we have set forth for hydrogen bonding. However, it has a much lower heat of vaporization than we would expect for a substance with strong intermolecular forces. Why should this be? We find that there is indeed hydrogen bonding in acetic acid. In fact, the hydrogen bonds are strong enough to produce *dimers* (double molecules). Even when acetic acid vaporizes, many of the dimers hold together; not all the hydrogen bonds are broken. It therefore takes less energy to convert a given quantity of liquid to vapor than we might expect, and the heat of vaporization is abnormally low. The structure of a dimer of acetic acid is shown in Figure 11.22.

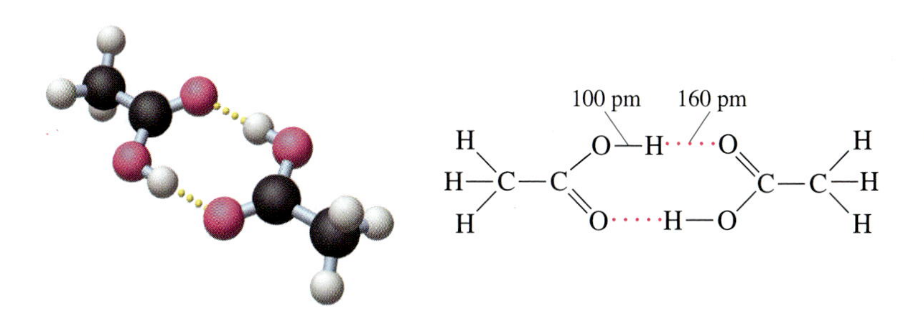

Figure 11.22 Hydrogen bonding in acetic acid
The two acetic acid molecules are joined through two intermolecular hydrogen bonds (dotted lines) into a "double" molecule, or dimer. Note that the hydrogen bond lengths (H···O) are longer than the H—O covalent bond lengths.

In some organic molecules a hydrogen bond may form between two nonmetal atoms *in the same molecule*. These molecules have an *intra*molecular hydrogen bond. An example is seen in salicylic acid (Figure 11.23), an analgesic (pain reliever) and antipyretic (fever reducer) related to aspirin.

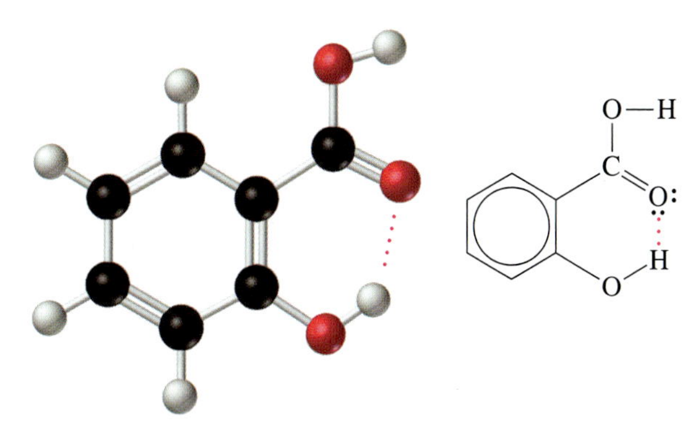

Figure 11.23 Hydrogen bonding in salicylic acid
An intramolecular hydrogen bond (dotted line) joins the OH group to the doubly bonded oxygen atom of the carboxyl group on the same molecule.

Hydrogen Bonding in Proteins

Proteins are polymers with molecular masses ranging from several thousand to several million atomic mass units. Nearly all proteins of all living species, from bacteria to humans, are constructed from a basic set of 20 amino acids as monomers. Almost all amino acids [$RCH(NH_2)COOH$] are characterized by an amino group ($-NH_2$) bonded to the alpha (α) carbon atom, that is, to the carbon atom adjacent to the carbon atom in a carboxyl group. The individual amino acid groups have different R groups.

Two amino acid molecules join by eliminating a molecule of water between them and forming a *peptide* bond ($-CO-NH-$).

$$H_2N-\overset{\alpha}{\underset{R}{CH}}-\overset{O}{\overset{\|}{C}}-O-H + H-\overset{H}{N}-\overset{\alpha}{\underset{R}{CH}}-COOH \longrightarrow H_2N-\underset{R}{CH}-\overset{O}{\overset{\|}{C}}-\underbrace{NH}_{\text{Peptide bond}}-\underset{R}{CH}-COOH + H_2O$$

The groups at the ends of this dipeptide can each join with another amino acid molecule, forming a chain that contains four amino acid units and three peptide bonds. Two more amino acid molecules can be added to the chain, and so on, until a long chain of amino acids forms.

$$\cdots NH-\underset{\displaystyle R}{\underset{|}{CH}}-\overset{\displaystyle O}{\overset{\|}{C}}-NH-\underset{\displaystyle R}{\underset{|}{CH}}-\overset{\displaystyle O}{\overset{\|}{C}}-NH-\underset{\displaystyle R}{\underset{|}{CH}}-\overset{\displaystyle O}{\overset{\|}{C}}-NH-\underset{\displaystyle R}{\underset{|}{CH}}-\overset{\displaystyle O}{\overset{\|}{C}}-NH-\underset{\displaystyle R}{\underset{|}{CH}}-\overset{\displaystyle O}{\overset{\|}{C}}\cdots$$

An organism's genetic code specifies the *sequence* of amino acids in each kind of protein (Chapter 23). This sequence is called the *primary structure* of the protein. The chains, in turn, can twist and fold to give a variety of shapes to protein molecules. These shapes are called the *secondary structure* of the protein. As suggested by Figure 11.24, hydrogen bonding is of great importance in determining the two main kinds of secondary structures found in all proteins.

Proteins differ in the *sequence* of the amino acids along the protein chain, the lengths of the chains, and in the number of chains. In some proteins, the chains are twisted about one another into larger cables and fibers, which are used for connections, support, and structure. These are the *fibrous proteins*; they are found, for example, in hair, skin, and muscle, and in insect fibers such as silk. In other proteins, the chains are folded back on themselves to make compact *globular proteins*. Hemoglobin, enzymes, and the gamma globulins that serve as protective antibodies are globular proteins. Enzymes are proteins that make nearly all the reactions in living cells possible.

A protein molecule functions only when it is in the proper configuration. Heat, ultraviolet light, and some chemical substances *denature* proteins, that is, change them in ways that destroy their function. They do so by disrupting the hydrogen bonds that hold the molecules in their proper arrangement.

Most proteins are denatured when heated above 50 °C. Heat, ultraviolet radiation, and organic compounds such as alcohols and phenols are used to disinfect things by denaturing the proteins in bacteria and thus killing them. We cook most of our protein-containing food, in part to kill harmful microorganisms, and also because denatured proteins are usually easier to chew and easier for digestive enzymes to break down.

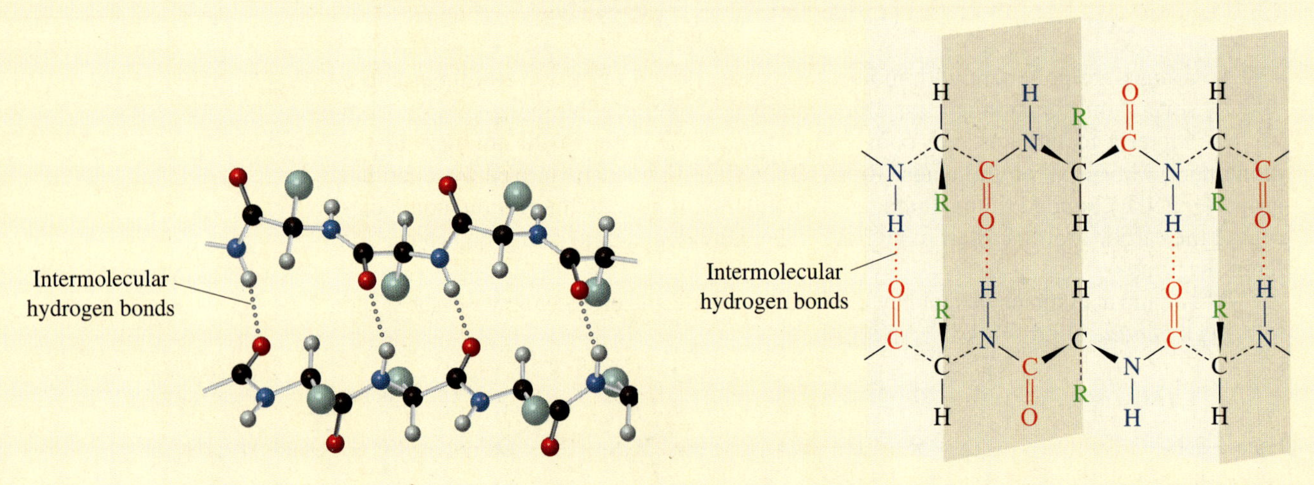

(a)

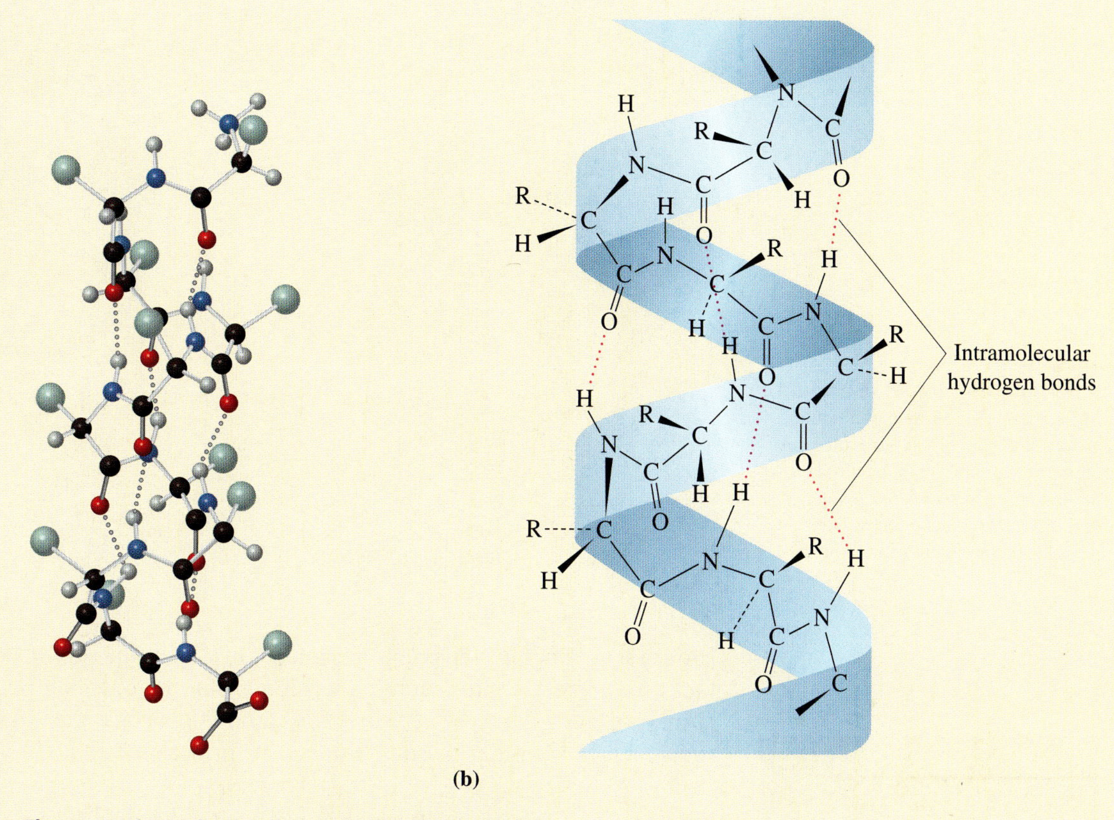

▲ **Figure 11.24 The two principal secondary structures of proteins**
(a) In the pleated sheet arrangement (page 486), protein chains run parallel and in alternating directions. The chains are held together by hydrogen bonds that join a NH group on one chain with a CO group on a neighboring chain. The hydrogen bonds are shown in both the molecular model and in the structural formula. In addition, the structural formula emphasizes the pleats (imagine that it is written on a pleated sheet of paper). (b) In the α-helical arrangement, the protein chain is coiled into a helix. Each NH group is hydrogen bonded to a CO group one helical turn (3.6 amino acid units) away in the same chain, giving a fairly rigid cylindrical structure with side chains on the outside.

11.7 Intermolecular Forces and Two Liquid Properties

So far, we have seen that the strengths of intermolecular forces affect densities, melting points, boiling points, and enthalpies of vaporization. Other physical properties are also affected. In this section, we look at two familiar ones that reflect intermolecular forces quite well: surface tension and viscosity.

Surface Tension

Regardless of the nature of the intermolecular forces in a liquid, molecules within the bulk of a liquid are attracted to more neighboring molecules than are surface molecules (Figure 11.25). Because they experience a greater net intermolecular attractive force, molecules in the bulk of a liquid are in a lower energy state than are surface molecules. Molecules crowd into the interior of a liquid to the greatest extent possible, leaving a minimum of surface area. A sphere has a smaller ratio of surface area to volume than any other three-dimensional figure, and free-falling liquids therefore tend to form spherical drops.

It takes energy to increase the surface area of a liquid because to do so requires that some molecules pass from the lower-energy bulk liquid to the higher-energy

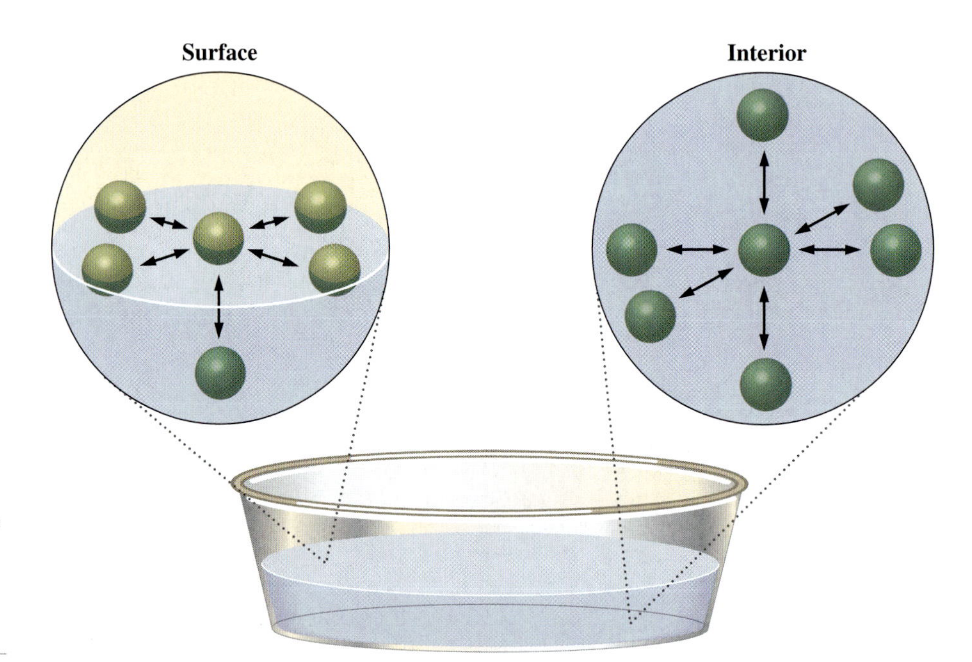

▶ **Figure 11.25**
Intermolecular forces in a liquid
Molecules at the surface of a liquid are attracted only by other molecules at the surface and by molecules below the surface. Molecules in the interior of a liquid experience forces from neighboring molecules in all directions.

surface state. **Surface tension** is the amount of work required to extend a liquid surface. We usually express surface tension (γ) in the unit joule per square meter (J/m^2)—that is, the work required to extend the surface area of a liquid by one square meter. Consider these two values of surface tension at 20 °C

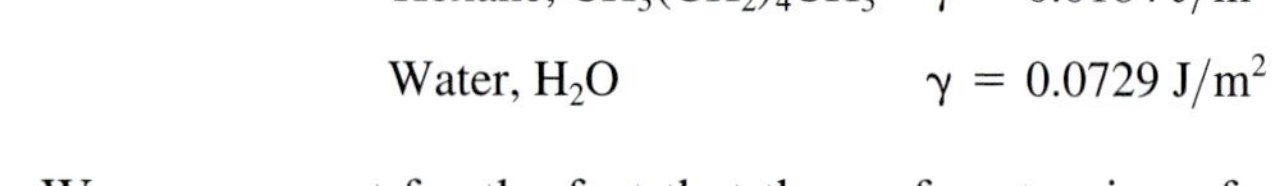

$$\text{Hexane, } CH_3(CH_2)_4CH_3 \quad \gamma = 0.0184\ J/m^2$$
$$\text{Water, } H_2O \quad \gamma = 0.0729\ J/m^2$$

We can account for the fact that the surface tension of water is significantly greater than that of hexane by comparing intermolecular forces. Strong intermolecular hydrogen bonding makes it more difficult to extend the surface of liquid water than that of hexane, where the intermolecular forces are dispersion forces only. When liquids are heated, the intermolecular forces are more easily overcome by the increased thermal energy of the molecules. Surface tension therefore *decreases* with an increase in temperature.

The tendency of molecules to be drawn toward the interior of a liquid makes the surface act as if it were covered with a tight "skin." This skin allows the steel needle in Figure 11.26 to float on water, even though the needle is much denser than water and apparently should sink. The force of gravity on the needle (its weight) is not enough to break the skin, that is, to spread the surface of the water over the needle.

When a drop of liquid spreads across a surface, we say that the liquid "wets" the surface. The energy required to spread the drop comes from the collapse of the drop under the force of gravity. However, we need to consider two other factors in determining whether a liquid will spread across a surface: the strengths of adhesive and cohesive forces. **Adhesive forces** are intermolecular forces between unlike molecules, and **cohesive forces** are those between like molecules. If the adhesive forces between a liquid and a surface are stronger than the cohesive forces within the liquid, the liquid will wet the surface. If the adhesive forces are weaker, the liquid will not wet the surface.

In order to clean a surface with water, the water must wet the surface. Glass and certain fabrics, for example, are easily wet by water. If glass is coated with a film

▲ **Figure 11.26**
Surface tension of water
The surface tension of water enables the steel needle, though denser than water, to float on the surface of the water. It also supports the water strider.

▲ **Figure 11.27 Adhesive and cohesive forces**
Strong adhesive forces cause a thin film of water to spread on a clean glass surface (left). When glass is coated with an oil film (right), adhesive forces between oil and water cannot overcome the cohesive forces in water. Drops of water stand on the glass.

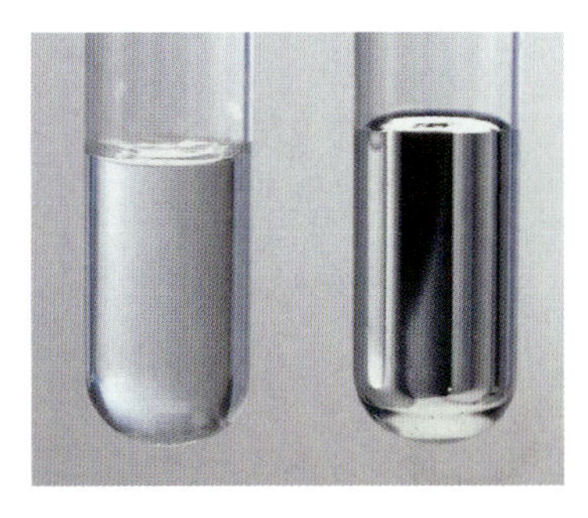

▲ **Figure 11.28 Meniscus formation**
Because it wets glass, water forms a *concave* meniscus in a glass container (left). Mercury does not wet glass (right) and forms a *convex* meniscus.

of grease or oil, however, water won't wet the surface. That is, water does not wet the oily film; instead, water droplets bead on the greasy film (Figure 11.27). A detergent acts to disperse the grease in water. To waterproof a fabric (a canvas tent, for example), we purposely coat the material with an oil that prevents water from wetting the fabric.

If a liquid wets the surface of its container, the liquid is drawn up the walls of the container to a slight extent. The interface between the liquid and the air above it, called a **meniscus,** has a *concave*, or saucerlike, shape (◡). In contrast, if a liquid does not wet the walls of its container, the liquid pulls away from the wall and exhibits a *convex* meniscus, like an inverted saucer (◠). Water in a glass container produces a concave meniscus, and mercury forms a convex one (Figure 11.28). Meniscus formation is greatly exaggerated in tubes of very small diameter, called *capillary* tubes. We call this phenomenon *capillary* action. As shown in Figure 11.29, water is drawn up several centimeters into a small glass capillary tube. The rise of water in a sponge also occurs through capillary action.

▲ **Figure 11.29 Capillary action**
The spread of a thin film of water up the capillary walls produces a slight drop in pressure below the meniscus. Atmospheric pressure pushes a column of water up the capillary to offset the pressure difference. The rise of water in the capillary is opposed by the force of gravity on the liquid column, limiting the height to which the water rises.

Viscosity

When a liquid flows, one portion of the liquid moves with respect to neighboring portions, and cohesive forces within the liquid create an "internal friction" that reduces the rate of flow. We use the term **viscosity** to describe a liquid's resistance to flow. Liquids such as molasses, honey, and heavy motor oil do not flow readily; they have a high viscosity and are said to be *viscous*. Liquids like hexane, water, and ethanol flow easily and are said to be *mobile*.

At least in part, the viscosity of a liquid is related to the strengths of intermolecular forces—the stronger these forces, the greater the viscosity. We can see this rather clearly in the following comparison of three alcohols.

H	H	H
|	|	|
H—C—O—H	H—C—O—H	H—C—O—H
|	|	|
H—C—H	H—C—O—H	H—C—O—H
|	|	|
H—C—H	H—C—H	H—C—O—H
|	|	|
H	H	H
Propyl alcohol	Propylene glycol	Glycerol
1-propanol	1,2-propanediol	1,2,3-propanetriol
0.02 P	0.5 P	10 P

▲ **Figure 11.30**
Measuring viscosity
The flow of automotive motor oil is started at the same time from the two identical funnels. The highly viscous oil (SAE 40) on the left flows much more slowly than the less viscous oil (SAE 10) on the right.

The unit of viscosity shown above is called the *poise* (P).* The effect of increasing molar mass should be to produce a gradual increase in viscosity in a comparison from left to right. However, the dramatic increases that we see are attributed to hydrogen bonding. As the number of —OH groups in an alcohol increases, the possibilities for hydrogen bonding increase, and the resistance to flow (the viscosity) also increases.

The viscosity of a liquid can be determined by timing its flow under carefully controlled conditions. The flows of two motor oils of widely different viscosity are compared in Figure 11.30. Intermolecular forces become less effective with an increase in temperature, and as is the case with surface tension, viscosity *decreases* with an increase in temperature.

The Structures of Solids

As we have noted, the constituent atoms, ions, or molecules of a solid are in close contact. If these structural units are in disorganized clusters with no long-range order, a solid is *amorphous*. In many solids, however, the structural units exist in highly ordered assemblies called *crystals*. Recall that we described the formation of a crystal of NaCl in Section 9.3 (page 363). Table 11.5 lists some characteristics of the major types of crystalline solids. In the remainder of this chapter, we will present more information about several of these types of solids.

11.8 Network Covalent Solids

In most covalently bonded substances, bonds between atoms in molecules, *intra*molecular forces, are quite strong. In these same substances, attractions between molecules, *inter*molecular forces, are much weaker, perhaps only a few percent as strong. As a result, many molecular substances exist as gases at room temperature, and the rest are mostly liquids or solids with low to moderate melting points. However, in a few covalently bonded substances, called **network covalent solids,** a network of covalent bonds extends throughout a crystalline solid, holding it together by exceptionally strong forces. Two prime examples are diamond and graphite, the two principal forms of carbon.

Diamond

Knowing that the carbon-to-carbon bonds in diamond are *single* covalent, we can attempt to write a Lewis structure.

```
        ·C̣·
         |
   ·Ċ—C—Ċ·
    ·    |   ·
        ·C·
         ·
```

An unsatisfactory Lewis structure for diamond

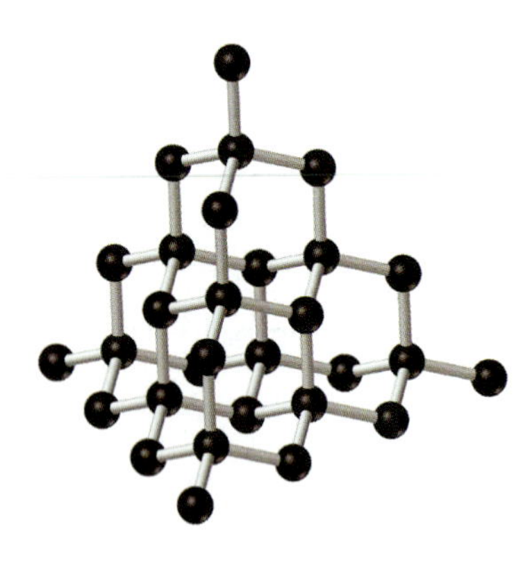

▲ **Figure 11.31**
The crystal structure of diamond
Each carbon atom is bonded to four others in a tetrahedral fashion.

In this structure, only one of the five C atoms has a valence-shell octet. Even though the Lewis structure will still be unsatisfactory, we can increase the proportion of C atoms with valence-shell octets by adding more carbon atoms. In this way, we can build up a giant "molecule" that consists of all the atoms in the diamond crystal. The C atoms of a diamond form a network covalent solid.

As shown in the tiny portion of a diamond crystal represented in Figure 11.31, each carbon atom is bonded to four other carbon atoms in a tetrahedral arrangement. We associate this bonding arrangement with sp^3 hybridization of the carbon atoms.

* The poise (named for J. L. Poiseuille) is a non-SI unit. The SI unit of viscosity is $1\ N \cdot s \cdot m^{-2} = 10$ P.

To scratch or break a diamond crystal, we have to break many covalent bonds. This is quite hard to do, and as a result, diamond is the hardest substance known. No other substance can scratch a diamond, but a diamond can scratch other solids such as glass. Diamond is the best abrasive available. To melt a diamond, we must also break covalent bonds; this accounts for its exceptionally high melting point of more than 3500 °C.

Because silicon is in the same group as carbon (Group 4A), we might expect Si atoms to be able to substitute for some of the C atoms in the diamond structure. This actually occurs in the compound silicon carbide, SiC. Silicon carbide is best known as the abrasive Carborundum®, widely used to make grindstones.

Diamond does not conduct electricity. The two requirements for electrical conductivity are (1) charged particles must be present, and (2) the particles must be free to move in an electric field. In diamond, the charged particles (electrons) are present, but they are all *localized* in covalent bonds. They are not set in motion in an electric field, and consequently diamond is a *nonconductor* of electricity.

Graphite

Another bonding scheme for carbon that uses all of the valence electrons has each C atom bonded to three other C atoms in the same plane. According to this model, three of the four valence electrons of each C atom are *localized* in sp^2 hybrid orbitals, but the fourth is in a $2p$ orbital perpendicular to the plane of the sp^2 orbitals. These electrons are located between planar layers of C atoms, and they are *delocalized*. The bonding is similar to that in benzene, except that the delocalized electrons are spread throughout the regions between planes of C atoms rather than being

TABLE 11.5 Some Characteristics of Crystalline Solids

Type	Structural Particles	Intermolecular Forces	Typical Properties	Examples
Molecular				
Nonpolar	Atoms or nonpolar molecules	Dispersion forces	Extremely low to moderate melting points; soluble in nonpolar solvents	Ar, H_2, I_2, CCl_4, CH_4, CO_2
Polar	Polar molecules	Dispersion forces, dipole–dipole and dipole-induced dipole attractions	Low to moderate melting points; soluble in some polar and some nonpolar solvents	HCl, H_2S, $CHCl_3$, $(CH_3)_2O$, $(CH_3)_2CO$
Hydrogen-bonded	Molecules with H bonded to N, O, or F	Hydrogen bonds	Low to moderate melting points; soluble in some hydrogen-bonded and some polar liquids	H_2O, HF, NH_3, CH_3OH, CH_3COOH
Network Covalent	Atoms	Covalent bonds	Most are very hard; sublime or melt at very high temperatures; most are nonconductors of electricity	C(diamond), C(graphite) SiC, SiO_2, BN
Ionic	Cations and anions	Electrostatic attractions	Hard; brittle; moderate to very high melting points; nonconductors as solids, but electrical conductors as liquids; many are soluble in water	NaCl, CaF_2, K_2S, MgO
Metallic	Cations and delocalized electrons	Metallic bonds	Hardness varies from soft to very hard; melting points vary from low to very high; lustrous; ductile; malleable; good to excellent conductors of heat and electricity	Na, Mg, Al, Fe, Cu, Zn, Mo, Ag, Cd, W, Pt, Hg, Pb

confined to separate hexagonal rings of C atoms. The crystal structure of this modification of carbon is shown in Figure 11.32.

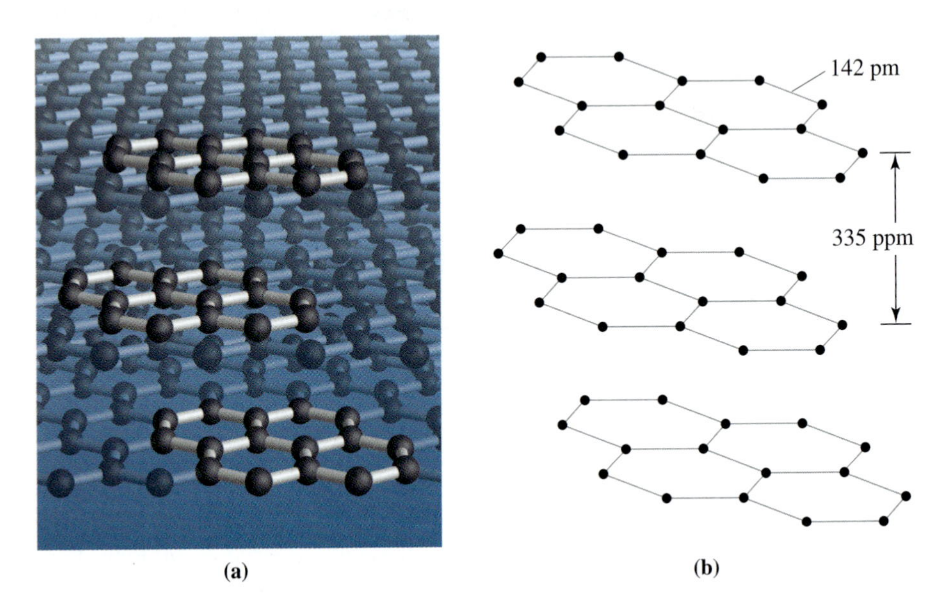

▶ **Figure 11.32**
The crystal structure of graphite
(a) A ball-and-stick model. (b) A schematic showing bond distances.

Graphite has some interesting properties that are consistent with the bonding scheme just described:

1. The carbon-to-carbon bond lengths *within* layers (142 pm) are comparable to those in benzene (139 pm). The carbon-to-carbon distances *between* layers, however, are much greater (335 pm).
2. The large distances and weak bonding between layers allows the layers to glide over one another rather easily. This makes graphite a good lubricant* and is also what makes graphite pencils work.
3. Graphite is a good electrical conductor because the delocalized *p* electrons between layers of carbon atoms can be set in motion by an external electric field. Graphite is extensively used as electrodes in batteries and in electrolysis reactions.

The solids diamond and graphite are *polymorphic* forms of carbon, but they are more than that. Two or more forms of an *element* that differ in their basic *molecular* structure are called **allotropes.** Diamond and graphite are allotropes of carbon.

Other Allotropes of Carbon

In 1985, a number of previously unknown carbon molecules were discovered in the products formed when graphite was vaporized. The predominant species had a molecular mass of 720 u, corresponding to the formula C_{60}. The structure proposed for the molecule is a roughly spherical collection of atoms in the shape of hexagons and pentagons, very much like a soccer ball, as seen in Figure 11.33. The C_{60} molecule was named "buckminsterfullerene" because it resembles the geodesic-domed structures pioneered by architect R. Buckminster "Bucky" Fuller. The general name

* A major factor in the lubricating properties of graphite appears to be the presence of oxygen molecules between the layers of carbon atoms. After being strongly heated in a vacuum, graphite becomes a much poorer lubricant.

fullerenes is now used for C_{60} and similar molecules discovered subsequently, such as C_{70}, C_{74}, and C_{82}. They are often colloquially called "buckyballs."

The fullerenes are *allotropic* forms of carbon in that they have a molecular structure different from diamond or graphite. More recently, allotropic forms of carbon called *nanotubes* have been discovered; they are also network covalent solids. We can visualize a nanotube as a fullerene that has been stretched out into a hollow cylinder by the insertion of many many more C atoms. Another way is to picture a two-dimensional array of hexagonal rings of carbon atoms, rather like ordinary chicken wire. The array, called a graphene sheet, is then rolled into a cylinder and capped at each end by half of a C_{60} molecule (see Figure 11.34). These nanotubes have unusual mechanical and electrical properties that are of great interest in current research.

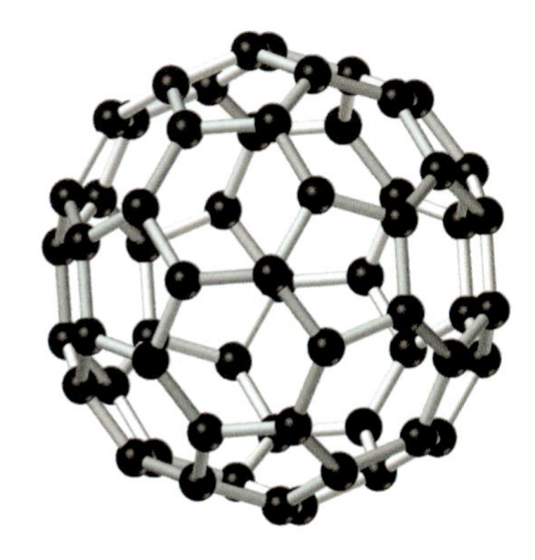

▲ **Figure 11.33**
A ball-and-stick model of C_{60}, a "buckyball"

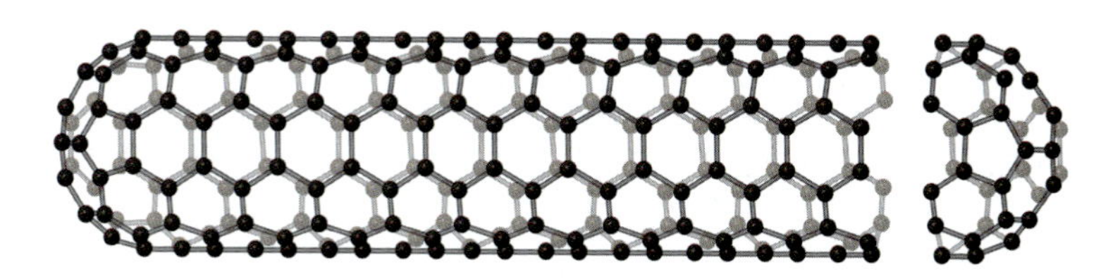

◀ **Figure 11.34**
A ball-and-stick model of a carbon nanotube
As indicated by the break in the structure, the molecules vary in length from a few nanometers to a micrometer or more. Their diameters are generally in the range of a few nanometers.

11.9 Ionic Bonds as "Intermolecular" Forces

When we introduced ionic compounds in Chapter 2 and when we described ionic bonding in Chapter 9, we noted that there are no "molecules" of a solid ionic compound, and therefore there can't be intermolecular forces. There are simply interionic attractions in which each ion is simultaneously attracted to several ions of the opposite charge. These interionic attractions—ionic bonds—extend throughout an ionic crystal.

In Chapter 9, we identified the *lattice energy* as a property that measures the strength of interionic attractions. We can make qualitative comparisons, however, without using actual values of lattice energies. The following generalization, illustrated in Figure 11.35, describes fairly well how lattice energy is related to atomic properties.

> *The attractive force between a pair of oppositely charged ions increases as the charges on the ions increase and as the ionic radii decrease. Lattice energies increase accordingly.*

◀ **Figure 11.35**
Interionic forces of attraction
Because of the higher charges on the ions and the reduced interionic distance, the attractive force between Mg^{2+} and O^{2-} is about seven times as great as between Na^{+} and Cl^{-}. The interionic distances are the sums of the ionic radii listed in Figure 8.14.

Most lattice energies are high enough that ionic solids do not readily sublime. We can melt ionic solids, however, if we supply enough thermal energy to break down the crystalline lattice. In general, the greater its lattice energy, the higher the melting point of an ionic solid is.

Solid ionic compounds meet only one of the two requirements for electrical conductivity cited on page 491. They have charged particles (ions), but they do not meet the second requirement because the ions are fixed in a crystalline lattice. Consequently, ionic solids do not conduct electricity. However, when we melt an ionic solid or dissolve it in a suitable solvent, such as water, the ions are free to move. Solutions of ionic compounds are good electrical conductors, as we noted in Chapter 4.

EXAMPLE 11.8

Arrange the following three ionic solids in the expected order of increasing melting point: MgO, NaBr, and NaCl.

SOLUTION

We need these two ideas: (1) The greater the lattice energy of an ionic compound, the higher its melting point, and (2) lattice energy is related to ionic charges and sizes as described earlier in this section. Mg^{2+} and O^{2-} ions have higher charges than do Na^+, Cl^-, and Br^-. Also, Mg^{2+} is a smaller ion than Na^+, and O^{2-} is smaller than both Cl^- and Br^-. As a result, we expect the lattice energy of MgO to be much greater than that of either NaCl or NaBr. Because it has the highest lattice energy, MgO should have the *highest* melting point of the three compounds.

Now we need to determine whether NaCl or NaBr has the greater lattice energy. We see that the compounds have the same cation and that their anions have the same charge. The only difference is in the anion radii. The Cl^- ion is smaller than the Br^- ion, and therefore the interionic attractions in NaCl are stronger than in NaBr. We expect NaCl to have a higher lattice energy and a higher melting point than NaBr. The expected order is NaBr $<$ NaCl $<$ MgO. The observed melting points—747 °C for NaBr, 801 °C for NaCl, and 2832 °C for MgO—support our reasoning.

EXERCISE 11.8

Arrange the following ionic solids in the expected order of increasing melting point: CsBr, KCl, KI, and MgF_2.

11.10 The Structure of Crystals

In a macroscopic view, a **crystal** is a solid substance with a regular shape having plane surfaces and sharp edges that intersect at fixed angles. The microscopic view has a small number of atoms, ions, or molecules making up a unit that is repeated over and over. An essential feature of a crystal is that we must be able to figure out its entire structure from this tiny unit.

Crystal Lattices

Most of us have dealt with a repeating pattern at one time or another. Usually, we need to deal with a pattern in only one dimension, as in stringing beads, or in two dimensions, as in laying floor tiles. These two everyday examples are illustrated in Figure 11.36.

In stringing beads (Figure 11.36a), we can single out a group of four beads, say red-red-green-blue, as a repeating unit. By repeating this unit many times, we could generate strings of beads of any length, all of which would have the same pattern. The situation with the hexagonal floor tiles (Figure 11.36b) is a bit more complex. If we use the red hexagon as a repeating unit, we can generate its neighbors in the same row by *single* displacements to the *left* and to the *right*. To generate its neighbors in the row below, however, requires *two* displacements: *down*, and then to the *left* or to the *right*. However, we can select a different repeating unit that will gen-

Repeating unit

(a) A string of beads represents a one-dimensional pattern.

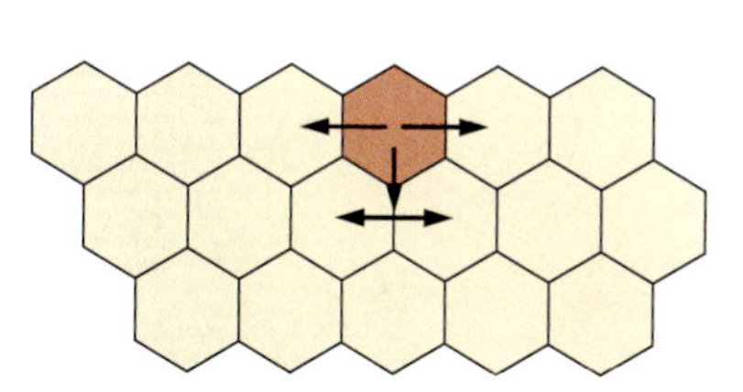

(b) A hexagonal floor tile layout is a two-dimensional pattern.

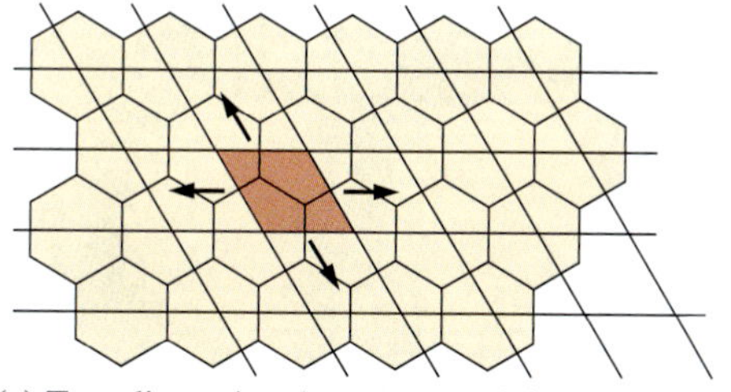

(c) Two-dimensional pattern overlain with a grid of parallel lines. The repeating unit is in red.

◀ **Figure 11.36 One- and two-dimensional patterns**

erate the entire pattern with just *single* displacements in all directions and without overlapping any of the pattern previously formed as the unit is moved. This is the parallelogram shown in red in Figure 11.36c.

To describe crystals, we need to work with *three*-dimensional patterns. The framework on which we outline the pattern is a *lattice*. We would need 14 different lattices to describe all crystalline solids, but we will limit our discussion to the types called *cubic*.

The lattice shown in Figure 11.37 consists of three sets of equidistant, mutually perpendicular planes. A geometric figure called a *parallelepiped* is outlined in color. It has *six* faces formed by the intersection of three pairs of parallel planes. This special kind of parallelepiped is a *cube*, because the distance between each pair of parallel planes is the same and the intersecting planes form 90° angles. A single parallelepiped that can be used to generate the entire lattice by simple, straight-line displacements—the repeating unit of the lattice—is a **unit cell.** The cube in color is a unit cell of the lattice in Figure 11.37.

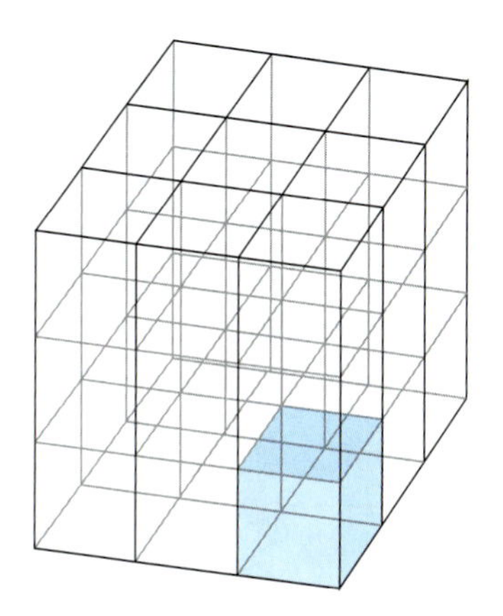

▲ **Figure 11.37 The cubic lattice**

The entire lattice can be generated by straight-line displacements (left and right, front and back, up and down) of the unit cell shaded in blue.

The simplest unit cell has structural particles (atoms, ions, or molecules) only at its corners; it is a **simple cubic cell** (Figure 11.38a). However, sometimes we can better describe a crystal structure by using a unit cell that has more structural particles. The **body-centered cubic (bcc)** structure has an additional structural particle at the center of the cube (see Figure 11.38b). The **face-centered cubic (fcc)** structure has an additional particle at the center of each face (Figure 11.38c).

An important concept in describing crystals is the *coordination number*. This is the number of neighboring particles with which a given structural particle of the crystal is in contact. For example, in the body-centered unit cell, we can clearly see from Figure 11.38b that the atom in the center of the cell is in contact with each of the eight corner atoms; its coordination number is *eight*. The coordination number in the simple cubic cell (Figure 11.38a) is just a bit harder to assess. However, if we visualize several unit cells, as in Figure 11.39, we see that the coordination number is *six*. The coordination number in a face-centered cubic cell (Figure 11.38c) is *twelve*, as is most readily established in the manner illustrated in the next section.

Close-Packed Structures

In a metallic crystal, we view the metal atoms as a collection of identical spheres, much like a collection of marbles in a box. We can't pack marbles in a box in a way that will fill *all* the space. Rather, there must always be some open spaces, or *voids*,

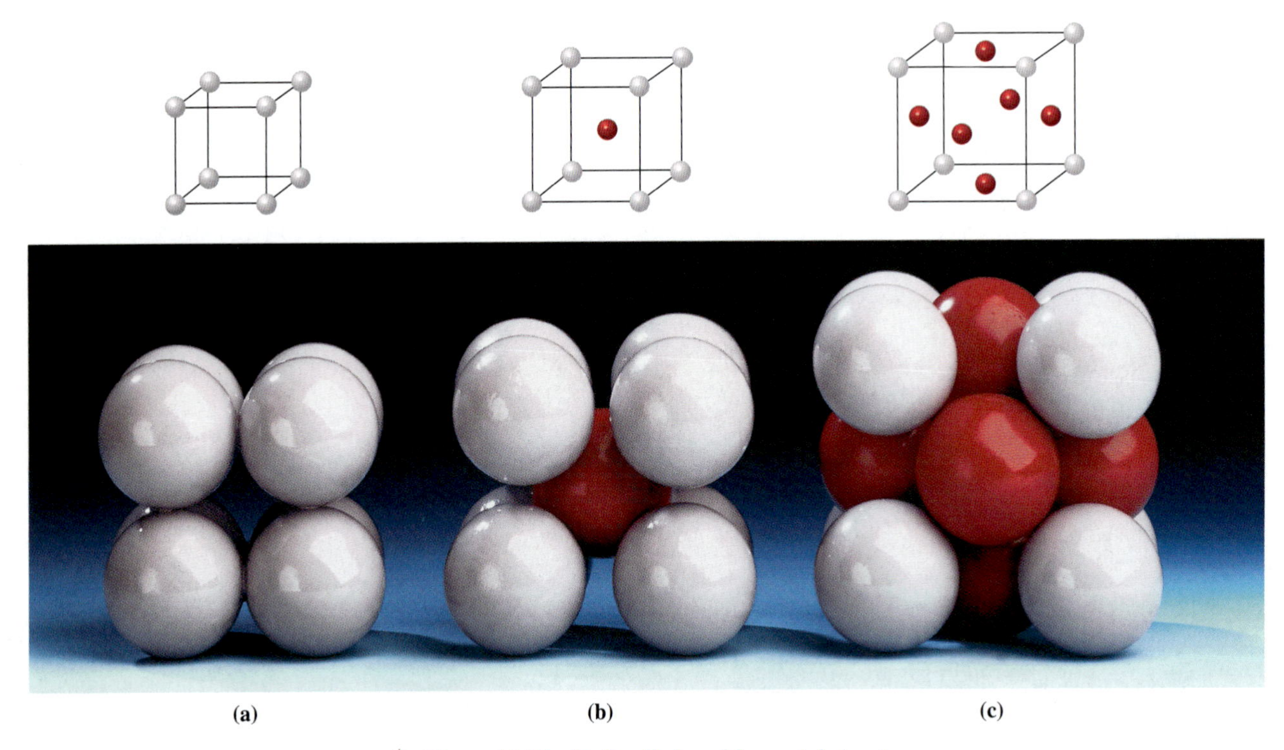

▲ **Figure 11.38 Unit cells in cubic crystal structures**
Only the centers of the spheres (atoms) are shown in the top row. The space-filling models in the bottom row show that certain of the spheres are in direct contact.

▲ **Figure 11.39 Coordination number 6 in a simple cubic cell**
The atoms shared by the four unit cells shown (tan) has six other atoms as nearest neighbors (shown in red).

among the marbles. Certain arrangements of identical spheres, called *close-packed*, are more efficient in filling an available volume than others.

To illustrate, let's first consider arranging spheres in a single layer on a table top. Two possible arrangements, both viewed from above, are suggested by Figure 11.40. In the "open" arrangement, we see that each sphere is in contact with *four* neighboring spheres (in red) and that the amount of space among the spheres is greater than in the close-packed arrangement. In the latter arrangement, each sphere touches *six* neighboring spheres (in red). Now let's focus on the close-packed arrangement and begin to build multiple layers of spheres. An example of a multi-layered, close packing is visualized in Figure 11.41.

The voids among the spheres in the close-packed arrangement in Figure 11.40(b) are all alike. This layer is also shown in Figure 11.41 as the bottom layer

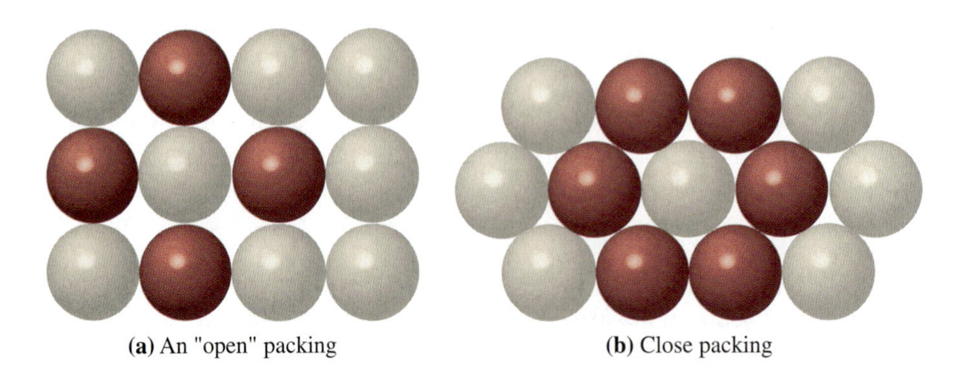

▶ **Figure 11.40 Packing spheres in two dimensions**

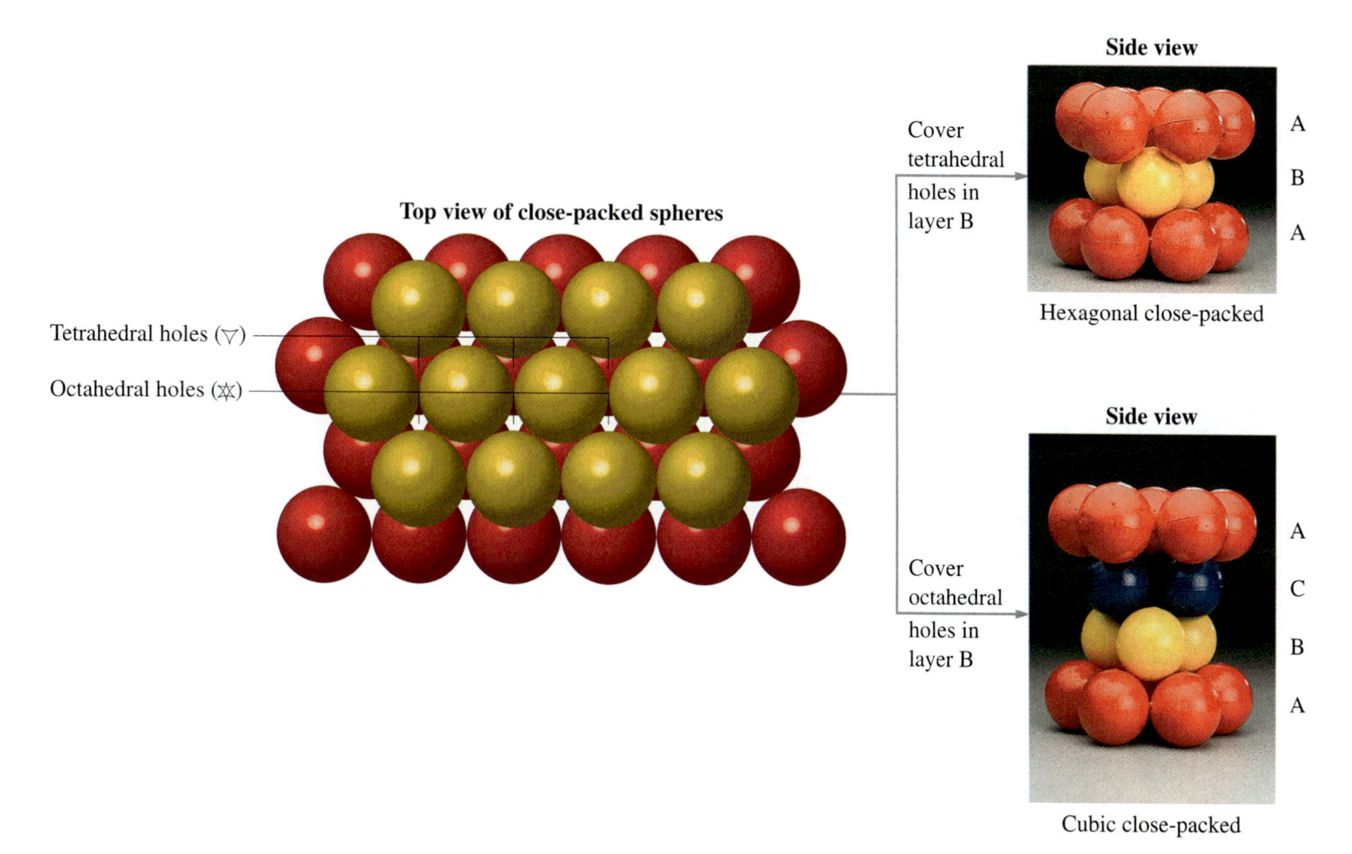

▲ **Figure 11.41 Close-packing of spheres in three dimensions**

(Layer A). It makes no difference where we begin the process of adding spheres in a second layer (Layer B) because each added sphere will rest in the hollow above a void in the bottom layer. When the second layer of spheres is in place, we see that there are now *two* types of voids. One type of void, called a *tetrahedral hole*, falls above a sphere in the bottom layer. The second type of void is called an *octahedral hole*. It falls above a void in the bottom layer. In adding a third layer of spheres (Layer C), there are *two* possibilities. If we cover all the tetrahedral holes, the third layer is identical to the bottom layer and the pattern begins to repeat itself. This is called the **hexagonal close-packed (hcp)** arrangement. In contrast, if we cover the octahedral holes the third layer is *not* identical with the bottom layer. It is only when we add a fourth layer that the pattern begins to repeat itself. This is called the **cubic close-packed (ccp)** arrangement. Figure 11.41 also helps us to assess the coordination numbers in the close-packed arrangements. Within any single layer of the structure, a given atom is in contact with *six* others. It is also in contact with *three* atoms in the layer below and *three* in the layer above, for a total of *twelve*.

If we now return to the three unit cells shown in Figure 11.38, we see that the simple cubic structure is not a close-packed structure; in fact, 47.64% of the volume in this structure is in voids. The body-centered cubic structure is a more tightly packed arrangement with only 31.98% of the total volume in voids. The face-centered cubic structure is the one that results from the cubic close-packed arrangement just described; it has 25.96% of its volume in voids (as does the hcp arrangement). Figure 11.42 shows that the ccp arrangement does indeed correspond to a face-centered cubic unit cell.

Many metals crystallize in the close-packed arrangements. For example, Cu, Ag, and Au solidify in the ccp arrangement. Mg, Zn, and Cd solidify in the hcp arrangement. However, some metals do not use the closest packed arrangements possible. Fe and Cr, as well as the alkali metals—Li, Na, K, Rb, and Cs—form body-centered cubic crystals.

▲ **Figure 11.42**
Cubic close packing of spheres and the face-centered cubic unit cell
The 14 spheres on the top have been taken from a larger array in the cubic close-packed arrangement. The two middle layers have six atoms each, and the top and bottom layers have one, accounting for the 14 atoms in the fcc unit cell. Can you see that rotating the group of 14 spheres reveals the fcc unit cell (bottom)?

Apportioning the Atoms in a Unit Cell

We will shortly introduce some simple calculations that are possible with such data as atomic radii and unit-cell dimensions. In these calculations, we will need to know the number of atoms in a unit cell, but this is *not* the same as the number we use when we draw a picture of a unit cell.

Consider the simple cubic cell in Figure 11.38(a). Our picture shows *eight* atoms, but we need to recognize that all eight are shared with neighboring unit cells. Look again at Figure 11.39. The tan atom with six red atoms as nearest neighbors is shared by the four unit cells pictured and by four more that are not pictured. (Can you sketch in these four?) We can allot only *one-eighth* of this atom to any particular unit cell. By considering that one-eighth of each of the eight corner atoms belongs to a given unit cell, we conclude that the simple cubic unit cell contains the equivalent of *one* atom, that is, $1/8 \times 8 = 1$.

As we show in Figure 11.43, only one-eighth of each of the eight corner atoms in the body-centered unit cell belongs to the unit cell. The atom in the center of the cube, however, belongs entirely to the unit cell. The equivalent number of atoms in a bcc unit cell is $(1/8 \times 8) + 1 = 2$. Example 11.9 illustrates how we determine the number of atoms in a fcc unit cell.

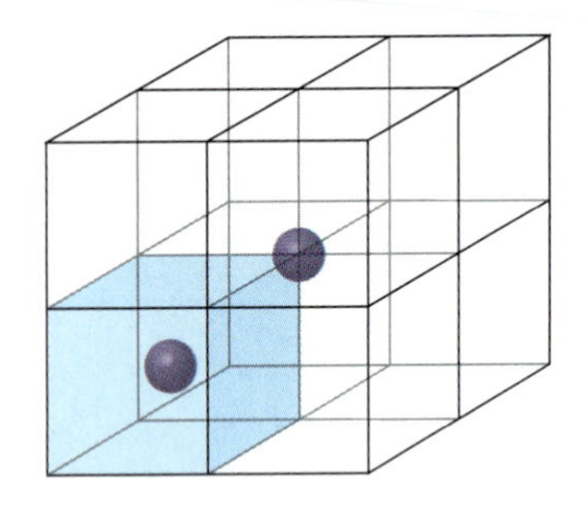

▲ **Figure 11.43**
Apportioning atoms in a unit cell
The eight unit cells outlined here are body-centered cubic. Two atoms are shown. One is in the exact center of the unit cell in color. The other, a corner atom, is shared among all eight unit cells. The effective number of atoms in a bcc unit cell is *two*.

Calculations Based on Atomic Radii and Unit-Cell Dimensions

Now that we have seen the different ways that atoms are arranged in crystal lattices and their unit cells, we can perform a variety of calculations. We will illustrate some calculations here and present additional possibilities in the end-of-chapter problems.

EXAMPLE 11.9

Copper crystallizes in the cubic close-packed arrangement. The atomic (metallic) radius of a Cu atom is 127.8 pm.

a. What is the length of a unit cell of Cu, in picometers?

b. What is the volume of a unit cell of Cu, in cubic centimeters?

c. How many atoms belong to the unit cell?

SOLUTION

We first need to recognize that the ccp arrangement (Figure 11.41) has a face-centered cube (Figures 11.38 and 11.42) as its unit cell. We can then construct a simple sketch to help answer the specific questions. The sketch in Figure 11.44 is a cross section of the front face of the unit cell.

a. Each face of a cube is a square. The length of the unit cell is the length of the square, marked l in Figure 11.44. The atomic radius of Cu is r in the sketch. The diagonal of the face, that is the diagonal of the square, has a length of $4r$. Notice that the diagonal divides the square into two right triangles and is the hypotenuse of these triangles. According to the theorem of Pythagoras, the square of the hypotenuse of a right triangle is the sum of the squares of the legs of the triangle: $c^2 = a^2 + b^2$. Each of the legs of the triangle is the length of the unit cell, l. That is,

$$(4r)^2 = l^2 + l^2$$

$$16r^2 = 2l^2$$

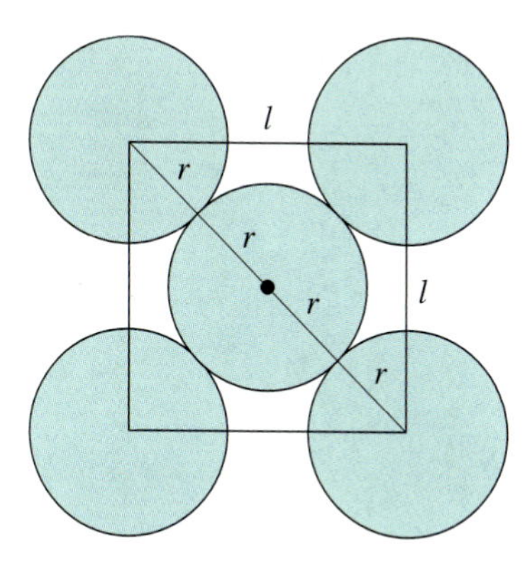

▲ **Figure 11.44**
Example 11.9 illustrated

$$l^2 = 8r^2$$

$$l = r\sqrt{8} = 127.8 \text{ pm} \times \sqrt{8} = 127.8 \text{ pm} \times 2.828 = 361.5 \text{ pm}$$

b. We can determine the length, l, in centimeters, by using conversion factors.

$$l = 361.5 \cancel{\text{pm}} \times \frac{10^{-12} \cancel{\text{m}}}{1 \cancel{\text{pm}}} \times \frac{100 \text{ cm}}{1 \cancel{\text{m}}} = 3.615 \times 10^{-8} \text{ cm}$$

The volume of the unit cell, V, is given by the formula $V = l^3$.

$$V = (3.615 \times 10^{-8})^3 \text{ cm}^3 = 4.724 \times 10^{-23} \text{ cm}^3$$

c. Let's consider the drawing in Figure 11.44 again. The atoms at the corners of the square are also at the corners of the unit cell. There are *eight* corner atoms in the cube, and as in the simple cubic and body-centered cubic cells, these eight atoms are shared by eight unit cells. They contribute the equivalent of *one* atom to the unit cell. The atom in the center of a face is shared by just two unit cells. There are six faces to the cube, and the total contribution of the six atoms in the faces is $1/2 \times 6 =$ *three* atoms. The total number of atoms in the unit cell is thus $1 + 3 = 4$.

EXERCISE 11.9

The unit cell of crystalline iron is body-centered cubic. The atomic radius of iron is 124.1 pm. Determine **(a)** the length, in picometers; and **(b)** the volume, in cubic centimeters, of a unit cell of iron.

EXAMPLE 11.10

Use the results of Example 11.9, the molar mass of copper, and Avogadro's number to calculate the density of metallic copper.

SOLUTION

The density of copper is the mass of the unit cell, in grams, divided by the volume of the unit cell, in cubic centimeters. We determined the volume of the unit cell in Example 11.9b. We now need to determine the mass of the unit cell. The molar mass, 63.546 g/mol, is the mass of Avogadro's number of copper atoms. We can calculate the mass of a single Cu atom.

$$\frac{63.546 \text{ g}}{1 \cancel{\text{mol Cu}}} \times \frac{1 \cancel{\text{mol Cu}}}{6.0221 \times 10^{23} \text{ Cu atoms}} = 1.0552 \times 10^{-22} \text{ g/Cu atom}$$

In Example 11.9c, we established that an equivalent of *four* copper atoms belong to the unit cell. Thus, the mass of the unit cell is four times the mass of a single Cu atom.

$$4 \cancel{\text{Cu atoms}} \times \frac{1.0552 \times 10^{-22} \text{ g}}{1 \cancel{\text{Cu atom}}} = 4.2208 \times 10^{-22} \text{ g}$$

Finally, we can calculate the density of copper.

$$d = \frac{m}{V} = \frac{4.2208 \times 10^{-22} \text{ g}}{4.724 \times 10^{-23} \text{ cm}^3} = 8.935 \text{ g/cm}^3$$

PROBLEM-SOLVING NOTE
We can use this type of calculation to determine Avogadro's number if we have a precisely measured density of a crystal (see Problems 77 and 78).

EXERCISE 11.10

Use the results of Exercise 11.9, the molar mass of iron, and Avogadro's number to calculate the density of metallic iron.

Ionic Crystal Structures

Ionic crystal structures are somewhat more complicated than those of metals for two reasons: (1) Ionic crystals have two different types of structural particles, cations and anions, rather than atoms of a single kind; and (2) the cations and anions have different sizes. One approach to ionic crystals is to consider that some voids in a close-packed arrangement of anions are filled by the smaller cations. In this arrangement, cations and anions can be in contact, but the structure would not be stable if the anions were also in direct contact. The cations must be of a size to fill voids and also force some separation between the anions. The particular packing arrangement of anions in an ionic crystal, then, depends on the ratio of the cation radius (r_c) to the anion radius (r_a), that is r_c/r_a.

Recall that in a close-packed arrangement of spheres there are two type of voids: tetrahedral holes and octahedral holes. Tetrahedral holes are the smaller holes and can accommodate only certain small cations in contact with larger anions. The range of radius ratios that lead to the filling of tetrahedral holes is shown below.

$$\textit{Tetrahedral}: 0.225 < r_c/r_a < 0.414$$

The larger octahedral holes can accommodate somewhat larger cations in an array of anions. The range of radius ratios for the filling of octahedral holes is the following.

$$\textit{Octahedral}: 0.414 < r_c/r_a < 0.732$$

If the cations and anions are more nearly of equal size, there isn't enough room for the cations in the voids in a close-packed arrangement of anions. The anions are forced into a more open cubic structure.

$$\textit{Cubic}: r_c/r_a > 0.732$$

In this case, think of a sphere resting in one of the hollows of the "open" structure of spheres in Figure 11.40(a), and then surmounted by a layer of spheres identical to the bottom layer shown.

To define a unit cell of an ionic crystal, we must choose a portion of the crystal that (1) generates the entire crystal by straight-line displacements in three dimensions, (2) indicates the coordination numbers of the ions, and (3) gives the correct formula of the compound.

The radius ratio for cesium chloride is $r_{Cs^+}/r_{Cl^-} = 169 \text{ pm}/181 \text{ pm} = 0.933$. The unit cell of CsCl is the body-centered cubic structure shown in Figure 11.45. Notice how this unit cell is consistent with the formula CsCl. The unit cell has one Cs^+ ion in the center of the cube and the equivalent of one Cl^- ion among the eight corner ions (that is, $1/8 \times 8 = 1$).

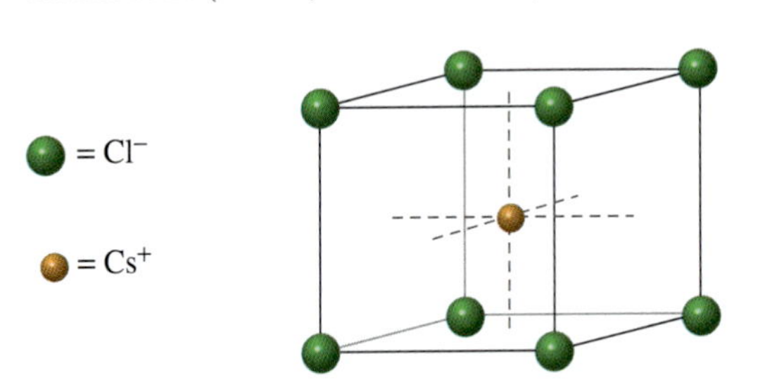

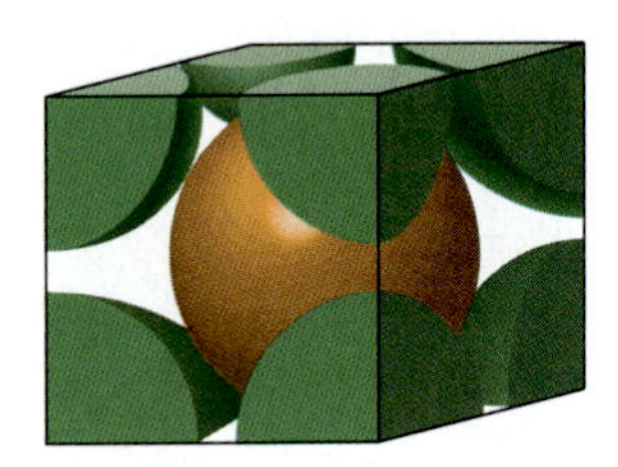

▲ **Figure 11.45 A unit cell of cesium chloride**
Left: For clarity, only the centers of the ions are shown, although the Cs^+ ion at the center of the cell is in contact with the eight Cl^- ions at the corners of the cube. The coordination number of Cs^+ is *eight*. Because we can draw the unit cell with a Cl^- ion at the center and Cs^+ ions at the corners, the coordination number of Cl^- is also *eight*. Right: Space filling models of ions in the unit cell.

The radius ratio for sodium chloride is $r_{Na^+}/r_{Cl^-} = 99 \text{ pm}/181 \text{ pm} = 0.55$. The Na^+ ions fill the octahedral holes in a close-packed array of Cl^- ions, producing the unit cell shown in Figure 11.46. To show that this unit cell is consistent with the formula NaCl, notice that one Na^+ ion belongs entirely to the cell and that each of the 12 Na^+ ions at the centers of the edges is shared among four unit cells. The number of Na^+ ions in the cell is $1 + (1/4 \times 12) = 4$. One-eighth of each of the eight Cl^- ions at the corners, and one-half of each of the six Cl^- ions in the centers of the faces belong to the unit cell. The number of Cl^- ions in the cell is $(1/8 \times 8) + (1/2 \times 6) = 4$. The unit cell contains the equivalent of *four* formula units of NaCl.

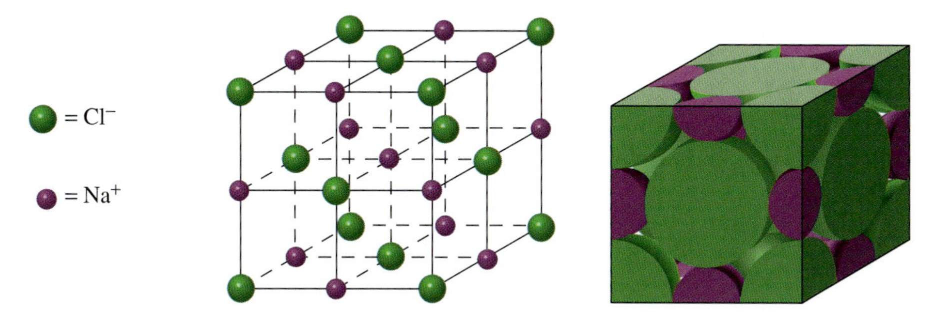

▲ Figure 11.46 A unit cell of sodium chloride
Only the centers of the ions are shown, although oppositely charged ions are in contact along the edges of the unit cell. Here, Cl^- ions are at the corners and in the centers of the faces. Na^+ ions are at the centers of the edges and at the center of the cell. The Na^+ ion at the center is in contact with four Cl^- ions in the same plane, with one Cl^- ion in the plane above, and one in the plane below; the coordination number of Na^+ is *six*. The Cl^- ion in the front face is in contact with four Na^+ ions in the same face, with the Na^+ ion in the center, and with one Na^+ ion directly in front of the unit cell; the coordination number of the Cl^- is also *six*.

Experimental Determination of Crystal Structures

We cannot see the patterns of atoms, ions, or molecules in crystalline solids, even with the help of ordinary light microscopes, because these particles are much too small. To figure out these patterns, we must use radiation with wavelengths comparable to the dimensions of unit cells. X rays are ideal for this purpose.

Figure 11.47 suggests the interaction of X rays with a crystal and a geometric analysis of the results. In the illustration, wave *a* is reflected by one plane of atoms or ions and wave *b*, by the next plane below. Wave *b* travels farther than does wave *a*. To achieve a maximum intensity of the reflected radiation, waves *a* and *b* must reinforce each other—that is, their crests and troughs must line up. For this to happen, the additional distance traveled by wave *b* must be an integral multiple, *n*, of the wavelength, λ, of the X rays:

$$n\lambda = 2d \sin \theta$$

By measuring the angle θ at which scattered X rays of known wavelength have their greatest intensity, one can calculate the spacing *d* between atomic planes. If the process is repeated for different orientations of the crystal, eventually the complete crystal structure can be worked out.

Summary

Vaporization is the passage of molecules from the liquid to the gaseous (vapor) state; condensation is the reverse process. Dynamic equilibrium is established when vaporization and condensation occur at the same rate in a closed system. The pressure exerted by the vapor

Key Terms

adhesive force (11.7)
allotrope (11.8)
body-centered cubic (bcc) (11.10)
boiling point (11.2)
cohesive force (11.7)
condensation (11.2)
cooling curve (11.3)
critical point (11.2)
critical pressure (11.2)
critical temperature (11.2)
crystal (11.10)
cubic close-packed (ccp) (11.10)
dispersion force (11.5)
dynamic equilibrium (11.2)
enthalpy (heat) of fusion (11.3)
enthalpy (heat) of sublimation (11.3)
enthalpy (heat) of vaporization (11.2)
face-centered cubic (11.10)
freezing point (11.3)
fusion (melting) (11.3)
heating curve (11.3)
hexagonal close-packed (hcp) (11.10)
hydrogen bond (11.6)
intermolecular force (page 476)
melting point (11.3)
meniscus (11.7)
network covalent solid (11.8)
normal boiling point (11.2)
normal melting point (11.3)
phase change (page 460)
phase diagram (11.4)
polarizability (11.5)
polymorphism (11.4)
simple cubic cell (11.10)
states of matter (11.1)
sublimation (11.3)
sublimation curve (11.3)
supercooling (11.3)
supercritical fluid (11.2)
surface tension (11.7)
triple point (11.3)
unit cell (11.10)
vaporization (11.2)
vapor pressure (11.2)
vapor pressure curve (11.2)
viscosity (11.7)

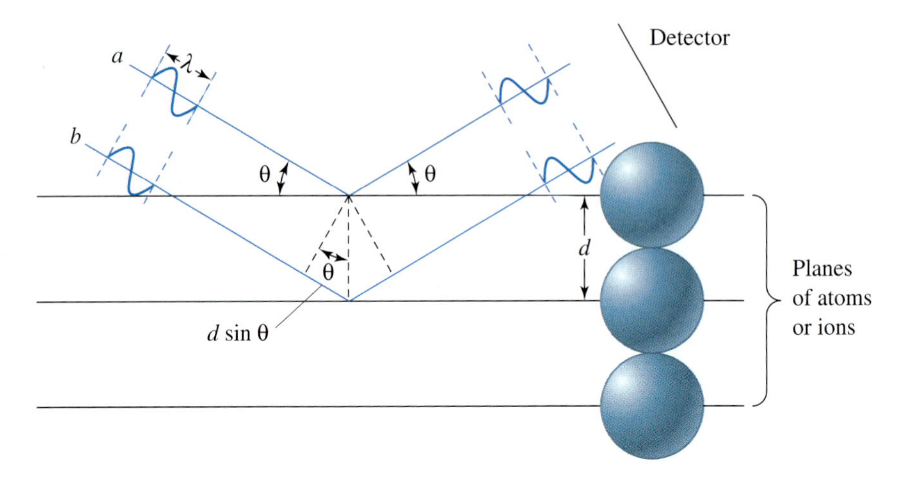

▲ **Figure 11.47 X-ray determination of crystal structure**
The hypotenuse of each triangle is equal to the interatomic distance, d. The side opposite the angle θ has a length of $d \sin \theta$. Wave b travels farther than a by the distance $2d \sin \theta$, and this distance is an integral multiple of the X-ray wavelength λ. Thus, $n\lambda = 2d \sin \theta$, where $n = 1, 2, 3, \ldots$

is called the vapor pressure. A graph of vapor pressure as a function of temperature is called a vapor pressure curve. The normal boiling point of a liquid is the temperature at which its vapor pressure is 1 atm. The critical point is the highest pressure-temperature point on a vapor pressure curve, and the critical temperature is the highest temperature at which a vapor can be liquefied by applying pressure alone.

A phase diagram is a pressure-temperature plot that shows the conditions under which a substance exists as a solid, a liquid, a gas, or some combination of these phases. Two-phase equilibria are represented by curves in the diagram: the vaporization (vapor pressure) curve, the sublimation curve, and the fusion curve. The triple point is the unique temperature and pressure at which the solid, liquid, and vapor states of a substance coexist.

Fluctuations in electron charge density in a molecule produce an instantaneous dipole, which in turn induces dipoles in neighboring molecules. Attractions between instantaneous and induced dipoles, called dispersion forces, are the intermolecular forces found in nonpolar substances. Polar substances consist of permanent dipoles and have additional intermolecular forces of the dipole–dipole and dipole-induced dipole types. Collectively known as van der Waals forces, dispersion forces, dipole–dipole forces, and dipole-induced dipole forces determine physical properties such as melting points and boiling points.

A strong intermolecular force—a hydrogen bond—forms when a H atom attached to an O, N, or F atom of one molecule is simultaneously attracted to an O, N, or F atom of another molecule. Hydrogen bonding accounts for some unusual properties of water (for example, an unusually high boiling point and a greater density as a liquid than as a solid). It also governs aspects of the behavior of biologically active molecules such as proteins and the nucleic acids.

The intermolecular forces in a liquid determine its resistance to flow, its viscosity. Differences in the intermolecular forces between molecules in the bulk of a liquid and at the surface give rise to the property of surface tension. Several familiar natural phenomena are related to surface tension.

In some solids, covalent bonds extend throughout a crystal. Such solids generally are harder, have much higher melting points, and are less volatile than other molecular solids. In ionic solids, interionic attractions bind all the ions together in a crystal. The strengths of these attractions depend primarily on ionic charges and sizes and partly determine physical properties like melting points.

A three-dimensional pattern called a lattice is used to describe the structure of a crystal. There are three types of cubic lattices: simple cubic, body-centered cubic, and face-centered cubic. An important concept relating to crystal lattices is that of the unit cell. Among the quantities that can be calculated from measured unit cell dimensions are atomic radii and the densities of crystalline substances. Measurements on crystalline substances can be used to evaluate Avogadro's number. The crystal structures of metals can be described as the packing of spheres. The close-packing of spheres leads to a minimum in the fraction of the volume occupied by holes or voids. The crystal structures of ionic substances can be approximated by a model in which cations occupy the voids present among a close-packed array of anions.

Review Questions

1. What is meant by a condition of *dynamic equilibrium*?
2. What do we mean when we say that a liquid is *volatile*? *nonvolatile*?
3. What is the distinction in the terms *vaporization* and *boiling*?
4. What is the distinction between these terms: boiling point of a liquid and *normal* boiling point of a liquid?
5. Which of the following variables do you expect to affect the vapor pressure of a liquid?
   (a) temperature
   (b) volume of liquid in the liquid-vapor equilibrium
   (c) volume of vapor in the liquid-vapor equilibrium
   (d) area of contact between liquid and vapor
   Explain.
6. Although the terms gas and vapor are often used interchangeably, sometimes a distinction is made between the two terms. What is this distinction?
7. What is meant by the *critical point* of a substance?
8. What is a *phase diagram*?
9. In a phase diagram, how many different phases are present in portions of the diagram represented by areas and curves?
10. What is a *triple point*? Is it the same as a *normal* melting point? Explain.
11. How are the following represented in phase diagrams: **(a)** polymorphic solids and **(b)** supercritical fluids?
12. Do all phase diagrams have a critical point? a normal boiling point? Can they have more than one triple point? Explain.
13. What is the difference between an *intra*molecular force and an *inter*molecular force? Can they ever be the same? Explain.
14. What is the difference between an *instantaneous* and an *induced* dipole?
15. What is a dispersion force?
16. What is the difference in meaning of the terms *dipole moment* and *polarizability* of a molecule?
17. Why does a polar liquid generally have a higher normal boiling point than a nonpolar liquid of the same molar mass?
18. State the principal reasons why CH_4 is a gas at room temperature whereas H_2O is a liquid.
19. Diamond and graphite are both network covalent solids. Explain why their properties are so dissimilar.
20. What is a meniscus, and what determines its shape (that is, whether convex or concave)?
21. Describe the general way in which ionic charges and sizes affect lattice energy, and the general way that lattice energy affects the melting point of an ionic solid.
22. What is meant by the terms *crystal lattice* and *repeating unit* in a crystal lattice?
23. What is the difference between a formula unit and a unit cell of an ionic compound?
24. What is the coordination number of an atom or ion in a crystal structure of atoms or ions?
25. What is the meaning of the terms *tetrahedral holes* and *octahedral holes* in the close packing of atoms? What is the meaning of the abbreviations *ccp* and *hcp*?
26. What is the meaning of the abbreviations *bcc* and *fcc* when referring to the unit cells of crystals?

Problems

Heat of Vaporization

27. Refer to Table 11.1, and determine how many kilojoules of heat are required to vaporize 1.00 kg $CS_2(l)$.
28. Refer to Table 11.1, and determine how many joules of heat are required to vaporize 0.25 mL $CCl_4(l)$ (d = 1.59 g/mL).
29. How many kilojoules of heat are required to convert 25.0 g H_2O from liquid at 18.0 °C to vapor at 25.0 °C? (*Hint:* What is the specific heat of water, and what is its heat of vaporization?)
30. How many kilojoules of heat are released when 1.25 mol $CH_3OH(g)$ at 25.0 °C is condensed and cooled to 15.5 °C? The specific heat of $CH_3OH(l)$ is 2.53 J g^{-1} $°C^{-1}$. (*Hint:* What is the heat of vaporization of CH_3OH?)
31. How many grams of propane must be burned to supply the heat required to vaporize 0.750 L of water at 298 K?

$$C_3H_8(g) + 5\ O_2(g) \longrightarrow 3\ CO_2(g) + 4\ H_2O(l)$$
$$\Delta H = -2.22 \times 10^3\ kJ$$

$$H_2O(l) \longrightarrow H_2O(g) \qquad \Delta H_{vapn} = 44.0\ kJ$$

32. The combustion of 1.25 g of pentane produces enough heat to vaporize 165 g of hexane. What is the molar enthalpy of vaporization of hexane?

$$C_5H_{12}(l) + 8\ O_2(g) \longrightarrow 5\ CO_2 + 6\ H_2O(l)$$
$$\Delta H = -3.51 \times 10^3\ kJ$$

$$C_6H_{14}(l) \longrightarrow C_6H_{14}O(g) \qquad \Delta H_{vapn} = ?$$

33. You can hold your hand in an oven at 100 °C for some time without feeling great discomfort. However, you can hold your hand a centimeter or two above rapidly boiling water in a kettle for only a few seconds. Explain why the effects are different.

34. A double boiler is a device sometimes used in cooking. Cooking takes place in an inner container in contact with the vapor from boiling water in an outside container. (A related laboratory device is called a steam bath.) **(a)** How is energy conveyed to the food being cooked? **(b)** What is the maximum temperature that can be reached in the inside container? Explain.

Vaporization, Vapor Pressure

35. By referring to Figure 11.4, estimate the approximate value of **(a)** the vapor pressure of carbon disulfide at 30 °C; **(b)** the boiling point of ethanol when the barometric pressure is 720 mmHg.

36. With reference to Figure 11.4, estimate the approximate value of **(a)** the vapor pressure of aniline at 100 °C; **(b)** the boiling point of methanol when the barometric pressure is 680 mmHg.

37. How many silver atoms are present in a vapor volume of 486 mL when Ag(l) and Ag(g) are in equilibrium at 1360 °C? (Vapor pressure of silver at 1360 °C = 1.00 mmHg.)

38. Even though the vapor pressure of liquid mercury is quite low, enough atoms are present in the vapor to present a significant health hazard when inhaled over time. Determine the number of Hg atoms in a tightly sealed room of 27.5 m^3 volume when equilibrium is established between Hg(l) and Hg(g) at 22.0 °C? (Vapor pressure of mercury at 22.0 °C = 0.00143 mmHg.)

39. Equilibrium is established between a small quantity of CCl_4(l) and its vapor at 40.0 °C in a flask having a volume of 285 mL. The total mass of vapor present is 0.480 g. What is the vapor pressure of CCl_4, in mmHg, at 40.0 °C?

40. The density of acetone vapor in equilibrium with liquid acetone, $(CH_3)_2CO$, at 32 °C is 0.876 g/L. What is the vapor pressure of acetone, in mmHg, at 32 °C?

41. A 1.82-g sample of H_2O is injected into a 2.55-L flask at 30.0 °C. In what form(s) will the H_2O be present, that is, solid and/or liquid and/or gas? Explain.

42. A 0.625-g sample of H_2O is injected into a 178.5-L flask at 0 °C. In what form(s) will the H_2O be present, that is, solid and/or liquid and/or gas? Explain.

43. How would you explain to a young schoolchild the phenomenon depicted in the photograph: Water is boiled in a paper cup over an open flame.

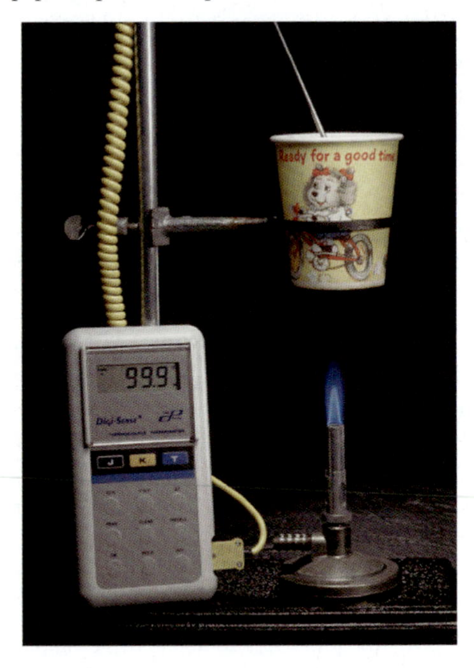

44. How would you explain to a young schoolchild the following phenomenon: A small quantity of water is boiled in a can; the can is capped off and cooled; the can collapses.

The Critical Point

45. Can a gas always be liquefied by applying sufficient pressure? by sufficiently reducing the temperature? by a combination of a pressure and temperature change? Explain.

46. The states of matter in bottled propane at room temperature are liquid and gas. Would the same be true for bottled methane? Explain.

Phase Changes

47. How many kilojoules of heat are required to melt an ice cube having the dimensions 3.5 cm × 2.6 cm × 2.4 cm? The density of ice is 0.92 g/cm^3. (Refer to Table 11.4.)

48. An "ice" calorimeter measures quantities of heat by the amount of ice melted. How many grams of ice would be melted by the heat released in the combustion of 875 mL of $CH_4(g)$, measured at 25 °C and 748 mmHg?

$$CH_4(g) + 2\,O_2(g) \longrightarrow CO_2(g) + 2\,H_2O(l) \quad \Delta H = -890.3\ \text{kJ}$$

$$H_2O(s) \longrightarrow H_2O(l) \quad \Delta H_{\text{fusion}} = 6.01\ \text{kJ}$$

49. A 0.506-kg chunk of ice at 0.0 °C is added to an insulated container holding 315 mL of water at 20.2 °C. Will any ice remain after thermal equilibrium is established?

50. An ice cube weighing 25.5 g at a temperature of 0.0 °C is added to 125 mL of water at 26.5 °C in an insulated container. What will be the final temperature after the ice has melted?

51. *Without doing detailed calculations*, and with data from this chapter and Chapter 6, determine which of the following would require the greatest input of heat: (a) melting 3.0 mol of ice at 0 °C; (b) evaporating 10.0 g $H_2O(l)$ at 298 K; (c) melting 2.0 mol of ice at 0 °C and heating the resulting $H_2O(l)$ to 10 °C; (d) subliming 1.0 mol of ice at 0 °C.

52. *Without doing detailed calculations*, and with data from this chapter and Chapter 6, determine which of the following would evolve the greatest quantity of heat: (a) freezing 2.00 kg Hg(l) at −38.9 °C; (b) condensing 0.50 mol of steam [$H_2O(g)$] at 100 °C; (c) forming 10.0 g of frost [$H_2O(s)$] from water vapor at −2.0 °C; (d) cooling 100.0 mL of $H_2O(l)$ from 51.0 °C to 1.0 °C.

Phase Diagrams

53. The figure provided is a phase diagram for iodine. **(a)** Indicate the phases present in the portions of the diagram marked (?). **(b)** Use the letters *A*, *B*, *C*, and *D* to represent the triple point, the normal melting point, the normal boiling point, and the critical point, respectively. **(c)** Describe the phase changes that occur as the temperature of a sample is raised, at constant pressure, from point X to point Y.

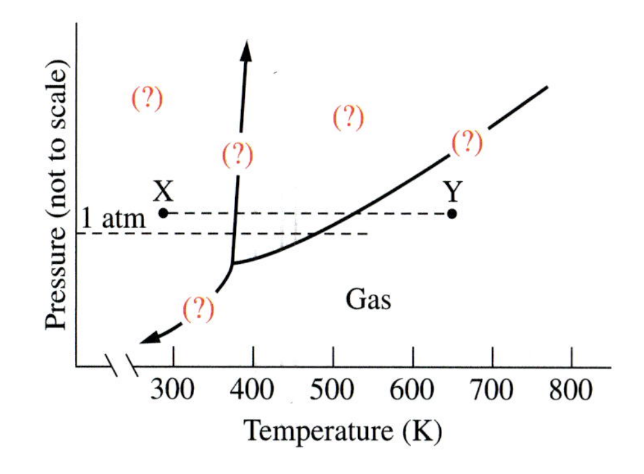

54. The drawing provided is a portion of the phase diagram for sulfur. The stable form of solid sulfur at room temperature is rhombic sulfur, S_α; at the normal melting point, it is monoclinic sulfur, S_β. **(a)** Indicate the phases present in the portions of the diagram marked (?). **(b)** Identify the *three* triple points, and indicate the phases at equilibrium at each one. **(c)** Describe the phase changes that occur as the pressure of a sample is raised, at constant temperature, from point X to point Y.

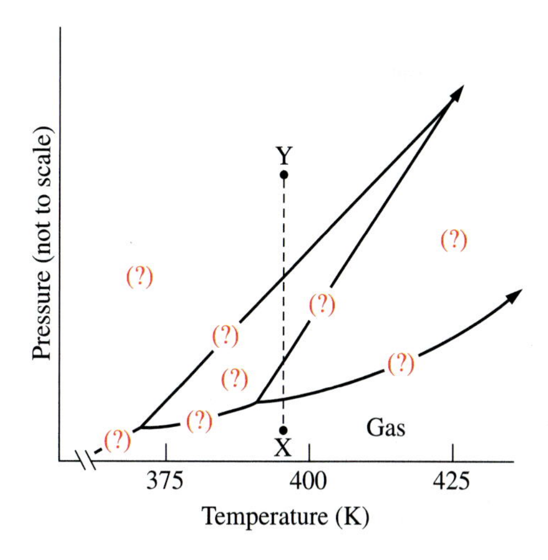

55. Sketch a phase diagram for hydrazine, and locate these points: the triple point (2.0 °C and 3.4 mmHg), the normal boiling point (113.5 °C), and the critical point (380 °C and 145 atm).

56. Sketch a phase diagram for benzene and locate these points: the triple point (5.5 °C and 35.8 mmHg), the normal boiling point (80.1 °C), the critical point (288.5 °C and 47.7 atm).

57. Indicate where each of the following points should fall on the phase diagram for water, that is, whether in a one-phase region (solid, liquid, or gas) or on a two-phase curve (vaporization, sublimation, or fusion). Use information supplied with Figure 11.13 and elsewhere in the chapter.

(a) 88.15 °C and 0.954 atm pressure

(b) 25.0 °C and 0.0313 atm pressure

(c) 0 °C and 2.50 atm pressure

(d) −10 °C and 0.100 atm pressure

58. Describe the phase changes that occur when a sample of solid carbon dioxide is heated in a device similar to that pictured in Figure 11.14 at a constant pressure of 10.0 atm from −100 °C to 100 °C. Also represent these phase changes through a heating curve.

Intermolecular Forces

59. Which of the following would you expect to have the *lower* boiling point, carbon disulfide or carbon tetrachloride? Why?

60. Which of the following would you expect to have the *higher* boiling point, hexane, C_6H_{14}, or 2,2-dimethylbutane, $CH_3CH_2C(CH_3)_3$? Why?

61. Which of the following would you expect to have the *higher* melting point, pentane, $CH_3CH_2CH_2CH_2CH_3$, or diethyl ether, $(CH_3CH_2)_2O$? Why?

62. Which of the following would you expect to have the *higher* melting point, 1-pentanol, $CH_3CH_2CH_2CH_2CH_2OH$, or 3,3-dimethylpentane, $CH_3CH_2C(CH_3)_2CH_2CH_3$? Why?

63. Only one of these substances is a solid at STP: C_6H_5COOH, $CH_3(CH_2)_8CH_3$, C_6H_{14}, $(CH_3CH_2)_2O$. Which do you think it is, and why?

64. Only one of these substances is a gas at STP: NI_3, BF_3, PCl_3, CH_3COOH. Which do you think it is, and why?

65. Arrange the following substances in the expected order of *increasing* boiling point: C_4H_9OH, NO, C_6H_{14}, N_2, $(CH_3)_2O$. Give the reasons for your ranking.

66. Arrange the following substances in the expected order of *increasing* boiling point: H_2O, NH_3, CH_4, CH_3CH_3. Give the reasons for your ranking.

67. Arrange the following substances in the expected order of *increasing* melting point: Cl_2, CsCl, CCl_4, $MgCl_2$. Give the reasons for your ranking.

68. Arrange the following substances in the expected order of *increasing* melting point: NaOH, CH_3OH, LiOH, C_6H_5OH. Give the reasons for your ranking.

Surface Tension

69. Which of the following liquids would you expect to have the higher surface tension, octane, C_8H_{18}, or 1-octanol, $C_8H_{17}OH$? Explain.

70. Which of the following liquids would you expect to have the higher surface tension, isopropyl alcohol, $(CH_3)_2CH_2OH$, or ethylene glycol, CH_2OHCH_2OH? Explain.

71. What is meant by the statement that water wets glass? Does water "wet" all materials? Can additives to water make it wetter, as is sometimes claimed in advertisements for cleaning agents? Explain.

72. Consider a typical sponge with a rectangular shape of 4 cm × 15 cm and a thickness of 2 cm. It can soak up more than its own mass of water. If the water-filled sponge is held in air with its 2-cm dimension vertical, all the water is retained. With its 4-cm dimension held vertically, the sponge also retains all the water. However, when its 15-cm dimension is held vertically, water drains from the sponge. Explain these observations in terms of the natural phenomena involved.

Crystal Structures

73. In the two-dimensional pattern pictured, **(a)** can both of the outlined portions be repeating units (in the same sense that a unit cell is the repeating unit of a crystal structure)? If not, why not, and which one is a repeating unit? **(b)** What are the numbers of ♥, ♦, and ♣ that should be assigned to the repeating unit?

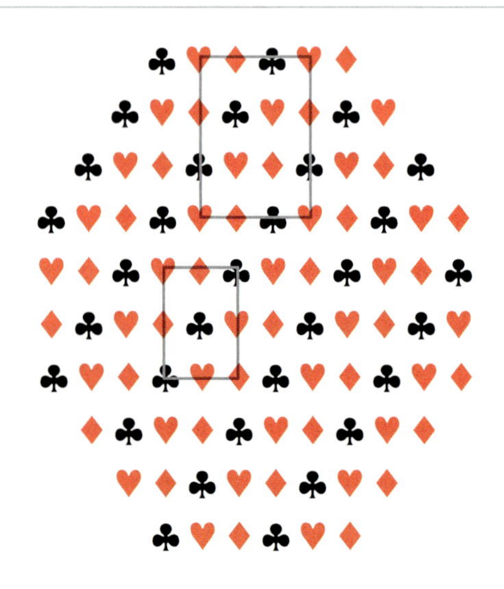

74. In the two-dimensional pattern pictured, **(a)** can both of the outlined portions be repeating units (in the same sense that a unit cell is the repeating unit of a crystal structure)? If not, why not, and which one is a repeating unit? **(b)** What are the numbers of ♦, ♣, and ♥ that should be assigned to the repeating unit?

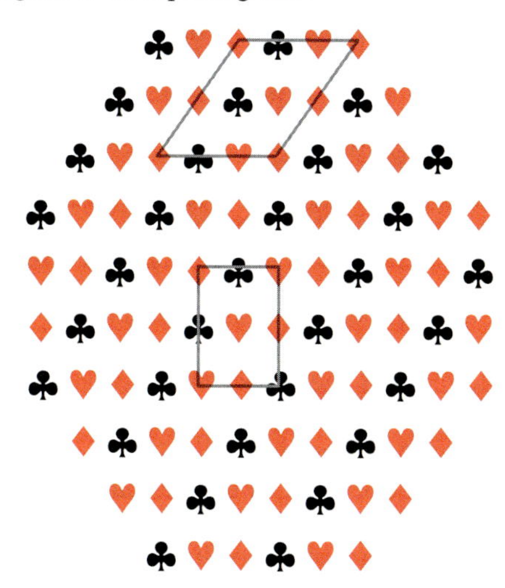

75. Silver has an atomic radius of 144.4 pm, and it has an *fcc* crystal structure. **(a)** What is the length of the unit cell, in picometers? **(b)** What is the volume of the unit cell, in cubic centimeters? **(c)** Calculate the density of silver.

76. Chromium has an atomic radius of 124.9 pm, and it has a *bcc* crystal structure. **(a)** What is the length of the unit cell, in picometers? **(b)** What is the volume of the unit cell, in cubic centimeters? **(c)** Calculate the density of chromium.

77. Use 169 pm for the radius of Cs^+, 181 pm for Cl^-, the unit cell pictured in Figure 11.45, and a density of CsCl(s) of 3.988 g/cm^3 to determine **(a)** the length of the unit cell, and **(b)** an estimate of Avogadro's number, N_A.

78. Potassium chloride has the same type of crystal structure as sodium chloride. The internuclear distance between K^+ and Cl^- is found to be 314.54 pm. The density of KCl(s) is 1.9893 g/cm^3. Determine **(a)** the length of the unit cell; and **(b)** the value of Avogadro's number, N_A.

79. Show that the unit cell of calcium fluoride is consistent with the formula CaF_2. What are the coordination numbers of Ca^{2+} and F^-? Would you expect them to be the same? Explain.

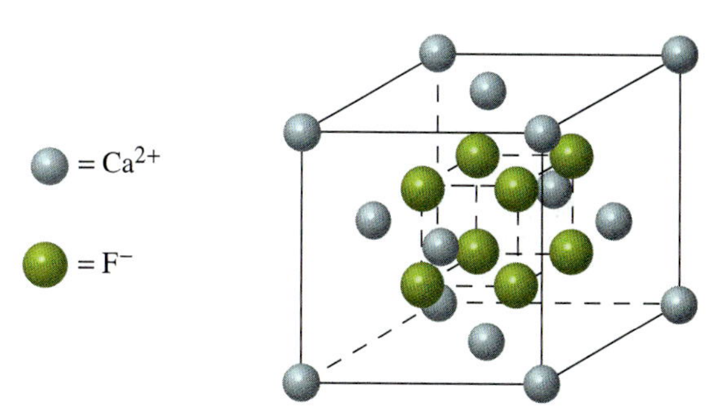

80. Show that the unit cell of titanium dioxide is consistent with the formula TiO_2. What are the coordination numbers of Ti^{4+} and O^{2-}? Would you expect them to be the same? Explain.

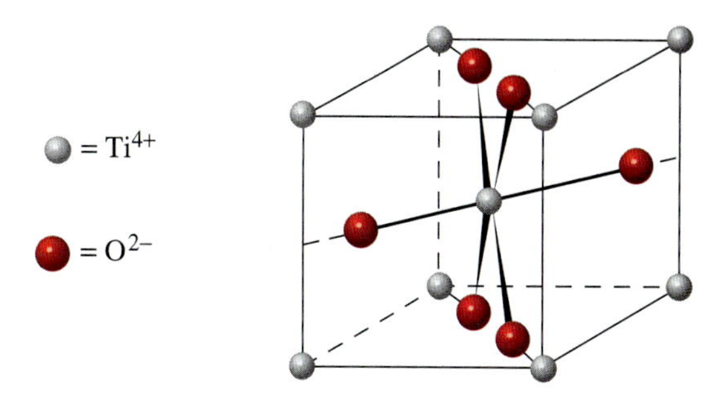

Additional Problems

81. A 150.0-mL sample of N_2(g) at 25.0 °C and 750.0 mmHg is passed through benzene, C_6H_6(l), until the gas becomes saturated with C_6H_6(g). The new volume of the gas is 172 mL at a total pressure of 750.0 mmHg. What is the vapor pressure of benzene at 25.0 °C?

82. The following data are given for CCl_4: normal melting point, −23 °C; normal boiling point, 77 °C; density of liquid, 1.59 g/mL; heat of fusion, 3.28 kJ/mol; vapor pressure at 25.0 °C, 110 mmHg. **(a)** How much heat must be absorbed to convert 10.0 g of solid CCl_4 to liquid at −23 °C? **(b)** What is the volume occupied by 1.00 mol of the saturated vapor at 77 °C? **(c)** What phases—solid, liquid, and/or vapor—are present if 3.5 g CCl_4 is kept in an 8.21-L volume at 25.0 °C?

83. State *several* reasons why you would expect the boiling point of hexanoic acid, $CH_3(CH_2)_4COOH$, to be *higher* than that of 2-methylbutane, $CH_3CH(CH_3)CH_2CH_3$.

84. A newspaper article describing the unusual properties of water includes the statement, "Water is the only kind of matter that can exist as solid, liquid, and gas all at the same temperature." Is this an accurate statement? Explain.

85. Over a certain small temperature range, liquid water can have the same density at *two* different temperatures. Explain how this is possible. At approximately what temperatures would you expect this to occur?

86. What is the *lowest* temperature at which liquid water can be made to boil? Explain how this can be done.

87. The vapor pressure of liquid ammonia is given by the following equation, where p is the vapor pressure in mmHg and t is the Celsius temperature.

$$\log p = 7.3605 - \frac{1617.9}{t + 240.17}$$

(a) What is the vapor pressure of $NH_3(l)$ at -75.0 °C?
(b) What is the normal boiling point of $NH_3(l)$?
(c) The critical temperature of ammonia is $T_c = 405.6$ K. Calculate the critical pressure, P_c.

88. The vapor pressure of liquid chlorine is given by the following equation

$$\log p = 6.9379 - \frac{861.34}{t + 246.33}$$

and that of solid chlorine, by the equation

$$\log p = 9.7051 - \frac{1444.2}{t + 267.13}$$

where p is the vapor pressure in mmHg and t is the Celsius temperature. Estimate the temperature and pressure at the triple point of chlorine.

89. The graphs show the normal boiling points of hydrides of the elements in Groups 4A–7A. Explain the trends represented in these graphs.

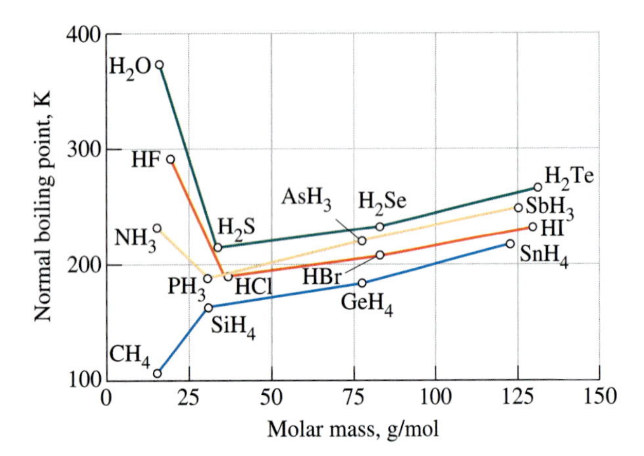

90. Following are several characteristic enthalpies of transition: ΔH_{vapn}, ΔH_{fusion}, ΔH_{condn}, ΔH_{subln}. Arrange them in order of increasing magnitude (that is, increasing in numerical value *without* regard for signs). In general, how do these values compare to the enthalpy changes of chemical reactions? Explain your reasoning. Would you expect exceptions to this generalization? If so, cite some.

91. In the final condition reached in Example 11.5, that is, condition (c), what mass of water is present as liquid and what mass as vapor?

92. A mixture of 1.00 g H_2 and 10.00 g O_2 is ignited in a 3.15-L flask. What is the pressure in the flask when it is cooled to 25 °C? (*Hint:* What is the reaction that occurs? What are the contents of the flask?)

93. A 1.00-g sample of $H_2O(g)$ is injected into a 40.0-L flask at a temperature of 35.0 °C and cooled. To the nearest degree Celsius, determine the temperature at which the first liquid water will condense.

94. The phenomenon illustrated in the two photographs is called the *regelation* of ice. How is the thin wire pulled through the block of ice without cutting the block in two? Would you expect the same phenomenon to occur with dry ice [$CO_2(s)$]? Explain.

95. Intramolecular bonding in salicylic acid was illustrated on page 485. Salicylic acid actually forms both intramolecular and intermolecular hydrogen bonds. Draw the structure of a dimer of salicylic acid that shows four hydrogen bonds altogether.

96. Sketch a phase diagram for tin, including the labeling of significant points in the diagram, given that there are three polymorphic forms of the solid: α, gray tin below 19 °C; β, white tin between 19 and 161 °C; and γ, brittle tin from 161 °C to the melting point at 232 °C. The normal boiling point of tin is 2623 °C. Sketch the cooling curve you would expect to obtain if a sample of liquid tin at 250 °C is slowly cooled to 0 °C.

97. Would you expect the crystalline structure of magnesium oxide to be of the CsCl type or NaCl type? Explain. (*Hint:* Use data from Chapter 8, as necessary.)

98. In the text, we stated that the percent voids in the simple cubic, the *bcc*, and the *fcc* unit cells are 47.64%, 31.98%, and 25.96%, respectively. Show by calculation that this is the case. [*Hint:* This is a geometry problem. Choose some simple radius of a sphere (for example, 1 cm) and calculate the volume of a unit cell and of the spheres contained in the cell.]

99. The unit cell of diamond is shown below. Use the methods of Examples 11.9 and 11.10 and a carbon-to-carbon bond length of 154.45 pm to calculate the density of diamond.

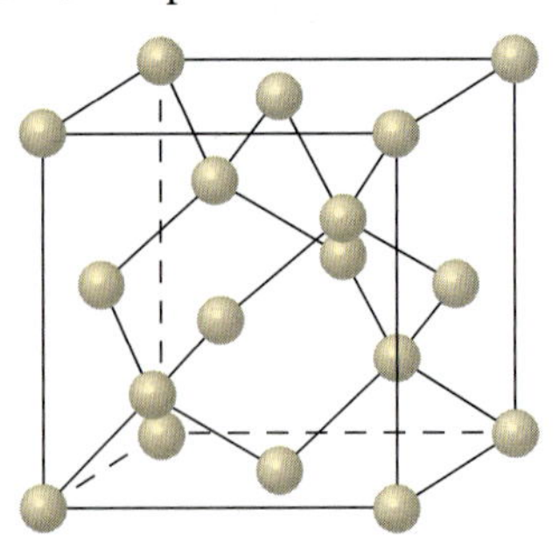

100. A heat pump is a device that is used to provide cooling during hot weather and heating during cold weather. It is based on the same condensation/evaporation cycle as a refrigerator. Describe how the heat pump works. Why do you think heat pumps are of limited value in very cold climates?

Physical Properties of Solutions

CHAPTER

12

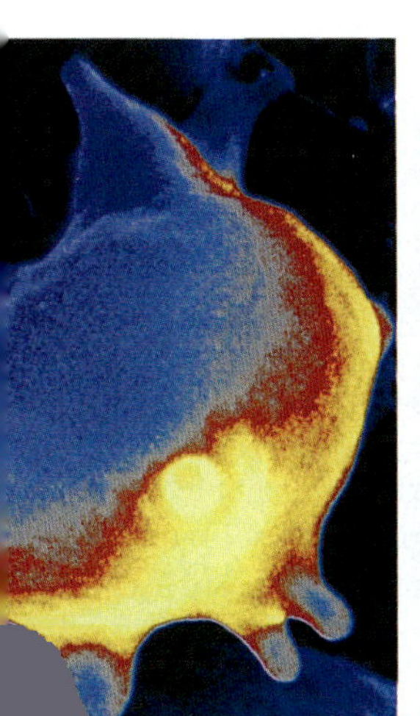
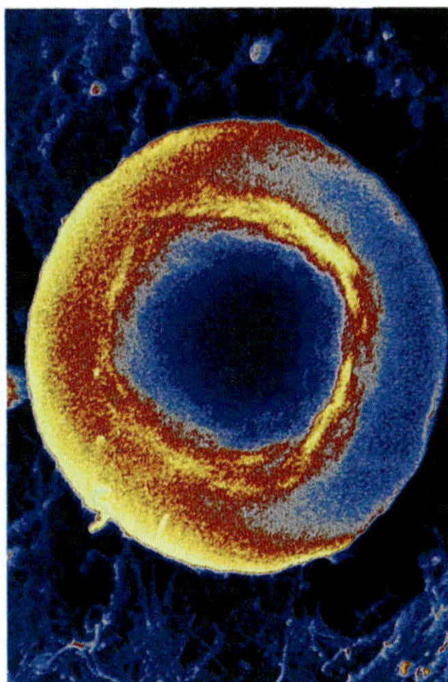
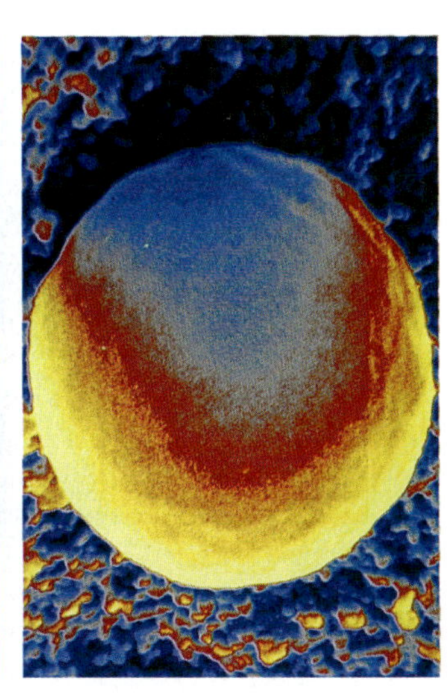

The red blood cell on the left is shrunken in shape because it is suspended in a solution that has a greater osmotic pressure than that in a normal red blood cell (center). The red blood cell on the right is swollen because it is in a solution that has a lower osmotic pressure than that in a normal cell. The cells are shown in false color at a magnification of about 2000×. Osmotic pressure is one of the physical properties of solutions presented in this chapter.

CONTENTS

One of the broad categories of matter that we described in Chapter 1 was homogeneous mixtures, or *solutions*, and we have often encountered solutions in the chapters that followed. In Chapter 3, we considered some basic terminology, and we introduced the concentration unit molarity and used it in stoichiometric calculations. We then devoted all of Chapter 4 to an overview of the basic types of chemical reactions that occur in aqueous solutions.

In this chapter, we will examine some *physical* properties of solutions and phenomena related to them. Consider the coolant in an automobile engine. We could use water alone to cool the engine, but we usually prefer a solution of ethylene glycol ($HOCH_2CH_2OH$) in water because this solution has both a higher *boiling point* and a lower *freezing point* than does water. The solution is less likely than water to boil away in a hot engine or to freeze at low temperatures; it acts as both a coolant and an antifreeze.

Consider petroleum, a solution composed of hundreds of compounds, many of them hydrocarbons. To make gasoline from petroleum, we separate hydrocarbons that boil over different temperature ranges from one another by fractional distilla-

● **ChemCDX**
When you see this icon, there are related animations, demonstrations, and exercises on the CD accompanying this text.

tion. This process is related to the *vapor pressures* of the solution components. Also consider how crucial a knowledge of solutions is to health care professionals. They inject fluids intravenously to restore fluids to the body, but they cannot use pure water for this purpose. The water would enter blood cells, causing them to swell and burst. Instead, they need aqueous solutions with the correct value of a property known as *osmotic pressure*.

Boiling points, freezing points, vapor pressures, and osmotic pressures are interrelated physical properties of solutions that we will explore in this chapter. We will also consider some fundamental questions about solution formation; for example, which substances dissolve in one another and why?

12.1 Some Types of Solutions

This mouse survives by breathing a solution of perfluorodecalin ($C_{10}F_{18}$) saturated with $O_2(g)$—a solution of a gas in a liquid. These and similar solutions are widely used in Japan as temporary substitutes for blood during surgery. In the United States, such solutions are used to treat premature babies whose lungs are not fully developed.

We generally label as the *solvent* the solution component that is present in the greatest amount and determines the state of matter of the solution. The other component—*or* components—is the *solute*. In a solution of sucrose (cane sugar) in water, solid sucrose is the solute, and liquid water is the solvent. This is true whether the amount of sucrose is small (dilute solution) or large (concentrated solution). In carbonated water, gaseous CO_2 is the solute, and again water is the solvent. However, solutions are by no means limited to the liquid state of matter. For example, all gaseous mixtures are solutions, and even some mixtures of solids are solutions. The several possibilities for types of solutions and their solutes and solvents are listed in Table 12.1.

TABLE 12.1 Some Common Types of Solutions

Solute	Solvent	Solution	Examples [solute(s) listed prior to solvent]
Gas	Gas	Gas	Air (O_2, Ar, CO_2, ... in N_2) Natural gas (C_2H_6, C_3H_8, ... in CH_4)
Gas	Liquid	Liquid	Club soda (CO_2 in H_2O) Blood substitute (O_2 in perfluorodecalin)
Liquid	Liquid	Liquid	Vodka (CH_3CH_2OH in H_2O) Vinegar (CH_3COOH in H_2O)
Solid	Liquid	Liquid	Saline solution (NaCl in H_2O) Racing fuel (naphthalene in gasoline)
Gas	Solid	Solid	Hydrogen (H_2) in palladium (Pd)
Solid	Solid	Solid	14-Karat gold (Ag in Au) Yellow brass (Zn in Cu)

To study the physical properties of solutions, we need several different concentration units. In the next section, we review some of the units that we have seen before and introduce some new ones.

12.2 Solution Concentration

The *molarity* of a solution relates an amount of solute in moles and a solution volume in liters.

$$\text{Molarity (M)} = \frac{\text{amount of solute (mol)}}{\text{volume of solution (L)}}$$

Thus, if we measure the volume of a solution and know its molarity, we can readily calculate the number of moles of solute present. This makes molarity the ideal concentration unit to use in titrations and other stoichiometric calculations, as we noted in Section 3.11. However, molarity is not as generally useful for some of the applications in this chapter.

Percent by Mass, Percent by Volume, and Mass/Volume Percent

For many practical applications, we can simply express solution concentration through percentage composition. Then, if we require a precise quantity of solute, we just measure out a mass or volume of solution. For use in storage batteries, for example, sulfuric acid is supplied as a solution that is 35.7% H_2SO_4 *by mass*. Every 100.0 g of this solution contains 35.7 g H_2SO_4 as solute and 64.3 g H_2O as solvent. If we want a quantity of solution containing 100.0 g H_2SO_4, we can measure out 280 g of the battery acid.

$$\text{Mass} = 100.0\ \cancel{\text{g H}_2\text{SO}_4} \times \frac{100.0\ \text{g battery acid}}{35.7\ \cancel{\text{g H}_2\text{SO}_4}} = 280\ \text{g battery acid}$$

EXAMPLE 12.1

How would you prepare 750 g of an aqueous solution that is 2.5% NaOH by mass?

SOLUTION

Our goal is to calculate how many grams of NaOH as solute and how many grams of H_2O as solvent we need to make 750 g of solution. It's easiest to determine the mass of NaOH first because we can use the percent composition of the solution as a conversion factor from mass of solution to mass of solute.

$$?\ \text{g NaOH} = 750\ \cancel{\text{g solution}} \times \frac{2.5\ \text{g NaOH}}{100\ \cancel{\text{g solution}}} = 19\ \text{g NaOH}$$

Because the combined mass of the solute and solvent must be 750 g, we can get the mass of water by difference.

$$?\ \text{g H}_2\text{O} = 750\ \text{g solution} - 19\ \text{g NaOH} = 731\ \text{g H}_2\text{O}$$

To make the 750 g of solution, weigh out 19 g NaOH, and dissolve it in 731 g H_2O.

EXERCISE 12.1A

Glucose is a sugar found abundantly in nature—for example, in ripe grapes. What is the percent by mass of a solution made by dissolving 163 g glucose in 755 g water? Do you need to know the formula of glucose? Explain.

EXERCISE 12.1B

Most of us know sucrose as table sugar. What is the mass percent sucrose in the solution obtained by mixing 225 g of an aqueous solution that is 6.25% sucrose by mass with 135 g of an aqueous solution that is 8.20% sucrose by mass?

Because it is so easy to measure liquid volumes if both the solute and solvent are liquids, a solution concentration is often expressed as *percent by volume*. Ethanol, CH_3CH_2OH, for medicinal purposes is generally known as USP* ethanol;

* USP is an abbreviation of *United States Pharmacopoeia*, the official publication of standards for pharmaceutical products.

it is 95% CH_3CH_2OH, by volume. That is, it consists of 95 mL $CH_3CH_2OH(l)$ per 100 mL of aqueous solution.

In tests for intoxication, blood alcohol levels are expressed as percent by volume. A blood alcohol level of 0.08% by volume means 0.08 mL CH_3CH_2OH per 100 mL of blood, and is considered proof of intoxication in many states. The Breathalyzer® test, shown here, measures ethanol in expired air. The reaction is complex, but it involves reducing yellow-orange $K_2Cr_2O_7$ to blue-green chromium(III) compounds. Readings from the Breathalyzer® are calibrated against actual blood alcohol levels.

EXAMPLE 12.2

USP ethanol is an aqueous solution containing 95.0% ethanol by volume. At 20 °C, pure ethanol has a density of 0.789 g/mL and the USP ethanol, 0.813 g/mL. What is the mass percent ethanol in USP ethanol?

SOLUTION

Because percent by volume means "parts per hundred" by volume, a convenient sample of the USP ethanol on which to base our calculation is 100.0 mL. We can get the mass of the 100.0 mL of solution by using its density to convert from milliliters to grams.

$$\text{Mass of solution} = 100.0\ \cancel{\text{mL}} \times \frac{0.813\ \text{g}}{1\ \cancel{\text{mL}}} = 81.3\ \text{g}$$

Next, we need to determine the mass of ethanol in the 100.0 mL of USP ethanol. To do this, we first use the volume percent composition to determine the number of milliliters of *pure* ethanol, and then we use the density of ethanol.

$$?\ \text{g ethanol} = 100.0\ \cancel{\text{mL solution}} \times \frac{95.0\ \cancel{\text{mL ethanol}}}{100.0\ \cancel{\text{mL solution}}} \times \frac{0.789\ \text{g ethanol}}{1\ \cancel{\text{mL ethanol}}}$$
$$= 75.0\ \text{g ethanol}$$

Now, to express the percent composition in the usual manner, we form the mass ratio of ethanol to solution and multiply by 100%.

$$\text{Mass percent ethanol} = \frac{75.0\ \text{g ethanol}}{81.3\ \text{g solution}} \times 100\% = 92.3\%$$

EXERCISE 12.2A

Assume that the volumes are additive, and determine the volume percent toluene, $C_6H_5CH_3$, in a solution made by mixing 40.0 mL of toluene with 75.0 mL of benzene, C_6H_6.

EXERCISE 12.2B

At 20 °C, the density of benzene is 0.879 g/mL and that of toluene is 0.866 g/mL. For the solution described in Exercise 12.2A, what is **(a)** the mass percent toluene and **(b)** the density?

Another concentration unit widely used in medicine is *mass/volume percent*. For example, a solution of sodium chloride used in intravenous injections has the composition 0.92% (mass/vol) NaCl; that is, it contains 0.92 g NaCl per 100 mL solution. A volume of 100 mL is also one-tenth of a liter, or one *deciliter*. If we express the mass of solute in *milligrams*, we then have a mass/volume concentration unit in *milligrams per deciliter* (mg/dL). A blood cholesterol reading of 187, for example, means 187 mg cholesterol/dL blood.

Parts per Million, Parts per Billion, and Parts per Trillion

For solutions that are extremely dilute, we often express concentrations in parts per million (**ppm**), parts per billion (**ppb**), or even parts per trillion (**ppt**). Thus, fluoridated drinking water has a fluoride ion concentration of about 1 *ppm*. A typical level of the contaminant chloroform ($CHCl_3$) in a municipal drinking water sample taken from the lower Mississippi River is 8 *ppb*.

We get a sense of the meanings of these terms by considering that one person in the city of San Diego, California, represents about one ppm, and one person in China, about one ppb. Figure 12.1 demonstrates the extreme dilution implied by the units ppm, ppb, and ppt.

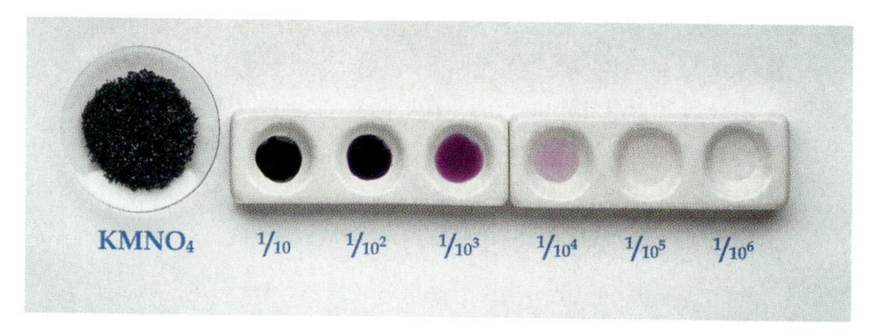

▲ **Figure 12.1 Visualizing one part per million**
The intensely colored solid is potassium permanganate ($KMnO_4$). In the first solution, a sample of $KMnO_4$ is dissolved in 9 times its mass of water, making its concentration 1 part in 10 by mass. The second solution is a tenfold dilution of the first; it has 1 part of solute per 100. The third solution has 1 part per 1000, and so on. The color of the solution is no longer discernible when the concentration has been reduced to 1 part per 1,000,000 (1 ppm).

For liquid solutions, ppm, ppb, and ppt are generally based on mass. Thus, 1 ppm of solute in a solution is the same as 1 g solute per 1×10^6 g solution. For gaseous solutions, such as air, ppm, ppb, and ppt are generally based on numbers of molecules or on volumes. Thus, 1 ppm of some substance in air is 1 molecule of the substance per 1×10^6 molecules of air or 1 L of that substance per 1×10^6 L of air. When applied to dilute aqueous solutions, the units ppm, ppb, and ppt have the following useful equivalents (see Example 12.3).

$$1 \text{ ppm} = 1 \text{ mg/L} \quad 1 \text{ ppb} = 1 \ \mu\text{g/L} \quad 1 \text{ ppt} = 1 \text{ ng/L}$$

EXAMPLE 12.3

The maximum allowable level of nitrates in drinking water set by the State of California is 45 mg NO_3^-/L. What is this level expressed in parts per million (ppm)?

SOLUTION

The density of water, even if it contains traces of dissolved substances, is essentially 1.00 g/mL. One liter of water has a mass of 1000 g. We can therefore express the allowable nitrate level as a ratio.

$$\frac{45 \text{ mg } NO_3^-}{1000 \text{ g water}}$$

To convert this nitrate level to ppm, we need to have the numerator and denominator in the same unit. Let's use milligrams. This makes the denominator one million milligrams. The numerator will then express the parts per million of solute, that is, ppm NO_3^-.

Setting Environmental Standards

Generally our physical senses cannot detect substances in the ppb or ppt range (see again Figure 12.1). A few substances can be detected at these levels through their odor. For example, hydrogen sulfide (H_2S) in air is toxic at concentrations of about 10 ppm, but we can detect it through its foul odor at much lower levels. Most toxic substances, however, are detected only by sophisticated analytical instruments. As technology advances even further, our ability to detect minute quantities of materials increases. This increase in the sensitivity of analytical techniques raises important questions for which there are no set answers. For example, what should we think about a substance detected for the first time because our instruments are now sensitive in the ppb, or even ppt, range? Should we assume it to be a contaminant that's new to the environment? Or should we assume that the substance has always been there, but at levels that were previously undetectable? What is the relationship between the level at which a substance can be detected and the level at which it is injurious to individuals or the environment?

$$\frac{45 \text{ mg } NO_3^-}{1000 \text{ } \cancel{\text{g water}}} \times \frac{1 \text{ } \cancel{\text{g water}}}{1000 \text{ mg water}} = \frac{45 \text{ mg } NO_3^-}{1{,}000{,}000 \text{ mg water}} = 45 \text{ ppm } NO_3^-$$

EXERCISE 12.3A

What is the concentration in **(a)** ppb and **(b)** ppt corresponding to a maximum allowable level in water of 0.1 μg/L of the pesticide chlordane?

EXERCISE 12.3B

What is the concentration of Na^+, in ppm by mass, in 0.00152 M Na_2SO_4?

Molality

Figure 12.2 illustrates an observation: The *molarity* of a solution varies with temperature. Because the volume of the solution increases as its temperature increases from 20 °C to 25 °C, the molarity of the solution *decreases*. If we need a concentration unit that is *independent* of temperature, it must be based on mass only, not volume. Percent by mass is such a unit; percent by volume and mass/volume percent are not. Molality is another unit independent of temperature. The **molality (*m*)** of a solution is the number of moles of solute per *kilogram of solvent* (not of solution).

$$\text{Molality } (m) = \frac{\text{amount of solute (mol)}}{\text{mass of solvent (kg)}}$$

For example, a solution of 0.2 mol ethanol in 500 g of water is 0.4 *m* ethanol.

$$\frac{0.2 \text{ mol ethanol}}{0.5 \text{ kg water}} = 0.4 \, m \text{ ethanol}$$

(a)

(b)

Figure 12.2
The effect of temperature on molarity
(a) A 0.1000 M HCl solution prepared at 20.0 °C. (b) The solution expands on being warmed to 25.0 °C. At this temperature, because the same amount of solute is present in a larger volume of solution than at 20.0 °C, the molarity is slightly *less than* 0.1000 M HCl.

EXAMPLE 12.4

What is the molality of a solution prepared by dissolving 5.05 g naphthalene [$C_{10}H_8(s)$] in 75.0 mL of benzene, C_6H_6 ($d = 0.879$ g/mL)?

SOLUTION

Molality refers to *mol solute/kg solvent.* We therefore need to convert 5.05 g $C_{10}H_8$ to mol $C_{10}H_8$ and 75.0 mL C_6H_6 to kg C_6H_6. Let's first perform each of these conversions and then take the appropriate ratio of solute to solvent.

$$? \text{ mol } C_{10}H_8 = 5.05 \cancel{\text{ g } C_{10}H_8} \times \frac{1 \text{ mol } C_{10}H_8}{128.2 \cancel{\text{ g } C_{10}H_8}} = 0.0394 \text{ mol } C_{10}H_8$$

$$? \text{ kg } C_6H_6 = 75.0 \cancel{\text{ mL } C_6H_6} \times \frac{0.879 \cancel{\text{ g } C_6H_6}}{1 \cancel{\text{ mL } C_6H_6}} \times \frac{1 \text{ kg } C_6H_6}{1000 \cancel{\text{ g } C_6H_6}} = 0.0659 \text{ kg } C_6H_6$$

The molality of the solution is

$$\frac{0.0394 \text{ mol } C_{10}H_8}{0.0659 \text{ kg } C_6H_6} = 0.598\ m\ C_{10}H_8$$

EXERCISE 12.4

What is the molality of a solution prepared by dissolving 225 mg glucose ($C_6H_{12}O_6$) in 5.00 mL ethanol ($d = 0.789$ g/mL)?

EXAMPLE 12.5

How many grams of benzoic acid, C_6H_5COOH, must be dissolved in 50.0 mL of benzene, C_6H_6 ($d = 0.879$ g/mL), to produce 0.150 m C_6H_5COOH?

SOLUTION

Here we use the molality as a conversion factor between the mass of solvent (in kg) and the number of moles of solute. As shown in the outline that follows, we first need two conversions to get the quantity of solvent in kilograms. Then we apply molality as the central factor (in red). Finally, we convert moles to grams of solute.

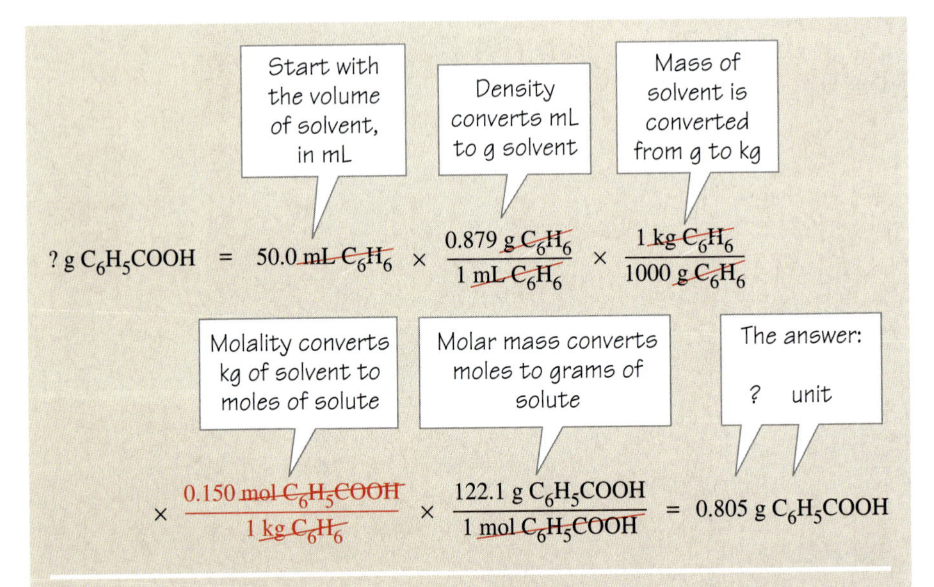

EXERCISE 12.5A

How many milliliters of ethanol (CH_3CH_2OH, $d = 0.789$ g/mL) must be mixed with 125 mL of benzene (C_6H_6, $d = 0.879$ g/mL) to produce a solution that is 0.0652 m CH_3CH_2OH?

EXERCISE 12.5B

How many milliliters of water ($d = 0.998$ g/mL) are required to dissolve 25.0 g of urea and thereby produce a 1.65 m solution of urea, $CO(NH_2)_2$?

Mole Fraction and Mole Percent

Later in the chapter, we will consider vapor pressure, a solution property that requires concentrations to be expressed in mole fractions. Similar to our description for mixtures of gases (page 206), the **mole fraction** (x_i) of a solution component i is the fraction of all the molecules in the solution that are molecules of i.

$$x_i = \frac{\text{amount of component } i \text{ (mol)}}{\text{total amount of solution components (mol)}} = \frac{n_i}{n_{total}}$$

Note that mole fraction has no unit; like all fractions, it is dimensionless. Note also that the sum of the mole fractions of all the components in a solution is equal to 1.

$$x_1 + x_2 + x_3 + \cdots = 1$$

The **mole percent** of a solution component is its mole fraction multiplied by 100%. Thus, a solution made by mixing 4 mol methanol and 6 mol ethanol has the mole fractions: $x_{CH_3OH} = 4/(4 + 6) = 0.40$ and $x_{CH_3CH_2OH} = 6/(4 + 6) = 0.60$. The corresponding mole percents are 40 mol% CH_3OH and 60 mol% CH_3CH_2OH.

The calculations of solution properties that we will do later in the chapter require that solution concentrations be expressed in a particular unit. This may require us to do a preliminary calculation in which we convert from one concentration unit to another. Some typical conversions are illustrated in Example 12.6.

EXAMPLE 12.6

An aqueous solution of ethylene glycol used as an automobile engine coolant is 40.0% $HOCH_2CH_2OH$ by mass and has a density of 1.05 g/mL. What are the **(a)** molarity, **(b)** molality, and **(c)** mole fraction of $HOCH_2CH_2OH$ in this solution?

SOLUTION

To find the molarity in (a), we need the number of moles of solute in a known *volume*. To find the molality in (b), we also need a number of moles of solute, but in a known *mass* of solvent. And to find the mole fractions in (c), we need the number of moles of both the solute and the solvent. When we are not given a particular quantity of solution to consider, we can pick a convenient one, usually 1.00 L or 1.00 kg. Let's choose 1.00 L for this problem. We will need conversion factors based on density, mass percent composition, and molar masses.

a. We need to determine the mass of 1.00 L of the solution, the mass of $HOCH_2CH_2OH$ in this 1.00 L, and then the number of moles of $HOCH_2CH_2OH$.
The mass of solution is simply the product of the solution volume in milliliters and the density.

$$? \text{ g solution} = 1.00 \cancel{\text{L}} \times \frac{1000 \cancel{\text{mL}}}{1 \cancel{\text{L}}} \times \frac{1.05 \text{ g}}{1 \cancel{\text{mL}}} = 1050 \text{ g}$$

We can use the mass percent composition to convert from mass of solution to mass of $HOCH_2CH_2OH$. Then we can use the molar mass of $HOCH_2CH_2OH$ to convert from grams to moles of $HOCH_2CH_2OH$.

$$? \text{ g } HOCH_2CH_2OH = 1050 \cancel{\text{g solution}} \times \frac{40.0 \text{ g } HOCH_2CH_2OH}{100 \cancel{\text{g solution}}}$$

$$= 420 \text{ g } HOCH_2CH_2OH$$

$$? \text{ mol } HOCH_2CH_2OH = 420 \cancel{\text{g } HOCH_2CH_2OH} \times \frac{1 \text{ mol } HOCH_2CH_2OH}{62.07 \cancel{\text{g } HOCH_2CH_2OH}}$$

$$= 6.77 \text{ mol } HOCH_2CH_2OH$$

Finally, we calculate the molarity of the solution, based on our chosen 1.00-L volume.

$$\frac{6.77 \text{ mol } HOCH_2CH_2OH}{1 \text{ L}} = 6.77 \text{ M } HOCH_2CH_2OH$$

b. From part (a), we have the number of moles of $HOCH_2CH_2OH$ in 1.00 L of solution (6.77 mol) and the mass of solution (1050 g) and solute (420 g). The mass of water, the *solvent*, is the difference in mass between that of the solution and the solute. Expressed in kilograms, the mass of water is

$$? \text{ kg } H_2O = (1050 - 420) \cancel{\text{g } H_2O} \times \frac{1 \text{ kg } H_2O}{1000 \cancel{\text{g } H_2O}} = 0.630 \text{ kg } H_2O$$

Now we can express the molality as moles of $HOCH_2CH_2OH$ per kg H_2O.

$$\frac{6.77 \text{ mol } HOCH_2CH_2OH}{0.630 \text{ kg}} = 10.7 \; m \; HOCH_2CH_2OH$$

c. To get the mole fractions of the solution components, we need the number of moles of each. We have the number of moles of $HOCH_2CH_2OH$ (6.77 mol) from part (a), and we can get the number of moles of H_2O in the 630 g H_2O calculated in part (b).

$$? \text{ mol } H_2O = 630 \cancel{\text{g } H_2O} \times \frac{1 \text{ mol } H_2O}{18.02 \cancel{\text{g } H_2O}} = 35.0 \text{ mol } H_2O$$

Now we can apply the definition of mole fraction.

$$x_{HOCH_2CH_2OH} = \frac{6.77 \text{ mol } HOCH_2CH_2OH}{6.77 \text{ mol } HOCH_2CH_2OH + 35.0 \text{ mol } H_2O} = 0.162$$

EXERCISE 12.6A

Calculate **(a)** the molality of CH_3OH in an aqueous solution that is 7.50% CH_3OH in CH_3CH_2OH by mass and **(b)** the mole percent of urea, $CO(NH_2)_2$, in an aqueous solution that is 1.05 *m* $CO(NH_2)_2$.

EXERCISE 12.6B

A 2.90 *m* solution of methanol (CH_3OH) in water has a density of 0.984 g/mL. What are the **(a)** mass percent, **(b)** molarity, and **(c)** mole percent of methanol in this solution?

EXAMPLE 12.7—An Estimation Example

Without doing detailed calculations, determine which of the following aqueous solutions has the greatest mole fraction of CH_3OH: (a) 1.0 *m* CH_3OH; (b) 10.0% CH_3OH by mass; (c) $x_{CH_3OH} = 0.10$.

SOLUTION

The mole fraction of solution (c) is given: 0.10. The ratio of moles of CH_3OH to moles of H_2O in solution (c) is 1/9. Let's estimate the corresponding ratios for the other solutions and compare them to 1/9. We will need to use the molar masses of H_2O and CH_3OH, but these can be just rough estimates: 18 g H_2O/mol and 32 g CH_3OH/mol.

Solution (a) contains 1.0 mol CH_3OH per *kilogram* of H_2O. One kilogram of water is 1000 g H_2O, and 1000 g H_2O is about 50 mol H_2O (1000/18). The ratio: mol CH_3OH/mol H_2O is about 1/50.

Solution (b) has 10.0 g CH_3OH for every 90.0 g H_2O. The 90.0 g H_2O is almost exactly 5.0 mol H_2O, and the 10.0 g CH_3OH is about $10/32 \approx 0.3$ mol CH_3OH. The ratio mol CH_3OH/mol H_2O in solution (b) is about 0.3/5.0. If we multiply both factors in this ratio by 3, we get 0.9/15, which we can round to 1/15. Thus, there is about 1 mol CH_3OH for each 15 mol H_2O.

Solution (c) has the highest ratio (1/9) of mol CH_3OH to mol H_2O, and thus the greatest mole fraction CH_3OH.

EXERCISE 12.7

Without doing detailed calculations, determine which of the following aqueous solutions has the *greatest* mole percent CH_3CH_2OH: (a) 0.50 M CH_3CH_2OH (d = 0.994 g/mL); (b) 5.0% CH_3CH_2OH by mass; (c) 0.50 *m* CH_3CH_2OH; (d) 5.0% CH_3CH_2OH by volume (d = 0.991 g/mL). (The density of pure liquid CH_3CH_2OH is 0.789 g/mL.)

12.3 Energetics of Solution Formation

Why do some substances dissolve in a given solvent, but others do not? For example, why does ethanol dissolve in water, but the natural gas component ethane does not? Why does sodium carbonate, but not magnesium carbonate, dissolve in water? Why does grease dissolve in kerosene, but not in water? To answer questions like these, let's consider two significant factors in solution formation: the enthalpy of solution and intermolecular forces in mixtures.

Enthalpy of Solution

We can gain some insight into solution formation by considering it to take place in three *hypothetical* steps: (1) Move the molecules of solvent apart to make room for the solute molecules. This requires work to overcome the intermolecular forces, and the enthalpy of the solvent increases: $\Delta H_1 > 0$. (2) Separate the molecules of solute to the distances found between them in the solution. Again, work is required and $\Delta H_2 > 0$. (3) Allow the separated solute and solvent molecules to mix randomly. For this process, we expect energy to be released, ΔH_3, because now there are forces of attraction between the solute and solvent molecules. These steps and the net change are summarized below.

(1) Pure solvent $\longrightarrow$ separated solvent molecules $\quad \Delta H_1$

(2) Pure solute $\longrightarrow$ separated solute molecules $\quad \Delta H_2$

(3) Separated solvent and solute molecules $\longrightarrow$ solution $\quad \Delta H_3$

Net: Pure solvent + pure solute $\longrightarrow$ solution $\quad \Delta H_{soln} = \Delta H_1 + \Delta H_2 + \Delta H_3$

Whether the formation of a solution is an endothermic process ($\Delta H_{soln} > 0$) or an exothermic process ($\Delta H_{soln} < 0$), then, depends on the relative values of the enthalpy changes of the three hypothetical steps. Figure 12.3 presents a generalized enthalpy diagram for an endothermic solution process, an exothermic process, and a special case in which $\Delta H_{soln} = 0$.

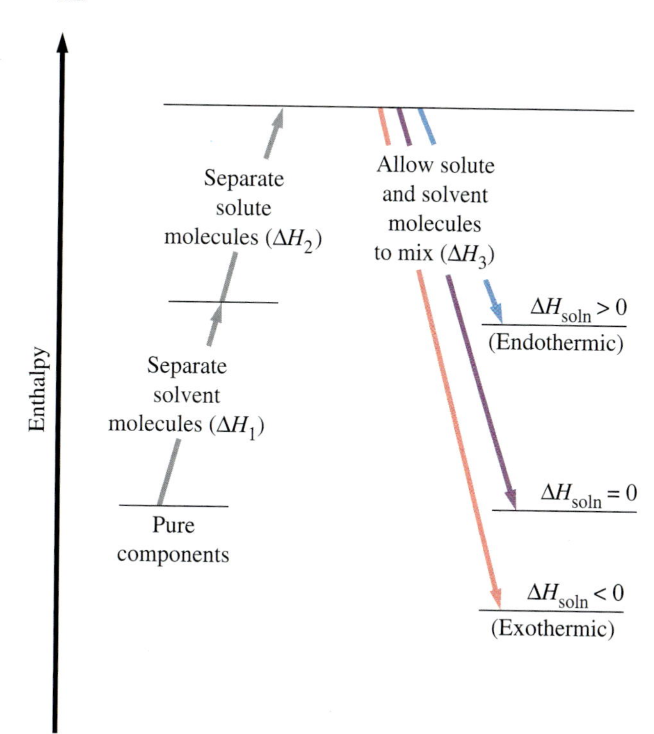

Figure 12.3 Visualizing an enthalpy of solution with an enthalpy diagram

As we shall see next, we can relate the enthalpy changes in the three-step hypothetical solution process to intermolecular forces of attraction.

Intermolecular Forces in Solution Formation

Figure 12.4 suggests three types of intermolecular forces in a solution. The intermolecular forces A–A in Figure 12.4 are between solvent molecules; their strength determines the enthalpy change ΔH_1 in the hypothetical three-step solution process.

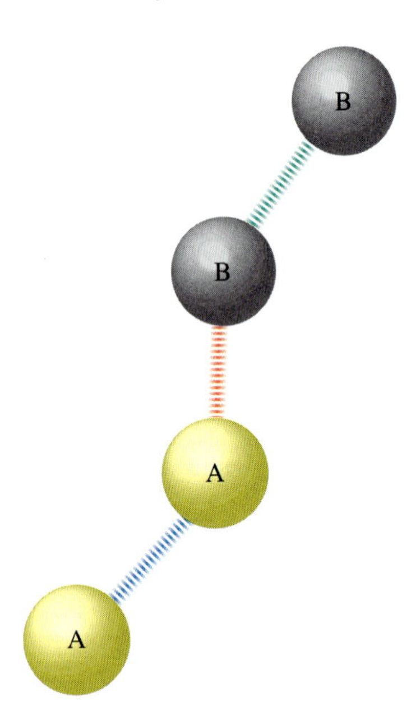

▲ **Figure 12.4**
Intermolecular forces in solution
The intermolecular force between two solvent molecules, A, is represented in blue; between two solute molecules, B, in green; and between a solvent molecule, A, and a solute molecule, B, in red.

The intermolecular forces B–B are between solute molecules and determine the value of ΔH_2. The intermolecular forces A–B are between molecules of solvent and solute; these forces establish ΔH_3.

The comparative strengths of these three types of intermolecular forces are an important factor in determining whether a solution forms between a given solvent and solute. Let's consider the following four possibilities.

1. *All intermolecular forces are of comparable strength.* If the three different intermolecular forces depicted in Figure 12.4, A–A, B–B, and A–B, are of the same type and comparable in strength, we expect the molecules of solute and solvent to intermingle randomly into a solution. Moreover, this situation might well lead to the case in Figure 12.3 in which $\Delta H_{soln} = 0$. There is no net energy change in the formation of the solution, and we expect the volume of the solution to be the sum of the volumes of the solvent and the solute ($\Delta V_{soln} = 0$). A solution that meets these requirements is called an **ideal solution**.

Mixtures of ideal gases are ideal solutions. There is probably no liquid solution that meets the above requirements perfectly, but some come close enough that we consider them to be ideal. One example is the mixture of hydrocarbons that make up gasoline. Another is a solution of toluene in benzene. Because of the similarities in their molecular structures (Figure 12.5), we expect intermolecular forces between benzene molecules and between toluene molecules to be quite similar in magnitude.

▶ **Figure 12.5**
Two molecules with similar structures
The structure of benzene is a six-carbon ring with an H atom attached to each C atom. Toluene differs only in that a $-CH_3$ group substitutes for one of the H atoms.

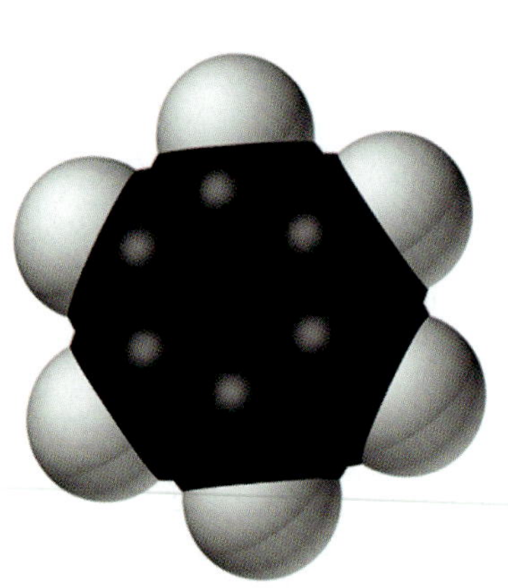
Benzene, C_6H_6

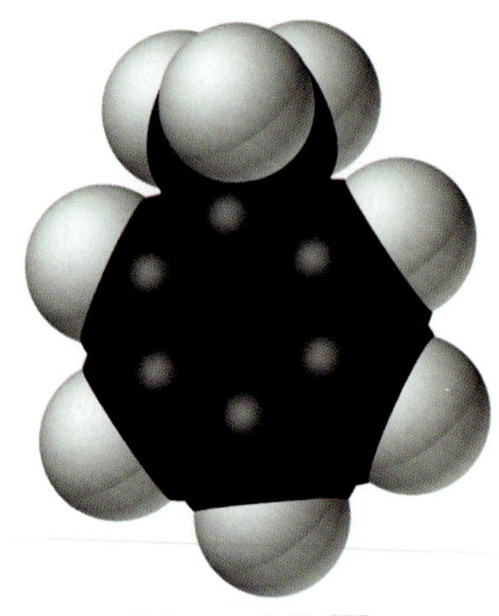
Toluene, $C_6H_5CH_3$

2. *Intermolecular forces between solute and solvent molecules are* stronger *than other intermolecular forces.* When this condition is met (Figure 12.3; red arrow), the magnitude of ΔH_3 can exceed the sum of ΔH_1 and ΔH_2. We expect a solution to form with the evolution of energy as heat. For example, in forming a solution of ethanol in water, $\Delta H_3 > \Delta H_1 + \Delta H_2$, and solution formation is exothermic. Moreover, as shown in Figure 12.6, in ethanol-water solutions, the volume of the solution is *less than* the sum of the volumes of the ethanol and water. We expect solutions of this type to be *nonideal* because $\Delta H_{soln} < 0$ and $\Delta V_{soln} < 0$.
3. *Intermolecular forces between solute and solvent molecules are* weaker *than other intermolecular forces.* This is a situation in which the magnitude of ΔH_3 in Figure 12.3 is less than the sum of ΔH_1 and ΔH_2. A solution may form, but it will be nonideal; $\Delta H_{soln} > 0$, and the solution process is endothermic.

Figure 12.6
A nonideal solution: Ethanol and water
Identical volumetric flasks are filled to the 50.0-mL mark with ethanol and with water. When the two liquids are mixed, the volume is seen to be less than the expected 100.0 mL; it is only about 95 mL.

How can a solution form through an endothermic process, in which the solution is in a higher energy (enthalpy) state than the separate components? You might well suspect that an additional factor is involved, and there is: Systems have a natural inclination toward states of increased disorder at the molecular level. We can measure this tendency through the change in a thermodynamic property called *entropy*. We will introduce entropy in Chapter 17, and we will also describe the role of *entropy changes* in solution formation there.

4. *Intermolecular forces between solute and solvent molecules are* much weaker *than other intermolecular forces*. In this case, the magnitude of ΔH_3 in Figure 12.3 may be so much smaller than the sum of ΔH_1 and ΔH_2 that the solution state is at too high an energy to be attained. The solute does not dissolve in the solvent; the two remain as separate phases in a *heterogeneous* mixture. For example, the dominant intermolecular forces among the nonpolar hydrocarbon molecules in gasoline are dispersion forces; among water molecules, they are hydrogen bonds. However, no hydrogen bonds form between water and hydrocarbon molecules, and gasoline and water do not form solutions.

This oil slick produced by the Exxon *Valdez* oil spill in Alaska in 1989 reminds us that hydrocarbon oils and water do not mix.

Two rules of thumb help us summarize three of these four possibilities. The rule that *like dissolves like* works well for the first two cases, mixtures in which (1) all intermolecular forces are of comparable strength and (2) those in which intermolecular forces between solute and solvent molecules are *stronger* than other intermolecular forces. The old saying, "*Oil and water don't mix*," is another way of stating (4): that substances whose molecules are structurally dissimilar, such as gasoline and water, tend not to dissolve in one another. Neither rule of thumb helps with possibility (3), cases in which intermolecular forces between solute and solvent molecules are *weaker* than other intermolecular forces. As we noted, the controlling factor in this case is the tendency toward increased disorder.

Aqueous Solutions of Ionic Compounds

Now let's consider ionic compounds dissolving in water. Interionic attractions hold ions together in an ionic solid. The forces causing the solid to dissolve are *ion–dipole* forces—the attraction of water dipoles for cations and anions. As suggested by Figure 12.7, the attraction of water dipoles for ions pulls the ions out of a crystalline lattice and into aqueous solution. Thus, ion–dipole forces break down the crystal of a soluble ionic compound. The ion–dipole forces in the solution lessen

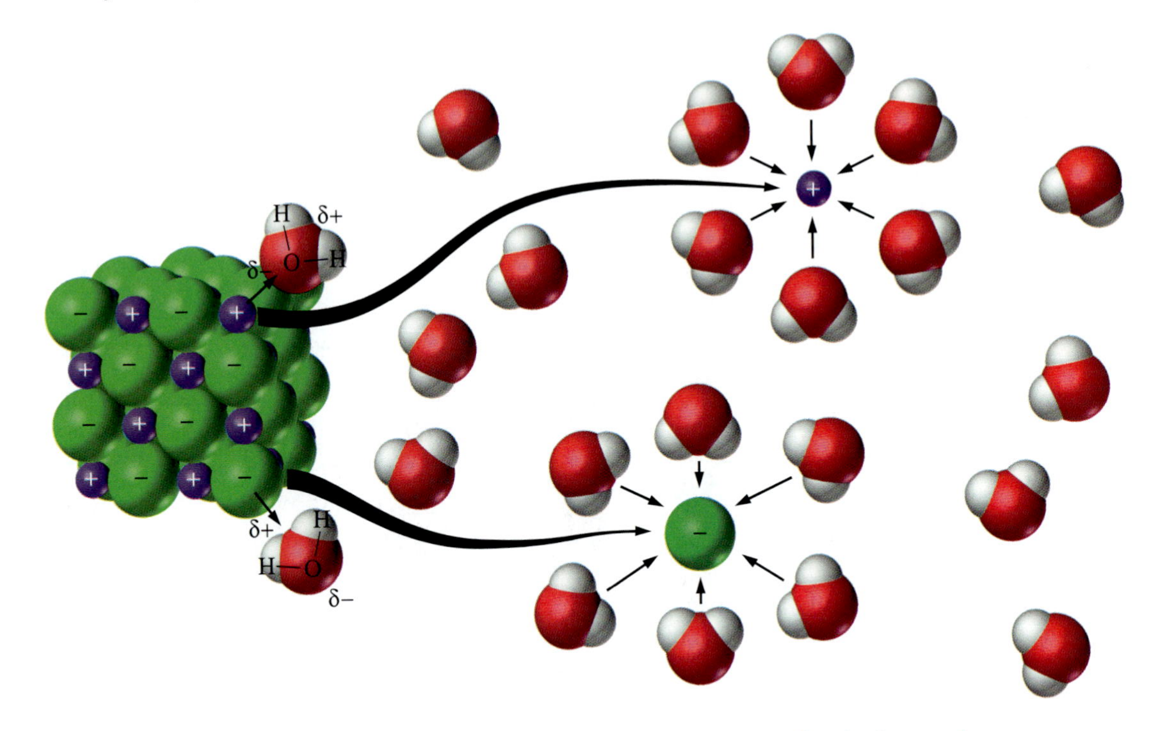

▲ **Figure 12.7 Ion–dipole forces in the dissolving of an ionic crystal**
The attraction of water dipoles for ions pulls ions out of a crystalline lattice and into aqueous solution. The ion–dipole forces exist in the solution as well, lessening the tendency for ions to return to the crystalline state. The combination of an ion in solution and the neighboring water dipoles to which it is attracted is called a *hydrated ion*.

● **ChemCDX** Dissolution of NaCl in Water

the tendency for ions to return to the crystalline state. All ions in an aqueous solution have a certain number of water molecules associated with them. The ions are *hydrated*.

The extent to which an ionic solid dissolves in water is determined largely by the competition between interionic attractions that hold ions in a crystal and the ion–dipole attractions that pull them into solution. If the ion–dipole attractions predominate, the solid is water soluble. This is the case with the common compounds of the Group 1A metals and ammonium compounds (Table 4.3, page 149). If interionic attractions are the dominant factor, as is the case with many carbonates, hydroxides, and sulfides, the solid is insoluble.

● **ChemCDX** Dissolution of $KMnO_4$

It is often difficult to predict the water solubility of a specific compound. For example, we would expect the interionic attractions to be strong between the highly charged Al^{3+} and SO_4^{2-} ions, leading to a limited water solubility. Yet $Al_2(SO_4)_3$ is quite soluble in water. Ion–dipole attractions must be especially strong in the solution, possibly because the negative charge centers of water dipoles can closely approach the small, highly charged Al^{3+} ions. In general, we can most readily predict the solubilities of ionic compounds by using the solubility rules in Chapter 4. There we see that the high solubility of $Al_2(SO_4)_3$ in water conforms to the solubility rule that most sulfates are soluble.

EXAMPLE 12.8

Predict whether each of the following is likely to be a solution or a heterogeneous mixture.

a. methanol, CH_3OH, and water, HOH

b. pentane, $CH_3(CH_2)_3CH_3$, and octane, $CH_3(CH_2)_6CH_3$

c. sodium chloride, NaCl, and carbon tetrachloride, CCl_4

d. 1-decanol, $CH_3(CH_2)_8CH_2OH$, and water, HOH

SOLUTION

a. The most important structural feature of the molecules is the OH group. This group is attached to a H atom in water and to a CH_3 group in methanol. The molecules are similar in that hydrogen bonding is the major intermolecular force in both pure liquids and in their mixtures. We expect methanol and water to form a solution.

b. The hydrocarbons pentane and octane are quite similar; they differ only by three CH_2 groups. Intermolecular forces are all of the dispersion type and of similar magnitude. We expect pentane and octane to form a solution (like dissolves like).

c. Sodium chloride and carbon tetrachloride are both chlorides, but there, the similarity ends. NaCl is an ionic compound, but CCl_4 is molecular. Moreover, the tetrahedral CCl_4 molecule is *nonpolar*, so there are no ion–dipole forces available to separate the ions in NaCl(s). There is no dissolving, and we expect the mixture to be *heterogeneous*.

d. This case may seem similar to (a)—a hydrocarbon chain substituted for an H atom in HOH— but here, the hydrocarbon chain is ten carbon atoms long. This long hydrocarbon chain is the principal structural feature of 1-decanol, and dispersion forces are its dominant molecular forces. We expect 1-decanol and water mixtures to be *heterogeneous* (no significant dissolving occurs). This is a case of "oil and water don't mix," with 1-decanol being the "oil."

PROBLEM-SOLVING NOTE

Alcohols become less and less soluble in water as the number of carbon atoms increases beyond three.

EXERCISE 12.8A

In which solvent would you expect nitrobenzene, $C_6H_5NO_2$, to be more soluble: H_2O or C_6H_6? Explain.

EXERCISE 12.8B

Figure 12.8 shows molecular models of four substances. Identify the substances, and arrange them in the expected order of increasing solubility in water.

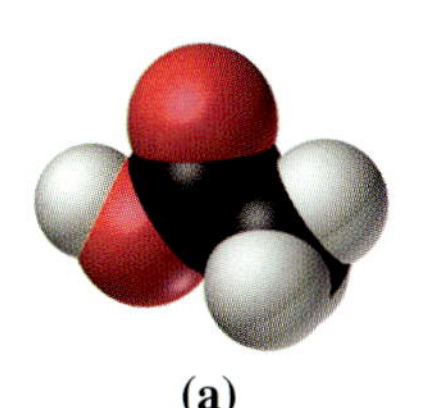
(a)

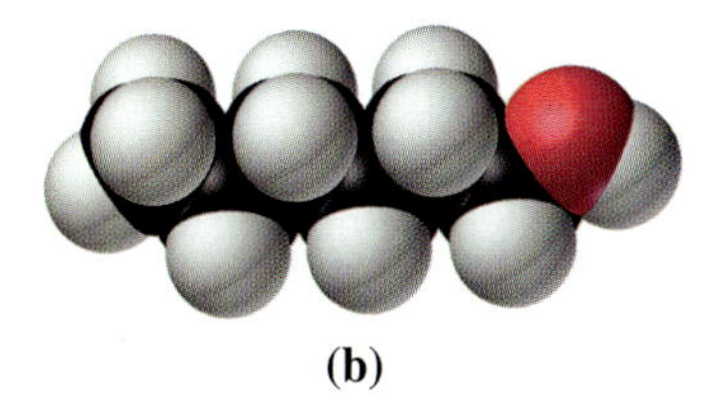
(b)

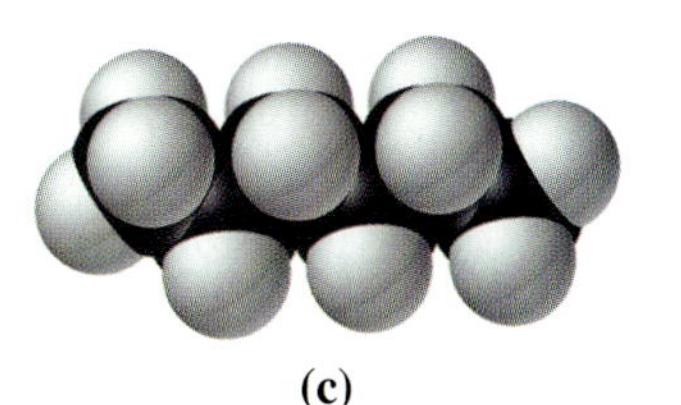
(c)

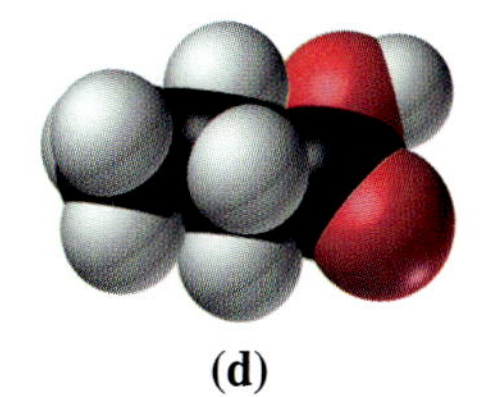
(d)

▲ **Figure 12.8 Exercise 12.8B illustrated**

12.4 Equilibrium in Solution Formation

Some substances can dissolve in each other without limit. Liquids that mix in all proportions are said to be *miscible*. For example, we can prepare an ethanol-water solution with a mass percent ethanol ranging from almost 0 to nearly 100% ethanol. Ethanol and water are completely miscible. For most substances, though, there are limits to their solubilities. This limit varies with the nature of the solute and of the solvent, and with temperature.

What happens when we place 40 g of NaCl in 100 g water at 20 °C? Initially many of the Na^+ and Cl^- ions are plucked from the surface of crystals by ion–dipole forces, and they wander about at random through the solution; some of the NaCl *dissolves*. In their wanderings, some of the ions pass near a crystal surface and are attracted back to the surface by ions of the opposite charge; some NaCl *crystallizes*. As more and more NaCl dissolves, the number of "wanderers" that return to the crystals increases. Eventually (when 36 g NaCl has dissolved) the number of ions returning to the crystals equals the number of ions leaving the crystals. At this point, the rate of crystallization is equal to the rate of dissolving, and a condition of *dynamic equilibrium* is established. The net amount of NaCl in solution remains the same, despite the fact that there is a lot of activity as ions come and go from the surface of the crystals. The net quantity of undissolved NaCl also remains constant (in this example, 4 g), even though individual crystals may change in size and shape. Some small crystals may disappear as others grow larger. This process is illustrated in Figure 12.9.

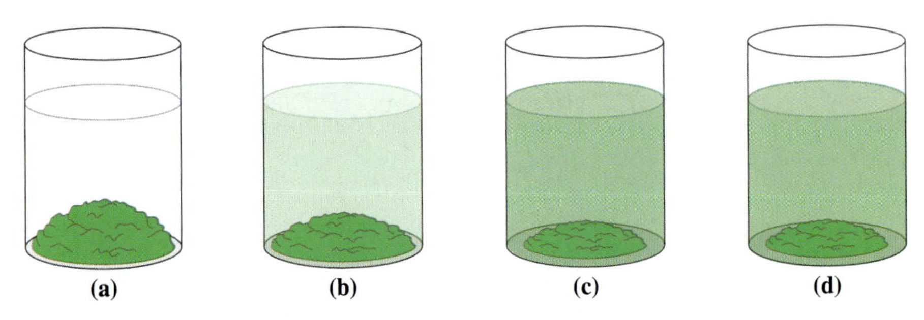

▲ **Figure 12.9 Formation of a saturated solution**
(a) A solid solute is added to a fixed quantity of water. **(b)** After a few minutes, the solution is colored due to the dissolved solute, and there is less undissolved solute than in (a). **(c)** After a longer time, the solution color has deepened, and the quantity of undissolved solute is further diminished from that in (b). The solution in (b) must be unsaturated. **(d)** Still later, the solution color and the quantity of undissolved solute appear to be the same as that in (c). Dynamic equilibrium must have been attained in (c) and persists in (d). In both (c) and (d), the solution is saturated.

When dynamic equilibrium is established between an undissolved solute and a solution, the solution is **saturated**. We refer to the concentration of the saturated solution as the **solubility** of the solute. The solubility of NaCl in water at 20 °C is 36 g NaCl/100 g H_2O. Any solution containing less solute than can be held at equilibrium is **unsaturated**. At 20 °C, a solution with 24 g NaCl/100 g H_2O is unsaturated, as is one with 32 g NaCl/100 g H_2O.

The terms unsaturated and saturated are not related to whether a solution is dilute or concentrated. For example, a saturated aqueous solution of $Ca(OH)_2$ at 20 °C is 0.023 M, quite a dilute solution. In contrast, at 20 °C, a 10 M NaOH solution, though quite concentrated, is still unsaturated.

Solubility as a Function of Temperature

Solubility varies with temperature, so when we cite solubility data, we should indicate the temperature. Later in the text, we will comment further on the relationship of solubility to temperature, but for now, let's just use this generalization regarding the solubilities of ionic compounds in water:

> *About 95% of all ionic compounds have aqueous solubilities that increase significantly with temperature. Most of the remainder have solubilities that change little with temperature. A very few—for example, some sulfates—even have solubilities that decrease with temperature.*

A graph of solubility as a function of temperature is called a **solubility curve**. Figure 12.10 shows the solubility curves of several ionic compounds in water in a unit often used in reference books: g solute/100 g solvent.

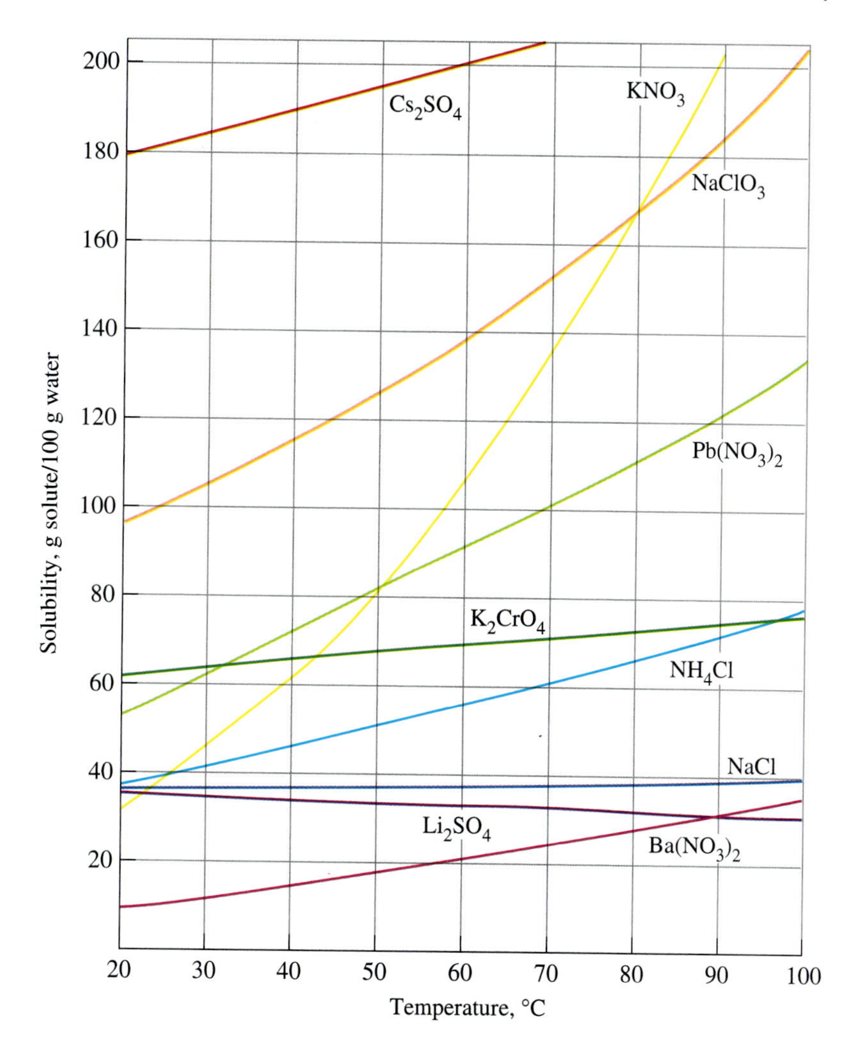

Figure 12.10 **The solubility curves for several salts in water**

If we cool a saturated solution of lead nitrate in contact with excess $Pb(NO_3)_2(s)$, some solute crystallizes from solution until equilibrium is once again established at the lower temperature. For example, referring to Figure 12.10, consider a saturated solution of lead nitrate at 90 °C. The solution contains 122 g of $Pb(NO_3)_2$ for each 100 g water. If we cool the solution to 20 °C, the solution can contain only 54 g $Pb(NO_3)_2$ per 100 g water. The excess of 68 g crystallizes out.

Now consider what would happen if we started to cool a saturated solution of lead nitrate with no excess solute present. Would lead nitrate crystallize? It should. Then again, there is no dynamic equilibrium because there are no crystals to capture wandering ions. If we are able to cool the solution without the occurrence of crystallization, it will contain more solute than it would contain at equilibrium. The solution is **supersaturated**. Supersaturated solutions are unstable, and addition of a "seed" crystal of solute will generally cause all of the excess solute to suddenly crystallize (Figure 12.11).

Crystallizing a solute from a concentrated solution by lowering the temperature can be an effective method of purifying a substance. In this technique, the impure solute is dissolved in a minimum quantity of a hot solvent. The concentrated solution is then cooled to a temperature at which the solubility of the solute is much smaller than in the hot solvent. The excess solute crystallizes from solution, but the impurities, being present in much smaller quantities, remain in solution. The

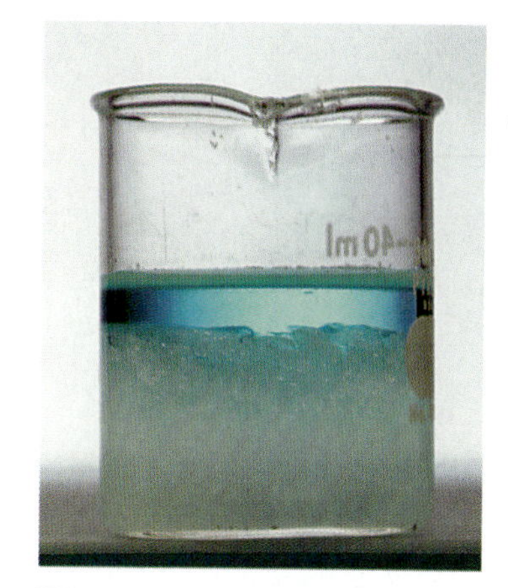

When $KNO_3(s)$ is crystallized from an aqueous solution of KNO_3 containing $CuSO_4$ as an impurity, $CuSO_4$ (blue) remains in the solution.

Some Supersaturated Solutions in Nature

Supersaturated solutions are not just laboratory curiosities; they occur naturally. Honey is one example; the principal solute is the sugar glucose, and the solvent is water. If honey is left to stand, the glucose eventually crystallizes. We say, not very scientifically, that the honey has "turned to sugar." Supersaturated sucrose (cane sugar) solutions are fairly common in cooking. Jellies are one example. Sucrose often crystallizes from jelly that has been standing for a long time.

Some wines have high concentrations of potassium hydrogen tartrate, $KHC_4H_4O_6$. When chilled, the solution may become supersaturated. After a time, crystals may form and settle out if the wine is stored in a refrigerator. Modern wineries solve this problem—and render the wine less acidic—by a process known as cold stabilization. The wine is chilled to near 0 °C, a temperature below usual refrigerator temperatures. Tiny seed crystals of $KHC_4H_4O_6$ are added to the supersaturated wine. Crystallization is complete after a period of time, and the excess crystals are filtered off. At one time, winemaking was the principal source of potassium hydrogen tartrate, the cream of tartar used in baking.

Imagine lunch without the supersaturated solution we call jelly!

Figure 12.11 Seeding a supersaturated solution
A tiny seed crystal is added to a supersaturated solution of sodium acetate (left). Crystallization begins (center) and continues until all the excess solute has crystallized from the solution (right).

crystals are filtered off and washed with small quantities of the cold solvent to remove any adhering solution.

12.5 The Solubilities of Gases

Solutions of gases in liquids, especially in water, are quite common. Consider the familiar case of carbonated beverages. All are solutions of $CO_2(g)$ in water, usually with added flavors and sweeteners. Other examples of solutions of gases include blood, which contains dissolved $O_2(g)$ and $CO_2(g)$; formalin, an aqueous solution of formaldehyde gas (HCHO) that is used as a biological preservative; and a variety of household cleaners that are aqueous solutions of $NH_3(g)$. In addition, *all* natural waters contain dissolved $O_2(g)$ and $N_2(g)$ and traces of other gases.

As with other solutes, the solubilities of gases depend on temperature, but they depend even more significantly on pressure.

The Effect of Temperature

It is difficult to make satisfactory generalizations about the effect of temperature on the solubilities of gases in liquids. Most gases become *less* soluble in water as the temperature increases. However, the situation is often the reverse for gases in organic solvents: Gas solubilities tend to increase with increased temperature.

The solubility of air in water at standard atmospheric pressure and different temperatures is shown in graphic form in Figure 12.12. Nitrogen and oxygen together make up about 99% of air. Note that the solubility curve for air (black) is essentially the sum of the curves for N_2 in air (red) and O_2 in air (blue). Although oxygen makes up only about 23% of air by mass, it constitutes about 35% of the air dissolved in water because $O_2(g)$ is more water soluble than is $N_2(g)$.

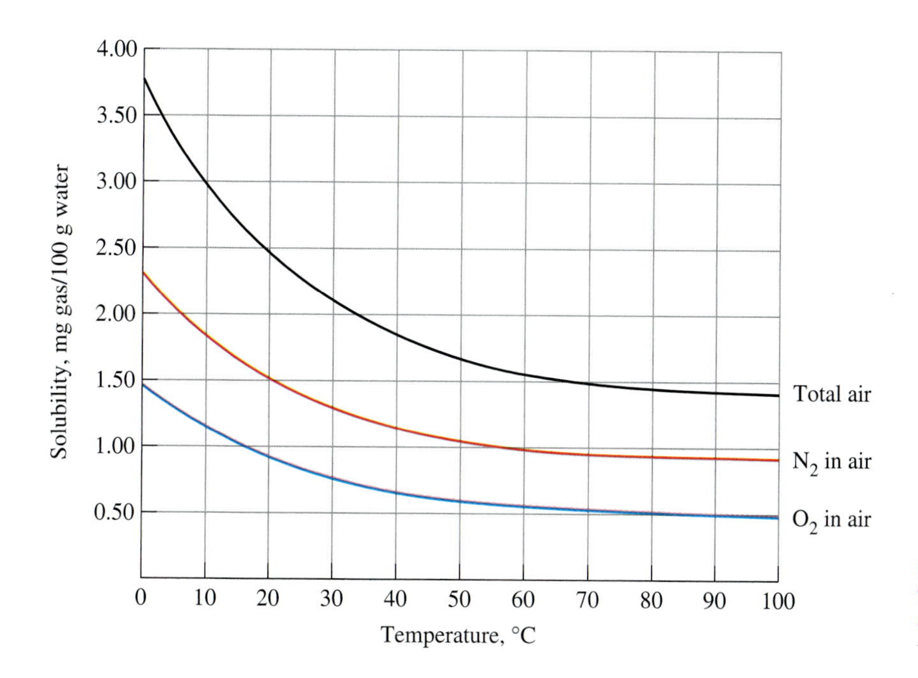

◀ **Figure 12.12 Solubility of air in water as a function of temperature at 1 atm pressure**

The solubility of air in water, as limited as it is, is essential to aquatic life. Fish depend on dissolved air for $O_2(g)$. Moreover, the fact that the solubility *decreases* with temperature explains why many fish (trout, for example) cannot survive in warm water; they don't get enough oxygen. At 30 °C, the amount of dissolved $O_2(g)$ in water is only about one-half of that at 0 °C.

APPLICATIONS NOTE

The world's major ocean fisheries are located in *cold* regions such as the Bering Sea off the coast of Alaska and the Grand Banks off Newfoundland.

The Effect of Pressure

At a constant temperature, the solubility (S) of a gas is directly proportional to the pressure of the gas (P_{gas}) in equilibrium with the solution. Thus, doubling the pressure of a gas doubles its solubility; a tenfold increase in pressure produces a tenfold increase in solubility; and so on. This relationship is expressed through the following equation of a straight line, having the slope k.

$$S = k\,P_{gas}$$

The value of k depends on the particular gas and the solvent.

● **ChemCDX** Henry's Law

The effect of pressure on the solubility of a gas was stated by William Henry in 1803 and is known as **Henry's law**. Figure 12.13 suggests a way to rationalize Henry's law: To dissolve in a liquid, gas molecules must first collide with the liquid surface. Increasing the pressure of a gas in contact with a saturated solution in-

▲ **Figure 12.13**
The effect of pressure on the solubility of a gas
As the gas is compressed into a smaller volume, increasing the number of molecules per unit volume, the number of dissolved molecules per unit volume—the concentration of the solution—also increases.

creases the number of molecules per unit volume in the gas. This increases the collision rate with the liquid surface. More gas molecules dissolve and their concentration in solution increases. However, the number of molecules returning to the gaseous state from this more concentrated solution is also increased. In the new dynamic equilibrium at the higher gas pressure, the concentrations of molecules in both the gas and solution phases are greater than in the original saturated solution. The solubility increases with increased gas pressure. If a gaseous solute reacts with the solvent, as do HCl(g) and NH_3(g), the situation is more complicated, and Henry's law may not accurately describe the solubility of the gas.

The straight-line plots in Figure 12.14 represent the solubilities in water of different gases as a function of gas pressure. We use data from this figure in the application of Henry's law in Example 12.9.

EXAMPLE 12.9

A 225-g sample of pure water is shaken with air under a pressure of 0.95 atm at 20 °C. How many milligrams of Ar(g) will be present in the water when solubility equilibrium is reached? Use data from Figure 12.14 and the fact that the mole fraction of Ar in air is 0.00934.

SOLUTION

Let's plan an approach in which we use the Henry's law equation, $S = k\, P_{gas}$. To do so, we need values of P_{gas} and k. We can get P_{gas} by using a relationship from Chapter 5 (page 206); we can get k from data in Figure 12.14.

Determining P_{gas}:
The mole fraction and partial pressure of argon are related by this expression for mixtures of gases.

$$x_{Ar} = \frac{P_{Ar}}{P_{total}} = \frac{P_{Ar}}{P_{air}} = \frac{P_{Ar}}{0.95\ \text{atm}} = 0.00934$$

$$P_{Ar} = 0.0089\ \text{atm}$$

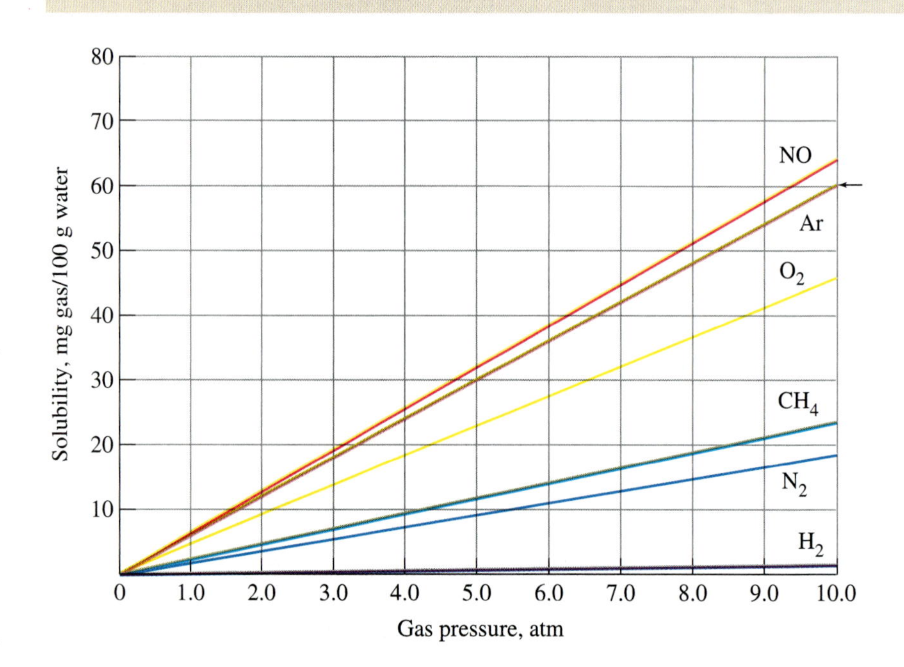

▶ **Figure 12.14**
The effect of gas pressure on aqueous solubilities of gases at 20 °C
We refer to the small arrow in Example 12.9.

Determining k:
To derive a value of k, we can choose a point on the argon solubility curve in Figure 12.14 for which we can read the relevant data with some precision. Let's choose the point identified by the arrow: $P_{Ar} = 10.0$ atm and $S = 60$ mg Ar/100 g water. We can solve Henry's law equation for the constant k, and substitute these data into the equation.

$$k = \frac{S}{P_{gas}} = \frac{S}{P_{Ar}} = \frac{60 \text{ mg Ar}/100 \text{ g water}}{10.0 \text{ atm}} = \frac{0.060 \text{ mg Ar}}{\text{g water atm}}$$

Determining the solubility of argon at 0.0089 atm:
Now that we have values for k and P_{Ar}, we can solve for the solubility, S.

$$S = k\,P_{Ar} = \frac{0.060 \text{ mg Ar}}{\text{g water } \cancel{\text{atm}}} \times 0.0089 \cancel{\text{ atm}} = \frac{5.3 \times 10^{-4} \text{ mg Ar}}{\text{g water}}$$

Determining the mass of argon in 225 g of water:
The value of S just derived is expressed "per gram of water." We just need to multiply S by the mass of water, 225 g.

$$\text{Mass of Ar} = 225 \cancel{\text{ g water}} \times \frac{5.3 \times 10^{-4} \text{ mg Ar}}{\cancel{\text{g water}}} = 0.12 \text{ mg Ar}$$

EXERCISE 12.9A

At 25 °C and 1 atm gas pressure, the solubility of $CO_2(g)$ is 149 mg/100 g water. When air at 25 °C and 1 atm is in equilibrium with water, what is the quantity of dissolved CO_2, in mg/100 g water? Air contains 0.036 mole % $CO_2(g)$.

EXERCISE 12.9B

How many milligrams of a mixture containing equal numbers of moles of CH_4 and N_2 at 10.0 atm total pressure will dissolve in 1.00 L of water at 20 °C?

A moderate pressure of $CO_2(g)$ above the beverage in a soft-drink bottle (right) keeps a significant quantity of the gas dissolved in the water. When the bottle is opened (left), this pressure is released and dissolved $CO_2(g)$ escapes, causing the familiar fizzing.

Colligative Properties

In the next four sections, we will consider **colligative properties**, physical properties of solutions that depend on the *number* of solute particles present but *not* on the identity of the solute. For example, aqueous 0.10 m sucrose ($C_{12}H_{12}O_{11}$) and 0.10 m urea $[CO(NH_2)_2]$ have the same number of molecules (0.10 mol) in the same mass of solvent (1.00 kg H_2O). The solute molecules are quite different in composition and structure, but the solutions have the same colligative properties, such as freezing point and boiling point. All of the solutes and solvents in the solutions that we describe in Sections 12.6, 12.7, and 12.8 exist in molecular form. That is, they are solutions of *nonelectrolytes*. In Section 12.9, we will consider solutions in which the solutes are electrolytes.

12.6 Vapor Pressures of Solutions

In 1887, F. M. Raoult reported the results of his studies on the vapor pressures of solutions. He found that the presence of a solute *lowers* the vapor pressure of the solvent in a solution. We begin with a modern statement of **Raoult's law**.

Deep-Sea Diving: Applications of Henry's Law

People who dive deeply into the sea must carry their own supply of air. They normally breathe about 800 L of air per hour. To have enough air for an hour or so of underwater exploring, divers carry air that is compressed to a much smaller volume. Moreover, high-pressure air is needed to keep the lungs inflated at the high pressures that prevail under depths of water. As we should expect from Henry's law, compressed air is much more soluble in blood and other body fluids than is air at normal pressures. One of the dissolved gases, nitrogen, can cause two kinds of problems: nitrogen narcosis and decompression sickness.

Nitrogen narcosis is experienced by divers breathing air at depths of 100 ft (30 m) or more. Ordinarily, the nitrogen we breathe in air has no physiological effect, but at the high pressures at these depths, nitrogen acts as a narcotic. It often produces pleasurable sensations similar to those of narcotics such as morphine, accounting for the alternate name, "rapture of the deep." Nitrogen narcosis impairs judgment and often causes serious diving accidents. Divers can minimize the risk by using a compressed helium-oxygen mixture as a substitute for compressed air. Helium is considerably less soluble than is nitrogen. Its absorption into the bloodstream is quite limited, and it has no narcotic effect. Excess oxygen presents no problem because it is consumed in metabolism.

Decompression sickness, caused by bubbles of nitrogen gas in the blood and other tissues, occurs when a person is exposed to a sudden drop in atmospheric pressure. Tiny bubbles block blood flow in capillaries, and the person suffers severe pains in joints and muscles that often cause the victim to curl up, a reaction that explains the common name of the illness, "the bends." Divers must be careful to return to the surface slowly or to spend considerable time in a decompression chamber where the pressure is gradually lowered. If they don't, excess $N_2(g)$ quickly escapes from its solution in blood as bubbles that can block blood flow through capillaries. The divers may faint or suffer deafness, paralysis, or even death. People who work in deep mines or tunnels, where compressed air is used to keep water from infiltrating, have similar problems, as do passengers in an airplane at high altitude that suddenly loses pressure.

If the return to normal pressures is slow enough, the excess gases leave the blood gradually. Excess O_2 is used in metabolism, and excess N_2 is removed to the lungs and expelled by normal breathing. For each atmosphere of pressure above normal that the diver experiences, about 20 minutes of slow decompression is needed.

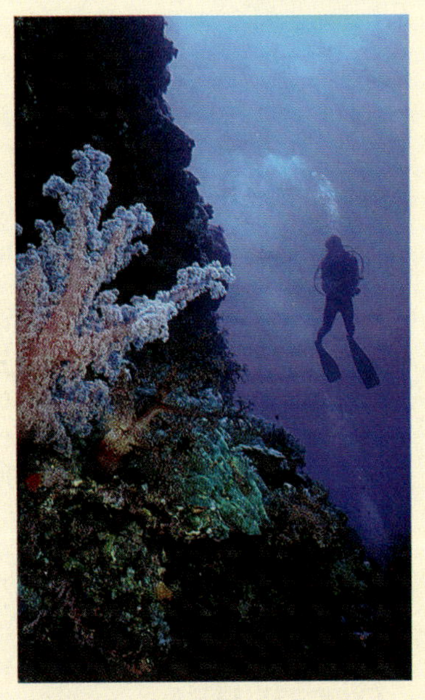

Underwater divers who surface too quickly may experience decompression sickness, a painful and dangerous condition also known as "the bends."

The vapor pressure of the solvent above a solution (P_{solv}) is the product of the vapor pressure of the pure solvent (P°_{solv}) and the mole fraction of the solvent in the solution (x_{solv}):

$$P_{solv} = x_{solv} \cdot P^\circ_{solv}$$

Note that this expression conforms to Raoult's observation of the lowering of vapor pressure. The presence of a solute in a solvent means that x_{solv} is less than one, and consequently P_{solv} is less than P°_{solv}. Thus, the solute has lowered the vapor pressure of the solvent. Raoult's law is strictly followed only in an *ideal* solution, but it often works well for nonideal solutions that are sufficiently *dilute*. A solution in which $x_{solv} > 0.95$, for example, might meet this requirement. Although various "molecular" explanations of Raoult's law have been given, we will present a more satisfactory explanation in Chapter 17.

EXAMPLE 12.10

The vapor pressure of pure water at 20.0 °C is 17.5 mmHg. What is the vapor pressure of water above a solution of 1.00 *m* $CO(NH_2)_2$ (urea) at 20.0 °C?

SOLUTION

We have one of the terms required in Raoult's law, $P^{\circ}_{solv} = 17.5$ mmHg. The other term we need is $x_{solv} = x_{H_2O}$. To get x_{H_2O}, we need to know the number of moles of both the water and urea in a solution that is 1.00 *m* $CO(NH_2)_2$. The proportions of solute and solvent in that solution are as follows.

$$1.00 \text{ mol } CO(NH_2)_2 \quad \text{and} \quad 1 \text{ kg } H_2O = 1000 \text{ g } H_2O$$

We can convert from mass to number of moles of water in the usual way.

$$? \text{ mol } H_2O = 1000 \text{ g } H_2O \times \frac{1 \text{ mol } H_2O}{18.02 \text{ g } H_2O} = 55.5 \text{ mol } H_2O$$

Now we can determine the mole fraction of water.

$$x_{H_2O} = \frac{55.5 \text{ mol } H_2O}{1.00 \text{ mol } CO(NH_2)_2 + 55.5 \text{ mol } H_2O} = 0.982$$

Finally, we use Raoult's law to calculate the vapor pressure of water above the solution.

$$P_{H_2O} = x_{H_2O} \cdot P^{\circ}_{H_2O} = 0.982 \times 17.5 \text{ mmHg} = 17.2 \text{ mmHg}$$

EXERCISE 12.10

The vapor pressure of benzene (C_6H_6) at 25 °C is 95.1 mmHg. What is the vapor pressure of benzene above a solution in which 5.05 g C_6H_5COOH (benzoic acid) is dissolved in 245 g C_6H_6?

Raoult's law is not limited to the solvent if a solution also contains volatile *solutes*. We can write a Raoult's law expression for *any* volatile component in a solution, whether it is the solvent or a solute, as long as the solution is nearly ideal. The benzene-toluene solution in Examples 12.11 and 12.12 is an essentially ideal solution in which both components are volatile. In Example 12.11, we calculate the vapor pressures of the two components in the solution, and in Example 12.12, we calculate the composition of the vapor in equilibrium with the solution.

EXAMPLE 12.11

The vapor pressures of pure benzene (C_6H_6) and toluene (C_7H_8) at 25 °C are 95.1 and 28.4 mmHg, respectively. A solution is prepared that has equal mole fractions of C_6H_6 and C_7H_8. Determine the vapor pressures of C_6H_6 and C_7H_8 and the total vapor pressure above this solution. Consider the solution to be ideal.

SOLUTION

We need to apply Raoult's law twice, once for benzene and once for toluene. We are given the vapor pressures of the pure liquids, and we are also told that their mole fractions in solution are equal. Because $x_{C_6H_6} = x_{C_7H_8}$ and because $x_{C_6H_6} + x_{C_7H_8} = 1$, it follows that $x_{C_6H_6} = x_{C_7H_8} = 0.500$. Now we can apply Raoult's law.

$$P_{\text{benzene}} = x_{\text{benzene}} P^{\circ}_{\text{benzene}} = 0.500 \times 95.1 \text{ mmHg} = 47.6 \text{ mmHg}$$

$$P_{\text{toluene}} = x_{\text{toluene}} P^{\circ}_{\text{toluene}} = 0.500 \times 28.4 \text{ mmHg} = 14.2 \text{ mmHg}$$

$$P_{\text{total}} = P_{\text{benzene}} + P_{\text{toluene}} = 47.6 \text{ mmHg} + 14.2 \text{ mmHg} = 61.8 \text{ mmHg}$$

PROBLEM-SOLVING NOTE
It doesn't matter what particular mass of the components you choose to work with.

EXERCISE 12.11

What are the partial pressures of benzene and toluene and the total pressure at 25 °C above a solution containing equal *masses* of the two components?

EXAMPLE 12.12

What is the composition, expressed as mole fractions, of the vapor in equilibrium with the benzene-toluene solution of Example 12.11?

SOLUTION

The vapor is a mixture of gaseous benzene and toluene. We can use the following basic expression from Chapter 5 (page 206) that describes a mixture of gases.

$$x_i = \frac{n_i}{n_{\text{total}}} = \frac{P_i}{P_{\text{total}}}$$

We can get the partial pressures of the two gases (P_i) and the total pressure (P_{total}) directly from Example 12.11.

$$x_{\text{benzene vapor}} = \frac{47.6 \cancel{\text{mmHg}}}{61.8 \cancel{\text{mmHg}}} = 0.770$$

$$x_{\text{toluene vapor}} = \frac{14.2 \cancel{\text{mmHg}}}{61.8 \cancel{\text{mmHg}}} = 0.230$$

Note that, as expected, $x_{\text{benzene vapor}} + x_{\text{toluene vapor}} = 0.770 + 0.230 = 1.000$

EXERCISE 12.12

Above which of these solutions does the vapor have the greater mole fraction of benzene at 25 °C: a solution with equal masses of benzene and toluene, or a solution with equal numbers of moles of benzene and toluene?

(*Hint:* You will need information from Examples 12.11 and 12.12, but you do not need to do detailed calculations.)

Consider the following: Pure benzene, C_6H_6, has a higher vapor pressure (95.1 mmHg at 25 °C) than does pure toluene, C_7H_8 (28.4 mmHg at 25 °C). In benzene-toluene solutions, benzene is the more volatile component. In Example 12.11, we started with a benzene-toluene solution that had $x_{C_6H_6} = 0.500$. In Example 12.12, we found that the vapor in equilibrium with that solution had $x_{C_6H_6} = 0.770$. These observations lead to an important idea about liquid-vapor equilibrium in a solution of two volatile components.

The vapor in equilibrium with an ideal solution of two volatile components has a higher mole fraction of the more volatile component than is found in the liquid.

If vapor having the mole fraction $x_{C_6H_6} = 0.770$ were condensed to a liquid and again vaporized, the new vapor would be richer still in benzene. (It would have $x_{C_6H_6} = 0.918$.) In a series of vaporizations and condensations, it should be possible ultimately to obtain pure benzene from the vapor. In practice, rather than to carry out the vaporizations and condensations at a constant temperature (here, 25 °C), we would use a constant pressure (atmospheric pressure) and boil the benzene-toluene solution in an apparatus like that shown in Figure 12.15. In this method, called **fractional distillation**, pure benzene would be condensed from the vapor at the top of the column, and the less volatile, pure toluene would be retained as a liquid in the warmer flask at the bottom of the column.

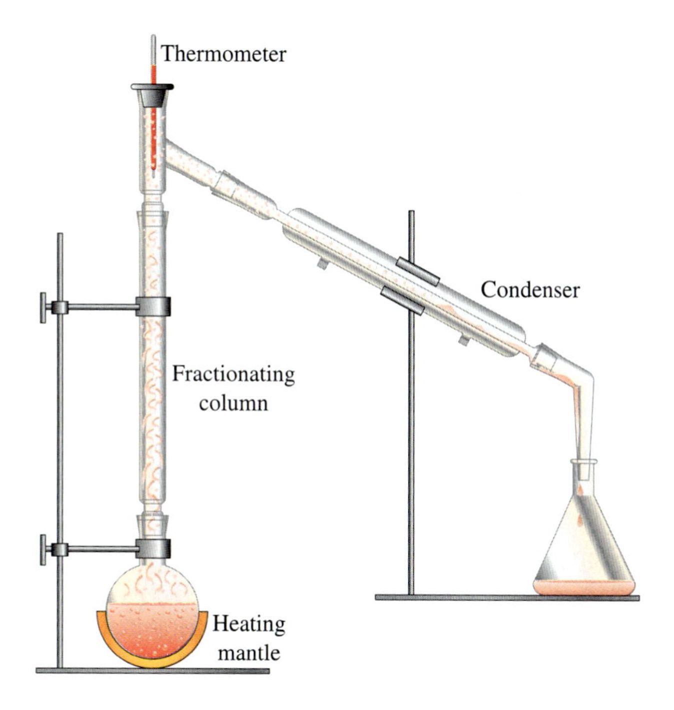

Figure 12.15
Fractional distillation
The upright column is packed with an inert material such as glass beads. As the solution boils, vapor rises up the column and condenses on the cooler beads. As more hot vapor rises, it heats the beads and revaporizes the condensed liquid. This vapor then condenses farther up the column. The process is repeated many times as the vapor moves to the top of the column, becoming successively richer in the more volatile, lower-boiling component. If the column is tall enough, the pure lower-boiling substance will leave the top of the column and enter the water-cooled condenser, where it is converted to a liquid and collected in a receiving flask.

EXAMPLE 12.13—A Conceptual Example

Figure 12.16 shows two different *aqueous* solutions placed in the same enclosure. After a time, the solution level has risen in container A and dropped in container B. Explain how and why this happens.

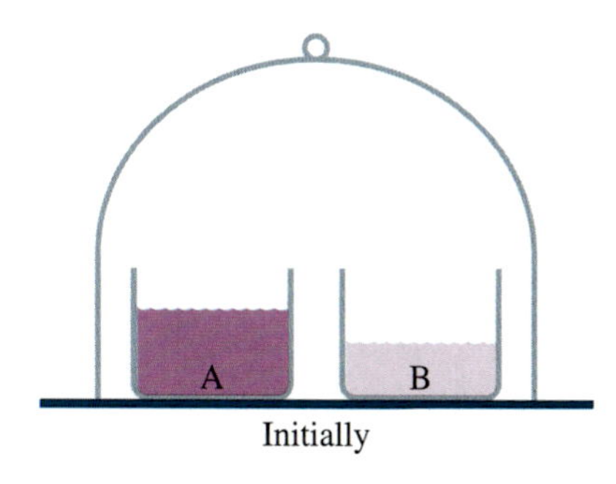

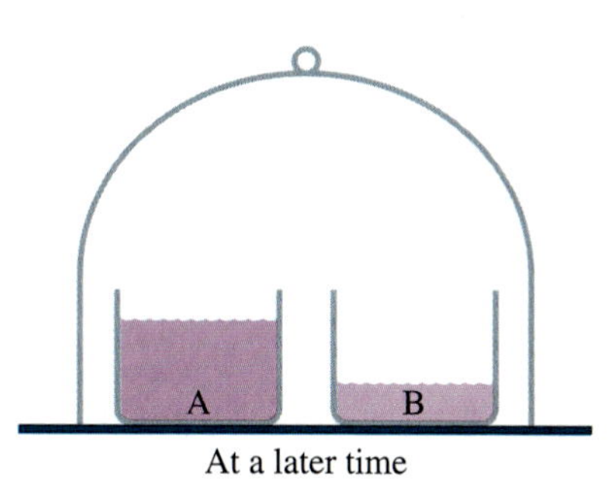

Figure 12.16
A phenomenon related to the vapor pressures of solutions

SOLUTION

The only way that water can move from the one container to the other is through the vapor. There must be a net vaporization from container B and a net condensation into container A. This requires that the vapor pressure of H_2O above solution B be greater than that above solution A. According to Raoult's law, the vapor pressure of water above both solutions is given by the following expression.

$$P_{H_2O} = x_{H_2O} \cdot P^{\circ}_{H_2O}$$

Thus, the mole fraction of water must be greater in the solution in container B than in the solution in container A. The solution in B must be more dilute. Water passes, as a vapor, from the more dilute to the more concentrated solution.

It doesn't matter what the solute is in solutions A and B; the solutions can even have *different* solutes. As long as the solutions are dilute and have the same solvent (water), vapor pressure lowering depends only on the *concentration* of the solute and not on other characteristics of the solute molecules. Vapor pressure lowering is a prime example of a colligative property.

EXERCISE 12.13

Will the process depicted in Figure 12.16 continue until the solution in container B evaporates to dryness? Explain.

12.7 Freezing Point Depression and Boiling Point Elevation

In this section, we focus on solutions with a volatile solvent and a solute that is: (a) *nonvolatile*; (b) a *nonelectrolyte*; and (c) soluble in the liquid solvent but not in the solid solvent. In solutions with these characteristics, the vapor pressure of the solution is simply that of the solvent in the solution, and at all temperatures, this vapor pressure is *lower* than that of the pure solvent. This statement is illustrated in Figure 12.17, in which a partial phase diagram for the solution (in red) is superimposed onto the phase diagram of the pure solvent (in blue).

Let's examine Figure 12.17. The dotted line at $P = 1$ atm intersects the fusion curve (red) of the solution at a *lower* temperature (fp) than it does the fusion curve

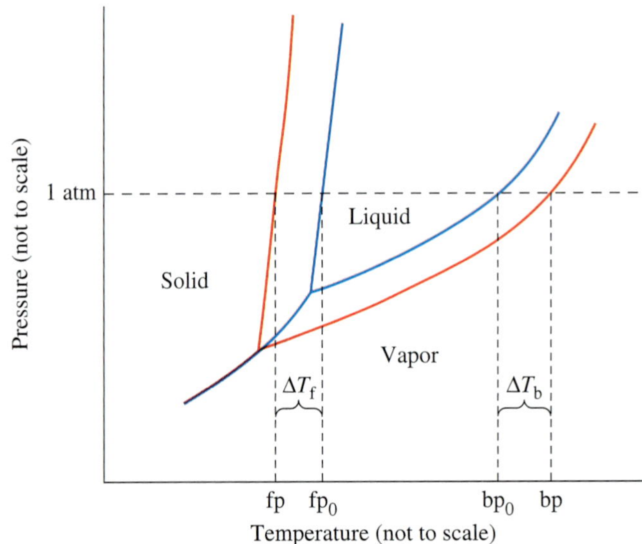

Figure 12.17
Vapor pressure lowering by a nonvolatile solute
In a solution, the vapor pressure of the solvent is lowered and the fusion curve is displaced to lower temperatures (red curves). As a consequence, the freezing point of the solvent is lowered by the amount ΔT_f, and the boiling point is raised by ΔT_b. Because the solute is assumed to be insoluble in the solid solvent, the sublimation curve of the solid solvent is unaffected by the presence of a solute in the liquid solution.

(blue) of the pure solvent (fp_0). The presence of the solute lowers or *depresses* the freezing point by the amount ΔT_f. The dotted line at $P = 1$ atm intersects the vapor pressure curve (red) for the solution at a *higher* temperature (bp) than it does the vapor pressure curve of the pure solvent (bp_0). The presence of a solute raises, or *elevates*, the boiling point by the amount ΔT_b.

The extent of the vapor pressure lowering depends on the *mole fraction* of solute present in a solution. So do the freezing point depression and boiling point elevation. However, in *dilute* solutions, the molality of solute is proportional to its mole fraction, and because of this we can write

In *dilute aqueous* solutions, molarity and molality are nearly equal and *both* are proportional to the mole fraction of solute.

$$\Delta T_f = T_f(\text{solution}) - T_f(\text{solvent}) = -K_f \times m$$

$$\Delta T_b = T_b(\text{solution}) - T_b(\text{solvent}) = K_b \times m$$

In these equations, ΔT_f and ΔT_b are the *freezing point depression* and *boiling point elevation*, respectively, and m is the molality of solute. K_f and K_b are proportionality constants. The constants K_f and K_b represent the freezing point depression and boiling point elevation when a solution is one *molal* (1 m). (For example, $\Delta T_b = K_b \times m = K_b \times 1 = K_b$.) For this reason, K_f and K_b are often referred to as the molal freezing point depression constant and the molal boiling point elevation constant, respectively (see Table 12.2)* Notice that the freezing point depression equation has a minus sign because ΔT_f must be a negative quantity.

TABLE 12.2 Molal Freezing Point Depression and Molal Boiling Point Elevation Constants

Solvent	Normal Freezing Point, °C	K_f, °C m^{-1}	Normal Boiling Point, °C	K_b °C m^{-1}
Acetic acid	16.63	3.90	117.90	3.07
Benzene	5.53	5.12	80.10	2.53
Cyclohexane	6.55	20.0	80.74	2.79
Nitrobenzene	5.8	8.1	210.8	5.24
Water	0.00	1.86	100.00	0.512

For the solutions we are describing, only the solvent freezes, and only the solvent escapes as vapor during boiling; the solute(s) remains in solution. As a result, the solute concentration increases as freezing or boiling occur. This causes the freezing point to drop still more and the boiling point to rise higher. The freezing and boiling points of solutions are *not* constant, as they are for pure liquids. When we speak of the freezing point of a solution, we mean the temperature at which the first bit of solid freezes from solution. In the case of boiling point, we mean the temperature at which boiling first occurs.

EXAMPLE 12.14

What is the freezing point of an aqueous sucrose solution that has 25.0 g $C_{12}H_{22}O_{11}$ per 100.0 g H_2O?

SOLUTION

The freezing point of the solution is that of pure water (0.00 °C) plus the freezing point depression (ΔT_f). We need to find ΔT_f. We can get a value of K_f for water from Table

* These constants are also called the *cryoscopic* (K_f) and *ebullioscopic* (K_b) constants.

12.2, and so our main problem is to determine the molality of the solution. We can proceed as we did in Example 12.4.

First, let's determine the number of moles of sucrose in 25.0 g.

$$? \text{ mol } C_{12}H_{22}O_{11} = 25.0 \text{ g } C_{12}H_{22}O_{11} \times \frac{1 \text{ mol } C_{12}H_2O_{11}}{342.3 \text{ g } C_{12}H_2O_{11}} = 0.0730 \text{ mol } C_{12}H_{22}O_{11}$$

The quantity of H_2O is 100.0 g = 0.100 kg, and the molality of the sucrose is

$$? \, m = \frac{0.0730 \text{ mol } C_{12}H_{22}O_{11}}{0.1000 \text{ kg } H_2O} = 0.730 \, m$$

We can use this molality to determine the depression of the freezing point, ΔT_f.

$$\Delta T_f = -K_f \times m = -1.86 \text{ °C } m^{-1} \times 0.730 \, m = -1.36 \text{ °C}$$

To obtain the freezing point of the solution, we add the freezing point depression, ΔT_f, to the freezing point of the pure solvent, fp_0.

$$fp_0 + \Delta T_f = 0.00 \text{ °C} - 1.36 \text{ °C} = -1.36 \text{ °C}$$

EXERCISE 12.14A

What is the freezing point of a solution in which 10.0 g of naphthalene $[C_{10}H_8(s)]$ is dissolved in 50.0 g of benzene $[C_6H_6(l)]$?

EXERCISE 12.14B

What mass of sucrose, $C_{12}H_{22}O_{11}$, should be added to 75.0 g H_2O to raise the boiling point to 100.35 °C?

(*Hint:* What must the molality of the solution be? What mass of sucrose is required to produce a solution of this molality? Use data from Table 12.2.)

EXAMPLE 12.15

Sorbitol is a sweet substance found in fruits and berries and sometimes used as a sugar substitute. An aqueous solution containing 1.00 g sorbitol in 100.0 g water is found to have a freezing point of −0.102 °C.

a. What is the molar mass of sorbitol?

b. Elemental analysis indicates that sorbitol consists of 39.56% C, 7.75% H, and 52.70% O by mass. What is the molecular formula of sorbitol?

SOLUTION

a. We can use the freezing point depression equation to determine the molality of the sorbitol solution. Once we have the molality, we can determine the number of moles of sorbitol. And from the number of moles of sorbitol and its mass (1.00 g), we can get the molar mass.

The freezing point depression is the difference between the freezing point of the solution and that of the pure solvent.

$$\Delta T_f = -0.102 \text{ °C} - 0.000 \text{ °C} = -0.102 \text{ °C}$$

We can get the molality of the solution from the equation, $\Delta T_f = -K_f m$. That is,

$$\text{Molality} = \frac{\Delta T_f}{-K_f} = \frac{-0.102 \text{ °C}}{-1.86 \text{ °C } m^{-1}} = 0.0548 \, m$$

Now we use the molality to determine the number of moles of sorbitol in the solution.

$$? \text{ mol sorbitol} = 0.1000 \text{ kg } H_2O \times \frac{0.0548 \text{ mol sorbitol}}{\text{kg } H_2O} = 0.00548 \text{ mol sorbitol}$$

The molar mass is simply the ratio of the number of grams of sorbitol to the number of moles of sorbitol in the same sample.

$$\text{Molar mass} = \frac{1.00 \text{ g sorbitol}}{0.00548 \text{ mol sorbitol}} = 182 \text{ g/mol sorbitol}$$

b. We can determine the molecular formula from the mass percent data and the molar mass by using the methods of Examples 3.9 and 3.11. Let's consider an alternative approach that works equally well. Start with a one-mole sample of sorbitol (182 g), and use the mass percent data to determine the mass of each element in the sample. Convert these masses to amounts in moles, and use those amounts as the subscripts in a molecular formula.

$$? \text{ mol C} = 182 \text{ g sorbitol} \times \frac{39.56 \text{ g C}}{100 \text{ g sorbitol}} \times \frac{1 \text{ mol C}}{12.01 \text{ g C}} = 5.99 \text{ mol C}$$

$$? \text{ mol H} = 182 \text{ g sorbitol} \times \frac{7.75 \text{ g H}}{100 \text{ g sorbitol}} \times \frac{1 \text{ mol H}}{1.008 \text{ g H}} = 13.99 \text{ mol H}$$

$$? \text{ mol O} = 182 \text{ g sorbitol} \times \frac{52.70 \text{ g O}}{100 \text{ g sorbitol}} \times \frac{1 \text{ mol O}}{16.00 \text{ g O}} = 5.99 \text{ mol O}$$

The molecular formula of sorbitol is $C_6H_{14}O_6$.

EXERCISE 12.15

A 1.065-g sample of an unknown substance is dissolved in 30.00 g of benzene; the freezing point of the solution is 4.25 °C. The compound is found to consist of 50.69% C, 4.23% H, and 45.08% O by mass. Use these data and data from Table 12.2 to determine the molecular formula of the substance.

Freezing point depression and boiling point elevation have practical applications. Both are involved, for example, in the action of an antifreeze such as ethylene glycol ($HOCH_2CH_2OH$) in an automotive cooling system. If water alone were used as the engine coolant, it might boil over when the engine overheats in hot weather, and it might freeze in the depths of northern winters. Addition of antifreeze raises the boiling point of the coolant and also lowers its freezing point. With the proper ratio of ethylene glycol and water, it is possible to protect an automotive cooling system to temperatures as low as −48 °C, and from boil-over to temperatures of about 113 °C.

Citrus growers are keenly aware of the role that freezing point depression plays in protecting their crop in cold weather. If the temperature drops a degree or two below the freezing point of water for just a few hours, citrus fruit is well protected from freezing. Solutes in the fruit juice, mainly sugars, lower the freezing point of the juice below 0 °C. Also, growers know that lemons are more prone to freezing than oranges because lemon juice is less concentrated in solutes than is orange juice.

Salts such as NaCl and $CaCl_2$ are scattered on icy sidewalks and streets to melt ice. The ice melts as long as the outdoor temperature is above the lowest freezing point of a salt-and-water mixture. Below this temperature, the ice will not melt.

Alcohols whose molecules contain multiple hydroxy groups (OH) are commonly used for de-icing aircraft.

Sodium chloride can melt ice at temperatures as low as −21 °C (−6 °F), but no lower. Calcium chloride is effective at temperatures as low as −55 °C (−67 °F). We will shortly see why $CaCl_2$ is more effective than NaCl at lowering the freezing point (Section 12.9).

12.8 Osmotic Pressure

Everyday experience tells us that we can separate coffee grounds from brewed coffee by passing the mixture through filter paper. Paper is *permeable* to water and other solvents, but it is *impermeable* to the insoluble coffee grounds. However, filter paper will not separate the caffeine from brewed coffee. Paper is *permeable* both to water, the solvent, and caffeine, a solute. Are there, perhaps, materials that are *semipermeable*, that allow solvent but not solute molecules to pass? Yes, there are.

Semipermeable membranes usually are sheets or films of a material containing a network of microscopic holes or pores through which small solvent molecules can pass, but not larger solute molecules. These membranes may be composed of natural materials of animal or vegetable origin, such as pig's bladder or parchment. Or they may be made of synthetic materials such as cellophane. In the human body, cell membranes, the lining of the digestive tract, and the walls of blood vessels are all semipermeable.

Figure 12.18 illustrates a phenomenon first observed in 1748 and called *osmosis*. The flared end of a funnellike tube is covered with a semipermeable membrane. An aqueous solution is introduced into the tube, and the tube is placed in a beaker of pure water. Some time later, the liquid level in the tube is seen to have risen to a greater height than the water in the beaker. Figure 12.18 also gives a molecular-level explanation. The larger solute molecules are unable to pass through the holes in the semipermeable membrane, but the small molecules of water can pass in both directions. However, because the concentration of H_2O molecules is greater in the pure water than in the aqueous solution, there is a net flow of water molecules into the tube. This net flow causes the solution level to rise in the tube.

Osmosis is the net flow of solvent molecules from pure solvent through a semipermeable membrane into a solution. For a 20% aqueous sucrose solution, by mass, osmosis could lift a column of solution to a height of 150 m. Osmosis can also occur between two solutions. The net flow of solvent molecules is from the more dilute solution to the more concentrated solution. The driving force behind osmosis is the tendency of solvent molecules to equalize their concentrations in two solutions in contact with each other. The concentration of *solvent* molecules is greater in a dilute solution than in a concentrated one.

We can reduce the net flow of solvent molecules through a semipermeable membrane into a solution by increasing the flow of solvent molecules *out* of the solution. We can do this by applying pressure to the solution. At a sufficiently high pressure, there is no net flow of solvent through the semipermeable membrane. The pressure required to stop osmosis is called the **osmotic pressure** of the solution. For a 20% aqueous sucrose solution, the osmotic pressure is about 15 atm.

To understand that osmotic pressure is a colligative property, think of the solute only in terms of reducing the mole fraction of solvent in a solution. The lower the mole fraction of solvent, the greater will be the net flow of solvent molecules into the solution and the greater the osmotic pressure of the solution. The identity of the solute is immaterial.

An expression that works quite well for calculating osmotic pressures of *dilute* solutions of nonelectrolytes bears a striking resemblance to the ideal gas equation.

$$\pi = \frac{n}{V}RT$$

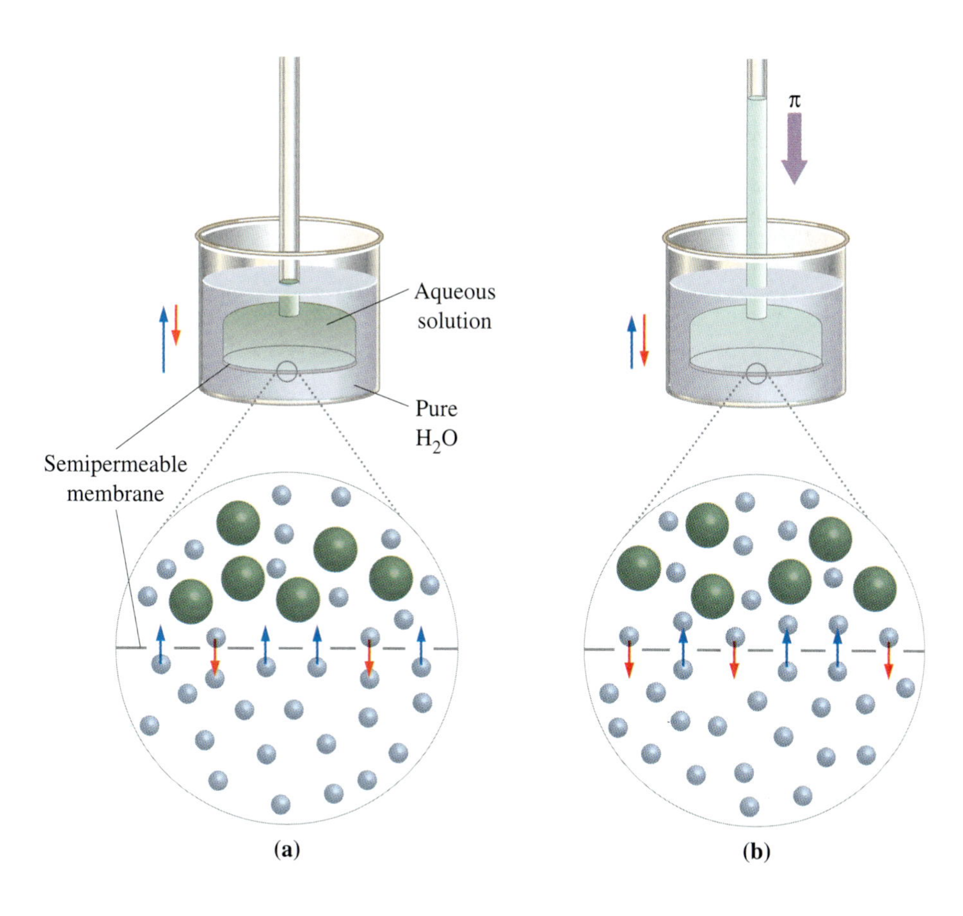

Figure 12.18
Osmosis and osmotic pressure
(a) An aqueous solution (green) is separated from pure water by a membrane permeable to H_2O molecules but not to solute particles. (b) A net flow of water through the semipermeable membrane dilutes the green solution somewhat and causes the liquid level in the tube to rise. The liquid column now exerts a downward pressure (a *hydrostatic* pressure) that helps push H_2O molecules through the membrane and into the beaker. When the flow of H_2O through the membrane is the same in both directions, there is no further change, and the hydrostatic pressure at that point is called the *osmotic pressure*, π.

In this equation, π is the osmotic pressure in atm, n is the number of moles of solute in a solution with a volume of V (liters), R is the universal gas constant, and T is the Kelvin temperature. Because the ratio of number of moles of solute to a solution volume (L) is the *molarity* (M), we can simply write

$$\pi = \text{M}\,RT$$

As the above equation suggests, we can use osmotic pressure measurements to determine molar masses of substances. However, for most solutions, osmotic pressures are so high that they are difficult to measure. In contrast, for solutions of *macromolecular* substances (polymers), osmotic pressures are fairly low and are easily measured. Osmotic pressure and freezing point depression measurements complement each other nicely for molar mass determinations. Generally, in cases where an osmotic pressure measurement works well, freezing point depression doesn't, and vice versa, as we illustrate in Example 12.16.

Crabs have blood that contains the protein hemocyanin.

EXAMPLE 12.16

An aqueous solution is prepared by dissolving 1.50 g of hemocyanin, a protein obtained from crabs, in 0.250 L of water. The solution has an osmotic pressure of 0.00342 atm at 277 K. (**a**) What is the molar mass of hemocyanin? (**b**) What should the freezing point of the solution be?

SOLUTION

a. We can obtain the molarity of the solution after rearranging the equation for osmotic pressure, $\pi = \text{M}\,RT$, to the form $\text{M} = \pi/RT$.

$$M = \frac{0.00342 \text{ atm}}{0.08206 \text{ L atm mol}^{-1} \text{ K}^{-1} \times 277 \text{ K}} = 1.50 \times 10^{-4} \text{ mol L}^{-1}$$

From the molarity and volume of solution, we can determine the number of moles of hemocyanin.

$$? \text{ mol hemocyanin} = 0.250 \text{ L} \times \frac{1.50 \times 10^{-4} \text{ mol hemocyanin}}{\text{L}}$$

$$= 3.75 \times 10^{-5} \text{ mol hemocyanin}$$

And from the number of moles and mass (1.50 g) of hemocyanin, we can establish its molar mass.

$$\text{Molar mass of hemocyanin} = \frac{1.50 \text{ g}}{3.75 \times 10^{-5} \text{ mol}} = 4.00 \times 10^{4} \text{ g/mol}$$

b. Because the solution is so dilute, the molarity and molality will have the same numerical value.

$$\Delta T_f = -K_f m = -1.86 \text{ °C } m^{-1} \times 1.50 \times 10^{-4} m = -2.79 \times 10^{-4} \text{ °C}$$

The freezing point of the solution should be −0.000279 °C. This freezing point would be extremely difficult to measure with precision and accuracy. Freezing point depression is not a good method for determining large molar masses.

PROBLEM-SOLVING NOTE

Low osmotic pressures are often measured and expressed in mmH_2O. Use the method of Example 5.2 to relate a pressure in atmospheres to millimeters of mercury and then to millimeters of H_2O.

EXERCISE 12.16A

An aqueous solution is prepared by dissolving 1.08 g of human serum albumin, a protein obtained from blood plasma, in 50.0 mL of water. The solution has an osmotic pressure of 0.00770 atm at 298 K. What is the molar mass of the albumin?

EXERCISE 12.16B

What is the osmotic pressure, expressed in the unit mmH_2O, of a solution having 125 μg of vitamin B_{12} ($C_{63}H_{88}CoN_{14}O_{14}P$) in 2.50 mL H_2O at 25 °C?

Practical Applications of Osmosis

We find examples of osmosis in living organisms everywhere: Cells are much like semipermeable bags filled with solutions of ions, small and large molecules, and still larger cell components. If we place red blood cells in pure water, a net flow of water into the cells causes them to burst. In contrast, if we place the cells in a solution that is 0.92% NaCl (mass/vol), there is no net flow of water through the cell membranes, and the cells are stable. The fluids inside the cells have the same osmotic pressure as the sodium chloride solution. A solution having the same osmotic pressure as body fluids is an **isotonic solution**. If the concentration of NaCl in a saline solution is greater than 0.92%, a net flow of water *out* of the cells causes them to shrink. The saline solution is a **hypertonic solution**; it has a higher osmotic pressure than red blood cells. If the concentration of NaCl in the solution surrounding the cells is less than 0.92%, water flows *into* the cells. We call the saline solution a **hypotonic solution**; it has a lower osmotic pressure than red blood cells.

One modern application of osmosis, called **reverse osmosis**, is based on *reversing* the normal net flow of water molecules through a semipermeable membrane. That is, by applying to a solution a pressure *exceeding* the osmotic pressure, water can be driven from a solution into pure water. In this way, pure water can be

Medical Applications of Osmosis

The rupture of a cell by a *hypotonic* solution is called *plasmolysis*. If the cell is a red blood cell, the more specific term is *hemolysis*. The shrinkage of a cell in a *hypertonic* solution, called *crenation*, can lead to the death of a cell.

When replacing body fluids or providing nourishment intravenously, it is important to use an isotonic fluid. Otherwise, hemolysis or crenation results, and the patient's well-being is jeopardized. As we have already described, an 0.92% NaCl (mass/vol) solution, called physiological saline, is isotonic with the fluid in red blood cells. The "D5W" so often referred to by television's doctors and paramedics is a 5.5% (mass/vol) solution of glucose (also called dextrose, D) in water (W). It, too, is isotonic with the fluid in red blood cells. The 0.92% NaCl is about 0.16 M, and the 5.5% glucose solution is approximately 0.31 M. (In Section 12.9, we will see why these two solutions have different molarities.)

There are shortcomings in the use of D5W for intravenous feeding. A patient can accommodate only about 3 L of water in a day. If an isotonic solution of 5.5% glucose is used, 3.0 L of this solution supplies only about 160 g glucose. The glucose yields about 640 kcal (640 food Calories) per day, a woefully inadequate amount of energy. Even a resting patient requires about 1400 kcal/day. And for a person suffering from serious burns, for example, requirements may be as high as 10,000 kcal/day. With carefully formulated solutions containing other vital nutrients as well as glucose, the feeding of a patient can be increased to about 1200 kcal/day, but this still falls short of the requirements of many seriously ill people.

One possibility is to use solutions that are about six times as concentrated as isotonic solutions. However, instead of administering them through a vein in an arm or a leg, the solution is infused directly through a tube into the superior vena cava, a large blood vessel leading to the heart. The large volume of blood flowing through this vein quickly dilutes the solution to levels that do not damage the blood. Using this technique, patients have been given 5000 kcal/day and have even gained weight.

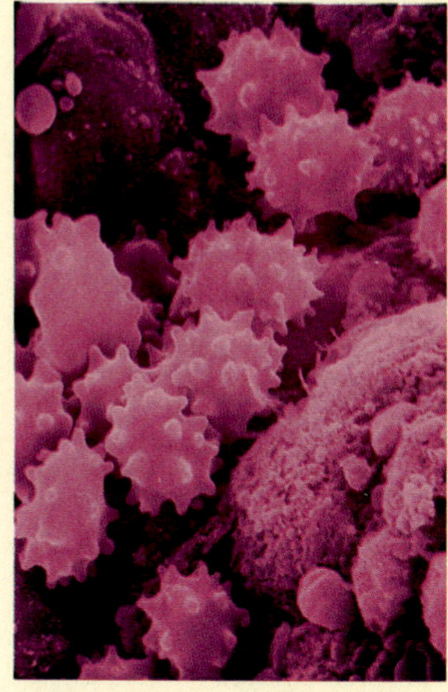

Red blood cells in a hypertonic solution, as seen through an electron microscope.

extracted from brackish water, seawater, or industrial wastewater. For reverse osmosis to succeed, membranes that can withstand high pressures are required. Reverse osmosis is widely used in ships at sea and in arid nations of the Middle East.

12.9 Solutions of Electrolytes

In our introduction to colligative properties (page 529), we noted that our discussion would first focus on solutions of nonelectrolytes. What about electrolyte solutions? Do they have colligative properties? We have already implied that they do by noting the osmotic properties of isotonic saline solutions and the use of NaCl and $CaCl_2$ in deicing roads. The major difference in treating solutions of electrolytes is in how we assess solute concentrations.

Nonelectrolyte solutions have *molecules* as their solute particles. Strong electrolyte solutions have *ions* as solute particles. Weak electrolyte solutions have *both molecules and ions* as solute particles. When we use the formula of a solute in establishing the molarity (M) or molality (m) of a solution, we generally disregard

Jacobus Henricus Van't Hoff (1852–1911) received the first Nobel Prize in chemistry in 1901 for his work on the physical properties of solutions. However, he is also known for proposing (simultaneously with the French chemist, Joseph Le Bel) the tetrahedral geometries of carbon compounds (for example, CH_4). Van't Hoff was 22 years old at the time.

these facts. To assess colligative properties, however, we must base the solute concentration on the particles that are actually in solution (ions and/or molecules). For example, 0.010 M NaCl(aq) is a strong electrolyte solution and has $[Na^+] = 0.010$ M, $[Cl^-] = 0.010$ M, and a total ion concentration of 0.020 M. In 0.010 M CH_3COOH(aq), a weak electrolyte solution, most of the CH_3COOH molecules remain intact, but a small fraction of them do ionize to form H^+ and CH_3COO^- ions. The total particle concentration is slightly greater than 0.010 M but less than 0.020 M.

We can modify the equations for colligative properties by accounting for the possible presence of ions in solution through a factor called the **van't Hoff factor** (*i*). The modified equations are

Freezing point depression: $\Delta T_f = -i\, K_f\, m$

Boiling point elevation: $\Delta T_b = i\, K_b\, m$

Osmotic pressure: $\pi = i\, M\, RT$

For nonelectrolyte solutions, $i = 1$. That means that the molarity or molality based on the solute formula is the actual concentration of solute particles. For a solution of the strong electrolyte NaCl, we should expect $i = 2$; *two* moles of ions for each mole of solute. For Na_2SO_4 and $CaCl_2$ we should expect $i = 3$. For a solution of a weak electrolyte, *i* is somewhat greater than 1 but less than 2 because a weak electrolyte is only partially ionized. The extent to which a weak electrolyte ionizes depends on the concentration of the solution, which means that *i* also depends on the concentration. For example, $i \approx 1.04$ for 0.010 M CH_3COOH, which signifies that about 4% of the molecules ionize; in 0.10 M CH_3COOH, $i \approx 1.01$.

The value of *i* for strong electrolytes also depends on the concentration. Only if the solution is quite dilute will *i* be an integer (that is, 2, 3, ...); otherwise it will be a noninteger. The values of *i* for various solutions of NaCl(aq) are: 1.0 *m*, $i = 1.81$; 0.10 *m*, $i = 1.87$; 0.010 *m*, $i = 1.94$; and 0.0010 *m*, $i = 1.97$. In concentrated solutions of strong electrolytes, attractive forces between cations and anions can cause some of them to associate into *ion pairs* and to otherwise behave as if the solute were not completely dissociated. In this text, we will assume solutions are dilute enough that strong electrolytes are completely dissociated into ions.

EXAMPLE 12.17—A Conceptual Example

Without doing detailed calculations, place the following solutions in order of *decreasing* osmotic pressure: (a) 0.01 M $C_{12}H_{22}O_{11}$(aq) at 25°C; (b) 0.01 M $C_6H_{12}O_6$(aq) at 37 °C; (c) 0.01 *m* KNO_3(aq) at 25°C; (d) a solution of 1.00 g polystyrene (molar mass: 3.5×10^5 g/mol) in 100 mL of benzene at 25 °C.

SOLUTION

We begin by writing the most general equation for the osmotic pressure of a solution, $\pi = i\, M\, RT$, and then we apply it to each of the solutions.

a. Let's designate the osmotic pressure of 0.01 M $C_{12}H_{22}O_{11}$(aq), a *nonelectrolyte* solution ($i = 1$), as π_a. Then we will compare the other osmotic pressures to π_a.

b. 0.01 M $C_6H_{12}O_6$ is also a solution of a *nonelectrolyte* ($i = 1$). The same number of particles are present per unit volume as in (a). However, because the temperature is *slightly higher*, the osmotic pressure of this solution should be *slightly* greater than that in (a), by a factor of 310/298: $\pi_b > \pi_a$.

c. The solution is 0.01 *m* KNO_3 rather than 0.01 M KNO_3. However, in dilute aqueous solutions, because $d \approx 1.0$ g/mL, molality and molarity are essentially equal. This is a solution of an *electrolyte* for which $i \approx 2$. The osmotic pressure is about twice as great as that of solution (a) and also almost twice that of solution (b).

$$\pi_c > \pi_b > \pi_a.$$

d. Each of the other three solutions has at least 0.001 mol of solute particles in a 100-mL sample. The *mass* of 0.001 mol of the polystyrene is 0.001 mol $\times$ 3.5 $\times$ 10^5 g/mol = 350 g, but the mass of polystyrene in the given solution is only 1.00 g. The solution is much more dilute than 0.01 M. The fact that the solute is a polymer and the solvent is an aromatic hydrocarbon has no bearing on the situation. Solution (d) has the lowest osmotic pressure by far; solution (c) has the highest

$$\pi_c > \pi_b > \pi_a > \pi_d.$$

EXERCISE 12.17

Which of the following solutions has the *lowest* and which has the *highest* freezing point: 0.010 *m* $C_6H_{12}O_6$(aq), 0.0080 M HCl(aq), 0.0050 *m* $MgCl_2$(aq), 0.0030 M $Al_2(SO_4)_3$(aq)?

Mixtures: Solutions, Colloids, and Suspensions

Once a solute and solvent are thoroughly mixed, the atoms, ions, or molecules of the solute remain randomly distributed in the solvent; the *solute does not settle out*. The mixture is *homogeneous*—a *solution*. If we try to dissolve sand (silica, SiO_2) in water, the best we can achieve is a temporary dispersion of the sand in water, called a *suspension*. Grains of sand quickly settle to the bottom of the container. The sand-water mixture is *heterogeneous*. Are there mixtures that lie between true solutions and suspensions? There are such mixtures, and they are called *colloids*.

12.10 Colloids

A mixture is a colloid not because of the kind of matter present but because of the *size* of the dispersed particles. True solutions have solute and solvent particles measuring less than about 1 nm in diameter. Ordinary suspensions have particle dimensions of about 1000 nm or more. Such particles are generally visible to the eye or through an ordinary microscope. We define a **colloid** as a dispersion in an appropriate medium of particles having sizes that range from about 1 nm to 1000 nm. Colloidal particles generally cannot be seen through an ordinary microscope.

The shapes of the particles in a colloid can vary widely, but in every case, at least one dimension (length, width, thickness) is in the range of about 1 to 1000 nm. The particles in an aqueous colloid of silica (SiO_2) are spheres. Some particles in colloids are rod-shaped, for example, certain viruses. Still other colloidal particles are disk-shaped, such as the blood plasma protein gamma globulin. Some natural and synthetic polymers form long, randomly coiled filaments.

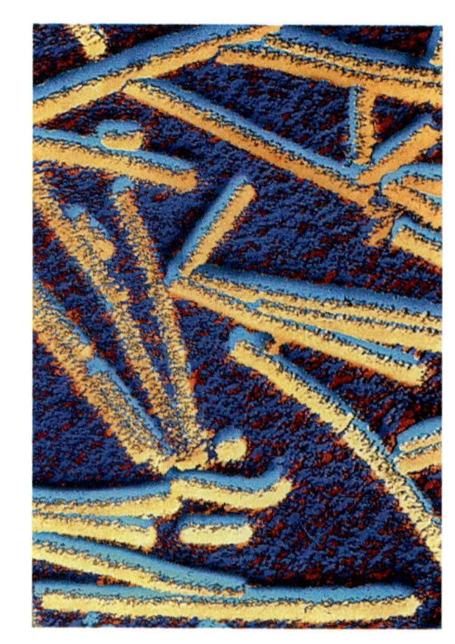

These rod-shaped tobacco mosaic viruses are colloidal-sized particles.

We can classify colloids according to the states of matter of the particles (the dispersed phase) and of the medium in which they are dispersed. In all, eight possibilities are summarized in Table 12.3, and each is given a different name. For example, a *sol* is a dispersion of solid particles in a liquid medium; an *emulsion* is a

TABLE 12.3 Some Common Types of Colloids

Dispersed Phase	Dispersion Medium	Type	Examples
Solid	Liquid	Sol	Starch solution, clay sols, jellies
Liquid	Liquid	Emulsion	Milk, mayonnaise
Gas	Liquid	Foam	Whipped cream, meringues
Solid	Gas	Aerosol	Fine dust or soot in air
Liquid	Gas	Aerosol	Fog, hair sprays
Solid	Solid	Solid sol	Ruby glass, black diamond
Liquid	Solid	Solid emulsion	Pearl, opal, butter
Gas	Solid	Solid foam	Floating soap, pumice, lava

dispersion of liquid particles (microscopic droplets) in a liquid medium; and an *aerosol* is a dispersion of either solid or liquid particles in a gas (usually air).

Much of living matter is in the form of sols and emulsions. Substances with high molecular masses, such as starches and proteins, generally form colloidal mixtures with water, not true solutions. Smog is an aerosol in part, and its suspended particles are both solid (smoke) and liquid (fog): *sm*oke + f*og* = smog. However, other constituents of smog are of molecular size, for example, SO_2, CO, NO, and O_3.

As you might expect, the properties of colloidal dispersions are different from those of true solutions and suspensions. Colloids often appear milky or cloudy, and even those that appear clear reveal their colloidal nature by scattering a beam of light passing through them. This phenomenon, first studied by John Tyndall in 1869, is known as the **Tyndall effect** (see Figure 12.19). The spectacular sunsets often seen in desert areas are caused, at least in part, by the preferential scattering of blue light as sunlight passes through dust-laden air (an aerosol). The transmitted light is deficient in the color blue, and thus has a reddish color.

Figure 12.19 The Tyndall effect
The light beam is not visible as it passes through a true solution (left), but it is readily visible as it passes through colloidal iron (III) oxide in water.

We have already noted that the properties of surfaces are in some ways different from those of bulk matter (recall our discussion of surface tension on page 487). A special property of surfaces is the ability to *adsorb*, or attach, ions from a solution. This ability has important consequences for colloids. The particles in a colloid generally adsorb certain cations or anions in preference to others. For example, the SiO_2 particles in colloidal silica adsorb OH^- (see Figure 12.20). As a result, even though the mixture as a whole is electrically neutral, the particles carry a net *negative* charge. Because all the silica particles are negatively charged, they repel one another. These repulsions are strong enough to overcome the force due to gravity, and the particles do not settle. However, a high concentration of an electrolyte can cause a colloid to *coagulate*, or *precipitate* (see Figure 12.21), by neutralizing the charges on colloidal particles.

Removing excess ions can prevent coagulation and make a colloid more stable. Excess ions can be removed through *dialysis*, a process similar to osmosis. In dialysis, ions pass from the colloidal dispersion through a semipermeable membrane into pure water, but the much larger colloidal particles do not. In *electrodialysis*, the process is facilitated by the attractions of ions to an electrode having the opposite charge. A human kidney acts as a dialyzer by removing excess electrolytes from blood, which is a colloid. In some diseases, the kidney loses its dialyzing ability, but fortunately this function can be served to a degree by a dialysis machine.

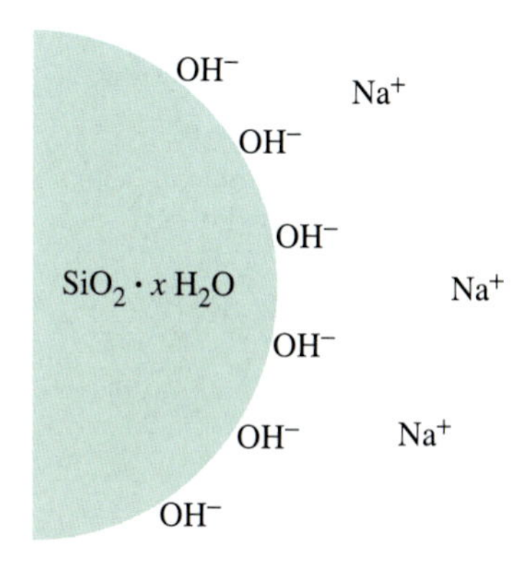

▲ **Figure 12.20**
Surface of an SiO_2 particle in colloidal silica
The silica in the colloid is hydrated, as noted by the formula $SiO_2{\cdot}xH_2O$. Hydroxide ions are preferentially *adsorbed* at the surface. Cations are present in the solution as counter ions to balance the charge. In the immediate vicinity of the colloidal particle, however, the negative charges outnumber the positive charges, and the particle carries a net negative charge.

◀ **Figure 12.21**
Formation and coagulation of a colloid
The red colloidal hydrous iron(III) oxide, $Fe_2O_3{\cdot}xH_2O$ (left), is produced by adding concentrated $FeCl_3(aq)$ to boiling water. When a few drops of $Al_2(SO_4)_3(aq)$ are added, the suspended particles rapidly coagulate into a precipitate of $Fe_2O_3{\cdot}xH_2O$ (right).

Summary

Several concentration units are used in describing solutions. Molarity is expressed as moles of solute per liter of solution. Molality is expressed in moles of solute per kilogram of solvent. The concentration of a component can also be specified as the fraction of all the molecules in solution that are of that component; such a concentration is called a mole fraction. Compositions can also be given on a percent basis, such as grams of solute per 100 grams of solution. Some related terms used for very dilute solutions are parts per million (ppm), parts per billion (ppb), and parts per trillion (ppt).

An ideal solution is one in which intermolecular forces between all molecules are of the same type and are essentially equal in magnitude. Where the types and magnitudes of solute-solvent intermolecular forces differ appreciably from solute-solute and solvent-solvent intermolecular forces, either the solution is nonideal or the components remain segregated in a heterogeneous mixture.

The solubility of a solute is the concentration of a saturated solution of the solute. In most cases, the solubility of a solid solute in water increases with temperature, but a few solid solutes become less soluble with an increase in temperature. The solubilities of gases in

Key Terms

colligative property (page 529)
colloid (12.10)
fractional distillation (12.6)
Henry's law (12.5)
hypertonic solution (12.8)
hypotonic solution (12.8)
ideal solution (12.3)
isotonic solution (12.8)
molality (m) (12.2)
mole fraction (x) (12.2)
mole percent (12.2)
osmosis (12.8)
osmotic pressure (12.8)
parts per billion (ppb) (12.2)
parts per million (ppm) (12.2)

parts per trillion (ppt) (12.2)
Raoult's law (12.6)
reverse osmosis (12.8)
saturated solution (12.4)
semipermeable membrane (12.8)
solubility (12.4)
solubility curve (12.4)
supersaturated (12.4)
Tyndall effect (12.10)
unsaturated (12.4)
Van't Hoff factor (i) (12.9)

water generally decrease with an increase in temperature and increase with an increase in the pressure of the gas above the solution.

The presence of a solute(s) lowers the vapor pressure of the solvent in a solution. For ideal solutions and dilute nonideal solutions, the lowering of vapor pressure follows Raoult's law: $P_{solv} = x_{solv} \cdot P^{\circ}_{solv}$. The vapor in equilibrium with a solution of two volatile components has a higher mole fraction of the more volatile component than does the solution. This behavior is the basis of the separation method known as fractional distillation.

Vapor pressure lowering, freezing point depression, boiling point elevation, and osmotic pressure are colligative properties. They depend on the particular solvent and the number of solute particles present, but not on the identity of the solute. Osmotic pressure is the pressure that must be applied to a solution to prevent the flow of solvent molecules from the pure solvent, through a semipermeable membrane, into the solution. Osmotic pressure measurements can be used to establish molar masses, especially of macromolecular materials (polymers).

Equations relating to colligative properties must be modified for solutes that are strong or weak electrolytes. This is done through the van't Hoff factor, i. The value of i is determined by the extent to which a solute ionizes and the ions interact in solution.

In colloids, the particles of dispersed matter range in size from about 1 to 1000 nm. Although colloidal dispersions often resemble solutions in outward appearance, they differ in important ways. For example, colloids scatter light and true solutions do not. Colloids can exist in any of the three states of matter, and many common materials, ranging from blood and other body fluids to food products to atmospheric smog, are colloidal.

Review Questions

1. Define or explain and, where possible, illustrate:
(a) percent by mass **(b)** percent by volume **(c)** molality **(d)** mole fraction.

2. Explain why molarity is temperature-dependent, whereas molality is not. Is mole fraction temperature-dependent? Is volume percent? Is mass percent?

3. Explain the meaning of each of these terms in describing a solution:
(a) dilute **(b)** concentrated **(c)** saturated **(d)** unsaturated **(e)** supersaturated

4. In a dynamic equilibrium involving a saturated solution, describe the two processes for which the rates are equal.

5. Precipitation is induced in a supersaturated solution by adding a seed crystal. When no more solid crystallizes, is the solution saturated, unsaturated, or supersaturated? Explain.

6. List some of the characteristics of an *ideal* solution. Would you expect a solution of ethanol (CH_3CH_2OH) in water to be ideal? Explain.

7. Does NaCl dissolve in benzene, C_6H_6? Explain.

8. Does motor oil dissolve in water? In benzene (C_6H_6)? Explain.

9. When the cap is removed from a bottle of carbonated beverage, carbon dioxide gas escapes from the liquid. Explain why this is so.

10. The vapor in equilibrium with a pure liquid has the same composition as the liquid. Is this statement also true for solutions? Explain.

11. What is a colligative property? What are the chief colligative properties discussed in this chapter? Is the density of a solution a colligative property? Explain.

12. Explain why a nonvolatile solute *raises* the boiling point of a solvent, whereas it *lowers* the freezing point of the same solvent.

13. Which laboratory measurement gives a more accurate result for the molar mass of a polymeric substance: freezing point depression or osmotic pressure? Explain.

14. How do the terms *osmosis* and *osmotic pressure* differ in meaning?

15. Define the terms *isotonic*, *hypotonic*, and *hypertonic* solution. Do all *isotonic* solutions have the same concentration? Explain.

16. The net direction in which gaseous molecules diffuse is from *higher* to lower pressure. In osmosis, the net direction in which molecules diffuse is from *lower* to higher solution concentration. Do these statements seem incompatible? Explain.

17. Contrast a true solution and a colloid with regard to **(a)** the size of the solute particles; **(b)** the nature of the distribution of solute and solvent particles; **(c)** the color and clarity of the solution; and **(d)** the Tyndall effect.

18. Compare and contrast a colloid and a suspension.

19. Is it possible to have a colloidal dispersion of one gas in another? Explain.
20. Compare and contrast osmosis and dialysis with respect to the kinds of particles that pass through a semipermeable membrane.
21. What is meant by the term *reverse osmosis*? Give an example of its use.
22. For each pair of solutions at 25 °C, indicate which member has the higher osmotic pressure.
   (a) 0.1 M $NaHCO_3$, 0.05 M $NaHCO_3$
   (b) 1 M NaCl, 1 M glucose
   (c) 1 M NaCl, 1 M $CaCl_2$
   (d) 1 M NaCl, 3 M glucose

Problems

Percent Concentration

23. Describe how you would prepare 5.00 kg of an aqueous solution that is 10.0% NaCl *by mass*.
24. Describe how you would prepare exactly 2.00 L of an aqueous solution that is 2.00% acetic acid *by volume*.
25. What is the mass percent of solute in each of the following solutions?
   (a) 4.12 g NaOH in 100.0 g water
   (b) 5.00 mL ethanol (d = 0.789 g/mL) in 50.0 g water
   (c) 1.50 mL glycerol (d = 1.324 g/mL) in 22.25 mL H_2O (d = 0.998 g/mL)
26. What is the mass percent of solute in each of the following solutions?
   (a) 175 mg NaCl/g solution
   (b) 0.275 L methanol (d = 0.791 g/mL)/kg water
   (c) 4.5 L ethylene glycol (d = 1.114 g/mL) in 6.5 L propylene glycol (d = 1.036 g/mL)
27. What is the volume percent of the first named component in each of the following solutions?
   (a) 35.0 mL water in 725 mL of an ethanol-water solution
   (b) 10.00 g acetone (d = 0.789 g/mL) in 1.55 L of an acetone-water solution
   (c) 1.05% by mass 1-butanol (d = 0.810 g/mL) in ethanol (d = 0.789 g/mL) [*Hint:* Assume the volumes are additive.]
28. What is the volume percent of the first named component in each of the following solutions?
   (a) 58.0 mL water in 625 mL of an ethanol-water solution
   (b) 10.00 g methanol (d = 0.791 g/mL) in 75.00 g ethanol (d = 0.789 g/mL) [*Hint:* Assume the volumes are additive.]
   (c) 24.0% by mass ethanol (d = 0.789 g/mL) in an aqueous solution with d = 0.963 g/mL
29. On average, glucose makes up about 0.10% of human blood, by mass. What is the approximate concentration of glucose in blood in milligrams per deciliter?
30. A vinegar sample has a density of 1.01 g/mL and contains 5.88% acetic acid by mass. What mass of acetic acid is contained in a one-pint bottle of the vinegar? (8 pt = 1 gal; 1 gal = 3.785 L.)

Parts per Million, Parts per Billion, Parts per Trillion

31. *Without doing detailed calculations*, arrange *aqueous* solutions with the following concentrations in the order of increasing mass percent of solute. **(a)** 1% by mass; **(b)** 1 mg solute/dL solution; **(c)** 1 ppb; **(d)** 1 ppm; **(e)** 1 ppt.
32. *Without doing detailed calculations*, determine which has the greater mass percent of solute: 0.0010 M $MgCl_2(aq)$ or $MgCl_2(aq)$ with 105 ppm Mg^{2+}.
33. Express the following aqueous concentrations in the unit indicated.
   (a) 1 μg benzene/L water, as ppb benzene
   (b) 0.0035% NaCl, by mass, as ppm of NaCl
   (c) 2.4 ppm F^-, as molarity of fluoride ion, $[F^-]$
34. Express the following aqueous concentrations in the unit indicated.
   (a) 5 μg trichloroethylene/L water, as ppb trichloroethylene
   (b) 0.0025 g KI/L water, as ppm of KI
   (c) 37 ppm SO_4^{2-}, as molarity of sulfate ion, $[SO_4^{2-}]$

Molarity and Molality

35. Calculate the *molality* of a solution that has 18.0 g glucose, $C_6H_{12}O_6$, dissolved in 80.0 g water.
36. Calculate the *molality* of a solution prepared by dissolving 125 mL pure methanol, CH_3OH, (d = 0.791 g/mL) in 275 g ethanol.
37. A typical commercial-grade phosphoric acid is 75% H_3PO_4, by mass, and has d = 1.57 g/mL. What are the molarity and the molality of H_3PO_4 in this acid?
38. An aqueous solution is prepared by diluting 3.30 mL acetone, CH_3COCH_3, (d = 0.789 g/mL) with water to a final volume of 75.0 mL. The density of the solution is 0.993 g/mL. What are the molarity and molality of acetone in this solution?
39. A dilute sulfuric acid solution that is 3.39 m H_2SO_4 has a density of 1.18 g/mL. How many moles of H_2SO_4 are there in 375 mL of this solution?
40. How many moles of ethylene glycol are present in 2.30 L of 6.27 m $HOCH_2CH_2OH$ (d = 1.035 g/mL)?

Mole Fraction and Mole Percent

41. What is the mole fraction of naphthalene, $C_{10}H_8$, in a solution prepared by dissolving 23.5 g $C_{10}H_8(s)$ in 315 g $C_6H_6(l)$?

42. What is the mole fraction of urea, $CO(NH_2)_2$, in a solution prepared by dissolving 25.2 g $CO(NH_2)_2$ in 125 mL of water at 20 °C (d = 0.998 g/mL)?

43. What is the mole fraction of naphthalene, $C_{10}H_8$, in a 0.250 m solution of $C_{10}H_8$ in benzene, C_6H_6?

44. What is the mole fraction of ethanol in an aqueous solution that is 22.4% ethanol by mass?

45. *Without doing detailed calculations*, determine which of the following *aqueous* solutions has the greatest mole fraction of solute: **(a)** 1.00 m CH_3OH; **(b)** 5.0% CH_3CH_2OH, by mass; **(c)** 10.0% $C_{12}H_{22}O_{11}$, by mass. Explain your reasoning.

46. *Without doing detailed calculations*, determine which of the following *aqueous* solutions has the greater mass percent urea, $CO(NH_2)_2$: 0.10 m urea or x_{urea} = 0.010? Explain your reasoning.

Solubilities

47. Predict which of the following substances is highly soluble in water, which is slightly soluble, and which is insoluble.

(a) chloroform, $CHCl_3$

(b) benzoic acid, C_6H_5COOH

(c) propylene glycol, $CH_3CHOHCH_2OH$

48. One of the following substances is soluble in *both* water and in benzene, C_6H_6. Which is it? Explain.

(a) chlorobenzene, C_6H_5Cl

(b) ethylene glycol, $HOCH_2CH_2OH$

(c) phenol, C_6H_5OH.

Saturated, Unsaturated, and Supersaturated Solutions

49. Refer to Figure 12.10, and determine whether a mixture of 25 g NH_4Cl and 55 g H_2O at 60 °C is saturated, unsaturated, or supersaturated. Explain your reasoning.

50. Refer to Figure 12.10, and determine whether an aqueous solution can be prepared that is 30.0% by mass of KNO_3 at 20 °C. Explain your reasoning.

51. Use data from Figure 12.10 to determine **(a)** the additional mass of water that should be added to a mixture of 35 g K_2CrO_4 and 35 g H_2O to dissolve all the solute at 25 °C; and **(b)** the approximate temperature to which a mixture of 50.0 g KNO_3 and 75.0 g water must be heated to complete the dissolving of the KNO_3.

52. Use data from Figure 12.10 to determine **(a)** the molality of saturated $Li_2SO_4(aq)$ at 50 °C; and **(b)** the additional mass of $Pb(NO_3)_2$ that can be dissolved in an aqueous solution containing 325 mg $Pb(NO_3)_2$/g solution at 30 °C.

53. Can you think of a way in which solute can be crystallized from an unsaturated solution without changing the solution temperature? Explain.

54. Are there any exceptions to the general rule that a supersaturated solution can be made to deposit excess solute by cooling? Explain.

Solubilities of Gases

55. The solubility of $O_2(g)$ in water is 4.43 mg O_2/100 g H_2O at 20 °C when the gas pressure is maintained at 1 atm. **(a)** What is the molarity of the saturated solution? **(b)** What pressure of $O_2(g)$ would be required to produce a saturated solution that is 0.010 M O_2?

56. When 250 g of water saturated with air is heated from 20 °C to 80 °C, estimate the quantity of air released in terms of its **(a)** mass, in mg; **(b)** volume, in mL (at STP). Use data from Figure 12.12, and assume 28.96 g/mol as the molar mass of air.

Liquid-Vapor Equilibrium

57. A solution has a 1:4 *mole* ratio of pentane to hexane. The vapor pressures of the pure hydrocarbons at 20 °C are 441 mmHg for pentane and 121 mmHg for hexane. **(a)** What are the partial pressures of the two hydrocarbons above the solution? **(b)** What is the mole fraction composition of the vapor?

58. A solution has a 2:3 *mass* ratio of toluene (C_7H_8) to benzene (C_6H_6). The vapor pressures of toluene and benzene at 25 °C are 28.4 mmHg and 95.1 mmHg, respectively. **(a)** What are the partial pressures of the two hydrocarbons above the solution? **(b)** What is the mole fraction composition of the vapor?

59. What is the vapor pressure of water above an 0.20 m glucose ($C_6H_{12}O_6$) solution at 20 °C? Use vapor pressure data from Table 11.2.

60. What is the vapor pressure of water above 3.49 M glycerol ($C_3H_8O_3$) solution (d = 1.073 g/mL) at 20 °C? Use vapor pressure data from Table 11.2.

Freezing Point Depression and Boiling Point Elevation (Use data from Table 12.2, as necessary.)

61. Determine the freezing points of the following solutions.

(a) 0.25 *m* urea in water

(b) 5.0% *para*-dichlorobenzene, $C_6H_4Cl_2$, by mass in benzene

62. Determine the boiling points of the following solutions.

(a) 0.44 *m* naphthalene in benzene

(b) 1.80 M sucrose ($C_{12}H_{22}O_{11}$) in water ($d = 1.23$ g/mL)

63. What is the molality of a naphthalene-in-benzene solution with the same freezing point as 0.55 *m* sucrose in water?

64. How many grams of naphthalene, $C_{10}H_8$, would you add to 50.0 g of benzene, C_6H_6(l), to produce a solution that has the same freezing point as pure water?

65. A 2.11-g sample of naphthalene ($C_{10}H_8$) dissolved in 35.00 g of *para*-xylene has a freezing point of 11.25 °C. The pure solvent has a freezing point of 13.26 °C. What is the K_f of *para*-xylene?

66. A 1.45-g sample of an unknown compound is dissolved in 25.00 mL benzene ($d = 0.879$ g/mL). The solution freezes at 4.25 °C. What is the molar mass of the unknown?

67. An unknown compound consists of 33.81% C, 1.42% H, 45.05% O, and 19.72% N by mass. A 1.505-g sample of the compound, when dissolved in 50.00 mL benzene ($d = 0.879$ g/mL), lowers the freezing point of the benzene to 4.70 °C. What is the molecular formula of the compound?

68. Coniferin, a sugar derivative found in conifers such as fir trees, has a composition of 56.13% C, 6.48% H, and 37.39% O, by mass. A 2.216-g sample is dissolved in 48.68 g of H_2O, and the solution is found to have a boiling point of 100.068 °C. What is the molecular formula of coniferin?

Osmotic Pressure

69. Pickles are made by soaking cucumbers in a salt solution. Which has the higher osmotic pressure, the salt solution or the liquid in the cucumber? Explain.

70. Describe how the phenomenon discussed in Example 12.13 is related to osmosis.

71. What is the osmotic pressure at 37 °C of an aqueous solution that is 1.80% CH_3CH_2OH (mass/vol)? Is this solution isotonic, hypotonic, or hypertonic? [*Hint:* Compare the osmotic pressure to that of a 5.5% (mass/vol) solution of glucose, $C_6H_{12}O_6$.]

72. The osmotic pressure exerted by a 0.325-g sample of the polymer polystyrene in 50.00 mL of benzene at 25 °C is capable of supporting a column of the solution ($d = 0.88$ g/mL) 5.3 mm in height. What is the molar mass of the polystyrene? (*Hint:* What is the height of mercury ($d = 13.6$ g/mL) equivalent to that of the polymer solution?)

73. In the illustration, if aqueous solution A is 0.15 M sucrose ($C_{12}H_{22}O_{11}$) and solution B has 5.15 g urea [$CO(NH_2)_2$] in 75.0 mL of H_2O, in what direction will there be a net flow of water through the semipermeable membrane—to the left or to the right? Explain.

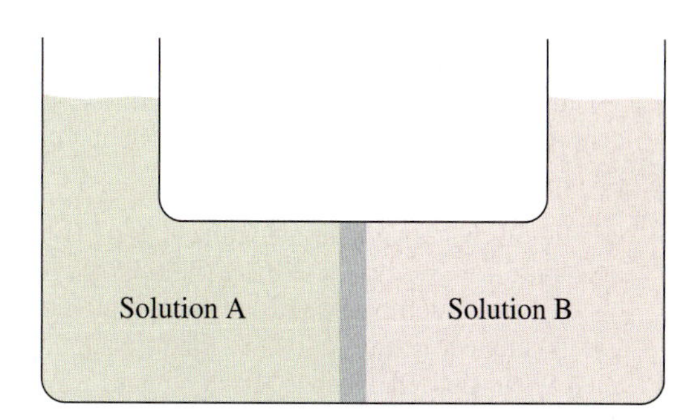

74. In the illustration above, if aqueous solution A is 0.18 M sucrose ($C_{12}H_{22}O_{11}$) and solution B is 0.22 M glucose ($C_6H_{12}O_6$), what pressure must be applied and to which arm of the apparatus—left or right—so that there is no net flow of water through the semipermeable membrane?

Colligative Properties of Strong Electrolytes, Weak Electrolytes, and Nonelectrolytes

75. Hydrogen chloride is soluble both in water and in benzene. The freezing point *depression* in 0.01 *m* HCl(aq) is about 0.04 °C, and in 0.01 *m* HCl(in benzene), it is about 0.05 °C. Are these the results you would expect for the two solutions? Explain.

76. Predict approximate freezing points of the following aqueous solutions.

(a) 0.10 *m* glucose ($C_6H_{12}O_6$)

(b) 0.10 *m* $CaCl_2$

(c) 0.10 *m* CH_3COOH

(d) 0.10 *m* KI

Which of these predictions is probably the most accurate? Explain.

77. The text describes an isotonic solution of NaCl as being about 0.16 M, whereas an isotonic solution of glucose is about 0.31 M. Explain why the concentrations of these isotonic solutions are not the same.

78. What would be the effect on red blood cells if they were placed in **(a)** 5.5% (mass/vol) NaCl(aq); **(b)** in 0.92% (mass/vol) glucose(aq). Explain.

79. Arrange the following aqueous solutions in order of *decreasing* freezing point, and state your reasons: **(a)** 0.15 *m* CH_3COOH; **(b)** 0.15 *m* $CO(NH_2)_2$ (urea); **(c)** 0.10 *m* H_2SO_4; **(d)** 0.10 *m* $Mg(NO_3)_2$; **(e)** 0.10 *m* NaBr.

80. Arrange the following five solutions in the order of *increasing* boiling point elevation, ΔT_b, and state your reasons: **(a)** 0.25 *m* sucrose(aq); **(b)** 0.15 *m* KNO_3(aq); **(c)** 0.048 *m* $C_{10}H_8$ (naphthalene) in benzene; **(d)** 0.15 *m* CH_3COOH(aq); **(e)** 0.15 *m* H_2SO_4(aq).

81. The boiling point of water when the barometric pressure is 744 mmHg is 99.4 °C. What approximate mass of NaCl would you add to 3.50 kg of the boiling water to raise the boiling point to 100.0 °C?

82. The text lists −21 °C as the lowest temperature at which NaCl can melt ice. What is the approximate mass percent of NaCl in an aqueous solution having this freezing point? Why is the calculation only approximate?

Colloids

83. Which 0.005 M solution would be most effective in coagulating the colloidal silica represented in Figure 12.20: KCl(aq), $MgCl_2$(aq), or $AlCl_3$(aq)? Explain.

84. Aluminum sulfate is commonly used to coagulate or precipitate colloidal suspensions of clay particles in municipal water-treatment plants. Why do you suppose aluminum sulfate is more effective than sodium chloride for this purpose?

Additional Problems

85. An aqueous solution with density 0.980 g/mL at 20 °C is prepared by dissolving 11.3 mL CH_3OH (d = 0.793 g/mL) in enough water to produce 75.0 mL of solution. What is the percent CH_3OH, expressed as **(a)** volume percent; **(b)** mass percent; **(c)** mass/volume percent; and **(d)** mole percent?

86. The liquids water and triethylamine $[(CH_3CH_2)_3N]$ are only partially miscible at 20 °C. That is, triethylamine dissolves in water to some extent, and water dissolves in triethylamine to some extent. In a mixture of 50.0 g water and 50.0 g triethylamine, 40.0 g of a phase consisting of 84.5% water and 15.5% triethylamine is obtained—a saturated solution of triethylamine in water. What is the percent by mass of water in the second phase—a saturated solution of water in triethylamine?

87. A solution is prepared by mixing 25.00 mL methanol, CH_3OH, (d = 0.791 g/mL) and 25.00 mL water (d = 0.998 g/mL). The solution can be described as **(a)** the solute methanol in water or **(b)** the solute water in methanol. Explain why the molality is different depending on which of these two descriptions is used. In which case, **(a)** or **(b)**, is the molality greater? (*Hint:* An actual calculation of molalities is not required.)

88. What mass of sucrose, $C_{12}H_{22}O_{11}$, must be dissolved per liter of water (d = 0.998 g/mL) to obtain a solution with 2.50 mole percent $C_{12}H_{22}O_{11}$? (*Hint:* Think of the required number of moles of sucrose as an unknown, x.)

89. Example 12.2 shows that the mass percent ethanol in an ethanol-water solution is less than the volume percent. Would you expect this to be the case for all ethanol-water solutions? for all solutes in water solution? Explain.

90. Use data from Figure 12.12 to establish the solubility of air in water at 20 °C in the units mL air (STP)/100 g water. This is the unit commonly used in listing gas solubilities in handbooks. (*Hint:* Use 28.96 g/mol as the apparent molar mass of air.)

91. We wish to compare three different solutions: saturated NaCl(aq) in contact with NaCl(s); 0.10 *m* NaCl(aq); and 0.20 *m* $(CH_3)_2CO$(aq). [$(CH_3)_2CO$ is acetone, a liquid with a boiling point of 56.5 °C.]

    (a) Which solution would you expect to have the greatest total vapor pressure? Explain.

    (b) Which solution would you expect to have the lowest freezing point? Explain.

    (c) When in an open container, which solution(s) has a total vapor pressure that changes with time, and which solution(s), a vapor pressure that does not? Explain.

92. Instructions on a container of Prestone® (ethylene glycol, mp −13.5 °C) give the following volumes of Prestone® to be used in protecting a 12-qt cooling system against freeze-up to the following temperatures (the remaining liquid is water): 10 °F, 3 qt; 0 °F, 4 qt; −15 °F, 5 qt; −34 °F, 6 qt. Why do you suppose the instructions do not include adding even more Prestone® to achieve a greater protection against freezing?

93. Given that the vapor pressures of benzene and toluene at 100.0 °C are 1351 mmHg and 556.3 mmHg, respectively, calculate the composition of a benzene-toluene solution that has a boiling point of 100.0 °C.

94. The photograph suggests that when the hydrate $CaCl_2{\cdot}6H_2O$(s) is transferred from a closed bottle to a

watch glass and exposed to the atmosphere, the material becomes wet with liquid. Explain what is happening. Do you think all solids behave in this way? (*Hint:* What do you think the liquid is?)

95. An aqueous, isotonic solution can be described in terms of freezing point depression as well as osmotic pressure. It has a freezing point of −0.52 °C. Show that this definition describes isotonic NaCl(aq) reasonably well [that is, 0.92% NaCl (mass/vol)].

96. In Chapter 4, we described a solution that is used in oral rehydration therapy (ORT) (Exercise 4.1B, page 137). Show that this solution is essentially isotonic. Use the definition of an isotonic solution given in Exercise 95.

97. The solubility of $CO_2(g)$ is 149 mg CO_2/100 g H_2O at 20 °C when the $CO_2(g)$ pressure is maintained at 1 atm. What is the concentration of CO_2 in water that is saturated with air at 20 °C? Express this concentration as mL CO_2 (STP)/100 g water. Use the fact that the mole percent of CO_2 in air is 0.036%.

98. A 1.684-g sample of an unknown oxygen derivative of a hydrocarbon yields 3.364 g CO_2 and 1.377 g H_2O upon complete combustion. A 0.605 g sample of the same compound dissolved in 34.89 g water lowers the freezing point of the water to −0.244 °C. What is the molecular formula of the compound?

99. Suppose that, initially, solution A in Figure 12.16 is 200.0 g of an aqueous solution with a mole fraction of urea, $CO(NH_2)_2$, of 0.100 and solution B is 100.0 g of an aqueous solution with a mole fraction of urea of 0.0500. What will be the mass and mole fraction concentration of each solution when equilibrium is established, that is, when there is no longer a net transfer of water between the two solutions?

100. The illustration suggests the successive subdivision of a cube of material, 1.0 cm on edge. That is, first the cube is cut in half; then, each of the two sections is cut in half; and the four sections are again cut in half. In this first subdivision, there are eight cubes, each 0.50 cm on edge. Now the subdivision into cubes is repeated many times, as suggested by the series of arrows in the bottom row.

(a) How many successive subdivisions are required before the material enters the colloidal domain, that is, with a particle dimension less than 1000 nm?

(b) At the point described in part (a), what are the volume and total surface area of the particles of material, and how do these compare with the volume and surface area of the starting material? Explain the idea that colloidal particles have a much larger surface area-to-volume ratio than does bulk matter.

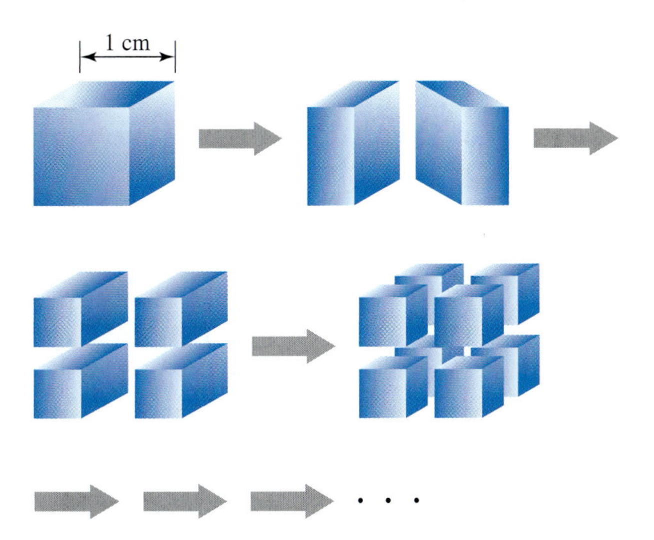

CHAPTER

13

Chemical Kinetics: Rates and Mechanisms of Chemical Reactions

CONTENTS

The rate of chirping of crickets is affected by temperature in much the same way that the rates of ordinary chemical reactions are affected, indicating that the underlying physiological processes are of a chemical nature. (See also Problem 93 on page 593.)

As we have seen, we can write chemical equations and use them to relate quantities of reactants and products in a chemical reaction (Chapter 3). We also have given special consideration to three important types of reactions—acid-base, precipitation, and oxidation-reduction (Chapter 4). And we have studied the heat effects that accompany chemical reactions (Chapter 6). You might wonder what more there is to learn about chemical reactions.

Actually, there are some important questions we have yet to answer, and we consider some of them in this chapter: How fast does a reaction go, that is, how much reactant is consumed or how much product is formed in a given time? Are there ways we can speed up or slow down a chemical reaction? How do reactions occur at the molecular level, that is, what is the step-by-step mechanism that leads from reactants to products?

● **ChemCDX**
When you see this icon, there are related animations, demonstrations, and exercises on the CD accompanying this text.

13.1 Chemical Kinetics—A Preview

The questions we posed in the introduction are the essence of **chemical kinetics**, a study of the rates of chemical reactions, the factors that affect these rates, and the sequences of molecular events—called reaction mechanisms—by which reactions occur.

Reactions proceed at different rates. Some require a very long time to consume the reactants; they are exceedingly slow. The disintegration of an aluminum can by atmospheric oxidation or of a plastic bottle by the action of sunlight can take years, decades, or even centuries. Some reactions occur so fast that the blink of an eye is a long time by comparison. A neutralization reaction, for example, occurs just as rapidly as we can mix the acid and base. Moreover, some reactions take place at widely different rates depending on conditions. Iron rusts rather rapidly in a humid environment, but in a desert region, iron corrodes so slowly that objects discarded half a century ago may show only slight signs of rust. Hydrogen and fluorine at room temperature form hydrogen fluoride in a highly exothermic and extremely rapid reaction. The reaction of hydrogen and oxygen to form water is also highly exothermic, but the reaction is immeasurably slow at room temperature. When the hydrogen-oxygen mixture is ignited at a high temperature, however, the reaction occurs with explosive speed (Figure 13.1). Reaction rates often increase dramatically as the temperature increases.

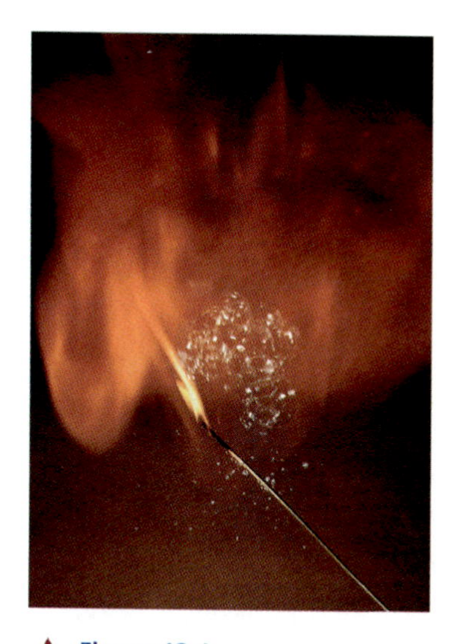

▲ **Figure 13.1**
The reaction of hydrogen and oxygen to form water
The hydrogen-filled soap bubbles, in contact with oxygen in the air, explode when they are ignited.

Carbon monoxide and nitrogen monoxide, both serious air pollutants, are produced in large quantities in automobile engines. Interestingly, these gases can react to form carbon dioxide and nitrogen, which are generally safer products.

$$2\,CO(g) + 2\,NO(g) \longrightarrow 2\,CO_2(g) + N_2(g)$$

Unfortunately, this reaction is extremely slow, even at high temperatures. However, we can speed up the reaction by using a catalyst, and that is what the catalytic converter in an automobile does. A **catalyst** is a substance that speeds up a reaction, but the catalyst emerges from the reaction unchanged. Obviously, a catalyst is not just a "spectator" (like the spectator ions in an ionic reaction). Later in the chapter, we will consider how catalysts work, including the action of enzymes, perhaps the most important catalysts of all.

Two goals of chemical kinetics that we will discuss in the first several sections of this chapter are to *measure* and to *predict* the rates of chemical reactions. For example, through experiments in laboratory smog chambers, atmospheric scientists have acquired enough data to predict pollution levels during smog episodes. Scientists also use rates of reactions to establish plausible step-by-step pathways or *mechanisms* of chemical reactions, an equally important topic. For example, scientists used data on the rates of a wide variety of atmospheric reactions to develop a plausible mechanism for the ozone depletion observed in the stratosphere in polar regions. This mechanism indicates that chlorine atoms from chlorofluorocarbons (CFCs) act as catalysts in ozone destruction. Confidence in this proposed mechanism has led the nations of the world to take measures to reduce and eventually eliminate their use of CFCs.

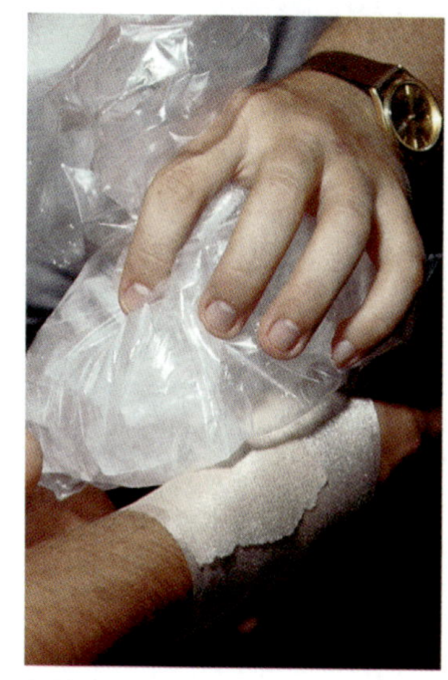

Athletic injuries often are treated by applying an ice pack as soon as possible. The biochemical reactions that cause inflammation are slower at the lower temperature, thus presumably lessening damage to the tissues. After a day or so, heat packs are sometimes applied to the injured area. Perhaps the higher temperature speeds up the reactions of the healing process.

If we can predict reaction rates, often we also can control them. We can speed up or slow down reactions by adjusting certain variables, such as the following.

- *Concentrations of reactants*. Reaction rates generally increase as we increase the concentrations of the reactants.
- *Temperature*. Reaction rates generally increase rapidly as we increase the temperature.

- *Surface area*. For reactions that occur on a surface rather than in a solution, the rate increases as the surface area is increased (Figure 13.2).
- *Catalysis*. Catalysis is the use of catalysts to speed up reactions.

13.2 The Meaning of the Rate of a Reaction

Rate or *speed* refers to how much something changes in a unit of time. Consider a runner moving at a speed of 16 km/h. The person's position changes by 16 km in one h. In a chemical reaction, the change is in the concentration of a reactant or product, which we can express as moles per liter (mol L^{-1}) or molarity (M). Thus, the rate of reaction has the units: *moles per liter per (unit of) time*. If we choose 1 s as the unit of time, the rate of reaction is expressed as $mol\ L^{-1}\ s^{-1}$ or $M\ s^{-1}$.

Consider the reaction in which sucrose (cane sugar) is broken down into the simpler sugars glucose and fructose. The reaction takes place in an aqueous solution with H^+ as a catalyst.

$$\underset{\text{Sucrose}}{C_{12}H_{22}O_{11}} + H_2O \xrightarrow{H^+} \underset{\text{Glucose}}{C_6H_{12}O_6} + \underset{\text{Fructose}}{C_6H_{12}O_6}$$

APPLICATION NOTE
Honeybees use this reaction to produce honey, a mixture of sucrose, glucose, and fructose. The bees use the enzyme *invertase* as a catalyst.

We can express the rate of reaction as the change in concentration of the *reactant* sucrose in a unit of time, or we can use the change in concentration of one of the *products*, that is, glucose or fructose.

In the expression below, $[\text{sucrose}]_1$ is the molarity of sucrose at an initial time, t_1, and $[\text{sucrose}]_2$ is the molarity at a *later* time, t_2. Using the delta (Δ) symbol to represent a change, the change in concentration is $\Delta[\text{sucrose}] = [\text{sucrose}]_2 - [\text{sucrose}]_1$. The change in time is $\Delta t = t_2 - t_1$. The rate at which sucrose is consumed, that is, the rate at which sucrose disappears is shown in the following equation.

$$\text{Rate of disappearance of sucrose} = \frac{\text{change in concentration of sucrose}}{\text{change in time}}$$
$$= \frac{[\text{sucrose}]_2 - [\text{sucrose}]_1}{t_2 - t_1} = \frac{\Delta[\text{sucrose}]}{\Delta t}$$

▲ **Figure 13.2**
The effect of surface area on the rate of a reaction
Finely divided and dispersed flour dust undergoes rapid combustion because of the large surface area on which the combustion reaction occurs. This swift flaming of flour has been responsible for many flour mill explosions.

Because sucrose is *consumed* in the reaction, $[\text{sucrose}]_2$ is *smaller* than $[\text{sucrose}]_1$. Thus, $\Delta[\text{sucrose}]$ is a *negative* quantity. By convention, we express rates of reactions as *positive* quantities, and so the rate of the reaction is the *negative* of the rate of disappearance of sucrose.

$$\text{Rate} = -\ \text{rate of disappearance of sucrose} = -\frac{\Delta[\text{sucrose}]}{\Delta t}$$

Now let's consider glucose, a product of the reaction. The concentration of glucose *increases* with time, and $[\text{glucose}]_2$ is larger than $[\text{glucose}]_1$. The rate of formation of glucose is a *positive* quantity.

$$\text{Rate} = \text{rate of formation of glucose} = \frac{\Delta[\text{glucose}]}{\Delta t}$$

To summarize, the **rate of a reaction** is the change in concentration of a product per unit of time (rate of formation of product) or the *negative* of the change in concentration of a reactant per unit of time (−rate of disappearance of reactant).

The General Rate of Reaction

Now let's consider the rate of the reaction by which the common household antiseptic hydrogen peroxide, $H_2O_2(aq)$, decomposes.

$$2\ H_2O_2(aq) \longrightarrow 2\ H_2O(l) + O_2(g)$$

We can base the rate on the disappearance of H_2O_2 or formation of O_2.

$$\text{Rate} = -\text{ rate of disappearance of } H_2O_2 = -\frac{\Delta[H_2O_2]}{\Delta t}$$

$$\text{Rate} = \text{rate of formation of } O_2 = \frac{\Delta[O_2]}{\Delta t}$$

However, these two rates are not the same.

From the balanced equation, we note that *two* moles of H_2O_2 are consumed to produce *one* mole of O_2; H_2O_2 disappears twice as fast as O_2 forms. Thus, the reaction rate based on the disappearance of H_2O_2 is *twice* the rate based on the formation of O_2. In describing the rate of a reaction, we may need to be specific about which reactant or product the rate refers to.

Alternatively, we can define a *general* rate of reaction that has the same value, regardless of which reactant or product we study. In the decomposition of hydrogen peroxide, to say that H_2O_2 disappears twice as fast as O_2 forms is the same as saying that O_2 forms only half as fast as H_2O_2 disappears.

$$\text{Rate} = -\frac{1}{2}\frac{\Delta[H_2O_2]}{\Delta t} = \frac{\Delta[O_2]}{\Delta t}$$

This expression is called the *general* rate of reaction for

$$2\ H_2O_2(aq) \longrightarrow 2\ H_2O(l) + O_2(g)$$

We get a general rate of reaction by *dividing* the rate of disappearance of a reactant or the rate of formation of a product by the stoichiometric coefficient of that reactant or product in the balanced chemical equation.

Applied to the more general reaction

$$a\,A + b\,B \longrightarrow c\,C + d\,D$$

the general rate of reaction is

$$\text{Rate} = -\frac{1}{a}\frac{\Delta[A]}{\Delta t} = -\frac{1}{b}\frac{\Delta[B]}{\Delta t} = \frac{1}{c}\frac{\Delta[C]}{\Delta t} = \frac{1}{d}\frac{\Delta[D]}{\Delta t}$$

To avoid confusion, in this chapter we will write chemical equations so that the general rate of reaction is the same as the rate based on the reactant or product whose concentration we are monitoring. For example, in the decomposition of hydrogen peroxide, we will monitor the concentration of hydrogen peroxide as a function of time, and we will therefore write the chemical equation in a form that shows 1 mol H_2O_2.

$$H_2O_2(aq) \longrightarrow H_2O(l) + \tfrac{1}{2}\ O_2(g)$$

We can then express the rate of reaction in the following form.

$$\text{Rate} = -\frac{\Delta[H_2O_2]}{\Delta t}$$

The Average Rate of Reaction

The decomposition of $H_2O_2(aq)$ starts off rapidly, but the rate slows down as more and more of the reactant decomposes. For this reason, when we use the expression, rate $= -\Delta[H_2O_2]/\Delta t$, what we calculate is an *average* rate of reaction during a time interval Δt. At the beginning of the interval, the rate is faster than this average rate; and at the end of the interval, it is slower. The situation is rather like taking your foot off the accelerator when driving an automobile at 50 mph and coming to a stop at a red light. Your average speed in this time interval may be 25 mph, but the actual speed in the interval would decrease through the range from 50 mph to 0 mph.

We calculate an average rate of reaction in Example 13.1 and illustrate different ways of expressing the rate of a reaction in Exercises 13.1A and 13.1B. We will determine more exact rates of reaction in the next section.

EXAMPLE 13.1

Consider the hypothetical reaction: $A + 2\,B \longrightarrow 3\,C + 2\,D$. Suppose that at some point during this reaction, $[A] = 0.4658$ M, and suppose that 125 s later $[A] = 0.4282$ M. **(a)** What is the average rate of reaction during this time period, based on the reactant A and expressed in M s^{-1}? **(b)** What is the rate of formation of C, expressed in M min^{-1}.

SOLUTION

a. Let's relate the rate of reaction to the rate of disappearance of A. Then we can substitute the change in concentration and the time interval into this expression.

$$\text{Rate} = -\text{ rate of disappearance of A} = -\frac{[A]_2 - [A]_1}{\Delta t}$$

$$= -\frac{0.4282\text{ M} - 0.4658\text{ M}}{125\text{ s}} = -\frac{(-0.0376\text{ M})}{125\text{ s}}$$

$$= 3.01 \times 10^{-4}\text{ M s}^{-1}$$

b. From the stoichiometric coefficients in the chemical equation, we see that *three* moles of C are produced for every mole of A that is consumed. This means that C is formed *three* times as fast as A disappears. In part (a), we described the rate of reaction in terms of the disappearance of A, so the rate of formation of C is simply three times the reaction rate.

$$\text{Rate of formation of C} = 3.01 \times 10^{-4}\,\frac{\cancel{\text{mol A}}}{\text{L s}} \times \frac{3\text{ mol C}}{1\,\cancel{\text{mol A}}} = 9.03 \times 10^{-4}\text{ M s}^{-1}$$

However, to express the rate of formation of C in the unit M min^{-1}, we must multiply this result by the conversion factor 60 s/1 min.

$$\text{Rate of formation of C} = 9.03 \times 10^{-4}\text{ M }\cancel{\text{s}^{-1}} \times \frac{60\,\cancel{\text{s}}}{1\text{ min}} = 5.42 \times 10^{-2}\text{ M min}^{-1}$$

EXERCISE 13.1A

Consider the hypothetical reaction $2\,A + B \longrightarrow 2\,C + D$. Suppose that at some point during the reaction $[D] = 0.2885$ M and that 2.55 min (2 min, 33 s) later $[D] = 0.3546$ M. **(a)** What is the average rate of reaction during this time period, based on the reactant D and expressed in M min^{-1}? **(b)** What is the rate of formation of C, expressed in M s^{-1}?

EXERCISE 13.1B

In the reaction $2\,A + B \longrightarrow 3\,C + D$, $-\Delta[A]/\Delta t$ is found to be 2.10×10^{-5} M s^{-1}. Using the idea of a general rate of reaction, what is **(a)** the rate of reaction, and **(b)** the rate of formation of C?

13.3 Measuring Reaction Rates

Over a period of time, the 3% aqueous solution of hydrogen peroxide many of us keep in our medicine cabinets loses its effectiveness because the H_2O_2 decomposes to oxygen gas and water.

$$H_2O_2(aq) \longrightarrow H_2O(l) + \tfrac{1}{2}\,O_2(g)$$

Figure 13.3 suggests a simple method of determining the rate at which $H_2O_2(aq)$ decomposes: Allow $O_2(g)$ to escape and determine the mass of the reaction mixture at various times. For example, the difference in mass for the 60-s time interval between Figure 13.3a and 13.3b is the mass of oxygen produced, 2.960 g. Some typical data are given in Table 13.1 on page 559. Footnote (b) in the table outlines the calculations by which we obtain the molarity of H_2O_2 remaining in solution from the mass of $O_2(g)$ evolved. We have plotted the molarities of H_2O_2 from Table 13.1 at the corresponding times, giving the black curve in Figure 13.4. The red, blue, and purple lines convey information about the rate of decomposition of H_2O_2 that we explore next.

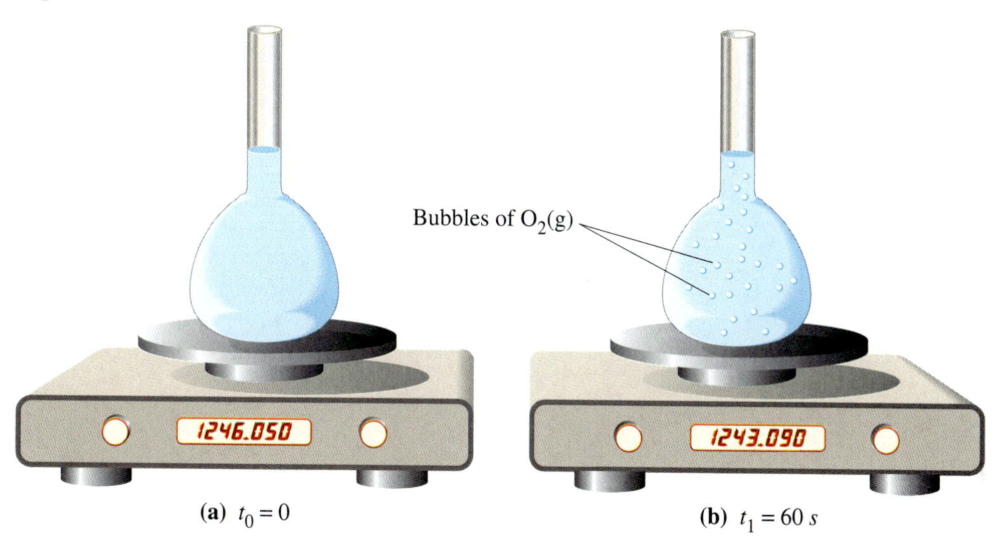

▲ **Figure 13.3 An experimental setup for determining the rate of decomposition of H_2O_2**
The difference between (a) the initial mass and (b) the mass at a later time, t_1 is the mass of $O_2(g)$ that was produced in the time interval. Because the reaction would otherwise occur too slowly, I^- (aq) has been added as a catalyst. The mechanism by which I^- catalyzes the reaction is discussed on page 579.

- *In general, the greater the concentration of a reactant, the faster the reaction goes.* We can confirm this statement by comparing data points in Table 13.1 and also by inspecting the black curve in Figure 13.4. For example, in the first 60 s, $[H_2O_2]$ drops by 0.185 M (from 0.882 M to 0.697 M). In contrast, in the 60-s interval from 540 s to 600 s, $[H_2O_2]$ drops by only 0.026 M (from 0.120 M to 0.094 M).
- The *average* rate of reaction during the experiment (from $t = 0$ to $t = 600$ s) is the *negative* of the slope of the dashed purple line. That is, the slope of the line is the average rate of disappearance of H_2O_2, and we change the sign to get a positive rate of reaction.

$$\text{Average rate} = -\ \text{average rate of disappearance of } H_2O_2 = -\frac{(0.094 - 0.882)\ \text{M}}{(600 - 0)\ \text{s}}$$
$$= 1.31 \times 10^{-3}\ \text{M s}^{-1}$$

- Because the rate constantly decreases during the reaction, to get a precise value at a particular point in the reaction, we must use a very short time-interval in the rate calculation. The rate of reaction at a given time is called an **instantaneous rate of reaction**.

$$\text{Instantaneous rate} = -\ \text{instantaneous rate of disappearance of } H_2O_2 = -\frac{\Delta[H_2O_2]}{\Delta t}$$

(where Δt is very small, that is, $\Delta t \longrightarrow 0$).

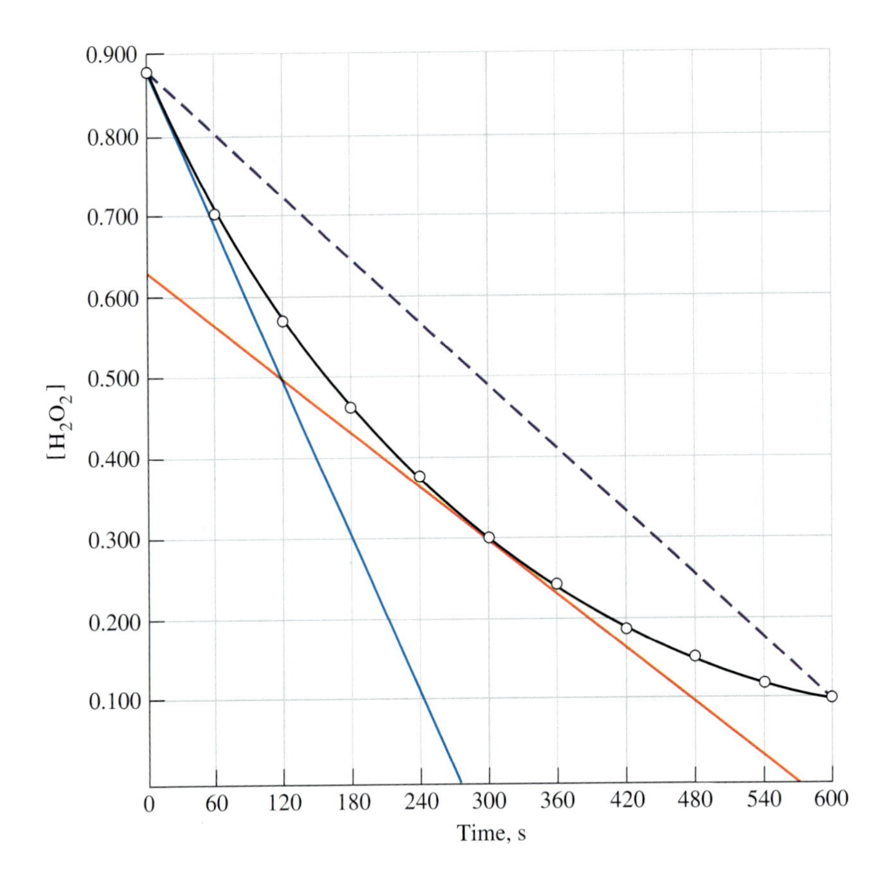

Figure 13.4
Kinetic data for the reaction: $H_2O_2(aq) \longrightarrow H_2O(l) + \frac{1}{2}O_2(g)$
Reaction rates are obtained from the slopes of the straight lines: an average rate from the purple line, the instantaneous rate at $t = 300$ s from the red line, and the initial rate from the blue line.

Alternatively, we can use the red line, a tangent line to the concentration-time graph (black). The *slope* of the tangent is the instantaneous rate of disappearance of H_2O_2, and the instantaneous reaction rate is the *negative of the slope of the tangent*.

- The instantaneous rate at the beginning of a reaction is called the **initial rate of reaction**. The initial rate is the negative of the slope of the blue tangent line in Figure 13.4. Note that the blue line and the black curve essentially coincide in the earliest stages of the reaction. At the very start of the reaction, the average and instantaneous rates are practically the same.
- When interpreting a plot of the concentration of a *product* versus time, all the preceding points still apply, with one important exception. Average, instantaneous, and initial rates are *equal* to the slopes of tangent lines. That is, no change of sign is required, because the slopes of the tangent lines are *positive*.

EXAMPLE 13.2

Use data from Table 13.1 and/or Figure 13.4, to **(a)** determine the initial rate of reaction and **(b)** calculate $[H_2O_2]$ at $t = 30$ s.

SOLUTION

a. Because the tangent line and concentration-time curve coincide at the start of a reaction, let's determine the initial rate in two ways.
Method One. Use the first two data points in Table 13.1.

$$\text{Initial rate} = -\frac{\Delta[H_2O_2]}{\Delta t} = -\frac{(0.697\text{ M} - 0.882\text{ M})}{(60 - 0)\text{ s}} = 3.08 \times 10^{-3}\text{ M s}^{-1}$$

TABLE 13.1 Decomposition of H_2O_2[a]

Time, s	Accumulated Mass O_2, g	$[H_2O_2]$, M[b]
0	0	0.882
60	2.960	0.697
120	5.056	0.566
180	6.784	0.458
240	8.160	0.372
300	9.344	0.298
360	10.336	0.236
420	11.104	0.188
480	11.680	0.152
540	12.192	0.120
600	12.608	0.094

[a] The decomposition of H_2O_2 in 1.00 L of 0.882 M, with I^- as the catalyst.
[b] Values of $[H_2O_2]$ are calculated in the manner outlined below for $t = 60$ s.

Number of moles of O_2 produced:

$$2.960\ \cancel{\text{g O}_2} \times \frac{1\ \text{mol O}_2}{32.00\ \cancel{\text{g O}_2}} = 0.09250\ \text{mol O}_2$$

Number of moles of H_2O_2 consumed:

$$0.09250\ \cancel{\text{mol O}_2} \times \frac{1\ \text{mol H}_2\text{O}_2}{\frac{1}{2}\ \cancel{\text{mol O}_2}} = 0.1850\ \text{mol H}_2\text{O}_2$$

Number of moles of H_2O_2 present at $t = 0$:

$$1.00\ \cancel{\text{L}} \times \frac{0.882\ \text{mol H}_2\text{O}_2}{1\ \cancel{\text{L}}} = 0.882\ \text{mol H}_2\text{O}_2$$

Number of moles of H_2O_2 remaining at $t = 60$ s:

$$0.882\ \text{mol H}_2\text{O}_2\ \text{initially} - 0.1850\ \text{mol H}_2\text{O}_2\ \text{consumed} = 0.697\ \text{mol H}_2\text{O}_2$$

$[H_2O_2]$ at $t = 60$ s:

$$[\text{H}_2\text{O}_2] = \frac{0.697\ \text{mol H}_2\text{O}_2}{1.00\ \text{L}} = 0.697\ \text{M}$$

Method Two. To obtain the initial rate from the blue tangent line in Figure 13.4, we need the slope of the line. Let's base the slope on the points where the line intersects the two axes. The intercept on the y axis is 0.882 M at 0 s. The intercept on the x axis is 0 M at an estimated time of 275 s.

$$\text{Slope of tangent} = \frac{y_2 - y_1}{x_2 - x_1} = \frac{(0 - 0.882)\ \text{M}}{(275 - 0)\ \text{s}} = -3.21 \times 10^{-3}\ \text{M s}^{-1}$$

$$\text{Initial rate} = -\text{slope of tangent line} = -(-3.21 \times 10^{-3}\ \text{M s}^{-1}) = 3.21 \times 10^{-3}\ \text{M s}^{-1}$$

The results agree rather well, but of the two, the initial rate based on the slope of the tangent line is likely to be the better one. Notice that in Figure 13.4 the blue and black lines begin to diverge before $t = 60$ s.

b. We can use the same equation for the initial rate as in part (a), but in this case $\Delta[H_2O_2]$ is the unknown in that equation. First, we need to solve for $\Delta[H_2O_2]$.

$$\text{Initial rate} = -\frac{\Delta[\text{H}_2\text{O}_2]}{\Delta t}$$

$$3.21 \times 10^{-3}\ \text{M s}^{-1} = -\frac{\Delta[\text{H}_2\text{O}_2]}{30\ \text{s}}$$

$$\Delta[\text{H}_2\text{O}_2] = -3.21 \times 10^{-3}\ \text{M s}^{-1} \times 30\ \text{s} = -0.096\ \text{M}$$

PROBLEM SOLVING NOTE

In general, we can use a calculation of the type $\Delta[H_2O_2]/\Delta t$ if the elapsed time (Δt) is less than a few percent of the total reaction time. Here it is 10%.

We can find a value for $[H_2O_2]_{(0\ s)}$ (0.882 M) in Table 13.1, so we can solve the following relationship for $[H_2O_2]_{(30\ s)}$ and then substitute the other two quantities.

$$\Delta[H_2O_2] = [H_2O_2]_{(30\ s)} - [H_2O_2]_{(0\ s)}$$

$$[H_2O_2]_{(30\ s)} = \Delta[H_2O_2] + [H_2O_2]_{(0\ s)}$$

$$= -0.096\ M + 0.882\ M = 0.786\ M$$

EXERCISE 13.2A

From Figure 13.4, **(a)** determine the instantaneous rate of reaction at $t = 300$ s. **(b)** Then use the result of (a) to calculate a value of $[H_2O_2]$ at $t = 310$ s.

EXERCISE 13.2B

Estimate the time at which the instantaneous rate of reaction is the same as the average rate given by the purple line in Figure 13.4.

(*Hint:* Draw a tangent line that has the same slope as the purple line. Is more than one tangent possible?)

13.4 The Rate Law of a Chemical Reaction

In Section 13.3, we used experimental data to obtain reaction rates during the decomposition of H_2O_2. We used both calculation and a graphical method (that is, from the slope of a tangent line), but there is a serious limitation: Our results applied only to a solution that was initially 0.882 M H_2O_2. What if we want to know the initial rates for the decomposition of 0.225 M H_2O_2 or 0.500 M H_2O_2 or some other concentration of H_2O_2? Do we have to do an experiment with each of these solutions to gather the necessary data?

On page 553, we noted that, in general, the greater the reactant concentrations the faster a reaction goes. This suggests that the initial rates for 0.225 M H_2O_2 and 0.500 M H_2O_2 aqueous solutions should both be smaller than the one in Example 13.2 because both of these solutions are less concentrated than 0.882 M H_2O_2. We can deal with this and other matters more definitively by using the *rate law* for the decomposition of H_2O_2.

The Rate Law and its Meaning

The **rate law** for a chemical reaction relates the rate of reaction to the concentrations of reactants. Consider again a general reaction of the type introduced on page 555.

$$a\,A + b\,B \cdots \longrightarrow c\,C + d\,D \cdots$$

The dots suggest that there may be additional reactants and products that are not indicated. We can express the general rate of reaction in terms of the rate of disappearance of the reactants.

$$\text{Rate} = -\frac{1}{a}\frac{\Delta[A]}{t} = -\frac{1}{b}\frac{\Delta[B]}{t} = \cdots$$

The rate law for this reaction is of the form

$$\text{Rate} = k[A]^m[B]^n \cdots$$

In the rate law, [A], [B], $\cdots$ are molarities of the reactants at a particular time. The exponents m, n, $\cdots$ are generally small positive integers (0, 1, 2), but they may be negative and occasionally nonintegral.

The exponents in a rate law must be determined by experiment. They are not derived from the stoichiometric coefficients in an overall chemical equation, but in some instances—for example, in some simple, one-step reactions (Section 13.9)—they may be the same.

The values of the exponents in a rate law establish the **order of a reaction**. If $m = 1$, the reaction is *first order* in A. If $n = 2$, the reaction is *second order* in B, and so on. The *overall order* of a reaction is the sum of the exponents in a rate law: $m + n + \cdots$.

The proportionality constant, k, is the **rate constant**. The numerical value of k depends on (1) the particular reaction, (2) the temperature, and (3) the presence of a catalyst (if any). As we shall see shortly, the units of k depend on the values of $m, n, \cdots$.

Now let's consider again the decomposition of H_2O_2. The rate law is a simple one—the reaction is first order.

$$\text{Rate} = k[H_2O_2]^1 = k[H_2O_2]$$

Notice that the exponent $m = 1$ is equal to the coefficient in the chemical equation when the equation is written as

$$H_2O_2(aq) \longrightarrow H_2O(l) + \tfrac{1}{2}O_2(g)$$

but not when it is written as

$$2\,H_2O_2(aq) \longrightarrow 2\,H_2O(l) + O_2(g)$$

Nevertheless, the reaction is first order, no matter how we choose to write the chemical equation.

We usually write the rate law in a form to calculate a rate, but we can also solve it for another quantity. For example, when a rate has been determined experimentally, we can use the measured rate and the corresponding concentration(s) to calculate the rate constant k. In Example 13.2, we found the initial rate to be $3.21 \times 10^{-3}\ \text{M s}^{-1}$ when $[H_2O_2] = 0.882$ M, and we can calculate the rate constant from those values.

$$k = \frac{\text{Rate}}{[H_2O_2]} = \frac{3.21 \times 10^{-3}\ \cancel{\text{M}}\ \text{s}^{-1}}{0.882\ \cancel{\text{M}}} = 3.64 \times 10^{-3}\ \text{s}^{-1}$$

The unit s^{-1} is what we expect for k in a first-order reaction. The product of s^{-1} and the molarity unit M is M s^{-1}, the units for a reaction rate.

With the rate law and a value of k, we can calculate initial rates for *any* initial concentration of H_2O_2, which is what we originally set out to do. Thus, for an initial $[H_2O_2] = 0.775$ M, the following holds true.

$$\text{Rate} = k[H_2O_2] = 3.64 \times 10^{-3}\ \text{s}^{-1} \times 0.775\ \text{M} = 2.82 \times 10^{-3}\ \text{M s}^{-1}$$

More About the Rate Constant *k*

We just confirmed that s^{-1} is an appropriate unit for k for a first-order reaction. What are the units for reactions of other orders? For the hypothetical zero-order reaction, $A \rightarrow B$, the rate law is as follows.

$$\text{Rate} = k[A]^0 = k$$

Because any quantity raised to the zero power is equal to one, the rate is equal to the rate constant k. For zero-order reactions, the rate constant has the same units as a rate of reaction, M s^{-1}.

The reaction of hydrogen and iodine monochloride to produce iodine and hydrogen chloride is first order in H_2, first order in ICl, and second order overall.

$$H_2(g) + 2\,ICl(g) \longrightarrow I_2(g) + 2\,HCl(g)$$
$$\text{Rate} = k[H_2][ICl]$$

Note that the units of k for a second-order reaction must be $M^{-1}\ s^{-1}$ and that the product of units in the rate law gives $M\ s^{-1}$ as the units for the reaction rate.

$$\text{Rate} = \overbrace{k}^{M^{-1}\,s^{-1}} \times \overbrace{[H_2]}^{M} \times \overbrace{[ICl]}^{M} = M\ s^{-1}$$

Overall reaction order	Units for k
Zero	$M\ s^{-1}$
First	s^{-1}
Second	$M^{-1}\ s^{-1}$
Third	$M^{-2}\ s^{-1}$

From these examples and the brief summary in the margin, we see that the unit of k for a reaction of *any* overall order is $M^{(1-\text{overall order})}\ s^{-1}$.

Sometimes the distinction between the *rate* of a reaction and the *rate constant* of a reaction can be confusing. It helps to remember these points.

- The two terms have quite different meanings. The *rate* of a reaction is the change in concentration with time, whereas the *rate constant* is the proportionality constant relating reaction rate to the concentrations of reactants.
- The rate constant remains *constant* throughout a reaction, regardless of the initial concentrations of the reactants. Except for a few cases, the rate of a reaction *varies* as concentrations vary.
- The rate and rate constant have the same numerical values *and* units only in *zero*-order reactions.
- For reaction orders other than zero, the rate and rate constant are numerically equal *only* when the concentrations of all reactants are 1 M. Even then, their units are different.

The Method of Initial Rates

To determine the rate law for a reaction, it is essential to devise experiments to determine the *exponents* ($m, n, \cdots$) in the rate law. One way to do this is to determine the initial rate for different initial concentrations of the reactants, a procedure called the **method of initial rates**. In this method we set up a series of experiments in which the initial concentrations of some reactants are held constant and others are varied in convenient multiples. We equate the initial rate to the average rate at the start of the reaction, at which point the two are essentially equal. Also, we know the exact concentrations of reactants with certainty at the beginning of the reaction. If there is any tendency for a reverse reaction to occur, that is, for products to react to reform the reactants, the effect is minimal at the very start of the reaction. We illustrate the method using the following reaction.

$$2\ NO(g) + Cl_2(g) \longrightarrow 2\ NOCl(g)$$

Table 13.2 lists data for three experiments. Each experiment differs from the others only in the initial concentration of one reactant. For example, the initial $[Cl_2]$ is the same for Experiments 1 and 3, but the initial [NO] in Experiment 3 is *twice* that of Experiment 1. In contrast, the initial rate in Experiment 3 is *four* times that of Experiment 1. Let's put these facts in mathematical form by writing the rate law for each experiment and obtaining a ratio of the two initial rates. In the expression below, the subscript numbers 1 and 3 refer to Experiments 1 and 3, respectively. Quantities that are the same for each experiment cancel.

$$\frac{(\text{Initial rate})_3}{(\text{Initial rate})_1} = \frac{\cancel{k}[NO]_3^m \cancel{[Cl_2]_3^n}}{\cancel{k}[NO]_1^m \cancel{[Cl_2]_1^n}} = \frac{(2 \times 0.0125)^m}{(0.0125)^m} = \frac{2^m \times \cancel{(0.0125)^m}}{\cancel{(0.0125)^m}} = 2^m$$

TABLE 13.2 Initial Rates of the Reaction: $2\ NO(g) + Cl_2(g) \longrightarrow 2\ NOCl(g)$

Experiment	Initial [NO]	Initial $[Cl_2]$	Initial Rate, $M\ s^{-1}$
1	0.0125 M	0.0255 M	2.27×10^{-5}
2	0.0125 M	0.0510 M	4.55×10^{-5}
3	0.0250 M	0.0255 M	9.08×10^{-5}

Thus, the ratio of the initial rates is

$$\frac{(\text{Initial rate})_3}{(\text{Initial rate})_1} = 2^m$$

However, we can also calculate the actual value of this ratio.

$$\frac{(\text{Initial rate})_3}{(\text{Initial rate})_1} = \frac{9.08 \times 10^{-5}\ \cancel{M\ s^{-1}}}{2.27 \times 10^{-5}\ \cancel{M\ s^{-1}}} = 4$$

The left sides of these two expressions are the same, and therefore $2^m = 4$. The value of $m = 2$, that is, $2^2 = 4$, and thus the reaction is *second* order in NO.

We can determine the exponent n by comparing Experiments 1 and 2.

$$\frac{(\text{Initial rate})_2}{(\text{Initial rate})_1} = \frac{\cancel{k}\ \cancel{[NO]_2^m}\ [Cl_2]_2^n}{\cancel{k}\ \cancel{[NO]_1^m}\ [Cl_2]_1^n} = \frac{(2 \times 0.0255)^n}{(0.0255)^n}$$

$$= \frac{2^n \times \cancel{(0.0255)^n}}{\cancel{(0.0255)^n}} = 2^n = \frac{4.55 \times 10^{-5}\ \cancel{M\ s^{-1}}}{2.27 \times 10^{-5}\ \cancel{M\ s^{-1}}} = 2$$

The value of $n = 1$, that is, $2^1 = 2$, and the reaction is *first* order in Cl_2.

Thus, the rate law for the reaction is *third* order overall.

$$\text{Rate} = k[NO]^2[Cl_2]$$

This is an example of a reaction in which—by coincidence—reaction orders and stoichiometric coefficients coincide.

We will obtain the value of k in Example 13.3.

The following list summarizes the effects on the initial rate caused by doubling the concentration of one reactant while the concentrations of the others are held constant. If the reaction is

- *Zero* order in the reactant, there is *no effect* on the rate
- *First* order in the reactant, the rate *doubles*
- *Second* order in the reactant, the rate *quadruples*
- *Third* order in the reactant, the rate increases *eightfold*

EXAMPLE 13.3

Refer to the reaction described in the text and in Table 13.2. **(a)** What would be the initial rate in Experiment 4, which has $[NO] = 0.0500$ M and $[Cl_2] = 0.0255$ M? **(b)** What is the value of k for the reaction?

SOLUTION

To answer both questions, we need the rate law for the reaction, which we established as rate $= k[NO]^2[Cl_2]$. We can most easily answer part (a) by comparing the initial rates in Experiments 3 and 4. To answer part (b), we can use the initial rate and initial concentrations for any of the four experiments to solve the rate law for k. Because k is a rate *constant*, we expect the same value of k in each case.

a. The value of $[Cl_2]$ is the same in Experiments 3 and 4, and the value of [NO] in Experiment 4 (0.0500 M) is *twice* that in Experiment 3 (0.0250 M). Because the reaction is *second* order in NO, doubling its initial concentration causes a *fourfold* increase in the initial rate.

$$(\text{Initial rate})_4 = 4 \times (\text{initial rate})_3 = 4 \times 9.08 \times 10^{-5}\ \text{M s}^{-1}$$
$$= 3.63 \times 10^{-4}\ \text{M s}^{-1}$$

b. We can solve the rate law for k and then substitute data from Table 13.2, for example, for Experiment 1.

$$k = \frac{\text{Initial rate}}{[\text{NO}]^2[\text{Cl}_2]} = \frac{2.27 \times 10^{-5}\ \text{M s}^{-1}}{(0.0125\ \text{M})^2(0.0255\ \text{M})} = 5.70\ \text{M}^{-2}\ \text{s}^{-1}$$

EXERCISE 13.3A

Refer to Example 13.3 and calculate the initial rate if the initial concentrations are $[\text{NO}] = 0.200$ M and $[\text{Cl}_2] = 0.400$ M.

(*Hint:* Recall that we now know a value of k.)

EXERCISE 13.3B

In the thermal decomposition of acetaldehyde,

$$CH_3CHO(g) \longrightarrow CH_4(g) + CO(g)$$

the rate of reaction is found to increase by a factor of about 2.8 when the initial concentration of acetaldehyde is doubled. What is the order of the reaction?

(*Hint:* Can the order be integral?)

13.5 First-Order Reactions

In this section, we focus our discussion on **first-order reactions** in which a single reactant yields products—that is, reactions of the type

$$\text{A} \longrightarrow \text{products}$$

The rate law for such a reaction is written as follows.

$$\text{Rate} = k[\text{A}]^1 = k[\text{A}]$$

Concentration As a Function of Time: The Integrated Rate Law

In the preceding section, we noted that decomposition of $H_2O_2(aq)$ is a first-order reaction with the rate law, rate $= k[H_2O_2]$, and we used this rate law to evaluate k and to calculate the rate of decomposition at various concentrations. However, such calculations will not answer this rather basic question: How long can 3% $H_2O_2(aq)$ be stored before decomposition progresses to the point at which the solution loses its antiseptic qualities? In this and many other practical situations, what we want to know is this:

What will be the concentration of a reactant at a later time if we know its concentration initially?

In the language of calculus, a rate law is called a *differential* equation. The equation is solved by a technique called *integration*, which we apply to a first-order rate law in Appendix A.

We can answer this question by using an equation derived from the rate law. The **integrated rate law** is an equation that describes the concentration of a reactant as a function of time. The form of the equation depends on the overall order of the reaction. For a first-order reaction, the integrated rate law has the following form.

$$\ln \frac{[\text{A}]_t}{[\text{A}]_0} = -kt$$

Refer to Appendix A for a review of the characteristics of logarithms.

In this equation, k has the same meaning as in the rate law; it is the rate constant. The elapsed time in the reaction is represented by t. The concentrations of A are $[\text{A}]_t$ at time t and $[\text{A}]_0$ at $t = 0$. The initial concentration is $[\text{A}]_0$. We express the concentration terms as a ratio, and we take the natural logarithm of the ratio, denoted by the symbol ln*.

The integrated rate law for a first-order reaction is the equation of a *straight line*. We can get the more familiar form of a straight line by replacing $\ln [A]_t/[A]_0$ by the equivalent term, $\ln [A]_t - \ln [A]_0$, and solving for $\ln [A]_t$.

$$\ln [A]_t - \ln [A]_0 = -kt$$

Integrated rate equation for first-order reaction: $\underbrace{\ln [A]_t}_{y} = \underbrace{(-k)}_{m}\underbrace{t}_{x} + \underbrace{\ln [A]_0}_{b}$

Equation of straight line: $y = mx + b$

If we plot $\ln [A]_t$ against time, t, and the result is a straight line, the reaction is first order and the slope of the line is the *negative* of k. If the plot is not a straight line, the reaction is not first order.

Data for the first-order decomposition of H_2O_2 from Table 13.1 are presented again in Table 13.3, this time also listing $\ln [H_2O_2]$. The straight-line plot of $\ln[H_2O_2]$ versus time in Figure 13.5 indicates that the reaction is first order. We can derive the value of the rate constant, k, from the slope (m) of the straight-line plot as follows.

$$m = (y_2 - y_1)/(x_2 - x_1) = -1.50/410 \text{ s} = -3.66 \times 10^{-3} \text{ s}^{-1}$$

$$k = -(\text{slope}) = -(-3.66 \times 10^{-3} \text{ s}^{-1}) = 3.66 \times 10^{-3} \text{ s}^{-1}$$

TABLE 13.3 Decomposition of H_2O_2: Data Required to Test for a First-Order Reaction

Time, s	$[H_2O_2]$, M	$\ln [H_2O_2]$
0	0.882	−0.126
60	0.697	−0.361
120	0.566	−0.569
180	0.458	−0.781
240	0.372	−0.989
300	0.298	−1.21
360	0.236	−1.44
420	0.188	−1.67
480	0.152	−1.88
540	0.120	−2.12
600	0.094	−2.36

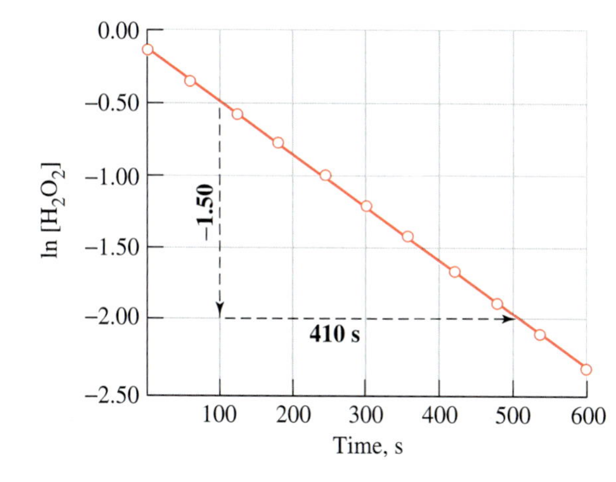

◀ **Figure 13.5 Test for a first order reaction: decomposition of $[H_2O_2]$**
The data plotted are from Table 13.3. The straight-line plot of $\ln[H_2O_2]$ versus time proves the reaction to be first order.

Example 13.4 and Exercise 13.4A illustrate two types of calculations that are possible with the integrated rate law for a first-order reaction. Exercise 13.4B applies the integrated rate law in a broader context.

EXAMPLE 13.4

Given a value of $k = 3.66 \times 10^{-3} \text{ s}^{-1}$ and initial $[H_2O_2] = 0.882$ M, determine the following for the first-order decomposition of $H_2O_2(aq)$: **(a)** the time at which $[H_2O_2] = 0.600$ M; **(b)** $[H_2O_2]$ after 225 s.

SOLUTION

For both parts of the question, we must use the *integrated* rate law for this first-order reaction: $\ln [H_2O_2]_t/[H_2O_2]_0 = -kt$. In each part, we know three of the four quantities in the equation and can solve for the fourth.

* The units of molarity cancel in the numerator and denominator of the ratio $[A]_t/[A]_0$, yielding a dimensionless quantity. We can take the logarithms only of pure numbers.

a. Because we know both $[H_2O_2]_t$ and $[H_2O_2]_0$, we can evaluate the left side of the equation and solve for t on the right side.

$$\ln \frac{0.600\ \cancel{M}}{0.882\ \cancel{M}} = -3.66 \times 10^{-3}\ s^{-1} \times t$$

$$\ln 0.680 = -0.385 = -3.66 \times 10^{-3}\ s^{-1} \times t$$

$$t = \frac{-0.385}{-3.66 \times 10^{-3}\ s^{-1}} = 105\ s$$

b. We know k and t and can solve for the left side of the integrated rate law.

$$\ln \frac{[H_2O_2]_t}{[H_2O_2]_0} = -kt = -\{3.66 \times 10^{-3}\ \cancel{s^{-1}} \times 225\ \cancel{s}\} = -0.824$$

Next we need to find the number whose natural logarithm is -0.824 and which is equal to the ratio of concentration terms.

$$\frac{[H_2O_2]_t}{[H_2O_2]_0} = e^{-0.824} = 0.439$$

Finally, we can use the known value of $[H_2O_2]_0$ and calculate $[H_2O_2]_t$.

$$[H_2O_2]_t = 0.439 \times [H_2O_2]_0 = 0.439 \times 0.882\ M = 0.387\ M$$

PROBLEM-SOLVING NOTE

To obtain ln 0.680 using a simple electronic calculator, enter "0.680", followed by the "ln" key. To find the number having -0.824 as its logarithm means to evaluate $e^{-0.824}$. That is, a number, N, and $\ln N$, are related as follows: $N = e^{\ln N}$. Enter the number "-0.824" into an electronic calculator, followed by the "e^x" key (or the "INV" key, then the "ln" key). If you use a graphing calculator, a slightly different procedure may be required.

EXERCISE 13.4A

The decomposition of nitramide, NH_2NO_2, is a first-order reaction.

$$NH_2NO_2(aq) \longrightarrow H_2O(l) + N_2O(g)$$

The rate law is rate $= k[NH_2NO_2]$, with $k = 5.62 \times 10^{-3}\ min^{-1}$ at 15 °C. Starting with 0.105 M NH_2NO_2, **(a)** at what time will $[NH_2NO_2] = 0.0250$ M and **(b)** what is $[NH_2NO_2]$ after 6.00 h?

EXERCISE 13.4B

Refer to the reaction in Exercise 13.4A. Starting with a 0.0750 M NH_2NO_2, what is the *rate* of the reaction after 35.0 min?

(*Hint:* You will need to use both the rate law and the integrated rate law.)

Variations of the Integrated Rate Law

At times, it is convenient to replace molarities in an integrated rate law by quantities that *are proportional to concentration*. For example, if we multiply the molarity (M) by the volume of a reaction mixture (V) we get the number of moles (n) of reactant. Further multiplication by the molar mass gives the mass of the reactant. Thus, mass is proportional to molarity.

In gas-phase reactions, it is easy to measure pressures. In a constant-volume mixture at a constant temperature, the partial pressure of a gas is proportional to its molarity. We can see this by rearranging the ideal gas equation. That is, rearrange $PV = nRT$ to $P = (n/V) \times RT$. Replace n/V by the molarity, M, and, because R and T are constants, P = constant $\times$ M.

We will make use of these small variations by substitutions in the integrated rate law in the next section.

Half-life of a Reaction

We often describe a first-order reaction in terms of the time required to reduce a reactant concentration (or mass or partial pressure) to a *fraction* of its initial value. A convenient fraction is one half ($\frac{1}{2}$), for then we can define a *half-life*. The **half-life** ($t_{1/2}$) of a reaction is the time required to consume one-half of the reactant orig-

is between *three* and *four* half-life periods. If we use a half-life of about 120 s, which we found in Example 13.5, we see that our answer is as follows.

$$3 \times 120 \text{ s} < time, t < 4 \times 120 \text{ s}$$
$$360 \text{ s} < time, t < 480 \text{ s}$$

Of the values we have to choose from, only (c) is probable: 400 s.

EXERCISE 13.6

Without doing detailed calculations or extrapolating the graph, estimate $P_{N_2O_5}$ at 370 s in the graph of Figure 13.6.

PROBLEM-SOLVING NOTE

You don't need to start at $t = 0$ in a graph such as Figure 13.6 when applying the half-life concept. The difference in time between *any* two points which differ in molarity or pressure by a factor of 2 is a half-life, $t_{1/2}$. For example, between $P = 600$ mmHg at about 50 s and $P \approx 300$ mmHg at about 170 s, one half-life has elapsed (about 120 s).

13.6 Reactions of Other Orders

The reactions that we will consider in this section are zero order and second order.

Zero-Order Reactions

The idea that the rates of some reactions are *independent* of the concentrations of the reactants may seem strange at first. However, sometimes factors other than reactant concentrations control how fast reactants can enter into a reaction. For example, in a reaction that requires the absorption of light, the intensity of light determines the reaction rate. In a surface-catalyzed reaction, the available surface area determines the rate. In the rate law for a reaction that is **zero order** overall, the sum of the exponents is zero: $m + n + \cdots = 0$.

The reaction by which ammonia, NH_3 (g), decomposes on a tungsten (W) surface is zero order.

$$NH_3(g) \xrightarrow{W} \tfrac{1}{2} N_2(g) + \tfrac{3}{2} H_2(g)$$
$$\text{Rate} = k[NH_3]^0 = k$$

In Figure 13.7, the concentration of $NH_3(g)$ is plotted as a function of time for two different initial concentrations: 2.00×10^{-3} M (blue) and 1.00×10^{-3} M (red). Both experiments were carried out at 1100 °C. Note the following.

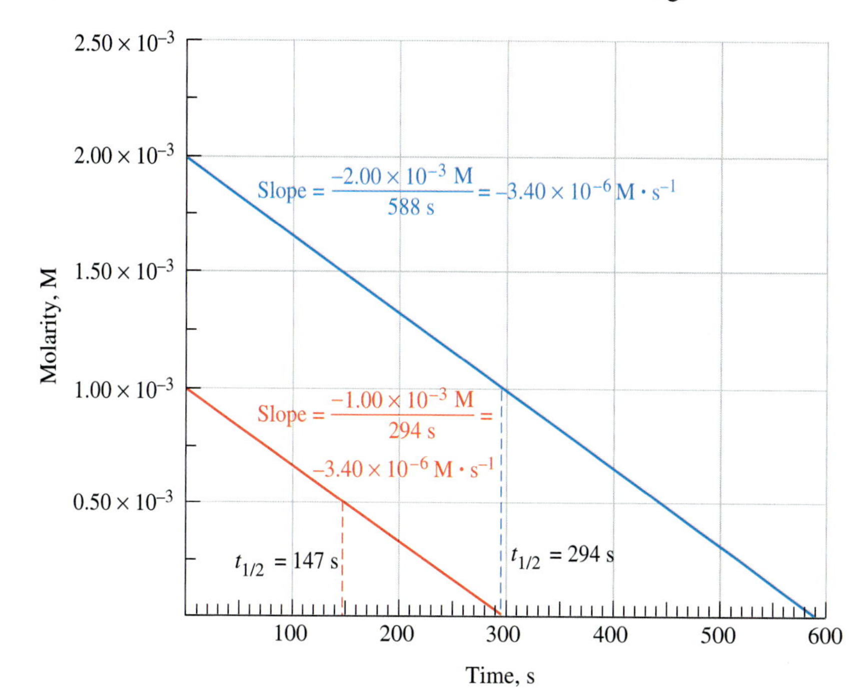

Figure 13.7
The decomposition of ammonia on a tungsten surface at 1100 °C: a zero-order reaction
The concentration of $NH_3(g)$ is plotted as a function of time for two different initial concentrations. The parallel, straight-line graphs show this to be a zero-order reaction. The slopes of the lines are indicated, as are the half-lives for the two experiments.

- The graph of concentration versus time for each experiment is a *straight line* with a *negative* slope. The blue and red lines are *parallel*. They have the same slope: -3.40×10^{-6} M s^{-1}.
- The rate of reaction, which remains *constant* throughout the reaction, is equal to the rate constant k and to the *negative* of the slope.

$$\text{Rate} = k = -(-3.40 \times 10^{-6}\ \text{M s}^{-1}) = 3.40 \times 10^{-6}\ \text{M s}^{-1}.$$

- Because the reaction rate has the same value at all points in a *zero-order* reaction, the rate is also independent of the initial concentration of the reactant.
- The half-life is *different* for the two experiments; it is directly proportional to the initial concentration. In Figure 13.7, we see that the half-life of 147 s when $[NH_3]_0 = 1.00 \times 10^{-3}$ M doubles to 294 s when $[NH_3]_0$ is doubled, to 2.00×10^{-3} M.

Second-Order Reactions

A **second-order reaction** has a rate law in which the sum of the exponents, $m + n + \cdots = 2$. One example is the reaction of NO(g) and O_3(g).

$$NO(g) + O_3(g) \longrightarrow NO_2(g) + O_2(g)$$

$$\text{Rate} = k[NO][O_3]$$

A simpler type of second-order reaction depends on the concentration of a single reactant: A $\longrightarrow$ products.

$$\text{Rate} = k[A]^2$$

The integrated rate law, which expresses [A] as a function of time, has the following form.

A brief derivation of this integrated rate law is given in Appendix A.

$$\frac{1}{[A]_t} = kt + \frac{1}{[A]_0}$$

From this equation, we see that a graph of 1/[A] versus time is a straight line. The slope of the straight line is equal to the rate constant, k, and the intercept ($t = 0$) is $1/[A]_0$. We can obtain the *half-life*, by substituting $[A]_t = \frac{1}{2}[A]_0$ into the integrated rate law and simplifying.

$$t_{1/2} = \frac{1}{k[A]_0}$$

As with a zero-order reaction, *the half life of a second-order reaction depends on the initial concentration as well as on the rate constant k.*

EXAMPLE 13.7

The second-order decomposition of HI(g) at 700 K is represented in Figure 13.8.

$$HI(g) \longrightarrow \tfrac{1}{2}H_2(g) + \tfrac{1}{2}I_2(g);\ \text{Rate} = k[HI]^2$$

What are the **(a)** rate constant and **(b)** half life of the decomposition of 1.00 M HI(g) at 700 K?

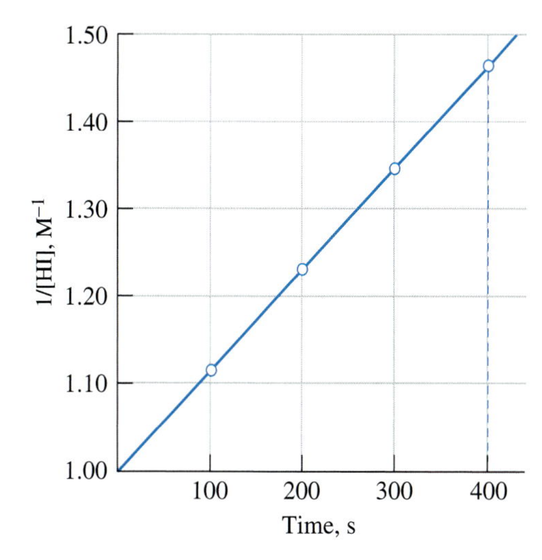

Figure 13.8
The decomposition of hydrogen iodide at 700 K: a second-order reaction
The *reciprocal* of the concentration of HI(g), 1/[HI], is plotted as a function of time. The rate constant for this second-order reaction is equal to the slope of the line: $k = 1.2 \times 10^{-3}\ M^{-1}\ s^{-1}$.

SOLUTION

In this problem, $[HI]_0 = 1.00$ M. This makes $1/[HI]_0 = 1.00\ M^{-1}$ at $t = 0$ s. For a second-order reaction, we can obtain k from the slope of the straight-line graph of $1/[A]_t$ versus t, and $t_{1/2}$ from the equation, $t_{1/2} = 1/k[A]_0$.

a. Let's base the slope of the line in Figure 13.8 on the points (1.46 M^{-1}, 400 s) and (1.00 M^{-1}, 0 s). In the equation for the slope, Δy is the length of the broken line in Figure 13.8: $1.46\ M^{-1} - 1.00\ M^{-1} = 0.46\ M^{-1}$; Δx is the difference, 400 s − 0 s = 400 s. The rate constant, k, is *equal* to the slope.

$$k = \text{slope} = \frac{\Delta y}{\Delta x} = \frac{0.46\ M^{-1}}{400\ s} = 1.2 \times 10^{-3}\ M^{-1}\ s^{-1}$$

b. We now have values of $[HI]_0$ and k to substitute in the half-life equation.

$$t_{1/2} = \frac{1}{k[A]_0} = \frac{1}{1.2 \times 10^{-3}\ M^{-1}\ s^{-1} \times 1.00\ M} = 8.3 \times 10^2\ s$$

PROBLEM-SOLVING NOTE
An equation for a half-life must be of a form that yields a unit of time for the half-life, regardless of the order of the reaction.

EXERCISE 13.7A

If it takes 55 s for [A] to fall to 0.40 M from an initial [A] = 0.80 M in a reaction with Rate = $k[A]^2$, what is the rate constant k for the reaction?

EXERCISE 13.7B

In the reaction described in Exercise 13.7A, how long would it take for [A] to fall to 0.20 M? 0.10 M? Describe how the half-life varies from one half-life period to the next in the reaction.

EXAMPLE 13.8—A Conceptual Example

Shown on page 572 are graphs of [A] versus time for two different experiments dealing with the reaction: A $\longrightarrow$ products. What is the order of this reaction?

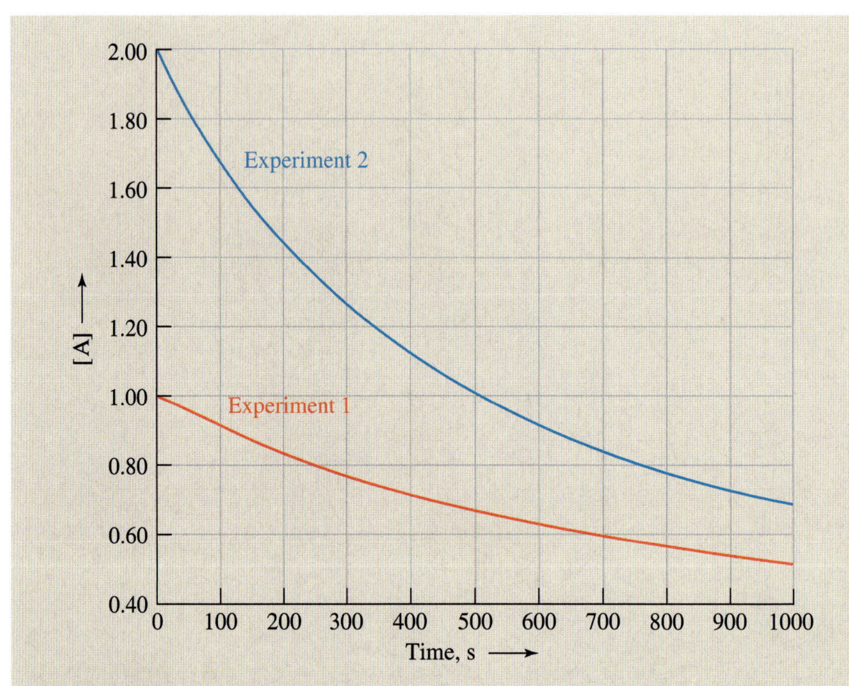

SOLUTION

Our task is to find the value of the exponent m in the rate law

$$\text{Rate} = k[\text{A}]^m$$

To assist in our task, let's refer to Table 13.4, a compilation of facts about zero-, first-, and second-order reactions of the type: A $\longrightarrow$ products.

Test for Zero-Order Reaction If the reaction were zero order, the rate of reaction would be constant (rate $= k$), and the graphs would be straight lines. The reaction is not zero order.

Test for First-Order Reaction One way to test for first order is to see if the half-life is constant. In Experiment 1, [A] falls from 1.00 M to one-half this value, 0.50 M, in 1000 s. In Experiment 2, [A] falls from 2.00 M to 1.00 M in 500 s. The half-life is *not* constant; the reaction is *not* first order.

Test for Second-Order Reaction There are *three* methods that we might use: (1) Obtain the *initial* rates of reaction, either from the slopes of tangent lines or by evaluating $-(\Delta[\text{A}]/\Delta t)$. Then apply the method of initial rates. (2) Determine if the graph 1/[A] versus time is a straight line. (3) Obtain the value of k from the half-life in each of the two experiments, using the equation: $t_{1/2} = 1/k[\text{A}]_0$. If the value of k is the same for the two, the reaction is second order.

TABLE 13.4 A Summary of Kinetic Data for Reactions of the Type: A $\longrightarrow$ Products

Order	Rate Law	Integrated Rate Law	Straight-Line Plot	$k =$	Units of k	Half-life
0	Rate $= k$	$[\text{A}]_t = -kt + [\text{A}]_0$	[A] vs. t	−slope	M s^{-1}	$\frac{[\text{A}]_0}{2k}$
1	Rate $= k[\text{A}]$	$\ln \frac{[\text{A}]_t}{[\text{A}]_0} = -kt$	ln [A] vs. t	−slope	s^{-1}	$\frac{0.693}{k}$
2	Rate $= k[\text{A}]^2$	$\frac{1}{[\text{A}]_t} = kt + \frac{1}{[\text{A}]_0}$	1/[A] vs. t	slope	$\text{M}^{-1}\,\text{s}^{-1}$	$\frac{1}{k[\text{A}]_0}$

Let's use the third method and begin by relating k to $t_{1/2}$, using the half-lives we found when applying the test for a first order reaction.

$$k = \frac{1}{t_{1/2}[A]_0}$$

Experiment 1

When $[A]_0 = 1.00\ M$, $t_{1/2} = 1000\ s$: $k = \dfrac{1}{1000\ s \times 1.00\ M} = 1.00 \times 10^{-3}\ M^{-1}\ s^{-1}$

Experiment 2

When $[A]_0 = 2.00\ M$, $t_{1/2} = 500\ s$: $k = \dfrac{1}{500\ s \times 2.00\ M} = 1.00 \times 10^{-3}\ M^{-1}\ s^{-1}$

The two values of k are the same, indicating that the reaction is second order.

EXERCISE 13.8

Use data from the graphs in Example 13.8 to show that the reaction is second order by using the first two of the three methods outlined.

13.7 Theories of Chemical Kinetics

So far, we have explored a number of practical matters related to the rates of chemical reactions, none of which required us to look at what happens at the molecular level. However, you may still be wondering, why are some reactions first order, others second order, and so on? Why do some reactions go so fast and others so slow? How do catalysts work? To answer such questions, we need to shift our focus to the molecular level. And we will do so in much of the rest of the chapter.

Collision Theory

Before atoms, molecules, or ions can react, they must first come together—they must *collide*. However, if all collisions between molecules produced a chemical reaction, reaction rates would be much greater than what we generally observe. Some of the collisions must be ineffective. The reaction rate therefore is proportional to the product of the frequency of molecular collisions *and* the fraction of these collisions that are effective ones.

An effective collision between two molecules puts enough energy into certain key bonds to break them. Such a collision may occur between two fast-moving molecules or perhaps when an especially fast molecule collides with a slower one. A collision between two slow-moving molecules will most likely fail to break bonds. The **activation energy (E_a)** is the minimum energy that must be supplied by collisions for a reaction to occur.

The kinetic-molecular theory allows us to calculate the fraction of all the molecules in a collection that possess a certain kinetic energy. Figure 13.9 represents the situation at two different temperatures. As the graph shows, the average kinetic energy at T_1, the lower temperature, is less than the average kinetic energy at T_2. Seen another way, T_1 has a *large* fraction of molecules at *low* kinetic energies, compared with T_2. If we assume that molecules require kinetic energies in excess of the value marked by the red arrow to react, we are led to two conclusions:

- Only a small fraction of the molecules at either temperature are energetic enough to react.
- This energetic fraction becomes larger as the temperature increases.

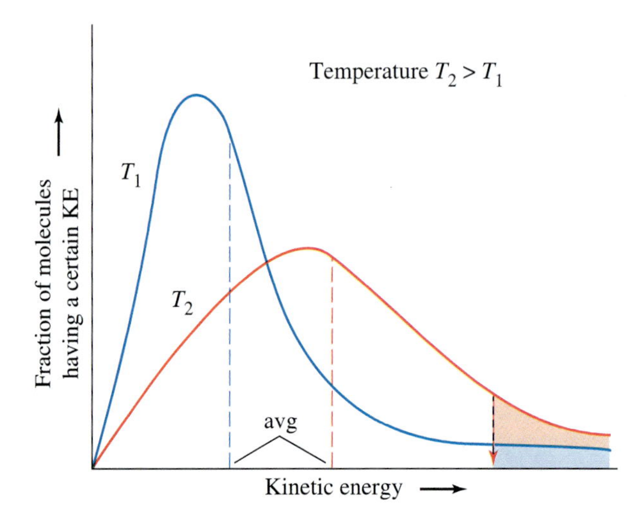

▶ **Figure 13.9 Distribution of kinetic energies of molecules**
The fraction of the molecules having energies in excess of the value marked by the red arrow is small compared to the total number of molecules. However, this fraction increases rapidly with temperature.

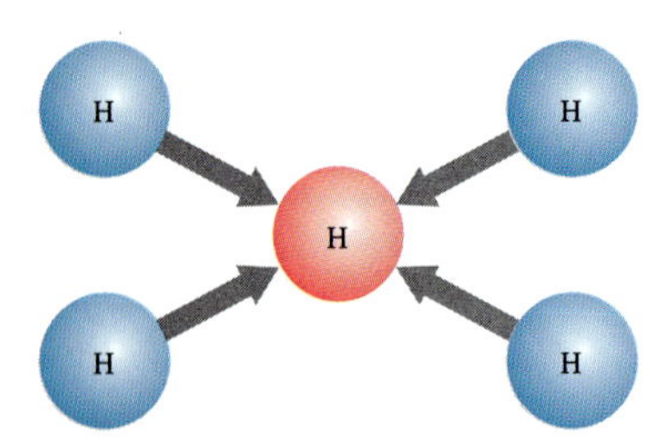

▲ **Figure 13.10 A reaction in which the orientation of colliding molecules is unimportant**
No matter from which direction an oncoming hydrogen atom (blue) approaches the "target" hydrogen atom (red), its "view" of the impending collision is the same. There is no preferred direction for one hydrogen atom to approach another when they react to form a hydrogen molecule.

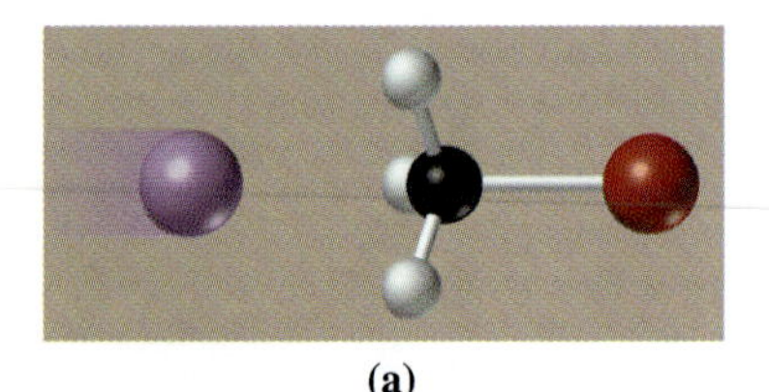
(a)

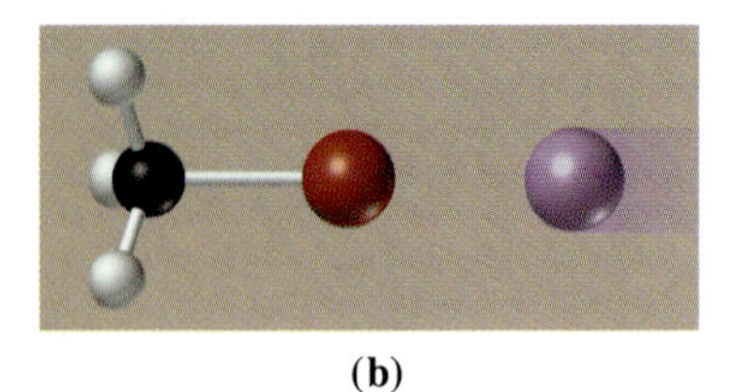
(b)

▲ **Figure 13.11 The importance of orientation of colliding molecules**
The I^- ion in (a) collides with the C atom of CH_3Br, a favorable collision for reaction to occur. The I^- ion in (b) collides with the Br atom of CH_3Br, an unfavorable collision for chemical reaction.

In many cases, we cannot account for a rate of reaction just by considering the collision frequency and the fraction of activated species. The *orientation* of the colliding species also affects a reaction rate. As shown in Figure 13.10, because of the symmetrical distribution of the electron cloud of a hydrogen atom, every approach of one H atom to another (front, back, top, bottom, and so forth) is the same. The orientation of the colliding atoms is not a factor in the rate of the following reaction.

$$H\cdot + \cdot H \longrightarrow H_2$$

In most cases, however, the orientation of the colliding species is a factor. Consider the reaction of iodide ion with methyl bromide to produce methyl iodide and bromide ion.

$$I^- + CH_3Br \longrightarrow CH_3I + Br^-$$

As shown in Figure 13.11, a collision of an I^- ion with the C atom of CH_3Br can be effective in leading to a reaction. A collision of an I^- ion with the Br atom of CH_3Br is not.

Transition State Theory

We can picture a chemical reaction as more than just the result of a collision. We can imagine, in slow motion, the gradual breaking of bonds in the reactants and formation of bonds in the products during the collision. However, during the collision there is, so to speak, a point of no return. Before this point is reached, the colliding species rebound with the reactants unchanged. Beyond this point, the colliding species proceed to products. The configuration of the atoms of the colliding species at this crucial point is called the **transition state**, and the transitory species having this configuration is called the **activated complex**.

For the reaction in which iodide ion displaces bromide ion from methyl bromide, we can represent the progress of the reaction in this way:

$$\underset{\text{Reactants}}{I^- + CH_3{-}Br} \longrightarrow \underset{\text{Activated complex}}{\overset{\delta^-}{I}\cdots\cdots CH_3 \cdots\cdots \overset{\delta^-}{Br}} \longrightarrow \underset{\text{Products}}{I{-}CH_3 + Br^-}$$

In the reactants, there is no bond between the I^- ion and the C atom and a full bond between the C and Br atoms. In the activated complex, a partial bond (·····) is formed between the I and C atoms, the bond between the C and Br atoms is partially broken (·····), and a unit of negative charge is divided into partial charges (δ^-) on the I and Br atoms. When the activated complex decomposes, there is a full bond between the I and C atoms and no bond between the C atom and the Br^- ion.

Another way of thinking about a reaction is illustrated in Figure 13.12. This representation, called a **reaction profile,** shows potential energy plotted as a function of a parameter called the progress of the reaction. Think of the progress of the reaction as a representation of how far the reaction has gone. That is, the reaction begins with reactants on the left, passes through a transition state, and ends with products on the right.

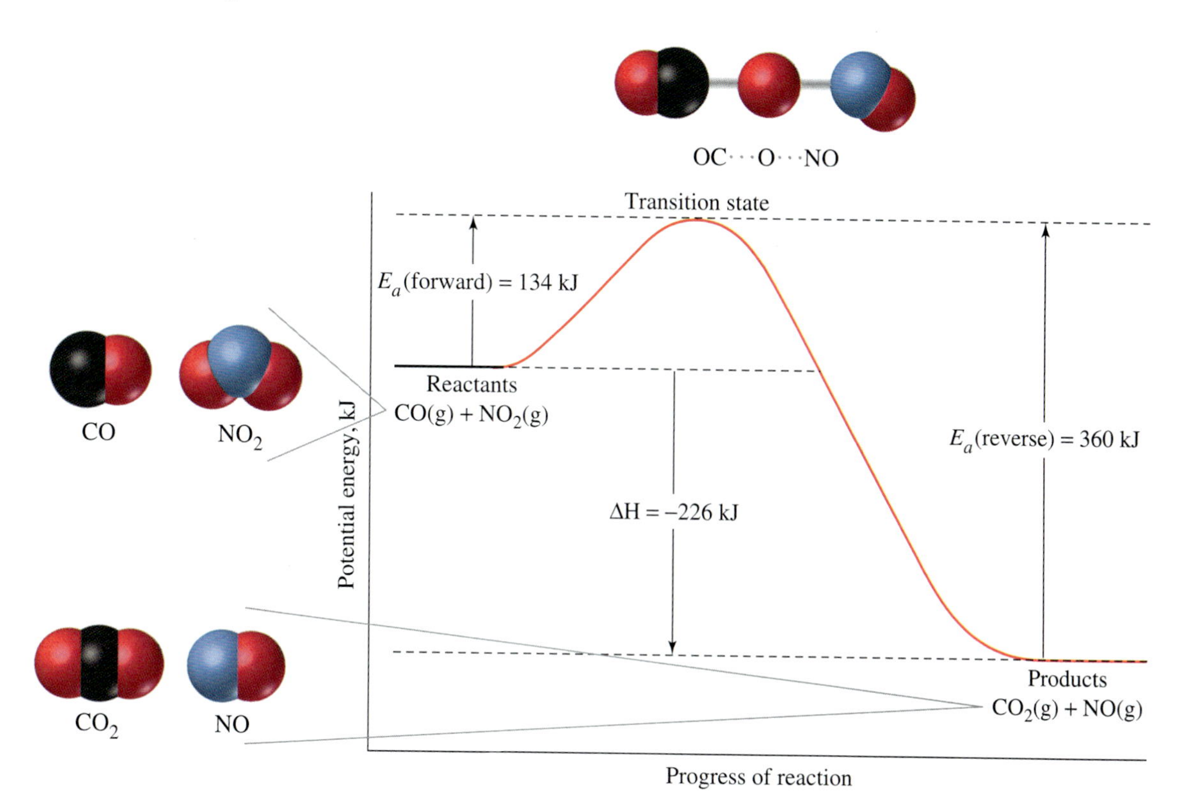

▲ **Figure 13.12 A reaction profile for the reaction $CO(g) + NO_2(g) \longrightarrow CO_2(g) + NO(g)$**
This reaction profile follows energy changes during the course of the reaction. From left to right, we proceed from reactants through the transition state to products.

Figure 13.12 is a reaction profile of the following reaction at temperatures above 600 K.

$$CO(g) + NO_2(g) \longrightarrow CO_2(g) + NO(g) \qquad \Delta H = -226 \text{ kJ}$$

The difference in potential energies between the products and reactants is ΔH for the reaction,* which is exothermic. An energy barrier separates the products from the reactants, and reactant molecules must have enough energy to surmount this barrier if a reaction is to occur. In transition state theory, the activation energy (E_a) is the difference in potential energy between the top of the barrier—the transition state—and that of the reactants. The value of E_a signifies that, collectively, one

* The difference in potential energy between reactants and products is ΔU of a reaction. For this reaction, $\Delta H = \Delta U$, but as we learned in Section 6.4, even for reactions in which these terms are not equal, the difference between ΔH and ΔU is usually quite small.

mole of CO molecules and one mole of NO_2 molecules must bring 134 kJ of energy into their collisions to form one mole of activated complex. The activated complex then dissociates into product molecules.

Figure 13.12 also describes the reverse reaction in which CO_2 and NO react to form CO and NO_2. The activation energy of the reverse reaction, $E_a(\text{reverse}) = 360$ kJ, is greater than that of the forward reaction.

Two important ideas are illustrated in Figure 13.12. (1) The enthalpy of a reaction, ΔH, is equal to the difference in activation energies of the forward and reverse reactions.

$$\Delta H = E_a(\text{forward}) - E_a(\text{reverse})$$

(2) For an *endothermic* reaction, the activation energy—E_a(reverse) in Figure 13.12—must always be equal to or greater than the enthalpy of reaction, ΔH. Figure 13.13 suggests an analogy to a reaction profile and activation energy.

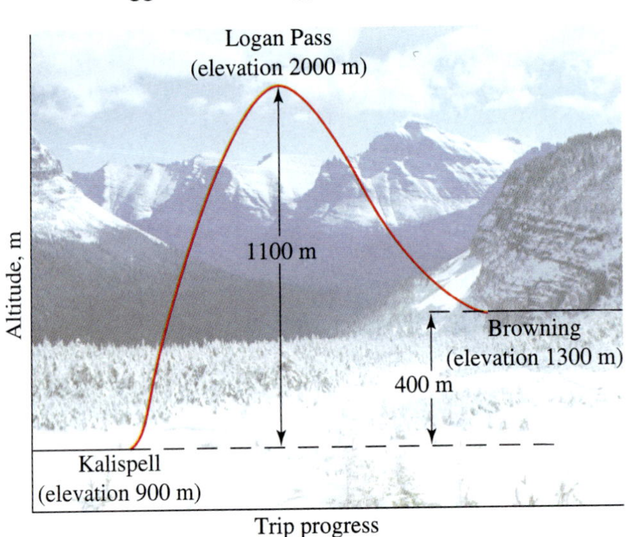

Figure 13.13 An analogy to a reaction profile and activation energy
If you were in Kalispell, Montana, (the reactants) and wished to travel to Browning (the products), you could choose to drive the scenic Going-to-the-Sun Highway (the reaction profile) through Glacier National Park. To do this you would have to cross the continental divide at Logan Pass (the transition state). First, you would have to climb 1100 m (the activation energy), but then it would be downhill the rest of the way.

13.8 Effect of Temperature on the Rates of Reactions

Charcoal (carbon) reacts so slowly with oxygen at room temperature that we see no change in it at all. When we raise the temperature of the charcoal with flames from a lighter fluid, the charcoal begins to burn more rapidly. As it does so, the heat of combustion raises the temperature even more, and the charcoal burns faster still. To slow down reactions, we generally lower the temperature. We refrigerate butter to slow down the reactions that cause it to go rancid. In the laboratory, we usually store $H_2O_2(aq)$ in the refrigerator to slow down its decomposition.

From a theoretical point of view, it seems reasonable that raising the temperature should speed up a reaction. The average kinetic energies of molecules increase, leading to more frequent collisions. But increased collision frequency is not the most important factor. As we saw in Figure 13.9, higher temperatures mean that more of the molecules are energetic enough to create a reaction. So not only are there more collisions, but the percentage of *effective* collisions is also greater, and the reaction rate increases. Reaction rate increases as the temperature increases for both exothermic and endothermic reactions. The reaction rate is governed by the height of the potential energy barrier (E_a) and not by the potential energy difference between reactants and products (ΔH).

In 1889, Svante Arrhenius proposed the following mathematical expression for the effect of temperature on the rate constant, k.

$$k = Ae^{-E_a/RT}$$

In the *Arrhenius equation*, the term "e" is the base number for natural logarithms, E_a is the activation energy, and $e^{-E_a/RT}$ represents the fraction of molecular collisions producing a reaction. (R is the gas constant, 8.3145 J mol^{-1} K^{-1}, and T is the Kelvin temperature.) The constant A, called the *frequency factor*, is the product of the collision frequency (Z) and a probability factor (p) that takes into account the orientations required for effective molecular collisions.

We can represent the Arrhenius equation in a form in which we take the logarithms of both sides.

$$\ln k = \ln A + \ln e^{-E_a/RT}$$

$$\ln k = \ln A - \frac{E_a}{RT}$$

$$\ln k = -\frac{E_a}{RT} + \ln A$$

This final equation is that of a straight line in the form $y = \text{m}\,x + \text{b}$ where $y = \ln k$ and $x = 1/T$. The slope of the line, $\text{m} = -E_a/R$ and the intercept $\text{b} = \ln A$. A typical, straight-line plot of $\ln k$ versus $1/T$ is shown in Figure 13.14, and the activation energy, E_a, is evaluated from the slope.

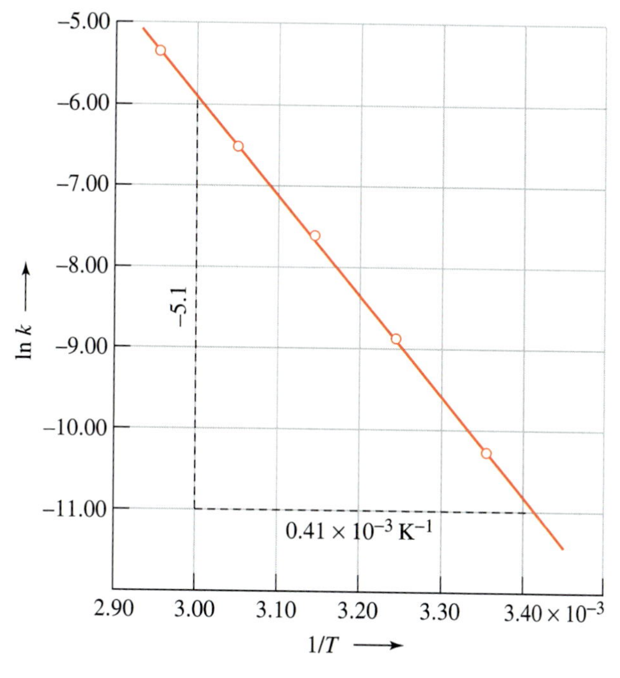

Figure 13.14
A plot of $\ln k$ versus $1/T$ for the decomposition of dinitrogen pentoxide:

$$N_2O_5(g) \longrightarrow 2\,NO_2(g) + O_2(g)$$

Evaluation of E_a:

slope $= -1.2 \times 10^4$ K
$E_a = -(\text{slope}) \times R$
$E_a = 1.2 \times 10^4 \text{ K} \times 8.3145 \text{ J mol}^{-1}\text{ K}^{-1} = 1.0 \times 10^5$ J/mol

An alternative method for determining E_a that requires no graphing is to measure the rate constant at two different temperatures and apply the Arrhenius equation in the following form.

$$\ln \frac{k_2}{k_1} = \frac{E_a}{R}\left[\frac{1}{T_1} - \frac{1}{T_2}\right]$$

In this equation, k_1 and k_2 are the rate constants at the Kelvin temperatures T_1 and T_2, and E_a is the activation energy in joules per mole. We can use the equation not only to calculate E_a, but any one of the five quantities from known values of the other four. The required form of the gas constant R is 8.3145 J mol^{-1} K^{-1}.

The derivation of this form of the Arrhenius equation is given in Appendix A. For problem solving some prefer the following form of the Arrhenius equation:

$$\ln \frac{k_2}{k_1} = \frac{E_a}{R}\left[\frac{T_2 - T_1}{T_1 \cdot T_2}\right]$$

EXAMPLE 13.9

Estimate a value of k at 375 K for the decomposition of dinitrogen pentoxide illustrated in Figure 13.14, given that $k = 2.5 \times 10^{-3}$ s^{-1} at 332 K.

SOLUTION

First, notice that we can't answer this question by just reading off a value of ln k from the straight-line graph of Figure 13.14. The point corresponding to 375 K ($1/T = 2.67 \times 10^{-3}\ K^{-1}$) is far outside the range shown. Instead, we can use the Arrhenius equation.

$$\ln\frac{k_2}{k_1} = \frac{E_a}{R}\left[\frac{1}{T_1} - \frac{1}{T_2}\right]$$

It doesn't matter whether we denote the unknown rate constant as k_1 or k_2, but typically we have used 1 for the starting condition, and 2 for the final or unknown condition. Let's do the same here; that is, let's call the unknown rate constant k_2. The data we need are

$$k_2 = ? \qquad T_2 = 375\ \text{K} \qquad E_a = 1.0 \times 10^5\ \text{J mol}^{-1}\ \text{(from Figure 13.14)}$$

$$k_1 = 2.5 \times 10^{-3}\ \text{s}^{-1} \qquad T_1 = 332\ \text{K} \qquad R = 8.3145\ \text{J mol}^{-1}\ \text{K}^{-1}$$

$$\ln\frac{k_2}{k_1} = \frac{1.0 \times 10^5\ \cancel{\text{J mol}^{-1}}}{8.3145\ \cancel{\text{J mol}^{-1}}\ \text{K}^{-1}}\left[\frac{1}{332\ \text{K}} - \frac{1}{375\ \text{K}}\right]$$

$$= 1.2 \times 10^4\ \cancel{\text{K}}(0.00301 - 0.00267)\ \cancel{\text{K}^{-1}} = 4.1$$

$$\frac{k_2}{k_1} = e^{4.1} = 60$$

$$k_2 = 60 \times k_1 = 60 \times 2.5 \times 10^{-3}\ \text{s}^{-1} = 1.5 \times 10^{-1}\ \text{s}^{-1}$$

PROBLEM-SOLVING NOTE

This result shows the strong effect of temperature on a reaction rate. A temperature increase from 59 °C to 102 °C causes the reaction to go 60 times faster. As a rough rule of thumb, at about room temperature, a reaction rate doubles for about a 10 °C increase in temperature.

EXERCISE 13.9A

Use data from Example 13.9 to determine the temperature at which the rate constant $k = 1.0 \times 10^{-5}\ s^{-1}$ for the decomposition of dinitrogen pentoxide.

EXERCISE 13.9B

Di-*tert*-butyl peroxide (DTBP) is used as a catalyst in the manufacture of polymers. In the gaseous state, DTBP decomposes to acetone and ethane by a first-order reaction.

$$C_8H_{18}O_2(g) \longrightarrow 2\ (CH_3)_2CO(g) + C_2H_6(g)$$

The half-life of DTBP is 17.5 h at 125 °C and 1.67 h at 145 °C. What is the activation energy, E_a, of the decomposition reaction?

13.9 Reaction Mechanisms

A few simple reactions consist only of the single step suggested by the balanced equation for the reaction. The reaction between methyl bromide and iodide ion (page 574) is an example. That reaction occurs as the direct result of one effective collision: I^- and CH_3Br.

$$I^- + CH_3Br \longrightarrow CH_3I + Br^-$$

Now consider the reaction between nitrogen monoxide and oxygen, a contributing reaction in the formation of smog.

$$2\ NO(g) + O_2(g) \longrightarrow 2\ NO_2(g)$$

It is highly unlikely that two NO molecules and one O_2 molecule would collide *simultaneously*, just as it is extremely unlikely that *three* basketballs might simultaneously collide in midair as basketball players take warmup shots at the basket. Instead, the overall reaction is accomplished in simpler steps.

A **reaction mechanism** is a series of simple steps that ultimately lead from the initial reactants to the final products of a reaction. An **elementary reaction** represents, at the molecular level, a single stage in the progress of the overall reaction. An elementary step involves an altering of the energy or geometry of starting molecules, or the formation of new molecules. Two requirements must be met by a plausible reaction mechanism.

- The mechanism must account for the *experimentally determined* rate law.
- The mechanism must be consistent with the stoichiometry of the overall or net reaction.

In proposing a reaction mechanism, we attempt to describe how countless molecules behave to produce the observable changes in the reaction. We do this by tracing the likely path of a few molecules through a series of elementary reactions. We assume—as seems quite likely—that all the reactant molecules follow the same pathway. In contrast, we cannot begin to predict the behavior of an entire human population by observing just a few people. And even at the molecular level, we cannot establish a reaction mechanism with total certainty. We can determine only that a reaction mechanism is *plausible*. Occasionally, two or more equally plausible mechanisms have been proposed for the same reaction.

Elementary Reactions

Before turning to specific reaction mechanisms, let's consider a few ideas about the elementary reactions that make up a reaction mechanism. The **molecularity** of an elementary reaction refers to the number of free atoms, ions, or molecules that enter into the reaction. An elementary reaction in which a single molecule dissociates is a **unimolecular reaction**; one in which two molecules collide effectively is a **bimolecular reaction**. A **termolecular reaction**, requiring the simultaneous collision of three molecules, is much less likely than unimolecular or bimolecular reactions.

Unimolecular

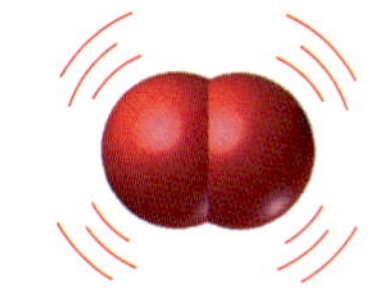

Bimolecular

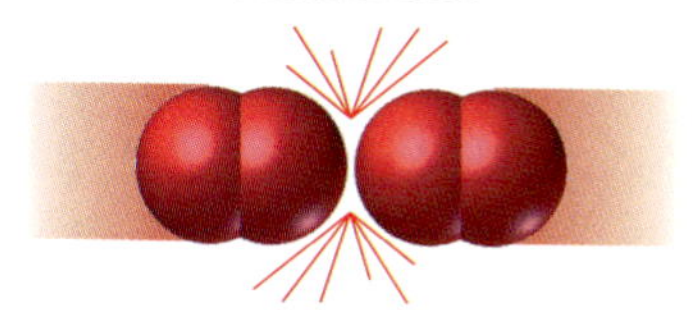

Termolecular

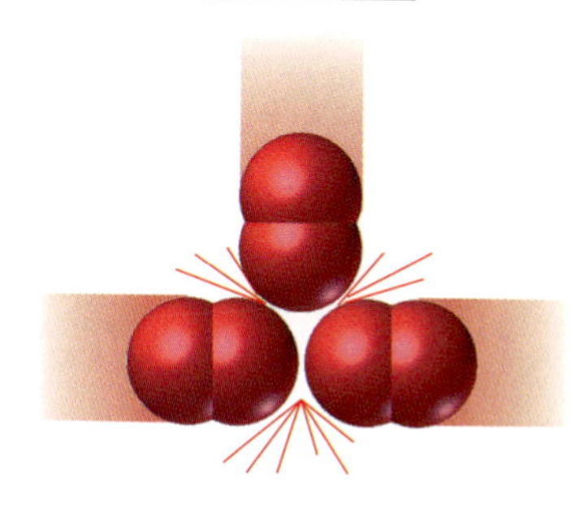

The exponents in the rate law for an *elementary reaction* are the *same* as the stoichiometric coefficients in the chemical equation for the reaction. (Recall that this is usually *not* the case with the rate law for the *overall* reaction.) Elementary reactions are reversible, that is, the forward and reverse reactions occur simultaneously. Some reach a state of equilibrium in which the rates of forward and reverse reactions are equal. Also, one elementary reaction may be much slower than all the others. In many cases this is the **rate-determining step**, the crucial step in establishing the rate of the overall reaction.

A Mechanism with a Slow Step Followed by a Fast Step

In presenting data on the decomposition of hydrogen peroxide (Figure 13.3), we pointed out that to speed up the reaction to a more easily measurable rate we usually use a catalyst, such as $I^-(aq)$.

$$2\,H_2O_2(aq) \xrightarrow{I^-} 2\,H_2O + O_2(g)$$

A reaction mechanism must give a description at the molecular level that is consistent with the facts we acquire in studying the rate of a reaction. The facts in the decomposition of hydrogen peroxide are: (1) The rate of decomposition of H_2O_2 is first order in H_2O_2 *and* in I^-, or second order overall. (2) The reactant I^- is unchanged in the reaction and hence does not appear in the equation for the net reaction. A plausible reaction mechanism consistent with these facts is the following.

Slow step: $H_2O_2 + I^- \longrightarrow H_2O + OI^-$

Fast step: $H_2O_2 + OI^- \longrightarrow H_2O + O_2 + I^-$

Net equation: $2\,H_2O_2 \longrightarrow 2\,H_2O + O_2$

The OI^- ion formed in the first step is consumed in the second step. As a result, OI^- does not appear in the equation for the overall reaction; OI^- is an *intermediate*. In contrast, I^- ion, which also does not appear in the net equation, is consumed in the first step and produced in the second step—it is recycled through the mechanism; I^- ion is a *catalyst*. The rate of the reaction is determined almost exclusively by that of the first step, which is the *rate-determining step*. In the rate law below, we set the reaction rate equal to that of the slow step.

$$\text{Rate} = \text{rate of slow step} = k[H_2O_2][I^-]$$

Earlier in the chapter, we treated the catalyzed decomposition of H_2O_2 as a first-order reaction. We were able to do this because $[I^-]$ remains constant throughout the reaction and the product $k \times [I^-]$ is itself a constant, noted below as k'.

$$\text{Rate} = k'[H_2O_2]$$

Because the value of k' depends on $[I^-]$, the more of the catalyst used, the faster the reaction goes.

As an analogy to a slow first step followed by a fast second step, consider driving to a market 500 m away. To get there, however, one has to cross a bridge under construction, where traffic is limited to one lane controlled by a road worker. The average waiting time is 15 minutes. The trip to the market, on average, will take just over 15 minutes. The time to complete the trip is determined almost entirely by the rate of the slow step: crossing the bridge.

A Mechanism with a Fast Reversible Step Followed by a Slow Step

Many reactions occur by a mechanism that has a fast reversible step followed by a slow step. The first step of the mechanism is an elementary reaction in which the reactants form an intermediate (rate constant k_1), but the intermediate readily decomposes back into the initial reactants in a reverse reaction (rate constant k_{-1}). We assume that the rates of the forward and reverse reactions quickly become equal, in a condition called a *rapid equilibrium*. However, a small quantity of the intermediate is removed by reacting in a slow second step, the *rate-determining step*, to form the final products (rate constant k_2).

Consider the reaction of NO(g) and O_2(g) to form NO_2(g), a smog-forming reaction we mentioned earlier.

$$2\,NO(g) + O_2(g) \longrightarrow 2\,NO_2(g)$$

The rate law for the reaction is experimentally determined.

$$\text{Rate} = k[NO]^2[O_2]$$

A simple mechanism with only one step would account for the observed rate law, but this would be a highly unlikely *termolecular* reaction. A more plausible mechanism is outlined below.

Fast step: $$2\,NO \underset{k_{-1}}{\overset{k_1}{\rightleftharpoons}} \cancel{N_2O_2}$$

Slow step: $$\cancel{N_2O_2} + O_2 \xrightarrow{k_2} 2\,NO_2$$

Net equation: $$2\,NO + O_2 \longrightarrow 2\,NO_2$$

Note that by canceling out the intermediate, N_2O_2, we obtain the net or overall equation for the reaction.

So far the mechanism looks good, but what does it predict for a rate law? If we base the rate law on the slow rate-determining step, we get

$$\text{Rate} = k_2[N_2O_2][O_2]$$

Remember, though, that we can state the experimentally determined rate law only in terms of substances in the net equation. We must eliminate $[N_2O_2]$. To do this, we begin with the equation for the first elementary reaction. Then we assume a rapid equilibrium, that is, that the forward and reverse reactions occur at the same rate. Next we describe those rates in terms of the formation and disappearance of N_2O_2 and set the rate laws equal. Then we solve for $[N_2O_2]$.

$$2\,NO \underset{k_{-1}}{\overset{k_1}{\rightleftharpoons}} N_2O_2$$

$$\text{Forward rate} = \text{reverse rate}$$

$$\text{Rate of formation of } N_2O_2 = -(\text{rate of disappearance of } N_2O_2)$$

$$k_1[NO]^2 = k_{-1}[N_2O_2]$$

$$\frac{k_1[NO]^2}{k_{-1}} = [N_2O_2]$$

We can substitute this expression for $[N_2O_2]$ into the rate law for the rate-determining step.

$$\text{Rate} = k_2[N_2O_2][O_2] = k_2\frac{k_1[NO]^2}{k_{-1}}[O_2]$$

$$= \frac{k_2k_1}{k_{-1}}[NO]^2[O_2] = k[NO]^2[O_2]$$

Thus, the proposed mechanism is consistent with the observed rate law. The observed rate constant, k, is a combination of three elementary-step rate constants, that is, $k = k_1k_2/k_{-1}$.

EXAMPLE 13.10

A proposed mechanism for the reaction: $H_2(g) + I_2(g) \longrightarrow 2\,HI(g)$ is

Fast step: $I_2 \underset{k_{-1}}{\overset{k_1}{\rightleftharpoons}} 2\,I$

Slow step: $2\,I + H_2 \xrightarrow{k_2} 2\,HI$

a. What is the net equation based on this mechanism?

b. What is the order of the reaction according to this mechanism?

SOLUTION

a. We can see at a glance that when we add the two steps of the proposed mechanism, the terms "2 I" cancel, leading to the anticipated net equation:

$$H_2 + I_2 \longrightarrow 2\,HI$$

b. To establish the order of the reaction, we must examine the rate law for the net reaction. To get this rate law, we begin with the rate law for the slow rate-determining step.

$$\text{Rate} = k_2[H_2][I]^2$$

Then we eliminate the concentration of the intermediate, [I], by assuming rapid equilibrium in the first step. That is, in the first elementary reaction, we assume that the rates of the forward and reverse reactions are equal.

$$k_1[I_2] = k_{-1}[I]^2$$

Then we solve for the term $[I]^2$.

$$[I]^2 = \frac{k_1}{k_{-1}}[I_2]$$

The rate law for the net reaction is

$$\text{Rate} = k_2[H_2][I]^2 = k_2\frac{k_1}{k_{-1}}[H_2][I_2]$$

$$= k[H_2][I_2]$$

The reaction is first order in H_2, first order in I_2, and second order overall.

EXERCISE 13.10A

The decomposition of nitrosyl chloride is found to be a first-order reaction.

$$2\ NOCl(g) \longrightarrow 2\ NO(g) + Cl_2(g)$$

Propose a mechanism for this reaction consisting of one fast step and one slow step.

EXERCISE 13.10B

An alternate mechanism to the one described in the text for the reaction

$$2\ NO(g) + O_2(g) \longrightarrow 2\ NO_2(g)$$

involves NO_3 as an intermediate instead of N_2O_2. **(a)** Write the steps for this alternate mechanism, and **(b)** show that the alternate mechanism is also consistent with the observed rate law.

● **ChemCDX** Bimolecular Reactions

13.10 Catalysis

We have seen that raising the temperature generally speeds up a reaction, sometimes dramatically. But raising the temperature in a living organism can have a disastrous effect; it can kill the organism. Living things generally rely on catalysis, not an increase in temperature, to speed up reactions. Many industrial and laboratory processes require catalysis as well.

The decomposition of potassium chlorate is sometimes used to produce small quantities of oxygen in the laboratory.

$$2\ KClO_3(s) \longrightarrow 2\ KCl(s) + 3\ O_2(g)$$

Without a catalyst, the $KClO_3(s)$ must be heated to over 400 °C to produce $O_2(g)$ at a useful rate. However, if we add a small quantity of $MnO_2(s)$, we can get the same rate of oxygen evolution at just 250 °C. Further, after the reaction is complete, we can recover essentially all the manganese dioxide, unchanged. As we have noted, a *catalyst* increases the reaction rate without itself being changed by the reaction. In general, a catalyst works by changing the mechanism of a chemical reaction. The pathway of a catalyzed reaction has a lower activation energy than that of an uncatalyzed reaction. If the activation energy is lowered, then more molecules have sufficient energies to engage in effective collisions. Figure 13.15 extends the analogy of Figure 13.13 to describe the difference in reaction profiles of a catalyzed and an uncatalyzed reaction.

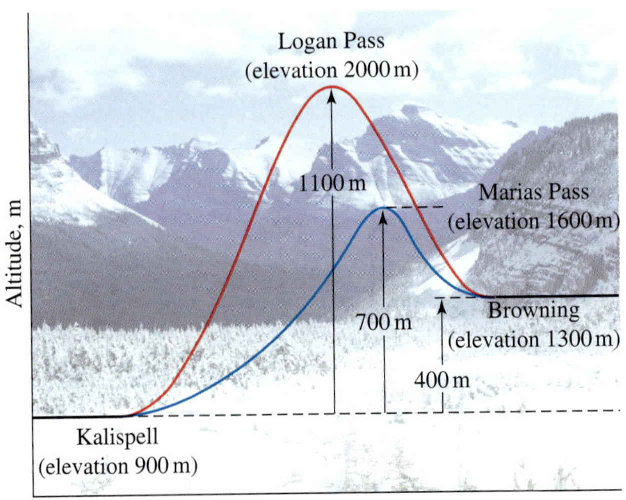

▶ **Figure 13.15**
An analogy to the reaction profile and activation energy in a catalyzed reaction
To return to our analogy of the trip from Kalispell to Browning (Figure 13.13), it is possible to take an alternate route. U. S. Highway 2 crosses the continental divide through Marias Pass (catalyzed reaction). This route involves a climb of only 700 m (lower activation energy), compared with 1100 m via Logan Pass (uncatalyzed reaction). This alternate route is analogous to the different pathway provided by a catalyst in a chemical reaction.

Homogeneous Catalysis

If a catalyzed reaction occurs in a *homogeneous* mixture, it involves *homogeneous* catalysis. Shown below is one simple mechanism for homogeneous catalysis in a reaction in which two reactants—A and B—yield two products: C and D.

Step 1:	$\text{A} + \cancel{\text{catalyst}} \longrightarrow \cancel{\text{intermediate}} + \text{C}$
Step 2:	$\text{B} + \cancel{\text{intermediate}} \longrightarrow \text{D} + \cancel{\text{catalyst}}$
Net equation:	$\text{A} + \text{B} \longrightarrow \text{C} + \text{D}$

As expected, neither the intermediate nor the catalyst appears in the overall, or net, equation.

The case shown below is an example of this mechanism. Here, O_3 and O are the reactants A and B, respectively, and O_2 is the product (both C and D). Cl is the catalyst, and ClO is an intermediate.

Step 1:	$O_3 + \cancel{Cl} \longrightarrow \cancel{ClO} + O_2$	$E_a = 2.1\ \text{kJ}$
Step 2:	$\cancel{ClO} + O \longrightarrow \cancel{Cl} + O_2$	$E_a = 0.4\ \text{kJ}$
Net equation:	$O_3 + O \longrightarrow 2\ O_2$	

APPLICATION NOTE

The seasonal depletion of ozone observed in the stratosphere over the polar regions is thought to occur by this mechanism involving Cl atoms as catalysts. We will explore this matter further in Chapter 25.

Figure 13.16 presents reaction profiles for the uncatalyzed and catalyzed decomposition of ozone. The two activation energies for the catalyzed route correspond to the two steps in the mechanism, but even the higher of the two (2.1 kJ) is much lower than that of the uncatalyzed reaction (17.1 kJ). As a consequence, the rate constant for the catalyzed reaction is thousands of times greater than of the uncatalyzed reaction. Notice that the catalyst (Cl) appears in both the initial and final states, but otherwise these two states are the same for the two reaction pathways. Notice also that the enthalpy of reaction, ΔH is identical in the catalyzed and uncatalyzed reaction.

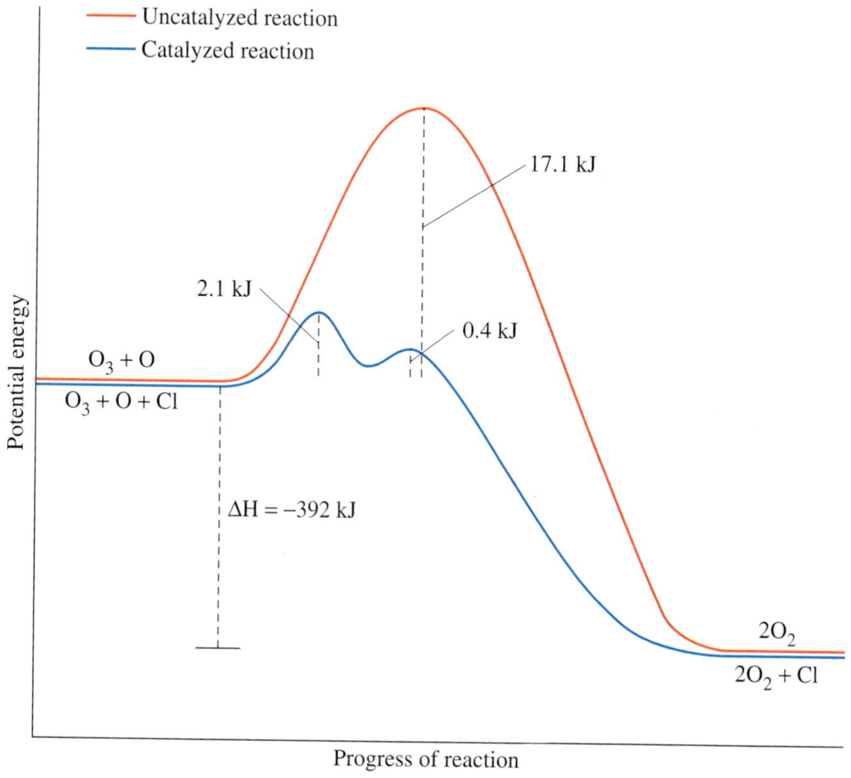

Uncatalyzed reaction
Catalyzed reaction
17.1 kJ
2.1 kJ
0.4 kJ
$O_3 + O$
$O_3 + O + Cl$
$\Delta H = -392$ kJ
$2O_2$
$2O_2 + Cl$
Potential energy
Progress of reaction

Figure 13.16 Reaction profile for the uncatalyzed and catalyzed decomposition of ozone
The net reaction in both cases is $O_3 + O \longrightarrow 2\ O_2$. In the catalyzed reaction, Cl is the catalyst.

The decomposition of hydrogen peroxide, $H_2O_2(aq)$, to $H_2O(l)$ and $O_2(g)$ is a highly exothermic reaction that is strongly catalyzed by the surface of the platinum metal in this wire mesh electrode.

Heterogeneous Catalysis

Reactions often can be catalyzed by the surfaces of appropriate solids. An essential aspect of this catalytic activity is the ability of surfaces to *adsorb* (bind) reactant molecules from the gaseous or liquid state. Because a surface-catalyzed reaction occurs in a *heterogeneous* mixture, the catalytic action is called *heterogeneous* catalysis. There are four basic steps in heterogeneous catalysis.

1. Reactant molecules are *adsorbed.*
2. Reactant molecules diffuse along the surface.
3. Reactant molecules react to form product molecules.
4. Product molecules are *desorbed* (that is, released from the surface).

Figure 13.17 shows a hypothetical reaction profile for a surface-catalyzed reaction and compares it with an uncatalyzed, homogeneous gas-phase reaction.

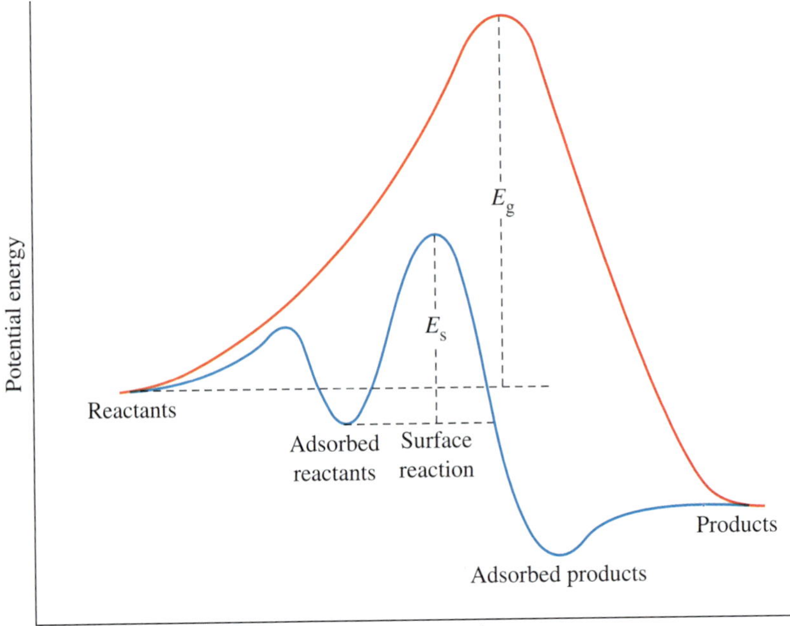

Figure 13.17
Reaction profile for a surface-catalyzed reaction
In the reaction profile (blue) for the surface-catalyzed reaction, the activation energy for the reaction step, *Es*, is considerably less than in the reaction profile (red) for the uncatalyzed gas-phase reaction, *Eg*.

The zero-order decomposition of ammonia on tungsten described on page 569 is a surface-catalyzed reaction. So is the hydrogenation of food oils to produce solid or semisolid fats that we described in Section 9.11. A simplified mechanism of the surface-catalyzed conversion of oleic to stearic acid is suggested in Figure 13.18.

● **ChemCDX** Surface Reactions—Hydrogenation

$$\underset{\text{Oleic acid}}{CH_3(CH_2)_7CH{=}CH(CH_2)_7COOH} + H_2 \xrightarrow{\text{Ni}} \underset{\text{Stearic acid}}{CH_3(CH_2)_7CH_2CH_2(CH_2)_7COOH}$$

13.11 Enzyme Catalysis

Substances such as platinum and nickel are able to catalyze a variety of reactions, but the catalysts for the chemical reactions in living organisms are usually quite specific in their catalytic action. Almost all of these catalysts are high-molecular-mass proteins called **enzymes**. An enzyme usually catalyzes one specific reaction and no others. In living organisms, even a simple reaction like the conversion of carbon dioxide to carbonic acid is catalyzed by an enzyme, called *carbonic anhydrase*. Each enzyme molecule can convert 100,000 CO_2 molecules per second.

$$CO_2(g) + H_2O \xrightleftharpoons{\text{carbonic anhydrase}} H_2CO_3(aq)$$

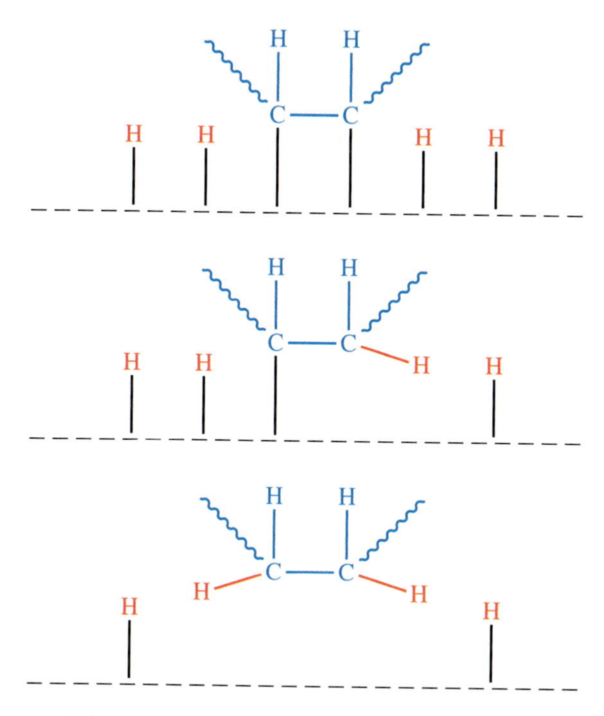

Figure 13.18 Heterogeneous catalysis on a nickel surface
A molecule of oleic acid (blue) adsorbs to the surface through C atoms at the double bond. The double bond is converted to a single bond. H_2 molecules dissociate into H atoms when they adsorb. First one, and then a second H atom attach to the surface-bonded C atoms. The stearic acid molecule formed detaches itself from the surface and escapes.

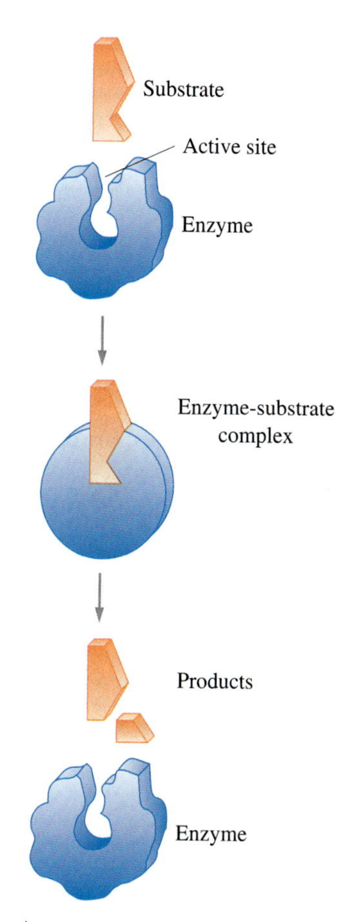

Figure 13.19 Model of enzyme action
The *induced-fit model* holds that the shapes of the substrate and the active site are not perfectly complementary, but the active site adapts to fit the substrate, much like a glove molds to fit the hand that is inserted into it.

Biochemists often use the model pictured in Figure 13.19, called the *induced-fit* model, to explain enzyme action. The reacting substance, called the **substrate (S)**, attaches itself to an area on the enzyme (E) called the **active site,** to form an enzyme-substrate complex (ES). The enzyme-substrate complex decomposes to form products (P), and the enzyme is regenerated. Using the symbolism that biochemists often use, we can represent the process as follows.

$$\text{E} + \text{S} \longrightarrow \text{ES} \longrightarrow \text{E} + \text{P}$$

Factors Influencing Enzyme Activity

The rates of enzyme-catalyzed reactions are influenced by several factors, including concentration of substrate,[S], concentration of enzyme, [E], acidity of the medium, and temperature.

Figure 13.20 shows how the reaction rate changes as the concentration of substrate increases while the enzyme concentration remains constant. The reaction rate increases as the concentration of substrate increases, but only up to a point. Eventually, a concentration is reached that saturates all the active sites on the enzyme molecules. The rate levels off, and a further increase in substrate concentration leaves the rate unchanged.

At low substrate concentrations, the reaction is first order in the substrate because the rate of enzyme-substrate complex formation is proportional to [S].

$$\text{Rate} = k[\text{S}]$$

At high substrate concentrations, adding more substrate cannot accelerate the reaction. The rate of the reaction is independent of substrate concentration, and the reaction is zero order.

$$\text{Rate} = k'[\text{S}]^0 = k''$$

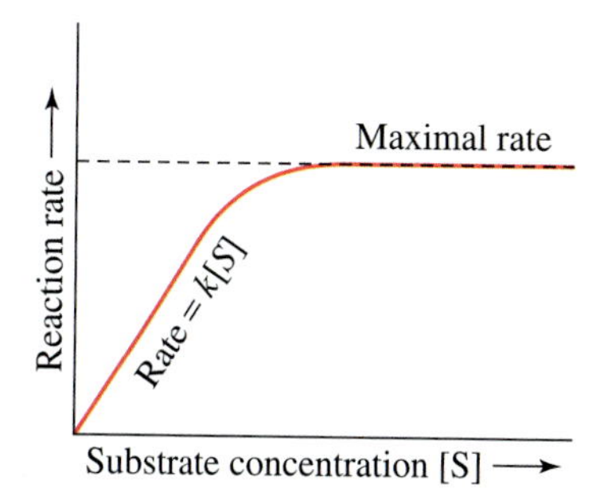

Figure 13.20 The effect of substrate concentration [S] on reaction rate with enzyme concentration held constant
When [S] is low, the reaction rate increases with [S]. At high values of [S], the reaction rate is independent of [S], and a maximal rate is observed.

It's as if 10 taxis (enzyme molecules) were waiting at a taxi stand to take people (substrate) on a 10-minute trip to a concert hall, one passenger at a time (corresponding to one active site per enzyme molecule). If only 5 people are present at the stand, the rate of their arrival at the concert hall is 5 people in 10 minutes. If the number of people at the stand is increased to 10, the rate increases to 10 arrivals in

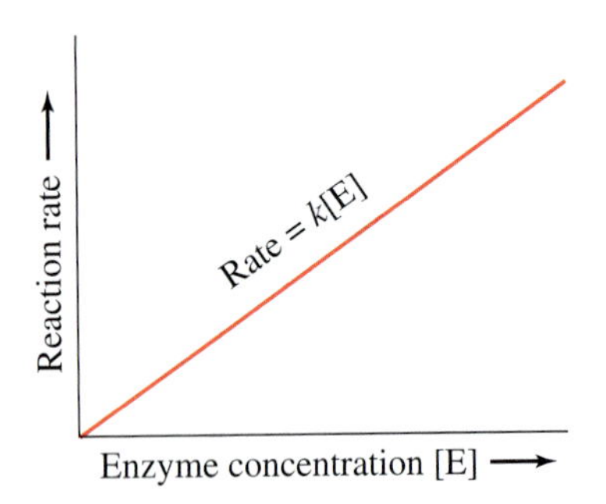

▲ **Figure 13.21 The effect of enzyme concentration on reaction rate, with the concentration of substrate held constant and in excess**

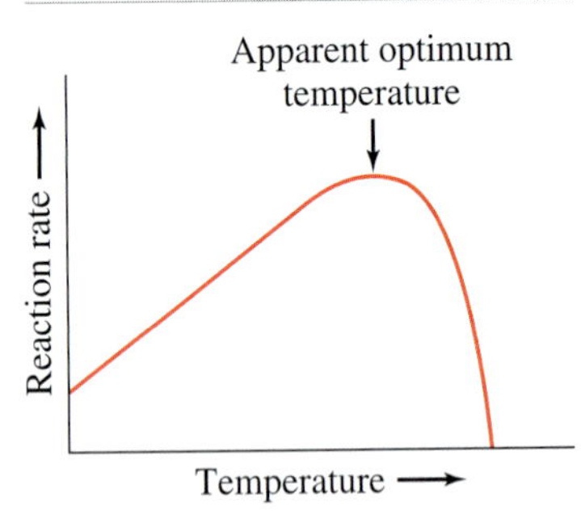

▲ **Figure 13.22 Enzyme activity as a function of temperature**
If the hypothetical reaction whose rate is suggested here occurs in humans, the optimum temperature is likely to be 37 °C—normal body temperature.

10 minutes. With 20 people at the taxi stand, the rate would still be 10 arrivals in 10 minutes. The taxis have been "saturated."

If the concentration of substrate is held constant and *in excess* (if we always have more people than taxis), the rate of a reaction is proportional to the concentration of enzyme. The more enzyme there is, the faster the reaction goes (the more taxis, the more people transferred). This relationship holds over a wide range of enzyme concentration (Figure 13.21).

Enzymes are proteins with acidic groups (such as —COOH) and basic groups (such as $-NH_2$). Enzyme activity depends on the concentration of H^+ ions present in the medium surrounding the enzyme. Each enzyme has its own *optimum acidity*. Those that operate in the stomach work best in highly acidic media. Those that catalyze reactions in the small intestine, in contrast, show optimum activity in slightly basic media.

As an example, consider *lysozyme*, an enzyme that breaks certain bonds in complex structural molecules found in the cell walls of bacteria. Lysozyme is active only when two conditions are met simultaneously.

1. Aspartic acid, one of the amino acid components of lysozyme, must be ionized. That is, the free —COOH group of aspartic acid must be converted to $-COO^-$.
2. Glutamic acid, another amino acid component of lysozyme, must *not* be ionized. That is, the —COOH group of glutamic acid must be intact.

When the medium is slightly acidic, both conditions are met, and lysozyme is active. However, in either highly acidic or highly basic solutions, the enzyme cannot function. In the former case, both acid groups are present in their —COOH form. In the latter case, both groups are present as $-COO^-$.

Enzyme action is sensitive to temperature changes. For living cells, there is often a rather narrow range of optimum temperatures. Both higher and lower temperatures can disable an enzyme or even kill an organism. A change in temperature changes the shape of the protein molecule so that the substrate no longer fits the active site. The temperature range for the activity of enzymes generally is 10 to 50 °C. However, some enzymes found in *thermophilic* (heat-loving) bacteria operate well near 100 °C. The *optimum temperature* for many enzymes in the human body is 37 °C—normal body temperature (Figure 13.22).

Summary

The rate law for a reaction is the equation: rate $= k[A]^m[B]^n \cdots$. The constant k is the rate constant, m is the order of the reaction in A, n is the order in B, and $m + n + \cdots$ is the overall order. Values of $m, n, \cdots$ must be obtained by experiment. The rate law can be used to calculate a value of k once the order of the reaction is known.

An integrated rate law relates concentration and time. For a first-order reaction, $\ln [A]_t/[A]_0 = -kt$. The rate constant, k, is the negative of the slope of the straight line graph: $\ln [A]_t$ versus t. In a zero-order reaction, the rate is independent of the concentration of reactant, and a graph of reactant concentration as a function of time is a straight line. In a second-order reaction, the sum of the exponents in the rate law is *two*. In the second-order reaction with the rate law, Rate $= k[A]^2$, a plot of $1/[A]$ versus t yields a straight line. The half-life of a reaction is the time in which one-half of the reactant initially present is consumed. For a first-order reaction the half-life is constant: $t_{1/2} = 0.693/k$. For both zero-order and second-order reactions, the half-life depends on concentration as well as on the value of k.

Chemical reactions occur when sufficiently energetic molecules collide while in the proper orientation. The point in the collision where the reactants are converted to products is the transition state. A reaction profile charts the progress of a reaction through changes in the potential energy of the reaction mixture. The potential energy rises from that of the reactants to that of the transition state, and then falls back to the potential energy of the products. The excess energy of the transition state over that of the reactants is the activation

Key Terms

activated complex (13.7)
activation energy (E_a)(13.7)
active site (13.11)
bimolecular reaction (13.9)
catalyst (13.10)
elementary reaction (13.9)
enzyme (13.11)
first-order reaction (13.5)
half-life (13.5)
initial rate of reaction (13.3)
instantaneous rate of reaction (13.3)
integrated rate law (13.5)
method of initial rates (13.4)

Enzyme Inhibition

An *inhibitor* is a substance that makes an enzyme less active or completely inactive. Some inhibitors bind at the active site of the enzyme and block out the substrate. Others bind elsewhere in the enzyme molecule but change the shape of the active site or otherwise hinder a substrate molecule's access to the site. Inhibition is an important, naturally occurring process that is essential in growth and metabolism, but some substances that we ingest—in food or otherwise—also act as inhibitors. Let's look at some inhibitors that kill—poisons—and others that sustain life—drugs.

Enzymes are proteins. Many proteins contain the amino acid *cysteine*, which has a sulfhydryl group (—SH) attached to each unit. Heavy metal ions such as Hg^{2+} and Pb^{2+} can deactivate enzymes by reacting with sulfhydryl groups to form sulfides. This poisoning can occur at a position removed from the active site, distorting or destroying the site (Figure 13.23).

Enzymes play a key role in the chemistry of the nervous system. When an electric signal from the brain reaches the end of a nerve cell, the substance *acetylcholine* is liberated and carries the signal across a tiny gap to a second nerve cell. This second cell acts as a receptor for the signal. Once acetylcholine has carried the signal to the next cell, the enzyme *cholinesterase* catalyzes its rapid breakdown into acetic acid and choline.

$$\underset{\text{Acetylcholine}}{CH_3COOCH_2CH_2N^+(CH_3)_3} + H_2O \xrightleftharpoons[\text{acetylase}]{\text{cholinesterase}} \underset{\text{Acetic acid}}{CH_3COOH} + \underset{\text{Choline}}{HOCH_2CH_2N^+(CH_3)_3}$$

In effect, the breakdown of acetylcholine resets the receptor cell to the "off" position, making it ready to receive the next impulse. Other enzymes, such as acetylase, convert the acetic acid and choline back to acetylcholine, completing the cycle.

Anticholinesterase poisons block the action of cholinesterase. The insecticides *malathion* and *parathion* and the warfare agents *tabun* and *sarin*, are well-known examples of such nerve poisons. The molecule of poison binds tightly to the cholinesterase (Figure 13.24), preventing the enzyme from performing its normal function. If the breakdown of acetylcholine is blocked, the concentration of this messenger compound builds up, causing the receptor nerve cell to "fire" repeatedly, that is, to be continuously "on." This overstimulates the muscles, glands, and organs. The heart beats wildly and irregularly. The victim goes into convulsions and dies quickly.

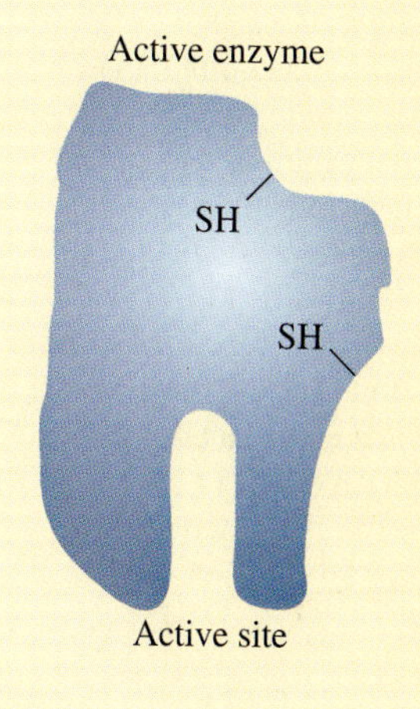

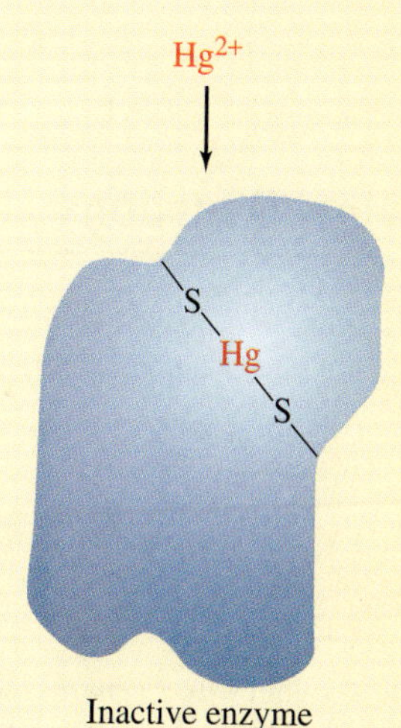

▲ **Figure 13.23**
Mercury poisoning as an example of enzyme inhibition
Mercury(II) ions react with sulfhydryl groups to change the conformation of the enzyme and destroy the active site.

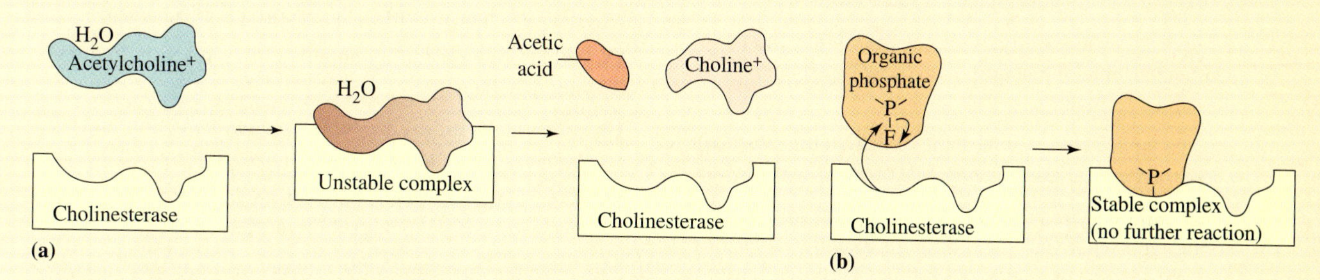

▲ **Figure 13.24 Action of cholinesterase and its inhibition**
(a) Cholinesterase catalyzes the hydrolysis of acetylcholine into acetic acid and choline. (b) An organic phosphate (as in the nerve gas, Sarin) binds to cholinesterase, preventing it from breaking down acetylcholine.

molecularity (13.9)
order of a reaction (13.4)
rate constant (13.4)
rate-determining step (13.9)
rate law (rate equation) (13.4)
reaction mechanism (13.9)
reaction profile (13.7)
second-order reaction (13.6)
substrate (S) (13.11)
termolecular reaction (13.9)
transition state (13.7)
unimolecular reaction (13.9)
zero-order reaction (13.6)

energy of the reaction. The difference in potential energy between reactants and products, the enthalpy change (ΔH) of a reaction, is also shown in the reaction profile.

Reactions generally go faster at higher temperatures because of the large increase in the fraction of molecular collisions effective in producing products. The effect of temperature and activation energy on the rate constant of a reaction can generally be expressed through the Arrhenius equation, $k = Ae^{-E_a/RT}$, and a straight-line graph of ln k versus $1/T$ has a slope $\frac{-E_a}{R}$.

A plausible reaction mechanism generally consists of a series of elementary reactions. The mechanism must yield the observed net equation, and the rate law deduced from the mechanism must be the same as that found experimentally.

A catalyst speeds up a reaction by changing the mechanism to one of lower activation energy; the catalyst is regenerated in the reaction. Some catalysts function in a homogeneous mixture (homogeneous catalysis); others provide a surface on which the reaction occurs (heterogeneous catalysis). In enzyme catalysis, the reactants (substrate) bind to an active site on the enzyme to form a complex that dissociates into product molecules. Enzyme activity depends on the concentrations of substrate and enzyme, acidity of the medium, and temperature.

Review Questions

1. State two quantities that must be measured to establish the rate of a chemical reaction.
2. Cite three or four factors that affect the rate of a chemical reaction.
3. Explain why the rate of disappearance of NO and the rate of formation of N_2 are not the same in the reaction, $2\,CO(g) + 2\,NO(g) \longrightarrow 2\,CO_2(g) + N_2(g)$.
4. Explain why a *general* rate of reaction has the same value regardless of which reactant or product we monitor. Use the reaction, $N_2(g) + 3\,H_2(g) \longrightarrow 2\,NH_3(g)$ to illustrate your explanation.
5. At what point in a reaction is the rate of reaction usually greatest? Are there any exceptions? Explain.
6. What is the difference between the average rate and the instantaneous rate of a reaction? Can the average rate ever be the same as the instantaneous rate? Explain.
7. What is the difference between the initial rate and an instantaneous rate of reaction? Can these two quantities be the same? Explain.
8. Explain the difference in meaning of these terms when applied to a chemical reaction: *rate*, *rate constant*, and *rate law*.
9. What variables are related through the *rate law* of a reaction, and what variables, through the *integrated rate law*? Write these two equations for the first-order reaction: $N_2O_5(g) \longrightarrow 2\,NO_2(g) + \frac{1}{2}\,O_2(g)$.
10. What is meant by the half-life of a reaction? Is the half-life constant for a zero-order reaction? first-order reaction? second-order reaction? Explain.
11. What is the relationship, if any, between the average kinetic energy of the molecules in a reaction mixture and the activation energy of the reaction?
12. With regard to the collision theory, what factor is most responsible for the fact that the rate of a chemical reaction generally increases sharply with a rise in temperature?
13. Why is the orientation of the colliding molecules an important factor in determining the rate of a reaction in some reactions but not in all reactions?
14. What is the reaction profile of a chemical reaction? In relation to the reaction profile, what are the transition state and the activated complex?
15. What variables does the Arrhenius equation relate? Explain each of the terms in the equation.
16. What plot of experimental data can you use to evaluate the activation energy, E_a, of a reaction? How is E_a related to this plot?
17. What are the chief requirements that must be met by a plausible reaction mechanism? Why do we say "plausible" mechanism rather than "correct" mechanism?
18. What do we mean by an elementary reaction in a reaction mechanism?
19. What is the difference between a *unimolecular* and a *bimolecular* elementary reaction in a reaction mechanism?
20. What is the difference between an *activated complex* and an *intermediate* in a reaction mechanism?
21. What do we mean by the rate-determining step in a reaction mechanism? Which elementary reaction in a reaction mechanism is often the rate-determining step?
22. State *two* important requirements of a substance in order for it to be considered a catalyst in a chemical reaction.
23. Neither a catalyst nor an intermediate appears in the net equation for a chemical reaction. Are they the same thing? If not, how do they differ?

24. What is the difference between *homogeneous* and *heterogeneous* catalysis?

25. To what class of macromolecular substances do enzymes belong?

26. What is the substrate in an enzyme-catalyzed reaction?

27. Describe the induced-fit model of an enzyme-catalyzed reaction.

28. Besides substrate concentration, what other factors affect the rate of an enzyme-catalyzed reaction?

Problems

Rate of Reaction and Reaction Order

29. In the reaction, $H_2O_2(aq) \longrightarrow H_2O(l) + \frac{1}{2} O_2(g)$, the initial concentration of H_2O_2 is 0.1108 M, and 12 s later the concentration is 0.1060 M. What is the initial rate of this reaction expressed in **(a)** $M\ s^{-1}$ and **(b)** $M\ min^{-1}$?

30. In the reaction $H_2O_2(aq) \longrightarrow H_2O(l) + \frac{1}{2} O_2(g)$, the initial concentration of H_2O_2 is 0.2546 M, and the initial rate of reaction is $9.32 \times 10^{-4}\ M\ s^{-1}$. What will be $[H_2O_2]$ at $t = 35$ s?

31. In the reaction $2\ A + 2\ B \longrightarrow C + 2\ D$, the rate of disappearance of A is $-2.2 \times 10^{-4}\ M\ s^{-1}$. **(a)** What is the rate of disappearance of B? **(b)** What is the rate of formation of C? **(c)** What is the general rate of reaction as described on page 555?

32. In the reaction $A + 2\ B \longrightarrow C + 3\ D$, the rate of disappearance of B is $-6.2 \times 10^{-4}\ M\ s^{-1}$. **(a)** What is the rate of disappearance of A? **(b)** What is the rate of formation of D? **(c)** What is the general rate of reaction as described on page 555?

33. For the reaction, $A \longrightarrow$ products, a graph of [A] versus time is a straight line. What is the order of this reaction?

34. The rate of a reaction and the rate constant, k, have the same units only for reactions of a particular overall order. Verify this statement, and identify the reaction order.

35. Following are two statements pertaining to the reaction $2\ A + B \longrightarrow 2\ C$, for which the rate law is *rate* $= k[A][B]$. Identify which statement is true and which is false, and explain your reasoning.

(a) The value of k is *independent* of the initial concentrations, $[A]_0$ and $[B]_0$.

(b) The unit of the rate constant for this reaction can be expressed either as s^{-1} or min^{-1}.

36. Following are two statements pertaining to the reaction $2\ A \longrightarrow B + C$. One of these statements is true, and the other may be false. Identify which, is which and explain your reasoning.

(a) The half-life of this reaction can be determined with the expression $t_{1/2} = 1/k[A]_0$, where k is the rate constant and $[A]_0$ is the initial concentration of A.

(b) The rate at which B is produced is not the same as the rate at which A is consumed.

37. The rate of the following reaction in aqueous solution is monitored by measuring the number of moles of Hg_2Cl_2 that precipitate per liter per minute. The data obtained are listed in the table.

$$2\ HgCl_2 + C_2O_4^{2-} \longrightarrow 2\ Cl^- + 2\ CO_2(g) + Hg_2Cl_2(s)$$

Experiment	$[HgCl_2]$, M	$[C_2O_4]^{2-}$, M	Initial rate, mol L^{-1} min^{-1}
1	0.105	0.15	1.8×10^{-5}
2	0.105	0.30	7.1×10^{-5}
3	0.052	0.30	3.5×10^{-5}
4	0.052	0.15	8.9×10^{-6}

(a) Determine the order of the reaction with respect to $HgCl_2$, with respect to $C_2O_4^{2-}$ and overall.

(b) What is the value of the rate constant k?

(c) What would be the initial rate of reaction if $[HgCl_2] = 0.020$ M and $[C_2O_4^{2-}] = 0.22$ M?

38. The rate of the following reaction in aqueous solution is monitored by measuring the rate of formation of I_3^-. The data obtained are listed in the table.

$$S_2O_8^{2-} + 3\ I^- \longrightarrow 2\ SO_4^{2-} + I_3^-$$

Experiment	$[S_2O_8^{2-}]$, M	$[I^-]$, M	Initial rate, $M\ s^{-1}$
1	0.038	0.060	1.4×10^{-5}
2	0.076	0.060	2.8×10^{-5}
3	0.076	0.120	5.6×10^{-5}

(a) Determine the order of the reaction with respect to $S_2O_8^{2-}$, with respect to I^-, and overall.

(b) What is the value of the rate constant k?

(c) What would be the initial rate of reaction if $[S_2O_8^{2-}] = 0.025$ M and $[I^-] = 0.045$ M?

39. In the reaction $A \longrightarrow$ products, we find that when [A] has fallen to half its initial value, the reaction proceeds at only one-quarter of its initial rate. Is the reaction zero order, first order, or second order? Explain.

40. In the reaction, $A \longrightarrow$ products, with the initial concentration $[A]_0 = 1.512$ M, [A] is found to be 1.496 M at $t = 30$ s. With the initial concentration $[A]_0 = 2.584$ M, [A] is found to be 2.552 M at $t = 1$ min. What is the order of this reaction?

41. Use a value of $k = 3.66 \times 10^{-3}\ s^{-1}$ to establish the instantaneous rate of the first-order decomposition of 2.05 M $H_2O_2(aq)$.

42. What should be the instantaneous rate of the first-order decomposition of 3% $H_2O_2(aq)$, by mass? Assume that the solution has a density of 1.00 g/mL and that $k = 3.66 \times 10^{-3}\ s^{-1}$.

First-Order Reactions

43. A first-order reaction, A $\longrightarrow$ products, has a rate of reaction of 0.00250 M s^{-1} when [A] = 0.484 M. What is the rate constant, k, for this reaction?

44. A first-order reaction, A $\longrightarrow$ products, has a rate constant, k, of 0.0462 min^{-1}. What is [A] at the time when the reaction is proceeding at a rate of 0.0150 M min^{-1}?

45. For the first-order reaction, A $\longrightarrow$ products, does $t_{3/4}$ depend on the initial concentration? Does $t_{4/5}$? Explain.

46. It takes 12 minutes for the first-order reaction, A $\longrightarrow$ products, to go to 12% completion and 48 minutes to go to 56% completion. *Without doing detailed calculations*, determine how long it should take the reaction to go to 78% completion. (*Hint:* What is the percent remaining at each of these times?)

47. In the first-order decomposition of dinitrogen pentoxide at 335 K,

$$N_2O_5(g) \longrightarrow 2\ NO_2 + \frac{1}{2}O_2$$

if we start with a 2.50-g sample of N_2O_5 at 335 K and have 1.50 g remaining after 109 s, **(a)** What is the value of the rate constant k? **(b)** What is the half-life of the reaction? **(c)** What mass of N_2O_5 will remain after 5.0 min?

48. The half-life for the first-order decomposition of sulfuryl chloride at 320 °C is 8.75 h.

$$SO_2Cl_2(g) \longrightarrow SO_2(g) + Cl_2(g)$$

(a) What is the value of the rate constant k, in s^{-1}? **(b)** What is the pressure of sulfuryl chloride 3.00 h after the start of the reaction, if its initial pressure is 722 mmHg? **(c)** How long after the start of the reaction will it be before the pressure of sulfuryl chloride becomes 125 mmHg?

49. The smog constituent peroxyacetyl nitrate (PAN) dissociates into peroxyacetyl radicals and $NO_2(g)$ in a first-order reaction with a half-life of 32 min.

$$\underset{\text{PAN}}{CH_3\overset{\overset{O}{\|}}{C}OONO_2} \longrightarrow \underset{\text{Peroxyacetyl radical}}{CH_3\overset{\overset{O}{\|}}{C}OO\cdot} + NO_2$$

If the initial concentration of PAN in an air sample is 5.0×10^{14} molecules/L, what will be the concentration 1.50 h later?

50. The decomposition of dimethyl ether at 504 °C is a first-order reaction with a half-life of 27 min.

$$(CH_3)_2O(g) \longrightarrow CH_4(g) + H_2(g) + CO(g)$$

(a) What will be the partial pressure of $(CH_3)_2O(g)$ after 1.00 h if its initial partial pressure was 626 mmHg?

(b) What will be the total gas pressure after 1.00 h? [*Hint:* $(CH_3)_2O$ and its decomposition products are the only gases present in the reaction vessel.]

Reactions of Various Orders

51. Refer to Figure 13.7, which describes the decomposition of ammonia on a tungsten surface at 1100 °C, and determine $[NH_3]$ at t = 335 s if initially $[NH_3]$ = 0.0452 M.

52. Consider Problem 51: What is the half-life for the reaction in which $[NH_3]$ = 0.0452 M?

53. The rate constant for the second-order decomposition of hydrogen iodide at 700 K is $k = 1.2 \times 10^{-3}\ M^{-1}\ s^{-1}$. In a reaction in which $[HI]_0$ = 0.56 M, what will [HI] be at t = 2.00 h?

54. Consider Problem 53: At what time will [HI] = 0.28 M for the reaction described?

55. The half-lives of zero-order and second-order reactions both depend on the initial concentration as well as on the rate constant k. In one case, the half-life gets longer as the initial concentration increases, and in the other case, it gets shorter. Which is which, and why isn't the situation the same for both?

56. Consider the reaction, A $\longrightarrow$ products, and three hypothetical reactions all having the same numerical value for the rate constant k: one that is zero order; one, first order; and one, second order. What must be the initial concentration of reactant, $[A]_0$, if **(a)** the zero- and first-order; **(b)** the zero- and second-order; **(c)** the first- and second-order reactions are to have the same half-life?

57. The following data were obtained in two separate experiments in the reaction: A $\longrightarrow$ products. Determine the rate law for this reaction, including the value of k.

Experiment 1		Experiment 2	
[A], M	Time, s	[A], M	Time, s
0.800	0	0.400	0
0.775	40	0.390	64
0.750	83	0.380	132
0.725	129	0.370	203
0.700	179	0.360	278

58. Listed below are initial rates, expressed in terms of the rate of decrease of partial pressure of a reactant in the following reaction at 826 °C. Determine the rate law for this reaction, including the value of k.

$$NO(g) + H_2(g) \longrightarrow \frac{1}{2} N_2(g) + H_2O(g)$$

With initial P_{H_2} = 400 mmHg		With initial P_{NO} = 400 mmHg	
Initial P_{NO}, mmHg	Rate, mmHg/s	Initial P_{H_2}, mmHg	Rate, mmHg/s
359	0.750	289	0.800
300	0.515	205	0.550
152	0.125	147	0.395

Collision Theory; Activation Energy

59. Chemical reactions occur as a result of molecular collisions, and the frequency of molecular collisions can be calculated with the kinetic-molecular theory. Calculations of the rates of chemical reactions, however, are generally not very successful. Explain why this is so.

60. A temperature increase of 10 °C causes an increase in collision frequency of only a few percent, yet it can cause the rate of a chemical reaction to increase by a factor of two or more. How do you explain this apparent discrepancy?

61. A mixture of hydrogen and oxygen gases is indefinitely stable at room temperature, but if struck by a spark, the mixture immediately explodes. What explanation can you offer for this observation?

62. The activation energy for an endothermic reaction is never less than the enthalpy change for the reaction. Can we make a similar statement about exothermic reactions? Explain.

63. Rate constants for the first-order decomposition of acetonedicarboxylic acid

$$\underset{\text{Acetonedicarboxylic acid}}{CO(CH_2COOH)_2(aq)} \longrightarrow \underset{\text{Acetone}}{CO(CH_3)_2(aq)} + 2\,CO_2$$

are $k = 4.75 \times 10^{-4}\ s^{-1}$ at 293 K and $k = 1.63 \times 10^{-3}\ s^{-1}$ at 303 K. What is the activation energy, E_a, of this reaction?

64. The decomposition of di-*tert*-butyl peroxide (DTBP) is a first-order reaction with a half-life of 320 min at 135 °C and 100 min at 145 °C. Calculate E_a for this reaction.

$$\underset{\text{DTBP}}{C_8H_{18}O_2(g)} \longrightarrow \underset{\text{Acetone}}{2\,(CH_3)_2CO(g)} + \underset{\text{Ethane}}{C_2H_6(g)}$$

65. The decomposition of ethylene oxide at 652 K

$$(CH_2)_2O(g) \longrightarrow CH_4(g) + CO(g)$$

is a first-order reaction with $k = 0.0120\ min^{-1}$. The activation energy of the reaction is 218 kJ/mol. Calculate the rate constant of the reaction at 525 K.

66. For the reaction in Problem 65, calculate the temperature at which the rate constant $k = 0.0100\ min^{-1}$.

Reaction Mechanisms

67. It is easy to see how a bimolecular elementary step in a reaction mechanism can occur as a result of a collision between two molecules. How do you suppose a *unimolecular* process is able to occur?

68. Why shouldn't we necessarily expect the rate law of a net reaction to be the same as the rate law of one of the elementary reactions in a plausible reaction mechanism? Cite *two* situations, however, in which this may indeed be the case.

69. The following is proposed as a plausible reaction mechanism:

$$A + B \longrightarrow I \quad \text{(slow)}$$
$$I + B \longrightarrow C + D \quad \text{(fast)}$$

(a) What is the net reaction described by this mechanism?

(b) What is a plausible rate law for the reaction?

70. The reaction $A + 2\,B \longrightarrow C + 2\,D$ is found to be first order in A and first order in B. A proposed mechanism for the reaction involves the following first step:

$$A + B \longrightarrow I + D \quad \text{(slow)}$$

(a) Write a plausible second step in a two-step mechanism.

(b) Is the second step slow or fast? Explain.

71. At temperatures below 600 K, the following reaction exhibits the rate law: Rate $= k[NO_2]^2$.

$$NO_2(g) + CO(g) \longrightarrow NO(g) + CO_2(g)$$

Propose a two-step mechanism involving one fast step and one slow step that is consistent with the net equation and the observed rate law.

72. Explain why the following mechanism is *not* plausible for the reaction in Problem 71.

Fast: $2\,NO_2 \rightleftharpoons N_2O_4$

Slow: $N_2O_4 + 2\,CO \longrightarrow 2\,NO + 2\,CO_2$

73. The following reaction exhibits the rate law: Rate $= k[NO]^2[Cl_2]$.

$$2\,NO(g) + Cl_2(g) \longrightarrow 2\,NOCl(g)$$

Explain why the following mechanism is *not* plausible for this reaction.

Fast: $NO + Cl_2 \rightleftharpoons NOCl + Cl$

Slow: $NO + Cl \longrightarrow NOCl$

74. Propose a two-step mechanism involving one fast reversible step and one slow step that is consistent with the net equation and the observed rate law for the reaction in Problem 73.

75. Show that the proposed mechanism is consistent with the rate law for the following reaction in aqueous solution,

$$Hg_2^{2+} + Tl^{3+} \longrightarrow 2\,Hg^{2+} + Tl^{+}$$

for which the observed rate law is

$$\text{Rate} = k \times \frac{[Hg_2^{2+}][Tl^{3+}]}{[Hg^{2+}]}$$

Proposed mechanism:

Fast: $Hg_2^{2+} \rightleftharpoons Hg^{2+} + Hg$

Slow: $Hg + Tl^{3+} \longrightarrow Hg^{2+} + Tl^{+}$

76. Show that the proposed mechanism is consistent with the rate law for the following reaction

$$2\,H_2(g) + 2\,NO(g) \longrightarrow N_2(g) + 2\,H_2O(g)$$

for which the rate law is: Rate $= k[H_2][NO]^2$.

Proposed mechanism:

$$2\,NO \rightleftharpoons N_2O_2$$

$$N_2O_2 + H_2 \longrightarrow N_2O + H_2O$$

$$N_2O + H_2 \longrightarrow N_2 + H_2O$$

Which must be the rate-determining step in this mechanism? Explain.

Catalysis

77. The decomposition of $H_2O_2(aq)$ is usually studied in the presence of $I^-(aq)$. The reaction is first order in both H_2O_2 and in I^-. Why can we treat the reaction as if it were only first order overall rather than second order?

78. Describe in a general way how the reaction profile for the surface-catalyzed reaction of SO_2 and O_2 to form SO_3 differs from the reaction profile for the noncatalyzed, homogeneous gas-phase reaction.

79. Describe ways in which an enzyme inhibitor may function.

80. A bacterial enzyme has an optimum temperature of 35 °C. Will the enzyme be more or less active at the normal human body temperature? Will it be more active or less active if the patient has a fever of 40 °C?

81. The kinetics of some surface-catalyzed reactions are similar to enzyme-catalyzed reactions. Explain this connection.

82. What happens to the rate of an enzyme-catalyzed reaction if the concentration of substrate X is doubled, when **(a)** the concentration of X is low, **(b)** the concentration of X is very high?

Additional Problems

83. In the first-order reaction, A $\longrightarrow$ products, the following concentrations were found at the indicated times: $t = 0$ s, [A] $= 0.88$ M; 25 s, 0.74 M; 50 s, 0.62 M; 75 s, 0.52 M; 100 s, 0.44 M; 125 s, 0.37 M; 150 s, 0.31 M. Calculate the instantaneous rate of reaction at $t = 125$ s.

84. The decomposition of $H_2O_2(aq)$ can be followed by removing samples from the reaction mixture and titrating them with $MnO_4^-(aq)$.

$$2\,MnO_4^- + 5\,H_2O_2 + 6\,H^+ \longrightarrow 2\,Mn^{2+} + 8\,H_2O + 5\,O_2(g)$$

Assume that at each of the times listed in Table 13.1, a 5.00-mL sample of the remaining $H_2O_2(aq)$ is removed and titrated with 0.0500 M $KMnO_4$. Add a column to the table listing the volume of $KMnO_4$ required for the titration, initially, at $t = 60$ s, at $t = 120$ s, and so on.

85. Refer to Problem 84. Because the volumes of 0.0500 M $KMnO_4$ required for the titrations are proportional to the remaining $[H_2O_2]$, you can plot ln (mL $KMnO_4$) versus time and determine a value of k for the decomposition of H_2O_2(aq). Show that the same value is obtained as in Figure 13.5.

86. Benzenediazonium chloride decomposes in water yielding $N_2(g)$.

$$C_6H_5N_2Cl \longrightarrow C_6H_5Cl + N_2(g)$$

The data tabulated below were obtained for the decomposition of an 0.071 M solution at 50 °C ($t = \infty$ corresponds to the completed reaction). To obtain $[C_6H_5N_2Cl]$ as a function of time, note that during the first 3 min, the volume of $N_2(g)$ produced was 10.8 mL of a total 58.3 mL, corresponding to this fraction of the total reaction: 10.8 mL/58.3 mL = 0.185. This same fraction of the available $C_6H_5N_2Cl$ was consumed during the same time.

Time, min	$N_2(g)$, mL	Time, min	$N_2(g)$, mL
0	0	18	41.3
3	10.8	21	44.3
6	19.3	24	46.5
9	26.3	27	48.4
12	32.4	30	50.4
15	37.3	∞	58.3

(a) Plot graphs showing the disappearance of $C_6H_5N_2Cl$ and the formation of $N_2(g)$ as a function of time.

(b) What is the initial rate of formation of $N_2(g)$?

(c) What is the rate of disappearance of $C_6H_5N_2Cl$ at $t = 20$ min?

(d) What is the half-life, $t_{1/2}$, of the reaction?

(e) Write the rate law for this reaction, including a value for k.

87. Refer to Problem 64. At 147 °C, the decomposition of DTBP is a first-order reaction with a half-life of 80.0 min. If a 4.50-g sample of DTBP is introduced into a 1.00-L flask at 147 °C, **(a)** what is the initial gas pressure in the flask? **(b)** What is the *total* gas pressure in the flask after 80.0 min? **(c)** the *total* gas pressure after 125 min?

88. Refer to Problem 49. The half-life for the decomposition of PAN at 25 °C is 32 min, and the activation energy of the reaction is 113 kJ/mol.**(a)** What is the half-life at 35 °C? **(b)** What is the initial rate of decomposition of PAN if 6.0×10^{14} molecules of PAN are injected into an air sample at 35 °C?

89. The rate constant for the first-order dissociation of cyclobutane to ethylene,

$$cyclo\text{-}C_4H_8(g) \longrightarrow 2\,C_2H_4(g)$$

can be represented as $k = 4.0 \times 10^{16}\ s^{-1}\ (e^{-E_a/RT})$, where E_a = 262 kJ/mol. Determine the temperature at which the half-life of the reaction is 1.00 min.

90. The conversion of *tert*-butyl bromide to *tert*-butyl alcohol is achieved in the first-order reaction

$$(CH_3)_3CBr + H_2O \longrightarrow (CH_3)_3COH + HBr$$

The half-life of this reaction is 14.1 h at 25 °C and 48.8 min at 50 °C. How long will it take for this conversion to go to 90% completion at 65 °C?

91. A rule of thumb in chemical kinetics states that for many reactions, the rate of reaction approximately doubles for a temperature rise of 10 °C. What must be the activation energy of a reaction if the rate is indeed found to double between 25 °C and 35 °C?

92. For which type of reaction would you expect the rate to increase more rapidly with increasing temperature, one with high or one with low activation energy? Explain.

93. It has been shown that the chirping of tree crickets (*Oceanthus*) can be described by the Arrhenius rate equation. Use the experimental measurement of 179 chirps/min at 25.0 °C and 142 chirps/min at 21.7 °C to determine **(a)** the activation energy for the chirping process; **(b)** the number of chirps/min at 20.0 °C; and **(c)** how closely the result in (b) conforms to the rule of thumb that the Fahrenheit temperature is "40 plus the number of chirps in 15 s."

94. The observed rate law for the reaction: $2\,O_3 \longrightarrow 3\,O_2$ is: Rate = $k[O_3]^2/[O_2]$. Propose a two-step mechanism for this reaction.

95. Hydroxide ion is involved in the mechanism but not consumed in this reaction in aqueous solution.

$$OCl^- + I^- \xrightarrow{OH^-} OI^- + Cl^-$$

(a) From the data in the table, determine the order of the reaction with respect to OCl^-, I^-, and OH^-, and the overall order.

$[OCl^-]$, M	$[I^-]$, M	$[OH^-]$, M	Rate of formation of OI^-, mol $L^{-1}\,s^{-1}$
0.0040	0.0020	1.00	4.8×10^{-4}
0.0020	0.0040	1.00	5.0×10^{-4}
0.0020	0.0020	1.00	2.4×10^{-4}
0.0020	0.0020	0.50	4.6×10^{-4}
0.0020	0.0020	0.25	9.4×10^{-4}

(b) Write the rate law, and determine a value of the rate constant, k.

(c) Show that the following mechanism is consistent with the net equation and with the rate law. Which is the rate-determining step?

$$OCl^- + H_2O \rightleftharpoons HOCl + OH^-$$
$$I^- + HOCl \longrightarrow HOI + Cl^-$$
$$HOI + OH^- \longrightarrow H_2O + OI^-$$

(d) Is it appropriate to refer to OH^- as a catalyst in this reaction? Explain.

96. The enzyme acetylcholinesterase has —COOH, —OH, and —NH_2 groups on side chains of amino acids at the active site. The enzyme is inactive in acidic solution, but the activity increases as the solution becomes more basic. Explain.

CHAPTER

14 Chemical Equilibrium

CONTENTS

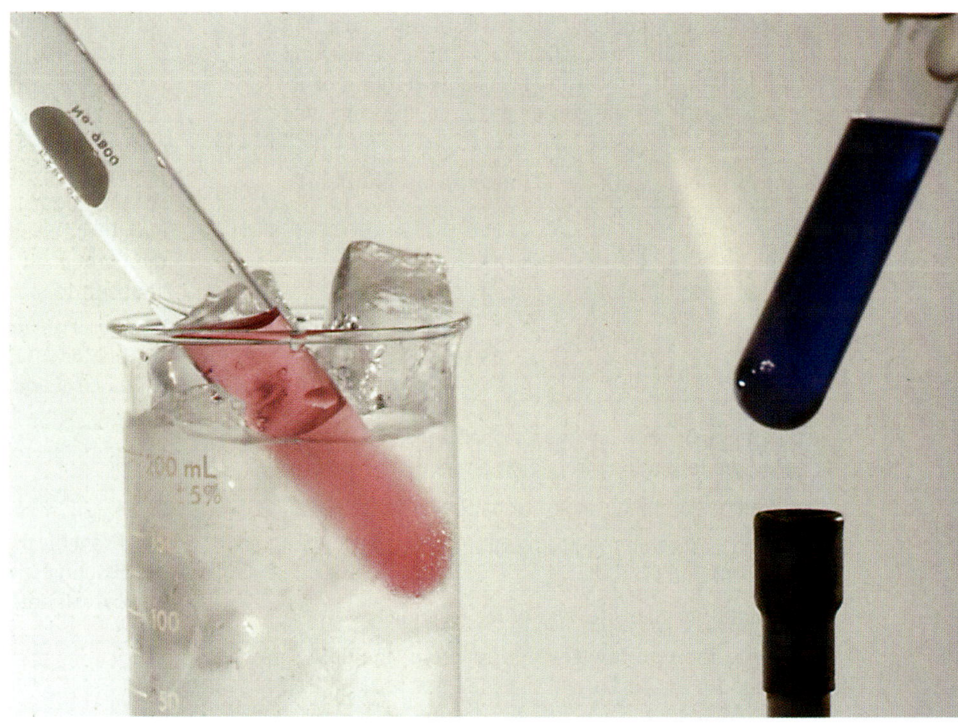

The ions $[Co(H_2O)_6]^{2+}$ (pink) and $[CoCl_4]^{2-}$ (blue) exist at equilibrium in aqueous solutions.

$$[Co(H_2O)_6]^{2+} + 4\,Cl^- \rightleftharpoons [CoCl_4]^{2-} + 6\,H_2O$$

At room temperature, both ions exist in significant concentrations, giving a violet colored solution. Heating the solution shifts the equilibrium to the right, forming more blue $[CoCl_4]^{2-}$. Cooling the solution shifts the equilibrium to the left, forming more pink $[Co(H_2O)_6]^{2+}$. In this chapter, we discuss principles of chemical equilibrium, including the effect of temperature on equilibrium.

Up to now, we have limited our calculations concerning chemical reactions to those that can be described by stoichiometry alone—that is, to reactions that go to completion. But not all reactions go to completion. Reversible reactions reach an equilibrium state when forward and reverse reactions proceed at the same rate. To calculate amounts of reactants and products at equilibrium, we need a new quantity called the equilibrium constant.

In our study of equilibrium, which extends to some degree over the next five chapters, we will address a number of fundamental questions. How do we determine equilibrium constants by experiment? How can we *calculate* them from other quantities? How do changes in pressure, volume, temperature, and amounts of reactants affect the equilibrium state? What effect does a catalyst have on equilibrium? As we proceed through these chapters, you will see that the principles of dynamic

● **ChemCDX**
When you see this icon, there are related animations, demonstrations, and exercises on the CD accompanying this text.

equilibrium are encountered in the laboratory, in chemical industry, in living organisms, and in other natural phenomena.

14.1 The Dynamic Nature of Equilibrium

As we have noted before, equilibrium involves opposing processes occurring at equal rates. In vapor pressure equilibrium, the rate of evaporation of a liquid is equal to the rate of condensation of its vapor. In solubility equilibrium, the rate of dissolution of the solid is equal to its rate of crystallization from solution. Moreover, the equilibria that we examine are *dynamic* (not *static* like a teaspoon balanced on the lip of a teacup). We can use radioactivity to demonstrate dynamic equilibrium, as is illustrated in Figure 14.1. If we add a small amount of NaCl(s) containing a trace of radioactive sodium-24 to saturated NaCl(aq), radioactivity shows up immediately in the saturated solution as well as in the undissolved solid, indicating that some solid dissolves. And, because the concentration of a saturated solution remains constant, the amount of additional NaCl(s) that dissolves must be just matched by the amount of additional NaCl(s) that crystallizes from solution.

Figure 14.1 Dynamic equilibrium in saturated solution formation. (a) A trace of radioactive NaCl(s) (red) is added to saturated NaCl(aq). (b) Radioactivity immediately appears in the solution phase. This proves that the dissolution process does not stop when a solution becomes saturated.

Figure 14.2 shows how concentrations vary with time in the decomposition of HI(g) to $H_2(g)$ and $I_2(g)$, at 698 K. The curves are similar to some of the concentration versus time graphs in Chapter 13, but note an important difference: After the time marked t_e, the curves level off. In any reaction in which reactants and products attain constant, nonzero concentrations, we know that the reaction is reversible and that the forward reaction does not go to completion. We use a double arrow to denote a reversible reaction when writing a chemical equation.

$$2\,HI(g) \rightleftharpoons H_2(g) + I_2(g)$$

In chemical kinetics, we focus on the portion of the concentration versus time graph before t_e. In chemical equilibrium, we focus on what transpires *after* t_e.

> *In a condition of* **equilibrium**, *a forward and a reverse reaction proceed at equal rates, and the concentrations of reactants and products remain constant.*

We could show that equilibrium in this reaction is dynamic by introducing some $I_2(g)$ containing a trace of radioactive iodine-131 into the equilibrium mixture. The radioactivity would soon show up in the HI(g) as well as in the $I_2(g)$.

As an example of dynamic equilibrium, think of bailing out a leaking rowboat. Water leaking into the boat is analogous to a forward reaction and pouring buckets

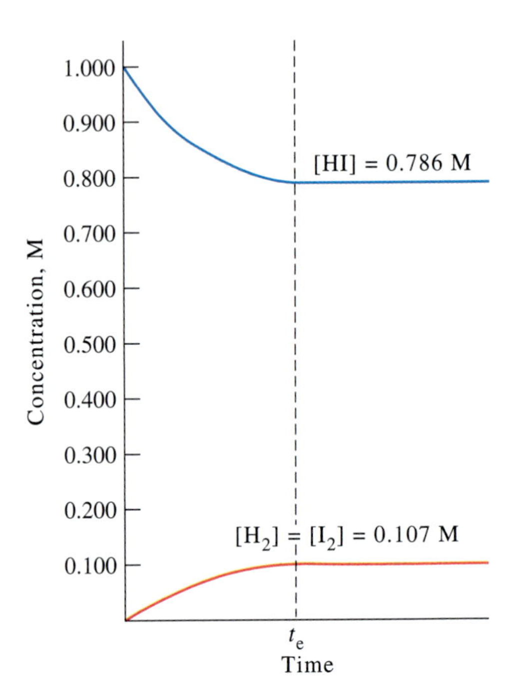

▶ **Figure 14.2**
Concentration versus time graph for the reversible reaction $2\ HI(g) \rightleftharpoons H_2(g) + I_2(g)$ at 698 K
After the time, t_e, the reaction is at equilibrium, and the concentrations of reactant and products undergo no further change. The data shown here, together with those for two other experiments, are listed in Table 14.1.

of water overboard is analogous to the reverse reaction. If we bail water out of the boat just as fast as it leaks in, the pool of water at the bottom of the boat is kept at a constant depth, analogous to the constant concentrations of reactants and products at equilibrium.

14.2 The Equilibrium Constant Expression

In the experiment described in Figure 14.2, when HI(g) is initially introduced into the reaction vessel, only the forward reaction occurs because neither H_2 nor I_2 is present. However, as soon as some of these products form, the reverse reaction begins. As time goes on, the forward reaction slows down because the concentration of HI(g) decreases. The reverse reaction speeds up as more $H_2(g)$ and $I_2(g)$ accumulate. Eventually, the forward and reverse reactions go at the same rate, and the reaction mixture is at equilibrium.

Figure 14.2 presents data for Experiment 1 in Table 14.1. Experiments 2 and 3 involve different initial concentrations for the same reaction. Consider the data in the third column of Table 14.1, the equilibrium concentrations of HI, H_2, and I_2. Note that these data have nothing in common for the three experiments. Let's use trial and error (called a heuristic approach) to try to find a common feature of the equilibrium state. For example, consider the following ratios of equilibrium concentrations listed in the fourth and fifth columns.

$$\frac{[H_2][I_2]}{[HI]} \quad \text{and} \quad \frac{[H_2][I_2]}{2[HI]}$$

Neither ratio has a common feature; their values are different for the three experiments. In the sixth column of Table 14.1, we raise the concentration factors to pow-

TABLE 14.1 Three Experiments Involving the Reaction, $2\ HI(g) \rightleftharpoons H_2(g) + I_2(g)$ at 698 K

Experiment Number	Initial Concentrations, M		Equilibrium Concentrations, M	$\frac{[H_2][I_2]}{[HI]}$	$\frac{[H_2][I_2]}{2[HI]}$	$\frac{[H_2][I_2]}{[HI]^2}$
1	[HI]:	1.000	0.786	0.0146	0.00728	0.0185
	$[H_2]$:	0.000	0.107			
	$[I_2]$:	0.000	0.107			
2	[HI]:	0.000	1.573	0.0288	0.0144	0.0183
	$[H_2]$:	1.000	0.213			
	$[I_2]$:	1.000	0.213			
3	[HI]:	1.000	2.360	0.0434	0.0217	0.0184
	$[H_2]$:	1.000	0.320			
	$[I_2]$:	1.000	0.320			

ers given by the stoichiometric coefficients. Now, allowing for slight variations due to experimental errors, we see that the ratios do have the same value. The ratio of *equilibrium* concentrations shown below is called the **equilibrium constant expression**. It has a constant value regardless of the initial concentrations of reactants and products. This constant is denoted by the symbol $\boldsymbol{K_c}$ and is called the **concentration equilibrium constant**.

The *inverse* of the equilibrium constant expression also has a constant value. We'll consider its significance in Section 14.3.

$$K_c = \frac{[H_2][I_2]}{[HI]^2} = 1.84 \times 10^{-2} \text{ (at 698 K)}$$

The subscript c in K_c signifies that concentrations (molarities) are used. We note the temperature because equilibrium constants are temperature-dependent. Thus, the value $K_c = 1.84 \times 10^{-2}$ (at 698 K) applies only to the reaction: $2\ HI(g) \rightleftharpoons H_2(g) + I_2(g)$ and only at 698 K.

Keep in mind that the equilibrium state can be approached from an initial condition in which only the reactant(s) are present (Experiment 1 in Table 14.1). It can also be approached from the product side (Experiment 2). Or, reactants and products may all be present initially (Experiment 3). In every case, the value of K_c is the same.

Another example of a reversible reaction is the oxidation of NO(g) to $NO_2(g)$, a reaction that contributes to the formation of smog.

$$2\ NO(g) + O_2(g) \rightleftharpoons 2\ NO_2(g)$$

Here, the following ratio of equilibrium concentrations has a constant value.

$$K_c = \frac{[NO_2]^2}{[NO]^2[O_2]}$$

From these two examples, we can begin to see the general nature of an equilibrium constant expression:

- Concentrations of the products appear in the numerator, and concentrations of the reactants appear in the denominator. (Concentrations are expressed as molarities, but *units* are omitted in the K_c expression.)
- The exponents of the concentrations are identical to the stoichiometric coefficients in the chemical equation. Consider the following hypothetical reaction and the equilibrium constant expression we write for it.

$$a\,\mathrm{A} + b\,\mathrm{B} + \cdots \rightleftharpoons g\,\mathrm{G} + h\,\mathrm{H} + \cdots$$

$$K_c = \frac{[\mathrm{G}]^g[\mathrm{H}]^h\cdots}{[\mathrm{A}]^a[\mathrm{B}]^b\cdots}$$

We can get a sense of the significance of an equilibrium constant expression by considering a case in which the decomposition of HI(g) at 698 K produces equilibrium concentrations of H_2 and I_2 of 0.0250 M. To find the equilibrium concentration of HI, we first write the equilibrium constant expression and substitute these concentrations.

One method of determining $[I_2]$ in the equilibrium mixture is to extract a sample from the mixture, followed by titration with $Na_2S_2O_3(aq)$.

$$I_2(aq) + 2\,S_2O_3^{2-}(aq) \longrightarrow S_4O_6^{2-}(aq) + 2\,I^-(aq)$$

$$K_c = \frac{[\mathrm{H_2}][\mathrm{I_2}]}{[\mathrm{HI}]^2} = \frac{(0.0250)(0.0250)}{[\mathrm{HI}]^2} = 1.84 \times 10^{-2}$$

Then we solve for [HI].

$$[\mathrm{HI}]^2 = \frac{(0.0250)(0.0250)}{1.84 \times 10^{-2}}$$

$$[\mathrm{HI}] = \sqrt{\frac{(0.0250)(0.0250)}{1.84 \times 10^{-2}}} = 0.184\ \mathrm{M}$$

We will consider more calculations based on equilibrium constant expressions later in the chapter.

EXAMPLE 14.1

If the equilibrium concentrations of Cl_2 and $COCl_2$ are found to be the same at 395 °C, find the equilibrium concentration of CO in the reaction that follows.

$$\mathrm{CO(g)} + \mathrm{Cl_2(g)} \rightleftharpoons \mathrm{COCl_2(g)} \quad K_c = 1.2 \times 10^3 \text{ at } 395\ ^\circ\mathrm{C}$$

SOLUTION

If asked to determine the concentration of a substance involved in a reversible reaction at equilibrium, we must use the equilibrium constant expression.

$$K_c = \frac{[\mathrm{COCl_2}]}{[\mathrm{CO}][\mathrm{Cl_2}]} = 1.2 \times 10^3$$

Because $[Cl_2]$ equals $[COCl_2]$ in the equilibrium at 395 °C, we can see that these two terms cancel.

$$\frac{\cancel{[COCl_2]}}{[CO]\cancel{[Cl_2]}} = \frac{1}{[CO]} = K_c = 1.2 \times 10^3$$

We can then solve for [CO].

$$[CO] = \frac{1}{K_c} = \frac{1}{1.2 \times 10^3} = 8.3 \times 10^{-4}\ \text{M}$$

Thus, there is only *one* possible value of [CO] for the condition of equal concentrations of Cl_2 and $COCl_2$, regardless of what those concentrations are.

EXERCISE 14.1A

If [CO] equals [Cl_2] at equilibrium for the reaction in Example 14.1, is there just one possible value of [$COCl_2$]? Explain.

EXERCISE 14.1B

Suppose [O_2] is fixed at a certain constant value when equilibrium is reached in the following reversible reaction.

$$2\ SO_2(g) + O_2(g) \rightleftharpoons 2\ SO_3(g) \quad K_c = 1.00 \times 10^2$$

Do [SO_2] and [SO_3] have unique values? Does the ratio [SO_2]/[SO_3] have a unique value? Explain. Describe the equilibrium state when [O_2] = 1.00 M.

The Condition of Equilibrium—A Kinetics View

Consider the following reaction.

$$2\ HI(g) \rightleftharpoons H_2(g) + I_2(g)$$

At equilibrium, the rate of the forward reaction equals the rate of the reverse reaction. Also, from Chapter 13 (pages 570 and 581), we have rate laws for both of these reactions.

$$\text{Rate of forward reaction} = k_f[HI]^2$$

$$\text{Rate of reverse reaction} = k_r[H_2][I_2]$$

At equilibrium, the rates are equal, so we can set the right sides of the two equations equal.

$$k_f[HI]^2 = k_r[H_2][I_2]$$

We can then gather the two rate constants on the same side of the equation.

$$\frac{k_f}{k_r} = \frac{[H_2][I_2]}{[HI]^2} = K_c$$

Thus, the ratio k_f/k_r is equal to the equilibrium constant, K_c.

Because the exponents of the concentrations in K_c are based on stoichiometric coefficients, the concentration units in k_f/k_r will match those of K_c *only* if they, too, are based on stoichiometric coefficients. This match is certain to occur only when the forward and reverse reactions proceed by a simple one-step mechanism.* The kinetics approach to chemical equilibrium is of theoretical interest, but because it, too, requires experimental data, we usually just evaluate equilibrium constants directly by experiments.

The Condition of Equilibrium—The Thermodynamic View

In Chapter 17, we will show that the equilibrium constant can be related to other fundamental thermodynamic properties, and we will then call it the *thermodynamic equilibrium constant*, K_{eq}. Moreover, we will learn how to use tabulated thermodynamic data to *predict* values of equilibrium constants.

In anticipation of that switch to K_{eq}, we have written equilibrium constants as dimensionless numbers because the thermodynamic equilibrium constant expression uses dimensionless quantities known as *activities*. In this chapter, we will use molarities and partial pressures (in atm) in equilibrium constant expressions. In both cases we will omit the units. In Chapter 17 we will see why it is permissible to do so.

14.3 Modifying Equilibrium Constant Expressions

Sometimes we need to modify an equilibrium constant expression to make it applicable to a particular situation. We consider a few important modifications in this section.

Modifying the Chemical Equation

The following equation is one way to describe the *formation* of $NO_2(g)$ at 298 K.

$$2\,NO(g) + O_2(g) \rightleftharpoons 2\,NO_2(g)$$

Using appropriate experimental data like those in Table 14.1, we could establish the following numerical value of K_c.

$$K_c = \frac{[NO_2]^2}{[NO]^2[O_2]} = 4.67 \times 10^{13} \text{ (at 298 K)}$$

If we were interested in the *decomposition* of $NO_2(g)$ at 298 K, we would likely write the chemical equation as the *reverse* of that for its formation.

$$2\,NO_2(g) \rightleftharpoons 2\,NO(g) + O_2(g) \quad K'_c = \text{? (at 298 K)}$$

However, we don't need to do another set of experiments to establish the value of the new equilibrium constant, designated K'_c. The equilibrium constant for the decomposition of $NO_2(g)$ is the *inverse* of the equilibrium constant for its formation.

$$K'_c = \frac{[NO]^2[O_2]}{[NO_2]^2} = \frac{1}{\dfrac{[NO_2]^2}{[NO]^2[O_2]}} = \frac{1}{K_c} = \frac{1}{4.67 \times 10^{13}} = 2.14 \times 10^{-14}$$

The preceding modification illustrates a general rule.

> *When we reverse a chemical reaction with the equilibrium constant, K_c, we invert the equilibrium constant. That is, the reverse reaction has the equilibrium constant, $1/K_c$.*

* If we base the relationship between rate constants and the equilibrium constant on an actual reaction mechanism, we are not limited to this condition of one-step mechanisms (see Problem 86).

Suppose we decide to describe the decomposition of $NO_2(g)$ based on *one* mole of reactant instead of two.

$$NO_2(g) \rightleftharpoons NO(g) + \tfrac{1}{2}O_2(g) \quad K''_c = ? \quad \text{(at 298 K)}$$

Again, we don't need any additional experimental data because we can see the following relationship.

$$K''_c = \frac{[NO][O_2]^{1/2}}{[NO_2]} = \left[\frac{[NO]^2[O_2]}{[NO_2]^2}\right]^{1/2} = (K'_c)^{1/2} = \left[\frac{1}{K_c}\right]^{1/2} = 1.46 \times 10^{-7}$$

The preceding illustrates another general rule.

> *When the coefficients of an equation are multiplied by a common factor,* n, *to produce a new equation, we raise the original* K_c *value to the power* n *to obtain the new equilibrium constant.*

In the preceding example, $n = \frac{1}{2}$. If we double the coefficients in an equation, the factor $n = 2$; and so on.

In summary, the form of an equilibrium constant expression and the value of K_c depend on exactly how the chemical equation for a reversible reaction is written. *Thus, we must write the balanced chemical equation when citing a value for* K_c.

EXAMPLE 14.2

The equilibrium constant for the reaction

$$\tfrac{1}{2}H_2(g) + \tfrac{1}{2}I_2(g) \rightleftharpoons HI(g)$$

at 718 K is 7.07. **(a)** What is the value of K_c at 718 K for the reaction $HI(g) \rightleftharpoons \frac{1}{2}H_2(g) + \frac{1}{2}I_2(g)$? **(b)** What is the value of K_c at 718 K for the reaction $H_2(g) + I_2(g) \rightleftharpoons 2\,HI(g)$?

SOLUTION

a. Because the reaction in question is the *reverse* of the one for which the equilibrium constant is 7.07, the equilibrium constant we seek is the *inverse* of 7.07.

$$K_c = \frac{1}{7.07} = 0.141$$

b. In this chemical equation, the coefficients of the original equation have been *doubled.* Thus, we need to raise the value of the original equilibrium constant to the power of *two*; that is, *square* it.

$$K_c = (7.07)^2 = 50.0$$

EXERCISE 14.2A

The equilibrium constant for the reaction $SO_2(g) + \frac{1}{2}O_2(g) \rightleftharpoons SO_3(g)$ is 20.0 at 973 K. Calculate K_c at 973 K for the following reaction.

$$2\,SO_3(g) \rightleftharpoons 2\,SO_2(g) + O_2(g)$$

EXERCISE 14.2B

The equilibrium constant for the reaction $\frac{1}{2}\,N_2(g) + \frac{3}{2}\,H_2O(g) \rightleftharpoons NH_3(g) + \frac{3}{4}\,O_2(g)$ at 900 K is 1.97×10^{-20}. Calculate K_c at 900 K for the following reaction.

$$4\,NH_3(g) + 3\,O_2(g) \rightleftharpoons 2\,N_2(g) + 6\,H_2O(g)$$

The Equilibrium Constant for a Net Reaction

In Section 6.6, we combined the equations for individual reactions to obtain a net equation. At the same time, we used Hess's law to combine the enthalpy changes for the individual reactions to obtain the enthalpy change for the net reaction. We use a similar approach to obtain the equilibrium constant for a net reaction.

Suppose we want to know the equilibrium constant at 298 K for this reaction.

$$(1) \qquad N_2O(g) + \tfrac{3}{2}O_2(g) \rightleftharpoons 2\,NO_2(g)$$

If we know values at 298 K for reactions (2) and (3), we can add these equations to get the equation for reaction (1) as the net equation.

$$(2) \qquad N_2O(g) + \tfrac{1}{2}O_2(g) \rightleftharpoons \cancel{2\,NO(g)} \qquad K_c(2) = 1.7 \times 10^{-13}$$

$$(3) \qquad \cancel{2\,NO(g)} + O_2(g) \rightleftharpoons 2\,NO_2(g) \qquad K_c(3) = 4.67 \times 10^{13}$$

$$\textit{Net:}\ (1) \qquad N_2O(g) + \tfrac{3}{2}O_2(g) \rightleftharpoons 2\,NO_2(g) \qquad K_c(1) = ?$$

Now we can find the relationship between the unknown $K_c(1)$ and the known $K_c(2)$ and $K_c(3)$.

$$\frac{\cancel{[NO]^2}}{[N_2O][O_2]^{1/2}} \times \frac{[NO_2]^2}{\cancel{[NO]^2}[O_2]} = \frac{[NO_2]^2}{[N_2O][O_2]^{3/2}}$$

$$K_c(2) \times K_c(3) = K_c(1)$$

$$1.7 \times 10^{-13} \times 4.67 \times 10^{13} = 7.9$$

The preceding illustrates another general rule.

> *When we* add *the equations for individual reactions to obtain a net equation, we* multiply *their equilibrium constants to obtain the equilibrium constant for the net reaction.*

Equilibria Involving Gases

In reactions involving gases, we often find it convenient to measure partial pressures rather than molarities. Consider the following general gas-phase reaction.

$$a\,A(g) + b\,B(g) + \cdots \rightleftharpoons g\,G(g) + h\,H(g) + \cdots$$

We can define a **partial pressure equilibrium constant, K_p**, as follows.

$$K_p = \frac{(P_G)^g (P_H)^h \cdots}{(P_A)^a (P_B)^b \cdots}$$

At times, we have a value of K_c for a reaction and need to know K_p, or vice versa. We can derive a relationship between K_c and K_p as we show for the following reaction at 298 K.

$$2\,NO(g) + O_2(g) \rightleftharpoons 2\,NO_2(g) \quad K_c = 4.67 \times 10^{13}$$

Suppose we apply the ideal gas law ($PV = nRT$) to NO_2 and solve it for P_{NO_2}. Then let's replace n_{NO_2}/V by its equivalent, the molarity of NO_2, that is, $[NO_2]$.

$$P_{NO_2} = \frac{n_{NO_2}}{V} \times RT = [NO_2]RT$$

Imagine doing the same for NO and O_2 and then writing the K_p expression for the reaction.

$$K_p = \frac{(P_{NO_2})^2}{(P_{NO})^2(P_{O_2})} = \frac{([NO_2]RT)^2}{([NO]RT)^2[O_2]RT} = \frac{[NO_2]^2\cancel{(RT)^2}}{[NO]^2\cancel{(RT)^2}[O_2](RT)} = \frac{[NO_2]^2}{[NO]^2[O_2]} \times \frac{1}{RT}$$

The expression shown in red above is simply K_c for the reaction, and the relationship between K_p and K_c is therefore the following.

$$K_p = \frac{K_c}{RT} = K_c(RT)^{-1}$$

Now let's consider again a general reaction.

$$a\,A(g) + b\,B(g) + \cdots \rightleftharpoons g\,G(g) + h\,H(g) + \cdots$$

In a manner similar to that for the reaction $2\,NO(g) + O_2(g) \rightleftharpoons 2\,NO_2(g)$, we could derive an equation for the general reaction.

$$K_p = K_c(RT)^{\Delta n_{gas}}$$

The exponent Δn_{gas} is the *change in number of moles of gas as the reaction occurs in the forward direction*. That is, $\Delta n_{gas} = (g + h + \cdots) - (a + b + \cdots)$.

Returning to the reaction: $2\,NO(g) + O_2(g) \rightleftharpoons 2\,NO_2(g)$, we see that $\Delta n_{gas} = 2 - (2 + 1) = 2 - 3 = -1$, and $K_p = K_c(RT)^{-1}$. To evaluate K_p from the known value of K_c (4.67×10^{13}), we use $R = 0.08206$ and $T = 298$.

$$K_p = K_c(RT)^{-1} = 4.67 \times 10^{13} \times (0.08206 \times 298)^{-1} = 1.91 \times 10^{12}$$

Unless otherwise indicated, K_p values are based on pressures expressed in atm. In most respects, we can deal with K_p expressions as we did with K_c expressions, as illustrated in Example 14.3.

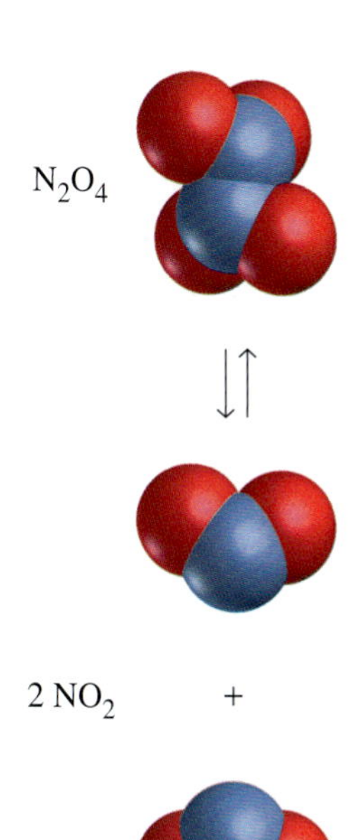

EXAMPLE 14.13

Consider equilibrium between dinitrogen tetroxide and nitrogen dioxide.

$$N_2O_4(g) \rightleftharpoons 2\,NO_2(g) \qquad K_p = 0.660 \text{ at } 319\text{ K}$$

(a) What is the value of K_c for this reaction? **(b)** What is the value of K_p for the reaction, $2\,NO_2(g) \rightleftharpoons N_2O_4(g)$? **(c)** If the equilibrium partial pressure of $NO_2(g)$ is found to be 0.332 atm, what is the equilibrium partial pressure of $N_2O_4(g)$?

SOLUTION

a. To find K_c, we need the expression that relates K_c and K_p.

$$K_p = K_c(RT)^{\Delta n_{gas}}$$

$\Delta n_{gas} = \Sigma$ stoichiometric coefficients of product(s) $- \Sigma$ stoichiometric coefficients of reactant(s). In this example, $\Delta n_{gas} = 2 - 1 = 1$.

We can now solve for K_c by using the known values of K_p, T, and R.

PROBLEM-SOLVING NOTE

Note that we omit units because equilibrium constants are written without units.

$$K_c = \frac{K_p}{(RT)^1} = \frac{0.660}{0.08206 \times 319} = 0.0252$$

b. Because the reaction for which we seek the K_p value, $2\ NO_2(g) \rightleftharpoons N_2O_4(g)$, is the *reverse* of the reaction for which we are given a K_p value, we must *invert* the known value (0.660).

$$K_p = \frac{1}{0.660} = 1.52$$

c. To find the partial pressure of $N_2O_4(g)$, we can use either the original K_p expression or the one derived in part (b). Let's use the original one for the reaction: $N_2O_4(g) \rightleftharpoons 2\ NO_2(g)$.

PROBLEM-SOLVING NOTE

Note that although the K_p expression does not include units, the partial pressure values that we use are in atmospheres.

$$K_p = \frac{(P_{NO_2})^2}{(P_{N_2O_4})} = 0.660$$

$$= \frac{(0.332)^2}{(P_{N_2O_4})} = 0.660$$

$$P_{N_2O_4} = \frac{(0.332)^2}{0.660} = 0.167 \text{ atm}$$

EXERCISE 14.3A

Given $K_c = 1.8 \times 10^{-6}$ for the reaction $2\ NO(g) + O_2(g) \rightleftharpoons 2\ NO_2(g)$ at 457 K, derive the value of K_p at 457 K for the reaction $NO_2(g) \rightleftharpoons NO(g) + \frac{1}{2}\ O_2(g)$.

EXERCISE 14.3B

For the reaction, $NO(g) + \frac{3}{2}\ H_2O(g) \rightleftharpoons NH_3(g) + \frac{5}{4}\ O_2(g)$, $K_p = 2.6 \times 10^{-16}$ at 900 K. What is K_c at 900 K for the reaction $4\ NH_3(g) + 5\ O_2(g) \rightleftharpoons 4\ NO(g) + 6\ H_2O(g)$?

Equilibria Involving Pure Solids and Liquids

Reactions with only gaseous reactants and products—the kind we have considered to this point in this chapter—are *homogeneous reactions*. The reaction medium is a single gaseous phase. In *heterogeneous* reactions, the reactants and products do *not* coexist in the same phase, and we need to make some accommodations in equilibrium constant expressions.

A general feature of equilibrium constant expressions is that they conform to the following idea.

The equilibrium constant expression does not include terms for pure solid and liquid phases because their concentrations do not change in a reaction.

Although the *amounts* of pure solid and liquid phases change during a reaction, these phases remain pure and their *concentrations* do not change.

Consider the *reversible* decomposition of calcium carbonate (the chief constituent of limestone).

$$CaCO_3(s) \rightleftharpoons CaO(s) + CO_2(g)$$

If we heat pure $CaCO_3(s)$ in a closed container, it decomposes to $CaO(s)$ and $CO_2(g)$. The $CaO(s)$ and $CO_2(g)$ also recombine to form $CaCO_3(s)$. When we say that the concentration of $CaCO_3(s)$ does not change, we mean that at all times, the solid is pure (100% $CaCO_3$), even as its amount decreases from what was originally present. Similarly, the $CaO(s)$ formed is pure. In fact, the $CO_2(g)$ is also pure. However, as the amount of $CO_2(g)$ increases, so do the concentration and pressure of CO_2 in the closed container. Thus, CO_2 appears in the equilibrium constant expressions and $CaCO_3$ and CaO do not.

$$K_c = [CO_2(g)] \quad \text{and} \quad K_p = P_{CO_2}$$

We can write similar expressions for equilibrium between pure liquid water and vapor.

$$H_2O(l) \rightleftharpoons H_2O(g)$$

$$K_c = [H_2O(g)] \quad \text{and} \quad K_p = P_{H_2O}$$

Note that K_p for a liquid-vapor equilibrium is simply the *vapor pressure* of the liquid.

EXAMPLE 14.4

The reaction of steam and coke (carbon) produces a mixture of carbon monoxide and hydrogen called water-gas. This reaction has long been used to make combustible gases from coal.

$$C(s) + H_2O(g) \rightleftharpoons CO(g) + H_2(g)$$

Write an equilibrium constant expression, K_c, for this reaction.

SOLUTION

The products CO and H_2 and the reactant H_2O are all gases and are represented in the equilibrium constant expression, but C(s), a solid, is not.

$$K_c = \frac{[CO][H_2]}{[H_2O]}$$

EXERCISE 14.4A

Write the partial pressure equilibrium constant expression for the reaction in Example 14.4.

EXERCISE 14.4B

The reaction of steam with iron is an old method of producing hydrogen gas.

$$3\ Fe(s) + 4\ H_2O(g) \rightleftharpoons Fe_3O_4(s) + 4\ H_2(g)$$

Write the equilibrium constant expressions K_c and K_p for this reaction.

Equilibrium Constants: When Do We Need Them and When Don't We?

In principle, every reaction is reversible, at least to some extent, and can be described through an equilibrium constant expression. In many cases, however, we don't need to use equilibrium constants in calculations. How can this be? Let's answer this question by considering three specific cases.

Consider the reaction of hydrogen and oxygen gases at 298 K.

$$2\,H_2(g) + O_2(g) \rightleftharpoons 2\,H_2O(l)$$

$$K_p = \frac{1}{(P_{H_2})^2(P_{O_2})} = 1.4 \times 10^{83}$$

Starting with a 2:1 mole ratio of hydrogen to oxygen, the equilibrium partial pressures of $H_2(g)$ and $O_2(g)$ must become extremely small—approaching zero—in order for the K_p value to be so large. For all practical purposes, the hydrogen and oxygen are totally consumed in the reaction. We say that a reaction *goes to completion* if one or more reactants is totally consumed, and we can do calculations just by using the principles of stoichiometry (Chapter 3).

A very large numerical value of K_c or K_p signifies that a reaction goes to completion, or essentially so. (In effect, the reaction is not reversible.)

(It is difficult to say exactly what we mean by "very large," but K values with double-digit powers of ten generally meet the requirement.)

Now let's consider a reaction with a very different outcome: the decomposition of limestone at 298 K.

$$CaCO_3(s) \rightleftharpoons CaO(s) + CO_2(g) \qquad K_p = P_{CO_2} = 1.9 \times 10^{-23}$$

The enduring quality of Michelangelo's marble sculpture "David," finished in 1504, is a testimony to the very, very limited decomposition of $CaCO_3(s)$ that occurs at 298 K. Many types of marble are nearly pure $CaCO_3(s)$.

Intuitively, we know that limestone, which is mainly $CaCO_3(s)$, doesn't decompose to any great extent at normal temperatures. The K_p value tells us that the partial pressure of $CO_2(g)$ in equilibrium with $CaCO_3(s)$ and $CaO(s)$ is exceedingly small. $P_{CO_2} = K_p = 1.9 \times 10^{-23}$ atm.

A very small numerical value of K_c or K_p signifies that the forward reaction, as written, does not occur to any significant extent.

(Double-digit *negative* powers of ten generally meet the requirement for "very small.") In fact, in many such cases, we say that the forward reaction doesn't take place at all. Therefore, we sometimes describe a reaction with a very small equilibrium constant as follows.

$$CaCO_3(s) \xrightarrow{298\text{ K}} \text{"no reaction"}$$

The situation is quite different when we consider the decomposition of limestone at about 1300 K.

$$CaCO_3(s) \rightleftharpoons CaO(s) + CO_2(g) \qquad K_p \approx 1$$

Here, the forward and reverse reactions are both significant, and we do indeed need to use the K_p expression in calculations.

Finally, we should keep in mind that an equilibrium constant expression applies only to a reversible reaction *at equilibrium*. Rates of reaction determine how long it takes to reach equilibrium, and thus, indirectly, when the equilibrium constant expression can be used. Although K_p for the reaction of $H_2(g)$ and $O_2(g)$ at 298 K is very large, suggesting that the reaction should go to completion, the reaction proceeds at an immeasurably slow rate because of its high activation energy. The reaction never reaches equilibrium at 298 K. It is only when the mixture is ignited by a spark, heated or catalyzed that the reaction occurs at an explosive speed. Chemists say that the reaction of $H_2(g)$ and $O_2(g)$ at 298 K is *thermodynamically favorable* (meaning that K_p is large) but is *kinetically controlled* (meaning that the exceedingly slow rate of reaction prevents any significant reaction from occurring).

EXAMPLE 14.5

Is the reaction $CaO(s) + CO_2(g) \rightleftharpoons CaCO_3(s)$ likely to occur to any appreciable extent at 298 K?

SOLUTION

This reaction is the *reverse* of that describing the decomposition of limestone. Its K_p value is the *reciprocal* of that for the decomposition of limestone: $K_p = 1/(1.9 \times 10^{-23}) = 5.3 \times 10^{22}$. The large value of K_p leads us to expect the forward reaction to occur to a very significant extent. In fact, over time, the reaction should go essentially to completion.

EXERCISE 14.5

Refer to Example 14.1, and determine if we can assume that the reaction $CO(g) + Cl_2(g) \longrightarrow COCl_2(g)$ goes essentially to completion at 395 °C. Explain your reasoning.

The Reaction Quotient, Q: Predicting the Direction of Net Change

As we have noted, *at equilibrium*, only certain concentrations or partial pressures of reactants and products are possible for a given reaction. That is, concentrations must be compatible with the equilibrium constant expression K_c, and partial pressures must be in agreement with K_p. However, initially we can bring together reactants and/or products in just about any concentrations or partial pressures. For these *nonequilibrium* conditions, the expression having the same form as K_c or K_p is called the **reaction quotient,** Q_c or Q_p. The reaction quotient is *not* constant for a reaction, but it is quite useful because it allows us to predict the direction in which a net reaction must occur to establish equilibrium. To illustrate, let's turn again to the decomposition of HI(g) and the data at 698 K listed in Table 14.1.

$$2\,HI(g) \rightleftharpoons H_2(g) + I_2(g)$$

The Value of Q_c for the Initial Conditions in Experiment 1 In this reaction, we start with only the reactant. Because there are no products initially, a net reaction must occur in the *forward* direction (to the right). When we substitute the initial concentrations [HI] = 1.000 M and $[H_2] = [I_2] = 0.000$ M into the reaction quotient expression, we find that

$$Q_c = \frac{[H_2][I_2]}{[HI]^2} = \frac{(0) \times (0)}{(1.000)^2} = 0$$

The initial value of Q_c is 0, but as the reaction proceeds in the forward direction, the numerator of this ratio—$[H_2][I_2]$—increases in value, and the denominator—$[HI]^2$—gets smaller. Both of these changes cause the value of Q_c to increase. Equilibrium is reached when $Q_c = K_c$. This analysis suggests the following criterion.

> *If $Q_c < K_c$, a net reaction proceeds in the forward direction, that is, from left to right. (The rate of the forward reaction exceeds that of the reverse reaction until equilibrium is reached.)*

The Value of Q_c for the Initial Conditions in Experiment 2 Here, we start with products only, and know that a net reaction must occur in the reverse direction (to the left). We can calculate the value of Q_c for the concentrations [HI] = 0.000 M and $[H_2] = [I_2] = 1.000$ M.

$$Q_c = \frac{[H_2][I_2]}{[HI]^2} = \frac{(1.000) \times (1.000)}{(0.000)^2} \longrightarrow \infty$$

As the reaction proceeds in the reverse direction, the numerator of this ratio decreases in value and the denominator gets larger, both of which cause the value of Q_c to decrease. Again, equilibrium is reached when $Q_c = K_c$. This analysis suggests another criterion.

> *If $Q_c > K_c$, a net reaction proceeds in the reverse direction, that is, from right to left. (The rate of the reverse reaction exceeds that of the forward reaction until equilibrium is reached.)*

For both of the preceding cases, we were able to predict the direction of a net reaction, without having to evaluate Q_c. Sometimes, however, we need to compare the reaction quotient and the equilibrium constant to predict the direction of a net reaction. Experiment 3 of Table 14.1 is an example of such a situation, as illustrated in Example 14.6. Figure 14.3 summarizes the relationship between the reaction quotient and equilibrium constant.

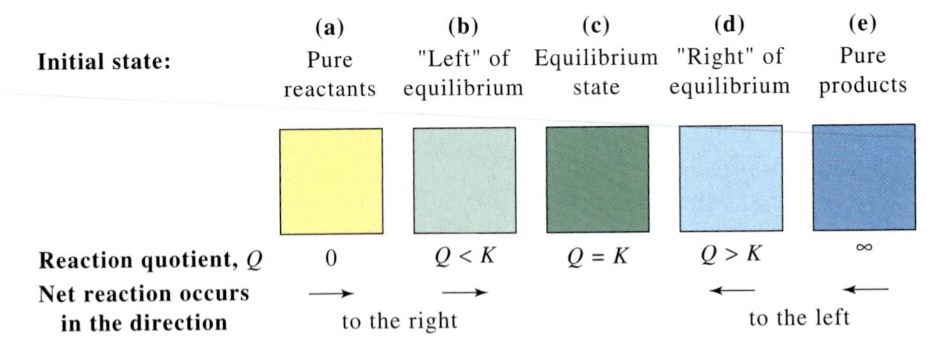

▶ Figure 14.3
Relating Q and K and predicting the direction of net reaction
For the first (a) and last (e) of the five situations shown, the reaction quotient (Q_c or Q_p) has zero and infinitely large values, respectively. For the other three situations, the value of Q_c or Q_p is in relation to K_c or K_p. A net reaction proceeds to the right if $Q < K$, and to the left $Q > K$.

EXAMPLE 14.6

Predict the direction of net change for Experiment 3 in Table 14.1 for the reaction, $2\,HI(g) \rightleftharpoons H_2(g) + I_2(g)$.

SOLUTION

Let's proceed as we did for Experiments 1 and 2. That is, we substitute the initial concentrations for Experiment 3, $[HI] = [H_2] = [I_2] = 1.000$ M, into the expression for the reaction quotient, Q_c. Then we can compare the value of Q_c with that of K_c from page 597: $K_c = 1.84 \times 10^{-2}$.

$$Q_c = \frac{[H_2][I_2]}{[HI]^2} = \frac{(1.000) \times (1.000)}{(1.000)^2} = 1.000$$

Because $Q_c > K_c$ ($1.000 > 1.84 \times 10^{-2}$), we conclude that a net reaction occurs in the *reverse* direction, from *right to left*. Notice how our prediction is supported by the equilibrium data for Experiment 3 in Table 14.1: [HI] = 2.360 M and $[H_2] = [I_2] =$ 0.320 M. The equilibrium concentration of HI is greater than its initial concentration, whereas the equilibrium concentrations of H_2 and I_2 are less than their initial concentrations. These changes in concentration correspond to a net reaction from right to left.

EXERCISE 14.6A

In which direction would a net reaction occur if the initial conditions in the reaction of Example 14.6 were [HI] = 1.00 M and $[H_2] = [I_2] = 0.100$ M?

EXERCISE 14.6B

For the reaction $H_2S(g) + I_2(s) \rightleftharpoons 2\ HI(g) + S(s)$, $K_p = 1.34 \times 10^{-5}$ at 60 °C. Initially, we bring together the following species: $H_2S(g)$ at a partial pressure of 0.010 atm, HI(g) at 0.0010 atm, $I_2(s)$, and S(s). When equilibrium is established, which gas will have a partial pressure that is greater than its initial partial pressure? Which gas will have a partial pressure that is less than its initial partial pressure? Which solid will increase in amount, and which will decrease?

14.4 Qualitative Treatment of Equilibrium: Le Châtelier's Principle

In working with the condition of equilibrium, sometimes nonnumerical answers or simple ballpark estimates serve our needs as well as precise numerical results do. A useful qualitative guide to equilibrium, called **Le Châtelier's Principle**, was framed by Henri Le Châtelier in 1888. Le Châtelier stated his principle in a rather lengthy manner, but the following rough paraphrase will serve our purposes.

When a change (that is, a change in concentration, temperature, pressure, or volume) is imposed on a system at equilibrium, the system responds by attaining a new equilibrium condition that minimizes the impact of the imposed change.

You can best acquire a sense of Le Châtelier's principle as we apply it to some specific situations.

The distinctive aroma and flavor of oranges are due to the ester, octyl acetate, $CH_3(CH_2)_6CH_2OCOCH_3$. This compound can easily be synthesized to produce artificial orange flavorings.

Changing the Amounts of Reacting Species

Let's look at the reaction for the formation of octyl acetate from 1-octanol and acetic acid.

$$\underset{\text{1-Octanol}}{CH_3(CH_2)_6CH_2OH} + \underset{\text{Acetic acid}}{CH_3COOH} \xrightleftharpoons{H^+} \underset{\text{Octyl acetate}}{CH_3(CH_2)_6CH_2OCOCH_3} + H_2O$$

An equilibrium mixture of the four components is homogeneous; it exists in a single liquid phase. Let's start with a mixture that is *at equilibrium*, at which point the reaction quotient, Q_c, is equal to the equilibrium constant, K_c.

$$Q_c = \frac{[\text{octyl acetate}][H_2O]}{[\text{1-octanol}][CH_3COOH]} = K_c$$

The Significance of Chemical Equilibrium, in the Words of Henri Le Châtelier

Henri Le Châtelier (1850–1936), a French chemist, was one of the first to appreciate the power of thermodynamics in dealing with chemical problems. For example, he fully understood the difference between a reaction that goes to completion and one that can only reach a state of equilibrium. He stated this distinction rather nicely in a journal article in 1888, from which we have taken the following quotation. In reading Le Châtelier's account, think of a *limited reaction* as a reversible reaction at equilibrium.

"It is known that in the blast furnace the reduction of iron oxide is produced by carbon monoxide, according to the reaction $Fe_2O_3(s) + 3\ CO(g) \rightleftharpoons 2\ Fe(s) + 3\ CO_2(g)$, but the gas leaving the chimney contains a considerable proportion of carbon monoxide, ... Because this incomplete reaction was thought to be due to an insufficiently prolonged contact between carbon monoxide and the iron ore, the dimensions of the furnaces have been increased. In England they have been made as high as thirty meters. But the proportion of carbon monoxide escaping has not diminished, thus demonstrating, by an experiment costing several hundred thousand francs, that the reduction of iron oxide by carbon monoxide is a limited reaction. Acquaintance with the laws of chemical equilibrium would have permitted the same conclusion to be reached more rapidly and far more economically."

Then let's disturb the equilibrium by adding more acetic acid, CH_3COOH. We indicate the resulting increase in concentration by using red type for acetic acid in the expression below.

$$Q_c = \frac{[\text{octyl acetate}][H_2O]}{[\text{1-octanol}][CH_3COOH]} < K_c$$

Because we have increased the denominator, the ratio of concentrations, Q_c, is now smaller than K_c, but this condition exists only temporarily. The concentrations must change in such a way as to make Q_c once again equal to K_c. This requires the numerator to become larger, which happens if some of the added CH_3COOH is consumed in a net forward reaction. Additional octyl acetate and water are produced. At the same time, however, 1-octanol is consumed. When equilibrium is reestablished, the concentrations of acetic acid, octyl acetate, and water will all be greater than in the original equilibrium (red); that of 1-octanol will be less (blue).

$$Q_c = \frac{[\text{octyl acetate}][H_2O]}{[\text{1-octanol}][CH_3COOH]} = K_c$$

We arrive at the same conclusion from Le Châtelier's principle, but without having to work through the reaction quotient. To counter the effect of an added reactant, the reaction that can consume some of that reactant is stimulated. In this case, acetic acid is consumed in the *forward* reaction. In the new equilibrium, the reaction has gone farther in the forward direction, or we might say that a net reaction occurs to the right. Le Châtelier's principle predicts the following results for each species.

- *Acetic acid*, CH_3COOH. *Some*, but not all, of the added acetic acid is consumed in the forward reaction. There will still be *more* acetic acid in the new equilibrium mixture than in the original equilibrium state.
- *1-Octanol*. There will be *less* 1-octanol than in the original equilibrium state. Some of the 1-octanol present in the original equilibrium mixture reacts with some of the added acetic acid.
- *Octyl acetate and water*. There will be more of each of these products in the new equilibrium state. They are formed as the forward reaction is stimulated.

Organic chemists often use an excess of acetic acid, as much as 10 mol acetic acid to 1 mol of the more expensive 1-octanol. This drives the equilibrium toward the product side, giving a good yield of octyl acetate. Another method of improving the equilibrium yield of octyl acetate is to remove water, as illustrated in Example 14.7.

EXAMPLE 14.7

Describe the effect on equilibrium in the reaction of 1-octanol and acetic acid if water is removed from the equilibrium mixture.

$$\underset{\text{1-Octanol}}{CH_3(CH_2)_6CH_2OH} + \underset{\text{Acetic acid}}{CH_3COOH} \overset{H^+}{\rightleftharpoons} \underset{\text{Octyl acetate}}{CH_3(CH_2)_6CH_2OCOCH_3} + H_2O$$

SOLUTION

As water is *removed*, the reverse reaction slows down compared to the forward reaction, which tends to form more H_2O. Not all of the water that is removed is replaced, however, so that in the new equilibrium, the amount of water is somewhat *less* than in the original equilibrium. The amount of octyl acetate in the new equilibrium is *greater* than in the original equilibrium, and the amounts of both 1-octanol and acetic acid are *less*.

EXERCISE 14.7

What should be the effect of each of the following changes on a constant-volume equilibrium mixture of N_2, H_2, and NH_3?

$$N_2(g) + 3\,H_2(g) \rightleftharpoons 2\,NH_3(g)$$

(a) Adding $H_2(g)$ **(b)** removing $N_2(g)$ **(c)** removing $NH_3(g)$

When we add or remove a reacting species from a *homogeneous* equilibrium mixture, we change its concentration. If the concentration of one reactant changes, so too must all the others to reestablish the constant value of K_c. If the component added or removed is a pure solid or liquid in a *heterogeneous* equilibrium mixture, there is *no change* in the equilibrium condition. As we have seen, liquids and solids do not appear in the equilibrium constant expression. Thus, the pressure of the $CO_2(g)$ in equilibrium with $CaO(s)$ and $CaCO_3(s)$ is unaffected by the amounts of the two solids present (Figure 14.4).

$$CaCO_3(s) \rightleftharpoons CaO(s) + CO_2(g) \qquad K_p = P_{CO_2}$$

Similarly, the vapor pressure of water is not affected by the addition or removal of liquid water in the following equilibrium.

$$H_2O(l) \rightleftharpoons H_2O(g) \qquad K_p = P_{H_2O}$$

▲ **Figure 14.4**
A heterogeneous equilibrium
At the same temperature, the pressure of $CO_2(g)$ in equilibrium with $CaO(s)$ and $CaCO_3(s)$ is unaffected by the amounts of the two solids that are present.

Changing External Pressure or Volume in Gaseous Equilibria

We can *increase* the partial pressure of a gaseous component in a constant-volume equilibrium mixture by *adding* more of it, or we can *decrease* its partial pressure by *removing* some of it. A net reaction occurs to the left or to the right in the manner that we have already described (see Exercise 14.7).

We can *increase* the partial pressures of *all* the gases in an equilibrium mixture by *increasing* the external pressure and thereby *reducing* the reaction volume. Likewise, we can *reduce* the partial pressures of all the gases by *reducing* the external pressure and thereby *increasing* the reaction volume. We can also reduce partial pressures by transferring the reaction mixture into an evacuated container of larger volume.

Let's consider the decomposition of $N_2O_4(g)$ to $NO_2(g)$ at 298 K.

$$N_2O_4(g) \rightleftharpoons 2\,NO_2(g) \qquad K_p = 0.145$$

Figure 14.5a depicts an equilibrium mixture under an external pressure of 1 atm. The molecules shown—17 in all—are in about their actual proportions: 12 N_2O_4 molecules to 5 NO_2 molecules. Now suppose we quickly increase the external pressure to 2 atm. We can predict what should happen by comparing Q_p and K_p.

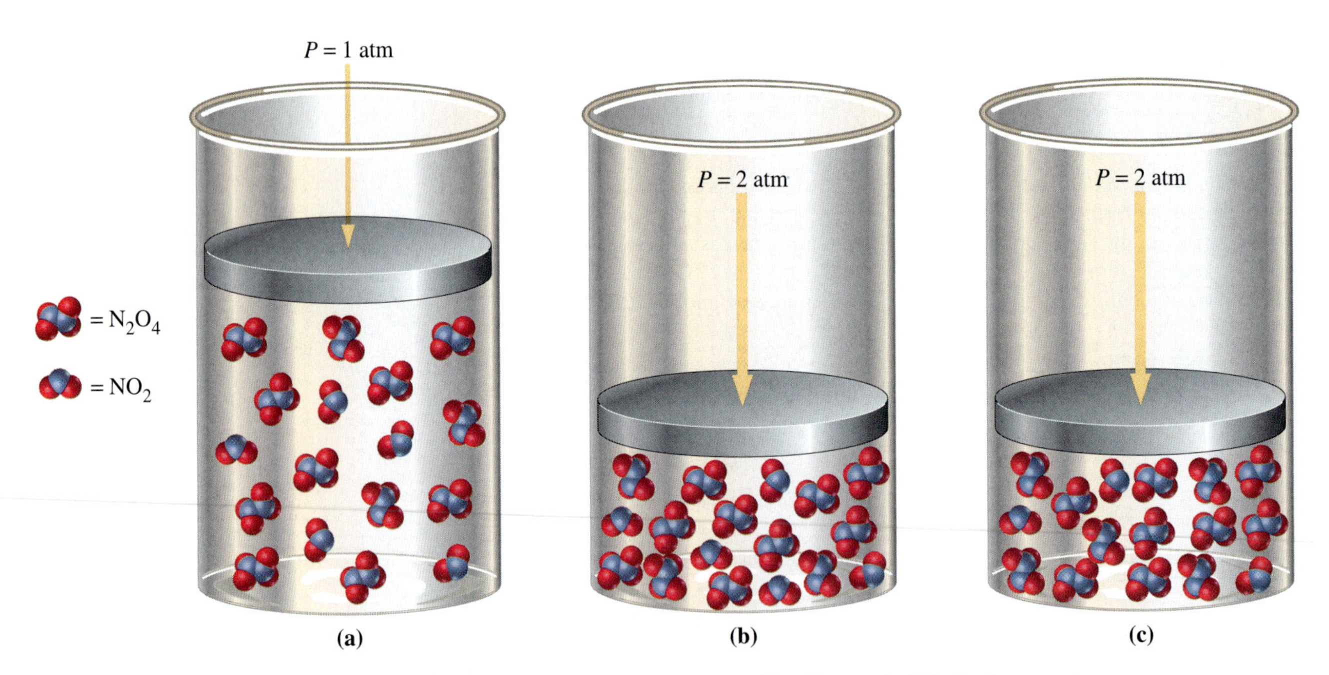

▲ **Figure 14.5 Illustrating Le Châtelier's principle in the reaction $N_2O_4(g) \rightleftharpoons 2\,NO_2(g)$ at 298 K**

(a) Equilibrium is established at a total pressure of 1 atm. For every *17* molecules, *five* are NO_2 and *twelve* are N_2O_4. (b) The total pressure is increased to 2 atm. Momentarily, the same *17* molecules are present. (c) The system accommodates to the reduced volume. Two NO_2 molecules combine to form one N_2O_4 molecule. The new equilibrium has *16* molecules in place of the original *17*. Of these, *three* are NO_2 and *thirteen* are N_2O_4. Notice that in every case the same total number of atoms is present: 29 N atoms and 58 O atoms.

Initial Equilibrium Mixture If $P_{N_2O_4}$ and P_{NO_2} are the partial pressures, as in Figure 14.5a, then the initial equilibrium is described by the following expression.

$$Q_p = K_p = \frac{(P_{NO_2})^2}{P_{N_2O_4}}$$

Disturbed Equilibrium Mixture If the amounts of N_2O_4 and NO_2 were to remain unchanged, as in Figure 14.5b, each partial pressure would double because the volume has been reduced to one-half its initial value.

$$Q_p = \frac{2P_{NO_2} \times \cancel{2}P_{NO_2}}{\cancel{2}P_{N_2O_4}} = 2 \times \frac{(P_{NO_2})^2}{P_{N_2O_4}} = 2 \times K_p > K_p$$

New Equilibrium Mixture Because Q_p is greater than K_p in Figure 14.5b, a net reaction should occur in the *reverse* direction. This will reduce the numerator, increase the denominator, and reduce Q_p so that it once again is equal to K_p. The new equilibrium mixture will have more N_2O_4 and less NO_2 than initially. Figure 14.5c represents the new equilibrium, now consisting of 13 N_2O_4 molecules for every 3 NO_2 molecules.

Now let's apply Le Châtelier's principle. When we decrease the volume of an equilibrium mixture by increasing the external pressure, we "crowd" the molecules more closely together. The reverse reaction is stimulated because *one* mole of gaseous reactant (N_2O_4) replaces *two* moles of gaseous product (NO_2). In Figure 14.5c, we see that *two* of the NO_2 molecules from Figures 14.5a and 14.5b have been replaced by *one* N_2O_4 molecule. As a result, the new equilibrium has 16 molecules for every 17 that were present in the initial equilibrium. The smaller number of molecules accommodate better to the more crowded conditions in the reduced volume.

The following statements summarize the effect of changes in external pressure (or system volume) on an equilibrium involving gases.

- When the external *pressure is increased* (or system *volume is reduced*), an equilibrium shifts in the direction producing the *smaller number of moles of gas*.
- When the external *pressure is decreased* (or system *volume is increased*), an equilibrium shifts in the direction producing the *larger number of moles of gas*.
- If there is *no change* in the number of moles of gas in a reaction, changes in external pressure (or system volume) have *no effect* on an equilibrium.

If we produce changes in gas pressures or volumes by adding an inert gas to an equilibrium mixture, the effects are somewhat different. If the inert gas is added at a *constant external pressure*, the volume expands to accommodate the added gas. This has the same effect as transferring the mixture to a container of larger volume. If the inert gas is added to a *constant volume* mixture, the concentrations and partial pressures of reactants and products do not change, and the inert gas does not affect the equilibrium.

EXAMPLE 14.8

An equilibrium mixture of $SO_2(g)$, $O_2(g)$, and $SO_3(g)$ is transferred from a 1.00-L flask to a 2.00-L flask. In which direction does a net reaction proceed to restore equilibrium?

$$2\ SO_3(g) \rightleftharpoons 2\ SO_2(g) + O_2(g)$$

SOLUTION

Transferring the mixture to a larger flask increases the volume to be filled with molecules. We expect the equilibrium to shift in the direction that produces the larger number of moles of gas—in this case, the forward reaction. Some of the $SO_3(g)$ is converted to $SO_2(g)$ and $O_2(g)$, and equilibrium shifts to the right.

EXERCISE 14.8A

Consider the reaction $H_2(g) + I_2(g) \rightleftharpoons 2\ HI(g)$. How is the equilibrium amount of HI(g) changed by compressing an equilibrium mixture into a smaller volume? Explain.

EXERCISE 14.8B

In which direction will a net reaction occur if additional $NO_2(g)$ is added to an equilibrium mixture at the same time that the mixture is transferred from a 1.00-L to a 1.50-L flask? Explain.

$$2\,NO(g) + O_2(g) \rightleftharpoons 2\,NO_2(g)$$

Changing the Equilibrium Temperature

The changes we have described thus far do not change the value of the equilibrium constant, but *changing the temperature* of an equilibrium mixture *does* change the value of K_p or K_c. If the equilibrium constant becomes larger, the forward reaction is favored, and equilibrium shifts to the right. If it becomes smaller, the reverse reaction is favored, and equilibrium shifts to the left. We can use Le Châtelier's principle to assess, qualitatively, the effect of temperature on equilibrium.

To change the temperature of a reaction mixture, we must add heat to raise the temperature or remove heat to lower the temperature. Adding heat to an equilibrium mixture will stimulate the reaction that can absorb some of the heat—the endothermic reaction. The removal of heat stimulates the exothermic reaction. These effects are summarized as follows.

> Raising *the temperature of an equilibrium mixture shifts equilibrium in the direction of the* endothermic reaction; *lowering the temperature shifts equilibrium in the direction of the* exothermic reaction.

In Example 14.9, we will show that we can also consider heat as if it were a "reactant" or a "product" of a reaction. Then we can reason as we would for changing the amount of a reacting species.

EXAMPLE 14.9

Is the amount of NO(g) formed from given amounts of $N_2(g)$ and $O_2(g)$ greater at high or low temperatures?

$$N_2(g) + O_2(g) \rightleftharpoons 2\,NO(g) \qquad \Delta H^\circ = +180.5\text{ kJ}$$

SOLUTION

As written, the ΔH° value given is for the forward reaction. Because ΔH° is positive, the forward reaction is endothermic. Thus, the equilibrium state shifts to the right as the temperature is raised. The conversion of $N_2(g)$ and $O_2(g)$ to NO(g) is favored at high temperatures.

Alternatively, we can rewrite the equation in this way.

$$N_2(g) + O_2(g) + \text{heat} \rightleftharpoons 2\,NO(g)$$

We can then reason as follows: Raising the temperature, that is, adding heat (a "reactant") stimulates the forward reaction. Thus, equilibrium is shifted to the right.

APPLICATION NOTE

Because conversion of $N_2(g)$ and $O_2(g)$ to NO(g) is favored at high temperatures, NO(g) is found in the exhaust of high-compression automobile engines—they operate at high temperatures.

EXERCISE 14.9

Is the conversion of $SO_2(g)$ to $SO_3(g)$ more nearly complete at high or low temperatures?

$$2\,SO_2(g) + O_2(g) \rightleftharpoons 2\,SO_3(g) \qquad \Delta H^\circ = -198\text{ kJ}$$

Recall from Section 12.4 that the majority of solid solutes (95% or more) have aqueous solubilities that *increase* with temperature. In these cases, solution formation is an *endothermic* process, and an endothermic process is favored at *higher* temperatures. This means that at higher temperatures, more of the solute dissolves before the solution becomes saturated, that is, before equilibrium is reached.

Adding a Catalyst

The reaction of $SO_2(g)$ with $O_2(g)$ to produce $SO_3(g)$ is greatly accelerated by a catalyst (such as platinum metal). However, the reverse reaction—the decomposition of $SO_3(g)$ to $SO_2(g)$—is also greatly speeded by the catalyst.

$$2\,SO_2(g) + O_2(g) \xrightleftharpoons{Pt} 2\,SO_3(g) \quad K_c = 2.8 \times 10^2 \text{ at } 1000\text{ K}$$

Because the rates of the forward and reverse reactions are increased to the same extent, the proportion of $SO_3(g)$ in the equilibrium mixture is the same as if no catalyst were present at all. That is, a catalyst does not shift an equilibrium to the right or left, nor does it affect the value of the equilibrium constant.

The role of a catalyst is to change the mechanism of a reaction to one of lower activation energy. Because a catalyst does not affect an equilibrium state, we can conclude that this state is a function only of the states of the reactants and products and not the reaction path. We will explore this matter further in Chapter 17.

EXAMPLE 14.10—A Conceptual Example

Flask A, pictured below, contains an equilibrium mixture in the reaction represented by the following equation.

$$CO(g) + H_2O(g) \rightleftharpoons CO_2(g) + H_2(g) \qquad \Delta H = -41\text{ kJ};\ K_c = 9.03 \text{ at } 698\text{ K}$$

Flask A is connected to flask B, and a new equilibrium is established when the valve between the two is opened. Describe, qualitatively, how the amounts of CO, H_2O, CO_2, and H_2 in the new equilibrium compare to their amounts in the initial equilibrium when the contents of flask B are those listed below. If you are uncertain, explain why.

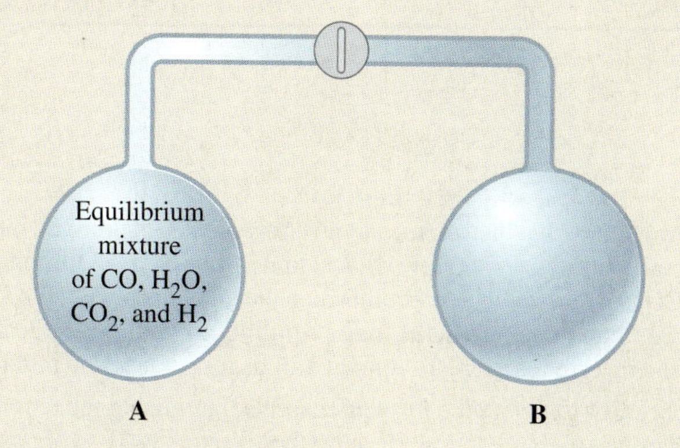

a. Flask *B* contains Ar(g) at 1 atm pressure.

b. Flask *B* contains 1.0 mol CO_2.

c. Flask *B* contains 1.0 mol CO, and the temperature of the *A-B* assembly is raised by 100 °C.

SOLUTION

a. Ar(g) is an inert gas and has no effect on the reaction. Because the reaction involves the same number of reactant and product molecules, the equilibrium is unaffected by the change in volume. The amounts of CO, H_2O, CO_2, and H_2 are all unchanged.

Chemical Equilibrium and the Synthesis of Ammonia

Ammonia is an important industrial chemical, ranking about sixth in quantity of the chemicals produced in the United States. It is used directly as a fertilizer, to make other nitrogen-containing fertilizers, and in the production of explosives and plastics. Following are typical conditions used in its manufacture.

$$N_2(g) + 3\,H_2(g) \rightleftharpoons 2\,NH_3(g)$$

$$\Delta H° = -92.22\text{ kJ}$$

Reactants: 3:1 mol ratio of H_2 to N_2
Temperature: 400–600 °C
Pressure: 140–340 atm
Catalyst: Fe_3O_4 with small amounts of Al_2O_3, MgO, CaO, and K_2O. The Fe_3O_4 is reduced to metallic iron before use.

Based on the chemical equation and Le Châtelier's principle, we would conclude that high yields of $NH_3(g)$ are favored by the following conditions.

1. *Low temperatures* because the forward reaction is exothermic.
2. *High pressures* because the forward reaction is accompanied by a decrease in number of moles of gas.
3. *Continuous removal of NH_3* because the removal of a product stimulates the forward reaction to form additional product (see Figure 14.6).

The synthesis reaction is indeed carried out at high pressures. The temperatures used, however, are moderately high, not low. The theoretical percent conversion of a 3 mol H_2:1 mol N_2 mixture to NH_3 is over 90% at high pressures and room temperature, but with these conditions, it would take far too long to reach equilibrium. The ammonia synthesis reaction is kinetically controlled (page 600). Even though only 20% conversion of the reactants to NH_3 is achieved in an equilibrium mixture at 500 °C and 200 atm, with these conditions and a catalyst, equilibrium is reached in less than one minute.

The equilibrium mixture of gases is cooled to the point at which the ammonia liquefies. The $NH_3(l)$ is removed, and the unreacted $H_2(g)$ and $N_2(g)$, still in a 3:1 mole ratio, are recycled through the process.

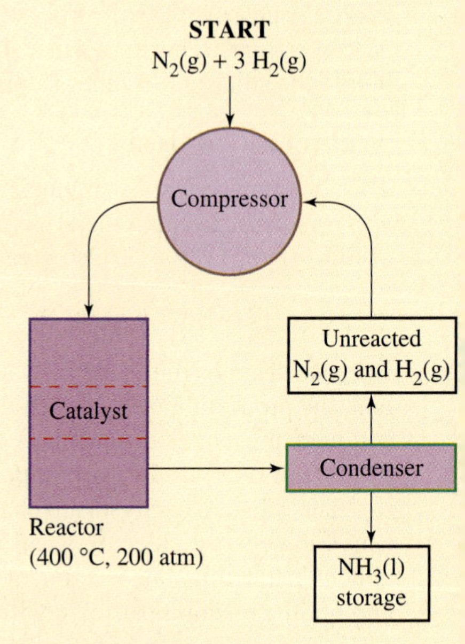

▲ **Figure 14.6**
The synthesis of ammonia: The Haber process
The N_2/H_2 mixture is brought into a reactor at a high temperature and pressure. The equilibrium mixture is removed from the reactor and cooled in a condenser. Liquid NH_3 is removed, and the N_2/H_2 mixture is returned to the reactor and mixed with additional reactant gases.

b. As in part (a), increasing the volume has no effect on the equilibrium, but having more $CO_2(g)$ present stimulates the *reverse* reaction. Some of the additional CO_2 is converted to CO and H_2. In the new equilibrium, the amounts of CO, H_2O, and CO_2 will all be *greater* than they were in the initial equilibrium in flask A. Because H_2 is consumed in the reverse reaction, its amount will be *less* than in the initial equilibrium.

c. Adding more CO(g) favors the *forward* reaction, but raising the temperature favors the endothermic *reverse* reaction. Because these factors work in opposition, we cannot make a qualitative prediction.

EXERCISE 14.10

Respond as directed in Example 14.10 to these additional conditions.

a. Flask *B* contains 1.0 mol $H_2(g)$ at 1 atm pressure.

b. Flask *B* contains 1.0 mol $H_2(g)$ and 1.0 mol $H_2O(g)$.

c. Flask *B* contains 1.0 mol H_2O, and the temperature of the *A*-*B* assembly is lowered by 100 °C.

14.5 Some Illustrative Equilibrium Calculations

We will conclude this chapter by learning how to use equilibrium constants to solve problems. We will find the knowledge gained here quite useful in solving many more such problems in the next two chapters. For convenience, we will divide the problems into two basic types: those in which we use experimental data to determine equilibrium constants, and those in which equilibrium constants are used to calculate equilibrium concentrations or partial pressures.

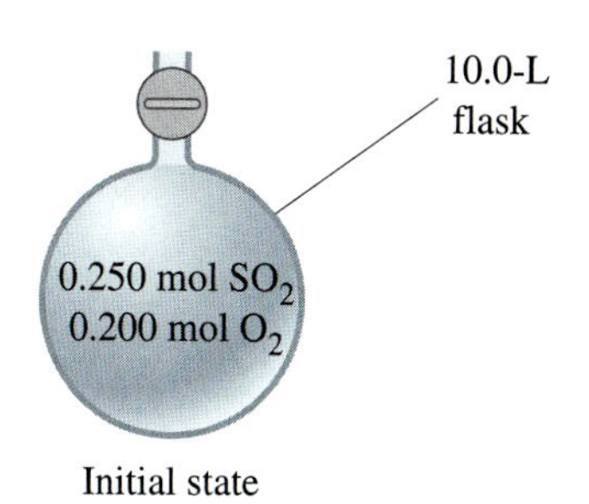

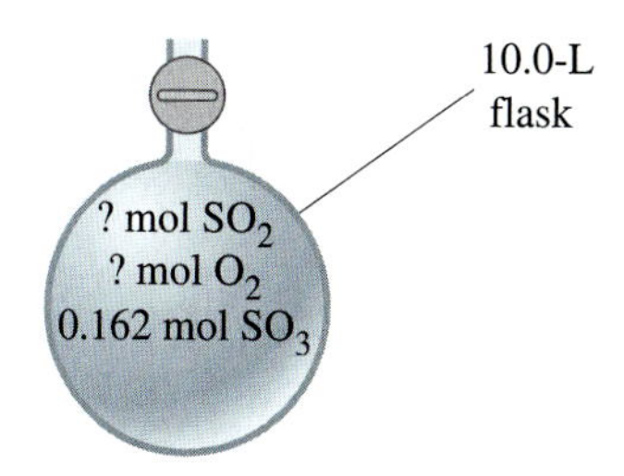

▲ **Figure 14.7**
Example 14.11 illustrated
The key in Example 14.11 is to determine the amounts of SO_2 and O_2 *consumed* to reach equilibrium.

Determining Values of Equilibrium Constants from Experimental Data

In Example 14.11, we seek a K_c value. As suggested by Figure 14.7, we are given the *initial* amounts of the reactants and the *equilibrium* amount of the product. From these data, we can establish the amounts of *all* the species in the equilibrium state and then the equilibrium concentrations. Finally, we can calculate K_c from those concentrations.

A useful general approach is to tabulate under the chemical equation: (a) the concentrations of substances present initially, (b) the changes in these concentrations that occur in reaching equilibrium, and (c) the equilibrium concentrations. Often the key step is (b), in which we identify the changes that occur and determine their relationship to one another.

EXAMPLE 14.11

In a 10.0-L vessel at 1000 K, 0.250 mol SO_2 and 0.200 mol O_2 react to form 0.162 mol SO_3 at equilibrium. What is K_c, at 1000 K, for the reaction?

$$2\,SO_2(g) + O_2(g) \rightleftharpoons 2\,SO_3(g)$$

SOLUTION

Let's begin with the *initial* concentrations of the three gases.

$$[SO_2] = \frac{0.250\text{ mol}}{10.0\text{ L}} = 0.0250\text{ M} \quad [O_2] = \frac{0.200\text{ mol}}{10.0\text{ L}} = 0.0200\text{ M} \quad [SO_3] = 0$$

From the given information, we can also calculate the *equilibrium* concentration of SO_3.

$$[SO_3] = \frac{0.162\text{ mol}}{10.0\text{ L}} = 0.0162\text{ M}$$

Next let's arrange these quantities in the following format.

The reaction:	$2\,SO_2(g)$ +	$O_2(g)$	$\rightleftharpoons$ $2\,SO_3(g)$
Initial concentrations, M:	0.0250	0.0200	0
Changes, M:	?	?	?
Equilibrium concentrations, M	?	?	0.0162

Now let's fill in the blanks. Because we started with no SO_3 and produced an equilibrium concentration of 0.0162 M, the *change* in $[SO_3]$ must be +0.0162 M. The *positive* sign signifies that something is formed. From the chemical equation, we see that the same number of moles per liter of SO_2 must be consumed as moles per liter of SO_3 produced. The *change* in $[SO_2]$ equals −0.0162 M; the *negative* sign signifies that something is consumed. Because only *one* mole per liter of O_2 is required for every *two* moles per liter of SO_3 produced, the *change* in $[O_2]$ is $-\frac{1}{2} \times 0.0162$ M, which equals -0.0081 M. Now we can complete the table by *adding* these changes to the initial concentrations to get equilibrium concentrations.

The reaction:	$2\ SO_2(g)$ +	$O_2(g)$ $\rightleftharpoons$	$2\ SO_3(g)$
Initial concentrations, M:	0.0250	0.0200	0
Changes, M:	−0.0162	−0.0081	+0.0162
Equilibrium concentrations, M:	0.0088	0.0119	0.0162

Finally, we substitute the equilibrium concentrations into the equilibrium constant expression.

$$K_c = \frac{[SO_3]^2}{[SO_2]^2[O_2]} = \frac{(0.0162)^2}{(0.0088)^2(0.0119)} = 2.8 \times 10^2$$

EXERCISE 14.11A

Initially we start with 1.00 mol each of PCl_3 and Cl_2 in a 1.00-L flask. When equilibrium is established at 250 °C in the reaction, $PCl_3(g) + Cl_2(g) \rightleftharpoons PCl_5(g)$, the amount of PCl_5 present is 0.82 mol. What is K_c for this reaction?

PROBLEM-SOLVING NOTE

We can use partial pressures in a tabular format under a chemical equation in the same way as we used concentrations in Example 14.11. Also, we can calculate K_c and use the equation $K_p = K_c(RT)^{\Delta n_{gas}}$ to convert K_c to K_p.

EXERCISE 14.11B

A 1.00-kg sample of $Sb_2S_3(s)$ and a 10.0-g sample of $H_2(g)$ are allowed to react in a 25.0-L container at 713 K. At equilibrium, 72.6 g $H_2S(g)$ is present. What is the value of K_p at 713 K for the following reaction?

$$Sb_2S_3(s) + 3\ H_2(g) \rightleftharpoons 2\ Sb(s) + 3\ H_2S(g)$$

Calculating Equilibrium Quantities from K_c and K_p Values

One of the most common types of equilibrium calculations is illustrated in Example 14.12. We start with initial reactants and no products and with the known value of the equilibrium constant. Then we use those data to calculate the amounts of substances present at equilibrium. Typically we use a symbol such as x to identify one of the changes in concentration that occurs in establishing equilibrium. Then, we relate all the other concentration changes to x, substitute appropriate terms into the equilibrium constant expression, and solve it for x.

EXAMPLE 14.12

Consider the following reaction.

$$H_2(g) + I_2(g) \rightleftharpoons 2\ HI(g) \qquad K_c = 54.3 \text{ at } 698\text{ K}$$

If we start with 0.500 mol H_2 and 0.500 mol $I_2(g)$ in a 5.25-L vessel at 698 K, how many moles of each gas will be present at equilibrium?

SOLUTION

First let's calculate the initial concentrations in the 5.25-L flask.

$$[H_2] = [I_2] = \frac{0.500\text{ mol}}{5.25\text{ L}} = 0.0952\text{ M} \quad [HI] = 0$$

If we let $-x$ represent the changes in concentration of H_2 and I_2, the change in [HI] is $+2x$ because two moles of HI are formed for every mole of H_2 and I_2 that react. We enter these changes, together with the initial and equilibrium concentrations, into the following tabular format.

The reaction:	$H_2(g)$	+	$I_2(g)$	$\rightleftharpoons$	$2\ HI(g)$
Initial concentrations, M:	0.0952		0.0952		0
Changes, M:	$-x$		$-x$		$+2x$
Equilibrium concentrations, M:	$(0.0952 - x)$		$(0.0952 - x)$		$2x$

Next we can enter the *equilibrium* concentrations into the K_c expression.

$$K_c = \frac{[HI]^2}{[H_2][I_2]} = \frac{(2x)^2}{(0.0952 - x)(0.0952 - x)} = \frac{(2x)^2}{(0.0952 - x)^2} = 54.3$$

From this point, the most direct approach is to take the square root of each side of the equation and solve for x.

$$\left[\frac{(2x)^2}{(0.0952 - x)^2}\right]^{1/2} = (54.3)^{1/2}$$

$$\frac{2x}{(0.0952 - x)} = (54.3)^{1/2}$$

$$2x = (54.3)^{1/2} \times (0.0952 - x)$$

$$2x = 7.37 \times (0.0952 - x)$$

$$2x = 0.702 - 7.37x$$

$$9.37x = 0.702$$

$$x = 0.0749$$

We can now calculate the equilibrium concentrations.

$$[H_2] = [I_2] = 0.0952 - x = 0.0952 - 0.0749 = 0.0203\ M$$

$$[HI] = 2x = 2 \times 0.0749 = 0.150\ M$$

To determine the equilibrium *amounts*, we multiply the equilibrium concentrations by the volume.

$$\text{mol } H_2 = \text{mol } I_2 = 5.25\ \cancel{L} \times 0.0203\ \text{mol}/\cancel{L} = 0.107\ \text{mol}$$

$$\text{mol HI} = 5.25\ \cancel{L} \times 0.150\ \text{mol}/\cancel{L} = 0.788\ \text{mol}$$

PROBLEM-SOLVING NOTE

A useful check on this result is to substitute the calculated equilibrium concentrations into the K_c expression to see what value of K_c they yield:

$$K_c = \frac{[HI]^2}{[H_2][I_2]} = \frac{(0.150)^2}{(0.0203)(0.0203)} = 54.6$$

This is close enough to the given value of K_c (54.3) to confirm that our calculation is correct.

EXERCISE 14.12A

Starting with 0.100 mol each of CO and H_2O in a 5.00-L flask, equilibrium is established in the following reaction at 600 K.

$$CO(g) + H_2O(g) \rightleftharpoons CO_2(g) + H_2(g) \qquad K_c = 23.2 \text{ at } 600\ K$$

What will be **(a)** the number of moles and **(b)** the partial pressure of $H_2(g)$ when equilibrium is established?

EXERCISE 14.12B

Show that for the reaction in Example 14.12, $H_2(g) + I_2(g) \rightleftharpoons 2\ HI(g)$, the equilibrium amounts of reactants and products are *independent* of the volume of the reaction flask. Would you expect this to be the case for all reversible reactions at equilibrium? Explain.

In Example 14.13, the initial amounts of the reactants are *unequal*. The setup of the problem is similar to that in Example 14.12, but the algebraic solution uses the quadratic formula, which is commonly needed in equilibrium calculations.

EXAMPLE 14.13

Suppose that in the reaction of Example 14.12, the initial amounts are 0.800 mol H_2 and 0.500 mol I_2. What will be the amounts of reactants and products when equilibrium is attained?

SOLUTION

As in Example 14.12, let's first determine the *initial* concentrations.

$$[H_2] = \frac{0.800 \text{ mol}}{5.25 \text{ L}} = 0.152 \text{ M} \quad [I_2] = \frac{0.500 \text{ mol}}{5.25 \text{ L}} = 0.0952 \text{ M} \quad [HI] = 0$$

Then we can set up a tabular format as usual.

The reaction:	$H_2(g)$	+	$I_2(g)$	$\rightleftharpoons$	2 HI(g)
Initial concentrations, M:	0.152		0.0952		0
Changes, M:	$-x$		$-x$		$+2x$
Equilibrium concentrations, M:	$(0.152 - x)$		$(0.0952 - x)$		$2x$

Next we can enter the equilibrium concentrations into the K_c expression.

$$K_c = \frac{[HI]^2}{[H_2][I_2]} = 54.3$$

$$\frac{(2x)^2}{(0.152 - x)(0.0952 - x)} = 54.3$$

We cannot extract the square root of each side of this equation because the two terms in the denominator on the left are not identical. Instead, we have to multiply both sides of the equation by the factor $(0.152 - x)(0.0952 - x)$.

$$4x^2 = 54.3 \times (0.152 - x)(0.0952 - x)$$

$$4x^2 = 54.3 \times (0.0145 - 0.247x + x^2)$$

$$4x^2 = 54.3x^2 - 13.4x + 0.787$$

Now we can gather the terms into the form $ax^2 + bx + c = 0$. We note that $a = 50.3$, $b = -13.4$, and $c = 0.787$.

$$50.3x^2 - 13.4x + 0.787 = 0$$

The solutions of this equation are given by the quadratic formula.

See also the discussion on quadratic equations in Appendix A.

$$x = \frac{-b \pm \sqrt{b^2 - 4ac}}{2a}$$

We then substitute for a, b, and c.

$$x = \frac{-(-13.4) \pm \sqrt{(-13.4)^2 - (4 \times 50.3 \times 0.787)}}{2 \times 50.3}$$

$$= \frac{13.4 \pm \sqrt{21.22}}{100.6} = \frac{13.4 \pm 4.61}{100.6}$$

Now we are faced with a choice; the solution gives two values for x.

$$x = \frac{13.4 + 4.61}{100.6} = 0.179$$

$$x = \frac{13.4 - 4.61}{100.6} = 0.0874$$

Can you see that the correct answer must be $x = 0.0874$ and not 0.179? The decrease in $[I_2]$ must be *less than* 0.0952 M, the initial concentration. We can now calculate the equilibrium concentrations.

$$[H_2] = 0.152 - x = 0.152 - 0.0874 = 0.065 \text{ M}$$

$$[I_2] = 0.0952 - x = 0.0952 - 0.0874 = 0.0078 \text{ M}$$

$$[HI] = 2x = 2 \times 0.0874 = 0.175 \text{ M}$$

Then we determine the equilibrium *amounts* by multiplying the equilibrium concentrations by the volume.

$$? \text{ mol } H_2 = 5.25 \cancel{\text{L}} \times 0.065 \text{ mol } H_2/\cancel{\text{L}} = 0.34 \text{ mol } H_2$$

$$? \text{ mol } I_2 = 5.25 \cancel{\text{L}} \times 0.0078 \text{ mol } I_2/\cancel{\text{L}} = 0.041 \text{ mol } I_2$$

$$? \text{ mol HI} = 5.25 \cancel{\text{L}} \times 0.175 \text{ mol HI}/\cancel{\text{L}} = 0.919 \text{ mol HI}$$

EXERCISE 14.13A

Starting with 0.100 mol CO and 0.200 mol Cl_2 in a 25.0-L flask, how many moles of $COCl_2$ will be present *at equilibrium* in the following reaction?

$$CO(g) + Cl_2(g) \rightleftharpoons COCl_2(g) \qquad K_c = 1.2 \times 10^3 \text{ at } 668 \text{ K}$$

EXERCISE 14.13B

Starting with 0.78 mol N_2 and 0.21 mol O_2 (their proportions in one mole of air), what will be the *mole fraction* of NO(g) *at equilibrium* in this reaction?

$$N_2(g) + O_2(g) \rightleftharpoons 2\,NO(g) \qquad K_c = 2.1 \times 10^{-3} \text{ at } 2500 \text{ K}$$

(*Hint:* Do you need to know the volume of the reaction mixture? Recall Exercise 14.12B.)

If we start with a mixture in which a reactant or product is absent, we know that a net reaction must occur in the direction in which some of that species is produced. This was the case in Examples 14.12 and 14.13. If the initial mixture contains *all* of the reactants and products, we don't immediately know in which direction a net reaction will occur, but we can find out easily enough. We can evaluate the *reaction quotient* for the initial conditions and compare its value to that of the equilibrium constant (page 608).

EXAMPLE 14.14

Carbon monoxide and chlorine react to form phosgene, $COCl_2$, which is used in the manufacture of pesticides, herbicides, and plastics.

$$CO(g) + Cl_2(g) \rightleftharpoons COCl_2(g) \qquad K_c = 1.2 \times 10^3 \text{ at } 668 \text{ K}$$

What will be the amount of each substance when equilibrium is established in a reaction mixture that *initially* has 0.0100 mol CO, 0.0100 mol Cl_2, and 0.100 mol $COCl_2$ in a 10.0-L flask?

SOLUTION

Let's first determine the initial concentrations, and evaluate Q_c.

$$[CO]_{initial} = [Cl_2]_{initial} = \frac{0.0100 \text{ mol}}{10.0 \text{ L}} = 0.00100 \text{ M}$$

$$[COCl_2]_{initial} = \frac{0.100 \text{ mol}}{10.0 \text{ L}} = 0.0100 \text{ M}$$

$$Q_c = \frac{[COCl_2]_{initial}}{[CO]_{initial}[Cl_2]_{initial}} = \frac{0.0100}{(0.00100)(0.00100)} = 1.0 \times 10^4$$

Because $Q_c > K_c$ (that is, $1.0 \times 10^4 > 1.2 \times 10^3$), a net reaction must occur in the *reverse* direction. At equilibrium, the concentrations of CO and Cl_2 will be greater than initially, and the concentration of $COCl_2$ will be less. Now we are ready to calculate the equilibrium concentrations.

Direction of net change:	←		
The reaction:	CO(g) +	Cl_2(g) ⇌	$COCl_2$(g)
Initial concentrations, M:	0.00100	0.00100	0.0100
Changes, M:	$+x$	$+x$	$-x$
Equilibrium concentrations, M:	$(0.00100 + x)$	$(0.00100 + x)$	$(0.0100 - x)$

$$K_c = \frac{[COCl_2]}{[CO] \times [Cl_2]} = \frac{(0.0100 - x)}{(0.00100 + x)(0.00100 + x)} = 1.2 \times 10^3$$

$$0.0100 - x = 1.2 \times 10^3 \times (0.00100 + x)(0.00100 + x)$$

$$0.0100 - x = 1.2 \times 10^3 \times (1.00 \times 10^{-6} + 0.00200x + x^2)$$

$$0.0100 - x = 1.2 \times 10^{-3} + 2.4x + (1.2 \times 10^3)x^2$$

Now let's put the equation in the form $ax^2 + bx + c = 0$.

$$(1.2 \times 10^3)x^2 + 3.4x - 0.0088 = 0$$

Only one of the two possible solutions to this quadratic equation is acceptable.

$$x = \frac{-3.4 \pm \sqrt{(3.4)^2 - [4 \times 1.2 \times 10^3 \times (-0.0088)]}}{2(1.2 \times 10^3)}$$

$$x = \frac{-3.4 \pm \sqrt{53.8}}{2.4 \times 10^3} = \frac{-3.4 + 7.33}{2.4 \times 10^3} = 0.0016$$

We can now calculate the equilibrium concentrations.

$$[CO] = [Cl_2] = 0.00100 + x = 0.00100 + 0.0016 = 0.0026 \text{ M}$$

$$[COCl_2] = 0.0100 - x = 0.0100 - 0.0016 = 0.0084 \text{ M}$$

To determine the equilibrium *amounts*, we multiply the equilibrium concentration of each species by the volume of the equilibrium mixture.

PROBLEM-SOLVING NOTE

In evaluating $\pm\sqrt{53.8}$, we use only the positive root. Using the negative root leads to a negative value of x and *negative* concentrations of CO and Cl_2, an obviously impossible solution.

PROBLEM-SOLVING NOTE

In checking for the correctness of our answer, as we did in Example 14.12, we find

$$K_c = [COCl_2]/[CO][Cl_2] = 0.0084/(0.0026)^2 = 1.2 \times 10^3$$

$$\text{mol CO} = \text{mol Cl}_2 = 10.0\ \cancel{\text{L}} \times 0.0026\ \text{mol}/\cancel{\text{L}} = 0.026\ \text{mol}$$

$$\text{mol COCl}_2 = 10.0\ \cancel{\text{L}} \times 0.0084\ \text{mol}/\cancel{\text{L}} = 0.084\ \text{mol}$$

EXERCISE 14.14A

How many moles of each reactant and product will be present when equilibrium is established in a gaseous mixture that initially contains 0.0100 mol H_2 and 0.100 mol HI in a 5.25-L volume at 698 K?

$$H_2(g) + I_2(g) \rightleftharpoons 2\ HI(g) \qquad K_c = 54.3 \text{ at } 698\ K$$

EXERCISE 14.14B

How many moles of each reactant and product will be present when equilibrium is established in the reaction of Exercise 14.14A if the mixture initially contains 0.0100 mol H_2, 0.0100 mol I_2, and 0.100 mol HI?

As shown in Figure 14.8 and Example 14.15, in a calculation based on K_p, we emphasize the partial pressures of the various gases and the total gas pressure. The pressure of a gas in a mixture at constant temperature and constant volume is proportional to the concentration of gas. Thus, we can use changes in partial pressures (P in Example 14.15) for the changes in concentrations, x, that we've used in previous examples.

$P_{tot} = P_{CO} + P_{Cl_2} + P_{COCl_2}$
$P_{CO} = P$
$P_{Cl_2} = P$
P_{COCl_2} =
0.351 − P
Equilibrium state

▲ **Figure 14.8**
Example 14.15 illustrated
Because two moles of gaseous products are produced for every mole of reactant in the reaction $COCl_2(g) \rightleftharpoons CO(g) + Cl_2(g)$, we expect a higher total pressure at equilibrium than initially. Also, because $COCl_2$ must be consumed to produce CO and Cl_2, we expect P_{COCl_2} at equilibrium to be less than the initial P_{COCl_2}.

EXAMPLE 14.15

A sample of phosgene, $COCl_2(g)$, is introduced into a constant-volume vessel at 395 °C and observed to exert an initial pressure of 0.351 atm. When equilibrium is established, what will be the partial pressure of each gas and the total gas pressure?

$$CO(g) + Cl_2(g) \rightleftharpoons COCl_2(g) \qquad K_p = 22.5$$

SOLUTION

Although we can work with the equation as given and describe a net reaction *to the left*, let's think of the partial dissociation of $COCl_2$ in terms of the *reverse* equation. This way a net reaction proceeds *to the right*. Of course, when we reverse the equation, the value of K_p is the reciprocal of the given value, that is, 1/22.5 or 0.0444.

The reaction:	$COCl_2(g)$	$\rightleftharpoons CO(g)$	$+\ Cl_2(g)$	$K_p = 0.0444$
Initial pressures, atm:	0.351	0	0	
Changes, atm:	$-P$	$+P$	$+P$	
Equilibrium pressures, atm:	$(0.351 - P)$	P	P	

Now substitute into the K_p expression and solve for P.

$$K_p = \frac{(P_{CO})(P_{Cl_2})}{P_{COCl_2}} = 0.0444$$

$$= \frac{P \times P}{0.351 - P} = 0.0444$$

$$P^2 = 0.0156 - 0.0444P$$

$$P^2 + 0.0444P - 0.0156 = 0$$

$$P = \frac{-0.0444 \pm \sqrt{(0.0444)^2 - [4 \times 1 \times (-0.0156)]}}{2}$$

$$= \frac{-0.0444 \pm 0.254}{2} = 0.105 \text{ or } -0.149$$

PROBLEM-SOLVING NOTE
We reject the negative root of the quadratic equation because *negative* partial pressures have no physical significance.

We can calculate the equilibrium partial pressures.

$$P_{CO} = P_{Cl_2} = 0.105 \text{ atm} \quad \text{and} \quad P_{COCl_2} = 0.351 - 0.105 = 0.246 \text{ atm}$$

We then find the total pressure.

$$P_{total} = P_{CO} + P_{Cl_2} + P_{COCl_2} = 0.105 + 0.105 + 0.246 = 0.456 \text{ atm}$$

PROBLEM-SOLVING NOTE
We can check the correctness of these partial pressures in the usual way.

$$K_p = (P_{CO})(P_{Cl_2})/P_{COCl_2}$$
$$= (0.105)^2/0.246 = 0.0448$$

EXERCISE 14.15A

What are the equilibrium partial pressures for each species and the total pressure for the reaction in Example 14.15 if initially $P_{CO} = 1.00$ atm and $P_{Cl_2} = 1.00$ atm?

EXERCISE 14.15B

Ammonium hydrogen sulfide dissociates readily, even at room temperature. What is the total pressure of the gases in equilibrium with $NH_4HS(s)$ at 25 °C?

$$NH_4HS(s) \rightleftharpoons NH_3(g) + H_2S(g) \qquad K_p = 0.108 \text{ at } 25\ °C$$

We have carefully chosen our Examples and Exercises so that we could solve them with the quadratic formula, and we will continue to do so in the next two chapters. However, you may occasionally encounter situations that require solving more complex equations. Scientists and engineers routinely use graphing calculators or computers to solve such equations, but the equations can also be solved by an approximation method that is illustrated in Appendix A. Even with the availability of computers, scientist still look for opportunities to make valid assumptions that will simplify problems.

Summary

In a reversible reaction at equilibrium, the concentrations of all reactants and products remain constant with time. For the general reaction represented by the equation

$$a\,A + b\,B + \cdots \rightleftharpoons g\,G + h\,H + \cdots$$

the concentrations of reactants and products must conform to the expression

$$K_c = \frac{[G]^g[H]^h \cdots}{[A]^a[B]^b \cdots}$$

In this expression, terms in the numerator are for the products and those in the denominator are for the reactants. The exponents correspond to the coefficients in the chemical equation. A K_p expression can be written for equilibria involving gases. It has the same form as K_c but uses partial pressures in place of concentrations. The relationship between the two is $K_p = K_c(RT)^{\Delta n_{gas}}$, where Δn_{gas} is the difference between the number of moles of gaseous products and gaseous reactants.

If the chemical equation for a reversible reaction is modified, the equilibrium constant expression must also be modified. If the equation is reversed, the K_c or K_p expression is inverted. If the coefficients of the equation are multiplied by a common factor, the K_c or K_p expression is raised to the corresponding power. Two other important ideas about equilibrium constants are that (1) K_c and K_p are expressed as dimensionless numbers, and (2) pure solid and liquid phases are not represented in equilibrium constant expressions.

In general, if K_c or K_p for a reaction is very large, the forward reaction goes to completion; if K_c or K_p is very small, the forward reaction occurs to a very limited extent. Usually, calculations based on the equilibrium constant expression are necessary only when K_c or K_p values lie between these extremes.

The reaction quotient is a ratio of concentrations (Q_c) or partial pressures (Q_p) having the same form as an equilibrium constant expression but using *nonequilibrium* concentrations or partial pressures. By comparing the values of Q and K, we can determine whether a net reaction will occur in the forward direction ($Q < K$) or in the reverse direction ($Q > K$) to establish equilibrium.

Qualitative predictions about the effect of various changes—amounts of reactants or products, volume, external pressure, or temperature—can be based on Le Châtelier's principle: When a change is imposed on a system at equilibrium, the system responds by attaining a new equilibrium in which the impact of the change is minimized. Catalysts speed the attainment of equilibrium but have no effect on the equilibrium state.

The most common types of equilibrium calculations are (a) determining the value of an equilibrium constant from initial and equilibrium conditions, and (b) using initial conditions and the equilibrium constant to determine equilibrium conditions. We need to solve algebraic equations to carry out some of these calculations. Several techniques for doing so are introduced in Section 14.5.

Key Terms

concentration equilibrium constant (14.2)
equilibrium (14.1)
equilibrium constant expression (14.2)
K_c (14.2)
K_p (14.3)
Le Châtelier's principle (14.4)
partial pressure equilibrium constant (14.3)
reaction quotient (Q_c or Q_p) (14.3)

Review Questions

1. What is a reversible reaction, and what is the condition of equilibrium in such a reaction?

2. Why is the condition of equilibrium in a reversible reaction a *dynamic* one? Cite some experimental evidence to support this idea.

3. What is meant by the equilibrium constant expression and the equilibrium constant of a reversible chemical reaction?

4. What is the difference between K_c and K_p for a chemical reaction?

5. What is the reaction quotient of a reaction, and what is the relationship between Q_c and K_c?

6. Describe how the equilibrium constant for a net reaction is related to the equilibrium constants for the individual reactions that are combined into the net reaction.

7. What is the difference between a *homogeneous* and a *heterogeneous* reaction?

8. Why can we say that the vapor pressure of a liquid at a given temperature is an equilibrium constant?

9. What role does a catalyst play in a reversible chemical reaction? How does a catalyst affect the value of the equilibrium constant for a reaction?

10. In principle, all chemical reactions are reversible. Why is it that we are able to consider that some of them go to completion?

11. Why does the chemical equation alone supply sufficient information for calculating the yield of products in a reaction that goes to completion, but not in a reversible reaction? What additional information is needed to calculate the yield of product(s) in a reversible reaction?

12. Write equilibrium constant expressions, K_c, for the following reactions.

(a) $C(graphite) + CO_2(g) \rightleftharpoons 2\,CO(g)$

(b) $H_2S(g) + I_2(s) \rightleftharpoons 2\,HI(g) + S(s)$

(c) $4\,CuO(s) \rightleftharpoons 2\,Cu_2O(s) + O_2(g)$

13. Balance the following equations, and write equilibrium constant expressions, K_c, for the reactions they represent.

(a) $H_2S(g) \rightleftharpoons H_2(g) + S_2(g)$

(b) $CS_2(g) + H_2(g) \rightleftharpoons CH_4(g) + H_2S(g)$

(c) $CO(g) + H_2(g) \rightleftharpoons CH_4(g) + H_2O(g)$

14. Write equilibrium constant expressions, K_p, for the following reactions.

(a) $CO(g) + H_2O(g) \rightleftharpoons CO_2(g) + H_2(g)$

(b) $N_2(g) + 3\,H_2(g) \rightleftharpoons 2\,NH_3(g)$

(c) $NH_4HS(s) \rightleftharpoons NH_3(g) + H_2S(g)$

15. Write an equilibrium constant expression, K_p, based on the formation of one mole of each of the following gaseous compounds from its elements at 25 °C.

(a) $NO(g)$ **(b)** $NH_3(g)$ **(c)** $NOCl(g)$

16. Write an equilibrium constant expression, K_c, for each of the following reversible reactions.

(a) Carbon monoxide gas reduces nitrogen monoxide gas to gaseous nitrogen; carbon dioxide gas is the other product.

(b) Oxygen gas oxidizes gaseous ammonia to gaseous nitrogen monoxide; water vapor is the other product.

(c) Solid sodium hydrogen carbonate decomposes to form solid sodium carbonate, water vapor, and carbon dioxide gas.

17. Describe Le Châtelier's principle, and indicate how it allows us to make qualitative predictions about chemical equilibria.

18. Describe how you might be able to drive a reaction having a small value of K_c to completion.

19. Equilibrium is established in the reversible reaction.

$$4\,HCl(g) + O_2(g) \rightleftharpoons 2\,H_2O(g) + 2\,Cl_2(g) \qquad \Delta H^\circ = -114.4\text{ kJ}$$

Describe four changes that can be made to this mixture to increase the amount of $Cl_2(g)$ at equilibrium.

20. In each of the following reactions, will the amount of product at equilibrium increase if the total gas pressure is raised from 1 atm to 10 atm? Explain.

(a) $SO_2(g) + Cl_2(g) \rightleftharpoons SO_2Cl_2(g)$

(b) $N_2(g) + O_2(g) \rightleftharpoons 2\,NO(g)$

(c) $SO_2(g) + \frac{1}{2}\,O_2(g) \rightleftharpoons SO_3(g)$

21. The extent of dissociation that occurs in one of the following reactions depends on the volume of the reaction vessel, and in the other, it does not. Identify the situation for each reaction, and explain why they are not the same.

(a) $2\,NO(g) \rightleftharpoons N_2 + O_2(g)$

(b) $2\,NOCl(g) \rightleftharpoons 2\,NO(g) + Cl_2(g)$

22. The extent of dissociation that occurs in one of the following reactions increases with an increase in temperature, and with the other, it decreases with an increase in temperature. Identify the situation for each reaction, and explain why they are not the same.

(a) $2\,NO(g) \rightleftharpoons N_2(g) + O_2(g) \qquad \Delta H^\circ = -180.5\text{ kJ}$

(b) $2\,SO_3(g) \rightleftharpoons 2\,SO_2(g) + O_2(g) \qquad \Delta H^\circ = 197.8\text{ kJ}$

Problems

Equilibrium Constant Relationships

23. Determine the values of K_p that correspond to the following values of K_c.

(a) $CO(g) + Cl_2(g) \rightleftharpoons COCl_2(g)$, $K_c = 1.2 \times 10^3$ at 668 K

(b) $2\,NO(g) + Br_2(g) \rightleftharpoons 2\,NOBr(g)$, $K_c = 1.32 \times 10^{-2}$ at 1000 K

(c) $2\,COF_2(g) \rightleftharpoons CO_2(g) + CF_4(g)$, $K_c = 2.00$ at 1000 °C

24. Determine the values of K_c that correspond to the following values of K_p.

(a) $SO_2Cl_2(g) \rightleftharpoons SO_2(g) + Cl_2(g)$, $K_p = 2.9 \times 10^{-2}$ at 303 K

(b) $2\,NO_2(g) \rightleftharpoons 2\,NO(g) + O_2(g)$, $K_p = 0.275$ at 700 K

(c) $CO(g) + Cl_2(g) \rightleftharpoons COCl_2(g)$, $K_p = 22.5$ at 395 °C

25. For the reaction $N_2(g) + O_2(g) \rightleftharpoons 2\,NO(g)$, $K_c = 4.08 \times 10^{-4}$ at 2000 K. What is the value of K_c at 2000 K for the reaction $NO(g) \rightleftharpoons \frac{1}{2}\,N_2(g) + \frac{1}{2}\,O_2(g)$?

26. For the reaction $SO_3(g) \rightleftharpoons SO_2(g) + \frac{1}{2}\,O_2(g)$, $K_c = 16.7$ at 1000 K. What is the value of K_c at 1000 K for the reaction, $2\,SO_2(g) + O_2(g) \rightleftharpoons 2\,SO_3(g)$?

27. Given that

$$2\,NH_3(g) \rightleftharpoons N_2(g) + 3\,H_2(g) \qquad K_c = 6.46 \times 10^{-3}\text{ at 300 °C}$$

determine K_p at 300 °C for the reaction

$$\tfrac{1}{3}\,N_2(g) + H_2(g) \rightleftharpoons \tfrac{2}{3}\,NH_3(g)$$

28. Given that

$$4\,HCl(g) + O_2(g) \rightleftharpoons 2\,Cl_2(g) + 2\,H_2O(g) \qquad K_p = 27.3\text{ at 450 °C}$$

determine K_c at 450 °C for the reaction

$$\tfrac{1}{2}\,Cl_2(g) + \tfrac{1}{2}\,H_2O(g) \rightleftharpoons HCl(g) + \tfrac{1}{4}\,O_2(g)$$

29. What is the value of K_p at 298 K for the reaction $\frac{1}{2}\,CH_4(g) + H_2O(g) \rightleftharpoons \frac{1}{2}\,CO_2(g) + 2\,H_2(g)$, given the following data at 298 K?

$$CO_2(g) + H_2(g) \rightleftharpoons CO(g) + H_2O(g) \qquad K_p = 9.80 \times 10^{-6}$$

$$CH_4(g) + H_2O(g) \rightleftharpoons CO(g) + 3\,H_2(g) \qquad K_p = 1.25 \times 10^{-25}$$

30. What is the value of K_p at 298 K for the reaction $N_2(g) + 2\,O_2(g) \rightleftharpoons N_2O_4(g)$, given the following data at 298 K?

$$\tfrac{1}{2}\,N_2(g) + \tfrac{1}{2}\,O_2(g) \rightleftharpoons NO(g) \qquad K_p = 6.9 \times 10^{-16}$$

$$NO_2(g) \rightleftharpoons NO(g) + \tfrac{1}{2} O_2(g) \qquad K_p = 6.7 \times 10^{-7}$$

$$N_2O_4(g) \rightleftharpoons 2\,NO_2(g) \qquad K_p = 0.15$$

31. Determine K_c at 298 K for the reaction $\frac{1}{2} N_2(g) + \frac{1}{2} O_2(g) + \frac{1}{2} Cl_2(g) \rightleftharpoons NOCl(g)$, given the following data at 298 K.

$$\tfrac{1}{2} N_2(g) + O_2 \rightleftharpoons NO_2(g) \qquad K_p = 1.0 \times 10^{-9}$$

$$NOCl(g) + \tfrac{1}{2} O_2(g) \rightleftharpoons NO_2Cl(g) \qquad K_p = 1.1 \times 10^{2}$$

$$NO_2(g) + \tfrac{1}{2} Cl_2(g) \rightleftharpoons NO_2Cl(g) \qquad K_p = 0.3$$

(*Hint:* How are K_c and K_p for the reaction related?)

32. Determine K_c at 298 K for the reaction $2\,CH_4(g) \rightleftharpoons C_2H_2(g) + 3\,H_2(g)$, given the following data at 298 K.

$$CH_4(g) + H_2O(g) \rightleftharpoons CO(g) + 3\,H_2(g) \qquad K_p = 1.2 \times 10^{-25}$$

$$2\,C_2H_2(g) + 3\,O_2(g) \rightleftharpoons 4\,CO(g) + 2\,H_2O(g) \qquad K_p = 1.1 \times 10^{2}$$

$$H_2(g) + \tfrac{1}{2} O_2(g) \rightleftharpoons H_2O(g) \qquad K_p = 1.1 \times 10^{40}$$

(*Hint:* How are K_c and K_p for the reaction related?)

33. The reversible reaction $N_2O_4(g) \rightleftharpoons 2\,NO_2(g)$ has a value of $K_p = 0.145$ at 25 °C. *Without doing a calculation*, explain whether the numerical value of K_p at 25 °C for the reaction, $\frac{1}{2} N_2O_4(g) \rightleftharpoons NO_2(g)$ is greater than, equal to, or less than 0.145.

34. Is K_c for the reaction $2\,ICl(g) \rightleftharpoons I_2(g) + Cl_2(g)$ necessarily greater than K_c' for the reaction $ICl(g) \rightleftharpoons \frac{1}{2} I_2(g) + \frac{1}{2} Cl_2(g)$? Explain.

35. If the equilibrium concentrations found in the reaction $A(g) + 2\,B(g) \rightleftharpoons 2\,C(g)$ are [A] = 0.025 M, [B] = 0.15 M, and [C] = 0.55 M, calculate the value of K_c.

36. If the equilibrium concentrations found in the reaction $2\,A(g) + B(g) \rightleftharpoons C(g)$ are $[A] = 2.4 \times 10^{-2}$ M, $[B] = 4.6 \times 10^{-3}$ M, and $[C] = 6.2 \times 10^{-3}$ M, calculate the value of K_c.

37. In the reaction $CO(g) + Cl_2(g) \rightleftharpoons COCl_2(g)$, $K_c = 1.2 \times 10^3$ at 395 K. What is the equilibrium value of $[COCl_2]$ if at equilibrium, $[CO] = 2\,[Cl_2] = \frac{1}{2}\,[COCl_2]$?

38. In the reaction $2\,H_2(g) + S_2(g) \rightleftharpoons 2\,H_2S(g)$, $K_c = 6.28 \times 10^3$ at 900 K. What is the equilibrium value of $[H_2]$ if at equilibrium $[H_2S] = [S_2]^{1/2}$?

39. For the reaction $C(s) + CO_2(g) \rightleftharpoons 2\,CO(g)$, $K_p = 63$. What will be the *total* pressure of the gases above an equilibrium mixture if $P_{CO} = 10\,P_{CO_2}$?

40. For the reaction $H_2S(g) + I_2(s) \rightleftharpoons S(s) + 2\,HI(g)$, $K_p = 1.33 \times 10^{-5}$ at 333 K. What will be the *total* pressure of the gases above an equilibrium mixture if $P_{HI} = 0.010 \times P_{H_2S}$?

41. In the reaction $N_2O_4(g) \rightleftharpoons 2\,NO_2(g)$, equilibrium is reached at a temperature at which $P_{NO_2} = 3\,(P_{N_2O_4})^{1/2}$. What must be the value of K_p at this temperature?

42. For the reaction $CO(g) + H_2O(g) \rightleftharpoons CO_2(g) + H_2(g)$, $K_p = 23.2$ at 600 K. Is it possible to have an equilibrium mixture at 600 K in which **(a)** $P_{CO} = P_{H_2O} = P_{CO_2} = P_{H_2}$? **(b)** $P_{H_2}/P_{H_2O} = P_{CO_2}/P_{CO}$? **(c)** $(P_{CO_2})(P_{H_2}) = (P_{CO})(P_{H_2O})$? **(d)** $P_{CO_2}/P_{H_2O} = P_{H_2}/P_{CO}$? Explain.

43. *Without doing detailed calculations*, determine which of the following statement(s) is(are) correct in describing the equilibrium established when an initial mixture containing an equal number of moles of SO_2 and O_2 is brought to equilibrium in a constant-volume vessel at a temperature of 1030 °C in the following reaction.

$$2\,SO_2(g) + O_2(g) \rightleftharpoons 2\,SO_3(g) \qquad K_c = 1.98 \text{ at } 1030\ °C$$

(a) The pressure of the mixture increases; **(b)** the number of moles of SO_3 at equilibrium is twice the number of moles of O_2 present initially; **(c)** the number of moles of SO_3 at equilibrium depends on the volume of the reaction vessel; **(d)** the equilibrium mixture no longer has equal numbers of moles of SO_2 and O_2; **(e)** K_p for the reaction is greater than K_c.

44. *Without doing detailed calculations*, determine which of the following statement(s) is(are) correct in describing the equilibrium established when equal numbers of moles of H_2, N_2, and HCN and an excess of graphite are allowed to come to equilibrium in a constant-volume reaction vessel at 2025 K in the following reaction.

$$H_2(g) + N_2(g) + C(\text{graphite}) \rightleftharpoons 2\,HCN(g) \qquad K_c = 3.43 \times 10^{-3} \text{ at } 2025\text{ K}$$

(a) The equilibrium amount of HCN can be increased by reducing the reaction volume; **(b)** the amounts of H_2 and N_2 increase from their initial amounts but remain in equimolar proportions; **(c)** the amount of graphite at equilibrium is the same as initially; **(d)** K_p for the reaction is less than K_c; **(e)** the change in partial pressure of HCN from its initial value is twice as great as the changes in partial pressures of H_2 and N_2.

Le Châtelier's Principle

45. Of the N_2O_4 molecules present in a sample of the gas, 12.5% are dissociated into NO_2 when equilibrium is established at 25 °C.

$$N_2O_4(g) \rightleftharpoons 2\,NO_2(g) \qquad \Delta H = 57.2\text{ kJ}$$

Will the percent dissociation be greater or less than 12.5% if **(a)** the reaction mixture is transferred to a vessel of twice the volume? **(b)** the temperature is raised to 50 °C? **(c)** a catalyst is added to the reaction vessel? **(d)** neon gas is added to the reaction mixture in a constant-volume flask?

46. The water-gas reaction is used to produce combustible gases from carbon (coal) and steam.

$$C(s) + H_2O(g) \rightleftharpoons CO(g) + H_2(g)$$

$$K_p = 9.7 \times 10^{-17} \text{ at } 298\text{ K} \qquad \Delta H^\circ = 131\text{ kJ}$$

What will be the effect on the final equilibrium amount of $H_2(g)$ if a gaseous mixture originally at equilibrium with a large excess of C(s) at 298 K is subjected to the following changes: **(a)** more $H_2O(g)$ is added; **(b)** a catalyst is added; **(c)** the mixture is transferred to a reaction vessel of greater volume; **(d)** more CO(g) is added; **(e)** an inert gas is added to the reaction vessel to increase the total pressure; **(f)** the temperature is raised to 1000 K; **(g)** a small amount of C(s) is removed?

47. Explain why the extent of dissociation of diatomic molecules of an element into atoms of the element must increase with an increase in temperature, for example, $Cl_2(g) \rightleftharpoons 2\,Cl(g)$, $S_2(g) \rightleftharpoons 2\,S(g)$, $H_2(g) \rightleftharpoons 2\,H(g)$, and so on.

48. Is the statement made in Problem 47 equally valid for the dissociation of molecules of *compounds* into the constituent elements, for example, $2\,H_2O(g) \rightleftharpoons 2\,H_2(g) + O_2(g)$, $2\,NO(g) \rightleftharpoons N_2(g) + O_2(g)$, and so on? Explain. (*Hint:* Use data from Appendix C if necessary.)

49. In the formation of the following gaseous compounds from their gaseous elements, which reactions occur to a greater extent at high pressures, and which reactions are unaffected by the total pressure: (a) NO(g) from $N_2(g)$ and $O_2(g)$, (b) $NH_3(g)$ from $N_2(g)$ and $H_2(g)$, (c) HI(g) from $H_2(g)$ and $I_2(g)$, (d) $H_2S(g)$ from $H_2(g)$ and $S_2(g)$?

50. Use Le Châtelier's principle to develop an explanation of why the application of a high pressure causes ice to melt.

The Reaction Quotient and the Direction of Net Change

51. If a 2.50-L reaction vessel at 1000 °C contains 0.525 mol CO_2, 1.25 mol CF_4, and 0.75 mol COF_2, in what direction will a net reaction occur to reach equilibrium? Explain.

$$CO_2(g) + CF_4(g) \rightleftharpoons 2\,COF_2(g)$$

$$K_c = 0.50 \text{ at } 1000\text{ °C}$$

52. If a gaseous mixture at 588 K contains 20.0 g each of CO and H_2O and 25.0 g each of CO_2 and H_2, in what direction will a net reaction occur to reach equilibrium? Explain.

$$CO(g) + H_2O(g) \rightleftharpoons CO_2(g) + H_2(g)$$

$$K_c = 31.4 \text{ at } 588\text{ K}$$

53. The following amounts of substances are added to a 7.25-L reaction vessel at 773 K: 0.103 mol CO, 0.205 mol H_2, 2.10 mol CH_4, 3.15 mol H_2O. In what direction will a net reaction occur to reach equilibrium? Explain.

$$CO(g) + 3\,H_2(g) \rightleftharpoons CH_4(g) + H_2O(g)$$

$$K_p = 102 \text{ at } 773\text{ K}$$

54. The following quantities of substances are added to a 15.5-L reaction vessel at 773 K: 25.2 g CO, 15.1 g H_2, 130.2 g CH_4, 125.0 g H_2O. In what direction will a net reaction occur to reach equilibrium? Explain.

$$CO(g) + 3\,H_2(g) \rightleftharpoons CH_4(g) + H_2O(g)$$

$$K_p = 102 \text{ at } 773\text{ K}$$

Experimental Determination of Equilibrium Constants

55. Equilibrium is established in a sealed 1.75-L vessel at 250 °C in the reaction, $PCl_5(g) \rightleftharpoons PCl_3(g) + Cl_2(g)$. The quantities found at equilibrium are 0.562 g PCl_5, 1.950 g PCl_3, and 1.007 g Cl_2. What is the value of K_c for this reaction? What is the value of K_p?

56. Equilibrium is established in a sealed 10.5-L vessel at 184 °C in the reaction, $2\,NO_2(g) \rightleftharpoons 2\,NO(g) + O_2(g)$. The quantities found at equilibrium are 1.353 g NO_2, 0.0960 g NO, and 0.0512 g O_2. What is the value of K_c for this reaction? What is the value of K_p?

57. In the dissociation of ammonium hydrogen sulfide at 25 °C, if we start with a sample of pure $NH_4HS(s)$, the *total* pressure of the gases is 0.658 atm when equilibrium is established. What is the value of K_p for the reaction $NH_4HS(s) \rightleftharpoons NH_3(g) + H_2S(g)$?

58. In the dissociation of ammonium carbamate at 30 °C, if we start with a sample of pure $NH_2COONH_4(s)$, the *total* pressure of the gases is 0.164 atm when equilibrium is established. What is the value of K_p for the reaction $NH_2COONH_4(s) \rightleftharpoons 2\,NH_3(g) + CO_2(g)$?

59. In the reaction $2\ ICl(g) \rightleftharpoons I_2(g) + Cl_2(g)$ at 682 K, a 0.682-g sample of ICl(g) is placed in a 625-mL reaction vessel. When equilibrium is reached, 0.0383 g I_2 is found in the mixture. What is K_c for this reaction?

60. A mixture of 5.25 g I_2 and 2.15 g Br_2 is brought to equilibrium in a 3.15-L reaction vessel at 115 °C. At equilibrium, 1.98 g of I_2 is present. What is K_c at 115 °C for the reaction $I_2(g) + Br_2(g) \rightleftharpoons 2\ IBr(g)$?

Calculations Based on K_c

61. The equilibrium constant for the isomerization of butane at 25 °C is $K_c = 7.94$.

$$\underset{\text{Butane}}{CH_3CH_2CH_2CH_3} \rightleftharpoons \underset{\text{Isobutane}}{CH_3\overset{\overset{CH_3}{|}}{C}HCH_3}$$

If 5.00 g butane is introduced into a 12.5-L flask at 25 °C, what mass of isobutane will be present when equilibrium is reached?

62. At 25 °C, the following equilibrium can be established in a liquid solution.

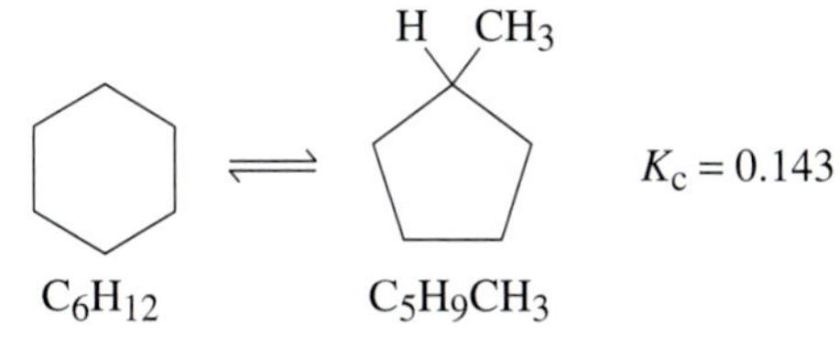

If equilibrium is established in a mixture initially consisting of 1.00×10^2 g cyclohexane, what mass of methylcyclopentane will be present? (*Hint:* Does the volume of solution matter?)

63. For the reaction $CO(g) + H_2O(g) \rightleftharpoons CO_2(g) + H_2(g)$, $K_c = 23.2$ at 600 K. If 0.250 mol each of CO and H_2O are introduced into a reaction vessel and equilibrium is established, how many moles each of CO_2 and H_2 will be present? (*Hint:* Does the volume of the reaction mixture matter?)

64. For the water-gas reaction, $C(s) + H_2O(g) \rightleftharpoons CO(g) + H_2(g)$, $K_c = 0.111$ at about 1100 K. If 0.100 mol $H_2O(g)$ and 0.100 mol $H_2(g)$ are mixed with excess C(s) at this temperature and equilibrium is established, how many moles of CO(g) will be present? No CO(g) is present initially.

65. For the synthesis of phosgene at 395 °C, $CO(g) + Cl_2(g) \rightleftharpoons COCl_2(g)$, $K_c = 1.2 \times 10^3$. If 20.0 g CO and 35.5 g Cl_2 are placed in an 8.05-L reaction vessel at 395 °C and equilibrium is established, how many grams of $COCl_2$ will be present?

66. For the decomposition of carbonyl fluoride, $2\ COF_2(g) \rightleftharpoons CO_2(g) + CF_4(g)$, $K_c = 2.00$ at 1000 °C. If 0.500 mol $COF_2(g)$ is placed in a 3.23-L reaction vessel at 1000 °C, how many moles of $COF_2(g)$ will remain *undissociated* when equilibrium is reached?

67. To establish equilibrium in the following reaction at 250 °C,

$$PCl_3(g) + Cl_2(g) \rightleftharpoons PCl_5(g)$$

$$K_c = 26 \text{ at } 250\ °C$$

0.100 mol each of PCl_3 and Cl_2 and 0.0100 mol PCl_5 are introduced into a 6.40-L reaction flask. How many moles of each of the gases should be present when equilibrium is established?

68. If 0.250 mol SO_2Cl_2, 0.150 mol SO_2, and 0.0500 mol Cl_2 come to equilibrium in a 12.0-L reaction vessel at 102 °C, how many moles of each gas will be present?

$$SO_2Cl_2(g) \rightleftharpoons SO_2(g) + Cl_2(g)$$

$$K_c = 7.77 \times 10^{-2} \text{ at } 102\ °C$$

69. If an initial 1.00 mol SO_3 is brought to equilibrium in a 5.00-L reaction vessel at 1030 °C, what percent of the SO_3 will remain *undissociated*?

$$2\ SO_3(g) \rightleftharpoons 2\ SO_2(g) + O_2(g)$$

$$K_c = 0.504 \text{ at } 1030\ °C$$

(*Hint:* Use the method of successive approximations in Appendix A.)

70. If an initial 1.00 mol NH_3 is brought to equilibrium in a 10.0-L reaction vessel at 300 °C, what percent of the NH_3 will remain *undissociated*?

$$2\ NH_3(g) \rightleftharpoons N_2(g) + 3\ H_2(g)$$

$$K_c = 6.46 \times 10^{-3} \text{ at } 300\ °C$$

(*Hint:* Use the method of successive approximations in Appendix A.)

Calculations Based on K_p

71. In the reaction $C(s) + S_2(g) \rightleftharpoons CS_2(g)$, $K_p = 5.60$ at 1009 °C. If, at equilibrium, $P_{CS_2} = 0.152$ atm, what must be **(a)** P_{S_2} and **(b)** the total gas pressure, P_{total}?

72. In the reaction $Sb_2S_3(s) + 3\ H_2(g) \rightleftharpoons 2\ Sb(s) + 3\ H_2S(g)$, $K_p = 0.429$ at 713 K. If, at equilibrium, $P_{H_2S} = 0.200$ atm, what must be **(a)** P_{H_2} and **(b)** the total gas pressure, P_{total}?

73. For the reaction $C(s) + 2\ H_2(g) \rightleftharpoons CH_4(g)$, $K_p = 0.263$ at 1000 °C. Calculate the total pressure when 0.100 mol CH_4 and an excess of C(s) are brought to equilibrium at 1000 °C in a 4.16-L reaction vessel.

74. Solid molybdenum is kept in contact with $CH_4(g)$, initially at a pressure of 1.00 atm, in a reaction at 973 K. Calculate the total pressure when equilibrium is established.

$$2\ Mo(s) + CH_4(g) \rightleftharpoons Mo_2C(s) + 2\ H_2(g)$$

$$K_p = 3.55 \text{ at } 973\text{ K}$$

75. For the synthesis of methanol at 483 K, $CO(g) + 2\ H_2(g) \rightleftharpoons CH_3OH(g)$, $K_p = 9.23 \times 10^{-3}$. A gaseous mixture containing H_2 and CO in a 2:1 mole ratio and at an initial total pressure of 12.0 atm is allowed to come to equilibrium at 483 K. What will be the partial pressure of $CH_3OH(g)$ in the equilibrium mixture? (*Hint:* Use the method of successive approximations of Appendix A.)

76. A gaseous mixture is prepared at 900 K in which initially the partial pressures are $P_{SO_3} = 1.50$ atm, $P_{SO_2} = 1.00$ atm, $P_{O_2} = 0.500$ atm. What is the partial pressure of SO_3 after the reaction comes to equilibrium?

$$2\ SO_3(g) \rightleftharpoons 2\ SO_2(g) + O_2(g)$$

$$K_p = 0.023 \text{ at } 900\text{ K}$$

(*Hint:* Use the method of successive approximations of Appendix A.)

Additional Problems

77. At 500 °C, an equilibrium mixture in the reaction $CO(g) + H_2O(g) \rightleftharpoons CO_2(g) + H_2(g)$ is found to contain 0.021 mol CO, 0.121 mol H_2O, 0.179 mol CO_2, and 0.079 mol H_2. If an additional 0.100 mol H_2 is added to the mixture, **(a)** in what direction must a net reaction occur to restore equilibrium, and **(b)** what will be the amounts of the four substances when equilibrium is reestablished?

78. For the isomerization of butane to isobutane, $K_c = 7.94$ (see Problem 61). Sketch a graph of concentration of butane and of isobutane as a function of time, in the manner of Figure 14.2. Start with any initial concentration of butane (but no isobutane), and show the equilibrium concentrations in their expected relative proportions.

79. An analysis of the gaseous phase [$S_2(g)$ and $CS_2(g)$] present at equilibrium at 1009 °C in the reaction $C(s) + S_2(g) \rightleftharpoons CS_2(g)$ shows it to be 13.71% C and 86.29% S, by mass. What is K_c for this reaction?

80. Equilibrium is established in the homogeneous reaction of acetic acid and ethanol by starting with 1.51 mol CH_3COOH and 1.66 mol CH_3CH_2OH.

$$\underset{\text{Acetic acid}}{CH_3COOH} + \underset{\text{Ethanol}}{CH_3CH_2OH} \rightleftharpoons \underset{\text{Ethyl acetate}}{CH_3COOCH_2CH_3} + H_2O \qquad K_c = ?$$

After equilibrium is reached, exactly one-hundredth of the equilibrium mixture is titrated with $Ba(OH)_2(aq)$ to determine the amount of acetic acid present.

$$2\ CH_3COOH + Ba(OH)_2(aq) \longrightarrow Ba(CH_3COO)_2(aq) + 2\ H_2O$$

The volume of 0.1025 M $Ba(OH)_2(aq)$ required for the titration is 22.44 mL. Use these data to calculate K_c for the formation of ethyl acetate.

81. A 0.100-mol sample of N_2O_4 is placed in a 2.50-L flask and brought to equilibrium at 348 K. The mole fraction of NO_2 at equilibrium is 0.746. Determine K_p at 348 K for the reaction $N_2O_4(g) \rightleftharpoons 2\ NO_2(g)$.

82. Consider the dissociation of $H_2S(g)$ at 750 °C.

$$2\ H_2S(g) \rightleftharpoons 2\ H_2(g) + S_2(g)$$

$$K_c = 1.1 \times 10^{-6}$$

If 1.00 mol H_2S is placed in a 7.50-L reaction vessel and heated to 750 °C, how many moles of H_2 and S_2 will be present when equilibrium is reached? (*Hint:* Look for an assumption that can greatly simplify the algebraic solution.)

83. The decomposition of calcium sulfate, $2\ CaSO_4(s) \rightleftharpoons 2\ CaO(s) + 2\ SO_2(g) + O_2(g)$, has $K_p = 1.45 \times 10^{-5}$ at 1625 K. A sample of $CaSO_4(s)$ is introduced into a reaction vessel that is filled with air at 1.0000 atm pressure at 1625 K. What will be the partial pressure of $SO_2(g)$ when equilibrium is established? Recall that the mole percent O_2 in air is 20.95%.

84. If, at 25 °C, a large quantity of anhydrous (dry) $Na_2HPO_4(s)$ is added to a 3.45-L container containing 42 mg $H_2O(g)$, will any of the hydrate, $Na_2HPO_4 \cdot 2H_2O(s)$ form?

$$Na_2HPO_4(s) + 2\ H_2O(g) \rightleftharpoons Na_2HPO_4 \cdot 2H_2O(s)$$
$$K_p = 6.01 \times 10^3$$

If so, how many moles will form? (*Hint:* What is the equilibrium pressure of water vapor above a mixture of the hydrate and the anhydrous compound?)

85. The photograph shows saturated $I_2(aq)$ floating on colorless $CCl_4(l)$ (left). After the mixture is vigorously shaken and equilibrium is established, most of the iodine has passed from the water layer to the CCl_4 layer, which now has a purple color (right). For this equilibrium, $I_2(aq) \rightleftharpoons I_2(\text{in } CCl_4)$,

$$K_c = \frac{[I_2]_{CCl_4}}{[I_2]_{aq}} = 85.5$$

The concentration of saturated $I_2(aq)$ is 1.33×10^{-3} M.

(a) If 25.0 mL of saturated $I_2(aq)$ is shaken with 25.0 mL of $CCl_4(l)$, how many milligrams of I_2 remain in the aqueous layer at equilibrium?

(b) If the 25.0-mL sample of saturated $I_2(aq)$ is shaken with 50.0 mL of $CCl_4(l)$ instead of 25.0 mL, would you expect the mass of I_2 remaining in the aqueous layer to be the same, more, or less than that calculated in (a)? Explain, but without doing a calculation.

(c) If the aqueous layer at equilibrium in (a) is brought to equilibrium with a second 25.0-mL sample of $CCl_4(l)$, would you expect the mass of I_2 remaining in the aqueous layer in this second equilibrium to be the same, more, or less than that in (b)? Explain, but without doing a calculation.

86. Show that the equilibrium constant expression for the reaction $H_2(g) + I_2(g) \rightleftharpoons 2\ HI(g)$ is consistent with this proposed two-step mechanism for the reaction.

Fast: $$I_2(g) \underset{k_{-1}}{\overset{k_1}{\rightleftharpoons}} 2\ I(g)$$

Slow: $$2\ I(g) + H_2(g) \underset{k_{-2}}{\overset{k_2}{\rightleftharpoons}} 2\ HI(g)$$

87. Benzene can be produced from hexane in the reversible reaction $C_6H_{14}(g) \rightleftharpoons C_6H_6(g) + 4\ H_2(g)$. The partial pressure equilibrium constant, K_p, for this reaction has been found to vary with Kelvin temperature according to the following equation.

$$\log_{10} K_p = 23.45 - 13{,}941/T$$

Equilibrium is established by starting with pure $C_6H_{14}(g)$. What must be the temperature if the *initial* gas pressure is 1.00 atm and the equilibrium partial pressure of C_6H_6 is 0.100 atm?

88. What is the mole percent $COCl_2$ in the equilibrium gaseous mixture when $COCl_2(g)$ dissociates at 395 °C and the *total* equilibrium pressure is 3.00 atm?

$$COCl_2(g) \rightleftharpoons CO(g) + Cl_2(g)$$
$$K_p = 4.44 \times 10^{-2} \text{ at } 395\ °C$$

89. Suppose that in the reaction $PCl_5(g) \rightleftharpoons PCl_3(g) + Cl_2(g)$, the fraction of the PCl_5 molecules that dissociate to reach equilibrium is α and the *total* gas pressure at equilibrium is P.

(a) Show that

$$K_p = \frac{\alpha^2 P}{1 - \alpha^2}$$

(*Hint:* Start with 1.00 mol PCl_5 in a container of volume V. Use the ideal gas law, as needed.)

(b) At 250 °C, $K_p = 1.78$. What is the fraction of PCl_5 that dissociates at this temperature when the total equilibrium pressure is 2.50 atm?

(c) Under what total pressure must the gaseous mixture be maintained to limit dissociation of PCl_5 to 10.0% at 250 °C?

CHAPTER

15

Acids, Bases, and Acid-Base Equilibria

CONTENTS

We use acid-base chemistry in every aspect of our lives, including the preparation and enjoyment of food. The acid-base reaction that occurs when we use lemon juice on fish is explained on page 649.

Acids and bases are such important compounds that we introduced them in Chapter 2. We have used acids and bases by name and formula often in following chapters. In Chapter 4, we classified acids and bases in aqueous solutions as *strong* or *weak* according to their ability to conduct electricity. We also presented acid-base reactions as one of the three main types of reactions that occur in aqueous solutions, and we introduced the stoichiometry of acid-base titrations.

In this chapter, we will examine factors that affect acid and base strength. We will describe what we mean by the pH of a solution—such as that of an acid rain sample with a pH of 4.2. We will show how to calculate the pH of a solution and how a constant pH can be maintained—such as a normal pH of 7.4 in blood. Also, we will discuss acid-base indicators and take another look at acid-base neutralization reactions carried out by titration.

● **ChemCDX**
When you see this icon, there are related animations, demonstrations, and exercises on the CD accompanying this text.

15.1 The Brønsted-Lowry Theory of Acids and Bases

The first successful theory of acids and bases was that of Arrhenius, which we introduced in Section 2.9. According to the Arrhenius theory, an acid in aqueous solution produces hydrogen ions, $H^+(aq)$, and a base produces hydroxide ions, $OH^-(aq)$. Later extensions of the theory also distinguished between *strong* acids and bases and *weak* acids and bases. A strong acid ionizes essentially completely into $H^+(aq)$ and accompanying anions, and a strong base dissociates nearly completely into $OH^-(aq)$ and accompanying cations. (Recall that Table 4.1, page 139, lists the common strong acids and strong bases.) The ionizations of weak acids and bases are reversible. They reach a state of equilibrium in which only a small percent of the acid or base exists as ions.

The Arrhenius theory is limited, however, in that it applies only to *aqueous* solutions and it does not adequately explain why such compounds as ammonia (NH_3) are bases. Arrhenius's theory seems to require that a base *contain* OH^-, or at least an —OH group that can become OH^-. There is no such ion or group in NH_3, but as we saw in Chapter 4, ammonia is a weak base. In fact, because Arrhenius and his contemporaries thought that a base had to contain OH^-, aqueous ammonia was long known as ammonium hydroxide, and its formula written NH_4OH. That name and formula are still sometimes used today, but it is unlikely that discrete molecules of NH_4OH exist in an aqueous solution. It is not possible, for example, to write a Lewis structure for NH_4OH.

Solutions of $NH_3(aq)$ are often labeled NH_4OH and called ammonium hydroxide.

The Brønsted-Lowry Theory

To a large degree, the shortcomings of the Arrhenius theory are overcome by a theory proposed independently by J. N. Brønsted in Denmark and T. M. Lowry in Great Britain in 1923. According to this theory, an acid is a **proton donor** and a base is a **proton acceptor**. By a "proton," we mean an ionized hydrogen atom, that is, H^+. However, a substance cannot simply shed a proton, because the proton cannot exist in a solution as a free, independent particle. The positively charged proton attaches itself to a center of negative charge, such as a lone pair of electrons on an atom in some other species. Thus, for every acid, there must be an available base. The reaction between them is an *acid-base* reaction. We can represent the ionization of hydrochloric acid in this way.

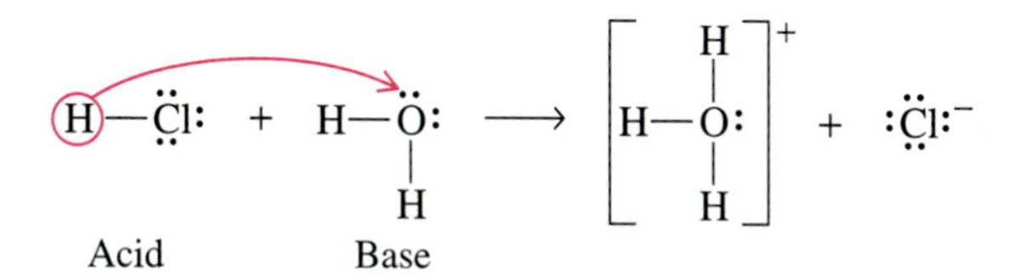

In the Brønsted-Lowry theory, H_3O^+ takes the place of the H^+ of the Arrhenius theory. The H_3O^+ ion, called the *hydronium ion*, forms when the O atom of an H_2O molecule uses a lone pair of electrons to form a coordinate covalent bond with a proton, H^+.*

Brønsted-Lowry theory is not limited to reactions in aqueous solutions. For example, HCl is an acid in liquid ammonia, $NH_3(l)$, just as it is in water. The difference, of course, is that the base that accepts a proton from HCl is the ammonia molecule.

*The proton, H^+, is probably associated with *several* H_2O molecules, for example, four H_2O molecules in $H(H_2O)_4^+$. However, in applying the Brønsted-Lowry theory, we will assume that the simple hydrated proton, H_3O^+, is the species present in aqueous solutions.

$$\underset{\text{Acid}}{H{-}\ddot{\underset{\cdot\cdot}{Cl}}:} + \underset{\text{Base}}{H{-}\underset{\substack{| \\ H}}{\ddot{N}}{-}H} \longrightarrow \left[H{-}\underset{\substack{| \\ H}}{\overset{\substack{H \\ |}}{N}}{-}H\right]^+ + :\ddot{\underset{\cdot\cdot}{Cl}}:^-$$

Also, the Brønsted-Lowry theory successfully explains how ammonia acts as a base in water, which the Arrhenius theory fails to do. In the ionization of NH_3, water is the acid.

$$\underset{\text{Base}}{H{-}\underset{\substack{| \\ H}}{\ddot{N}}{-}H} + \underset{\text{Acid}}{H{-}\underset{\substack{| \\ H}}{\ddot{O}}:} \longrightarrow \left[H{-}\underset{\substack{| \\ H}}{\overset{\substack{H \\ |}}{N}}{-}H\right]^+ + \left[:\ddot{\underset{\cdot\cdot}{O}}{-}H\right]^-$$

Recall that NH_3 is a *weak* base; it does not ionize completely. That leads us to the conclusion that the equation we have just written for the ionization of NH_3 is incomplete. The ionization is a *reversible* reaction and reaches an equilibrium state. In the reverse reaction, NH_4^+ is the acid and OH^- is the base. This reversible acid-base reaction is represented below and in Figure 15.1.

$$\underset{\text{Base(1)}}{H{-}\underset{\substack{| \\ H}}{\ddot{N}}{-}H} + \underset{\text{Acid(2)}}{H{-}\underset{\substack{| \\ H}}{\ddot{O}}:} \rightleftharpoons \underset{\text{Acid(1)}}{\left[H{-}\underset{\substack{| \\ H}}{\overset{\substack{H \\ |}}{N}}{-}H\right]^+} + \underset{\text{Base(2)}}{\left[:\ddot{\underset{\cdot\cdot}{O}}{-}H\right]^-}$$

The labels in the above equation designate the acid and base for the forward reaction *and* for the reverse reaction. Notice that the acid on one side of the equation differs from the base on the other side only by a proton, that is, H^+. The combinations acid(1)/base(1) and acid(2)/base(2) are called **conjugate acid-base pairs**. The **conjugate acid** of a base is the base *plus* an attached proton, H^+. Thus, acid(1), NH_4^+, is the conjugate acid of base(1), NH_3, and the combination NH_4^+/NH_3 is a conjugate acid-base pair. The **conjugate base** of an acid is the acid *minus* a proton, H^+. Thus, base(2), OH^-, is the conjugate base of acid(2), H_2O, and the combination H_2O/OH^- is also a conjugate acid-base pair.

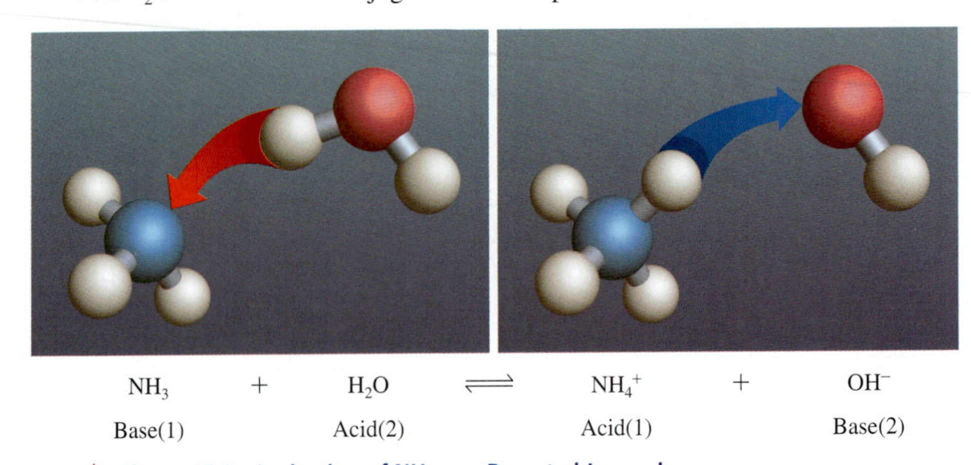

▲ **Figure 15.1 Ionization of NH_3 as a Brønsted-Lowry base**
Proton transfer is represented by a red arrow in the forward reaction and a blue arrow in the reverse reaction. Because ammonium ion is a stronger acid than water and because hydroxide ion is a stronger base than ammonia, the equilibrium condition lies quite far to the left. The relative strengths of some acids and bases are presented in Table 15.1 on page 638.

The ionization of acetic acid is also a reversible reaction, which we can represent as follows.

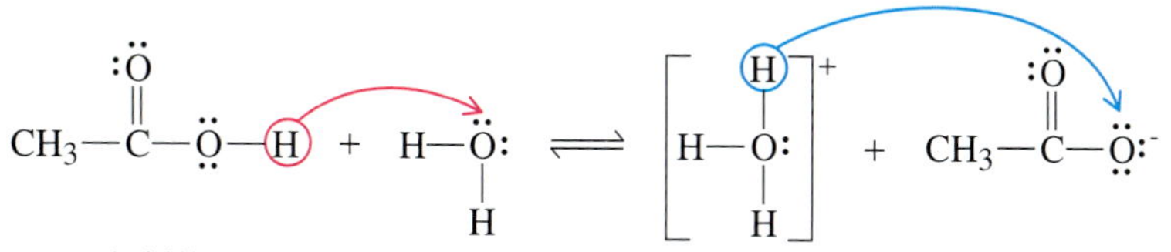

$$CH_3COOH + H_2O \rightleftharpoons [H_3O]^+ + CH_3COO^-$$

Acid(1) Base(2) Acid(2) Base(1)

The conjugate acid-base pairs are CH_3COOH/CH_3COO^- and H_3O^+/H_2O. Notice that in this reaction, H_2O is a base, whereas in the ionization of ammonia, H_2O was an acid. This dual role of H_2O is suggested below.

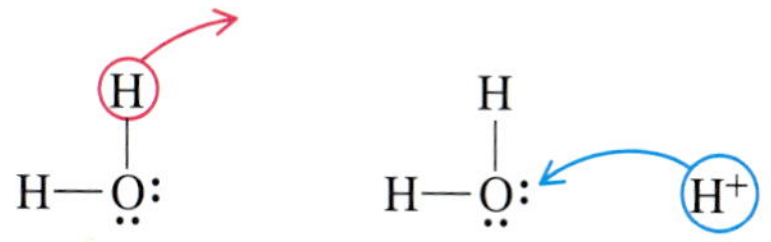

H_2O as an acid H_2O as a base

A substance like water that can act either as an acid or a base is **amphiprotic**. We identify Brønsted-Lowry acids and bases and amphiprotic species in Example 15.1 and Exercises 15.1A and B.

EXAMPLE 15.1

Identify the Brønsted-Lowry acids and bases and their conjugates in each of the following reactions.

a. $H_2S + NH_3 \rightleftharpoons NH_4^+ + HS^-$

b. $OH^- + H_2PO_4^- \rightleftharpoons H_2O + HPO_4^{2-}$

SOLUTION

To identify Brønsted-Lowry acids and bases, we look for the proton donors and proton acceptors in each reaction.

a. Because H_2S is converted to HS^- by donating a proton, H_2S is an acid; HS^- is its conjugate base. Because NH_3 accepts the proton lost by the H_2S, NH_3 is a base; NH_4^+ is its conjugate acid.

$$\underset{\text{Acid(1)}}{H_2S} + \underset{\text{Base(2)}}{NH_3} \rightleftharpoons \underset{\text{Acid(2)}}{NH_4^+} + \underset{\text{Base(1)}}{HS^-}$$

b. OH^- accepts a proton from $H_2PO_4^-$. Therefore OH^- is a base; H_2O is its conjugate acid. $H_2PO_4^-$ donates a proton to OH^-. Thus, $H_2PO_4^-$ is an acid; HPO_4^{2-} is its conjugate base.

$$\underset{\text{Base(1)}}{OH^-} + \underset{\text{Acid(2)}}{H_2PO_4^-} \rightleftharpoons \underset{\text{Acid(1)}}{H_2O} + \underset{\text{Base(2)}}{HPO_4^{2-}}$$

EXERCISE 15.1A

Identify the Brønsted-Lowry acids and bases and their conjugates in each of the following reactions.

a. $NH_3 + HCO_3^- \rightleftharpoons NH_4^+ + CO_3^{2-}$

b. $H_3PO_4 + H_2O \rightleftharpoons H_2PO_4^- + H_3O^+$

EXERCISE 15.1B

Identify the *amphiprotic* species in each reaction in Exercise 15.1A, and tell whether it acts as an acid or a base in that reaction. Then write an equation for a reaction in which it acts in the opposite manner. Specifically, if the amphiprotic species was an acid in Exercise 15.1A, use it as a base, and write an equation for its reaction with HCl. If the amphiprotic species was a base in Exercise 15.1A, now use it as an acid, and write an equation for its reaction with NH_3.

We can write equilibrium constant expressions for the reversible ionizations of $CH_3COOH(aq)$ and $NH_3(aq)$ as we did for other reversible reactions in Chapter 14. In each case, we treat the solvent, $H_2O(l)$, as if it were a pure liquid because its concentration does not change appreciably. Also, instead of calling the equilibrium constants K_c, we use the symbols K_a, called the **acid ionization constant**, and K_b, the **base ionization constant**.

$$CH_3COOH(aq) + H_2O(l) \rightleftharpoons H_3O^+(aq) + CH_3COO^-(aq)$$

$$K_a = \frac{[H_3O^+][CH_3COO^-]}{[CH_3COOH]} = 1.8 \times 10^{-5}$$

$$NH_3(aq) + H_2O(l) \rightleftharpoons NH_4^+(aq) + OH^-(aq)$$

$$K_b = \frac{[NH_4^+][OH^-]}{[NH_3]} = 1.8 \times 10^{-5}$$

Values of K_a and K_b are determined by experiments of a type that we will describe a bit later. The fact that K_a for acetic acid and K_b for ammonia have the same value is of no significance; it is purely coincidental.

Strengths of Conjugate Acid-Base Pairs

We chose HC1 as our first example of a Brønsted-Lowry acid and represented its ionization in an equation with a *single* arrow.

$$\underset{\text{Acid(1)}}{HCl} + \underset{\text{Base(2)}}{H_2O} \longrightarrow \underset{\text{Acid(2)}}{H_3O^+} + \underset{\text{Base(1)}}{Cl^-}$$

Recall (Chapter 4) that electrical conductivity experiments and other evidence indicate that HCl is a *strong* acid; it is essentially completely ionized in aqueous solution. Similar evidence shows that acetic acid is a *weak* acid. Only a fraction of the CH_3COOH molecules ionize in an acetic acid solution, and we represent its ionization as a reversible reaction.

$$\underset{\text{Acid(1)}}{CH_3COOH} + \underset{\text{Base(2)}}{H_2O} \rightleftharpoons \underset{\text{Acid(2)}}{H_3O^+} + \underset{\text{Base(1)}}{CH_3COO^-}$$

We can describe the difference in acid strengths of HCl and CH_3COOH two ways. First, because HCl is a much *stronger* acid than CH_3COOH, HCl has a much greater tendency to transfer protons to water molecules than CH_3COOH does. Second, we can consider the possible reverse reactions. Because Cl^- is a much *weaker* base than CH_3COO^-, Cl^- shows much less tendency to accept a proton from H_3O^+ than does CH_3COO^-. As a consequence, the reverse of the ionization of HCl is negligible, whereas the reverse of the ionization of CH_3COOH is quite significant. We can summarize these observations in the following generalizations about conjugate acid-base pairs.

- *The stronger an acid, the weaker is its conjugate base.*
- *The stronger a base, the weaker is its conjugate acid.*
- *An acid-base reaction is favored in the direction from the stronger member to the weaker member of each conjugate acid-base pair.*

Let's use these ideas to determine the direction of net change in the following reaction.

$$CH_3COOH + Br^- \overset{?}{\rightleftharpoons} HBr + CH_3COO^-$$

Like HCl, HBr is a strong acid. In contrast, its conjugate base, Br^-, is a very weak base. The strength of HBr favors the reverse reaction, and the weakness of Br^- severely limits the forward reaction. As a result, the reaction occurs almost exclusively in the *reverse* reaction. The direction of net change is from the strong to the weak member of the HBr/Br^- conjugate acid-base pair. We arrive at the same conclusion when we consider the conjugate pair CH_3COOH/CH_3COO^-. Acetate ion is stronger as a base than is acetic acid as an acid, and the *reverse* reaction is also the direction from the stronger to the weaker member of this pair.

To apply the above line of reasoning requires that we have a relative ranking of acid and base strengths, such as that shown in Table 15.1. The strongest acids (left) and the weakest conjugate bases (right) are at the top of the table. At the bottom of the table, we find the weakest acids (left) and the strongest conjugate bases (right).

In constructing a ranking such as that of Table 15.1, we can use K_a values to compare the strengths of weak acids. Thus, hydrofluoric acid ($K_a = 6.6 \times 10^{-4}$) is a somewhat stronger proton donor than acetic acid ($K_a = 1.8 \times 10^{-5}$). For strong acids, water has a *leveling effect*. *All* strong acids appear to donate essentially all of their protons to water molecules. To compare the strengths of strong acids, we need to study them in a solvent that is a weaker base than water, for example, in diethyl ether or acetone. Ionization in these solvents is no longer complete, and so we can rank strong acids according to their K_a values in certain nonaqueous solutions.

15.2 Molecular Structure and Strengths of Acids and Bases

As we have noted before, a central challenge in chemistry is to explain observable properties of a substance in terms of its molecular structure. Referring to Table 15.1, we would like to explain, for example, why HF is a weak acid whereas HCl is strong, why HNO_3 is a stronger acid than HNO_2, and so on. Let's now consider several factors that affect acid strength and thus help explain these facts, looking first at binary acids.

A binary acid is composed of hydrogen and another nonmetal, X. The formula of the acid is HX if X is in Group 7A, H_2X if X is in Group 6A, and H_3X if X is in Group 5A. Let's see what factors affect the acid strengths of binary acids within the same group of the periodic table and then within the same period.

TABLE 15.1 Relative Strengths of Some Brønsted-Lowry Acids and Their Conjugate Bases

Acid	Conjugate Base
HI (hydroiodic acid)	I^- (iodide ion)
HBr (hydrobromic acid)	Br^- (bromide ion)
HCl (hydrochloric acid)	Cl^- (chloride ion)
H_2SO_4 (sulfuric acid)	HSO_4^- (hydrogen sulfate ion)
HNO_3 (nitric acid)	NO_3^- (nitrate ion)
HSO_4^- (hydrogen sulfate ion)	SO_4^{2-} (sulfate ion)
HNO_2 (nitrous acid)	NO_2^- (nitrite ion)
HF (hydrofluoric acid)	F^- (fluoride ion)
CH_3COOH (acetic acid)	CH_3COO^- (acetate ion)
H_2CO_3 (carbonic acid)	HCO_3^- (hydrogen carbonate ion)
NH_4^+ (ammonium ion)	NH_3 (ammonia)
HCO_3^- (hydrogen carbonate ion)	CO_3^{2-} (carbonate ion)
CH_3OH (methanol)	CH_3O^- (methoxide ion)

Increasing acid strength (upward) — Increasing base strength (downward)

Variations in Strength of Binary Acids in a Group of the Periodic Table

As an example, let's consider the binary acids in Group 7A: the hydrogen halides. First, we can write this general expression.

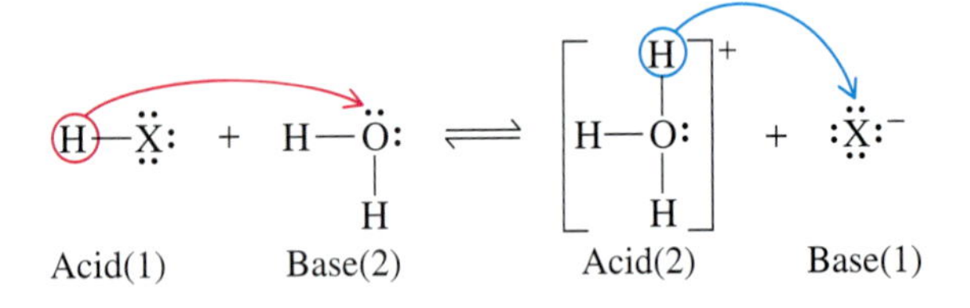

The strength of base(2)—H_2O—and its conjugate acid(2)—H_3O^+—do not depend on the particular acid we consider. Any differences in acid strengths must be related to the molecule HX and the ion X^-.

In the expression above, the red arrow shows the transfer of a proton from HX to H_2O. The greater the tendency for this transfer, the more the *forward* reaction is favored and the *stronger* the acid(1). The blue arrow shows the transfer of a proton from H_3O^+ to X^-. The greater the tendency for this transfer to occur, the more the *reverse* reaction is favored and the *weaker* the acid(1).

We can consider the proton transfer indicated by the red arrow to be somewhat related to the process of breaking a bond in a gaseous molecule to produce gaseous atoms, even though in aqueous solutions ions are formed, not atoms. The bond-dissociation energy (Chapter 9) is a measure of the tendency for bond breakage. The lower the bond-dissociation energy, the more easily the H—X bond is broken. Then, as suggested below, the more easily the H—X bond is broken, the more readily a proton is transferred and the stronger the acid.

Bond-dissociation energy (kJ/mol) decreases	569	>	431	>	368	>	297
	HF		**HCl**		**HBr**		**HI**
Acid strength increases K_a	6.6×10^{-4}	<	$\approx 10^6$	<	$\approx 10^8$	<	$\approx 10^9$

We must also evaluate the tendency for proton transfer in the reverse of the ionization reaction (blue arrow). Protons (H^+) are attracted to the negatively charged ion X^-. The smaller the X^- ion, the stronger is this attraction and the greater is the tendency for the reverse reaction to occur. Conversely, the *larger* the X^- ion, the *smaller* the tendency for the reverse reaction to occur and the *stronger* is the acid HX.

Anion radius (pm) increases	136	<	181	<	195	<	216
	HF		**HCl**		**HBr**		**HI**
Acid strength increases K_a	6.6×10^{-4}	<	$\approx 10^6$	<	$\approx 10^8$	<	$\approx 10^9$

Thus, both factors, bond energy and anion size, lead to the same conclusion.

In a group of the periodic table, the strengths of binary acids increase from top to bottom in the group.

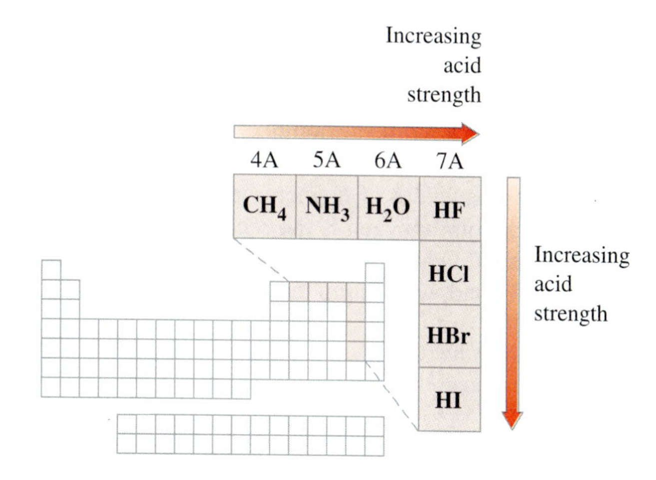

Representative trends in strengths of binary acids.

Variations in Strength of Binary Acids in a Period of the Periodic Table

Variations in acid strengths in the same period correlate well with the *electronegativity difference* (ΔEN) between H and the element to which it is bonded. Recall that a small ΔEN signifies a bond that is largely covalent in character, whereas a large difference indicates separation into partial charges. An acid with partial ionic charges loses H^+ to a base more readily, and is thus a stronger acid, than an acid without partial charges. We can use a ranking for the second period binary compounds of hydrogen to make the generalization on the next page.

In a period of the periodic table, the strengths of binary acids increase from left to right.

ΔEN increases	0.4	<	0.9	<	1.4	<	1.9
Acid strength increases	$\mathbf{CH_4}$	<	$\mathbf{NH_3}$	<	$\mathbf{H_2O}$	<	**HF**

Methane (CH_4) does not ionize as an acid. In aqueous solution, NH_3 ionizes weakly, but as a base not as an acid. The ability of H_2O to ionize as an acid is quite limited, as we shall see in Section 15.3. In contrast, HF is a weak acid with $K_a = 6.6 \times 10^{-4}$.

Strengths of Oxoacids

Most acids contain oxygen as well as hydrogen and a third nonmetal. In oxoacids, there is at least one H atom bonded to an O atom.

H—O—E

The broken lines in this structure suggest that other groups may also be bonded to the E atom. The strength of the O—H bond is affected by the attraction of the E atom for electrons in that bond. If the E atom strongly attracts electrons, then to some extent, electrons are withdrawn from the O—H bond. We would expect this electron-withdrawing effect to increase as the electronegativity of E increases. This weakens the O—H bond and increases the acid strength. The data below show the effect of the electronegativity of the E atom (I, Br, and Cl) on the strengths of three comparable acids.

Electronegativity increases:	2.5	<	2.8	<	3.0
	HOI		**HOBr**		**HOCl**
Acid strength increases	$K_a = 2.3 \times 10^{-11}$	<	2.5×10^{-9}	<	2.9×10^{-8}

For a given E atom, the number of terminal O atoms in the molecule profoundly influences the strength of an oxoacid. In the following comparison, where E is Cl, we see that the values of K_a increase by several powers of ten for each additional terminal O atom. Oxygen is the second most electronegative of all the elements. Thus, terminal O atoms act together with the E atom to withdraw electrons from the O—H bond, weakening the bond and increasing the strength of the acid.

Number of terminal O atoms (red) increases	0		1		2		3
	H—O—Cl		H—O—Cl—O		H—O—Cl(=O)—O		H—O—Cl(=O)(—O)—O
Acid strength increases	$K_a = 2.9 \times 10^{-8}$	<	1.1×10^{-2}	<	≈ 1000	<	$\approx 10^8$

Strengths of Carboxylic Acids

As with other acids, the strength of a carboxylic acid depends on the ease with which electrons can be withdrawn from an O—H bond to facilitate the liberation of a proton (H^+). Because all carboxylic acids share the —COOH group, we must look to differences in the R groups to explain variations in acid strength.

$$R-\overset{\overset{\large O}{\|}}{C}-O-H$$

If the R group is simply a hydrocarbon chain, it has little effect on acid strength, as seen in the K_a values of the following two- and five-carbon acids.

CH_3COOH $K_a = 1.8 \times 10^{-5}$	$CH_3(CH_2)_3COOH$ $K_a = 1.5 \times 10^{-5}$
Acetic acid (Ethanoic acid)	Valeric acid (Pentanoic acid)

If the R groups contain atoms of high electronegativity, these atoms can withdraw electrons from the O—H bond, thereby weakening the bond and increasing the acid strength. In these cases, we need to consider two factors when predicting the strength of the carboxylic acid: (1) How many highly electronegative atoms are present in the R group? (2) How close are these atoms to the —COOH group? In the comparison below, the condensed structural formulas identify the electronegative substituents and indicate their numbers and distances from the —COOH group. Also, recall that Cl (EN = 3.0) is more electronegative than I (EN = 2.5).

$I-CH_2CH_2COOH$		$Cl-CH_2CH_2COOH$		$CH_3CHClCOOH$		CH_3CCl_2COOH
3-Iodopropanoic acid		3-Chloropropanoic acid		2-Chloropropanoic acid		2,2-Dichloropropanoic acid
$K_a = 8.3 \times 10^{-5}$	$<$	$K_a = 1.0 \times 10^{-4}$	$<$	$K_a = 1.4 \times 10^{-3}$	$<$	$K_a = 8.7 \times 10^{-3}$

Increasing acid strength →

EXAMPLE 15.2

Select the stronger acid in each of the following pairs.

a. nitrous acid, HNO_2, or nitric acid, HNO_3 **b.** $BrCH_2COOH$ and CCl_3COOH

SOLUTION

a. We need to replace the conventional formulas by Lewis structures to assess the numbers of terminal atoms in the two acids. From the structures in the margin, we expect HNO_3 to be the stronger acid. It has *two* terminal O atoms, compared to *one* in HNO_2. (In fact, HNO_3 is a strong acid, whereas HNO_2 is a weak acid.)

b. We should expect $BrCH_2COOH$ to be a somewhat stronger acid than CH_3COOH (acetic acid) because of the presence of the electronegative Br atom on the C atom adjacent to the —COOH group. However, CCl_3COOH, with *three* of the somewhat more electronegative Cl atoms adjacent to the C atom of the —COOH group, should be much stronger than both CH_3COOH and $BrCH_2COOH$. (The tabulated K_a values are 1.8×10^{-5}, for CH_3COOH; 1.3×10^{-3} for $BrCH_2COOH$; and 3.0×10^{-1} for CCl_3COOH.)

$$H-\ddot{\underset{..}{O}}-\ddot{N}=\underset{..}{O}:$$

Nitrous acid

$$H-\ddot{\underset{..}{O}}-\overset{\overset{\ddot{O}:}{\|}}{N}-\ddot{\underset{..}{O}}:$$

Nitric acid

EXERCISE 15.2A

Select the stronger acid in each of the following pairs.

a. H_2S or H_2Te **b.** $CH_3CH_2CH_2CHBrCOOH$ or $ClCH_2CH_2CH_2CH_2COOH$

EXERCISE 15.2B

Arrange the following acids in order of increasing strength (that is, from the acid you expect to be weakest to the acid you expect to be strongest). Explain the basis of your order.

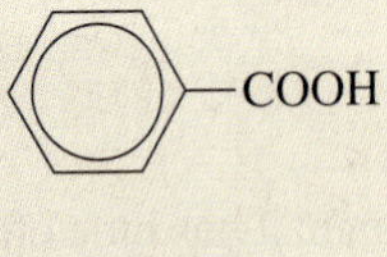

COOH (a)

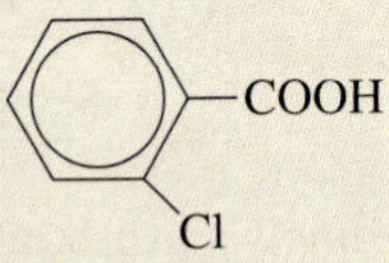

COOH, Cl (b)

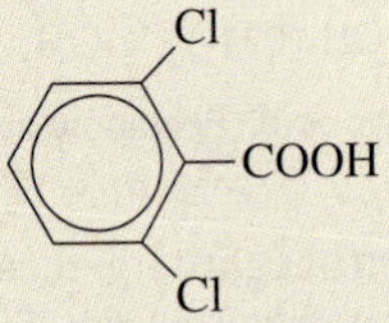

Cl, COOH, Cl (c)

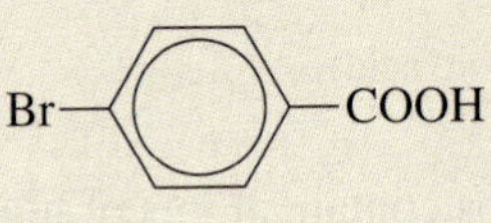

Br, COOH (d)

Strengths of Amines as Bases

Let's now take a brief look at molecular structure and base strength. We have just seen that any atom or group that withdraws electrons from the bond where a proton (H^+) is to be released makes the acid *stronger*. As we might expect, any atom or group that withdraws electrons from the atom to which H^+ bonds makes the base *weaker*. For example, due to the presence of the Br atom, which is more electronegative than an H atom, $BrNH_2$ ($K_b = 2.5 \times 10^{-8}$) is a weaker base than NH_3 ($K_b = 1.8 \times 10^{-5}$).

Aromatic amines are much weaker bases than aliphatic (nonaromatic) amines. Aniline, $C_6H_5NH_2$, for example, has a $K_b = 7.4 \times 10^{-10}$. Recall from Chapter 10 that the π electrons of the benzene ring are *delocalized* (spread out) through resonance over the entire ring. In aniline, the lone pair of electrons on the N atom are also delocalized through resonance involving structures III-V. (Structures I and II are the usual Kekulé structures for benzene rings.)

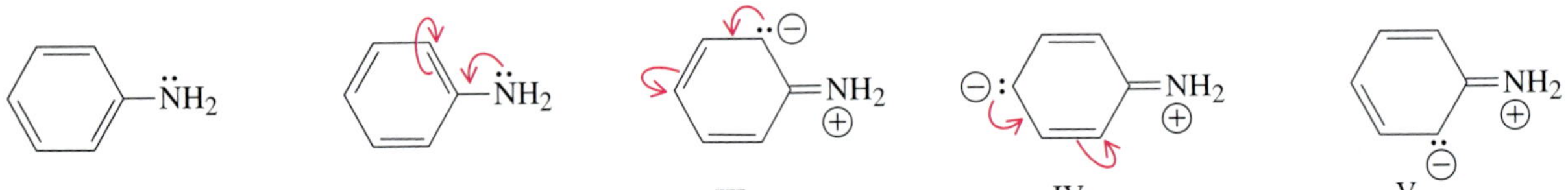

$\ddot{N}H_2$ I $\ddot{N}H_2$ II $=NH_2^{\oplus}$ III $=NH_2^{\oplus}$ IV $=NH_2^{\oplus}$ V

Electron-withdrawing groups on the ring further diminish the basicity of aromatic amines relative to aniline, as is illustrated in Example 15.3.

EXAMPLE 15.3

Select the weaker base in each of the following pairs.

a. $HOCH_2CH_2NH_2$ and $CH_3CH_2NH_2$

b. 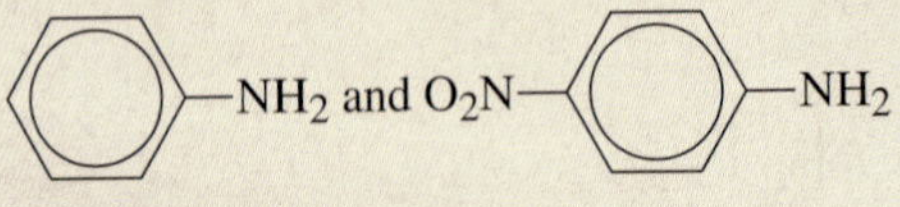 C_6H_5–NH_2 and O_2N–C_6H_4–NH_2

SOLUTION

a. Note that $HOCH_2CH_2NH_2$ has an electronegative oxygen atom on the second carbon atom from the $-NH_2$ group. Thus, we expect $HOCH_2CH_2NH_2$ to be a somewhat weaker base than $CH_3CH_2NH_2$ because of the presence of the oxygen atom. (The tabulated K_b values are 2.5×10^{-5} for $HOCH_2CH_2NH_2$ and 4.3×10^{-4} for $CH_3CH_2NH_2$.)

b. The NO_2 group, with three electronegative atoms, withdraws electrons from the NH_2 group, making $p\text{-}NO_2C_6H_4NH_2$ less likely to attract an H^+ ion and therefore a weaker base than $C_6H_5NH_2$.

EXERCISE 15.3

Arrange the following bases in order of increasing strengths (that is, from the base you expect to be weakest to the base you expect to be strongest). Explain the basis of your order.

a. Cl—C_6H_4—NH_2 (para) **b.** $CH_3(CH_2)_3NH_2$

c. $CH_3CHCl(CH_2)_2NH_2$ **d.** Cl—C_6H_3(Cl)—NH_2

15.3 Self-Ionization of Water—The pH Scale

Even the purest water conducts electricity, though it takes exceptionally sensitive measurements to detect this. Electrical conductivity requires the presence of ions, but where do they come from in pure water? The Brønsted-Lowry theory helps us to see how they form. Recall that water is *amphiprotic*. So, to a slight extent, water molecules can transfer protons among themselves. In the self-ionization of water, for every H_2O molecule that loses a proton, another H_2O molecule gains it. In the equation below, acid(2)—H_3O^+—is much stronger than acid(1)—H_2O. Also, base(1)—OH^-—is much stronger than base(2)—H_2O. As a result, the *reverse* reaction is strongly favored, and the equilibrium state lies *far to the left*, as suggested by the unequal arrow lengths.

$$\underset{\text{Acid(1)}}{H_2O} + \underset{\text{Base(2)}}{H_2O} \rightleftharpoons \underset{\text{Acid(2)}}{H_3O^+} + \underset{\text{Base(1)}}{OH^-}$$

We can write the equilibrium constant expression for the self-ionization of water in the usual manner. We call the equilibrium constant the **ion product of water** and represent it by the symbol K_w.

$$2\,H_2O(l) \rightleftharpoons H_3O^+(aq) + OH^-(aq)$$

$$K_w = [H_3O^+][OH^-]$$

At 25 °C, the experimentally determined equilibrium concentrations in pure water are as follows.

$$[H_3O^+] = [OH^-] = 1.0 \times 10^{-7}\,M$$

We can then calculate the value of K_w.

Like other equilibrium constants, the value of K_w depends on temperature.

$$K_w = [H_3O^+][OH^-] = (1.0 \times 10^{-7})(1.0 \times 10^{-7}) = 1.0 \times 10^{-14} \text{ (at 25 °C)}$$

The equilibrium constant K_w is extremely important because it applies to *all aqueous solutions*—that is, to solutions of acids, bases, salts, or nonelectrolytes—not just to pure water. Consider, for example, a solution that is 0.00015 M HCl. Because HCl is a strong acid, its ionization is complete.

$$HCl + H_2O \longrightarrow H_3O^+ + Cl^-$$

And because the ionization is complete, it produces a concentration of H_3O^+ of 0.00015 M. This is over 1000 times greater than the 1×10^{-7} mol H_3O^+ per liter found in pure water, and we can state the value of $[H_3O^+]$ in 0.00015 M HCl as follows.

$$[H_3O^+] = 0.00015 \text{ M} = 1.5 \times 10^{-4} \text{ M}$$

With the K_w expression, we can now calculate $[OH^-]$ in the solution.

$$[OH^-] = \frac{K_w}{[H_3O^+]} = \frac{1.0 \times 10^{-14}}{1.5 \times 10^{-4}} = 6.7 \times 10^{-11} \text{ M}$$

pH and pOH

In 1909, the Danish biochemist Søren Sørenson proposed a convenient convention that is still used today. He let the term pH refer to "the power of the hydrogen ion." By this, he meant the hydrogen ion concentration expressed as a negative power of ten: $[H^+] = 10^{-pH}$. Using $[H_3O^+]$ rather than $[H^+]$ and a logarithm (log) in place of a power of ten, we define the **pH** of a solution as the *negative* of the logarithm of $[H_3O^+]$.

PROBLEM-SOLVING NOTE
Note that pH is defined in terms of the common logarithm, "log" (the base 10 logarithm), and *not* the natural logarithm, "ln" (the base *e* logarithm). (Also, see Appendix A.)

$$pH = -\log [H_3O^+]$$

To determine the pH of 0.00015 M HCl, we first determine $[H_3O^+]$ in the solution. As we saw earlier, $[H_3O^+] = 1.5 \times 10^{-4}$ M.

$$pH = -\log [H_3O^+] = -\log (1.5 \times 10^{-4}) = -[\log 1.5 + \log 10^{-4}]$$
$$= -[0.18 - 4.00] = -[-3.82] = 3.82$$

PROBLEM-SOLVING NOTE
Note that we define pH as the *negative* of log $[H_3O^+]$ in order to make a typical pH value a *positive* number.

The number 1.5×10^{-4} has *two* significant figures; the exponent 4 is simply the decimal-point locator in the decimal equivalent: 0.00015. Similarly, in log $1.5 \times 10^{-4} = 3.82$, the digit 3 does not enter into the assessment of significant figures; only the *two* digits following the decimal point are significant.

It is equally important to be able to calculate $[H_3O^+]$ corresponding to a given pH. Such problems require an inverse calculation. We can calculate $[H_3O^+]$ in a solution with pH = 2.19 as follows.

$$-\log [H_3O^+] = pH = 2.19$$
$$\log [H_3O^+] = -2.19$$
$$[H_3O^+] = \text{antilog} (-2.19) = 10^{-2.19} = 6.5 \times 10^{-3}$$

We can express the concentration of any ion in solution with a logarithmic expression similar to that used for pH. In particular, we can define **pOH** as follows.

$$\text{pOH} = -\log [\text{OH}^-]$$

A 2.5×10^{-3} M NaOH solution has $[\text{OH}^-] = 2.5 \times 10^{-3}$ M and

$$\text{pOH} = -\log (2.5 \times 10^{-3}) = -(-2.60) = 2.60$$

We can also use logarithmic expressions to replace exponential numbers in equilibrium constants. Thus, we can define $\mathbf{p}\boldsymbol{K}_\mathbf{w}$ as the negative logarithm of K_w. At 25 °C, $\text{p}K_w = 14.00$. We introduce $\text{p}K_w$ in part in order to derive a simple relationship between the pH and pOH of a solution.

$$K_w = [\text{H}_3\text{O}^+][\text{OH}^-] = 1.0 \times 10^{-14}$$

$$-\log K_w = -\log ([\text{H}_3\text{O}^+][\text{OH}^-]) = -\log (1.0 \times 10^{-14})$$

$$\text{p}K_w = -\log [\text{H}_3\text{O}^+] - \log [\text{OH}^-] = 14.00$$

$$\text{p}K_w = \text{pH} + \text{pOH} = 14.00$$

If we have a value of either pH or pOH in a solution, we can calculate the value of the other. Recall that we found the pOH of 2.5×10^{-3} M NaOH to be 2.60. We can readily calculate the pH of this solution as follows.

$$\text{pH} = \text{p}K_w - \text{pOH} = 14.00 - \text{pOH} = 14.00 - 2.60 = 11.40$$

In pure water, the concentrations of H_3O^+ and OH^- are equal: $[\text{H}_3\text{O}^+] = [\text{OH}^-] = 1.0 \times 10^{-7}$ M at 25°C. Thus, pH and pOH are both 7.00. Pure water and all aqueous solutions with pH = 7.00 at 25°C are *neutral*. If the pH is less than 7.00, a solution is *acidic*; if the pH is above 7.00, a solution is *basic*, or *alkaline*. As a solution becomes more acidic, $[\text{H}_3\text{O}^+]$ increases and pH decreases. As a solution becomes more basic, $[\text{H}_3\text{O}^+]$ decreases and pH increases. These ideas are summarized in the diagram below.

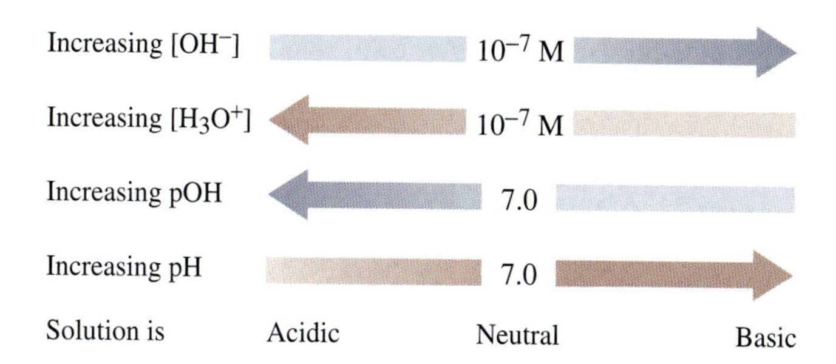

Figure 15.2 gives the pH values of a number of familiar materials. In viewing Figure 15.2, keep in mind that because pH is on a logarithmic scale, every unit change in pH represents a *tenfold* change in $[\text{H}_3\text{O}^+]$. Thus, lemon juice (pH $\approx$ 2.3) is approximately ten times as acidic as orange juice (pH $\approx$ 3.5) and about 100 times more acidic than tomato juice (pH $\approx$ 4.5).

EXAMPLE 15.4

By the method suggested in Figure 15.3, a student determines the pH of milk of magnesia, a suspension of solid magnesium hydroxide in its saturated aqueous solution, and obtains a value of 10.52. What is the molarity of $Mg(OH)_2$ in its saturated aqueous solution? The suspended, undissolved $Mg(OH)_2(s)$ does not affect the measurement.

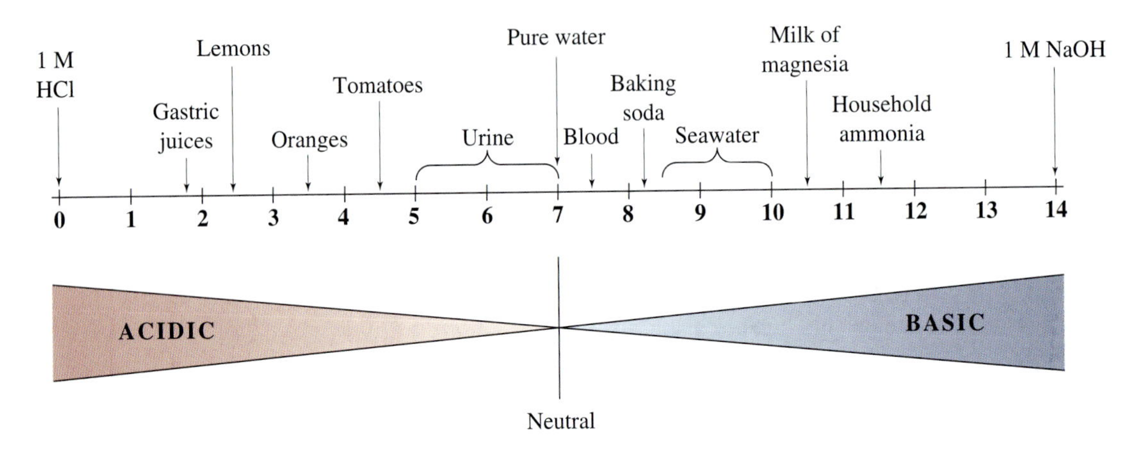

▲ **Figure 15.2 The pH scale**
The pH values of common substances range from about 0 to 14. Negative numbers are occasionally encountered ($[H_3O^+] = 10$ M corresponds to pH $= -1$), as are values somewhat greater than 14 ($[OH^-] = 10$ M corresponds to pH $= 15$). However, accurate pH measurements are difficult to make below about pH 1. They are also difficult to make above about pH 13.

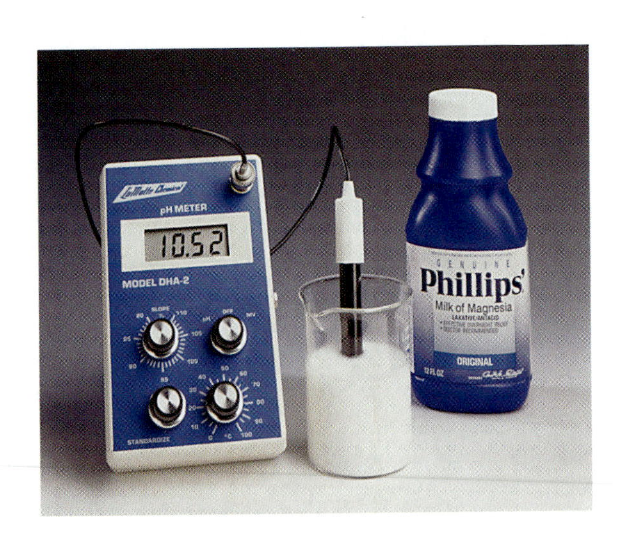

▶ **Figure 15.3 Measurement of pH with a pH meter—Example 15.3 illustrated**
No modern laboratory is complete without an electrical measuring instrument known as a pH meter.

SOLUTION

We will assume that the dissolved magnesium hydroxide is a strong base and therefore completely dissociated into ions.

$$Mg(OH)_2(aq) \longrightarrow Mg^{2+}(aq) + 2\,OH^-(aq)$$

We have a simple way of obtaining $[OH^-]$ from the pOH; and, in turn, we can get the pOH from the pH.

$$pOH = 14.00 - pH = 14.00 - 10.52 = 3.48$$

$$\log [OH^-] = -3.48$$

$$[OH^-] = 10^{-3.48} = 3.3 \times 10^{-4}\ M$$

Now we need only to relate the molarity of $Mg(OH)_2$ to the hydroxide ion concentration.

$$\text{Molarity} = \frac{3.3 \times 10^{-4}\ \cancel{\text{mol OH}}}{1\ \text{L}} \times \frac{1\ \text{mol Mg(OH)}_2}{2\ \cancel{\text{mol OH}}}$$

$$= \frac{1.7 \times 10^{-4}\ \text{mol Mg(OH)}_2}{1\ \text{L}} = 1.7 \times 10^{-4}\ \text{M Mg(OH)}_2$$

EXERCISE 15.4A

What is the pH of a solution prepared by dissolving 0.0105 mol HNO_3 in 225 L of water?

EXERCISE 15.4B

What is the pH of a solution containing 2.65 g $Ba(OH)_2$ in 735 mL of aqueous solution? Assume that the $Ba(OH)_2$ is completely dissociated.

EXAMPLE 15.5—A Conceptual Example

Is the solution 1.0×10^{-8} M HCl acidic, basic, or neutral?

SOLUTION

Because the solute is a strong acid, our first thought is that the solution should be acidic. However, let's check our assumption by calculating the pH.

$$HCl + H_2O \longrightarrow H_3O^+ + Cl^-$$

$$[H_3O^+] = 1.0 \times 10^{-8}\ \text{M} \quad \text{and} \quad \text{pH} = -\log [H_3O^+] = -\log (1.0 \times 10^{-8}) = 8.00$$

Because the pH is greater than 7.00, the solution appears to be basic.

The quandary here is that the HCl(aq) is so dilute that the self-ionization of water produces more H_3O^+ than does the strong acid. Still, the total $[H_3O^+]$ from the *two* sources is somewhat greater than 1.0×10^{-7} M. The pH is somewhat less than 7.00; the solution is *acidic*.

We can generally ignore the slight self-ionization of water if other ionization processes predominate. Usually, however, we cannot ignore it if the pH is within one unit or so of pH = 7.

PROBLEM-SOLVING NOTE

An exact calculation of the pH is the subject of Problem 111.

EXERCISE 15.5

Is a solution that is 1.0×10^{-8} M NaOH acidic, basic, or neutral? Explain.

15.4 Equilibrium in Solutions of Weak Acids and Weak Bases

To calculate the pH value for a solution of a weak acid or a weak base, we first must determine $[H_3O^+]$ or $[OH^-]$, using an equilibrium calculation based on the acid or base ionization constant, K_a and K_b. Table 15.2 is a brief listing of K_a and K_b values, and Appendix C includes a more extensive listing. Table 15.2 also includes listings of pK_a and pK_b, which are logarithmic expressions of ionization constants.

$$pK_a = -\log K_a \qquad \text{and} \qquad pK_b = -\log K_b$$

TABLE 15.2 Ionization Constants of Some Weak Acids and Weak Bases in Water at 25 °C

	Ionization Equilibrium	Ionization Constant, K	pK
Inorganic Acids		$K_a =$	
Chlorous acid	$HClO_2 + H_2O \rightleftharpoons H_3O^+ + ClO_2^-$	1.1×10^{-2}	1.96
Nitrous acid	$HNO_2 + H_2O \rightleftharpoons H_3O^+ + NO_2^-$	7.2×10^{-4}	3.14
Hydrofluoric acid	$HF + H_2O \rightleftharpoons H_3O^+ + F^-$	6.6×10^{-4}	3.18
Hypochlorous acid	$HOCl + H_2O \rightleftharpoons H_3O^+ + OCl^-$	2.9×10^{-8}	7.54
Hypobromous acid	$HOBr + H_2O \rightleftharpoons H_3O^+ + OBr^-$	2.5×10^{-9}	8.60
Hydrocyanic acid	$HCN + H_2O \rightleftharpoons H_3O^+ + CN^-$	6.2×10^{-10}	9.21
Carboxylic Acids		$K_a =$	
Chloroacetic acid	$CH_2ClCOOH + H_2O \rightleftharpoons H_3O^+ + CH_2ClCOO^-$	1.4×10^{-3}	2.85
Formic acid	$HCOOH + H_2O \rightleftharpoons H_3O^+ + HCOO^-$	1.8×10^{-4}	3.74
Benzoic acid	$C_6H_5COOH + H_2O \rightleftharpoons H_3O^+ + C_6H_5COO^-$	6.3×10^{-5}	4.20
Acetic acid	$CH_3COOH + H_2O \rightleftharpoons H_3O^+ + CH_3COO^-$	1.8×10^{-5}	4.74
Inorganic Bases		$K_b =$	
Ammonia	$NH_3 + H_2O \rightleftharpoons NH_4^+ + OH^-$	1.8×10^{-5}	4.74
Hydrazine	$H_2NNH_2 + H_2O \rightleftharpoons H_2NNH_3^+ + OH^-$	8.5×10^{-7}	6.07
Hydroxylamine	$HONH_2 + H_2O \rightleftharpoons HONH_3^+ + OH^-$	9.1×10^{-9}	8.04
Amines		$K_b =$	
Dimethylamine	$(CH_3)_2NH + H_2O \rightleftharpoons (CH_3)_2NH_2^+ + OH^-$	5.9×10^{-4}	3.23
Ethylamine	$CH_3CH_2NH_2 + H_2O \rightleftharpoons CH_3CH_2NH_3^+ + OH^-$	4.3×10^{-4}	3.37
Methylamine	$CH_3NH_2 + H_2O \rightleftharpoons CH_3NH_3^+ + OH^-$	4.2×10^{-4}	3.38
Pyridine	$C_5H_5N + H_2O \rightleftharpoons C_5H_5NH^+ + OH^-$	1.5×10^{-9}	8.82
Aniline	$C_6H_5NH_2 + H_2O \rightleftharpoons C_6H_5NH_3^+ + OH^-$	7.4×10^{-10}	9.13

Small values of pK_a and pK_b correspond to large values of K_a and K_b, just as a low value of pH corresponds to a high $[H_3O^+]$. We arranged the entries in Table 15.2 by increasing value of pK_a or pK_b, that is, in the order of decreasing acid strength and base strength.

Organic Bases

As we noted in Section 2. 10, amines are compounds in which one or more of the H atoms in NH_3 is replaced by a hydrocarbon group, R. Amines are bases, and water-soluble amines can accept a proton from water. For example, trimethylamine, $(CH_3)_3N$, reacts with water as follows.

$$(CH_3)_3N(aq) + H_2O(l) \rightleftharpoons [(CH_3)_3N{-}H]^+(aq) + OH^-(aq) \qquad K_b = 6.3 \times 10^{-5}$$

Nearly all amines, including those that are not very soluble in water, react with strong acids to form water-soluble salts. For example, only 3.5 g of aniline ($C_6H_5NH_2$) dissolves in 100 mL of water at 25 °C. However, when aniline reacts with hydrochloric acid, it forms the salt commonly called aniline hydrochloride. The salt is soluble to the extent of about 100 g per 100 mL of water.

$$\underset{\text{Aniline}}{C_6H_5NH_2(l)} + HCl(aq) \rightleftharpoons \underbrace{C_6H_5NH_3^+(aq) + Cl^-(aq)}_{\text{Aniline hydrochloride}}$$

Many medicines, both synthetic and naturally occurring, contain the amine functional group. *Alkaloids* are basic nitrogen-containing compounds produced by plants. Familiar alkaloids include caffeine, cocaine, morphine, nicotine, and strychnine (Figure 15.4). Many alkaloids have important physiological effects: They are medicines, poisons, and drugs of abuse.

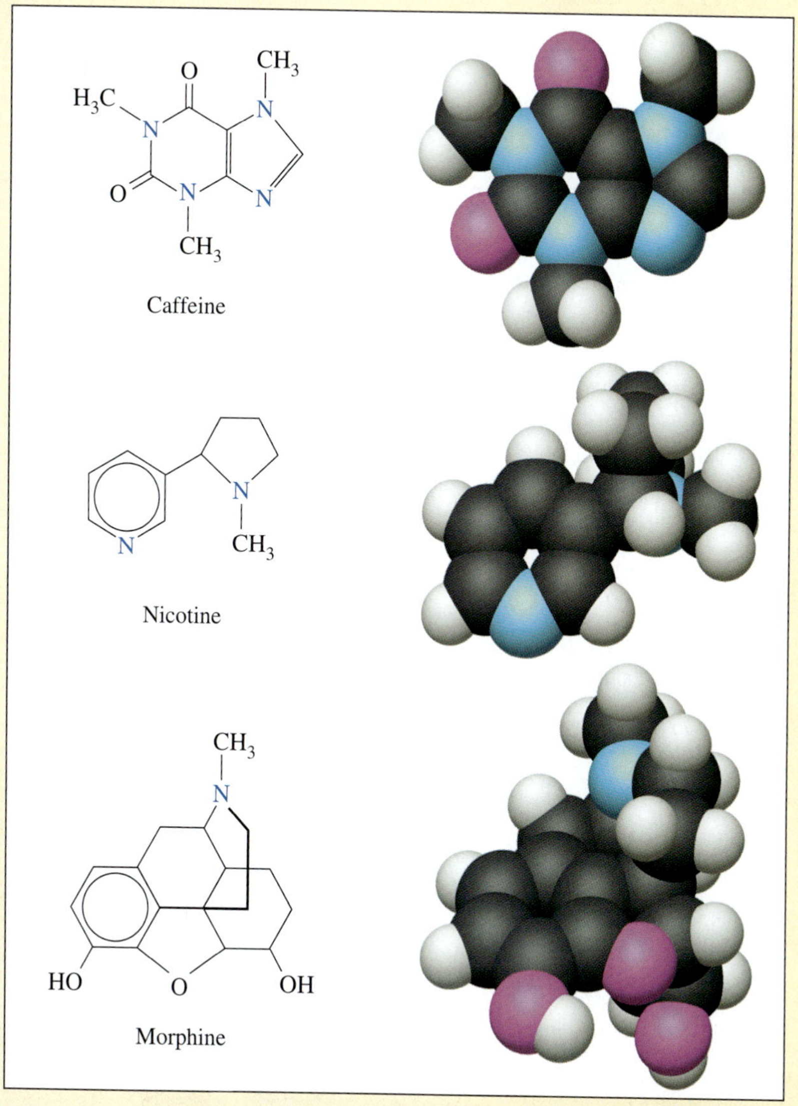

Caffeine

Nicotine

Morphine

Figure 15.4
Some common alkaloids

Medicines that are amines are often converted to salts to enhance their water solubility. For instance, procaine (Figure 15.5) is soluble only to the extent of 0.5 g in 100 g of water. In contrast, its hydrochloride salt is soluble to the remarkable degree of 100 g in 100 g of water. Procaine hydrochloride, often called by the trade name Novocain®, is a well-known local anesthetic.

We also make use of the chemistry of amines when we put lemon juice on fish. The unpleasant fishy odor is due to amines. The citric acid in the lemon juice converts the amines to nonvolatile salts, thus reducing the odor.

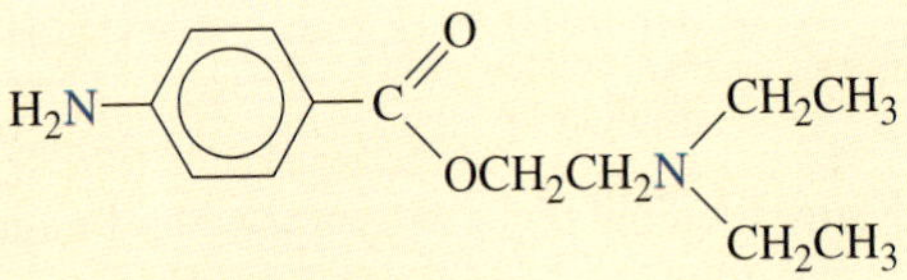

$H_2N-C_6H_4-C(=O)-OCH_2CH_2N(CH_2CH_3)_2$

Procaine

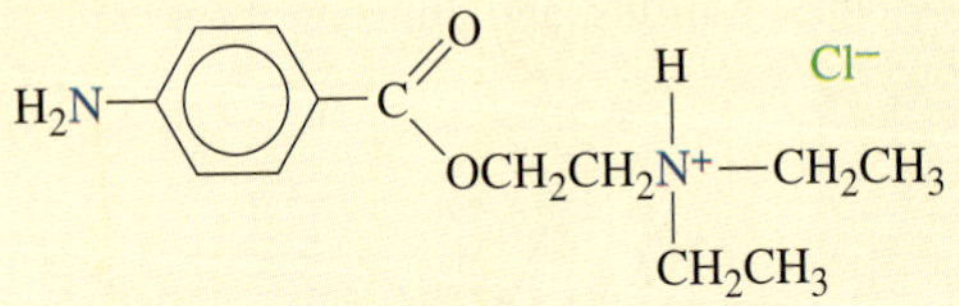

$H_2N-C_6H_4-C(=O)-OCH_2CH_2N^{+}H(CH_2CH_3)_2\ \ Cl^-$

Procaine hydrochloride (Novocain®)

Figure 15.5
An amine and its salt

Some Acid-Base Equilibrium Calculations

The equilibrium calculations we do in this chapter are much like those of the preceding chapter, and much of what we did there applies here as well. We write an equation for the reversible reaction, organize data under the equation, assess the changes that occur in establishing equilibrium, and then calculate the equilibrium concentrations.

EXAMPLE 15.6

Ordinary vinegar is approximately 1 M CH_3COOH, and as shown in Figure 15.6, it has a pH of about 2.4. Calculate the expected pH of 1.00 M $CH_3COOH(aq)$, and show that the calculated and measured pH values are in good agreement.

Figure 15.6
Demonstrating that acetic acid is a weak acid: Measuring the pH of vinegar
The experimentally determined pH of vinegar—an aqueous solution of acetic acid—is considerably higher than that of a strong acid of the same molarity (about 1 M).

SOLUTION

To calculate the pH of a solution, we must first calculate $[H_3O^+]$. To calculate $[H_3O^+]$, we need to identify the sources of H_3O^+. In 1.00 M $CH_3COOH(aq)$, H_3O^+ comes from

(1) the ionization of CH_3COOH and (2) the self-ionization of H_2O. We can assess the relative importance of these two sources by comparing K_a of acetic acid and K_w of water.

$$CH_3COOH + H_2O \rightleftharpoons H_3O^+ + CH_3COO^- \quad K_a = 1.8 \times 10^{-5}$$

$$H_2O + H_2O \rightleftharpoons H_3O^+ + OH^- \quad K_w = 1.0 \times 10^{-14}$$

Because K_w is so much smaller than K_a of acetic acid, the self-ionization of water is a negligible source of H_3O^+, and we can neglect it in the equilibrium calculation. We need to consider only the ionization of acetic acid. The initial concentrations listed below are those before the ionization of acetic acid begins. There is no acetate ion in solution, but there is a trace of H_3O^+ from the water itself. So we write $[CH_3COO^-] = 0$, but $[H_3O^+] \approx 0$.

PROBLEM-SOLVING NOTE

The symbol $\approx$ means "approximately equal to."

The reaction:	CH_3COOH + H_2O $\rightleftharpoons$	H_3O^+ +	CH_3COO^-
Initial concentrations, M:	1.00	≈ 0	0
Changes, M:	$-x$	$+x$	$+x$
Equilibrium concentrations, M:	$(1.00 - x)$	x	x

$$K_a = \frac{[H_3O^+][CH_3COO^-]}{[CH_3COOH]} = \frac{x \cdot x}{1.00 - x} = \frac{x^2}{1.00 - x} = 1.8 \times 10^{-5}$$

Note that this is a quadratic equation; the highest power is x^2. We can rearrange the equation to the familiar form shown here.

$$x^2 + 1.8 \times 10^{-5}x - 1.8 \times 10^{-5} = 0$$

Then we can use the quadratic formula to solve for x, as we did several times in Chapter 14. However, the equation is much simpler to solve if we make the following assumption, *provided the assumption is valid.*

Assume that x is *much* smaller than 1.00, that is, that $x \ll 1$. Then, replace the term $(1.00 - x)$ by 1.00. The equation we must solve becomes

PROBLEM-SOLVING NOTE

The symbol $\ll$ means "much smaller than."

$$\frac{x^2}{1.00 - x} = \frac{x^2}{1.00} = 1.8 \times 10^{-5}$$

$$x^2 = 1.8 \times 10^{-5}$$

$$x = [H_3O^+] = (1.8 \times 10^{-5})^{1/2} = 4.2 \times 10^{-3}\ M$$

$$pH = -\log\ [H_3O^+] = -\log\ (4.2 \times 10^{-3}) = 2.38$$

When we make an assumption, we should always test its validity. Here we can do so by using the calculated value of x to see if the value of $(1.00 - x)$ is indeed very nearly equal to 1.00. We find that $1.00 - x = 1.00 - 4.2 \times 10^{-3} = 1.00 - 0.0042 = 0.9958 = 1.00$ (to two decimal places). Our assumption is valid for the precision allowed by the calculation—two decimal places.

The agreement between the calculated and experimentally determined pH is remarkably good—both are pH = 2.38. This excellent agreement is partly fortuitous because the acetic acid concentration in commercial vinegar is somewhat variable.

EXERCISE 15.6A

Determine the pH of 0.250 M CH_3CH_2COOH (propionic acid). Obtain the K_a value from Appendix C.

EXERCISE 15.6B

The solubility of *o*-nitrophenol in water is 2.1 g/L. What is the pH of this saturated aqueous solution?

$$C_6H_4(NO_2)OH + H_2O \rightleftharpoons H_3O^+ + C_6H_4(NO_2)O^- \qquad K_a = 6.0 \times 10^{-8}$$

PROBLEM-SOLVING NOTE

Review the discussion of aromatic compounds (page 451) if you need help in interpreting the formula representing *o*-nitrophenol.

The solution we considered in Example 15.6 is 1.00 M $CH_3COOH(aq)$. Let's call 1.00 M the *nominal* molarity of the solution. It describes how to prepare the solution—dissolve 1.00 mole of pure CH_3COOH (60.05 g) in enough water to make one liter of solution. Once the acetic acid is in solution, some of it ionizes, so that concentration of CH_3COOH molecules, that is, $[CH_3COOH]$ is no longer 1.00 M. Instead, $[CH_3COOH] = (1.00 - x)$ M, where x is equal to $[H_3O^+] = [CH_3COO^-]$. Recall that we made the assumption that $x \ll 1$, so that $(1.00 - x) \approx 1.00$. This assumption implies that so little of the CH_3COOH ionizes that the concentration of nonionized acetic acid $[CH_3COOH]$ is equal to the nominal concentration: 1.00 M.

In general, this type of assumption is acceptable if the following condition is met.

$$x \text{ is less than } 5\% \text{ of } M_{acid}$$

In this statement, $x = [H_3O^+]$ and M_{acid} is the nominal molarity of the acid. We can see that this was the case in Example 15.6.

$$\frac{x}{M_{acid}} \times 100\% = \frac{4.2 \times 10^{-3}\ \cancel{M}}{1.00\ \cancel{M}} \times 100\% = 0.42\%$$

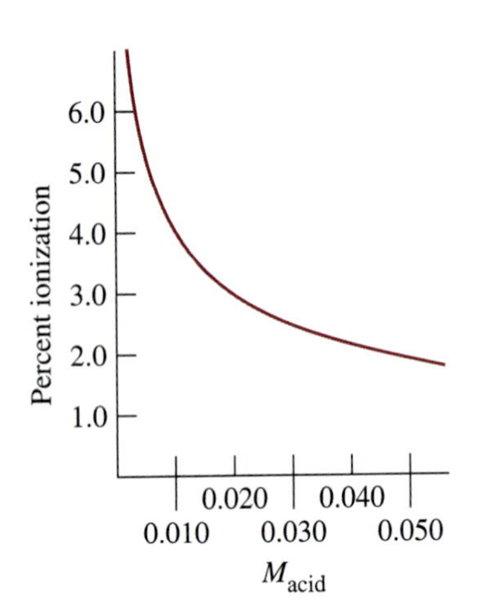

▲ **Figure 15.7**
The effect of molarity on the percent ionization of acetic acid

Whether the "5% rule" is met depends on both the nominal molarity of the acid and the value of K_a. We can generally expect it to be met if the following is true.

$$\text{The ratio } M_{acid}/K_a \text{ is greater than } 100$$

Figure 15.7 suggests that acetic acid follows the rule if M_{acid} is greater than about 0.005 M.

EXAMPLE 15.7

What is the pH of 0.00200 M $CH_2ClCOOH(aq)$?

SOLUTION

As in Example 15.6, we can tabulate data as follows

The reaction:	$CH_2ClCOOH + H_2O \rightleftharpoons$	H_3O^+ +	CH_2ClCOO^-
Initial concentrations, M:	0.00200	≈ 0	0
Changes, M:	$-x$	$+x$	$+x$
Equilibrium concentrations, M:	$(0.00200 - x)$	x	x

We find the acid ionization constant for chloroacetic acid, $CH_2ClCOOH$, in Table 15.2 and use it in the setup below.

$$K_a = \frac{[H_3O^+][CH_2ClCOO^-]}{[CH_2ClCOOH]}$$

$$1.4 \times 10^{-3} = \frac{x \cdot x}{0.00200 - x} = \frac{x^2}{0.00200 - x}$$

At this point, we could try the assumption that $x \ll 0.00200$ and that $(0.00200 - x) \approx 0.00200$. Then we could complete the solution as in Example 15.6. In this case, however, let's first evaluate the ratio M_{acid}/K_a.

$$\frac{M_{acid}}{K_a} = \frac{0.00200}{1.4 \times 10^{-3}} = 1.4$$

Because this ratio is much less than 100, we expect the assumption to fail. Therefore, we must use the quadratic formula to solve the equation below.

$$x^2 = (0.00200 - x) \times 1.4 \times 10^{-3}$$

$$x^2 + 1.4 \times 10^{-3}x - 2.8 \times 10^{-6} = 0$$

$$x = \frac{-1.4 \times 10^{-3} \pm \sqrt{(1.4 \times 10^{-3})^2 - 4 \times 1 \times (-2.8 \times 10^{-6})}}{2}$$

$$x = \frac{-1.4 \times 10^{-3} \pm \sqrt{1.3 \times 10^{-5}}}{2}$$

$$x = [H_3O^+] = \frac{-1.4 \times 10^{-3} \pm 3.6 \times 10^{-3}}{2} = \frac{2.2 \times 10^{-3}}{2} = 1.1 \times 10^{-3}$$

Thus, $x = [H_3O^+] = 1.1 \times 10^{-3}$ M. We can now determine the pH.

$$pH = -\log [H_3O^+] = -\log (1.1 \times 10^{-3}) = 2.96$$

PROBLEM-SOLVING NOTE

Note that using the simplifying assumption that $x \ll 0.00200$ M and $0.00200 \text{ M} - x = 0.00200$ M would have introduced a serious error because $0.00200 \text{ M} - 0.0011 \text{ M} = 0.0009$ M. Also, as with other equilibrium calculations, we can check our answer by substituting $x = 1.1 \times 10^{-3}$ into the expression, $x^2/(0.00200 - x)$. We get 1.3×10^{-3}, compared to $K_a = 1.4 \times 10^{-3}$.

EXERCISE 15.7A

Repeat the equilibrium calculation in Example 15.7 by making the assumption that $(0.00200 - x) \approx 0.00200$ and show that the assumption fails.

EXERCISE 15.7B

Calculate the pH of 0.0100 M chlorous acid. Obtain K_a from Table 15.2.

We can use the methods applied in Examples 15.6 and 15.7 to calculate pH values of solutions of weak bases if we make the following adjustments:

- Use a K_b expression to calculate $[OH^-]$.
- Calculate pOH from $[OH^-]$.
- Calculate pH by using the expression: pH + pOH = 14.00.
- Use the following as the criterion governing simplifying assumptions in the solution of a quadratic equation: M_{base}/K_b must be greater than 100.

EXAMPLE 15.8

What is the pH of 0.500 M $NH_3(aq)$?

SOLUTION

Following the steps outlined above, we first calculate $[OH^-]$ in this solution. Note that the self-ionization of water makes a negligible contribution to $[OH^-]$, because K_w is so much smaller than K_b. We can therefore set up the relevant data as follows.

The reaction:	$NH_3 + H_2O \rightleftharpoons$		NH_4^+ +	OH^-
Initial concentrations, M:	0.500		0	≈ 0
Changes, M:	$-x$		$+x$	$+x$
Equilibrium concentrations, M:	$(0.500 - x)$		x	x

We find the base ionization constant for ammonia in Table 15.2 and use it in the setup below.

$$K_b = \frac{[NH_4^+][OH^-]}{[NH_3]}$$

$$1.8 \times 10^{-5} = \frac{x \cdot x}{0.500 - x} = \frac{x^2}{0.500 - x}$$

To determine whether we can make the assumption that $x \ll 0.500$, we can evaluate the following ratio.

$$\frac{M_{base}}{K_b} = \frac{0.500}{1.8 \times 10^{-5}} = 2.8 \times 10^4$$

Because this ratio is much greater than 100, the assumption should be valid. We can solve the simplified equation.

$$\frac{x^2}{0.500 - x} \approx \frac{x^2}{0.500} = 1.8 \times 10^{-5}$$

$$x^2 = 0.500 \times 1.8 \times 10^{-5} = 9.0 \times 10^{-6}$$

$$x = [OH^-] = (9.0 \times 10^{-6})^{\frac{1}{2}} = 3.0 \times 10^{-3}\ M$$

PROBLEM-SOLVING NOTE

Can you see at this point that our assumption is justified? The quantity $0.500 - x = 0.500 - 0.0030 = 0.497$.

From this $[OH^-]$ we can calculate the pOH.

$$pOH = -\log [OH^-] = -\log (3.0 \times 10^{-3}) = 2.52$$

Finally, we can calculate the pH.

$$pH + pOH = 14.00$$

$$pH = 14.00 - pOH = 14.00 - 2.52 = 11.48$$

EXERCISE 15.8A

What is the pH of a solution that is 0.200 M in methylamine, CH_3NH_2? Obtain a value of K_b from Table 15.2.

EXERCISE 15.8B

What is the pH of the solution obtained by diluting 5.00 mL of 0.0100 M $NH_3(aq)$ to 1.000 L?

In Chapter 14, we learned how to calculate an equilibrium constant from experimental data. The K_a of a weak acid or K_b of a weak base can be established in several ways. Perhaps the simplest way is measure the pH and calculate $[H_3O^+]$ or $[OH^-]$ from the pH value. In Example 15.9, we determine K_b of dimethylamine from the pH of an aqueous solution.

EXAMPLE 15.9

The pH of a 0.164 M aqueous solution of dimethylamine is found to be 11.98. What are the values of K_b and pK_b?

$$\underset{\text{Dimethylamine}}{(CH_3)_2NH} + H_2O \rightleftharpoons \underset{\text{Dimethylammonium ion}}{[(CH_3)_2NH_2]^+} + OH^- \qquad K_b = ?$$

SOLUTION

Let's follow the procedure that we first illustrated in Example 14.11 (page 617). That is, let's start with this preliminary setup.

The reaction:	$(CH_3)_2NH + H_2O \rightleftharpoons$	$[(CH_3)_2NH_2]^+ +$	OH^-
Initial concentrations, M:	0.164	0	≈0
Changes, M:	?	?	?
Equilibrium concentrations, M:	?	?	?

In Example 14.11, the *equilibrium* concentration of one of the products was given. Here, we are given an equilibrium concentration indirectly. We can get $[OH^-]$ (marked ? above) from the pH of the solution.

$$pOH = 14.00 - pH = 14.00 - 11.98 = 2.02$$

$$-\log [OH^-] = pOH = 2.02$$

$$\log [OH^-] = -2.02$$

$$[OH^-] = 10^{-2.02} = 9.5 \times 10^{-3}\ M$$

Now let's go back and substitute a numerical value for each of the blank items in the tabulated data.

The reaction:	$(CH_3)_2NH + H_2O \rightleftharpoons$	$[(CH_3)_2NH_2]^+ +$	OH^-
Initial concentrations, M:	0.164	0	≈0
Changes, M:	-9.5×10^{-3}	$+9.5 \times 10^{-3}$	$+9.5 \times 10^{-3}$
Equilibrium concentrations, M:	$(0.164 - 9.5 \times 10^{-3})$	9.5×10^{-3}	9.5×10^{-3}

Finally, we substitute the equilibrium concentrations into the K_b expression and solve for K_b. Notice that no assumptions are needed here; we know each of the concentration terms in the K_b expression.

$$K_b = \frac{[(CH_3)_2NH_2^+][OH^-]}{[(CH_3)_2NH]} = \frac{(9.5 \times 10^{-3})(9.5 \times 10^{-3})}{(0.164 - 0.0095)} = 5.8 \times 10^{-4}$$

$$pK_b = -\log K_b = -\log (5.8 \times 10^{-4}) = 3.24$$

EXERCISE 15.9A

Suppose you discovered a new acid, HZ, and found that the pH of a 0.0100 M solution is 3.12. What are K_a and pK_a for HZ?

$$HZ + H_2O \rightleftharpoons H_3O^+ + Z^- \qquad K_a = ?$$

EXERCISE 15.9B

What is the molarity of $NH_3(aq)$ if the solution has the same pH as 0.200 M $(CH_3)_2NH(aq)$? Use data from Table 15.2.

EXAMPLE 15.10—A Conceptual Example

Without doing detailed calculations, indicate which solution has the greater $[H_3O^+]$, 0.030 M HCl or 0.050 M CH_3COOH.

SOLUTION

HCl(aq) is a strong acid. It is essentially 100% ionized, and in 0.030 M HCl, $[H_3O^+] =$ 0.030 M. Acetic acid, CH_3COOH, is a weak acid. The 0.050 M CH_3COOH would have to be more than 50% ionized to have $[H_3O^+]$ equal to that in 0.030 M HCl. On page 652 and in Figure 15.7, we saw that a solution as dilute as 0.005 M CH_3COOH is only about 5% ionized; the percent ionization in 0.050 M CH_3COOH is much smaller still. Also, just by reflecting on the magnitude of K_a for acetic acid (1.8×10^{-5}), we expect the ionization not to go very far before equilibrium is reached, certainly not to the point at which more than half the molecules are ionized. The 0.030 M HCl has the greater $[H_3O^+]$.

EXERCISE 15.10

Without doing detailed calculations, determine which of these solutions has the *higher* pH: 0.025 M $NH_3(aq)$ or 0.030 M methylamine, $CH_3NH_2(aq)$.

(*Hint:* Refer to Table 15.2.)

15.5 Polyprotic Acids

A **monoprotic acid** has *one* ionizable H atom per molecule. Hydrochloric acid, HCl, is a monoprotic acid. Acetic acid, CH_3COOH, has four H atoms per molecule, but only one of them is ionizable; it is also a monoprotic acid. A **polyprotic acid** has *more than one* ionizable H atom per molecule. Carbonic acid, H_2CO_3, has two H atoms, both ionizable; it is a *diprotic* acid.

Phosphoric acid, H_3PO_4, has three H atoms, all ionizable; it is a *triprotic* acid. In this section, we will consider three polyprotic acids—phosphoric acid, carbonic acid, and sulfuric acid. Additional polyprotic acids are listed in Appendix C.

Phosphoric Acid

APPLICATION NOTE

The principal use of phosphoric acid is in the manufacture of phosphate fertilizers. Phosphoric acid and its salts are also widely used in the food industry.

A key feature of polyprotic acids is that ionizations of the ionizable H atoms occur separately. For example, a molecule of H_3PO_4 does not give up its three ionizable H atoms in one single action. Instead, the complete ionization takes place in three distinct steps. Each step involves a reversible reaction with its own distinctive K_a value.

$$(1)\quad H_3PO_4 + H_2O \rightleftharpoons H_3O^+ + H_2PO_4^- \qquad K_{a_1} = \frac{[H_3O^+][H_2PO_4^-]}{[H_3PO_4]} = 7.1 \times 10^{-3}$$

$$(2)\quad H_2PO_4^- + H_2O \rightleftharpoons H_3O^+ + HPO_4^{2-} \qquad K_{a_2} = \frac{[H_3O^+][HPO_4^{2-}]}{[H_2PO_4^-]} = 6.3 \times 10^{-8}$$

$$(3)\quad HPO_4^{2-} + H_2O \rightleftharpoons H_3O^+ + PO_4^{3-} \qquad K_{a_3} = \frac{[H_3O^+][PO_4^{3-}]}{[HPO_4^{2-}]} = 4.3 \times 10^{-13}$$

It is easy to explain why the second ionization constant, K_{a_2}, is much smaller than the first, K_{a_1}. In the first ionization, as a proton leaves an H_3PO_4 molecule, it must overcome the attraction of the $H_2PO_4^-$ ion left behind. In the second ionization, the proton that leaves must overcome the attraction of the HPO_4^{2-} ion left behind. Because HPO_4^{2-} has twice the charge of $H_2PO_4^-$, the departing proton experiences a stronger interionic attraction than in the first ionization, and proton separation occurs much less readily. In the final ionization, proton separation from the ion PO_4^{3-} is more difficult still, and this means that K_{a_3} is much smaller than K_{a_2}. Thus, the first ionization constant for a polyprotic acid is the largest, and subsequent ionization constants are progressively smaller: $K_{a_1} > K_{a_2} > K_{a_3} \ldots$. Often, each K_a is thousands of times smaller than the preceding one, and in these cases we can make two additional generalizations.

Phosphoric acid

- Because K_{a_2}, K_{a_3}, ... are so small, few of the anions produced in the first ionization step ionize further.
- In all but extremely dilute solutions, essentially all the H_3O^+ ions come from the first ionization step alone.

In the first ionization step of phosphoric acid, H_3O^+ and $H_2PO_4^-$ ions are produced in equal number. If few of the $H_2PO_4^-$ ions undergo further ionization, we see that $[H_3O^+] \approx [H_2PO_4^-]$. Note the interesting result when we incorporate this idea into the expression for K_{a_2}.

$$K_{a_2} = \frac{\cancel{[H_3O^+]}[HPO_4^{2-}]}{\cancel{[H_2PO_4^-]}} = 6.3 \times 10^{-8}$$

The molarities of H_3O^+ and $H_2PO_4^-$ cancel.

$$[HPO_4^{2-}] = K_{a_2} = 6.3 \times 10^{-8}$$

This relationship holds regardless of the values of $[H_3O^+]$ and $[H_2PO_4^-]$ and, thus, of the molarity of the phosphoric acid solution itself.

EXAMPLE 15.11

Calculate the following concentrations in an aqueous solution that is 5.0 M H_3PO_4: **(a)** $[H_3O^+]$; **(b)** $[H_2PO_4^-]$; **(c)** $[HPO_4^{2-}]$; **(d)** $[PO_4^{3-}]$.

SOLUTION

We can obtain these ion concentrations by calculations using the general ideas about polyprotic acids that we developed at the beginning of this section.

a. Because K_{a_1} exceeds K_{a_2} by more than 10^5, when phosphoric acid ionizes, essentially all the H_3O^+ ions come from the first ionization step. This ionization is summarized below.

	$H_3PO_4 + H_2O \rightleftharpoons$	H_3O^+ +	$H_2PO_4^-$
The reaction:	$H_3PO_4 + H_2O$	H_3O^+	$H_2PO_4^-$
Initial concentrations, M:	5.0	≈ 0	0
Changes, M:	$-x$	$+x$	$+x$
Equilibrium concentrations, M:	$(5.0 - x)$	x	x

$$K_{a_1} = \frac{[H_3O^+][H_2PO_4^-]}{[H_3PO_4]} = \frac{x \cdot x}{(5.0 - x)} = 7.1 \times 10^{-3}$$

If we assume that $x \ll 5.0$, so that $(5.0 - x) \approx 5.0$, we can write

$$x^2 = 5.0 \times 7.1 \times 10^{-3} = 3.6 \times 10^{-2}$$

$$x = [H_3O^+] = 0.19\text{ M}$$

Let's use the "5% rule" to check the validity of the assumption that $x \ll 5.0$.

$$\frac{x}{\text{M}_{\text{acid}}} \times 100\% = \frac{0.19\text{ M}}{5.0\text{ M}} \times 100\% = 3.8\%$$

The assumption meets the requirement of the rule.

b. Because K_{a_2} is so small, little of the $H_2PO_4^-$ produced in the first ionization undergoes further ionization. For this reason, $[H_2PO_4^-] \approx [H_3O^+]$ and $[H_2PO_4^-] = 0.19$ M.

A typical cola drink contains phosphoric acid (see Exercise 15.11B).

c. We established on page 657 that $[HPO_4^{2-}] = K_{a_2}$ regardless of the molarity of a phosphoric acid solution. Thus, $[HPO_4^{2-}] = K_{a_2} = 6.3 \times 10^{-8}$ M.

d. PO_4^{3-} is formed only through the very limited third ionization. We now have all the data needed to calculate $[PO_4^{3-}]$.

$$\frac{[H_3O^+][PO_4^{3-}]}{[HPO_4^{2-}]} = K_{a_3} = 4.3 \times 10^{-13}$$

In this expression, $[H_3O^+] = 0.19$ and $[HPO_4^{-}] = 6.3 \times 10^{-8}$.

$$\frac{0.19 \times [PO_4^{3-}]}{6.3 \times 10^{-8}} = 4.3 \times 10^{-13}$$

$$[PO_4^{3-}] = 1.4 \times 10^{-19}\ \text{M}$$

EXERCISE 15.11A

What is the pH of a solution that is 0.125 M in maleic acid, an additive used to retard rancidity in fats and oils?

$$\text{HOOCCH{=}CHCOOH} + H_2O \rightleftharpoons H_3O^+ + \text{HOOCCH{=}CHCOO}^- \qquad K_{a_1} = 1.2 \times 10^{-2}$$

$$\text{HOOCCH{=}CHCOO}^- + H_2O \rightleftharpoons H_3O^+ + {}^-\text{OOCCH{=}CHCOO}^- \qquad K_{a_2} = 4.7 \times 10^{-7}$$

EXERCISE 15.11B

Acids are added to cola drinks to lower the pH to about 2.5 to impart tartness. Show that a cola that contains from 0.057 to 0.084% of 75% phosphoric acid (H_3PO_4), by mass, meets this pH requirement.

Carbonic Acid

When carbon dioxide dissolves in water this reaction occurs, producing weak, diprotic carbonic acid.

$$CO_2(aq) + H_2O(l) \rightleftharpoons H_2CO_3(aq)$$

The reaction is reversible. Carbonic acid is unstable and readily reverts to $CO_2(g)$ and H_2O. In an open vessel, $CO_2(g)$ escapes and the reaction goes to completion *to the left*, just as we would expect from Le Châtelier's principle.

The two ionization steps of H_2CO_3 and their K_a values are given below. Because K_{a_1} is much greater than K_{a_2}, carbonic acid conforms to our generalizations about polyprotic acids (page 657): Few of the HCO_3^- ions produced in the first ionization step ionize further.

$$(1)\quad H_2CO_3 + H_2O \rightleftharpoons H_3O^+ + HCO_3^- \qquad K_{a_1} = \frac{[H_3O^+][HCO_3^-]}{[H_2CO_3]} = 4.2 \times 10^{-7}$$

$$(2)\quad HCO_3^- + H_2O \rightleftharpoons H_3O^+ + CO_3^{2-} \qquad K_{a_2} = \frac{[H_3O^+][CO_3^{2-}]}{[HCO_3^-]} = 4.7 \times 10^{-11}$$

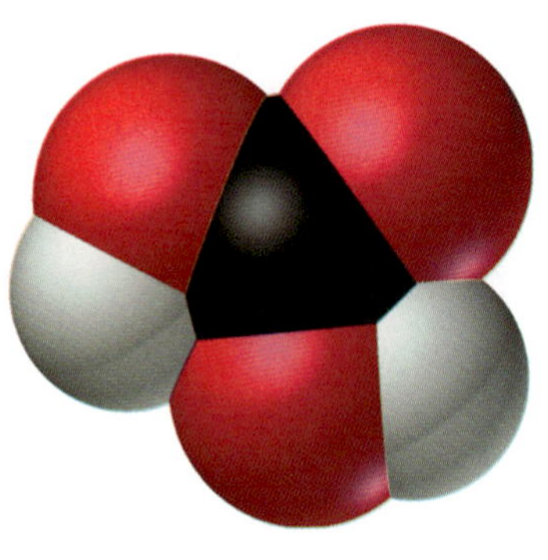

Carbonic acid

Neutralization of H_2CO_3 with 1 mol of OH^- in the first step produces salts such as $NaHCO_3$, sodium hydrogen carbonate. Neutralization with 2 mol OH^- produces carbonate salts, such as Na_2CO_3, sodium carbonate.

Equilibria based on the ionization of carbonic acid are important in several natural phenomena, such as the formation of hard water and limestone caves. These equilibria are also essential in maintaining the proper pH of blood.

Sulfuric Acid

Sulfuric acid is an unusual diprotic acid in that its first ionization step goes essentially to completion, but its second step does not.

$$(1)\quad H_2SO_4 + H_2O \longrightarrow H_3O^+ + HSO_4^- \quad K_{a_1} \approx 10^3$$

$$(2)\quad HSO_4^- + H_2O \rightleftharpoons H_3O^+ + SO_4^{2-} \quad K_{a_2} = \frac{[H_3O^+][SO_4^{2-}]}{[HSO_4^-]} = 1.1 \times 10^{-2}$$

APPLICATION NOTE
Sulfuric acid is produced in greater quantity than any other manufactured chemical. Its main use is in the production of fertilizers.

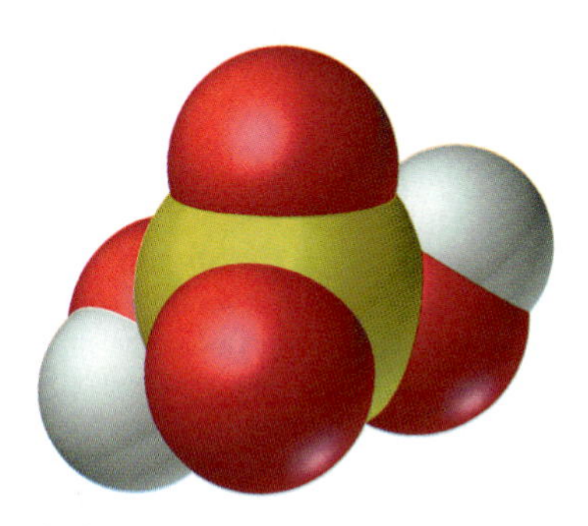

Sulfuric acid

For purposes of calculations, we can consider three categories of sulfuric acid solutions, described below in increasing order of complexity.

1. *Concentrated solutions* (greater than 0.50 M H_2SO_4). In these solutions, essentially all the H_3O^+ is produced in the first ionization because it goes to completion, whereas the second ionization is reversible. Thus, in 1.00 M H_2SO_4, we expect $[H_3O^+] \approx 1.00$ M. (See also Example 15.12)
2. *Very dilute solutions* (less than 0.0010 M H_2SO_4). Although K_{a_2} is small compared to K_{a_1}, it is large enough that in sufficiently dilute solutions, the second ionization goes to completion as well as the first. We predict that two H_3O^+ ions are produced for every H_2SO_4 molecule present originally and also that there are essentially no H_2SO_4 molecules or HSO_4^- ions in solution. For example, in 0.0010 M H_2SO_4, $[H_3O^+] \approx 0.0020$ M and $[SO_4^{2-}] \approx 0.0010$ M. (See Exercise 15.12A.)
3. *"Intermediate concentrations"* (between 0.0010 M H_2SO_4 and 0.50 M H_2SO_4). Both ionization steps must be considered when calculating ion concentrations. (See Exercise 15.12B.)

EXAMPLE 15.12

What is the approximate pH of 0.71 M H_2SO_4?

SOLUTION

This solution fits the first of the three categories listed above: Essentially all the H_3O^+ comes from the first step in the ionization of H_2SO_4, and that step goes to completion.

$$H_2SO_4 + H_2O \longrightarrow H_3O^+ + HSO_4^-$$

Thus, 0.71 M H_2SO_4 has $[H_3O^+] = 0.71$ M.

$$pH = -\log [H_3O^+] = -\log 0.71 = 0.15$$

EXERCISE 15.12A

What is the approximate pH of 8.5×10^{-4} M H_2SO_4?

EXERCISE 15.12B

Without doing detailed calculations, indicate which of the following is most likely to be closest to the measured $[H_3O^+]$ in 0.020 M H_2SO_4: (a) 0.020 M; (b) 0.025 M; (c) 0.039 M; (d) 0.045 M? Explain your reasoning.

15.6 Ions as Acids and Bases

A package of an everyday wall cleaner carries a common warning found on products that are either rather strongly acidic or strongly basic: Avoid contact with eyes and prolonged contact with skin. The principal component of this cleaner is listed as sodium carbonate, Na_2CO_3. Our first thought might be that Na_2CO_3 is neither an acid nor a base, because we see no H atoms, no OH groups, and no N atoms with lone-pair electrons. Yet, the washing soda shown in Figure 15.8, 1 M $Na_2CO_3(aq)$, is quite basic.

Figure 15.8
Hydrolysis of carbonate ion
This sodium carbonate solution contains a few drops of thymolphthalein indicator, which is colorless below pH 9.4 and blue above pH 10.6. The blue color of the indicator shows that the pH of the solution is greater than 10.6. The solution is rather strongly basic as a result of the hydrolysis of CO_3^{2-} as a base.

When the ionic compound $Na_2CO_3(s)$ dissolves, Na^+ and CO_3^{2-} ions enter the solution.

$$Na_2CO_3(s) \xrightarrow{H_2O} 2\,Na^+(aq) + CO_3^{2-}(aq)$$

The Brønsted-Lowry theory explains how carbonate ions react to produce OH^- ions.

$$\underset{\text{Base(1)}}{CO_3^{2-}} + \underset{\text{Acid(2)}}{H_2O} \rightleftharpoons \underset{\text{Acid(1)}}{HCO_3^-} + \underset{\text{Base(2)}}{OH^-}$$

This reaction raises $[OH^-]$ to a value much higher than 1.0×10^{-7} M, and $[H_3O^+]$ decreases accordingly. Thus, the pH rises well above 7.0. Because sodium ions do not react with water, they do not affect the pH of a solution.

$$Na^+(aq) + H_2O \longrightarrow \text{no reaction}$$

Although acid-base reactions of ions with water are fundamentally no different from other acid-base reactions, they are often called **hydrolysis** reactions. In $Na_2CO_3(aq)$, we say that CO_3^{2-} hydrolyzes and Na^+ does not. Cations of Group 1A and Group 2A do not hydrolyze, but many other metal cations do hydrolyze, particularly those with a small size and high charge.

Now let's consider some useful generalizations about hydrolysis. In the photographs on the next page, the solutions contain a few drops of bromthymol indicator solution. The color of the indicator depends on pH as follows:

$pH < 7$	$pH = 7$	$pH > 7$
Yellow	Green	Blue

In the numbered items that follow, certain salts are indicated in red type, and the photographs of their solutions are also identified through red type.

1. *Salts of strong acids and strong bases form* neutral *solutions* (pH = 7). The strong bases are the ionic hydroxides of the Group 1A and 2A metals. Examples of salts of strong acids and strong bases are NaCl, KNO_3, and BaI_2. The anions Cl^-, NO_3^-, and I^- are conjugate bases of strong acids and are therefore all very weak bases. They do not hydrolyze, nor do Group 1A and Group 2A cations.
2. *Salts of weak acids and strong bases form* basic *solutions* (pH > 7). The anion hydrolyzes as a base. Examples: Na_2CO_3, KNO_2, CH_3COONa. The anions CO_3^{2-}, NO_2^-, and CH_3COO^- are the conjugate bases of weak acids and are therefore considerably stronger bases than Cl^-, NO_3^-, and I^- (recall Table 15.1). Again, Group 1A and Group 2A cations do not hydrolyze.
3. *Salts of strong acids and weak bases form* acidic *solutions* (pH < 7). The cation hydrolyzes as an acid. Examples: NH_4Cl, NH_4NO_3, and NH_4Br. The cation NH_4^+ is the conjugate acid of the weak base NH_3. As in the first case, anions such as Cl^-, NO_3^-, and Br^- are the conjugate bases of strong acids and do not hydrolyze.

NaCl(aq)

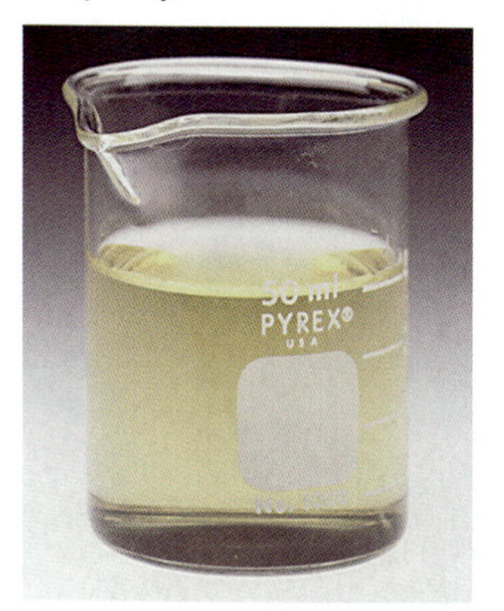

$NH_4Cl(aq)$

$CH_3COONH_4(aq)$

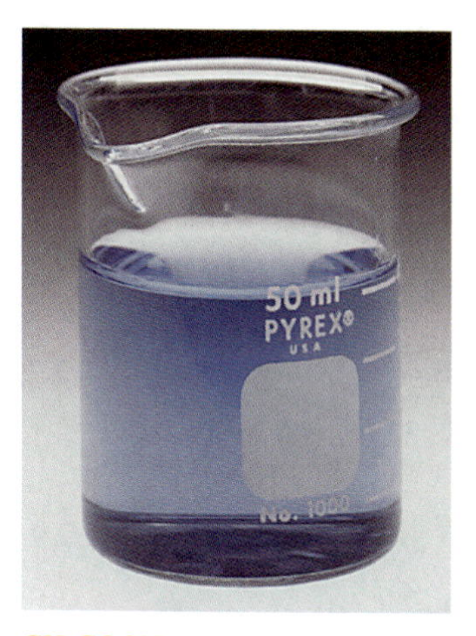

$CH_3COONa(aq)$

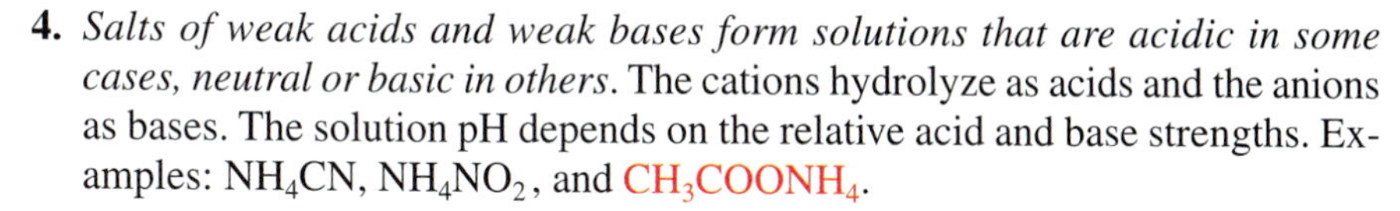

4. *Salts of weak acids and weak bases form solutions that are acidic in some cases, neutral or basic in others*. The cations hydrolyze as acids and the anions as bases. The solution pH depends on the relative acid and base strengths. Examples: NH_4CN, NH_4NO_2, and CH_3COONH_4.

We can summarize all four of these cases in a single statement:

> *Only ions that are* the conjugates *of weak acids or weak bases hydrolyze appreciably.*

EXAMPLE 15.13

Indicate whether the following solutions are acidic, basic, or neutral.

a. $NH_4I(aq)$ **b.** $CH_3COONH_4(aq)$

SOLUTION

a. NH_4I is the salt of a strong acid, HI, and a weak base, NH_3, an example of case 3 above. The cation NH_4^+ hydrolyzes, and the solution is *acidic*.

$$NH_4^+ + H_2O \rightleftharpoons NH_3 + H_3O^+$$

The anion, I^-, a very weak base, does not hydrolyze appreciably.

b. Ammonium acetate is the salt of a weak acid, CH_3COOH, and a weak base, NH_3. It is an example of case 4; both ions hydrolyze.

$$NH_4^+ + H_2O \rightleftharpoons NH_3 + H_3O^+$$

$$CH_3COO^- + H_2O \rightleftharpoons CH_3COOH + OH^-$$

From the observation that the pH of aqueous ammonium acetate is 7 (photograph on page 661), we should expect the equilibrium constants for the two hydrolysis reactions to have about the same value.

EXERCISE 15.13A

Indicate whether the following solutions are acidic, basic, or neutral. Explain.

a. $NaNO_3(aq)$ **b.** $CH_3CH_2CH_2COOK(aq)$

EXERCISE 15.13B

Arrange the following solutions in the expected order of *increasing pH*, and state your reasoning in doing so: NaCl(aq), HCl(aq), NaOH(aq), $KNO_2(aq)$, $NH_4I(aq)$.

To make a quantitative prediction of the pH of a solution in which hydrolysis occurs, we need an equilibrium constant for the hydrolysis reaction. Consider, for example, the hydrolysis of acetate ion.

$$CH_3COO^- + H_2O \rightleftharpoons CH_3COOH + OH^- \qquad K_b = \frac{[CH_3COOH][OH^-]}{[CH_3COO^-]} = ?$$

Two of the concentration terms in the K_b expression are the same as in the K_a expression for the ionization of acetic acid, the conjugate acid of CH_3COO^-. It seems, then, that K_b for CH_3COO^- and K_a for CH_3COOH should be related to each other. Suppose we multiply both the numerator and denominator of the K_b expression by $[H_3O^+]$.

$$K_b = \frac{[CH_3COOH][OH^-][H_3O^+]}{[CH_3COO^-][H_3O^+]} = \frac{K_w}{K_a} = \frac{1.0 \times 10^{-14}}{1.8 \times 10^{-5}} = 5.6 \times 10^{-10}$$

Note that the terms in blue are equivalent to K_w. Those in red give the *inverse* of K_a for acetic acid, that is, $1/K_a$. The result is commonly written as follows.

$$K_a \times K_b = K_w$$

That is, the product of K_a and K_b of a conjugate acid-base pair equals the ion product of water, which at 25 °C is $K_w = 1.0 \times 10^{-14}$. Many tables of ionization constants list only pK_a values, but we can get a K_b value by calculating it from the expression $pK_a + pK_b = pK_w = 14.00$.

Once we obtain ionization constants in this way, we can calculate the pH of solutions of salts that hydrolyze, as is illustrated in Example 15.14.

EXAMPLE 15.14

Calculate the pH of a sodium acetate solution that is 0.25 M $CH_3COONa(aq)$.

SOLUTION

Acetate ion hydrolyzes as a base, and we just derived a numerical value of K_b for this hydrolysis. We present the data under the equation for the hydrolysis reaction in the usual way.

The reaction:	CH_3COO^-	+ H_2O $\rightleftharpoons$	CH_3COOH	+	OH^-
Initial concentrations, M:	0.25		0		≈ 0
Changes, M:	$-x$		$+x$		$+x$
Equilibrium concentrations, M:	$(0.25 - x)$		x		x

$$K_b = \frac{[CH_3COOH][OH^-]}{[CH_3COO^-]} = \frac{K_w}{K_a} = \frac{1.0 \times 10^{-14}}{1.8 \times 10^{-5}} = 5.6 \times 10^{-10}$$

$$\frac{x \cdot x}{0.25 - x} = 5.6 \times 10^{-10}$$

Let's assume that $x \ll 0.25$, so that $(0.25 - x) = 0.25$.

$$\frac{x^2}{0.25} = 5.6 \times 10^{-10}$$

$$x^2 = 1.4 \times 10^{-10}$$

$$x = [OH^-] = (1.4 \times 10^{-10})^{1/2} = 1.2 \times 10^{-5}\ M$$

$$pOH = -\log [OH^-] = -\log (1.2 \times 10^{-5}) = 4.92$$

$$pH = 14.00 - pOH = 14.00 - 4.92 = 9.08$$

PROBLEM-SOLVING NOTE

We see that the assumption is justified, since $[OH^-]$ (1.2×10^{-5}) is much smaller than M_{base} (0.25).

EXERCISE 15.14A

Calculate the pH of a 0.052 M NH_4Cl solution.

EXERCISE 15.14B

A 50.00-mL sample of 0.120 M CH_3COOH is neutralized with 18.75 mL of 0.320 M NaOH. What is the pH of the neutralized solution?

(*Hint:* What is the neutralization reaction, and what are the concentrations of its products?)

EXAMPLE 15.15

What is the molarity of a NH_4NO_3 solution that has a pH = 4.80?

SOLUTION

Ammonium nitrate is the salt of a strong acid (HNO_3) and a weak base (NH_3). In NH_4NO_3(aq), NH_4^+ hydrolyzes and NO_3^- does not. The hydrolysis equilibrium and equilibrium constant for NH_4^+ are as follows.

$$NH_4^+ + H_2O \rightleftharpoons H_3O^+ + NH_3 \quad K_a = \frac{K_w}{K_b} = \frac{1.0 \times 10^{-14}}{1.8 \times 10^{-5}} = 5.6 \times 10^{-10}$$

PROBLEM-SOLVING NOTE

Because K_a for the hydrolysis of NH_4^+ and K_b for the hydrolysis of CH_3COO^- are equal, we can see why CH_3COONH_4(aq) should be pH neutral, as predicted in part (b) of Example 15.13.

As usual, we can calculate $[H_3O^+]$ from the pH of the solution.

$$\log [H_3O^+] = -pH = -4.80$$

$$[H_3O^+] = 10^{-4.80} = 1.6 \times 10^{-5}\ M$$

This $[H_3O^+]$ is much larger than the hydronium ion concentration associated with pure water $(1.0 \times 10^{-7}\ M)$. We should therefore be able to assume that all the hydronium ion comes from the hydrolysis reaction. We do so in the setup below, in which x represents the unknown initial concentration of NH_4^+.

The reaction:	NH_4^+	+ H_2O $\rightleftharpoons$	H_3O^+	+	NH_3
Initial concentrations, M:	x		≈ 0		0
Changes, M:	-1.6×10^{-5}		$+1.6 \times 10^{-5}$		$+1.6 \times 10^{-5}$
Equilibrium concentrations, M:	$(x - 1.6 \times 10^{-5})$		1.6×10^{-5}		1.6×10^{-5}

We can now substitute equilibrium concentrations into the ionization constant expression for the hydrolysis reaction.

$$K_a = \frac{[H_3O^+][NH_3]}{[NH_4^+]} = \frac{(1.6 \times 10^{-5})(1.6 \times 10^{-5})}{(x - 1.6 \times 10^{-5})} = 5.6 \times 10^{-10}$$

We can make the usual simplifying assumption here. That is, if we assume that ammonium ion is mostly *nonhydrolyzed*, the change in $[NH_4^+]$ should be much smaller than the initial $[NH_4^+]$. That is, if $1.6 \times 10^{-5} \ll x$, we can replace $(x - 1.6 \times 10^{-5})$ by x. Then we can solve for x.

$$\frac{(1.6 \times 10^{-5})^2}{x} = 5.6 \times 10^{-10}$$

$$x = \frac{(1.6 \times 10^{-5})^2}{5.6 \times 10^{-10}} = [NH_4^+] = 0.46 \text{ M}$$

The solution is 0.46 M NH_4NO_3. Note that our assumption is valid: $x - 1.6 \times 10^{-5} = 0.46 - 1.6 \times 10^{-5} = 0.46$.

EXERCISE 15.15A

What is the molarity of a $CH_3COONa(aq)$ solution that has a pH = 9.10?

EXERCISE 15.15B

Without doing detailed calculations, determine which of the following solutions should have the *higher* pH: 0.10 M NH_4NO_2 or 0.10 M NH_4CN. Explain your reasoning.

15.7 The Common Ion Effect

The two solutions pictured in Figure 15.9 both contain acetic acid, CH_3COOH, at the same molarity. Both also contain an indicator known as bromophenol blue. The solution at the right, however, also contains sodium acetate, CH_3COONa, as a second solute. As is evident from their different colors, the two solutions have *different* pH values. The solution having both acetic acid *and* sodium acetate as solutes has a pH that is greater by about *two* units. $[H_3O^+]$ in this solution is only about one-hundredth of that in the solution with acetic acid as the only solute. There is no mystery here, however. This is a classic example of Le Châtelier's principle.

1.00 M CH_3COOH 1.00 M CH_3COOH-1.00 M CH_3COONa

▶ **Figure 15.9**
The common ion effect
Both solutions contain bromophenol blue indicator (yellow below pH 3.0; blue above pH 4.6). The yellow color indicates that the pH < 3.0 in 1.00 M CH_3COOH, and the blue-violet color that pH > 4.6 in the solution that is 1.00 M in both CH_3COOH and CH_3COONa.

When we add sodium acetate to an acetic acid solution, the sodium acetate dissociates completely to yield $CH_3COO^-(aq)$ and $Na^+(aq)$. Thus, the concentration of $CH_3COO^-(aq)$, one of the products of the ionization of acetic acid, increases. According to Le Châtelier's principle, if we increase the concentration of one of the products of a reversible reaction, in this case CH_3COO^-, equilibrium shifts to the *left*.

When a salt supplies CH_3COO^-, equilibrium shifts to the *left*.

$$\underset{\text{Acid(1)}}{CH_3COOH} + \underset{\text{Base(2)}}{H_2O} \rightleftharpoons \underset{\text{Acid(2)}}{H_3O^+} + \underset{\text{Base(1)}}{CH_3COO^-}$$

As CH_3COO^-, a base, is consumed in the reverse reaction, so too is H_3O^+, an acid. $[H_3O^+]$ decreases, and the pH increases accordingly. Because it is found both in aqueous acetic acid and in sodium acetate, acetate ion is a *common ion*.

The **common ion effect** is the suppression of the ionization of a weak acid or a weak base by the presence of a common ion from a strong electrolyte.

In Example 15.16, we calculate the pH of the acetic acid–sodium acetate solution we have just described. In Exercise 15.16A, we see the effect of NH_4^+ as a common ion on the ionization of $NH_3(aq)$, and in Exercise 15.16B, we explore the common ion effect in a solution containing both a strong acid and a weak acid.

EXAMPLE 15.16

Calculate the pH of a solution that is both 1.00 M CH_3COOH and 1.00 M CH_3COONa.

SOLUTION

First, the pH of the solution does not depend on the method of preparing it. That is, we can dissolve sodium acetate in an aqueous solution of acetic acid, or we can dissolve acetic acid in an aqueous solution of sodium acetate. Suppose we begin with 1.00 M $CH_3COONa(aq)$. In this solution, the ionic compound CH_3COONa is completely dissociated into its ions.

$$CH_3COONa(s) \xrightarrow{H_2O} CH_3COO^-(aq) + Na^+(aq)$$

Then we can add the required amount of acetic acid. In this way, the initial concentrations of both CH_3COO^- and CH_3COOH are 1.00 M.

The reaction:	CH_3COOH + H_2O $\rightleftharpoons$	H_3O^+ +	CH_3COO^-
Initial concentrations, M:	1.00	≈ 0	1.00
Changes, M:	$-x$	$+x$	$+x$
Equilibrium concentrations, M:	$(1.00 - x)$	x	$(1.00 + x)$

$$K_a = \frac{[H_3O^+][CH_3COO^-]}{[CH_3COOH]}$$

$$1.8 \times 10^{-5} = \frac{x(1.00 + x)}{1.00 - x}$$

If x is very small, we can assume that $(1.00 - x) \approx (1.00 + x) \approx 1.00$.

$$\frac{x(\cancel{1.00})}{\cancel{1.00}} = 1.8 \times 10^{-5} \qquad x = [H_3O^+] = 1.8 \times 10^{-5}\,M$$

$$pH = -\log [H_3O^+] = -\log (1.8 \times 10^{-5}) = 4.74$$

Compare this value of the pH (4.74) with that found in Example 15.6, where there was no common ion (pH 2.38), and you will see just how effectively acetate ion suppresses the ionization of acetic acid.

Note that the assumption that x is very small is valid.

$$(1.00 - 1.8 \times 10^{-5}) = 1.00 \quad \text{and} \quad (1.00 + 1.8 \times 10^{-5}) = 1.00.$$

EXERCISE 15.16A

Calculate the pH of a solution that is 0.15 M NH_3 and 0.35 M NH_4NO_3.

$$NH_3 + H_2O \rightleftharpoons NH_4^+ + OH^- \qquad K_b = 1.8 \times 10^{-5}$$

EXERCISE 15.16B

What is the common ion in a solution that is 0.10 M in HCl and 0.10 M in CH_3COOH? What is $[H_3O^+]$ in this solution? What is $[CH_3COO^-]$?

15.8 Buffer Solutions

The acetic acid–sodium acetate solution of Example 15.16 is a **buffer solution**, a solution that changes pH *only slightly* when small amounts of a strong acid or a strong base are added.

Buffer solutions have many important applications in industry, in the laboratory, and in living organisms. Some chemical reactions consume acids, others produce acids, and many are catalyzed by H_3O^+. If we want to study the kinetics of these reactions or simply to control their reaction rates, we often need to control the pH. We can minimize pH changes by conducting the reactions in buffered solutions. Enzyme-catalyzed reactions are particularly sensitive to pH changes. Studies that involve proteins are usually performed in buffered media because the magnitude and kind of electric charge carried by the protein molecules depend on the pH.

Figure 15.10 shows that pure water totally lacks the ability to buffer. Even tiny amounts of acid or base produce large changes in the pH of water. To be a buffer, a solution usually must have one of the following compositions.

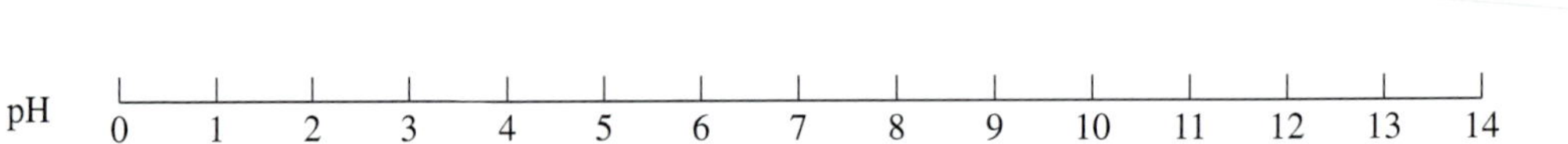

▲ **Figure 15.10 Representing buffer action**
The pH of pure water undergoes very large changes in pH when a small amount of either an acid or a base is added. Water has no buffering ability. In contrast, the corresponding pH changes in a buffer solution that is 1.00 M in CH_3COOH and 1.00 M in CH_3COONa are almost imperceptible.

A solution containing *a weak acid and its salt (conjugate base).*
A solution containing *a weak base and its salt (conjugate acid).*

The acid component of the buffer can neutralize small added amounts of OH^-, and the basic component can neutralize small added amounts of H_3O^+.

How a Buffer Solution Works

Let's look more closely at the buffer solution of Example 15.16, which is 1.00 M in both CH_3COOH and CH_3COONa We can represent it through the following equation and equilibrium constant expression.

$$CH_3COOH + H_2O \rightleftharpoons H_3O^+ + CH_3COO^-$$

$$K_a = \frac{[H_3O^+][CH_3COO^-]}{[CH_3COOH]} = 1.8 \times 10^{-5}$$

At equilibrium, $[H_3O^+]$ is given by the following expression.

$$[H_3O^+] = K_a \times \frac{[CH_3COOH]}{[CH_3COO^-]}$$

$$= 1.8 \times 10^{-5} \times \frac{[CH_3COOH]}{[CH_3COO^-]}$$

Now let's substitute the equilibrium concentrations from Example 15.16 into this expression, including the simplifying assumption that $x \ll 1.00$.

$$[H_3O^+] = K_a \times \frac{[CH_3COOH]}{[CH_3COO^-]}$$

$$= 1.8 \times 10^{-5} \times \frac{(1.00 - x)}{(1.00 + x)} = 1.8 \times 10^{-5} \times \frac{1.00}{1.00} = 1.8 \times 10^{-5}\text{ M}$$

From that value of $[H_3O^+]$, we can get the pH of the buffer solution.

$$pH = -\log [H_3O^+] = -\log (1.8 \times 10^{-5}) = 4.74.$$

Now suppose we add enough of a strong base to the buffer solution to neutralize 2% of the acetic acid. This will reduce its concentration from 1.00 M to 0.98 M and raise that of acetate ion from 1.00 M to 1.02 M. A neutralization reaction in a buffer solution always converts some of one buffer component to the other.

The buffer reaction:	CH_3COOH	+ OH^- ⟶	H_2O + CH_3COO^-
Initial buffer:	1.00 M	≈0	1.00 M
Add:		0.02 M	
Changes:	−0.02 M	−0.02 M	+0.02 M
After neutralization:	0.98 M	≈0 M	1.02 M

Now let's calculate $[H_3O^+]$ and pH in a solution that is 0.98 M CH_3COOH and 1.02 M CH_3COONa. We can do this in the same way that we determined $[H_3O^+]$ in the solution of Example 15.16, which was 1.00 M in both CH_3COOH and CH_3COONa. However, let's simplify our calculation by anticipating that $[H_3O^+]$ (or x) is very small compared both to 0.98 M, the $[CH_3COOH]$, and to 1.02 M, the $[CH_3COO^-]$. We can then use the expression for $[H_3O^+]$ in the buffer.

$$[H_3O^+] = K_a \times \frac{[CH_3COOH]}{[CH_3COO^-]}$$

$$= 1.8 \times 10^{-5} \times \frac{(0.98 - x)}{(1.02 + x)} = 1.8 \times 10^{-5} \times \frac{0.98}{1.02} = 1.7 \times 10^{-5}\ M$$

$$pH = -\log [H_3O^+] = -\log (1.7 \times 10^{-5}) = 4.77.$$

Note that the $[H_3O^+]$ and pH of the buffer solution have hardly been affected.

When we add an acid to a buffer solution, the situation is quite similar. If 2% of the conjugate base (acetate ion) is neutralized by adding H_3O^+, we can make the following table.

The buffer reaction:	CH_3COO^-	$+ H_3O^+$	$\longrightarrow H_2O +$	CH_3COOH
Initial buffer:	1.00 M	≈0 M		1.00 M
Add:		0.02 M		
Changes:	−0.02 M	−0.02 M		+0.02 M
After neutralization:	0.98 M	≈0 M		1.02 M

To calculate $[H_3O^+]$ and pH in a solution that is 1.02 M CH_3COOH and 0.98 M CH_3COONa, we can proceed as before.

$$[H_3O^+] = K_a \times \frac{[CH_3COOH]}{[CH_3COO^-]}$$

$$= 1.8 \times 10^{-5} \times \frac{(1.02 - x)}{(0.98 + x)} = 1.8 \times 10^{-5} \times \frac{1.02}{0.98} = 1.9 \times 10^{-5}\ M$$

$$pH = -\log [H_3O^+] = -\log (1.9 \times 10^{-5}) = 4.72.$$

Again, the $[H_3O^+]$ and pH are hardly affected.

We have shown that a buffer solution resists a change in pH with the addition of either an acid or base. However, we should note that when an *acid* is added, the pH is *lowered* very slightly, and when a *base* is added the pH is *raised* a tiny bit. We can use this fact as a helpful check on buffer calculations.

An Equation for Buffer Solutions

In certain applications, we need to repeatedly calculate the pH of buffer solutions. We can do this with a single, simple equation, *provided that we are aware of its limitations*.

To establish such an equation, we begin with the equation for $[H_3O^+]$ in a buffer containing acetic acid and sodium acetate—an equation that we have used before.

$$[H_3O^+] = K_a \times \frac{[CH_3COOH]}{[CH_3COO^-]}$$

Now, we take the *negative* logarithm of each side of the equation.

$$-\log [H_3O^+] = -\log K_a - \log \frac{[CH_3COOH]}{[CH_3COO^-]}$$

Next, let's substitute pH for $-\log [H_3O^+]$ and pK_a for $-\log K_a$. Also, let's replace the logarithm of the ratio of concentrations by the equivalent expression: $\log [CH_3COOH] - \log [CH_3COO^-]$.

$$pH = pK_a - \{\log [CH_3COOH] - \log [CH_3COO^-]\}$$

$$pH = pK_a - \log [CH_3COOH] + \log [CH_3COO^-]$$

PROBLEM-SOLVING NOTE

The net effect of the manipulations with the log terms is to invert the original ratio and change the sign of its log from negative to positive.

Now, we replace the difference in logarithms so that we again have the logarithm of a ratio.

$$pH = pK_a + \log \frac{[CH_3COO^-]}{[CH_3COOH]}$$

This equation for the pH of an acetic acid–acetate ion buffer is a special case of a more general equation, known as the **Henderson-Hasselbalch equation**.

$$pH = pK_a + \log \frac{[\text{conjugate base}]}{[\text{weak acid}]}$$

The Henderson-Hasselbalch equation has an important limitation. It works only if we can substitute the measured or nominal molarities for equilibrium molarities of weak acids and conjugate bases. We treat weak acids and conjugate bases *as if* their molarities were unchanged by the ionization equilibrium. This is the same assumption that we discussed on page 652. It results in substitutions of this sort: M_{acid} for $(M_{acid} - x)$, M_{base} for $(M_{base} + x)$, and so on.

The net effect of the limitations of the Henderson-Hasselbalch equation is that it works only for *buffer* solutions that fit the following criteria.

- The ratio, [conjugate base]/[weak acid] has a value *between 0.10 and 10.*
- Both [conjugate base] and [weak acid] exceed K_a by a factor of *100 or more*.

EXAMPLE 15.17

A buffer solution is prepared that is 0.24 M NH_3 and 0.20 M NH_4Cl. **(a)** What is the pH of this buffer? **(b)** If 0.0050 mol NaOH is added to 0.500 L of this solution, what will be the pH?

SOLUTION

To determine the pH in part (a), we will use the same method that we used in introducing the common ion effect, although we could use the Henderson-Hasselbalch equation instead. We will use that equation in part (b).

a. We begin with the usual format describing the equilibrium.

	NH_3	$+ H_2O \rightleftharpoons$	NH_4^+	$+ OH^-$
The reaction:	NH_3	$+ H_2O \rightleftharpoons$	NH_4^+	$+ OH^-$
Initial concentrations, M:	0.24		0.20	≈ 0
Changes, M:	$-x$		$+x$	$+x$
Equilibrium. concentrations, M:	$(0.24 - x)$		$(0.20 + x)$	x

In substituting equilibrium concentrations into the K_b expression, let's assume that $x \ll 0.20$. This allows us to substitute 0.20 for $(0.20 + x)$ and 0.24 for $(0.24 - x)$.

$$K_b = \frac{[NH_4^+][OH^-]}{[NH_3]}$$

$$1.8 \times 10^{-5} = \frac{(0.20 + x)x}{(0.24 - x)}$$

$$1.8 \times 10^{-5} = \frac{0.20x}{0.24}$$

Now let's solve for x, which is equal to $[OH^-]$.

$$0.20x = (1.8 \times 10^{-5}) \times 0.24$$

$$x = [OH^-] = 1.8 \times 10^{-5} \times \frac{0.24}{0.20} = 2.2 \times 10^{-5}\ M$$

PROBLEM-SOLVING NOTE

Note that our assumptions are valid: $0.20 + x = 0.20 + 2.2 \times 10^{-5} = 0.20$ and $0.20 - x = 0.20 - 2.2 \times 10^{-5} = 0.20$.

We can use the $[OH^-]$ to calculate the pOH and the pH of the solution.

$$pOH = -\log [OH^-] = -\log (2.2 \times 10^{-5}) = 4.66$$

$$pH = 14.00 - pOH = 14.00 - 4.66 = 9.34$$

b. First, we need to calculate the result of neutralizing the added NaOH. Adding 0.005 mol OH^- to 0.500 L of buffer produces an immediate $[OH^-]$ = 0.005 mol OH^-/0.500 L = 0.010 M solution. This $[OH^-]$ is reduced to near zero by the neutralization.

The buffer reaction:	NH_4^+	+ OH^-	$\rightleftharpoons$ H_2O +	NH_3
Initial buffer:	0.20 M	≈0 M		0.24 M
Add:		0.01 M		
Changes:	−0.01 M	−0.01 M		+0.01 M
After neutralization:	0.19 M	≈0 M		0.25 M

We now need to calculate $[H_3O^+]$ and pH using the new molarities of NH_3 and NH_4^+. *Solution based on the equilibrium constant expression*: This is the same method used in part (a) for the original buffer solution. Let's begin by substituting equilibrium concentrations into the K_b expression and assuming that $x \ll 0.19$. This allows us to substitute 0.19 for $(0.19 + x)$ and 0.25 for $(0.25 - x)$.

$$K_b = \frac{[NH_4^+][OH^-]}{[NH_3]}$$

$$1.8 \times 10^{-5} = \frac{(0.19 + x)x}{(0.25 - x)} = \frac{0.19x}{0.25}$$

Now we solve for x, which is equal to $[OH^-]$.

$$0.19x = (1.8 \times 10^{-5}) \times 0.25$$

$$x = [OH^-] = 1.8 \times 10^{-5} \times \frac{0.25}{0.19} = 2.4 \times 10^{-5}\ M$$

We first calculate pOH and then pH.

$$pOH = -\log [OH^-] = -\log (2.4 \times 10^{-5}) = 4.62$$

$$pH = 14.00 - pOH = 14.00 - 4.62 = 9.38$$

Solution based on the Henderson-Hasselbalch equation: Here, we simply begin with the Henderson-Hasselbalch equation and substitute the appropriate information into the right side of the equation.

$$pH = pK_a + \log \frac{[\text{conjugate base}]}{[\text{weak acid}]}$$

The weak acid in this buffer is NH_4^+, derived from the salt NH_4Cl. The conjugate base is NH_3. We need the pK_a value for NH_4^+. We can get this from the K_b value of NH_3 by recalling that $pK_a + pK_b = 14.00$.

$$pK_a = 14.00 - pK_b = 14.00 - [-\log(1.8 \times 10^{-5})]$$
$$= 14.00 - 4.74 = 9.26$$

The concentrations are 0.25 M for $[NH_3]$, the conjugate base, and 0.19 M for NH_4^+, the weak acid.

$$pH = 9.26 + \log \frac{0.25}{0.19} = 9.26 + 0.12 = 9.38$$

EXERCISE 15.17A

What is the final pH if 0.03 mol HCl is added to 0.500 L of a buffer solution that is 0.24 M NH_3 and 0.20 M NH_4Cl?

EXERCISE 15.17B

In Example 15.17, we saw that the addition of 0.0050 mol NaOH to 0.500 L of a solution that is 0.24 M NH_3 and 0.20 M NH_4Cl raised the pH from 9.34 to 9.38. How many moles of NaOH should have been added to raise the pH to 9.50?

(*Hint:* What must be the value of the ratio $[NH_3]/[NH_4^+]$?)

Quite often in the laboratory, we need to prepare a buffer solution of a particular pH. In Example 15.18 we use the Henderson-Hasselbalch equation to do the necessary calculation, although we could use the unmodified equilibrium constant expression instead.

EXAMPLE 15.18

What concentration of acetate ion in 0.500 M CH_3COOH produces a buffer solution with pH = 5.00?

SOLUTION

We begin with the Henderson-Hasselbalch equation for an acetic acid–acetate ion buffer solution.

$$pH = pK_a + \log \frac{[CH_3COO^-]}{[CH_3COOH]}$$

We can replace pH by 5.00, pK_a by $-\log K_a$, and $[CH_3COOH]$ by 0.500. Then we can solve for $[CH_3COO^-]$.

$$5.00 = -\log(1.8 \times 10^{-5}) + \log \frac{[CH_3COO^-]}{0.500}$$

$$5.00 = 4.74 + \log \frac{[CH_3COO^-]}{0.500}$$

$$\log \frac{[CH_3COO^-]}{0.500} = 5.00 - 4.74 = 0.26$$

$$\frac{[CH_3COO^-]}{0.500} = 10^{0.26} = 1.82$$

$$[CH_3COO^-] = 0.500 \times 1.82 = 0.910 \text{ M}$$

EXERCISE 15.18A

What concentration of acetic acid in 0.250 M CH_3COONa is needed to produce a buffer solution with pH = 4.50?

EXERCISE 15.18B

What mass of NH_4Cl must be present in 0.250 L of 0.150 M NH_3 to produce a buffer solution with pH = 9.05?

Buffer Capacity and Buffer Range

In Example 15.17, we examined the buffering ability of a solution of 0.24 M NH_3 and 0.20 M NH_4Cl. We saw that when 0.0050 mol OH^- was added to 0.500 L of this solution, the pH increased only from 9.34 to 9.38. If we had considered the addition of 0.050 mol OH^- (ten times as much), we would have found the buffer less effective; the pH would have increased to 9.79.

There is a limit to the capacity of a buffer solution to neutralize added acid or base, and this limit is reached before all of one of the buffer components has been consumed. In the case of the 0.24 M NH_3/0.20 M NH_4Cl buffer, the limits are just under 0.20 mol OH^-/L (which would neutralize all the NH_4^+) and just less than 0.24 mol H_3O^+/L (which would neutralize all the NH_3).

In general, the more concentrated the buffer components in a solution, the more added acid or base the solution can neutralize. And, as a rule, a buffer is most effective if the concentrations of the buffer acid and its conjugate base are *equal*. In the Henderson-Hasselbalch equation, the equal concentrations cancel, and because $\log 1 = 0$, $pH = pK_a$. For an acetic acid–sodium acetate buffer, for example, we get $pH = pK_a = 4.74$.

$$pH = pK_a + \log \frac{[CH_3COO^-]}{[CH_3COOH]}$$

$$pH = 4.74 + \log 1$$

$$pH = 4.74$$

In general, the pH range over which a buffer solution is effective is about one pH unit on either side of $pH = pK_a$. This corresponds to the following ratios.

Notice that the conditions that govern a high buffer capacity are the same as those required in the Henderson-Hasselbalch equation.

$$0.10 < \frac{[\text{conjugate base}]}{[\text{weak acid}]} < 10.0$$

$$-1.0 < \log \frac{[\text{conjugate base}]}{[\text{weak acid}]} < 1.0$$

Thus, for an acetic acid–acetate ion buffer, the effective range is $3.74 < pH < 5.74$. For an ammonia–ammonium ion buffer, where pK_a of NH_4^+ is 9.26, the effective range is $8.26 < pH < 10.26$.

15.9 Acid-Base Indicators

An **acid-base indicator** is a weak acid having one color and the conjugate base of the acid having a different color. (One of the colors may be "colorless.") The color of a solution containing the indicator depends on the relative proportions of the weak acid and conjugate base; in turn, these proportions depend on the pH.

In Figure 15.11, when a few drops of phenol red indicator are added to pure water, the color is *orange*. A few drops of lemon juice—an acid—change the color to *yellow*, whereas a few drops of household ammonia—a base—change the color to *red*. To explain these color changes, let's call the yellow form of phenol red the indicator acid, HIn, and the red form, the indicator base, In^-. The following equilibrium exists between the two.

$$\underset{\text{Yellow}}{\mathrm{HIn}} + \mathrm{H_2O} \rightleftharpoons \mathrm{H_3O^+} + \underset{\text{Red}}{\mathrm{In^-}}$$

Figure 15.11
Phenol red—A pH indicator
Phenol red, when in an acidic solution, has a yellow color (left). In pure water, it is orange (center). In an alkaline solution, the indicator is red (right).

In an acidic solution, $[H_3O^+]$ is high. Because H_3O^+ is a common ion, it suppresses the ionization of the indicator acid. In accordance with Le Châtelier's principle, equilibrium shifts to the left, favoring the yellow form of the indicator. In a basic solution, such as $NH_3(aq)$, H_3O^+ from the indicator acid is neutralized. Equilibrium shifts to the right, favoring the red indicator anion. In a solution that is closer to neutral, the indicator acid and anion are present in about equal concentrations, and the observed color is orange—a mixture of red and yellow.

Phenol red indicator:	$pH < 6.6$	$6.6 < pH < 8.2$	$pH > 8.2$
	Yellow	Orange	Red

One of the most commonly used acid-base indicators in the introductory chemistry laboratory is litmus, a natural dye extracted from lichens. It is often used in the form of paper strips that have been impregnated with a water solution of litmus and allowed to dry. The litmus paper is moistened with the solution being tested.

Buffers in Blood

Buffers are of utmost importance in our blood and other body fluids. Cells in living organisms must maintain a proper pH in order to carry out essential life processes, primarily because enzyme function (Section 13.11) is sharply dependent on pH. The normal pH value of blood plasma is 7.4. Sustained variations of a few tenths of a pH unit can cause severe illness or death.

Acidosis, a condition in which the pH of blood decreases, can be brought on by heart failure, kidney failure, diabetes mellitus, persistent diarrhea, a long-term high-protein diet, or other factors. Prolonged, intense exercise can cause temporary acidosis.

Alkalosis, characterized by an increase in the pH of blood, may result from severe vomiting, hyperventilation (excessive breathing, sometimes caused by anxiety or hysteria), or exposure to high altitudes (altitude sickness). Arterial blood samples taken from climbers who reached the summit of Mount Everest (8848 m = 29,028 ft) without supplemental oxygen had pH values between 7.7 and 7.8. This alkalosis was caused by hyperventilation; to compensate for the very low partial pressures of O_2 at this altitude (about 43 mmHg compared to about 159 mmHg at sea level), the climber must breathe extremely rapidly.

A good measure of the buffer capacity of human blood is that the addition of 0.01 mol HCl to one liter of blood lowers the pH only from 7.4 to 7.2. In contrast, the same amount of HCl added to a saline (NaCl) solution isotonic with blood lowers the pH from 7.0 to 2.0. The saline solution has no buffer capacity.

Several buffers are involved in the control of the pH of blood. Perhaps the most important one is the HCO_3^- (bicarbonate ion) and H_2CO_3 (carbonic acid) system. [In this system, we treat $CO_2(g)$ as if it were completely converted to H_2CO_3, and we consider only the first ionization of the diprotic acid H_2CO_3.]

$$\text{(1)} \quad H_2CO_3 + H_2O \rightleftharpoons H_3O^+ + HCO_3^-$$

Carbon dioxide enters the blood from tissues as the by-product of metabolic reactions. In the lungs, $CO_2(g)$ is exchanged for $O_2(g)$, which is transported throughout the body by the blood.

The blood buffers must be able to neutralize excess acid, such as the lactic acid produced by exercise. A relatively high concentration of HCO_3^- helps in this regard; it reacts with excess acid to reverse the ionization reaction shown in equation (1).

Excess alkalinity is much less common than excess acidity. Should it occur, however, additional H_2CO_3 can be formed in the lungs by reabsorbing $CO_2(g)$ to build up the H_2CO_3 content of the blood.

$$CO_2(g) + H_2O(l) \rightleftharpoons H_2CO_3(aq)$$

The H_2CO_3 then ionizes as needed to neutralize the excess base.

Other blood buffers include the dihydrogen phosphate ($H_2PO_4^-$)/monohydrogen phosphate (HPO_4^{2-}) system.

$$\text{(2)} \quad H_2PO_4^- + H_2O \rightleftharpoons H_3O^+ + HPO_4^{2-}$$

The HPO_4^{2-} reacts with excess acid in the reverse of reaction (2). Excess alkali entering the blood can be neutralized in part by reaction with $H_2PO_4^-$.

Some plasma proteins also act as buffers. The $-COO^-$ groups on the protein molecules can react with excess acid.

$$-COO^- + H_3O^+ \longrightarrow -COOH + H_2O$$

The $-NH_3^+$ groups on the protein molecules can neutralize excess base.

$$-NH_3^+ + OH^- \longrightarrow -NH_2 + H_2O$$

The blood buffers do a remarkable job of maintaining the pH of blood at 7.4. They can be overwhelmed, however, if the body's metabolism goes badly amiss.

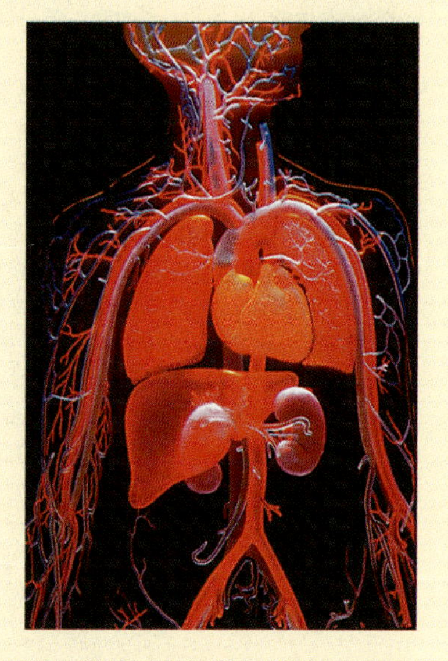

Blood is highly buffered. It is carried by the human circulatory system at a pH of 7.4.

The color change of litmus occurs over a much broader pH range than of other indicators, and it is used only to give a general indication of whether a solution is acidic or basic (see Figure 15.12).

Litmus indicator:	pH < 4.5	pH > 8.3
	Red	Blue

Figure 15.12
Litmus: A general-purpose indicator
We often use litmus paper as a rough measure to determine whether a material is acidic or basic. The soil sample on the left turns red litmus blue and is basic. The soil sample on the right turns blue litmus red and is acidic.

Figure 15.13 shows the pH ranges and colors for several other common acid-base indicators.

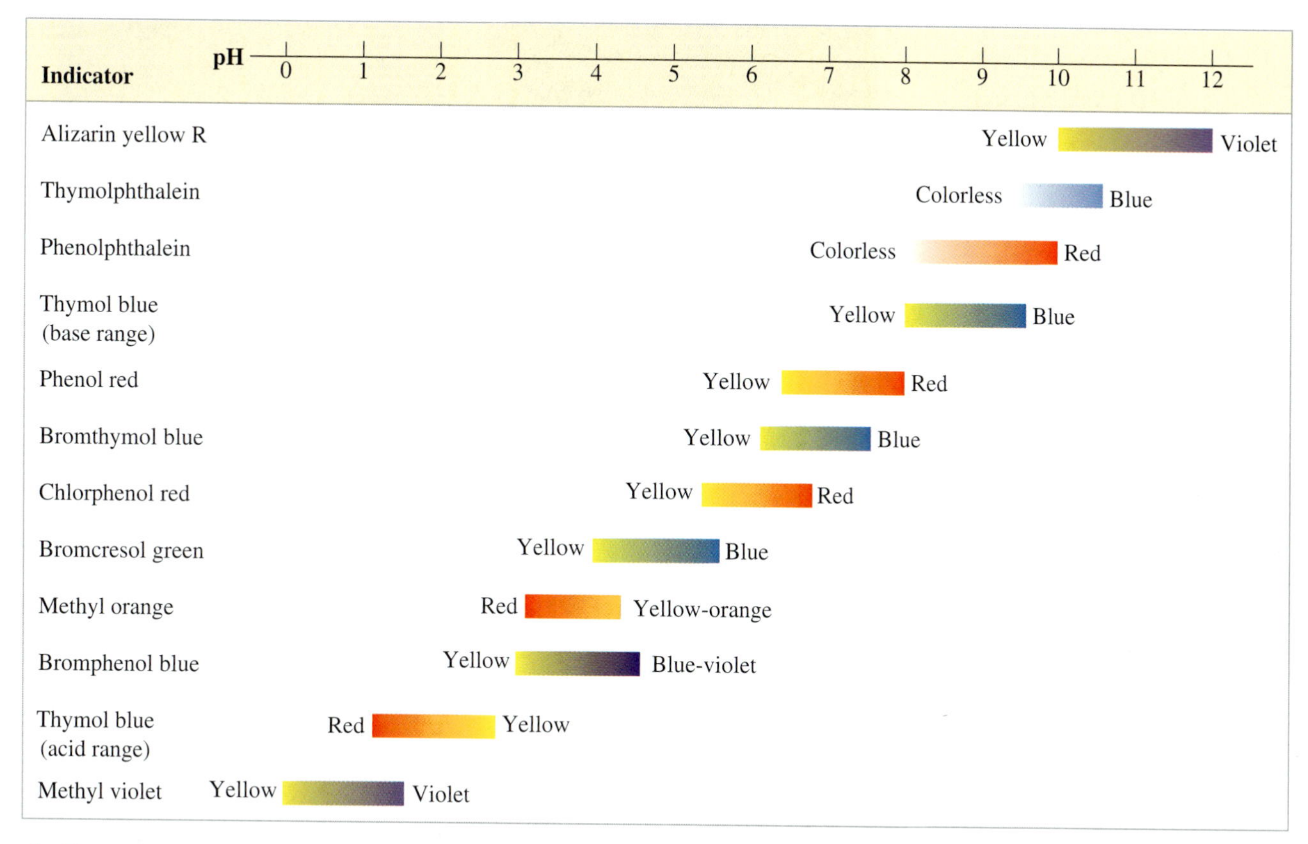

Figure 15.13 **pH ranges and colors of several common indicators**

Acid-base indicators are often used for applications in which a precise pH reading isn't necessary. A familiar use is in acid-base titrations, as we will describe in the next section.

EXAMPLE 15.19—A Conceptual Example

Explain the series of color changes of *thymol blue* indicator produced by the following actions pictured in the photographs.

(a) (b) (c) (d)

a. A few drops of thymol blue are added to HCl(aq). Solution color after addition: red.

b. A quantity of sodium acetate is added to solution (a). Solution color after addition: yellow.

c. A small quantity of sodium hydroxide is added to solution (b). Solution color after addition: yellow.

d. An additional, larger quantity of sodium hydroxide is added to solution (c). Solution color after addition: blue.

SOLUTION

a. The red color of the indicator shows the solution to be rather strongly acidic: pH < 1.2 (see Figure 15.13). The HCl is completely ionized.

$$HCl + H_2O \longrightarrow H_3O^+ + Cl^-$$

b. The yellow color shows that the pH has risen to a value greater than 3.0. This corresponds to $[H_3O^+] < 1 \times 10^{-3}$. A reaction between excess acetate ion and H_3O^+ (limiting reactant) goes nearly to completion.

$$\underset{\text{(From HCl)}}{H_3O^+} + \underset{\text{(From } CH_3COONa)}{CH_3COO^-} \longrightarrow CH_3COOH + H_2O$$

The resulting solution is one of a weak acid, CH_3COOH, and its conjugate base, CH_3COO^-. It is a buffer solution with a pH of about pH = pK_a = 4.74.

c. The small quantity of added NaOH is neutralized by the weak acid.

$$CH_3COOH + OH^- \longrightarrow H_2O + CH_3COO^-$$

This produces only a small change in the ratio $[CH_3COO^-]/[CH_3COOH]$ and correspondingly small changes in $[H_3O^+]$ and pH. The solution color remains yellow.

d. The quantity of NaOH added is enough to neutralize all the CH_3COOH in the buffer. The buffer capacity has been exceeded, and the buffer action is destroyed. Unreacted OH^- raises the pH to a value above 9.6, at which point thymol blue indicator has a blue color.

$$CH_3COOH + \underset{\text{(Excess)}}{OH^-} \longrightarrow H_2O + CH_3COO^-$$

EXERCISE 15.19

A solution is known to be one of the following four: (a) 1.00 M NH_4Cl; (b) 1.00 M NH_4Cl/1.00 M NH_3; (c) 1.00 M HCl/1.00 M HNO_3; (d) 1.00 M CH_3COOH/1.00 M CH_3COONa. A few drops of bromcresol green indicator are added to the solution, producing a green color (see Figure 15.13). Which solution(s) does this observation eliminate? What additional simple test involving a common laboratory acid and/or base could you perform to determine which of the four it is?

15.10 Neutralization Reactions and Titration Curves

In Section 4.2, we learned that *neutralization* is the reaction of an acid and a base, and we learned that *titration* is a commonly used technique for conducting a neutralization. The critical point in a titration is the **equivalence point**, at which the acid and base have been brought together in stoichiometric proportions and neither is in excess.

We can use the color change of an acid-base indicator to locate the equivalence point. The point in the titration at which the indicator changes color is called the **end point**, and the trick is to match the indicator end point and the equivalence point of the neutralization. Specifically, we need an indicator whose color change occurs over a pH range that includes the pH at the equivalence point.

We can best match the indicator end point and the equivalence point for a neutralization reaction by plotting a **titration curve**, a graph of pH versus volume of titrant. The **titrant** is the solution added from a buret. We can measure the pH with a pH meter. A recorder connected to the pH meter can automatically plot the titration curve. In this section, we will calculate the pH expected at some characteristic points on two types of titration curves. Through these calculations, we can also review some of the acid-base equilibria presented earlier in this chapter.

In a typical titration, the volume of titrant is less than 50 mL, and the molarity of the titrant is generally less than 1 M. The typical amount of H_3O^+ or OH^- delivered from the buret is only a few thousandths of a mole—for example,

APPLICATION NOTE

Medical doctors sometimes "titrate" a patient's disease symptoms with a medicine. They give increasing quantities of the drug until the desired response is achieved. Linus Pauling, in his book, *Vitamin C and the Common Cold*, recommended titrating a cold with vitamin C. The cold sufferer was supposed to take vitamin C in increasing dosages until the cold symptoms disappeared. Fortunately, end points in an acid-base titration are much easier to detect than those in a medical titration.

2.00×10^{-3} mol. We can avoid having to deal with a lot of exponential terms in calculations if we work with the unit millimole (mmol) rather than mole. Because 1 mmol is 0.001 mol, we can get millimoles by dividing moles by 1000. We also get milliliters by dividing liters by 1000, and as a result we can redefine molarity in this way:

$$M = \frac{\text{mol}}{\text{L}} = \frac{\text{mol}/1000}{\text{L}/1000} = \frac{\text{mmol}}{\text{mL}}$$

Thus, a solution that is 0.500 M HCl has 0.500 mol HCl per liter of solution or 0.500 mmol HCl per milliliter of solution.

Titration of a Strong Acid with a Strong Base

Suppose that we place 20.00 mL of 0.500 M HCl, a strong acid, in a small flask and slowly add to it 0.500 M NaOH, a strong base. To establish data for a titration curve, we can calculate the pH of the accumulated solution at different points in the titration. Then we can plot these pH values versus the volume of NaOH(aq) added. From the titration curve, we can establish the pH at the equivalence point and identify appropriate indicators for the titration.

APPLICATION NOTE
The photographs in Figure 4.5 (page 143) illustrate the technique of titration.

In Example 15.20, we calculate four representative points on the titration curve. In Exercise 15.20A, we focus on the region near the equivalence point to illustrate the very rapid rise in pH with volume of titrant added.

EXAMPLE 15.20

Calculate the pH at the following points in the titration of 20.00 mL of 0.500 M HCl with 0.500 M NaOH.

$$H_3O^+ + Cl^- + Na^+ + OH^- \longrightarrow Na^+ + Cl^- + 2\,H_2O$$

a. Before the addition of any NaOH (*the initial pH*).

b. After the addition of 10.00 mL of 0.500 M NaOH (*the half-neutralization point*). Half the original HCl is neutralized; half remains.

c. After the addition of 20.00 mL of 0.500 M NaOH (*the equivalence point*). Neither acid nor base is in excess.

d. After the addition of 20.20 mL of 0.500 M NaOH (*beyond the equivalence point*). Excess titrant is present.

SOLUTION

a. Because HCl is a strong acid, it ionizes completely. Therefore, the initial solution has $[H_3O^+] = 0.500$ M, and

$$\text{pH} = -\log\,[H_3O^+] = -\log\,(0.500) = 0.301$$

b. The total amount of H_3O^+ to be titrated is as follows.

$$20.00\ \cancel{\text{mL}} \times 0.500\ \text{mmol}\ H_3O^+/\cancel{\text{mL}} = 10.0\ \text{mmol}\ H_3O^+$$

The amount of OH^- in 10.00 mL of 0.500 M NaOH is

$$10.00\ \cancel{\text{mL}} \times 0.500\ \text{mmol}\ OH^-/\cancel{\text{mL}} = 5.00\ \text{mmol}\ OH^-$$

We can represent the progress of the neutralization reaction as follows.

The reaction:	H_3O^+	+ OH^-	$\longrightarrow$ $2\,H_2O$
Initial amounts, mmol:	10.0	≈ 0	
Add, mmol:		5.00	
Changes, mmol:	−5.00	−5.00	
After reaction, mmol:	5.0	≈ 0	

The total volume is 20.00 mL + 10.00 mL = 30.00 mL, and

$$[H_3O^+] = \frac{5.0 \text{ mmol } H_3O^+}{30.00 \text{ mL}} = 0.17 \text{ M}$$

$$pH = -\log [H_3O^+] = -\log 0.17 = 0.77$$

c. At the equivalence point, the solution is simply NaCl(aq). Because neither Na^+ nor Cl^- hydrolyzes, the solution pH = 7.00.

d. The amount of OH^- in 20.20 mL of 0.500 M NaOH is

$$20.20 \, \cancel{\text{mL}} \times \frac{0.500 \text{ mmol } OH^-}{1 \, \cancel{\text{mL}}} = 10.1 \text{ mmol } OH^-$$

Again, we can represent the progress of the neutralization as follows.

The reaction:	H_3O^+	+ OH^-	$\longrightarrow$ $2\,H_2O$
Initial amounts, mmol:	10.0	≈ 0	
Add, mmol:		10.1	
Changes, mmol:	−10.0	−10.0	
After reaction, mmol:	≈ 0	0.1	

We can calculate the concentration of OH^-, the pOH, and the pH as follows.

$$[OH^-] = \frac{0.1 \text{ mmol } OH^-}{(20.00 + 20.20) \text{ mL}} = 0.002 \text{ M}$$

$$pOH = -\log [OH^-] = -\log 0.002 = 2.7$$

$$pH = 14.00 - pOH = 14.00 - 2.7 = 11.3$$

EXERCISE 15.20A

For the titration described in Example 15.20, determine the pH after the addition of the following volumes of 0.500 M NaOH.

a. 19.90 mL **b.** 19.99 mL **c.** 20.01 mL **d.** 20.10 mL

EXERCISE 15.20B

In the titration of 25.00 mL of 0.220 M NaOH, what is the pH of the solution after 8.10 mL of 0.252 M HCl has been added?

Figure 15.14 illustrates these features of the titration curve for the titration of a *strong acid* with a *strong base*.

- The pH is *low* at the beginning of the titration.
- The pH changes slowly until just before the equivalence point.

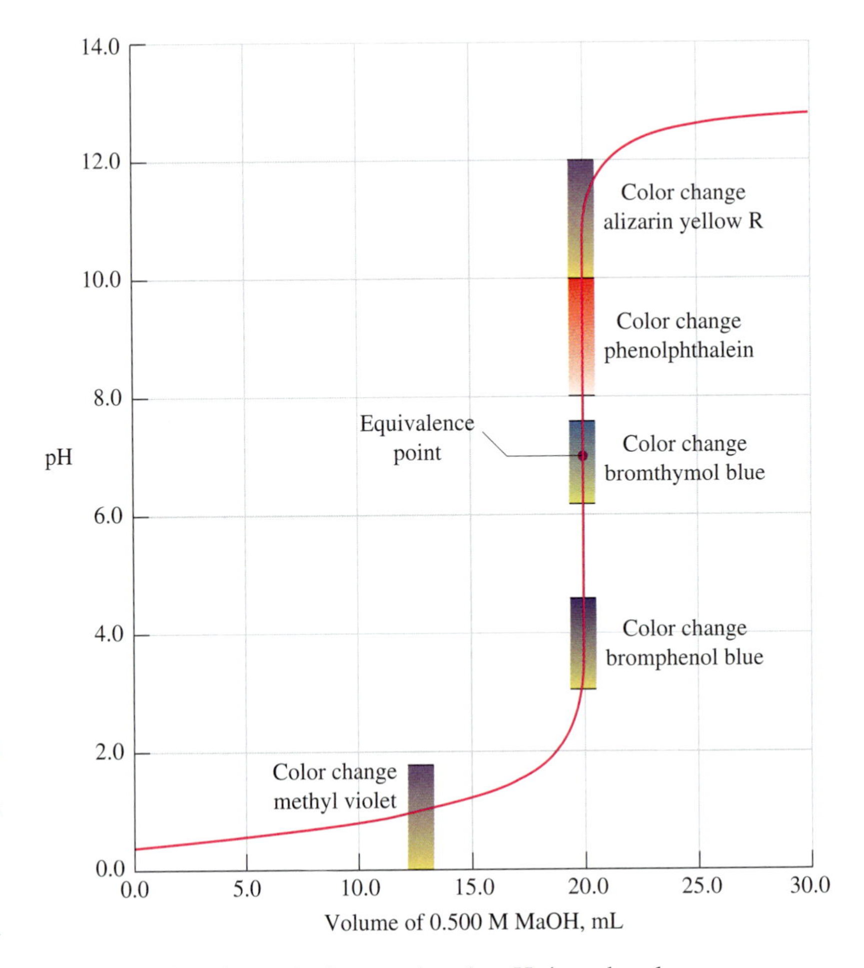

Figure 15.14 Titration curve for a strong acid by a strong base: 20.00 mL of 0.500 M HCl by 0.500 M NaOH
Any indicator that changes color along the steep portion of the titration curve is suitable for the titration. Methyl violet changes color too soon, and alizarin yellow R too late.

- Just before the equivalence point, the pH rises sharply.
- At the equivalence point, the pH is 7.00.
- Just past the equivalence point, the pH continues its sharp rise.
- Further beyond the equivalence point, the pH continues to increase, but much more slowly.
- Any indicator whose color changes in the pH range from about 4 to 10 can be used in the titration of a strong acid with a strong base.

Titration of a Weak Acid with a Strong Base

If we use the same strong base to titrate two different solutions of equal molarity—one a strong acid and the other, a weak acid—we get titration curves with two features in common: (1) The volume of base required to reach the equivalence point is the same in both cases. (2) The portions of the curves *after* the equivalence points are substantially the same. However, there are some important differences between the two cases, as we suggest in Figure 15.15.

In contrast to the titration of a strong acid with a strong base, in the titration of a weak acid with a strong base, we find these features.

- The initial pH is higher because the weak acid is only partially ionized.
- At the half-neutralization point, pH $= pK_a$. The solution at this point is a buffer solution in which the concentrations of the weak acid and its conjugate base (the anion) are equal.
- The pH is greater than 7 at the equivalence point because the anion of the weak acid hydrolyzes.

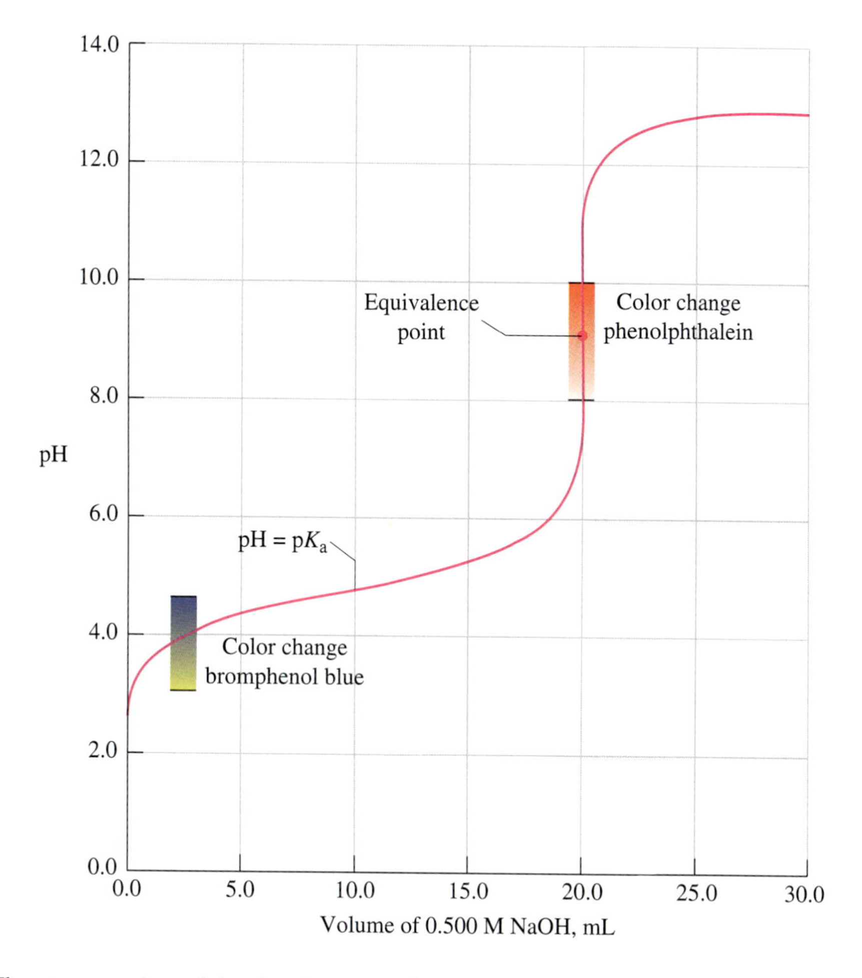

Figure 15.15 Titration curve for a weak acid by a strong base: 20.00 mL of 0.500 M CH_3COOH by 0.500 M NaOH
Phenolphthalein can be used as an indicator for this titration; bromphenol blue cannot. When exactly half the acid is neutralized, $[CH_3COOH] = [CH_3COO^-]$ and $pH = pK_a = 4.74$.

- The steep portion of the titration curve just prior to and just beyond the equivalence point is confined to a smaller pH range.
- The choice of indicators for the titration is more limited. The color change must occur in a basic solution. Generally, the midpoint of the pH range in which the indicator changes color must be well above pH 7.

Some of these points should be made clearer by the calculations in Example 15.21. As suggested by Figure 15.16, the calculations required for the titration curve of a weak acid with a strong base include most of the types considered throughout the chapter.

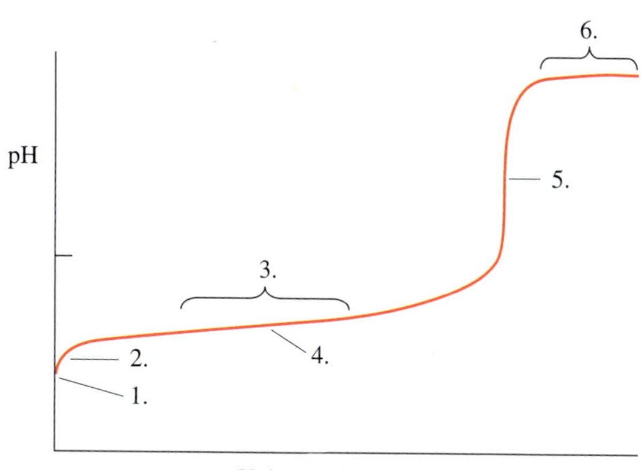

Figure 15.16 Summarizing types of equilibrium calculations in the titration curve for a weak acid with a strong base
The *six* types of calculations needed to establish the graph are:

1. Initial pH: Ionization of a weak acid.
2. Early stages of titration: Ionization of weak acid suppressed by common ion.
3. In the pH range, $pH = pK_a \pm 1$: Buffer solutions (Henderson-Hasselbalch equation).
4. At the half-neutralization point: $pH = pK_a$.
5. At the equivalence point: Hydrolysis of an anion (conjugate base of the weak acid).
6. Beyond the equivalence point: Strong base in aqueous solution.

EXAMPLE 15.21

Calculate the pH at the following points in the titration of 20.00 mL of 0.500 M CH_3COOH with 0.500 M NaOH.

$$CH_3COOH + Na^+ + OH^- \longrightarrow Na^+ + CH_3COO^- + H_2O$$

a. Before the addition of any NaOH (*the initial pH*).

b. After the addition of 8.00 mL of 0.500 M NaOH (*the buffer region*).

c. After the addition of 10.00 mL of 0.500 M NaOH (*the half-neutralization point*).

d. After the addition of 20.00 mL of 0.100 M NaOH (*the equivalence point*).

e. After the addition of 21.00 mL of 0.100 M NaOH (*beyond the equivalence point*).

SOLUTION

a. This calculation is similar to that of Example 15.6, except that the acid is 0.500 M.

	$CH_3COOH + H_2O \rightleftharpoons$	H_3O^+	$+ CH_3COO^-$
The reaction:	$CH_3COOH + H_2O \rightleftharpoons$	H_3O^+	$+ CH_3COO^-$
Initial concentrations, M:	0.500	≈ 0	0
Changes, M:	$-x$	$+x$	$+x$
Equilibrium concentrations, M:	$(0.500 - x)$	x	x

$$K_a = \frac{[H_3O^+][CH_3COO^-]}{[CH_3COOH]}$$

$$1.8 \times 10^{-5} = \frac{x \cdot x}{0.500 - x}$$

We can make the usual assumption, that is, that $x \ll 0.500$.

$$K_a = \frac{x^2}{0.500} = 1.8 \times 10^{-5}$$

$$x^2 = 9.0 \times 10^{-6}$$

$$x = [H_3O^+] = 3.0 \times 10^{-3}$$

$$pH = -\log [H_3O^+] = -\log (3.0 \times 10^{-3}) = 2.52$$

b. The addition of 8.00 mL of 0.500 M NaOH represents the addition of

$$8.00 \text{ mL} \times \frac{0.500 \text{ mmol OH}^-}{1 \text{ mL}} = 4.00 \text{ mmol OH}^-$$

At this point in the titration, we have the following data.

	CH_3COOH	$+ OH^- \longrightarrow$	$H_2O + CH_3COO^-$
The reaction:	CH_3COOH	$+ OH^- \longrightarrow$	$H_2O + CH_3COO^-$
Initial amounts., mmol:	10.0	≈ 0	≈ 0
Add, mmol:		4.00	
Changes, mmol:	-4.00	-4.00	$+4.00$
After reaction, mmol:	6.0	≈ 0	4.0

The simplest approach is to use the Henderson-Hasselbalch equation, with $pK_a = 4.74$, $[CH_3COO^-] = 4.0$ mmol/28.00 mL, and $[CH_3COOH] = 6.0$ mmol/28.00 mL.

$$pH = pK_a + \log \frac{[CH_3COO^-]}{[CH_3COOH]}$$

$$pH = 4.74 + \log \frac{(4.0/28.00)}{(6.0/28.00)}$$

$$pH = 4.74 + \log 0.67 = 4.74 - 0.17 = 4.57$$

PROBLEM-SOLVING NOTE

The units of the concentrations in the logarithmic expression are mmol/mL = M. There is no need to include them, however, because they appear in a ratio and cancel out.

c. At the point at which half the acid has been neutralized—the half-neutralization point—the titration has progressed to the following point.

The reaction:	CH_3COOH	+ OH^-	$\longrightarrow$ H_2O	+ CH_3COO^-
Initial amounts, mmol:	10.0	≈ 0		≈ 0
Add, mmol:		5.00		
Changes, mmol:	−5.00	−5.00		+5.00
After reaction, mmol:	5.0	≈ 0		5.0

Because the CH_3COOH and CH_3COO^- are present in equal amounts—5.0 mmol—and in the same 30.00 mL of solution, their concentrations are equal. This means that

$$pH = pK_a = 4.74$$

d. At the equivalence point, 10.0 mmol of CH_3COOH and 10.0 mmol of NaOH have reacted to produce 10.0 mmol of CH_3COONa in 40.00 mL of solution (20.00 mL acid + 20.00 mL base). The solution molarity is shown below.

$$\frac{10.0 \text{ mmol } CH_3COONa}{40.00 \text{ mL}} = 0.250 \text{ M } CH_3COONa$$

We calculated the pH of 0.250 M CH_3COONa in Example 15.14 in our discussion of hydrolysis and found pH = 9.08.

e. The addition of 21.00 mL of 0.500 M NaOH represents the addition of the following.

$$21.00 \text{ mL} \times \frac{0.500 \text{ mmol } OH^-}{1 \text{ mL}} = 10.5 \text{ mmol } OH^-$$

This point is beyond the equivalence point, where we have the following data.

The reaction:	CH_3COOH	+ OH^-	$\longrightarrow$ H_2O	+ CH_3COO^-
Initial amounts, mmol:	10.0	≈ 0		≈ 0
Add, mmol:		10.5		
Changes, mmol:	−10.0	−10.0		+10.0
After reaction, mmol:	≈ 0	0.5		10.0

Acetate ion is a weak base compared to OH^-. The hydroxide ion concentration is simply

$$[OH^-] = \frac{0.5 \text{ mmol}}{20.00 + 21.00 \text{ mL}} = 0.01 \text{ M}$$

$$pOH = -\log [OH^-] = -\log (1 \times 10^{-2}) = 2.0$$

$$pH = 14.00 - pOH = 14.00 - 2.0 = 12.0$$

EXERCISE 15.21

For the titration described in Example 15.21, determine the pH after the addition of the following volumes of 0.500 M NaOH.

a. 12.50 mL **b.** 20.10 mL

EXAMPLE 15.22—A Conceptual Example

The titration curve below involves 1.0 M solutions of an acid and a base. Identify the type of titration represented by this curve.

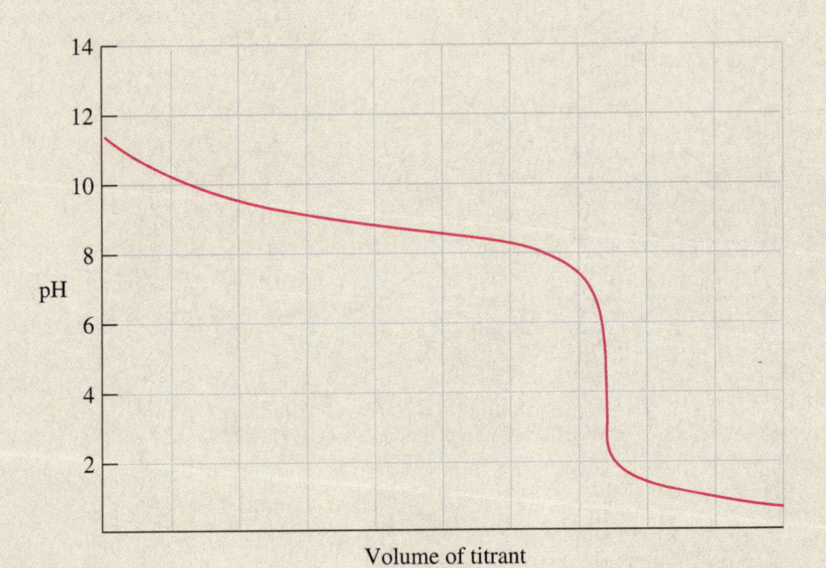

SOLUTION

● **ChemCDX** Acid-Base Titration.

We can identify these features of the titration curve and draw certain conclusions from them.

- The pH starts high and decreases during the titration. The solution being titrated is a *base*; the titrant is an *acid*.
- The base must be a *weak base*. The initial pH is about 11.5, but a 1.0 M strong base would dissociate completely so that $[OH^-] = 1.0$ M, $[H_3O^+] = 1.0 \times 10^{-14}$, and pH = 14.00.
- The pH at the equivalence point is *less than 7*. This is the pH expected for the hydrolysis of the salt of a *strong acid* and a *weak base*.
- The pH beyond the equivalence point drops to a low value (pH < 1), again suggesting that the titrant is a *strong acid*.

From these observations, we conclude that the curve represents the titration of a *weak base* with a *strong acid*.

EXERCISE 15.22

For the titration described in Example 15.22, **(a)** estimate the value of K_b of the weak base, and **(b)** obtain a value of the pH at the equivalence point, by calculation and by estimation from the titration curve.

(*Hint:* Is it necessary to know the actual volume of titrant?)

15.11 Lewis Acids and Bases

We have devoted most of this chapter to the Brønsted-Lowry theory of acids and bases, which works well for the reactions in aqueous solutions that we have emphasized here. However, there are reactions in nonaqueous solvents, in the gaseous state, and even in the solid state that can be thought of as acid-base reactions. Chemists have devised several more general acid-base theories. In this section, we will briefly consider one of the more important alternative theories.

In 1923, G. N. Lewis proposed an acid-base theory that focuses on the role of electron pairs, in contrast to the protons of the Brønsted-Lowry theory. A **Lewis acid** is a species that is an electron-pair *acceptor*. A **Lewis base** is a species that is an electron-pair *donor*. In a Lewis acid-base reaction, a covalent bond is formed between the acid and base. We actually encountered an example earlier in the text (page 386): the formation of a coordinate covalent bond between BF_3 and F^-.

$$:\ddot{F}-B(:\ddot{F}:)_2 \quad + \quad :\ddot{F}:^- \longrightarrow \left[:\ddot{F}-B(:\ddot{F}:)_2-\ddot{F}:\right]^-$$

Lewis acid Lewis base

This example suggests that in a Lewis acid-base reaction we should look for (1) a species that has an available empty orbital to accommodate an electron pair (such as the B atom in the BF_3 molecule) and (2) a species that has lone-pair electrons (such as the F^- ion).

This definition allows us to consider typical Brønsted-Lowry bases, such as OH^-, NH_3, and H_2O, also as Lewis bases. They all have electron pairs available for sharing.

$$\left[:\ddot{O}-H\right]^- \qquad H-\ddot{N}(H)-H \qquad H-\ddot{O}:(H)$$

The definition seems not to fit a Brønsted-Lowry acid like HCl as a Lewis acid. An HCl molecule cannot gain an electron pair. However, if we consider that HCl ionizes to produce H^+, the H^+ ion is a Lewis acid. It accepts an electron pair when it bonds to a lone pair of electrons in an H_2O molecule, forming the hydronium ion H_3O^+. We can think of HCl as a *source* of a Lewis acid.

Now let's consider a reaction that is an acid-base reaction according to the Lewis theory but not the Brønsted-Lowry theory—no hydrogen atoms are involved.

$$CaO(s) + SO_2(g) \longrightarrow CaSO_3(s)$$

The best way to see this as a Lewis acid-base reaction is to write Lewis structures and to identify an electron pair (in red) that is supplied by O^{2-} (the base) to SO_2 (the acid). A shift of an electron pair (in blue) out of the sulfur-to-oxygen bond to a lone pair position is also involved.

$$Ca^{2+} \; :\ddot{O}:^{2-} + \ddot{S}(=\ddot{O}:)(-:\ddot{O}:) \longrightarrow Ca^{2+}\left[:\ddot{O}-\ddot{S}(-:\ddot{O}:)(-:\ddot{O}:)\right]^{2-}$$

In organic chemistry, Lewis acids are often called *electrophiles*: species that seek electrons. Lewis bases are called *nucleophiles*, that is, species that seek a positive site. We consider applications to organic chemistry in Chapter 23.

Key Terms

acid-base indicator (15.9)
acid ionization constant, K_a (15.1)
amphiprotic (15.1)
base ionization constant, K_b (15.1)
buffer solution (15.8)
common ion effect (15.7)
conjugate acid (15.1)
conjugate acid-base pair (15.1)
conjugate base (15.1)
end point (15.10)
equivalence point (15.10)
Henderson-Hasselbalch equation (15.8)
hydrolysis (15.6)
ion product of water, K_w (15.3)
Lewis acid (15.11)
Lewis base (15.11)
monoprotic acid (15.5)
pH (15.3)
pK_a (15.4)
pK_b (15.4)
pK_w (15.3)
pOH (15.3)
polyprotic acid (15.5)
proton acceptor (15.1)
proton donor (15.1)
titrant (15.10)
titration curve (15.10)

Summary

In the Brønsted-Lowry theory, an acid is a proton donor and a base is a proton acceptor. Every acid has a conjugate base, and every base has a conjugate acid. In an acid-base reaction, the forward reaction is between an acid and a base; the reverse reaction is between a conjugate base and a conjugate acid.

$$\underset{\text{acid(1)}}{\text{H—A}} + \underset{\text{base(2)}}{\text{B:}} \rightleftharpoons \underset{\text{acid(2)}}{\text{H—B}} + \underset{\text{base(1)}}{\text{A}^-}$$

Brønsted-Lowry acids and bases can be ranked by relative strength. If an acid is strong, its conjugate base is weak; and if a base is strong, its conjugate acid is weak. An acid-base reaction is favored in the direction from the stronger member to the weaker member of each conjugate acid-base pair. Acid ionization requires bond breakage and release of a proton (H^+). The ease of bond breakage is affected by such factors as electronegativity, number of terminal O atoms in oxoacids, and the placement of substituent groups on carbon chains in carboxylic acids.

Water is amphiprotic; it can be either an acid or a base. A limited transfer of protons between H_2O molecules in pure water produces H_3O^+ and OH^-. The concentrations of H_3O^+ and OH^- are in accordance with the ion product.

$$K_w = [H_3O^+][OH^-] = 1.0 \times 10^{-14} \text{ at } 25\ °\text{C}$$

In the notation of negative logarithms:

$$\text{pH} = -\log\,[H_3O^+] \qquad \text{pOH} = -\log\,[OH^-] \qquad \text{p}K_w = -\log K_w$$

The pH in both pure water and in neutral solutions is 7. The pH in acidic solutions is less than 7, and the pH in basic solutions is greater than 7. In *all* aqueous solutions at 25 °C, pH + pOH = 14.00.

The strategies for calculations involving weak acids and weak bases are similar to those used in Chapter 14. We frequently make simplifying assumptions.

Ions, as well as neutral molecules, can act as acids and bases. Reactions of ions with water molecules, called hydrolysis reactions, cause certain salt solutions to be either acidic or basic. In calculating the pH of a salt solution, it is often necessary to establish the ionization constant of an acid or a base from that of its conjugate:

$$\text{p}K_a\,(\text{acid}) + \text{p}K_b\,(\text{conjugate base}) = 14.00.$$

A strong electrolyte that produces an ion common to the ionization equilibrium of a weak acid or a weak base suppresses the ionization of the weak electrolyte. An important application of this common ion effect is in buffer solutions. A buffer is a mixture of a weak acid and its conjugate base or a weak base and its conjugate acid. A buffer maintains an essentially constant pH when small amounts of a strong acid or strong base are added. One buffer component neutralizes small added amounts of the acid, and the other component neutralizes small added amounts of the base.

Another application of the common ion effect is in the action of acid-base indicators. H_3O^+ from the solution being tested affects the concentrations of the indicator acid, HIn, and the indicator base, In^-. In turn, this determines the color exhibited by the indicator. An appropriate indicator for an acid-base titration undergoes a color change at the pH of the equivalence point, the point in a titration where the acid and base have reacted in stoichiometric proportions and neither is in excess. A titration curve, a graph of pH versus volume of titrant solution, can be used to establish the equivalence point. To calculate the points on a titration curve, many of the concepts introduced throughout the chapter are used, especially if one of the reactants is a weak acid or a weak base.

In Lewis acid-base theory, a Lewis acid accepts an electron pair from a Lewis base, which is an electron-pair donor. In a Lewis acid-base reaction, new covalent bonds are formed.

Review Questions

1. Write equations to represent the ionization of HI as an acid in both the Arrhenius and Brønsted-Lowry theories.

2. Can a substance be a Brønsted-Lowry acid if it does not contain H atoms? Are there any characteristic atoms that must be present in a Brønsted-Lowry base?

3. Must every Brønsted-Lowry acid have a conjugate base? every base have a conjugate acid? Explain.

4. What is an *amphiprotic* species? Is H_3PO_4 amphiprotic? $H_2PO_4^-$? NH_4^+? H_3O^+? Explain.

5. Write equations for the ionizations and K_a expressions for each of the following as Brønsted-Lowry weak acids.

(a) HOClO **(b)** CH_3CH_2COOH **(c)** HCN **(d)** C_6H_5OH

6. Explain how the strengths of *binary* acids are affected by **(a)** bond energies and **(b)** the ionic radii of the anions they produce.

7. Explain how the strengths of *oxoacids* are affected by **(a)** the electronegativity of the central nonmetal atom, and **(b)** the number of terminal O atoms.

8. Describe **(a)** two factors that affect the acidic strength of a carboxylic acid, and **(b)** two factors that affect the basic strength of amines.

9. What is meant by the pH of a solution? What is the pOH? What is the relationship between the pH and pOH of a solution? What is the pH of pure water at 25 °C?

10. What is meant by the ion product constant of water? What is its value at 25 °C? How is the pH of pure water related to the ion product constant at 25 °C? What do the terms *acidic*, *basic*, and *neutral* mean in relation to the pH of water at 25 °C?

11. What is meant by the *percent ionization* of an acid? How is the percent ionization of a weak acid affected by the concentration of the acid?

12. What are the characteristic features of a *polyprotic* acid? Is C_6H_6 a polyprotic acid? Explain.

13. Explain why the K_{a_1} for a diprotic acid is larger than K_{a_2}.

14. Describe the difference between the *ionization* of a weak acid and the *hydrolysis* of a weak acid anion.

15. What is the relationship between K_a for a Brønsted-Lowry acid and K_b for its conjugate base?

16. Explain what is meant by the *common ion effect*.

17. What is a buffer solution? What are the necessary components of a buffer solution, and what role does each component play?

18. What is meant by the terms *capacity* and *range* in describing a buffer solution? Do either of these depend on the concentrations of the buffer components? Explain.

19. How does an acid-base indicator work? What is the role of the common ion effect in the functioning of an acid-base indicator?

20. Describe the difference between the *equivalence point* and the *end point* of an acid-base titration carried out with an acid-base indicator.

21. What is a titration curve? How can such a curve be used to determine **(a)** the equivalence point and **(b)** a suitable indicator for an acid-base titration?

22. If the titration of a *weak base* is carried out using a *strong acid* as the titrant, at what point on the titration curve will the pH be lowest and at what point will it be highest? At approximately what pH will the equivalence point be found?

23. The titration of a weak acid with a weak base generally proves to be unsatisfactory. Why do you suppose this is the case?

24. Describe the Lewis acid-base theory. Can a substance be a Lewis acid without being a Brønsted-Lowry acid? Explain.

Problems

[Unless otherwise indicated, assume a temperature of 25 °C.]

Brønsted-Lowry Acids and Bases

25. For each of the following, identify the conjugate acid-base pairs by using notation such as acid(1) and base(1).

(a) $HOClO_2 + H_2O \rightleftharpoons H_3O^+ + OClO_2^-$

(b) $HSeO_4^- + NH_3 \rightleftharpoons NH_4^+ + SeO_4^{2-}$

(c) $HCO_3^- + OH^- \rightleftharpoons CO_3^{2-} + H_2O$

(d) $C_5H_5NH^+ + H_2O \rightleftharpoons C_5H_5N + H_3O^+$

26. For each of the following, identify the conjugate acid-base pairs by using notation such as acid(1) and base(1). Then use Table 15.1 to rank the four reactions in order of increasing tendency for the reaction to go to completion (to the right).

(a) $HSO_4^- + F^- \rightleftharpoons HF + SO_4^{2-}$

(b) $NH_4^+ + Cl^- \rightleftharpoons NH_3 + HCl$

(c) $HCl + CH_3COO^- \rightleftharpoons CH_3COOH + Cl^-$

(d) $CH_3OH + Br^- \rightleftharpoons HBr + CH_3O^-$

27. Identify the species that is amphiprotic, and write one equation for its reaction with OH^-(aq) and another for its reaction with HBr(aq): H_2S, SO_2, HSO_3^-, H_2CO_3.

28. Identify the species that is amphiprotic and write two equations for its reaction with H_2O(l) that illustrate its amphiprotic character: HI, $H_2PO_4^-$, H_2SO_4, H_2CO_3, CO_3^{2-}.

29. With which of the following bases will the reaction of acetic acid, CH_3COOH, proceed furthest toward completion (to the right): (a) CO_3^{2-}; (b) F^-; (c) Cl^-; or (d) NO_3^-?

30. With which of the following acids will the reaction of sulfate ion, SO_4^{2-}, proceed furthest toward completion (to the right): HF, HCl, HCO_3^-, or CH_3COOH? Explain.

31. Liquid NH_3, like water, is an amphiprotic solvent. Write an equation for the self-ionization of NH_3(l). Draw Lewis structures for the reactants and products.

32. Aniline, $C_6H_5NH_2$, is a weak base in water, but it is a strong base in glacial acetic acid, CH_3COOH(l). Explain this difference in behavior.

Molecular Structure and the Strengths of Acids and Bases

33. Explain why perchloric acid, $HClO_4$, is a strong acid, whereas chlorous acid, $HClO_2$, is a weak acid.

34. Explain why perchloric acid, $HClO_4$, is a stronger acid than sulfuric acid, H_2SO_4.

35. Identify the stronger acid in each pair, and explain your choice.

(a) H_2S or H_2Se

(b) $HClO_3$ or HIO_3

(c) H_3AsO_4 or $H_2PO_4^-$

36. Identify the stronger acid in each pair, and explain your choice.

(a) H_2Se or HBr

(b) HN_3 or HCN

(c) HNO_3 or HSO_4^-

37. Refer to ideas and data presented on page 641 and Table 15.2, and estimate the pK_a of dichloroacetic acid. How would you expect pK_a of trichloroacetic acid to compare with that of dichloroacetic acid?

38. Refer to the ideas and data presented on page 641, and estimate a value of K_a for **(a)** 2,3-dichloropropanoic acid, $ClCH_2CHClCOOH$; and **(b)** 4-chlorobutanoic acid, $ClCH_2CH_2CH_2COOH$.

39. Phenol, C_6H_5OH, ionizes as an acid.

$$C_6H_5\text{-OH} + H_2O \rightleftharpoons H_3O^+ + C_6H_5\text{-O}^- \quad K_a = 1.0 \times 10^{-10}$$

Arrange the following substituted phenols in the order in which you would expect their K_a values to *increase*. Where would you expect phenol itself to fit into this ranking?

(a) phenol ring with OH and Cl substituents

(b) phenol ring with OH and two Cl substituents

(c) 2,4,6-trichlorophenol: Cl at positions 2, 4, 6 with —OH

(d) 2-chlorophenol: —OH with Cl

40. Aniline, $C_6H_5NH_2$, ionizes as a base.

$$C_6H_5NH_2 + H_2O \rightleftharpoons C_6H_5NH_3^+ + OH^- \quad K_b = 7.4 \times 10^{-10}$$

Arrange the following substituted anilines in the order in which you would expect their K_b values to *increase*. Where would you expect aniline itself to fit into this ranking?

(a) NH_2 with ortho NO_2 (b) O_2N—, NH_2, NO_2 (c) NO_2, NH_2, NO_2 (d) NH_2 with meta NO_2

The pH Scale

41. The pH of a cup of coffee at 25 °C is 4.32. What is the $[H_3O^+]$ in the coffee?

42. A detergent solution has a pH of 11.13 at 25 °C. What is the $[OH^-]$ in the solution?

43. What is the pH of each of the following aqueous solutions?

(a) 0.0025 M HCl **(c)** 0.015 M $Ba(OH)_2$

(b) 0.055 M NaOH **(d)** 1.6×10^{-3} M HBr

44. What is the pOH of each of the following aqueous solutions?

(a) 2.5×10^{-3} M NaOH **(c)** 3.6×10^{-4} M $Ca(OH)_2$

(b) 3.2×10^{-3} M HCl **(d)** 0.000220 M HNO_3

45. Which has the lower pH, 0.00048 M H_2SO_4 or a vinegar solution having pH = 2.42?

46. Which has the higher pH, 0.0062 M $Ba(OH)_2(aq)$ or an ammonia cleanser having a pH = 11.65?

47. Describe how you would prepare 2.00 L of an aqueous solution having a pH = 3.60, if you had available a supply of 0.100 M HCl.

48. Describe how you would prepare 5.00 L of an aqueous solution having a pH = 10.70, if you had available a supply of 0.250 M NaOH.

Equilibria in Solutions of Weak Acids and Weak Bases

(Where necessary, obtain K_a and K_b values from Table 15.2 or Appendix C.2.)

49. Formic acid, HCOOH ($K_a = 1.8 \times 10^{-4}$), is the irritant associated with ant stings (Latin, *formica*, ant). Calculate the pH of an aqueous solution containing 75.0 g HCOOH per liter.

50. Pyridine, C_5H_5N ($K_b = 1.5 \times 10^{-9}$), is an organic base used in the synthesis of vitamins, drugs, and fungicides. Calculate the pH of an aqueous solution having 1.25 g C_5H_5N in 125 mL of solution.

51. Hydrazoic acid, HN_3 ($pK_a = 4.72$), is perhaps best known through its sodium salt, sodium azide, NaN_3, the gas-forming substance in automobile air-bag systems. What molarity of HN_3 is required to produce an aqueous solution with pH = 3.10?

52. Phenol, C_6H_5OH ($pK_a = 10.00$), is widely used in synthesizing organic chemicals, but it is more familiarly known as a general disinfectant (carbolic acid). A saturated aqueous solution of phenol has a pH = 4.90. What is the molarity of phenol in this solution?

53. A 1.00-g sample of aspirin (acetylsalicylic acid) is dissolved in 0.300 L of water at 25 °C, and its pH is found to be 2.62. What is the K_a of aspirin?

$$o\text{-}C_6H_4(OCOCH_3)COOH + H_2O \rightleftharpoons H_3O^+ + o\text{-}C_6H_4(OCOCH_3)COO^- \quad K_a = ?$$

54. Codeine, $C_{18}H_{21}NO_3$, a commonly prescribed painkiller, is a weak base. A saturated aqueous solution contains 1.00 g codeine in 120 mL of solution and has a pH = 9.8. What is the K_b of codeine?

$$C_{18}H_{21}NO_3 + H_2O \rightleftharpoons [C_{18}H_{21}NHO_3]^+ + OH^- \qquad K_b = ?$$

55. Calculate the pH of 0.105 M CCl_3COOH (trichloroacetic acid, $pK_a = 0.52$).

56. Piperidine, $C_5H_{11}N$ ($pK_b = 2.89$), is a colorless liquid having the odor of pepper. Calculate the pH of 0.00250 M $C_5H_{11}N$.

57. What is the molarity of a formic acid solution, HCOOH(aq), that has the same pH as 0.150 M CH_3COOH(aq)?

58. What is the molarity of a methylamine solution, CH_3NH_2(aq), that has the same pH as 0.0850 M NH_3(aq)?

Polyprotic Acids

59. *Without doing detailed calculations*, determine which of the following polyprotic solutions will have the *lower pH*: 0.0045 M H_2SO_4 or 0.0045 M H_3PO_4.

60. *Without doing detailed calculations*, determine which of the following is the most likely pH for 0.010 M H_2SO_4: (a) 2.00, (b) 1.85, (c) 1.70, (d) 1.50?

61. Oxalic acid, HOOCCOOH, is a weak dicarboxylic acid found in the cell sap of many plants (for example, rhubarb leaves) as its potassium or calcium salt. The solubility of oxalic acid at 25 °C is about 83 g/L. What are the **(a)** pH and **(b)** $[^-OOCCOO^-]$ in the saturated solution?

$$HOOCCOOH + H_2O \rightleftharpoons H_3O^+ + HOOCCOO^- \qquad K_{a_1} = 5.4 \times 10^{-2}$$

$$HOOCCOO^- + H_2O \rightleftharpoons H_3O^+ + {}^-OOCCOO^- \qquad K_{a_2} = 5.3 \times 10^{-5}$$

62. Citric acid ($pK_{a_1} = 3.13$, $pK_{a_2} = 4.76$, $pK_{a_3} = 6.40$) is found in citrus fruit. Calculate the approximate pH of lemon juice, which is about 5% citric acid by mass.

```
        CH2COOH
        |
   HO—CCOOH
        |
        CH2COOH
```

Citric acid

63. For a solution that is 0.15 M H_3PO_4(aq), determine **(a)** pH, **(b)** $[H_3PO_4]$, **(c)** $[H_2PO_4^-]$, **(d)** $[HPO_4^{2-}]$, and **(e)** $[PO_4^{3-}]$.

64. In thiophosphoric acid (H_3PO_3S), an S atom substitutes for one of the O atoms in the H_3PO_4 molecule. The K_a values for this acid are

$$K_{a_1} = 1.6 \times 10^{-2}, K_{a_2} = 3.7 \times 10^{-6}, K_{a_3} = 8.3 \times 10^{-11}$$

For a solution that is 0.50 M H_3PO_3S(aq), determine **(a)** pH, **(b)** $[H_3PO_3S]$, **(c)** $[H_2PO_3S^-]$, **(d)** $[HPO_3S^{2-}]$, and **(e)** $[PO_3S^{3-}]$.

Hydrolysis

65. Write equations for the hydrolysis reactions that occur in the following solutions, and predict whether the solution is acidic, basic, or neutral.

(a) CH_3CH_2COOK(aq) **(c)** NH_4CN(aq)

(b) $Mg(NO_3)_2$(aq)

66. Write equations for the hydrolysis reactions that occur in the following solutions, and predict whether the solution is acidic, basic, or neutral.

(a) $RbClO_4$(aq) **(c)** $HCOONH_4$(aq)

(b) $CH_3CH_2NH_3Br$(aq)

67. Which of the following 0.100 M aqueous solutions has the *lowest* pH: (a) $NaNO_3$, (b) CH_3COOK, (c) NH_4I, or (d) Na_3PO_4?

68. Which of the solutions in Problem 67 has the *highest* pH?

69. For a solution that is 0.080 M NaOCl, **(a)** write an equation for the hydrolysis reaction that occurs, and determine **(b)** the equilibrium constant for this hydrolysis and **(c)** the pH.

70. For a solution that is 0.602 M NH_4Cl, **(a)** write an equation for the hydrolysis reaction that occurs, and determine **(b)** the equilibrium constant for this hydrolysis and **(c)** the pH.

71. What is the molarity of a sodium acetate solution that is found to have a pH = 9.05?

72. What is the molarity of an ammonium bromide solution that is found to have a pH = 5.75?

Common Ion Effect

73. Which of the following will *suppress* the ionization of formic acid, $HCOOH(aq)$: (a) NaCl; (b) $KOH(aq)$; (c) HNO_3; (d) $(HCOO)_2Ca$; (e) Na_2CO_3? Explain.

74. Which of the following will *increase* the ionization of formic acid, $HCOOH(aq)$: (a) KI; (b) $NaOH(aq)$; (c) HNO_3; (d) HCOONa; (e) NaCN? Explain.

75. Calculate $[NH_4^+]$ in a solution that is 0.15 M NH_3 *and* 0.015 M KOH.

76. Calculate $[C_6H_5COO^-]$ in a solution that is 0.015 M C_6H_5COOH and 0.051 M HCl.

77. Calculate the pH of a solution that is 0.350 M CH_3CH_2COOH and 0.0786 M in CH_3CH_2COOK. (For CH_3CH_2COOH, $K_a = 1.3 \times 10^{-5}$.)

78. Calculate the pH of a solution that is 0.132 M in diethylamine, $(CH_3CH_2)_2NH$, and 0.145 M in diethylammonium chloride, $(CH_3CH_2)_2NH_2Cl$. [For $(CH_3CH_2)_2NH$, $K_b = 6.9 \times 10^{-4}$]

More Acid-Base Equilibria

79. *Without doing detailed calculations*, determine which of the following solutions should have a pH greater than 7.00 and which, lower than 7.00: (a) 0.0050 M C_6H_5COOH; (b) 1.0×10^{-5} M NH_3; (c) 0.022 M CH_3NH_3Cl; (d) 0.10 M KH_2PO_4; (e) saturated $Ba(OH)_2(aq)$; (f) 0.25 M $NaNO_2$.

80. *Without doing detailed calculations*, arrange the following 0.10 M aqueous solutions in the order of *increasing* pH: (a) HCl, (b) CH_3COONa, (c) KCl, (d) H_3PO_4, (e) NH_3, (f) H_2SO_4, (g) NH_4NO_3, (h) KOH.

81. Write two equations—one to show the ionization of HPO_4^{2-} as an acid and one to show its ionization (hydrolysis) as a base. Would you expect an aqueous solution of Na_2HPO_4 to be acidic or basic? (*Hint:* What are the relevant K_a and K_b values?)

82. Predict whether an aqueous solution can be made up to have the following concentrations or whether a chemical reaction must occur.

(a) 0.25 M CH_3COONa and 0.15 M HI

(b) 0.050 M KNO_2 and 0.18 M KNO_3

(c) 0.0050 M $Ca(OH)_2$ and 0.0010 M HNO_3

(d) 0.20 M NH_4Cl and 0.35 M NaOH

Buffer Solutions

83. A solution is prepared that is 0.405 M HCOOH (formic acid) and 0.326 M HCOONa (sodium formate). What is the pH of this buffer solution?

84. What is the pH of a buffer solution that is 0.245 M CH_3NH_2 (methylamine) and 0.186 M CH_3NH_3Cl (methylammonium chloride)?

85. What mass of $(NH_4)_2SO_4(s)$ must be dissolved in 0.100 L of 0.350 M $NH_3(aq)$ to produce a buffer solution with pH = 10.05?

86. The only significant solute in white vinegar is acetic acid. Suppose you wanted to make a buffer solution from a particular white vinegar (d = 1.01 g/mL, 4.53% CH_3COOH by mass). What would be the pH of the buffer obtained by dissolving 10.0 g of sodium acetate in 0.250 L of the vinegar?

87. If 1.00 mL of 0.250 M HCl is added to 50.0 mL of the buffer solution described in Problem 83, what will be the pH of the final solution?

88. If 2.00 mL of 0.0850 M NaOH is added to 75.0 mL of the buffer solution described in Problem 84, what will be the pH of the final solution?

89. We have seen that a buffer solution contains an acid as one component and a base as the other. Can the two components be HCl and NaOH? Explain.

90. Aqueous solutions of acetic acid contain both an acid, CH_3COOH, and its conjugate base, CH_3COO^-. Can an acetic acid solution alone be considered a buffer? Explain.

Acid-Base Indicators

91. Why are so many more acid-base indicators suitable for the titration of a strong acid with a strong base than are suitable for the titration of a weak acid with a strong base?

92. Why can a single acid-base indicator suffice to determine the equivalence point in a titration, whereas more than one indicator is often needed to determine the pH of a solution?

93. Refer to Figure 15.13, and state the color you would expect for the following acid-base indicators in the given aqueous solution?

(a) bromthymol blue in 0.10 M NH_4Cl

(b) bromphenol blue in 0.10 M CH_3COOK

(c) phenolphthalein in 0.10 M Na_2CO_3

(d) methyl violet in 0.10 M CH_3COOH

94. A particular buffer solution is believed to have a pH greater than 9.5. Describe how you might use a pair of acid-base indicators to verify this. What two indicators from Figure 15.13 would you use? Explain.

95. A 0.100 M HCl(aq) solution contains thymol blue indicator and has a red color. A 0.100 M NaOH(aq) solution with phenolphthalein indicator also has a red color. What should be the color when equal volumes of the two solutions (with their indicators) are mixed?

96. When added to an unknown solution, bromcresol green indicator has a yellow color. When added to another sample of the same solution, thymol blue indicator also turns yellow. To within about one pH unit, what is the pH of the solution?

Neutralization Reactions and Titration Curves

97. Produce a listing comparable to that on pages 679-80 to describe the titration of a strong base (for example, NaOH) with a strong acid (for example, HCl).

98. Produce a listing comparable to that on pages 680-81 to describe the titration of a weak base (for example, NH_3) with a strong acid (for example, HCl).

99. Can the same indicator be used in the titration of a strong base with a strong acid as in the titration of a strong acid with a strong base? Explain.

100. Can the same indicator be used in the titration of a weak base with a strong acid as in the titration of a weak acid with a strong base? Explain.

101. In the titration of 20.00 mL of 0.500 M CH_3COOH by 0.500 M NaOH, calculate the pH at the point in the titration where 7.45 mL of 0.500 M NaOH has been added.

102. In the titration of 20.00 mL of 0.500 M HCl by 0.500 M NaOH, calculate the volume of 0.500 M NaOH required to reach a pH of 2.0. (*Hint:* Set up an equation in the unknown volume, V, and solve for V.)

103. In the titration of 25.00 mL of 0.324 M NaOH with 0.250 M HCl, how many milliliters of the HCl(aq) must be added to reach a pH of 11.50? (*Hint:* See Problem 102.)

104. In the titration of 10.00 mL of 1.050 M CH_3COOH with 0.480 M NaOH, how many milliliters of the NaOH(aq) must be added to reach a pH of 4.20? (*Hint:* What is the required value of $[CH_3COO^-]/[CH_3COOH]$? What volume of NaOH(aq), V, is required to produce this value?)

105. Construct, by calculation, a titration curve for the titration of 40.00 mL of 0.200 M NH_3 with 0.500 M HCl. Specifically, what is **(a)** the volume of titrant required to reach the equivalence point; and what is the pH **(b)** initially, **(c)** after the addition of 5.00 mL of 0.500 M HCl, **(d)** at the half-neutralization point, **(e)** after the addition of 10.00 mL of 0.500 M HCl, **(f)** at the equivalence point, and **(g)** after the addition of 20.00 mL of 0.500 M HCl?

106. Construct, by calculation, a titration curve for the titration of 25.00 mL of 0.250 M CH_3CH_2COOH (pK_a = 4.89) with 0.330 M NaOH. Specifically, what is **(a)** the volume of titrant required to reach the equivalence point; and what is the pH **(b)** initially, **(c)** after the addition of 2.00 mL of 0.330 M NaOH, **(d)** at the half-neutralization point, **(e)** after the addition of 12.00 mL of 0.330 M NaOH, **(f)** at the equivalence point, and **(g)** after the addition of 20.00 mL of 0.330 M NaOH?

Lewis Acids and Bases

107. Identify the Lewis acid and Lewis base in each of the following reactions.

(a) $OH^-(aq) + Al(OH)_3(s) \longrightarrow Al(OH)_4^-(aq)$

(b) $Cu^{2+}(aq) + 4\,NH_3(aq) \longrightarrow [Cu(NH_3)_4]^{2+}(aq)$

(c) $CO_2(g) + OH^-(aq) \longrightarrow HCO_3^-(aq)$

108. Identify the Lewis acid and Lewis base in each of the following reactions.

(a) $CaO(s) + CO_2(g) \longrightarrow CaCO_3(s)$

(b) $Ag^+(aq) + 2\,CN^-(aq) \longrightarrow Ag(CN)_2^-(aq)$

(c) $B(OH)_3(s) + OH^-(aq) \longrightarrow B(OH)_4^-(aq)$

Additional Problems

109. A batch of sewage sludge is dewatered to 28% solids; the remaining water has a pH of 6.0. The sludge is made quite alkaline to kill disease-causing microorganisms and make the sludge acceptable for spreading on farmland. How much quicklime (CaO) is needed to raise the pH of 1.0 ton of this sludge to 12.0?

$$CaO(s) + H_2O(l) \longrightarrow Ca(OH)_2(aq)$$

110. What volume of concentrated nitric acid (70.4% HNO_3 by mass; $d = 1.42$ g/mL) must be dissolved in water to prepare 0.500 L of a solution having pH = 2.20?

111. Calculate the pH of a solution that is 1.0×10^{-8} M HCl? (*Hint:* What are the two sources of H_3O^+?)

112. Can a solution have $[H_3O^+] = 2 \times [OH^-]$? Can a solution have pH = 2 × pOH? If so, will the two solutions be the same?

113. Refer to the series of photographs in Example 15.19. Suppose the solution in the photograph marked (b) is 100.0 mL of a buffer with $[CH_3COOH] = [CH_3COO^-] = 0.200$ M. Approximately how many moles of NaOH, at a minimum, must be added to get the color change in the photograph marked (d)? (*Hint:* Refer to Figure 15.13.)

114. Determine $[H_3O^+]$ in 0.020 M H_2SO_4(aq), and compare your result with that obtained in Exercise 15.12B. (*Hint:* What are the two sources of H_3O^+ that you must consider?)

115. Use data from Appendix C for the neutralization reaction: $H^+(aq) + OH^-(aq) \longrightarrow H_2O(l)$, to determine whether the ion product for water, K_w, increases or decreases with increasing temperature.

116. Explain why the *difference* in values between pK_{a_1} and pK_{a_2} is considerably greater for maleic acid than for fumaric acid.

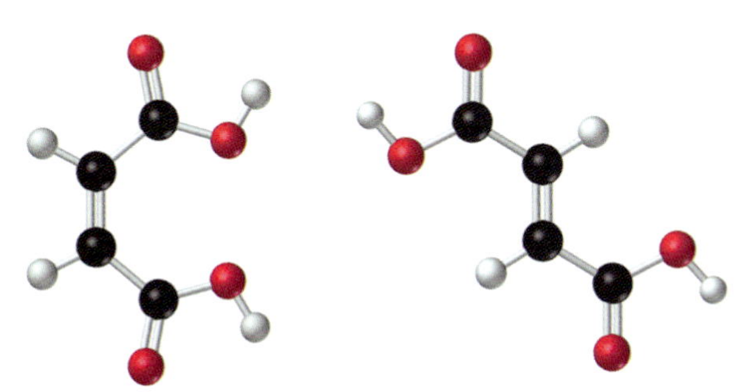

Maleic acid: $pK_{a_1} = 1.91$; $pK_{a_2} = 6.33$

Fumaric acid: $pK_{a_1} = 3.10$; $pK_{a_2} = 4.60$

117. A handbook lists for hydrazine, N_2H_4, that $pK_{b_1} = 6.07$ and $pK_{b_2} = 15.05$. Draw a structural formula for hydrazine, and write equations to show how hydrazine can ionize as a base in two distinctive steps. Explain why pK_{b_2} is so much larger than pK_{b_1}. Calculate the pH of 0.150 M N_2H_4.

118. A handbook gives the solubility of methylamine, CH_3NH_2, as 959 volumes of CH_3NH_2(g) per volume of water at 25 °C and under 1 atm pressure of CH_3NH_2(g). What is the pH of a saturated aqueous solution of methylamine under these conditions?

119. The titration curve in the accompanying figure involves 1.00 M solutions of an acid and a base. In the manner of Example 15.22 and Exercise 15.22, **(a)** identify the type of acid-base titration represented by this curve, **(b)** estimate the pK value of the *titrant*, and **(c)** obtain a value of the pH at the equivalence point, by calculation and by estimation from the titration curve.

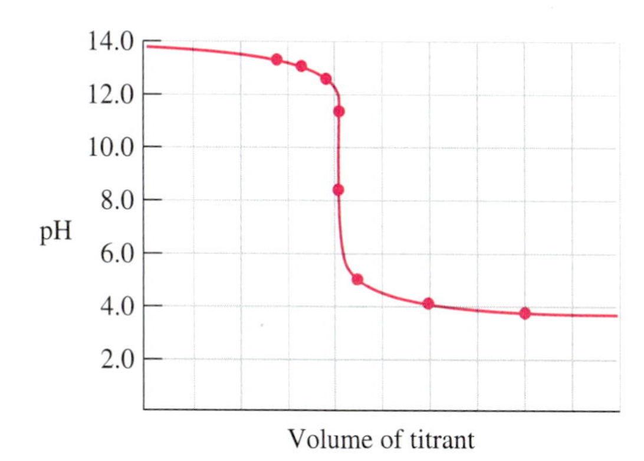

120. Assume that the color of an acid-base indicator is the "acid color" if at least 90% of the indicator is in the form of HIn, and the "base color" if at least 90% of the indicator is in the form of In^-. Show that the pH range for the indicator to change color is roughly pH = $pK_{HIn} \pm 1$

121. We have learned how to determine the pH of an aqueous solution in which hydrolysis of an ion occurs. Calculating the pH is more difficult if a given ion can undergo *both* hydrolysis and further ionization.

(a) Write one equation to represent the ionization of $H_2PO_4^-$ as an acid, and another to represent its hydrolysis (ionization as a base).

(b) Show that the pH of an aqueous solution containing $H_2PO_4^-$ is *independent* of the molarity, M, as long as the solution is sufficiently concentrated (greater than about 0.010 M). Show that the pH corresponds to the formula: $pH = \frac{1}{2}(pK_{a_1} + pK_{a_2})$.

122. *o*-Phthalic acid [*o*-$C_6H_4(COOH)_2$] is used to make phenolphthalein indicator. A saturated aqueous solution of *o*-phthalic acid has 0.6 g/100 mL and a pH = 2.33. A 0.10 M solution of potassium hydrogen *o*-phthalate has a pH = 4.19. Determine pK_{a_1} and pK_{a_2} for *o*-phthalic acid. (*Hint:* Refer to Problem 121.)

123. Normal rainfall is made slightly acidic from the dissolving of atmospheric CO_2 in the rainwater and ionization of the resulting carbonic acid, H_2CO_3. At a CO_2(g) partial pressure of 1 atm and a temperature of 25 °C, the solubility of CO_2(g) in water is 0.759 mL CO_2 at STP per mL H_2O. Given that air contains 0.036% CO_2 by volume, estimate the pH of rainwater that is saturated with CO_2(g). (*Hint:* Recall Henry's law [page 527], and assume that the dissolved CO_2 is present as carbonic acid.)

124. The accompanying titration curve is that of 10.0 mL of 0.100 M H_3PO_4 titrated with 0.100 M NaOH.

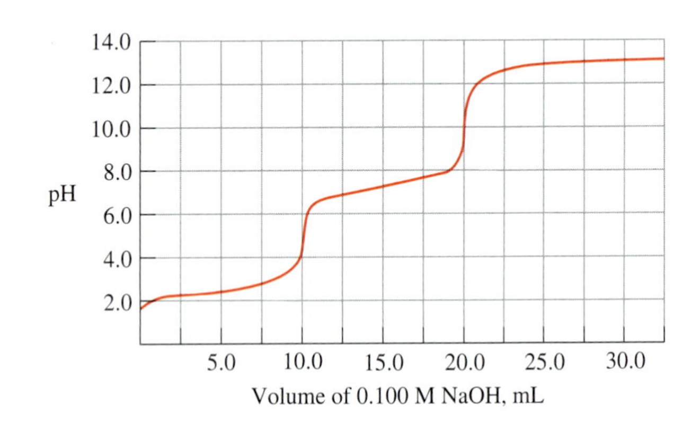

(a) Write a net ionic equation for the neutralization reaction that occurs prior to the first equivalence point. What is the substance present in solution at the first equivalence point?

(b) Write a net ionic equation for the neutralization reaction that occurs between the first and second equivalence points. What is the substance present in solution at the second equivalence point?

(c) Select an appropriate indicator for titration to each of the equivalence points.

(d) H_3PO_4 is a *tri*protic acid. Why do you suppose the titration curve is lacking a third equivalence point?

(e) In what way would the titration curve be altered if the solution being titrated were 0.100 M in *both* H_3PO_4 and HCl?

125. A requirement of buffer solutions is that they contain components capable of reacting with added acid or base. The dihydrogen phosphate ion is capable of doing this—it is amphiprotic. Why isn't $NaH_2PO_4(aq)$ alone a particularly effective buffer? Is a solution containing NaH_2PO_4 and Na_2HPO_4 an effective buffer? Explain. (*Hint:* Use the titration curve accompanying Problem 124.)

126. From the observation that 0.0500 M $H_2C{=}CHCH_2COOH$ (vinylacetic acid) has a freezing point of -0.096 °C, determine K_a for this acid.

127. Show that when $[H_3O^+]$ of a solution is reduced to half of its original value, the pH increases by 0.30, *regardless of the initial pH*. Is it also true that when any solution is diluted to half of its original concentration its pH increases by 0.30? Explain.

128. What is the pH of a solution that is 0.250 M CH_3COOH and also 0.150 M HCOOH? (*Hint:* You will need to solve two equations simultaneously.)

CHAPTER

16

More Equilibria in Aqueous Solutions: Slightly Soluble Salts and Complex Ions

A precipitate of lead(II) iodide forms when an aqueous solution of potassium iodide is added to one of lead(II) nitrate. Conditions that lead to the formation of precipitates are considered on page 703.

CONTENTS

In Chapter 14, we introduced general ideas about chemical equilibrium and applied them mostly to reactions involving gases. Much of Chapter 15 dealt with equilibria in aqueous solutions of weak acids and weak bases. In this chapter, we will focus on equilibria between slightly soluble salts and their ions in solution, and on equilibria that involve complex ions. As we shall see, though, these equilibria, in turn, are often interwoven with equilibria involving acids and bases.

Applications of the concepts in this chapter are as close as our teeth—tooth decay is related to the solubility of slightly soluble substances—and the film in our cameras—the developing of photographic film makes use of complex ion formation. We will consider both of these applications and many others as well.

● **ChemCDX**
When you see this icon, there are related animations, demonstrations, and exercises on the CD accompanying this text.

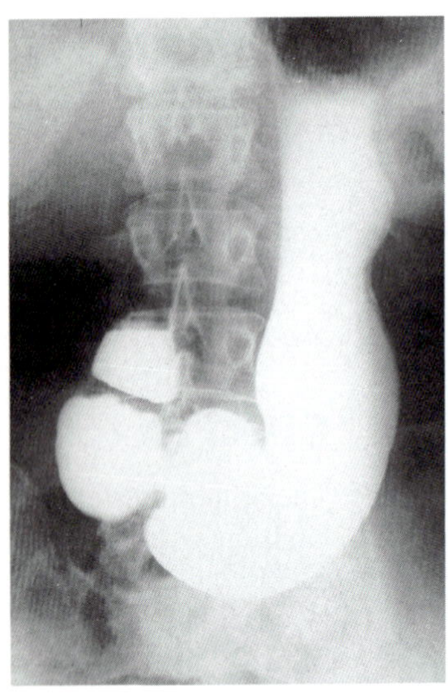

Barium sulfate is so very slightly soluble in water that it can be taken internally. It is also opaque to X rays; an outline of a stomach coated with $BaSO_4(s)$ shows up on this X-ray photograph. The $BaSO_4(s)$ is later eliminated from the digestive tract without harm to the patient.

16.1 The Solubility Product Constant, K_{sp}

Many important ionic compounds are only slightly soluble in water. In fact, we often use the term "insoluble" to describe such ionic compounds. For example, the solubility of barium sulfate, $BaSO_4$, is only 0.000246 g per 100 g water at 25 °C.

We can use the following equation to represent the equilibrium between $BaSO_4(s)$ and the ions present in a *saturated* aqueous solution.

$$BaSO_4(s) \rightleftharpoons Ba^{2+}(aq) + SO_4^{2-}(aq)$$

The symbol (aq) indicates that water is the solvent, but we do not write H_2O in the chemical equation. And, as we saw in Chapters 14 and 15, pure liquids and solids do not appear in equilibrium constant expressions (K_c). Also, as we did in several instances in Chapter 15, we use a special name and symbol for the K_c representing solubility equilibrium. We call it the solubility product constant and represent it as K_{sp}.

$$K_{sp} = [Ba^{2+}][SO_4^{2-}] = 1.1 \times 10^{-10} \text{ (at 25 °C)}$$

The **solubility product constant, K_{sp}**, is the product of the concentrations of the ions involved in a solubility equilibrium, each raised to a power equal to the stoichiometric coefficient of that ion in the chemical equation for the equilibrium. Like other equilibrium constants, K_{sp} depends on temperature. Table 16.1 lists some typical solubility equilibria and their solubility product constants at about room temperature. A more extensive listing of K_{sp} values is given in Appendix C.

As illustrated in Example 16.1, we must base a K_{sp} expression on a balanced equation, and the equation is always based on one mole of ionic solid. That is, the equation has a single term on the left—the formula of the ionic solid—and the coefficient of that term is *1*.

EXAMPLE 16.1

Write a solubility product constant expression for equilibrium in a saturated aqueous solution of each of the following slightly soluble salts.

a. iron(III) phosphate, $FePO_4$ **b.** chromium(III) hydroxide, $Cr(OH)_3$

SOLUTION

The chemical equations on which we must base the K_{sp} expression have one mole of solid represented on the left. The coefficients on the right side then represent the numbers of moles of cations and anions per mole of the compound. And, as with other equilibrium constant expressions, coefficients in the balanced equation appear as exponents in the K_{sp} expression.

a. $FePO_4(s) \rightleftharpoons Fe^{3+}(aq) + PO_4^{3-}(aq)$ $K_{sp} = [Fe^{3+}][PO_4^{3-}]$

b. $Cr(OH)_3(s) \rightleftharpoons Cr^{3+}(aq) + 3\,OH^-(aq)$ $K_{sp} = [Cr^{3+}][OH^-]^3$

EXERCISE 16.1A

Write a K_{sp} expression for equilibrium in a saturated aqueous solution of each of the following.

a. MgF_2 **b.** Li_2CO_3 **c.** $Cu_3(AsO_4)_2$

TABLE 16.1 Some Solubility Product Constants at 25 °C

Solute	Solubility Equilibrium	K_{sp}
Aluminum hydroxide	$Al(OH)_3(s) \rightleftharpoons Al^{3+}(aq) + 3\ OH^-(aq)$	1.3×10^{-33}
Barium carbonate	$BaCO_3(s) \rightleftharpoons Ba^{2+}(aq) + CO_3^{2-}(aq)$	5.1×10^{-9}
Barium sulfate	$BaSO_4(s) \rightleftharpoons Ba^{2+}(aq) + SO_4^{2-}(aq)$	1.1×10^{-10}
Calcium carbonate	$CaCO_3(s) \rightleftharpoons Ca^{2+}(aq) + CO_3^{2-}(aq)$	2.8×10^{-9}
Calcium fluoride	$CaF_2(s) \rightleftharpoons Ca^{2+}(aq) + 2\ F^-(aq)$	5.3×10^{-9}
Calcium sulfate	$CaSO_4(s) \rightleftharpoons Ca^{2+}(aq) + SO_4^{2-}(aq)$	9.1×10^{-6}
Calcium oxalate	$CaC_2O_4(s) \rightleftharpoons Ca^{2+}(aq) + C_2O_4^{2-}(aq)$	2.7×10^{-9}
Chromium(III) hydroxide	$Cr(OH)_3(s) \rightleftharpoons Cr^{3+}(aq) + 3\ OH^-(aq)$	6.3×10^{-31}
Copper(II) sulfide	$CuS(s) \rightleftharpoons Cu^{2+}(aq) + S^{2-}(aq)$	8.7×10^{-36}
Iron(III) hydroxide	$Fe(OH)_3(s) \rightleftharpoons Fe^{3+}(aq) + 3\ OH^-(aq)$	4×10^{-38}
Lead(II) chloride	$PbCl_2(s) \rightleftharpoons Pb^{2+}(aq) + 2\ Cl^-(aq)$	1.6×10^{-5}
Lead(II) chromate	$PbCrO_4(s) \rightleftharpoons Pb^{2+}(aq) + CrO_4^{2-}(aq)$	2.8×10^{-13}
Lead(II) iodide	$PbI_2(s) \rightleftharpoons Pb^{2+}(aq) + 2\ I^-(aq)$	7.1×10^{-9}
Magnesium carbonate	$MgCO_3(s) \rightleftharpoons Mg^{2+}(aq) + CO_3^{2-}(aq)$	3.5×10^{-8}
Magnesium fluoride	$MgF_2(s) \rightleftharpoons Mg^{2+}(aq) + 2\ F^-(aq)$	3.7×10^{-8}
Magnesium hydroxide	$Mg(OH)_2(s) \rightleftharpoons Mg^{2+}(aq) + 2\ OH^-(aq)$	1.8×10^{-11}
Magnesium phosphate	$Mg_3(PO_4)_2(s) \rightleftharpoons 3\ Mg^{2+}(aq) + 2\ PO_4^{3-}(aq)$	1×10^{-25}
Mercury(I) chloride	$Hg_2Cl_2(s) \rightleftharpoons Hg_2^{2+}(aq) + 2\ Cl^-(aq)$	1.3×10^{-18}
Mercury(II) sulfide	$HgS(s) \rightleftharpoons Hg^{2+}(aq) + S^{2-}(aq)$	2×10^{-53}
Silver bromide	$AgBr(s) \rightleftharpoons Ag^+(aq) + Br^-(aq)$	5.0×10^{-13}
Silver chloride	$AgCl(s) \rightleftharpoons Ag^+(aq) + Cl^-(aq)$	1.8×10^{-10}
Silver iodide	$AgI(s) \rightleftharpoons Ag^+(aq) + I^-(aq)$	8.5×10^{-17}
Strontium carbonate	$SrCO_3(s) \rightleftharpoons Sr^{2+}(aq) + CO_3^{2-}(aq)$	1.1×10^{-10}
Strontium sulfate	$SrSO_4(s) \rightleftharpoons Sr^{2+}(aq) + SO_4^{2-}(aq)$	3.2×10^{-7}
Zinc sulfide	$ZnS(s) \rightleftharpoons Zn^{2+}(aq) + S^{2-}(aq)$	1.6×10^{-24}

EXERCISE 16.1B

Write a K_{sp} expression for equilibrium in a saturated solution of **(a)** magnesium hydroxide, milk of magnesia; **(b)** scandium fluoride, ScF_3, used in the preparation of scandium metal; and **(c)** zinc(II) phosphate, used in dental cements.

16.2 The Relationship Between K_{sp} and Molar Solubility

The term "solubility" product constant suggests that K_{sp} is related to the solubility of an ionic solute. However, this does *not* mean that K_{sp} and molar solubility—the molarity of a solute in a saturated aqueous solution—are equivalent. It means only that we can determine K_{sp} from molar solubility and molar solubility from K_{sp}.

Although K_{sp} calculations are often easier than other equilibrium calculations, they are much more subject to error. This is so in part because some K_{sp} values are not precisely known, and in part because interionic attractions complicate equilibrium relationships when ion concentrations are high. We therefore do not ordinarily use the solubility product concept for moderately or highly soluble ionic solutes. Even for slightly soluble solutes, some calculations give only ballpark estimates.

Like other equilibrium calculations, those dealing with solubility equilibrium fall into two broad categories: determining a value of K_{sp} from experimental data and calculating equilibrium concentrations when a value of K_{sp} is known. The first type of calculation is illustrated by Example 16.2.

EXAMPLE 16.2

At 20 °C, a saturated aqueous solution of silver carbonate contains 32 mg of Ag_2CO_3 per liter of solution. Calculate the K_{sp} for Ag_2CO_3 at 20 °C.

$$Ag_2CO_3(s) \rightleftharpoons 2\,Ag^+(aq) + CO_3^{2-}(aq) \quad K_{sp} = ?$$

SOLUTION

To evaluate K_{sp}, we need the concentrations of the ions in the saturated solution, but we are given the solubility expressed in terms of the solute as a whole. We can use the chemical equation (above) to relate the ion concentrations to the given solubility. From the equation, we see that there are 2 mol Ag^+ and 1 mol CO_3^{2-} in solution for every 1 mol Ag_2CO_3 that dissolves. This fact provides us with the conversion factors (blue) in the following setups.

$$[Ag^+] = \frac{32\ \cancel{mg\ Ag_2CO_3}}{L} \times \frac{1\ \cancel{g\ Ag_2CO_3}}{1000\ \cancel{mg\ Ag_2CO_3}} \times \frac{1\ \cancel{mol\ Ag_2CO_3}}{275.8\ \cancel{g\ Ag_2CO_3}} \times \frac{2\ mol\ Ag^+}{1\ \cancel{mol\ Ag_2CO_3}}$$

$$= 2.3 \times 10^{-4}\ M$$

$$[CO_3^{2-}] = \frac{32\ \cancel{mg\ Ag_2CO_3}}{L} \times \frac{1\ \cancel{g\ Ag_2CO_3}}{1000\ \cancel{mg\ Ag_2CO_3}} \times \frac{1\ \cancel{mol\ Ag_2CO_3}}{275.8\ \cancel{g\ Ag_2CO_3}} \times \frac{1\ mol\ CO_3^{2-}}{1\ \cancel{mol\ Ag_2CO_3}}$$

$$= 1.2 \times 10^{-4}\ M$$

Now we can write the K_{sp} expression for Ag_2CO_3 and substitute in the equilibrium concentrations of the ions.

$$K_{sp} = [Ag^+]^2[CO_3^{2-}] = (2.3 \times 10^{-4})^2(1.2 \times 10^{-4}) = 6.3 \times 10^{-12}$$

EXERCISE 16.2A

In Example 15.4 (page 645), we determined the molar solubility of magnesium hydroxide (milk of magnesia) from the measured pH of its saturated solution. Use data from that example to determine the K_{sp} for $Mg(OH)_2$.

EXERCISE 16.2B

A saturated aqueous solution of silver(I) chromate is found to contain 14 ppm of Ag^+ by mass. Determine K_{sp} for silver(I) chromate. Assume that the solution has a density of 1.00 g/mL.

In a second type of calculation, illustrated in Example 16.3, we determine the molar solubility of a solute from a tabulated value of its K_{sp}. This is a practical type of calculation because many reference books list only K_{sp} values and not molar solubilities.

EXAMPLE 16.3

From the K_{sp} value for silver sulfate, calculate its molar solubility.

$$Ag_2SO_4(s) \rightleftharpoons 2\,Ag^+(aq) + SO_4^{2-}(aq) \qquad K_{sp} = 1.4 \times 10^{-5} \text{ at } 25\ °C$$

SOLUTION

The equation shows that when 1 mol $Ag_2SO_4(s)$ dissolves, 1 mol SO_4^{2-} and 2 mol Ag^+ appear in solution . If we let s represent the number of moles of Ag_2SO_4 that dissolve per liter of solution, the two ion concentrations are as shown here.

$$[SO_4^{2-}] = s \quad \text{and} \quad [Ag^+] = 2s$$

These concentrations must satisfy the K_{sp} expression. In substituting ion concentrations into the expression, it is important to note that (1) the factor 2 appears in $2s$ because the concentration of Ag^+ is *twice* that of SO_4^{2-}, and the exponent 2 appears in the term $(2s)^2$ because the concentration of Ag^+ must be raised to the *second* power in the K_{sp} expression.

$$K_{sp} = [Ag^+]^2[SO_4^{2-}]$$
$$1.4 \times 10^{-5} = (2s)^2(s) = 4s^3$$
$$s^3 = \frac{1.4 \times 10^{-5}}{4} = 3.5 \times 10^{-6}$$
$$s = (3.5 \times 10^{-6})^{1/3} = 1.5 \times 10^{-2}\text{ M}$$

Recall that s represents $[SO_4^{2-}]$ in the saturated solution and that the dissolution equation indicates that the molar solubility of Ag_2SO_4 will be the same as $[SO_4^{2-}]$. This is indicated by the conversion factor in blue.

$$\frac{1.5 \times 10^{-2}\ \cancel{\text{mol } SO_4^{2-}}}{\text{L}} \times \frac{1 \text{ mol } Ag_2SO_4}{1\ \cancel{\text{mol } SO_4^{2-}}} = 1.5 \times 10^{-2}\text{ M } Ag_2SO_4$$

EXERCISE 16.3A

Calculate the molar solubility of silver arsenate, given that

$$Ag_3AsO_4(s) \rightleftharpoons 3\,Ag^+(aq) + AsO_4^{3-}(aq) \qquad K_{sp} = 1.0 \times 10^{-22}$$

EXERCISE 16.3B

Determine the number of parts per million of I^- in a saturated aqueous solution of PbI_2 ($K_{sp} = 7.1 \times 10^{-9}$). Assume that the solution has a density of 1.00 g/mL.

At times, instead of calculating molar solubilities, we need only to make some qualitative judgments. Often, we can do this by comparing K_{sp} values, as illustrated in Example 16.4.

EXAMPLE 16.4—A Conceptual Example

Without doing detailed calculations, but using data from Table 16.1, establish the order of *increasing* solubility of these silver halides in water: AgCl, AgBr, AgI.

SOLUTION

The key to solving this problem is to recognize that all three solutes are of the same type; that is, the ratio of cation to anion in each is 1:1. As suggested by the following equation for the solubility equilibrium of AgX, the ion concentrations are equal to each other and to the molar solubility, s.

$$AgX(s) \rightleftharpoons \underset{s \text{ mol } Ag^+/\text{L}}{Ag^+(aq)} + \underset{s \text{ mol } X^-/\text{L}}{X^-(aq)}$$

$$K_{sp} = [Ag^+][X^-] = (s)(s) = s^2$$

The solute solubility $s = \sqrt{K_{sp}}$.

The order of increasing molar solubility is the same as the order of *increasing* K_{sp} values.

Increasing molar solubilities:	AgI	<	AgBr	<	AgCl
Increasing K_{sp}:	$K_{sp}: 8.5 \times 10^{-17}$	<	5.0×10^{-13}	<	1.8×10^{-10}

The method used here works only if we are comparing solutes that are *all* of the same type, that is, all MX, all MX_2, and so on.

EXERCISE 16.4

Without doing detailed calculations, and using data from Table 16.1, arrange the following solutes in order of increasing molar solubility: MgF_2, CaF_2, $PbCl_2$, PbI_2.

16.3 The Common-Ion Effect in Solubility Equilibria

The common-ion effect that we studied in Chapter 15 also operates in solubility equilibria. For example, we can add sodium sulfate to a saturated solution of silver sulfate. Sulfate ion from sodium sulfate,

$$Na_2SO_4(s) \xrightarrow{H_2O} 2\,Na^+(aq) + SO_4^{2-}(aq)$$

increases $[SO_4^{2-}]$ in the solubility equilibrium involving silver sulfate.

$$Ag_2SO_4(s) \rightleftharpoons 2\,Ag^+(aq) + SO_4^{2-}(aq)$$

This is a stress that, according to Le Châtelier's principle, is relieved as equilibrium shifts to the left. To establish a new equilibrium, several things happen.

- Some $Ag_2SO_4(s)$ precipitates.
- $[Ag^+]$ is *smaller than* its original equilibrium value.
- $[SO_4^{2-}]$ is *greater than* in the original equilibrium.

We can achieve a similar effect by adding $AgNO_3(aq)$ to saturated $Ag_2SO_4(aq)$. Now, we predict these effects:

- Some $Ag_2SO_4(s)$ precipitates.
- $[Ag^+]$ is *greater than* in the original equilibrium.
- $[SO_4^{2-}]$ is *smaller than* its original equilibrium value.

We can diagram the effect as follows.

When a salt supplies Ag^+ *or* SO_4^{2-}, equilibrium shifts to the *left*.

$$Ag_2SO_4(s) \rightleftharpoons 2\,Ag^+(aq) + SO_4^{2-}(aq)$$

And we can summarize the **common-ion effect** on a solubility equilibrium with the following statement.

The solubility of a slightly soluble ionic compound is lowered when a second solute that furnishes a common ion is added to the solution.

Figure 16.1 is a visualization of the effect of adding Na_2SO_4 to a clear, saturated solution of silver sulfate.

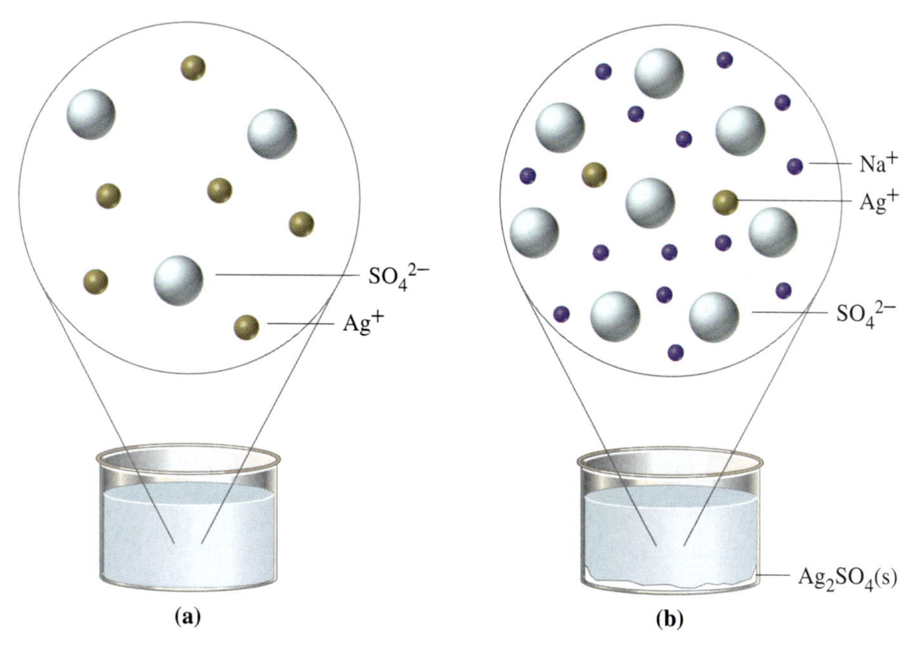

▲ **Figure 16.1 The common-ion effect in solubility equilibrium**
(a) Clear, saturated $Ag_2SO_4(aq)$ with undissolved solid removed. The solution contains two cations for every anion. (b) Following the addition of $Na_2SO_4(aq)$. There are still twice as many cations as anions, but most of the cations are Na^+. Some of the Ag^+ ions in the original equilibrium (four of the six) have precipitated as $Ag_2SO_4(s)$.

Qualitatively, both common ions, Ag^+ and SO_4^{2-}, have the same effect on the solubility of Ag_2SO_4; they reduce it. Quantitatively, however, the effect of Ag^+ is more pronounced than that of SO_4^{2-}, as can be seen by comparing the result of Exercise 16.5A with that of Example 16.5.

EXAMPLE 16.5

Calculate the molar solubility of Ag_2SO_4 in 1.00 M $Na_2SO_4(aq)$.

SOLUTION

Instead of water, let's use 1.00 M $Na_2SO_4(aq)$ as the solvent to prepare a saturated solution of Ag_2SO_4. We assume that the Na_2SO_4 is completely dissociated and that the presence of Ag_2SO_4 has no effect on this dissociation. Let's continue to use s to represent the number of moles of Ag_2SO_4 that dissolve per liter of saturated solution. From s mol Ag_2SO_4/L, we get s mol SO_4^{2-}/L and $2s$ mol Ag^+/L. We can tabulate this information, including the already present 1.00 mol SO_4^{2-}/L, in the usual format.

The reaction:	$Ag_2SO_4(s) \rightleftharpoons$	$2\,Ag^+(aq)$	$+\ SO_4^{2-}(aq)$	$K_{sp} = 1.4 \times 10^{-5}$
Initial concentrations, M:		0	1.00 (from the Na_2SO_4)	
From Ag_2SO_4:		$2s$	s	
Equilibrium concentrations, M:		$2s$	$(1.00 + s)$	

The usual K_{sp} relationship must be satisfied.

$$K_{sp} = [Ag^+]^2[SO_4^{2-}]$$
$$1.4 \times 10^{-5} = (2s)^2(1.00 + s)$$

PROBLEM-SOLVING NOTE

When we assume that $(1.00 + s) \approx 1.00$, we are saying that essentially all of the common ion comes from the added source, not from the slightly soluble solute. The assumption is valid here: $(1.00 + 1.9 \times 10^{-3}) = 1.00$ (to two decimal places).

To simplify the equation, let's assume that s is much smaller than 1.00 M, so that $(1.00 + s) \approx 1.00$.

$$(2s)^2(1.00) = 1.4 \times 10^{-5}$$
$$4s^2 = 1.4 \times 10^{-5}$$
$$s^2 = 3.5 \times 10^{-6}$$
$$s = \text{molar solubility} = (3.5 \times 10^{-6})^{1/2} = 1.9 \times 10^{-3} \text{ mol } Ag_2SO_4/L$$

In contrast, recall that in Example 16.3 we found the solubility of Ag_2SO_4 in pure water to be 1.5×10^{-2} mol Ag_2SO_4/L—about 8 times as great.

EXERCISE 16.5A

Calculate the molar solubility of Ag_2SO_4 in 1.00 M $AgNO_3$(aq).

EXERCISE 16.5B

How many grams of silver nitrate must be added to 250.0 mL of a clear, saturated solution of Ag_2SO_4(aq) to reduce the solubility of the Ag_2SO_4 to 1.0×10^{-3} M? (K_{sp} of $Ag_2SO_4 = 1.4 \times 10^{-5}$.)

● **ChemCDX** Common Ion Effect

The common-ion effect is frequently used in analytical chemistry, as is illustrated in the following method of determining the percent calcium in limestone. Limestone is principally calcium carbonate, $CaCO_3$, with small amounts of other constituents.

First, we obtain calcium ion in solution by dissolving the limestone sample in HCl(aq).

$$CaCO_3(\text{s, impure}) + 2\ H_3O^+(aq) \longrightarrow Ca^{2+}(aq) + 3\ H_2O\ (l) + CO_2(g)$$

After neutralizing excess, unreacted HCl(aq), we add an *excess* of ammonium oxalate solution, $(NH_4)_2C_2O_4$(aq). By an excess, we mean that we use enough $C_2O_4^{2-}$ so that no further precipitate forms and a significant concentration of $C_2O_4^{2-}$ remains in the saturated solution.

$$Ca^{2+}(aq) + C_2O_4^{2-}(aq) \longrightarrow CaC_2O_4(s)$$

The presence of excess oxalate ion, a common ion, greatly reduces the solubility of CaC_2O_4(s). Without this common ion, a small but significant percentage of Ca^{2+} remains in solution. We separate the precipitate by filtration and wash it to remove residual impurities. For the washing, we use $(NH_4)_2C_2O_4$(aq) rather than pure water—again so that the common ion, $C_2O_4^{2-}$, will keep the solubility very low.

Finally, we heat the precipitate to 500 °C to decompose the calcium oxalate to *pure* $CaCO_3$(s).

$$CaC_2O_4(s) \xrightarrow{\Delta} CaCO_3(\text{s, pure}) + CO(g)$$

We can then determine the mass of $CaCO_3$(s) obtained and calculate the percent Ca in the original limestone.

16.4 Will Precipitation Occur? Is It Complete?

Sometimes a student mistakenly uses municipal tap water instead of deionized water in preparing a solution. If the solute is $NaNO_3$, this may not cause a problem. However, if the solute is $AgNO_3$, a cloudy solution will probably result (see Figure 16.2). The cloudiness is due to a trace of a precipitate, principally AgCl. What concentration of Cl^- can be tolerated in water before a precipitate forms? Let's establish a criterion to answer this and similar questions.

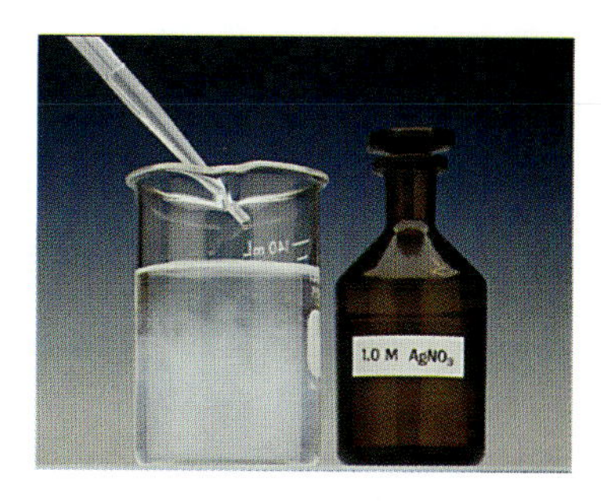

▲ **Figure 16.2**
Precipitation of anions from municipal tap water
When a few drops of $AgNO_3(aq)$ are added to municipal tap water, a white precipitate forms. The precipitate forms because certain ions present in the tap water (e.g., Cl^-, CO_3^{2-}, and SO_4^{2-}) form slightly soluble silver compounds.

To Determine Whether Precipitation Occurs

We can begin with the expressions that describe equilibrium between AgCl(s) and its ions, Ag^+ and Cl^-.

$$AgCl(s) \rightleftharpoons Ag^+(aq) + Cl^-(aq) \qquad K_{sp} = [Ag^+][Cl^-] = 1.8 \times 10^{-10}$$

Suppose we want to prepare 0.100 M $AgNO_3(aq)$ using municipal tap water as the solvent, and suppose the water has $[Cl^-] = 1 \times 10^{-6}$ M. Let's write a reaction quotient, an ion product denoted by the symbol Q_{ip}, for these initial conditions. Then, we can compare Q_{ip} to K_{sp}.

$$Q_{ip} = [Ag^+]_{initial} \times [Cl^-]_{initial} = (0.100) \times 1 \times 10^{-6} = 1 \times 10^{-7}$$

We see that Q_{ip} greatly exceeds the value of K_{sp} $(1 \times 10^{-7} > 1.8 \times 10^{-10})$. Recall that when a reaction quotient exceeds the value of the corresponding K, a net reaction should occur in the *reverse* direction; Ag^+ and Cl^- should combine to form AgCl(s). That is, *precipitation should occur*, and it should continue until Q_{ip} falls to a value equal to K_{sp}. At this point, the solid solute would be in equilibrium with its ions in a saturated solution.

Now suppose we prepare 0.100 M $AgNO_3$, using as the solvent a deionized water in which the chloride ion concentration has been lowered to 1×10^{-10} M. Again, we can calculate a Q_{ip} value.

$$Q_{ip} = [Ag^+]_{initial} \times [Cl^-]_{initial} = (0.100) \times 1 \times 10^{-10} = 1 \times 10^{-11}$$

This reaction quotient, Q_{ip}, is smaller than K_{sp} $(1 \times 10^{-11} < 1.8 \times 10^{-10})$. According to the criterion from Chapter 14, a net reaction should occur in the *forward* direction. However, this is not possible because there is no AgCl(s) present to dissolve. The solution is *unsaturated*, and certainly *no AgCl(s) will precipitate* from this solution.

Based on examples like these, we can formulate general rules that allow us to predict what should happen when we mix solutions containing ions capable of precipitating as a slightly soluble ("insoluble") ionic solid.

- Precipitation *should occur* if $Q_{ip} > K_{sp}$.
- Precipitation *cannot occur* if $Q_{ip} < K_{sp}$.
- A solution is *just saturated* if $Q_{ip} = K_{sp}$.

Example 16.6 illustrates that in applying these criteria we can proceed in three steps.

1 Determine the initial concentrations of ions.
2 Evaluate the reaction quotient Q_{ip}.
3 Compare Q_{ip} and K_{sp} to determine whether precipitation will occur.

EXAMPLE 16.6

If 1.00 mg of Na_2CrO_4 is added to 225 mL of 0.00015 M $AgNO_3(aq)$, will a precipitate form?

$$Ag_2CrO_4(s) \rightleftharpoons 2\,Ag^+(aq) + CrO_4^{2-}(aq) \qquad K_{sp} = 1.1 \times 10^{-12}$$

SOLUTION

Let's apply the three-step strategy outlined above.

1 *Initial concentrations of ions:* The initial concentration of Ag^+ is simply 1.5×10^{-4} M. To establish $[CrO_4^{2-}]$, we first need to determine the number of moles of CrO_4^{2-} placed in solution.

$$?\ \text{mol } CrO_4^{2-} = 1.00 \times 10^{-3}\ \text{g } Na_2CrO_4 \times \frac{1\ \text{mol } Na_2CrO_4}{162.0\ \text{g } Na_2CrO_4} \times \frac{1\ \text{mol } CrO_4^{2-}}{1\ \text{mol } Na_2CrO_4}$$

$$= 6.17 \times 10^{-6}\ \text{mol } CrO_4^{2-}$$

We have 6.17×10^{-6} mol CrO_4^{2-} in 225 mL (0.225 L) of solution, so the chromate ion concentration is

$$[CrO_4^{2-}] = \frac{6.17 \times 10^{-6}\ \text{mol } CrO_4^{2-}}{0.225\ \text{L}} = 2.74 \times 10^{-5}\ \text{M}$$

2 *Evaluation of Q_{ip}:*

$$Q_{ip} = [Ag^+]_{initial}^2[CrO_4^{2-}]_{initial} = (1.5 \times 10^{-4})^2(2.74 \times 10^{-5}) = 6.2 \times 10^{-13}$$

3 *Comparison of Q_{ip} and K_{sp}:* $Q_{ip} = 6.2 \times 10^{-13}$ and $K_{sp} = 1.1 \times 10^{-12}$. Because $Q_{ip} < K_{sp}$, we conclude that *no precipitation occurs.*

EXERCISE 16.6A

If 1.00 g $Pb(NO_3)_2$ and 1.00 g MgI_2 are both added to 1.50 L of H_2O, should a precipitate form? (K_{sp} for $PbI_2 = 7.1 \times 10^{-9}$.)

EXERCISE 16.6B

Will precipitation occur from a solution that has 2.5 ppm $MgCl_2$ and also has a pH of 10.35? Assume the solution density is 1.00 g/mL. [K_{sp} for $Mg(OH)_2 = 1.8 \times 10^{-11}$.]

EXAMPLE 16.7—A Conceptual Example

Pictured at the left is the result of adding a few drops of concentrated KI(aq) to a dilute solution of $Pb(NO_3)_2(aq)$. What is the yellow solid that first appears? Explain why it then disappears.

SOLUTION

Where the drops of concentrated KI(aq) first enter the $Pb(NO_3)_2$ solution, $[I^-]$ is large enough *immediately after addition* so that

$$Q_{ip} = [Pb^{2+}][I^-]^2 > K_{sp}$$

Because $Q_{ip} > K_{sp}$, yellow $PbI_2(s)$ precipitates. When the $PbI_2(s)$ that initially formed settles below the point of entry of the KI(aq), however, $[I^-]$ is no longer large enough to maintain a saturated solution for the concentration of Pb^{2+} present, and the solid redissolves. Based on a *uniform* distribution of I^- ion throughout the solution, the following holds true.

$$Q_{ip} = [Pb^{2+}][I^-]^2 < K_{sp}$$

EXERCISE 16.7

Describe how you might use observations of the type suggested by the photographs to make an estimate of the value of K_{sp} for PbI_2.

Example 16.7 provides a qualitative illustration that in applying the precipitation criteria, we must consider the effect of dilution when solutions are mixed. Example 16.8 demonstrates the dilution effect quantitatively.

EXAMPLE 16.8

If 0.100 L of 0.0015 M $MgCl_2$ and 0.200 L of 0.025 M NaF are mixed, should a precipitate of MgF_2 form?

$$MgF_2(s) \rightleftharpoons Mg^{2+}(aq) + 2\,F^-(aq) \quad K_{sp} = 3.7 \times 10^{-8}$$

SOLUTION

Let's assume that the volume of the mixture of solutions is 0.100 L + 0.200 L = 0.300 L. Then let's apply the three-step strategy as before.

1 *Initial concentrations of ions:* First, we need to determine $[Mg^{2+}]$ and $[F^-]$ as they initially exist in this mixed solution.

$$[Mg^{2+}] = \frac{0.100\,\cancel{L} \times \dfrac{0.0015\,\cancel{mol\,MgCl_2}}{\cancel{L}} \times \dfrac{1\,mol\,Mg^{2+}}{1\,\cancel{mol\,MgCl_2}}}{0.300\,L} = 5.0 \times 10^{-4}\,M$$

$$[F^-] = \frac{0.200\,\cancel{L} \times \dfrac{0.025\,\cancel{mol\,NaF}}{\cancel{L}} \times \dfrac{1\,mol\,F^-}{1\,\cancel{mol\,NaF}}}{0.300\,L} = 1.7 \times 10^{-2}\,M$$

2 *Evaluation of Q_{ip}:*

$$Q_{ip} = [Mg^{2+}][F^-]^2 = (5.0 \times 10^{-4})(1.7 \times 10^{-2})^2 = 1.4 \times 10^{-7}$$

3 *Comparison of Q_{ip} and K_{sp}:* $Q_{ip} = 1.4 \times 10^{-7}$ and $K_{sp} = 3.7 \times 10^{-8}$. Because $Q_{ip} > K_{sp}$, we conclude that *precipitation should occur.*

EXERCISE 16.8A

Equal volumes of 0.0010 M $MgCl_2(aq)$ and 0.020 M NaF(aq) are mixed. Should a precipitate of $MgF_2(s)$ form?

EXERCISE 16.8B

How many grams of KI(s) should be dissolved in 10.5 L of 0.00200 M $Pb(NO_3)_2(aq)$ in order that precipitation of $PbI_2(s)$ *will just begin*?

To Determine Whether Precipitation Is Complete

We have criteria for precipitation from solution, but how complete is the precipitation if it does occur? That is, what percentage of the ion in question is precipitated, and what percentage remains in solution? A slightly soluble solid never totally precipitates from solution, but we generally consider precipitation to be essentially complete if about 99.9% of the target ion is precipitated and only 0.1% or less is left in solution.

Three conditions that generally *favor completeness of precipitation* are as follows.

1. A very small value of K_{sp}
2. A high initial concentration of the target ion
3. A concentration of common ion that greatly exceeds that of the target ion

With these conditions, the solubility of the precipitate is suppressed. The concentration of the target ion remaining in solution is quite low; it is only a tiny percentage of the initial concentration.

Some types of kidney stones are mainly calcium oxalate, $CaC_2O_4(s)$.

EXAMPLE 16.9

To a solution with $[Ca^{2+}] = 0.0050$ M, we add sufficient solid ammonium oxalate, $(NH_4)_2C_2O_4(s)$, to make the initial $[C_2O_4^{2-}] = 0.0051$ M. Will the precipitation of Ca^{2+} as $CaC_2O_4(s)$ be complete?

$$CaC_2O_4(s) \rightleftharpoons Ca^{2+}(aq) + C_2O_4^{2-}(aq) \quad K_{sp} = 2.7 \times 10^{-9}$$

SOLUTION

First, let's use our usual three-step approach.

1 *Initial concentrations of ions*: These are given: $[Ca^{2+}] = 0.0050$ M, and $[C_2O_4^{2-}] = 0.0051$ M.

2 *Evaluation of Q_{ip}*:

$$Q_{ip} = [Ca^{2+}][C_2O_4^{2-}] = (5.0 \times 10^{-3})(5.1 \times 10^{-3}) = 2.6 \times 10^{-5}$$

3 *Comparison of Q_{ip} and K_{sp}*: Because Q_{ip} (2.6×10^{-5}) greatly exceeds K_{sp} for CaC_2O_4 (2.7×10^{-9}), precipitation should occur.

Next, to simplify the algebraic solution, let's use this two-step approach: (1) Assume that all the Ca^{2+} precipitates as $CaC_2O_4(s)$, and note the concentration of $C_2O_4^{2-}$ that remains in excess. (2) Calculate the solubility of $CaC_2O_4(s)$ in an aqueous solution containing this excess $C_2O_4^{2-}$ as a common ion. That is, the $[C_2O_4^{2-}]$ listed as the concentration *after* precipitation in step (1), 0.0001 M, becomes the initial concentration in the solubility equilibrium (step 2).

Step 1. The precipitation

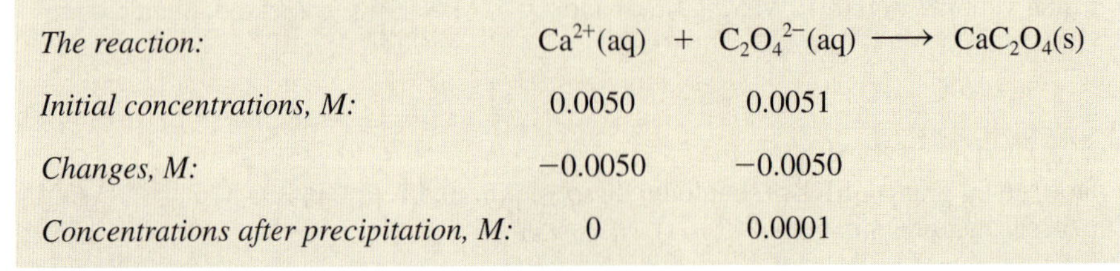

The reaction:	$Ca^{2+}(aq)$	+ $C_2O_4^{2-}(aq)$	$\longrightarrow CaC_2O_4(s)$
Initial concentrations, M:	0.0050	0.0051	
Changes, M:	−0.0050	−0.0050	
Concentrations after precipitation, M:	0	0.0001	

Step 2. The solubility equilibrium

The reaction:	$CaC_2O_4(s) \longrightarrow$	$Ca^{2+}(aq)$ +	$C_2O_4^{2-}(aq)$
Initial concentrations, M:		0	0.0001
Changes, M:		$+s$	$+s$
Equilibrium concentrations, M:		s	$(0.0001 + s)$

As usual, let's assume that $s \ll 0.0001$ and that $(0.0001 + s) \approx 0.0001$. We then obtain

$$K_{sp} = [Ca^{2+}][C_2O_4^{2-}] = s(0.0001 + s) \approx s \times 0.0001 = 2.7 \times 10^{-9}$$

$$s = [Ca^{2+}] = 3 \times 10^{-5}\ M$$

Of an initial 5×10^{-3} mol Ca^{2+}/L, 3×10^{-5} mol Ca^{2+}/L remains after the precipitation. The percentage of Ca^{2+} remaining in solution is

$$\frac{3 \times 10^{-5}\ \cancel{M}}{5 \times 10^{-3}\ \cancel{M}} \times 100 = 0.6\%$$

Because our rule for complete precipitation requires that less than 0.1% of an ion should remain in solution, we conclude that *precipitation is incomplete*.

PROBLEM-SOLVING NOTE

Without the simplifying assumption and by solving a quadratic equation, we would obtain $s = 2 \times 10^{-5}$. The percent Ca^{2+} remaining in solution would be 0.4%, but our conclusion that precipitation is incomplete is still valid.

EXERCISE 16.9

We have a solution in which $[Ca^{2+}] = 0.0050$ M. We add sufficient solid ammonium oxalate, $(NH_4)_2C_2O_4(s)$, to make the solution also $[C_2O_4^{2-}] = 0.0100$ M. Will the precipitation of Ca^{2+} as $CaC_2O_4(s)$ be complete? (K_{sp} for $CaC_2O_4 = 2.7 \times 10^{-9}$.)

Selective Precipitation: Precipitating One Ion and Not Another

In Example 16.4, we compared the solubilities of AgCl, AgBr, and AgI. We found AgI to be the least soluble and AgCl, most soluble. Now let's consider the series of questions in Example 16.10 to see how the difference in solubility of AgI and AgCl enables us to separate I^- from Cl^- in aqueous solution.

EXAMPLE 16.10

An aqueous solution that is 2.00 M in $AgNO_3$ is slowly added from a buret to an aqueous solution that is 0.0100 M in Cl^- and also 0.0100 M in I^- (see Figure 16.3).

$$AgCl(s) \rightleftharpoons Ag^+(aq) + Cl^-(aq) \qquad K_{sp} = 1.8 \times 10^{-10}$$

$$AgI(s) \rightleftharpoons Ag^+(aq) + I^-(aq) \qquad K_{sp} = 8.5 \times 10^{-17}$$

a. Which ion, Cl^- or I^-, is the first to precipitate from solution?

b. When the second ion begins to precipitate, what is the remaining concentration of the first ion?

c. Is separation of the two ions by selective precipitation feasible?

SOLUTION

Notice that the molarity of the silver nitrate solution is 200 times greater than $[Cl^-]$ or $[I^-]$. Only a very small volume of $AgNO_3(aq)$ is required in the precipitation. Because of this, we can neglect the slight dilution of the Cl^- and I^- that occurs upon addition of solution from the buret to that in the beaker.

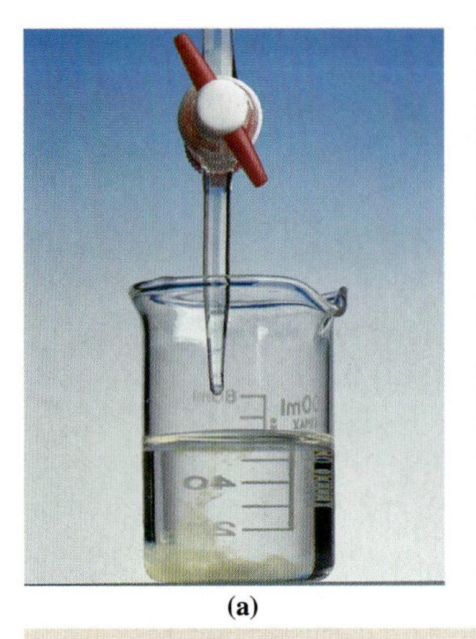

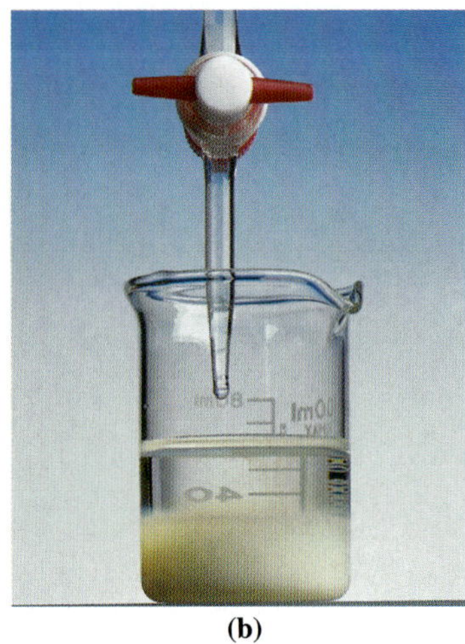

Figure 16.3 Selective precipitation—Example 16.10 illustrated
(a) The first precipitate to form when $AgNO_3(aq)$ is added to an aqueous solution containing Cl^- and I^- is yellow AgI(s). (b) Essentially all of the I^- has precipitated before the precipitation of white AgCl(s) begins.

a. Let's find the $[Ag^+]$ necessary to precipitate each of the anions. Precipitation will begin when the solution has just been saturated, a point at which $Q_{ip} = K_{sp}$.

To precipitate Cl^-. When Q_{ip} is equal to K_{sp} for silver chloride, $[Cl^-]$ in the solution is 0.0100 M, and we can calculate $[Ag^+]$ in the solution at this point.

$$Q_{ip} = [Ag^+][Cl^-] = K_{sp} = 1.8 \times 10^{-10}$$

$$[Ag^+] = \frac{K_{sp}}{[Cl^-]} = \frac{1.8 \times 10^{-10}}{0.0100} = 1.8 \times 10^{-8}\ M$$

To precipitate I^-. When Q_{ip} is equal to K_{sp} for silver iodide, $[I^-]$ in the solution is 0.0100 M, and we can calculate $[Ag^+]$ in the solution at this point.

$$Q_{ip} = [Ag^+][I^-] = K_{sp} = 8.5 \times 10^{-17}$$

$$[Ag^+] = \frac{K_{sp}}{[I^-]} = \frac{8.5 \times 10^{-17}}{0.0100} = 8.5 \times 10^{-15}\ M$$

As soon as the concentration of Ag^+ in the solution reaches 8.5×10^{-15} M, AgI(s) starts to precipitate. This concentration of Ag^+ is far below the 1.8×10^{-8} M Ag^+ needed to precipitate AgCl(s). I^- is the first ion to precipitate.

b. Next, let's determine $[I^-]$ remaining in solution when $[Ag^+] = 1.8 \times 10^{-8}$ M. This is the $[Ag^+]$ at which AgCl(s) begins to precipitate.

$[I^-]$ remaining when AgCl(s) begins to precipitate. We can rearrange the K_{sp} expression for silver iodide and solve for $[I^-]$.

$$[I^-] = \frac{K_{sp}}{[Ag^+]} = \frac{8.5 \times 10^{-17}}{1.8 \times 10^{-8}} = 4.7 \times 10^{-9}\ M$$

c. Now, let's see what percent of the I^- is still in solution when AgCl(s) begins to precipitate.

Percent of I^- remaining in solution.

$$?\% = \frac{4.7 \times 10^{-9}\ M}{0.0100\ M} \times 100\% = 4.7 \times 10^{-5}\%$$

This is far below the 0.1% suggested as a completeness-of-precipitation rule on page 706. The precipitation of I^- as AgI(s) is essentially complete before the precipitation of Cl^- as AgCl(s) begins. Separation of Cl^- and I^- by selective precipitation is feasible.

EXERCISE 16.10A

Answer the three questions posed in Example 16.10 if the solution has the concentrations $[Cl^-] = [Br^-] = 0.0100$ M. (K_{sp} for AgBr $= 5.0 \times 10^{-13}$.)

EXERCISE 16.10B

Use the results of Example 16.10 and Exercise 16.10A, and *without doing detailed calculations*, determine whether I^- and Br^- can be separated by selective precipitation from a solution in which $[I^-] = [Br^-] = 0.0100$ M. Explain your reasoning.

16.5 Effect of pH on Solubility

The solubility of an ionic solute is greatly affected by pH if an acid-base reaction also occurs as the solute dissolves. Let's see why this should be the case. If we add a strong acid to a solution containing fluoride ion, the F^- ion, a relatively strong base, accepts a proton from H_3O^+, and the weak acid HF is formed. This reaction is the reverse of the ionization of HF.

$$\underset{\text{Acid(1)}}{H_3O^+(aq)} + \underset{\text{Base(2)}}{F^-(aq)} \longrightarrow \underset{\text{Acid(2)}}{HF(aq)} + \underset{\text{Base(1)}}{H_2O(l)}$$

Now suppose that the solution containing F^- ion is a saturated solution of calcium fluoride in equilibrium with $CaF_2(s)$. According to Le Châtelier's principle, as $F^-(aq)$ is converted to HF(aq), the following equilibria shift to the right and a net reaction occurs in which the dissolution of $CaF_2(s)$ is stimulated.

Equilibria shift to the right: $\longrightarrow$

$$CaF_2(s) \rightleftharpoons Ca^{2+}(aq) + 2\,F^-(aq)$$

$$2\,H_3O^+(aq) + 2\,F^-(aq) \rightleftharpoons 2\,HF(aq) + 2\,H_2O(l)$$

Net reaction:

$$CaF_2(s) + 2\,H_3O^+(aq) \rightleftharpoons Ca^{2+}(aq) + 2\,HF(aq) + 2\,H_2O(l)$$

In contrast to calcium fluoride, the solubility of silver chloride is *independent* of pH. For example, AgCl(s) does not dissolve in dilute $HNO_3(aq)$. Chloride ion, the very weak conjugate base of the strong acid HCl, does not accept a proton from H_3O^+.

$$H_3O^+(aq) + Cl^-(aq) \longrightarrow \text{no reaction}$$

Thus, the solubility of AgCl(s) does not increase as the pH is lowered.

Carbonate ion is a moderately strong base, and the solubilities of carbonates therefore strongly depend on pH. When an acid is added to an insoluble carbonate, such as $CaCO_3$, carbonate ions are rapidly converted to hydrogen carbonate ions. These ions react further to form carbonic acid, which then dissociates into $CO_2(g)$ and water.

$$CO_3^{2-}(aq) + H_3O^+(aq) \longrightarrow HCO_3^-(aq) + H_2O(l)$$
$$HCO_3^-(aq) + H_3O^+(aq) \longrightarrow H_2CO_3(aq) + H_2O(l)$$
$$H_2CO_3(aq) \longrightarrow H_2O\,(l) + CO_2(g)$$

Because $CO_2(g)$ escapes, there is little opportunity for the reverse reactions to occur.

To say that the solubility of an "insoluble" carbonate increases as a solution is made more acidic means that the cation concentration in the solution increases. As we see from the equation written below for the reaction of $CaCO_3(s)$ and HCl(aq), a net chemical reaction occurs in which carbonate ions decompose. The solute in the acidic solution is not calcium carbonate; it is calcium chloride.

$$CaCO_3(s) + 2\,HCl(aq) \longrightarrow Ca^{2+}(aq) + 2\,Cl^-(aq) + H_2O(l) + CO_2(g)$$

Figure 16.4 shows a simple geological field test based on this reaction.

These photographs show the effect of acid rain on a marble statue of George Washington. The photo on top was made in 1935, and the one at the bottom in the mid-1990s. The primary constituent of marble is $CaCO_3(s)$.

EXAMPLE 16.11

What is the molar solubility of $Mg(OH)_2(s)$ in a buffer solution having $[OH^-] = 1.0 \times 10^{-5}$ M, that is, pH = 9.00?

$$Mg(OH)_2(s) \rightleftharpoons Mg^{2+}(aq) + 2\,OH^-(aq) \qquad K_{sp} = 1.8 \times 10^{-11}$$

SOLUTION

First, note that because the solution is a buffer, any OH^- coming from the dissolution of $Mg(OH)_2(s)$ is neutralized by the acidic component of the buffer. The pH will therefore remain at 9.00 and the $[OH^-]$ at 1.0×10^{-5} M. We can calculate $[Mg^{2+}]$ in the solution from the K_{sp} expression, using $[OH^-] = 1.0 \times 10^{-5}$ M.

$$K_{sp} = [Mg^{2+}][OH^-]^2 = 1.8 \times 10^{-11}$$
$$K_{sp} = [Mg^{2+}] \times (1.0 \times 10^{-5})^2 = 1.8 \times 10^{-11}$$
$$[Mg^{2+}] = \frac{1.8 \times 10^{-11}}{1.0 \times 10^{-10}} = 0.18\text{ M}$$

For every mole of $Mg^{2+}(aq)$ appearing in solution, one mole of $Mg(OH)_2(s)$ must have dissolved. The molar solubility of $Mg(OH)_2$ and the equilibrium concentration of Mg^{2+} are the same. The molar solubility of $Mg(OH)_2$ at pH 9.00 is 0.18 mol $Mg(OH)_2$/L.

EXERCISE 16.11A

What is the molar solubility of $Fe(OH)_2$ in a buffer solution with pH = 6.50? [K_{sp} for $Fe(OH)_2 = 8.0 \times 10^{-16}$.]

EXERCISE 16.11B

Can a buffer solution that is 0.520 M CH_3COOH and 0.180 M CH_3COONa also have $[Fe^{3+}] = 1.0 \times 10^{-3}$ M without any precipitate of $Fe(OH)_3(s)$ forming? Explain. [K_a for $CH_3COOH = 1.8 \times 10^{-5}$; K_{sp} for $Fe(OH)_3 = 4 \times 10^{-38}$.]

By the method we used in Example 16.11, we would calculate a molar solubility of $Mg(OH)_2$ at pH 3.00 of 1.8×10^{11} M—an impossible result. We can understand why it is impossible with the help of Figure 12.7 (page 522). There we see several H_2O molecules for every ion; the H_2O molecules provide the ion–dipole forces that keep the ions in solution. This suggests that there cannot be as many ions in solution as there are H_2O molecules, and so the maximum molar solubility of an ionic solute should be less than the 55.6 moles of H_2O per liter of pure water. If the pH of 3.00 is supplied by HCl(aq), the true molar solubility is that of the $MgCl_2$ formed in this reaction.

$$Mg(OH)_2(s) + 2\,H_3O^+(aq) + 2\,Cl^-(aq) \longrightarrow Mg^{2+}(aq) + 2\,Cl^-(aq) + 4\,H_2O(l)$$

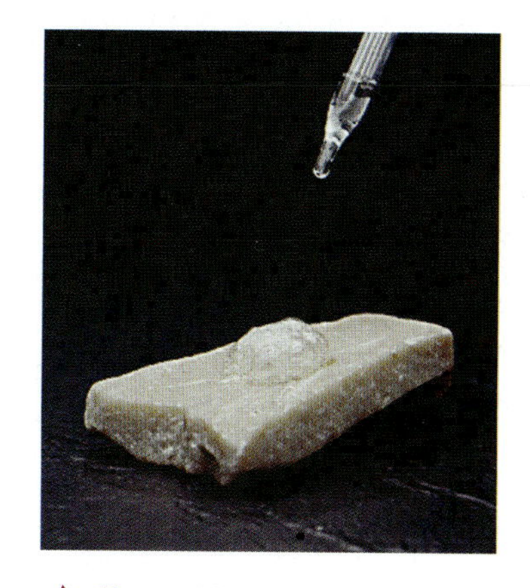

▲ **Figure 16.4**
The reaction of an acid and a carbonate
When a drop of HCl(aq) is placed on a piece of limestone, a characteristic fizzing occurs as $CO_2(g)$ is formed. This fizzing indicates that the limestone contains carbonates.

EXAMPLE 16.12—A Conceptual Example

Without doing detailed calculations, determine in which of the following solutions $Mg(OH)_2(s)$ is most soluble: **(a)** 1.00 M NH_3, **(b)** 1.00 M NH_3/1.00 M NH_4, **(c)** 1.00 M NH_4Cl?

$$Mg(OH)_2(s) \rightleftharpoons Mg^{2+}(aq) + 2\,OH^-(aq) \qquad K_{sp} = 1.8 \times 10^{-11}$$

SOLUTION

At high pH values, OH^- in the solution will act as a common ion and *reduce* the solubility of the $Mg(OH)_2$. At lower pH values, OH^- produced by the dissolution of $Mg(OH)_2$ is neutralized, and this causes more $Mg(OH)_2$ to dissolve, thus *increasing* the solubility. We need to compare the pH values of the three solutions.

a. We know that the pH of 1.00 M NH_3, a *weak base*, is considerably greater than 7.

b. Because $[NH_3] = [NH_4^+]$, the 1.00 M NH_3/1.00 M NH_4Cl buffer solution has a pH equal to the pK_a of NH_4^+. For NH_3, $K_b = 1.8 \times 10^{-5}$, and $pK_b = -\log(1.8 \times 10^{-5}) = 4.74$. For NH_4^+, $pK_a = 14.00 - pK_b = 14.00 - 4.74 = 9.26$. The buffer solution is *basic*.

c. In $NH_4Cl(aq)$, NH_4^+ hydrolyzes and Cl^- does not.

$$NH_4^+(aq) + H_2O(l) \rightleftharpoons H_3O^+(aq) + NH_3(aq) \qquad K_a = K_w/K_b$$

This solution should be acidic with a pH less than 7. The $NH_4Cl(aq)$, the solution with the lowest pH, should dissolve the greatest amount of $Mg(OH)_2$.

EXERCISE 16.12

Describe the dissolution of $Mg(OH)_2(s)$ in $NH_4Cl(aq)$ through a net acid-base reaction in which NH_4^+ is the acid.

● **ChemCDX** Dissolution of $Mg(OH)_2$ by Acid

16.6 Equilibria Involving Complex Ions

At first sight, the data in Table 16.2 on the next page are puzzling. They deal with the solubility of AgCl(s) in solutions containing the common ion Cl^-. The *predicted* solubilities, based on calculations such as in Example 16.5, are much smaller than the solubility in water, and they become progressively smaller as $[Cl^-]$ increases. In contrast, the *measured* solubilities are not nearly as small as predicted, and in solutions with $[Cl^-]$ greater than about 0.3 M, the solubility of AgCl(s) actually *increases* as $[Cl^-]$ *increases*. We need a new explanation for these unexpected data, and we find it in complex-ion formation.

Complex-Ion Formation

A **complex ion** is a polyatomic cation or anion consisting of a central metal atom or ion to which are bonded other groups called **ligands**. Four common ligands featured in this section are the anions Cl^- and OH^- and the molecules H_2O and NH_3.

TABLE 16.2 Solubility of Silver Chloride in NaCl(aq)

$[Cl^-]$, M	Predicted solubility mol AgCl/L × 10^5	Measured solubility mol AgCl/L × 10^5
0.000	1.3	1.3
0.0039	0.0046	0.072
0.036	0.00050	0.19
0.35	0.000051	1.7
1.4	0.000013	18
2.9	0.0000063	1000

Predicted solubility: Decreasing solubility (from top to bottom). Measured solubility: Decreasing solubility (from 0.0039 upward to 0.000); Decreasing solubility (from 2.9 upward to 0.036).

$:\ddot{\underset{..}{Cl}}:^-$ $\quad [:\ddot{\underset{..}{O}}-H]^-$ $\quad H-\ddot{O}:$ (with H below O) $\quad H-\ddot{N}-H$ (with H below N)

The Lewis structures of these ligands show that they all have in common at least one lone pair of electrons. One way to describe complex-ion formation is that the metal center is a *Lewis acid* that accepts electron pairs from the ligands, which are *Lewis bases*. The ligands and metal center are joined by coordinate covalent bonds.

pH, Solubility, and Tooth Decay

Tooth enamel, the hard outer part of human teeth, is composed principally of the mineral *hydroxyapatite*, $Ca_5(PO_4)_3OH$. Hydroxyapatite is insoluble in water but somewhat soluble in acidic solution, and therein lie the origins of tooth decay.

A carbohydrate-protein combination called *mucin* forms a film on teeth called *plaque*. If not removed by brushing and flossing, the buildup of plaque traps food particles. Bacteria ferment carbohydrates in these particles, producing lactic acid ($CH_3CHOHCOOH$). Saliva does not penetrate the plaque and thus cannot buffer against the buildup of acid. The pH drops to as low as 4.5. The H_3O^+ neutralizes OH^- and converts PO_4^{3-} of the hydroxyapatite to slightly ionized HPO_4^{2-}. The calcium salt of HPO_4^{2-} is water soluble, and some hydroxyapatite dissolves.

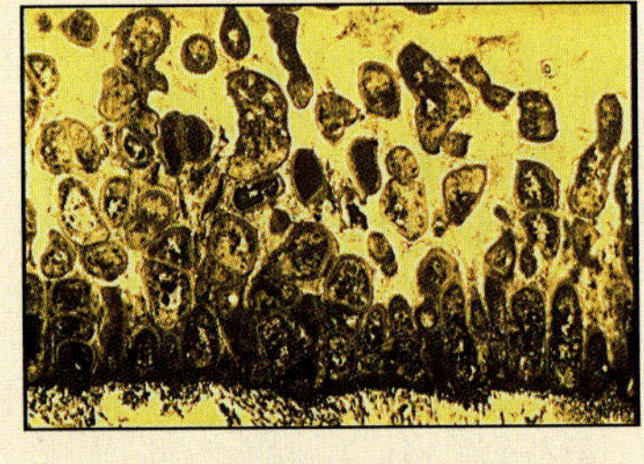

Plaque on tooth enamel, as seen with an electron microscope.

$$Ca_5(PO_4)_3OH(s) + 4\,H_3O^+(aq) \longrightarrow 5\,Ca^{2+}(aq) + 3\,HPO_4^{2-}(aq) + 5\,H_2O(l)$$

If unchecked, this dissolution of tooth enamel produces cavities. Tooth erosion is even more rapid in people with *bulimia*, an eating disorder characterized by binge eating followed by vomiting. Hydrochloric acid, vomited from the stomach, drops the pH in the mouth to as low as 1.5.

One way to fight tooth decay is to add fluorides to drinking water or to toothpaste. Fluorides convert some of the hydroxyapatite in enamel to *fluorapatite*.

$$\underset{\text{Hydroxyapatite}}{Ca_5(PO_4)_3OH(s)} + F^-(aq) \rightleftharpoons \underset{\text{Fluorapatite}}{Ca_5(PO_4)_3F(s)} + OH^-(aq)$$

Fluorapatite is less soluble in acids than is hydroxyapatite because F^- is a weaker base than OH^-. As a result, tooth decay is slowed.

The complex ion responsible for the increased solubility of AgCl(s) at high $[Cl^-]$ consists of a central Ag^+ ion and two Cl^- ions as ligands. The formation of this complex ion is represented in the equation below; the complex ion is the species within brackets. The structure of the complex ion is suggested in Figure 16.5.

$$Ag^+(aq) + 2\,Cl^-(aq) \rightleftharpoons [AgCl_2]^-(aq)$$

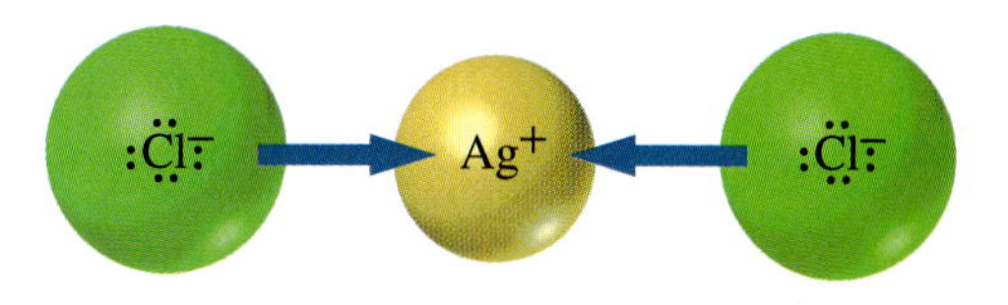

Figure 16.5 The structure of the complex ion $[AgCl_2]^-$.
This illustration suggests that lone pair electrons on the Cl^- ions are used to form coordinate covalent bonds with the central Ag^+ ion. The complex ion has a linear structure. Bonding and structures of complex ions are discussed in Chapter 22.

This complex ion is an *anion* with a net charge of 1−, because it is made of *one* unipositive cation (Ag^+) and *two* uninegative anions (Cl^-).

The complex ion formed between copper(II) ions and ammonia molecules is a complex *cation*—the central copper ion has a charge of 2+ and the ammonia molecules are neutral.

$$\underset{\text{Blue}}{Cu^{2+}(aq)} + 4\,NH_3(aq) \rightleftharpoons \underset{\text{Deep blue-violet}}{[Cu(NH_3)_4]^{2+}(aq)}$$

The formation of this complex ion is shown in Figure 16.6.

Figure 16.6 Complex-ion formation
The dilute $CuSO_4(aq)$ solution on the left owes its pale blue color to $Cu^{2+}(aq)$. When we add $NH_3(aq)$ (here labeled "conc ammonium hydroxide"), the solution color changes to a deep blue-violet (right) as $Cu^{2+}(aq)$ is converted to $[Cu(NH_3)_4]^{2+}(aq)$. The molecular view shows the ions $[Cu(NH_3)_4]^{2+}$ and SO_4^{2-}.

Not all of the Ag^+ in a solution of $Cl^-(aq)$ exists as $[AgCl_2]^-$, and not all of the Cu^{2+} in a solution of $NH_3(aq)$ exists as $[Cu(NH_3)_4]^{2+}$. The formation reaction of a complex ion is *reversible*, and we describe the equilibrium state in the reaction through an equilibrium constant called a **formation constant**, represented by the symbol $\boldsymbol{K_f}$.

$$Ag^+(aq) + 2\,Cl^-(aq) \rightleftharpoons [AgCl_2]^-(aq) \qquad K_f = \frac{[AgCl_2]^-}{[Ag^+][Cl^-]^2} = 1.2 \times 10^8$$

$$Cu^{2+}(aq) + 4\,NH_3(aq) \rightleftharpoons [Cu(NH_3)_4]^{2+}(aq) \qquad K_f = \frac{[Cu(NH_3)_4]^{2+}}{[Cu^{2+}][NH_3]^4} = 1.1 \times 10^{13}$$

Table 16.3 lists a few representative formation constants. A more extensive tabulation is given in Appendix C.

TABLE 16.3 Some Formation Constants for Complex Ions

Complex ion	Equilibrium reaction	K_f
$[Co(NH_3)_6]^{3+}$	$Co^{3+} + 6\,NH_3 \rightleftharpoons [Co(NH_3)_6]^{3+}$	4.5×10^{33}
$[Cu(NH_3)_4]^{2+}$	$Cu^{2+} + 4\,NH_3 \rightleftharpoons [Cu(NH_3)_4]^{2+}$	1.1×10^{13}
$[Fe(CN)_6]^{4-}$	$Fe^{2+} + 6\,CN^- \rightleftharpoons [Fe(CN)_6]^{4-}$	1×10^{37}
$[Fe(CN)_6]^{3-}$	$Fe^{3+} + 6\,CN^- \rightleftharpoons [Fe(CN)_6]^{3-}$	1×10^{42}
$[PbCl_3]^-$	$Pb^{2+} + 3\,Cl^- \rightleftharpoons [PbCl_3]^-$	2.4×10^{1}
$[Ag(NH_3)_2]^+$	$Ag^+ + 2\,NH_3 \rightleftharpoons [Ag(NH_3)_2]^+$	1.6×10^{7}
$[Ag(CN)_2]^-$	$Ag^+ + 2\,CN^- \rightleftharpoons [Ag(CN)_2]^-$	5.6×10^{18}
$[Ag(S_2O_3)_2]^{3-}$	$Ag^+ + 2\,S_2O_3^{2-} \rightleftharpoons [Ag(S_2O_3)_2]^{3-}$	1.7×10^{13}
$[Zn(NH_3)_4]^{2+}$	$Zn^{2+} + 4\,NH_3 \rightleftharpoons [Zn(NH_3)_4]^{2+}$	4.1×10^{8}
$[Zn(CN)_4]^{2-}$	$Zn^{2+} + 4\,CN^- \rightleftharpoons [Zn(CN)_4]^{2-}$	1×10^{18}
$[Zn(OH)_4]^{2-}$	$Zn^{2+} + 4\,OH^- \rightleftharpoons [Zn(OH)_4]^{2-}$	4.6×10^{17}

Complex-Ion Formation and Solubilities

Figure 16.7 shows that water-insoluble silver chloride *does dissolve* in aqueous ammonia. An Ag^+ ion from AgCl(s) combines with two NH_3 molecules to form the complex ion $[Ag(NH_3)_2]^+$. The Cl^- from AgCl(s) goes into solution as Cl^-(aq).This dissolution is best described as the net reaction that results by combining two equilibrium processes.

$$(1)\quad AgCl(s) \rightleftharpoons \cancel{Ag^+(aq)} + Cl^-(aq) \qquad K_{sp} = 1.8 \times 10^{-10}$$

$$(2)\quad \cancel{Ag^+(aq)} + 2\,NH_3(aq) \rightleftharpoons [Ag(NH_3)_2]^+(aq) \qquad K_f = 1.6 \times 10^{7}$$

$$\textit{Net reaction}: AgCl(s) + 2\,NH_3(aq) \rightleftharpoons [Ag(NH_3)_2]^+(aq) + Cl^-(aq)$$

$$K_c = K_{sp} \times K_f = 2.9 \times 10^{-3}$$

The equilibrium constant for the dissolution of AgCl(s) in NH_3(aq) is not large ($K_c = 2.9 \times 10^{-3}$). Nevertheless, in the presence of a high concentration of NH_3(aq), equilibrium lies far enough to the right so that a significant amount of AgCl(s) dissolves.

We can write similar equations and equilibrium constants, K_c, for the action of NH_3(aq) on AgBr(s) and AgI(s). In each case, $K_c = K_{sp} \times K_f$. The equilibrium constants, K_c, for the dissolution of AgCl(s), AgBr(s), and AgI(s) in NH_3(aq) are 2.9×10^{-3}, 8.0×10^{-6}, and 1.4×10^{-9}, respectively. These values are consis-

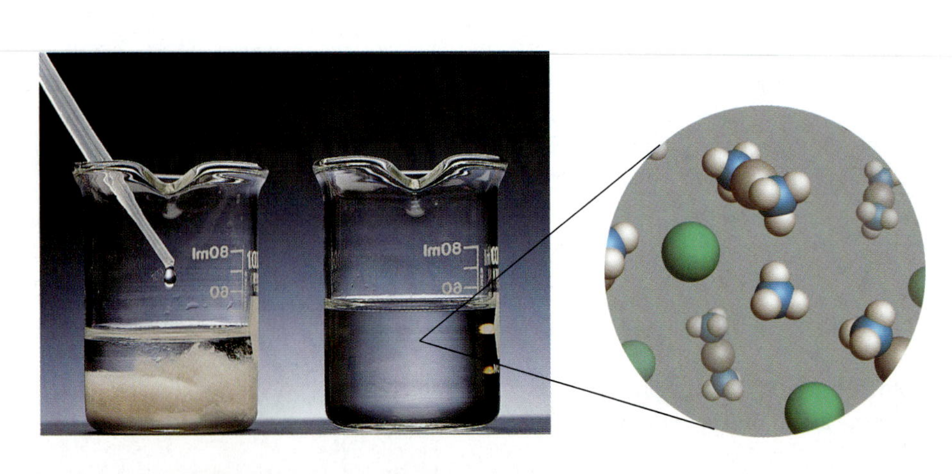

Figure 16.7 Complex-ion formation and solute solubility

A precipitate of AgCl(s) (left) readily dissolves in an aqueous solution of NH_3 because of the formation of the complex ion $[Ag(NH_3)_2]^+$ (right). The molecular view of the final solution shows $[Ag(NH_3)_2]^+$ and Cl^- ions and NH_3 molecules.

tent with the observation that AgCl(s) is moderately soluble, AgBr(s) is slightly soluble, and AgI(s) is only very slightly soluble in aqueous ammonia. The K_c values are directly proportional to values of K_{sp} because K_f is the same for each reaction. Thus, the solubilities of AgCl, AgBr, and AgI in $NH_3(aq)$ parallel their K_{sp} values.

Solubility in NH₃(aq) decreases:	AgCl	>	AgBr	>	AgI
K_{sp} *decreases:*	1.8×10^{-10}		5.0×10^{-13}		8.5×10^{-17}

Silver bromide, as an emulsion in gelatin, is used in photographic film. The photographic image is a deposit of metallic silver in the emulsion, produced by exposure of the film to light, followed by treatment with a mild reducing agent [for example, hydroquinone, $C_6H_4(OH)_2$]. After the film has been developed, it is necessary to "fix" the film—to remove the unexposed AgBr(s) so that the film will not darken on further exposure to light. For this purpose, we can use sodium thiosulfate (also called sodium hyposulfite, or "hypo").

The dissolving of AgBr(s) in $Na_2S_2O_3(aq)$, like that of AgBr(s) in $NH_3(aq)$, is the net result of two simultaneous equilibrium processes.

$$AgBr(s) \rightleftharpoons \cancel{Ag^+(aq)} + Br^-(aq) \qquad K_{sp} = 5.0 \times 10^{-13}$$

$$\cancel{Ag^+(aq)} + 2\,S_2O_3^{2-}(aq) \rightleftharpoons [Ag(S_2O_3)_2]^{3-}(aq) \qquad K_f = 1.7 \times 10^{13}$$

Net reaction: $AgBr(s) + 2\,S_2O_3^{2-}(aq) \rightleftharpoons [Ag(S_2O_3)_2]^{3-}(aq) + Br^-(aq)$

$$K_c = K_{sp} \times K_f = 5.0 \times 10^{-13} \times 1.7 \times 10^{13} = 8.5$$

Now we can see why sodium thiosulfate is used as a photographic fixer and ammonia is not. The equilibrium constant for the dissolution of AgBr(s) in $S_2O_3^{-2}(aq)$ ($K_c = 8.5$) is about a million times larger than for its dissolution in $NH_3(aq)$ ($K_c = 8.0 \times 10^{-6}$). Sodium thiosulfate solutions can dissolve the excess AgBr(s) from photographic film more completely than ammonia solutions can.

The three examples that follow illustrate three common types of calculations involving complex ions.

- In Example 16.13, we illustrate the use of a K_f expression to calculate equilibrium concentrations.
- In Example 16.14, we apply precipitation criteria to solutions containing complex ions.
- In Example 16.15, we illustrate calculating a solubility when complex ions are formed.

EXAMPLE 16.13

Calculate the concentration of free silver ion, $[Ag^+]$, in an aqueous solution that is prepared as 0.10 M $AgNO_3$ and 3.0 M NH_3.

$$Ag^+(aq) + 2\,NH_3(aq) \rightleftharpoons [Ag(NH_3)_2]^+(aq) \qquad K_f = 1.6 \times 10^7$$

SOLUTION

Because K_f is such a large number, let's begin by assuming that the formation reaction goes to completion.

The reaction:	Ag^+	$+\ 2\,NH_3$	$\longrightarrow$	$[Ag(NH_3)_2]^+$
Initial concentrations, M:	0.10	3.0		0
Changes, M:	−0.10	−0.20		+0.10
Final concentrations, M:	0	2.8		0.10

Now let's restate the question in this way: What is $[Ag^+]$ in a solution that is 0.10 M $[Ag(NH_3)_2]^+$ and 2.8 M NH_3?" To answer this question, we need to remember that equilibrium can be achieved from either direction, and we now consider the *reverse* of the formation reaction in solving for $[Ag^+]$.

The reaction:	Ag^+	$+\ 2\,NH_3$ $\rightleftharpoons$	$[Ag(NH_3)_2]^+$
Initial concentrations, M:	0	2.8	0.10
Changes, M:	$+x$	$+2x$	$-x$
Equilibrium concentrations, M:	x	$(2.8 + 2x)$	$(0.10 - x)$

Because the formation constant is so large, practically all the silver will be present as $[Ag(NH_3)_2]^+$ and very little as Ag^+. We can therefore assume that x is very small compared to 0.10 and that $2x$ is very small compared to 2.8.

$$\frac{[[Ag(NH_3)_2]^+]}{[Ag^+][NH_3]^2} = K_f$$

$$\frac{(0.10 - x)}{x(2.8 + 2x)^2} = 1.6 \times 10^7$$

$$\frac{0.10}{x(2.8)^2} = 1.6 \times 10^7$$

$$0.10 = x(2.8)^2 \times 1.6 \times 10^7$$

$$x = \frac{0.10}{(2.8)^2 \times 1.6 \times 10^7}$$

$$x = [Ag^+] = 8.0 \times 10^{-10}\ M$$

EXERCISE 16.13A

Calculate the concentration of free silver ion, $[Ag^+]$, in an aqueous solution that is 0.10 M $AgNO_3$ and 1.0 M $Na_2S_2O_3$.

$$Ag^+(aq) + 2\,S_2O_3^{2-}(aq) \rightleftharpoons [Ag(S_2O_3)_2]^{3-}(aq)\ K_f = 1.7 \times 10^{13}$$

EXERCISE 16.13B

What must be the *total* NH_3 concentration in a solution that is initially 0.050 M $AgNO_3$ if $[Ag^+]$ is to be 1.0×10^{-8} M? (K_f for $[Ag(NH_3)_2]^+ = 1.6 \times 10^7$.)

(*Hint:* The total NH_3 concentration refers to NH_3 that is free in solution and that present as ligands in $[Ag(NH_3)_2]^+$.)

EXAMPLE 16.14

If 1.00 g KBr is added to 1.00 L of the solution described in Example 16.13, should any AgBr(s) precipitate from the solution?

$$AgBr(s) \rightleftharpoons Ag^+(aq) + Br^-(aq) \qquad K_{sp} = 5.0 \times 10^{-13}$$

SOLUTION

The solution in Example 16.13 is 0.10 M in $AgNO_3$ and 3.0 M in NH_3. Our first task is to determine $[Ag^+]$, but we did this in Example 16.13: $[Ag^+] = 8.0 \times 10^{-10}$ M. Next we need to determine the $[Br^-]$ when 1.00 g KBr is dissolved in 1.00 L of the solution.

$$[Br^-] = \frac{1.00\ \cancel{g\ KBr} \times \dfrac{1\ \cancel{mol\ KBr}}{119.0\ \cancel{g\ KBr}} \times \dfrac{1\ mol\ Br^-}{1\ \cancel{mol\ KBr}}}{1\ L} = 8.40 \times 10^{-3}\ M$$

Now we must compare Q_{ip} and K_{sp} for AgBr.

$$Q_{ip} = [Ag^+]_{initial} \times [Br^-]_{initial} = (8.0 \times 10^{-10})(8.40 \times 10^{-3}) = 67.2 \times 10^{-13}$$

Because $Q_{ip} > K_{sp}$, we conclude that some AgBr(s) should precipitate from the solution.

EXERCISE 16.14A

If 1.00 g KI is added to 1.00 L of the solution described in Exercise 16.13A, should any AgI(s) precipitate from the solution?

$$AgI(s) \rightleftharpoons Ag^+(aq) + I^-(aq) \quad K_{sp} = 8.5 \times 10^{-17}$$

EXERCISE 16.14B

What is the maximum mass of KBr that can be added to 250.0 mL of a solution that has $[NH_3] = 1.00$ M and $[[Ag(NH_3)_2]^+] = 0.050$ M without a precipitate of AgBr(s) forming? (K_f for $[Ag(NH_3)_2]^+ = 1.6 \times 10^7$; K_{sp} for AgBr $= 5.0 \times 10^{-13}$.)

EXAMPLE 16.15

What is the molar solubility of AgBr(s) in 3.0 M NH_3?

$$AgBr(s) + 2\,NH_3(aq) \rightleftharpoons [Ag(NH_3)_2]^+(aq) + Br^-(aq) \qquad K_c = 8.0 \times 10^{-6}$$

SOLUTION

First, notice that the dissolution reaction is a net reaction. The equilibrium constant for this reaction, K_c, is the product of K_f for $[Ag(NH_3)_2]^+$ and K_{sp} for AgBr. If we had not been given a value of K_c, we would have had to establish it in the same way that we did for AgCl(s) in NH_3(aq) on page 714. Now notice that for every formula unit of AgBr(s) that dissolves, we get one $[Ag(NH_3)_2]^+$ and one Br^- ion. The molar solubility we seek is the same as the molarities of either of these ions at equilibrium. We can use our standard equilibrium format:

	AgBr(s) + $2\,NH_3$(aq) ⇌	$[Ag(NH_3)_2]^+$ +	Br^-(aq)
The reaction:	AgBr(s) + $2\,NH_3$(aq) ⇌	$[Ag(NH_3)_2]^+$ +	Br^-(aq)
Initial concentrations, M:	3.0	0	0
Changes, M:	$-2s$	$+s$	$+s$
Equilibrium concentrations, M:	$(3.0 - 2s)$	s	s

Now let's write the equilibrium constant expression for the reaction and substitute the above data.

$$K_c = \frac{[[Ag(NH_3)_2]^+][Br^-]}{[NH_3]^2} = \frac{s \cdot s}{(3.0 - 2s)^2} = 8.0 \times 10^{-6}$$

At this point, we could assume that $s \ll 3.0$ and proceed in the usual fashion, but we can simplify our calculation if we express the equation as follows.

$$\frac{s^2}{(3.0 - 2s)^2} = \left(\frac{s}{3.0 - 2s}\right)^2 = 8.0 \times 10^{-6}$$

Next, we can take the square root of each side of the equation.

$$\frac{s}{3.0 - 2s} = (8.0 \times 10^{-6})^{1/2} = 2.8 \times 10^{-3}$$

Then we can complete the calculation.

$$s = (3.0 - 2s)(2.8 \times 10^{-3}) = (8.4 \times 10^{-3}) - 5.6 \times 10^{-3}s$$

$$1.0056\,s = 8.4 \times 10^{-3}$$

$$s = [[Ag(NH_3)_2]^+] = [Br^-] = 8.4 \times 10^{-3}\text{ M}$$

The molar solubility is therefore 8.4×10^{-3} mol AgBr/L.

EXERCISE 16.15A

What is the molar solubility of AgBr(s) in 0.500 M $Na_2S_2O_3$(aq)?

$$AgBr(s) + 2\,S_2O_3^{2-}(aq) \rightleftharpoons [Ag(S_2O_3)_2]^{3-}(aq) + Br^-(aq) \qquad K_c = 8.5$$

EXERCISE 16.15B

Without doing detailed calculations, determine in which of the following 0.100 M solutions AgI(s) should be most soluble: NH_3(aq), $Na_2S_2O_3$(aq), or NaCN(aq). Explain.

EXAMPLE 16.16—A Conceptual Example

Figure 16.8 shows that a precipitate forms when HNO_3(aq) is added to the colorless solution obtained in Figure 16.7. Write an equation(s) to show what happens.

▶ **Figure 16.8**
Destruction of a complex ion
To the clear solution in Figure 16.7 (left) is added HNO_3(aq) (right).

SOLUTION

The principal solute species in the solution in Figure 16.7 are $[Ag(NH_3)_2]^+$(aq), Cl^-(aq), and free NH_3(aq). There is a trace of free Ag^+. The added HNO_3(aq) is a source of H_3O^+. A proton transfer occurs between H_3O^+, an acid, and the free NH_3, a base.

(a) $$H_3O^+(aq) + NH_3(aq) \longrightarrow NH_4^+(aq) + H_2O(l)$$

According to Le Châtelier's principle, the stress imposed by this removal of NH_3(aq) stimulates the *decomposition* of $[Ag(NH_3)_2]^+$(aq). This produces more NH_3(aq) and, simultaneously, more Ag^+(aq).

(b) $[Ag(NH_3)_2]^+(aq) \longrightarrow Ag^+(aq) + 2\ NH_3(aq)$

As $[Ag^+]$ increases, the ion product $[Ag^+][Cl^-]$ soon exceeds K_{sp}, and AgCl(s) precipitates.

(c) $Ag^+(aq) + Cl^-(aq) \longrightarrow AgCl(s)$

The net equation, then, is 2(a) + (b) + (c):

$$[Ag(NH_3)_2]^+(aq) + Cl^-(aq) + 2\ H_3O^+(aq) \longrightarrow AgCl(s) + 2\ NH_4^+(aq) + 2\ H_2O(l)$$

EXERCISE 16.16

Would you expect AgCl(s) to precipitate in Figure 16.8 if $NH_4NO_3(aq)$ were used instead of $HNO_3(aq)$? Explain.

Complex Ions in Acid-Base Reactions

In discussing hydrolysis reactions in Chapter 15, we stated that cations of Groups 1A and 2A do not hydrolyze. However, we also noted that many other metal cations do hydrolyze, particularly those with a small size and high charge (page 660). We can use complex-ion formation to explain why this is so.

Aqueous solutions of iron(III) compounds, for example, are about as acidic as acetic acid solutions. To explain how H_3O^+ ions might be produced in these solutions, let's begin by noting that most of the Fe(III) is present as the complex ion $[Fe(H_2O)_6]^{3+}$. Water molecules are quite commonly found as ligands in complex ions because H_2O is a Lewis base and there are many, many H_2O molecules in every aqueous solution.

Now, consider that the electron-withdrawing power of the small, highly charged Fe^{3+} ion weakens an O—H bond in a *ligand* water molecule, causing it to give up a proton to a water molecule in the solution. The result is the formation of an H_3O^+ ion in solution and the conversion of an H_2O ligand to OH^-. This process is suggested by Figure 16.9 and by the following equation.

$$[Fe(H_2O)_6]^{3+}(aq) + H_2O(l) \rightleftharpoons [FeOH(H_2O)_5]^{2+}(aq) + H_3O^+(aq) \quad K_a = 9 \times 10^{-4}$$

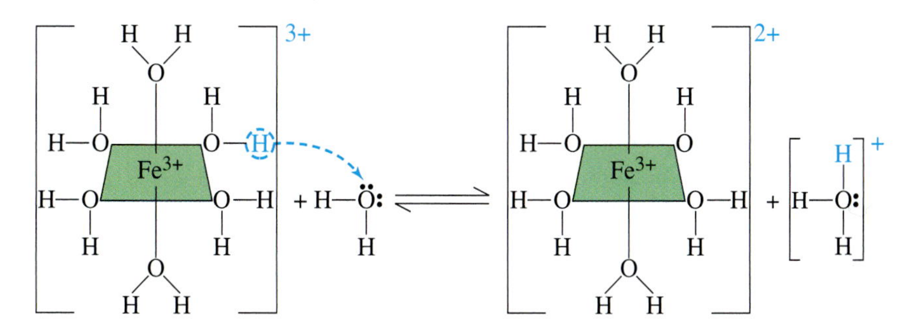

Figure 16.9
Ionization of $[Fe(H_2O)_6]^{3+}(aq)$ as an acid
The transfer of protons from ligand water molecules in $[Fe(H_2O)_6]^{3+}(aq)$ to solvent water molecules produces an increase in $[H_3O^+]$ in the solution.

In the iron(II) complex ion $[Fe(H_2O)_6]^{2+}$, the electron-withdrawing effect of the larger, less highly charged Fe^{2+} central ion is weaker. As a consequence, $[Fe(H_2O)_6]^{2+}$ does not ionize as extensively as $[Fe(H_2O)_6]^{3+}$.

$$[Fe(H_2O)_6]^{2+}(aq) + H_2O(l) \rightleftharpoons [Fe(H_2O)_5OH]^+(aq) + H_3O^+(aq) \quad K_a = 1 \times 10^{-7}$$

As shown in Figure 16.10, $Al(OH)_3(s)$, which is insoluble in water, reacts with and dissolves in *both* HCl(aq) and NaOH(aq). Complex-ion formation helps us to explain this behavior.

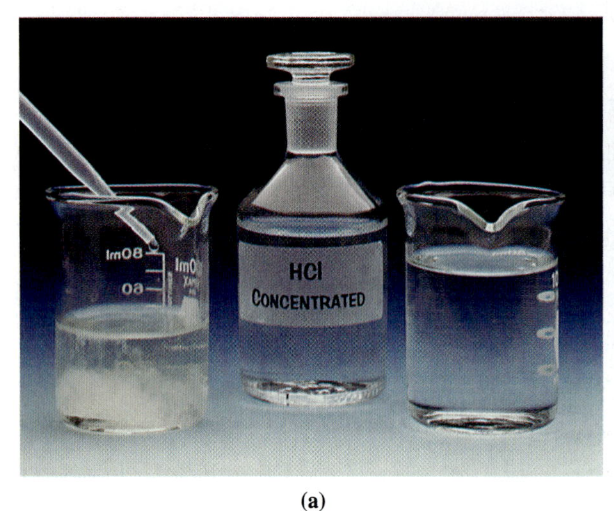

(a)

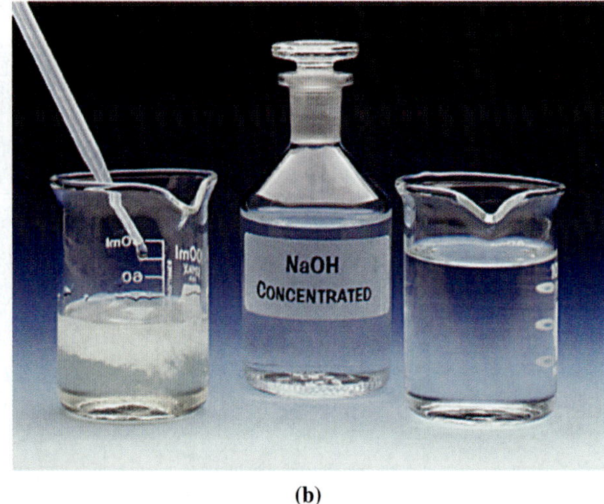

(b)

▲ **Figure 16.10 Amphoteric behavior of $Al(OH)_3(s)$**
(a) Aluminum hydroxide, $Al(OH)_3(s)$, reacts with HCl(aq). In the resulting colorless solution (right), the aluminum is present as the complex ion, $[Al(H_2O)_6]^{3+}(aq)$. (b) Freshly precipitated $Al(OH)_3(s)$ reacts with NaOH(aq) to form a solution (right) of the colorless aluminum-containing complex ion $[Al(OH)_4]^-(aq)$.

When $Al(OH)_3(s)$ reacts with HCl(aq) (Figure 16.10a), H_3O^+ from HCl(aq) reacts with OH^- from $Al(OH)_3(s)$ to form H_2O. The H_2O molecules become ligands in the complex ion $[Al(H_2O)_6]^{3+}$. The net result is the following.

$$Al(OH)_3(s) + 3\,H_3O^+(aq) \longrightarrow [Al(H_2O)_6]^{3+}(aq)$$

When $Al(OH)_3(s)$ reacts with NaOH(aq) (Figure 16.10b), OH^- from the NaOH(aq) and the OH^- ions in $Al(OH)_3(s)$ together become the ligands in the complex ion $[Al(OH)_4]^-$. The net result in this case is shown in the following equation.

$$Al(OH)_3(s) + OH^-(aq) \longrightarrow [Al(OH)_4]^-(aq)$$

Aluminum hydroxide is *amphoteric*. An **amphoteric** hydroxide can react with either an acid or a base. The hydroxides of zinc and chromium(III) are also amphoteric. The ions produced in acidic solution are $[Zn(H_2O)_4]^{2+}(aq)$ and $[Cr(H_2O)_6]^{3+}(aq)$, respectively. In basic solution, they are $[Zn(OH)_4]^{2-}$ and $[Cr(OH)_4]^-$. The hydroxide of iron(III), on the other hand, is not amphoteric. $Fe(OH)_3$ dissolves in acidic solutions to produce the cation $[Fe(H_2O)_6]^{3+}(aq)$, but it does not react in basic solutions.

The oxides Al_2O_3, ZnO, and Cr_2O_3 behave toward acids and bases much as their hydroxides do. For example, consider the reactions of aluminum oxide.

$$Al_2O_3(s) + 6\,H_3O^+(aq) + 3\,H_2O(l) \longrightarrow 2\,[Al(H_2O)_6]^{3+}(aq)$$

$$Al_2O_3(s) + 2\,OH^-(aq) + 3\,H_2O(l) \longrightarrow 2\,[Al(OH)_4]^-(aq)$$

Like hydroxides, oxides that can react with and dissolve in either an acid or a base are often said to be *amphoteric*.*

* The term *amphiprotic* (page 635) refers to ions or molecules that are able either to donate or accept a proton. It is best to use the term *amphoteric*, however, when describing the ability of substances, particularly oxides and hydroxides, to react with both acids and bases. For example, although Al_2O_3 is amphoteric, it can't be amphiprotic because it contains no H atoms.

16.7 Qualitative Inorganic Analysis

Many concepts that we have studied in this text—particularly in this and the preceding chapter—have important applications in analytical chemistry. For example, acid-base chemistry, precipitation reactions, oxidation-reduction, and complex-ion formation all come into sharp focus in an area of analytical chemistry called *classical qualitative inorganic analysis.*

"Qualitative" signifies that we are interested in determining what is present, but not how much (that would be *quantitative* analysis). "Inorganic" indicates that we are analyzing for inorganic constituents, usually ions. By "classical," we mean analyses involving chemical reactions and methods developed mostly in the nineteenth century. Although classical qualitative analysis is not widely used today, it is still a good vehicle for applying all the basic concepts of equilibria in aqueous solutions. Also, certain qualitative tests, such as that to distinguish between deionized water and municipal tap water in Figure 16.2, are so simple that we still favor them over more modern, sophisticated tests.

Figure 16.11 outlines a general scheme for analyzing an aqueous solution for the possible presence of any or all of about 25 common cations. We can use differences in solubilities of certain compounds of these cations to separate the cations into five groups. Then we do additional separations in each group. Ultimately we

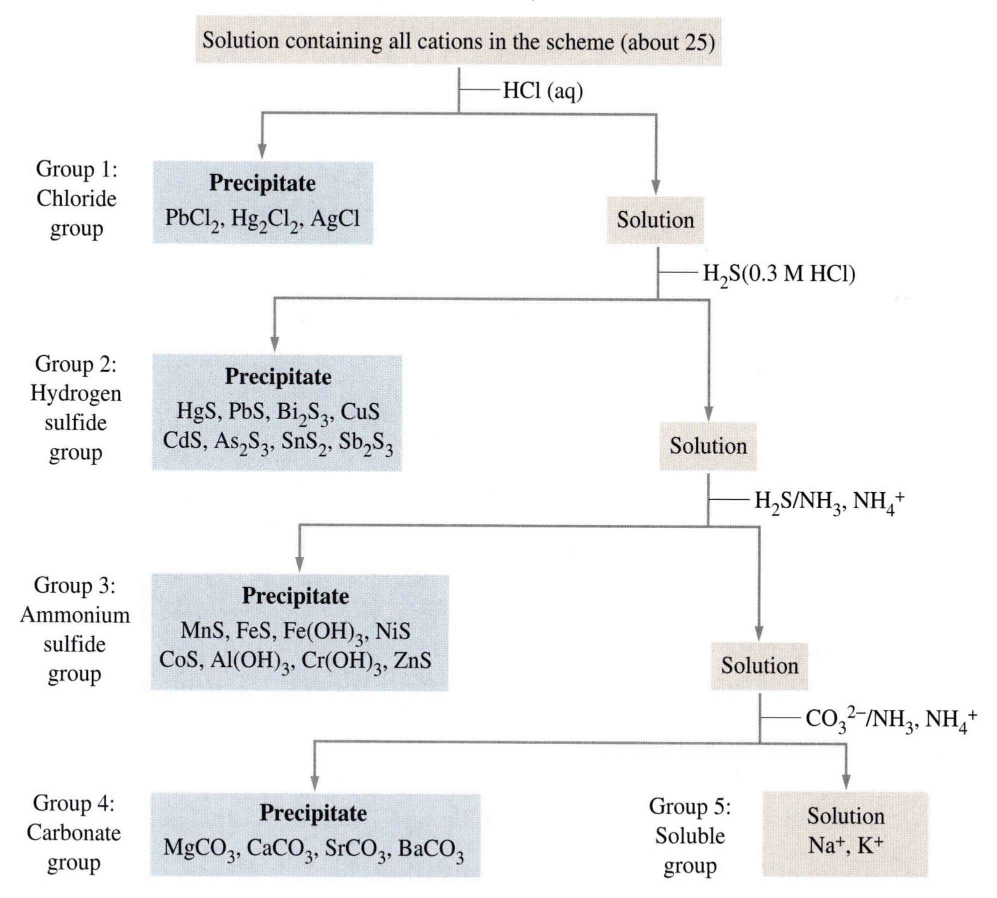

▲ **Figure 16.11 Outline of the qualitative analysis scheme for some common cations**
The text describes how we can use differences in solubilities of compounds to separate the cations into five groups, do additional separations in each group to obtain each cation that is present in a separate solution, and then test for that cation in the solution.

▲ **Figure 16.12**
Precipitates in cation group 1
Left. The cation group 1 precipitate: $PbCl_2$ (white), Hg_2Cl_2 (white), AgCl (white). *Middle.* Reaction product in test for Hg_2^{2+}: a mixture of Hg (black) and $HgNH_2Cl$ (white). *Right.* Reaction product in test for Pb^{2+}: $PbCrO_4$ (yellow), formed in the reaction of K_2CrO_4(aq) with saturated $PbCl_2$(aq).

The K_{sp} values of the precipitates for group 1 cations are: $PbCl_2$, 1.6×10^{-5}; Hg_2Cl_2, 1.3×10^{-18}; and AgCl, 1.8×10^{-10}.

obtain each cation, if present, in a solution from which other cations have been eliminated. We then perform a test for that cation in the solution. A separate scheme is used to test for anions. In this discussion, we will consider the group of cations called group 1 in some detail. We will consider the remainder of the cation analysis scheme more briefly.

Cation Group 1

To an aqueous solution that might contain up to 25 different cations, called the *unknown*, we add HCl(aq). If a precipitate forms, we know that the unknown contains one or more of these cations: Pb^{2+}, Hg_2^{2+}, Ag^+; they are the only ones that form insoluble chlorides. If there is no group 1 precipitate, we know that Pb^{2+}, Hg_2^{2+}, and Ag^+ are *absent* from the unknown. If there is a precipitate, we filter off this precipitate and save it. We save the solution separated from the precipitate for the analysis of other groups.

If we do get a group 1 precipitate, it could be any one or a mixture of these chlorides: $PbCl_2$, Hg_2Cl_2, AgCl. Of the three, $PbCl_2$(s) is the most water soluble. We can get some $PbCl_2$ to dissolve just by washing the precipitate with *hot* water. To test for the presence of Pb^{2+}, we add K_2CrO_4(aq) to the hot water washings. If Pb^{2+} is present, chromate ion combines with lead ion to form a precipitate of *yellow* lead chromate, $PbCrO_4$(s), which is less soluble than $PbCl_2$(s) (see Figure 16.12).

$$Pb^{2+}(aq) + CrO_4^{2-}(aq) \longrightarrow PbCrO_4(s) \qquad K_{sp} = 2.8 \times 10^{-13}$$

Next, we treat the undissolved portion of the precipitate with NH_3(aq). If AgCl(s) is present in this precipitate, it dissolves by the following reaction, which is illustrated in Figure 16.7 (page 714).

$$AgCl(s) + 2\,NH_3(aq) \longrightarrow [Ag(NH_3)_2]^+(aq) + Cl^-(aq)$$

We can confirm that the AgCl(s) has dissolved by treating the $[Ag(NH_3)_2]^+$(aq) with HNO_3(aq), with the result illustrated in Figure 16.8 (page 718).

$$[Ag(NH_3)_2]^+(aq) + Cl^-(aq) + 2\,H_3O^+(aq) \longrightarrow AgCl(s) + 2\,NH_4^+(aq) + 2\,H_2O(l)$$

At the same time that AgCl(s) dissolves in NH_3(aq), any Hg_2Cl_2(s) present undergoes an *oxidation-reduction* reaction. One of the products of the reaction is a dark gray mixture of elemental mercury and $HgNH_2Cl$(s) (see Figure 16.12).

$$Hg_2Cl_2(s) + 2\,NH_3(aq) \longrightarrow \underbrace{Hg(l) + HgNH_2Cl(s)}_{\text{dark gray}} + NH_4^+(aq) + Cl^-(aq)$$

The presence of a dark gray mixture at this point indicates that mercury(I) ions were present in the original unknown solution.

Hydrogen Sulfide in the Qualitative Analysis Scheme

Once the cation group 1 chlorides have been precipitated, we use hydrogen sulfide as the next reagent in the qualitative analysis scheme. H_2S is a *weak diprotic acid.*

$$H_2S(aq) + H_2O \rightleftharpoons H_3O^+(aq) + HS^-(aq) \qquad K_{a_1} = 1.0 \times 10^{-7}$$

$$HS^-(aq) + H_2O \rightleftharpoons H_3O^+(aq) + S^{2-}(aq) \qquad K_{a_2} = 1.0 \times 10^{-19}$$

From the extremely small value of K_{a_2}, we expect very little ionization of hydrogen sulfide ion, HS^-. This suggests that HS^- and not S^{2-} itself is the precipitating agent. The precipitation of a metal sulfide MS can be represented by the following net equation.

$$H_2S(aq) + H_2O(l) \rightleftharpoons H_3O^+(aq) + \cancel{HS^-(aq)}$$
$$M^{2+}(aq) + \cancel{HS^-(aq)} + H_2O(l) \rightleftharpoons MS(s) + H_3O^+(aq)$$

Net reaction: $$M^{2+}(aq) + H_2S(aq) + 2\,H_2O(l) \rightleftharpoons MS(s) + 2\,H_3O^+(aq)$$

Hydrogen sulfide gas has a familiar "rotten egg" odor that is especially noticeable in volcanic areas and near sulfur hot springs. At levels of 10 ppm, the gas can produce headaches and nausea, and it can be lethal at about 100 ppm. Because of its toxicity, $H_2S(g)$ is generally produced only in small quantities and directly in the solution where it is to be used. For example, it is slowly released when thioacetamide is heated in aqueous solution.

$$\underset{\text{Thioacetamide}}{CH_3\overset{\overset{S}{\|}}{C}NH_2(aq)} + H_2O(l) \longrightarrow \underset{\text{Acetamide}}{CH_3\overset{\overset{O}{\|}}{C}NH_2(aq)} + H_2S(aq)$$

Cation Groups 2, 3, 4, and 5

When we apply Le Châtelier's principle to the net reaction for the precipitation of metal sulfides, we see that the reaction to the *left* is favored in *acidic* solutions, where $[H_3O^+]$ is high, and to the *right* in *basic solutions*, where $[H_3O^+]$ is low. Thus, metal sulfides are more soluble in acidic solutions and less soluble in basic solutions.

$$M^{2+}(aq) + H_2S(aq) + 2\,H_2O(l) \underset{\text{acidic solution}}{\overset{\text{basic solution}}{\rightleftharpoons}} MS(s) + 2\,H_3O^+(aq)$$

The concentration of HS^- is so low in a strongly acidic solution of $H_2S(aq)$ that only the most insoluble sulfides precipitate. The sulfides in this category are the eight metal sulfides in group 2.* They precipitate from $H_2S(aq)$ that is also 0.3 M in HCl. Of the eight cations in group 3, five form sulfides that are soluble in acidic solution but insoluble in an alkaline NH_3/NH_4Cl buffer solution. The other three group 3 cations form hydroxide precipitates in the alkaline solution.

The cations of groups 4 and 5 form soluble sulfides, even in basic solution. Their hydroxides, except $Mg(OH)_2$, are also moderately or highly soluble. The group 4 cations are precipitated as carbonates from a buffered alkaline solution. The cations of group 5 remain soluble in the presence of all common reagents.

Within each of the qualitative analysis groups, we need to use additional reactions to dissolve group precipitates and to separate and selectively precipitate individual cations for identification and confirmation. These reactions include oxidation-reduction, complex-ion formation, and amphoteric behavior, as well as precipitation. Flame tests are a prominent feature for several of the ions in groups 4 and 5 (recall page 348).

Consider, for example, how we might detect the presence of Cr^{3+} in an unknown. When the group 3 cations are precipitated, groups 1 and 2 have already been removed, and cations of groups 4 and 5 remain in the filtrate. We dissolve the group 3 precipitate in a mixture of HCl(aq) and $HNO_3(aq)$, obtaining a solution in which the ions possibly present are

$$Fe^{3+},\ Mn^{2+},\ Co^{2+},\ Ni^{2+},\ Al^{3+},\ Zn^{2+},\ \text{and}\ Cr^{3+}.$$

Now we neutralize the solution, make it basic with NaOH(aq), and also add hydrogen peroxide, $H_2O_2(aq)$. Although all the group 3 hydroxides are insoluble in

* We find Pb^{2+} in both groups 1 and 2. $PbCl_2(s)$ is soluble enough that sufficient Pb^{2+} remains in the group 1 *filtrate* to precipitate again in group 2 as PbS(s), which is much less soluble than $PbCl_2(s)$.

water, three of them are *amphoteric*. In addition to being basic, they are acidic enough to dissolve in the alkaline solution.

$$Al(OH)_3(s) + OH^-(aq) \longrightarrow [Al(OH)_4]^-(aq)$$

$$Zn(OH)_2(s) + 2\,OH^-(aq) \longrightarrow [Zn(OH)_4]^{2-}(aq)$$

$$Cr(OH)_3(s) + OH^-(aq) \longrightarrow [Cr(OH)_4]^-(aq)$$

The hydrogen peroxide, an *oxidizing* agent, brings about several oxidations, including that of $[Cr(OH)_4]^-$ to CrO_4^{2-}.

$$3\,H_2O_2(aq) + 2\,[Cr(OH)_4]^-(aq) + 2\,OH^-(aq) \longrightarrow 2\,CrO_4^{2-}(aq) + 8\,H_2O(l)$$

Green Yellow

Group 3 is thus separated into two subgroups: a precipitate of hydroxides and a filtrate. The filtrate may or may not contain $[Al(OH)_4]^-$, $[Zn(OH)_4]^{2-}$, and CrO_4^{2-}. The appearance of a yellow color is a strong indication that CrO_4^{2-} is present; the other ions that may be present in the filtrate are colorless. The presence of CrO_4^{2-} can also be confirmed by other tests, such as the precipitation of yellow $BaCrO_4(s)$.

Key Terms

amphoteric (16.6)
common-ion effect (16.3)
complex ion (16.6)
formation constant, K_f (16.6)
ligand (16.6)
solubility product constant, K_{sp} (16.1)

Summary

The solubility product constant, K_{sp}, represents equilibrium between a slightly soluble ionic compound and its ions in a saturated aqueous solution. The K_{sp} and molar solubility are related in a way that either one can be established from a value of the other.

The solubility of a slightly soluble ionic compound is reduced, often significantly, in the presence of an excess of one of the ions involved in the solubility equilibrium. This *common-ion effect* is particularly useful in procedures involving precipitates in quantitative analysis.

We can determine what will happen when certain ions are brought together in solution by comparing the reaction quotient or ion product, Q_{ip}, with K_{sp}.

- If $Q_{ip} > K_{sp}$, a precipitate should form.
- If $Q_{ip} < K_{sp}$, a precipitate will not form.
- If $Q_{ip} = K_{sp}$, a solution is just saturated.

We assume that precipitation is complete if no more than 0.1% of the target ion remains in solution after precipitation has occurred. A small value of K_{sp}, a high initial ion concentration, and the presence of a common ion favor complete precipitation. A mixture of ions in solution can sometimes be separated by the addition of an ion with which similar precipitates are formed. For example, chloride and iodide ions can be separated using $Ag^+(aq)$ to precipitate $AgI(s)$ while Cl^- remains in solution. Selective precipitation requires that the K_{sp} values of the precipitates be widely different.

The solubilities of some slightly soluble compounds depend strongly on pH. For example, solutes become more soluble in acidic solutions if their anions are sufficiently basic to accept protons from $H_3O^+(aq)$. Hydroxides, carbonates, and fluorides all become more soluble as the solution pH is lowered.

Certain solutes become more soluble in the presence of species that can serve as ligands in complex ions. Thus, $AgCl(s)$ is soluble in $NH_3(aq)$ because Ag^+ can join with two NH_3 molecules to form The extent to which a solute dissolves in the presence of complexing ligands depends on the values of K_{sp} and the formation constant of a complex ion, K_f. K_f describes equilibrium between the complex ion, the free cation, and the free ligands.

The ability of H_2O ligand molecules to donate protons accounts for the acidic character of some complex ions. The formation of complex ions with OH^- ions as ligands accounts for the amphoterism of some oxides and hydroxides [for example, those of aluminum, chromium(III), and zinc]. An amphoteric oxide or hydroxide can react with either an acid or a base.

Precipitation, acid-base, and oxidation-reduction reactions, together with complex-ion formation and amphoteric behavior, are all used extensively in the classical scheme for the qualitative analysis of common cations.

Review Questions

1. Write the solubility product constant expression for equilibrium in a saturated solution of iron(III) hydroxide.

$$Fe(OH)_3(s) \rightleftharpoons Fe^{3+}(aq) + 3\ OH^-(aq)$$

2. K_{sp} for zinc phosphate, $Zn_3(PO_4)_2$, is 9.0×10^{-33}. Write the equation for the reversible reaction to which this equilibrium constant applies.
3. Explain why the solubility product concept is limited to ionic compounds that are only slightly soluble in water. That is, why isn't the concept useful in describing saturated solutions of NaCl or $NaNO_3$, for example?
4. What is the difference between the *solubility* and the *solubility product constant* of a slightly soluble ionic compound?
5. Which compound is more soluble, $BaSO_4$ ($K_{sp} = 1.1 \times 10^{-10}$) or $PbSO_4$ ($K_{sp} = 1.6 \times 10^{-8}$)? Explain.
6. Describe the *common-ion effect* on the equilibrium between a slightly soluble ionic solute and its cations and anions in solution. Is the magnitude of the effect the same whether the common ion is the anion or the cation? Explain.
7. What is the reaction quotient, Q_{ip}, and how is it related to K_{sp}? How can the comparative values of Q_{ip} and K_{sp} be used to determine whether a precipitate will form in a solution?
8. What are the conditions that favor the essentially complete precipitation of a slightly soluble solute?
9. Describe how ions in solution can be separated by *selective precipitation*. What conditions are necessary for a successful separation?
10. In quantitative analysis, a precipitate is often washed with a dilute salt solution rather than with pure water. Explain why this is done. What type of salt solution should be used?
11. When does pH affect the solubility of a slightly soluble solute, and when does it not? Give some examples.
12. Which of the following solutions could be used to increase the aqueous solubility of $Fe(OH)_3(s)$? That is, which would produce a solution with a greater $[Fe^{3+}]$ than found in saturated $Fe(OH)_3(aq)$: HCl(aq), NaOH(aq), $CH_3COOH(aq)$, $NH_3(aq)$? Explain.
13. What is a complex ion? What are the ligands in a complex ion? Describe complex-ion formation as a Lewis acid–base reaction.
14. What is the formation constant, K_f, of a complex ion? What are the concentration terms that appear in the K_f expression?
15. Explain why $PbCl_2(s)$ is less soluble in 1 M $Pb(NO_3)_2$ than in pure water, but somewhat more soluble in 1 M HCl(aq) than in pure water.
16. Explain why sodium thiosulfate, $Na_2S_2O_3(aq)$ ("hypo"), is used to remove excess AgBr from photographic film rather than $NH_3(aq)$, a cheaper reagent.
17. Which of the following solutions will dissolve an appreciable quantity of $Cu(OH)_2(s)$? That is, which would produce a solution with a greater total concentration of copper(II) ion than what is found in saturated $Cu(OH)_2(aq)$: $H_2O(l)$, $NH_3(aq)$, NaOH(aq), HCl(aq)? Explain.
18. Aqueous solutions of Al^{3+} are acidic. Write an equation involving $[Al(H_2O)_6]^{3+}$ to account for this fact.
19. What are the properties that characterize an *amphoteric* oxide or hydroxide? Give an example of an amphoteric hydroxide.
20. What is the reagent used to separate the group 1 cations from other cations in the qualitative analysis scheme? What reagents are used to separate the group 2 cations from those of groups 3, 4, and 5?
21. In qualitative analysis cation group 1, what reagent is used to separate $PbCl_2(s)$ from the other chlorides? What reagent is used to separate AgCl(s) from $Hg_2Cl_2(s)$?
22. Explain how amphoterism is used to separate the cations of qualitative analysis group 3 into two subgroups.

Problems

(Use data from Tables 16.1 and 16.3 or Appendix C, as necessary.)

The Solubility Product Constant, K_{sp}

23. Write a chemical equation representing solubility equilibrium and write the solubility product constant expression for **(a)** $Hg_2(CN)_2$, $K_{sp} = 5 \times 10^{-40}$; and **(b)** Ag_3AsO_4, $K_{sp} = 1.0 \times 10^{-22}$.
24. Write a chemical equation representing solubility equilibrium and write the solubility product constant expression for **(a)** YF_3, $K_{sp} = 6.6 \times 10^{-13}$; and **(b)** $Fe_4[Fe(CN)_6]_3$, $K_{sp} = 3.3 \times 10^{-41}$.

The Relationship Between Solubility and K_{sp}

25. Can the numerical values of the molar solubility and the solubility product constant of a slightly soluble ionic compound ever be the same? Which of the two is usually the larger value? Explain.

26. The K_{sp} values of $CuCO_3$ and $ZnCO_3$ are 1.4×10^{-10} and 1.4×10^{-11}, respectively. Does this mean that $CuCO_3$ has ten times the solubility of $ZnCO_3$? Explain.

27. A reference book lists the solubility of barium chromate as 0.0010 g $BaCrO_4$/100 mL H_2O. Calculate the K_{sp} of $BaCrO_4$.

28. A reference book lists the solubility of cadmium iodate as 0.097 g $Cd(IO_3)_2$/100 mL H_2O. Calculate the K_{sp} of $Cd(IO_3)_2$.

29. Determine which of these slightly soluble solutes has the greater molar solubility: AgCl, or Ag_2CrO_4.

30. Determine which of these slightly soluble solutes has the greater molar solubility: $Mg(OH)_2$ or $MgCO_3$.

31. A 250.0-mL sample of a saturated aqueous solution of lanthanum(III) hydroxide required 2.3 mL of 0.0010 M HCl(aq) for its neutralization.

$$La(OH)_3(\text{saturated aq}) + 3\,HCl(aq) \longrightarrow LaCl_3(aq) + 3\,H_2O(l)$$

Use this fact to determine K_{sp} for lanthanum hydroxide.

32. A 50.0-mL sample of a saturated aqueous solution of barium thiosulfate was treated with HCl(aq), and 6.4 mg of sulfur was obtained in the reaction that follows.

$$S_2O_3^{2-}(aq) + 2\,H_3O^+ \longrightarrow 3\,H_2O + SO_2(g) + S(s)$$

Use these data to determine K_{sp} for barium thiosulfate.

33. *Without doing detailed calculations*, indicate which of the following saturated aqueous solutions has the highest concentration of Ca^{2+} ion: $CaCO_3$, CaF_2, $CaSO_4$. Explain.

34. *Without doing detailed calculations*, indicate which of the following saturated aqueous solutions has the highest concentration of PO_4^{3-} ion: $Ca_3(PO_4)_2$, $FePO_4$, $Mg_3(PO_4)_2$. Explain.

35. Calculate the concentration of Cu^{2+} in parts per billion (ppb) in a saturated solution of copper(II) arsenate, $Cu_3(AsO_4)_2(aq)$. (*Hint:* Recall that 1 ppb signifies 1 g Cu^{2+} per 10^9 g solution.)

36. Fluoridated drinking water contains about 1 part per million (ppm) of F^-. Is MgF_2 sufficiently soluble in water to be used as the source of fluoride ion for this fluoridation? Explain. (*Hint:* Recall that 1 ppm signifies 1 g F^- per 10^6 g solution.)

37. Given the K_{sp} values for $PbCl_2$ of 1.6×10^{-5} at 25 °C and 3.3×10^{-3} at 80 °C, if 1.00 mL of saturated $PbCl_2(aq)$ at 80 °C is cooled to 25 °C, will a sufficient amount of $PbCl_2(s)$ precipitate to be visible? Assume that you can detect as little as 1 mg of the solid.

38. When a 1.00-L sample of a saturated aqueous solution of calcium tungstate is heated from 15 °C to 100 °C, 5.2 mg $CaWO_4$ deposits from solution. Given that

$$CaWO_4(s) \rightleftharpoons Ca^{2+}(aq) + WO_4^{2-}(aq)$$

$$K_{sp} = 4.9 \times 10^{-10} \text{ at } 15\ °C$$

what is the value of K_{sp} at 100 °C?

The Common-Ion Effect in Solubility Equilibria

39. In which of the following is the slightly soluble AgBr likely to be *most soluble*: pure water, 0.050 M KBr, or 1.25 M $AgNO_3$? Explain.

40. In which of the following is the slightly soluble Ag_2CrO_4 likely to be *least soluble*: pure water, 0.10 M K_2CrO_4, or 0.10 M $AgNO_3$? Explain.

41. Calculate the molar solubility of $Mg(OH)_2$ in 0.10 M $MgCl_2$.

42. Calculate the molar solubility of $Mg(OH)_2$ in 0.25 M NaOH(aq).

43. A 15.0-g sample of NaI is dissolved in water to make 0.250 L of solution. The solution is then saturated with PbI_2. What will be $[Pb^{2+}]$ and $[I^-]$ in the saturated solution?

44. A 1.05-g sample of LiCl is used to make 0.125 L of an aqueous solution. The solution is then saturated with Li_3PO_4. What will be $[Li^+]$ and $[PO_4^{3-}]$ in the saturated solution?

45. A solution is saturated with Ag_2SO_4. **(a)** Calculate $[Ag^+]$ in this saturated solution. **(b)** What mass of Na_2SO_4 must be added to 0.500 L of the solution to decrease $[Ag^+]$ to 4.0×10^{-3} M?

46. A solution is saturated with Ag_2CrO_4. **(a)** Calculate $[CrO_4^{2-}]$ in this saturated solution. **(b)** What mass of $AgNO_3$ must be added to 0.635 L of the solution to reduce $[CrO_4^{2-}]$ to 1.0×10^{-8} M?

47. Calculate the molar solubility of strontium chromate in 0.0025 M $Na_2CrO_4(aq)$. (*Hint:* Check any simplifying assumptions that you make.)

48. Calculate the molar solubility of barium thiosulfate in 0.0033 M $Na_2S_2O_3(aq)$. (*Hint:* Check any simplifying assumptions that you make.)

Precipitation Criteria

49. What $[CrO_4^{2-}]$ must be present in 0.00105 M $AgNO_3(aq)$ to just cause $Ag_2CrO_4(s)$ to precipitate?

50. What must be the pH of a solution that is 0.050 M Fe^{3+} to just cause $Fe(OH)_3(s)$ to precipitate?

51. Will a precipitate form if 0.0010 mol $Hg_2(NO_3)_2$ and 0.0010 mol NaCl are added to 20.00 L of water?

52. Will a precipitate form if 0.48 mg $MgCl_2$ and 12.2 mg Na_2CO_3 are added to 225 mL of water?

53. Will a precipitate form when 235 mL of 0.0022 M $MgCl_2$ and 485 mL of 0.0055 M NaF are mixed? If so, what is the precipitate?

54. Will a precipitate form when 136 mL of 0.00015 M $Pb(NO_3)_2$ and 234 mL of 0.00028 M Na_3AsO_4 are mixed? If so, what is the precipitate?

55. $[Ca^{2+}]$ in hard water is about 2.0×10^{-3} M. Water is fluoridated with 1.0 g F^- per 1.0×10^3 L of water. Will $CaF_2(s)$ precipitate from hard water upon fluoridation?

56. A certain municipal water sample contains 46.1 mg SO_4^{2-}/L and 30.6 mg Cl^-/L. How many drops (1 drop = 0.05 mL) of 0.0010 M $AgNO_3$ must be added to 1.00 L of this water to just produce a precipitate? What will this precipitate be?

Completeness of Precipitation and Selective Precipitation

57. The first step in the extraction of magnesium metal from seawater is the precipitation of Mg^{2+} as $Mg(OH)_2(s)$. $[Mg^{2+}]$ in seawater is about 0.059 M. If a seawater sample is treated so that its $[OH^-]$ is kept constant at 2.0×10^{-3} M, what will be $[Mg^{2+}]$ remaining in solution after precipitation has occurred? Is precipitation of $Mg(OH)_2(s)$ complete under these conditions?

58. Is precipitation of $CaSO_4$ complete if a solution that is 0.0250 M in Ca^{2+} is also made 0.500 M in $K_2SO_4(aq)$?

59. What pH must be maintained by a buffer solution so that no more than 0.010% of the Mg^{2+} present in 0.360 M $MgCl_2(aq)$ remains in solution following the precipitation of $Mg(OH)_2(s)$?

60. Nineteen centuries ago, the Romans added calcium sulfate to wine. It clarifies the wine and removes dissolved lead.

    (a) Calculate $[SO_4^{2-}]$ in wine saturated with $CaSO_4$.

    (b) What $[Pb^{2+}]$ remains in wine that is saturated with $CaSO_4$?

61. Concentrated NaF(aq) is slowly added to a solution that is 0.010 M $CaCl_2$ and 0.010 M $MgCl_2$.

    (a) What is the first precipitate to form?

    (b) What is the $[F^-]$ needed to start precipitating the second precipitate?

    (c) Can Ca^{2+} and Mg^{2+} be separated by selective precipitation of their fluorides? Explain.

62. Concentrated $Pb(NO_3)_2(aq)$ is slowly added to a solution that is 0.15 M $Na_2CrO_4(aq)$ and 0.15 M Na_2SO_4. **(a)** What is the first precipitate to form? **(b)** What is the $[Pb^{2+}]$ needed to start precipitating the second precipitate? **(c)** Can CrO_4^{2-} and SO_4^{2-} be separated by selective precipitation with Pb^{2+}? Explain.

63. A 10.00-mL portion of a 0.50 M $AgNO_3(aq)$ solution is added to 100.0 mL of a solution that is 0.010 M in Cl^- and 0.010 M in SO_4^{2-}. **(a)** Will AgCl(s) precipitate from this solution? If so, how many moles? **(b)** Will $Ag_2SO_4(s)$ precipitate? If so, how many moles?

64. A 10.00-mL portion of 0.25 M $AgNO_3(aq)$ solution is added to 100.0 mL of a solution that is 0.010 M in Cl^- and 0.010 M in Br^-.

    (a) Will AgCl(s) precipitate from this solution? If so, how many moles?

    (b) Will AgBr(s) precipitate from this solution? If so, how many moles?

Effect of pH on Solubility

65. Calculate the solubility of $Al(OH)_3(s)$ in a buffer solution with pH = 4.50.

66. Calculate the solubility of $Mg(OH)_2$ in a buffer solution that is 0.75 M NH_3 and 0.50 M NH_4Cl.

67. Which of the following would you add to a mixture of $CaCO_3(s)$ and its saturated solution to increase the molar solubility of $CaCO_3$: more water, Na_2CO_3, NaOH, $NaHSO_4$? Explain.

68. Which of the following solids are likely to be more soluble in acidic solution, and which in basic solution? Which are likely to have a solubility that is essentially independent of pH? Explain.

(a) C_6H_5COOH
(b) $BaCO_3$
(c) CaC_2O_4
(d) $LiNO_3$
(e) $CaCl_2$
(f) $Sr(OH)_2$

69. Boiler scale (mostly $CaCO_3$) is insoluble in water. Hydrochloric acid is sometimes used to remove the scale from boilers in commercial power plants. Write an equation to show what happens. Boiler scale is also produced in automatic coffeemakers. The recommended way to remove it is to run vinegar (acetic acid) through the machine. Write an equation to show what happens.

70. If a concentrated aqueous solution of sodium acetate is mixed with an aqueous solution of silver nitrate, a precipitate of silver acetate forms. The precipitate dissolves readily when nitric acid is added. Write equations that explain what happens.

Complex Ion Formation

71. A complex ion has Cr^{3+} as its central ion and four NH_3 molecules and two Cl^- ions as ligands. Write the formula of this complex ion.

72. A complex ion has Fe^{2+} as the central ion and six cyanide ions as ligands. Write the formula of this complex ion.

73. Which of the following complex ions would you expect to have the *lowest* $[Ag^+]$ in a solution that is 0.10 M in the complex ion and 1.0 M in the free ligand? Explain.

(a) $[Ag(NH_3)_2]^+$
(b) $[Ag(CN)_2]^-$
(c) $[Ag(S_2O_3)_2]^{3-}$

74. Which of the following reagents will increase significantly the concentration of free Zn^{2+} ions when it is added to a solution of the complex ion $[Zn(NH_3)_4]^{2+}$? Explain.

(a) $H_2O(l)$
(b) $NaHSO_4(aq)$
(c) $NH_3(aq)$
(d) $HCl(aq)$

75. When a few drops of concentrated $Na_2SO_4(aq)$ are added to a dilute solution of $AgNO_3(aq)$, a white precipitate forms. When a small quantity of concentrated $NH_3(aq)$ is added to the mixture, the precipitate redissolves, resulting in a colorless solution. When this solution is made acidic with $HNO_3(aq)$, a white precipitate again appears. Write equations to represent these three observations.

76. Figure 16.7 shows that AgCl(s) can be dissolved in $NH_3(aq)$, and Figure 16.8 shows that AgCl(s) can be reprecipitated by adding $HNO_3(aq)$. AgCl(s) can also be dissolved in concentrated HCl(aq), but it cannot be reprecipitated by the addition of $HNO_3(aq)$. Explain this difference.

77. Write plausible net ionic equations to represent what happens, if anything, in each of the following. If no reaction occurs, write N.R.

(a) $ZnCl_2(aq) + NaOH(conc.\ aq) \longrightarrow$

(b) $Al_2O_3(s) + HCl(conc.\ aq) \longrightarrow$

(c) $Fe(OH)_3(s) + NaOH(conc.\ aq) \longrightarrow$

78. Write plausible net ionic equations to represent what happens, if anything, in each of the following. If no reaction occurs, write N.R.

(a) $Fe(OH)_3(s) + HCl(conc.\ aq) \longrightarrow$

(b) $NaCl(aq) + NaOH(conc.\ aq) \longrightarrow$

(c) $CrCl_3(aq) + NaOH(conc.\ aq) \longrightarrow$

Complex-Ion Equilibria

79. What is $[Zn^{2+}]$ in a solution that is 0.25 M in $[Zn(NH_3)_4]^{2+}$ and has $[NH_3]$ equal to 1.50 M?

80. What must be $[NH_3]$ in a solution that is 0.15 M in $[Ag(NH_3)_2]^+$ if $[Ag^+]$ is to be kept at 1.0×10^{-8} M?

81. Should $PbI_2(s)$ precipitate if 2.00 mL of 2.00 M KI(aq) is added to 0.300 L of an aqueous solution that is 0.010 M in $[PbCl_3]^-$ and has $[Cl^-]$ equal to 1.50 M?

82. Should $BaS_2O_3(s)$ precipitate if 2.50 mL of 1.50 M $Ba(NO_3)_2(aq)$ is added to 0.275 L of an aqueous solution that is 0.066 M in $[Ag(S_2O_3)_2]^{3-}$ and has $[Ag^+]$ equal to 1.25 M?

83. How many milligrams of KBr can be added before AgBr(s) precipitates from 1.42 L of an aqueous solution that is 0.220 M in $[Ag(NH_3)_2]^+$ and has $[NH_3]$ equal to 0.805 M?

84. What is the minimum $[CN^-]$ required in 0.750 L of an aqueous solution that is 0.250 M in $[Ag(CN)_2]^-$ so that AgI(s) will not precipitate when 12.5 g KI is later added?

85. Calculate the molar solubility of AgBr(s) in 0.100 M $Na_2S_2O_3(aq)$. (*Hint:* Write an equation for the dissolution reaction, and determine its K_c value as shown in the text.)

86. Calculate the molar solubility of AgI(s) in 0.100 M NaCN(aq). (*Hint:* Write an equation for the dissolution reaction, and determine its K_c value as shown in the text.)

Qualitative Inorganic Cation Analysis

87. Both Pb^{2+} and Ag^+ form insoluble chlorides and sulfides. In the qualitative analysis scheme, Pb^{2+} appears both in cation group 1 and group 2, whereas Ag^+ appears only in group 1. Give a plausible explanation for this observation.

88. The test for NH_4^+ in the qualitative analysis scheme is suggested in the accompanying photograph. A sample of the original unknown is treated with concentrated NaOH(aq), and a strip of moistened red litmus paper is suspended in the vapor above the solution. In a positive test for NH_4^+, the color of the litmus turns to blue.

(a) Explain how this test for NH_4^+ works, by writing ionic equations for the reactions involved.

(b) Why is it possible to test for NH_4^+ on the *original* sample, whereas all the other cations must be separated into groups before testing?

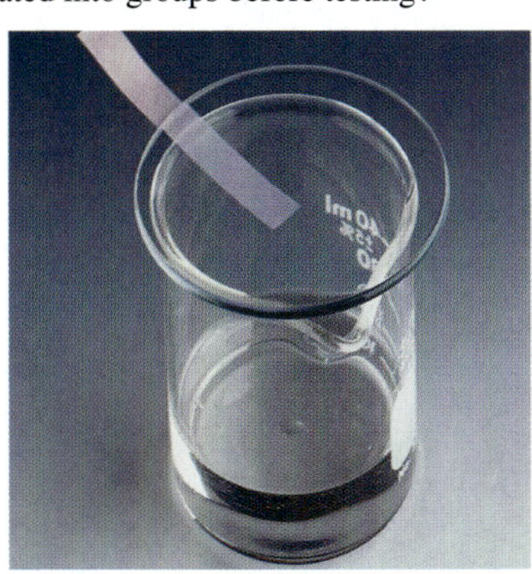

89. A cation group 1 unknown is treated with HCl(aq) and yields a white precipitate. When NH_3(aq) is added to the precipitate, it turns a gray color. According to these tests, indicate for each of the group 1 cations whether it is definitely present, definitely absent, or whether its presence remains uncertain.

90. A cation group 3 unknown is treated with NaOH(aq) and H_2O_2. No precipitate forms, but the solution becomes bright yellow in color. Indicate for each of the group 3 cations whether it is probably present, most likely absent, or its presence remains uncertain.

Additional Problems

(Use data from Tables 16.1 and 16.3 or Appendix C, as necessary.)

91. Ba^{2+}(aq) is poisonous when ingested. The lethal dosage in mice is about 12 mg Ba^{2+} per kg of body mass. Despite this fact, $BaSO_4$ is widely used in medicine to obtain X-ray photographs of the gastrointestinal tract.

(a) Explain why $BaSO_4$(s) is safe to take internally, even though Ba^{2+}(aq) is poisonous.

(b) What is the concentration of Ba^{2+}, in milligrams per liter, in saturated $BaSO_4$(aq)?

(c) $MgSO_4$ can be mixed with $BaSO_4$(s) in this medical procedure. What function does the $MgSO_4$ serve?

92. Write equations to show why $Zn(OH)_2$(s) is readily soluble in each of the following dilute solutions: HCl(aq), CH_3COOH(aq), NH_3(aq), and NaOH(aq).

93. Most ordinary soaps are sodium salts of long-chain fatty acids and are water soluble. Soaps of divalent cations, such as Ca^{2+}, are only slightly soluble and are often seen in the common soap scum formed in hard water. A typical calcium soap is calcium palmitate: $Ca[CH_3(CH_2)_{14}COO]_2$. A handbook lists the solubility of this soap as 0.003 g/100 mL at 25 °C. If sufficient sodium soap is used to produce a concentration of palmitate ion equal to 0.10 M in a water sample having 25 ppm Ca^{2+}, would you expect any soap scum to form? If so, how many grams of calcium palmitate would precipitate in a bowl containing 6.5 L of this water?

94. Will lead(II) oxalate precipitate if 50.0 mg $Pb(NO_3)_2$ is added to 0.500 L of 0.0100 M $H_2C_2O_4$? For PbC_2O_4, $K_{sp} = 4.8 \times 10^{-10}$; for $H_2C_2O_4$, $K_{a_1} = 5.4 \times 10^{-2}$, $K_{a_2} = 5.3 \times 10^{-5}$.

95. Show that if an aqueous solution is 0.350 M $MgCl_2$ and also 0.750 M NH_3, a precipitate of $Mg(OH)_2$(s) will form. The precipitate can be kept from forming if NH_4Cl is added to the solution. Explain why this is so. What minimum mass of NH_4Cl should be present in 0.500 L of the solution to prevent the precipitation of $Mg(OH)_2$(s)?

96. Calculate the number of moles of NH_3 that must be added to prevent the precipitation of AgCl(s) from 0.250 L of 0.100 M $AgNO_3$ when 0.100 mol NaCl is subsequently added.

97. A 0.100-L sample of saturated SrC_2O_4(aq) is analyzed for its oxalate ion content by titration with permanganate ion in acidic solution. The volume of 0.00500 M $KMnO_4$ required is 3.22 mL. Use these data to determine K_{sp} for SrC_2O_4. In the titration of oxalate ion with permanganate ion, Mn^{2+}(aq) and CO_2(g) are obtained as products.

98. The chief compound in marble is $CaCO_3$. Marble has been widely used for statues and ornamental work on buildings, but marble is readily attacked by acids. Determine the solubility of marble (that is, $[Ca^{2+}]$ in a saturated solution) in **(a)** normal rainwater with pH = 5.6, and **(b)** "acid" rainwater with pH = 4.22. Assume that the following is the net reaction that occurs.

$$CaCO_3(s) + H_3O^+(aq) \rightleftharpoons Ca^{2+}(aq) + HCO_3^-(aq) + H_2O \qquad K = 6.0 \times 10^1$$

99. Obtain K_a for hydrazoic acid, HN_3, and K_{sp} for lead azide, $Pb(N_3)_2$ from Appendix C, and use these to determine a value of K_c for the dissolution of $Pb(N_3)_2(s)$ in an acidic solution. Then calculate the molar solubility of $Pb(N_3)_2(s)$ in a buffer solution with pH = 2.85.

100. Exercise 16.12 (page 711) asks for an equation describing the dissolution of $Mg(OH)_2(s)$ in $NH_4Cl(aq)$. Determine a value of K_c for this reaction, and then calculate the molar solubility of $Mg(OH)_2$ in 0.500 M $NH_4Cl(aq)$.

101. The text gives the following equation for the precipitation of a metal sulfide in the qualitative analysis scheme.

$$M^{2+}(aq) + H_2S(aq) + 2\,H_2O \rightleftharpoons MS(s) + 2\,H_3O^+(aq)$$

Use the description of the solubility product constant of MS(s) given in a footnote to Table C.2C (Appendix C), together with other relevant data, to show that the equilibrium constant for the precipitation reaction has the form: $K_c = K_w K_{a_1}/K_{sp}$. What is the value of K_c if the metal sulfide is FeS?

102. The dissolution of a metal sulfide in an acidic solution can be represented by the reverse of the equation written in Problem 101. With the aid of such an equation and its equilibrium constant, determine the molar solubility of MnS(s) in a buffer solution that is 1.50 M in CH_3COOH and 0.45 M in CH_3COO^-.

103. How many milliliters of 0.0050 M NaBr(aq) can be added before any AgBr(s) precipitates from 165 mL of an aqueous solution that is 0.62 M in $[Ag(NH_3)_2]^+$ and has $[NH_3]$ equal to 0.50 M?

104. A *white* solid mixture consists of *two* compounds, and the two compounds contain different cations. When treated with water, part of the mixture dissolves and part does not. The solution obtained is treated with $NH_3(aq)$ and yields a *white* precipitate. The part of the original solid mixture that is insoluble in water dissolves in HCl(aq) with the evolution of a *gas*. The resulting solution is treated with $(NH_4)_2SO_4(aq)$ and produces a *white* precipitate. State which of the following cations *might* have been present and which *could not* have been present in the original solid mixture. Explain your reasoning.

(a) Mg^{2+} **(b)** Cu^{2+} **(c)** Ba^{2+} **(d)** Na^+ **(e)** NH_4^+

Thermodynamics: Spontaneity, Entropy, and Free Energy

CHAPTER

17

The development of the science of thermodynamics was closely associated with the invention of the steam engine, such as the one powering this locomotive.

CONTENTS

Thermodynamics deals with the relationship between heat and work. What can this possibly have to do with chemistry? Actually, quite a bit. In Chapter 6, we focused on thermochemistry, specifically on the enthalpy changes (heats) of chemical reactions. In subsequent chapters, we considered heat effects in relation to several concepts. And thermochemistry is just one branch of thermodynamics.

In this chapter, we will consider three main ideas: *spontaneity*—whether or not a process can take place unassisted; *entropy*—a measure of the degree of randomness among the atoms and molecules of a system; and *free energy*—a thermodynamic function that relates enthalpy and entropy to spontaneity. Perhaps most important of all, we will relate free energy change to equilibrium constants, and this makes it possible to *predict* equilibrium constants from tabulated data.

● **ChemCDX**
When you see this icon, there are related animations, demonstrations, and exercises on the CD accompanying this text.

17.1 Why Study Thermodynamics?

Two meanings of the word "spontaneous"
In "spontaneous combustion," a combustible material suddenly bursts into flame, as in the sudden ignition of this shipment of latex gloves that caused a fire at the Brooklyn Naval Yard in 1995. In common usage, we say that the combustion is "spontaneous" because it happens suddenly and without forewarning. In scientific usage, a spontaneous reaction is one that has a natural tendency to occur. However, to say that a reaction is spontaneous says nothing about whether the reaction will occur rapidly or be immeasurably slow.

Having the right kind of knowledge can bring great rewards. With a knowledge of thermodynamics and by making a few calculations before embarking on a new venture, scientists and engineers can save themselves a great deal of time, money, and frustration. Recall Le Châtelier's statement about the importance of understanding chemical equilibrium, a key thermodynamics concept (page 610).

The practical value of thermodynamics was well recognized early in this century by two American chemists, G. N. Lewis and Merle Randall, who noted in the introduction to their landmark textbook,*

> *"To the manufacturing chemist thermodynamics gives information concerning the stability of his substances, the yield which he may hope to attain, the methods of avoiding undesirable substances, the optimum range of temperature and pressure, the proper choice of solvent. . . ."*

With considerable foresight, they further offered the opinion that "... the greatest development of applied thermodynamics is still to come." Even Lewis and Randall may have been surprised at how widely thermodynamics is now applied. In our study, we will find that we can apply thermodynamics to systems as complex as living organisms, not just to reactions in beakers, batteries, and blast furnaces.

17.2 Spontaneous Change

We know from experience that some processes occur by themselves, without requiring us to do anything. The ice melts as a glass of iced tea stands at room temperature. An iron nail rusts when exposed to the oxygen and water vapor of moist air. A violent reaction occurs when sodium metal and chlorine gas come in contact (Figure 9.3, page 362).

$$2\,Na(s) + Cl_2(g) \longrightarrow 2\,NaCl(s)$$

These are all examples of spontaneous processes. A **spontaneous process** is one that can occur in a system left to itself; no action from outside the system is necessary to bring it about.

We know equally well that certain other processes don't occur by themselves. They are *nonspontaneous*. Water does *not* freeze to form ice at room temperature. The iron oxide of a rusty nail does *not* revert to iron metal and oxygen gas. Solid sodium chloride does *not* decompose into sodium metal and chlorine gas. A **nonspontaneous process** is one that cannot take place in a system left to itself. We can summarize these observations by the following statement.

> *If a process is spontaneous, the reverse process is nonspontaneous and vice versa.*

Some nonspontaneous processes are impossible, such as freezing liquid water at atmospheric pressure and a constant temperature of 25 °C. Others, like the conversion of sodium chloride to sodium metal and chlorine gas, are *not* impossible. To make a nonspontaneous process go, we need to impose some action from outside the system. We can melt solid sodium chloride, and we *can* convert the molten

* G. N. Lewis and M. Randall, *Thermodynamics and the Free Energy of Chemical Substances*, McGraw-Hill, 1923.

NaCl to sodium and chlorine. We can do this, for example, through the action of an electric current in a process called electrolysis.

The term spontaneous signifies nothing about how fast a process occurs. The reaction between sodium and chlorine is extremely fast; the rusting of iron is much slower. If we mix H_2 and O_2 gases at room temperature, we see no evidence of a chemical reaction. Yet, thermodynamic criteria indicate that this reaction is indeed spontaneous. Thus, thermodynamics can tell us if a process is *possible*, but only chemical kinetics (Chapter 13) can tell us *how fast* the process will occur. Chemical kinetics dictates that the reaction between H_2 and O_2 will be immeasurably slow at room temperature because of its high activation energy. At high temperatures, thermodynamics again predicts that the reaction will be spontaneous, and chemical kinetics forecasts that the reaction will be fast. In line with these predictions, hydrogen and oxygen combine with explosive speed at high temperatures. To state the matter another way, thermodynamics determines the equilibrium *state* of a system; kinetics determines the *pathway* by which equilibrium is reached. Both are important in answering the question: Will a reaction occur?

In Example 17.1, we show, even this early in our discussion, how we can identify some processes as spontaneous or nonspontaneous. However, there are other cases in which we can't be sure.

The sign "Rock Slide Area" warns of a spontaneous process. Moreover, the implication is that the barrier preventing a rockfall is weak (low activation energy). A heavy rainstorm or slight earth tremor may provide enough of a "push" to set some rocks in motion.

EXAMPLE 17.1

Indicate whether each of the following processes is spontaneous or nonspontaneous. Comment on cases in which a clear determination cannot be made.

a. The action of toilet bowl cleaner, HCl(aq), on "lime" deposits, $CaCO_3$(s).

b. The boiling of water at normal atmospheric pressure and 65 °C.

c. The reaction of N_2(g) and O_2(g) to form NO(g) at room temperature.

d. The melting of an ice cube.

SOLUTION

a. When we add the acid, the fizzing that occurs—the escape of a gas—indicates that the reaction occurs without any further action on our part. The net ionic equation is as shown here.

$$CaCO_3(s) + 2\,H_3O^+(aq) \longrightarrow Ca^{2+}(aq) + 3\,H_2O(l) + CO_2(g)$$

The reaction is *spontaneous*.

b. We know that the normal boiling point of a liquid is the temperature at which the vapor pressure is equal to 1 atm. For water, this temperature is 100 °C. Thus, the boiling of water at 65 °C and 1 atm pressure is *nonspontaneous* and not possible.

c. Nitrogen and oxygen gases occur mixed in air, and there is no evidence of their reaction to form toxic NO at room temperature. This makes their reaction to form NO(g) appear to be *nonspontaneous*. Or this could be an example of a spontaneous reaction that occurs extremely slowly, like that of H_2 and O_2 to form H_2O. We need to develop some additional criteria before we can answer this question.

d. To answer this question, we would have to know the temperature. We know that at 1 atm pressure ice melts spontaneously at temperatures above its normal melting point of 0 °C. Below this temperature, it does not.

EXERCISE 17.1

Indicate whether each of the following processes is spontaneous or nonspontaneous. Comment on cases in which a clear determination cannot be made.

a. The decay of a piece of wood buried in soil.

b. The formation of sodium, Na(s), and chlorine, Cl_2(g), by vigorously stirring an aqueous solution of sodium chloride, NaCl(aq).

c. The formation of lime, CaO(s), and carbon dioxide, CO_2(g) at 1 atm pressure, from limestone, $CaCO_3$(s), at 600 °C.

d. The ionization of hydrogen chloride when HCl(g) dissolves in liquid water.

The unresolved cases in Example 17.1 indicate that it would be helpful to have criteria that tell us whether a process proceeds spontaneously. Common knowledge can help. You know that if you leave your car in neutral gear and forget to set the parking brake, it will roll downhill. You also know that water in a stream always flows only one way—downhill. And you probably know why these things happen: If no counteracting forces are present, the force of gravity attracts objects toward the center of Earth, that is, in the direction of *decreasing* potential energy.

By similar reasoning, early chemists proposed that spontaneous chemical reactions should occur in the direction of decreasing energy. In Chapter 6, we defined internal energy, U, as the total energy content of a system and ΔU as the change in internal energy accompanying a process such as a chemical reaction. But because most chemical reactions are carried out in vessels open to the atmosphere, we found it convenient to use the closely related energy quantity, called *enthalpy*, H. The enthalpy change, ΔH, during a chemical reaction is equal to the heat of a reaction, q_p, the amount of heat that enters or leaves a system in which a reaction occurs at constant pressure.

Figure 17.1 compares a spontaneous mechanical process—the fall of water—with two chemical processes. The direction of *spontaneous* flow of water in the waterfall is to a lower elevation, to a lower potential energy. Internal energy in a chemical system is analogous to potential energy in a mechanical system. However, in place of internal energy, let's use the familiar and closely related enthalpy, H. By the waterfall analogy, we might expect reactions in which enthalpy decreases to be spontaneous. Reactions with $\Delta H < 0$ are *exothermic*. Similarly, we might expect reactions in which enthalpy increases to be nonspontaneous. Reactions with $\Delta H > 0$ are *endothermic*. Figure 17.1 shows that the reaction of H_2(g) and O_2(g) to form H_2O(l) is an *exothermic* reaction. Moreover, it is a *spontaneous* reaction. The figure shows that the vaporization of water is an *endothermic* process. However, at 25 °C and for pressures up to 0.0313 atm, the vaporization of water is also *spontaneous*.

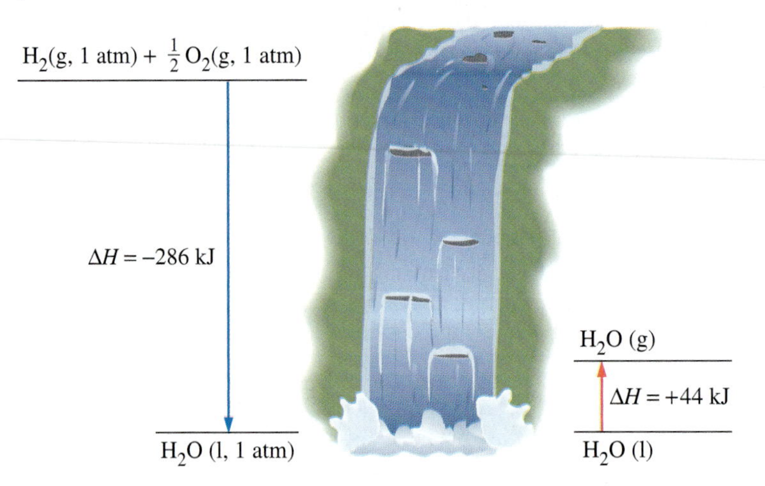

▶ Figure 17.1 The direction of decrease in energy: A criterion for spontaneous change?
Water in the waterfall (center) flows spontaneously to a lower elevation, where the potential energy is lower. The enthalpy diagram (left) shows the lowering of enthalpy when liquid water is spontaneously formed from its elements at 25 °C and 1 atm. The situations seem analogous. The enthalpy diagram (right) is for the vaporization of water to produce H_2O(g) at its equilibrium vapor pressure at 25 °C. This process is also spontaneous, but the enthalpy *increases*, so the waterfall analogy doesn't work in this case.

The idea that exothermic reactions are spontaneous, whereas endothermic reactions are not, works in many cases. However, as in the vaporization of water just cited, enthalpy change is not a sufficient criterion for predicting spontaneous change. We need to look at some additional factors.

17.3 Entropy: Disorder and Spontaneity

Suppose you have a new deck of cards arranged by suit and rank and spread out on a card table. Now suppose someone opens a window and the wind blows the cards off the table onto the floor. Which of the arrangements of cards pictured in Figure 17.2 would you expect to see? You would certainly *not* expect arrangement (a). Some of the cards are still in midair, and the potential energy has *not* been minimized. Arrangements (b) and (c) are both in the minimum energy state available to them. Considering their energy states alone, we would say that the two arrangements are equally likely. Yet you know that arrangement (c) is closer to what you would observe. A scrambled or *disordered* arrangement is considerably more likely than the uniform arrangement in (b), in which all the cards are face up. You might spend a lifetime scattering decks of cards and never observe the arrangement in (b). This example suggests that there are *two* natural tendencies behind spontaneous processes: the tendency to achieve a lower energy state and the tendency toward a more disordered state. We need to consider *two* questions when judging whether a change will be spontaneous.

- Does the enthalpy of the system increase or decrease, and by how much?
- Does the system become more or less disordered, and by how much?

When a process involves both a decrease in energy and an increase in disorder, as in Figure 17.2(c), we can easily predict the direction of spontaneous change. In some cases, one of the two factors is negligible and the other takes precedence. An example is shown in Figure 17.3—the formation of an ideal solution of benzene and toluene. Because no chemical reaction occurs and also because intermolecular forces are essentially the same in the solution and in the pure liquids, $\Delta H \approx 0$. Because enthalpy change is not important, the natural tendency toward increased disorder rules. A condition in which the two types of molecules remain segregated from each other is a highly organized one and cannot endure. A solution is formed in which the molecules are uniformly and randomly mixed. This is the condition of maximum possible *disorder*.

In many cases, however, the two factors work in opposition: The enthalpy decreases and the degree of disorder decreases, or the enthalpy increases and the de-

(a)

(b)

(c)

▲ **Figure 17.2 Entropy: The importance of randomness or disorder**
A newly opened deck of cards is highly ordered, with the cards arranged according to suit and rank. When the cards are blown off the table by a blast of air, we expect an arrangement like the one in (c).

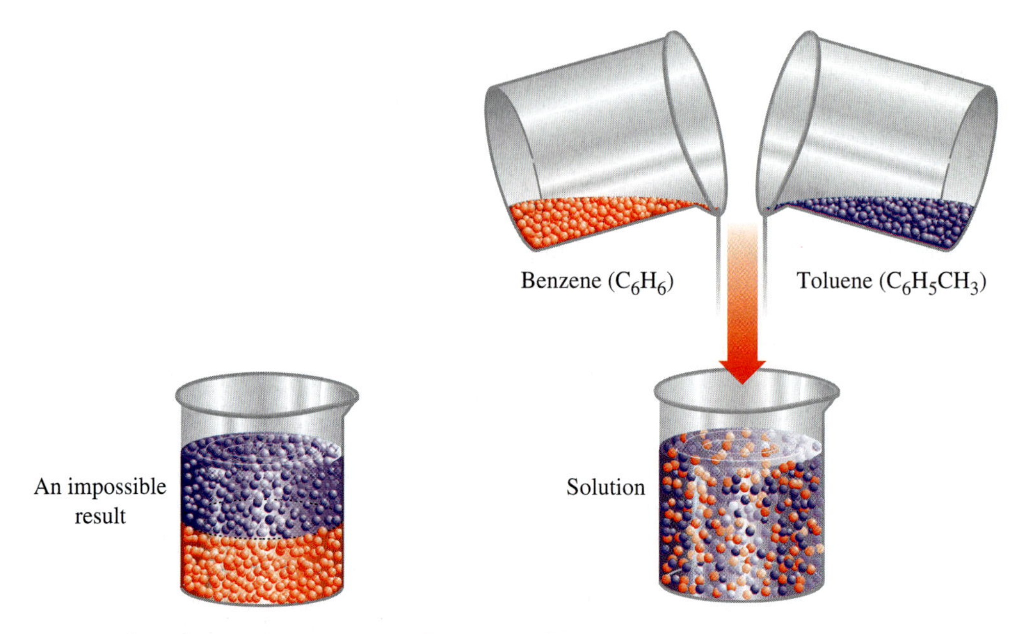

▲ **Figure 17.3 Formation of an ideal solution**
The situation on the left, in which the mixture of benzene and toluene molecules remains segregated, is impossible. It is more ordered than the solution with randomly mixed molecules. Formation of the solution is driven by a natural tendency to achieve the maximum state of disorder possible.

gree of disorder increases. In these cases, we need to determine which factor predominates. Soluble ionic substances that dissolve endothermically in water ($\Delta H > 0$) are a common example. The energy of the system increases, but when the solute dissolves, the system generally becomes more disordered.

Entropy

The thermodynamic property of a system that is related to its degree of randomness or disorder is called **entropy (S)**:

The greater the degree of disorder in a system, the greater is its entropy.

Recall (Chapter 6) that enthalpy and enthalpy change are *state functions*. We will demonstrate shortly that entropy is also a *state function*. That is, the entropy of a system has a unique value when the composition, temperature, and pressure of the system are specified. Also, the difference in entropy between two states, the **entropy change (ΔS)**, has a unique value.

We have noted that the ideal solution in Figure 17.3 is more disordered than the components from which it is formed. Entropy should increase in the formation of an ideal solution. We show this below in an expression in which we write S_{mixt} for the entropy of the solution, S_A for the entropy of the solvent (A), and S_B for the entropy of the solute (B).

$$\Delta S = S_{mixt} - [S_{A(l)} + S_{B(l)}] > 0$$

The increase in disorder and entropy that accompanies the vaporization of water is suggested by Figure 17.4. In general, entropy *increases* for the following processes.

- Solids melt to liquids.
- Solids or liquids vaporize to form gases.
- Solids or liquids dissolve in a solvent to form nonelectrolyte solutions.

▲ **Figure 17.4 Increase of disorder and entropy in the vaporization of water**
The photograph depicts the vaporization of water at the macroscopic level, even though nothing appears to be taking place. In the molecular view, we see that there are widely spaced molecules in the vapor state (high disorder) and closely spaced molecules in the liquid state (high order), and that all the molecules are in motion.

- A chemical reaction produces an increase in the number of molecules of *gases*.
- A substance is heated. (Increased temperature means increased atomic, ionic, or molecular motions. Increased motion means increased disorder.)

In applying these generalizations in Example 17.2, the following mental picture may help you to visualize a degree of disorder or randomness. Imagine a class of kindergarten students. Maximum *order* exists when they sit at their desks and listen to a story with rapt attention. When they tire and begin to fidget, there is less order and more disorder. There is still more disorder when they are free to pursue activities away from their desks. Maximum *disorder* results when they are released to the playground for recess.

EXAMPLE 17.2

Predict whether each of the following leads to an increase or decrease in entropy of a system. If in doubt, explain why.

a. The synthesis of ammonia.

$$N_2(g) + 3\,H_2(g) \longrightarrow 2\,NH_3(g)$$

b. Preparation of a sucrose solution.

$$C_{12}H_{22}O_{11}(s) \xrightarrow{H_2O(l)} C_{12}H_{22}O_{11}(aq)$$

c. Evaporation to dryness of a solution of urea, $CO(NH_2)_2$, in water.

$$CO(NH_2)_2(aq) \longrightarrow CO(NH_2)_2(s)$$

SOLUTION

a. Four moles of gaseous reactants produce two moles of gaseous product. Because two moles of gas—a highly disorganized state of matter—are *lost*, we predict a *decrease* in entropy.

b. The sucrose molecules are highly ordered in the solid state, whereas they are randomly distributed in water in an aqueous solution. We predict an *increase* in entropy.

c. In the evaporation of water, a liquid is converted to a gas, suggesting an *increase* in entropy. The urea, in contrast, passes from its random molecular distribution in the solution to a highly ordered solid, suggesting a *decrease* in entropy. Without further information, we can't say whether the entropy of the system increases or decreases.

EXERCISE 17.2A

Predict whether each of the following leads to an increase or decrease in the entropy of a system. If in doubt, explain why.

a. $NH_3(g) + HCl(g) \longrightarrow NH_4Cl(s)$

b. $2\ KClO_3(s) \longrightarrow 2\ KCl(s) + 3\ O_2(g)$

c. $CO(g) + H_2O(g) \longrightarrow CO_2(g) + H_2(g)$

EXERCISE 17.2B

The transformation of a liquid to a vapor is accompanied by increased disorder, as suggested by the molecular view in Figure 17.4. Explain why the liquid water does not completely disappear.

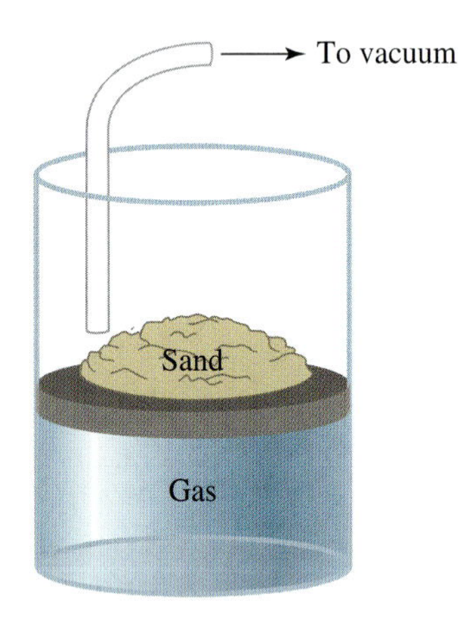

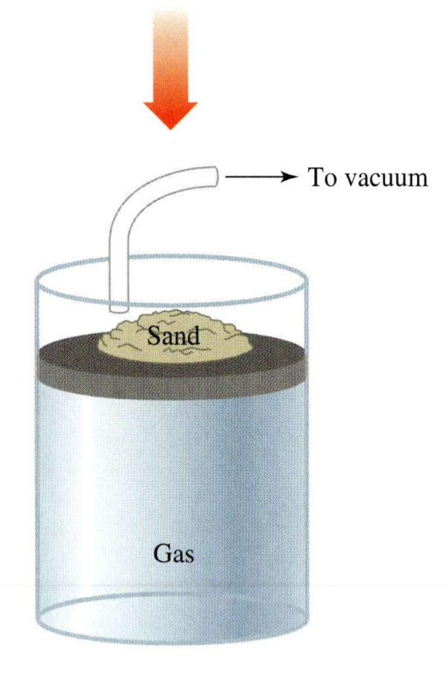

▲ **Figure 17.5**
A nearly reversible process
A *reversible* process can be made to reverse direction with just an infinitesimal change in a system variable. Here, grains of sand are being removed one at a time and the gas very slowly expands. If at any time a grain of sand is *added* to the pile, the piston will reverse direction and compress the gas. The process is not quite reversible because grains of sand have more than an infinitesimal mass.

Sometimes we need *quantitative* values of entropy changes. To get these values, we must relate the entropy change of a system, ΔS, to the quantity of heat, q_{rev}, that is exchanged in a reversible manner between a system and its surroundings at the Kelvin temperature, T.

$$\Delta S = \frac{q_{rev}}{T}$$

Just what do we mean by q_{rev}? As we noted on page 234, ΔU and ΔH are state functions but q and w are not. To make the entropy change ΔS a state function, we have to specify a particular path in order to give q a unique value. That path must be a *reversible* process. A reversible process is never more than an infinitesimal step from equilibrium. That is, it is a process that can be made to reverse its direction by just a minute change in some variable. For the reversible flow of heat, there can be only an infinitesimal difference in temperature between the system and its surroundings. Perhaps it is easier to imagine a process in which *work* is performed reversibly, as suggested in Figure 17.5.

We can rationalize the equation $\Delta S = q_{rev}/T$ by noting that heating a substance increases both atomic and molecular motion. This increased motion introduces greater disorder into the substance and its entropy increases. This is true regardless of whether the substance is a solid, liquid, or gas. The extent of the increased disorder depends on the quantity of heat, and hence ΔS is directly proportional to q_{rev}.

Finally, the *inverse* relationship between ΔS and the Kelvin temperature, T, reflects the fact that a quantity of heat produces more disorder in a system that is highly ordered than in one that is highly disordered. For example, a given quantity of heat produces more disorder in a solid at *low* temperatures than in a gas at *high* temperatures. Earthquakes provide an analogy: A tremor of 5.0 magnitude on the Richter scale might produce considerable disorder in the well-stocked shelves of a supermarket. However, a 5.0 aftershock following a 6.7 earthquake would probably add little to the disorder produced by the original quake (see Figure 17.6).

The enthalpy change for the transition between two phases of matter at their equilibrium temperature has a unique value. This enthalpy change is the q_{rev} we need in calculating the entropy change for the transition. For example, we can represent the fusion of ice at 273 K by the following equation.

$$H_2O(s) \longrightarrow H_2O(l) \qquad \Delta H_{fusion} = 6.01 \text{ kJ} \qquad \Delta S_{fusion} = ?$$

We can then calculate the entropy change, ΔS_{fusion}, as follows.

$$\Delta S_{fusion} = \frac{q_{rev}}{T} = \frac{\Delta H_{fusion}}{T} = \frac{6.01 \text{ kJ}}{273 \text{ K}} = 0.0220 \text{ kJ K}^{-1} = 22.0 \text{ J K}^{-1}$$

In this calculation, we see that the units of entropy change are those of energy divided by temperature: kJ K^{-1} or J K^{-1}.

We will shortly use entropy to formulate the *second law of thermodynamics*, and that will provide us with a criterion for spontaneous change. First, however, let's look more closely at evaluating entropy and entropy changes.

▲ **Figure 17.6**
An analogy to an entropy change
The 6.7-magnitude Northridge, California, earthquake in 1994 produced a great deal of disorder in this Los Angeles area supermarket. Smaller aftershocks produced little additional disorder, however.

Standard Molar Entropies

Because entropy is related to disorder, in a system with perfect order and no disorder, the entropy should be zero. An example of perfect order is that found in a crystalline substance at the absolute zero of temperature, where atomic and molecular motion cease. We can state the **third law of thermodynamics** as follows.

The entropy of a pure, perfect crystal can be taken to be zero at 0 K.

We can use the third law of thermodynamics to obtain the entropy of a substance from experimental data. Beginning at temperatures very near 0 K, heat is slowly added to a substance and the temperature is measured. Entropy changes are calculated in increments, using the expression $\Delta S = q_{rev}/T$. Because the entropy at 0 K is zero, when these incremental entropy changes are added together, the result is an *absolute* entropy.

Recall that with internal energy and enthalpy, we are not able to work with absolute values. We can work only with *differences*, that is, ΔU and ΔH.

Figure 17.7 plots absolute entropy as a function of Kelvin temperature for two substances, methyl chloride and hydrogen. Each graph has some interesting features. There is a sharp increase in entropy at the melting point, where the solid is converted to the more disordered liquid state. At the boiling point, there is an even sharper increase, suggesting that the greatest increase in disorder comes in converting a liquid to a gas.

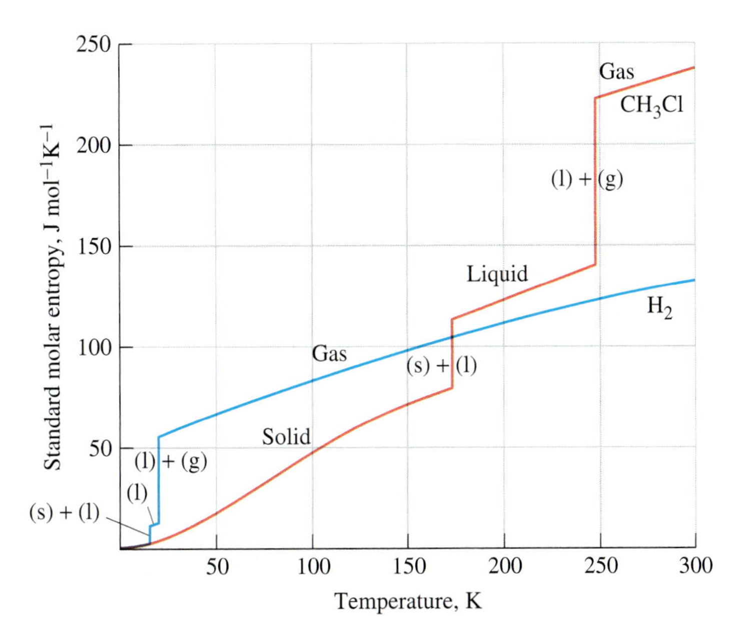

◀ **Figure 17.7**
Entropy as a function of temperature
The entropies of methyl chloride, CH_3Cl, and hydrogen, H_2, are plotted as a function of temperature from 0 K to 300 K. Data are for molar quantities. The phases present at various temperatures are noted. The first vertical line segment on each graph represents fusion, (s) + (l); the second represents vaporization, (l) + (g).

In comparing methyl chloride and hydrogen at 298 K, we see that the entropy of $CH_3Cl(g)$ is significantly greater than that of $H_2(g)$. One important difference is in their molecular structures: CH_3Cl molecules have five atoms, whereas H_2 molecules have two. And the five atoms in CH_3Cl have more ways of moving with respect to one another—more vibrational modes—than do the two atoms in H_2. The more ways that molecules can absorb energy, the greater their entropy. This observation suggests that we add this generalization about entropy to the list on pages 736–37.

In general, the more atoms in its molecules, the greater the entropy of a substance will be.

Another point implied by Figure 17.7 is that entropy, like internal energy and enthalpy, is an *extensive* property—its magnitude depends on the amount of substance. The data in Figure 17.7 are standard *molar* entropies. The **standard molar entropy**, $S°$, is the entropy of one mole of a substance in its standard state. Usually, standard molar entropies are tabulated for a temperature of 25 °C (298 K). A listing of typical data is given in Appendix C.

We will use standard molar entropies mainly to determine standard entropy changes for chemical reactions. We can relate the entropy change to the entropies of reactants and products in the same way that we relate enthalpy change to enthalpies of formation (page 258). That is, we can write the following expression.

$$\Delta S° = \Sigma\, \nu_p\, S°(\text{products}) - \Sigma\, \nu_r\, S°(\text{reactants})$$

In this equation, Σ refers to a sum and ν_p and ν_r are the stoichiometric numbers, the coefficients of the products and reactants in the equation for the reaction. We apply this expression and data from Appendix C in Example 17.3.

EXAMPLE 17.3

Use data from Appendix C to calculate the standard entropy change at 25 °C for the Deacon process. This is a high-temperature, catalyzed reaction used to convert hydrogen chloride, a by-product from organic chlorination reactions, into chlorine.

$$4\ HCl(g) + O_2(g) \longrightarrow 2\ Cl_2(g) + 2\ H_2O(g)$$

SOLUTION

Because five moles of gaseous reactants yield four moles of gaseous products, we expect from the generalization on page 737 that entropy *decreases* in this reaction ($\Delta S < 0$). Let's see if that is the case.

We relate $\Delta S°$ for the reaction to the standard molar entropies, $S°$, of reactants and products as shown below.

$$\begin{aligned}\Delta S° &= \Sigma\, \nu_p\, S°(\text{products}) - \Sigma\, \nu_r\, S°(\text{reactants}) \\ &= \left[2\ \text{mol} \times S°_{Cl_2(g)} + 2\ \text{mol} \times S°_{H_2O(g)}\right] \\ &\quad - \left[4\ \text{mol} \times S°_{HCl(g)} + 1\ \text{mol} \times S°_{O_2(g)}\right] \\ &= (2\ \cancel{\text{mol}} \times 223.0\ \text{J}\ \cancel{\text{mol}^{-1}}\ \text{K}^{-1}) + (2\ \cancel{\text{mol}} \times 188.7\ \text{J}\ \cancel{\text{mol}^{-1}}\ \text{K}^{-1}) \\ &\quad - (4\ \cancel{\text{mol}} \times 186.8\ \text{J}\ \cancel{\text{mol}^{-1}}\ \text{K}^{-1}) - (1\ \cancel{\text{mol}} \times 205.0\ \text{J}\ \cancel{\text{mol}^{-1}}\ \text{K}^{-1}) \\ &= -128.8\ \text{J K}^{-1}\end{aligned}$$

Just as we expected, $\Delta S°$ is indeed negative; entropy decreases.

EXERCISE 17.3A

Use data from Appendix C to calculate the standard molar entropy change at 25 °C for the following reaction. Note that this is a reaction in Exercise 17.2A for which we could not predict whether entropy increases or decreases.

$$CO(g) + H_2O(g) \longrightarrow CO_2(g) + H_2(g)$$

EXERCISE 17.3B

The following reaction is one of those that may occur in the high-temperature, catalyzed oxidation of $NH_3(g)$. Use data from Appendix C to calculate the standard molar entropy change for this reaction at 25 °C.

$$NH_3(g) + O_2(g) \longrightarrow N_2(g) + H_2O(g) \quad \text{(not balanced)}$$

The Second Law of Thermodynamics

We have seen that we can't use the enthalpy change in a reaction (ΔH) as the sole criterion for spontaneous change. Can we use the tendency toward increasing disorder ($\Delta S > 0$) as a sole criterion?

In fact, we can, but only if we consider the entropy change of a system *and* of the *surroundings*. We call this total entropy change the *entropy change of the universe*.

$$\Delta S_{\text{total}} = \Delta S_{\text{universe}} = \Delta S_{\text{system}} + \Delta S_{\text{surroundings}}$$

The **second law of thermodynamics** establishes that *all spontaneous or natural processes increase the entropy of the universe*. For a spontaneous process, we have the following relationship.

$$\Delta S_{\text{univ}} > 0$$

If a process results in an entropy *increase* in *both* the system and the surroundings, then surely it is *spontaneous*. Just as surely, if a process results in an entropy *decrease* in the system *and* the surroundings, the process is *nonspontaneous* and cannot occur. What will ΔS_{univ} be if one term is positive and the other is negative? Consider the freezing of liquid water at −15 °C.

$$H_2O(l) \longrightarrow H_2O(s)$$

We can see that $\Delta S_{\text{sys}} < 0$. Ice is a more ordered state than liquid water. Let's not forget, however, that when water freezes it gives off heat—the heat of fusion. And because this heat is absorbed by the surroundings, $\Delta S_{\text{surr}} > 0$. By calculation, it can be shown that the magnitude of ΔS_{surr} exceeds that of ΔS_{sys}.* The total entropy change, $\Delta S_{\text{univ}} > 0$, and the overall process is *spontaneous*, just as we intuitively know it should be.

Although ΔS_{univ} gives a single criterion for spontaneous change, we find it difficult to apply because we need to consider often complex interactions between a system and its surroundings. In Section 17.4, we will see that there is an easier approach.

* These entropy changes must be calculated for a hypothetical, *reversible* process having the same initial and final states as the freezing of supercooled water at −15 °C. However, such calculations are somewhat beyond the scope of the text.

Entropy and Probability

In general, the likelihood of winning a prize is greater in a local charity raffle than in a statewide lottery. There are fewer entrants and a greater probability of holding the winning ticket in the smaller event. Probability also plays a role in determining the likelihood of events in the molecular world.

At the microscopic level there are more possibilities, and hence a greater probability, for a disordered state than for a more ordered one. To illustrate this point, let's return to the spilling of cards described on page 735 but limit ourselves to just four cards. There are two possible ways a given card may fall—face up or face down—and the number of possible arrangements of the four cards is $2^4 = 16$. These 16 possibilities are pictured in the figure, where they are divided into five categories:

(a) all four cards with faces showing,

(b) three cards with faces showing and one with face down ( ),

(c) two cards with faces showing and two with faces down ( ),

(d) one card with face showing and three with faces down ( ), and

(e) four cards with faces down ( ).

From the figure we see that there is a probability of 1/16 that the cards will be in arrangement (a) and 1/16 in arrangement (e). These are the most ordered arrangements. The probability of arrangement (b) is 4/16, as is that of arrangement (d). The probability of arrangement (c) is the greatest, 6/16. This is the arrangement that we would expect to see the most if we observed the spilling of four cards over a large number of observations.

Early workers in thermodynamics developed the concept of entropy without regard to the ultimate nature of matter—some did not even subscribe to the existence of atoms as the building blocks of matter. Ludwig Boltzmann (1844-1906) was the first to promote the idea that macroscopic observations in thermodynamics are governed by mechanical laws and laws of chance applied to atoms and molecules. Toward this end, he derived an equation relating entropy and probability.

$$S = k \ln W$$

In this equation, S is the entropy of a system in a particular state and W is the number of different configurations of equal energy in which the microscopic particles can be found in that state. The proportionality constant, k, is called the Boltzmann constant and is equal to R/N_A, the universal gas constant divided by Avogadro's number.

Because there is only one possibility for arrangements (a) and (e) in the figure, Boltzmann's equation indicates that $W = 1$, $\ln 1 = 0$, and $S = 0$. Arrangements (a) and (e) represent perfect order and a zero of entropy. For arrangements (b) and (d), there are four possibilities each, and the entropy is proportional to ln 4 (1.39). For arrangement (c), with six possibilities, the entropy is proportional to ln 6 (1.79). Thus, the more possible arrangements—the more disordered a system—the greater is the entropy. Notice that Boltzmann's equation also predicts that the entropy of a pure perfect crystal at 0 K should be zero; in this state W = 1. Thus, Boltzmann's equation is consistent with the third law of thermodynamics.

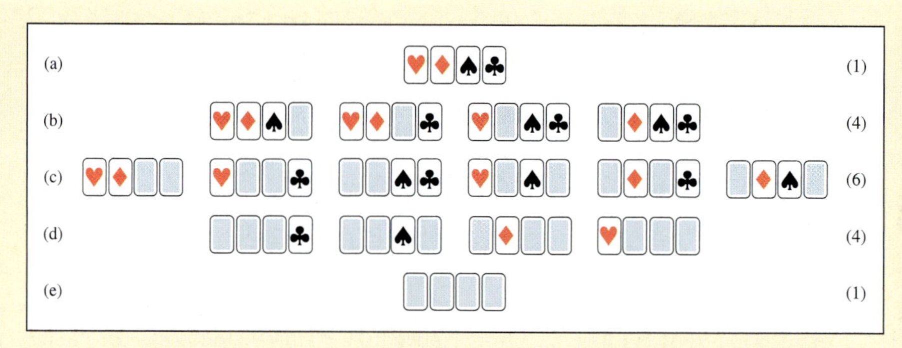

The 16 possible arrangements of four spilled playing cards
Four playing cards—a diamond, a heart, a spade, and a club—are allowed to fall to the floor. The cards may land face up or face down ( ). The number of possibilities for each arrangement is shown at the right.

17.4 Free Energy and Free Energy Change

Our objective in this section will be to introduce a new thermodynamic function and a new criterion for spontaneous change based *just on the system*, without regard for the surroundings.

To do so, imagine a process conducted at constant temperature and pressure and with work limited to pressure-volume work (page 240). For this process, $\Delta H_{sys} = q_p$. Because the process is at constant temperature, this heat must be exchanged with the surroundings, that is, $q_{surr} = -q_p = -\Delta H_{sys}$. We can use this information to evaluate ΔS_{surr}.*

$$\Delta S_{surr} = \frac{q_{surr}}{T} = \frac{-\Delta H_{sys}}{T}$$

Now let's turn to the expression

$$\Delta S_{univ} = \Delta S_{sys} + \Delta S_{surr}$$

and substitute the above expression written for ΔS_{surr}.

$$\Delta S_{univ} = \Delta S_{sys} - \frac{\Delta H_{sys}}{T}$$

Next we can multiply by T.

$$T\Delta S_{univ} = T\Delta S_{sys} - \Delta H_{sys} = -(\Delta H_{sys} - T\Delta S_{sys})$$

Then we can multiply by -1, changing signs.

$$-T\Delta S_{univ} = \Delta H_{sys} - T\Delta S_{sys}$$

The significance of this equation is that it relates the elusive ΔS_{univ} to two quantities, ΔH_{sys} and $-T\Delta S_{sys}$, that are based only on the system itself. We don't have to consider the surroundings at all.

Now let's introduce another thermodynamic function. We set the term on the left of the above equation equal to the change in a function called the **Gibbs free energy**, ***G***. That is,

$$\Delta G = -T\Delta S_{univ}$$

Then the **free energy change**, **ΔG**, for a process at constant temperature is given by the *Gibbs equation.*

$$\Delta G_{sys} = \Delta H_{sys} - T\Delta S_{sys}$$

Because each term is now based on the system, we can drop the subscripts.

$$\Delta G = \Delta H - T\Delta S$$

Because the criterion for spontaneous change is that $\Delta S_{univ} > 0$, and because $\Delta G = -T\Delta S_{univ}$, the criterion for spontaneous change is also that $\Delta G < 0$. That

J. Willard Gibbs (1839–1903), a professor of mathematical physics at Yale University, was not well known to his contemporaries, mostly because his ideas were quite abstract and published in obscure journals. Eventually, though, his ideas were understood and he was recognized as a giant in the emerging field of chemical thermodynamics.

* For this equation to be valid, heat must be gained or lost by the surroundings by a *reversible* path, that is, $q_{surr} = q_{rev}$.

is, the change in free energy must be *negative*. In summary, these are the criteria that apply to a process at constant temperature and pressure.

- If $\Delta G < 0$ (negative), a process is *spontaneous*.
- If $\Delta G > 0$ (positive), a process is *nonspontaneous*.
- If $\Delta G = 0$, neither the forward nor the reverse process is favored; there is no net change, and the process is at *equilibrium*.

Using ΔG as a Criterion for Spontaneous Change

We will first consider some qualitative applications of the Gibbs equation.

$$\Delta G = \Delta H - T\Delta S$$

In the equation, we see that if ΔH is *negative* and ΔS is *positive*, then ΔG must be *negative* and a reaction will be *spontaneous*. We can just as easily see that if ΔH is *positive* and ΔS is *negative*, then ΔG must be *positive* and a reaction will be *nonspontaneous*.

Situations in which ΔH and ΔS are *both positive* or *both negative* require more thought. In these cases, whether a reaction is spontaneous or not—that is, whether ΔG is negative or positive—depends on the temperature. Usually, though, the ΔH term will dominate at *lower* temperatures and the $T\Delta S$ term at *higher* temperatures, as suggested by Figure 17.8. Altogether, there are *four* possibilities for ΔG, depending on the signs of ΔH and ΔS. These four possibilities are outlined in Table 17.1 and illustrated in Example 17.4.

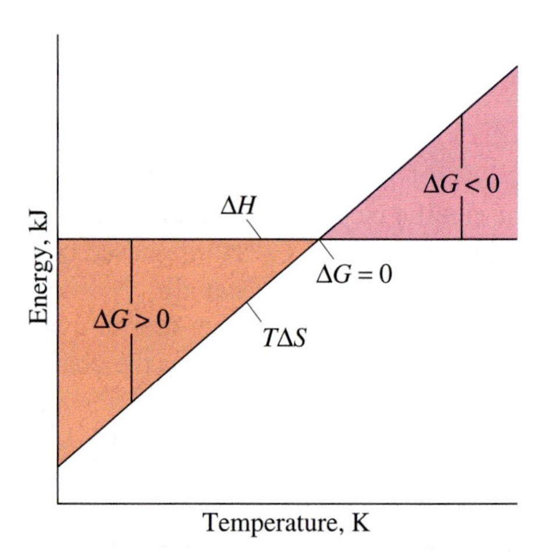

Figure 17.8
ΔG as a criterion for spontaneous change
ΔG, ΔH, and $T\Delta S$ all have the units of energy. The value of ΔG at any temperature is the value on the ΔH line *minus* the value on the $T\Delta S$ line. That is, ΔG is the distance between the two lines. At the temperature at which the lines intersect, this distance is *zero* ($\Delta G = 0$), and the system is at equilibrium. Below this temperature, $\Delta G > 0$ and the reaction is *nonspontaneous*. Above this temperature, $\Delta G < 0$, and the reaction is *spontaneous*. The situation described here is that of case 3 in Table 17.1.

TABLE 17.1 Criterion for Spontaneous Change: $\Delta G = \Delta H - T\Delta S$.

Case	ΔH	ΔS	ΔG	Result	Example
1	−	+	−	Spontaneous at all T	$2\ O_3(g) \longrightarrow 3\ O_2(g)$
2	−	−	−	Spontaneous toward low T	$N_2(g) + 3\ H_2(g) \longrightarrow 2\ NH_3(g)$
	−	−	+	Nonspontaneous toward high T	
3	+	+	+	Nonspontaneous toward low T	$2\ H_2O(g) \longrightarrow 2\ H_2(g) + O_2(g)$
	+	+	−	Spontaneous toward high T	
4	+	−	+	Nonspontaneous at all T	$2\ C(\text{graphite}) + 2\ H_2(g) \longrightarrow C_2H_4(g)$

EXAMPLE 17.4

Predict which of the four cases in Table 17.1 you expect to apply to the following reactions.

a. $C_6H_{12}O_6(s) + 6\,O_2(g) \longrightarrow 6\,CO_2(g) + 6\,H_2O(g)$ $\Delta H = -2540$ kJ

b. $Cl_2(g) \longrightarrow 2\,Cl(g)$

SOLUTION

a. The reaction is exothermic; $\Delta H < 0$. Because twelve moles of gaseous products replace six moles of gaseous reactant, we expect $\Delta S > 0$ for this reaction. Because $\Delta H < 0$ and $\Delta S > 0$, we expect this reaction to be spontaneous at all temperatures (Case 1 of Table 17.1).

b. One mole of gaseous reactant produces two moles of gaseous product, and we expect $\Delta S > 0$. No value is given for ΔH. However, we can determine the sign of ΔH by examining what occurs during the reaction. Bonds are broken, but no new bonds are formed. Because molecules must absorb energy for bonds to break, the reaction must be endothermic. Thus, ΔH must be *positive*. Because ΔH and ΔS are both positive, this reaction is an example of Case 3 in Table 17.1. We expect the reaction to be nonspontaneous at low temperatures and spontaneous at high temperatures.

EXERCISE 17.4A

Predict which of the four cases in Table 17.1 are likely to apply to each of the following reactions.

a. $N_2(g) + 2\,F_2(g) \longrightarrow N_2F_4(g)$ $\Delta H = -7.1$ kJ

b. $COCl_2(g) \longrightarrow CO(g) + Cl_2(g)$ $\Delta H = +110.4$ kJ

EXERCISE 17.4B

Predict which of the four cases in Table 17.1 is likely to apply to the following reaction. Use the value −93.85 kJ/mol for ΔH_f° for BrF(g) and additional data from Appendix C as necessary.

$$Br_2(g) + BrF_3(g) \longrightarrow 3\,BrF(g)$$

EXAMPLE 17.5—A Conceptual Example

Molecules exist over only a relatively short range of possible temperatures—from near 0 K to a few thousand K. Explain why this is to be expected.

SOLUTION

To cause a molecule to dissociate into its atoms, we must supply the molecule with enough energy to induce such vigorous vibrations that its atoms fly apart, an endothermic process ($\Delta H > 0$). However, a system of individual atoms is more disordered than the same atoms united into molecules ($\Delta S > 0$). The key factor is the temperature, T. At low temperatures, ΔH is the determining factor, and molecules are generally stable with respect to uncombined atoms. However, no matter how large the value of ΔH, eventually a temperature is reached at which the magnitude of $T\Delta S$ exceeds that of ΔH. Then ΔG is negative, and the dissociation becomes spontaneous.

$$\Delta H - T\Delta S = \Delta G < 0$$

For all known molecules, this high temperature limit is no more than a few thousand K.

EXERCISE 17.5

If the process described in Figure 17.8 is CO_2(s, 1 atm) $\longrightarrow$ CO_2(g, 1 atm), at what temperature will the two lines intersect? Explain.

(*Hint:* Use data from Chapter 11.)

17.5 Standard Free Energy Change, $\Delta G°$

The **standard free energy change**, $\Delta G°$, of a reaction is the free energy change when reactants and products are in their standard states. The convention for standard states is the same as that used for $\Delta H°$ on page 256: The standard state of a solid or liquid is the pure substance at 1 atm pressure* and at the temperature of interest. For a gas, the standard state is the pure gas behaving as an ideal gas at 1 atm pressure* and the temperature of interest. One way we can evaluate the standard free energy change of a reaction is to use standard enthalpy and entropy changes in the Gibbs equation.

$$\Delta G° = \Delta H° - T\Delta S°$$

In using this equation, we must be careful to use the same energy unit for $\Delta H°$ and $\Delta S°$. For example, tabulated data generally use *kilojoules* for enthalpy changes and *joules* per K for entropies.

Another way to evaluate $\Delta G°$ is to use tabulated standard free energies of formation. The **standard free energy of formation, $\Delta G_f°$** is the free energy change that occurs in the formation of 1 mol of a substance in its standard state from the reference forms of its elements in their standard states. The reference forms of the elements generally are their most stable forms. Like standard enthalpies of formation, the standard free energies of formation of the *elements* in their reference forms have values of *zero*. Appendix C contains a listing of the standard free energies of formation of many substances at 25 °C (298 K), the temperature usually used for tabulations.

To obtain a standard free energy of reaction from standard free energies of formation, we use the same type of relationship that we used for enthalpies (page 258).

$$\Delta G° = \Sigma\nu_p\,\Delta G_f°(\text{products}) - \Sigma\nu_r\,\Delta G_f°(\text{reactants})$$

Both methods of calculating a standard free energy change are illustrated in Example 17.6.

EXAMPLE 17.6

Calculate $\Delta G°$ at 298 K for the reaction

$$4\,HCl(g) + O_2(g) \longrightarrow 2\,Cl_2(g) + 2\,H_2O(g) \qquad \Delta H° = -114.4\text{ kJ}$$

(a) with the Gibbs equation, and **(b)** from standard free energies of formation.

SOLUTION

a. To use the Gibbs equation, we need values of both $\Delta H°$ and $\Delta S°$. The value of $\Delta H°$ is given, and we can obtain $\Delta S°$ from standard molar entropies. We determined $\Delta S°$

* As we noted before (page 256), the IUPAC has recommended that the standard state pressure be chosen as 1 bar (1×10^5 Pa) rather than 1 atm (101,325 Pa). Because the effect of changing the standard state pressure is very small, we will continue to use 1 atm.

in Example 17.3 and found a value of $-128.8\ \text{J K}^{-1}$. To express $\Delta H°$ and $\Delta S°$ in the same energy unit, we can convert $-128.8\ \text{J K}^{-1}$ to $-0.1288\ \text{kJ K}^{-1}$.

$$\Delta G° = \Delta H° - T\Delta S° = -114.4\ \text{kJ} - [298\ \cancel{\text{K}} \times (-0.1288\ \text{kJ}\ \cancel{\text{K}^{-1}})]$$
$$= -114.4\ \text{kJ} + 38.4\ \text{kJ} = -76.0\ \text{kJ}$$

b. We must look up standard free energies of formation in Appendix C and use them in the expression

$$\Delta G° = 2\ \Delta G_f°[Cl_2(g)] + 2\ \Delta G_f°[H_2O(g)] - 4\ \Delta G_f°[HCl(g)] - \Delta G_f°[O_2(g)]$$
$$= 2\ \cancel{\text{mol}} \times (0.00\ \text{kJ}/\cancel{\text{mol}}) + 2\ \cancel{\text{mol}}$$
$$\times (-228.6\ \text{kJ}/\cancel{\text{mol}}) - 4\ \cancel{\text{mol}} \times (-95.30\ \text{kJ}/\cancel{\text{mol}})$$
$$- 1\ \cancel{\text{mol}} \times (0.00\ \text{kJ}/\cancel{\text{mol}})$$
$$= (-457.2 + 381.2)\ \text{kJ} = -76.0\ \text{kJ}$$

EXERCISE 17.6A

Use the Gibbs equation to determine the standard free energy change at 25 °C for these reactions.

a. $2\ NO(g) + O_2(g) \longrightarrow 2\ NO_2(g)$
$\Delta H° = -114.1\ \text{kJ} \qquad \Delta S° = -146.2\ \text{J K}^{-1}$

b. $2\ CO(g) + 2\ NH_3(g) \longrightarrow C_2H_6(g) + 2\ NO(g)$
$\Delta H° = 409.0\ \text{kJ} \qquad \Delta S° = -129.1\ \text{J mol}^{-1}\ \text{K}^{-1}$

EXERCISE 17.6B

Use standard free energies of formation from Appendix C to determine the standard free energy change at 25 °C for these reactions:

a. $CS_2(l) + 2\ S_2Cl_2(g) \longrightarrow CCl_4(l) + 6\ S(s)$
b. $NH_3(g) + O_2(g) \longrightarrow N_2(g) + H_2O(g)$ (not balanced)

17.6 Free Energy Change and Equilibrium

So far, we have considered conditions in which $\Delta G < 0$ (spontaneous process) and in which $\Delta G > 0$ (nonspontaneous process). Now let's consider a process that is neither spontaneous nor nonspontaneous. In this case, there is no net change because a forward and reverse process occur at the same rate. This is the equilibrium condition, at which $\Delta G = 0$.

$$\Delta G = \Delta H - T\Delta S = 0 \quad \textit{(at equilibrium)}$$

and

$$\Delta H = T\Delta S \quad \textit{(at equilibrium)}$$

At the *equilibrium* temperature, we can use this expression to determine ΔH from a value of ΔS, or ΔS from a value of ΔH.

When we relate the enthalpy and entropy of vaporization at the *normal* boiling point of a liquid, we are dealing with *standard* enthalpy and entropy changes.

The liquid and vapor are both in the standard state: 1 atm pressure. Consider the boiling of benzene at 80.10 °C.

$$C_6H_6(l, 1\text{ atm}) \rightleftharpoons C_6H_6(g, 1\text{ atm}) \qquad \Delta H^\circ = 30.76\text{ kJ/mol}$$

$$\Delta S^\circ_{vapn} = \frac{\Delta H^\circ_{vapn}}{T_{bp}} = \frac{30.76\text{ kJ mol}^{-1}}{(80.10 + 273.15)\text{ K}} = 0.08708\text{ kJ mol}^{-1}\text{ K}^{-1}$$

We can also express the standard molar entropy of vaporization of benzene at 80.10 °C in the units J mol^{-1} K^{-1}.

$$\Delta S^\circ_{vapn} = 0.08708\text{ kJ mol}^{-1}\text{ K}^{-1} \times \frac{1000\text{ J}}{1\text{ kJ}} = 87.08\text{ J mol}^{-1}\text{ K}^{-1}$$

Chemists have noted that many liquids have about the same standard molar entropy of vaporization at their normal boiling points—approximately 87 J mol^{-1} K^{-1}—an observation summarized by **Trouton's rule**.

$$\Delta S^\circ_{vapn} = \frac{\Delta H^\circ_{vapn}}{T_{bp}} \approx 87\text{ J mol}^{-1}\text{ K}^{-1}$$

Trouton's rule implies that about the same amount of disorder is generated in the passage of one mole of substance from liquid to vapor when comparisons are made at the normal boiling point. The rule works best for *nonpolar* substances, such as those in Figure 17.9. It generally fails for liquids with a more ordered structure, such as the liquid structure produced by extensive hydrogen bonding. In these cases, a greater increase in disorder occurs during vaporization and ΔS°_{vapn} is greater than 87 J mol^{-1} K^{-1}. For example, for the vaporization of H_2O at 100 °C, ΔS°_{vapn} = 109 J mol^{-1} K^{-1}.

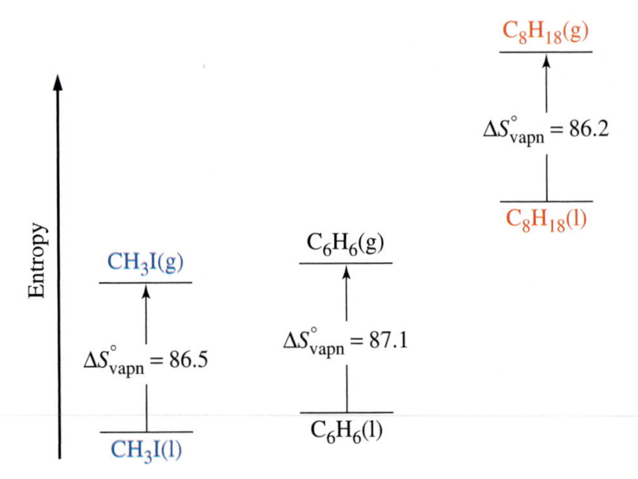

▲ **Figure 17.9 Illustrating Trouton's rule**
Entropies of vaporization are given in J K^{-1} per mole of liquid vaporized. The three liquids all have a standard molar entropy of vaporization of about 87 J mol^{-1} K^{-1}. However, the following tabulation suggests that the liquids have little else in common.

	Molecular mass, u	Molar entropy, $S^\circ_{298\,K}$, J mol^{-1} K^{-1}	At the normal boiling point: BP, K	ΔH°_{vapn}, kJ/mol	ΔS°_{vapn}, J mol^{-1} K^{-1}
CH_3I, iodomethane	142	163	315.6	27.3	86.5
C_6H_6, benzene	78	173	353.3	30.8	87.1
C_8H_{18}, octane	114	358	398.9	34.4	86.2

EXAMPLE 17.7

At its normal boiling point, the enthalpy of vaporization of pentadecane, $CH_3(CH_2)_{13}CH_3$, is 49.45 kJ/mol. What should its approximate normal boiling point temperature be?

SOLUTION

We have just seen that Trouton's rule provides an approximate value (87 J mol^{-1} K^{-1}) of the molar entropy of vaporization of a *nonpolar* liquid at its normal boiling point. Pentadecane, an alkane, is a nonpolar liquid, and when it is in equilibrium with its vapor at 1 atm pressure, we can apply Trouton's rule.

$$\Delta S^\circ_{\text{vapn}} = \frac{\Delta H^\circ_{\text{vapn}}}{T_{\text{bp}}} \approx 87 \text{ J mol}^{-1}\text{ K}^{-1}$$

$$T_{\text{bp}} \approx \frac{\Delta H^\circ_{\text{vapn}}}{87 \text{ J mol}^{-1}\text{ K}^{-1}} \approx \frac{49.45 \times 10^3 \cancel{\text{J}}\ \cancel{\text{mol}^{-1}}}{87 \cancel{\text{J}}\ \cancel{\text{mol}^{-1}}\text{ K}^{-1}} \approx 570 \text{ K}$$

How good is this approximation? The experimentally determined boiling point of pentadecane is 543.8 K. Our estimate is within about 5% of the true value.

EXERCISE 17.7A

The normal boiling point of styrene, $C_6H_5CH{=}CH_2$ is 145.1 °C. Estimate the molar enthalpy of vaporization of styrene at its normal boiling point.

EXERCISE 17.7B

Would you expect $\Delta S^\circ_{\text{vapn}}$ for methanol, CH_3OH, to be greater, less than, or about equal to 87 J mol^{-1} K^{-1}? Explain. Use the normal boiling point of 64.7 °C and data from Appendix C to estimate a value of $\Delta S^\circ_{\text{vapn}}$.

Raoult's Law Revisited

Recall Raoult's law from Section 12.6.

$$P_{\text{solv}} = x_{\text{solv}} \cdot P^\circ_{\text{solv}}$$

Because the mole fraction of solvent in a solution (x_{solv}) is less than 1, the vapor pressure of the solvent (P_{solv}) in an ideal solution is lower than that of the pure solvent (P°_{solv}). The concept of entropy gives us a way to explain Raoult's law, as illustrated in Figure 17.10.

Figure 17.10 shows that a solution is more disordered, and therefore at a higher entropy, than the pure solvent. Because the intermolecular forces in the solution and in the pure solvent are comparable, we expect the enthalpy of vaporization (ΔH_{vapn}) to be the same from the solution and from the pure solvent. The entropy of vaporization should also be the same because $\Delta S_{\text{vapn}} = \Delta H_{\text{vapn}}/T$. However, because the ideal solution is more disordered than the pure solvent, the vapor from the solution must also be more disordered than the vapor from the pure solvent. Disorder in a vapor increases if molecules can roam more freely, that is, if they are at a *lower* pressure. Thus, solutes in a solution reduce the vapor pressure of the solvent.

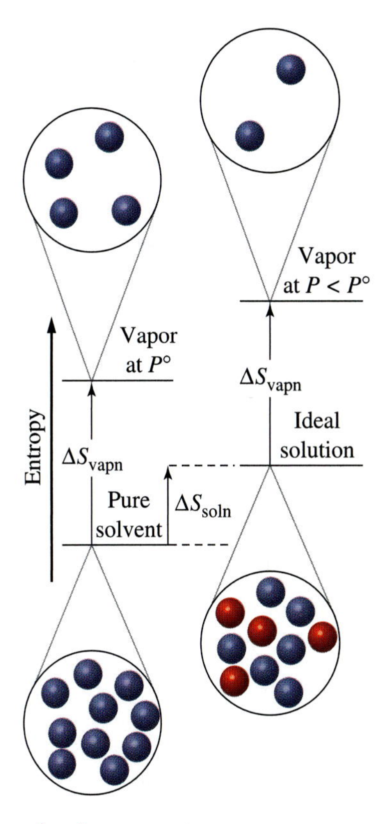

▲ **Figure 17.10**
An entropy-based explanation of Raoult's law
The features brought out by the diagram are:
- The entropy of the solution is greater than that of the pure solvent.
- The entropy change for vaporization of a quantity of the solvent, ΔS_{vapn}, is the same, whether from the pure solvent or from an ideal solution.
- The vapor phase above the solution is more disordered than above the pure solvent; it is at a lower pressure.

Relationship of ΔG° to the Equilibrium Constant, K_{eq}

Liquid water is in equilibrium with water vapor at 1 atm pressure and 100 °C. As we have seen, ΔG for this (or any) *equilibrium* process is *zero*. Moreover, because the liquid and vapor are both in their standard states (1 atm), we can write

$$H_2O(\text{l, 1 atm}) \rightleftharpoons H_2O(\text{g, 1 atm}) \qquad \Delta G^\circ_{373} = 0$$

We can write the same equation for the vaporization of water at 25 °C and determine ΔG°_{298} from tabulated standard free energies of formation.

$$H_2O(l, 1\ atm) \rightleftharpoons H_2O(g, 1\ atm) \qquad \Delta G^\circ_{298} = +8.590\ kJ$$

The *positive* value of ΔG°_{298} shows that the process is *nonspontaneous*. This does not mean that water will not vaporize at 25 °C, just that it will not produce a vapor *at 1 atm pressure*; equilibrium is displaced *to the left*. From Table 11.2 (page 464) we see that the *equilibrium* vapor pressure of water at 25 °C is 23.8 mmHg (or 0.0313 atm), a fact that we can represent by the following equation.

$$H_2O(l, 0.0313\ atm) \rightleftharpoons H_2O(g, 0.0313\ atm) \qquad \Delta G_{298} = 0$$

To summarize, $\Delta G^\circ = 0$ is a criterion for equilibrium at a single temperature, the one temperature at which the equilibrium state has all reactants and products in their standard states. At equilibrium at every other temperature, some or all of the reactants and products must be in a *nonstandard* state. For these *nonstandard* conditions, the criterion for equilibrium is that $\Delta G = 0$ (*not* $\Delta G^\circ = 0$). This seems to make ΔG° of limited value, but we will now see that it really is quite useful.

The quantities ΔG and ΔG° are related through the *reaction quotient*, Q, by the following equation.

$$\Delta G = \Delta G^\circ + RT \ln Q$$

Now consider a reaction at equilibrium, in which $\Delta G = 0$ and $Q = K_{eq}$. This leads to the expression

$$0 = \Delta G^\circ + RT \ln K_{eq} \quad \textit{(at equilibrium)}$$

and

$$\Delta G^\circ = -RT \ln K_{eq} \quad \textit{(at equilibrium)}$$

This is one of the most important equations in all of chemical thermodynamics. It says that if we have ΔG° for a reaction at a given temperature, we can *calculate* the equilibrium constant at that temperature. With the equilibrium constant, we can then calculate equilibrium concentrations or partial pressures, just as we have done in earlier chapters. In applying this equation, we should note the following.

- R, the gas constant, is generally expressed as 8.3145 J mol^{-1} K^{-1}.
- T, the temperature, is expressed in kelvins, K.
- K_{eq} must have the format described below.

The Equilibrium Constant, K_{eq}

Recall that in Chapter 14, we wrote a K_c expression if concentrations were used to describe reactants and products, and we used a K_p expression if reactants and products were described through their partial pressures. In the expression $\Delta G^\circ = -RT \ln K_{eq}$, we write the **equilibrium constant K_{eq}** rather than K_c or K_p, and K_{eq} may differ from both of them. Because the equation relating the standard free energy change and the equilibrium constant includes the term "ln K_{eq}," there is a special requirement of K_{eq}: It must be a *dimensionless* number. K_{eq} can have no units because we can't take the logarithm of a unit.

The dimensionless quantities needed in the equilibrium constant K_{eq} are called *activities*, but we can continue to use molarities and partial pressures if we use the following conventions regarding activities.

- *For pure solid and liquid phases*: The activity $a = 1$.
- *For gases*: Assume ideal gas behavior, and replace the activity by the *numerical* value of the gas partial pressure in atm.
- *For solutes in aqueous solution*: Assume that intermolecular or interionic attractions are negligible—that is, that the solution is *dilute*—and replace solute activity by the *numerical* value of the solute molarity.

EXAMPLE 17.8

Write the equilibrium constant expression, K_{eq}, for the oxidation of chloride ion by manganese dioxide in an acidic solution.

$$MnO_2(s) + 4\,H^+(aq) + 2\,Cl^-(aq) \rightleftharpoons Mn^{2+}(aq) + Cl_2(g) + 2\,H_2O(l)$$

SOLUTION

First, we can write the K_{eq} expression in terms of activities (a), and then we can substitute the appropriate quantities for these activities in accordance with the conventions just introduced.

$$K_{eq} = \frac{a_{Mn^{2+}} \times a_{Cl_2} \times (a_{H_2O})^2}{a_{MnO_2} \times (a_{H^+})^4 \times (a_{Cl^-})^2}$$

No term will appear for $MnO_2(s)$; it is a pure solid phase with $a = 1$. For Mn^{2+}, H^+, and Cl^-, ionic species in dilute aqueous solution, we substitute molarities for activities. For the activity of chlorine gas, we substitute its partial pressure. Because H_2O is the preponderant species in the dilute aqueous solution, the activity of H_2O is essentially the same as in pure liquid water: $a = 1$. Making these substitutions, we get the following.

$$K_{eq} = \frac{[Mn^{2+}](P_{Cl_2})}{[H^+]^4[Cl^-]^2}$$

Notice that this K_{eq} expression contains both molarities and a partial pressure. It is neither a K_c nor a K_p.

EXERCISE 17.8A

Write the K_{eq} expression for the following reaction.

$$2\,Al(s) + 6\,H^+(aq) \longrightarrow 2\,Al^{3+}(aq) + 3\,H_2(g)$$

EXERCISE 17.8B

Write an equation for the dissolution of magnesium hydroxide in an acidic solution, and then write the K_{eq} expression for this reaction.

Calculating Equilibrium Constants, K_{eq}

Now we are ready to combine several ideas to calculate an equilibrium constant, K_{eq}, and to consider its significance. Let's calculate K_{eq} for the vaporization of water at 25 °C, for which we have previously written

$$H_2O(l) \rightleftharpoons H_2O(g) \qquad \Delta G^\circ_{298} = +8.590 \text{ kJ}$$

We can begin by rearranging the equation, $\Delta G^\circ = -RT \ln K_{eq}$, to

$$\ln K_{eq} = \frac{-\Delta G^\circ}{RT}$$

Then we substitute values of ΔG°, R, and T. In making this substitution, we change the value of ΔG° first to 8.590 kJ/mol and then to 8.590×10^3 J/mol. The "mol^{-1}" portion of the unit kJ/mol signifies that the quantities in the chemical equation are on a mole basis.* The conversion from kJ/mol to J/mol leads to the proper cancellation of units in the calculation that follows.

$$\ln K_{eq} = \frac{-8.590 \times 10^3 \cancel{\text{J}}\ \cancel{\text{mol}^{-1}}}{8.3145 \cancel{\text{J}}\ \cancel{\text{mol}^{-1}}\ \cancel{\text{K}^{-1}} \times 298.15 \cancel{\text{K}}} = -3.465$$

$$K_{eq} = e^{-3.465} = 0.0313$$

According to the conventions regarding activities, the activity of $H_2O(l)$ is one, and the appropriate expression for K_{eq} is

$$K_{eq} = P_{H_2O(g)} = 0.0313$$

That is, the *equilibrium* vapor pressure of water at 25 °C is 0.0313 atm (23.8 mmHg). This calculated vapor pressure is the same as the experimentally measured value listed in Table 11.2.

EXAMPLE 17.9

Determine the value of K_{eq} at 25 °C for the reaction: $2\ NO_2(g) \rightleftharpoons N_2O_4(g)$.

SOLUTION

Our first task is to obtain a value of ΔG°, which we can do from tabulated standard free energies of formation in Appendix C.

$$\Delta G^\circ = \Delta G^\circ_f[N_2O_4(g)] - 2 \times \Delta G^\circ_f[NO_2(g)]$$

$$\Delta G^\circ = \left(1 \cancel{\text{mol } N_2O_4} \times \frac{97.82 \text{ kJ}}{1 \cancel{\text{mol } N_2O_4}}\right) - \left(2 \cancel{\text{mol } NO_2} \times \frac{51.30 \text{ kJ}}{1 \cancel{\text{mol } NO_2}}\right)$$

$$= -4.78 \text{ kJ}$$

Next, let's change the unit of ΔG° from kJ to J/mol, to signify that the quantities in the chemical equation are expressed in moles. Then we can rearrange the following expression and solve for K_{eq}, after substituting as follows.

$$\Delta G^\circ = -4780 \text{ J mol}^{-1}, R = 8.3145 \text{ J mol}^{-1} \text{ K}^{-1}, \text{ and } T = 298.15 \text{ K.}$$

$$\Delta G^\circ = -RT \ln K_{eq}$$

$$\ln K_{eq} = \frac{-\Delta G^\circ}{RT} = \frac{-(-4780 \cancel{\text{J}}\ \cancel{\text{mol}^{-1}})}{8.3145 \cancel{\text{J}}\ \cancel{\text{mol}^{-1}}\ \cancel{\text{K}^{-1}} \times 298.15 \cancel{\text{K}}} = 1.93$$

$$K_{eq} = e^{1.93} = 6.9$$

* Alternatively, we can say that the balanced equation represents "one mole of reaction," and that ΔG° = 8.590 kJ/mol signifies 8.590 kilojoules per mole of reaction. The free energy change for a reaction should always be linked to a balanced equation, and it is the equation that establishes one mole of reaction.

EXERCISE 17.9A

Use data from Appendix C to determine K_{eq} at 25 °C for the reaction: $2\ HgO(s) \rightleftharpoons 2\ Hg(l) + O_2(g)$.

EXERCISE 17.9B

Use data from Appendix C to determine K_{eq} at 25 °C for the reaction: $2\ NO(g) + Br_2(l) \rightleftharpoons 2\ NOBr(g)$. If NO(g) at 1 atm pressure is allowed to react with an excess of $Br_2(l)$, what will be the partial pressures of NO(g) and NOBr(g) at equilibrium?

The Sign and Magnitude of $\Delta G°$: Their Significance

In Chapter 14 (page 606), we found that we could use the magnitude of K_{eq} to decide, roughly, whether a reaction would (a) go essentially to completion, (b) occur hardly at all in the forward direction, or (c) reach an equilibrium condition that must be described through K_{eq}. Now we have an equation linking K_{eq} and $\Delta G°$.

$$\Delta G° = -RT \ln K_{eq}$$

We can use $\Delta G°$ alone to decide among these three possibilities. In Figure 17.11, we have plotted three reaction profiles, suggesting how free energy can change during a reaction, starting with reactants in their standard states on the left and proceeding to the products in their standard states on the right. The minimum point in a reaction profile represents chemical equilibrium. This minimum point can be approached from either direction, depending on the value of the reaction quotient Q. Once this minimum is reached, however, the reaction does not proceed any further along the profile.

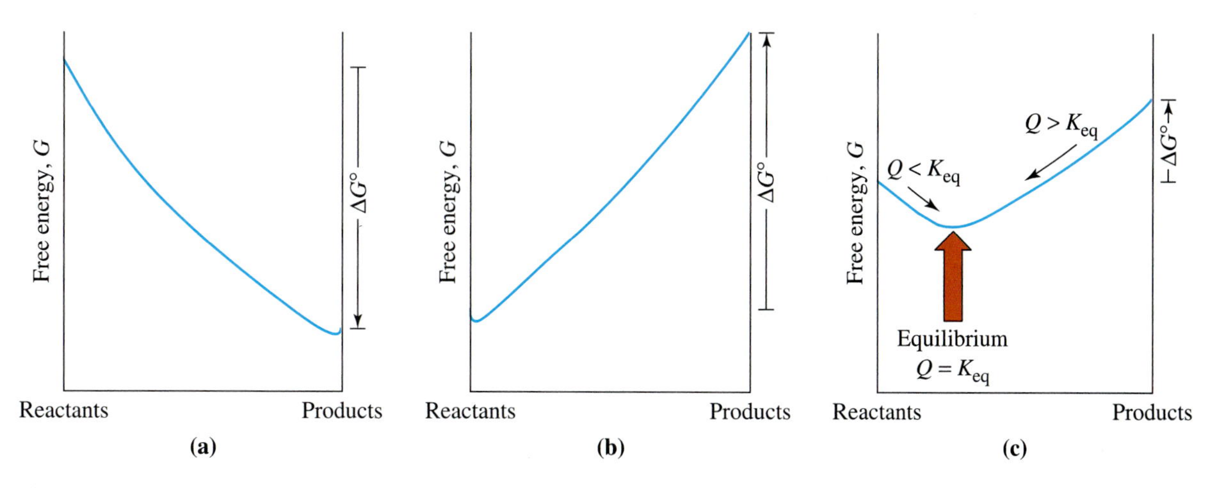

▲ **Figure 17.11 $\Delta G°$ and the direction and extent of spontaneous change**
(a) $\Delta G°$ *is large and negative*: Equilibrium lies far to the right. The reaction goes essentially to completion. (b) $\Delta G°$ *is large and positive*: Equilibrium lies far to the left. Reaction hardly occurs at all. (c) $\Delta G°$ *is neither very large nor very small and either positive or negative*: The equilibrium point lies well within the profile. If $Q < K_{eq}$, a net reaction proceeds in the forward direction; if $Q > K_{eq}$, in the reverse direction.

In case (a), the free energy of the products is much lower than that of the reactants. The difference in free energy between them—the standard free energy change $\Delta G°$—is a *large, negative* quantity and equilibrium is so far to the right that we say the reaction goes to completion. In case (b), the situation is reversed:

The free energy of the products is much higher than that of the reactants, the standard free energy change is a *large, positive* quantity, and equilibrium is so far to the left that we say no reaction occurs. In case (c), the difference in free energies of the reactants and products is small, and equilibrium lies more toward the interior of the reaction profile.

We cannot be highly specific as to what is a "large" value of ΔG° and what is a "small" one. As a general rule, however, if ΔG° is several hundred kilojoules per mole and negative in sign, it is quite likely that a reaction will go essentially to completion. If ΔG° is several hundred kilojoules per mole and positive in sign, it is just as likely that no significant reaction will occur. When the negative or positive value of ΔG° is much closer to zero, we usually have to do an equilibrium calculation.

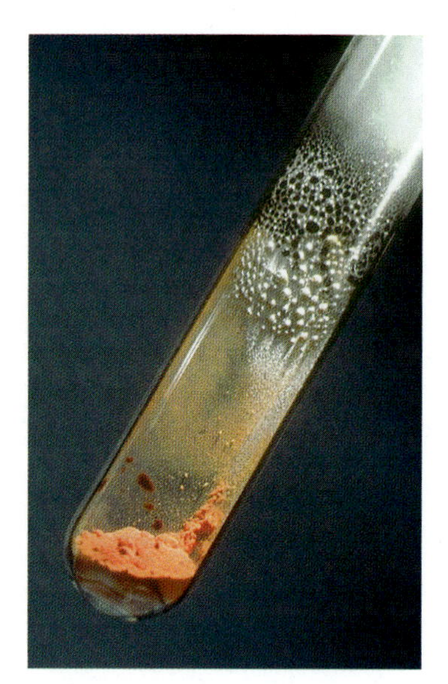

The decomposition of red mercury(II) oxide, HgO, by heating. Note the silvery droplets of liquid mercury.

Coupled Reactions

Consider the decomposition of mercury(II) oxide.

$$HgO(s) \longrightarrow Hg(l) + \tfrac{1}{2}\,O_2(g) \quad \Delta G^\circ_{298} = +58.56\text{ kJ} \qquad \Delta H^\circ_{298} = +90.83\text{ kJ}$$

Because the reaction produces a gas from a solid, entropy increases: $\Delta S^\circ > 0$. When $\Delta H^\circ > 0$ and $\Delta S^\circ > 0$, the forward reaction is favored at higher temperatures (case 3 of Table 17.1). Because $\Delta G^\circ{}_{298}$ is not particularly large, the reaction to produce $O_2(g)$ in its standard state (1 atm) becomes spontaneous at a moderate temperature of about 400 °C. The rather favorable thermodynamics of this reaction made it possible for Joseph Priestley (1733–1804) to isolate oxygen in 1774 just by heating mercury(II) oxide.

Now consider the reaction in which copper(I) oxide decomposes to produce metallic copper and oxygen gas.

$$Cu_2O(s) \longrightarrow 2\,Cu(s) + \tfrac{1}{2}\,O_2(g) \quad \Delta G^\circ_{298} = +149.9\text{ kJ} \qquad \Delta H^\circ_{298} = +170.7\text{ kJ}$$

Here, ΔG° is considerably more positive than in the reaction of mercury(II) oxide. $Cu_2O(s)$ must be heated to a much higher temperature than HgO(s)—to more than 2500 °C—before its decomposition under standard state conditions becomes spontaneous. It is not feasible to produce copper metal simply by heating the oxide.

Suppose, however, that we heat $Cu_2O(s)$ in the presence of carbon. The carbon combines with oxygen that comes from the $Cu_2O(s)$. Here we can think in terms of two reactions and an overall reaction. We obtain the overall equation by combining the equations for the two reactions, and we obtain ΔG for the overall reaction as the sum of two ΔG values. This is the same way we treat ΔH values when applying Hess's law (page 254).

$$\begin{array}{lr} Cu_2O(s) \longrightarrow 2\,Cu(s) + \cancel{\tfrac{1}{2}\,O_2(g)} & \Delta G^\circ_{298} = +149.9\text{ kJ} \\ C(\text{graphite}) + \cancel{\tfrac{1}{2}\,O_2(g)} \longrightarrow CO(g) & \Delta G^\circ_{298} = -137.2\text{ kJ} \\ \hline Cu_2O(s) + C(\text{graphite}) \longrightarrow 2\,Cu(s) + CO(g) & \Delta G^\circ_{298} = +12.7\text{ kJ} \end{array}$$

The overall reaction with reactants and products in their standard states is not spontaneous at 298 K, but it is much closer to being so than is the unaided decomposition of $Cu_2O(s)$. Although the commercial production of copper by this reaction is carried out at much higher temperatures, the overall reaction actually becomes spontaneous for standard state conditions at about 100 °C.

Any combination of two or more simpler reactions is a **coupled reaction**. However, the term is usually used to refer to a combination of a *spontaneous* reaction and a *nonspontaneous* reaction to produce an overall reaction that is *spontaneous*.

17.7 The Dependence of $\Delta G°$ and K_{eq} on Temperature

We have now established criteria for spontaneous change, and we have examined a number of qualitative and quantitative assessments that we can make with just these two equations.

$$\Delta G = \Delta H - T\Delta S$$

$$\Delta G° = -RT \ln K_{eq}$$

There appears to be one serious limitation, however. All our quantitative calculations have been at the single temperature of 25 °C because tabulated thermodynamic data are given mostly for that temperature. We know that for practical purposes, however, we must carry out reactions at a variety of temperatures.

To obtain equilibrium constants at different temperatures, we will assume that ΔH and ΔS do not change much with temperature. (Notice that we made this assumption in Figure 17.8.) In particular, we will assume that values of $\Delta H°$ and $\Delta S°$ at 25 °C will apply at other temperatures as well. This assumption generally works because standard enthalpies of formation and standard molar entropies of the products and reactants all change in roughly the same way with changes in temperature.

To obtain a value of $\Delta G°$, we can substitute the 25 °C values of $\Delta H°$ and $\Delta S°$ and the desired temperature into the expression

$$\Delta G° = \Delta H° - T\Delta S°$$

Then, to obtain K_{eq} at the desired temperature, we can use the equation

$$\Delta G° = -RT \ln K_{eq}$$

Because we most often wish to calculate K_{eq}, another approach is to *combine* the two equations into a single equation relating K_{eq} and temperature.

$$\Delta H° - T\Delta S° = \Delta G° = -RT \ln K_{eq}$$

We can then obtain the following equation.

$$\ln K_{eq} = \frac{-\Delta H°}{RT} + \frac{\Delta S°}{R}$$

If we assume that $\Delta H°$ and $\Delta S°$ remain essentially constant over a range of temperatures, we can replace the term $\Delta S°/R$ by a constant and then write the following equation.

$$\ln K_{eq} = \frac{-\Delta H°}{RT} + \text{constant}$$

This is the equation for a straight line if we plot $\ln K_{eq}$ against $1/T$. The slope of the line is $-\Delta H°/R$, and the intercept is $\Delta S°/R$. The best test of the assumption that $\Delta H°$ and $\Delta S°$ are independent of temperature is to plot some equilibrium data and see if they do yield a straight line. This is illustrated in Figure 17.12.

By the method shown in Appendix A, we can replace the above equation with the one that follows.

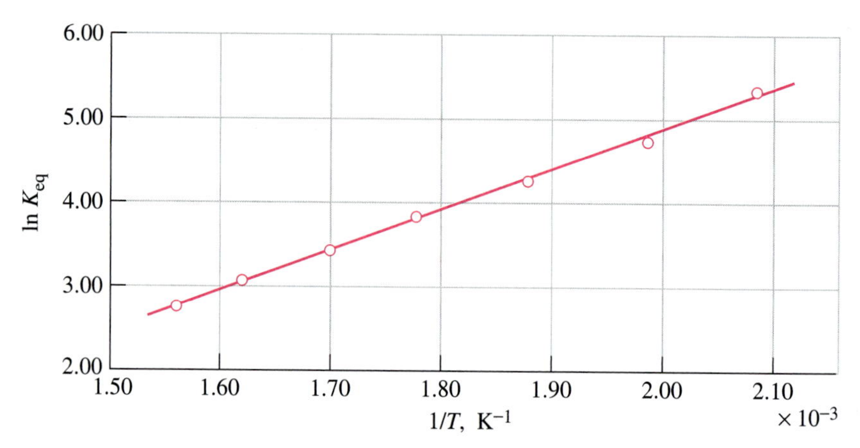

▲ **Figure 17.12 Temperature dependence of K_{eq} for the reaction, $CO(g) + H_2O(g) \rightleftharpoons CO_2(g) + H_2(g)$**
The value of $\Delta H°$ obtained from the slope of this graph is about −40 kJ, a value reasonably close to that of −41.2 kJ for $\Delta H°_{298}$, obtained from data in Appendix C. Some of the data used in the graph are shown here.

T, K	$1/T$	K_{eq}	ln K_{eq}	T, K	$1/T$	K_{eq}	ln K_{eq}
478	2.09×10^{-3}	210	5.35	588	1.70×10^{-3}	31	3.43
533	1.88×10^{-3}	73	4.29	643	1.56×10^{-3}	16	2.77

$$\ln \frac{K_2}{K_1} = \frac{\Delta H°}{R}\left(\frac{1}{T_1} - \frac{1}{T_2}\right)$$

This equation, called the *van't Hoff equation*, relates the equilibrium constant, K_{eq}, at two temperatures. The value of the equilibrium constant at temperature T_1 is K_1, and the value at temperature T_2 is K_2. $\Delta H°$ is the standard enthalpy change in the reaction, and R is the gas constant.

If the equilibrium is between a liquid or solid and its vapor, the equilibrium constant is simply the pressure of the vapor. Consequently, we can substitute P (vapor pressure) for K (equilibrium constant) and $\Delta H°_{vapn}$ or $\Delta H°_{subl}$ for $\Delta H°$. The equation is then called the *Clausius-Clapeyron equation*.

PROBLEM-SOLVING NOTE
For problem solving, some prefer one of the following forms of the van't Hoff equation.

$$\ln \frac{K_2}{K_1} = \frac{-\Delta H°}{R}\left(\frac{1}{T_2} - \frac{1}{T_1}\right)$$

$$\ln \frac{K_2}{K_1} = \frac{\Delta H°}{R}\left(\frac{T_2 - T_1}{T_1 \cdot T_2}\right)$$

Vaporization

$$\ln \frac{P_2}{P_1} = \frac{\Delta H°_{vapn}}{R}\left(\frac{1}{T_1} - \frac{1}{T_2}\right)$$

Sublimation

$$\ln \frac{P_2}{P_1} = \frac{\Delta H°_{subl}}{R}\left(\frac{1}{T_1} - \frac{1}{T_2}\right)$$

EXAMPLE 17.10

Consider this reaction at 298 K.

$$CO(g) + H_2O(g) \rightleftharpoons CO_2(g) + H_2(g) \qquad \Delta H°_{298} = -41.2 \text{ kJ}$$

Determine K_{eq} for the reaction at 725 K.

SOLUTION

This is the reaction for which K_{eq} is plotted as a function of $1/T$ in Figure 17.12. However, we can't simply use the graph to obtain ln K_{eq} when $1/T = 1/725$ K. This point falls outside the range of the data plotted. However, we can select one of the K_{eq} values listed in the caption of Figure 17.12. Suppose, for example, that we use these data in the van't Hoff equation.

$$K_2 = ? \quad T_2 = 725\text{ K} \quad \Delta H° = \Delta H°_{298} = -41.2\text{ kJ/mol}$$

$$K_1 = 16 \quad T_1 = 643\text{ K}$$

$$\ln\frac{K_2}{K_1} = \frac{\Delta H°}{R}\left(\frac{1}{T_1} - \frac{1}{T_2}\right) = \frac{(-41.2 \times 1000)\ \cancel{\text{J mol}^{-1}}}{8.3145\ \cancel{\text{J mol}^{-1}}\ \text{K}^{-1}}\left(\frac{1}{643\text{ K}} - \frac{1}{725\text{ K}}\right)$$

$$\ln\frac{K_2}{K_1} = -4.96 \times 10^3\ \cancel{\text{K}} \times \left(1.56 \times 10^{-3} - 1.38 \times 10^{-3}\right)\ \cancel{\text{K}^{-1}}$$

$$\ln\frac{K_2}{K_1} = -0.89$$

$$\frac{K_2}{K_1} = e^{-0.89} = 0.41$$

$$K_2 = 0.41 \times K_1 = 0.41 \times 16 = 6.6$$

PROBLEM-SOLVING NOTE

To avoid an inadvertent loss of significant figures, it is best to store the value of $(1/T_1 - 1/T_2)$ in your calculator rather than to write intermediate results, such as the 1.56×10^{-3} and 1.38×10^{-3} shown here.

EXERCISE 17.10A

In Example 17.9, we determined the value of K_{eq} at 25 °C for the reaction: $2\ NO_2(g) \rightleftharpoons N_2O_4(g)$. What is the value of K_{eq} for this reaction at 65 °C? (*Hint:* You will have to use tabulated data to establish the value of $\Delta H°$.)

EXERCISE 17.10B

At 25 °C the vapor pressure of water is 23.8 mmHg and the enthalpy of vaporization is 44.0 kJ/mol. Use the Clausius-Clapeyron equation to calculate the vapor pressure of water at 40.0 °C. Compare your result with data from Table 11.2.

EXAMPLE 17.11—A Conceptual Example

Estimate the value of $\Delta S°_{298}$ for the dissociation of copper(II) oxide,

$$4\ CuO(s) \rightleftharpoons 2\ Cu_2O(s) + O_2(g) \qquad \Delta H°_{298} = 283\text{ kJ}$$

given that the equilibrium pressure of $O_2(g)$ is 1.00 atm at 1395 K.

SOLUTION

Although we should not expect the enthalpy and entropy changes to remain constant over the broad temperature interval in this problem, for *estimation* purposes we can assume that their values do not change significantly. Thus, we assume that $\Delta S°_{298}$ is about the same as $\Delta S°_{1395}$. Also, $\Delta H°_{1395} \approx \Delta H°_{298}$. We can calculate $\Delta S°_{1395}$ from the equation

$$\Delta G°_{1395} = \Delta H°_{1395} - T\Delta S°_{1395}$$

$$\Delta S°_{298} \approx \Delta S°_{1395} = \frac{\Delta H°_{1395} - \Delta G°_{1395}}{T} \approx \frac{\Delta H°_{298} - \Delta G°_{1395}}{T}$$

$$\approx \frac{283\text{ kJ} - \Delta G°_{1395}}{1395\text{ K}}$$

Now all we need is the value of $\Delta G°_{1395}$, and this we can get from the following equation.

$$\Delta G°_{1395} = -RT \ln K_{eq}$$

Finally, then, it all comes down to the question, "What is K_{eq} at 1395 K?"

Because the activities of the solids are equal to 1, for this reaction, $K_{eq} = P_{O_2}$. At 1395 K, the equilibrium pressure of O_2 is 1 atm. Thus, at 1395 K,

$$K_{eq} = 1.00 \text{ and } \Delta G^\circ_{1395} = -RT \ln 1.00 = 0$$

The value we seek is

$$\Delta S^\circ_{298} \approx \frac{283 \text{ kJ} - 0}{1395\ K} = 0.203 \text{ kJ K}^{-1}$$

EXERCISE 17.11

Evaluate ΔS°_{298} for the reaction

$$4\,CuO(s) \rightleftharpoons 2\,Cu_2O(s) + O_2(g) \qquad \Delta H^\circ_{298} = 283 \text{ kJ}$$

in the following way. First, determine K_{eq} at 298 K by using the van't Hoff equation. Then, use the following equations.

$$\Delta G^\circ_{298} = -RT \ln K_{eq} \text{ and } \Delta G^\circ_{298} = \Delta H^\circ_{298} - T\,\Delta S^\circ_{298}$$

Show that the result is about the same as obtained in Example 17.11, and explain why you would expect it to be.

Key Terms

coupled reaction (17.6)
entropy, S (17.3)
entropy change, ΔS (17.3)
equilibrium constant, K_{eq} (17.6)
free energy change, ΔG (17.4)
Gibbs free energy, G (17.4)
nonspontaneous process (17.2)
second law of thermodynamics (17.3)
spontaneous process (17.2)
standard free energy change, ΔG° (17.5)
standard free energy of formation, ΔG°_f (17.5)
standard molar entropy, S° (17.3)
thermodynamics (page 731)
third law of thermodynamics (17.3)
Trouton's rule (17.6)

Summary

A key objective of this chapter is to establish criteria for spontaneous change. A spontaneous change in a system occurs by itself without outside intervention. To develop criteria for spontaneous change, it is necessary to use the concept of entropy, a measure of the degree of randomness or disorder in a system.

The direction of spontaneous change is that in which entropy (disorder) increases. The entropy change that must be assessed, however, is the *sum* of the entropy change of a system *and* of its surroundings. By the second law of thermodynamics, this so-called entropy of the universe, S_{univ}, must always increase for a spontaneous process, that is, $\Delta S_{univ} > 0$.

The third law of thermodynamics states that the entropy of a pure, perfect crystal at 0 K can be taken to be zero. This serves as a starting point for the experimental determination of standard molar entropies, which can be used to calculate entropy changes in chemical reactions.

The free energy change, ΔG, is equal to $-T\Delta S_{univ}$, and it applies *just to a system itself*, without regard for the surroundings. It is defined by the Gibbs equation.

$$\Delta G = \Delta H - T\Delta S$$

For a spontaneous process at constant temperature and pressure, ΔG must be negative; that is $\Delta G < 0$. In many cases, the sign of ΔG can be predicted just by knowing the signs of ΔH and ΔS.

The standard free energy change, ΔG°, can be calculated (1) by substituting standard enthalpies and entropies of reaction and a Kelvin temperature into the Gibbs equation, or (2) by combining standard free energies of formation through the expression

$$\Delta G^\circ = \left[\Sigma \nu_p\, \Delta G^\circ_f(\text{products})\right] - \left[\Sigma \nu_r\, \Delta G^\circ_f(\text{reactants})\right]$$

The condition of equilibrium is one for which $\Delta G = 0$. The *standard* free energy change is a particularly useful property for describing equilibrium because of its relationship to the equilibrium constant, K_{eq}.

$$\Delta G^\circ = -RT \ln K_{eq}$$

Thermodynamics and Living Organisms

In living organisms, the energy stored in food—potential energy—is changed into the kinetic energy associated with breathing, jogging, and twiddling thumbs. How do these energy changes come about? Do they obey the laws of thermodynamics?

A reaction with a decrease in free energy is *exergonic* ($\Delta G < 0$), and one with a free energy increase is *endergonic* ($\Delta G > 0$). Living organisms carry out exergonic reactions to maintain life processes, but there must also be free energy-storing (endergonic) processes to provide for the exergonic reactions.

In living organisms, complex molecules such as proteins and DNA are produced from simpler ones such as O_2, CO_2, and H_2O. These processes usually involve a *decrease* in entropy ($\Delta S < 0$), an *increase* in enthalpy ($\Delta H > 0$), and subsequently an *increase* in free energy ($\Delta G > 0$).

$$\Delta G = \Delta H - T\Delta S$$

We expect them to be *nonspontaneous*. How can nonspontaneous reactions essential to life proceed in directions that are not natural?

Consider this: If you drop a book, it falls to the floor—a spontaneous process. The book won't leap back into your hands—a nonspontaneous process. Suppose, however, the book on the floor is tied by a rope to another, larger book that you hold. Now, you can grasp the rope and let the larger book fall. As it falls, the smaller book rises. What you have is a primitive pulley (see Figure 17.13). The spontaneous process (fall of the larger book) drives the nonspontaneous process (rise of the smaller book). This is a *coupled* process, analogous to coupled chemical reactions (page 754).

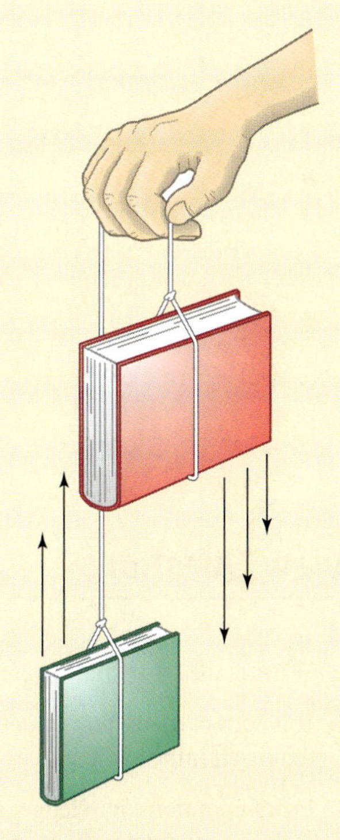

▲ **Figure 17.13**
An analogy to coupled reactions
The small book will not rise on its own; that is a nonspontaneous process. When coupled by being attached by a rope to a larger book that is allowed to fall (a spontaneous process), the small book rises. The coupled process is spontaneous.

The reaction by which plants combine the simple sugars glucose and fructose into sucrose (common table sugar) has a positive ΔG; it is endergonic. The reaction must be coupled to one that is exergonic, such as the hydrolysis of ATP (adenosine triphosphate) to ADP (adenosine diphosphate). The sum of the two reactions has $\Delta G < 0$.

$$\begin{array}{lr} \text{Glucose} + \text{fructose} \longrightarrow \text{sucrose} + \cancel{H_2O} & \Delta G = +29.3\text{ kJ} \\ \text{ATP} + \cancel{H_2O} \longrightarrow \text{ADP} + HPO_4^{2-} & \Delta G = -30.5\text{ kJ} \\ \hline \text{Glucose} + \text{fructose} + \text{ATP} \longrightarrow \text{sucrose} + \text{ADP} + HPO_4^{2-} & \Delta G = -1.2\text{ kJ} \end{array}$$

The reactions are coupled through the *intermediate* glucose 1-phosphate. First, ATP reacts with glucose to form this intermediate. Some of the energy stored in the bonds of ATP is transferred to bonds in the intermediate.

$$\text{(a)} \qquad \text{Glucose} + \text{ATP} \longrightarrow \text{glucose 1-phosphate} + \text{ADP}$$

When the intermediate reacts with fructose, the energy is transferred to the sucrose molecule.

$$\text{(b)} \qquad \text{Glucose 1-phosphate} + \text{fructose} \longrightarrow \text{sucrose} + HPO_4^{2-}$$

Glucose 1-phosphate, an intermediate, drops out of the equation for the overall reaction $\left[(a) + (b)\right]$.

$$\text{Glucose} + \text{fructose} + \text{ATP} \longrightarrow \text{sucrose} + \text{ADP} + HPO_4^{2-} \qquad \Delta G = -1.2\text{ kJ}$$

One of the requirements of this equation is that the K_{eq} expression be based on *dimensionless* quantities called activities. However, in place of activities, we can generally substitute the *numerical* values of molarities for solutes and partial pressures in atmospheres for gases.

The value of $\Delta G°$ is in itself often sufficient to determine whether a reaction is likely to (a) go to completion ($\Delta G°$ very large and negative), (b) occur hardly at all ($\Delta G°$ very large and positive), or (c) reach an equilibrium condition requiring an equilibrium calculation ($\Delta G°$ not very large and either positive or negative). Sometimes a nonspontaneous reaction can be made to occur by combining it with a spontaneous reaction. That is, two reactions can be *coupled* to produce a spontaneous overall reaction.

Tabulated values of $\Delta G_f°$, $\Delta H_f°$, and $S°$ are generally for 25 °C. To use these data to obtain values of K_{eq} at other temperatures, we generally assume that $\Delta H°$ and $\Delta S°$ are independent of temperature. With this assumption, the *van't Hoff equation* can be derived to relate the equilibrium constant and temperature.

$$\ln\frac{K_2}{K_1} = \frac{\Delta H°}{R}\left(\frac{1}{T_1} - \frac{1}{T_2}\right)$$

Review Questions

1. What is meant by the phrase "a spontaneous process"?
2. Is it correct to say that a nonspontaneous process in a system is an impossible process? Explain.
3. From common knowledge, predict which of the following are spontaneous changes: **(a)** the souring of milk; **(b)** obtaining copper metal from copper ore; **(c)** the rusting of a steel can in moist air. Explain.
4. Based on the relationship between entropy and disorder, indicate whether each of the following changes represents an increase or decrease in entropy of the system.
   (a) the freezing of acetic acid
   (b) the sublimation of the moth repellent, *para*-dichlorobenzene ($C_6H_4Cl_2$)
   (c) the burning of gasoline
5. Why is the notion of a reversible process so important to the concept of entropy?
6. Why is the entropy change in a system not always a reliable predictor of whether the process producing the change is spontaneous?
7. What does the second law of thermodynamics tell us about the concept of entropy?
8. What is the basic idea underlying the third law of thermodynamics?
9. Why do we tabulate standard entropies of substances, $S°$, rather than their entropies of formation, $\Delta S_f°$, as we do for values of $\Delta H_f°$ and $\Delta G_f°$?
10. Which would you expect to have the higher absolute molar entropy at 25 °C, NOF_3(g) or NO_2F(g)? Explain.
11. What are the two tendencies that act as driving forces in establishing the direction of spontaneous change?
12. Under what conditions of temperature—low or high—would you expect a reaction to proceed furthest in the forward direction if the reaction has $\Delta H < 0$ and $\Delta S < 0$?
13. Would you expect the following reaction to occur spontaneously at room temperature? Explain.

$$H_2O(l, 1\ atm) \longrightarrow H_2O(g, 1\ atm)$$

14. What is the special significance of a reaction for which **(a)** $\Delta G = 0$; **(b)** $\Delta G° = 0$?
15. For the mixing of ideal, inert gases, would you expect each of the following quantities to be positive, negative, or zero: ΔH, ΔS, ΔG? Explain.
16. What is Trouton's rule? How is it used? Why does it not work for all liquids?
17. What do we mean by the term *coupled reactions*? How are coupled reactions used?
18. How does the equilibrium constant K_{eq} resemble, and how does it differ from, K_c and K_p?
19. How does thermodynamics make it possible to *calculate* the value of an equilibrium constant, K_{eq}? What kind of data are needed for this calculation?
20. If a reaction has $\Delta G° = 0$, what is K_{eq} for the reaction? Explain.

Problems

Disorder, Entropy, and Spontaneous Change

21. For each of the following reactions, indicate whether you would expect the entropy of the system to increase or decrease. If you cannot tell just by inspecting the equation, explain why.

(a) $CH_3OH(l) \longrightarrow CH_3OH(g)$

(b) $N_2O_4(g) \longrightarrow 2\,NO_2(g)$

(c) $CO(g) + H_2O(g) \longrightarrow CO_2(g) + H_2(g)$

(d) $2\,KClO_3(s) \longrightarrow 2\,KCl(s) + 3\,O_2(g)$

22. For each of the following reactions, indicate whether you would expect the entropy of the system to increase or decrease. If you cannot tell just by inspecting the equation, explain why.

(a) $CH_3COOH(l) \longrightarrow CH_3COOH(s)$

(b) $N_2(g) + O_2(g) \longrightarrow 2\,NO(g)$

(c) $N_2H_4(l) \longrightarrow N_2(g) + 2\,H_2(g)$

(d) $2\,NH_3(g) + H_2SO_4(aq) \longrightarrow (NH_4)_2SO_4(aq)$

23. Compare the following processes, and decide which should have the *larger* value of ΔS. Explain your choice.

(a) $H_2O(s, 1\text{ atm}) \longrightarrow H_2O(l, 1\text{ atm})$

(b) $H_2O(s, 1\text{ atm}) \longrightarrow H_2O(g, 4.58\text{ mmHg})$

24. Compare the following processes, and decide which should have the *larger* value of ΔS. Explain your choice.

(a) $C(\text{diamond}, 1\text{ atm}) \longrightarrow C(\text{graphite}, 1\text{ atm})$

(b) $CO_2(s, 1\text{ atm}) \longrightarrow CO_2(g, 1\text{ atm})$

25. In the manner used to describe Figure 17.3, describe the situation if the two liquids are water and octane (C_8H_{18}, a component of gasoline). That is, what would you expect to find for the final condition of the mixture? Explain.

26. In the manner suggested by Figure 17.4, make sketches to represent entropy changes occurring in **(a)** the melting of a solid, and **(b)** the condensation of a vapor.

27. Three possible completions of a statement are given. Tell what is wrong with each one, and then complete the statement correctly. *For a process to occur spontaneously*,

(a) The entropy of the system must increase.

(b) The entropy of the surroundings must increase.

(c) Both the entropy of the system and of the surroundings must increase.

28. Three possible completions of a statement are given. Tell what is wrong with each one, and then complete the statement correctly. *The free energy change of a reaction indicates*

(a) whether the reaction is endothermic or exothermic

(b) whether the reaction is accompanied by an increase or decrease in molecular disorder

(c) whether equilibrium in the reaction is favored at high or low temperatures

29. For the following reaction

$$2\,S(s, \text{rhombic}) + Cl_2(g) \longrightarrow S_2Cl_2(g)$$

Comment on the difficulty in deciding whether $\Delta S°$ is positive or negative when using the generalizations on pages 736–37. Describe how the additional generalization on page 740 can help you decide. Calculate the actual value of $\Delta S°$ using data from Appendix C.

30. The values of $\Delta S°_{298}$ obtained in Example 17.11 and in Exercise 17.11 are only estimates. What data can you use to obtain a more exact value? Do so, and compare the results.

31. Based on a consideration of entropy changes, why is it so difficult to eliminate environmental pollution?

32. An environmental problem posed by the use of aluminum cans is that their disintegration under environmental conditions appears to take an almost endless period of time. Can we say that the environmental disintegration of aluminum is a nonspontaneous process? Explain.

Free Energy and Spontaneous Change

33. Explain why free energy change is easier to use as a criterion for spontaneous change in a system than is entropy change alone.

34. Explain why in a reversible reaction, both the forward and the reverse reactions can be spontaneous, but not under the same conditions.

35. Make a sketch similar to Figure 17.8 that corresponds to Case 2 of Table 17.1. Explain the significance of the plot in relation to the prediction in Table 17.1.

36. Make a sketch similar to Figure 17.8 that corresponds to Case 1 of Table 17.1. Explain the significance of the plot in relation to the prediction in Table 17.1.

37. Give an example of a phase change that is nonspontaneous at low temperatures and spontaneous at higher temperatures. What is the equilibrium temperature?

38. Give an example of a phase change that is spontaneous at low temperatures and nonspontaneous at higher temperatures. What is the equilibrium temperature?

39. Would you expect each of the following reactions to be spontaneous at low temperatures, high temperatures, all temperatures, or not at all? Explain.

(a) $PCl_3(g) + Cl_2(g) \longrightarrow PCl_5(g)$
$\Delta H° = -87.9$ kJ

(b) $2\ NH_3(g) \longrightarrow N_2(g) + 3\ H_2(g)$
$\Delta H° = +92.2$ kJ

(c) $2\ N_2O(g) \longrightarrow 2\ N_2(g) + O_2(g)$
$\Delta H° = -164.1$ kJ

40. Would you expect each of the following reactions to be spontaneous at low temperatures, high temperatures, all temperatures, or not at all? Explain.

(a) $H_2O(g) + \frac{1}{2}\ O_2(g) \longrightarrow H_2O_2(g)$
$\Delta H° = +105.5$ kJ

(b) $CH_4(g) + O_2(g) \longrightarrow CO_2(g) + 2\ H_2O(g)$
$\Delta H° = -802.3$ kJ

(c) $2\ CO(g) + O_2(g) \longrightarrow 2\ CO_2(g)$
$\Delta H° = -566.0$ kJ

41. The following reaction is nonspontaneous under standard state conditions at room temperature.

$$COCl_2(g) \longrightarrow CO(g) + Cl_2(g)$$

To make it a spontaneous reaction, would you raise or lower the temperature? Explain.

42. Would you expect the following reaction to be spontaneous under standard state conditions at room temperature?

$$3\ O_2(g) \longrightarrow 2\ O_3(g) \qquad \Delta H° = +285.4 \text{ kJ}$$

What effect will changing the temperature have on the spontaneity of this reaction? Explain.

Standard Free Energy Change

43. Use data from Appendix C to determine $\Delta G°$ values for the following reactions at 25 °C.

(a) $C_2H_4(g) + H_2(g) \longrightarrow C_2H_6(g)$

(b) $SO_3(g) + CaO(s) \longrightarrow CaSO_4(s)$

44. Use data from Appendix C to determine $\Delta G°$ values for the following reactions at 25 °C.

(a) $FeO(s) + H_2(g) \longrightarrow Fe(s) + H_2O(g)$

(b) $CdO(s) + 2\ HCl(g) \longrightarrow CdCl_2(s) + H_2O(l)$

45. Use data from Appendix C to determine $\Delta H°$ and $\Delta S°$, at 298 K, for the following reaction. Then determine $\Delta G°$ in two ways, and compare the results.

$$C(\text{graphite}) + H_2O(g) \longrightarrow CO(g) + H_2(g)$$

46. Use data from Appendix C to determine $\Delta H°$ and $\Delta S°$, at 298 K, for the following reaction. Then determine $\Delta G°$ in two ways, and compare the results.

$$CS_2(l) + 3\ O_2(g) \longrightarrow CO_2(g) + 2\ SO_2(g)$$

47. Why is the standard free energy change, $\Delta G°$, so important in dealing with the question of spontaneous change, even though the conditions in a chemical reaction are usually *nonstandard*?

48. Under what conditions will ΔG for a reaction be negative even though $\Delta G°$ is positive?

Free Energy Change and Equilibrium

49. Estimate the normal boiling point of heptane, C_7H_{16}, given that at this temperature $\Delta H°_{vapn} = 31.69$ kJ/mol.

50. A reference book lists the following values for chloroform (trichloromethane) at 298 K: $\Delta H_f°[CHCl_3(l)] = -132.3$ kJ/mol, $\Delta H_f°[CHCl_3(g)] = -102.9$ kJ/mol. Estimate the normal boiling point of chloroform.

51. The normal boiling point of $Br_2(l)$ is 59.47 °C. Estimate $\Delta H°_{vapn}$ of bromine. Compare your result with a value based on data from Appendix C.

52. The normal boiling point of sulfuryl chloride, $SO_2Cl_2(l)$, is 69.3 °C. Estimate $\Delta H°_{vapn}$ of sulfuryl chloride. Compare your result with a value based on data from Appendix C.

53. At 765 K, for the reaction, $H_2(g) + I_2(g) \rightleftharpoons 2\ HI(g)$, $K_p = 46.0$.

(a) What is ΔG for a mixture at equilibrium at 765 K?

(b) What is $\Delta G°$ for this reaction at 765 K?

54. At 850 K, for the reaction, $2\ SO_2(g) + O_2(g) \rightleftharpoons 2\ SO_3(g)$, $\Delta G° = -36.3$ kJ

(a) What is ΔG for a mixture at equilibrium at 850 K?

(b) What is K_p for this reaction at 850 K?

55. Write K_{eq} expressions for the following reactions. Which, if any, of these expressions correspond to equilibrium constants that we have previously denoted as K_c, K_p, K_a, and so on?

(a) $2\ NO(g) + O_2(g) \rightleftharpoons 2\ NO_2(g)$
(b) $MgSO_3(s) \rightleftharpoons MgO(s) + SO_2(g)$
(c) $HCN(aq) + H_2O(l) \rightleftharpoons H_3O^+(aq) + CN^-(aq)$

56. Write K_{eq} expressions for the following reactions. Which, if any, of these expressions correspond to equilibrium constants that we have previously denoted as K_c, K_p, K_a, and so on?

(a) $2\ NaHSO_3(s) \rightleftharpoons Na_2SO_3(s) + H_2O(g) + SO_2(g)$
(b) $Mg(OH)_2(s) \rightleftharpoons Mg^{2+}(aq) + 2\ OH^-(aq)$
(c) $CH_3COO^-(aq) + H_2O(l) \rightleftharpoons CH_3COOH(aq) + OH^-(aq)$

57. Use data from Appendix C to determine K_p at 298 K for these reactions.

(a) $2\ N_2O(g) + O_2(g) \rightleftharpoons 4\ NO(g)$
(b) $2\ NH_3(g) + 2\ O_2(g) \rightleftharpoons N_2O(g) + 3\ H_2O(g)$

58. Use data from Appendix C to determine K_p at 298 K for these reactions.

(a) $2\ SO_2(g) + O_2(g) \rightleftharpoons 2\ SO_3(g)$
(b) $CH_4(g) + 2\ H_2O(g) \rightleftharpoons CO_2(g) + 4\ H_2(g)$

59. The following are data for the vaporization of toluene at 298 K: $\Delta S° = 99.7$ J mol^{-1} K^{-1}; $\Delta H° = 38.0$ kJ mol^{-1}. Calculate the equilibrium vapor pressure of toluene at 298 K.

$$C_6H_5CH_3(l) \rightleftharpoons C_6H_5CH_3(g)$$

60. The following data are given for the sublimation of naphthalene at 298 K: $\Delta S° = 168.7$ J mol^{-1} K^{-1}; $\Delta H° = 73.6$ kJ mol^{-1}. Calculate the pressure of naphthalene vapor in equilibrium with solid naphthalene at 298 K.

$$C_{10}H_8(s) \rightleftharpoons C_{10}H_8(g)$$

61. In an *equilibrium* mixture in the following reaction at 345 °C,

$$CO(g) + H_2O(g) \rightleftharpoons CO_2(g) + H_2(g)$$

the mole fractions of the gases were found to be: $\chi_{CO_2} = \chi_{H_2} = 0.320$, $\chi_{CO} = 0.0133$, and $\chi_{H_2O} = 0.347$.

(a) What is $\Delta G°$ for this reaction at 345 °C?
(b) In what direction will a net reaction occur if one brings together 0.085 mol CO, 0.112 mol H_2O, 0.145 mol CO_2 and 0.226 mol H_2 and allows them to come to equilibrium?
(c) What is the composition of the equilibrium mixture obtained by the reaction in (b)?

62. At 440 °C, for the reaction $CO_2(g) + 4\ H_2(g) \rightleftharpoons CH_4(g) + 2\ H_2O(g)$, $\Delta G° = -32.3$ kJ.

(a) What is K_p for this reaction at 440 °C?
(b) In what direction will a net reaction occur to establish equilibrium in a mixture in which initially $P_{CO_2} = 0.25$ atm, $P_{H_2} = 0.65$ atm, $P_{CH_4} = 1.12$ atm, and $P_{H_2O} = 2.52$ atm?
(c) What are the partial pressures of the gases at equilibrium in (b)?

63. Use data from Appendix C to determine which of the following can be coupled with the reaction

$$CoO(s) \longrightarrow Co(s) + \tfrac{1}{2}\ O_2(g) \qquad \Delta G° = 237.9\ \text{kJ}$$

to make the reduction of CoO(s) to Co(s) spontaneous for standard state conditions at 298 K: (a) oxidation of C(graphite) to CO(g); (b) oxidation of $H_2(g)$ to $H_2O(g)$; (c) oxidation of CO(g) to $CO_2(g)$? Explain.

64. Use data from Appendix C to determine which of the following reactions,

(a) $FeO(s) \longrightarrow Fe(s) + \tfrac{1}{2}\ O_2(g)$
(b) $CaO(s) \longrightarrow Ca(s) + \tfrac{1}{2}\ O_2(g)$
(c) $MnO_2(s) \longrightarrow Mn(s) + O_2(g)$

can be coupled with the oxidation of Al(s) to $Al_2O_3(s)$ to give an overall reaction that is spontaneous for standard state conditions at 298 K? Explain.

$\Delta G°$ and K_{eq} as Functions of Temperature

65. The following reaction is carried out on an industrial scale for the production of thionyl chloride, a chemical used in the manufacture of pesticides.

$$SO_3(g) + SCl_2(l) \rightleftharpoons OSCl_2(l) + SO_2(g)$$

The thermodynamic data listed below are for 298 K. Use these data and data from Appendix C to determine the temperature at which $K_{eq} = 1.0 \times 10^{15}$ for the reaction.

	$\Delta H°_f$, kJ/mol	$S°$, J mol^{-1} K^{-1}
$SCl_2(l)$	−50.0	184
$OSCl_2(l)$	−245.6	121

66. The thermodynamic data listed below are for 298 K. Use these data and data from Appendix C to determine K_{eq} at 45 °C for the reaction

$$CO_2(g) + SF_4(g) \rightleftharpoons CF_4(g) + SO_2(g)$$

	$\Delta H°_f$, kJ/mol	$S°$, J mol^{-1} K^{-1}
$SF_4(g)$	−763	299.6
$CF_4(g)$	−925	261.6

67. The normal boiling point of 1-butanol, $CH_3(CH_2)_2CH_2OH(l)$, is 117.8 °C.

$$CH_3(CH_2)_2CH_2OH(l, 1\ atm) \rightleftharpoons CH_3(CH_2)_2CH_2OH(g, 1\ atm) \qquad \Delta H° = 43.82\ kJ$$

Calculate the boiling point of 1-butanol when the barometric pressure is 747 mmHg.

68. The normal boiling point of diethyl ether, $CH_3CH_2OCH_2CH_3(l)$, is 34.66 °C.

$$CH_3CH_2OCH_2CH_3(l, 1\ atm) \rightleftharpoons CH_3CH_2OCH_2CH_3(g, 1\ atm) \qquad \Delta H° = 26.70\ kJ$$

Calculate the vapor pressure of diethyl ether at 32.50 °C.

69. Use data from Appendix C to calculate K_p at 155 °C for the following reaction.

$$2\ NO(g) + O_2(g) \rightleftharpoons 2\ NO_2(g)$$

70. Use data from Appendix C to calculate K_{eq} at 375 °C for the reversible reaction in which hydrogen sulfide and carbon dioxide gases produce gaseous water and carbonyl sulfide, COS(g).

71. With data from Appendix C, estimate the normal boiling point of carbon tetrachloride.

72. With data from Appendix C, estimate the temperature at which the sublimation pressure of $P_4(g)$ above solid red phosphorus is 1 atm.

73. With data from Appendix C, determine the temperature to which $Ag_2O(s)$ must be heated to produce $O_2(g)$ at an equilibrium partial pressure of 0.21 atm, the same as found in the atmosphere.

$$2\ Ag_2O(s) \rightleftharpoons 4\ Ag(s) + O_2(g)$$

74. Equilibrium is established in the following reaction at 60.0 °C.

$$2\ BrCl(g) \rightleftharpoons Br_2(g) + Cl_2(g)$$

Starting with an initial pressure of BrCl(g) of 1.00 atm, what is the partial pressure of BrCl(g) when equilibrium is reached?

Additional Problems

75. If you drop fifty pennies on the floor, explain why it is unlikely that all the pennies will have "heads" up. What is more likely to happen?

76. The reversible expansion of a gas is suggested in Figure 17.5. Is the expansion of a gas pictured in Figure 6.8 (page 232) reversible? Explain.

77. On several occasions, we have made the assumption that $\Delta H°$ and $\Delta S°$ change very little with temperature. Why can't we make the same assumption about $\Delta G°$?

78. Which of the following substances would you expect to follow Trouton's rule most closely: **(a)** $CH_3(CH_2)_6CH_2OH$, **(b)** CH_3CH_2OH, **(c)** $CH_3(CH_2)_6CH_3$? Which would you expect to deviate most from the rule? Explain.

79. Explain the meaning of the celebrated remark attributed to Rudolf Clausius in 1865. "Die Energie der Welt ist konstant; die Entropie der Welt strebt einem Maximum zu." ("The energy of the world is constant; the entropy of the world increases toward a maximum.")

80. Show that the effect of temperature on equilibrium that we predicted qualitatively in Chapter 14 is consistent with the effect of temperature on K_{eq} discussed in Section 17.7.

81. The free energy of solution formation is related to concentration by the expression: $\Delta G = \Delta G_f^\circ + RT \ln c$, where c is the solute molarity. The concentration of creatine, an amino acid derivative formed during muscle contraction, is 2.0 mg/100 mL in blood and 75 mg/100 mL in urine. Calculate the free energy change for the transfer of creatine from blood to urine at 37 °C. (*Hint:* It is not necessary to convert concentrations from mg/100 mL to molarity. The proportionality constant between the two cancels out in the calculation.)

82. For the decomposition of $NaHCO_3(s)$, estimate the temperature at which the total pressure of the gases above the solids is 1 atm.

$$2\ NaHCO_3(s) \rightleftharpoons Na_2CO_3(s) + H_2O(g) + CO_2(g)$$

83. The vapor pressure of hydrazine, N_2H_4, at 35.0 °C is 25.7 mmHg; at 50.0 °C, it is 57.0 mmHg. Estimate a value of the standard free energy change at 25 °C for the vaporization, and determine a more exact value with data from Appendix C. Compare the results.

$$N_2H_4(l) \longrightarrow N_2H_4(g)$$

84. When ethane is passed over a catalyst at 900 K and 1 atm, it is partly decomposed to ethylene and hydrogen.

$$CH_3CH_3(g) \rightleftharpoons CH_2{=}CH_2(g) + H_2(g)$$

Calculate the mole percent $H_2(g)$ at equilibrium. (*Hint:* Use data from Appendix C.)

85. Following are values of K_p at different temperatures for the reaction, $2\ SO_2(g) + O_2(g) \rightleftharpoons 2\ SO_3(g)$. At 800 K, $K_p = 9.1 \times 10^2$; at 900 K, $K_p = 4.2 \times 10^1$; at 1000 K, $K_p = 3.2$; at 1100 K, $K_p = 0.39$; and at 1170 K,

$K_p = 0.12$. Plot $\ln K_p$ versus $1/T$, and determine $\Delta H°$ for this reaction.

86. An ultrahigh vacuum system is capable of reducing the pressure in a system to about 10^{-9} mmHg. Do you think it is possible to achieve such a high vacuum at 25 °C if a sample of $NH_4Cl(s)$ is present in the system (see illustration). If not, can the ultralow pressure be achieved by changing the temperature? Explain. [*Hint:* Consider the dissociation of $NH_4Cl(s)$ to $NH_3(g)$ and $HCl(g)$.]

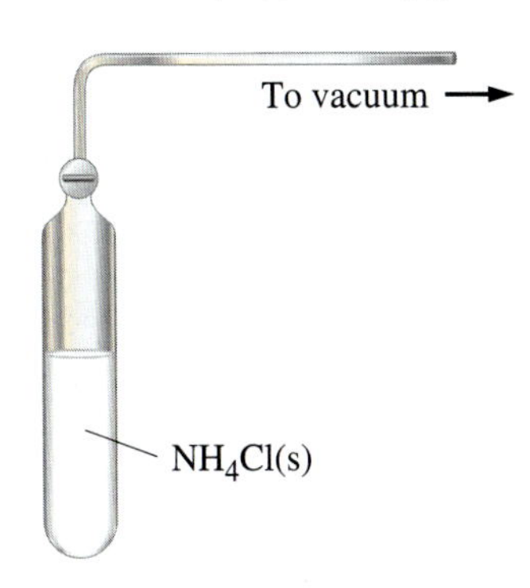

87. Use data from Appendix C, together with the molar enthalpy of fusion of mercury of 2.30 kJ/mol at the melting point of −38.86 °C, to estimate the vapor pressure of solid mercury at the sublimation temperature of dry ice $[CO_2(s)]$, −78.5 °C.

88. Hydrogen cyanide is produced in large quantities for use in the manufacture of plastics. One process for its manufacture involves the reaction of methane (natural gas) and ammonia at about 1200 °C.

$$CH_4(g) + NH_3(g) \rightleftharpoons HCN(g) + 3\,H_2(g)$$

A reaction vessel contains $CH_4(g)$ and $NH_3(g)$, both initially at 1.00 atm pressure, at 1200 °C. What will be the total gas pressure when the system is brought to equilibrium?

89. The decomposition of phosgene gas is represented by the following equation.

$$COCl_2(g) \rightleftharpoons CO(g) + Cl_2(g)$$

At what temperature will a sample of $COCl_2(g)$ be 10.0% dissociated at a total pressure of 1.00 atm? (*Hint:* Refer to page 631, Problem 89.)

90. A round-trip spaceship voyage to Mars carrying astronauts would require a great deal of fuel. One proposal to reduce the quantity of fuel is to take along a small quantity of hydrogen. On Mars, this hydrogen would be allowed to react with the abundant carbon dioxide in the Martian atmosphere to produce methane and water. The water would then be decomposed to hydrogen and oxygen by electrolysis. The hydrogen would be recycled into the first reaction, and the oxygen would be stored. On the return trip, the spacecraft would be powered by the reaction of the methane and oxygen produced on Mars. Do you think this proposal is thermodynamically feasible? Explain.

91. Use thermodynamic data from Appendix C to obtain a value of K_{sp} for Ag_2SO_4, and compare your result with the one found in the table of solubility product constants in Appendix C.

92. Use thermodynamic data from Appendix C to obtain K_{eq} for the following reaction.

$$Mg(OH)_2(s) + 2\,NH_4^+(aq) \rightleftharpoons Mg^{2+}(aq) + 2\,NH_3(aq) + 2\,H_2O(l)$$

Then obtain K_{eq} from other tabulated equilibrium constants in Appendix C, and compare the results.

93. Solid silver is added to a solution that is 0.100 M each in Ag^+, Fe^{2+}, and Fe^{3+}. What will be the molarities of the three ions after equilibrium is reached in the following reaction? (*Hint:* Use thermodynamic data from Appendix C.)

$$Ag^+(aq) + Fe^{2+}(aq) \rightleftharpoons Ag(s) + Fe^{3+}(aq)$$

CHAPTER

18

Electrochemistry

CONTENTS

● **ChemCDX**
When you see this icon, there are related animations, demonstrations, and exercises on the CD accompanying this text.

Electric current, a flow of electrons, can be used to bring about chemical change through a process called electrolysis. Conversely, oxidation-reduction reactions can be used to produce electricity in devices called voltaic cells. Corrosion, as in this rusting hulk of a ship, is also an electrochemical process involving voltaic cells.

No one needs to be reminded how useful electricity is. We use it to cook foods, heat or cool homes, run the motors in vacuum cleaners and other household devices, and power television sets and computers. In the most common method of generating electricity, the heat of combustion of a fuel is converted to mechanical work and then to electricity. However, the second law of thermodynamics imposes a restriction on the conversion of heat to work: Such conversions must always be accompanied by waste heat. In a typical power plant, less than half the heat of combustion is converted to electricity. In internal combustion engines of automobiles, the conversion of heat to work is even less efficient.

A much more efficient way to generate electricity is to avoid heat-to-work conversions and go directly from chemical energy to electric energy. **Electrochemistry** deals with the links between chemical reactions and electricity. Electricity involves the flow of electrons, so the types of chemical reactions associated with electricity are those in which electron transfers occur—*oxidation-reduction* reactions.

First we will consider a powerful alternative to the change in oxidation state method of balancing oxidation-reduction equations. However, we will focus on learning how *spontaneous* chemical reactions can produce electricity and how electricity can be used to produce *nonspontaneous* reactions. We will also examine practical applications that range from batteries and fuel cells as electric power sources, to methods of controlling corrosion, to the manufacture of key chemicals and metals.

Oxidation-Reduction: The Transfer of Electrons

You are probably familiar with the facts that an aqueous solution containing copper(II) ion has a blue color (recall Figure 16.6) and solid silver metal has a silvery color. And with these facts, we can easily describe the result of immersing a coil of copper wire in an aqueous solution containing Ag^+ ions (Figure 18.1). Our casual description might be, "Copper goes into solution, and silver comes out of solution." Let's try to explain what happens more scientifically.

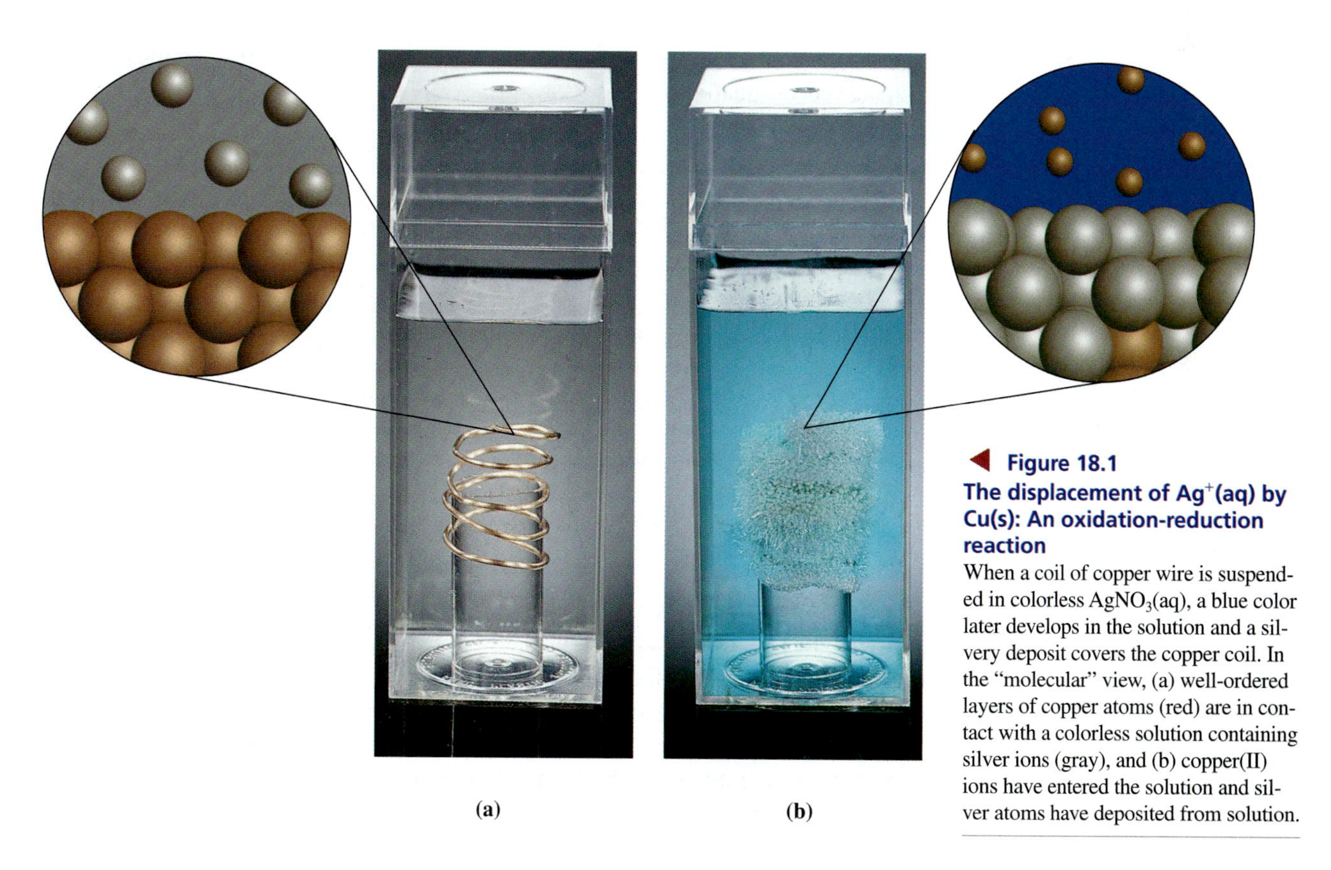

Figure 18.1
The displacement of $Ag^+(aq)$ by Cu(s): An oxidation-reduction reaction
When a coil of copper wire is suspended in colorless $AgNO_3(aq)$, a blue color later develops in the solution and a silvery deposit covers the copper coil. In the "molecular" view, (a) well-ordered layers of copper atoms (red) are in contact with a colorless solution containing silver ions (gray), and (b) copper(II) ions have entered the solution and silver atoms have deposited from solution.

18.1 Half-Reactions

What happens to the copper in Figure 18.1 is that Cu *atoms* give up electrons to become Cu^{2+} *ions*. The electrons remain on the copper wire, and the Cu^{2+} ions go into solution. Because the oxidation state of the copper increases from 0 to +2, this process is an oxidation.

$$\textit{Oxidation}: \mathrm{Cu(s)} \longrightarrow \mathrm{Cu^{2+}(aq)} + 2\,\mathrm{e^-}$$

Something else occurs at the same time: Ag^+ ions in the solution strike the copper wire, gain electrons, and deposit as Ag atoms. Because the oxidation state of the silver decreases from +1 to 0, this process is a reduction.

$$\textit{Reduction}: \mathrm{Ag^+(aq)} + \mathrm{e^-} \longrightarrow \mathrm{Ag(s)}$$

The two processes just described are called half-reactions. A **half-reaction** is either an oxidation or a reduction process. Taken together, two half-reactions—one an oxidation, one a reduction—constitute an oxidation-reduction reaction.

18.2 The Half-Reaction Method of Balancing Redox Equations

We can combine the equations for two half-reactions into an overall oxidation-reduction equation, but generally we can't just add the two together. We must make sure that the same number of electrons is involved in the oxidation half-reaction as in the reduction half-reaction. In the reaction between Cu(s) and Ag^+(aq), each Cu atom loses *two* electrons and each Ag^+ gains only *one*. We must multiply the reduction half-equation by the factor 2 before combining the half-equations into the overall redox equation.

$$\begin{array}{llcl} \textit{Oxidation}: & \mathrm{Cu(s)} & \longrightarrow & \mathrm{Cu^{2+}(aq)} + \cancel{2\,\mathrm{e^-}} \\ \textit{Reduction}: & 2\,\mathrm{Ag^+(aq)} + \cancel{2\,\mathrm{e^-}} & \longrightarrow & 2\,\mathrm{Ag(s)} \\ \hline \textit{Overall}: & \mathrm{Cu(s)} + 2\,\mathrm{Ag^+(aq)} & \longrightarrow & \mathrm{Cu^{2+}(aq)} + 2\,\mathrm{Ag(s)} \end{array}$$

We have just demonstrated a fundamental principle underlying the half-reaction method of balancing an oxidation-reduction equation:

> *All the electrons "lost" in an oxidation half-reaction must be "gained" in a reduction half-reaction. That is, although electrons can move around in a chemical reaction, they cannot be created or destroyed.*

The strategy we used to balance the equation for the reaction between Cu(s) and Ag^+(aq) is summarized below.

- Separate an oxidation-reduction equation into two half-equations, one for oxidation and one for reduction.
- Balance the atoms and the electric charge in each half-equation. Electrons appear on the *left* in the *reduction* half-equation and on the *right* in the *oxidation* half-equation.
- Adjust the coefficients in the half-equations so that the same number of electrons appears in each half-equation. This may require multiplying one or both half-equations by an appropriate factor.
- Add together the two adjusted half-equations to obtain an *overall* oxidation-reduction equation.

Redox Reactions in Acidic Solution

We can implement the basic strategy just outlined through a specific six-step procedure. First, we will consider redox reactions that occur in *acidic* solution. In these cases, we can expect H^+ and/or H_2O to appear in one or both half-equations and in the overall equation.

EXAMPLE 18.1

Permanganate ion, MnO_4^-, is used in the laboratory as an oxidizing agent and thiosulfate ion, $S_2O_3^{2-}$, as a reducing agent. Write a balanced equation for the reaction of these ions in an acidic aqueous solution to produce manganese(II) ion and sulfate ion.

SOLUTION

The reactants are MnO_4^- and $S_2O_3^{2-}$; the products are Mn^{2+} and SO_4^{2-}. Before turning to the six-step procedure, let's (a) write the equation to be balanced and (b) assign oxidation numbers. The oxidation numbers are shown above the equation.

$$\overset{+7\ -2}{MnO_4^-} + \overset{+2\ -2}{S_2O_3^{2-}} \longrightarrow \overset{+2}{Mn^{2+}} + \overset{+6\ -2}{SO_4^{2-}} \quad (\textit{not balanced})$$

Because the reaction occurs in an *acidic aqueous* solution, H^+ and H_2O are also likely to be involved. We will see where to put them as we balance the equation.

1 *Identify the species undergoing oxidation and reduction and write two "skeleton" half-equations.** MnO_4^- is *reduced* to Mn^{2+} (the oxidation number of Mn decreases), and $S_2O_3^{2-}$ is *oxidized* to SO_4^{2-} (the oxidation number of S increases).

$$MnO_4^- \longrightarrow Mn^{2+} \quad \text{(skeleton half-equation for } \textit{reduction}\text{)}$$
$$S_2O_3^{2-} \longrightarrow SO_4^{2-} \quad \text{(skeleton half-equation for } \textit{oxidation}\text{)}$$

2 *Balance the numbers of atoms in each half-equation.* In this step, the following order works best.

- First, balance all atoms *except* H and O.
- Second, balance O atoms by adding as many molecules of H_2O as needed.
- Third, balance H atoms by adding as many H^+ ions as needed.

Mn atoms are balanced in the skeleton half-equation for reduction. To balance the S atoms in the skeleton half-equation for oxidation, we place the coefficient 2 on the right side.

$$MnO_4^- \longrightarrow Mn^{2+} \quad (\textit{1 Mn atom on each side})$$
$$S_2O_3^{2-} \longrightarrow 2\,SO_4^{2-} \quad (\textit{2 S atoms on each side})$$

To balance O atoms, we add *four* H_2O on the *right* side of the reduction half-equation and *five* H_2O on the *left* side of the oxidation half-equation.

$$MnO_4^- \longrightarrow Mn^{2+} + 4\,H_2O \quad (\textit{4 O atoms on each side})$$
$$S_2O_3^{2-} + 5\,H_2O \longrightarrow 2\,SO_4^{2-} \quad (\textit{8 O atoms on each side})$$

To balance H atoms, we add *eight* H^+ to the *left* side of the reduction half-equation and *ten* H^+ to the *right* side of the oxidation half-equation.

$$MnO_4^- + 8\,H^+ \longrightarrow Mn^{2+} + 4\,H_2O \quad (\textit{8 H atoms on each side})$$
$$S_2O_3^{2-} + 5\,H_2O \longrightarrow 2\,SO_4^{2-} + 10\,H^+ \quad (\textit{10 H atoms on each side})$$

3 *Balance each half-equation for electric charge by adding electrons (e^-).* In adding electrons to a half-equation, we need to answer two questions: to which side, and

* The half-reaction method is not dependent on labeling half-equations as oxidation or reduction at this point. There are occasional cases, for example, some reactions involving organic compounds, in which this identification is more easily made in step **3**.

PROBLEM-SOLVING NOTE

To simplify the redox equations we write in this chapter, we will represent hydronium ion as H^+ rather than H_3O^+. Also, when balancing redox equations we will omit state designations such as (l) and (aq), except in the final balanced equation.

how many? The first answer is rather easy. Electrons are *gained* in a *reduction* process and must appear on the *left* side. Electrons are *lost* in an *oxidation* process and must appear on the *right* side. The second answer is that to whichever side we add electrons, we must add the number required to *make the net charge equal on the two sides.*

Thus, we add 5 electrons to the *left* side of the reduction half-equation. This reduces the net charge on the left from 7+ to 2+, the same as the 2+ net charge on the right. We add 8 electrons to the *right* side of the oxidation half-equation. This reduces the net charge on the right from 6+ to 2−, the same as the net charge on the left.

Charges:	1−	8+	5−		2+
Reduction:	$MnO_4^- + 8\,H^+ + 5\,e^- \longrightarrow Mn^{2+} + 4\,H_2O$				
Net charges:	*left*: −1 + 8 − 5 = 2+				*right*: 2+
Charges:	2−		4−	10+	8−
Oxidation:	$S_2O_3^{2-} + 5\,H_2O \longrightarrow 2\,SO_4^{2-} + 10\,H^+ + 8\,e^-$				
Net charges:	*left*: 2−		*right*: −4 + 10 −8 = 2−		

4 *Multiply coefficients by factors that make the number of electrons in the oxidation and reduction half-equations equal.* The reduction half-equation has 5 e^- and the oxidation half-equation, 8 e^-. The smallest common multiple of 5 and 8 is 40. We must multiply the reduction half-equation by 8 and the oxidation half-equation by 5.

Combine the adjusted half-equations into an *overall* equation.

Reduction: $8\,MnO_4^- + 64\,H^+ + \cancel{40\,e^-} \longrightarrow 8\,Mn^{2+} + 32\,H_2O$

Oxidation: $5\,S_2O_3^{2-} + 25\,H_2O \longrightarrow 10\,SO_4^{2-} + 50\,H^+ + \cancel{40\,e^-}$

Overall: $8\,MnO_4^- + 5\,S_2O_3^{2-} + 64\,H^+ + 25\,H_2O \longrightarrow 8\,Mn^{2+} + 10\,SO_4^{2-} + 50\,H^+ + 32\,H_2O$

5 *Simplify the overall equation.* To simplify means to eliminate from one side of the equation any species that appears on both sides of the equation. Thus, to eliminate H^+ from the *right* side of the above equation, we *subtract* 50 H^+ from each side. To eliminate H_2O from the *left* side, we *subtract* 25 H_2O from each side.

$$8\,MnO_4^- + 5\,S_2O_3^{2-} + 64\,H^+ + 25\,H_2O \longrightarrow 8\,Mn^{2+} + 10\,SO_4^{2-} + 50\,H^+ + 32\,H_2O$$
$$-\,50\,H^+ - 25\,H_2O \qquad\qquad -\,50\,H^+ - 25\,H_2O$$

At this point the equation is balanced:

$$8\,MnO_4^-(aq) + 5\,S_2O_3^{2-}(aq) + 14\,H^+(aq) \longrightarrow 8\,Mn^{2+}(aq) + 10\,SO_4^{2-}(aq) + 7\,H_2O(l)$$

6 *Verify that the equation is balanced.* Check to ensure that both atoms and electric charge are balanced. This check is shown below in a tabular format.

	Reactants (left)	*Products* (right)
Mn	8	8
S	10	10
O	47	47
H	14	14
Charge	(8 × 1−) + (5 × 2−) + (14 × 1+) = 4−	(8 × 2+) + (10 × 2−) = 4−

Because the charges are equal on both sides and the same number of atoms appear on each side, the equation is balanced.

EXERCISE 18.1A

Write a balanced equation for the reaction in which zinc is oxidized to zinc(II) ion by a dilute acidic solution containing nitrate ion. Dinitrogen monoxide gas is also formed.

EXAMPLE 18.1B

The oxidation of phosphorus by nitric acid is described as follows.

$$P_4(s) + H^+(aq) + NO_3^-(aq) \longrightarrow H_2PO_4^-(aq) + NO(g)$$

Write a balanced equation for this reaction.

Redox Reactions in Basic Solution

For a reaction in basic solution, $OH^-(aq)$ should appear instead of $H^+(aq)$ in the balanced equation. A method commonly used to balance such equations is to balance the equation *as if* the reaction occurs in acidic solution. Then, to *each* side of the net equation, add a number of OH^- ions equal to the number of H^+ appearing in the equation. As a result, one side of the equation will have H^+ and OH^- ions in equal number; they can be combined and replaced by H_2O molecules. The other side of the equation will have OH^- ions. We apply this method in Example 18.2.

EXAMPLE 18.2

In *basic* solution, Br_2 disproportionates to bromate and bromide ions. Use the half-reaction method to balance the equation for this reaction.

$$Br_2(l) \longrightarrow Br^-(aq) + BrO_3^-(aq)$$

SOLUTION

As in Example 18.1, let's begin by assigning oxidation numbers.

$$\overset{0}{Br_2} \longrightarrow \overset{-1}{Br^-} + \overset{+5}{BrO_3^-}$$

1 *Identify the species undergoing oxidation and reduction, and write the two half-equations.* Some of the Br_2 is *reduced* to Br^- (the oxidation number of Br decreases). At the same time, some of the Br_2 is *oxidized* to BrO_3^- (the oxidation number of Br increases). This means that Br_2 is a reactant in *both* half-reactions.

Reduction: $Br_2 \longrightarrow Br^-$

Oxidation: $Br_2 \longrightarrow BrO_3^-$

2 *Balance the numbers of atoms in each half-equation.* First, we balance the Br atoms.

$$Br_2 \longrightarrow 2\,Br^-$$

$$Br_2 \longrightarrow 2\,BrO_3^-$$

Then we balance the O atoms.

$$Br_2 \longrightarrow 2\,Br^-$$

$$Br_2 + 6\,H_2O \longrightarrow 2\,BrO_3^-$$

PROBLEM-SOLVING NOTE

In a *disproportionation* reaction, a substance is both oxidized and reduced in the same reaction. We considered a similar disproportionation reaction in Exercise 4.10B (page 159).

Finally, we balance the H atoms.

$$Br_2 \longrightarrow 2\,Br^-$$

$$Br_2 + 6\,H_2O \longrightarrow 2\,BrO_3^- + 12\,H^+$$

3 *Balance each half-equation for electric charge by adding electrons (e^-).*

Reduction: $Br_2 + 2\,e^- \longrightarrow 2\,Br^-$

Oxidation: $Br_2 + 6\,H_2O \longrightarrow 2\,BrO_3^- + 12\,H^+ + 10\,e^-$

4 *Multiply by factors that make the number of electrons in the oxidation and reduction half-equations equal. Combine the adjusted half-equations into an overall equation.* Multiply the reduction half-equation by *five*; leave the oxidation half-equation as is.

Reduction: $5\,(Br_2 + 2\,e^- \longrightarrow 2\,Br^-)$

$5\,Br_2 + \cancel{10\,e^-} \longrightarrow 10\,Br^-$

Oxidation: $Br_2 + 6\,H_2O \longrightarrow 2\,BrO_3^- + 12\,H^+ + \cancel{10\,e^-}$

Overall: $6\,Br_2 + 6\,H_2O \longrightarrow 10\,Br^- + 2\,BrO_3^- + 12\,H^+$

5 *Simplify*. All the coefficients in the equation in **4** are divisible by *two*.

$$3\,Br_2 + 3\,H_2O \longrightarrow 5\,Br^- + BrO_3^- + 6\,H^+$$

6 *Convert from acidic to basic solution.* Add *six* OH^- ions to each side of the equation in **5**.

$$3\,Br_2 + 3\,H_2O + 6\,OH^- \longrightarrow 5\,Br^- + BrO_3^- + 6\,H^+ + 6\,OH^-$$

On the right side, combine 6 H^+ and 6 OH^- into 6 H_2O.

$$3\,Br_2 + 3\,H_2O + 6\,OH^- \longrightarrow 5\,Br^- + BrO_3^- + 6\,H_2O$$

Simplify by subtracting *three* H_2O molecules from each side.

$$3\,Br_2(l) + 6\,OH^-(aq) \longrightarrow 5\,Br^-(aq) + BrO_3^-(aq) + 3\,H_2O(l)$$

7 *Verify*. The balanced equation has 6 atoms each of Br, H, and O on the left and on the right. The net charge on each side is 6–.

EXERCISE 18.2A

Cyanate ion in waste solutions from gold mining operations can be destroyed by treatment with hypochlorite ion in basic solution. Write a balanced oxidation-reduction equation for this reaction.

$$OCN^-(aq) + OCl^-(aq) + OH^-(aq) \longrightarrow CO_3^{2-}(aq) + N_2(g) + Cl^-(aq) + H_2O(l)$$

(*Hint:* Notice that in the oxidation half-reaction, OCN^- yields two products.)

EXERCISE 18.2B

In basic solution permanganate ion oxidizes ethanol to acetate ion and is itself reduced to solid manganese(IV) oxide. Write a balanced equation for this redox reaction.

Voltaic Cells

The "lemon battery" in Figure 18.2—strips of zinc and copper sticking into a lemon—is an example of a voltaic cell. As we will learn in this section, a voltaic cell uses a spontaneous oxidation-reduction reaction to produce electricity.

▲ **Figure 18.2**
A simple voltaic cell
Two dissimilar metal strips (copper and zinc) and an electrolyte (lemon juice) are the key components of a device that converts chemical energy to electricity—a *voltaic cell*. Electrons flow through the wires and the electric meter. Ions in the lemon juice flow between the metal strips. The electron and ion flows together constitute an electric current. The significance of the meter reading is described on page 775.

18.3 A Qualitative Description of Voltaic Cells

When we dip a strip of zinc metal into an aqueous solution of zinc sulfate, some zinc atoms are *oxidized* (Figure 18.3). Each Zn atom that is oxidized leaves behind two electrons and enters the solution as a Zn^{2+} ion.

$$\textit{Oxidation}: \mathrm{Zn(s)} \longrightarrow \mathrm{Zn^{2+}(aq)} + 2\,\mathrm{e^-}$$

At the same time, some Zn^{2+} ions in solution gain two electrons from the zinc strip and deposit as Zn atoms. They are *reduced* (Figure 18.3).

$$\textit{Reduction}: \mathrm{Zn^{2+}(aq)} + 2\,\mathrm{e^-} \longrightarrow \mathrm{Zn(s)}$$

The opposing oxidation and reduction processes quickly come to an equilibrium.

$$\mathrm{Zn(s)} \underset{\text{Reduction}}{\overset{\text{Oxidation}}{\rightleftharpoons}} \mathrm{Zn^{2+}(aq)} + 2\,\mathrm{e^-}$$

A strip of metal used in an electrochemical experiment is called an **electrode**. The equilibrium established at the surface of the electrode when it is immersed in a solution of its ions is called an *electrode equilibrium*. Another example of an electrode equilibrium is that between a copper electrode and an aqueous solution of Cu^{2+}.

$$\mathrm{Cu(s)} \underset{\text{Reduction}}{\overset{\text{Oxidation}}{\rightleftharpoons}} \mathrm{Cu^{2+}(aq)} + 2\,\mathrm{e^-}$$

In Chapter 4 we used the activity series of the metals (Figure 4.15) to show that zinc is a good reducing agent and thus rather easily oxidized. In contrast, copper is a poor reducing agent that is not so readily oxidized. We can demonstrate that the forward direction (oxidation) is more strongly favored in the electrode equilibrium at a zinc electrode than at a copper electrode with the voltaic cell pictured in Figure 18.4.

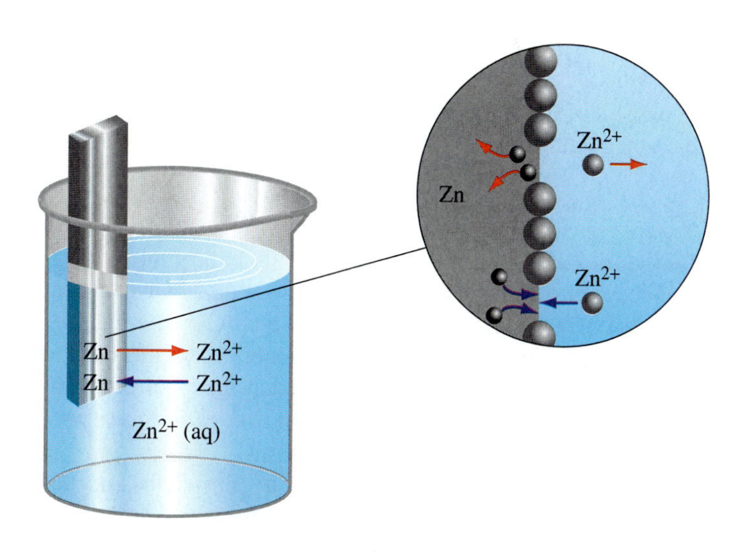

◄ **Figure 18.3**
Electrode equilibrium
The zinc metal strip, called an electrode, is partially immersed in a solution containing Zn^{2+} ions. Oxidation and reduction occur at the electrode until a condition of equilibrium is reached. (The anions needed to produce an electrically neutral solution are not shown.)

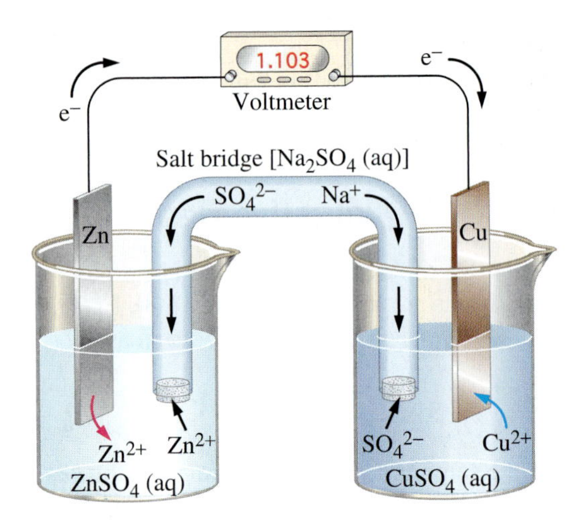

Figure 18.4
A zinc-copper voltaic cell
Zinc atoms lose electrons and enter the solution as Zn^{2+} ions (red arrow). Electrons flow through the external circuit from the Zn to the Cu electrode. At the copper electrode, Cu^{2+} ions gain electrons and deposit as Cu atoms (blue arrow). Ion migrations through the solutions are noted by the black arrows. Porous plugs prevent the bulk flow of solution but permit the ions to migrate through the salt bridge. The voltmeter reading is discussed on page 775. Voltaic cells similar to these (Daniell cells) were used to power telegraph lines in the early days of the telegraph.

A **half-cell** consists of a metal immersed in a solution of its ions. Figure 18.4 represents two half-cells: one with Zn(s) in $ZnSO_4$(aq) and the other, Cu(s) in $CuSO_4$(aq). The solutions in the two half-cells are joined by a **salt bridge**. This is an inverted U-shaped tube containing a salt solution, for example, Na_2SO_4(aq). Porous plugs at the ends of the tube prevent the bulk flow of solution, while still permitting the migration of ions. Metal wires connect the electrodes to the terminals of an electric meter called a voltmeter. The meter indicates that electrons flow continuously from the zinc to the copper electrode. (The voltmeter and the significance of its reading are discussed on page 775.) We can account for the observed electric current in this way.

At the zinc electrode, oxidation occurs. Zinc atoms lose electrons and enter the solution as Zn^{2+} ions.

$$\textit{Oxidation:}\quad \mathrm{Zn(s)} \longrightarrow \mathrm{Zn^{2+}(aq)} + 2\,\mathrm{e^-}$$

Electrons left behind on the zinc electrode by the departing Zn^{2+} ions do not just accumulate. They flow through the wires and the voltmeter to the copper electrode. At the copper electrode, the incoming flood of electrons overcomes any tendency for copper to be oxidized. Instead, Cu^{2+} ions from the $CuSO_4$(aq) gain electrons at the copper electrode and are reduced to copper atoms.

$$\textit{Reduction:}\quad \mathrm{Cu^{2+}(aq)} + 2\,\mathrm{e^-} \longrightarrow \mathrm{Cu(s)}$$

As a result of the half-reactions at the electrodes, the following oxidation-reduction reaction occurs.

$$\textit{Overall reaction:}\quad \mathrm{Zn(s)} + \mathrm{Cu^{2+}(aq)} \longrightarrow \mathrm{Zn^{2+}(aq)} + \mathrm{Cu(s)}$$

Our description to this point tells only part of the story, however. There can be no electric current unless electric charge is also conveyed through the solutions in the half-cells, and this electric charge *cannot be carried by electrons*. Electrons carry the electric current through the external circuit, but they cannot get beyond the electrodes. Electrons are exchanged at the electrode surfaces in the oxidation and reduction half-reactions, and *cations and anions carry electric charge through the solutions*.

As zinc is oxidized, the number of Zn^{2+} ions in the zinc half-cell increases, tending to build up a positive charge in the solution. Simultaneously, as Cu^{2+} is reduced, the number of Cu^{2+} ions decreases in the copper half-cell, tending to build up a negative charge in the solution. This buildup of charges is prevented by the salt bridge. SO_4^{2-} ions migrate out of the salt bridge into the zinc half-cell to neutralize the excess positive charge associated with the Zn^{2+} ions produced in the oxidation. Also, some SO_4^{2-} ions migrate out of the copper half-cell into the salt bridge. At the same time, cations migrate in the opposite direction from the anions. Na^+ ions from the salt bridge migrate into the copper half-cell to replace Cu^{2+} ions that have been reduced. Also, some Zn^{2+} ions migrate out of the zinc half-cell into the salt bridge.

Credit for the discovery of current electricity is generally given to Luigi Galvani (1737–1798). However, Alessandro Volta (1745–1827) first constructed cells similar to the one in Figure 18.4 for the production of electricity, hence the name *voltaic* cell. However, the same cells are sometimes referred to as *galvanic* cells.

Some Important Electrochemical Terms

We have established that a **voltaic cell** is an electrochemical device in which electric current is generated as a spontaneous oxidation-reduction reaction takes place in physically separated oxidation and reduction half-reactions. Let's consider a few other important terms used to describe voltaic cells.

The **anode** is the electrode at which oxidation occurs. The **cathode** is the electrode at which reduction occurs. In Figure 18.4, zinc is the anode and copper is the cathode. Because the anode is the source of electrons in a voltaic cell, it is also called the *negative* electrode. In turn, the cathode is the receiver of electrons and is the *positive* electrode. In fact, however, the buildup of electrons at the anode compared to the cathode is only very slight.

Another way to describe the difference between the anode and cathode is to say that the potential energy of the electrons at the anode is greater than at the cathode. Thus, when the two electrodes are joined in a voltaic cell, the flow of electrons is from the anode to the cathode. This is the same as describing the flow of water over a waterfall—from a higher to lower potential energy of the water. The change in potential energy depends on the height of the waterfall. The property in an electric circuit related to the height of a waterfall is the electric potential. The *electric potential* is the energy per unit of charge that flows. If we take *coulomb* as the unit of charge and *joule* as the unit of energy, an electric potential of 1 **volt (V)** is defined as 1 J/C.

The following mnemonic (memory) device may help you remember the difference between the cathode and anode.*

Car	Auto
a e	n x
t d	o i
h u	d d
o c	e a
d t	t
e i	i
o	o
n	n

* Kindly supplied by Professor Richard S. Treptow, Chicago State University.

A *voltmeter* measures a difference in electric potential between two points in an electric circuit. If the two points are the electrodes in a voltaic cell, the *potential difference* is the driving force that propels electrons from the anode to the cathode; it is also called the **cell potential** (E_{cell}). Because the measurements are in volts, the cell potential is also called the **cell voltage**. The cell voltage of the Zn-Cu voltaic cell of Figure 18.4 is 1.103 V. The voltage of the zinc/copper/lemon voltaic cell in Figure 18.2 is 0.914 V, or 914 *milli*volts. The overall oxidation-reduction reaction that occurs in a voltaic cell is called the **cell reaction**.

Cell Diagrams

It is more convenient to represent an electrochemical cell by writing a **cell diagram** than by making a drawing as in Figure 18.4. By international agreement, the following conventions are used for cell diagrams:

- Place the *anode* on the *left* side of the diagram.
- Place the *cathode* on the *right* side of the diagram.
- Use a *single* vertical line (|) to represent the boundary between different phases, such as between an electrode and a solution.
- Use a *double* vertical line (||) to represent a salt bridge or porous barrier separating two half-cells.

The cell diagram for Figure 18.4 is written as follows.

$$\underset{\text{Anode}}{\searrow} \quad \underset{\text{Salt bridge}}{\downarrow} \quad \underset{\text{Cathode}}{\swarrow}$$

$$Zn(s)\,|\,Zn^{2+}(aq)\,||\,Cu^{2+}(aq)\,|\,Cu(s)$$

Half-cell (oxidation) — Half-cell (reduction)

In many cases, we use an electrode that does not participate in an oxidation-reduction equilibrium. Rather, the inert electrode simply furnishes the surface on which an electric potential is established. For example, to establish electrode equilibrium between chlorine gas and chloride ions, we can immerse a strip of the inert metal platinum into a solution of $Cl^-(aq)$ and bubble chlorine gas over its surface. The following oxidation-reduction equilibrium is established on the surface of the platinum and imparts a characteristic electric potential to the metal.

$$2\,Cl^-(aq) \xrightleftharpoons{\text{on Pt}} Cl_2(g) + 2\,e^-$$

Shown below is the symbolism that we would use for half-cells having this Cl_2/Cl^- electrode.

As an anode (oxidation): $Pt, Cl_2(g)\,|\,Cl^-(aq)$

As a cathode (reduction): $Cl^-(aq)\,|\,Cl_2(g), Pt$

EXAMPLE 18.3

Describe the half-reactions and the overall reaction that occur in the following voltaic cell.

$$Pt(s)\,|\,Fe^{2+}(aq), Fe^{3+}(aq)\,||\,Cl^-(aq)\,|\,Cl_2(g), Pt(s)$$

SOLUTION

Because platinum is inert, the electrodes do not enter into the reaction. The half-reactions occur on the surface of the platinum, but they involve the other species in the cell diagram. According to the cell diagram conventions, the electrode on the *left* is the *anode*, where *oxidation* occurs. Fe^{2+} is oxidized to Fe^{3+}.

Oxidation: $Fe^{2+}(aq) \longrightarrow Fe^{3+}(aq) + e^-$

The electrode on the *right* is the *cathode*, where *reduction* occurs. Chlorine gas is reduced to chloride ion.

Reduction: $Cl_2(g) + 2\,e^- \longrightarrow 2\,Cl^-(aq)$

To combine these half-equations into an oxidation-reduction equation, we must multiply the oxidation half-equation by the factor 2. Then we can add it to the reduction half-equation, and the electrons will cancel.

Oxidation: $2\,Fe^{2+}(aq) \longrightarrow 2\,Fe^{3+}(aq) + \cancel{2\,e^-}$

Reduction: $Cl_2(g) + \cancel{2\,e^-} \longrightarrow 2\,Cl^-(aq)$

Cell reaction: $2\,Fe^{2+}(aq) + Cl_2(g) \longrightarrow 2\,Fe^{3+}(aq) + 2\,Cl^-(aq)$

EXERCISE 18.3A

Write the equation for the cell reaction that occurs in the following voltaic cell.

$$Al(s)\,|\,Al^{3+}(aq)\,||\,H^+(aq)\,|\,H_2(g), Pt(s)$$

EXERCISE 18.3B

Write a cell diagram for a voltaic cell in which the cell reaction would be the same as the reaction illustrated in Figure 18.1 on page 767.

As a final note, remember that the left/right (anode/cathode) convention in a cell diagram is *arbitrary*. When we assemble a voltaic cell of unknown half-cells, the one on the left won't necessarily be the anode. We can only say it is the anode if we determine *by experiment* that oxidation occurs in that half-cell. And even if the half-cell on the left is the anode, someone on the opposite side of the lab bench would see the same anode half-cell on their right. Left and right in the three-dimensional world are not as unambiguous as on a sheet of paper.

18.4 Standard Electrode Potentials

We have described how to construct a voltaic cell by combining two half-cells, and how to *measure* the cell voltage with a voltmeter. However, it is also possible to *calculate* cell voltages, without making specific reference to measurements. This is done by establishing characteristic potentials for individual half-cells and then determining differences in potential (cell voltages) when the half-cells are combined into voltaic cells. But how should we go about determining the electrode potentials for individual half-cells? The difficulty is that we cannot measure an individual electrode potential; we must always measure a potential difference. The situation is similar to determining the elevation of a point on Earth. We need to give the elevation relative to an arbitrarily assigned zero of elevation—mean sea level.

The zero point of electrode potentials is taken to be that of the half-cell pictured in Figure 18.5 and called the *standard hydrogen electrode*. Other electrode potentials are in relation to this *arbitrarily* assigned zero. In the **standard hydrogen electrode**, hydrogen gas at exactly 1 atm pressure is bubbled over an inert platinum electrode and into a hydrochloric acid solution. The concentration of the acid is adjusted so that the *activity* of H_3O^+ is exactly one ($a = 1$). This means that the hydrochloric acid solution is approximately 1 M HCl.* And, for simplicity, let's use the symbol H^+ in place of H_3O^+. Equilibrium between H_2 molecules and H^+ ions is established on the platinum surface; H_2 is oxidized to H^+, and H^+ is reduced to H_2. Finally, we will write the reversible half-equation with the *reduction* process as the forward direction (left to right) and the oxidation process as the reverse (right to left).

$$2\,H^+(a = 1) + 2\,e^- \xrightleftharpoons{\text{on Pt}} H_2(g, 1\text{ atm}) \quad E° = 0\text{ V } (\textit{exactly})$$

By international agreement, a **standard electrode potential, $E°$**, is based on the tendency for *reduction* to occur at the electrode. All solution species are present at unit activity ($a = 1$), which is about 1 M, and all gases are at 1 atm pressure. When no other metal is indicated as the electrode material, the potential is that for

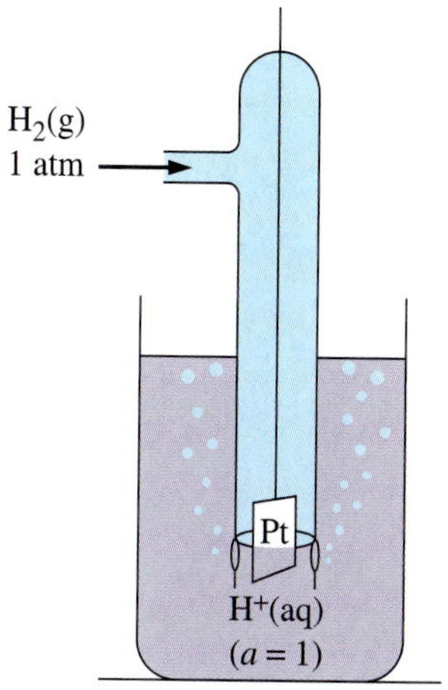

▲ **Figure 18.5**
The standard hydrogen electrode

The inert platinum strip acquires a potential that is determined by the equilibrium: $2\,H^+(a = 1) + 2\,e^- \rightleftharpoons H_2(g, 1\text{ atm})$. The condition of $a_{H^+} = 1$ can be approximated by $[H^+] = 1$ M.

* As we have noted before, activities should be used instead of molarities for precise work dealing with solution properties, equilibrium constants, and the like. We will continue to use molarity in place of activity, but we will report results *as if* we had used activities.

Because standard electrode potentials are based on reduction tendencies, they are also called standard *reduction* potentials.

equilibrium established on an inert surface, such as platinum metal. The electrode potential for the standard hydrogen electrode, as we have seen, is arbitrarily set at *exactly* zero volts.

To establish standard electrode potentials for other half-reactions, we can construct a voltaic cell like that shown in Figure 18.6. In that cell a standard hydrogen electrode is joined with a standard copper electrode. We find that electrons flow *from* the hydrogen electrode (anode) *to* the copper electrode (cathode) at a measured voltage of 0.340 V.

$$\underset{\text{(anode)}}{\text{Pt, } H_2(\text{g, 1 atm})\,|\,H^+(1\text{ M})}\,||\,\underset{\text{(cathode)}}{Cu^{2+}(1\text{ M})\,|\,Cu(\text{s})}$$

The cell voltage, called the **standard cell potential (E°_{cell})**, is the *difference* between the standard potential of the *cathode* and that of the *anode*.

$$E^\circ_{cell} = E^\circ(\text{cathode}) - E^\circ(\text{anode})$$

Or, if we set up the cell in the same manner as the conventional notation of cell diagrams, the anode is on the left and the cathode is on the right.

$$E^\circ_{cell} = E^\circ(\text{right}) - E^\circ(\text{left})$$

In Figure 18.6, E°_{cell} is *measured* and found to be +0.340 V. We will represent the potential of the electrode on the right—the standard Cu^{2+}/Cu electrode—in the form

$$E^\circ{}_{Cu^{2+}/Cu}$$

The electrode on the left is the standard hydrogen electrode, for which the defined potential is

$$E^\circ{}_{H^+/H_2} = 0.000\text{ V}$$

To find the standard electrode potential for the reduction of Cu^{2+} to Cu(s), we can write

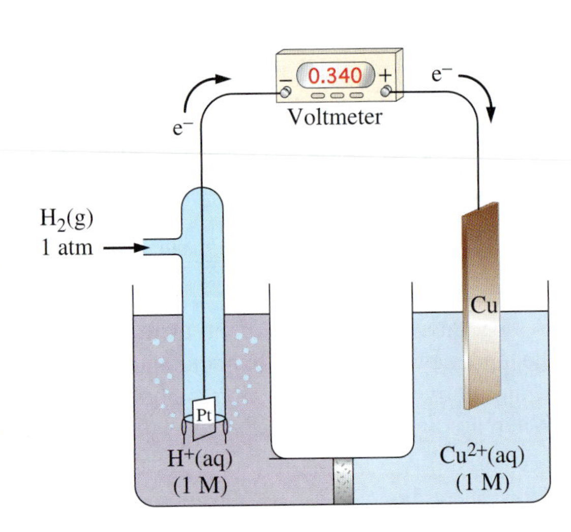

Figure 18.6 Measuring the standard potential of the Cu^{2+}/Cu electrode

The standard hydrogen electrode is the anode and the Cu^{2+}/Cu electrode is the cathode. Contact between the solutions in the two half-cells is through a porous plate, which permits the migration of ions but prevents bulk flow of the solutions. The direction of electron flow and the voltmeter reading are shown.

$$E^\circ_{cell} = E^\circ_{Cu^{2+}/Cu} - E^\circ_{H^+/H_2} = +0.340\ V$$

$$E^\circ_{Cu^{2+}/Cu} - 0.000\ V = +0.340\ V$$

$$E^\circ_{Cu^{2+}/Cu} = +0.340\ V$$

The *positive* value of the standard electrode potential means that Cu^{2+} ions are *more readily* reduced to Cu(s) than H^+ ions are reduced to $H_2(g)$.

Now consider what happens in the cell in Figure 18.7, where we replace the Cu^{2+}/Cu half-cell by a Zn^{2+}/Zn half-cell. The voltmeter in the circuit registers −0.763 V, and, the cell diagram is

$$\text{Pt, } H_2(\text{g, 1 atm})|H^+(1\ M)||Zn^{2+}(1\ M)|Zn(s) \quad E^\circ_{cell} = -0.763\ V$$

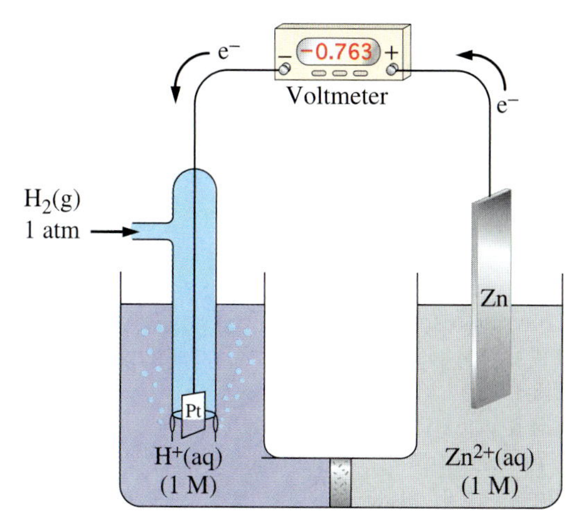

Figure 18.7 Measuring the standard potential of the Zn^{2+}/Zn electrode
As in Figure 18.6, the standard hydrogen electrode appears on the left and the metal electrode on the right. However, as signified by the direction of electron flow and the *negative* voltage, the standard hydrogen electrode is the *cathode*. The Zn^{2+}/Zn electrode is the *anode*. When voltaic cells are assembled as in Figures 18.6 and 18.7, correct magnitudes and signs of all standard electrode potentials can be established.

This standard cell potential and the standard electrode potentials are still related as follows.

$$E^\circ_{cell} = E^\circ(\text{right}) - E^\circ(\text{left})$$

$$E^\circ_{cell} = E^\circ_{Zn^{2+}/Zn} - E^\circ_{H^+/H_2} = -0.763\ V$$

$$E^\circ_{Zn^{2+}/Zn} - 0.000\ V = -0.763\ V$$

$$E^\circ_{Zn^{2+}/Zn} = -0.763\ V$$

The *negative* value of the standard electrode potential means that $Zn^{2+}(aq)$ is *less readily* reduced to Zn(s) than $H^+(aq)$ is reduced to $H_2(g)$.

What is the physical significance of the negative value of E°_{cell} in Figure 18.7? It has to do with how the electrodes are connected to the voltmeter. The proper connection for a *positive* reading is that the anode be connected to the (−) terminal of the voltmeter and the cathode to the (+) terminal. If the connections are reversed—anode to (+) and cathode to (−)—the voltmeter reading will be *negative*. In the cell that we have been describing, we assumed that the Zn electrode is the cathode, *but it is actually the anode.*

The three standard electrode potentials that we have been discussing are represented diagramatically in Figure 18.8. We have arranged them in terms of *decreasing* value of E°. The electrode potential for the easiest reduction, that of Cu^{2+}

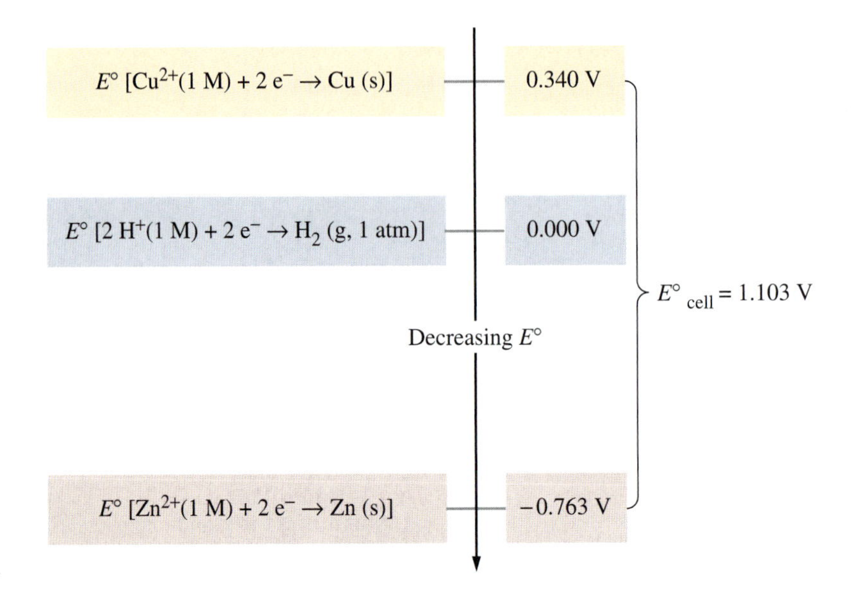

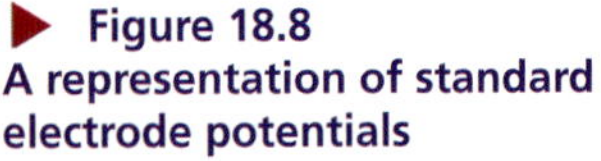

Figure 18.8
A representation of standard electrode potentials
The standard potentials for the copper and zinc electrodes are shown in relation to the standard hydrogen electrode potential. The 1.103-V potential *difference* between the copper and zinc electrodes—the cell voltage for the voltaic cell of Figure 18.4—is also shown.

to Cu, is at the top of the diagram. The electrode potential for the most difficult reduction, that of Zn^{2+} to Zn, is at the bottom.

Table 18.1 lists some standard electrode potentials and the reduction half-reactions that they describe. The arrangement is also in order of decreasing $E°$ values. A more extensive listing is given in Appendix C.

Viewed from the *left* side of the half-equations in Table 18.1, the best oxidizing agents, those most easily reduced, appear high in the table (for example, F_2, O_3, and so on). The poorest oxidizing agents, those most difficult to reduce, are at the bottom of the table. Viewed from the *right* side of the half-equations, thus for the reverse reactions, the best reducing agents are at the *bottom* of the table (Li, K, and so forth), and the poorest reducing agents are at the top of the table.

Tables of standard electrode potentials are sometimes written in the reverse order, with the best reducing agents—those most easily oxidized—at the top of the table.

We will find Table 18.1 very useful. First we will apply it to the rather simple task of determining a standard cell potential, E°_{cell}, from standard electrode potentials, $E°$. Consider how we can calculate E°_{cell} for the voltaic cell in Figure 18.4.

$$Zn(s)\,|\,Zn^{2+}(aq)\,||\,Cu^{2+}(aq)\,|\,Cu(s)$$

We need only to substitute standard electrode potentials from Table 18.1 into the expression that follows.

$$\begin{aligned} E^{\circ}_{cell} &= E^{\circ}(\text{right}) - E^{\circ}(\text{left}) \\ &= E^{\circ}(\text{cathode}) - E^{\circ}(\text{anode}) \\ &= E^{\circ}_{Cu^{2+}/Cu} - E^{\circ}_{Zn^{2+}/Zn} \\ &= 0.340\text{ V} - (-0.763\text{ V}) = 1.103\text{ V} \end{aligned}$$

In this type of calculation, three quantities will always be involved: E°_{cell}, $E°$(cathode), and $E°$(anode). If we have values for any two, we can calculate that of the third. In Example 18.4, we use this idea to determine the value of an unknown standard electrode potential.

TABLE 18.1 Some Selected Standard Electrode Potentials at 25 °C

Reduction Half-Reaction	$E°$, volt
Acidic Solution	
$F_2(g) + 2\,e^- \longrightarrow 2\,F^-(aq)$	+2.866
$O_3(g) + 2\,H^+(aq) + 2\,e^- \longrightarrow O_2(g) + H_2O(l)$	+2.075
$S_2O_8^{2-}(aq) + 2\,e^- \longrightarrow 2\,SO_4^{2-}(aq)$	+2.01
$H_2O_2(aq) + 2\,H^+(aq) + 2\,e^- \longrightarrow 2\,H_2O(l)$	+1.763
$MnO_4^-(aq) + 8\,H^+(aq) + 5\,e^- \longrightarrow Mn^{2+}(aq) + 4\,H_2O(l)$	+1.51
$PbO_2(s) + 4\,H^+(aq) + 2\,e^- \longrightarrow Pb^{2+}(aq) + 2\,H_2O(l)$	+1.455
$Cl_2(g) + 2\,e^- \longrightarrow 2\,Cl^-(aq)$	+1.358
$Cr_2O_7^{2-}(aq) + 14\,H^+(aq) + 6\,e^- \longrightarrow 2\,Cr^{3+}(aq) + 7\,H_2O(l)$	+1.33
$MnO_2(s) + 4\,H^+(aq) + 2\,e^- \longrightarrow Mn^{2+}(aq) + 2\,H_2O(l)$	+1.23
$O_2(g) + 4\,H^+(aq) + 4\,e^- \longrightarrow 2\,H_2O(l)$	+1.229
$2\,IO_3^-(aq) + 12\,H^+(aq) + 10\,e^- \longrightarrow I_2(s) + 6\,H_2O(l)$	+1.20
$Br_2(l) + 2\,e^- \longrightarrow 2\,Br^-(aq)$	+1.065
$NO_3^-(aq) + 4\,H^+(aq) + 3\,e^- \longrightarrow NO(g) + 2\,H_2O(l)$	+0.956
$Ag^+(aq) + e^- \longrightarrow Ag(s)$	+0.800
$Fe^{3+}(aq) + e^- \longrightarrow Fe^{2+}(aq)$	+0.771
$O_2(g) + 2\,H^+(aq) + 2\,e^- \longrightarrow H_2O_2(aq)$	+0.695
$I_2(s) + 2\,e^- \longrightarrow 2\,I^-(aq)$	+0.535
$Cu^{2+}(aq) + 2\,e^- \longrightarrow Cu(s)$	+0.340
$SO_4^{2-}(aq) + 4\,H^+(aq) + 2\,e^- \longrightarrow 2\,H_2O(l) + SO_2(g)$	+0.17
$Sn^{4+}(aq) + 2\,e^- \longrightarrow Sn^{2+}(aq)$	+0.154
$S(s) + 2\,H^+(aq) + 2\,e^- \longrightarrow H_2S(g)$	+0.14
$2\,H^+(aq) + 2\,e^- \longrightarrow H_2(g)$	0
$Pb^{2+}(aq) + 2\,e^- \longrightarrow Pb(s)$	−0.125
$Sn^{2+}(aq) + 2\,e^- \longrightarrow Sn(s)$	−0.137
$Co^{2+}(aq) + 2\,e^- \longrightarrow Co(s)$	−0.277
$Fe^{2+}(aq) + 2\,e^- \longrightarrow Fe(s)$	−0.440
$Zn^{2+}(aq) + 2\,e^- \longrightarrow Zn(s)$	−0.763
$Al^{3+}(aq) + 3\,e^- \longrightarrow Al(s)$	−1.676
$Mg^{2+}(aq) + 2\,e^- \longrightarrow Mg(s)$	−2.356
$Na^+(aq) + e^- \longrightarrow Na(s)$	−2.713
$Ca^{2+}(aq) + 2\,e^- \longrightarrow Ca(s)$	−2.84
$K^+(aq) + e^- \longrightarrow K(s)$	−2.924
$Li^+(aq) + e^- \longrightarrow Li(s)$	−3.040
Basic Solution	
$O_3(g) + H_2O(l) + 2\,e^- \longrightarrow O_2(g) + 2\,OH^-(aq)$	+1.246
$OCl^-(aq) + H_2O(l) + 2\,e^- \longrightarrow Cl^-(aq) + 2\,OH^-(aq)$	+0.890
$O_2(g) + 2\,H_2O(l) + 4\,e^- \longrightarrow 4\,OH^-(aq)$	+0.401
$2\,H_2O(l) + 2\,e^- \longrightarrow H_2(g) + 2\,OH^-(aq)$	−0.828

EXAMPLE 18.4

Determine $E°$ for the reduction half-reaction: $Ce^{4+}(aq) + e^- \longrightarrow Ce^{3+}(aq)$, given the cell voltage for the following voltaic cell.

$$Co(s)|Co^{2+}(1\text{ M})||Ce^{4+}(1\text{ M}), Ce^{3+}(1\text{ M})|Pt(s) \quad E°_{cell} = 1.887\text{ V}$$

SOLUTION

The reduction half-reaction occurs in the cathode half-cell, shown on the right in the cell diagram. We are seeking $E°$(cathode), that is, $E°_{Ce^{4+}/Ce^{3+}}$. The oxidation half-reaction: $Co(s) \longrightarrow Co^{2+} + 2\,e^-$ occurs in the anode half-cell on the left. From Table 18.1, we find that $E°_{Co^{2+}/Co} = -0.277$ V. In the same manner as illustrated for the Zn-Cu voltaic cell, we can write the following.

$$\begin{aligned} E°_{cell} &= E°(\text{right}) - E°(\text{left}) \\ &= E°_{cathode} - E°_{anode} \\ &= E°_{Ce^{4+}/Ce^{3+}} - E°_{Co^{2+}/Co} \\ 1.887\text{ V} &= E°_{Ce^{4+}/Ce^{3+}} - (-0.277\text{ V}) \\ 1.887\text{ V} - 0.277\text{ V} &= E°_{Ce^{4+}/Ce^{3+}} = 1.610\text{ V} \end{aligned}$$

EXERCISE 18.4A

Use the following information and data from Table 18.1 to determine $E°_{Sm^{2+}/Sm}$.

$$Sm(s)\,|\,Sm^{2+}(1\text{ M})\,||\,I^-(1\text{ M})\,|\,I_2(s), Pt(s) \qquad E°_{cell} = 3.21\text{ V}$$

EXERCISE 18.4B

Use the following information and data from Table 18.1 to determine $E°_{ClO_4^-/Cl_2}$.

$$Pt(s), Cl_2(g)\,|\,ClO_4^-(1\text{ M})\,||\,Cl^-(1\text{ M})\,|\,Cl_2(g), Pt(s) \qquad E°_{cell} = -0.034\text{ V}$$

Figure 18.9
$E°_{cell}$ is an intensive property
That $E°_{cell}$ is an intensive property can be seen in the constant voltage (1.5 V) of dry cell batteries, whether they are large D batteries or small AA penlite batteries. The voltage does not depend on the amounts of substances involved in the cell reaction.

Following are two additional important points about electrode potentials and cell voltages.

- Electrode potentials and cell voltages are *intensive* properties. Their magnitudes are fixed once the particular species and their concentrations are specified. The magnitudes do *not* depend on the total amounts of the species present, for example, not on the size of a half-cell or voltaic cell. Figure 18.9 illustrates this fact with a familiar example.
- Cell voltages can be ascribed to oxidation-reduction reactions without regard to voltaic cells. Specifically, we can calculate $E°_{cell}$ from the equation for a cell reaction without writing a cell diagram. This idea is illustrated in Example 18.5.

EXAMPLE 18.5

Balance the following oxidation-reduction equation and determine $E°_{cell}$ for the reaction.

$$O_2(g) + H^+(aq) + I^-(aq) \longrightarrow H_2O(l) + I_2(s)$$

SOLUTION

First we need to separate the redox equation into half-equations for oxidation and reduction. Then we can obtain a balanced equation by the half-reaction method of Section 18.1. The balanced half-equations and their combination into the balanced redox equation are as follows.

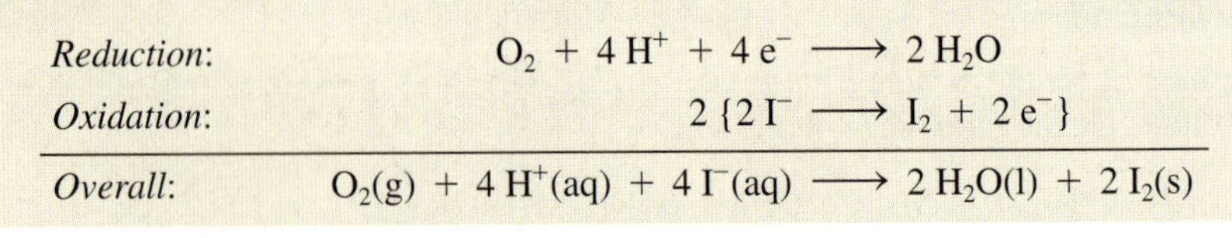

Reduction: $O_2 + 4\,H^+ + 4\,e^- \longrightarrow 2\,H_2O$

Oxidation: $2\,\{2\,I^- \longrightarrow I_2 + 2\,e^-\}$

Overall: $O_2(g) + 4\,H^+(aq) + 4\,I^-(aq) \longrightarrow 2\,H_2O(l) + 2\,I_2(s)$

We can complete our answer by obtaining $E°$ values for the two half-reactions from Table 18.1 and matching them to this expression.

$$E^\circ_{cell} = E^\circ(\text{cathode}) - E^\circ(\text{anode})$$

If the reaction were carried out in a voltaic cell, the half-reaction at the cathode would be a *reduction* and at the anode, an *oxidation*. That is, we can also write the following.

$$E^\circ_{cell} = E^\circ(\text{reduction}) - E^\circ(\text{oxidation})$$

Then we can substitute the $E°$ values.

$$E^\circ_{cell} = E^\circ_{O_2/H_2O} - E^\circ_{I_2/I^-} = +1.229\text{ V} - 0.535\text{ V} = +0.694\text{ V}$$

Notice that although we doubled the oxidation half-equation before combining it with the reduction half-equation, we did not alter the $E^\circ_{I_2/I^-}$ value in any way.

EXERCISE 18.5

Determine the E°_{cell} values for each of the following redox reactions.

a. $2\,Al(s) + 3\,Cu^{2+}(aq) \longrightarrow 2\,Al^{3+}(aq) + 3\,Cu(s)$

b. $S_2O_8^{2-}(aq) + Mn^{2+}(aq) + H_2O(l) \longrightarrow SO_4^{2-} + MnO_4^-(aq) + H^+(aq)$ (*not balanced*)

c. $2\,NO_3^-(aq) + 3\,Pb^{2+}(aq) + 2\,H_2O(l) \longrightarrow 2\,NO(g) + 3\,PbO_2(s) + 4\,H^+(aq)$

18.5 Electrode Potentials, Spontaneous Change, and Equilibrium

When a reaction takes place in a voltaic cell, it performs work. We can think of this electrical work as the work produced by electric charges in motion. The total work done is the product of (1) the cell voltage, (2) the number of moles of electrons, n, transferred between electrodes, and (3) the electric charge per mole of electrons—a quantity called the **Faraday constant**, F, and equal to 96,485 coulombs per mole.

$$\omega_{elec} = n \times F \times E_{cell}$$

The product $n \times F \times E_{cell}$ has the units volt × coulomb. From the definition of the volt: 1 V = 1 J/C (page 775), we see that the unit of electrical work is the joule: 1 J = 1 V C.

Electrical work is related to free energy change in the following way: The maximum amount of useful work that a system can do is $-\Delta G$, and this maximum amount of work can be realized as electrical work. As a consequence, the electrical work done by a voltaic cell is as follows.

$$-\Delta G = \omega_{elec} = n \times F \times E_{cell}$$

Thus, for an oxidation-reduction reaction we can write this equation.

$$\Delta G = -n \times F \times E_{cell}$$

Notice that in this equation we have not written the superscript "°". The term E_{cell} signifies that the conditions at the electrodes are *not* standard conditions—solute concentrations may *not* be 1 M and gas pressures may *not* be 1 atm. Similarly, the term ΔG also implies *nonstandard* conditions. The equation is perfectly general, however, and we can apply it to a cell in which all substances are in the standard state. In that case, we should use the superscript "°" and write the following.

$$\Delta G^\circ = -n \times F \times E^\circ_{cell}$$

Criteria for Spontaneous Change in Redox Reactions

In order for a reaction to proceed spontaneously, $\Delta G < 0$, and if ΔG is *negative*, then E_{cell} must be *positive*. This leads to some important new ideas concerning spontaneous change.

- If E_{cell} is *positive*, the reaction in the forward direction (from left to right) is *spontaneous*.
- If E_{cell} is *negative*, the reaction in the forward direction is *nonspontaneous*.
- If $E_{cell} = 0$, a system is at equilibrium.
- When a cell reaction is *reversed*, E_{cell} and ΔG change signs.

When we obtain an E_{cell} value by combining standard electrode potentials from Table 18.1, we get a *standard* cell potential, E°_{cell}. Any predictions we make will therefore be for reactions having reactants and products in their standard states. Usually, however, *qualitative* predictions based on standard-state conditions apply over a wide range of nonstandard conditions as well. Let's now apply these new criteria for spontaneous change.

EXAMPLE 18.6

Will copper metal displace silver ion from aqueous solution? That is, does this reaction occur spontaneously from left to right?

$$Cu(s) + 2\,Ag^+(1\,M) \longrightarrow Cu^{2+}(1\,M) + 2\,Ag(s)$$

SOLUTION

Let's separate the overall equation into its two half-equations, assign the appropriate E° values, and then recombine the half-equations to obtain E°_{cell}. If the resulting E°_{cell} is positive, then the reaction is spontaneous in the forward direction.

Reduction: $2\,\{Ag^+(1\,M) + e^- \longrightarrow Ag(s)\}$

Oxidation: $Cu(s) \longrightarrow Cu^{2+}(1\,M) + 2\,e^-$

Overall: $Cu(s) + 2\,Ag^+(1\,M) \longrightarrow Cu^{2+}(1\,M) + 2\,Ag(s)$

$$\begin{aligned} E^\circ_{cell} &= E^\circ(\text{cathode}) - E^\circ(\text{anode}) \\ &= E^\circ(\text{reduction}) - E^\circ(\text{oxidation}) \\ &= E^\circ_{Ag^+/Ag} - E^\circ_{Cu^{2+}/Cu} \\ &= 0.800\text{ V} - 0.340\text{ V} = 0.460\text{ V} \end{aligned}$$

Because E°_{cell} is positive, the forward direction should be the direction of spontaneous change. Copper metal should displace silver ions from solution.

EXERCISE 18.6A

Should the following reaction occur spontaneously as written for standard-state conditions?

$$Cu^{2+}(aq) + 2\,Fe^{2+}(aq) \longrightarrow 2\,Fe^{3+}(aq) + Cu(s)$$

EXERCISE 18.6B

Which is the direction of spontaneous change for the following reaction in which all reactants are in their standard states? Explain.

$$2\,Mn^{2+}(aq) + 2\,IO_3^-(aq) + 2\,H_2O(l) \rightleftharpoons 2\,MnO_4^-(aq) + I_2(s) + 4\,H^+(aq)$$

Note that the spontaneous reaction predicted in Example 18.6 is the same one we illustrated in Figure 18.1. There, we saw that the reaction occurs by simply immersing a coil of copper wire into an aqueous solution of silver nitrate. It is important to understand that although we use cell terminology to make predictions about the direction of spontaneous change in redox reactions, these predictions apply *even if the reactions are* not *carried out in voltaic cells.*

The Activity Series of the Metals Revisited

We can now give a theoretical explanation of the activity series of the metals introduced in Section 4.5. According to the relationship between E_{cell} and the direction of spontaneous change, a metal will displace from a solution of its ions any metal lying *above* it in a listing of electrode potentials arranged by *decreasing value*, such as in Table 18.1. And, treating the listing for hydrogen as just another entry in the table, a metal appearing *below* the standard hydrogen electrode (for example, Fe, Zn, and Al) will react with a mineral acid (a solution having H^+ as the only oxidizing agent) to produce $H_2(g)$. A metal appearing *above* the standard hydrogen electrode (for example, Cu and Ag) will not displace $H_2(g)$ from a solution of $H^+(aq)$. This assessment yields the same conclusions we previously stated for the activity series of metals (Figure 4.15).

EXAMPLE 18.7—A Conceptual Example

The photograph in Figure 18.10 shows strips of copper and zinc joined together and then dipped in HCl(aq). Explain what happens. That is, what are the gas bubbles on the zinc and on the copper, and how did they get there?

▲ **Figure 18.10**
Copper-zinc assembly in HCl(aq)

SOLUTION

A metal should react with HCl(aq), displacing $H_2(g)$ from solution, if the metal lies above $H_2(g)$ in the activity series of the metals (Figure 4.15) or below $H_2(g)$ in the listing of standard electrode potentials in Table 18.1. Zinc fits this requirement, as we can establish by using $E°$ data.

$$\begin{aligned}
&\textit{Reduction:} && 2\,H^+(1\,M) + \cancel{2e^-} \longrightarrow H_2(g,\,1\,atm)\\
&\textit{Oxidation:} && Zn(s) \longrightarrow Zn^{2+}(1\,M) + \cancel{2e^-}\\
\hline
&\textit{Overall:} && Zn(s) + 2\,H^+(1\,M) \longrightarrow Zn^{2+}(1\,M) + H_2(g,\,1\,atm)
\end{aligned}$$

$$\begin{aligned}
E°_{cell} &= E°(\text{reduction}) - E°(\text{oxidation})\\
&= E°_{H^+/H_2} - E°_{Zn^{2+}/Zn}\\
&= 0.000\text{ V} - (-0.763\text{ V}) = 0.763\text{ V}
\end{aligned}$$

The gas on the zinc is H_2 formed in the direct reaction of Zn with HCl(aq). But how does a gas form on the copper? Cu(s) lies *below* H_2(g) in the activity series of the metals and *above* H_2(g) in the listing of standard electrode potentials in Table 18.1. It should not displace H_2(g) from HCl(aq).

$$\begin{array}{lrcl} \textit{Reduction:} & 2\,H^+(1\,M) + \cancel{2\,e^-} & \longrightarrow & H_2(g,\,1\,atm) \\ \textit{Oxidation:} & Cu(s) & \longrightarrow & Cu^{2+}(1\,M) + \cancel{2\,e^-} \\ \hline \textit{Overall:} & Cu(s) + 2\,H^+(1\,M) & \longrightarrow & Cu^{2+}(1\,M) + H_2(g,\,1\,atm) \end{array}$$

$$\begin{aligned} E^\circ_{cell} &= E^\circ(\text{reduction}) - E^\circ(\text{oxidation}) \\ &= E^\circ_{H^+/H_2} - E^\circ_{Cu^{2+}/Cu} \\ &= 0.000\text{ V} - (0.340\text{ V}) = -0.340\text{ V} \end{aligned}$$

The reaction of Cu(s) with HCl(aq) is *nonspontaneous.*

Here is a plausible explanation for the appearence of a gas on the copper. The gas *is* H_2 but it is *not* formed by a reaction involving Cu(s). The electrons required for the reduction of H^+ to H_2 on the copper come from the oxidation of the *zinc*. Copper, an excellent electrical conductor, acts as just an inert extension of the zinc electrode. Thus, although H_2 will form on the zinc even if the copper is not present, H_2 will *not* form on the copper *unless* the copper is connected to a metal more active than hydrogen, such as zinc.

The device in Figure 18.10 is actually a voltaic cell. The portion of the reduction half-reaction that occurs on the copper is physically separated from the oxidation half-reaction that occurs only on the zinc. Electrons flow from the zinc to the copper.

EXERCISE 18.7

Offer a plausible explanation of what occurs in the "lemon battery" in Figure 18.2 (page 773).

Equilibrium Constants for Redox Reactions

In Chapter 17, we introduced an important relationship between ΔG° and K_{eq} of a reaction. A similar one exists between K_{eq} and E°_{cell}.

$$\Delta G^\circ = -RT \ln K_{eq} = -n \times F \times E^\circ_{cell}$$

This allows us to write the following equation.

$$E^\circ_{cell} = \frac{RT \ln K_{eq}}{nF}$$

In this equation, E°_{cell} is the standard cell potential, R is the gas constant expressed as 8.3145 J mol^{-1} K^{-1}, T is the Kelvin temperature, n is the number of moles of electrons involved in the reaction, and F is the Faraday constant. E°_{cell} values are generally obtained from tables of E° at 25 °C, and the equation is almost always used at 25 °C. As a result, the quantity RT/F has the same value from one calculation to another. It is easiest to replace it by its numerical equivalent, 0.025693 V. The ratio J/C (joule/coulomb) is the unit V (volt).

The equivalent expression using common logarithms is $E^\circ_{cell} = (0.059160\text{ V}/n) \log K_{eq}$

$$E^\circ_{cell} = \frac{RT}{nF} \ln K_{eq} = \frac{8.3145\text{ J } \cancel{\text{mol}^{-1}}\ \cancel{\text{K}^{-1}} \times 298.15\ \cancel{\text{K}}}{n \times 96485\text{ C } \cancel{\text{mol}^{-1}}} \ln K_{eq} = \frac{0.025693\text{ V}}{n} \ln K_{eq}$$

We can establish n from the half-cell reactions: It is the number of electrons in either the reduction or oxidation half-equation *after* the coefficients have been adjusted to balance the equation. That is, n is the number of electrons that cancel out of each side when the half-equations are combined to give the balanced redox equation.

EXAMPLE 18.8

Calculate the values of $\Delta G°$ and K_{eq} at 25 °C for the reaction

$$Cu(s) + 2Ag^{+}(1\ M) \rightleftharpoons Cu^{2+}(1\ M) + 2\ Ag(s)$$

SOLUTION

Look at the half-equations and the overall equation that we wrote for this reaction in Example 18.6. We had two electrons in each half-equation before we combined them, so in this case, $n = 2$. We found $E°_{cell}$ for the reaction to be +0.460 V.

We can get $\Delta G°$ by substituting $n = 2$ and known values of F and $E°_{cell}$ into the expression: $\Delta G° = -n \times F \times E°_{cell}$. Also, note that 1 J = 1 V C.

$$\Delta G° = -2 \times 96{,}485\ C \times 0.460\ V = -8.88 \times 10^{4}\ J\ (\text{or} -88.8\ kJ)$$

To obtain K_{eq}, we substitute into the expression

$$E°_{cell} = \frac{0.025693\ V}{n} \ln K_{eq}$$

$$0.460\ V = \frac{0.025693\ V}{2} \ln K_{eq}$$

$$\ln K_{eq} = \frac{2 \times 0.460\ \cancel{V}}{0.025693\ \cancel{V}} = 35.8$$

$$K_{eq} = e^{35.8} = 4 \times 10^{15}$$

EXERCISE 18.8A

Calculate the values of $\Delta G°$ and K_{eq} at 25 °C for the reaction that follows.

$$3\ Mg(s) + 2\ Al^{3+}(1\ M) \rightleftharpoons 3\ Mg^{2+}(1\ M) + 2\ Al(s)$$

EXERCISE 18.8B

Determine K_{eq} for the reaction of silver metal with nitric acid. The silver is oxidized to $Ag^{+}(aq)$, and nitrate ion is reduced to NO(g).

Thermodynamics, Equilibrium, and Electrochemistry: A Summary

Let's pause for a moment and use Figure 18.11 as an aid to review some important relationships. At the center of the figure we have grouped the three related properties: $\Delta G°$, K_{eq}, and $E°_{cell}$. We have joined them by double headed arrows to show that the quantitative relationships can be applied in either direction. Noted around the edges of the figure are the principal experimental methods used to establish values of the three properties.

Calorimetric measurements can be quite precise, and they are important in establishing $\Delta G°$ values. Electrical measurements are among the most precise that can

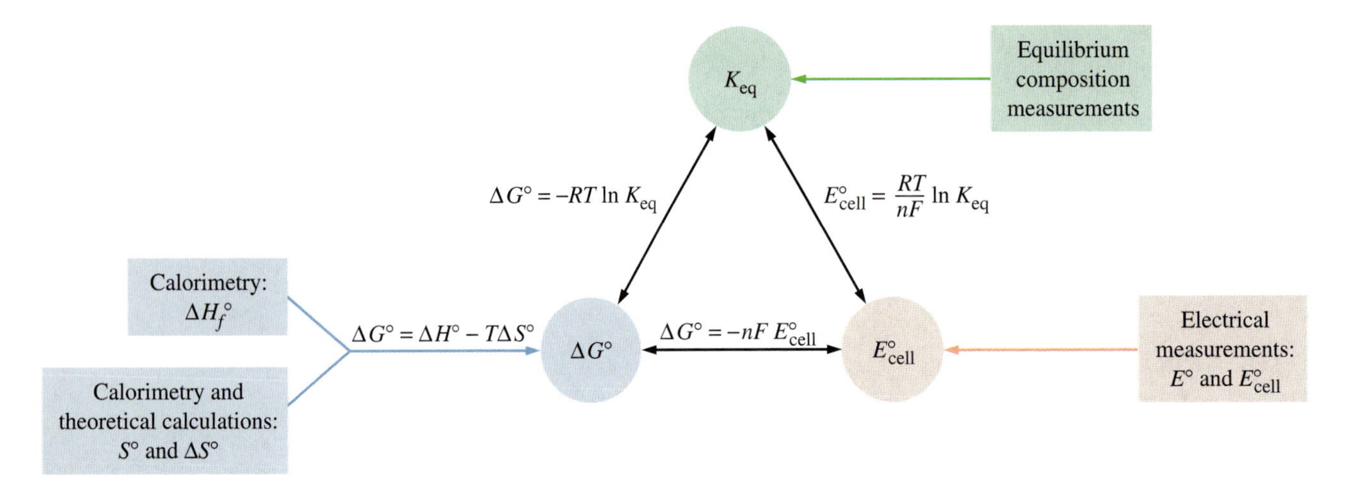

▲ **Figure 18.11 Summarizing important relationships from thermodynamics, equilibrium, and electrochemistry**

be made in the chemical laboratory, and they are an important source of $\Delta G°$ and K_{eq} values as well as $E°_{cell}$ values. In contrast, equilibrium composition measurements are used to establish some K_{eq} values, but they are less often used to generate $\Delta G°$ and $E°_{cell}$ values.

Table 18.2 gives additional comparative information for $\Delta G°$, K_{eq}, and $E°_{cell}$.

TABLE 18.2 Comparing $\Delta G°$, K_{eq}, and $E°_{cell}$

	$\Delta G°$	K_{eq}	$E°_{cell}$
Equilibrium under standard-state conditions	0	1	0
Spontaneous reaction under standard-state conditions	<0	>1	>0
Approximate range of values encountered	± several hundred kJ	$10^{\pm 100}$	± a few volts

18.6 Effect of Concentrations on Cell Voltage

We can do quite a bit with *standard* cell voltages, $E°_{cell}$, but most cell measurements—and many calculations—involve *nonstandard* conditions. First, let's look *qualitatively* at the relationship between $E°_{cell}$ and E_{cell}. To do this, we turn again to the voltaic cell of Figure 18.4.

According to Le Châtelier's principle, the *forward* reaction should be favored if we increase $[Cu^{2+}]$ and/or decrease $[Zn^{2+}]$ (represented on page 789, in red). When the forward reaction is favored, $-\Delta G$ and E_{cell} *increase*. The *reverse* reaction should be favored by decreasing $[Cu^{2+}]$ and/or increasing $[Zn^{2+}]$ (represented on page 789, in blue). When the reverse reaction is favored, $-\Delta G$ and E_{cell} *decrease*. Cell reactions and cell voltages for standard conditions and two nonstandard conditions are summarized as follows.

$$\begin{array}{llll}
 & \text{increase } [Cu^{2+}] \longrightarrow & & \\
 & \longrightarrow \text{decrease } [Zn^{2+}] & & \\
\textit{Nonstandard:} & Zn(s) + Cu^{2+}(1.5\text{ M}) \rightleftharpoons Zn^{2+}(0.075\text{ M}) + Cu(s) & & E_{cell} = 1.142\text{ V} \\
\textit{Standard:} & Zn(s) + Cu^{2+}(1.0\text{ M}) \rightleftharpoons Zn^{2+}(1.0\text{ M}) + Cu(s) & & E^{\circ}_{cell} = 1.103\text{ V} \\
\textit{Nonstandard:} & Zn(s) + Cu^{2+}(0.075\text{ M}) \rightleftharpoons Zn^{2+}(1.5\text{ M}) + Cu(s) & & E_{cell} = 1.064\text{ V} \\
 & \text{decrease } [Cu^{2+}] \longleftarrow & & \\
 & \longleftarrow \text{increase } [Zn^{2+}] & &
\end{array}$$

Now let's look at the matter *quantitatively* by combining several familiar expressions. We start with the equation introduced on page 750 that relates ΔG, ΔG°, and the reaction quotient Q.

$$\Delta G = \Delta G^{\circ} + RT \ln Q$$

Into this equation, we can substitute the expressions

$$\Delta G = -nFE_{cell} \qquad \text{and} \qquad \Delta G^{\circ} = -nFE^{\circ}_{cell}$$

This gives the equation

$$-nFE_{cell} = -nFE^{\circ}_{cell} + RT \ln Q$$

Next, we change signs

$$nFE_{cell} = nFE^{\circ}_{cell} - RT \ln Q$$

and then solve for E_{cell}.

$$\frac{\cancel{nF}E_{cell}}{\cancel{nF}} = \frac{\cancel{nF}E^{\circ}_{cell}}{\cancel{nF}} - \frac{RT}{nF}\ln Q$$

$$E_{cell} = E^{\circ}_{cell} - \frac{RT}{nF}\ln Q$$

If we limit our discussion to 25 °C, we can replace the term RT/F by 0.025693 V, as we did when relating E°_{cell} and $\ln K_{eq}$.

$$E_{cell} = E^{\circ}_{cell} - \frac{0.025693\text{ V}}{n}\ln Q$$

This equation has customarily been written in terms of common logarithms. To switch from natural to common logarithms, we have to multiply 0.025693 V by ln 10. That is, 0.025693 V $\times$ 2.3026 = 0.059161 V, usually rounded off to 0.0592 V. As a result, the form of the equation that we will use is as follows.

$$E_{cell} = E^{\circ}_{cell} - \frac{0.0592\text{ V}}{n}\log Q$$

Walther Nernst (1864–1941) is well known for several important achievements in chemistry. He formulated the Nernst equation in 1889, when he was only 25 years old. In the same year, he suggested the solubility product concept. In 1906, he proposed a hypothesis that we now know as the third law of thermodynamics.

The equation we have just derived, the **Nernst equation**, relates a cell voltage for *nonstandard* conditions, E_{cell}, to a standard cell voltage, E°_{cell}, and the concentrations of reactants and products expressed through the reaction quotient, Q. In the equation, n is the number of moles of electrons involved in the cell reaction.

The Nernst equation is especially useful for determining the concentration of a species in a voltaic cell through a measurement of E_{cell}. Also, the equation helps us to understand why a cell voltage does not remain constant as electric current is drawn from a voltaic cell. As the cell reaction proceeds, the concentrations of products increase and the concentrations of reactants decrease. The value of Q increases continuously, and correspondingly, the voltage falls. The voltage falls to zero when the reaction reaches equilibrium $(Q = K_{eq})$.

EXAMPLE 18.9

Calculate the expected voltmeter reading for the voltaic cell pictured in Figure 18.12.

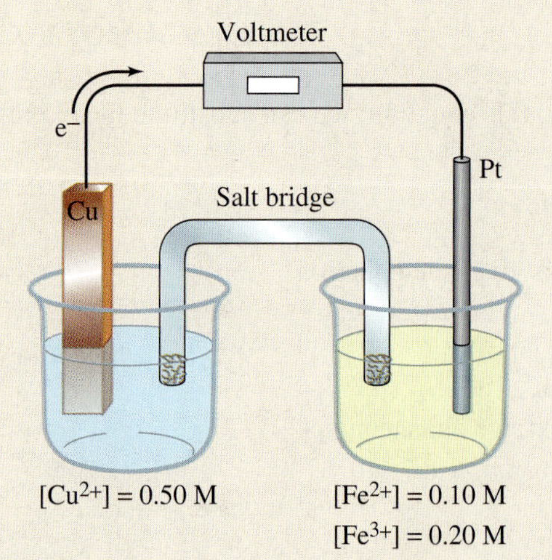

Figure 18.12
Example 18.9 Illustrated

SOLUTION

The expected voltmeter reading is E_{cell}, which we will determine with the Nernst equation. First we will use data from Table 18.1 to determine E°_{cell} for the cell reaction. Then we will formulate the reaction quotient Q and determine E_{cell}.

Determining E°_{cell}:

The direction of electron flow shown in Figure 18.12 enables us to identify the anode and cathode.

$$\begin{array}{lrl} \textit{Cathode (reduction):} & 2\,\{Fe^{3+}(1\,M) + e^- & \longrightarrow Fe^{2+}(1\,M)\} \\ \textit{Anode (oxidation):} & Cu(s) & \longrightarrow Cu^{2+}(1\,M) + 2\,e^- \\ \hline \textit{Overall:} & Cu(s) + 2\,Fe^{3+}(1\,M) & \longrightarrow Cu^{2+}(1\,M) + 2\,Fe^{2+}(1\,M) \end{array}$$

$$\begin{aligned} E^\circ_{cell} &= E^\circ(\text{cathode}) - E^\circ(\text{anode}) \\ &= E^\circ_{Fe^{3+}/Fe^{2+}} - E^\circ_{Cu^{2+}/Cu} \\ &= 0.771\text{ V} - 0.340\text{ V} = 0.431\text{ V} \end{aligned}$$

Determining E_{cell}:

We seek E_{cell} for a voltaic cell with the following cell reaction.

$$Cu(s) + 2\,Fe^{3+}(0.20\,M) \longrightarrow Cu^{2+}(0.50\,M) + 2\,Fe^{2+}(0.10\,M)$$

For this, we need to make the following substitutions into the Nernst equation: $E^\circ_{cell} = +0.431$ V; $n = 2$; $[Fe^{3+}] = 0.20$ M; $[Fe^{2+}] = 0.10$ M; $[Cu^{2+}] = 0.50$ M.

$$E_{cell} = E^\circ_{cell} - \frac{0.0592 \text{ V}}{n} \log Q$$

$$E_{cell} = 0.431 \text{ V} - \frac{0.0592 \text{ V}}{2} \log \left(\frac{[Cu^{2+}][Fe^{2+}]^2}{[Fe^{3+}]^2} \right)$$

$$E_{cell} = 0.431 \text{ V} - \frac{0.0592 \text{ V}}{2} \log \left(\frac{0.50(0.10)^2}{(0.20)^2} \right)$$

$$E_{cell} = 0.431 \text{ V} - (0.0296 \text{ V} \times \log 0.125) = 0.431 \text{ V} - (-0.027) \text{ V} = +0.458 \text{ V}$$

EXERCISE 18.9

Use the Nernst equation to determine E_{cell} at 25 °C for the following voltaic cells.

a. $Zn(s)|Zn^{2+}(2.0 \text{ M})||Cu^{2+}(0.050 \text{ M})|Cu(s)$

b. $Zn(s)|Zn^{2+}(0.050 \text{ M})||Cu^{2+}(2.0 \text{ M})|Cu(s)$

c. $Cu(s)|Cu^{2+}(1.0 \text{ M})||Cl^-(0.25 \text{ M})|Cl_2(g, 0.50 \text{ atm}), Pt$

Concentration Cells

The voltaic cell in Figure 18.13 differs from others that we have considered in that the two electrodes are *identical*. However, the solutions in the half-cells have different concentrations, and this creates a potential difference between the electrodes. Let's determine the source of this voltage and the nature of the cell reaction. First, for standard conditions,

Cathode (reduction): $Cu^{2+}(1 \text{ M}) + \cancel{2e^-} \longrightarrow \cancel{Cu(s)}$

Anode (oxidation): $\cancel{Cu(s)} \longrightarrow Cu^{2+}(1 \text{ M}) + \cancel{2e^-}$

Overall: $Cu^{2+}(1 \text{ M}) \longrightarrow Cu^{2+}(1 \text{ M})$

$$\begin{aligned} E^\circ_{cell} &= E^\circ(\text{cathode}) - E^\circ(\text{anode}) \\ &= E^\circ_{Cu^{2+}/Cu} - E^\circ_{Cu^{2+}/Cu} \\ &= 0.340 \text{ V} - 0.340 \text{ V} = 0.000 \text{ V} \end{aligned}$$

The *standard* cell voltage is *zero*, but this is what we should expect. One electrode reaction is simply the reverse of the other, and nothing happens. To calculate the cell voltage for the *nonstandard* conditions, we can use the Nernst equation.

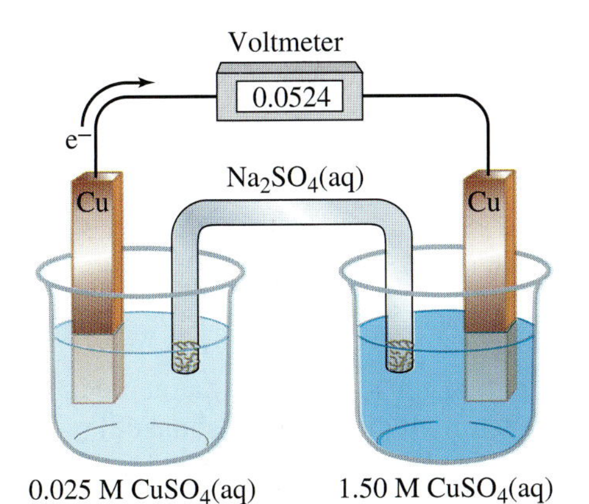

Figure 18.13
A concentration cell
The electrodes are identical, but the solution concentrations differ. The driving force for the cell reaction is the tendency for the solution concentrations to become equalized: $Cu^{2+}(1.50 \text{ M}) \longrightarrow Cu^{2+}(0.025 \text{ M})$.

$$Cu^{2+}(1.50\text{ M}) \longrightarrow Cu^{2+}(0.025\text{ M}) \quad E_{cell} = ?$$

$$E_{cell} = E^{\circ}_{cell} - \frac{0.0592\text{ V}}{n}\log Q$$

In this case, $E^{\circ}_{cell} = 0$, $n = 2$, and $Q = 0.025/1.50$.

$$E_{cell} = -\frac{0.0592\text{ V}}{2}\log\frac{0.025}{1.50} = -0.0296\text{ V} \times \log 0.017$$

$$= -0.0296\text{ V} \times (-1.77) = 0.0524\text{ V}$$

If the potential of a cell is determined solely by a difference in concentration of solutes in equilibrium with identical electrodes, that cell is called a **concentration cell**. The cell reaction is not a chemical reaction. It simply represents the migration of solute from a more concentrated to a more dilute solution. The more concentrated solution is diluted and the more dilute one becomes more concentrated, just as would happen if we simply brought two solutions of different concentration in contact. Disorder and entropy increase, and the process is spontaneous. A concentration cell is unusual in that it uses the natural tendency for solutions to mix as a basis of generating electricity.

pH Measurement

Suppose we construct a concentration cell consisting of two hydrogen electrodes. One is a standard hydrogen electrode, and the other is in a solution having an unknown pH; that is, with $[H^+] = x$. If $x < 1.0$ M—and this is generally the case—oxidation occurs at the nonstandard hydrogen electrode (the anode) and reduction occurs at the standard hydrogen electrode (the cathode).

Reduction:	$2\,H^+(1\text{ M}) + \cancel{2\,e^-} \longrightarrow \cancel{H_2(g, 1\text{ atm})}$
Oxidation:	$\cancel{H_2(g, 1\text{ atm})} \longrightarrow 2\,H^+(x\text{ M}) + \cancel{2\,e^-}$
Overall:	$2\,H^+(1\text{ M}) \longrightarrow 2\,H^+(x\text{ M})$

The voltage of this concentration cell is given by the Nernst equation, which takes the form

$$E_{cell} = -\frac{0.0592\text{ V}}{2}\log\frac{x^2}{1^2}$$

$$E_{cell} = -0.0296\text{ V}\log x^2$$

$$E_{cell} = -0.0296\text{ V}\,(2\log x) = -(0.0592\log x)\text{ V}$$

Because we often represent the acidity of a solution by the expression, $\text{pH} = -\log[H^+] = -\log x$, we can also write for the concentration cell

$$E_{cell} = 0.0592\text{ pH}$$

Thus, if we measure an E_{cell} value of 0.225 V in such a concentration cell at 25 °C, the unknown solution must have a pH of $0.225/0.0592 = 3.80$.

The pH Meter

A standard hydrogen electrode is difficult to construct and to maintain. In practice, to measure the pH of a solution we generally replace it with some other reference electrode having a precisely known E°. We also replace the second hydrogen elec-

The Electrochemistry of a Heartbeat

The heart is a pump made mostly of muscle, and like all muscles, it contracts to an electric signal. In muscle, electric current consists of the movement of ions across cell membranes. Sodium ion concentrations are high outside the cell and low inside; potassium ion concentrations are low outside the cell and high inside. These concentration gradients lead to a voltage difference, called the *membrane potential*.

Unlike other muscles, the heart has its own stimulator and built-in circuitry—it keeps on beating even if outside nerve connections are cut. Cut the nerve connections to the pectoral muscles, in contrast, and your arms are paralyzed for good.

The cells controlling the action of the heart valves, pacemaker cells, have a natural voltage that depends mainly on the concentrations of sodium and potassium ions. All cell membranes have a mechanism that moves these ions across the membranes, keeping K^+ inside the cell and expelling Na^+. Some K^+ leaks out, however, and this gives the interior of the cell a negative potential with respect to the outside. When the potential difference reaches a critical value, channels open up in the pacemaker cell membranes and Na^+ ions rush in. This results in an electric discharge, a current flow that spreads almost instantaneously to other heart muscle cells. The cells act in concert, and the heart beats.

Calcium ions also play a vital role in the heartbeat. Following the electric discharge, Ca^{2+} ions flow into the cells for a fraction of a second. When the flow of Ca^{2+} ions into the cells ceases, K^+ ions begin to leak out again, reestablishing a potential difference across the membrane.

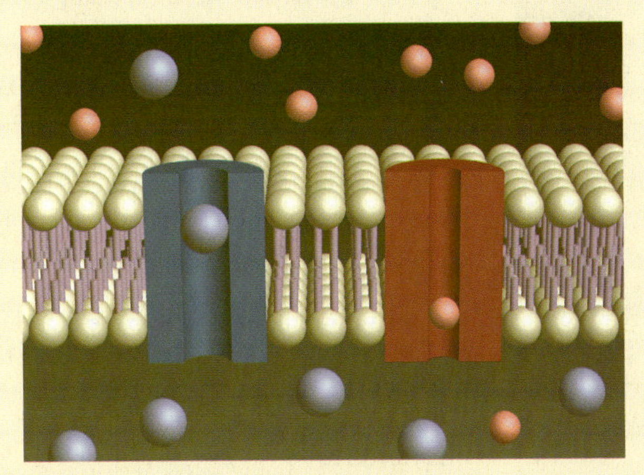

▲ This computer-generated model of a portion of a cell membrane shows a K^+ channel (blue) and an Na^+ channel (red). The area outside the cell (top) is rich in Na^+ and low in K^+. Inside the cell, the fluids are relatively rich in K^+ and low in Na^+.

The proper concentration of K^+ is crucial. Normal blood concentrations are 3.5 to 5.0 mmol K^+/L blood. The normal ratio of $[K^+]$ between the inside and outside of the pacemaker cells is about 30:1. If $[K^+]$ in the blood is too *low*, the ionic imbalance goes beyond the critical stage, and the pacemaker cells are overcharged; they remain cocked like broken pistols, but do not fire. The heart does not beat. If $[K^+]$ in the blood is too *high*, the potential difference never gets high enough to fire off a proper signal. The heartbeat diminishes to nothingness. Wasting of tissue from disease or starvation, loss of fluids through prolonged diarrhea or excessive vomiting, and long-term use of diuretic drugs all lead to dangerously low K^+ levels. Excessive dietary supplements and kidney diseases in which the body does not excrete enough potassium lead to K^+ levels that are too high. Both extremes—$[K^+]$ too low or too high—lead to cardiac arrest.

trode by a special electrode known as a *glass electrode*. This electrode, which is enclosed within a thin glass membrane, contains HCl(aq), into which is immersed a silver wire coated with AgCl(s). When the electrode is dipped into a solution of unknown pH, H^+ ions are exchanged across the membrane. The potential developed on the silver wire depends on $[H^+]$ in the unknown solution. The potential difference between the glass electrode and the reference electrode similarly depends on $[H^+]$ in the unknown solution (see Figure 18.14).

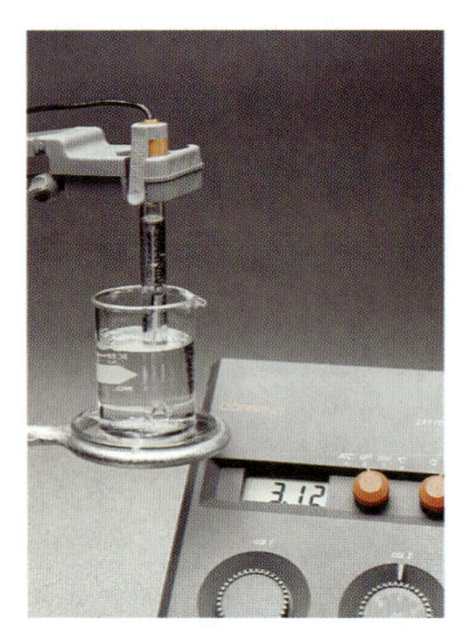

▲ **Figure 18.14**
The glass electrode and the pH meter
A reference electrode and a glass electrode are both enclosed in the combination electrode shown. The potential difference between the two is converted to the pH value of the solution (3.12) displayed by the meter.

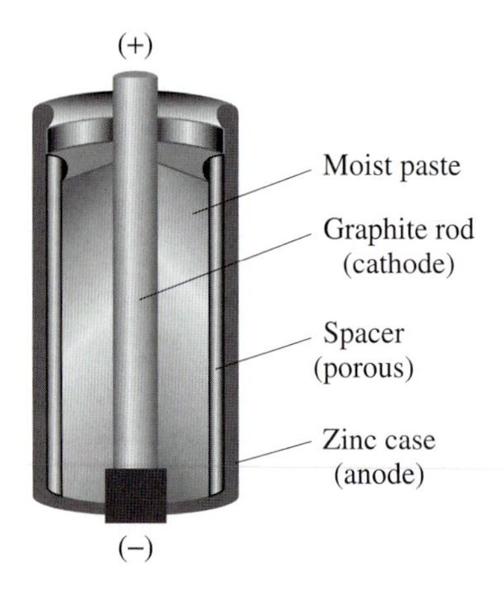

▲ **Figure 18.15**
Cross-section of a Leclanché (dry) cell
The moist paste consists of $NH_4Cl(aq)$ and $ZnCl_2(aq)$; carbon black (a finely divided form of carbon) and $MnO_2(s)$ are also present.

18.7 Batteries: Using Chemical Reactions to Make Electricity

In everyday life, we call a device that stores chemical energy for later release of electricity a **battery**. A flashlight battery consists of a single voltaic cell with two electrodes in contact with one or more electrolytes. Sometimes, a battery is defined as an assembly of two or more voltaic cells connected together. By this definition, an automobile battery is a true battery, but a flashlight battery is not. In this section, we consider three different types of cells or batteries.

The Dry Cell

In most flashlights and portable electronic devices, we use *primary* batteries. The cell reactions in primary batteries are irreversible. During use, reactants are converted to products, and when the reactants are used up, the battery is "dead." A typical example, the *Leclanché cell*—better known as a dry cell—is diagrammed in Figure 18.15.

A zinc container is the anode, and an inert carbon (graphite) rod in contact with $MnO_2(s)$ is the cathode. The electrolyte is a moist paste of NH_4Cl and $ZnCl_2$. The cell is called a dry cell because there is no free-flowing liquid. The concentrated electrolyte solution is thickened into a gellike paste by an agent such as starch. The cell diagram for a Leclanché cell is

$$Zn(s)|ZnCl_2(aq), NH_4Cl(aq)|MnO(OH)(s)|MnO_2(s)|C(graphite)$$

The reactions that occur when electric current is drawn from a dry cell are quite complex and not completely understood. We will give only a simplified description. As suggested by the cell diagram, zinc metal is oxidized to Zn^{2+} at the anode.

$$\textit{Anode}: Zn(s) \longrightarrow Zn^{2+}(aq) + 2\,e^-$$

The reduction reaction at the cathode appears to be

$$\textit{Cathode}: MnO_2(s) + H^+(aq) + e^- \longrightarrow MnO(OH)(s)$$

followed by

$$2\,MnO(OH)(s) \longrightarrow Mn_2O_3(s) + H_2O(l)$$

The Leclanché cell is cheap to make and has a maximum cell voltage of 1.55 V, but it has two disadvantages: (1) The cell voltage drops rapidly during use because of the build-up of Zn^{2+} near the anode and the change of pH at the cathode, and (2) the zinc metal slowly reacts with the electrolyte, even when the cell is disconnected.

Alkaline cells cost more but last longer than ordinary dry cells. They have a longer shelf life and can be kept in service longer. An alkaline cell has NaOH or KOH in place of NH_4Cl as the electrolyte. It uses essentially the same reduction half-reaction as the ordinary dry cell, but in an alkaline medium. Zinc ion produced in the oxidation half-reaction combines with $OH^-(aq)$ to form $Zn(OH)_2(s)$

$$\textit{Cathode}: \quad 2\,MnO_2(s) + H_2O(l) + \cancel{2e^-} \longrightarrow Mn_2O_3(s) + \cancel{2\,OH^-(aq)}$$

$$\textit{Anode}: \quad Zn(s) + \cancel{2\,OH^-(aq)} \longrightarrow Zn(OH)_2(s) + \cancel{2e^-}$$

$$\textit{Cell reaction}: \quad Zn(s) + 2\,MnO_2(s) + H_2O(l) \longrightarrow Zn(OH)_2(s) + Mn_2O_3(s)$$

$$\begin{aligned} E^\circ_{\text{cell}} &= E^\circ(\text{cathode}) - E^\circ(\text{anode}) \\ &= E^\circ_{\text{MnO}_2/\text{Mn}_2\text{O}_3} - E^\circ_{\text{Zn(OH)}_2/\text{Zn}} \\ &= 0.118\text{ V} - (-1.246\text{ V}) = 1.364\text{ V} \end{aligned}$$

In practice, because the conditions in the cell are *nonstandard*, the cell potential is not equal to 1.364 V—it is actually somewhat higher, about 1.5 V.

Another important advantage of the alkaline cell over the ordinary dry cell is that the voltage does not drop as rapidly when current is drawn. This is because the concentration of zinc ion does not build up at the anode and the pH remains more nearly constant at the cathode.

The Lead-Acid Storage Battery

The lead-acid storage battery used in automobiles is a *secondary* battery; it is rechargeable. The cell reaction can be reversed and the battery restored to its original condition. The battery can be used through repeated cycles of discharging and recharging. Figure 18.16 shows a portion of a cell in the lead-acid battery. Several anodes and several cathodes are connected together in each cell to increase its current-delivering capacity. Each cell has a voltage of 2 V, and six cells are connected together in series fashion, + to −, to form a 12-volt battery.

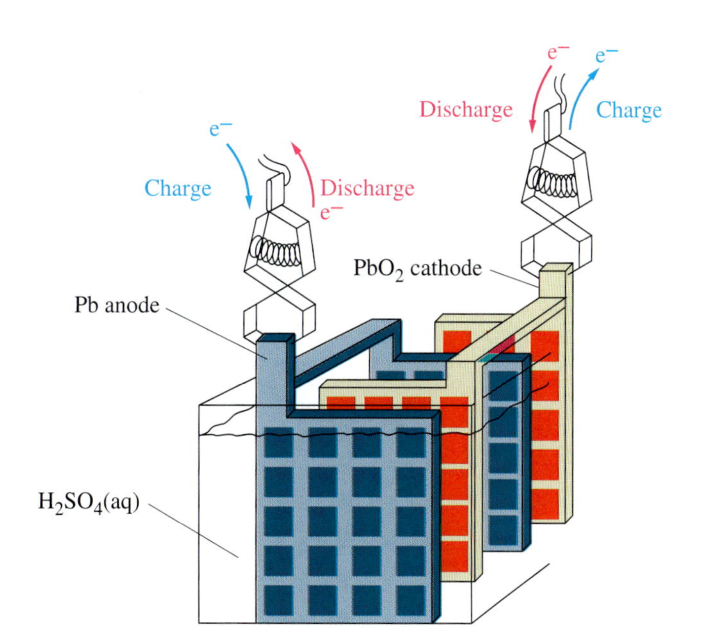

Figure 18.16
A lead-acid (storage) cell
The composition of the electrodes, the cell reaction, and the cell voltage are described in the text. Shown here are two anode plates and two cathode plates in parallel connections. This type of connection increases the surface area of the electrodes and the capacity of the cell to deliver current.

The anodes in the lead-acid storage cell are made of a lead alloy, and the cathodes are made of a lead alloy impregnated with red lead dioxide. The electrolyte is dilute sulfuric acid. The half-reactions and the cell reaction are as follows.

Cathode: $PbO_2(s) + 4H^+(aq) + SO_4^{2-}(aq) + \cancel{2e^-} \longrightarrow PbSO_4(s) + 2H_2O(l)$

Anode: $Pb(s) + SO_4^{2-}(aq) \longrightarrow PbSO_4(s) + \cancel{2e^-}$

Cell reaction: $Pb(s) + PbO_2(s) + 4\,H^+(aq) + 2\,SO_4^{2-}(aq) \longrightarrow 2\,PbSO_4(s) + 2\,H_2O(l)$

$$\begin{aligned} E^\circ_{\text{cell}} &= E^\circ(\text{cathode}) - E^\circ(\text{anode}) \\ &= E^\circ_{\text{PbO}_2/\text{PbSO}_4} - E^\circ_{\text{PbSO}_4/\text{Pb}} \\ &= 1.690\text{ V} - (-0.356\text{ V}) = 2.046\text{ V} \end{aligned}$$

Even though the conditions in a lead-acid cell are nonstandard, the voltage delivered by the cell is just about the same as the $E°_{cell}$ value.

As the cell reaction proceeds, $PbSO_4(s)$ precipitates and partially coats both electrodes; the water formed dilutes the $H_2SO_4(aq)$. In this condition, the cell is *discharged*. By connecting the cell to an external electric energy source we can force electrons to flow in the opposite direction. The half-cell reactions and the cell reaction are reversed and the battery is recharged.

Recharging reaction: $2\,PbSO_4(s) + 2\,H_2O(l)$

$$\longrightarrow Pb(s) + PbO_2(s) + 4\,H^+(aq) + 2\,SO_4^{2-}(aq) \quad E°_{cell} = -2.046\text{ V}$$

The negative value of $E°_{cell}$ signifies that the recharging reaction is *nonspontaneous*. The external source used to recharge the battery must provide a voltage in excess of 2.046 V for each cell in the battery, thus in excess of 6×2.046 V for a typical 12-volt battery.

In an automobile, the battery is discharged when the engine is started. While running, the engine powers the alternator, which produces electric energy required to recharge the battery. The battery is constantly recharged as the automobile is driven. When a dead battery is unable to start the engine, the engine can often be jump-started by connecting the weak battery to another fully charged battery. The engine gets its start from the good battery, and immediately the recharging of the weak battery begins. As a battery is discharged, the $H_2SO_4(aq)$ is diluted; it is reconcentrated during recharging. A simple method of determining the state of charge in a lead-acid battery is to measure the density of the acid. The more concentrated the acid, the greater its density is.

In principle, a lead-acid storage battery should last indefinitely, but it does not. During discharge of the battery, $PbSO_4(s)$ deposits on the electrodes in a finely divided form. If a discharged battery stands for a long time, the small grains of $PbSO_4(s)$ may grow into large crystals that fall from the electrodes. When the $PbSO_4(s)$ is in this form, the battery can no longer be recharged. This condition of "sulfating" is an important cause of battery failure.

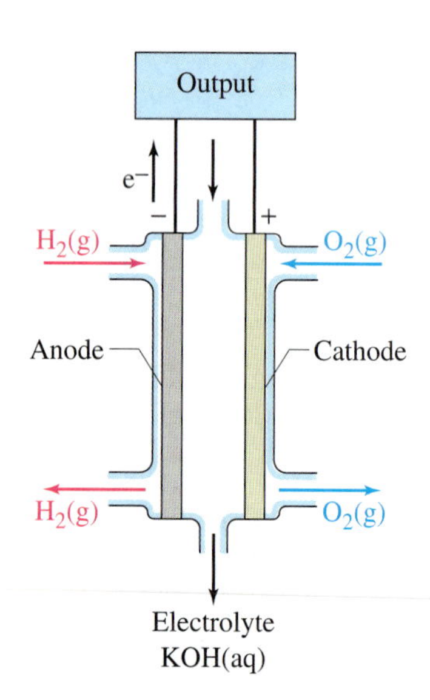

▲ **Figure 18.17**
A hydrogen-oxygen fuel cell
The electrodes are porous to allow easy access of the gaseous reactants to the electrolyte. The electrode material also catalyzes the electrode reactions.

A fuel cell is like a battery with the contents outside the battery case rather than inside. It is a flow battery. In principle, a fuel cell will continue to operate as long as reactants are fed into it.

Fuel Cells

Modern civilization runs mainly on fossil fuels, but as we have noted, combustion of fuels and conversion of the evolved heat to electricity is quite limited in efficiency. A voltaic cell, in contrast, is able to convert chemical energy to electricity with an efficiency of 90% or more. Because combustion reactions and voltaic cell reactions both involve oxidation and reduction, why not design a voltaic cell in which the cell reaction is equivalent to a combustion reaction? It's already been done. Scientists have been studying such devices for over 50 years. They are called *fuel cells*. The reaction within the cell is as follows.

$$\text{Fuel} + \text{oxygen} \longrightarrow \text{oxidation products}$$

In the hydrogen-oxygen cell, hydrogen and oxygen gases flow over separate inert electrodes in contact with an electrolyte such as KOH(aq) (see Figure 18.17). The reactions are shown below.

Cathode:	$O_2(g) + 2\,H_2O + 4\,e^- \longrightarrow 4\,OH^-$
Anode:	$2\{H_2(g) + 2\,OH^-(aq) \longrightarrow 2\,H_2O + 2\,e^-\}$
Overall:	$2\,H_2(g) + O_2(g) \longrightarrow 2\,H_2O(l)$

$$\begin{aligned} E^\circ_{\text{cell}} &= E^\circ(\text{cathode}) - E^\circ(\text{anode}) \\ &= E^\circ_{O_2/OH^-} - E^\circ_{H_2O/H_2} \\ &= 0.401\ \text{V} - (-0.828\ \text{V}) = 1.229\ \text{V} \end{aligned}$$

This E°_{cell} value is for 25 °C. Fuel cells of this type are generally operated at nonstandard conditions and at temperatures often considerably higher than 25 °C. Their operating voltages are about 1.0 to 1.1 V. Hydrogen-oxygen fuel cells have been widely used in space vehicles where, in addition to the electricity, the water formed is also a valuable product.

In principle, a fuel cell can run directly on a hydrocarbon fuel, such as methane.

$$\begin{array}{lr} \textit{Cathode:} & 2\,\{O_2(g) + 4\,H^+ + 4\,e^- \longrightarrow 2\,H_2O\} \\ \textit{Anode:} & CH_4(g) + 2\,H_2O \longrightarrow CO_2(g) + 8\,H^+ + 8\,e^- \\ \hline \textit{Overall:} & CH_4(g) + 2\,O_2(g) \longrightarrow CO_2(g) + 2\,H_2O(l) \end{array}$$

An indirect method that accomplishes the same objective involves steam reforming of the fuel to produce $H_2(g)$

$$CH_4(g) + H_2O(g) \longrightarrow CO(g) + 3\,H_2(g)$$

followed by use of $H_2(g)$ in a hydrogen-oxygen fuel cell.

Hydrogen-oxygen fuel cells show great promise for powering automobiles because they produce little in the way of undesirable products (beyond humid air). Government agencies classify such vehicles as zero-emission.

The Daimler-Benz NECAR 3 prototype is the world's first vehicle to produce hydrogen on board. Hydrogen and carbon dioxide gases are formed from gasoline and oxygen. Then, the hydrogen and oxygen are used in a fuel cell that produces electricity through the migration of protons (H^+) from one side of a membrane to the other. This technology could use existing petroleum refining and gasoline distribution systems.

Air Batteries

An air battery uses $O_2(g)$ from air as an oxidizing agent. Typically, the reducing agent is a metal, such as zinc or aluminum.

In an aluminum-air battery, oxidation occurs at an aluminum anode and reduction at a carbon cathode. The electrolyte circulated through the battery is NaOH(aq). Aluminum is oxidized to Al^{3+}, but because of the high concentration of OH^-, the complex anion $[Al(OH)_4]^-$ is formed (page 720).

$$\begin{array}{lr} \textit{Cathode:} & 3\,\{O_2(g) + 2\,H_2O(l) + 4\,e^- \longrightarrow 4\,OH^-(aq)\} \\ \textit{Anode:} & 4\,\{Al(s) + 4\,OH^-(aq) \longrightarrow [Al(OH)_4]^-(aq) + 3\,e^-\} \\ \hline \textit{Overall:} & 4\,Al(s) + 3\,O_2(g) + 6\,H_2O(l) + 4\,OH^-(aq) \longrightarrow 4\,[Al(OH)_4]^-(aq) \end{array}$$

$$\begin{aligned} E^\circ_{\text{cell}} &= E^\circ(\text{cathode}) - E^\circ(\text{anode}) \\ &= E^\circ_{O_2/OH^-} - E^\circ_{Al(OH)_4^-/Al} \\ &= 0.401\ \text{V} - (-2.310\ \text{V}) = 2.711\ \text{V} \end{aligned}$$

During operation, the battery is fed chunks of aluminum metal and water. The electrolyte is circulated outside the battery, where the $[Al(OH)_4]^-(aq)$ is precipitated as $Al(OH)_3(s)$. Aluminum is later recovered from the $Al(OH)_3(s)$ in an aluminum manufacturing facility. An aluminum-air battery can power an automobile for several hundred miles before requiring refueling and removal of $Al(OH)_3(s)$.

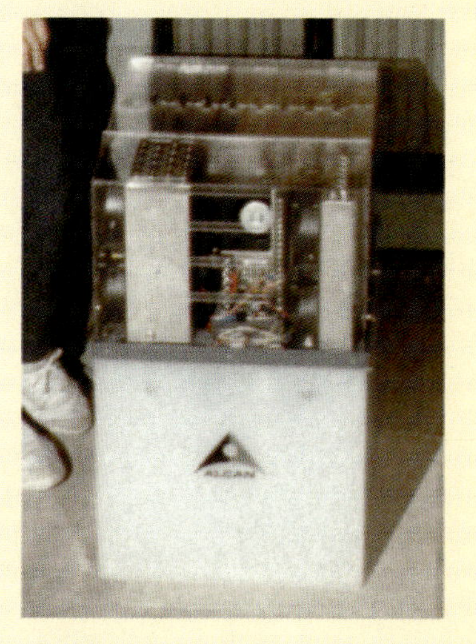

Aluminum-air batteries use oxygen gas from the air as an oxidizing agent and aluminum metal as the reducing agent. The battery consumes aluminum metal and water and produces aluminum hydroxide.

18.8 Corrosion: Metal Loss Through Voltaic Cells

Corrosion causes great economic loss, as shown by these rusted chains.

When carried out in voltaic cells, redox reactions are important sources of electricity. At the same time, some redox reactions produce electricity that underlies the destruction caused by corrosion. This is a matter of great economic importance. It is estimated that in the United States alone, corrosion costs $70 billion a year. Perhaps 20% of all the iron and steel produced in the United States each year goes to replace corroded material.

Let's look first at the corrosion of iron. In moist air, iron can be *oxidized* to Fe^{2+}, particularly at scratches, nicks, or dents. These regions, called *anodic* areas, are in a slightly higher energy state, and oxidation occurs a bit more readily here than elsewhere on the surface. The anodic areas become pitted. Electrons lost in the oxidation are conducted along the iron to other regions, called *cathodic* areas, where atmospheric $O_2(g)$ is reduced to OH^-.

$$\begin{array}{lrcl} \textit{Cathode:} & O_2(g) + 2\,H_2O(l) + 4\,e^- & \longrightarrow & 4\,OH^-(aq) \\ \textit{Anode:} & 2\,\{Fe(s) & \longrightarrow & Fe^{2+}(aq) + 2\,e^-\} \\ \hline \textit{Overall:} & 2\,Fe(s) + O_2(g) + 2\,H_2O(l) & \longrightarrow & 2\,Fe^{2+}(aq) + 4\,OH^-(aq) \end{array}$$

$$\begin{aligned} E^\circ_{cell} &= E^\circ(\text{cathode}) - E^\circ(\text{anode}) \\ &= E^\circ_{O_2/OH^-} - E^\circ_{Fe^{2+}/Fe} \\ &= 0.401\text{ V} - (-0.440\text{ V}) = 0.841\text{ V} \end{aligned}$$

The corrosion of iron, then, looks every bit like a spontaneous redox reaction occurring in a voltaic cell, and it is. Oxidation and reduction can occur at separate points on the metal. Electrons are conducted through the metal, and the circuit is completed by an electrolyte in aqueous solution. In the snow belt, this solution is often the slush from road salt and melting snow; in coastal areas, it may be ocean spray.

Iron(II) ions migrate from anodic areas to cathodic areas, where they combine with OH^- ions to form insoluble iron(II) hydroxide.

$$Fe^{2+}(aq) + 2\,OH^-(aq) \longrightarrow Fe(OH)_2(s)$$

Usually, iron(II) hydroxide is further oxidized to iron(III) hydroxide.

$$4\,Fe(OH)_2(s) + O_2(g) + 2\,H_2O(l) \longrightarrow 4\,Fe(OH)_3(s)$$

Iron(III) hydroxide is generally represented as a hydrated iron(III) oxide of variable composition: $Fe_2O_3 \cdot xH_2O$. This material is the familiar *iron rust*. The overall process described here is illustrated in Figure 18.18.

In order to emphasize the resemblance to a voltaic cell, we have focused on a situation in which the anodic and cathodic regions are some distance apart and intermediate stages of the corrosion are identifiable. Often, though, all the reactions occur simultaneously on the corroding portion of the iron. Then, all we really notice over time is the accumulation of rust.

Aluminum is more easily oxidized than iron. We might expect aluminum to corrode more rapidly than iron, but the occasional litter of beer and soda pop cans attests to the fact that it doesn't. How can this be? When freshly prepared aluminum is exposed to air, an oxide coating forms on its surface very rapidly. However, this thin, hard film of aluminum oxide is impervious to air and protects the underlying metal from further oxidation. Iron oxide (rust), in contrast, readily flakes off and constantly exposes fresh iron surface to further corrosion. This difference in be-

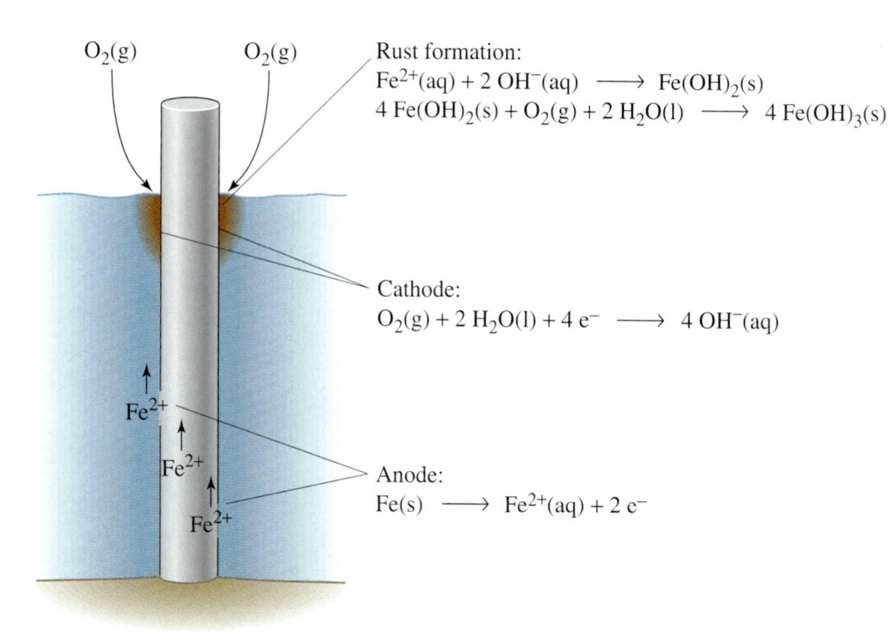

Figure 18.18 Corrosion of an iron piling: An electrochemical process
This schematic drawing illustrates the anodic and cathodic regions, their half-reactions, and the final formation of rust. The cathodic region is near the air-water interface, where the availability of $O_2(g)$ is greatest. The anodic region is at greater depths below the water's surface. $Fe^{2+}(aq)$ from the anodic region migrates to the cathodic region, where rust formation occurs.

havior of their corrosion products explains why steel cans eventually disintegrate in the environment whereas aluminum cans seem to last forever.

The simplest line of defense against the corrosion of iron is to paint it to exclude oxygen from the surface. Another approach is to coat the iron with a thin layer of another, less active metal, as in coating a steel can with tin. Either way, however, the surface is protected only as long as the coating does not crack, chip, or peel. A tin can may begin to rust rapidly if it is dented, and a car body usually rusts first where the paint is scratched or chipped.

An entirely different approach is to protect iron with a *more* active metal, as in the zinc-clad iron known as *galvanized iron*. The oxide coating on zinc is like that on aluminum—thin, hard, and impervious to air. And it doesn't matter if a break occurs in the zinc plating. The zinc continues to corrode rather than the iron because zinc is more easily oxidized than iron.

Sacrificial magnesium anodes attached to the steel structure of a ship provide cathodic protection against corrosion.

Another method that works on the same general principle as galvanized iron is known as *cathodic protection*. The iron or steel object to be protected—a ship, a pipeline, a storage tank, a water heater, or the plumbing system in a swimming pool—is connected to a chunk of an active metal such as magnesium, aluminum, or zinc, either directly or with a wire. Oxidation occurs at the active metal, and that metal slowly dissolves. The iron surface acquires electrons from the active metal and supports the *cathodic* or reduction half-reaction. As long as some of the active metal remains, the iron is protected. The active metal is called a *sacrificial anode*. One of the more important uses of magnesium metal is in sacrificial anodes (about 12 million tons annually in the United States).

Another example of the electrochemical control of corrosion is seen in a method for the removal of tarnish from silver. The tarnish is generally black insoluble silver sulfide formed when silver comes in contact with sulfur compounds, either in the atmosphere or in foods such as eggs. We can use a silver polish to remove the tarnish, but in doing so, we also lose part of the silver.

The concentration of Ag^+ in an aqueous solution in contact with $Ag_2S(s)$ is extremely low, yet aluminum is a sufficient reducing agent to reduce this Ag^+ back to silver metal. A precious metal is conserved at the expense of a cheaper one.

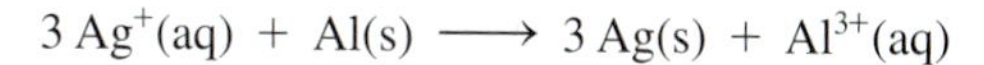

$$3\ Ag^+(aq) + Al(s) \longrightarrow 3\ Ag(s) + Al^{3+}(aq)$$

Sodium hydrogen carbonate ($NaHCO_3$) works well as an electrolyte to sustain this desired voltaic cell reaction. The tarnished silver is placed in contact with aluminum foil, covered with $NaHCO_3(aq)$, and heated.

EXAMPLE 18.10—A Conceptual Example

In Figure 18.19, iron nails are placed in a warm colloidal dispersion of agar in water. Phenolphthalein indicator and potassium ferricyanide, $K_3[Fe(CN)_6]$, are also present in the dispersion. When the faintly yellow-colored dispersion cools and stands for a few hours, it solidifies into a gel, and soon the colored regions begin to develop. Explain what is happening in Figure 18.19(a).

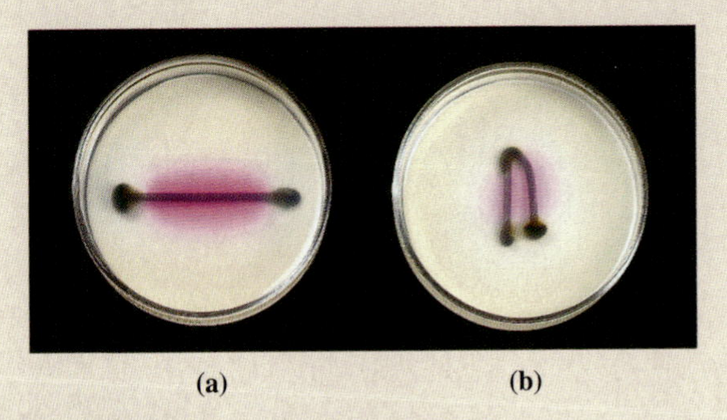

(a) (b)

▲ **Figure 18.19 Demonstrating the corrosion of iron**

SOLUTION

We learned in Chapter 15 that phenolphthalein, a pH indicator, is colorless in acidic and neutral solution and turns to a pink color above pH 10. The source of the OH^- ions responsible for the pink color is the cathodic half-reaction in the corrosion of iron.

$$\textit{Cathode:}\quad O_2(g) + 2\,H_2O(l) + 4\,e^- \longrightarrow 4\,OH^-(aq)$$

We can't have a cathodic half-reaction without an anodic half-reaction, and in the corrosion of iron the following half-reaction occurs.

$$\textit{Anode:}\quad Fe(s) \longrightarrow Fe^{2+}(aq) + 2\,e^-$$

The $Fe^{2+}(aq)$ reacts with the $[Fe(CN)_6]^{3-}(aq)$ to form a complex precipitate having a blue color (called Turnbull's blue).

The head and tip of the nail are *anodic* areas and are preferentially oxidized because the metal is more strained (and hence more energetic) in these regions. The body of the nail is a *cathodic* region.

EXERCISE 18.10

Explain what is happening in Figure 18.19(b) and why this differs in one detail from what is seen in Figure 18.19(a).

Electrolytic Cells

The general name for a combination of two half-cells, with the appropriate connections between electrodes and solutions, is an **electrochemical cell**. If the cell produces electric current through a *spontaneous* reaction, it is a *voltaic* cell. When electric current from an external source is passed through an electrochemical cell, a *nonspontaneous* reaction called **electrolysis** occurs; the cell is called an **electrolytic cell**. In the remainder of this chapter, we consider some electrolysis reactions and the electrolytic cells in which they are carried out.

If we use a battery or other source of direct electric current to pass electricity through molten sodium chloride, we observe that yellow-green chlorine gas forms at one electrode and silvery, liquid, metallic sodium forms at the other electrode and floats on the molten salt (see Figure 18.20). The external source of electricity acts like an "electron pump." It pulls electrons away from the *anode*, where the *oxidation* of Cl^- occurs.

$$2\,Cl^- \longrightarrow Cl_2(g) + 2\,e^-$$

Also, it pushes electrons onto the *cathode*, where *reduction* occurs.

$$Na^+ + e^- \longrightarrow Na(l)$$

The overall reaction is

$$2\,NaCl(l) \xrightarrow{\text{electrolysis}} 2\,Na(l) + Cl_2(g)$$

From this description, we see that in an electrolytic cell, as in a voltaic cell, oxidation occurs at the anode and reduction occurs at the cathode. However, the polarities of the electrodes are reversed from those in the voltaic cell because now the external source controls the flow of electrons. Because electrons are withdrawn from it, the anode of an electrolytic cell has a *positive* charge relative to the cathode. The cathode has a *negative* charge because of the electrons forced onto it.

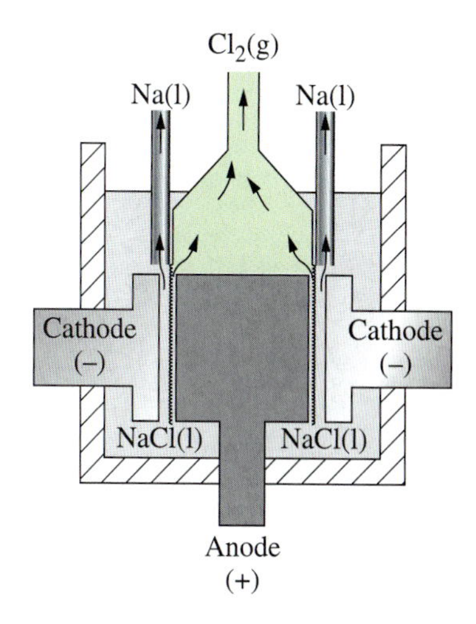

Figure 18.20
The electrolysis of molten sodium chloride
The electrolysis cell pictured here is called a *Downs cell*. The electrolyte is molten NaCl with a small amount of $CaCl_2$ added to lower its melting point. Liquid sodium forms at the steel cathode and gaseous chlorine at the graphite anode. A steel gauze diaphragm keeps the sodium and chlorine from recombining to form sodium chloride.

18.9 Predicting Electrolysis Reactions

As long as the potential difference applied to the electrodes in Figure 18.20 is sufficiently great (in excess of about 4 V), the electrolysis of NaCl(l) occurs and produces Na(l) and $Cl_2(g)$ as the sole products. If we apply the same voltage in an electrolytic cell containing an *aqueous* solution of NaCl, we don't get the same products, as we shall see shortly.

A useful approach to predicting the products of an electrolysis is to consider various oxidation and reduction half-reactions that might occur and then to select the oxidation and reduction that can be accomplished at the *lowest* voltage.

The Electrolysis of NaCl(aq)

Imagine an electrolytic cell in which two platinum electrodes are immersed in the same aqueous solution of NaCl and then connected to a battery. The battery will make one of the electrodes an anode and the other a cathode. Two possible half-reactions are essentially the ones observed in Figure 18.20. Two other possible half-reactions in an aqueous solution are the oxidation and reduction of water molecules. These four half-reactions are arranged below in order of decreasing value of the standard reduction potential, $E°$.

Anode (oxidation):

$$2\,Cl^-(aq) \longrightarrow Cl_2(g) + 2\,e^- \qquad E°_{Cl_2/Cl^-} = 1.358\text{ V}$$

$$2\,H_2O(l) \longrightarrow 4\,H^+(aq) + O_2(g) + 4\,e^- \qquad E°_{O_2/H_2O} = 1.229\text{ V}$$

Cathode (reduction):

$$2\,H_2O(l) + 2\,e^- \longrightarrow 2\,OH^-(aq) + H_2(g) \qquad E°_{H_2O/H_2} = -0.828\text{ V}$$

$$Na^+(aq) + e^- \longrightarrow Na(s) \qquad E°_{Na^+/Na} = -2.713\text{ V}$$

Suppose we were to determine E°_{cell} in the usual fashion, using the following equation.

$$E^\circ_{cell} = E^\circ(\text{cathode}) - E^\circ(\text{anode})$$

No matter what combination of cathode and anode half-reactions we consider, all will give a *negative* value of E°_{cell}. But this simply means that the electrolysis reaction won't occur spontaneously. Still, it is reasonable to expect that the reaction most likely to occur is the one with the *least* negative value of E°_{cell}. It would require the lowest applied voltage from the external electricity source. The favored electrolysis reaction should involve the reduction of H_2O ($E^\circ = -0.828$ V) rather than Na^+ ($E^\circ = -2.713$ V) at the cathode. It would also seem to involve the oxidation of H_2O ($E^\circ = 1.229$ V) rather than Cl^- ($E^\circ = 1.358$ V) at the anode.

There are other factors involved, however. One is the fact that the reactants and products may not be in their standard states. For example, in the industrial electrolysis of NaCl(aq) the pH is adjusted to about 4. This is not the pH of 0 associated with $[H^+] \approx 1$ M. An adjustment for this difference in pH gives $E = 0.99$ V as the *nonstandard* reduction potential of the half-reaction, $2\ H_2O(l) \longrightarrow 4\ H^+(aq) + O_2(g) + 4\ e^-$. Lowering this reduction potential from 1.229 V to 0.99 V makes it even more likely that the oxidation of H_2O to O_2 is favored over the oxidation of Cl^- to Cl_2.

A second factor works in the opposite direction, however. In many half-reactions, particularly those involving gases, various interactions at electrode surfaces make the required voltage for electrolysis higher than the voltage calculated from E° data. **Overvoltage** is the excess voltage above the voltage calculated from E° values that is required in electrolysis. The overvoltage for the formation of $H_2(g)$ at the cathode is almost zero, but the overvoltage for the formation of $O_2(g)$ at the anode is high, and it is significantly greater than for the formation of $Cl_2(g)$.

The net effect of the two factors we have been describing is that $O_2(g)$ is formed in the electrolysis of very dilute NaCl(aq), but in more concentrated NaCl(aq), the product at the anode is almost exclusively $Cl_2(g)$. In the commercial electrolysis of NaCl(aq), which we further describe on page 807, the overall electrolysis reaction is the following.

$$2\ Cl^-(aq) + 2\ H_2O(l) \xrightarrow{\text{electrolysis}} 2\ OH^- + H_2(g) + Cl_2(g)$$

When H_2O is oxidized at the anode and also reduced at the cathode, the overall reaction is simply the electrolysis of water.

$$2\ H_2O(l) \xrightarrow{\text{electrolysis}} 2\ H_2(g) + O_2(g)$$

However, the water must also contain ions to make it electrically conducting even though the ions are not involved in the electrode reactions.

Inert Electrodes and Active Electrodes

In the electrolysis of NaCl(l) and NaCl(aq), the electrodes used were *inert*. They provided a surface on which the exchange of electrons and hence the oxidation and reduction half-reactions could occur. Otherwise, they did not participate in the electrolysis reaction. In contrast, some electrodes are *active* electrodes; they directly participate in a half-reaction. Thus, in predicting electrolysis reactions, we must always consider the nature of the electrodes, as illustrated in Example 18.11.

● ChemCDX Electrolysis of Water

EXAMPLE 18.11

Predict the electrolysis reaction when $AgNO_3(aq)$ is electrolyzed **(a)** using platinum electrodes and **(b)** using a silver anode and a platinum cathode.

SOLUTION

a. The platinum electrodes are inert, so we need to consider only the species present in the solution; these are Ag^+ and NO_3^- ions and H_2O molecules. Silver ion is easily reduced, as we can see from the half-reaction in Table 18.1.

$$Ag^+(aq) + e^- \longrightarrow Ag(s) \qquad E°_{Ag^+/Ag} = 0.800\ V$$

This reduction should occur exclusively over that of water molecules.

$$2\ H_2O(l) + 2\ e^- \longrightarrow H_2(g) + 2\ OH^-(aq) \qquad E°_{H_2O/H_2} = -0.828\ V$$

Because NO_3^- has the N atom in its highest possible oxidation state (+5), it cannot be oxidized. The only likely oxidation is that of H_2O molecules.

$$2\ H_2O(l) \longrightarrow O_2(g) + 4\ H^+(aq) + 4\ e^- \qquad E°_{O_2/H_2O} = 1.229\ V$$

The calculated voltage required for the electrolysis is

$$E°_{cell} = E°(\text{cathode}) - E°(\text{anode})$$
$$= 0.800\ V - 1.229\ V = -0.429\ V$$

The electrolysis reaction is as follows.

$$4\ Ag^+(aq) + 2\ H_2O(l) \xrightarrow{\text{electrolysis}} 4\ Ag(s) + 4\ H^+(g) + O_2(g)$$

b. The reduction half-reaction at the cathode is the same as in part (a).

$$Ag^+(aq) + e^- \longrightarrow Ag(s) \qquad E°_{Ag^+/Ag} = 0.800\ V$$

But, in contrast to the platinum electrode, the silver anode is an *active* electrode; the silver undergoes oxidation.

$$Ag(s) \longrightarrow Ag^+(aq) + e^- \qquad E°_{Ag^+/Ag} = 0.800\ V$$

The calculated voltage required for the electrolysis is

$$E°_{cell} = E°(\text{cathode}) - E°(\text{anode})$$
$$= 0.800\ V - 0.800\ V = 0\ V$$

This required voltage of zero does not mean, however, that electrolysis can be accomplished just by placing the electrodes in the solution without using a battery or other source of electricity. A small voltage is required to drive electrons through the external circuit and ions through the solution. The overall electrolysis simply involves the transport of silver, as Ag^+, through the solution from the anode to the cathode.

$$Ag(s)_{anode} \xrightarrow{\text{electrolysis}} Ag(s)_{cathode}$$

EXERCISE 18.11A

Predict the electrolysis reaction when KBr(aq) is electrolyzed using platinum electrodes.

EXERCISE 18.11B

Predict the electrolysis reaction when $AgNO_3$(aq) is electrolyzed using a silver anode and a copper cathode. What is the approximate external voltage required for the electrolysis?

EXAMPLE 18.12—A Conceptual Example

Two electrochemical cells are connected as shown. Specifically, the zinc electrodes of the two cells are joined, as are the copper electrodes. Will there be **(a)** no current, **(b)** a flow of electrons in the direction of the red arrows, or **(c)** a flow of electrons in the direction of the blue arrows?

e^- Zn(s) | Zn^{2+}(1 M) || Cu^{2+}(0.10 M) | Cu(s) e^- | e^- Zn(s) | Zn^{2+}(0.10 M) || Cu^{2+}(1.0 M) | Cu(s) e^-

Cell A Cell B

SOLUTION

Notice that these are voltaic cells like in Figure 18.4 set up in opposition. That is, a stream of electrons coming out of the zinc electrode of one cell encounters a stream of electrons coming from the zinc electrode of the other cell. The net flow of electrons will be in the direction that has the greater "push" (voltage) behind it.

If the two cells were *identical*, their E_{cell} values would be the same and there would be no current flow. Because the cells are *not* identical, we can eliminate possibility (a).

When the cells operate as voltaic cells, the cell reaction, as we have seen before, is this one.

$$Zn(s) + Cu^{2+}(aq) \longrightarrow Zn^{2+}(aq) + Cu(s)$$

Cell A has $[Zn^{2+}] = 1.0$ M (standard state) and $[Cu^{2+}] = 0.10$ M (more dilute than in the standard state). From Le Châtelier's principle, we would expect the reaction to be *less* favorable than under standard state conditions. For Cell A: $E_{cell} < E^\circ_{cell}$.

Cell B has $[Zn^{2+}] = 0.10$ M (more dilute than the standard state) and $[Cu^{2+}] = 1.0$ M (standard state). From Le Châtelier's principle, we would expect the forward reaction to be *more* favorable than under standard state conditions. For Cell B: $E_{cell} > E^\circ_{cell}$.

We conclude that there is a net flow of electrons from the Zn electrode of Cell B to the Zn electrode of Cell A. That is, Cell B is a voltaic cell, forcing electrons into the Zn cathode of Cell A, an electrolytic cell. Also, electrons are withdrawn from the Cu anode of Cell A. The direction of electron flow is that of the red arrows.

EXERCISE 18.12

Write net ionic equations for the cell reactions in Cell A and Cell B when they are connected as shown in Example 18.12. Explain why electric current cannot continue to flow indefinitely.

Michael Faraday (1791–1867) is known for his discoveries in electricity and magnetism, as well as in electrochemistry. Faraday was unschooled in mathematics but had an unrivaled intuitive ability to visualize physical phenomena.

18.10 Quantitative Electrolysis

The quantitative basis of electrolysis was largely established by Michael Faraday. The quantity of a reactant consumed or product formed during electrolysis is related to (1) the molar mass of the substance, (2) the quantity of electric charge used, and (3) the number of electrons transferred in the electrode reaction.

The unit of electric charge is the **coulomb (C)**, and the charge carried by one electron is -1.6022×10^{-19} C. Electric *current*, expressed in **amperes (A)**, is the rate of flow of electric charge. One ampere of current is a flow of one coulomb per second.

$$1\ A = 1\ C/s$$

The total electric charge involved in an electrolysis reaction, then, is a product of the current and the time that it flows.

$$\text{Electric charge (C)} = \text{current (C/s)} \times \text{time (s)}$$

We need this expression and the Faraday constant (96,485 C/mol e^-) to calculate the quantitative outcomes of electrolysis reactions. Our usual strategy will be the following.

1 *Determine the amount of charge—the product of current and time.*

2 *Convert the amount of charge (C) to moles of electrons: 1 mol e^- = 96,485 C.*

3 *Use a half-equation to relate moles of electrons to moles of a reactant or product*. For example, the half-equation $Cu^{2+}(aq) + 2\ e^- \longrightarrow Cu(s)$ yields the factor

$$\frac{1\ \text{mol Cu}}{2\ \text{mol e}^-}$$

4 *Convert from moles of reactant or product to the final quantity desired*—mass, volume of a gas, and so on.

For some problems, we need to use variations of this four-step strategy. For example, in Exercise 18.13A, each step of the strategy is inverted, and the steps are performed in the reverse order: 4, 3, 2, 1.

EXAMPLE 18.13

We can use electrolysis to determine the gold content of a sample. The sample is dissolved, and all the gold converted to $Au^{3+}(aq)$. The reduction half-reaction is $Au^{3+}(aq) + 3\ e^- \longrightarrow Au(s)$. What mass of gold will be deposited at the cathode in 1.00 hour by a current of 1.50 A?

SOLUTION

We can apply the four-step procedure just outlined.

1 *Determine the amount of charge*. The total electric charge involved is

$$\frac{1.50\ \text{C}}{\text{s}} \times 1\ \text{hr} \times \frac{60\ \text{min}}{1\ \text{hr}} \times \frac{60\ \text{s}}{1\ \text{min}} = 5.40 \times 10^3\ \text{C}$$

2 *Convert the amount of charge (C) to moles of electrons*. The number of moles of electrons is

$$5.40 \times 10^3\ \text{C} \times \frac{1\ \text{mol e}^-}{96{,}485\ \text{C}} = 0.0560\ \text{mol e}^-$$

3 *Use a half-equation to relate moles of electrons to moles of the product*. According to the reduction half-equation, 3 mol e^- are required for every mole of Au deposited.

$$0.0560\ \text{mol e}^- \times \frac{1\ \text{mol Au}}{3\ \text{mol e}^-} = 0.0187\ \text{mol Au}$$

4 *Convert from moles to grams of product*. The mass of Au, in grams, is simply the product of the amount in moles and the molar mass.

$$0.0187\ \text{mol Au} \times \frac{197.0\ \text{g Au}}{1\ \text{mol Au}} = 3.68\ \text{g Au}$$

EXERCISE 18.13A

For how many minutes must the electrolysis of a solution of $CuSO_4$(aq) be carried out, at a current of 2.25 A, to deposit 1.00 g of Cu(s) at the cathode?

EXERCISE 18.13B

A coulometer is a device for measuring a quantity of electric charge. One type of coulometer is an electrolytic cell with platinum electrodes in $AgNO_3$(aq). If 2.175 g Ag(s) is deposited on the cathode in an electrolysis lasting 21 minutes and 12 seconds, **(a)** what is the amount of charge, in coulombs, passing through the cell, and **(b)** what is the magnitude of the current, in amperes?

EXAMPLE 18.14—An Estimation Example

Without doing detailed calculations, determine which of the following solutions will yield the greatest mass of metal at a platinum cathode during its electrolysis by a 1.50-A electric current for 30.2 min: $CuSO_4$(aq), $AgNO_3$(aq), or $AuCl_3$(aq).

SOLUTION

We can begin by comparing the cathode half-reactions:

$$Cu^{2+} + 2\,e^- \longrightarrow Cu(s) \qquad Ag^+ + e^- \longrightarrow Ag(s) \qquad Au^{3+} + 3\,e^- \longrightarrow Au(s)$$

From these half-equations, we see that for every 1 mol e^- we could obtain

$$1/2 \text{ mol Cu(s)} \quad \text{or} \quad 1 \text{ mol Ag(s)} \quad \text{or} \quad 1/3 \text{ mol Au(s)}$$

With just a glance at a table of atomic masses (Cu = 63.5 u; Ag = 108 u; Au = 197 u), we can conclude that the

$$\text{mass of } 1/2 \text{ mol Cu} < \text{mass of } 1/3 \text{ mol Au} < \text{mass of } 1 \text{ mol Ag}.$$

We see that the $AgNO_3$(aq) yields the greatest mass of metal. Notice that we didn't need to use the current and time data because we didn't have to calculate actual masses of the three metals.

EXERCISE 18.14

Without doing detailed calculations, determine which of the following solutions will yield the greatest volume of gas (at STP) at a platinum anode when the same amount of electric charge is passed through each of them: $Mg(NO_3)_2$(aq), NaCl(aq), or KI(aq)?

18.11 Applications of Electrolysis

In concluding this chapter, let's consider a few more of the many many practical applications of electrolysis.

Producing Chemicals by Electrolysis

Electrolysis plays an important role in the manufacture and purification of a host of substances: sodium, calcium, magnesium, aluminum, copper, zinc, silver, hydrogen, chlorine, fluorine, hydrogen peroxide, sodium hydroxide, potassium dichromate, and potassium permanganate, among others.

One of the most commercially important electrolyses is that of NaCl(aq). We wrote a net ionic equation for this electrolysis on page 802, but we can also include the spectator ion, Na^+.

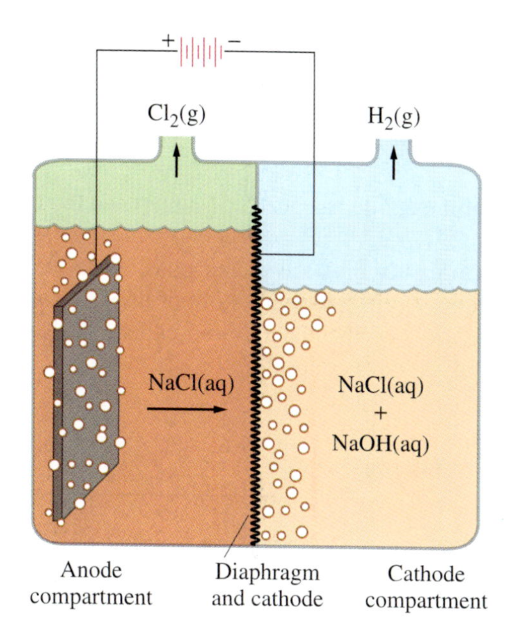

▲ **Figure 18.21**
A diaphragm chlor-alkali cell
The anode is a specially treated titanium metal. The diaphragm and cathode are a composite unit consisting of an asbestos-polymer mixture deposited on a steel wire mesh. A difference in the solution levels is maintained, so that NaCl(aq) moves through the diaphragm during the electrolysis.

$$2\,Na^+(aq) + 2\,Cl^- + 2\,H_2O(l) \longrightarrow \underset{\text{Sodium hydroxide}}{2\,Na^+(aq) + 2\,OH^-(aq)} + \underset{\text{Hydrogen}}{H_2(g)} + \underset{\text{Chlorine}}{Cl_2(g)}$$

The products most sought are the chlorine gas and sodium hydroxide solution. Hence, the electrolysis of NaCl(aq) is called the *chlor-alkali* process.

One common type of chlor-alkali electrolytic cell, called a diaphragm cell, is pictured in Figure 18.21. The major challenge in a chlor-alkali process is to keep $Cl_2(g)$ from coming into contact with the NaOH(aq). In an alkaline solution, Cl_2 disproportionates, producing ClO^- and Cl^-. The diaphragm prevents this contact. A chlor-alkali cell must also prevent the mixing of Cl_2 and H_2. Mixtures of these gases can be explosive.

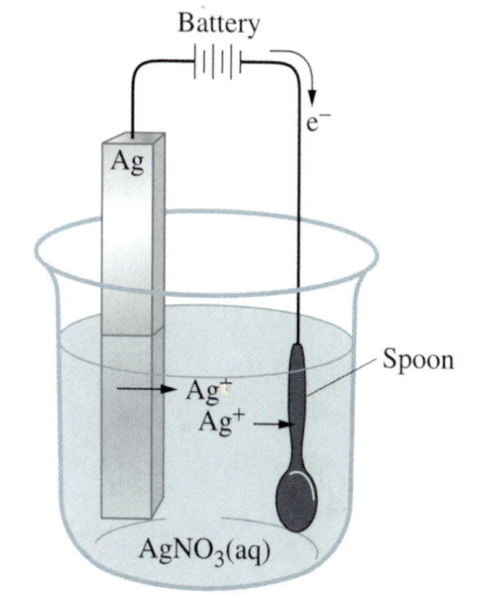

▲ **Figure 18.22**
An electrochemical cell for silver plating
The anode is a silver bar, and the cathode is an iron spoon.

Electroplating

Electrolysis can be used to coat one metal onto another, a process called *electroplating*. Usually, the object to be electroplated, such as a spoon, is cast of an inexpensive metal. It is then coated with a thin layer of a more attractive, more corrosion-resistant, and more expensive metal, such as silver or gold. The finished product costs far less than if the object had been made entirely of the more expensive metal.

A cell for electroplating silver is shown in Figure 18.22. A piece of pure silver is the anode, the spoon is the cathode, and a solution of silver nitrate is the electrolyte. Think of the spoon as replacing the inert platinum cathode, and you will see that the electrolysis is the same as that described in Example 18.11(b) on page 803.

$$Ag(s,\ anode) \longrightarrow Ag(s,\ cathode)$$

A similar, large-scale industrial use of electroplating is in the refining of copper. High-purity copper is required for applications that depend on copper's high electrical conductivity. A massive anode of impure copper and a thin cathode of pure copper are suspended in an electrolyte such as $CuSO_4(aq)$. Just as in the case of silver plating, the overall process is the transfer of copper atoms from the anode to the cathode.

Copper is refined by electroplating copper from an impure copper anode onto a pure copper cathode.

Anode:	$Cu(s, impure) \longrightarrow Cu^{2+}(aq) + \cancel{2e^-}$	
Cathode:	$Cu^{2+}(aq) + \cancel{2e^-} \longrightarrow Cu(s, pure)$	
Overall:	$Cu(s, impure) \longrightarrow Cu(s, pure)$	

Electrolysis is continued until most of the anode has been consumed and the cathode has built up into a thick sheet.

Because copper is oxidized more readily than silver and gold, these metals, which may be present in the impure copper, drop to the bottom of the electrolysis cell below the anode. The value of the precious metals in this so-called *anode mud* is often sufficient to pay for the cost of the electrorefining.

Key Terms

ampere, A (18.10)
anode (18.3)
battery (18.7)
cathode (18.3)
cell diagram (18.3)
cell reaction (18.3)
cell voltage (cell potential), E_{cell} (18.3)
concentration cell (18.6)
coulomb, C (18.10)
electrochemical cell (page 800)
electrochemistry (page 766)
electrode (18.3)
electrolysis (page 801)
electrolytic cell (page 801)
electromotive force (emf) (18.3)
Faraday constant (F) (18.5)
half-cell (18.3)
half-reaction (18.1)
Nernst equation (18.6)
overvoltage (18.9)
salt bridge (18.3)
standard cell potential, E°_{cell} (18.4)
standard electrode potential, E° (18.4)
standard hydrogen electrode (18.4)
volt, V (18.3)
voltaic cell (18.3)

Summary

An oxidation-reduction reaction can be separated into two half-reactions, one for oxidation and the other for reduction. Half-equations for half reactions can be balanced separately and then recombined to yield a balanced redox equation. In this recombination, all the electrons "lost" in the oxidation half-reaction must be "gained" in the reduction half-reaction.

Half-reactions can be conducted in half-cells. In a voltaic cell, the electrodes in two half-cells are joined by a wire, and electrolyte solutions by a salt bridge. Oxidation occurs in the anode half-cell and reduction in the cathode half-cell. The spontaneous cell reaction produces electricity.

A cell diagram for a voltaic cell is written with the anode on the left and the cathode on the right. Phase boundaries are represented by a vertical line ($|$), and a salt bridge or porous barrier by a double vertical line ($\|$). Solution concentrations may also be noted, as in the voltaic cell

$$Zn(s)\,|\,Zn^{2+}(0.10\text{ M})\,\|\,Cu^{2+}(0.50\text{ M})\,|\,Cu(s)$$

A standard hydrogen electrode has H^+ ion at unit activity ($\approx$1 M) in equilibrium with H_2(g, 1 atm) on an inert platinum electrode. Standard electrode potentials, E°, are written for reduction half-reactions. They are obtained by comparison to the standard hydrogen electrode, which is assigned a potential of zero. E°_{cell} for a reaction is the difference between the standard potentials of the cathode and anode.

$$E^\circ_{cell} = E^\circ(\text{cathode}) - E^\circ(\text{anode})$$

A redox reaction for which $E_{cell} > 0$ occurs spontaneously. If $E_{cell} < 0$, the reaction is nonspontaneous. Usually these criteria are applied for standard conditions, and E°_{cell} values are obtained from tables of standard electrode potentials. Another important use of E°_{cell} is in determining values of ΔG° and K_{eq}.

$$\Delta G^\circ = -nFE^\circ_{cell} = -RT \ln K_{eq}$$

The *Nernst equation* relates a cell voltage under nonstandard conditions to E°_{cell}, the number of electrons transferred in the cell reaction (n), and the reaction quotient (Q). At 25 °C,

$$E_{cell} = E^\circ_{cell} - \frac{0.0592}{n} \log Q$$

In concentration cells, the half-cells have identical electrodes and solutions of the same electrolyte but at different concentrations. The cell voltage depends only on these concentrations. That is, the Nernst equation reduces to

$$E_{cell} = \frac{-0.0592}{n} \log Q \quad \text{(at 25 °C)}$$

Batteries are voltaic cells, taken singly or joined together to yield higher voltages or higher currents. In primary batteries, the electrode reactions are irreversible; in secondary batteries, they are reversible. Thus, primary batteries cannot be recharged, but secondary batteries can. In a fuel cell, a fuel and oxygen (from air) combine to produce oxidation products, and chemical energy is released as electricity.

A corroding metal consists of anodic areas, at which dissolution of the metal occurs, and cathodic areas, where atmospheric oxygen is reduced to hydroxide ion. A metal can be protected against corrosion by plating it with a second metal that corrodes less readily. In another method, cathodic protection, an active metal is sacrificed to protect a less active metal to which it is joined.

In an electrolytic cell, direct electric current from an external source produces nonspontaneous changes. Electrode potential data can be used to predict the probable products of an electrolysis reaction, although other factors may have to be considered as well. Electrolysis is an important method of manufacturing chemicals and refining metals. The amount of chemical change produced in electrolysis depends on the amount of charge that is transferred at the electrodes. The amount of charge, in turn, depends on the magnitude of the electric current used and the time required for the electrolysis.

Review Questions

1. What is a *half-reaction* of an oxidation-reduction reaction? How do electrons participate in half-reactions?

2. What are the basic ideas involved in balancing a redox equation by the half-reaction method?

3. What is meant by the term *electrode*?

4. What is a half-cell and what is a voltaic cell?

5. What is meant by the terms *anode* and *cathode* in an electrochemical cell?

6. What is the function of a salt bridge in an electrochemical cell?

7. In what ways are voltaic cells and electrolytic cells similar and how do they differ?

8. Can a Brønsted-Lowry acid-base reaction be the basis of a half-reaction in an electrochemical cell? Explain.

9. What is a cell diagram? What aspects of a cell are incorporated in a cell diagram?

10. Which of the following pairs of terms have the same meaning and which have a different meaning? Explain.

(a) cell voltage and cell potential

(b) electrode potential and standard electrode potential

(c) half-cell and half-reaction

11. Some standard electrode potentials have positive values and some have negative values. Why don't all electrode potentials have the same sign?

12. Explain the distinction between the symbols $E°$ and $E°_{cell}$; between E_{cell} and $E°_{cell}$.

13. Describe the standard hydrogen electrode. What is the potential assigned to it?

14. Explain the relationship between the activity series of the metals introduced in Chapter 4 and the table of standard electrode potentials introduced in this chapter.

15. What is the criterion for spontaneous chemical change based on cell potentials? Explain.

16. Which two of the following metals do *not* react with HCl(aq): Mg, Ag, Zn, Fe, Au?

17. Explain why electrode potentials can be used to make predictions of spontaneous change even for reactions that are not carried out in electrochemical cells. Are there some reactions for which they cannot be used? Explain.

18. If $E°_{cell} < 0$, does this mean that the cell reaction will not occur spontaneously, regardless of the conditions? Explain.

19. How is $E°_{cell}$ related to $\Delta G°$ and to K_{eq}?

20. What is the Faraday constant, F? What are its numerical value and units? Describe some situations where it is used.

21. What is the purpose served by the Nernst equation in electrochemistry?

22. What is a concentration cell? What is the value of $E°_{cell}$ for a concentration cell? What is the overall change that occurs as electric current is drawn from the cell?

23. What is the difference between a primary and a secondary battery? Give an example of each.

24. What is a fuel cell? What is the basic principle involved in its operation?

25. Describe the electrochemical nature of corrosion. Use the corrosion of iron as an example.

26. Explain why iron corrodes much more readily than does aluminum, even though iron is the less active of the two metals.

27. What is cathodic protection? How does it provide protection against corrosion?

28. How do standard electrode potentials enter into determining the voltage required to carry out an electrolysis?

29. Should an object to be electroplated with a metal be made the anode or the cathode in an electrolytic cell? Explain.

30. What electrolysis is implied by the term chlor-alkali process? What are the products of the electrolysis?

Problems

Oxidation-Reduction Reactions

31. Complete and balance the following half-equations, and indicate whether oxidation or reduction is involved.

(a) $ClO_2(g) \longrightarrow ClO_3^-(aq)$ (acidic solution)

(b) $MnO_4^-(aq) \longrightarrow MnO_2(s)$ (acidic solution)

(c) $SbH_3(g) \longrightarrow Sb(s)$ (basic solution)

32. Complete and balance the following half-equations, and indicate whether oxidation or reduction is involved.

(a) $P_4(s) \longrightarrow H_3PO_4(aq)$ (acidic solution)

(b) $MnO_2(s) \longrightarrow MnO_4^-(aq)$ (basic solution)

(c) $CH_3CH_2OH(aq) \longrightarrow CO_2(g)$ (basic solution)

33. Use the half-reaction method to balance the following equations for redox reactions in acidic solution.

(a) $Ag(s) + NO_3^-(aq) + H^+(aq) \longrightarrow Ag^+(aq) + H_2O(l) + NO(g)$

(b) $H_2O_2(aq) + MnO_4^-(aq) + H^+(aq) \longrightarrow Mn^{2+}(aq) + H_2O(l) + O_2(g)$

(c) $Cl_2(g) + I^-(aq) + H_2O(l) \longrightarrow Cl^-(aq) + IO_3^-(aq) + H^+(aq)$

34. Use the half-reaction method to balance the following redox equations for reactions in acidic solution.

(a) $Fe^{2+}(aq) + Cr_2O_7^{2-}(aq) + H^+(aq) \longrightarrow Fe^{3+}(aq) + Cr^{3+}(aq) + H_2O(l)$

(b) $S_8(s) + O_2(g) + H_2O(l) \longrightarrow SO_4^{2-}(aq) + H^+(aq)$

(c) $Fe^{3+}(aq) + (NH_2OH_2)^+(aq) \longrightarrow Fe^{2+}(aq) + N_2O(g) + H^+(aq) + H_2O(l)$

35. Use the half-reaction method to balance the following equations for redox reactions in basic solution.

(a) $Fe(OH)_2(s) + O_2(g) + H_2O(l) \longrightarrow Fe(OH)_3(s)$

(b) $S_8(s) + OH^-(aq) \longrightarrow S_2O_3^{2-}(aq) + S^{2-}(aq) + H_2O(l)$

(c) $CrI_3(s) + H_2O_2(aq) + OH^-(aq) \longrightarrow CrO_4^{2-}(aq) + IO_4^-(aq) + H_2O(l)$

36. Use the half-reaction method to balance the following equations for redox reactions in basic solution.

(a) $CrO_4^{2-}(aq) + AsH_3(g) + H_2O(l) \longrightarrow Cr(OH)_3(s) + As(s) + OH^-(aq)$

(b) $CH_3OH(aq) + MnO_4^-(aq) \longrightarrow HCOO^-(aq) + MnO_2(s) + H_2O(l) + OH^-(aq)$

(c) $[Fe(CN)_6]^{3-}(aq) + N_2H_4(aq) + OH^-(aq) \longrightarrow [Fe(CN)_6]^{4-}(aq) + N_2(g) + H_2O(l)$

37. Write balanced equations for the following redox reactions.

(a) The reaction of oxalic acid (HOOCCOOH) and permanganate ion in acidic solution to produce manganese(II) ion and carbon dioxide gas

(b) The reaction of $Cr_2O_7^{2-}$ and UO^{2+} to produce UO_2^{2+} and Cr^{3+} in an acidic aqueous solution

(c) The reaction of nitrate ion and zinc in basic solution to produce zinc(II) ion and gaseous ammonia

38. Write balanced equations for the following redox reactions.

(a) The reaction of thiosulfate and permanganate ions in acidic solution to form sulfate and manganese(II) ions

(b) The action of permanganate ion on acetaldehyde (CH_3CHO) in basic solution to produce manganese(IV) oxide and acetate ion

(c) The disproportionation of manganate ion (MnO_4^{2-}) to permanganate ion and solid manganese(IV) oxide in a basic solution

Electrode Potentials and Voltaic Cells

Use data from Table 18.1 and Appendix C, as necessary.

39. For the voltaic cell pictured, write an equation for the cell reaction that occurs, and determine the voltmeter reading if the metal, M, is **(a)** Sn, **(b)** Zn, and **(c)** Cu. Note that the voltmeter can display both positive and negative voltages.

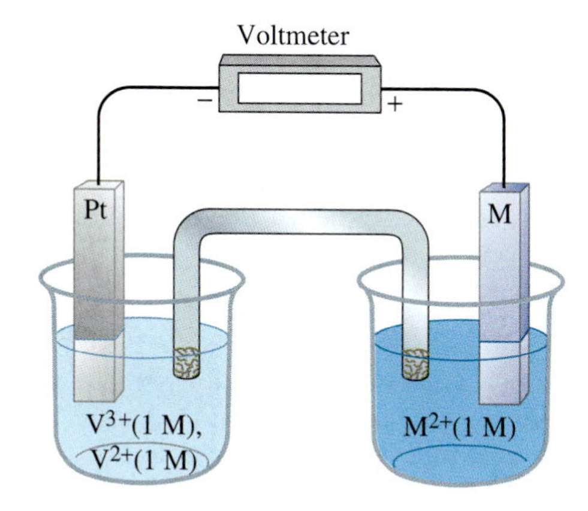

40. From the indicated voltages for the voltaic cell pictured, determine the standard electrode potential, $E°_{M^{3+}/M}$, if the metal, M, is **(a)** In, $E°_{cell} = 0.086$ V; **(b)** La, $E°_{cell} = -1.96$ V; **(c)** Tl, $E°_{cell} = 1.14$ V

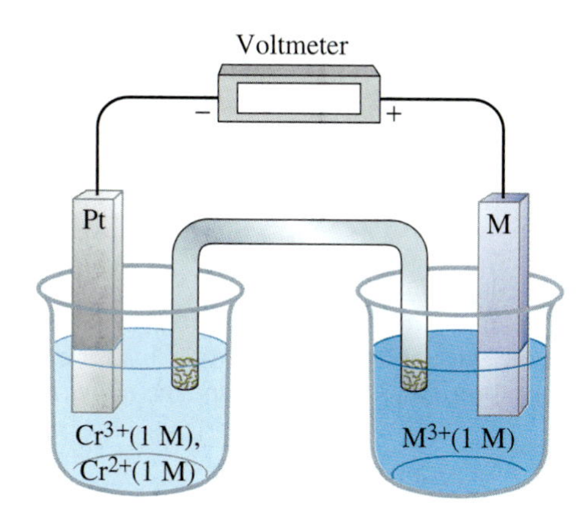

41. For the reaction

$$2\,CuI(s) + Cd(s) \longrightarrow Cd^{2+}(aq) + 2\,I^-(aq) + 2\,Cu(s) \qquad E°_{cell} = +0.23\text{ V}$$

Given that $E°_{Cd^{2+}/Cd} = -0.403$ V, determine the value of $E°$ for the half-reaction

$$2\,CuI(s) + 2\,e^- \longrightarrow 2\,Cu(s) + 2\,I^-(aq)$$

42. For the reaction

$$3\,V(s) + 2\,SbO^+(aq) + 4\,H^+(aq) \longrightarrow 3\,V^{2+}(aq) + 2\,Sb(s) + 2\,H_2O(l) \qquad E°_{cell} = 1.387\text{ V}$$

Given that $E°_{V^{2+}/V} = -1.175$ V, determine the value of $E°$ for the half-reaction

$$SbO^+(aq) + 2\,H^+(aq) + 3\,e^- \longrightarrow Sb(s) + H_2O(l)$$

43. $E°_{cell} = 1.47$ V for the voltaic cell

$$V(s)\,|\,V^{2+}(1\text{ M})\,\|\,Cu^{2+}(1\text{ M})\,|\,Cu(s)$$

Determine the value of $E°_{V^{2+}/V}$.

44. $E°_{cell} = 3.73$ V for the voltaic cell

$$Y(s)\,|\,Y^{3+}(1\text{ M})\,\|\,Cl^-(1\text{ M})\,|\,Cl_2(g, 1\text{ atm}), Pt$$

Determine the value of $E°_{Y^{3+}/Y}$.

45. Write equations for the half-reactions and the overall cell reaction, and calculate $E°_{cell}$ for each of the voltaic cells diagrammed below.

(a) $Pt, I_2(s)\,|\,I^-(aq)\,\|\,Cl^-(aq)\,|\,Cl_2(g), Pt$

(b) $Pt, PbO_2(s)\,|\,Pb^{2+}(aq), H^+(aq)\,\|\,S_2O_8^{\ 2-}(aq), SO_4^{\ 2-}(aq)\,|\,Pt$

46. Write equations for the half-reactions and the overall cell reaction, and calculate $E°_{cell}$ for each of the voltaic cells diagrammed below.

(a) $Pt\,|\,Fe^{2+}(aq), Fe^{3+}(aq)\,\|\,Cr_2O_7^{\ 2-}(aq), Cr^{3+}(aq)\,|\,Pt$

(b) $Pt, NO(g)\,|\,NO_3^{\ -}(aq), H^+(aq)\,\|\,H^+(aq), H_2O_2(aq)\,|\,Pt$

47. Each of the following reactions is made to take place in a voltaic cell. Write equations for the half-reactions and the overall cell reaction. Write a cell diagram for the voltaic cell, and calculate the value of $E°_{cell}$.

(a) $Zn(s) + Ag^+(aq) \longrightarrow Ag(s) + Zn^{2+}(aq)$

(b) $Fe^{2+}(aq) + O_2(g) + H^+(aq) \longrightarrow Fe^{3+}(aq) + H_2O(l)$

48. Each of the following reactions is made to take place in a voltaic cell. Write equations for the half-reactions and the overall cell reaction. Write a cell diagram for the voltaic cell, and calculate the value of $E°_{cell}$.

(a) $Fe^{3+}(aq) + Sn^{2+}(aq) \longrightarrow Fe^{2+}(aq) + Sn^{4+}(aq)$

(b) $Cu(s) + H^+(aq) + NO_3^{\ -}(aq) \longrightarrow Cu^{2+}(aq) + H_2O(l) + NO(g)$

E°_{cell} and the Spontaneity of Redox Reactions

Use data from Table 18.1 and Appendix C, as necessary.

49. Predict whether a spontaneous reaction will occur in the forward direction in each of the following. Assume that all reactants and products are in their standard states.

(a) $Sn(s) + Co^{2+}(aq) \longrightarrow Sn^{2+}(aq) + Co(s)$

(b) $6\,Br^-(aq) + Cr_2O_7^{2-}(aq) + 14\,H^+(aq) \longrightarrow 2\,Cr^{3+}(aq) + 7\,H_2O(l) + 3\,Br_2(l)$

50. Predict whether a spontaneous reaction will occur in the forward direction in each of the following. Assume that all reactants and products are in their standard states.

(a) $Sn^{4+}(aq) + 2\,I^-(aq) \longrightarrow Sn^{2+}(aq) + I_2(s)$

(b) $2\,MnO_2(s) + 3\,ClO^-(aq) + 2\,OH^-(aq) \longrightarrow 2\,MnO_4^-(aq) + 3\,Cl^-(aq) + H_2O(l)$

51. Predict whether each of the following processes will proceed in the forward direction to any appreciable extent.

(a) The displacement of $Cd^{2+}(aq)$ by $Al(s)$

(b) The oxidation of $Cl^-(aq)$ to $Cl_2(g)$ by $Br_2(l)$

(c) The oxidation of $Cl^-(aq)$ to $ClO_3^-(aq)$ by H_2O_2 in basic solution

52. Predict whether each of the following processes will proceed in the forward direction to any appreciable extent.

(a) The reduction of $Sn^{4+}(aq)$ to $Sn^{2+}(aq)$ by $Cu(s)$

(b) The oxidation of $I_2(s)$ to $IO_3^-(aq)$ by $O_3(g)$ in acidic solution

(c) The oxidation of $Cr(OH)_3(s)$ to $CrO_4^{2-}(aq)$ by H_2O_2 in basic solution

53. Silver does not react with HCl(aq), but it does react with $HNO_3(aq)$. **(a)** Explain the difference in the behavior of silver toward these two acids. **(b)** Write a plausible net ionic equation for the reaction of silver with $HNO_3(aq)$.

54. Can we use sodium metal to displace $Mg^{2+}(aq)$ from aqueous solution? If the reaction does occur, write the half-equations and the net equation. If the displacement reaction does *not* occur, write the equation for the reaction that does occur.

55. Rhodium is a rare metal used as a catalyst. The metal does not react with HCl(aq), but it does react with $HNO_3(aq)$, producing $Rh^{3+}(aq)$ and NO(g). Copper will displace Rh^{3+} from aqueous solution, but silver will not. Estimate a value of $E^\circ_{Rh^{3+}/Rh}$.

56. Palladium is a rare metal used as a catalyst. Copper and silver will both displace Pd^{2+} from aqueous solution. The metal itself will react with $HNO_3(aq)$ producing $Pd^{2+}(aq)$ and NO(g). Estimate a value of $E^\circ_{Pd^{2+}/Pd}$.

E°_{cell}, ΔG°, K_{eq}

57. Determine the values of E°_{cell} and ΔG° for the following reactions.

(a) $Al(s) + 3\,Ag^+(aq) \longrightarrow Al^{3+}(aq) + 3\,Ag(s)$

(b) $4\,IO_3^-(aq) + 4\,H^+(aq) \longrightarrow 2\,I_2(s) + 2\,H_2O(l) + 5\,O_2(g)$

58. Determine the values of E°_{cell} and ΔG° for the following reactions.

(a) $O_2(g) + 4\,I^-(aq) + 4\,H^+(aq) \longrightarrow 2\,H_2O(l) + 2\,I_2(s)$

(b) $Cr_2O_7^{2-}(aq) + 3\,Cu(s) + 14\,H^+(aq) \longrightarrow 2\,Cr^{3+}(aq) + 3\,Cu^{2+}(aq) + 7\,H_2O(l)$

59. Write the equilibrium constant expression for each of the following reactions, and determine the numerical value of K_{eq} at 25 °C.

(a) $Ag^+(aq) + Fe^{2+}(aq) \rightleftharpoons Fe^{3+}(aq) + Ag(s)$

(b) $MnO_2(s) + 4\,H^+(aq) + 2\,Cl^-(aq) \rightleftharpoons Mn^{2+}(aq) + 2\,H_2O(l) + Cl_2(g)$

(c) $2\,OCl^-(aq) \rightleftharpoons 2\,Cl^-(aq) + O_2(g)$ (basic solution)

60. Write the equilibrium constant expression for each of the following reactions, and determine the numerical value of K_{eq} at 25 °C.

(a) $PbO_2(s) + 4\,H^+(aq) + 2\,Cl^-(aq) \rightleftharpoons Pb^{2+}(aq) + 2\,H_2O(l) + Cl_2(g)$

(b) $3\,O_2(g) + 2\,Br^-(aq) \rightleftharpoons 2\,BrO_3^-(aq)$ (basic solution)

61. A strip of tin metal is immersed in 1.00 M $Pb^{2+}(aq)$ at 25 °C. What is the reaction that occurs? What will be the concentrations of the cations in solution when equilibrium is reached?

62. A strip of copper metal is immersed in 1.00 M $Ag^+(aq)$ at 25 °C. What is the reaction that occurs? What will be the concentrations of the cations in solution when equilibrium is reached?

Effect of Concentration on E_{cell}

Use data from Table 18.1 and Appendix C, as necessary.

63. What is the value of E_{cell} of each of the following reactions when carried out in a voltaic cell?

(a) $Fe(s) + 2\,Ag^{+}(0.0015\ M) \longrightarrow Fe^{2+}(1.33\ M) + 2\,Ag(s)$

(b) $4\,VO^{2+}(0.050\ M) + O_2(g, 0.25\ atm) + 2\,H_2O(l) \longrightarrow 4\,VO_2^{+}(0.75\ M) + 4\,H^{+}(0.30\ M)$

64. What is the value of E_{cell} of each of the following reactions when carried out in a voltaic cell?

(a) $Pb(s) + 2\,H^{+}(0.0025\ M) \longrightarrow Pb^{2+}(0.85\ M) + H_2(g, 0.95\ atm)$

(b) $ClO_3^{-}(0.65\ M) + 3\,Mn^{2+}(0.25\ M) + 3\ H_2O(l) \longrightarrow Cl^{-}(1.50\ M) + 3\ MnO_2(s) + 6\,H^{+}(1.25\ M)$

65. What is E_{cell} for the voltaic cell diagrammed below?

$$Pt, H_2(g, 1\ atm)\,|\,0.0025\ M\ HCl\,||\,H^{+}(1\ M)\,|\,H_2(g, 1\ atm), Pt$$

66. What is E_{cell} for the voltaic cell diagrammed below?

$$Pt, H_2(g, 1\ atm)\,|\,0.0675\ M\ HCl\,||\,0.0250\ M\ KOH\,|\,H_2(g, 1\ atm), Pt$$

67. The voltaic cell diagrammed below registers $E_{cell} = 0.108\ V$

$$Pt, H_2(g, 1\ atm)\,|\,H^{+}(x\ M)\,||\,H^{+}(1\ M)\,|\,H_2(g, 1\ atm), Pt$$

What is the pH of the unknown solution?

68. A voltaic cell represented by the following cell diagram has $E_{cell} = -0.015$ V. Calculate $[Ag^{+}]$ in the cell.

$$Pt\,|\,Fe^{2+}(0.125\ M), Fe^{3+}(0.068\ M)\,||\,Ag^{+}(x\ M)\,|\,Ag(s)$$

69. *Without doing detailed calculations*, determine which of the following voltaic cells should have the higher cell voltage. Explain your reasoning.

(a) $Cu(s)\,|\,Cu^{2+}(1.25\ M)\,||\,Ag^{+}(0.55\ M)\,|\,Ag(s)$

(b) $Cu(s)\,|\,Cu^{2+}(0.12\ M)\,||\,Ag^{+}(0.60\ M)\,|\,Ag(s)$

70. *Without doing detailed calculations*, determine which of the following voltaic cells has the highest molarity of Cu^{2+}(? M). Explain your reasoning.

(a) $Zn(s)\,|\,Zn^{2+}(1.00\ M)\,||\,Cu^{2+}(?\ M)\,|\,Cu(s)$ $E_{cell} = 1.06\ V$

(b) $Zn(s)\,|\,Zn^{2+}(0.10\ M)\,||\,Cu^{2+}(?\ M)\,|\,Cu(s)$ $E_{cell} = 1.15\ V$

(c) $Zn(s)\,|\,Zn^{2+}(1.0 \times 10^{-3}\ M)\,||\,Cu^{2+}(?\ M)\,|\,Cu(s)$ $E_{cell} = 1.16\ V$

Batteries

71. Describe plausible electrode reactions and the cell reaction for a battery system having a Zn anode and a Cl_2 cathode. Determine its E°_{cell}.

72. Describe plausible electrode reactions and the cell reaction for a battery system having a Mg anode and an O_2 cathode. Determine its E°_{cell}.

73. Silver-zinc cells, called button batteries, are tiny cells used in watches, electronic calculators, hearing aids, and cameras. The battery has the following cell diagram.

$$Zn(s), ZnO(s)\,|\,KOH(sat'd\ aq)\,|\,Ag_2O(s), Ag(s)$$

Its storage capacity is about six times that of a lead-acid cell of the same size. Write equations for the half-reactions and the overall reaction that occur when the cell is discharged.

74. A mercury battery, once widely used, is being phased out because of the problem of disposing of the toxic mercury. A simplified cell diagram for the battery is

$$Zn(s), ZnO(s)\,|\,KOH(sat'd\ aq)\,|\,HgO(s), Hg(l)$$

Write equations for the half-reactions and the overall reaction that occur when the cell is discharged.

Corrosion

75. Why are water, an electrolyte, and oxygen all required for the corrosion of iron?

76. Iron can be protected from corrosion by copper plating or by zinc plating. Both methods are effective so long as the plating remains intact. If a break occurs in the plating, however, zinc plating proves to be far superior to copper plating in providing protection to the underlying iron. Explain why this should be so.

77. Explain why the term "sacrificial anode" is an appropriate one in describing one method of protecting iron from corrosion.

78. Describe how silver tarnish can be removed without the loss of silver.

79. Example 18.10 and Exercise 18.10 dealt with interpreting parts **(a)** and **(b)** of Figure 18.19. The metal shown with the iron nail in part **(c)** of the same photograph is zinc. Explain how and why part **(c)** of this photograph differs from Figure 18.19(a).

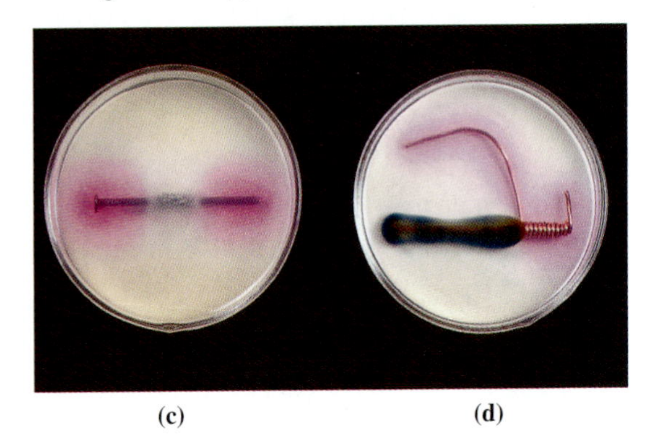

(c) **(d)**

80. Explain how and why part **(d)** of the photograph shown above differs from Figure 18.19(a). The metal shown with the iron nail is copper.

Electrolysis

81. Write a net ionic equation for the expected reaction when the electrolysis of $NiSO_4$(aq) is conducted using **(a)** a nickel anode and an iron cathode; **(b)** a nickel anode and an inert platinum cathode; **(c)** an inert platinum anode and a nickel cathode.

82. Write a net ionic equation for the expected reaction when the electrolysis of $Cu(NO_3)_2$(aq) is conducted using **(a)** a copper anode and a copper cathode; **(b)** an inert platinum anode and an iron cathode; **(c)** inert platinum for both electrodes.

83. Use electrode potential data to predict the probable products and the minimum voltage required in the electrolysis with inert platinum electrodes of each of the following.

(a) $BaCl_2$(l) **(c)** $NaNO_3$(aq)

(b) HBr(aq)

84. Use electrode potential data to predict the probable products and the minimum voltage required in the electrolysis with inert platinum electrodes of each of the following.

(a) $ZnSO_4$(aq) **(c)** $NiCl_2$(aq)

(b) $MgBr_2$(l)

85. How many grams of silver are deposited at a platinum cathode in the electrolysis of $AgNO_3$(aq) by 1.73 A of electric current in 2.05 hours?

86. How many mL of H_2(g), measured at 23.5 °C and 749 mmHg, are produced at a platinum cathode in the electrolysis of H_2SO_4(aq) by 2.45 A of electric current in 5.00 minutes?

87. How many coulombs of electric charge are required to deposit 25.0 g Cu(s) at the cathode in the electrolysis of $CuSO_4$(aq)?

88. What is the electric current, in amperes, if 212 mg Ag(s) is deposited at the cathode in 1435 seconds in the electrolysis of $AgNO_3$(aq)?

89. *Without doing detailed calculations*, determine which of the following will yield the greatest mass of metal deposit on a platinum cathode when the solution is electrolyzed for exactly one hour with a current of 1.00 A. Explain your choice.

(a) $Cu(NO_3)_2$(aq) **(c)** $AgNO_3$(aq)

(b) $Zn(NO_3)_2$(aq) **(d)** $NaNO_3$(aq)

90. *Without doing detailed calculations*, determine which of the following solutions will take the longest time to electrolyze under these conditions: Equal volumes of the solutions are used, and electrolysis is carried out with inert platinum electrodes and a current of 1.00 A. Electrolysis is stopped when the solution concentration falls to half its initial value. Explain your choice.

(a) 0.50 M $Cu(NO_3)_2$ **(c)** 0.80 M $AgNO_3$

(b) 0.75 M HNO_3 **(d)** 0.30 M $Zn(NO_3)_2$

Additional Problems

91. When $P_4(s)$ is heated with water, it disproportionates to phosphine, $PH_3(g)$, and phosphoric acid. Write a balanced equation for this reaction.

92. Cyanide wastes can be detoxified by adding chlorine gas to a basic solution of the wastes. The cyanide ion is converted to cyanate ion (OCN^-), and the chlorine is reduced to chloride ion. Following the addition of some acid so that the solution is not quite so basic, further reaction with chlorine gas converts cyanate ion to hydrogen carbonate (bicarbonate) ion and nitrogen gas. Write balanced equations for the two reactions just described.

93. Balance the following redox equations by the half-reaction method.

(a) $B_2Cl_4 + OH^- \longrightarrow BO_2^- + Cl^- + H_2O + H_2(g)$

(b) $CH_3CH_2ONO_2 + Sn + H^+ \longrightarrow CH_3CH_2OH + NH_2OH + Sn^{2+} + H_2O$

(c) $F_5SeOF + OH^- \longrightarrow SeO_4^{2-} + F^- + H_2O + O_2(g)$

(d) $As_2S_3 + OH^- + H_2O_2 \longrightarrow AsO_4^{3-} + SO_4^{2-} + H_2O$

(e) $XeF_6 + OH^- \longrightarrow XeO_6^{4-} + F^- + H_2O + Xe(g) + O_2(g)$

94. Although we have used the half-reaction method only to balance equations for redox reactions occurring in aqueous solution, the method can actually be used more widely. Balance the following equation by the half-reaction method, and explain why the method works.

$$NO(g) + H_2(g) \longrightarrow NH_3(g) + H_2O(g)$$

95. Figure 18.3 describes what happens when a zinc electrode is partially immersed in $ZnSO_4(aq)$. Will a similar description apply to a sodium electrode partially immersed in NaCl(aq)? Explain.

96. Consider the silver plating of an iron spoon in Figure 18.22. Show that, in principle, Ag^+ should be reduced to Ag(s) simply by immersing the spoon in $AgNO_3(aq)$. Why do you suppose that electroplating is used rather than displacement of Ag^+ from solution?

97. How many milliliters of $O_2(g)$, collected over water and measured at 20.0 °C and a barometric pressure of 761.5 mmHg, should be liberated at a platinum anode at the same time that 1.02 g Ag(s) is deposited at a platinum cathode in the electrolysis of $AgNO_3(aq)$?

98. The electrolysis of 0.100 L of 0.785 M $AgNO_3(aq)$ using platinum electrodes is carried out with a current of 1.75 A. What is the molarity of the $AgNO_3(aq)$ 25.0 minutes after electrolysis is begun?

99. Use standard electrode potential data to show that $MnO_2(s)$ should not react with HCl(aq) to liberate $Cl_2(g)$. Yet when $MnO_2(s)$ and concentrated HCl(aq) are heated together, $Cl_2(g)$ does form. In fact, this is a common laboratory method of generating small quantities of $Cl_2(g)$. Explain why the reaction occurs.

100. What minimum voltage is required to recharge a lead-acid storage battery? If a voltage much greater than this minimum is used, a danger exists that a potentially explosive mixture of gases could accumulate in the battery. Explain why this is so.

101. Calculate E_{cell} for the following voltaic cell.

$$Pt, H_2(g, 1\ atm)|CH_3COOH(0.45\ M) \| H^+(0.010\ M)|H_2(g, 1\ atm), Pt$$

102. Calculate E_{cell} for the following voltaic cell.

$$Pt, H_2(g, 1\ atm)|NH_3(0.45\ M), NH_4^+(0.15\ M) \| H^+(0.010\ M)|H_2(g, 1\ atm), Pt$$

103. What is E_{cell} of the following voltaic cell?

$$Cu(s)|Cu^{2+}(0.10\ M)\|Ag_2CrO_4(sat'd\ aq)|Ag(s)$$

(*Hint:* What is $[Ag^+]$ in saturated $Ag_2CrO_4(aq)$?)

104. What $[Cl^-]$ should be maintained in the anode half-cell if the following voltaic cell is to have $E_{cell} = 0.100$ V?

$$Ag(s), AgCl(s)|Cl^-(x\ M)\|Cu^{2+}(0.25\ M)|Cu(s)$$

105. What happens to the voltage of the concentration cell in Figure 18.13 as it operates over a period of time? That is, does the voltage increase, decrease, or remain constant? If the cell operates continuously, does it stop producing electricity at some point? If so, what is the condition in each half-cell compartment at this point?

106. The *efficiency value*, ϵ, of a fuel cell reaction is $\epsilon = \Delta G°/\Delta H°$. What is the efficiency value of the methane-oxygen fuel cell described on page 797? What is the theoretical voltage, $E°_{cell}$, of this fuel cell?

107. Many tabulated solubility product constants have been obtained from electrochemical measurements. The voltage of the following concentration cell is $E_{cell} = 0.417$ V.

$$Ag(s)|Ag^+(sat'd\ AgI(aq))\|Ag^+(0.100\ M)|Ag(s)$$

Use these data to obtain a value of K_{sp} for AgI.

108. Refer to the discussion of the aluminum-air battery on page 797.

(a) How many grams of aluminum are consumed if a 10.0-A electric current is drawn from the battery for 4.00 hours?

(b) Use data from page 797 and Appendix C to obtain a value of $\Delta G_f^\circ\{[Al(OH)_4]^-\}$.

109. A 250.0-mL sample of 0.1000 M $CuSO_4(aq)$ is electrolyzed with a current of 3.512 A for 1368 seconds. Sufficient NH_3 is added to complex any remaining Cu^{2+} and to maintain a free $[NH_3] = 0.10$ M. If the blue color of $[Cu(NH_3)_4]^{2+}$ is detectable at concentrations as low as 1×10^{-5} M, will the blue color be seen in this case?

110. Refer to Example 18.12 and the electrochemical cells described there through cell diagrams. What will be the concentrations of the ions in each half-cell of each electrochemical cell when electric current no longer flows between them?

111. Consider the following reversible reaction with the indicated initial concentrations. What will the ion concentrations be when equilibrium is reached?

$$Hg^{2+}(0.250\ M) + 2\ Fe^{2+}(0.180\ M) \rightleftharpoons 2\ Fe^{3+}(0.210\ M) + Hg(l)$$

112. When AgCl(s) is added to a solution of $Br^-(aq)$, this reversible reaction occurs.

$$AgCl(s) + Br^-(aq) \rightleftharpoons AgBr(s) + Cl^-(aq)$$

If the initial $[Br^-] = 0.4000$ M, what will be the concentration of Br^- at equilibrium?

113. An acid-base titration is carried out in an electrochemical cell in which 100.00 mL of 0.0350 M NaOH(aq) and a hydrogen electrode are the anode half-cell and a standard hydrogen electrode is the cathode half-cell. Determine the pH in the anode half-cell and E_{cell} after the following volumes of 0.150 M HCl(aq) are added: **(a)** 0.00 mL; **(b)** 5.00 mL, **(c)** 10.00 mL; **(d)** 15.00 mL; **(e)** 22.00 mL; **(f)** 23.00 mL; **(g)** 24.00 mL; **(h)** 25.00 mL. Sketch two titration curves, one based on pH and the other on E_{cell}. How do they compare?

114. The electrolysis of $Na_2SO_4(aq)$ by a 6-V battery is conducted in two separate half-cells joined by a salt bridge also containing $Na_2SO_4(aq)$. The cell diagram for the electrolysis is

$$Pt|Na_2SO_4(aq)||Na_2SO_4(aq)|Pt$$

Phenolphthalein indicator is added to each half-cell.

(a) What are the likely half-reactions occurring at the anode and at the cathode of the electrolysis cell? Describe any color changes occuring in the half-cell compartments.

(b) Could the electrolysis also be carried out with a 1.5-V dry cell battery? Explain.

(c) After electrolysis is stopped, the solutions of the two half cells are mixed. Describe and explain any color changes that occur.

(d) Explain why your answer in **(c)** does not depend on the concentration of the $Na_2SO_4(aq)$, the volumes of the solutions in the half-cells, the amount of current used in the electrolysis, or on how long the electrolysis is carried out.

(e) In one experiment, a 10.00-mL sample of HCl(aq) of unknown concentration is added to the cathode compartment along with phenolphthalein. Electrolysis is carried out with a 23.2-mA (milliampere) current, and the solution color becomes pink after 8 minutes and 22 seconds. What is the molarity of the HCl(aq)?

Nuclear Chemistry

CHAPTER

19

The wood in the haft of this stone axe found in the Hebrides off northern Scotland was dated to around 3200 B.C. using radioactive carbon-14. This is one of several applications of nuclear chemistry discussed in this chapter.

CONTENTS

How old is the object in the photograph? Nature left an imprint on the wood from which it was made that is almost as definite as a date stamp, but we have to learn how to read it. Reading this particular date stamp requires that we determine the quantity of carbon-14 present in the wood. There is no practical way to distinguish the very rare carbon-14 from the far more abundant carbon-12 and carbon-13 by *chemical* tests. The three isotopes behave the same chemically. However, we *can* distinguish between the three through their *nuclear* properties. The most distinctive difference in nuclear properties is that carbon-14 is *radioactive* and carbon-12 and carbon-13 are not. **Radioactivity**, or **radioactive decay**, is the spontaneous decay of the nuclei of certain atoms, accompanied by the emission of subatomic particles and/or high-frequency electromagnetic radiation. In this chapter, we will learn how the radioactivity of carbon-14 is used to determine the ages of objects in the method called radiocarbon dating (page 826).

● ChemCDX
When you see this icon, there are related animations, demonstrations, and exercises on the CD accompanying this text.

Our main concern in this chapter will be radioactivity, including its applications to chemistry, the life sciences, and medicine. However, we will also consider topics such as nuclear energy and the effects of ionizing radiation on matter.

19.1 Radioactivity and Nuclear Equations

There are five ways in which atomic nuclei may display radioactivity; that is, there are five types of radioactive decay. We can describe them in terms of the radiation emitted. In our discussion, we will write *nuclear* equations to represent nuclear changes. We will call a nucleus with a specified number of protons and neutrons a **nuclide**.* In a nuclear equation, nuclides and emitted particles are represented in the form: ${}^{A}_{Z}E$. Z is the atomic number—-the number of protons in the nucleus of an atom of the element, E. A is the mass number—the total of the number of protons and neutrons in the nucleus. Together, protons and neutrons are called **nucleons**, and the basic principle in writing a nuclear equation is that nucleons are conserved in a nuclear reaction (just as atoms are conserved in a chemical reaction).

The two sides of a nuclear equation must have the same totals of atomic numbers and mass numbers.

Some of the points in the following discussion are summarized in Table 19.1.

TABLE 19.1 Types of Radioactive Decay: A Summary

		Changes in nucleus	
Mode of decay	**Radiation emitted**	**Atomic number**	**Mass number**
Alpha emission (α)	${}^{4}_{2}He$	−2	−4
Beta emission (β^-)	${}^{0}_{-1}e$	+1	0
Gamma emission (γ)	${}^{0}_{0}\gamma$	0	0
Positron emission (β^+)	${}^{0}_{1}e$	−1	0
Electron capture (EC)	X rays	−1	0

Alpha-particle emission: An **alpha (α) particle** has the same composition as a helium nucleus: two protons and two neutrons. Thus an α particle has a mass of 4 u and a charge of 2+. Because they carry a positive charge, α particles are deflected in electric and magnetic fields. Their deflection in an electric field is illustrated in Figure 19.1. The penetrating power of alpha particles through matter is low; the particles can generally be stopped by a sheet of paper. The symbol for an alpha particle is ${}^{4}_{2}He$. We can represent α-particle emission with a nuclear equation, as in the radioactive decay of uranium-238.

$$\begin{array}{lccccccc} \textit{Sum of mass numbers:} & 238 & & 234 & + & 4 & = & 238 \\ & {}^{238}_{92}U & \longrightarrow & {}^{234}_{90}Th & + & {}^{4}_{2}He & & \\ \textit{Sum of atomic numbers:} & 92 & & 90 & + & 2 & = & 92 \end{array}$$

When a nucleus emits an alpha particle, *its atomic number decreases by 2 and its mass number decreases by 4*. The new nuclide is that of a different element than the decaying nuclide.

* We have previously noted that two or more nuclides of the same element are called isotopes, for example, ${}^{12}_{6}C$ and ${}^{13}_{6}C$.

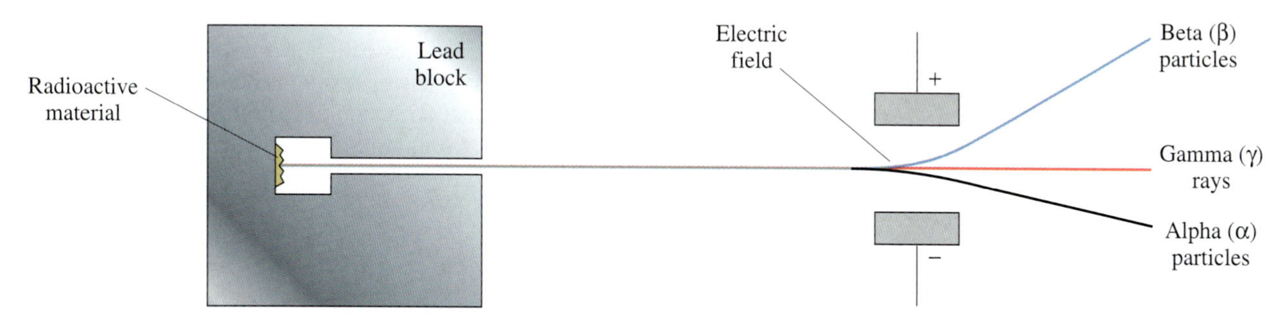

▲ **Figure 19.1 Three types of radiation from radioactive materials**

Beta-particle emission: **Beta (β^-) particles** are electrons. Like all electrons, they have very little mass and carry a charge of 1−. Beta particles are deflected in electric and magnetic fields, but in the opposite direction from alpha particles (see Figure 19.1). Beta particles are more penetrating than α particles. They can pass through aluminum foil 2 to 3 mm thick. The symbol for a beta particle is ${}^{0}_{-1}\text{e}$.

Although an atomic nucleus contains the protons and neutrons that make up an alpha particle, a nucleus does not contain electrons. Instead, a neutron is converted to a proton and an electron, represented in the following nuclear equation.

$${}^{1}_{0}\text{n} \longrightarrow {}^{1}_{1}\text{p} + {}^{0}_{-1}\text{e}$$

Because the atomic number represents the positive charge on a particle, the neutron has an atomic number of 0 (no charge). The electron has the equivalent of an atomic number of −1; it carries the same charge as a proton, but negative in sign. An example of a radioactive decay that produces beta particles is shown below.

$$\begin{array}{llcccccc}
\textit{Sum of mass numbers:} & 234 & & 234 & + & 0 & = & 234 \\
& {}^{234}_{90}\text{Th} & \longrightarrow & {}^{234}_{91}\text{Pa} & + & {}^{0}_{-1}\text{e} & & \\
\textit{Sum of atomic numbers:} & 90 & & 91 & + & -1 & = & 90
\end{array}$$

When a nucleus emits a beta particle, *its atomic number increases by 1 and its mass number is unchanged*. The new nuclide is that of a different element than the decaying nuclide.

Gamma-ray emission: **Gamma (γ) rays** are a highly penetrating form of electromagnetic radiation. They consist of photons, and thus they are not particles of matter. They are emitted by energetic nuclei as a means of reaching a lower energy state. In a nuclear equation for gamma-ray emission, we represent the energetic nucleus by affixing the symbol m to its mass number. For example, in the radioactive decay of uranium-238 by alpha-particle emission, 23% of the thorium-230 nuclei formed are in an excited state: ${}^{230\text{m}}_{90}\text{Th}$. These nuclei then emit energy as gamma rays.

$$\begin{array}{llcccccc}
\textit{Sum of mass numbers:} & 230 & & 230 & + & 0 & = & 230 \\
& {}^{230\text{m}}_{90}\text{Th} & \longrightarrow & {}^{230}_{90}\text{Th} & + & \gamma & & \\
\textit{Sum of atomic numbers:} & 90 & & 90 & + & 0 & = & 90
\end{array}$$

When a nucleus emits a gamma ray, *both its atomic number and mass number remain constant*. The new and old nuclide are of the same element. As we would expect for a form of electromagnetic radiation, gamma rays are unaffected by electric and magnetic fields (see Figure 19.1).

Positron emission: **Positrons** are particles having the same mass as electrons but carrying a charge of 1+. They are sometimes called positive electrons and referred to as β^+ particles. Their penetrating power through matter is very limited because when a positron comes into contact with an electron, the two particles annihilate each other and are converted to two gamma rays. Positrons are formed in the nucleus through the conversion of a proton to a neutron and a positron.

$${}^{1}_{1}\text{p} \longrightarrow {}^{1}_{0}\text{n} + {}^{0}_{1}\text{e}$$

Positrons are most commonly emitted in the radioactive decay of certain nuclides of the lighter elements. The radioactive decay of aluminum-26 is 82% by positron emission.

Sum of mass numbers:	26		26	+	0	=	26
	${}^{26}_{13}\text{Al}$	$\longrightarrow$	${}^{26}_{12}\text{Mg}$	+	${}^{0}_{1}\text{e}$		
Sum of atomic numbers:	13		12	+	1	=	13

When a decaying nucleus emits a positron, *its atomic number decreases by 1 and its mass number is unchanged*. The new nuclide is that of a different element from the decaying nuclide.

Electron capture: **Electron capture (EC)** is a process in which the nucleus absorbs an electron from an inner electronic shell, usually the first or second. An X ray is released when an electron from a higher quantum level drops to fill the level vacated by the captured electron. Once inside the nucleus, the captured electron combines with a proton to form a neutron.

$${}^{0}_{-1}\text{e} + {}^{1}_{1}\text{p} \longrightarrow {}^{1}_{0}\text{n}$$

Nuclear equations for electron capture usually show the captured electron as a reactant. Iodine-125, used in medicine to diagnose pancreatic function and intestinal fat absorption, decays by electron capture.

Sum of mass numbers:	125		0		125
	${}^{125}_{53}\text{I}$	+	${}^{0}_{-1}\text{e}$	$\longrightarrow$	${}^{125}_{52}\text{Te}$
Sum of atomic numbers:	53		−1		52

The result of electron capture is the same as positron emission. *The atomic number of the nucleus decreases by 1, and the mass number is unchanged*. The new nuclide is that of a different element than the decaying nuclide.

EXAMPLE 19.1

Radon-222 is enclosed in capsules as a radiation source for treatment of some types of cancer; phosphorus-32 is used to label red blood cells for blood volume determinations; technetium-99m is used in a host of medical imaging and diagnostic procedures. Write nuclear equations for **(a)** α-particle emission by radon-222, **(b)** β^- decay of phosphorus-32, **(c)** γ decay of technetium-99m.

SOLUTION

We need to recognize that nucleons are conserved in a nuclear reaction and that the changes nuclei undergo in radioactive decay are those summarized in Table 19.1. In this way, if we have data on all but one species in the nuclear equation—reactant or product—we can deduce data for the unknown (?).

a. We identify two of the species from the information given—radon-222 and ^4_2He.

$$^{222}_{86}Rn \longrightarrow ? + ^4_2He$$

The missing nuclide must have $Z = 86 - 2 = 84$ and $A = 222 - 4 = 218$. The element with $Z = 84$ is polonium.

$$^{222}_{86}Rn \longrightarrow ^{218}_{84}Po + ^4_2He$$

b. The atomic number increases by one unit and the mass number remains constant in β^- emission. The two species described in the statement are phosphorus-32 and $^{0}_{-1}e$.

$$^{32}_{15}P \longrightarrow ? + ^{0}_{-1}e$$

The missing nuclide must have $Z = 15 + 1 = 16$ and $A = 32$. The element with $Z = 16$ is sulfur.

$$^{32}_{15}P \longrightarrow ^{32}_{16}S + ^{0}_{-1}e$$

c. Both the mass and atomic numbers remain constant in gamma decay. On the right side of the equation, we simply remove the symbol m from Tc and add a γ ray.

$$^{99m}_{43}Tc \longrightarrow ^{99}_{43}Tc + \gamma$$

EXERCISE 19.1A

Write a nuclear equation for the decay of **(a)** radon-212 by α-particle emission, **(b)** argon-37 by electron capture, and **(c)** cobalt-60 by γ-ray emission.

EXERCISE 19.1B

Write a nuclear equation for each of the following:

a. The α-particle decay of an isotope to produce lead-214.

b. The production of sulfur-36 through positron emission.

19.2 Naturally Occurring Radioactivity

Let us first consider a few of the naturally occurring radioactive nuclides among the lighter elements. Hydrogen-3 (tritium) and carbon-14 are formed by cosmic radiation entering Earth's upper atmosphere. They are found in trace amounts in the atmosphere and in compounds that incorporate H and C atoms—including living matter. Both are β^- emitters.

● **ChemCDX** Separation of Alpha, Beta, and Gamma Rays

$$^3_1H \longrightarrow ^3_2He + ^{0}_{-1}e$$

$$^{14}_6C \longrightarrow ^{14}_7N + ^{0}_{-1}e$$

Potassium-40 makes up 0.0118% of naturally occurring potassium atoms. There are three modes of radioactive decay of this nuclide: β^- emission, positron emission, and electron capture.

$$^{40}_{19}K \longrightarrow ^{40}_{20}Ca + ^{0}_{-1}e \qquad ^{40}_{19}K \longrightarrow ^{40}_{18}Ar + ^{0}_{1}e \qquad ^{40}_{19}K + ^{0}_{-1}e \longrightarrow ^{40}_{18}Ar$$

When Earth was young, potassium-40 was much more abundant than it is today. Practically all the argon in Earth's atmosphere is argon-40, and scientists think that it has come from the radioactive decay of potassium-40.

Radioactive Decay Series

Most of the naturally occurring nuclides of the lighter elements have stable nuclei; that is, they are nonradioactive. In contrast, *all* nuclides of the heaviest elements are radioactive. A dividing line is found in the element bismuth ($Z = 83$). All nuclides with atomic number greater than 83 are radioactive. We'll have more to say about this difference in nuclear properties between the light and heavy elements when we discuss nuclear stability in Section 19.6.

Even though they are radioactive, many nuclides of high atomic number are found in natural sources. And a few of them decay very slowly, as reflected in their half-lives. The half-life for radioactive decay is similar to the half-life of a chemical reaction. The **half-life ($t_{1/2}$)** of a radioactive nuclide is the time required for one-half the nuclei in a sample of the nuclide to decay. The decay of uranium-238 by alpha-particle emission has a half-life of 4.51×10^9 years.

The decay of a uranium-238 atom is followed by other decays that continue until the stable nucleus, $^{206}_{82}Pb$, is reached. A series of radioactive decays beginning with a long-lived radioactive nuclide and ending with a nonradioactive one is called a **radioactive decay series**. Nuclear equations for the first few steps in the uranium-238 series are written at the left, and the entire series is summarized in Figure 19.2.

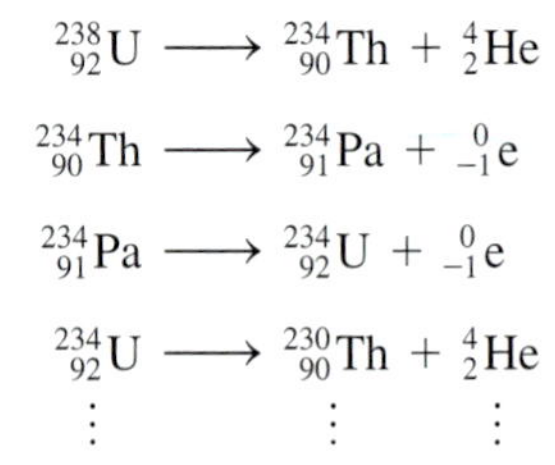

$$^{238}_{92}U \longrightarrow {}^{234}_{90}Th + {}^{4}_{2}He$$
$$^{234}_{90}Th \longrightarrow {}^{234}_{91}Pa + {}^{0}_{-1}e$$
$$^{234}_{91}Pa \longrightarrow {}^{234}_{92}U + {}^{0}_{-1}e$$
$$^{234}_{92}U \longrightarrow {}^{230}_{90}Th + {}^{4}_{2}He$$
$$\vdots \qquad \vdots \qquad \vdots$$

This uranium-238 decay series helps us to understand a number of interesting observations. Even though some of the nuclides in the decay series have very short half-lives, all the nuclides in the series exist because they are constantly being formed. The 1600-year half-life of radium-226 gives an expectation of less than 1 gram of radium in the several tons of uranium ore processed by Marie Curie in 1898. Thus, we can truly marvel at her feat of extracting 120 milligrams of $RaCl_2$. For the nuclide polonium-210 with a half-life of only 138 days, the expectation is

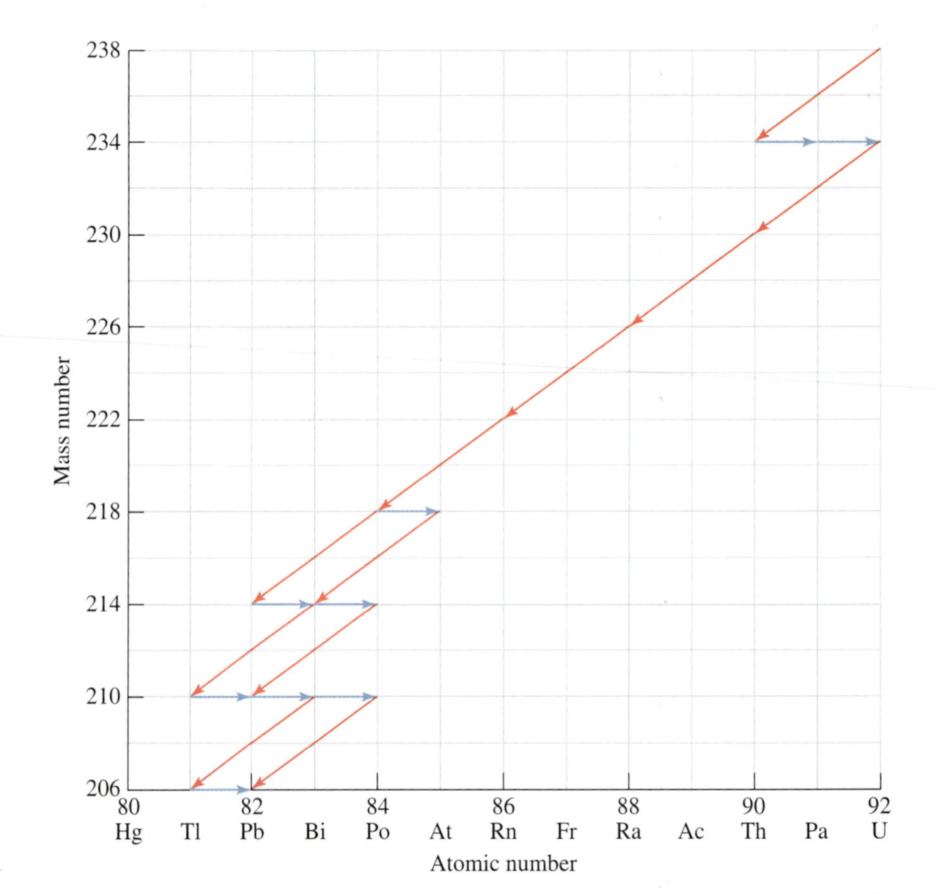

▶ **Figure 19.2**
The natural radioactive decay series for $^{238}_{92}U$
The long red arrows pointing down and to the left correspond to α-particle emissions. The short blue horizontal arrows represent β^--particle emissions.

that there was only a fraction of a milligram of polonium present, and it is understandable that she could not isolate it.

Marie Sklodowska Curie (1867–1934) was born in Poland and went to Paris to do graduate work. There she met and married the physicist Pierre Curie. The Curies and Henri Becquerel shared in the Nobel Prize in physics in 1903. Madame Curie won the Nobel Prize in chemistry in 1911 for her prodigious feat in discovering radium and polonium and in isolating radium. She is the only scientist to have won Nobel Prizes in both physics and chemistry. Interestingly, most of the work leading to these prizes was from her doctoral thesis.

The Age of Earth

Radioactive decay schemes give us a way to approach an intriguing question: How old is Earth? The parent nuclides of the four natural decay series and their half-lives are as follows.

	$^{232}_{90}Th$	$^{238}_{92}U$	$^{235}_{92}U$	$^{237}_{93}Np$
$t_{1/2}$ =	1.4×10^{10} y	4.5×10^{9} y	7.1×10^{8} y	2.2×10^{6} y

Let's assume that the lowest detectable level of radioactivity occurs after about 30 half-life periods. At this time, only about $(1/2)^{30}$ of the atoms of the radioactive nuclide originally present remain, about one atom for an original one billion atoms. For neptunium-237, this would be about 7×10^7 y (that is, $30 \times 2.2 \times 10^6$ y). Because there is essentially no neptunium-237 to be found, Earth must be older than 7×10^7 years. However, because uranium-235 can still be found, Earth is probably not as old as 2×10^{10} years (that is, $30 \times 7.1 \times 10^8$ y).

Our best estimate of the age of Earth comes from comparing the percent natural abundance of uranium-238 (99.28%) with that of uranium-235 (0.72%). It would have taken about 6 billion years for these percent relative abundances to develop if the two isotopes had initially been present in equal abundance. There are reasons to believe that the initial proportion of uranium-235 might not have been as great as that of uranium-238, and so 6 billion years is an upper limit. Other dating methods suggest a probable age of about 5 billion years for Earth.

19.3 Radioactive Decay Rates

In chemical kinetics, we can relate the half-life of a reaction to the rate constant and rate of reaction if we know the order of the reaction (Chapter 13.) We can do something similar for radioactive decay.

In a radioactive sample, there is no way to predict when a *particular* atom will decay. It could happen within the next second or not for a million years. The decay process is random. However, if we have a sample with a large number of atoms, we can measure how many of the atoms decay in a unit of time, and that number is quite reproducible. Moreover, in a sample with twice as many atoms, we find that twice as many decay in a unit of time. This observation is stated through the **radioactive decay law**:

A metamorphic rock from Nuuk, Greenland, dated by radioactivity at over three billion years old.

The rate of disintegration of a radioactive nuclide, called the decay rate or activity, A, is directly proportional to the number of atoms present.

$$\text{Rate of radioactive decay} = A = \lambda N$$

Note the resemblance of this equation to the rate law for the first-order reaction: A $\longrightarrow$ products.

$$\text{Rate of reaction} = k[\text{A}]$$

Radioactive decay is a first-order process. The rate of decay or activity is analogous to a rate of reaction; the *decay constant*, λ, is analogous to k; and the number of atoms, N, is analogous to the concentration [A]. The unit of λ is $(\text{time})^{-1}$, for example, s^{-1}, d^{-1}, or y^{-1}; and the unit for the rate of decay is atoms/time, for example, atoms/s.

Consider a 5,500,000-atom sample with a decay rate of 85 atoms per second. $N = 5.5 \times 10^6$ atom and $A = 85$ atom s^{-1}. The value of λ is

$$\lambda = \frac{A}{N} = \frac{85 \text{ atom s}^{-1}}{5.5 \times 10^6 \text{ atom}} = 1.5 \times 10^{-5} \text{ s}^{-1}$$

As with other first-order processes, we can also write

$$\ln \frac{N_t}{N_0} = -\lambda t \quad \text{and} \quad t_{1/2} = \frac{0.693}{\lambda}$$

In these equations, N_0 represents the initial number of atoms at $t = 0$, that is, at the time we begin our measurement. N_t is the number of atoms at a later time t; λ is the decay constant; and $t_{1/2}$ is the half-life of the decay process.

Iodine-131, a radioactive nuclide used in studies of the thyroid gland, has a half-life of eight days. Thus, half the atoms in a sample undergo decay in an eight-day period; the rate of decay, the activity, A, falls to half its initial value in eight days. The activity is down to one-fourth of its initial value in 16 days, to one eighth in 24 days, and so on.

The shorter the half-life, the larger the value of λ and the faster the decay proceeds. Half-lives range from microseconds—radium-218 has a half-life of 14 μs—to billions of years (like uranium-238). Table 19.2 lists several radioactive nuclides, their half-lives, and some uses.

TABLE 19.2 Half-Lives of Representative Radioactive Nuclides

Nuclide	Half-life[a]	Typical Use
Hydrogen-3	12.26 y	Biochemical tracer
Carbon-11	20.39 min	PET scans
Carbon-14	5730 y	Dating of artifacts
Sodium-24	14.659 h	Tracer, cardiovascular system
Phosphorus-32	14.3 d	Biochemical tracer
Potassium-40	1.25×10^9 y	Dating of rocks
Iron-59	44.496 d	Tracer, red blood cell lifetime
Cobalt-60	5.271 y	Radiation treatment of cancer
Strontium-90	28.5 y	No present use (component of radioactive fallout)
Iodine-131	8.040 d	Tracer, thyroid studies
Radium-226	1.60×10^3 y	Radiation therapy for cancer
Uranium-238	4.51×10^9 y	Dating of rocks and Earth's crust

[a] s = second, m = minute, h = hour, d = day, y = year.

● **ChemCDX** First-Order Decay

EXAMPLE 19.2

The nuclide sodium-24 (Table 19.2) is used to detect constrictions and obstructions in the human circulatory system. It emits β^- particles. **(a)** What is the decay constant, in s^{-1}, for sodium-24? **(b)** What is the activity of a freshly synthesized 1.00-mg (1.00×10^{-3} g) sample of sodium-24? **(c)** What will be the rate of decay of the 1.00-mg sample after one week (168 h)?

SOLUTION

a. We can begin by finding $t_{1/2}$ for sodium-24 in Table 19.2; it is 14.659 h. So that the unit of λ will be s^{-1}, let's first express the half-life in seconds.

$$t_{1/2} = 14.659 \text{ h} \times \frac{60 \text{ min}}{1 \text{ h}} \times \frac{60 \text{ s}}{1 \text{ min}} = 5.2772 \times 10^4 \text{ s}$$

Now we can relate the decay constant to $t_{1/2}$.

$$\lambda = \frac{0.693}{t_{1/2}} = \frac{0.693}{5.2772 \times 10^4\,\text{s}} = 1.31 \times 10^{-5}\,\text{s}^{-1}$$

b. The decay rate or activity (A) is related to the decay constant (λ) and the number of atoms (N) through the radioactive decay law: $A = \lambda N$. We have the decay constant, and we need to determine the number of atoms in 1.00 mg of sodium-24.

$$N = 1.00 \times 10^{-3}\,\text{g}\,{}^{24}\text{Na} \times \frac{1\,\text{mol}\,{}^{24}\text{Na}}{24.0\,\text{g}\,{}^{24}\text{Na}} \times \frac{6.022 \times 10^{23}\,\text{atoms}\,{}^{24}\text{Na}}{1\,\text{mol}\,{}^{24}\text{Na}}$$

$$= 2.51 \times 10^{19}\,\text{atoms}\,{}^{24}\text{Na}$$

The decay rate is

$$A = \lambda N = 1.31 \times 10^{-5}\,\text{s}^{-1} \times 2.51 \times 10^{19}\,\text{atoms} = 3.29 \times 10^{14}\,\text{atoms/s}$$

c. For this part of the question we need to use the integrated rate law: $\ln N_t/N_0 = -\lambda t$. We have a value of λ from part (a), and we know N_0 and A_0 from part (b). The value of t is 1 week, but since the unit of the decay constant is s^{-1}, we need to express t in seconds.

$$t = 1\,\text{week} \times \frac{7\,\text{d}}{1\,\text{week}} \times \frac{24\,\text{h}}{1\,\text{d}} \times \frac{60\,\text{min}}{1\,\text{h}} \times \frac{60\,\text{s}}{1\,\text{min}} = 6.05 \times 10^5\,\text{s}$$

We can substitute activities for numbers of atoms in the integrated rate law. That is, if $A = \lambda N$, we can also say that $N = A/\lambda$. Thus

$$\ln \frac{N_t}{N_0} = -\lambda t$$

$$\ln \frac{A_t/\lambda}{A_0/\lambda} = -\lambda t$$

$$\ln \frac{A_t}{A_0} = -\lambda t$$

$$\ln \frac{A_t}{A_0} = -1.31 \times 10^{-5}\,\text{s}^{-1} \times 6.05 \times 10^5\,\text{s} = -7.93$$

$$A_t/A_0 = e^{-7.93} = 3.6 \times 10^{-4}$$

$$A_t = 3.6 \times 10^{-4} \times A_0 = 3.6 \times 10^{-4} \times 3.29 \times 10^{14}\,\text{atoms/s}$$

$$= 1.2 \times 10^{11}\,\text{atoms/s}$$

EXERCISE 19.2A

Refer to Table 19.2. If the current rate of decay of a sample of phosphorus-32 is 2.50×10^{10} atoms/s, what will be the decay rate one year from now?

EXERCISE 19.2B

The half-life of plutonium-239 is 2.411×10^4 y. How long would it take for a sample of plutonium-239 to decay to 1.00% of its present activity?

PROBLEM-SOLVING NOTE

An effective way to detect and count radioactive atoms is through their activity, A. For this reason, it is often necessary to substitute A/λ for N.

EXAMPLE 19.3—An Estimation Example

The half-life of iodine-131 is 8.040 days. Following the release of large quantities of iodine-131 during the Chernobyl nuclear disaster of 1986, approximately how long did it take for the activity of this iodine-131 to fall to 1% of its initial value?

SOLUTION

We could do an exact calculation with the methods of Example 19.2, but note that 1% represents 1/100th of the initial activity. During one half-life period, the activity would fall to half its initial value; during the next half-life period, to one quarter; and so on.

Time:	$t_{1/2}$	$2(t_{1/2})$	$3(t_{1/2})$	$4(t_{1/2})$	$5(t_{1/2})$	$6(t_{1/2})$	$7(t_{1/2})$
Fraction remaining:	1/2	1/4	1/8	1/16	1/32	1/64	1/128

It took between 6 and 7 half-lives for the activity to fall to 1/100th of its initial value. If we estimate 6.5 half-lives, the time required was $6.5 \times 8\text{ d} \approx 52\text{ d}$.

EXERCISE 19.3

Refer to Table 19.2. *Without doing detailed calculations*, determine which of these samples has the greatest rate of decay: **(a)** 1 μmol sodium-24, **(b)** 1 μg carbon-11, **(c)** 1 g uranium-238. Explain your reasoning.

Radiocarbon Dating

Carbon-14 is formed at a nearly constant rate in the upper atmosphere by the bombardment of nitrogen-14 with neutrons from cosmic radiation. When a neutron is absorbed by the nucleus of a nitrogen-14 atom, a proton is ejected. The result is that the nitrogen-14 atom is converted to a carbon-14 atom. We can show this through a nuclear equation, in which we represent the neutron as ${}^{1}_{0}\text{n}$ and the proton as ${}^{1}_{1}\text{H}$.

$$ {}^{14}_{7}\text{N} + {}^{1}_{0}\text{n} \longrightarrow {}^{14}_{6}\text{C} + {}^{1}_{1}\text{H} $$

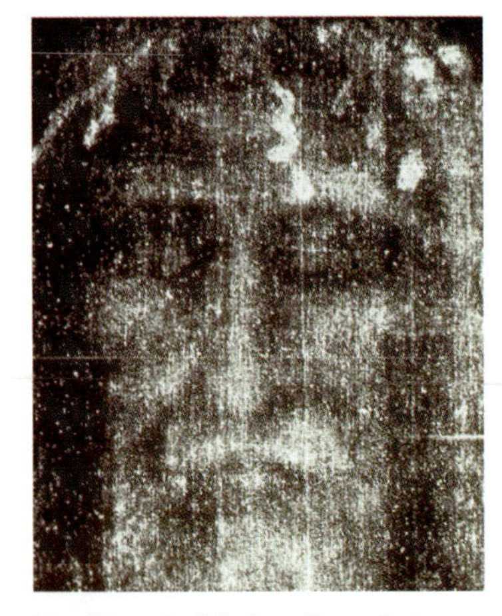

The Shroud of Turin, a linen cloth over 4 meters long, shows the faint image of a man. The image is best visualized in a photographic negative, as shown here. The shroud was thought by some to be the burial shroud of Jesus Christ, in which case it would be about 2000 years old. The cloth was shown by carbon-14 dating in three separate laboratories to be, at most, about 700 years old. This corresponds well to records that indicate that the shroud appeared first in the 1300s.

The carbon-14 is eventually incorporated into atmospheric carbon dioxide. At the same time that carbon-14 is being formed, carbon-14 in the environment is undergoing decay, resulting in an essentially constant concentration of about one atom of carbon-14 for every 10^{12} atoms of carbon-12. A living plant consumes carbon dioxide, and animals consume plants. Plants and animals incorporate carbon-14 into their tissues as readily as carbon-12, and therefore in the same proportions as found in the environment. Thus, as carbon-14 atoms in living organisms undergo decay, they are constantly replaced with "fresh" carbon-14 atoms.

Carbon-14 in living matter decays by β^- emission at a rate of about 15 disintegrations per minute (dis min^{-1}) per gram of carbon. But consider what happens when a tree is cut down. It no longer takes up carbon dioxide. The carbon-14 that decays is no longer replaced, and as the concentration of carbon-14 falls, the disintegration or decay rate falls as well. We can use the reduced decay rate at some later time to estimate the age of an object made of wood from the tree.

Because carbon-14 has a half-life of 5730 years, radiocarbon dating does not work well for objects less than a few hundred years old. The carbon-14 decay rate in the object will be too close to what it was initially. Nor does the method work well for objects more than about 50,000 years old. The level of radioactivity will have fallen to the point at which it is not much greater than the background radiation level found in the surroundings everywhere. Between these limits, however, radiocarbon dating has had some remarkable successes.

EXAMPLE 19.4

A wooden object from an Egyptian tomb is subjected to radiocarbon dating. The decay rate observed for its carbon-14 content is 7.2 dis min^{-1} per g carbon. What is the age of the wood in the object (and, presumably, of the object itself)? The half-life of carbon-14 is 5730 years, and the decay rate for carbon-14 in living organisms is 15 dis min^{-1} per g carbon.

SOLUTION

First, we can determine the decay constant, λ.

$$\lambda = \frac{0.693}{t_{1/2}} = \frac{0.693}{5730 \text{ y}} = 1.21 \times 10^{-4} \text{ y}^{-1}$$

As we did in Example 19.2(c), let's work with activities (A) instead of numbers of atoms (N). To do so, we can rewrite the radioactive decay law in the form shown here.

$$N_0 = A_0/\lambda \quad \text{and} \quad N_t = A_t/\lambda$$

We need the integrated rate law in the form below.

$$\ln\frac{N_t}{N_0} = \ln\frac{A_t/\lambda}{A_0/\lambda} = \ln\frac{A_t}{A_0} = -\lambda t$$

Finally, we can substitute the known data and solve for the unknown time, t.

$$\ln\frac{A_t}{A_0} = \ln\frac{7.2}{15} = -\lambda t = -(1.21 \times 10^{-4} \text{ y}^{-1}) \times t$$

$$t = \frac{-\ln 7.2/15}{1.21 \times 10^{-4} \text{ y}^{-1}} = 6.1 \times 10^3 \text{ y}$$

EXERCISE 19.4A

What will be the decay rate of the carbon-14 in the object described in Example 19.4 at a time 1500 years in the future?

EXERCISE 19.4B

Tritium (^3H), a β^- emitting hydrogen nuclide, can be used to determine the age of items up to about 100 years. A sample of brandy, stated to be 25 years old and offered for sale at a premium price, has tritium with half the activity of that found in new brandy. Is the claimed age of the beverage authentic? Use data from Table 19.2, and assume that the natural abundance of tritium is a fixed quantity.

19.4 Synthetic Nuclides

For hundreds of years, alchemists tried—without success—to change one element into another, a process called *transmutation*. Modern scientists have learned how to do so. Ernest Rutherford brought about the first transmutation in a laboratory experiment in 1919. By bombarding nitrogen-14 nuclei with α particles, he obtained oxygen-17 and protons as products. We can represent the process through the following nuclear equation.

$$^{14}_{7}\text{N} + ^{4}_{2}\text{He} \longrightarrow ^{17}_{8}\text{O} + ^{1}_{1}\text{H}$$

Irène Curie, daughter of Pierre and Marie Curie, met and married Frédéric Joliot when both were assistants in Marie Curie's laboratory. They just missed out on two great discoveries—the neutron and the positron—but were awarded the Nobel Prize in chemistry in 1935 for their discovery of artificially induced radioactivity.

Oxygen-17, which is *not* radioactive, is a naturally occurring oxygen nuclide; its natural abundance is 0.037%.

Rutherford's experiment was especially important because it established the existence of protons *outside* the nuclei of atoms. Fifteen years later, in an experiment similar to Rutherford's, Irène Curie and Frédéric Joliot got a surprisingly different result. They bombarded aluminum-27 with α particles and observed the emission of *two* types of particles: neutrons and positrons. When they stopped the bombardment, the emission of neutrons stopped, but that of positrons continued. They hypothesized that the nuclear bombardment produces phosphorus-30, which then decays by the emission of positrons.

$$^{27}_{13}\text{Al} + {}^{4}_{2}\text{He} \longrightarrow {}^{30}_{15}\text{P} + {}^{1}_{0}\text{n}$$

$$^{30}_{15}\text{P} \longrightarrow {}^{30}_{14}\text{Si} + {}^{0}_{+1}\text{e}$$

Phosphorus-30 was the first *synthetic* radioactive nuclide. Since its discovery, scientists have synthesized over a thousand others. The number of known radioactive nuclides now greatly exceeds the number of nonradioactive ones. We will examine some of the uses of synthetic radioactive nuclides in later sections of this chapter.

EXAMPLE 19.5

When a magnesium-24 nucleus is bombarded with a deuteron, the sodium-22 nucleus is formed. What other particle is produced?

SOLUTION

To learn the identity of the particle, let's apply the principles of nuclear equation balancing. (The deuteron is the nucleus of a deuterium atom, $^{2}_{1}\text{H}$.)

$$^{24}_{12}\text{Mg} + {}^{2}_{1}\text{H} \longrightarrow {}^{22}_{11}\text{Na} + ?$$

The unknown particle must have $A = 26 - 22 = 4$ and $Z = 13 - 11 = 2$. It is therefore an α particle, and the complete nuclear equation is as shown below.

$$^{24}_{12}\text{Mg} + {}^{2}_{1}\text{H} \longrightarrow {}^{22}_{11}\text{Na} + {}^{4}_{2}\text{He}$$

EXERCISE 19.5A

Write a nuclear equation to represent the bombardment of a chlorine-35 nucleus with a neutron to produce the nucleus of sulfur-35.

EXERCISE 19.5B

Write a nuclear equation to represent the bombardment of a californium-249 nucleus with a nitrogen-15 nucleus to produce the nucleus of dubnium-260 and neutrons as products.

19.5 Transuranium Elements

In 1940, the first of the **transuranium elements**—elements with $Z > 92$—were synthesized. This was accomplished by bombarding uranium-238 nuclei with neutrons. First, unstable uranium-239 is formed, and then this nuclide decays by beta emission, producing neptunium ($Z = 93$).

$$^{238}_{92}\text{U} + {}^{1}_{0}\text{n} \longrightarrow {}^{239}_{92}\text{U}$$

$$^{239}_{92}\text{U} \longrightarrow {}^{239}_{93}\text{Np} + {}^{0}_{-1}\text{e}$$

Neptunium-239 decays to plutonium ($Z = 94$).

$$^{239}_{93}\text{Np} \longrightarrow {}^{239}_{94}\text{Pu} + {}^{0}_{-1}\text{e}$$

Neutrons are especially effective projectiles for nuclear bombardment because they have no charge and thus are not repelled as they approach a nucleus. However, neutron bombardment produces only small changes in atomic number. To produce large changes in atomic number, rather massive positive ions are needed. For example, the discovery of element 111 in 1994 involved producing three atoms of the new element by bombarding bismuth-209 with nickel-64 nuclei.

Considerable energy must be imparted to a positive ion in order for it to overcome repulsion by a positively charged nucleus. Only in this way can the ion collide with the nucleus and induce a nuclear reaction. Energetic ions are usually produced by accelerating them to high speeds. One type of *charged-particle accelerator*, called a cyclotron, is described in Figure 19.3.

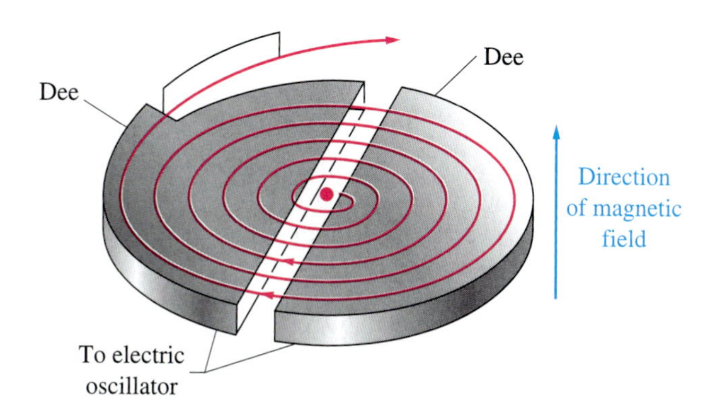

Figure 19.3
The cyclotron—A charged-particle accelerator
The two hollow, flat boxes, called dees, are held in a magnetic field and charged electrically. Positive ions are produced between the dees, attracted into the negatively charged dee, and forced into a circular path by the magnetic field. As the ions leave the first dee, the electric charge is reversed, and the ions are attracted into the second dee and are again accelerated. The process is repeated until the ions have achieved the desired energy. The ions are then brought out of the accelerator. A playground swing provides an analogy: The rider is given a push at the end of each swing, and the swing travels through an ever-increasing arc.

19.6 Nuclear Stability

At some point, you may have wondered, "If positive charges repel one another, how is it possible for protons to be packed so closely in the nuclei of atoms?" The answer is that there are attractive *nuclear* forces that are much stronger than electrostatic forces. The strengths of these forces are closely related to the numbers of protons and neutrons in a nucleus. We begin with some observations about the naturally occurring *stable* nuclides.

- About 160 stable nuclides have an *even* number of protons and an *even* number of neutrons, for example, $^{12}_{6}\text{C}$.
- About 50 stable nuclides have an *even* number of protons and an *odd* number of neutrons, for example, $^{25}_{12}\text{Mg}$.
- About 50 stable nuclides have an *odd* number of protons and an *even* number of neutrons, for example, $^{19}_{9}\text{F}$.
- Only 4 stable nuclides have an *odd* number of protons and an *odd* number of neutrons. They are $^{2}_{1}\text{H}$, $^{6}_{3}\text{Li}$, $^{10}_{5}\text{B}$, and $^{14}_{7}\text{N}$.

One theoretical approach to nuclear stability is the *nuclear shell theory*. In simplest terms, this theory proposes that the protons and the neutrons each exist in shells within the nucleus. This is much like the existence of electrons in shells outside the nucleus. The similarity extends to the special stability associated with the

Maria Goeppert-Mayer (1906–1972) followed her chemist husband to several academic locations, working mostly in temporary positions. While at the University of Chicago, stimulated by questions posed by Enrico Fermi, she worked out the shell model of nuclear structure (supposedly in 10 minutes). Her work earned her a share of the 1963 Nobel Prize in physics.

closing of shells, similar to what is seen with the noble gases in electron configurations. In the nuclear shell theory, a special stability is associated with nuclei that have any of the following numbers as numbers of protons or neutrons.

$$2, 8, 20, 28, 50, 82, 126$$

These numbers are called **magic numbers** because scientists recognized their significance in relation to nuclear stability before they had developed a theory to explain them.

One observation consistent with these magic numbers is the fact that alpha particles are very stable; they have 2 protons and 2 neutrons and are "doubly magic." Another observation is that tin ($Z = 50$) has *ten* naturally occurring stable nuclides, more than any other element. The magic number of protons (50) seems to allow for a greater variation in the number of neutrons in the tin nucleus than in others. Also, the uranium-238 radioactive decay series terminates in the nuclide $^{206}_{82}Pb$; the uranium-235 series, in $^{207}_{82}Pb$; and the thorium-232 series, in $^{208}_{82}Pb$. All these terminating nuclides have the magic number 82 in lead ($Z = 82$); $^{208}_{82}Pb$ is doubly magic, with 82 protons and 126 neutrons.

A crucial factor in the stability of a nucleus is the ratio of the neutron number (N) to the proton number (Z). Some nuclides of the lightest elements have an N/Z ratio of 1, and as a group these nuclides have an average ratio slightly greater than 1. Examples of nuclides in this group are $^{4}_{2}He$, $^{16}_{8}O$, $^{27}_{13}Al$, $^{39}_{19}K$, and $^{40}_{20}Ca$. Nuclides with $Z > 20$ require a larger number of neutrons than protons to moderate the effect of increasing proton repulsions. For example, the N/Z ratio in $^{56}_{26}Fe$ is $30/26 = 1.15$; in $^{133}_{55}Cs$ it is $78/55 = 1.42$, and in $^{209}_{83}Bi$, $126/83 = 1.52$. For nuclides with $Z > 83$, the proton repulsions are too large to be overcome by proton-neutron interactions, and the nuclides are all radioactive.

The general pattern for nuclear stability in terms of neutron and proton numbers is shown in Figure 19.4, a graph of neutron number (N) versus proton number (Z). All of the naturally occurring *stable* nuclides are indicated by dots within the belt labeled the *belt of stability*. However, other nuclides in this belt that are not shown are radioactive. Also, *all* nuclides falling *outside* the belt are radioactive, and their mode of decay is one that brings the nuclides formed in the decay process into the belt. Nuclides above the belt decay by beta emission, and those below the belt by positron emission and electron capture. Many of the nuclides in the upper right corner decay by alpha emission. If the numbers 114 and 184 are also magic numbers, as some scientists think they are, there might be a small island of stability centered on the nuclide with $Z = 114$ and $N = 184$. This nuclide may someday be synthesized, and there have even been unsuccessful attempts to find element 114 in natural sources.

As illustrated in Examples 19.6 and 19.7, we can use data from Figure 19.4 and the list of items on page 829 to answer questions about nuclear stability and radioactive decay.

EXAMPLE 19.6

Which of the following would you expect to be radioactive: $^{118}_{50}Sn$, $^{234}_{91}Pa$, $^{54}_{25}Mn$, $^{74}_{30}Zn$?

SOLUTION

$^{118}_{50}Sn$: This nuclide has 50 protons and $(118 - 50) = 68$ neutrons. This is an even-even combination, the most common for stable nuclides. Also, the neutron:proton ratio of 68:50 lies within the belt of stability in Figure 19.4. $^{118}_{50}Sn$ is *nonradioactive*.

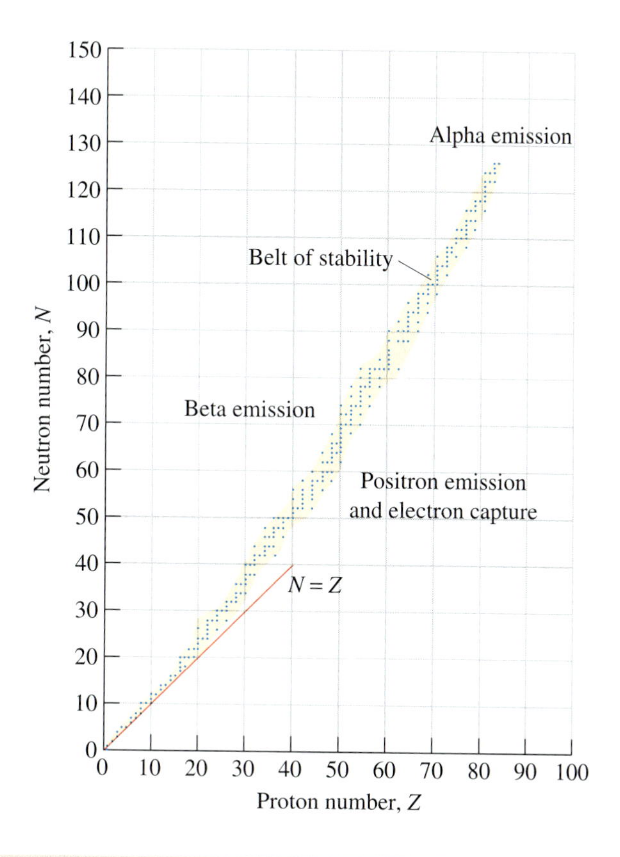

Figure 19.4
Neutron-to-proton ratio and the stability of nuclides
All the stable nuclides lie within the belt of stability (as do some radioactive ones). At low atomic numbers, the neutron:proton ratios are 1:1 or slightly greater; the belt borders the line $N = Z$. At higher atomic numbers, the neutron:proton ratios rise to about 1.5:1. The belt ends at $Z = 83$. Nuclides outside the belt tend to undergo radioactive decay by the modes indicated.

$^{234}_{91}Pa$: Atomic number 91 exceeds the limit for the naturally occurring stable nuclides ($Z > 83$). $^{234}_{91}Pa$ is *radioactive*.

$^{54}_{25}Mn$: This nuclide has 25 protons and $(54 - 25) = 29$ neutrons. This is an odd-odd combination found only in four stable nuclides of low atomic numbers. We should expect that it is *radioactive*.

$^{74}_{30}Zn$: This nuclide has 30 protons and $(74 - 30) = 44$ neutrons. From Figure 19.4, we see that at $Z = 30$, the upper limit of the belt of stability is at about $N = 40$. The nuclide $^{74}_{30}Zn$ lies above the belt and is *radioactive*.

EXERCISE 19.6

Using information like that in Example 19.6, cite two isotopes of calcium that you are quite certain are *nonradioactive* and one isotope that is *radioactive*. Explain your reasoning.

EXAMPLE 19.7—A Conceptual Example

What kind of radioactive decay would you expect the nuclide $^{84}_{40}Zr$ to undergo?

SOLUTION

When we check the belt of stability in Figure 19.4 at $Z = 40$, we see that a nuclide with $N = 44$ lies below the belt. This confirms that the nuclide is radioactive. We would expect a decay that moves the neutron:proton ratio closer to the belt. This means converting a proton to a neutron. The atomic number goes down by one, and the mass number remains the same. These changes are achieved either by positron emission or electron capture.

Positron emission: $$^{84}_{40}Zr \longrightarrow ^{84}_{39}Y + ^{0}_{1}e$$

Electron capture: $$^{84}_{40}Zr + ^{0}_{-1}e \longrightarrow ^{84}_{39}Y$$

Notice that in each case, the neutron:proton ratio increases from 44/40 in $^{84}_{40}Zr$ to 45/39 in $^{84}_{39}Y$.

EXERCISE 19.7

The nuclide $^{84}_{39}Y$ is radioactive. What kind of decay would you expect it to undergo? Is the product of this decay likely to be nonradioactive? Explain.

19.7 Energetics of Nuclear Reactions

The details of nuclear phenomena may be a great mystery to many people, but almost everyone is keenly aware that nuclear processes are potential sources of enormous quantities of energy. We will consider nuclear energy in this and the following section.

In 1905, while working out his theory of special relativity, Albert Einstein derived the equation for the equivalence of mass and energy.

$$E = mc^2$$

The constant c^2, the square of the speed of light, relates energy in joules to mass in kilograms.

In a typical spontaneous nuclear reaction, a small quantity of matter is destroyed and replaced by a corresponding quantity of energy. Presumably this is also true of ordinary chemical reactions, but there, the energy changes are so small that the corresponding mass changes are undetectable. This is why we can use the principle of conservation of mass (total mass is unchanged in a chemical reaction) as the basis of stoichiometric calculations.

Nuclear energies are generally expressed in the unit MeV (megaelectronvolt). The **electronvolt** is the energy that an electron acquires as it moves through a potential difference of 1 volt. Given the charge on an electron, 1.6022×10^{-19} C, the product of this charge and a 1 V potential difference is that shown below.

$$1\text{ eV} = 1.6022 \times 10^{-19}\text{ C} \times 1\text{ V} = 1.6022 \times 10^{-19}\text{ V C} = 1.6022 \times 10^{-19}\text{ J}$$

The electronvolt is such an extremely small energy unit that we usually use the unit MeV instead.

$$1\text{ MeV} = 1 \times 10^6\text{ eV} = 1 \times 10^6 \times 1.6022 \times 10^{-19}\text{ J} = 1.6022 \times 10^{-13}\text{ J}$$

Another useful relationship is the energy equivalent to 1 atomic mass unit (u). To obtain this, we first find the SI mass equivalent of 1 u, and then we find the energy equivalent of this mass. Because the mass of a carbon-12 atom is exactly 12 u, we have the following.

$$1\text{ u} = \frac{1}{12} \times (\text{mass of one }^{12}\text{C atom})$$

Given that the molar mass of carbon-12 is exactly 12 g, the mass of one ^{12}C atom is

$$\frac{12.00000\text{ g}}{6.0221 \times 10^{23}\ {}^{12}\text{C atoms}} = 1.9927 \times 10^{-23}\text{ g}/{}^{12}\text{C atom}$$

and the mass of 1 u is

$$1\ \text{u} = \frac{1.9927 \times 10^{-23}\ \text{g}}{12} = 1.6606 \times 10^{-24}\ \text{g} = 1.6606 \times 10^{-27}\ \text{kg}$$

Now we can apply Einstein's equation, $E = mc^2$, to determine the energy equivalent to 1 u.

$$E = mc^2 = 1.6606 \times 10^{-27}\ \text{kg} \times (2.9979 \times 10^8)^2\ \text{m}^2\ \text{s}^{-2}$$
$$= 1.4924 \times 10^{-10}\ \text{kg m}^2\ \text{s}^{-2} = 1.4924 \times 10^{-10}\ \text{J}$$

Finally, expressed in MeV, this energy is

$$E = 1.4924 \times 10^{-10}\ \cancel{\text{J}} \times \frac{1\ \text{MeV}}{1.6022 \times 10^{-13}\ \cancel{\text{J}}} = 931.5\ \text{MeV}$$

In Example 19.8, where we determine the energy change in a nuclear reaction, we use *nuclear* masses. These are easily obtained from atomic masses.

Nuclear mass = atomic mass − mass of extranuclear electrons

In dealing with energy changes in nuclear reactions, we use the principle that the total mass/energy of the products is equal to the total mass/energy of the reactants. This suggests that

- Mass *lost* in a nuclear reaction must be replaced by kinetic energy in the products.
- Mass *gained* in a nuclear reaction must come from the kinetic energy of the reactants.

EXAMPLE 19.8

Given the nuclear masses

$$^{241}_{95}\text{Am} = 241.0046\ \text{u} \qquad ^{237}_{93}\text{Np} = 236.9970\ \text{u} \qquad ^{4}_{2}\text{He} = 4.0015\ \text{u}$$

calculate the energy associated with the α decay of americium-241, in MeV.

SOLUTION

First, let's write a nuclear equation and note the masses of the species involved.

$$^{241}_{95}\text{Am} \longrightarrow {}^{237}_{93}\text{Np} + {}^{4}_{2}\text{He}$$

241.0046 u | 236.9970 u | 4.0015 u

Next, we determine the mass change in the decay of $^{237}_{93}\text{Np}$, by subtracting the initial mass from the final mass.

$$236.9970\ \text{u} + 4.0015\ \text{u} - 241.0046\ \text{u} = -0.0061\ \text{u}$$

In megaelectronvolts, the energy of the α decay is

$$-0.0061\ \cancel{\text{u}} \times \frac{931.5\ \text{MeV}}{1\ \cancel{\text{u}}} = -5.7\ \text{MeV}$$

The negative sign signifies that energy is given off.

PROBLEM-SOLVING NOTE

In determining the change in mass in a nuclear reaction, we must have available precise values of the nuclear masses because the mass changes are quite small.

EXERCISE 19.8A

Neptunium-237 decays by α-particle emission. Determine the energy in MeV associated with this decay. The relevant *nuclear* masses are $^{237}_{93}Np$, 236.9970 u; $^{233}_{91}Pa$, 232.9901 u; and $^{4}_{2}He$, 4.0015 u.

EXERCISE 19.8B

Determine the energy requirement in MeV for the following nuclear reaction.

$$^{27}_{13}Al + ^{4}_{2}He \longrightarrow ^{30}_{15}P + ^{1}_{0}n$$

The *atomic* masses are $^{27}_{13}Al$, 26.9815 u; $^{4}_{2}He$, 4.0026 u; and $^{30}_{15}P$, 29.9783 u. Use 1.0087 u for the mass of a neutron. Determine this energy in two ways. First, use *nuclear* masses derived from the given atomic masses, and then use the *atomic* masses directly.

PROBLEM-SOLVING NOTE

In Exercise 19.8B, we see that the calculated energy of a nuclear reaction is the same, whether we use nuclear masses or atomic masses.

Nuclear Binding Energy

We say the mass number of a helium-4 nucleus is *four* because it is made up of *two* protons and *two* neutrons. However, the *mass* of the nucleus is slightly *less* than the sum of the nucleon masses. Figure 19.5 depicts the combination of two protons and two neutrons into a nucleus. There is a mass loss of 0.0305 u, a quantity called the *mass defect* of the nucleus. This lost mass is liberated as energy.

$$0.0305\ \cancel{u} \times \frac{931.5\ \text{MeV}}{1\ \cancel{u}} = 28.4\ \text{MeV}$$

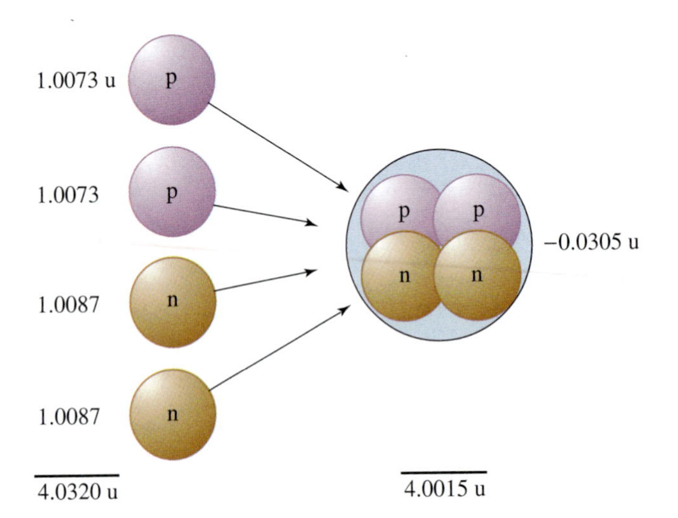

▶ Figure 19.5
Nuclear binding energy in $^{4}_{2}He$
The mass of a helium nucleus, 4.0015 u, is less than that of two protons and two neutrons, 4.0320 u. The energy equivalent of the 0.0305 u mass defect is the nuclear energy that binds the nucleons together.

The energy released in forming a nucleus from its protons and neutrons is called the **nuclear binding energy** and is expressed as a positive quantity. Alternatively, nuclear binding energy is the quantity of energy necessary to separate a nucleus into individual protons and neutrons. If we apportion the nuclear binding energy of $^{4}_{2}He$ among the four nucleons, the binding energy per nucleon is 28.4 MeV/4 = 7.10 MeV. Figure 19.6 is a graph of binding energy per nucleon as a

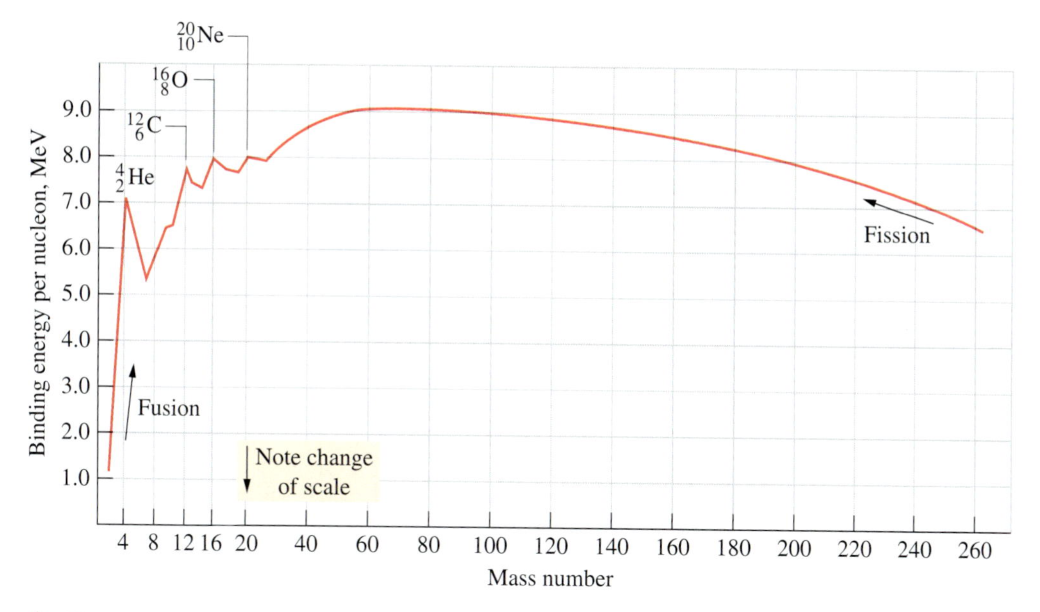

▲ **Figure 19.6 Average binding energy per nucleon as a function of atomic number**

function of atomic number. This graph is useful in explaining nuclear fission and nuclear fusion, discussed in the next section.

19.8 Nuclear Fission and Nuclear Fusion

Figure 19.6 shows that the nuclide $^{56}_{26}Fe$ has about the maximum in binding energy per nucleon. We can draw two conclusions from the existence of this maximum.

- If light nuclei combine to form a heavier nucleus, mass is converted to energy and the binding energy per nucleon in the product nucleus increases as A increases, reaching a maximum at about $A = 56$. The process of combining light nuclei into a heavier one is called **nuclear fusion**.
- If very heavy nuclei split into lighter ones, mass is also converted to energy and the binding energies per nucleon in the product nuclei increase as A decreases, again reaching a maximum at about $A = 56$. The breakup of a heavy nucleus into two lighter fragments is called **nuclear fission**.

Nuclear fission and nuclear fusion are so important in modern life that we will now look at each one in some detail.

The Discovery of Nuclear Fission

In 1934, Enrico Fermi and Emilio Segrè bombarded uranium-238 with slow neutrons, called *thermal neutrons*. The target material emitted β^- particles, but Fermi and Segrè did not firmly establish the source of the radiation. In 1938, the chemists Otto Hahn and Fritz Strassman showed that the products of the bombardment of uranium with thermal neutrons were not elements with $Z > 92$, as Fermi and Segrè had expected. They were radioactive nuclides of much lighter elements, such as strontium and barium. In 1939, Lise Meitner became the first to publish the idea that nuclear fission occurs in the bombardment of uranium with thermal neutrons. Actually, the nuclear fission was not that of uranium-238, but of the minor isotope uranium-235. The fission process is depicted in Figure 19.7.

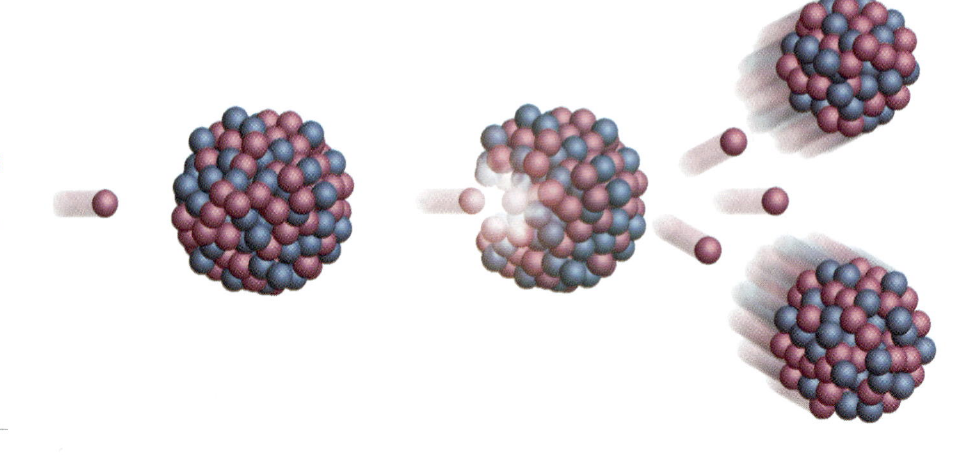

Figure 19.7
Nuclear fission of a uranium-235 nucleus with thermal neutrons
A uranium-235 nucleus is struck by a neutron with ordinary thermal energy. An unstable uranium-236 nucleus is formed, but it breaks into two fragments with the release of several neutrons. The neutrons can induce the fission of other uranium-235 nuclei.

When a nucleus undergoes fission, some mass is converted to energy. On average, the energy release is about 3.2×10^{-11} J (200 MeV) per fission event. This seems like a tiny quantity of energy, but it looks much bigger on a per gram basis: about 8×10^{7} kJ/g. We would have to burn nearly three tons of coal to get the same amount of energy.

Nuclear Reactors

The fission of each ^{235}U nucleus yields additional neutrons—an average of about 2.5 per fission event. These neutrons are important fission products indeed. Those from the first fission event, on average, cause two more ^{235}U nuclei to split. The neutrons from the second round of fission can cause another four or five ^{235}U nuclei to split, and so on. A chain reaction is set off. If the reaction is not controlled, the rapidly released energy causes an explosion. The early atomic bombs were nuclear fission devices. Actually, some of the neutrons produced by fission escape without inducing further fission. In order for an explosion to occur, a **critical mass** of ^{235}U is needed—a large enough quantity to sustain a chain reaction.

A nuclear reactor is designed to tame the nuclear fission process so that energy is released in a controlled manner. In the nuclear reactor pictured in Figure 19.8, fission energy is used to generate steam, the steam powers a steam turbine, and the turbine turns an electric generator. In the core of the reactor, rods of uranium-235–enriched fuel are immersed in liquid water maintained under a pressure of from 70 to 150 atm. The water serves two purposes.

Control rods such as those shown here are inserted into a nuclear reactor to regulate the rate of the chain reaction by absorbing some of the neutrons released during fission.

- Water acts as a *moderator* to slow down neutrons. Slower neutrons are the ones most likely to induce fission.
- Water is a heat-transfer medium. Water under high pressure and superheated to about 300 °C by the fission reaction is pumped to a heat exchanger, where it converts colder water to steam.

The nuclear reaction is regulated by a set of *control rods*, which readily absorb neutrons; the rods are usually made of cadmium. When the rods are lowered into the reactor, they absorb enough neutrons to slow the fission process. When the rods are raised, more neutrons enter the uranium fuel and the rate of fission increases.

A nuclear reactor used to generate electric power "burns" perhaps 1 to 3 kg ^{235}U per day, and a variety of highly radioactive waste products with long half-lives accumulate, for example, $^{239}_{94}Pu$ with a half-life of 24,400 years. The long-term disposal of radioactive wastes is a controversial issue. At present, most nuclear wastes

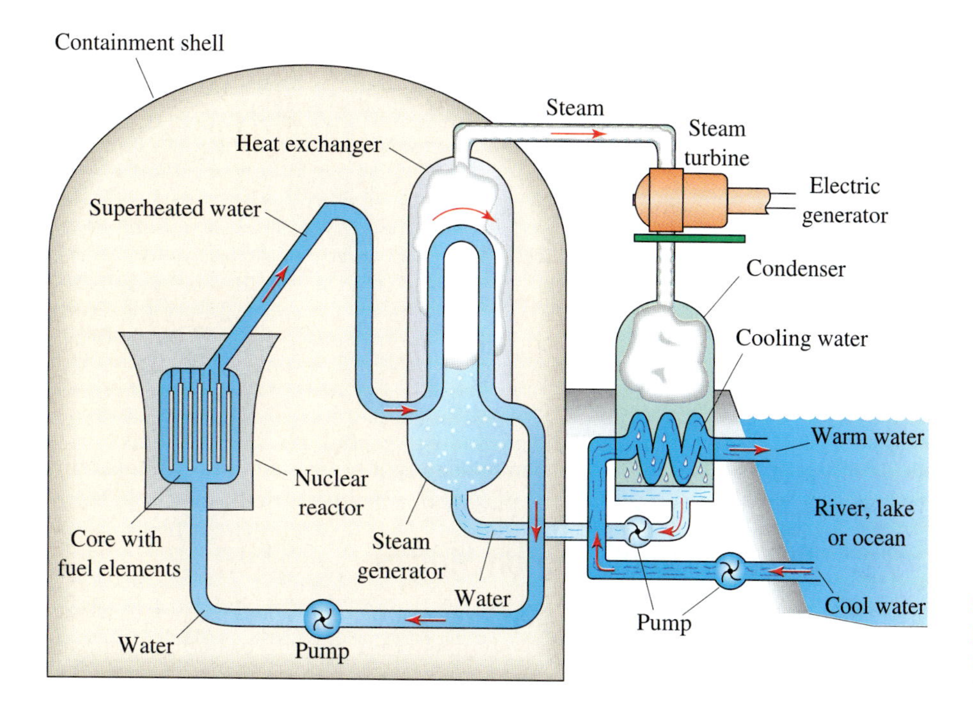

◀ **Figure 19.8**
Schematic diagram of a nuclear power plant used to generate electricity

are kept at the reactor sites, but they cannot remain there permanently. Scientists are studying a proposed burial site in Nevada for the ultimate disposal of radioactive wastes, but the environmental isolation of the site is still uncertain, and the plan has many opponents.

Nuclear Fusion

What would you think of constructing one giant nuclear reactor out in space that would transmit abundant energy to Earth almost forever? Well, that's exactly what we earthlings have in our sun, 92 million miles away. The sun is powered by the *fusion* of atomic nuclei, and its fuel supply—mostly ${}^{1}_{1}H$—will last for billions of years.

On Earth, scientists have unleashed the extraordinary energy of uncontrolled fusion reactions in hydrogen bombs. In a hydrogen bomb, nuclear fusion is initiated by the fission reaction in a fission (atomic) bomb. However, such a totally uncontrolled fusion reaction cannot be used for practical purposes. Control of fusion reactions as energy sources is probably still decades away.

Scientists face a daunting challenge in developing a fusion energy source. The most promising nuclear reaction is the deuterium-tritium reaction.

$$ {}^{2}_{1}H + {}^{3}_{1}H \longrightarrow {}^{4}_{2}He + {}^{1}_{0}n $$

Before they will fuse, however, the nuclei of deuterium and tritium must be forced extremely close together. And because the positively charged nuclei repel one another so strongly, close approach requires that the nuclei have enormously high thermal energies. At the required temperatures, gases are completely ionized into a mixture of atomic nuclei and electrons known as *plasma*. A temperature of over 40,000,000 K is necessary to initiate *self-sustaining* fusion—a nuclear reaction that releases more energy than it takes to get it started. Another requirement is that the plasma be confined at an enormously high density long enough for the fusion to occur. Moreover, this confinement must be done without the plasma contacting the

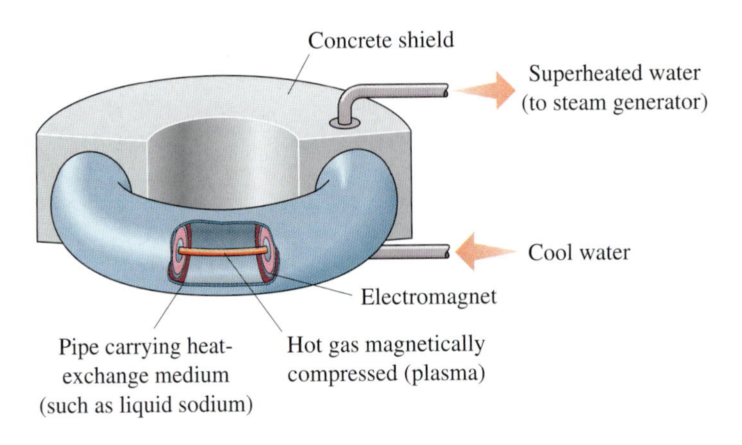

Figure 19.9
A nuclear fusion device
Nuclear fusion requires that plasma be confined at exceedingly high temperatures and pressures. A giant doughnut-shaped electromagnet called a tokamak is one device designed to contain the plasma.

walls of the reactor, where it would immediately lose heat and thus its capability to fuse. One method is to confine the plasma in a magnetic field (Figure 19.9).

19.9 Effect of Radiation on Matter

Ionizing radiation (α, β^-, γ) dislodges electrons from atoms and molecules to produce ions (Figure 19.10). The relatively massive α particle has a high ionizing power and thus produces a large number of ions in passing through matter. The much less massive β^- particle produces far fewer ions, and γ rays have the lowest ionizing power of all. Electrons freed by ionizing radiation are called *primary* electrons. Some primary electrons are energetic enough to ionize atoms and molecules, producing *secondary* electrons. Also, ionizing radiation can raise some electrons to higher energy levels without dislodging them. When these excited electrons return to their ground states, they give off electromagnetic radiation (X rays, ultraviolet or visible light). In the encounter of a β^+ particle (positron) and an electron, the two particles annihilate each other, and the mass that is destroyed is converted into energy. The energy appears as two γ rays originating at the point of annihilation and traveling in opposite directions.

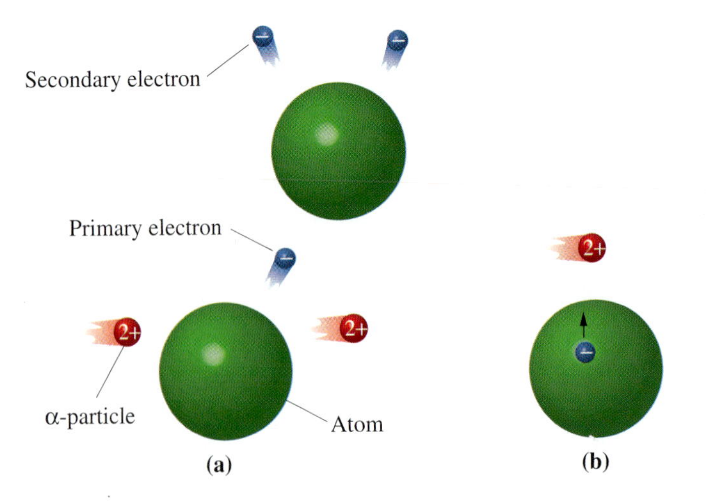

Figure 19.10
Some effects of ionizing radiation
(a) An α-particle passes through an atom and dislodges a *primary* electron. The primary electron in turn dislodges a *secondary* electron from another atom. The atoms that lose electrons become positive ions. **(b)** An α-particle passing near an atom, but not colliding with it, may excite electrons within the atom to higher energy levels.

Radiation Detectors

One of the simplest and oldest ways to detect ionizing radiation is to observe the clouding it produces on photographic film. This is the method that Becquerel used in discovering radioactivity, and it is the principle behind the film badges used by those who work with X rays or radioactive materials.

Summary

The five types of decay of radioactive nuclides involve emission of alpha (α) particles, ${}^{4}_{2}He$; beta (β^-) particles, ${}^{0}_{-1}e$; positrons, ${}^{0}_{1}e$; gamma (γ) rays—high-frequency electromagnetic radiation; and electron capture (EC) followed by the emission of X rays. Nuclear equations can be used to represent these modes of radioactive decay. Both sides of a nuclear equation must have the same total of atomic numbers and of mass numbers.

All known nuclides with $Z > 83$ are radioactive, and many of them occur naturally as members of four radioactive decay series. There are also a few important naturally occurring radioactive nuclides of lower atomic numbers, such as ${}^{3}_{1}H$, ${}^{14}_{6}C$, and ${}^{40}_{19}K$. The majority of radioactive nuclides of the lighter elements can be obtained synthetically only through appropriate nuclear reactions. Transuranium elements can also be produced synthetically. With massive, highly charged ions as projectiles in charged particle accelerators, atoms of all the elements up to $Z = 112$ have been produced. Among the factors that establish whether a nuclide is stable or radioactive are the ratio of the neutron number (N) to the proton number (Z), whether the proton and neutron numbers are even or odd, and whether either of them is a "magic number."

In the formation of an atomic nucleus from its protons and neutrons, a quantity of mass is converted to an equivalent quantity of energy, called the nuclear binding energy. The fusing of lighter nuclei to produce heavier ones is the basis of nuclear fusion energy, and the splitting of heavier nuclei to produce lighter fragments is the basis of nuclear fission energy. Nuclear fission occurs in atomic bombs and in nuclear power reactors. Nuclear fusion is the basis of the hydrogen bomb, but it has not yet been harnessed for practical use.

Radiation from radioactive materials interacts with matter, principally by forming ions and breaking chemical bonds. Less energetic radiation excites electrons to higher energy states, leading to the emission of electromagnetic radiation. The operation of radiation detectors involves detecting and analyzing the interactions of radiation and matter. The interaction of radiation with living matter can cause birth defects and various cancers. These effects are thought to be closely linked to the dosage of radiation received.

Key Terms

alpha (α) particle (19.1)
beta (β^-) particle (19.1)
critical mass (19.8)
electron capture (EC) (19.1)
electronvolt (19.7)
gamma (γ) ray (19.1)
half-life ($t_{1/2}$) (19.6)
magic numbers (19.6)
nuclear binding energy (19.7)
nuclear fission (19.8)
nuclear fusion (19.8)
nucleon (19.1)
nuclide (19.1)
positron (β^+) (19.1)
rad (19.9)
radioactive decay law (19.3)
radioactive decay series (19.2)
radioactive tracer (19.10)
radioactivity (radioactive decay) (page 817)
rem (19.9)
transuranium elements (19.5)

Review Questions

1. Describe the basic nature of α, β^-, and γ radiation. Which has the greatest ability to penetrate through matter? Which has the greatest ability to ionize matter?
2. What is the main difference between a γ ray and an X ray?
3. What type of radiation is emitted when the atomic number of a nucleus increases by one unit? decreases by two units?
4. What change in atomic number occurs when a nucleus emits a β^- particle? a positron?
5. Describe the relationship between the following terms: *nucleus*, *nuclide*, *nucleon*.
6. Why does radioactive decay by electron capture produce X rays?
7. What principles underlie the balancing of nuclear equations?
8. What is a radioactive decay series?
9. Distinguish among the terms *activity*, *decay constant*, and *half-life*, all of which are used in describing radioactive decay rates.
10. How is the activity of a radioactive sample related to the amount of sample? How is the activity of the sample related to time?
11. How does radiocarbon dating work? What are its limitations in establishing the ages of objects?
12. How can a nuclide that does not occur naturally be synthesized?
13. What is a charged-particle accelerator and how is it used?
14. What is meant by the term *magic numbers*? What does it mean for a nucleus to be "doubly magic"?
15. What is a *nuclear binding energy*? What is a nuclear binding energy per nucleon? Where in the periodic table would you expect to find the elements with the highest nuclear binding energy per nucleon?
16. What is the meaning of the term *mass defect*, when applied to a nucleus?
17. Explain why nuclear reactions liberate so much more energy than do chemical reactions.
18. What is nuclear fission, and how does it differ from nuclear fusion?
19. What is the meaning of the term *critical mass*? To what type of nuclear process does it apply?
20. In a nuclear reactor, what do *control rods* control, and what does the *moderator* moderate?

mass of radium-229 is 228.9866 u, and the mass of an electron is 5.486×10^{-4} u.

62. Thorium-228 decays by α-particle emission. In this decay, 5.50 MeV of energy is given off. What are **(a)** the *nuclear* mass and **(b)** the *atomic* mass of thorium-228? The nuclear masses of radium-224 and helium-4 are 223.9719 u and 4.00150 u, respectively.

63. What is the energy, in megaelectronvolts (MeV), required to bring about the following nuclear reaction?

$$^{232}_{90}\text{Th} + ^{4}_{2}\text{He} \longrightarrow ^{235}_{92}\text{U} + ^{1}_{0}\text{n}$$

The *atomic* masses are ^{232}Th, 232.0382 u; ^{4}He, 4.0026 u; ^{235}U, 235.0439 u. The mass of a neutron is 1.0087 u.

64. Given that the atomic masses of ^{235}U and ^{236}U are 235.0439 u and 236.0456 u, respectively, explain the point made on page 835 that fission can be initiated with slow neutrons, that is, neutrons with only the kinetic energy associated with ordinary temperatures.

Nuclear Stability

65. Of these two fluorine nuclides, $^{17}_{9}$F and $^{22}_{9}$F, one decays by β^- emission and the other by β^+ emission. Which does which? Explain.

66. Which of the following nuclides do you think are radioactive and which do you think are stable? Explain.

(a) $^{10}_{6}$C **(b)** $^{15}_{7}$N **(c)** $^{30}_{14}$Si **(d)** $^{36}_{15}$P

67. In some cases, we can identify the most abundant nuclide of an element by rounding off the weighted average atomic mass listed on the inside front cover. Thus, the principal nuclide of sodium (weighted average atomic mass, 22.9898 u) is sodium-23, and that of potassium (weighted average atomic mass, 39.0983 u) is potassium-39. In some cases, though, this method won't work. Cite some examples, and explain why the method fails.

68. In general, the elements with *even* atomic numbers have a greater number of naturally occurring isotopes than those with *odd* atomic numbers. Explain why you would expect this to be the case.

69. There are six naturally occurring nuclides of calcium. They have the mass numbers 40, 42, 43, 44, 46, 48. Why do you suppose that calcium-40 is by far the most abundant of these nuclides?

70. For what neutron number N, would you expect to find the greatest number of stable nuclides? Explain.

Fission and Fusion

71. Suppose that one of the reactions in the fission of uranium-235 is

$$^{235}_{92}\text{U} + ^{1}_{0}\text{n} \longrightarrow ^{139}_{54}\text{Xe} + ^{95}_{38}\text{Sr} + 2\,^{1}_{0}\text{n}$$

Calculate the energy in megaelectronvolts (MeV) released in this fission. The nuclear masses are: ^{235}U, 234.9934 u; ^{139}Xe, 138.8891; ^{95}Sr, 94.8985 u; ^{1}n = 1.0087 u.

72. Calculate the energy in megaelectronvolts (MeV) released in the nuclear fusion reaction:

$$^{2}_{1}\text{H} + ^{3}_{1}\text{H} \longrightarrow ^{4}_{2}\text{He} + ^{1}_{0}\text{n}.$$

The nuclear masses are: ^{2}H, 2.0135 u; ^{3}H, 3.0156; ^{4}He, 4.0015 u; ^{1}n, 1.0087 u. How does the energy per gram of ^{4}He produced in this reaction compare with the energy per gram of ^{235}U that undergoes fission, described on page 836.

Radiation Dosage and the Effect of Radiation on Matter

73. Explain why radioactive nuclides with half-lives of intermediate value are generally more hazardous than those with extremely short or extremely long half-lives.

74. Explain why some radioactive substances are hazardous from a distance, whereas others must be taken internally to constitute a hazard.

75. Estimates of the total release of iodine-131 from the Chernobyl nuclear accident of 1986 range from 40 to 50 million curies (Ci). Given the definition of a curie in Table 19.3 and that $t_{1/2}$ = 8.040 d for iodine-131, approximately how many grams of iodine-131 were released in the accident?

76. An evaluation of the iodine-131 contamination caused by the Chernobyl nuclear accident placed the activity close to the site at greater than 37,000 kBq/m^2. Given the definition of the becquerel (Bq) in Table 19.3 and that $t_{1/2}$ = 8.040 d for iodine-131, what was the minimum level of contamination with iodine-131 near the plant site, expressed in milligrams of iodine-131 per square kilometer?

Additional Problems

77. Refer to Figure 19.1 and indicate the type of deflection expected for a positron. Why is the deflection of the α-particle not as great as that of β^-? Draw a sketch of the expected deflections of α, β^-, β^+, and γ rays in a magnetic field, rather than an electric field.

78. Francium is such a rare element that it has been estimated that less than about 30 grams are present in Earth's crust at any one time. How do you think such an estimate of the natural abundance of francium is made? Would you expect to find francium in the same natural sources as the other Group 1A (alkali) metals? Explain.

79. Argon is by far the most abundant of the noble gases, making up almost 1% of the atmosphere by volume. What is the most likely reason for this?

80. An aqueous solution of HCl(aq) is prepared incorporating some ^3H as a tracer in HCl. Then the labeled HCl(aq) is used in the following reactions, which are carried out in succession:

$$NH_3(aq) + HCl(aq) \longrightarrow NH_4Cl(aq)$$

$$NH_4Cl(aq) + NaOH(aq) \longrightarrow NaCl(aq) + H_2O(l) + NH_3(g).$$

Would you expect radioactivity to appear **(a)** in the $NH_3(g)$, **(b)** in the NaCl(aq), **(c)** in NaCl(s) crystallized from the solution? Explain.

81. In the decay of uranium-238, some mass is converted to energy carried away by the α particle. What is the energy of the α particle, in MeV? The *atomic* masses are ^{238}U = 238.0508 u, ^{234}Th = 234.0436 u, and ^4He = 4.0026 u.

82. In the γ-ray decay of $^{230m}_{90}Th$, the energy of the γ rays is 0.050 MeV. What is the wavelength of these γ rays?

83. An electric power plant burns natural gas that is 92.0 % $CH_4(g)$ and 8.0 % $C_2H_6(g)$ by volume. Assume complete combustion of the gas, yielding $CO_2(g)$ and $H_2O(l)$ as products. How many liters of this natural gas, measured at 22 °C and 744 mmHg, would have to be burned to liberate as much energy as that produced in the fission of 1.00 mg ^{235}U? Assume the energy release in the fission is that cited on page 836.

84. Calculate the minimum kinetic energy, in megaelectronvolts (MeV), and the velocity in meters per second (m/s), that deuterons must have in order to bring about this nuclear reaction:

$$^{75}_{33}As + ^{2}_{1}H \longrightarrow ^{75}_{34}Se + 2\,^{1}_{0}n.$$

The *atomic* masses are: ^{75}As, 74.9216 u; ^2H, 2.0140 u; ^{75}Se, 74.9225 u. The mass of a neutron is 1.0087 u.

85. Determine the mass, in grams, equivalent to the energy released in the combustion of 1.00 kg of carbon to produce carbon dioxide.

86. The curie, described in Table 19.3, is the rate of radioactive decay of 1.00 g of radium-226. The atomic mass of radium-226 is 226.0254 u. Use this information and the radioactive decay law to obtain a value of Avogadro's number. Would you expect this method to be as accurate as the one based on X-ray crystallography described in Chapter 11 (Problems 77 and 78)? Explain.

87. Tritium, ^3_1H, is a β^- emitter with a 12.26 year half-life. Some HCl with ^3H as a tracer is added to HCl(g), which is then used to prepare 50.0 mL of 1.00 M HCl(aq). The activity, *A*, of the solution is found to be 2.5×10^{10} disintegrations per second. Then a 0.455-g sample of zinc is allowed to react with the 50.0 mL of 1.00 M HCl. What will be the activity of the $H_2(g)$ produced?

88. The percent natural abundance of radioactive potassium-40 ($t_{1/2} = 1.25 \times 10^9$ y) is 0.0117%. The decay of potassium-40 takes place principally by β^- emission. Approximately how many β^- particles are produced per minute by a 50.0-mg sample of the mineral *carnallite*, $KMgCl_3 \cdot 6H_2O$?

89. Use the radioactive decay law to verify the statement on page 823 that the percent natural abundances of ^{238}U and ^{235}U correspond to an age of Earth of about 6 billion years, assuming that the two nuclides were originally present in equal abundances. (*Hint:* You will need to solve two equations in two unknowns.)

90. Problem 89 suggests how the ratio of two uranium nuclides can be used to establish an approximate age of Earth. Another ratio that can be used to establish the age of certain rocks in Earth's crust is the that of ^{206}Pb to ^{238}U. Suppose that the molten material that froze to become a rock originally contained 1.00 g ^{238}U. Assuming that no lead was present initially and that none entered or left the rock, what mass of ^{206}Pb would you expect to find in the rock after one half-life period of the ^{238}U? (*Hint:* What are the ultimate products of the radioactive decay series that starts with ^{238}U?)

CHAPTER

20

CONTENTS

● **ChemCDX**
When you see this icon, there are related animations, demonstrations, and exercises on the CD accompanying this text.

The *s*-Block Elements

A characteristic reaction of most of the *s*-block elements is with water to produce hydrogen gas. If the reaction is sufficiently vigorous, as here between potassium metal and water, the hydrogen will ignite.

Early chemical knowledge was mostly of a practical nature, a how-to of chemical processes. Chemists mainly used trial and error to transform the materials around them, accumulating a wealth of factual information while doing so. In contrast, modern chemical knowledge is based largely on principles that explain facts and answer the "why" as well as the how-to of chemical change. So far in this text, we have focused on principles, although we have often made references to their application. In the remaining chapters, we will emphasize facts and applications, but we will refer to underlying principles repeatedly.

The periodic table provides us an organizational scheme for a systematic study of the elements and their compounds. In this study, generally called *descriptive chemistry*, we describe the behavior of the elements and their compounds and, where possible, we attempt to rationalize this behavior by using fundamental ideas about atomic and molecular structure, intermolecular forces, thermodynamics, kinetics, oxidation-reduction, acids and bases, and so on. This chapter deals with the *s*-block elements, and the next two deal with the *p*-block and *d*-block elements, respectively.

Four of the 14 elements that comprise the *s*-block are somewhat unusual cases, although not all for the same reason. We discussed the difficulty in placing hydrogen in the periodic table in Chapter 8 (page 332). We tend to place hydrogen in Group 1A based on its electron configuration of $1s^1$. On the other hand, because hydrogen is a nonmetal with little resemblance to the active metals of Group 1A, we can place it in a special location at the top and near the center of the table, as is done in some periodic tables.

Because the helium atom has electrons only in its 1*s* orbital ($1s^2$), it is an *s*-block element. In its properties, however, helium closely resembles only the other noble gases, which have the outer-shell electron configuration ns^2np^6. They are *p*-block elements, and we will consider helium with the *p*-block elements despite its placement in the *s* block.

In their physical and chemical properties, francium and radium do resemble the other members of Group 1A and 2A, respectively. However, because all their isotopes are radioactive, our main interest is in their nuclear properties. We said something about these elements in Chapter 19 but will not discuss them further in this chapter.

1671 Boyle: Flammable gas in metal-acid reaction.

1766 Cavendish: Discovery of H_2.

1803 Dalton: H chosen as atomic weight standard.

1810 Davy: H shown to be common to all acids.

1885 Balmer: Equation describing emission spectrum of H.

1913 Bohr: Quantum mechanical treatment of H atom.

1926 Schrödinger: Wave mechanics developed for H atom.

1927 Heitler & London: Wave mechanics applied to H_2.

▲ **Figure 20.1 Important developments linked to the element hydrogen**

Hydrogen

Because hydrogen atoms are the simplest of all the atoms, hydrogen has always been considered rather special. For example, hydrogen was Dalton's choice as an atomic weight standard, the element found to be common to all acids, and the focus of early quantum mechanical studies (see Figure 20.1).

A *molecule* of H_2 is lower in energy than two H atoms by 436 kJ/mol, and hydrogen is therefore found as H_2 molecules and not H atoms. The bond between the two H atoms in H_2 is a single covalent bond. Intermolecular forces in a sample of hydrogen are weak, and hydrogen is a gas that conforms reasonably well to the ideal gas law over a range of temperatures and pressures. Hydrogen can be liquefied at 1 atm pressure at 20.39 K, and it freezes to a solid at 13.98 K.

20.1 Occurrence and Preparation of Hydrogen

Hydrogen makes up 0.9% of the mass and 15.1% of the atoms in Earth's crust. (Earth's crust is the layer of rock covering the planet to an average depth of 17 km.) If we look beyond our home planet, H atoms are thought to make up 89% of the atoms on the sun, and 85 to 95% of the atoms in the atmospheres of the outer planets. In the universe as a whole, about 90% of all atoms are H (the rest are mainly He).

Optical image of the emission nebula NGC 6357, located about 5500 light-years from Earth. It is a cloud of hydrogen (60 light-years or 3.5×10^{14} mi across) ionized and lit up by young, hot, blue stars embedded in it.

Only trace amounts of free hydrogen (H_2) are found on Earth; it must be obtained from its compounds. Although hydrogen occurs in more compounds than does any other element, only a few compounds are economically viable sources of the element. Water, the most abundant hydrogen-containing compound, is usually the first choice. We can obtain H_2 in reactions of steam [$H_2O(g)$] with

- *carbon* (as coal or coke)
- *carbon monoxide*
- *hydrocarbons* (for example, methane or natural gas)

These reactions are carried out at high temperatures, and some require catalysts.

Water-gas reaction: $$C(s) + H_2O(g) \xrightarrow{1000\ ^\circ C} CO(g) + H_2(g)$$

$$\textit{Water-gas shift reaction:} \quad CO(g) + H_2O(g) \xrightarrow[1000\ °C]{\text{catalyst}} CO_2(g) + H_2(g)$$

$$\textit{Reforming of methane:} \quad CH_4(g) + H_2O(g) \xrightarrow[1000\ °C]{\text{catalyst}} CO(g) + 3\ H_2(g)$$

The first reaction produces water gas, a combustible mixture that consists mainly of CO and H_2. The second reaction—the water-gas shift reaction—is used to increase the yield of H_2. The third reaction—the reforming of methane—is the principal commercial source of hydrogen. (The term *reforming* is used to signify the restructuring of a hydrocarbon.)

Hydrogen is an important by-product of petroleum refining operations. For example, in the process of *catalytic reforming*, an alkane hydrocarbon of low octane rating (page 264), such as hexane, is converted to a hydrocarbon of higher octane rating, such as benzene. Hydrogen is also produced.

$$\underset{\text{Hexane}}{C_6H_{14}} \xrightarrow{\text{catalyst}} \underset{\text{Benzene}}{C_6H_6} + 4\ H_2$$

As recently as the 1930s, before the advent of inexpensive natural gas, there were over 11,000 coal gasifiers operating in the United States. They supplied homes and industry with water gas and related gaseous products. These gases were variously called producer gas, town gas, or city gas and were stored in large tanks.

● **ChemCDX** Electrolysis of Water

The most direct method of producing hydrogen (and oxygen as well) is the decomposition of water. Some compounds can be decomposed into their elements simply by heating strongly enough to break chemical bonds. Even at 2000 °C, though, water is only about 1% decomposed into H_2 and O_2. When decomposition by heating is not feasible, chemists often use *electrolysis*. Recall that water is a nonelectrolyte and can conduct electric current only when ions of a solute are present (Section 4.1). We learned how to predict the probable products of an electrolysis in Section 18.9. Specifically, we saw that if no other substances are present that are more readily oxidized and reduced than H_2O, the products of the electrolysis are H_2 and O_2.

In dilute sulfuric acid, H^+ ions are reduced to $H_2(g)$ at the cathode, and H_2O molecules are oxidized to $O_2(g)$ at the anode (Figure 20.2).

$$\textit{Reduction (cathode):} \quad 2\ \{2\ H^+(aq) + 2\ e^- \longrightarrow H_2(g)\}$$

$$\textit{Oxidation (anode):} \quad 2\ H_2O(l) \longrightarrow 4\ H^+(aq) + O_2(g) + 4\ e^-$$

$$\textit{Net:} \quad 2\ H_2O(l) + \cancel{4\ H^+(aq)} + \cancel{4\ e^-} \longrightarrow 2\ H_2(g) + \cancel{4\ H^+(aq)} + O_2(g) + \cancel{4\ e^-}$$

$$E^\circ_{cell} = E^\circ(\text{cathode}) - E^\circ(\text{anode}) = 0.000\ V - 1.229\ V = -1.229\ V$$

In summary,

$$2\ H_2O(l) \xrightarrow{\text{electrolysis}} 2\ H_2(g) + O_2(g) \qquad E^\circ_{cell} = -1.229\ V$$

The value of E°_{cell} tells us two things about this reaction: (1) It is nonspontaneous as it is written (That is why we need an external agent—in this case, electrolysis.) (2) The required voltage for the electrolysis must exceed 1.229 V. We should be able to electrolyze water with an ordinary flashlight battery (1.5 V).

The electrolysis of water is an expensive way to make hydrogen. About 0.1 kilowatt hour (kWh) of electrical energy is required to produce 1 mol $H_2(g)$. It takes as much energy to obtain just two grams of hydrogen by the electrolysis of water as to operate a 100-watt electric lightbulb for one hour. Electrolysis of water is economically feasible only where hydroelectric power is cheap (as in Canada and Norway).

In the laboratory, we often can prepare chemicals by methods that are not economically feasible for commercial production. A common laboratory method for

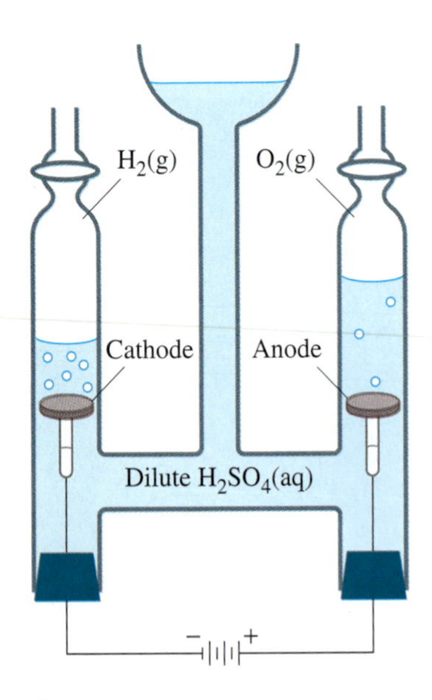

▲ **Figure 20.2**
The electrolysis of water
Hydrogen gas forms at the negative electrode (cathode), and oxygen gas at the positive electrode (anode). The volume of hydrogen produced is exactly twice that of oxygen.

preparing small quantities of hydrogen involves the displacement of $H^+(aq)$ by metals such as zinc that lie above hydrogen in the activity series of metals or below hydrogen in a table of standard electrode potentials (Table 18.1). For example

$$Zn(s) + 2\,H^+(aq) \longrightarrow Zn^{2+}(aq) + H_2(g)$$

The most active of the metals, those of Group 1A and the heavier members of Group 2A, displace $H_2(g)$ even from pure water, where the concentration of H^+ is very low. Here, H_2O is the reactant, rather than H^+, and ionic hydroxides are formed together with $H_2(g)$

$$2\,M(s) + 2\,H_2O(l) \longrightarrow 2\,MOH(aq) + H_2(g) \qquad \text{(where M is any Group 1A metal)}$$

$$M(s) + 2\,H_2O(l) \longrightarrow M(OH)_2(aq) + H_2(g) \qquad \text{(where M is Ca, Sr, Ba, or Ra)}$$

20.2 Binary Compounds of Hydrogen

Hydrogen reacts with other nonmetals to form molecular compounds. We can call these compounds *molecular hydrides*. Two familiar ones are hydrogen chloride and ammonia.

$$H_2(g) + Cl_2(g) \longrightarrow 2\,HCl(g) \qquad \Delta H^\circ = -184.62\ \text{kJ};\ K_p = 2.5 \times 10^{33}\ \text{at 298 K}$$

$$3\,H_2(g) + N_2(g) \rightleftharpoons 2\,NH_3(g) \qquad \Delta H^\circ = -92.22\ \text{kJ};\ K_p = 6.2 \times 10^{5}\ \text{at 298 K}$$

The value of K_p for the first reaction is so large that the reaction goes essentially to completion. In fact, the reaction occurs with explosive violence when a mixture of hydrogen and chlorine is exposed to light. The second reaction is reversible. According to Le Châtelier's principle (Section 14.4), we should expect the production of ammonia to be favored at high pressures of $H_2(g)$ and $N_2(g)$ and at low temperatures. However, because of the high activation energy, equilibrium is reached very slowly at low temperatures. The synthesis of ammonia is carried out at elevated temperatures (page 616).

Among the most important molecular hydrides are those of carbon and hydrogen. *Hydrocarbons* are the customary entry point to the field of organic chemistry (Sections 2.9 and 2.10), and they are the fuels derived from petroleum.

Hydrogen reacts with the most active metals to form *ionic hydrides*, in which hydrogen exists as the *hydride* ion, H^-. The formation of many ionic hydrides is energetically favorable.

$$Na(s) + \tfrac{1}{2}\,H_2(g) \longrightarrow NaH(s) \qquad \Delta H_f^\circ = -56.3\ \text{kJ}$$

$$Ca(s) + H_2(g) \longrightarrow CaH_2(s) \qquad \Delta H_f^\circ = -186\ \text{kJ}$$

Ionic hydrides react with water to liberate $H_2(g)$ and are sometimes used as a source of hydrogen for weather-observation balloons.

$$CaH_2(s) + 2\,H_2O(l) \longrightarrow Ca^{2+}(aq) + 2\,OH^-(aq) + 2\,H_2(g)$$

When hydrides are formed from transition elements, the products, called *metallic hydrides*, retain some metallic properties such as electrical conductivity. Many of the metallic hydrides are *nonstoichiometric*—the ratio of H atoms to metal atoms is variable. For example, the formula of titanium hydride can vary from $TiH_{1.8}$ to TiH_2. We can picture H atoms in metallic hydrides as filling the voids or holes

The effervescence in this beaker is caused by hydrogen escaping from the reaction of calcium hydride and water. The deep pink color is that of phenolphthalein indicator in the presence of $OH^-(aq)$, which is also produced in the reaction. The fog above the beaker is condensed water vapor, indicating that the contents of the beaker are much warmer than the surroundings. The reaction is exothermic.

among metal atoms in the solid—much like fitting cherries into the holes among oranges in a box (Figure 20.3).* Because some holes fill while others do not, the ratio of H to metal atoms is variable.

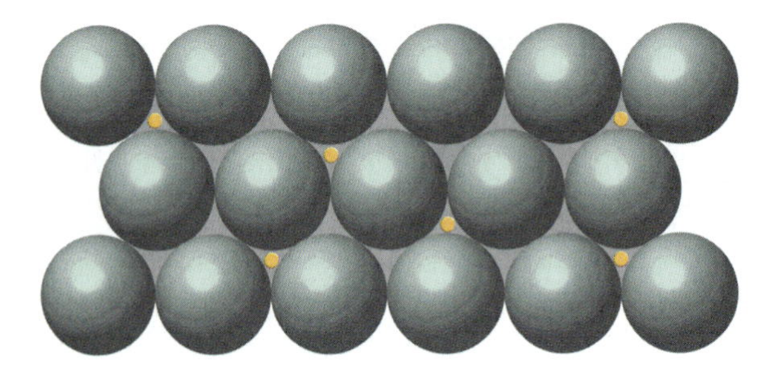

Figure 20.3
Representing a metal hydride
The small yellow spheres represent H atoms interspersed in the voids of a close-packed collection of larger metal atoms.

20.3 Uses of Hydrogen

Nearly half of the hydrogen gas produced is used in the manufacture of ammonia (NH_3). Ammonia, in turn, is used in the manufacture of fertilizers, plastics, and explosives. The second most important use of hydrogen is in the petrochemical industry. For example, hydrogen is used to convert benzene to cyclohexane, a cyclic hydrocarbon used as an intermediate in the production of nylon.

Another chemical manufacturing process that uses hydrogen is the synthesis of methanol, an industrial solvent and a raw material for making other organic compounds.

$$CO(g) + 2\,H_2(g) \xrightarrow{\text{catalyst}} CH_3OH(g)$$

In metallurgy, the extraction of pure metals from their mineral sources generally involves a *reduction* process. The reducing agent most commonly used is some form of carbon (charcoal, coke, or coal). In some cases, however, carbon is not a sufficient reducing agent, and in others it may react with the metal to form unwanted metal carbides. In such instances, hydrogen is often used, as in the reduction of WO_3 to tungsten metal.

$$WO_3(s) + 3\,H_2(g) \xrightarrow{850\ ^\circ\text{C}} W(s) + 3\,H_2O(g)$$

Liquid hydrogen is used as a rocket fuel. The space shuttle rocket engines, for example, use the reaction of hydrogen and oxygen.

$$H_2(g) + \tfrac{1}{2}\,O_2(g) \longrightarrow H_2O(l) \qquad \Delta H^\circ = -285.8\ \text{kJ}$$

Both reactants are stored as liquids. The fuel tank holds 1.5×10^6 L of liquid hydrogen. The oxygen tank carries 5.4×10^5 L of liquid oxygen. During liftoff, these propellants power the shuttle's main engines for about 8.5 minutes; liquid hydrogen is consumed at a rate of nearly 3000 L/s.

When hydrogen is passed through an electric arc, H_2 molecules absorb energy and dissociate into H atoms. The H atoms then recombine into H_2 molecules, in a highly exothermic process. This recombination, coupled with the combustion of H_2, creates an extremely high flame temperature.

* The actual situation may be somewhat more complicated because the arrangement of metal atoms in metallic hydrides is not always the same as in the pure metal.

A Possible Future: The Hydrogen Economy

The prospect of dwindling supplies and higher prices for coal, natural gas, and petroleum and the threat of global warming mean that we may need a substitute for fossil fuels in future centuries. Hydrogen is often mentioned as a candidate.

Hydrogen is an attractive fuel for several reasons. For example, an automobile engine burning hydrogen is 25 to 50% more energy-efficient than is an engine burning gasoline. Because the only significant combustion product is H_2O, the exhaust from a hydrogen engine is far lower in pollutants than the exhaust of a gasoline engine. The heat of combustion per gram of liquid hydrogen is more than twice that of jet fuel. Aircraft using liquid hydrogen could fly much farther than those using conventional jet fuel.

In addition to taking the place of gasoline and jet fuels for transportation, hydrogen could also replace natural gas for heating homes and other buildings. Because it is an excellent reducing agent, hydrogen could largely replace carbon in metallurgy. Finally, with hydrogen readily available, the cost of producing ammonia and products derived from it would remain low. If we adopt the widespread use of hydrogen, as outlined here, major changes in our way of life would follow, producing what is called a hydrogen economy.

Hydrogen looks good for a future fuel, but tough problems must be solved before we adopt a hydrogen economy. There are no hydrogen mines, nor can we pump it from the ground by drilling holes. A hydrogen economy requires a cheap way of making hydrogen and storing it. The most likely future source of hydrogen will be water. The decomposition of water might be accomplished as the net reaction of a series of thermochemical reactions, or by electrolysis if this can be done cheaply enough. Of course, as much energy is required to break down water into hydrogen and oxygen as is obtained in burning hydrogen as a fuel. What is actually accomplished is a transfer of energy from one form—nuclear, hydroelectric, solar—into another highly portable and versatile form—hydrogen.

Because of the large volumes involved, storing hydrogen as a gas is not feasible. Hydrogen, when liquefied, occupies a much smaller volume, but then it must be maintained at extremely low temperatures (below −240 °C). And, in either case, hydrogen must be kept from contact with oxygen (air), with which it forms explosive mixtures. One promising possibility is to dissolve the hydrogen gas in a metal, such as an iron-titanium alloy. The gas can be released from the metal by mild heating. In a hydrogen-burning automobile engine, for example, the required heat might come from the engine exhaust.

Planning for a possible hydrogen economy is a source of fascinating challenges for chemists and other natural and social scientists as well.

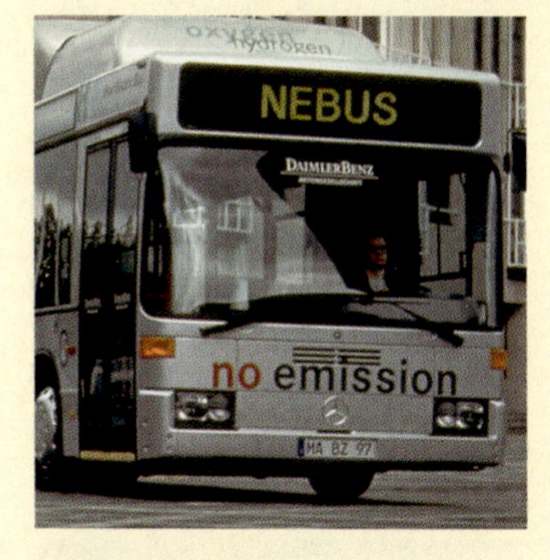

The Daimler-Benz fuel-cell bus NEBUS in front of the company headquarters in Stuttgart-Mohringen. NEBUS is a genuine zero-emission vehicle: It produces no pollutants whatsoever; only pure water vapor emerges from the exhaust pipe. This bus is propelled by quiet electric motors. The necessary power (250 kW) is provided by fuel cells in the rear.

$$\begin{array}{lll} 2\,H(g) \longrightarrow H_2(g) & & \Delta H^\circ = -436\text{ kJ} \\ H_2(g) + \tfrac{1}{2}\,O_2(g) \longrightarrow H_2O(g) & & \Delta H^\circ = -241.8\text{ kJ} \\ \hline 2\,H(g) + \tfrac{1}{2}\,O_2(g) \longrightarrow H_2O(g) & & \Delta H^\circ = -678\text{ kJ} \end{array}$$

An oxyhydrogen welding torch using atomic hydrogen readily cuts through steel and can be used to melt tungsten, which has a melting point of about 3400 °C.

Flame tests for Li, Na, and K were shown in Chapter 8. Those for Rb (top) and Cs are shown here.

Group 1A: The Alkali Metals

Table 20.1 lists some properties of the alkali metals. We introduced most of these properties in earlier chapters, for example, electron configurations in Section 8.4, atomic and ionic radii and ionization energies in Section 8.7, flame colors in Section 8.9, electronegativity in Section 9.7 and electrode potentials in Section 18.4. Table 20.1 also includes data on two new properties: electrical conductivity and hardness. The numerical values listed for these properties are based on comparisons between substances rather than on particular sets of units.

TABLE 20.1 Some Properties of the Group 1A Metals

	Li	Na	K	Rb	Cs
Atomic number, Z	3	11	19	37	55
Valence-shell electron configuration	$2s^1$	$3s^1$	$4s^1$	$5s^1$	$6s^1$
Atomic (metallic) radius, pm	152	186	227	248	265
Ionic (M^+) radius, pm	59	99	138	148	169
First ionization energy (I_1), kJ/mol	520	496	419	403	376
Electronegativity	1.0	0.9	0.8	0.8	0.7
Flame color	Carmine	Yellow	Violet	Bluish red	Blue
Melting point, °C	180.54	97.81	63.65	39.1	28.40
Density, g/cm^3 at 20 °C	0.534	0.971	0.862	1.532	1.873
Electrode potential, $E°$, V $M^+(aq) + e^- \longrightarrow M(s)$	−3.040	−2.713	−2.924	−2.924	−2.923
Electrical conductivity[a]	18.6	37.9	25.9	12.7	8.0
Hardness[b]	0.6	0.4	0.5	0.3	0.2

[a] On a scale relative to silver, 100; copper, 95.0; gold, 67.7.
[b] On a scale (Mohs) from 0 to 10. Each item on this scale is able to scratch only materials with hardness less than its own. Examples: talc (0), wax (0.2), asphalt (1–2), fingernail (2.5), copper (2.5–3), iron (4–5), glass (5–6), steel (5–8.5), diamond (10).

20.4 Properties and Trends in Group 1A

The Group 1A metals exhibit regular trends for a number of properties. Just as we should expect, the radii of the alkali metal atoms and ions *increase* regularly from Li to Cs, because the number of electron shells increases in proceeding down the group. The regular *decrease* in first ionization energies occurs because the larger atoms can be stripped of their ns^1 electron more easily than can smaller atoms. The *decrease* in electronegativities from top to bottom in Group 1A signifies an increase in metallic character, which we can also associate with increasing atomic radii and decreasing ionization energies. A continuous increase or decrease of a property—a regular trend—suggests that a single factor is primarily responsible for that property.

Irregular trends suggest that factors are working against each other in determining a property. Consider, for example, the fact that the density of K is lower than that of the element preceding it (Na) and also lower than that of the one following it (Rb). The factors we need to examine are mass and volume. Let's work with the molar mass and molar volume.

$$d\ (\text{g/cm}^3) = \frac{\text{molar mass (g/mol)}}{\text{molar volume (cm}^3\text{/mol)}}$$

Sodium, a soft metal, can be cut with a knife.

Molar mass increases as atoms become heavier, and molar volume generally increases as atoms become larger.* Eight elements separate Na ($Z = 11$) and K ($Z = 19$). The added molar mass in the short third period is less significant than the jump in atomic size in going from three to four electron shells. The molar volume (denominator) increases by a greater factor than the molar mass (numerator) and the density *decreases*. Eighteen elements separate K ($Z = 19$) and Rb ($Z = 37$). The added molar mass in the long fourth period is more significant then the jump in atomic size in going from four to five electron shells. The molar mass (numerator) increases by a greater factor than the molar volume (denominator) and the density *increases*.

Two notable physical properties of the alkali metals are that the solids are soft and have low melting points. All are soft enough to be scratched with a toothpick and cut with a knife (even one made of plastic). The melting points of cesium and rubidium are low enough that both would be liquids on a very hot day.

When freshly cut, the alkali metals are bright and shiny—typical metallic properties. The metals quickly tarnish, however, as they react with oxygen in the atmosphere. Another characteristic property is their ability to conduct electric current. In a ranking of electrical conductivity, silver, copper, and gold are number 1, 2, and 3, respectively, but sodium is not far behind at number 7. In Chapter 24 we will discuss a theory of bonding in metals that explains electrical conductivity and other metallic properties.

Diagonal Relationships: The Special Case of Lithium

The first member of a group of the periodic table often differs in some significant ways from the other members of its group. The eccentric member of Group 1A is lithium. Following are some of the ways that lithium and its compounds differ from the other Group 1A metals.

- Lithium carbonate, fluoride, hydroxide, and phosphate are much less soluble in water than are corresponding salts of the other alkali metals.
- Lithium forms a nitride (Li_3N), the only alkali metal to do so.
- When it burns in air, lithium forms a normal oxide (Li_2O) rather than a peroxide (M_2O_2) or a superoxide (MO_2).
- Lithium carbonate and lithium hydroxide decompose to form the oxide on heating, while the carbonates and hydroxides of the other Group 1A metals are thermally stable.

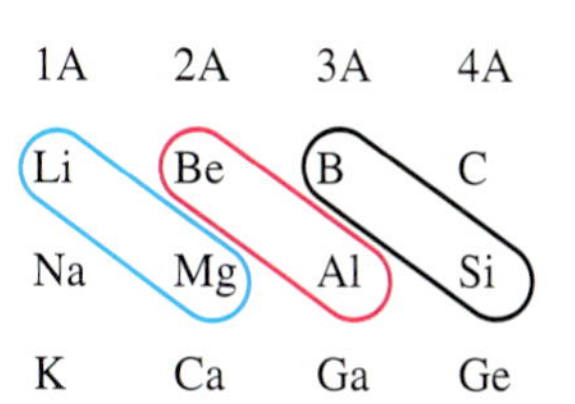

▲ **Figure 20.4**
The diagonal relationship in the periodic table
The elements in each encircled pair have a number of similar properties.

In large part, these differences can be attributed to the high charge density of Li^+ compared to that of the other Group 1A cations. The charge density is the ratio of the ionic charge to the ionic volume.

In some of its properties, lithium and its compounds resemble magnesium and its compounds. This is an example of the **diagonal relationship**, depicted in Figure 20.4. The similarity between Li and Mg probably results from the roughly equal sizes of the Li and Mg atoms and of the Li^+ and Mg^{2+} ions. The diagonal relationship between beryllium and aluminum is described in Section 20.8.

* The way in which atoms pack is also involved in relating molar volume to atomic size. However, the Group 1A metals all have the same crystal structure—body-centered cubic.

20.5 Occurrence, Preparation, Uses, and Reactions of Group 1A Metals

The natural abundances of sodium and potassium in Earth's solid crust are 2.27% and 1.84% by mass, respectively; the other alkali metals are much scarcer, ranging from 78 to 18 to 2.6 parts per million (ppm) of Rb, Li, and Cs, respectively. Francium is exceptionally rare. It is formed by the radioactive decay of heavier elements, and there is probably no more than 15 g of francium in the top 1 km of Earth's crust. Because francium is both rare and highly radioactive, few of its properties have been determined.

Preparation of the Alkali Metals

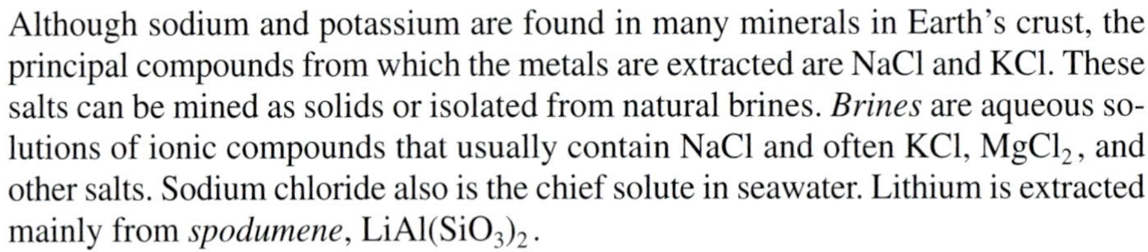

Although sodium and potassium are found in many minerals in Earth's crust, the principal compounds from which the metals are extracted are NaCl and KCl. These salts can be mined as solids or isolated from natural brines. *Brines* are aqueous solutions of ionic compounds that usually contain NaCl and often KCl, $MgCl_2$, and other salts. Sodium chloride also is the chief solute in seawater. Lithium is extracted mainly from *spodumene*, $LiAl(SiO_3)_2$.

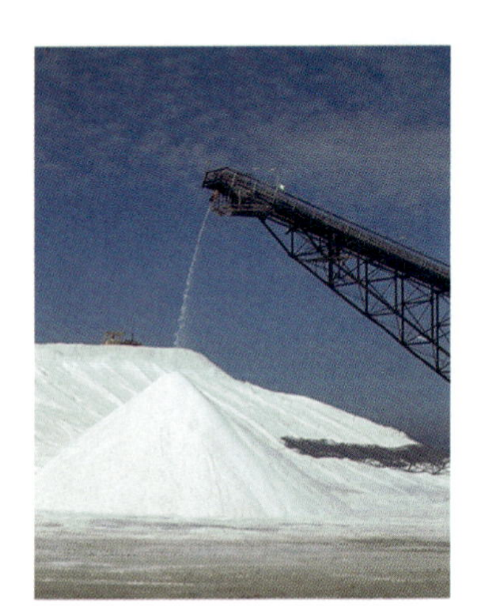

Sodium chloride (salt) is mined, pumped from brine wells, and in some seaside locations, harvested from seawater.

To convert an alkali metal ion to an alkali metal atom, the ion must be forced to take on an electron—a process of *reduction*. These reductions are not easily accomplished with spontaneous reactions because the alkali metals are themselves excellent reducing agents. Reduction can be achieved by the electrolysis of molten salts, usually the chlorides.

Potassium metal was the first to be prepared by electrolysis; Humphry Davy electrolyzed molten KOH in 1807. The usual industrial method involves a chemical reaction between sodium metal and potassium chloride.

$$\mathrm{KCl(l) + Na(l) \xrightarrow{850\ ^\circ C} NaCl(l) + K(g)}$$

In this reaction, Na atoms lose their valence electrons and K^+ ions acquire them. The resulting gaseous potassium escapes from the reaction mixture and is converted, first to a liquid and then to a solid, by cooling. The reverse reaction is actually favored because K atoms lose electrons more readily than do Na atoms (they have a lower ionization energy). The method works because sodium is a liquid at 850 °C whereas potassium is a gas. The escape of gaseous potassium stimulates the forward reaction and causes it to go to completion (Le Châtelier's principle).

The alkali metals have a few important uses. *Liquid* sodium is used as a heat transfer medium in some types of nuclear reactors. Liquid sodium is especially good for this purpose because it has the following characteristics.

- A higher specific heat than most liquid metals—it takes a relatively large quantity of heat per gram to change its temperature.
- Good thermal conductivity—heat is readily conveyed through the liquid and delivered to other media.
- A low density and low viscosity—liquid sodium is easy to pump.
- A low vapor pressure—the tendency of liquid sodium to vaporize is limited, even at high temperatures (550 °C).

Small quantities of sodium, as sodium vapor, are used in lamps for outdoor lighting. Perhaps the most important use of sodium metal is as a reducing agent in producing refractory (high melting point) metals, such as titanium, zirconium, and hafnium.

$$MCl_4 + 4\,Na \longrightarrow M + 4\,NaCl \quad (\text{where } M = Ti, Zr, \text{ or } Hf)$$

The use of potassium is limited to a few applications in which sodium, the cheaper metal, cannot be used. One application is its conversion to potassium superoxide, KO_2, which is used in life-support systems, where it acts both to absorb CO_2 and to produce O_2.

$$4\,KO_2(s) + 2\,CO_2(g) \longrightarrow 2\,K_2CO_3(s) + 3\,O_2(g)$$

(Sodium superoxide, NaO_2, can be prepared only under very carefully controlled conditions.)

Lithium is used in lightweight electrical batteries of the type found in clocks and watches, hearing aids, and heart pacemakers. Lithium is a desirable battery electrode because (1) the mass of Li required to release 1 mole of electrons is so small (6.941 g) and (2) high voltages can be achieved by combining the oxidation of Li(s) ($E^\circ_{Li^+/Li} = -3.040$ V) with an appropriate reduction half-reaction. Lithium is also used in alloys with other light metals; it imparts high-temperature strength to aluminum and ductility to magnesium when it is added in small quantities. A silver-lithium alloy is used for brazing (welding together) metals. A possible future use of lithium is in the production of tritium (^3_1H) for use in nuclear fusion reactors.

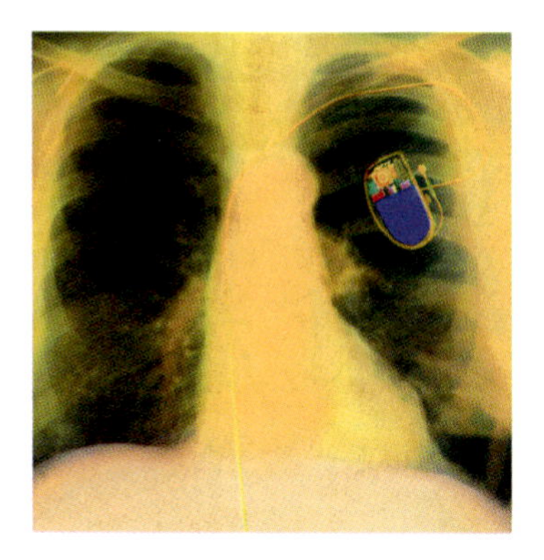

A heart pacemaker powered by a lithium battery is seen in this X-ray photograph.

Some Reactions of Li, Na, and K

The chemistry of the Group 1A metals reflects the relative ease of removal of the ns^1 electron to form the metal ions. The alkali metals (M) react directly with elements of Group 7A, the halogens (X_2), to form ionic binary halides (MX). As we saw in Section 20.2, they also react with hydrogen to form ionic hydrides (MH). The reaction of $O_2(g)$ with the metals produces various products. Table 20.2 summarizes some of the reaction chemistry of the Group 1A metals.

TABLE 20.2 Some Typical Reactions of the Alkali Metals, M

With halogens (Group 7A), X_2

$2\,M(s) + X_2 \longrightarrow 2\,MX(s)$ (e.g., LiF, NaCl, KBr, CsI)

With hydrogen, H_2:

$2\,M(s) + H_2(g) \longrightarrow 2\,MH(s)$ (e.g., LiH, NaH)

With excess oxygen, O_2*

$4\,Li(s) + O_2(g) \longrightarrow 2\,Li_2O(s)$ (plus some Li_2O_2)

$2\,Na(s) + O_2(g) \longrightarrow Na_2O_2(s)$ (plus some Na_2O)

$M(s) + O_2(g) \longrightarrow MO_2(s)$ (where M = K, Rb, or Cs)

With water, H_2O

$2\,M(s) + 2\,H_2O(l) \longrightarrow 2\,MOH(aq) + H_2(g)$

*Li_2O is a *normal* oxide; Na_2O_2 is a *peroxide*; MO_2 is a *superoxide*. These oxides are described further in Section 21.9. Under appropriate conditions all the alkali metals can form M_2O, M_2O_2, and MO_2.

Lithium, sodium, and potassium are such active metals (as are also rubidium and cesium) that they will displace $H_2(g)$ from water itself. Some of these reactions occur with explosive violence. The liberated hydrogen is raised to a high temperature by the heat of reaction and may spontaneously ignite, combining with oxygen in air to form water (see page 848). In their reactions with water, the alkali metals produce aqueous solutions of ionic hydroxides, MOH, as well as hydrogen.

$$2\,M(s) + 2\,H_2O(l) \longrightarrow 2\,MOH(aq) + H_2(g)$$

It is because Group 1A hydroxides are *basic*, or *alkaline*, that the Group 1A metals have long been called the alkali metals.

20.6 Important Compounds of Li, Na, and K

Lithium carbonate is the usual starting material for making other lithium compounds. For example, lithium hydroxide is prepared by the following reaction.

$$Li_2CO_3(aq) + Ca(OH)_2(aq) \longrightarrow CaCO_3(s) + 2\,LiOH(aq)$$

Solid $CaCO_3$ is removed by filtration, leaving an aqueous solution of lithium hydroxide. The LiOH can be obtained as a solid by evaporation of water from the solution.

An interesting use of lithium hydroxide is to remove $CO_2(g)$ from expired air in confined quarters such as submarines and space vehicles.

$$2\,LiOH(s) + CO_2(g) \longrightarrow Li_2CO_3(s) + H_2O$$

LiOH is preferred over other ionic hydroxides because, per mole of CO_2 removed from expired air, a smaller mass of LiOH is required. Similarly, in its reaction with water, LiH produces a greater volume of $H_2(g)$ per unit mass of hydride than do other, cheaper hydrides.

With an annual production in the United States of about 50 million tons, sodium chloride is easily the most important sodium compound. In fact, it is the most used of all minerals for the production of chemicals. Table 20.3 lists several of these NaCl-based chemicals and their estimated annual production in the United States.

TABLE 20.3 Some Chemicals Produced from NaCl

Chemical	U.S. Production, 1997, billions of pounds
Sodium hydroxide, NaOH	22.74
Chlorine, Cl_2	25.99
Hydrochloric acid, HCl	8.44[a]
Sodium sulfate, Na_2SO_4	1.28[b]

[a] Most HCl is produced as a by-product of the chlorination of hydrocarbons; for example, in the chlorination of methane: $CH_4(g) + Cl_2(g) \longrightarrow CH_3Cl(g) + HCl(g)$
[b] Includes both Na_2SO_4 and $Na_2SO_4 \cdot 10H_2O$.

Figure 20.5 is a diagram that shows how chemists sometimes outline important reactions. A central compound is chosen (NaCl), and the reactant(s) required to convert it to other compounds is(are) noted along arrows. Additional information may be placed along the arrows to indicate reaction conditions, such as whether heating (Δ) or catalysts are required. In Example 20.1, we use a reaction diagram as the basis of writing a series of equations for the reactions that might collectively be used in a laboratory or industrial process.

EXAMPLE 20.1

Use Figure 20.5 to write chemical equations for the Leblanc process—the conversion of NaCl to Na_2CO_3.

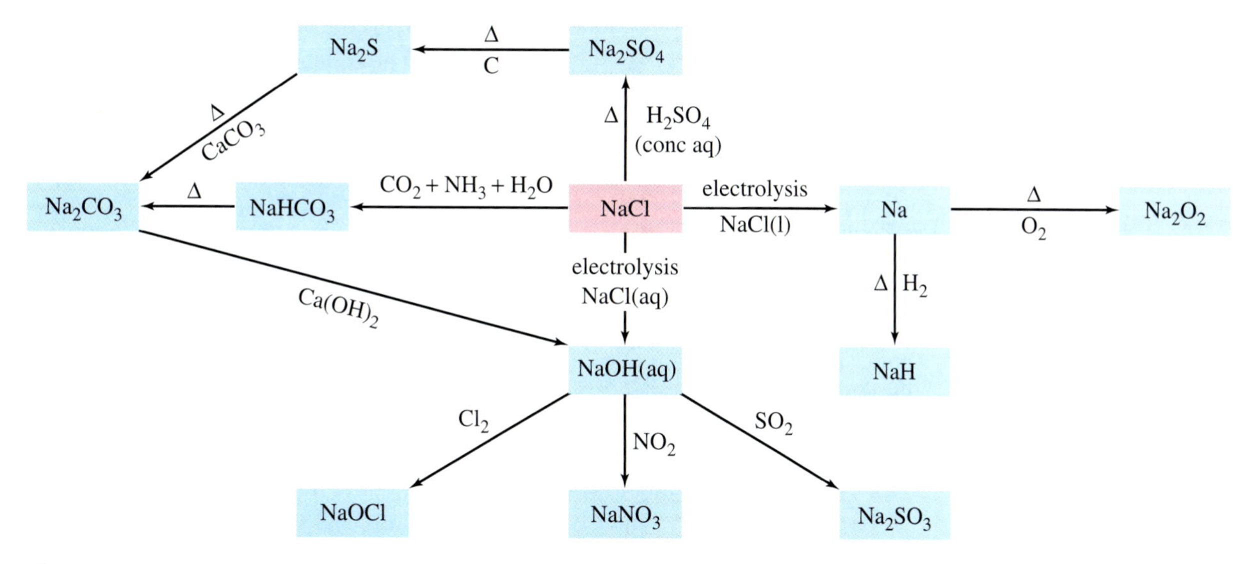

▲ **Figure 20.5 Preparation of sodium compounds from NaCl**
The methods of preparation suggested by this diagram are not necessarily the preferred industrial methods.

SOLUTION

Figure 20.5 describes the necessary transformations in an abbreviated form. To translate this information into chemical equations, we need only identify the reactants in each reaction and predict plausible products. For example, to convert NaCl(s) to Na_2SO_4(s) requires reaction with H_2SO_4(conc aq). The other plausible product is HCl(g).

$$2\ NaCl(s) + H_2SO_4(conc\ aq) \longrightarrow Na_2SO_4(s) + 2\ HCl(g)$$

The reaction of Na_2SO_4(s) with carbon yields Na_2S(s), with CO(g) as another plausible product.

$$Na_2SO_4(s) + 4\ C(s) \longrightarrow Na_2S(s) + 4\ CO(g)$$

The final step is the reaction of Na_2S(s) with $CaCO_3$(s), yielding Na_2CO_3(s) and CaS(s).

$$Na_2S(s) + CaCO_3(s) \longrightarrow Na_2CO_3(s) + CaS(s)$$

Water-soluble Na_2CO_3 is extracted from the mixed solid.

EXERCISE 20.1

Write plausible chemical equations for each of the following conversions outlined in Figure 20.5.

a. NaCl to NaH **b.** NaCl to NaOCl

Sodium hydroxide is produced by the electrolysis of concentrated NaCl(aq). In the electrolysis, Na^+ is unchanged, Cl^- is converted to Cl_2(g), and H_2O is converted to H_2(g) and OH^-.

$$2\ Na^+(aq) + 2\ Cl^-(aq) + 2\ H_2O(l) \xrightarrow{\text{electrolysis}} 2\ Na^+(aq) + 2\ OH^-(aq) + H_2(g) + Cl_2(g)$$

Gunpowder is a mixture of potassium nitrate (white), sulfur (yellow), and charcoal (black).

The gases are collected separately. If *solid* NaOH is desired (rather than an aqueous solution), water is evaporated from the solution after the electrolysis is completed.

Sodium sulfate is obtained from both natural sources and by a synthetic method introduced by J. R. Glauber more than 300 years ago.

$$H_2SO_4(\text{conc aq}) + 2\ NaCl(s) \xrightarrow{\Delta} Na_2SO_4(s) + 2\ HCl(g)$$

The principle of this reaction is that a volatile acid (HCl) is produced by heating one of its salts (NaCl) with a nonvolatile acid (H_2SO_4). Other volatile acids can be produced by similar reactions. The major use of sodium sulfate (Na_2SO_4) is in the paper industry (about 70% of the annual U.S. consumption). In making kraft paper, undesirable lignin is removed from wood by treating the wood with an alkaline solution of sodium sulfide (Na_2S). The Na_2S is produced by the reaction of Na_2SO_4 with carbon. About 100 lb of Na_2SO_4 is used in the production of one ton of paper.

Potassium compounds have some uses similar to their sodium counterparts (for example, K_2CO_3 in glass and ceramics), but their most important use by far is in fertilizers, which account for 95% of the commercial use of potassium compounds. Potassium (as K^+) is one of the three main nutrients required by plants (nitrogen and phosphorus are the other two). KCl is commonly used as a fertilizer because this is the form in which most potassium is obtained from natural sources.

Glass bottles being manufactured. The principal current use of sodium carbonate is in glassmaking.

Figure 20.6 shows another way that chemists and chemical engineers diagram the industrial production of a chemical. You can think of Figure 20.6 as an elaboration of the part of Figure 20.5 dealing with the conversion of NaCl to $NaHCO_3$. The diagram identifies the raw materials (limestone, brine, and ammonia) and illustrates how they are used in the main reaction. The diagram also shows how substances are recycled. The only ultimate by-product of the Solvay process is $CaCl_2$. The demand for $CaCl_2$ is limited, however, and in the past, much of the $CaCl_2$ was dumped, generally into lakes or streams. Such dumping is now prohibited by environmental regulations. Partly for this reason and partly because of the discovery of abundant natural sources of Na_2CO_3 in Wyoming, the Solvay process is now obsolete in the United States. It continues in use elsewhere in the world, however.

The case of sodium carbonate emphasizes the importance of economic and environmental factors in establishing a kind of natural progression for industrial chemical processes—advent, modification, and decline.

20.7 The Alkali Metals and Living Matter

Potatoes are a good dietary source of K^+.

Hydrogen, oxygen, carbon, and nitrogen are the most abundant elements in the human body, in the order listed. They account for 99% of all the atoms in the body. Sodium and potassium, in the form of Na^+ and K^+, are in a second tier of seven elements that account for about 0.9% of the atoms. Life appears to have evolved using the elements that are most available in the environment, and sodium and potassium have comparatively high abundances in Earth's crust and in freshwater and seawater.

Sodium ions are found primarily in fluids outside cells, for example, in blood plasma. Potassium ions are abundant in fluids within cells. Na^+ and K^+ ions are involved in the transmission of nerve signals and in regulating and controlling body fluids. We commented on their role in the electrochemistry of a heartbeat in Chapter 18 (page 793).

Ordinary salt (NaCl), which is naturally present in most foods and is often added to foods as table salt, supplies the body with Na^+ and also with chloride ions, which are necessary for the production of stomach acid, HCl(aq). Bananas and orange juice are good sources of K^+. The recommended daily intake of K^+ is variously given as 2000 mg up to 5600 mg.

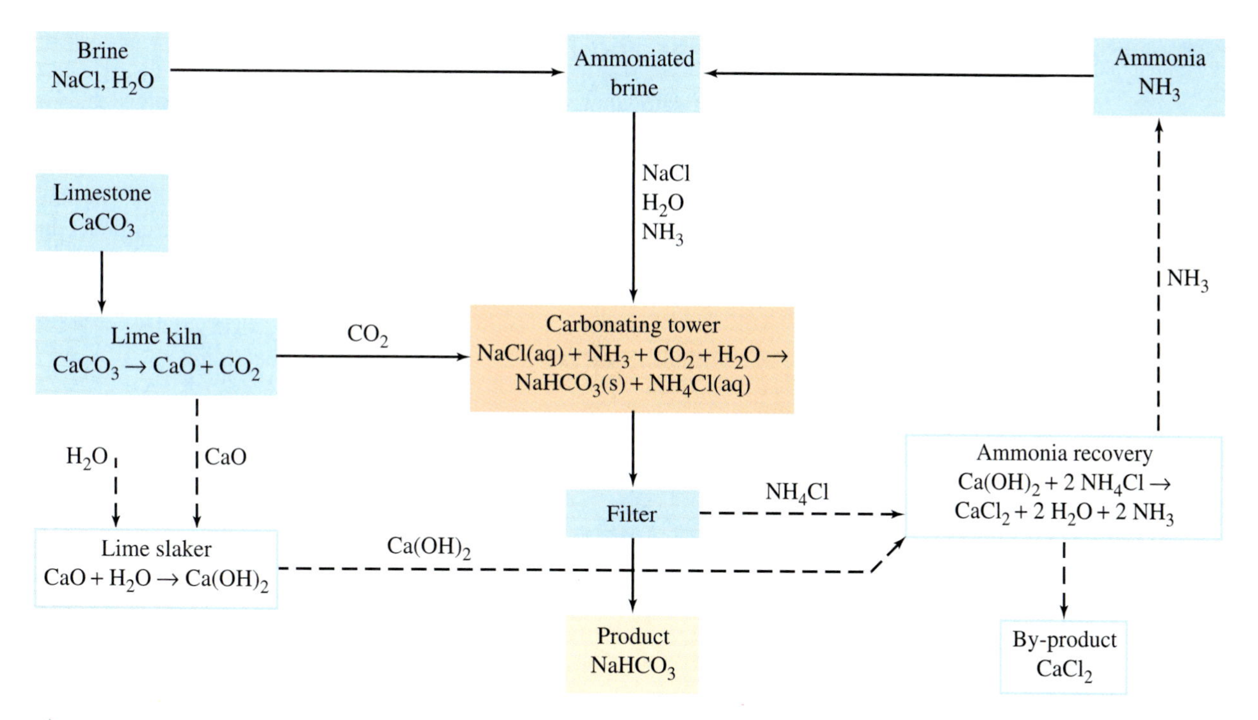

▲ **Figure 20.6 The Solvay process**
The solid boxes and arrows outline the steps essential to the main reaction and recovery of the product. The broken-line boxes and arrows indicate the recycling steps and formation of the by-product.

Potassium ion is an essential nutrient for plants. It is generally readily available to plants, except in soil depleted by high-yield agriculture. The usual form of potassium in commercial fertilizers is potassium chloride, KCl.

Because most alkali metal compounds are water soluble, many acidic drugs are administered in the form of their sodium or potassium salts. For example, "free" penicillin G (benzylpenicillic acid) is only sparingly soluble in water, whereas potassium penicillin G is freely soluble in water.

Lithium carbonate is used in medicine to level out the dangerous manic "highs" that occur in manic-depressive psychoses. Some practitioners also recommend lithium carbonate for the depression stage of the cycle. It appears to affect the transport of chemical substances across cell membranes in the brain.

Sodium sulfate decahydrate, $Na_2SO_4 \cdot 10H_2O$, was one of the first synthetic chemicals used in medicine. Still known today as *Glauber's salt*, it acts as a cathartic (a substance that purges the bowels). Glauber's synthesis of this salt helped turn the focus of alchemists from attempts to convert base metals into gold to the synthesis of medicines.

Group 2A: The Alkaline Earth Metals

The Group 2A oxides, MO, and hydroxides, $M(OH)_2$, are basic, or *alkaline*, though none of these compounds is soluble in water to any great extent. In the early days of chemistry, the term "earth" described substances that are insoluble or only slightly soluble in water and that are not decomposed by heating. These properties account for the group family name: the *alkaline earth* metals. Table 20.4 lists data for the Group 2A metals.

20.8 Properties and Trends in Group 2A

Table 20.4 shows the same general trends of increasing atomic and ionic sizes and decreasing ionization energies from top to bottom of Group 2A that we saw in Group 1A. Table 20.4 lists both the first and second ionization energies because an alkaline earth metal atom loses two electrons when it forms the ion M^{2+}. The ionization energies of the first two members, Be and Mg, are noticeably higher than those of the other three, Ca, Sr, and Ba. A parallel observation is that the first two do not exhibit a flame color, whereas the latter three do. The electronegativity values are those that we expect of active metals, and the decreasing values down the group are in line with the increasing metallic character of the atoms.

TABLE 20.4 Some Properties of the Group 2A Metals

	Be	Mg	Ca	Sr	Ba
Atomic number, Z	4	12	20	38	56
Valence-shell electron configuration	$2s^2$	$3s^2$	$4s^2$	$5s^2$	$6s^2$
Atomic (metallic) radius, pm	111	160	197	215	217
Ionic (M^{2+}) radius, pm	31	65	99	113	135
Ionization energy, kJ/mol					
I_1	900	738	590	550	503
I_2	1757	1451	1145	1064	965
Electronegativity	1.5	1.2	1.0	1.0	0.9
Flame color	None	None	Orange-red	Scarlet	Green
Melting point, °C	1278	649	839	769	729
Density, g/cm^3 at 20 °C	1.848	1.738	1.550	2.540	3.594
Electrode potential, $E°$, V $M^{2+}(aq) + 2e^- \longrightarrow M(s)$	−1.70	−2.356	−2.84	−2.89	−2.92
Electrical conductivity[a]	40	36	46	6.9	3.2
Hardness[b]	≈5	2.0	1.5	1.8	≈2

[a] On a scale relative to silver, 100; copper, 95.0; gold, 67.7.
[b] On a scale (Mohs) from 0 to 10. Each item on this scale is able to scratch only materials with hardness less than its own. Examples: talc (0), wax (0.2), asphalt (1–2), fingernail (2.5), copper (2.5–3), iron (4–5), glass (5–6), steel (5–8.5), diamond (10).

The higher densities of the Group 2A metals relative to Group 1A are mainly a consequence of the large difference in atomic sizes between the two groups. Molar volumes are significantly smaller in Group 2A than in Group 1A, and the molar masses are only slightly larger. The standard electrode potentials ($E°_{M^{2+}/M}$) are large and negative, showing that the alkaline earth metal ions M^{2+} are reduced with difficulty. This means that the alkaline earth metals themselves are readily oxidized; they are good reducing agents.

An interesting trend related to data in Table 20.4 is in the solubilities of the metal hydroxides in water. The molar solubility of $M(OH)_2$ at 20 °C increases in the following manner in the group.

$Mg(OH)_2$	$Ca(OH)_2$	$Sr(OH)_2$	$Ba(OH)_2$
0.0002 M	0.021 M	0.066 M	0.23 M

As the cation size increases from left to right in this series, the interionic attractions that hold the crystalline solid together decrease in strength. As the interionic attractions weaken, the solubilities of the compounds in water increase.

The Special Case of Beryllium

Like lithium in Group 1A, beryllium—the first member in Group 2A—differs in some ways from other members of the group. Some of its distinctive properties are described below.

- BeO does not react with water. [The other Group 2A metal oxides do react with water: $MO + H_2O \longrightarrow M(OH)_2$.]
- Be and BeO dissolve in strongly basic solutions to form the ion BeO_2^-. The oxide BeO has *acidic* properties. The other alkaline earth metal oxides are *basic*.
- $BeCl_2$ and BeF_2 in the molten state are poor conductors of electricity; they are molecular substances (see Figure 20.7). In contrast, the other alkaline earth metal compounds are almost entirely ionic.

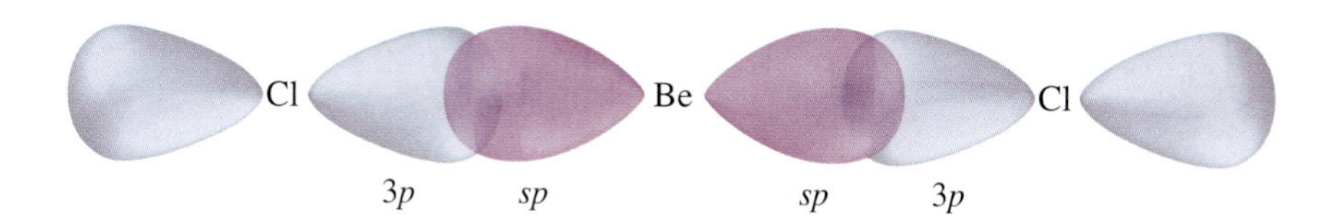

Figure 20.7 Molecular structure of $BeCl_2$
The hybridization scheme for the Be atom is

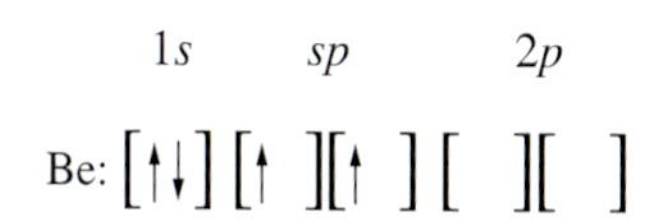

The *sp* hybrid orbitals are directed outward from the Be atom along a straight line. The $BeCl_2$ molecule is a linear triatomic molecule. The molecular structure in the solid state is more complex, however.

In Chapter 21, we will see that in some ways Al, Al_2O_3, and $AlCl_3$ resemble their beryllium counterparts. This is another example of a *diagonal relationship* (see again, Figure 20.4).

20.9 Occurrence, Preparation, Uses, and Reactions of Group 2A Metals

The natural abundances of calcium and magnesium in Earth's crust are 4.66% and 2.76% by mass, respectively. They rank fifth and sixth among the elements in crustal abundance, just ahead of sodium and potassium. Calcium and magnesium minerals occur in many types of rocks. Limestone rock is mainly $CaCO_3$, and dolomite is a mixed carbonate, $CaCO_3 \cdot MgCO_3$. Strontium and barium are both present at about 400 parts per million, and beryllium is found at 2 ppm. An important mineral source of beryllium is *beryl*, $3BeO \cdot Al_2O_3 \cdot 6SiO_2$. Some familiar gemstones, including aquamarine and emerald, are based on the mineral beryl.

Preparation of the Alkaline Earth Metals

To obtain beryllium metal, beryl is first converted to BeF_2. Then the BeF_2 is reduced to beryllium, using magnesium as the reducing agent.

$$BeF_2(g) + Mg(l) \xrightarrow{1000\ °C} Be(s) + MgF_2(s)$$

Calcium is generally obtained by electrolysis of molten calcium chloride. Strontium and barium can also be obtained by electrolysis, but more commonly they are obtained by the high-temperature reduction of their oxides, using aluminum as the reducing agent.

Magnesium is obtained by the electrolysis of molten $MgCl_2$, in a process that begins with seawater or natural brines. There is about 1.3 g Mg^{2+} per kilogram of seawater. Magnesium can be extracted from seawater because Mg^{2+} is the only cation present that forms an insoluble hydroxide.

The White Cliffs of Dover, England, are made of chalk, a soft form of limestone, $CaCO_3$.

The Dow Process for the Production of Magnesium

In the Dow process, calcium carbonate, as either limestone or seashells, is decomposed to CaO(s) and CO_2(g) by heating. Treatment of CaO with water produces $Ca(OH)_2$, the source of OH^- for the precipitation of $Mg(OH)_2$(s).

$$Mg^{2+}(aq) + 2\,OH^-(aq) \longrightarrow Mg(OH)_2(s)$$

The precipitated $Mg(OH)_2$(s) is washed, filtered, and dissolved in HCl(aq).

$$Mg(OH)_2(s) + 2\,HCl(aq) \longrightarrow MgCl_2(aq) + 2\,H_2O(l)$$

The resulting solution is evaporated to dryness. The $MgCl_2$ is then melted and electrolyzed, yielding pure Mg metal and Cl_2(g).

$$MgCl_2(l) \xrightarrow{\text{electrolysis}} Mg(l) + Cl_2(g)$$

The Cl_2(g) is converted to HCl and recycled. The Dow process is outlined in Figure 20.8.

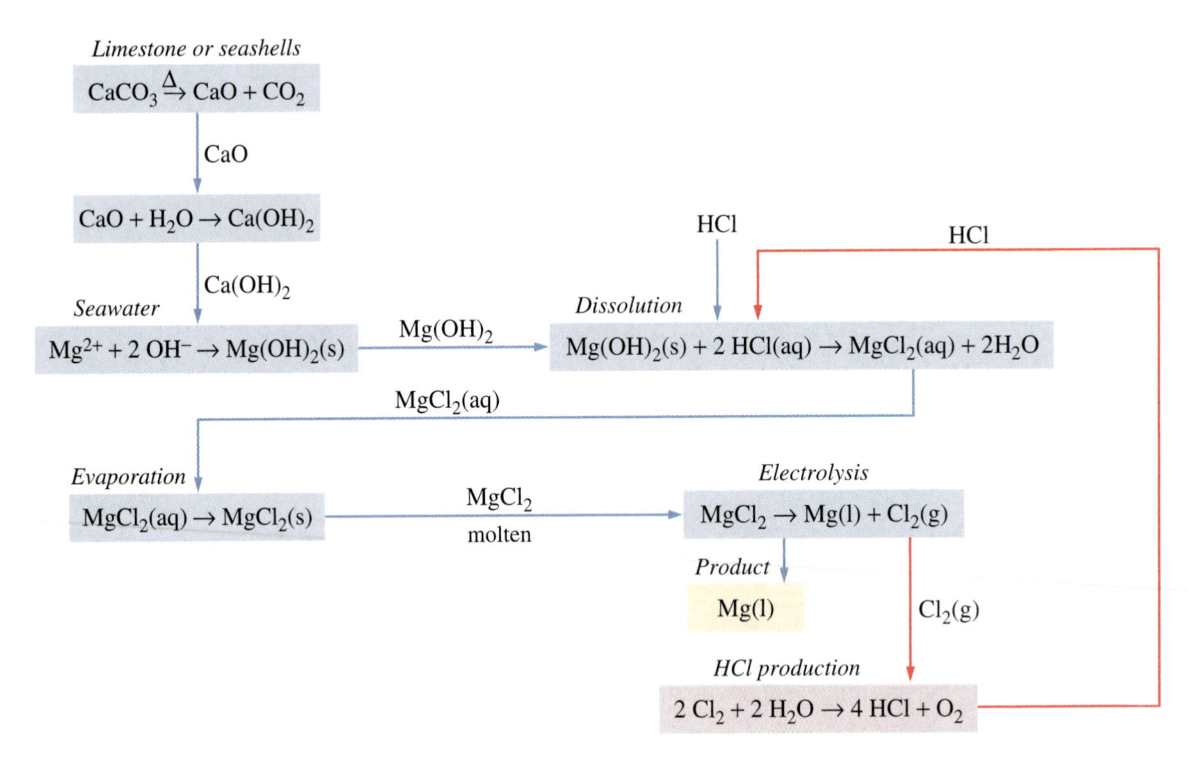

▲ **Figure 20.8 The Dow process for the production of magnesium**
The steps in the process are traced by the blue arrows. The red box and arrow show that Cl_2 is recycled by conversion to HCl.

Electrolyses and the heating of materials to high temperatures are among the industrial operations that consume the most energy. The Dow process involves both. The production of 1 kg Mg requires about 300 MJ of energy (300 MJ = 300×10^6 J). In contrast, it takes only about 7 MJ of energy to melt and recast 1 kg of recycled Mg. Recycling of magnesium is especially cost-effective.

Uses of the Alkaline Earth Metals

Alloys of beryllium with other metals have many applications. An alloy with copper having about 2% beryllium is used in springs, clips, and electrical contacts because it is able to withstand metal fatigue, the cracking or rupture of a metal subjected to repeated stress. Because of their low densities, other Be alloys are used in lightweight structural materials. The small Be atom has little stopping power for X rays or neutrons; this makes beryllium useful for "windows" in X-ray tubes and for components in nuclear reactors. Mixtures of beryllium and radium compounds have long been used as neutron sources. Alpha particles from the radium induce the nuclear reaction

$$^{9}_{4}\mathrm{Be} + ^{4}_{2}\mathrm{He} \longrightarrow ^{12}_{6}\mathrm{C} + ^{1}_{0}\mathrm{n}$$

Beryllium and its compounds are extremely hazardous when ingested or inhaled, causing lung cancer. Special precautions are necessary when working with them.

Magnesium has a lower density than any other structural metal. Lightweight aircraft parts are manufactured from magnesium alloyed with aluminum and other metals; these parts are easily fabricated with standard machining and metallurgical methods. Also, magnesium is an important metallurgical reducing agent, for example, in the production of beryllium (page 863) and titanium.

$$\mathrm{TiCl_4} + 2\,\mathrm{Mg} \longrightarrow \mathrm{Ti} + 2\,\mathrm{MgCl_2}$$

Magnesium is also used in batteries, fireworks, flash photography, and in the synthesis of important reagents for organic chemical reactions.

Calcium is used to reduce the oxides or fluorides of less common metals (Sc, W, Th, U, Pu, and most of the lanthanides) to the free metals. For example,

$$\mathrm{UO_2} + 2\,\mathrm{Ca} \longrightarrow \mathrm{U} + 2\,\mathrm{CaO}$$

$$2\,\mathrm{ScF_3} + 3\,\mathrm{Ca} \longrightarrow 2\,\mathrm{Sc} + 3\,\mathrm{CaF_2}$$

Calcium is also used in the manufacture of batteries and in forming alloys with aluminum, silicon, and lead.

Reactions of the Alkaline Earth Metals

To illustrate the trends in chemical properties found in Group 2A of the periodic table, consider the reaction of the heavier Group 2A metals, M, with water.

$$\mathrm{M(s)} + 2\,\mathrm{H_2O(l)} \longrightarrow \mathrm{M(OH)_2} + \mathrm{H_2(g)}$$

If M = Ca, slow reaction with cold water
= Sr, reaction more rapid than Ca
= Ba, reaction more rapid than Sr

The slow reaction of calcium with water was illustrated in Figure 8.21 (page 349). In the case of magnesium, an impervious film of $Mg(OH)_2$ covers the surface and immediately stops the reaction. Magnesium does react with steam, however, but MgO is formed rather than $Mg(OH)_2$.

$$\mathrm{Mg(s)} + \mathrm{H_2O(g)} \longrightarrow \mathrm{MgO(s)} + \mathrm{H_2(g)}$$

Beryllium fails to react with either cold water or steam.

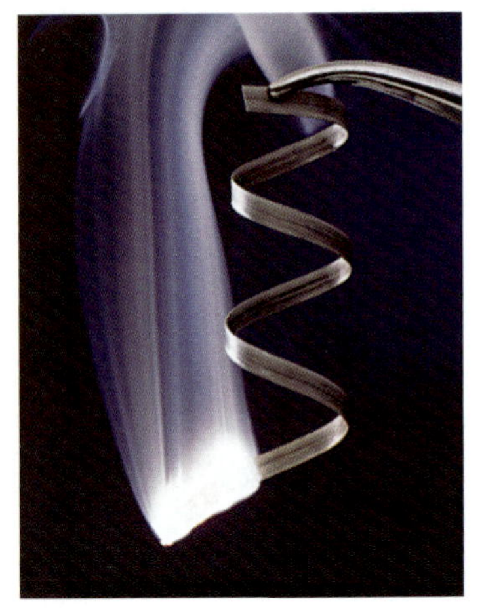

Magnesium burns in air with a bright flame.

All the alkaline earth metals react with dilute acids to displace hydrogen.

$$M(s) + 2\,H^+(aq) \longrightarrow M^{2+}(aq) + H_2(g)$$

Some typical reactions of the alkaline earth metals, illustrated below for magnesium, are with the halogens (X_2), with oxygen, and with nitrogen.

$$Mg + X_2 \longrightarrow MgX_2$$
$$(\text{where } X = F, Cl, Br, I)$$
$$2\,Mg + O_2 \longrightarrow 2\,MgO$$
$$3\,Mg + N_2 \longrightarrow Mg_3N_2$$

EXAMPLE 20.2—A Conceptual Example

A 1.000-g strip of magnesium is heated in air. Is it possible to calculate with certainty the mass of the product obtained?

SOLUTION

If we know the reactant(s) and product(s) of a reaction, we can write an equation and calculate a theoretical yield. If the reaction has a 100% yield, then we might expect to obtain the theoretical yield. The difficulty here is that two reactions can occur simultaneously.

$$2\,Mg(s) + O_2(g) \longrightarrow 2\,MgO(s)$$
$$3\,Mg(s) + N_2(g) \longrightarrow Mg_3N_2(s)$$

We can't be sure how much of the product is MgO(s) and how much is $Mg_3N_2(s)$. In air, there is a plentiful supply of both $N_2(g)$ and $O_2(g)$, and Mg(s) is the limiting reactant in either reaction. We can make an estimate of the amount of product, but we cannot do a precise calculation.

EXERCISE 20.2

What is the estimated answer to Example 20.2 just referred to above? That is, what is the maximum and what is the minimum mass of product that might be obtained?

20.10 Important Compounds of Magnesium and Calcium

Several magnesium compounds occur naturally, either in mineral form or in brines. These include the carbonate, chloride, hydroxide, and sulfate. Other magnesium compounds can be prepared from these. A few compounds and some of their uses are listed in Table 20.5.

TABLE 20.5 Important Magnesium Compounds

Compound	Uses
$MgCO_3$	Manufacture of refractory bricks; glass, inks, rubber reinforcing agent; dentifrices, cosmetics, antacids, and laxatives
$MgCl_2$	Manufacture of magnesium metal, textiles, and paper; fireproofing agents, cements; refrigeration brine
MgO	Refractories (furnace linings), ceramics syntheses, cements, SO_2 removal from stack gases
$MgSO_4$	Fireproofing, textile manufacturing, ceramics, fertilizers, cosmetics, dietary supplements

Limestone is a naturally occurring form of calcium carbonate, containing some clay and other impurities. Calcium carbonate is the most widely used calcium compound. Let's examine the special role of limestone in the chemical industry.

The decomposition (calcination) of limestone is carried out in a long rotary kiln, producing quicklime, CaO, and carbon dioxide. In a cement kiln, the limestone is mixed with clay and sand to produce the complex mixture of calcium silicates and aluminates known as portland cement.

Limestone: Building Stone and Chemical Raw Material

Some limestone is used as building stones. However, most limestone is used to manufacture other building materials. *Portland cement* is a complex mixture of calcium and aluminum silicates obtained by heating limestone, clay, and sand in a high-temperature rotary kiln. When the cement is mixed with sand, gravel, and water, it solidifies into the familiar material *concrete*. Ordinary *soda-lime glass* (used to make bottles and windows, for example) is a mixture of sodium and calcium silicates formed by heating together limestone, sand, and sodium carbonate.

Limestone is used in the metallurgy of iron and steel to produce an easily liquefied mixture of calcium silicates called *slag* that carries away impurities from the molten metal. And, through chemical reactions producing CaO and $Ca(OH)_2$, limestone is the basis of a large part of the inorganic chemical industry.

In practical applications, often the first step is decomposition of limestone by heating, a process called *calcination*.

$$\textit{Calcination:}\quad CaCO_3(s) \xrightarrow{\Delta} \underset{\text{Quicklime}}{CaO(s)} + CO_2(g)$$

Calcination is carried out in a high-temperature kiln (about 1000 °C) with continuous removal of $CO_2(g)$ to promote the forward reaction. The product formed, CaO(s), is called *lime* or *quicklime*. In the process of *hydration*, the reaction of quicklime with water produces $Ca(OH)_2$, known as *slaked lime*.

$$\textit{Hydration:}\quad CaO(s) + H_2O(l) \longrightarrow \underset{\text{Slaked lime}}{Ca(OH)_2(s)}$$

In the process of *carbonation*, $CO_2(g)$ is bubbled through a suspension of $Ca(OH)_2(s)$ in water and $CaCO_3(s)$ is formed once more.

$$\textit{Carbonation:}\quad Ca(OH)_2(s) + CO_2(g) \longrightarrow CaCO_3(s) + H_2O(l)$$

The three steps just described can be combined and used to prepare chemically pure $CaCO_3(s)$ from limestone, an impure material. This chemically pure $CaCO_3(s)$, called *precipitated* calcium carbonate, is extensively used as a filler, providing bulk to such materials as paint, plastics, printing inks, and rubber. It is also used in toothpastes, food, cosmetics, and antacids. Added to paper, calcium carbonate makes the paper bright, opaque, smooth, and capable of absorbing ink well.

Quicklime and slaked lime are the cheapest and most widely used bases. Because they are cheap, they are usually the first choice anytime someone wants to neutralize unwanted acids. Thus, lime is used to neutralize acidic soils of lawns, gardens, and farmland, and to treat excess acidity in lakes.

$$CaO(s) + 2\,H^+(aq) \longrightarrow Ca^{2+}(aq) + H_2O(l)$$

Another application of quicklime is in air pollution control. When coal is burned in an electric power plant, sulfur in the coal is converted to sulfur dioxide gas, a major culprit in the formation of acid rain. When powdered limestone is mixed with powdered coal before combustion, the limestone decomposes to CaO(s), which reacts with $SO_2(g)$ that would otherwise escape.

APPLICATION NOTE

When heated to a high temperature CaO (mp 2614 °C) emits light. Before the days of electric lighting, lime was heated in a gas flame and the emitted light was focused into a beam for spotlighting, hence the expression, "in the limelight." Limelight is a blackbody radiation (page 291).

$$CaO(s) + SO_2(g) \longrightarrow CaSO_3(s)$$

By reaction with oxygen, the calcium sulfite is converted to calcium sulfate (gypsum), which has a number of uses. Quicklime is also used in treating wastewater effluents and sewage.

Slaked lime, $Ca(OH)_2$, is the cheapest commercial base and is used in all applications where high water solubility is not essential. Slaked lime is used in the manufacture of other alkalis and bleaching powder, in sugar refining, in tanning hides, and in water softening.

A mixture of slaked lime, sand, and water is the familiar mortar used in bricklaying. In the initial setting of the mortar, bricks absorb excess water, which is then lost through evaporation. In the final setting, $CO_2(g)$ from air reacts with $Ca(OH)_2$ and converts it back to $CaCO_3$.

$$Ca(OH)_2(s) + CO_2(g) \longrightarrow CaCO_3(s) + H_2O(g)$$

Hydrates

These dunes at White Sands National Monument in New Mexico are composed of gypsum.

The mineral gypsum has the formula $CaSO_4 \cdot 2H_2O$. Recall that a compound that incorporates water molecules into its crystal structure is called a *hydrate* (Section 2.7). In gypsum, *two* water molecules are present for every formula unit of $CaSO_4$; the chemical name is calcium sulfate *dihydrate*.

Another hydrate of calcium sulfate has one water molecule for every *two* formula units of $CaSO_4$. We could write its formula as $2CaSO_4 \cdot H_2O$, but more commonly, we write $CaSO_4 \cdot \frac{1}{2} H_2O$ and call the compound calcium sulfate *hemihydrate*. This hydrate is called plaster of paris. It is obtained by heating gypsum.

$$\underset{\text{Gypsum}}{CaSO_4 \cdot 2H_2O(s)} \longrightarrow \underset{\text{Plaster of paris}}{CaSO_4 \cdot \tfrac{1}{2} H_2O(s)} + \tfrac{3}{2} H_2O(g)$$

When mixed with water, plaster of paris reverts to gypsum, expanding slightly as it does so. A mixture of plaster of paris and water is used to make castings where sharp details of an object must be retained, as in dental work and jewelry making. The most important use is in making gypsum board, which has largely supplanted plaster in the construction industry.

Hydrate formation occurs infrequently among alkali metal compounds, but it is commonly found in alkaline earth metal compounds. Typical hydrates have the formula $MX_2 \cdot 6H_2O$ (where M = Mg, Ca, or Sr, and X = Cl or Br).

20.11 The Group 2A Metals and Living Matter

Persons of average size have approximately 25 g of magnesium in their bodies. The magnesium serves a variety of functions in metabolic processes and nerve action. The recommended daily intake of magnesium for adults is 280–350 mg. In plant life, magnesium plays a role in photosynthesis, the process by which plants convert carbon dioxide and water into sugars. The magnesium is present in the green pigment *chlorophyll*, which is necessary for the capture of energy from sunlight. This solar energy is the ultimate energy source that makes photosynthesis possible.

Calcium is essential to all living matter. The human body typically contains from 1 to 1.5 kg of calcium. About 99% of it is in the bones and teeth, principally as the mineral *hydroxyapatite*, $Ca_5(PO_4)_3OH$. The other 1% is involved in several processes: the clotting of blood, maintenance of a regular heartbeat, the proper functioning of nerves that control muscles, and in various aspects of metabolism.

Children require an adequate daily intake of calcium for the proper development of bones and teeth. Adults—especially older women—require calcium to prevent osteoporosis, a condition in which the bones become porous, brittle, and easily broken. The recommended daily intake of calcium for adults is at least 800 mg.

Strontium is not essential to living matter, but it is of interest because of its chemical similarity to calcium. Strontium can follow some of the same chemical pathways in organisms as calcium does. Because of this, the body easily ingests and absorbs the dangerously radioactive isotope strontium-90, a product of the fallout from the nuclear fission that occurs in nuclear explosions.

Barium also has no known function in organisms; in fact the Ba^{2+} ion is toxic. Despite this fact, water suspensions of $BaSO_4(s)$ are used in X-ray imaging of the gastrointestinal tract: a barium "milkshake" for the upper tract or a "barium enema" for the lower tract. Barium ions are good absorbers of X rays and make the tract visible in an X-ray photograph. Because it is insoluble ($K_{sp} = 1.1 \times 10^{-10}$), $BaSO_4(s)$ is eliminated by the body with no significant absorption of Ba^{2+} ions.

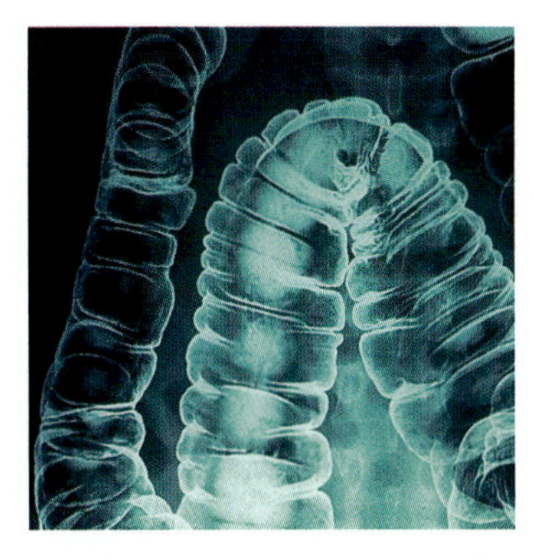

The gastrointestinal tract is rendered visible through the X-ray–absorbing ability of a barium sulfate coating on the tract walls.

Chemistry of Groundwater

Carbonates, especially $CaCO_3$, are involved in a number of natural phenomena. One process begins when rainwater dissolves atmospheric $CO_2(g)$. The $CO_2(g)$ reacts with the water to form unstable carbonic acid, H_2CO_3, which ionizes to form hydrogen carbonate ion, HCO_3^-. We can write an equation for each of these reactions and then combine them into an overall equation that represents rainwater containing dissolved CO_2.

$$CO_2(g) + H_2O(l) \rightleftharpoons \cancel{H_2CO_3(aq)}$$

$$\cancel{H_2CO_3(aq)} + H_2O(l) \rightleftharpoons H_3O^+(aq) + HCO_3^-(aq)$$

$$CO_2(g) + 2\,H_2O(l) \rightleftharpoons H_3O^+(aq) + HCO_3^-(aq) \quad \text{(rainwater containing } CO_2\text{)}$$

Rainwater with dissolved CO_2 is *acidic*. When this water seeps through limestone, *insoluble* $CaCO_3$ is converted to *soluble* $Ca(HCO_3)_2$. In the equations below, the first represents rainwater charged with CO_2, and the second, an acid-base reaction with H_3O^+ as the acid and CO_3^{2-} (from $CaCO_3$) as the base. The net dissolution equation is the sum of the two.

$$CO_2(g) + 2\,H_2O(l) \rightleftharpoons \cancel{H_3O^+(aq)} + HCO_3^-(aq)$$

$$CaCO_3(s) + \cancel{H_3O^+(aq)} \rightleftharpoons Ca^{2+}(aq) + HCO_3^-(aq) + H_2O(l)$$

$$CaCO_3(s) + H_2O(l) + CO_2(g) \rightleftharpoons Ca^{2+}(aq) + 2\,HCO_3^-(aq)$$

APPLICATION NOTE

The reactions that produce limestone caves also account for the deterioration of limestone buildings and marble statues. The reactions become even more significant under conditions of acid rain (Section 25.7).

Over time, the dissolving action, represented by the forward direction of this equation, can produce a large cavity in a limestone bed—a cave. Evaporation of $Ca(HCO_3)_2(aq)$, represented by the reverse direction of the equation, leads to a loss of both H_2O and CO_2. As a result, $Ca(HCO_3)_2(aq)$ is converted back to $CaCO_3(s)$. This process is extremely slow, but over a period of many years, as $Ca(HCO_3)_2(aq)$ drips from the ceiling of a cave, it is converted to icicle-like deposits of $CaCO_3(s)$

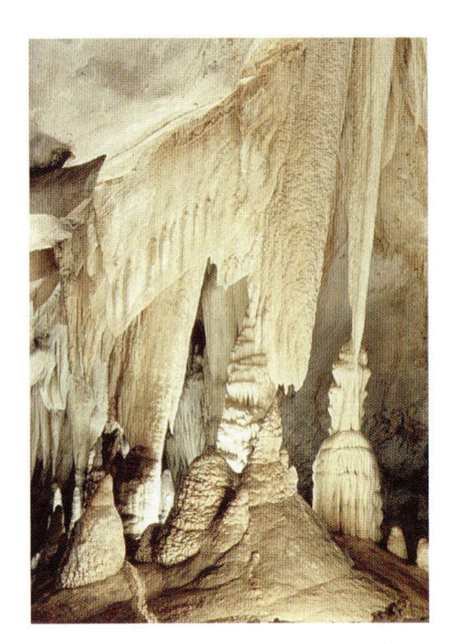

▲ **Figure 20.9**
Stalactites, stalagmites, and columns in a limestone cavern

called *stalactites*. When some of the dripping $Ca(HCO_3)_2(aq)$ hits the floor of the cave, decomposition occurs there, and limestone ($CaCO_3$) deposits build up from the floor in formations called *stalagmites*. Some of the stalactites and stalagmites grow together into limestone columns (see Figure 20.9).

20.12 Hard Water and Water Softening

We have just seen how rainwater can become infused with calcium hydrogen carbonate. **Hard water** is groundwater that contains significant concentrations of ions from natural sources, principally Ca^{2+}, Mg^{2+}, and sometimes Fe^{2+}, along with associated anions. If the primary anion is the hydrogen carbonate ion, HCO_3^-, the hardness is said to be **temporary hardness**. If the predominant anions are other than HCO_3^-, for example, Cl^- or SO_4^{2-}, the hardness is called **permanent hardness**.

When water with temporary hardness is heated, hydrogen carbonate ions decompose.

$$2\,HCO_3^-(aq) \xrightarrow{\Delta} CO_3^{2-}(aq) + H_2O(l) + CO_2(g)$$

The CO_3^{2-} formed in this way reacts with $M^{2+}(aq)$ (that is, Ca^{2+}, Mg^{2+} and Fe^{2+}) to form solid carbonates.

$$M^{2+}(aq) + CO_3^{2-} \longrightarrow MCO_3(s)$$

When water with temporary hardness is heated, $CaCO_3(s)$ is deposited.

The mixed precipitate of $CaCO_3$, $MgCO_3$, and $FeCO_3$, together with any other undissolved solids, is commonly called *boiler scale*. Formation of boiler scale is a serious problem associated with hard water. A boiler clogged with boiler scale heats unevenly, and overheating in parts of the boiler can lead to an explosion. Hard water also interferes with the action of soap (p. 873). For many of its uses, then, hard water must be softened.

Water Softening

Water softening is the removal of objectionable cations and anions in water with either temporary or permanent hardness. Water with temporary hardness can be softened by boiling, but this produces boiler scale. The water can also be softened by addition of a base, which converts HCO_3^- to CO_3^{2-}. The CO_3^{2-} then combines with M^{2+} ions to form precipitates. Metal carbonates are thus removed from the water, but remain as a gritty solid that can cause problems in washing machines and other devices.

$$HCO_3^-(aq) + OH^-(aq) \longrightarrow CO_3^{2-}(aq) + H_2O(l)$$

$$CO_3^{2-}(aq) + M^{2+}(aq) \longrightarrow MCO_3(s)$$

In this method of water softening, the usual source of $OH^-(aq)$ is slaked lime, $Ca(OH)_2$.

Water with permanent hardness cannot be softened by boiling. Addition of washing soda (Na_2CO_3) softens the water by precipitating cations such as Ca^{2+}, Mg^{2+}, and Fe^{2+} as carbonates, but salts such as NaCl and Na_2SO_4 remain in solution.

Ion Exchange

An effective way to soften water is through **ion exchange**. In this process, the undesirable ions in hard water are exchanged for ions that are less objectionable. The ion-exchange medium may be a synthetic resin or a natural or synthetic sodium aluminosilicate called a *zeolite*. In either case, the ion-exchange materials consist of macromolecular (polymer) particles. In contact with water, these materials ionize to produce two types of ions:

1. *Fixed* ions that remain attached to the polymer surface.
2. Free or mobile *counterions*. The counterions have signs opposite to those of the fixed ions. The total charge of the counterions must equal that of the fixed ions, so that the resin as a whole is electrically neutral.

Zeolites have an open three-dimensional structure through which ions can migrate rather freely, as suggested by this molecular model.

In the ion-exchange resin pictured in Figure 20.10, the fixed ions are negatively charged and the counterions are cations. In a fully charged resin, all the counterions at the beginning of the process are Na^+. When hard water is passed through, the more highly charged Ca^{2+}, Mg^{2+}, and Fe^{2+} ions in the water attach to the negatively charged surfaces of the ion-exchange resin and displace the Na^+ ions as counterions. Water with Na^+ ions is a "soft" water in that no precipitates form when the water is boiled; nor do the Na^+ ions interfere with the action of soap as do Ca^{2+}, Mg^{2+}, and Fe^{2+}. To regenerate the resin, concentrated NaCl(aq) is passed through the bed. In *high* concentration, through their sheer numbers, the Na^+ ions are able to dislodge the 2+ cations and restore the resin to its original condition. The ion-exchange material has an indefinite lifetime. To represent the ion-exchange process through simple chemical equations, we can write

$$Na_2Z + M^{2+}(aq) \longrightarrow MZ + 2\,Na^+(aq) \quad \text{(where Z represents a zeolite)}$$

$$Na_2R + M^{2+}(aq) \longrightarrow MZ + 2\,Na^+(aq) \quad \text{(where R represents a synthetic resin)}$$

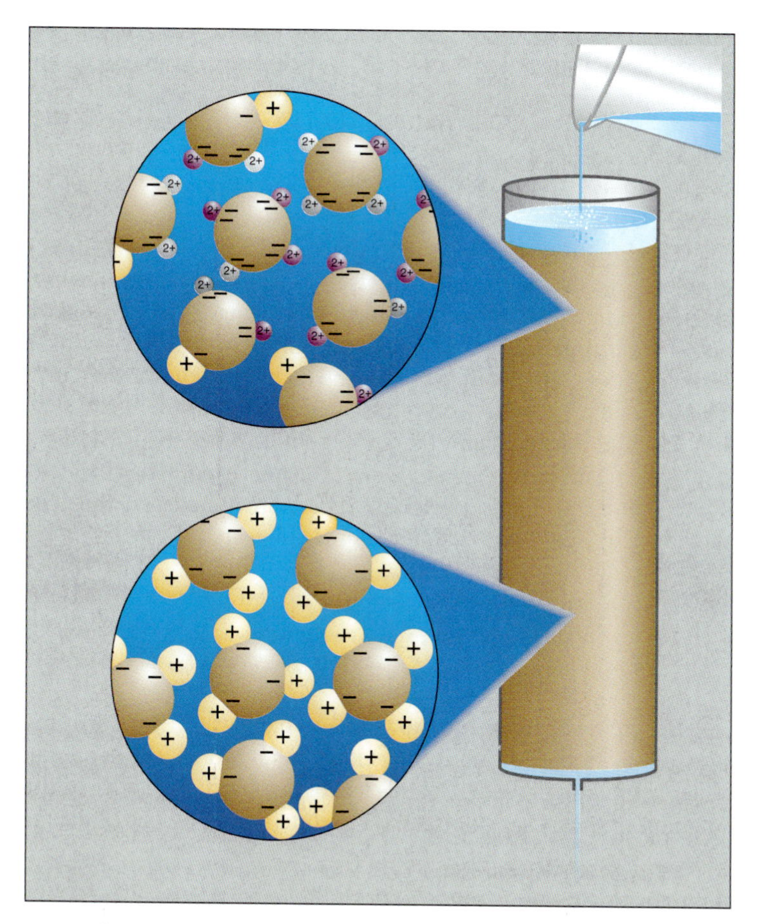

Figure 20.10 Water softening by ion exchange
The anions R^-—the fixed ions—are attached to the surface of the ion-exchange resin. The mobile counterions are Na^+. At the top of the column, Ca^{2+}, Mg^{2+}, and/or Fe^{2+} in hard water replace Na^+ as counterions. By the time the water reaches the bottom of the column, the more highly charged cations have been removed and only Na^+ ions are present. The ion-exchange material in this column is in the form of tiny beads that are very much larger than the cations; relative sizes are distorted in the microscopic views.

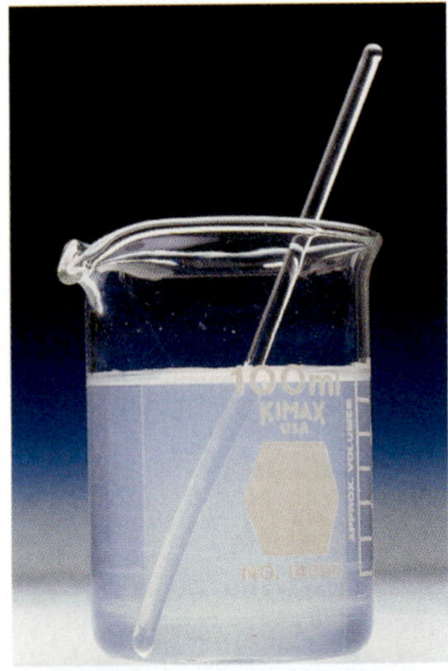

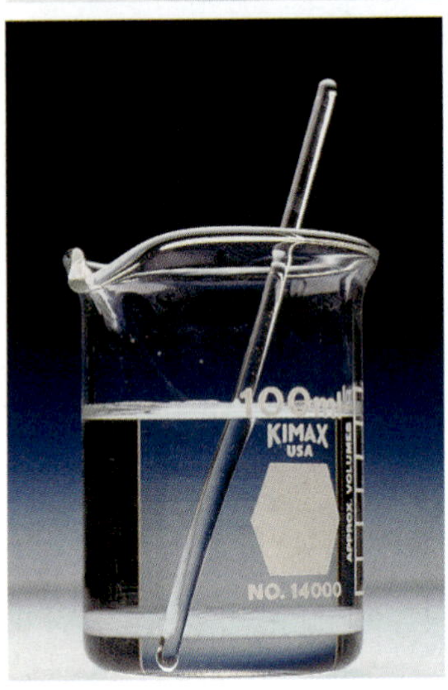

▲ **Figure 20.11**
Deionized water in the chemical laboratory
When a solution of $AgNO_3$ is prepared in tap water, a white precipitate forms (top). Most likely this precipitate is AgCl(s), formed by the reaction of Ag^+(aq) with traces of Cl^-(aq) in the water. No precipitate forms when $AgNO_3$ is dissolved in *deionized* water (bottom).

Suppose instead of NaCl, we use concentrated HCl(aq) to recharge an ion-exchange resin. This leaves H^+ as the counterions. Now when a hard water sample is passed through the resin, the exchange reaction is

$$H_2R + M^{2+}(aq) \longrightarrow MR + 2\,H^+(aq)$$

The water becomes acidic, and we can do the following.

1. Determine the amount of acid in a solution by titration. If we titrate the H^+ eluted (washed) from an ion-exchange column, we can establish the level of hardness in the water (see Problem 79).
2. Pass the effluent from the cation-exchange resin, which contains only H^+ as cations, through an anion-exchange resin to replace anions by OH^-. As a result of the following reaction, the water is freed of all its ions.

$$H^+(aq) + OH^-(aq) \longrightarrow H_2O(l)$$

The product is called **deionized water.** We use deionized water rather than ordinary tap water in the laboratory because ions present in tap water may interfere with chemical reactions, for example, by forming unwanted precipitates (see Figure 20.11).

20.13 Soaps and Detergents

Recall from earlier discussions that fats are esters of glycerol and long-chain carboxylic acids known as *fatty acids* (pages 70 and 265). **Soaps** are salts of fatty acids formed in the reaction of fats with a base such as NaOH(aq). Sodium palmitate, a typical soap, is a salt of the 16-carbon palmitic acid. It can be represented as

$$\underset{\text{Sodium palmitate (a soap)}}{CH_3(CH_2)_{14}COO^-\,Na^+} \quad \text{or} \quad RCOO^-\,Na^+ \quad [\text{where } R = CH_3(CH_2)_{14}]$$

The function of a soap is to disperse grease and oil films into microscopic droplets. The droplets detach themselves from the surfaces being cleaned, become suspended in water, and are removed by rinsing. As suggested by Figure 20.12, at an oil-water interface, soap molecules orient themselves with their hydrocarbon residues, R—, dissolved in the oil and their ionic ends, $-COO^-Na^+$, in the water. The soap emulsifies or "solubilizes" the oil.

The alkali metal soaps are water soluble,* but soaps having an M^{2+} cation are not. This means that if an alkali metal soap is used in hard water, calcium, magnesium, and/or iron soaps precipitate. These precipitates are the major components of the familiar "soap scum." For example,

$$\underset{\text{Sodium soap}}{2\,RCOO^-Na^+} + Ca^{2+} \longrightarrow \underset{\text{Calcium soap (soap scum)}}{Ca(RCOO)_2(s)} + 2\,Na^+(aq)$$

Before a soap can function well in hard water, part of it must be used up to precipitate all the M^{2+} ions present. In other words, the soap must first soften the water before it will work. But this leaves an objectionable, dirty, grayish film of soap scum on clothes. One alternative is to soften water by one of the methods previously

* Except when very dilute, alkali metal soaps do not form true solutions in water. Rather, they form colloidal dispersions (Section 12.10).

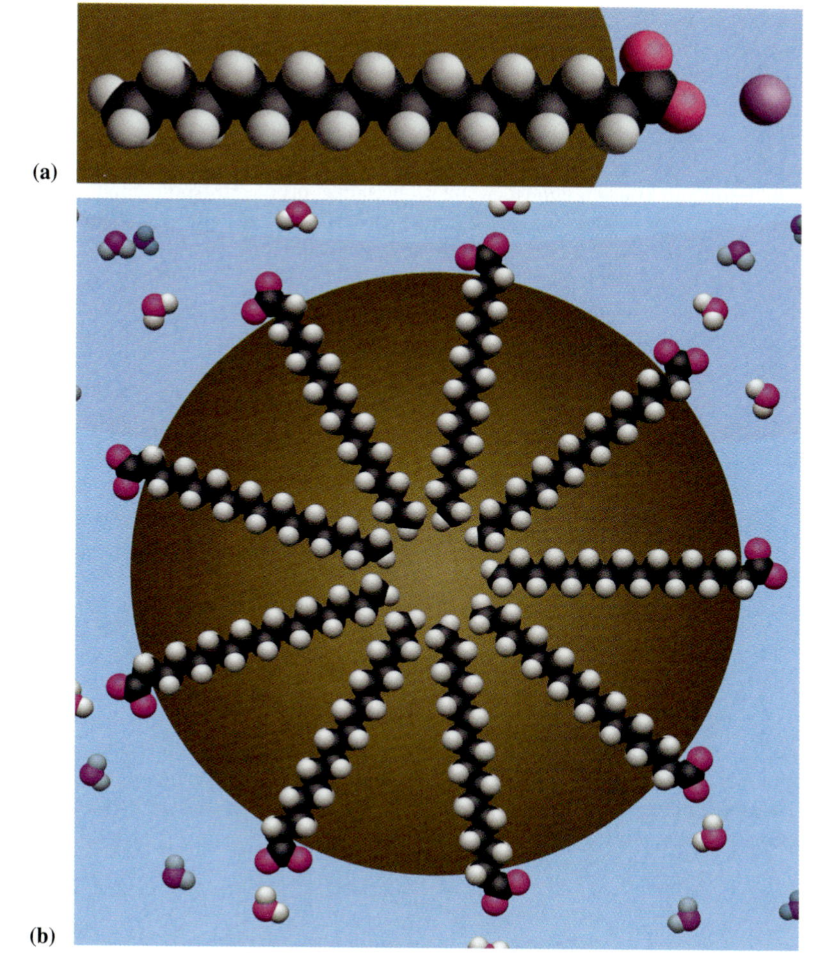

Figure 20.12 Cleaning action of a soap visualized
(a) Sodium palmitate, $CH_3(CH_2)_{14}COO^-Na^+$, a typical soap. (b) A microscopic view of soap action. In an oil droplet suspended in water, the hydrocarbon residues (R groups) of soap molecules are immersed in the oil, and the ionic ends extend into the water. Attractive forces between water molecules and the ionic ends cause the oil droplet to be "solubilized."

● **ChemCDX** Surfactant Molecules

described before using it with soap. Another alternative is to use a synthetic detergent in place of a soap.

A synthetic detergent functions in much the same way as a soap, but it does not form precipitates with M^{2+} cations. The detergent can therefore be used in hard water. Early detergents, called *alkylbenzene sulfonates* (*ABS*), had highly branched alkyl groups. ABS detergents are *not* biodegradable. The detergent type represented by the following structural formula is called a *linear alkylbenzene sulfonate* (*LAS*).

$$CH_3(CH_2)_nCH(CH_3){-}C_6H_4{-}SO_3^-Na^+ \quad \text{(where } n \text{ is usually 9 to 13)}$$

LAS detergents, like soaps, are biodegradable. Bacteria break them down to carbon dioxide, water, and inorganic ions such as sulfate ion.

Summary

The main sources of the element hydrogen are the electrolysis of water, the water-gas reactions, and the reforming of hydrocarbons. Small quantities of the gas can be produced by the reaction of very active metals (Group 1A and Ca, Sr, or Ba) with water, or less active metals (for example, Zn and Fe) with mineral acids. Among the uses of hydrogen are (a) the synthesis of ammonia, (b) in hydrogenation reactions, (c) as a reducing agent in metallurgy, and (d) as a fuel.

Key Terms

deionized water (20.12)
diagonal relationship (20.4)
hard water (20.12)
ion exchange (20.12)
permanent hardness (20.12)
soap (20.13)
temporary hardness (20.12)

Of all the elements, the Group 1A metals (the alkali metals) have the largest atomic radii and lowest ionization energies; they also have low densities and low melting points. They form ionic solids with nonmetals and react with water to produce ionic hydroxides and hydrogen gas. In some of its physical and chemical behavior, lithium resembles magnesium. This resemblance is called a diagonal relationship. Sodium is formed in the electrolysis of molten sodium chloride. Sodium hydroxide, chlorine, and hydrogen form in the electrolysis of aqueous sodium chloride. Other chemicals produced from sodium chloride include sodium carbonate, sodium sulfate, and hydrochloric acid. The vast majority of Group 1A compounds are water soluble. Sodium and potassium ions are essential to living organisms, and many Group 1A compounds have medicinal value.

The Group 2A metals (the alkaline earth metals) have smaller atomic radii and greater ionization energies, densities, and melting points than do the Group 1A metals. Beryllium is an exception to general trends found in Group 2A. Calcium and magnesium are found in many minerals, for example, in limestone ($CaCO_3$) and dolomite ($CaCO_3 \cdot MgCO_3$). The principal method used to obtain the metals is by electrolysis of the molten chloride. The heavier Group 2A metals react with water to liberate hydrogen; magnesium does so with steam, but beryllium does not react with water. Among the important Group 2A compounds are the carbonates, chlorides, hydroxides, oxides, and sulfates of calcium and magnesium. Many alkaline earth compounds are insoluble or only slightly soluble in water. Also, many occur as hydrates, as in $CaCl_2 \cdot 6H_2O$. Magnesium and calcium are essential to all living organisms.

The slight dissolution of minerals (for example, $CaCO_3$, $CaSO_4$) in acidified rainwater introduces ions into groundwater. This dissolving action is responsible for natural caverns, limestone formations, and "hardness" in water. Hard water is objectionable because it yields mineral deposits when heated and interferes with the cleaning action of soaps. Water can be "softened" by chemical reactions that remove the offending cations Ca^{2+}, Mg^{2+}, and Fe^{2+}. Another water-softening method is ion exchange, in which offending ions are replaced by innocuous ones. Ion exchange can also be used to eliminate virtually all ions from water, yielding deionized water. An alternative that permits the use of hard water for laundry purposes is to substitute detergents for soaps.

Review Questions

1. Which is the most abundant alkali metal in Earth's crust? Which is the most abundant alkaline earth metal?
2. Name several naturally occurring alkali metal and alkaline earth metal compounds.
3. Which of the Group 1A elements are known to be essential to living organisms? Which of the Group 2A elements?
4. Explain why the percent abundance of hydrogen atoms in Earth's crust is greater when based on percent by number of atoms than percent by mass. Would you expect the same relationship to hold for all the elements? Explain.
5. In the electrolysis of water pictured in Figure 20.2, does the molarity of the H_2SO_4 change during the electrolysis? If so, in what way? Explain.
6. Helium is an *s*-block element, but we don't consider it to be a member of Group 2A. Explain.
7. Which of the Group 1A elements react with cold water to produce $H_2(g)$? Which of the Group 2A elements?
8. Name some elements that form ionic hydrides and some that form binary molecular hydrides.
9. What is a *nonstoichiometric* compound? Give an example of a type of compound that might be nonstoichiometric.
10. What is meant by the reforming of a hydrocarbon? Give a commercially important example.
11. Explain the function of the metallurgical process known as *reduction*.
12. What is meant by the term *calcination* (for example, the calcination of limestone)?
13. What is meant by the *diagonal relationship* among elements? Which elements display this relationship?
14. Describe several ways in which **(a)** lithium differs from the other Group 1A elements; **(b)** beryllium differs from the other Group 2A elements.
15. What are the principal raw materials required for the manufacture of each of the following products?
    (a) portland cement **(c)** mortar
    (b) soda-lime glass **(d)** plaster of paris
16. Describe how each of the following is used in medicine.
    (a) lithium carbonate **(b)** barium sulfate
17. Describe a function served in the human body by **(a)** sodium, **(b)** potassium, **(c)** magnesium, and **(d)** calcium.

18. Explain the difference between *temporary* and *permanent* hardness in water.

19. What is "boiler scale"? How is it formed?

20. Describe some ways in which hard water can be softened.

21. Describe how deionized water can be prepared from ordinary tap water through the use of ion-exchange resins.

22. What are the structural features of a soap? How does a calcium soap differ in structure from a sodium soap?

23. What are the structural features of a detergent? What do the symbols ABS and LAS mean when applied to a detergent?

24. How do soaps and detergents function; that is, how do they emulsify oil or grease?

Problems

Group Trends

25. Arrange the following compounds in the expected order of increasing solubility in water, and give the basis for your arrangement: Li_2CO_3, Na_2CO_3, $MgCO_3$.

26. Arrange the following compounds in the expected order of increasing melting point, and give the basis for your arrangement: MgO; $CaCl_2$; $SrBr_2$.

Nomenclature

27. Supply a name or formula for each of the following:
    (a) Li_2CO_3
    (b) magnesium nitride
    (c) calcium bromide hexahydrate
    (d) $KHSO_4$

28. Supply a name or formula for each of the following:
    (a) LiH
    (b) magnesium iodide
    (c) barium hydroxide octahydrate
    (d) $Ca(HCO_3)_2$.

29. Give formulas and acceptable scientific names of the substances known by these common names: **(a)** quicklime; **(b)** Glauber's salt; **(c)** gypsum; **(d)** dolomite?

30. Which of the following are pure substances, and which are mixtures? For those that are mixtures, what are the main components?
    (a) plaster of paris
    (b) water gas
    (c) soap scum
    (d) slaked lime

Industrial Processes

31. For the Dow process for the production of magnesium, describe the raw materials required, the main product, any by-products, and substances that are recycled. Outline the steps in the process by writing chemical equations.

32. For the Solvay process for the production of sodium hydrogen carbonate, describe the raw materials required, the main product, any by-products, and substances that are recycled. Outline the steps in the process by writing chemical equations.

33. What is Glauber's salt? Write an equation for the reaction by which it is produced. What is the underlying principle of this reaction?

34. Write equations to show how methanol (CH_3OH) can be synthesized from coke (carbon) and water.

35. Sodium was first obtained by electrolysis of sodium hydroxide, which has a lower melting point (322 °C) than sodium chloride (801 °C). Why do you suppose that the primary industrial method of producing sodium today is the electrolysis of NaCl rather than of NaOH?

36. According to ionization energies and electrode potentials in Table 20.1, K^+ is more difficult to reduce than Na^+. Does this mean that there is no possible reaction in which sodium metal can reduce a potassium compound to metallic potassium? Explain.

37. In the manner used in Figure 20.5, complete the diagram outlined below. Specifically, indicate the reactants and any special conditions needed to produce each substance, using $MgCO_3$ as the starting material.

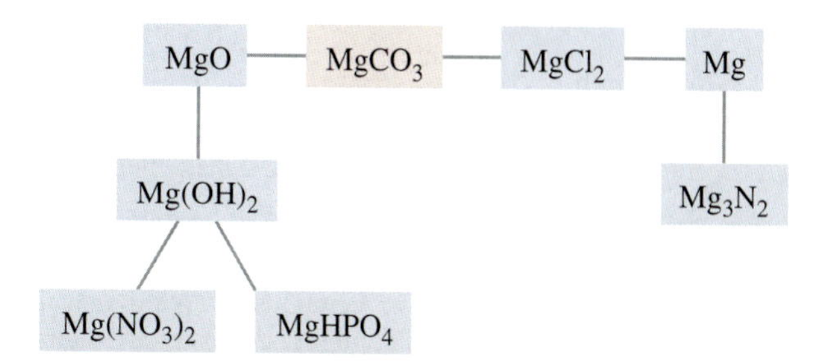

38. In the manner used in Figure 20.5, complete the diagram outlined. Specifically, indicate the reactants and any special conditions needed to produce each substance, using $BaCl_2$ as the starting material.

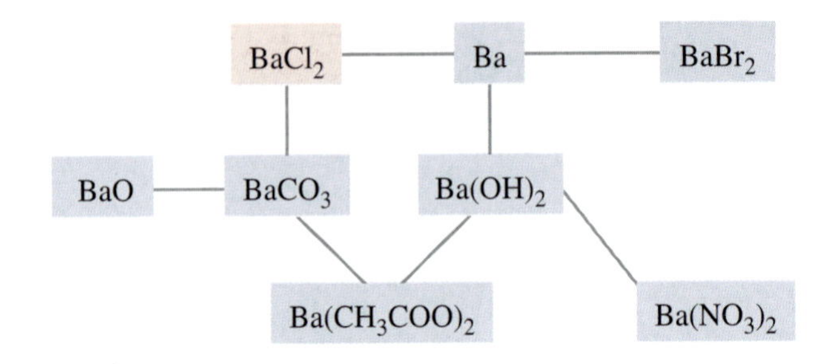

Chemical Equations

39. Complete the following equations representing the reactions of substances with water.

(a) $Sr(s) + H_2O(l) \longrightarrow$

(b) $LiH(s) + H_2O(l) \longrightarrow$

(c) $BaO(s) + H_2O(l) \longrightarrow$

(d) $C(s) + H_2O(g) \xrightarrow{\Delta}$

40. Complete the following equations representing the reactions of substances with acids.

(a) $Ca(s) + HCl(aq) \longrightarrow$

(b) $KHCO_3(s) + HCl(aq) \longrightarrow$

(c) $NaF(s) + H_2SO_4(\text{conc aq}) \xrightarrow{\Delta}$

41. Write chemical equations to represent each of the following.

(a) the displacement of $H_2(g)$ from HCl(aq) by Al(s)

(b) the *reforming* of ethane gas (C_2H_6) with steam

(c) the complete hydrogenation of methylacetylene, $CH_3C{\equiv}CH$

(d) the reduction of $MnO_2(s)$ to Mn(s) with $H_2(g)$

42. Write chemical equations to represent each of the following.

(a) the reaction of lithium metal with chlorine gas

(b) the reaction of potassium metal with water

(c) the reaction of cesium metal with liquid bromine

(d) the combustion of potassium metal to form potassium superoxide

43. Write chemical equations to represent each of the following.

(a) the reaction of BeF_2 with metallic sodium to produce metallic beryllium

(b) the reaction of calcium metal with dilute acetic acid, $CH_3COOH(aq)$

(c) the reaction of plutonium(IV) oxide with calcium to produce metallic plutonium

(d) the calcination of dolomite, a mixed calcium magnesium carbonate ($CaCO_3 \cdot MgCO_3$)

44. Write equations for the reactions you would expect to occur when **(a)** $BaCO_3(s)$ is heated to a high temperature; **(b)** $CaCl_2(l)$ is electrolyzed; **(c)** Ca(s) is added to cold dilute HCl(aq); and **(d)** an excess of slaked lime is added to aqueous sulfuric acid.

45. Write an equation for the reaction that you would expect to occur when **(a)** $MgCO_3(s)$ is added to HCl(aq); **(b)** $CO_2(g)$ is bubbled into KOH(aq); and **(c)** KCl(s) is heated with concentrated $H_2SO_4(aq)$.

46. Write equations to show how each of the following substances can be used in the preparation of $H_2(g)$:

(a) $H_2O(l)$

(b) HI(aq)

(c) Mg(s)

(d) $CH_4(g)$

Use other reactants as necessary—water, acids or bases, metals, and so forth.

47. Write plausible equations to show how you would convert **(a)** NaCl(s) to $Na_2SO_4(s)$; and **(b)** $Mg(HCO_3)_2(aq)$ to MgO(s).

48. Write plausible equations to show how you would convert **(a)** $BaCO_3(s)$ to $Ba(OH)_2(s)$; and **(b)** NaCl(s) to $Na_2O_2(s)$.

Reaction Stoichiometry

(*Hint*: Write chemical equations, as necessary.)

49. How many grams of $CaH_2(s)$ are required to generate sufficient $H_2(g)$ to fill a 126-L weather observation balloon at 746 mmHg and 15 °C?

50. How many liters of $H_2(g)$, measured at 22 °C and 15.5 atm pressure, are required to convert 175 g propylene ($CH_3CH{=}CH_2$) to propane (C_3H_8)? What mass of propane is obtained?

51. How many liters of 6.0 M HCl are required to neutralize 55.6 kg $Ca(OH)_2(s)$?

52. How many cubic meters of $CO_2(g)$, measured at 748 mmHg and 22 °C, would be produced in the calcination of 1.00×10^3 kg of the mineral dolomite ($CaCO_3 \cdot MgCO_3$)?

53. Assume that the magnesium content of seawater is 1270 g Mg^{2+}/ton seawater and that the density of seawater is

1.03 g/mL. What minimum volume of seawater, in liters, must be used in the Dow process to obtain 1.00 kg of magnesium? Why is the actual volume required greater than this calculated minimum volume?

54. In the Solvay process, for every 1.00 kg $NaHCO_3(s)$ produced in the main reaction, how many kilograms of $CaCl_2$ are obtained as a by-product of the ammonia recovery step?

55. *Without doing detailed calculations*, determine which of the following reactions produces the greatest mass of hydrogen per 100 grams of the limiting reactant indicated: **(a)** Zn(s) with an excess of HCl(aq); **(b)** Na(s) with an excess of $H_2O(l)$; **(c)** $CaH_2(s)$ with an excess of $H_2O(l)$.

56. *Without doing detailed calculations*, determine which of the following hydrates has the greatest mass percent H_2O: **(a)** $Ba(OH)_2 \cdot 8H_2O$; **(b)** $CaCl_2 \cdot 6H_2O$; **(c)** $MgSO_4 \cdot 7H_2O$.

Hard Water

57. Describe how water with *temporary* hardness can be softened by the addition of $NH_3(aq)$. (*Hint*: What ions are present in the water? What chemical reactions occur?)

58. With reference to Problem 57, do you think that water with *permanent* hardness can be softened with $NH_3(aq)$? Explain.

59. A particular water sample has a hardness expressed as 115 ppm HCO_3^-. Assuming that Ca^{2+} is the only cation present, how many milligrams of "boiler scale" would you expect to deposit when 725 mL of the water is evaporated to dryness? (*Hint*: Think of parts per million as g $HCO_3^-/10^6$ g water $\approx$ g $HCO_3^-/10^3$ L water.)

60. A water sample has a hardness of 126 ppm HCO_3^-. How many kilograms of $Ca(OH)_2$ are required to soften 1.50×10^7 L of the water? (*Hint*: Refer to Problem 59 and the equations on page 870.)

61. Describe the chemical reactions that lead to the formation of limestone caves and stalactites and stalagmites in those caves. How are these reactions related to the formation of hard water?

62. To soften water with temporary hardness, Ca^{2+} is precipitated as $CaCO_3$. One way to do this is to add CaO (quicklime) to the water. Why doesn't adding a calcium compound simply increase $[Ca^{2+}]$ in the water?

63. Describe how a water sample containing the ions Fe^{3+}, Ca^{2+}, Cl^-, and SO_4^{2-} can be *deionized* by the use of ion-exchange resins.

64. Write chemical equations to represent the preparation of deionized water from the following solutions.

(a) $Ca(HCO_3)_2(aq)$

(b) NaCl(aq)

(c) NaOH(aq)

Additional Problems

65. Write chemical equations for the following reactions referred to on page 855.

(a) the formation of lithium oxide

(b) the reaction of lithium with nitrogen gas to form lithium nitride

(c) the calcination of lithium carbonate

66. The use of potassium superoxide in life-support systems is described on page 857. Write a balanced equation to show how sodium peroxide can be used instead.

67. When stored lime is exposed to air for a long time it loses the properties of CaO, for example, it no longer reacts with water. What reaction(s) must have occurred in the lime?

68. The world's leading producer of cesium and its compounds is a German company called Chemetall Gmbh. Assuming the availability of water, common reagents (acids, bases, salts), and simple laboratory equipment, give a practical method that could be used to prepare each of the following substances from cesium chloride (CsCl)

(a) Cs metal

(b) CsOH

(c) $CsNO_3$

(d) Cs_2SO_4

(e) $CsHCO_3$

(f) Cs_2CO_3

69. A 1.00-kg sample of spent uranium and an excess of water and nitric acid react to produce $UO_2(NO_3)_2$ and $H_2(g)$. How many liters of hydrogen gas are formed if the gas is measured at 25 °C and 745 torr?

70. In Example 20.2, we saw that when magnesium is burned in air, the product is a mixture of MgO and Mg_3N_2. When a small quantity of water is added to the mixture, a strong ammonia smell is detected. When the mixture is then strongly heated, pure MgO is obtained. Write equations for the reactions that occur. Do you think this is an economically feasible method of obtaining MgO? If not, why not, and what method would be more feasible?

71. Suppose the 1.000-g strip of magnesium described in Example 20.2 yields 1.537 g of product. What is the mass percent of MgO in this product?

72. Use electrode potential data from Appendix C to show that the following metals should all be able to displace H_2 from water (as mentioned on page 851): Li, Na, K, Rb, Cs, Ca, Sr, Ba.

73. A particular lithium battery has a voltage of 3.0 V. The capacity of the battery is 0.5 ampere • hour (A • h). Assume that to regulate the heartbeat requires 5 microwatts (μW). How many years can the implanted battery operate? What is the minimum mass of lithium that must be present in the battery? (*Hint*: Refer to Appendix B.)

74. When a 5.00-g sample of impure lime containing some $CaCO_3$ is dissolved in HCl(aq), 143 mL of a gas, measured at 746 mmHg and 22 °C, was liberated. Calculate the percent of $CaCO_3$ in the sample.

75. Determine whether liquid hydrogen or jet fuel has the greater heat of combustion **(a)** on a mass basis and **(b)** on a volume basis. Jet fuel is a mixture of hydrocarbons, but assume $C_{12}H_{26}$ is a representative formula. The densities of $H_2(l)$ and $C_{12}H_{26}(l)$ are 0.0708 g/mL and 0.749 g/mL, respectively. $\Delta H_f^\circ[C_{12}H_{26}(l)] = -350.9$ kJ/mol.

76. A particular water sample has a hardness of 117 ppm SO_4^{2-}. The cations present are Ca^{2+}.

(a) Show how this water can be softened with Na_2CO_3 (washing soda).

(b) How many grams of Na_2CO_3 are required to soften 162 L of this water?

77. How many grams of the soap sodium palmitate are consumed in softening 5.00 L of a hard water with 135 ppm HCO_3^-. Assume that the only cations in the water are Ca^{2+}.

78. A sample of water whose hardness is expressed as 185 ppm Ca^{2+} is passed through an ion-exchange column in which the Ca^{2+} is replaced by Na^+. What is the molarity of Na^+ in the treated water?

79. A 100.0-mL sample of hard water is passed through a column of the ion-exchange resin H_2R. The water coming off the column requires 15.17 mL of 0.02650 M NaOH for its titration. What is the hardness of the water, expressed as ppm Ca^{2+}?

80. Workers who deal with plaster and stucco are aware that considerable heat is given off when slaked lime is produced from quicklime. Determine the maximum temperature reached if 1.00 kg each of quicklime and water are mixed at 25 °C. Use data from Appendix C, and assume specific heats of 0.75 J g^{-1} °C^{-1} for quicklime, 1.18 J g^{-1} °C^{-1} for slaked lime, and 4.18 J g^{-1} °C^{-1} for water.

81. Use atomic radii from Table 20.1 and the fact that sodium, potassium, and rubidium all crystallize in a body-centered cubic structure to calculate the densities of these three metals. Show that your results lead to the same trend in densities as found in Table 20.1 and discussed on page 855.

82. Consider the following sequence of actions and observations: (1) A small chunk of solid CO_2 (dry ice) is added to 0.005 M $Ca(OH)_2(aq)$. (2) Initially, a white precipitate forms. (3) After a short time, the precipitate redissolves. Write chemical equations to explain these observations.

83. Refer to Problem 82. If the 0.005 M $Ca(OH)_2(aq)$ is replaced by 0.005 M $CaCl_2(aq)$, would you expect a precipitate to form? If the 0.005 M $Ca(OH)_2(aq)$ is replaced by 0.010 M $Ca(OH)_2$, a precipitate does form, but it does not redissolve. Explain why. (*Hint*: What equilibrium expressions apply, and what are the equilibrium constants describing them?)

84. Comment on the feasibility of using the following reaction as a means of converting $Na_2CO_3(aq)$ to NaOH(aq).

$$Ca(OH)_2(s) + Na_2CO_3(aq) \longrightarrow CaCO_3(s) + 2\,NaOH(aq)$$

For example, assume that the $Na_2CO_3(aq)$ is 2.00 M, and determine the molarity of the NaOH(aq) that can be produced. (*Hint*: Write an ionic equation. What is K for the reaction?)

85. Use data from elsewhere in the text to obtain an equilibrium constant for the reversible reaction involved in the dissolution of $CaCO_3(s)$ in CO_2-charged rainwater described on page 869.

The p-Block Elements

CHAPTER

21

An image of Jupiter's moon Io. The bright colors are due to sulfur compounds, and the dark features are probably silicate lava flows. Sulfur and silicon are two important p-block elements.

In this chapter, we continue our discussion of descriptive chemistry begun with the s-block elements in Chapter 20. Here, we look at the elements of periodic table Groups 3A through 8A—the p-block elements. The p-block includes all the noble gases except helium, all the nonmetals except hydrogen, all the metalloids, and even a few metals, such as aluminum, tin, and lead.

Three of the p-block elements—O, Si, and Al—are the most abundant in Earth's crust, six of them—C, N, O, P, S, and Cl—are among the elements making up the bulk of living matter, and five others—B, F, Si, Se, and I—are required in trace amounts by most plant and animal life. Two of the p-block elements—C and S—can occur in the free state and have been known from prehistoric times, and three of them—Sn, Sb, and Pb—have been known for several thousand years. Most of the p-block elements, however, were not discovered until much later.

In our discussion of the p-block elements, we will relate the properties of the elements to their positions in the periodic table. We will describe important compounds of the elements, comment on their uses, and discover some ways in which

CONTENTS

● **ChemCDX**
When you see this icon, there are related animations, demonstrations, and exercises on the CD accompanying this text.

B	C	N	O	F	Ne
Al	Si	P	S	Cl	Ar
Ga	Ge	As	Se	Br	Kr
In	Sn	Sb	Te	I	Xe
Tl	Pb	Bi	Po	At	Rn

The *p*-block elements

we encounter them in daily life. In this discussion, we will use several principles introduced earlier in the text: bonding theory, enthalpy and free-energy changes, electrode potentials, and equilibrium constants. In doing so, we will once more see the power of fundamental principles to explain chemical phenomena.

GROUP 3A

The first two members of Group 3A—boron, a metalloid; and aluminum, a metal—are the most important of the five group members. We will describe only a few aspects of the chemistry of boron but deal more extensively with aluminum. In the following section, we first take a brief look at the group as a whole.

21.1 Properties and Trends in Group 3A

Several atomic and physical properties of the Group 3A elements are listed in Table 21.1. Let's consider the valence-shell electron configuration, ns^2np^1, and how it changes when Group 3A atoms form compounds. Boron atoms are small with high ionization energies. Boron therefore tends to use its $2s^22p^1$ valence electrons to form covalent bonds with other nonmetal atoms. Aluminum also forms some covalent bonds, but Al atoms have a tendency to lose the $3s^23p^1$ valence electrons. This happens, for example, in the reaction of aluminum with acids.

$$2\,Al(s) + 6\,H^+(aq) \longrightarrow 2\,Al^{3+}(aq) + 3\,H_2(g)$$

Because Al^{3+} is a small, highly charged ion, it is usually coordinated with other species. Thus, $Al^{3+}(aq)$ referred to in the above equation is mostly $[Al(H_2O)_6]^{3+}$.

TABLE 21.1 Some Properties of the Group 3A Elements

	B	Al	Ga	In	Tl
Atomic number	5	13	31	49	81
Valence-shell electron configuration	$2s^22p^1$	$3s^23p^1$	$4s^24p^1$	$5s^25p^1$	$6s^26p^1$
Atomic radius, pm[a]	88	143	122	163	170
Ionic radius, pm[b]	23	50	62	81	95
First ionization energy, kJ/mol	801	577	579	558	589
Electronegativity	2.0	1.6	1.8	1.8	2.0
Density at 20°C, g/cm^3	2.34	2.70	5.91	7.31	11.85
Melting point, °C	2300	660	30	157	304
Electrical conductivity[c]	9×10^{-11}	59.9	5.9	19.0	8.8

[a] For B, the single covalent radius; for the others, the metallic radius.
[b] For the ion M^{3+}.
[c] On a scale relative to silver as 100.

Gallium, indium, and thallium also form tripositive ions, but the electron configurations of these ions are not those of noble-gas atoms. The electron configuration of Ga^{3+}, for example, is $[Ar]3d^{10}$. Unlike aluminum, which forms only a 3+ ion, gallium, indium, and thallium also form 1+ ions. Thus, when a gallium atom loses its $4p^1$ electron but retains its $4s^2$ electrons, it becomes the Ga^+ ion with the electron configuration $[Ar]3d^{10}4s^2$. The heavier elements have an even greater tendency to form 1+ ions. In fact, Tl^+ is more stable in aqueous solution than Tl^{3+}. This is reflected in the large *positive* potential for the reduction of Tl^{3+} to Tl^+, and in the *negative* potential for the reduction of Tl^+ to Tl(s).

$$Tl^{3+}(aq) + 2\,e^- \longrightarrow Tl^+(aq) \qquad E° = +1.25\ V$$
$$Tl^+(aq) + e^- \longrightarrow Tl(s) \qquad E° = -0.336\ V$$

Thus, $Tl^{3+}(aq)$ in contact with Tl(s) is readily reduced to $Tl^+(aq)$.

$$Tl^{3+}(aq) + 2\,Tl(s) \longrightarrow 3\,Tl^+(aq) \qquad E°_{cell} = 1.25\ V - (-0.336\ V) = 1.59\ V$$

The valence-shell pair of electrons retained in the Tl^+ ion ($6s^2$) is called an **inert pair**. This retention of the ns^2 pair is common in the metals that follow a transition series.

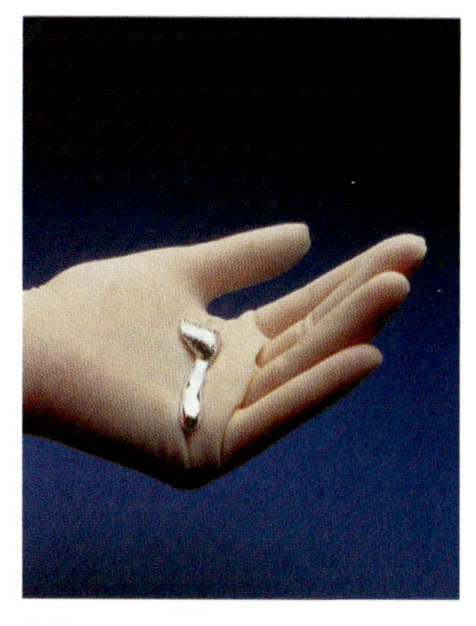

Gallium is a liquid from 30 °C (just above room temperature) to about 2400 °C. This is one of the longest temperature ranges for the liquid state of any substance. Gallium finds some use in high-temperature thermometers.

One physical property in Table 21.1 that suggests boron's nonmetallic character is its high melting point, which can be attributed to the network covalent bonding in the solid. The other four members of Group 3A are good electrical conductors, but boron is a poor conductor. However, because boron displays the electrical properties of a *semiconductor* (Section 24.8), we classify it as a metalloid.

Boron, the second period member of Group 3A, has a *diagonal relationship* with silicon, the third period member of Group 4A. Recall our discussion of diagonal relationships in Chapter 20 (page 855), where we noted some similarities of Li to Mg and of Be to Al. One similarity between B and Si, for example, is that both are semiconductors.

21.2 Boron

Much of the chemistry of boron compounds is based on the *lack* of an octet of electrons about the central boron atom. These compounds are electron-deficient, and this deficiency causes them to exhibit some unusual bonding features that we explore in our look at boron hydrides.

Boron Hydrides Carbon atoms have four valence electrons, and the simplest hydrocarbon is methane, CH_4. By analogy, because boron atoms have three valence electrons, we might expect the simplest boron-hydrogen compound to be BH_3, *borane*. However, the boron atom in BH_3 lacks a valence-shell octet. It has only *six* electrons in its valence shell.

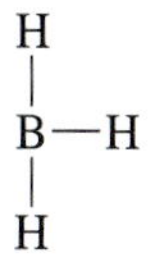

Borane does not exist as a stable compound.

To complete the octet of the boron atom, the BH_3 group can become part of a more extensive structure by forming a coordinate covalent bond with another atom that has a lone pair of electrons. In the structure that follows, this bond is between the B atom of BH_3 and the O atom of dimethyl ether. The compound, which results from the "addition" of one structure to another, is called an **adduct**.

$$CH_3{-}\ddot{O}{:}(CH_3) + BH_3 \longrightarrow CH_3{-}\ddot{O}{:}BH_3(CH_3)$$

Dimethyl ether Borane Dimethyl ether–borane adduct

The simplest boron hydride that can be isolated is *diborane*, B_2H_6. For many years following its isolation, diborane presented a bonding puzzle. What holds the two BH_3 units together?

$$H_3B \; ? \; BH_3$$

The problem is that the structure has a total of only *12* valence electrons, and the number required would seem to be *14* (as in ethane, C_2H_6).

The difficulty is resolved through a type of bonding we have not seen before: a bond in which a single pair of electrons is able to join *three* atoms rather than the usual two. As shown in Figure 21.1, there are two of these so-called *three-center* bonds in diborane. The H atoms in the three-center bonds are simultaneously bonded to *two* atoms rather than the usual one. We can think of these H atoms as bridging the gap between the two B atoms.

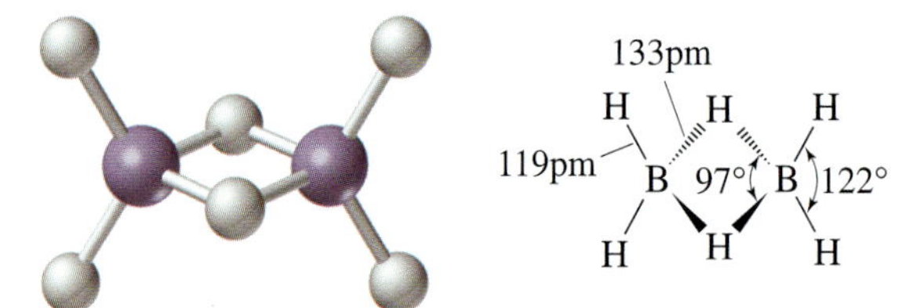

▶ **Figure 21.1**
Structure of diborane, B_2H_6

We can explain the fact that the B—H bond lengths are greater in the three-center bonds than in the other B—H bonds by noting that two electrons cannot bind three atoms as tightly as they bind two atoms. The orientation of the four H atoms around each B atom is roughly tetrahedral, which we can describe through sp^3 hybridization of the bonding orbitals of the boron atoms. However, the bond angles are not tetrahedral (109.5°). The most satisfactory explanation of bonding in B_2H_6 comes from modern molecular orbital theory. Recall that this theory is able to account for orbitals that are delocalized among several atoms, as in benzene (page 449).

In all, about two dozen different boron-hydrogen compounds (boranes) are now known, and their structures have been determined by combining molecular orbital theory and experimental measurements. Boranes such as diborane and organic derivatives of borane are frequently used as reagents in the synthesis of organic compounds.

Borax, Boric Oxide, Boric Acid, and Borates Boron compounds are fairly widely distributed on Earth. Boron comprises about 9 ppm of Earth's crust and 4.8 ppm of seawater, and it is an essential element for some organisms. However, concentrated mineral deposits of boron compounds are found only in a few locations, such as Turkey and the desert regions of California. *Borax*, $Na_2B_4O_7 \cdot 10H_2O$, a hydrated borate, is the primary source of boron.

The first step in the production of boron and boron compounds is the conversion of borax to boric acid, $B(OH)_3$.

$$Na_2B_4O_7 \cdot 10H_2O + H_2SO_4 \longrightarrow 4\,B(OH)_3 + Na_2SO_4 + 5\,H_2O$$

In the 1880s, borax was hauled from salt flats in Death Valley, California, by a team of 20 mules pulling a pair of wagons—plus a vital 1200-gallon water tank. The borax was hauled 165 miles across an almost waterless desert and mountains to a railroad terminal in Mojave. Borax is still sold today as a laundry aid under the brand name, 20 Mule Team Borax®.

Following this, the compound boric oxide, B_2O_3, is prepared by the dehydration of $B(OH)_3$ by heating.

$$2\,B(OH)_3(s) \xrightarrow{\Delta} B_2O_3(s) + 3\,H_2O(g)$$

Elemental boron and boron compounds are prepared from B_2O_3, as illustrated in Example 21.1.

EXAMPLE 21.1

Write chemical equations to represent **(a)** the high-temperature reduction of B_2O_3 to elemental boron with magnesium as a reducing agent; **(b)** the preparation of boron trichloride by heating boric oxide with carbon and chlorine gas. (The carbon is oxidized to carbon monoxide.)

SOLUTION

Remember that the key to writing equations for chemical reactions is to have a complete description of the initial reactants, the final products, and the reaction conditions. Most of this information is provided in the statement of the problem.

a. $B_2O_3(s) + 3\,Mg(s) \xrightarrow{\Delta} 2\,B(s) + 3\,MgO(s)$

b. $B_2O_3(s) + 3\,C(s) + 3\,Cl_2(g) \xrightarrow{\Delta} 2\,BCl_3(g) + 3\,CO(g)$

EXERCISE 21.1

Write an equation to represent the preparation of pure boron by the reduction of $BCl_3(g)$ with hydrogen gas.

Boric oxide, B_2O_3, reacts with water to form boric acid, $B(OH)_3$.

$$B_2O_3(s) + 3\,H_2O(l) \longrightarrow 2\,B(OH)_3(s)$$

We write the formula of boric acid as $B(OH)_3$ rather than H_3BO_3 because boric acid is an extremely weak *monoprotic* acid, not triprotic as the formula H_3BO_3 would indicate. Further, its acidity is displayed differently from most other weak acids. Instead of donating a proton, the acid accepts a hydroxide ion, forming the complex ion $[B(OH)_4]^-$. We can think of a hydroxide ion produced in the self-ionization of water as attaching itself to the B atom of $B(OH)_3$, an electron-deficient structure, through a coordinate covalent bond.

$$OH^- + B(OH)_3 \rightleftharpoons \left[HO{-}B(OH)_2{-}OH \right]^-$$

The net ionization reaction is the following.

$$B(OH)_3(aq) + 2\,H_2O(l) \rightleftharpoons H_3O^+(aq) + [B(OH)_4]^-(aq) \quad K_a = 5.6 \times 10^{-10}$$

Boric acid is quite toxic if taken internally, but dilute solutions can be used externally as a mild antiseptic—for example, in eyewash solutions. Boric acid is also used as an insecticide against cockroaches and black carpet beetles.

Solutions of the salts of boric acid—borate solutions—are generally quite complex because they contain polymers of the borate anions as well as the simple anions $[B(OH)_4]^-$, BO_3^{3-}, and BO_4^{5-}. Borates hydrolyze in water to give basic solutions, and this is why borax is used in some cleaning agents. Sodium perborate, $NaBO_3 \cdot 4H_2O$, crystallizes from an aqueous solution of hydrogen peroxide and borax. It is used in denture cleansers and as a color-safe "oxygen" bleach for clothes that would be harmed by a "chlorine" bleach (sodium hypochlorite). The bleaching action is actually that of H_2O_2 released in the hydrolysis of sodium perborate.

$$[B_2(O_2)_2(OH)_4]^{2-}(aq) + 4\,H_2O(1) \longrightarrow 2\,H_2O_2(aq) + 2\,[B(OH)_4]^-(aq)$$

The oxidation number of boron in $NaBO_3 \cdot 4H_2O$ is *not* +5, as this empirical formula suggests. A more representative formula is $Na_2[B_2(O_2)_2(OH)_4] \cdot 6H_2O$, with the "$O_2$" signifying peroxides linkages, and with a +3 oxidation number for boron. The structure of the perborate ion is suggested below.

21.3 Aluminum

The most important metal of Group 3A is aluminum.* Over 5 million tons of the metal are produced per year in the United States; most of it is used in lightweight alloys. In this section, we consider the commercial production of the metal, its properties and uses, and some of its important compounds.

Production of Aluminum Earth's crust is 8.3% by mass of aluminum. This makes aluminum the third most abundant element and the most abundant metal. Aluminum metal was not isolated until 1825, when Hans Oersted produced it in impure form. For the next several decades it remained a semiprecious metal used mostly in jewelry and artwork. It was still rare and expensive in 1884 when an aluminum cap was placed atop the newly completed Washington Monument. Just a couple of years later, however, the situation changed completely. Charles Martin Hall, in the United States, and Paul Heroult, in France, discovered an inexpensive way to make aluminum by electrolysis.

Charles Martin Hall (1863–1914), was motivated by a professor at Oberlin College who remarked that anyone discovering a cheap method of producing aluminum would become rich and famous. Hall's discovery, in his home laboratory within eight months of his graduation, was the foundation of the aluminum industry in the United States.

The Hall-Heroult process involves two key features: First, it uses *bauxite*, a mixture of hydrates of aluminum oxide, as the source of aluminum, and it takes advantage of amphoterism to remove the principal impurity, Fe_2O_3, from the bauxite. Al_2O_3 is amphoteric; it can accept protons, thus acting as a base, or—like boric acid—it can react with hydroxide ions, thus acting as an acid. Fe_2O_3 is not amphoteric; it has only basic properties. Thus, when the ore is treated with hot, concentrated NaOH(aq), only the Al_2O_3 reacts.

$$Al_2O_3(s) + 2\,OH^-(aq) + 3\,H_2O(l) \longrightarrow 2\,[Al(OH)_4]^-(aq)$$

$$Fe_2O_3(s) + OH^-(aq) \longrightarrow \text{no reaction}$$

* In much of the world, the element of atomic number 13, Al, is spelled alumin*i*um (pronounced al-you-MIN-ee-um). In the United States, though, it is usually spelled aluminum (pronounced a-LOO-min-um).

The solution containing $[Al(OH)_4]^-$ is separated from the undissolved "red mud" (mostly Fe_2O_3), diluted with water, and slightly acidified, causing $Al(OH)_3$ to precipitate.

$$[Al(OH)_4]^-(aq) + H_3O^+(aq) \longrightarrow Al(OH)_3(s) + 2\,H_2O(l)$$

The $Al(OH)_3(s)$ is heated to about 1200 °C and decomposes to pure $Al_2O_3(s)$.

$$2\,Al(OH)_3(s) \longrightarrow Al_2O_3(s) + 3\,H_2O(g)$$

The second key feature of the Hall-Heroult process deals with the fact that the melting point of Al_2O_3 (2045 °C) is much too high to permit the electrolysis of molten Al_2O_3. Moreover, molten Al_2O_3 is not a particularly good conductor. Hall and Heroult made a crucial innovation by using molten *cryolite*, Na_3AlF_6, as the electrolyte, with a few percent dissolved Al_2O_3. The mixture is a good electrical conductor and remains liquid even at about 950 °C.

The electrolysis cell pictured in Figure 21.2 uses a carbon-lined steel cell as the cathode and large chunks of carbon as the anodes. $Al_2O_3(s)$ is continually added to the bath as liquid aluminum is drawn off. The electrode reactions are complex, but the overall electrolysis reaction is

Reduction:	$4\,\{Al^{3+} + 3\,e^- \longrightarrow Al(l)\}$
Oxidation:	$3\,\{C(s) + 2\,O^{2-} \longrightarrow CO_2(g) + 4\,e^-\}$
Overall:	$3\,C(s) + 4\,Al^{3+} + 6\,O^{2-} \longrightarrow 4\,Al(l) + 3\,CO_2(g)$

Paul Heroult (1863–1914), a student of Le Châtelier's, was, like Hall, 23 years old when he discovered the same method of producing aluminum. Heroult's discovery was the foundation of the aluminum industry in Europe.

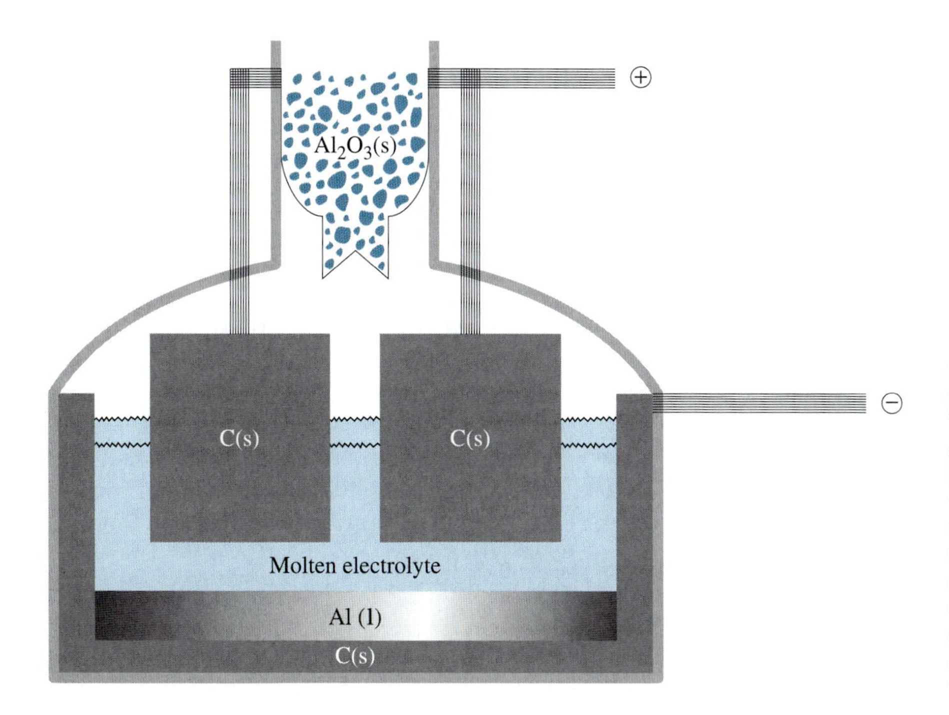

Figure 21.2
Electrolysis cell for aluminum production
The cathode is the carbon lining of a steel tank. The anodes are made of blocks of carbon. Liquid aluminum, which is more dense than the electrolyte, collects at the bottom of the tank and is removed. The electrolyte is alumina (Al_2O_3) dissolved in molten cryolite (Na_3AlF_6). Fresh alumina is continuously added from a hopper above the cell.

The production of aluminum requires a great deal of energy, about 15,000 kWh per ton of Al. To produce a ton of steel requires only about one-fifth this much energy. On the other hand, because the density of aluminum (2.70 g/cm^3) is much less than that of iron (7.87 g/cm^3), less mass of aluminum may be required for a particular application where aluminum can replace iron. Moreover, because it takes only about one-twentieth as much energy to recycle aluminum as it does to produce it from bauxite, much of the aluminum produced in the United States is recycled.

Properties and Uses of Aluminum The reduction of $Al^{3+}(aq)$ to Al(s) occurs only with difficulty.

$$Al^{3+}(aq) + 3\,e^- \longrightarrow Al(s) \quad E^\circ = -1.676\ V$$

This means that the reverse process, the oxidation of Al(s), occurs readily. Thus, Al(s) is a good reducing agent. One interesting reaction based on the reducing power of aluminum is the highly exothermic *thermite* reaction (recall Figure 4.12). The liquid iron produced in the reaction can be used to weld large iron objects.

$$Fe_2O_3(s) + 2\,Al(s) \longrightarrow 2\,Fe(l) + Al_2O_3(s)$$

As we noted in Section 18.8, the oxide that readily forms on aluminum is a thin impervious film that protects the underlying metal from corrosion. The oxide film can be made thicker by making aluminum the *anode* in an electrolytic cell with dilute $H_2SO_4(aq)$ as the electrolyte. Equations for the half-reactions and overall reaction are written below.

$$\begin{array}{llcl}
\textit{Anode:} & 2\,Al(s) + 3\,H_2O(l) & \longrightarrow & Al_2O_3(s) + 6\,H^+ + 6\,e^- \\
\textit{Cathode:} & 3 \times \{2\,H^+(aq) + 2\,e^- & \longrightarrow & H_2(g)\} \\
\hline
\textit{Overall:} & 2\,Al(s) + 3\,H_2O(l) & \longrightarrow & Al_2O_3(s) + 3\,H_2(g)
\end{array}$$

This *anodized aluminum* can be colored by adding dyes to the electrolyte solution. Bronze and brown anodized aluminum are especially popular in modern buildings and as window frames in homes.

Anodized aluminum.

As an active metal, aluminum readily reacts with acids to produce hydrogen gas.

$$2\,Al(s) + 6\,H^+(aq) \longrightarrow 2\,Al^{3+}(aq) + 3\,H_2(g)$$

However, aluminum also dissolves in *basic* solutions. We can think of the reaction as a two-step process. First, the Al_2O_3 film on the metal dissolves in $OH^-(aq)$, just as we described for the purification of bauxite.

$$Al_2O_3(s) + 3\,H_2O(l) + 2\,OH^-(aq) \longrightarrow 2\,[Al(OH)_4]^-(aq)$$

Once the oxide film is removed from the aluminum, the metal is then free to display its true activity. It reacts with water in the alkaline solution, displacing hydrogen gas.

$$2\,Al(s) + 6\,H_2O(l) + 2\,OH^-(aq) \longrightarrow 2\,[Al(OH)_4]^-(aq) + 3\,H_2(g) \quad E^\circ_{cell} = 1.482\ V$$

Some drain cleaners make use of this reaction. They consist of a mixture of solid sodium hydroxide and granules of aluminum metal. When added to water, the heat of solution of the NaOH(s) and the heat of the above reaction help to melt fat and grease. Agitation resulting from the formation of the hydrogen gas helps to dislodge obstructions and unplug the drain. (Often, though, a plumber's "snake" is more effective than chemical drain cleaners.)

Because its combustion is a highly exothermic reaction, powdered aluminum is used as a component in rocket fuels, explosives, and fireworks.

$$2\,Al(s) + \tfrac{3}{2}\,O_2(g) \longrightarrow Al_2O_3(s) \quad \Delta H^\circ = -1676\ kJ$$

Perhaps most familiar is aluminum's use in beverage cans, cookware, and as a foil for wrapping foods. Most aluminum, though, is consumed in structural materials, usually alloyed with other metals to impart greater strength. Most modern aircraft use aluminum alloys, as do some automobile engines and the exterior trim of modern buildings. Because aluminum has good electrical conductivity (about 63.5% that of an equal volume of copper) and a low density, it is also widely used in the electrical industry. High-power transmission lines in the United States are now made mostly of aluminum alloys.

Aluminum Compounds Among the aluminum halides, AlF_3 is rather different than the others. For example, although AlF_3 and $AlCl_3$ both have crystal structures based on Al atoms surrounded by six halogen atoms, only AlF_3 seems to have the properties normally associated with ionic substances. It is nonvolatile, very slightly soluble in water, has a high melting point (1090 °C), and is a good electrical conductor in the liquid state. In contrast, solid $AlCl_3$ sublimes, is soluble in water and also in some organic solvents, such as ethanol, diethyl ether, and carbon tetrachloride, has a low melting point (193 °C), and is a poor electrical conductor in the liquid state. In the liquid and gaseous states, aluminum chloride exists as *dimers* of $AlCl_3$, that is, as Al_2Cl_6 molecules. In Figure 21.3, the Lewis structure shows that two of the Cl atoms in Al_2Cl_6 are bonded to both Al atoms; they form bridges between the two $AlCl_3$ units. Two of the bonds are coordinate covalent bonds in which the Cl atoms supply both electrons. The bonding scheme shows that the orbital overlaps in the molecules involve sp^3 hybrid orbitals of Al and $3p$ orbitals of Cl. The coordinate covalent bonds are shown as arrows. The chlorine-atom bridges between the two $AlCl_3$ units are also seen in the space-filling model.

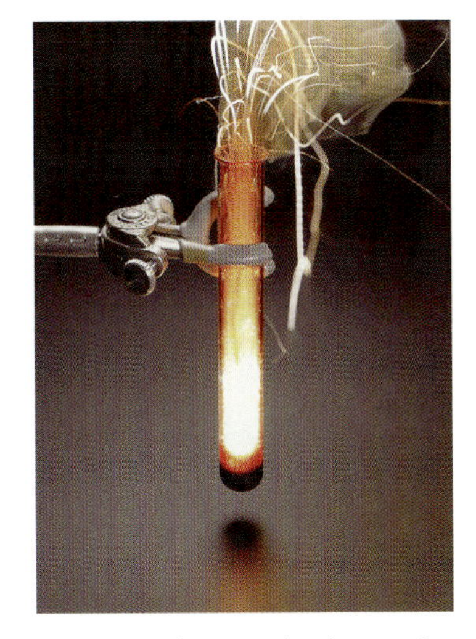

The reaction between aluminum and liquid bromine.

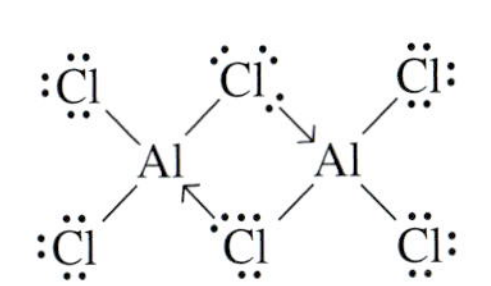

Lewis structure

Cl Cl Cl Al Al Cl Cl Cl $\sigma(sp^3, 3p)$

Bonding scheme

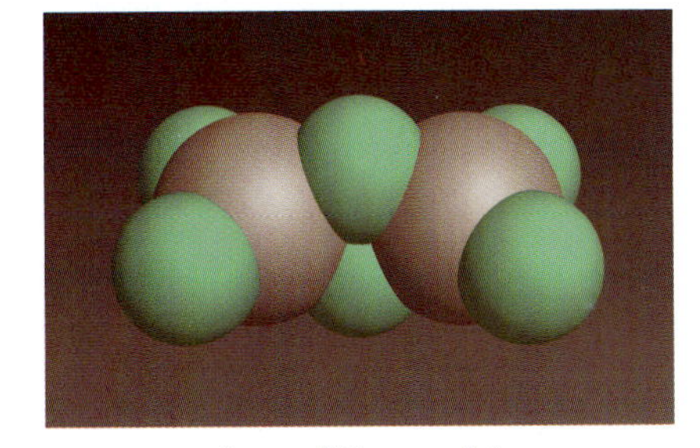

Space-filling model

▲ Figure 21.3 Bonding in Al_2Cl_6

Aluminum chloride is an important catalyst in organic chemistry (Chapter 23), where its electron-deficient structure enables it to form complexes with most oxygen-containing organic compounds and other species having lone-pair electrons.

Lithium aluminum hydride, $LiAlH_4$, is used as a reducing agent in organic chemistry. If we think of hydride ion ($H:^-$) as a *pseudohalide* ion (some periodic tables place hydrogen at the top of Group 7A, as well as at the top of Group 1A), then the AlH_4^- ion can be considered an adduct of AlH_3 and H^-. An example of its use in organic chemistry is the reduction of carboxylic acids, such as those from fats, to alcohols. For palmitic acid, the reduction proceeds as follows.*

$$\underset{\substack{\text{Palmitic acid}\\\text{(from palm oil)}}}{CH_3(CH_2)_{14}COOH} + LiAlH_4 \longrightarrow \underset{\text{1-Hexadecanol}}{CH_3(CH_2)_{14}CH_2OH}$$

APPLICATION NOTE

Fatty alcohols such as 1-hexadecanol are used to make specialty detergents for toothpastes and shampoos.

Aluminum sulfate is the most important industrial aluminum compound. It is prepared by the action of hot, concentrated $H_2SO_4(aq)$ on $Al_2O_3(s)$. The product that crystallizes from solution is $Al_2(SO_4)_3 \cdot 18H_2O$. About half of the 1 million tons of aluminum sulfate produced annually in the United States is used in water treatment. In this application, the pH is raised to the point at which $Al(OH)_3(s)$ precipitates.

*The mechanism of this reaction is quite complex. It proceeds through four steps, and the final mixture must be acidified to obtain the alcohol. In the course of the reaction, $LiAlH_4$ is decomposed to Li^+, Al^{3+}, and $H_2(g)$.

Corundum and emery are widely used in abrasive materials, as in the abrasive belt of this sander.

$$Al^{3+}(aq) + 3\,OH^-(aq) \longrightarrow Al(OH)_3(s)$$

As it settles, the gelatinous $Al(OH)_3(s)$ traps and removes suspended solids from the water.

We have described the use of aluminum oxide in the manufacture of aluminum. Another important use is in refractory materials for high-temperature furnaces. A refractory is a solid that retains its chemical identity and physical dimensions even when subjected to high temperatures; the melting point of Al_2O_3 is 2045 °C. Aluminum oxide is also used in the manufacture of ceramic materials (page 892). The mineral *corundum* is a pure form of Al_2O_3, while *emery* is corundum contaminated with iron oxides (Fe_2O_3 and/or Fe_3O_4) and silica (SiO_2). Both of these materials have a hardness value of 9 on a ten-point scale, with diamond as the hardest substance, and both are used in the manufacture of abrasive materials such as grinding wheels and sandpaper. Many gemstones are naturally occurring impure forms of aluminum oxide.

Iron and titanium impurities in aluminum oxide produce blue sapphires.

Group 4A

The members of Group 4A all have the valence-shell electron configuration ns^2np^2, where n is the period number. Carbon, the first member, is a nonmetal. Carbon atoms use their four valence electrons to form four covalent bonds in nearly all carbon compounds. The next two members, silicon and germanium, also form covalent bonds for the most part. Both are metalloids and have interesting properties as semiconductors (Section 24.8). Tin and lead are more metallic in their behavior. Both form 2+ and 4+ ions.

21.4 Carbon

The ground-state electron configuration of carbon is $1s^22s^22p^2$. As we learned in Sections 10.4 and 10.5, hybridization produces three different sets of orbitals: (1) four sp^3 orbitals, as in ethane, C_2H_6, (2) three sp^2 orbitals plus one p orbital, as in ethene (ethylene), C_2H_4, or (3) two sp orbitals plus two p orbitals, as in ethyne (acetylene), C_2H_2. These hybridization possibilities permit the formation of carbon chain and ring structures having multiple bonds between C atoms, as well as the more common single bonds. In these structures, C atoms also bond to H, O, N, S, halogens, and several other types of atoms. Taken together, these factors lead to a myriad of *organic* compounds. We have frequently referred to organic compounds since introducing them in Chapter 2, and we will consider them in more detail in Chapter 23. Our emphasis here will be more on the inorganic chemistry of carbon.

Elemental carbon exists in nature mainly as the two allotropes diamond and graphite, although recently the allotropic forms known as fullerenes and nanotubes

Gemstones: Natural and Artificial

Gemstones are found only rarely in nature. They must contain just the right combination of substances, brought together under the appropriate conditions of temperature, pressure, concentrations, and so on, in order to form the proper crystals. One class of gemstones has aluminum oxide (corundum) as the principal constituent and tiny amounts of transition metal oxides as impurities. The particular impurities present determine the colors of the gem (see Table 21.2). Corundum gemstones with pink to dark red colors are *rubies*; those of all other colors are *sapphires*.

Natural rubies (bottom left) and a synthetic ruby (top right).

TABLE 21.2 Common Gemstones Based on Aluminum Oxide

Gem	Impurity
Ruby	Cr
Sapphires	
white	none
orange	Ni, Cr
yellow	Ni
green	Co, V, and/or Ni
blue	Fe, Ti
star	Ti

Scientists have even learned how to mimic nature by synthesizing rubies and sapphires in high-temperature furnaces. A mixed powder of Al_2O_3 and the appropriate transition metal oxide(s) is sprayed into the upper region of the furnace. The powder melts in the hottest regions of the furnace, forming a liquid layer that solidifies. The solidified material is gradually withdrawn from the furnace as layer after layer is added to it. Gemstones produced in this way have many industrial applications. The ruby is rare in nature and one of the most valuable of all gemstones. Rubies can now be manufactured for use in jewelry, but good quality natural rubies are still more valuable than synthetic ones. The natural gems are identified by their slight imperfections; the synthetic ones—as unbelievable as it might seem—are too near to perfection to pass as "natural."

have been discovered. We described the structures of all the important allotropic forms of carbon in Section 11.8. Graphite is perhaps most familiar as the writing material in pencils. Because it conducts electric current, graphite is used for electrodes in batteries and industrial electrolysis. It can also withstand high temperatures, leading to its use in foundry molds, furnaces, and other high-temperature devices.

This Beech Starship jet aircraft is made largely of composites.

Another modern use of graphite is in the form of fibers. When carbon-based fibers such as rayon are heated to a high temperature, other elements are driven off in gaseous products, leaving behind graphite fibers. These fibers are imbedded in plastic materials to make high-strength, lightweight composites. The composites are used in products as diverse as tennis rackets, canoes, fishing rods, and airplanes.

Diamonds are used in jewelry, but they also have industrial uses. Because they have a high thermal conductivity (they dissipate heat quickly) and are extremely hard, diamonds are used as abrasives and in drill bits for cutting steel and other

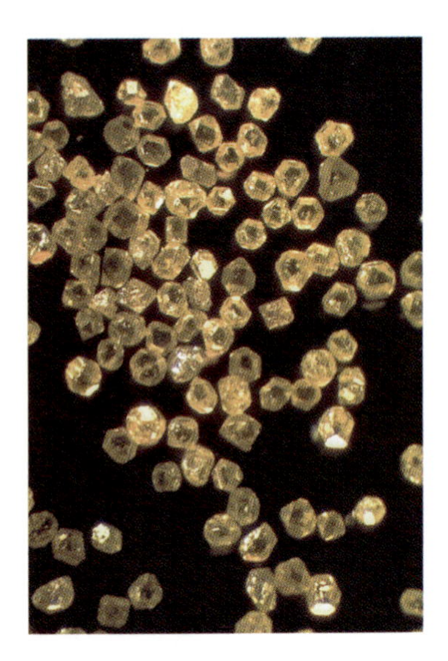

▲ **Figure 21.4**
Synthetic diamonds
Synthetic diamonds are made from graphite. Graphite is the more stable form of carbon at room temperatures and pressures and remains so up to temperatures of 3000 °C and pressures of 10^4 atm or more. The conversion of graphite to diamonds requires temperatures of 1000 °C to 2000 °C and pressures of 10^5 atm or more.

hard materials. In general, natural diamonds are used as gemstones and synthetic diamonds (Figure 21.4) for industrial purposes.

Carbon also exists in amorphous forms. The term *amorphous* implies a noncrystalline solid, but amorphous carbon is probably microcrystalline graphite. When coal is heated in the absence of air, volatile substances are driven off, leaving a high-carbon residue called *coke*. Coke is the principal metallurgical reducing agent; it is used in the reduction of iron oxide to iron in a blast furnace, for example. A similar *destructive distillation* of wood produces charcoal. As you've probably observed in an improperly adjusted Bunsen burner, incomplete combustion of natural gas produces a smoky flame. This smoke can be deposited as a powdery soot called *carbon black*. Carbon black is used as a filler in rubber tires, as a pigment in printing inks, and as the transfer material in carbon paper, typewriter ribbons, and photocopying machines.

Activated carbon is formed by heating carbon black to 800–1000 °C in the presence of steam to expel all volatile matter. This form of carbon is highly porous, like sponges or honeycombs. Because of its high ratio of surface area to volume, activated carbon has a great capacity to adsorb substances from liquids and gases. Activated carbon is used in gas masks to adsorb poisonous gases from air, in water filters to remove organic contaminants from water, in the recrystallization of sugar solutions to remove colored impurities, in air-conditioning systems to control odors, and in industrial plants for the control and recovery of vapors.

Inorganic Carbon Compounds

We have encountered carbon monoxide and carbon dioxide, the two principal oxides of carbon, many times throughout the text. In Table 21.3, we summarize some sources and uses of these gases. In Chapter 25, we will consider some environmental issues concerning CO(g) and CO_2(g).

TABLE 21.3 Sources and Uses of Oxides of Carbon

Sources	Uses
Carbon Monoxide, CO	***Carbon Monoxide, CO***
Incomplete combustion of hydrocarbons: $2\,CH_4(g) + 3\,O_2(g) \longrightarrow 2\,CO(g) + 4\,H_2O(l)$	Manufacture of methanol and other organic compounds from synthesis gas (CO/H_2 mixture)
Steam reforming of natural gas: $CH_4(g) + H_2O(g) \longrightarrow CO(g) + 3\,H_2(g)$	Metallurgical reducing agent, as in the blast furnace reaction: $Fe_2O_3(s) + 3\,CO(g) \longrightarrow 2\,Fe(l) + 3\,CO_2(g)$
Carbon Dioxide, CO_2	***Carbon Dioxide, CO_2***
Complete combustion of hydrocarbons: $CH_4(g) + 2\,O_2(g) \longrightarrow CO_2(g) + 2\,H_2O(l)$	Refrigerant in freezing, storage, and transport of foods
Decomposition (calcination) of limestone at about 900 °C: $CaCO_3(s) \longrightarrow CaO(s) + CO_2(g)$	Production of carbonated beverages
	Petroleum recovery in oil fields
Fermentation by-product in the production of ethanol	Fire extinguisher systems

Carbon combines with most metals to form compounds called carbides. With active metals, the carbides are ionic. For example, calcium carbide is formed in the high-temperature reaction of lime and coke.

$$CaO(s) + 3\,C(s) \xrightarrow{2000\ °C} CaC_2(s) + CO(g)$$

Calcium carbide, an ionic compound with a carbon-to-carbon triple bond in the C_2^{2-} anion, reacts with water to produce acetylene, $H{-}C{\equiv}C{-}H$.

$$CaC_2(s) + 2\,H_2O(l) \longrightarrow Ca(OH)_2(s) + \underset{\text{Acetylene}}{C_2H_2(g)}$$

Calcium carbide is convenient to use. It can be transported as a solid, and the gaseous fuel acetylene can be generated simply by adding water to the solid.

Two other binary compounds of carbon are carbon disulfide, CS_2, and carbon tetrachloride, CCl_4. Carbon disulfide is prepared by the reaction of methane with sulfur vapor in the presence of a catalyst.

$$CH_4(g) + 4\,S(g) \longrightarrow CS_2(l) + 2\,H_2S(g)$$

CS_2 is a flammable, volatile liquid that dissolves sulfur, phosphorus, bromine, iodine, fats, and oils. Its toxicity limits its uses as a solvent, however. Carbon disulfide is an important intermediate in the manufacture of rayon and cellophane (Section 24.9).

Carbon tetrachloride can be prepared by the direct chlorination of methane.

$$CH_4(g) + 4\,Cl_2(g) \longrightarrow CCl_4(l) + 4\,HCl(g)$$

Once extensively used as a solvent, dry-cleaning agent, and fire extinguisher, CCl_4 is declining in importance because it causes liver and kidney damage, and it is a suspected carcinogen.

Cyanide ion, CN^-, is similar to halide ions, X^-, in several ways: It forms an insoluble silver salt, AgCN, and an acid, HCN, though this acid (hydrocyanic acid) is quite weak. Cyanide ions differ from halide ions in that they are quite toxic. Despite its toxicity, HCN, a liquid that boils at about room temperature, is widely used in the manufacture of plastics. It is also used—carefully, by well-trained personnel—as a fumigant to kill rodents and insects on ships.

Just as two Cl atoms combine to form the molecule Cl_2, two CN groups can combine to form the *cyanogen* molecule, $(CN)_2$ or C_2N_2. In basic solution, cyanogen disproportionates just as does chlorine.

$$(CN)_2(g) + 2\,OH^-(aq) \longrightarrow CN^-(aq) + OCN^-(aq) + H_2O(l)$$
$$Cl_2(g) + 2\,OH^-(aq) \longrightarrow Cl^-(aq) + OCl^-(aq) + H_2O(l)$$

$:N{\equiv}C{-}C{\equiv}N:$

Cyanogen

Cyanogen is used as a reagent in organic synthesis, as a fumigant, and as a rocket propellant.

21.5 Silicon

We have seen important differences between the second- and third-period members of Groups 1A (Li/Na), 2A (Be/Mg), and 3A (B, Al). In Group 4A, the difference is perhaps greatest of all. The most distinctive feature of C atoms—their tendency to bond together into chain and ring structures—is much less significant in Si atoms. The $Si{-}Si$ bond (bond energy = 226 kJ/mol) and $Si{-}H$ bond (318 kJ/mol) are weaker than the corresponding $C{-}C$ (347 kJ/mol) and $C{-}H$ (414 kJ/mol) bonds, but this is probably not the primary reason why chains and rings of Si atoms lack the stability of those of C atoms. Instead, the activation energies of reactions of silicon chain and ring compounds are much lower than those of the corresponding carbon compounds; their rates of reaction are also correspondingly greater. Thus, disilane, Si_2H_6, spontaneously ignites on contact with oxygen (producing SiO_2 and H_2O), whereas the ignition of ethane, C_2H_6, requires a spark or open flame. In any

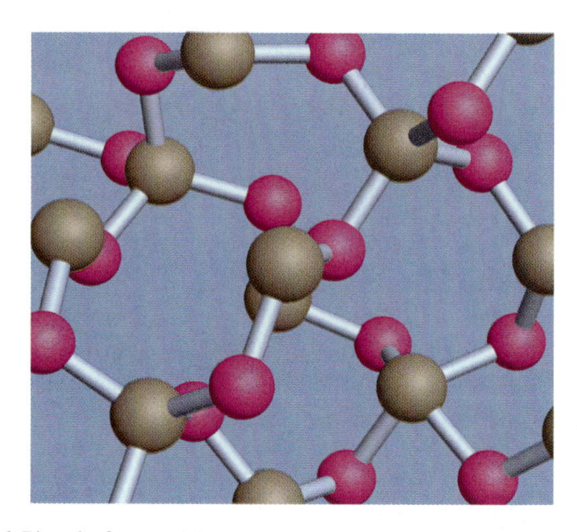

Figure 21.5
The structure of silica, SiO_2
Each Si atom forms bonds to four O atoms and each O atom forms bonds to two Si atoms in this three-dimensional network covalent structure.

event, strong Si—O bonds (464 kJ/mol) favor silicates as the predominant naturally occurring compounds of silicon. Thus, silicon is the key element of the mineral world just as carbon is the key element of the living world.

A silicon atom, like a carbon atom, forms four bonds in almost all cases. In elemental silicon, each atom uses its four valence electrons ($3s^23p^2$) in an sp^3 hybridization scheme. Silicon crystallizes in an *fcc* covalent network similar to diamond. There is no allotrope of silicon equivalent to graphite.

Silica, SiO_2

Unlike carbon, silicon forms few multiple covalent bonds. The sidewise overlap of a $3p$ orbital of a Si atom with a $2p$ orbital of an O atom, for example, is too limited to form a strong π bond. Considerably less energy is released in forming two S—O double bonds than in forming four Si—O single bonds. As a result, SiO_2 is not made up of discrete molecules (as is CO_2); it is a network covalent solid (Figure 21.5). Quartz, a form of pure silica, is quite hard (a hardness of 7, compared to diamond's hardness of 10), has a high melting point (about 1700 °C), and is a nonconductor of electricity. Silica is the basic raw material of the glass, ceramics, and refractory materials industries.

Ceramic insulators of the type commonly seen in electric power substations.

Ceramics

A **ceramic** is an inorganic solid generally produced at high temperatures and characterized by such physical properties as hardness, brittleness, stability at high temperatures, and a high melting point. Some ceramics are crystalline materials and others are amorphous solids. The range of ceramic products is broad and includes structural clay products (bricks and tiles), whiteware (dinnerware and porcelain items), abrasives, refractories (furnace linings), and glass. Silica and silicates are common constituents of ceramics, and glass is the most widely produced ceramic product.

Certain molten materials become viscous when they cool; eventually they cease to flow entirely. Such a material is called a *glass*. A glass is amorphous; that is, it does not have the long-range order found in crystalline solids. Unlike a crystalline solid, which has a definite melting point, a glass softens and becomes liquid over a range of temperatures. When a mixture of sodium carbonate, calcium carbonate, and sand (silica) is heated to about 1500 °C, the molten product is a water-insoluble mixture of sodium and calcium silicates; it is called *soda-lime* glass.

Because soda-lime glass expands and contracts with changes in temperature, it is subject to shattering when its temperature is changed rapidly When the calcium carbonate in glass is replaced by boric oxide, a borosilicate glass is formed. This glass has a low thermal expansion and is not as subject to thermal shock as is soda-lime glass. It is extensively used for laboratory glassware and ovenware, and is commonly known by its trade name—Pyrex®.

Some new ceramic materials have specially designed electrical, magnetic, or optical properties. Also new are some methods of preparing ceramic materials. In the *sol-gel* process, the particle size of solid particles can be carefully controlled by forming the particles as a colloidal dispersion (sol). The sol is then converted to a rigid gel, which is the starting point of the fabrication process. Some exceptionally lightweight ceramic materials can be produced by this method. Ceramics with desirable high-temperature mechanical and structural properties are being developed for gas turbines. Another possibility is the development of a ceramic automobile engine. In Section 24.7, we will consider some ceramic materials with unusual electrical properties—they are superconductors. The promise of ceramic materials has caused some to speak of this as the dawn of a "new stone age."

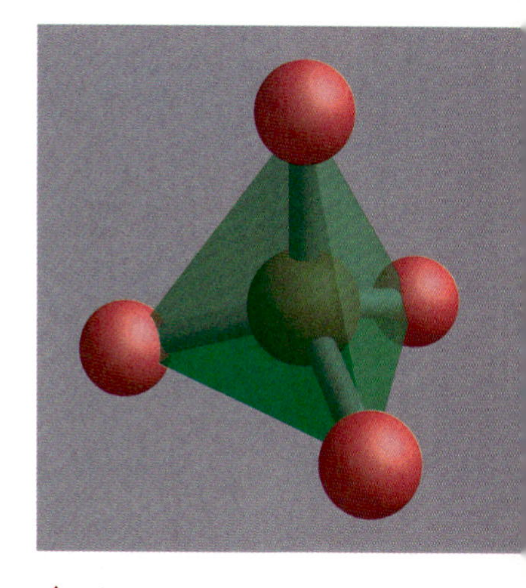

▲ **Figure 21.6**
The silicate anion, SiO_4^{4-}
The Si atom is at the center of the tetrahedron and an O atom is at each of the four corners. The shaded figure (green) shows how the tetrahedron is typically represented in a mineral structure.

Silicate Minerals

Silicon is the second most abundant element (after oxygen) in Earth's crust, accounting for 27.2% of its mass. Silicon occurs primarily as silica and silicates. Silicate anions are often quite complex, but most have as a basic structural unit a tetrahedron with a Si atom at the center and O atoms at the four corners. Silicon has the oxidation number +4 and oxygen, −2; the tetrahedral anion has a charge of 4−, SiO_4^{4-} (see Figure 21.6). In silicate minerals, SiO_4 tetrahedra may be arranged in a variety of ways, leading to a host of mineral forms. Some of the possibilities are

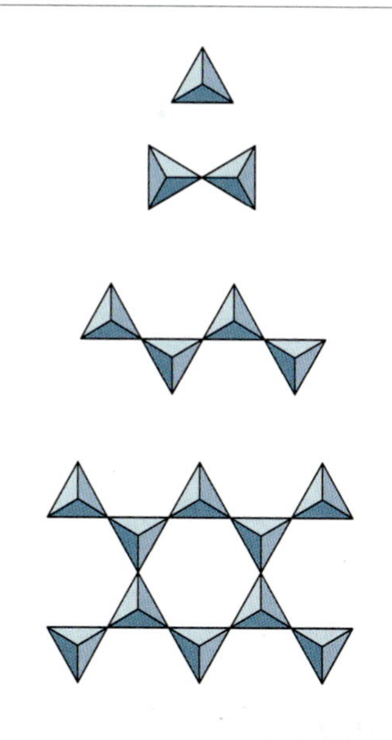

- The anions may be simple SiO_4^{4-} tetrahedra. Typical minerals and their compositions are *thorite* ($ThSiO_4$) and *zircon* ($ZrSiO_4$).
- The anions may be combinations of two SiO_4 tetrahedra. In the ion $Si_2O_7^{6-}$, the Si atoms of two tetrahedra share an O atom between them. A typical mineral is *thortveitite* ($Sc_2Si_2O_7$).
- SiO_4 tetrahedra may be joined together into long Si—O chains. Each Si atom shares two O atoms with Si atoms in adjacent tetrahedra. A typical mineral is *spodumene*, which has the empirical formula $LiAl(SiO_3)_2$; it is the principal natural source of lithium and lithium compounds.
- SiO_4 tetrahedra may be joined together into double Si—O chains. Half of the SiO_4 tetrahedra share three O atoms, while the other half share two. In *chrysotile asbestos*, the double chains are bound together by cations, chiefly Mg^{2+}. The mineral has a fibrous appearance. The empirical formula is $Mg_3(Si_2O_5)(OH)_4$.
- SiO_4 tetrahedra may share three of their four O atoms with adjacent tetrahedra in two-dimensional sheets, as suggested by Figure 21.7. In *muscovite mica*, the counterions to the silicate anions are mostly K^+ and Al^{3+}; its empirical formula is $KAl_2(AlSi_3O_{10})(OH)_2$. Because bonding within sheets is stronger than between sheets, mica flakes easily. Vermiculite, used as loose-fill insulation, has sheets of mica separated by double water layers.
- SiO_4 tetrahedra may share all four of their O atoms with adjacent tetrahedra in a three-dimensional array. This is the most common arrangement in silicate minerals in Earth's crust. It is the arrangement found in quartz itself, and in mineral forms such as
  - *amethyst:* a purple quartz containing iron as an impurity
  - *agate:* an impure silica that crystallizes in colored bands
  - *petrified wood:* very old wood in which colored agate replaces organic matter but in which the microscopic structure of the wood is retained

Petrified wood.

Muscovite mica.

▶ **Figure 21.7**
A two-dimensional sheet in the structure of mica
Three of the four O atoms in each tetrahedron are shared with another tetrahedron in a two-dimensional array of tetrahedra. The cations found between the planes of silicate anions are mostly K^+ and Al^{3+}.

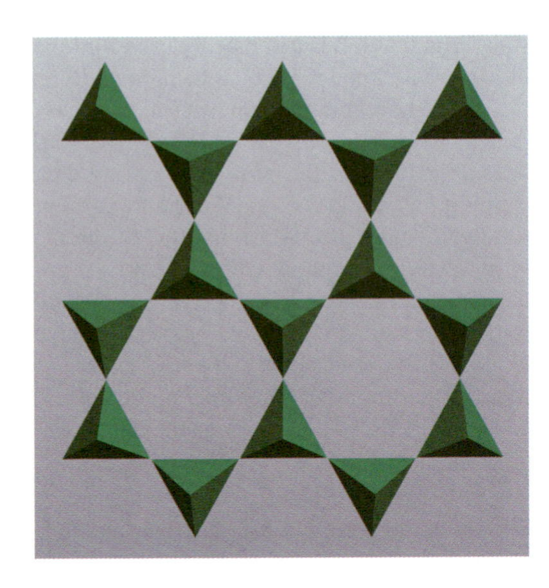

Organosilicon Compounds

Although we have stressed the inorganic chemistry of silicon, there is also a chemistry that emulates the chemistry of carbon—an organic chemistry—although it is not nearly as rich as that of carbon. Silicon can form Si — Si bonds in chains of up to about a dozen Si atoms.

```
    H           H    H                 H    H    H
    |           |    |                 |    |    |
H—Si—H     H—Si—Si—H           H—Si—Si—Si—H ...
    |           |    |                 |    |    |
    H           H    H                 H    H    H
  Silane       Disilane               Trisilane
```

The *silanes* are thermally unstable. When heated, the silanes with larger numbers of silicon atoms decompose to silanes with fewer silicon atoms or to the elements. Like the hydrocarbons, the silanes are combustible; the combustion products are $SiO_2(s)$ and H_2O. As we have noted, however, unlike hydrocarbons, the silanes burst into flame in air. In this regard, silanes are like boranes (Section 21.2), a fact that illustrates the diagonal relationship of the two elements.

Other atoms can be rather easily substituted for H atoms in silanes. For example, if we substitute Cl for H atoms in SiH_4, we obtain successively SiH_3Cl, SiH_2Cl_2, $SiHCl_3$, and $SiCl_4$. The reaction of $(CH_3)_2SiCl_2$ with water produces dimethylsilanol, $(CH_3)_2Si(OH)_2$, a starting material for the production of a class of polymers called *silicones*.

$$(CH_3)_2SiCl_2 + 2\,H_2O \longrightarrow \underset{\text{Dimethylsilanol}}{(CH_3)_2Si(OH)_2} + 2\,HCl$$

21.6 Tin and Lead

Tin and lead are quite similar to each other. They are soft and malleable and melt at low temperatures for metals. Their standard electrode potentials for the reduction $M^{2+}(aq) + 2\,e^- \longrightarrow M(s)$ are both slightly negative. Tin and lead are placed just above hydrogen in the activity series of metals (Figure 4.15).

Lead has only one solid form, but tin exists in two allotropic forms. The α (gray) or nonmetallic form is stable below 13 °C, and the β (white) or metallic form is stable above 13 °C. When held below 13 °C for a long time, white tin changes to gray tin. The tin expands and crumbles to a powder. This transformation, called "tin disease," has led to the disintegration of organ pipes, buttons, medals, and other objects made of tin that are kept below 13 °C.

Nearly half of all the tin metal produced is used in tin plate, mostly in plating steel for use in cans to store foods. The next most important use (about 25%) is in the manufacture of solders—alloys with low melting points that are used to join wires or pieces of metal. Other important alloys of tin are bronze (90% Cu, 10% Sn) and pewter (85% Sn, 7% Cu, 6% Bi, and 2% Sb).

Over half of the lead produced is used in lead-acid (storage) batteries. Other uses include the manufacture of solder and other alloys, lead shot, and radiation shields (to protect against X rays and γ rays).

A significant difference between tin and lead is that the +4 oxidation state is much more stable in tin than in lead. For example, when tin is heated in air, the product is $SnO_2(s)$; whereas the air oxidation of lead produces $PbO(s)$. Metallic tin reacts with $Cl_2(g)$ to form $SnCl_4$ (bp, 115 °C), a reaction that is used to recover tin from scrap tin plate. Lead reacts with $Cl_2(g)$ to form $PbCl_2$ (mp, 501 °C). Tin reacts with sulfur to form SnS, but the reaction can proceed further to produce SnS_2. Lead forms only the sulfide PbS. A few uses of tin and lead compounds are listed in Table 21.4.

Tin and lead, together with bismuth and cadmium, are common constituents of low-melting-point alloys. The alloy pictured here has a melting point below the boiling point of water.

TABLE 21.4 Common Compounds of Tin and Lead

Compound	Use(s)
$SnCl_2$	Reducing agent in the laboratory [reduces Fe(III) to Fe(II), Hg(II) to Hg(I), Cu(II) to Cu(I)]; tin galvanizing; catalyst
SnF_2	Cavity preventive in toothpastes (stannous fluoride)
SnO_2	Jewelry abrasive, ceramic glazes, preparation of perfumes and cosmetics, catalyst
SnS_2	Pigment, imitation gilding (tin bronze)
$Pb(NO_3)_2$	Preparation of other lead compounds
PbO	Used in glass, ceramic glazes, and cements
PbO_2	Oxidizing agent, lead-acid battery electrodes, matches and explosives
$PbCrO_4$	A yellow pigment (chrome yellow) for industrial paints, plastics, and ceramics

Group 5A

All the members of Group 5A have the valence-shell electron configuration ns^2np^3, but their atoms can alter this electron configuration in different ways when forming compounds. In a few cases they can gain three electrons to produce a 3− ion with the noble-gas electron configuration ns^2np^6. Far more often, though, they acquire this configuration by sharing three electrons. This is especially likely for the smaller atoms N and P. The larger atoms—As, Sb, and Bi—can give up the p^3 set of electrons. This leads to an electron configuration with a next-to-outermost shell of 18 and an outer shell of 2. This "18 + 2" configuration is found in several ionic species derived from metals following a transition series. In still other cases, all five valence electrons are involved in compound formation, leading to an oxidation number of +5.

Lead Poisoning

The Latin name of lead, *plumbum*, is reflected in its longtime use in plumbing. Pipes made from this easily forged metal have been used to carry water from the time of the Romans until present times. Some older homes still have lead plumbing systems, and many more have copper pipes fitted with lead solder. However, it was not until well into the twentieth century that drinking water was identified as a potential source of lead poisoning. Now, the U.S. Environmental Protection Agency requires municipal water utilities to test for the lead content in water. The maximum permitted level is 50 mg/L (50 ppb). In addition to contamination in drinking water, we are also exposed to lead through cooking and eating utensils, leaded crystal glass, and pottery glazes. In the past, the chief sources of lead contamination in the environment have been tetraethyllead, an antiknock additive for gasoline, and lead-based paints.

Mild forms of lead poisoning produce nervousness and mental depression. More severe cases can lead to permanent nerve, brain, and kidney damage, especially in children. Lead also interferes with the biochemical reactions that produce the iron-containing heme group in hemoglobin. Fortunately, average blood lead levels have dropped dramatically since leaded gasoline was phased out. However, lead contamination still persists in certain soils and in painted surfaces in some older homes.

The ancient Romans drank wine from lead-containing metal goblets. Many of them seemed to suffer from lead poisoning. Some historians speculate that lead poisoning may have contributed to the decline and fall of the Roman Empire.

Among the Group 5A elements, we note the usual decrease of first ionization energy with increasing atomic number: 1402 kJ/mol for N and 1012, 947, 834, and 703 kJ/mol for P, As, Sb, and Bi, respectively. Electronegativities decrease from 3.0 for N to 2.1, 2.0, 1.9, and 1.9 for P, As, Sb, and Bi, respectively. These data suggest virtually no metallic character for nitrogen and a progressively increasing metallic character for the heavier members of Group 5A. None of the elements in Group 5A is highly metallic, however. That is, none has a metallic character comparable to the elements of Groups 1A and 2A.

Liquid nitrogen can be used for the cryogenic preservation of the genetic material of endangered species. This technique aids the New York Zoological Society in efforts to breed the endangered snow leopard.

21.7 Nitrogen

Nitrogen is found in greater abundance in the atmosphere than anywhere else. The mass percent of nitrogen in Earth's atmosphere is 76.8%, but that in Earth's solid crust is only 0.002%. There are only two important mineral sources of nitrogen: KNO_3 (niter or saltpeter) and $NaNO_3$ (soda niter or Chile saltpeter). Because both of these compounds are water soluble, they are found on Earth's surface mainly in a few desert regions. Nitrogen compounds occur in all living matter, and some organic nitrogen compounds can be extracted from plant or animal sources or from the fossilized remains of ancient plant life, such as coal.

Nitrogen is used as an inert blanketing atmosphere in industrial operations, such as in metallurgy and the manufacture of electronics components. Liquid nitrogen is used in low-temperature applications, such as the fast freezing of foods. The only important commercial method of producing nitrogen is the fractional distillation of liquid air (see Figure 21.8).

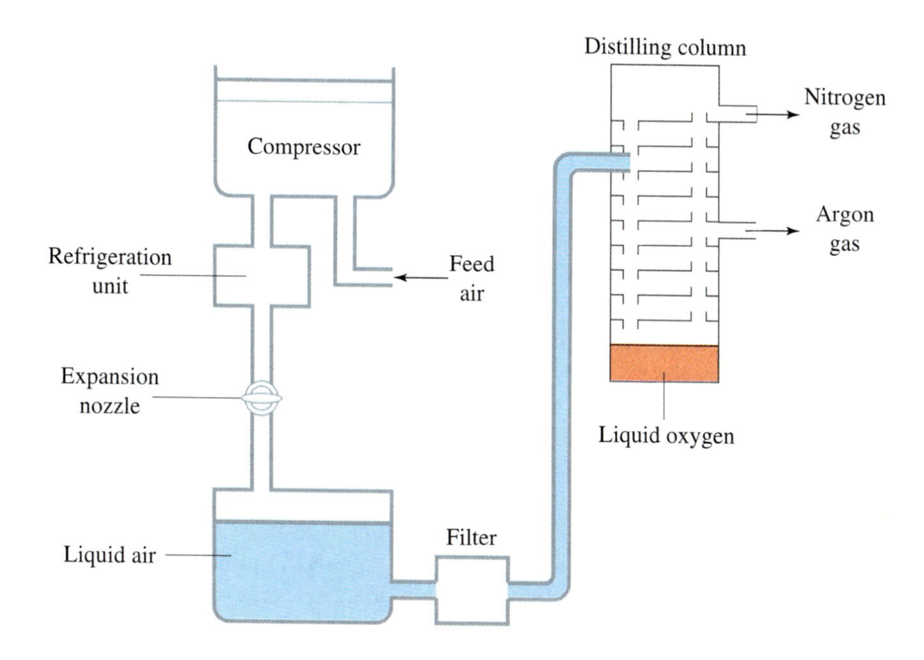

Figure 21.8 The fractional distillation of liquid air
Clean air is compressed and then cooled by refrigeration. Upon expanding, the air cools. Following a series of compressions and expansions, the air cools to a temperature at which it liquefies. The liquid air is filtered to remove $CO_2(s)$ and then distilled. Nitrogen is the most volatile component, with a normal boiling point of 77.4 K; it comes off as a gas. Argon, which boils at 87.5 K, is removed from the middle of the column, and liquid oxygen, the least volatile component with a normal boiling point of 90.2 K, collects at the bottom of the column.

Bonding in the N_2 Molecule

When two nitrogen atoms combine to form a molecule of N_2, they acquire valence-shell octets by forming a *triple* covalent bond between them. The nitrogen-to-nitrogen triple bond is one of the strongest chemical bonds known. The rupture of one mole of these bonds requires the absorption of 945.4 kJ of energy.

$$N{\equiv}N(g) \longrightarrow 2\,N(g) \qquad \Delta H = +945.4\text{ kJ}$$

Chemical reactions in which strong bonds are replaced by weaker bonds are endothermic. This is because the energy required to break bonds is not offset by the energy released in forming new bonds. As a result, many nitrogen compounds have positive enthalpies of formation, for example, NO(g).

$$\tfrac{1}{2}\,N_2(g) + \tfrac{1}{2}\,O_2(g) \longrightarrow NO(g) \qquad \Delta H_f^\circ = +90.25\text{ kJ}$$

Generally, highly endothermic reactions do not occur to any appreciable extent, and this is why $N_2(g)$ and $O_2(g)$ can coexist so peacefully in the atmosphere. If the formation of NO(g) from $N_2(g)$ and $O_2(g)$ occurred to any appreciable extent at room temperature, which we would expect if the formation reaction were exothermic, the amount of this highly noxious gas in the atmosphere would be significant. At the same time, the amount of life-sustaining $O_2(g)$ would be reduced. There would be no life as we know it on Earth.

The Synthesis of Ammonia

In Section 14.4, we learned that the basic problem in ammonia synthesis is that under most conditions, the following reaction does not go to completion. It is a reversible reaction.

$$N_2(g) + 3\,H_2(g) \rightleftharpoons 2\,NH_3(g)$$

The N_2 for the Haber-Bosch process is obtained from air. In some cases, pure N_2 obtained by the fractional distillation of air is used; and in others, ordinary air (78.1

Anhydrous liquid ammonia being applied directly into the soil.

mol% N_2) is used directly. Typically, the source of H_2 is the water-gas reaction or the steam reforming of hydrocarbons (page 850). The reaction conditions used to obtain a good yield of NH_3 and an outline of the synthesis reaction were presented on page 616.

Fertilizers: Ammonia and Its Compounds

Ammonia ranks fifth or sixth, by mass, among chemicals produced in the United States. Worldwide, annual production is about 140 million tons. About 85% of this production goes into the manufacture of fertilizers. Pure liquid ammonia, in a form called *anhydrous ammonia*, is the most used nitrogen-based fertilizer, because of its exceptionally high (82%) mass percent of nitrogen and ease of application.

Urea, $CO(NH_2)_2$, is another nitrogen-based fertilizer, often made at the site of the ammonia synthesis plant. The required reactants are ammonia and carbon dioxide, and the products are urea and water. About 80% of the urea produced in the United States is used as a fertilizer. With 47% N by mass, urea has a greater nitrogen content than any other solid fertilizer. The urea is usually applied as a solid or in aqueous solution with ammonia and ammonium nitrate. Another agricultural use of urea is as a source of supplemental nitrogen in cattle feed.

```
        O
        ‖
H—N—C—N—H
  |     |
  H     H
```

Urea

The nitrogen-based fertilizer most familiar to the home gardener is probably ammonium sulfate. It has only 21% N by mass, but it is easily handled. Also, it is not highly soluble in water, so it remains available to the plants for a relatively long time and doesn't readily leach into the groundwater. Furthermore, it also provides sulfur, another necessary plant nutrient. Ammonium sulfate can be made by the reaction of ammonia and sulfuric acid, but mostly it is obtained as a by-product of other chemical manufacturing processes.

The ammonium phosphates are also dual-purpose fertilizers, supplying plants with both nitrogen and phosphorus. They represent one of the fastest growing segments of the fertilizer industry, and now account for about two-thirds of the total production of phosphate fertilizers in the United States.

Ammonium phosphate fertilizers are formed by neutralizing aqueous ammonia with phosphoric acid. The two principal products are ammonium dihydrogen phosphate, $NH_4H_2PO_4$, commonly called monoammonium phosphate (MAP), and ammonium hydrogen phosphate, $(NH_4)_2HPO_4$, commonly called diammonium phosphate (DAP). Together, these two products are the world's leading fertilizers.

A platinum-rhodium catalyst for the oxidation of ammonia. It is made in the form of a gauze to increase the surface area for the reaction. A mixture of $NH_3(g)$ and air is passed through the gauze very quickly (about 1 millisecond) to minimize side reactions. Nearly 100% conversion of $NH_3(g)$ to NO(g) is achieved.

Nitric Acid and Nitrates

Another industrial process in nitrogen chemistry, one that is perhaps second in importance only to the ammonia synthesis reaction, is the catalyzed oxidation of $NH_3(g)$ to NO(g).

$$4\,NH_3(g) + 5\,O_2(g) \xrightarrow{\text{Pt/Rh catalyst}} 4\,NO(g) + 6\,H_2O(g)$$

This is the first step in the commercial preparation of nitric acid. The additional steps involve the oxidation of NO(g) to $NO_2(g)$ and the reaction of $NO_2(g)$ with water.

$$2\,NO(g) + O_2(g) \longrightarrow 2\,NO_2(g)$$

$$3\,NO_2(g) + H_2O(l) \longrightarrow 2\,HNO_3(aq) + NO(g)$$

The NO(g) formed in the second reaction is recycled in the first reaction.

Some of the industrial uses of nitric acid, such as the manufacture of ammonium nitrate, are based on its acidic properties—namely its ability to neutralize bases.

$$NH_3(aq) + HNO_3(aq) \longrightarrow NH_4NO_3(aq)$$

Most of the nitric acid produced in the United States goes into the manufacture of ammonium nitrate and other fertilizers.

Nitric acid is also used as an *oxidizing agent*. The actual agent is nitrate ion in an acidic solution. When a metal reacts with nitric acid, the metal is oxidized to aqueous metallic ions, but hydrogen gas is rarely obtained as a product. The reduction product has nitrogen atoms with an oxidation number lower than +5: usually $NO(g)$ or $NO_2(g)$, but even $N_2O(g)$ or NH_4^+ in some instances. The reaction of copper with nitric acid was the subject of Example 4.11 (page 162).

$$Cu(s) + 4\,H^+(aq) + 2\,NO_3^-(aq) \longrightarrow Cu^{2+}(aq) + 2\,NO_2(g) + 2\,H_2O(l)$$

Ammonium nitrate has nitrogen with oxidation numbers −3 (in NH_4^+) and +5 (in NO_3^-). It can therefore participate in oxidation-reduction reactions all by itself; no other agent is required to react with it. Nitrate ion is an oxidizing agent, and ammonium ion is a reducing agent. When heated gently at about 200 °C, ammonium nitrate forms dinitrogen monoxide and water.

$$NH_4NO_3(l) \xrightarrow{\Delta} N_2O(g) + 2\,H_2O(g)$$

Stronger heating or mechanical shock can induce a much more vigorous reaction. This reaction is the basis of the use of ammonium nitrate as an explosive.

$$2\,NH_4NO_3(l) \xrightarrow{\Delta} 2\,N_2(g) + 4\,H_2O(g) + O_2(g)$$

APPLICATION NOTE

The explosive potential of ammonium nitrate in contact with organic materials was dramatically revealed in the ammonium nitrate-fuel oil explosion of a cargo ship in Texas City, Texas, in 1947. Nearly 600 people were killed and several thousand injured. More recently, explosive mixtures of ammonium nitrate fertilizer and fuel oil have been used in terrorist attacks.

Oxides of Nitrogen

The common oxides of nitrogen include examples of nitrogen with every oxidation number from +1 to +5. Three of these oxides—N_2O, NO, NO_2—were known before Dalton's time. In fact, Dalton cited them in establishing his law of multiple proportions (page 38).

NO—A Messenger Molecule

In 1992, the simple molecule NO, notorious as an air pollutant, was shown to be a messenger molecule that carries signals between cells in the body. All of the previously known messenger molecules were complex substances that act by fitting specific receptors in cell membranes. Nitric oxide is essential to maintaining blood pressure and establishing long-term memory. It also aids in the immune response to foreign invaders in the body, and it mediates the relaxation phase of intestinal contractions in the digestion of food.

The discovery of the physiological role of NO by Louis Ignarro, Robert F. Furchgott, and Ferid Murad was recognized in a 1998 Nobel Prize. This award has a fortuitous link back to Alfred Nobel, whose invention of dynamite provided the financial basis of the Nobel Prizes. Nitroglycerin, the explosive ingredient of dynamite, has long been used to relieve the chest pain of heart disease. In his later years, Nobel refused to take nitroglycerin for his heart disease because it causes headaches and he did not think it would relieve his chest pain. Now we know that nitroglycerin acts by releasing NO.

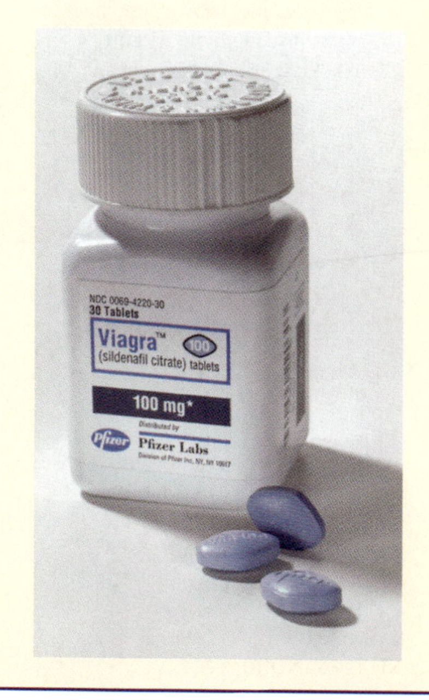

NO dilates the blood vessels that allow blood flow into the penis to cause an erection. Research on this role of NO led to the development of the anti-impotence drug Viagra®. Related research has led to drugs for treating shock and a drug for treating high blood pressure in newborn babies.

Earlier in this section, we considered methods of producing three of the oxides of nitrogen: $N_2O(g)$ by the mild decomposition of ammonium nitrate, and NO(g) and $NO_2(g)$ by the reduction of nitrate ion in acidic solution (such as in the reactions of metals with nitric acid). NO(g) is produced commercially by the catalytic oxidation of ammonia as we have already noted. The role of oxides of nitrogen in the production of photochemical smog is well established and is discussed in Chapter 25.

Other Nitrogen Compounds

A nitrogen-hydrogen compound related to ammonia is *hydrazine*, NH_2NH_2. Because of its *two* N atoms, hydrazine can ionize in *two* steps.

$$NH_2NH_2(aq) + H_2O(l) \rightleftharpoons NH_2NH_3^+(aq) + OH^-(aq) \qquad K_{b_1} = 8.5 \times 10^{-7}$$

$$NH_2NH_3^+(aq) + H_2O(l) \rightleftharpoons [NH_3NH_3]^{2+}(aq) + OH^-(aq) \qquad K_{b_2} = 8.9 \times 10^{-16}$$

Some of hydrazine's most important uses, however, are not as a base, but as a reducing agent. For example, it is used to remove dissolved $O_2(g)$ from boiler water. In the acidic medium in the redox reaction below, hydrazine exists in its cationic form, $NH_2NH_3^+$ (just as ammonia exists as NH_4^+ in acidic solutions).

$$NH_2NH_3^+(aq) + O_2(g) \longrightarrow 2\,H_2O(l) + H^+(aq) + N_2(g) \qquad E^\circ_{cell} = +1.46\text{ V}$$

The direct reaction of liquid hydrazine with oxygen gas can be used in a fuel cell and also for rocket propulsion.

$$N_2H_4(l) + O_2(g) \longrightarrow N_2(g) + 2\,H_2O(l) \qquad \Delta H^\circ = -622.2\text{ kJ}$$

The oxidation of hydrazine in acidic solution by nitrite ion produces hydrogen azide, HN_3.

$$NH_2NH_3^+(aq) + NO_2^-(aq) \longrightarrow HN_3(aq) + 2\,H_2O(l)$$

In aqueous solution, HN_3 is a weak acid ($K_a = 1.9 \times 10^{-5}$) called hydrazoic acid; its salts are called azides. Azides are extremely unstable; lead azide, for example, is used to make detonators. Sodium azide, NaN_3, also decomposes readily, releasing $N_2(g)$.

$$2\,NaN_3(s) \longrightarrow 2\,Na(l) + 3\,N_2(g)$$

Sodium azide is used in air-bag safety systems in automobiles.

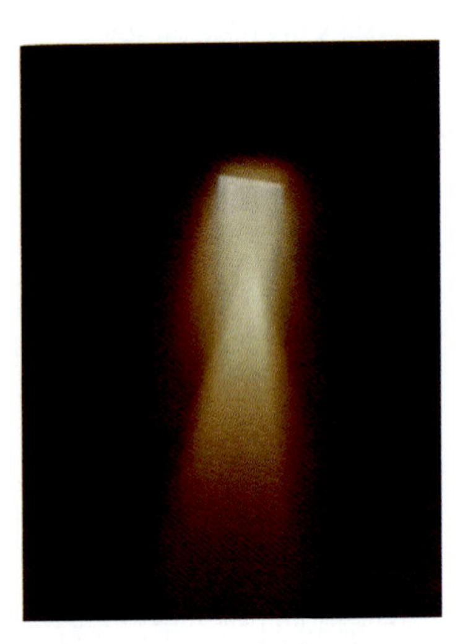

The upfiring thrusters of the space shuttle *Columbia*, shown here in an in-flight test, use methylhydrazine, CH_3NHNH_2, as a fuel.

21.8 Phosphorus

Phosphorus is the eleventh most abundant element in Earth's crust. It occurs mainly in phosphate rock, produced in nature from the remains of marine animals deposited millions of years ago. Phosphate rock is a member of the class of minerals called *apatites*. Fluorapatite, $3Ca_3(PO_4)_2 \cdot CaF_2$, is an example. Elemental phosphorus is prepared by heating phosphate rock, silica, and coke in an electric furnace.

$$2\,Ca_3(PO_4)_2(s) + 10\,C(s) + 6\,SiO_2(s) \longrightarrow 6\,CaSiO_3(l) + 10\,CO(g) + P_4(g)$$

The $P_4(g)$ is collected, condensed to a solid, and stored under water.

The solid phosphorus prepared in this way is a waxy, white solid. This *white* phosphorus can be cut with a knife. It melts at 44.1 °C, is a nonconductor of electricity, and ignites spontaneously in air (that's why it is stored under water). White phosphorus is insoluble in water but soluble in nonpolar solvents like carbon disulfide (CS_2).

As shown in Figure 21.9, white phosphorus is made up of tetrahedral P_4 molecules, with a P atom at each corner of the tetrahedron. The P—P bonds in P_4 appear to involve the overlap of $3p$ orbitals almost exclusively. Normally such overlap produces 90° bond angles, but in P_4 the P—P—P bond angles are 60°. The bonds are said to be *strained*, and species with strained bonds, like P_4, are generally quite reactive.

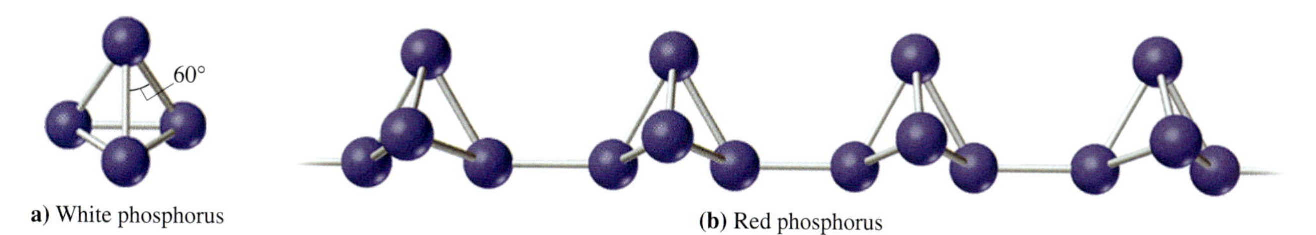

▲ Figure 21.9 The molecular structures of white and red phosphorus
(a) White phosphorus is composed of P_4 molecules. The molecules have a pyramidal shape, with 60° P—P—P bond angles. **(b)** In red phosphorus, one P—P bond breaks in each P_4 molecule, and the P_4 fragments join together in long chains.

When white phosphorus is heated to about 300 °C in the absence of air, it is transformed to *red* phosphorus. At the molecular level, it seems that one P—P bond per P_4 molecule breaks, and the fragments join together into long chains, as shown in Figure 21.9. Red phosphorus and white phosphorus are *allotropes*; they have different basic structural units. The two allotropes differ appreciably in their properties. For example, red phosphorus is less reactive than white phosphorus.

The glow of white phosphorus. Vapor from the sublimation of white phosphorus reacts slowly with atmospheric oxygen. Energy is evolved as light. The name phosphorus is derived from Greek (*phos*, light; *phoros*, bringing).

Phosphorus forms two important oxides. In the oxide with the empirical formula P_2O_3, P has the oxidation number +3. In the other, with empirical formula P_2O_5, P has the oxidation number +5. Although these oxides have been commonly called phosphorus *tri*oxide and phosphorus *pent*oxide, respectively, their true molecular formulas are double the empirical formulas, that is, tetraphosphorus hexoxide, P_4O_6, and tetraphosphorus decoxide, P_4O_{10}. The structures of these oxide molecules are shown in Figure 21.10. P_4O_6 forms in the reaction of P_4 with a limited quantity of $O_2(g)$, and P_4O_{10}, with excess $O_2(g)$. An oxide that reacts with water to produce an acid as the sole product is often called an *acid anhydride*. P_4O_6 and P_4O_{10} are the acid anhydrides of phosphorous acid and phosphoric acid, respectively.

$$P_4O_6 + 6\,H_2O \longrightarrow \underset{\text{Phosphorous acid}}{4\,H_3PO_3}$$

$$P_4O_{10} + 6\,H_2O \longrightarrow \underset{\text{Phosphoric acid}}{4\,H_3PO_4}$$

:Ö:
|
H—Ö—P—Ö—H
|
H

Phosphorous acid

:Ö:
|
H—Ö—P—Ö—H
|
:Ö—H

Phosphoric acid

Phosphoric acid is a *tri*protic acid. In contrast, phosphorous acid is a *di*protic acid. One of the H atoms in H_3PO_3 is bonded directly to the central P atom. It is not ionizable.

Although we represent aqueous phosphoric acid as $H_3PO_4(aq)$, the solution generally contains a complex mixture of species. The simplest phosphoric acid, called *orthophosphoric* acid, can react to form *pyrophosphoric* acid.

$$2\,H_3PO_4 \longrightarrow H_4P_2O_7 + H_2O$$

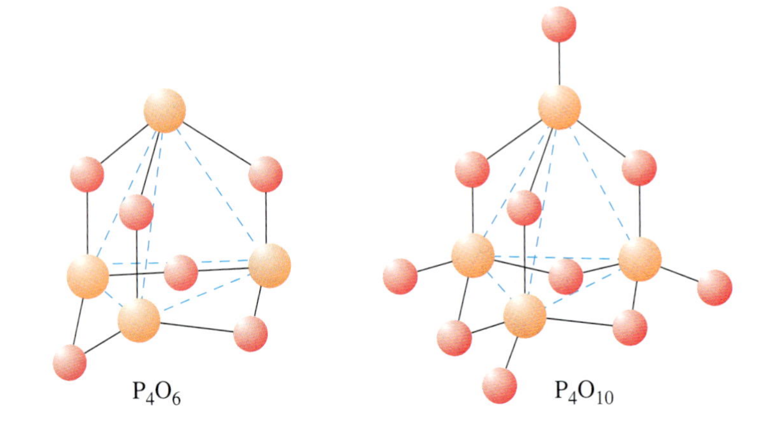

▶ **Figure 21.10**
Molecular structures of P_4O_6 and P_4O_{10}
In P_4O_6, an O atom is inserted between each pair of P atoms of the basic P_4 structure, leading to a total of six O atoms in the molecule. In P_4O_{10}, an additional O atom is bonded to each P atom, yielding a total of ten O atoms per molecule.

As shown through the structural formulas below, the essential reaction is the elimination of a molecule of H_2O from two molecules of orthophosphoric acid.

$$\underset{\text{Orthophosphoric acid}}{\begin{array}{c}\text{O}\\ |\\ \text{H—O—P—O—H}\\ |\\ \text{O—H}\end{array}} + \underset{\text{Orthophosphoric acid}}{\begin{array}{c}\text{O}\\ |\\ \text{H—O—P—O—H}\\ |\\ \text{O—H}\end{array}} \longrightarrow \underset{\text{Pyrophosphoric acid}}{\begin{array}{c}\text{O}\quad\text{O}\\ |\quad\;|\\ \text{H—O—P—O—P—O—H}\\ |\quad\;|\\ \text{O—H}\;\;\text{O—H}\end{array}} + H_2O$$

A pyrophosphoric acid molecule can react with an additional orthophosphoric acid molecule to form triphosphoric acid.

$$\begin{array}{c}\text{O}\quad\text{O}\\ |\quad\;|\\ \text{H—O—P—O—P—O—H}\\ |\quad\;|\\ \text{O—H}\;\;\text{O—H}\end{array} + \begin{array}{c}\text{O}\\ |\\ \text{H—O—P—O—H}\\ |\\ \text{O—H}\end{array} \longrightarrow \underset{\text{Triphosphoric acid}}{\begin{array}{c}\text{O}\quad\text{O}\quad\text{O}\\ |\quad\;|\quad\;|\\ \text{H—O—P—O—P—O—P—O—H}\\ |\quad\;|\quad\;|\\ \text{O—H}\;\;\text{O—H}\;\;\text{O—H}\end{array}} + H_2O$$

The reaction can continue to produce still longer-chain *polyphosphoric* acids. We will consider the environmental problem of *eutrophication* in Chapter 25, to which the salts of phosphoric acid and polyphosphoric acids are important contributors.

APPLICATION NOTE
In the food industry, phosphoric acid is used to impart tartness to soft drinks. Phosphate salts are used as dietary supplements, in baking powders, in dairy products, and as polishing agents in toothpastes.

Arsenic, Antimony, and Bismuth

Arsenic and antimony are generally classified as metalloids, and bismuth as a metal. All are used in the manufacture of alloys. When As and Sb are added to lead, the resulting alloy has more desirable properties for use in electrodes in lead-acid storage batteries than does lead alone. Because of their relatively low melting points, antimony (mp 631 °C) and bismuth (mp 271 °C) are used in low-melting-point alloys for solders and fire-protection systems. Bismuth is one of only three elements (Ga and Ge are the other two) that shares an unusual property with water: it expands on freezing from the molten state.

An important arsenic compound is the oxide As_4O_6, known as *white arsenic*. It can be made by burning arsenic in air, and it is the acid anhydride of arsenious acid, H_3AsO_3.

$$As_4O_6(s) + 6\,H_2O(l) \longrightarrow 4\,H_3AsO_3(aq)$$

Both inorganic and organic compounds derived from As_4O_6 have found extensive use as insecticides. However, perhaps the most important arsenic compound currently is gallium arsenide, GaAs. Gallium arsenide has some semiconductor and optical properties that make it an important material for supercomputers and communication satellites. (Semiconductors are discussed in Section 24.8.)

Group 6A

The elements of Group 6A have the valence-shell electron configuration ns^2np^4. Oxygen and sulfur are clearly nonmetallic, but selenium is less so; tellurium is a metalloid; and polonium is generally considered a metal.

21.9 Oxygen

Oxygen is one of the most active nonmetals and one of the most important. It forms compounds with all the elements except the light noble gases (He, Ne, and Ar). In general, oxygen forms ionic compounds with metals, and it forms covalent compounds with other nonmetals. Oxygen ranks first in abundance in Earth's crust, about 45.5% by mass.

Oxygen and its compounds are ubiquitous, and we have encountered them many times in this text; for example, in combustion reactions and in the reactions of acids and bases in aqueous solutions. Also, oxygen atoms are key components of many functional groups in organic compounds: alcohols, ethers, aldehydes, ketones, carboxylic acids, esters, and so forth. Furthermore, whenever we discuss the chemistry of another element, we discuss its oxygen compounds.

In most of its compounds oxygen has the oxidation number −2. This is true both in molecular substances such as H_2O and CO_2 and in ionic substances such as Li_2O and MgO. Among the ionic oxides of active metals, there are three possibilities illustrated by the Lewis structures below. In these structures, notice the relationship between the formal charges and the ionic charges, and notice also that the superoxide ion is paramagnetic.

(−2)	(−1) (−1)	(−1) (0)
$[:\ddot{\underset{..}{O}}:]^{2-}$	$[:\ddot{\underset{..}{O}}-\ddot{\underset{..}{O}}:]^{2-}$	$[:\ddot{\underset{..}{O}}-\ddot{\underset{..}{O}}\cdot]^{-}$
Oxide ion	*Peroxide* ion	*Superoxide* ion

Oxygen has the oxidation number −2 in oxides, −1 in peroxides, and $-\frac{1}{2}$ in superoxides. In peroxides such as hydrogen peroxide, H_2O_2, the oxidation number of the oxygen is also −1.

The chief reactions of elemental, atmospheric oxygen are oxidation processes: combustion (rapid burning), rusting and other forms of corrosion, and respiration. Oxygen reacts rapidly with active metals. Magnesium, for example, burns with a brilliant white flame when ignited in air (page 866).

$$2\ Mg(s) + O_2(g) \longrightarrow 2\ MgO(s) + \text{heat} + \text{light}$$

This same reaction occurs more slowly on a freshly prepared surface of magnesium at room temperature. Similar reactions occur with aluminum and titanium. The oxides formed in these reactions are thin, transparent coatings that are impervious to air, preventing further oxidation of the metal. Iron, especially in the presence of moisture and electrolytes, typically forms iron(III) oxide (rust). This oxide flakes off the surface of the metal, and oxidation (rusting) continues. We discussed the corrosion of iron and methods of preventing it in Section 18.8.

Preparation and Uses of Oxygen

Oxygen gas, obtained by the fractional distillation of liquid air (recall Figure 21.8), is an important commercial chemical. Typically, it ranks third in quantity produced in the United States, following sulfuric acid and nitrogen. Some important uses of oxygen are listed in Table 21.5.

TABLE 21.5 Common Uses of Oxygen

Manufacture of iron, steel, and other metals
Metal welding and cutting by torch
Manufacture of chemicals
Water treatment
Oxidizer of rocket fuels
Respiration therapy and other medical uses

Occasionally, small quantities of oxygen are prepared by the decomposition of compounds containing oxoanions, usually in the presence of a catalyst. For example, the decomposition of potassium chlorate is catalyzed by MnO_2.

$$2\ KClO_3(s) \xrightarrow{MnO_2(s)} 2\ KCl(s) + 3\ O_2(g)$$

The decomposition of certain oxides also yields oxygen gas. Joseph Priestley discovered oxygen in 1774 by heating mercury(II) oxide.

$$2\ HgO(s) \longrightarrow 2\ Hg(l) + O_2(g)$$

Lavoisier used this decomposition reaction as the source of oxygen in his early studies of combustion.

APPLICATION NOTE
The enzyme *peroxidase*, found in blood, catalyzes the decomposition of hydrogen peroxide. Thus, the formation of O_2 is accelerated when hydrogen peroxide is used as an antiseptic on an open wound.

Oxygen can also be produced by the decomposition of aqueous solutions of hydrogen peroxide, as we noted in Chapter 13.

$$2\ H_2O_2(aq) \longrightarrow 2\ H_2O(l) + O_2(g)$$

The reaction is slow but can be speeded up greatly by using a catalyst, such as $Br^-(aq)$ or $I^-(aq)$ (page 579). Another oxygen-producing reaction uses potassium superoxide.

$$4\ KO_2(s) + 2\ CO_2(g) \longrightarrow 2\ K_2CO_3(s) + 3\ O_2(g)$$

This reaction, because it consumes CO_2 while producing O_2, is used in emergency breathing apparatus and in recycling the air in submarines and spacecraft.

Oxygen is also produced, together with hydrogen, during the electrolysis of water (Section 20.1).

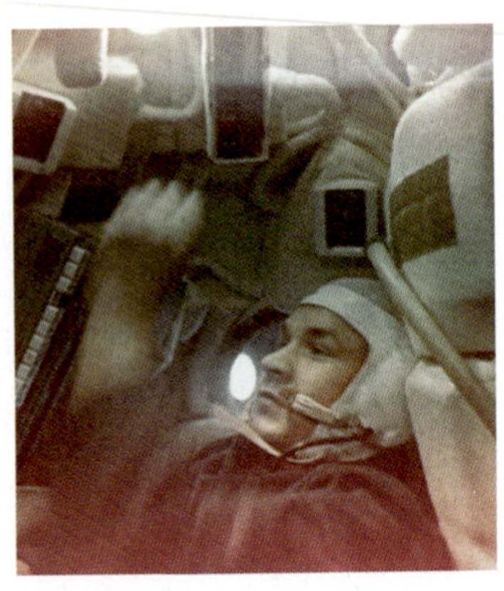

The Soviet Union used potassium superoxide to replenish the air breathed by astronaut V. Kubasov in the cabin of this 1970 space flight of Soyuz 6.

Ozone

In the past few decades *ozone*, O_3, an allotrope of oxygen, has become something of a "celebrity" chemical. This came first through recognition of its role in photochemical smog and then in the observation of seasonal depletions ("ozone holes") in stratospheric ozone levels over the polar regions. We will consider environmental issues involving ozone in Chapter 25, but ozone is also of chemical interest.

In Chapter 9, we described the ozone molecule as the hybrid of two resonance structures, and with the VSEPR method, we predict for this AX_2E molecule the trigonal planar shape that is observed by experiment (O—O—O bond angle: 116.8°).

$$:\ddot{\underset{..}{O}}{-}\ddot{O}{=}\ddot{O}: \longleftrightarrow :\ddot{O}{=}\ddot{O}{-}\ddot{\underset{..}{O}}:$$

Ozone is a powerful oxidizing agent, especially in acidic solution.

$$O_3(g) + 2\ H^+(aq) + 2\ e^- \longrightarrow O_2(g) + H_2O(l) \quad E° = +2.075\ V$$

It is used as an oxidizing agent in organic reactions and in the purification of drinking water, treatment of industrial wastes, and bleaching of paper and textiles.

The production of $O_3(g)$ directly from $O_2(g)$ is a highly endothermic reaction and occurs only rarely in the lower atmosphere, chiefly during electrical storms.

$$3\,O_2(g) \longrightarrow 2\,O_3(g) \quad \Delta H^\circ = +285\text{ kJ}$$

Ozone, which can be identified by its pungent odor, is occasionally formed around heavy-duty electrical equipment and xerographic office copiers.

To produce ozone, $O_2(g)$ must be maintained in a high-energy environment by passing either an electric discharge or ultraviolet radiation through it. Because it is unstable and decomposes back to $O_2(g)$, ozone is usually generated at the point where it is to be used.

Ozone from this portable generator is used to sanitize foods.

21.10 Sulfur

Sulfur and oxygen are similar in several ways, as we expect from the electron configurations of their atoms. Both form ionic compounds with active metals, and they form some similar covalent compounds, for example,

$$H_2O \text{ and } H_2S \quad CO_2 \text{ and } CS_2 \quad CH_3CH_2OH \text{ and } CH_3CH_2SH$$

However, oxygen and sulfur compounds also differ in important ways. Let's consider two examples.

- *Hydrogen bonding is an important intermolecular force in H_2O, but not in H_2S.* As an indication of this difference, H_2S (34 g/mol) is a gas at room temperature; it boils at −61 °C. In contrast, H_2O (18 g/mol) is a liquid at room temperature; it boils at 100 °C.
- *Sulfur can employ an expanded valence shell, but oxygen cannot.* As a consequence, sulfur can form compounds such as SF_4, CH_2SF_4, and SF_6, but there are no comparable compounds of oxygen.

Elemental sulfur exists as several molecular species, some of which are pictured in Figure 21.11. Some familiar forms of sulfur are described below, and a few of these are pictured in Figure 21.12.

- *Rhombic sulfur* (S_α) is a solid made up of S_8 molecules. At 95.5 °C S_α converts to
- *monoclinic sulfur* (S_β). Monoclinic sulfur, also made up of S_8 molecules, melts at 119 °C, yielding
- *liquid sulfur* (S_λ). This yellow, transparent, mobile liquid is made up of S_8 molecules. At 160 °C, the S_8 rings open up and join into long spiral-chain molecules, resulting in

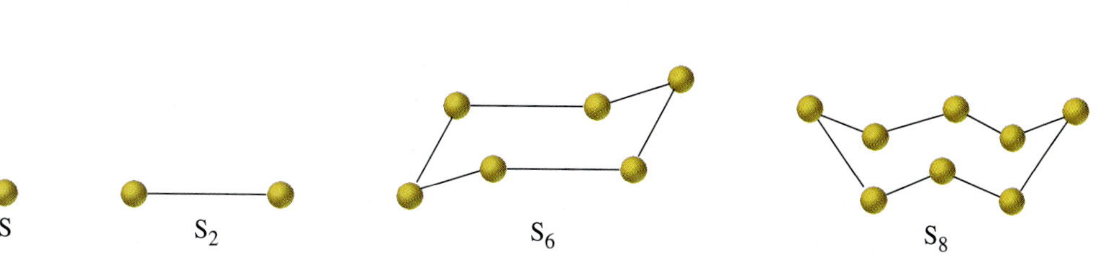

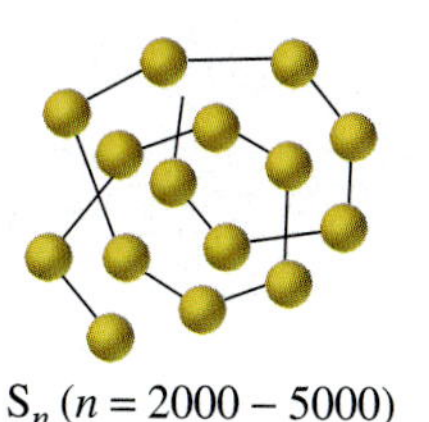

▲ **Figure 21.11 Several molecular forms of sulfur**

(a)

(b)

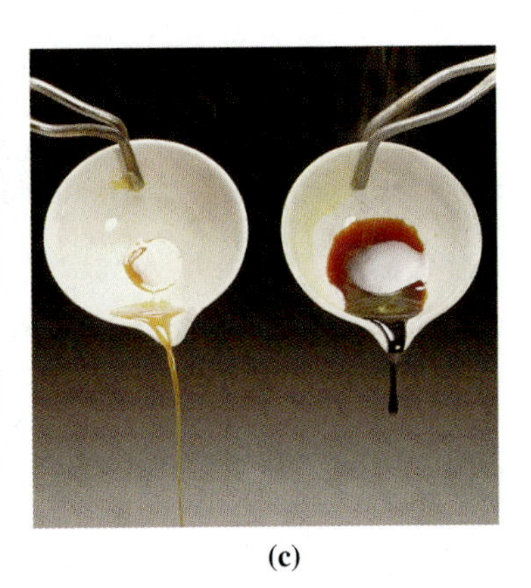

(c)

(d)

▲ **Figure 21.12 Several modifications of sulfur**
(a) Rhombic sulfur. **(b)** Monoclinic sulfur. **(c)** At its melting point of 119 °C, (left) sulfur is a straw-colored, mobile liquid. At temperatures above 160 °C (right), the liquid becomes dark colored and viscous (Sμ). **(d)** Liquid sulfur (Sμ) is poured into water to form plastic sulfur.

- *liquid sulfur* (S_μ). This is a dark, viscous liquid. If liquid S_μ is poured into cold water, the solid first formed is a rubber-like material called *plastic sulfur*. On standing, plastic sulfur slowly reverts to rhombic sulfur. At higher temperatures the chain molecules in liquid S_μ break up and the liquid flows more freely. At 445 °C liquid S_μ boils, producing
- *sulfur vapor*. The molecules in the vapor range from S_2 to S_{10}; S_8 predominates at the boiling point and S_2 at higher temperatures.

These forms of sulfur and the changes they undergo are summarized in the temperature profile below.

$$S_\alpha \xrightarrow{95.5\ °C} S_\beta \xrightarrow{119\ °C} S_\lambda \xrightarrow{160\ °C} S_\mu \xrightarrow{445\ °C} S_8(g) \longrightarrow S_6(g) \longrightarrow$$

$$S_4(g) \xrightarrow{1000\ °C} S_2(g) \xrightarrow{2000\ °C} S(g)$$

Sulfur is abundant in Earth's crust. It occurs as elemental sulfur, as mineral sulfides and sulfates, as $H_2S(g)$ in natural gas, and as organic sulfur compounds in oil and coal. Extensive deposits of elemental sulfur are found in Texas and Louisiana, some in offshore sites. This sulfur is mined in an unusual way known as the Frasch process (Figure 21.13). A mixture of superheated water and steam (at about 160 °C and 16 atm) is forced down the outermost of three concentric pipes into an underground bed of sulfur-containing rock. The sulfur melts and forms a liquid pool. Compressed air (at 20–25 atm) is pumped down the innermost pipe and forces liquid sulfur up the middle pipe. (The superheated water and steam descending in the outermost pipe help to keep the ascending liquid sulfur in the middle pipe from solidifying.) The Frasch process is no longer the most important way of producing sulfur in the United States. Because of the need to control emissions of oxides of sulfur, much of the sulfur now in use is recovered as a by-product of industrial operations.

A small amount of elemental sulfur is used directly in vulcanizing rubber and as a pesticide (for example, in dusting grapevines). Most, however, is burned to $SO_2(g)$, and the greatest proportion of this is converted by way of $SO_3(g)$ to H_2SO_4.

Sulfur dioxide reacts with water to produce a solution of sulfurous acid, $H_2SO_3(aq)$, but pure H_2SO_3 is unstable; it has never been isolated. Salts of sulfurous acid, called sulfites, are good reducing agents in reactions in which they are oxidized to sulfates, as in the following reaction.

$$Cl_2(g) + SO_3^{2-}(aq) + H_2O(l) \longrightarrow 2\,Cl^-(aq) + SO_4^{2-}(aq) + 2\,H^+(aq)$$

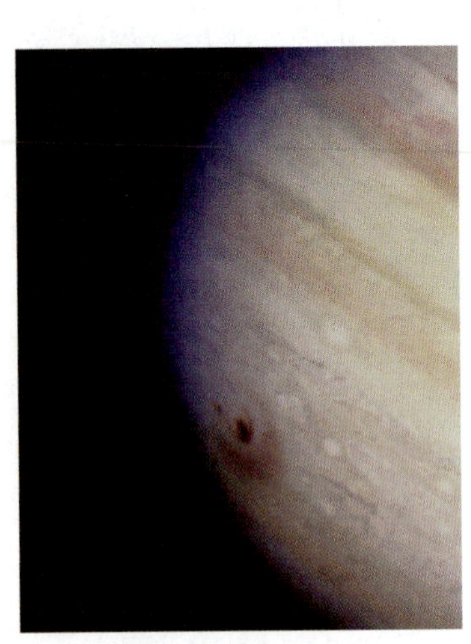

The 1994 impact of comet Shoemaker-Levy 9 with Jupiter (left center). The impact expelled gases, including $S_2(g)$, from the depths of Jupiter's atmosphere. The molecules were identified by their characteristic spectra.

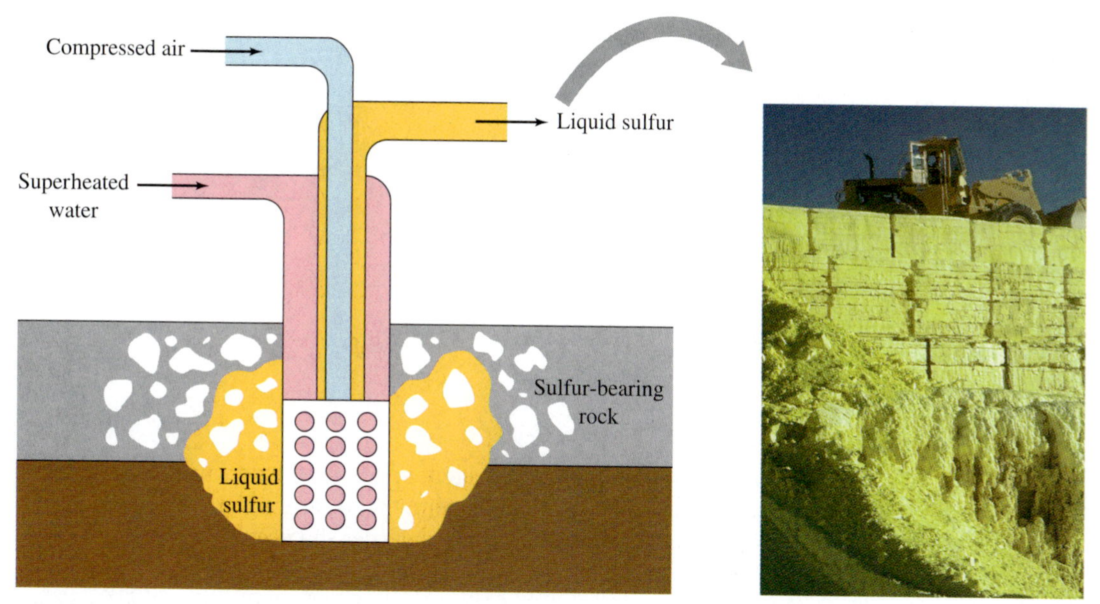

▲ **Figure 21.13 The Frasch process for mining sulfur**
Sulfur is melted by superheated water forced down the outermost of three concentric pipes. Compressed air is blown down the innermost of the pipes, and the liquid sulfur is forced up the middle pipe. The liquid sulfur is pumped to collection pools where it is allowed to freeze into large solid formations (right).

However, sulfites can also act as oxidizing agents, as in the reaction

$$2\,H_2S(g) + 2\,H^+(aq) + SO_3^{2-}(aq) \longrightarrow 3\,H_2O(l) + 3\,S(s)$$

Sulfur dioxide and sulfites are widely used in the food industry as decolorizing agents and preservatives. The main use of sulfites, however, is in the pulp and paper industry. When wood is digested in an aqueous sulfite solution, chemical reactions occur that separate the cellulosic constituents of wood from unwanted materials such as lignin. The wood pulp thus formed is used to make paper or rayon.

Concentrated sulfuric acid is added to sucrose (cane sugar) on the left. The carbon produced in the reaction is seen on the right.

Sulfuric acid, $H_2SO_4(aq)$, is a strong acid. Dilute aqueous solutions neutralize bases, react with metals to form $H_2(g)$, and react with carbonates to liberate $CO_2(g)$. Concentrated sulfuric acid, on the other hand, has some rather distinctive properties. For one, it is a fairly good oxidizing agent. It will react with copper metal, for example, producing $Cu^{2+}(aq)$ and $SO_2(g)$. But also, it is a very strong dehydrating agent, so strong that it will extract H and O atoms in the proportion H_2O from certain compounds. Concentrated sulfuric acid extracts water from a carbohydrate leaving a carbon residue.

$$\underset{\text{Sucrose}}{C_{12}H_{22}O_{11}(s)} \xrightarrow{H_2SO_4(\text{concd})} 12\,C(s) + 11\,H_2O\,(l)$$

Reactions of this type account for the the damage to clothing and the severe burns that are caused by the concentrated acid.

Salts of sulfuric acid—sulfates—have many important uses. We discussed the use of *gypsum*, $CaSO_4 \cdot 2H_2O$, in the construction industry in Section 20.10, and aluminum sulfate in water treatment earlier in this chapter (page 887). Copper(II) sulfate is used in electroplating, and it is also an effective fungicide and algicide.

Thiosulfates are related to sulfates in that they have an S atom replacing one of the O atoms in SO_4^{2-} (see Figure 21.14). Thiosulfates can be prepared by boiling elemental sulfur in an alkaline solution of sodium sulfite. The sulfur is oxidized, and the sulfite ion is reduced, both to thiosulfate ion.

$$SO_3^{2-}(aq) + S(s) \longrightarrow S_2O_3^{2-}(aq)$$

SO_4^{2-} $S_2O_3^{2-}$

S O

▶ Figure 21.14
Structures of the sulfate and thiosulfate ions
(a) Sulfate ion. **(b)** Thiosulfate ion, formed by replacing one of the terminal O atoms of the sulfate ion with an S atom.

The thiosulfate ion is a good reducing agent and is used extensively in analytical chemistry. It is also a good complexing agent, that is, as a ligand in complex ions. This is the property of thiosulfate ion that underlies the use of sodium thiosulfate as a fixer in photography (page 715).

21.11 Selenium, Tellurium, and Polonium

Selenium and tellurium are similar to sulfur but more metallic. For example, sulfur is an electrical insulator, whereas selenium and tellurium are semiconductors. Polonium is an electrical conductor and has other metallic properties, such as the ability to form cations. Because it is highly radioactive, however, polonium has few uses. Polonium-218—an α particle emitter—has been implicated in the environmental problems associated with radon gas.

Se and Te occur as selenides and tellurides mixed in sulfide ores (for example, as Cu_2Se and Cu_2Te). The free elements are obtained mainly from the anode mud in the electrolytic refining of copper (page 808). Both Se and Te find some use in alloys, and their compounds are used as additives to control the color of glass. Selenium is an important element in modern technology, primarily because it is a *photoconductor*; its electrical conductivity increases in the presence of light. Photoconductivity is the property that makes selenium useful in photocells for cameras and as the light-sensitive element in photocopying machines.

Selenium-coated light-sensitive element from a photocopier.

Group 7A

Atoms of the Group 7A elements—the halogens—have the valence-shell electron configuration ns^2np^5. They have high ionization energies, large negative electron affinities, and high to moderately high electronegativities. This means that they gain electrons readily and lose electrons only with difficulty—just as we expect for a family of nonmetals.

The halogen elements exist as diatomic molecules, X_2 (where X = F, Cl, Br, or I). Much of their reaction chemistry involves redox reactions, and standard electrode potentials are the best guide to their reactivity in aqueous solutions. From the standard electrode potentials listed in Table 21.6, we see clearly that F_2 is the most easily reduced; it is the strongest oxidizing agent of the halogens. In fact, F_2 is the strongest oxidizing agent of all the elements. Because they are so easily reduced, fluorine, chlorine, and bromine occur naturally only as the anions F^-, Cl^-, and Br^-. Iodine also occurs as iodide ion, I^-, but it is found in positive oxidation states as well (for example, as $NaIO_3$).

TABLE 21.6 Standard Electrode Potentials of the Halogens

Reduction half-reaction	$E°$, Volts
$F_2(g) + 2\,e^- \rightarrow 2\,F^-(aq)$	2.866
$Cl_2(g) + 2\,e^- \rightarrow 2\,Cl^-(aq)$	1.358
$Br_2(l) + 2\,e^- \rightarrow 2\,Br^-(aq)$	1.065
$I_2(s) + 2\,e^- \rightarrow 2\,I^-(aq)$	0.535

Astatine, at the bottom of Group 7A, is a radioactive element. All its isotopes are short-lived, and only about 0.05 mg has ever been prepared at one time. These samples rapidly undergo radioactive decay, so that few of the properties of astatine have been determined with certainty. We will not consider the element further here.

21.12 Sources and Uses of the Halogens

Fluorine

Fluorine has been known to exist since early in the nineteenth century, but Henri Moissan was the first to succeed in preparing $F_2(g)$, in 1886. Moissan's method involves the electrolysis of HF dissolved in molten potassium fluoride.

$$2\,H^+ + 2\,F^- \xrightarrow{\text{electrolysis}} H_2(g) + F_2(g)$$

Because F_2 is so easily reduced to F^-, the reverse process—oxidation of F^- to F_2—occurs only with extreme difficulty. This is yet another example of electrolysis being used to accomplish oxidations or reductions when chemical reactions are not feasible.

Fluorine is used in the production of $UF_6(g)$ for separation of the uranium-235 and uranium-238 isotopes by gaseous diffusion (page 216). Uranium enriched in the uranium-235 isotope is used in nuclear fuels and weapons.

This frosted design was etched on glass by hydrofluoric acid.

Fluorine is also used to make $SF_6(g)$, which forms when sulfur is burned in fluorine gas. $SF_6(g)$ is a nonflammable, unreactive gas of low toxicity that is used as an insulating gas in high-voltage electrical equipment. An **interhalogen** is a compound of two (or even three) different halogens. Interhalogen compounds containing fluorine, such as ClF_3 and BrF_3, are often used as substitutes for fluorine in reactions because they are easier to handle than elemental fluorine.

Like oxygen, fluorine forms compounds with all the elements except the light noble gases. Hydrogen fluoride is the starting reagent for preparing most other fluorine compounds. An unusual property of HF is its ability to etch glass.

$$SiO_2(s) + 4\,HF(aq) \longrightarrow 2\,H_2O(l) + SiF_4(g)$$

Ultimately, the glass is eroded away, and for this reason HF(aq) must be stored in special containers made of wax or Teflon®.

Chlorine

Chlorine was first produced by Carl Wilhelm Scheele in 1774, using the reaction of $MnO_2(s)$ and HCl(aq).

$$MnO_2(s) + 4\,HCl(aq) \xrightarrow{\Delta} MnCl_2(aq) + 2\,H_2O(l) + Cl_2(g)$$

This reaction is occasionally used in the laboratory, but the only significant commercial methods of preparing $Cl_2(g)$ today involve electrolysis. Some chlorine is made by the electrolysis of molten NaCl or $MgCl_2$, but most comes from the electrolysis of an aqueous solution of NaCl (page 807).

Elemental chlorine, with an annual U.S. production of about 12 million tons, generally ranks among the top ten chemicals. It has three main commercial uses: About 70% is used to produce chlorinated organic compounds; about 20% is used as a bleach in the paper and textile industries and for the treatment of municipal water, sewage, and swimming pools; and about 10% is used to produce chlorine-containing inorganic compounds.

Chlorine forms stable compounds with most of the elements except the noble gases. Especially important are chlorine-containing carbon compounds—that is, organic chlorine compounds. Chlorine reacts with hydrocarbons to produce chlorinated hydrocarbons. For example, in the direct reaction of $Cl_2(g)$ with $CH_4(g)$, chlorine atoms substitute for hydrogen atoms, and products ranging from CH_3Cl to CCl_4 are formed, with HCl as a by-product of the reactions. Similarly, the reaction of ethane, C_2H_6, with chlorine yields products with one to six Cl atoms substituted for H atoms. Several chlorinated hydrocarbons are used as solvents for nonpolar or slightly polar solutes.

The chlorofluorocarbons (CFCs) have F and Cl atoms bonded to carbon. The most widely used have been $CFCl_3$ and CF_2Cl_2. Because these substances are volatile liquids or easily condensable gases, they work well as refrigerants and as blowing agents to form the pores in foam plastics. We will discuss the role of CFCs in the destruction of stratospheric ozone in Chapter 25.

Bromine and Iodine

Bromine is extracted from subterranean brines. Brines from Arkansas contain 3800–5000 parts per million (ppm) of bromine (as Br^-), and waters in the Dead Sea, about 4500–5000 ppm. The water is treated with $Cl_2(g)$, which oxidizes Br^- to Br_2.

$$Cl_2(g) + 2\,Br^-(aq) \longrightarrow 2\,Cl^-(aq) + Br_2(l) \quad E^\circ_{cell} = +0.293\text{ V}$$

We first described halogen displacement reactions like this one in Chapter 8 (page 347).

APPLICATION NOTE
Iodide ion is used in the thyroid gland to synthesize thyroxin, a substance that helps to regulate metabolism. An iodide ion deficiency leads to retardation in children. An estimated 10 million children in China have been affected. Iodide ion deficiency causes an enlargement of the thyroid gland in adults, a condition called goiter. Both goiter and retardation can be prevented by the use of iodized salt, a product that is mainly NaCl but with a small quantity of added NaI or KI.

Iodine was once obtained in small quantities from dried seaweed. (Certain marine plants concentrate I^- selectively in the presence of Cl^- and Br^-.) Nowadays, iodine is also obtained from brines. Wells in Michigan, Oklahoma, and Japan yield brines with 100 ppm I^-. Elemental iodine is obtained from these brines by a process similar to that used for bromine.

Organic bromine compounds are used as pharmaceuticals, dyes, fumigants, and pesticides. Bromine compounds are also used as fire extinguishers and fire retardants. An important inorganic bromine compound is AgBr, the primary light-sensitive agent used in photographic film.

Iodine is of less commercial importance than chlorine and bromine, although iodine and its compounds do have applications as catalysts, in medicine, and in photographic emulsions (AgI). Iodine and its compounds also have important uses in analytical chemistry.

21.13 Hydrogen Halides

In aqueous solution, the hydrogen halides are acids. Except for HF, they are strong acids. Strong hydrogen bonding in HF accounts for the fact that HF(l) has the highest boiling point among the liquid hydrogen halides (Section 11.6), even though it has the lowest molecular mass. Hydrogen bonding may also play a role in making HF(aq) weak rather than strong as an acid.

The hydrogen halides can be prepared by the direct combination of the elements.

$$H_2(g) + X_2(g) \longrightarrow 2\,HX(g)$$

The reaction between H_2 and F_2 occurs with explosive violence. Mixtures of $H_2(g)$ and $Cl_2(g)$ are stable in the dark, but react explosively in the presence of light. $Br_2(g)$ and $I_2(g)$ react more slowly with $H_2(g)$; a catalyst is required to get a reasonable reaction rate.

Hydrogen halides can also be made by heating a halide salt (for example, CaF_2 or NaCl) with a concentrated, *nonvolatile* acid such as H_2SO_4.

$$CaF_2(s) + H_2SO_4(\text{concd aq}) \longrightarrow CaSO_4(s) + 2\,HF(g)$$
$$2\,NaCl(s) + H_2SO_4(\text{concd aq}) \longrightarrow Na_2SO_4(s) + 2\,HCl(g)$$

However, we can't use H_2SO_4 to prepare HBr(g) or HI(g) because H_2SO_4 is a sufficiently good oxidizing agent to oxidize Br^- to Br_2 and I^- to I_2. To get around this difficulty, we can use a *nonoxidizing* acid, such as phosphoric acid.

$$NaBr(s) + H_3PO_4(\text{concd aq}) \longrightarrow NaH_2PO_4(s) + HBr(g)$$

EXAMPLE 21.2—A Conceptual Example

We have discussed that HBr(g) cannot be prepared by heating a bromide salt with concentrated sulfuric acid because bromine gas is obtained instead.

a. Show that the reaction

$$2\,NaBr(s) + 2\,H_2SO_4(l) \longrightarrow Na_2SO_4(s) + 2\,H_2O(g) + SO_2(g) + Br_2(g)$$

is not spontaneous at 25 °C and standard-state conditions.

b. Why does this reaction occur at higher temperatures?

SOLUTION

a. We need to evaluate the standard free energy change for the reaction. We can do this with standard free energies of formation from Appendix C and the expression

$$\Delta G^\circ = [\sum \nu_p\, \Delta G_f^\circ(\text{products})] - [\sum \nu_r\, \Delta G_f^\circ(\text{reactants})].$$

$$2\,NaBr(s) + 2\,H_2SO_4(l) \longrightarrow Na_2SO_4(s) + 2\,H_2O(g) + SO_2(g) + Br_2(g)$$

ΔG_f°, kJ/mol −349.0 −690.1 −1270 −228.6 −300.2 3.14

$$\begin{aligned}\Delta G^\circ &= \{[1\ \text{mol} \times -1270\ \text{kJ/mol}] + [2\ \text{mol} \times -228.6\ \text{kJ/mol}]\\ &\quad + [1\ \text{mol} \times -300.2\ \text{kJ/mol}]\\ &\quad + [1\ \text{mol}\ (3.14\ \text{kJ/mol})]\} - \{[2\ \text{mol} \times -349.0\ \text{kJ/mol}]\\ &\quad + [2\ \text{mol} \times -690.1\ \text{kJ/mol}]\}\\ &= -2024\ \text{kJ} - (-2078\ \text{kJ}) = 54\ \text{kJ}\end{aligned}$$

The fact that $\Delta G^\circ = +54$ kJ tells us that the reaction will not occur spontaneously with reactants and products in their standard states at 25 °C. This value is so large and positive that we wouldn't expect the reaction to be spontaneous at 25 °C even for most nonstandard conditions.

b. The simplest approach to showing that the reaction is spontaneous at higher temperatures is to use the Gibbs equation: $\Delta G = \Delta H - T\Delta S$. We don't need to do calculations, however. Just by inspecting the balanced equation we can see that ΔS is large and *positive* for this reaction: Two moles each of a solid and a liquid yield one mole of solid and *four* moles of gas. A reaction with a large, positive ΔS is favored at higher temperatures because this ensures a large, negative value of $-T\Delta S$. At some temperature $-T\Delta S$ is certain to be large enough to overcome a positive value of ΔH. This situation, in which $\Delta H > 0$ and $\Delta S > 0$, corresponds to case 3 in Table 17.1.

EXERCISE 21.2

Use standard electrode potentials to determine whether this reaction should occur spontaneously as written.

$$Br^-(aq) + 4\,H^+(aq) + SO_4^{2-}(aq) \longrightarrow 2\,H_2O(l) + SO_2(g) + Br_2(l)$$

Is your result consistent with the results obtained in parts (a) and (b) of Example 21.2? Explain.

21.14 Oxoacids and Oxoanions of the Halogens

In its compounds, fluorine always has the oxidation number −1. The other halogens, however, can have positive oxidation numbers: +1, +3, +5, and +7. These oxidation numbers are found in the oxoacids of Cl, Br, and I listed in Table 21.7. Chlorine forms a complete set of oxoacids, but bromine and iodine do not.

TABLE 21.7 Oxoacids of the Halogens[a]

Oxidation Number of Halogen	Chlorine	Bromine	Iodine
+1	HOCl	HOBr	HOI
+3	HOClO	—	—
+5	$HOClO_2$	$HOBrO_2$	$HOIO_2$
+7	$HOClO_3$	$HOBrO_3$	$HOIO_3$[b]

[a] In all these acids, H atoms are bonded to an O atom, not to the central halogen atom. Nevertheless, the formulas of oxoacids are often written in the form: HClO (for HOCl), $HClO_2$ (for HOClO), and so on.
[b] $(HO)_5IO$ is also a component of periodic acid solutions. It is formed in the addition and rearrangement of two H_2O molecules and one of $HOIO_3$.

Hypochlorous acid, HOCl, is formed to some extent by the disproportionation of Cl_2 in water.

$$Cl_2(g) + H_2O(l) \rightleftharpoons HOCl(aq) + H^+(aq) + Cl^-(aq) \qquad E^\circ_{cell} = -0.27\text{ V}$$

APPLICATION NOTE
Alkaline solutions of sodium hypochlorite can be used as a ready source of $Cl_2(g)$. All that is required is to neutralize the excess alkali; the two reactions described here are then reversed.

HOCl(aq) is an effective germicide, often used to disinfect water. HOCl exists only in solution; it cannot be isolated in the pure state.

If we dissolve Cl_2 in a basic solution, such as NaOH(aq), equilibrium in the above reaction is displaced far to the right as HOCl is neutralized. An aqueous solution of a hypochlorite salt is formed.

$$Cl_2(g) + 2\,OH^-(aq) \longrightarrow OCl^-(aq) + Cl^-(aq) + H_2O(l) \qquad E^\circ_{cell} = +0.84\text{ V}$$

Common household bleaches are aqueous alkaline solutions of NaOCl, as are some drain cleaners and a solution commonly used to chlorinate home swimming pools.

Chlorine dioxide, ClO_2, is a bleach for paper, fibers, and flour. When reduced by peroxide ion in aqueous solution, $ClO_2(g)$ is converted to chlorite ion.

$$2\,ClO_2(g) + O_2^{2-}(aq) \longrightarrow 2\,ClO_2^-(aq) + O_2(g)$$

APPLICATION NOTE
Some emergency oxygen generators are activated by a reaction between powdered iron and sodium chlorate. The heat of this reaction raises the temperature and causes the rest of the sodium chlorate to decompose into sodium chloride and oxygen. The accidental activation of a cargo of oxygen generators caused the Valujet crash in Florida in 1996.

Sodium chlorite is used to bleach textiles.

When $Cl_2(g)$ is added to *hot* alkaline solutions, chlorate salts are formed.

$$3\,Cl_2(g) + 6\,OH^-(aq) \longrightarrow 5\,Cl^-(aq) + ClO_3^-(aq) + 3\,H_2O(l)$$

Chlorates are good oxidizing agents. Solid chlorates are used in matches and fireworks. When heated, chlorates decompose to produce oxygen gas, which supports the combustion of the other ingredients.

Electrolysis of chlorate salt solutions yields perchlorates; oxidation of ClO_3^- occurs at a Pt anode in the following half-reaction.

$$ClO_3^-(aq) + H_2O(l) \longrightarrow ClO_4^-(aq) + 2\,H^+(aq) + 2\,e^-$$

An important laboratory use of perchlorate salts is in solution studies; ClO_4^- has the least tendency of any anion to act as a ligand in complex ion formation. (Recall that a ligand is a group attached to the central ion of a complex ion.) In the presence of readily oxidizable substances, such as most organic compounds, perchlorates often react explosively. Mixtures of ammonium perchlorate and aluminum powder are used as solid propellants for rockets. Ammonium perchlorate is dangerous to handle: an oxidizing agent, ClO_4^-, and a reducing agent, NH_4^+, occur in the same compound.

Potassium chlorate, an oxidizing agent, and sucrose, a reducing agent, react vigorously, producing a bright flame. The violet color of the flame is caused by K^+ ions (please see Figure 8.19).

Group 8A

We conclude our survey of *p*-block elements with another brief look at the unusual elements of Group 8A, the noble gases. For many years after their discovery, these elements were thought to be inert. The Lewis theory of bonding was based in part on this apparent inertness. We still use the Lewis theory, but we no longer regard the noble gases as inert. The change in thinking came about mainly as a result of some experiments in the 1960s.

In 1962, Neil Bartlett and D. H. Lohman identified a compound of O_2 and PtF_6 in a 1:1 mole ratio. Properties of the compound indicate that it is $(O_2)^+(PtF_6)^-$. It takes 1177 kJ/mol of energy to extract an electron from O_2 to form the ion O_2^+. Noting that this is almost identical to the first ionization energy of Xe (1170 kJ/mol), and that the size of the Xe atom is roughly the same as that of the O_2 molecule, Bartlett was able to substitute Xe for O_2 and obtain a yellow crystalline solid, apparently $XePtF_6$. (We now know the formula to be $Xe(PtF_6)_n$, where n is between 1 and 2.) Before long, chemists around the world had synthesized several additional noble-gas compounds. For example, depending on the reaction conditions, XeF_2, XeF_4, and XeF_6 can be prepared by the direct reaction of Xe(g) and F_2(g). Some compounds of Kr and Rn exist, but no compounds of He, Ne, or Ar are known.

In Chapter 10, we used the VSEPR method to make qualitative predictions about the molecular geometry of two xenon compounds, XeF_2 and XeF_4. In Table 10.1 (page 413), we saw that XeF_2 has the VSEPR notation AX_2E_3 and is *linear*, whereas XeF_4 has the notation AX_4E_2 and is *square planar*. It is difficult to describe bonding in noble-gas compounds in a way that is consistent with all the experimental facts, and we won't attempt to do so. In fact, the discovery of noble-gas compounds has done much to stimulate the development of modern bonding theories.

Crystals of xenon tetrafluoride.

21.15 Occurrence of the Noble Gases

Except for helium and radon, the noble gases are found only in the atmosphere. Helium is found in some natural gas deposits, particularly those underlying the Great Plains of the United States. This helium is formed by the α-particle decay of naturally occurring radioactive isotopes. The α-particles ($^4He^{2+}$) acquire two electrons, becoming helium atoms. As we noted in Figure 19.2 (page 822), radon, a radioactive element, is formed in the radioactive decay series that originates with ^{238}U.

Most of the noble gases, except argon, have escaped from the atmosphere since Earth was formed. The fact that argon is much more abundant than the other noble gases can be explained rather easily. It is a product of the radioactive decay of potassium-40, a fairly abundant naturally occurring isotope. We can show this decay through a nuclear equation.

$$^{40}_{19}\text{K} \longrightarrow {}^{40}_{20}\text{Ar} + {}^{0}_{-1}\text{e}$$

Although helium is also constantly being formed through α-particle emissions by radioactive isotopes, it escapes from the atmosphere into outer space at a higher rate than argon because the molar mass of helium is only one-tenth that of argon.

21.16 Properties and Uses

Helium is used to fill balloons and dirigibles. Its lifting power is nearly as good as that of hydrogen, but it has the important advantage of being nonflammable. Helium is also used to provide an inert atmosphere for the welding of metals that otherwise might be attacked by oxygen in air.

"And Justice for All." A motif of randomly assigned geometric patterns illuminated by multicolored neon lights, by Lili Lakich c. 1990.

Liquid helium is used to achieve extremely low temperatures for scientific and commercial purposes. It boils at −268.9 °C (only 4.2 K), and it exists as a liquid to temperatures approaching 0 K. All other substances freeze to solids at temperatures well above 0 K; even hydrogen has a freezing point of 14 K. Large quantities of liquid helium are used in *cryogenics*, the study of materials at extremely low temperatures. Some metals become superconductors at liquid helium temperatures (that is, they essentially lose their resistance to the flow of electric current). Powerful magnets can be created by immersing the metal-wire coils of electromagnets in liquid helium. These magnets are the key components in nuclear magnetic resonance (NMR) instruments in research laboratories and magnetic resonance imaging (MRI) devices in hospitals.

Another interesting use of helium is in helium–oxygen breathing mixtures for deep-sea divers (page 530). Similar mixtures are also used in the treatment of asthma, emphysema, and other conditions involving respiratory obstruction. The same low molar mass that gives helium its lifting power also permits it to diffuse into partially obstructed areas of the lungs more rapidly than nitrogen does. The helium–oxygen mixture puts less strain on the muscles involved in breathing than air does.

Neon is used in lighted advertising signs. A tube with imbedded electrodes is shaped into letters or symbols and filled with neon at low pressure. An electric current passed through the gas causes the neon atoms to emit light of characteristic wavelengths. With a spectroscope, we would see individual spectral lines, but with the unaided eye, we merely see the familiar orange-red glow. Other colors can be obtained with mixtures of neon and argon or mercury vapor.

Argon, the most plentiful of the noble gases, can be separated from air rather inexpensively (recall Figure 21.8). Like helium, argon gas provides an inert atmosphere. In the laboratory, an argon atmosphere is an ideal medium for reactions in which reactants or products are sensitive to air oxidation or to reaction with nitrogen. In industry, argon is used to blanket materials that need to be protected from nitrogen and oxygen, such as certain types of welding, and in the preparation of ultrapure semiconductor materials like silicon and germanium. A mixture of argon and nitrogen is used to fill incandescent lightbulbs to increase their efficiency and life.

APPLICATION NOTE
A major problem in aluminum recycling is the formation of dross, a solid material that is mainly aluminum oxide formed in the melting furnace when the molten aluminum reacts with oxygen from the air. Argon provides a nonoxidizing atmosphere that reduces dross formation by 60%.

Krypton and xenon are too expensive to have many important commercial applications. Both are used in lasers and in flash lamps in photography. Radon can be collected from the radioactive decay of radium-226.

$$^{226}_{88}\text{Ra} \longrightarrow {}^{222}_{86}\text{Rn} + {}^{4}_{2}\text{He}$$

The radon can be sealed in small vials and used for radiation therapy of certain malignancies.

Summary

Because they have only three valence electrons, boron atoms tend to form electron-deficient compounds, and this leads to some unusual bonding patterns. The chemistry of boron focuses on borax, boric oxide, boric acid, borates, and boron hydrides.

Aluminum, an active metal, is protected against corrosion by a film of $Al_2O_3(s)$. Both $Al_2O_3(s)$ and Al react with acids and strong bases. The production of aluminum is based on the amphoterism of $Al_2O_3(s)$ and its electrolysis in molten cryolite. The electron-deficiency of aluminum chloride is a useful property in organic syntheses.

Carbon is the key element of organic chemistry, but the free element also has uses. Diamond is prized for hardness and thermal conductivity; graphite for electrical conductivity and refractory properties. Coke, carbon black, and activated carbon are amorphous forms of carbon.

Silicon forms some noncarbon organic compounds, but these do not occur naturally. Silicon is the key element of the mineral world, as silica, SiO_2, and as various minerals based on the silicate anion, SiO_4^{4-}.

Tin and lead are metals that are slightly more active than hydrogen. They can display both the +2 and +4 oxidation numbers, with Sn(II) being a good reducing agent and Pb(IV), a good oxidizing agent.

Some of the nitrogen compounds described in the chapter are ammonia, urea, nitric acid, ammonium salts, hydrazine, and hydrazoic acid and azides. The structure of phosphorus is based on the pyramidal molecule, P_4, both in the white and red modifications. The structures of the oxides P_4O_6 and P_4O_{10} are related to that of the P_4 molecule. The principal compounds of phosphorus are the phosphates and polyphosphates.

Oxygen forms compounds with all elements except the lighter noble gases. Most oxygen is obtained, together with nitrogen and argon, by the fractional distillation of liquid air. Oxygen forms three types of anions when combined with active metals—oxide (O^{2-}), peroxide (O_2^{2-}), and superoxide, (O_2^-). Ozone, O_3, is an allotrope of the much more common O_2. Current interest in ozone is due to certain environmental issues (discussed in Chapter 25). However, ozone is also useful as an oxidizing agent, both in the laboratory and in chemical industry. Sulfur differs from oxygen in important ways, such as in its variety of allotropic forms. Its important compounds are the oxides, oxoacids, sulfites, sulfates, and thiosulfates, and many of their reactions are oxidation-reduction reactions.

The halogens (Group 7A) are nonmetals, with fluorine being the most nonmetallic of all elements. Fluorine and chlorine are prepared by electrolysis, and bromine and iodine by displacement reactions. Two different halogens can react to form an interhalogen compound. Halogen atoms can substitute for H atoms in hydrocarbons and other organic compounds. Hydrogen halides form by the direct combination of the elements or by the reaction of a halide salt with a nonvolatile acid. An important class of halogen compounds are the oxoacids and their salts. The chemical reactions of these compounds are mostly oxidation-reduction reactions.

Interest in the noble gases centers on their physical properties and inertness. In contrast, the ability of the heavier noble gases to form some chemical compounds provides important insights into bonding theory.

Key Terms

adduct (21.2)
ceramic (21.5)
inert pair (21.1)
interhalogen compound (21.12)

Review Questions

1. What is the primary feature that characterizes a *p*-block element?
2. Which two *p*-block elements have some similar properties governed by the *diagonal relationship*?

3. Three of the p-block elements are the three most abundant elements in Earth's crust. What are they?
4. Which is the most active metal and which is the most active nonmetal among the p-block elements?
5. What is a three-center or bridge bond? Describe the bridge bonds in B_2H_6.
6. Describe the bridge bonds in Al_2Cl_6. How do they differ from those in B_2H_6?
7. What is *anodized* aluminum? What purposes are served in anodizing aluminum?
8. What is the purpose served by each of these two common ingredients of common drain cleaners: pellets of NaOH(s) and granules of Al(s)?
9. A given copper wire is a better electrical conductor than an aluminum wire of the same diameter and length. Why is it, then, that most high-voltage transmission lines in the United States are made of aluminum rather than copper?
10. In 1866, aluminum metal cost \$250 to \$750 per kilogram. In 1924, it cost about \$0.50/kg. Account for this dramatic difference in price.
11. What is *carbon black*? How is it prepared, and what are some of its uses?
12. What is *activated carbon*? How does it differ from carbon black, and what are some of its uses?
13. What is the basic composition of soda-lime glass and how is this composition modified in Pyrex® glass?
14. What is the basic structural unit of the silicate anion, and what are some of the ways in which these units are combined in silicate minerals?
15. Cite *one* similarity and *two* differences between the peroxide and superoxide ions.
16. Cite one example of allotropy and one example of polymorphism in the element sulfur.
17. Explain why H_2S is a gas at room temperature, whereas H_2O, with only about half the molar mass of H_2S, is a liquid.
18. Explain why copper does not react with HCl(aq), but does react with H_2SO_4(concd aq).
19. What is an *adduct*? Name two elements closely associated with adduct formation.
20. What is the *inert pair* effect? Name some elements in which this effect is seen.
21. What type of compound is suggested by the term interhalogen compound? Give an example.
22. What type of compounds are suggested by the terms chlorinated hydrocarbons and chlorofluorocarbons? Give two examples of each.
23. Although we have not specifically considered these compounds in the text, name them by using information from this chapter or principles discussed elsewhere in the text.

    (a) AgN_3 **(b)** KSCN **(c)** At_2O **(d)** H_2TeO_4

24. Although we have not specifically considered these compounds in the text, write plausible formulas for them by using information from this chapter or principles discussed elsewhere in the text.

    (a) selenic acid **(b)** hydrotelluric acid **(c)** lead azide **(d)** silver astatide

25. What are the formulas of the following compounds mentioned in this chapter: bauxite, boric oxide, corundum, cyanogen, hydrazine, silica?
26. Name key elements found in the following minerals mentioned in this chapter: fluorapatite, borax, corundum, quartz?

Problems

Boron

27. Elemental boron can be prepared by heating magnesium metal and boric oxide. Write an equation for this reaction.
28. Boron trifluoride reacts with lithium fluoride to form an ionic compound called lithium tetrafluoborate. Write a plausible equation for this reaction. (*Hint:* The tetrafluoborate ion is related to the borate ion but contains no O atoms.)
29. Although they have similar formulas, the compound BF_3 exists and the compound BH_3 does not. Offer a plausible explanation of this fact.
30. Although they have similar formulas, phosphoric acid, H_3PO_4, is a weak *triprotic* acid, whereas boric acid, H_3BO_3, is a weak, *monoprotic* acid. Explain this difference.
31. Use the diagram outlined at the right to write equations for the conversion of borax to BF_3.

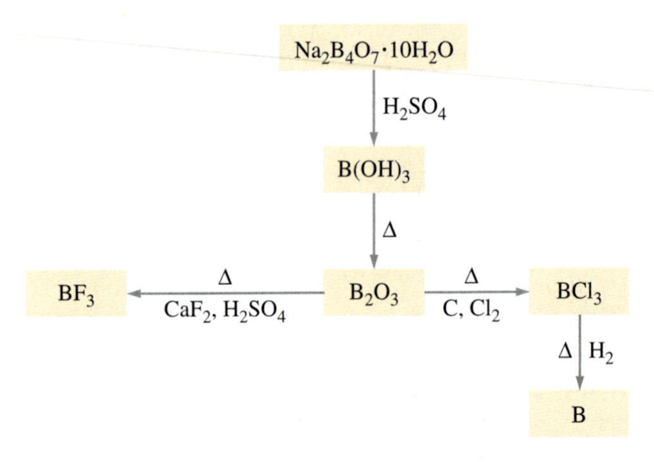

32. Use the diagram outlined above to write equations for the conversion of borax to pure boron.

Aluminum

33. Write equations for a series of reactions that could be used to make the following conversions of aluminum compounds:

Aluminum oxide $\longrightarrow$ aluminum sulfate $\longrightarrow$

aluminum hydroxide $\longrightarrow$ aluminum chloride

34. A typical baking powder contains baking soda ($NaHCO_3$) and alum [$NaAl(SO_4)_2$] as its active ingredients. During baking, the baking powder undergoes a reaction that yields $CO_2(g)$ and $Al(OH)_3(s)$ as two of its products. Write a plausible equation for this reaction.

35. What are the two key features that make possible the low-cost production of aluminum from bauxite?

36. What is the principle that underlies the fact that thallium forms both 1+ and 3+ ions, whereas aluminum forms only a 3+ ion?

37. Use data from Appendix C to estimate the enthalpy change for the thermite reaction given on page 886. Why is your value only an estimate?

38. Verify the value of E°_{cell} at 298 K for the reaction of aluminum with water in basic solution given on page 886.

39. Why cannot aluminum cookware be used in cooking strongly acidic foods? Should one scour an aluminum pot using a typical oven cleaner containing NaOH? Explain.

40. In the use of aluminum sulfate in water treatment (see page 887), the water to be treated is usually kept between pH 5 and pH 8. Why do you suppose this is the case?

Carbon

41. Sodium forms a carbide similar to calcium carbide—that is, containing the C_2^{2-} ion. Write the formula of sodium carbide and an equation for its reaction with water.

42. Unlike calcium carbide, which yields acetylene, $Al_4C_3(s)$ yields methane when it reacts with water. Write an equation for this reaction.

43. Write plausible Lewis structures for CS_2, CCl_4, and C_2N_2.

44. Write plausible Lewis structures for COS, CN^-, and OCN^-.

Silicon

45. Sketch the geometrical structure and write a Lewis structure for the anion $Si_2O_7^{6-}$.

46. Write a formula and Lewis structure, and sketch the geometric structure for an anion that consists of three SiO_4 units joined through O atoms.

47. Show that the empirical formula given on page 893 is consistent with the expected oxidation states of the elements in chrysotile asbestos.

48. Show that the empirical formula given on page 893 is consistent with the expected oxidation numbers of the elements in muskovite mica.

49. In many respects, the naming of organosilicon compounds follows the same rules as organic carbon compounds. Write formulas for the following compounds: **(a)** tetramethylsilane, **(b)** dichlorodimethylsilane, **(c)** triethylsilane.

50. What is the formula of tetrasilane? Write an equation for its combustion in air.

Tin and Lead

51. Use data from Appendix C to determine if $PbO_2(s)$ is a sufficiently strong oxidizing agent to carry out the following oxidations in acidic solution.

(a) $ClO_3^-(aq)$ to $ClO_4^-(aq)$

(b) $H_2O(l)$ to $H_2O_2(aq)$

(c) $Ag^+(aq)$ to $Ag^{2+}(aq)$

(d) $Sn^{2+}(aq)$ to $Sn^{4+}(aq)$

52. Use data from Appendix C to determine if $Sn^{2+}(aq)$ is a sufficiently strong reducing agent to carry out the following reductions.

(a) $Cu^{2+}(aq)$ to $Cu^+(aq)$

(b) $V^{3+}(aq)$ to $V^{2+}(aq)$

(c) $Ag^+(aq)$ to $Ag(s)$

(d) $Cr^{3+}(aq)$ to $Cr^{2+}(aq)$

53. Write plausible equations for the production of tin compounds described in a chemical dictionary as follows.
 (a) Stannous chloride [tin(II) chloride]: "By dissolving tin in hydrochloric acid."
 (b) Stannic chloride [tin(IV) chloride]: "Treatment of stannous chloride [tin(II) chloride] with chlorine."
 (c) Stannic oxide [tin(IV) oxide]: "Precipitated from stannic chloride [tin(IV) chloride] solution by ammonium hydroxide [$NH_3(aq)$]."

54. Write plausible equations for the production of lead compounds described in a chemical dictionary as follows.
 (a) Lead acetate: "By the action of acetic acid on litharge (PbO)."
 (b) Red lead oxide, Pb_3O_4: "By gently heating litharge (PbO) in a furnace in a current of air."
 (c) Lead dioxide: "By action of an alkaline solution of calcium hypochlorite on lead(II) hydroxide."

Nitrogen

55. *Without doing detailed calculations*, place the following in order of increasing mass percent nitrogen: **(a)** dinitrogen monoxide, **(b)** ammonia, **(c)** nitrogen monoxide, **(d)** ammonium chloride, **(e)** hydrazine, **(f)** ammonium sulfate.

56. *Without doing detailed calculations*, determine which of the following can produce the greater mass of nitrogen dioxide when 1.00 mole of the metal reacts with an excess of concentrated nitric acid: copper or silver.

57. Complete and balance the following equations based on plausible reaction products.
 (a) $NH_2NH_2(aq) + HCl(aq) \longrightarrow$
 (b) $Cu(s) + H^+(aq) + NO_3^-(aq) \longrightarrow$
 (c) $NO(g) + O_2(g) \longrightarrow$

58. Complete and balance the following equations based on plausible reaction products.
 (a) $NO_2(g) + H_2O(l) \longrightarrow$
 (b) $NH_3(g) + O_2(g) \longrightarrow$
 (c) $NH_4NO_3(l) \xrightarrow{200\ °C}$

59. In acidic solution, hydrazine reduces Fe^{3+} to Fe^{2+}. The hydrazine, which in acidic solution is present as $NH_2NH_3^+$, is oxidized to $N_2(g)$. Write half-equations and a redox equation for this reaction. If $E°_{cell}$ for this reaction is found to be 1.00 V, what is $E°$ for the reduction of $N_2(g)$ to $NH_2NH_3^+(aq)$?

60. The text mentions the combustion of liquid hydrazine as a reaction that can be used for rocket propulsion or as the reaction occurring in a fuel cell (page 900). Use data from Appendix C and Chapter 18 to determine the value of $E°_{cell}$ for this fuel cell.

Phosphorus

61. What are the two principal allotropic forms of phosphorus? Which is the more reactive? How do they differ in molecular structure?

62. What are the formulas of the oxides in which phosphorus has oxidation numbers of +3 and +5, respectively? What are their molecular structures?

63. The industrial preparation of phosphine, $PH_3(g)$, involves the reaction of white phosphorus with KOH(aq). The other reaction product is $KH_2PO_2(aq)$. Write a balanced equation for this redox reaction.

64. Phosphorus trichloride is used to produce many other phosphorus compounds. Write a balanced equation to show how phosphoryl chloride, $POCl_3$, can be made by the reaction of phosphorus trichloride, chlorine, and tetraphosphorus decoxide.

Oxygen

65. Write an equation to represent **(a)** the action of oxygen on a freshly prepared aluminum surface; **(b)** the formation of oxygen by the decomposition of potassium chlorate; **(c)** the formation of oxygen by the action of water on sodium peroxide (sodium hydroxide is the other product); **(d)** the oxidation of $Pb^{2+}(aq)$ to $PbO_2(s)$ by ozone in an acidic solution (oxygen gas is the other product).

66. Write an equation to represent **(a)** the formation of ozone by the passage of an electric discharge through oxygen; **(b)** the formation of oxygen by the thermal decomposition of mercury(II) oxide; **(c)** the formation of oxygen by the action of water on potassium superoxide (potassium hydroxide is the other product); **(d)** the oxidation of $Cl^-(aq)$ to $ClO_3^-(aq)$ by ozone in an acidic solution (oxygen gas is the other product).

Sulfur

67. Describe the physical properties and physical principles that make possible the Frasch process for mining sulfur.

68. Outline the phase changes and changes in molecular structure that occur when a sample of rhombic sulfur is heated from room temperature to about 450 °C.

69. The equation given on page 907 for the formation of thiosulfate ion from sulfite ion and sulfur in an alkaline solution is for an oxidation-reduction reaction. Write equations for the half-reactions that occur. Derive the overall equation from these half-equations.

70. When an aqueous solution of sodium thiosulfate is acidified, unstable thiosulfuric acid is formed. The acid decomposes immediately to sulfurous acid and sulfur. Write equations for the two reactions. (*Hint:* What is the formula for thiosulfuric acid?)

The Halogens

71. Which of the following would you add to an aqueous solution containing Br^- to oxidize $Br^-(aq)$ to $Br_2(l)$: I_2, I^-, Cl_2, Cl^-, F^-? Explain your choice or choices.

72. Can you think of a reagent that can be used to displace F^- (aq) and produce $F_2(g)$? Explain.

73. If Br^- and I^- occur together in an aqueous solution, I^- can be oxidized to IO_3^- with $Cl_2(aq)$. Simultaneously, Br^- is oxidized to Br_2, which is extracted with $CS_2(l)$. Write chemical equations for the reactions that occur.

74. Write equations to illustrate the oxidation of $Fe^{2+}(aq)$ to $Fe^{3+}(aq)$ by the halogen, X_2. (*Hint:* How many different equations can you write?)

75. One of the following is an *interhalogen*, one is a *halide*, one is a *halate*, and one fits none of these categories: ICl, NaCl, $(CN)_2$, $KBrO_3$. Identify each.

76. One of the following is a *halide*, one is a *halate*, and one fits neither of these categories: BrF_3, $MgBr_2$, $NaClO_3$. Identify each.

77. Write equations to represent the reaction of Cl_2 with each of the following.

(a) $H_2O(l)$
(b) cold NaOH(aq)
(c) hot NaOH(aq)

78. Write equations to represent the reaction of KI with each of the following.

(a) H_2SO_4(concd aq)
(b) H_3PO_4(concd aq)

Noble Gases

79. Use VSEPR theory to predict the shapes of these molecules.

(a) XeO_3

(b) XeO_4

80. Use VSEPR theory to predict the shapes of these molecules.

(a) XeF_4

(b) $XeOF_4$

81. Why is it that helium is formed in many natural radioactive decay processes, whereas argon is formed in only one radioactive decay?

82. Argon was discovered in 1894 through the observation that the density of nitrogen obtained from air was different from the density of nitrogen obtained from nitrogen compounds. Why did the nitrogen from these two sources appear to have different densities? Which of the two sources do you think yielded the gas with the greater density? Explain.

Additional Problems

83. Predict the likely geometric structures of these interhalogen molecules.

(a) BrF_3
(b) IF_5

84. Predict the likely geometric structures of these polyatomic ions.

(a) ICl_2^-
(b) I_3^-

85. The reaction of borax, calcium fluoride, and concentrated sulfuric acid yields sodium hydrogen sulfate, calcium sulfate, water, and boron trifluoride as products. Write a balanced equation for this reaction.

86. To prepare potassium aluminum alum, $KAl(SO_4)_2 \cdot 12H_2O$, aluminum foil is dissolved in KOH(aq). The solution thus obtained is treated with $H_2SO_4(aq)$, and the alum is crystallized from the resulting solution. Write plausible equations for these reactions.

87. In 1986, fluorine was prepared by a chemical method (that is, not involving electrolysis). The reactions used were that of hexafluoromanganate(IV) ion, MnF_6^{2-} with antimony pentafluoride to produce manganese(IV) fluoride and SbF_6^-, followed by the dissociation of manganese(IV) fluoride to manganese(II) fluoride and fluorine gas. Write equations for these two reactions.

88. To prevent air oxidation of $Sn^{2+}(aq)$ to $Sn^{4+}(aq)$, metallic tin can be kept in contact with $Sn^{2+}(aq)$. Use electrode potential data to explain the role of the metallic tin.

89. $E° = +2.32$ V for the reduction half-reaction

$$XeF_2(aq) + 2\,H^+(aq) + 2\,e^- \longrightarrow Xe(g) + 2\,HF(aq)$$

Show that XeF_2 decomposes in aqueous solution, producing $O_2(g)$. Write an equation for this reaction and calculate $E°_{cell}$. Would you expect this decomposition to be favored in acidic or in basic solution? Explain.

90. The gemstone mineral beryl consists of SiO_4^{4-} tetrahedra arranged in six-membered rings with Be^{2+} and Al^{3+} in a 3:2 ratio as counterions. Make a sketch of this structure, and deduce the empirical formula beryl.

91. In a reaction described on page 907, sulfite ion in acidic solution oxidizes $H_2S(g)$ to free sulfur. What volume of air, at 25 °C and 755 mmHg and containing 1.5 mole percent H_2S, could be purged of H_2S by 2.50 L of 1.75 M Na_2SO_3?

92. What masses of C(s) and $Al_2O_3(s)$, in kilograms, are consumed to produce 1.00×10^3 kg Al by the electrolysis reaction of page 885? How many coulombs of electric charge are required for the electrode reactions?

93. Polonium is the only element known to crystallize in the simple cubic structure. In this structure, the interatomic distance is 335 pm. Use this description of the crystal structure to estimate the density of polonium. (*Hint:* How many atoms are in a unit cell? What are the mass and volume of this unit cell?)

94. In aqueous solution, $O_3(g)$ is rapidly decomposed to $O_2(g)$. Write equations for two half-reactions that combine to give the decomposition of $O_3(g)$ as the overall redox reaction. What is the value of $E°_{cell}$ for this reaction?

95. Magnesium can reduce NO_3^- to $NH_3(g)$ in a basic solution.

$$Mg(s) + NO_3^-(aq) + H_2O(l) \longrightarrow Mg(OH)_2(s) + OH^-(aq) + NH_3(g) \quad \textit{(not balanced)}$$

The gaseous NH_3 can be neutralized with excess HCl(aq), and the unreacted HCl(aq) can then be titrated with NaOH(aq). A 25.00-mL sample of a solution with unknown molarity of NO_3^- was treated in this way, and the $NH_3(g)$was passed into 50.00 mL of 0.1500 M HCl. The excess HCl required 32.10 mL of 0.1000 M NaOH for its titration. What was $[NO_3^-]$ in the sample?

96. It has been estimated that if all the O_3 in the atmosphere were brought to sea level at STP, the gas would form a layer 0.3 cm thick. Estimate the number of O_3 molecules in Earth's atmosphere. (Assume that Earth's radius is 4000 mi.)

97. Methane and sulfur vapor react to form carbon disulfide and hydrogen sulfide. Carbon disulfide reacts with chlorine gas to form carbon tetrachloride and disulfur dichloride. Further reaction of carbon disulfide and disulfur dichloride produces additional carbon tetrachloride and solid sulfur. Write equations for these three reactions.

98. The composition of a phosphate mineral can be expressed as % P, % P_2O_5, or % BPL [bone phosphate of lime, $Ca_3(PO_4)_2$].

(a) Show that % P = 0.436 × (% P_2O_5), and % BPL = 2.185 × (% P_2O_5).

(b) What is the significance of a % BPL greater than 100%?

(c) What is the % BPL of a typical phosphate rock?

99. You are given these bond-dissociation energies in kJ/mol at 298 K: O_2, 498; N_2, 946; F_2, 159; Cl_2, 243; ClF, 251; OF (in OF_2), 187; OCl (in Cl_2O), 205; and NF (in NF_3), 280. Calculate ΔH_f° at 298 K for 1 mole of **(a)** ClF(g); **(b)** $OF_2(g)$; **(c)** $Cl_2O(g)$; **(d)** $NF_3(g)$.

100. You have available elemental sulfur, chlorine gas, metallic sodium, and water. Show how you would use these substances to produce aqueous solutions of **(a)** Na_2SO_3; **(b)** Na_2SO_4; and **(c)** $Na_2S_2O_3$.

101. Write plausible half-equations and a balanced redox equation for the reaction of XeF_4 and water to form Xe, XeO_3, and HF. Xe and XeO_3 are produced in a 2:1 mole ratio, and $O_2(g)$ is also formed.

102. Nitramide and hyponitrous acid both have the formula $H_2N_2O_2$. Hyponitrous acid is a weak diprotic acid; nitramide contains the $-NH_2$ group. Based on this information, write plausible Lewis structures for these two substances.

The *d*-Block Elements and Coordination Chemistry

CHAPTER

22

Stained-glass windows in the Spiral Church of Dallas, Texas. Several of the *d*-block metal ions impart characteristic colors to glass: Cr^{3+} (green), Mn^{3+} (purple), Fe^{2+} (blue green), Co^{2+} (violet blue), Ni^{2+} (brown). The origin of color in complex ions is one of the topics discussed in this chapter.

CONTENTS

Our primary focus in this chapter will be on the *d*-block elements. As in the preceding two chapters, we will relate the physical and chemical properties and uses of this large and varied group of elements to principles of bonding, structure, and reactivity.

Some of the *d*-block elements have many uses and are among the most important of all the elements. Others, because they are rare, have fewer uses. Yet, even some of the rare ones meet needs that only they can serve. Nine of the *d*-block elements have biological functions that are essential to human life.

One special characteristic of the *d*-block elements is their ability to form a variety of complex ions. We devote the latter half of this chapter to complex ions and compounds containing them, called coordination compounds. We will examine the nature of bonding in complex ions, the structures of complex ions, and properties related to their structures.

● **ChemCDX**
When you see this icon, there are related animations, demonstrations, and exercises on the CD accompanying this text.

The *d*-Block Elements

All the *d*-block elements are metals. Because they provide a *transition* from the highly metallic character of the *s*-block metals to the less metallic character of the *p*-block metals, the *d*-block (and *f*-block) metals are known as *transition metals*. A more precise definition, however, states that a **transition metal** is an element following Groups 1A and 2A in the periodic table and having *d*-orbital vacancies in the next-to-outermost electron shell of either its atoms or cations. In this sense, Zn, Cd, and Hg are not transition metals because they do not have the required *d*-orbital vacancies in their atoms or cations. Our treatment in this chapter, then, is of the transition metals plus Zn, Cd, and Hg.

22.1 General Properties and Their Trends

We begin our survey of the *d*-block elements with a look at some general properties. We will consider trends in these properties, and we will compare the properties of the *d*-block elements to those of the *s*-block and *p*-block elements.

Some Properties of the *d*-Block Elements of the Fourth Period

The fourth period elements with atomic numbers 21 to 30 are the first series of *d*-block elements. Table 22.1 lists some of their properties. These elements have the valence electron shell $n = 4$ and a $3d$ subshell that is partially filled, or filled in the case of Cu and Zn atoms.

One indication that these *d*-block elements are metals is seen in their electronegativity values, which range from 1.4 to 1.9. As we saw in Figure 9.8 (page 373), as a rough generalization, metalloids have electronegativities of about 2. Nonmetals have electronegativities greater than 2, and metals, less than 2. Note also the gradual increase in electronegativity in moving from left to right across the table, a general trend described in Section 9.7. Finally, from the relationship between the percent ionic character and the electronegativities of the bonded atoms, we expect bonds between metal and nonmetal atoms to have somewhat less ionic character for *d*-block metals than for *s*-block metals.

Most of the cations of the fourth-period *d*-block atoms are produced by the loss of valence-shell electrons and one or more $3d$ electrons. Only zinc atoms fail to lose $3d$ electrons. The variability of oxidation number is typical for *d*-block elements, as is shown in the fifth row in Table 22.1. For the fourth-period elements up to and including manganese (Group 7B), the highest oxidation number corresponds to the periodic table group number: +3 for Sc (Group 3B), +4 for Ti (Group 4B), and so on. With the later members of the series (beyond iron), the energy of the $3d$ subshell has fallen far enough that it effectively becomes part of the electron core (Figure 22.1). Electrons in the $3d$ subshell are less likely to participate in chemical bonding, and the range of oxidation numbers for elements beyond iron is more limited.

The negative standard electrode potentials show that the $M^{2+}(aq)$ cations are more difficult to reduce than is $H^{+}(aq)$, with $Cu^{2+}(aq)$ as the only exception.

$$M^{2+}(aq) + 2\,e^- \longrightarrow M(s) \qquad E^\circ < 0\text{ V}$$
$$2\,H^{+}(aq) + 2\,e^- \longrightarrow H_2(g) \qquad E^\circ = 0\text{ V}$$
$$Cu^{2+}(aq) + 2\,e^- \longrightarrow Cu(s) \qquad E^\circ = 0.340\text{ V}$$

Most of the *s*-block metals react with water to displace $H_2(g)$, but only scandium of the fourth-period *d*-block is active enough to do so.

TABLE 22.1 Selected Properties of the *d*-Block Elements of the Fourth Period

	Sc	Ti	V	Cr	Mn	Fe	Co	Ni	Cu	Zn
Atomic number	21	22	23	24	25	26	27	28	29	30
Electron configuration[a]	$3d^14s^2$	$3d^24s^2$	$3d^34s^2$	$3d^54s^1$	$3d^54s^2$	$3d^64s^2$	$3d^74s^2$	$3d^84s^2$	$3d^{10}4s^1$	$3d^{10}4s^2$
Electronegativity	1.4	1.5	1.6	1.7	1.6	1.8	1.9	1.9	1.9	1.7
Common cations	3+	2+, 3+	2+, 3+	2+, 3+	2+, 3+	2+, 3+	2+, 3+	2+	1+, 2+	2+
Common oxidation numbers[b]	+3	+2, +3, +4	+2, +3, +4, +5	+2, +3, +6	+2, +3, +4, +6, +7	+2, +3, +6	+2, +3	+2, +3	+1, +2	+2
Atomic radius, pm	161	145	132	125	124	124	125	125	128	133
$E°$, V[c]	−2.03	−1.63	−1.13	−0.90	−1.18	−0.440	−02.77	−0.257	+0.340	−0.763
Melting point, °C	1397	1672	1710	1900	1244	1530	1495	1455	1083	420
Density, g/cm^3	3.00	4.50	6.11	7.14	7.43	7.87	8.90	8.91	8.95	7.14
Electrical conductivity[d]	3	4	6	12	1	16	25	23	95	27
Thermal conductivity[d]	4	5	7	22	2	19	23	21	93	27

[a] Each atom has an argon inner core configuration.
[b] The most important oxidation numbers are printed in red.
[c] For the reduction $M^{2+}(aq) + 2e^- \longrightarrow M(s)$ [except for Sc, where the ion is $Sc^{3+}(aq)$].
[d] Electrical and thermal conductivities are on an arbitrary scale relative to 100 for silver, the best metallic conductor.

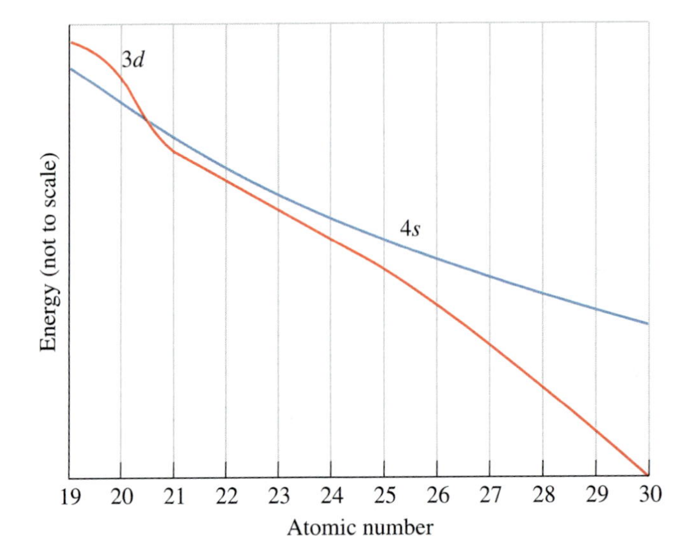

Figure 22.1
Relative orbital energies in the fourth-period transition series
The energy level of the $3d$ subshell crosses and then drops below the energy level of the $4s$ subshell with increasing atomic number in the fourth period of the periodic table.

The physical properties of the fourth-period *d*-block elements are also consistent with those expected of metals: moderate to high melting points and moderately high densities. (By comparison, sulfur, a typical solid nonmetal, has a melting point of 119 °C and a density at 25 °C of 2.07 g/cm^3.)

Perhaps the most distinctive physical property of the *d*-block metals is their ability to conduct electricity and heat. Although only copper comes close to matching silver in these abilities, the values for the other metals in Table 22.1 are still significantly high compared to nonmetals. For example, the electrical conductivity of the nonmetal white phosphorus is only $1.6 \times 10^{-17}\%$ of that for silver, and its thermal conductivity is only 0.055% of that for silver.

Other Trends Among the *d*-Block Elements

Recall an earlier generalization (page 335) that atoms become larger from the top to the bottom of a group in the periodic table. In comparing atomic radii in groups of *d*-block metals, we see exceptions to this rule. Between the fourth and fifth periods, atomic radii do increase as expected; but between the fifth and sixth periods, they remain nearly constant or even decrease slightly (Figure 22.2).

The sixth period is an especially long one. It has 32 members because the $4f$ subshell fills before the $5d$. The $4f$ subshell fills in the lanthanide series. The shapes of f orbitals are such that electrons in these orbitals are not very good at screening valence electrons from the nucleus. As a result, the increase in effective nuclear charge with increasing atomic number is greater than we would otherwise anticipate, and consequently, so is the attraction of valence electrons to the nucleus. Atom-

TABLE 22.2 Oxidation Numbers of Some Transition Metals

	Common oxidation numbers[a]		
Period	**Group 4B**	**Group 6B**	**Group 8B**
4	Ti: +3, +4	Cr: +2, +3, +6	Fe: +2, +3, +6
5	Zr: +4	Mo: +2, +3, +4, +5, +6	Ru: +2, +3, +4, +6, +8
6	Hf: +4	W: +2, +3, +4, +5, +6	Os: +2, +3, +4, +6, +8

[a] The most important oxidation numbers are printed in red.

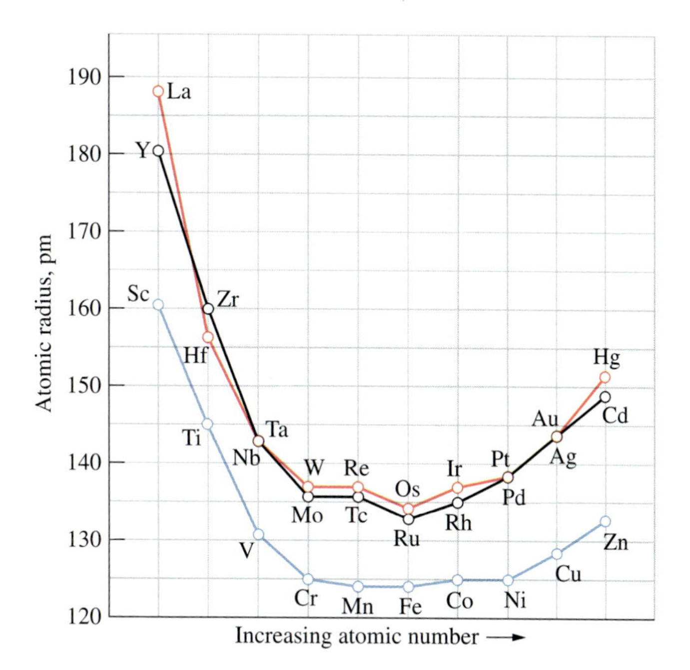

Figure 22.2 Atomic radii of the *d*-block elements The general trend for each period of the *d*-block is similar. The radii of the fourth-period atoms (blue) are smaller than those in the succeeding periods. However, because of the lanthanide contraction, the radii of the fifth-period atoms (black) and the corresponding sixth-period atoms (red) are nearly the same.

ic size does not increase between the fifth- and sixth-period members of a group of transition elements, a phenomenon known as the **lanthanide contraction**.

A comparison within groups of transition elements (Table 22.2) shows that the heavier group members display a greater range of oxidation numbers in their compounds than the lighter members. In its oxides, iron has only the oxidation numbers +2 (in FeO) and +3 (in Fe_2O_3). In contrast, ruthenium and osmium have oxidation numbers +4 (in RuO_2 and OsO_2) and +8 (in RuO_4 and OsO_4). Moreover, compounds in higher oxidation states exhibit a greater degree of covalency in their bonding. In line with this tendency, the melting points of FeO and Fe_2O_3 are 1337 and 1462 °C, respectively, whereas those of RuO_4 and OsO_4 are only 25 and 41 °C. RuO_4 and OsO_4 are molecular compounds; their melting points depend on the strengths of intermolecular forces rather than interionic attractions.

Comparing Main-Group and *d*-Block Elements

Among the main-group elements, particularly those of the first and second periods, the nature of the chemical bonding is established by the *s* and *p* electrons of the valence shell. For *p*-block elements of the third period and beyond, *d* orbitals may also be involved in forming chemical bonds, but presumably only in compounds in which the central atom has an expanded valence shell, such as SF_4, SF_6, and PF_5. With most of the *d*-block elements, in contrast, *d* orbitals are important in establishing physical and chemical properties. Differences in behavior between main-group and *d*-block elements regarding complex-ion formation, color, magnetic properties, and catalytic activity reflect the differing roles played by *s*, *p*, and *d* orbitals.

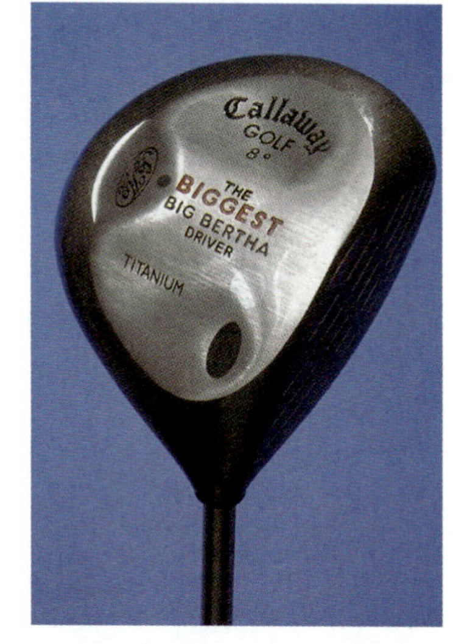

Golf clubs made of titanium metal are so popular that they have become a prime target for thieves.

22.2 The Elements Scandium Through Manganese

Now we shift our focus from a general overview to a closer look at specific elements. We begin with the first five *d*-block elements, one element at a time.

Scandium

Scandium is rather widely distributed in Earth's crust, but there is only one mineral from which it is usually extracted: *thortveitite*, $Sc_2Si_2O_7$. This scarcity of scan-

White $TiO_2(s)$, mixed with other components to produce the desired color, is the leading pigment used in paints.

dium minerals accounts in part for scandium's relatively recent history. Scandium oxide was not isolated until 1879, and the metal itself was not prepared until 1937. Scandium metal can be obtained by the electrolysis of a mixture of molten chlorides that includes $ScCl_3$.

Despite its relative rarity, the physical and chemical properties of scandium have been well characterized. Its chemistry is based mostly on the ion Sc^{3+}. In many ways, scandium resembles not the transition metals but the main-group metals, particularly aluminum. For example, scandium reacts with either acidic or basic solutions, and even with water itself, to liberate $H_2(g)$. Another similarity to aluminum is that scandium hydroxide is amphoteric. $Sc(OH)_3(s)$ reacts with acidic solutions to produce Sc^{3+} and with alkaline solutions to form $[Sc(OH)_6]^{3-}$.

Titanium

Titanium is the ninth most abundant element in Earth's solid crust and the second most abundant transition metal (after iron). Its chief mineral sources are rutile, TiO_2, and ilmenite, $FeTiO_3$.

Most titanium is produced in a two-step process. First, $TiO_2(s)$ is heated with carbon and chlorine at about 800 °C to form gaseous $TiCl_4$. Then the $TiCl_4(g)$ is reduced with magnesium in an argon atmosphere at about 1000 °C to produce pure Ti(s).

Three desirable properties make titanium a highly useful metal: (1) low density (4.5 g/cm^3), (2) high structural strength, even at high temperatures, and (3) corrosion resistance. The first two properties account for its use in aircraft, racing bicycles, and gas-turbine engines. The third property accounts for its use in the chemical industry, especially for reaction vessels to hold mineral acids, alkaline solutions, chlorine, and a host of organic reagents.

The most important compound of titanium is the oxide, TiO_2, and its most important use is as a white pigment in paints, papers, and plastics. Naturally occurring rutile is impure TiO_2. It is usually colored because of impurities, and it requires extensive processing to yield the pure white powder required as a pigment.

Vanadium

Although vanadium is reasonably abundant in Earth's solid crust, ranking nineteenth among the elements, it is mostly obtained as a by-product in the production of uranium from the mineral *carnotite*, $K_2(UO_2)_2(VO_4)_2 \cdot 3H_2O$. Vanadium metal can be prepared in high purity, but mostly it is produced as an iron-vanadium alloy, *ferrovanadium*.

Vanadium is of interest for three principal reasons: (1) as an alloying element in steel, (2) the catalytic activity of some of its compounds, principally V_2O_5, and (3) the range of oxidation numbers in its ions.

The inclusion of vanadium in steel imparts strength and toughness, making the steel ideal for use in springs and high-speed machine tools. The catalytic activity of vanadium pentoxide probably involves the reversible loss of O_2 that occurs when V_2O_5 is heated. V_2O_5 is the principal catalyst for the conversion of $SO_2(g)$ to $SO_3(g)$ in the manufacture of sulfuric acid.

Vanadium exhibits the oxidation numbers +2, +3, +4, and +5 in solid compounds and in aqueous solutions. It displays its lower oxidation numbers in the monatomic cations V^{2+} and V^{3+}. In higher oxidation numbers, the vanadium is part of a polyatomic cation (VO^{2+} and VO_2^+) or a polyatomic anion (VO_4^{3-}). The distinctive colors of vanadium species are shown in Figure 22.3.

▲ Figure 22.3
Some vanadium species in solution
The yellow solution contains VO_2^+, in which the oxidation number of vanadium is +5. The blue solution contains VO^{2+}, with an oxidation number of +4 for vanadium. The green solution contains V^{3+}, and the violet solution contains V^{2+}.

Chromium

The most important source of chromium is the mineral *chromite*, $FeCr_2O_4$ or $FeO \cdot Cr_2O_3$. An iron-chromium alloy, *ferrochrome*, is made by reducing chromite ore with carbon.

$$FeCr_2O_4 + 4\,C \xrightarrow{\Delta} \underbrace{Fe + 2\,Cr}_{\text{Ferrochrome}} + 4\,CO(g)$$

Ferrochrome and other alloying elements can be added directly to iron to produce stainless steel. When needed, pure chromium can be prepared by reducing Cr_2O_3 in the thermite reaction.

$$Cr_2O_3(s) + 2\,Al(s) \longrightarrow Al_2O_3(s) + 2\,Cr(l)$$

In addition to its use in alloys, chromium can be plated onto other metals, generally by electrolysis from a solution containing CrO_3 in $H_2SO_4(aq)$. Because this electrolysis involves reducing Cr(VI) to Cr(0), it takes 6 moles of electrons to produce 1 mole of chromium plate: Chromium plating consumes more electric energy than do other types of plating. (Silver plating, for instance, requires 1 mole of electrons per mole of Ag plated.) Chromium plating is desirable because the metal is hard and bright. Moreover, the chromium is covered by a transparent oxide coating that gives it some protection from corrosion.

Chromium plating for decorative purposes is generally applied in an extremely thin layer—about 10 nm thick. The metal is first plated with a copper or nickel layer about 100 times thicker than the chromium layer. The function of the chromium is to provide an unusually bright surface.

When chromite ore is heated in molten sodium hydroxide in the presence of air, oxygen in the air oxidizes $FeCr_2O_4$ to sodium chromate, Na_2CrO_4. Most other chromium compounds are prepared from sodium chromate.

Chromate ion in basic or neutral aqueous solutions is bright yellow (Figure 22.4, left). If the solution is made acidic, the color changes to a bright orange, because of the formation of dichromate ion, $Cr_2O_7^{2-}$ (Figure 22.4, right). Chromate and dichromate ions participate in a reversible reaction.

$$2\,CrO_4^{2-}(aq) + 2\,H^+(aq) \rightleftharpoons Cr_2O_7^{2-}(aq) + H_2O(l)$$

In acidic solutions, the forward reaction is favored. In basic solutions, H^+ is removed by reaction with OH^- to form water. This favors the reverse reaction, the conversion of $Cr_2O_7^{2-}$ to CrO_4^{2-}.

Figure 22.4
Chromate and dichromate ions
Chromium(VI) exists as CrO_4^{2-} in basic solution (left) and as $Cr_2O_7^{2-}$ in acidic solution (right).

Dichromate ion is a common oxidizing agent in organic chemistry. In an acidic solution, it oxidizes alcohols to aldehydes, ketones, and carboxylic acids (page 164). It is also used extensively in the analytical chemistry laboratory; its standard electrode potential is large and *positive*.

Chromium(IV) oxide, CrO_2, has electrical and magnetic properties that make it useful in magnetic recording tape.

$$Cr_2O_7^{2-}(aq) + 14\,H^+(aq) + 6\,e^- \longrightarrow 2\,Cr^{3+}(aq) + 7\,H_2O(l) \qquad E° = 1.33\text{ V}$$

In contrast, chromate ion in basic solution is not a good oxidizing agent; its standard electrode potential is slightly *negative*.

$$CrO_4^{2-}(aq) + 4\,H_2O(l) + 3\,e^- \longrightarrow Cr(OH)_4^-(aq) + 4\,OH^-(aq) \qquad E° = -0.13\text{ V}$$

Instead, $CrO_4^{2-}(aq)$ is a precipitating agent. For example, lead chromate and barium chromate precipitates are encountered in the qualitative analysis scheme (Section 16.7). Insoluble chromates, particularly $PbCrO_4$ and $ZnCrO_4$, find some use as paint pigments. These inorganic pigments are less affected by the action of sunlight and chemical agents than are organic dyes.

The main oxides of chromium are Cr_2O_3 and CrO_3. Both Cr_2O_3 and the corresponding hydroxide, $Cr(OH)_3$, are amphoteric. The amphoterism of $Cr(OH)_3$ is illustrated in Figure 22.5. The ability of $Cr(OH)_3$ to dissolve in an alkaline solution, and the ease with which $[Cr(OH)_4]^-$ can be oxidized to CrO_4^{2-}, play a role in the qualitative analysis scheme (page 724).

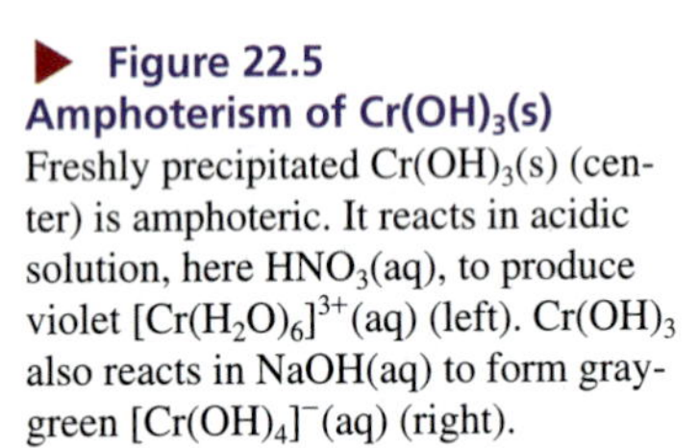

Figure 22.5 Amphoterism of $Cr(OH)_3(s)$
Freshly precipitated $Cr(OH)_3(s)$ (center) is amphoteric. It reacts in acidic solution, here $HNO_3(aq)$, to produce violet $[Cr(H_2O)_6]^{3+}(aq)$ (left). $Cr(OH)_3$ also reacts in NaOH(aq) to form gray-green $[Cr(OH)_4]^-(aq)$ (right).

The oxide CrO_3, in contrast, has only acidic properties. It dissolves in water to produce a strongly acidic solution.

$$2\,CrO_3(s) + 3\,H_2O(l) \longrightarrow 2\,H_3O^+(aq) + Cr_2O_7^{2-}(aq)$$

The name chromium is based on the Greek word *chroma*, which means color. The element is so named because its compounds exhibit many colors (recall Figures 22.4 and 22.5). The simple ions $Cr^{2+}(aq)$ and $Cr^{3+}(aq)$ form many complex ions that extend the variety of colors considerably. We will explore this matter later in the chapter.

EXAMPLE 22.1—A Conceptual Example

Write a plausible equation to explain the reaction in Figure 22.6. Pure ammonium dichromate (left) is ignited with a match, producing *pure* chromium(III) oxide (right).

SOLUTION

The oxidation number of Cr in ammonium dichromate, $(NH_4)_2Cr_2O_7(s)$, is +6. In chromium(III) oxide, the oxidation number of Cr is +3. Dichromate ion is *reduced*. If a reduction occurs, an oxidation must also occur.

▲ **Figure 22.6 Decomposition of ammonium dichromate**

In the ammonium ion, NH_4^+, the oxidation number of N is −3, the lowest possible for N. Ammonium ion can be *oxidized* to any of several higher oxidation states. Because $Cr_2O_3(s)$ is the only product found in the reaction container, any other product(s) must be gaseous. One of the gases is probably $H_2O(g)$, to account for the H and O atoms of the initial reactant. So far, we have accounted for all the Cr and O atoms in the reactants in the products.

$$(NH_4)_2Cr_2O_7(s) \longrightarrow Cr_2O_3(s) + 4\,H_2O(g) \quad \textit{(equation incomplete)}$$

Another product must contain nitrogen. Although we might postulate it to be N_2, N_2O, NO, or NO_2, we see that no O atoms from $(NH_4)_2Cr_2O_7$ are available to form oxides of nitrogen, so the additional product is probably $N_2(g)$. (If any oxides were to form, the required oxygen would have to come from the air.)

$$(NH_4)_2Cr_2O_7(s) \longrightarrow Cr_2O_3(s) + 4\,H_2O(g) + N_2(g)$$

PROBLEM-SOLVING NOTE

You can also obtain this equation by using the half-reaction method (page 768) for the oxidation of NH_4^+ to $N_2(g)$ and the reduction of $Cr_2O_7^{2-}$ to $Cr_2O_3(s)$. If you assume any other products, such as N_2O or NO, you cannot keep NH_4^+ and $Cr_2O_7^{2-}$ in the required ratio of 2:1.

EXERCISE 22.1

As noted on page 927, most chromium compounds are made from sodium chromate. The sodium chromate is produced by heating chromite ore in molten NaOH in the presence of air. Write a plausible equation for this reaction.
[*Hint*: Both Cr(III) and Fe(II) are oxidized.]

Manganese

Manganese is obtained mainly from the mineral *pyrolusite*, MnO_2. Relatively pure manganese can be made by directly reducing MnO_2. However, because most manganese is used in steel alloys, common practice is to reduce a mixture of MnO_2 and Fe_2O_3 to obtain an iron-manganese alloy called *ferromanganese*.

$$MnO_2 + Fe_2O_3 + 5\,C \xrightarrow{\Delta} \underbrace{Mn + 2\,Fe}_{\text{Ferromanganese}} + 5\,CO(g)$$

Ferromanganese alloys are wear-resistant and shock-resistant and are used for railroad tracks, bulldozers, and road scrapers.

We can explain the chemical behavior of manganese and its compounds with the help of its electron configuration: $[Ar]3d^54s^2$.

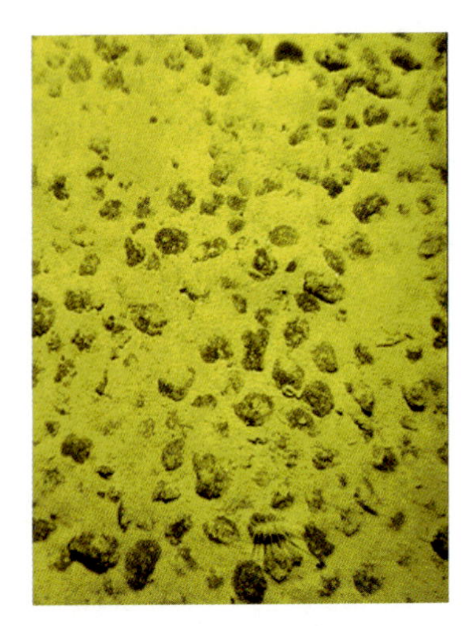

Manganese nodules are composed of layers of oxides of Mn and Fe, with small quantities of Co, Cu, and Ni. The metal oxides are washed into the ocean, where they coalesce into rocks that grow a few millimeters each million years. Marine organisms may play a role in the formation of the nodules. The nodules are a potential source of several elements, but the technological problems involved in bringing them up from the sea floor are daunting.

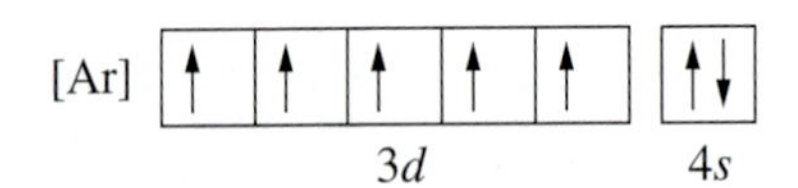

First, by giving up its two valence electrons, a manganese atom acquires the oxidation number +2. Then, by making use of its unpaired 3*d* electrons, the manganese atom can exhibit all additional oxidation numbers from +3 to +7.

Manganese dioxide is the starting point for making most other manganese compounds. To obtain the higher oxidation numbers of manganese, MnO_2 is first oxidized to a manganate salt, such as K_2MnO_4, in which the oxidation number of Mn is +6. A strong oxidizing agent ($KClO_3$) in a basic medium (KOH) is required.

$$3\,MnO_2(s) + ClO_3^-(aq) + 6\,OH^-(aq) \longrightarrow 3\,MnO_4^{2-}(aq) + Cl^-(aq) + 3\,H_2O(l)$$

To obtain Mn(VII), K_2MnO_4 is oxidized one step further to $KMnO_4$, with Cl_2 as the oxidizing agent, for instance.

Potassium permanganate is an important oxidizing agent, used in both the analytical and organic chemistry laboratory. Its oxidizing power also makes it useful as a medical disinfectant and as a substitute for Cl_2 in water purification. In oxidation-reduction reactions in acidic solutions, MnO_4^- is usually reduced to Mn^{2+}; in basic solutions, it is reduced to $MnO_2(s)$.

The lower oxidation states of manganese can be obtained by reducing $MnO_2(s)$. For example, in the cell reaction in a dry cell (page 794), Zn is oxidized to Zn^{2+}; and MnO_2 is reduced to Mn_2O_3. The following reaction is the starting point for the preparation of Mn(II) compounds.

$$MnO_2(s) + 4\,H^+(aq) + 2\,Cl^-(aq) \longrightarrow Mn^{2+}(aq) + 2\,H_2O(l) + Cl_2(g)$$

EXAMPLE 22.2

Use data from Appendix C to demonstrate that permanganate ion in acidic solution is a more powerful oxidizing agent than either $O_2(g)$ or $Cr_2O_7^{2-}(aq)$.

SOLUTION

An oxidizing agent undergoes *reduction* in a redox reaction. To compare the strengths of oxidizing agents, we can compare the standard electrode potentials for the half-reactions in which the oxidizing agents are reduced. We find these values in Appendix C:

$$MnO_4^-(aq) + 4\,H^+(aq) + 3\,e^- \longrightarrow MnO_2(s) + 2\,H_2O(l) \qquad E° = 1.70\text{ V}$$

$$MnO_4^-(aq) + 8\,H^+(aq) + 5\,e^- \longrightarrow Mn^{2+}(aq) + 4\,H_2O(l) \qquad E° = 1.51\text{ V}$$

$$O_2(g) + 4\,H^+(aq) + 4\,e^- \longrightarrow 2\,H_2O(l) \qquad E° = 1.229\text{ V}$$

$$Cr_2O_7^{2-}(aq) + 14\,H^+(aq) + 6\,e^- \longrightarrow 2\,Cr^{3+}(aq) + 7\,H_2O(l) \qquad E° = 1.33\text{ V}$$

The larger values of $E°$ for the reduction half-reactions with $MnO_4^-(aq)$ suggest that in acidic solution, MnO_4^- (aq) is a stronger oxidizing agent than $O_2(g)$ and $Cr_2O_7^{2-}(aq)$.

EXERCISE 22.2

Show that $MnO_4^-(aq)$ is a good oxidizing agent in basic solution. (*Hint:* What are some oxidations that it is able to bring about?)

APPLICATION NOTE

A brown ring is often found on the inside surfaces of bottles in which permanganate solutions are stored. The ring is $MnO_2(s)$ formed by the reduction of $MnO_4^-(aq)$. Some $H_2O(l)$ is simultaneously oxidized to $O_2(g)$.

22.3 The Iron Triad: Fe, Co, and Ni

Iron is the fourth most abundant element in Earth's crust. Cobalt and nickel are not nearly as common, but they are still sufficiently abundant that their annual production is in the thousands of tons. Cobalt is used primarily in alloys with other metals. About 80% of the U.S. production of nickel goes into alloys, and about 15% is used for electroplating. Smaller quantities of nickel are used as electrodes in batteries and fuel cells and as a catalyst. Another familiar use is in coinage, such as the U.S. five-cent coin, the nickel.

Ferromagnetism

Most of the transition elements—including iron, cobalt, and nickel—have some unpaired *d* electrons. These unpaired electrons account for *paramagnetism* (page 333), but Fe, Co, and Ni exhibit a much stronger magnetic effect known as **ferromagnetism**.

In a paramagnetic solid, magnetic moments are associated with atoms or ions that have unpaired electrons. In the absence of a magnetic field, the magnetic moments are randomly oriented, as shown in Figure 22.7. (Think of a magnetic moment as being like a microscopic bar magnet having a north pole at one end and a south pole at the other.) Figure 22.7 shows that when a paramagnetic solid is placed in a magnetic field, most of the magnetic moments align in the direction of the field. When the magnetic field is removed, however, the directions of the magnetic moments again become random.

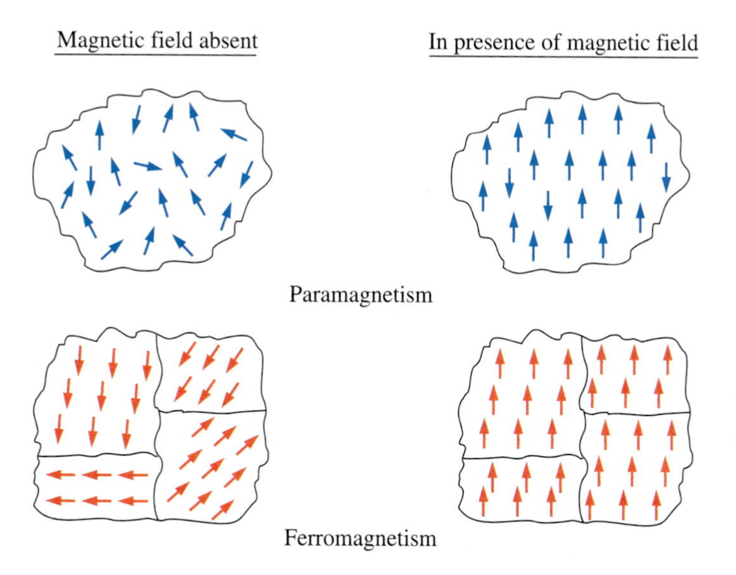

Figure 22.7 Ferromagnetism and paramagnetism compared A paramagnetic solid in the absence and presence of a magnetic field is represented at the top. The magnetic moments associated with the spins of unpaired electrons in individual atoms are shown as arrows. A ferromagnetic solid in the absence and presence of a magnetic field is represented at the bottom. Although only a few magnetic moments are shown, the four regions, called *domains*, contain many atoms.

A ferromagnetic solid consists of regions called *domains*. A domain contains a large number of atoms that have their individual magnetic moments aligned—that is, pointed in the same direction. However, this direction varies from one domain to another, as shown in Figure 22.7. When the ferromagnetic solid is placed in a magnetic field, all the magnetic moments in the domains line up in the direction of the field. The solid becomes magnetized. Moreover, when the field is removed, the orientation of the magnetic moments persists and the solid remains magnetized.

Ordinary paramagnetism becomes ferromagnetism only when the interatomic distances are just right, so that atoms can be arranged into domains. Iron, cobalt, and nickel meet the requirements, as do certain alloys of other metals, such as manganese in the combinations Al-Cu-Mn, Ag-Al-Mn, and Bi-Mn.

Color-enhanced image of magnetic domains in a ferromagnetic garnet film.

Oxidation States

All three iron triad elements form 2+ ions.

$$Fe^{2+}\ [Ar]\ 3d^6 \qquad Co^{2+}\ [Ar]\ 3d^7 \qquad Ni^{2+}\ [Ar]\ 3d^8$$

For cobalt and nickel, the oxidation number +2 is the most common, but for iron it is +3. When an iron atom loses its third electron, forming Fe^{3+}, the electron configuration that results is $[Ar]3d^5$. A half-filled $3d$ subshell with five unpaired electrons is a very stable electron configuration. As a result, iron(II) is oxidized to iron(III) without difficulty.

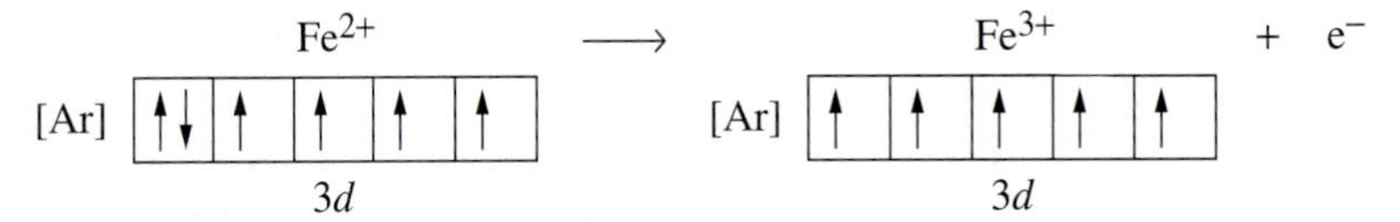

Iron(III) can be oxidized further to FeO_4^{2-}, but only with great difficulty. The conversions of Co(II) and Ni(II) to Co(III) and Ni(III) are also difficult because their +3 oxidation states do not have half-filled $3d$ subshells. The +3 oxidation number of cobalt can be stabilized in complex ions, such as $[Co(NH_3)_6]^{3+}$, but the +3 oxidation number is rare in nickel compounds.

22.4 Group 1B: Cu, Ag, and Au

From earliest times, copper, silver, and gold have been used to make coins. Their use in coins is based on the comparative lack of reactivity of the metals, in keeping with their positive electrode potentials (Table 22.3). Thus, the metal ions are easily reduced to the free metals, and, in turn, the metals are difficult to oxidize.

Another insight from Table 22.3 concerns atomic radii and densities. That silver and gold have the same atomic radius is another example of the lanthanide contraction (page 925). Although the two atoms are about the same size, gold atoms have nearly twice the mass of silver atoms. This largely accounts for the much higher density of gold compared to silver.

The Group 1B elements are excellent metals, as is indicated by their physical properties. They are exceptionally malleable and ductile, and they have the highest electrical and thermal conductivities of all metals. Like transition elements in general, the Group 1B metals display variable oxidation numbers; but the variability is limited, just as it is in the iron triad. Additional similarities to other transition elements include paramagnetism and color in some of their compounds and the ability to form complex ions.

Several ancient coins. The Group 1B metals are called "coinage metals" because copper, silver, and gold have long been used in coins.

Because of their durability, metallic luster, malleability, and ductility, the Group 1B metals are valued in the decorative arts and jewelry making. Electrical wiring for residential purposes is almost always made of copper. Silver and gold are used in electronic components. The resistance of copper to corrosion accounts for its use in plumbing, whereas its good thermal conductivity leads to the use of copper in cookware. The primary use of gold is a rather unusual one: It serves as a monetary reserve for individuals and nations.

We have noted that the Group 1B metals do not displace $H^+(aq)$ from solutions; for example, they do not react with HCl(aq). Copper and silver do react with concentrated $H_2SO_4(aq)$ and $HNO_3(aq)$. These reactions produce Cu^{2+} and Ag^+, along with SO_2 and oxides of nitrogen, respectively, but no H_2. Gold does not react with any single acid, but it will react with aqua regia, a mixture of 1 part HNO_3 and 3 parts HCl. In this reaction, HNO_3 oxidizes gold to Au^{3+}, and HCl furnishes the Cl^- necessary to form the complex ion, $[AuCl_4]^-$. Formation of this stable complex ion drives the reaction in the forward direction by the removal of Au^{3+} ions.

d-Block Elements in Living Matter

As indicated in Figure 22.8, relatively few elements are essential to living matter. Elements with low atomic numbers ($Z < 21$) make up the bulk of living matter. Most plant and animal life requires only trace amounts of some elements, and most of these trace elements are members of the fourth-period *d*-block. An adult human body that weighs 70 kg, for example, contains only about 4 g of iron, but those 4 g are essential for good health.

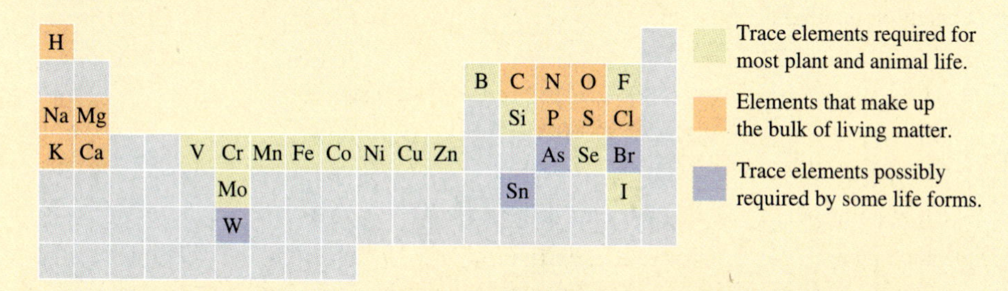

Figure 22.8
The elements in living matter
Most of the elements essential to life are relatively abundant in the natural environment. Nine of the 11 most essential elements are also the most abundant in seawater: H, O, Cl, Na, Mg, S, K, Ca, and C. The other two, nitrogen and phosphorus, are also among the 20 most abundant elements in seawater. It is believed that the life forms on Earth developed from the elements readily available to them.

In biological matter, the transition metals are often incorporated into large organic molecules. The protein hemoglobin, found in red blood cells and responsible for the color of blood, consists of iron-containing heme units attached to polypeptide molecules called globins (Figure 22.9). Hemoglobin transports O_2 molecules from the lungs, through the arteries, to all parts of the body for use in the metabolism of glucose. It then carries CO_2 produced in this metabolism, through the veins, back to the lungs.

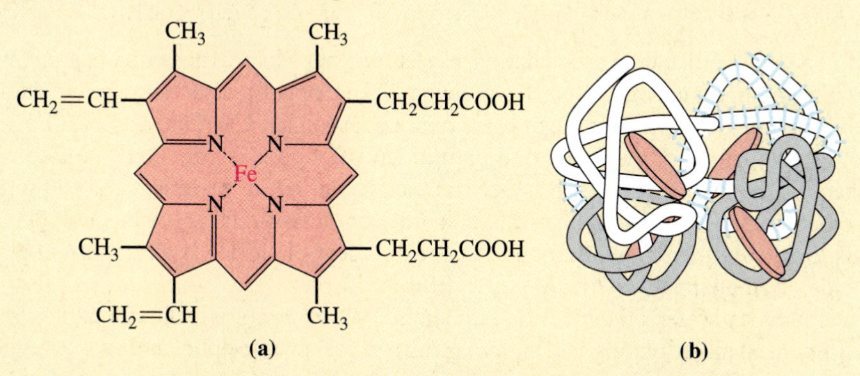

Figure 22.9
Heme and hemoglobin
(a) The structure of heme. (b) Four heme units and four coiled polypeptide chains are bonded together in a molecule of hemoglobin.

To perform its essential function, hemoglobin must have iron present as Fe^{2+}. If the iron is oxidized to Fe^{3+}, the resulting substance, called methemoglobin, can't transport oxygen. This oxidation occurs in the disease called *methemoglobinemia*. High concentrations of nitrate ion in groundwater used for domestic purposes can cause this condition.

Another important biochemical process involving both iron and molybdenum is the fixation of atmospheric nitrogen by certain bacteria through the action of the enzyme *nitrogenase*. Nitrogen molecules are apparently held to Mo atoms in the enzyme, as seen at the right.

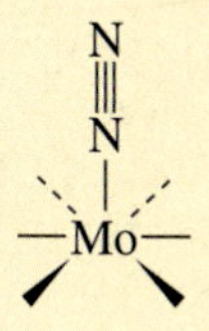

The enzyme acts by facilitating the transfer of six electrons to the $N{\equiv}N$ molecule, reducing the oxidation numbers of the N atoms from 0 to -3. The N atoms are then released as NH_3 molecules. The electron transfers appear to occur through changes in the oxidation numbers of the Fe and Mo atoms. Thus, the variability of oxidation numbers of *d*-block elements plays an important role in the biological activity of the transition elements.

TABLE 22.3 Some Properties of Copper, Silver, and Gold

	Cu	Ag	Au
Atomic number	29	47	79
Electron configuration	$[Ar]3d^{10}4s^1$	$[Kr]4d^{10}5s^1$	$[Xe]4f^{14}5d^{10}6s^1$
Atomic radius, pm	128	144	144
Electronegativity	1.9	1.9	2.4
Oxidation numbers[a]	+1, +2	+1, +2, +3	+1, +3
Electrode potential, V			
$M^+(aq) + e^- \longrightarrow M(s)$	+0.522	+0.800	+1.83
$M^{2+}(aq) + 2\,e^- \longrightarrow M(s)$	+0.340	+1.39	—
$M^{3+}(aq) + 3\,e^- \longrightarrow M(s)$	—	—	+1.52
Melting point, °C	1083	962	1064
Density, g/cm^3	8.96	10.5	19.3
Electrical conductivity[b]	95	100	68
Thermal conductivity[b]	93	100	74

[a] The most common oxidation number is printed in red.
[b] Electrical and thermal conductivities are on an arbitrary scale relative to 100 for silver, the best metallic conductor.

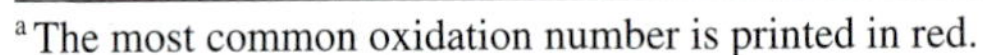

$$Au(s) + 4\,H^+(aq) + NO_3^-(aq) + 4\,Cl^-(aq) \longrightarrow [AuCl_4]^-(aq) + 2\,H_2O(l) + NO(g)$$

Basic copper carbonate, $Cu_2(OH)_2CO_3$, imparts a characteristic green color to copper roofs and bronze statues.

Gold is resistant to air oxidation, as is silver; but silver tarnishes by reacting with sulfur compounds to produce Ag_2S (page 799). Copper resists corrosion in dry air, but in moist air, it forms green basic copper carbonate, $Cu_2(OH)_2CO_3$.

The principal manufactured compound of copper is copper(II) sulfate pentahydrate, $CuSO_4 \cdot 5H_2O$. It is used in electroplating, in batteries, and to prepare other copper salts. Although essential in trace amounts to living organisms, Cu^{2+} ion is toxic at higher concentrations. As a result, copper sulfate is an important pesticide, used against bacteria, algae, and fungi.

Silver nitrate is the chief manufactured compound of silver. Among its varied uses are in silver plating and silvering mirrors, as an antiseptic, and as a reagent in analytical chemistry laboratories. Its most important use, however, is in making silver halides (AgCl, AgBr, AgI) for photography. Gold compounds are used in electroplating, photography, medicinal chemistry, and in ruby glass and ceramics.

22.5 Group 2B: Zn, Cd, and Hg

Zinc, cadmium, and mercury are *d*-block metals, but—strictly speaking—they are not transition metals. They resemble the transition metals in some ways, for example, in forming many complex ions. In other ways, they resemble the alkaline earth metals of Group 2A. (Mendeleev combined Groups 2A and 2B into Group II of his periodic table.) For example, zinc and cadmium form only the 2+ cation, and the cations are diamagnetic and colorless in solution.

Mercury differs from zinc and cadmium in at least six significant ways. Mercury is a liquid at room temperature. It does not displace H_2 from acidic solutions. Mercury displays the oxidation number +1 in the diatomic ion Hg_2^{2+}, which features an Hg—Hg covalent bond. It forms many molecular compounds (for example, $HgCl_2$ is only slightly ionized in aqueous solution). Mercury forms few water-soluble compounds, and most of its compounds are not hydrated. Mercury shows little tendency to oxidize, and its oxide, HgO, readily decomposes to the elements upon heating.

Zinc has many uses. Large quantities are used in alloys. For example, *brass* is a copper alloy having from 20 to 45% zinc and small quantities of tin, lead, and iron. Brass has an attractive yellow metallic sheen, is corrosion-resistant, and is a good electrical conductor. Zinc oxidizes in air to form a thin, adherent oxide coating that protects the underlying metal from further corrosion. This property explains the use of zinc in making galvanized (zinc-coated) iron. Also, the fact that zinc is more easily oxidized than iron underlies its use as sacrificial anodes that offer cathodic protection to iron (page 799).

Cadmium can substitute for zinc in coating metals for certain applications. Its primary uses are in alloys and as electrodes in batteries. Because of its capacity to absorb neutrons, cadmium is used in control rods in nuclear reactors. The uses of cadmium are severely limited, however. Whereas zinc in trace amounts is an essential element for humans, cadmium is quite toxic. Its effect may be to substitute for zinc in certain enzymes.

Brass is used extensively on sailing ships because it resists the corrosive action of seawater.

The physical properties of mercury, especially its metallic and liquid properties and its high density, determine many of its uses. Thus, mercury is used in thermometers, barometers, manometers, electric relays and switches, and as electrodes for certain batteries and electrolysis cells. Some uses of mercury are related to its ability to form alloys called *amalgams* with most metals. Dental amalgam used to fill tooth cavities consists primarily of silver and tin with a small amount of mercury. The thermal properties of expansion and contraction of the amalgam closely match those of teeth. Another widespread use of mercury is as mercury vapor in fluorescent tubes and streetlamps. The quantity in one lamp is so small, however, that the total amount of mercury used for this purpose is quite small.

Liquid mercury was once treated as a laboratory plaything, but we now know that long-term exposure can present a serious health hazard. Mercury vapor is toxic, and levels exceeding 0.05 mg Hg/m^3 are unsafe. Liquid mercury has a low vapor pressure, but even so, the concentration of mercury in the saturated vapor at room temperature is well above the safe limit. Unsafe mercury vapor levels have been found in various locales in which the element is used—chlor-alkali plants, thermometer factories, smelters, and dental laboratories. In chlor-alkali plants that use mercury, the loss to the environment has been reduced to about 200 mg Hg per ton of Cl_2 produced. This is about one-thousand-fold less than decades ago. In dentistry, various metal powders, porcelain, and plastics are supplanting dental amalgam in many applications.

Mercury poisons the body's systems, in part, by interfering with sulfur-containing enzymes. Hg^{2+} reacts with $-SH$ groups in an enzyme to change the shape of the enzyme and render it inactive. Some microorganisms can convert inorganic mercury to methylmercury (CH_3Hg^+) compounds, which are readily absorbed by organisms. These compounds concentrate in the food chain, and can lead to unsafe mercury levels in fish.

APPLICATION NOTE

In humans, mercury affects the nervous system and causes brain damage. Hatter's disease (which probably afflicted the Mad Hatter in *Alice's Adventures in Wonderland*) was a form of chronic mercury poisoning. Mercury compounds were once used to convert fur to felt for felt hats.

22.6 The Lanthanide Elements (Rare Earths)

The elements cerium ($Z = 58$) through lutetium ($Z = 71$) comprise the first series of the f-block. We can think of the f-block series as an insertion in the sixth-period d-block. The series is preceded by a d-block element, lanthanum, and followed by a d-block element, hafnium. Because the 14 elements in the series immediately follow lanthanum, they are often called the lanthanide elements. They are also called *inner-transition elements* because the subshell that is being filled is *two* principal quantum levels below the valence shell; in the sixth period, the 4*f* subshell fills before the 5*d*. These elements are frequently called "rare earths," a name of historical origin but a clear misnomer because several are not rare at all. Cerium (Ce) is the twenty-fifth most abundant element in Earth's crust. Both cerium and neodymium (Nd) are more abundant than lead. All the lanthanides except promethium (Pm) are more abundant than such well-known elements as Cd, Ag, Hg, and Au. All promethium isotopes are radioactive. The most striking feature of the lanthanide elements, which are often referred to by the general symbol Ln, is a similarity to each other and to La and the Group 3B metals.

In spite of their relative abundance, the term rare earths seems appropriate for two reasons: (1) For many years after their discovery, there was only one known source of the elements, *monazite*, a mixed phosphate of La, Th, and the lanthanides. (2) The individual lanthanides are difficult to separate and isolate. (In the nineteenth century, it was extremely difficult to separate them.)

When ions differ sufficiently in a property such as solubility, we can often separate them in a one-step process. For example, we can readily separate Ag^+ and Cu^{2+} in an aqueous solution by adding Cl^- to the solution; Ag^+ completely precipitates as AgCl(s) and Cu^{2+} remains in solution. If we try to use differences in solubility to separate individual lanthanide cations (Ln^{3+}), we get only a slight separation. We have to go through many, many cycles of precipitation and redissolving to achieve total separation. Researchers took about 70 years, from 1839 to 1907, to isolate all of the lanthanide elements. (Promethium, found only as a product of nuclear fission, was not isolated until 1945.) Recall that in 1913 H. G. J. Moseley related the frequency of X rays to the atomic number of the target element in an X-ray tube (page 328). In just a few days, he was able to identify by atomic number all the lanthanide elements in a sample of monazite. Today, the lanthanides are generally separated by complex-ion formation and ion-exchange techniques.

For many uses, the lanthanides do not have to be separated from one another. A mixture of the lanthanide metals with about 25% La is used in certain steel and magnesium-based alloys. Lanthanide-cobalt alloys are used to make permanent magnets. A number of lanthanide compounds also have commercial uses. Several of the oxides are used as phosphors in the screens of color television sets, as colorants for glass and ceramic glazes, and as catalysts. Cerium(IV) compounds are important oxidizing agents in analytical chemistry.

Coordination Chemistry

The two substances in Figure 22.10 have similar formulas, but they are strikingly different in color. They are both *coordination compounds:* Each is derived from two simpler compounds that are somehow coordinated, or joined together. The following formulas suggest that the two simpler compounds are $CoCl_3$ and NH_3.

$$CoCl_3 \cdot 6NH_3 \qquad CoCl_3 \cdot 5NH_3$$

Golden orange — Purple

The two coordination compounds differ in more than color. Each substance has three moles of Cl^- per mole of compound. As we might expect, when one mole of the golden orange compound is dissolved in water and treated with $AgNO_3$(aq), three moles of AgCl(s) are obtained. However, only two moles of AgCl(s) are obtained from one mole of the purple compound. In unraveling the mystery of these two compounds, and others of a similar nature, the Swiss chemist Alfred Werner established the modern basis of coordination chemistry.

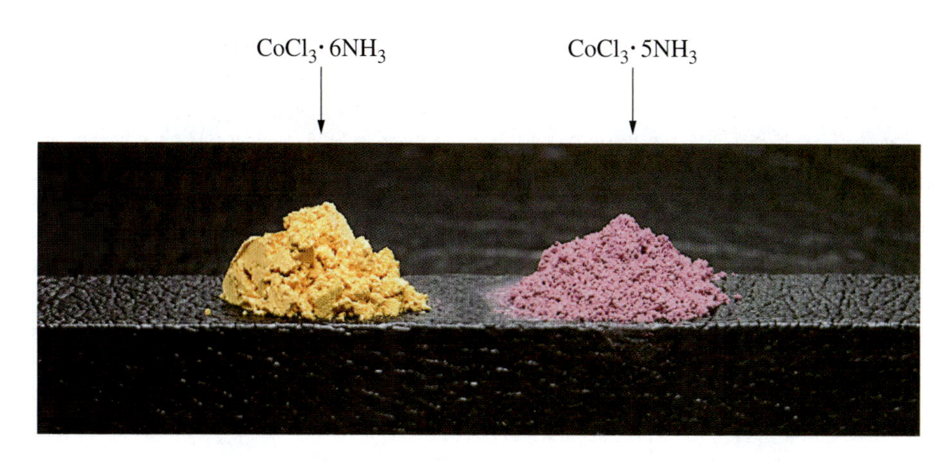

Figure 22.10 Two coordination compounds of cobalt(III)

22.7 Werner's Theory of Coordination Chemistry

In 1893, Werner explained the difference between the golden orange and purple compounds by representing their dissociation in water as follows.

$$[Co(NH_3)_6]Cl_3(s) \xrightarrow{H_2O(l)} [Co(NH_3)_6]^{3+}(aq) + 3\,Cl^-(aq)$$

Golden orange

$$[CoCl(NH_3)_5]Cl_2(s) \xrightarrow{H_2O(l)} [CoCl(NH_3)_5]^{2+}(aq) + 2\,Cl^-(aq)$$

Purple

Thus, the purple compound yields only two moles of AgCl(s) when treated with an excess of $AgNO_3(aq)$.

Alfred Werner (1866–1919) claimed that the basic idea of coordination theory came to him, at age 25, during his sleep. He arose at 2 A.M. and worked out the essentials by 5 A.M. Werner was awarded the 1913 Nobel Prize in chemistry for his work in coordination chemistry.

Some Terminology for Complexes

We can use the two compounds studied by Werner, $[Co(NH_3)_6]Cl_3$ and $[CoCl(NH_3)_5]Cl_2$, to illustrate some of the terms used to describe complexes. An entity within brackets, for example, $[Co(NH_3)_6]^{3+}$, is properly called a *coordination entity*, but more commonly, it is called a complex. A **complex** consists of a central atom, which is usually a metal atom or ion (for example, Co^{3+}), and attached groups called **ligands**. The ligands may be neutral molecules, such as NH_3, or anions, for example, Cl^-. The ligands may all be of the same type, as in $[Co(NH_3)_6]^{3+}$, or of different types, as in $[CoCl(NH_3)_5]^{2+}$.

The region surrounding the central atom or ion and containing the ligands is called the *coordination sphere*. The **coordination number** is the total number of points at which a central atom or ion attaches ligands. Both in $[Co(NH_3)_6]^{3+}$ and $[CoCl(NH_3)_5]^{2+}$, the coordination number is six. The most common coordination numbers seen in complexes are 2, 4, and 6, and the geometric structures they produce are shown in Figure 22.11. Depending on the charges on the metal and on the ligands, the overall charge on a complex may be positive, negative, or zero. If a complex carries a net electric charge, it is called a **complex ion**. $[Co(NH_3)_6]^{3+}$ and $[CoCl(NH_3)_5]^{2+}$ are complex ions with charges of 3+ and 2+, respectively. A substance consisting of one or more complexes is called a **coordination compound**. Golden orange $[Co(NH_3)_6]Cl_3$ and purple $[CoCl(NH_3)_5]Cl_2$ are both coordination compounds. Several of the terms introduced here are illustrated in the margin.

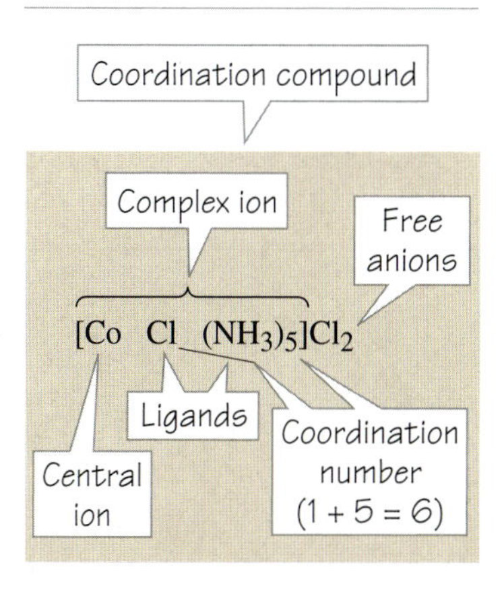

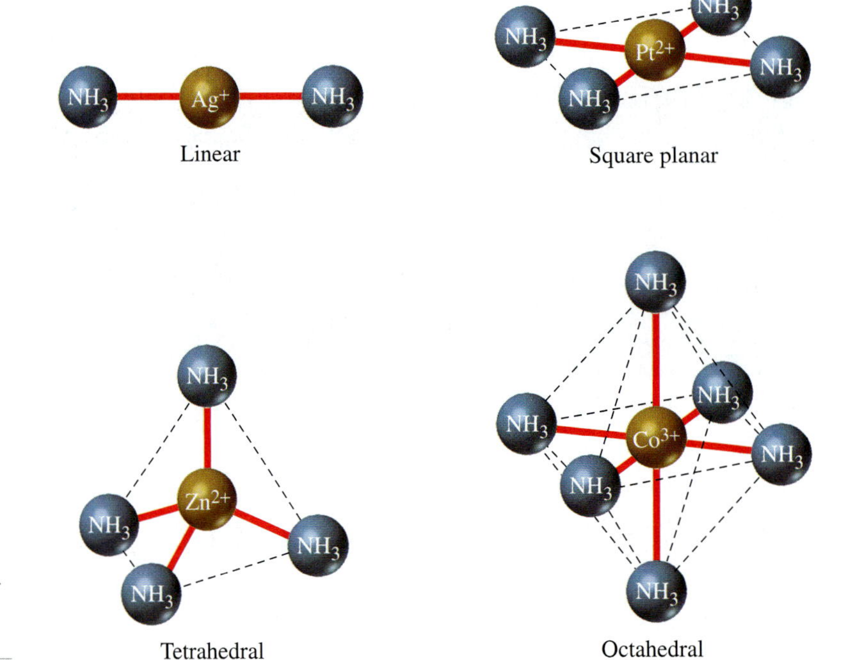

Figure 22.11
Four common structures of complex ions
In these complex ions, the NH_3 molecules use their lone-pair electrons to attach to the central metal ion.

EXAMPLE 22.3

What are the coordination number and the oxidation number of the central atom in the complexes **(a)** $[CoCl_4(NH_3)_2]^-$ and **(b)** $[Ni(CO)_4]$?

SOLUTION

a. The metal at the center of this complex ion is cobalt, and six ligands—four Cl^- ions and two NH_3 molecules—are attached to it. The coordination number is 6. The complex is an anion with a net charge of 1−. The NH_3 ligands carry no charge, and four Cl^- ions carry a total charge of 4−. If we let x equal the charge on the cobalt, then $x - 4 = -1$, and $x = +3$. The central ion is Co^{3+}, and the oxidation number of Co is +3.

b. The central metal in this complex is nickel, and four ligands—all CO molecules—are attached to it. The complex is electrically neutral, as are the CO molecules. The oxidation number of Ni is 0.

EXERCISE 22.3

What are the coordination number and the oxidation number of the central atom in the complexes **(a)** $[Co(SO_4)(NH_3)_5]^+$ and **(b)** $[Fe(CN)_6]^{4-}$?

Ligands

Many metal atoms and ions, particularly those of the transition metals, have empty orbitals that can accommodate electron pairs. The central metal of a complex is a Lewis acid (electron-pair acceptor). Ligands are characterized by lone-pair electrons; they are Lewis bases (electron-pair donors).

The atom in a ligand that furnishes an electron pair is called a *donor* atom. Ligands with one donor atom have just one point of attachment to the central metal atom or ion. They are **monodentate** (one-toothed) ligands. Chloride ions, water molecules, and ammonia molecules are monodentate ligands. When anions are ligands, their names are given an “o” ending. There is no distinctive ending used in naming neutral molecules as ligands.

	$[:\ddot{\underset{..}{Cl}}:]^-$	$[:\ddot{\underset{..}{O}}-H]^-$	$H-\ddot{\underset{..}{O}}:$ (with H above O)	$H-\underset{..}{N}-H$ (with H above N)
Ligand name:	*Chloro*	*Hydroxo*	*Aqua*	*Ammine* *

Another common monodentate ligand is nitrite ion, NO_2^-, but here the donor atom can be either the N atom or one of the O atoms. The ligand is given a different name for each case. The true structure is a resonance hybrid of two resonance structures, but we can see the lone-pair electrons on the donor atoms by writing just one of the structures.

$$\left[:\ddot{O}{=}\underset{..}{N}{-}\ddot{\underset{..}{O}}:\right]^-$$

Ligand name: *Nitro*, if N is the donor atom
Nitrito, if O is the donor atom

Nitrite ion is monodentate because it can bond through only one of its donor atoms at a time. Ethylenediamine also has two donor atoms that are far enough apart that they can bond independently. The ligand molecule can wrap itself around the central metal atom or ion and attach to it through both donor atoms at the same time. It is a **bidentate** (two-toothed) ligand.

$$H-\ddot{N}(H)-CH_2-CH_2-\ddot{N}(H)-H$$

Ligand name: *ethylenediamine* (*en*)

Ligands with multiple points of attachment are **polydentate** (many-toothed) ligands. Table 22.4 lists some common monodentate and polydentate ligands.

One of the complex ions whose structure is shown in Figure 22.11 is $[Pt(NH_3)_4]^{2+}$. Suppose we replace each pair of NH_3 molecules along an edge of the square with the bidentate ligand ethylenediamine. The resulting complex ion, $[Pt(en)_2]^{2+}$, has the structure shown in Figure 22.12, where the dashed lines outline the square planar structure.

Note that this structure has two five-membered rings (pentagons), outlined by the solid red lines (chemical bonds). Each ring consists of one Pt, two N, and two C atoms. When a five- or six-membered ring is produced by the attachment of a polydentate ligand, the complex is a **chelate** (pronounced KEY-late). The process of attachment of a polydentate ligand to a central metal atom or ion is called *chelation*, and the ligand is referred to as a *chelating agent*. Chelate comes from the Greek word for claw. In the $[Pt(en)_2]^{2+}$ complex ion, it's as if the central Pt^{2+} ion is being pinched by two sets of molecular claws.

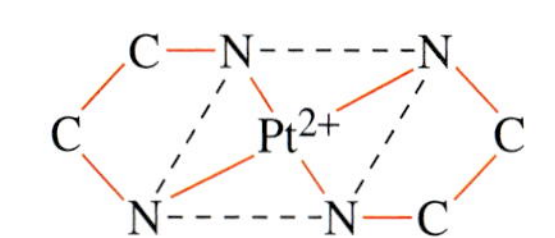

▲ **Figure 22.12**
The chelate $[Pt(en)_2]^{2+}$
The N atoms of ethylenediamine (en) bond through lone-pair electrons. They attach at the corners along an edge of the square (black dashed lines), but they *do not* bridge the square by attaching at opposite ends of a diagonal. For simplicity, H atoms are not shown.

22.8 Naming Complex Ions and Coordination Compounds

The collection of rules needed to name all possible complexes is extensive and has been revised many times over the years. We will use a simplified list that will help us match an acceptable name with a formula and an appropriate formula with a name.

* The spelling of a*mm*ine with a double m signifies that the ligand is a*mm*onia, also spelled with a double m. In contrast, the functional group $-NH_2$ in organic compounds is called the a*m*ine group and is spelled with a single m.

TABLE 22.4 Some Common Ligands

MONODENTATE					
Formula[a]	Name as Ligand[b]	Formula[a]	Name as Ligand[b]	Formula[a]	Name as Ligand[b]
Neutral Molecules					
NH_3	Ammine	NO	Nitrosyl	H_2O	Aqua
CH_3NH_2	Methylamine	CO	Carbonyl	C_5H_5N	Pyridine
Anions					
F^-	Fluoro	NO_2^-	Nitro	Cl^-	Chloro
ONO^-	Nitrito	Br^-	Bromo	SCN^-	Thiocyanato
I^-	Iodo	NCS^-	Isothiocyanato	OH^-	Hydroxo
OSO_3^{2-}	Sulfato	CN^-	Cyano	SSO_3^{2-}	Thiosulfato

POLYDENTATE		
Name of Ligand[b]	Abbreviation	Formula[a]
Ethylenediamine	en	$H_2NCH_2CH_2NH_2$
Oxalato	ox	$[OOCCOO]^{2-}$
Ethylenediaminetetraacetato	EDTA	$[(OOCCH_2)_2NCH_2CH_2N(CH_2COO)_2]^{4-}$

[a] Donor atoms are shown in red.

[b] Most neutral ligands carry the unmodified name, except for aqua, ammine, carbonyl, and nitrosyl. Anion ligand names end in "o," which requires changing the terminal *-e* to *-o* (e.g., sulfate ⟶ sulfato). With many common anions, an entire *−ide* ending is changed to *-o* (e.g., cyanide ⟶ cyano).

1. *In naming a complex*, first name the ligands and then the central metal atom or ion as a single compound word. *Example:* $[Cu(NH_3)_4]^{2+}$ is tetraammine-copper(II) ion.
2. *When writing a name from a formula*, name the ligands in alphabetical order, based on the first letters of their names and without regard for prefixes. *Example:* $[CuCl(NH_3)_3]^+$ has the name triamminechlorocopper(II) ion.

   When writing a formula from a name, place anions *before* neutral ligands. The order of the anions and neutral ligands is alphabetical based on the first letters of their symbols or formulas. *Example:* Diamminediaquadichloro-cobalt(III) ion has the formula $[CoCl_2(H_2O)_2(NH_3)_2]^+$.
3. Designate the number of ligands with a prefix: *mono* = 1 (usually omitted), *di* = 2, *tri* = 3, *tetra* = 4, and so on. *Example:* $[Ag(NH_3)_2]^{2+}$ is *di*ammine silver(I) ion.

   If the ligand name itself includes a prefix, place parentheses around the name, and precede it with *bis* = 2, *tris* = 3, *tetrakis* = 4, and so on. *Example:* $[Pt(en)_2]^{2+}$ is *bis*(ethylene*di*amine)platinum(II) ion. Bis, tris, and so on may also be used, when needed, to avoid ambiguity.
4. Use the unmodified name of the central metal in a complex cation. In a complex anion, add the ending *-ate* to the name of the central metal. For certain common metals in complex anions, use a Latin-based name (Table 22.5). In all cases, denote the oxidation number of the central metal by a Roman numeral in parentheses. *Examples:* $[Zn(H_2O)_4]^{2+}$ is tetraaquazinc(II) ion. $[Zn(OH)_4]^{2-}$ is tetrahydroxozinc*ate* (II) ion. $[CuCl_4]^{2-}$ is tetrachloro*cuprate* (II) ion.
5. In writing the formula or name of a coordination compound, place the ions in the usual order: cation followed by anion. *Example:* $Na_2[Zn(OH)_4]$ is sodium tetrahydroxozincate(II).

Some coordination compounds have long been known and are still referred to by common names. Two examples are potassium ferr*o*cyanide, $K_4[Fe(CN)_6]$, and potassium ferr*i*cyanide, $K_3[Fe(CN)_6]$. The "o" and "i" indicate that the central ions are ferr*o*us (Fe^{2+}) and ferr*i*c (Fe^{3+}), respectively. With this information and the coordination number 6, it's not hard to relate the common names to the formulas of these two coordination compounds. Nevertheless, their systematic names are even more indicative of the formulas: potassium hexacyanoferrate(II) for $K_4[Fe(CN)_6]$ and potassium hexacyanoferrate(III) for $K_3[Fe(CN)_6]$.

TABLE 22.5 Names for Some Metals in Complex Anions

Copper	*Cuprate*
Gold	*Aurate*
Iron	*Ferrate*
Lead	*Plumbate*
Silver	*Argentate*
Tin	*Stannate*

EXAMPLE 22.4

Name the following: **(a)** $[CrCl_2(NH_3)_4]^+$; **(b)** $K[PtBrCl_2NH_3]$.

SOLUTION

a. The ligands are four ammonia molecules (ammine) and two chloride ions (chloro). We refer to them as *tetra*ammine and *di*chloro, and we list them in alphabetical order. The two Cl^- ligands carry *two* units of negative charge, and the net charge on the complex ion is *one plus*. The central Cr must therefore be Cr^{3+}.

The name is tetraamminedichlorochromium(III) ion.

b. A formula unit of this coordination compound consists of a K^+ cation and a complex anion, $[PtBrCl_2(NH_3)]^-$. In alphabetical order, the ligands in the complex are one NH_3 molecule (ammine), one Br^- ion (bromo), and two Cl^- ions (dichloro). The central ion must be Pt^{2+}, to account for the net charge of 1− on the complex: $+2 - 1 - 2 = 1-$.

The name of the coordination compound is potassium amminebromodichloroplatinate(II).

EXERCISE 22.4

Name the following: **(a)** $[Co(NH_3)_6]^{2+}$; **(b)** $[AuCl_4]^-$; **(c)** $[CoBr(NH_3)_5]Br_2$.

EXAMPLE 22.5

Write formulas for the following:

a. triamminechlorodinitroplatinum(IV) ion

b. sodium hexanitrocobaltate(III)

SOLUTION

a. This is a complex cation of Pt^{4+} having three (tri) ammonia (ammine) molecules, one chloride (chloro) ion, and two (di) nitrite ions as ligands. The net charge on the complex ion is $4 - 1 - 2 = 1$. In writing the formula of the complex ion, the name nitro signifies that the donor atom of the nitrite ligand is the N atom and that the ligand should be represented as NO_2. Note also that we place the anions before the neutral ammonia molecules in the formula.

$$[PtCl(NO_2)_2(NH_3)_3]^+$$

b. This is a coordination compound made up of monatomic Na^+ ions and complex anions. The name cobaltate(III) tells us that the central ion in the anion is Co^{3+}. This central ion and six NO_2^- ions as ligands produce a complex anion with a net charge of $+3 - 6 = 3-$ and the formula $[Co(NO_2)_6]^{3-}$. Because there must be three Na^+ cations for each anion, we write the following formula for the coordination compound.

$$Na_3[Co(NO_2)_6]$$

EXERCISE 22.5

Write formulas for the following:

a. (tris)ethylenediaminecobalt(III) ion

b. diamminetetrachlorochromate(III) ion

c. dichlorobis(ethylenediamine)platinum(IV) sulfate

22.9 Isomerism in Complex Ions and Coordination Compounds

Recall that isomers are compounds with the same molecular formula but different structures and properties. We find several types of isomers among complex ions and coordination compounds.

Structural Isomers

Structural isomers differ in the ligands that are attached to the central atom or in the donor atoms through which the ligands are bonded. Thus, although the following complex ions have the same ligands in the same numbers, they are isomers because a different atom acts as donor in the NO_2^- ligand—N in one case and O in the other. The donor atoms are shown in color in the formulas written below.

$[Co(NO_2)(NH_3)_5]^{2+}$	$[Co(ONO)(NH_3)_5]^{2+}$
Pentaamminenitrocobalt(III) ion	Pentaamminenitritocobalt(III) ion

The two coordination compounds below have the same percent composition by mass and the same molar mass, but they are isomers because they differ in the placement of the anions. One compound has SO_4^{2-} as a ligand attached to the Cr^{3+} central ion and Cl^- as a free anion outside the coordination sphere. The other compound has Cl^- as the ligand and SO_4^{2-} as the free anion.

$[Cr(SO_4)(NH_3)_5]Cl$	$[CrCl(NH_3)_5]SO_4$
Pentaamminesulfatochromium(III) chloride	Pentaamminechlorochromium(III) sulfate

EXAMPLE 22.6

For each pair, indicate whether the species are isomers.

a. $[Co(NH_3)_6][Cr(CN)_6]$ and $[Cr(NH_3)_6][Co(CN)_6]$

b. $[Cr(H_2O)_5(NH_3)]^{3+}$ and $[Cr(NH_3)(H_2O)_5]^{3+}$

c. $[Co(H_2O)_5(NH_3)]^{3+}$ and $[Co(H_2O)(NH_3)_5]^{3+}$

SOLUTION

a. These two coordination compounds have the same overall compositions and molar masses. They also have the same numbers and types of ligands. They differ in the distribution of ligands between the complex cation and complex anion. The cation in one compound is hexaamminecobalt(III) ion, whereas in the other, it is hexaamminechromium(III) ion. The two compounds are isomers.

b. Each of these complex ions has one NH_3 and five H_2O molecules as ligands and Cr^{3+} as the central ion. The formulas are written differently, but the structures are identical. These two formulas do *not* represent isomers; they represent the same complex ion.

c. Each complex ion has Co^{3+} as the central ion and a total of six ligands. However, one ion has five H_2O and one NH_3 molecule, and the other has five NH_3 and one H_2O molecule. These two formulas represent *different* complex ions. This is not a case of isomerism.

EXERCISE 22.6A

Indicate whether the species in each pair are isomers. Explain.

a. $[FeSCN(H_2O)_5]^{2+}$ and $[FeNCS(H_2O)_5]^{2+}$

b. $[Zn(NH_3)_4][CuCl_4]$ and $[Cu(NH_3)_4][ZnCl_4]$

c. $[PtCl_2(NH_3)_2]$ and $[Pt(NH_3)_4]Cl_2$

EXERCISE 22.6B

Are different isomers of the following possible? Explain.

a. $[Zn(H_2O)(NH_3)_3]^{2+}$ **b.** $[Cu(NH_3)_4][PtCl_4]$

In another type of isomerism, known as **stereoisomerism**, the number and types of ligands and their mode of attachment are all the same, but the manner in which the ligands occupy the space around the central atom differs. We will consider two important types of stereoisomers.

Geometric Isomerism

Recall the complex ion $[Pt(NH_3)_4]^{2+}$ pictured in Figure 22.11. Suppose we replace two of the NH_3 ligands by Cl^- ions. Figure 22.13 shows that we can do this in two ways: The two Cl^- ions can be either along the same edge of the square (cis) or on opposite corners (trans). This is similar to the cis-trans isomerism of alkenes (Section 10.5).

NH_3

Cl^-—Pt^{2+}—NH_3

Cl^-

cis-Diamminedichloroplatinum(II)
(cisplatin)

Cl^-

H_3N—Pt^{2+}—NH_3

Cl^-

trans-Diamminedichloroplatinum(II)
(transplatin)

Figure 22.13 Geometric isomerism in a square planar complex

The two geometric isomers of diamminedichloroplatinum(II) offer a striking example of the critical relationship between structure and properties. The cis isomer, known as cisplatin, is a powerful antitumor drug used in chemotherapy (see Figure 22.13, left). Apparently, the two Cl^- ligands are just the right distance apart so that the complex can attach to the DNA in cancerous cells and inhibit further cell growth. In the trans isomer, the Cl^- ligands are too far apart for the complex to be effective.

Figure 22.14 presents an example of cis and trans isomerism in an octahedral complex of cobalt(III), $[CoCl_2(en)_2]^+$. The cis isomer has two Cl^- ions along the same edge of the octahedron. In the trans isomer, the two Cl^- ions are on opposite corners of the octahedron.

Look again at Figure 22.13. If we replace the two NH_3 ligands of cisplatin by Cl^- or the two Cl^- by NH_3, there is no isomerism in the product ($[PtCl_4]^{2-}$ or $[Pt(NH_3)_4]^{2+}$). Now consider a substitution in $[CoCl_2(en)_2]^+$ in Figure 22.14. If we replace the two Cl^- ligands of the cis isomer by a third ethylenediamine (en), it might appear that there is no isomerism. In fact, however, there are *two* different tris(ethylenediamine)cobalt(III) ions, as we shall see next.

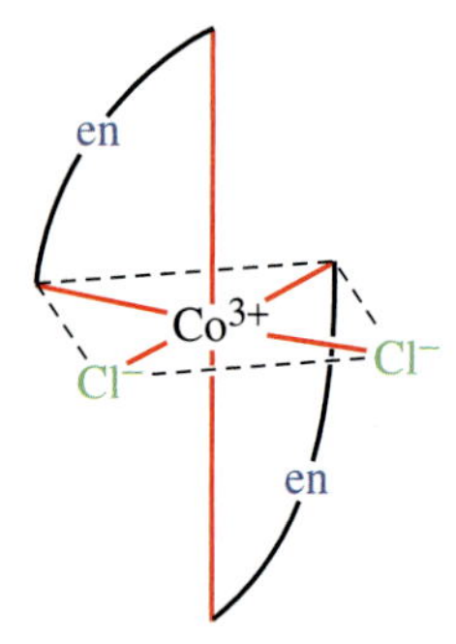

(a) *cis*-Dichlorobis-(ethylenediamine)cobalt(III) ion

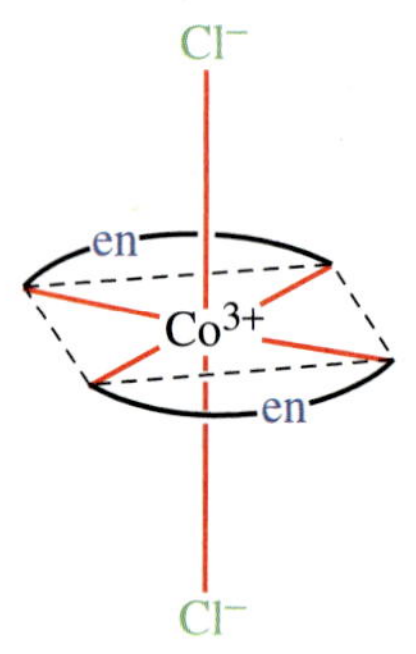

(b) *trans*-Dichlorobis-(ethylenediamine)cobalt(III) ion

▲ **Figure 22.14 Geometric isomerism in an octahedral complex**

Optical Isomers

To understand optical isomerism, we need to consider the relationship between an object and its mirror image. Think of a plain rubber ball and its image in a mirror. If it were possible to lift the image from the mirror, we would be able to place the image over the ball. The real object and its image would be indistinguishable. In contrast, the mirror image of a right hand is a left hand, and one is not superimposable on the other. As shown in Figure 22.15, two *nonsuperimposable* structures are possible for the tris(ethylenediamine)cobalt(III) ion. Molecules or ions with structures that are nonsuperimposable in the same way as an object and its nonsuperimposable mirror image are called **enantiomers**.

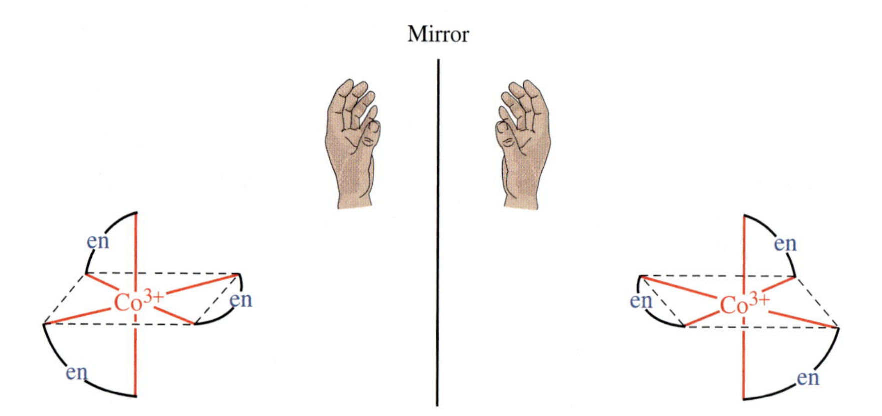

▶ **Figure 22.15 Optical isomers**
The two structures are nonsuperimposable mirror images. They are like a right hand and a left hand.

Enantiomers have identical physical and chemical properties except in a few situations that depend on "handedness" at the molecular level. Think of handedness in this way: Whether persons are right-handed or left-handed, they can all use the same pencils, hammers or screwdrivers. However, a left-handed person has difficulty using a pair of right-handed scissors. One physical property that is dependent on the handedness of enantiomers is *optical activity*.

In Figure 7.10, we pictured an electromagnetic wave as an oscillation of electric and magnetic fields. In normal light, these oscillations occur in all directions perpendicular to the line along which the light is propagated. The light is *unpolarized*. When the oscillations are limited to a single direction, the wave motion occurs in only one plane, and the light is *plane polarized*. Some materials, called polarizers, have the ability to filter out all the oscillations except those in a single plane. They transmit plane-polarized light. Figure 22.16 shows a light source, the unpolarized light that it produces, and the plane-polarized light transmitted by a polarizer. Optically active materials have the ability to *rotate the plane of polarized light*. The optically active substance in the sample tube of Figure 22.16 rotates the plane of polarized light to the right. The angle of rotation is measured by the required orientation of an analyzer (a second polarizer) to absorb the polarized light.

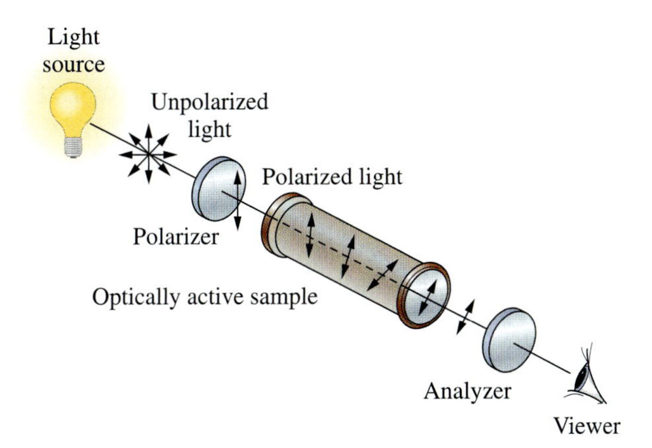

Figure 22.16
Optical activity
Ordinary light consists of electromagnetic waves that vibrate in all directions perpendicular to the line along which the light is propagated; it is unpolarized. Light having all vibrations in the same plane is *polarized*. Optically active substances are able to rotate the plane of polarized light.

Optical isomers are isomers that differ in their ability to rotate the plane of polarized light. Enantiomers are optical isomers that rotate the plane of polarized light to the same degree but in opposite directions, that is, one to the right and the other to the left. Measurements of optical activity can sometimes help to establish a correct structure from several hypothetical structures, as illustrated in Example 22.7.

EXAMPLE 22.7—A Conceptual Example

At one time, a triangular prism was considered as a possible geometric structure for coordination number 6. Assuming that ethylenediamine (en) molecules can link only to adjacent coordination sites, determine the number of possible geometric and/or optical isomers of $[Co(en)_3]^{3+}$ based on this structure.

SOLUTION

Four structures are sketched below, arranged as two pairs. If we look first at structures I and II, we see that each has (en) ligands along the three vertical edges of the prism. The two structures are identical. Structures III and IV have one (en) ligand along a vertical edge and the remaining two (en) ligands in the upper and lower triangular faces. Structures III and IV are mirror images.

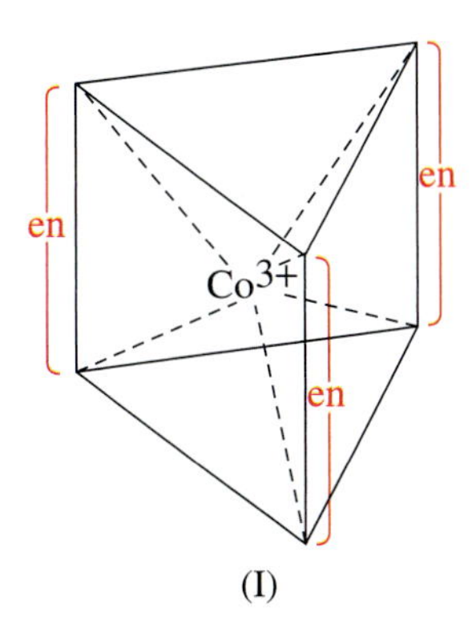

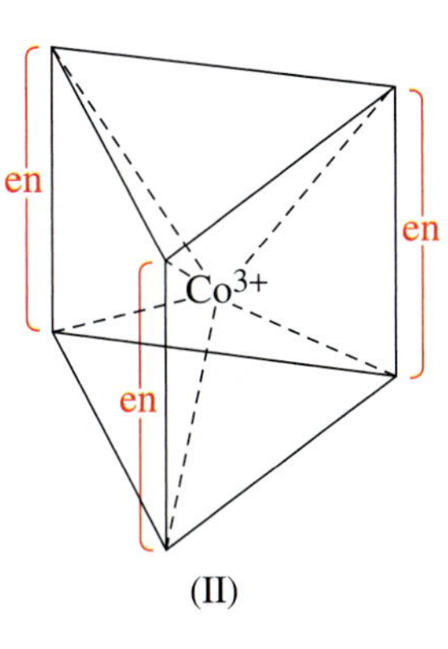

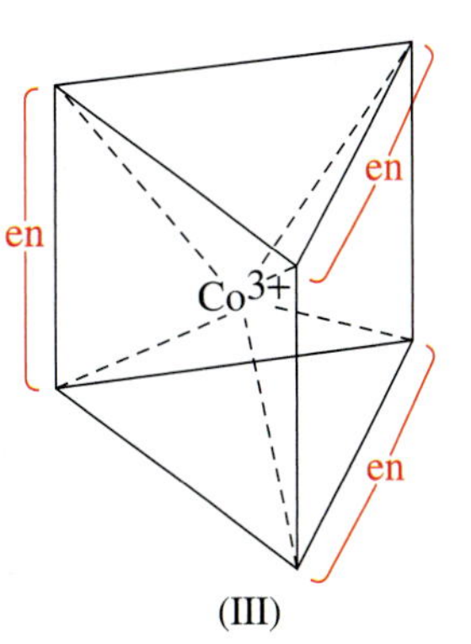

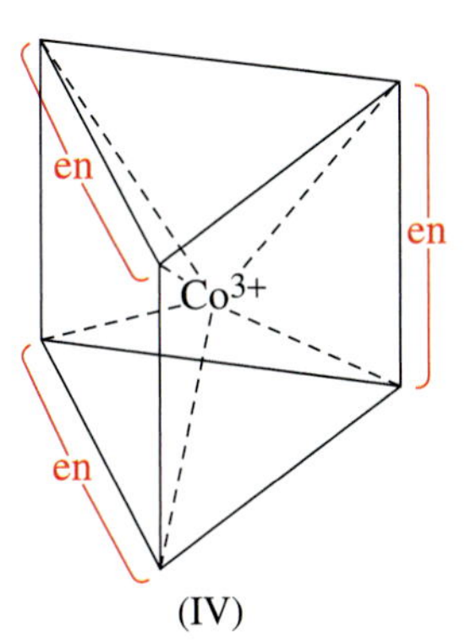

Structure I can be superimposed on structure II, and, following the appropriate rotation, structure III can be superimposed on its mirror image, structure IV. As a result, there is no optical isomerism. In contrast, structure I and structure III are different; they cannot be superimposed on one another. Adjacent coordination sites along a vertical edge of the prism are different from adjacent sites along an edge of a triangular face. Structures I and III are *geometric isomers.*

In summary, if the structure of $[Co(en)_3]^{3+}$ were a triangular prism, there would be *two* geometric isomers and *no* optical isomers. Instead, as we have seen in Figure 22.15, there are *two* optical isomers and *no* geometric isomers.

EXERCISE 22.7

Another early suggestion for coordination number 6 was that of a planar hexagon with the central atom or ion at its center.

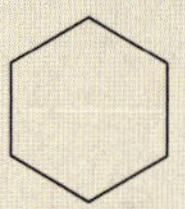

How many geometric and/or optical isomers would you predict for $[Co(en)_3]^{3+}$ based on the hexagonal structure?

When an optically active complex is synthesized, the amounts of the two enantiomers obtained are the same, and the rotation of the plane of polarized light produced by one isomer is just offset by the opposite rotation of the other. As a result, the mixture, called a *racemic mixture*, does not rotate the plane of polarized light at all. Separating the enantiomers in a racemic mixture requires some reaction that distinguishes between a structure and its mirror image. This is the molecular equivalent of separating a bin filled with gloves into separate piles for right- and left-handed gloves. We will encounter optical activity again in Chapter 23.

22.10 Bonding in Complexes: Crystal Field Theory

The Lewis acid-base theory helps explain how a metal ion and ligands join to form a complex ion. But we need more than the Lewis theory to explain certain properties of complex ions, especially their characteristic colors and magnetic properties. A theory that works well is the *crystal field theory*.

According to the crystal field theory, the attractions between a central atom or ion and its ligands are largely electrostatic. Lone-pair electrons on the ligands are attracted to the positively charged nucleus of the central metal. In addition, however, there are repulsions between the ligand electrons and *d* electrons of the central atom or ion. Crystal field theory focuses on the effect of these repulsions.

We first pictured the *five d* orbitals in Figure 7.27. We show them again in Figure 22.17, this time as a group in yellow and red. Six ligands are shown approaching the set of *d* orbitals along the *x*, *y*, and *z* axes. This direction of approach produces an octahedral structure for a complex. From this figure, we see that the ligands approaching along the *z* axis encounter most directly the lobes of the d_{z^2} orbitals (yellow), and those approaching along *x* and *y* axes encounter the lobes of the $d_{x^2-y^2}$ orbitals (yellow). As a result, the strongest repulsions occur between these two orbitals and the approaching ligands. This raises the energies of these *d* orbitals above what they would be in the free ion with no ligands present. The energies of the other *d* orbitals, d_{xy}, d_{xz}, and d_{yz}, are not raised as much. As a result,

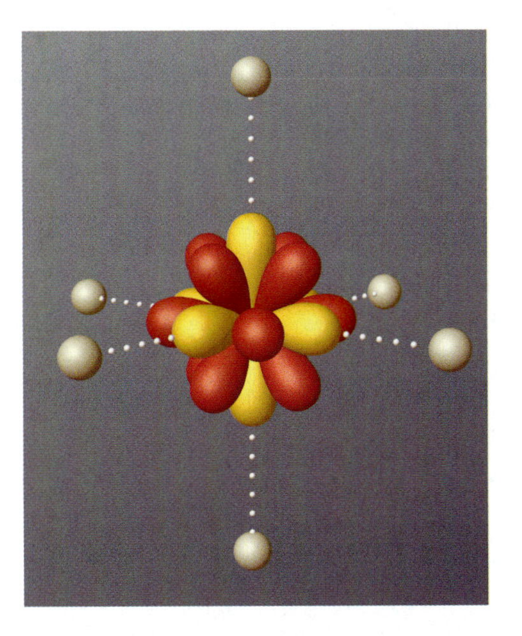

Figure 22.17
Ligand approach leading to formation of an octahedral complex
In the formation of an octahedral complex, ligands approach the central atom or ion along the *x*, *y*, and *z* axes. Maximum interference occurs with the d_{z^2} and $d_{x^2-y^2}$ orbitals (shown in yellow). The energies of these orbitals are raised with respect to those of the d_{xy}, d_{xz}, and d_{yz} orbitals (shown in red).

the *d* energy level of the central atom or ion is split into two groups. We will use the symbol Δ to designate the energy difference between the two groups.

In a similar manner, we could describe how ligands approach a central metal atom or ion to produce tetrahedral or square planar complexes. The observations we would make, together with that for the octahedral complex, are summarized in Figure 22.18. From left to right in this schematic representation, we see (1) the energy levels of *d* orbitals in a hypothetical central atom or ion free of ligands, (2) the *average* energy level to which the *d* orbitals are raised by the ligands, (3) the splitting of the energy levels that occurs in the case of (a) tetrahedral, (b) octahedral, and (c) square planar complexes.

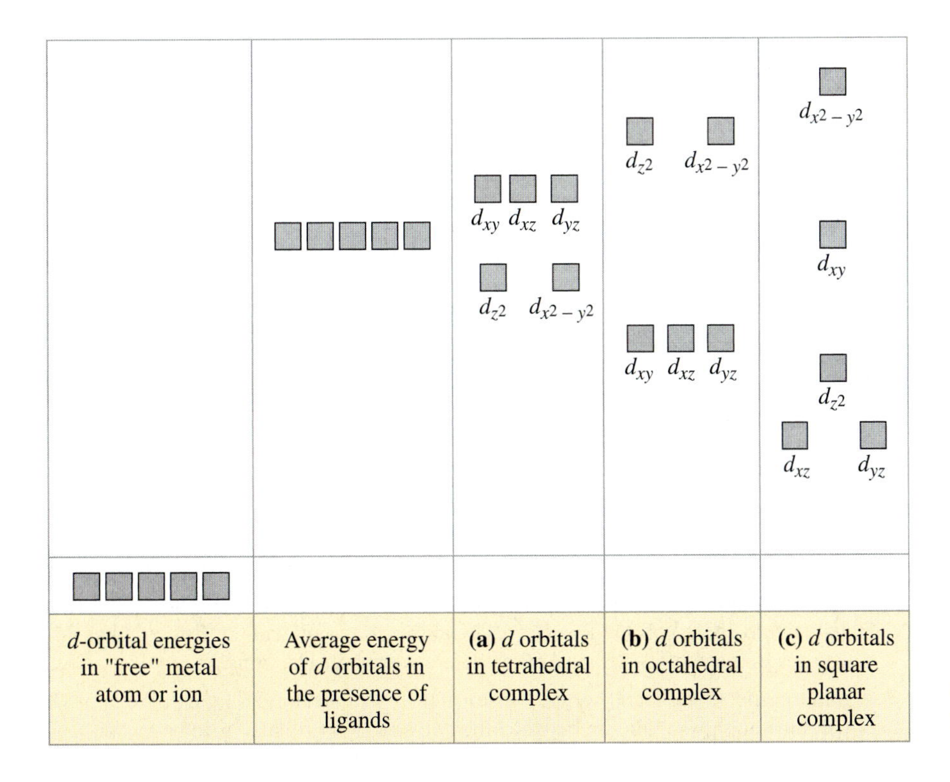

Figure 22.18
Schematic representation of *d*-level splitting

Of what importance is all this? To see, let's consider how the five 3*d* electrons in Fe^{3+} are distributed among the 3*d* orbitals in two of its complex ions. (Remember that iron has $Z = 26$ and the Fe^{3+} ion therefore has 23 electrons in the configuration $[Ar]3d^5$.)

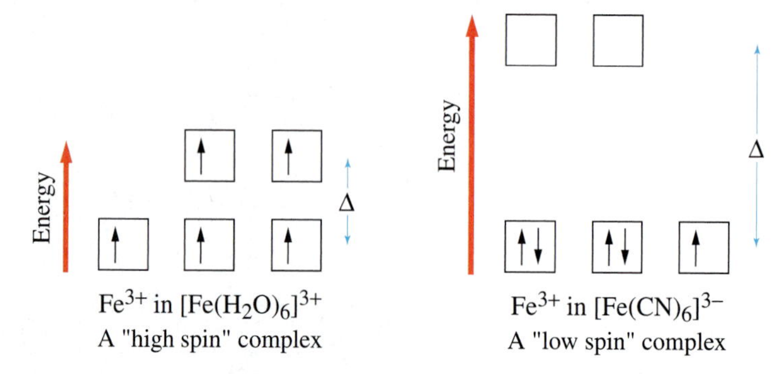

Because the energy separation (Δ) between the two groups of *d* orbitals is small in $[Fe(H_2O)_6]^{3+}$, the five 3*d* electrons of Fe^{3+} distribute themselves in the usual way. They appear as *five* unpaired electrons in the 3*d* orbitals. Because these unpaired electrons have parallel spins, we call $[Fe(H_2O)_6]^{3+}$ a "high spin" complex. In contrast, in $[Fe(CN)_6]^{3-}$ the energy separation (Δ) between the two groups of *d* orbitals is large compared to the energy benefit derived from unpaired electrons. The five electrons are all found in the lower energy *d* orbitals. Four are paired, and only *one* is unpaired. $[Fe(CN)_6]^{3-}$ is a "low spin" complex.

To predict the magnetic properties of a complex ion, we need to know its structure (tetrahedral, octahedral, or square planar) and the ability of its ligands to split the *d*-orbital energy levels. The following general arrangement, called the **spectrochemical series**, shows the relative abilities of some common ligands to split the *d*-orbital energy levels. Ligands that produce large energy separations between groups of *d* orbitals are called "strong field" ligands, and those that produce small energy separations are called "weak field" ligands.

Field strength: Strong → Weak

$CN^- > NO_2^- > en > NH_3 > H_2O > OH^- > F^- > Cl^- > Br^- > I^-$

d-Level splitting, Δ: Large → Small

The energy-level diagrams that we drew for the cyano and aqua complex ions of Fe^{3+} are consistent with the order suggested by the spectrochemical series.

EXAMPLE 22.8

How many unpaired electrons would you expect for the octahedral complex ion $[CoF_6]^{3-}$?

SOLUTION

We must do three things: (1) determine the number of 3*d* electrons in the central cobalt ion, (2) assess whether the energy difference between the two groups of *d* orbitals, Δ, is likely to be large or small, and (3) use the aufbau principle to distribute electrons among the *d* orbitals.

1. The complex ion has six F^- ions as ligands and a net charge of 3−. The central ion must carry a charge of 3+; it is Co^{3+}. Cobalt has $Z = 27$, and the Co^{3+} ion has 24 electrons in the configuration $[Ar]3d^6$. There are six 3*d* electrons.
2. From the spectrochemical series, we see that F^- is a weak field ligand. We should expect the energy separation between the two sets of *d* orbitals, Δ, to be small.

3. Because of this small energy separation, we should assign electrons to all five of the *d* orbitals singly before forming any pairs. There should be *four* unpaired electrons.

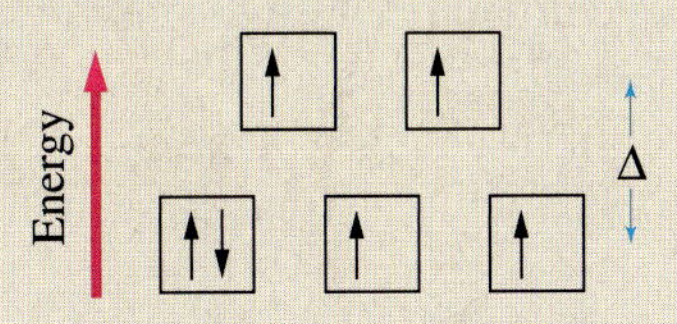

EXERCISE 22.8

How many unpaired electrons would you expect to find in each of the following complex ions: **(a)** the octahedral complex ion, $[Co(CN)_6]^{3-}$; **(b)** the tetrahedral complex ion, $[NiCl_4]^{2-}$?
(*Hint*: Refer to Figure 22.18.)

22.11 Color in Complex Ions and Coordination Compounds

A substance absorbs photons of light if the energies of the photons match the energies required to excite electrons to higher energy levels. If the absorbed photons are of visible light, the light *transmitted* by the substance is colored. That is, the substance absorbs one or more colors of light and transmits the rest. The color of the transmitted light is the *complementary* color of the absorbed light. This idea is illustrated in Figure 22.19. The colors of five solutions encountered elsewhere in the text are shown in Figure 22.20 in relation to the color of light they absorb and the color of light they transmit.

Ions having a noble-gas electron configuration (for example, Na^+ or Cl^-), an outer shell with 18 electrons (for example, Zn^{2+}), or an "18 + 2" configuration

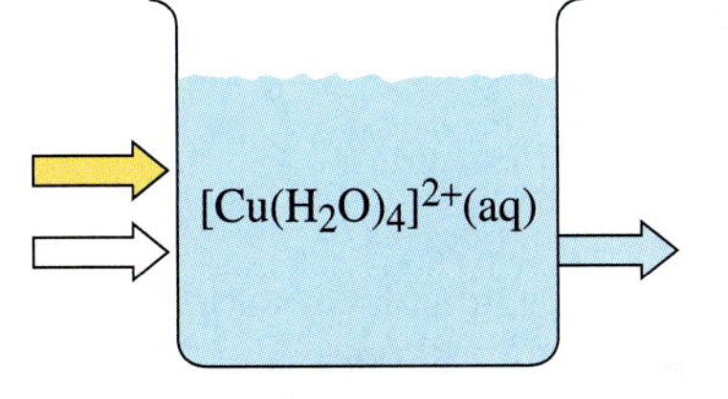

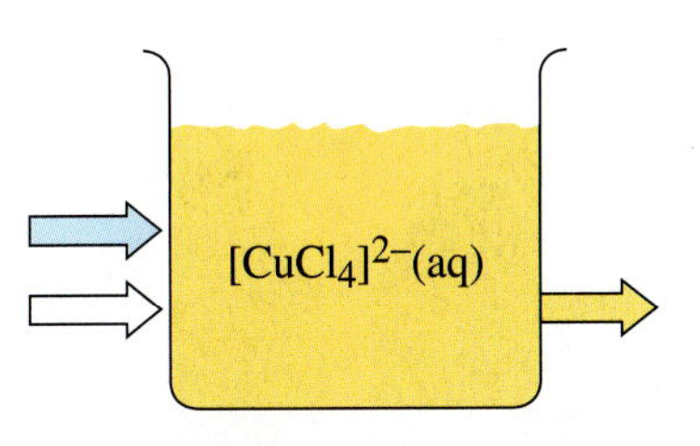

▲ **Figure 22.19**
Light absorption and the origin of color
In $[Cu(H_2O)_4]^{2+}(aq)$, the yellow component of white light is absorbed, and the *complementary* color of yellow—*blue*—is transmitted. Yellow light is not transmitted. In $[CuCl_4]^{2-}(aq)$, the situation is reversed: Blue light is absorbed, and only its complementary color—*yellow*—is transmitted. Blue light is not transmitted.

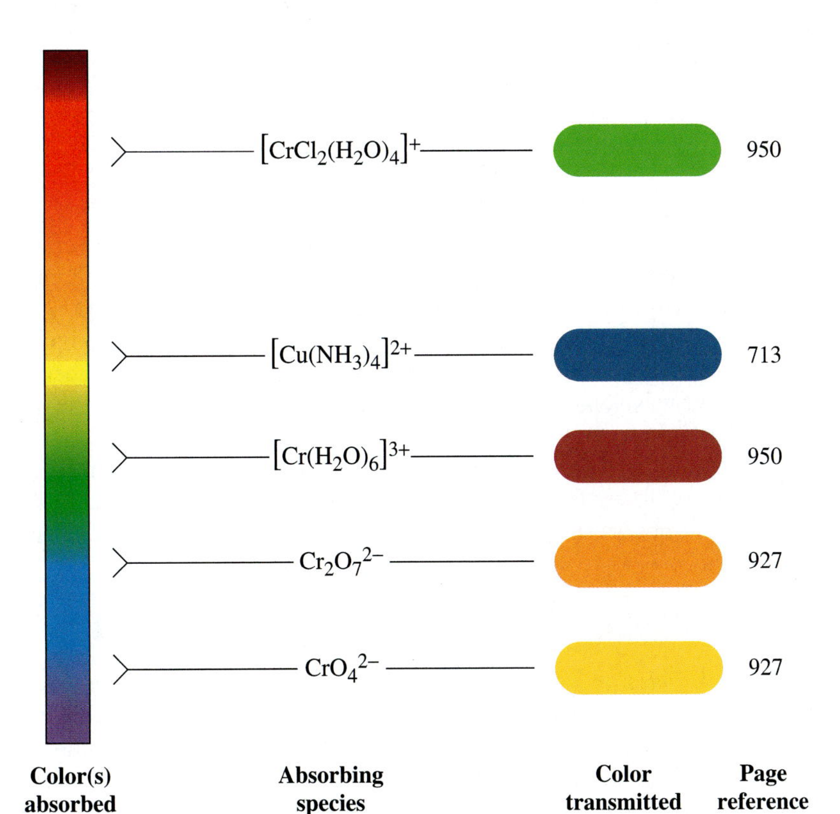

◀ **Figure 22.20**
The relationship between absorbed and transmitted light

(for example, Sn^{2+}) have *no* electron transitions in the energy range of visible light. These ions do not absorb visible light, and their aqueous solutions are colorless.

Many complex ions are colored because the energy differences between *d* orbitals, Δ, match the energies of components of visible light. Substituting one ligand for another can produce subtle changes in the energy levels of the *d* orbitals and striking changes in the colors of complex ions, as seen in Figures 22.21 and 22.22.

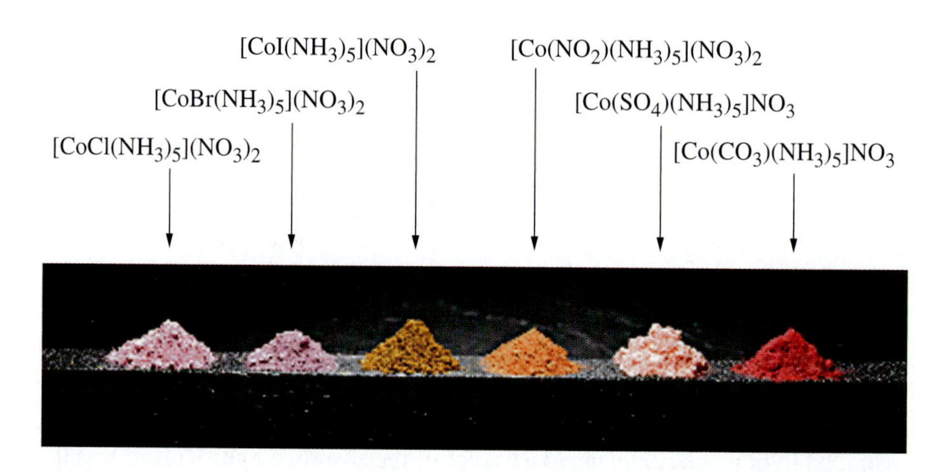

▶ **Figure 22.21**
Effect of ligands on colors of coordination compounds

▲ **Figure 22.22**
Colors of chromium(III) complex ions
The green solid $CrCl_3 \cdot 6H_2O$ dissolves in water to form a green solution. The green color is due to $[CrCl_2(H_2O)_4]^+$ (left). A slow exchange of H_2O molecules for Cl^- ions as ligands leads to a reddish solution of $[Cr(H_2O)_6]^{3+}$ in one or two days (right).

We can use crystal field theory to explain the colors of complex ions. For example, to explain why solutions of $[Cr(H_2O)_6]^{3+}$ are *violet* whereas those of $[Cr(NH_3)_6]^{3+}$ are *yellow*, let's construct a *d*-orbital energy-level diagram for these octahedral complex ions. We begin by noting that the electron configuration of Cr is $[Ar]3d^54s^1$ and that of Cr^{3+} is $[Ar]3d^3$. The three electrons are assigned to 3*d* orbitals as indicated below.

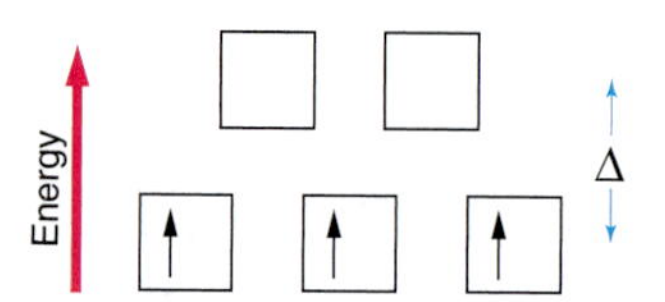

The color of light absorbed in promoting an electron from a lower energy to a higher energy *d* orbital depends on the magnitude of the energy difference, Δ. Because NH_3 is a stronger field ligand than H_2O, Δ is larger for $[Cr(NH_3)_6]^{3+}$ than for $[Cr(H_2O)_6]^{3+}$. Compared to the aqua complex ion, the ammine complex ion absorbs light of higher energy, and thus higher frequency. Light of higher frequency is of shorter wavelength. We should expect $[Cr(NH_3)_6]^{3+}$ to absorb toward the violet end of the spectrum and transmit toward the red end. Thus, the yellow color of $[Cr(NH_3)_6]^{3+}$ is a reasonable expectation. The light absorbed by $[Cr(H_2O)_6]^{3+}$ is of lesser energy, lower frequency, and longer wavelength. Figure 22.20 shows that $[Cr(H_2O)_6]^{3+}$ absorbs in the green region of the spectrum and displays a violet color.

22.12 Chelates: Complexes of Special Interest

Recall that *chelation* (page 939) is the formation of five- or six-membered rings of atoms through the attachment of polydentate ligands to central metal atoms or ions. We gave an example of a chelate in Figure 22.12. Now we will focus on a group of common chelating agents—the salts of *ethylenediaminetetraacetic acid* (EDTA). The actual ligand in these salts is the ethylenediaminetetraacetate ion, $EDTA^{4-}$.

The formula of $EDTA^{4-}$ is shown in Table 22.4. Figure 22.23 shows the structure of a metal-EDTA complex. The five-membered rings (five) are outlined in red.

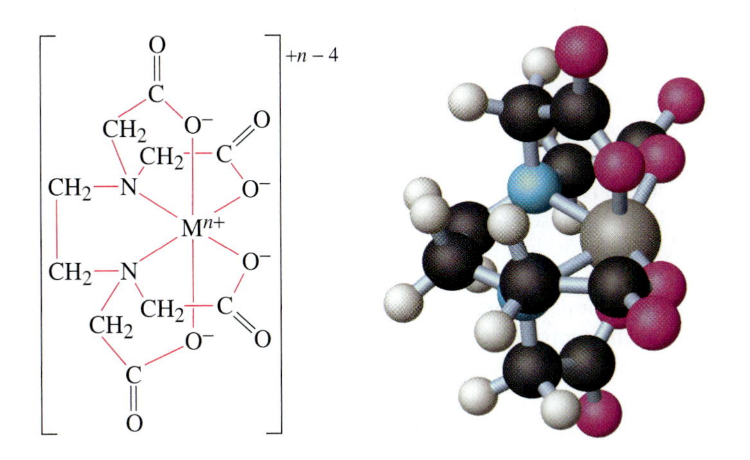

Figure 22.23 Structure of a metal-EDTA complex
The central metal ion M^{n+} (gray) can be any of many different cations: Ca^{2+}, Mg^{2+}, Fe^{2+}, Fe^{3+}, and so on. The charge on the EDTA anion is 4−, and the net charge of the complex ion is that of the central ion ($n+$) plus that of the anion (4−), that is, $+n - 4$.

Stabilities of Chelates

Chelates generally are much more stable than complexes with monodentate ligands. Consider, for example, the net reaction in which an $EDTA^{4-}$ ion replaces the six NH_3 molecules as ligands in the complex ion $[Ni(NH_3)_6]^{2+}$.

$$[Ni(NH_3)_6]^{2+}(aq) + EDTA^{4-}(aq) \rightleftharpoons [NiEDTA]^{2-}(aq) + 6\,NH_3(aq) \quad K = 8 \times 10^9$$

In Chapter 17, we learned that (1) an equilibrium constant is related to a standard free energy change: $\Delta G^\circ = -RT \ln K$, and (2) that the free energy change depends on the changes in enthalpy and entropy: $\Delta G^\circ = \Delta H^\circ - T\Delta S^\circ$. The large value of K and the corresponding negative value of ΔG° indicate that this reaction is thermodynamically favorable. As a consequence, we say that $[NiEDTA]^{2-}$ is more stable than $[Ni(NH_3)_6]^{2+}$. The negative value of ΔG° can be attributed in large part to the *increase* in entropy in the reaction. For every two particles of reactant—the ions $[Ni(NH_3)_6]^{2+}$ and $EDTA^{4-}$—seven particles are produced—the ion $[NiEDTA]^{2-}$ and six NH_3 molecules. The final state in the reaction is more disordered than the initial state. The increase in entropy during chelation, then, is a contributing factor to the stabilities of chelates.

Sequestering Metal Ions

To sequester a metal ion in solution means to tie it up in a form that effectively removes it from solution. $EDTA^{4-}$ is one of the best ligands for sequestering metal ions.

EDTA is sometimes added to boiler water to sequester metal ions. When they are the central ions in chelate structures, Ca^{2+}, Mg^{2+}, and Fe^{3+} are no longer able to form boiler scale (page 870). Another application is to prevent the growth of certain bacteria in liquid soaps, shampoos, and other personal products. Chelation with EDTA removes Ca^{2+} and Mg^{2+} ions, which are important constituents of the cell walls of these bacteria. The cell walls disintegrate, and the bacteria die.

This iron chelate plant food is an iron–EDTA complex.

Because of its hexadentate character and high negative charge, EDTA has the interesting effect of converting a simple cation into the central ion of an *anion*, for example, $Fe^{3+}(aq)$ to $[FeEDTA]^-$. For use as a plant food, iron can be effectively transported through soils as $[FeEDTA]^-(aq)$. Clay particles in soils have anions on their surfaces, and $Fe^{3+}(aq)$ cannot easily migrate through such a soil; rather it is immobilized by combining with surface anions. However, complex anions containing iron(III) are not held back by these surface anions, and the iron(III) is thus available to plants.

Key Terms

bidentate (22.7)
chelate (22.7)
complex (22.7)
complex ion (22.7)
coordination compound (22.7)
coordination number (22.7)
crystal field theory (22.10)
enantiomers (22.9)
ferromagnetism (22.3)
lanthanide contraction (22.1)
ligand (22.7)
monodentate (22.7)
optical isomers (22.9)
polydentate (22.7)
spectrochemical series (22.10)
stereoisomerism (22.9)

Summary

The elements of the *d*-block—all metals—exist in several oxidation states and form many complex ions and colored compounds. The early members in a period in the *d*-block are active metals but later members are less active. Within the B-groups, fifth- and sixth-period members have a greater tendency to exist in their higher oxidation states than do fourth-period members. Also, because of the poor shielding of valence electrons by electrons in the *f* subshell, atomic radii of the members of the fifth and sixth periods are about the same (the lanthanide contraction).

Scandium resembles aluminum but is not widely used. The uses of titanium depend on its high strength, low density, and corrosion resistance. TiO_2 is used as a white pigment. Vanadium exists in several oxidation states, and its ions display a variety of colors. Vanadium, chromium, and manganese are all used in the manufacture of steel. Dichromate and permanganate ions ($Cr_2O_7^{2-}$ and MnO_4^-) are widely used oxidizing agents. Dichromate ion converts to chromate ion (CrO_4^{2-}) in a pH-dependent equilibrium. Chromate ion is a precipitating agent. Iron, cobalt, and nickel have similar properties, one being that they are ferromagnetic; they can be permanently magnetized.

Copper, silver, and gold differ from earlier members of their periods by being much less active metals. They are unable to displace H_2 from acidic solutions. Uses of these metals are based mostly on their resistance to corrosion and their exceptional abilities to conduct heat and electricity. Zinc, cadmium, and mercury are not transition elements; their atoms and ions all have filled *d* subshells. Mercury differs from zinc and cadmium in several ways, for example, in its inability to displace H_2 from acidic solutions.

The central metal atom or ion of a metal complex is a Lewis acid. It forms coordinate covalent bonds by accepting lone-pair electrons from ligands, which are Lewis bases. A monodentate ligand attaches at a single coordination site of the central metal, and a polydentate ligand attaches at two or more sites.

Isomerism among complexes is of two general types. Structural isomers differ in their ligands and/or the donor atoms through which the ligands are bonded to the metal center. Stereoisomers differ in the manner in which ligands occupy the space around the central atom. Geometric isomerism (cis-trans) is one type of stereoisomerism and optical isomerism is another. Enantiomers, a type of optical isomers, bear the same relationship to one another as an object and its nonsuperimposable mirror image. The only physical difference between a pair of enantiomers is in the direction in which they rotate the plane of polarized light. One isomer rotates it to the right, and the other rotates it to the left.

Interactions between lone-pair electrons on ligands and electrons in the *d* orbitals of the central metal atom or ion produce a splitting of the *d*-orbital energy level. Electron transitions between *d* orbitals of different energy provide a way for a complex to absorb some wavelength components of visible light and transmit others, giving rise to color. Color is a common property of transition metal complexes, and it is strongly influenced by the particular ligands present. Explanations of the colors and magnetic properties of complexes are facilitated by a listing of common ligands called the spectrochemical series.

Review Questions

1. Why are *d*-block elements first found in the fourth period and *f*-block elements in the sixth period?
2. One atom has the electron configuration $[Kr]4d^{10}5s^25p^1$, and another has the electron configuration $[Ar]3d^74s^2$. What are these atoms, and which is a transition element? Explain.
3. Name three properties in which transition metals differ from main-group metals.
4. What is the lanthanide contraction, and among which elements is it observed?
5. What is ferromagnetism, and among which elements is it observed?
6. Name one or more *d*-block metal hydroxides having only basic properties, and name one or more that is amphoteric.
7. Give the chemical composition of the material known as **(a)** galvanized iron, **(b)** chromite ore, **(c)** basic copper carbonate.
8. Give a common name that is often used for the material that is **(a)** an iron-manganese alloy obtained in the met-

allurgy of manganese, **(b)** a copper-zinc alloy, **(c)** a mixture of nitric and hydrochloric acids.

9. Why is the most stable oxidation state of Fe +3, whereas that of Co and Ni is +2?

10. Write a name that adequately describes each of the following: **(a)** $ScCl_3$, **(b)** Fe_2SiO_4, **(c)** Na_2MnO_4, **(d)** CrO_3.

11. Write a formula that adequately describes each of the following: **(a)** barium dichromate, **(b)** chromium(III) oxide, **(c)** mercury(I) bromide, **(d)** vanadium pentoxide.

12. What happens when copper is added to concentrated HCl(aq)? to concentrated HNO_3(aq)? Why are the results different in the two cases?

13. What are the component parts of a complex ion?

14. What is the relationship between a complex ion and a coordination compound?

15. Explain how it is possible that a complex containing a cation may itself be an anion. Could the complex be an electrically neutral molecule?

16. Explain how it is possible that the coordination number and the number of ligands in a complex ion may not be the same.

17. How is the Lewis acid-base theory used to describe complex-ion formation? How is it possible that ammonia is both a Brønsted-Lowry base and a Lewis base?

18. Explain the meaning and give an example of a monodentate ligand and of a polydentate ligand.

19. What is a chelate, and what conditions are required for its formation?

20. What are the ligands represented by the following names or symbols?

(a) aqua **(c)** ammine **(e)** ox

(b) chloro **(d)** en **(f)** EDTA

21. What do the terms bis and tris mean when they are used in the name of a complex ion?

22. What does the ending -ate signify when referring to the metal atom in a complex ion, for example, cobaltate(III)?

23. What is the difference between a nitro and a nitrito ligand in a complex ion?

24. What is the difference between a cis and a trans isomer of a complex ion?

25. How does stereoisomerism differ from structural isomerism? How does optical isomerism differ from geometric isomerism?

26. What is meant by the following terms?

(a) "high spin" and "low spin" complexes

(b) "strong field" and "weak field" ligands

(c) the spectrochemical series

Problems

Properties of the Transition Elements

27. Sc and Ca both have two valence shell electrons ($4s^2$), yet Ca has only the oxidation number +2 in its compounds whereas Sc has the oxidation number +3. Explain this difference.

28. In comparing some metal atoms and their ions, we find that Sc is paramagnetic and Sc^{3+} is diamagnetic, whereas both Ti and Ti^{2+} are paramagnetic and both Zn and Zn^{2+} are diamagnetic. Explain these differences.

29. The two adjacent elements in the periodic table, K and Ca, have atomic (metallic) radii of 227 and 197 pm, respectively. Another two adjacent elements, Mn and Fe, each has an atomic radius of 124 pm. Explain why **(a)** the atomic radius of Ca is smaller than that of K; **(b)** the atomic radius of Mn is smaller than that of Ca; and **(c)** the atomic radii of Mn and Fe are the same.

30. In Group 5B, Nb and Ta each has an atomic (metallic) radius of 143 pm, whereas that of V, also a member of Group 5B, is 131 pm. **(a)** Why are all three radii not the same? **(b)** If there is a reason why all three should not be the same, why are two of the radii still found to be the same?

Scandium Through Manganese

31. Write plausible equations for the following.

(a) the reaction of Sc(s) with HCl(aq)

(b) the reaction of $Sc(OH)_3$(s) with HCl(aq)

(c) the reaction of $Sc(OH)_3$(s) with excess NaOH(aq)

32. Write plausible equations for the following.

(a) the reaction of Mn(s) with HCl(aq)

(b) the reaction of $Mn(OH)_2$(s) with HCl(aq)

(c) the reaction of $Cr(OH)_3$(s) with NaOH(aq)

33. Suggest reactions with common chemicals by which the following syntheses can be accomplished.

(a) $BaCrO_4$(s), starting with barium metal and K_2CrO_4(aq)

(b) MnO_2(s), starting with $KMnO_4$(aq)

34. Suggest reactions with common chemicals by which the following syntheses can be accomplished.

(a) $CrCl_3$(s), starting with $K_2Cr_2O_7$(aq)

(b) $MnCO_3$(s), starting with $KMnO_4$(aq)

35. A qualitative analysis test for Mn^{2+} involves its oxidation to MnO_4^- in acidic solution by sodium bismuthate, $NaBiO_3$. The bismuthate ion is reduced to Bi^{3+}. Write an equation for this redox reaction.

36. The text suggests a two-step process for obtaining $KMnO_4$ from $MnO_2(s)$ (page 930). Write a pair of balanced equations for the process.

37. When *colorless* $Pb(NO_3)_2(aq)$ is added to *orange* $K_2Cr_2O_7(aq)$, *yellow* $PbCrO_4(s)$ precipitates. Explain how this comes about.

38. When *colorless* HCl(aq) is added to *orange* $K_2Cr_2O_7(aq)$, a *green* solution and a *yellowish green* gas are formed. Explain how this comes about.

The Iron Triad

39. Which of the following would you expect to be the best *reducing* agent: Fe(s), $Fe^{2+}(aq)$, $Co^{3+}(aq)$, or Co(s)? Explain.

40. Which would you expect to be the better *oxidizing* agent: $Fe^{2+}(aq)$ or $Co^{3+}(aq)$? Explain.

41. Although iron does not commonly occur with an oxidation number higher than +3, $Fe_2O_3(s)$ can be oxidized to ferrate ion, FeO_4^{2-}, by $Cl_2(g)$ in a strongly basic solution. Write a balanced redox equation for this reaction.

42. Show that an acidic solution of $Co^{3+}(aq)$ is unstable, and write a plausible balanced redox equation for the reaction that occurs. (For the half-reaction $Co^{3+}(aq) + e^- \longrightarrow Co^{2+}(aq)$, $E° = 1.92$ V.)

Group 1B

43. Write plausible net ionic equations for the following reactions that are described on page 932.

(a) $Cu(s) + H_2SO_4(\text{concd aq}) \longrightarrow$

(b) $Cu(s) + HNO_3(\text{concd aq}) \longrightarrow$

44. Write plausible equations for the following reactions that are described on page 932.

(a) $Ag(s) + H_2SO_4(\text{concd aq}) \longrightarrow$

(b) $Ag(s) + HNO_3(\text{concd aq}) \longrightarrow$

45. Explain why it is that silver reacts with HNO_3(concd aq) and gold does not.

46. Explain why gold, which reacts with neither HCl(aq) nor $HNO_3(aq)$, does react with a mixture of the two acids. Write a net ionic equation.

Group 2B

47. Describe some important ways in which the Group 2B elements differ from those in Group 2A.

48. Describe some important ways in which mercury differs chemically from zinc and cadmium.

49. Write plausible net ionic equations to represent the following reactions.

(a) Hg(l) with HNO_3(concd aq)

(b) ZnO(s) with $CH_3COOH(aq)$

50. Write plausible net ionic equations to represent the following reactions.

(a) Cd(s) with H_2SO_4(concd aq)

(b) $ZnSO_4(aq) \xrightarrow[\text{Pt electrodes}]{\text{Electrolysis}}$

Coordination Chemistry

51. What is the coordination number of the central metal atom in the following complexes?

(a) $[PtCl_3(NH_3)_3]^+$ **(c)** $[Ni(CO)_4]$

(b) $[Cu(en)_2]^{2+}$ **(d)** $[FeEDTA]^{2-}$

52. How many ligands of the following type would be found in an octahedral complex with Cr^{3+} as the central metal ion?

(a) CN^- **(c)** $^-OOCCOO^-$

(b) $H_2NCH_2CH_2NH_2$ **(d)** $EDTA^{4-}$

53. Indicate the oxidation number of the central metal atom in each of the following.

(a) $[Pt(en)_2]^{2+}$ **(c)** $[Mn(CN)_6]^{3-}$

(b) $[CoCl_2(NH_3)_4]^+$ **(d)** $[CrCl_4(H_2O)_2]^-$

54. Indicate the oxidation number of the central metal atom in each of the following.

(a) $[NiBr_4]^{2-}$ **(c)** $[PtCl_2(NH_3)_4]^{2+}$

(b) $[FeF_5(H_2O)]^{2-}$ **(d)** $[Fe(CO)_5]$

Nomenclature

55. Name the following complex ions.
   (a) $[Cu(NH_3)_4]^{2+}$
   (b) $[FeF_6]^{3-}$
   (c) $[PtCl_2(NH_3)_4]^{2+}$
   (d) $[Cr(en)_3]^{3+}$

56. Name the following complex ions.
   (a) $[Fe(OH)(H_2O)_5]^+$
   (b) $[Co(NO_2)_2(en)_2]^+$
   (c) $[Ag(S_2O_3)_2]^{3-}$
   (d) $[Fe(ox)_3]^{3-}$

57. Write the formula of each of the following.
   (a) hexaaquairon(II) ion
   (b) tetraaminebromochlorochromium(III) ion
   (c) trioxalatoaluminate(III) ion

58. Write the formula of each of the following.
   (a) hexafluorocobaltate(III) ion
   (b) tetraaquadichlorochromium(III) ion
   (c) dibromobis(ethylenediamine)cobalt(III) ion

59. Name the following coordination compounds.
   (a) $K_4[Cr(CN)_6]$
   (b.) $K_3[Cr(ox)_3]$

60. Name the following coordination compounds.
   (a) $K_2[PtCl_6]$
   (b) $NH_4[Cr(NCS)_4(NH_3)_2]$

61. Write formulas for the following coordination compounds.
   (a) sodium tetrahydroxozincate(II)
   (b) tris(ethylenediamine)chromium(III) sulfate
   (c) dipotassium sodium hexanitrocobaltate(III)

62. Write formulas for the following coordination compounds.
   (a) tetraamminecopper(II) tetrachloroplatinate(II)
   (b) tris(ethylenediamine)chromium(III) hexacyanocobaltate(III)
   (c) tetraammineplatinum(II) tetrachloroplatinate(II)

63. Each of the following names is in error. Point out the error, and give the correct name.
   (a) tetrahydroxozinc(II) ion
   (b) iron(III)hexafluoride ion

64. Each of the following names is in error. Point out the error, and give the correct name.
   (a) tetraaquacuprate(II) ion
   (b) pentaamminesulfatocobalt ion.

Isomerism

65. Which of the following pairs must be isomers? Explain your conclusion in each case.
   (a) $[Cu(NH_3)_2(H_2O)_2]Cl_2$ and $[Cu(H_2O)_2(NH_3)_2]Cl_2$
   (b) $[PtCl_2(NH_3)_4]Br_2$ and $[PtBr_2(NH_3)_4]Cl_2$
   (c) $[Co(NH_3)_6]Cl_3$ and $[Co(NH_3)_6]Cl_2$
   (d) $[Co(NO_2)(NH_3)_5]^{2+}$ and $[Co(ONO)(NH_3)_5]^{2+}$

66. Which of the following pairs must be isomers? Explain your conclusion in each case.
   (a) $[Co(NCS)(NH_3)_5]^+$ and $[Co(SCN)(NH_3)_5]^+$
   (b) $[PtCl(NH_3)_3]_2[PtCl_4]$ and $[Pt(NH_3)_4][PtCl_3(NH_3)]_2$
   (c) $K_3[Fe(CN)_6]$ and $K_4[Fe(CN)_6]$
   (d) $[Co(NO_3)(NH_3)_5]SO_4$ and $[Co(SO_4)(NH_3)_5]NO_3$

67. Draw appropriate sketches to show that the complex ion $[CrCl_4(en)]^-$ *does not* exhibit geometric isomerism.

68. Draw appropriate sketches to show that the complex $[CrCl_3(NH_3)_3]$ *does* exhibit geometric isomerism.

69. Would you expect the octahedral complex ion $[Cr(en)(ox)_2]^-$ to exist as optical isomers? Explain.

70. Would you expect the tetrahedral complex ion $[ZnCl(NH_3)_3]^+$ to exist as optical isomers? Explain.

Magnetism and Color in Complexes

71. Use an orbital diagram to show the expected distribution of electrons in the 3*d* orbitals of the central metal ion in each of the following complex ions. If more than one distribution seems possible, indicate whether you expect the low-spin or the high-spin state to be favored. Also. indicate the number of unpaired electrons in each ion.
   (a) octahedral $[Cr(H_2O)_6]^{3+}$
   (b) octahedral $[FeCl_6]^{3-}$
   (c) octahedral $[Mn(CN)_6]^{3-}$
   (d) tetrahedral $[CoCl_4]^{2-}$

72. Use an orbital diagram to show the expected distribution of electrons in the 3*d* orbitals of the central metal ion in each of the following complex ions. If more than one distribution seems possible, indicate whether you expect the low-spin or the high-spin state to be favored. Also, indicate the number of unpaired electrons in each ion.
   (a) octahedral $[PtCl_6]^{2-}$
   (b) tetrahedral $[Zn(H_2O)_4]^{2+}$
   (c) octahedral $[Fe(CN)_6]^{4-}$
   (d) square planar $[PtCl_4]^{2-}$

73. One of the following compounds has a yellow color, and the other is violet: $[Cr(H_2O)_6]Cl_3$ and $[Cr(NH_3)_6]Cl_3$. Which do you think is the yellow compound, and which is the violet? Explain.

74. One of the following compounds has a green color, and the other has a yellow color: $Fe(NO_3)_2 \cdot 6H_2O$ and $K_4[Fe(CN)_6] \cdot 3H_2O$. Which do you think is the green compound, and which is the yellow? Explain.

Additional Problems

75. The text states that chromate ion is not a particularly good oxidizing agent. Yet, when $Na_2CrO_4(s)$ is added to hot concentrated HCl(aq), chlorine gas is formed. Explain why, in fact, this is what you should expect to observe.

76. Write a plausible balanced equation to represent the air oxidation of copper to basic copper carbonate, $Cu_2(OH)_2CO_3$.

77. Manganate ion, MnO_4^{2-}, is unstable in acidic solution and disproportionates. Write a plausible ionic equation for the disproportionation reaction. (*Hint:* What are the likely products?)

78. Show that the oxidation of $Fe^{2+}(aq)$ to $Fe^{3+}(aq)$ by atmospheric oxygen in acidic solution is a spontaneous reaction when all reactants and products are in their standard states. Is the reaction also spontaneous if the partial pressure of $O_2(g) = 0.20$ atm, $[Fe^{3+}] = 1.0$ M, $[Fe^{2+}] = 0.10$ M, and pH $= 5$? (*Hint:* Either use the Nernst equation or determine K_{eq} for the reaction.)

79. You have available as reducing agents $Sn^{2+}(aq)$, $V^{2+}(aq)$, and $Fe^{2+}(aq)$. For each one, determine whether it is capable of carrying out the following reductions in acidic solution.

(a) $MnO_2(s)$ to $Mn^{2+}(aq)$ (c) $H^+(aq)$ to $H_2(g)$

(b) $I_2(s)$ to $I^-(aq)$

80. You have available as oxidizing agents: $Fe^{3+}(aq)$, $NO_3^-(aq)$, and $V^{3+}(aq)$. For each one, determine whether it is capable of carrying out the following oxidations in acidic solution.

(a) $H_2(g)$ to $H^+(aq)$ (c) $Hg(l)$ to $Hg^{2+}(aq)$

(b) $Sn^{2+}(aq)$ to $Sn^{4+}(aq)$

81. What mass of chromium can be electrodeposited in 3.25 h with a current of 2.57 A from a chromium-plating bath of the type described on page 927?

82. The vapor pressure of Hg as a function of temperature is given by the following equation.

$$\log P\,(\text{mmHg}) = \frac{-a}{T} + b$$

In the equation, $a = 3{,}236$, $b = 8.118$, and $T =$ Kelvin temperature. Show that the concentration of Hg(g) in equilibrium with Hg(l) at 25 °C greatly exceeds the maximum permissible level of 0.05 mg Hg/m^3.

83. Explain why one of the cis-trans isomers of $[CoCl_2(en)_2]$ exhibits optical isomerism and the other does not. Which is which?

84. Would you expect to find optical isomerism in either of the following hypothetical tetrahedral complexes: $[ZnA_2B_2]^{2+}$ and $[ZnABCD]^{2+}$ (where the ligands A, B, C, and D are different neutral molecules)? Explain.

85. The complex $[PtCl_2(NH_3)_2]$ displays cis-trans isomerism, but $[ZnCl_2(NH_3)_2]$ does not. Why do you suppose these two cases are different?

86. Does the chelate $[M(EDTA)]^{+n-4}$ pictured in Figure 22.23 display optical isomerism? Explain.

87. The magnetic properties of the octahedral complex ion $[Cr(L)_6]^{3+}$ are independent of the identity of the ligands (L). How do you account for this fact?

88. In some of its complex ions, the paramagnetism of Cr^{2+} corresponds to two unpaired electrons and in others to four. How do you account for this difference?

89. A compound known before Werner's time was Magnus's green salt, having the *empirical* formula $PtCl_2 \cdot 2NH_3$. It is actually a coordination compound comprised of a dipositive complex cation and a dinegative complex anion. Propose a formula that is consistent with this information.

90. Zeise's salt has the *empirical* formula $PtCl_2 \cdot KCl \cdot C_2H_4$, but it is actually a coordination compound consisting of a unipositive cation and a complex anion. Propose a formula that is consistent with this information.

91. The complex ion $[Ni(CN)_4]^{2-}$ is *diamagnetic*. Use crystal field theory to determine whether the structure of this complex ion is octahedral, square planar, or tetrahedral.

92. Four structures are shown in the accompanying sketch. Indicate whether any of these structures are identical, whether any are geometric isomers, and whether any are optical isomers.

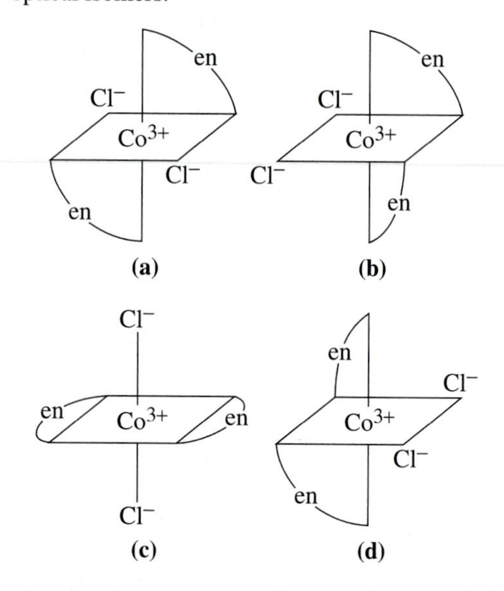

Chemistry and Life: More On Organic, Biological, and Medicinal Chemistry

CHAPTER

23

The chemistry of carbon compounds pervades almost every aspect of our lives. The fragrant odor of jasmine is largely due to benzyl acetate, $C_6H_5CH_2OCOCH_3$, an ester synthesized by various species of *Jasminum*. Chemists have determined the structure of the ester and it is now synthesized in factories for use in fragrances.

CONTENTS

In this chapter and the two that follow, we extend our study of chemistry into more complex systems. For example, we will see that the principles of chemistry apply to reactions in living cells (Sections 23.7–23.10), to the materials that contribute so much to our modern way of life (Chapter 24), and to the environment in which we live (Chapter 25). In doing so, we demonstrate that science is a unified whole, and not just a collection of isolated facts.

We begin this chapter by recapitulating some of the organic chemistry discussed in earlier chapters. Then we apply some of the principles that we learned subsequently to still more organic compounds. With this background, we move on to the more complex compounds that make up living matter. Then we consider some techniques for determining molecular structures. Finally, we examine some of the chemistry involved in the design and action of common drugs.

Organic compounds do not differ in any fundamental way from inorganic compounds. Indeed, we have used both types to illustrate molecular structures, acid-base chemistry, thermodynamic principles, and other key chemical concepts. However,

● **ChemCDX**
When you see this icon, there are related animations, demonstrations, and exercises on the CD accompanying this text.

because there are so many of them, because they can be conveniently organized into families, and for historical reasons, organic compounds are usually studied as a separate branch of chemistry.

Millions of carbon compounds are already known, and new ones are being discovered every day. Why so many carbon compounds? Recall that carbon atoms are unique in their ability to bond to each other to form long chains, the chains can have branches, and they can form rings of various sizes. Add to this the fact that carbon atoms also bond strongly to other elements, such as hydrogen, oxygen, and nitrogen, and that these atoms can be arranged in many different ways, and it soon becomes obvious that carbon can form an almost infinite number of molecules of various shapes, sizes, and compositions.

We use thousands of carbon compounds every day without even realizing it because they are silently carrying out important chemical reactions within our bodies. Many of these carbon compounds are so vital that we literally could not live without them.

Hydrocarbons

Many of the millions of organic compounds contain only two elements, carbon and hydrogen. These **hydrocarbons** are further classified as follows.

- *Alkanes*, or *saturated hydrocarbons*, first introduced in Section 2.9, have the general formula C_nH_{2n+2}. All the bonds between carbon atoms are single bonds.
- *Alkenes* (Section 9.11) have one or more double bonds between carbon atoms. Those with one double bond have the general formula C_nH_{2n}.
- *Alkynes* (Section 9.11) have one or more triple bonds between carbon atoms. Those with one triple bond have the general formula C_nH_{2n-2}. Alkenes and alkynes collectively are called *unsaturated hydrocarbons*.
- *Aromatic hydrocarbons* have the special kind of bonding noted in Section 10.8.

The systematic nomenclature of alkanes, alkenes, and alkynes is presented in Appendix D.

23.1 Alkanes

Alkane molecules are essentially nonpolar. For a given molecular mass, alkanes are among the lowest boiling of all the organic compounds (Section 11.5). They are also essentially insoluble in water (Section 12.3). Chemically, alkanes are unreactive toward most ionic compounds. There is no center of either positive or negative charge to attract an ion. Alkanes generally do not react with aqueous acids, bases, or oxidizing agents. Consider butane as an example.

Paraffin wax, a mixture of high molecular mass alkanes, is more dense as a solid than as a liquid (left). In contrast, ice is less dense than liquid water (right).

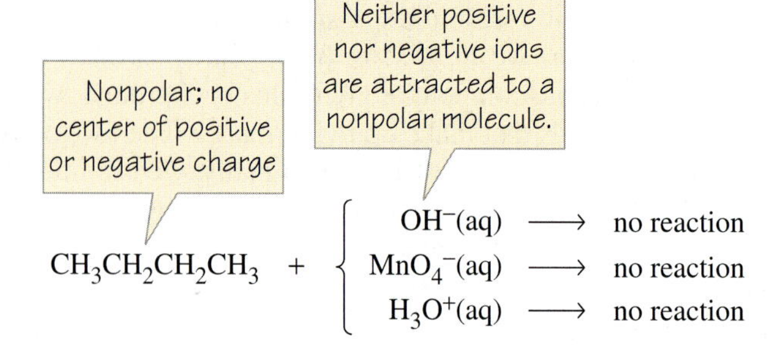

Nonpolar; no center of positive or negative charge
Neither positive nor negative ions are attracted to a nonpolar molecule.
$CH_3CH_2CH_2CH_3$ +
$OH^-(aq)$ ⟶ no reaction
$MnO_4^-(aq)$ ⟶ no reaction
$H_3O^+(aq)$ ⟶ no reaction

This lack of reactivity toward ionic reagents makes alkanes useful solvents for a variety of organic reactions. Alkanes also are good solvents for other substances of low polarity. Thus, hexane is used to extract food oils from soybeans, cottonseed, corn, and other seeds.

Although unreactive toward ionic substances, alkanes react readily at high temperatures with oxygen and with fluorine, chlorine, and bromine. These reactions do not involve ions, but rather proceed by steps that involve free radicals (Section 9.9) as intermediates. The most familiar and important reaction of alkanes is combustion (Section 6.8). We burn alkanes as fuels in automobiles, home heating furnaces, electric power plants, and many other places. An example is the combustion of butane.

$$2\ CH_3CH_2CH_2CH_3(g) + 13\ O_2(g) \longrightarrow 8\ CO_2(g) + 10\ H_2O(l)$$

This gas turbine burns natural gas, a mixture of alkanes (mostly methane), to generate electricity.

Halogens, like oxygen, are generally good oxidizing agents and they react with hydrocarbons. The reactions, however, are quite different from combustion. In combustion, the carbon skeleton is completely destroyed. In halogenation, the reaction can be controlled to provide varying degrees of halogenation and leave the carbon skeleton intact. Further, the trend in reactivities of halogens with alkanes is the same as the trend in the strengths of the halogens as oxidizing agents: $F_2 > Cl_2 > Br_2 > I_2$. Fluorine reacts explosively. Thus chemists generally use fluorine compounds rather than $F_2(g)$ to fluorinate organic hydrocarbons. Chlorination proceeds slowly in the dark at room temperature, but quite rapidly at elevated temperatures or in strong light. Bromine is still less reactive, also requiring high temperatures or strong light. Iodine is essentially unreactive toward alkanes.

The reaction of an alkane and a halogen is a **substitution reaction**; a halogen atom substitutes for one or more of the hydrogen atoms of the hydrocarbon. Consider, for example, the reaction of methane and chlorine. With an excess of methane, the main organic product is chloromethane.

$$CH_4(g) + Cl_2(g) \xrightarrow[\text{light}]{\text{Heat or}} CH_3Cl(g) + HCl(g)$$

The mechanism is that of a *chain reaction* initiated by heat or light. In the following scheme we show only the electrons involved in bond breakage or bond formation.

Initiation: $$Cl{:}Cl \xrightarrow[\text{light}]{\text{Heat or}} 2\ Cl\cdot$$

Propagation: $$H_3C{:}H + Cl\cdot \longrightarrow H_3C\cdot + H{:}Cl$$
$$H_3C\cdot + Cl{:}Cl \longrightarrow H_3C{:}Cl + Cl\cdot$$

Termination: $$2\ Cl\cdot \longrightarrow Cl{:}Cl \quad \text{or} \quad H_3C\cdot + Cl\cdot \longrightarrow H_3C{:}Cl \quad \text{or} \quad 2\ H_3C\cdot \longrightarrow H_3C{:}CH_3$$

The rapid reaction is *initiated* when Cl_2 molecules absorb enough energy to dissociate into chlorine atoms ($Cl\cdot$). The reaction is then *propagated* in two steps that are repeated many times: Chlorine atoms collide with CH_4 molecules, extracting an H atom to form the product HCl(g), and leaving a methyl radical, $H_3C\cdot$. The methyl radicals then collide with Cl_2 molecules, extracting a Cl atom to form the product $CH_3Cl(g)$*, and leaving a Cl atom. The Cl atom attacks another CH_4 molecule, and

APPLICATION NOTE

Interestingly, although nearly all organic compounds are combustible, many halogen-containing organic compounds, such as the halons (p. 3), are used to put out fires. Carbon tetrachloride, CCl_4, was once widely used in fire extinguishers, but has been replaced by presumably safer substances.

*The chlorination of methane usually produces a mixture of products. In addition to CH_3Cl, some CH_2Cl_2, $CHCl_3$, and CCl_4 are also formed. Formation of CH_3Cl can be maximized, however, by using a large excess of methane. Traces of ethane, CH_3CH_3, are also formed, just as one of the termination steps of the mechanism predicts.

the propagation steps are repeated. The chain reaction finally ends with a *termination* reaction, halting the production of further free radicals.

Free radicals are often used to initiate polymerization reactions, such as that of ethylene to form polyethylene (Chapter 24). Free radical reactions are also involved in the destruction of the ozone layer that protects life on Earth from deadly ultraviolet radiation. We will examine this issue further when we study environmental chemistry in Chapter 25.

23.2 Alkenes and Alkynes

Like the alkanes, alkenes and alkynes are essentially nonpolar. All three kinds of hydrocarbons have similar physical properties. They are essentially insoluble in water. Boiling points of alkenes and alkynes are quite close to those of alkanes with the same number of carbon atoms. Chemically, the main similarity among the hydrocarbons is that they all burn. Otherwise, the unsaturated hydrocarbons (alkenes and alkynes) are quite different from the saturated hydrocarbons (alkanes). The difference in reactivity reflects differences in structure. Compare the three two-carbon hydrocarbons:

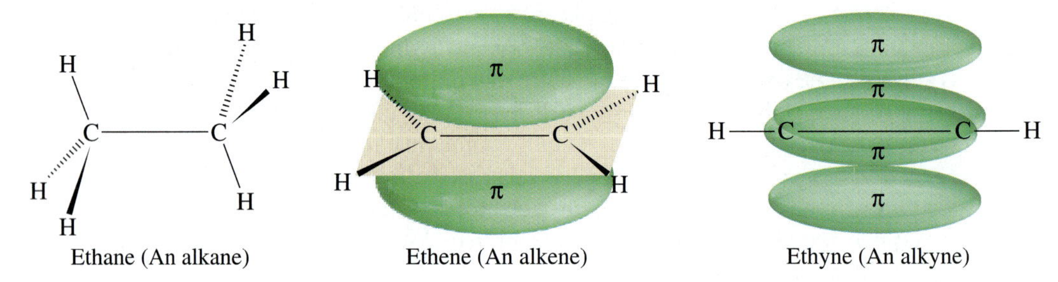

In ethane, all the bonds are sigma (σ) bonds, and the bonding electrons are localized *between* the atoms that form the bond. In addition to a σ-bond framework, ethene and ethyne have pi (π) bonds. The electrons in π bonds are delocalized and readily accessible to **electrophilic reagents**, that is, to reagents that are deficient in electrons and thus are attracted to electrons in other molecules. The π bonds in alkenes and alkynes make them much more chemically reactive than alkanes.

In contrast to the substitution reactions of alkanes, the typical reactions of alkenes are **addition reactions**: two or more reactant molecules combine to give a single product molecule. These reactions generally involve ions; we say they proceed by *ionic* mechanisms. Unlike free-radical reactions, addition reactions usually occur readily even at room temperature and in the dark. An electrophile (electron-seeking molecule or ion) uses the π electrons of the unsaturated hydrocarbon to attach itself to one of the carbon atoms of a multiply-bonded pair. A second atom or group then links to the other carbon atom, completing the addition reaction. Consider the addition of HI to ethene.

APPLICATION NOTE
Although alkenes and alkynes burn, they are not generally used as fuels. Rather, alkenes are starting materials for the synthesis of many valuable materials, especially polymers, as we saw in Chapter 9 and will revisit in Chapter 24. Alkynes have fewer uses. Some serve as starting materials for synthesis. You may recall that ethyne (acetylene) is used as fuel for the oxyacetylene torch.

$$\underset{\text{Ethene}}{CH_2{=}CH_2} + HI \longrightarrow \underset{\text{Iodoethane}}{CH_3CH_2I}$$

Because the product molecule no longer has a double bond, it is no longer an *alkene* but rather a halogenated *alkane*. The mechanism for the reaction involves *ions* as intermediates. Somewhat simplified, the mechanism has the H^+ ion using the π electrons to attach itself to one of the carbon atoms, leaving the second carbon atom with a positive charge.

$$\text{Step 1: } H^+ + CH_2{=}CH_2 \longrightarrow CH_3CH_2^+$$

(Organic chemists use a curved arrow to indicate the movement of a pair of electrons.) The intermediate, $CH_3CH_2^+$, is called a **carbocation** because it has a positive charge on a carbon atom. In the second step, the carbocation reacts with I^- to form the product.

$$\text{Step 2: } CH_3CH_2^+ + I^- \longrightarrow CH_3CH_2I$$

Addition reactions in which the first step involves an electrophile are called *electrophilic additions*. Many other reactions take place by a similar mechanism. For example, ethene reacts with chlorine to give 1,2-dichloroethane.

$$H_2C{=}CH_2 + Cl{-}Cl \longrightarrow H{-}\underset{Cl}{\overset{H}{C}}{-}\underset{Cl}{\overset{H}{C}}{-}H$$

Ethene 1,2-Dichloroethane

In this reaction, it may not be apparent at first that Cl_2 is an electrophile. Recall, however, that chlorine has a high (negative) electron affinity. We can envision the reaction as follows.

$$\text{Step 1: } Cl{-}Cl + CH_2{=}CH_2 \longrightarrow Cl^- + ClCH_2CH_2^+$$
$$\text{Step 2: } ClCH_2CH_2^+ + Cl^- \longrightarrow ClCH_2CH_2Cl$$

Bromine reacts much like chlorine. Br_2 forms brownish-red solutions in carbon tetrachloride, CCl_4. When such a solution is added to an alkene, the color disappears because the bromine reacts with the alkene to form a colorless dibromoalkane.

$$CH_3(H)C{=}CH_2 + Br{-}Br \longrightarrow CH_3{-}\underset{Br}{\overset{H}{C}}{-}\underset{Br}{\overset{H}{C}}{-}H$$

Propene (Colorless) Bromine (Brownish red) 1,2-Dibromopropane (Colorless)

The decolorization of bromine solutions is frequently used as a simple test for the presence of alkenes.

In Chapter 9 we described the reaction of alkenes and alkynes with H_2 in the presence of a catalyst. A common example is the hydrogenation of oils. Unsaturated fats are converted to saturated fats as double bonds are replaced by single bonds.

Another important reaction of alkenes is the addition of a water molecule. This reaction, called *hydration*, requires H^+ as a catalyst. The H^+ is usually furnished by sulfuric acid.

$$CH_2{=}CH_2 + H_2O \xrightarrow{H_2SO_4} CH_3CH_2OH$$

Ethene Ethanol

APPLICATION NOTE

Food chemists use the addition of iodine to the double bonds in unsaturated fats to determine the degree of unsaturation. The *iodine number* of a fat is the number of grams of I_2 consumed by 100 g of the fat. Polyunsaturated oils, such as sunflower oil, have high iodine numbers. Highly saturated fats, such as butter, have relatively low iodine numbers. The addition of bromine to the double bonds in unsaturated fat was illustrated on p. 398.

APPLICATION NOTE

Large quantities of ethanol are made in this way for use as an industrial solvent. Although the alcohol is identical to that used in alcoholic beverages, federal law requires that all beverage alcohol be produced by the natural process of fermentation.

23.3 Conjugated and Aromatic Compounds

Molecules can have more than one multiple bond. Compounds with molecules that have two carbon-to carbon double bonds are called *dienes*; with three double bonds, *trienes*; and so on.

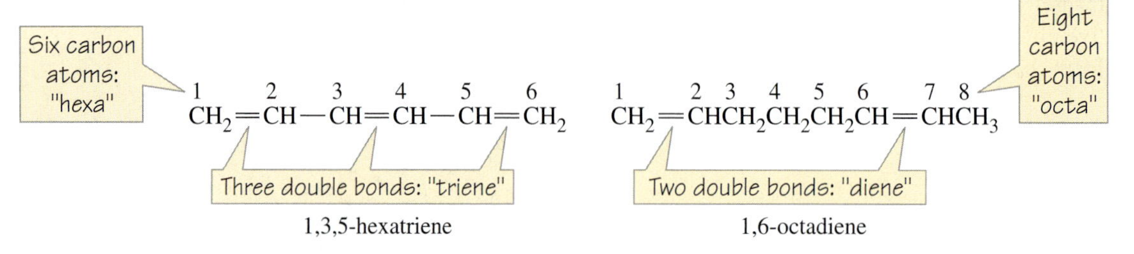

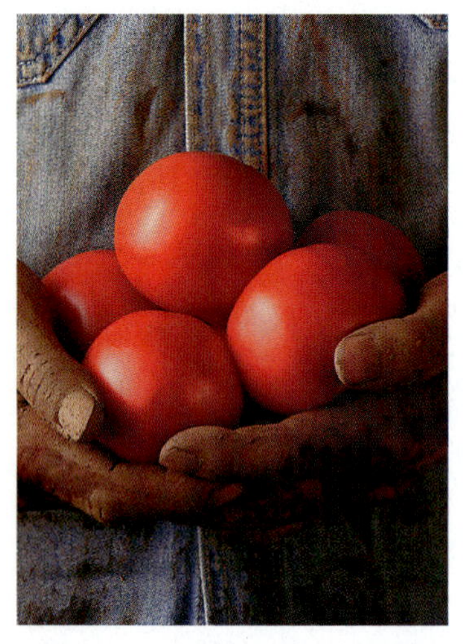

The bright red color of these tomatoes is due to lycopene, a molecule with an extended conjugated bonding system.

Conjugated Bonding Systems

Note that 1,3,5-hexatriene has alternate double and single bonds between the carbon atoms. A structure that has a series of alternate double and single bonds is said to have a **conjugated bonding system**. Note that 1,6-octadiene is *not* conjugated; the double bonds are separated by more than one single bond. *Aromatic* compounds have conjugated bonding in a ring system, and the most common ones have six π electrons, as does benzene.

Aromatic Compounds

In Section 10.9 we divided organic compounds into two main categories. Compounds that are related in the broadest sense to the benzene molecule, C_6H_6, are called *aromatic* compounds. Recall from Chapter 10 that the benzene (C_6H_6) molecule consists of a planar ring of six carbon atoms, each with one hydrogen atom attached. Also, *six* delocalized π electrons are found in π molecular orbitals (Figure 10.29). As a reminder, some representations of benzene are shown below, and a computer-generated model is shown in the margin.

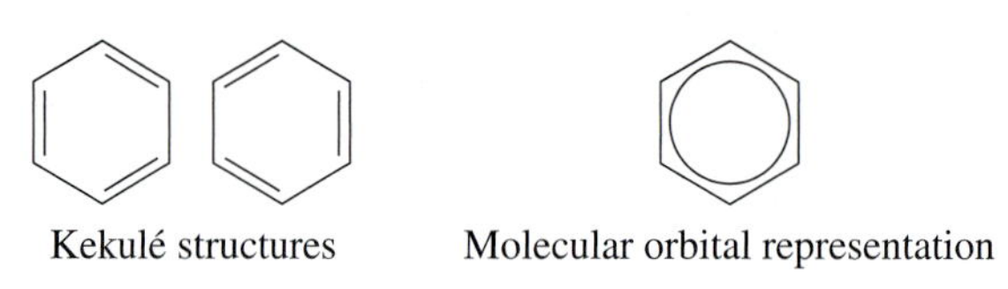

Many common aromatic compounds—toluene ($C_6H_5CH_3$), aniline ($C_6H_5NH_2$), phenol (C_6H_5OH), and benzoic acid (C_6H_5COOH), for example—are considered to be derivatives of benzene.

All open-chain organic compounds are *aliphatic* compounds. Ring compounds that are not benzenelike are also aliphatic. Cycloalkanes have properties much like open-chain alkanes and are aliphatic.

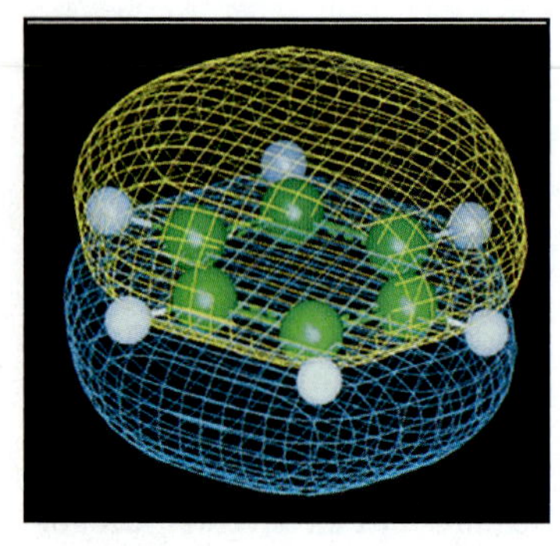

A computer-generated molecular orbital model of benzene.

We defined aromatic compounds as those that resemble benzene. But just which properties of benzene must a compound possess to be aromatic? What properties do all aromatic compounds have in common?

Experimentally, aromatic compounds have molecular formulas that would lead us to expect a high degree of unsaturation. However, aromatic compounds are resistant to the addition reactions generally characteristic of unsaturated compounds. Instead of addition reactions, we often find that these aromatic compounds undergo *substitution* reactions. Presumably, the cause of the aromatic compounds' resistance toward addition reactions is an unusual stability. Low heats of hydrogenation and low heats of combustion provide evidence of this stability. Let's now contemplate some of the theory of aromaticity.

- Aromatic molecules are cyclic, consisting of rings of atoms, *and* planar (flat).
- Aromatic molecules have a *conjugated bonding system* that produces cyclic clouds of delocalized π electrons above and below the plane of the molecule. The conjugated system must extend completely around the ring(s) and the π clouds must contain a total of $(4n + 2)$ *electrons*, where $n = 0, 1, 2, \ldots$

Most common aromatic compounds have $n = 1$ so that $4n + 2 = 4 \times 1 + 2 = 6$.

EXAMPLE 23.1

Which of the following compounds are aromatic and which are aliphatic? Explain.

$CH_2{=}CH{-}CH{=}CH{-}CH{=}CH_2$

(a)

N

(b) **(c)** **(d)**

SOLUTION

a. This compound has six π electrons and a conjugated bonding system, but it is not *cyclic*. Like *all* open chain compounds, this one is aliphatic.

b. This compound has six π electrons and a cyclic, conjugated bonding system; it is aromatic.

c. This compound has 10 π electrons ($n = 2; 4 \times 2 + 2 = 10$) and a cyclic, conjugated bonding system; it is aromatic.

d. This compound has a cyclic, conjugated bonding system, but it does not extend all the way around the ring and has only four π electrons. The compound is aliphatic.

EXERCISE 23.1

Which of the following compounds are aromatic and which are aliphatic? Explain.

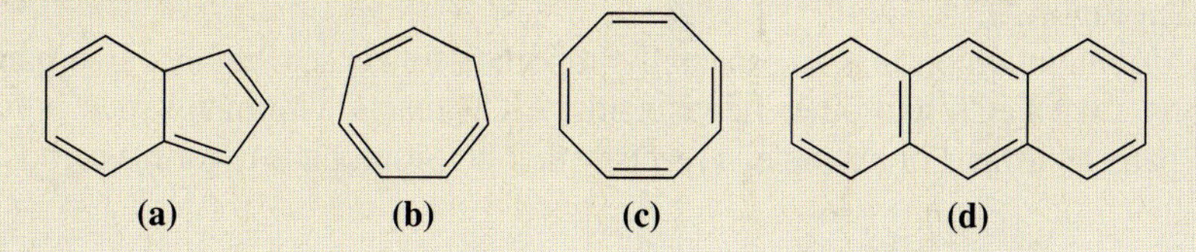

(a) **(b)** **(c)** **(d)**

Reactions of Benzene

Its molecular formula suggests that benzene is an unsaturated hydrocarbon. Six carbon atoms could bond as many as $2n + 2 = 14$ hydrogen atoms, as in hexane, C_6H_{14}. Even removing two H atoms to form a ring would leave 12 hydrogen atoms, as in cyclohexane, C_6H_{12}. Although the benzene molecule is unsaturated, it does not readily undergo the addition reactions typical of alkenes (page 960). Instead, its reactions are mainly *substitution* reactions. For example, with $FeCl_3$ as a catalyst, benzene reacts with chlorine to form chlorobenzene.

$$C_6H_6 + Cl_2 \xrightarrow{FeCl_3} C_6H_5Cl + HCl$$

To gain a better understanding of this and other aromatic substitution reactions, let's look at the mechanism of the reaction.

The first step is a Lewis acid-base reaction between the catalyst, $FeCl_3$, and Cl_2.

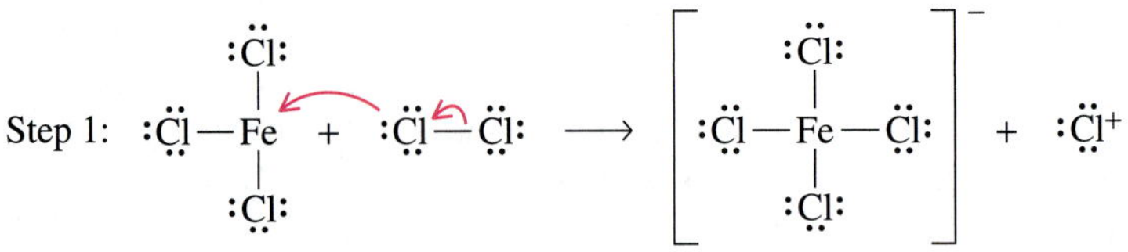

Step 1: $FeCl_3 + Cl{-}Cl \longrightarrow [FeCl_4]^- + Cl^+$

Note that in this case *both* bonding electrons of Cl_2 go with one of the chlorine atoms. This is quite unlike the free radical chlorination of methane, in which strong light or high temperatures split Cl_2 into two chlorine atoms, each with an unpaired electron. The electron-deficient chlorine atom* in this case has a positive charge; it is an electrophile. The Cl^+ attracts the π electrons of the benzene ring and uses a pair of them to attach itself to a carbon atom of the ring. Below, we use one of the Kekulé structures to represent benzene.

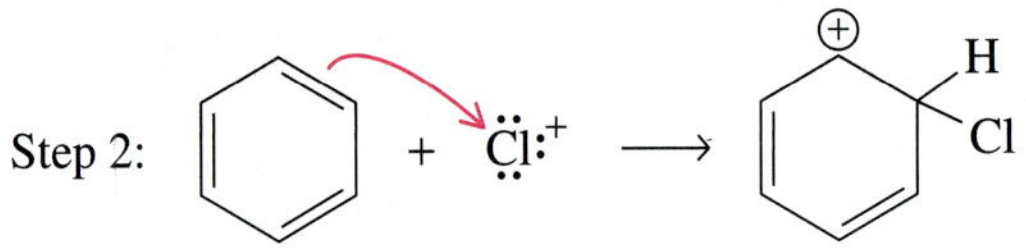

Step 2: Benzene + Cl^+ $\longrightarrow$ Carbocation intermediate

The intermediate carbocation gives up a hydrogen ion to $[FeCl_4]^-$, forming the *product*, chlorobenzene, the *by-product*, HCl, and regenerating the *catalyst*, $FeCl_3$.

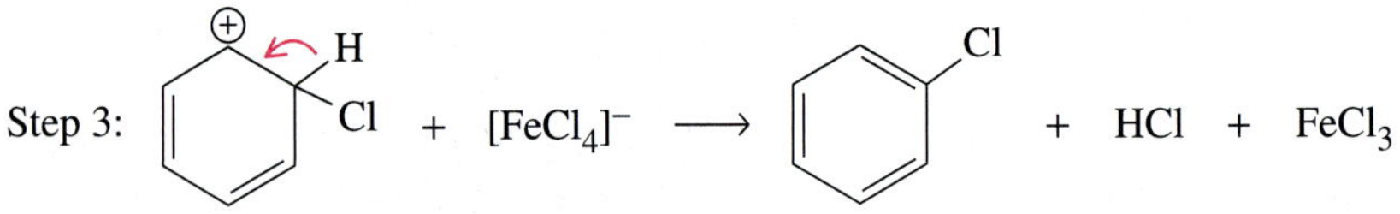

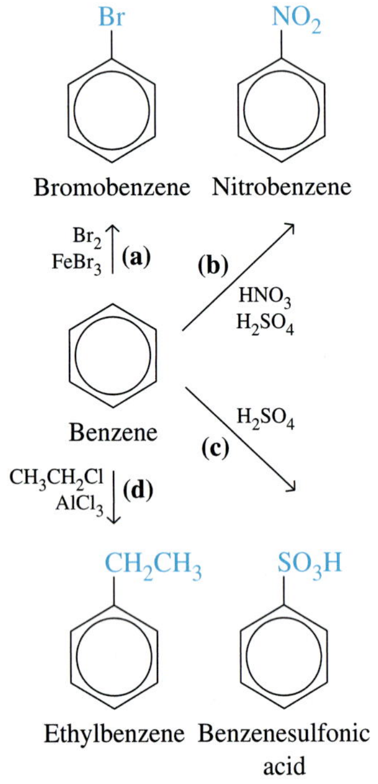

Step 3: Carbocation intermediate + $[FeCl_4]^-$ $\longrightarrow$ Chlorobenzene + HCl + $FeCl_3$

This type of reaction, involving the replacement of one of the hydrogen atoms of an aromatic ring by an electrophile (in this case, Cl^+), is called an **electrophilic aromatic substitution** reaction. Four other examples of such reactions are outlined in Figure 23.1.

Aluminum chloride is an especially important Lewis acid catalyst in organic reactions. Like $FeCl_3$ and other Lewis acids, $AlCl_3$ reacts with many species having lone-pair electrons. For example, it reacts with 2-chloropropane to form a carbocation.

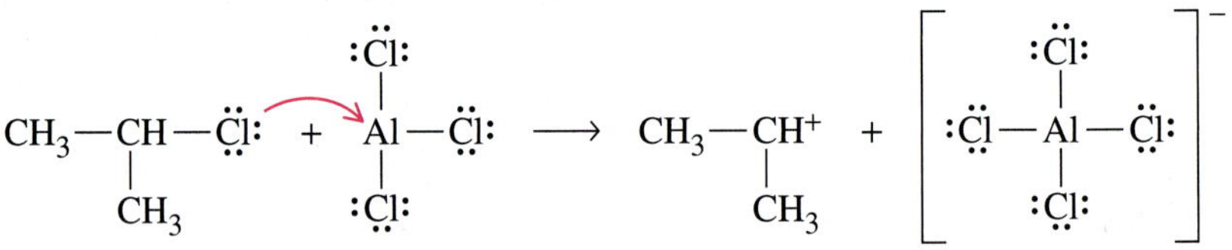

$$CH_3{-}CH(CH_3){-}Cl + AlCl_3 \longrightarrow CH_3{-}CH^+(CH_3) + [AlCl_4]^-$$

This carbocation is an electrophile that can then attach itself to a benzene ring. The net alkylation reaction is one in which new carbon-to-carbon bonds are formed in a molecule.

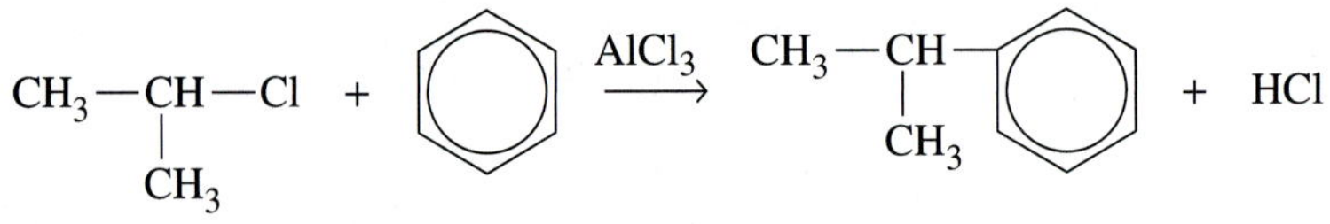

$$CH_3{-}CH(CH_3){-}Cl + C_6H_6 \xrightarrow{AlCl_3} CH_3{-}CH(CH_3){-}C_6H_5 + HCl$$

2-Chloropropane (Isopropyl chloride) · Benzene · Isopropylbenzene

Bromobenzene · Nitrobenzene

(a) Br_2, $FeBr_3$ · (b) HNO_3, H_2SO_4 · (c) H_2SO_4 · (d) CH_3CH_2Cl, $AlCl_3$

Benzene

Ethylbenzene · Benzenesulfonic acid

▲ **Figure 23.1 Some electrophilic aromatic substitution reactions of benzene**
(a) halogenation, (b) nitration, (c) sulfonation, and (d) alkylation.

APPLICATION NOTE

Electrophilic substitution reactions enter into the manufacture of detergents. The linear alkylbenzenesulfonate (LAS) detergents described in Section 20.13 are made by alkylation of benzene, followed by sulfonation of the alkylbenzene, and then neutralization of the alkylbenzenesulfonic acid with sodium carbonate.

* It is unlikely that an independent Cl^+ ion exists. Rather, as $FeCl_3$ begins to bond to Cl_2, the chlorine atom with a *developing* positive charge ($Cl_3Fe^{\delta-}{-}Cl{-}Cl^{\delta+}$) begins to form a bond to the benzene ring.

Reactions of Other Aromatic Compounds

Some substituted benzenes undergo the same types of reactions as their aliphatic analogs. For example, toluene ($C_6H_5CH_3$) reacts with chlorine at high temperatures, undergoing free radical substitution in the *side chain*; the CH_3 group reacts like an alkane.

$$C_6H_5{-}CH_3 + Cl_2 \xrightarrow[\text{light}]{\text{heat or}} C_6H_5{-}CH_2Cl + HCl$$

Like benzene, toluene and other substituted benzenes also undergo electrophilic aromatic substitution reactions. With chlorine and $FeCl_3$, a chlorine atom substitutes for a *ring* hydrogen atom. There is a complication, however. Whereas all six positions on an unsubstituted benzene ring are equivalent, on a substituted benzene ring they are not. Thus, in the reaction of benzene and chlorine, a chlorine atom can substitute for an H atom at any of the six positions and there is only one possible product, chlorobenzene. In the reaction of toluene and chlorine, however, the substitution of a Cl atom for an H atom can yield *three* possible products.

$$\underset{\text{Toluene}}{C_6H_5CH_3} + Cl_2 \xrightarrow{FeCl_3} \underset{o\text{-Chlorotoluene}}{o\text{-}ClC_6H_4CH_3} + \underset{m\text{-Chlorotoluene}}{m\text{-}ClC_6H_4CH_3} + \underset{p\text{-Chlorotoluene}}{p\text{-}ClC_6H_4CH_3} + HCl$$

Recall from page 451 that the two positions adjacent to a substituent on a benzene ring are *ortho*. The position directly opposite the substituent is *para*, and the remaining two positions are *meta*.

In practice, chlorination of toluene gives a mixture composed of mainly *o*- and *p*-chlorotoluene; very little *m*-chlorotoluene is formed. In general, groups present on a substituted benzene ring act like a traffic director, sending an incoming electrophile preferentially either to the meta position or to ortho and para positions. The CH_3 group directs the incoming chlorine atoms preferentially to the ortho and para positions. We will not consider the directing influence of substituent effects further here. Nitration of toluene with excess HNO_3 produces 2,4,6-trinitrotoluene (TNT), an explosive.

$$C_6H_5{-}CH_3 + HNO_3 \xrightarrow{H_2SO_4} O_2N{-}C_6H_2(NO_2)_2{-}CH_3 + H_2O$$

Aromatic carboxylic acids, such as benzoic acid, C_6H_5COOH, react much like aliphatic acids, forming salts with bases and esters with alcohols (Section 23.6). Aromatic amines, such as aniline, $C_6H_5NH_2$, undergo typical reactions of amines, such as salt formation in reactions with acids and amide formation with carboxylic acids (Section 23.6).

Functional Groups

Recall that many organic compounds have two parts, a hydrocarbon portion that is generally unreactive and a *functional group* where most of the reactions of the molecule take place. The functional group is usually polar, with electron-rich and electron-poor sites where ions are likely to attack. *Electrophiles*, often H^+ or another

Lewis acid, are attracted to a partially negative site. **Nucleophiles**, ions or molecules attracted to a partially positive carbon atom, are typically negative ions or other Lewis bases. The hydrocarbon portion is essentially nonpolar, an unlikely location for ionic attack. The carbonyl group is common to many families of organic compounds (Table 2.7, page 68). Consider, for example, a molecule such as the ketone 2-octanone.

$$\overset{\displaystyle O\,\delta^-}{\underset{\delta^+}{\overset{\|}{CH_3CCH_2CH_2CH_2CH_2CH_2CH_3}}}$$

Positive ions (electrophiles), often H^+, attack here.

Ions are unlikely to attack the nonpolar hydrocarbon chain.

Negative ions (nucleophiles), such as CN^-, attack here.

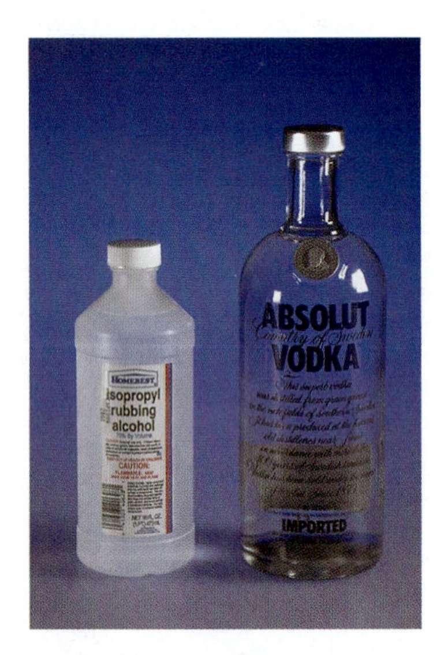

Two familiar alcohols are ethanol (vodka is about 40 to 50% ethanol) and 2-propanol (rubbing alcohol is about 70% 2-propanol).

23.4 Alcohols and Ethers

In Section 2.10, we introduced alcohols as compounds with a hydroxyl group (OH) bonded to a carbon atom of an alkyl group (R), giving them the general formula ROH. Ethers have two hydrocarbon groups joined through an oxygen atom: ROR′. An alcohol molecule can act as a Brønsted base. One of the lone pairs on the oxygen atom can accept a proton (H^+), just as a water molecule does in forming a hydronium ion.

$$R{-}\ddot{\underset{..}{O}}{-}H \;+\; H{-}A \;\rightleftharpoons\; \left[R{-}\ddot{\underset{\substack{|\\H}}{O}}{-}H\right]^+ \;+\; A^-$$

An alcohol — An acid — A protonated alcohol

The reaction forms the conjugate acid of the alcohol, and we say that the alcohol has been *protonated*. Many reactions of alcohols occur in acidic solution, and the first step is often one in which the alcohol acts as a Brønsted base.

If the alcohol is 1-propanol and the acid is HI, the overall reaction forms 1-iodopropane and water.

$$\underset{\text{1-Propanol}}{CH_3CH_2CH_2OH} + HI \longrightarrow \underset{\text{1-Iodopropane}}{CH_3CH_2CH_2I} + H_2O$$

In this reaction, I substitutes for the OH. Direct substitutions are rare, however, because the OH group is difficult to displace. 1-Propanol does *not* react with I^-.

$$CH_3CH_2CH_2OH + I^- \longrightarrow \text{no reaction}$$

Only after protonation of the alcohol does the I^- enter into the reaction; it displaces the OH_2 as water (H_2O).

To explain these facts, organic chemists propose reaction mechanisms in the manner we described in our study of chemical kinetics (Chapter 13). Consider the following mechanism consisting of two elementary reactions, the first of which is the protonation reaction.

$$\text{Acid-base step (fast): } \underset{\text{(a base)}}{CH_3CH_2CH_2{-}\ddot{\underset{..}{O}}{:}H} + \underset{\text{(an acid)}}{H{-}I} \rightleftharpoons \left[CH_3CH_2CH_2{-}\underset{\displaystyle H}{\overset{..}{\underset{|}{O}}}{:}H\right]^+ + I^-$$

$$\text{Substitution step (slow): } I^- + \left[CH_3CH_2\overset{\delta+}{CH_2}{-}\underset{\displaystyle H}{\overset{\delta-}{\underset{|}{O}}}{:}H\right]^+ \longrightarrow CH_3CH_2CH_2I + H_2O$$

The first step is a fast step—an acid-base reaction between HI and 1-propanol. Note that the intermediate, $CH_3CH_2CH_2OH_2^+$, has a polar C-to-O bond. Electrons are displaced toward the more electronegative O atom ($\delta-$) and away from the C atom ($\delta+$).

In the slow, second step the *negative* ion I^- attacks the partially *positive* C atom from the backside, that is, the side opposite the OH_2 end of the intermediate (Figure 23.2). The I^- is a *nucleophile*, a seeker of positive charge. Because this second elementary reaction involves two species, the I^- ion and the protonated alcohol, it is a *second-order* reaction. This reaction mechanism is called a *nucleophilic substitution, second order* and is symbolized as S_N2.

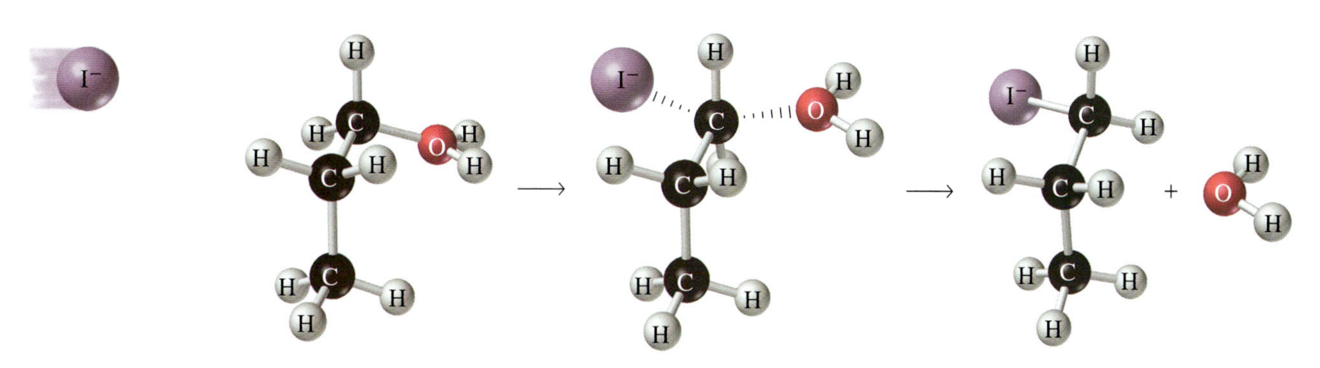

Figure 23.2 An S_N2 Reaction
The I^- ion displaces OH_2 as a water molecule in the second step of the conversion of 1-propanol to 1-iodopropane.

Alcohols can also undergo *dehydration* reactions. During dehydration, the elements of water—an OH group and an H atom—are removed from the alcohol. In concentrated H_2SO_4 and at high temperatures, dehydration leads to an alkene. Thus, dehydration of ethanol produces ethene.

$$CH_3CH_2OH \xrightarrow[\Delta]{H_2SO_4} CH_2{=}CH_2 + H_2O$$

In contrast, if less H_2SO_4 is used and the reaction is carried out at a lower temperature, diethyl ether is the main product.

$$CH_3CH_2OH + HOCH_2CH_3 \xrightarrow{H_2SO_4} \underset{\text{Diethyl ether}}{CH_3CH_2OCH_2CH_3} + H_2O$$

Notice that in alkene formation an H atom and an OH group both are removed from one alcohol molecule. In ether formation, however, an H atom is removed from the

OH group of one alcohol molecule, and the OH group is removed from a second alcohol molecule.

In both reactions, the first step is protonation of the alcohol.

$$\textit{Acid-base step (fast):}\quad CH_3CH_2OH + H_2SO_4 \rightleftharpoons CH_3CH_2OH_2^+ + HSO_4^-$$

What happens next depends on the reaction conditions.

Alkene formation: With excess sulfuric acid, nearly all the alcohol is converted to $CH_3CH_2OH_2^+$. The protonated alcohol then reacts with a base such as water. Attack of the water molecule at the carbon atom that bears the OH_2^+ group simply leads to the reversal of the protonation reaction. However, if it attacks a hydrogen atom on the adjacent carbon atom, we get an elimination reaction.

$$\textit{Acid–base step (slow):}\quad H_2\ddot{O}\!: \longrightarrow H{-}CH_2{-}CH_2{-}\overset{+}{O}H_2$$

This step involves extracting a proton from a *carbon* atom and is much slower than the fast acid-base first step. It occurs readily only at relatively high temperatures. The slow, rate-determining step involves both H_2O and $HCH_2CH_2OH_2^+$. It is therefore a second-order *elimination* reaction, designated E2.

Ether formation: If the temperature is relatively low and the ethanol concentration relatively high, an ethanol molecule acts as a nucleophile, displacing water from a protonated alcohol ion.

$$\textit{Displacement reaction (slow):}\quad CH_3CH_2OH + CH_3CH_2{-}OH_2^+ \longrightarrow \left[CH_3CH_2\underset{\displaystyle H}{\underset{|}{O}}CH_2CH_3\right]^+ + H_2O$$

The product of this step is the conjugate acid of diethyl ether. This protonated ether molecule acts as an acid, readily giving up a proton to a base such as CH_3CH_2OH or H_2O.

$$\textit{Acid-base step (fast):}\quad \left[CH_3CH_2\underset{\displaystyle H}{\underset{|}{O}}CH_2CH_3\right]^+ + H_2O \longrightarrow CH_3CH_2OCH_2CH_3 + H_3O^+$$

The slow step is bimolecular, and the ether-forming reaction is a second-order *substitution* reaction (S_N2).

Like water (HOH), alcohols (ROH) undergo self ionization.

$$2ROH \rightleftharpoons ROH_2^+ + RO^-$$

In general, alcohols are weaker acids than water, and the pH of a solution of an alcohol in water would be determined mainly by the self ionization of water. Alcohols are acidic enough, however, to undergo some reactions that resemble those of water. For example, alcohols react with sodium metal to form hydrogen gas and ionic compounds called sodium *alkoxides*. In this oxidation-reduction reaction, only the H atom of the OH group is replaced.

$$2\,ROH(l) + 2\,Na(s) \longrightarrow 2\,RO^-Na^+(\text{in ROH solution}) + H_2(g)$$

The concentration of H^+ produced in the self-ionization of ROH is large enough that $E_{cell} > 0$: the reduction of H^+ to $H_2(g)$ is a spontaneous reaction.

S_N2 Displacement Reactions in Organic and Biochemistry

Many organic and biochemical reactions fit the category of S_N2 displacement reactions. In organic chemistry, S_N2 reactions can be used to convert alkyl halides to amines, to sulfur-containing compounds called thiols, and to hundreds of other products.

$$\underset{\text{Bromoethane}}{CH_3CH_2Br} + 2\,NH_3 \longrightarrow \underset{\text{Ethylamine}}{CH_3CH_2NH_2} + NH_4Br$$

$$\underset{\text{1-Bromobutane}}{CH_3CH_2CH_2CH_2Br} + NaSH \longrightarrow \underset{\text{1-Butanethiol}}{CH_3CH_2CH_2CH_2SH} + NaBr$$

Interestingly, 1-butanethiol has a potent skunk odor. A related compound, ethanethiol, CH_3CH_2SH, is used as an odorant in natural gas so that we can detect leaks of this otherwise odorless material.

In biochemistry, many reactions involve reaction of an NH_2 group on DNA with some *alkylating agent*, such as an alkyl halide. The alkylated DNA is *mutated*, and if not repaired, can lead to cancer or a disease that can be passed on to one's descendants. Fortunately, our cells have special enzymes that can ordinarily cut out and replace segments of damaged DNA.

Skunks use thiols similar to 1-butanethiol as a defensive weapon.

Other typical oxidation-reduction reactions of alcohols are those in which they are oxidized to aldehydes and ketones. Aldehydes are then further oxidized to carboxylic acids.

$$\underset{\text{An alcohol}}{R-CH_2OH} \xrightarrow{\text{oxidation}} \underset{\text{An aldehyde}}{R-\overset{\overset{\large O}{\|}}{C}-H} \xrightarrow{\text{oxidation}} \underset{\text{A carboxylic acid}}{R-\overset{\overset{\large O}{\|}}{C}-OH}$$

$$\underset{\text{An alcohol}}{R-\overset{\overset{\large OH}{|}}{CH}-R'} \xrightarrow{\text{oxidation}} \underset{\text{A ketone}}{R-\overset{\overset{\large O}{\|}}{C}-R'}$$

We described these reactions in some detail in Section 4.6.

23.5 Aldehydes and Ketones

Above and in Chapter 4 we noted that aldehydes and ketones are formed by the oxidation of alcohols. If the OH group of an alcohol is on a terminal carbon atom, an *aldehyde* is formed, and if the OH group is bonded to an interior carbon atom, a *ketone* is produced. Both aldehydes and ketones have a *carbonyl* (C$=$O) functional group.

$$\underset{\text{An aldehyde}}{R-\overset{\overset{\large O}{\|}}{C}-H} \qquad \text{Carbonyl group} \qquad \underset{\text{A ketone}}{R-\overset{\overset{\large O}{\|}}{C}-R'}$$

Large quantities of formaldehyde are used to make phenol-formaldehyde resins for bonding plywood and other building materials.

An aldehyde has a hydrogen atom attached to the carbonyl carbon atom; a ketone has *two* hydrocarbon groups attached to the carbonyl carbon atom. These structural formulas are often written on one line as RCHO and RCOR′, in which cases the carbon-oxygen double bond is understood.

Cinnamaldeyde is the main flavor component in cinnamon.

The delightful aroma of butter is largely due to 2,3-butanedione.

APPLICATION NOTE

For the oxidation of ethanol, the half-reaction is

$$CH_3CH_2OH \longrightarrow CH_3CHO + 2\,H^+ + 2\,e^-$$

In living cells, the oxidizing agent for this oxidation is a complex substance, nicotinamide adenine dinucleotide. It is reduced from NAD^+ to NADH.

$$NAD^+ + H^+ + 2\,e^- \longrightarrow NADH$$

The overall redox reaction is

$$CH_3CH_2OH + NAD^+ \longrightarrow CH_3CHO + NADH + H^+$$

The carbonyl group is also a part of more complicated functional groups, such as those of carboxylic acids, esters, and amides. We also encounter the carbonyl group in carbohydrates, fats, proteins, nucleic acids, hormones, vitamins, and the host of organic compounds critical to living systems. Right now, we will focus on the carbonyl group in aldehydes and ketones.

Formaldehyde, HCHO, is familiar to many as a biological tissue preservative, as in embalming fluid. It is used industrially in huge quantities, mainly to make phenol-formaldehyde resins (Chapter 24) for bonding plywood. Formaldehyde and other simple aldehydes have rather foul, irritating odors. Some higher aldehydes, however, are used in flavors and fragrances. The following are familiar examples. The carbonyl group is highlighted in each compound.

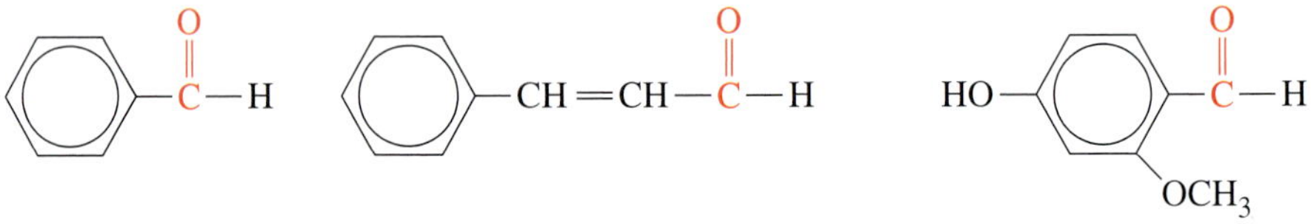

Benzaldehyde (Almond flavor) | Cinnamaldehyde (Cinnamon flavor) | Vanillin (Vanilla flavor)

The simplest ketone, acetone (2-propanone; CH_3COCH_3), is a common solvent used in applications as varied as varnishes and lacquers, rubber cement, and fingernail polish remover. The following are familiar examples of other ketones.

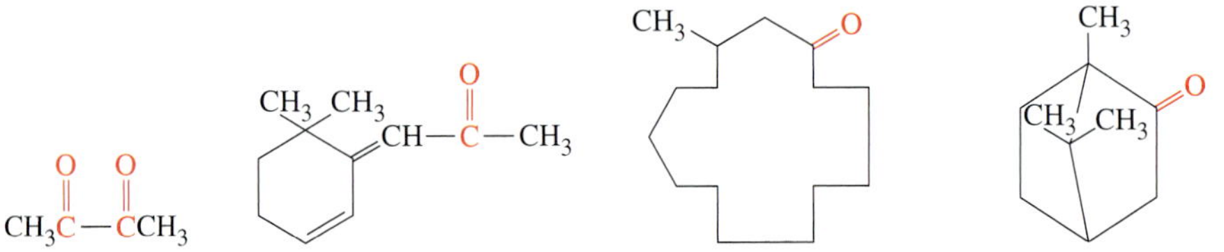

2,3-Butanedione (Butter flavoring) | Irone (Odor of violets) | Muscone (Musk oil, an ingredient in perfumes) | Camphor (An ingredient in some insect repellants)

Aldehydes and ketones undergo reactions at the carbonyl group, C=O. Many of these reactions start with a simple addition to the C=O group, but this addition is often followed by more steps. Consider the addition of HCN. In an acidic solution, the H^+ ion adds to the partially negative O atom of the carbonyl group, and then CN^- adds to the partially positive C of the C=O group.

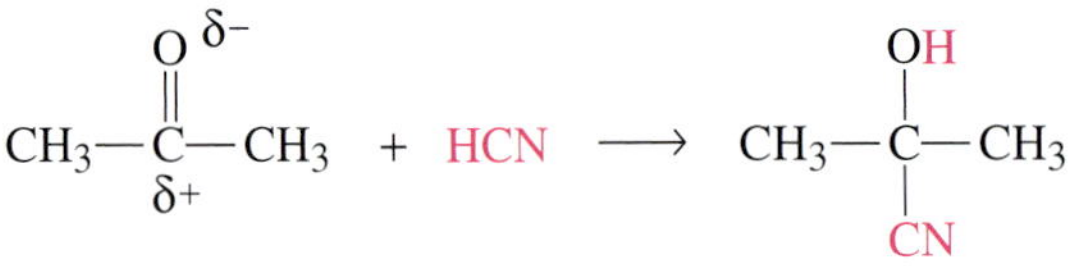

23.6 Carboxylic Acids, Esters, and Amides

We first introduced carboxylic acids (RCOOH) in Chapter 2. We used acetic acid and other members of the family as examples of weak acids in Chapter 4 and in acid-base equilibria calculations in Chapter 15. Carboxylic acids and some compounds derived from them are common in nature and quite important in biochemistry. We will consider two classes of carboxylic acid derivatives here. The *esters* (RCOOR′) also include fats (Section 23.7). The *amides* ($RCONH_2$) are also carboxylic acid derivatives. Proteins (Section 23.9) are perhaps the most spectacular example of amides. Here, we look at simple carboxylic acids, esters, and amides. Later, we turn to the more complex worlds of lipids and proteins.

Carboxylic Acids

The simplest carboxylic acids are widely known by their common names. For example, HCOOH is formic acid and CH_3COOH is acetic acid. Systematic nomenclature is discussed in Appendix D.

By far the most common reaction of the carboxylic acids is as acids. Benzoic acid, the simplest aromatic acid, is a solid only slightly soluble in water. However, it reacts with and dissolves in a solution of an aqueous base.

$$C_6H_5\text{—COOH(s)} + \text{NaOH(aq)} \longrightarrow C_6H_5\text{—COONa(aq)} + H_2O(l)$$

Two other important reactions are the formation of esters and amides.

Ester formation closely resembles ether formation (Section 23.4). In ester formation, the OR from an alcohol molecule replaces the OH of a carboxylic acid molecule. A molecule of water is eliminated. As in ether formation, a strong acid such as H_2SO_4 catalyzes the reaction. We can write general and specific reactions as follows.

$$\underset{\text{An acid}}{R\text{—}\overset{\overset{\displaystyle O}{\|}}{C}\text{—OH}} + \underset{\text{An alcohol}}{HO\text{—}R'} \xrightleftharpoons{H_2SO_4} \underset{\text{An ester}}{R\text{—}\overset{\overset{\displaystyle O}{\|}}{C}\text{—O—}R'} + H_2O$$

$$\underset{\text{Butyric acid}}{CH_3CH_2CH_2COOH} + \underset{\text{Ethanol}}{HOCH_2CH_3} \xrightleftharpoons{H_2SO_4} \underset{\text{Ethyl butyrate}}{CH_3CH_2CH_2COOCH_2CH_3} + H_2O$$

The reactions are reversible and reach an equilibrium condition.

Just as an ester is derived from a carboxylic acid and an alcohol, an **amide** is derived from a carboxylic acid and ammonia ($RCONH_2$) or an amine (RCONHR and $RCONR_2$). For example, the following amide is made from butyric acid (butanoic acid) and ammonia.

$$CH_3CH_2CH_2\text{—}\overset{\overset{\displaystyle O}{\|}}{C}\text{—}NH_2$$

The amide functional group.

The OH group of the carboxylic acid molecule is replaced by an NH_2 group from an ammonia molecule. Amides are named by changing the ending of the name of the carboxylic acid from *-ic acid* to *-amide*. The amide above is named butyramide (common) or butanamide (IUPAC).

Amides can be made by heating ammonium (or amine) salts of carboxylic acids. For example, acetamide is formed by heating ammonium acetate.

Carboxylic acids occur in many common household items. Vinegar contains acetic acid (Chapter 2); aspirin is acetylsalicylic acid (Chapter 1); vitamin C is ascorbic acid, an antioxidant (Chapter 4), lemons contain citric acid, a triprotic acid (Chapter 15), and both spinach and Zud cleanser contain oxalic acid.

APPLICATION NOTE
Nylon (Chapter 24), a synthetic polyamide (a polymer with amide functional groups), is made by heating the salt of a dicarboxylic acid and a diamine. Proteins (Section 23.9) are natural polyamides.

$$\underset{\substack{\text{Ammonium acetate}\\\text{(Ammonium ethanoate)}}}{CH_3{-}\overset{\overset{\displaystyle O}{\|}}{C}{-}O^-NH_4^+} \xrightarrow{\Delta} \underset{\substack{\text{Acetamide}\\\text{(Ethanamide)}}}{CH_3{-}\overset{\overset{\displaystyle O}{\|}}{C}{-}NH_2} + H_2O$$

This method is usually carried out in two steps: First, a carboxylic acid is neutralized with ammonia or an amine. Then the resulting salt is heated. We can illustrate with the following reactions.

$$\underset{\text{Acetic acid}}{CH_3{-}\overset{\overset{\displaystyle O}{\|}}{C}{-}OH} + \underset{\text{Ethylamine}}{CH_3CH_2NH_2} \longrightarrow \underset{\text{Ethylammonium acetate}}{CH_3{-}\overset{\overset{\displaystyle O}{\|}}{C}{-}O^-H_3N^+{-}CH_2CH_3}$$

$$CH_3{-}\overset{\overset{\displaystyle O}{\|}}{C}{-}O^-H_3N^+{-}CH_2CH_3 \xrightarrow{\Delta} \underset{\textit{N}\text{-Ethylacetamide}}{CH_3{-}\overset{\overset{\displaystyle O}{\|}}{C}{-}NHCH_2CH_3} + H_2O$$

Many esters are used as fragrances and flavors. Ethyl butyrate, made by the reaction of ethyl alcohol with butyric acid, is a pineapple flavoring. Pentyl acetate (also called amyl acetate), made by the reaction of 1-pentanol with acetic acid, is a banana flavoring.

Reactions of Esters and Amides

The principal reaction of esters is hydrolysis, a process that converts the ester back to the alcohol and the carboxylic acid from which it was made.

$$\underset{\text{An ester}}{R{-}\overset{\overset{\displaystyle O}{\|}}{C}{-}O{-}R'} + H_2O \underset{}{\overset{H_2SO_4}{\rightleftharpoons}} \underset{\text{A carboxylic acid}}{R{-}\overset{\overset{\displaystyle O}{\|}}{C}{-}OH} + \underset{\text{An alcohol}}{HO{-}R'}$$

The reaction is reversible and reaches an equilibrium. Like ester formation, an acid catalyzes the hydrolysis. Esters can also be hydrolyzed in basic solution, and basic hydrolysis goes to completion.

$$\underset{\text{An ester}}{R{-}\overset{\overset{\displaystyle O}{\|}}{C}{-}O{-}R'} + NaOH(aq) \xrightarrow{\Delta} \underset{\text{A carboxylate salt}}{R{-}\overset{\overset{\displaystyle O}{\|}}{C}{-}O^-Na^+} + \underset{\text{An alcohol}}{HO{-}R'}$$

The hydrolysis of esters is the basis for soap making (Section 23.7).

The principal reaction of amides is also hydrolysis, a process that converts the amide back to the carboxylic acid and ammonia (or amine) from which it was made. These reactions are catalyzed by either an acid or a base. In basic solution, $NH_3(g)$ is formed, and can be detected either by its odor or by the fact that it turns moist red litmus paper blue.

$$\underset{\text{An amide}}{R{-}\overset{\overset{\displaystyle O}{\|}}{C}{-}NH_2} + NaOH(aq) \xrightarrow{\Delta} \underset{\text{A carboxylate salt}}{R{-}\overset{\overset{\displaystyle O}{\|}}{C}{-}O^-Na^+} + NH_3(g)$$

The digestion of proteins (Section 23.9) involves the hydrolysis of amide linkages, breaking the protein molecule down into individual amino acids that are absorbed into the bloodstream.

Biological Molecules

Our bodies are collections of chemicals—exquisitely organized chemicals, to be sure, but chemicals nonetheless. Fats, carbohydrates, proteins, and nucleic acids, together with water, make up most of the mass of the human body. Countless crucial chemical reactions take place within each cell every minute of every day. Living organisms depend on the reactions of fats and carbohydrates to furnish energy for our every activity. Proteins provide the building materials that help us to grow and develop. Perhaps most intriguing of all are the complex molecules called nucleic acids, macromolecules involved in the processes of heredity.

To place our discussion in context, we need to describe briefly the cell, the basic structural unit of living matter where biomolecules—proteins, nucleic acids, fats, carbohydrates, and so on—are found. Each cell of a plant or animal organism is enclosed in a cell membrane through which the cell gains nutrients and gets rid of wastes. Inside the cell are a number of structures, called *organelles*, that serve a variety of functions. The largest organelle usually is the cell nucleus, which contains the nucleic acids that control heredity. Protein synthesis, which involves other nucleic acids, occurs in ribosomes. The energy-producing activities of the cell occur in mitochondria.

23.7 Lipids

We classify most organic compounds in terms of functional groups, and we represent them by general formulas such as ROH for alcohols and RCOOR′ for esters. Lipids, however, are classified on the basis of their *solubilities*. Thus, we can give no general structural formula for all lipids. Recall our rule (Chapter 12) that "like dissolves like." A **lipid** is any component of plant or animal tissue that is insoluble in water but soluble in solvents of low polarity. Diethyl ether, hexane, carbon tetrachloride, and benzene are familiar examples of solvents with low polarity. Lipids include fats and oils (triglycerides), cholesterol, sex hormones, some vitamins (A, D, E, and K), and components of cell membranes called phospholipids.

In Chapter 6, we discussed the heat of combustion for a common source of energy in our diets: fats. You may recall that fats yield more energy per gram than other types of biomolecules. You are familiar with triglycerides such as beef fat, olive oil, and other foods. Here we will take a closer look at them, and briefly examine a few other lipids as well.

Triglycerides: Fats and Oils

Triglycerides* are esters of the trihydroxy alcohol glycerol, $HOCH_2CHOHCH_2OH$, with long-chain carboxylic acids (fatty acids). Table 23.1 lists some common fatty acids. If the three fatty acids in a triglyceride are the same, the molecule is a *simple triglyceride*. Usually, though, the fatty acids are different, and the molecule is a *mixed triglyceride*.

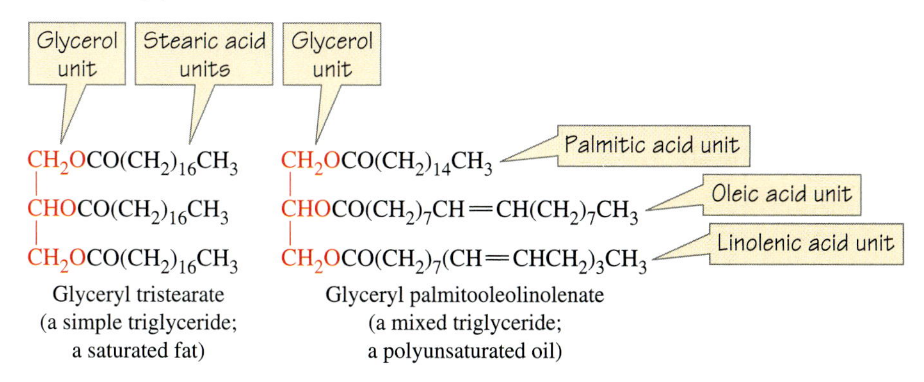

Glyceryl tristearate (a simple triglyceride; a saturated fat)

Glyceryl palmitooleolinolenate (a mixed triglyceride; a polyunsaturated oil)

* The systematic name for these compounds is *triacylglycerols*, but the common name *triglycerides* is widely used.

The fatty acid components of safflower oil, a polyunsaturated cooking oil, are about 6-7% palmitic, 2-3% stearic, 12-14% oleic, 75-80% linoleic, and 0.5-1.5% linolenic acid.

TABLE 23.1 Some Common Fatty Acids

Common Name	IUPAC Name	Condensed Structural Formula
Saturated acids		
Lauric acid	dodecanoic acid	$CH_3(CH_2)_{10}COOH$
Myristic acid	tetradecanoic acid	$CH_3(CH_2)_{12}COOH$
Palmitic acid	hexadecanoic acid	$CH_3(CH_2)_{14}COOH$
Stearic acid	octadecanoic acid	$CH_3(CH_2)_{16}COOH$
Monounsaturated acids		
Oleic acid	9-octadecenoic acid	$CH_3(CH_2)_7CH{=}CH(CH_2)_7COOH$
Polyunsaturated acids		
Linoleic acid	9,12-octadecadienoic acid	$CH_3(CH_2)_3(CH_2CH{=}CH)_2(CH_2)_7COOH$
Linolenic acid	9,12,15-octadecatrienoic acid	$CH_3(CH_2CH{=}CH)_3(CH_2)_7COOH$

Like other esters, a principal reaction of triglycerides is hydrolysis (page 972). Our digestion of triglycerides involves hydrolysis catalyzed by enzymes. In the laboratory, hydrolysis is usually done in basic solution, and is called **saponification** (Latin, *sapon*, soap; *facere*, to make). Saponification of the mixed triglyceride, glyceryl palmitooleolinolenate, gives glycerol and the salts of the three carboxylic acids from which the triglyceride was made.

$$\begin{array}{l} CH_2OCO(CH_2)_{14}CH_3 \\ | \\ CHOCO(CH_2)_7CH{=}CH(CH_2)_7CH_3 \\ | \\ CH_2OCO(CH_2)_7(CH{=}CHCH_2)_3CH_3 \\ \text{Glyceryl palmitooleolinolenate} \end{array} + 3\,NaOH \longrightarrow \begin{array}{l} CH_2OH \\ | \\ CHOH \\ | \\ CH_2OH \\ \text{Glycerol} \end{array} + \begin{array}{l} CH_3(CH_2)_{14}COONa \\ \text{Sodium palmitate} \\ CH_3(CH_2)_7CH{=}CH(CH_2)_7COONa \\ \text{Sodium oleate} \\ CH_3(CH_2CH{=}CH)_3(CH_2)_7COONa \\ \text{Sodium linolenate} \end{array}$$

The mixture of carboxylate salts is called **soap**. (We discussed the cleansing action of soap in Section 20.13.)

Phospholipids

Another important class of lipids is the *phospholipids*. These phosphorus-containing compounds are especially important constituents of cell membranes. Many phospholipids are esters of glycerol with fatty acids and with phosphoric acid, with another group also attached to the phosphoric acid unit. A fairly simple type of phospholipid called a *phosphatide*, for example, has glycerol combined with two fatty acid units and one phosphoric acid unit. The phosphoric acid unit in turn is esterified with an aminoalcohol such as ethanolamine, $HOCH_2CH_2NH_2$.

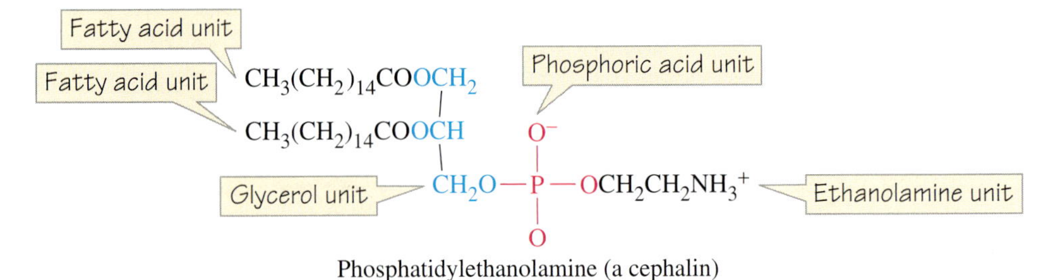

Phosphatidylethanolamine (a cephalin)

Like soap molecules (Section 20.13), the phospholipids are *amphipathic*: The phosphoric acid-ethanolamine portion corresponds to a hydrophilic "head," and there are *two* hydrophobic fatty acid "tails," whereas soap molecules have only one hydrophobic tail.

Cephalins are important components of brain and nerve cells. Related compounds called lecithins occur in all living cells. They have a choline group ($-OCH_2CH_2N(CH_3)_3^+$) in place of the ethanolamine unit of cephalins. Lecithin, isolated from soybeans, is used as an emulsifying agent in salad dressings and other foods that combine oil and water components.

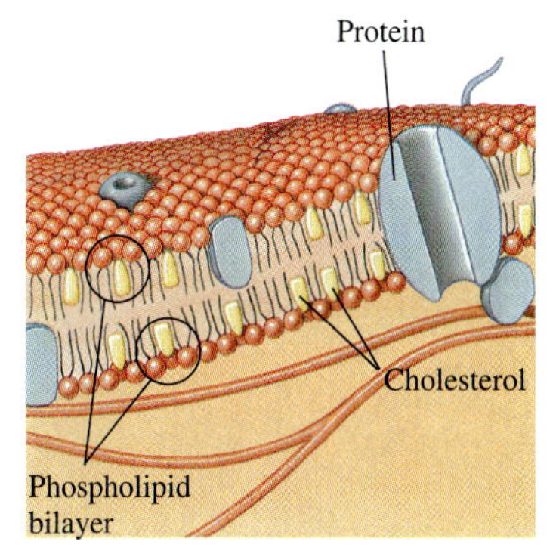

As shown in this schematic diagram, a cell membrane is a double layer (bilayer) of phospholipid molecules with cholesterol and proteins embedded in it. Some of the protein molecules provide channels through which polar substances can enter and leave the cell. Nonpolar and slightly polar substances such as diethyl ether can diffuse directly through the phospholipid bilayer.

23.8 Carbohydrates

Carbohydrates supply most of the energy needs of our bodies. Carbohydrates were so named because they were once thought to be hydrates of carbon. Indeed, many have formulas of the type $C_x(H_2O)_y$.

Carbohydrate	Molecular formula	"Hydrate" formula
Glucose	$C_6H_{12}O_6$	$C_6(H_2O)_6$
Sucrose	$C_{12}H_{22}O_{11}$	$C_{12}(H_2O)_{11}$
Starch	$(C_6H_{10}O_5)_n$	$[C_6(H_2O)_5]_n$

In fact, carbohydrates have complex molecular structures, whereas simple hydrates (Section 2.7) have discrete H_2O molecules associated with other molecules or ions. **Carbohydrates** are polyhydroxy aldehydes or polyhydroxy ketones or compounds that are derived from, or that can be hydrolyzed to, polyhydroxy aldehydes or polyhydroxy ketones. Carbohydrates that have aldehyde groups are called *aldoses* and those with ketone groups, *ketoses*. Carbohydrates with five-carbon atom molecules are *pentoses* and those with six-carbon atom molecules are *hexoses*. For example, we can classify the two familiar compounds, glucose and fructose, as follows.

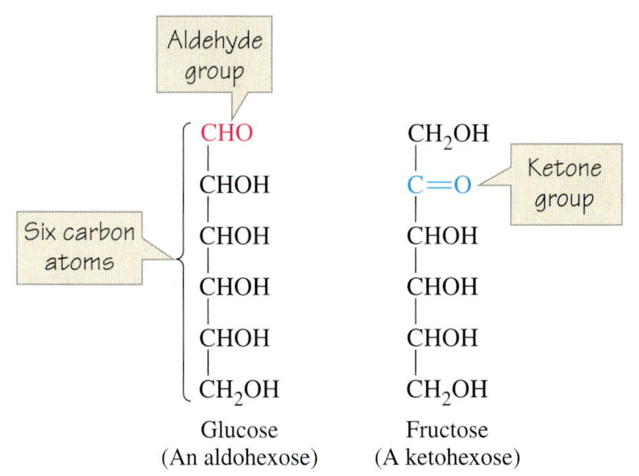

Glucose (An aldohexose)

Fructose (A ketohexose)

Actually, glucose is only one of 16 possible isomers, all of which have the same molecular formula and the same groups in the same order of attachment on the carbon chain. They differ in the arrangement of the atoms in space, and this makes them *stereoisomers* (page 943).

Simple carbohydrates taste sweet and are called *sugars*. Sugars that cannot be hydrolyzed to simpler carbohydrates are **monosaccharides**. Glucose and fructose are familiar examples. Most monosaccharides have from three to six carbon atoms in their molecules. The two three-carbon monosaccharides are widely known by common names, and they occur as important intermediates in the metabolism of carbohydrates in living cells.

Aldehyde group

Three carbon atoms

CHO
|
CHOH
|
CH_2OH

Glyceraldehyde (An aldotriose)

CH_2OH
|
C=O
|
CH_2OH

Ketone group

Dihydroxyacetone (A ketotriose)

Emil Fischer (1852-1919) was awarded the Nobel Prize in 1902 for his research on the structure of sugar molecules. He later figured out how amino acids join together to form proteins.

Glyceraldehyde exists as a pair of *enantiomers* (page 944) that can be distinguished experimentally by their action on polarized light (see again Figure 22.15). Enantiomers are stereoisomers that are related like a right and left hand; one is the nonsuperimposable mirror image of the other. Enantiomers are optically active: One isomer rotates the plane of polarized light to the right (clockwise) and is said to be **dextrorotatory** (designated +). The other isomer rotates the plane of polarized light to the left (counterclockwise) and is the **levorotatory** (−) isomer. Most organic compounds that exhibit optically activity have one or more carbon atoms with *four different groups attached*. Such a substituted carbon atom is called a **chiral carbon** atom.

Look again at the structures of the two trioses. Note that the middle carbon atom of glyceraldehyde bears four different groups, CHO, H, OH, and CH_2OH; it is a chiral carbon atom. (There is no chiral carbon atom in dihydroxyacetone; the top and bottom carbon atoms both have *two* H atoms, and the middle one has only *three* groups attached.)

The arrangement of the groups about a chiral carbon atom is called its **absolute configuration**. Biochemists use a system devised by Emil Fischer over a century ago to specify the absolute configurations of carbohydrates. The *Fischer projection* of glyceraldehyde has the carbon atom chain vertically and places the most oxidized group (CHO) at the top and the least oxidized group (CH_2OH) at the bottom. The H and OH are written to the sides.

CHO
|
H—C—OH
|
CH_2OH

D-(+)-Glyceraldehyde

CHO
|
HO—C—H
|
CH_2OH

L-(−)-Glyceraldehyde

H, O, C, OH, C, C, OH, H, H, H

(a)

CHO
|
H—C—OH
|
CH_2OH

(b)

▲ **Figure 23.3**
Two representations of the structure of D-(+)-glyceraldehyde
(a) The three-dimensional structure, and (b) a Fischer projection.

The Fischer projection of glyceraldehyde in Figure 23.3(b) places the chiral carbon atom in the plane of the page, the top and bottom groups attached to the chiral carbon atom *behind* the plane of the page, and the side groups on the chiral atom *out* from the page toward the viewer. Fischer arbitrarily assigned the configuration that has the H on the left and the OH on the right of the chiral carbon atom to the dextrorotatory form of glyceraldehyde. He called this the D configuration. He assigned the arrangement with the H on the right and the OH on the left to the levorotatory form and called it the L configuration. The actual assignments were confirmed decades later by X-ray diffraction studies. Just by chance, Fischer had guessed correctly. The configurations of all other carbohydrates are related to those of D- and L-glyceraldehyde. We assume the larger molecules are built by adding carbon atoms at the top end of a glyceraldehyde molecule, and we make a D or L assignment based on the configuration at the bottom chiral carbon atom (blue)—the one assumed to have come from D- or L-glyceraldehyde. Nature seems to favor D configurations for carbohydrates, as we see in the four most common hexoses.

$$\begin{array}{c} CHO \\ | \\ H-C-OH \\ | \\ HO-C-H \\ | \\ H-C-OH \\ | \\ H-C-OH \\ | \\ CH_2OH \end{array} \quad \begin{array}{c} CH_2OH \\ | \\ C{=}O \\ | \\ HO-C-H \\ | \\ H-C-OH \\ | \\ H-C-OH \\ | \\ CH_2OH \end{array} \quad \begin{array}{c} CHO \\ | \\ HO-C-H \\ | \\ HO-C-H \\ | \\ H-C-OH \\ | \\ H-C-OH \\ | \\ CH_2OH \end{array} \quad \begin{array}{c} CHO \\ | \\ H-C-OH \\ | \\ HO-C-H \\ | \\ HO-C-H \\ | \\ H-C-OH \\ | \\ CH_2OH \end{array}$$

D-(+)-Glucose D-(−)-Fructose D-(+)-Mannose D-(+)-Galactose

Note that there is no correlation between the absolute configuration (D or L) and the sign of the rotation of polarized light (+ or −). The absolute configuration is assigned from the projection formula; the sign of rotation must be determined by experiment.

EXAMPLE 23.2

Classify as fully as possible and give the configuration (D or L) and the sign of the rotation (+ or −) for ribose, a sugar found in ribonucleic acid (RNA).

$$\begin{array}{c} CHO \\ | \\ H-C-OH \\ | \\ H-C-OH \\ | \\ H-C-OH \\ | \\ CH_2OH \end{array}$$

Ribose

SOLUTION

The ribose molecule has an aldehyde group (CHO) and five carbon atoms; it is an *aldopentose*. The configuration about the bottom chiral carbon atom (blue) has the OH on the right—the D configuration. There is no way to tell the sign of the rotation from the structure.

EXERCISE 23.2

Classify as fully as possible and give the configuration (D or L) and the sign of the rotation (+ or −) for (**a**) gulose and (**b**) erythrulose.

$$\begin{array}{c} CHO \\ | \\ HO-C-H \\ | \\ HO-C-H \\ | \\ H-C-OH \\ | \\ HO-C-H \\ | \\ CH_2OH \end{array} \qquad \begin{array}{c} CH_2OH \\ | \\ C{=}O \\ | \\ HO-C-H \\ | \\ CH_2OH \end{array}$$

Gulose Erythrulose

Enantiomers differ in only one physical property: the *direction* (not the extent) of rotation of polarized light. They also differ in only one kind of chemical property: their reaction with other chiral molecules. This may seem like a minor difference, but it is all important in biochemistry. The enzymes that catalyze nearly all

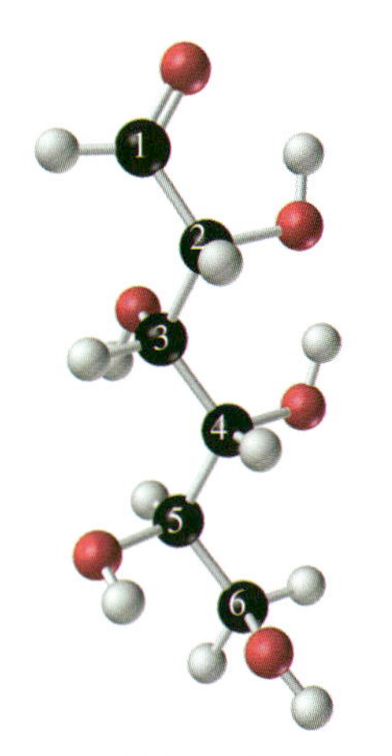

Extended

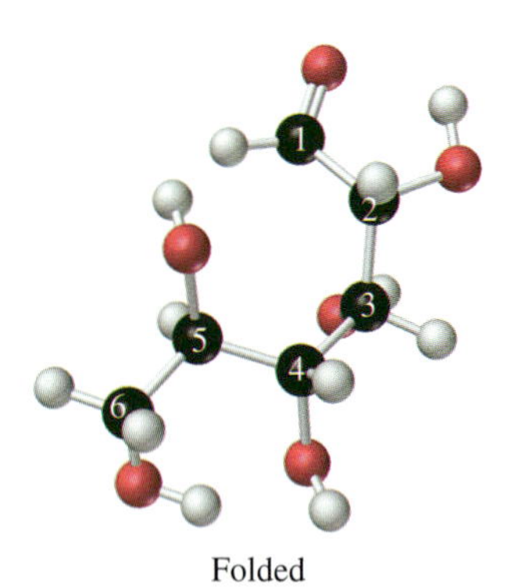

Folded

▲ **Figure 23.4**
Models of two arrangements of the free aldehyde form of glucose
Note that the O atom on C-5 in the folded model is much nearer the carbonyl carbon than is the O atom on C-5 in the extended model.

The liquid interior of a chocolate-covered cherry is a result of the hydrolysis of sucrose. The cherry is covered with a paste of sucrose containing the enzyme invertase. When the paste hardens, the cherries are dipped in chocolate and stored for a week or two. Invertase catalyzes the hydrolysis of sucrose to a syrupy mixture of glucose and fructose (*invert sugar*).

biochemical reactions are chiral; the reactions that they catalyze involve a reactant (substrate) that must fit the active site on the enzyme. The substrate must have the proper "handedness" to fit the active site, just as it takes a right handed glove to fit a right hand. The enzymes that catalyze the metabolism of glucose fit only the "right-handed" D-glucose that occurs in nature. Our bodies can derive no energy whatsoever from L-glucose.

Cyclic Structures of Monosaccharides

Although we have represented monosaccharides as straight-chain structures, most exist mainly in cyclic forms. For example, less than 0.02% of D-(+)-glucose exists in solution in the straight-chain form. In glucose, the OH group on the fifth carbon (designated C-5) adds to the carbonyl group (C-1), forming a ring with five C atoms and one O atom (Figure 23.4). Ring closure results in a new OH group that can extend up from the ring or down below the ring and makes C-1 a new chiral carbon atom. Therefore there are *two* cyclic forms of D-glucose. The form in which the OH extends downward is designated α-glucose and that with the OH pointing up is β-glucose.

Disaccharides

The carbohydrates we have studied so far are all monosaccharides. The monosaccharide units can join together in chains. Two monosaccharide molecules can join together by an OH group on one molecule adding to the carbonyl group of another, and then eliminating a water molecule to form a **disaccharide**. There are three common disaccharides. They are listed below by name and by the monosaccharide units they form upon hydrolysis.

$$\text{Maltose} \xrightarrow{H_2O} \text{glucose} + \text{glucose}$$

$$\text{Lactose} \xrightarrow{H_2O} \text{glucose} + \text{galactose}$$

$$\text{Sucrose} \xrightarrow{H_2O} \text{glucose} + \text{fructose}$$

In characterizing disaccharides, however, we must also consider *how* the monosaccharide units are joined together. Structures of maltose, lactose, and sucrose are shown in Figure 23.5.

In maltose (malt sugar), C-1 of one glucose unit is joined to C-4 of the other through an α linkage. The configuration at C-1 of the second glucose unit can be either α or β. Lactose (milk sugar) has a galactose unit joined C-1 to C-4 of a glucose unit. The glucose ring on the right in lactose can have either an α or a β configuration at C-1. Sucrose (cane sugar or beet sugar) has C-1 of a glucose unit joined to C-2 of a fructose unit. Looking at the molecule from the glucose end, the linkage is α; from the fructose unit, the linkage is β.

A disaccharide can be hydrolyzed to monosaccharide units in a reaction catalyzed by acids or by the appropriate enzyme. As we have indicated, sucrose is hydrolyzed to glucose and fructose. Sucrose is dextrorotatory. Of the products, glucose is also dextrorotatory (accounting for its alternate name dextrose), but fructose is levorotatory (accounting for the alternate name levulose). For equal concentrations, fructose rotates plane-polarized light much farther to the left than glucose does to the right. Therefore when sucrose is hydrolyzed in solution, the optical rotation changes from (+) to (−), and the process is known as the *inversion* of sucrose. The product, a mixture of glucose and fructose, is known as *invert sugar*.

Maltose (α-form)

Lactose (β-form)

(glucose unit)

(fructose unit)

Sucrose

▲ **Figure 23.5 Three disaccharides**

Polysaccharides

Complex carbohydrates having more than ten monosaccharide units in a chain are called **polysaccharides** (Figure 23.6). *Starches*, polymers of glucose, are the principal energy storage materials of plants, especially cereal grains and potatoes. A simple starch called *amylose* has roughly 60 to 300 glucose units joined C-1 of one unit to C-4 of the next through α linkages, just as in the disaccharide maltose. *Amylopectin* has an amylose-like chain structure but with branching on the chains. Animals store glucose in the form of *glycogen*, a high molecular mass polysaccharide with even more chain branching than amylopectin. Glycogen is stored in animal muscle and liver cells and can be converted back to glucose as needed for energy.

Starch forms water-insoluble granules, such as those that make up the bulk of the cells in a potato.

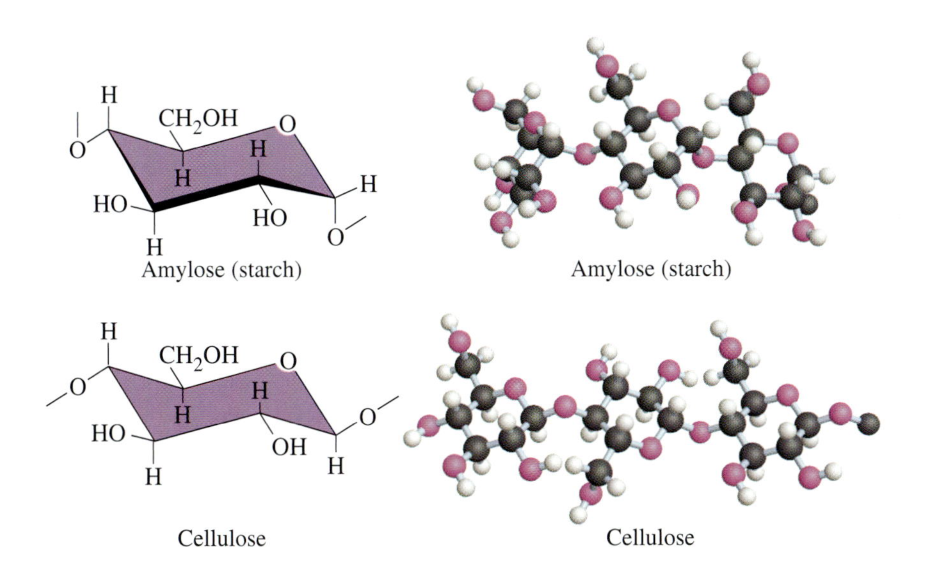

◀ **Figure 23.6 Two polysaccharides**
Structural formulas and molecular models.

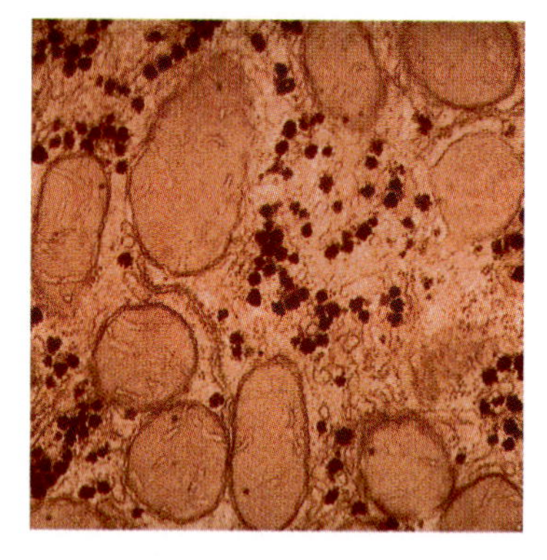

Electron micrograph of glycogen granules in a rat liver cell.

Electron micrograph of the cell wall of an alga. The wall consists of successive layers of cellulose fibers in parallel arrangement.

Cellulose, also a polymer of glucose, is the main structural material of plants. It is the main component of cotton, flax, wood pulp, and other plant fibers. Cellulose is composed of chains of about 1800–3000 glucose units, joined C-1 of one unit to C-4 of the next through β linkages. Humans, like most higher animals, lack enzymes to catalyze the hydrolysis of cellulose; the enzymes do not fit the β linkages that join the glucose units. We get no food energy from cellulose. Ruminants (cattle, goats, deer, sheep) have bacteria in their digestive tracts that can hydrolyze cellulose to glucose. The animals can use the glucose for energy. Termites also have bacteria in their guts that can digest cellulose, allowing the termites to live on a diet of wood.

23.9 Proteins

As we noted in Chapter 11, proteins are polymers of α-amino acids. Recall that an α-amino acid is a carboxylic acid with an amino group on the carbon atom *next to* the carboxyl group.

$$R{-}\underset{\displaystyle NH_2}{\underset{|}{CH}}{-}COOH$$

The α carbon atom

Carboxyl group

Amino group

Twenty different amino acids serve as monomer units in the protein polymers. Some typical amino acids are listed in Table 23.2.

As they occur in nature, 19 of the 20 amino acids are optically active and have the L configuration. As with carbohydrates, the configuration is related to that of glyceraldehyde.

$$H_2N{-}\overset{\displaystyle COOH}{\overset{|}{\underset{\displaystyle R}{\underset{|}{C}}}}{-}H \qquad HO{-}\overset{\displaystyle CHO}{\overset{|}{\underset{\displaystyle CH_2OH}{\underset{|}{C}}}}{-}H$$

An L amino acid L-Glyceraldehyde

Glycine is not optically active because the glycine molecule, H_2NCH_2COOH, has no chiral carbon atom.

Our bodies are unable to synthesize certain of the amino acids needed to build functioning protein molecules; these **essential amino acids**, indicated by footnote *b* in Table 23.2, must be included in our diet. The essential amino acids act as limiting reactants. If there is a shortage of any of these building blocks, the body can't make enough of the proper protein molecules.

Amino acids are amphoteric; they react either with acids or with bases. In highly acidic aqueous solutions (at low pH), the amino acids accept a proton from the aqueous acid and are *cations*. In highly basic solutions, the amino acids donate a proton to the aqueous base and are *anions*. At some intermediate point, they exist as dipolar ions called *zwitterions*.

$$H_3N^+{-}\underset{\displaystyle R}{\underset{|}{CH}}{-}COOH \underset{H^+}{\overset{OH^-}{\rightleftharpoons}} H_3N^+{-}\underset{\displaystyle R}{\underset{|}{CH}}{-}COO^- \underset{H^+}{\overset{OH^-}{\rightleftharpoons}} H_3N{-}\underset{\displaystyle R}{\underset{|}{CH}}{-}COO^-$$

Cation Zwitterion (Dipolar ion) Anion

TABLE 23.2 The 20 Amino Acids Commonly Found in Proteins

Name	Symbol	Condensed Structural Formula	p*I*[a]
Neutral amino acids			
Glycine	Gly	$HCH(NH_2)COOH$	6.03
Alanine	Ala	$CH_3CH(NH_2)COOH$	6.10
Valine[b]	Val	$(CH_3)_2CHCH(NH_2)COOH$	6.04
Leucine[b]	Leu	$(CH_3)_2CHCH_2CH(NH_2)COOH$	6.04
Isoleucine[b]	Ile	$CH_3CH_2CH(CH_3)CH(NH_2)COOH$	6.04
Phenylalanine[b]	Phe	$C_6H_5CH_2CH(NH_2)COOH$	5.74
Serine	Ser	$HOCH_2CH(NH_2)COOH$	5.70
Threonine[b]	Thr	$CH_3CHOHCH(NH_2)COOH$	5.6
Methionine[b]	Met	$CH_3SCH_2CH_2CH(NH_2)COOH$	5.71
Cysteine[b]	Cys	$HSCH_2CH(NH_2)COOH$	5.05
Tyrosine	Tyr	*p*-$HOC_6H_4CH_2CH(NH_2)COOH$	5.70
Tryptophan[b]	Trp	$C_8H_6NCH_2CH(NH_2)COOH$	5.89
Asparagine	Asn	$H_2NCOCH_2CH(NH_2)COOH$	5.4
Glutamine	Gln	$H_2NCOCH_2CH_2CH(NH_2)COOH$	5.7
Proline[c]	Pro	C_4H_8NCOOH	6.21
Acidic amino acids			
Aspartic acid	Asn	$HOOCCH_2CH(NH_2)COOH$	2.96
Glutamic acid	Glu	$HOOCCH_2CH_2CH(NH_2)COOH$	3.22
Basic amino acids			
Lysine[b]	Lys	$H_2N(CH_2)_4CH(NH_2)COOH$	9.74
Arginine	Arg	$H_2NC(=NH)NH(CH_2)_3CH(NH_2)COOH$	10.73
Histidine[b]	His	$C_3H_3N_2CH_2CH(NH_2)COOH$	7.58

[a] pH of at the *isoelectric point*.
[b] Essential amino acids.
[c] Proline has a *secondary* amine functional group (—NH—) making it an *imino* acid.

C_8H_6N = (indole ring with CH_2— and NH)

C_4H_8N = (five-membered ring with HN, positions 2, 3, 4, 5)

$C_3H_3N_2$ = (imidazole ring with CH_2—, N, and NH)

The pH at which the amino acid exists mainly as the dipolar ion is called the **isoelectric point (p*I*)**. Recall that in an electric field, cations migrate to the cathode and anions move toward the anode. At the isoelectric point, the amino acid is overall electrically neutral, and it does not migrate in an electric field. At pH values near its isoelectric point, an amino acid acts very much like the salt of a weak acid and a weak base. As such, it acts as a buffer. Added acid reacts with the $—COO^-$ group and added base reacts with the $—NH_3^+$ group.

Amino acids can join together through an amide linkage by eliminating a molecule of water. In biochemistry, the amide linkage is known as a *peptide linkage* and the short chains of amino acid units are called **peptides**. A *dipeptide* has two amino acid units, a *tripeptide* has three, and so on. With a large number of amino acid units, the molecule is called a **polypeptide**.

Peptides are described by using the three-letter abbreviations in Table 23.2 to specify the amino acid units, starting at the end with the free amino group (the *N-terminal*) and proceeding to the end having a free carboxyl group (the *C-terminal*). We can represent the following pentapeptide as Ala-Ser-Val-Ala-Met.

$$H_2N—CH(CH_3)—CO—NH—CH(CH_2OH)—CO—NH—CH(CH_2(CH_3)_2)—CO—NH—CH(CH_3)—CO—NH—CH(CH_2CH_2SCH_3)—COOH$$

Alanyl   Seryl   Valyl   Alanyl   Methionine

The boundary between polypeptides and proteins is quite arbitrary, often drawn at a molecular mass of about 10,000 u. Thus, we often call molecules with 50 or more amino acid units **proteins** and those with fewer than 50 amino acid units, polypeptides.

APPLICATION NOTE
The primary structure of β-endorphin (page 421), the opiate-like substance produced in our bodies, is Tyr-Gly-Gly-Phe-Met-Thr-Ser-Glu-Lys-Ser-Gln-Thr-Pro-Leu-Val-Thr-Leu-Phe-Lys-Asn-Ala-Ile-Ile-Lys-Asn-Ala-Tyr-Lys-Lys-Gly-Glu.

As we noted in Chapter 11, the **primary structure** of a protein is the exact sequence of amino acid units in the protein molecule. The primary structure of the above pentapeptide is therefore simply Ala-Ser-Val-Ala-Met. We also noted in Chapter 11 that the **secondary structure** of a protein is the structure or shape of an entire protein chain, and that the two principal secondary structures are the α-helix and the β-pleated sheet (see again Figure 11.24).

Many proteins exhibit further structural features. They may have domains of both helical and pleated sheet structures and these may be twisted and tangled in various arrangements. Figure 23.7 shows the *tertiary structure* of triose phosphate isomerase, an enzyme involved in sugar metabolism. The **tertiary structure** of a protein is its complete three-dimensional shape. The folding of a protein chain brings otherwise distant parts of it primary structure close together. Interactions between side chains of the amino acids in these regions stabilize the tertiary structure. Note that the molecule in Figure 23.7 has regions of β strands and α helices connected by turns and folds called *random coils* that do not have repeating structures.

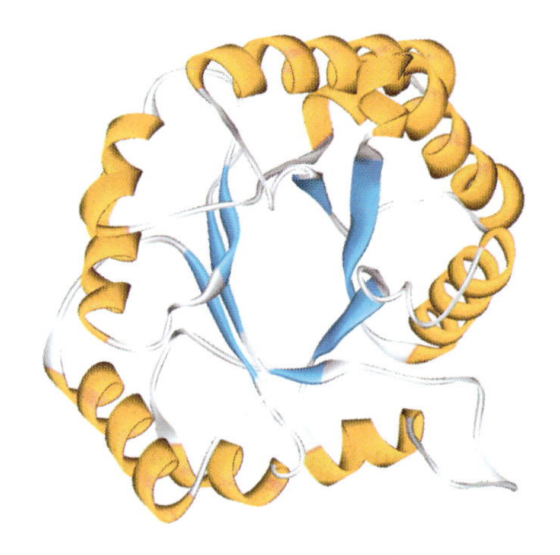

Figure 23.7 ▶ Tertiary structure of a protein
This representation of triose phosphate isomerase, an enzyme involved in sugar metabolism, has a cylindrical arrangement of β strands (blue) surrounded by a garland of α helices (orange). These are connected by turns and folds (white and gray) that lack a repeating structure.

Some protein molecules associate in discrete clusters called a **quaternary structure**. Perhaps the most familiar example of a quaternary structure is hemoglobin. This oxygen-carrying blood protein is an aggregate of four folded chains, as shown in Figure 22.9.

23.10 Nucleic Acids: Molecules of Heredity

Nucleic acids serve as the information and control centers of the cell. There are two kinds of nucleic acid: **deoxyribonucleic acid (DNA)**, found primarily in the cell nucleus, and **ribonucleic acid (RNA)**, found in all parts of the cell. Both are polymers of repeating units (monomers) called **nucleotides**. Each nucleotide, in turn, consists of three parts: a sugar, a phosphate unit, and a cyclic amine base. We can represent a nucleotide schematically as shown in the margin.

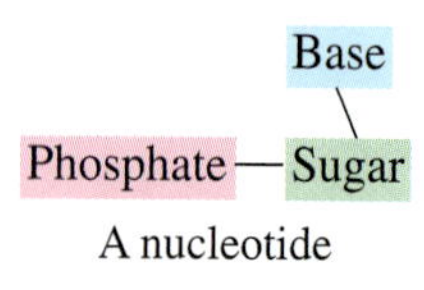

A nucleotide

The Sugars Two sugars occur in nucleic acids (Figure 23.8). Ribose is found in ribonucleic acid (RNA), and 2-deoxyribose is the sugar found in deoxyribonucleic acid (DNA). (The 2-*deoxy* indicates that this sugar lacks an oxygen atom on C-2).

The Phosphate Ester Group A phosphate ester group replaces the OH group on C-5 of the sugar unit. Think of this replacement as resulting from the reaction of an OH group of a phosphoric acid molecule, $OP(OH)_3$, with the H atom of the OH group on C-5 of the sugar. Accordingly, an H_2O molecule is eliminated.

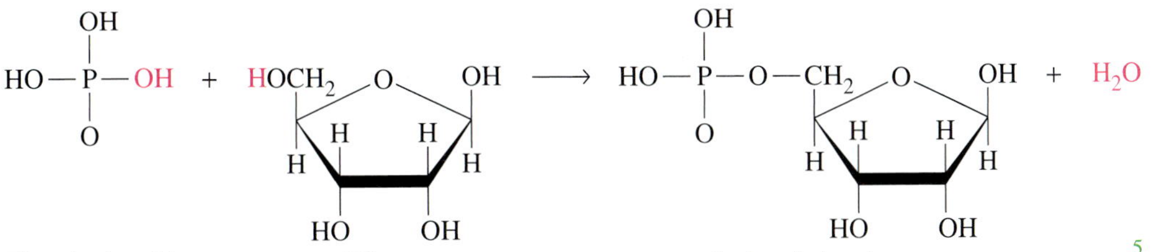

Phosphoric acid + Ribose ⟶ A phosphate ester + H_2O

The Bases In the nucleotide, one of *five* possible heterocyclic amine bases replaces the hydroxyl group on C-1 of the sugar. These bases, shown in Figure 23.9, are the single-ring units cytosine, thymine, and uracil, called *pyrimidines*, and the fused double-ring units adenine and guanine, called *purines*.

Nucleotides are joined to one another through the phosphate groups to form nucleic acid chains. The phosphate unit on one nucleotide is joined to the hydroxyl group on C-3 of the sugar unit in a second nucleotide through an ester linkage. A third nucleotide is attached in the same way, and the process is repeated to build up a long nucleotide chain (Figure 23.10). Schematically, we can represent the polymer as follows.

```
              Base              Base               Base
                \                 \                  \
... Phosphate — Sugar — Phosphate — Sugar — Phosphate — Sugar ...
```

The backbone of the chain consists of alternating sugar and phosphate units. The heterocyclic bases, which are branched off this backbone, can be present in various combinations. In DNA nucleotides, the bases are adenine, guanine, cytosine, and thymine. The bases in RNA nucleotides are adenine, guanine, cytosine, and uracil. Table 23.3 summarizes the compositions of DNA and RNA.

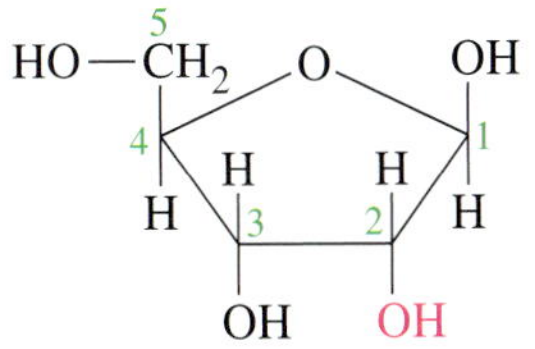

Ribose

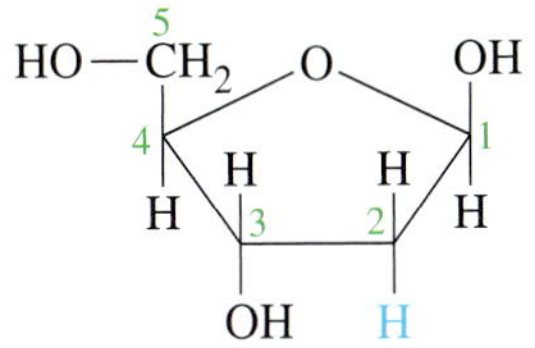

2-Deoxyribose

▲ **Figure 23.8**
The sugars found in nucleic acids
Ribose and 2-deoxyribose differ in that 2-deoxyribose lacks an oxygen atom ("deoxy") on the second carbon atom. That is, ribose has an H atom and an OH group attached to the second carbon atom and 2-deoxyribose has two H atoms.

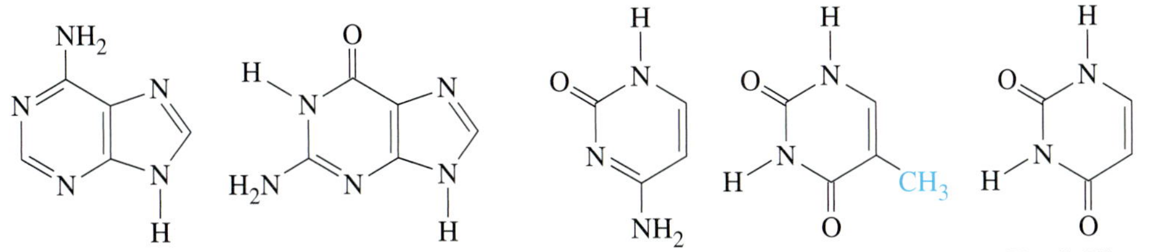

Adenine (A) Guanine (G) — Purine bases

Cytosine (C) Thymine (T) Uracil (U) — Pyrimidine bases

▲ **Figure 23.9 Heterocyclic bases found in nucleic acids**

▶ **Figure 23.10 The backbone of a deoxyribonucleic acid molecule**

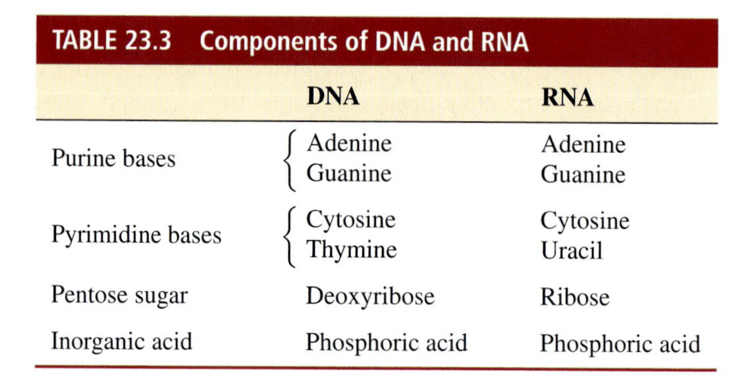

TABLE 23.3 Components of DNA and RNA

	DNA	RNA
Purine bases	Adenine Guanine	Adenine Guanine
Pyrimidine bases	Cytosine Thymine	Cytosine Uracil
Pentose sugar	Deoxyribose	Ribose
Inorganic acid	Phosphoric acid	Phosphoric acid

James D. Watson (1928–) (right) and Francis Crick (1916–) (seated) worked out the double helix model of DNA in 1953 and were awarded the Nobel Prize for this work in 1968.

Base Sequence in Nucleic Acids

A crucial feature of a nucleic acid molecule is the sequence of the four bases along the strand, called the *base sequence*. The molecules are huge, with molecular masses ranging into the billions for mammalian DNA, so the four bases may be arranged in an essentially infinite number of variations. The specific sequence of the bases along the chain is the information storage system needed to build organisms. But before we examine that aspect of nucleic acid chemistry, let's consider one more important feature of nucleic acid structure.

Base Pairing in DNA

In DNA the molar amount of adenine (A) is equal to the molar amount of thymine (T). Similarly, the molar amount of guanine (G) is equal to that of cytosine (C). To account for this balance, the bases must be paired, A to T and G to C. But how? James D. Watson and Francis Crick worked out the structure of DNA in 1953. They determined that DNA is composed of two helixes coiled around one another. The sugar and phosphate backbone of the polymer chains forms the outside of the structure, rather like a spiral staircase. The heterocyclic amines are paired on the inside, with guanine on one helix always opposite cytosine on the other. Similarly, adenine on one helix is always opposite thymine on the other. In our staircase analogy, these base pairs are the steps (Figure 23.11).

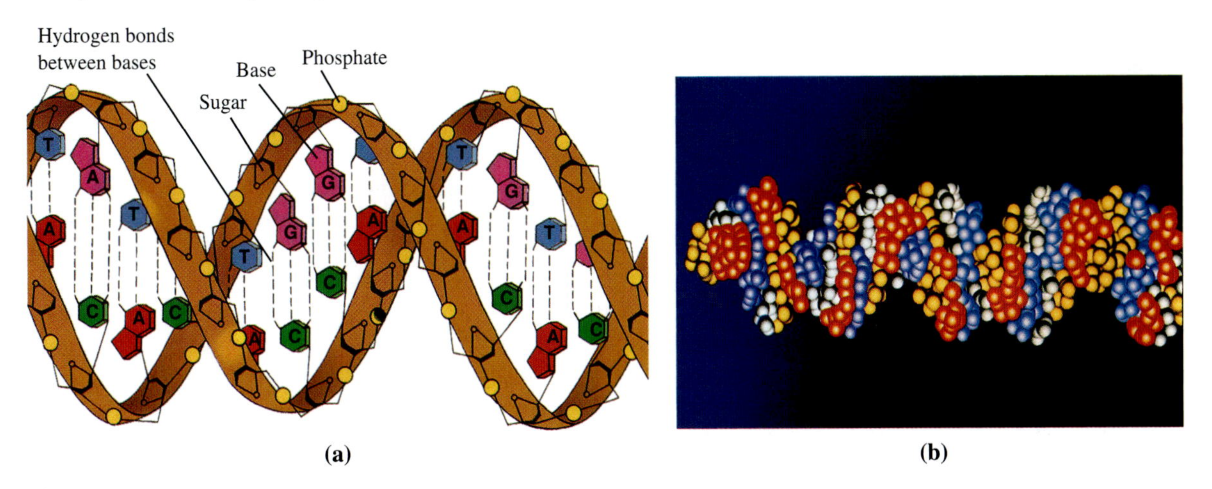

▲ **Figure 23.11 The DNA double helix**
(a) A schematic representation. Two strands of DNA are wound around each other in a double helix. Hydrogen bonds between complementary bases hold the strands together. (b) A space-filling model of the DNA molecule.

The two strands of a DNA double helix are not identical. Rather, they are *complementary*. The base A in one chain is always paired to T in the other, and G in one chain to C in the other. This is the only pairing in which hydrogen bonds can effectively join the two strands of nucleotides at the proper internuclear distance for maximum hydrogen bond formation. Figure 23.12 shows the two sets of base pairs. Note that the matching of these particular pairs, a pyrimidine on one chain and a purine on the other, leads to hydrogen bonding at a fixed interchain distance (1.085 nm). The base pairs fit like pieces of a jigsaw puzzle (but the puzzle is only two pieces wide). Other pairing of bases, such as pyrimidine–pyrimidine or purine–purine would not produce hydrogen bonds of the correct length.

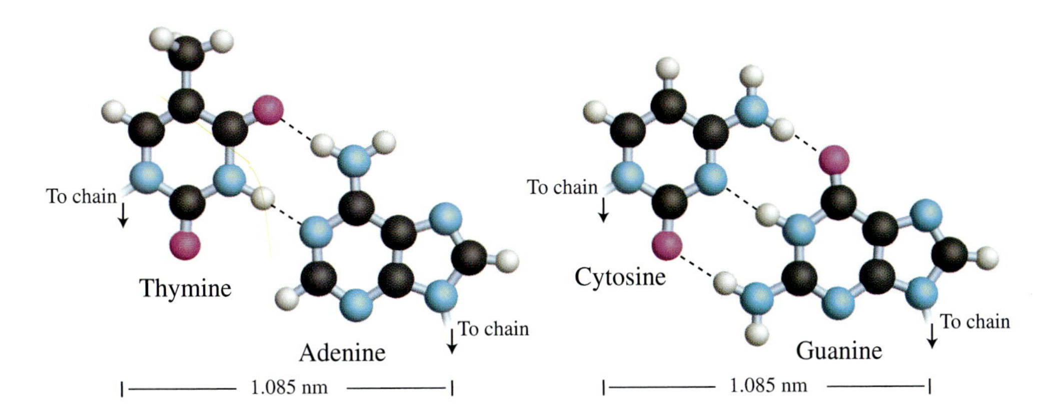

▲ **Figure 23.12 The pairing of bases in the DNA double helix**

The Watson–Crick structure answers many crucial questions. It explains how cells are able to divide and go on functioning, how genetic data are passed to new generations, and even how proteins are built to required specifications. It all depends on base pairing.

Nucleic Acids and Heredity

Lions have cubs that grow up to be lions. Eagles lay eggs that produce eaglets that grow up to be eagles. How is it that each species reproduces after its own kind? How does a fertilized egg "know" that it should develop into a kangaroo and not a koala?

The physical basis of heredity has been known for a long time. Humans and most other higher organisms reproduce sexually. A sperm cell from the male unites with an egg cell from the female. For human beings, that single fertilized egg cell must carry the information for making legs, liver, lungs, heart, head, hair, and hands—in short, all the instructions needed for growth and maintenance of the individual.

Hereditary material is found in the nuclei of all cells, concentrated in elongated, threadlike bodies called *chromosomes*. The number of chromosomes varies with the species. Most human cells have 46 chromosomes, while egg and sperm cells carry half that number. In sexual reproduction, an egg and sperm combine to form the entire complement of chromosomes. A new individual receives half its hereditary material from each parent.

Chromosomes are made of DNA and proteins. The DNA is the primary hereditary material. Arranged along the chromosomes are the basic units of heredity, the genes. Structurally, **genes** are sections of the DNA molecule (although some viral genes contain only RNA).

APPLICATION NOTE

Scientists are now embarked on a huge scientific venture, called the Human Genome Project, to determine the base sequence of the complete set, or *genome*, of human genes, and to map the location of each of the 100,000 or so genes on the chromosomes. They expect to complete the project in the first decade of the twenty-first century.

Replication During cell division, each chromosome produces a duplicate of itself. Transmission of genetic information therefore requires *replication* (copying or duplication) of DNA molecules. The Watson–Crick double helix provides a ready model for replication.

As shown in Figure 23.13, the process of replication begins with the separation of the double helix (Figure 23.13a). This separation is much like the unzipping of a zipper (Figure 23.13b). The exposed single strands pick up appropriate nucleotides from the cellular fluid, and enzymes catalyze their attachment to the sugar-phosphate backbone of a new complementary chain (Figure 23.13c). Ultimately, *two* double helixes replace the original one (Figure 23.13d). The information encoded in the original DNA double helix is now contained in each of the two replicates. When a cell divides, each daughter cell gets one of the DNA molecules, and with the DNA it gets all the information available to the parent cell.

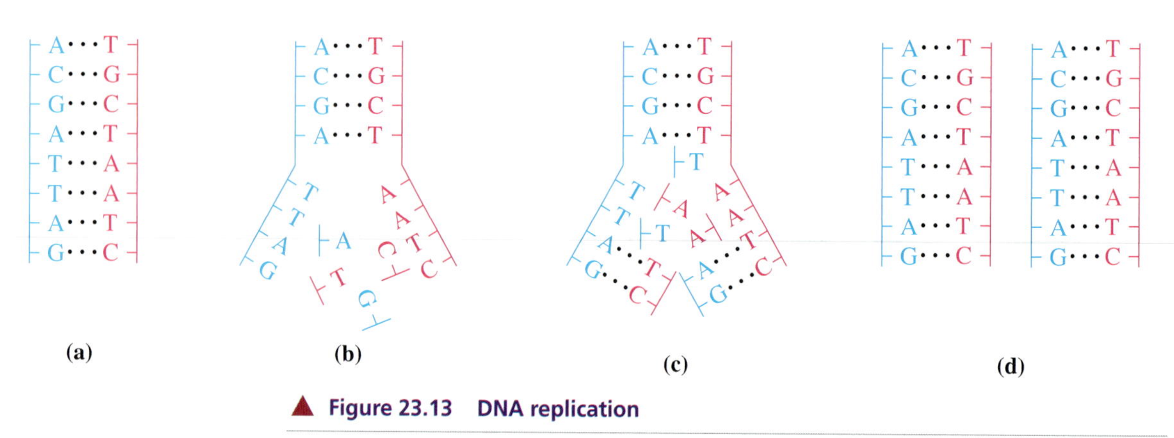

▲ **Figure 23.13 DNA replication**

The sequence of bases along the DNA chain encodes the directions for building an organism. Just as in English the sequence of letters *art* means one thing and the same letters in the order *rat* means another, the sequence of bases CGT means one thing, and GCT means something else. Although there are only four "letters" in the genetic code of DNA—the four bases—their sequence along the long strands can vary so widely that they can convey essentially unlimited information. Each cell carries in its DNA all the information it needs to determine all the hereditary characteristics of even the most complex organism.

Transcription The message carried by DNA in the cell nucleus is relayed by various kinds of RNA to other parts of the cell. In the first step of protein synthesis, called **transcription**, DNA transfers its information to *messenger RNA* (*mRNA*). The base sequence of DNA specifies the base sequence of mRNA. The only difference in transcription and DNA replication is that in transcription, adenine (A) requires uracil (U) rather than thymine (T) (Table 23.4). Each mRNA contains all the information necessary to make one particular protein.

TABLE 23.4 DNA Bases and Their Complementary RNA Bases

DNA base	Complementary RNA base
Adenine (A)	Uracil (U)
Thymine (T)	Adenine (A)
Cytosine (C)	Guanine (G)
Guanine (G)	Cytosine (C)

Translation The next step in creating a protein involves deciphering the code copied by mRNA and **translation** of that code into a specific protein structure. The deciphering occurs when the mRNA travels from the nucleus and attaches itself to a ribosome in the cytoplasm of the cell. Ribosomes are constructed of RNA and proteins.

Transfer RNA (tRNA) translates the base sequence of mRNA into a specific amino acid sequence in a protein. The tRNA has the looped structure shown in Figure 23.14. At the head of the molecule is a set of three base units, called a *base triplet*, that pairs with a set of three complementary bases on mRNA. This triplet determines which amino acid is carried at the tail of the tRNA. To illustrate, the base triplet GUA on a segment of mRNA pairs with the base triplet CAU on a tRNA molecule. All tRNA molecules with the base triplet CAU always carry the amino acid valine.

Each of the 61 different tRNA molecules carries a specific amino acid into place on a growing peptide chain. The protein chain gradually built up in this way is released from the tRNA and mRNA as it is formed. Three tRNAs carry *stop* signals that call for termination of the protein chain. Figure 23.15 provides an overall summary of protein synthesis.

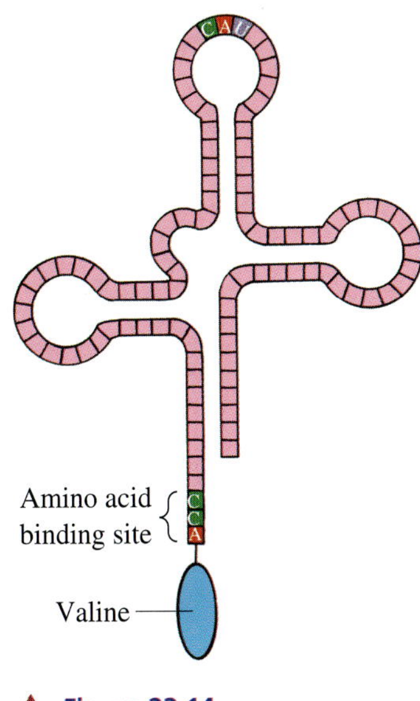

▲ **Figure 23.14 Transfer RNA (tRNA)**

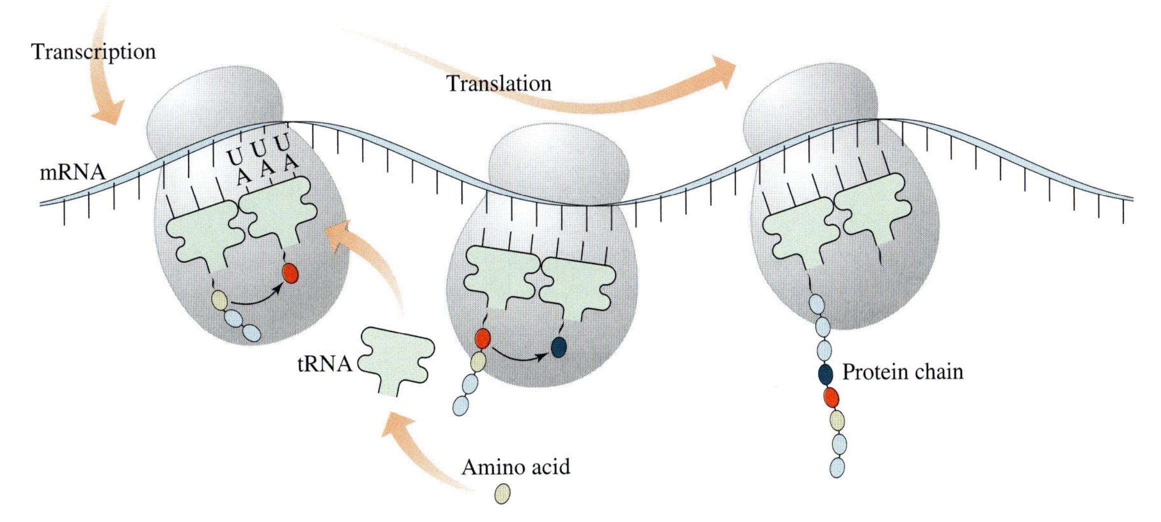

▲ **Figure 23.15 Protein synthesis visualized: From DNA through mRNA and tRNA to protein**

Spectroscopy in Organic Chemistry

We often use the structures of molecules to explain properties of matter, and for this reason we have focused much of our study on molecular structure. Molecular structures are determined by **spectroscopy**, a set of methods based on the interaction of electromagnetic radiation with matter. In the sections that follow, we discuss briefly three of the most important methods, especially as they pertain to organic compounds. Each method supplies particular kinds of information, but combining results of several methods gives us the most complete description of a molecular structure. For example, when applied to organic molecules:

Infrared (IR) spectroscopy indicates the functional groups that are present.

Ultraviolet-visible (UV-VIS) spectroscopy reveals whether the structure has a series of alternate single and double bonds (conjugated π electron systems).

Nuclear magnetic resonance (NMR) spectroscopy describes the carbon-hydrogen framework of the molecule.

23.11 The Interaction of Matter with Electromagnetic Radiation

We introduced the electromagnetic spectrum and explored the basis of light emission in Chapter 7. We also noted the basis of emission spectroscopy: that the frequencies of emitted radiation depend on energy differences between excited and ground states of the emitting atoms or molecules. Ground-state atoms or molecules can *absorb* radiation to become excited. This is the basis of *absorption spectroscopy*, the focus of our discussion here.

Table 23.5 lists several types of electromagnetic radiation and the kinds of changes that their absorption causes in molecules. For example, ultraviolet (UV) light has the shortest wavelength and greatest energy of the forms listed in Table 23.5, and UV radiation can excite the electrons in π bonds in a molecule. Infrared light, having a much lower energy than UV, generally can only affect the vibrations of molecules, the stretching or bending of bonds.

In general, a *spectrometer* (Figure 23.16) measures the extent to which radiation of a particular wavelength is absorbed by a sample. Light from the radiation source is limited to a narrow beam of certain wavelengths and is then passed through a sample. If the radiation does not interact with molecules in the sample, the light beam passes through with undiminished intensity. At a wavelength where the molecules do absorb radiation, the transmitted radiation will be diminished in intensity or completely lacking.

TABLE 23.5 Electromagnetic Radiation and Its Interactions with Molecules

Type of Radiation	Wavelength[a]	Type of Interaction
Vacuum ultraviolet	180–200 nm	Interacts with σ electrons
Ultraviolet	200–400 nm	Interacts with π electrons
Visible	400–800 nm	Interacts with conjugated π electrons
Near infrared	800–2000 nm	Stretches bonds
Infrared	2.0–30.0 μm	Stretches and bends bonds
Radio	1.0–5.0 m	Changes nuclear spin states

[a] m = meter; μm = micrometer (1×10^{-6} m); nm = nanometer (1×10^{-9} m)

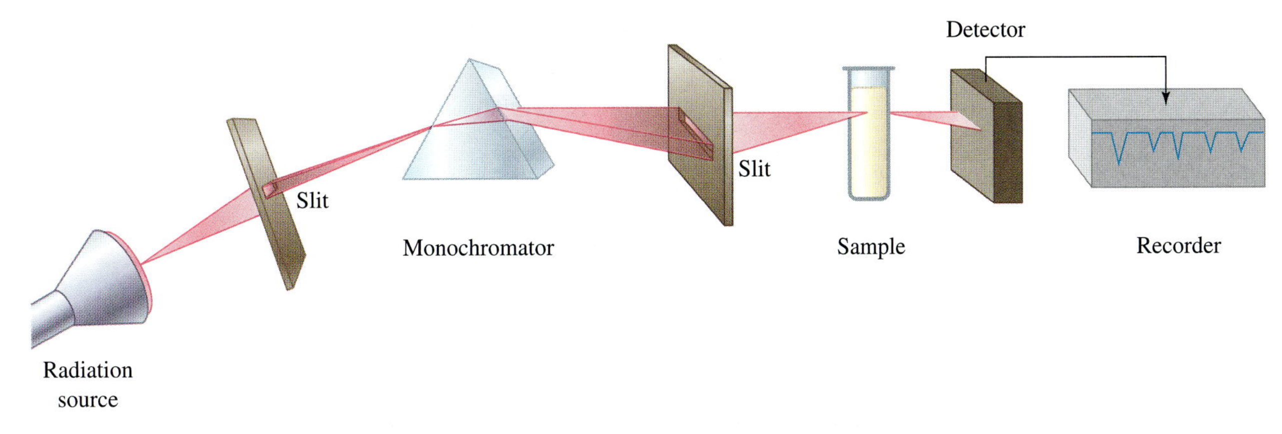

▲ **Figure 23.16 Schematic diagram of a simple spectrometer**
The essential components that comprise the spectrometer are (1) a source of radiation for the portion of the electromagnetic spectrum required; (2) a monochromator—a device (shown here as a prism of glass, quartz, rock salt, or other suitable material) that disperses the incoming radiation into a spectrum; (3) a slit that allows the desired narrow wavelength band to pass while other wavelengths are screened out; (4) a detector that responds to the power or intensity of the radiation striking it by producing an electric signal of a proportionate magnitude; (5) a recorder that plots the strength of the electric signal; peaks occur at wavelengths of the radiation absorbed, and the heights of the peaks indicate the quantity of radiation absorbed.

Through the years, chemists have assembled tables of the structural features in molecules responsible for absorption at particular wavelengths. With appropriate experimental data and the help of these tables we can determine the molecular structure of almost any substance.

23.12 Infrared Spectroscopy

Infrared radiation (IR) typically changes the intensity of the vibrations of atoms or groups of atoms about the bonds that connect them. Like electronic transitions, the vibrational motion of atoms is quantized: Infrared radiation of a specific wavelength is required to activate a molecule from one vibrational energy level to a higher one. Figure 23.17 illustrates vibrational changes called stretching and bending modes.

An IR spectrometer records *peaks* at wavelengths at which absorption occurs. The location of an absorption peak is usually specified by its frequency, but stated in a special way. The **wavenumber** ($\tilde{\nu}$) is the number of cycles of the wave in a 1-cm length of the IR beam. Its unit is reciprocal centimeters, (cm^{-1}).

$$\tilde{\nu} = \frac{1}{\lambda} \text{ (with } \lambda \text{ in centimeters)}$$

Consider the oxygen-to-hydrogen bond, O—H. IR radiation of frequency 3250 to 3450 cm^{-1} is absorbed as the bonded atoms vibrate more intensely. Any molecule having an O—H group absorbs radiation in the 3250–3450 cm^{-1} range. This range varies from one alcohol to another, depending on the overall structure of the molecule. If we find an absorption peak in this range we know that a compound could have an O—H group. We can't be certain, however, because other groups also absorb in this region. For example, the absorption frequency range for the N—H group overlaps somewhat that of the O—H group. We need other evidence to distinguish between the two possibilities. If an elemental analysis of the

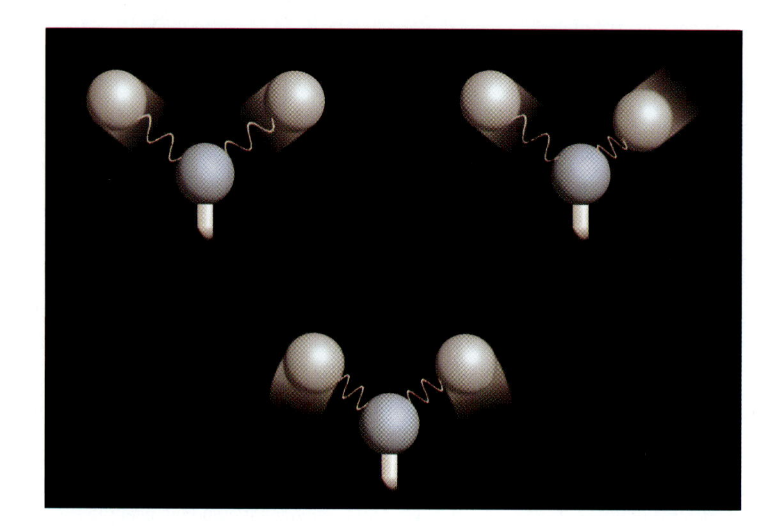

Figure 23.17 Stretching and bending of covalent bonds

A hypothetical functional group of three atoms is attached to a larger molecule, and the different modes of vibration are illustrated. Two different modes of stretching (top) and the bending mode (bottom) are shown.

compound shows the absence of nitrogen, we can be more confident that an absorption peak at 3250–3450 cm^{-1} is due to O—H. The infrared absorption spectrum of ethanol in Figure 23.18 shows prominent peaks due to C—H and O—H bonds.

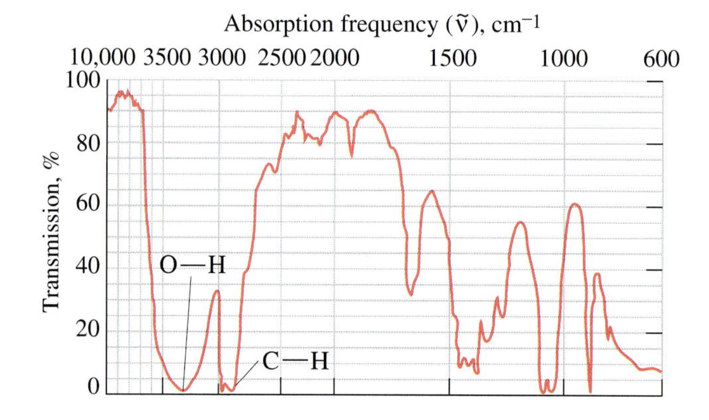

Figure 23.18 The infrared spectrum of ethanol, CH_3CH_2OH

The absorption peaks due to O—H and C—H stretching frequencies are labeled.

The carbonyl group, C=O, absorbs IR radiation in the range 1630–1780 cm^{-1}, causing the atoms in the bond to vibrate more intensely. Aldehydes, ketones, carboxylic acids, and esters all have carbonyl groups and absorption peaks in the 1630–1780 cm^{-1} range.

R—C(=O)—H	R—C(=O)—R	R—C(=O)—OH	R—C(=O)—OR
Aldehyde	Ketone	Carboxylic acid	Ester

Note that the carboxylic acid has *both* a C=O group and an OH group. However, the OH absorption peak is somewhat different from that in alcohols. Figure 23.19 presents the infrared absorption spectrum of 2-hexanone, a ketone. The C—H and C=O peaks are marked.

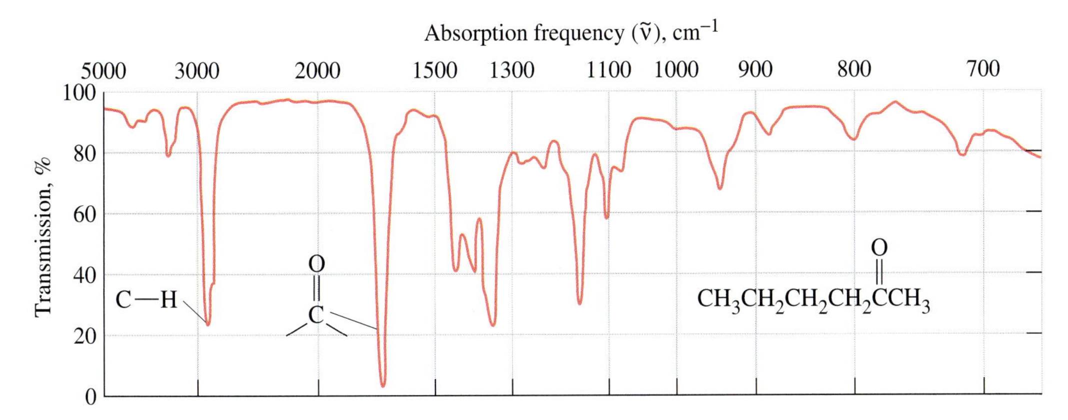

▲ **Figure 23.19 The infrared spectrum of 2-hexanone**
The absorption peaks due to C—H and C=O stretching frequencies are labeled.

Infrared absorption peaks of typical functional groups are given in Table 23.6. To illustrate how these data can be used to identify a compound from its infrared absorption spectrum, consider the case of lidocaine (Figure 23.20), a local anesthetic. The groups of lidocaine that we expect to have distinctive absorption peaks are the alkyl and aromatic C—H bonds, the N—H bond, and the C=O bond. Accordingly, the lidocaine spectrum shows the following peaks and the groups of atoms from which they arise.

- At about 3200 cm^{-1}, caused by the amide N—H bond
- At 2940 cm^{-1}, caused by the aromatic C—H bonds (that is, those of the benzene ring).
- At 2800 cm^{-1}, caused by the alkyl C—H bonds (that is, those of the methyl and ethyl groups).
- At 1680 cm^{-1}, caused by the amide C=O bond.

TABLE 23.6 Infrared Absorptions of Bonds in Characteristic Functional Groups

Functional Group	Bond	Frequency Range, cm^{-1}
Alkyl	C—H	2850–2960
Aromatic	C—H	~3030
Alcohol or phenol	O—H	3250–3450
Amine	N—H	3300–3500
Ketone	C=O	1680–1750
Ester	C=O	1735–1750
Carboxylic acid	C=O O—H	1710–1780 2500–3000
Amide	C=O N—H	1630–1690 3140–3500

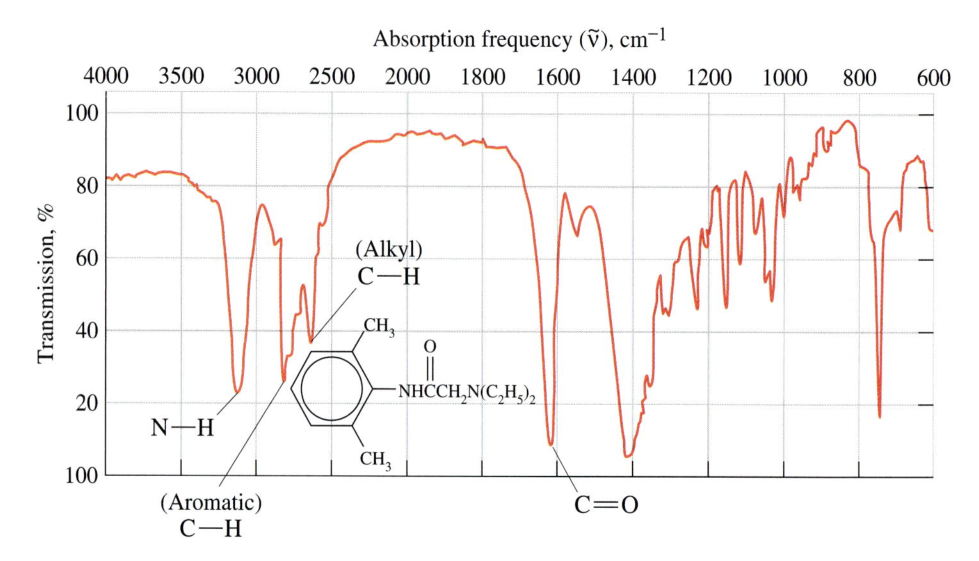

▲ **Figure 23.20 The infrared spectrum of lidocaine**
The spectrum was taken on a sample formed into a KBr pellet. The absorption peaks due to N—H, two kinds of C—H (aromatic and alkyl), and C=O stretching frequencies are labeled.

Infrared spectra are often called the "fingerprints" of organic compounds. We can say with confidence that two pure samples that give identical IR spectra, matched peak for peak, are the same compound. If the spectra do not match, the samples must be different substances.

23.13 Ultraviolet-Visible Spectroscopy

Ultraviolet (UV) and visible radiation have sufficient energy to boost valence electrons of molecules to higher energy levels. The visible region of the electromagnetic spectrum extends from 390 nm to 760 nm in wavelength. The ultraviolet ranges from 390 nm down to about 40 nm, but only the portion from 200 nm to 390 nm is commonly used in laboratory studies. UV light in this range is especially effective in producing energy-level changes in *delocalized* π electrons.

As a case in point, let's consider benzene, C_6H_6. Recall that we can represent benzene as the resonance hybrid of two Kekulé structures of alternate double and single bonds.

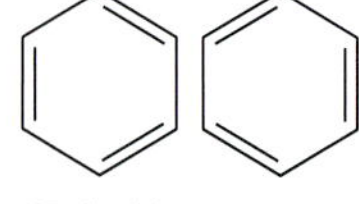

Kekulé structures of benzene

Molecular orbital theory indicates that the six π electrons reside in three bonding molecular orbitals and that three vacant antibonding molecular orbitals lie at higher energies (Figure 10.29). Ultraviolet radiation can boost electrons from the bonding orbital of highest energy, called the *highest occupied molecular orbital (HOMO)*, to the antibonding orbital of lowest energy, called the *lowest unoccupied molecular orbital (LUMO)*. Similar absorptions of energy occur in other molecules with a *conjugated* bonding arrangement.

We record a UV spectrum by irradiating a sample with UV light that is continuously varied in wavelength. When the wavelength corresponds to the difference in energy between the HOMO and the LUMO orbitals of a conjugated system, some of the radiation is absorbed by the sample. The extent of this absorption is measured and plotted on a chart as wavelength versus the percent of the radiation that is absorbed. UV spectra often feature a single broad peak, and we label the wavelength at the top of the peak as λ_{max}. For benzene, the most prominent peak is at λ_{max} = 203 nm.

The ultraviolet spectrum of a typical conjugated straight-chain molecule, 1,3-butadiene, is shown in Figure 23.21, and that of a typical aromatic molecule, styrene, is shown in Figure 23.22. The λ_{max} value for styrene, 245 nm, comes at a longer wavelength than does that of benzene, 203 nm, because in the styrene molecule, the π electrons of the benzene ring are conjugated with those of the double bond in the side chain.

—CH=CH$_2$

The π electrons of styrene require less energy to jump from the HOMO to the LUMO than do those of benzene. This lesser energy can be supplied by radiation of lower frequency. Thus, as the length of conjugation increases in a molecule, λ_{max} of the absorption peak shifts to longer wavelengths. We also see this effect in the following three molecules.

$CH_2{=}CH{-}CH{=}CH_2$ $\lambda_{max} = 217$ nm
1,3-Butadiene

$CH_2{=}CH{-}CH{=}CH{-}CH{=}CH_2$ $\lambda_{max} = 258$ nm
1,3,5-Hexatriene

$CH_2{=}CH{-}CH{=}CH{-}CH{=}CH{-}CH{=}CH_2$ $\lambda_{max} = 290$ nm
1,3,5,7-Octatetraene

An especially long conjugated system absorbs visible light, and compounds with absorption peaks in the visible region are colored. The transmitted light is the complementary color of that which is absorbed (see Table 23.7). Substances with no absorption peaks in the visible region are colorless. From Table 23.7 we see, for example, that a compound that absorbs blue light (475 nm) will appear orange, the complementary color of blue. β-Carotene, a pigment prominent in carrots, pumpkins, and some other vegetables, has 11 conjugated double bonds. It absorbs at 466 and 497 nm and is orange-red in color.

TABLE 23.7 The Wavelengths of Visible Colors[a]

Approximate Wavelength, nm	Color of Light	Complementary Color
425	Violet	Yellow
475	Blue	Orange
500	Blue-green	Red
525	Green	Purple
575	Yellow	Violet
625	Orange	Blue
700	Red	Blue-green

[a] The relationship between the colors of absorbed and transmitted light is also illustrated, in a different way, in Figure 22.20.

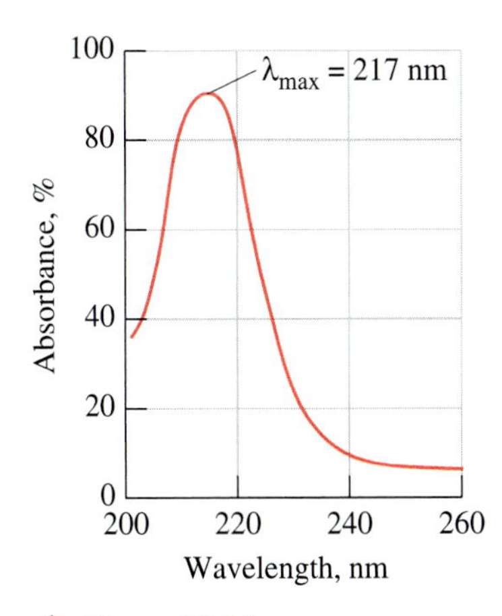

▲ **Figure 23.21**
The ultraviolet spectrum of 1,3-butadiene

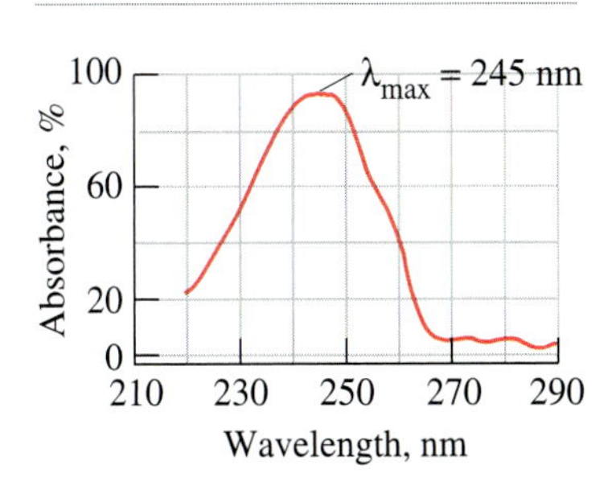

▲ **Figure 23.22**
The ultraviolet spectrum of styrene, $C_6H_5-CH{=}CH_2$
Styrene is the monomer from which polystyrene plastics (see Chapter 24) are made. The π electrons of the benzene ring are conjugated with those of the double bond in the side chain.

APPLICATION NOTE
Conjugated bonding systems are important in ultraviolet-visible spectroscopy (Section 23.13). They are also a factor in producing color in organic compounds such as lycopene, the red pigment of tomatoes, and β-carotene, the orange pigment of carrots. There are other reasons for substances to exhibit visible colors, though, as we saw in Chapter 22.

Finally, note that no matter how many double bonds are present in a molecule, if they are not part of a *conjugated* system, no absorption will occur in the UV region above 200 nm or in the visible region of the spectrum. Thus, neither 1-hexene nor 1,5-hexadiene has a UV-VIS absorption peak above 200 nm.

$CH_2{=}CHCH_2CH_2CH_2CH_3$ (1-Hexene) $\qquad$ $CH_2{=}CHCH_2CH_2CH{=}CH_2$ (1,5-Hexadiene)

23.14 Nuclear Magnetic Resonance Spectroscopy

The carbon-hydrogen framework is the backbone of nearly all organic molecules. Nuclear magnetic resonance (NMR) spectroscopy provides better information about this framework than either UV-VIS or IR spectroscopy.

NMR is based on the absorption of energy by the *nuclei* of certain atoms in a molecule. If an atomic nucleus has an odd number for either the atomic number Z or the mass number A, the nucleus will be active in NMR spectroscopy. The most important nuclei for NMR studies of organic compounds are hydrogen-1 (^1_1H) and carbon-13 ($^{13}_{6}C$). We will consider only ^1_1H spectra.

NMR-active atomic nuclei spin about their axes like tiny tops. Because they are electrically charged, spinning nuclei act like miniature magnets. In the absence of an external magnetic field to line them up, the tiny magnetic fields of the spinning nuclei are randomly oriented. In the presence of a strong external magnetic field of magnitude H_0, the nuclei assume preferred orientations, much as compass needles align themselves with Earth's magnetic field (see Figure 23.23). When a hydrogen-containing compound is placed in a strong magnetic field, the tiny fields of the nuclei can align either in the same direction (*parallel*) as the external magnetic field or in the opposite direction (*antiparallel*). The parallel orientation is a slightly lower energy state than the antiparallel alignment, and slightly more nuclei are found in the parallel than in the antiparallel state.

If electromagnetic radiation of the proper energy is focused on a sample within a magnetic field, some of the hydrogen nuclei in the lower energy state absorb the radiation and are boosted to the higher energy level. These nuclei undergo a "spin flip," changing their orientation in the field from parallel to antiparallel. The required radiation is of the low energies found in the radiofrequency (RF) portion of the electromagnetic spectrum. When the radiation has just the right energy to

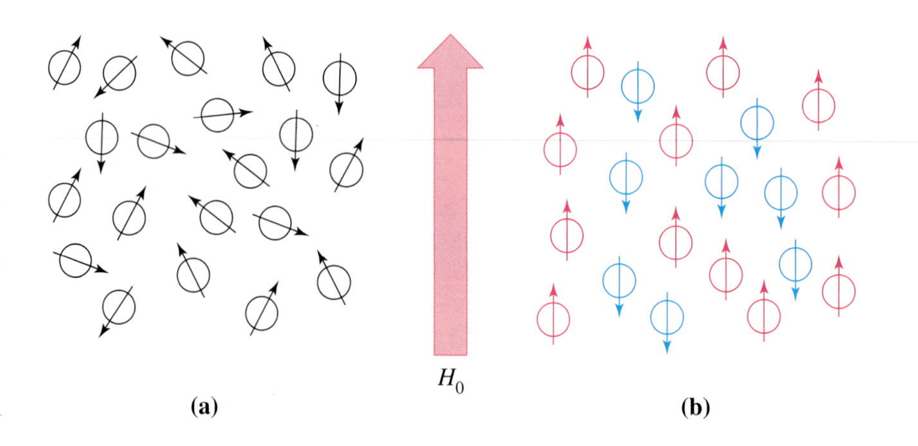

Figure 23.23 Nuclear spins illustrated
(a) In the absence of a strong external field, nuclear spins are oriented at random. (b) In the presence of a strong applied field, there are only two possible orientations. Some nuclear spins are aligned in the direction of the field (red) and some are aligned in the opposite direction (blue). Those with spins in the direction of (parallel to) the applied field are in the lower energy state.

cause a spin flip, the nuclei are said to be in resonance with the applied frequency; hence the name *nuclear magnetic resonance*.

The frequency of RF energy required* to produce a resonance condition depends on the strength of the applied magnetic field and on the electronic environment of the nuclei. Electrons are involved because they, too, are spinning charged particles and have their own tiny magnetic fields. Electron fields in the vicinity of a hydrogen nucleus usually act in opposition to the applied field. The *effective* field that actually acts on a nucleus is therefore a bit weaker than the applied field.

$$H_{\text{actual}} = H_0 - H_{\text{local}}$$

Just how much weaker the field is depends on the electrons that bond hydrogen to another atom and on electrons in adjacent groups in the molecule.

In NMR spectroscopy, we detect the absorption of energy and correlate it with the environment of particular hydrogen atoms. If two hydrogen atoms in the same molecule have a different electronic environment, one hydrogen nucleus will absorb radiation of slightly different energy than the other hydrogen nucleus. For example, in the toluene molecule, the electronic environment of the hydrogen atoms in the methyl group (red) is different from the electronic environment of the hydrogen atoms in the benzene ring (blue).

The methyl H atoms and the ring H atoms have different electronic environments. The nuclei of these two "kinds"† of hydrogen atoms absorb different frequencies of RF energy in a constant magnetic field. The two kinds of H atoms produce two separate peaks in the NMR spectrum of toluene (Figure 23.24).

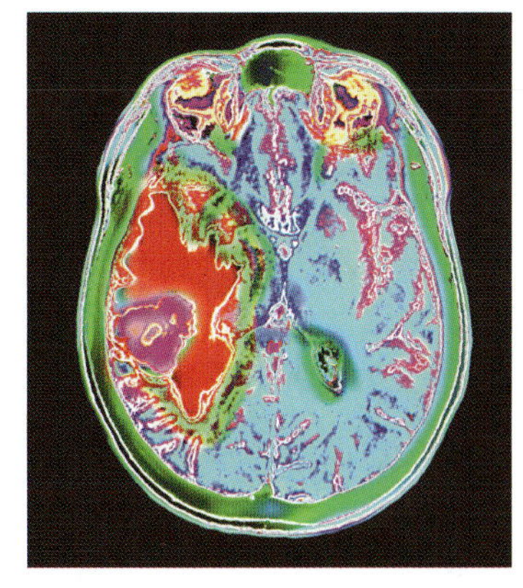

Magnetic resonance imaging (MRI), a medical technique that was developed from basic research in nuclear magnetic resonance, is now widely used to make images of cross-sections of the human body. The MRI signal originates mainly from the hydrogen atoms of water molecules. Because tissues vary in water content and in the way the water is held in the tissues, signals vary from one type of tissue to another. An MRI scan can therefore precisely locate a tumor or injured tissues without surgery. This MRI of an axial section of a brain shows a tumor (purple). The growing tumor has damaged the surrounding tissue (red), which has filled with fluid.

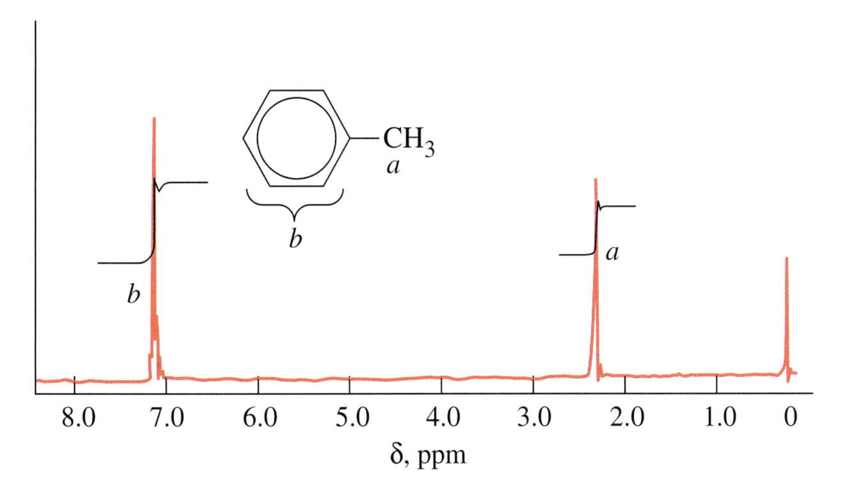

Figure 23.24
The NMR spectrum of toluene
There are only two signals. The height of each integral curve (thin black curves in the peak region) is a measure of the area of that peak. These areas are in a ratio of 5:3, just as we expect from the formula $C_6H_5CH_3$.

* Rather than varying the radiofrequency, we usually obtain an NMR spectrum by holding the radio frequency constant and varying the magnetic field. It is easier to operate an NMR spectrometer this way, and it yields equivalent results.

† In some aromatic compounds, the ortho-, meta-, and para-hydrogen atoms show up as slightly "different" in the NMR spectra. In toluene, the difference is so slight that we can ignore it here.

TABLE 23.8 Positions of Absorption of Hydrogen Atoms in NMR Spectroscopy[a]

Type of hydrogen	Chemical shift, δ (ppm)
$(CH_3)_4Si$ (TMS)	0
$R{-}\mathbf{CH_3}$	0.8–1.0
$R{-}\mathbf{CH_2}{-}R$	1.2–1.4
$R{-}\mathbf{CH}(R){-}R$	1.4–1.7
$R{-}C({=}O){-}\mathbf{CH_3}$	2.1–2.6
$R{-}C({=}O){-}\mathbf{H}$	9.5–9.6
$R_2C{=}\mathbf{CH_2}$	4.6–5.0
$R\mathbf{CH_2}OH$	3.3–4.0
Ar**H**	6.0–9.5

[a] R— stands for an aliphatic (nonaromatic) hydrocarbon group and Ar— represents an aromatic hydrocarbon group.

NMR spectra are displayed on charts with the applied field strength increasing from left to right. The hydrogen nuclei of the compound tetramethylsilane (TMS), $(CH_3)_4Si$, absorb at a higher field strength than those of nearly all other hydrogen-containing compounds. Chemists have adopted TMS as a standard for NMR studies, and its absorption peak appears at the extreme right of an NMR spectrum. The position of an absorption peak for a certain kind of hydrogen nucleus relative to that of the TMS peak is called the **chemical shift** for that kind of hydrogen nucleus. Chemical shifts are stated in units of frequency on an arbitrary *delta* scale, where one delta unit (δ) is equal to one part per million of the operating frequency of the NMR spectrometer. (The chemical shift of TMS, of course, is zero.) Table 23.8 lists typical positions of absorption peaks of several kinds of hydrogen atoms.

Other features of NMR spectra are helpful in determining the structures of molecules. One attribute is that the area under each peak is proportional to the number of hydrogen nuclei giving rise to that peak. In Figure 23.24, the ratio of the peak areas of aromatic H to methyl H is 5:3. This is just the ratio that we expect from the formula of toluene, $C_6H_5CH_3$.

Another vital feature of NMR spectra (but one that we will not explore in any detail) is the fact that often an absorption peak is split into a collection of several smaller peaks. This splitting occurs whenever the hydrogen nucleus responsible for a peak is adjacent to a carbon atom having other hydrogen atoms bonded to it. For example, in the ethyl group

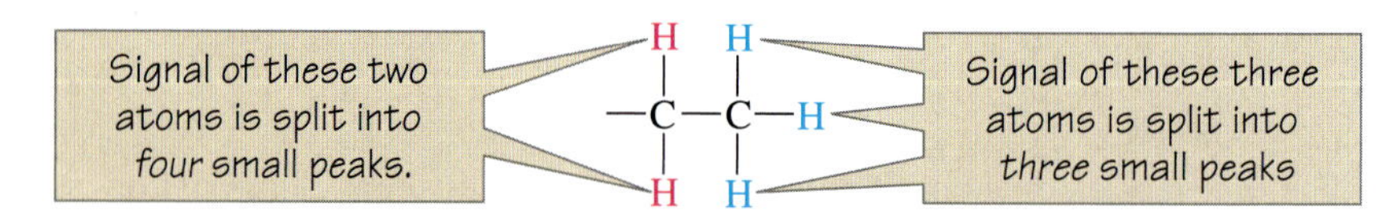

The splitting of signals into smaller peaks is illustrated in the NMR spectrum of ethylbenzene in Figure 23.25. It arises from a phenomenon called *spin-spin splitting*, caused by the nuclear spin of one H atom interacting with those of nearby H atoms. Spin-spin splitting does not occur when chemically equivalent (the same kind) H atoms are found on the same or adjacent C atoms. Neither does it occur very often when nonequivalent (different kinds) H atoms are two or more C atoms apart, as in toluene, $C_6H_5CH_3$.

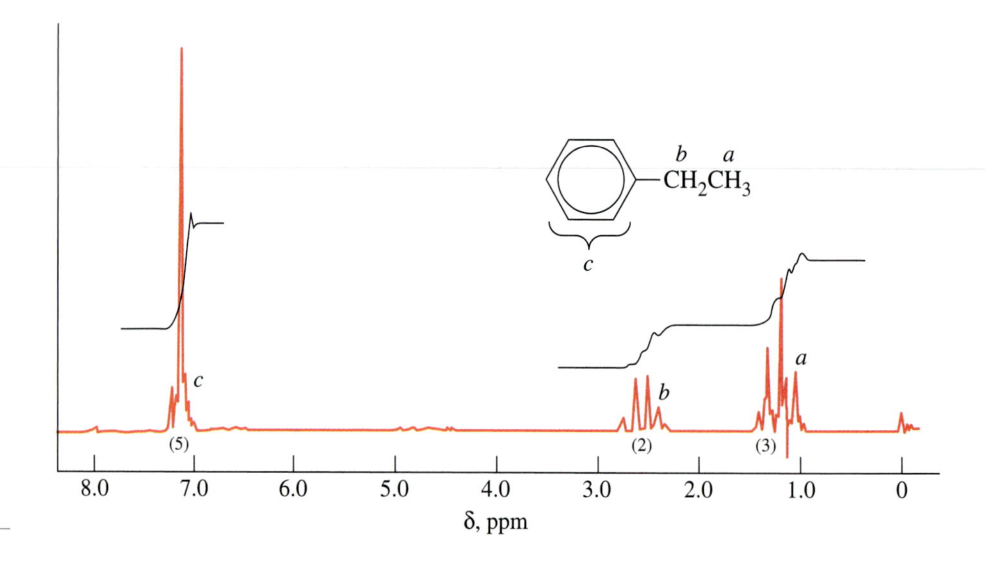

▶ **Figure 23.25 The NMR spectrum of ethylbenzene**

Some Organic and Biochemistry of Drugs

People have used medicines in attempts to relieve pain and cure illnesses since prehistoric times. However, for most of human history the variety of drugs was limited, and they were discovered mainly through trial and error. Most societies used ethyl alcohol. Some used marijuana, some knew the narcotic effect of the opium poppy, and the native tribes of the Andes Mountains chewed the leaves of the coca plant for the stimulating effect of the cocaine in the leaves.

Only at the beginning of the twentieth century was a scientific rationale for the use of chemical substances to treat diseases developed. In 1904, Paul Ehrlich (1854–1915), a German chemist, realized that certain chemicals were more toxic to disease organisms than to human cells. These chemicals could be used to control or cure infectious diseases. Ehrlich coined the term *chemotherapy*, a shorter version of the term "chemical therapy." Ehrlich found that certain dyes used to stain bacteria to make them more visible under a microscope also could be used to kill the bacteria. He used dyes against the organism that causes African sleeping sickness. He also synthesized an arsenic compound effective against the organism that causes syphilis. Ehrlich was awarded the Nobel Prize in 1908.

23.15 Molecular Shapes and Drug Action

For most of the twentieth century, scientists found useful drugs mainly by chance. After identifying an effective drug and determining its structure, chemists would synthesize dozens of related compounds. Medical research workers would then use laboratory animals to test each of these compounds for toxicity and effectiveness. Those that passed these tests were then tested on humans. This process takes years and is very expensive.

In Chapter 10, we noted that many drugs act by fitting specific receptor sites on cell membranes. In Chapter 13, we described how enzymes act by fitting a substrate molecule into an active site. Drugs and poisons act by blocking or otherwise changing the shapes of certain active sites. It is quite difficult to determine the precise shape of a receptor site and to determine what natural molecule the body produces that activates the receptor. Scientists have had to learn the normal biochemistry of cells before they could develop drugs that treat abnormal conditions. Gertrude Elion and George Hitchings of Burroughs Wellcome Research Laboratories in North Carolina and James Black of Kings College in London made notable contributions to the basic biochemistry, leading to the development of antiviral drugs. Such drugs are used to treat herpes, AIDS, and other conditions caused by viruses. Their work also led to the development of many anticancer drugs. They and other scientists determined the shapes of cell membrane receptors, and they learned much about how normal cells work. With that knowledge, scientists were able to design drugs to block receptors. Today scientists use powerful computers to design molecules to fit receptors. Although drug design often was hit or miss in its early decades, it is now becoming a more precise science. We will consider only a few examples here.

Gertrude Elion shared the 1988 Nobel prize for physiology and medicine with Hitchings and Black (not shown).

Penicillins: The Enzyme-Inhibition Wars

Penicillin, the first antibiotic, was discovered in 1928 by Alexander Fleming (1881–1955), a Scottish microbiologist then working at the University of London. Antibiotics are water–soluble substances derived from molds or bacteria that inhibit the growth of other microorganisms. Fleming was studying an infectious bacterium, *Staphylococcus aureus*, when one of his cultures became contaminated with

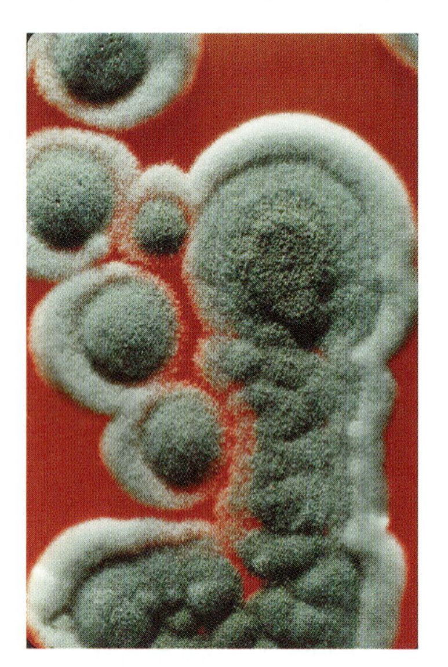

These symmetrical colonies of mold are *Penicillium chrysogenum*, a mutant form that now produces nearly all the commercial penicillin.

blue mold. Contaminated cultures generally are useless, but Fleming noted that the bacterial colonies had been destroyed in the vicinity of the mold. Fleming was able to make crude extracts of the active substance, later called penicillin. It was further purified and improved by Howard Florey (1898–1968), an Australian, and Ernst Boris Chain (1906–1979), a refugee from Nazi Germany. Fleming, Florey, and Chain shared the 1945 Nobel prize in physiology and medicine for their work on penicillin.

Penicillin is not a single compound but a group of compounds with related structures. All have in common a four-membered ring, called a β lactam; they differ in the nature of the R groups. By varying the R groups, chemists could design penicillin molecules with different properties. The penicillins vary in effectiveness. Bacteria resistant to one penicillin may be killed by another. The structures of several common penicillins are shown in Figure 23.26.

Penicillin (General formula): R—C(=O)NH—HC—CH—S—C(CH_3)(CH_3)—CH(COO$^-$)—N—C(=O)

R = C_6H_5—CH_2— Penicillin G

= HO—C_6H_4—CH(NH_2)— Amoxicillin

= C_6H_5—O—CH_2— Penicillin V

▲ Figure 23.26 Some common penicillins
The β-lactam ring is shown in red. Amoxicillin was the most widely prescribed drug in the United States in 1988, but the development of resistant bacterial strains has caused a dramatic decline in its use.

Penicillins act by inhibiting the enzyme that catalyzes the synthesis of bacterial cell walls. In a reaction analogous to the hydrolysis of an amide, a nucleophilic group on the enzyme attacks the carbonyl carbon atom of the penicillin molecule. The net effect is that the penicillin molecule attaches to the enzyme, changing the active site and inactivating the enzyme.

Active enzyme —ÖH + R—C(=O)NH—CH—CH—S—C(CH_3)(CH_3)—CH(COO$^-$)—N—C(=O) ⟶ Inactive enzyme —O—C(=O)—CH(NH—C(=O)R)—CH—S—C(CH_3)(CH_3)—CH(COO$^-$)—NH

Human cells do not have cell *walls*. Their external *membranes* differ from the cell walls of bacteria and are not affected by penicillin. Thus, penicillin can destroy bacteria without harming human cells.

Bacteria become resistant to penicillin through mutations. The resistant bacteria produce an enzyme, β-lactamase, that catalyzes the hydrolysis of the amide linkage of the β-lactam.

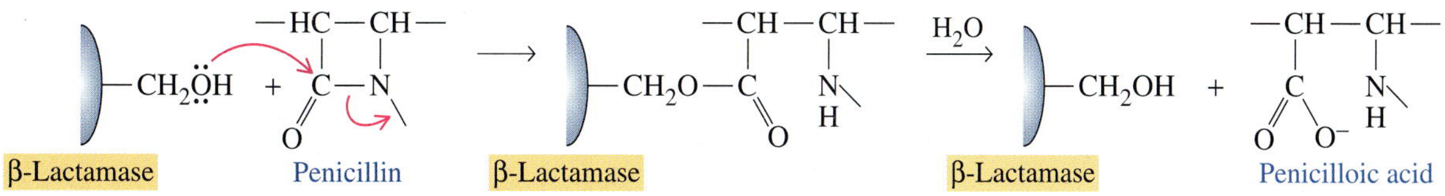

β-Lactamase—$CH_2\ddot{O}H$ + Penicillin → β-Lactamase—CH_2O—C(=O)… $\xrightarrow{H_2O}$ β-Lactamase—CH_2OH + Penicilloic acid

β-Lactamase Penicillin β-Lactamase β-Lactamase Penicilloic acid

The hydrolyzed penicillin, called penicilloic acid, is unable to interfere with bacterial cell wall synthesis.

In the continuing fight against infectious bacteria, chemists have developed drugs that inhibit β-lactamase. These drugs can be given along with penicillin, and with β-lactamase inhibited, the penicillin can inhibit the enzyme that catalyzes bacterial cell wall synthesis. Thus we see how simple organic reactions, such as nucleophilic addition to a carbonyl group and hydrolysis of an amide, become important in the fight against infectious diseases.

Brain Amines: Depression and Mania

Drugs are used to treat diseases of the mind as well as those of the body. The nervous system is made up of about 1.2×10^{10} neurons (nerve cells) with 10^{13} connections between them. Each neuron is connected to other cells though a long fiber called an *axon*. An axon on a given nerve cell may be up to 60 cm long, but there is no continuous pathway to the next cell. Rather, messages must be transmitted across tiny, fluid-filled gaps, called *synapses*. When an electrical signal from the brain reaches the end of an axon, specific chemicals called *neurotransmitters* are liberated and carry the impulse across the synapse to the next cell. There are perhaps a hundred or so neurotransmitters. Each has a specific function. Messages are carried to other nerve cells, to muscles, and to the endocrine glands (such as the adrenal glands). Each neurotransmitter fits one or more receptor sites on the receptor cell. Many drugs (and some poisons) act by mimicking the action of the neurotransmitter. Others act by blocking the receptor and preventing the neurotransmitter from acting on it. Several of the neurotransmitters are amines, as are some of the drugs that affect the chemistry of our brains.

We all have mood swings. These ups and downs probably result from multiple causes, but it is likely that a variety of chemical compounds formed in the brain are involved. Before we consider the brain chemicals, however, let's take a look at epinephrine, an amine formed in the adrenal glands and sometimes called adrenaline.

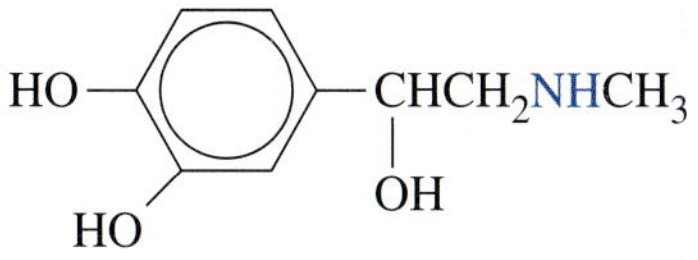

$(HO)_2C_6H_3$—$CH(OH)CH_2NHCH_3$

Epinephrine

A tiny amount of epinephrine causes a great increase in blood pressure. When a person is under stress or is frightened, the flow of adrenaline prepares the body for fight or flight. Because culturally imposed inhibitions prevent fighting or fleeing in many modern situations, the adrenaline-induced supercharge is not used. In some cases, repeated episodes of this sort of frustration has been implicated in some forms of mental illness.

Biochemical theories of mental illness involve brain amines. One is norepinephrine (NE), a relative of epinephrine.

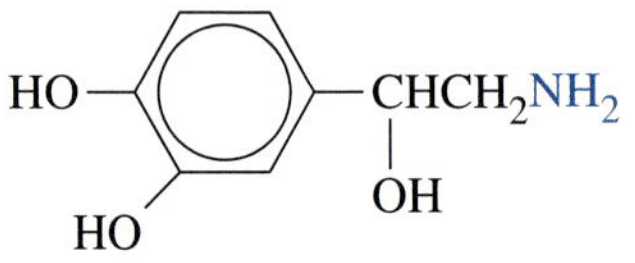

HO—C6H3(HO)—CH(OH)CH$_2$NH$_2$

Norepinephrine (NE)

NE is a neurotransmitter formed in the brain. When formed in excess, NE causes the person to be elated—perhaps even hyperactive. In large excess, NE induces a manic state. A deficiency of NE, in contrast, could cause depression.

Another brain amine is serotonin, also a neurotransmitter.

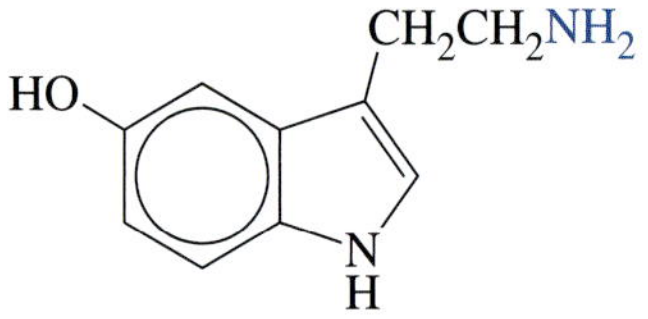

HO—(indole, N—H)—CH$_2$CH$_2$NH$_2$

Serotonin

Serotonin is involved in sleep, sensory perception, and the regulation of body temperature. In the body, serotonin is oxidized to 5-hydroxyindoleacetic acid (5-HIAA).

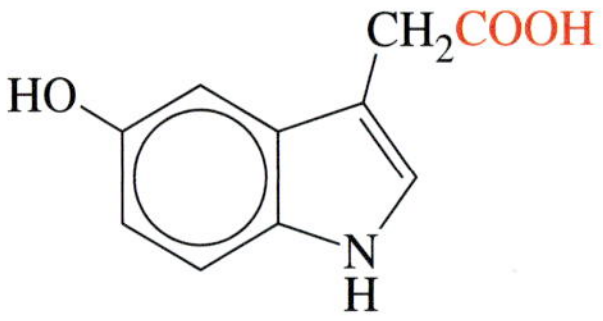

HO—(indole, N—H)—CH$_2$COOH

5-HIAA

Using partial structures, we can represent the oxidation as follows.

$$-CH_2CH_2NH_2 \longrightarrow -CH_2COOH$$

5-HIAA is found in unusually low levels in the spinal fluid of violent suicide victims. This indicates that abnormal serotonin metabolism plays a role in depression. Recent research indicates that a low flow of serotonin through the synapses in the frontal lobe of the brain causes depression.

Our human cells have at least six different receptors that are activated by NE and related compounds. NE agonists (drugs that enhance or mimic its action) are stimulants. NE antagonists (drugs that block the action of NE) slow down various processes. Drugs called beta blockers reduce the stimulant action of epinephrine and NE on various kinds of cells. Propranolol (Inderal), a well-known beta blocker, is used to treat cardiac arrhythmias, angina, and hypertension by lessening slightly the force of the heartbeat. Unfortunately, it also causes lethargy and depression. Another beta blocker, metoprolol (Lopressor) acts selectively on a different receptor found on the cells of the heart. It can be used by hypertensive patients who have asthma because it does not act on receptors in the bronchi.

Serotonin agonists are used to treat depression and anxiety. Agonist drugs are used to treat obsessive–compulsive disorder. Serotonin antagonists are used to treat migraine headaches and to relieve the nausea caused by cancer chemotherapy.

Richard Wurtman of the Massachusetts Institute of Technology has found a relationship between diet and serotonin levels in the brain. Serotonin is produced in the body from the amino acid tryptophan. The synthesis involves several steps, omitted here for simplicity.

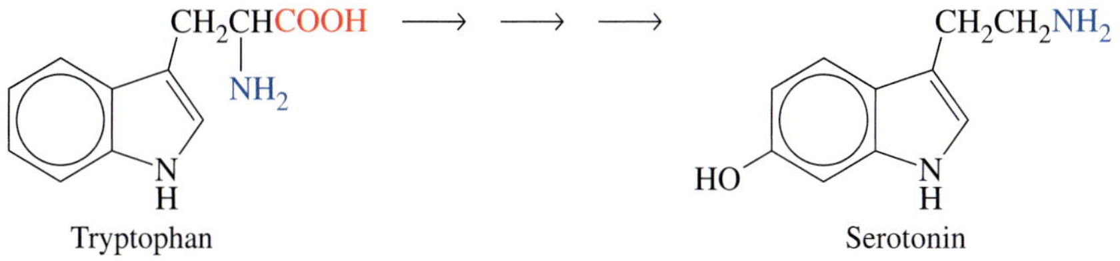

$CH_2CHCOOH$ (with NH_2) → → → $CH_2CH_2NH_2$ (HO on ring)

Tryptophan Serotonin

Wurtman found that diets high in carbohydrates lead to high levels of serotonin. High protein diets lower the serotonin concentration.

Norepinephrine also is synthesized in the body from an amino acid, tyrosine. Because tyrosine is also a component of our diets, it may well be that our mental state depends somewhat on our diet. The synthesis of tyrosine is complex and proceeds through several intermediates, only a few of which we show in the following scheme.

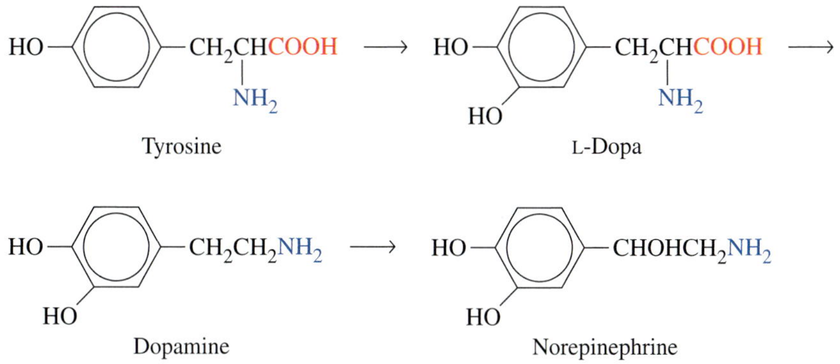

HO–C_6H_4–$CH_2CHCOOH$ (NH_2) → HO(HO)–C_6H_3–$CH_2CHCOOH$ (NH_2) →

Tyrosine L-Dopa

HO(HO)–C_6H_3–$CH_2CH_2NH_2$ → HO(HO)–C_6H_3–$CHOHCH_2NH_2$

Dopamine Norepinephrine

L-Dopa, the left-handed form of the compound, is used to treat Parkinson's disease, which is characterized by rigidity and stiffness of muscles. Parkinson's results from inadequate dopamine production, but dopamine cannot be used directly to treat the disease because it is not absorbed into the brain. Dopamine has been used to treat hypertension. It also is a neurotransmitter; schizophrenia has been attributed to an overabundance of dopamine in nerve cells.

Stimulant drugs, such as amphetamine and methamphetamine, are related to epinephrine and norepinephrine.

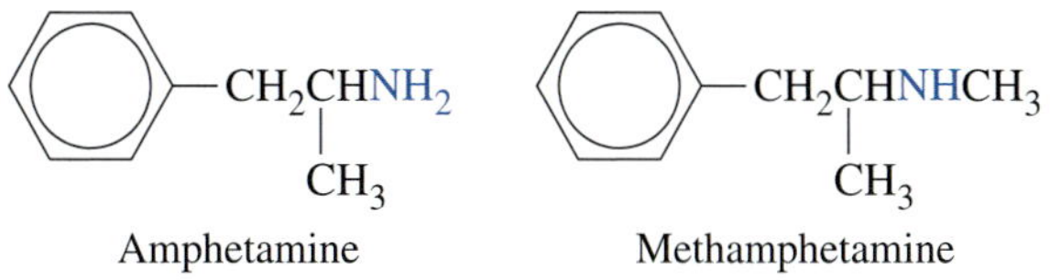

C_6H_5–CH_2CHNH_2 (CH_3) C_6H_5–$CH_2CHNHCH_3$ (CH_3)

Amphetamine Methamphetamine

These two synthetic amines are derivatives of β-phenylethylamine. They probably act as stimulants by mimicking the natural brain amines. Methamphetamine has a more pronounced psychological effect than amphetamine.

Some people seek stimulants. Others seek relief for anxiety. Benzodiazepines, such as diazepam (Valium), are used to treat anxiety. Certain benzodiazepines are also used to treat insomnia. Triazolam (Halcion) is a familiar example.

Love: A Chemical Connection

Can love be chemical in origin? An unsettling idea, perhaps. However, it is possible that the emotions that trigger romantic relationships are governed in part by a chemical called β-phenylethylamine (PEA).

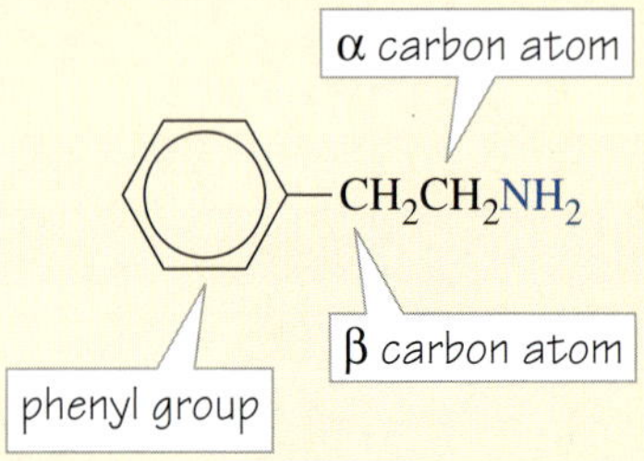

α carbon atom

$C_6H_5-CH_2CH_2NH_2$

β carbon atom

phenyl group

PEA functions as a neurotransmitter in the human brain. It appears to create excited, alert feelings and moods. Increased levels of PEA produce a "high" feeling identical to the feeling that people describe as "being in love." Not surprisingly, the chemical structure of PEA resembles that of norepinephrine.

How much PEA does it take to get back that old feeling? Levels of PEA in the brain can be estimated by measuring levels of its metabolite, phenylacetic acid ($C_6H_5CH_2COOH$), in the urine. Low levels of urinary phenylacetic acid correlate with depression. This has prompted researchers to investigate factors that increase PEA levels in the brain. There are no food sources of PEA, but protein-rich foods contain phenylalanine (Section 23.9), an amino acid precursor of PEA. Perhaps a steak dinner is a way to your true love's heart after all.

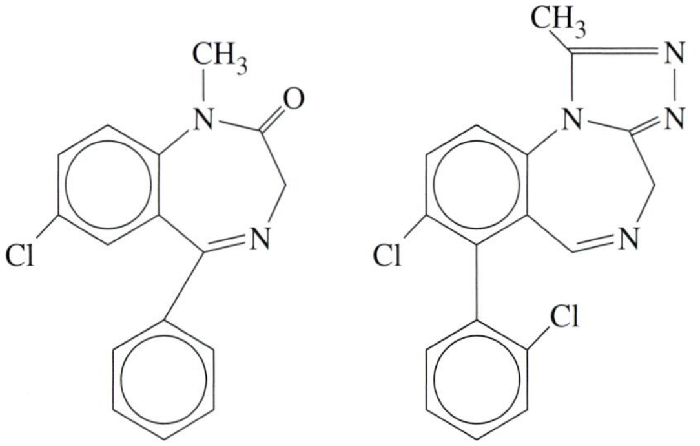

Diazepam (Valium) Triazolam (Halcion)

Like most other mind-altering drugs, the benzodiazepines act by fitting specific receptors. Presumably our bodies produce compounds that fit these receptors. To date, no such compound has been found. Rather, scientists have found compounds, called β-carbolines, that act on the brain's anxiety receptors to produce terror. There is still much to learn about the chemistry of the brain.

A class of drugs called phenothiazines, such as chlorpromazine (Thorazine), is used to treat psychotic patients.

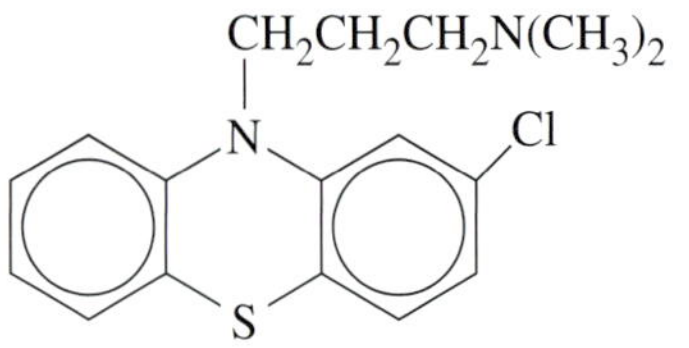

$CH_2CH_2CH_2N(CH_3)_2$

Chlorpromazine (Thorazine)

These drugs calm mentally ill people. Chlorpromazine is quite effective in controlling the symptoms of schizophrenia. These drugs and related ones have truly revolutionized mental illness therapy, serving to reduce greatly the number of patients confined to mental hospitals. Today, 95% of all people with schizophrenia no longer need hospitalization.

The phenothiazines are dopamine antagonists. Dopamine is important in the control of detailed motion (such as grasping small objects), in memory and emotions, and in exciting the cells of the brain. Some researchers think schizophrenic patients produce too much dopamine, others that they have too many dopamine receptors. In either case, blocking the action of dopamine relieves some of the symptoms of schizophrenia.

Some drugs related to promazine have a very different effect, indicating that slight changes in structure can result in profound changes in properties. Replacing the sulfur atom of promazine with a CH_2CH_2 group produces imipramine (Tofranil), an *antidepressant*.

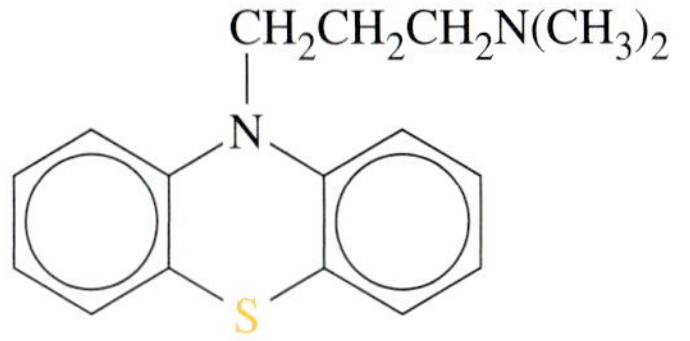

$CH_2CH_2CH_2N(CH_3)_2$ N S

Promazine
(A tranquilizer)

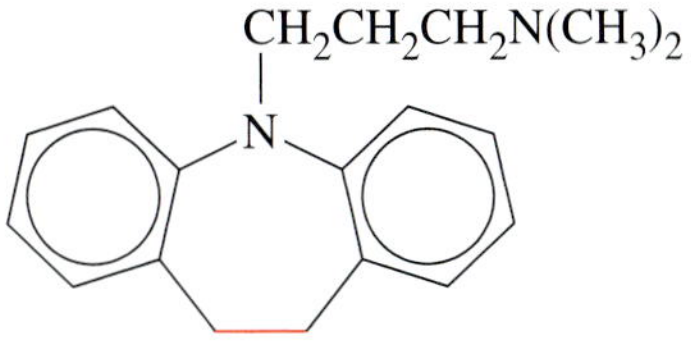

$CH_2CH_2CH_2N(CH_3)_2$ N

Imipramine
(An antidepressant)

These tricyclic antidepressants are only mildly successful. There is a narrow range in which the dose is both safe and effective.

Newer antidepressants are now available, including fluoxetine (Prozac). Doctors prescribe Prozac to help people cope with gambling problems, obesity, fear of public speaking, or premenstrual syndrome (PMS). The drug works by enhancing the effect of serotonin, blocking its reabsorption by the cells. It seems to be safer than the older antidepressants and more easily tolerated.

The most widely used antianxiety drug, usually self-prescribed, is most likely ethyl alcohol, found in beer, wine, and other alcoholic beverages. Excessive ethanol consumption causes intoxication, and long term consumption can lead to addiction. Just how ethanol intoxicates is still somewhat of a mystery. Researchers have found that ethanol disrupts a receptor for γ-aminobutyric acid (GABA), a substance that the body uses to shut off nerve cell activity. Ethanol thus disrupts the brain's control over muscle activity, causing the drunk to stagger and fall.

Nearly one out of every ten people in the United States suffers from some degree of mental illness. Over half the patients in hospitals are there because of mental problems. However, many mental illnesses can be alleviated by drugs. When we more fully understand the biochemistry of the brain, such illnesses may be cured by administration of drugs.

23.16 Acid-Base Chemistry and Drug Action

Acid-base chemistry is widely encountered in organic and biochemistry. As we have seen, acid-base steps are a part of many organic reaction mechanisms. Control of pH is vital in many biochemical reactions because most enzymes work most efficiently at an optimal pH (Chapter 13). The solubility and volatility of chemical compounds, including drugs, are also affected by pH. We will look at a few examples here.

Coca leaves.

Cocaine

Cocaine was first used in medicine as a local anesthetic, but it also is a powerful stimulant.

Cocaine

The drug, an alkaloid, is obtained from the leaves of a shrub that grows on the eastern slopes of the Andes Mountains. Many of the natives living in that area chew coca leaves—mixed with lime and ashes—for their stimulant effect. Cocaine used to arrive in the United States as the salt, cocaine hydrochloride. Now much of it comes in the form of broken lumps of the free base, a form called crack cocaine. We can use the molecular formula to write equations for acid-base reactions.

$$\underset{\text{Cocaine (free base)}}{C_{17}H_{21}O_4N} + HCl \longrightarrow \underset{\text{Cocaine hydrochloride}}{C_{17}H_{21}O_4NH^+Cl^-}$$

Cocaine hydrochloride can be converted back to the free base by reaction with a strong base.

$$\underset{\text{Cocaine hydrochloride}}{C_{17}H_{21}O_4NH^+Cl^-} + OH^- \longrightarrow \underset{\text{Cocaine (free base)}}{C_{17}H_{21}O_4N} + H_2O + Cl^-$$

Cocaine hydrochloride, a salt, is quite soluble in water (100 g in 40 mL H_2O) and fairly high melting (~195 °C). It is therefore readily absorbed through the watery mucous membrane of the nose. It is the form used by those who snort cocaine.

Pure cocaine is only somewhat polar and therefore only slightly soluble in water (0.17 g in 100 mL H_2O) but fairly soluble in nonpolar solvents (8.3 g in 100 mL olive oil.). Cocaine melts at 98 °C, but is quite volatile above 90 °C. Those who smoke cocaine use the free base (crack), which readily vaporizes at the temperature of a burning cigarette. When smoked, cocaine is readily absorbed through the fatty cells that line the lungs, reaching the brain in 15 seconds. Cocaine acts by preventing the neurotransmitter dopamine from being reabsorbed after it is released by nerve cells. High levels of dopamine are therefore available to stimulate the pleasure centers of the brain. After the binge, dopamine is depleted in less than an hour. This leaves the user in a pleasureless state, often craving more cocaine.

Nicotine

Another common stimulant is nicotine.

Nicotine

Like cocaine, nicotine is an alkaloid. A high-boiling (247 °C), oily liquid, nicotine is quite soluble in both water and nonpolar solvents. Therefore, nicotine is readily absorbed both through the watery mucous membrane of the mouth from chewing tobacco and through the fatty cells that line the lungs when taken by smoking. Nicotine, which is highly toxic to animals, has been used in agriculture as a contact insecticide. It is especially deadly when injected; the lethal dose for a human is estimated to be about 50 mg.

Because nicotine is toxic, the body must get rid of some of it lest concentrations build up to dangerous levels. The principal route of excretion is in the urine. Excretion is more efficient when the urine is acidic because the nicotine at acidic pH is mainly in the form of its conjugate acid.

$$\underset{\text{Nicotine}}{C_{10}H_{14}N_2} + H^+ \longrightarrow \underset{\text{Conjugate acid of nicotine}}{C_{10}H_{14}N_2H^+}$$

The conjugate acid, with its positive charge, is more soluble in water and much *less* soluble in fatty body tissues than nicotine is. Therefore the lower the pH of the urine, the more rapidly the nicotine is excreted, and the sooner the smoker craves another cigarette. When does the pH of urine drop? After a meal, acids enter the urine from the metabolism of foods. After drinking ethanol, some of the ethanol is oxidized to acetic acid which enters the urine, making it more acidic. When under stress, the body breaks down tissues, producing acids. When does a smoker just have to have another cigarette? After a meal. While drinking alcoholic beverages. When under stress.

Nicotine seems to have a rather transient effect as a stimulant. This initial response is followed by depression, but smokers generally keep a near-constant level of nicotine in their bloodstreams by indulging frequently.

Is nicotine addictive? Consider the 1972 memorandum by a tobacco company scientist who noted that "no one has ever become a cigarette smoker by smoking cigarettes without nicotine." He suggested that the company "think of the cigarette as a dispenser for a dose unit of nicotine."

Summary

Mostly, alkanes are burned as fuels. Alkanes also undergo substitution reactions with halogens. Methane and chlorine undergo a chain reaction involving free-radical intermediates. Alkenes typically undergo addition reactions: they add halogens, water, and other substances across the carbon-to-carbon double bond.

Most of the reactions of organic compounds with functional groups take place at or near the functional group. The hydrocarbon portion is generally less reactive. Alcohols often react by another group substituting for the OH group, for example, substitution of a halogen atom for the OH group. Alcohols can be dehydrated to either alkenes or ethers. Aldehydes and ketones undergo many reactions that start with an addition to the carbonyl group. Carboxylic acids react as acids, but they also react with alcohols to form esters and with ammonia or amines to form amides. Esters and amides can be hydrolyze in either aqueous acids or aqueous bases.

Benzene and related aromatic compounds react mainly by electrophilic aromatic substitution. Lewis acids such as aluminum chloride catalyze several of these reactions.

The principal substances in living organisms are lipids, carbohydrates, proteins, and nucleic acids. The most familiar lipids are triglycerides, esters of fatty acids and glycerol. The esters of saturated fatty acids predominate in fats and unsaturated fatty acids in oils. The hydrolysis of fats and oils in basic solutions yields soaps—metal salts of fatty acids.

Carbohydrates include simple sugars, starches, and cellulose. Two monosaccharide units combine to form a disaccharide; for example, glucose and fructose join to form sucrose. Polysaccharides are polymers having monosaccharides as monomers, for example, glucose is the monomer in starch and cellulose. Optical activity is an important property of carbohydrates.

Key Terms

absolute configuration (23.8)
addition reaction (23.2)
amide (23.6)
carbocation (23.2)
carbohydrate (23.8)
chemical shift (23.14)
chiral carbon (23.8)
conjugated bonding system (23.3)
deoxyribonucleic acid (DNA) (23.10)
dextrorotatory (+) (23.8)
disaccharide (23.8)
electrophilic reagent (23.2)
electrophilic aromatic substitution (23.3)
essential amino acids (23.9)
gene (23.10)
hydrocarbon (page 958)
isoelectric point (p*I*) (23.9)
levorotatory (−) (23.8)
lipid (23.7)
monosaccharide (23.8)
nucleophile (page 966)
nucleotide (23.10)
peptide (23.9)
polypeptide (23.9)
polysaccharide (23.8)
primary structure (23.9)
protein (23.9)
quaternary structure (23.9)
ribonucleic acid (RNA) (23.10)
saponification (23.7)
secondary structure (23.9)
soap (23.7)
spectroscopy (page 988)
stereoisomers (23.8)
substitution reaction (23.1)
tertiary structure (23.9)
transcription (23.10)
translation (23.10)
wavenumber ($\tilde{\nu}$) (23.12)

Proteins are polymers of about 20 different α-amino acids. Individual amino acids join through peptide bonds. Protein molecules have 50 or more amino acid units. The primary structure of a protein refers to the amino acid sequence in a protein molecule.

Nucleic acids have nucleotides as their monomers. Nucleotides consist of a sugar, a phosphate group, and a cyclic amine base. The structure of DNA is that of two long chains of nucleotides joined into a double helix. In protein synthesis, the amino acid sequence in the protein is dictated by DNA through a code carried by an mRNA molecule and interpreted by tRNA molecules.

Three types of spectroscopy are described in this chapter. In infrared spectroscopy (IR), the interaction of radiation and matter causes changes in the vibrational modes of molecules. In ultraviolet-visible (UV-VIS) spectroscopy, electrons in bonds are promoted to higher energy states. In organic compounds, these transitions occur most readily among π electrons in conjugated systems. In nuclear magnetic resonance (NMR) spectroscopy, nuclei with a spin absorb electromagnetic radiation and change their orientation in a magnetic field. When ^1H is the target nucleus, a NMR spectrum pinpoints different "kinds" of H atoms for different locations and electronic environments in a molecule.

Chemotherapy is the use of chemical substances to control or cure diseases. Many drugs act by fitting specific sites on cell membranes or by inhibiting a specific enzyme. Penicillin, an antibiotic, acts by inhibiting the synthesis of bacterial cell walls.

The brain synthesizes amines that affect our mental state. Synthetic stimulant drugs, such as the amphetamines, mimic the action of norepinephrine. Other drugs act as either agonists or antagonists to various brain amines.

Acid-base chemistry is important in organic and biochemical reactions. Cocaine is a powerful stimulant. As the salt cocaine hydrochloride, it is readily absorbed through the watery mucous membranes of the nose. The free base is absorbed through the fatty cells that line the lungs. Nicotine has a transient stimulant effect. The rate of excretion of nicotine depends on the pH of the urine.

Review Questions

1. Briefly identify the important distinctions between:
   (a) alkane and alkyl
   (b) aliphatic and aromatic
   (c) hydrocarbon and carbohydrate
   (d) addition and substitution reaction
   (e) electrophile and nucleophile
2. Using appropriate examples, distinguish between:
   (a) an aldehyde and a ketone
   (b) an ester and an ether
   (c) an amide and an amine
   (d) straight-chain alkane and branched-chain alkane
3. Define and illustrate each of the following.
   (a) esterification
   (b) acid hydrolysis
   (c) saponification
4. Define, describe, illustrate, or explain each of the following terms.
   (a) triglyceride
   (b) aldose
   (c) hexose
   (d) disaccharide
   (e) polysaccharide
   (f) aldopentose
   (g) ketotetrose
   (h) tripeptide
   (i) zwitterion
5. Indicate whether each of the following compounds is aromatic, aliphatic, or both.

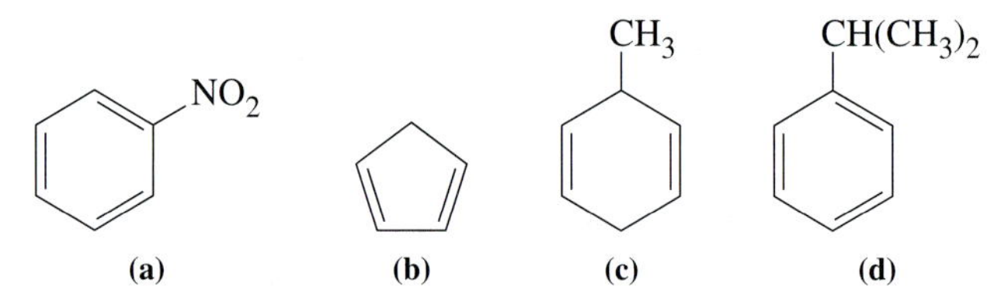

6. Distinguish between terms in the following pairs:
   (a) fat and oil
   (b) simple triglyceride and mixed triglyceride
   (c) polypeptide and protein
   (d) N-terminal and C-terminal amino acid
   (e) primary and secondary structure of a protein
7. What products are formed when a triglyceride is saponified?
8. Draw formulas for D-glyceraldehyde and L-glyceraldehyde. What do the prefixes mean?
9. Why are (+)-glucose and (−)-fructose both classified as D sugars?
10. How do amylose and cellulose differ from each other? How are they similar?

11. What is the general structure for an α-amino acid? Do most naturally occurring amino acids have a D or an L configuration? Explain.
12. Name the two kinds of nucleic acids. State two differences between them. Which one is found in the cell nucleus?
13. What is the difference between a pyrimidine base and a purine base? What is meant by base pairing and what kind of intermolecular force is involved?
14. Describe the process of DNA replication.
15. We say that DNA controls protein synthesis, yet most DNA resides within the cell nucleus while protein synthesis occurs outside the nucleus. How does DNA exercise its control?
16. Describe the process of protein synthesis.
17. What are the similarities and differences between emission and absorption spectroscopy?
18. Describe the meaning of a conjugated system of double bonds and illustrate with three specific examples.
19. What are the types of changes in molecules produced by the absorption of IR radiation? of UV-VIS radiation?
20. In which of these forms of spectroscopy—UV, VIS, IR, NMR—is the absorbed radiation of greatest energy content? longest wavelength?
21. Describe the relationship between the frequency of electromagnetic radiation, ν, and its wavenumber,
22. Describe the meaning of the terms HOMO and LUMO and their relationship to the absorption of electromagnetic radiation.
23. What is the type of change brought about by the absorption of electromagnetic radiation in NMR spectroscopy?
24. What is the meaning of each of the following terms, and in what spectroscopic method is it encountered?
    (a) chemical shift
    (b) λ_{max}
    (c) parallel and antiparallel
25. What is the essential structural feature of a penicillin molecule? How does penicillin kill bacteria without killing human cells?
26. How do bacteria become resistant to penicillin? How can resistant bacteria still be killed by penicillin?
27. How do each of the following affect our mental state?
    (a) norepinephrine
    (b) serotonin
    (c) dopamine
28. What are beta blockers? How do they work?
29. In what way might our mental state be related to our diet?
30. How do antianxiety drugs work? Are they a cure for schizophrenia? Explain.
31. How does cocaine act as a stimulant?
32. How does the pH of urine affect a smoker's desire for a cigarette?

Problems

Nomenclature

[You should be familiar with the material in Appendix D before attempting this section.].

33. Draw the structural formulas for each of the following.
    (a) 4-methyl-2-hexene
    (b) 3-methylpentane
    (c) 2,2,5-trimethylhexane
    (d) 4-ethyl-3-methyloctane
    (e) 3-isopropyl-1-hexyne
    (f) 2,3-dimethyl-2-butene
34. Draw the structural formulas for each of the following.
    (a) 2-ethyl-1-butene
    (b) 4-ethyl-2-methylhexane
    (c) 2,4,6,6-tetramethyl-2-heptene
    (d) 4-ethyl-3-isopropyloctane
    (e) ethylcyclobutane
    (f) 3-ethyl-2-pentene
35. Name the following compounds by the IUPAC system.
    (a) $CH_3CH_2CH(CH_3)CH_2CH_3$ (b) $(CH_3)_2CHC{\equiv}CCH_2CH(CH_3)_2$
    (c) $CH_3CH(CH_2CH_3)_2$ (d) $CH_3CH_2CH{=}C(CH_3)_2$
36. Name the following compounds by the IUPAC system.
    (a) $CH_3C(CH_3){=}CHCH_3$ (b) $CH_3CHClCH_2C{\equiv}CCH_2CHCl_2$
    (c) $(CH_3)_3CCH{=}C(CH_3)CH_2CH_3$ (d) $CH_2{=}C(CH_3)CH_2CH_2CH_3$
37. Name the following compounds by the IUPAC system.
    (a) $CH_3CH_2CH_2CH_2CH_2CH_2OH$
    (b) $CH_3CH_2CH_2CH_2CHOHCH_3$

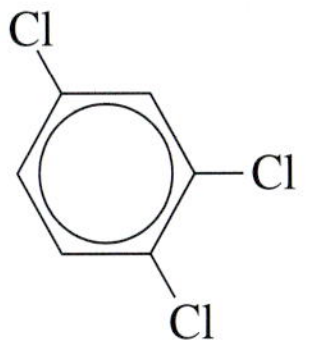

    (c) benzene ring with Cl, Cl, Cl substituents
    (d) benzene ring with Br, Br, CH_3 substituents
38. Name the following compounds by the IUPAC system.
    (a) $CH_3CH_2CHOHC(CH_3)_3$
    (b) $(CH_3)_2CHCH_2OH$

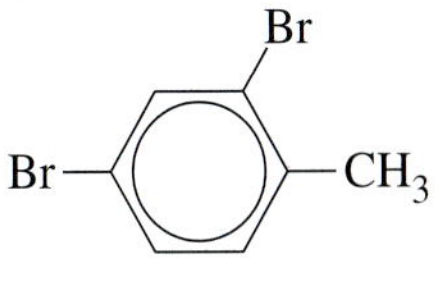

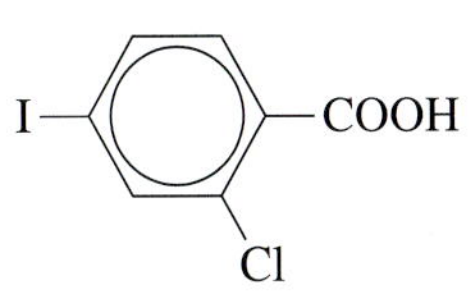

    (c) benzene ring with I, COOH, Cl substituents
    (d) benzene ring with NO_2, Cl, NH_2 substituents

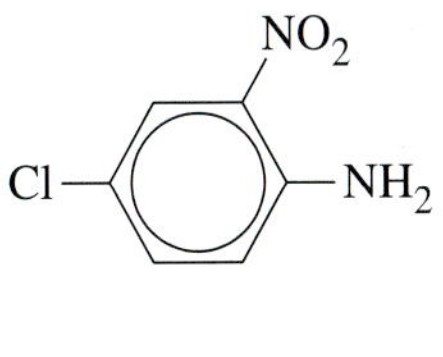

39. Give structural formulas for each of the following compounds.

(a) 3-hexanol
(b) 3,3-dimethyl-2-butanol
(c) 4,5-dimethyl-3-heptanol
(d) 2-ethyl-1-butanol
(e) 2,3-dibromobenzoic acid
(f) *m*-isopropylbenzoic acid

40. Give structural formulas for each of the following compounds.

(a) cyclopentanol
(b) 4-methyl-2-hexanol
(c) heptanoic acid
(d) 3-methylbutanoic acid
(e) *o*-nitrobenzoic acid
(f) *p*-chlorobenzoic acid

41. Write structural formulas for compounds **(a)**, **(b)**, and **(c)**, whose models are shown below. Give IUPAC names to each of the compounds.

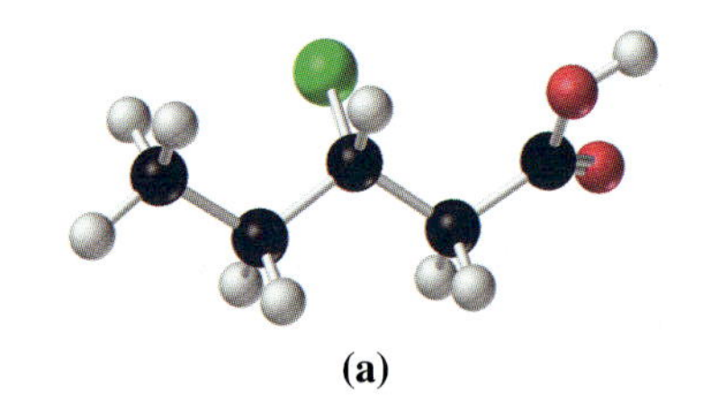

(a)

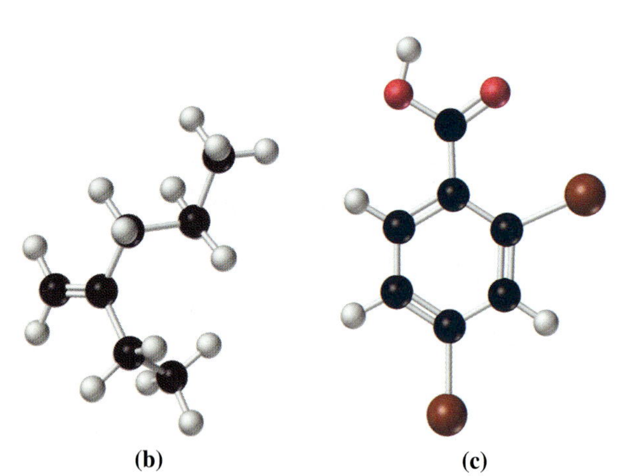

(b) **(c)**

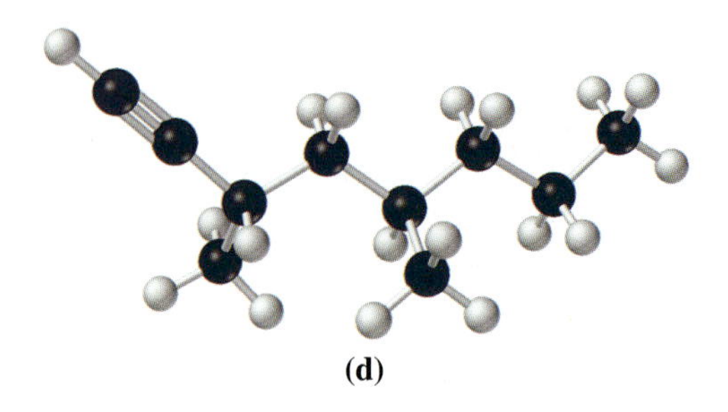

(d)

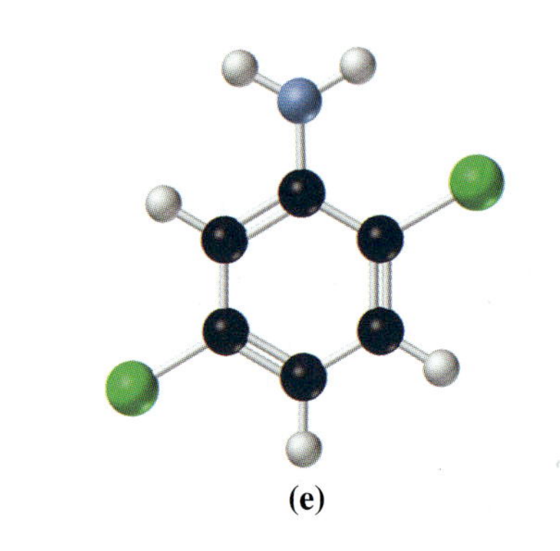

(e)

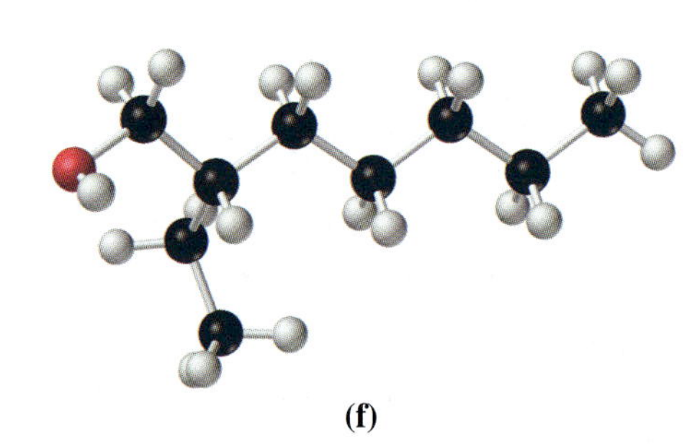

(f)

42. Write structural formulas for compounds **(d)**, **(e)**, and **(f)**, whose models are shown above. Give IUPAC names to each of the compounds.

Organic Reactions

43. Classify each of the following conversions as oxidation, dehydration, or hydration. (Only the organic starting material and the product are shown.)

$$CH_3CH_2\underset{\displaystyle OH}{\underset{|}{C}}HCH_3 \longrightarrow CH_3CH{=}CHCH_3$$

(a)

$$CH_3CH_2CH_2CH_2OH \longrightarrow CH_3CH_2CH_2\overset{\displaystyle O}{\overset{\|}{C}}OH$$

(b)

44. Classify each of the following conversions as oxidation, dehydration, or hydration. (Only the organic starting material and the product are shown.)

(a) $CH_3CH_2CH_2\underset{\displaystyle OH}{\underset{|}{C}}HCH_3 \longrightarrow CH_3CH_2CH_2\underset{\displaystyle O}{\underset{\|}{C}}CH_3$

(b) (cyclopentene) $\longrightarrow$ (cyclopentane)—OH

45. Complete the following equations by giving the structural formula for the main organic product.

$$(CH_3)_2C{=}CH_2 \ + \ Br_2 \longrightarrow$$
(a)

$$CH_2{=}C(CH_3)CH_2CH_3 \ + \ H_2 \xrightarrow{Ni}$$
(b)

46. Complete the following equations by giving the structural formula for the main organic product.

$$CH_2{=}CHCH{=}CH_2 \ + \ 2\,H_2 \xrightarrow{Ni}$$
(a)

$$(CH_3)_2C{=}C(CH_3)_2 \xrightarrow[H_2SO_4]{H_2O}$$
(b)

47. Write an equation for the dehydration of 1-propanol to yield **(a)** an alkene and **(b)** an ether.

48. Write an equation for the dehydration of cyclohexanol to an alkene.

49. Give the structure of the alkene from which each of the following alcohols is made by reaction with water in acidic solution.

(a) $CH_3CH(OH)CH_3$ (b) $CH_3C(CH_3)(OH)CH_3$

50. Give the structure of the alkene from which each of the following alcohols is made by reaction with water in acidic solution.

(a) C_6H_5–$CH(OH)CH_3$ (b) cyclohexane ring–OH

51. Write an equation for **(a)** the reaction of butyric acid with $NaHCO_3(aq)$ and **(b)** the acid-catalyzed hydrolysis of ethyl acetate.

52. Write equations for **(a)** the reaction of benzoic acid with NaOH(aq) and **(b)** the base-catalyzed hydrolysis of butyl acetate.

Lipids

53. Which of these fatty acids are saturated and which are unsaturated? How many carbon atoms are there in each?

(a) decanoic acid

(b) 9-tetradecenoic acid

(c) 9,12-hexadecadienoic acid

54. Which of these fatty acids are saturated, and which are unsaturated? How many carbon atoms are there in each?

(a) octanoic acid

(b) 6-decenoic acid

(c) 9,11,13-octadecatrienoic acid

55. Give structural formulas for the following.

(a) stearic acid **(c)** glyceryl trioleate

(b) potassium myristate **(d)** glyceryl tripalmitate

56. Give structural formulas for the following.

(a) sodium oleate

(b) glyceryl oleopalmitostearate

(c) linolenic acid

(d) a highly unsaturated oil

Carbohydrates

57. Identify each of the following as **(a)** an aldose or a ketose and **(b)** as a triose, tetrose, pentose, or hexose.

(a) D-glyceraldehyde **(d)** D-fructose

(b) D-ribose **(e)** L-glucose

(c) L-deoxyribose

58. Identify each of the following as **(a)** an aldose or a ketose and **(b)** as a triose, tetrose, pentose, or hexose.

(a) D-glucose **(d)** L-mannose

(b) L-fructose **(e)** D-deoxyribose

(c) D-galactose **(f)** L-glyceraldehyde

59. Specify whether each of the following is a D sugar or an L sugar.

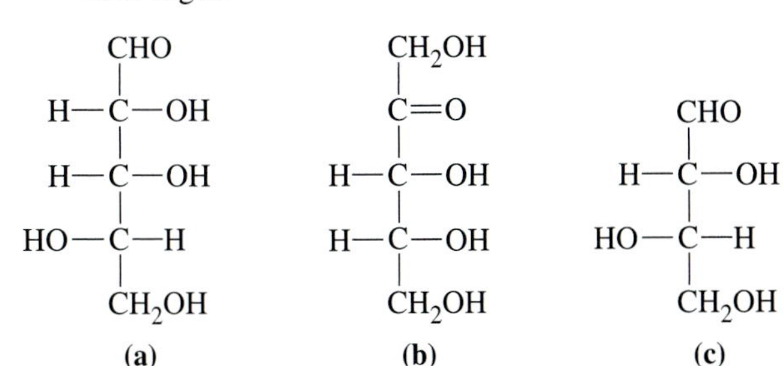

60. Specify whether each of the following is a D sugar or an L sugar:

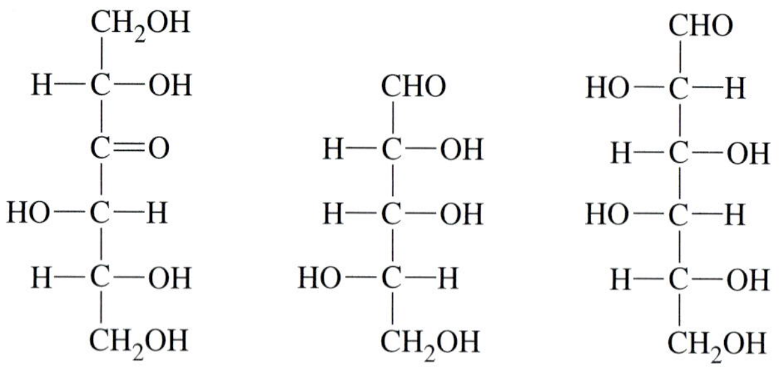

(a) CH_2OH; H—C—OH; C=O; HO—C—H; H—C—OH; CH_2OH

(b) CHO; H—C—OH; H—C—OH; HO—C—H; CH_2OH

(c) CHO; HO—C—H; H—C—OH; HO—C—H; H—C—OH; CH_2OH

61. Draw cyclic structures for **(a)** α-D-glucose and **(b)** β-D-fructose.

62. Draw a cyclic structure for **(a)** β-D-glucose. **(b)** Use that structure as a reference to draw the structure for β-D-galactose.

63. For each of these abbreviated sugar formulas, indicate whether the linkage between units is α or β.

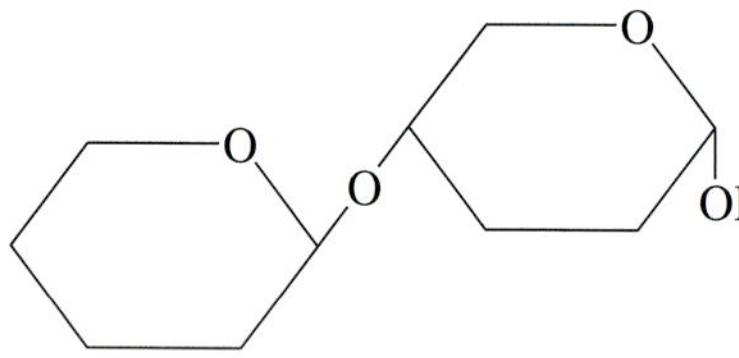

O O O OH

(a)

O O O OH

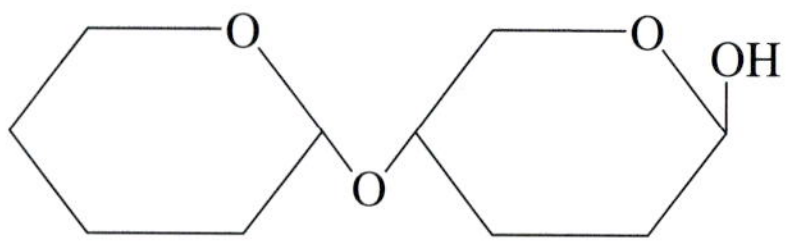

(b)

64. For each of these abbreviated sugar formulas, indicate whether the linkage between units is α or β.

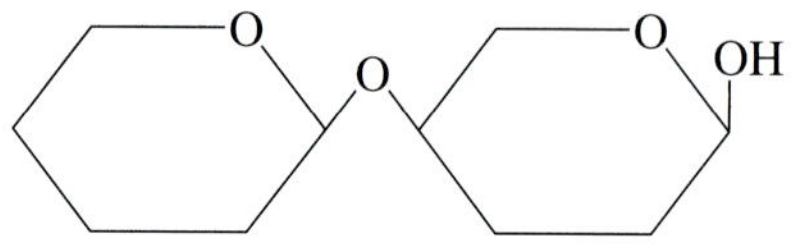

O O O OH

(a)

O O O CH_2OH

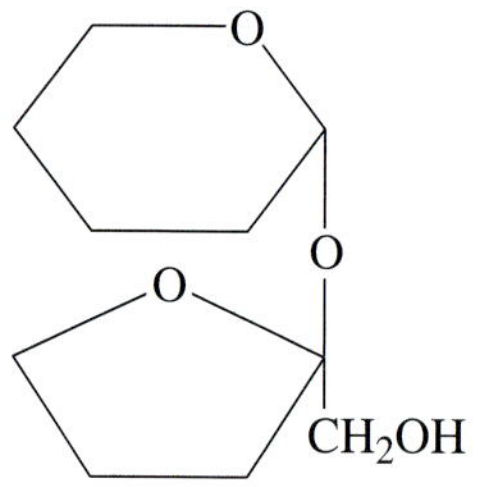

(b)

Amino Acids and Proteins

65. Draw the side chains of each of the following amino acids.

(a) alanine **(c)** tyrosine
(b) lysine **(d)** isoleucine

66. Draw the side chains of each of the following amino acids.

(a) aspartic acid **(d)** serine
(b) cysteine **(e)** phenylalanine
(c) valine

67. Write a structural formula for an amino acid that has an acidic side chain, and give its name.

68. Write a structural formula for an amino acid that has a basic side chain, and give its name.

69. Write a structural formula for the anion formed when glycine reacts with a base.

70. Write a structural formula for the cation formed when phenylalanine reacts with an acid.

71. Under what conditions does a protein have **(a)** a net positive charge, **(b)** a net negative charge, and **(c)** a net zero charge?

72. How do the properties of a protein at its isoelectric pH differ from its properties in solutions at other pH values?

Nucleic Acids

73. Identify the sugar and the base in the following nucleotide.

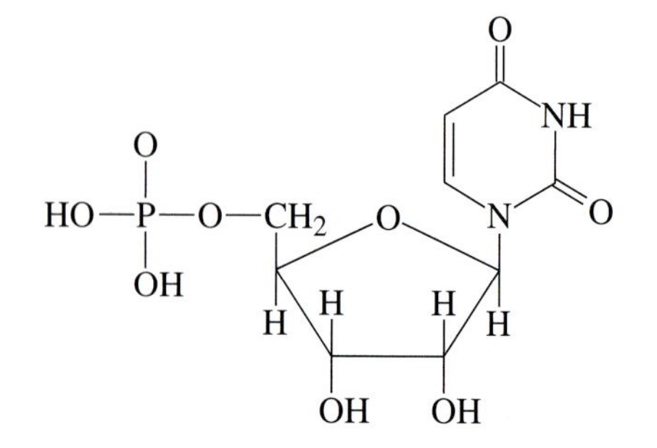

74. Identify the sugar and the base in the following nucleotide.

NH_2; N; OH; O—P—O—CH_2; O; N; O; OH; H H H H; OH OH

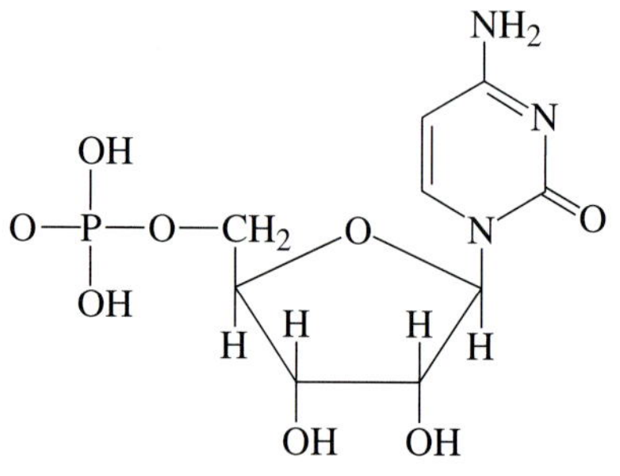

75. Is the molecule shown a nucleotide? Is the base a purine or a pyrimidine? Would the compound be incorporated in DNA or RNA?

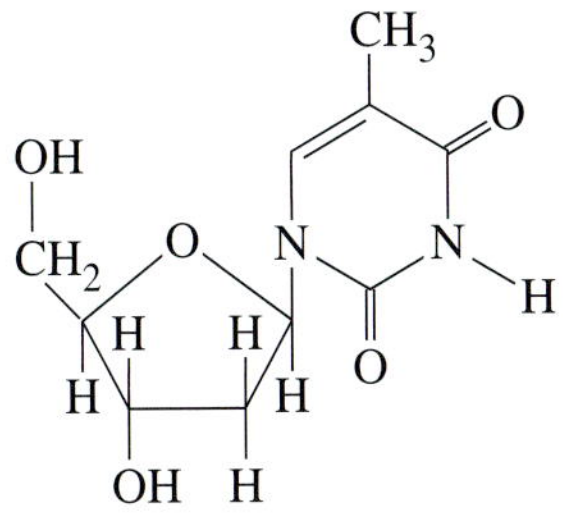

76. Is the molecule shown a nucleotide? Is the base a purine or a pyrimidine? Would the compound be incorporated in DNA or RNA?

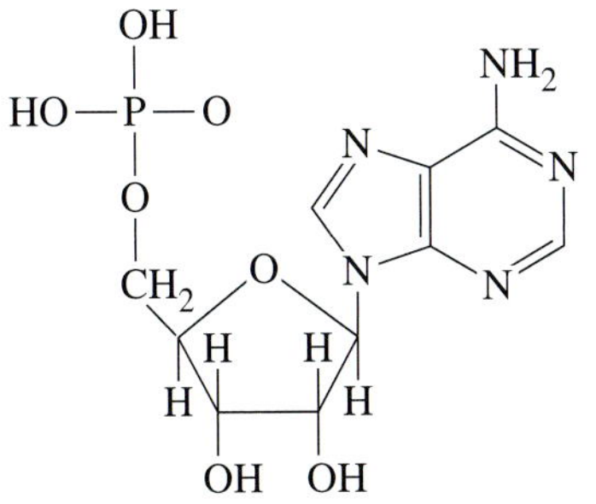

77. The base sequence along one strand of DNA is ATTCGC. What would be the sequence of the complementary strand of DNA? What sequence of bases would appear in the messenger RNA molecule copied from the original DNA strand?

78. If the sequence of bases along a messenger RNA strand is UCCGAU, what was the sequence along the DNA template?

79. What are the base triplets on tRNA that pair with the following triplets on mRNA?

(a) UUU **(c)** AGC

(b) CAU **(d)** CCG

80. What are the base triplets on mRNA for the following triplets on tRNA molecules?

(a) UUG **(b)** GAA

Spectroscopy of Organic Compounds

81. Could you distinguish between acetic acid, CH_3COOH, and methyl acetate, CH_3COOCH_3, through IR spectroscopy? Explain.

82. Could you distinguish between phenol, C_6H_5OH, and benzoic acid, C_6H_5COOH, through UV spectroscopy? through visible spectroscopy? Explain.

83. We have noted that if an organic molecule lacks a conjugated system of double bonds it will not have an absorption peak in the 200–400 nm range of the UV or in the visible region of the spectrum. Will it have an IR spectrum? Explain.

84. Which of the following compounds would you expect to absorb UV radiation between about 215 and 400 nm: **(a)** 2-pentene, **(b)** 1,3-pentadiene, **(c)** 2,4-dimethylpentane, **(d)** 1,5-hexadiene? Explain.

85. The substance *o*-nitrophenol is yellow in color. In what region of the visible spectrum does it absorb light?

86. The aldehyde $CH_3(CH{=}CH)_8CHO$ has $\lambda_{max} = 436$ nm. What color would you expect it to be?

Additional Problems

87. You find an unlabeled jar containing a solid that melts at 48 °C. It ignites readily and burns cleanly. The substance is insoluble in water and floats on the surface of the water. Is the substance likely to be organic or inorganic? Explain.

88. From what alcohol might each acid be prepared via oxidation with acidic dichromate?

(a) CH_3CH_2COOH **(c)** HCOOH

(b) HOOCCOOH **(d)** $(CH_3)_2CHCH_2COOH$

89. A lactone is a cyclic ester. What product is formed in each of the following reactions?

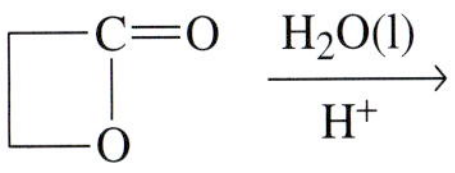

$H_2O(l)$, H^+

(a)

C=O + NaOH(aq) ⟶

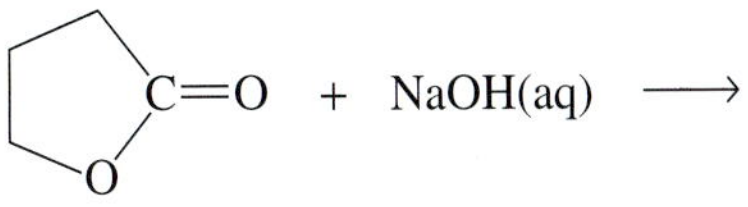

(b)

90. Pellagra is a vitamin-deficiency disease. Corn contains the antipellagra factor nicotinamide, which is not readily absorbed in the digestive tract. When corn is treated with slaked lime (calcium hydroxide) to make hominy, the vitamin is rendered more available. What reaction takes place? Write the equation.

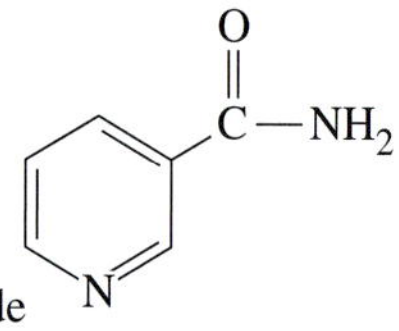

O
‖
C—NH_2
N

Nicotinamide

91. For neutralization, 5.10 g of a monocarboxylic acid requires 125 mL of a 0.400 M NaOH solution. Write all possible structural formulas for the acid.

92. Each of the four isomeric butyl alcohols is treated with potassium dichromate in acid. Draw the product (if any) expected from each reaction.

93. Draw the structural formula of the ether formed by the *intra*molecular dehydration of the molecule shown below.

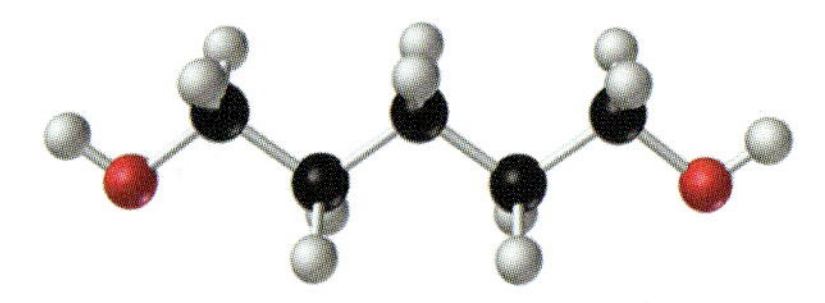

94. If one strand of a DNA molecule has the base sequence AGC, what must be the sequence on the opposite strand? Sketch the structure of this portion of the double helix, showing all hydrogen bonds.

95. What is the energy, in kJ/mol, associated with IR radiation having a wavenumber of 3590 cm^{-1}?

96. The structures of two aromatic compounds are shown below.

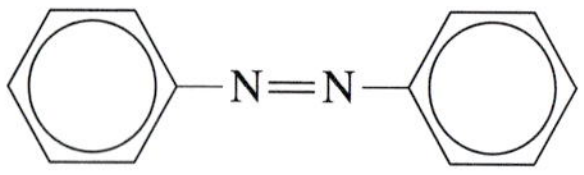

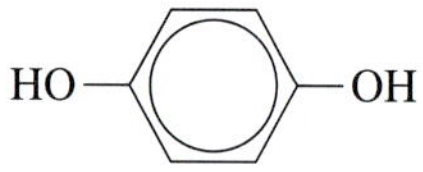

—N=N—   HO— —OH

One of these has an orange-red color and the other is white. Indicate which is which, and explain your reasoning.

97. State how infrared absorption spectra would enable you to distinguish between the following pairs of compounds. (*Hint:* Use data from Table 23.6.)

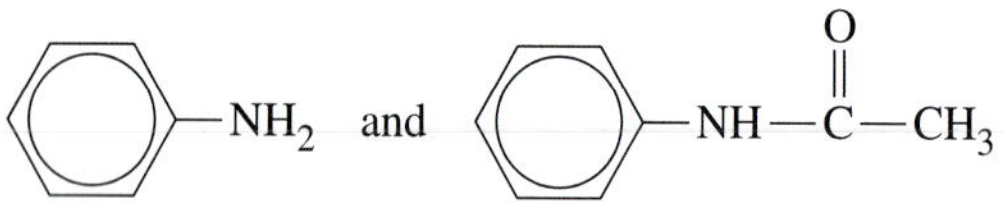

—NH_2 and —NH—C(=O)—CH_3

(a)

$CH_3CH_2CHCH_3$ (OH) and $CH_3CH_2CCH_3$ (=O)

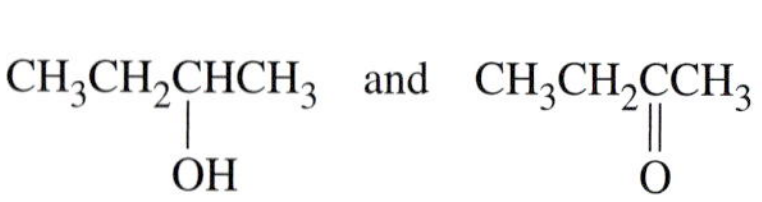

(b)

98. How many different "kinds" of H atoms are there in the following molecule? Explain.

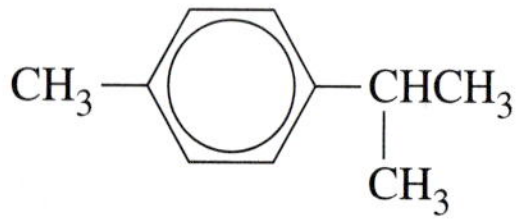

CH_3— —$CHCH_3$ (CH_3)

99. Describe how one of the types of spectroscopy (IR, UV-VIS, or NMR) could be used to distinguish between the compounds in each of the following pairs.

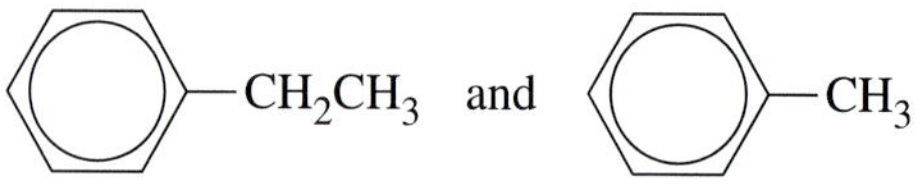

—CH_2CH_3 and —CH_3

(a)

CH_3CHCH_3 (OH) and CH_3CCH_3 (=O)

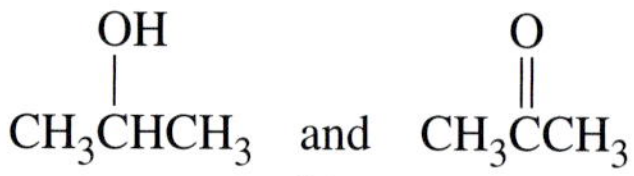

(b)

100. Use information from Table 23.8 to predict the approximate position of NMR absorption peaks (δ values) of the H atoms in boldface type.

$CH_3CCH_2\mathbf{CH_3}$ (=O) $CH_3CHC\mathbf{H_2}CH_3$ (OH)

(a)   **(b)**

101. What is the meaning of "spin-spin splitting," and in what spectroscopic method is it observed?

102. The "$n + 1$" rule in NMR spectroscopy applies to the spin-spin splitting of an NMR absorption into several small peaks: The signal from hydrogen nuclei with n neighboring H atoms is always split into $n + 1$ peaks. Show how this rule applies to NMR absorption of the ethyl group described on page 996.

103. Use information from Problem 102 to predict the NMR splitting pattern for each type of H atom in **(a)** 1,1-dibromoethane; **(b)** $CH_3OCH_2CH_2Cl$.

104. Use information from Table 23.8 to predict the approximate position of NMR absorption peaks (δ values) of the H atoms in bold-face type, and from Problem 102 to indicate the multiplicity of the signal, that is, whether it will be unsplit (a singlet), split in two (a doublet), in three (a triplet), in four (a quartet), ...

$\mathbf{CH_3}CH_2CH{=}CH_2$ $CH_3\mathbf{CH_2}C{\equiv}CH$

(a)   **(b)**

$\mathbf{CH_3}CCH_2CH_3$ (=O)   —$C\mathbf{H}CH_3$ (CH_3)

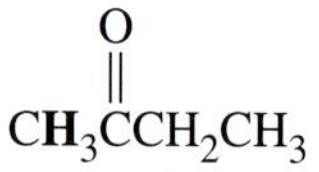

(c) 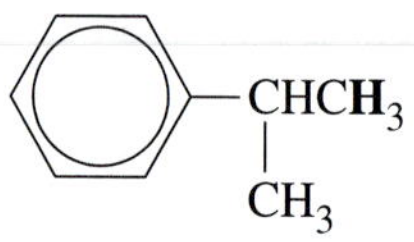   **(d)**

105. The pH of urine varies about 4.5 to 8.0. The K_b for nicotine is 1.05×10^{-6}. Use the Henderson-Hasselbalch equation to calculate the ratio of $[C_{10}H_{14}N_2H^+]$ to $[C_{10}H_{14}N_2]$ at intervals of 0.5 pH unit over the indicated range.

106. Aldehydes are named much like the carboxylic acids (Appendix D). Write the structure of **(a)** *cis*-3-hexenal, a compound with an herbal odor, and **(b)** *trans*-2-*cis*-6-nonadienal.

Chemistry of Materials: Bronze Age to Space Age

CHAPTER

24

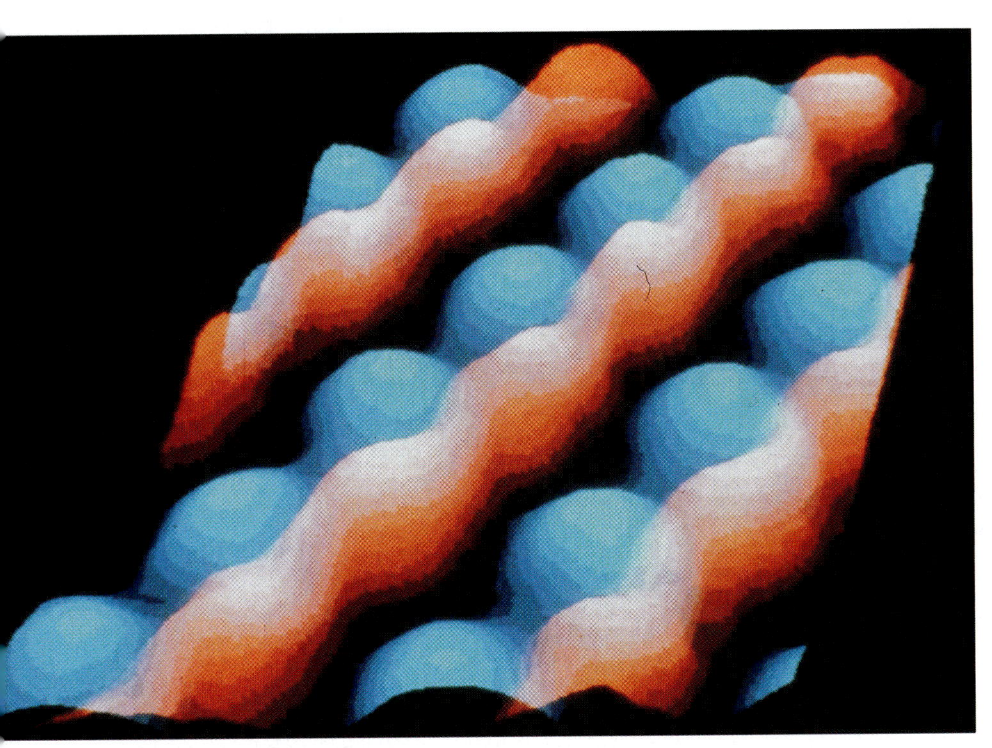

The arrangement of atoms in gallium arsenide is indicated in this false-color image from a scanning tunneling microscope (page 308). GaAs is a semiconductor, one of several important materials that we discuss in this chapter. Ga atoms are shown in red and As atoms in blue.

CONTENTS

Prehistoric people were limited to natural materials that they could use without much alteration. However, the discovery of fire gave them an effective way of transforming natural materials into new materials. For example, they developed primitive methods of heating minerals to extract tin, lead, and other metals. When they fused tin with naturally occurring copper, they obtained a hard metallic material—bronze—that craftsmen could fashion into tools. With this discovery about 7000 years ago, humankind passed from the Stone Age to the Bronze Age. The use of iron and the advent of the Iron Age came somewhat later. Metals, which were so important to ancient people that the metal names are associated with various periods of prehistory, are still materials of great importance. We consider metals first in this chapter.

In contrast, many of the materials that are critical to modern society were all but unknown until well into the twentieth century. We focus on two types of new materials in the remainder of the chapter: semiconductors and polymers. Semiconductors have revolutionized the electronics industry, which in turn is the driving

● ChemCDX
When you see this icon, there are related animations, demonstrations, and exercises on the CD accompanying this text.

force behind much of our high-tech world. And in many of the materials of our daily lives—ranging from clothing to housing materials to automobiles—polymers have replaced natural fibers, wood, and metals. Fully half of all industrial chemists in the United States work in some way with polymers.

Cinnabar, a mercury ore, principally HgS.

Metallurgy

In this section, we will first consider general approaches for obtaining metals from natural sources. Then we will illustrate these approaches for a few common metals. Also, we will look at some newer and more specialized methods.

24.1 Some General Considerations

Most metals are found in *minerals*—crystalline inorganic compounds in Earth's solid crust. For example, titanium is obtained from the mineral *rutile*, TiO_2, and we get the metalloid antimony from *stibnite*, Sb_2S_3. An **ore** is a solid deposit containing a sufficiently high percentage of a mineral to make extraction of a metal economically feasible. A titanium ore, for example, has a significant percentage of rutile. The ores in the following list are of commercial importance.

- *Native ores (free metal)*: Gold, silver, platinum, and copper.
- *Oxides*: Iron, manganese, aluminum, tin, and titanium.
- *Sulfides*: Copper, nickel, zinc, lead, mercury, antimony, bismuth, and silver.
- *Carbonates*: Sodium, potassium, magnesium, calcium, manganese, iron, and zinc.
- *Chlorides (often in aqueous solution)*: Sodium, potassium, magnesium, and calcium.

Rutile, a titanium ore, mainly TiO_2, in a matrix of quartz.

Because silicon is such an abundant element (exceeded only by oxygen in Earth's crust), many metals occur as silicates. For example, aluminum is widely distributed in kaolinite clay, $Al_2Si_2O_5(OH)_4$. However, it is too expensive to extract aluminum from this source. In general, silicates are not used as ores.

Extractive Metallurgy

Metallurgy is the general study of metals, and *extractive metallurgy* focuses on the activities required to obtain a pure metal from one of its ores. Although there is no single method of extractive metallurgy that applies to all metals, the five processes described next are all commonly encountered.

- *Mining*: Ores are mined in many ways. Generally, we think of a mine as a hole dug into Earth's crust. Deep mines are the method of choice when the ore is far below the surface. Ores near the surface, however, can be dug up in open-pit mines. Most copper ore in the western United States has been mined in this way. Some mining is done by pumping seawater or brines from salt lakes or subsurface wells to obtain the metals they contain.
- *Concentration*: Often, the metal-containing mineral makes up only a small fraction of the material mined. To eliminate as much extraneous material as possible in the metallurgical scheme, ores are generally *concentrated*, physically separating the ore from waste rock.

  Figure 24.1 illustrates the **flotation** method of concentrating an ore. In this process, a metal ore is ground into a powder. The powder and certain additives are placed in water in a large vat. The mixture is agitated with air. Particles of ore become attached to air bubbles and rise to the top of the mixture as a froth,

Gold (Au) embedded in quartz.

which is allowed to overflow. Particles of the undesired waste rock, called *gangue*, fall to the bottom of the vat. The success of the flotation method depends upon some critical additives. One additive, called a frother, produces a stable foam. Another additive, called the collector, coats the particles of ore but does not "wet" particles of gangue.

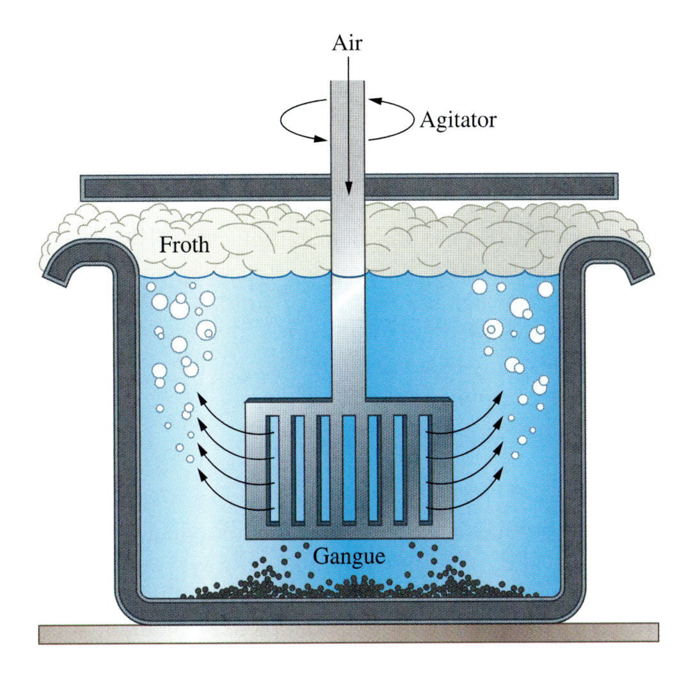

◀ **Figure 24.1 Concentration of an ore by flotation**
Powdered ore, additives, and water are agitated with air. Air bubbles carry ore particles into the froth. Particles of waste rock, called gangue, fall to the bottom of the vat.

An open-pit copper mine near Salt Lake City, Utah.

- *Roasting*: Following concentration, often ores are heated to a high temperature in a process called *roasting*. Roasting converts many metal compounds to metal oxides. For example, zinc carbonate yields zinc oxide when heated strongly.

$$ZnCO_3(s) \longrightarrow ZnO(s) + CO_2(g)$$

When strongly heated in air, sulfides liberate $SO_2(g)$. Thus, when roasted in air, lead(II) sulfide produces lead(II) oxide and $SO_2(g)$.

$$2\,PbS(s) + 3\,O_2(g) \longrightarrow 2\,PbO(s) + 2\,SO_2(g)$$

- *Reduction*: This term can have two somewhat different meanings in extractive metallurgy. In one sense, it refers to the removal of oxygen from a metal oxide. In principle, this can be done by heating the metal oxide to a sufficiently high temperature. In practice, though, this method of reduction works only for a couple of oxides that decompose at relatively low temperatures. Ag_2O decomposes rapidly at 250 to 300 °C and HgO at about 500 °C.

$$2\,HgO(s) \longrightarrow 2\,Hg(l) + O_2(g)$$

Usually, a reducing agent is required for the removal of oxygen from a metal oxide. The most widely used and inexpensive reducing agent is coke (page 902). In most cases, the partial oxidation of carbon produces CO(g), which serves as the actual reducing agent.

$$SnO_2(s) + 2\,C(s) \longrightarrow Sn(l) + 2\,CO(g)$$
$$SnO_2(s) + 2\,CO(g) \longrightarrow Sn(l) + 2\,CO_2(g)$$

In a few specialized cases, an especially strong metallic reducing agent, such as aluminum or calcium, is required. This is the case in the thermite reaction (Section 21.3).

APPLICATION NOTE
The iron ore *magnetite*, Fe_3O_4, can be concentrated by the process of magnetic beneficiation. The Fe_3O_4 is attracted to a magnet and thus separated from waste rock which is composed of nonmagnetic materials.

APPLICATION NOTE
Sulfur dioxide produced in metallurgical processes is generally converted to sulfuric acid. Environmental restrictions preclude venting $SO_2(g)$ into the atmosphere (Section 25.2).

$$Cr_2O_3(s) + 2\,Al(s) \longrightarrow Al_2O_3(s) + 2\,Cr(l)$$

The term reduction can also refer to *electrolytic* reduction of an ionic form of a metal to the free metal. Thus, we have described the reduction of Na^+ to Na(l) in molten NaCl (Section 18.9), Mg^{2+} to Mg(l) in molten $MgCl_2$ (Section 20.9), and Al^{3+} to Al(l) from Al_2O_3 dissolved in molten Na_3AlF_6 (Section 21.3).

APPLICATION NOTE

The glassy slag is sometimes formed into fibers and used as *rock wool* for insulation.

- *Slag Formation*: During the reduction process, high-melting-point impurities such as sand (SiO_2; melting point about 1720 °C) and aluminum oxide (Al_2O_3; melting point 2072 °C) are often removed as a **slag**, a lower-melting (about 1200 °C), glassy product. In slag formation, a basic oxide (often CaO) reacts with acidic oxides (for example, SiO_2) and amphoteric oxides (for example, Al_2O_3). Slag formation plays a crucial role in the metallurgy of iron.
- *Refining*: If a metal is obtained by electrolytic reduction, it often requires no further processing before use. A metal obtained by chemical reduction, for example, through the reaction of a metal oxide with CO(g), is often too impure for its intended use. **Refining** is the process of removing impurities from a metal by any of a variety of chemical or physical means. In a number of instances, refining is done by electrolysis. We described the electrolytic refining of copper in Section 18.11.

Alloys

The final product sought in many metallurgical processes is not the pure metal itself but an alloy. An **alloy** is a mixture of two or more metals, or of a metal with a nonmetal, formulated to produce properties that make the mixture more desirable than the pure metal itself. For example, a small amount of copper added to gold increases its hardness, and the presence of nickel and chromium in iron greatly improves its corrosion resistance.

Some alloys are heterogeneous mixtures, like the familiar lead-tin alloy *solder*. Although the alloy appears homogeneous to the eye, the separate solid phases are visible under a microscope. Other alloys are solid solutions, and among solid solutions there are two possibilities. In *substitutional* solid solutions, atoms of one metal may substitute for some of the atoms in the crystal structure of another metal. This type of solid solution generally requires a close match in the atomic radii of the metals in the alloy. Thus, silver and gold, both with an atomic radius of 144 pm, form a series of solid solutions whose concentrations can range from pure silver to pure gold (recall, Figure 1.3). In *interstitial* solid solutions, small atoms of one constituent occupy voids among the larger atoms of the major component. An example of this type of solid solution is one with about 1% carbon in iron. A small number of alloys are actually intermetallic compounds, such as the amalgam $NaHg_2$.

In Section 24.2, we will consider important alloys that have iron as the major component. A few other common alloys, their compositions, and previous references to them are listed in Table 24.1.

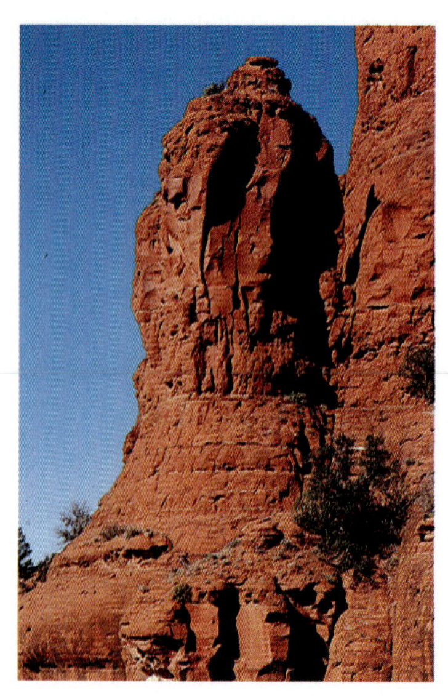

Iron compounds give soil and rocks a characteristic red color, as in this rock formation near Sedona, Arizona.

24.2 Iron and Steel

Among all the metals, iron is first in commercial importance and second only to aluminum in natural abundance. Iron-containing minerals commonly are present in red soil and red-rock formations.

Pig Iron

For the last 700 years or more, people have used a device called a *blast furnace* to reduce iron ore to iron. A modern blast furnace is pictured in Figure 24.2. Chemical reactions occur as the solid reactants settle from the top of the furnace and

TABLE 24.1 Several Common Alloys

Name	Composition	Page of previous reference
Battery plate	94% Pb, 6% Sb	902
Gun metal bronze	90% Cu, 10% Sn	895
Magnalium	70–90% Al, 10–30% Mg	865
Pewter	85% Sn, 7% Cu, 6% Bi, 2% Sb	895
Plumber's solder	67% Pb, 33% Sn	895
Sterling silver	92.5% Ag, 7.5% Cu	932
Yellow brass	67% Cu, 33% Zn	935

Molten slag being dumped.

gaseous reactants rise from the bottom. The reactions are complex, but we can divide them into four basic categories.

- Formation of reducing agents, principally CO(g) and H_2(g)
- Reduction of iron oxide
- Slag formation
- Impurity sources

Representative reactions are noted in Figure 24.2, keyed to the regions of their main occurrence in the blast furnace.

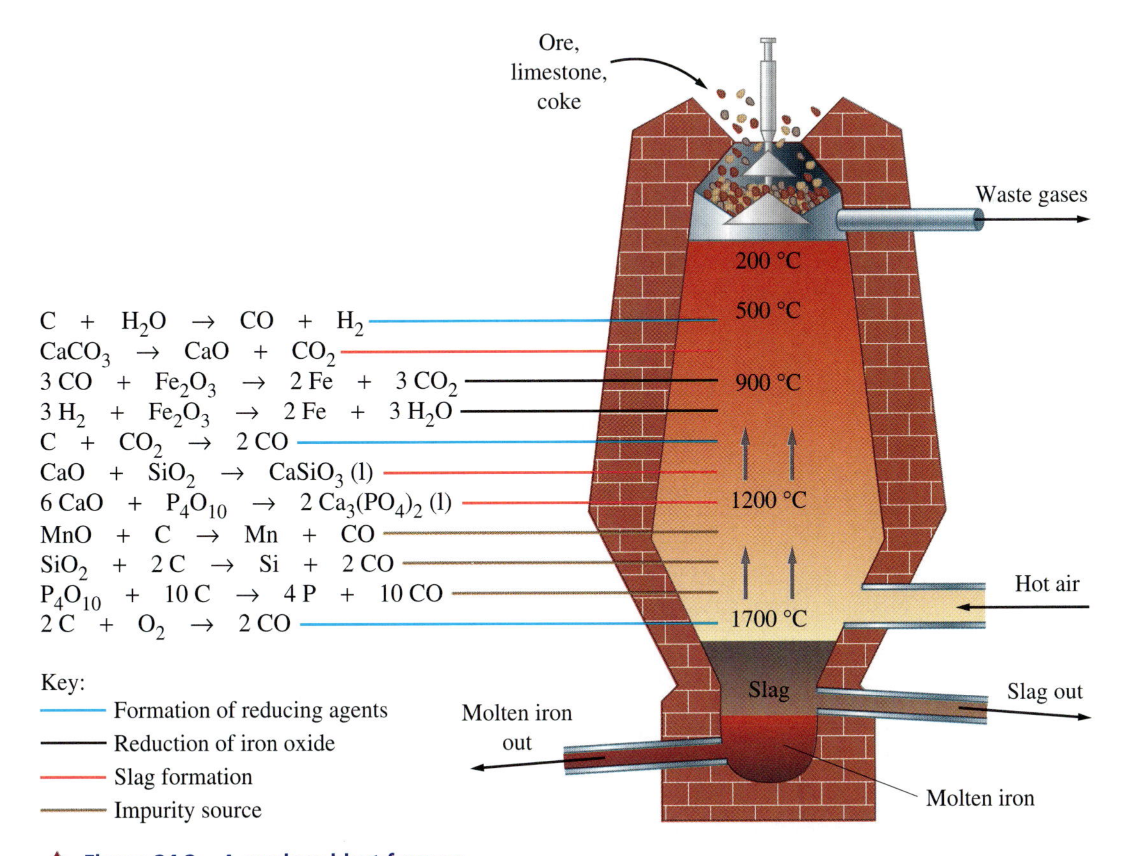

▲ **Figure 24.2 A modern blast furnace**
A modern blast furnace stands up to 90 m high, is computer controlled, and is equipped with environmental control devices.

APPLICATION NOTE
Long ago, steel swords were made by heating wrought iron swords in charcoal. This was followed by rapid quenching of the hot metal in a cold liquid. The sword cooled so fast that the atoms were trapped in an irregular array that left the steel strong and flexible. Some metalworkers preferred biological fluids such as urine—a readily available resource—as a quenching medium.

APPLICATION NOTE
In 1912, the ocean liner *Titanic* struck an iceberg and sank. In 1985, the wreckage was found, and scientists sought to discover why this "unsinkable" ship sank so quickly. Metallurgical analysis showed that wrought iron rivets in the ship's hull contained a high amount of slag. This made the rivets brittle at ice-water temperatures, causing the hull to rip apart rapidly. Heat treatment would have made the rivets less brittle, a fact presumably unknown to the ship's builders.

Acidic oxides such as SiO_2 predominate as impurities in most iron ores, so iron producers commonly use basic oxides to react with them to form slags. Typically, the basic oxide is CaO formed by the decomposition of limestone ($CaCO_3$). Conversely, if the iron ore contains carbonates as impurities, sand (SiO_2) is used to form the slag.

The iron formed in a blast furnace is called **pig iron**. It is impure iron, generally containing 3 to 4% C, 0.5 to 3.5% Si, 0.5 to 1% Mn, 0.05 to 2% P, and 0.05 to 0.15% S. Most pig iron is transferred directly to steelmaking furnaces as a liquid.

The solid metal obtained from liquid pig iron is called **cast iron**. Crude cast iron is brittle when cold, but malleable above 250 °C. Its properties are modified by remelting, reprocessing, and cooling at a controlled rate. Some of the uses of cast irons include automotive engines, boilers, stoves, and cookware.

Cast iron is *wrought* by hammering at 800 to 900 °C. This mechanically squeezes out some of the remaining slag and solid impurities and burns out most of the remaining carbon. Wrought iron has many uses, including decorative fences, gates, and grills.

Steel

Most iron is converted to alloys known collectively as **steel**. In general, a steel has more desirable properties (strength, malleability, corrosion resistance) than iron itself. Steel manufacturers make the following fundamental changes in pig iron to convert it to steel.

- They reduce the carbon content to less than 1.5% C.
- They remove major impurities (Si, Mn, P, and S) and some minor ones.

The resulting steel with carbon as the principal alloying element is called *carbon steel*. Carbon steels account for the greatest volume of steel production.

The remainder of steel production is in the form of *alloy steel*. An alloy steel has some element(s)—Cr, Ni, Mn, V, Mo, W—in addition to carbon as a major component. Table 24.2 lists a few carbon and alloy steels and their uses.

Although other steelmaking methods have been used in the past, most steelmakers now use the basic oxygen process (Figure 24.3). Carbon and sulfur are burned off as gaseous oxides. Silicon, manganese, and phosphorus also are oxidized, but their oxides form slags rather than gases. SiO_2 and MnO combine to form $MnSiO_3(l)$; P_4O_{10} combines with CaO (from $CaCO_3$ added to the furnace

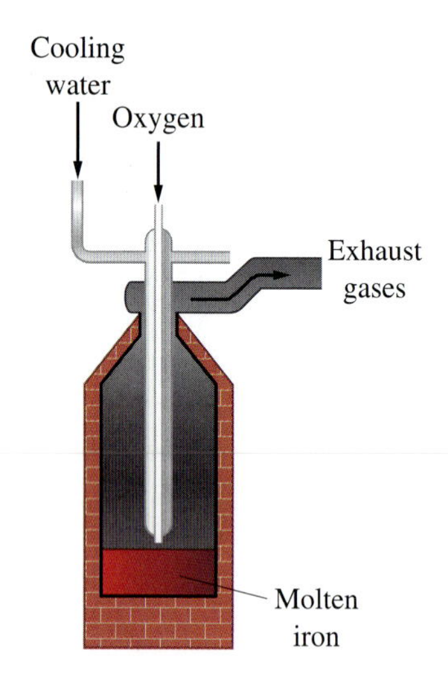

▲ **Figure 24.3**
A basic oxygen furnace
Oxygen at about 10 atm pressure and limestone are discharged into molten pig iron with 15 to 30% scrap iron and steel.

TABLE 24.2 Common Types of Steel

Type	Alloying Element(s)	Typical Uses
Low-carbon	<0.25% C	Beams, steel rods for reinforcing concrete
Medium-carbon	0.25–0.70% C	Machine components requiring high strength and fatigue resistance
High-carbon	>0.70% C	Railroad rails and other uses requiring wear resistance
High-speed tool	0.6–0.8% C, 4.0% Cr, 10.0% W, 0.8–1.0% Mo	Cutting tools, boring tools, saws
Stainless N	0.18% C, 18% Cr, 8% Ni	Chemical plant construction, flatware and cutlery, ornamental use in architecture
Silicon	0.6–5.0% Si	Transformers, motors, generators

charge) to form $Ca_3(PO_4)_2(l)$. The liquid slag floats on top of the liquid iron and is poured off. Alloying metals are added as a final step.

24.3 Tin and Lead

The methods used to extract tin and lead from their ores are good illustrations of the basic metallurgical processes described in Section 24.1. Their relatively uncomplicated extractive metallurgies help to account for the discovery of these metals in ancient times.

Tin

Tin occurs in nature mainly as the ore *cassiterite*, SnO_2, which can be concentrated by flotation. Roasting of the ore oxidizes metallic impurities and drives off sulfur and arsenic as volatile oxides. The SnO_2 that remains is then reduced with coke.

$$SnO_2(s) + 2\,C(s) \longrightarrow Sn(l) + 2\,CO(g)$$

The metal is first solidified and then remelted. Tin is fairly low melting (mp 232 °C), and the molten tin is poured off, leaving behind unmelted impurities. Impurities that are soluble in liquid tin are oxidized with air, and the oxide film is skimmed off.

Recycling is also an important source of tin. In one recycling method, scrap tin plate is treated with chlorine gas, which converts the tin to $SnCl_4(l)$ without affecting the underlying metal (usually steel). $SnCl_4$ is then converted to SnO_2, and the SnO_2 is reduced to metallic tin. The tin plate is mainly in the form of "tin" cans—steel cans with a protective layer of tin (page 799).

Lead

Lead is found chiefly as *galena*, PbS. The ore is concentrated by flotation and then roasted.

$$2\,PbS(s) + 3\,O_2(g) \longrightarrow 2\,PbO(s) + 2\,SO_2(g)$$

Reduction is carried out with coke.

$$PbO(s) + C(s) \longrightarrow Pb(l) + CO(g)$$
$$PbO(s) + CO(g) \longrightarrow Pb(l) + CO_2(g)$$

The lead at this point contains several possible impurities, such as Cu, Ag, Au, Sn, As, and Sb. Lead producers take advantage of physical properties such as melting points and solubilities to remove these impurities. When the lead is melted (mp 327 °C), copper rises to the top of the liquid as an insoluble solid and is skimmed off. When the temperature is raised further, Sn, As, and Sb are oxidized, and the oxide film is skimmed off. At this point, zinc is added to the molten lead, and the zinc melts. Silver and gold in the molten lead pass into the molten zinc, in which they are more soluble. When the molten mixture is cooled below 420 °C (the melting point of zinc), the zinc solidifies to a crust containing most of the Au and Ag; then this crust is skimmed off. Combined with electrolytic refining, this treatment can produce lead with a purity of 99.99%.

The recycling of used lead is an important alternative to the production of new lead. Currently, about 70% of manufactured lead is recycled lead.

The stainless steel in these sinks (Type 302) contains 17–19% Cr, 8–10% Ni, and a maximum of 0.15% C.

These tin-coated steel cans can be recycled. The tin metal is recovered through the chemical reactions described in the text.

24.4 Copper and Zinc

The metallurgy of copper is somewhat more complicated than the general procedure described in Section 24.1, primarily because sulfide ores of copper generally contain appreciable quantities of iron sulfides. The metallurgy of zinc is more straightforward, but it also illustrates some interesting variations on the general procedure.

Copper

As we have noted, copper ores commonly contain iron as well. Thus, copper made by the general metallurgical scheme outlined in Section 24.1 would probably be contaminated with significant amounts of iron. To avoid this contamination, copper producers generally follow the four-step procedure outlined below.

1. The copper ore is concentrated, generally by flotation.
2. The ore is partially roasted so that iron sulfides are mostly converted to iron oxides but copper sulfides are left largely unchanged.

$$2\,FeS(s) + 3\,O_2(g) \longrightarrow 2\,FeO(s) + 2\,SO_2(g)$$

Pouring molten copper into anode casts prior to electrolytic refining.

3. The roasted ore is heated in a furnace at about 1100 °C, at which point the materials melt and separate into two layers. The bottom layer is *copper matte*, a mixture of molten sulfides of copper and iron (mostly Cu_2S and FeS). The upper layer is a silicate slag formed by the reaction of oxides of Fe, Ca, and Al with SiO_2 added to the furnace charge.
4. In a process called *conversion*, air is blown through the molten copper matte. Any remaining iron sulfide is converted to iron oxide, and the iron oxide is removed as an iron silicate slag. Other reactions also occur:

$$2\,Cu_2S + 3\,O_2(g) \longrightarrow 2\,Cu_2O + 2\,SO_2(g)$$
$$2\,Cu_2O + Cu_2S \longrightarrow 6\,Cu(l) + SO_2(g)$$

The product, called *blister copper*, is about 97 to 99% Cu with entrapped bubbles of $SO_2(g)$. It is sufficiently pure for some applications, such as plumbing. Where high-purity copper is required, as in electrical applications, refining can be done electrolytically (Section 18.11).

A copper ore typically used to produce copper generally has only about 0.5% Cu. This means that enormous quantities of finely ground waste rock are generated in concentrating the copper ore by flotation. Disposal of this waste rock in environmentally acceptable ways poses serious problems. The recycling of used copper is now an important alternative to the production of new copper. Currently, nearly half of manufactured copper is recycled copper.

Zinc

Zinc occurs mainly as ZnS (sphalerite) and $ZnCO_3$ (smithsonite). During roasting, sphalerite is converted to zinc oxide and $SO_2(g)$; smithsonite, to zinc oxide and $CO_2(g)$. Zinc oxide is reduced with coke or powdered coal. The reduction is carried out at about 1100 °C, well above the boiling point of zinc (907 °C). Zinc *vapor* is condensed to a liquid, and the impurities, mostly cadmium and lead, can be removed by fractional distillation of the liquid zinc.

Alternatively, zinc oxide from the roasting step can be dissolved in $H_2SO_4(aq)$. Addition of powdered zinc to the zinc sulfate solution displaces less active metals such as cadmium.

$$Zn(s) + Cd^{2+}(aq) \longrightarrow Zn^{2+}(aq) + Cd(s)$$

Then the $ZnSO_4(aq)$ is electrolyzed, and pure zinc metal is deposited at the cathode.

EXAMPLE 24.1—A Conceptual Example

Consider the electrolytic method of zinc metallurgy just described. Why must the ions of metals less active than zinc (for example, Cd^{2+}) be removed before the electrolysis of $ZnSO_4(aq)$ is carried out?

SOLUTION

Recall from Section 18.4 that the value of E°_{cell} for a redox reaction is given by the following relationship.

$$E^\circ_{cell} = E^\circ(\text{reduction}) - E^\circ(\text{oxidation})$$

For electrolysis reactions, which are *nonspontaneous*, the value of E°_{cell} will be *negative*. And, the minimum voltage required to bring about the electrolysis is given by this expression.

$$E^\circ_{electrolysis} = -E^\circ_{cell}$$

In the electrolysis we are considering, a metal ion is reduced to the metal at the cathode, and an oxidation occurs at the anode. If the metal ion reduced is $Zn^{2+}(aq)$,

$$\begin{aligned} E^\circ_{cell} &= E^\circ_{Zn^{2+}/Zn} - E^\circ(\text{oxidation}) \\ &= -0.763\text{ V} - E^\circ(\text{oxidation}) \end{aligned}$$

If the metal ion reduced is $Cd^{2+}(aq)$,

$$\begin{aligned} E^\circ_{cell} &= E^\circ_{Cd^{2+}/Cd} - E^\circ(\text{oxidation}) \\ &= -0.403\text{ V} - E^\circ(\text{oxidation}) \end{aligned}$$

Regardless of the oxidation that occurs at the anode [actually it is the oxidation of $H_2O(l)$ to $O_2(g)$], the minimum voltage required for the reduction of $Cd^{2+}(aq)$ at the cathode is given by this expression.

$$E^\circ_{electrolysis} = -E^\circ_{cell} = 0.403\text{ V} + E^\circ(\text{oxidation})$$

This voltage is *less than* that required for the reduction of $Zn^{2+}(aq)$.

$$E^\circ_{electrolysis} = -E^\circ_{cell} = 0.763\text{ V} + E^\circ(\text{oxidation})$$

Thus, the voltage that we must use to obtain Zn(s) at the cathode is more than adequate to produce Cd(s) as well. Consequently, we must remove $Cd^{2+}(aq)$ and any other metal ions that are more easily reduced than Zn^{2+} *before* we electrolyze the $ZnSO_4(aq)$.

EXERCISE 24.1

Consider the displacement of metal ions discussed in Example 24.1. Why do you suppose that zinc is the metal used rather than some other active metal, for example, aluminum?

24.5 Hydrometallurgy

Metallurgical methods that use ore concentration, roasting, chemical or electrolytic reduction, and slag formation are often called **pyrometallurgy** (*pyro* means fire or heat). Pyrometallurgical methods are energy intensive and usually require expensive controls to avoid polluting the environment. In some cases, these methods are being replaced by others, collectively called **hydrometallurgy**, that involve processing aqueous solutions of metallic compounds. For example, copper ores can be treated with $H_2SO_4(aq)$. This converts the copper compounds in the ore to $CuSO_4(aq)$, which is then electrolyzed to produce copper. Following are the essential operations of hydrometallurgy.

Large spherical autoclaves are used to leach ilmenite ore, $FeTiO_3$, with HCl(aq) at 120 °C and 3.5 atm. This is a step in the production of pure TiO_2, for use as a pigment or in the metallurgy of titanium.

- *Leaching*: Metal ions are extracted from the ore with a liquid—water, acids, bases, or salt solutions. Sometimes an oxidation-reduction reaction is involved. In some instances, leaching can be done in place, without even mining the ore.
- *Purification and/or concentration*: Impurities are removed from the leach solution and the solution is concentrated by evaporation to facilitate further processing. Two methods used for the removal of impurities are ion exchange and adsorption on activated carbon (page 890).

- *Precipitation and reduction*: In some processes, the desired metal ions are precipitated as an insoluble ionic solid. In a subsequent step, the metal ions are reduced to the free metal either by displacement from solution by a more active metal or by electrolysis.

Hydrometallurgy has several advantages over pyrometallurgy. For example, it is a better process for low-grade ores, and it is more energy-efficient. It is also less polluting of the atmosphere. In the pyrometallurgy of zinc, for example, there are inevitably some emissions of $SO_2(g)$ and Hg(g) and mercury compounds. In the hydrometallurgy of zinc, compounds of sulfur and mercury are retained in solution. Hydrometallurgy presents some difficult problems of its own, however, such as how to contain and dispose of liquid solutions and solid waste. The mercury compounds in the waste solution from the hydrometallurgy of zinc can be an environmental hazard if the solution is not properly treated before disposal.

In the hydrometallurgy of zinc, ZnS ore is leached with sulfuric acid solution and $O_2(g)$ at 7 atm pressure at 150 °C.

$$ZnS(s) + H_2SO_4(aq) + \tfrac{1}{2}\,O_2(g) \longrightarrow ZnSO_4(aq) + S(s) + H_2O(l)$$

Zinc is obtained at the cathode when purified $ZnSO_4(aq)$ is electrolyzed. $H_2SO_4(aq)$ is also a product of the electrolysis; it is recovered for use in the leaching process. The production of zinc from zinc oxide by electrolytic reduction, described on page 1020, is another example of hydrometallurgy.

The purification of bauxite ore in the metallurgy of aluminum (Section 21.3) is a hydrometallurgical process based on the amphoterism of $Al(OH)_3$. Hydrometallurgy has long been used in extracting silver and gold from their ores.

Silver and gold are both found free in nature, but all easily accessible known deposits have been mined. A typical gold ore today contains only about 10 g Au per ton. In one older, environmentally damaging method, low-grade gold ores were treated with mercury. Gold dissolved in the liquid mercury to form an *amalgam*. (Amalgam is the general name for mercury alloys.) The mercury in the amalgam was driven off by heating, leaving behind pure gold.

A method more widely used today is *cyanidation*, a special case of hydrometallurgy. In this method $O_2(g)$ in air oxidizes the free metal to Au^+, which is then complexed with CN^-.

$$4\,Au(s) + 8\,CN^-(aq) + O_2(g) + 2\,H_2O(l) \longrightarrow 4\,[Au(CN)_2]^-(aq) + 4\,OH^-(aq)$$

Gold can be oxidized in the above reaction only because $[Au(CN)_2]^-$ is such a stable complex ion. Without $CN^-(aq)$ present, the oxidation of Au to Au^+ hardly occurs at all. Gold is displaced from $[Au(CN)_2]^-(aq)$ by an active metal such as zinc.

$$2\,[Au(CN)_2]^-(aq) + Zn(s) \longrightarrow 2\,Au(s) + [Zn(CN)_4]^{2-}(aq)$$

The cyanidation of ores poses substantial environmental problems. Aqueous solutions of cyanides are poisonous and are workplace hazards. Equally important, waste solutions containing cyanides must be held in containment ponds. The ponds must be lined to prevent solutions from entering goundwater, and migratory birds and other wildlife must be kept away from the ponds. Environmental protection is at least as important a consideration as technical and economic matters when planning any new mining operation.

The metallurgy of silver resembles that of gold in some respects, but there are differences. Silver is found mostly in mineral form, principally as Ag_2S. And, because $Ag_2S(s)$ is highly insoluble in water, sulfide ion must be eliminated from the solution in which complex-ion formation is to occur. This is accomplished by blowing air through a suspension of $Ag_2S(s)$ in a solution of $CN^-(aq)$. Sulfide ion is ox-

Waste solution from the leaching operation at a gold-mining facility in the Mojave Desert of California is stored in a large containment pond.

idized to sulfate ion, and silver sulfate is readily soluble in solutions of $CN^-(aq)$. The complex ion $[Ag(CN)_2]^-$ has a very large formation constant, $K_f = 5.6 \times 10^{18}$.

$$Ag_2S(s) + 4\,CN^-(aq) + 2\,O_2(g) \longrightarrow 2\,[Ag(CN)_2]^-(aq) + SO_4^{2-}(aq)$$

Free silver is displaced from solution by an active metal such as zinc, just as in the metallurgy of gold. Generally, small quantities of other metals deposit with the silver, and further processing is required to refine the silver.

Bonding in Metals and Semiconductors

In Section 11.10, we used a microscopic view of a metal—its crystal structure—to predict a macroscopic property—the density of the metal. Other macroscopic properties of metals, however, such as electrical conductivity, ductility, and malleability, are determined by the nature of the bonding between metal atoms. In the sections that follow, we will consider two bonding theories for metals. We will find that one bonding theory (band theory) works well for metals, and it also accounts for the interesting electrical properties of semiconductors, displayed by such metalloids as silicon and germanium.

24.6 The Free-Electron Model of Metallic Bonding

From the electron configuration $1s^2 2s^1$, we can write a Lewis structure that shows a bond between two Li atoms.

$$Li{:}Li$$

The molecule Li_2 does exist in the gaseous state, but this fact doesn't help us to explain how a Li atom is bonded to *eight* nearest neighbors in the solid metal. This example highlights the basic problem in explaining bonding in metals: The atoms don't seem to have enough valence electrons to form all the bonds needed.

Let's assume for a moment that the bonding is ionic. Suppose half the Li atoms in a crystal were to lose their $2s^1$ valence electron, forming Li^+ ions, and the other half were to gain an electron to form Li^- ions. Then, much as in the ionic crystal Na^+Cl^-, each ion could surround itself with ions of the opposite charge. There are two good reasons, however, why this cannot be the case: (1) We do not expect one Li atom to transfer an electron to another Li atom because all Li atoms have the same electronegativity. (2) Solid ionic compounds, unlike metals, do not conduct electricity—all the electrons are bound to individual ions, and the ions are essentially immobile.

The key to metallic bonding is that certain electrons must be *delocalized*. These delocalized electrons are not bound to individual atoms, and they can therefore serve to bind large numbers of metal atoms together.

According to one theory of metallic bonding, each atom in a metallic crystal loses its valence electrons, releasing them to the crystal as a whole. Because these valence electrons are freed from individual atoms, this theory is known as the **free-electron model** of metallic bonding. We picture the metallic crystal as a lattice of positive ions immersed in a "gas" made up of electrons (see Figure 24.4).

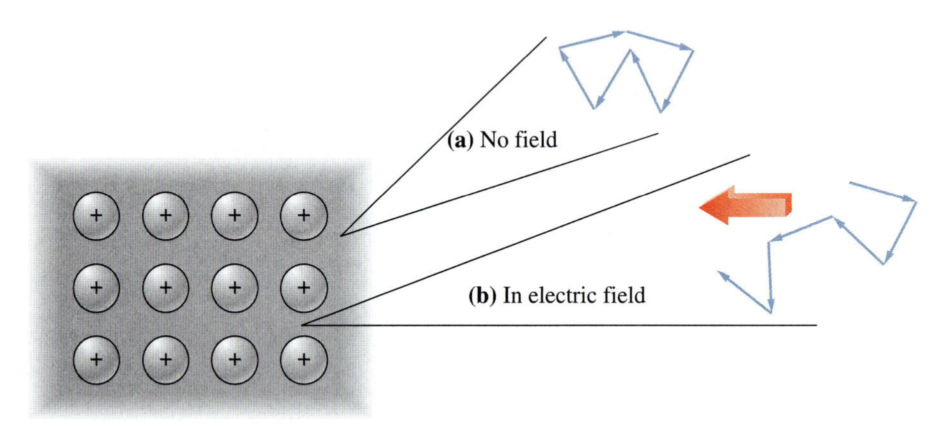

Figure 24.4 The free-electron model
A lattice of cations is immersed in a cloud of negative electric charge made up of the free valence electrons of the metal atoms. The high-speed, random motion of an electron is suggested by the series of connected arrows. (a) In the absence of an electric field, on average, an electron returns to its starting point. (b) In an electric field, there is a net drift of electrons (as suggested by the red arrow).

The comparatively massive metal ions undergo lattice vibrations but are otherwise immobile. The valence electrons, in contrast, are highly mobile, zipping about much like atoms or molecules in a gas. The cloud of negative charge associated with the free electrons envelopes the positive-ion lattice and is the "glue" that holds the metallic crystal together.

Figure 24.4 suggests how the free-electron model accounts for electrical conductivity. In the absence of an electric field, the motion of the free electrons is completely random. Over a period of time, although electrons may have traveled great distances along zigzag paths, their distribution within a metal remains unchanged. However, if the metal is connected to the terminals of a battery, electrons drift toward the positive terminal, even as they continue their zigzag motion. Some electrons leave the metal under the influence of the electric field, and others enter the metal to take their place. An electric current flows.

Figure 24.5 shows how the free-electron model accounts for the malleability generally found in metals, in contrast to the brittleness of an ionic crystal. When a force is applied to the top layer of ions in the metallic crystal in Figure 24.5(a), the environment of the highlighted ion (red) is unchanged. The deformation is easily accommodated; the metal is malleable. When a similar force is applied to a layer of ions in the ionic crystal in Figure 24.5(b), like-charged ions are brought into proximity. Repulsive forces cause the crystal to rupture; the ionic solid is brittle.

The electron-gas model is less successful in explaining the effect of temperature on the electrical conductivities of metals. We expect a gas to flow more readily as its temperature and molecular speeds increase. Yet, the electrical resistance of a metal increases with temperature; the metal's ability to conduct electric current *decreases*. One possibility, of course, is that the vibrations of the cations in-

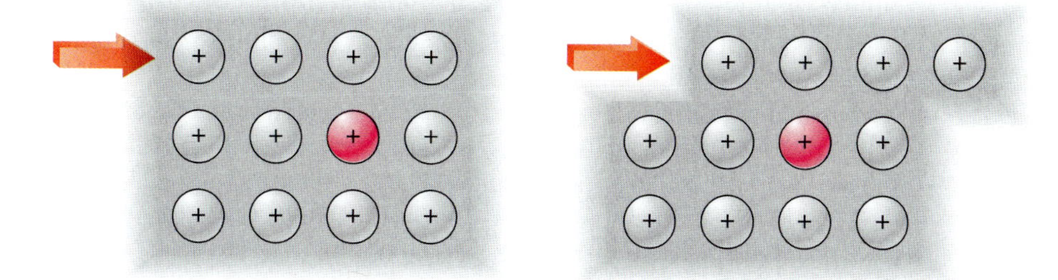
(a) Metal

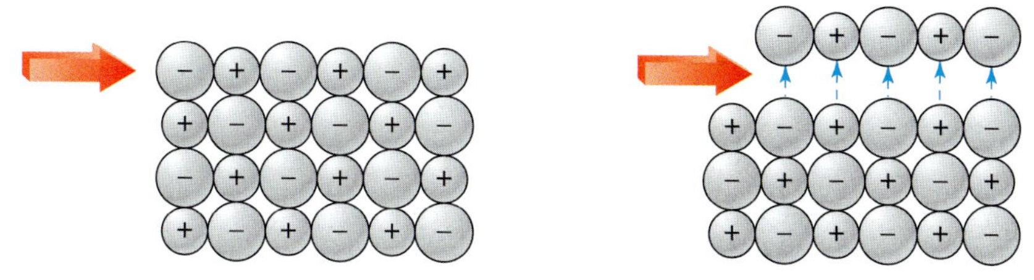
(b) Ionic crystal

▲ **Figure 24.5 Deformation of a metal compared to an ionic solid**
(a) A metal accommodates a distorting force without breaking; it is malleable. (b) An ionic crystal breaks when distorted; it is brittle.

terfere with the migration of electrons in an electric field, but, in any case, the electron-gas model cannot account for the details of electrical conductivity.

The fundamental shortcoming of the free-electron theory is that it is a classical theory patterned after the kinetic theory of gases. It allows for the specification of an electron's position and momentum more precisely than is permitted by Heisenberg's uncertainty principle (Section 7.8). This suggests that a quantum mechanical treatment of bonding in metals should be more satisfactory, as we shall see in the next section.

24.7 Band Theory

In Section 10.6, we used molecular orbital theory to describe covalent bonding. We noted that a combination of two atomic orbitals, one from each atom in a bonded pair, produces two molecular orbitals. One of the molecular orbitals, a *bonding* orbital, lies at an energy below that of the atomic orbitals. The other molecular orbital, an *antibonding* orbital, lies at a higher energy. Valence electrons distribute themselves between the molecular orbitals in accordance with the aufbau principle. Figure 24.6 shows the formation of molecular orbitals in Li_2 from the $2s$ orbitals of two Li atoms. Also shown is the situation in Li_3, in which three $2s$ orbitals combine; and in Li_4, in which four $2s$ orbitals combine. In each case, the number of molecular orbitals is equal to the number of atomic orbitals that combine.

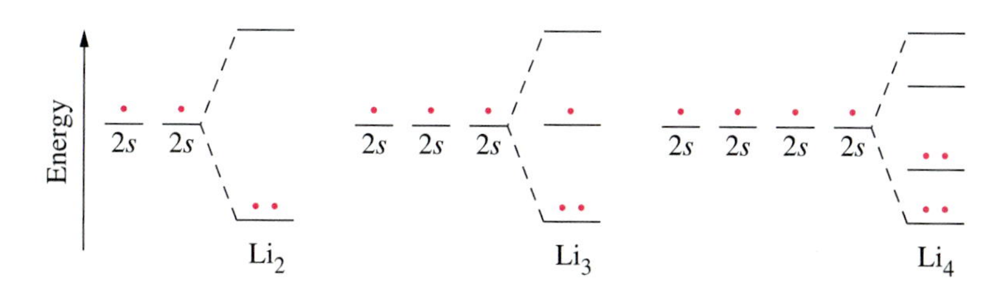

◀ **Figure 24.6 Molecular orbitals for Li_2, Li_3, and Li_4**
Each Li atom contributes a $2s$ orbital to the formation of molecular orbitals. Li_2 has one bonding and one antibonding molecular orbital. In Li_3, the additional molecular orbital is at about the same energy as the isolated atomic orbitals; it is essentially a nonbonding molecular orbital. In Li_4, there are two bonding and two antibonding molecular orbitals. Electrons are represented by bold dots.

The examples in Figure 24.6 give us a fairly good idea of what to expect for a very large collection of Li atoms, say all the atoms in a crystal of the metal. Figure 24.7 depicts molecular orbital energy levels for Li_n, where n is a very large number. Here, we see that a huge number of energy levels, n, are contained within an energy range not much greater than the energy difference between the bonding and antibonding molecular orbitals in Li_2. The spacing between the energy levels is so minute that the levels essentially merge into a **band** (much as the individual wavelength components of visible light merge into a continuous spectrum). And because this band is occupied by the valence electrons of the lithium atoms, it is called a **valence band**.

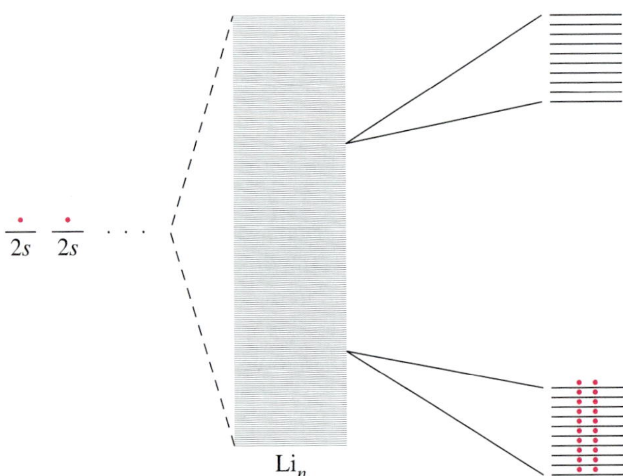

▶ **Figure 24.7**
The 2*s* band in lithium metal
The atomic orbitals of a very large number (n) of Li atoms are combined into a band of molecular orbitals having closely spaced energies. At 0 K, the levels in the bottom half of the band are occupied by electron pairs; those in the top half are empty.

At 0 K, the n valence electrons associated with n Li atoms would occupy the lower half ($n/2$) of the energy levels—two per level. However, the energy difference between levels in the valence band is so small that electrons are easily excited from the highest filled levels to the unfilled levels immediately above them. This excitation can be brought on by heating the metal or applying a small voltage. Thus, in the presence of an electric field, electrons are stimulated to move into empty levels within the band, and this motion creates a net drift of electrons in the metal, an electric current.

The band theory description brings out an important requirement for electrical conductivity: the presence of a **conduction band**, a *partially* filled band of energy levels. The valence band in lithium meets this requirement for a conduction band; it is half filled. But this requirement seems to present problems when applied to certain metals. Consider magnesium. In Mg_n the 3*s* band formed from the 3*s* orbitals of Mg atoms should be filled: *two* valence electrons ($3s^2$) per atom and $2n$ electrons for n molecular orbitals. The band formed by combining 3*p* atomic orbitals, in contrast, should be empty because magnesium has no 3*p* electrons. If an energy band is empty, there are no electrons to jump between energy levels, and if a band is filled, there is no place for the electrons to go. It would seem that magnesium should not conduct an electric current at all, but it does, and quite well.

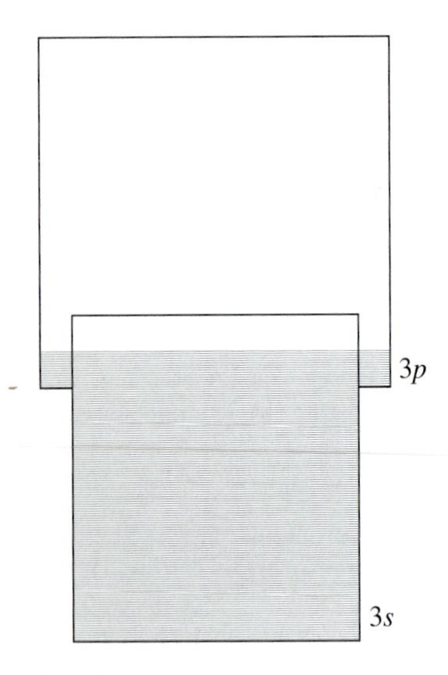

▲ **Figure 24.8**
Band overlap in magnesium
Because the 3*s* and 3*p* bands overlap in energy, some of the lower energy levels in the 3*p* band are filled, and some of the higher energy levels in the 3*s* band are empty. Both the 3*s* and 3*p* bands are partially filled and function as conduction bands. Magnesium is an electrical conductor.

The resolution of this paradox, as illustrated in Figure 24.8, is that the lowest energy bonding molecular orbitals in the 3*p* band lie at a *lower* energy than do the highest energy antibonding molecular orbitals in the 3*s* band—the bands overlap. As a result, some of the valence electrons in magnesium that we would expect to occupy the top of the 3*s* band are instead found in the lower levels of the 3*p* band. This means that both the 3*s* and 3*p* bands are only partially filled, and this meets the requirement for electrical conductivity.

Band theory provides a good explanation of metallic luster and metallic colors. Because the energy levels in bands are so closely spaced, there are electronic transitions in a partially filled band that match in energy every component of visible light. Metals absorb the light that falls on them and are therefore opaque. At the same time, electrons that have absorbed energy from incident light are very effective in reradiating light of the same frequency. This is why metals are good reflectors of light and have a shiny, often mirrored, appearance.

Most metals—magnesium, aluminum, and silver, for example—are equally effective in reradiating (or reflecting) light of all wavelengths, and all have a typical metallic or silvery color. In the case of copper and gold, however, the metal absorbs and reflects light of some wavelengths better than others. Being richer in some wavelength components, the reflected light (and hence the metal) is colored (Figure 24.9).

▲ **Figure 24.9**
Metallic luster and color
Gold and copper have the "metallic" sheen or luster associated with reflected light. However, because they absorb and reflect more effectively in the longer wavelength regions of the visible spectrum than elsewhere, they have characteristic colors: copper (reddish) and gold (yellow).

24.8 Semiconductors

As suggested by Figure 24.10, an electrical insulator has a conduction band that lies at a much higher energy level than does the valence band. Few electrons can acquire enough energy to jump the **energy gap (E_g)** separating the two bands. Thus, insulators are extremely poor electrical conductors. Also shown in Figure 24.10 is a situation in which the energy gap between the two bands is much smaller. Here, some electrons are able to jump the energy gap, resulting in a limited electrical conductivity. This is the energy band picture for a **semiconductor**. For a metal, as we have seen, there is no corresponding energy gap separating a partially filled band and an empty one. Either the valence band is itself a conduction band (as in lithium) or conduction is made possible by the overlap of the valence band and an empty energy band (as in magnesium).

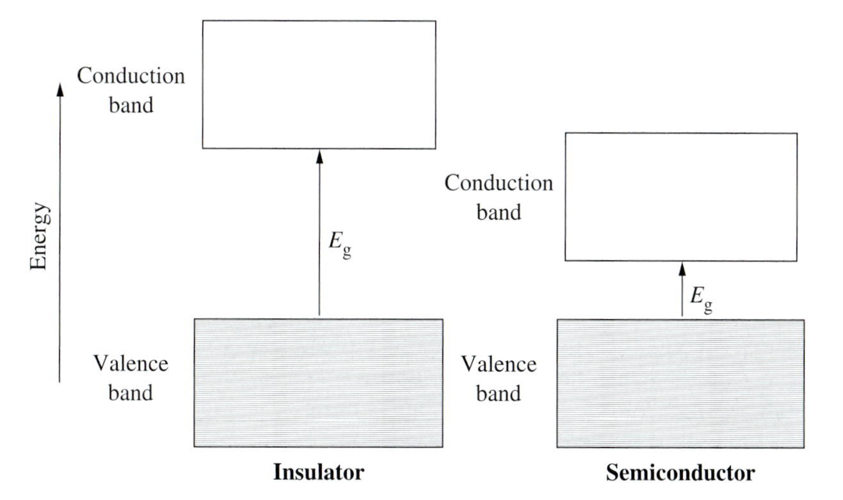

◄ **Figure 24.10**
Band structure of insulators and semiconductors
Insulators and semiconductors have an empty conduction band above a filled valence band. In insulators, the energy gap, E_g, is large. Hardly any electrons can jump the gap between the bands. In semiconductors, the energy gap is much smaller, and small numbers of electrons are able to jump the gap, especially as the temperature is increased.

Figure 24.11 suggests two ways of looking at the electrical conductivity of silicon, a typical semiconductor. In the localized picture, an electron is shown leaving a bond between two Si atoms and entering the crystal as a free electron. At the site of the bond rupture, a vacancy forms which tends to be filled by a free electron from a neighboring bond. This converts the one-electron bond back to a two-electron bond. By attracting an electron, the vacancy acts like a positively charged center; it is called a **positive hole**. Electric current is carried through the semiconductor by both free electrons and positive holes.

In the delocalized picture, every electron that leaves a bond jumps from the valence band to the conduction band. A positive hole is left behind in the valence band. Electric current consists of the movement of electrons between levels in the conduction band and of positive holes between levels in the valence band.

Superconductors

Atoms in a metal lattice vibrate about fixed positions. These vibrations interfere with the flow of electrons and produce an electrical resistance. The more intense the vibrations, the greater the resistance is. As a result, electric current flows more easily at low temperatures than at high temperatures. Theoretically, the electrical resistance of a metal should be zero at 0 K.

In 1911, Kamerlingh Onnes found that some metals abruptly lose their electrical resistance at temperatures still well above 0 K; they become superconductors. A **superconductor** is a material that allows electric current to flow indefinitely, with no loss of energy, when the temperature is held below a critical value. The highest critical temperature among the metals is 23 K. This temperature is far below room temperature, but above the boiling point of liquid helium (4.2 K). Superconducting metals in liquid helium find use in powerful electromagnets in NMR spectrometers (Chapter 23), charged-particle accelerators, nuclear fusion research, and magnetic resonance imaging (MRI).

In 1986, a class of materials was discovered that become superconductors at much higher temperatures than metals. For example, the compound $YBa_2Cu_3O_7$ becomes a superconductor at 90 K. Thus, it is a superconductor at the boiling point of liquid nitrogen (77 K), as shown in the photograph.

Although these new superconductors contain metal atoms, they are not metals. They are ceramic materials (page 892). The structures of ceramic superconductors are complex, but are known to involve sheets of copper and oxygen atoms separated by other constituent atoms. An ultimate goal of superconductor research is to create a superconductor that will function at room temperature and above. Electric transmission lines made of high-temperature superconductors would be able to transmit electricity with no loss of energy. Trains could be propelled at high speed and low energy cost by using the levitation principle illustrated in the photograph.

The highest critical temperature for a ceramic superconductor is still far below room temperature. Add to this the fact that ceramic superconductors are brittle and not easily made into wires, and that their current carrying capacity is limited. As a result, practical applications of ceramic superconductors still seem rather distant.

Superconductivity has also been observed among some fullerene compounds (page 492). For example, the compound K_3C_{60} exhibits metallic conductivity at room temperature and becomes superconducting below 19 K. One or two other fullerenes are superconducting at even higher temperatures, and it seems likely that the upper temperature limit will keep rising as more discoveries are made.

When a small magnet is dropped above a superconductor in liquid nitrogen, the falling magnet induces an electric current in the superconductor. In turn, the electric current induces a magnetic field around the superconductor that opposes the field of the falling magnet. The magnet remains suspended at the point where the upward force of the repulsive magnetic field is just matched by the downward force due to Earth's gravitational field. The magnet remains suspended as long as the current in the superconductor persists. And this continues as long as the superconductor is maintained at the boiling point of liquid nitrogen (77 K).

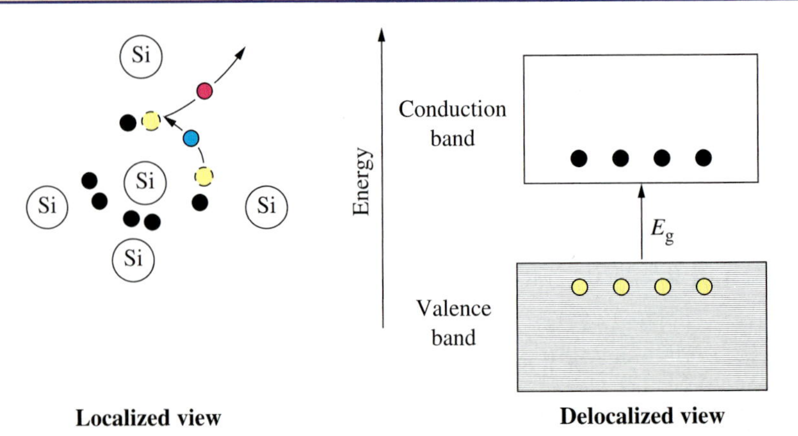

Figure 24.11
Two views describing electrical conductivity in semiconductors
In the localized view, an electron (•) can escape from the bond between two Si atoms and move into the crystal at large. The vacancy it leaves at the bond site (•) is a positive hole. If an electron (•) from a neighboring bond moves into this positive hole, a new positive hole (•) forms there, and so on. The positive hole migrates. In the delocalized view, a positive hole (•) is produced in the valence band for every electron (•) that jumps from the valence band to the conduction band.

The electrical conductivity of a semiconductor can be *increased* by raising its temperature. With an increase in temperature, electrons acquire more energy and more of them are able to jump the energy gap. This behavior is an important distinction between semiconductors and metals. The electrical conductivity of a metal, as we noted previously, *decreases* as its temperature is raised.

The conductivity of a semiconductor can also be increased by **doping**. A semiconductor is doped by adding small, carefully controlled amounts of impurities. Two different types of semiconductors are produced by doping.

n-Type and *p*-Type Semiconductors

Suppose we dope a crystal of silicon with a trace of arsenic. Arsenic atoms have five valence electrons, and silicon has four. To be accommodated into the silicon structure, each As atom must give up one electron. As shown in the left part of Figure 24.12, the energy level of the As atoms, called *donor* atoms, lies quite close to the conduction band. Thermal energy alone is enough to cause the "extra" valence electrons to be lost to the conduction band; the As atoms become immobile positive ions, As^+. Electrical conductivity in this type of semiconductor involves primarily the movement of electrons in the conduction band, with the majority of the electrons coming from the donor atoms. A semiconductor with these characteristics is called an ***n*-type semiconductor**; the *n* stands for the negative charge carried by an electron.

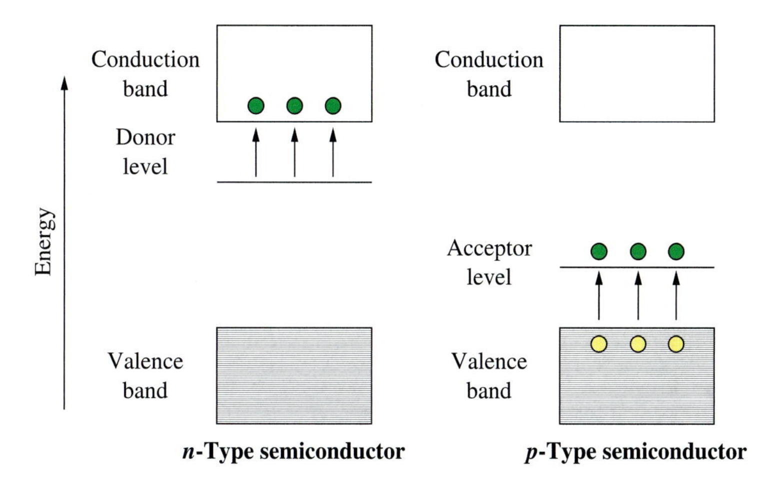

Figure 24.12
***p*- and *n*-type semiconductors**
In a semiconductor with donor atoms, the donor level lies just beneath the conduction band. Electrons (•) are easily promoted into the conduction band. The semiconductor is of the *n*-type. In a semiconductor with acceptor atoms, the acceptor level lies just above the valence band. Electrons (•) are easily promoted to the acceptor level, leaving positive holes (•) in the valence band. The semiconductor is of the *p*-type.

Now consider a silicon semiconductor doped with aluminum. Because it has only three valence electrons, an Al atom can form regular two-electron bonds with three neighboring Si atoms but only a one-electron bond with the fourth Si atom. Figure 24.12 shows that the energy level of the aluminum atoms, called *acceptor* atoms, is only slightly above the valence band. Electrons are easily promoted from the valence band into the acceptor level, where they are gained by Al atoms to form immobile negative ions, Al^-. Meanwhile, positive holes have been created in the valence band. Because electrical conductivity in this type of semiconductor involves primarily the migration of positive holes, it is called a ***p*-type semiconductor**. The *p* stands for the positive charge of a hole.

A Semiconductor Device: The Photovoltaic Cell

Semiconductor devices have revolutionized the field of electronics. The list of modern technological devices that use semiconductors is almost endless—electronic calculators, computers, radios, television sets, stereo equipment, and cellular telephones, to name a few.

Zone Refining: Obtaining Pure Semiconductor Materials

The characteristics of a semiconductor are critically dependent on the concentration of dopant atoms present. For example, a *p*-type semiconductor may require only one boron atom for every several million silicon atoms. Furthermore, any impurity atoms in the silicon itself must be kept at levels far below one part per million. A semiconductor manufacturing process may require levels as low as 10 impurity atoms per billion. There are not many purification methods that yield such ultrapure materials. Zone refining is one that does.

In **zone refining**, as illustrated in Figure 24.13, heating coils move along the length of a cylindrical rod of the material to be purified. Sections of the rod are alternately melted and refrozen. Impurities are much more soluble in the molten phase than in the solid, and they concentrate in the liquid zone. Ultimately, almost all the impurity atoms are swept into the liquid zone at the end of the rod, which is cut off and discarded.

Figure 24.13
Zone refining

Solar cells provide electricity to an outdoor telephone.

The semiconductor device pictured in Figure 24.14 converts light (solar energy) to electricity. It is a **photovoltaic cell** or a *solar cell*. The cell consists of a thin layer of *p*-type semiconductor, such as Si doped with Al, in contact with an *n*-type semiconductor, such as Si doped with P. When the cell is in the dark, charge does not flow across the *p–n* junction. For charge to flow, positive holes crossing from the *p* side would have to move away from Al^- ions, and electrons crossing from the *n* side would have to move away from P^+ ions. Separating oppositely charged particles requires energy and is therefore not favored.

Some of the electrons in Si — Si bonds in the *p*-type semiconductor, however, absorb energy from light and are promoted to the conduction band. This leaves behind positive holes in the valence band (recall Figure 24.11). Unlike positive holes, conduction electrons can freely cross the *p–n* junction and leave the cell as an electric current. Returning electrons from the electric circuit simply neutralize the positive holes previously formed. Electricity is generated continuously as long as light strikes the solar cell.

The *p*-type semiconductor in the solar cell must be very thin—about 1×10^{-4} cm. This is to reduce the tendency for conduction electrons produced by sunlight to be captured by positive holes and immobilized in covalent bonds.

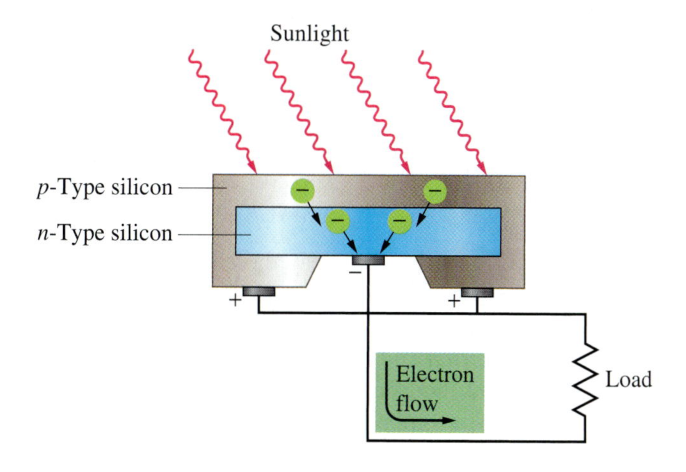

Figure 24.14
A photovoltaic (solar) cell

Polymers

Polymers are the "super molecules" of chemistry. Chemists also call them **macromolecules**, a term derived from the Greek *makros*, which means large or long. Macromolecules are made from smaller molecules, much as a brick wall is constructed from individual bricks. The small building-block molecules are called **monomers**. Macromolecular materials are called **polymers**.

Synthetic polymers are a mainstay of modern life, but nature also makes polymers; they are found in all living matter. In the remaining sections of this chapter, we will explore the wide range of types and uses of polymers.

24.9 Natural Polymers

In Chapter 23, we discussed three types of natural polymers: polysaccharides (starch and cellulose), proteins, and nucleic acids. The first attempts to manufacture polymers involved chemical modifications of natural polymers to improve their properties.

Consider cellulose, with the structure shown in Figure 24.15. It is a polymer with β linkages between glucose monomers (page 978). (The corresponding polymer with α linkages is starch.) The molecular masses of cellulose macromolecules range from about 500,000 to 2,400,000 u. Cellulose can be modified by reactions involving its many hydroxyl (—OH) groups.

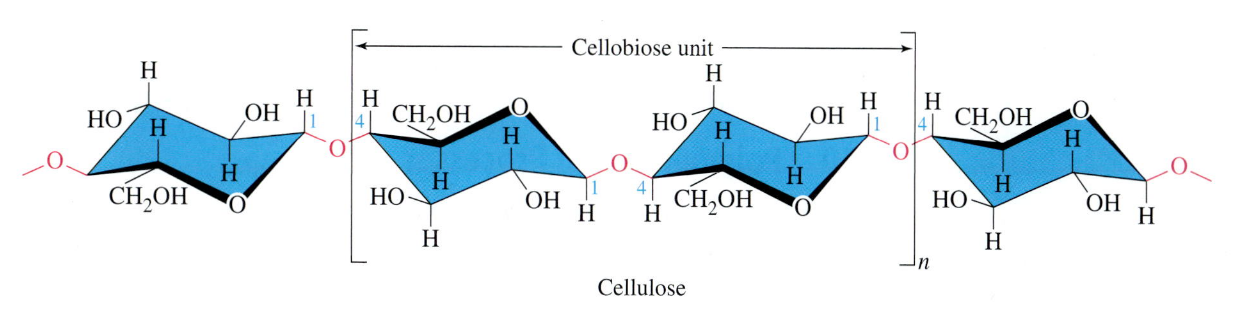

Figure 24.15 The molecular structure of cellulose
Cellulose is a polymer of the monomer glucose. To bring out the β linkages between monomer units (page 978), the disaccharide cellobiose is chosen as the repeating unit in the structure of cellulose shown here.

APPLICATION NOTE

When dissolved in camphor and ethanol, cellulose nitrate forms *celluloid*. Celluloid was originally developed as a substitute for ivory in billiard balls. However, the balls of celluloid sometimes exploded on impact. Celluloid was also the material used in the film of early movies, but the dangerous flammability of celluloid led to its replacement by safer materials. Celluloid survives today only as Ping-Pong® balls.

Recall (page 70) that alcohols and acids react to form *esters*. Cellulose, with its many alcohol functional groups, readily forms a variety of cellulose esters. With nitric acid, the ester formed is cellulose nitrate. In the representation below, —OH represents one of the many hydroxyl groups in the cellulose molecule.

$$\underset{\text{Cellulose}}{\text{cell—OH}} + \underset{\text{Nitric acid}}{HONO_2} \longrightarrow \underset{\text{Cellulose nitrate}}{\text{cell—ONO}_2} + H_2O$$

The ONO_2 group is properly called nitrate, but sometimes the first O is ignored, leaving just NO_2 and leading to an alternative name for cellulose nitrate, *nitrocellulose*. When the cellulose nitrate is made from cotton (nearly pure cellulose), it is called *guncotton* and is used in gunpowder and other explosives.

Several other modifications of cellulose are of economic importance. We will consider two. When cellulose is treated with sodium hydroxide and carbon disulfide (CS_2), a water-soluble intermediate called cellulose xanthate is formed.

$$\underset{\text{Cellulose}}{\text{cell—OH}} + Na^+OH^- + CS_2 \longrightarrow \underset{\text{Cellulose xanthate}}{\text{cell—O—}\underset{\underset{\displaystyle S}{\|}}{C}\text{—}S^-Na^+} + H_2O$$

Viscose can be regenerated by mixing it with crystals of $Na_2SO_4 \cdot 10H_2O$ and heating. The crystals melt to form $Na_2SO_4(aq)$, producing holes in the regenerated cellulose. The product is cellulose sponge.

A viscous aqueous solution of cellulose xanthate is called *viscose*. When viscose is forced through fine holes in a spinneret into dilute $H_2SO_4(aq)$, the cellulose is regenerated as fine, continuous, cylindrical threads called *rayon*.

$$\underset{\text{Cellulose xanthate}}{\text{cell—O—}\underset{\underset{\displaystyle S}{\|}}{C}\text{—}S^-} + H_3O^+ \longrightarrow \underset{\text{Regenerated cellulose}}{\text{cell—OH}} + CS_2 + H_2O$$

If viscose is forced through a narrow slit, a thin transparent film of the regenerated cellulose is obtained, a product called *cellophane*. Cellophane was once widely used as a wrapping material for consumer goods. Now, it is used as the dialyzing membrane in artificial kidneys.

Cellulose reacts with acetic anhydride, a substance derived from acetic acid, to form the ester *cellulose acetate* (also called *rayon acetate* or simply *acetate*).

$$\underset{\text{Cellulose}}{\text{cell—OH}} + \underset{\text{Acetic anhydride}}{CH_3\underset{\underset{\displaystyle O}{\|}}{C}\text{—O—}\underset{\underset{\displaystyle O}{\|}}{C}CH_3} \longrightarrow \underset{\text{Cellulose acetate}}{\text{cell—O—}\underset{\underset{\displaystyle O}{\|}}{C}\text{—}CH_3} + \underset{\text{Acetic acid}}{CH_3COOH}$$

The cellulose acetate is dissolved in acetone and forced through the tiny holes in a spinneret. Warm air causes the solvent to evaporate, leaving fine, lustrous threads of rayon acetate. The difference in appearance of the natural cellulose in cotton fabrics and this semisynthetic modification is striking (Figure 24.16).

▲ **Figure 24.16**
The wrinkled garment is made of unmodified cellulose. The blue cellulose acetate garment is sleek and silky.

24.10 Polymerization Processes

In general, whether in a laboratory or in a living system, polymers are made by hooking together many smaller molecules. But the result is not just like hooking box cars together in a train. The polymer is as different from the monomers as long strips of spaghetti are from the particles of flour that make up the spaghetti. For example, polyethylene, the familiar solid, waxy "plastic" of plastic bags, is a polymer prepared from the gaseous monomer, ethylene. There are two general types of reactions for producing polymers from monomers—*addition polymerization* and *condensation polymerization*.

Addition Polymerization

The key feature of **addition polymerization** is that monomers *add* to one another in such a way that the polymeric product contains all the atoms of the starting monomers. We can represent the polymerization of ethylene to polyethylene as follows.

$$n\ CH_2{=}CH_2 \longrightarrow {+}CH_2CH_2{+}_n$$

Recall that in Section 9.12 we gave a general overview of the formation of the addition *polymer* polyethylene from the *monomer* ethylene, and in Section 23.2 we discussed other addition reactions. Now, let's briefly consider some of the steps in the mechanism of addition polymerization.

Initiation The reaction is started by an *initiator*, usually a free radical, R• (page 385). The radical R• bonds to one of the C atoms of ethylene when its unpaired electron pairs with one of the electrons from the C=C double bond. This converts the double bond to a single bond and leaves one unpaired electron. The product of this reaction is still a free radical.

$$R\cdot + H_2C{=}CH_2 \longrightarrow R{-}CH_2{-}CH_2\cdot$$

Propagation The radical formed in the initiation step joins with another ethylene molecule to form a larger radical.

$$R{-}CH_2{-}CH_2\cdot + H_2C{=}CH_2 \longrightarrow R{-}CH_2{-}CH_2{-}CH_2{-}CH_2\cdot$$

The larger radical forms a still larger one by reacting with another ethylene molecule; and so on, through hundreds of steps.

Termination The propagation finally ends when a molecule is produced that no longer has an unpaired electron. One possible termination step involves the combination of two radicals through an electron pair bond.

$$R{+}CH_2)_nCH_2\cdot + \cdot CH_2(CH_2{+}_{\overline{m}}R' \longrightarrow R{+}CH_2)_nCH_2CH_2(CH_2{+}_{\overline{m}}R'$$

As a result of the polymerization, a low-molecular-mass gaseous *alkene* is converted to a high-molecular-mass solid *alkane*.

Figure 24.17 shows two molecular models of a small portion of a polyethylene molecule. Real polyethylene molecules have huge numbers of carbon atoms—from a few hundred to several thousand.

Alkenes commonly serve as the monomers in addition polymerization. By substituting various groups for one or more hydrogen atoms of the simple ethylene molecule, chemists can produce a fantastic array of synthetic polymers. (See, again, the listing in Table 9.2.)

Condensation Polymerization

In **condensation polymerization**, a small portion of the monomer molecule is *not* incorporated in the final polymer. As an example, we can consider the formation of a type of *nylon* that has 6-aminohexanoic acid as its monomer. The polymerization involves the reaction of the OH portion of the carboxyl group of one monomer with an H atom of the $-NH_2$ (amine) group of another. The monomers are held

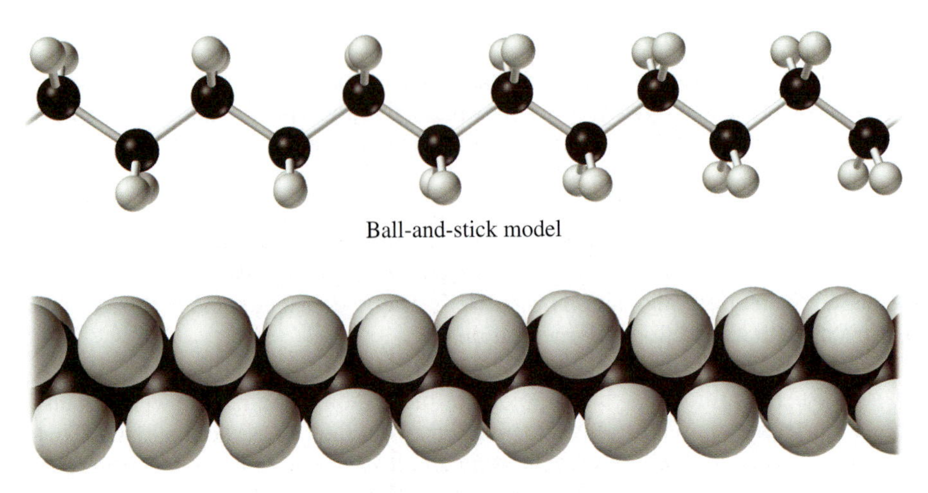

▲ **Figure 24.17 Molecular models of a segment of a polyethylene molecule**

together by the newly formed bond—an amide bond—and a molecule of water is also produced.

$$n\text{HO}-\overset{\overset{\text{O}}{\|}}{\text{C}}-\text{CH}_2\text{CH}_2\text{CH}_2\text{CH}_2\text{CH}_2-\overset{\overset{\text{H}}{|}}{\text{N}}-\text{H} + n\text{HO}-\overset{\overset{\text{O}}{\|}}{\text{C}}-\text{CH}_2\text{CH}_2\text{CH}_2\text{CH}_2\text{CH}_2-\overset{\overset{\text{H}}{|}}{\text{N}}-\text{H}$$

$$\longrightarrow \left[\overset{\overset{\text{O}}{\|}}{\text{C}}-\text{CH}_2\text{CH}_2\text{CH}_2\text{CH}_2\text{CH}_2-\overset{\overset{\text{H}}{|}}{\text{N}}-\overset{\overset{\text{O}}{\|}}{\text{C}}-\text{CH}_2\text{CH}_2\text{CH}_2\text{CH}_2\text{CH}_2-\overset{\overset{\text{H}}{|}}{\text{N}}\right]_n + 2n\text{H}_2\text{O}$$

Amide bond

Following is a list of the key features of condensation polymerization.

- Each monomer molecule contains at least *two* functional groups.
- The monomers are linked through the functional groups.
- Small molecules are formed as by-products as the monomers are linked.

Several of the commercially important polymers used in fibers, fabrics, and plastics are formed by condensation polymerization, as we will see shortly.

EXAMPLE 24.2

Write a condensed structural formula for polypropylene, made by the polymerization of propylene ($CH_2{=}CHCH_3$).

SOLUTION

First we need to identify the type of polymerization involved. The double bond indicates that the monomer units can join together with no loss of atoms, and we therefore expect polypropylene to be an addition polymer. We also must recognize that linkage occurs through the carbon atoms in the monomers that are joined by the double bond. The methyl groups are attached along the polymer chain as substituent groups. Thus, of the following formulas, the top one is incorrect, and the bottom one is correct.

$$n\,CH_2{=}CH{-}CH_3 \longrightarrow +CH_2{-}CH{-}CH_3+_n \quad (incorrect)$$

$$n\;\underset{H}{\overset{H}{}}C{=}C\underset{CH_3}{\overset{H}{}} \longrightarrow \left[\begin{array}{cc} H & H \\ | & | \\ -C- & -C- \\ | & | \\ H & CH_3 \end{array}\right]_n \quad (correct)$$

Polypropylene

EXERCISE 24.2

Describe the error(s) in the condensed structural formula, $+CH_2{-}CH{-}CH_3+_n$, for polypropylene from Example 24.2.

EXAMPLE 24.3

Write the condensed structural formula for the polymer formed from dimethylsilanol, $(CH_3)_2Si(OH)_2$.

SOLUTION

We can start by writing the structural formula of the monomer.

$$\begin{array}{c} CH_3 \\ | \\ HO{-}Si{-}OH \\ | \\ CH_3 \end{array}$$

There are no double bonds in this molecule, so we do not expect addition polymerization. Rather, we expect a condensation reaction, in which an —OH group of one molecule and an H atom of another combine to form a molecule of water. Moreover, because there are two —OH groups per molecule, each monomer can form bonds with the neighbors on both sides. This is a key requirement for polymerization. We can represent the reaction as follows.

$$\begin{array}{c} CH_3 \\ | \\ HO{-}Si{-}OH \\ | \\ CH_3 \end{array} + \begin{array}{c} CH_3 \\ | \\ HO{-}Si{-}OH \\ | \\ CH_3 \end{array} + \begin{array}{c} CH_3 \\ | \\ HO{-}Si{-}OH \\ | \\ CH_3 \end{array} + \ldots \longrightarrow \left[\begin{array}{c} CH_3 \\ | \\ -Si{-}O- \\ | \\ CH_3 \end{array}\right]_n + n\,H_2O$$

EXERCISE 24.3

Write the condensed structural formula for the polymer formed from 3-hydroxypropanoic acid ($HOCH_2CH_2COOH$).

The polymer in Example 24.3 is a *poly*siloxane, but is more commonly known as a *silicone*. Silicones contain carbon atoms, but they have no carbon-to-carbon bonds. The —Si—O—Si— skeleton of a silicone gives it greater high-temperature stability and resistance to oxidation than observed in carbon-based polymers.

One of the many uses of silicone polymers is in waterproof caulking compounds for use in the installation of window frames and plumbing fixtures.

24.11 Physical Properties of Polymers

Polymers are classified not only according to their method of synthesis, but also according to their response to heating. To the general public, synthetic polymers are often known as "plastics." However, *plastic* has a more restricted meaning to chemists. A **thermoplastic polymer** is one that can be softened by heating and then formed into desired shapes by applying pressure. In contrast, **thermosetting polymers** become permanently hard at elevated temperatures and pressures. After setting, they cannot be softened and remolded. We can relate properties such as these to specific structural features of macromolecules.

▲ **Figure 24.18**
The effect of heat on polyethylene bottles
These bottles were heated in the same oven for the same length of time. The one that melted has branched-chain polyethylene molecules; the other is composed of straight-chain molecules.

We can illustrate how variations in structure affect the properties of polymers by examining two basic kinds of polyethylene plastics. Of the two, *high-density polyethylene (HDPE)* has a higher density, greater rigidity, greater strength, and a higher melting point. It is used for such things as threaded bottle caps, radio and television cabinets, toys, and large-diameter pipes. In contrast, *low-density polyethylene (LDPE)* is a waxy, semirigid, translucent material with a low melting point. LDPE is used in insulation for electric wiring, plastic bags, refrigerator dishes, squeeze bottles, and many other common household articles. An essential difference between the two types of polyethylene is shown in Figure 24.18.

What aspects of their structures can account for the differences in properties of HDPE and LDPE? High-density polyethylene consists primarily of linear molecules, that is, of long unbranched chains. The chains can run alongside one another in close contact over relatively great distances. This permits strong intermolecular forces of attraction between the chains. The forces are *dispersion forces* (recall Section 11.5). The overall effect is to produce an ordered (crystalline) structure that imparts rigidity, strength, and a higher melting temperature to the polymeric material.

Low-density polyethylenes have branched chains that prevent the macromolecules from assuming a crystalline structure.* The branches get in the way when two chains try to come into close contact (much as logs with short protruding branch stems are hard to arrange in a neat pile). This decreases the dispersion forces, weakens the attractions between chains, and produces a lower melting, more flexible material than HDPE. These two different polymer chain arrangements are suggested in Figure 24.19.

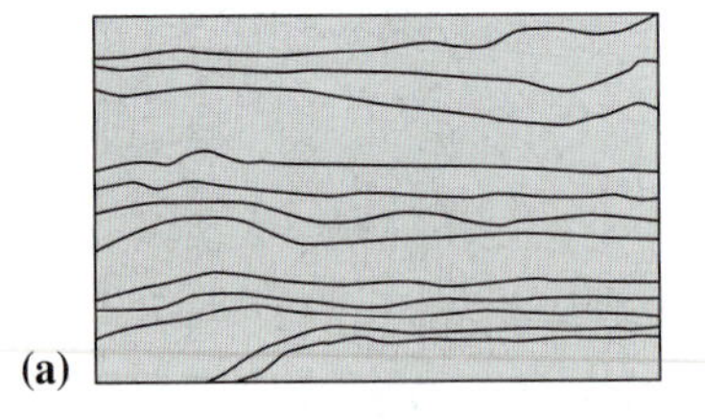
(a)

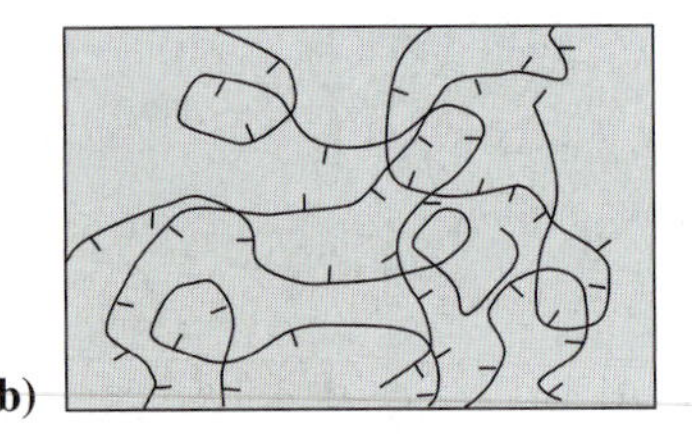
(b)

▶ **Figure 24.19**
Organization of polymer molecules
(a) Polymer chains in high-density polyethylene (HDPE). (b) Polymer chains in low-density polyethylene (LDPE).

What if we want a really tough material, something quite rigid and strong, something that won't melt or soften on heating? The answer could well be *Bakelite®*, the oldest synthetic thermosetting polymer. Bakelite® is a condensation polymer prepared by combining *two different* monomers, phenol and formaldehyde. Bakelite® is more generally referred to as a phenol-formaldehyde resin. The polymerization involves substituted phenols that are formed as intermediates in the reaction of phenol and formaldehyde.

*One form of branching occurs if the free-radical end of a growing chain coils and attacks a carbon atom several atoms back on the chain. A short chain branch is formed and new growth continues from the point of attack.

Phenol Formaldehyde

The substituted phenols then interact by splitting out water molecules. Water is driven off by heat liberated while the polymer sets. The final product is a huge, complex, three-dimensional network (see Figure 24.20). The extensive cross-linking of polymer chains results in rigidity. The final polymer has great strength without having great mass, a useful combination of properties. About 50% of the United States production of phenol-formaldehyde resins is used as a binder in plywood. Other uses include varnishes for electric coils, molding compounds for the manufacture of electric plugs and switches, and automobile parts such as steering wheels.

Figure 24.20
A small segment of Bakelite®, a phenol-formaldehyde resin

Bakelite was used in the 1920s and 1930s to make a variety of products, such as camera boxes, pool balls, and telephones.

24.12 Elastomers

Hardness and rigidity are not the only properties we seek in plastics. Frequently we want flexibility and, more particularly, elasticity. *Flexibility* is the ability of a material to yield to forces—to bend or "give"—but without breaking. *Elasticity* is the

Conducting Polymers

In the polymerization of ethylene, an alkene is converted to an alkane. What kind of product is formed when an *alkyne* is polymerized? As we did for ethylene on page 397, let's consider a few steps in the polymerization of acetylene.

$$\text{R}\cdot + \text{H—C}\equiv\text{C—H} \longrightarrow \text{R—C(H)=C(H)}\cdot$$

$$\text{R—C(H)=C(H)}\cdot + \text{H—C}\equiv\text{C—H} \longrightarrow \text{R—C(H)=C(H)—C(H)=C(H)}\cdot$$

$$\text{R—C(H)=C(H)—C(H)=C(H)}\cdot + \text{H—C}\equiv\text{C—H} \longrightarrow \text{R—C(H)=C(H)—C(H)=C(H)—C(H)=C(H)}\cdot \quad \text{... and so on}$$

We can represent a section of a polyacetylene molecule with a condensed structural formula.

$$\cdots\text{—CH=CH—CH=CH—CH=CH—CH=CH—}\cdots$$

A ball-and-stick model is presented in Figure 24.21. Both clearly show that the bonding in the backbone of the molecule is a *conjugated* system (Section 23.13). Single and double carbon-to-carbon bonds alternate throughout the molecule. The π electrons in a conjugated system are *delocalized*. In effect, an electron can move the entire length of a polyacetylene molecule, and this means that the molecule is an electrical conductor.

To make polyacetylene an effective conductor, however, there must be a way for electrons to get from one molecule to another. This is accomplished by doping the polyacetylene with an additive such as iodine. Explanations of conducting polymers are similar to the descriptions given for bonding in metals and semiconductors in Section 24.7. Two important differences, however, are that (1) electron bands are formed from molecular orbitals of polymer molecules rather than the atomic orbitals of the atoms in the polymer chain and (2) the dopants reside between polymer chains rather than substituting for atoms in the chains.

Conducting polymers were discovered accidentally in the 1970s. They are now beginning to appear in commercial products such as polymer batteries.

Batteries made of foil-like sheets of conducting polymers are light and flexible. They save considerable weight compared to lead-acid batteries and can be crammed into almost any available space.

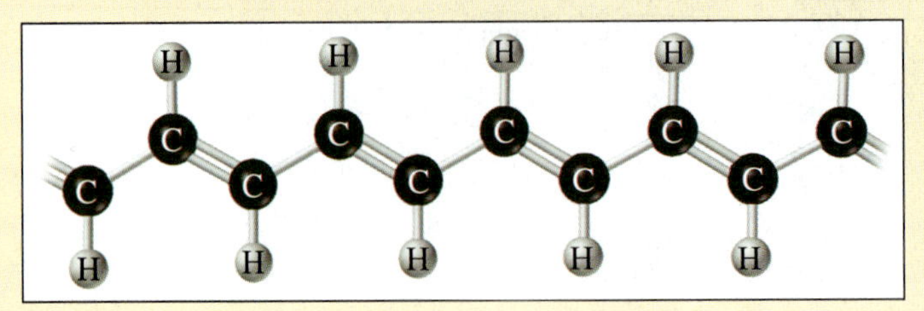

Figure 24.21 A ball-and-stick model of polyacetylene

ability of a material to regain its former shape after a distorting force is removed. **Elastomers** are flexible, elastic materials. The natural polymer *rubber* is the prototype for this kind of material.

Natural rubber is a polymer of the simple hydrocarbon monomer *isoprene* (2-methyl-1,3-butadiene).

```
CH2    CH2
  \\   //
   C—C
  /     \
CH3      H
```

The polymer has the following structure. The repeating unit is shown in red.

```
···CH2     CH2—CH2     CH2—CH2     CH2—CH2     CH2···
     \    /      \    /      \    /      \    /
      C=C         C=C         C=C         C=C
     /   \       /   \       /   \       /   \
  CH3     H   CH3     H   CH3     H   CH3     H
```

Notice that the $—CH_2—CH_2—$ units are all cis, that is, on the same side of the C=C double bonds. The polymer chains in natural rubber are coiled, twisted, and intertwined with one another. The stretching of rubber involves straightening out the coiled macromolecules.

Natural rubber is soft and tacky when hot. It can be made harder in a reaction with sulfur, called *vulcanization*. In this process, sulfur atoms form cross-links between hydrocarbon chains, as shown in Figure 24.22. As with Bakelite®, this cross-linked, three-dimensional structure makes vulcanized rubber a harder, stronger substance than natural rubber. Thus, vulcanized rubber makes excellent automobile tires, whereas natural rubber is totally unsuited for this purpose.

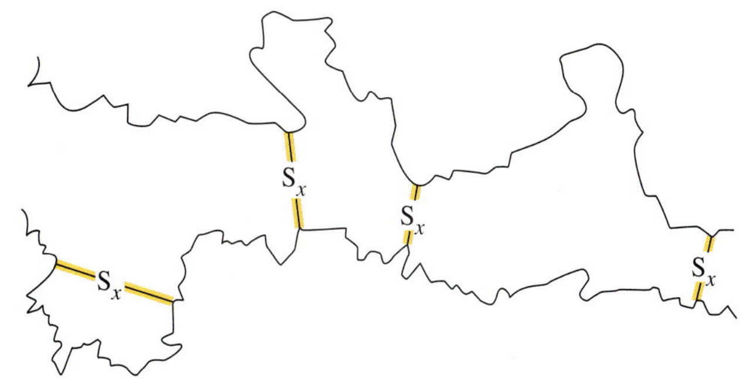

Figure 24.22 Vulcanized rubber
The subscript x indicates a variable number of S atoms, usually 1 to 4.

Table 24.2 lists several kinds of rubber and their degree of cross-linking. The smaller the average number of intervening monomer units on each chain between successive cross-links, the greater the degree of cross-linking is. Notice that the harder items in Table 24.2 have more extensive cross-linking. Just the right degree of cross-linking can even improve elasticity over that of natural rubber. The individual chains are still relatively free to uncoil and stretch, but when a stretched piece of the rubber is released, the cross-links serve to pull the chains back to their original prestretched arrangement (Figure 24.23). Rubber bands owe their "snap" to this sort of molecular structure.

TABLE 24.3 Extent of Cross-linking in Rubber Products

Product	Monomer Units Between Cross-links
Surgical gloves	100–150
Kitchen gloves	50–80
Artificial heart membrane	30–40
Bicycle inner tube	20–30
Bicycle tire	10–20

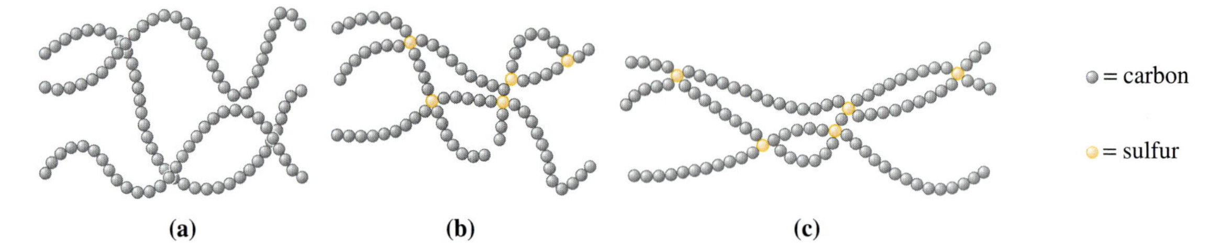

▲ **Figure 24.23 Vulcanization of rubber produces cross-links between molecular chains**
(a) In unvulcanized rubber, the chains slip past one another when the rubber is stretched. (b) Vulcanization involves the addition of sulfur cross-linkages between the chains. (c) When the vulcanized rubber is stretched, the sulfur cross-linkages prevent the chains from slipping past one another. Vulcanized rubber is stronger than unvulcanized rubber.

Several kinds of synthetic rubber were developed during and after World War II. Some of them bear a striking molecular resemblance to nature's own elastomer. For example, polychloroprene (neoprene) is made from a monomer, chloroprene (2-chloro-1,3-butadiene), that is similar to isoprene but has a chlorine atom in place of a methyl group.

```
CH2     CH2
  \\    //
   C — C
  /      \
Cl        H
```

Another kind of synthetic rubber illustrates the principle of *copolymerization*. In this process, a mixture of two different monomers forms a product in which the chain contains both monomers as building blocks. The product is called a *copolymer*. Styrene-butadiene rubber (SBR) is a copolymer of styrene (25%) and butadiene (75%). A segment of an SBR macromolecule might look something like this.

```
···CH2     CH2—CH2     CH2—CH2      CH2       CH2···
     \     /     \     /     \     /   \     /
      C=C         C=C         C         C=C
     /   \       /   \   (C6H5) \      /   \
   CH3    H    CH3    H          H   H      H

  Butadiene unit  Butadiene unit  Styrene unit  Butadiene unit
```

This synthetic rubber is more resistant to oxidation and abrasion than natural rubber, but it has less satisfactory mechanical properties.

APPLICATION NOTE
Chemists have even learned to make polyisoprene, the structural equivalent of natural rubber, except that it's not harvested in rubber tree plantations.

24.13 Fibers and Fabrics

A *fiber* is a natural or synthetic material obtained in long, threadlike structures that can be woven into fabrics. One factor affecting the quality of a fiber is its *tensile strength*, a measure of the extent to which a material can be subjected to a stretching force without breaking.

Cotton, wool, and silk are natural fibers of great tensile strength that have long been spun and woven into cloth. Now, synthetic fibers have been developed that match the natural fibers in tensile strength and outdo them in their resistance to stretching and shrinking—and to attack by moths. Synthetic polymers have revolutionized the clothing industry.

Polyacrylonitrile (Acrilan®, Creslan®, and the like) is an addition polymer.

$$n\ \mathrm{CH_2{=}CH(CN)} \longrightarrow \left[\!\!\begin{array}{c}\mathrm{H\quad H}\\ |\quad\ |\\ \mathrm{-C-C-}\\ |\quad\ |\\ \mathrm{H\quad CN}\end{array}\!\!\right]_n$$

Polyacrylonitrile

Polyesters are condensation polymers. Dacron® polyester is made from the condensation of ethylene glycol with terephthalic acid.

$$n\,\mathrm{HO{-}CH_2{-}CH_2{-}OH} + n\,\mathrm{HO{-}\overset{O}{\overset{\|}{C}}{-}C_6H_4{-}\overset{O}{\overset{\|}{C}}{-}OH} \longrightarrow \left[\mathrm{CH_2CH_2O{-}\overset{O}{\overset{\|}{C}}{-}C_6H_4{-}\overset{O}{\overset{\|}{C}}}\right]_n + 2n\,\mathrm{H_2O}$$

Ethylene glycol — Terephthalic acid — Dacron

This polymer can be extruded as a film. Coated with a magnetic material, the film is the "tape" used in tape recorders and videotape machines. Perhaps the most familiar use of this polyester is in plastic bottles for soft drinks.

Polyamides are also condensation polymers. The protein fibers in silk and wool are polyamides. The most common synthetic polyamide is nylon-66. It is made by the condensation of 1,6-hexanediamine and adipic acid.

$$n\,\mathrm{H{-}\overset{H}{\overset{|}{N}}{-}CH_2CH_2CH_2CH_2CH_2CH_2{-}\overset{H}{\overset{|}{N}}{-}H} + n\,\mathrm{HO{-}\overset{O}{\overset{\|}{C}}{-}CH_2CH_2CH_2CH_2{-}\overset{O}{\overset{\|}{C}}{-}OH} \longrightarrow$$

1,6-Hexanediamine — Adipic acid

$$\left[\mathrm{\overset{H}{\overset{|}{N}}{-}CH_2CH_2CH_2CH_2CH_2CH_2{-}\overset{H}{\overset{|}{N}}{-}\overset{O}{\overset{\|}{C}}{-}CH_2CH_2CH_2CH_2{-}\overset{O}{\overset{\|}{C}}}\right]_n + 2n\,\mathrm{H_2O}$$

Nylon-66

Silk, wool, Dacron®, and nylon-66, like high-density polyethylene, owe their strength to the ordered, relatively rigid arrangement of their long molecules. However, unlike nonpolar polyethylene, the polyesters and polyamides have polar functional groups. Interactions among these groups give the polymers their unique tensile strengths.

Chemists can now design materials with seemingly impossible combinations of properties in a single substance. For example, spandex fibers, which are used for stretch fabrics (Lycra®) in ski pants, exercise wear, and bathing suits, combine the elasticity of rubber and the tensile strength of a fiber. How can the coiled, flaccid molecular chains required for elasticity and the fairly rigid, highly ordered chains needed for tensile strength be combined into a single polymer? It can be done by grafting two molecular structures into a polymer chain. Thus, a single macromolecule can have blocks of elastomer components and blocks of fiber components. The fabric made from these macromolecules exhibits flexibility *and* rigidity.

Velcro, the familiar material used in fasteners for shoes, clothes, watch bands, and the like, has hooks on one surface and loops on the other. The hooks and loops are made of nylon. This image was made by a scanning electron microscope.

24.14 Biomedical Polymers

One of the most interesting uses of polymers has been in replacements for diseased, worn-out, or missing parts of the human body. Artificial ball-and-socket hip joints made of stainless steel (the ball) and plastic (the socket) are now being installed at a rate of 130,000 per year. People crippled by arthritis or injuries are not only free of pain but are given much more freedom of movement. Patients with heart and circulatory problems can enter a hospital for a "valve job" or the replacement of worn-out or damaged parts. Pyrolytic carbon heart valves derived from a polymer are widely used. Knitted Dacron® tubes can replace arteries blocked or damaged by atherosclerosis.

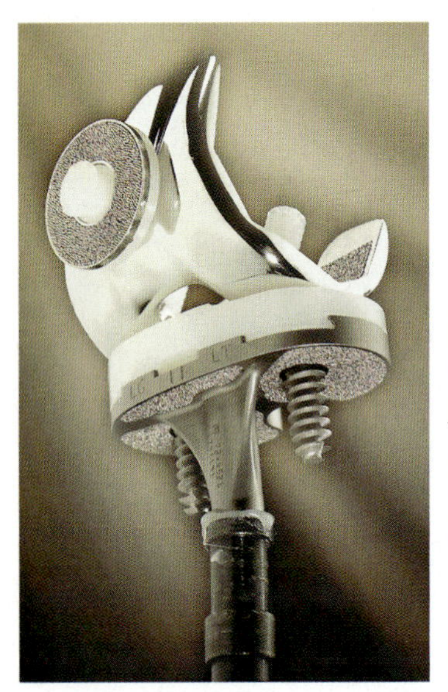

Synthetic polymers are an important part of this hip joint replacement.

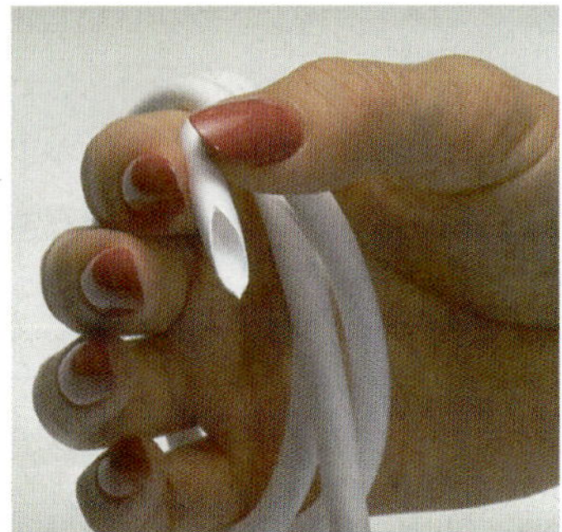

A synthetic artery.

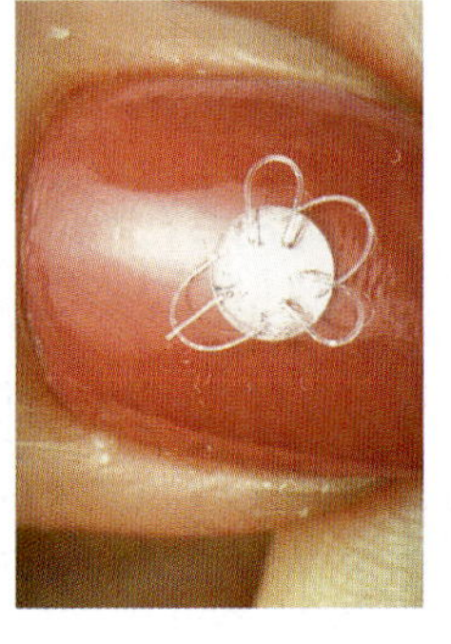

A plastic artificial eye lens.

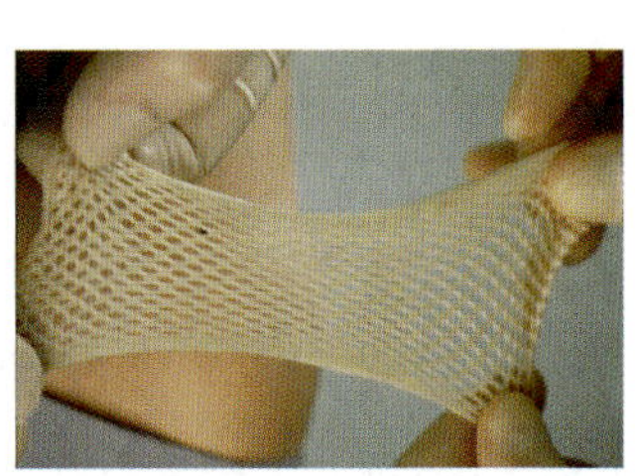

Synthetic skin.

Blood begins to clot when it comes in contact with most foreign substances. To prevent this clotting, synthetic polymers generally must be chemically treated and coated with an anticoagulant before implantation. Another approach that circumvents the problem is to use naturally occurring substances to construct biomedical polymers. For example, polymers of glycolic acid and lactic acids have been used in synthetic films for covering burn wounds.

$$\underset{\text{Glycolic acid}}{HOCH_2COOH} \qquad \underset{\text{Lactic acid}}{CH_3\underset{\substack{|\\ OH}}{C}HCOOH}$$

Ordinarily, to prevent infection and excessive loss of fluids, burns have to be covered with human skin from donors or with specially treated pigskin. These covers must be changed frequently because the body tends to reject foreign tissues. In contrast, polymer films made from natural substances are absorbed and metabolized rather than rejected.

The development of biomedical polymers has barely begun. In the future lies the prospect of the replacement of entire organs of the body. Already, artificial hearts made of synthetic elastomers are used to keep people alive while awaiting a heart transplant.

Summary

Some of the general operations of extractive metallurgy are the mining, concentration, and roasting of an ore; reduction of a metal compound to the free metal, either chemically or electrolytically; and refining of the metal. Removal of impurities as a slag may also be required.

Roasting is not required in the metallurgy of iron because the principal ores are oxides. Coke (carbon) is used to reduce the iron oxides. The slag is mainly calcium silicate. The impure iron (pig iron) is in most cases converted to iron-carbon alloys called steel.

Traditional extractive metallurgy, called pyrometallurgy, is based on high-temperature reactions involving solids. An alternative, hydrometallurgy, uses lower temperature reactions involving solutions. The cyanidation process for the extraction of gold and silver from their ores is an example.

In the free-electron model of metals, valence electrons of the atoms in a crystal are combined into an "electron gas" surrounding a network of positive ions. The model accounts for such properties as malleability and ductility, as well as electrical and thermal conductivity.

Band theory describes bonding both in metals and semiconductors in terms of bands of closely-spaced molecular orbitals. The presence of a band that is only partly filled with electrons is required for electrical conductivity. In some metals, the band obtained by combining the valence orbitals of the metal atoms is only partly occupied. Here, the valence band is also a conduction band. In other metals, an empty conduction band and a filled valence band overlap.

In an electrical insulator, the energy gap between the valence band and a conduction band is so great that very few electrons can make the transition. In semiconductors, the energy gap is much smaller, and a more significant number of electrons can jump the gap. The rate at which electrons jump the gap increases with temperature, explaining why the conductivity of a semiconductor increases with temperature. A semiconductor becomes a much better conductor when it is doped. In an *n*-type semiconductor, the energy level of electron donor atoms is close to the conduction band, and electric current is carried by electrons in the conduction band. In a *p*-type semiconductor, the energy level of electron acceptor atoms is close to the valence band, and electric current is carried by positive holes in the valence band.

Polymers, both natural and synthetic, are macromolecules that are made up of large numbers of smaller molecules called monomers. Two basic methods are used to produce polymers from monomers. In addition polymerization, monomer units add to one another and the final polymer contains all the atoms of the monomers. In condensation polymerization, small molecules such as H_2O are liberated when monomers join through certain functional groups. Each monomer must have two or more of the appropriate functional groups.

Polymers can be classified in several ways. One classification divides polymers into three basic groups: fibers, elastomers, and plastics. Another classification is according to the effect of temperature on the polymer properties: thermoplastic and thermosetting. Still other classifications are based on the degree of branching on polymer chains and on the long-range order or crystallinity found in the polymer chains.

Key Terms

addition polymerization (24.10)
alloy (24.1)
band (24.7)
cast iron (24.2)
condensation polymerization (24.10)
conduction band (24.7)
doping (24.8)
elastomer (24.12)
energy gap (E_g) (24.8)
flotation (24.1)
free-electron model (24.6)
hydrometallurgy (24.5)
macromolecule (page 1031)
monomer (page 1031)
n-type semiconductor (24.8)
ore (24.1)
photovoltaic cell (24.8)
pig iron (24.2)
polymer (page 1031)
positive hole (24.8)
p-type semiconductor (24.8)
pyrometallurgy (24.5)
refining (24.1)
semiconductor (24.8)
slag (24.1)
steel (24.2)
superconductor (page 1028)
thermoplastic polymer (24.11)
thermosetting polymer (24.11)
valence band (24.7)
zone refining (24.8)

Review Questions

1. Name some metals that can be found free or uncombined in nature. Where do you expect to find them in the activity series of the metals?
2. What is an ore? What are the principal compounds of metals found in ores?
3. What is the flotation process in metallurgy? What is its purpose?
4. In pyrometallurgy, what is the purpose of "roasting" an ore? Must all ores be roasted? Explain.
5. What is accomplished in the metallurgical operation called reduction? What is the reducing agent most commonly employed? Is it always necessary to use a reducing agent? Explain.
6. Cite two methods that can be used to purify a metal, and give an example of each.
7. What is an alloy? Can an alloy be a pure substance? Are all alloys homogeneous? Explain.
8. What functions are served by a blast furnace in the metallurgy of iron?
9. What is pig iron? What are its principal impurities? What are carbon steel and alloy steel? In what ways are these two types of steel similar, and how do they differ?
10. Cite several differences between pyrometallurgy and hydrometallurgy.
11. What is meant by the terms amalgamation and cyanidation? Name an element that is obtained by these processes.
12. Briefly explain why the ionic and covalent bond descriptions do not work for metals.
13. Briefly describe the meaning of the term "electron gas" when applied to electrons in a metal? Does this term apply to all the electrons or only some? Explain.
14. What is wrong (or incomplete) in the following description of electrical conductivity in the free-electron theo-

ry? "The random motion of the electrons in the 'electron gas' in a wire creates an electric current."

15. Briefly describe the band theory of metals.

16. Briefly describe the difference between a valence band and a conduction band in a metal. Can they be the same? Under what circumstances?

17. What is a doped semiconductor?

18. What is an *n*-type semiconductor? What is a *p*-type semiconductor?

19. What is a positive hole? How do positive holes move through a semiconductor?

20. Describe the meaning of the terms monomer, polymer, and macromolecule.

21. In general terms, describe the features that characterize the three broad classes of polymers: plastics, elastomers, and fibers. Can a single polymer be synthesized to fit two of these categories? Explain.

22. Describe the essential features of each of the two general polymerization processes: addition polymerization and condensation polymerization.

23. What are thermosetting and thermoplastic polymers? Give an example of each, and tell how they differ in structure and properties?

24. How do low-density and high-density polyethylene differ in structure and properties?

25. What is meant by the vulcanization of rubber? What purpose does it serve?

26. Describe the general features of the following polymers: **(a)** polyesters; **(b)** polyamides. Give a common example of each.

Problems

Metallurgy

27. An ore containing the highest percentage of a metal may not be the best industrial source of the metal. State some reasons why this is so.

28. Is it correct to say that hydrometallurgy is more environmentally "friendly" than pyrometallurgy? Explain.

29. In the metallurgy of zinc, why is it reasonable to expect to find cadmium as an impurity? Indicate one simple way in which cadmium can be removed from zinc.

30. In the metallurgy of copper, why is it reasonable to expect to find silver and gold as impurities? How can they be removed from the copper?

31. Roasting of a metal sulfide generally converts it to the metal oxide. Explain why this is not the case with cinnabar, HgS.

32. In the metallurgy of silver and gold, often an alloy of the two metals is obtained. The metals can be separated by treating the alloy with nitric acid in a process called *parting*. Explain how this method works.

33. Write the equation for a reaction in which silver metal is displaced from an aqueous solution of $[Ag(CN)_2]^-$.

34. When silver oxide is heated, it decomposes to silver metal and oxygen gas. Write an equation for this reaction.

35. Write chemical equations to describe the following hydrometallurgy of zinc: Zinc oxide is dissolved in $H_2SO_4(aq)$; powdered zinc is added to displace less active metals; and the solution is electrolyzed. What are the electrode half-reactions in the electrolysis? Show how $H_2SO_4(aq)$ is recycled.

36. Write chemical equations showing how pure zinc can be obtained from smithsonite ore, $ZnCO_3$.

37. In one method for recovering tin from scrap tin plate, the metal is treated with $Cl_2(g)$ to form $SnCl_4(l)$. The $SnCl_4(l)$ is hydrolyzed in water to form the hydrated oxide $SnO_2 \cdot xH_2O$. The hydrated tin(IV) oxide is dehydrated by heating, and the product is reduced to metallic tin with carbon. Write plausible equations to represent this method.

38. As an alternative to the method in Problem 37, tin plate is treated with $O_2(g)$ in a basic solution to produce hexahydroxostannate(IV) ion. Then the solution is acidified to precipitate $SnO_2 \cdot xH_2O$. The hydrated oxide is dehydrated by heating, and the product is reduced to metallic tin with carbon. Write plausible equations to represent this method.

39. A blast furnace produces 1.0×10^7 kg of pig iron per day. Assume the pig iron is 95% Fe by mass. How many kilograms of iron ore are consumed in this furnace per day? Assume that the ore is 82% by mass hematite, Fe_2O_3.

40. The blast furnace described in Problem 39 produces 0.5 kg of slag per kg of pig iron. What is the minimum daily requirement of limestone in this furnace? Assume that the limestone is 91% $CaCO_3$, that the slag is exclusively $CaSiO_3$, and that the limestone is the only source of calcium in the slag.

Band Theory

41. Calcium is a somewhat better electrical conductor than potassium, even though the 4*s* atomic orbitals are filled in Ca atoms and only half filled in K atoms. Explain how this can be.

42. Propose a plausible band structure for aluminum that is consistent with its electrical conductivity.

43. How does band theory account for the lustrous appearance of a clean metal surface? How does it account for the fact that a few metals are colored?

44. According to the band theory, can a bulk metal be transparent to visible light? Explain.

45. How many energy levels are there in the valence band of a 35.0-mg single crystal of sodium? How many electrons are in the band?

46. How many energy levels are there in the $3p$ conduction band of a 55.5-mg single crystal of magnesium? What is the total number of electrons in the $3s$ and $3p$ bands of this crystal?

Semiconductors

47. How does one distinguish between a metallic conductor and a semiconductor in band theory?

48. How does one distinguish between a semiconductor and an insulator in band theory?

49. Classify the following semiconductors as n-type or p-type: **(a)** Ge doped with As; **(b)** Si doped with B.

50. Classify the following semiconductors as n-type or p-type: **(a)** Ge doped with Al; **(b)** Si doped with P.

51. In silicon and germanium, conduction electrons and positive holes are present in equal numbers. Is this also true of n-type and p-type semiconductors based on silicon and germanium? Explain.

52. Would you expect silicon and phosphorus-doped silicon to have about the same electrical conductivity? Explain.

53. Some metals become superconductors at temperatures approaching 0 K. Would you expect the same behavior in semiconductors? Explain.

54. The electrical conductivity of a pure semiconductor material, such as silicon, is strongly dependent on temperature. Would you expect the temperature dependence of the conductivity of a boron-doped silicon to be about the same, greater, or less than that of silicon? Explain.

55. Certain compounds having the same average number of valence electrons as silicon or germanium are good semiconductors. Which of the following compounds meet this requirement: **(a)** CuS, **(b)** ZnSe, **(c)** PbO, **(d)** GaP?

56. As the percent ionic character of a semiconductor increases, so does the energy gap E_g. Which of the following would you expect to have the greatest energy gap: Ge, CuBr, or GaAs? Explain.

Polymers

57. What is the chemical makeup of each of the following materials, and how is each one made?

(a) cellulose nitrate
(b) rayon
(c) HDPE

58. What is the chemical makeup of each of the following materials, and how is each one made?

(a) cellophane
(b) rayon acetate
(c) LDPE

59. Why is rubber elastic? How does vulcanization improve the elasticity of rubber?

60. In which of these rubber products is a greater degree of cross-linking required, surgical gloves or automobile tires? Explain.

61. Write the condensed general formula for the polymer made from *cis*-1,2-dichloroethene: ClCH═CHCl.

62. Write the condensed general formula of neoprene, a trans polymer of chloroprene (2-chloro-1,3-butadiene) (page 1040).

63. Represent the structures of addition polymers made from the following monomers by showing at least four repeating units.

Acrylonitrile, $CH_2{=}CH{-}CN$
(a)

Vinyl acetate, $CH_2{=}CH{-}O{-}C({=}O){-}CH_3$
(b)

64. Represent the structures of addition polymers made from the following monomers by showing at least four repeating units.

Styrene, $CH_2{=}CH{-}C_6H_5$
(a)

1-Hexene, $CH_2{=}CHCH_2CH_2CH_2CH_3$
(b)

65. Represent the structures of the condensation polymers made from the following monomers by showing at least four repeating units.

Glycolic acid, $HOCH_2C({=}O){-}OH$
(a)

Terephthalic acid, $HO{-}C({=}O){-}C_6H_4{-}C({=}O){-}OH$,

and 1,4 phenylenediamine, $H_2N{-}C_6H_4{-}NH_2$
(b)

66. Represent the structures of the condensation polymers made from the following monomers by showing at least four repeating units.

(a) Lactic acid, $CH_3CH(OH)-C(=O)-OH$

(b) Adipic acid, $HO-C(=O)(CH_2)_4C(=O)-OH$, and 1,4-diaminobutane, $H_2NCH_2CH_2CH_2CH_2NH_2$

67. Based on the partial structural formula of Quiana® given below, determine **(a)** the structures of the monomer(s) and **(b)** whether the polymer is a polyester or a polyamide.

$$-NH-C_6H_{10}-CH_2-C_6H_{10}-NH-C(=O)-(CH_2)_6-C(=O)-NH-C_6H_{10}-CH_2-C_6H_{10}-NH-$$

68. Based on the partial structural formula of Kevlar® given below, determine **(a)** the structures of the monomer(s) and **(b)** whether the polymer is a polyester or a polyamide.

$$-C(=O)-C_6H_4-C(=O)-NH-C_6H_4-NH-C(=O)-C_6H_4-C(=O)-NH-C_6H_4-NH-$$

Additional Problems

69. A sample of the alloy *stainless steel A* is to be analyzed for its chromium content. By suitable treatment, a 5.000-g sample of the steel is used to produce 250.0 mL of a solution containing $Cr_2O_7^{2-}$. A 10.00-mL portion of this solution is added to $BaCl_2(aq)$. When the pH of the solution is properly adjusted, 0.1387 g $BaCrO_4(s)$ precipitates. What is the %Cr, by mass, in the steel sample?

70. A sample of the alloy *manganese steel* is to be analyzed for its manganese content. By suitable treatment, a 1.250-g sample of the steel is used to produce 100.0 mL of a solution containing MnO_4^-. A 10.00-mL portion of this solution requires 17.66 mL of 0.0826 M $Fe^{2+}(aq)$ for its titration in an acidic medium. What is the %Mn, by mass, in the steel sample. (*Hint:* What are the likely oxidation and reduction products in the titration?)

71. The electrical conductivity of a semiconductor increases if either donor or acceptor atoms are incorporated into the structure. However, if both are added in equal number, there is little or no effect on the conductivity. Explain how this can be.

72. The energy gap, E_g, for silicon is 110 kJ/mol. What is the minimum wavelength light that can promote an electron from the valence band to the conduction band? In what region of the electromagnetic spectrum is this light? Does silicon absorb all *visible* light, some of it, or none of it? Explain.

73. Isobutylene, $(CH_3)_2C{=}CH_2$, polymerizes to form polyisobutylene, a sticky polymer used as an adhesive. Copolymerized with 1,3-butadiene, isobutylene forms butyl rubber. Represent a segment of **(a)** the polyisobutylene macromolecule and **(b)** the butyl rubber macromolecule. In each case, show at least four monomer units.

74. The microorganism *Alcaligenes eutrophus* produces a natural macromolecular material called polyhydroxybutyrate. The material is a polymer of 3-hydroxybutanoic acid. Represent a segment of the polymer showing at least four monomer units. To what class of polymers does polyhydroxybutyrate belong?

75. The monomers in the polyester Kodel® are terephthalic acid and 1,4-cyclohexanedimethanol. Represent a segment of the Kodel® macromolecule containing at least two of each monomer unit. (*Hint:* The methanol group is $-CH_2OH$.)

76. Draw a structure of the likely product(s) if ethanol is substituted for ethylene glycol in the reaction used to make Dacron® (page 1041).

77. The monomers 1,2-ethanediol and 1,2-benzenedicarboxylic acid form a soft, tacky polymer. If 1,2,3-propanetriol is substituted for the 1,2-ethanediol, a hard, brittle resin is formed. Explain this difference in behavior.

78. Generally speaking, the molecular masses of polymers formed by addition polymerization are greater than those formed by condensation polymerization. Give a plausible explanation of this observation.

79. When speaking of the molecular mass of a polymer, chemists have to deal with averages. The *number average molecular mass* of a polymer is analogous to the weighted average atomic mass of an element. What is the number average molecular mass of a sample of polyethylene if 28% of the molecules have the mass 786 u, 25% have a mass of 702 u, 15% have a mass of 814 u, and 32% have a mass of 758?

Environmental Chemistry

CHAPTER
25

Planet Earth is mostly solid, but it has liquid water on its surface and a gaseous atmosphere that makes it hospitable to higher forms of life. In this chapter, we examine Earth's atmosphere and waters and some of the effects of human activities on our environment. In doing so, we apply some of the chemistry that we learned in preceding chapters. We will find that a knowledge of chemistry is essential to understanding environmental problems and to their solution.

CONTENTS

When astronomers speculate about life elsewhere in the universe, they look for the possibility of planets within a size and temperature range that will allow for liquid water and a gaseous atmosphere. Most of the other worlds in our solar system are barren and airless, like Mercury and Earth's moon, or they have crushing atmospheres with pressures thousands of times greater than that of Earth, like Jupiter, Saturn, Uranus, and Neptune.

The nature of life makes it dependent on water because only water has the unique properties required to sustain life. Thus, the presence of large amounts of liquid water makes our planet unusual in the solar system, probably the only one capable of supporting higher forms of life. Scientists who look for life elsewhere in our solar system place their scant hope on the presence of water beneath the dry surface of Mars or the existence of liquid water beneath the frozen seas of Jupiter's

● **ChemCDX**
When you see this icon, there are related animations, demonstrations, and exercises on the CD accompanying this text.

moon, Europa. If we should someday discover higher forms of life on distant planets of other suns, it will likely be on another watery planet similar to our own.

In this chapter, we look at the composition and properties of Earth's atmosphere and water supplies, and we look at some of the ways that human activities affect our air and water. Most of all, though, we focus on how a knowledge of chemistry can illuminate environmental issues and how chemistry can often be used to alleviate environmental problems. We can then use our knowledge to protect the only planet that we know is hospitable to humans.

The Atmosphere

Let's look first at Earth's atmosphere. How unique is the air we breathe? A look at other worlds in our solar system gives us a clue. Astronauts have walked on the dry, dusty, airless surface of the moon. Scientists have sent robotic probes down through clouds of sulfuric acid and a thick blanket of carbon dioxide to land on the hot, inhospitable surface of Venus. Other space probes have descended through the sparse atmosphere of Mars and used a robot to analyze its rocks. Still other spacecraft have examined the crushing, turbulent, toxic atmospheres of Jupiter, Saturn, Uranus, and Neptune. We don't know much about the atmosphere of distant Pluto because it has not been examined except from vast distances.

Of all the worlds in the solar system, Earth's atmosphere is unique in its ability to support human life. In the sections that follow, first we consider the normal composition of the atmosphere and then some substances introduced into the atmosphere by human activities.

25.1 Composition, Structure, and Natural Cycles

Without food, we can live about a month; without water, only a few days. But without air, we would die within minutes. Air is vital because it contains free oxygen (O_2), an element essential to the basic processes of respiration and metabolism. However, life as we know it could not exist in an atmosphere of pure oxygen because oxidation processes would be greatly accelerated by the increased concentration of O_2. We would burn out in a lot less than our biblically allotted three score and ten years. The oxygen in air is diluted with nitrogen, thus lessening the tendency for everything in contact with air to become oxidized. Carbon dioxide and water vapor are but minor components in air, yet they are the primary raw materials of the plant kingdom, and plants produce the food on which we and all other animal life depend. And even ozone, a gas present only in trace quantities, plays vital roles in shielding Earth's surface from harmful ultraviolet (UV) radiation and in maintaining a proper energy balance in the atmosphere.

On a mole percent basis, dry air in the lower atmosphere consists of about 78% N_2, 21% O_2, and 1% Ar. Among the minor constituents of dry air, the most abundant is carbon dioxide. The concentration of CO_2 in air has increased from about 275 ppm in 1880 to its present value of more than 360 ppm. The concentration most likely will continue to rise as more and more fossil fuels (coal, oil, and natural gas) are burned. The composition of dry air is summarized in Table 25.1.

Altogether, the thin blanket of gases that makes up the atmosphere is spread over a surface area of 5.0×10^8 km^2 and has a mass of about 5.2×10^{15} metric tons (1 metric ton = 1000 kg). That is about 10 million tons of air over each square kilometer of surface, 10 tons over each square meter, or about 1 ton (nearly the mass of one small automobile) over each square foot.

TABLE 25.1 Composition of Dry Air (Near Sea Level)

Component	Mole Percent[a]
Nitrogen (N_2)	78.084
Oxygen (O_2)	20.946
Argon (Ar)	0.934
Carbon dioxide (CO_2)	0.0360
Neon (Ne)	0.001818
Helium (He)	0.000524
Methane (CH_4)	0.0002
Krypton (Kr)	0.000114
Hydrogen (H_2)	0.00005
Dinitrogen monoxide (N_2O)	0.00005
Xenon (Xe)	0.000009
	Plus traces of: Ozone (O_3); Sulfur dioxide (SO_2); Nitrogen dioxide (NO_2); Ammonia (NH_3); Carbon monoxide (CO); Iodine (I_2)

[a] The compositions of gaseous mixtures are often expressed in percentages by volume. Volume percent and mole percent compositions are the same.

How deep is the atmosphere? It's hard to say, for the atmosphere does not end abruptly. It gradually fades away as the distance from Earth's surface increases. We know, though, that 99% of the mass of the atmosphere lies within 30 km of Earth's surface—a thin layer of air indeed, akin to the peel of an apple, only relatively thinner. The density of air is about 1.3 grams per *liter* at sea level, but air gets "thinner" (less dense) with altitude. Above the stratosphere, the density drops into the range of micrograms and even nanograms per liter.

Rather arbitrarily, we divide the atmosphere into the layers shown in Figure 25.1. The layer nearest Earth's surface, the *troposphere*, extends about 12 km above the surface and contains about 90% of the mass of the atmosphere. Weather and nearly all human activity occur in the troposphere. Temperatures generally decline as altitude increases in the troposphere, ranging from a maximum of about 320 K at Earth's surface to a minimum of about 220 K in the upper troposphere.

The next layer, the *stratosphere*, extends from about 12 to 55 km above Earth's surface. The ozone layer that shields living creatures from deadly UV radiation lies within the stratosphere, from about 15 to 30 km above Earth's surface. Supersonic aircraft fly in the lower part of the stratosphere. Temperatures in the stratosphere are fairly constant at about 220 K from 12 to 25 km, then rise with increasing altitude to about 280 K at 50 km.

The next layer above the stratosphere, ranging from about 55 to 80 km, is called the *mesosphere*. Here the temperature falls continuously to about 180 K as altitude increases. The layer above the mesosphere is known as the *thermosphere* or *ionosphere*. In this region molecules can absorb such highly energetic electromagnetic radiation from the sun that they dissociate into atoms; some of the atoms are further broken down into positive and negative ions and free electrons.

Temperatures in the thermosphere rise to about 1500 K as the altitude increases,* but high temperatures in this region don't have the same significance as on Earth's surface. The temperatures are high because the average kinetic energy

APPLICATION NOTE

Mt. Cayambe in Ecuador is only a few kilometers from the equator, but its peak is covered with snow, illustrating the decrease in temperature with increased altitude in the troposphere. At an altitude of 5.79 km, the top of the mountain extends about halfway through the troposphere. (See Figure 25.1, p. 1050.)

* The temperature in the thermosphere depends on the amount of solar radiation and varies between night and day and with sunspot activity.

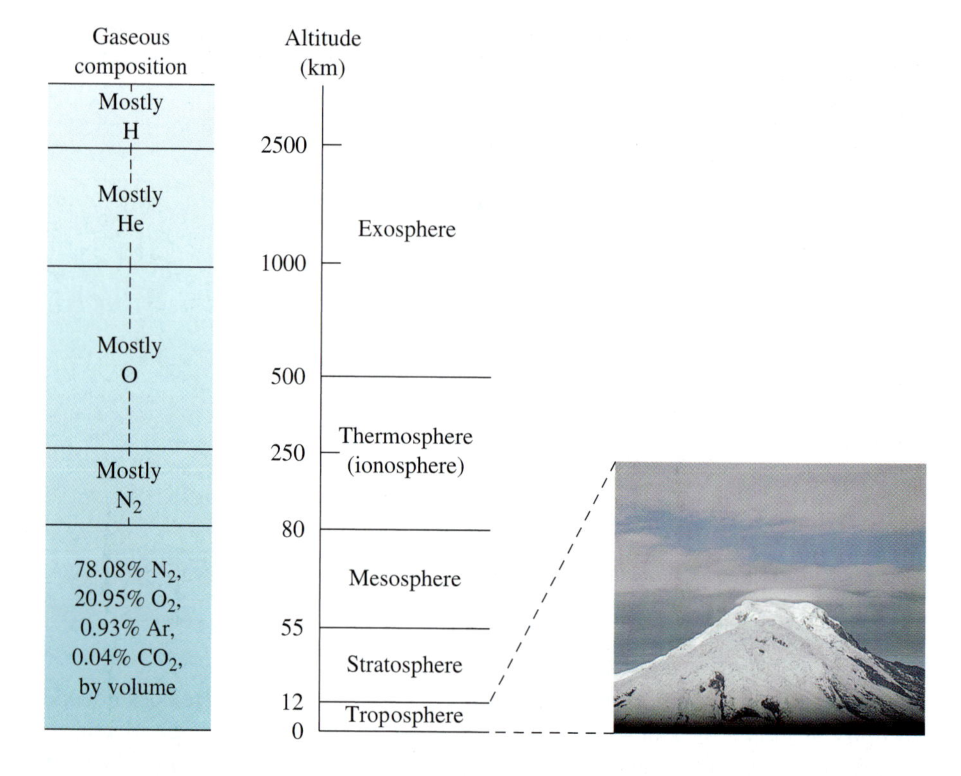

Figure 25.1 Layers of the atmosphere
The altitudes of the several layers of the atmosphere are only approximate. For example, the height of the troposphere varies from about 8 km at the poles to 16 km at the equator. The approximate temperatures in these layers are cited in the text.

of the gaseous particles is high. However, because there are relatively few of these particles per unit volume, little energy is transferred through collisions. A cold object brought into this region does not get hot; there are far too few collisions to bring the object to temperature equilibrium with the gas molecules.

This meteor seen against a starry sky at dusk is an extraterrestrial chunk of matter that has entered Earth's atmosphere. It emits light because it is heated to a high temperature. This heating does not occur in the high-temperature thermosphere, however. It occurs through the frictional resistance the object encounters as it passes through gases lower in the atmosphere (about 80 to 100 km).

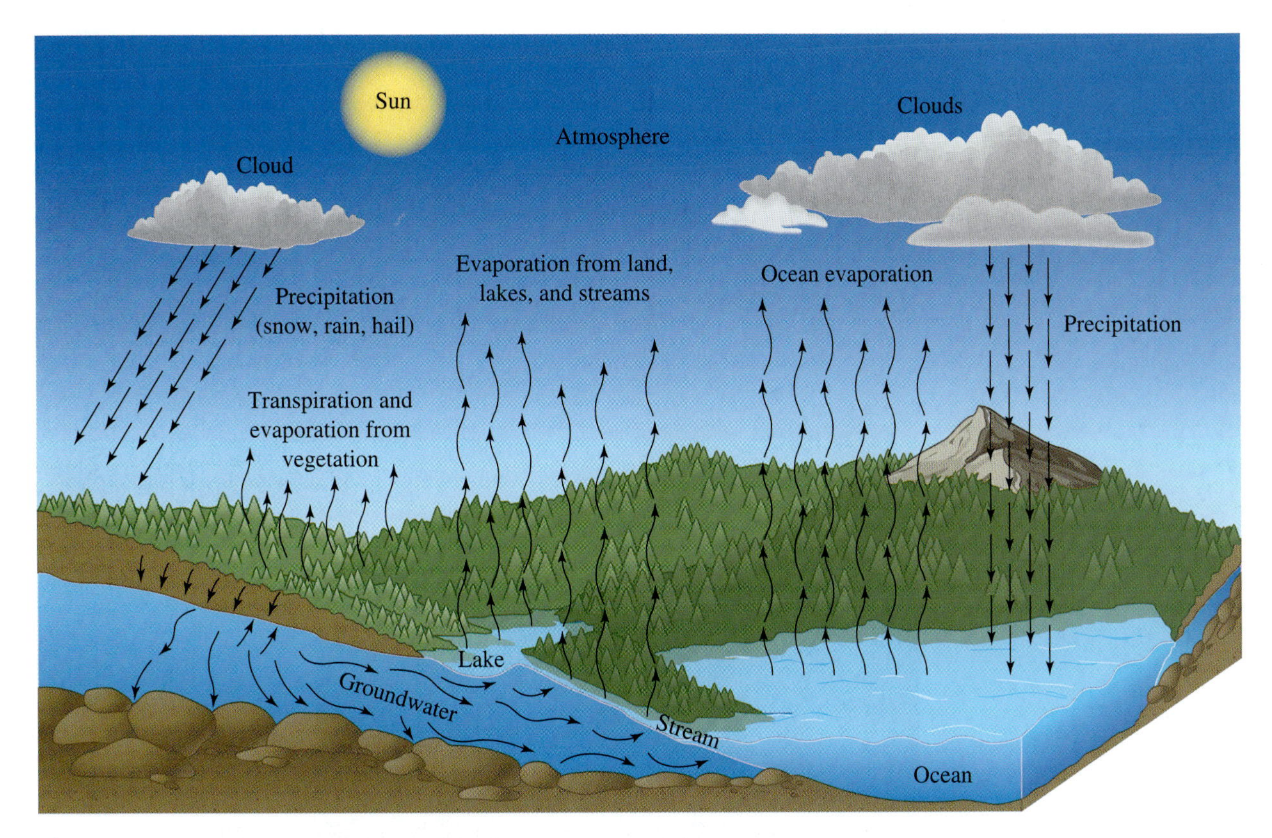

▲ **Figure 25.2 The hydrologic (water) cycle**
The oceans on Earth are vast reservoirs of water. Water evaporates from the oceans, producing moist air masses that move over the land. When the moist air is cooled, the water vapor forms clouds and the clouds produce rain. Rainwater replenishes groundwater and forms lakes and streams. The water eventually returns to the ocean. Water is also returned to the atmosphere by evaporation throughout the cycle.

Water Vapor in the Atmosphere

Unless it has been specially dried, air invariably contains water vapor. Atmospheric water vapor plays one of the key roles in the **hydrologic (water) cycle**—the series of natural processes by which water is recycled through the environment (see Figure 25.2).

The proportion of water vapor in air is quite variable, ranging from trace amounts to about 4% by volume. *Humidity* is a general term describing the water vapor content of air. The *absolute humidity* is the actual quantity of water vapor present in an air sample, usually expressed in g H_2O/m^3 air. The *relative humidity* of air is a measure of water vapor content as a percentage of the maximum possible; it compares the actual partial pressure of water vapor in an air sample to the maximum partial pressure that could exist at the given temperature—the vapor pressure of water.

$$\text{Relative humidity} = \frac{\text{partial pressure of water vapor}}{\text{vapor pressure of water}} \times 100\%$$

A number of different experimental methods exist for determining the relative humidity of air. A crude but colorful method is illustrated in Figure 25.3.

▲ **Figure 25.3 A measure of relative humidity**
The strips of filter paper were impregnated with an aqueous solution of cobalt(II) chloride and allowed to dry. In dry air, the strip is blue, the color of anhydrous $CoCl_2$. In more humid air, the strip acquires the red color of the hexahydrate, $CoCl_2 \cdot 6H_2O$.

EXAMPLE 25.1

The partial pressure of water vapor in a certain air sample at 20.0 °C is 12.8 mmHg. What is the relative humidity of this air?

SOLUTION

To establish the relative humidity of the air sample, we need (1) the partial pressure of water vapor in the sample and (2) the vapor pressure of water at the given temperature. The first quantity is given (12.8 mmHg). For the second, we need the vapor pressure of water at 20.0 °C. We can find it in a tabulation such as Table 11.2, page 464.

$$\text{Relative humidity} = \frac{12.8 \cancel{\text{mmHg}}}{17.5 \cancel{\text{mmHg}}} \times 100\% = 73.1\%$$

EXERCISE 25.1

If an air sample at 20.0 °C has a relative humidity of 38.5%, what is the partial pressure of water vapor in this sample?

Dill covered with morning dew. When the temperature drops to the point that the absolute humidity of the air is greater than the vapor pressure of water, water vapor condenses to the familiar liquid known as dew.

If we warm the air sample described in Example 25.1, the relative humidity *decreases* because the vapor pressure of water increases sharply with temperature, whereas the measured partial pressure of water vapor in the air sample changes more slowly as the temperature increases. In contrast, if we cool the air sample, the relative humidity *increases*. If the vapor pressure of the water remained at 12.8 mmHg, the relative humidity at 15.0 °C would be 100%. The air would then be saturated with water vapor. At temperatures below 15.0 °C, the relative humidity would exceed 100%, and the air would be supersaturated with water vapor. This is an unstable situation that is not at equilibrium; it cannot persist. Some of the water vapor would condense as droplets of liquid water that we call *dew*. The highest temperature at which condensation of water vapor from an air sample can occur is known as the *dew point*. When the dew point is below the freezing point of water (0 °C), the water condenses as *frost* without going through the liquid state.

▲ **Figure 25.4**
Deliquescence of calcium chloride
Water vapor from the air condenses onto the solid $CaCl_2 \cdot 6H_2O$ and produces a solution of $CaCl_2(aq)$. Here, the solution is saturated, but eventually all the solid would dissolve and the solution would become unsaturated. The deliquescence of $CaCl_2 \cdot 6H_2O$ occurs only when the relative humidity exceeds 32%. Other water soluble solids deliquesce under other conditions of relative humidity.

The water vapor that condenses from air at the dew point is pure water. The liquid in Figure 25.4 also results from the condensation of atmospheric water vapor, but it is not pure water. Rather, it is a solution of calcium chloride, formed in this way: If the partial pressure of water vapor in the air exceeds the vapor pressure of a saturated solution of $CaCl_2 \cdot 6H_2O$, water vapor condenses on the solid $CaCl_2 \cdot 6H_2O$ and dissolves some of it, producing some saturated solution. This condensation of water vapor on a solid followed by solution formation is called **deliquescence**; it continues until all the solid has dissolved and the vapor pressure of the solution (now unsaturated) equals the partial pressure of water vapor in the air.

Nitrogen Fixation: The Nitrogen Cycle

Although it exists in large quantities in the atmosphere, nitrogen, an element essential to life, cannot be used directly by higher plants or animals. The N_2 molecules first must be "fixed," that is, converted to compounds that are more readily usable by living organisms. This conversion of atmospheric nitrogen into nitrogen compounds is called **nitrogen fixation**.

Certain bacteria, such as cyanobacteria (blue-green algae) found in water and a variety of bacteria that live in root nodules of specific plants, are able to fix atmospheric nitrogen by converting it to ammonia. These nitrogen-fixing bacteria are concentrated in the roots of leguminous plants such as clover, soybeans, and peas. Other plants take up nitrogen atoms in the form of nitrate ion or ammonium ion. Nitrogen atoms in plants, combined with carbon compounds from photosynthesis, form amino acids, the building blocks of proteins (page 980). The food chain for animals originates with plant life. The decay of plant and animal life returns nitrogen to the environment as nitrates and ammonia. Eventually nitrogen is returned to the atmosphere as N_2 by the action of denitrifying bacteria on nitrates.

Nitrogen fixation occurring during electrical storms is an important part of the natural nitrogen cycle.

Lightning also "fixes" some atmospheric nitrogen by creating a high-energy environment in which nitrogen and oxygen can combine. Nitrogen monoxide and nitrogen dioxide are formed in this way.

$$N_2(g) + O_2(g) \xrightarrow{\text{Lightning}} 2\,NO(g)$$

$$2\,NO(g) + O_2(g) \longrightarrow 2\,NO_2(g)$$

Nitrogen dioxide reacts with water to form nitric acid.

$$3\,NO_2(g) + H_2O(l) \longrightarrow 2\,HNO_3(aq) + NO(g)$$

The nitric acid falls in rainwater, adding to the available nitrates in sea and soil. The net effect of all these natural activities is a constant recycling of nitrogen atoms in the environment in a **nitrogen cycle** (see Figure 25.5).

Industrial fixation of nitrogen by the manufacture of inorganic nitrogen fertilizers has modified the nitrogen cycle. These fertilizers have greatly increased the

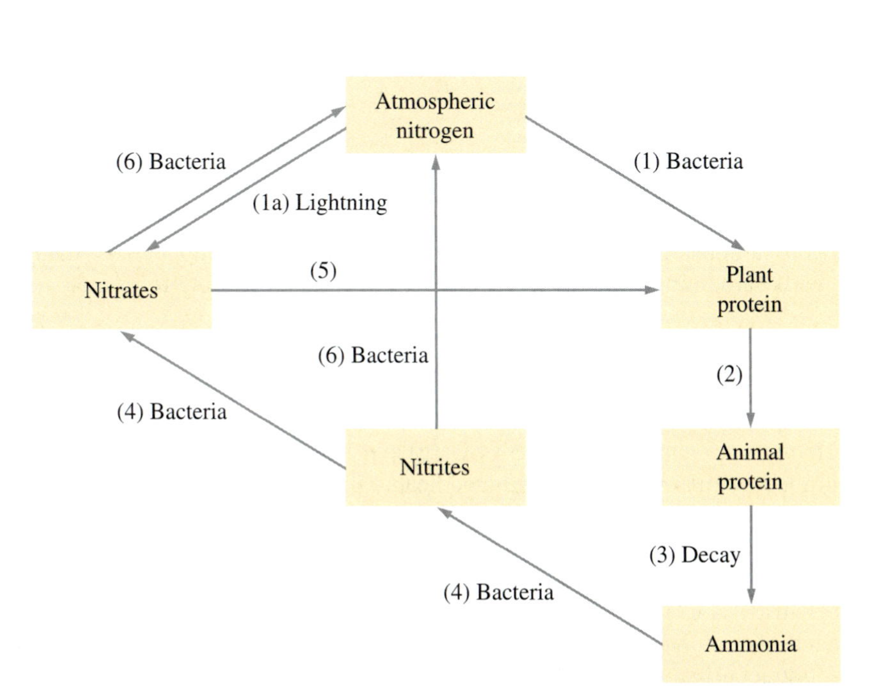

Figure 25.5
The nitrogen cycle
Bacteria fix atmospheric nitrogen and convert it through chemical reactions into plant proteins (1). Plants also convert nitrates to proteins (5). Animals feed on plants and other animals (2). The decay of plant and animal proteins produces ammonia (3). Through a series of bacterial actions, ammonia is converted to nitrites and nitrates (4). Denitrifying bacteria decompose nitrites and nitrates, returning N_2O and N_2 to the atmosphere (6). Some atmospheric nitrogen is converted to nitrates during electric storms (1a). A significant amount of nitrogen fixation and denitrification occurs in oceans. Modern manufacturing and agricultural processes also play important roles in the cycle.

world's food supply because the availability of fixed nitrogen is often the limiting factor in the production of food. Not all the consequences of this interference have been favorable, however. Excessive runoff of dissolved nitrogen fertilizers has led to serious water pollution in some areas, but modern methods of high-yield farming seem to demand synthetic fertilizers.

The Carbon Cycle

Carbon atoms are cycled throughout Earth's solid crust, oceans, and atmosphere by natural processes. In the process of photosynthesis, atmospheric CO_2 is converted to carbohydrates, the chief structural material of plants. For example, the photosynthesis of glucose, one of the simplest carbohydrates, occurs through dozens of sequential steps leading to the following net change.

$$6\,CO_2(g) + 6\,H_2O(l) \longrightarrow C_6H_{12}O_6(s) + 6\,O_2(g)$$

Animals acquire carbon compounds as they consume plants or other animals, and they return CO_2 to the atmosphere as a product of respiration. The decay of plant and animal matter also returns CO_2 to the air. Most photosynthesis occurs in the oceans, where algae and related plants convert CO_2 into organic compounds. Some of Earth's carbon is locked up in fossilized forms—in coal, petroleum, and natural gas from decaying organic matter and in limestone from the shells of decayed mollusks from ancient seas. Figure 25.6 shows a simplified carbon cycle.

Note in Figure 25.6 that human activities today play a key role in the carbon cycle by releasing carbon atoms as CO and CO_2 when we burn wood and fossil fuels.

25.2 Air Pollution

Human activities affect the atmosphere in many ways. In addition to the release of carbon dioxide into Earth's atmosphere, the combustion of fossil fuels causes regional problems with high levels of carbon monoxide and particles of ash, dust, and soot, and with sulfur dioxide pollution and acid rain. Oxides of nitrogen, formed by high-temperature combustion in air, are involved in the production of photochemical smog. In general, an **air pollutant** is a substance found in air in greater abundance than normally occurs naturally and having one or more harmful effects on human health or the environment. Let's start our study of air pollution with carbon monoxide.

Carbon Monoxide

Carbon monoxide and carbon dioxide are formed in varying quantities when fossil fuels are burned. Coal is mostly carbon; natural gas and petroleum are mainly hydrocarbons. The combustion of methane, the major component of natural gas, yields carbon monoxide and carbon dioxide by the following reactions.

$$2\,CH_4(g) + 3\,O_2(g) \longrightarrow 2\,CO(g) + 4\,H_2O(l)$$

$$CH_4(g) + 2\,O_2(g) \longrightarrow CO_2(g) + 2\,H_2O(l)$$

With an excess of O_2, as is the case when air is readily available, the combustion products are almost exclusively $CO_2(g)$ and $H_2O(l)$. If the quantity of air is more limited, as in a dust- or dirt-clogged heater, CO(g) is also formed.

The chief source of CO(g) in polluted air is the incomplete combustion of gasoline hydrocarbons in automobile engines. Millions of tons of this invisible but deadly gas are poured into the atmosphere each year, about 75% of it from automobile exhausts. The United States government has set danger levels at 9 ppm CO averaged over an 8-hour period and 35 ppm averaged over a 1-hour period. Even in off-street urban areas, levels often reach 8 ppm or more. On streets and in parking

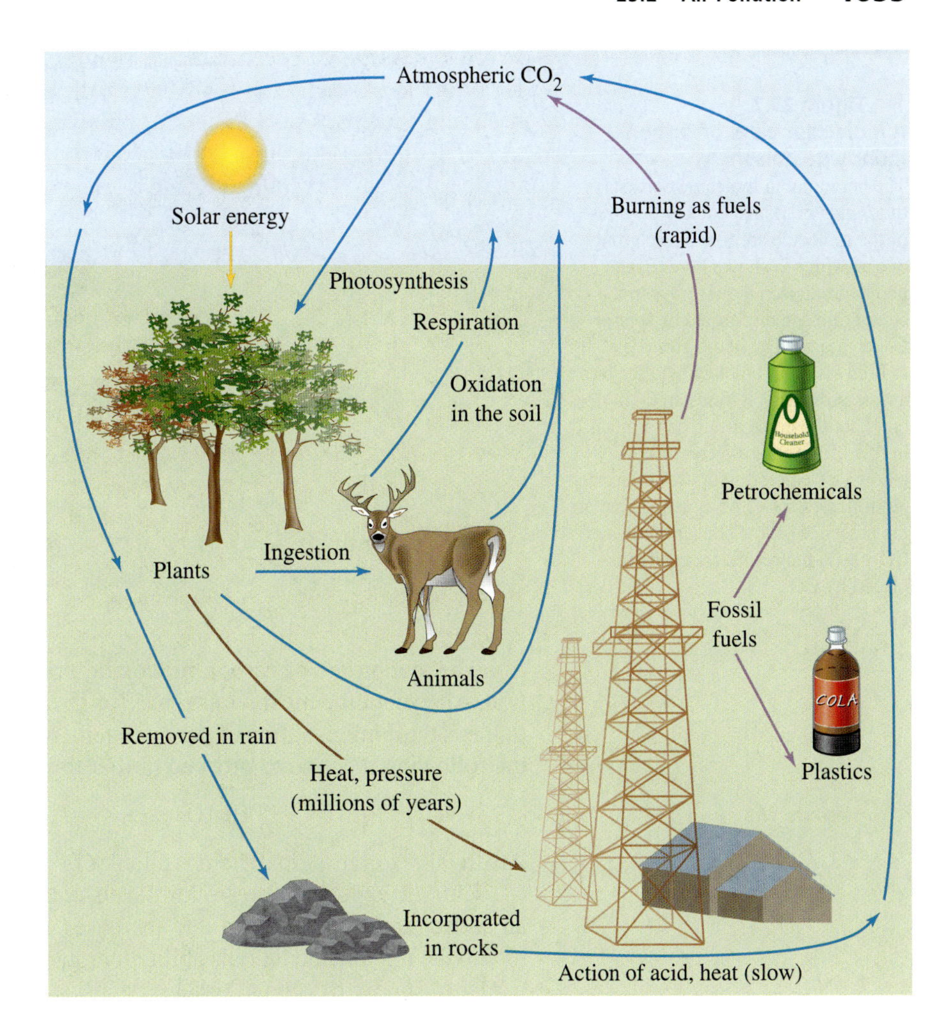

▲ **Figure 25.6 The carbon cycle**
The natural main cycle is indicated by blue arrows. Some carbon atoms are locked up by the formation of fossil fuels and limestone deposits, so-called fossilization tributaries (brown arrows). Disruption of the cycle by human activities is becoming increasingly important (purple arrows).

garages, danger levels are exceeded much of the time. Such levels do not cause immediate death, but exposure over a long period can cause physical and mental impairment.

Because CO is an invisible, odorless, tasteless gas, we can't tell when it is around without using test reagents or instruments. Drowsiness is usually the only symptom of carbon monoxide poisoning, and drowsiness is not always unpleasant. How many auto accidents are caused by drowsiness or sleep induced by CO(g) escaping into the car from a faulty exhaust system? No one knows for sure.

Carbon monoxide exerts its insidious effect because CO molecules replace O_2 molecules normally bonded to Fe atoms in hemoglobin in blood. We represented the heme units and polypeptide chains in hemoglobin in Figure 22.9. The form of hemoglobin carrying carbon monoxide is shown in Figure 25.7. The reversible uptake of O_2 and CO by hemoglobin (Hb) to form HbO_2 and HbCO is represented by the equations below. CO is highly effective in displacing O_2 from HbO_2 because reaction (2) proceeds much farther toward products than does reaction (1).

$$(1)\quad Hb + O_2 \rightleftharpoons HbO_2$$

$$(2)\quad Hb + CO \rightleftharpoons HbCO$$

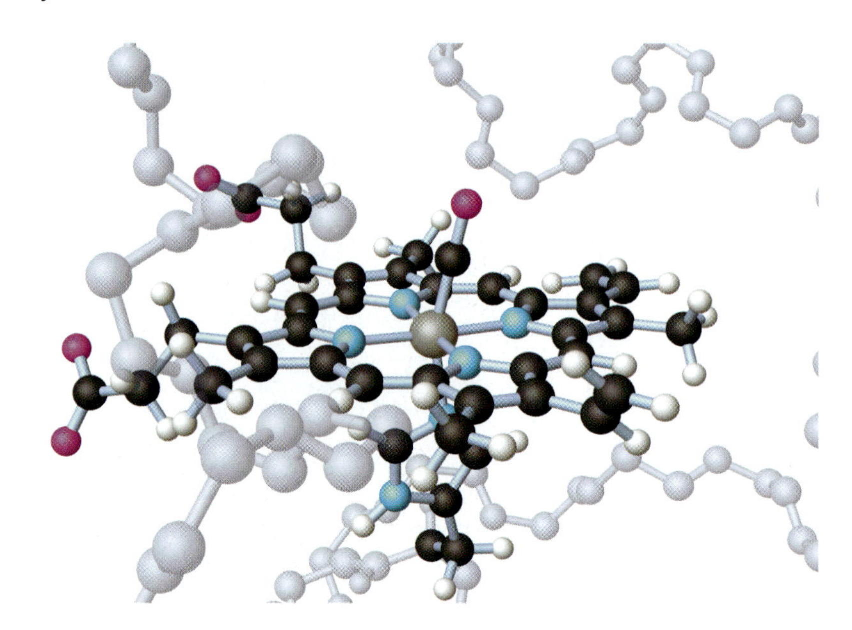

▶ **Figure 25.7**
A molecular view of carbon monoxide poisoning
The hemoglobin molecule consists of thousands of atoms, but key portions of the molecule are four heme groups, one of which is shown here. Each heme has an iron atom (gray) in the center of a square formed by four nitrogen atoms. The heme group is able to bind one small molecule to the iron atom. Ordinarily, it binds an O_2 molecule, but the heme group has a much higher affinity for a CO molecule (shown here pointing up from the iron atom) than for O_2. As a result, even in the presence of low concentrations of CO(g), O_2 molecules are easily displaced by CO.

The symptoms of carbon monoxide poisoning are those of oxygen deprivation. All except the most severe cases of the poisoning are reversible. The best antidote is the administration of pure oxygen. At high concentrations, the O_2 can cause the following reaction to proceed farther toward products.

$$HbCO + O_2 \rightleftharpoons HbO_2 + CO$$

Artificial respiration may help if a tank of oxygen is not available.

Chronic exposure, even to low levels of CO, as through cigarette smoking, puts an added strain on the heart and increases the chances of a heart attack. Carbon monoxide impairs the blood's ability to transport oxygen, and the heart has to work harder to supply oxygen to tissues.

Carbon monoxide is a local pollution problem: Its greatest threat is in urban areas with heavy traffic. In laboratory tests, CO molecules survive about 3 years in contact with air. But nature is able to prevent the buildup of CO in the atmosphere, despite the large amounts entering the environment. Bacteria in the soil presumably convert CO to CO_2. In fact, it is estimated that on a global basis, up to 80% of the CO in the atmosphere comes from natural sources. Except for highly localized situations that can be quite severe, nature seems to have carbon monoxide under control.

Photochemical Smog

We usually think of sunlight as something pleasant. However, when sunlight falls on air containing a mix of nitrogen oxides, hydrocarbons, and other pollutants, it produces a mix of pollutants called **photochemical smog**. This smog usually has a considerably higher than normal concentration of ozone. Photochemical smog production starts with the formation of nitrogen monoxide (NO). The reaction of $N_2(g)$ and $O_2(g)$ at ordinary temperatures is obviously extremely slow because these gases coexist in the atmosphere. However, at high temperatures, such as those attained by the combustion of a fuel in air, some NO is formed.

$$N_2(g) + O_2(g) \longrightarrow 2\,NO(g)$$

This reaction takes place in power plants that burn fossil fuels and in incinerators. The greatest source of NO, though, is in the exhaust fumes of automobile engines.

In sufficiently high concentrations, NO can react with hemoglobin in blood and rob it of its oxygen-carrying ability, just as CO does. However, these concentrations are rarely reached in polluted air. The main role of NO as an air pollutant is its participation in various reactions that yield several other pollutants.

Nitrogen dioxide, the gas that gives the red-brown color often seen in polluted air in major urban centers, is formed by the oxidation of NO. Nitrogen dioxide is an irritant to the eyes and respiratory system. Tests with laboratory animals indicate that chronic exposure to levels of NO_2 in the range of 10 to 25 ppm might lead to emphysema or other degenerative lung diseases. However, as with NO, we are not so much concerned with the direct effects of NO_2 but rather with its chemical reactions.

In the presence of sunlight ($h\nu$),* NO_2 decomposes.

$$NO_2(g) + h\nu \longrightarrow NO(g) + O(g)$$

Oxygen *atoms* produced by the photochemical (light-induced) decomposition of NO_2 are highly reactive. The O atoms react with many substances that are generally available in polluted air. For example, they react with O_2 molecules to form ozone (O_3).

$$O(g) + O_2(g) \longrightarrow O_3(g)$$

A normal radish plant (left) and one damaged by air pollution (right).

Ozone is the main cause of the breathing difficulties that some people experience during smog episodes. Another effect of ozone is that it causes rubber to crack and deteriorate.

In addition to ozone, nitrogen oxides, and hydrocarbons, photochemical smog contains *peroxyacetyl nitrate* (PAN). PAN is an organic compound formed by the combination of two free radicals.

$$CH_3\overset{\displaystyle O}{\overset{\|}{C}}{-}O{-}O\cdot \; + \; \cdot NO_2 \longrightarrow CH_3\overset{\displaystyle O}{\overset{\|}{C}}{-}O{-}ONO_2$$

(Recall from Chapter 9 that NO_2 is an odd-electron molecule; it has an unpaired electron and is therefore a free radical.) PAN is a powerful *lacrimator*. It makes the eyes form tears.

In addition to their role in smog formation, NO and NO_2 (collectively called NO_x) contribute to the fading and discoloration of fabrics. By forming nitric acid, they contribute to the acidity of rainwater, which accelerates the corrosion of metals and building materials. They also produce crop damage, although specific effects of these gases are difficult to separate from those of other pollutants. Finally, the components of photochemical smog generally reduce visibility (Figure 25.8).

▲ **Figure 25.8**
Photochemical smog
Photochemical smog is characterized by amber haze, like that seen in this street scene in Mexico City.

The reactions that form photochemical smog are exceedingly complex and are still not completely understood. We will point out only a few of the more important features of this reaction chemistry.

We have already noted two reactions: the photochemical decomposition of NO_2, followed by the production of ozone. If ozone formation is to continue, there must be a continuous source of NO_2. We have previously noted one reaction for the formation of NO_2.

$$2\,NO(g) + O_2(g) \longrightarrow 2\,NO_2(g)$$

However, at the low concentrations of NO in a smoggy atmosphere and at normal air temperatures, this reaction is too slow to produce much NO_2. The NO appears

* The expression $h\nu$ is derived from the equation for the energy of a photon of light, $E = h\nu$ (Chapter 7). Here, it represents a photon of sunlight. One photon of the proper energy can split one molecule of $NO_2(g)$ into a molecule of NO(g) and an oxygen atom, represented as O(g).

to be converted to NO_2 in another—much faster—way that involves hydrocarbons, which come mostly from automotive exhaust.

For example, a hydrocarbon molecule, RH, can react with an oxygen atom to produce two free radicals, and the OH radical can react with another hydrocarbon molecule.

$$RH + O \longrightarrow R\cdot + \cdot OH$$
$$RH + \cdot OH \longrightarrow R\cdot + H_2O$$

The hydrocarbon radicals, $R\cdot$, can react with O_2 to produce new radicals, called *peroxyl* radicals.

$$R\cdot + O_2 \longrightarrow RO_2\cdot$$

And then the peroxyl radicals can react with NO to form NO_2.

$$RO_2\cdot + NO \longrightarrow RO\cdot + NO_2$$

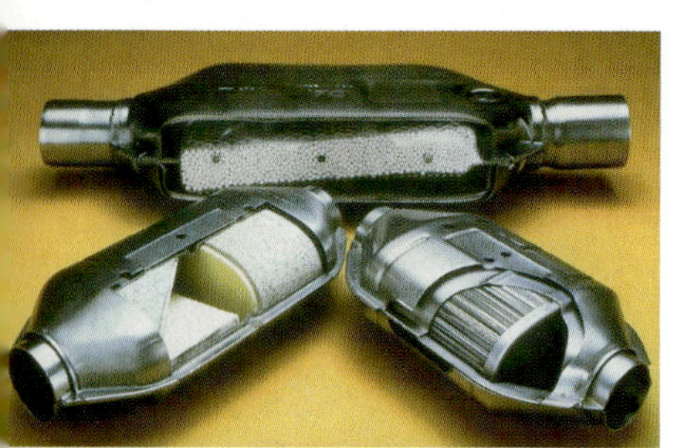

A cutaway view of a dual-bed automobile catalytic converter.

Automotive exhaust is a crucial contributor to the production of photochemical smog, but geographic factors are also important. Smog is especially likely to occur in a region like the Los Angeles basin, which is surrounded by mountains. In the absence of strong winds in the basin, the only direction in which mixing and dilution of air pollutants can occur is vertically into the atmosphere. However, at times, the region may also experience a temperature inversion—a mass of warm air overlaying a body of colder air below. This inversion layer acts like a lid on a reaction vessel, but it does let sunlight through. The sunlight acts on the primary pollutants to form smog, which is then trapped in the cooler stagnant layer of air. The most severe smog episodes usually occur during periods of strong temperature inversions.

Control of Photochemical Smog

Most measures to reduce the levels of photochemical smog focus on automobiles, but potential sources of smog precursors range from power plants to lawn mowers to charcoal lighter fluid. In many parts of the world, automobiles are now equipped with catalytic converters. The first function of these converters is to catalyze the oxidation of carbon monoxide and unburned hydrocarbons to CO_2 and H_2O. The catalyst is usually palladium (Pd) or platinum (Pt) or a mixture of the two. To remove NO from automotive exhaust by reducing it to N_2 requires a reduction catalyst, which is different from the Pt/Pd oxidation catalyst. Some automobiles therefore use a dual-catalyst system. The NO content of automotive exhaust can also be lowered by using a fuel-rich mixture of fuel and air. This results in some unburned hydrocarbons and CO(g), which can reduce NO to N_2. For example, CO reduces NO, forming N_2 and CO_2.

$$2\,CO(g) + 2\,NO(g) \longrightarrow 2\,CO_2(g) + N_2(g)$$

Excess unburned hydrocarbons and CO(g) are then oxidized to CO_2 and H_2O in the catalytic converter.

This copper smelter emitted 900 tons of SO_2 daily before it ceased operation in January of 1987.

Industrial Smog

Photochemical smog occurs mainly in warm, sunny weather and is characterized by high levels of hydrocarbons, nitrogen oxides (NO and NO_2) and ozone. The other major kind of smog is usually associated with industrial activities and is therefore called **industrial smog**. Industrial smog occurs mainly in cool, damp weather and is usually characterized by high levels of sulfur oxides (SO_2 and SO_3, collectively called SO_x) and of particulate matter (dust, smoke, and so on).

A major source of SO_x in some areas is smelters, where sulfide ores are roasted as the first step in the production of metals such as copper, lead, and zinc. For example, the smelting of zinc ore to produce ZnO also yields SO_2.

$$2\ ZnS(s) + 3\ O_2(g) \longrightarrow 2\ ZnO(s) + 2\ SO_2(g)$$

Coal, especially soft coal from the eastern United States, has a relatively high sulfur content. When this coal is burned, sulfur compounds in the coal also burn, forming SO_2.

Sulfur dioxide, a choking, acrid gas, is readily absorbed in the respiratory system. It is a powerful irritant and is known to aggravate the symptoms of people who suffer from asthma, bronchitis, emphysema, and other lung diseases.

Some of the sulfur dioxide reacts further with oxygen in air to form sulfur trioxide.

$$2\ SO_2(g) + O_2(g) \longrightarrow 2\ SO_3(g)$$

Sulfur trioxide then reacts with water to form sulfuric acid.

$$SO_3(g) + H_2O(l) \longrightarrow H_2SO_4(aq)$$

Fine droplets of this acid form an aerosol mist that is even more irritating to the respiratory tract than sulfur dioxide.

Particulate matter consists of solid and liquid particles of greater than molecular size (Figure 25.9). The largest particles often are visible in air as dust and smoke. Smaller particles, 1 μm or less in diameter, are called *aerosols* and are often invisible to the naked eye.

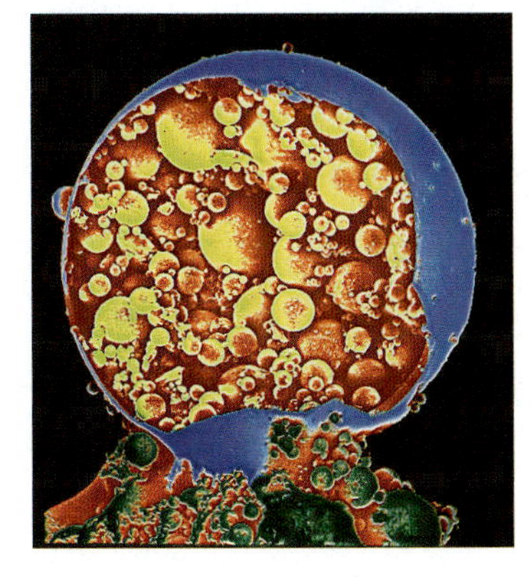

▲ **Figure 25.9**
False-color scanning electron micrograph of fly ash from a coal-burning power plant

Particulate matter consists in part of *soot* (unburned carbon). A larger portion is made up of the mineral matter that occurs in coal but does not burn. In the roaring fire of a huge boiler in a factory or power plant, some of this solid mineral matter is left behind as *bottom ash*. However, a lot of solid matter is carried aloft in the tremendous draft created by the fire. This *fly ash* settles over the surrounding area, covering everything with dust. It is also inhaled, contributing to respiratory problems in animals and humans.

Perhaps the most insidious form of particulate matter is the sulfates. Some of the sulfuric acid in smog reacts with ammonia to form solid ammonium sulfate.

$$2\ NH_3(g) + H_2SO_4(aq) \longrightarrow (NH_4)_2SO_4(s)$$

The solid ammonium sulfate and minute liquid droplets of sulfuric acid are trapped in the lungs, where they can cause considerable damage.

The harmful effects of sulfur dioxide and particulate matter may be magnified by their interaction. A certain level of sulfur dioxide, without the presence of particulate matter, might be reasonably safe. A certain level of particulate matter, without sulfur dioxide around, might be fairly harmless. But take these same levels of the two together, and the effect might be considerably enhanced, aggravating respiratory problems such as bronchitis or triggering acute asthma attacks. *Synergistic effects* such as this are quite common whenever certain chemicals are brought together: The effect of the two together is much greater than the sum of their individual effects. For example, some forms of asbestos are carcinogenic, and about 35 or 40 of the chemicals in cigarette smoke are carcinogens. Asbestos workers who smoke develop cancer at a much greater rate than do people who are exposed to one of these carcinogens but not the other.

When inhaled deeply into the lungs, the pollutants in industrial smog break down the cells of the tiny air sacs, called alveoli, where oxygen and carbon dioxide exchange ordinarily occurs. The alveoli lose their resilience; it becomes difficult for them to expel carbon dioxide. Such lung damage contributes to pulmonary

APPLICATION NOTE

According to the World Health Organization, Mexico City has the world's worst air pollution. Pollution levels are at least double those considered safe. One five-day episode in 1996 caused 400,000 pollution-related hospital and clinic contacts. The pollution originates from automobiles, industries, and propane used in home heating and cooking. The problem is compounded by location; Mexico City is in a valley surrounded by mountains.

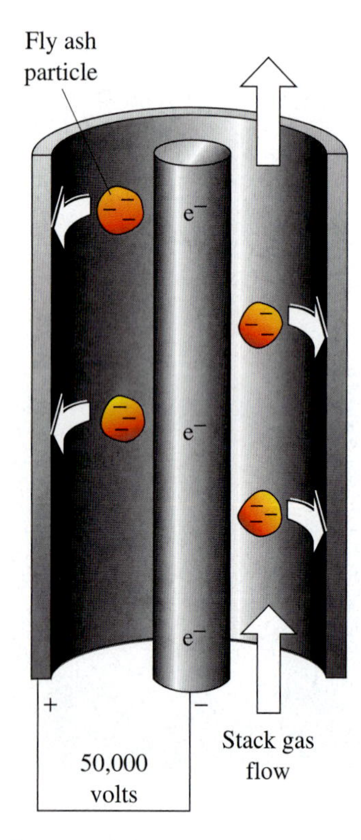

▲ **Figure 25.10**
An electrostatic precipitator
Electrons emitted from the negatively charged electrode in the center attach themselves to the particles of fly ash, giving them a negative charge. The negatively charged particles are attracted to the positively charged cylindrical collector plate and are deposited there.

emphysema, a condition characterized by an increasing shortness of breath. Emphysema is one of the fastest growing causes of death in the United States. Although the principal factor in the rise of emphysema is cigarette smoking, air pollution is also a factor.

Not only are the oxides of sulfur and the aerosol mists of sulfuric acid damaging to humans and other animals, but they damage plants as well. Leaves become bleached and mottled when exposed to sulfur oxides. The yield and quality of farm crops can be severely affected. These compounds are also major contributors to acid rain.

Controlling Industrial Smog

Soot and fly ash can be removed from smokestack gases in several ways. One method, the use of electrostatic precipitators, is illustrated in Figure 25.10. The particles in the smokestack gases are given an electric charge, and the charged particles deposit on the oppositely charged collector plate. The energy requirement of this method is high. About 10% of the electricity produced in a power plant is required to operate the electrostatic precipitators. And the ash has to be put somewhere. Ash production in the United states is about 70 million metric tons per year. Some of the ash is used in making cement, and some is melted and formed into fibers called mineral wool for insulation. Most of the ash is simply stored on the ground or buried in landfills.

Emissions of SO_x can be reduced by removing sulfur-containing minerals before burning coal. For example, FeS_2 (*pyrite*) can be concentrated and removed by flotation (page 1015). Another way is to convert the coal to gaseous or liquid hydrocarbons by reaction with $H_2(g)$, leaving minerals such as pyrite behind. Both methods are expensive, however.

An alternative is to remove SO_2 from stack gases after the coal has been burned. In the method described by the following equations, powdered coal and limestone are burned together. The limestone decomposes, and SO_2 reacts with the CaO.

$$CaCO_3(s) \longrightarrow CaO(s) + CO_2(g)$$
$$CaO(s) + SO_2(g) \longrightarrow CaSO_3(s)$$

Sulfur dioxide can also be allowed to react with hydrogen sulfide, producing elemental sulfur that is easily recovered.

$$2\,H_2S(g) + SO_2(g) \longrightarrow 3\,S(s) + 2\,H_2O(l)$$

Still another possibility is to convert sulfur dioxide to sulfuric acid. However, if all the SO_2 in smokestack gases were converted to sulfuric acid, the quantity produced would greatly exceed current demand. Then the sulfuric acid would present a disposal problem.

25.3 The Ozone Layer

The **ozone layer** is a band of the *stratosphere* about 20 km thick, centered at about 25 to 30 km altitude. It has a much higher ozone concentration than the rest of the atmosphere. Ozone absorbs harmful ultraviolet (UV) radiation, and the ozone layer thus protects life on Earth.

UV radiation, invisible to human eyes, has profound effects on living matter. Proteins, nucleic acids, and other cell components are broken down by UV radiation with wavelengths shorter than about 290 nm. Of this radiation, O_2 and other atmospheric components effectively filter out wavelengths below 230 nm. Ozone alone absorbs UV radiation in the wavelength range from 230 to 290 nm. Radiation with wavelengths in the range of 290 to 320 nm, called UV-B radiation, produces sunburns and can also cause eye damage and skin cancer. UV-B radiation is only partially absorbed by ozone. Thus the amount of UV radiation reaching Earth's surface is critically dependent on the concentration of $O_3(g)$ in the ozone layer. Society therefore faces a challenge to maintain the appropriate concentration of ozone in the ozone layer.

When two opposing processes occur, one producing and the other consuming a substance, the result is a fairly constant concentration. Let's consider the processes that lead to a natural balanced concentration of about 8 ppm of ozone in the ozone layer. Ozone is produced in the upper atmosphere in a sequence of two reactions. First, an O_2 molecule absorbs UV radiation and dissociates into two O atoms.

$$O_2 + h\nu \longrightarrow O + O$$

Then atomic and molecular oxygen react to form ozone. In this reaction, a "third body," M, often an N_2 molecule, carries off excess energy. Otherwise the energetic O_3 would simply decompose back to O_2 and O.

$$O_2 + O + M \longrightarrow O_3 + M$$

Ozone is decomposed in a sequence of two reactions. First, an ozone molecule decomposes when it absorbs UV radiation.

$$O_3 + h\nu \longrightarrow O_2 + O$$

Then, an atom of oxygen reacts with another ozone molecule, forming two O_2 molecules and releasing heat to the environment.

$$O + O_3 \longrightarrow 2\,O_2 \qquad \Delta H = -391.9\text{ kJ}$$

The heat released in this reaction accounts for the characteristic temperature increase with altitude in the stratosphere (page 1049).

Other naturally occurring species also contribute to the decomposition of ozone. Atmospheric NO, produced mainly from N_2O released by soil bacteria, decomposes ozone through this sequence of reactions.

$$\begin{array}{ll} (1) & \cancel{NO} + O_3 \longrightarrow \cancel{NO_2} + O_2 \\ (2) & \cancel{NO_2} + O \longrightarrow \cancel{NO} + O_2 \\ \hline \textit{Overall reaction:} & O_3 + O \longrightarrow 2\,O_2 \end{array}$$

Note that NO consumed in the first reaction is regenerated in the second. A little NO goes a long way. Moreover, we should expect that the injection of any additional NO into the stratosphere would increase the destruction of ozone and reduce its steady-state concentration. This additional NO can come, for example, from combustion processes in supersonic jets operating in the stratosphere.

Of all human activities that affect the ozone layer, release of chlorofluorocarbons (CFCs) is thought to be the most significant. CFCs have a long lifetime in the atmosphere. Some of the molecules eventually rise and appear in low concentrations in the stratosphere. There, they absorb UV radiation and decompose to produce atomic and molecular fragments (radicals).

$$CCl_2F_2 + h\nu \longrightarrow \cdot CClF_2 + Cl\,\cdot$$

The ozone-destroying cycles set up beyond this point are quite complex, but we can give a simplified representation as follows.

$$\begin{array}{ll} (1) & \cancel{Cl\cdot} + O_3 \longrightarrow \cancel{ClO\cdot} + O_2 \\ (2) & \cancel{ClO\cdot} + O \longrightarrow \cancel{Cl\cdot} + O_2 \\ \hline \textit{Overall reaction:} & O_3 + O \longrightarrow 2\,O_2 \end{array}$$

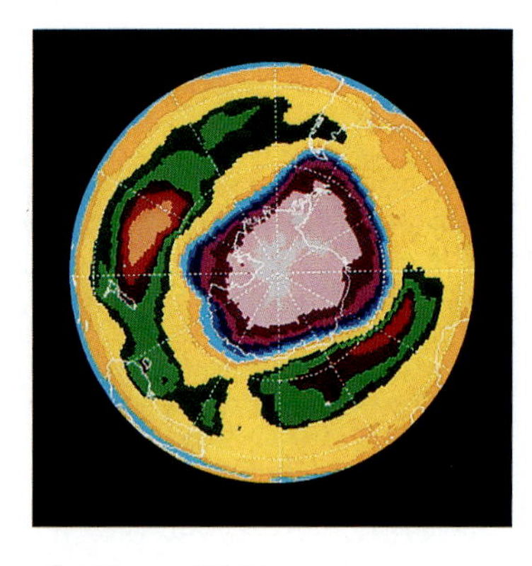

▲ **Figure 25.11**
The ozone hole over Antarctica
In this measurement, recorded in 1994, the different colors represent different ozone concentrations. The gray, pink, and purple areas at the center constitute the ozone hole, with the ozone concentration being lowest in the gray region. Higher concentrations of ozone are found in the yellow, green, and brown areas.

The chlorine atom that reacts in step (1) is regenerated in step (2); one chlorine atom can therefore destroy thousands of ozone molecules. (As in polymerization, the reaction proceeds until ended by the combination of two radicals.) Perhaps the best evidence for the depletion of stratospheric ozone is that obtained from studies in Antarctica (Figure 25.11).

Currently, CFCs and other chlorine- or bromine-containing compounds that might diffuse into the stratosphere and contribute to ozone depletion are being replaced with more benign substances. Fluorocarbons, such as CH_2FCH_3, have no Cl or Br to form radicals and are one kind of replacement. Another kind are hydrochlorofluorocarbons (HCFCs), such as CH_3CCl_2F. These molecules break down more readily in the troposphere, and thus fewer ozone-destroying molecules reach the stratosphere.

25.4 Global Warming: Carbon Dioxide and the Greenhouse Effect

We all exhale carbon dioxide with every breath; it is a normal product of respiration. Consequently, we generally do not think of CO_2 as an air pollutant. Low levels of CO_2 are not toxic. At a concentration of 360 ppm CO_2 (0.0360 mol %), carbon dioxide is a minor component of Earth's atmosphere. However, even this relatively small quantity of CO_2 plays a role in determining Earth's climate. Small increases in the concentration of CO_2 could have a profound effect on the environment by producing a significant increase in the average global temperature, an effect called **global warming**.

APPLICATION NOTE
Trees consume water vapor and carbon dioxide gas in the process of photosynthesis. The deforestation of tropical rain forests, as in "slash-and-burn" clearing for agriculture, is a contributing factor in the increase of $CO_2(g)$ levels in the atmosphere.

When electromagnetic radiation from the sun reaches Earth, some is reflected back into space, some is absorbed by substances in the atmosphere, and some reaches Earth's surface and is absorbed there. The surface gets rid of some of the solar energy by emitting infrared radiation toward outer space. Certain atmospheric gases, principally $CO_2(g)$ and $H_2O(g)$, absorb some of this infrared radiation. This radiant energy is retained in the atmosphere and warms it. The process, known as the **greenhouse effect** because it resembles the retention of heat in a greenhouse, is summarized in Figure 25.12. The greenhouse effect is natural, and it is crucial to maintaining the proper temperature for life on Earth. Without it, Earth would be an icehouse, permanently covered with snow and ice. Scientists are concerned, however, with the effects of continued increases in the concentrations of greenhouse-enhancing gases.

From 1880 to 1995, the average CO_2 content of the atmosphere increased from 275 to 359 ppm. The 1996 level was about 360 ppm, and it currently increases about 1 ppm each year. These increases come from combustion of carbon-containing fuels—wood, coal, natural gas, gasoline—and the cutting of trees that would otherwise consume atmospheric CO_2 through photosynthesis.

The number of huge icebergs that break off the continental ice shelf in Antarctica may increase as a result of global warming.

Computer models of the atmosphere indicate that a CO_2 buildup is likely to cause an increase in Earth's average temperature. There are many uncertainties, however. It is impossible to identify all the factors that should be included in computer models and to know how heavily to weight each factor. For example, warming of the atmosphere could lead to the increased evaporation of water and an accompanying increase in cloud cover. Because clouds reflect some incoming radiation into space, this could cause global cooling rather than global warming. Still, most models predict warming, with some of them forecasting that a doubling of the CO_2 content in air from its preindustrial values will cause a global temperature increase of from 1.5 to 4.5 °C. Major effects could be experienced as early as the middle of the twenty-first century.

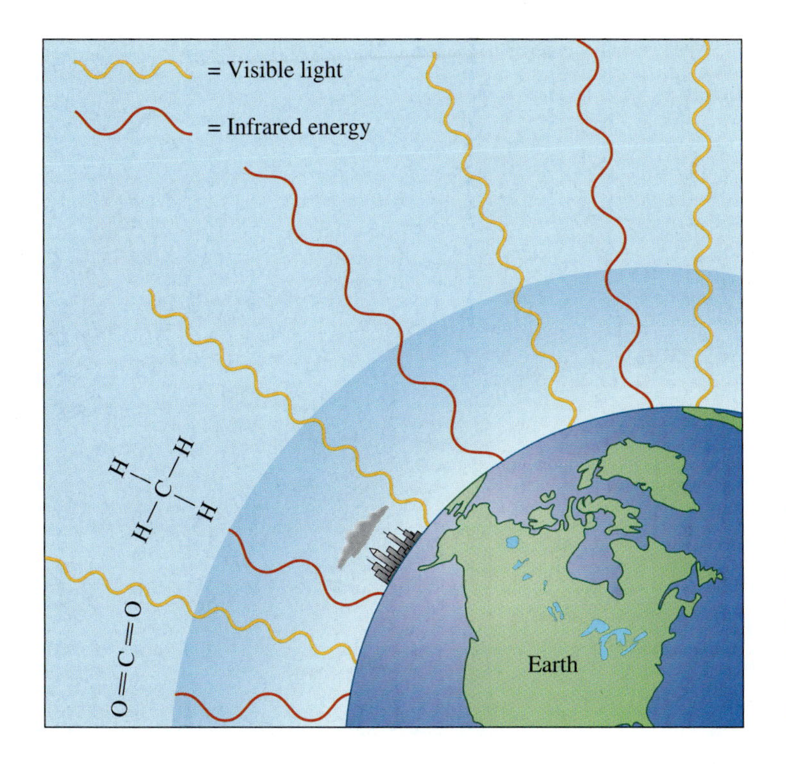

Figure 25.12
The greenhouse effect
Sunlight passing through the atmosphere is absorbed, warming Earth's surface. The warm surface emits infrared radiation. Some of this radiation is absorbed by CO_2, H_2O, and other gases and retained in the atmosphere as thermal energy.

Even a small increase in average global temperature could have profound effects. Two probable effects are:

- Local climate changes: For example, the Great Plains states might become much drier with hotter summers.
- A rise in mean sea level caused by increased melting of polar icecaps and thermal expansion of seawater: The sea level could rise as much as several meters, flooding coastal cities and causing increased erosion of beaches.

Scientists are not limited to computer modeling to assess the likelihood of global warming. Direct experimental evidence also exists. For example, the ice in the Greenland and Antarctic icecaps is laid down in layers much like the annual growth rings of trees. Analyses of tiny air bubbles trapped in these layers show a strong correlation between atmospheric CO_2 content and estimated global temperatures over the past 160,000 years—lower CO_2 levels correlate with lower temperatures, and higher levels with higher temperatures. Thus it seems reasonable to expect that global temperatures will continue to rise as CO_2 levels increase.

Most atmospheric scientists think that global warming is already under way. As if in confirmation, 1997 was the warmest year since accurate temperature records have been kept, and nine of the ten warmest years on record occurred between 1987 and 1997. Others still question the reality of global warming, noting large swings in average temperatures over the years. The uncertainty arises from the fact that the increase in average temperature is only a few tenths of a degree Celsius, whereas annual variations in certain regions are often as much as 4 or 5 °C.

The main strategy for countering a possible global warming is to curtail the use of fossil fuels, but this may not be enough. For example, several gases—methane, ozone, nitrous oxide (N_2O), and CFCs—are even better absorbers of infrared radiation than carbon dioxide is. At an international meeting in Kyoto, Japan, in 1998,

APPLICATION NOTE

Motion pictures such as *Deep Impact* and *Armageddon* feature a large comet or asteroid threatening Earth. In reality, if such a comet or asteroid landed in an ocean, it would blast vast quantities of water vapor into the atmosphere. Water vapor is a powerful greenhouse gas, and one result could be severe global warming. Even before impact, the frictional heat generated by the object would cause nitrogen and oxygen in the atmosphere to form vast quantities of NO. The NO could lead to substantial destruction of the ozone layer.

Natural Pollutants

Even before there were people, there was pollution. Volcanoes erupted, spewing ash and poisonous gases into the atmosphere. They still do. Kilauea in Hawaii emits 200 to 300 tons of SO_2 per day. Acid rain downwind from this volcano has created a barren region called the Kau desert.

The 1991 eruption of Mount Pinatubo in the Philippines injected such vast quantities of particulates into the stratosphere that the reflection of incoming solar radiation by these particles apparently produced a temporary reversal of the trend toward global warming. Earth's average surface temperature fell from 15.47 °C in 1990 to 15.13 °C in 1992.

Dust storms, especially in arid regions, add massive amounts of particulate matter to the atmosphere. Smoke and dust from forest fires in Mexico and Central America drift across the United States and into Canada. Dust from the Sahara Desert reaches the Caribbean area and South America. Swamps and marshes emit noxious gases such as hydrogen sulfide, a toxic gas with the odor of rotten eggs.

An important source of particulates in the atmosphere is the sea. Wave action causes seawater droplets to be suspended in the air, and when these droplets evaporate, salt particles are left behind. Tropical thunderstorms even carry chloride ions into the stratosphere, where they can be converted to chlorine atoms and perhaps contribute to the destruction of ozone. Thus, the greatest contributor to all the particulate matter in the atmosphere is ordinary salt, and it comes from a perfectly natural source. Nature isn't always benign.

An eruption at Kilauea volcano on the island of Hawaii.

the industrial nations agreed to limit emissions of CO_2 over the next several decades. There is still much debate about the need for immediate action and how drastic our efforts should be. The debate is likely to continue for years.

The Hydrosphere

Earth is a water world. Most of its surface is covered with oceans, seas, lakes, and streams, waters collectively called the *hydrosphere*. The human body is mostly water, too—about two thirds by mass. The water in blood is similar to water in the ocean. In fact, we are much like walking sacks of seawater. Although primitive life forms are found in some seemingly inhospitable places on Earth and could possibly exist elsewhere in the solar system, Earth alone has the large quantity of liquid water necessary for higher life forms as we know them.

Some of the properties of water that make it able to support life, however, also make it easy to pollute. Many chemical substances are soluble in water. In fact, much of the chemistry we have studied is that of aqueous solutions. Soluble substances are easily dispersed and eventually enter our water supplies. Removing them, once they are there, is often quite difficult. A daily supply of clean drinking water is essential to life and good health. Contaminated water can cause disease and in some severe cases, death.

25.5 Earth's Natural Waters

We have noted the unusual properties of water in previous chapters. The following list provides a quick review of some of the properties and their consequences.

- Water commonly occurs as a liquid, the only prevalent naturally occurring liquid on Earth's surface.
- The solid form of water (ice) is less dense than the liquid; water expands when it freezes.
- Water has a higher density than most other familiar liquids; hydrocarbons and other organic compounds that are insoluble in water and less dense than water float on its surface.
- Water has a high heat capacity and a high heat of vaporization. Thus, a given quantity of heat produces a much greater temperature increase in a landmass than in a body of water covering the same area. Bodies of water tend to make daily and seasonal temperature variations more moderate along their shores.

Although three fourths of Earth's surface is covered with water, nearly 98% is salty seawater, unfit for drinking and unsuitable for most industrial purposes. Something less than 2% of Earth's water is frozen in the polar icecaps, and less than 1% is available as freshwater. Freshwater falls on Earth in enormous amounts as rain and snow, but most of it falls into the sea or in areas otherwise inaccessible. Thus, available water is not always where the people are. Some areas with adequate rainfall are unsuitable for human habitation because of extreme cold climates or steep mountain slopes. Some areas have adequate *average* rainfall but have periods of drought and spans of flooding. In some places with adequate freshwater supplies, the water is too polluted for many uses.

Natural waters such as rainwater and groundwater are not pure H_2O. Rainwater carries dust particles from the atmosphere and dissolves a little oxygen, nitrogen, and carbon dioxide as it falls through the atmosphere. During electrical storms, traces of nitric acid are found in rainwater as well. Groundwater dissolves minerals from rocks and soil as it moves along on or beneath Earth's surface. Groundwater also dissolves matter from decaying plants and animals. The principal cations in natural groundwater are Na^+, K^+, Ca^{2+}, Mg^{2+}, and sometimes Fe^{2+} or Fe^{3+}. The anions are usually SO_4^{2-}, HCO_3^-, and Cl^-. Table 25.2 provides a summary of substances found in natural waters.

25.6 Water Pollution

Early people did little to pollute the water and the air, if only because their numbers were few. The industrial revolution and a concurrent large increase in population led to serious pollution of the environment. Even then, the pollution was mostly local and largely biological. Human wastes were dumped on the ground or into the nearest stream. Disease organisms were transmitted through food, water, and direct contact.

TABLE 25.2 Some Substances Found in Natural Waters

Substance	Formula	Source
Carbon dioxide	CO_2	Atmosphere
Dust	—	Atmosphere
Nitrogen	N_2	Atmosphere
Oxygen	O_2	Atmosphere
Nitric acid (thunderstorms)	HNO_3	Atmosphere
Sand and soil particles	—	Soil and rocks
Sodium ions	Na^+	Soil and rocks
Potassium ions	K^+	Soil and rocks
Calcium ions	Ca^{2+}	Limestone rocks
Magnesium ions	Mg^{2+}	Dolomite rocks
Iron(II) ions	Fe^{2+}	Soil and rocks
Chloride ions	Cl^-	Soil and rocks
Sulfate ions	SO_4^{2-}	Soil and rocks
Bicarbonate ions	HCO_3^-	Soil and rocks

Contamination of water supplies by microorganisms from human wastes was a severe problem throughout the world until about 100 years ago. During the 1830s, severe epidemics of cholera swept the Western world. Typhoid fever and dysentery were common. In 1900, for example, there were more than 35,000 deaths from typhoid in the United States. Today, as a result of chemical treatment, municipal water supplies in the more developed nations are generally safe. However, waterborne diseases are still quite common in much of Asia, Africa, and Latin America. Worldwide, an estimated 80% of all the world's sickness is caused by contaminated water. People with waterborne diseases fill half the world's hospital beds and die at a rate of 25,000 people per day. Less than 10% of the people of the world have access to sufficient clean water. We still have epidemics of cholera, typhoid, and dysentery in many parts of the world.

How much water does one really need? We need only about 1.5 L per day for drinking. The average United States resident uses about 7 L for drinking and cooking, but overall consumes directly about 400 L (Table 25.3). The combined resi-

TABLE 25.3 Average Daily per Person Use of Water in the United States

Use	Amount (L)	
Direct use		
Drinking and cooking	7	
Flushing toilets	80	
Swimming pools and lawn watering	85	
Dish washing	14	
Bathing	70	
Laundry	35	
Miscellaneous	90	
Total direct use		400
Indirect use		
Industrial	3800	
Irrigation (agriculture)	2150	
Municipal water (nonindustrial)	550	
Total indirect use		6500
Total overall use		6900

dential, industrial, commercial, and agricultural use of water adds up to an average of 6900 L per person per day; and the rate of use is rapidly increasing. Much of this water is used *indirectly* in agriculture and industry to produce food and other materials. For example, it takes 800 L of water to produce 1 kg of vegetables and 13,000 L of water to produce a steak from beef cattle fed from irrigated croplands. We also use water for recreation (for example, swimming, boating, and fishing). For most of these purposes, we need water that is free from bacteria, viruses, and parasitic organisms.

The threat of biological contamination has not been totally eliminated from the developed nations. An estimated 30 million people in the United States are at risk because of bacterial contamination of drinking water. Hepatitis A, a viral disease spread through drinking water and contaminated food, at times threatens to reach epidemic proportions, even in the most developed nations. Biological contamination also lessens the recreational value of water, leading to a ban of swimming in many areas.

Chemical Contamination: From Farm, Factory, and Home

In the past, factories often were built on the banks of streams, and wastes were dumped into the water to be carried away. Currently, fertilizers and pesticides used in agriculture and on lawns and recreational areas such as golf courses have found their way into the water system and further contaminated it. Transportation of petroleum results in oil spills in oceans, estuaries, and rivers. Acids enter waterways from mines and factories and from acid precipitation. Household chemicals also contribute to water pollution when detergents, solvents, and other chemicals are dumped down drains.

About half the people in the United States drink surface water (from streams and lakes). The other half get their drinking water from groundwater. Toxic chemicals from various sources have been found in both kinds of water supplies. For example, people living close to the Rocky Mountain Arsenal, near Denver, have found their wells contaminated by wastes from the production of pesticides. Wells near Minneapolis, Minnesota, are contaminated with creosote, a chemical used as a wood preservative. Water wells in Wisconsin and on Long Island have been contaminated with aldicarb, a pesticide used on potato crops. Community water supplies in New Jersey have been shut down because of contamination with industrial wastes.

Chemicals buried in dumps—often years ago, before there was much awareness of environmental problems—have now infiltrated groundwater supplies. Often, as at the Love Canal site in Niagara Falls, New York, people built schools and houses on or near old dump sites. Common contaminants are hydrocarbon solvents, such as benzene and toluene, and chlorinated hydrocarbons, such as carbon tetrachloride (CCl_4), chloroform ($CHCl_3$), and methylene chloride (CH_2Cl_2). Especially common is trichloroethylene ($CCl_2{=}CHCl$), widely used as a dry-cleaning solvent and as a degreasing compound. These organic compounds dissolve in water only in trace quantities, often in the ppm or ppb range, and it is these tiny amounts that are found in groundwater. However, these chlorinated hydrocarbons are unwanted even in trace amounts, for most of them are suspected carcinogens. Unfortunately, the compounds are notably lacking in reactivity. They react so slowly that they are likely to be around for a long time.

Another major source of groundwater contamination is leaking underground storage tanks. Gasoline at service stations has traditionally been stored in buried steel tanks. There are perhaps 2.5 million such tanks in the United States. The tanks last an average of about 15 years before they rust through and begin to leak. As many as 200,000 may now be leaking, many of them at stations that went out of business

Ironing Out Pollutants

Contaminated groundwater from a leaking landfill often moves in a plume toward a nearby lake or stream. Is there an effective way to remove contaminants as the water moves through the ground? One way that is relatively inexpensive and quite effective for removing chlorinated compounds is to place a pit containing scrap iron filings as a barrier in the path of the flow (Figure 25.13).

Recall that iron corrodes through a redox reaction in which dissolved oxygen is the oxidizing agent and iron is the reducing agent.

Oxidation half-reaction: $2\{Fe \longrightarrow Fe^{2+} + 2e^-\}$

Reduction half-reaction: $O_2 + 4H^+ + 4e^- \longrightarrow 2H_2O$

Overall reaction: $2Fe + O_2 + 4H^+ \longrightarrow 2Fe^{2+} + 2H_2O(l)$

In a reaction with chlorinated compounds, Fe atoms give electrons to the chlorine atoms, converting them to chloride ions. In the reduction half-reaction, a chlorine-containing compound, designated RCl, is converted to a hydrocarbon, RH.

Reduction half-reaction: $RCl + H^+ + 2e^- \longrightarrow RH + Cl^-$

The overall reaction can be written as follows.

$$Fe + RCl + H^+ \longrightarrow Fe^{2+} + RH + Cl^-$$

The hydrocarbons formed are generally less toxic and more easily degraded by microorganisms than the chlorine-containing compounds.

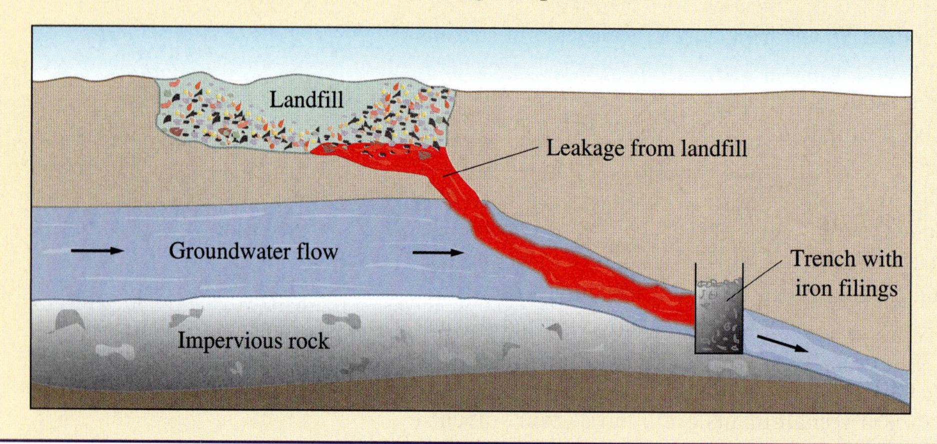

Figure 25.13
Using iron filings to remove chlorinated hydrocarbons from contaminated groundwater
As contaminated groundwater from a leaking landfill moves toward a nearby lake or stream, it passes through a barrier of iron filings. The iron converts the chlorinated compound to a hydrocarbon and chloride ions.

during the fuel shortages of the 1970s. Gasoline is now found in wells near these tanks in many areas of the country. Laws now require replacement of old gasoline storage tanks and proper cleanup of any contaminated ground.

Groundwater contamination is vexing because, once contaminated, an underground aquifer may remain unusable for decades or longer. There is no easy way to remove the contaminants. Pumping out the water and purifying it could take years and cost billions of dollars.

Groundwater pollution is often overdramatized by the news media. The amounts of contaminants are often truly minute. For example, the U.S. Environmental Protection Agency has set a safety limit of 10 ppb for aldicarb—that's 10 mg of aldicarb in 1000 L of water. You would have to drink 32,000 L of water to get as much aldicarb as there is aspirin in one tablet. Aldicarb is moderately toxic to mammals, but it breaks down rather quickly in the environment. Contamination of groundwater by toxic substances is serious, but biological contamination is much more widespread and often more deadly than chemical contamination.

Industries in the United States have eliminated a considerable proportion of the water pollution they once produced. Most industries are now in compliance with the Water Pollution Control Act, which requires that they use the best practicable technology. Let's consider a specific case in pollution control.

Waste chromium, in the form of chromate ions (CrO_4^{2-}), and cyanide ions (CN^-) are products of the chromium plating of steel. In the past, these toxic substances were often dumped into waterways. Nowadays, they generally are removed to a large extent by chemical treatment.

Cyanide ions are removed by a reaction with chlorine in basic solution to form nitrogen gas, bicarbonate ions, and chloride ions.

$$10\ OH^-(aq) + 2\ CN^-(aq) + 5\ Cl_2(g) \longrightarrow N_2(g) + 2\ HCO_3^-(aq) + 10\ Cl^-(aq) + 4\ H_2O(l)$$

Bicarbonate and chloride ions are generally not considered toxic at the levels formed in this reaction.

Chromate ions are removed by reduction with sulfur dioxide, forming Cr^{3+} ion and sulfate ions.

$$2\ CrO_4^{2-}(aq) + 3\ SO_2(g) + 2\ H_2O(l) \longrightarrow 2\ Cr^{3+}(aq) + 3\ SO_4^{2-}(aq) + 4\ OH^-(aq)$$

Cr^{3+} can be precipitated and removed as $Cr(OH)_3(s)$. However, proper control of the pH is essential because $Cr(OH)_3$ is amphoteric; it redissolves in a solution that is either too strongly acidic or too basic. Sulfate ion is generally not a serious pollutant.

Many industries have contributed to water pollution. Wastes from the textile industries include conditioners, dyes, bleaches, oils, dirt, and other organic debris. Most of these can be removed by conventional sewage treatment. Wastes from meatpacking plants include blood and various animal parts. These and other food industry wastes also are usually treated by regular sewage treatment plants.

TABLE 25.4 Some Substances Added to Water by the Breakdown of Organic Matter

Substance	Formula
Aerobic conditions	
Carbon dioxide	CO_2
Nitrate ions	NO_3^-
Phosphate ions	PO_4^{3-}
Sulfate ions	SO_4^{2-}
Bicarbonate ions	HCO_3^-
Anaerobic conditions	
Methane	CH_4
Ammonia	NH_3
Amines	RNH_2
Hydrogen sulfide	H_2S
Methanethiol	CH_3SH

Some Chemistry and Biology of Sewage

Dumping of human sewage into waterways spreads *pathogenic* (disease-causing) microorganisms, but disease is not the only problem. The organic matter in sewage is decomposed by bacteria, but this depletes dissolved oxygen [($O_2(aq)$)] in the water and enriches the water with plant nutrients. A flowing stream can handle a small amount of waste without difficulty, but large quantities of raw sewage cause undesirable changes.

Most organic material can be degraded (broken down) by microorganisms. This biodegradation can be either *aerobic* or *anaerobic*. **Aerobic oxidation** occurs in the presence of dissolved oxygen. The **biochemical oxygen demand (BOD)** measures the quantity of oxygen, in milligrams, needed for the oxidation of the organic compounds in 1 liter of water. If the BOD is high enough, oxygen is depleted and higher life forms, such as fish, can no longer survive in the water. However, rapidly flowing streams can regenerate themselves downstream as oxygen is dissolved from the air by the moving water.

With adequate dissolved oxygen, aerobic bacteria (those that require oxygen) oxidize the organic matter to carbon dioxide, water, and a variety of inorganic ions (Table 25.4). The water is relatively clean, but the ions, particularly the nitrates and phosphates, may serve as nutrients for the growth of algae, which also cause problems. When the algae die, they become organic waste and increase the BOD through a process called **eutrophication**. Algal bloom and die-off are also stimulated by the

An algal bloom leads to oxygen depletion in the water of this pond.

runoff of fertilizers from farms and lawns and seepage from feedlots. This combination leads to dead and dying streams and lakes that nature cannot purify nearly as quickly as we can pollute them.

When the dissolved oxygen in a body of water is depleted by too much organic matter—whether from sewage, dying algae, or other sources—**anaerobic decay** processes take over. Instead of oxidizing the organic matter, anaerobic bacteria reduce it. Methane (CH_4) is formed. Sulfur is converted to hydrogen sulfide (H_2S) and other foul-smelling organic compounds. Nitrogen is reduced to ammonia and odorous amines. The foul odors indicate that the water is overloaded with organic wastes. No life, other than a few anaerobic organisms, can survive in such water.

Methane, sometimes called marsh gas, is formed by the anaerobic decay of organic matter. This painting, *Dalton Collecting Marsh Fire Gas*, shows John Dalton, developer of the atomic theory (Chapter 2).

Water Treatment: A Drop to Drink

About half of us drink water that comes from reservoirs, lakes, and rivers. Many cities use water that has been used by other cities upstream. Such water may be polluted with chemicals and pathogenic microorganisms. To make the water safe and palatable involves several steps of physical and chemical treatment (Figure 25.14).

The water to be purified usually is placed in a settling basin, where it is treated with slaked lime and a flocculating agent such as aluminum sulfate. These materials react to form a gelatinous mass (flocs) of aluminum hydroxide that carries down dirt particles and bacteria.

$$\underset{\text{Slaked lime}}{3\,Ca(OH)_2(aq)} + Al_2(SO_4)_3(aq) \longrightarrow 2\,Al(OH)_3(s) + 3\,CaSO_4(s)$$

The water is then filtered through sand and gravel.

The next step is usually *aeration*: Water is sprayed into the air to remove odorous compounds and to improve its taste (water without dissolved air tastes flat). Sometimes the water is filtered through charcoal, which adsorbs colored and odorous compounds. In the final step, water is *chlorinated*: Chlorine is added to kill any remaining bacteria. In some communities that use lake or river water, a lot of chlorine is needed to kill all the bacteria, and the residual chlorine imparts a taste to the water.

Some pollutants are difficult to remove from water. For example, nitrates get in the groundwater from the use of fertilizers on farms and lawns, from decomposition of organic wastes in sewage treatment, and from runoff from animal feedlots.

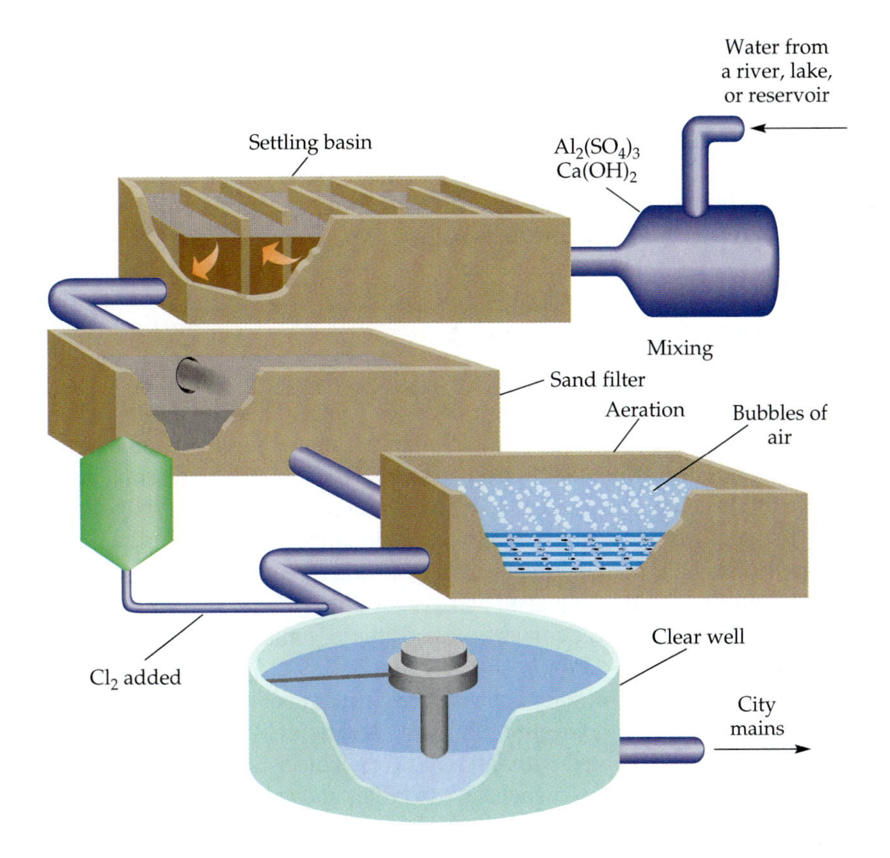

Figure 25.14
A diagram of a municipal water purification plant

Nitrates are highly soluble, and only expensive treatment can remove nitrate compounds from water once they are there.

Many people drink bottled water to avoid real or perceived problems with public water supplies. Bottled water was largely unregulated until 1995, when the Food and Drug Administration set standards to ensure that its minimum quality was equal to that of public water supplies. Much of it comes from the same municipal supplies as tap water.

In addition to chlorination to kill bacteria , many communities add fluorides to drinking water to prevent dental caries (tooth decay). Tooth decay was once considered to be the leading chronic disease of childhood, but it is rarer today, thanks in part to fluoridation of drinking water. While only 29% of the nine-year-olds in the United States were cavity-free in 1971, over two thirds of that age group are cavity-free today.

Recall that tooth enamel is a complex calcium phosphate called hydroxyapatite (page 712). Fluoride ions replace some of the hydroxide ions, forming a harder mineral called fluorapatite.

$$Ca_5(PO_4)_3OH + F^- \longrightarrow Ca_5(PO_4)_3F + OH^-$$

Enough fluoride (usually as H_2SiF_6 or Na_2SiF_6) is added to the water to give concentrations of 0.7 to 1.1 ppm (by mass) of fluoride. Evidence indicates that such fluoridation results in a reduction in the incidence of dental caries by as much as two thirds in some areas.

There is some concern about cumulative effects of consuming fluorides in drinking water, in the diet, in toothpaste, and from other sources. Excessive fluoride consumption during early childhood can cause mottling of the tooth enamel.

The enamel becomes brittle in certain areas and gradually discolors. At higher concentrations than those found in drinking water, fluorides may interfere with calcium metabolism, with kidney action, with thyroid function, and with the actions of other glands and organs. In moderate to high concentrations, fluorides are acute poisons. Indeed, sodium fluoride is used as a poison for roaches and rats. However, there is little or no evidence that fluoridation at the levels now used causes any health problems. Nevertheless, fluoridation of public water supplies remains a controversial matter.

Wastewater Treatment Plants

For many years, most communities simply held sewage in settling ponds for a while before discharging it into a stream, lake, or ocean. This constitutes what now is called **primary sewage treatment**. Primary treatment removes some of the solids as *sludge*, but the effluent still has a huge BOD. Often all the dissolved oxygen in the pond is used up, and anaerobic decomposition—with its resulting odors—takes over. Effluent from a primary treatment plant contains considerable dissolved and suspended organic matter. A **secondary sewage treatment** plant passes effluent from a settling tank through sand and gravel filters. There is some aeration in this step, and aerobic bacteria convert most of the organic matter to stable inorganic materials.

A combination of primary and secondary treatment methods, known as the **activated sludge method** (Figure 25.15), is frequently employed. The sewage is placed in tanks and aerated with large blowers. This causes the formation of large, porous clumps called *flocs*, which serve to filter and absorb contaminants. The aerobic bacteria further convert the organic material to sludge. A part of the sludge is recycled to keep the process going, but huge quantities must be removed for disposal. This sludge is stored on land (where it requires large areas), dumped at sea (where it pollutes the ocean), or burned in incinerators (where it requires energy—such as natural gas—and contributes to air pollution). Sometimes the sludge is processed for use as fertilizer.

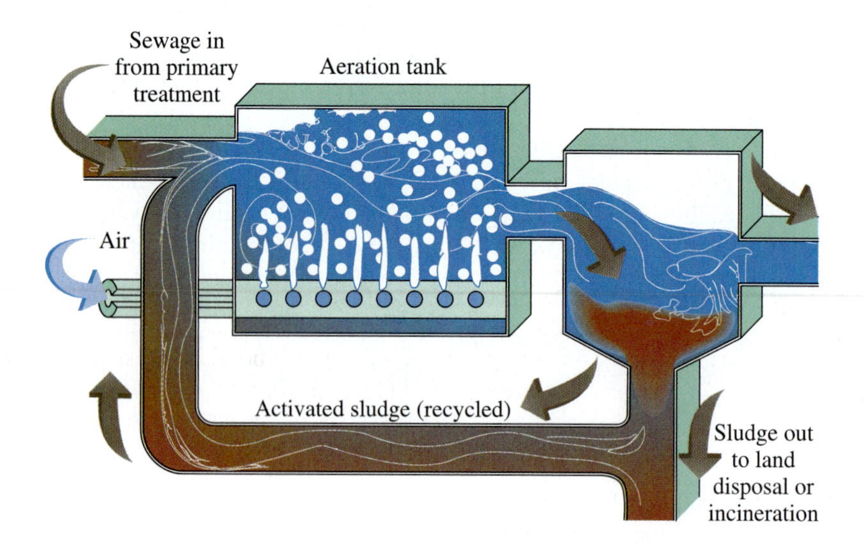

Figure 25.15 **A diagram of a secondary sewage treatment plant that uses the activated sludge method of treatment**

In many areas, secondary treatment of wastewater is inadequate. **Advanced treatment**, sometimes called *tertiary treatment*, is increasingly required. Several advanced processes are now in use. One process that is becoming more important is charcoal filtration. Charcoal adsorbs organic molecules that are difficult to remove

TABLE 25.5 Summary of Wastewater Treatment Methods

Method	Cost	Material Removed	Percent Removal
Primary			
Sedimentation	Low	Dissolved organics	25–40
		Suspended solids	40–70
Secondary			
Trickling filters	Moderate	Dissolved organics	80–95
		Suspended solids	70–92
Activated sludge	Moderate	Dissolved organics	85–95
		Suspended solids	85–95
Advanced (tertiary)			
Carbon bed with regeneration	Moderate	Dissolved organics	90–98
Ion exchange	High	Nitrates and phosphates	80–92
Chemical precipitation	Moderate	Phosphates	88–95
Filtration	Low	Suspended solids	50–90
Reverse osmosis	Very high	Dissolved solids	65–95
Electrodialysis	Very high	Dissolved solids	10–40
Distillation	Extremely high	Dissolved solids	90–98

by other methods. There are federal mandates for more communities to treat their water with charcoal, but some cities have not yet complied because of the expense. Even more costly processes, such as reverse osmosis (page 540), may be needed in some cases. Table 25.5 provides a summary of wastewater treatment methods.

The effluent from sewage plants usually is treated with chlorine before it is returned to a waterway. Chlorine is added in an attempt to kill any pathogenic microorganisms that might remain. Such treatment has been quite effective in preventing the spread of waterborne infectious diseases such as typhoid fever. Further, some chlorine remains in the water, providing residual protection against pathogenic bacteria. However, chlorination is not effective against viruses, such as those that cause hepatitis.

25.7 Acid Rain and Acid Waters

We have seen how sulfur oxides are converted to sulfuric acid and nitrogen oxides to nitric acid in the atmosphere. These acids fall to Earth as acid rain or acid snow or are deposited from acid fog or adsorbed on (attached to) particulate matter. When rainfall is more acidic than it would be just by dissolving atmospheric $CO_2(g)$, it is called **acid rain**. Some rainfall has been reported that is even more acidic than vinegar or lemon juice. Acid rain comes mainly from sulfur oxides emitted from power plants and smelters and from nitrogen oxides from automobiles. These acids may be carried for hundreds of kilometers before falling as rain or snow (Figure 25.16).

Acid Waters: Dead Lakes

Acid rain corrodes metals, limestone, and marble, and even ruins the finishes on our automobiles. Acids also flow into streams from abandoned mines.

Acid water is detrimental to life in lakes and streams. More than 1000 bodies of water in the eastern United States are acidified, and 11,000 others have only a limited ability to neutralize the acids that enter them. In the Canadian province of Ontario, 48,000 lakes are threatened, and more than 100 lakes are so acidic that

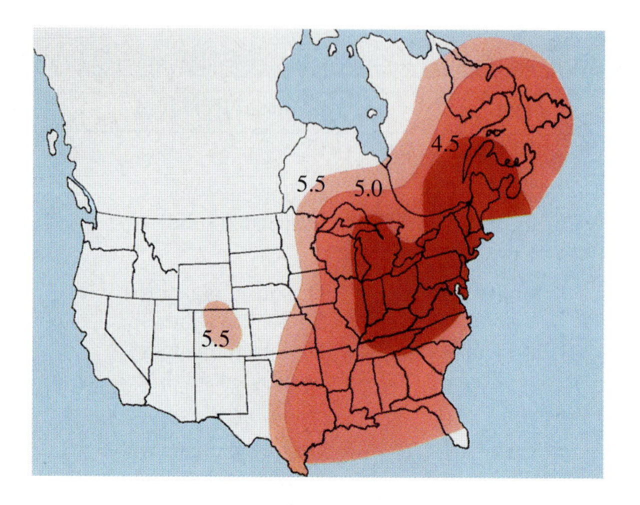

Figure 25.16
Acid rain
Acids formed from oxides of sulfur, SO_x, mainly from coal-burning power plants, and oxides of nitrogen, NO_x, from power plants and automobiles, may fall in rain hundreds of kilometers from their sources. They cross oceans and international boundaries, causing political as well as environmental problems. This map shows that acid deposition is most severe over the northeastern United States and eastern Canada. The numbers indicate average pH values.

they are devoid of life. The acid rain falling on these areas is thought to originate mainly in the Ohio River Valley and Great Lakes regions.

Acid rain has been linked to declining crop and forest yields. The effects of acid waters on living organisms are hard to pin down precisely. Probably the greatest effect of acidity is that it causes the release of toxic ions from rocks and soil. For example, aluminum ions, which are tightly bound in clays and other minerals, are released in acidic solution.

$$\underset{\text{A clay}}{Al_2Si_2O_5(OH)_4(s)} + 6\,H_3O^+(aq) \longrightarrow 2\,Al^{3+}(aq) + \underset{\text{Sand}}{2\,SiO_2(s)} + 11\,H_2O(l)$$

Aluminum ions have low toxicity to humans, but they seem to be deadly to young fish. Many of the dying lakes have only old fish; none of the young fish survive. Ironically, lakes destroyed by excess acidity are often quite beautiful. The water is clear and sparkling—quite a contrast to those in which fish are killed by oxygen depletion following algal blooms.

Acids are no threat to lakes and streams in areas where the rock is limestone, which can neutralize excess acid. Where rock is principally granite, however, no such neutralization occurs. Acidic waters can be neutralized by adding lime or pulverized limestone. A few such attempts have been carried out, but the process is costly and the results last only a few years. An obvious way to alleviate the problem is to remove the sulfur from the coal before combustion or to scrub the sulfur oxides from the smokestack gases. Considerable progress has been made in this area, but these remedies are expensive, and they add to the cost of electricity.

Toxic Substances in the Biosphere

In concluding our study of chemistry, we briefly consider the most important "sphere" on Earth: The **biosphere** is the relatively thin film of air, water, and soil in which almost all life on Earth exists. Nearly all the processes in the biosphere are driven by energy from the sun. About 23% of the solar energy that reaches Earth drives the water cycle. A tiny portion is absorbed by green plants, which use it to power photosynthesis, directly supplying the energy that sustains life. Let's now turn our attention to toxic substances that sometimes seem to threaten living things.

One of the main concerns about certain pollutants is their toxicity. Toxic substances, often called poisons, have always been with us. However, our knowledge of them is greater, and more of them are available than ever before. Industrial accidents, such as that at Bhopal, India, in 1984, which killed 2500 people and injured perhaps 100,000 others, have made the public acutely aware of problems with toxic substances. People are also concerned about long-term exposure to toxic substances in the air, in their drinking water, and in their food. Chemists can detect exceedingly tiny quantities of such substances. It is still quite difficult, however, to determine the effects of these trace amounts of toxic materials on human health. **Toxicology** is the study of the effects of poisons, their identification or detection, and the development and use of antidotes.

Many common plants and plant parts including hydrangeas (top), holly berries (middle), and the philodendron (bottom) are poisonous.

25.8 Poisons

What is a poison? Perhaps a better question would be, How much is a poison? A substance may be harmless—or even a necessary nutrient—in one amount, and injurious, or even deadly, in another. Even a common substance such as table salt can be poisonous when eaten in abnormally large amounts. Too much salt—sodium chloride—can induce vomiting. There have even been cases of fatal poisoning when salt was accidentally substituted for lactose (milk sugar) in formulas for infants. Some substances are obviously more toxic than others, however. It would take a massive dose of salt to kill the average healthy adult, whereas only a few micrograms of some of the nerve poisons can be fatal. Toxicity largely depends on the chemical nature of the substance.

People also respond differently to the same chemical. To cite an extreme case, 10 to 20 grams of sugar would cause no acute symptoms in most people but might be dangerous to a diabetic. Excessive amounts of salt would be especially serious to a person with edema (swelling due to excessive amounts of fluid in the tissues).

Still another complicating factor is that chemicals behave differently when administered in different ways. Nicotine is more than 50 times as toxic when applied intravenously as when taken orally. Good, fresh water is delightful when we drink it, but even water can be deadly when *inhaled* in sufficient quantity. Further complications arise from the fact that even closely related animal species can react differently to a given chemical. Even individuals within a species may react to different degrees.

Poisons Around the House and in the Garden

Many household chemicals are poisonous. Drain cleaners, oven cleaners, and toilet bowl cleaners are highly corrosive to tissues. Some insecticides and rodenticides are quite toxic. Laundry bleach and ammonia are both toxic, and when they are mixed together, they can be deadly. Sodium hypochlorite in the bleach reacts with NH_3 to form highly toxic chloramines (NH_2Cl, $NHCl_2$, and NCl_3) and nitrosyl chloride ($NOCl$).

Even seemingly harmless products around the house can be dangerous if a young child happens to get into them. A bottle of cough syrup can trigger an emergency visit to the hospital. Toxic substances are also used in home gardens. Herbicides and insecticides are not the only poisons you are likely to find in a garden. Sometimes the plants themselves are toxic. Iris are beautiful, and so are azaleas, hydrangeas, and oleander, but all of these popular perennials are poisonous. Holly berries, wisteria seeds, and the leaves and berries of privet hedges are also among the more poisonous products of the home garden. Some houseplants, such as the philodendron, are also toxic.

Corrosive Poisons

Strong acids and bases and strong oxidizing agents are highly corrosive to human tissue. These chemicals indiscriminately destroy living cells. Both acids and bases, even in dilute solutions, catalyze the hydrolysis of protein molecules in living cells. These reactions involve the breaking of the amide (peptide) linkages in the molecules. In cases of severe exposure, the fragmentation continues until the tissue is completely destroyed.

Acidic air pollutants, such as sulfuric acid aerosols or acids formed in the incineration of plastics and other wastes, are particularly destructive of lung tissue. Other air pollutants also damage living cells. Ozone, peroxyacetyl nitrate (PAN), and the other oxidizing components of photochemical smog probably do their main damage through the deactivation of enzymes. The active sites of enzymes often incorporate —SH groups, and ozone can oxidize these groups to sulfonic acid groups, $—SO_3H$. This change renders the enzyme inactive and halts vital processes in the cell. No doubt, oxidizing agents can also break bonds in many of the chemical substances in a cell. Such powerful agents as ozone are more likely to make an indiscriminate attack than to react in a highly specific way.

Agents That Block Oxygen Transport and Use

Certain chemical substances block the transport of oxygen in the bloodstream and prevent the oxidation of metabolites by oxygen in the cells. All act on the iron atoms in complex protein molecules. Carbon monoxide binds tightly to the iron atom in hemoglobin, blocking the transport of oxygen (Section 25.2). Nitrates also diminish the ability of hemoglobin to carry oxygen (Section 25.6). Nitrates are reduced to nitrites by microorganisms in the digestive tract, and in turn the iron atoms in hemoglobin are oxidized from Fe^{2+} to Fe^{3+}. The resulting *methemoglobin* is incapable of carrying oxygen.

Cyanides are among the most notorious poisons in both fact and fiction. Sodium cyanide is used to extract gold and silver from ores and in electroplating baths, and cyanides can enter the environment from these processes. Hydrogen cyanide is used (with great care by specially trained experts) to exterminate insects and rodents in the holds of ships, in warehouses, in railway cars, and on citrus and other fruit trees. Hydrogen cyanide is generated easily enough from a cyanide salt by treatment with an acid.

$$CN^-(aq) + H_3O^+(aq) \longrightarrow HCN(g) + H_2O(l)$$

NaCN can be accidentally or deliberately ingested, and HCN can be inhaled. In either case, cyanides act almost instantaneously, and it takes only tiny quantities to kill. The average fatal dose for an adult is only 50 or 60 mg.

Cyanide blocks the oxidation of glucose inside the cell by forming a stable complex with iron- and copper-containing enzymes called *cytochrome oxidases*. The enzymes normally act by providing electrons for the reduction of oxygen in the cell. Cyanide ties up these mobile electrons, rendering them unavailable for the reduction process and bringing an abrupt end to cellular respiration.

Sodium thiosulfate ($Na_2S_2O_3$) is an antidote for cyanide poisoning, but it must be administered quickly. A sulfur atom is transferred from the thiosulfate ion to the cyanide ion, converting cyanide to relatively innocuous thiocyanate ions (SCN^-).

$$S_2O_3^{2-} + CN^- \longrightarrow SCN^- + SO_3^{2-}$$

Unfortunately, few victims of cyanide poisoning survive long enough to be treated.

Heavy Metal Poisons

Most metals and their compounds show some toxicity when ingested in large amounts. Even the essential mineral nutrients can be toxic when taken in excessive amounts (Figure 25.17). In many cases, too little of a metal (deficiency) can be as dangerous as too much (toxicity). For example, the average adult requires 10 to 18 mg of iron every day. If less is taken in, the person suffers from anemia. Yet an overdose can cause vomiting, diarrhea, shock, coma, and even death. As few as 10 to 15 tablets containing 325 mg each of iron (as $FeSO_4$) have been fatal to children.

▲ **Figure 25.17**
The effect of copper ions on the height of oat seedlings
From left to right, the concentrations of Cu^{2+} are 0, 3, 6, 10, 20, 100, 500, 2000, and 3000 mg/L. Plants on the left show varying degrees of deficiency; those on the right show copper ion toxicity. The optimum level of Cu^{2+} for oat seedlings is therefore about 100 mg/L.

We don't know exactly how iron poisoning works. However, the heavy metals—those near the bottom of the periodic table—exert their action primarily by inactivating enzymes (Section 13.11). Both mercury (Hg^{2+}) and lead (Pb^{2+}) act in this way. The symptoms of mercury poisoning, which include loss of equilibrium, sight, feeling, and hearing, often do not show up for several weeks. By the time the symptoms become recognizable, extensive damage has already been done to the brain and the rest of the nervous system. This damage is largely irreversible. Lead poisoning can be treated if detected early enough. As we saw in Chapter 22, it usually is treated by intravenous administration of the calcium salt of EDTA. Pb^{2+} is exchanged for Ca^{2+}, and the lead-EDTA complex is excreted. As with mercury poisoning, the neurological damage done by lead compounds is essentially irreversible. Treatment must be performed early to be effective.

Cadmium (as Cd^{2+}) is also toxic (page 935). Its mode of action is different from that of lead and mercury. Cadmium poisoning leads to loss of calcium ions (Ca^{2+}) from the bones, leaving them brittle and easily broken. It also causes severe abdominal pain, vomiting, diarrhea, and a choking sensation.

Nerve Poisons

To understand how these poisons work, let's consider the action of acetylcholine as a neurotransmitter with the help of the following equation.

$$\underset{\text{Acetylcholine}}{CH_3COOCH_2CH_2N^+(CH_3)_3} + H_2O \underset{\text{Acetylase}}{\overset{\text{Cholinesterase}}{\rightleftharpoons}} \underset{\text{Acetic acid}}{CH_3COOH} + \underset{\text{Choline}}{HOCH_2CH_2N^+(CH_3)_3}$$

Acetylcholine carries a signal from one nerve cell to another. After doing so, it hydrolyzes into acetic acid and choline through the action of the enzyme *cholinesterase*. Another enzyme *acetylase* converts acetic acid and choline back to acetylcholine, which is now ready to carry the next nerve impulse.

Nerve poisons can disrupt the acetylcholine cycle in three different ways.

- Botulin, the deadly toxin given off by the anaerobic bacterium *Clostridium botulinum*, found in improperly processed canned food, blocks the synthesis of acetylcholine. No messenger is formed and no messages are carried between nerve cells. Paralysis sets in, and death occurs, usually by respiratory failure.

- Curare, atropine, and some local anesthetics block the acetylcholine receptor sites. In this case, the message is sent but not received. In the case of local anesthetics, this can be good for pain relief in a limited area, but anesthetics, too, can be lethal in sufficient quantity.

- The third category, called *anticholinesterase poisons*, inhibits the action of cholinesterase. This keeps the level of acetylcholine high; the message is turned on continuously, overstimulating the receptor cells. Anticholinesterase poisons include the organic phosphorus insecticides and chemical warfare compounds such as tabun, sarin, and soman.

The nerve poisons are among the most toxic synthetic chemicals known. They kill when they are inhaled or absorbed through the skin, resulting in the complete loss of muscular coordination and subsequent death by cessation of breathing. The usual antidote is atropine injection and artificial respiration. Without the antidote, death may occur in 2 to 10 minutes.

Chlorinated hydrocarbon pesticides such as DDT are relatively safe to mammals, but they are also nerve poisons. Acute DDT poisoning causes tremors, convulsions, and cardiac or respiratory failure. Chronic exposure to DDT leads to the degeneration of the central nervous system. Other chlorinated compounds such as the PCBs act in a similar manner.

Despite their tremendous potential for death and destruction, the nerve poisons have helped us gain an understanding of the chemistry of the nervous system. It is that knowledge that enables scientists to design antidotes for the nerve poisons. In addition, our increased understanding should contribute to progress along more positive lines—in the control of pain, for example.

Detoxification and Potentiation of Poisons

The human body can handle moderate amounts of some poisons. The liver is able to detoxify some compounds by oxidation, reduction, or coupling with amino acids or other normal body chemicals. Perhaps the most common route is oxidation. Ethanol is detoxified by oxidation to acetaldehyde, which in turn is oxidized to acetic acid, a normal constituent of cells. The acetic acid is then oxidized to carbon dioxide and water.

Highly toxic nicotine from tobacco is detoxified by oxidation to cotinine.

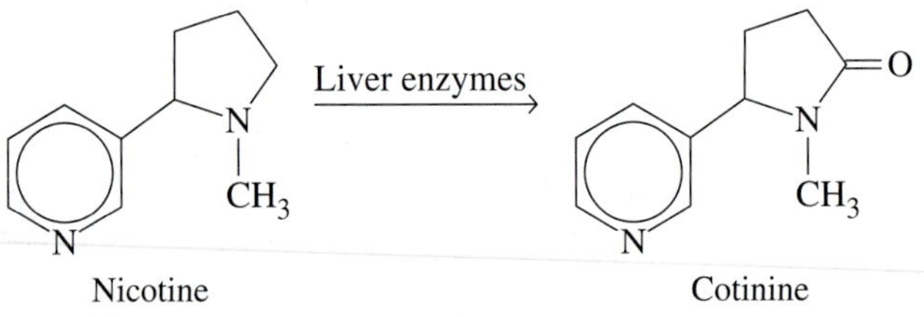

Liver enzymes →

N, CH_3, N — Nicotine

O, N, CH_3, N — Cotinine

Cotinine is less toxic than nicotine. The added oxygen atom also makes cotinine more water soluble than nicotine, and thus more readily excreted in the urine.

The liver is equipped with a system of enzymes, called P-450, that oxidize fat-soluble substances. The P-450 system converts these fat-soluble compounds, which are likely to be retained in the body, into water-soluble ones that are readily excreted. It can also join molecules to amino acids. For example, toluene is essentially insoluble in water. The P-450 enzymes oxidize toluene to more soluble benzoic acid. The latter is then coupled with the amino acid glycine to form hippuric acid, which is still more soluble and is readily excreted.

$$C_6H_5-CH_3 \xrightarrow{\text{Oxidation}} C_6H_5-\overset{O}{\overset{\|}{C}}-OH \xrightarrow[\text{(Glycine)}]{H_2N-CH_2-\overset{O}{\overset{\|}{C}}-OH} C_6H_5-\overset{O}{\overset{\|}{C}}-NH-CH_2-\overset{O}{\overset{\|}{C}}-OH$$

Toluene — Benzoic acid — Hippuric acid

The liver enzymes simply oxidize, reduce, or join molecules together. The end product is not necessarily less toxic. For example, methanol is oxidized to a more toxic form, formaldehyde. It is probably the reaction of formaldehyde with the protein in cells that causes blindness, convulsions, respiratory failure, and death. Also, the same enzymes that oxidize the alcohols deactivate the male hormone testosterone. The buildup of these enzymes in a chronic alcoholic leads to a more rapid destruction of testosterone. This appears to be the mechanism for alcoholic impotence, a well-known characteristic of alcoholism.

Benzene, because of its general inertness in the body, is not acted on until it gets to the liver. There it is slowly oxidized to an epoxide.

$$C_6H_6 \xrightarrow[\text{the liver}]{\text{Oxidation in}} C_6H_6O \text{ (epoxide)}$$

The epoxide is a highly reactive molecule that can attack certain key proteins. This type of reaction, called *potentiation*, converts a relatively harmless chemical into a much more toxic one. The damage done by this epoxide sometimes results in leukemia.

Carbon tetrachloride, CCl_4, is also quite inert in the body. But when it reaches the liver, it is converted to the reactive trichloromethyl free radical $Cl_3C\cdot$, which in turn attacks the unsaturated fatty acids in the body. This action can trigger cancer.

25.9 Carcinogens and Anticarcinogens

Tumors, abnormal growths of new tissue, may be either benign or malignant. Benign tumors are characterized by slow growth; they often regress spontaneously, and they do not invade neighboring tissues. Malignant tumors, often called *cancers*, may grow slowly or rapidly, but their growth is generally irreversible. Malignant growths invade and destroy neighboring tissues. Cancer is not a single disease, but rather a catchall term for over 200 different afflictions. Many are not even closely related to each other.

What causes cancer? The mechanisms of action of chemical and physical factors are probably quite varied. Some chemicals modify DNA, thus scrambling the code for replication and for the synthesis of proteins. For example, aflatoxin B is known to bind to guanine residues in DNA. Just how this initiates cancer, however, is not known for sure.

There is a genetic component to the development of many forms of cancer. Certain genes, called *oncogenes*, seem to trigger or sustain the processes that convert normal cells to cancerous ones. Oncogenes arise from ordinary genes that regulate cell growth and cell division. These oncogenes can be activated by chemical carcinogens, radiation, or perhaps some viruses. It seems that more than one oncogene must be turned on, perhaps at different stages of the process, before a cancer develops. We also have suppressor genes that ordinarily prevent the development of cancers. These genes must be inactivated before a cancer develops. Suppressor gene inactivation can occur through mutation, alteration, or loss of the gene. In all, 10 or 15 mutations may be required in a cell before it turns cancerous. Thus, our bodies have some natural protection against cancer.

A **carcinogen** is a material that causes cancer. Many people seem to believe that chemicals are a major cause of cancer, but most cancers are caused by lifestyle factors. Nearly two thirds of all cancer deaths in the United States are linked to tobacco, our diets, and the lack of exercise with resulting obesity. Even among the chemicals that are most suspect, such as pesticides, many cannot be shown to be carcinogenic. Only about 30 chemical compounds have been identified as human carcinogens. Another 300 or so have been shown to cause cancer in laboratory animals, but it is often difficult to equate the results of tests on laboratory animals with risks to humans. Some of these 300 are widely used and are therefore a subject of some concern.

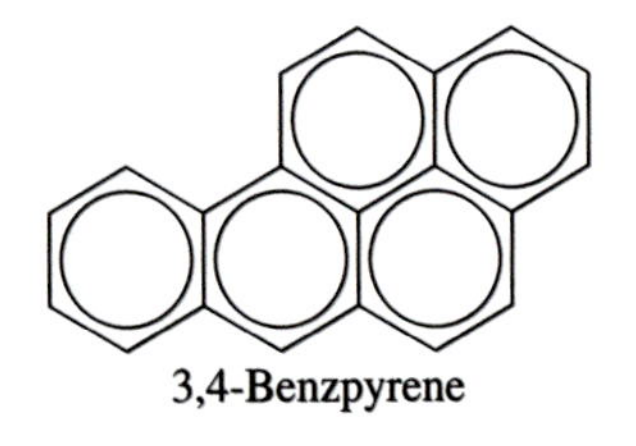
3,4-Benzpyrene

Some of the more notorious carcinogens are polycyclic aromatic hydrocarbons, of which 3,4-benzpyrene is perhaps the best known. Carcinogenic hydrocarbons are formed during the incomplete burning of nearly any organic material. They have been found in charcoal-grilled meats, cigarette smoke, automobile exhausts, coffee, burnt sugar, and many other materials. Not all polycyclic aromatic hydrocarbons are carcinogenic. There are strong correlations between carcinogenicity and certain molecular sizes and shapes.

Another important class of carcinogens is the aromatic amines. Two prominent ones are β-naphthylamine and benzidine. These compounds once were used widely in the dye industry. They were responsible for a high incidence of bladder cancer among the workers whose jobs brought them into prolonged contact with the compounds.

Not all carcinogens are aromatic compounds. Prominent among the aliphatic (nonaromatic) carcinogens are dimethylnitrosamine [$(CH_3)_2N{=}N{=}O$] and vinyl chloride, the monomer from which the polymer polyvinyl chloride is made. Other carcinogens include three- and four-membered heterocyclic rings containing nitrogen or oxygen. Many epoxides, cyclic ethers with three-membered rings, are carcinogenic.

$CH_2-CH-CH-CH_2$ (each pair bridged by O)

Bis(epoxy)butane

H_2C, H_2C ring with $N-(CH_2)_{11}CH_3$

N-Laurylethyleneimine

H_2C-CH_2, $O-C{=}O$

β-Propiolactone

Three small-ring heterocyclic carcinogens.

Few of the known carcinogens are synthetic chemicals. Some, such as safrole in sassafras and the aflatoxins produced by molds on foods, occur naturally. Some researchers estimate that 99.99% of all carcinogens that we ingest are natural ones. Plants produce compounds to protect themselves from fungi, insects, and higher animals, including humans. Some of the compounds are carcinogens that are found in mushrooms, basil, celery, figs, mustard, pepper, fennel, parsnips, and citrus oils—almost every place that a curious chemist looks. Carcinogens are also produced during cooking and as products of normal metabolism.

There are many natural carcinogens in food. So why don't we all get cancer? Some substances in food act as **anticarcinogens**. Fiber is believed to protect against colon cancer. The food additive BHT (butylated hydroxytoluene) may give protection against stomach cancer. And certain vitamins have been shown to have anticarcinogenic effects.

A diet rich in cruciferous vegetables (cabbage, broccoli, brussels sprouts, kale, and cauliflower) has been shown to reduce the incidence of cancer both in animals

How Cigarette Smoking Causes Cancer

The association between cigarette smoking and cancer has been known for decades, but the precise mechanism was not known until 1996. The research scientists who found the link focused on a tumor-suppressor gene called P53, a gene that is mutated in about 60% of all lung cancers. They also focused on a metabolite of benzpyrene, a carcinogen found in tobacco smoke.

In the body, benzpyrene is oxidized into an active carcinogen (an epoxide) that binds to specific nucleotides in the gene—sites called hot spots—where mutations frequently occur. It seems quite likely that the benzpyrene metabolite causes many of those mutations in the P53 gene.

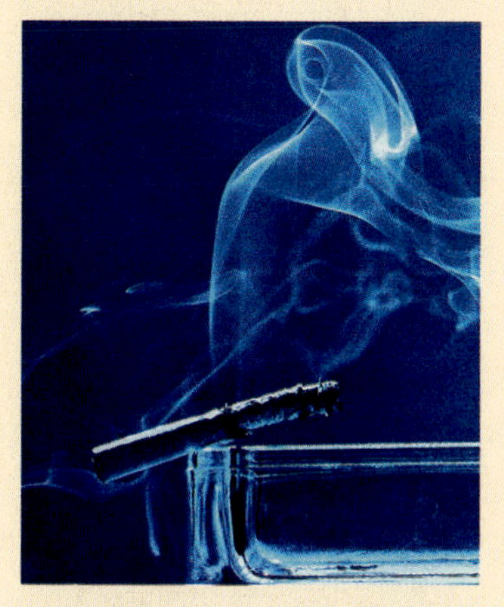

Cigarette smoke contains at least 40 carcinogens, including 3,4-benzpyrene.

and in human population groups. We still don't know for sure what components of these foods protect against cancer. Perhaps it is a combination of substances, rather than just one or two. The vitamins that are antioxidants (vitamin C, vitamin E, and β-carotene, a precursor to vitamin A) seem to exhibit the strongest anticancer properties. Some studies with vitamins A, C, and E, used separately or in combination, have confirmed the fact that each of these vitamins has some ability to lower the incidence of cancer. There probably are many other anticarcinogens in our food that have not yet been identified.

OH

$(CH_3)_3C$ $C(CH_3)_3$

CH_3

BHT

25.10 Hazardous Materials

Stories about toxic substances have caused increasing concern about the problems involving hazardous materials in the environment. Problems with chemical dumps have made household words out of Love Canal, in New York, and Valley of the Drums, in Kentucky. Although often overblown in the news media, serious problems do exist.

The U.S. Environmental Protection Agency categorizes **hazardous materials** on the basis of their properties.

- **Ignitable materials** are substances such as gasoline and other hydrocarbons that catch fire readily.
- **Corrosive materials** are substances such as strong acids that corrode storage containers and other equipment.
- **Reactive materials** are substances such as explosives and materials such as powdered bleach (calcium hypochlorite) that react with water to produce toxic fumes.
- **Toxic materials** are substances such as chlorine, ammonia, formaldehyde, and pesticides that are injurious when inhaled or ingested.

Broccoli, cauliflower, and brussels sprouts are among the cruciferous vegetables that have been shown to reduce the incidence of cancer in human population groups.

(a)

(b)

(a) A waste dump in 1970 at Malkins Bank, Cheshire, England, with drums leaking chemical wastes. (b) The same site, cleaned up and restored, is now a municipal golf course.

From this list, we see that hazardous materials can cause fires or explosions, pollute the air, contaminate our food and water, and occasionally poison by direct contact. As long as we want the products our industries produce, however, we will have to deal with the problems of hazardous wastes (Table 25.6).

Many hazardous materials can be rendered less harmful by chemical treatment. For example, acid wastes can be neutralized with inexpensive bases such as lime. However, the best way to handle hazardous wastes is not to produce them in the first place. Many industries have modified processes to minimize the amount of wastes. Some wastes can be reprocessed to recover energy or materials. Hydrocarbon solvents such as hexane can be purified and reused, or burned as fuels. Sometimes, one industry's waste can be a raw material for another industry. For example, waste nitric acid from the metals industry can be converted to fertilizer. Finally, if a hazardous waste cannot be used, incinerated, or treated to render it less hazardous, it must be stored in a secure landfill. Unfortunately, landfills often leak, contaminating the groundwater. We clean up one toxic waste dump and move the materials to another, playing a rather macabre shell game. The best technology at present for treating organic wastes, including chlorinated compounds, is incineration. For example, at 1260 °C, more than 99.9999% destruction of chlorinated compounds, such as PCBs, is achieved.

Perhaps *biodegradation* of the wastes will be the way of the future. Microorganisms can degrade hydrocarbons such as those in gasoline. Other bacteria, when provided with proper nutrients, can degrade chlorinated hydrocarbons.

Weighing Risks and Benefits

Increasingly, as citizens we are having to decide whether the benefits we gain from hazardous substances are worth the risks we assume by using them. Many issues involving toxic chemicals are emotional, and most of the decisions regarding them are political. Nevertheless, possible solutions to problems posed by toxic chemicals often lie in the field of chemistry. We hope that the chemistry you have learned in this text will help you make intelligent decisions. Most of all, we hope that you will continue to learn more about chemistry, because chemistry affects nearly everything you do. We wish you success and happiness, and may the joy of learning go with you always.

TABLE 25.6 Industrial Products and Hazardous Waste By-products

Product	Associated Waste
Plastics	Organic chlorine compounds
Pesticides	Organic chlorine compounds, organophosphate compounds
Medicines	Organic solvents and residues, heavy metals (e.g., mercury and zinc)
Paints	Heavy metals, pigments, solvents, organic residues
Oil, gasoline	Oil, phenols and other organic compounds, heavy metals, ammonium salts, acids, caustics
Metals	Heavy metals, fluorides, cyanides, acidic and alkaline cleaners, solvents, pigments, abrasives, plating salts, oils, phenols
Leather	Heavy metals, organic solvents
Textiles	Heavy metals, dyes, organic chlorine compounds, solvents

Summary

Starting at Earth's surface, the primary regions of the atmosphere are the troposphere, stratosphere, mesosphere, and thermosphere (ionosphere). The chief components of dry air are N_2, O_2, and Ar. The humidity of air is a measure of its water vapor content. Dew and frost formation and deliquescence are phenomena related to relative humidity. Water vapor is an important participant in the hydrologic (water) cycle.

Nitrogen is involved in a natural cycle that has been modified by chemical fixation of nitrogen to make fertilizers. Oxides of nitrogen, in the presence of unburned hydrocarbons and sunlight, lead to photochemical smog. Temperature inversions also contribute to smog conditions. Smog-control measures focus on catalytic converters for automobiles and the control of combustion processes to reduce emissions.

Ozone, O_3, in the stratosphere protects living organisms by absorbing ultraviolet radiation. The integrity of the ozone layer, however, is threatened by human activities, such as the release of chlorofluorocarbons (CFCs) into the atmosphere.

Atmospheric CO_2 is the carbon source for carbohydrate synthesis in the carbon cycle. Some carbon is locked out of the cycle in fossil fuels (coal, natural gas, and petroleum), but the combustion of these fuels returns CO and CO_2 to the cycle. CO is an air pollutant, and the continuing buildup of CO_2 in the atmosphere may result in global warming.

As atmospheric pollutants, SO_2 and SO_3 contribute to the problems of industrial smog and acid rain. SO_2 is oxidized to SO_3, which reacts with water to form H_2SO_4. Particulate matter is also present in industrial smog. Control measures for industrial smog are designed to reduce the emissions of SO_2 and particulate matter, both of which have serious health effects.

Water covers three fourths of Earth's surface, but only about 1% of it is available as fresh water. Water has a higher density, heat capacity, and heat of vaporization than most other liquids. Unlike most liquids, water expands when it freezes.

Freshwater can readily become contaminated by various kinds of chemicals and microorganisms, some from natural processes and others from human activities. Dumping sewage into water increases its biochemical oxygen demand (BOD). Water pollutants that are nutrients for the growth of algae can lead to eutrophication of a lake or stream. Also, acid rain can cause lakes and streams to become so acidic that they damage fish populations and other aquatic life.

Groundwater and surface water each provide drinking water for about half the U.S. population. Municipal water supplies are treated in several physical and chemical steps, including settling, filtration, aeration, and chlorination. Fluoridation appears to have greatly reduced the incidence of dental caries.

Wastewater treatment usually includes sludge removal, sand and gravel filtration, and chlorination. Advanced treatment methods include charcoal filtration, ion exchange, and reverse osmosis.

Toxicology is the study of the responses of living organisms to poisons. The statement, "The dose makes the poison" suggests that anything can be a poison if the dose is large enough. Strong acids and bases are corrosive poisons. Substances such as ozone are toxic because they are strong oxidizing agents. Carbon monoxide and nitrites are toxic because they interfere with the blood's transport of oxygen. Cyanide is a poison because it shuts down cell respiration. Heavy metal poisons, such as lead and mercury, inactivate enzymes by tying up their —SH groups. Nerve poisons, such as organophosphates, interfere with the acetylcholine cycle. Carcinogens are slow poisons that trigger the growth of malignant tumors. Hazardous materials include all kinds of industrial products and by-products that can cause illness or death. The four classes of hazardous materials are based on their properties: ignitability, corrosivity, reactivity, and toxicity.

Key Terms

acid rain (25.7)
activated sludge method (25.6)
advanced treatment (25.6)
aerobic oxidation (25.6)
air pollutant (25.2)
anaerobic decay (25.6)
anticarcinogen (25.9)
biochemical oxygen demand (BOD) (25.6)
biosphere (page 1074)
carcinogen (25.9)
corrosive material (25.10)
deliquescence (25.1)
eutrophication (25.6)
global warming (25.4)
greenhouse effect (25.4)
hazardous material (25.10)
hydrologic (water) cycle (25.1)
ignitable material (25.10)
industrial smog (25.2)
nitrogen cycle (25.1)
nitrogen fixation (25.1)
ozone layer (25.3)
particulate matter (25.2)
photochemical smog (25.2)
primary sewage treatment (25.6)
reactive material (25.10)
secondary sewage treatment (25.6)
toxic material (25.10)
toxicology (page 1075)

Review Questions

1. Which layer of the atmosphere **(a)** lies nearest the surface of Earth? **(b)** contains the ozone layer?
2. List the three major components of dry air, and give the approximate (nearest whole number) mole percent of each.
3. Explain the difference between the absolute and the relative humidity of an air sample.
4. Describe the formation of dew and frost from air.
5. What is the nitrogen cycle? How has industrial fixation of nitrogen to make fertilizers affected the nitrogen cycle?
6. Briefly describe each of the following terms dealing with a natural phenomenon.
   (a) deliquescence **(b)** the greenhouse effect
7. What specific materials are implied by these terms for atmospheric pollutant(s)?
   (a) PAN **(b)** SO_x **(c)** fly ash
8. By name and/or formula, indicate **(a)** two gases able to displace O_2 in blood hemoglobin; **(b)** two "greenhouse" gases, in addition to CO_2 and H_2O; and **(c)** a constituent of acid rain.
9. List two uses of chlorofluorocarbons (CFCs). How are CFCs implicated in the depletion of the ozone layer?
10. What is photochemical smog? What is the role of sunlight in its formation?
11. What is industrial smog? How is it formed?
12. What conditions favor the formation of carbon monoxide during the combustion of gasoline in an automobile engine?
13. How does carbon monoxide exert its poisonous effect?
14. What is synergism? Indicate one specific example of a synergistic effect concerning air pollution.
15. What are the health effects associated with ozone in the stratosphere and at ground level? Why are they not the same?
16. Trace the steps by which the burning of high-sulfur coal leads to acid rain.
17. How is each of the following used to reduce air pollution?
    (a) electrostatic precipitator
    (b) catalytic converter
18. Which of the following are important contributors to the formation of photochemical smog, and which are not? Explain.
    (a) NO **(c)** hydrocarbon vapors
    (b) CO **(d)** SO_2
19. What is a temperature inversion? How does a temperature inversion contribute to air pollution problems?
20. How are the following terms related to one another regarding air pollution: aerosol, fly ash, and particulate matter?
21. Describe measures that can be used to control the emission of nitrogen oxides in automotive exhaust, and explain why these are not the same measures used to control emissions of hydrocarbons and carbon monoxide.
22. Is all air pollution the result of human activity? Explain.
23. What proportion of Earth's water is seawater? Why are the seas salty?
24. List some waterborne diseases. Why are these diseases no longer common in developed countries?
25. What impurities are present in rainwater?
26. List four cations and three anions present in groundwater.
27. What is BOD? Why is a high BOD undesirable?
28. What are the products of the breakdown of organic matter by aerobic bacteria? by anaerobic bacteria?
29. List some ways in which groundwater is contaminated. What are some common industrial contaminants of groundwater?
30. Why do chlorinated hydrocarbons remain in groundwater for such a long time?
31. List two ways by which lakes and streams have become acidic. Why is acidic water especially harmful to fish?
32. List several ways by which the acidity of rain can be reduced. What kind of rocks tend to neutralize acidic waters? How can we restore (at least temporarily) lakes that are too acidic?
33. List two toxic compounds found in wastes from the chromium plating process. How is each removed?
34. Describe a primary sewage treatment plant. What impurities are removed by primary sewage treatment? What impurities remain in wastewater after primary treatment?
35. Describe a secondary sewage treatment plant. What impurities are removed by secondary sewage treatment? What substances remain in wastewater after effective secondary treatment?
36. Describe the activated sludge method of sewage treatment. Why is wastewater chlorinated before it is returned to a waterway?
37. Why is drinking water chlorinated?
38. What is meant by advanced treatment of wastewater? What kinds of substances are removed from wastewater by charcoal filtration? Why is it so difficult to remove nitrate ions from water?
39. Why are municipal water supplies **(a)** treated with aluminum sulfate and slaked lime? **(b)** aerated?
40. What is toxicology? Is sodium chloride (table salt) poisonous? Explain your answer fully.
41. Give an example that shows how the toxicity of a substance depends on the route of administration.
42. List three corrosive poisons. How do dilute solutions of acids and bases damage living cells? How does ozone damage living cells?
43. How does cyanide exert its toxic effect? How does sodium thiosulfate act as an antidote for cyanide poisoning?

44. Iron (as Fe^{2+}) is a necessary nutrient. What are the effects of too little Fe^{2+}? Of too much?

45. What is acetylcholine? Describe its action.

46. How do **(a)** botulin, **(b)** sarin, and **(c)** organophosphorus compounds affect the acetylcholine cycle?

47. What is the P-450 system? What is its function? Does it always detoxify foreign substances?

48. List two ways that the conversion of nicotine to cotinine in the liver lessens the risk of nicotine poisoning.

49. List two steps in the detoxification of ingested toluene. What is the effect of these steps?

50. What is a tumor? How are benign and malignant tumors different?

51. What are **(a)** oncogenes and **(b)** suppresser genes? How is each involved in the development of cancer?

52. List some conditions under which polycyclic hydrocarbons are formed.

53. What is a hazardous material?

54. Define and give an example of **(a)** a reactive material, **(b)** an ignitable material, **(c)** a corrosive material.

Problems

The Atmosphere

55. The text states that 99% of the mass of the atmosphere lies within 30 km of the surface of Earth. Which of the following is a reasonable estimate of air pressure at an altitude of 30 km: **(a)** 0.1 mmHg, **(b)** 1 mmHg, **(c)** 10 mmHg, or **(d)** 100 mmHg? Explain your reasoning. (*Hint:* Recall the basic ideas relating to pressure from Section 5.3.)

56. When present in a very small proportion in air, the concentration of a gas is customarily indicated in parts per million (ppm) rather than in mole percent or volume percent. Use data in Table 25.1 to determine the parts per million in air of the noble gases that are listed there.

Water Vapor in the Atmosphere

57. What are the mole percent and ppm of H_2O in an air sample at STP in which the partial pressure of water vapor is 2.00 mmHg?

58. What is the relative humidity of a sample of air at 25 °C in which the partial pressure of water vapor is 10.5 mmHg? (*Hint:* Use data from Table 11.2.)

59. What is the partial pressure of water vapor in a sample of air having a relative humidity of 75.5% at 20 °C? (*Hint:* Use data from Table 11.2.)

60. A parcel of air has an absolute humidity, expressed as a partial pressure of water vapor, of 18.0 mmHg. At which of the following temperatures does the air have the greatest relative humidity: 25 °C, 30 °C, or 40 °C? Explain.

61. What is the dew point of the parcel of air described in Problem 60? (*Hint:* Use data from Table 11.2.)

62. Why is it that condensed water vapor (steam) can be seen above a kettle of boiling water even in a hot kitchen, whereas you can see your breath (steam) only on a cold day?

Carbon, CO, and CO_2

63. The combustion of a hydrocarbon, especially if the quantity of oxygen is limited, produces a mixture of carbon dioxide and carbon monoxide. The decomposition of a metal carbonate by an acid produces only carbon dioxide, even if the quantity of acid is limited. Explain this difference in behavior.

64. Indicate a natural process or processes by which carbon atoms are **(a)** removed from the atmosphere; **(b)** returned to the atmosphere; **(c)** effectively withdrawn from the carbon cycle.

65. Write an equation that represents the complete combustion of the hydrocarbon hexane, $C_6H_{14}(l)$. Explain why it is not possible to write a unique equation to represent its incomplete combustion.

66. Carbon monoxide is a poisonous gas, even in low concentrations, whereas carbon dioxide is not. Yet, except in some local situations, there is less environmental concern over carbon monoxide than over carbon dioxide. Explain why this is so.

67. The United States leads the world in per capita emissions of $CO_2(g)$, with 19.8 metric tons (t) per person per year (1 metric ton = 1000 kg). What mass, in metric tons, of each of the following fuels would yield this quantity of CO_2?

(a) CH_4

(b) C_8H_{18}

(c) coal that is 94.1% C by mass

68. Tabulations on carbon dioxide emissions often list cement manufacture as one of the sources. Describe two ways in which the manufacture of cement injects carbon dioxide into the atmosphere.

Oxides of Sulfur

69. Write equations for the following reactions.
    (a) Sulfur burns in air forming sulfur dioxide.
    (b) Zinc sulfide, heated in air, yields zinc oxide and sulfur dioxide.
    (c) Sulfur dioxide reacts with oxygen, forming sulfur trioxide.
    (d) Sulfur trioxide reacts with water, forming sulfuric acid.
    (e) Sulfuric acid is completely neutralized by aqueous ammonia.

70. Per ton of material consumed, which of the following would you expect to produce the greatest quantity of $SO_2(g)$: **(a)** smelting zinc sulfide, **(b)** smelting lead sulfide, **(c)** burning coal, or **(d)** burning natural gas? Explain.

Particulate Matter

71. Describe how the following particulate matter may be produced.
    (a) sodium chloride from seawater
    (b) sulfate particles in an industrial smog

72. The average person takes 15 breaths per minute, inhaling 0.50 L of air with each breath. What mass of particulates, in milligrams, would the person breathe in a day, if the particulate level in air were 75 $\mu g/m^3$?

Water and Water Pollution

73. Give the equation for the neutralization of acidic water by limestone.

74. Wastewater disinfected with chlorine must be dechlorinated before it is returned to sensitive bodies of water. The dechlorinating agent often is sulfur dioxide. Write the equation for the reaction. Is the chlorine oxidized or reduced? Identify the oxidizing agent and reducing agent in the reaction.

Additional Problems

75. Assume that a typical urban air contains 100 μg of suspended particles per m^3 of air. Assume that the average particle is spherical in shape, with a diameter of 1 μm and a density of 1 g/cm^3. Estimate the number of particles per cm^3 of the air.

76. The text states that 5.2×10^{15} metric tons of atmospheric gases are spread over a surface area of 5.0×10^8 km^2. Use these facts, together with data from Appendix B, to estimate a value of standard atmospheric pressure.

77. At 20 °C, the vapor pressure of a saturated solution of $CaCl_2 \cdot 6H_2O$ is 5.67 mmHg. If a quantity of this solution is placed in a large sealed container at 20 °C and the solution kept saturated by the presence of excess solid, what relative humidity will be maintained in the air in the container? How effective is $CaCl_2 \cdot 6H_2O$ in dehumidifying air?

78. A 12.012-L sample of air is saturated with water vapor at 25.0 °C. The air is then cooled to 20.0 °C. What mass of water (dew) will deposit on the walls of the container?

79. There are different ways to assess how much the combustion of various fuels contributes to the buildup of CO_2 in the atmosphere. One relates the mass of CO_2 formed to the mass of fuel burned; another relates the mass of CO_2 to the quantity of heat evolved in the combustion. Which of the three fuels C(graphite), $CH_4(g)$, or $C_4H_{10}(g)$ produces the smallest mass of CO_2 **(a)** per gram of fuel; **(b)** per kJ of heat evolved? (*Hint:* Use data from Appendix C, and assume that all the products of each combustion are gases.)

80. A large coal-fired electric plant burns 2500 tons of coal per day. The coal that is burned contains 0.65% S by mass. Assume that all of the sulfur is converted to SO_2 and, because of a thermal inversion, remains trapped in a parcel of air that is 45 km $\times$ 60 km $\times$ 0.40 km. Will the level of SO_2 in this air exceed the primary national air quality standard of 365 $\mu g\ SO_2/m^3$ air?

81. Use the following and other data from the text to show that if all the sulfur in coal used in electric power plants were converted to sulfuric acid, the quantity of acid produced would exceed current demand. (1) Annual U.S. coal consumption by electric power plants: approximately 8.7×10^8 ton. (2) Average SO_2 formation in the combustion of coal: 2 mg SO_2/kJ heat evolved. (3) Typical annual U.S. production of sulfuric acid: approximately 80×10^9 lb. [*Hint:* You need to estimate the heat of combustion of coal. To do this, assume that coal is 100% C (graphite) and use data from Appendix C.]

82. It has been estimated that if all the ozone in the atmosphere were brought to sea level at STP, the gas would form a layer 0.3 cm thick.
    (a) Estimate the number of O_3 molecules in Earth's atmosphere.
    (b) Comment on the feasibility of a caller's suggestion to a radio science program that depleted ozone in the

stratosphere be replaced by transporting unwanted ozone from low altitudes into the stratosphere.

83. A small gasoline-powered engine with a 1-L storage tank is inadvertently left running overnight in a large warehouse, 95 m × 38 m × 16 m. When workers arrive in the morning, are they likely to enter an environment in which the level of CO exceeds the danger level of 35 ppm? (*Hint:* Use C_8H_{18} as a representative formula of gasoline, and make other reasonable assumptions.)

84. For the complete combustion of gasoline, the mass ratio of air to fuel should be about 14.5 to 1. Use ideas from this chapter and elsewhere in the text to show that this is about the ratio that you would predict. (*Hint:* Assume that C_8H_{18} is a representative formula of gasoline, and use the composition of air given in Table 25.1. You may find it useful to work with the concept of a mole of air.)

85. An important variable in the combustion of gasoline in an internal-combustion engine is the air/fuel ratio. The accompanying figure shows how the emission of pollutants is related to the air/fuel ratio. Provide a plausible interpretation of this figure. (*Hint:* The stoichiometric ratio is verified in Problem 84. Also, recall that RH represents hydrocarbons.)

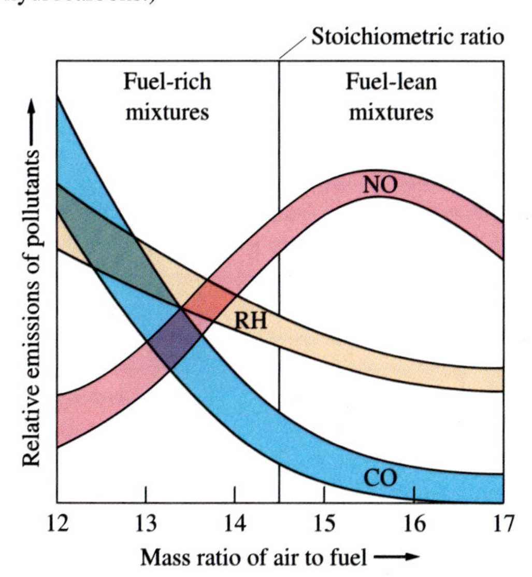

86. The workplace standard for $SO_2(g)$ in air is 5 ppm. Approximately what mass of sulfur could be burned in an enclosed workplace, 10.5 m × 5.4 m × 3.6 m, before this limit is exceeded?

87. The decomposition of ozone by chlorine atoms can be described by the rate law: rate $= k[Cl][O_3]$.

$$Cl(g) + O_3(g) \longrightarrow ClO(g) + O_2(g)$$
$$k = 7.2 \times 10^9\ M^{-1}\ s^{-1} \text{ at } 298\ K$$

What would be the effect on the rate of ozone destruction of doubling the concentration of chlorine atoms?

88. A newspaper article states the following: "Never dump oil down a sewer or water main. One quart of motor oil can make 250,000 gallons of water undrinkable. That's more water than 30 people drink in a lifetime." Estimate the concentration of the oil in the contaminated water, assuming thorough mixing. What other assumptions must you make?

89. The concentration of trichloroethane, CH_3CCl_3, in a sample of groundwater is 34 ppb. What is the concentration in nanomoles per liter?

90. To prevent tooth decay, drinking water is usually made 1.60 ppm in F^- ion. A cylindrical water tank has a diameter of 5.00 m and a depth of 12.0 m. What mass of NaF(s) is needed to establish an F^- ion concentration of 1.60 ppm in a tankful of water? What volume of 0.150 M NaF would be needed to achieve 1.60 ppm of F^-?

91. A leaking tank spills 875 kg of 2-propanol into a lake with a volume of 1.8×10^8 L. How much is the BOD (in milligrams per liter) increased by the spill? Assume that the following reaction occurs.

$$C_3H_8O + \tfrac{9}{2}O_2 \longrightarrow 3\ CO_2 + 4\ H_2O$$

92. To grow properly, common species of algae need C, N, and P in the approximate atom ratio 106 C : 16 N : 1 P. Which is the limiting nutrient in a lake that has the following concentrations: C, 25 mg/L; N, 98 mg/L; and P, 21 mg/L?

93. Chlorine, even at very low concentrations, can be quite toxic to aquatic organisms. It also reacts with organic substances in the water to form toxic chlorinated organic compounds such as chloroform and chlorinated phenols. Wastewater that contains chlorine is therefore often dechlorinated before it is discharged. At low pH (below pH 1), chlorine exists principally as $Cl_2(aq)$. At high pH (above pH 8.5), the chlorine largely disproportionates to OCl^- and Cl^-. Two principal reagents are used to dechlorinate wastewater, sulfur dioxide (or salts that release SO_2, such as $NaHSO_3$) and hydrogen peroxide. Write equations for the reaction of OCl^- in wastewater with a pH of 12.3 with **(a)** $NaHSO_3$ and **(b)** H_2O_2.

94. Many industrial wastewater streams are contaminated with metal ions. Many of these ions can be effectively removed by precipitation as insoluble salts. As^{3+}, Cd^{2+}, Cr^{3+}, Cu^{2+}, Fe^{2+}, Mn^{2+}, Ni^{2+}, Pb^{2+}, and Zn^{2+} can be precipitated as hydroxides. Cd^{2+}, Co^{2+}, Cu^{2+}, Fe^{2+}, Hg^{2+}, Mn^{2+}, Ni^{2+}, Ag^+, Sn^{2+}, and Zn^{2+} can be precipitated as sulfides. Cd^{2+}, Ni^{2+}, and Pb^{2+} can be precipitated as carbonates. All these reactions are sensitive to the pH of the solution. For each group, write an equation for a representative reaction and state whether the reaction should be carried out at high pH or low pH. Explain.

95. Various regulations often require the reduction of phosphate levels before wastewater can be discharged to waterways. This can be done with any of several different precipitating agents such as **(a)** iron(III) chloride, **(b)** aluminum sulfate, or **(c)** CaO. Write the equation for the precipitation reaction that occurs with each of the reagents. In each case, calculate the cation concentration that must be present if the concentration of PO_4^{3-} is to be reduced to 10 ppm.

96. Propose a disposal method for each of the following wastes. Be as specific as possible, and justify your choice.

(a) hydrochloric acid contaminated with iron salts

(b) picric acid (an explosive)

(c) pentane contaminated with residues from penicillin production

97. Describe the mechanism by which benzpyrene causes lung cancer in cigarette smokers.

98. The world's termite population is estimated to be 2.4×10^{17}. Annually, these termites produce an estimated 4.6×10^{16} g CO_2. The atmosphere contains 5.2×10^{15} metric tons of air (1 metric ton = 1000 kg). On a number basis, the current CO_2 level in the atmosphere is 360 ppm. What percent increase in this CO_2 level would the termites cause if none of the CO_2 they produce were removed by natural processes?

99. In investigating an old dump, a student finds a metal container filled with 12 ounces of liquid. The label indicates that the can contains a substance that a reference book lists as a carcinogen. The concentration of the substance is given as 8.2 mg/fluid oz. On further investigation, he finds that 20 billion such containers were once filled and distributed in the United States each year.

(a) How many milligrams of the carcinogen are in each can?

(b) How many metric tons of the carcinogen were distributed in this manner each year?

Appendix A

Some Mathematical Operations

A.1 Exponential Notation

A number is in exponential form when it is written as the product of a coefficient—usually with a value between 1 and 10—and a power of ten. Following are two examples of *exponential notation*, a form usually employed by scientists and sometimes called *scientific notation.*

$$4.18 \times 10^3 \quad \text{and} \quad 6.57 \times 10^{-4}$$

We generally express numbers in exponential form for two reasons: (1) We can write very large or very small numbers in a minimum of printed space and with a reduced chance of typographical error. (2) Numbers in exponential form convey explicit information about the precision of measurements: The number of significant figures in a measured quantity is stated unambiguously.

In the expression 10^n, n is the exponent of 10, and we say that the number 10 is raised to the nth power. If n is a *positive* quantity, 10^n has a value *greater than 1*. If n is a *negative* quantity, 10^n has a value *less than 1*. We are particularly interested in cases where n is an integer, as in the following examples.

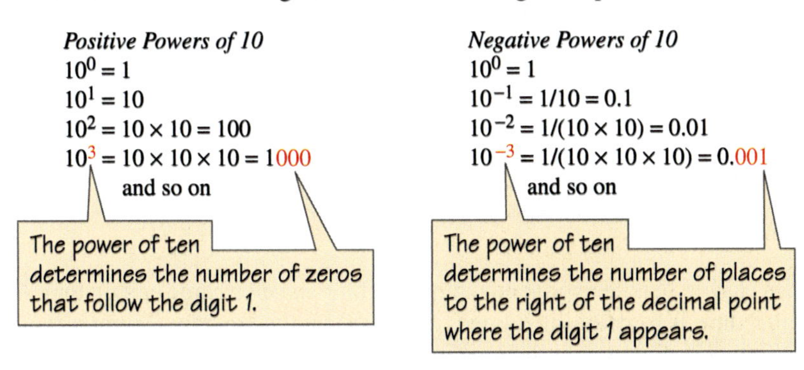

Positive Powers of 10
10⁰ = 1
10¹ = 10
10² = 10 × 10 = 100
10³ = 10 × 10 × 10 = 1000
and so on
The power of ten determines the number of zeros that follow the digit 1.
Negative Powers of 10
10⁰ = 1
10⁻¹ = 1/10 = 0.1
10⁻² = 1/(10 × 10) = 0.01
10⁻³ = 1/(10 × 10 × 10) = 0.001
and so on
The power of ten determines the number of places to the right of the decimal point where the digit 1 appears.

We express (a) 612,000 and (b) 0.000505 in exponential form as follows.

$$\text{(a)} \quad 612{,}000 = 6.12 \times 100{,}000 = 6.12 \times 10^5$$

$$\text{(b)} \quad 0.000505 = 5.05 \times 0.0001 = 5.05 \times 10^{-4}$$

The following steps provide a more direct approach to converting numbers to the exponential form.

- Count the number of places a decimal point must be moved to produce a coefficient having a value between 1 and 10.
- The number of places counted then becomes the power of 10.
- The power of 10 is *positive* if the decimal point is moved to the *left*.

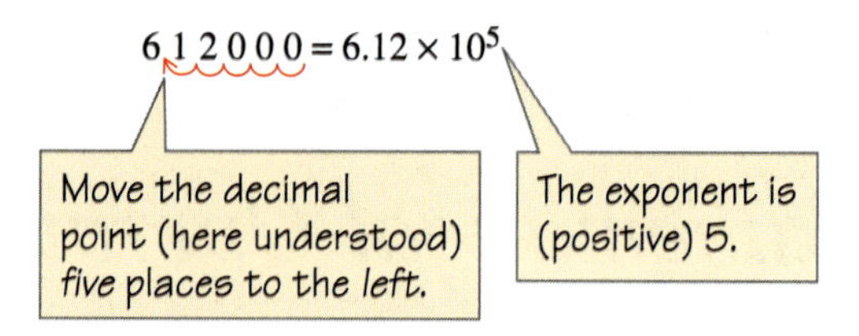

- The power of 10 is *negative* if the decimal point is moved to the *right*.

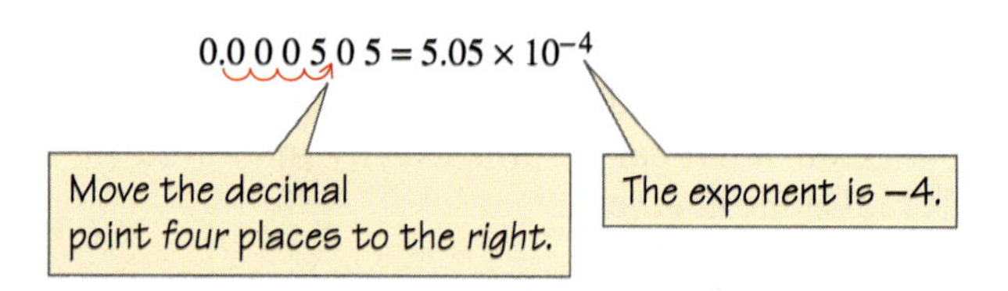

To convert a number from exponential form to the conventional form, move the decimal point in the opposite direction.

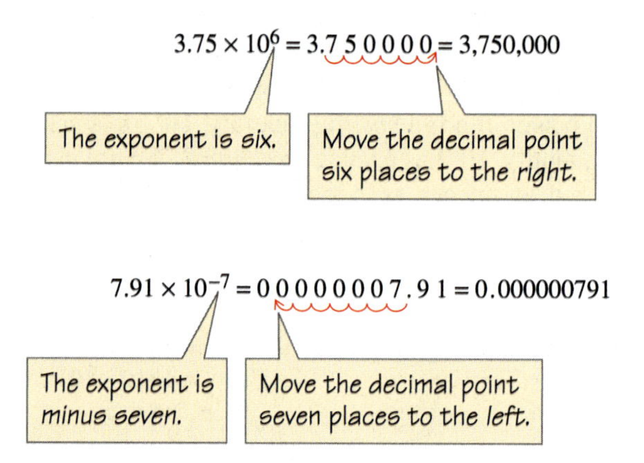

The key strokes required with your calculator may be different from those shown here. Check the specific instructions in the manual supplied with the calculator.

It is easy to handle exponential numbers on most electronic calculators. A typical procedure is to enter the number, followed by the key EXP. To enter the number 2.85×10^7, the key strokes required are [2] [.] [8] [5] [EXP] [7], and the result is displayed as $\left[2.85^{07}\right]$.

For the number 1.67×10^{-5}, the key strokes are [1] [.] [6] [7] [EXP] [5] [±], and the result is displayed as $\left[1.67^{-05}\right]$.

Many calculators can be set to convert all numbers and calculated results to the exponential form, regardless of the form in which the numbers are entered. Generally, the calculator can also be set to display a fixed number of significant figures in results.

Addition and Subtraction

To add or subtract numbers in exponential notation with pencil and paper only, we must express each quantity as the same power of ten. This treats the power of ten in the same way as a unit—it is simply "carried along" in the calculation. In the following, each quantity is expressed with the power 10^{-3}.

$$(3.22 \times 10^{-3}) + (7.3 \times 10^{-4}) - (4.8 \times 10^{-4})$$
$$= (3.22 \times 10^{-3}) + (0.73 \times 10^{-3}) - (0.48 \times 10^{-3})$$
$$= (3.22 + 0.73 - 0.48) \times 10^{-3}$$
$$= 3.47 \times 10^{-3}$$

In contrast, most calculators perform these operations automatically, and you generally will not need to convert the numbers to the same power of ten when using a calculator.

Multiplication and Division

To multiply numbers expressed in exponential form, *multiply* all coefficients to obtain the coefficient of the result, and *add* all exponents to obtain the power of ten in the result.

$$0.0803 \times 0.0077 \times 455 = (8.03 \times 10^{-2}) \times (7.7 \times 10^{-3}) \times (4.55 \times 10^{2})$$
$$= (8.03 \times 7.7 \times 4.55) \times 10^{(-2-3+2)}$$
$$= (2.8 \times 10^{2}) \times 10^{-3} = 2.8 \times 10^{-1}$$

To divide two numbers in exponential form, *divide* the coefficients to obtain the coefficient of the result, and *subtract* the exponent in the denominator from the exponent in the numerator to obtain the power of ten. The example below combines multiplication and division. First we apply the rule for multiplication separately to the numerator and to the denominator, and then we use the rule for division.

$$\frac{0.015 \times 0.0088 \times 822}{0.092 \times 0.48} = \frac{(1.5 \times 10^{-2})(8.8 \times 10^{-3})(8.22 \times 10^{2})}{(9.2 \times 10^{-2})(4.8 \times 10^{-1})}$$
$$\frac{1.1 \times 10^{-1}}{4.4 \times 10^{-2}} = 0.25 \times 10^{-1-(-2)} = 0.25 \times 10^{1}$$
$$= 2.5 \times 10^{-1} \times 10^{1} = 2.5$$

As with addition and subtraction, most electronic calculators perform multiplication, division, and combinations of the two with no need to record intermediate results.

Raising a Number to a Power and Extracting the Root of an Exponential Number

To raise an exponential number to a given power, raise the coefficient to that power, and multiply the exponent by that power. For example, we can *cube* a number (that is, raise it to the *third* power) in the following manner.

$$(0.0066)^3 = (6.6 \times 10^{-3})^3 = (6.6)^3 \times 10^{3 \times (-3)}$$

Rewrite in exponential form.

Cube the coefficient.

Multiply the exponent by 3.

$$= (2.9 \times 10^{2}) \times 10^{-9} = 2.9 \times 10^{-7}$$

To extract the root of an exponential number, we raise the number to a *fractional* power: one-half power for a square root, one-third power for a cube root, and so on. Most calculators have keys designed for extracting square roots and cube roots. Thus, to extract the square root of 1.57×10^{-5}, enter the number 1.57×10^{-5} into the calculator, and use the [$\sqrt{\ }$] key.

$$\sqrt{1.57 \times 10^{-5}} = 3.96 \times 10^{-3}$$

To extract the cube root of 3.18×10^{10}, enter the number 3.18×10^{10} into an electronic calculator, and use $[\sqrt[3]{\ }]$ key.

$$\sqrt[3]{3.18 \times 10^{10}} = 3.17 \times 10^{3}$$

Some calculators allow you to extract roots by keying in the root as a fractional exponent.

$$(2.75 \times 10^{-9})^{1/5} = 1.94 \times 10^{-2}$$

We can also extract the roots of numbers by using logarithms.

A.2 Logarithms

The common logarithm (log) of a number (N) is the exponent (x) to which the base 10 must be raised to yield the number.

$\log N = x$ ← means that → $N = 10^x$ ← or that → $N = 10^{\log N}$

In the expressions below, the numbers N are printed in blue and their logarithms ($\log N$) are printed in red.

$\log 1 = \log 10^0 = 0$	$\log 1 = \log 10^0 = 0$
$\log 10 = \log 10^1 = 1$	$\log 0.1 = \log 10^{-1} = -1$
$\log 100 = \log 10^2 = 2$	$\log 0.01 = \log 10^{-2} = -2$
$\log 1000 = \log 10^3 = 3$	$\log 0.001 = \log 10^{-3} = -3$

Most of the numbers that we commonly encounter, of course, are not integral powers of ten, and their logarithms are not integral numbers. From the above pattern, though, we do have a general idea of what their logarithms might be. Consider, for example, the numbers 655 and 0.0078.

$$100 < 655 < 1000 \qquad 0.001 < 0.0078 < 0.01$$
$$2 < \log 655 < 3 \qquad -3 < \log 0.0078 < -2$$

We can see that log 655 is between 2 and 3 and that log 0.0078 is between -3 and -2. To get a more exact value, however, we must use a table of logarithms or the [LOG] key on a calculator.

$$\log 655 = 2.816 \qquad \log 0.0078 = -2.11$$

In working with logarithms, we often need to find the number that has a certain value for its logarithm. The number is sometimes called an *antilogarithm*, and we can think of it in the following terms.

If $\log N = 3.076$, then $N = 10^{3.076} = 1.19 \times 10^3$.
If $\log N = -4.57$, then $N = 10^{-4.57} = 2.7 \times 10^{-5}$.

With a calculator, we simply enter the value of the logarithm (that is, 3.076 or -4.57) and then use the $[10^x]$ key.

Significant Figures in Logarithms

At first sight, $\log N = 3.076$ appears to have four significant figures, and $N = 1.19 \times 10^3$ to have only three, but in reality both values have only *three*. Digits to the *left* of the decimal point in a logarithm simply relate to the power of ten in the exponential form of a number. The only significant digits in a logarithm are those to the *right* of the decimal point. The coefficient of the exponential form of the number should have this same number of digits. Thus, to express the logarithm of 2.5×10^{-12} to two significant figures, we would write: $\log 2.5 \times 10^{-12} = -11.60$.

Some Relationships Involving Logarithms

We can use the definition of logarithms to write: $M = 10^{\log M}$ and $N = 10^{\log N}$. For the product $(M \times N)$, we can write either of the following.

$$(M \times N) = 10^{\log M} \times 10^{\log N} = 10^{(\log M + \log N)}$$

$$(M \times N) = 10^{\log(M \times N)}$$

This means that the logarithm of the product of several terms is equal to the sum of the logarithms of the individual terms. Thus,

$$(1) \quad \log (M \times N) = (\log M + \log N)$$

We can establish two other relationships in a similar manner.

$$(2) \quad \log \frac{M}{N} = (\log M - \log N)$$

$$(3) \quad \log N^{a} = a \log N$$

You may find explicit use for relationship (3) in calculations because it affords a simple method of extracting the roots of number. For example, to determine $(2.75 \times 10^{-9})^{1/5}$, write

$$\log (2.75 \times 10^{-9})^{1/5} = 1/5 \times \log (2.75 \times 10^{-9})$$

$$= 1/5 \times (-8.561) = -1.712$$

$$(2.75 \times 10^{-9})^{1/5} = 10^{-1.712} = 0.0194$$

Natural Logarithms

Choosing *10* as the base for common logarithms is arbitrary. Other choices can be made as well. For example, to the base 2, $\log_2 8 = 3$. This simply means that $2^3 = 8$. And $\log_2 10 = 3.322$ means that $2^{3.322} = 10$.

Several of the relationships in this text involve *natural logarithms*. The base for natural logarithms (ln) is the quantity e, which has the value, $e = 2.71828\cdots$. We encounter the ln function in circumstances in which the rate of change of a variable is proportional to the value of that variable at the time the rate is measured. Such circumstances are common in physical science, including, for example, the rate of decay of a radioactive material.

Generally we can work entirely within the natural logarithm system by using the calculator keys [ln] and $[e^x]$ rather than [LOG] and $[10^x]$. However, if we need to convert between natural and common logarithms, we can use the following conversion factor based on the relationship $\log_e 10 = 2.303$

$$\ln N = 2.303 \log N$$

A.3 Algebraic Operations

To solve an algebraic equation requires that we isolate one quantity—the unknown—on one side of the equation and the known quantities on the other side. This generally requires rearranging terms in the equation, and in these rearrangements, the guiding principle is that *whatever we do to one side of the equation we must do to the other side as well.* Consider the equation

$$\frac{(5x^2 - 12)}{(x^2 + 4)} = 3$$

1. Multiply both sides of the equation by $(x^2 + 4)$.

$$\cancel{(x^2 + 4)} \frac{(5x^2 - 12)}{\cancel{(x^2 + 4)}} = 3 \times (x^2 + 4)$$

$$5x^2 - 12 = 3x^2 + 12$$

2. Subtract $3x^2$ from each side of the equation.

$$5x^2 - 3x^2 - 12 = \cancel{3x^2} - \cancel{3x^2} + 12$$

$$2x^2 - 12 = 12$$

3. Add 12 to each side of the equation.

$$2x^2 - \cancel{12} + \cancel{12} = 12 + 12 = 24$$

4. Divide each side of the equation by 2.

$$\frac{\cancel{2}x^2}{\cancel{2}} = \frac{24}{2} = 12$$

5. Extract the square root of each side of the equation.

$$\sqrt{x^2} = \pm\sqrt{12} = \pm\sqrt{4} \times \sqrt{3}$$

$$x = \pm 2\sqrt{3}$$

$$x = \pm 3.464$$

Quadratic Equations

A quadratic equation has 2 as the highest power of the unknown x. At times, quadratic equations are of this form.

$$(x + n)^2 = m^2$$

To solve for x, extract the square root of each side.

$$(x + n) = \sqrt{m^2} = \pm m$$

and

$$x = m - n \quad \text{or} \quad x = -m - n$$

You will find an example of a quadratic equation of this type in Example 14.12 (page 618).

More often, however, the quadratic equation will be of the form

$$ax^2 + bx + c = 0$$

where a, b, and c are constants. To solve this equation for x, we can use the *quadratic formula.*

$$x = \frac{-b \pm \sqrt{b^2 - 4ac}}{2a}$$

Consider the solution of the following quadratic equation.

$$50.3x^2 - 13.4x + 0.787 = 0$$

$$x = \frac{-(-13.4) \pm \sqrt{(-13.4)^2 - (4 \times 50.3 \times 0.787)}}{2 \times 50.3}$$

$$x = \frac{13.4 \pm \sqrt{21.22}}{100.6} = \frac{13.4 \pm 4.61}{100.6}$$

A.5 Some Key Equations

On several occasions in the text, we refer to this appendix for details on the derivations of key equations or their manipulation into more useful forms. Abbreviated treatments follow. The first two require some prior knowledge of calculus.

Integrated Rate Equation for First-Order Reaction (page 564)

For the reaction

$$\text{A} \longrightarrow \text{products}$$

having the rate law

$$\text{Rate of reaction} = -(\text{rate of disappearance of A}) = k[\text{A}]$$

1. Replace the rate of disappearance of A by the derivative $d[\text{A}]/dt$.

$$-\frac{d[\text{A}]}{dt} = k[\text{A}]$$

2. Rearrange this expression to the form

$$\frac{d[\text{A}]}{[\text{A}]} = -kdt$$

3. Integrate between the limits A_0 at time $t = 0$ and A_t at time t.

$$\int_{[\text{A}]_0}^{[\text{A}]_t} \frac{d[\text{A}]}{[\text{A}]} = -k\int_0^t dt$$

4. The result obtained is

$$\ln\frac{[\text{A}]_t}{[\text{A}]_0} = -kt$$

Integrated Rate Equation for Second-Order Reaction (page 570)

For the reaction

$$\text{A} \longrightarrow \text{products}$$

having the rate law

$$\text{Rate of reaction} = -(\text{rate of disappearance of A}) = k[\text{A}]^2$$

1. Replace the rate of disappearance of A by the derivative $d[\text{A}]/dt$.

$$-\frac{d[\text{A}]}{dt} = k[\text{A}]^2$$

2. Rearrange this expression to the form

$$\frac{d[\text{A}]}{[\text{A}]^2} = -kdt$$

3. Integrate between the limits A_0 at time $t = 0$ and A_t at time t.

$$\int_{[\text{A}]_0}^{[\text{A}]_t} \frac{d[\text{A}]}{[\text{A}]^2} = -k\int_0^t dt$$

4. The result obtained is

$$-\frac{1}{[\text{A}]_t} + \frac{1}{[\text{A}]_0} = -kt \quad \text{or} \quad \frac{1}{[\text{A}]_t} = kt + \frac{1}{[\text{A}]_0}$$

The Arrhenius Equation (page 577)

Our goal is to convert the equation for the straight-line graph of Figure 13.14,

$$\ln k = \frac{-E_a}{RT} + \ln A$$

into an equation that eliminates the constant term, $\ln A$.

1. Write the equation for two different temperatures, T_1 and T_2, at which the rate constants are k_1 and k_2. (E_a and R are constants.)

$$\ln k_2 = \frac{-E_a}{RT_2} + \ln A \quad \ln k_1 = \frac{-E_a}{RT_1} + \ln A$$

2. Subtract $\ln k_1$ from $\ln k_2$.

$$\ln k_2 - \ln k_1 = \frac{-E_a}{RT_2} + \ln A - \left(\frac{-E_a}{RT_1} + \ln A\right)$$

3. Replace $\ln k_2 - \ln k_1$ by $\ln \frac{k_2}{k_1}$, and eliminate $\ln A$.

$$\ln \frac{k_2}{k_1} = \frac{E_a}{RT_1} - \frac{E_a}{RT_2} + \cancel{\ln A} - \cancel{\ln A}$$

4. Rearrange the equation to the final form.

$$\ln \frac{k_2}{k_1} = \frac{E_a}{R}\left(\frac{1}{T_1} - \frac{1}{T_2}\right)$$

The van't Hoff Equation (page 756)

Our goal is to convert the equation for the straight-line graph of Figure 17.12,

$$\ln K_{eq} = \frac{-\Delta H^\circ}{RT} + \text{constant}$$

into an equation that eliminates the term "constant," represented below as A.

1. Write the equation for two different temperatures, T_1 and T_2, at which the equilibrium constants are K_1 and K_2. (ΔH° and R are constants.)

$$\ln K_2 = \frac{-\Delta H^\circ}{RT_2} + \ln A \quad \ln K_1 = \frac{-\Delta H^\circ}{RT_1} + \ln A$$

2. Subtract $\ln K_1$ from $\ln K_2$.

$$\ln K_2 - \ln K_1 = \frac{-\Delta H^\circ}{RT_2} + \ln A - \left(\frac{-\Delta H^\circ}{RT_1} + \ln A\right)$$

3. Replace $\ln K_2 - \ln K_1$ by $\ln \frac{K_2}{K_1}$, and eliminate $\ln A$.

$$\ln \frac{K_2}{K_1} = \frac{\Delta H^\circ}{RT_1} - \frac{\Delta H^\circ}{RT_2} + \cancel{\ln A} - \cancel{\ln A}$$

4. Rearrange the equation to the final form.

$$\ln \frac{K_2}{K_1} = \frac{\Delta H^\circ}{R}\left(\frac{1}{T_1} - \frac{1}{T_2}\right)$$

Appendix B
Some Basic Physical Concepts

B.1 Velocity and Acceleration

The speed of an object is the distance it travels per unit time. An automobile with a speedometer that reads 105 km/h will, if it continues at this constant speed for exactly one hour, travel a distance of 105 km. For scientific work, *velocity* is a more appropriate term. Velocity has two components: a *magnitude* (speed) and a *direction* (up, down, east, southwest, and so forth). The SI units of velocity are distance $\times$ time^{-1}, that is, m s^{-1}.

The velocity of an object changes if its speed changes or if the direction of its motion changes. The rate of change of velocity is called *acceleration*, which has the units of velocity $\times$ time^{-1}, that is, m s^{-1} $\times$ s^{-1} = m s^{-2}. For an object under a constant acceleration (a), its velocity (v) as a function of time (t) is shown below.

$$v = at \qquad \text{(B.1)}$$

The distance (d) traveled is given by the following equation, which can be established by the methods of calculus.

$$d = \tfrac{1}{2}at^2 \qquad \text{(B.2)}$$

The *constant acceleration due to gravity* (g) experienced by a freely falling body is 9.8066 m s^{-2}.

B.2 Force and Work

According to Newton's *first law* of motion, an object has a natural tendency—called *inertia*—to remain in motion at a constant velocity if it is moving or to remain at rest if it is not moving. A *force* is required to overcome the inertia of an object—that is, to give motion to an object at rest or to change the velocity of a moving object. Because a change in velocity is an acceleration, we can say that *a force is required to provide acceleration to an object.*

Newton's *second law* of motion describes the force (F) required to produce an acceleration (a) in an object of mass (m).

$$F = ma \qquad \text{(B.3)}$$

The SI unit of force is the *newton* (N). It is the force required to produce an acceleration of 1 m s^{-2} in a 1 kg mass.

$$1\ \text{N} = 1\ \text{kg} \times 1\ \text{m s}^{-2} = 1\ \text{kg m s}^{-2} \qquad \text{(B.4)}$$

The weight W of an object is the force of gravity on the object. It is the mass of the object multiplied by the acceleration due to gravity.

$$W = F = mg$$

Work is done when a force acts through a distance.

$$\text{Work } (w) = \text{force } (F) \times \text{distance } (d)$$

A joule (J) is the work done when a force of one newton acts through a distance of one meter. When we combine this definition and the SI units of the newton from Equation B.4, we obtain the SI units of the joule.

$$1\ \text{J} = 1\ \text{N} \times 1\ \text{m} = 1\ \text{N m}$$
$$1\ \text{J} = 1\ \text{kg} \times \text{m s}^{-2} \times 1\ \text{m} = 1\ \text{kg m}^2\ \text{s}^{-2}$$

B.3 Energy

Energy is the capacity to do work. A moving object has *kinetic* energy as a result of its motion. The work associated with the moving object is given by the previous expressions:

$$w = F \times d = ma \times d$$

From equation B.2, we can substitute the expression shown in color for the distance, d.

$$w = ma \times \tfrac{1}{2}at^2 = \tfrac{1}{2} \times m(at)^2$$

Now, from equation B.1, we can substitute velocity (v) for the term at.

$$w = \tfrac{1}{2} \times m \times v^2$$

This is the work required to provide an object of mass m with a velocity v. This quantity of work appears as the kinetic energy (E_k) of the moving object.

$$E_k = \tfrac{1}{2}mv^2$$

In addition to kinetic energy associated with motion, an object may possess *potential energy*. Potential energy is stored energy that can be released under appropriate circumstances. Think of it as energy that stems from the condition, position, or composition of an object. In principle, equations can be written for the various ways in which potential energy is stored in an object, but we do not specifically use such equations in the text.

B.4 Magnetism

Attractive and repulsive forces associated with magnets are centered in regions of the magnets called *poles*. A magnet has a north pole and a south pole. If two magnets are arranged so that the north pole of one magnet is brought near the south pole of another magnet, there is an attractive force between the magnets. If like poles are brought close together—either both north poles or both south poles—there is a repulsive force. *Unlike poles attract, and like poles repel.*

A magnetic field exists in the region surrounding a magnet in which the influence of the magnet can be felt. For example, a magnetic field can be detected through deflections of a compass needle, or the field can be visualized through the attractive forces that cause a characteristic alignment of iron filings.

▲ **Visualizing a magnetic field of a bar magnet**
The sprinkling of iron filings outlines the magnetic field of a bar magnet.

B.5 Electricity

Electricity is a phenomenon closely related to magnetism. Ultimately, all bulk matter contains electrically charged particles: protons and electrons. However, an object displays a net electric charge—positive or negative—only when the numbers of electrons and protons in the object are unequal. The basic expression dealing with stationary electrically charged particles—static electricity—is Coulomb's law: The magnitude of the force (F) between electrically charged objects is directly proportional to the magnitudes of the charges (Q) and inversely proportional to the *square* of the distance (r) between them.

$$F \propto \frac{Q_1 \times Q_2}{r^2}$$

Like charges repel. Whether both charges are positive or both are negative, their product is a positive quantity. A *positive* force is a *repulsive force. Unlike charges attract.* If one charge is positive and the other negative, their product is a negative quantity. A *negative* force is an *attractive* force.

An *electric field* exists in the region surrounding an electrically charged object in which the influence of the electric charge is felt. If an uncharged object is brought into the field of a charged object, an electric charge of the opposite sign may be *induced* in the previously uncharged object. This leads to a force of attraction between the two. (See Figure 11.17, page 478).

Electric current is a flow of charged particles—electrons in metallic conductors and positive and negative ions in molten salts and in aqueous salt solutions. The unit of electric charge is the *coulomb* (C). The unit of electric current is the *ampere* (A). A current of one ampere is the flow of one coulomb of electric charge per second.

$$1\ \text{A} = 1\ \text{C}/1\ \text{s} = 1\ \text{C s}^{-1}$$

Electric potential, or voltage, is the energy per unit of charge in an electric current. With coulomb as the unit of charge and joule as the unit of energy, the unit of electrical potential, 1 *volt* (V), is

$$1\ \text{V} = \frac{1\ \text{J}}{1\ \text{C}}$$

Electric *power* is the rate of production (or consumption) of electric energy. The electric power unit, the *watt* (W), signifies the production (or consumption) of one joule of energy per second.

$$1\ \text{W} = 1\ \text{J s}^{-1}$$

Because electric energy in joules is the product (volts × coulombs) and because coulombs per second (C s^{-1}) represents a current in amperes (A), we can also write the following expressions.

$$1\ \text{W} = 1\ \text{V C s}^{-1}$$
$$= 1\ \text{V} \times 1\ \text{A}$$

As an example, the electric power associated with the passage of 10.0 amp through a 110-volt electric circuit is

$$110\ \text{V} \times 10.0\ \text{A} = 1100\ \text{W}$$

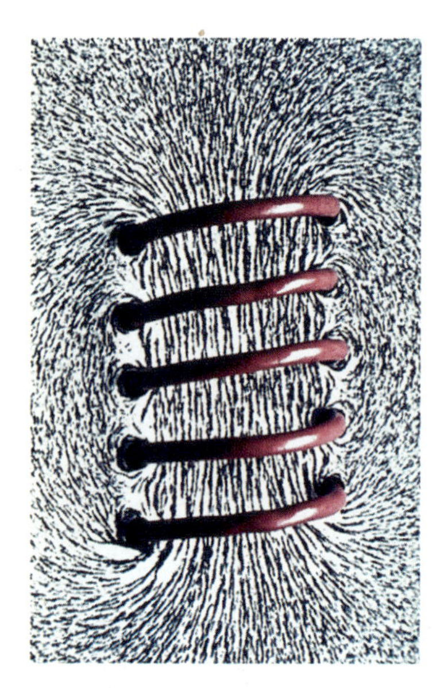

▲ **Visualizing a magnetic field of an electromagnet**
An electric current flowing through a coil of wire produces a magnetic field, outlined here by a sprinkling of iron filings.

B.6 Electromagnetism

A variety of relationships between electricity and magnetism, collectively called *electromagnetism*, underlie some important practical applications: (1) Magnetic fields are associated with the flow of electrons, as in *electromagnets* (see the photograph below). (2) Forces are experienced by current-carrying conductors in a magnetic field, as in *electric motors*. (3) Electric currents are induced when electric conductors are moved through a magnetic field, as in *electric generators*. Several phenomena described in this text are electromagnetic effects.

▲ **An electromagnet**
Electric current from the battery passes through the coil of wire wrapped around an iron bar. The electric current induces a magnetic field and causes the bar to act as a magnet, attracting small iron objects. When the electric current is cut off, the magnetic field dissipates and the bar loses its magnetism.

Appendix C

Data Tables

C.1 Thermodynamic Properties of Substances at 298.15 K

Substances are at 1 atm pressure.[a] For aqueous solutions, solutes are at unit activity (≈1 M).

Inorganic Substances

	ΔH_f°, kJ mol^{-1}	ΔG_f°, kJ mol^{-1}	S°, J mol^{-1} K^{-1}
Aluminum			
Al(s)	0	0	28.3
Al^{3+}(aq)	−531	−485	−321.7
$AlCl_3$(s)	−705.6	−630.1	109.3
Al_2Cl_6(g)	−1291	−1221	490
AlF_3(s)	−1504	−1425	66.48
Al_2O_3(α, solid)	−1676	−1582	50.92
$Al(OH)_3$(s)	−1276	—	—
$Al_2(SO_4)_3$(s)	−3441	−3100	239
Barium			
Ba(s)	0	0	62.3
Ba^{2+}(aq)	−537.6	−560.8	9.6
$BaCO_3$(s)	−1216	−1138	112
$BaCl_2$(s)	−858.1	−810.4	123.7
BaF_2(s)	−1209	−1159	96.40
BaO(s)	−548.1	−520.4	72.09
$Ba(OH)_2$(s)	−946.0	−859.4	107
$Ba(OH)_2 \cdot 8H_2O$(s)	−3342	−2793	427
$BaSO_4$(s)	−1473	−1362	132
Beryllium			
Be(s)	0	0	9.54
$BeCl_2$(s)	−496.2	−449.5	75.81
BeF_2(s)	−1027	−979.5	53.35
BeO(s)	−608.4	−579.1	13.77
Bismuth			
Bi(s)	0	0	56.74
$BiCl_3$(s)	−379	−315	177
Bi_2O_3(s)	−573.9	−493.7	151
Boron			
B(s)	0	0	5.86
BCl_3(l)	−427.2	−387	206

Inorganic Substances (continued)

	ΔH_f°, kJ mol^{-1}	ΔG_f°, kJ mol^{-1}	S°, J mol^{-1} K^{-1}
BF_3(g)	−1137	−1120.3	254.0
B_2H_6(g)	36	86.6	232.0
B_2O_3(s)	−1273	−1194	53.97
Bromine			
Br(g)	111.9	82.43	174.9
Br^-(aq)	−121.6	−104.0	82.4
Br_2(g)	30.91	3.14	245.4
Br_2(l)	0	0	152.2
BrCl(g)	14.6	−0.96	240.0
BrF_3(g)	−255.6	−229.5	292.4
BrF_3(l)	−300.8	−240.6	178.2
Cadmium			
Cd(s)	0	0	51.76
Cd^{2+}(aq)	−75.90	−77.61	−73.2
$CdCl_2$(s)	−391.5	−344.0	115.3
CdO(s)	−258	−228	54.8
Calcium			
Ca(s)	0	0	41.4
Ca^{2+}(aq)	−542.8	−553.6	−53.1
$CaBr_2$(s)	−682.8	−663.6	130.
$CaCO_3$(s)	−1207	−1128	88.70
$CaCl_2$(s)	−795.8	−748.1	105
CaF_2(s)	−1220	−1167	68.87
CaH_2(s)	−186	−147	42
$Ca(NO_3)_2$(s)	−938.4	−743.2	193
CaO(s)	−635.1	−604.0	39.75
$Ca(OH)_2$(s)	−986.1	−898.6	83.39
$Ca_3(PO_4)_2$(s)	−4121	−3885	236
$CaSO_4$(s)	−1434	−1322	106.7
Carbon (See also the table of organic substances.)			
C(g)	716.7	671.3	158.0
C (diamond)	1.90	2.90	2.38

[a] The IUPAC standard-state pressure is 1 bar (10^5 Pa). The values given here are for 1 atm, but they do not differ significantly from those at 1 bar. For example, for CO_2(g), the values of ΔH_f° and ΔG_f° are the same at 1 atm and 1 bar; the value of S° = 213.6 J mol^{-1} K^{-1} at 1 atm and 213.8 J mol^{-1} K^{-1} at 1 bar.

Inorganic Substances (continued)	ΔH_f°, kJ mol^{-1}	ΔG_f°, kJ mol^{-1}	S°, J mol^{-1} K^{-1}
C (graphite)	0	0	5.74
CCl_4(g)	−102.9	−60.63	309.7
CCl_4(l)	−135.4	−65.27	216.2
C_2N_2(g)	308.9	297.2	242.3
CO(g)	−110.5	−137.2	197.6
CO_2(g)	−393.5	−394.4	213.6
CO_3^{2-}(aq)	−677.1	−527.8	−56.9
C_3O_2(g)	−93.72	−109.8	276.4
C_3O_2(l)	−117.3	−105.0	181.1
$COCl_2$(g)	−220.9	−206.8	283.8
COS(g)	−138.4	−165.6	231.5
CS_2(l)	89.70	65.27	151.3
Chlorine			
Cl(g)	121.7	105.7	165.1
Cl^-(aq)	−167.2	−131.2	56.5
Cl_2(g)	0	0	223.0
ClF_3(g)	−163.2	−123.0	281.5
ClO_2(g)	102.5	120.5	256.7
Cl_2O(g)	80.33	97.49	267.9
Chromium			
Cr(s)	0	0	23.66
Cr_2O_3(s)	−1135	−1053	81.17
CrO_4^{2-}(aq)	−881.2	−727.8	50.21
$Cr_2O_7^{2-}$(aq)	−1490	−1301	261.9
Cobalt			
Co(s)	0	0	30.0
CoO(s)	−237.9	−214.2	52.97
$Co(OH)_2$ (pink solid)	−539.7	−454.4	79
Copper			
Cu(s)	0	0	33.15
Cu^{2+}(aq)	64.77	65.49	−99.6
$CuCO_3 \cdot Cu(OH)_2$(s)	−1051	−893.7	186
Cu_2O(s)	−168.6	−146.0	93.14
CuO(s)	−157.3	−129.7	42.63
$Cu(OH)_2$(s)	−450.2	−373	108
$CuSO_4 \cdot 5H_2O$(s)	−2279.6	−1880.1	300.4
Fluorine			
F(g)	78.99	61.92	158.7
F^-(aq)	−332.6	−278.8	−13.8
F_2(g)	0	0	202.7
Helium			
He(g)	0	0	126.0
Hydrogen			
H(g)	218.0	203.3	114.6
H^+(aq)	0	0	0
H_2(g)	0	0	130.6
HBr(g)	−36.40	−53.43	198.6
HCl(g)	−92.31	−95.30	186.8
HCl(aq)	−167.2	−131.3	56.48
HCN(g)	135	125	201.7
HF(g)	−271.1	−273.2	173.7

Inorganic Substances (continued)	ΔH_f°, kJ mol^{-1}	ΔG_f°, kJ mol^{-1}	S°, J mol^{-1} K^{-1}
HI(g)	26.48	1.72	206.5
HNO_3(l)	−173.2	−79.91	155.6
HNO_3(aq)	−207.4	−113.3	146.4
H_2O(g)	−241.8	−228.6	188.7
H_2O(l)	−285.8	−237.2	69.91
H_2O_2(g)	−136.1	−105.5	232.9
H_2O_2(l)	−187.8	−120.4	110
H_2S(g)	−20.63	−33.56	205.7
H_2SO_4(l)	−814.0	−690.1	156.9
H_2SO_4(aq)	−909.3	−744.6	20.08
Iodine			
I(g)	106.8	70.28	180.7
I^-(aq)	−55.19	−51.57	111.3
I_2(g)	62.44	19.36	260.6
I_2(s)	0	0	116.1
IBr(g)	40.84	3.72	258.7
ICl(g)	17.78	−5.44	247.4
ICl(l)	−23.89	−13.60	135.1
Iron			
Fe(s)	0	0	27.28
Fe^{2+}(aq)	−89.1	−78.90	−137.7
Fe^{3+}(aq)	−48.5	−4.7	−315.9
$FeCO_3$(s)	−740.6	−666.7	92.88
$FeCl_3$(s)	−399.5	−334.1	142.3
FeO(s)	−272	−251.5	60.75
Fe_2O_3(s)	−824.2	−742.2	87.40
Fe_3O_4(s)	−1118	−1015	146
$Fe(OH)_3$(s)	−823.0	−696.6	107
Lead			
Pb(s)	0	0	64.81
Pb^{2+}(aq)	−1.7	−24.43	10.5
PbI_2(s)	−175.5	−173.6	174.8
PbO_2(s)	−277	−217.4	68.6
$PbSO_4$(s)	−919.9	−813.2	148.6
Lithium			
Li(s)	0	0	29.12
Li^+(aq)	−278.5	−293.3	13.4
LiCl(s)	−408.6	−384.4	59.33
Li_2O(s)	−597.94	−561.18	37.57
LiOH(s)	−484.9	−439.0	42.80
$LiNO_3$(s)	−483.1	−381.1	90.0
Magnesium			
Mg(s)	0	0	32.69
Mg^{2+}(aq)	−466.9	−454.8	−138.1
$MgCl_2$(s)	−641.3	−591.8	89.62
$MgCO_3$(s)	−1096	−1012	65.7
MgF_2(s)	−1124	−1071	57.24
MgO(s)	−601.7	−569.4	26.94
$Mg(OH)_2$(s)	−924.7	−833.9	63.18
$MgSO_4$(s)	−1285	−1171	91.6
Manganese			
Mn(s)	0	0	32.0
Mn^{2+}(aq)	−220.8	−228.1	−73.6

Inorganic Substances (continued)

	ΔH_f°, kJ mol^{-1}	ΔG_f°, kJ mol^{-1}	S°, J mol^{-1} K^{-1}
MnO_2(s)	−520	−465.2	53.05
MnO_4^-(aq)	−541.4	−447.2	191.2
Mercury			
Hg(g)	61.32	31.85	174.9
Hg(l)	0	0	76.02
HgO(s)	−90.83	−58.56	70.29
Nitrogen			
N(g)	472.7	455.6	153.2
N_2(g)	0	0	191.5
NF_3(g)	−124.7	−83.2	260.7
NH_3(g)	−46.11	−16.48	192.3
NH_3(aq)	−80.29	−26.57	111.3
NH_4^+(aq)	−132.5	−79.31	113.4
NH_4Br(s)	−270.8	−175	113.0
NH_4Cl(s)	−314.4	−203.0	94.56
NH_4F(s)	−464.0	−348.8	71.96
NH_4HCO_3(s)	−849.4	−666.1	121
NH_4I(s)	−201.4	−113	117
NH_4NO_3(s)	−365.6	−184.0	151.1
NH_4NO_3(aq)	−339.9	−190.7	259.8
$(NH_4)_2SO_4$(s)	−1181	−901.9	220.1
N_2H_4(g)	95.40	159.3	238.4
N_2H_4(l)	50.63	149.2	121.2
NO(g)	90.25	86.57	210.6
N_2O(g)	82.05	104.2	219.7
NO_2(g)	33.18	51.30	240.0
N_2O_4(g)	9.16	97.82	304.2
N_2O_4(l)	−19.6	97.40	209.2
N_2O_5(g)	11.3	115.1	355.7
NO_3^-(aq)	−205.0	−108.7	146.4
NOBr(g)	82.17	82.4	273.5
NOCl(g)	51.71	66.07	261.6
Oxygen			
O(g)	249.2	231.7	160.9
O_2(g)	0	0	205.0
O_3(g)	142.7	163.2	238.8
OH^-(aq)	−230.0	−157.2	−10.75
OF_2(g)	24.5	41.8	247.3
Phosphorus			
P (α white)	0	0	41.1
P (red)	−17.6	−12.1	22.8
P_4(g)	58.9	24.5	279.9
PCl_3(g)	−287.0	−267.8	311.7
PCl_3(l)	−319.7	−272.3	217.1
PCl_5(g)	−374.9	−305.0	364.5
PCl_5(s)	−443.5	—	—
PH_3(g)	5.4	13.4	210.1
P_4O_{10}(s)	−2984	−2698	228.9
PO_4^{3-}(aq)	−1277	−1019	−222
Potassium			
K(g)	89.24	60.63	160.2
K(l)	2.28	0.26	71.46
K(s)	0	0	64.18
K^+(aq)	−252.4	−283.3	102.5
KBr(s)	−393.8	−380.7	95.90

Inorganic Substances (continued)

	ΔH_f°, kJ mol^{-1}	ΔG_f°, kJ mol^{-1}	S°, J mol^{-1} K^{-1}
KCN(s)	−113	−101.9	128.5
KCl(s)	−436.7	−409.2	82.59
$KClO_3$(s)	−397.7	−296.3	143
$KClO_4$(s)	−432.8	−303.2	151.0
KF(s)	−567.3	−537.8	66.57
KI(s)	−327.9	−324.9	106.3
KNO_3(s)	−494.6	−394.9	133.1
KOH(s)	−424.8	−379.1	78.87
KOH(aq)	−482.4	−440.5	91.63
K_2SO_4(s)	−1438	−1321	175.6
Silicon			
Si(s)	0	0	18.8
SiH_4(g)	34	56.9	204.5
Si_2H_6(g)	80.3	127	272.5
SiO_2(quartz)	−910.9	−856.7	41.84
Silver			
Ag(s)	0	0	42.55
Ag^+(aq)	105.6	77.11	72.68
AgBr(s)	−100.4	−96.90	107
AgCl(s)	−127.1	−109.8	96.2
AgI(s)	−61.84	−66.19	115
$AgNO_3$(s)	−124.4	−33.5	140.9
Ag_2O(s)	−31.0	−11.2	121
Ag_2SO_4(s)	−715.9	−618.5	200.4
Sodium			
Na(g)	107.3	76.78	153.6
Na(l)	2.41	0.50	57.86
Na(s)	0	0	51.21
Na^+(aq)	−240.1	−261.9	59.0
Na_2(g)	142.0	104.0	230.1
NaBr(s)	−361.1	−349.0	86.82
Na_2CO_3(s)	−1131	−1044	135.0
$NaHCO_3$(s)	−950.8	−851.0	102
NaCl(s)	−411.1	−384.0	72.13
NaCl(aq)	−407.3	−393.1	115.5
$NaClO_3$(s)	−365.8	−262.3	123
$NaClO_4$(s)	−383.3	−254.9	142.3
NaF(s)	−573.7	−543.5	51.46
NaH(s)	−56.27	−33.5	40.02
NaI(s)	−287.8	−286.1	98.53
$NaNO_3$(s)	−467.9	−367.1	116.5
$NaNO_3$(aq)	−447.4	−373.2	205.4
Na_2O_2(s)	−510.9	−447.7	94.98
NaOH(s)	−425.6	−379.5	64.48
NaOH(aq)	−469.2	−419.2	48.1
NaH_2PO_4(s)	−1537	−1386	127.5
Na_2HPO_4(s)	−1748	−1608	150.5
Na_3PO_4(s)	−1917	−1789	173.8
$NaHSO_4$(s)	−1125	−992.9	113
Na_2SO_4(s)	−1387	−1270	149.6
Na_2SO_4(aq)	−1390	−1268	138.1
$Na_2SO_4 \cdot 10H_2O$(s)	−4327	−3647	592.0
$Na_2S_2O_3$(s)	−1123	−1028	155

Inorganic Substances (continued)	ΔH_f°, kJ mol^{-1}	ΔG_f°, kJ mol^{-1}	S°, J mol^{-1} K^{-1}
Sulfur			
S (rhombic)	0	0	31.8
$S_8(g)$	102.3	49.16	430.2
$S_2Cl_2(g)$	−18.4	−31.8	331.5
$SF_6(g)$	−1209	−1105	291.7
$SO_2(g)$	−296.8	−300.2	248.1
$SO_3(g)$	−395.7	−371.1	256.6
$SO_4^{2-}(aq)$	−909.3	−744.5	20.1
$S_2O_3^{2-}(aq)$	−648.5	−522.5	67
$SO_2Cl_2(g)$	−364.0	−320.0	311.8
$SO_2Cl_2(l)$	−394.1	−314	207
Tin			
Sn (white)	0	0	51.55
Sn (gray)	−2.1	0.1	44.14
$SnCl_4(l)$	−511.3	−440.2	259
$SnO(s)$	−286	−257	56.5
$SnO_2(s)$	−580.7	−519.7	52.3

Inorganic Substances (continued)	ΔH_f°, kJ mol^{-1}	ΔG_f°, kJ mol^{-1}	S°, J mol^{-1} K^{-1}
Titanium			
$Ti(s)$	0	0	30.6
$TiCl_4(g)$	−763.2	−726.8	355
$TiCl_4(l)$	−804.2	−737.2	252.3
$TiO_2(s)$	−944.7	−889.5	50.33
Uranium			
$U(s)$	0	0	50.21
$UF_6(g)$	−2147	−2064	378
$UF_6(s)$	−2197	−2069	228
$UO_2(s)$	−1085	−1032	77.03
Zinc			
$Zn(s)$	0	0	41.6
$Zn^{2+}(aq)$	−153.9	−147.1	−112.1
$ZnCl_2(s)$	−415.1	−369.4	111.5
$ZnO(s)$	−348.3	−318.3	43.64

Organic Substances

Formula	Name	ΔH_f°, kJ mol^{-1}	ΔG_f°, kJ mol^{-1}	S°, J mol^{-1} K^{-1}
$CH_4(g)$	Methane(g)	−74.81	−50.75	186.2
$C_2H_2(g)$	Acetylene(g)	226.7	209.2	200.8
$C_2H_4(g)$	Ethylene(g)	52.26	68.12	219.4
$C_2H_6(g)$	Ethane(g)	−84.68	−32.89	229.5
$C_3H_8(g)$	Propane(g)	−103.8	−23.56	270.2
$C_4H_{10}(g)$	Butane(g)	−125.7	−17.15	310.1
$C_6H_6(g)$	Benzene(g)	82.93	129.7	269.2
$C_6H_6(l)$	Benzene(l)	48.99	124.4	173.3
$C_6H_{12}(g)$	Cyclohexane(g)	−123.1	31.8	298.2
$C_6H_{12}(l)$	Cyclohexane(l)	−156.2	26.7	204.3
$C_{10}H_8(g)$	Naphthalene(g)	149	223.6	335.6
$C_{10}H_8(s)$	Naphthalene(s)	75.3	201.0	166.9
$CH_2O(g)$	Formaldehyde(g)	−117.0	−110.0	218.7
$CH_3OH(g)$	Methanol(g)	−200.7	−162.0	239.7
$CH_3OH(l)$	Methanol(l)	−238.7	−166.4	126.8
$CH_3CHO(g)$	Acetaldehyde(g)	−166.1	−133.4	246.4
$CH_3CHO(l)$	Acetaldehyde(l)	−191.8	−128.3	160.4
$CH_3CH_2OH(g)$	Ethanol(g)	−234.4	−167.9	282.6
$CH_3CH_2OH(l)$	Ethanol(l)	−277.7	−174.9	160.7
$C_6H_5OH(s)$	Phenol(s)	−165.0	−50.42	144.0
$(CH_3)_2CO(g)$	Acetone(g)	−216.6	−153.1	294.9
$(CH_3)_2CO(l)$	Acetone(l)	−247.6	−155.7	200.4
$CH_3COOH(g)$	Acetic acid(g)	−432.3	−374.0	282.5
$CH_3COOH(l)$	Acetic acid(l)	−484.1	−389.9	159.8
$CH_3COOH(aq)$	Acetic acid(aq)	−488.3	−396.6	178.7
$C_6H_5COOH(s)$	Benzoic acid(s)	−385.1	−245.3	167.6
$CH_3NH_2(g)$	Methylamine(g)	−23.0	32.3	242.6
$C_6H_5NH_2(g)$	Aniline(g)	86.86	166.7	319.2
$C_6H_5NH_2(l)$	Aniline(l)	31.6	149.1	191.3
$C_6H_{12}O_6(s)$	Glucose(s)	−1273.3	−910.4	212.1

C.2 Equilibrium Constants

A. Ionization Constants of Weak Acids at 25 °C

Name of Acid	Formula	K_a
Acetic	$HC_2H_3O_2$	1.8×10^{-5}
Acrylic	$HC_3H_3O_2$	5.5×10^{-5}
Arsenic	H_3AsO_4	6.0×10^{-3}
	$H_2AsO_4^-$	1.0×10^{-7}
	$HAsO_4^{2-}$	3.2×10^{-12}
Arsenous	H_3AsO_3	6.6×10^{-10}
Benzoic	$HC_7H_5O_2$	6.3×10^{-5}
Bromoacetic	$HC_2H_2BrO_2$	1.3×10^{-3}
Butyric	$HC_4H_7O_2$	1.5×10^{-5}
Carbonic	H_2CO_3	4.4×10^{-7}
	HCO_3^-	4.7×10^{-11}
Chloroacetic	$HC_2H_2ClO_2$	1.4×10^{-3}
Chlorous	$HClO_2$	1.1×10^{-2}
Citric	$H_3C_6H_5O_7$	7.4×10^{-4}
	$H_2C_6H_5O_7^-$	1.7×10^{-5}
	$HC_6H_5O_7^{2-}$	4.0×10^{-7}
Cyanic	$HOCN$	3.5×10^{-4}
Dichloroacetic	$HC_2HCl_2O_2$	5.5×10^{-2}
Fluoroacetic	$HC_2H_2FO_2$	2.6×10^{-3}
Formic	$HCHO_2$	1.8×10^{-4}
Hydrazoic	HN_3	1.9×10^{-5}
Hydrocyanic	HCN	6.2×10^{-10}
Hydrofluoric	HF	6.6×10^{-4}
Hydrogen peroxide	H_2O_2	2.2×10^{-12}
Hydroselenic	H_2Se	1.3×10^{-4}
	HSe^-	1×10^{-11}
Hydrosulfuric	H_2S	1.0×10^{-7}
	HS^-	1×10^{-19}
Hydrotelluric	H_2Te	2.3×10^{-3}
	HTe^-	1.6×10^{-11}
Hypobromous	$HOBr$	2.5×10^{-9}
Hypochlorous	$HOCl$	2.9×10^{-8}
Hypoiodous	HOI	2.3×10^{-11}
Hyponitrous	$HON{=}NOH$	8.9×10^{-8}
	$HON{=}NO^-$	4×10^{-12}
Iodic	HIO_3	1.6×10^{-1}
Iodoacetic	$HC_2H_2IO_2$	6.7×10^{-4}
Malonic	$H_2C_3H_2O_4$	1.5×10^{-3}
	$HC_3H_2O_4^-$	2.0×10^{-6}
Nitrous	HNO_2	7.2×10^{-4}
Oxalic	$H_2C_2O_4$	5.4×10^{-2}
	$HC_2O_4^-$	5.3×10^{-5}
Phenol	HOC_6H_5	1.0×10^{-10}
Phenylacetic	$HC_8H_7O_2$	4.9×10^{-5}
Phosphoric	H_3PO_4	7.1×10^{-3}
	$H_2PO_4^-$	6.3×10^{-8}
	HPO_4^{2-}	4.2×10^{-13}
Phosphorous	H_3PO_3	3.7×10^{-2}
	$H_2PO_3^-$	2.1×10^{-7}
Propionic	$HC_3H_5O_2$	1.3×10^{-5}
Pyrophosphoric	$H_4P_2O_7$	3.0×10^{-2}
	$H_3P_2O_7^-$	4.4×10^{-3}
	$H_2P_2O_7^{2-}$	2.5×10^{-7}
	$HP_2O_7^{3-}$	5.6×10^{-10}
Selenic	H_2SeO_4	strong acid
	$HSeO_4^-$	2.2×10^{-2}
Selenous	H_2SeO_3	2.3×10^{-3}
	$HSeO_3^-$	5.4×10^{-9}
Succinic	$H_2C_4H_4O_4$	6.2×10^{-5}
	$HC_4H_4O_4^-$	2.3×10^{-6}
Sulfuric	H_2SO_4	strong acid
	HSO_4^-	1.1×10^{-2}
Sulfurous	H_2SO_3	1.3×10^{-2}
	HSO_3^-	6.2×10^{-8}
Thiophenol	HSC_6H_5	3.2×10^{-7}
Trichloroacetic	$HC_2Cl_3O_2$	3.0×10^{-1}

B. Ionization Constants of Weak Bases at 25 °C

Name of Base	Formula	K_b
Ammonia	NH_3	1.8×10^{-5}
Aniline	$C_6H_5NH_2$	7.4×10^{-10}
Codeine	$C_{18}H_{21}O_3N$	8.9×10^{-7}
Diethylamine	$(C_2H_5)_2NH$	6.9×10^{-4}
Dimethylamine	$(CH_3)_2NH$	5.9×10^{-4}
Ethylamine	$C_2H_5NH_2$	4.3×10^{-4}
Hydrazine	NH_2NH_2	8.5×10^{-7}
	$NH_2NH_3^+$	8.9×10^{-16}
Hydroxylamine	NH_2OH	9.1×10^{-9}
Isoquinoline	C_9H_7N	2.5×10^{-9}
Methylamine	CH_3NH_2	4.2×10^{-4}
Morphine	$C_{17}H_{19}O_3N$	7.4×10^{-7}
Piperidine	$C_5H_{11}N$	1.3×10^{-3}
Pyridine	C_5H_5N	1.5×10^{-9}
Quinoline	C_9H_7N	6.3×10^{-10}
Triethanolamine	$C_6H_{15}O_3N$	5.8×10^{-7}
Triethylamine	$(C_2H_5)_3N$	5.2×10^{-4}
Trimethylamine	$(CH_3)_3N$	6.3×10^{-5}

C. Solubility Product Constants[a]

Name of Solute	Formula	K_{sp}
Aluminum hydroxide	$Al(OH)_3$	1.3×10^{-33}
Aluminum phosphate	$AlPO_4$	6.3×10^{-19}
Barium carbonate	$BaCO_3$	5.1×10^{-9}
Barium chromate	$BaCrO_4$	1.2×10^{-10}
Barium fluoride	BaF_2	1.0×10^{-6}
Barium hydroxide	$Ba(OH)_2$	5×10^{-3}
Barium sulfate	$BaSO_4$	1.1×10^{-10}
Barium sulfite	$BaSO_3$	8×10^{-7}
Barium thiosulfate	BaS_2O_3	1.6×10^{-5}
Bismuthyl chloride	$BiOCl$	1.8×10^{-31}
Bismuthyl hydroxide	$BiOOH$	4×10^{-10}
Cadmium carbonate	$CdCO_3$	5.2×10^{-12}
Cadmium hydroxide	$Cd(OH)_2$	2.5×10^{-14}
Cadmium sulfide[b]	CdS	8×10^{-28}
Calcium carbonate	$CaCO_3$	2.8×10^{-9}
Calcium chromate	$CaCrO_4$	7.1×10^{-4}
Calcium fluoride	CaF_2	5.3×10^{-9}
Calcium hydrogen phosphate	$CaHPO_4$	1×10^{-7}
Calcium hydroxide	$Ca(OH)_2$	5.5×10^{-6}
Calcium oxalate	CaC_2O_4	2.7×10^{-9}
Calcium phosphate	$Ca_3(PO_4)_2$	2.0×10^{-29}
Calcium sulfate	$CaSO_4$	9.1×10^{-6}
Calcium sulfite	$CaSO_3$	6.8×10^{-8}
Chromium(II) hydroxide	$Cr(OH)_2$	2×10^{-16}
Chromium(III) hydroxide	$Cr(OH)_3$	6.3×10^{-31}
Cobalt(II) carbonate	$CoCO_3$	1.4×10^{-13}
Cobalt(II) hydroxide	$Co(OH)_2$	1.6×10^{-15}
Cobalt(III) hydroxide	$Co(OH)_3$	1.6×10^{-44}
Copper(I) chloride	$CuCl$	1.2×10^{-6}
Copper(I) cyanide	$CuCN$	3.2×10^{-20}
Copper(I) iodide	CuI	1.1×10^{-12}
Copper(II) arsenate	$Cu_3(AsO_4)_2$	7.6×10^{-36}
Copper(II) carbonate	$CuCO_3$	1.4×10^{-10}
Copper(II) chromate	$CuCrO_4$	3.6×10^{-6}
Copper(II) ferrocyanide	$Cu_2[Fe(CN)_6]$	1.3×10^{-16}
Copper(II) hydroxide	$Cu(OH)_2$	2.2×10^{-20}
Copper(II) sulfide[b]	CuS	6×10^{-37}
Iron(II) carbonate	$FeCO_3$	3.2×10^{-11}
Iron(II) hydroxide	$Fe(OH)_2$	8.0×10^{-16}
Iron(II) sulfide[b]	FeS	6×10^{-19}
Iron(III) arsenate	$FeAsO_4$	5.7×10^{-21}
Iron(III) ferrocyanide	$Fe_4[Fe(CN)_6]_3$	3.3×10^{-41}
Iron(III) hydroxide	$Fe(OH)_3$	4×10^{-38}
Iron(III) phosphate	$FePO_4$	1.3×10^{-22}
Lead(II) arsenate	$Pb_3(AsO_4)_2$	4.0×10^{-36}
Lead(II) azide	$Pb(N_3)_2$	2.5×10^{-9}
Lead(II) bromide	$PbBr_2$	4.0×10^{-5}
Lead(II) carbonate	$PbCO_3$	7.4×10^{-14}
Lead(II) chloride	$PbCl_2$	1.6×10^{-5}
Lead(II) chromate	$PbCrO_4$	2.8×10^{-13}
Lead(II) fluoride	PbF_2	2.7×10^{-8}
Lead(II) hydroxide	$Pb(OH)_2$	1.2×10^{-15}

C. Solubility Product Constants (continued)

Name of Solute	Formula	K_{sp}
Lead(II) iodide	PbI_2	7.1×10^{-9}
Lead(II) sulfate	$PbSO_4$	1.6×10^{-8}
Lead(II) sulfide[b]	PbS	3×10^{-28}
Lithium carbonate	Li_2CO_3	2.5×10^{-2}
Lithium fluoride	LiF	3.8×10^{-3}
Lithium phosphate	Li_3PO_4	3.2×10^{-9}
Magnesium ammonium phosphate	$MgNH_4PO_4$	2.5×10^{-13}
Magnesium carbonate	$MgCO_3$	3.5×10^{-8}
Magnesium fluoride	MgF_2	3.7×10^{-8}
Magnesium hydroxide	$Mg(OH)_2$	1.8×10^{-11}
Magnesium phosphate	$Mg_3(PO_4)_2$	1×10^{-25}
Manganese(II) carbonate	$MnCO_3$	1.8×10^{-11}
Manganese(II) hydroxide	$Mn(OH)_2$	1.9×10^{-13}
Manganese(II) sulfide[b]	MnS	3×10^{-14}
Mercury(I) bromide	Hg_2Br_2	5.6×10^{-23}
Mercury(I) chloride	Hg_2Cl_2	1.3×10^{-18}
Mercury(I) iodide	Hg_2I_2	4.5×10^{-29}
Mercury(II) sulfide[b]	HgS	2×10^{-53}
Nickel(II) carbonate	$NiCO_3$	6.6×10^{-9}
Nickel(II) hydroxide	$Ni(OH)_2$	2.0×10^{-15}
Scandium fluoride	ScF_3	4.2×10^{-18}
Scandium hydroxide	$Sc(OH)_3$	8.0×10^{-31}
Silver acetate	CH_3COOAg	2.0×10^{-3}
Silver arsenate	Ag_3AsO_4	1.0×10^{-22}
Silver azide	AgN_3	2.8×10^{-9}
Silver bromide	$AgBr$	5.0×10^{-13}
Silver chloride	$AgCl$	1.8×10^{-10}
Silver chromate	Ag_2CrO_4	1.1×10^{-12}
Silver cyanide	$AgCN$	1.2×10^{-16}
Silver iodate	$AgIO_3$	3.0×10^{-8}
Silver iodide	AgI	8.5×10^{-17}
Silver nitrite	$AgNO_2$	6.0×10^{-4}
Silver sulfate	Ag_2SO_4	1.4×10^{-5}
Silver sulfide[b]	Ag_2S	6×10^{-51}
Silver sulfite	Ag_2SO_3	1.5×10^{-14}
Silver thiocyanate	$AgSCN$	1.0×10^{-12}
Strontium carbonate	$SrCO_3$	1.1×10^{-10}
Strontium chromate	$SrCrO_4$	2.2×10^{-5}
Strontium fluoride	SrF_2	2.5×10^{-9}
Strontium sulfate	$SrSO_4$	3.2×10^{-7}
Thallium(I) bromide	$TlBr$	3.4×10^{-6}
Thallium(I) chloride	$TlCl$	1.7×10^{-4}
Thallium(I) iodide	TlI	6.5×10^{-8}
Thallium(III) hydroxide	$Tl(OH)_3$	6.3×10^{-46}
Tin(II) hydroxide	$Sn(OH)_2$	1.4×10^{-28}
Tin(II) sulfide[b]	SnS	1×10^{-26}
Zinc carbonate	$ZnCO_3$	1.4×10^{-11}
Zinc hydroxide	$Zn(OH)_2$	1.2×10^{-17}
Zinc oxalate	ZnC_2O_4	2.7×10^{-8}
Zinc phosphate	$Zn_3(PO_4)_2$	9.0×10^{-33}
Zinc sulfide[b]	ZnS	2×10^{-25}

[a] Data are at various temperatures around room temperature, from 18 to 25 °C.

[b] For a solubility equilibrium of the type: $MS(s) + H_2O \rightleftharpoons M^{2+}(aq) + HS^-(aq) + OH^-(aq)$.

C. Complex-Ion Formation Constants[a]

Formula	K_f
$[Ag(CN)_2]^-$	5.6×10^{18}
$[Ag(EDTA)]^{3-}$	2.1×10^{7}
$[Ag(en)_2]^+$	5.0×10^{7}
$[Ag(NH_3)_2]^+$	1.6×10^{7}
$[Ag(SCN)_4]^{3-}$	1.2×10^{10}
$[Ag(S_2O_3)_2]^{3-}$	1.7×10^{13}
$[Al(EDTA)]^-$	1.3×10^{16}
$[Al(OH)_4]^-$	1.1×10^{33}
$[Al(ox)_3]^{3-}$	2×10^{16}
$[Cd(CN)_4]^{2-}$	6.0×10^{18}
$[Cd(en)_3]^{2+}$	1.2×10^{12}
$[Cd(NH_3)_4]^{2+}$	1.3×10^{7}
$[Co(EDTA)]^{2-}$	2.0×10^{16}
$[Co(en)_3]^{2+}$	8.7×10^{13}
$[Co(NH_3)_6]^{2+}$	1.3×10^{5}
$[Co(ox)_3]^{4-}$	5×10^{9}
$[Co(SCN)_4]^{2-}$	1.0×10^{3}
$[Co(EDTA)]^-$	10^{36}
$[Co(en)_3]^{3+}$	4.9×10^{48}
$[Co(NH_3)_6]^{3+}$	4.5×10^{33}
$[Co(ox)_3]^{3-}$	10^{20}
$[Cr(EDTA)]^-$	10^{23}
$[Cr(OH)_4]^-$	8×10^{29}
$[CuCl_3]^{2-}$	5×10^{5}
$[Cu(CN)_4]^{3-}$	2.0×10^{30}
$[Cu(EDTA)]^{2-}$	5×10^{18}
$[Cu(en)_2]^{2+}$	1×10^{20}
$[Cu(NH_3)_4]^{2+}$	1.1×10^{13}
$[Cu(ox)_2]^{2-}$	3×10^{8}
$[Fe(CN)_6]^{4-}$	10^{37}
$[Fe(EDTA)]^{2-}$	2.1×10^{14}

C. Complex-Ion Formation Constants (continued)

Formula	K_f
$[Fe(en)_3]^{2+}$	5.0×10^{9}
$[Fe(ox)_3]^{4-}$	1.7×10^{5}
$[Fe(CN)_6]^{3-}$	10^{42}
$[Fe(EDTA)]^-$	1.7×10^{24}
$[Fe(ox)_3]^{3-}$	2×10^{20}
$[Fe(SCN)]^{2+}$	8.9×10^{2}
$[HgCl_4]^{2-}$	1.2×10^{15}
$[Hg(CN)_4]^{2-}$	3×10^{41}
$[Hg(EDTA)]^{2-}$	6.3×10^{21}
$[Hg(en)_2]^{2+}$	2×10^{23}
$[HgI_4]^{2-}$	6.8×10^{29}
$[Hg(ox)_2]^{2-}$	9.5×10^{6}
$[Ni(CN)_4]^{2-}$	2×10^{31}
$[Ni(EDTA)]^{2-}$	3.6×10^{18}
$[Ni(en)_3]^{2+}$	2.1×10^{18}
$[Ni(NH_3)_6]^{2+}$	5.5×10^{8}
$[Ni(ox)_3]^{4-}$	3×10^{8}
$[PbCl_3]^-$	2.4×10^{1}
$[Pb(EDTA)]^{2-}$	2×10^{18}
$[PbI_4]^{2-}$	3.0×10^{4}
$[Pb(OH)_3]^-$	3.8×10^{14}
$[Pb(ox)_2]^{2-}$	3.5×10^{6}
$[Pb(S_2O_3)_3]^{4-}$	2.2×10^{6}
$[PtCl_4]^{2-}$	1×10^{16}
$[Pt(NH_3)_6]^{2+}$	2×10^{35}
$[Zn(CN)_4]^{2-}$	1×10^{18}
$[Zn(EDTA)]^{2-}$	3×10^{16}
$[Zn(en)_3]^{2+}$	1.3×10^{14}
$[Zn(NH_3)_4]^{2+}$	4.1×10^{8}
$[Zn(OH)_4]^{2-}$	4.6×10^{17}
$[Zn(ox)_3]^{4-}$	1.4×10^{8}

[a] The ligands referred to in this table are monodentate: Cl^-, CN^-, I^-, NH_3, OH^-, SCN^-, $S_2O_3^{2-}$; bidentate: ethylenediamine, en; oxalate ion, ox ($C_2O_4^{2-}$); tetradentate: ethylenediaminetetraacetato ion, $EDTA^{4-}$.

C.3 Standard Electrode (Reduction) Potentials at 25 °C

Reduction Half-reaction	$E°$, V
$F_2(g) + 2e^- \rightarrow 2F^-(aq)$	+2.866
$OF_2(g) + 2H^+(aq) + 4e^- \rightarrow H_2O(l) + 2F^-(aq)$	+2.1
$O_3(g) + 2H^+(aq) + 2e^- \rightarrow O_2(g) + H_2O(l)$	+2.075
$S_2O_8^{2-}(aq) + 2e^- \rightarrow 2SO_4^{2-}(aq)$	+2.01
$Ag^{2+}(aq) + e^- \rightarrow Ag^+(aq)$	+1.98
$H_2O_2(aq) + 2H^+(aq) + 2e^- \rightarrow 2H_2O(l)$	+1.763
$MnO_4^-(aq) + 4H^+(aq) + 3e^- \rightarrow MnO_2(s) + 2H_2O(l)$	+1.70
$PbO_2(s) + SO_4^{2-}(aq) + 4H^+(aq) + 2e^- \rightarrow PbSO_4(s) + 2H_2O(l)$	+1.69
$Au^{3+}(aq) + 3e^- \rightarrow Au(s)$	+1.52
$MnO_4^-(aq) + 8H^+(aq) + 5e^- \rightarrow Mn^{2+}(aq) + 4H_2O(l)$	+1.51
$2BrO_3^-(aq) + 12H^+(aq) + 10e^- \rightarrow Br_2(l) + 6H_2O(l)$	+1.478
$PbO_2(s) + 4H^+(aq) + 2e^- \rightarrow Pb^{2+}(aq) + 2H_2O(l)$	+1.455
$ClO_3^-(aq) + 6H^+(aq) + 6e^- \rightarrow Cl^-(aq) + 3H_2O(l)$	+1.450
$Au^{3+}(aq) + 2e^- \rightarrow Au^+(aq)$	+1.36
$Cl_2(g) + 2e^- \rightarrow 2Cl^-(aq)$	+1.358
$Cr_2O_7^{2-}(aq) + 14H^+(aq) + 6e^- \rightarrow 2Cr^{3+}(aq) + 7H_2O(l)$	+1.33
$MnO_2(s) + 4H^+(aq) + 2e^- \rightarrow Mn^{2+}(aq) + 2H_2O(l)$	+1.23
$O_2(g) + 4H^+(aq) + 4e^- \rightarrow 2H_2O(l)$	+1.229
$2IO_3^-(aq) + 12H^+(aq) + 10e^- \rightarrow I_2(s) + 6H_2O(l)$	+1.20
$ClO_4^-(aq) + 2H^+(aq) + 2e^- \rightarrow ClO_3^-(aq) + H_2O(l)$	+1.19
$ClO_3^-(aq) + 2H^+(aq) + e^- \rightarrow ClO_2(g) + H_2O(l)$	+1.175
$NO_2(g) + H^+(aq) + e^- \rightarrow HNO_2(aq)$	+1.07
$Br_2(l) + 2e^- \rightarrow 2Br^-(aq)$	+1.065
$NO_2(g) + 2H^+(aq) + 2e^- \rightarrow NO(g) + H_2O(l)$	+1.03
$[AuCl_4]^-(aq) + 3e^- \rightarrow Au(s) + 4Cl^-(aq)$	+1.002
$VO_2^+(aq) + 2H^+(aq) + e^- \rightarrow VO^{2+}(aq) + H_2O(l)$	+1.000
$NO_3^-(aq) + 4H^+(aq) + 3e^- \rightarrow NO(g) + 2H_2O(l)$	+0.956
$Cu^{2+}(aq) + I^-(aq) + e^- \rightarrow CuI(s)$	+0.86
$Hg^{2+}(aq) + 2e^- \rightarrow Hg(l)$	+0.854
$Ag^+(aq) + e^- \rightarrow Ag(s)$	+0.800
$Fe^{3+}(aq) + e^- \rightarrow Fe^{2+}(aq)$	+0.771
$O_2(g) + 2H^+(aq) + 2e^- \rightarrow H_2O_2(aq)$	+0.695
$2HgCl_2(aq) + 2e^- \rightarrow Hg_2Cl_2(s) + 2Cl^-(aq)$	+0.63
$MnO_4^-(aq) + e^- \rightarrow MnO_4^{2-}(aq)$	+0.56
$I_2(s) + 2e^- \rightarrow 2I^-(aq)$	+0.535
$Cu^+(aq) + e^- \rightarrow Cu(s)$	+0.520
$H_2SO_3(aq) + 4H^+(aq) + 4e^- \rightarrow S(s) + 3H_2O(l)$	+0.449
$Cu^{2+}(aq) + 2e^- \rightarrow Cu(s)$	+0.340
$C_2N_2(g) + 2H^+(aq) + 2e^- \rightarrow 2HCN(aq)$	+0.37
$[Fe(CN)_6]^{3-}(aq) + e^- \rightarrow [Fe(CN)_6]^{4-}(aq)$	+0.361
$VO^{2+}(aq) + 2H^+(aq) + e^- \rightarrow V^{3+}(aq) + H_2O(l)$	+0.337
$PbO_2(s) + 2H^+(aq) + 2e^- \rightarrow PbO(s) + H_2O(l)$	+0.28
$Hg_2Cl_2(s) + 2e^- \rightarrow 2Hg(l) + 2Cl^-(aq)$	+0.2676
$HAsO_2(aq) + 3H^+(aq) + 3e^- \rightarrow As(s) + 2H_2O(l)$	+0.240
$AgCl(s) + e^- \rightarrow Ag(s) + Cl^-(aq)$	+0.2223
$SO_4^{2-}(aq) + 4H^+(aq) + 2e^- \rightarrow 2H_2O(l) + SO_2(g)$	+0.17
$Cu^{2+}(aq) + e^- \rightarrow Cu^+(aq)$	+0.159
$Sn^{4+}(aq) + 2e^- \rightarrow Sn^{2+}(aq)$	+0.154
$S(s) + 2H^+(aq) + 2e^- \rightarrow H_2S(g)$	+0.14
$AgBr(s) + e^- \rightarrow Ag(s) + Br^-(aq)$	+0.071
$2H^+(aq) + 2e^- \rightarrow H_2(g)$	0
$Pb^{2+}(aq) + 2e^- \rightarrow Pb(s)$	−0.125
$Sn^{2+}(aq) + 2e^- \rightarrow Sn(s)$	−0.137
$AgI(s) + e^- \rightarrow Ag(s) + I^-(aq)$	−0.152
$V^{3+}(aq) + e^- \rightarrow V^{2+}(aq)$	−0.255
$Ni^{2+}(aq) + 2e^- \rightarrow Ni(s)$	−0.257
$H_3PO_4(aq) + 2H^+(aq) + 2e^- \rightarrow H_3PO_3(aq) + H_2O(l)$	−0.276
$Co^{2+}(aq) + 2e^- \rightarrow Co(s)$	−0.277

C.3 Standard Electrode (Reduction) Potentials at 25 °C (continued)

Reduction Half-reaction	$E°$, V
$PbSO_4(s) + 2\,e^- \rightarrow Pb(s) + SO_4^{2-}(aq)$	−0.356
$Cd^{2+}(aq) + 2\,e^- \rightarrow Cd(s)$	−0.403
$Cr^{3+}(aq) + e^- \rightarrow Cr^{2+}(aq)$	−0.424
$Fe^{2+}(aq) + 2\,e^- \rightarrow Fe(s)$	−0.440
$2\,CO_2(g) + 2\,H^+(aq) + 2\,e^- \rightarrow H_2C_2O_4(aq)$	−0.49
$Zn^{2+}(aq) + 2\,e^- \rightarrow Zn(s)$	−0.763
$Cr^{2+}(aq) + 2\,e^- \rightarrow Cr(s)$	−0.90
$Mn^{2+}(aq) + 2\,e^- \rightarrow Mn(s)$	−1.18
$Ti^{2+}(aq) + 2\,e^- \rightarrow Ti(s)$	−1.63
$U^{3+}(aq) + 3\,e^- \rightarrow U(s)$	−1.66
$Al^{3+}(aq) + 3\,e^- \rightarrow Al(s)$	−1.676
$Mg^{2+}(aq) + 2\,e^- \rightarrow Mg(s)$	−2.356
$Na^+(aq) + e^- \rightarrow Na(s)$	−2.713
$Ca^{2+}(aq) + 2\,e^- \rightarrow Ca(s)$	−2.84
$Sr^{2+}(aq) + 2\,e^- \rightarrow Sr(s)$	−2.89
$Ba^{2+}(aq) + 2\,e^- \rightarrow Ba(s)$	−2.92
$Cs^+(aq) + e^- \rightarrow Cs(s)$	−2.923
$K^+(aq) + e^- \rightarrow K(s)$	−2.924
$Rb^+(aq) + e^- \rightarrow Rb(s)$	−2.924
$Li^+(aq) + e^- \rightarrow Li(s)$	−3.040
Basic Solution	
$O_3(g) + H_2O(l) + 2\,e^- \rightarrow O_2(g) + 2\,OH^-(aq)$	+1.246
$ClO^-(aq) + H_2O(l) + 2\,e^- \rightarrow Cl^-(aq) + 2\,OH^-(aq)$	+0.890
$HO_2^-(aq) + H_2O(l) + 2\,e^- \rightarrow 3\,OH^-(aq)$	+0.88
$BrO^-(aq) + H_2O(l) + 2\,e^- \rightarrow Br^-(aq) + 2\,OH^-(aq)$	+0.766
$ClO_3^-(aq) + 3\,H_2O(l) + 6\,e^- \rightarrow Cl^-(aq) + 6\,OH^-(aq)$	+0.622
$2\,AgO(s) + H_2O(l) + 2\,e^- \rightarrow Ag_2O(s) + 2\,OH^-(aq)$	+0.604
$MnO_4^-(aq) + 2\,H_2O(l) + 3\,e^- \rightarrow MnO_2(s) + 4\,OH^-(aq)$	+0.60
$BrO_3^-(aq) + 3\,H_2O(l) + 6\,e^- \rightarrow Br^-(aq) + 6\,OH^-(aq)$	+0.584
$Ni(OH)_3(s) + e^- \rightarrow Ni(OH)_2(s) + OH^-(aq)$	+0.48
$2\,BrO^-(aq) + 2\,H_2O(l) + 2\,e^- \rightarrow Br_2(l) + 4\,OH^-(aq)$	+0.455
$2\,IO^-(aq) + 2\,H_2O(l) + 2\,e^- \rightarrow I_2(s) + 4\,OH^-(aq)$	+0.42
$O_2(g) + 2\,H_2O(l) + 4\,e^- \rightarrow 4\,OH^-(aq)$	+0.401
$Ag_2O(s) + H_2O(l) + 2\,e^- \rightarrow 2\,Ag(s) + 2\,OH^-(aq)$	+0.342
$Co(OH)_3(s) + e^- \rightarrow Co(OH)_2(s) + OH^-(aq)$	+0.17
$NO_3^-(aq) + H_2O(l) + 2\,e^- \rightarrow NO_2^-(aq) + 2\,OH^-(aq)$	+0.01
$CrO_4^{2-}(aq) + 4\,H_2O(l) + 3\,e^- \rightarrow [Cr(OH)_4]^-(aq) + 4\,OH^-(aq)$	−0.13
$HPbO_2^-(aq) + H_2O(l) + 2\,e^- \rightarrow Pb(s) + 3\,OH^-(aq)$	−0.54
$HCHO(aq) + 2\,H_2O(l) + 2\,e^- \rightarrow CH_3OH(aq) + 2\,OH^-(aq)$	−0.59
$SO_3^{2-}(aq) + 3\,H_2O(l) + 4\,e^- \rightarrow S(s) + 6\,OH^-(aq)$	−0.66
$AsO_4^{3-}(aq) + 2\,H_2O(l) + 2\,e^- \rightarrow AsO_2^-(aq) + 4\,OH^-(aq)$	−0.67
$AsO_2^-(aq) + 2\,H_2O(l) + 3\,e^- \rightarrow As(s) + 4\,OH^-(aq)$	−0.68
$2\,H_2O(l) + 2\,e^- \rightarrow H_2(g) + 2\,OH^-(aq)$	−0.828
$OCN^-(aq) + H_2O(l) + 2\,e^- \rightarrow CN^-(aq) + 2\,OH^-(aq)$	−0.97
$As(s) + 3\,H_2O(l) + 3\,e^- \rightarrow AsH_3(g) + 3\,OH^-(aq)$	−1.21
$[Zn(OH)_4]^{2-}(aq) + 2\,e^- \rightarrow Zn(s) + 4\,OH^-(aq)$	−1.285
$Sb(s) + 3\,H_2O(l) + 3\,e^- \rightarrow SbH_3(g) + 3\,OH^-(aq)$	−1.338
$Al(OH)_4^-(aq) + 3\,e^- \rightarrow Al(s) + 4\,OH^-(aq)$	−2.310
$Mg(OH)_2(s) + 2\,e^- \rightarrow Mg(s) + 2\,OH^-(aq)$	−2.687

Appendix D
Organic Nomenclature

In the early days of organic chemistry, compounds were given common names, often based on natural sources. For example, butyric acid ($CH_3CH_2CH_2COOH$) was so named from the Latin *butyrum*, meaning butter. (The *but-* part of the name survives in the systematic names of compounds with four carbon atoms per molecule: $CH_3CH_2CH_2CH_3$, butane; $CH_3CH_2CH_2CH_2OH$, 1-butanol; and so on). Today, there are millions of organic compounds, and we could no more memorize individual names for all of them than we could memorize all the listings in the New York City telephone directories. To bring some order to the haphazard naming of the rapidly increasing roster of organic compounds, an international assembly of chemists met in 1892 for the first of many meetings on nomenclature—a system for assigning names. This organization exists today as the International Union of Pure and Applied Chemistry (IUPAC). Here, we will examine some simple rules for naming a few families of organic compounds.

D.1 Alkanes

An IUPAC name of an organic compound often consists of three basic parts: one or more prefixes, a stem, and an ending. The following rules will enable us to name most alkanes.

1 Use the ending *-ane* to indicate that the compound is an alkane.

2 The longest continuous chain (LCC) of carbon atoms is the *parent chain*; it provides the *stem* name. For example, the following alkane is named as a derivative of hexane because there are six carbon atoms in the LCC.

$$CH_3CH_2\underset{\displaystyle CH_3}{\underset{|}{C}}HCH_2CH_2CH_3$$

The stem *hex-* and ending *-ane* combine to give *hexane*. (Recall that the names of the stems for chains with up to 10 carbon atoms were given in Table 2.6 on page 62.)

3 The prefixes in a name indicate the groups attached to the parent chain. If the group contains only carbon and hydrogen with no double or triple bonds, it is called an *alkyl* group. Alkyl groups are derived from the corresponding alkane by removing one H atom, and the alkyl group is named after the alkane. For example, the *methyl* group, $CH_3—$, is derived from *methane*, CH_4. There is only one alkyl derived from methane and one from ethane. For the general alkane C_nH_{2n+2} with $n = 3$ (propane) or higher, different alkyl groups are formed depending on which H atom is removed. For example, removal of an end H atom from propane gives a *propyl* group, whereas removal of an H atom from the middle carbon atom of propane yields an *isopropyl* group. The names and formulas of the four simplest alkyl groups are as follows.

$CH_3—$	$CH_3CH_2—$	$CH_3CH_2CH_2—$	$CH_3\underset{|}{C}HCH_3$
Methyl group	Ethyl group	Propyl group	Isopropyl group

4 The smallest possible Arabic numerals are used to indicate the position(s) on the longest carbon chain at which the substituents (alkyl groups, in this case) are attached. Thus, to complete the name of the alkane in item 2, we again identify the LCC (red).

$$CH_3CH_2\underset{\displaystyle CH_3}{\underset{|}{C}}HCH_2CH_2CH_3$$

Next we note that the methyl group (CH_3—) is attached to the third carbon atom from the left end. Thus, the compound is *3-methylhexane*. Notice that if we number the carbon atoms from the right end, the methyl group is on the *fourth* carbon atom. The name "4-methylhexane" is incorrect, however, because we must use the *lowest* possible numbers for constituent groups.

5 We use the prefixes *di-* for two, *tri-* for three, and *tetra-* for four to denote two, three, or four identical groups attached to the parent chain. If two identical groups are bonded to the same carbon atom, we must repeat the position number for each group.

$$\begin{array}{l} \quad CH_3 \\ \quad\ | \\ CH_3CCH_2CH_2CH_2CH_2CH_3 \\ \quad\ | \\ \quad CH_3 \end{array}$$

2,2-Dimethylheptane

$$\begin{array}{l} \qquad\quad CH_3 \qquad CH_3 \\ \qquad\quad\ | \qquad\qquad\ | \\ CH_3-C-CH_2CHCH_3 \\ \qquad\quad\ | \\ \qquad\quad CH_3 \end{array}$$

2,2,4-Trimethylpentane

Notice that we use commas to separate numbers from each other and hyphens to separate numbers from words.

6 Groups are listed in alphabetical order. Thus, the proper name for the compound

$$\begin{array}{l} CH_3-CHCH_2CH-CH_2CH_3 \\ \qquad\quad | \qquad\ \ | \\ \qquad CH_3 \quad CH_2CH_3 \end{array}$$

is 4-ethyl-2-methylhexane, not "2-methyl-4-ethylhexane."

As a final example, let's name the following compound.

$$\begin{array}{l} \ 5 \quad\ 4 \quad\ 3 \qquad\quad 2 \quad 1 \\ CH_3CH_2CH-CHCH_3 \\ \qquad\qquad | \qquad\ \ | \\ \qquad\quad CH_3 \ \ CH_3 \end{array}$$

The LCC has five carbon atoms; the compound is named as a derivative of pentane. We see that there are two methyl groups that are attached to the second and third carbon atoms. The correct name is 2,3-dimethylpentane. Note that we numbered the carbon atoms so that the substituent groups would have the lowest numbers possible.

EXAMPLE D.1

Give an appropriate IUPAC name for the compound

$$\begin{array}{l} CH_3CH-CH_2CHCH_3 \\ \qquad | \qquad\qquad\ | \\ \quad CH_2 \qquad\ CH_3 \\ \qquad | \\ \quad CH_3 \end{array}$$

SOLUTION

Let's begin by identifying the LCC. This one is a little tricky. The parent compound is the longest continuous chain, not necessarily the chain drawn straight across the page. The LCC (red) contains *six* carbon atoms.

$$\begin{array}{l} \qquad 4 \qquad\quad 3 \quad\ 2 \quad 1 \\ CH_3CH-CH_2CHCH_3 \\ \qquad | \qquad\qquad\ | \\ \ 5\,CH_2 \qquad\ CH_3 \\ \qquad | \\ \ 6\,CH_3 \end{array}$$

The substituents (black) are two methyl groups, one on the second carbon atom of the LCC and one on the fourth. Again, we use the lowest combination of numbers, in this case counting from the left end. The correct name is 2,4-dimethylhexane, not "2-ethyl-4-methylpentane."

EXAMPLE D.2

Draw the structure of 5-isopropyl-2-methyloctane.

SOLUTION

The compound is derived from octane, so we start with a chain of eight carbon atoms.

$$-C-C-C-C-C-C-C-C-$$

Next, starting on the left, we add a methyl group to the second C atom.

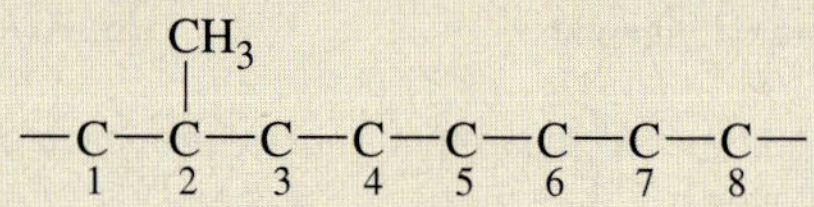

$$\begin{array}{cccccccc} & CH_3 & & & & & & \\ & | & & & & & & \\ -C- & C- & C- & C- & C- & C- & C- & C- \\ 1 & 2 & 3 & 4 & 5 & 6 & 7 & 8 \end{array}$$

Then, continuing to count from the left, we add an isopropyl group to the fifth C atom. (Remember that an isopropyl group is a three-carbon chain attached by the *middle* carbon atom.)

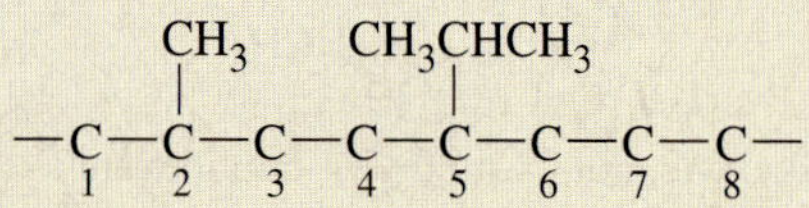

$$\begin{array}{cccccccc} & CH_3 & & & CH_3CHCH_3 & & & \\ & | & & & | & & & \\ -C- & C- & C- & C- & C- & C- & C- & C- \\ 1 & 2 & 3 & 4 & 5 & 6 & 7 & 8 \end{array}$$

Finally, we add enough hydrogen atoms to give each carbon atom four bonds. The structure is

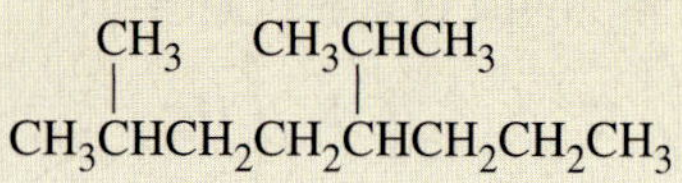

$$\begin{array}{l} \quad CH_3 \quad CH_3CHCH_3 \\ \quad\ | \qquad\quad | \\ CH_3CHCH_2CH_2CHCH_2CH_2CH_3 \end{array}$$

D.2 Alkenes and Alkynes

A few simple alkenes are best known by common names (Section 9.11), but we need systematic names for the many isomers of higher alkenes. Some of the IUPAC rules for alkenes are as follows.

1. All alkenes have names ending in *-ene*.
2. The longest chain of carbon atoms *containing the double bond* is the parent chain. The stem name is the same as that of the alkane with the same number of carbon atoms. The compound $CH_2{=}CH_2$ has the same number of C atoms as ethane, CH_3CH_3; it is called *ethene*. Similarly, $CH_3CH{=}CH_2$ has the same number of C atoms as propane, $CH_3CH_2CH_3$; it is called *propene*.
3. For chains of four or more carbon atoms, we must indicate the location of the double bond. To do this, number the carbon atoms from whichever end of the parent chain gives the first carbon atom in the double bond the lowest possible number. For example, the compound $CH_3CH{=}CHCH_2CH_3$, has the double bond between the second and third carbon atoms. Its name is *2-pentene*.
4. Substituent groups are named as in alkanes, and their positions are indicated by a number. Thus, the following alkene is 2-ethyl-5-methyl-1-hexene.

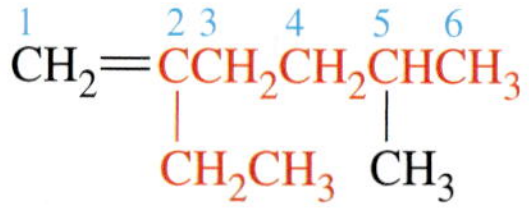

$$\begin{array}{l} \ 1 \qquad\quad 2\ 3 \quad\ 4 \quad\ 5 \quad 6 \\ CH_2{=}CCH_2CH_2CHCH_3 \\ \qquad\quad | \qquad\qquad | \\ \qquad\ CH_2CH_3 \quad CH_3 \end{array}$$

Notice that we did not choose the LCC of seven carbon atoms (red) as the parent chain. We chose the *six-carbon chain containing the double bond*.

To name *alkynes* we use a set of rules almost identical to those for the alkenes. The only difference is that alkynes have a carbon-to-carbon triple bond instead of a double bond, and we use the ending *-yne* rather than *-ene*. The IUPAC name of $HC{\equiv}CH$, is *ethyne*, and that of $CH_3C{\equiv}CH$ is *propyne*. The compounds $CH_3CH_2C{\equiv}CH$ and $CH_3C{\equiv}CCH_3$ are *1-butyne* and *2-butyne*, respectively.

EXAMPLE D.3

Name the compound

$$\begin{array}{l} CH_3C{\equiv}CCH_2\underset{\displaystyle CH_3}{\underset{|}{C}}H{-}\underset{\displaystyle CH_3}{\underset{|}{C}}HCH_3 \end{array}$$

SOLUTION

The LCC containing the triple bond has seven carbon atoms; the seven-carbon alkyne is heptyne. Keep in mind that in naming alkenes and alkynes, we give the lowest position number to the double bond or triple bond, not to the substituent groups. To do this, we number the C atoms as shown below.

$$\overset{1}{C}H_3\overset{2}{C}{\equiv}\overset{3}{C}\overset{4}{C}H_2\underset{\displaystyle CH_3}{\underset{|}{\overset{5}{C}}}H{-}\underset{\displaystyle CH_3}{\underset{|}{\overset{6}{C}}}H\overset{7}{C}H_3$$

The parent compound is 2-heptyne. There are methyl substituents on the fifth and sixth carbon atoms. The name of the compound is 5,6-dimethyl-2-heptyne.

EXAMPLE D.4

Give the structural formula for (**a**) 2-methyl-2-pentene, and (**b**) 4-methyl-2-hexyne.

SOLUTION

a. The stem *pent* and ending *-ene* tell us that the LCC containing the double bond has five carbon atoms. The *2* indicates that the double bond is between the second and third carbon atoms.

$$C{-}C{=}C{-}C{-}C$$

The *2-methyl* locates the $-CH_3$ substituent on the second carbon atom.

$$C{-}\underset{\displaystyle CH_3}{\underset{|}{C}}{=}C{-}C{-}C$$

Adding enough hydrogen atoms to give each carbon atom four bonds gives us the structural formula.

$$CH_3{-}\underset{\displaystyle CH_3}{\underset{|}{C}}{=}CH{-}CH_2{-}CH_3$$

b. The stem *hex-* and ending *-yne* tell us that the LCC containing a *triple* bond has six carbon atoms. The 2 tells us that the triple bond is between the second and third carbon atoms, and the *4-methyl* locates the $-CH_3$ substituent on the fourth carbon atom. Adding hydrogen atoms gives the structural formula.

$$CH_3C{\equiv}C\underset{\displaystyle CH_3}{\underset{|}{C}}HCH_2CH_3$$

D.3 Alcohols

In Section 2.10, we introduced alcohols as compounds with a hydroxyl group (—OH) bonded to a carbon atom of an alkyl group (R—), giving them the general formula ROH. The —OH group is a substituent on the LCC, but we generally don't name it as a substituent. Rather, we indicate its presence by an ending:

- The presence of the —OH functional group is denoted by an *-ol* ending on an alkane stem name (not through the prefix, hydroxyl).
- In numbering the carbon atoms of the LCC, the position of the —OH group is given first priority, that is, it is given the lowest number possible.

The two simplest alcohols are known by the common names methyl alcohol and ethyl alcohol. The IUPAC name of CH_3OH is based on the alkane methane, CH_4. We drop the *-e* of methane and add the ending *-ol*; its name is *methanol*. Similarly, the name of CH_3CH_2OH is based on the alkane ethane, CH_3CH_3; its name is *ethanol*. There are two propyl alcohols. Their structures and names are given below.

$$CH_3CH_2CH_2OH$$

1-Propanol

$$CH_3\underset{\displaystyle OH}{\underset{|}{C}}HCH_3$$

2-Propanol

Recall that their common names are propyl alcohol and isopropyl alcohol, respectively.

EXAMPLE D.5

The compound commonly known as *tert*-butyl alcohol, $(CH_3)_3COH$, and its methyl ether, $(CH_3)_3COCH_3$, are used as octane boosters in gasoline. What is the IUPAC name of the alcohol?

SOLUTION

Let's begin by converting the condensed formula to a more complete structural formula. There are three methyl groups and a hydroxyl group, all attached to a single C atom.

$$CH_3-\overset{\displaystyle CH_3}{\overset{|}{\underset{\displaystyle OH}{\underset{|}{C}}}}-CH_3$$

The LCC is three carbon atoms long, and the hydroxyl group is on the second carbon of this chain, yielding, for the moment, 2-propanol. There is also a methyl group attached to the second carbon of the parent chain, giving the IUPAC name *2-methyl-2-propanol*.

D.4 Carboxylic Acids

The simplest carboxylic acids are widely known by their common names. For example, HCOOH is called formic acid and CH_3COOH is called acetic acid. The IUPAC names are based on alkanes with the same number of carbon atoms. The *-e* ending of the alkane name is replaced by *-oic acid*. Thus, HCOOH is methanoic acid, and CH_3COOH is ethanoic acid. The carboxylic acid with an LCC of eight carbon atoms is octanoic acid. For locating substituents, numbering begins with the carboxylic carbon atom as number 1, as illustrated in Example D.6.

EXAMPLE D.6

Give the structural formula for 4-ethyl-6-methyloctanoic acid.

SOLUTION

Octanoic tells us that the compound has an LCC of eight carbon atoms, with an end C atom as part of a carboxyl group.

$$\overset{8}{C}-\overset{7}{C}-\overset{6}{C}-\overset{5}{C}-\overset{4}{C}-\overset{3}{C}-\overset{2}{C}-\overset{1}{COOH}$$

The substituents are an ethyl group on the fourth carbon atom (counting *from the carboxyl end*) and a methyl group on the sixth carbon atom. Adding these substituents and enough H atoms to give each C atom four bonds, we get

$$CH_3CH_2\underset{\displaystyle CH_3}{\underset{|}{C}}HCH_2\underset{\displaystyle CH_2CH_3}{\underset{|}{C}}HCH_2CH_2COOH$$

There are many other families of organic compounds. We listed several of the most common functional groups and examples of compounds that contain them in Table 2.7 (page 68). With what we have presented here and in Chapters 2 and 23, you should be able to relate the names and formulas of many types of organic compounds.

Appendix E
Glossary

Absolute configuration is the three-dimensional arrangement about a chiral center in a molecule.

Absolute zero is the lowest possible temperature: $-273.15\ °C = 0\ K$.

The **accuracy** of a set of measurements refers to the closeness of the average of the set to the "correct" or most probable value.

An **acid** is (1) a hydrogen-containing compound that, under appropriate conditions, can produce hydrogen ions, H^+ (Arrhenius theory); (2) a proton donor (Brønsted–Lowry theory); (3) an atom, ion, or molecule that can accept a pair of electrons to form a covalent bond (Lewis theory).

An **acid-base indicator** is a substance added to the reaction mixture in a titration that changes color at or near the equivalence point.

An **acidic oxide** is a nonmetal oxide whose reaction with water produces a ternary acid as its sole product.

The **acid ionization constant (K_a)** is the equilibrium constant describing equilibrium in the reversible ionization of a weak acid.

Acid rain is rainfall that is more acidic than is water in equilibrium with atmospheric carbon dioxide.

The **actinide** elements constitute the portion of the f-block of the periodic table in which the $5f$ subshell fills in the aufbau process.

An **activated complex** is an aggregate of atoms in the transition state of a reaction formed by a favorable collision. (*See also* **transition state**.)

The **activated sludge method** of sewage treatment is a process in which sludge from the secondary stage is put into aeration tanks to facilitate decomposition by aerobic microorganisms.

The **activation energy (E_a)** of a reaction refers to the minimum total kinetic energy that molecules must bring into their collisions so that a chemical reaction may occur.

The **active site** on an enzyme is the region of the enzyme molecule where the substrate attaches and a chemical reaction occurs. (*See also* **enzyme** and **substrate**.)

The **activity series of the metals** is a listing of the metals in terms of their ability to displace one another from solutions of their ions or to displace H^+ as $H_2(g)$ from acidic solutions (see Figure 4.15).

The **actual yield** is the measured quantity of a desired product obtained in a chemical reaction. (*See also* **theoretical yield** and **percent yield**.)

Addition polymerization is a type of polymerization reaction in which monomers add to one another to produce a polymeric product that contains all the atoms of the starting monomers.

In an **addition reaction**, substituent groups join to hydrocarbon molecules at points of unsaturation—double or triple bonds.

An **adduct** is a compound that results from the addition, through a coordinate covalent bond, of one structure to another.

Adhesive forces are intermolecular forces between unlike molecules, for example, between those in a liquid and those in a surface over which the liquid is spread.

Advanced treatment (*tertiary treatment*) is a third stage of sewage treatment in which nitrates, phosphates, and (sometimes) dissolved organic substances are removed from the effluent from a secondary treatment process.

Aerobic oxidation is an oxidation process that occurs in the presence of oxygen.

An **air pollutant** is a substance that is found in air in greater abundance than normally occurs naturally and that has some harmful effect(s) on the environment.

An **alcohol (ROH)** is an organic substance whose molecules contain the hydroxyl group, $-OH$, attached to an alkyl group.

An **aldehyde (RCHO)** is an organic compound whose molecules have a carbonyl functional group ($C{=}O$) with a hydrogen atom attached to the carbonyl carbon.

Aliphatic compounds are those organic compounds with open chains of carbon atoms, or rings of carbon atoms that are similar in structure and properties to the open-chain compounds. (*See* **aromatic compounds**.)

An **alkane** is a saturated hydrocarbon having the general formula C_nH_{2n+2}. (*See also* **saturated hydrocarbon**.)

An **alkene** is a hydrocarbon (carbon–hydrogen compound) whose molecules contain at least one carbon-to-carbon double bond.

An **alkyl group ($R-$)** is a substituent group in organic molecules that is derived from an alkane molecule by removal of one hydrogen atom.

An **alkyne** is a hydrocarbon (carbon–hydrogen compound) whose molecules contain at least one carbon-to-carbon triple bond.

An **allotrope** is one of two or more forms of an element that differ in their basic molecular structure.

An **alloy** is a metallic material consisting of two or more elements.

An **alpha (α) particle** consists of two protons and two neutrons. It is identical to a doubly ionized helium ion, He^{2+}.

An **amide ($RCONH_2$)** is an organic compound in which a nitrogen atom is bonded to the carbon atom of a carbonyl group.

An **amine** is an organic substance in which one or more H atoms of an ammonia molecule is replaced by a hydrocarbon residue.

The **ampere (A)** is the basic unit of electric current. One ampere is a current of one coulomb per second: $1\ A = 1\ C/s$.

An **amphiprotic** substance can ionize either as a Brønsted–Lowry acid or base, depending on the acid–base properties of other species in the solution.

An **amphoteric** substance (usually an oxide or hydroxide) can react either with an acid or a base. The central element of the oxide or hydroxide appears in a cation in acidic solutions and in an anion in basic solutions.

Anaerobic decay is decomposition in the absence of oxygen.

Analytical chemistry is the branch of chemistry that deals with determining the composition of substances and mixtures.

An **anion** is a negatively charged ion.

An **anode** is an electrode at which an oxidation half-reaction occurs. It is the negative electrode in a voltaic cell and the positive electrode in an electrolysis cell.

An **antibonding molecular orbital** places a high electron charge density (electron probability) away from the region between bonded atoms. (*See also* **bonding molecular orbital** and **molecular orbital**.)

An **anticarcinogen** is a substance that opposes the action of a carcinogen; it prevents or retards the development of cancer.

An **aqueous solution** is a solution in which water is the solvent.

Aromatic compounds are organic compounds with benzenelike structures. To describe their electronic structures (Lewis structures) one must use resonance theory.

The **atmosphere (atm)** is a unit used to measure gas pressure. It is equal to the pressure of a column of mercury having a height of exactly 760 mm. That is, 1 atm = 760 mmHg.

The **atomic mass** of an element is the weighted average of the masses of the atoms of the naturally occurring isotopes of the element.

An **atomic mass unit (u)** is *exactly* one-twelfth the mass of an atom of carbon-12.

The **atomic number (Z)** of an atom is the number of protons in the atomic nucleus.

An **atomic orbital** is a wave function for an electron corresponding to the assignment of specific values to the n, l, and m_l quantum numbers in a wave equation.

The **atomic radius** is a measure of the size of an atom based on the measurement of internuclear distances. (*See also* **covalent radius**, **ionic radius**, and **metallic radius**.)

Atoms are the smallest distinctive units of a sample of matter. Atoms of one element differ from atoms of all other elements. (*See also* **element**.)

The **aufbau principle** is a hypothetical process for building up an atom from the atom of preceding atomic number by adding a proton and the requisite number of neutrons to the nucleus and one electron to the appropriate atomic orbital.

Average bond energy is the average of the bond dissociation energies for a number of different molecular species containing the particular bond. (*See also* **bond dissociation energy**.)

Avogadro's hypothesis states that equal numbers of molecules of different gases, when compared at the same temperature and pressure, occupy equal volumes.

Avogadro's law states that at a fixed temperature and pressure the volume of a gas is directly proportional to the amount of gas.

Avogadro's number (N_A) is the number of elementary units in a mole—6.0221367×10^{23} mol^{-1}.

A **band**, in describing bonding in metals and semiconductors, is a collection of a large number of closely-spaced molecular orbitals, obtained by combining atomic orbitals of many atoms.

A **barometer** is a device used to measure the pressure exerted by the atmosphere.

A **base** is (1) a compound that produces hydroxide ions, OH^-, in water solution (Arrhenius theory); (2) a proton acceptor (Brønsted–Lowry theory); (3) an atom, ion, or molecule that can donate a pair of electrons to form a covalent bond (Lewis theory).

The **base ionization constant (K_b)** is the equilibrium constant describing equilibrium in the reversible ionization of a weak base.

A **basic oxide** is a metal oxide whose reaction with water produces a base.

A **battery** is a device that stores chemical energy for later release as electricity as a result of a chemical reaction.

A **beta (β^-) particle** is identical to an electron and is emitted by the nuclei of certain radioactive atoms as they undergo decay. In the decaying nucleus, the atomic number increases by one unit and the mass number remains unchanged.

A **bidentate** ligand has two points of attachment to a metal center in a complex.

A **bimolecular reaction** is an elementary reaction in a reaction mechanism that involves the collision of two molecules.

Biochemical oxygen demand (BOD) is the amount of oxygen needed by aerobic microorganisms to metabolize the organic wastes in water.

The **biosphere** is the part of Earth occupied by living organisms.

A **body-centered cubic (bcc)** crystal structure has as its unit cell a cube with a structural unit at each corner and one in the center of the cell. (*See also* **unit cell**.)

The **boiling point** of a liquid is the temperature at which the liquid boils—the temperature at which the vapor pressure of the liquid is equal to the prevailing atmospheric pressure.

The **bond dissociation energy (D)** of a particular covalent bond between two atoms is the quantity of energy required to break one mole of bonds of that type in a gaseous species.

A **bonding molecular orbital** places a high electron charge density (electron probability) in the region between two bonded atoms. (*See also* **antibonding molecular orbital** and **molecular orbital**.)

A **bonding pair** is a pair of electrons shared between two atoms in a molecule.

The **bond length** of a particular covalent bond is the distance between the nuclei of two atoms joined by that type of bond.

A **bond moment** describes the extent to which a separation of positive ($\delta+$) and negative (δ^-) charges exists in a covalent bond between two atoms.

Bond order is a term used to indicate the nature of a covalent bond—that is, whether single (bond order = 1), double (bond order = 2), or triple (bond order = 3). In molecular orbital theory it is one half the difference between the number of electrons in bonding molecular orbitals and in antibonding molecular orbitals.

Boyle's law states that for a given amount of gas at a constant temperature, the volume of a gas varies inversely with its pressure. That is, $V \propto 1/P$, or PV = constant.

A **buffer solution** is a solution containing a weak acid and its conjugate base, or a weak base and its conjugate acid. Small quantities of added acid are neutralized by one buffer component and small quantities of added base by the other. As a result the solution pH is maintained nearly constant.

A **buret** is a graduated, long glass tube constructed to deliver precise volumes of a liquid solution through a stopcock valve.

A **calorie (cal)** is the amount of energy needed to raise the temperature of 1 g of water by 1 °C (more precisely, from 14.5 to 15.5 °C). 1 cal = 4.184 J.

A **calorimeter** is a device in which quantities of heat are measured.

A **carbocation** is an intermediate species in certain reactions of organic compounds in which a positive charge is centered on a carbon atom in the species.

A **carbohydrate** is a compound consisting of carbon, hydrogen, and oxygen; a starch, sugar, or cellulose.

The **carbon cycle** refers to the totality of activities in which carbon atoms are cycled through the environment.

A **carboxylic acid (RCOOH)** is an organic substance whose molecules contain the carboxyl group, —COOH.

A **carcinogen** is an agent that causes cancer.

Cast iron is a carbon-iron alloy cast to shape and containing 1.8 to 4.5% C.

A **catalyst** is a substance that increases the rate of a reaction without itself being consumed in the reaction. A catalyst changes a reaction mechanism to one with a lower activation energy.

A **cathode** is an electrode at which a reduction half-reaction occurs. It is the positive electrode in a voltaic cell and the negative electrode in an electrolysis cell.

A **cathode ray** is a beam of electrons that travels from the cathode to the anode when an electric discharge is passed through an evacuated tube.

A **cation** is a positively charged ion.

A **cell diagram** is a schematic representation of an electrochemical cell.

Cell voltage (E_{cell}) (cell potential) refers to the potential difference between the electrodes in a voltaic cell.

A **cell reaction** is the overall oxidation-reduction reaction that occurs in a voltaic cell.

A **central atom** in a molecule or polyatomic ion is an atom bonded to two or more other atoms.

Charles's law states that the volume of a fixed amount of a gas at a constant pressure is directly proportional to its Kelvin temperature. That is, $V \propto T$, or $V = \text{constant} \times T$.

A **chelate** is a five- or six-membered ring structure(s) produced in a complex through the attachment of a polydentate ligand(s) to a metal center.

A **chemical bond** is a force that holds atoms together in molecules or ions in crystals.

Chemical change. (*See* **chemical reaction**).

A **chemical equation** is a description of a chemical reaction that uses symbols and formulas to represent the elements and compounds involved in the reaction. Numerical coefficients preceding each symbol or formula and indicating molar proportions may be needed to balance a chemical equation.

A **chemical formula** indicates the composition of a compound through symbols of the elements present and subscripts to indicate the relative numbers of atoms of each element.

Chemical nomenclature is a systematic way of relating the names and formulas of chemical compounds.

A **chemical property** is a characteristic that a sample of matter displays as it undergoes a change in composition.

A **chemical reaction** is a process in which a sample of matter undergoes a change in composition and/or structure of its molecules. One or more original substances (reactants) is changed into one or more new substances (products).

Chemical shift is a term used in nuclear magnetic resonance (NMR) spectroscopy to indicate the location of an absorption peak relative to a standard. The magnitudes of the chemical shifts can be used to determine structural features of a molecule.

A **chemical symbol** is a representation of an element made up of one or two letters derived from the English name of the element (or, sometimes, from the Latin name of the element or one of its compounds).

A **chiral carbon** is a carbon atom that is attached to four different groups.

Chemistry is a study of the composition, structure, and properties of matter and of the changes that occur in matter.

The term **cis** is used to describe isomers in which two substituent groups are attached on the same side of a double bond in an organic molecule, or along the same edge of a square planar or octahedral complex ion. (*See also* **geometric isomerism**.)

Cohesive forces are intermolecular forces between like molecules.

A **colligative property** is a physical property—such as vapor pressure lowering, freezing point depression, boiling point elevation, and osmotic pressure—that depends on the concentration of solute in the solution but not on the identity of the solute.

A **colloid** is a dispersion in which the dispersed matter has one or more dimensions (length, width, or thickness) in the range from about 1 nm to 1000 nm.

The **common-ion effect** refers to the ability of ions from a strong electrolyte to (a) suppress the ionization of a weak acid or weak base or (b) to reduce the solubility of a slightly soluble ionic compound.

A **complex** consists of a central atom, which is usually a metal ion, and attached groups called ligands.

A **complex ion** is a complex that carries a net electric charge, either positive (a complex cation) or negative (a complex anion).

Composition refers to the types of atoms and their relative proportions in a sample of matter.

A **compound** is a substance made up of atoms of two or more elements, with the different atoms joined in fixed proportions.

A **concentration cell** is a voltaic cell having identical electrodes in contact with solutions of different concentrations.

A **concentration equilibrium constant (K_c)** is the numerical value of an equilibrium constant expression in which molarities of products and reactants are used.

Condensation is the conversion of a gas (vapor) to a liquid.

In **condensation polymerization**, monomers with at least two functional groups link together by eliminating small-molecule by-products.

A **conduction band** is a partially filled band of very closely spaced energy levels.

A **conjugate acid-base pair** is an acid and its conjugate base or a base and its conjugate acid.

A **conjugate acid** is formed when a Brønsted–Lowry base accepts a proton. Every base has a conjugate acid.

A **conjugate base** is formed when a Brønsted–Lowry acid donates a proton. Every acid has a conjugate base.

A **conjugated bonding system** refers to a molecular structure having a series of alternate single and double bonds. Substances having this feature can absorb UV and/or visible light.

Contributing structure. (*See* **resonance structure**.)

A **conversion factor** is a ratio of terms—equivalent to the number one—used to change the unit(s) in which a quantity is expressed.

A **cooling curve** is a graph of temperature as a function of time obtained as a substance is cooled. Constant-temperature segments of the curve correspond to phase changes, for example, condensation and freezing.

A **coordinate covalent bond** is a linkage between two atoms in which one atom provides both of the electrons of the shared pair.

A **coordination compound** is a substance made up of one or more complexes.

The **coordination number** of the metal center in a complex is the total number of points around a central atom at which bonding to ligands can occur.

Core electrons are electrons found in the inner electronic shells of atoms. (*See also* **valence electrons**.)

A **corrosive material** is one that degrades a metal or alloy by a chemical reaction. A corrosive material requires a special container because it corrodes conventional container materials.

The **coulomb (C)** is the SI unit of electric charge. The electric charge on an electron, for example, is -1.602×10^{-19} C.

A **coupled reaction** is one that involves two separate processes that can be combined to give a single reaction. In most cases, a thermodynamically unfavorable reaction is combined with another reaction to give an overall reaction that is thermodynamically favorable.

A **covalent bond** is a bond formed by a pair of electrons shared between atoms.

The **covalent radius** of an atom is one-half the distance between the nuclei of two like atoms joined into a molecule.

Critical mass is the minimum mass of a fissionable element that must be present to sustain a chain reaction. This is the mass required to produce an explosion of a nuclear bomb.

The **critical point** refers to the condition at which the liquid and gaseous (vapor) states of a substance become identical. It is the highest temperature point on a vapor pressure curve.

The **critical pressure** of a substance is the pressure at its critical point.

The **critical temperature** of a substance is the temperature at its critical point.

A **crystal** is a structure having plane surfaces, sharp edges, and a regular geometric shape. The fundamental units—atoms, ions, or molecules—are assembled in a regular, repeating manner extending in three dimensions through the crystal.

Crystal field theory is a theory of bonding in complexes that focuses on the abilities of ligands to produce a splitting of a d-subshell energy level of the metal center in a complex.

A **cubic close-packed (ccp)** structure has units (atoms, ions, or molecules) arranged in one of the two ways that minimize the voids between the units. The layers are stacked in the arrangement ABCABC···. (*See also* **hexagonal close-packed**.)

Dalton's law of partial pressures states that in a mixture of gases each gas expands to fill the container and exerts its own pressure, called a partial pressure. The total pressure of the mixture is the sum of the partial pressures exerted by the separate gases.

Data are the facts of science obtained by careful observations and measurements made during experiments.

The ***d*-block** is the portion of the periodic table in which the $(n - 1)d$ subshell (the d subshell of the next-to-outermost shell) fills in the aufbau process. The d-block comprises the B-group elements in the periodic table.

The **debye (D)** is the unit used to express the dipole moments of polar molecules. One debye is equal to 3.34×10^{-30} C m.

Degenerate orbitals are two or more orbitals that are at the same energy level.

Deionized water is water that has been freed of ions through ion exchange processes.

Deliquescence is the condensation of water vapor on a solid followed by solution formation.

The **density (*d*)** of a sample of matter is its mass per unit volume, that is, the mass of the sample divided by its volume: $d = m/V$.

Deoxyribonucleic acid (DNA) is a polymer of nucleotides. The nucleotides consist of the sugar deoxyribose, a phosphate ester, and a cyclic amine base (adenine, guanine, thymine, or cytosine).

A **dextrorotatory (+)** substance rotates the plane of polarized light to the right.

A **diagonal relationship** refers to the similarity of certain second-period elements in one group of the periodic table with third-period elements of the next group.

Diamagnetism is the weak repulsion in a magnetic field by a substance in which all electrons are paired.

Diffusion is the process by which one substance mixes with one or more other substances as a result of the movement of molecules.

Dilution is a process of producing a more dilute solution from a more concentrated one by the addition of an appropriate quantity of solvent.

The **dipole moment (μ)** of a polar molecule is the product of the magnitude of the charges (δ) and the distance that separates them.

A **disaccharide** is a carbohydrate with molecules that can be hydrolyzed to two monosaccharide units.

A **dispersion force** is an attractive force between an instantaneous dipole and an induced dipole.

A **disproportionation reaction** is an oxidation–reduction reaction in which the same substance is both oxidized and reduced.

Doping refers to the addition of trace amounts of certain elements to a semiconductor to change the semiconducting properties. (*See also* ***n*-type semiconductor** and ***p*-type semiconductor**.)

A **double bond** is a covalent linkage in which two atoms share two pairs of electrons between them.

Dynamic equilibrium occurs when two opposing processes occur at exactly the same rate, with the result that no net change occurs.

The **effective nuclear charge (Z_{eff})** acting on an electron in an atom is the actual nuclear charge less the screening effect of other electrons in the atom.

Effusion is a process in which a gas escapes from its container through a tiny hole (an orifice). (*See also* **Graham's law of effusion**.)

An **elastomer** is a polymeric material that can be stretched by a relatively low stress to at least twice its length and upon release of the stress will return to its original length.

An **electrochemical cell** is a combination of two half-cells in which metal electrodes are joined by a wire and the solutions are brought into contact through a salt bridge or by other means. (*See also* **electrolytic cell**, **half-cell**, and **voltaic cell**.)

Electrochemistry is a study of the use of electricity to cause a nonspontaneous chemical reaction (electrolysis) and the use of a spontaneous chemical reaction to produce electricity (voltaic cell).

An **electrode** is a metal strip or carbon rod dipped into a solution or molten electrolyte to carry electricity to or from the liquid. (*See also* **anode** and **cathode**.)

Electrode potential is a property related to the electric charge density on an electrode in an oxidation–reduction equilibrium with ions in solution.

Electrolysis is the decomposition of compounds by passing electricity through an ionic solution or a molten salt. A nonspontaneous chemical change is produced.

An **electrolyte** is a compound that conducts electricity when molten or in a liquid solution.

An **electrolytic cell** is an electrochemical cell in which electricity is passed through a solution or a molten salt to produce a nonspontaneous chemical reaction.

The **electromagnetic spectrum** is the range of wavelengths and frequencies found for electromagnetic waves, extending from very long wavelength radio waves to the shortest gamma rays.

An **electromagnetic wave** originates in the vibrations of electrically charged objects and is propagated through oscillations of electric and magnetic fields.

Electromotive force (emf). (*See* **cell voltage**.)

An **electron** is a particle carrying the fundamental unit of negative electric charge. Electrons have a mass of 0.0005486 u and are found outside of the nuclei of atoms.

Electron affinity is the energy change that occurs when an electron is added to an atom in the gaseous state.

Electron capture (EC) is a type of radioactive decay in which a nucleus absorbs an electron from the first or second electronic shell.

The **electron configuration** of an atom describes the distribution of electrons among atomic orbitals in the atom.

The **electronegativity (EN)** of an element is a measure of the tendency of its atoms in molecules to attract electrons to themselves.

An **electron group** is a collection of valence electrons localized in a region around a central atom that exerts repulsions on other groups of valence electrons. It may be a bonding pair of electrons in a single bond, two pairs of electrons in a double bond, three pairs of electrons in a triple bond, a lone pair of electrons, or even a lone unpaired electron.

The **electron-group geometry** of a molecule or ion is the arrangement of all the electron groups—both bonding and nonbonding—about a central atom.

The **electron spin quantum number (m_s)** is a fourth quantum number (in addition to the three required by the Schrödinger wave equation) necessary to characterize an electron in an orbital. The two possible values of the spin quantum number are $+\frac{1}{2}$ and $-\frac{1}{2}$.

The **electronvolt** is a unit of energy equal to that acquired by an electron as it passes through a potential difference of 1 V in a vacuum.

An **electrophilic reagent** is an electron-deficient molecule or ion that accepts an electron pair from a molecule, forming a covalent bond.

Electrophilic aromatic substitution is a reaction type in which an electrophile attaches to an aromatic ring such as benzene, replacing a hydrogen atom on the ring.

An **element** is a substance that cannot be broken down into simpler substances by chemical reactions. All atoms of a given element have the same atomic number.

An **elementary reaction** represents, at the molecular level, a single stage in the overall mechanism by which a chemical reaction occurs. (*See also* **unimolecular**, **bimolecular**, and **termolecular**.)

An **emission (line) spectrum** is a dispersion of electromagnetic radiation into a discrete set of wavelength components. These components can be rendered as images of a slit (lines) in light from a spectroscope.

An **empirical formula** is the *simplest* formula describing the elements in a compound and the smallest integral (whole number) ratio in which their atoms are combined.

Enantiomers are pairs of mirror-image isomers that differ only in the direction in which they rotate the plane of polarized light. One isomer rotates the plane to the right, and the other rotates the plane to the same degree, but to the left.

An **endothermic reaction** is a reaction in which thermal energy is converted to chemical energy. In an endothermic process a temperature decrease occurs in an isolated system, or, in a nonisolated system, heat is absorbed from the surroundings.

The **end point** is the point in a titration at which an added indicator changes color. An indicator is chosen so that its end point matches the equivalence point of the reaction. (*See also* **equivalence point**.)

An **energy gap (E_g)** is the energy separation between a valence band and a conduction band that lies above it. In a semiconductor the gap is relatively small and in an insulator it is very large. (*See also* **valence band** and **conduction band**.)

An **energy level (shell)** is the state of an atom determined by the location of its electrons among the various principal shells and subshells.

Enthalpy (H) is a thermodynamic function defined as the sum of the internal energy and the pressure–volume product: $H = E + PV$.

The **enthalpy change (ΔH)** in a chemical reaction is equal to the heat of reaction at constant temperature and pressure, q_P.

An **enthalpy diagram** is a graphical representation of the change in enthalpy that occurs in a chemical reaction.

The **enthalpy (heat) of fusion (ΔH_{fusion})** is the quantity of heat required to melt a given quantity of a solid.

The **enthalpy (heat) of sublimation (ΔH_{subl})** is the quantity of heat required to vaporize a given quantity of solid at a constant temperature. It is equal to the sum of the enthalpies of fusion and vaporization.

The **enthalpy (heat) of vaporization (ΔH_{vapn})** is the quantity of heat required to vaporize a given quantity of liquid at a constant temperature.

Entropy (S) measures the degree of randomness or disorder in a system. The greater the degree of disorder in a system, the greater is its entropy.

Entropy change (ΔS) is the difference in entropy between two states of a system, as between the products and reactants of a chemical reaction.

An **enzyme** is a protein that catalyzes reactions occurring in living organisms.

Equilibrium is a condition that is reached when two opposing processes occur at equal rates. As a result, the concentrations (or partial pressures) of the reacting species remain constant with time.

The **equilibrium constant (K_{eq})** is the form of the equilibrium constant needed in thermodynamic relationships, such as $\Delta G° = -RT \ln K_{eq}$. In the K_{eq} expression, species in solution are represented through their molarities and gases by their partial pressures in atm.

An **equilibrium constant expression** is a particular ratio of concentrations (or partial pressures) of products to reactants in a chemical reaction at equilibrium. The expression has a constant value that is independent of the manner in which equilibrium is reached. (*See also* K_c and K_p.)

The **equivalence point** of a titration is the point at which two reactants have been introduced into a reaction mixture in their stoichiometric proportions.

An **essential amino acid** is an amino acid that cannot be synthesized in the body and must therefore be included in the diet.

An **ester (R'COOR)** is a compound derived from a carboxylic acid and an alcohol. The —OH of the acid is replaced by an —OR group.

An **ether (R'OR)** is a compound having two hydrocarbon groups joined through an oxygen atom.

Eutrophication describes a process in which an overabundance of nutrients leads to an overgrowth of algae in a body of water. The algae then die, and their decay depletes the dissolved oxygen in the water.

An **excited state** of an atom is one in which one or more electrons has been promoted to a higher energy level than in the ground state. (*See also* **ground state**.)

An **exothermic reaction** is a reaction in which chemical energy is converted to thermal energy. In an exothermic process a temperature increase occurs in an isolated system, or, in a nonisolated system, heat is given off to the surroundings.

The term **expanded valence shell** refers to a situation in which the central atom in a Lewis structure is able to accommodate more than the usual octet (8) of electrons in its valence shell.

An **experiment** is a carefully controlled procedure devised to test a hypothesis or a theory.

An **extensive property** is a physical property, such as mass or volume, that depends on the size or quantity of the sample of matter being considered.

A **face-centered cubic (fcc)** crystal structure has as its unit cell a cube with a structural unit at each of the corners and in the center of each face of the cube. (*See also* **unit cell**.)

The **Faraday constant (F)** is the electric charge, in coulombs, per mole of electrons—96,485 C/mol.

The ***f*-block** is the portion of the periodic table in which the $(n - 2)f$ subshell (the f subshell of the second-from-outermost shell) fills in the aufbau process. The f-block consists of the lanthanides and actinides.

Ferromagnetism is a magnetic effect much stronger than paramagnetism and associated with iron, cobalt, and nickel, and certain alloys. It requires that atoms be both paramagnetic and of the right size to be able to form magnetic domains.

The **first law of thermodynamics** states that the internal energy of an isolated system is constant, or if a system interacts with its surroundings by exchanging heat and/or work, the exchange must occur in such a way that no energy is created or destroyed. (*See also* **law of conservation of energy**.)

A **first-order reaction** has a rate equation in which the sum of the exponents, $m + n + \cdots = 1$.

Flotation is a metallurgical method by which an ore is separated from waste rock based on selective wetting of the ore by a surface-active agent.

Formal charge, a concept used in writing Lewis structures, is the number of valence electrons in an isolated atom minus the number of electrons assigned to that atom in a Lewis structure.

The **formation constant (K_f)** describes equilibrium between a complex ion and the cation and ligands from which it is formed.

Formula mass is the mass of a formula unit relative to that of a carbon-12 atom; it is the sum of the masses of the atoms or ions represented by the formula.

A **formula unit** is the simplest combination of atoms or ions consistent with the formula of a compound. In an ionic compound, it is the smallest possible electrically neutral collection of ions.

Fractional crystallization is a method of purifying a solid by dissolving it in a suitable solvent and changing the solution temperature to a value where the solute solubility is lower (usually a lower temperature). Excess solute crystallizes as pure solid and soluble impurities remain in solution.

Fractional distillation is a method of separating the volatile components of a solution having different vapor pressure and boiling points. It involves repeated vaporizations and condensations occurring continuously in a distillation column.

The **free electron model of metals** considers the metal to be composed of positive ions surrounded by a "sea" of mobile electrons.

The **free energy change (ΔG)** is the difference in free energy between two states of a system, as between the free energies of the products and reactants of a chemical reaction. It is given by the equation $\Delta G = \Delta H - T\Delta S$.

A **free radical** is a highly reactive atom or molecular fragment characterized by having an unpaired electron(s). Free radicals are encountered as intermediates in some chemical reactions.

Freezing point is the temperature at which a liquid freezes—that is, the liquid comes into equilibrium with solid. For a pure substance the freezing point and melting point are the same.

The **frequency (ν)** of a wave is the number of cycles of the wave (the number of wavelengths) that pass through a point in a unit time.

A **functional group** is an atom or grouping of atoms attached to or within a hydrocarbon chain or ring that confers characteristic properties to the molecule as a whole.

Fusion (melting) is the process of changing a solid to a liquid.

Galvanic cell. *See* **voltaic cell**.

A **gamma (γ) ray** is a highly penetrating form of electromagnetic radiation emitted by the nuclei of certain radioactive atoms as they undergo decay.

A **gene** is a section of a DNA molecule found in the chromosomes of cells; genes are the basic units of heredity.

Geometric isomers in organic compounds are isomers (cis, trans) that differ in the positions of attachment of substituent groups at a double bond. In complexes the isomers differ in the positions of attachment of ligands to the central metal ion.

Gibbs free energy (G), a thermodynamic function used in establishing criteria for equilibrium and for spontaneous change, is defined as $G = H - TS$, where H is the enthalpy; T, the Kelvin temperature; and S, the entropy of a system.

Global warming refers to the anticipated increase in Earth's average temperature resulting from the injection of CO_2 and other infrared-absorbing gases into the atmosphere.

Graham's law of effusion states that the rates of effusion of gas molecules are inversely proportional to the square roots of their molar masses.

The **greenhouse effect** refers to the ability of $CO_2(g)$ and certain other gases to absorb and trap energy radiated by Earth's surface as infrared radiation.

The **ground state** of an atom is the atom at its lowest energy level. (*See also* **excited state**.)

Groups of the periodic table are the vertical columns of elements having similar properties.

A **half-cell** is a metal electrode partially immersed in a solution of ions that participate in an oxidation–reduction equilibrium at the electrode. (*See also* **electrochemical cell**.)

The **half-life** of a chemical reaction is the time required to consume one-half of the initial quantity of a reactant. For radioactive decay, it is the time in which one-half of the atoms of a radioactive nuclide disintegrate.

A **half-reaction** is a portion of an oxidation–reduction reaction that represents either the oxidation process or the reduction process.

Hard water is groundwater containing significant concentrations of doubly-charged cations derived from natural sources, such as Ca^{2+}, Mg^{2+}, and Fe^{2+}, and associated anions.

A **hazardous material** is one that, when improperly managed, can cause or contribute to death or illness or threaten human health or the environment.

Heat (q) is an energy transfer into or out of a system caused by a difference in temperature between a system and its surroundings.

A **heating curve** is a graph of temperature as a function of time obtained by gradually heating a substance. Constant-temperature segments of the curve correspond to phase changes. (*See also* **cooling curve**.)

The **heat of reaction (q_{rxn})** is the quantity of heat exchanged between a system and its surroundings when a chemical reaction occurs at a constant temperature and pressure.

The **Henderson–Hasselbalch equation** is used to relate the pH of a solution of a weak acid and its conjugate base to the pKa of the weak acid and the stoichiometric concentrations of the weak acid and of the conjugate base: pH = pKa + log [conjugate base]/[weak acid].

Henry's law states that the solubility of a gas is directly proportional to the pressure maintained in the gas in equilibrium with the solution.

Hess's law states that the enthalpy change of a reaction is constant, whether the reaction is carried out directly in one step or indirectly through a number of steps.

A **heterogeneous mixture** is a mixture in which the composition and/or properties vary from one region to another within the mixture.

A **hexagonal close-packed (hcp)** structure is one of the two crystal arrangements in which the structural units are close-packed. The layers are stacked in the arrangement ABABAB$\cdots$. (*See also* **cubic close-packed**.)

A **homogeneous mixture** is a mixture having the same composition and properties throughout the given mixture.

A **homologous series** is a series of organic compounds whose formulas and structures vary in a regular manner and whose properties are predictable based on this regularity.

Humidity is a measure of the water vapor content of air. The *absolute humidity* is the actual quantity of water vapor present in an air sample, and the *relative humidity* of air is a measure of water vapor content as a percentage of the maximum possible quantity.

Hund's rule states that electrons occupy atomic orbitals of identical energy singly before any pairing of electrons occurs. Furthermore, the electrons in the singly occupied orbitals have parallel spins.

Hybridization is a hypothetical process in which pure atomic orbitals are combined to produce a set of new orbitals called hybrid orbitals to describe covalent bonding by the valence bond method. (*See also* ***sp***, ***sp***^{2}, ***sp***^{3}, ***sp***^{3}***d***, ***sp***^{3}***d***^{2}.)

Hybrid orbitals are formed by a combination of atomic orbitals to produce a set of new orbitals.

A **hydrate** is a compound that incorporates water molecules into its basic solid structure. The formula unit of a hydrate includes a fixed number of water molecules.

A **hydrocarbon** is a compound containing only hydrogen and carbon atoms.

In a **hydrogenation reaction**, $H_2(g)$ is a reactant and H atoms are added to C atoms at a carbon-to-carbon double or triple bond.

A **hydrogen bond** is a type of intermolecular force in which a hydrogen atom covalently bonded in one molecule is simultaneously attracted to a nonmetal atom in a neighboring molecule. In most cases, both the atom to which the hydrogen atom is bonded and the one to which it is attracted must be small atoms of high electronegativity: N, O, or F.

The **hydrologic (water) cycle** is the series of natural processes by which water is recycled through the environment—Earth's solid crust, oceans and freshwater bodies, and the atmosphere.

In a general sense, **hydrolysis** is the reaction of a substance with water in which both the substance and the water molecules split apart. In a more limited sense, it is an acid-base reaction between an ion and water that usually results in the solution becoming either somewhat acidic or somewhat basic.

Hydrometallurgy is the extraction of a metal from its ores by processes that involve water and aqueous solutions.

A **hypertonic solution** is a solution having an osmotic pressure greater than that of body fluids (blood, tears). A hypertonic solution has a greater osmotic pressure than an isotonic solution.

A **hypothesis** is a *tentative* explanation or prediction concerning some phenomenon.

A **hypotonic solution** is a solution having an osmotic pressure less than that of body fluids (blood, tears). A hypotonic solution has a lower osmotic pressure than an isotonic solution.

An **ideal gas** is a gas that strictly obeys the simple gas laws and the ideal gas law.

Ideal gas equation. *See* **ideal gas law**.

The **ideal gas law** states that the volume of a gas is directly proportional to the amount of a gas and its Kelvin temperature and inversely proportional to its pressure: $PV = nRT$.

An **ideal solution** is one for which the heat of solution is zero and the volume of solution is the total of the volumes of the solution components. In general, the physical properties of an ideal solution can be predicted from the properties of its components.

An **ignitable material** is one that burns readily on ignition, presenting a fire hazard.

Industrial smog is polluted air associated with industrial activities. The principal pollutants are oxides of sulfur and particulate matter.

An **inert pair** refers to the ns^2 electrons in the valence shell of the posttransition elements of Groups 3A, 4A, and 5A. These electrons may remain in the valence shell following the loss of the np electrons, as in Tl^+, Sn^{2+}, Pb^{2+}, and Bi^{3+}.

The **initial rate of reaction** is the rate of a reaction immediately after the reactants are brought together. The rate is generally expressed in terms of the rate of change with time of the concentration of one of the reactants or one of the products.

An **instantaneous rate of reaction** is the rate of a reaction at some particular time in the course of a reaction. It is established through a tangent line to a concentration versus time graph at the time in question.

An **integrated rate law** is an equation derived from the rate law for a reaction that expresses the concentration of a reactant as a function of time. (*See also* **rate law**.)

An **intensive property** is a property of a sample of matter, such as temperature or density, that is independent of the quantity of matter being considered.

An **interhalogen compound** is a compound of two halogen elements (even three in a few cases).

An **intermediate** is a substance that is produced in one elementary step in a reaction mechanism and consumed in another. The intermediate does not appear in the chemical equation for the overall reaction.

An **intermolecular force** is a force *between* molecules.

The **internal energy (*U*)** is the total amount of energy contained in a thermodynamic system. The components of internal energy are energy associated with random molecular motion (thermal energy) and that associated with chemical bonds and intermolecular forces (chemical energy).

An **ion** is an electrically charged particle comprised of one or more atoms.

Ion exchange is the replacement in solution of ions carrying a single unit of charge for other ions, usually multiply charged.

Ionic bonds are attractive forces between positive and negative ions, holding them together in solid crystals.

An **ionic compound** is a compound that consists of oppositely-charged ions held together by electrostatic attractions.

Ionic radii are measures of the size of a cation or anion based on the distance between the centers of ions in an ionic compound.

Ionization energy is the energy required to remove the least tightly bound electron from a ground-state atom (or ion) in the gaseous state.

The **ion product of water (K_w)** is the product of the concentration of hydronium ion, $[H_3O^+]$, and the concentration of hydroxide ion, $[OH^-]$, in pure water or a water solution. At 25 °C, its value is 1.0×10^{-14}.

The **isoelectric point (p*I*)** is the pH value at which an amino acid exists as a zwitterion.

Isoelectronic species (atoms, ions, molecules) have the same number of electrons in the same electron configuration.

Isomers are compounds having the same molecular formula but different structural formulas.

An **isotonic solution** is one that has the same osmotic pressure as body fluids (blood, tears).

Isotopes are atoms that have the same number of protons in their nuclei—the same atomic number—but different numbers of neutrons and, therefore, different mass numbers.

The **joule (J)** is the basic unit of energy in SI. It is the work done by a force of 1 newton (N) acting over a distance of 1 meter. That is, $1\ J = 1\ N\ m = 1\ kg\ m^2\ s^{-2}$.

K_c is the numerical value of an equilibrium constant expression in which molarities of products and reactants are used.

K_p is the numerical value of an equilibrium constant expression in which the partial pressures (usually in atm) of gaseous products and reactants are used.

A **kelvin (K)** is the SI base unit of temperature. An interval of one Kelvin on the Kelvin temperature scale is the same as one degree on the Celsius scale.

The **Kelvin scale** is an absolute temperature scale with its zero at −273.15 °C; its relationship to the Celsius scale is $T\ (K) = t\ (°C) + 273.15$.

A **ketone** is an organic substance whose molecules have a carbonyl group (C=O) between two other C atoms.

The **kilogram (kg)** is the SI base unit of mass.

A **kilopascal (kPa)** is 1000 pascals (Pa). (*See also* **pascal**.)

Kinetic energy (E_k) is energy of motion, given by the expression $E_k = \frac{1}{2}mu^2$.

The **kinetic–molecular theory of gases** is a theory based on a small number of postulates concerning gas molecules from which simple gas laws, the ideal gas law, and equations dealing with temperature and molecular speeds can be derived.

The **lanthanide** elements constitute the portion of the f-block of the periodic table in which the $4f$ subshell fills in the aufbau process.

The **lanthanide contraction** describes the general downward trend in the radii of lanthanide atoms and ions with increasing atomic number.

Lattice energy is the enthalpy change that accompanies the formation of one mole of an ionic solid from its gaseous ions.

The **law of combining volumes** states that, when gases measured at the same temperature and pressure are allowed to react, the volumes of gaseous reactants and products are in small whole-number ratios.

The **law of conservation of energy** states that in a physical or chemical change energy can neither be created nor destroyed.

The **law of conservation of mass** states that the total mass remains constant during a reaction. That is, the mass of the products of a reaction is always equal to the total mass of the reactants consumed.

The **law of constant composition** states that all samples of a particular compound have the same composition. That is, all samples have the same proportions by mass of the elements present.

The **law of definite proportions.** *See* **law of constant composition**.

The **law of multiple proportions** states that when two or more different compounds of the same two elements are compared, the masses of one element that combine with a fixed mass of the second element are in the ratio of small whole numbers.

Le Châtelier's principle is a statement that permits qualitative predictions about the effects produced by changes (amounts of reactants or products, reaction volume, temperature, ···) imposed on a system at equilibrium. (*See* page 609 for a statement of the principle.)

A **levorotatory (−)** substance rotates the plane of polarized light to the left.

Lewis acid. *See* **acid**.

Lewis base. *See* **base**.

A **Lewis structure** is a representation of covalent bonding through Lewis symbols, shared electron pairs, and lone-pair electrons.

A **Lewis symbol** is a representation of an element in which the chemical symbol stands for the core of the atom and dots placed around the symbol for its valence electrons.

A **ligand** is a species (atom, molecule, anion, or, rarely, cation) that is bonded to a metal center in a complex.

The **limiting reactant** (reagent) is the reactant that is completely consumed in a chemical reaction, thereby limiting the amounts of products formed.

The **line spectrum** of an element is the pattern of lines produced by the light emitted by the element. (*See also* **emission spectrum**.)

A **liquid crystal** is a physical form of a substance that has the fluid properties of a liquid and the optical properties of a crystalline solid.

A **lipid** is a cellular component that is insoluble in water but soluble in solvents of low polarity such as hexane, diethyl ether, and benzene.

A **liter (L)** is a metric unit of volume equal to one cubic decimeter or 1000 cubic centimeters: $1\ L = 1\ dm^3 = 1000\ cm^3$.

Lone pairs are electron pairs assigned exclusively to one of the atoms in a Lewis structure. They are not shared and hence are not involved in the chemical bonding.

Macromolecules are giant molecules (polymers) having small molecules (monomers) as their building blocks.

Magic numbers are numbers of protons and neutrons in stable nucleon shells that make up especially stable atomic nuclei.

The **magnetic quantum number (m_l)** is the last of three parameters that must be assigned a specific value to achieve a solution of Schrödinger's wave equation for the hydrogen atom: m_l is an integer between $-l$ and $+l$ (including 0). (*See also* **angular momentum quantum number** and **principal quantum number**.)

A **main-group element** is an element in which the subshell being filled in the aufbau process is either an s or p subshell of the principal shell of highest principal quantum number (the outermost shell). Main-group elements are located in the s- and p-blocks of the periodic table.

A **manometer** is a device used to measure a gas pressure. Generally a manometer must be used in combination with a barometer.

Mass is the quantity of matter in an object. It is related to the force required to move the object or to change its velocity if the object is already in motion.

The **mass percent composition** of a substance is the proportion, by mass, of each element in the substance expressed as a percentage.

The **mass number (A)** is the sum of the number of protons and neutrons in the nucleus of an atom. Also called the *nucleon number*.

A **mass spectrometer** is a device that separates ions according to their mass-to-charge ratios.

The **melting point** of a solid is the temperature at which it melts, that is, the temperature at which it comes into equilibrium with the liquid phase.

A **meniscus** is the interface between a liquid and the air above it.

A **meta director** is a substituent already on a benzene ring, such as $-COOH$ or $-NO_2$, that causes an incoming electrophile to substitute mainly in the meta position.

A **metal** is an element having a distinctive set of properties: luster, good heat and electrical conductivity, malleability, and ductility. Metal atoms generally have small numbers of valence electrons and a tendency to form cations. Metals are found to the left of the stepped diagonal line in the periodic table. All s-block (except hydrogen and helium), d-block, and f-block elements are metals, as are a few in the p-block.

The **metallic radius** is one-half the distance between the nuclei of adjacent atoms in a solid metal.

A **metalloid** is an element that has the physical appearance of a metal but some nonmetallic properties as well. Metalloids are located along the stepped diagonal line in the periodic table.

The **meter (m)** is the SI base unit of length.

The **method of initial rates** is an experimental method of establishing the rate law of a reaction. To establish the order of the reaction with respect to one of the reactants, the initial rates are compared for two different concentrations of that reactant, with the concentrations of all other reactants held constant. (*See also* **initial rate of reaction**, **rate law**, and **order of a reaction**.)

A **millimeter of mercury (mmHg)** is a unit used to express gas pressure: 1 mmHg = 1/760 atm (exactly). (*See also* **atmosphere**.)

A **mixture** is a type of matter that has a composition and properties that may vary from one sample to another. *See also* **heterogeneous mixture** and **homogeneous mixture.**

The **molality (m)** of a solution is the amount of solute, in moles, per kilogram of solvent (not of solution).

Molar concentration. *See* **molarity**.

Molar heat capacity is the quantity of heat required to change the temperature of one mole of a substance by 1 °C (or 1 K); it is the heat capacity of one mole of substance.

The **molarity (M)** of a solution is the amount of solute, in moles, per liter of solution.

The **molar mass** of a substance is the mass of 1 mol of that substance. It is numerically equal to the atomic mass, molecular mass, or formula mass, and expressed as g/mol.

The **molar volume of a gas** refers to the volume occupied by 1 mol of gas at a fixed temperature and pressure; it is essentially independent of the identity of the gas. At standard temperature and pressure, the molar volume of an ideal gas is 22.4141 L.

A **mole (mol)** is an amount of substance that contains as many elementary units (atoms, molecules, formula units) as there are atoms in exactly 12 g of the isotope carbon-12.

A **molecular compound** has molecules as its smallest characteristic entities, and these molecules determine the properties of the compound.

A **molecular formula** gives the symbol and *exact* numbers of each kind of atom found in a molecule.

Molecular geometry describes the geometric figure formed when appropriate atomic nuclei in a molecule or polyatomic ion are joined by straight lines. Molecular geometry refers to the geometric shape of a molecule or polyatomic ion.

The **molecularity** of a reaction is the number of molecules (or ions) that come together to form the activated complex.

Molecular mass is the average mass of a molecule of a substance relative to that of a carbon-12 atom; it is the sum of the masses of the atoms represented in the molecular formula.

A **molecular orbital** is a region in a molecule where there is a high electron charge density or a high probability of finding an electron(s). (*See also* **antibonding molecular orbital** and **bonding molecular orbital**.)

A **molecule** is a group of two or more atoms held together in a definite arrangement by forces called covalent bonds.

The **mole fraction (x)** of a component in a homogeneous mixture (a solution) is the fraction of all the molecules in the mixture contributed by that component.

The **mole percent** of a component in a homogeneous mixture (a solution) is the percentage of all the molecules in the mixture contributed by that component.

A **monodentate ligand** attaches to the metal center in a complex through one pair of electrons on a donor atom.

Monomers are small molecules that are capable of independent existence, but which under appropriate conditions can join together to form a giant molecule called a polymer. (*See also* **polymerization**.)

A **monoprotic acid** has one ionizable hydrogen atom per molecule.

A **monosaccharide** is a carbohydrate that cannot be hydrolyzed into simpler sugars.

A **multiple bond** is a covalent linkage in which two atoms share either two pairs (double bond) or three pairs (triple bond) of electrons between them.

The **Nernst equation** relates a cell voltage under nonstandard conditions, E_{cell}, to the standard cell potential, E°_{cell}, and the concentrations of reactants and products of a redox reaction. Its form at 25 °C is $E_{cell} = E^{\circ}_{cell} - (0.0592/n) \log Q$, where n is the number of moles of electrons transferred in the oxidation and reduction half-reaction of a redox reaction and Q is the reaction quotient.

A **net ionic equation** is an equation that represents the actual molecules or ions that participate in a chemical reaction, eliminating all nonparticipating species (so-called "spectator" ions).

In a **network covalent solid**, covalent bonds extend throughout the crystalline solid.

A **neutralization** reaction is one in which an acid and a base react in such a manner that there is neither excess acid or base in the final solution. The products of the reaction are water and a salt.

The **neutron** is a fundamental particle of matter found in the nuclei of atoms. Neutrons have a mass of 1.0087 u and no electric charge.

A **newton (N)** is the basic unit of force in SI. It is the force required to give a 1 kg mass an acceleration of 1 m/s^2. That is, $1\ N = 1\ kg \cdot m \cdot s^{-2}$.

The **nitrogen cycle** refers to the totality of activities in which nitrogen atoms are cycled through the environment.

Nitrogen fixation refers to the conversion of atmospheric nitrogen (N_2) into nitrogen compounds. This occurs naturally in the nitrogen cycle or artificially, as in the synthesis of ammonia. (*See also* **nitrogen cycle**.)

A **noble gas** is an element in Group 8A of the periodic table. Noble gases have the valence-shell electron configuration ns^2np^6 (except helium, $1s^2$).

A **nonelectrolyte** is a substance that exists exclusively or almost exclusively in molecular form, whether in the pure state or in solution.

A **nonmetal** is an element that lacks metallic properties. Nonmetals are generally poor conductors of heat and electricity and brittle when in the solid state. Nonmetal atoms generally have larger numbers of valence electrons than do metals and a tendency to form anions. Nonmetal atoms are confined to the p-block of the periodic table (plus hydrogen).

In a **nonpolar** bond there is an equal sharing of the electrons between the bonded atoms. The electrons are not any closer to one atom than to the other and so there is no charge separation.

A **nonspontaneous process** will not occur in a system left to itself. It can only be made to occur through intervention from outside the thermodynamic system.

The **normal boiling point** of a liquid is the temperature at which the liquid boils at 1 atm pressure.

The **normal melting point** of a solid is the temperature at which the solid melts at 1 atm pressure.

An ***n*-type semiconductor** is a semiconductor doped with donor atoms that can lose electrons to the conduction band. Electric current in this type of semiconductor is carried primarily by these donor electrons. (*See also* **doping**.)

Nuclear binding energy is the energy associated with the nucleons bound together into the nucleus of an atom. It is the energy equivalent of the mass lost in creating a nucleus from its individual protons and neutrons.

Nuclear fission is the splitting of a large unstable nucleus into two lighter fragments and two or more neutrons. Mass destroyed in this process is converted to an equivalent quantity of energy, which is evolved.

Nuclear fusion is the joining together or fusing of lighter nuclei into a heavier one. In the process some matter is converted to energy, which is released.

Nucleon is the general term for the nuclear particles protons and neutrons.

A **nucleophile** is a molecule or ion that donates a lone pair of electrons to another molecule, forming a covalent bond.

Nucleotides are the structural units, consisting of a sugar, a phosphate ester group, and a cyclic amine base, that make up deoxyribonucleic acid (DNA) and ribonucleic acid (RNA).

Nuclide is a term used to signify an atomic species having a particular atomic number and mass number, such as ${}^{12}_{6}C$. (*See also* **isotopes**.)

The **octet rule** states that most covalently bonded atoms represented in a Lewis structure have eight electrons in their outermost (valence) shells. In the formation of ionic compounds, the ions of the main-group elements also tend to follow the octet rule.

Optical isomers are nonsuperimposable mirror images; they differ only in the way they rotate the plane of polarized light. (*See also* **enantiomers**.)

The **orbital angular momentum quantum number (l)** is the second of three parameters that must be assigned a specific value to achieve a solution of Schrödinger's wave equation for the hydrogen atom: $l = 0, 1, 2, 3, \ldots, n - 1$. The value of l establishes a particular sublevel or subshell within a principal energy level.

An **orbital diagram** is a method of denoting an electron configuration in which parentheses or boxes are used to represent orbitals within subshells and arrows are used to represent electrons in the orbitals.

The **order of a reaction** is determined by the exponents of the concentration terms in the rate law for a reaction: Rate of reaction = $kA^mB^n \ldots$. The order of the reaction with respect to A is m; with respect to B it is n; and so on. The overall order of the reaction is $m + n + \ldots$.

An **ore** is a naturally occurring mineral containing a metal in a form and concentration that makes extraction of the metal feasible.

Osmosis is the net flow of a solvent through a semipermeable membrane, from pure solvent into a solution or from a solution of a lower concentration into one of a higher concentration.

The **osmotic pressure** of a solution is the pressure that must be applied to a solution to prevent the flow of solvent molecules into the solution when the solution and pure solvent are separated by a semipermeable membrane.

The **overvoltage** of an electrode reaction is the excess voltage above that calculated from $E°$ values required to bring about the reaction.

Oxidation is a process in which the oxidation number of an element increases. It is the half-reaction of an oxidation–reduction reaction in which electrons are "lost."

The **oxidation number** of an element in a compound is a means of designating the number of electrons that its atoms have lost, gained, or shared in forming that compound.

An **oxidizing agent** (oxidant) is a substance that makes possible the oxidation that occurs in an oxidation–reduction reaction. The oxidizing agent itself is reduced.

The **ozone layer** is a band of the stratosphere, about 20 km thick and centered at an altitude of about 25 to 30 km, that has a much higher concentration of ozone than the rest of the atmosphere.

Paramagnetism is the attraction into an external magnetic field of substances that have unpaired electrons.

Partial pressure. *See* **Dalton's law of partial pressure**.

A **partial pressure equilibrium constant (K_p)** is the numerical value of an equilibrium constant expression in which the partial pressures (usually in atm) of gaseous products and reactants are used.

Particulate matter is an air pollutant consisting of solid and liquid particles of greater than molecular size but small enough to remain suspended in air.

Parts per billion (ppb) expresses the composition of a mixture as the number of parts of one component per billion parts of the mixture as a whole, usually on a mass basis for liquid solutions and a mole basis for gaseous mixtures.

Parts per million (ppm) expresses the composition of a mixture as the number of parts of one component per million parts of the mixture as a whole, usually on a mass basis for liquid solutions and a mole basis for gaseous mixtures.

Parts per trillion (ppt) expresses the composition of a mixture as the number of parts of one component per trillion parts of the mixture as a whole, usually on a mass basis for liquid solutions and a mole basis for gaseous mixtures.

A **pascal (Pa)** is the basic unit of pressure in SI. It is a pressure of 1 newton per square meter, $1\ N \cdot m^{-2}$.

The **Pauli exclusion principle** states that no two electrons in an atom may have all four quantum numbers alike. A consequence of this principle is that there may be no more than two electrons in an orbital, and that the two electrons must have opposing spins.

The ***p*-block** is the portion of the periodic table in which the np subshell (the p subshell of the outer shell) fills in the aufbau process. The p-block elements are all main-group elements.

A **peptide** is composed of a chain of two or more amino acids joined through peptide (amide) linkages in chains of proteins, polypeptides, and peptides.

The **percent yield** is the ratio of the actual yield to the theoretical yield of a chemical reaction, expressed as a percentage.

The **periodic law** states that certain sets of physical and chemical properties recur at regular intervals (periodically) when the elements are arranged according to increasing atomic number.

A **periodic table** is a tabular arrangement of the elements according to increasing atomic number that places elements having similar properties into the same vertical columns. (Mendeleev's original periodic table was arranged according to atomic weights, not atomic numbers.)

Permanent hardness in water is the condition which the predominant anions are other than HCO_3^-. (*See also* **hard water**.)

The **pH** is the negative of the logarithm of the hydronium ion concentration in a solution: $pH = -\log [H_3O^+]$.

A **phase change** is a change from one phase to another, as in solid to liquid or liquid to gas.

A **phase diagram** is a pressure–temperature plot indicating the conditions under which a substance exists as a solid phase(s), a liquid, or a gas, or some combination of these in equilibrium.

Photochemical smog is air that is polluted with oxides of nitrogen and unburned hydrocarbons, together with ozone and several other components produced by the action of sunlight.

The **photoelectric effect** refers to the emission of electrons from the surface of certain materials when they are struck by light of the appropriate frequency.

A **photon** is a quantum of energy in the form of light. The energy of the photon is given by the expression $E = h\nu$.

A **photovoltaic cell** is a device that uses semiconductors to convert solar energy (light) into electricity.

In a **physical change**, a sample of matter undergoes a change in phase or state or other property that is observable but does not involve a change in composition.

A **physical property** is a characteristic that a sample of matter displays without undergoing a change in composition.

A **pi (π) bond** forms by the overlap in a parallel or side-by-side fashion of p orbitals of the bonded atoms. A double bond consists of one σ and one π bond; a triple bond, of one σ and two π bonds.

Pig iron is crude, high-carbon iron produced by reduction of iron ore in a blast furnace.

pK_a is the negative of the logarithm of the ionization constant of an acid: $pK_a = -\log K_a$.

pK_b is the negative of the logarithm of the ionization constant of a base: $pK_b = -\log K_b$.

pK_w is the negative of the logarithm of the ion product of water: $pK_w = -\log K_w = -\log (1.0 \times 10^{-14}) = 14.00$ (at 25° C). (*See also* **ion product of water**.)

Planck's constant (h) is the numerical constant relating the energy of a photon of light and its frequency: $E = h\nu$. Its value is $6.6260755 \times 10^{-34}$ J C s.

pOH is the negative of the logarithm of the hydroxide concentration in an aqueous solution: $pOH = -\log [OH^-]$.

In a **polar** bond between two atoms, electrons are drawn closer to the more electronegative atom, creating a separation of charge. One end of the bond has a small negative charge, $\delta-$, and the other end, a small positive charge, $\delta+$.

Polarizability is a measure of the ease with which electron charge density in an atom or molecule is distorted by an external electric field. It measures the ease with which a dipole can be induced in an atom or molecule.

In a **polar molecule**, depending on the electronegativities of bonded atoms and on the molecular geometry, a small separation of positive ($\delta+$) and negative ($\delta-$) charge exists.

A **polyatomic** ion is a charged particle containing two or more covalently bonded atoms.

A **polydentate ligand** attaches to a metal center in a complex at more than one point.

A **polymer** is a giant molecule formed by the combination of smaller molecules (monomers) in a repeating manner.

Polymerization is a type of reaction in which small repeating units (monomers) combine to form giant molecules (polymers).

Polymorphism is the property of a substance crystallizing in two or more forms, such as sulfur in its rhombic and monoclinic forms.

A **polypeptide** is a polymer of amino acids; usually of lower molecular mass than protein.

A **polyprotic acid** has more than one ionizable H atom per molecule. The ionization of a polyprotic acid occurs in distinctive steps.

A **polysaccharide** is carbohydrate each molecule of which can be hydrolyzed into many monosaccharide units.

A **positive hole** is a "missing electron" in a semiconductor; the vacancy acts like a positive ion.

A **positron (β^+)** is a positively charged particle having the same mass as a β^- particle.

The **potential difference**, measured in volts, is the difference in electric potential between two points in an electric circuit, for example, between the electrodes in an electrochemical cell.

Potential energy is energy due to position or arrangement. It is the energy associated with forces of attraction and repulsion between objects.

A **precipitate** is an insoluble ionic compound formed by a reaction in solution.

A **precipitation reaction** is a chemical reaction between ions in solution that produces an insoluble solid—a precipitate.

The **precision** of a set of measurements refers to how closely members of a set of measurements agree with one another. It reflects the degree of reproducibility of the measurements.

Pressure (P) is a force per unit area—that is, $P = F/A$.

Primary sewage treatment involves treatment of sewage in a holding pond intended to remove some of the sewage solids as sludge by simply the settling (sedimentation) of solids.

The **primary structure** is the amino acid sequence in a protein or of nucleotides in a nucleic acid.

The **principal quantum number (n)** is the first of three quantum numbers that must be assigned a specific numerical value to achieve a solution to Schrödinger's wave equation for the hydrogen atom: $n = 1, 2, 3, \cdots$. Its value designates the principal energy level of an electron in an atom.

A **principal shell (level)** refers to the collection of orbitals having the same principal quantum number.

A **product** is a substance that is produced in a chemical reaction. The formulas of products appear on the right side of a chemical equation.

A **protein** is a high-molecular-mass polymer of amino acids.

A **proton** is a nuclear particle carrying the fundamental unit of positive charge and having a mass of 1.0073 u.

A **proton acceptor** is a Brønsted–Lowry base. (*See also* **base**.)

A **proton donor** is a Brønsted–Lowry acid. (*See also* **acid**.)

Pseudohalogens are certain groupings of atoms, such as CN and OCN, that mimic the characteristics of a halogen atom.

A ***p*-type semiconductor** is a semiconductor that has been doped with acceptor atoms that extract electrons from chemical bonds in the semiconductor, producing positive holes in the valence band. Electric current in these semiconductors is carried primarily by positive holes. (*See also* **doping**.)

A **quantum** is the smallest quantity of energy that can be emitted or absorbed in a process, as given by the expression $E = h\nu$.

Quantum (wave) mechanics is the mathematical description of atomic structure based on the wave properties of subatomic properties.

Quantum numbers are certain integral values assigned to three parameters in a wave equation to obtain acceptable solutions to the equation.

The **quaternary structure** of a protein is the arrangement of protein subunits in geometric shapes.

A **rad** (radiation absorbed dose) corresponds to the absorption of 1×10^{-2} J of energy per kilogram of matter.

The **radioactive decay law** states that the rate of disintegration of a radioactive isotope, called the decay rate or activity, is directly proportional to the number of atoms present.

A **radioactive decay series** is a sequence of nuclear processes involving α and β emissions by which an initial long-lived radioactive nucleus is eventually converted to a stable nonradioactive nucleus.

A **radioactive tracer** is a radionuclide that can be used to follow a physical or chemical process through the ionizing radiation that it emits.

Radioactivity (radioactive decay) is the spontaneous emission of ionizing radiation by the atomic nuclei of certain isotopes.

Raoult's law states that the addition of a solute lowers the vapor pressure of the solvent, and that the fractional lowering of the vapor pressure is equal to the mole fraction of the solute.

The **rate constant (k)** of a reaction is a numerical constant that relates the rate of the reaction to the concentrations of the reactants. Rate constants are functions of temperature. (*See also* **rate law**.)

The **rate-determining step** in a reaction mechanism is the step (usually the slowest) that is crucial in establishing the rate of an overall reaction.

The **rate law (rate equation)** of a chemical reaction is an expression relating the rate of the reaction to the concentrations of the reactants.

A **reactant** is a starting material or substance consumed in a chemical reaction. The formulas of reactants appear on the left side of a chemical equation.

A **reactive material** is one that tends to react spontaneously or to react vigorously with air or water.

A **reaction mechanism** is a detailed representation of a chemical reaction consisting of a series of elementary reactions. A plausible mechanism must be consistent with the stoichiometry and the rate law of the net reaction.

A **reaction profile** is a schematic representation of changes in energy during the course of a reaction. The profile shows activation energies and enthalpies of reaction and identifies the energies of reactants, transition state(s), and products.

A **reaction quotient (Q_c or Q_p)** has the same format as an equilibrium constant (K) but uses initial concentrations rather than equilibrium concentrations.

A **reducing agent** (reductant) is a substance that makes possible the reduction that occurs in an oxidation–reduction reaction. The reducing agent itself is oxidized.

Reduction is a process in which the oxidation number of an element decreases. It is the half-reaction of an oxidation–reduction reaction in which electrons are "gained."

Refining is the process of removing impurities from a metal by any of a variety of chemical or physical means.

A **rem** (roentgen equivalent for man) is a *rad* (*which see*) multiplied by a factor that takes into account the fact that different types of radiation of the same energy have different effects on people.

Resonance is a term used to describe a situation in which several plausible Lewis structures can be written to represent a species but in which the true structure cannot be written. The plausible structures are called contributing structures or resonance structures, and the true structure, which is a composite of the contributing structures, is called the resonance hybrid.

A **resonance hybrid** is a composite of two or more plausible contributing Lewis structures. The resonance hybrid represents the true structure of a molecule or ion.

A **resonance structure** is one of two or more plausible Lewis structures that can be written to represent a molecule or ion.

The **resultant dipole moment** is the dipole moment of a molecule as a whole based on an assessment of bond moments and the molecular geometry. (*See also* **bond moment** and **dipole moment**.)

Reverse osmosis refers to the net flow of solvent through a semipermeable membrane in the opposite direction from that expected for osmosis. It is produced by applying a pressure to a solution that exceeds its osmotic pressure. (*See also* **osmosis**.)

Ribonucleic acid (RNA) is a polymer of nucleotides. The nucleotide consists of the sugar ribose, a phosphate ester, and a cyclic amine base (adenine, guanine, uracil, or cytosine).

The **root-mean-square speed (u_{rms})** of the molecules of a gas is the square root of the average of the squares of the molecular speeds.

A **salt** is an ionic compound in which hydrogen atoms of an acid are replaced by metal ions. Salts are produced in the reaction of an acid and a base.

A **salt bridge** is an inverted U-shaped tube filled with a salt solution used to connect the two *solutions* in a voltaic cell.

Saponification is soap making; the alkaline hydrolysis of a fat or other ester.

A **saturated hydrocarbon** has molecules that contain the maximum number of hydrogen atoms for the carbon atoms present. All bonds in the molecules are single covalent bonds.

A **saturated solution** is one in which dynamic equilibrium exists between undissolved solute and a solution. The solution contains the maximum amount of solute that can be dissolved in a particular quantity of solvent at the given temperature.

The ***s*-block** is the portion of the periodic table in which the *ns* subshell (the *s* subshell of the outer shell) fills in the aufbau process.

A **scientific law** is a brief statement, sometimes in mathematical terms, used to summarize and describe patterns in large collections of scientific data.

The **second (s)** is the SI base unit of time.

Secondary sewage treatment consists of passing the effluent from a primary treatment plant through gravel and sand filters to aerate the water and remove suspended solids.

The **secondary structure** of a protein is the arrangement of the protein chains with respect to the nearest neighbor amino acid units, for example, a helix or pleated sheet.

One statement of the **second law of thermodynamics** is that all natural or spontaneous processes are accompanied by an increase in entropy of the universe. That is,

$$\Delta S_{univ} = \Delta S_{system} + \Delta S_{surroundings} > 0.$$

A **second-order reaction** has a rate equation in which the sum of the exponents $m + n + \cdots = 2$.

A **semiconductor** is a substance in which there is only a small energy gap between the valence and conduction band. The electrical conductivity of a semiconductor is not nearly as good as that of a metal, but still much better than that of an insulator.

A **semipermeable membrane** is a material that permits the flow of solvent molecules but severely restricts the flow of solute molecules of a solution.

The **significant figures** in a measured quantity are all the digits known with certainty plus the first uncertain digit.

A **sigma (σ) bond** results from the end-to-end overlap of pure or hybridized atomic orbitals between the bonded atoms. A σ bond exists along a line joining the nuclei of the bonded atoms.

A **simple cubic cell** has an atom, molecule, or ion at each corner of a cube.

A **single bond** is a covalent linkage in which two atoms share one pair of electrons between them.

The **skeletal structure** of a polyatomic species (molecule, ion) indicates the order in which atoms are attached to one another.

Slag is a metallurgical term for a relatively low-melting-point product of the reaction of an acidic oxide and a basic oxide.

Soaps are salts of long-chain carboxylic acids called fatty acids (because they are derived from fats).

The **solubility** of a solute in a particular solvent refers to the concentration of the solute in a saturated solution.

A **solubility curve** is a graph of the solubility of a solute as a function of temperature.

The **solubility product constant (K_{sp})** describes the equilibrium that exists between a slightly soluble ionic solute and its ions in a saturated aqueous solution.

Solubility rules are a set of generalizations used to describe classes of substances that are either soluble or insoluble in water.

A **solute** is a solution component that is dissolved in a solvent. A solution may have several solutes, which are generally present in lesser amounts than is the solvent.

A **solution** is a homogeneous mixture of two or more substances. The composition and properties are uniform throughout a solution.

The **solvent** is the solution component (usually present in greatest amount) in which one or more solutes are dissolved to form the solution.

sp describes a hybridization scheme in the valence bond method in which an *s* and one *p* orbital are combined into two *sp* hybrid orbitals oriented in a linear fashion.

spdf notation uses numbers to designate a principal shell and the letters *s*, *p*, *d* , and *f* to identify a subshell.

sp^2 describes a hybridization scheme in the valence bond method in which an *s* and two *p* orbitals are combined into three sp^2 hybrid orbitals oriented in a trigonal planar fashion.

sp^3 describes a hybridization scheme in the valence bond method in which an *s* and three *p* orbitals are combined into four sp^3 hybrid orbitals oriented in a tetrahedral fashion.

sp^3d describes a hybridization scheme in the valence bond method in which an *s*, three *p*, and a *d* orbital are combined into five sp^3d hybrid orbitals oriented in a trigonal bipyramidal fashion.

sp^3d^2 describes a hybridization scheme in the valence bond method in which an *s*, three *p*, and two *d* orbitals are combined into six sp^3d^2 hybrid orbitals oriented in an octahedral fashion.

The **specific heat** of a substance is the quantity of heat required to raise the temperature of one gram of substance by 1 °C (or 1 K).

The **spectrochemical series** is a listing of ligands in terms of their abilities to produce a splitting of a *d*-subshell energy level in a complex. Ligands in the series are referred to as strong field, intermediate field, or weak field depending on the degree of splitting they produce. (*See also* **crystal field theory**.)

Spectroscopy is a study, using various instrumental methods, of atomic and molecular structures through the absorption or emission of electromagnetic radiation by the substances.

A **spontaneous process** is one that occurs in a system left to itself. Once started, no action from outside the system is required to keep the process going.

A **standard cell potential** (E°_{cell}) (standard cell voltage) is the potential difference (in volts) between the electrodes in a voltaic cell when all species are present in their standard states.

The **standard electrode potential** (E°) measures the tendency of a reduction process to occur when all species in a half-cell are present in their standard states. This tendency is measured in volts, relative to an assigned value of zero for the standard hydrogen electrode.

The **standard enthalpy of formation** (ΔH°_f) of a substance is the enthalpy change that occurs in the formation of 1 mol of the substance in its standard state from the reference forms of its elements in their standard states. The reference forms of the elements are generally their most stable forms at the given temperature and 1 atm pressure.

The **standard enthalpy of reaction** (ΔH°) is the enthalpy change for a reaction in which all reactants and products are in their standard states.

The **standard free energy change** (ΔG°) is the free energy change of a process in which the reactants and products are all in their standard states.

The **standard free energy of formation** (ΔG°_f) of a substance is the free energy change that occurs in the formation of 1 mol of the substance in its standard state from the reference forms of its elements in their standard states. The reference forms of the elements are generally their most stable forms at the given temperature and 1 atm pressure.

A **standard hydrogen electrode (SHE)** has hydrogen gas at 1 atm pressure and hydronium ion at unit activity (about 1 M) in oxidation–reduction equilibrium on an inert platinum electrode. The potential arbitrarily assigned to this electrode is zero.

The **standard molar entropy** (S°) of a substance is its entropy at standard pressure and a specified temperature.

The **standard state** of a solid or liquid substance is the pure element or compound at 1 atm pressure and at the temperature of interest. For a gaseous substance the standard state is the (hypothetical) pure gas behaving as an ideal gas at 1 atm pressure and the temperature of interest.

Standard temperature and pressure (STP) for a gas are 273.15 K (0 °C) and 1 atm (760 mmHg).

A **state** of a system is its exact condition, determined by the kinds and amounts of matter present, the structure of this matter at the molecular level, and the prevailing temperature and pressure.

A **state function** is a property that has a unique value that depends only on the present state of a system and not on how that state was reached.

The **states of matter** are the *three* fundamental conditions in which samples of matter may be obtained: solid, liquid, and gas.

A **steel** is an alloy of iron containing small amounts of carbon and usually containing other metals such as manganese, nickel, and chromium.

Stereoisomerism is a type of isomerism in which the number and types of groups and their mode of attachment in a molecule or complex are the same, but in which their spatial arrangements differ in the isomers.

Stereoisomers are isomers that have the same molecular formulas but differ in the arrangement of atoms in three-dimensional space.

A **stoichiometric coefficient** is a coefficient placed in front of a formula in a chemical equation to balance the equation.

A **stoichiometric factor** is a conversion factor relating molar amounts of two species involved in a chemical reaction (that is, a reactant to a product, one reactant to another, and so on). The numerical values used in formulating the factor are the stoichiometric coefficients.

Stoichiometric proportions refer to relative amounts of reactants that are in the mole ratios corresponding to the coefficients in a balanced equation.

Stoichiometry refers to quantitative measurements and relationships involving substances and mixtures of chemical interest.

A **strong acid** is an acid that is essentially completely ionized in solution, that is, an acid that is a strong electrolyte. (*See also* **acid**.)

A **strong base** is a base that is essentially completely ionized in solution, that is, a base that is a strong electrolyte. (*See also* **base**.)

A **strong electrolyte** is a substance that exists exclusively or almost exclusively in ionic form in solution.

A **structural formula** is a chemical formula that shows how the atoms in a molecule are attached to one another.

Sublimation is the direct passage of molecules from the solid state to the vapor state.

A **sublimation curve** is a graph of the vapor pressure (sublimation pressure) of a solid as a function of temperature. It is analogous to the vapor pressure curve of a liquid.

A **subshell (sublevel)** is the collection of orbitals of a given type present in a principal shell. For example, the three $2p$ orbitals constitute the $2p$ subshell.

Subshell (sublevel) notation is a method of denoting an electron configuration that uses numbers to represent the principal shells and the letters s, p, d, and f for subshells. A superscript number following the letter indicates the number of electrons in the subshell.

A **substance** is a type of matter having a definite, or fixed, composition and fixed properties that do not vary from one sample to another. All substances are either elements or compounds.

In a **substitution reaction**, a substituent group replaces a hydrogen atom in a hydrocarbon molecule. This type of reaction is characteristic of alkane and aromatic hydrocarbons.

The **substrate** in an enzyme-catalyzed reaction is the reactant species that attaches to the active site on an enzyme molecule and undergoes chemical reaction.

A **superconductor** is a metal, alloy, or ceramic material whose resistance to the flow of electricity vanishes at a sufficiently low temperature.

Supercooling is a condition in which a liquid is cooled below its freezing point without the appearance of any solid. A supercooled liquid usually begins to freeze when the temperature is lowered sufficiently. The temperature of the freezing liquid then rises back to the normal freezing point.

A **supercritical fluid** is a fluid at a temperature above its critical temperature and at a pressure above its critical pressure.

A **supersaturated solution** contains more solute than is present in a saturated solution in equilibrium with undissolved solute.

Surface tension is the amount of work required to extend a liquid surface, usually expressed in joules per square meter, $J\ C\ m^{-2}$.

The **surroundings** refer to that part of the universe with which a system interacts by exchanging heat and/or work and/or matter.

A **system** is that part of the universe chosen for a thermochemical or thermodynamic study. (*See also* **surroundings**.)

Temperature is a physical property related to the kinetic energies of the atoms or molecules in a substance that indicates the direction of heat flow. Kinetic energy, as heat, is transferred from more energetic (higher temperature) to less energetic (lower temperature) atoms or molecules.

Temporary hardness is present in hard water that has hydrogen carbonate (bicarbonate) ion, HCO_3^- as its primary anion. (*See also* **hard water**.)

A **terminal atom** in a polyatomic species (molecule, ion) is bonded to just one other atom.

A **termolecular reaction** in a reaction mechanism involves the simultaneous collision of three molecules.

The **tertiary structure** of a protein is the folds, bends, and twists in protein or nucleic acid structure.

The **theoretical yield** is the calculated quantity of a product expected in a chemical reaction.

A **theory** provides explanations of observed phenomena and predictions that can be tested by experimentation. It is the intellectual framework for explaining scientific data and scientific laws.

Thermochemistry is the study of energy changes associated with chemical reactions or physical processes, especially energy changes that appear as heat.

Thermodynamics is the science dealing with the relationship between heat and motion (work) and with transformations of energy from one form to another. Thermochemistry is a subfield within thermodynamics.

A **thermoplastic polymer** is one that can be softened by heating and formed into desired shapes by applying pressure.

A **thermosetting polymer** becomes permanently hard at elevated temperatures and pressures.

The **third law of thermodynamics** states that the entropy of a pure, perfect crystal at 0 K is zero. This is the starting point for the experimental determination of absolute molar entropies.

A **titrant** is a solution of known concentration and composition used in analytical titrations. Also called a standard solution.

Titration is a laboratory procedure in which two reactants in solution are made to react in their stoichiometric proportions.

A **titration curve** in a neutralization reaction is a graph of pH versus volume of titrant added from a buret.

A **Torr** is a unit used to express gas pressure: 1 Torr = 1 mmHg. (*See also* **millimeter of mercury**.)

A **toxic material** is one that contains or releases poisonous substances in amounts large enough to threaten human health or the environment.

Toxicology is a study of the effects of poisons on the body, their identification and detection, and remedies against them.

The term **trans** is used to indicate geometric isomers in which two groups are attached to opposite sides of a double bond in an organic molecule, or at opposite corners of a square in a square planar complex, or above and below the central plane in an octahedral complex.

Transcription is the process by which DNA directs the synthesis of an mRNA molecule during protein synthesis.

A **transition element** is one in which the subshell being filled in the aufbau process is in a principal shell of less than the highest quantum number (an inner shell). Transition elements are located in the *d*- and *f*-blocks of the periodic table.

A **transition state** is a state that lies between the reactants and products of a chemical reaction. It is produced as a result of collisions between especially energetic molecules.

Translation is the process by which the information contained in a base triplet of an mRNA molecule is converted to a protein structure.

The **transuranium elements** are those with atomic number (Z) greater than 92.

A **triglyceride** is an ester formed by the chemical combination of glycerol with three fatty acids. Also called a *triacylglycerol*.

A **triple bond** is a covalent linkage in which two atoms share three pairs of electrons between them.

A **triple point** is a particular temperature and pressure at which three phases of a pure substance are at equilibrium—solid, liquid, and vapor; or two solid phases and the liquid; or two solid phases and the vapor.

Trouton's rule is that the entropy of vaporization of a nonpolar liquid at its normal boiling point is approximately 87 $J \cdot K^{-1} \cdot mol^{-1}$.

The **Tyndall effect** is the scattering of light by colloidal particles, which makes a colloidal dispersion distinguishable from a true solution.

Heisenberg's **uncertainty principle** states that the product of the uncertainty in the position of an object and the uncertainty in its momentum (mass, m, times speed, u) cannot be less than $h/4\pi$. Thus, it is not possible to know with certainty both the position of a subatomic particle and details of its motion.

A **unimolecular reaction** in a reaction mechanism is one in which a single molecule undergoes rearrangement or decomposition.

The **unit cell** of a crystal structure is the simplest parallelepiped that can be used to generate the entire crystalline lattice through straight-line displacements in all three dimensions.

The **universal gas constant (*R*)** is the numerical constant required to relate pressure, volume, amount, and temperature of a gas in the ideal gas equation, $PV = nRT$. Its numerical value is 0.082057 $L \cdot atm \cdot mol^{-1} \cdot K^{-1}$ or 8.3145 $J \cdot mol^{-1} \cdot K^{-1}$.

An **unsaturated hydrocarbon** is a carbon–hydrogen compound having one or more multiple bonds (double, triple) between carbon atoms.

An **unsaturated solution** contains less of a solute in a given quantity of solution than is present in a saturated solution. It is a solution having a concentration less than the solubility limit.

A **valence band** is formed by combining atomic orbitals of the valence electrons of a large number of atoms into a set of molecular orbitals very closely spaced in energy. If the band is only partially filled with electrons it is also a conduction band. (*See also* **band** and **conduction band**.)

The **valence bond method** describes a covalent bond as a region of high electron charge density that results from the overlap of atomic orbitals between the bonded atoms and the sharing of a pair of electrons in the region of overlap.

Valence electrons are electrons with the highest principal quantum number. They are found in the outermost electronic shells of atoms. (*See also* **core electrons**.)

The **valence shell** is the outer shell of electrons of an atom; the electrons with the highest principal quantum number.

The **valence-shell electron-pair repulsion (VSEPR) theory** is an approach to describing the geometric shapes of molecules and polyatomic ions in terms of the geometrical distribution of electron groups in the valence shell(s) of central atom(s).

The **van't Hoff factor (*i*)** is a correction factor that must be incorporated in equations for colligative properties so that the equations may be applied to solutions of strong or weak electrolytes.

A **vapor** is a gas at a temperature below its critical temperature. A vapor can be liquefied by application of pressure without lowering the temperature.

Vaporization or evaporation is the process of conversion of a liquid to a gas (vapor).

The **vapor pressure** of a liquid is the pressure exerted by the vapor in dynamic equilibrium with the liquid at a constant temperature.

A **vapor pressure curve** is a graph of the vapor pressure of a liquid as a function of temperature. The curve is the boundary between the liquid and vapor areas in a phase diagram.

The **viscosity** of a liquid substance is a measure of its resistance to flow.

Volt (V) is the unit used to measure electrode potentials and electrical potential differences.

A **voltaic cell** is an electrochemical cell that produces electricity through an oxidation–reduction reaction.

A **wave** is a progressive, repeating disturbance propagated from a point of origin to a more distant point.

The **wavelength** is the distance between any two identical points in consecutive cycles of a wave, for example, the distance between the peaks or crests of the wave.

The **wavenumber** ($\tilde{\nu}$) of radiation expresses the frequency of radiation as the number of cycles per cm of the wave: $\tilde{\nu} = 1/\lambda$

A **weak acid** is an acid that exists partly in ionic form and partly in molecular form in solution, that is, an acid that is a weak electrolyte. (*See also* **acid**.)

A **weak base** is a base that exists partly in ionic form and partly in molecular form in solution, that is, a base that is a weak electrolyte. (*See also* **base**.)

A **weak electrolyte** is a substance that is present partly in molecular form and partly in ionic form in its solutions.

Work is (1) the result of a force acting through a distance, for example, $1\ J = 1\ N \times 1\ m$, or (2) an energy transfer into or out of a thermodynamic system that can be expressed as the product of a force and a distance.

An **X-ray** is a type of electromagnetic radiation produced by the impact of cathode rays (electrons) on a solid, such as on a dense metal anode (a target) in a cathode-ray tube.

A **zero-order reaction** has a rate that is independent of the concentration of reactant(s). The sum of the exponents in its rate equation, $m + n + \ldots = 0$.

Zone refining is a process of purification in which a rod of material is subjected to repeated cycles of melting and freezing. This sweeps impurities in a molten zone to the end of the rod, which is then cut off.

Appendix F

Answers to Selected Problems

These answers are for in-text Exercises and selected Review Questions, Problems, and Additional Problems. *Note*: Some of your answers may differ slightly from those given here, depending on the number of digits used in expressing atomic masses and other precisely known quantities and whether intermediate results were rounded off.

Chapter 1

Exercises: 1.1 (a) 7.42 ms; **(b)** 5.41 μm; **(c)** 1.19 ng; **(d)** 5.98 km. **1.2 (a)** 4.75×10^{-7} m; **(b)** 2.25×10^{-7} s; **(c)** 1.415×10^{6} m; **(d)** 2.26×10^{3} kg. **1.3 (a)** 185 °F; **(b)** 10.0 °F; **(c)** 179 °C; **(d)** −29.3 °C. **1.4** 8.9×10^{-3} m^3. **1.5** 925 g zinc. **1.6A (a)** 100.5 m; **(b)** 1.50×10^{2} g; **(c)** 415 g; **(d)** 6.3 L. **1.6B (a)** 1.80×10^{3} m^2; **(b)** 2.33 g/mL; **(c)** 72 kg/m^3; **(d)** 0.63 g/cm^3. **1.7 (a)** 0.0763 m; **(b)** 8.56×10^{4} mg; **(c)** 1.82×10^{3} ft; **(d)** 1.38 kg; **(e)** 448 fl oz. **1.8 (a)** 25.0 m/s; **(b)** 1.53 km/h; **(c)** 15.0 kg/h; **1.9A (a)** 73.8 in.^{2}; **(b)** 0.256 m^3. **1.9B** 1.034×10^{4} kg/m^2. **1.10A** 0.568 g/cm^3. **1.10B** 13.6 g/cm^3. **1.11A** 3.34 gal. **1.11B** 8.72 gal. **1.12** 500 °C. **1.13** 20 kg. **1.14** The hull can be a "honeycomb" of air-filled chambers. If a few chambers are punctured, only these will fill with water. If too many chambers fill, however, the overall density exceeds that of seawater and the ship sinks. **1.15** Assumptions: (1) all envelopes have the same mass; (2) all scale readings precise to ±3 g. Sources of failure: (1) masses of envelopes too close to 28.35; (2) scale incorrectly calibrated against known masses. **Review Questions: 2.** (a), (b), (c), (e). **11.** (c) The burning of natural gas is a chemical change; composition changes. **12.** Physical change: (a), (e); chemical change: (b), (c), (d). **13.** Elements: (a), (c), (e); not elements: (b), (d), (f). **14.** Substances: (a), (d); mixtures: (b), (c). **Problems: 25 (a)** 4.54 ng; **(b)** 3.76 km; **(c)** 6.34 μg. **27 (a)** 74.3 °C; **(b)** 78.83 °C; **(c)** −144 °F. **29 (a)** 57.8 °C; **(b)** −184 °F. **31 (a)** 5.00×10^{4} m; **(b)** 0.0479 L; **(c)** 0.578 ms; **(d)** 1.55×10^{8} mg; **(e)** 8.74×10^{3} mm^2; **(f)** 1.60 m/s. **33.** (4) $<$ (3) $<$ (1) $<$ (2). **35 (a)** four; **(b)** two; **(c)** five; **(d)** three; **(e)** four; **(f)** four. **37 (a)** 2.804×10^{3} m; **(b)** 9.01×10^{2} s; **(c)** 9.0×10^{-4} cm; **(d)** 2.210×10^{2} s. **39 (a)** 505.5 m; **(b)** 2120 s, zero is not significant; **(c)** 0.00610 g; **(d)** 40,000 mL, final two zeros not significant. **41 (a)** 37.8 m; **(b)** 155 g; **(c)** 111 mL; **(d)** 2.4 cm. **43 (a)** 2.32×10^{3}; **(b)** 4.80×10^{3}; **(c)** 4.6×10^{4}; **(d)** 1.92×10^{-4}. **45.** 3.12 g/mL. **47.** 5.23 g/cm^3. **49.** 11.3 g/cm^3. **51.** 39.6 g. **53.** 1.16×10^{5} cm^3; 7.73×10^{4} cm. **55.** 0.0042 g/cm^3. **57.** (2). **59.** 7.1 g/cm^3. **Additional Problems: 61.** Sample A weighs more than sample B because of the greater force of gravity on Earth. Because of the weaker force of gravity on the moon, sample C must have a greater mass than sample A. **63.** Actual area, 1.41 m^2; error, 0.02 m^2. **65.** 36.8 °C. **67.** 7.92 in./link. **69.** 2.47 acres/hectare. **71.** 0.251 mm. **73.** 1.8×10^{4} cm^3. **75.** 4.9 mg m^{-2} h^{-1}. **77.** 7.134 g/cm^3. **79.** Vessel on the right. **81 (a)** 5.178 ± 0.008 m; **(b)** 1.392 ± 0.004 m^2, 0.3% error; **(c)** By carrying four significant figures in the area—1.392 m^2—the implied percent error is the expected 0.3%. The extra significant figure is justified. It would not be justified if the dimensions were 1.250 m × 0.762 m, for example. **83 (a)** 26.9%; **(b)** 4.7%; **(c)** the person in (b).

Chapter 2

Exercises: 2.1 7.500 g. The mass of the product (magnesium oxide) is equal to the sum of the masses of the magnesium and oxygen consumed. The masses of the materials not involved in the reaction are unchanged. **2.2A** 3.778 g. **2.2B** 10.361 g. **2.3A** ${}^{116}_{50}Sn$ **2.3B** $Z = 48$; $A = 114$; 48 protons and 48 electrons. **2.4A** 20.18 u. **2.4B** 69.17% copper-63, 30.83% copper-65. **2.5** ${}^{24}Mg$ is most abundant; its mass is closest to 24.305 u. The proportions of ${}^{25}Mg$ and ${}^{26}Mg$ depend on the percent abundance of ${}^{24}Mg$, which is not given. **2.6** N_2F_4; dinitrogen tetrafluoride. **2.7 (a)** P_4O_{10}; **(b)** iodine heptafluoride. **2.8 (a)** AlF_3; **(b)** K_2S; **(c)** Ca_3N_2; **(d)** Li_2O. **2.9A (a)** calcium bromide; **(b)** lithium sulfide; **(c)** iron(II) bromide; **(d)** copper(I) iodide. **2.9B** copper(I) sulfide, Cu_2S. **2.10 (a)** $(NH_4)_2CO_3$; **(b)** $Ca(ClO)_2$; **(c)** $Cr_2(SO_4)_3$. **2.11 (a)** potassium hydrogen carbonate; **(b)** iron(III) phosphate; **(c)** magnesium dihydrogen phosphate. **2.12 (a)** No, different formulas; **(b)** Yes. **2.13** $C_{11}H_{24}$ to about $C_{15}H_{32}$. **Review Questions: 2 (a)** 104.3 g; **(b)** 40.3 g MgO; **(c)** conservation of mass and constant composition; **(d)** 80.6 g MgO (law of constant composition). **3.** 1.5 : 1 or 3 : 2 (law of multiple proportions). **10.** isotopes, (b); isobars, (a). **11 (a)** ${}^{8}_{5}B$; **(b)** ${}^{14}_{6}C$; **(c)** ${}^{235}_{92}U$; **(d)** ${}^{60}_{27}Co$. **17.** Systematic: iron(II) chloride and iron(III) chloride; common: ferrous chloride and ferric chloride. **18.** Substance: (a), (d), (f); ion or group: (b), (c), (e). **25.** Molecular formula sufficient: (a), (b), (d); structural formula needed: (c), (e). **Problems: 27.** Yes, the total mass remains 1.200 g. Before reaction: 1.000 g Zn + 0.200 g S; after reaction, 0.608 g ZnS + 0.592 g Zn.

29. Yes, each sample is 62.5% C and 4.2% H. **31.** 2.61 g SO_2. **33.** (a) and (c). **35 (a)** 20 p, 20 e^-; **(b)** 11 p, 11 e^-; **(c)** 9 p, 9 e^-; **(d)** 18 p, 18 e^-; **(e)** 4 p, 4 e^-. **37 (a)** 30 p, 32 n; **(b)** 94 p, 147 n; **(c)** 43 p, 56 n; **(d)** 42 p, 57 n. **39.** The neutral atoms are $^{40}_{20}Ca$, $^{48}_{22}Ti$, and $^{48}_{20}Ca$. Calcium-40 and calcium-48 are isotopes. **41.** There must be at least two isotopes, one or more with mass number greater than 80 and one or more with mass number less than 80. **43.** 69.7 u. **45.** 20.18 u. **47.** 72.164% ^{85}Rb and 27.836% ^{87}Rb. **49 (a)** nonmetal, Group 4A, Period 2; **(b)** metal, Group 2A, Period 4; **(c)** nonmetal, Group 6A, Period 3; **(d)** metal, Group 4A, Period 5; **(e)** metal, Group 4B, Period 4; **(f)** nonmetal, Group 7A, Period 4; **(g)** metal, Group 5A, Period 6; **(h)** metal, Group 3A, Period 5; **(i)** metal, Group 1B, Period 6; **(j)** metal,Group 6B, Period 5. **51 (a)** O_2; **(b)** Br_2; **(c)** H_2; **(d)** N_2. **53.** Binary molecular compounds: (b) and (d). Compounds (a) and (e) each consists of *three* elements, and (c) is a binary *ionic* compound. **55 (a)** N_2O; **(b)** SF_6; **(c)** P_4S_3; **(d)** carbon disulfide; **(e)** diboron tetrachloride; **(f)** dichlorine heptoxide. **57 (a)** potassium ion; **(b)** oxide ion; **(c)** copper(II) ion; **(d)** Al^{3+}; **(e)** N^{3-}; **(f)** Cr^{3+}. **59 (a)** carbonate ion; **(b)** sulfate ion; **(c)** hydroxide ion; **(d)** dihydrogen phosphate ion; **(e)** NH_4^+; **(f)** NO_2^-; **(g)** CN^-; **(h)** HCO_3^-. **61 (a)** sodium oxide; **(b)** magnesium bromide; **(c)** iron(II) chloride; **(d)** aluminum oxide; **(e)** lithium iodide trihydrate; **(f)** potassium sulfide; **(g)** calcium hydroxide; **(h)** ammonium nitrate; **(i)** chromium(III) sulfate; **(j)** sodium hydrogen sulfite; **(k)** potassium permanganate; **(l)** magnesium perchlorate; **(m)** copper(II) hydroxide; **(n)** ammonium oxalate; **(o)** iron(III) phosphate dihydrate. **63 (a)** $FeCO_3$; **(b)** $BaI_2 \cdot 2H_2O$; **(c)** $Al_2(SO_4)_3$; **(d)** $KHCO_3$; **(e)** $NaBrO_3$; **(f)** $CaCl_2 \cdot 6H_2O$; **(g)** $Cu(NO_3)_2 \cdot 3H_2O$; **(h)** $LiHSO_4$; **(i)** $Mg(CN)_2$; **(j)** $Fe_2(SO_4)_3$; **(k)** $(NH_4)_2Cr_2O_7$; **(l)** $Mg(ClO_4)_2$. **65.** Correct: (b). Incorrect: (a) Calcium hypochlorite is $Ca(ClO)_2$; (c) the prefix "di" is unnecessary; (d) calcium chlorate is $Ca(ClO_3)_2$; (e) calcium oxychloride is a nonspecific name. **67 (a)** HI; **(b)** H_2SO_4; **(c)** LiOH; **(d)** HNO_2; **(e)** periodic acid; **(f)** calcium hydroxide; **(g)** hydrobromic acid; **(h)** phosphorous acid.

69

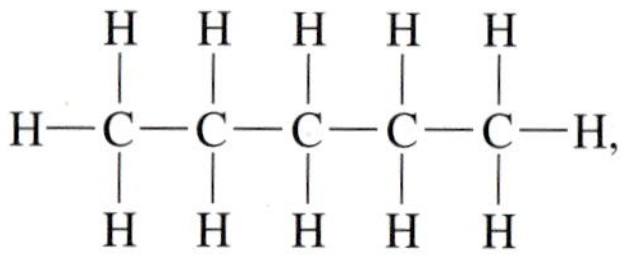

```
   H  H  H  H  H
   |  |  |  |  |
H—C—C—C—C—C—H, CH3(CH2)3CH3;
   |  |  |  |  |
   H  H  H  H  H
```

(a)

```
   H  H  H  O
   |  |  |  ||
H—C—C—C—C—O—H, CH3(CH2)2COOH;
   |  |  |
   H  H  H
```

(b)

```
   H  H     H  H
   |  |  H  |  |
H—C—C—N—C—C—H, (CH3CH2)2NH;
   |  |     |  |
   H  H     H  H
```

(c)

```
   H  H
   |  |
H—C—C—H, CH2—CH2
   |  |      |    |
H—C—C—H  CH2—CH2
   |  |
   H  H
```

(d)

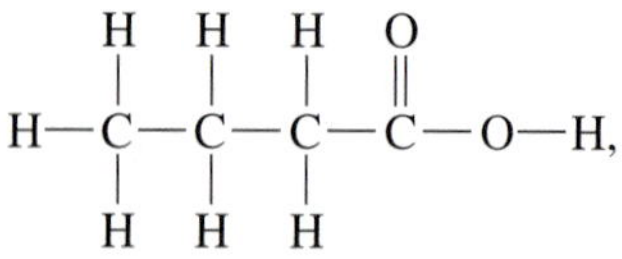

```
      H
      |
   H—C—H
   H  |  H
   |  |  |
H—C—C—C—H, CH(CH3)3;
   |  |  |
   H  H  H
```

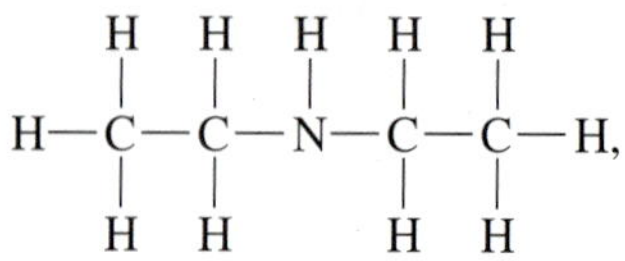

(e)

```
   H  H  O
   |  |  ||
H—C—C—C—O—H, CH3CH2COOH;
   |  |
   H  H
```

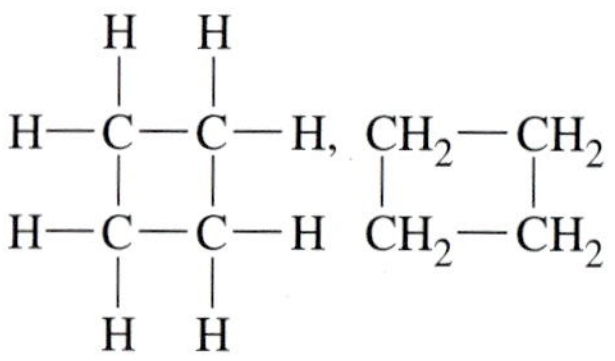

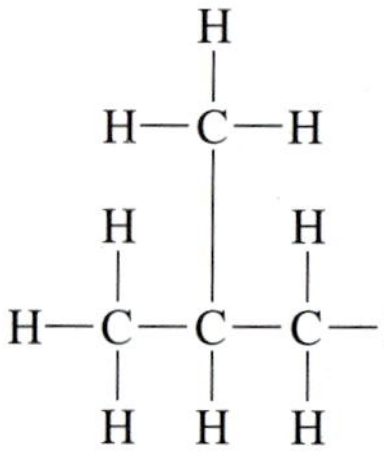

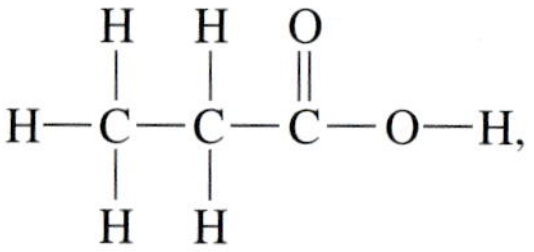

(f)

```
   H     H
   |     |
H—C—O—C—H, (CH3)2O;
   |     |
   H     H
```

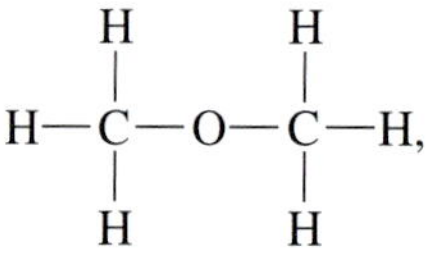

(g)

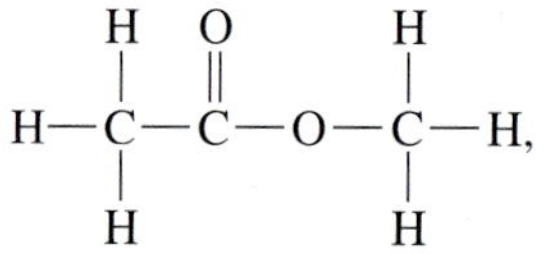

```
   H  O     H
   |  ||    |
H—C—C—O—C—H, CH3COOCH3.
   |        |
   H        H
```

(h)

71 (a) butyric acid, $CH_3CH_2CH_2\overset{O}{\overset{\|}{C}}OH$;

isobutyric acid, $CH_3\underset{}{\overset{CH_3}{\overset{|}{C}}}H—\overset{O}{\overset{\|}{C}}OH$;

(b) propyl butyrate, $CH_3CH_2CH_2\overset{O}{\overset{\|}{C}}OCH_2CH_2CH_3$;

propyl isobutyrate, $CH_3\overset{CH_3}{\overset{|}{C}}H—\overset{O}{\overset{\|}{C}}OCH_2CH_2CH_3$.

73. Isomers of the given structure: (a) and (b). Structure (c) is identical to (b); (d) is a different structure. **75.** $CH_3OCH_2CH_2CH_3$; $CH_3OCH(CH_3)_2$; $CH_3CH_2OCH_2CH_3$. **77 (a)** straight-chain alkane; **(b)** alcohol; **(c)** hydrocarbon (could be cyclic alkane); **(d)** hydrocarbon; **(e)** carboxylic acid; **(f)** inorganic compound; **(g)** ester; **(h)** ether. **79.** (b). **Additional Problems: 81.** The total mass remains constant; before reaction: 10.00 g + 114.8 g = 124.8 g; after reaction: 120.40 g + 4.39 = 124.79 g. **83.** Based on the constant percentages of carbon (46.59%) and hydrogen (8.81%), the data are consistent with the law of constant composition. However, we cannot be certain whether the other elements—N and O—also have fixed percentages. **85.** The first compound has a 25 : 1 mass ratio of Hg to O; the second, 12.5 : 1. The ratio, 25 : 12.5 = 2 : 1. **87.** The ratio of masses of Fe per gram of O for the given compound and compound (a) are 1.33 : 1 or 4 : 3. For (b), (c), and (d), the ratios are not those of small whole numbers. **89.** $^{40}_{20}Ca$. **91.** 78.99% ^{24}Mg, 10.00% ^{25}Mg, 11.01% ^{26}Mg. **93.** $C_{11}H_{24}$, 196 °C; $C_{12}H_{26}$, 217; $C_{13}H_{28}$, 236; $C_{14}H_{30}$, 254; $C_{15}H_{32}$, 271; $C_{21}H_{44}$, 355; $C_{22}H_{46}$, 367. **95 (a)** No, the structures are identical; **(b)** isomers; **(c)** different compounds; **(d)** identical structures. **97 (a)** $C_nH_{(2n+1)}OH$; **(b)** $C_{n-1}H_{(2n-1)}COOH$; **(c)** $C_nH_{(2n-2)}(OH)_2$. **99.** (f): $CH_3CHOHCOOH$. **101.** C_2H_6N and $C_4H_{12}N_2$. **103.** (f).

Chapter 3

Exercises: 3.1A (a) 257 u; **(b)** 32.1 u; **(c)** 98.0 u; **(d)** 72.1 u. **3.1B (a)** 125.97 u; **(b)** 92.011 u; **(c)** 86.177 u; **(d)** 74.079 u. **3.2A (a)** 29.9 u; **(b)** 148 u; **(c)** 234 u; **(d)** 295 u. **3.2B (a)** 104.06 u; **(b)** 117.49 u; **(c)** 392.18 u; **(d)** 249.69 u. **3.3A (a)** 6.83 g C_3H_8; **(b)** 7.37 mg C_2H_6; **(c)** 2.84×10^3 mol C_4H_{10}; **(d)** 7.76×10^{-4} mol H_3PO_4. **3.3B (a)** 17 mol Al; **(b)** 134 mL CCl_4. **3.4A (a)** 3.470×10^{-22} g; **(b)** 1.529×10^{-22} g; **(c)** 2.15×10^{20} N_2 molecules; **(d)** 1.70×10^{25} atoms. **3.4B (a)** 1.47×10^{24} Br_2 molecules; **(b)** 0.970 L CH_3CH_2OH. **3.5** (d). The sample contains about 1/6 mol Mg, weighing about 4.0 g. **3.6A (a)** 21.20% N, 6.10% H, 24.27% S, 48.43% O; **(b)** 20.00% C, 26.64% O, 46.65% N, 6.71% H; urea has the greatest mass percent nitrogen. **3.6B** 9.39% N. **3.7A** 1.37×10^3 mg Na^+. **3.7B** 119 g N. **3.8** LiH_2PO_4. **3.9A** $C_6H_{12}O$. **3.9B** $C_5H_{10}NO_2$. **3.10A** C_7H_5. **3.10B** $C_7H_5N_3O_6$. **3.11** ethylene, C_2H_4; cyclohexane, C_6H_{12}, 1-pentene, C_5H_{10}. **3.12 (a)** 80.24% c, 9.62% H, 10.14% O; **(b)** $C_{21}H_{30}O_2$. **3.13 (a)** $SiCl_4 + 2\ H_2O \longrightarrow SiO_2 + 4\ HCl$; **(b)** $PCl_3 + 3\ H_2O \longrightarrow H_3PO_3 + 3\ HCl$; **(c)** $6\ CaO + P_4O_{10} \longrightarrow 2\ Ca_3(PO_4)_2$. **3.14 (a)** $2\ C_4H_{10} + 13\ O_2 \longrightarrow 8\ CO_2 + 10\ H_2O$; **(b)** $CH_3CH_2CH_2CHOHCH_2OH + 7\ O_2 \longrightarrow 5\ CO_2 + 6\ H_2O$. **3.15 (a)** $FeCl_3 + 3\ NaOH \longrightarrow Fe(OH)_3 + 3\ NaCl$; **(b)** $3\ Ba(NO_3)_2 + Al_2(SO_4)_3 \longrightarrow 3\ BaSO_4 + 2\ Al(NO_3)_3$; **(c)** $3\ Ca(OH)_2 + 2\ H_3PO_4 \longrightarrow Ca_3(PO_4)_2 + 6\ H_2O$. **3.16** $2\ C_6H_{14}O_4 + 15\ O_2 \longrightarrow 12\ CO_2 + 14\ H_2O$. **3.17 (a)** 1.59 mol CO_2; **(b)** 305 mol H_2O; **(c)** 0.6060 mol CO_2. **3.18A** 21.4 g Mg. **3.18B** 26.4 g N_2 and 15.1 g O_2. **3.19** 774 mL H_2O. **3.20A** 4.77 g H_2S; 0.9 g FeS in excess. **3.20B** 1.40 g H_2. **3.21A** 26.8 g. **3.21B** 374 g. **3.22** 1.35 kg $(NH_4)_2HPO_4$. **3.23A (a)** 1.26 M; **(b)** 0.0242 M; **(c)** 0.0349 M. **3.23B (a)** 6.99×10^{-3} M; **(b)** 7.19 M; **(c)** 0.34 M. **3.24A (a)** 673 g KOH; **(b)** 0.0561 g KOH; **(c)** 4.91 g KOH. **3.24B** 23.2 mL. **3.25A** 23.5 M HCOOH. **3.25B** 70.4% $HClO_4$. **3.26A** 466 mL. **3.26B** 17.0 mL. **3.27A** 375 mL. **3.27B (a)** 11.9 g CO_2; **(b)** 0.662 M NaCl. **Review Questions: 3.** 6.02×10^{23} O_2 molecules; 1.20×10^{24} O atoms. **4.** 6.02×10^{23} Ca^{2+} ions; 1.20×10^{24} Cl^- ions. **8 (a)** HO; **(b)** CH_2; **(c)** C_5H_4; **(d)** $C_6H_{16}O$. **13. (a)** $2\ Mg(s) + O_2(g) \longrightarrow 2\ MgO(s)$; **(b)** $NH_4NO_3(s) \longrightarrow N_2O(g) + 2\ H_2O(l)$; **(c)** $C_7H_{16}(l) + 11\ O_2(g) \longrightarrow 7\ CO_2(g) + 8\ H_2O(l)$. **19.** The units are the same; 1000 mg per gram and 1000 mL per liter. **Problems: 21.** Molecular mass: (a) 157.0 u, (c) 98.00 u; formula mass: (b) 162.1 u, (d) 294.2 u, (e) 666.4 u, (f) 399.2 u. **23 (a)** 324.4 u; **(b)** 152.1 u. **25 (a)** 47.1 g; **(b)** 22.0 g; **(c)** 35.1 g. **27 (a)** 1.56 mol; **(b)** 0.112 mol; **(c)** 0.0677 mol; **(d)** 6.80×10^{-3} mol. **29 (a)** 2.82×10^{24}; **(b)** 4.55×10^{23}; **(c)** 1.055×10^{-22} g. **31.** Answer: (c). Reason: (a), 2 mol N atoms; (b), slightly more than 2 mol Na atoms; (c), slightly less than 3 mol atoms; (d), 2 mol Mg atoms. **33 (a)** 64.35% Ba, 13.16% Si, 22.49% O; **(b)** 58.54% C, 4.09% H, 11.38% N, 25.99% O; **(c)** 16.61% Mg, 1.38% H, 16.42% C, 65.60% O; **(d)** 4.71% Al, 41.85% Br, 50.28% O, 3.17% H. **35.** 48.44% O. **37.** 5.03% Be; 50.3 g Be from 1.00 kg beryl. **39.** $C_6H_4Cl_2$. **41.** $C_{18}H_{21}NO_3$. **43.** $C_6H_6O_2$. **45.** $LiClO_4 \cdot 3H_2O$. **47.** NH_4NO_2. It has a lower formula mass than other compounds with 2 mol N/mol compound, and the formula mass of NH_4Cl is more than half that of NH_4NO_2. **49.** 48.67% C, 8.22% H, 43.11% O, $C_3H_6O_2$. **51.** C_4H_4S. **53 (a)** MTBE $<$ ethanol $<$ methanol (one O atom per molecule and decreasing order of molecular masses); **(b)** yes; **(c)** 15% MTBE. **55 (a)** $Cl_2O_5 + H_2O \longrightarrow 2\ HClO_3$; **(b)** $V_2O_5 + 2\ H_2 \longrightarrow V_2O_3 + 2\ H_2O$; **(c)** $4\ Al + 3\ O_2 \longrightarrow 2\ Al_2O_3$; **(d)** $2\ C_4H_{10} + 13\ O_2 \longrightarrow 8\ CO_2 + 10\ H_2O$; **(e)** $Sn + 2\ NaOH \longrightarrow Na_2SnO_2 + H_2$; **(f)** $PCl_5 + 4\ H_2O \longrightarrow H_3PO_4 + 5\ HCl$; **(g)** $2\ CH_3OH + 3\ O_2 \longrightarrow 2\ CO_2 + 4\ H_2O$; **(h)** $3\ Zn(OH)_2 + 2\ H_3PO_4 \longrightarrow Zn_3(PO_4)_2 + 6\ H_2O$. **57 (a)** $2\ CO(g) + 2\ NO(g) \longrightarrow 2\ CO_2(g) + N_2(g)$; **(b)** $C_3H_8(g) + 3\ H_2O(g) \longrightarrow 3\ CO(g) + 7\ H_2$; **(c)** $Mg_3N_2(s) + 6\ H_2O(l) \longrightarrow 3\ Mg(OH)_2(s) + 2\ NH_3(g)$; **(d)** $Pb(s) + PbO_2(s) + 2\ H_2SO_4(aq) \longrightarrow 2\ PbSO_4\ (s) + 2\ H_2O(l)$. **59.** $3\ Fe_2O_3(s) + H_2(g) \longrightarrow 2\ Fe_3O_4(s) + H_2O(g)$. **61 (a)** 1.4×10^5 mol CO_2; **(b)** 5.5×10^5 mol O_2. **63.** $3\ PbO + 2\ NH_3 \longrightarrow 3\ Pb + N_2 + 3\ H_2O$ **(a)** 3.82 g NH_3; **(b)** 2.54 g N_2. **65.** $CaCN_2$. Both produce 2 mol NH_3 per mole of compound, but $CaCN_2$ has the smaller molar mass and greater number of moles per kilogram of compound. **67.** 8.97×10^3 g CO_2. **69.** 1.60 g CO_2. **71.** 35.0 mL. **73.** Limiting reactant: LiOH; 0.0750 mol Li_2CO_3 forms. **75.** Answer: (d). Incorrect: (a)—some Hg(l) must be consumed to produce HgO(s); (b)—cannot have both Hg(l) and $O_2(g)$ in excess; (c)—Hg(l) and $O_2(g)$ are not in stoichiometric proportions. **77 (a)** 528 g KI; **(b)** 74 g HI in excess. **79.** Theoretical yield, 0.727 g ZnS; percent yield, 83.4%. **81.** 51.3 g NH_4HCO_3. **83. (a)** 2.40 M HCl; **(b)** 0.0700 M Li_2CO_3; **(c)** 0.9079 M H_2SO_4; **(d)** 1.95 M $C_6H_{12}O_6$; **(e)** 0.9462 M $C_3H_8O_3$; **(f)** 1.83 M $CH_3CHOHCH_3$. **85 (a)** 0.0294 mol NaOH; **(b)** 7.66 g $C_6H_{12}O_6$; **(c)** 75.55 mL $N(CH_2CH_2OH)_3$; **(d)** 87.9 mL $CH_3CHOHCH_2CH_3$. **87 (a)** 403 mL; **(b)** 2.49×10^3 mL; **(c)** 104 mL. **89.** 14.9 M HNO_3. **91.** 0.0520 M Na_2CO_3. **93 (a)** 165.2 mL; **(b)** 16.20 mL. **95.** Answer: (c). Concentration must be between 0.100 M and 0.200 M, but closer to 0.200 M—a greater volume of 0.200 M NH_3 is used than of 0.100 M NH_3. **97.** Prepare 85.00 mL of 0.100 M $AgNO_3$, as follows: Transfer three 25.00-mL and one 10.00-mL samples of 0.04000 M $AgNO_3$ to a beaker and dissolve 0.8664 g $AgNO_3$ in the solution. **99.** 46.5

g $BaSO_4$. **101.** 2.2 M HCl. **103.** 1.91 g CO_2. **105.** 0.2 cm^2. **Additional Problems: 107 (a)** 2.19 g; **(b)** 4.76 g; **(c)** 9.40 g; **(d)** 3.63 g. **109.** 894 u. **111.** M is Fe; atomic mass, 55.85 u. **113.** The molecular formula is $C_4H_8O_2$. There are several possible structures; one is $CH_3CH_2CH_2COOH$. **115.** 89.8% Ag_2O. **117.** Theoretical yield: 24.0 g C_4H_9Br; 70.0% yield. **119.** 1.81 g Na. **121.** 0.931 M $CO(NH_2)_2$.

Chapter 4

Exercises: 4.1A $[Na^+] = 0.438$ M; $[Mg^{2+}] = 0.0512$ M; $[Cl^-] = 0.540$ M. **4.1B** $[C_6H_{12}O_6] = 0.111$ M; $[C_6H_5O_7^{3-}] = 0.0112$ M; $[K^+] = 0.0201$ M; $[Cl^-] = 0.0800$ M; $[Na^+] = 0.0935$ M. **4.2 (a)** $Ca(OH)_2(aq) + 2\ HCl(aq) \longrightarrow CaCl_2(aq) + 2\ H_2O(l)$; **(b)** $Ca^{2+}(aq) + 2\ OH^-(aq) + 2\ H^+(aq) + 2\ Cl^-(aq) \longrightarrow Ca^{2+}(aq) + 2\ Cl^-(aq) + 2\ H_2O(l)$; **(c)** $OH^-(aq) + H^+(aq) \longrightarrow H_2O(l)$. **4.3A** 74.53 mL. **4.3B** 110. mL. **4.4A** 0.05040 M H_2SO_4. **4.4B** 12.6 mL. **4.5** The bulb will be dimly lit with CH_3NH_2 (weak base) and brightly lit with HNO_3 (strong acid). When the solutions are mixed, the $CH_3NH_3^+NO_3^-$ (salt) formed maintains the brightly lit bulb. **4.6 (a)** $Mg^{2+}(aq) + 2\ OH^-(aq) \longrightarrow Mg(OH)_2(s)$; **(b)** $2\ Fe^{3+}(aq) + 3\ S^{2-}(aq) \longrightarrow Fe_2S_3(s)$; **(c)** No reaction. $CaCO_3(s)$ is insoluble in H_2O and in NaCl(aq). **4.7** H^+ from the acid combines with OH^- from $Fe(OH)_3$ to form $H_2O(l)$: $Fe(OH)_3(s) + 3\ H^+(aq) \longrightarrow Fe^{3+}(aq) + 3\ H_2O(l)$. **4.8A** 42.20% NaCl. **4.8B (a)** 17.4 g AgCl(s); **(b)** 14.9 g $Mg(OH)_2(s)$. **4.9 (a)** Al, +3; O, −2; **(b)** P, 0; **(c)** Na, +1; Mn, +7, O, −2; **(d)** Cl, +1; O, −2; **(e)** H, +1, As, +5, O, −2; **(f)** H, +1, Sb, +5; F, −1; **(g)** Cs, +1; O, −1/2; **(h)** C, −2; H, +1, F, −1; **(i)** C, +2; H, +1, Cl, −1; **(j)** C, 0; H, +1; O, −2. **4.10A** $2\ MnO_4^-(aq) + 5\ C_2O_4^{2-}(aq) + 16\ H^+(aq) \longrightarrow 2\ Mn^{2+}(aq) + 8\ H_2O(l) + 10\ CO_2(g)$. **4.10B** $3\ Cl_2(g) + 6\ OH^-(aq) \longrightarrow ClO_3^-(aq) + 5\ Cl^-(aq) + 3\ H_2O(l)$. **4.11** $Cr_2O_7^{2-}(aq)$ and $HNO_3(aq)$ are oxidizing agents and do not react with each other. HCl(aq) is a reducing agent; $Cl^-(aq)$ is oxidized, probably to $Cl_2(g)$. The $Cr_2O_7^{2-}$ is reduced to Cr^{3+}, accounting for the green color. **4.12** 22.04 mL. **Review Questions: 1.** Strong electrolyte: (b), (c), (e), (g); weak electrolyte: (d), (f); nonelectrolyte: (a), (h). **2.** Strong acid: (e); weak acid: (d); strong base: (b), (h); weak base: (f); salt: (a), (c), (g). **3.** Highest: (c); lowest: (b). **4.** (d). **5.** (a). NaCl(aq) is the only strong electrolyte. **6.** Highest $[H^+]$: (b). H_2SO_4 is strong in its first ionization and weak in the second, producing more H^+ ions than the other strong acid (a) and the weak acid (c). **9.** An acid-base reaction occurs in the first mixture; no reaction in the second—mixture of a strong base and a weak base. **10.** $Ca(OH)_2(s) + 2\ H^+(aq) \longrightarrow Ca^{2+}(aq) + 2\ H_2O(l)$. **14.** No reaction occurs in the first mixture; $Zn(OH)_2(s)$ precipitates from the second. **15.** (d). The other three compounds fall in the categories soluble or mostly soluble in Table 4.3. **16.** (c) $CO_3^{2-}(aq) + Mg^{2+}(aq) \longrightarrow MgCO_3(s)$. **21 (a)** 0; **(b)** +3; **(c)** −2; **(d)** +6; **(e)** −3; **(f)** +4; **(g)** +4; **(h)** +5; **(i)** +2.5; **(j)** −1. **22 (a)** −3; **(b)** +2; **(c)** +1; **(d)** −2; **(e)** +3. **25.** Reaction occurs in the first mixture because Zn is above H_2 in the activity series: $Zn(s) + 2\ CH_3COOH(aq) \longrightarrow Zn^{2+}(aq) + 2\ CH_3COO^-(aq) + H_2(g)$. **26.** Only the first equation is correctly balanced for electric charge. **Problems: 27 (a)** $[Li^+] = 0.0385$ M; **(b)** $[Cl^-] = 0.070$ M; **(c)** $[Al^{3+}] = 0.0224$ M; **(d)** $[Na^+] = 0.24$ M. **29.** $[Na^+] = 0.0844$ M; $[Cl^-] = 0.0554$ M; $[SO_4^{2-}] = 0.0145$ M. **31.** $[NO_3^-] = 6.99 \times 10^{-3}$ M. **33.** 1.91×10^4 mg Cl^-/L. **35.** 67.5 mL. **37.** (c). $[Cl^-] > 0.01$ M in this solution. $[Cl^-] \ll 0.01$ M in the other two solutions. **39 (a)** $HI(g) \xrightarrow{H_2O} H^+(aq) + I^-(aq)$; **(b)** $KOH(s) \xrightarrow{H_2O} K^+(aq) + OH^-(aq)$; **(c)** $HNO_2(aq) \rightleftharpoons H^+(aq) + NO_2^-(aq)$; **(d)** $H_2PO_4^-(aq) \rightleftharpoons H^+(aq) + HPO_4^{2-}(aq)$; **(e)** $CH_3NH_2 + H_2O(l) \rightleftharpoons CH_3NH_3^+(aq) + OH^-(aq)$; **(f)** $CH_3CH_2COOH(aq) \rightleftharpoons H^+(aq) + CH_3CH_2CH_2COO^-(aq)$. **41 (a)** $RbOH(aq) + HCl(aq) \longrightarrow RbCl(aq) + H_2O(l)$; **(b)** $Rb^+(aq) + OH^-(aq) + H^+(aq) + Cl^-(aq) \longrightarrow Rb^+(aq) + Cl^-(aq) + H_2O(l)$; **(c)** $OH^-(aq) + H^+(aq) \longrightarrow H_2O(l)$. **43.** $2\ CH_3COOH(aq) + CaCO_3(s) \longrightarrow Ca^{2+}(aq) + 2\ CH_3COO^-(aq) + H_2O(l) + CO_2(g)$. **45 (a)** 46.8 mL; **(b)** 11.9 mL; **(c)** 23.1 mL. **47.** 0.8051 M. **49 (a)** 4.61 M H_2SO_4; **(b)** 35.7% H_2SO_4. **51.** Tums®; it neutralizes 2 mol H^+ per mole of compound, compared to 1 mol H^+ for Alka-Seltzer®. The difference in their molar masses is minor. **53.** (d). All the CH_3COOH has been neutralized, and the excess $OH^-(aq)$ is one-tenth the amount of $CH_3COO^-(aq)$. **55 (a)** $2\ I^-(aq) + Pb^{2+}(aq) \longrightarrow PbI_2(s)$; **(b)** no reaction; **(c)** $Cr^{3+}(aq) + 3\ OH^-(aq) \longrightarrow Cr(OH)_3(s)$; **(d)** no reaction; **(e)** $OH^-(aq) + H^+(aq) \longrightarrow H_2O(l)$; **(f)** $HSO_4^-(aq) + OH^-(aq) \longrightarrow H_2O(l) + SO_4^{2-}(aq)$. **57 (a)** $MgO(s) + 2\ H^+(aq) \longrightarrow Mg^{2+}(aq) + H_2O(l)$; **(b)** $HCOOH(aq) + NH_3(aq) \longrightarrow NH_4^+(aq) + HCOO^-(aq)$; **(c)** no reaction; **(d)** $Cu^{2+}(aq) + CO_3^{2-}(aq) \longrightarrow CuCO_3(s)$; **(e)** no reaction. **59.** No. $MgSO_4$ dissolves in HCl(aq) because it is water soluble and $Mg(OH)_2$ as a result of an acid-base reaction. Use water as the solvent; $MgSO_4$ will dissolve and $Mg(OH)_2$ will not. **61.** $Cu^{2+}(aq) + CO_3^{2-}(aq) \longrightarrow CuCO_3(s)$. **63.** Add $BaCl_2(aq)$ to one sample. If a precipitate ($BaSO_4$) forms, the solution is $Na_2SO_4(aq)$. If there is no precipitate, add $Na_2SO_4(aq)$ to a second sample. Here, a precipitate ($BaSO_4$) indicates the solution is $Ba(NO_3)_2(aq)$, and no precipitate, that it is $NH_3(aq)$. **65 (a)** oxidation (increase in oxidation number); **(b)** neither (no changes in oxidation number); **(c)** neither (no changes in oxidation number). **67 (a)** $4\ HCl + O_2 \longrightarrow 2\ Cl_2 + 2\ H_2O$; **(b)** $2\ NO + 5\ H_2 \longrightarrow 2\ NH_3 + 2\ H_2O$; **(c)** $CH_4 + 4\ NO \longrightarrow 2\ N_2 + CO_2 + 2\ H_2O$. **69 (a)** not possible; **(b)** $3\ Ag + 4\ H^+ + NO_3^- \longrightarrow 3\ Ag^+ + 2\ H_2O + NO$; **(c)** $5\ H_2O_2 + 2\ MnO_4^- + 6\ H^+ \longrightarrow 2\ Mn^{2+} + 8\ H_2O + 5\ O_2$; **(d)** not possible; **(e)** $IO_4^- + 7\ I^- + 8\ H^+ \longrightarrow 4\ I_2 + 4\ H_2O$. **71 (a)** $3\ Zn + Cr_2O_7^{2-} + 14\ H^+ \longrightarrow 3\ Zn^{2+} + 2\ Cr^{3+} + 7\ H_2O$; **(b)** $SeO_3^{2-} + 4\ I^- + 6\ H^+ \longrightarrow Se + 2\ I_2 + 3\ H_2O$; **(c)** $2\ Mn^{2+} + 5\ S_2O_8^{2-} + 8\ H_2O \longrightarrow 2\ MnO_4^- + 10\ SO_4^{2-} + 16\ H^+$; **(d)** $5\ CaC_2O_4 + 2\ MnO_4^- + 16\ H^+ \longrightarrow 5\ Ca^{2+} + 2\ Mn^{2+} + 8\ H_2O + 10\ CO_2$; **(e)** $CrO_4^{2-} + AsH_3 + H_2O \longrightarrow Cr(OH)_3 + As + 2\ OH^-$; **(f)** $3\ S_2O_4^{2-} + 2\ CrO_4^{2-} + 2\ H_2O + 2\ OH^- \longrightarrow 2\ Cr(OH)_3 + 6\ SO_3^{2-}$; **(g)** $P_4 + 3\ H_2O + 3\ OH^- \longrightarrow 3\ H_2PO_2^- + PH_3$. **73 (a)** $5\ HOOCCOOH(aq) + 2\ MnO_4^-(aq) + 6\ H^+(aq) \longrightarrow 2\ Mn^{2+}(aq) + 8\ H_2O(l) + 10\ CO_2(g)$; **(b)** $3\ CH_3CHO(aq) + 2\ MnO_4^-(aq) + OH^-(aq) \longrightarrow 3\ CH_3COO^-(aq) + 2\ MnO_2(s) + 2\ H_2O(l)$; **(c)** $S_2O_3^{2-}(aq) + 4\ Cl_2(g) + 5\ H_2O(l) \longrightarrow 2\ HSO_4^-(aq) + 8\ Cl^-(aq) + 8\ H^+(aq)$. **75.** Oxidizing agent: (a) MnO_4^-; (b) MnO_4^-; (c) Cl_2; reducing agent: (a) HOOCCOOH;

(b) CH_3CHO; (c) $S_2O_3^{2-}$. **77 (a)** $Zn(s) + 2\ H^+(aq) \longrightarrow Zn^{2+}(aq) + H_2(g)$; **(b)** no reaction; **(c)** $Fe(s) + 2\ Ag^+(aq) \longrightarrow Fe^{2+}(aq) + 2\ Ag(s)$; **(d)** no reaction. **79 (a)** 12.39 mL; **(b)** 47.67 mL. **81.** $3\ Mn^{2+} + 2\ MnO_4^- + 4\ OH^- \longrightarrow 5\ MnO_2 + 2\ H_2O$; $[Mn^{2+}] = 0.1226$ M. **83.** $6\ Fe^{2+} + Cr_2O_7^{2-} + 14\ H^+ \longrightarrow 6\ Fe^{3+} + 2\ Cr^{3+} + 7\ H_2O$; 47.39% Fe. **Additional Problems: 85.** $[Cl^-] = 0.235$ M. **87.** $MgSO_4$. **89.** $H_2SO_4 + Na_2CO_3 \longrightarrow Na_2SO_4 + H_2O + CO_2$; 1.5×10^3 kg Na_2CO_3. **91.** $[OH^-] = 1.2$ M. **93.** 31.7% H_2SO_4. **95 (a)** $2\ CrI_3 + 27\ H_2O_2 + 10\ OH^- \longrightarrow 2\ CrO_4^{2-} + 6\ IO_4^- + 32\ H_2O$; **(b)** $As_2S_3 + 14\ H_2O_2 + 12\ OH^- \longrightarrow 2\ AsO_4^{3-} + 3\ SO_4^{2-} + 20\ H_2O$; **(c)** $I_2 + 5\ H_5IO_6 \longrightarrow 7\ IO_3^- + 9\ H_2O + 7\ H^+$; **(d)** $2\ XeF_6 + 16\ OH^- \longrightarrow Xe + XeO_6^{4-} + 12\ F^- + O_2 + 8\ H_2O$; **(e)** $24\ As_2S_3 + 80\ H^+ + 80\ NO_3^- + 32\ H_2O \longrightarrow 48\ H_3AsO_4 + 9\ S_8 + 80\ NO$. **97. (a)** $13\ C_6H_{12}O_6 + 6\ H_2O \longrightarrow 36\ C_2H_4O_2 + 6\ CO_2 + 12\ H_2$; **(b)** $2\ CO_2 + 4\ H_2 \longrightarrow C_2H_4O_2 + 2\ H_2O$; **(c)** $C_2H_4O_2 + 2\ O_2 \longrightarrow 2\ CO_2 + 2\ H_2O$. **99.** $CH_3OH + 6\ H^+ + 6\ ClO_3^- \longrightarrow 6\ ClO_2 + CO_2 + 5\ H_2O$; 1.25×10^4 mL CH_3OH. **101.** 0.02401 M $KMnO_4$.

Chapter 5

Exercises: 5.1A (a) 720. mmHg; **(b)** 737 Torr; **(c)** 760.7 Torr; **(d)** 1.01 atm. **5.1B** 2.00×10^2 kPa; 1.97 atm. **5.2A** 6.50 m. **5.2B** 2.90 m due to the water; 3.90 atm total. **5.3** 735 mmHg < 745 mmHg < 750 mmHg < 101 kPa < 103 kPa; 103 kPa = 773 mmHg, but it is uncertain how this compares to the pressure "above 762 mmHg" in Example 5.3. **5.4A** 503 Torr. **5.4B** 17.8 mL. **5.5** 400 mmHg. The volume increases three-fold, and the pressure decreases by this same amount. **5.6A** 234 L. **5.6B** 338 K (65 °C). **5.7** 7.5 L. The Kelvin temperature increases about three-fold. **5.8A** 98.4 g C_3H_8. **5.8B** 3.75×10^3 L CO_2. **5.9A** about 670 °C. **5.9B** $P_2 = (n_2/n_1) \times P_1$. Pressure is directly proportional to amount of gas. With more molecules in the same volume, the frequency of molecular collisions, and hence the pressure, increases. **5.10** 1.58 mol N_2. **5.11A** −139 °C. **5.11B** 21.6 g N_2. **5.12** 74.3 g/mol. **5.13A** 126 u. **5.13B** $C_4H_6O_2$. **5.14** 1.25 g/L. **5.15A** 94 °C. **5.15B** 92 °C. **5.16A** 2.78 L $O_2(g)$. **5.16B** 207 L $O_2(g)$. **5.17A** 4.16×10^4 L CO_2. **5.17B** 746 L $C_5H_{10}(l)$. **5.18A** Partial pressures: O_2, 0.209 atm; Ar, 0.00932 atm; CO_2, 0.0005 atm; total pressure: 0.999 atm. **5.18B** 12.90 atm. **5.19A** Partial pressures: N_2 0.741 atm; O_2, 0.150 atm; H_2O , 0.060 atm; Ar, 0.009 atm; CO_2, 0.040 atm. **5.19B** Mole fractions: CH_4, 0.636; C_2H_6, 0.253; C_3H_8, 0.054; C_4H_{10}, 0.056. **5.20** If only H_2 were added, its partial pressure would be 2.50 atm instead of 2.00 atm. A gas other than H_2 is needed at a partial pressure of 0.50 atm; it can be additional He, but does not have to be. **5.21A** Total mass of wet gas: 0.0235 g; mass of H_2, 0.0197 g. **5.21B** 0.502 g $KClO_3$. **5.22** 5000 K. $T > 273 \times$ (molar mass O_2/molar mass H_2). **5.23A** N_2 effuses faster by the factor $(39.95/28.01)^{1/2} = 1.194$. **5.23B** NO effuses faster by the factor $(46.07/30.01)^{1/2} = 1.239$, or 23.9% faster. **5.24A** 60 g/mol. **5.24B** 166 s. **Review Questions: 14 (a)** decrease; **(b)** decrease; **(c)** increase. **15 (a)** increase; **(b)** increase; **(c)** increase. **16 (a)** temperature decreases; **(b)** pressure decreases. **17.** The pressures are the same. **22 (a)** greater density in A because of higher pressure; **(b)** densities the same because number of molecules per unit volume is the same; **(c)** greater density in B because more molecules are present. **Problems: 31 (a)** 749 mmHg; **(b)** 1.12 atm; **(c)** 85.6 kPa; **(d)** 801 mmHg. **33.** 121 atm. **35 (a)** 3.31 m; **(b)** 2.14 m. **37.** 758 Torr. **39.** Height requirement depends on pressure to be measured, that is, 760 mm for pressures up to 1 atm, 1520 mm for pressures up to 2 atm, and so on. **41 (a)** 1.21 L; **(b)** 338 mL; **(c)** 973 mL. **43.** 46.4 m^3. **45.** 748 Torr. **47 (a)** 9.12×10^3 L; **(b)** 19.0 h. **49.** 117 mL. **51.** 160 °C. **53.** 150 K. **55.** 4.10 kg Ne. **57.** 575 mL. **59.** (d). Response (c) is less than 2 mol; (b) is slightly more than 2 mol; (a) is 2.5 mol; (d) is about 3 mol. **61.** 2.49×10^3 Torr. **63.** 24.5 L. **65.** 0.489 m^3. **67 (a)** 22.3 L; **(b)** 27.9 atm; **(c)** 3.10 mg H_2; **(d)** 1.68×10^3kPa. **69.** 4.93 L. **71.** 7.26×10^8 mol. **73.** 4.1×10^{-17} atm. **75.** 153 u. **77.** C_7H_8. **79 (a)** 1.25 g CO/L; **(b)** 1.03 g Ar/L. **81.** 365 K. **83.** S_8. **85.** (c). Density of the H_2 (a) is lowest. He (b) has twice the molar mass of H_2, and STP differs only slightly from T and P for H_2; density of He is about twice that of H_2. CH_4 (c) is at a lower T and higher P than He, and has four times the molar mass; it is more than four times as dense as the He. C_2H_6 (d) has almost twice the molar mass of CH_4, but is at half the P and at T only 60 K higher than CH_4; it is not as dense as the CH_4. **87.** 1.15 L $SO_3(g)$. **89.** 122 L. **91.** 28.2 mg Mg. **93.** (c). The two gases are not uniformly distributed in (a); in (b) equal numbers of molecules of the two gases are shown, but the number of H_2 molecules must be twice that of He. **95.** Partial pressures: N_2, 585 mmHg; O_2, 153 mmHg; CO_2, 24 mmHg. **97 (a)** mole fractions: helium, 0.205; oxygen, 0.795; **(b)** partial pressures: helium, 2.40 atm; oxygen, 9.30 atm; **(c)** total pressure: 11.70 atm. **99.** Oxygen: partial pressure, 710 Torr; mole fraction, 0.957. **101.** 0.154 g O_2; 0.145 g $C_6H_{12}O_6$. **103.** 233 L O_2. **105.** (b). In (a) no He effuses. In (c) there are 7 H_2 molecules for every He atom, but H_2 does not effuse seven times faster than He. In (d) there are equal numbers of H_2 molecules and He atoms, but the two gases do not effuse at equal rates. H_2 effuses faster than He by the factor $\sqrt{4.003/2.016} = 1.41$—about the ratio of H_2 to He shown in (b). **107.** The partial pressures depend on the number of moles of each gas; they can easily be different. However, there can be only one e_k of the gas molecules in the mixture. Because temperature depends on e_k, the two gases must be at the same temperature. **109 (a)** 81 u; **(b)** 26 u. **Additional Problems: 111 (a)** equilateral hyperbola similar to Figure 5.6; V inversely proportional to P, that is, $V = a/P$, where $a = nRT$; **(b)** straight line through the origin; n directly proportional to P, that is $n = b \times P$, where $b = V/RT$; **(c)** straight line through the origin; that is, $T = c \times P$, where $c = V/nR$; **(d)** equilateral hyperbola similar to Figure 5.6; n inversely proportional to T, that is, $n = d/T$, where $d = PV/R$. **113.** The amount of each gas is within 1% of 0.00400 mol, and all gases occupy 100.0 mL at the same T and P. This is consistent with Avogadro's hypothesis: "equal volumes-equal numbers of molecules." **115.** $C_2Cl_2F_4$. **117 (a)** C_6H_{14};

```
       CH3                          CH3
 H      |   H   H   H         H   H  |   H   H
 |      |   |   |   |         |   |  |   |   |
H—C—C—C—C—C—H        H—C—C—C—C—C—H
 |   |   |   |   |            |   |   |   |   |
 H   H   H   H   H            H   H   H   H   H
```

(b)

119. Air pressure must exceed P_{bar} by 1.1×10^2 psi to lift the 73-ton object. **121.** 4.1×10^4 L CO_2. **123.** $\bar{u} = 6.52 \times 10^3$ m/s; $u_{rms} = 7.15 \times 10^3$ m/s. **125 (a)** 9.78 atm; **(b)** 9.38 atm; **(c)** Intermolecular forces of attraction cause the gas to exert less pressure than expected for an ideal gas. Van der Waals calculation (b) takes this into account. **127.** 1.2 atm. **129.** Somewhat higher than 30 km, but not as high as 40 km.

Chapter 6

Exercises: 6.1A $\Delta U = -478$ J. **6.1B** The system absorbs 122.6 J of heat. **6.2 (a)** gas does work ($w < 0$); **(b)** internal energy decreases ($\Delta U < 0$); **(c)** temperature decreases ($\Delta T < 0$). **6.3A** $\Delta H = 142.7$ kJ. **6.3B** $\Delta H = -148.9$ kJ. **6.4** $N_2O_3(g) \longrightarrow NO(g) + NO_2(g)$ $\Delta H = 40.5$ kJ. **6.5A** $\Delta H = -1.17 \times 10^3$ kJ. **6.5B** 2.80×10^4 L $CH_4(g)$. **6.6A** 11 J/°C. **6.6B** $q = -153$ kJ. **6.7A** $q = 6.67 \times 10^4$ cal (66.7 kcal). **6.7B** 4.39×10^5 kg H_2O. **6.8A** 95.1 °C. **6.8B** 1.57×10^3 g Cu. **6.9A** 0.13 J g^{-1} °C^{-1}. **6.9B** 1.64 J g^{-1} °C^{-1}. **6.10** 60 °C. **6.11** $q_{rxn} = -2.81$ kJ. **6.12A** $\Delta H = -56.2$ kJ. **6.12B** 29.9 °C. **6.13A** $\Delta H = -395$ kJ. **6.13B** 5.99 kJ/°C. **6.14A** $\Delta H = -1215$ kJ. **6.14B** $\Delta H = -137.0$ kJ. **6.15A** $\Delta H° = -44.2$ kJ. **6.15B** $2\ C_4H_{10}(g) + 13\ O_2(g) \longrightarrow 8\ CO_2(g) + 10\ H_2O(l)$ $\Delta H° = -5755$ kJ. **6.16A** $\Delta H_f° = -52.3$ kJ/mol $C_2Cl_4(l)$. **6.16B** $C_4H_4S(l) + 6\ O_2(g) \longrightarrow 4\ CO_2(g) + SO_2(g) + 2\ H_2O(l)$; $\Delta H_f° = 81$ kJ/mol $C_4H_4S(l)$. **6.17** $\Delta H_f°$ of CH_3CH_2OH and CH_3OH differ only by 39 kJ/mol. In contrast, the enthalpies of the combustion products are much lower for CH_3CH_2OH than for CH_3OH [by one mole each of $CO_2(g)$ and $H_2O(l)$]. On a mole basis, combustion of CH_3CH_2OH gives off the greater amount of heat. **6.18A** $\Delta H_f°$ [$BaSO_4$] $= -1473$ kJ/mol. **6.18B** $\Delta H° = 2.4$ kJ [per mole of $Mg(OH)_2(s)$ formed]. **Review Questions: 19.** $2\ Fe(s) + 3/2\ O_2(g) \longrightarrow Fe_2O_3(s)$. **21.** The values of $\Delta H_f°$ of compounds are relative to $\Delta H_f° = 0$ for the reference form of each element. $\Delta H_f°$ values of ions must be related to one ion assigned a value of $\Delta H_f° = 0$; $H^+(aq)$ has been chosen for this purpose. **Problems: 25.** $\Delta U = 130$ J. **27.** $w = 2.4 \times 10^2$ J. **29.** A system can do work and also absorb heat. If the system absorbs exactly as much heat as the work it does, $\Delta U = 0$. **31.** Endothermic. Because the reaction is at constant pressure, $q = q_P = \Delta H = +241$ kJ. **33.** $CaCO_3(s) \longrightarrow CaO(s) + CO_2(g)$ $\Delta H = 178$ kJ. **35. (a)** $\Delta H = -15$ kJ; **(b)** $\Delta H = 23$ kJ. **37.** $1/4\ P_4(s) + 3/2\ Cl_2(g) \longrightarrow PCl_3(g)$ $\Delta H = -287$ kJ. **39.** $q = -581$ kJ. **41.** $q = -276$ kJ. **43.** 1.59×10^4 L $C_2H_6(g)$. **45.** 32.2 g CaO. **47.** Per mole, C_4H_{10} evolves 1.85 times as much heat as $C_2H_6(g)$ (-5.76×10^3 kJ/-3.12×10^3 kJ). However, the molar mass of C_4H_{10} (58 g/mol) is 1.93 times the molar mass of C_2H_6 (30 g/mol), so C_2H_6 evolves more heat on a per gram basis. **49.** $C = 4.1$ J/°C. **51 (a)** $q = 6.36$ kJ; **(b)** $q = 16.2$ kJ. **53.** 42.0 °C. **55.** 0.373 J g^{-1}°C^{-1}. **57.** 131 °C. **59.** 2.25 g copper. The mass of lead is twice that of copper, but its specific heat is only about one-third that of copper. The mass and specific heat of silver are both less than for copper. **61.** $\Delta H = -53.7$ kJ. **63.** $\Delta H = 25.8$ kJ. **65.** $q = -27.4$ kJ/g coal. **67.** 11.6 kJ/°C. **69.** -5.64×10^3 kJ/mol $C_{10}H_{14}O$. **71.** $\Delta H = -24.0$ kJ. **73.** $\Delta H = -1370.9$ kJ. **75.** $\Delta H = -1170.$ kJ. **77. (a)** $\Delta H° = -176.0$ kJ; **(b)** $\Delta H° = -146.3$ kJ; **(c)** $\Delta H° = -154.5$ kJ; **(d)** $\Delta H° = -11$ kJ. **79.** $\Delta H_f° = -206$ kJ/mol ZnS(s). **81.** $\Delta H_f° = -628$ kJ/mol $C_4H_{10}O_3(l)$. **83.** $\Delta H° = 30.6$ kJ. **85. (a)** The graph becomes essentially linear for C_6H_{14}, C_7H_{16}, and C_8H_{18}. **(b)** By extrapolation of the straight line, the enthalpies of combustion of $C_{10}H_{22}$ and $C_{12}H_{26}$ appear to be 6700 kJ/mol and 7900 kJ/mol, respectively. **87.** 14%. **Additional Problems: 89.** (c). The e_k of a single molecule (a) is orders of magnitude smaller than the others. The e_k of the BB shot (b) is $1/2\ mv^2 = 5.00$ J. The energy required to raise the temperature of 10 g H_2O by 1 °C (c) is 42 J. **91.** The water should warm slightly as potential energy is converted to kinetic energy and then to thermal energy. **93.** 29.5 °C. **95 (a)** constant $= 25.2$ J mol^{-1} °C^{-1}. **(b)** Plot specific heat versus 1/molar mass for a straight-line graph. **(c)** 59.9 u; **(d)** 83 mL H_2O. **97 (a)** 1.80×10^3 g CH_4; **(b)** 3.78×10^3kJ. **99.** The liberated heat would raise the temperature of the products to about 5000 °C if no melting occurred. Thus, when $T = 1530$ °C the iron melts. **101.** ≈80 °C. **103.** 33.54 °C. **105 (a)** Because ΔH_f is a state function, ΔH_f values of $H_2SO_4(aq)$ depend on concentration. **(b)** T increases because heat is evolved on dilution of $H_2SO_4(aq)$. **107.** The species on each line, starting at the highest level, are $3\ H_2(g) + 3/2\ O_2(g) + 2\ NO(g)$; $N_2(g) + 3\ H_2(g) + 5/2\ O_2(g)$; $2\ NH_3(g) + 5/2\ O_2(g)$; $2\ NO(g) + 3\ H_2O(l)$. The $\Delta H°$ values of the downward pointing arrows, from the highest level, are -180.5 kJ; -92.22 kJ; -584.7 kJ. The $\Delta H°$ of the upward pointing arrow is $+857.4$ kJ.

Chapter 7

Exercises: 7.1A 2.80×10^{11} Hz. **7.1B** 3.07×10^3 nm. **7.2** Microwave radiation has a longer wavelength than the visible light. **7.3A** 1.91×10^{-23} J/photon. **7.3B** 8.46×10^{-19} J/photon. **7.4A** 299 kJ/mol photons. **7.4B** infrared (≈1200 nm). **7.5** $E_6 = -6.053 \times 10^{-20}$J. **7.6** $\Delta E_{level} = 4.086 \times 10^{-19}$ J. **7.7A** 3.083×10^{15} s^{-1}. **7.7B** 434.1 nm (visible region). **7.8** (a). ΔE_{level} between $n = 1$ and $n = 2$ is $3/4\ B$. The only transitions that would require more energy are from $n = 1$ to $n \geqq 3$. **7.9** 0.105 nm. **7.10** Possible: (b), (c); not possible: (a), (d); m_l cannot be less than $-l$ nor greater than $+l$. **7.11 (a)** three p orbitals in a p subshell; **(b)** 16 orbitals in the $n = 4$ shell (*one* 4*s*, *three* 4*p*, *five* 4*d*, *seven* 4*f*); **(c)** f subshells consist of seven orbitals, that is, the subshells 4*f*, 5*f*,.... **Review Questions: 7.** The order of the elements by atomic number is the same as by atomic mass (a), except in a few instances. For example, Co ($Z = 27$) has atomic mass 58.9332 u, and Ni ($Z = 28$), 58.698 u. There are no exceptions to (b). **14.** A quantum of energy is very small, but of great significance at the microscopic level, where overall energies are small. It is not of significance at the macroscopic level, where energies are much larger. **19.** E_n must approach zero when n^2 is very large; n^2 must appear in the denominator. The negative sign signifies a negative energy and an electrostatic attraction of a bound electron to the nucleus. **22.** In the Schrödinger H atom, 52.9 pm is the most probable distance from the nucleus to an electron in the 1*s* orbital. The Bohr atom requires 52.9 pm as a unique, fixed distance for the orbit $n = 1$. **26 (a)** 3*d*; **(b)** 2*s*; **(c)** 4*p*; **(d)** 4*f*. **27 (a)** *three* subshells: 3*s*, 3*p*, 3*d*; **(b)** *two* subshells: 2*s*, 2*p*; **(c)** *four* subshells: 4*s*, 4*p*, 4*d*, 4*f*. **Problems: 31.** Charge: $+1.602 \times 10^{-19}$ C; mass: $1.672 \times$

10^{-27} kg. **33.** 1.6×10^{-19} C; all droplets carry a charge that is an integral multiple of this number. **35.** Values are atomic units of mass and charge, followed by SI units: **(a)** $-80:1 = -8.29 \times 10^{-7}$ kg/C; **(b)** $-9:1 = -9.33 \times 10^{-8}$ kg/C; **(c)** $+40:1 = 4.15 \times 10^{-7}$ kg/C. Values are approximate because mass numbers were used instead of actual atomic masses. **37.** ≈ 72.5 u. **39.** 1.5×10^4 s (4.2 h). **41.** 302 nm. **43.** 4.61×10^{14} s^{-1} (orange). **45.** 700 nm; 4.29×10^{14} s^{-1}; red. **47.** 4.92×10^{-19} J; this is greater than that of 655-nm wavelength light ($E = 3.03 \times 10^{-19}$ J). **49.** 2.55×10^{-19} J/photon; 154 kJ/mol photons. **51.** 276 nm. **53.** 627 nm (orange). **55 (a)** 2.092×10^{-18} J; **(b)** 4.90×10^{-20} J. **57 (a)** 4.567×10^{14} s^{-1}; **(b)** 3.083×10^{15} s^{-1}. **59.** No. $E = -2.179 \times 10^{-21}$ J, is *exactly* $-0.001\,B$, that is, $-B/n^2 = -B/1000$. This would require $n = \sqrt{1000} = 31.6$—a nonintegral number. **61.** 1312 kJ/mol H. **63.** 1.56×10^{-4} nm. **65.** 8.62×10^3 m/s. **67 (a)** $n = 2$; **(b)** $n = 4$. **69.** $l = 0, 1, 2, 3$, or 4, if $n = 5$; $m_l = -3, -2, -1, 0, 1, 2$, or 3, if $l = 3$. **71.** Permissible: (a), (c); not permissible: (b), because l cannot equal n, and (d), because n cannot be zero. **73 (a)** $n = 3, l = 0, m_l = 0$; **(b)** $n = 5, l = 3, m_l = -3, -2, -1, 0, 1, 2$, or 3; **(c)** $3p$. **75 (a)** n is $\geqq 3$, $m_s = \pm\frac{1}{2}$; possible orbitals: $3d, 4d, 5d, \ldots$; **(b)** $l = 1$; $2p$ orbital; **(c)** $l = 3$ or 2; $m_s = \pm\frac{1}{2}$; possible orbitals: $4f$ or $4d$; **(d)** $n = 1, 2, 3, \ldots$; $m_l = 0$; possible orbitals: $1s, 2s, 3s, \ldots$. **Additional Problems: 77.** 200.6 u. This value is only approximate because only mass numbers (not actual atomic masses) are given. **79.** Period and frequency are inversely related: $1/60\ s^{-1} = 0.017$ s. **81.** In the microscopic world "quantum jumps" are the only incremental energy changes that can occur. There are no more gradual changes. **83.** (c). Between $n = 1$ and $n = 3$, $\Delta E_{level} = (3/4)B + B(1/4 - 1/9) = B - (B/9) = (32/27)(27/36)B = (32/27)(3/4)B$. **85.** maximum: 121.6 nm; minimum: 91.16 nm. **87.** $n = 4 \longrightarrow n = 2$. **89.** 8.716×10^{-18} J. **91.** $E = h\nu = hc/\lambda = mc^2$; $\lambda = h/mc$ (substitute particle velocity v for the speed of light c). **93.** 7.66×10^3 m/s. **95.** 7×10^2 photons/s. **97.** Probability based on a single volume element is greatest at the nucleus, but combined probability for all equivalent volume elements is greatest in spherical shell of 52.9-nm radius.

Chapter 8

Exercises: 8.1 (a) P: $1s^2\,2s^2\,2p^6\,3s^2\,3p^3$; [Ne]$3s^2\,3p^3$;

[Ne] [↑↓] [↑ | ↑ | ↑]
$3s$ $3p$

8.1 (b) Cl: $1s^2 2s^2 2p^6 3s^2 3p^5$; [Ne]$3s^2 3p^5$;

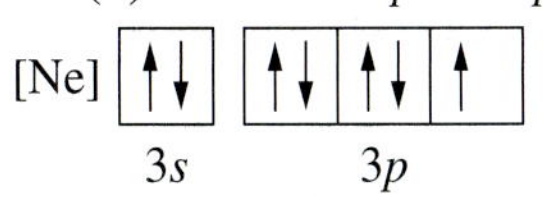

8.2A (a) Mo: $1s^2 2s^2 2p^6 3s^2 3p^6 3d^{10} 4s^2 4p^6 4d^5 5s^1$; [Kr]$4d^5 5s^1$; **(b)** Bi: $1s^2 2s^2 2p^6 3s^2 3p^6 3d^{10} 4s^2 4p^6 4d^{10} 4f^{14} 5s^2 5p^6 5d^{10} 6s^2 6p^3$; [Xe]$4f^{14} 5d^{10} 6s^2 6p^3$. **8.2B (a)** Sn: [Kr]$4d^{10} 5s^2 5p^2$;

(b) Zr: [Kr] [↑ | ↑ | | |] [↑↓]
$4d$ $5s$

8.3A Se^{2-}, [Ar]$3d^{10} 4s^2 4p^6$; Pb^{2+}: [Xe]$4f^{14} 5d^{10} 6s^2$. **8.3B** I^-: [Kr]$4d^{10} 5s^2 5p^6$; Cr^{3+}: [Ar]$3d^3$. **8.4** The paramagnetic species and odd number of electrons are (a), 1; (d), 3; (f) 2; (g) 1. **8.5 (a)** F < N < Be; **(b)** Be < Ca < Ba; **(c)** F < Cl < S; **(d)** Mg < Ca < K. **8.6A** $Y^{3+} < Sr^{2+} < Rb^+ < Br^- < Se^{2-}$. **8.6B** $Cr^{3+} < Cr^{2+} < Ca^{2+} < K^+ < Cs^+ < Cl^-$. **8.7 (a)** Be < N < F; **(b)** Ba < Ca < Be; **(c)** S < P < F; **(d)** K < Ca < Mg. **8.8** +400 kJ/mol; Se^{2-} is larger than S^{2-}, for which $EA_2 \approx 450$ kJ/mol. **8.9** More nonmetallic: (a) O; (b) S; (c) F. **8.10 (a)** $Z = 52$ (Te); **(b)** $Z = 21$ (Sc); **(c)** lowest I_1, $Z = 39$ (Y); highest I_1, $Z = 48$ (Cd); **(d)** N. **Review Questions: 3.** identical energies for H atom: (a) $2s$ and $2p$, (b) $3s$, $3p$, and $3d$, (c) $3p_x, 3p_y, 3p_z$; identical energies for multielectron atom: (c) $3p_x, 3p_y$, and $3p_z$. **6 (a)** F; **(b)** Cu; **(c)** N; **(d)** Na; **(e)** Fe. **10 (a)** ns, **(b)** np, **(c)** $(n-1)d$, **(d)** $(n-2)f$, where n is the valence shell. **12 (a)** 4; **(b)** 8; **(c)** 7; **(d)** 3; **(e)** 2. **14.** Discovery of $Z = 113$—the element after $Z = 112$—is likely. An element between Mg ($Z = 12$) and Al ($Z = 13$) appears impossible; Z is an *integral* number. **15.** 4A: ns^2np^2; 6A: ns^2np^4. **18.** Atoms with a half-filled orbital(s) are paramagnetic; this includes *all* atoms with odd Z and *some* with even Z, such as C and O. **Problems: 29.** Possible: (c), (d), (f); not possible: (a)—one orbital has two electrons with same m_s; (b)—one orbital has three electrons; (e) the three unpaired electrons do not have parallel spins. **31. (a)** $2p$ subshell fills before $3s$; **(b)** $2p$ subshell fills completely before $3s$; **(c)** there is no $2d$ subshell. **33 (a)** Maximum of two electrons in the $2s$ subshell. Also, $2p$ fills before $3s$. **(b)** A $2p$ orbital cannot accommodate three electrons, as required by $2p^7$. **(c)** There is no $2d$ subshell. **35 (a)** Al, $1s^2 2s^2 2p^6 3s^2 3p^1$; **(b)** Cl, $1s^2 2s^2 2p^6 3s^2 3p^5$; **(c)** Na, $1s^2 2s^2 2p^6 3s^1$; **(d)** B, $1s^2 2s^2 2p^1$; **(e)** He, $1s^2$; **(f)** O, $1s^2 2s^2 2p^4$; **(g)** C, $1s^2 2s^2 2p^2$; **(h)** Li, $1s^2 2s^1$; **(i)** Si, $1s^2 2s^2 2p^6 3s^2 3p^2$. **37 (a)** Ba, [Xe]$6s^2$; **(b)** Rb, [Kr]$5s^1$; **(c)** As, [Ar]$3d^{10} 4s^2 4p^3$; **(d)** F, [He]$2s^2 2p^5$; **(e)** Se, [Ar]$3d^{10} 4s^2 4p^4$; **(f)** Sn, [Kr]$4d^{10} 5s^2 5p^2$.

39 C, [↑↓] [↑↓] [↑ | ↑ |] ;
$1s$ $2s$ $2p$
(a)

O, [↑↓] [↑↓] [↑↓ | ↑ | ↑] ;
$1s$ $2s$ $2p$
(b)

K, [↑↓] [↑↓] [↑↓ | ↑↓ | ↑↓] [↑↓] [↑↓ | ↑↓ | ↑↓] [↑] ;
$1s$ $2s$ $2p$ $3s$ $3p$ $4s$
(c)

Al, [↑↓] [↑↓] [↑↓ | ↑↓ | ↑↓] [↑↓] [↑ | |] ;
$1s$ $2s$ $2p$ $3s$ $3p$
(d)

S, [↑↓] [↑↓] [↑↓ | ↑↓ | ↑↓] [↑↓] [↑↓ | ↑ | ↑] ;
$1s$ $2s$ $2p$ $3s$ $3p$
(e)

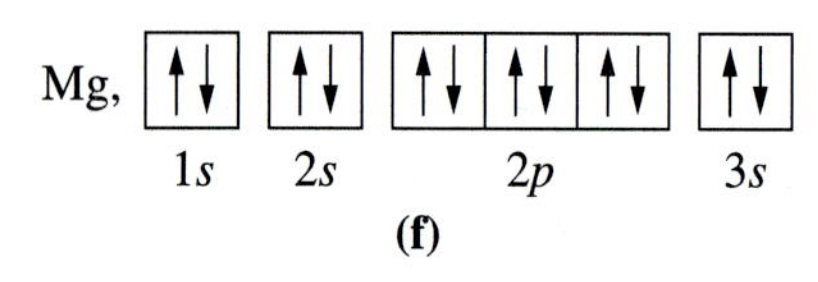

(f)

41. Hf is the second element in the *d* block in the sixth period:

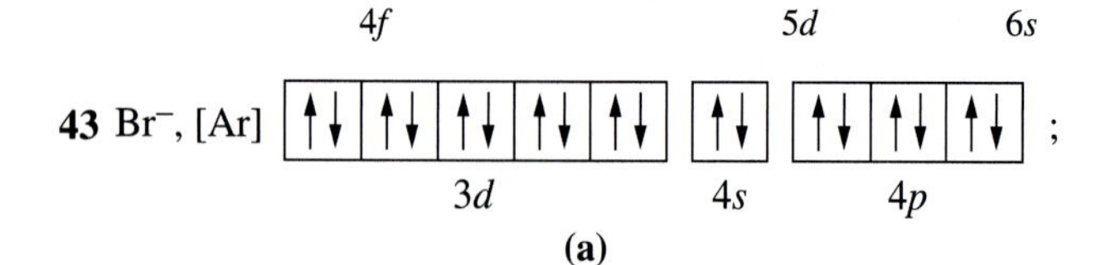

43 Br^-, [Ar] | ↑↓ | ↑↓ | ↑↓ | ↑↓ | ↑↓ | (3*d*) | ↑↓ | (4*s*) | ↑↓ | ↑↓ | ↑↓ | (4*p*);

(a)

Ni^{2+}, [Ar] | ↑↓ | ↑↓ | ↑↓ | ↑ | ↑ | (3*d*);

(b)

Sb^{3+}, [Kr] | ↑↓ | ↑↓ | ↑↓ | ↑↓ | ↑↓ | (4*d*) | ↑↓ | (5*s*);

(c)

Te^{2-}, [Kr] | ↑↓ | ↑↓ | ↑↓ | ↑↓ | ↑↓ | (4*d*) | ↑↓ | (5*s*) | ↑↓ | ↑↓ | ↑↓ | (5*p*).

(d)

45 (a) Period 2, Group 8A; **(b)** Period 3, Group 4A; **(c)** Period 3, Group 1A; **(d)** Period 2, Group 2A; **(e)** Period 2, Group 5A; **(f)** Period 3, Group 3A. **47 (a)** 5; **(b)** 32 ($4s^2 4p^6 4d^{10} 4f^{14}$); **(c)** 5; **(d)** 2; **(e)** 10. **49.** diamagnetic: (b), (d); paramagnetic: (a), (c), (e). **51 (a)** S; **(b)** S^{2-}; **(c)** Mg; **(d)** F^-. **53 (a)** Al < Mg < Na (inverse of their order in the third period); **(b)** Mg < Ca < Sr (their descending order in Group 2A). **55.** The largest atoms are in the lower left corner of the periodic table (Cs, Fr, Ra). **57 (a)** $r_{Ca} > r_{Cl}$, because $r_{Ca} > r_{Mg}$, and $r_{Mg} > r_{Cl}$. **(b)** $r_{K^+} < r_{Cl^-}$, but r_{F^-} is also smaller than r_{Cl^-}; a prediction is not possible. **59 (a)** Ba < Ca < Mg (I_1 increases from bottom to top of group); **(b)** Al < P < Cl (I_1 increases from left to right in period); **(c)** Na < Fe < Cl < F < Ne (I_1 increases from left to right in third period, bottom to top in Group 7A, and left to right in second period). **61.** Successive ionization energies increase because of increasing positive charge on the ion from which the electron is ionized. $I_4 \gg I_3$ because the fourth electron must be removed from the filled 2*p* subshell of Al^{3+}. **63.** Group 7A atoms are the smallest atoms in their periods and gain an electron most easily (most negative electron affinities). **65.** The electron gained by Si goes into an empty 3*p* orbital, resulting in a half-filled 3*p* subshell. This is energetically more favorable than for electrons to pair up in a 3*p* orbital of the half-filled 3*p* subshell of P. **67.** The graph has peaks for the Group 1A elements, resembling Figure 8.9. **69.** Metallic character is related to I_1, atomic radii, and location in the periodic table (and designation as a metal, nonmetal or metalloid). Based on these factors, the probable order is P < Ge < Bi < Al < Ca < K < Rb. **71 (a)** is Ge; **(b)** S; **(c)** Tl; **(d)** Ne; **(e)** Se. Increasing I_1: Tl < Ge < Se < S < Ne. **73 (a)** $Cl_2(g) + 2\,Br^-(aq) \longrightarrow 2\,Cl^-(aq) + Br_2(l)$; **(b)** $I_2(s) + 2\,F^-(aq) \longrightarrow$ no reaction; **(c)** $Br_2(l) + 2\,I^-(aq) \longrightarrow 2\,Br^-(aq) + I_2(s)$. **75 (a)** $N_2O_5(s) + 2\,H_2O(l) \longrightarrow 2\,HNO_3(aq)$; **(b)** $MgO(s) + 2\,CH_3COOH(aq) \longrightarrow Mg(CH_3COO)_2(aq) + H_2O(l)$; **(c)** $Li_2O(s) + H_2O(l) \longrightarrow 2\,LiOH(aq)$. **Additional Problems: 77 (a)** Period 4, Group 8A (Kr); **(b)** Period 5, Group 5A (Sb); **(c)** Period 4, Group 3B (Sc); **(d)** several elements along the stepped diagonal line in the periodic table; **(e)** a Group 3A metal oxide, such as Al_2O_3, and other possibilities in the *d* block. **79 (a)** Photoelectric effect with visible light: probable for Cs, Rb, K, and not for Zn, Cd, Hg. **(b)** similar to others in the *f* block: ≈600 kJ/mol; **(c)** Based on Group 1A metals (Figure 8.9): ≈275 pm. **81 (a)** Rb, $1s^3 2s^3 2p^9 3s^3 3p^9 3d^7 4s^3$; Na, $1s^3 2s^3 2p^5$; with three possibilities for m_s, Rb would be in the *d* block and Na, in the *p* block; **(b)** Rb, $1s^2 1p^6 2s^2 2p^6 2d^{10} 3s^2 3p^6 3d^1 4s^2$; Na, $1s^2 1p^6 2s^2 2p^1$; again, Rb would be in the *d* block and Na, in the *p* block. **83 (a)** $I_1 = 1.312 \times 10^3$ kJ/mol is the difference between E_∞ and E_1; **(b)** The difference in energy between the shortest wavelength line in the Balmer series ($n = \infty \longrightarrow n = 2$) and the longest in the Lyman series ($n = 2 \longrightarrow n = 1$) corresponds to the transition $n = \infty \longrightarrow n = 1$. The same magnitude but opposite sign to (a). **85.** $Cl(g) + e^- \longrightarrow Cl^-(g)$ $\Delta H = -349$ kJ [EA_1 for Cl(g)]. Overall process is exothermic. **87.** Obtain volume of 1 mol Na atoms and divide by N_A to obtain the volume of one atom; use the formula $V = 4/3\,\pi r^3$ to obtain *r*. The method is inaccurate because spherical atoms cannot pack to eliminate empty space. The value obtained is too large. **89 (a)** From left to right, the peaks are 2*p*, 2*s*, and 1*s*. **(b)** For $1s^2 2s^2 2p^1$, the 1*s* and 2*s* peaks are of similar size—each orbital has two electrons; the 2*p* peak is half as large. **(c)** Ionization energies from photoelectron spectroscopy are a first ionization energy, but from every subshell rather than just from the one at the highest energy. Ionization from the 2*p* subshell gives I_1. No ionizations in the photoelectron spectrum correspond to I_2, I_3, and so on. **(d)** 6.3 nm **(e)** The spectrum has five peaks, corresponding to the subshells in $1s^2 2s^2 2p^6 3s^2 3p^1$.

Chapter 9

Exercises: 9.1 (a) :Ar: **(b)** Ca· **(c)** :Br: **(d)** ·As: **(e)** K· **(f)** ·Se:

9.2A Barium iodide, BaI_2, $Ba^{2+} + 2\,[:I:]^-$ **9.2B** Aluminum oxide, Al_2O_3, $2\,Al^{3+} + 3\,[:O:]^{2-}$ **9.3A** $\Delta H_f^\circ = -615$ kJ. **9.3B** Lattice energy = −863 kJ/mol LiCl(s). **9.4 (a)** Ba < Ca < Be; **(b)** Ga < Ge < Se; **(c)** Te < S < Cl; **(d)** Bi < P < S. **9.5** C—S < C—H < C—Cl < C—O < C—Mg.

9.6A H—N(H)—N(H)—H (each N with a lone pair) **9.6B** H—C(H)(H)—C(H)(H)—Cl:

9.7A :S=C=O: **9.7B** :N—O(=O:)—F:

9.8A $[:C\equiv N:]^-$ **9.8B** $\left[\begin{array}{c} H \\ | \\ H-P-H \\ | \\ H \end{array}\right]^+$

9.9A $:\ddot{O}=\ddot{N}-\ddot{Cl}:$ **9.9B** H–C(H)(H)–C(=O:)–Ö–C(H)(H)–H

9.10 The resonance hybrid involves equal contributions from the following three structures:

$$\left[:\ddot{O}=N(-\ddot{O}:)-\ddot{O}:\right]^- \longleftrightarrow \left[:\ddot{O}-N(-\ddot{O}:)=\ddot{O}:\right]^- \longleftrightarrow \left[:\ddot{O}-N(=O:)-\ddot{O}:\right]^-$$

9.11A $:\ddot{Cl}-\ddot{P}(-\ddot{Cl}:)-\ddot{Cl}:$ **9.11B** (a) $:\ddot{F}-\ddot{Cl}(-\ddot{F}:)-\ddot{F}:$ (b) SF_4: $:\ddot{F}-\ddot{S}(-\ddot{F}:)_2-\ddot{F}:$

9.12 **(a)** Incorrect. ClO_2 has 19 valence electrons, but the Lewis structure has 20. **(b)** Correct. **(c)** Incorrect. The structure has 26 electrons rather than the 24 available. The correct structure is $:\ddot{F}-\ddot{N}=\ddot{N}-\ddot{F}:$ **9.13A** 144 pm. **9.13B** 145 pm. **9.14** $\Delta H = -113$ kJ. **Review Questions: 6** **(a)** Na· **(b)** $\cdot\ddot{O}:$ **(c)** $\cdot\dot{Si}\cdot$ **(d)** $:\ddot{Br}:$ **(e)** Ca· (f) $\cdot\dot{As}:$ **8** **(a)** K^+, $[Ne]3s^2 3p^6$, K^+ **(b)** S^{2-}, $[Ne]3s^2 3p^6$, $[:\ddot{S}:]^{2-}$ **(c)** Al^{3+}, $1s^2 2s^2 2p^6$, Al^{3+} **(d)** Cl^-, $[Ne]3s^2 3p^6$, $[:\ddot{Cl}:]^-$ **(e)** Mg^{2+}, $1s^2 2s^2 2p^6$, Mg^{2+} **(f)** N^{3-}, $1s^2 2s^2 2p^6$, $[:\ddot{N}:]^{3-}$ **9** **(a)** $Ca\cdot + 2:\ddot{Br}: \longrightarrow Ca^{2+} + 2[:\ddot{Br}:]^-$ **(b)** $Ba\cdot + \cdot\ddot{O}: \longrightarrow Ba^{2+} + [:\ddot{O}:]^{2-}$ **(c)** $2\cdot\dot{Al}\cdot + 3\cdot\ddot{S}: \longrightarrow 2Al^{3+} + 3[:\ddot{S}:]^2$ **10.** The only such compound is lithium hydride: $Li^+ [:H]^-$. **12.** $:\ddot{I}-\ddot{I}:$, where the dash represents a bonding pair of electrons and the pairs of dots, lone pair electrons. **13.** $\overset{\delta+}{H}-\overset{\delta-}{\ddot{F}}:$ **15.** The stronger attraction of electrons is by the more electronegative atom: **(a)** N; **(b)** Cl; **(c)** F; **(d)** O. **16** **(a)** F; **(b)** Br; **(c)** Cl; **(d)** N. **17.** ionic: (a), (c), (e), (f); polar covalent: (b), (d), (i); nonpolar covalent: (g), (h). **18** **(a)** 1; **(b)** 4; **(c)** 2; **(d)** 1, **(e)** 3, **(f)** 1. **20.** C, N, O, and S form double covalent bonds, and C and N form triple covalent bonds. **21.** Odd electron species, e.g., NO, NO_2, ClO_2; incomplete octets, e.g., BF_3, BCl_3; expanded valence shells, e.g., SF_6, PCl_5, ClF_3. **22.** Nonmetal atoms having valence shells with $n \geqq 3$ can employ expanded valence shells in their compounds. **25.** Unsaturated hydrocarbon: (a) $HC\equiv CCH_3$, (b) $CH_3CH_2CH=CH_2$, (c) $HC\equiv CCH_2CH_3$. **26.** Alkene: (b); alkynes (a), (c). **Problems: 29.** Noble-gas configurations (b), (d), (e): **(a)** Cr^{3+}, $[Ar]3d^3$; **(b)** Sc^{3+}, $[Ne]3s^2 3p^6$; **(c)** Zn^{2+}, $[Ar]3d^{10}$; **(d)** Te^{2-}, $[Kr]4d^{10} 5s^2 5p^6$; **(e)** Zr^{4+}, $[Ar]3d^{10} 4s^2 4p^6$, **(f)** Cu^{2+}, $[Ar]3d^{10}$. **31** **(a)** $K^+ + [:\ddot{I}:]^-$; **(b)** $Ba^{2+} + 2[:\ddot{F}:]^-$; **(c)** $2\ Rb^+ + [:\ddot{S}:]^{2-}$; **(d)** $2\,Al^{3+} + 3[:\ddot{O}:]^{2-}$. **33.** $\Delta H_f^\circ = -559$ kJ/mol (Appendix C value: -573.7 kJ/mol). **35.** Lattice energy $= -645$ kJ/mol CsCl(s) [I_1 for Cs from Table 8.4; EA_1 for Br from Table 8.5; for the dissociation: $\frac{1}{2}Br_2(g) \longrightarrow Br(g)$, $\Delta H = 96.4$ kJ, from ΔH_f° data in Appendix C]. **37.** Lattice energy $= -2809$ kJ/mol $Li_2O(s)$ [I_1 for Li from Table 8.4; EA_1 and EA_2 for O from page 342; for the dissociation: $\frac{1}{2}O_2(g) \longrightarrow O(g)$, $\Delta H = 249.2$ kJ, and $\Delta H_f^\circ[Li_2O(s)]$ from Appendix C].

39 (a) $H-\ddot{P}(-H)-H$ (b) $:\ddot{F}-C(-\ddot{F}:)_2-\ddot{F}:$

41 (a) $H-C(H)(H)-\ddot{O}-H$ (b) $H-C(=\ddot{O}:)-H$ (c) $H-\ddot{N}(H)-\ddot{O}-H$ (d) $H-\ddot{N}(H)-\ddot{N}(H)-H$ (e) $:\ddot{F}-C(=\ddot{O}:)-\ddot{F}:$ (f) $:\ddot{Cl}-\ddot{P}(-\ddot{Cl}:)-\ddot{Cl}:$

43 **(a)** B < N < F; **(b)** Ca < As < Br; **(c)** Ga < C < O.

45 **(a)** $F-F < \overset{\delta+}{Cl}-\overset{\delta-}{F} < \overset{\delta+}{Br}-\overset{\delta-}{F} < \overset{\delta+}{I}-\overset{\delta-}{F} < \overset{\delta+}{H}-\overset{\delta-}{F}$;

(b) $H-H < \overset{\delta+}{H}-\overset{\delta-}{I} < \overset{\delta+}{H}-\overset{\delta-}{Br} < \overset{\delta+}{H}-\overset{\delta-}{Cl} < \overset{\delta+}{H}-\overset{\delta-}{F}$.

47. The formal charges in SO_2 are S, +1; =O, 0; —O, −1. In CO, formal charges are C, −1; O, +1. In HO_2^-, formal charges are H, 0; —O—, 0; —O, −1. In HCN, formal charges are H, 0; C, 0; N, 0. **49.** For the structure $[:\ddot{C}=N=\ddot{O}:]^-$, formal charges are C, −2; N, +1, O, 0. For $[:N\equiv C-\ddot{O}:]^-$, formal charges are N, 0; C, 0; O, −1. The second structure is the preferred structure because of the smaller formal charges. **51.** The resonance hybrid has carbon-to-oxygen bonds to the terminal O atoms that are between single and double, as indicated by the resonance structures:

$$\left[H-\ddot{O}-C(=\ddot{O}:)-\ddot{O}:\right]^- \longleftrightarrow \left[H-\ddot{O}-C(-\ddot{O}:)=\ddot{O}\right]^-$$

53 (a) $H-\ddot{O}-\ddot{N}=\ddot{O}:$ (b) $H-\ddot{O}-N(-\ddot{O}:)=\ddot{O}: \longleftrightarrow H-\ddot{O}-N(=\ddot{O}:)-\ddot{O}:$

A single structure works for nitrous acid (HONO), but two equivalent resonance structures are needed for nitric acid ($HONO_2$) to show that the nitrogen-to-oxygen bonds to terminal O atoms are between single and double. **55 (a)** odd-electron molecule, ·N═O: **(b)** expanded valence shell, :F—Cl—F: with :F: bonded below Cl

(c) incomplete octet, :Cl—B—Cl: with :Cl: bonded below B

(d) expanded valence shell, :F—Se—F: with :F: bonded above and :F: bonded below Se

57 (a) A structure with no formal charges and an expanded valence shell is :S═S—F: (with :F: bonded below the second S). Another structure without valence shell expansion is :S—S—F:. (with :F: bonded below the second S). It has formal charges of +1 for one S atom and −1 for the other. Both structures are likely to contribute to a resonance hybrid. The molecule :F—S—S—F:, on the other hand, is not a resonance structure but an isomer.

(b) [:I—I—I:]$^-$ **(c)** H—O—C—O—H (with C═O: below C)

(d) [:C≡N:]$^-$ **(e)** [:F—S—F:]$^-$ (with two :F: above S and :F: below S) **(f)** [:O—Br—O:]$^-$ (with :O: bonded above Br)

Several possible resonance structures involving one and two bromine-to-oxygen double bonds can be written. However, as discussed on pages 387-388, it is questionable whether these structures based on an expanded valence shell are important contributors to the resonance hybrid.

59

(a) H—C(H)(H)—C(H)(—CH$_2$H...) structure:

```
    H   H
    |   |
H—C—C—O—H
    |   |
    H   |
  H—C—H
      |
      H
```
(a)

```
    :O:
     ||
 H—C—O—H
```
(b)

```
   H       H
   |       |
H—C—O—C—H
   |       |
   H       H
```
(c)

61 (a) The C atom has an incomplete octet. The carbon-to-sulfur bond is a double bond: [:S═C═N:]$^-$. **(b)** There are only 16 electrons shown instead of the required 17; :O═N—O: ↔ :O—N═O: **(c)** Nitrogen cannot have an expanded valence shell, and all nitrogen-to-chlorine bonds are single: :Cl—N—Cl: (with :Cl: bonded below N)

63. Empirical formula: C_2H_5O. This is an odd-electron species (19 valence electrons) and is not likely to be a stable molecule. A structure based on a molecular formula of $C_4H_{10}O_2$ has 38 valence electrons.

unlikely:
```
        H   H
        |   |
H—O—C—C·
        |   |
        H   H
```

likely:
```
        H   H   H   H
        |   |   |   |
H—O—C—C—C—C—O—H   (plus isomers)
        |   |   |   |
        H   H   H   H
```

65 (a) 233 pm; **(b)** 149 pm. Estimates are likely to be too high because the bonds are polar. In general, the more polar the bond, the more the bond length is shortened from calculated value. **67.** The smallest percent difference should be for the least polar bond—the bond with the smallest electronegativity difference—HI. **69.** The nitrogen-to-nitrogen bond is a double bond, and the expected structure is :F—N═N—F: **71.** $\Delta H = -535$ kJ. **73.** Energy released in forming new bonds is greater than that expended in breaking old bonds; reaction is exothermic ($\Delta H = -113$ kJ). **75.** The oxygen-to-oxygen bond energy in O_3 is 302 kJ/mol. **77.** Bond dissociation energy of NO(g): 632 kJ/mol. **79 (a)** $H_2C{=}CH_2$; **(b)** $H_2C{=}CHCH_2CH_3$; **(c)** $HC{\equiv}CCH_3$; **(d)** $H_3CC{\equiv}CCH_2CH_3$. **81 (a)** $H_2C{=}CH_2 + H_2 \longrightarrow H_3C{-}CH_3$; **(b)** $HC{\equiv}CH + 2\,H_2 \longrightarrow H_3C{-}CH_3$. **83 (a)** $\Delta H = -128$ kJ; **(b)** $\Delta H = -136.94$ kJ.

85.
```
  H  H  H  H  H  H  H  H
  |  |  |  |  |  |  |  |
—C—C—C—C—C—C—C—C—
  |  |  |  |  |  |  |  |
  H  H  H  H  H  H  H  H
```
(a)

```
  H  H  H  H  H  H  H  H
  |  |  |  |  |  |  |  |
—C—C—C—C—C—C—C—C—
  |  |  |  |  |  |  |  |
  H  Cl H  Cl H  Cl H  Cl
```
(b)

```
  H   H    H   H    H   H    H   H
  |   |    |   |    |   |    |   |
—C—C—C—C—C—C—C—C—
  |   |    |   |    |   |    |   |
  H  CH3 H  CH3 H  CH3 H  CH3
```

(c)

Additional Problems: 87. In alkanes, all valence electrons are bonding pair electrons represented as dashes. There are no lone-pair electrons represented by dots. The structural formula and the Lewis structure are identical. This is not true for organic compounds containing O or N atoms, for example, where lone pair electrons will appear on O and N atoms.

89. dimethyl ether:

```
    H       H
    |       |
H—C—Ö—C—H
    |   ··  |
    H       H
```

ethanol:

```
    H  H
    |  |
H—C—C—Ö—H
    |  |  ··
    H  H
```

91.

```
                                                        H  H  H  H
                                                        |  |  |  |
                                                        C=C—C=C
                                                        |        |
                                                        H        H
H      H  H              H  H          H          H        H  H
|      |  |              |  |          |          |        |  |
C=C=C—C—H   H—C≡C—C—C—H   H—C—C≡C—C—H   C—C—H
|         |              |  |          |          |      ||  ||
H         H              H  H          H          H      C—C—H
                                                          |  |
                                                          H  H
```

93. $\mathrm{H{-}\ddot{N}{-}N{\equiv}N:} \longleftrightarrow \mathrm{H{-}\ddot{N}{=}N{=}\ddot{N}:}$

95 (a) From the percent composition, the empirical formula is C_3O_2 (formula mass 68 u). Because the *sole* product in the reaction of the oxide with water has a molar mass of 104 g/mol, the reaction product must be $H_4C_3O_4$; the reaction is $C_3O_2 + 2\ H_2O \longrightarrow H_4C_3O_4$. **(b)** From the titration data, we find 2 mol ionizable H atoms per mole of compound. Two of the four H atoms in $H_4C_3O_4$ are bonded in a different way than the other two. Plausible Lewis structures are

(reactant) $\mathrm{:\ddot{O}{=}C{=}C{=}C{=}\ddot{O}:}$

```
             :O: H  :O:
              ||  |   ||
(product) H—Ö—C—C—C—Ö—H
                   |
                   H
```

97. Based on ΔH_f° data and Table 9.1, bond energy = 299 kJ/mol; based on Lewis structure and Table 9.1, bond energy = 320 kJ/mol. **99.** $\Delta H_f^\circ = -15$ kJ/mol MgCl(s). **101.** Electron affinity: −65 kJ/mol H(g).

Chapter 10

Exercises: 10.1A (a) $SiCl_4$: VSEPR notation, AX_4: electron-group geometry and molecular geometry both tetrahedral. **(b)** $SbCl_5$: VSEPR notation, AX_5: electron-group geometry and molecular geometry both trigonal bipyramidal. **10.1B (a)** BF_4^-: VSEPR notation, AX_4: electron-group geometry and molecular geometry both tetrahedral (Lewis structure on page 386). **(b)** N_3^-: Lewis structure, $[\mathrm{:\ddot{N}{=}N{=}\ddot{N}:}]^- \longleftrightarrow [\mathrm{:N{\equiv}N{-}\ddot{\underset{..}{N}}:}]^-$, VESPR notation, AX_2; electron-group geometry and molecular geometry both linear. **10.2A** SF_4: VSEPR notation, AX_4E. Electron-group geometry is trigonal bipyramidal. In the seesaw structure (Table 10.1), two LP-BP repulsions at 90° and two at 120°; in the pyramidal structure given, three LP-BP at 90° and one at 180°. Because 90° LP-BP interactions are especially unfavorable, the seesaw structure is adopted. **10.2B** ClF_3: VSEPR notation AX_3E_2. Electron group geometry is trigonal bipyramidal. In the T-shaped structure (Table 10.1), four LP-BP repulsions at 90°; in the trigonal planar structure, six LP-BP repulsions at 90°. The T-shaped structure is observed. **10.3A** For Lewis structure of $(CH_3)_2O$, see answer to Chapter 9, Problem 59(c). Each C atom bonded to four atoms in tetrahedral fashion (AX_4). The O atom (AX_2E_2) has tetrahedral electron-group geometry, and the C—O—C linkage is angular or bent (bond angle ≈ 109°). **10.3B** For structural formula of CH_3CH_2OH, see page 67. Each C atom has four atoms or groupings of atoms in a tetrahedral arrangement (AX_4). The O atom (AX_2E_2) displays angular or bent molecular geometry (C—O—H bond angle ≈ 109°). **10.4A** Polar: SO_2, angular or bent (AX_2E); BrCl, linear but with a small ΔEN between atoms. Nonpolar: BF_3, symmetrical, trigonal planar (AX_3); N_2, linear, no polarity in bond. **10.4B** SO_3, symmetrical, trigonal planar (AX_3)—nonpolar; SO_2Cl_2, symmetrical, tetrahedral (AX_4), but with different EN for terminal O and Cl atoms—polar; ClF_3 nonsymmetric, T-shaped (AX_3E_2)—polar; BrF_5, nonsymmetric, square pyramidal shape (AX_5E)—polar. **10.5** NOF, 110°; NO_2F, 118°. LP-BP repulsions from lone-pair electrons on the N atom, force O and F atoms closer together than in NO_2F, with no lone-pair electrons on N atom. **10.6A** $SiCl_4$ is a tetrahedral molecule (AX_4); hybridization scheme for Si: sp^3. **10.6B** I_3^-, a linear ion with trigonal bipyramidal electron group geometry; sp^3d hybridization for central I atom. **10.7A (a)** Tetrahedral molecular geometry around C atom and angular or bent around O atom. **(b)** Hybridization: sp^3 for both C and O. **(c)** 3 σ C(sp^3)-H(1s); σ C(sp^3)-O(sp^3); σ O(sp^3)-H(1s). **10.7B (a)** Linear molecule, :N≡C—C≡N: **(b)** Hybridization: sp for both C and N. **(c)** σ C(sp)-C(sp); 2 σ C(sp)-N(sp) + 4 π C(2p)-N(2p). **10.8 (a)** Planar $H_2C{=}CH_2$ is nonpolar, but replacing one H with F makes $H_2C{=}CHF$ polar. **(b)** Because of the very small Δ EN between C and H and the symmetrical shape of *trans*-2-butene (see page 441), the molecule is *nonpolar*. **(c)** Acetylene, H—C≡C—H, is a symmetrical, linear *nonpolar* molecule. **(d)** Substitute Cl atoms for the two H atoms at the double bond in *cis*-2-butene, and EN differences make the resulting nonsymmetric molecule *polar*. **10.9** H_2^- should be somewhat stable; two electrons in bonding MO and one, in antibonding MO—bond order is 1/2. **10.10** The 11 valence electrons are distributed in the MO diagram of either N_2 or O_2. In either case, bond order $= (8 - 3)/2 = 2.5$. **Review Questions: 6 (a)** linear, 180°; **(b)** trigonal planar, 120°; **(c)** tetrahedral, 109.5°. **7.** Possible with diatomic molecules. HCl is a linear molecule with tetrahedral electron group geometry for the Cl atom. **10.** No, a bond dipole exists only if there is an EN difference between the bonded atoms. **11.** No, $\mu = 0$ for a molecule with no bond dipoles or with a symmetrical shape in which bond dipoles cancel. **17.** Trigonal planar, sp^2; octahedral, sp^3d^2. **20.** Yes. π bonds are found only in molecules with multiple bonds, and all bonds in alkanes are single bonds (σ). **21.** No. π bonds are found only in multiple bonds, and all the single bonds in alkenes (C—C and C—H) are σ.

30. [Chlorobenzene structures: two Kekule structures and a resonance hybrid]

Kekule structures Resonance hybrid

Problems: 31 **(a)** H—Ö(H): (AX_2E_2); **(b)** [:Ö—Cl:]$^-$ (AXE_3), with either O or Cl as the central atom; **(c)** [:I—I—I:]$^-$ (AX_2E_3). **33 (a)** Linear geometry not possible for a triatomic molecule, AX_2E_2 (see Problem 31a). **(b)** Electron-group geometry and molecular geometry both trigonal planar for AX_3 molecule, such as SO_3. **(c)** PH_3—an AX_3E molecule with trigonal pyramidal molecular geometry; it cannot be planar. **35** BF_3 (Table 10.1) has 3 bonding pairs of electrons on the central B atom (AX_3); it must be trigonal planar. ClF_3 (see Chapter 9, Problem 55b) has five electron pairs around the central Cl atom—three bonding and two lone pairs—in trigonal bipyramidal distribution (AX_3E_2); smaller electron-pair repulsions in T-shaped than in trigonal planar structure (see Exercise 10.2B). **37 (a)** AX_3E, trigonal pyramidal; **(b)** AX_4, tetrahedral; **(c)** AX_4E_2, square planar; **(d)** AX_2, linear; **(e)** AX_5E, square pyramidal. **39 (a)** H_2O_2: The O—O—H bonds are bent; overall shape is a nonplanar "zigzag". **(b)** C_3O_2: The C atoms are central atoms, all bonds are double, and the molecule is linear. **(c)** OSF_4: A trigonal bipyramidal shape with S as the only central atom. Picture replacing the LP electrons in the SF_4 structure of Table 10.1 with an O atom. **(d)** N_2O: One central N atom with two electron groups (AX_2)—linear triatomic molecule. **41.** CH_4. All valence electrons in CH_4 are in bonds (AX_4), and terminal atoms are identical; we expect the exact bond angles of 109.5°. In $COCl_2$ (AX_3), the terminal atoms are different, two Cl and one O. LP-BP interactions involving LP electrons of the O atom force the Cl—O—Cl bond angle to be <120°. **43 (a)** linear, symmetric, *nonpolar*; **(b)** bent, nonsymmetric, *polar*; **(c)** square planar, symmetric, *nonpolar*; **(d)** T-shaped, nonsymmetric, *polar*. **45.** Both H_2O and OF_2 are bent molecules with a similar bond angle (AX_2E_2); both are polar because of EN differences. H_2O should have the greater dipole moment because ΔEN is greater in H_2O than in OF_2. **47. (a)** NF_3 is a trigonal pyramidal molecule (AX_3E), with electrons displaced toward the F atoms. However, their effect is diminished by LP electrons on the N atom. **(b)** SF_4 has a seesaw shape (Table 10.1). Bond dipoles are directed toward F atoms. Two are in opposite directions along a straight line and effectively cancel. The other two are in the trigonal planar central plane. The molecule is polar, but LP electrons on the S atom tend to counteract the two S—F bond dipoles in the central plane. **49.** More extensive overlap of $2p$ orbitals in F_2 than $2s$ orbitals in Li_2 give F_2 the greater bond energy. **51 (a)** bent shape (AX_2E_2), sp^3 hybridization of O; **(b)** tetrahedral shape (AX_4), sp^3 hybridization of N; **(c)** linear shape (AX_2), sp hybridization of C; **(d)** trigonal planar shape (AX_3), sp^2 hybridization of C. **53 (a)** Molecule is linear; both central C atoms are sp hybridized. **(b)** H—N—C portion is trigonal planar, and N atom is sp^2; N=C=O portion is linear, and C atom is sp. **(c)** H_2N—O portion is pyramidal, and N atom is sp^3; the N—O—H portion is bent, and O atom is sp^3 hybridized. **(d)** The C of H_3C— group is sp^3; —COOH portion has sp^2 hybridization of C and sp^3 hybridization of O atom in C—O—H linkage.

55 :Cl—N(=O:)—Ö: ⟷ :Cl—N(—Ö:)=Ö: trigonal planar, bonding

(a)

scheme: 1 σ Cl($3p$)-N(sp^2), 2 σ N(sp^2)-O($2p$), 1 π N($2p$)-O($2p$). **(b)** :F—Ö—F: bent, bonding scheme: 2 σ F($2p$)-O(sp^3).

[:Ö—C(=O:)—Ö:]$^{2-}$ ⟷ [:Ö—C(—Ö:)=Ö:]$^{2-}$ ⟷ [:Ö=C(—Ö:)—Ö:]$^{2-}$

(c)

trigonal planar, bonding scheme: 3 σ C(sp^2)-O($2p$), 1 π C($2p$)-O($2p$). **57.** [:Cl—I—Cl:]$^+$ has bent shape (AX_2E_2) and sp^3 hybridization of I. [:Cl—I—Cl:]$^-$ is linear (AX_2E_3) and has sp^3d hybridization of I. **59.** The 157 pm bond length indicates C-to-C single bond. The 125 pm bond length indicates C-to-O bonds intermediate between single and double bonds and suggests a resonance hybrid.

[:Ö—C(=O:)—C(=O:)—Ö:]$^{2-}$ ⟷ [:Ö=C(—Ö:)—C(=O:)—Ö:]$^{2-}$ ⟷

[:Ö—C(=O:)—C(—Ö:)=Ö:]$^{2-}$ ⟷ [:Ö=C(—Ö:)—C(—Ö:)=Ö:]$^{2-}$

In each resonance structure, the two C atoms are sp^2 hybridized, and the bonding scheme is 1 σ C(sp^2)-C(sp^2), 4 σ C(sp^2)-O($2p$), 2 π C($2p$)-O($2p$).

61 H—N(H)—O—H sp^3 hybridization of both N and O atoms;

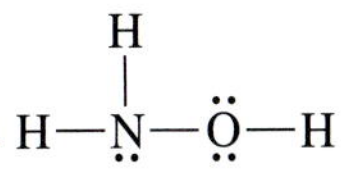

(a)

bonding scheme: 2 σ N(sp^3)-H(1s), 1 σ N(sp^3)-O(sp^3), 1 σ O(sp^3)-H(1s). **(b)** H—N̈=C=Ö: sp^2 hybridization of N and sp hybridization of C; bonding scheme: 1 σ H(1s)-N(sp^2), 1 σ N(sp^2)-C(sp), 1 π N(2p)-C(2p), 1 σ C(sp)-O(2p), 1 π C(2p)-O(2p). H—C(H)(H)—C(=O)—N(H)—H sp^3 hybridization of first C,

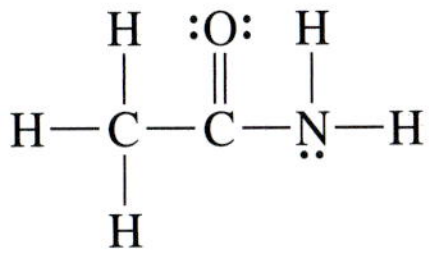

(c)

sp^2 hybridization of second C, sp^3 hybridization of N; bonding scheme: 3 σ C(sp^3)-H(1s), 2 σ N(sp^3)-H(1s), 1 σ C(sp^3)-C(sp^2), 1 σ C(sp^2)-N(sp^3), 1 σ C(sp^2)-O(2p), 1 π C(2p)-O(2p). **63.** (a) and (c) have cis-trans isomers; one H atom at each C atom of double bond, and two distinctly different structures possible. No cis-trans isomerism in (b); two H atoms on one C atom at the double bond allows one structure to be flipped over into the other. **65.** Neither exhibits cis-trans isomerism; each has two identical groups on one C atom at the double bond. Substituting Cl for one H on the first C atom produces cis-trans isomerism in 1-butene, but not in isobutylene, because one C atom at the double bond still has two identical groups (CH_3). **67.** The electron lost by F_2 to become F_2^+ comes from an antibonding MO, thereby strengthening the F—F bond. The electron gained by F_2 goes into an antibonding MO, thereby weakening the F—F bond in F_2^- (see Figure 10.27). **69.** By the octet rule, a Lewis structure for C_2 has all 8 valence electrons in a *quadruple* bond, C≣C. In the MO diagram, two of 8 valence electrons are in an antibonding MO, reducing the bond order to 2 (see Figure 10.27). **71.** MO diagrams are CN^+: $\sigma_{2s}^2\,\sigma_{2s^*}^2\,\pi_{2p}^4$ (same as C_2) and CN^-: $\sigma_{2s}^2\,\sigma_{2s^*}^2\,\pi_{2p}^4\,\sigma_{2p}^2$ (same as N_2). **(a)** The stronger C-to-N bond is in CN^- (triple compared to double). **(b)** Species are diamagnetic; all electrons are paired. **73.** Based on the MO diagram of O_2, superoxide ion, O_2^-, has a bond order of $(8 - 5)/2 = 1.5$ and is paramagnetic; peroxide ion, O_2^{2-}, has a bond order of $(8 - 6)/2 = 1$ and is diamagnetic.

75 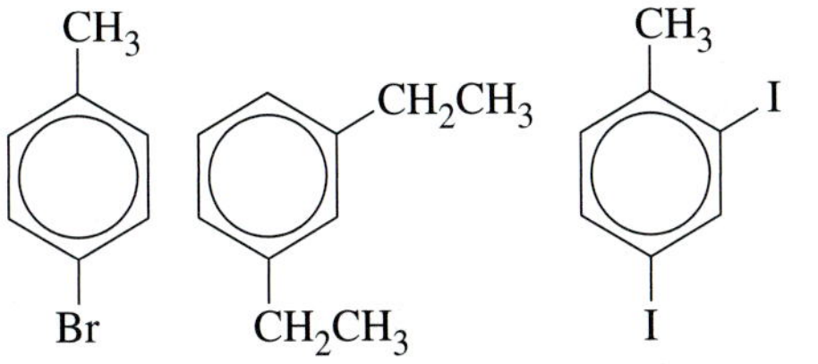 (a) 4-bromotoluene structure (CH_3 and Br on ring); (b) ring with CH_2CH_3, CH_2CH_3; (c) ring with CH_3, I, I

(a) (b) (c)

77 (a) 2,6-dichlorophenol; **(b)** 1,3,5-trinitrobenzene; **(c)** 2,5-dibromotoluene.

79 cis: N=N with both phenyl rings on same side; trans: N=N with phenyl rings on opposite sides

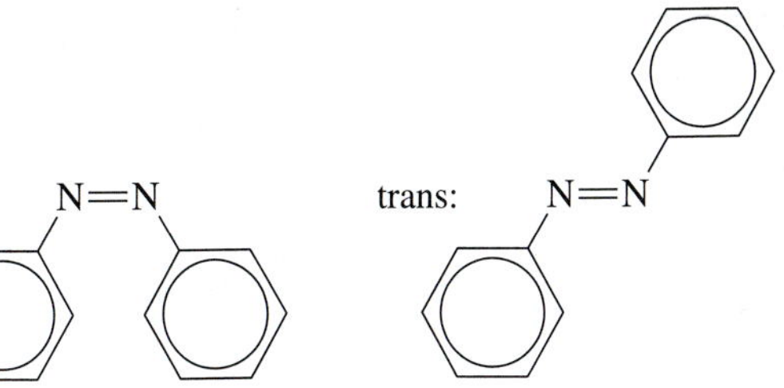

(a)

(b) All C and N atoms are sp^2 hybridized; additionally each benzene ring has delocalized electrons in π molecular orbitals. **(c)** All C—C—C, H—C—C, and C—N—N angles are about 120°. **Additional Problems: 81.** Statement is valid only for diatomic molecules. For polyatomic molecules, individual bond dipoles may be oriented in a way as to cancel, resulting in a nonpolar molecule. **83.** The terminal atoms in NO_2F are not identical—two O and one F. Because of differences in EN between O and F, one bond dipole differs from the other two. There should be a displacement of electrons toward F. **85.** VSEPR notation, AX_5E: square pyramidal shape. **87.** PCl_5: sp^3d hybridization of P; PCl_4^+: sp^3 hybridization of P; PCl_6^-: sp^3d^2 hybridization of P. All P—Cl bonds involve the overlap of a hybrid orbital from P with a 3p orbital of Cl.

89. For CH_3CNO: **(a)** Lewis structure: H—C(H)(H)—N̈=C=Ö:

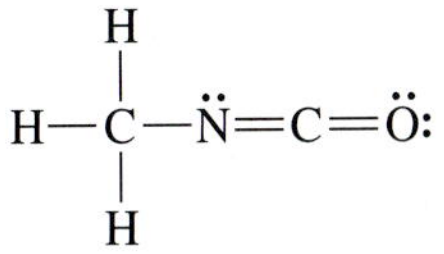

(b) From left to right in the Lewis structure, C (sp^3), N (sp^2), C(sp). (c) Approximate bond angles: H—C—H, 109.5°; H—C—N, 109.5°; C—N—C, 120°; N—C—O, 180°. **(d)** Molecular shape: The H_3CN portion is tetrahedral (much like CH_4, with N substituting for an H). The NCO portion is linear, tilted at a 120° angle from the H_3C—N bond. **91.** Lewis structure of C_3O_2, :Ö=C=C=C=Ö:, hybridization scheme: sp for all three C atoms; bonding scheme: C-to-O bonds consist of σ: C(sp)-O(2p) and π: C(2p)-O(2p), C-to-C bonds consist of σ: C(sp)-C(sp) and π: C(2p)-C(2p).

93. Lewis structure: :F̤—S(F)(F)—S̤—F̤: Hybridization of S: left, sp^3d, right, sp^3 Molecular shape: Refer to the seesaw structure (AX_4E) of SF_4 in Table 10.1. Replace one of the F atoms in the central plane by S and reattach the fourth F atom to this S atom, producing a S—S—F bond angle of about 109.5°.

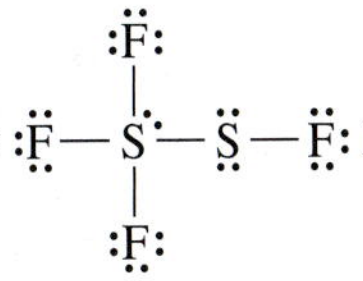

Chapter 11

Exercises: 11.1A $\Delta H_{vapn} = 34.0$ kJ/mol C_6H_6. **11.1B** $\Delta H_{total} = 92.0$ kJ. **11.2** CCl_4. A relatively low ΔH_{vapn} and a relatively high molar mass lead to the smallest heat requirement per kilogram. **11.3A** 4.00×10^2 mmHg. **11.3B** 5.33×10^{-3}g H_2O. **11.4** Room temperature is far above T_c of methane; a pressure gauge can be used to measure its amount. **11.5** If only gas were present, the pressure would be 10 atm—far in excess of the vapor pressure of H_2O at 30.0 °C. Most of the H_2O condenses to liquid. (It cannot all be liquid, because the liquid volume is only about 20 mL and the system volume is 2.61 L.) The final condition is a point on the vapor pressure curve at 30.0 °C. **11.6A** IBr and BrCl have about the same polarity, but IBr has a greater molar mass. IBr has stronger intermolecular forces and is the solid. BrCl is the gas. **11.6B (a)** The molar masses are comparable, but aniline is somewhat polar and toluene is nonpolar. Toluene has the lower boiling point. **(b)** The lower boiling point is that of nonpolar *trans*-1,2-dichloroethene; the cis isomer is polar. **(c)** The symmetrical placement of Cl atoms in *para*-dichlorobenzene makes it nonpolar and gives it a lower boiling point; the ortho isomer is polar. **11.7** Hydrogen bonding is an important intermolecular force in NH_3, C_6H_5OH, CH_3COOH, H_2O_2. **11.8.** $CsBr < KI < KCl < MgF_2$. **11.9 (a)** 286.6 pm; **(b)** 2.354×10^{-23} cm^3. **11.10.** d = 7.880 g/cm^3. **Review Questions: 5.** Of the variables given, only temperature (a) affects the vapor pressure. The other variables may affect the amount of vapor or the rate at which it forms, but not the vapor pressure. **9.** Each area represents one pure phase; each curve represents two phases in equilibrium, that is, liquid-gas, solid-liquid, solid-gas, $solid_1$-$solid_2$, and so on. **12.** All phase diagrams have a critical point. They have a normal boiling point only if all pressures on the sublimation curve are below 1 atm. Any T and P at which three phases coexist in equilibrium is a triple point: the more polymorphic forms of a solid substance, the more triple points. **Problems: 27.** 3.60×10^2 kJ. **29.** 61.7 kJ. **31.** 36.4 g C_3H_2. **33.** The transfer of heat from air in the oven to the hand occurs slowly, and so the hand warms slowly. Above the boiling water there is an almost instantaneous transfer of a large quantity of heat (the heat of condensation) from condensed steam directly into the hand. **35 (a)** about 425 mmHg; **(b)** about 77 °C. **37.** 2.87×10^{18} Ag atoms. **39.** 214 mmHg. **41.** The sample cannot be liquid only; 1.82 g $H_2O(l)$ occupies less than 2 mL volume and the container holds 2.55 L. It cannot be vapor only because the $H_2O(g)$ would exert a pressure (749 mmHg) far in excess of the vapor pressure at 30.0 °C. It is a mixture of liquid and gas. **43.** Heat goes into vaporizing water—the water boils at a constant temperature. Because its ignition temperature is greater than 100 °C, the paper does not burn. Boiling occurs at 99.9 °C rather than 100 °C because barometric pressure is slightly below 1 atm. **45.** A gas cannot be liquefied above T_c, regardless of the pressure applied. A gas can be either liquefied or solidifed by a sufficient lowering of the temperature, regardless of its pressure. A gas can always be liquefied by an appropriate combination of pressure and temperature changes. **47.** 6.7 kJ. **49.** Not enough heat is liberated in cooling 315 mL H_2O to 0 °C to melt 0.506 kg of ice. An ice-water mixture at 0 °C results. **51.** (d)—subliming 1 mol $H_2O(s)$ requires about 50 kJ ($\Delta H_{vapn} + \Delta H_{fusion}$). Melting 3.0 mol of ice (a) takes only 18 kJ; melting 2.0 mol ice and heating the liquid to 10 °C (c) takes even less heat. Evaporating 10.0 g $H_2O(l)$ (a) requires only about 25 kJ. **53 (a)** The area above "Gas" on the left is *solid*; on the right, *liquid*. The curve separating solid and gas is the sublimation curve (solid + gas). The curve separating liquid and gas is the vapor pressure curve (liquid + vapor). The line separating solid and liquid is the fusion curve (solid + liquid). **(b)** Point A is the point common to the solid, liquid, and gas areas; point B is the intersection of the line at 1 atm and the fusion curve; point C is the intersection of the line at 1 atm and the vapor pressure curve; point D is the highest T on the vapor pressure curve. **(c)** The sample remains solid until the fusion curve is reached, when melting begins. After melting is complete, T rises until the vapor pressure curve is reached; here, vaporization occurs. T remains constant until all the liquid vaporizes. Then T increases, but no further phase changes occur. **55.** The diagram consists of three one-phase areas (S, L, G) and three two-phase curves (S-G, S-L, L-G) that come together at the triple point (2.0 °C and 3.4 mmHg). The normal boiling point is the point—113.5 °C, 1 atm—on the vapor pressure curve. The critical point is the final point on the vapor pressure curve. **57 (a)** From Table 11.2, vapor pressure of water at 88.15 °C is about 500 mmHg. The point (88.15 °C, 0.954 atm) is in the *liquid* area. **(b)** P = 0.0313 atm = 23.8 mmHg. This is the vapor pressure of water at 25 °C (Table 11.2); the point is on the vapor pressure curve (L-G). **(c)** The point (0 °C, 1.00 atm) is on the fusion curve. Because the fusion curve slopes to the left, the point (0 °C, 2.50 atm) is in the *liquid* area. **(d)** The point (−10 °C, 0.100 atm) is below the triple point temperature, and above the triple point pressure. The point lies in the *solid* area. **59.** CS_2 has the lower boiling point. Both substances are nonpolar, but CCl_4 has a higher molecular mass, stronger dispersion forces, and a higher boiling point. **61.** The two have comparable molecular masses, but because $(CH_3CH_2)_2O$ is polar, the added dipole-dipole intermolecular forces give it the higher melting point. **63.** C_6H_5COOH (122 u) has both dipole-dipole interactions and hydrogen bonds in addition to dispersion forces; it is a solid at STP. $C_{10}H_{22}$ has a slightly higher molecular mass (142 u) but is nonpolar. CH_3OH (32 u) would be a gas were it not for hydrogen bonding, but the strength of these bonds is only enough to make it a liquid at STP, not a solid. $(CH_3CH_2)_2O$ (74 u) is a polar liquid. **65.** $N_2 < NO < (CH_3)_2O < C_6H_{14} < C_4H_9OH$. NO has a slightly higher molar mass than N_2 and is somewhat polar; $(CH_3)_2O$ (46 u) is polar, but its molecular mass is considerably smaller than for C_6H_{14} (86 u); hydrogen bonding is important enough in C_4H_9OH (74 u) to give it the highest boiling point. **67.** $Cl_2 < CCl_4 < CsCl < MgCl_2$. Cl_2 and CCl_4 are nonpolar molecular substances. CsCl and $MgCl_2$ are ionic and have higher melting points than the molecular substances. The smaller size and higher charge of Mg^{2+} compared to Cs^+ gives $MgCl_2$ the highest melting point. **69.** 1-Octanol. The two have comparable molar masses, but 1-octanol has hydrogen bonding. Its greater intermolecular forces give 1-octanol the higher surface tension. **71.** Water wets glass because adhesive forces between water and glass exceed cohesive forces in water. The reverse is true with substances that water does not wet, for example, Teflon®. Wetting agents lower the surface tension of water and improve its wetting ability. **73 (a)** The larger unit is a unit cell, but not the smaller one. The larger cell can be shifted left, right, up, and down with no gaps. That is, the bottom edge of one cell coincides with the top edge of the identical cell below it. In contrast, the bottom edge of the small cell does not coincide with the top edge of the identical cell below it; there is a gap between them. **(b)** There are 3 hearts, 3 diamonds, and 3 clubs in a unit cell. **75 (a)** 408.4 pm; **(b)** 6.812×10^{-23} cm^3; **(c)** $d = 10.52$ g Ag/cm^3. **77 (a)** 404 pm; **(b)** $N_A \approx 6.4 \times 10^{23}$. **79.** There are four Ca^{2+} and eight F^- ions per unit cell, corresponding to the formula CaF_2. Coordination numbers are 4 for F^- and 8 for Ca^{2+}. Because there are twice as many F^- ions as Ca^{2+}, we expect the coordination of Ca^{2+} to be twice that of F^-. **Additional Problems: 81.** 96 mmHg. **83.** Hexanoic acid has dipole-dipole interactions and hydrogen bonding that are absent in 2-methylbutane; also it has a higher molecular mass and, therefore, greater dispersion forces. **85.** The density of $H_2O(l)$ rises from its value at the melting point to a maximum at 3.98 °C. Above this temperature its density falls with temperature in a customary fashion. Thus, for every density in the temperature range from 0 °C to 3.98 °C, there is another temperature in a range extending a few degrees above 3.98 °C at which the same density is observed. **87 (a)** 56.9 mmHg; **(b)** −33.36 °C; **(c)** 108.9 atm. **89.** The Group 4A hydrides show the expected trend: dispersion forces and boiling points increase with increasing molar mass. For the Group 5A, 6A, and 7A hydrides of periods 3, 4, and 5, boiling points also increase with molar mass. Three anomalies are seen for NH_3, H_2O, HF. The unusually high boiling points for NH_3, H_2O, and HF result from hydrogen bonding in these liquids. **91.** 120. g $H_2O(l)$ and 870. g $H_2O(g)$. **93.** The pressure is equal to the vapor pressure of water between 26 °C and 27 °C, probably close to 26.5 °C. **95.** Two *intra*molecular and two *inter*molecular hydrogen bonds are seen below:

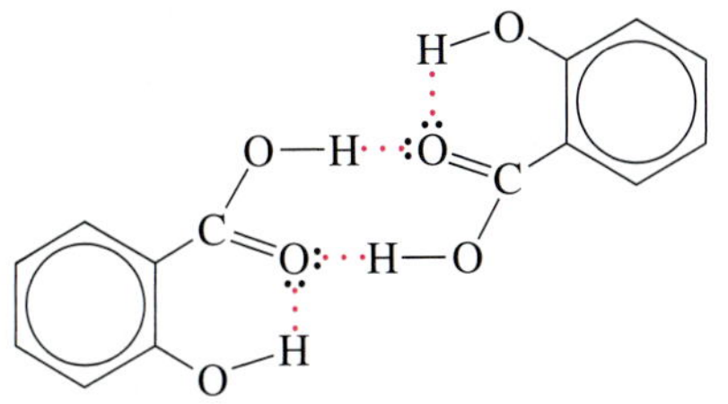

97. MgO should have the NaCl-type of crystal structure (*fcc*). The ratio, $r_{Mg^{2+}}/r_{Cl^-}$, is 0.46, which is similar to the ratio, $r_{Na^+}/r_{Cl^-} = 0.55$, and falls in the range for filling octahedral holes. **99.** $d = 3.516$ g/cm^3.

Chapter 12

Exercises: 12.1A 17.8% glucose, by mass. **12.1B** 7.00% sucrose, by mass. **12.2A** 34.8% toluene, by volume. **12.2B (a)** 34.4% toluene, by mass; **(b)** $d = 0.874$ g/mL. **12.3 (a)** 0.1 ppb; **(b)** 100 ppt. **12.3B** 69.9 ppm Na^+. **12.4** 0.317 m $C_6H_{12}O_6$. **12.5A** 0.418 mL CH_3CH_2OH. **12.5B** 253 mL H_2O. **12.6A (a)** 2.53 m CH_3OH; **(b)** 1.86 mol% $CO(NH_2)_2$. **12.6B (a)** 8.50% CH_3OH, by mass; **(b)** 2.61 M CH_3OH; **(c)** 4.97 mol% CH_3OH. **12.7** (b). The solutions are all rather dilute and have densities of approximately 1.0 g/mL. The mole percents of CH_3CH_2OH are not large, and we expect the largest to be in the solution having the greatest quantity of CH_3CH_2OH per liter or kilogram of solution. Solution (a) has 0.5 mol CH_3CH_2OH per liter; (b) has slightly more than 1 mol CH_3CH_2OH per kilogram of solution; (c) has slightly less than 0.5 mol CH_3CH_2OH per kilogram of solution; (d) has somewhat less than 1 mol CH_3CH_2OH per liter. **12.8A** Because of the similarity of its structure to that of benzene, nitrobenzene is more soluble in benzene than in water. **12.8B** The molecular models: (a) acetic acid, (b) 1-hexanol, (c) hexane, (d) butanoic acid; the order of increasing solubility in water: $c < b < d < a$. **12.9A** 5.4×10^{-2} mg CO_2/100 g H_2O. **12.9B** 2.1×10^2 mg. **12.10** 93.9 mmHg. **12.11** Partial and total pressures: $P_{benz} = 51.4$ mmHg; $P_{tol} = 13.0$ mmHg; $P_{total} = 64.4$ mmHg. **12.12** The solution with equal masses has a greater mole fraction of the more volatile benzene (78 g/mol) than of toluene (92 g/mol); it produces vapor with the greater $x_{benzene}$. **12.13** No. Water vapor passes from the more dilute (B) to the more concentrated solution (A) until the two solutions have the same concentration. Then, there is no further net transfer of water between the solutions. **12.14A** −2.46 °C. **12.14B** 18 g sucrose. **12.15** $C_6H_6O_4$. **12.16A** 6.86×10^4 g/mol. **12.16B** 9.33 mmH_2O. **12.17** Lowest fp: 0.0080 M HCl; highest fp: 0.010 m $C_6H_{12}O_6$. **Review Questions: 2.** Because they are based on volume, which varies with T, molarity and volume percent are dependent on T. **10.** Usually not. If only the solvent is volatile, the vapor cannot have the same composition as the solution—the vapor is pure solvent. Even if all components are volatile, the vapor and solution compositions are likely to differ because of differences in the volatilities of the components. **16.** Osmotic flow is of solvent molecules from a *dilute* solution, which has a higher vapor pressure, into a more concentrated one. This is similar to gases diffusing from higher to lower pressure. **22 (a)** 0.1M $NaHCO_3$; **(b)** 1 M NaCl; **(c)** 1 M $CaCl_2$; **(d)** 3 M glucose. **Problems: 23.** Dissolve 0.500 kg NaCl(s) in 4.50 kg $H_2O(l)$. **25 (a)** 3.96% NaOH; **(b)** 7.31% ethanol; **(c)** 8.21% glycerol. **27 (a)** 4.61% H_2O; **(b)** 0.811% acetone; **(c)** 1.02% 1-butanol. **29.** 1.0×10^2 mg glucose per dL. **31.** $e < c < d < b < a$; that is, $1 \times 10^{-12} < 1 \times 10^{-9} < 1 \times 10^{-6} < 1 \times 10^{-5} < 1 \times 10^{-2}$. **33 (a)** 1 ppb benzene; **(b)** 35 ppm NaCl; **(c)** $[F^-] = 1.3 \times 10^{-4}$ M. **35.** 1.25 m $C_6H_{12}O_6$. **37.** 12 M H_3PO_4 and 31 m H_3PO_4. **39.** 1.13 mol H_2SO_4. **41.** mole fraction $C_{10}H_8$: 0.0435. **43.** mole fraction $C_{10}H_8$: 0.0192. **45.** The solution with the greatest amount of solute per unit mass of solvent has the greatest mole fraction of solute—(b). Solution (a) has 1 mol solute per 1000 g H_2O; (b) has slightly more than 1 mol solute in 950 g H_2O; (c) has about 0.3 mol solute per 900 g H_2O. **47 (a)** $CHCl_3$, insoluble; **(b)** C_6H_5COOH, slightly soluble; **(c)** $CH_3CHOHCH_2OH$, highly soluble. **49.** Unsaturated. **51 (a)** 21 g H_2O; **(b)** ≈44 °C. **53.** Allow solvent to evaporate. When the solution becomes saturated, crystallization begins to occur. **55 (a)** 1.38×10^{-3} M O_2; **(b)** 7.2 atm. **57 (a)** Partial pressures: pentane, 88.2 mmHg; hexane, 96.8 mmHg; **(b)** Vapor composition: $x_{pent} = 0.477$; $x_{hex} = 0.523$. **59.** 17.4 mmHg. **61 (a)** −0.47 °C; **(b)** 3.70 °C. **63.** 1.28 m. **65.** $K_f = 4.27$ °C m^{-1}. **67.** $C_6H_3N_3O_6$. **69.** The salt solution has the higher osmotic pressure. As it shrinks into a pickle, the cucumber loses water to NaCl(aq). **71.** 9.94 atm. The solution is hypertonic. **73.** To the right (A → B). **75.** They are the results expected for the expression: $\Delta T_f = -i \cdot K_f \cdot m$, with $i = 1$ for HCl dissolved in C_6H_6, and $i = 2$, in H_2O. **77.** For NaCl, $i = 2$, and for glucose, $i = 1$. An isotonic solution of NaCl(aq) has only half the molarity of an isotonic solution of glucose. **79.** Order of decreasing freezing points:

b > a > e > c > d. **81.** 1.2×10^2 g NaCl. **83.** The charge on the particles is negative. Al^{3+} is the most highly charged cation of the three, making $AlCl_3(aq)$ the most effective coagulant. **Additional Problems: 85 (a)** 15.1% by volume; **(b)** 12.2% by mass; **(c)** 11.9% mass/volume; **(d)** 7.25 mole %. **87.** The 25.00 mL H_2O in 25.00 mL CH_3OH has the greater molality. The no. mol H_2O > no. mol CH_3OH, and the number kg solvent is smaller in 25.00 mL CH_3OH than in 25.00 mL H_2O. **89.** All ethanol-water solutions have a lower mass percent than volume percent; d of ethanol is less than that of water. For other liquid solutes, if $d_{solute} > d_{water}$, mass percent solute > volume percent solute. **91 (a)** 0.20 m $(CH_3)_2CO$ has the highest total vapor pressure because x_{H_2O} is as great as in the other solutions and $(CH_3)_2CO$ is volatile; **(b)** NaCl(satd. aq) has the highest concentration of solute particles and the lowest freezing point. **(c)** The vapor pressure of NaCl(satd. aq) remains constant because its concentration does not change as $H_2O(g)$ is lost. The concentrations and vapor pressures of the other solutions do change. **93.** $x_{benzene} = 0.256$; $x_{toluene} = 0.744$. **95.** 0.92% NaCl (mass/vol) is 0.157 m, corresponding to a freezing point of -0.58 °C if $i = 2$. Actually $i < 2$ and agreement with -0.52 °C is good. **97.** 0.027 mL CO_2(STP)/100 g H_2O. **99.** At equilibrium, each solution has a mass fraction of urea of 0.230 and a mole fraction of 0.0822.

Chapter 13

Exercises: 13.1A (a) 0.0259 M min^{-1}: **(b)** 8.63×10^{-4} M s^{-1}. **13.1B (a)** 1.05×10^{-5} M s^{-1}; **(b)** 3.15×10^{-5} M s^{-1}. **13.2A (a)** 1.11×10^{-3} M s^{-1}: **(b)** $[H_2O_2]_{310s} = 0.287$ M. **13.2B** about 270 s. **13.3A** 0.0912 M s^{-1}. **13.3B** 3/2 order. **13.4A (a)** $t = 255$ min; **(b)** $[NH_2NO_2] = 0.0139$ M. **13.4B** 3.46×10^{-4} M min^{-1}. **13.5 (a)** $t \approx 480$ s; **(b)** 0.149 g N_2O_5. **13.6** The time is about 10 s longer than $3 \times t_{1/2}$. The pressure of N_2O_5 should be slightly less than 100 mmHg, perhaps about 95 mmHg. **13.7A** $k = 0.023$ M^{-1} s^{-1}. **13.7B** The half-life doubles for each succeeding half-life period—55 s for the first $t_{1/2}$, 110 s for the second $t_{1/2}$, 220 s for the third $t_{1/2}$, and so on. [A] = 0.20 M at t = (55 + 110) = 160 s; [A] = 0.10 M at t = (160 s + 220) = 380 s. **13.8** $Rate_2 = 3.2 \times 10^{-3}$ M s^{-1}, $rate_1 = 8.0 \times 10^{-4}$ M s^{-1}, $rate_2/rate_1 = 2^n = 4$, and $n = 2$. For straight-line graphs, plot 1/[A] versus t with data from the graphs given. **13.9A** T = 288 K (15 °C). **13.9B** E_a = 163 kJ/mol. **13.10A** First step: (slow, rate-determining), $NOCl \longrightarrow NO + Cl$; second step (fast), $NOCl + Cl \longrightarrow NO + Cl_2$. Rate of reaction = k[NOCl]. **13.10B** Mechanism consists of a fast, reversible first step: $NO + O_2 \underset{k_{-1}}{\overset{k_1}{\rightleftharpoons}} NO_3$, followed by a slow, rate-determining step: $NO_3 + NO \xrightarrow{k_2} 2\,NO_2$. Overall reaction is $2\,NO + O_2 \longrightarrow 2\,NO_2$. Assume rapid equilibrium in the first step to establish the rate law: rate = $k[NO]^2[O_2]$. **Review Questions: 5.** Except for zero-order reactions, whose rates are independent of concentration, the rate is greatest at the start of the reaction. **6.** The average rate is based on a change in concentration measured over a finite time interval; the instantaneous rate is based on an infinitesimally small time period. **7.** An instantaneous rate can be determined at any precise point in a reaction; the initial rate is the instantaneous rate at the very beginning of the reaction. The two are equal at other times only for zero-order reactions. **10.** $t_{1/2}$ is the time required for half the amount of reactant present initially to be consumed; $t_{1/2}$ is a constant (depending only on k) only for first-order reactions. For zero-order and second-order reactions, $t_{1/2}$ depends on concentration as well as on k. **12.** The sharp increase in the fraction of molecules energetic enough to react mostly accounts for increased reaction rates with temperature. **Problems: 29 (a)** 4.0×10^{-4} M s^{-1}; **(b)** 2.4×10^{-2} M min^{-1}. **31 (a)** rate of disappearance of B = -2.2×10^{-4} M s^{-1}; **(b)** rate of formation of C = 1.1×10^{-4} M s^{-1}; **(c)** general rate of reaction = 1.1×10^{-4} M s^{-1}. **33.** Zero-order reaction: $[A]_t = -kt + [A]_0$. **35 (a)** True. The rate law depends on the exponents, $m, n, \ldots$, not on concentrations; k is a constant determined by experiment. **(b)** Not true. For a second-order reaction, the units of k are M^{-1} s^{-1} and M^{-1} min^{-1}, not s^{-1} or min^{-1}. **37 (a)** First order in $HgCl_2$; second order in $C_2O_4^{2-}$; third order overall. **(b)** $k = 7.6 \times 10^{-3}$ M^{-2} min^{-1}; **(c)** 7.4×10^{-6} M min^{-1}. **39.** Second order reaction. The drop in initial rate by 1/4 when the initial concentration is reduced by 1/2, corresponds to $(1/2)^n = 1/4$, and $n = 2$. **41.** rate = 7.50×10^{-3} M s^{-1}. **43.** $k = 5.17 \times 10^{-3}$ s^{-1}. **45.** Just as $\ln 1/2 = -kt_{1/2}$, $\ln 1/4 = -kt_{3/4}$, $\ln 1/5 = -kt_{4/5}$, and so on. Thus, $t_{1/2}, t_{3/4}, t_{4/5}, \ldots$ are all constants, independent of initial concentration. **47 (a)** $k = 4.69 \times 10^{-3}$ s^{-1}; **(b)** $t_{1/2} = 148$ s; (c) 0.61 g N_2O_5. **49.** 7.1×10^{13} molecules PAN/L. **51.** $[NH_3] = 0.0441$ M. **53.** [HI] = 0.096 M. **55.** $t_{1/2}$ gets longer as $[A]_0$ increases for zero-order—in a reaction with a constant rate, the more molecules present, the longer it takes to consume half of them. $t_{1/2}$ gets shorter with increased $[A]_0$ for second-order—the greater $[A]_0$, the faster the initial rate, and the sooner half the molecules are consumed. **57.** Initial rates: rate(1) = 6.25×10^{-4} M s^{-1}; rate (2) = 1.56×10^{-4} M s^{-1}. Rate law: rate = $k[A]^2$. Rate constant, $k = 9.77 \times 10^{-4}$ M^{-1} s^{-1}. **59.** The fraction of colliding molecules with the proper orientation and sufficient energy to react cannot be precisely calculated—a calculation of overall collision frequency is insufficient. **61.** E_a is very high. Very few molecules can react at room temperature. In the high-temperature region of the spark, the number of energetic molecules is greatly increased. Reaction starts here; liberated heat raises the temperature and the entire mixture reacts with explosive speed. **63.** E_a = 91.0 kJ/mol. **65.** $k = 7.14 \times 10^{-7}$ min^{-1}. **67.** In a unimolecular process, a molecule acquires the necessary energy through molecular collisions, but a collision is not required in the actual dissociation process. **69 (a)** $A + 2\,B \longrightarrow C + D$; **(b)** rate = k[A][B]. **71.** Plausible mechanism: a slow step, $2\,NO_2 \longrightarrow NO_3 + NO$, followed by the fast step, $NO_3 + CO \longrightarrow NO_2 + CO_2$. **73.** The mechanism gives the rate law, rate = $k[NO]^2[Cl_2]/[NOCl]$, rather than rate = $k[NO]^2[Cl_2]$. **75.** Use the rapid equilibrium condition in the fast step to obtain [Hg] = $k_1[Hg_2^{2+}]/k_{-1}[Hg^{2+}]$. Substitute this value into the rate equation for the slow step: rate = $k_2[Hg][Tl^{3+}]$ to obtain the rate law. **77.** Because $[I^-]$ remains constant (it is a catalyst), the rate law, rate = $k[I^-][H_2O_2]$, simplifies to rate = $k'[H_2O_2]$. The value of k' depends on $[I^-]$ chosen, but once $[I^-]$ is fixed, the value of

k' is fixed. **79.** An inhibitor may block active sites on an enzyme, or it may distort the sites so that they can no longer accept substrate molecules. **81.** In both cases, molecules must attach to active sites and react, followed by the release of products. **Additional Problems: 83.** From three different pairs of data, $t_{1/2}$ is a constant—100 s. Determine k, and use the rate law: rate $= k[A]$, with $[A] = 0.37$ M at 125 s. Rate $= 2.6 \times 10^{-3}$ M s^{-1}. **85.** Graph ln (mL $KMnO_4$) versus t with data from Problem 84 (e.g., 0 s, 35.3 mL; 120 s, 22.6 mL; 240 s, 14.9 mL; 360 s, 9.44 mL; 480 s, 6.08 mL; 600 s, 3.76 mL). Determine k from the graph ($k = 3.63 \times 10^{-3}$ s^{-1}). **87 (a)** $P = 1.06$ atm; **(b)** $P = 2.12$ atm; **(c)** $P = 2.46$ atm. **89.** $T = 738$ K (465 °C). **91.** $E_a = 53$ kJ/mol. **93 (a)** $E_a = 51.3$ kJ/mol; **(b)** 126 chirps/min; **(c)** estimated: 71.4 °F; actual: 68 °F. **95 (a)** Rate is first order in OCl^-, first order in I^-, negative first order in OH^-, and first order overall. **(b)** rate $= k \times [OCl^-][I^-]/[OH^-]$. **(c)** The first step is a fast, reversible reaction; the rapid equilibrium assumption leads to $[HOCl] = (k_1/k_{-1}) \times [OCl^-][H_2O]/[OH^-]$. The second step is rate determining: rate $= k_2[I^-][HOCl] = (k_1 k_2[H_2O]/k_{-1}) \times [I^-][OCl^-]/[OH^-] = k[I^-][OCl^-]/[OH^-]$ (note that $[H_2O]$ is a constant). The fast third step—neutralization of HOI—removes OH^- and forms the product OI^-. **(d)** OH^- is an inhibitor rather than a catalyst. It slows down the reaction.

Chapter 14

Exercises: 14.1A No. $[COCl_2] = K_C[CO][Cl_2] = K_C[CO]^2$, but there are many possible values for $[CO] = [Cl_2]$. **14.1B** $[SO_2]$ and $[SO_3]$ do not have unique values, but the following ratios do: $[SO_3]^2/[SO_2]^2$, $[SO_2]^2/[SO_3]^2$, $[SO_3]/[SO_2]$, $[SO_2]/[SO_3]$. If $[O_2] = 1.00$ M, $[SO_3] = 10.0[SO_2]$. **14.2A** $K_C = 2.5 \times 10^{-3}$. **14.2B** $K_C = 6.64 \times 10^{78}$. **14.3A** $K_P = 4.6 \times 10^3$. **14.3B** $K_C = 3.0 \times 10^{60}$. **14.4A** $K_P = (P_{CO})(P_{H_2})/P_{H_2O}$. **14.4B** $K_C = [H_2]^4/[H_2O]^4$; $K_P = (P_{H_2})^4/(P_{H_2O})^4$. **14.5** Reaction probably does not go to completion; $K_C = 1.2 \times 10^3$ is not a particularly large value. **14.6A** Net reaction goes to the right. **14.6B** Compared to initial values, at equilibrium: P_{H_2S} increases, P_{HI} decreases, amount of I_2(s) increases, amount of S(s) decreases. **14.7** Compared to the original equilibrium, there will be **(a)** more NH_3 and H_2, less N_2 (equilibrium shifts to the right); **(b)** less NH_3 and N_2, more H_2 (to the left); **(c)** less NH_3, N_2, and H_2 (to the right). **14.8A** There is no change because no. mol gaseous products = no. mol gaseous reactants. **14.8B** Both changes shift equilibrium to the left. The amounts of NO and O_2 increase as well as the amount of NO_2. **14.9** At low temperatures, because the forward reaction is exothermic. **14.10 (a)** More CO, H_2O, and H_2 and less CO_2 (equilibrium shifts to the left). **(b)** Prediction not possible. Added H_2 favors the reverse reaction; added H_2O, the forward reaction. **(c)** Both changes shift equilibrium to the right. There will be more CO_2 and H_2, and less CO. Whether the amount of H_2O increases or decreases depends on the original equilibrium condition. **14.11A** $K_C = 25$. **14.11B** $K_P = 0.429$. **14.12A** 0.0828 mol H_2, and $P_{H_2} = 0.815$ atm. **14.12B** If the equilibrium concentrations ([ ]) are written as [ ] = mol/L, then the unit, L, cancels if it appears as many times in the numerator and the denominator. This only occurs if the same total number of moles of gas appear on each side of the reaction equation. **14.13A** 0.085 mol $COCl_2$. **14.13B** $x_{NO} = 0.018$. **14.14A** 1.76×10^{-2} mol H_2, 7.6×10^{-3} mol I_2, 0.085 mol HI. **14.14B** 0.0128 mol H_2; 0.0128 mol I_2; 0.094 mol HI. **14.15A** $P_{CO} = P_{Cl_2} = 0.19$ atm; $P_{COCl_2} = 0.81$ atm; $P_{total} = 1.19$ atm. **14.15B** $P_{total} = 0.658$ atm. **Review Questions: 12 (a)** $K_C = [CO]^2/[CO_2]$; **(b)** $K_C = [HI]^2/[H_2S]$; (c) $K_C = [O_2]$. **13 (a)** $K_C = [S_2][H_2]^2/[H_2S]^2$; **(b)** $K_C = [CH_4][H_2S]^2/[CS_2][H_2]^4$; **(c)** $K_C = [CH_4][H_2O]/[CO][H_2]^3$. **14 (a)** $K_P = (P_{CO_2})(P_{H_2})/(P_{CO})(P_{H_2O})$; **(b)** $K_P = (P_{NH_3})^2/(P_{N_2})(P_{H_2})^3$; **(c)** $K_P = (P_{NH_3})(P_{H_2S})$. **15 (a)** $K_P = (P_{NO})/(P_{N_2})^{1/2}(P_{O_2})^{1/2}$; **(b)** $K_P = (P_{NH_3})/(P_{N_2})^{1/2}(P_{H_2})^{3/2}$; **(c)** $K_P = (P_{NOCl})/(P_{N_2})^{1/2}(P_{O_2})^{1/2}(P_{Cl_2})^{1/2}$. **16 (a)** $K_C = [N_2][CO_2]^2/[CO]^2[NO]^2$; **(b)** $K_C = [NO]^4[H_2O]^6/[NH_3]^4[O_2]^5$; **(c)** $K_C = [H_2O][CO_2]$. **Problems: 23 (a)** $K_P = 22$; **(b)** $K_P = 1.61 \times 10^{-4}$; **(c)** $K_P = 2.00$. **25.** $K_C = 49.5$. **27.** $K_P = 0.412$. **29.** $K_P = 1.13 \times 10^{-10}$. **31.** $K_C = 1 \times 10^{-11}$. **33.** $K_P > 0.145$. For numbers smaller than 1, square roots are larger than the numbers themselves. **35.** $K_C = 5.4 \times 10^2$. **37.** $[COCl_2] = 6.7 \times 10^{-3}$ M. **39.** $P_{total} = 6.9$ atm. **41.** $K_P = 9$. **43 (a)** False; P_{total} decreases. **(b)** False. The maximum amount of SO_3 can't exceed the amount of SO_2 present initially. **(c)** True. The forward reaction is favored by a smaller volume. **(d)** True. Two moles of SO_3 are consumed for every mole of O_2. **(e)** False. $K_P < K_C$ because $\Delta n_{gas} < 0$. **45 (a)** $> 12.5\%$; **(b)** $> 12.5\%$; **(c)** 12.5%; **(d)** 12.5%. **47.** Because bonds are broken without new ones being formed, the reactions are endothermic. An increase in temperature favors the forward reaction—dissociation. **49 (a)** unaffected; **(b)** occurs to a greater extent; **(c)** unaffected; **(d)** occurs to a greater extent. **51.** Net reaction occurs to the left: $Q_C = 0.86 > K_C$. **53.** A slight net reaction occurs to the right, because Q_P (98.4) is slightly less than K_P (102). **55.** $K_C = 4.27 \times 10^{-2}$; $K_P = 1.83$. **57.** $K_P = 0.108$. **59.** $K_C = 1.49 \times 10^{-3}$. **61.** 4.44 g isobutane. **63.** 0.207 mol CO_2; 0.207 mol H_2. **65.** 48.2 g $COCl_2$. **67.** 0.082 mol PCl_3, 0.082 mol Cl_2, 0.028 mol PCl_5. **69.** 28% *undissociated.* **71 (a)** $P_{S_2} = 0.0271$ atm; **(b)** $P_{total} = 0.179$ atm. **73.** $P_{total} = 3.65$ atm. **75.** Partial pressure of CH_3OH(g) = 1.00 atm. **Additional Problems: 77 (a)** Equilibrium shifts to the left. **(b)** $K_C = 5.6$ for the reaction. Amounts when equilibrium is reestablished: 0.036 mol CO, 0.136 mol H_2O, 0.164 mol CO_2, 0.164 mol H_2. **79.** $K_C = 5.60$. **81.** $K_P = 4.0$. **83.** $P_{SO_2} = 8.32 \times 10^{-3}$ atm. **85 (a)** 9.77×10^{-2} mg I_2. **(b)** Additional I_2 would be present in the CCl_4, and $[I_2]_{aq}$ would be smaller than in (a). **(c)** $[I_2]_{aq}$ would be reduced more by using *two* 25.0-mL extractions with CCl_4 rather than a single one with 50.0 mL. With two extractions, $[I_2]_{aq}$ is reduced by about $1/(80 \times 80)$; with a single extraction using 50.0 mL CCl_4, the reduction is by about $1/(2 \times 80)$. **87.** $T = 536$ K (263 °C). **89. (a)** Starting with 1.00 mol PCl_5, at equilibrium: α mol PCl_3, α mol Cl_2, (1.00 − α) mol PCl_5, and (1.00 + α) mol, total. Determine the partial pressures of the three gases at temperature T and volume V, and substitute them into the K_P expression. Note also, that $RT/V = P_{total}/n_{total}$. **(b)** $\alpha = 0.645$; **(c)** $P = 176$ atm.

Chapter 15

Exercises: 15.1A (a) base: NH_3, conjugate acid, NH_4^+; acid: HCO_3^-, conjugate base, CO_3^{2-}; **(b)** acid: H_3PO_4, conjugate base, $H_2PO_4^-$; base: H_2O, conjugate acid, H_3O^+. **15.1B** Amphiprotic species in 15.1A: HCO_3^- acts as an acid in (a), and as a base in $HCO_3^- + H_3O^+ \rightleftharpoons H_2CO_3 + H_2O$; H_2O acts as a base in (b), and as an acid in $H_2O + NH_3 \rightleftharpoons OH^- + NH_4^+$. **15.2A (a)** H_2Te. The Te atom is larger than S, and the H—Te bonds are weaker. **(b)** $CH_3CH_2CH_2CHBrCOOH$. Although Cl is somewhat more electronegative than Br, it is located much farther from the —COOH group. **15.2B** a < d < b < c. The Cl and Br atoms are electron-withdrawing. The effect is weakest with Br opposite the —COOH group, but stronger with Cl adjacent to the group, and stronger still with two Cl atoms adjacent. **15.3** d < a < c < b. Aromatic amines are much weaker than aliphatic ones. Cl atoms weaken amine bases because they are electron-withdrawing, and more so with more Cl atoms closer to the $—NH_2$ group. **15.4A** pH = 4.33. **15.4B** pH = 12.624. **15.5** Basic. OH^- from the NaOH raises $[OH^-]$ above the 1.0×10^{-7} M found in water. **15.6A** pH = 2.74. **15.6B** pH = 4.52. **15.7A** $[H_3O^+] = 1.7 \times 10^{-3}$ M compared to 1.1×10^{-3} M without the assumption—an error of 55%. **15.7B** pH = 2.20. **15.8A** pH = 11.96. **15.8B** pH = 9.34. **15.9A** $K_a = 6.3 \times 10^{-5}$; $pK_a = 4.20$. **15.9B** $[NH_3] = 6.56$ M. **15.10** Methylamine. A larger K_a and higher molarity make $[OH^-]$ and pH of CH_3NH_2 greater than of $NH_3(aq)$. **15.11A** pH = 1.48. **15.11B** The molarity of H_3PO_4 in the cola is between about 4.4×10^{-3} M and 6.5×10^{-3} M. Assume the first ionization of H_3PO_4 produces all the $[H_3O^+]$. (The assumption that $[H_3PO_4] \gg K_{a_1}$ doesn't hold, however.) For 0.0044 M H_3PO_4, pH = 2.51. **15.12A** Assume complete ionization: pH = 2.77. **15.12B** (b)—complete ionization in the first step and limited in the second. Response (a) has no ionization in the second step and (c), nearly complete ionization. In (d), $[H_3O^+]$ cannot exceed 0.040 M. **15.13 (a)** Neutral—salt of strong acid and strong base; **(b)** basic—salt of weak acid and strong base. **15.13** $HCl(aq) < NH_4I(aq) < NaCl(aq) < KNO_2(aq) < NaOH(aq)$. **15.14A** pH = 5.27. **15.14B** Neutralization produces 0.0873 M CH_3COONa; pH = 8.84. **15.15A** 0.28 M $CH_3COONa(aq)$. **15.15B** 0.10 M NH_4CN. HCN is a much weaker acid than HNO_2—the weaker the acid, the stronger the anion as a base. **15.16A** pH = 8.89. **15.16B** H_3O^+ is the common ion; $[H_3O^+] = 0.10$ M; $[CH_3COO^-] = 1.8 \times 10^{-5}$ M. **15.17A** pH = 9.10. **15.17B** 0.020 mol NaOH. **15.18A** $[CH_3COOH] = 0.43$ M. **15.18B** 3.2 g NH_4Cl. **15.19** The pH ≈ 5. This eliminates (b)—a basic buffer—and (c)—strongly acidic. Add a small quantity of acid or base. If there is a color change, the solution is (a), if not, it is the buffer (d). **15.20A (a)** pH = 2.90; **(b)** pH = 3.90; **(c)** pH = 10.10; **(d)** pH = 11.10. **15.20B** pH = 13.02. **15.21 (a)** pH = 4.96; **(b)** pH = 11.10. **15.22** $K_b \approx 1 \times 10^{-5}$; pH at the equiv. point: ≈ 5 (estimated from graph); 4.7 (calculated, assuming $K_b = 1 \times 10^{-5}$). **Review Questions: 1.** Arrhenius: $HI \xrightarrow{H_2O} H^+ + I^-$; Brønsted-Lowry: $HI + H_2O \longrightarrow H_3O^+ + I^-$. **5 (a)** $HClO_2 + H_2O \rightleftharpoons H_3O^+ + ClO_2^-$; $K_a = [H_3O^+][ClO_2^-]/[HClO_2]$; **(b)** $CH_3CH_2COOH + H_2O \rightleftharpoons H_3O^+ + CH_3CH_2COO^-$; $K_a = [H_3O^+][CH_3CH_2COO^-]/[CH_3CH_2COOH]$; **(c)** $HCN + H_2O \rightleftharpoons H_3O^+ + CN^-$; $K_a = [H_3O^+][CN^-]/[HCN]$; **(d)** $C_6H_5OH + H_2O \rightleftharpoons H_3O^+ + C_6H_5O^-$; $K_a = [H_3O^+][C_6H_5O^-]/[C_6H_5OH]$. **15.** K_a(acid) × K_b(conjugate base) = K_w. **20.** Neutralization is just complete at the equivalence point. The end point is reached when the indicator changes color. **22.** The pH is highest at the start of the titration and lowest at the end. The pH < 7 at the equivalence point. **23.** The slope of the titration curve is too gradual to assess the equivalence point precisely. **Problems: 25** The pairs, acid(1)/base(1), followed by base(2)/acid(2), are **(a)** $HOClO_2/OClO_2^-$, H_2O/H_3O^+; **(b)** $HSeO_4^-/SeO_4^{2-}$, NH_3/NH_4^+; **(c)** HCO_3^-/CO_3^{2-}, OH^-/H_2O; **(d)** $C_5H_5NH^+/C_5H_5N$, H_2O/H_3O^+. **27.** $HSO_3^-(aq) + OH^-(aq) \longrightarrow H_2O(l) + SO_3^{2-}(aq)$; $HSO_3^-(aq) + HBr(aq) \longrightarrow H_2SO_3(aq) + Br^-(aq)$. **29.** (a). CO_3^{2-} is the strongest base: $CH_3COOH(aq) + CO_3^{2-}(aq) \longrightarrow HCO_3^-(aq) + CH_3COO^-(aq)$.

31. $H_3N: + H_3N: \rightleftharpoons [NH_4]^+ + [H—\ddot{N}—H]^-$ (Lewis structures: H—N(H)—H with lone pair + H—N(H)—H with lone pair ⇌ [H—N(H)(H)—H]$^+$ + [H—N̈—H]$^-$)

33. The molecule, $HOClO_3$, has three terminal O atoms, whereas HOClO has only one. **35. (a)** H_2Se; bond strengths: H—Se < H—S. **(b)** $HClO_3$; $HOClO_2$ and $HOIO_2$ have the same number of terminal O atoms, but Cl has a greater EN than I. **(c)** H_3AsO_4; the loss of a proton by $H_2PO_4^-$ is more difficult than by the neutral molecule H_3AsO_4. **37.** pK_a of $ClCH_2COOH$, 2.85; expected pK_a of $Cl_2CHCOOH$, ≈ 2 (compare 2-chloropropanoic acid and 2,2-dichloropropanoic acids); expected pK_a of Cl_3COOH, significantly below 2. **39.** a < d < b < c; based on number of Cl atoms and their proximity to the —OH group. Phenol, C_6H_5OH, should be weaker than (a). **41.** $[H_3O^+] = 4.8 \times 10^{-5}$ M. **43 (a)** pH = 2.60; **(b)** pH = 12.74; **(c)** pH = 12.48; **(d)** pH = 2.80. **45.** The vinegar solution. **47.** Dilute 5.0 mL of 0.100 M HCl to 2.00 L with water. **49.** pH = 1.77. **51.** 0.033 M HN_3. **53.** $K_a = 3.6 \times 10^{-4}$. **55.** pH = 1.08. **57.** 0.017 M HCOOH. **59.** 0.0045 M H_2SO_4. H_2SO_4 is completely ionized in its first step and even its $K_{a_2} > K_{a_1}$ of H_3PO_4, **61 (a)** pH = 0.70; **(b)** $[^-OOCCOO^-] = 5.3 \times 10^{-5}$ M. **63 (a)** pH = 1.53; **(b)** $[H_3PO_4] = 0.12$ M; **(c)** $[H_2PO_4^-] = 0.029$ M; **(d)** $[HPO_4^{2-}] = 6.3 \times 10^{-8}$ M; **(e)** $[PO_4^{3-}] = 9.3 \times 10^{-19}$ M. **65 (a)** basic: $CH_3CH_2COO^- + H_2O \rightleftharpoons CH_3CH_2COOH + OH^-$; **(b)** neutral: salt of strong acid and strong base; **(c)** basic: $NH_4^+ + H_2O \rightleftharpoons H_3O^+ + NH_3$ and $CN^- + H_2O \rightleftharpoons HCN + OH^-$, but CN^- ionizes more strongly as a base than NH_4^+ as an acid. **67.** (c). NH_4^+ is the only ion of the group that ionizes as an acid. **69 (a)** $OCl^- + H_2O \rightleftharpoons HOCl + OH^-$; **(b)** $K = K_w/K_a = 3.4 \times 10^{-7}$; **(c)** pH = 10.22. **71.** 0.23 M CH_3COONa. **73.** (c) and (d). $HNO_3(aq)$ supplies H_3O^+ as a common ion, and HCOONa(aq) supplies $HCOO^-$. **75.** $[NH_4^+] = 1.8 \times 10^{-4}$ M. **77.** pH = 4.24. **79.** pH > 7: (b) NH_3, a weak base, (e) $Ba(OH)_2$, a strong base, (f) $NaNO_2$, salt of a weak acid and strong base; pH < 7: (a) C_6H_5COOH, a weak acid, (c) CH_3NH_3Cl, salt of a strong acid and weak base, (d) $H_2PO_4^-$ ionizes more extensively as an acid ($K_a = 6.3 \times 10^{-8}$) than as a base ($K_b = 1.4 \times 10^{-12}$). **81.** $HPO_4^{2-} + H_2O \rightleftharpoons$

$H_3O^+ + PO_4^{3-}$; $K_{a_3} = 4.3 \times 10^{-13}$; $HPO_4^{2-} + H_2O \rightleftharpoons H_2PO_4^- + OH^-$; $K_b = K_w/K_{a_2} = 1.6 \times 10^{-7}$. Hydrolysis of HPO_4^{2-} exceeds its further ionization as an acid; $Na_2HPO_4(aq)$ is basic. **83.** pH $= 3.65$. **85.** 0.38 g $(NH_4)_2SO_4$. **87.** pH $= 3.63$. **89.** No, HCl and NaOH react to form NaCl(aq), which has no buffer capacity. If either reactant is in excess, the solution is either acidic or basic, not a buffer. **91.** The rapid change in pH at the equivalence point occurs over a broader range in a strong acid/strong base titration than in a weak acid/strong base titration. **93 (a)** yellow; **(b)** blue; **(c)** red; **(d)** violet. **95.** The solution will be 0.0500 M NaCl(aq) with pH $= 7$; the indicator will be yellow. **97.** The pH starts high, changes slowly until pH ≈ 11, then falls steeply to pH ≈ 3 with the addition of only a drop or two of titrant. The equivalence point (pH $= 7$) is in the middle of the region of steep decline. Beyond pH ≈ 3, the pH decreases very slowly. Use the same indicators as in the titration of a strong acid with a strong base. **99.** Yes (see answer to Problem 97). **101.** pH $= 4.51$. **103.** 31.7 mL. **105 (a)** 16.00 mL to equivalence point; **(b)** initial pH $= 11.28$; **(c)** after 5.00 mL, pH $= 9.60$; **(d)** at half-neutralization, pOH $= pK_b = 4.74$, and pH $= 9.26$; **(e)** after 10.00 mL, pH $= 9.04$; **(f)** at equivalence point, pH $= 5.05$; **(g)** after 20.00 mL, pH $= 1.48$. **107.** Lewis acids: (a) $AlOH_3(s)$, (b) $Cu^{2+}(aq)$, (c) $CO_2(g)$; Lewis bases: (a) $OH^-(aq)$, (b) $NH_3(aq)$, (c) $OH^-(aq)$. **Additional Problems: 109.** 1.8×10^2 g CaO. **111.** pH $= 6.98$. **113.** 0.0200 mol NaOH converts all the CH_3COOH to CH_3COO^-. The pH is ≈ 9 (recall Example 15.15), about the pH at which thymol blue changes color. **115.** K_w increases with T. The neutralization reaction is exothermic, and the reverse reaction is endothermic.

117. $H_2N{-}NH_2 + H_2O \rightleftharpoons [H_3N{-}NH_2]^+ + OH^-$, $pK_{b_1} = 6.07$;

$[H_3N{-}NH_2]^+ + H_2O \rightleftharpoons [H_3N{-}NH_3]^{2+} + OH^-$, $pK_{b_2} = 15.05$;

$pK_{b_2} \gg pK_{b_1}$ because the second proton must attach to a positive ion instead of a neutral molecule. For 0.150 M N_2H_4 pH $= 10.55$. **119 (a)** Strong base titrated with a weak acid; **(b)** $pK_a \approx 3.8$; **(c)** at equivalence point, $8 < \text{pH} < 9$ (from the graph), and pH $= 8.7$ (by calculation). **121 (a)** [1] $H_2PO_4^- + H_2O \rightleftharpoons H_3O^+ + HPO_4^{2-}$, K_{a_2}; [2] $H_2PO_4^- + H_2O \rightleftharpoons H_3PO_4 + OH^-$, $K_b = K_w/K_{a_1}$. **(b)** Assume $[H_3O^+] = [HPO_4^{2-}]$ in [1] and $[OH^-] = [H_3PO_4]$ in [2]. Also, assume H_3O^+ from reaction [1] neutralizes nearly all the OH^- from [2]. This means that $[H_3O^+] = [HPO_4^{2-}] - [H_3PO_4]$, the key equation to be solved. Substitute for $[HPO_4^{2-}]$ and $[H_3PO_4]$ using K values for [1] and [2], respectively. For $[OH^-]$, substitute $K_w/[H_3O^+]$. Solve the equation—now with the terms $[H_3O^+]$, $[H_2PO_4^-]$, K_{a_1} and K_{a_2},—for $[H_3O^+]$. **123.** pH $= 5.68$. **125.** Addition of small amounts of H_3O^+ or OH^- cause a large change in pH. The pH of $NaH_2PO_4(aq)$ is at the midpoint of the first steeply rising portion of the titration curve in the titration of H_3PO_4 with NaOH. The aqueous solution, NaH_2PO_4/NaH_2PO_4, is a buffer; it corresponds to the very slowly rising portion between the first and second equivalence points on the titration curve. **127.** When $[H_3O^+]$ is reduced by a factor of two, the pH change is log 2 $= 0.3010$. This effect of dilution on pH is true only for solutions of strong acids and strong bases. For example, the pH of neutral salt solutions and buffer solutions do not change on dilution.

Chapter 16

Exercises: 16.1A (a) $MgF_2(s) \rightleftharpoons Mg^{2+}(aq) + 2\,F^-(aq)$, $K_{sp} = [Mg^{2+}][F^-]^2$; **(b)** $Li_2CO_3(s) \rightleftharpoons 2\,Li^+(aq) + CO_3^{2-}(aq)$, $K_{sp} = [Li^+]^2[CO_3^{2-}]$; **(c)** $Cu_3(AsO_4)_2(s) \rightleftharpoons 3\,Cu^{2+}(aq) + 2\,AsO_4^{3-}(aq)$, $K_{sp} = [Cu^{2+}]^3[AsO_4^{3-}]^2$. **16.1B (a)** $Mg(OH)_2(s) \rightleftharpoons Mg^{2+}(aq) + 2\,OH^-(aq)$, $K_{sp} = [Mg^{2+}][OH^-]^2$; **(b)** $ScF_3(s) \rightleftharpoons Sc^{3+}(aq) + 3\,F^-(aq)$, $K_{sp} = [Sc^{3+}][F^-]^3$; **(c)** $Zn_3(PO_4)_2(s) \rightleftharpoons 3\,Zn^{2+}(aq) + 2\,PO_4^{3-}(aq)$, $K_{sp} = [Zn^{2+}]^3[PO_4^{3-}]^2$. **16.2A** $K_{sp} = 2.0 \times 10^{-11}$. **16.2B** $K_{sp} = 1.1 \times 10^{-12}$ M. **16.3A** 1.4×10^{-6} M. **16.3B** 3.1×10^2 ppm of I^-. **16.4** The trend in molar solubilities is the same as in K_{sp} values: CaF_2 ($K_{sp} = 5.3 \times 10^{-9}$) $< PbI_2$ (7.1×10^{-9}) $< MgF_2$ (3.7×10^{-8}) $< PbCl_2$ (1.6×10^{-5}). **16.5A** 1.4×10^{-5} M. **16.5B** 5.0 g $AgNO_3$. **16.6A** Yes, Q_{ip} (1.2×10^{-8}) $> K_{sp}$. **16.6B** No, Q_{ip} (1.3×10^{-12}) $< K_{sp}$. **16.7** Add KI(aq) dropwise from a buret to a known volume of solution of known $[Pb^{2+}]$. Stir after each drop is added, observing first the appearance and then disappearance of $PbI_2(s)$. Continue until a single drop produces a lasting precipitate. Now, $Q_{ip} = K_{sp}$. Calculated K_{sp} from the $[Pb^{2+}]$ and $[I^-]$. **16.8A** Yes, Q_{ip} (5×10^{-8}) $> K_{sp}$. **16.8B** 3.3 g KI. **16.9** Yes, only $\approx 0.01\%$ of Ca^{2+} remains in solution. **16.10A (a)** Br^- precipitates first; **(b)** $[Br^-] = 2.8 \times 10^{-5}$ M when Cl^- begins to precipitate; **(c)** precipitation of Br^- is not quite complete. **16.10B** I^- and Br^- can be separated. K_{sp} for AgI and AgBr differ by nearly 6000-fold, much greater than 360-fold between AgBr and AgCl. **16.11A** 0.80 M. **16.11B** No, Q_{ip} (7×10^{-33}) $> K_{sp}$. **16.12** $Mg(OH)_2(s) + 2\,NH_4^+(aq) \longrightarrow Mg^{2+}(aq) + 2\,H_2O(l) + 2\,NH_3(aq)$. **16.13A** $[Ag^+] = 9.2 \times 10^{-15}$ M. **16.13B** $[NH_3]_{total} = 0.66$ M. **16.14A** No, Q_{ip} (5.5×10^{-17}) $< K_{sp}$ of AgI. **16.14B** 4.8×10^{-3} g KBr. **16.15A** 0.22 M. **16.15B** NaCN(aq). K_f of $[Ag(CN)_2]^-$ is much larger than that of $[Ag(S_2O_3)_2]^{3-}$ or $[Ag(NH_3)_2]^+$. **Review Questions: 1.** $K_{sp} = [Fe^{3+}][OH^-]^3$. **2.** $Zn_3(PO_4)_2(s) \rightleftharpoons 3\,Zn^{2+}(aq) + 2\,PO_4^{3-}(aq)$. **5.** The compounds are of the same type (MX); $PbSO_4$ has the larger K_{sp} and is the more soluble. **10.** A solution with a common ion is used, because less precipitate will dissolve than in pure water. **12.** HCl and CH_3COOH react with OH^-, increasing $[Fe^{3+}]$. The other solutions reduce $[Fe^{3+}]$ by supplying the common ion OH^-. **15.** $Pb(NO_3)_2(aq)$ reduces the solubility of $PbCl_2$ through the common-ion effect; HCl(aq) increases it through the formation of the complex ion $[PbCl_3]^-$. **17.** HCl increases the solubility of $Cu(OH)_2$ through an acid-base reaction, and NH_3, through complex-ion formation. NaOH reduces the solubility through the common ion, OH^-. More water does not affect the solubility. **18.** $[Al(H_2O)_6]^{3+} + H_2O \rightleftharpoons H_3O^+ + [AlOH(H_2O)_5]^{2+}$. **Problems: 23 (a)** $Hg_2(CN)_2(s) \rightleftharpoons Hg_2^{2+}(aq) + 2\,CN^-(aq)$, $K_{sp} = [Hg_2^{2+}][CN^-]^2 = 5 \times 10^{-40}$; **(b)** $Ag_3AsO_4(s) \rightleftharpoons 3\,Ag^+(aq) + AsO_4^{3-}(aq)$,

$K_{sp} = [Ag^+]^3[AsO_4^{3-}] = 1.0 \times 10^{-22}$. **25.** Molar solubility and K_{sp} cannot have the same value. Molar solubility is equal to an ion concentration or some fraction of it; K_{sp} is a *product* of two ion concentrations raised to powers. Molar solubilities are generally larger than K_{sp} because the roots of a number < 1 are larger than the number. **27.** $K_{sp} = 1.6 \times 10^{-9}$. **29.** Despite its smaller K_{sp}, Ag_2CrO_4 is more soluble than AgCl. **31.** $K_{sp} = 2.4 \times 10^{-21}$. **33.** $CaSO_4$ is more soluble than $CaCO_3$; it has a larger K_{sp}. $[Ca^{2+}]$ is greater in $CaSO_4$(satd aq) than in CaF_2(satd aq), because $(9.1)^{1/2}$, is larger than $1/4 \times (5.3)^{1/3}$. **35.** 7.1 ppb Cu^{2+}. **37.** Yes, 22 mg $PbCl_2$ should precipitate. **39.** Pure water. The other two solutions contain common ions and reduce the solubility. **41.** 6.7×10^{-6} M. **43.** $[Pb^{2+}] = 4.4 \times 10^{-8}$ M; $[I^-] = 0.400$ M. **45 (a)** $[Ag^+] = 3.0 \times 10^{-2}$ M; **(b)** 62 g Na_2SO_4. **47.** 3.6×10^{-3} M. **49.** $[CrO_4^{2-}] = 1.0 \times 10^{-6}$ M. **51.** Yes, Q_{ip} $(1.25 \times 10^{-13}) > K_{sp}$ of Hg_2Cl_2. **53.** No, Q_{ip} $(9.9 \times 10^{-9}) < K_{sp}$ of MgF_2. **55.** No, Q_{ip} $(5.5 \times 10^{-12}) < K_{sp}$ of CaF_2. **57 (a)** $[Mg^{2+}] = 4\ 5 \times 10^{-6}$ M; **(b)** Yes, only 0.0076% of Mg^{2+} remains in solution. **59.** pH = 10.85. **61 (a)** CaF_2(s); **(b)** $[F^-] = 1.9 \times 10^{-3}$; **(c)** No, 15% of the Ca^{2+} remains in solution when MgF_2(s) begins to precipitate. **63 (a)** Essentially complete precipitation of 1.0×10^{-3} mol AgCl(s) occurs. **(b)** Because $[Ag^+]$ is too low following precipitation of AgCl(s), Ag_2SO_4 does not precipitate. **65.** 4.1×10^{-5} M. **67.** $NaHSO_4$. The reaction, $HSO_4^- + CO_3^{2-} \longrightarrow HCO_3^- + SO_4^{2-}$, replaces insoluble $CaCO_3$ by the more soluble $Ca(HCO_3)_2$. **69.** $CaCO_3(s) + 2\ H_3O^+(aq) \longrightarrow Ca^{2+}(aq) + 3\ H_2O(l) + CO_2(g)$; $CaCO_3(s) + 2\ CH_3COOH(aq) \longrightarrow Ca^{2+}(aq) + 2\ CH_3COO^-(aq) + H_2O(l) + CO_2(g)$. **71.** $[CrCl_2(NH_3)_4]^+$. **73.** $[Ag^+]$ is lowest in $[Ag(CN)_2]^-$. **75.** $2\ Ag^+(aq) + SO_4^{2-}(aq) \longrightarrow Ag_2SO_4(s)$; $Ag_2SO_4(s) + 2\ NH_3(aq) \longrightarrow [Ag(NH_3)_2]^+(aq) + SO_4^{2-}(aq)$; $[Ag(NH_3)_2]^+(aq) + 2\,H_3O^+(aq) \longrightarrow Ag^+(aq) + 2\,NH_4^+(aq) + 2\,H_2O(l)$. **77 (a)** $Zn^{2+}(aq) + 4\ OH^-(aq) \longrightarrow [Zn(OH)_4]^{2-}(aq)$; **(b)** $Al_2O_3(s) + 3\,H_2O(l) + 6\ H_3O^+(aq) \longrightarrow 2\ [Al(H_2O)_6]^{3+}(aq)$; **(c)** $Fe(OH)_3(s) + OH^-(aq) \longrightarrow$ N.R. **79.** $[Zn^{2+}] = 1.2 \times 10^{-10}$ M. **81.** A trace of PbI_2(s) should precipitate; Q_{sp} (2.1×10^{-8}) is slightly larger than K_{sp}. **83.** 4.0 mg KBr. **85.** 0.043 M. **87.** (PbS) $\ll K_{sp}$ ($PbCl_2$). Enough Pb^{2+} remains from cation group 1 that K_{sp} of PbS is exceeded. $[Ag^+]$ in equilibrium with AgCl(s) is so low that Ag^+ does not later precipitate as Ag_2S. **89.** Hg_2^{2+} is definitely present; the presence of Ag^+ and Pb^{2+} is uncertain—no tests were performed. **Additional Problems: 91 (a)** $[Ba^{2+}]$ in saturated $BaSO_4$(aq) is too low to be hazardous. **(b)** 1.4 mg Ba^{2+}/L. **(c)** SO_4^{2-} from $MgSO_4$ reduces the solubility of $BaSO_4$. **93.** 2.3 g calcium palmitate. **95.** NH_4^+ from NH_4Cl suppresses the ionization of NH_3, reducing $[OH^-]$. When $[OH^-]$ is low enough, $Mg(OH)_2$(s) no longer precipitates; 50 g NH_4Cl is required. **97.** $K_{sp} = 1.62 \times 10^{-7}$. **99.** $Pb(N_3)_2(s) + 2\ H_3O^+(aq) \rightleftharpoons Pb^{2+}(aq) + 2\,HN_3(aq) + 2\,H_2O(l)$; $K_c = 6.9$. Solubility at pH 2.85 is 0.015 mol $Pb(N_3)_2$/L. **101.** Combine these equations and their K values: (1) $M^{2+}(aq) + HS^-(aq) + OH^-(aq) \rightleftharpoons MS(s) + H_2O(l)$, $K = 1/K_{sp}$; (2) $2\ H_2O(l) \rightleftharpoons H_3O^+(aq) + OH^-(aq)$, $K = K_w$; (3) $H_2S(g) + H_2O(l) \rightleftharpoons H_3O^+(aq) + HS^-(aq)$, $K = K_{a_1}$. Overall: $K_c = K_w \times K_{a_1}/K_{sp}$. For FeS, $K_c = 2 \times 10^{-3}$. **103.** 0.10 mL 0.0050 M NaBr.

Chapter 17

Exercises: 17.1 (a) Spontaneous. Cellulose decomposes into simpler molecules, such as CO_2 and H_2O, through the action of microorganisms. **(b)** Nonspontaneous. A compound cannot be decomposed through a physical change. **(c)** Uncertain. All compounds decompose at very high temperatures, but it is uncertain whether this decomposition will produce CO_2(g) at 1 atm at 600 °C. **(d)** Spontaneous. HCl(g) dissociates completely simply by dissolving in water. **17.2A (a)** Decrease. Two moles of gas are converted to one of solid. **(b)** Increase. Two moles of solid are converted to two moles of solid and 3 moles of gas. **(c)** Uncertain. Two moles of gaseous reactants produce two moles of gaseous products. **17.2B** Vaporization produces an increase in entropy (disorder), but enthalpy also increases. In condensation, entropy and enthalpy both decrease. At equilibrium vaporization and condensation occur at equal rates. **17.3A** $\Delta S° = -42.1$ J/K. **17.3B** $\Delta S° = 131.0$ J/K. **17.4A (a)** $\Delta S < 0$, $\Delta H < 0$, case 2; **(b)** $\Delta S > 0$, $\Delta H > 0$, case 3. **17.4B** $\Delta S > 0$, $\Delta H° = -56.9$ kJ, case 1. **17.5** The phase diagram of CO_2 (Figure 11.12) shows CO_2(s) and CO_2(g) are in equilibrium at 1 atm at −78.5 °C. **17.6A (a)** $\Delta G° = -70.5$ kJ; **(b)** $\Delta G° = 447.5$ kJ. **17.6B (a)** $\Delta G° = -66.9$ kJ; **(b)** $\Delta G° = -1305.7$ kJ. **17.7A** $\Delta H°_{vap} = 36$ kJ mol^{-1}. **17.7B** Because of extensive hydrogen bonding in CH_3OH(l), $\Delta S°_{vapn} > 87$ J mol^{-1} K^{-1}. $\Delta S° = 112.9$ J mol^{-1} K^{-1} is based on $S°$ values in Appendix C; it is 112.5 J mol^{-1} K^{-1} based on $\Delta H_f°$ values and $\Delta S° = \Delta H°/T_{bp}$. **17.8A** $K_{eq} = [Al^{3+}]^2(P_{H_2})^3/[H^+]^6$. **17.8B** $Mg(OH)_2(s) + 2\ H_3O^+(aq) \longrightarrow Mg^{2+}(aq) + 4\ H_2O(l)$; $K_{eq} = [Mg^{2+}]/[H_3O^+]^2$. **17.9A** $K_{eq} = 3.02 \times 10^{-21}$. **17.9B** $K_{eq} = 27$; $P_{NO} = 0.16$ atm, $P_{NOBr} = 0.84$ atm. **17.10A** $K_{eq} = 0.45$. **17.10B** With the Clausius-Clapeyron equation, $P_{H_2O} = 55.7$ mmHg; from Table 11.2, $P_{H_2O} = 55.3$ mmHg. **17.11** $\Delta S° = 0.201$ kJ/K, about the same result as in Example 17.11. Example 17.11 and Exercise 17.11 are based on the same equations and on the same assumption that ΔH and ΔS are independent of T. We expect them to give the same result. **Review Questions: 3 (a)** Spontaneous. Microorganisms lead to the souring of milk without outside intervention. **(b)** Nonspontaneous. External action is required to convert copper ores into copper metal. **(c)** Spontaneous. No external action is required to get steel (iron) to corrode in moist air. **4 (a)** Decrease (liquid $\longrightarrow$ solid); **(b)** increase (solid $\longrightarrow$ gas); **(c)** increase (liquid + oxygen gas $\longrightarrow$ large volume of gaseous products). **10.** NOF_3 has a greater number of atoms and vibrational modes, and hence a greater entropy, than NO_2F. **12.** Low temperatures. The ΔH term dominates in the Gibbs equation, offsetting a positive value of $-T\Delta S$ and making $\Delta G < 0$. **13.** No. Water vaporizes spontaneously but produces vapor at $P \leqq$ vapor pressure. The vapor pressure of H_2O is 1 atm only at 100 °C. **15.** The mixture of gases is more disordered than the individual gases, $\Delta S > 0$. Intermolecular forces are the same in the separated gases and in the mixture, $\Delta H = 0$. From the Gibbs equation, $\Delta G = \Delta H - T\Delta S$, $\Delta G < 0$. **20.** If $\Delta G° = 0$, $K_{eq} = 1$, because $\Delta G° = -RT \ln K_{eq}$. **Problems: 21 (a)** Increase—gas is less ordered than liquid. **(b)** Increase—process produces increased number of molecules of gas. **(c)** Indeterminate—same number of moles of gas on each side of equation. **(d)** Increase—large amount of gas produced from a solid. **23.** (b). Sublimation yielding a gas at low pressure produces more disorder than melting of a solid at a higher pressure. **25.** The final state: heterogeneous mixture of octane floating on water. Solution formation would require $\Delta H > 0$ to break hydrogen bonds in the water. The increase in entropy would not be sufficient to overcome the large ΔH. **27.** Correct: For a process to occur spontaneously, the total entropy—S_{univ}—must increase. Errors in other statements: (a) entropy of the system may increase in some cases and decrease in others; (b) entropy of the surroundings may also increase or decrease; (c) entropy of the system and surroundings need not both increase, as long as the increase in one exceeds the decrease in the other. **29.** The first generalizations do not help because one mole of gaseous reactant produces one mole of gaseous product (discounting the small entropy of the solid). The additional generalization suggests that the more complex S_2Cl_2 has a higher entropy than Cl_2. With data from Appendix C, $\Delta S° = 44.9$ J/K. **31.** In pollution, complex substances break down into simpler ones and become widely dispersed—$\Delta S_{univ} > 0$. To clean up pollution, dispersed materials must be brought together and processed into other substances—$\Delta S_{syst} < 0$. This requires external actions, and $\Delta S_{surr} > 0$. **33.** A criterion based on ΔS requires assessing both ΔS_{syst} and ΔS_{surr}; the free energy change requires only measurements in the system: $\Delta G = \Delta H - T\Delta S$. **35.** The line representing ΔH as a function of T is essentially parallel to the T axis. The $T\Delta S$ line has a negative slope. It starts high at low T and becomes smaller as T increases. The $T\Delta S$ line lies above the ΔH line at low T values ($\Delta G < 0$) and below it at high T values ($\Delta G > 0$). At some intermediate T the lines cross and $\Delta G = 0$. **37.** Melting of a solid is nonspontaneous below its melting point and spontaneous above its melting point—0 °C for water. **39 (a)** Low temperatures—$\Delta S < 0$, $\Delta H < 0$. **(b)** High temperatures—$\Delta S > 0$, $\Delta H > 0$. (c) All temperatures—$\Delta S > 0$, $\Delta H < 0$. **41.** Because two moles of gaseous product are formed from one of gaseous reactant, $\Delta S > 0$. Because the reaction is nonspontaneous at room temperature, $\Delta H > 0$. The reaction becomes spontaneous by raising T. **43 (a)** $\Delta G° = -101.01$ kJ; **(b)** $\Delta G° = -346.9$ kJ. **45.** $\Delta H° = 131.3$ kJ; $\Delta S° = 133.8$ J/K; $\Delta G° = 91.4$ kJ (by both methods). **47.** With $\Delta G°$, we can evaluate K_{eq}: $\Delta G° = -RT \ln K_{eq}$. With K_{eq}, we can determine an equilibrium condition. To evaluate ΔG for nonstandard conditions we use $\Delta G = \Delta G° + RT \ln Q$ (where Q is reaction quotient). However, we cannot obtain ΔG without knowing $\Delta G°$. **49.** 364 K (91 °C). **51.** Estimate using Trouton's rule: $\Delta H° \approx 29$ kJ/mol. With data from Appendix C: $\Delta H° = 30.91$ kJ/mol. **53 (a)** $\Delta G = 0$; **(b)** $\Delta G° = -24.4$ kJ. **55 (a)** $K_{eq} = K_p = (P_{NO_2})^2/(P_{NO})^2(P_{O_2})$; **(b)** $K_{eq} = K_p = P_{SO_2}$; **(c)** $K_{eq} = K_a = [H_3O^+][CN^-]/[HCN]$. **57 (a)** $K_p = 6.5 \times 10^{-25}$; **(b)** $K_p = 1 \times 10^{96}$. **59.** 27 mmHg. **61 (a)** $\Delta G° = -15.9$ kJ. **(b)** Direction of net reaction: $\longrightarrow$. **(c)** 0.037 mol CO, 0.064 mol H_2O, 0.193 mol CO_2, 0.274 mol H_2. **63.** The reaction to be coupled with the given reaction must have $\Delta G° < -237.9$ kJ. Only (c) will work; it has $\Delta G° = -257.2$ kJ. **65.** $T = 269$ K (−4 °C). **67.** 390.4 K. **69.** $K_p = 2.0 \times 10^6$. **71.** Estimated boiling point: 348 K. Use $\Delta H°_{298}$ and $\Delta G°_{298}$ from Appendix C; calculate K_p at 298 K; use the van't Hoff equation to determine T at which $K_p = 1.00$ atm. **73.** 425 K. **Additional Problems: 75.** The "all heads" and "all tails" arrangements are highly improbable. There are many possibilities of "mixed up" arrangements, and one of these is far more likely to be seen. **77.** Even if $\Delta S°$ and $\Delta H°$ are independent of temperature, $T\Delta S°$ cannot be, nor can $\Delta G°$ ($\Delta G° = \Delta H° - T\Delta S°$) **79.** Conservation of energy (1st law) states that energy cannot be created or destroyed—the energy of the universe is constant. The 2nd law of thermodynamics states that all spontaneous change requires $\Delta S_{univ} > 0$. Thus, entropy constantly increases. **81.** $\Delta G = 9.34$ kJ/mol. **83.** Estimate: $\Delta G° = 9.8$ kJ/mol. (Use the Clausius–Clapeyron equation twice—to obtain $\Delta H°$ and then the vapor pressure at 298 K—followed by $\Delta G° = -RT \ln K_p$, at 298 K.) From data in Appendix C: $\Delta G° = 10.1$ kJ/mol. **85.** $\Delta H° \approx -190$ kJ/mol. **87.** Sublimation pressure of Hg(s) at −78.5 °C: $\approx 10^{-9}$ mmHg. **89.** $T = 631$ K. **91.** $K_{sp} = 1.1 \times 10^{-5}$ (close to the listed value of 1.4×10^{-5}). **93.** At equilibrium: $[Ag^+] = [Fe^{2+}] = 0.138$ M, $[Fe^{3+}] = 0.062$ M ($K_{eq} = 3.24$).

Chapter 18

Exercises: 18.1A $4\,Zn(s) + 2\,NO_3^-(aq) + 10\,H^+(aq) \longrightarrow 4\,Zn^{2+}(aq) + N_2O(g) + 5\,H_2O(l)$. **18.1B** $3\,P_4(s) + 20\,NO_3^-(aq) + 8\,H_2O(l) + 8\,H^+(aq) \longrightarrow 12\,H_2PO_4^-(aq) + 20\,NO(g)$. **18.2A** $2\,OCN^-(aq) + 3\,OCl^-(aq) + 2\,OH^-(aq) \longrightarrow 2\,CO_3^{2-}(aq) + N_2(g) + H_2O(l) + 3\,Cl^-(aq)$. **18.2B** $4\,MnO_4^-(aq) + 3\,CH_3CH_2OH(aq) \longrightarrow 4\,MnO_2(s) + 3\,CH_3COO^-(aq) + OH^-(aq) + 4\,H_2O(l)$. **18.3A** $2\,Al(s) + 6\,H^+(aq) \longrightarrow 2\,Al^{3+}(aq) + 3\,H_2(g)$. **18.3B** $Cu(s)|Cu^{2+}(aq)||Ag^+(aq)|Ag(s)$. **18.4A** $E°_{Sm^{2+}/Sm} = -2.67$ V. **18.4B** $E°_{ClO_4^-/Cl_2} = 1.392$ V. **18.5 (a)** $E°_{cell} = 2.016$ V; **(b)** $E°_{cell} = 0.50$ V; **(c)** $E°_{cell} = -0.499$ V. **18.6A** No, $E°_{cell} = -0.431$ V. **18.6B** The reverse reaction is spontaneous; for the forward reaction, $E°_{cell} = -0.31$ V. **18.7** Similar to Example 18.7, Zn is oxidized, and electrons pass through the voltmeter to the Cu electrode, where H^+ (from lemon juice) is reduced to $H_2(g)$. **18.8A** $\Delta G° = -394$ kJ; $K_{eq} = 1 \times 10^{69}$. **18.8B** For the reaction: $3\,Ag(s) + 4\,H^+(aq) + NO_3^-(aq) \longrightarrow 3\,Ag^+(aq) + NO(g) + 2\,H_2O(l)$, $E°_{cell} = 0.156$ V; $K_{eq} = 8.0 \times 10^7$. **18.9 (a)** $E_{cell} = 1.056$ V; **(b)** $E_{cell} = 1.150$ V; **(c)** $E_{cell} = 1.045$ V. **18.10** The bend, like the head and tip, is strained and more energetic than the body of the nail. It is an anodic area (blue precipitate). **18.11A** $2\,Br^-(aq) + 2\,H_2O(l) \longrightarrow Br_2(l) + 2\,OH^-(aq) + H_2(g)$; $E°_{cell} = -1.893$ V. **18.11B** To force the electrolysis, Ag(s,anode) $\longrightarrow$ Ag(s,cathode), the external voltage > 0.460 V. Otherwise, this reaction can occur: $Cu(s) + 2\,Ag^+(aq) \longrightarrow Cu^{2+}(aq) + 2\,Ag(s)$, $E°_{cell} = 0.460$ V. **18.12** Cell A: $Cu(s) + Zn^{2+}(1.0\text{ M}) \longrightarrow Cu^{2+}(0.10\text{ M}) + Zn(s)$; cell B: $Zn(s) + Cu^{2+}(1.0\text{ M}) \longrightarrow Zn^{2+}(0.10\text{ M}) + Cu(s)$. As current flows, $[Zn^{2+}]$ and $[Cu^{2+}]$ change in both cells; when concentrations in the two cells are the same, current stops. **18.13A** 22.5 min. **18.13B (a)** 1945 C; **(b)** 1.529 A. **18.14** NaCl(aq). The oxidation $H_2O \longrightarrow O_2$ produces 0.25 mol O_2 per mol e^-; $Cl^- \longrightarrow Cl_2$ produces 0.5 mol Cl_2 per mol e^-. Oxidation of I^- produces solid I_2. **Review Questions: 8.** No, an electrochemical cell is based on a reaction in which electrons are transferred—a redox reaction. **11.** $E°$ values compare other reduction half-reactions to the reduction of H^+(1 M) to H_2(g, 1 atm), which is assigned a value of zero. Thus $E°$ may be greater than or less than zero. **12.** $E°$ is for a reduction process with reactants and products in their standard states. $E°_{cell}$ is the *difference* in potential of two standard electrodes. E_{cell} is also a *difference* in the potentials of two electrodes, but for nonstandard conditions. **16.** Ag, Au. **17.** $E°_{cell}$ values are based on $E°$ values and related to $\Delta G°$ values for *redox* reactions. Predictions based on $E°_{cell}$ are the same as if based on $\Delta G°$. They apply regardless of how redox reactions are carried out. **18.** $E°_{cell} < 0$ means that a reaction is nonspontaneous when reactants and products are in their standard states. The reaction may be spontaneous for specific nonstandard conditions. **28.** $E°$ data are used to determine $E°_{cell}$ for the nonspontaneous reaction occurring in electrolysis; the applied voltage must exceed this $E°_{cell}$. **29.** An object to be electroplated is the cathode, and the metal to be plated is in solution as cations. **Problems: 31 (a)** $ClO_2(g) + H_2O(l) \longrightarrow ClO_3^-(aq) + 2\ H^+(aq) + e^-$ (oxidation); **(b)** $MnO_4^-(aq) + 4\ H^+(aq) + 3\ e^- \longrightarrow MnO_2(s) + 2\ H_2O(l)$ (reduction); **(c)** $SbH_3(g) + 3\ OH^-(aq) \longrightarrow Sb(s) + 3\ H_2O(l) + 3\ e^-$ (oxidation). **33 (a)** $3\ Ag(s) + NO_3^-(aq) + 4\ H^+(aq) \longrightarrow 3\ Ag^+(aq) + NO(g) + 2\ H_2O(l)$; **(b)** $2\ MnO_4^-(aq) + 5\ H_2O_2(aq) + 6\ H^+(aq) \longrightarrow 2\ Mn^{2+}(aq) + 5\ O_2(g) + 8\ H_2O(l)$; **(c)** $3\ Cl_2(g) + I^-(aq) + 3\ H_2O(l) \longrightarrow 6\ Cl^-(aq) + IO_3^-(aq) + 6\ H^+(aq)$. **35 (a)** $4\ Fe(OH)_2(s) + O_2(g) + 2\ H_2O(l) \longrightarrow 4\ Fe(OH)_3(s)$; **(b)** $S_8(s) + 12\ OH^-(aq) \longrightarrow 2\ S_2O_3^{2-}(aq) + 4\ S^{2-}(aq) + 6\ H_2O(l)$; **(c)** $2\ CrI_3(s) + 27\ H_2O_2(aq) + 10\ OH^-(aq) \longrightarrow 2\ CrO_4^{2-}(aq) + 6\ IO_4^-(aq) + 32\ H_2O(l)$. **37 (a)** $5\ H_2C_2O_4(aq) + 2\ MnO_4^-(aq) + 6\ H^+(aq) \longrightarrow 2\ Mn^{2+}(aq) + 10\ CO_2(g) + 8\ H_2O(l)$; **(b)** $Cr_2O_7^{2-}(aq) + 3\ UO^{2+}(aq) + 8\ H^+(aq) \longrightarrow 2\ Cr^{3+}(aq) + 3\ UO_2^{2+}(aq) + 4\ H_2O(l)$; **(c)** $4\ Zn(s) + NO_3^-(aq) + 6\ H_2O(l) \longrightarrow 4\ Zn^{2+}(aq) + NH_3(g) + 9\ OH^-(aq)$. **39 (a)** 0.118 V; **(b)** −0.508 V; **(c)** 0.595 V. **41.** $E° = -0.17$ V. **43.** $E°_{V^{2+}/V} = -1.13$ V. **45 (a)** $2\ I^-(aq) + Cl_2(g) \longrightarrow I_2(s) + 2\ Cl^-(aq)$; $E°_{cell} = 0.823$ V; **(b)** $Pb^{2+}(aq) + S_2O_8^{2-}(aq) + 2\ H_2O(l) \longrightarrow PbO_2(s) + 2\ SO_4^{2-}(aq) + 4\ H^+(aq)$; $E°_{cell} = 0.55$ V. **47 (a)** $Zn(s)|Zn^{2+}(aq)||Ag^+(aq)|Ag(s)$, $Zn(s) + 2\ Ag^+(aq) \longrightarrow Zn^{2+}(aq) + 2\ Ag(s)$, $E°_{cell} = 1.563$ V; **(b)** $Pt|Fe^{2+}(aq), Fe^{3+}(aq)||H_2O(l), H^+(aq)|O_2(g)$, Pt; $4\ Fe^{2+}(aq) + O_2(g) + 4\ H^+(aq) \longrightarrow 4\ Fe^{3+}(aq) + 2\ H_2O(l)$. $E°_{cell} = 0.458$ V. **49 (a)** No, $E°_{cell} = -0.140$ V; **(b)** Yes, $E°_{cell} = 0.26$ V. **51 (a)** Yes, $E°_{cell} = 1.273$ V; **(b)** No, $E°_{cell} = -0.293$ V; **(c)** Yes, $E°_{cell} = 0.26$ V (Note: In basic solution H_2O_2 is present as HO_2^-) **53 (a)** H^+ is not a good enough oxidizing agent to oxidize Ag(s) to Ag^+; $2\ Ag(s) + 2\ H^+(aq) \longrightarrow 2\ Ag^+(aq) + H_2(g)$, $E°_{cell} = -0.800$ V. **(b)** $NO_3^-(aq)$ oxidizes Ag(s) to Ag^+, $3\ Ag(s) + 4\ H^+(aq) + NO_3^-(aq) \longrightarrow 3\ Ag^+(aq) + NO(g) + 2\ H_2O(l)$, $E°_{cell} = 0.156$ V. **55.** $0.340 < E°_{Rh^{3+}/Rh} < 0.800$ V. **57 (a)** $E°_{cell} = 2.476$ V, $\Delta G° = -716.7$ kJ; **(b)** $E°_{cell}$−0.03 V, $\Delta G° = 6 \times 10^1$ kJ. **59 (a)** $K_{eq} = [Fe^{3+}]/[Fe^{2+}][Ag^+] = 3.1$; **(b)** $K_{eq} = [Mn^{2+}](P_{Cl_2})/[H^+]^4[Cl^-]^2 = 4 \times 10^{-5}$. **(c)** $K_{eq} = [Cl^-]^2(P_{O_2})/[OCl^-]^2 = 1 \times 10^{33}$ **61.** Reaction: $Sn(s) + Pb^{2+}(aq) \rightleftharpoons Sn^{2+}(aq) + Pb(s)$, $K_{eq} = 2.5$; equilibrium: $[Sn^{2+}] = 0.71$ M, $[Pb^{2+}] = 0.29$ M. **63 (a)** $E_{cell} = 1.069$ V; **(b)** $E_{cell} = 0.181$ V. **65.** $E_{cell} = 0.15$ V. **67.** pH = 1.82. **69.** The cell reaction is $Cu(s) + 2\ Ag^+(aq) \longrightarrow Cu^{2+}(aq) + 2\ Ag(s)$. Cell (b) has a higher E_{cell} because it has a lower ratio of $[Cu^{2+}]/[Ag^+]^2$. **71.** Anode: $Zn(s) \longrightarrow Zn^{2+}(aq) + 2\ e^-$; cathode: $Cl_2(g) + 2\ e^- \longrightarrow 2\ Cl^-(aq)$; cell reaction: $Zn(s) + Cl_2(g) \longrightarrow Zn^{2+}(aq) + 2\ Cl^-(aq)$; $E°_{cell} = 2.121$ V. **73.** Anode: $Zn(s) + 2\ OH^-(aq) \longrightarrow ZnO(s) + H_2O(l) + 2\ e^-$; cathode: $Ag_2O(s) + H_2O(l) + 2\ e^- \longrightarrow 2\ Ag(s) + 2\ OH^-(aq)$; cell reaction: $Zn(s) + Ag_2O(s) \longrightarrow ZnO(s) + 2\ Ag(s)$. **75.** Oxygen oxidizes Fe(s) to $Fe^{2+}(aq)$ and $Fe^{3+}(aq)$. Water is involved in the reduction half reaction and in the conversion of $Fe(OH)_2$ to $Fe(OH)_3$. The electrolyte completes the electric circuit between anodic and cathodic regions. **77.** The sacrificial anode is oxidized. Its sacrifice protects from oxidation the metal to which it is attached. **79.** Zinc is oxidized instead of iron (it is a sacrificial anode). There is a white precipitate of zinc ferricyanide but no Turnbull's blue. **81 (a)** Ni(s,anode) $\longrightarrow$ Ni(s,cathode); **(b)** Ni(s,anode) $\longrightarrow$ Ni(s,cathode); **(c)** $2\ Ni^{2+}(aq) + 2\ H_2O(l) \longrightarrow 2\ Ni(s) + O_2(g) + 4\ H^+(aq)$. **83 (a)** $BaCl_2(l) \longrightarrow Ba(l) + Cl_2(g)$, required voltage > 4.28 V; **(b)** $2\ HBr(aq) \longrightarrow H_2(g) + Br_2(l)$, required voltage > 1.065 V; **(c)** $2\ H_2O(l) \longrightarrow 2\ H_2(g) + O_2(g)$, required voltage > 2.057 V. **85.** 14.3 g Ag. **87.** 7.59×10^4 C. **89.** $AgNO_3$. $Ag^+(aq)$ yields 1 mol Ag/mol e^-, and $Cu^{2+}(aq)$ and $Zn^{2+}(aq)$ yield 0.5 mol/mol e^-. Also, the molar mass of Ag exceeds that of Cu and Zn. **Additional Problems: 91.** $2\ P_4(s) + 12\ H_2O(l) \longrightarrow 5\ PH_3(g) + 3\ H_3PO_4(aq)$. **93 (a)** $B_2Cl_4 + 6\ OH^- \longrightarrow 2\ BO_2^- + 4\ Cl^- + 2\ H_2O + H_2$; **(b)** $CH_3CH_2ONO_2 + 3\ Sn + 6\ H^+ \longrightarrow CH_3CH_2OH + NH_2OH + 3\ Sn^{2+} + H_2O$; **(c)** $2\ F_5SeOF + 16\ OH^- \longrightarrow 2\ SeO_4^{2-} + 12\ F^- + 8\ H_2O + O_2$; **(d)** $As_2S_3 + 14\ H_2O_2 + 12\ OH^- \longrightarrow 2\ AsO_4^{3-} + 3\ SO_4^{2-} + 20\ H_2O$; **(e)** $2\ XeF_6 + 16\ OH^- \longrightarrow Xe + XeO_6^{4-} + 12\ F^- + 8\ H_2O + O_2$. **95.** Instead of an electrode equilibrium, a redox reaction occurs: $2\ Na(s) + 2\ H_2O(l) \longrightarrow 2\ Na^+(aq) + 2\ OH^-(aq) + H_2(g)$. **97.** 58.1 mL $O_2(g)$. **99.** $MnO_2(s) + 4\ H^+(aq) + 2\ Cl^-(aq) \longrightarrow Mn^{2+}(aq) + Cl_2(g) + 2\ H_2O(l)$; $E°_{cell} = -0.13$ V. By using HCl(conc. aq) and allowing $Cl_2(g)$ to escape, the reaction proceeds in the forward direction, despite $E°_{cell} < 0$. **101.** $E_{cell} = 0.032$ V. **103.** $E_{cell} = 0.260$ V. **105.** E_{cell} decreases as $[Cu^{2+}]$ increases at the anode and decreases at the cathode. When $[Cu^{2+}] = 0.76$ M in each half-cell, $E_{cell} = 0$ and the current stops. **107.** K_{sp} of AgI $= 8.2 \times 10^{-17}$. **109.** The blue color will be seen. The Cu^{2+} is converted to $[Cu(NH_3)_4]^{2+}$, $[[Cu(NH_3)_4]^{2+}] = 4 \times 10^{-4}$ M. **111.** $[Hg^{2+}] = 0.177$ M; $[Fe^{2+}] = 0.034$ M; $[Fe^{3+}] = 0.356$ M. **113.** pH in anode half-cell and E_{cell}: **(a)** pH = 12.54, $E_{cell} = -0.742$ V; **(b)** 12.42, −0.735 V; **(c)** 12.26, −0.726 V; **(d)** 12.04, −0.713 V; **(e)** 11.21, −0.664 V; **(f)** 10.61, −0.628 V; **(g)** 3.09, −0.183 V; **(h)** 2.70, −0.160 V. The two curves are very similar if pH and $-E_{cell}$ are plotted versus mL HCl(aq).

Chapter 19

Exercises: 19.1A (a) ${}^{212}_{86}Rn \longrightarrow {}^{208}_{84}Po + {}^{4}_{2}He$; **(b)** ${}^{37}_{18}Ar + {}^{0}_{-1}e \longrightarrow {}^{37}_{17}Cl$; **(c)** ${}^{60m}_{27}Co \longrightarrow {}^{60}_{27}Co + \gamma$. **19.1B (a)** ${}^{218}_{84}Po \longrightarrow {}^{214}_{82}Pb + {}^{4}_{2}He$; **(b)** ${}^{36}_{17}Cl \longrightarrow {}^{36}_{16}S + {}^{0}_{1}e$. **19.2A** 5.20×10^2 atoms s^{-1}. **19.2B** $t = 1.60 \times 10^5$ y. **19.3** Answer: (b). There are about 10 times more ${}^{24}Na$ atoms than ${}^{11}C$, but ${}^{11}C$ atoms disintegrate about 50 times as fast. There are about

$10^4 - 10^5$ times as many ^{238}U atoms as ^{11}C, but $t_{1/2}$ of ^{238}U exceeds that of ^{11}C by a factor of more than 10^9. **19.4A** 6.0 dis min^{-1} per g carbon. **19.4B** The activity falls to one-half its initial value in 12.26 y ($t_{1/2}$ for tritium). The brandy is only about 12 years old, not 25. **19.5A** $^{35}_{17}Cl + ^{1}_{0}n \longrightarrow ^{35}_{16}S + ^{1}_{1}H$. **19.5B** $^{249}_{98}Cf + ^{15}_{7}N \longrightarrow ^{260}_{105}Db + 4\,^{1}_{0}n$. **19.6** "Magic numbers" of protons and neutrons in ^{40}Ca and ^{48}Ca make them stable. (Isotopes with $Z = 20$ and $A = 42$, 44, and 46 are also stable.) ^{39}Ca is below the belt of stability in Figure 19.4 and is radioactive. **19.7** ^{84}Y has 39 protons and 45 neutrons and lies outside the belt of stability (Figure 19.4). It should decay by electron capture or positron emission to yield ^{84}Sr, which is within the belt of stability and could be stable. **19.8A** $\Delta E = -5.0$ MeV. **19.8B** Energy requirement $= \Delta E = 2.7$ MeV. To convert from atomic to nuclear masses subtract the mass of 15 electrons from each side of the equation. The same result is obtained by using either atomic or nuclear masses. **Review Questions: 1.** γ rays have the greatest ability to penetrate matter, and alpha particles to ionize matter. **2.** γ rays are produced by changes within a nucleus, whereas X rays involve displacements in inner electron shells. Both are electromagnetic radiation, but γ rays generally are more energetic. **6.** Electron capture creates a vacancy in an inner electron shell that is filled by an electron from a higher numbered shell, producing an X ray. **17.** Chemical reactions involve bond breakage and formation, which involve changes in electrostatic forces. Nuclear reactions involve changes in the much stronger nuclear forces that bind nucleons together. **Problems: 29 (a)** $^{32}_{16}S \longrightarrow ^{32}_{17}Cl + ^{0}_{-1}e$; **(b)** $^{14}_{8}O \longrightarrow ^{14}_{7}N + ^{0}_{1}e$; **(c)** $^{238}_{92}U \longrightarrow ^{234}_{90}Th + ^{4}_{2}He$. **31 (a)** $^{206}_{82}Pb$; **(b)** $^{59}_{28}Ni$; **(c)** $^{208}_{82}Pb$. **33.** The coordinates of the graph are expressed as (A, Z) and the line segments, by the dash lines between the pairs of coordinates that follow. (232, 90)—(228, 88)—(228, 89)—(228, 90)—(224, 88)—(220, 86)—(216, 84)—(212, 82)—(212, 83)—(212, 84)

| |

(208, 81)—(208, 82).

35. Chemical reactions occur as a result of collisions, and their rates depend on the reaction mechanism. Different mechanisms lead to different overall reaction orders. The rate at which the nuclei of a particular nuclide decay depends on how many of them are present. Thus, the process is first order. **37 (a)** ^{15}O has the shortest half-life and the greatest activity. **(b)** A 75% reduction represents two half-life periods. The required $t_{1/2}$ is one day, and the nuclide having about this half-life is ^{28}Mg. **39.** For equal masses, the number of atoms (N) is about four times as great for ^{32}P as for ^{131}I. The decay constant (λ) of ^{32}P is about one-half that of ^{131}I. Because A is proportional to both λ and N, A of ^{32}P is about twice that of ^{131}I $(4 \times 1/2 = 2)$. **41.** $A = 8.11 \times 10^4$ atoms h^{-1} (22.5 atoms s^{-1}). **43.** $t_{1/2} = 88.5$ d. **45 (a)** t is slightly greater than $3 \times t_{1/2} \approx 24$ d. **(b)** $t = 26.7$ d. **47.** 11 dis min^{-1} per g carbon. **49.** The missing items are **(a)** $^{10}_{4}Be$; **(b)** $^{1}_{1}H$; **(c)** $^{4}_{2}He$. **51 (a)** $2\,^{2}_{1}H \longrightarrow ^{3}_{2}He + ^{1}_{0}n$; **(b)** $^{241}_{95}Am + ^{4}_{2}He \longrightarrow ^{243}_{97}Bk + 2\,^{1}_{0}n$; **(c)** $^{238}_{92}U + ^{4}_{2}He \longrightarrow ^{239}_{94}Pu + 3\,^{1}_{0}n$. **53.** $^{99}_{43}Tc + ^{1}_{0}n \longrightarrow ^{100}_{43}Tc \longrightarrow ^{100}_{44}Ru + ^{0}_{-1}e$. **55.** 0.5301 u. **57.** 8.57 MeV/nucleon. **59.** $\Delta E = -5.8$ MeV. **61.** nuclear mass: 228.9848 u; atomic mass: 229.0336 u. **63.** 11.0 MeV. **65.** ^{17}F, β^+ emission; ^{22}F, β^- emission. **67.** ^{36}Cl (17p, 19n) and ^{64}Cu (29p, 35n) do not occur naturally, even though their atomic masses are close to the weighted average atomic mass of the element (35.4527 u for Cl and 63.546 u for Cu). They are "odd-odd" nuclides and are radioactive. **69.** ^{40}Ca is a "doubly magic" nuclide (20p, 20 n) with a neutron:proton ratio of 1 : 1. These factors contribute to the stability of a nuclide of relatively low atomic number. **71.** $\Delta E = -183.6$ MeV. **73.** A nuclide with a very short half-life decays so quickly that it is not a significant hazard, whereas one with a very long half-life has a very low activity and is less hazardous for that reason. **75.** 80 g $<$ mass of ^{131}I $<$ 400 g. **Additional Problems: 77.** The deflection of β^+ should be the same as that for β^- but in the opposite direction; gamma rays would not be deflected. The β^+ and β^- particles would be deflected into tight circles and the more massive α particle into a circle of greater radius. (Note that a β^+ particle is annihilated on encountering an electron, producing two gamma rays.) **79.** ^{40}Ar is produced from naturally occurring ^{40}K by positron emission and electron capture. Although ^{4}He is also produced continuously (as α-particle emissions) some tends to be lost to outer space because of its low atomic mass. **81.** α particle energy: 4.3 MeV. **83.** 2.1×10^3 L of the natural gas. **85.** 3.65×10^{-7} g. **87.** 7.0×10^9 dis s^{-1}. **89.** Assume equal numbers of atoms initially: $^{235}N_0 = {}^{238}N_0$. Obtain the decay constants $^{235}\lambda$ and $^{238}\lambda$ from page 823. Use the radioactive decay law to obtain the expression: $\ln(^{235}N_t/^{235}N_0) - \ln(^{238}N_t/^{238}N_0) = -(^{235}\lambda - {}^{238}\lambda)\,t$. Obtain the value of $\ln(^{235}N_t/^{238}N_t)$ from the current ratio of the natural abundances: $^{235}N_t/^{238}N_t = 0.72/99.28$. Solve for t (about 6×10^9 y).

Chapter 20

Exercises: 20.1 (a) $2\,NaCl(l) \xrightarrow{\text{electrolysis}} 2\,Na(l) + Cl_2(g)$, followed by $2\,Na(s) + H_2(g) \longrightarrow 2\,NaH(s)$; **(b)** $2\,NaCl(aq) + 2\,H_2O(l) \xrightarrow{\text{electrolysis}} 2\,NaOH(aq) + Cl_2(g) + H_2(g)$, followed by $2\,NaOH(aq) + Cl_2(g) \longrightarrow NaCl(aq) + NaOCl(aq) + H_2O(l)$. **20.2** Maximum mass: 1.658 g (exclusively MgO); minimum mass: 1.384 (exclusively Mg_3N_2). **Review Questions: 2.** Some common naturally occurring compounds: NaCl, Na_2CO_3, $NaNO_3$; $CaCl_2$, $CaCO_3$, $CaSO_4$. **5.** There is no change in amount of H_2SO_4 because only $H_2O(l)$ is electrolyzed. If significant amounts of water are electrolyzed, the concentration of H_2SO_4 increases. **15 (a)** limestone ($CaCO_3$), clay (Al_2O_3), sand (SiO_2); **(b)** limestone ($CaCO_3$), sand (SiO_2), sodium carbonate (Na_2CO_3); **(c)** slaked lime [$Ca(OH)_2$], sand (SiO_2), water (H_2O); **(d)** gypsum ($CaSO_4 \cdot 2H_2O$). **Problems: 25.** Order of increasing solubility: $MgCO_3 < Li_2CO_3 < Na_2CO_3$. Interionic attractions are strongest between the small, highly charged Mg^{2+} and CO_3^{2-}, and are greater for Li_2CO_3 than Na_2CO_3 because Li^+ is a smaller ion than Na^+. **27 (a)** lithium carbonate; **(b)** Mg_3N_2; **(c)** $CaBr_2 \cdot 6H_2O$; **(d)** potassium hydrogen sulfate. **29 (a)** CaO, calcium oxide; **(b)** $Na_2SO_4 \cdot 10H_2O$, sodium sulfate decahydrate; **(c)** $CaSO_4 \cdot 2H_2O$, calcium sulfate dihydrate; **(d)** $CaCO_3 \cdot MgCO_3$, calcium magnesium carbonate. **31.** Raw materials: seawater or natural brine (source of Mg^{2+}), limestone or seashells [$CaCO_3$, for conversion to CaO and $Ca(OH)_2$], HCl(aq), water (for conversion of Cl_2 to HCl). Main product: Mg(l). By-products: $CO_2(g)$ (from calcination of

$CaCO_3$) and $O_2(g)$ (from reaction of steam and Cl_2). Recycled: $Cl_2(g)$ and HCl(g) [Cl_2 is converted to HCl(g), which is recycled]. Equations are given in Figure 20.8. **33.** H_2SO_4(conc aq) + 2 NaCl(s) $\xrightarrow{\Delta}$ $Na_2SO_4(s)$ + 2 HCl(g); $Na_2SO_4(s)$ + 10 $H_2O(l)$ $\longrightarrow$ $Na_2SO_4 \cdot 10H_2O(s)$ (Glauber's salt). Heating the salt of a volatile acid with a nonvolatile acid drives off the volatile acid. **35.** NaOH(s) has to be prepared from NaCl(aq) by electrolysis. To obtain Na from NaOH requires an additional electrolysis. This is a more expensive route than to electrolyze NaCl(l), despite the requirement of a higher temperature.

37. MgO $\longleftarrow$ $MgCO_3$ $\xrightarrow{HCl}$ $MgCl_2$ $\xrightarrow{electrolysis}$ Mg $\xrightarrow{\Delta,\ N_2}$ Mg_3N_2; $MgCO_3$ $\xrightarrow{H_2O}$ $Mg(OH)_2$ $\xrightarrow{HNO_3}$ $Mg(NO_3)_2$; MgO $\xrightarrow{H_2O}$ $Mg(OH)_2$ $\xrightarrow{H_3PO_4}$ $MgHPO_4$

39 (a) Sr(s) + 2 $H_2O(l)$ $\longrightarrow$ $Sr(OH)_2(aq)$ + $H_2(g)$; **(b)** LiH(s) + $H_2O(l)$ $\longrightarrow$ LiOH(aq) + $H_2(g)$; **(c)** BaO(s) + $H_2O(l)$ $\longrightarrow$ $Ba(OH)_2(aq)$; **(d)** C(g) + $H_2O(g)$ $\xrightarrow{\Delta}$ CO(g) + $H_2(g)$. [*Note:* Both $Sr(OH)_2(s)$ and $Ba(OH)_2(s)$ are only slightly soluble in water.] **41 (a)** 2 Al(s) + 6 HCl(aq) $\longrightarrow$ 2 $AlCl_3(aq)$ + 3 $H_2(g)$; **(b)** $CH_3CH_3(g)$ + 2 $H_2O(g)$ $\xrightarrow{\Delta}$ 2 CO(g) + 5 $H_2(g)$; (c) $CH_3C{\equiv}CH$ + 2 $H_2(g)$ $\longrightarrow$ $CH_3CH_2CH_3(g)$; **(d)** $MnO_2(s)$ + 2 $H_2(g)$ $\xrightarrow{\Delta}$ Mn(s) + 2 $H_2O(l)$. **43 (a)** $BeF_2(s)$ + 2 Na(l) $\xrightarrow{\Delta}$ Be(s) + 2 NaF(s); **(b)** Ca(s) + 2 $CH_3COOH(aq)$ $\longrightarrow$ $Ca(CH_3COO)_2(aq)$ + $H_2(g)$; **(c)** $PuO_2(s)$ + 2 Ca(s) $\xrightarrow{\Delta}$ Pu(s) + 2 CaO(s); **(d)** $CaCO_3 \cdot MgCO_3(s)$ $\xrightarrow{\Delta}$ CaO(s) + MgO(s) + 2 $CO_2(g)$ **45 (a)** $MgCO_3(s)$ + 2 HCl(aq) $\longrightarrow$ $MgCl_2(aq)$ + $H_2O(l)$ + $CO_2(g)$; **(b)** $CO_2(g)$ + 2 KOH(aq) $\longrightarrow$ $K_2CO_3(aq)$ + $H_2O(l)$; **(c)** 2 KCl(s) + H_2SO_4(conc aq) $\xrightarrow{\Delta}$ $K_2SO_4(s)$ + 2 HCl(g). **47 (a)** 2 NaCl(s) + H_2SO_4(conc aq) $\xrightarrow{\Delta}$ $Na_2SO_4(s)$ + 2 HCl(g); **(b)** $Mg(HCO_3)_2(aq)$ $\xrightarrow{\Delta}$ $MgCO_3(s)$ + $H_2O(l)$ + $CO_2(g)$, followed by $MgCO_3(s)$ $\xrightarrow{\Delta}$ MgO(s) + $CO_2(g)$. **49.** 1.10×10^2 g CaH_2. **51.** 2.5×10^2 L HCl(aq). **53.** 694 L seawater. **55.** (c). From the balanced equations, determine the moles of reactant to produce 1 mol H_2: (a) 1 mol Zn/1 mol H_2, (b) 2 mol Na/1 mol H_2, (c) 1/2 mol CaH_2/1 mol H_2. The mass of 1/2 mol CaH_2 is much less than the masses of 1 mol Zn or 2 mol Na. **57.** NH_3 (a base) reacts with HCO_3^- (an acid): $NH_3(aq)$ + $HCO_3^-(aq)$ $\longrightarrow$ $NH_4^+(aq)$ + $CO_3^{2-}(aq)$. This is followed by $M^{2+}(aq)$ + $CO_3^{2-}(aq)$ $\longrightarrow$ $MCO_3(s)$, where M^{2+} = Mg^{2+}, Ca^{2+}, or Fe^{2+}. **59.** 68.4 mg boiler scale ($CaCO_3$). **61.** A limestone cave and hard water both form through the forward direction of the reaction: $CaCO_3(s)$ + $H_2O(l)$ + $CO_2(g)$ $\rightleftharpoons$ $Ca(HCO_3)_2(aq)$. Stalactites and stalagmites form in the reverse of this reaction, as does boiler scale when hard water is boiled. **63.** In a cation exchange, all cations are replaced by H^+. Then, in an anion exchange, all anions are replaced by OH^-. The H^+ and OH^- combine to form H_2O, and the water is essentially free of all ions. **Additional Problems: 65 (a)** 4 Li(s) + $O_2(g)$ $\longrightarrow$ 2 $Li_2O(s)$; **(b)** 6 Li(s) + $N_2(g)$ $\longrightarrow$ 2 $Li_3N(s)$; **(c)** Li_2CO_3 $\xrightarrow{\Delta}$ $Li_2O(s)$ + $CO_2(g)$. **67.** The CaO(s) is converted to $CaCO_3(s)$ by one or both of these routes: CaO(s) + $CO_2(g)$ $\longrightarrow$ $CaCO_3(s)$, or CaO(s) + $H_2O(l)$ $\longrightarrow$ $Ca(OH)_2(s)$, followed by $Ca(OH)_2(s)$ + $CO_2(g)$ $\longrightarrow$ $CaCO_3(s)$ + $H_2O(g)$. **69.** 314 L $H_2(g)$. **71.** 60.2% MgO(s). **73.** The battery operates for 34 years and consumes 0.13 g Li(s). **75 (a)** H_2 has greater ΔH_{comb} on a mass basis: -141.8 kJ/g H_2 compared to -47.5 kJ/g $C_{12}H_{26}$; **(b)** $C_{12}H_{26}$ has greater ΔH_{comb} on a volume basis: -35.6 kJ/mL $C_{12}H_{26}(l)$ compared to -10.0 kJ/mL $H_2(l)$. **77.** 3.08 g soap. **79.** 80.56 ppm Ca^{2+}. **81.** Use methods of Exercises 11.9 and 11.10 to calculate densities: 0.96 g Na/cm^3, 0.90 g K/cm^3, 1.51 g Rb/cm^3. This is the order found in Table 20.1. **83.** No precipitation from 0.005 M $CaCl_2$. $Q_{ip} = [Ca^{2+}][CO_3^{2-}] = 5 \times 10^{-3} \times K_{a_2}$ of $H_2CO_3 = 2 \times 10^{-13} < K_{sp}$ of $CaCO_3$. In 0.010 M $Ca(OH)_2$, $[CO_3^{2-}]$ is much greater than in saturated $CO_2(aq)$ because of the reaction between H_2CO_3 and OH^-. The $CaCO_3(s)$ precipitate does not redissolve because $CO_2(g)$ is insufficiently soluble to drive the dissolution reaction to completion (see also, answer to Problem 85). **85.** The reaction is $CaCO_3(s)$ + $CO_2(aq)$ + $H_2O(l)$ $\rightleftharpoons$ $Ca^{2+}(aq)$ + 2 $HCO_3^-(aq)$, with $K = (K_{sp} \times K_{a_1})/K_{a_2} = [Ca^{2+}][HCO_3^-]^2/[CO_2] = 2.6 \times 10^{-5}$. In this expression, K_{sp} is for $CaCO_3$, K_{a_1} and K_{a_2} are for H_2CO_3, and $[CO_2]$ is substituted for $[H_2CO_3]$.

Chapter 21

Exercises: 21.1 2 $BCl_3(g)$ + 3 $H_2(g)$ $\longrightarrow$ 2 B(s) + 6 HCl(g). **21.2** The reaction is nonspontaneous for standard-state conditions: $E^\circ_{cell} = -0.90$ V and $\Delta G^\circ = 174$ kJ. The ΔG° differs from Example 22.2 because of different standard states, such as 1 M $Br^-(aq)$ instead of NaBr(s) and 1 M $H^+(aq)$ and 1 M $SO_4^{2-}(aq)$ instead of $H_2SO_4(l)$. **Review Questions: 4.** Most active *p*-block metal: Al; nonmetal, F. **6.** The bridge bonds in B_2H_6 are three-center bonds (B—H—B) with a single pair of electrons. Those in Al_2Cl_6 are coordinate covalent bonds between lone pair electrons of Cl in one $AlCl_3$ unit with Al of a second unit. **8.** NaOH(s) gives off heat when it dissolves, and Al(s) liberates $H_2(g)$ when it reacts with NaOH(aq). **9.** Even though a larger cable of Al is required to carry the same amount of current, the weight of the cable is less because of the considerably lower density of A1 compared to Cu. **10.** In 1866, Al was only obtained by expensive chemical reduction. In 1924, Al was being produced electrolytically. **15.** A similarity: both O_2^{2-} and O_2^- have two O atoms per ion; some differences: electric charges (2– and 1–, respectively), oxidation numbers (–1 and –1/2, respectively), magnetic properties (O_2^{2-}, diamagnetic, and O_2^-, paramagnetic). **17.** Because of hydrogen bonding, intermolecular attractions in H_2O are stronger than in some substances of much higher molar masses that lack hydrogen bonding (e.g., H_2S). **18.** Cu cannot be oxidized by H_3O^+, but can by concentrated SO_4^{2-} in acidic solution. **23 (a)** silver azide; **(b)** potassium thiocyanate; **(c)** astatine oxide, **(d)** telluric acid. **24 (a)** H_2SeO_4; **(b)** H_2Te; **(c)** $Pb(N_3)_2$; **(d)** AgAt. **25.** bauxite, Al_2O_3; boric oxide, B_2O_3; corundum, Al_2O_3; cyanogen, $(CN)_2$; hydrazine, N_2H_4; silica, SiO_2.

26 Key elements: fluorapatite: Ca, P, O, F; borax: Na, B, O, H; corundum: Al, O; quartz: Si, O. **Problems: 27.** $3\ Mg(s) + B_2O_3 \xrightarrow{\Delta} 2\ B(s) + 3\ MgO(s)$. **29.** Both are electron-deficient species, but the possibility of resonance structures with B-to-F double bonds results in a resonance hybrid of BF_3 that is a stable molecule. **31.** $Na_2B_4O_7 \cdot 10H_2O + H_2SO_4 \longrightarrow 4\ B(OH)_3 + Na_2SO_4 + 5\ H_2O$; $2\ B(OH)_3 \xrightarrow{\Delta} B_2O_3 + 3\ H_2O$; $B_2O_3 + 3\ CaF_2 + 3\ H_2SO_4 \xrightarrow{\Delta} 2\ BF_3 + 3\ CaSO_4 + 3\ H_2O$. **33.** $Al_2O_3(s) + 3\ H_2SO_4(conc\ aq) + 15\ H_2O(l) \xrightarrow{\Delta} Al_2(SO_4)_3 \cdot 18H_2O(s)$; $Al_2(SO_4)_3(aq) + 6\ OH^-(aq) \longrightarrow 2\ Al(OH)_3(s) + 3\ SO_4^{2-}(aq)$; $Al(OH)_3(s) + 3\ HCl(aq) \longrightarrow AlCl_3(aq) + 3\ H_2O(l)$. **35.** (1) The amphoterism of Al_2O_3 allows removal of Fe_2O_3 impurity from bauxite ore. (2) The use of molten cryolite, $Na_3AlF_6(l)$, as a solvent for Al_2O_3 permits electrolysis from a better electrically conducting medium and at a lower temperature than with $Al_2O_3(l)$ alone. **37.** $2\ Al(s) + Fe_2O_3(s) \longrightarrow Al_2O_3(s) + 2\ Fe(l)$; $\Delta H \approx -852$ kJ. The ΔH value is only an estimate because it is based on data at 298 K, and because it assumes the product Fe(s) rather than Fe(l). **39.** Strongly acidic food could dissolve the $Al_2O_3(s)$ coating on aluminum cookware, and even the underlying metal itself, The same effect might occur with an alkaline oven cleaner, but $[Al(OH)_4]^-(aq)$ would be produced rather than $Al^{3+}(aq)$. **41.** $Na_2C_2(s) + 2\ H_2O(l) \longrightarrow 2\ NaOH(aq) + C_2H_2(g)$.

43. $:\ddot{S}{=}C{=}\ddot{S}:$ CCl_4: C bonded to four $:\ddot{Cl}:$ atoms $:N{\equiv}C{-}C{\equiv}N:$

45. $[\,:\ddot{O}{-}Si(:\ddot{O}:)_2{-}\ddot{O}{-}Si(:\ddot{O}:)_2{-}\ddot{O}:\,]^{6-}$ Geometric structure: two tetrahedra sharing a corner.

47. $Mg_3(Si_2O_5)(OH)_4$ Oxidation numbers: Mg (3×2); Si (2×4); H (4×1); O (9×-2); total $= 18 - 18 = 0$. **49** **(a)** $Si(CH_3)_4$; **(b)** $SiCl_2(CH_3)_2$; **(c)** $SiH(CH_2CH_3)_3$. **51.** Reduction: $PbO_2(s) + 4\ H^+(aq) + 2\ e^- \longrightarrow Pb^{2+}(aq) + 2\ H_2O(l)$, $E° = 1.455$ V **(a)** Yes, $E°_{cell} = 0.27$ V; **(b)** No, $E°_{cell} = -0.308$ V; **(c)** $E°_{cell} = -0.53$ V; **(d)** Yes, $E°_{cell} = -1.301$ V. **53** **(a)** $Sn(s) + 2\ HCl(aq) \longrightarrow SnCl_2(aq) + H_2(g)$; **(b)** $SnCl_2 + Cl_2(g) \longrightarrow SnCl_4$; **(c)** $SnCl_4(aq) + 4\ NH_3(aq) + 2\ H_2O(l) \longrightarrow SnO_2(s) + 4\ NH_4^+(aq) + 4\ Cl^-(aq)$. **55.** f $<$ d $<$ c $<$ a $<$ b $<$ e. **57** **(a)** $NH_2NH_2(aq) + HCl(aq) \longrightarrow NH_2NH_3^+(aq) + Cl^-(aq)$, followed by $NH_2NH_3^+(aq) + HCl(aq) \longrightarrow NH_3NH_3^{2+}(aq) + Cl^-(aq)$; **(b)** $3\ Cu(s) + 8\ H^+(aq) + 2\ NO_3^+(aq) \longrightarrow 3\ Cu^{2+}(aq) + 2\ NO(g) + 4\ H_2O(l)$; **(c)** $2\ NO(g) + O_2(g) \longrightarrow 2\ NO_2(g)$. **59.** Oxidation: $NH_2NH_3^+(aq) \longrightarrow N_2(g) + 5\ H^+(aq) + 4\ e^-$; reduction: $Fe^{3+}(aq) + e^- \longrightarrow Fe^{2+}(aq)$; redox reaction: $4\ Fe^{3+}(aq) + NH_2NH_3^+(aq) \longrightarrow 4\ Fe^{2+}(aq) + N_2(g) + 5\ H^+(aq)$; $E°_{N_2/NH_2NH_3^+} = -0.23$ V. **61.** Principal allotropes: white P and red P; structures: white P consists of individual P_4 tetrahedra and red P consists of P_4 units joined together into long chains. **63.** $P_4(s) + 3\ KOH(aq) + 3\ H_2O(l) \longrightarrow 3\ KH_2PO_2(aq) + PH_3(g)$. **65** **(a)** $4\ Al(s) + 3\ O_2(g) \longrightarrow 2\ Al_2O_3(s)$; **(b)** $2\ KClO_3(s) \xrightarrow{\Delta} 2\ KCl(s) + 3\ O_2(g)$; **(c)** $2\ Na_2O_2(s) + 2\ H_2O(l) \longrightarrow 4\ NaOH(aq) + O_2(g)$; **(d)** $Pb^{2+}(aq) + H_2O(l) + O_3(g) \longrightarrow PbO_2(s) + 2\ H^+(aq) + O_2(g)$. **67.** Because of its low melting point (119 °C) S_8 is melted by superheated water. The water and $S_8(l)$ do not form a solution or react, and the mixture can be brought to the surface with compressed air. The $S_8(s)$ is in a pure state. **69.** Oxidation: $2\ S(s) + 6\ OH^- \longrightarrow S_2O_3^{2-} + 3\ H_2O + 4\ e^-$; reduction: $2\ SO_3^{2-} + 3\ H_2O + 4\ e^- \longrightarrow S_2O_3^{2-} + 6\ OH^-$; overall: $SO_3^{2-}(aq) + S(s) \longrightarrow S_2O_3^{2-}(aq)$. **71.** Only $Cl_2(g)$ will oxidize $Br^-(aq)$ to $Br_2(l)$. $I_2(s)$ is too poor an oxidizing agent for this, and I^-, Cl^-, and F^- can only act as reducing agents. **73.** (1) $I^-(aq) + 3\ Cl_2(g) + 3\ H_2O(l) \longrightarrow IO_3^-(aq) + 6\ Cl^-(aq) + 6\ H^+(aq)$; (2) $2\ Br^-(aq) + Cl_2(g) \longrightarrow 2\ Cl^-(aq) + Br_2(l)$; (3) When the mixture, $IO_3^-(aq) + Cl^-(aq) + Br_2(l)$, is extracted with $CS_2(l)$, only the $Br_2(l)$ dissolves in the $CS_2(l)$. **75.** Interhalogen: ICl; halide, NaCl; halate, $KBrO_3$, none of these categories: $(CN)_2$. **77** **(a)** $Cl_2(g) + H_2O(l) \longrightarrow HOCl(aq) + H^+(aq) + Cl^-(aq)$; **(b)** $Cl_2(g) + 2\ NaOH(aq) \longrightarrow NaOCl(aq) + NaCl(aq) + H_2O(l)$; **(c)** $3\ Cl_2(g) + 6\ NaOH(aq) \xrightarrow{\Delta} 5\ NaCl(aq) + NaClO_3(aq) + 3\ H_2O(l)$. **79** **(a)** XeO_3: 26 valence electrons, VSEPR notation AX_3E, trigonal pyramidal shape; **(b)** XeO_4: 32 valence electrons, VSEPR notation AX_4, tetrahedral shape. **81.** Argon is produced only by the decay of potassium-40, whereas helium is derived from alpha particles. Alpha particles are emitted by many radioactive isotopes, particularly those of high atomic and mass numbers. **Additional Problems: 83** **(a)** BrF_3: 28 valence electrons; VSEPR notation AX_3E_2; electron-group geometry, trigonal bipyramidal; molecular shape, T-shaped; **(b)** IF_5: 42 valence electrons; VSEPR notation AX_5E; electron-group geometry, octahedral; molecular shape, square pyramidal. **85.** $Na_2B_4O_7 \cdot 10H_2O + 6\ CaF_2 + 8\ H_2SO_4 \xrightarrow{\Delta} 4\ BF_3 + 2\ NaHSO_4 + 6\ CaSO_4 + 17\ H_2O$. **87.** (1) $MnF_6^{2-} + 2\ SbF_5 \longrightarrow MnF_4 + 2\ SbF_6^-$, (2) $MnF_4 \longrightarrow MnF_2 + F_2$. **89.** The oxidation half-reaction is $2\ H_2O \longrightarrow 4\ H^+ + O_2 + 4\ e^-$; the overall redox reaction is $2\ XeF_2 + 2\ H_2O \longrightarrow 2\ Xe(g) + 4\ HF + O_2(g)$, $E°_{cell} = 1.09$ V. The forward reaction is favored in basic solution, where the HF is neutralized. **91.** 1.4×10^4 L air. **93.** $d = 9.23\ g/cm^3$. **95.** $[NO_3^-] = 0.1716$ M. **97.** (1) $2\ CH_4 + S_8 \longrightarrow 2\ CS_2 + 4\ H_2S$; (2) $CS_2 + 3\ Cl_2 \longrightarrow CCl_4 + S_2Cl_2$; (3) $4\ CS_2 + 8\ S_2Cl_2 \longrightarrow 4\ CCl_4 + 3\ S_8$. **99** **(a)** $\Delta H_f°[ClF] \approx -50$ kJ/mol; **(b)** $\Delta H_f°[OF_2] \approx 34$ kJ/mol; **(c)** $\Delta H_f°[Cl_2O] \approx 82$ kJ/mol; **(d)** $\Delta H_f°[NF_3] \approx -129$ kJ/mol. **101.** Oxidation half-reactions: (1) $XeF_4 + 3\ H_2O \longrightarrow XeO_3 + 4\ HF + 2\ H^+ + 2\ e^-$, and (2) $2\ H_2O \longrightarrow O_2 + 4\ H^+ + 4\ e^-$; reduction half reaction: $XeF_4 + 4\ e^- \longrightarrow Xe + 4\ F^-$. To obtain overall redox equation, multiply half-equation (1) by 1, half-equation (2) by 3/2, and the reduction half-equation by 2; then add them together. Result: $3\ XeF_4 + 6\ H_2O \longrightarrow 2\ Xe + XeO_3 + 12\ HF + 3/2\ O_2$.

Chapter 22

Exercises: 22.1 $4\ FeCr_2O_4 + 16\ NaOH + 7\ O_2 \xrightarrow{\Delta} 8\ Na_2CrO_4 + 4\ Fe(OH)_3 + 2\ H_2O$. **22.2** $MnO_4^-(aq) + 2\ H_2O(l) + 3\ e^- \longrightarrow MnO_2(s) + 4\ OH^-(aq)$; $E° = 0.60$ V. In basic solution, MnO_4^- will oxidize any species having an $E° < 0.60$ V, including $Br^-(aq)$ to $BrO_3^-(aq)$, $Br_2(l)$ to $BrO^-(aq)$, $Ag(s)$ to $Ag_2O(s)$, and $NO_2^-(aq)$ to $NO_3^-(aq)$. **22.3** The coordination and oxidation numbers are **(a)** 6 and +3; **(b)** 6 and +2. **22.4 (a)** hexaamminecobalt(II) ion; **(b)** tetrachloroaurate(III) ion; **(c)** pentaamminebromocobalt(III) bromide. **22.5 (a)** $[Co(en)_3]^{3+}$; **(b)** $[CrCl_4(NH_3)_2]^-$; **(c)** $[PtCl_2(en)_2]SO_4$. **22.6A (a)** Isomers. NCS^- is bonded to Fe^{3+} through the S atom in one structure and the N atom in the other. **(b)** Isomers. The ligands, NH_3, are banded to Zn^{2+} in one structure and Cu^{2+} in the other, similarly with Cl^-. **(c)** Not isomers. One compound has two NH_3 ligands and the other, four. **22.6B (a)** No isomerism. The H_2O molecule can take up any one of four equivalent positions, and the NH_3 molecules, the other three. **(b)** Yes, ligands can be interchanged between the complex cation and complex anion, as in $[Pt(NH_3)_4][CuCl_4]$. **22.7** Assuming en ligands attach only at adjacent points on the hexagonal ring, all structures are identical to the following:

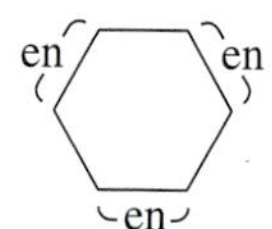

22.8 (a) CN^- is a strong field ligand, and the energy separation (Δ) is large. The six d electrons of $[Ar]3d^6$ are found as three pairs in the lower-energy set of $3d$ orbitals. **(b)** In a tetrahedral field, the eight $3d$ electrons of $[Ar]3d^8$ fill the two lower-level and one of the upper-level d orbitals. The other two upper level d orbitals have single electrons. There are two unpaired electrons. **Review Questions: 2.** $[Kr]4d^{10}5s^25p^1$ is the electron configuration of indium, a fifth-period, p-block, main-group element; $[Ar]3d^74s^2$ is that of cobalt, a fourth-period, d-block, transition element. **7 (a)** zinc-coated iron; **(b)** $FeCr_2O_4$; **(c)** $Cu_2(OH)_2CO_3$. **8 (a)** ferromanganese; **(b)** brass; **(c)** aqua regia. **10 (a)** scandium chloride; **(b)** iron(II) silicate; **(c)** sodium manganate; **(d)** chromium(VI) oxide. **11 (a)** $BaCr_2O_7$; **(b)** Cr_2O_3; **(c)** Hg_2Br_2; **(d)** V_2O_5. **16.** The number of ligands is the same as the coordination number if all the ligands attach at a single site (monodentate), but not if some ligands attach at more than one site (polydentate). **20 (a)** water; **(b)** chloride ion; **(c)** ammonia; **(d)** ethylenediamine; **(e)** oxalate ion; **(f)** ethylenediaminetetraacetate ion. **Problems: 27.** To acquire the argon electron configuration, a Ca atom loses two electrons ($4s^2$), whereas Sc loses three ($3d^14s^2$). **29 (a)** Ca < K because $Z_{eff} \approx 2$ for Ca and ≈ 1 for K. The nucleus attracts the valence electrons more strongly in Ca. **(b)** Mn < Ca because it has a larger nuclear charge but the same number of valence electrons as Ca. **(c)** The Mn and Fe atoms are the same size because they have the same Z_{eff} and the same number of valence electrons. **31 (a)** $2\ Sc(s) + 6\ HCl(aq) \longrightarrow 2\ ScCl_3(aq) + 3\ H_2(g)$; **(b)** $Sc(OH)_3(s) + 3\ HCl(aq) \longrightarrow ScCl_3(aq) + 3\ H_2O(l)$; **(c)** $Sc(OH)_3(s) + 3\ Na^+(aq) + 3\ OH^-(aq) \longrightarrow 3\ Na^+(aq) + [Sc(OH)_6]^{3-}(aq)$. **33 (a)** $Ba(s) + 2\ HCl(aq) \longrightarrow BaCl_2(aq) + H_2(g)$, followed by $BaCl_2(aq) + K_2CrO_4(aq) \longrightarrow BaCrO_4(s) + 2\ KCl(aq)$; **(b)** Combine $MnO_4^- + 2\ H_2O + 3\ e^- \longrightarrow MnO_2(s) + 4\ OH^-$, $E° = 0.60$ V and a half-reaction with $E° < 0.60$ V. For example, $3\ NO_2^- + 2\ MnO_4^- + H_2O \longrightarrow 3\ NO_3^- + 2\ MnO_2(s) + 2\ OH^-$, $E°_{cell} = E°_{MnO_4^-/MnO_2} - E°_{NO_3^-/NO_2^-} = 0.60\ V - 0.01\ V = 0.59$ V. **35.** $2\ Mn^{2+}(aq) + 5\ BiO_3^-(aq) + 14\ H^+(aq) \longrightarrow 2\ MnO_4^-(aq) + 5\ Bi^{3+}(aq) + 7\ H_2O(l)$. **37.** CrO_4^{2-} in equilibrium with $Cr_2O_7^{2-}$ combines with Pb^{2+} to form $PbCrO_4(s)$; the net reaction goes to completion. **39.** Fe(s); it is most easily oxidized. That is, $Fe^{2+}(aq) + 2\ e^- \longrightarrow Fe(s)$ has the lowest value of $E°$ (−0.440 V). **41.** $Fe_2O_3(s) + 3\ Cl_2(g) + 10\ OH^-(aq) \longrightarrow 6\ Cl^-(aq) + 2\ FeO_4^{2-}(aq) + 5\ H_2O(l)$. **43 (a)** $Cu(s) + SO_4^{2-}(aq) + 4\ H^+(aq) \longrightarrow Cu^{2+}(aq) + SO_2(g) + 2\ H_2O(l)$; **(b)** $3\ Cu(s) + 2\ NO_3^-(aq) + 8\ H^+(aq) \longrightarrow 3\ Cu^{2+}(aq) + 2\ NO(g) + 4\ H_2O(l)$. **45.** Compare three reduction half-reactions: (1) $NO_3^- + 4\ H^+ + 3\ e^- \longrightarrow NO(g) + 2\ H_2O$, $E° = 0.956$ V; (2) $Ag^+ + e^- \longrightarrow Ag(s)$, $E° = 0.800$ V; (3) $Au^{3+} + 3\ e^- \longrightarrow Au(s)$, $E° = 1.52$ V. The combination of (1) and the reverse of (2) has $E°_{cell} > 0$; (1) and the reverse of (3) has $E°_{cell} < 0$. **47.** Compared to Group 2A, the 2B elements form many complex ions and have smaller atomic radii, higher ionization energies, and $E°$ values that are less negative. **49 (a)** $3\ Hg(l) + 2\ NO_3^-(aq) + 8\ H^+(aq) \longrightarrow 3\ Hg^{2+}(aq) + 2\ NO(g) + 4\ H_2O(l)$; **(b)** $ZnO(s) + 2\ CH_3COOH(aq) \longrightarrow Zn^{2+}(aq) + 2\ CH_3COO^-(aq) + H_2O(l)$. **51 (a)** 6; **(b)** 4; **(c)** 4; **(d)** 6. **53 (a)** +2; **(b)** +3; **(c)** +3; **(d)** +3. **55 (a)** tetraamminecopper(II) ion; **(b)** hexafluoroferrate(III) ion; **(c)** tetraamminedichloroplatinum(IV) ion; **(d)** tris(ethylenediamine)chromium(III) ion. **57 (a)** $[Fe(H_2O)_6]^{2+}$; **(b)** $[CrBrCl(NH_3)_4]^+$; **(c)** $[Al(ox)_3]^{3-}$. **59 (a)** potassium hexacyanochromate(II); **(b)** potassium trioxalatochromate(III). **61 (a)** $Na_2[Zn(OH)_4]$; **(b)** $[Cr(en)_3]_2(SO_4)_3$; **(c)** $K_2Na[Co(NO_2)_6]$. **63 (a)** Zn^{2+} and four OH^- ions form a complex anion; its name is tetrahydroxozincate ion. **(b)** The ligand F^- is fluoro, and it must precede the name of the central ion. Also, the complex is an anion. The correct name is hexafluoroferrate(III) ion. **65 (a)** Identical compounds: the ligands are listed in a different order, but this does not produce isomerism. **(b)** Isomers: The compositions are identical, but one complex has Cl^- as ligands and the other Br^-. **(c)** Different compounds: One compound has Co^{3+} as the central ion, and the other, Co^{2+}; this also makes the compositions different. **(d)** Isomers: NO_2^- is linked to Co^{3+} through the N atom in one complex and the O atom in the other. **67.** The sketch should have Cr^{3+} as the central ion in an octahedron. No matter which pair of adjacent vertices is chosen for linkage of the en ligand, the remaining four vertices for linkage of Cl^- are equivalent. **69.** Yes. Refer to Figure 22.15. Replace any two en ligands by ox, both in the structure on the left and in its mirror image. The resulting structures are still nonsuperimposable mirror images.

71. (a) Cr^{3+} $[Ar]3d^3$; [↑][↑][↑] [ ][ ]; only one spin state is possible; it has three unpaired e^-.

(b) Fe^{3+} $[Ar]3d^5$; [↑][↑][↑] [↑][↑]; because Cl^- is a weak-field ligand, we expect the high-spin state with all five e^- unpaired.

(c) Mn^{3+} $[Ar]3d^4$; [↑↓][↑][↑] [ ][ ]; because CN^- is

overall they are half-filled with 9.17×10^{20} e^-. **47.** In a semiconductor, there is an energy gap between the valence and conduction bands. In a metal, either the valence band is itself a conduction band or it overlaps one; there is no energy gap. **49 (a)** *n*-type; **(b)** *p*-type. **51.** See Figure 24.12. In *n*-type, electrons in the donor level are readily promoted. Positive holes left in the donor level are immobilized. Electrons predominate as charge carriers. In *p*-type, electrons are promoted to and immobilized in the low-lying acceptor level. Positive holes predominate as charge carriers. **53.** No, the number of electrons that jump the gap *decreases* as the temperature is lowered. **55.** ZnSe and GaP. **57 (a)** Cellulose nitrate has $—ONO_2$ replacing $—OH$ groups, and is made by treating cellulose with nitric acid. **(b)** In rayon, cellulose is first converted to cellulose xanthate (in which $—OH$ groups are replaced by $—O—CS_2^-Na^+$) and then regenerated in an acid bath. **(c)** HDPE has the formula $\text{+}CH_2—CH_2\text{+}_n$. It is a polymer of $H_2C{=}CH_2$, and the polymer chains have little chain branching and pack tightly. **59.** Rubber is elastic because its coiled polymer chains are straightened out during stretching but return to their coiled state when relaxed. Vulcanization forms crosslinks between chains which pull the chains back into their original shape after stretching.

61 $\left[\text{—CHCl—CHCl—}\right]_n$ (each C bearing one H and one Cl)

63 (a) $—CH_2—CH(CN)—CH_2—CH(CN)—CH_2—CH(CN)—CH_2—CH(CN)—$

(b) $—CH_2—CH(OCOCH_3)—CH_2—CH(OCOCH_3)—CH_2—CH(OCOCH_3)—CH_2—CH(OCOCH_3)—$

65 (a) $—O—CH_2—C(=O)—O—CH_2—C(=O)—O—CH_2—C(=O)—O—CH_2—C(=O)—$

(b) $\left[—C(=O)—C_6H_4—C(=O)—NH—C_6H_4—NH—\right]_n$

67. Quiana, a polyamide. Monomers: $H_2N—C_6H_{10}—CH_2—C_6H_{10}—NH_2$ and $HOOC(CH_2)_6COOH$;

polymer formula: $\left[—NH—C_6H_{10}—CH_2—C_6H_{10}—NH—C(=O)—(CH_2)_6—C(=O)—\right]_n$

Additional Problems: 69. 14.23% Cr. **71.** Donor and acceptor atoms exchange electrons. All atoms have four valence electrons, and conductivity requires electrons to jump the energy gap from valence to conduction band, just as in the basic semiconductor itself.

73 (a) $—C(CH_3)_2—CH_2—C(CH_3)_2—CH_2—C(CH_3)_2—CH_2—C(CH_3)_2—CH_2—$

(b) $—C(CH_3)_2—CH_2—CH_2—CH{=}CH—CH_2—C(CH_3)_2—CH_2—CH_2—CH{=}CH—CH_2—$

75. $—C(=O)—C_6H_4—C(=O)—O—CH_2—C_6H_{10}—CH_2—O—C(=O)—C_6H_4—C(=O)—O—CH_2—C_6H_{10}—CH_2—O—$

77. 1,2-ethanediol has two —OH groups, whereas 1,2,3-propanetriol has three. The polymer of 1,2,3-propanetriol and 1,2-benzenedicarboxylic acid has extensive crosslinking of polymer chains and is much more rigid than the polymer with 1,2-ethanediol. **79.** 760 u.

Chapter 25

Exercise: 25.1 6.74 mmHg. **Review Questions: 1 (a)** troposphere; **(b)** stratosphere. **2.** 78 mol% N_2, 21 mol% O_2, 1 mol% Ar. **7 (a)** peroxyacetyl nitrate, $CH_3COOONO_2$; **(b)** SO_2 and SO_3; **(c)** air-borne incombustible mineral matter in coal carried by the wind from coal-burning facilities. **8 (a)** carbon monoxide, CO, and nitrogen oxide, NO; **(b)** CH_4 and CCl_2F_2; **(c)** HNO_3 and/or H_2SO_4. **12.** A combustion mixture that is rich in fuel and lean in $O_2(g)$ promotes the formation of CO(g). **16.** (1) S(in coal) + O_2(in air) $\longrightarrow$ $SO_2(g)$; (2) 2 SO_2 + O_2(in air) $\longrightarrow$ 2 $SO_3(g)$; (3) $SO_3(g)$ + $H_2O(l)$ $\longrightarrow$ $H_2SO_4(aq)$; (4) $H_2SO_4(l)$ + rain $\longrightarrow$ acid rain. **18.** NO and hydrocarbons are involved in the production of NO_2, O_3, PAN, and other photochemical smog components. CO is an air pollutant, but not a constituent of photochemical smog. SO_2 is an important component of industrial smog, but not of photochemical smog. **26.** Cations: Na^+, Ca^{2+}, Mg^{2+}, Fe^{2+}; anions: Cl^-, SO_4^{2-}, HCO_3^-. **39 (a)** Aluminum sulfate and slaked lime react to form $Al(OH)_3(s)$ which adsorbs impurities as the colloidal material settles. **(b)** Aeration of water removes odorous and volatile organic compounds; it also introduces dissolved air to improve the flavor of the water. **Problems: 55.** (c). Air pressure at 30 km is produced by the 1% of air that is at altitudes beyond 30 km. $P_{bar} \approx 0.01 \times 760$ mmHg ≈ 7.6 mmHg ≈ 10 mmHg. **57.** 0.263 mol% H_2O; 2.36 $\times$ 10^3 ppm H_2O. **59.** 13.2 mmHg. **61.** The dew point is about 20.4 °C, the temperature at which $P_{H_2O(g)} = 18.0$ mmHg. **63.** Combustion is a redox reaction. The oxidation numbers of the C atoms can be increased either to +4 (CO_2) or +2 (CO). The reaction of an acid and a carbonate is an acid-base reaction. The oxidation numbers of all C atoms is +4. Only CO_2 can form. **65.** 2 $C_6H_{14}(l)$ + 19 $O_2(g)$ $\longrightarrow$ 12 $CO_2(g)$ + 14 $H_2O(l)$. If combustion is incomplete, some C atoms combine with two O atoms (CO_2), some with only one (CO), and some with none (C); we cannot write a unique equation. **67.** Masses in metric tons: **(a)** 7.22 t CH_4; **(b)** 6.42 t C_8H_{18}; **(c)** 5.74 t coal. **69 (a)** $S_8(s)$ + 8 $O_2(g)$ $\longrightarrow$ 8 $SO_2(g)$; **(b)** 2 ZnS(s) + 3 $O_2(g)$ $\longrightarrow$ 2 ZnO(s) + 2 $SO_2(g)$; **(c)** 2 $SO_2(g)$ + $O_2(g)$ $\longrightarrow$ 2 $SO_3(g)$; **(d)** $SO_3(g)$ + $H_2O(l)$ $\longrightarrow$ $H_2SO_4(aq)$ **(e)** $H_2SO_4(aq)$ + 2 $NH_3(aq)$ $\longrightarrow$ $(NH_4)_2SO_4(aq)$. **71 (a)** Water evaporates from sea spray and leaves particles of NaCl(s) suspended in the air. **(b)** Sulfur dioxide emissions are converted to $SO_3(g)$, which reacts with water vapor to form $H_2SO_4(aq)$. Droplets of $H_2SO_4(aq)$ are neutralized by a basic substance (e.g., NH_3) to form solid sulfate particles. **73.** 2 $H_3O^+(aq)$ + $CaCO_3(s)$ $\longrightarrow$ $Ca^{2+}(aq)$ + 3 $H_2O(l)$ + $CO_2(g)$. **Additional Problems: 75.** 2 $\times$ 10^2 particles per cm^3. **77.** R.H. = 32.4%. $CaCl_2 \cdot 6H_2O$(satd aq) will remove water vapor from the air as long as the relative humidity of the air exceeds 32.4%. **79.** Per gram of fuel: $CH_4(g)$ gives 2.74 g CO_2; C_4H_{10}, 3.03 g CO_2; C, 3.66 g CO_2. Per kJ of heat evolved: $CH_4(g)$ gives 0.0494 g CO_2; C_4H_{10}, 0.0611 g CO_2; C, 0.111 g CO_2. Methane is best by both measures. **81.** If all the S in coal were converted to H_2SO_4, about 1.8 $\times$ 10^{11} lb H_2SO_4 would be produced, about twice the current annual U.S. production. **83.** If *all* the C_8H_{18} were converted to CO—an unlikely happening—the CO level would be about 24 ppm CO, well below the danger level of 35 ppm. **85.** Combustion of a fuel-rich mixture is incomplete and leads to high levels of CO and unburned hydrocarbons (RH). The greater the air/fuel mass ratio, the more nearly complete is the combustion, leading to lower levels of CO and RH. In fuel-lean mixtures, the additional air (and thus N_2) results in the production of more NO. Also, a fuel-lean mixture burns at a higher temperature, which favors the formation of NO. **87.** The rate of ozone destruction would double. **89.** 2.5 $\times$ 10^2 nmol CH_3CCl_3/L. **91.** 12 mg O_2/L. **93 (a)** OCl^- + HSO_3^- + OH^- $\longrightarrow$ Cl^- + SO_4^{2-} + H_2O; **(b)** OCl^- + H_2O_2 $\longrightarrow$ Cl^- + H_2O + O_2. **95 (a)** $Fe^{3+}(aq)$ + $PO_4^{3-}(aq)$ $\longrightarrow$ $FePO_4(s)$, $K_{sp} = 1.3 \times 10^{-22}$, and $[Fe^{3+}] = 1.2 \times 10^{-18}$ M; **(b)** $Al^{3+}(aq)$ + $PO_4^{3-}(aq)$ $\longrightarrow$ $AlPO_4(s)$, $K_{sp} = 6.3 \times 10^{-19}$, and $[Al^{3+}] = 5.7 \times 10^{-15}$ M; **(c)** CaO(s) reacts with water to produce $Ca(OH)_2(s)$, which is sufficiently soluble to supply Ca^{2+} required in the reaction 3 $Ca^{2+}(aq)$ + 2 $PO_4^{3-}(aq)$ $\longrightarrow$ $Ca_3(PO_4)_2(s)$, $K_p = 2.0 \times 10^{-29}$, and $[Ca^{2+}] = 1.2 \times 10^{-7}$ M. **97.** Enzymes catalyze the oxidation of benzpyrene to an epoxide that binds to specific nucleotides in a suppressor gene (P53). The gene is thus mutated to an inactive form and abnormal cell growth (lung cancer) develops. **99 (a)** 98 mg; **(b)** 2.0 $\times$ 10^3 metric tons.

Index

License Agreement

YOU SHOULD CAREFULLY READ THE FOLLOWING TERMS AND CONDITIONS BEFORE BREAKING THE SEAL ON THE PACKAGE. AMONG OTHER THINGS, THIS AGREEMENT LICENSES THE ENCLOSED SOFTWARE TO YOU AND CONTAINS WARRANTY AND LIABILITY DISCLAIMERS. BY BREAKING THE SEAL ON THE PACKAGE, YOU ARE ACCEPTING AND AGREEING TO THE TERMS AND CONDITIONS OF THIS AGREEMENT. IF YOU DO NOT AGREE TO THE TERMS OF THIS AGREEMENT, DO NOT BREAK THE SEAL. YOU SHOULD PROMPTLY RETURN THE PACKAGE UNOPENED.

License

Subject to the provisions contained herein, Prentice-Hall, Inc. ("PH") hereby grants to you a non-exclusive, non-transferable license to use the object code version of the computer software product ("Software") contained in the package on a single computer of the type identified on the package.

Software and Documentation.

PH shall furnish the Software to you on media in machine-readable object code form and may also provide the standard documentation ("Documentation") containing instructions for operation and use of the Software.

License Term and Charges.

The term of this license commences upon delivery of the Software to you and is perpetual unless earlier terminated upon default or as otherwise set forth herein.

Title.

Title, and ownership right, and intellectual property rights in and to the Software and Documentation shall remain in PH and/or in suppliers to PH of programs contained in the Software. The Software is provided for your own internal use under this license. This license does not include the right to sublicense and is personal to you and therefore may not be assigned (by operation of law or otherwise) or transferred without the prior written consent of PH. You acknowledge that the Software in source code form remains a confidential trade secret of PH and/or its suppliers and therefore you agree not to attempt to decipher or decompile, modify, disassemble, reverse engineer or prepare derivative works of the Software or develop source code for the Software or knowingly allow others to do so. Further, you may not copy the Documentation or other written materials accompanying the Software.

Updates.

This license does not grant you any right, license, or interest in and to any improvements, modifications, enhancements, or updates to the Software and Documentation. Updates, if available, may be obtained by you at PH's then current standard pricing, terms, and conditions.

Limited Warranty and Disclaimer.

PH warrants that the media containing the Software, if provided by PH, is free from defects in material and workmanship under normal use for a period of sixty (60) days from the date you purchased a license to it.

THIS IS A LIMITED WARRANTY AND IT IS THE ONLY WARRANTY MADE BY PH. THE SOFTWARE IS PROVIDED 'AS IS' AND PH SPECIFICALLY DISCLAIMS ALL WARRANTIES OF ANY KIND, EITHER EXPRESS OR IMPLIED, INCLUDING, BUT NOT LIMITED TO, THE IMPLIED WARRANTY OF MERCHANTABILITY AND FITNESS FOR A PARTICULAR PURPOSE. FURTHER, COMPANY DOES NOT WARRANT, GUARANTY OR MAKE ANY REPRESENTATIONS REGARDING THE USE, OR THE RESULTS OF THE USE, OF THE SOFTWARE IN TERMS OF CORRECTNESS, ACCURACY, RELIABILITY, CURRENTNESS, OR OTHERWISE AND DOES NOT WARRANT THAT THE OPERATION OF ANY SOFTWARE WILL BE UNINTERRUPTED OR ERROR FREE. COMPANY EXPRESSLY DISCLAIMS ANY WARRANTIES NOT STATED HEREIN. NO ORAL OR WRITTEN INFORMATION OR ADVICE GIVEN BY PH, OR ANY PH DEALER, AGENT, EMPLOYEE OR OTHERS SHALL CREATE, MODIFY OR EXTEND A WARRANTY OR IN ANY WAY INCREASE THE SCOPE OF THE FOREGOING WARRANTY, AND NEITHER SUBLICENSEE OR PURCHASER MAY RELY ON ANY SUCH INFORMATION OR ADVICE. If the media is subjected to accident, abuse, or improper use; or if you violate the terms of this Agreement, then this warranty shall immediately be terminated. This warranty shall not apply if the Software is used on or in conjunction with hardware or programs other than the unmodified version of hardware and programs with which the Software was designed to be used as described in the Documentation.

Limitation of Liability.

Your sole and exclusive remedies for any damage or loss in any way connected with the Software are set forth below. UNDER NO CIRCUMSTANCES AND UNDER NO LEGAL THEORY, TORT, CONTRACT, OR OTHERWISE, SHALL PH BE LIABLE TO YOU OR ANY OTHER PERSON FOR ANY INDIRECT, SPECIAL, INCIDENTAL, OR CONSEQUENTIAL DAMAGES OF ANY CHARACTER INCLUDING, WITHOUT LIMITATION, DAMAGES FOR LOSS OF GOODWILL, LOSS OF PROFIT, WORK STOPPAGE, COMPUTER FAILURE OR MALFUNCTION, OR ANY AND ALL OTHER COMMERCIAL DAMAGES OR LOSSES, OR FOR ANY OTHER DAMAGES EVEN IF PH SHALL HAVE BEEN INFORMED OF THE POSSIBILITY OF SUCH DAMAGES, OR FOR ANY CLAIM BY ANY OTHER PARTY. PH'S THIRD PARTY PROGRAM SUPPLIERS MAKE NO WARRANTY, AND HAVE NO LIABILITY WHATSOEVER, TO YOU. PH's sole and exclusive obligation and liability and your exclusive remedy shall be: upon PH's election, (i) the replacement of your defective media; or (ii) the repair or correction of your defective media if PH is able, so that it will conform to the above warranty; or (iii) if PH is unable to replace or repair, you may terminate this license by returning the Software. Only if you inform PH of your problem during the applicable warranty period will PH be obligated to honor this warranty. You may contact PH to inform PH of the problem as follows:

SOME STATES OR JURISDICTIONS DO NOT ALLOW THE EXCLUSION OF IMPLIED WARRANTIES OR LIMITATION OR EXCLUSION OF CONSEQUENTIAL DAMAGES, SO THE ABOVE LIMITATIONS OR EXCLUSIONS MAY NOT APPLY TO YOU. THIS WARRANTY GIVES YOU SPECIFIC LEGAL RIGHTS AND YOU MAY ALSO HAVE OTHER RIGHTS WHICH VARY BY STATE OR JURISDICTION.

Miscellaneous.

If any provision of this Agreement is held to be ineffective, unenforceable, or illegal under certain circumstances for any reason, such decision shall not affect the validity or enforceability (i) of such provision under other circumstances or (ii) of the remaining provisions hereof under all circumstances and such provision shall be reformed to and only to the extent necessary to make it effective, enforceable, and legal under such circumstances. All headings are solely for convenience and shall not be considered in interpreting this Agreement. This Agreement shall be governed by and construed under New York law as such law applies to agreements between New York residents entered into and to be performed entirely within New York, except as required by U.S. Government rules and regulations to be governed by Federal law.

YOU ACKNOWLEDGE THAT YOU HAVE READ THIS AGREEMENT, UNDERSTAND IT, AND AGREE TO BE BOUND BY ITS TERMS AND CONDITIONS. YOU FURTHER AGREE THAT IT IS THE COMPLETE AND EXCLUSIVE STATEMENT OF THE AGREEMENT BETWEEN US THAT SUPERSEDES ANY PROPOSAL OR PRIOR AGREEMENT, ORAL OR WRITTEN, AND ANY OTHER COMMUNICATIONS BETWEEN US RELATING TO THE SUBJECT MATTER OF THIS AGREEMENT.

U.S. Government Restricted Rights.

Use, duplication or disclosure by the Government is subject to restrictions set forth in subparagraphs (a) through (d) of the Commercial Computer-Restricted Rights clause at FAR 52.227-19 when applicable, or in subparagraph (c) (1) (ii) of the Rights in Technical Data and Computer Software clause at DFARS 252.227-7013, and in similar clauses in the NASA FAR Supplement.

Installation Instructions

Please see the README file for CD-ROM launching instructions.

Minimum System Requirements

Windows

Windows 95 or higher
(Software is not compatible with Windows 3.1)
486, 66 MHz processor or higher
16 MB RAM
4x CD-ROM drive
SVGA graphics card
Color monitor
Sound card
QuickTime v3.0X

Macintosh

System 7.x or higher
Power Macintosh
16 MB RAM
4x CD-ROM
Color monitor
QuickTime v3.0X

Selected Physical Constants

Acceleration due to gravity	g	$9.8066\ \text{m}\cdot\text{s}^{-2}$
Speed of light (in vacuum)	c	$2.99792458 \times 10^{8}\ \text{m}\cdot\text{s}^{-1}$
Gas constant	R	$0.0820584\ \text{L}\cdot\text{atm}\cdot\text{mol}^{-1}\cdot\text{K}^{-1}$
		$8.314510\ \text{J}\cdot\text{mol}^{-1}\cdot\text{K}^{-1}$
Electron charge	e^-	$-1.60217733 \times 10^{-19}\ \text{C}$
Electron rest mass	m_e	$9.1093897 \times 10^{-31}\ \text{kg}$
Planck's constant	h	$6.6260755 \times 10^{-34}\ \text{J}\cdot\text{s}$
Faraday constant	$\mathfrak{F}$	$9.6485309 \times 10^{4}\ \text{C}\cdot\text{mol}^{-1}$
Avogadro number	N_A	$6.0221367 \times 10^{23}\ \text{mol}^{-1}$

Some Common Conversion Factors

Length

1 meter (m) = 39.37007874 inches (in.)

1 in. = 2.54 centimeters (cm) (exact)

Mass

1 kilogram (kg) = 2.2046226 pounds (lb)

1 lb. = 453.59237 grams (g)

Volume

1 liter (L) = 1000 mL = 1000 cm^3 (exact)

1 L = 1.056688 quart (qt)

1 gallon (gal) = 3.785412 L

Force

1 newton (N) = 1 $\text{kg}\cdot\text{m}\cdot\text{s}^{-2}$

Energy

1 joule (J) = 1 $\text{N}\cdot\text{m}$ = 1 $\text{kg}\cdot\text{m}^2\cdot\text{s}^{-2}$

1 calorie (cal) = 4.184 J (exact)

1 electronvolt (eV) = 1.60218×10^{-19} J

1 eV/atom = 96.485 $\text{kJ}\cdot\text{mol}^{-1}$

1 kilowatt hour (kWh) = 3600 kJ (exact)

Mass-energy equivalence:

1 unified atomic mass unit (u)

= $1.6605402 \times 10^{-27}$ kg

= 931.4874 MeV

Some Useful Geometric Formulas

Perimeter of a rectangle = $2l + 2w$

Circumference of a circle = $2\pi r$

Area of a rectangle = $l \times w$

Area of a triangle = $\frac{1}{2}$ (base × height)

Area of a circle = πr^2

Area of a sphere = $4\pi r^2$

Volume of a parallelepiped = $l \times w \times h$

Volume of a sphere = $\frac{4}{3}\pi r^3$

Volume of a cylinder or prism = (area of base) × height

$\pi = 3.14159$